D.W. Pearson, N. C. Steele,
and R. F. Albrecht

Artificial Neural Nets and Genetic Algorithms

Proceedings of the International Conference
in Alès, France, 1995

Springer-Verlag Wien New York

Dr. David W. Pearson
Laboratoire de Génie Informatique et Ingénierie de Production
Ecole pour les Etudes et la Recherche en Informatique et Electronique (EMA-EERIE), Nîmes, France

Dr. Nigel C. Steele
Division of Mathematics
School of Mathematical and Information Sciences
Coventry University, Coventry, U.K.

Dr. Rudolf F. Albrecht
Institut für Informatik
Universität Innsbruck, Innsbruck, Austria

Camera-ready copies provided by authors and editors

Printed on acid-free and chlorine-free bleached paper

With 378 Figures

ISBN-13: 978-3-211-82692-8 e-ISBN-13: 978-3-7091-7535-4
DOI: 10.1007/978-3-7091-7535-4

Preface

This conference is the second in the series devoted to the topics of artificial neural networks and genetic algorithms. The first conference took place in Innsbruck, Austria, in April 1993 when about 100 papers were presented covering the theoretical, implementational and practical aspects. As well as providing a forum for the presentation of new results, the conference served to bring together people working in related areas and a number of productive collaborations have been established.

Delegates at the conference insisted that this particular blend of topics should form a basis for a series of biannual meetings, serving as a target for future publication. Largely due to the quality of the conference proceedings, awareness of the conference has spread widely through the artificial neural network and genetic algorithm communities and thus this volume contains both new contributions from those who took part in the first conference and a significant number from those who are participating for the first time.

In this conference it is pleasing to observe the number of contributions reporting successful applications of the technology to the solution of industrial/commercial problems. This may well reflect the maturity of the technology, notably in the sense that 'real' users of modelling/prediction techniques are prepared to accept neural networks as a valid paradigm. Theoretical issues also receive attention, notably in connection with the radial basis function neural network which is currently receiving considerable attention in the literature.

Rapid though progress has been in artificial neural networks it has been at least as rapid, arguably more so, in the field of genetic algorithms. Contributions herein reflect the wide range of current applications, including for example portfolio selection, filter design, frequency assignment, tuning of nonlinear PID controllers and so on. These techniques are also used extensively for combinatorial optimisation problems, as is evidenced within this volume.

The success of neural networks in modelling and data analysis has prompted a re-examination of classical statistical methods. In particular, attention has focused closely on the problem data providing in the first instance better training data for networks, but of more importance, a better understanding of the data themselves.

The success of genetic algorithms in a host of optimisation problems has provoked interest in heuristic methods in general. Thus this volume contains contributions discussing evolutionary strategies and tabu search methods.

In conclusion, it is interesting to note that the themes of this conference have their origins in attempts to obtain understanding of natural processes. Indeed, Nature itself seems capable of providing a source of inspiration both for new research areas and new methodologies.

David W. Pearson
Alès

Nigel C. Steele
Coventry

Rudolf F. Albrecht
Innsbruck

Contents

SESSION B4: Genetic Algorithms Theory

SESSION C4: Neural Networks Theory

SESSION A5: Time Series, Sequences and Filters

SESSION B5: Genetic Algorithms and Combinatorial Optimisation

SESSION C5: Parallelism and Neural Networks

Radial Basis Function Networks

SESSION A6: Multimedia. CAD and Software Tools

SESSION B6: Genetic and Neural Theory Combined

Genetic Algorithms Theory

SESSION C6: Optimisation

SESSION A7: Self-Organizing Maps and Auto-Associative Memory

SESSION B7: Parallelism and Genetic Algorithms

Optimisation

SESSION C7: Control Theory Applications

SESSION A9: Training Data Selection

SESSION B9: Parallelism and Genetic Algorithms

Genetic Algorithms and Combinatorial Optimisation

SESSION C9: Fuzzy Logic and Uncertainty

Time Series, Sequences and Filters

ICANNGA 95
International Conference on Artificial Neural Networks and Genetic Algorithms
Alès, France, April 19-21, 1995

Advisory Committee

Programme Committee

Workshop Summary

EMA-EERIE, Nîmes, France, April 18, 1995

The conference was preceded by a one day workshop. The morning was devoted to neural networks and the afternoon to genetic algorithms. General introductions, theoretical and implementational issues and specific applications were presented to the participants. We include here a brief summary of each workshop.

1 Introduction to Artificial Neural Networks

Nigel Steele
Coventry University

This session is aimed at those with limited knowledge of the field of Artificial Neural Networks, rather than for experienced practitioners.

An introduction to the field will be given, focussing on both the multilayer perceptron and the radial-basis function networks. As well as discussing training issues, performance of trained networks will be discussed together with an account of some application areas in for example, classification tasks and as an inference engine in a fuzzy-logic based control system.

It is intended that some "hands-on" experience will be available.

2 Source Separation: A Simple Example of an Application of Artificial Neural Networks

Anne Johannet
Ecole des Mines d'Alès

Source separation is a problem very often found in signal processing and in measurement. It consists of the non supervised separation of several superposed signals after their capture by sensors.

After the presentation of the problem and the notation, the exposition, starting from the biological fundamentals of the method, will procede with different proposed approaches. An electronic prototype will be presented.

3 Hardware Implementation of Neural Nets

Kevin Warwick
University of Reading

In this session some of the principles and practice of implementing ANNs in hardware will be presented. Following an introduction to some hardware analogues of ANNs, the basis and structure of some digital and analogue artificial networks will be described.

Issues involved in their implementation will be discussed, along with examples and case studies drawn from a wide range of applications including image processing, speech recognition and robotics.

Finally, some ways in which hardware ANNs can be expected to develop will be outlined.

4 Optimization with GAs

Colin Reeves
Coventry University

Genetic Algorithms (GAs) have aroused intense interest in the past few years because of their flexibility and versatility in solving problems which traditional methods of optimization find difficult. Reported applications cover the fields of Operational Research, Engineering, Biology, Medicine, Control Theory and Robotics - to mention only a few areas.

In this session, the basic concepts of Genetic Algorithms will be introduced, using a worked example for emphasis. Some of the many ways in which GAs have been modified and extended will then be reviewed. GAs need no simplifying assumptions of linearity, continuity etc., and thus can solve highly complex real-world problems, some of which will be discussed, including problems of optimization, sequencing, design and control.

Research in GAs is a burgeoning field, but despite their very real successes, they also have some disadvantages. We will look at some of these, and suggest ways in which they might be overcome. Chief among these may well be the development of hybrid procedures incorporating such techniques as Tabu Search; a brief introduction to the latter topic will also be included.

It is intended to have some GA software available for participants to use.

5 Genetic Algorithms: Experience Drawn from Case Studies

George Smith
University of East Anglia

This session looks in detail at a small number of case studies in the use of genetic algorithms. The applications presented here will be taken from the following set: the concentrator location problem, traffic routing/ channel assignment in telecommunications, the capital budgeting problem, the convoy routing (with delays) problem, the radio-link frequency assignment problem and the Steiner Tree problem in graphs, some of which are the result of collaborative work with national and international companies and organisations.

The emphasis of the session will be on the major (decision) issues of representation and fitness evaluation within straight GAs. However, we also illustrate the wealth of hybrid approaches possible and describe their strengths.

In addition, the participants will have the opportunity to "solve" some of the problems under study using the Genetic Algorithm toolkit, GAmeter, which has been developed by the GA group at the University of East Anglia.

KOHONEN NEURAL NETWORKS FOR MACHINE AND PROCESS CONDITION MONITORING

Dr Tom Harris BSc CertEd PhD MWeldI

Neural Applications Group, Brunel University
Runnymede Campus, Egham, Surrey UK

Abstract

Kohonen feature map neural networks have been used for a variety of successful applications since their introduction in the late eighties. They have one major advantage over their more common peers in that they are capable of unsupervised learning. This property makes them ideal for machine health monitoring situations.

Unsupervised learning allows the network to represent single or multiple classes according to distribution and density. Novelty detection is then possible on-line or classes can be labelled to give diagnosis. This presentation explains the special nature of machine monitoring applications in data availability and desired diagnosis information and provides examples of such systems working in different environments.

Presentation Summary

A neural network application is only ever as good as the training data that one can collect to train it with. Machine and Process monitoring tasks present special problems to neural networks in that training data can be very difficult to collect. There is nearly always a large quantity of data collected from the machine when it is running under normal conditions in good working order but very little of the fault condition data which the system needs to detect.

When a fault develops in a machine the normal priority is to shut it down, not collect data. It is also very rare to be able to run a machine with a deliberately induced fault, solely to collect data. Synthesised data can be of some help but does not give the confidence required in an on-line system.

Kohonen feature map neural networks have a two layer structure which optimises reference vector positions of the inputs in the training set. The optimisation is based on the position and density of the training examples. It is unsupervised during this training stage which means that no desired output or classification is given with the training example.

In the case of condition monitoring this means that a network can be trained on just good condition data and then used to detect variations from that, and hence warn of a problem. In such a way the network is acting like a normal operation template. The networks advantage over traditional modelling methods for this come from its ability to represent the whole machine in a fused fashion and cope with variations in the machine's speed and load.

One such example of this has been developed by the author to detect failure of tyres under high speed tests. It is more useful for the engineer to inspect a tyre when a fault has developed but before the tyre fails completely. One of the difficulties in this problem is that each tyre test is different and so are the sound and vibration characteristics. The Kohonen network learns the normal tyre sound at the beginning of each test and then warns of novelty as the sound changes indicating a developed fault.

Kohonen networks can also be used to classify data, and hence give diagnosis, if training examples are available. In this case the network is initially trained unsupervised on all the available data of all classes. When the network has trained each training example is presented again in a supervised manner and the winning neuron labelled as being of the same class as the input example. When running on-line, the network can then output a class as well as a novelty value.

The author has been involved with developing a system for monitoring food process pumps using this method at Leatherhead Food Research Association in the UK. In this application the condition of the machine is less important than the process. The network was trained to diagnose blockages and cavitation by monitoring vibrations in the pump casing. Faults were induced to collect the data and the system was developed to give diagnosis to an accuracy of 99.6%

An hybrid Kohonen network and expert system is also being developed by the author which allows for diagnosis without training data for a Kohonen network. Expert system have been developed to diagnose machinery faults by calculating key frequency components in vibration spectra. This hybrid system then uses these frequencies as its inputs. Novelties can then be attributed to specific inputs which tie back to specific components in the machine thus giving a diagnosis of the novelty. This system is currently being developed as part of a UK Government initiative to promote applications of neural networks.

Conclusion

Kohonen neural networks provide a variety of excellent tools in machine and process monitoring, both where fault data is available and not. Their use can contribute towards successful implementation of predictive maintenance strategies that can dramatically reduce costs and extend the use of machine components.

PROCESS MODELLING AND CONTROL WITH NEURAL NETWORKS: PRESENT STATUS AND FUTURE DIRECTIONS

Prof. Dr. Bernd Schürmann

Siemens AG,
Corporate Research & Development,
D-81730 München, Germany.

Abstract

Even though there exist numerous traditional approaches for solving nonlinear control tasks their realization in practice often proves to be difficult, mainly because of i) insufficient analytical knowledge of the system to be controlled, and ii) because of incomplete knowledge of the physical parameters of the system.

Neural networks can cope with these problems because of their capability to realize multivariate, nonlinear transformations and to learn these merely by empirical information, i.e. by representative training examples *(data driven modeling).*

In this contribution we discuss neural methods for modeling and control and illustrate their use in real world applications. In order to improve the corresponding conventional solutions, it is crucial to i) incorporate prior knowledge (analytical and/or rule based) into the neural network and ii) to optimize its architecture by refined learning methods. We point out the future need for neural control of the *complete* process rather than controlling its parts independently.

Further, we discuss the necessity to combine neural methods with the modern techniques of Nonlinear Dynamics for the modeling and control of processes with highly nonlinear dynamic behavior.

A GENETIC ALGORITHM FOR MULTICRITERIA INVENTORY CLASSIFICATION

H. Altay Güvenir

Department of Computer Engineering and Information Science,
Bilkent University, Bilkent, 06533 Ankara
guvenir@cs.bilkent.edu.tr

Abstract

One of application areas of the genetic algorithms is parameter optimization. This paper addresses the problem of optimizing a set of parameters that represent the weights of criteria, where the sum of all weights is 1. A chromosome represents the values of the weights, possibly along with some cut-off points. A new crossover operation, called *continuous uniform crossover*, is proposed, such that it produces valid chromosomes given that the parent chromosomes are valid. The new crossover technique is applied to the problem of multicriteria inventory classification. The results are compared with the classical inventory classification technique using Analytical Hierarchy Process.

INTRODUCTION

Management and control of inventories consisting of a large number of items is usually done by classifying these items in three groups, called the *ABC classification.* The items that require closest attention are called class A items. The class C items are given the least importance. All other items are classified as class B. In the traditional ABC classification, items are classified according to their total annual dollar usage. It has been suggested by Flores and Whybark (1986) that multiple criteria ABC classification can provide a more comprehensive managerial approach. Flores et. al. (1992) proposed the use of the Analytic Hierarchy Process (AHP) to reduce these multiple criteria to a univariate and consistent measure to consider multiple inventory management objectives. Through subjective pairwise comparisons of the criteria, a weight vector is derived. The elements of this vector represent the absolute weights of each criteria. The inventory items are then sorted according to this weight vector. The items on top of this sorted list are classified as class A, and the ones in the bottom (usually lower half) as class C. The remaining items are classified as class B. The difficulty with this approach is that it is quite difficult to reflect the relative weight of criteria in the form of pairwise comparisons.

In order to remedy this problem we propose a method, which uses direct classifications of a small set of training items. The method we propose here uses a genetic algorithm to learn a weight vector, along with the cut-off points between class A and B, and class B and C. The remaining set of items are then classified using this weight vector and cut-off points. Note that the approach proposed here is applicable to any multicriteria classification problem with any number of classes, provided that it is possible to reduce the problem to learning a weight vector along with the cut-off points between classes.

We compared the pairwise comparison (AHP) technique and the genetic algorithm approach on a sample inventory. We have noticed that are much more positive in classifying items than subjectively comparing two criteria. The classification made by our technique was closer to the desired classification than the one made by the AHP technique.

An important characteristic of the weight vector is that the elements are real values representing absolute weights of the criteria whose sum is 1. In order apply a genetic algorithm to the weight learning problem, we proposed a new crossover operator that guarantees the generation of off-spring that are valid representations of weight vectors. This approach is better than using the standard crossover operations and controlling the search through penalty functions only.

In the rest of the paper we describe the AHP technique, then define our approach. Next we compare both techniques on a sample inventory. The paper concludes with general remarks on the use of genetic algorithms in multicriteria classification.

MULTICRITERIA INVENTORY CLASSIFICATION USING AHP

AHP is proposed by Saaty (1980) as a general theory of measurement. Application of AHP to multiple criteria classification requires a pairwise comparison of all criteria by an expert. The decision maker is usually asked to assign a value in a scale 1 through 9 for each pair of criteria. These pairwise comparisons are combined in a square matrix called *pairwise comparison matrix*. Given a pairwise comparison matrix A, the entry A_{ij} represents the relative strength of criterion C_i when compared to C_j. That is, A_{ij} is expected to be w_i/wj, where w_i and wj are the relative strengths of the criteria C_i and C_j, respectively. Therefore, the pairwise comparison matrix is a symmetric matrix and its diagonal is equal to 1. If the expert was not consistent in the pairwise comparisons the resulting matrix will be inconsistent. In that case, the expert is asked to form the pairwise comparison matrix again. This process repeated until the matrix is consistent. The next step consists of the computation of a vector of priorities from the constructed matrix. According to the AHP methodology, the eigenvector of the comparison matrix with the largest eigenvalue provides the priority ordering of the criteria, that is their weights. The values of this vector are between 0 and 1, and their sum is 1.

For each inventory item, the values of each criteria are, then, organized in a way that the higher values increase the probability of an item being in Class A. The criteria values are normalized to the range of 0 to 1 for each criteria. The weighted score of each item in the inventory is computed as the sum of the product of weight and normalized value for each criteria. The items are sorted in descending order according to their weighted score values. Finally, the first 20% of the items are classified as class A, the next 20% as class B, and the remaining items as class C.

Although this technique is sound, the pairwise comparisons are based on the subjective judgments of an expert. Usually, the expert does not agree with the resulting classification of the inventory. In that case, the expert changes the pairwise comparisons and obtains a new weight vector. This process continues until there are no obvious misclassification of the inventory items. This difficulty arises because it is not easy to compare two criteria and assign a numerical value. However, experts are able to judge the appropriateness of the resulting classification.

MULTICRITERIA INVENTORY CLASSIFICATION USING GA

Here we propose an alternative method to learn the weight vector along with the cut-off values for multicriteria inventory classification. The method proposed here, called GAMIC (for Genetic Algorithm for Multicriteria Inventory Classification), uses a genetic algorithm to learn the weights of criteria along with AB and BC cut-off points from preclassified items. Once the criteria weights are obtained the weighted scores of the items in the inventory are computed as in the previous method. Then the items with score greater than AB cut-off value are classified as A, those with values between AB and BC as class B, and the remaining items are classified as class C.

Encoding

A chromosome encodes the weight vector along with two cut-off points. The values of the genes are real values between 0 and 1. The total value of the elements of the weight vector is always 1. Also the AB cut-off value (x_{AB}) is always greater than the BC cut-off value (x_{BC}). Therefore, if the classification is based on k criteria, a chromosome $\mathbf{c}$ is a vector defined as

$$\mathbf{c} = < w_1, w_2, \ldots w_k, x_{AB}, x_{BC} > .$$

Here, w_j represents the weight of jth criterion, $\sum_{j=1}^{k} w_j = 1$, and $x_{AB} < x_{BC}$. In this representation the relative weight of a criteria can be encoded as an absolute value in a chromosome. That is, an allele represents the weight of the corresponding criterion, independent of the other alleles.

Given a chromosome $\mathbf{c}$, classification of an inventory item i is done by computing its weighted sum, $ws_{c,i}$ as follows:

$$ws_{c,i} = \sum_{j=1}^{k} w_j \frac{i_j - min_j}{max_j - min_j}.$$

Here i_j is the value of the item i for the criterion j, max_j and min_j are maximum and minimum values of criterion j among all inventory items. The classification of an inventory item i according to chromosome $\mathbf{c}$ is

$$classification(i, \mathbf{c}) = \begin{cases} A & \text{if } ws_{\mathbf{c},i} \leq x_{AB} \\ B & \text{if } x_{BC} \leq ws_{\mathbf{c},i} < x_{AB} \\ C & \text{otherwise} \end{cases}$$

Given this encoding scheme, GAMIC applies the standard genetic operators (reproduction, crossover, and mutation) to the chromosomes in the population (Goldberg, 1989). GAMIC applies fitness proportionate roulette wheel selection in reproduction. GAMIC also uses the elitist approach, that is the best chromosome is always copied to the next generation. The evaluation of the fitness of a chromosome, the crossover, and the mutation operations employed by GAMIC are described below.

Fitness function: The fitness of a chromosome reflects its ability to classify the training set correctly. Therefore, any misclassified item should introduce a penalty. However, due the the linear ordering among the classes, we have to distinguish the error made by classifying an A as B than as C. In our implementation the fitness of a chromosome **c** was computed as

$$fitness(\mathbf{c}) = \frac{\sum_i^t p_i}{t},$$

$$p_i = \begin{cases} 1 & classifcation(i, c) = class(i), \\ 0.4 & |classifcation(i, c) - class(i)| = 1, \\ 0 & otherwise, \end{cases}$$

where t is the size of the training set. Note that this fitness function prefers a chromosome making two mistakes with difference 1 to a chromosome making a single mistake with difference of 2.

Crossover

Although the initial population is setup in a way that all chromosomes represent legal codings (each allele is between 0 and 1, the sum of all weight values is 1, and the AB cut-off is less than the BC cut-off), standard crossover operations are bound to result in illegal codings. Here we propose a new form of uniform crossover operation, called *continuous uniform crossover*, that guarantees the legality of the offsprings.

Continuous Uniform Crossover: Given two chromosomes $\mathbf{x} = < x_1, x_2, \ldots, x_n >$, and $\mathbf{y} = < y_1, y_2, \ldots, y_n >$, the offsprings are defined as $\mathbf{x}' = < x_1', x_2', \ldots, x_n' >$, and $\mathbf{y}' = < y_1', y_2', \ldots, y_n' >$, where

$$\begin{aligned} x_i' &= sx_i + (1-s)y_i \\ y_i' &= (1-s)x_i + sy_i \end{aligned}$$

Here s, called *stride*, is constant through a single crossover operation. This crossover preserves the sum of any subset of genes. In the case of inventory classification, if the genes 1 through m encode the criteria weights, $\sum_{i=1}^m x_i = 1$. After the crossover operation,

$$\sum_{i=1}^m x_i' = s\sum_{i=1}^m x_i + (1-s)\sum_{i=1}^m y_i = s + (1-s) = 1.$$

This is true for both offsprings; $\sum_{i=1}^m x_i' = 1$, and $\sum_{i=1}^m y_i' = 1$, Further, this crossover preserves the greater-than relation between genes, as well. That is, if $x_{AB} > x_{BC}$, then $x_{AB}' > x_{BC}'$. Therefore, continuous uniform crossover preserves the legality of chromosomes for the multicriteria ABC classification.

The choice of the stride (s) is an important issue. If $s = 0$, the off-springs are the same as the parents. If $0 \leq s$ then alleles remain to be between 0 and 1, however the alleles get closer to each other through generations. If $s = 0.5$, then both off-springs are the same, and the alleles are equal to the average values of the respective alleles in the parents. On the other hand, if $s < 0$, the alleles diverge from the respective values in the parents. However, if $s < 0$, then alleles may be outside of the limits; that is an allele may get a negative value of a value greater than 1, although their sum is still 1. In that case, a normalization of the chromosome is needed. In our implementation we chose s randomly from [-0.5, 0.5] for each crossover operation.

Mutation: Mutation operation in our implementation sets the value of a gene to 0 or 1 with equal probability. If a chromosome is modified by the mutation operation, it has to be normalized.

EXPERIMENT

We have compared these two techniques (AHP and GAMIC) for multicriteria classification on an inventory of 115 items used by a construction company in rock excavation jobs. The items are to be classified according to ten criteria. We first asked an expert to name five items from each of the three classes. This set of 15 items formed our training set for the GA. The genetic algorithm was run on a population of 100 chromosomes, with $p_c = 0.7$ and $p_m = 0.001$. Our genetic algorithm converged on the training set in the 340th generation after 24003 function evaluations. The best and average fitness values in each generations are shown in Figure 1.

We classified the remaining 100 items using the weights and cut-off values learned by the GA. The expert also was asked to classify these test items. The expert's classification did not agree with GA method on 7 items. In order to compare with the AHP technique, the expert was asked to pairwise compare the criteria. In the second try the expert obtained a consistent matrix. All 115 items were then classified according to the weights learned by the AHP technique. The

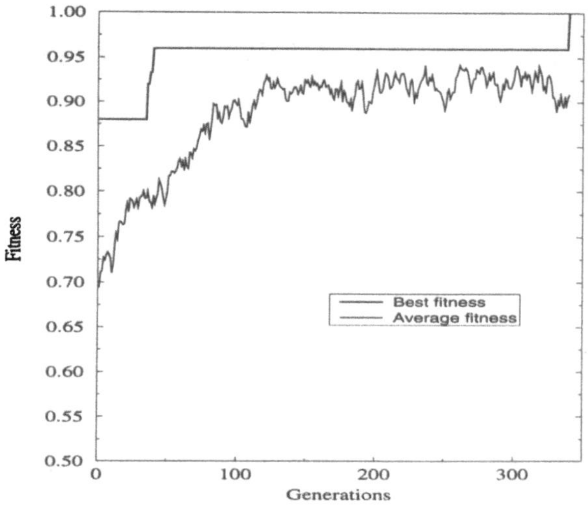

Figure 1: Best and average fitness values through generations.

expert did not agree with the classification of 13 items. As a result the genetic algorithm method proposed here performed better than the AHP technique in our test set.

CONCLUSION

We presented a new approach using genetic algorithms to multicriteria classification. This approach is based on learning a weight vector of absolute weights of each criteria along with a set of cut-off points. In order apply a genetic algorithm to the weight learning problem, we proposed a new crossover operator that guarantees the generation of off-spring that are valid representations of weight vectors.

We have compared our approach with classical AHP technique based on subjective pairwise comparisons of criteria on a sample inventory classification task. The approach, implemented in a program called GAMIC, is applicable to any multicriteria classification problem with any number of classes, provided that it is possible to reduce the problem to learning a weight vector along with the cut-off points between classes. For problems where this condition is not satisfied a more complex techniques using genetic programming techniques can be used.

References

[1] B.E. Flores, D.L. Olson and V.K. Dorai, Management of Multicriteria Inventory Classification, *Mathl. Comput. Modelling* Vol. 16, No. 12, pp. 71-82, 1992.

[2] B.E. Flores and D.C. Whybark, Multiple Criteria ABC Analysis, *International Journal of Operations and Production Management* Vol. 6, No. 3, pp. 38-46, 1986.

[3] D. E. Goldberg, Genetic Algorithms in Search, Optimization & Machine Learning, Reading, Massachusetts, Addison-Wesley, 1989.

[4] T.L. Saaty, The Analytic Hierarchy Process, New York, McGraw-Hill, 1980.

OPTIMIZING CLASSIFIERS FOR HANDWRITTEN DIGITS BY GENETIC ALGORITHMS

J. Schäfer, H. Braun

Institut für Logik, Komplexität und Deduktionssysteme
Universität Karlsruhe*

***Abstract*—We present the first large real-world application for the neural network optimizing genetic algorithm ENZO. Nets had several thousands links and the training data up to over 200,000 patterns. We evolved nets for a classification task that have an order of magnitude free parameters less than commonly used polynomial classifiers while maintaining the same performance.**

To achieve this we implemented some significant enhancements and minor improvements of the original algorithm.

It is also shown how to use ENZO as an efficient tool to create nets satisfying task-specific constraints.

I. Introduction

One of the unsolved problems in neural network research remains the design of a topology well suited to a specific task. Mostly the decision is based on past experience and intuition.

Our approach to solve this problem is called ENZO, an algorithm optimizing network topology by means of a hybrid genetic algorithm (GA). Experiments on small to medium sized problems proved the algorithm to be very efficient (cf. [1, 2]). The success encouraged us to apply ENZO to a big real-world benchmark, for which we chose a handwritten digit recognition task.

Our decision based on the following two considerations:

1. this problem is well understood and known to be solvable by neural nets,
2. a huge database of handwritten digits is publicly available.

The database was created by the US National Institute for Standardization and Technology (NIST) for a world-wide competition which took place in 1992. The data consists of a training set with more than 220,000 and a test set including more than 50,000 labeled pictures. All details about the data and the competition may be found in their technical report [3].

Most important for the digit classification task is to achieve fast response times during application – the time needed for training is of minor interest. Therefore it is essential to find a small net with the best generalization performance possible.

Depending on the application the use of varying cost-models is necessary which take into account constraints on time, accuracy, rejection rate, etc. As will be shown, ENZO provides a natural and convenient way for doing this.

II. ENZO

ENZO is a hybrid system utilizing both: efficient local optimization techniques and a robust global search. The local optimization is done with RPROP (cf. [5]), the fastest known gradient descent for neural nets, and a genetic algorithm for global searching. The former guarantees fast adjustment of the weights to a local minimum whereas the latter optimizes the topology. Additionally the GA enables the nets to escape from local minima by broadening the search space.

The genetic operators of ENZO are defined on a direct encoding which means that they manipulate the weight matrix directly. Network structure is put under user control by specifying a reference net, this may include restrictions on the depth of the network or if shortcut connections will be used. All nets generated by ENZO will conform to the topology given by this reference net, i.e. the search space for the genetic algorithm is restricted to subnets of it.

The principal line of evolution as done by ENZO is summarized in the following picture (fig. 1).

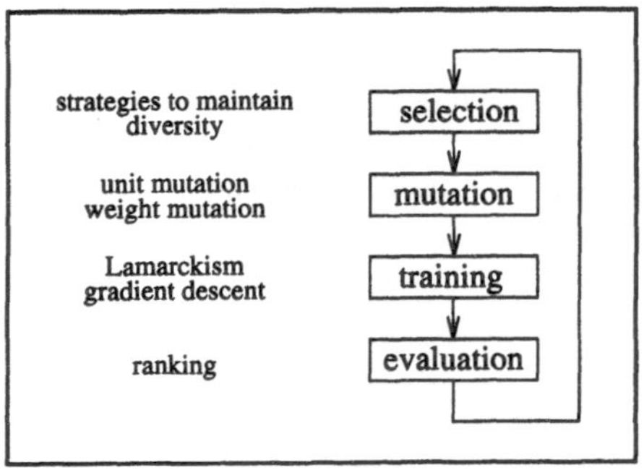

Figure 1: *Principal structure and features of* ENZO.

[1]ILKD, Universität Karlsruhe, D-76128 Karlsruhe;
e-mail: schaefr@ira.uka.de, braun@ira.uka.de

The most important operator is mutation, which encompasses link and unit mutation. These are applied independently to each net.

Link mutation For each link a decision based on a fixed probability p_{del} is made if it is to be deleted or inserted. Link mutation is supplemented with simple pruning, that deletes small weights.

Unit mutation The decision process for the unit mutation is twofold: the first question is if unit mutation takes place at all and the second is if a unit is inserted or deleted. Units are deleted with higher probability for those having less connections with the remaining net.

In case of deletion the unit may be 'bypassed': the flow of information is preserved through connecting former input units to the deleted unit with its targets. For inserted units all possible connections are established.

Note that no crossover is used.

A feature fundamental for the success of ENZO is called *Lamarckism*. This means that attributes acquired by the parents are inherited by the offsprings—realized in ENZO by using the parental weights as initialization for the offsprings. Furthermore using the parental weights speeds up training of an offspring considerably because the structure of the network didn't change that much so learning starts in the neighborhood of a good solution. As a consequence ENZO creates small trained nets, for which it might even be impossible to be trained starting with random weights.

III. Extensions

The idea of Lamarckism may not only be applied to the weights of the network but also to the training pattern set.

Accounting for the huge data set we used a simple pattern selection scheme: a net is first trained with a random subset. Every k epochs (k constant) it is replaced by patterns out of the whole set which aren't yet close enough to the target. The selected subset thus resembles the salient patterns, while redundant ones are ignored.

Instead of (re-)training the offsprings initially with a random subset, patterns selected by the parents are used and replaced again during learning.

The decision which patterns are to be selected depends on a threshold for the squared error being dynamically adapted. If the number of patterns exceeding the threshold is greater than a user-specified maximum number m it is increased otherwise decreased. Changing the threshold is don by multiplying it with constants $\sigma^+ > 1$ and $\sigma^- < 1$. So the training set consists of approximately m patterns with worst error, without any need for an time consuming sort.

As can be seen from figure 2 the error on the selected patterns becomes stable after a short time and the error for the nets trained with the selected patterns decreases much faster. Although the final error on the test set is greater than the error achieved by training with the whole pattern set the classification performance is about the same, which proves this simple technique to be quite effective.

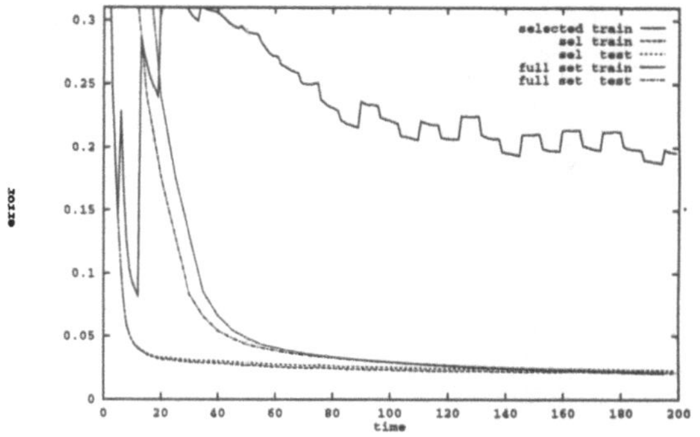

Figure 2: *Train an test errors for training with pattern selection (sel) and without (full set). The error is measured on the whole training and test set. For the subset of selected pattern the error is given as 'selected train'. The x-axis gives a normalized time measurement.*

For our task we chose $m = 5,000$ patterns out of a total 50,000. Selection took place after every $k = 7$ epochs. Since offsprings are re-trained up to twelve epochs this means once per offspring. If we compare the number of patterns examined by retraining the offspring we get: $12 \cdot 5,000 + 50,000 = 110,000$ and $12 \cdot 50,000 = 600,000$ pattern evaluations with and without selection respectively. So pattern selection enables faster learning by more than five times.

In summary an offspring starts its career in an optimal environment consisting of good initial weights, fine-tuned learning parameters and a subset including the most important patterns. Combining these features, learning time is speeded up by several orders of magnitude.

Moreover the simple pruning is replaced by a weighted link mutation taking the size of the weights into account. This fits much better into the GA-framework used by ENZO and avoids pruning and therefor another retraining.

Weighted Link mutation For each link the probability to be deleted is $p_{del} * N_\sigma(0, w)$ where p_{del} and σ are set by the user, $N_\sigma(0, w)$ means a normal distributed value with mean 0 and variance σ und w denotes the weight of the considered link. Since the weight of the link is covered by $N_\sigma(0, w)$ pruning can be abandoned.

We may call this soft pruning, since not only the weights below the threshold σ are pruned, but very small weights ($|w| << \sigma$) with probability about p_{del}, big weights ($|w| >> \sigma$) with probability about 0 and a soft interpolation in between (see fig. 3). Moreover, by the factor $N_\sigma(0, w)$ we get the effect that for each weight the probability of deletion is proportional (cum grano salis) to the probability of effecting no significant deterioration of the performance.

If we choose p_{del} such that only a few links are deleted by each weight mutation we get the following heuristic:

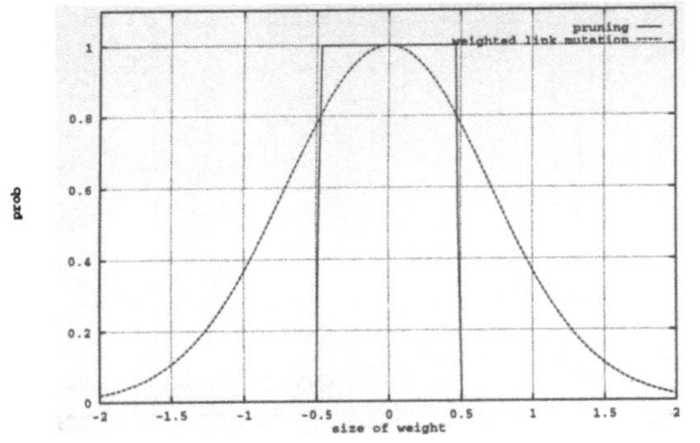

Figure 3: *Deletion probabilities for pruning and weighted link mutation. Pruning deletes all weights smaller than a given bound with no regard to their importance whereas weighted mutation allows some small links to remain in the net and deleting bigger ones.*

Softpruning Test a few weights for pruning preferring small weights. Whenever this pruning effects no significant deterioration, this variant will survive and will be subject to more pruning, - else this offspring is classified as failure and therefore not inserted in the population.

Therefore it is not necessary to estimate the effects of the deletion of a weight as is done by other heuristics (e.g. optimal brain damage)—just try and test it. Of course, for thinning out neural networks by hundreds of weights as done in our application, we cannot delete just one weight per offspring but we have to adjust p_{del} big enough. On the other hand, if p_{del} is too big, we may prune too many weights such that the offspring will not survive.

Other improvements are:

- A small perturbation of the offspring weights is made prior to retraining. This change made ENZO even more robust to local minima.

- Unit mutation is now allowed to mutate input units (*Input mutation*), too, which leads to consequent minimization of the number of inputs.

By these improvements ENZO creates even smaller nets for problems tested so far. As an example the net for the TC-Problem[1] that originally had nearly 300 links was already impressively reduced to a 10-2-1 net with 27 links by ENZO [2]. We could improve this even further to 9-2-1 with only 19 links left. See table 1 for details.

[1]The TC-Problem is to tell two patterns ('T' and 'C') on a 16x16 retina apart no matter what their orientation is. It was first proposed by Rumelhart et al [6] and further investigated by McDonnell and Waagen [4].

Reference	topology	no. of links
Original net	16-16-1	288
McDonnell, Waagen	15-7-1	60
Braun, Zagorski	10-2-1	27
	11-1-1	22
This paper	9-2-1	19

Table 1: *Nets generated for the TC-Problem. We could even improve the best known solution. Note especially the further decreased number of input units due to input-mutation.*

IV. Results

The training patterns from NIST were split randomly into four subsets consisting of 50,000 patterns and one subset with the remaining 20,000 patterns. In favor of more experiments we used one set of 50,000 pictures for training, one for evaluation the fitness and the remaining ones for conclusive testing.[2]

The best net we obtained has 3,295 links and classified 99.16% of our test patterns correctly. All nets range from 900 to 3,300 links and from 98.05 up to 99.16% correct.

Table 2 illustrates the use of the fitness-function for generating specialized nets for a given application, too. This can be done by adjusting the ratio Q for weighting of the number of links to the number of misclassified patterns.

links	correct	ppw : miss
892	98.05	1 : 2
1,648	98.71	1 : 5
2,047	98.86	1 : 10
3,295	99.16	1 : 1500

Table 2: *Some nets obtained from* ENZO *illustrating the ability to adapt to specific tasks by means of the fitness function.*

If the number of weights in the evaluated net is called W and the number of misclassified or rejected patterns M, the fitness function to be minimized is given by $f = w \cdot \text{ppw} + M \cdot \text{miss}$ with ppw (penalty per weight) and miss as user specified weighting factors. These are used to adjust the ratio Q mentioned above.

If evaluation time is of major concern, you can create nets with only a few links by greater values for Q. If you're interested only in classification rates you may decrease Q in order to get better but bigger nets.

So the fitness function corresponds directly to the desired cost model. Even more complicated models may be realized by alternative implementations of to the fitness function.

Training the nets obtained with ENZO from scratch results in slightly decreased performance: 98.72% for the 2,047-net and 98.54% for the one with 1,648 links (each best out of ten).

[2]We decided not to use the NIST-test-patterns, because they differ significantly from the digits in the training set.

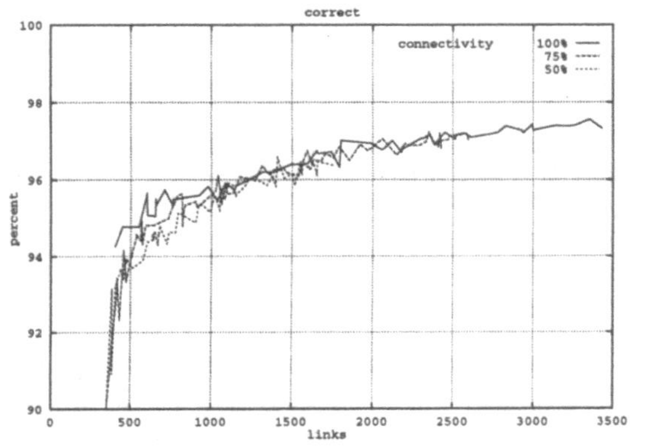

Figure 4: *Net size vs. classification performance for randomly created topologies. (Note that the performance is not as good as mentioned in the text because we used a fixed 5,000 pattern subset to conduct this experiment.)*

For a randomly created topology with the same number of links this is even worse (significantly below 98%). This shows that ENZO finds a good architecture as well as an optimal set of weights.

It may be interesting to test how much the performance depends on the chosen topology. In order to get a thorough statistic we limited the training set to 5,000 patterns and trained about 1,000 networks with different topologies each (see fig. 4). The results show that the performance depends strongly on the number of links, whereas the topology as such seems to be of minor importance. Therefore, it was a difficult and challenging task to diminish the network size without loosing performance.

V. Comparison to a Polynomial Classificator

We compared our neural network classification performance with that of a commercially used polynomial classifier of degree two. Trained on the same 50,000 pattern subset using the same features as the neural net the polynomial classifier achieves a classification rate of 99.06% correct. So both classification schemes perform equally well.

The major difference is in the number of free parameters: while a 2nd degree polynomial classifier uses 8.610 coefficients our nets range from less than 1,600 to 3,300 links. As a consequence the classification time in practical application is reduced to 20% of time needed by a PC.

Another advantage is the possibility to obtain a specialized net for a given time-accuracy tradeoff. By means of the fitness-function the user can support the evolution of either nets with few links, risking a small drop in performance, or more powerful nets with some links more.

VI. Conclusion

Through systematic use of Lamarckism we could extend ENZO to handle large tasks successfully and show that network design done with ENZO surpasses the commonly used trial-and-error method by far.

Especially important for large problems is the speed-up in training time achieved for offsprings by consistent use of Lamarckism:

- offsprings inheriting the weights from the parent and
- selecting the salient patterns out of the training set.

The two striking advantages of network creation done with ENZO are

1. generation of nets with outstanding topology and weights and
2. the possibility to adapt nets easily to various task-specific constraints.

These advantages are not limited to the classification task considered throughout this paper but carry over to all problems with a huge and dense training set. Application in neuro-control will be presented in a forthcoming paper.

References

[1] Heinrich Braun and Joachim Weisbrod. Evolving feedforward neural networks. *Proceedings of the International Conference on Artificial Neural Nets and Genetic Algorithms*, 1993.

[2] Heinrich Braun and Peter Zagorski. ENZO-M: A hybrid approach for optimizing neural networks by evolution and learning. In *Proceedings of the third parallel problem solving from Nature*, Jerusalem, Israel, 1994.

[3] R. Allan Wilkinson *et al.* The first census optical character recognition systems conference. NIST ir4912. Available at the NIST-Archive: **sequoya.ucsl.nist.gov.**

[4] John McDonell and Don Waagen. Neural structure design by evolutionary programming, 1993. NCCOSC, RDT&E Division, San Diego, CA 92152.

[5] Martin Riedmiller and Heinrich Braun. A direct adaptive method for faster backpropagation learning: The RPROP algorithm. In *Proceedings of the ICNN 93*, San Francisco, 1993.

[6] D. E. Rumelhart, G. Hinton, and R. Williams. Learning internal representations by error propagation. In D. E. Rumelhart and J.L. McClelland, editors, *Parallel Distributed Processing, Vol. I Foundations*, pages 318–362. MIT Press, Cambridge, MA, 1986.

DCS: A PROMISING CLASSIFIER SYSTEM

Philippe Collard Cathy Escazut

Laboratory I3S — CNRS-UNSA, Bât. 4
250, av. Albert Einstein, Sophia Antipolis, 06560 Valbonne, FRANCE

A classifier system is a machine learning system that learns syntactically simple string rules called classifiers. Such systems combine learning and evolution processes. The Bucket Brigade algorithm implements the first one, while the second one often use a genetic algorithm. Unfortunately, this kind of genetics-based machine learning systems suffers from a lot of problems yielding system instability, often resulting in poor performance. The main difficulty consists in maintaining good classifiers in the population during the evolution process.

In this paper we proposed an original classifier system, DCS, that manipulates pairs of doubles in the population. The mechanism of DCS is inspired from natural observations concerning the editing of RNA. *With this new characteristic, different classifiers may be interpreted in the same way. We report experiments demonstrating that the introduction of these special individuals in the population can lead to a significant increase in success rate of artificial learning, and a diminution of the rules specificity.*

1 Introduction

Learning in classifier systems (CS) is performed thanks to payoffs sent by the environment [4, 3] . With this mechanism, each classifier is evaluated according to its role played in obtaining the searched solution. This measure is computed by updating a coefficient for each classifier.

But, without an evolution mechanism, CSs cannot explore the whole search space. Thus, in order to inject new classifiers in the system, the learning algorithm is combined with an evolution process: a genetic algorithm (GA).

In this paper, we present an original CS using non standard classifiers. We illustrate our proposition with experimental results showing significant improvements.

2 Classifier Systems

Three main components enable the working of a CS:

- The rule and message system: The messages are the basic tokens of information exchange in a CS and between the system and its environment. Messages are most often represented as bit strings. The rules, called *classifiers*, are a kind of production rules. They are usually implemented as simple strings defined over the alphabet {0,1,#}, where the "don't care" symbol # represents either a '0' or a '1'.
- The apportionment algorithm: The environment manages the attribution of rewards or sanctions to the set of classifiers, called the *population* in order to encourage the best rules. The Bucket Brigade algorithm is generally used to achieve this step: it allows the measure of the reliability of each classifier updating its coefficient called *strength*.
- The genetic algorithm: The use of a GA enables the system to evolve and to react towards the modifications of its environment. Moreover, the evolution of the basic set of rules leads to the exploration of new areas of the search space.

The Bucket Brigade algorithm and the GA are both activated on classifiers and messages. Since the apportionment of credit mechanism is beyond the scope of this paper, we omit further description of the Bucket Brigade algorithm (Interested readers can refer to [6].). However, we can say that the Bucket Brigade algorithm has advantages of simplicity, but is not able to assure the learning by itself. Thus, we must use a mechanism inserting new classifiers in the population: a GA. It is based on the mechanics of natural selection and natural genetics: it uses the application of a reproductive technique and the application of operators such as crossover and mutation [3, 1].

3 DCS Framework

3.1 New Approach

Molecular biologists believed for a long time that all the informations contained in the molecule of DNA was used for the protein creation: they thought the elaboration was divided in two steps: first DNA became RNA, then the RNA enabled the making of the protein. The study of organisms more complex than simple bacteria has destroyed this basic idea of classic molecular biology. Indeed, several sequences of DNA, called *introns*, seem to be useless and thus are ignored during the transcription, while new sequences, usefull for the transcription and called *exons*, may be inserted: this phenomenon is called the *editing of* RNA These natural phenomenons may provide inspiration for GA mechanisms. Thus, the paper [5] reports experiments demonstrating that the insertion of introns into bit strings can lead to increases in success rates of artificial evolution. Our approach is based on the notion of exons: the condition part of a classifier contains two sorts of informations: the first one contains information needed to understand the second one. Thus, individuals are submitted to a particular treatment depending on a part of themselves.

We propose a quite simple implementation of this idea through a new encoding of the binary string. The first bit, called *head-bit*, manages the interpretation of the rest of the string. When the head-bit is set to '0', the rest of the string remains unchanged. If the head-bit is set to '1' the string is interpreted as the binary complement of the rest of the string. For instance the string 0101 becomes 101, and 1110 becomes 001. We call *transcription* this interpretation of the classifier condition part.

It's worth noticing that different classifiers may be interpreted in the same way. These strings behave as twins in the population since they have the same phenotype. Thus, we proposed to called them *doubles* with a special meaning: same behaviour but different appearance.

3.2 DCS behaviour

Of course, we must add a bit to the condition string, but this additional work is all the more negligible as the strings are long. And indeed, most of the CSs use long strings. Holland's schema theorem is fundamental to the theory of GA. The algorithm manipulates individual strings. Most of the theoretical work focuses on the implicit processing of *schemata* identifying a subset of individuals of the population. For instance, the schema 01⋆01 groups the individuals 01001 and 01101 together. The character '⋆' is a "don't care" symbol. In CSs this notion of schema is explicit thanks to the symbol #.

Our new CS leads to consider connections between the bits of the condition. When the head-bit is fixed, the transcription is easy. For instance, the condition 01#1 becomes 1#1, 11#0 becomes 0#1. However, when the head-bit is undetermined (its value is '#'), new individuals emerge . These new templates can only be expressed thanks to the use of two complementary variables. For instance, #01#0 becomes after transcription either 01#0 if the head-bit is considered as a '0', or 10#1 if it represents a '1'. This example shows that the introduction of a head-bit combined with the use of a binary alphabet allows an implicit manipulation of individuals as XX'#X where X' denotes the binary complement of X. Indeed, XX'#X represents either the condition 01#0 or 10#1. They can both be represented by the string #01#0. Since, X and X' are binary complements, they can also be represented the string #10#1.

The introduction of doubles in the population allows the expression of generalizations, thanks to the emergence of new hyperplanes. The figure 1represents classical 2-bits conditions ; there are only seven individuals possible inducing the four partitions: ##, #d, d# and dd, where the character 'd' means "defined bit". After the transcription of the doubles corresponding to this example, we note the appearance of a fifth partition corresponding to the order-2 doubles which head-bit is undetermined. Indeed the condition #00 and #11 are transcripted in XX, while #01 and #10 are both decoded as X'X .

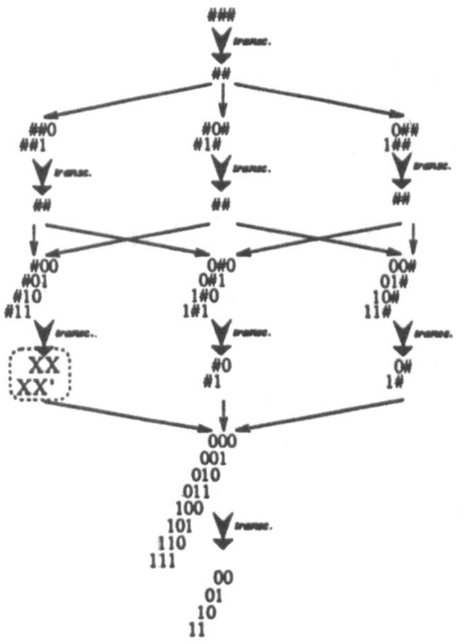

Fig. 1: Emergence of new schemata.

Using implicit variables, we treat the classifiers at an upper level of abstraction. New partitions are explored during the resolution. Moreover whatever the complexity of the problem may be, only one head-bit is needed because strings with variables represent all the hyperplanes missing in the traditional approach. Doubles have other interesting features. The first one is that though the search space is larger, the use of such classifiers allows an improvement in the learning. The other one is that the partitions manipulated by the AG are more likely to remain stable after the loss of one element in each pair of doubles.

For instance, consider the subset of conditions $H = \{00, 11\}$ (or a subset that includes these individuals). This subset is unstable: the result of crossover of distinct members of H does not belong to H (These aspects of stability/instability are developed in [8]). Such a subset may be devastated by crossover. We can observe that with DCS, the subset H becames $H' = \{000, 111, 011, 100\}$. In each pair of doubles, one disapears in order to maintain stability. For instance, if 000 and 100 are lost, the new subset $H'' = \{111, 011\}$ becames stable since crossover between 111 and 011 always gives new offsprings belonging to H''. Thus, thanks to our new encoding, we can avoid the destruction of information.

In the next section, we present experimental results which demonstrate that doubles can be successfully used in a CS. The messages sent by the environment have the classic aspect, while classifiers must have an added bit in their condition to play the head-bit role. Thus they have the following form: <*head − bit*><*condition*>:<*action*>. Let us note that the head-bit influence is only on the condition part of the classifier, so the action will never be altered. For instance, the message 00110 is matched by the classifiers 00#1#0:10, 11100#:10, ##0#10:10 or #11#01:10. Whatever the classifier activated, the action will be 10.

4 Empirical Study

In order to illustrate our proposition, we applied DCS to the learning of two boolean functions: the XOR problem, and the multiplexer function. We limit ourselves to a *stimulus-response* CS.

4.1 The XOR function

The XOR problem, famous in neural nets domain, consists in learning the boolean function: $f(0,0) = 0$, $f(0,1) = 1$, $f(1,0) = 1$ and $f(1,1) = 0$. In a standard CS, the condition part represents the arguments x and y of the function f, and the action part represents $f(x,y)$. So two bits are needed for the condition part, and one for the action part. For instance, possible classifiers are 00:0, 0#:1. The first one translate $f(0,0) = 0$, and the second one is interpreted as $f(0,0) = 1$ and $f(0,1) = 1$. Of course, in DCS we must add the head-bit.

In both cases, the initial population is randomly created with 22 classifiers, the GA is activated every 2500 iterations. A graph of correct answer proportion versus iteration is presented in figure 2. It presents a moving average over the last 50 iterations as well as a performance average over al time. One can note that our system behaves in a better way than a standard CS: approximately beyond cycle 2000, the system is able to give the right answer. When the standard CS is concerned, the final population contains the four totally specialized classifiers—i.e. 00:0, 01:1, 10:1, 11:0— from the 5000th cycle. The system has no other solution than learning the XOR solution by heart because it cannot generalized without making any error. While with DCS, only two classifiers are needed to assure the convergence: #00:0 and #01:1. The undetermined head-bit leads to the use of variables (XX:0 and XX':1 cf 3.2) and so the use of less specialized classifiers. The learning is realized at an upper level of abstraction.

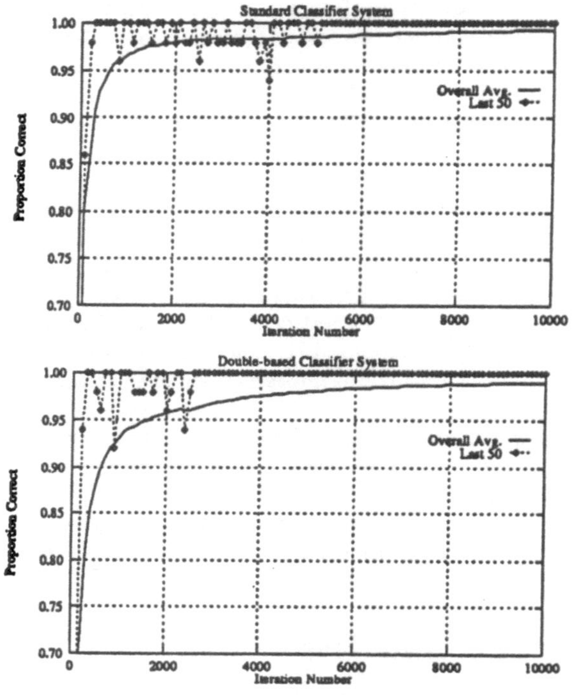

Fig. 2: The XOR function: Proportion Correct

4.2 The eleven-line multiplexer

The multiplexer function to be learned is an extension of the one Goldberg described in [3]. Eleven lines come into the multiplexer, the signals of the three first lines are decoded as an unsigned binary integer, the address value thus obtained, indicates which of the eight last lines is chosen. For instance, if the address signals are 101 (i.e. 5), the signal on line 5 is passed through to the multiplexer output.

In this case, two parts compose the condition of the classifier: three bits for the three address lines, and eight bits for the data. A third part is added in DCS for the head-bit. The action part represents the system's answer.

The initial population, in any case, is randomly created with 300 classifiers. The GA is invoked every 5000 iterations, and 20 percent of the current population undergoes reproduction, crossover, mutation and replacement following the De Jong's crowding model [2]. Thus, in 50,000 iterations, 600 new offspings, the equivalent of two entire populations, are generated.

The performance average of the two systems increases from the beginning, but the one of DCS evolves faster (Cf. figure 3). Indeed, since the cycle 35,000 the average is above 80%, against 70% for the standard classifier system. After 50,000 iterations the correct answer proportion reaches 83% with our system, while it is only near 72% with the standard one.

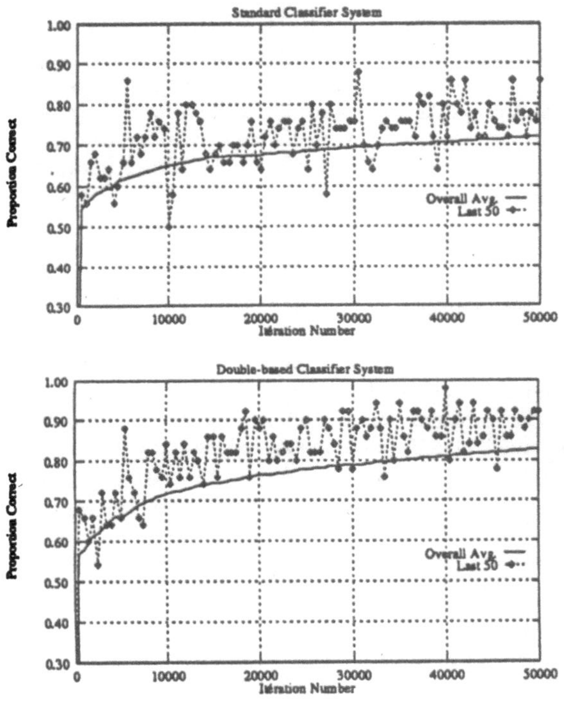

Fig. 3: The multiplexer function: Proportion Correct

5 Conclusions

In this paper we explored the power of an original CS. The presence of a head-bit in the binary strings leads to significant improvements. This simple and powerful mechanism is worth noticing. Of course, our experimental study is not exhaustive but the results are attractive and full of promise. Indeed, we propose a new CS allowing a more succinct representation of the solution space than the one obtained with a classical system, increasing the GA's ability. The implicit use of a couple of complementary variables in strings permits the potential expression of all the hyperplanes of the search space, without adding new basic symbol in the definition alphabet: the classifier representation keeps its simplicity, the genetic operators do not need to be altered.

For the moment, only the conditions are concerned by the variables. The next step in the evolution of DCS deals with the use of variables in the action part of the rules. Such a variable would take its value in the condition if it was assigned there, or would be randomly instancied if it does not appear before. This would even more increase the expressiveness of classifiers.

References

[1] L. D. Davis. *Handbook of Genetics algorithms*. VNR Computer Library, 1991.

[2] K. A. De Jong. *An analysis of the behavior of a class of genetic adaptive systems*. PhD thesis, University of Michigan, 1975.

[3] D. E. Goldberg. *Genetic algorithms in search, optimization, and machine learning*. Reading, MA : Addison-Wesley, 1989.

[4] J. H. Holland. Escaping brittleness : The possibilities of general purpose learning algorithms applied to parallel rule-based systems. In R.S. Michalski, J.G. Carbonell, and T.M. Mitchell, editors, *Machine Learning II*, pages 593–623. Morgan Kaufmann, 1986.

[5] James R. Levenick. Inserting introns improves genetic algorithms success rate: Taking a cue from biology. In R. K. Belew and L. B. Booker, editors, *Genetic algorithms and their applications : Proceedings of the Fourth International Conference on Genetic Algorithms*, pages 123–127, San Mateo, CA, 1991. Morgan Kaufmann.

[6] R. L. Riolo. Bucket brigade performance: 1. long sequences of classifiers, 2. default hierarchies. In J. J. Grefenstette, editor, *Genetic algorithms and their applications : Proceedings of the Second International Conference on Genetic Algorithms*, pages 184–201, Hillsdale, NJ, 1987. Lawrence Erlbaum Associates.

[7] L. Shu and J. Schaeffer. Vcs: Variable classifier systems. In J. D. Schaffer, editor, *Genetic algorithms and their applications : Proceedings of the Third International Conference on Genetic Algorithms*, pages 334–339, San Mateo, CA, 1989. Morgan Kaufmann.

[8] M. Vose. Generalizing the notion of schema in genetic algorithms. *Artificial Intelligence*, 50:385–396, 1991.

Application of neural networks for classification of temperature distribution patterns

Henrik Saxén*, Abhay Bulsari*, Leif Karilainen*,‡ and Kalevi Raipala‡

* Dept. Chem. Eng., Åbo Akademi, Biskopsg. 8, FIN–20500 Åbo, Finland
‡ Fundia, Koverhar Works, FIN–10820 Lappvik, Finland

Abstract

Classification of spatially distributed measurements in industrial processes is an important but difficult task. Interpretation of the signals from horizontal probes in blast furnaces is a typical example. A neural network-based classification of such temperature measurements is described in this paper. Using temperature profiles first classified by the team, training and test sets were formed, and feedforward neural networks of different size were trained and evaluated on the material. The network size was found to clearly affect the quality of the classifications. By a judicious choice of the number of hidden nodes, a reliable neural classifier was obtained, which was incorporated in a supervision and control system for a Finnish blast furnace.

Introduction

In many industrial processes, measurements of variables which show spatial distribution are available. Typical examples are temperature measurements or chemical analyses sampled at different locations in reactors. Since the measurements reflect the state of the process, they provide valuable information for process analysis and control purposes, but the interpretation of the patterns is often difficult. In this work, a set of measurements from an ironmaking blast furnace is studied. The furnace is an extremely complex industrial process, the internal conditions of which are characterised by multi-phase flows (solid, liquids, and gas), high temperatures (> 2000°C) and pressure. Often, the only internal measurements available are radial gas temperature and/or composition distributions obtained from horizontal probes in the shaft. These measurements are usually interpreted manually, because of complicated interrelations between the variables in the process. Another common approach is to compute gas flow indices [1, 2], but also statistical methods, such as discriminant analysis, have been used to classify probe data [3]. Recently, this problem has also been tackled by neural networks [4, 5].

The blast furnace studied in this work is equipped with a horizontal above-burden probe, where gas temperatures are measured frequently. In our approach, feedforward neural networks were used to classify temperature profiles [6] using patterns classified manually as teacher signals. The resulting neural classifier has been incorporated in a blast furnace supervision and control system [7, 8].

The data set

We consider a set of 11 temperature measurements, T_i, $i = 1, \ldots, 11$, from a horizontal probe located above the burden surface (Figure 1) in the blast furnace of Fundia Koverhar, Lappohja, Finland. The dimension of the problem was reduced — both to simplify the "manual" classification and to reduce the numerical effort in training the neural networks — by making the profiles symmetrical with respect to the furnace central axis, using the transformation $\tilde{T}_i = (T_i + T_{12-i})/2$; $i = 1, \ldots, 6$. Moreover, 3-hour mean values of the measurements were used to filter out noise. The resulting profiles were classified

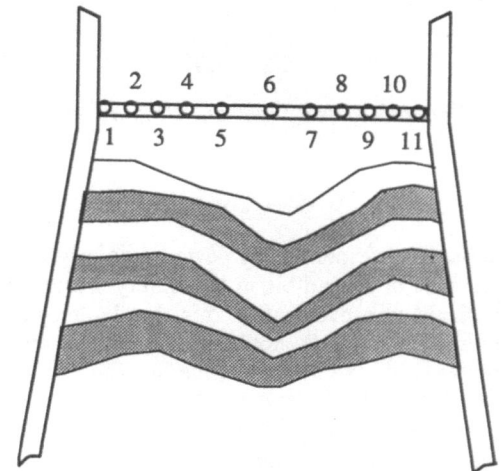

Fig. 1 The probe with its eleven measurement points.

by the authors [6] into six typical temperature classes: 1) flat profiles, 2) inverse v- or u-shape, 3) w-shape (with some peripheral gas flow), 4) W-shape (with considerable peripheral gas flow), 5) Sharp central flow (with some wall flow), and 6) Λ-shape with little peripheral flow.

The appearances of the above classes were depicted graphically, and a large number of temperature patterns taken from four different periods of normal operation of the furnace was classified. For each temperature pattern, i, the six classes were devised degrees of membership, $t_{i,j}, j = 1, \ldots, 6$, satisfying the constraint $\sum_j t_{i,j} = 1$. However, when the resulting classifications were analysed, several inconsistencies were found, which were caused by the fact that some of the six classes were strongly related. Especially, remarkable subjectivity was involved in classifications of patterns which were considered as fuzzy combinations of several classes. It was therefore considered desirable to rearrange the classes, so that neighbouring ones would show similarities, *e.g*, as in Fig. 2. It was decided that each pattern in the training and test sets be classified either as a *crisp* class ($t_j = 1,\ t_i = 0, \forall i \neq j$) or as a combination of two (vertical or horizontal) neighbour classes. This measure was found to simplify the work and it yielded improved consistency in the classification.

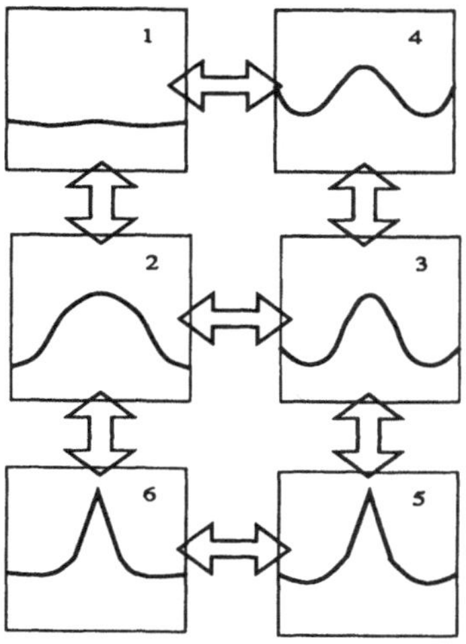

Fig. 2 Spatial organization of the classes. Arrows denote feasible transitions in the human classification.

After classifying 476 temperature profiles using the "rules" outlined above, 200 instances were selected for training and 276 for testing of the neural networks. Table 1 presents some information about the $M_{tr} = 200$ training and $M_{te} = 276$ test patterns.

Table 1 Statistics on the data sets.

Class j	$t_j \leq \frac{1}{4}$	$\frac{1}{4} < t_j \leq \frac{3}{4}$	$t_j > \frac{3}{4}$	$t_j > 0$
Train				
1	0	17	8	25
2	3	23	13	39
3	4	42	40	86
4	1	28	11	40
5	0	21	36	57
6	1	13	20	34
Test				
1	0	6	1	7
2	2	9	1	12
3	17	59	48	124
4	0	8	0	8
5	0	58	148	206
6	1	10	3	14

The neural classification

The reason for tackling the classification problem at hand by neural networks was that the relations between the inputs and the classifications are expected to be both nonlinear and complex. In the networks of this study, except for one in a reference run, both hidden and output nodes used sigmoidal activation: The outputs were squashed through nonlinearities since the memberships in the classes should obey $t_j \in (0,1)$. The network weights were estimated by the Marquardt method [9,10], which has proved to be efficient and reliable for training feedforward networks [11].

An important problem associated with the use of feedforward neural networks is how to determine the number of hidden nodes. Even though several heuristic or statistical methods have been suggested, the application of such tests on the material in this work was problematic, since the (human) classification of the patterns was known to be quite subjective, and sometimes inconsistent. Thus, it was desired that the networks show some major residuals after training, since learning erroneous classes would result in poor generalisation performance.

Various feedforward networks with one or two hidden layers were trained. Table 2 shows the square sums, E, and root mean square errors (rms), $\sqrt{E/M}$, on the training and test sets for some of the networks studied. As expected, the errors on the test set are considerable higher than the training errors.

Figure 3 shows the square sum of the networks reported in Table 2 versus number of parameters (weights); the diamond refers to the linear model, squares to networks with one hidden layer and crosses to networks with two hidden layers.

Table 2 Square sums and rms errors on the training and test sets. "L" denotes linear output activation.

Network	E_{tr} (rms)	E_{te} (rms)
$(6,6)_L$	76.84 (0.620)	103.91 (0.614)
(6,6)	57.65 (0.537)	86.50 (0.560)
(6,5,6)	25.01 (0.354)	74.23 (0.519)
(6,8,6)	18.22 (0.302)	72.44 (0.512)
(6,10,6)	14.92 (0.273)	73.73 (0.517)
(6,12,6)	10.99 (0.234)	103.19 (0.611)
(6,15,6)	6.79 (0.184)	147.63 (0.731)
(6,18,6)	6.58 (0.181)	113.49 (0.641)
(6,5,5,6)	17.48 (0.296)	71.31 (0.508)
(6,7,7,6)	11.49 (0.240)	118.31 (0.655)

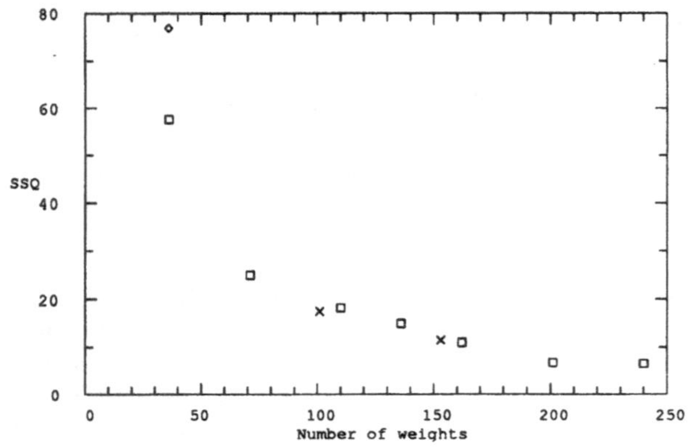

Fig. 3 Square sum vs. number of weights. ◇: linear, □: 1-hidden-layer, ×: 2-hidden-layer networks.

Based on these findings, the (6,8,6), (6,10,6) and (6,12,6) as well as (6,5,5,6) and (6,7,7,6) networks were considered more seriously. The smaller networks, where the number of hidden nodes is less than or equal to the number of outputs (classes), $(6,N\leq 6,6)$, did not seem to have enough capacity to express the underlying "classification rules", while the larger networks, $(6,N>12,6)$, clearly tended to "memorise" training data including all exceptions and errors, and did not generalise well.

A deeper analysis of the performance of the models on the test set revealed that the (6,8,6) and (6,10,6) networks showed several common large errors, while the number of additional large errors for the (6,12,6) network was considerable. The common residuals can possibly be attributed to errors in the human classification, while a majority of the latter errors arose from too sharp classification boundaries produced by the (6,12,6) network. Based on these findings, the (6,8,6) and (6,10,6) were considered more seriously. By studying the decision boundaries (between the classes) it was found that (6,8,6) tended to give multiple classes.

As for the two-hidden layer networks, (6,5,5,6) showed slightly diffuse class boundaries, probably because of insufficient capacity, while (6,7,7,6) had problems to recognize some classes (*e.g., Class 6*) and also suggested multiple classes. The (6,10,6) network yielded reasonable classification of the test set patterns, smooth transitions and less class overlapping. Also, this network produced outputs which usually summed up to unity, and for most patterns there was only one non-zero output. Based on these observations, the (6,10,6) was selected for further development and application. An illustration of the classifier on a set of test problems is given in ref. [6].

In order to improve the network's generalisation properties, the inputs were normalised by subtracting from each of the six temperatures the minimum temperature of the pattern. This makes the classification independent of the level of the profile. Further, since some tests on independent material suggested that the network had problems to classify patterns where the temperatures at the wall were low, about 30 extra patterns were included in the training set, and the (6,10,6) network was retrained. It was found that the network could incorporate the added patterns without any increase in classification error.

The application

A user-friendly tool for classification of temperature distributions in the blast furnace was developed based on the neural classifier, using a graphics software package [12]. The tool, which has been implemented as a module in a knowledge-based supervision and control system for the Koverhar blast furnace [7,8], automatically classifies the probe temperature distributions every 10th minute. Figure 4 shows the appearance of the interface for a situation where the gas profile has gradually changed from *Class 6* through *Class 2* to *Class 3*. In the right part of the figure, the colours of the stereotype patters indicate the classification of the latest profile, while the adjacent bars show the time evolution of the membership functions. The large panel to the left shows the measured temperature profile and the profile after the transformation (which makes it symmetrical).

Recent changes in the profiles can be examined by scrolling backwards in time, focusing on certain events, such as periods where charging control actions (which affect the gas distribution) have been taken. Moreover, some of the measurements on the probe can be set invalid, if considered unreliable.

These features have proved to be important for the acceptance of the classifier in the practical operation of the blast furnace. The feedback from the personnel is also important: the operators have already suggested new desired features, such as possibilities to present the classification as verbal statements including recommendation of possible control actions.

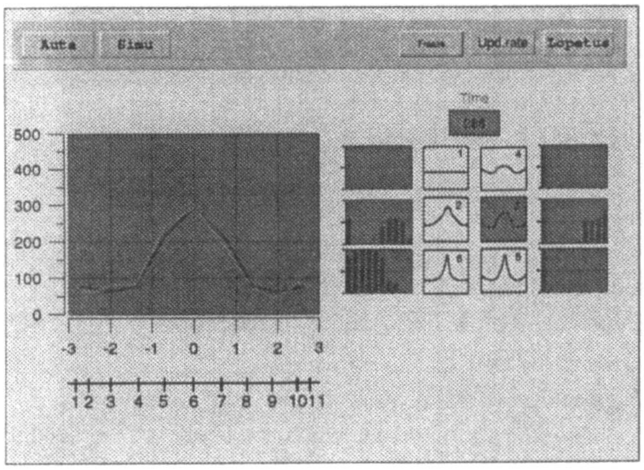

Fig. 4 The graphical interface of the neural classifier.

Concluding remarks

A method for classification of gas temperatures in a blast furnace, measured by a horizontal probe, has been presented. A large set of distributions was first transformed by making the profiles symmetric about the furnace center by averaging, and then classified into six possible classes based on process know-how. It was found that a network with about 10 hidden nodes could mimic the human way of classifying the temperature patterns, still being more consistent. A prototype of the classifier has been implemented as a part of a supervision and control system at a Finnish iron works. The performance of the classifier is presently studied, and it is expected that the tool will be equipped with a mechanism for presenting verbal evaluations of the gas distribution and changes therein to the furnace operators.

Acknowledgements

This work has been carried out within the Finnish research project SULA 2 — Energy in steel and base metal production. The financial support by the research programme is gratefully acknowledged.

References

1. Poveromo, J.J., Milanoski, S.V. and Dwelly, M.J., Proc. *42th Ironmaking Conference*, Atlanta, USA, pp. 491–501, ISS-AIME, 1983.

2. Nishio, H., Ariyama,T., Saito, H., Niwa, Y. and Yoshida, H., Proc. *41th Ironmaking Conference*, Pittsburgh, USA, pp. 174–181, ISS-AIME, 1982.

3. Nagai, M., Saino, M., Tamura, S., Kaneka, K., Okabe, K., Fukutake, K. and Taguchi, S., Proc. *36th Ironmaking Conference*, ISS-AIME, 1977.

4. Hirata, T., Yamamura, K., Morimoto, S. and Takada, H., Proc. *6th International Iron and Steel Congress*, Nagoya, Japan, pp. 23–27, ISIJ 1990.

5. Otsuka, Y., Matsuda, K., Hanaoka, K., Kadoguchi, K. and Maki, T., *Kobe Steel Engineering Reports* **41** (4) 20–23, 1991.

6. Bulsari, A.B. and Saxén, H., Proc. *Tenth International Conference on Systems Engineering (ICSE94)*, Coventry, England, pp. 157–163, 1994.

7. Karilainen, L., Saxén, H., Raipala, K. and Hinnelä, J., Proc. *53rd Ironmaking Conference*, Chicago, USA, pp. 365–371, 1994.

8. Karilainen, L. and Saxén, H., Preprint *2nd IFAC Workshop on Computer Software Structures Integrating AI/KBS Systems in Process Control*, Lund, Sweden, pp. 183–187, 1994.

9. Press, W.H., Flannery, B.P., Teukolsky, S.A. and Vetterling, W.T., *Numerical Recipes*, Cambridge University Press, Cambridge, USA, 1986.

10. Battiti, R., *Neural Computation* **4** (1992) 141–166.

11. Bulsari, A.B. and Saxén, H., *Neurocomputing* **3** (1991) 125–133.

12. *SL-GMS Reference Manual* (Version 4.0), Sherrill-Lubinski Corporation, USA, 1991.

COMBINATION OF GENETIC ALGORITHMS AND CLP IN THE VEHICLE-FLEET SCHEDULING PROBLEM

E. Stefanitsis, N. Christodoulou & J. Psarras

Management Systems Unit, Department of Electrical and Computer Engineering, National Technical University of Athens, 42 Patission Str., GR 10682 Athens, GREECE

Abstract: Constraint Logic Programming (CLP) is a technique which presents several advantages in dealing with combinatorial optimization problems as it combines the declarative aspects of Logic Programming with the efficiency of Constraint Techniques. However, the attempts to solve the Vehicle Scheduling Problem (VSP) through the application of CLP[1] has not been successful, due to the huge complexity of the problem and the scarcity of constraints able to drive the solution.

This paper presents a new approach to the problem which attempts to combine the advantages of CLP with the benefits of Genetic Algorithms in order to obtain satisfactory results with respect to execution time and solution quality.

1. Introduction

One of the most commonly occurring problems in transport management is Vehicle-Fleet Scheduling (VSP), which can be described as the design of minimum cost routes originating and terminating at a central depot, for a fleet of vehicles that service a set of customers with known demands.

The VSP is an NP-hard problem which means that it can not be solved optimally in polynomial time. It was initially introduced by Dantzig and Ramser[2] in 1959 as the truck dispatching problem. In 1964 Clarke and Wright[3] solved the problem using a clever heuristic based on the savings algorithm.

Since then, several heuristic algorithms have been developed for solving specific variations of the problem. But each one of the algorithms used was applicable only to a specific variation; a slight difference in the structure of the problem rendered the algorithm inefficient. This urged the researchers towards the creation of generalized schemes versatile enough to handle more than one variations of the problem.[4]

The first attempts, based on algorithmic languages (procedural programming) did not prove very efficient. On the contrary, descriptive programming was considered more suitable and further attempts were devoted to this methodology.

2. An Overview of the Solution Approach

In this paper a new approach facing the problem is introduced. It draws upon the combination of the advantages of Constraint Logic Programming (CLP) with the benefits of Genetic Algorithms in order to obtain satisfactory results with respect to execution time. CLP is used in generating some initial feasible solutions and then for checking every intermediate solution obtained from Genetic Algorithms so that it is in accordance with the constraints of the problem. Genetic Algorithms are implemented for the minimization of the cost of the initial solutions.

In order to keep the computational time as low as possible, and without much loss in cost with respect to the optimal solution, several operators were tested, such as exchanging points or parts of routes.

The previously referred operators are normally applied in problems consisting of only one route (Traveling Salesman Problem - TSP) and not of many (VSP), Thus, in order to apply these algorithms to VSP a formula corresponding each VSP problem to a TSP and vice versa was developed.

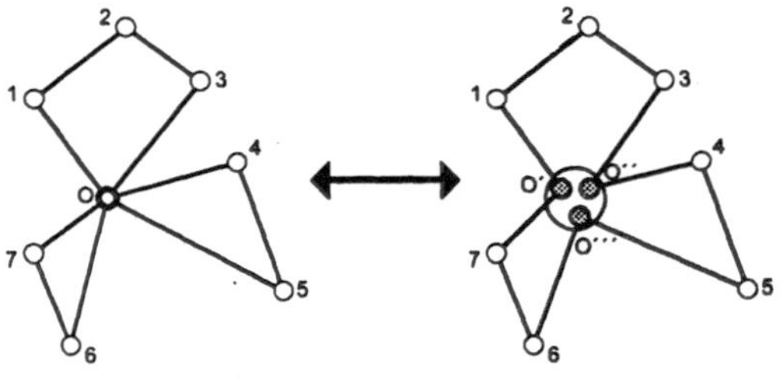

Figure 1: Schematic transformation of VSP to TSP

3. Transformation of VSP to TSP

A method, that could provide only one route (TSP) from many routes (VSP), would be to substitute the initial depot(s) of the problem e.g. the node 0 with multiple artificial nodes 0, with infinite distance from each other, while the distance between every one of them and the other nodes of the problem is equal to the real distance of the nodes from node 0. In other words, if in a vehicle scheduling problem all the routes are combined in one route, this route visits the depot (node 0) as many times as the total number of the trucks used[5]. The whole procedure is illustrated in Fig. 1.

Another issue that should be taken into serious consideration is the capacity of the trucks. Hence, the load of each truck resulting as the sum of all the loads between two artificial nodes-depots, should in noneway exceed the maximum allowable capacity of the truck.

4. Genetic Algorithms

Before implementing the genetic algorithm, a procedure that generates few initial feasible solutions in accordance with the constraints of the problem is applied. This procedure is identical for any operator tested, and produces the initial population that will be used thereafter by the operators in a later stage.

4.1. Initial Population

In order to obtain the initial population, the lower bound N of the number of the trucks is calculated. Then, the procedure attempts to construct a solution with as many routes as the already calculated value of N, by setting successively nodes to be served and interposing node-artificial depot, only if the capacity of a truck is violated. Simultaneously, all the rest of the problem constraints which have to be satisfied are taken into consideration.
If either one of these constraints, or the capacity of a truck are violated and there is no other possibility of rearrangement of the nodes, the whole procedure is repeated increasing the value of N by 1, until feasible solutions are obtained.

4.2. Operators

Below we describe briefly these operators that proved to be more efficient.

4.2.1. Re-insertion Given one tour and two nodes i, j, we insert node i between nodes j and j+1 and then we place node i-1 before node i+1. See Fig. 2a.

4.2.2. Nodes-2-exchange Given one tour and two nodes i, j, we exchange the nodes, i.e. we insert i in the position of j and vice versa. See Fig. 2b.

4.2.3. Sections-2-exchange Given one tour and two nodes i, j, we dissect the initial solution between points i, i+1 and j, j+1 and then we connect point i with j and point i+1 with point j+1.[6][7]. See Fig. 2c.

4.3. GA implementation

Below the 3 different approaches to the problem are described:

4.3.1. The standard approach We applied the above mentioned operators to the initial population of m solutions k times. We used random values for the variables i, j. From the k x m set of the produced solutions we kept only the best m as the initial population for the next step.

4.3.2. The improved approach This one is the same with the above approach, but the solutions with worse cost from their parent are rejected. There is also a variation where solutions with cost more than this of their parent, but less than a prespecified limit defined by their parent cost can be kept

4.3.3. Combination with local search This approach combines GA with local search. This can be easily explained as for every solution of the initial population, the operators were applied many times for every value of i and j and the best of the obtained solutions are kept as initial population for the next step propagation.
In all the approaches above and for every operator, all problem constraints are a-priori imposed using in the best way CLP and rejecting all unsuitable solutions. Simultaneously the developed in CLP schema remains accessible to most modifications and in that way, any new variation or constraint of the problem formulation can be easily handled.

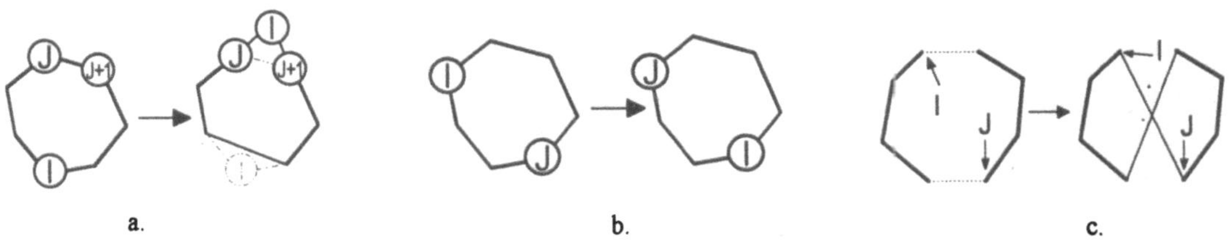

Figure 2: The three different operators tested

5. Results

All the above methods and operators were tested in various examples. The results given below are all mean values obtained by more than one executions for the same problem, in order to be more representative.
First we compare the different ways of implementing the genetic algorithm using the same operator and fixed population. The results are given on Table 1. The values of this Table were acquired after 10 executions for each version of each problem.
Then, Table 2 provides comparative results for problems with different size with the same operator and implementation of genetic algorithm concerning the population size this time. Values were taken as mean values after 5 executions.
Finally, Table 3 presents comparative results concerning the different operators. These results are separated in 2 cases according to the number of trucks in the solution.

6. Conclusions

Finally, an evaluation of the implementation of the genetic algorithms presented, is performed. More specifically, one can conclude that no single genetic algorithm can provide the best results for all problems. Various factors determine the most appropriate method for each problem, such as: size and type of the problem, required level of accuracy and execution time.
From the presented tables, many conclusions can be drawn. Firstly is the lack of existence of any combination of operators, population and genetic algorithm, that can provide us with the best results for all problems.
Concerning the genetic algorithm implementation, the case in which we accept children with cost 5% worse than their parents, seems to be the best. Our attempt to combine genetic algorithms with local search did not prove to be effective enough. This came inevitably as the good results are obtained after extremely long execution. Therefore this method should be rejected.

Table 1: Cost and time results obtained after different ways of implementing genetic algorithms

10 executions	Standard approach	better than parents	better than parents *1.05	better than parents *1.1	combination with local search
10 points	190/6sec	175/8sec	170/8sec	170/8sec	168/5sec
20 points	770/70sec	630/100sec	560/70sec	590/75sec	577/50sec
25 points	850/80sec	718/110sec	672/100sec	675/110sec	680/120sec
30 points	470/102sec	400/120sec	380/100sec	384/90sec	390/170sec
40 points	680/130sec	512/180sec	475/150sec	492/165sec	480/320sec
50 points	808/250sec	610/510sec	514/400sec	550/460sec	531/1100sec
60 points	1302/380sec	754/625sec	580/508sec	612/540sec	620/1210sec

Table 2: Cost and Time results obtained by different population sizes

5 executions	n=10	n=30	n=90
10 points	277/2sec	283/9sec	290/9sec
20 points	520/30sec	525/20sec	522/42sec
30 points	789/100sec	713/90sec	708/140sec
40 points	866/150sec	800/180sec	810/280sec
50 points	858/450sec	750/700sec	742/890sec
60 points	970/700sec	890/820sec	876/1200sec

Table 3 Cost and Time results obtained by different operators tested

	Few Routes			Many Routes		
No of Points	Re-insertion	Nodes-2-exchange	Section-2-exchange	Re-insertion	Nodes-2-exchange	Section-2-exchange
15	410/5sec	455/5sec	408/7sec	1255/7sec	1206/7sec	1240/12sec
25	739/34sec	790/55sec	724/90sec	1060/111sec	1145/70sec	1124/220sec
35	815/120sec	973/100sec	798/190sec	1280/241sec	1342/110sec	1245/430sec
45	890/250sec	1062/220sec	870/420sec	1381/400sec	1286/290sec	1358/800sec
55	830/410sec	982/380sec	817/580sec	734/652sec	744/495sec	745/1005sec

As far as the population size is concerned, we observed that the production of better results, as an outcome of the increase of the population, comes accordingly. It should be noted however, that after a particular point and onwards the improvement obtained, is little compared to the raise of the execution time. After many experiments in many problems, we noticed that an initial population equal to 30, which remains fixed during the procedure, can give the best results in problems with points lying between 20 and 150.

Finally, concerning the different operators tested, we have to admit that none of them could provide the best results for any problem category. Although the "section-2-exchange" method produces good cost results, it consumes much execution time.

Moreover, even though the "nodes-2-exchange" operator is faster than the other operators for all problem categories, the efficiency of each method varies according to the number of routes that appear in the final solution. According to the tables shown above, for cases with many routes in the final solution, the "nodes-2-exchange" is faster and provides results with comparable cost. For few routes, this gain in time when compared with the re-insertion operator produces loss in cost.

In the overall, the combination of CLP and Genetic Algorithms proved to be an applicable and promising approach towards a generic solution strategy for VSP.

The challenge, that also indicates future work directions, is to integrate automatic solution strategy selection to a tool that identifies the inherent semantics at specific problems, builds a proper set of solving algorithms and applies it to the best of the results quality.

References

1. Christodoulou, N., Stefanitsis, E., Kaltsas, E., & Assimakopoulos, V., "A Constraint Logic Programming Approach to the Vehicle - Fleet Scheduling Problem", Proceedings of the Second International Conference on the Practical Application of Prolog, , London, p137-149 (1994).

2.Dantzig, G.B., & Ramser, J.H., "The Truck dispatching Problem", Management Science 6, 81 (1959).

3. Clarke, G., & Wright, J.W., "Scheduling of vehicles from a central depot to a number of delivery points", Op. Res. 12, 568 (1964)

4. Christodoulou, N., "Implementation of Constraint Logic Programming (CLP) for the solution of the Vehicle Scheduling Problem (VSP), Phd Thesis, Athens, National Technical University of Athens (1994).

5. N., Christofides, and S., Eilon, "An algorithm for the vehicle dispatching problem", Op. Res. Quart. 20, 318 (1969)

6. Lin, S., and Kernigham, B.W., "An effective heuristic algorithm for the travelling salesman problem", Op. Res. 21, 498 (1973)

7.Lin, S., "Computer Solutions to the travelling salesman problem", Bel System Technical Journal 44, 2245 (1965)

MINIMUM COST TOPOLOGY OPTIMISATION OF THE COST 239 EUROPEAN OPTICAL NETWORK

M.C.Sinclair

Dept. of Electronic Systems Engineering, University of Essex,
Wivenhoe Park, Colchester, C04 3SQ, UK

A genetic algorithm is presented for the minimum-cost topology optimisation of the COST 239 European Optical Network (EON). The algorithm makes use of real traffic data where possible, and produces a network amongst twenty given nodes with two-shortest-node-disjoint routing between node pairs. Both routes are fully resourced, guaranteeing the network will survive a single component failure. The results are compared with the earlier 'hand-crafted' EON topology.

The Background

The work described in this paper was undertaken as part of COST Action 239. The COST framework (European Cooperation in the field of Scientific and Technical Research) is run by the European Commission, aiming to promote pre-competitive R&D cooperation between industry, universities and national research centres. COST 239, "Ultra-high Capacity Optical Transmission Networks", is studying the feasibility of a transparent optical overlay network capable of carrying all the international traffic between the main centres of Europe.

At the start of the project, twenty such centres were identified, each acting as the gateway for all (or half) of the European international traffic for their country.

The initial network design and analysis of a proposed European Optical Network based on those nodes was presented in [1,2]. The initial topology design was carried out by making what were presumed to be reasonable assumptions about the possible traffic distribution, given the node populations, and incorporating suitable structures to enhance reliability. The topological design was then analysed, using existing software [3,4], and incorporating the PD (population-distance) model for node-to-node traffic, two-shortest-link-disjoint-path routing and a simple model for link availabilities. This resulted in comparative link capacities, which were scaled to realistic levels for the immediate, medium and long term.

Subsequently, an improved model for European international telephony traffic, the PFD (population-factor-distance) model, has been developed [5].

The Problem

The problem this paper seeks to address is that of producing a minimum-cost topology for the COST 239 European Optical Network (EON), given the existing choice of twenty nodes. The traffic applied is the real traffic data (scaled to the long-term level estimated by COST 239, and completed, where not available, using the PFD model [5]); two-shortest-node-disjoint-path routing is used between node pairs; and a reliability constraint is employed. The latter is to ensure that there are two fully-resourced node-disjoint routes between node pairs, thereby guaranteeing that the network will survive the failure of any single component (node or link).

The size of the problem space is simply the number of different topologies. If only the nine central nodes are considered, then the space is 6.9×10^{10}, but for the full EON, this rises to 1.6×10^{57}.

Cost Models

To determine the cost of a given topology, separate models for both links and nodes are required.

Link First, the two-shortest-node-disjoint routes are determined for all node pairs. The 'length' of each path is taken to be the sum of the contributing link weights, with the weight of the bi-directional link between nodes i and j given by:

$$W_{ij} = 0.5N_i + L_{ij} + 0.5N_j \qquad (1)$$

where N_i and N_j are the Node Effective Distances of nodes i and j respectively (see below) and L_{ij} is the length of link (i,j) in km.

Then, the link carried traffic is determined for each link by summing the contributions from all the primary and alternative routes that make use of it. The alternative routes are fully-resourced, and thus are assumed to carry the same traffic as the corresponding primary routes i.e. the traffic they would be required to carry if their primary route failed.

The capacity of link (i,j) is taken to be:

$$V_{ij} = \text{ceil}_{2.5}[1.4T_{ij}] \tag{2}$$

where T_{ij} is the carried traffic in Gbit/s on the link, and $\text{ceil}_{2.5}$ rounds its argument up to the nearest 2.5 - the assumed granularity of the transmission links. The factor of 1.4 is to allow for stochastic effects in the traffic.

The cost of link (i,j) is then taken to be:

$$C_{ij} = L_{ij}V_{ij} \tag{3}$$

<u>Node</u> Node Effective Distance was used as a way of representing the cost of nodes in an optical network in equivalent distance terms. It can be regarded as the effective distance added to a path as a result of traversing a node. By including it in link weights (Equation 1) for the calculation of shortest paths, path weights reflect the cost of the nodes traversed (and half the costs of the end nodes). As a result, a longer geographical path may have a lower weight if it traverses fewer nodes, thereby reflecting the relatively high costs of optical switching. The Node Effective Distance of node i was taken to be:

$$N_i = E + n_i F \tag{4}$$

where n is the degree of node i, i.e. the number of bi-directional links attached to it. The constants E and F were taken to be 200km and 100km respectively. Node Effective Distance thus increases as the switch grows more complex.

Node capacity is simply the sum of the capacities of all the attached links, and node cost was taken to be:

$$C_i = 0.5N_iV_i \tag{5}$$

where V_i is the capacity of node i in Gbit/s. The cost is thus derived as if the node was a star of links, each of half the Node Effective Length, and each having the same capacity as the node itself.

<u>Network</u> The network cost is then taken to be the sum of the costs of all the individual links and nodes comprising the network.

<u>Initial Topology</u> Using the above cost models, the initial EON topology [1,2] has a cost of 7.22 million (in kmGbit/s), comprising 3.27M for the links and 3.95M for the nodes (see Fig. 1). It does not, however, meet the reliability constraint, as Luxembourg (LUX) has only one link, and cannot therefore support two node-disjoint routes.

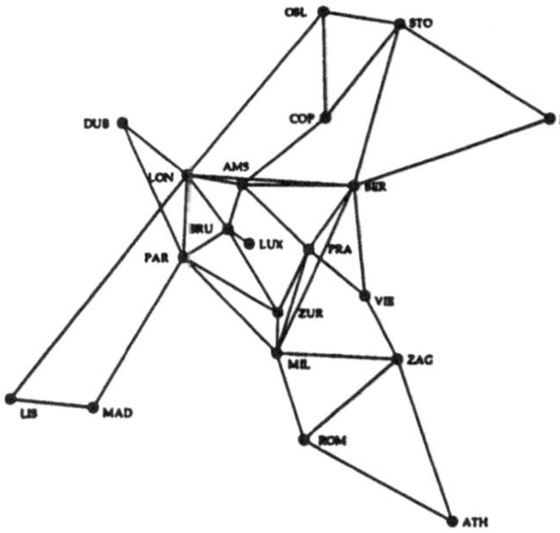

Fig. 1 Initial Topology (Cost: 7.22M)

<u>Topology Optimisation</u>

To search the vast problem space efficiently, a genetic algorithm (GA) was employed [6,7]. The objective was a topology of minimal network cost, although the fitness of a particular topology was not taken to be network cost alone, but network cost modified by a penalty function to ensure the reliability constraint was met. The penalty function adds P_A to the network cost for every node pair that has no alternative path, and P_N for every node pair with no routes at all between them. This avoids the false minimum of the topology with no links at all, whose cost, under the link and node cost models employed, would be zero.

GENESIS v5.0 was used [7], with topologies encoded as a simple binary string, each bit corresponding to a potential link in the topology.

<u>Results for Central Nodes</u>

For illustrative purposes, the GA was run for the central nine nodes (London, Amsterdam, Berlin, Paris, Brussels, Luxembourg, Prague, Zurich and Milan). Structure length was $(9 \times 8)/2 = 36$, and the algorithm parameters are given in Table 1. The penalty constants P_A and P_N were 0.1M and 0.2M respectively. These values were selected using a few trial runs to obtain values sufficiently high to ensure that the reliability constraint was met in the best topologies found.

Table 1 GA Parameters for Central Nodes

Elitist	
Population size	50
Crossover rate	0.6
Mutation rate	0.001
Generation gap	1.0
Scaling window	5

Five runs were carried out, all of which 'spun' within 3,000 trials. The best run was selected for detailed examination, and six of the topologies found are shown in Fig. 2.

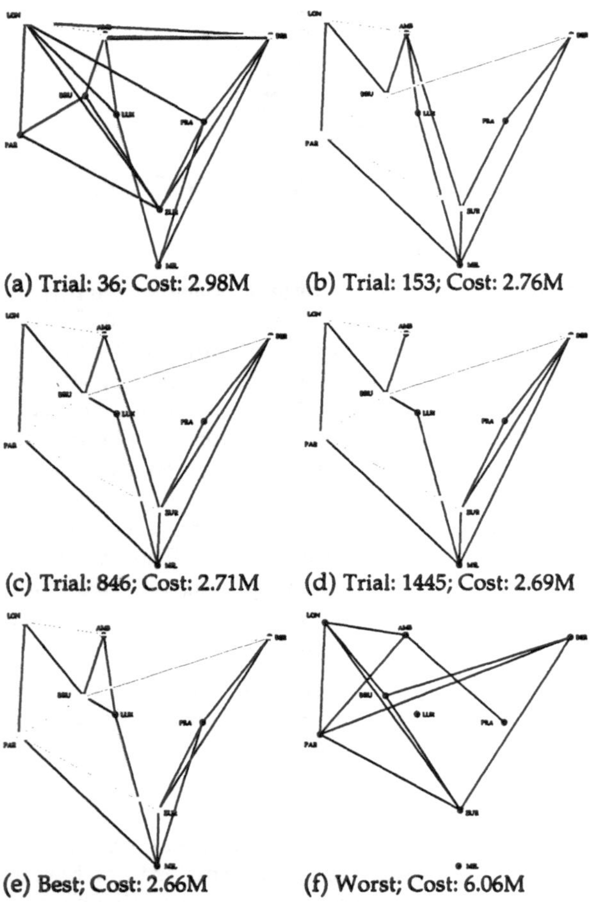

(a) Trial: 36; Cost: 2.98M (b) Trial: 153; Cost: 2.76M

(c) Trial: 846; Cost: 2.71M (d) Trial: 1445; Cost: 2.69M

(e) Best; Cost: 2.66M (f) Worst; Cost: 6.06M

Fig. 2 Results for Central Nodes

The first five topologies in Fig. 2, (a) to (e), represent the best individuals of their respective generations - 0, 4, 30, 54, and 83. There is a clear progression from (a), with far too many links, and consequently, higher cost, to (e), which was the best of the run. Of particular interest is the development of the triangle of links between Amsterdam, Brussels and Luxembourg, combined with the need for only a single direct link between those three nodes and Milan/Zurich.

By way of contrast, topology (f) is the worst individual from generation 0. In addition to the underlying cost of 2.46M for the links and nodes, it incurred a total penalty 3.6M for the two disconnected nodes (Luxembourg and Milan) and the one single-parented node (Prague). Thus without the addition of the penalty, this topology would have had a lower cost than (e), despite its obvious deficiencies.

Results for EON

The full European Optical Network includes an additional eleven nodes (Oslo, Stockholm, Dublin, Copenhagen, Moscow, Vienna, Zagreb, Lisbon, Madrid, Rome and Athens). The structure length was $(20 \times 19)/2 = 190$, the penalty constants P_A and P_N were 0.5M and 1.0M respectively, and the algorithm parameters were the same as those given in Table 1, except that several different population sizes were tried.

Five runs were made with a population size of 50. All of the runs spun in less than 8,000 trials, with the best result obtained as 7.00M.

Ten runs were carried out with a population size of 100. All of the runs spun in less than 48,000 trials, with the best result obtained as 6.85M.

Finally, five runs were carried out with a population size of 200. Three of the runs spun in less than 200,000 trials, with the best result obtained as 6.89M.

To further improve the results obtained by the GA alone, a simple hill-climbing algorithm was used on the best topology for each run. The hill climber removed that link which reduced the network cost most, and then iterated until no further reduction could be made.

The best topology overall is given in Fig. 3, and was found after 47,308 trials using a population size of 100. No links could be removed by the hill climber, and therefore the topology shown was that found by the GA alone.

The total cost of 6.85 million (in kmGbit/s) is made up of 3.09M for the links and 3.76M for the nodes.

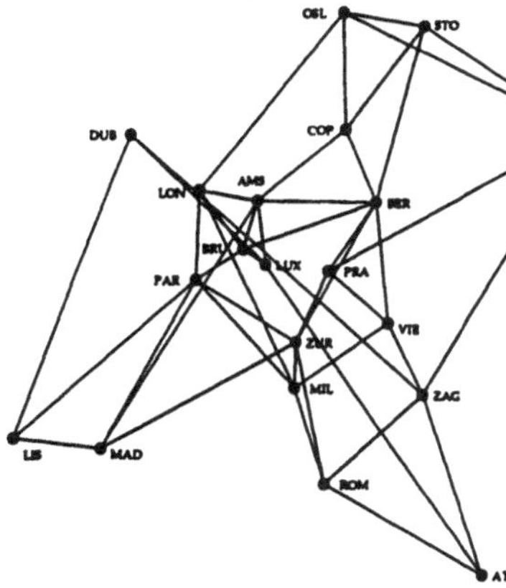

Fig. 3 Best Overall Topology (Cost: 6.85M)

Conclusions and Further Work

Although the initial EON topology [1,2], developed by hand, has a reasonably low cost (7.22M ignoring the penalty), it does not meet the reliability constraint. A single link failure could disconnect Luxembourg's international traffic.

In contrast, the topologies obtained so far using GA-based optimisation show both cost improvements (up to 5.1%) and superior reliability (in terms of the reliability constraint).

The work presented here could be extended by:

- Developing more rigorous cost models; for example [8] suggests link cost may only be proportional to the root of link capacity.
- Including maximising the network availability in the overall objective.
- Including wavelength allocation (assuming a multi-wavelength optical network), and the minimisation of the number of wavelengths used, in the overall objective.

References

1. Sinclair, M.C. & O'Mahony, M.J.: Proc. 36th RACE Concertation Meeting, Brussels 89 (1993)
2. O'Mahony, M., Sinclair, M.C. & Mikac, B.: Information, Telecommunications, Automata Journal 12 33 (1993)
3. Sinclair, M.C.: Proc. 8th UK Teletraffic Symposium, Beeston 8/1 (1991)
4. Sinclair, M.C.: Proc. IEE Colloquium on Resilience in Optical Networks, London 3/1 (1992)
5. Sinclair, M.C.: Electronics Letters 30 1468 (1994)
6. Goldberg, D.E.: Genetic Algorithms in Search, Optimization and Machine Learning. Addison-Wesley 1989.
7. Grefenstette, J.J.: A User's Guide to GENESIS Version 5.0 (1990)
8. Ishio, H.: NTT Review 4 62 (1992)

TIMETABLING USING GENETIC ALGORITHMS

Wilhelm Erben

Department of Computer Science
Fachhochschule Konstanz, D-78462 Konstanz, Germany

Abstract

This paper is concerned with the automated construction of timetables for a university. Timetabling essentially involves the assignment of course modules to periods and lecture rooms. It is a class of scheduling problems which is highly constrained. Several approaches including graph-theoretic and heuristic methods have been applied to it in the past. The timetabling system presented in this paper is based on Genetic Algorithms techniques. Hence it aims not only at finding a feasible solution to the problem; a timetable produced should also be of good quality, i.e. satisfy as many secondary constraints as possible. A corresponding objective function, and a Genetic Algorithm using problem-specific chromosome representations and operators have been developed and implemented in C and PROLOG. Testing was begun in a real-life environment with promising results.

1. Introduction

Scheduling problems belong to the most difficult but at the same time most interesting problems in the areas of Operations Research and Artificial Intelligence. They are known to be NP-complete, in general. Hence, the time to solve such a problem should be expected to increase exponentially with its size.

Timetabling is a special scheduling problem which is highly constrained. Search techniques that use heuristics have been applied to it with reasonable success, but they do not guarantee to find an optimal solution. By contrast, *Genetic Algorithms* techniques promise to come close to an optimal solution by means of a more exploitative search through the space of feasible timetables.

We started to apply the Genetic Algorithms approach to timetabling hoping to get deeper insight into handling constraint based scheduling tasks, in general. Since there is quite a lot of versions of the timetabling problem, differing from one school to the next, we focus on constructing course timetables at our own university rather than trying to develop a universal program.

Since new courses have been introduced at our university, we are confronted with increasing student numbers. As a consequence, there is a shortage of lecture rooms, laboratories and other resources to be shared. Moreover, according to our modular course structures a student may choose optional modules offered by his own or by other departments, rather than follow a fixed curriculum. Therefore we have a large number of conflicting constraints, and timetables are very difficult to schedule in this environment.

2. Weekly Course Timetabling

Our weekly course-timetabling problem involves scheduling *teachers*, *classes*, *subjects* (to be taught) and *rooms* to a number of *periods* of a week in such a way that no clashes will arise. A subject (or *course module*) may consist of more than one lecture (or laboratory or group-exercise class) to be delivered at distinct periods.

In addition, there are a lot of requirements which have to be fulfilled (e.g. some teachers may be unavailable at specific times, some classes must be taught in special computing labs), as well as second-order constraints which should be taken into account if possible (e.g. teachers as well as students do not like to have many „holes" in their timetables, some lectures should be spread uniformly over the whole week, some teachers wish to have a special furnitured classroom).

There is a number of requirements and constraints with respect to *time scheduling* (some classes have to be divided into course-work groups; some lectures are to be scheduled over two or more consecutive time periods; students may take a lot of options; etc.), as well as *room assignment* (to avoid much movement, allocated classrooms should be close to the host department; some rooms may be shared by different departments, others not; etc.)

As a consequence, we decided not to separate room assignment from time scheduling because one of the main goals consists in achieving efficient room utilization rates. This is accomplished in a more promising way by trying to do (or modify) time and room assignments simultaneously. So, apart from a large number of constraints, the input to our timetabling algorithm essentially consists of a set of subject-teacher and a set of subject-classes pairs, as well as of a list of classrooms of varying types.

Our *course-timetabling problem* covers most of the aspects of *exam scheduling* (see [2] for reference): For instance, in order to avoid a student having examinations (or lectures) in two adjacent periods, a corresponding (second-order) constraint can be established. If we wish to assign more than one room to a specific examination, to give another example, this can be done in the same way as dividing a class into course-work groups.

As mentioned above, each student may select a certain number of optional course modules. In order to give a real choice, conflicts where 2 lectures chosen by a student are scheduled at the same period of the week should be avoided. Again, a similar situation arises in exam scheduling.

3. The Genetic Algorithm

We agree with Michalewicz [5]: Problem-specific knowledge should be incorporated in both, the chromosome representation and the genetic operators.

Hence, the *chromosome representation* we chose is natural, and an extension of the one presented by de Werra [6]: A timetable is a map

$$f: C \times T \times S \times R \times P \rightarrow \{0,1\},$$

where $f(c,t,s,r,p)=1$ iff *class c* and *teacher t* have to meet for a lesson on *subject s* in *room r* at *period p*.

A timetable is *feasible* iff all required lessons are scheduled, there are no clashes at all, all allocated rooms are large enough to hold the class, and preassignments, teacher unavailabilities and specific room requirements are also taken into account.

The *initialization procedure* creates at random a population of feasible solutions. A number of necessary conditions for the existence of solutions must have been checked before. One of the arguments of the procedure is the number of periods available within a week. For a timetable to be generated the lectures which have to be placed are selected in random order, and each lecture is assigned to a randomly chosen period and lecture room without violating any feasibility constraints. The rooms are selected according to a list of preferences.

The initial population suffers from poor *fitness*, in general, because courses are allowed, for example, to run on Friday late evening, or some teachers might be employed fully on some day, and not at all on others.

One of the biggest problems appears to be the definition of an *objective function* for evaluating a given timetable. In our first approach it consists of several components, so-called „happiness functions“ measuring to what extent teachers, classes and rooms might be satisfied by the given timetable. (Rooms are „satisfied“ if they are just large enough to contain all the students.)

However, further tests must be done to find out how to weight these components, and what kind of second-order constraints should have preference. Some teachers, for example, are „happy“ if all their lectures take place within two days of the week, others might prefer to spread them over the whole week. Students may be exhausted after 3 hours of maths, whereas a teacher might insist in having his lectures in consecutive periods.

We do not need to use *penalties* in our approach because the concept of our domain-specific genetic operators is to produce only feasible solutions. The price we have to pay for this is that the definitions of the operators are not quite simple: Even small modifications of period or room assignments are likely to produce invalid timetables. Some points in the search space may be „isolated“, i.e. there might be no feasible timetable in a local neighbourhood of a given one.

The main difficulty in constructing a feasible *weekly* course timetable is that there is only a fixed number of periods within a week. In the exam scheduling problem, in contrast, this number may be left unspecified; hence, a free period can always be found for an unscheduled exam, and we do not run the risk of leaving the search space.

Mutation is defined in the following way: A number of time periods $p_1, p_2, \ldots, p_n$ is selected at random; the given timetable is modified only at the selected points, using the initialization procedure. As special cases of mutation we get operations like swapping time assignments between teachers or classes, or merely changing rooms, as long as the feasibility conditions are not violated. (The kernel of the mutation algorithm may also be used if we want to modify a given timetable slightly rather than develop a new one from the scratch.)

The *crossover* operator we defined resembles, in a way, standard path representation based operators for the Travelling Salesman Problem (see [5], e.g.). But things are more complicated in our timetabling problem. Having selected two parent timetables, parts which are exchangeable without violating the feasibility conditions are identified. In addition to some swapping operations which make the offspring look similar to the parents we need a nondeterministic algorithm to fill certain time periods at random, obeying the feasibility conditions. Again, the initialization procedure does the job.

4. Results and Conclusion

The basic algorithm has been implemented using C language on standard UNIX workstations. The initialization program and the domain-specific genetic operators have been written in PROLOG. Testing was begun using a small sample of data, but incorporating all types of possible constraints. The first results are promising, but a number of refinements still has to be done.

In order to ease this job, we will add a component to the system which enables us to adapt various parameters at run-time. A graphical X-Windows user interface for displaying the timetables produced by the genetic algorithm has already been implemented.

Next steps will be tests with full data (about 150 teachers, more than 60 classes, approximately 500 subjects, nearly 100 rooms, 30 time periods a week) including experimentation on parallelizing the algorithm. We will have to improve and extend the heuristics used in our genetic operators. In order to get accepted by potential users, the algorithm has to be made transparent. The timetabler should be enabled to modify, to a certain extent, the objective function according to his own understanding of a „good“ timetable.

References

[1] Abramson, D.: *Constructing School Timetables Using Simulated Annealing: Sequential and Parallel Algorithms*, Management Science 37/1, 98 (1991)

[2] Burke, E., Elliman, D., Weare, D.: *A Genetic Algorithm for University Timetabling*, AISB Workshop on Evolutionary Computing, Leeds

[3] Colorni, A., Dorigo, M., Maniezzo, V.: *Genetic Algorithms and Highly Constrained Problems: The Time-Table Case*, Proc. of the First Internat. Conf. on Parallel Problem Solving from Nature, 55 (1990)

[4] Le Kang, White, G.M.: *A Logic Approach to the Resolution of Constraints in Timetabling*, European Journal of Operational Research 61, 306 (1992)

[5] Michalewicz, Z.: *Genetic Algorithms+Data Structures=Evolution Programs*, Berlin Heidelberg New York: Springer 1992

[6] Werra, D. de, *An Introduction to Timetabling*, European Journal of Operational Research 19, 151 (1985)

RESOLUTION OF CARTOGRAPHIC LAYOUT PROBLEM BY MEANS OF IMPROVED GENETIC ALGORITHMS

Yassine DJOUADI

L.I.S.I. INSA/UCBL, Bât 502
20, Avenue Albert Einstein, 69621 Villeurbanne Cedex, France

Abstract. In this paper, a system for spatial layout and its application to cartographic domain is proposed. Genetic algorithms have been used as a way to deal with several kinds of constraints (fuzzy and Boolean), by means of the fitness function.
When a representation other than bit strings is used, it is often necessary to redefine the genetic operators. This led us to consider specific mutation and crossover operators. Specific probabilities are also needed in order to realise the combined goals of maintaining diversity in the population and sustaining the convergence capacity of the resolution process. Mutation process needs probabilistic model and is performed by means of the Gibbs-Boltzman probability. In this way, our system profits by the simulated annealing efficiency. This mixed method permits to examine solutions space without falling into a local minimum. The search process decreases a fitness function. This function describes the sum of costs inherent to different constraints. In order to explore entirely the solutions space, crossover process has been set up using adaptive probability. This probability varies during the resolution process. This variation is defined in order to process all regions of the cartographic map considering that, mutation operator is more dedicated to a given region of the map.

INTRODUCTION

Our aim is to design a system which displays the results of queries in cartographic databases. A query may be of the form: `{retrieve Q objects satisfying spatial-constraints}`, where `Q` is a quantifier and `spatial-constraints` represent typical geometrical constraints. In the retrieval process, a flexible form of constraints has been considered [1] and it may arises some cases where two (or more) retrieved objects are being lightly overlapped. It may also occurs cases where retrieved objects do not satisfy completely the rules of cartographic placement. Retrieved objects have their original placement however, some geometrical transformations (translation) are probably needed to enhance the quality of the cartographic restitution. Also, the system presented in this paper, attempts to find an optimal representation on a given medium of communication (map, screen,...) regarding to aesthetics, readability, etc.
Many studies of automatic layout have been carried out from the beginning of 80's. In [2], a mathematical approach using integer programming was proposed to perform label placement (label means textual data associated with cartographic object). The author argued that artificial intelligence should be used only when mathematical methods fail, and that cartographic layout problem has a mathematical composition. In our opinion, it seems that actual applied linear programming tools are not suited for integer programming (with 0 and 1 variables).
Another approach consists of using expert systems. Autonap [3] is a rule based system which performs also name placement. It uses a rule processor to measure the aesthetics quality of the placements. After each object is placed, a quality-evaluation rule is used to provide feedback to the rule processor. These rules supply a normalised "quality" number between 0.0 and 1.0. If expert systems are easy to run and set up, it seems particularly difficult in cartography to acquire and get expert knowledge.
In [4], an expandable system which uses neural networks is proposed. For the authors, it is possible to "learn" the rules used by the cartographer when placing labels, by examining an existing placement location and the conditions surrounding the choice of that placement location, by assessing the probability that the location is the optimum location and, by storing this probability as a weight in the neural network. The authors propose ten different locations for the placement of each label. The goal of the neural network is to examine the map and to assign a weight to each of those ten locations. All weights should initially be unset. As each point is processed, the process computes weight for location and average weight for that point text placement location in the following manner:

$$avg(n+1) = \frac{(avg(n) * n) + wt(n+1)}{n+1}$$

where *avg* is the current average weight for a given location, *n* is the total number of weights that have been used to calculate the average weight, and *wt(n+1)* is the newly computed weight of the feature being processed. Our system does not allow a priori locations, also the use of neural networks seems difficult to implement. Indeed our problem presents a dynamic aspect and, the weights would be continuously changed.
Genetic placement problem has been first addressed in [5], another works were performed afterwards ([6], etc...) and most of them were intended for electronic cell or VLSI placement. They did not take into consideration the topological aspect such (object *A* must be near object *B*). Topological aspect attempts to find relative layout from relationships between objects.
In cartographic layout process, Boolean constraints (e.g. no-overlap) coexist with fuzzy constraints (e.g. proximity). Genetic algorithms have been used in [7] as a method which processes them in a unified way. For this first proposition, crossover was heuristically performed and no probability was associated with it.

Also, our system did not profit fully from the power of GA's which arises especially from crossover operation. In this paper, adaptive probabilities for crossover and mutation are then proposed.

GEOMETRICAL ASPECTS

The two kinds of cartographic objects (i.e. linear, area) are modelled respectively as polylines and polygons. Such geometrical representation is more suited for translation operations. To simplify the problem (without loss of generality), some suppositions have been made:

- Object coordinates are considered as screen coordinates (in this way, the geometrical space is discretized unto pixel resolution),
- the convex hull for an object is considered instead of its natural shape,
- linear objects (e.g. roads, rivers,...) are considered as surface (a ribbon with unitary width).

Thus, we determine for each object its minimum bounding convex hull, its control point, and its target point. Target point corresponds to the original coordinates of the object transformed as screen coordinates. Control point corresponds to the gravity centre. Distance between target and control points is used to calculate the displacement cost. After that, the chosen method attempts to translate a given object and its corresponding control point towards the target point.

Two kinds of constraints (see Fig. 1) are considered in our problem. These constraints reflect the geometrical and topological aspect of the domain. A cost is associated for each kind of constraint as described in the next chapter.

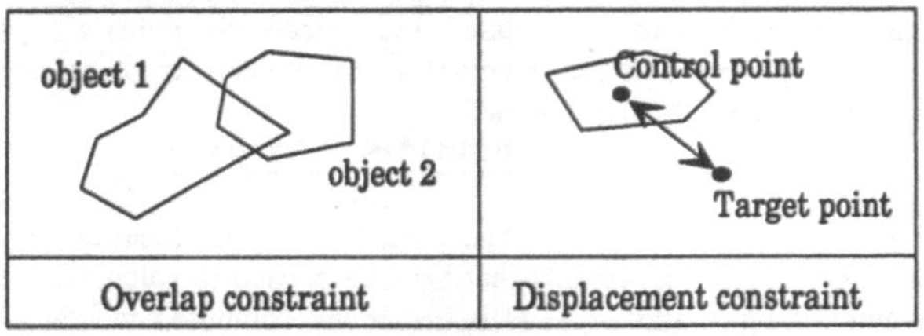

Fig. 1: The different constraints

GENETIC IMPLEMENTATION

Practical implementation for systems which use another representation than bit string is somewhat more complicated [8]. One of the most important aspect of such implementation is the choice of the appropriate encoding representation and suited operators.

Individual encoding method

The chosen encoding method is the more natural one. Each retrieved object consists of an allele whereas an individual consists of a solution (a layout of retrieved objects according to previously mapped fixed objects). Alleles may appear in different orders. Thematic groups (see Table 1) may be used to group these alleles. In the same manner, quadtrees or equivalent hierarchical data structure (see Table 2) used for geometrical data may also be used to represent a given individual.

Topics 1			....		Topics l		
$f_{t_1,1}$	...	$f_{t_1,k}$	$f_{t_2,1}$ ...		$f_{t_l,1}$	...	$f_{t_l,n}$

Table 1

North-West			North-East			South-West			South-East		
$f_{1,1}$	...	$f_{1,k}$	$f_{2,1}$	...	$f_{2,l}$	$f_{3,1}$	...	$f_{3,n}$	$f_{4,1}$	...	$f_{4,m}$

Table 2

Resolution method

The implementation of genetic algorithm which has been chosen is given as follows:

```
P←initial population obtained with constructor operator C;
p←|P| ;
t←initial_temperature ;
for ( i=0 ; i<Number_of_Generations ; i++) {
    for ( j=0 ; j<Number_of_Crossover ; j++) {
Offspring←{∅} ;
(parent1, parent2) = choice_2(P) ;
ind = Ψ(parent1, parent2);
with probability Pcros, do : Offspring←Offspring ∪ ind }
    P←selection(P, Offspring) ;
    for (j=0; j<Number_of_Mutations ; j++) {
ind = choice_1(P)
ind'←Θ(ind);
with probability Pmut, accept ind' }
t←t*r }
return individual with lowest fitness value
```

where:

P:	population; $P=\{ind_i\}_{i=1,p}$
p:	number of considered individuals
ind:	individual; $ind=\{f\}$
f:	cartographic object
$Pmut$:	probability of mutation
$Pcros$:	probability of crossover
t:	temperature used for the Boltzman mutation probability
r:	temperature decreasing rate
$Offspring$:	result of the crossover process.

Fitness function F: It sums costs inherent to geometrical and topological constraints being considered. In this way, Boolean overlapping constraint will be integrated.

$$F = \frac{1}{2}\sum_{f_i}\sum_{f_j} O_{f_i,f_j} + \sum_{f_i} T_{f_i}$$

The search of "good" solution is performed by means of minimising fitness function through the previous algorithm.

Overlap cost O: It is given by the half of the overlap area measured by pixels.

O_{f_i,f_j}: Overlap cost between the ith object and the jth object.

Displacement cost T: It represents the "disorder" of a given configuration. It is computed as the squared Euclidean distance between target point P_t and control point P_c.

$$T = \|P_c - P_t\|^2$$

Description of the operators

Two characteristics are needed in order to improve a given set of solutions. The first characteristics is the capacity to converge to an optimum in a given region of the solution space. The second is the capacity to explore new regions of the solution space. The balance between these characteristics of the considered algorithm is guided by the values of *Pmut* and *Pcros*.

Population constructor C: Two population constructors C_1 and C_2 yielding respectively the population sets P_1 and P_2 were considered. We obtain P_1 by applying random displacement to most constrained objects and P_2 by applying "force-directed displacement" [9] (see Fig. 2) to the most constrained objects. This heuristics is inspired from robots motion throughout many obstacles. Inside a given environment (called here a neighbourhood), an obstacle acts a reactive force F_r on the object to place whereas, its original location (i.e. target point) acts an attractive force F_a. The resulting force F is computed as the sum and gives the direction of the displacement.

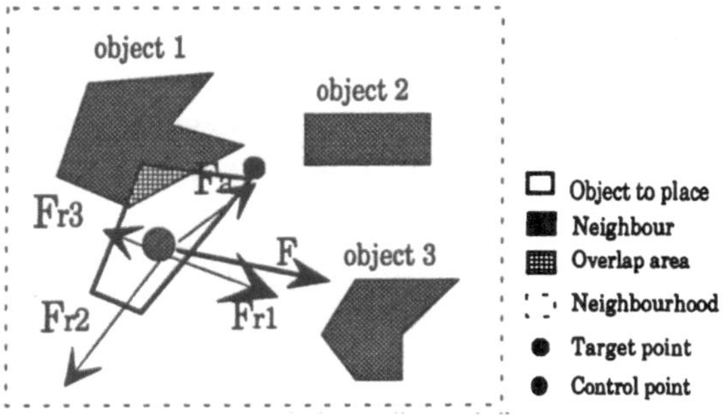

Fig. 2: Principle of "force-directed displacement"

The goal of generating population set is both good initial performance (in terms of fitness) and enough diversity (good distribution law) [5]. P_1 performs well the first goal and P_2 the second one. Also P uses a mixed population of both P_1 and P_2: $P = p_1.P_1 + p_2.P_2$.Where p_1 and p_2 are experimentally determined. For example p_1=50%, p_2=50%.

Choice function: The function *choice_2* is used for the crossover process and *choice_1* for the mutation process. Best scoring individual and another randomly chosen individual are taken to perform the crossover function whereas random individual is taken for the mutation process.

Selection function: It consists of selecting p individuals through the generated population P. The chosen function takes the half of individuals in P with best fitness value and the other half is randomly chosen.

Crossover operator (Ψ): Crossover operation is performed at the beginning of each new generation. The resulting offspring should resemble its parents but also should provide potentially meaningful new combinations [10]. The chosen method takes the best individual (i.e. with lowest fitness value) and another randomly chosen and crosses them as follows:

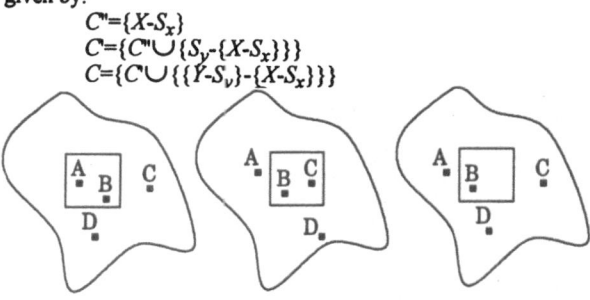

let us assume X the best parent ("basic"" parent), Y the other selected parent ("passing" parent) and, C the resulting offspring. X is taken to be the basis of C. A $k \times k$ square is cut from Y (where k is a random number in the range [1, $L/2$] and L is the side of the map). The set S_x (resp. S_y) corresponds to the objects contained in the square of the "basic" (resp. "passing") parent. Let S_x-S_y (resp. S_y-S_x) be the set of objects in S_x (resp. S_y) but not in S_y (resp. S_x). The result C of the crossover (see Fig. 3) is given by:

$$C' = \{X - S_x\}$$
$$C = \{C' \cup \{S_y - \{X - S_x\}\}\}$$
$$C = \{C \cup \{\{Y - S_y\} - \{X - S_x\}\}\}$$

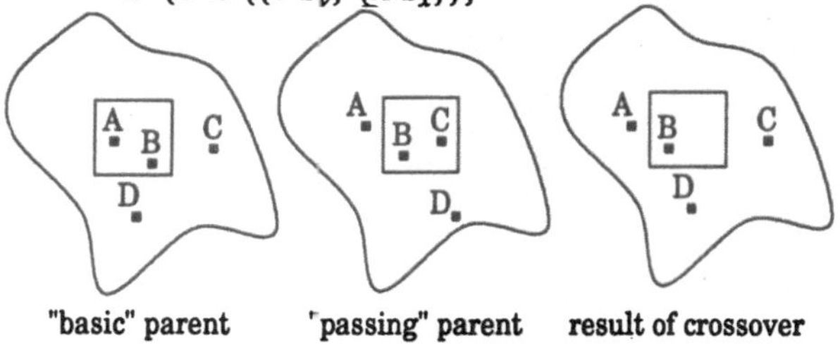

Fig. 3: Description of crossover operation

Since *Pcros* has to be increased when the difference for a given population decreases, it appears that *Pcros* is of the form:

$$P_{cros} = \frac{k}{F_{max} - F_{min}}$$

In the same manner, the probability of crossover has to be increased when the difference $F(X)$-$F(Y)$ between the two chosen parents increases. The probability *Pcros* is finally given by:

$$P_{cros} = \frac{F(X) - F(Y)}{F_{max} - F_{min}}$$

Mutation operator (Θ): Unlike VLSI layout problem where crossover is randomly performed and consists of swapping elements between them, in cartography there is no association (relation) between features that are long away and random crossover cannot be performed. Mutation arises for object which do not satisfy constraints. Asked questions for this operator are: how does it proceed to perform mutation operation and what is the probability associated with each operation. We do not allow a priori placements for objects and mutation consists to translate those objects. Some geometrical heuristics are used to perform this translation. On the other hand, Θ mutates m alleles, where m is a random number in the interval [0, $l/2$], l is the number of objects being overlapped.

The probability *Pmut* of mutation from a state i to a state i' is chosen as the Gibbs-Boltzman probability [11] which is given by:

$$P_{mut} = \begin{cases} \exp - \dfrac{(F_{i'} - F_i)}{t} & \text{if } F_{i'} > F_i \\ 1 & \text{otherwise} \end{cases}$$

EXPERIMENTAL RESULTS AND CONCLUSION

Our system is now implemented and experiments have showed interesting results. Contribution of the Boltzman probability as mutation probability is of a certain interest for avoiding local minima. Fig. 4 shows the fitness variation across iterations and generations with a random mutation probability. Two typical local minima are encountered during the resolution process. Fig. 5 shows more smooth and regular decreasing process by using the Boltzman probability.

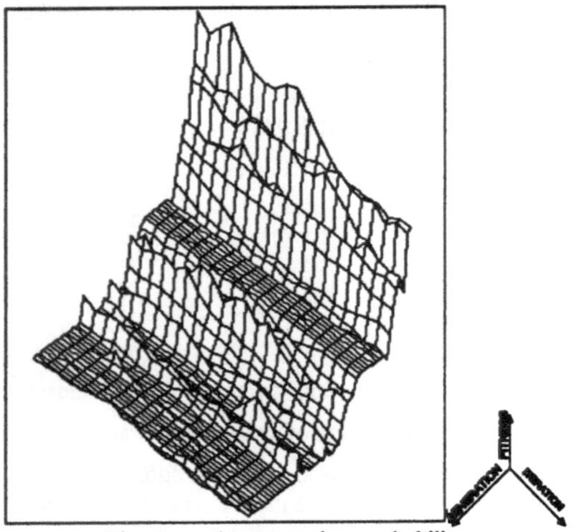

Fig. 4: random mutation probability

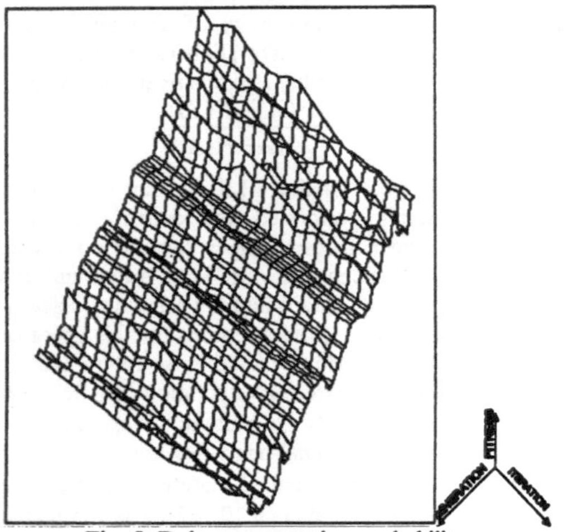

Fig. 5: Boltzman mutation probability

It has been also remarked (see table 4) that computation time and number of generated individuals are not really dependent from the number of objects but from goodness of the initial solutions and this may be useful for concrete applications. This tells us the power of the GA's as a combinatorial tool for the specific considered domain of the cartography. Ongoing works will consist to improve the heuristics for crossover and mutation.

Parameter	Value
initial population size *(p)*	10
initial temperature *(t)*	8
temperature decreasing rate *(r)*	0.9
total iterations	1000
Number of mutations by iteration	10
Number of crossover by iteration	5

Table 3: Values of parameters

Number of objects	Lowest fitness	Highest fitness	Computation time (μs)	Generated individuals
5	0	12,727922	4,969987	1834
50, with bad initial quality	60,455844	1970,414214	39,826297	2159
50, with good initial quality	9,485281	1084,414214	19,348824	1980
100	313,213203	2105,414214	60,366370	2141

Table 4: Experimental results

REFERENCES

1. Djouadi, Y.: Fuzzy modelling of spatial relationships for automated Placement/Retrieval System. Eurocarto conference, IX.1 (1994).
2. Zoraster, S.: Integer Programing Applied to the Map Label Placement Problem. Cartographica Review, Vol 23, N° 3, 16 (1986).
3. Freeman, H., Doerschler, J.S.: A Rule-Based System for Dense-map Name Placement. Communications of the ACM, Vol 35, No 1, 68 (1992).
4. Basoglu, U., Johnson, D.S.: The Use of Artificial Intelligence in the Automated Placement of Cartographic Names. Ninth Int. Sym. on Computer-Assisted Cartography, 225 (1989).
5. Cohoon, J.P., Paris, W.D.: Genetic Placement. Proc of the 23rd Design Automation Conference, 422 (1986).
6. Shahookar, K., Mazumder, P.: GASP - A Genetic Algorithm for Standard Cell Placement. EDAC Conference, 660 (1990).
7. Djouadi, Y.: Cartage - A Cartographic Layout System based on Genetic Algorithms. EGIS Conference, 48 (1994).
8. Grefenstette, J.J.: Genetic Algorithms. IEEE Review on Intelligent Systems and their Applications, 5 (1993).
9. Khatib, O. (1987), A Unified Approach for Motion and Force Control of Robot Manipulators : The Operational Space Formulation. IEEE Journal of Robotics and Automation, vol. RA-3, No 1, pp. 43-53.
10. Srinivas, M., Patnik, L.M.: Adaptive Probabilities of Crossover and Mutation in Genetic Algorithms. IEEE Trans. on SMC, Vol 24, N° 4, 656 (1994).
11. Metropolis, N., Rosenbluth, A., Teller, A.: Equations of State Calculations by fast Computing Machines. Journal Chem. Phys., vol 21, 1087 (1953).

USING GENETIC ALGORITHMS TO SOLVE THE RADIO LINK FREQUENCY ASSIGNMENT PROBLEM

A. Kapsalis, V. J. Rayward-Smith and G. D. Smith

School of Information Systems, University of East Anglia, Norwich, NR4 7TJ, UK

Abstract

The Radio Link Frequency Assignment Problem (RLFAP) is that of assigning frequencies to a number of radio links in such a manner as to simultaneously satisfy a large number of constraints and use as few distinct frequencies as possible. This problem is known to be NP-complete. We describe the application of a genetic algorithm to the solution of the RLFAP. The standard crossover operators were found to be of limited use due to the highly epistatic nature of the problem and a range of new (crossover and mutation) operators are described. Dynamically modifying the weights used in the fitness function has also proved to be effective in improving the performance of the standard genetic algorithm. This work is being undertaken as part of the EUCLID *CALMA Project - RTP 6.4*, Combinatorial Algorithms for Military Applications.

Introduction

This paper describes the results obtained by applying a genetic algorithm, *(GA)*, to the Radio Link Frequency Assignment Problem, *(RLFAP)*. This work forms part of a multinational comparative study into the various exact and heuristic techniques applied to combinatorial optimisation problems commonly found in military applications. A formulation of the *RLFAP* is given in the following section, followed by a description of the particular data sets used by all members of the consortium. The standard *GA* and actual representation used are then presented, followed by a description of some specialised genetic operators incorporated into the standard *GA* to improve the performance. The remainder of the paper addresses the results obtained.

RLFAP

The RLFAP can be stated as follows: We are given a set of links that have to be assigned frequencies. Typically, there may be of the order of 200 to 1000 links and approximately 50 discrete frequency values, so the latter are assigned to many different links. As a result, interference considerations generate many thousands of constraints. We have to find a **feasible** assignment if possible, i.e. one which satisfies all given constraints. In some cases, we have to find an **optimal**, feasible assignment, i.e. a feasible assignment that minimises some objective function such as the distinct number of frequency values used.

Formally, each link $i \in L = \{1, 2, 3, ..., n\}$ is assigned a frequency value f_i from a given domain D_i, which is a subset of the set of all available frequency values. Typically, the domains have non-null intersections. In the data sets with which we are working, there are six domains, each containing between 7 and 50 frequency values.

For the particular problem type we are studying, there exist a large number of interference constraints of the following types: for a given pair of links i and j, it is necessary that:

$$|f_i - f_j| > \varepsilon_{ij} \qquad (1)$$

or

$$|f_i - f_j| = \varepsilon_{ij} \qquad (2)$$

for some frequency separation ε_{ij}. In our examples, there are between 200 to 500 equality constraints and 1000 to 5000 inequality constraints.

There may be some instances with no feasible solutions. In such cases we find an assignment which satisfies as many of the constraints as possible. In some problems, each constraint is given a value corresponding to its importance. The objective function must include terms which give more weight to the higher priority constraints. The CELAR data sets described below contain objectives of this nature.

The *RLFAP* is easily shown to be NP-complete, see [2, 6], and generates an enormous solution space. For example, a particular problem with 50 frequency values and 200 links (the smallest problem in the CELAR data set) has a search space of 50^{200}.

In order to effect a proper comparison of the various search techniques, the members of the consortium have been primarily developing and testing their models using one set of problems, viz. the CELAR data sets, see [3]. There are eleven problems within the CELAR sets, and each problem is specified by a set of variables, a set of binary constraints of the type shown in equations (1) and (2), various domains of frequency values and an objective. The primary objective for each problem is to find a feasible solution minimising the number of different values assigned; if there is no feasible solution, find a solution which minimises the cost function with associated penalties. We initially present results for CELAR Problem 2, with 200 variables and 1235 constraints. This is the smallest, and hence easiest of the CELAR problems to solve. It is known to possess many feasible solutions and the optimum number of distinct frequencies is 14. The final section shows the results for the full set using a combination of selected mechanisms. This set of problems can be obtained through electronic mail or ftp to mscmga.ms.ic.ac.uk (155.198.66.4).

The Standard GA Approach

In function and combinatorial optimisation, the basic genetic algorithm starts with a population of (randomly) generated initial solutions. From these, a selection process decides which of these solutions will be used to create the next generation of solutions, using standard genetic operators, such as

Crossover which combines characteristics of two parent structures to form two similar offspring, and of which there exist many different forms, such as *Single Point Crossover*, *Multi-point crossover* and *Uniform crossover.*

Mutation which acts as a background operator and introduces new genetic material into the population of structures/solutions.

In general, the GA is a global search mechanism, capable of converging to good regions of the search space, and in many test problems it converges very quickly to the known optimal, see for example [5, 1, 4]. The software GAmeter was used to develop the genetic algorithm application, see [5].

Representation and Evaluation

Let N be the number of variables in the problem. An assignment is represented by a string of frequencies f_i,

f_1	f_2	f_3	$\cdots$	$\cdots$	f_N

where f_i is a frequency in Domain D_i, i.e. the domain associated with link/variable i.
Thus for CELAR problem 1, with 7 domains and 916 variables, the representation consists of a string of 916 indices, each index pointing to a frequency in one of the 7 domains. Since the domains can have as few as 6 frequency values and as many as 44, an integer in the range 1 to 64 (6 bits) will suffice to represent a frequency within a domain, with obvious redundancy.
For each given assignment, it is then possible to determine, among other things,

- the number of distinct frequency values assigned, D
- the number of equality constraints violated, EC
- the number of inequality constraints violated, IC

The resulting fitness function used to determine the suitability of a particular assignment is made up of weighted terms drawn from the various objectives:

$$fitness = w_1 * D + w_2 * EC + w_3 * IC + Penalties \quad (3)$$

In equation (3), the value of the term *Penalties* is determined from, among other things, the number and type of constraints violated.

Specialised Operators

In this section, we introduce specialised crossover-based and mutation-based operators which have been incorporated into the standard GA, in an attempt to improve the performance.

Constraint Crossover Select a constraint and check whether it is satisfied by any of two chosen parents. If it is satisfied, exchange the assignments to associated links between the two parents. Repeat these steps a specified number of times to create two new children.

Frequency Replacement Select a link. Find all other links with the same frequency assigned as the chosen link and belonging to the same domain. Select another frequency value at random from the associated domain and assign it to all links found. This is used in place of the standard mutation operator.

Swapper Select a pair of links randomly from the same domain. Swap the frequencies assigned to the selected links. Repeat these steps a specified number of times: It is used in addition to chosen mutation operator.

Constraint-Driven Mutation Randomly select a constraint to identify two links. Starting at two random positions in the respective domains, selectively search for and select two frequencies that satisfy the constraint. Replace the selected values in the assignment. Repeat these steps a specified number of times: This has the effect of forcing the constraint to be satisfied.

Another crossover operator, based on the variable grouping representation used in [2], was tried but led to no great improvement. Full details are given in [3].

Results using Standard GA and Specialised Operators

All the results that follow, unless otherwise stated, are for CELAR problem 2, the smallest and easiest to solve. This problem was chosen to provide a good comparison of the various developments to the basic genetic algorithm.
Table 1 gives details of the mechanism(s) in use for each experiment, while Table 2 shows a summary of the results.

For each experiment, there are two rows in Table 2. The first row represents the best (B) run from 5 runs while the second row represents the average (A) performance over 5 runs. The column headed **Fitness** shows the best/average fitness value

Exp. No.	*Description*
4	Standard GA -with 1-point crossover
5	*SGA* without crossover (mutation only)
7	*(5)* with multi-point crossover (random)
8	*(5)* with uniform crossover
9	*(5)* plus *Constraint Crossover*
12	*SGA* : with *Frequency Replacement*
13	*(9)* plus *Swapper*
15	*(9)* plus *Constraint-driven Mutation*

Table 1: Description of experiments for the Standard GA with and without specialised operators

Exp.		Fitness	EC	IC	D	Evaluations	Time
4	B	43	0	7	36	45663	119.7
	A	48.8	0.2	7.4	41.2	50405	132.2
5	B	47	0	3	44	26662	65.3
	A	48.6	0.2	5.6	42.8	42785	104.8
7	B	40	0	2	38	51545	266.7
	A	47	0	6.6	40.4	63358	327.9
8	B	43	0	5	38	54126	275.5
	A	48.6	0	8.2	40.4	56568	287.9
9	B	46	0	4	42	35057	142.8
	A	48.8	0	8	40.8	43442	176.9
12	B	44	0	6	38	77893	239.4
	A	44.2	0	9.8	34.4	82942	254.9
13	B	38	0	2	36	92965	283.6
	A	46.8	0	6.4	40.4	94234	287.5
15	B	36	0	0	36	68586	243.8
	A	39.2	0	0	39.2	35828	127.3

Table 2: Results for Standard GA with and without specialised operators

achieved by the experiment, i.e. equation (3). The columns headed **EC**, **IC** and **D** contain the number of equality constraints violated, the number of inequality constraints violated and the number of distinct frequencies, respectively. The columns headed **Evals** and **Time** respectively denote the number of evaluations and the elapsed time required to reach the best solution found - on a DEC 3000.

The main conclusions to be drawn from this initial set of experiments are that there is very little to choose between the different (standard) crossover operators, a constructive crossover operator such as *Constraint Crossover* appears to give no real advantage, and *Constraint-Driven Mutation*, (CDM), has a slight edge over the other mutation-like operators in that it is marginally faster and gives better quality solutions, in terms of the number of constraints satisfied. Also, *Frequency Replacement* improves the quality of solution obtained, in terms of the number of distinct frequencies used.

Modifications to the Objective Function

The following mechanisms were incorporated into the objective function (3) in an attempt to improve the search process.

Number of links 'not assigned' - Whenever, a constraint is violated, the two variables associated with it are marked as "not assigned". The weighted total of the number of links "not assigned" is added to the fitness function.

Constraint weighting by degree - For each constraint, a positive weight is determined, based on the number of times each of the links appears in the list of constraints. When evaluating the fitness of a solution, a term equal to the sum of the weights for the violated constraints is added to the fitness as a penalty.

Constraint weighting by frequency pair - This worked in the same way as the previous method, but in this case, the weights were determined by the number of frequency pairs that satisfy each of the constraints.

Dynamic fitness weights - After m generations for which no better solution is found, adjust one of the weights w_1, w_2 and w_3 in equation (3). Immediately, the entire population is re-evaluated and the optimisation continues.

Dynamic constraint violation weights - Every time a new solution is evaluated, a set of counters, one for each constraint, is updated according to which constraints were violated. After r generations, the counters are used to determine weights for all constraints, the constraint that is violated most being assigned the highest weight. The fitness function is augmented by a term in the same manner as in the *static constraint weighting.*

The set of experiments for this group of mechanisms is described in Table 3, while Table 4 shows a summary of the results.

Exp.	*Description*
32	*(15)* with *Number of links 'Not Assigned'*
20	*(9)* with *Constraint Weighting by degree* highest degree given the largest weight
21	*(9)* with *Constraint Weighting by degree* smallest degree given the largest weight
22	*(9)* with *Constraint Weighting by Frequency Pair* smallest number of pairs given the largest weight
23	*(9)* with *Constraint Weighting by Frequency Pair* largest number of pairs given the largest weight
27	*(9)* with *Dynamic Fitness weights* concentrating minimising *D*
25	*(9)* with *Dynamic Constraint Violation weights*

Table 3: Description of experiments for *Modifications to the Objective Function*

Exp.		Fitness	EC	IC	D	Evaluations	Time
32	B	42	0	0	42	4736	31.1
	A	43.4	0.2	0	43.2	5719.8	37.6
20	B	42	0	0	42	51193	162.2
	A	43.6	0.4	0.8	42.4	76579.4	242.7
21	B	45	0	1	44	65125	210.6
	A	46.8	2.8	0.4	43.6	102968	333.0
22	B	44	0	2	42	109501	354.5
	A	47	0.8	3.4	42.8	96445	312.2
23	B	35	0	1	34	193581	669.8
	A	35.8	0.8	3	32	119295	412.8
27	B	20	0	0	20	28325	122.0
	A	23.6	0	0	23.6	78079.6	336.4
25	B	42	0	0	42	137479	442.2
	A	43.2	0	0	43.2	134315	432.0

Table 4: Results for *Modifications to the Objective Function*

In summary of this section, we note that: use of the operator *Number of links 'not assigned'* dramatically improves the time to find feasible solutions. Also, using *Dynamic fitness weights* in such a way as to prioritise the minimisation of D (experiment 27), finds a near optimal solution (20 frequencies compared to optimal of 14).

Results for all CELAR problems

The following table shows the best results obtained to date, for all problems in the CELAR data set, using a combination of the most effective of the afore-mentioned mechanisms. In particular, we use *Constraint Crossover*, (Experiment 9) with *Constraint-Driven Mutation*, (Experiment 15), and incorporate the *Number of links 'not assigned'* mechanism, (Experiment 32) and the *Dynamic fitness weights* mechanism of Experiment 27.

In addition, an observed symmetry of the data set is utilised to simplify the problem. Specifically, in each of the problems, it is noted that the variables occur in pairs, in sequence starting with the pair (1,2) and continuing until (N-1,N); each pair is linked through an equality constraint, the right hand side of which is always the same value, namely 238, and no equality constraints exist other than these mutually exclusive, pairwise constraints. It is possible, therefore, for this type of data set, to reduce the problem. The representation used here, made up of half of the previous representation (for the odd numbered variables) plus N/2 bits, simultaneously assigns frequency values to the odd numbered variables and specifies whether 238 should be added or subtracted to those values to assign frequencies to the even numbered variables. We note particularly that an optimal solution has been obtained for Problem 2, while near optimal solutions are obtained for some others. A full description of the data sets and the results is given in [3]. The problems marked with an asterisk denote those for which no feasible solutions exist. In addition, for these problems, each constraint is associated with a penalty which is added to the fitness if the constraint is violated. Some of these penalties are of the order of 10^6, accounting for the large fitness values for Problem 7.

This work is ongoing. In particular, we are looking to use a *GA* at a meta level to control the parameters for a low-level search technique such as *simulated annealing*. The results of this will be reported at a later stage.

	Size		Fitness	IC	D	Time
Scen.	N	Co				
1	916	5548	22	0	22	4288
2	200	1235	16	0	16	95
3	400	2760	20	0	20	726
4	680	3967	54	8	46	2216
4a	680	3967	46	0	46	12
5	400	2598	187	147	40	1689
5a	400	2598	53	11	42	144
6*	200	1322	7978	238	44	652
7*	400	2865	8948446	331	44	4193
8*	916	5744	733	343	46	4100
9*	680	4103	54882	189	46	2026
10*	680	4103	53679	161	46	312
11	680	4103	34	2	32	2409

Table 5: Best results for full CELAR data set - N is the number of variables and Co is the number of constraints. The other columns have the same meaning as in previous tables.

References

[1] R.H. Berry and G.D. Smith. Using a genetic algorithm to investigate taxation-induced interactions in capital budgeting. In *Proceedings of the International Conference on Neural Networks and Genetic Algorithms*, Innsbruck, 1993. Springer-Verlag.

[2] W. Crompton, S. Hurley, and N. M. Stephens. Frequency assignment using a parallel genetic algorithm. In *Proceedings of Natural Algorithms in Signal Processing.* IEE, 1993.

[3] UEA CALMA Group. Genetic algorithm approaches to solving the radio link frequency assignment problem. Technical report, University of East Anglia, 1994.

[4] A. Kapsalis, V. J. Rayward-Smith, and G. D. Smith. Solving the graphical Steiner tree problem using genetic algorithms. *J. Oper. Res. Soc.*, 44(4):397–406, 1993.

[5] A. Kapsalis, V.J. Rayward-Smith, and G.D. Smith. Fast sequencial and parallel implementation of genetic algorithms using the gameter toolkit. In *Proceedings of the Int. Conf. on Artificial Neural Nets and Genetic Algorithms.* Springer Verlang, 1993.

[6] T. A. Lanfear. Graph theory and radio frequency assignment. Technical report, NATO EMC Analysis Programme, 1989. Project No. 5.

A TRANSFORMATION FOR IMPLEMENTING EFFICIENT DYNAMIC BACKPROPAGATION NEURAL NETWORKS

George L. Rudolph and Tony R. Martinez

Computer Science Department, Brigham Young University, Provo, Utah 84602
e-mail: george@axon.cs.byu.edu, martinez@cs.byu.edu

Abstract

Most Artificial Neural Networks (ANNs) have a fixed topology during learning, and often suffer from a number of short-comings as a result. Variations of ANNs that use dynamic topologies have shown ability to overcome many of these problems. This paper introduces Location-Independent Transformations (LITs) as a general strategy for implementing distributed feed-forward networks that use dynamic topologies (dynamic ANNs) efficiently in parallel hardware. A LIT creates a set of location-independent nodes, where each node computes its part of the network output independent of other nodes, using local information. This type of transformation allows efficient support for adding and deleting nodes dynamically during learning. In particular, this paper presents an LIT for standard Backpropagation with two layers of weights, and shows how dynamic extensions to Backpropagation can be supported.

1. INTRODUCTION

This paper proposes a strategy for implementing standard and dynamic backpropagation neural networks in parallel hardware: LIT BP (*Location-Independent Transformation for Backpropagation Networks*). LIT BP is part of an effort to develop a general strategy (called LIT) for implementing ANNs. This strategy supports general classes of ANNs *and* efficiently supports ANNs with dynamic topologies [5-7]. A *dynamic topology* is one which allows adding and deleting both nodes and weighted connections during learning.

Hardware support for ANNs is important for handling large, complex problems in real time. Learning times for complex applications can exceed tolerable limits using conventional computing schemes. Furthermore, hardware is becoming cheaper and easier to design. LIT overcomes several weaknesses of current hardware implementation methods.

Most ANNs use static topologies—the topology is fixed initially, and remains the same throughout learning. ANNs with static topologies often suffer from the following shortcomings: 1) sensitivity to user-supplied parameters—learning rate(s), momentum, etc., 2) local error minima during learning, and 3) there is no mechanism for deciding on an effective initial topology (number of nodes, number of layers, etc.) Current research is demonstrating the use of dynamic topologies in overcoming these problems [1], [3], [4].

Early ANN hardware implementations were model-specific, and are intended to support only static topologies. More recent *neurocomputer* systems have specialized neural hardware, and seek to support more general classes of ANNs [2]. Although some neurocomputers could potentially support dynamic topologies more directly in hardware, rather than in software, they currently do not. Of course, general parallel machines, like the Connection Machine, can simulate the desired dynamics in software, but these machines are not optimized for neural computation. LIT supports general classes of ANNs *and* dynamic topologies in an efficient parallel hardware implementation.

LIT redesigns the original network into a hierarchical, parallel network in which each node contains enough information locally to compute its part of the network output, independent of any other node. A network whose nodes have this property is said to be *location-independent*. The nodes also are location-independent—regardless of the physical location of any node in the network, the relative order in which they compute results, or the order in which those results are gathered, *the individual computations are the same, and the network output is the same*. Furthermore, because a node's information is local, adding or deleting nodes from the network can be done without affecting any other nodes. Thus, location-independence allows efficient support for dynamic topologies. A prototype VLSI chip set has been fabricated as proof-of-concept of the overall LIT strategy [8].

This paper presents the transformation for standard backpropagation (BP) networks with a single hidden layer, and shows how dynamic extensions to BP can be supported—the same basic transformation applies to networks with an arbitrary number of layers. Many dynamic extensions to BP have been proposed in the literature. Although any of these can be supported by LIT, one algorithm that allows addition and deletion of both nodes and weights [3] is chosen as an example for this paper.

Section 2 discusses the mapping from an original (either standard or with a dynamic extension) BP network to a transformed (LIT BP) network. Section 3 gives the algorithms for execution and learning as adapted to LIT, demonstrating the equivalence of the original and transformed networks. For brevity, the steps required for the dynamic topology have been included with the standard BP algorithm, rather than described separately. Section 4 is the conclusion.

2. CONSTRUCTION OF A BP NETWORK

In this paper, the term *Control Unit* refers to a mechanism that broadcasts inputs to a network, and reads the results from the network. The term *original* refers to a network before it is transformed, and the term *transformed* refers to a network after it has been transformed. These terms apply similarly to the nodes as well. The number of layers refers to the *number of weight layers*.

LIT is a two-step process:

1. Construct a set of location-independent nodes based on the original model.
2. Embed the nodes in a tree.

The original BP network has a distributed internal representation of learned information. Furthermore, the model requires an efficient backward phase of learning as well as forward. The key issues for transforming the BP network are to localize enough information at a node to satisfy location-independence and allow for efficient backpropagation, while essentially preserving the distributed character of the original model. A node structure that meets these goals is given below.

A binary tree topology is used for parallelism and communication in the transformed network, in this paper. Trees and other hierachical topologies allow adding or deleting nodes with time logarithmic in the number of hidden nodes in the original BP network. The connectivity among nodes in the transformed network does not explode because each node is connected to one parent and zero to two children. The algorithms for the transformed network show explicitly how communication can take place.

The heart of a transformation is the construction step: It defines the mapping between the original network and the set of LI-nodes, which in turn affects how the behavior of the original network is modeled in the transformed network. The original BP equations and structures reveal a symmetry both in the information used for computation, and in the behavior during forward and backpropagation phases. The construction given here reflects that symmetry, as well as achieving the goals of LIT.

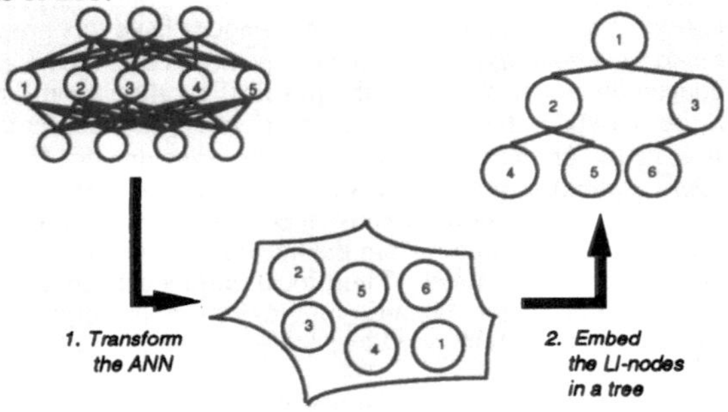

Figure 2. Transforming a 2-layer BP Network.

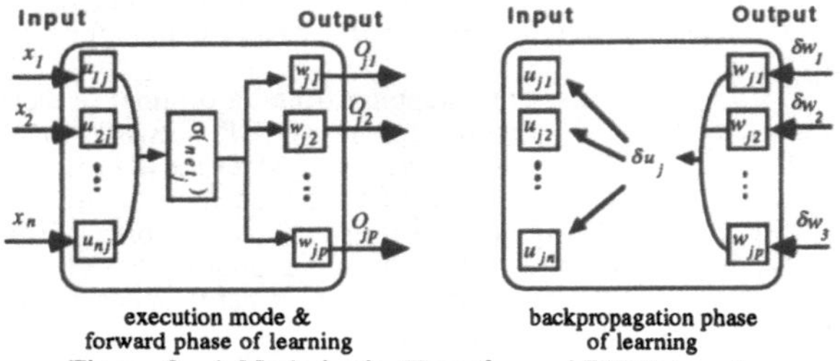

Figure 3. A Node in the Transformed BP Network

Figure 2 shows an original 2-layer BP network transformed to a LIT network with six nodes. There is one node in the transformed network corresponding to each original hidden node, plus one bias node. Figure 3 shows a general transformed BP node in detail. A transformed BP node (except the bias node) stores two vectors (layers) of weights—one on inputs (hereafter $\boldsymbol{u}_j$) and one on outputs (hereafter $\boldsymbol{w}_j$). Each element, u_{ij} or w_{jk}, of these vectors corresponds to a weight connected to the respective original hidden node. The original node 2, in figure 2, has four weights on inputs, and three weights on outputs. The transformed node 2 has four weights in $\boldsymbol{u}_j$ and three weights in $\boldsymbol{w}_j$. Each transformed node also stores a bias value θ_j, which corresponds to the bias value in the respective original hidden node. For purposes of learning, θ_j can be treated as a component of $\boldsymbol{u}_j$.

The bias node is like the other transformed nodes, with some exceptions. It needs no $\boldsymbol{u}_j$ because its activation is always 1. The elements of its $\boldsymbol{w}_j$ each correspond to one of the biases from the original output nodes. It has no additional bias values of its own. In all other respects, the bias node behaves like the other transformed nodes.

The reason each node stores two weight vectors has to do with backpropagation learning. The computation of δ error values at an original hidden node requires access to the weight values connected to *that* node from the subsequent layer. This computation is thus localized by storing the needed weights at each node. This difference in structure causes some differences in behavior between the original and transformed networks, though they remain quite similar. These differences are highlighted here—formal descriptions of the algorithms for the transformed network are given below in sections 3.1 and 3.2.

During execution, each transformed node computes a *vector* of output values $\boldsymbol{O}_j$ as follows: Each node recieves the input vector $\boldsymbol{X}$ and performs the sum-of-products with $\boldsymbol{u}_j$, and θ_j is added, to produce net_j. Each component of $\boldsymbol{w}_j$ is then multiplied by $f(net_j)$ to produce $\boldsymbol{O}_j$. The corresponding elements of each vector from all the nodes are summed together to produce a single vector of activation values. This summation is accomplished in the transformed network as each node sums its vector with those of its children, and sends the result to its parent. The parent of the root node is the Control Unit. The final output of the network is computed by applying the sigmoid function f to each element of this activation vector. Each element of the final output corresponds to one of the output nodes in the original network. As the root node sends its result to the Control Unit, the sigmoid function is applied to each element, thus the final output is held at the Control Unit.

During learning, the error values for the second layer of weights (output layer in the original) are calculated at the end of the forward phase, as given in the algorithm below. The resulting vector of error values is broadcast to all the nodes. This vector is used at each transformed node to compute weight changes for the second layer of weights (including the bias node) and the δ value for the first layer of weights just as in the original network. Weight changes for the first layer of weights are computed, also as in the original.

The preceeding discussion shows how the standard BP model is supported. This paragraph describes the dynamic extension to learning. The dynamic BP algorithm of [3] allows adding and deleting hidden nodes, but not input or output nodes (original network). Node addition is accomplished through probabilistic node division. The probability that a particular node will divide is a function of the global time since any node last divided, how much this node's error corrections have oscillated over the last epoch of training, and a probability threshold value. A node which oscillates wildly will eventually divide so that one node covers one hyperplane, and one node covers the other. Node deletion is accomplished through probabilistic weight deletion. A node self-deletes whenever all the weights in $\boldsymbol{u}_j$ are 0, or when all the weights in $\boldsymbol{w}_j$ are 0. This is a probabilistic event, since each weight computes a probability that it will set itself to 0 after each epoch of learning. This probability is a function of the global time since the last weight was deleted in the network and how much the weight has contributed correctly to its node's overall activity in the last epoch.

The learning algorithm below includes the steps of the dynamic algorithm based on [3], but the equations and further details are not included for reasons of space. More detail is available in [3] and [7].

3. ALGORITHMS

The following definitions and equations describe execution mode and learning for the 2-layer LIT BP. They are essentially the original equations, but altered to show the behavior of the LIT model. The learning algorithm also includes steps for the dynamic extension based on [3].

The main steps of the respective algorithms are in italics. The sub-steps (1.1, 2.1, etcetera) specify the details of how the LIT network accomplishes these steps.

3.1 Execution Algorithm

- q number of weight layers in the original network. $q = 2$.
- n number of nodes in the original input layer, size of $\mathbf{u}_j$.
- m number of nodes in the original hidden layer, number of nodes in the transformed network.
- p number of nodes in the original output layer, size of $\mathbf{w}_j$.
- a real-valued number, it will be either net_j or y_k.
- $\mathbf{O}_j$ output vector of new node j.
- O_j intermediate activation value for new node j, it corresponds to O_j for an original node j.
- net_j real-valued sum of the inputs to node j [equation (4)].
- f sigmoid function.
- $\mathbf{X}$ pattern clamped on inputs, has size n.
- x_i ith real-valued component of $\mathbf{X}$.
- $\mathbf{u}_j$ vector of weights on inputs in a new node j, has size n.
- u_{ij} real-valued ith component of $\mathbf{u}_j$. It corresponds in the original network to the weight from input node i to hidden node j.
- o_{jk} real-valued kth component of $\mathbf{O}_j$. It corresponds in the original network to the weight from hidden node j to output node k, multiplied by the original O_j.
- $\mathbf{w}_j$ vector of weights on output in a new node j, has size p.
- w_{jk} real-valued kth component of $\mathbf{w}_j$. It corresponds in the original network to the weight from hidden node j to output node k.
- $\mathbf{Y}$ vector of intermediate activation values, the sum of all $\mathbf{O}_j$.
- y_k real-valued kth component of $\mathbf{Y}$. It corresponds in the original network to the net_j of output node k.
- $\mathbf{Z}$ vector of network output values.
- z_k real-valued (always between 0 and 1) kth component of $\mathbf{Z}$. It corresponds in the original network to the output O_k of output node k.

$$f(a) = \frac{1}{1+e^{-a}} \tag{1}$$

$$O_j = f(net_j) \tag{2}$$

$$net_j = \sum_{i=1}^{n} x_i u_{ij} + \theta_j \tag{3}$$

$$o_{jk} = O_j w_{jk} \tag{4}$$

$$y_k = \sum_{j=1}^{m} o_{jk} \tag{5}$$

$$z_k = f(y_k) \tag{6}$$

The algorithm is as follows:

(1) Clamp an input pattern on the input nodes.
 1.1 Broadcast input vector X to the nodes.

(2) Each node takes the sigmoid of the sum-of-products of its weights and inputs.
 2.1 Each node computes its net_j and then computes O_j. O_j for the bias node is 1.
 2.2 Each node's weight vector $\mathbf{w}_j$ is multiplied by its (scalar) O_j to produce its output vector $\mathbf{O}_j$.
 2.3 Each node computes the vector sum of its $\mathbf{O}_j$ with the result from each child, and the sum (a vector) is sent to its parent. (The Control Unit is the parent of the root node.) The result is the vector $\mathbf{Y}$.

(3) Wait until the activation values on the output nodes are valid.
 3.1 $\mathbf{Y}$ is sent (through the sigmoid pipe) to the Control Unit. The result $\mathbf{Z}$ is the output of the network.

3.2 Learning Algorithm

Below are additional definitions and equations that describe the learning mode for the LIT BP. The forward phase of learning is almost identical to execution mode, and so follows the equations given above.

- α weight change momentum constant.
- $\delta\mathbf{w}$ vector of error values for all $\mathbf{w}_j$. It has size p.
- δw_k real-valued jth component of $\delta\mathbf{w}$. It corresponds in the original network to δ_k for output node k.
- δu_j real-valued error value for $\mathbf{u}_j$ in node j. It corresponds in the original network to δ_j for hidden node j.
- $\mathbf{T}$ vector of target values for the current pattern being presented. It has size p.
- t_k target value for z_k. It corresponds in the original network to the target value for O_k for output node k.
- $psse$ pattern sum-squared error for the current pattern being presented.
- $tsse$ total sum-squared error over all patterns in the current learning epoch.
- $ecrit$ maximum error tolerance for learning.
- s number of patterns in the training set (epoch).
- r simple index, used to avoid confusion with i, j, k, etc.
- c current training cycle number.
- η a real-valued learning constant.

$$f'(a) = \frac{\partial f(a)}{\partial a} = f(a)(1-f(a)) \tag{7}$$

$$\Delta w_{jk}(c) = \eta O_j \delta w_k + \alpha \Delta w_{jk}(c-1)$$

$$\Delta u_{ij}(c) = \eta x_i \delta u_j + \alpha \Delta u_{ij}(c-1) \tag{8}$$

$$\delta w_k = (t_k - z_k) f'(y_k) \tag{9}$$

$$\delta u_j = \left(\sum_{k=1}^{p} (\delta w_k w_{jk})\right) f'(O_j) \tag{10}$$

$$tsse = \sum_{r=1}^{s} psse_r \tag{11}$$

$$psse_r = \sum_{k=1}^{p} \delta w_k^{\,2} \tag{12}$$

As with the execution algorithm above, the main steps of the learning algorithm are in italics, and the sub-steps give the details of how LIT accomplishes each step. The learning algorithm is as follows:

Until Convergence ($tsse < ecrit$ at the end of an epoch)

(1) Present a training pattern to the network
 1.1 The forward phase of learning—this is same as execution above, up to (and including) step 2.3.

(2) Calculate the error δ of the output units (T - O)

2.1 As each y_k is sent to the Control Unit, the following occurs:

2.1.1 z_k is computed and received by the Control Unit.

2.1.2 z_k and t_k are used to compute δw_k, which is received by the Control Unit.

2.1.3 δw_k is used to compute δw_k^2.

2.1.4 Each δw_k^2 is summed with the value of *psse* so far.

(3) For each hidden layer Calculate δ_j using δ_k from the subsequent layer

3.1 The vector δw is broadcast to the nodes.

3.2 Calculate δu_j.

(4) Update the weights

4.1 Calculate Δw_{jk} and Δu_{ij}.

4.2 Change the weights.

(5) Update tsse

(6) Perform probabilistic weight deletion

6.1 Each weight computes a probability of self-deletion in the range 0-1.

6.2 Each weight generates a random value in the range 0-1.

6.3 If the probability is greater than the random value, then the weight is deleted.

(7) Perform node deletion on the hidden nodes (the number of input and output nodes remains fixed)

7.1 If $u_j=0$, or $w_j=0$, then node j self-deletes.

7.2 In the case that $u_j=0$, and the bias $\theta_j \neq 0$, then adjust the biases of the remaining nodes to compensate.

(8) Check for node addition (by probabilistic node division).

8.1 Each node computes a probability of dividing, in the range 0-1.

8.2 Each weight generates a random value in the range 0-1.

8.3 If the probability is less than the random value, then the node does not divide—skip step 9.

(9) Do node division (only if at least one node divides)

9.1 Two descendant nodes are created from the node: The u_j for both remain the same as the parent, but w_j are chnged so that one node covers one hyperplane and the other node covers the other hyperplane.

end

Step 5 may actually be computed any time after the end of step 2—independent of steps 3 or 4. Most algorithms assume that *psse* and *tsse* can be maintained, but give no explicit mechanism for doing so—LIT provides a mechanism.

Conceptually, step 9 could be done at the end of step 8. However, for algorithmic purposes, it is more convenient to check for the possibility of node division separate from actually performing node division. Node division is *potentially* the most expensive of all steps in terms of complexity and performance. However, it is important to realize that node division is a *rare* event in the operation of the network; when it does occur, it is highly unlikely to tax the resources of the network—particularly large networks. First, only highly unstable nodes have high probabilities of dividing. Second, as training time increases and the network becomes more stable, it is less likely for any particular node to become unstable, and correspondingly more remote that even a few nodes will simultaneously divide.

[7] shows that complexity of both learning and execution algorithms is $O(n+p+\log m)$ for a single pattern, where n is the number of inputs, p is the number of outputs, and m is the number of hidden nodes in the original network.

4. CONCLUSION

ANNs that use a static topology, i.e. a topology that remains fixed throughout learning, suffer from a number of short-comings. Current research is demonstrating the use of dynamic topologies in overcoming some of these problems. The Location-Independent Transformation (LIT) is a general strategy for implementing variations of ANNs that use dynamic topologies during learning.

This paper introduced ideas for LITs for static and dynamic feedforward, distributed neural networks. An LIT with associated execution and learning algorithms for backpropagation were formally defined, as an example.

Current work includes the following:

- LITs for other ANNs, aside from those for which transformations have been developed.
- VLSI design and fabrication of LIT models.

REFERENCES

[1] Fahlmann, Scott, C. Lebiere. The Cascade-Correlation Learning Architechture. in *Advances in Neural Information Processing 2*. pp. 524-532. Morgan Kaufmann Publishers: Los Altos, CA.

[2] Hammerstrom, D., W. Henry, M. Kuhn. Neurocomputer System for Neural-Network Applications. In *Parallel Digital Implementations of Neural Networks*. K. Przytula, V. Prasanna, Eds. Prentice-Hall, Inc. 1991.

[3] Odri, S.V., D.P. Petrovacki, G.A. Krstonosic. Evolutional Development of a Multilevel Neural Network. *Neural Networks, Vol. 6, #4*. pp. 583-595. Pergamon Press Ltd.: New York. 1993.

[4] Reilly, D.L., L.N. Cooper, C. Elbaum. Learning Systems Based on Multiple Neural Networks. (Internal paper). Nestor, Inc. 1988.

[5] Rudolph G., and T.R. Martinez. An Efficient Static Topology for Modeling ASOCS. *International Conference on Artificial Neural Networks,* Helsinki, Finland. In *Artificial Neural Networks,* Kohonen et al, pp. 279-734. North Holland: Elsevier Publishers, 1991.

[6] Rudolph G., and T.R. Martinez. A Transformation for Implementing Localist Neural Networks. Submitted, 1994.

[7] Rudolph G., Martinez, T. R. An Efficient Transformation for Implementing Two-Layer FeedForward Neural Networks. To appear in the *Journal of Artificial Neural Networks*, 1995.

[8] Stout, M., G. Rudolph, T.R. Martinez, L. Salmon, A VLSI Implementation of a Parallel, Self-Organizing Learning Model, Proceedings of the *International Conference on Pattern Recognition* (1994) 373-376.

AA1*: A DYNAMIC INCREMENTAL NETWORK THAT LEARNS BY DISCRIMINATION

Christophe Giraud-Carrier
Department of Computer Science
University of Bristol
Bristol, BS8 1TR , U.K.

Tony Martinez
Department of Computer Science
Brigham Young University
Provo, UT 84602, U.S.A.

Abstract

An incremental learning algorithm for a special class of self-organising, dynamic networks is presented. Learning is effected by adapting both the function performed by the nodes and the overall network topology, so that the network grows (or shrinks) over time to fit the problem. Convergence is guaranteed on any arbitrary Boolean dataset and empirical generalisation results demonstrate promise.

Introduction

Most connectionist models are based on static networks of compu ting nodes, where learning is effected by changin g the weights of the connections between nodes and/or by modifying the nodes' functions. Adaptive Self-Organising Concurrent Systems (ASOCS) [5] are a class of dynamic networks that learn by adapting not only the functio n performed by their nodes, but also their overall topology. Hence, these networks grow (or shrink) over time to fit the problem.

ASOCS have two basic modes of operation: learning and execution. In exec ution mode, the system functions as a logic network. It receives inputs and produces an output as the data flow asynchronously and in parallel through the network. In learning mode, the system is incrementally presented a set of Boolean instances. As each instance is presented, the network first executes and compares its computed output with that of the training instance. If the two are identical, no change is made and the next training instance is presented, otherwise, the network adapts to learn the new instance, while preserv ing consistency with previously acquired knowled ge.

Several ASOCS learning algorithms have been developed. They include Adaptive Algorithm (AA) 1 [10], AA2 [8] and AA3 [9]. AA2 ind uctively constructs a logic circuit consistent with the training instances. Though its convergence is evident, AA2 does not generalise to prev iously unseen instances. AA3 inductively cons tructs a binary decision tree. A formal proof of AA3's convergence is in [1]. However, AA3's generalisation is little more than memorisation in that the predicted output of any previously unseen instance is that of the first training instance. In most cases, this bias is too strong, dependent upon ordering of the training instances, and thus scarcely better than merely guessing. AA1 learns by discrimination and constructs a network in which

$$
\begin{array}{llcll}
A & B & \rightarrow & Z & (1) \\
B' & C & \rightarrow & Z & (2) \\
A & C' & \rightarrow & Z' & (3)
\end{array}
$$

Figure 1: Sample Instances

knowledge is distributed across all of the nodes. AA1 has also been shown to converge on any arbitrary Boolean instance set [3]. However, unlike AA2 and AA3, empirical studies show that AA1 also meaningfully generalises to previously unseen instances [7].

This paper presents AA1*, an extension to AA1 grounded in the *Occam's Razor Principle.* AA1* focuses on minimising network's growth to achieve greater predictive accuracy with smaller networks. The memory requirement at each node is also reduced by one third, and first-order features are available for discrimination during Node Selection. Results on several datasets demonstrate the validity of the extensions.

AA1* Learning

Only an overview of AA1* is possible here. Emphasis is placed on the differences with AA1. Further details and background on AA1 may be found in [3, 5, 10].

Instances

An *instance* is a conjunction of Boolean input values together with the (also Boolean) output value it implies. By convention, V is used to denote the fact that "(variable) V is true or positive", and V' to denote "V is false or negati ve". Fig. 1 contains three sample instances. The left-hand side of an instance needs not contain an input value for each input variable. Missing input variables are treated as though their value was *don't-care* and can thus be replaced by either true or false without affecting the output value. Formally, if I is an instance in which no value is specified for some input variable V, then I represents both instances obtained from I by adding V and V', respectively to the left-hand sid e of I. In Fig. 1, if A, B and C are the only three input variables, then instance (1) represents both instances $ABC \rightarrow Z$ and $ABC' \rightarrow Z$. In other words, instance (1) means that when A and B are both true, then Z must become true, regardless of the values of any other input variables. Henc e, instances can be viewed as if-then rules.

P	N
?	1
0	0
	1

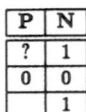

Figure 2: Sample NT

The *polarity* of an instance is the value of its implied output. Instances (1) and (2) of Fig. 1 are positive instances, while instance (3) is a negative instance. If two instances have the same polarity, then they are said to be *concordant* with respect to each other. Otherwise, they are *discordant*.

Nodes

Nodes are the basic architectural units of AA1*. However, their structure and connectivity differ from that of classical connectionist models. In AA1*,

- Each node has exactly two inputs and a node's output is the AND or OR of its (possibly inverted) inputs. Logical functions are extended in a natural way to accommodate missing input values denoted by ? (e.g., 0 AND ? implies 0, 0 OR ? implies ?).
- Each node records output information about previous training instances in its *node table* (NT).
- Each node's output is sent to 0 or more other nodes' input on weightless connections.
- The overall network structure is a rooted graph and the output of the root or *top node* is the network's output.

The NTs are ignored during execution. During learning, they are used to control discrimination as discussed below. An NT has two columns, P and N, holding the node's output values for each positive and negative training instance, respectively. There is, of course, a consistent one-to-one correspondence between training instances and NT cells. Fig. 2 shows a sample NT.

In AA1, input variables are generally not nodes. Rather, they combine in pairs and become inputs to nodes. Only when an odd number of input variables are recruited, does AA1 produce a node for one of the input variables. That node is simply constructed by having both inputs be the selected variable and setting the function to AND. Unlike AA1, AA1* produces one such node for each input variable. The motivation for this extension is given later.

Discrimination

If a cell in a node's NT contains the value 0 or 1, then that node *discriminates* the corresponding instance from all discordant instances whose cells contain the opposite value. For example, assume that the NT of Fig. 2 belongs to some node K and consider the second cell of the P column, whose corresponding instance is I. If K outputs 0, there is no way to tell whether it is I that is matched or the instance corresponding to the second cell of the N column. However, if K outputs 1, then clearly I is not matched. But K outputs 1 for the instances corresponding to the first and third cells of the N column. Hence, K discriminates I from these two instances. Clearly, cells that contain ? cannot participate in discrimination.

A node that discriminates every positive instance from every negative instance is a ***complete discriminant node.*** A complete discriminant node thus asserts one value (0 or 1) for all negative instances and the opposite value for all positive instances. A node that asserts one value for either all positive or all negative instances, and the opposite value for at least one discordant instance is a ***one-sided discriminant node.*** Given an instance I and a set S of discordant instances from which I must be discriminated, a node N's *discriminant count* (with respect to I and S), denoted by $DC_N(I, S)$, is the number of instances in S that N discriminates from I.

In AA1, information about which instances a node discriminates from another given instance is stored in the NT as a binary D column. AA1* computes D on-the-fly, thereby reducing the memory requirement at each node by one third, while not increasing computation time.

Learning Algorithm

As training instances are presented to the network, AA1* incrementally learns to discriminate each new training instance from all previously observed discordant instances. Essentially, the network constantly adapts so that its top node is complete discriminant with value 1 in the P column. Pseudocode for the learning algorithm is given in Fig. 3. It is executed as each new training instance is presented.

An instance set is *inconsistent* if it contains two discordant

```
(1) Compute add-list & delete-list by
    maintaining current instance set
    consistent and minimal
(2) For each instance in delete-list
      Empty corresponding cell in NTs
(3) For each instance in add-list
   (a) Do Execution with NT update
   (b) If (output != target value)
      (i) Do Node Selection
     (ii) If (more nodes are needed)
            Do Node Addition
    (iii) Do Node Combination
     (iv) Perform Self-Deletion
```

Figure 3: AA1* Learning Algorithm

instances whose conjunctive parts could be true simultaneously. Step (1) remedies such inconsistencies by removing one of the offending instances or adding variables sufficient to discriminate between them. Details of which instance to remove and how to add discriminant variables are in [5, 6]. Informally, an instance set is *minimal* if it can not be made smaller (i.e., less instances and/or variables) while retaining its functionality. Complete minimisation is not practical, but step (1) achieves partial minimisation through pairwise comparison of the trainin g instance and instances in the current instance set [5, 6]. Step (1) produces a delete-list and an add-list containing the instances to remove from and add to the current instance set to keep it consistent and somewhat minimal.

Step (2) processes the delete-list by removing instances from the current instance set and causing each node to empty the corresponding cells in its NT. Step (3) adds the instances of the add-list to the current instance set and presents each in turn to the network. Note that the maintenance of a *current instance set* in steps (2) and (3) is necessary only to step (1). The stored instance set is ignored during execution and network modi fication.

Step (3) is the heart of the learning algorithm. It is performed for each instance in the add-list. In step(3)(a), the network executes based on the values of the inst ance's input variables. Each node applies its stored function to its two inputs and passes the result to all of its successors. Since execution is with update, each node also stores its output value for the new instance in a corresponding cell of its NT. The output of the top node is then checked in step (3)(b). If it is concordant with that of the new instance, then no changes need be made and the next instance is processed. If, on the other hand, it is different from that of the new instance (i.e., either dis cordant or ?), then the current top node is no longer complete discriminant. Adjustments are made to the network in steps (3)(b)(i)-(iii) to create a new complete discriminant top node. Since the current top node is complete discriminant for all instances except the new one, the idea is to construct a one-sided discriminant node that discriminates the new instance from all discordant instances and to combine it with the current top node to obtain the new desired complete discriminant top node. Steps (3)(b)(i)- (iii) are detailed in the next section. In step (3)(b)(iv), the network undergoes a phase of self-deletion in which nodes that are no longer needed (e.g., non-discriminant nodes) are removed from the network. Self-deletion increases parsimony. Details of the various self-deletion mechanisms are beyond the scope of this paper. They may be found in [5, 10].

Network Modification

In steps (3)(b)(i)-(iii), AA1* recruits nodes that together suffice to produce a one-sided discriminant node that discriminates the new instance, NI, from the set, D, of all discordant instances. If N and M are nodes that discri minate NI from $D_1 \subseteq D$ and $D_2 \subseteq D$, respectively, then it is a result of [3] that the node obtained by combining N and M discriminates NI from $D_1 \cup D_2$. This result extends naturally to show that, given an appropriate set of nodes, step (3)(b)(iii) produces a new network whose top node is complete discriminant. Hence, the goal of the system is to compute a set $\{N_1, N_2, ..., N_n\}$ of nodes such that, if for each $1 \leq i \leq n$, N_i discriminates NI from $D_i \subseteq D$, then $\bigcup_{i=1}^{n} D_i = D$.

The set $\{N_i\}$ is naturally constructed by selecting existing nodes (step (3)(b)(i)), and addin g new nodes if necessary (step (3)(b)(ii)). In AA1, Node Selection successively recr uits the node that discriminates the new instance from the largest number of remaining, non-discriminated, discordant instances. That is, the *ith* node selected, N_i, is such that $DC_{N_i}(NI, D - \bigcup_{k=1}^{i-1} D_k) = \max_{N \in Net}\{DC_N(NI, D - \bigcup_{k=1}^{i-1})\}$. Selection goes on until the value of i such that $\max_{N \in Net}\{DC_N(NI, D - \bigcup_{k=1}^{i-1} D_k)\} = 0$, is reached (i.e., no further dis crimination can be achieved by existing nodes). The set of selected nodes produced by this procedure is clearly not guaranteed to be minimal. Hence, the resulting network may be larger than it needs be. Following the *Occam's Razor Principle*, AA1* extends Node Selection as follows. Let W_1 be the first node selected by AA1 and S_1 be the set of subsequently selected nodes. Let W_2 be the first node selected by AA1 when W_1 is left out of the competition and S_2 be the set of subsequently selected nodes. In general, let W_i be the first node selected by AA1 when W_1, ...,W_{i-1} are left out of the competition and S_i be the set of subsequently selec ted nodes. AA1 chooses the smallest of the S_i. In the current implementation, AA1* computes only S_1 through S_5 as empirical studies show that in most cases S_2 or S_3 are the smallest. This extension increases learning time only by a constant multiplicative factor. Execution time is reduced as the networks are smaller.

Another way in which AA1* improves on AA1 is by creating individual nodes for input variables, as mentioned above. These nodes represent first-order features that may be helpful in early discrimination. In AA1, candidate nodes for selection are combin ations of input variables, i.e., higher-order feature s. Recruitment of single variables is delayed until Node Addition. However, given a new instance which did not cause the combination of two inputs, the logical combination of these inputs is not guaranteed to have a better discriminant count than one of the input alone. Hence, smaller networks may be obtained by recruiting existing individual input variables during Node Selection.

Once selection is completed, if some instances in D remain to be discriminated from, then Node Addit ion takes place. A set of candidate input variables is computed and a sufficient subset is extracted by applying the aforementioned procedure used for Node Selection by AA1. The process is completed by Node Combination. Details are in [5, 10].

Simulation Results

Several datasets from [11] were used to test AA1 and AA1*'s generalisation. In both cases, generalisation is possible due to the bias towards parsimony, manifest in three ways. First, if the current network correctly predicts the value of a training instance, no unnecessary changes are made. Second, during node selection, the system uses heuristics to minimise the number of recruited nodes. And third, after a new instance has caused network modification, unne cessary nodes are deleted. Empirical results confirm these findings

Table 1: Simulation Results

	AA1		AA1*		
Application	**PA**	**IR**	**PA**	**IR**	**HC**
Monk1	95.5	0.63	98.4	0.43	1.8
Monk2	73.4	1.52	71.4	1.45	9
Monk3	92.0	0.64	91.3	0.61	1.5
hepatitis	72.8	0.88	71.8	0.88	2.1
shuttle-l-c	100	1.33	100	1.40	0
shuttle-exp	98.7	0.18	98.0	0.17	0.3
Averages	88.7	0.86	88.5	0.82	2.5

as shown in Table 1. The PA column shows predictive accuracy on the test set. The IR column gives the ratio of the number of nodes in the final network to the number of training instances and serves as a measure of parsimony. The HC column under AA1* shows the average number of times, in one pass over the training set, that a smalle r set of nodes was found using the above extension.

All non-Boolean inputs have their values translated into their equivalent binary form. Only one form of self-deletion, namely, complete discriminant deletion [10], is implemented in the simulations. The dataset *shuttle-exp* consists of the complete set of 278 instances resulting from expanding the 15 rules of the shuttle-landing-control (*shuttle-l-c*) datas et. Reported results for *hepatitis* and *shuttle-exp* were gathered using 10-way cross validation. Results for the Monk problems used the provided training and test sets, and results for *shuttle-l-c* were obtained by training on the 15 rules and testing on the 278 corres ponding instances. The outcome of the algorithm s being order-dependent, several random orderings of the training set were used and averaged.

In all cases (except *shuttle-l-c* which use s only 1 extra node), AA1* produces smaller networks. Surprisingly, predictive accuracy seems to suffer some, except in the case of *Monk1*. These preliminary result s demonstrate the potential of the proposed extensions, since on average, network size is reduced by 4.7% and predictive accuracy only by 0.2%. They also show that much can still be done as HC is small for some of the datasets. One would expect AA1* to produce a network of same size for both *shuttle-l-c* and *shuttle-exp*, and given the simple expressions of *Monk1* and *Monk3*, smaller networks could be anticipated. Also, HC is comparatively high for *hepatitis* and *Monk2* while the network's size is only slightly reduced and predictive accuracy not improved. Both *hepatitis* and *Monk2* are known to be difficult problems.

Conclusion

This paper presents an extension to the AA1 learning algorithm for ASOCS. The hierarchical nature of the topology guarantees that executi on is essentially O(log$n$), where n is the number of nodes in the network. Convergence on any arbitrary Boolean datase t is guaranteed and empirical generalisation result s demonstrate promise. The algorithm's ability to incrementally handle both examples and rules, and to integrate symbolic and connectionist repre sentations, are relevant to current effort s in neural network and machine learning research [2, 4, 12, 13].

AA1* extends the early AA1 [10]. More informed heuristics may be considered, as well as their impact on noisy datasets. Insisting on the instanc e set being consistent may not always be appropriate as nature is replete with inconsistencies. Moreover, the current handling of incons istencies introduces chronology as a strong lea rning bias, and makes the algorithm order-dependent. Ways to remove or appropriately use the order dependency may be considered. Introducing explicit forms of generalisation (e.g., ***dropping-condition*** rule), is an interesting avenue to pursue. Of particular concern is the interaction between such generalisations and the requirement for consistency. It still remains unclear how to do away with, or effectively replace, the memory-expensive node tables used by AA1*. The algorithm PDLA of [2] is one attempt at overcoming some of AA1*'s limitations.

References

[1] C. Barker and T.R. Martinez. Proof of cor rectness for ASOCS AA3 networks. *IEEE Transactions on Systems, Man, and Cybernetics*, 24(3):503-510, 1994.

[2] C. Giraud-Carrier and T. Martinez. Using precepts to augment training set learning. In *Proc. ANNES'93*, 46-51.

[3] C. Giraud-Carrier and T. Martinez. Anal ysis of the convergence and generalization of AA1. *JPDC*, 1994 (to appear).

[4] L.O. Hall and S.G. Romaniuk. A hybrid connectionist, symbolic learning system. In *Proc. AAAI'90*, 783-788.

[5] T.R. Martinez . *Adaptive Self-Organizing Networks*. PhD thesis, University of California, Los Angeles, 1986.

[6] T.R. Martinez. Consistency and gener alization in incrementally trained connectionist networks. In *Proc. International Symposium on Circuits and Systems*, 706-709, 1990.

[7] T.R. Martinez, J.C. Barker, C. Giraud-Carrier. A generalizing adaptive discriminant network. In *Proc. WCNN'93*, I:613-616 .

[8] T.R. Martinez and D.M. Campbell. A sel f-adjusting dynamic logic module. *JPDC*, 11(4):303-313, 1991.

[9] T.R. Martinez and D.M. Campbell. A self-organizing binary decision tree for incrementally defined rule based systems. *IEEE Transactions on Systems, Man, and Cybern etics*, 21(5):1231-1238, 1991.

[10] T.R. Martinez and J.J. Vidal. Adaptive parallel logic networks. *JPDC*, 5(1):26-58, 1988.

[11] P.M. Murphy and D.W. Aha. UCI reposit ory of machine learning databases. University of California, Irvine, Department of Inform ation and Computer Science, 1992.

[12] D. Ourston and R. Mooney. Changing the rules: A comp rehensive approach to theory refinement. In *Proc. AAAI'90*, 815-820.

[13] R. Sun. A connectionist model for commensense reasoning incorporating rules and similarities. *Knowledge Acquisition*, 4:293-321, 1992.

NON-SUPERVISED SENSORY-MOTOR AGENTS LEARNING

Raul Sidnei Wazlawick Dr. Eng.
EDUGRAF - INE - Universidade Federal de Santa Catarina - Brazil
raul@edugraf.ufsc.br

Antonio Carlos da Rocha Costa Dr. Sc.
Instituto de Informatica - Universidade Federal do Rio Grande do Sul - Brazil
rocha@inf.ufrgs.br

Abstract

This text discusses a proposal for creation and destruction of neurons based on the sensory-motor activity. This model, called *sensory-motor schema,* is used to define a sensory-motor agent as a collection of activity schemata. The activity schema permits a useful distribulion of neurons in a conceptual space, creating concepts based on action and sensation. Such approach is inspired in the theory of the Swiss psychologist and epistemologist Jean Piaget, and intends to make explicit the account of the processes of continuous interaction between sensory-motor agents and their environments when agents are producing cognitive structures.

1. Introduction

The notion of an *autonomous agent* plays a central role in contemporaneous research on Artificial Intelligence [3]. *Cognitive agents* are based on symbolic processing mechanisms. *Reactive agents* are based on alternative computational mechanisms like neural networks analogic processing etc. The alternative approach using autonomous agents based on hybrid mechanisms has not been extensively explored and the question of how to proceed for combining symbolic and non-symbolic mechanisms is an up-to-date research topic.

Our work [6] [1] is oriented towards Jean Piaget's Genetic Psychology and Epistemology [5]. According to that perspective, the two kinds of mechanisms are coordinated by the so called *equilibration mechanism.* The chracteristic of this kind of coordination is that the symbolic mechanism results from an elaborated construction, called *reflexive abstraction* [5] based on elements given by non-symbolic mechanism.

The sensory-motor mechanism is based on Kohonen's non-supervised neural networks [2] and on genetic algorithms [4] that were combined for reproducing the main structure of the sensory-motor level, called *sensory-motor schema* [5]. This combined mechanism is used to build neural networks capable of self-adaptation in changing environments.

2. Activity Schemata

A *signal* is defined as a rational value between () and 1. A *generalized .signal* is defined as a rational value between -1 and 1. An *alphabet* of *generalized signals* with a *p* digits mantissa is denoted by G_p. A sensorial entry e^n is defined as a signal string S_p^n. The i-th signal of e^n will be denoted by e_i^n.

A genetic alphahet A is defined as $A = S_p \cup \{\#\}$,where # denotes an improper element. The operation *difGene* $: Sp \times A \rightarrow \mathbf{Nat}$, is defined by *difGene* $= \lambda s.\lambda g.$*(if g = # then 1/3 else ABS(s-g)).* The operation *sumGene* $: AxGp \rightarrow A$ is defined by *sumGene* $= \lambda g.\lambda s.$*(if g=# then # else g+s).* The operation that adjuts a gene $g \in A$ to a signal $s \in S_g$, rated by a learning value $\propto$ is given by : *adjustGene:* $A \times S_p \times S_p \rightarrow A$, where *adjustGene* $= \lambda g.\lambda s.\lambda\propto$.*(if g=# then # else g + (g-s)*/$\propto$2).*

A chromosome is a string $<a_l, a_2, ..., a_n> \in A^n$. The *sirnilarity* between a sensorial entry $S^n \in S_p^n$ and a chromosome $c^n \in A^n$ produces a value that corresponds to the average of the application of the function *difGene* to the genes of c^n and to the signals at the corresponding positions at s^n. The function *similarity* $: S_n^p \times A^n \rightarrow \mathbf{Nat}$ is given by : *similarity* $= \lambda s^n . \lambda c^n .[\sum_{i=1}^{n} difGene(s_i^n, c_i^n)] /n..$ The operation*crossover* $: A^n x A^n x \mathbf{Nat} \rightarrow A^n x A^n$ is defined by *crossover* $= \lambda x^n.\lambda y^n.\lambda r.($ *if* $r = n$ *then* (x^n, y^n) *else* <

$x_1^n, ...x_r^n, y_{r+1}^n, ..., y_n^n>, < y_1^n, ..., y_r^n, x_{r+1}^n, ..., X_n^n >$). *A vector of mutation is* defined as a string of generalized signals from G_p. The operation *vmut* : $S_p x S_p x S_p \rightarrow$ ***Nat*** creates a vector of mutation respecting the following restrictions : *(i1 *t)* values are generated by a random function on the interval [- i2,+i2] and ((1-i1)*t) values are generated on the interval [-i3,+i3]. Therefore, *vmut* = $\lambda i1..\lambda i2.\lambda i3.\lambda t..$ < *randomSignal(i1,i2,i3)1 ,...,randomSignal(i1,i2,i3)t,* >. Mutation consists in the application of *sumGene* to the genes, of a chromosome $c^n \in A^n$ with the values of a vector of mutation $v^n \in G_p^n$.

Formally, *mutation* : $A^n \times G_p^n \rightarrow A^n$ is defined by *mutation* = $\lambda c^n \lambda v^n$.< sumGene(c_1^n, v_1^n), ...,sumGene (c_2^n, v_2^n),..., sumGene (c_n^n, v_n^n)>. The *reproduction of a pair of chromosomes* is an operation that generates a new pair of chromosomes. It's given by the function *reproduction* : $A^n \times A^n \rightarrow A^n \times A^n$ where reproduction = $\lambda c1^n . \lambda c2^n$.(let i = *random*n*, let $v1^n$ = *vmut*((0.1,1.0, 0.2), let $v2^n$=*vmut* (0.1,1.0,0.2) in (let ($p1^n, p2^n$) = *crossover* ($c1^n$,$c2^n$,i) in (*mutation*($p1^n$,$v1^n$), *mutation* ($p2^n$,$v2^n$))))

A neuron is defined as a pair *<c^n,1>*, where $c^n \in A^n$ and $I \in S_p$. The element I corresponds to an activation threshold for the neuron, that is, an excitation level above that the neuron triggers. The domain of neurons of size *n* will be denoted by : *Neuron*^{n} = A^n x S_p. The operation *application* : S_p^n x *Neuron*$^n \rightarrow$ is defined by *application* = $\lambda s^n..\lambda r^n$.*max((0, similarity(S^n, chromosome(r^n)) - threshold(r^n))*. The *adjustment* of a neuron to a sensorial entry is given by the adjustment of each of its genes. This function is given by : *adjust : Neuron^n* x S_p^n x $S_p \rightarrow$ *Neuron^n*, where *adjust* = λ*<c^n,I>..λe^n..$\lambda\alpha$.<<adjustGene* (c_1^n, e_1^n, α), ..., *adjustGene* (c_n^n, e_n^n, α)>,I>.

A schema consists of a tuple <<c_E^n,., I_E>, <c_0^n,I_0>, a^m, α, v>, where <c_E^n, I_E > is a sensorial entry neuron, <c_0^n,I_0> is a goal neuron, $a^m \in A^m$ is an action chromosome, $\alpha \in S_p$ is a learning value, and $v \in S_p$ is an evaluation value. The schemata domain will be denoted by *Schema^{n,m} = Neuron^{n}x Neuron^{n}xA^{m}x S_p x S_p*. The *training* of a schema is given by *training* = λesq^{n,m}..$\lambda S_E^n.\lambda S_0^n$.<*adjust(entry(esq^{n,m}),* S_E^n, learning(esq^{n,m})), adjust(goal(esq^{n,m}), S_0^n, learning(esq^{n,m})), action(esq^{n,m}), learning(esq^{n,m}), *evaluation(esq^{n,m})*>. The function *attenuation* : *Schema^{n,m}* x $S_p \rightarrow$ *Schema^{n,m}* is defined by : *attenuation* =λesq^{n,m}.$\lambda\varepsilon$.<*entry(esq^{n,m}), goal(esq^{n,m}), action(esq^{n,m}),* ε x *learning(esq^{n,m}), evaluation(esq^{n,m})*>. The function *re-evaluation : Schema^{n,m}* x S_p^n x $S_p^n \rightarrow$ *Schema^{n,m}* is given by *re-evaluation* = λesq^{n,m}..λ. $S_E^n.\lambda S_0^n$< *entry*(esq^{n,m}),goal (esq^{n,m}), *action*(esq^{n,m}), *learning*(esq^{n,m}), (*evaluation*(esq^{n,m}) x(1 - *similarity* (S_E^n,*entry* (esq^{n,m}))+ *similarity*(S_0^n, *goal*(esq^{n,m}))**similarity* (S_E^n,entry(esq^{n,m})))>.

Let $e1^{n,m}$= <<c_E^n,$I1_E$,>,<$c1_0^n$,$I1_0$ >,$a1^m$,α1,v1>, and $e2^{n,m}$= <<$c2_E^n$,$I2_E$ >,<$c2_0^n$,$I2_0$ >,$a2^m$,α2,v2> be schemata. The reproduction of the schemata $e1^{n,m}$ and $e2^{n,m}$ must originate a new pair of schemata $e3^{n,m}$ = <<$c3_E^n$,$I3_E$ >,<$c3_0^n$,$I3_0$ >,$a3^m$,α3,v3> and $e4^{n,m}$ = <<$c4_E^n$,$I4_E$ >,<$c4_0^n$,$I4_0$ >,$a4^m$,α4,v4> on the following way : (a) ($c3_E^n$,$c4_E^n$) = *reproduction* ($c1_E^n$,$c2_E^n$) ; (b) ($c3_0^n$,$c4_0^n$) = *reproduction* ($c1_0^n$,$c2_0^n$) ; (c) ($a3^m$,$a4^m$) = *reproduction* ($a1^m$,$a2^m$) ; (d) The pairs of activation threshold *($I3_E$,$I4_E$)* and ($I3_0$,$I4_0$) can be generated from *($I1_E$,$I2_E$)* and *($I1_0$,$I2_0$)* by the usual genetic process ; (e) The values $\alpha3$ and $\alpha4$ must be fixed near to their upper bound ; (f) The values of *v3* and *v4* must be fixed at an arbitrary value lower than 1/3.

A *population* consists of a set of schemata. The function *selection : Population^{n,m}* x $S_p^n \rightarrow$*Schema^{n,m}*, *is* given by *selection* = $\lambda Q^{n,m}$. λs^n.*(e^{n,m} such that e^{n,m} is selected among the schemata of Q^{n,m} with probability given by :*

$$application(e^{n,m}, s^n) / \sum_{j=1}^{n} application(e_j^{n,m}, S^n)).$$

<u>3. Non-Supervised Sensory-Motor Agent</u>

A non-.supervised sensory-motor agent will be defined as a set of sensors, effectors and a population $Q^{n,m}$. An agent lives in an environment $E^{n,m}$, that provides *n* signals to the agent's sensors, and it suffers actions composed by *m* signals coming from agent's effectors. An agent may be denoted by $A^{n,m}$ = <$Q^{n,m}$,e^m >, where $Q^{n,m}$ is a population, and e^m are signals present in the agent's effectors. The agent's environment may be denoted by $E^{m,n}$ = <$R^{m,n}$,s^n>, where s^n is a set of entry signals for the agent, and $R^{m,n}$ is a set of environment schemata. Let $S_{[t]}^n \in S_p^n$ and $e_{[t]}^m \in S_p^m$ be representations of signals at time *t*. Then, <$Q_{[t]}^{n,m}$,$e_{[t]}^m$Q> *is an instant description of a cognitive agent*, and <$R_{[t]}^{n,m}$,$S_{[t]}^m$> is an *instant description of environment.* The working of a non-supervised sensory-motor agent $A^{n,m} = <Q^{n,m}, e^{m}>$ in an environment $E^{m,n} = <R^{m,n}, s^{n}>$ is given by the following algorithm:

Step 1 : Selection. The agent observes the sensorial entry $s^{n}_{[t]}$ provided by the environment. Then the agent selects a winner schema for activation by $winner^{n,m}_{[t]} := selection(Q^{n,m}_{[t]}, s^{n}_{[t]})$.

Step 2 : Activation. The agent activates $winner^{n,m}_{[t]}$ by putting on the correspondent positions in $e^{m}_{[t+1]}$ the proper signals (those different of #) of the chromosome action($winner^{n,m}_{[t]}$).

Step 3 : Adjust. The agent adjusts the weights of the entry and goal neurons accordingly to the data provided by the environment : $entry(winner^{n,m}_{[t]}) := adjust(winnerl^{n,m}_{[t]}, s^{n}_{[t]}, learning(winner^{n,m}_{[t]})$; . and $goal(winner^{n,m}_{[t]}) := adjust(winner^{n,m}_{[t]}, s^{n}_{[t+1]}, learning(winner^{n,m}_{[t]})$.

Step 4 : Attenuation. The agent decrements the learning value to attenuating the adjusting rate of the schema: $winner^{n,m}_{[t]} := attenuation(winner^{n,m}_{[t]}, 0.95)$.

Step 5 : New evaluation. The agent observes again the sensorial entry after the effectors activation. Let this observation be represented by $s^{n}_{[t+1]}$. The agent calculates the new evaluation of $winner^{n,m}_{[t]}$ by $winnerl^{n,m}_{[t]} := re\text{-}evaluation(winnerl^{n,m}_{[t]}, s^{n}_{[t]}, s^{n}_{[t+1]})$.

Step 6 : Reproduction. If the re-evaluation of $winner^{n,m}_{[t]}$ reduced the value of $evaluation(winner^{n,m}_{[t]})$, then the agent makes the reproduction of $winner^{n,m}_{[t]}$ with another schema given by .$selection(Q^{n,m}_{[t+1]}, s^{n}_{[t]})$. The agent adds the new pair of schemata to $Q^{n,m}_{[t+1]}$ and eliminates from $Q^{n,m}_{[t+1]}$ the two schemata with the smallest values for v, and then returns to the step 1.

The constant adjustment of goal neurons increments gradually the evaluation of many schemata. In some moment, however, the action a^{m} applied to a certain entry may not product the desired effect, lowering the schema evaluation, and performing the role of a *perturbation,* as defined by Piaget [5].

Those perturbations begin schemata reproduction (Piaget has called this process *differentiation).* The reproduction doesn't destroy the schema that has suffered the perturbation, but the schemata with the lowest evaluation. Thus, the perturbed schema is maintained, and a pair of new schemata is created by reproduction. These new schemata will try to compensate the perturbation. In the case of reproduction with the perturbed schema, the weighted selection of schemata of highest value is in accord with the genetic algorithm process. This makes the agent to search some good schemata in order to try to accommodate the perturbation.

The interaction process between agent and environment may be thought of as a process going on through the evolution of the system *Agent x Environment,* represented by $<A^{n,m}, E^{m,n}>$. The evolution of this pair is denoted by a sequence of cycles of the form $<< Q^{n,m}_{[t]}, e^{m}_{[t]} >, < R^{n,m}_{[t]}, s^{n}_{[t]} >>$

$\rightarrow_A << Q^{n,m}_{[t]}, e^{m}_{[t+1]} >, < R^{n,m}_{[t]}, s^{n}_{[t]} >>$

$\rightarrow_E << Q^{m}_{[t]}, e^{m}_{[t+1]} >, < R^{m,n}_{[t+1]}, s^{n}_{[t+1]} >>$

$\rightarrow_A << Q^{m}_{[t+1]}, e^{m}_{[t+1]} >, < R^{m,n}_{[t+1]}, s^{n}_{[t+1]} >>$ where $\rightarrow_A$ represent steps of the agent's working, and $\rightarrow_E$ represent steps of the environment's working.

The sequence of those steps describes the join working of the pair agent/environment. It indicates how agent's actions modify its environment, and how these modifications affects the agent's schemata, and consequently it's sensory-motor capacity. Thus, it represents the agent's cognitive development, in interaction with its environment, and captures, in a certain sense, some processes identified by genetic epistemology.

4. Conclusions

The schema mechanism approximates a certain kind of human classification method. Concepts are associated to schemata, and schemata are built from observables, actions and goals. If the action reaches the schema's goal, then the observable belongs to the concept associated to the schema; otherwise the schema must be differentiated. The differentiation occurs when a couple of schemata is genetically reproduced. In this case, the action, observables and goals are recombined for creating new schemata, which try to better represent the activity structure related to the environmental objects. The genetic mechanism is responsible for finding the best configurations for the tuple *observahle x action x goal.*

5. References

[1] COSTA, Anlonio Carlos da R(xha *Machine Intelligence: Sketch of a Construtivist Approach.* Ph.D. Thesis, Porlo Alegre : CPGCC-UFRGS, 1993.
[2] DAYHOFF, Judilh E. *Neural Network Architectures - an introduction.* New York: Van Nostrand Reinhold, 1990.
[3] DEMAZEAU, Yvcs, MULLER, Jean-Pierre *Decen~ralized Artificial Inlelligence 1.* North-Holland, Elscvier Science Publishers, 1990.
[4] HOLLAND, *John Adaptation in Natural and Artificial Systems.* University of Michigan Press, 1975.
[5] PIAGET, Jean *The Equilibration of Cognitive Structures - The Central Problem of Cognitive Development.* Rio de Janeiro: Zahar, 1976.
[6] WAZLAWICK, Raul S. *An operatory Model for Knowledge Construction,* Ph.D. Thesis, Florianopolis: PGEP-UFSC, I 993 .

FUNCTIONAL EQUIVALENCE AND GENETIC LEARNING OF RBF NETWORKS

Roman Neruda*

Institute of Computer Science, Czech Academy of Sciences,
Pod vodárenskou věží 2, 182 07 Prague 8, Czech Republic
e-mail: roman@uivt.cas.cz

Abstract

In this paper a functional equivalence property of feedforward networks is introduced and studied for the case of radial basis function networks with Gaussian activation function and metrics induced by an inner product. The description of functional equivalent parameterizations is used for proposition of new genetic learning rules that operate only on a small part of the whole weight space.

1. Introduction

Feedforward network can be seen as a device computing a real function of several arguments with respect to the set of parameters (usually called weights). This view is particularly suitable if one tries to investigate the class of functions that can be approximated by a given type of network. This subject has been extensively studied in recent years and gradually a so called *universal approximation* property has been proven for both multilayer perceptrons and radial basis function networks with quite general activation functions. This property means that a given network is capable to approximate any continuous or measurable function with arbitrary precision.

The essential criterion for judging the suitability of a particular network architecture—its ability to learn all reasonable functions—thus holds for both most frequently used architectures. That is the reason for looking for another, "finer" criteria that can tell us which network architecture is more suitable in each particular case. The reasonable selection could be time or space complexity of the network. The space complexity means the number of network parameters, which is mostly affected by the number of (hidden) units in the network. Time complexity in this context means the learning time. Practical results show that the learning time represents a cruciaal factor in using feedforward networks for real world applications.

*This work was partially supported by Grant Agency of the Czech Republic under grants number 201/93/0427, 201/94/0729 and 201/95/0976.

In this paper we present an algorithm that should reduce the time complexity of learning by reduction of the parameter space that is necessary for learning.

Hecht-Nielsen in [5] proposed to study parameters of the network that determine the same input/output network function—so called functionally equivalent network parameterizations. His idea was that the knowledge about these parameterizations could speed up some learning algorithms. Several authors have studied functional equivalent parameterizations for multilayer perceptron networks with various activation functions. Sussmann [10], Chen et al. [3] studied perceptron-type networks with hyperbolic tangent. Kůrková and Kainen [8], Kainen et al. [7], Albertini and Sontag [1] extended the results for logistic sigmoid, polynomials, and other activation functions. Kůrková and Neruda in [9] characterized functional equivalent parameterizations for Gaussian radial basis function networks.

If the functional equivalent parameterizations can be described, one can divide the whole parameter space into the classes of functional equivalence. Furthermore, it is possible to choose one parameterization in each class (so called *canonical parameterization*) and to restrict the learning algorithm only to these representatives. Thus, we are still able to cover the whole output space by using only a subset of the parameter space. The cardinality of this subset depends on the structure of canonical parameterizations and varies depending on the network architecture and activation function type. Usual gradient learning algorithms, such as back propagation can not be restricted to a certain subset of the parameter space, but other learning schemes as genetic learning are able to operate only on canonical parameterizations.

In this paper we characterize canonical parameterizations of radial basis function networks with Gaussian activation function and propose a genetic learning algorithm that can take advantage of this description. The organization of the paper is as follows: necessary definitions and preliminaries are gathered in section 2 and the characterization of functional equivalent parameterizations is given in section 3.

Section 4 contains the proposed learning algorithm, while the conclusions and discussion appear in section 5.

2. Preliminaries

We study networks with n input units, one hidden layer containing k RBF-units with a radial function $\varphi : \mathcal{R}_+ \to \mathcal{R}$, a norm $\|\cdot\|$ on $\mathcal{R}^n$ and with a single linear output unit. Such networks compute functions $f(\mathbf{x})\colon \mathcal{R}^n \to \mathcal{R}$ of the form:

$$f(\mathbf{x}) = \sum_{i=1}^{k} w_i \varphi\left(\frac{\|\mathbf{x} - \mathbf{c}_i\|}{b_i}\right),$$

where $w_i, b_i \in \mathcal{R}, b_i > 0$, and $\mathbf{c}_i \in \mathcal{R}^n (i = 1, \dots, k)$ are parameters (w_i are called weights, b_i widths and $\mathbf{c}_i$ centroids of RBF units). The set of parameters $\mathbf{P} = \{w_i, b_i, \mathbf{c}_i; i = 1, \dots, k\}$ is called a $(\varphi, \|\cdot\|)$ RBF *network parameterization.*

Two $(\varphi, \|\cdot\|)$ RBF network parameterizations $\mathbf{P} = \{w_i, b_i, \mathbf{c}_i; i = 1, \dots, k\}$ and $\mathbf{P}' = \{w_i', b_i', \mathbf{c}_i'; i = 1, \dots, k'\}$ are called *functionally equivalent* if they determine the same input-output function, i.e.

$$\sum_{i=1}^{k} w_i \varphi\left(\frac{\|\mathbf{x} - \mathbf{c}_i\|}{b_i}\right) = \sum_{i=1}^{k'} w_i' \varphi\left(\frac{\|\mathbf{x} - \mathbf{c}_i'\|}{b_i'}\right).$$

Two network parameterizations are called *interchange equivalent*, if $k = k'$ and there exists a permutation π of the set $\{1, \dots, k\}$, such that $w_i = w'_{\pi(i)}$ and $b_i = b'_{\pi(i)}$ and $\mathbf{c}_i = \mathbf{c}'_{\pi(i)}$, for each $i \in \{1, \dots, k\}$.

Network parameters are set during a learning process with respect to a particular *training set* T—the collection of z pairs representing the input of the network $\mathbf{x}_j$ and the desired output y_j. $T = \{(\mathbf{x}_j, y_j); j = 1, \dots, z\}$. The elements of the training set are called *training patterns.*

3. Functional equivalence

The standard choice of a radial function is Gaussian and the most popular norms are those induced by various inner products. The following theorem states that in this case an input-output function determines the network parameterization uniquely up to a permutation of hidden units.

Theorem 1 (Kůrková and Neruda, [9]) *Let n be a positive integer, $\|\cdot\|$ norm on $\mathcal{R}^n$ induced by an inner product, $\gamma(t) = \exp(-t^2)$. Then any two reduced $(\gamma, \|\cdot\|)$ RBF network parameterizations are functionally equivalent if and only if they are interchange equivalent.*

This theorem enables us to describe easily a canonical representation of a network computing a particular function. Since no parameter transformations other than interchanges leave the input-output function unchanged, we only have to choose a canonical representation of a set of networks with permuted hidden units.

One of the possible choices is to impose a condition on a parameterization that weight vectors corresponding to hidden units are increasing in a lexicographic ordering. Represent a parameterization $\mathbf{P} = \{w_i, k_i, \mathbf{c}_i; i = 1, \dots, k\}$ as a vector $\mathbf{P} = \{\mathbf{p}_i, \dots \mathbf{p}_k\} \in \mathcal{R}^{k(n+2)}$, where $\mathbf{p}_i = \{w_i, b_i, c_{i1}, \dots, c_{in}\} \in \mathcal{R}^{n+2}$ is a weight vector corresponding to the i-th hidden unit.

Let $\prec$ denotes the lexicographic ordering on $\mathcal{R}^{n+2}$. For $\mathbf{p}, \mathbf{q} \in \mathcal{R}^{n+2}$ we say that $\mathbf{p}$ preceeds $\mathbf{q}$ and write $\mathbf{p} \prec \mathbf{q}$ if there exists an index $m \in \{1, \dots, n+2\}$ such that $p_j = q_j$ for $j < m$ and $p_m < q_m$.

For a norm $\|\cdot\|$ induced by an inner product, we call a $(\gamma, \|\cdot\|)$ RBF network parameterization $\mathbf{P}$ *canonical* if

$$\mathbf{p}_1 \prec \mathbf{p}_2 \prec \dots \prec \mathbf{p}_k. \tag{1}$$

Canonical parameterizations create subset of all parameterizations containing exactly one representative from each class of functional equivalence. From this point of view, the set of canonical parameterizations corresponds to the *minimal search set* proposed by Chen and Hecht-Nielsen in [2].

4. Genetic rules

In order to take advantage of the previous result a learning algorithm is needed that can operate only on canonical parameterizations. Unfortunately, neither back propagation nor more complicated three step learning algorithms of RBF networks (cf. Hlaváčková and Neruda, [6]) are suitable, since the analytical solution obtained in each step of the iterative process cannot in principle be limited to certain weight space subset. This is not so with genetic algorithm whose operations can be changed to preserve the property of being canonical.

The proposed algorithm works with the strings of network parameterizations (see Fig. 1). In the beginning a popula-

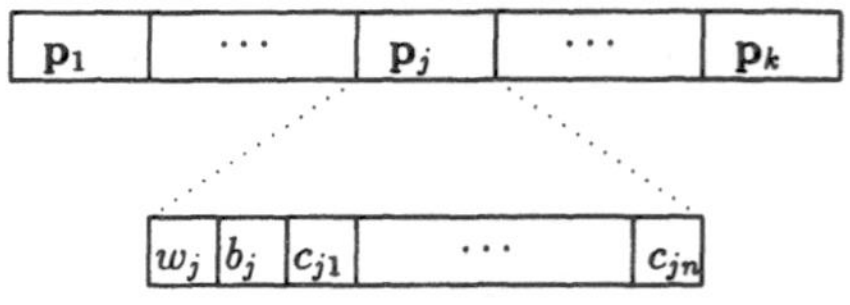

Figure 1: The string of the network parameterization $\mathbf{P}$

tion of m canonical parameterizations

$$\mathcal{P}_0 = \{\mathbf{P}_1, \dots, \mathbf{P}_m\}$$

is generated at random. Having population $\mathcal{P}_i$, the successive population $\mathcal{P}_{i+1}$ is generated by means of three basic genetic operations: *reproduction*, *crossover* and *mutation* that again generate only canonical parameterizations.

Reproduction

The reproduction operator represents a probabilistic selection of parameterization $\mathbf{P} \in \mathcal{P}_i$ according to the values of objective function $G(\mathbf{P})$. The simplest way how to compute values of G is to use the usual error function $E(\mathbf{P})$ (i.e. the sum of distances between the actual and desired output of the network over all patterns from the training set) and subtract it from a constant C. This constant can be either computed or estimated maximum of function realized by the network or it can simply be set according to the maximal error in each population.

Thus, we have

$$G(\mathbf{P}) = C - \sum_{j=1}^{z} ||f_{\mathbf{P}}(\mathbf{x}_j) - y_j||^2,$$

where y_j is the desired network output and $f_{\mathbf{P}}(\mathbf{x}_j)$ represents the actual response of the network with parameterization $\mathbf{P}$ when the input $\mathbf{x}_j$ is presented.

For each parameterization $\mathbf{P}_l \in \mathcal{P}_i$ the value of its objective function $G(\mathbf{P}_l)$ is computed and then normalized by dividing by the sum of objective function values over all parameterizations in the population:

$$p_l = \frac{G(\mathbf{P}_l)}{\sum_{r=1}^{m} G(\mathbf{P}_r)}.$$

The number p_l then represents the probability with which the parameterization is selected at random.

Mutation

The mutation operates at two levels—first the parameterization $\mathbf{P}$ is considered as a vector $(\mathbf{p}_1, \ldots, \mathbf{p}_k)$ and an element $\mathbf{p}_s$ is chosen randomly as a candidate for mutation. Because we have to preserve the canonical property, $\mathbf{p}_s$ can not be changed arbitrarily. It is necessary that the condition

$$\mathbf{p}_{s-1} \prec \mathbf{p}_s \prec \mathbf{p}_{s+1}$$

remains satisfied. Thus, the neighbours $\mathbf{p}_{s-1}$ and $\mathbf{p}_{s+1}$ determine the lower and upper border of the range in which the mutation can change $\mathbf{p}_s$. This change is done randomly again. The resulting parameterization $\mathbf{P}' \in \mathcal{P}_{i+1}$ has the form

$$\mathbf{P}' = (\mathbf{p}_1, \ldots, \mathbf{p}_{s-1}, \mathbf{p}'_s, \mathbf{p}_{s+1}, \ldots, \mathbf{p}_k),$$

where $\mathbf{p}'_s$ is modified $\mathbf{p}_s$ (see Fig. 2).

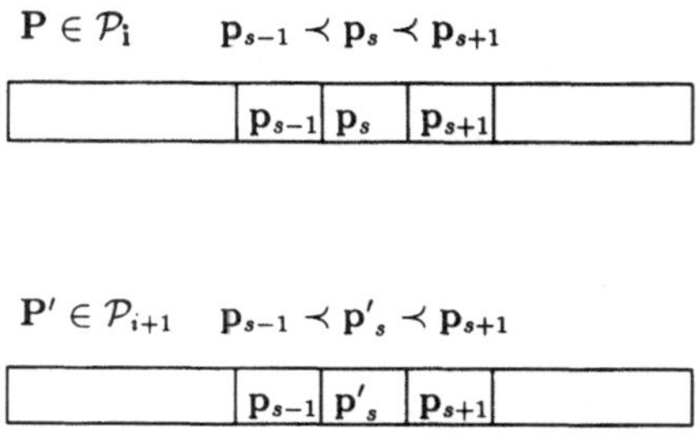

Figure 2: The mutation operator creates $\mathbf{P}' \in \mathcal{P}_{i+1}$ from $\mathbf{P} \in \mathcal{P}_i$.

Crossover

The crossover operation chooses two parameterizations $\mathbf{P} = (\mathbf{p}_1, \ldots, \mathbf{p}_k)$ and $\mathbf{Q} = (\mathbf{q}_1, \ldots, \mathbf{q}_k)$ in $\mathcal{P}_i$ that will provide a new offspring $\mathbf{P}' \in \mathcal{P}_{i+1}$, first. After that a position s is found at random such that the parameterization

$$\mathbf{P}' = (\mathbf{p}_1, \ldots, \mathbf{p}_s, \mathbf{q}_{s+1}, \ldots, \mathbf{q}_k)$$

still satisfies (1). In particular, it means that the condition

$$\mathbf{p}_s \prec \mathbf{q}_{s+1}$$

has to be verified. Fig. 3 illustrates the function of the crossover operator.

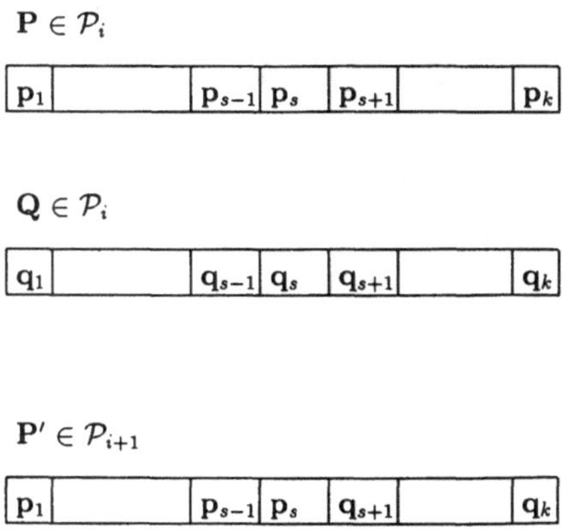

Figure 3: The crossover operator creates $\mathbf{P}' \in \mathcal{P}_{i+1}$ by combining $\mathbf{P}$ and $\mathbf{Q}$ from Π_i.

5. Discussion

We have presented a modification of genetic operators of reproduction, crosssover and mutation such that the genetic learning algorithm works only with the canonical network parameterizations. Time complexity of the above described operators is not constant as in the case of classical genetic learning. The crossover operator requires $O(k)$ operations and the time complexity of mutation is $O(k+n)$ (recall that k is the number of hidden units and n the input dimension). On the other hand, the reduction of search space is much greater—since there are $k!$ permutations of k hidden units, the algorithm works on the $1/k!$ of the whole weight space. For large k this allows to achieve a considerable time reduction.

We believe that the major drawback in using any genetic algorithm in neural networks learning is the necessity to present all training patterns to the network when computing the objective function values. But this can be overcome by the parallel nature of genetic learning. Genetic algorithms are more suitable for parallel implementation than back propagation since thay require less communication between parallel processes.

Such a parallel implementation of the proposed algorithm will be one of the directions of our further research. Currently, we are also extending our results concerning functional equivalence of radial basis function networks. Another extension leads to proposal of a similar genetic rules for other architectures such as multilayer perceptrons, where the structure of cannonical parameterizations is sometimes more complicated.

References

1. Albertini, F., Sontag, E.D.: For neural networks, function determines form, *Neural Networks*, **6**, 975-990, (1993).
2. Chen, A. M., Hecht-Nielsen, R.: On the geometry of feedforward neural network weight spaces. In *Proceedings of the 2nd IEE Conference on Artificial Neural Networks* 1-4, London: IEE Press, 1991.
3. Chen, A. M., Haw-minn, L., Hecht-Nielsen, R.: On the geometry of feedforward neural network error surfaces, *Neural Computation*, **5**, 910-927, (1993).
4. Goldberg, D. E.: Genetic algorithms in search, optimization, and machine learning. New York: Addison–Wesley, 1989.
5. Hecht-Nielsen, R.: On the algebraic structure of feedforward network weight spaces. In *Advanced Neural Computers*, 129-135, Elsevier, 1990.
6. Hlaváčková, K., Neruda, R.: Radial Basis Function Networks, *Neural Network World*, **3**, 1, 93-101 (1993).
7. Kainen, P. et al: A New Criterion for Choosing an Activation Function: Uniqueness Leads to Faster Learning, *Neural Parallel and Scientific Computations* (in press)
8. Kůrková, V., Kainen, P.: Functionally Equivalent Feedforward Networks, *Neural Computation*, **6**, 543-558, (1993).
9. Kůrková, V., Neruda, R. Uniqueness of Functional Representation by Gaussian Basis Function Networks. In *Proceedings of ICANN'94*, 417-474, Springer: London, 1994.
10. Sussmann, H.J.: Uniqueness of the weights for minimal feedforward nets with a given input-output map, *Neural Networks,* **5**, 4, 589-594, (1992).

TEACHING RELAXATION LABELING PROCESSES USING GENETIC ALGORITHMS

Marcello Pelillo **Fabio Abbattista** **Angelo Maffione**

Dipartimento di Informatica, Università di Bari
Via E. Orabona 4, 70126 Bari, Italy

Abstract – In a recent work, a learning procedure for relaxation labeling algorithms has been introduced which involves minimizing a certain cost function with classical gradient methods. The gradient-based learning algorithm suffers from some inherent drawbacks that could prevent its application to real-world problems of practical interest. Essentially, these include the inability to escape from local minima and its computational complexity. In this paper, we propose using genetic algorithms to solve the relaxation labeling learning problem to overcome the difficulties with the gradient algorithm. Experiments are presented which demonstrate the superiority of the proposed approach both in terms of quality of solutions and robustness.

1. INTRODUCTION

Relaxation labeling processes are a broad class of popular techniques within the pattern recognition and machine vision domains [1]-[3]. They are parallel iterative procedures that attempt to combine local and contextual information in order to remove, or at least reduce, labeling ambiguities in classification problems where local measurements may be noisy or unreliable. In (continuous) relaxation labeling models, contextual information is embedded in a set of real-valued *compatibility coefficients*, which quantitatively express the degree of agreement of label configurations. Over the past years, several heuristic statistical-based interpretations such as correlation [1] or mutual information [4] have been formulated.

Recently, a novel approach for determining the compatibilities of relaxation labeling procedures has been introduced which views the problem as one of *learning* [5]. This amounts to minimizing a certain cost function which quantifies the degree of "goodness" of a given set of compatibility strengths, so that the learning task is formulated in terms of an optimization problem. In previous work [5], classical gradient techniques were used to accomplish this. However, gradient-based learning algorithms exhibit some inherent limitations that could prevent them from being applied to high-dimensional problems of practical interest. These include the inability to escape from local minima and their high computational complexity – which is of the order of the fourth power of the number of labels (or classes) of the problem at hand [5]. In addition, we note that some relaxation schemes are even non-differentiable (see, e.g., [6]) and this completely prevents the gradient algorithm from being applied.

In this paper we attempt to overcome the limitations of gradient-based relaxation labeling learning procedures by proposing the use of genetic algorithms (GAs) [7]. We found GAs to be advantageous not only because are able to find nearly globally optimal solutions without being trapped into local optima, but also because are less computationally expensive, requiring on each step a time roughly proportional to the square of the number of labels.

2. RELAXATION LABELING AND THE LEARNING PROBLEM

Relaxation labeling processes involve a set of objects $B = \{b_1, \cdots, b_n\}$ and a set of labels $\Lambda = \{1, \cdots, m\}$. The purpose is to label each object of B with one label of Λ. By means of some local measurement it is generally possible to construct, for each object b_i, a vector $p_i^{(0)} = (p_{i1}^{(0)}, \cdots, p_{im}^{(0)})^T$ such that $p_{i\lambda}^{(0)} \geq 0$ (all i and λ) and $\sum_\lambda p_{i\lambda}^{(0)} = 1$ (all i). Each $p_i^{(0)}$ can therefore be interpreted as the *a priori* (non contextual) probability distribution of labels for b_i. By simply concatenating $p_1^{(0)}, p_2^{(0)}, \cdots, p_n^{(0)}$, we obtain an initial weighted labeling assignment for the objects of B that will be denoted by $p^{(0)} \in \mathbb{R}^{nm}$. The compatibility model is represented by a four-dimensional matrix of real-valued nonnegative compatibility coefficients R: the element $r_{ij}(\lambda,\mu)$ measures the strength of compatibility between the hypotheses "λ is on object b_i" and "μ is on object b_j." In what follows, we will find it convenient to "linearize" the compatibility matrix and consider it as a column vector $\boldsymbol{r}$.

The relaxation labeling algorithm accepts as input the initial labeling assignment $\boldsymbol{p}^{(0)}$ and updates it iteratively taking into account the compatibility model, in order to achieve global consistency. At the tth step ($t = 0,1,2,\cdots$) the labeling is updated according to the following formula [1]:

$$p_{i\lambda}^{(t+1)} = p_{i\lambda}^{(t)} q_{i\lambda}^{(t)} \Big/ \sum_{\mu=1}^{m} p_{i\mu}^{(t)} q_{i\mu}^{(t)} \qquad (1)$$

where

$$q_{i\lambda}^{(t)} = \sum_{j=1}^{n} \sum_{\mu=1}^{m} r_{ij}(\lambda,\mu) p_{j\mu}^{(t)} . \qquad (2)$$

The process is stopped when some termination condition is satisfied (e.g., when the distance between two successive labelings becomes negligible) and the final labeling is usually used to label the objects of $\boldsymbol{B}$ according to a maxima selection criterion [6].

Now, let us focus on the learning problem. Suppose that a set of learning samples $L = \{L_1,\cdots,L_N\}$ is given, where each sample L_γ is a set of labeled objects of the form

$$L_\gamma = \left\{ (b_i^\gamma, \lambda_i^\gamma) : 1 \le i \le n_\gamma,\ b_i^\gamma \in \boldsymbol{B},\ \lambda_i^\gamma \in \boldsymbol{\Lambda} \right\} .$$

For each $\gamma = 1\ldots N$, let $\boldsymbol{p}^{(L_\gamma)} \in \mathbf{R}^{n_\gamma m}$ denote the unambiguous labeling assignment for the objects of L_γ, i.e.,

$$p_{i\alpha}^{(L_\gamma)} = \begin{cases} 0, & \text{if } \alpha \ne \lambda_i^\gamma; \\ 1, & \text{if } \alpha = \lambda_i^\gamma. \end{cases}$$

Also, suppose that we have some mechanism for constructing an initial labeling $\boldsymbol{p}^{(I_\gamma)}$ on the basis of the objects in L_γ, and let $\boldsymbol{p}^{(F_\gamma)}$ denote the labeling produced by the relaxation algorithm when $\boldsymbol{p}^{(I_\gamma)}$ is given as input.

Broadly speaking, the learning problem for relaxation labeling is to determine a compatibility vector $\boldsymbol{r}$ so that the final labeling $\boldsymbol{p}^{(F_\gamma)}$ be as close as possible to the desired labeling $\boldsymbol{p}^{(L_\gamma)}$, for each γ. To do this, we can define a cost function measuring the loss incurred when $\boldsymbol{p}^{(F_\gamma)}$ is obtained instead of $\boldsymbol{p}^{(L_\gamma)}$, and attempt to minimize it. Here, we use the following information-theoretic divergence measure recently proposed by Lin [8]:

$$E_\gamma = n_\gamma - \sum_{i=1}^{n_\gamma} \log_2\left(1+p_{i\lambda_i^\gamma}^{(F_\gamma)}\right) . \qquad (3)$$

The total error over L can therefore be defined as

$$E = \sum_{\gamma=1}^{N} E_\gamma . \qquad (4)$$

In conclusion, the learning problem for relaxation labeling can be stated as the problem of minimizing the function E with respect to $\boldsymbol{r}$. In [5], this problem is solved by means of a gradient method which begins with an initial point $\boldsymbol{r}_0$ and produces a sequence $\{\boldsymbol{r}_k\}$ as follows: $\boldsymbol{r}_{k+1} = \boldsymbol{r}_k - \alpha_k \boldsymbol{u}_k$, where $\boldsymbol{u}_k$ is a direction vector determined from the gradient of E, and α_k is a suitable step size.

3. LEARNING COMPATIBILITY COEFFICIENTS WITH GENETIC ALGORITHMS

Genetic algorithms are parallel search procedures largely inspired from the mechanisms of evolution in natural systems [7]. They work with a constant-size population of chromosomes or individuals, each associated with a fitness value that determines its probability of surviving at the next generation. In the present application, each chromosome represents a compatibility vector $\boldsymbol{r}$; each coefficient $r_{ij}(\lambda,\mu)$ is mapped into a fixed-length string of bits, and the whole chromosome is then obtained by simply concatenating these strings.

The GA starts out with an initial population of S members generally chosen at random and, in its simplest version, makes use of three basic operators: reproduction, crossover, and mutation. The most popular way of implementing the reproduction operator, commonly referred to as *roulette-wheel* selection [7], consists of choosing the chromosomes that are to be copied in the next generation according to a probability proportional to its fitness. One problem with this mechanism is that the best individuals need not survive in future generations and this can slow down the convergence of the algorithm. To overcome this drawback, we made use of an *elitist* reproduction mechanism [9] which consists of copying deterministically the best individual of each generation into the succeeding one, the other members being copied according to the usual roulette-wheel strategy. Once that the best individuals have been selected, the crossover operator is applied between pairs of individuals in order to produce new offsprings. The operator proceeds in two steps. First, two members of the mating pool are chosen at random; next, a cut point is determined (again) randomly and the corresponding right-hand segments are swapped. The frequency with which crossover is applied is controlled by a parameter P_c. Recall that in our application each chromosome represents a vector of compatibilities; hence, the crossover point is allowed to fall only at the boundary between two succeeding coefficients. Doing so, crossover does not create new compatibility coefficients but simply exchanges coefficients between vectors. The task of producing new compatibility coefficients is assigned to the mutation operator which consists of reversing the value of every bit within a chromosome with fixed probability P_m.

To map the function E defined in (4) into a fitness function, we used the following formula proposed in [10]:

$$F = \tan\left\{ \frac{\pi}{2}\left(1 - E/E_{max}\right) \right\}, \tag{7}$$

where E_{max} is the maximum value of E, and to avoid premature convergence toward mediocre individuals, the fitness was scaled according to the scheme suggested in [7].

In contrast with the gradient algorithm, the GA-based learning procedure is much less computationally demanding as it requires, on each generation, a number of operations roughly proportional to $S \cdot m^2$ (this should be compared with the $O(m^4)$ operations required by the gradient procedure [5]). However, many generations are typically needed for the GA to find an optimal solution and this suggests that, as far as learning time is concerned, the GA-based learning procedure cannot be regarded as a serious competitor of the gradient method when $S \approx m^2$.

4. RESULTS

To assess the effectiveness of the proposed evolutionary learning algorithm, experiments over a simple computer vision application were carried out [1]. The problem consists of labeling the sides of a triangle over a background, according to a 3-D interpretation. Each side can be interpreted as a convex edge (+), a concave edge (−), or as one of two types of occluding edges (> or <). There are 64 possible labeling configurations but only eight of them correspond to meaningful interpretations [1].

In the experiments presented here we idealized a very unfavorable as well as unlikely scenario for the learning algorithm in that the initial labeling assignments were strongly biased toward incorrect yet meaningful interpretations. Fig. 1 shows the learning set used in the study; the goal was to train the relaxation process to produce the desired labelings, given the input labelings, after a predetermined number of iterations. In particular, to study how the number of relaxation labeling steps affects the learning abilities of the competing algorithms, three series of simulations were carried out by fixing them at 1, 5, and 10. It is readily seen that standard statistical compatibilities perform poorly on such a task.

Case	Initial labeling				Desired labeling			
	>	<	-	+	>	<	-	+
1)	.30	.70	.00	.00	1	0	0	0
	.30	.70	.00	.00	1	0	0	0
	.30	.70	.00	.00	1	0	0	0
2)	.70	.30	.00	.00	0	1	0	0
	.70	.30	.00	.00	0	1	0	0
	.70	.30	.00	.00	0	1	0	0
3)	.30	.70	.00	.00	1	0	0	0
	.00	.00	.30	.70	0	0	1	0
	.30	.70	.00	.00	1	0	0	0
4)	.30	.70	.00	.00	1	0	0	0
	.30	.70	.00	.00	1	0	0	0
	.00	.00	.30	.70	0	0	1	0
5)	.00	.00	.30	.70	0	0	1	0
	.30	.70	.00	.00	1	0	0	0
	.30	.70	.00	.00	1	0	0	0
6)	.70	.30	.00	.00	0	1	0	0
	.00	.00	.70	.30	0	0	0	1
	.70	.30	.00	.00	0	1	0	0
7)	.70	.30	.00	.00	0	1	0	0
	.70	.30	.00	.00	0	1	0	0
	.00	.00	.70	.30	0	0	0	1
8)	.00	.00	.70	.30	0	0	0	1
	.70	.30	.00	.00	0	1	0	0
	.70	.30	.00	.00	0	1	0	0

Fig. 1. Training set used in the experiments.

We made a comparison between the genetic and the gradient learning algorithm. The GA was run using the following parameters: $S = 50$, $P_c = 0.5$, $P_m = 0.01$, and each compatibility coefficient was encoded into a 10-bit string. The step size α_k in the gradient algorithm was kept fixed at 0.1, and the direction vector u_k was normalized to avoid unacceptable oscillations. For each predetermined number of relaxation labeling steps, ten independent runs of both the gradient and the genetic algorithm were performed, each started from randomly chosen initial compatibility vectors. The performance of the algorithms is shown in

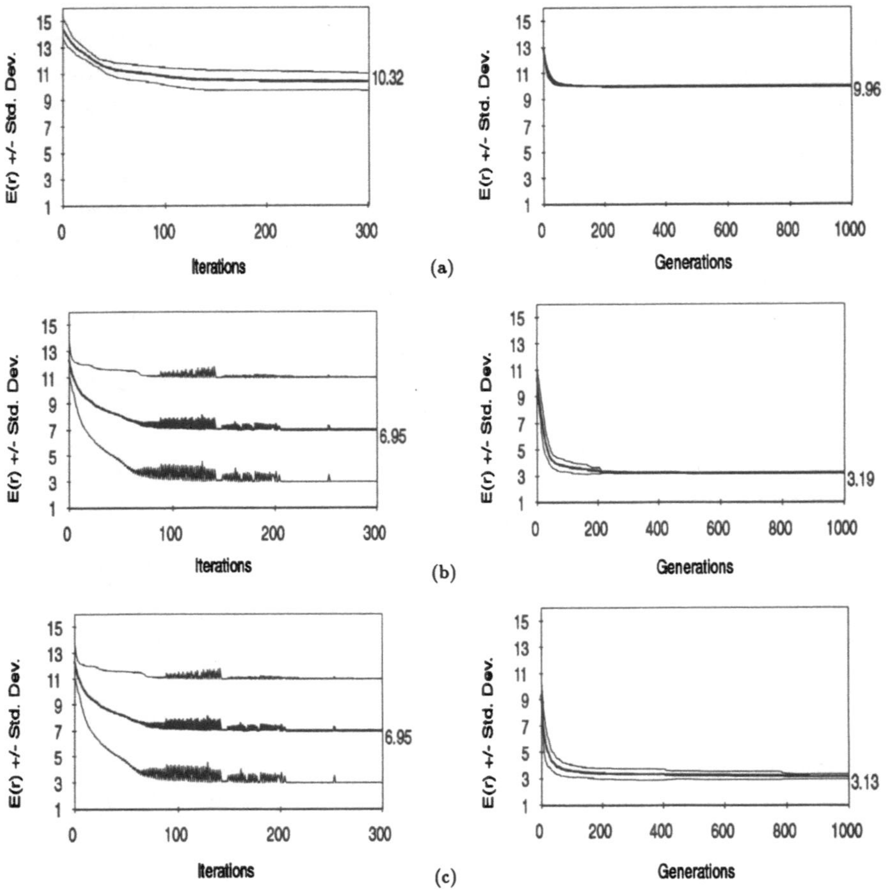

Fig. 2. Behavior of the mean error function E ($\pm$ standard deviation) during training. Left: gradient algorithm; Right: GA. The graphs from (a) to (c) correspond to 1, 5, and 10 relaxation labeling steps, respectively.

Fig. 2, where the mean learning curves along with their standard deviations (based on the ten trials) are plotted – in the case of the GA the graphs represent the average behavior of the *best* population members. As can be seen, in all the three series of experiments the GA performed significantly better than the gradient algorithm. The major point worth noting is that the GA not only determined on the average better compatibility coefficients but, in contrast with the gradient procedure, was also able to do so irrespective of the starting conditions, thereby exhibiting a much more robust behavior. It is interesting to notice how this superiority becomes more and more marked as the number of relaxation labeling steps increases. This behavior has an intuitive explanation. Observe, in fact, that allowing the relaxation process to perform more iterations amounts to increasing the degree of nonlinearity in the error function E; thus, the surface defined by E becomes much more complex and this emphasizes the tendency of the gradient algorithm to get stuck in local minima.

REFERENCES

1. A. Rosenfeld, *et al.*: *IEEE-SMC* **6**, 420 (1976).
2. R. Hummel, S. Zucker: *IEEE-PAMI* **5**, 267 (1983).
3. L. Davis, A. Rosenfeld: *Artif. Intell.* **17**, 245 (1981).
4. S. Peleg, A. Rosenfeld: *IEEE-SMC* **8**, 548 (1978).
5. M. Pelillo, M. Refice: *IEEE-PAMI* **16**, 933 (1994).
6. S. Zucker, *et al.*: *IEEE-PAMI* **3**, 117 (1981).
7. D. Goldberg: *Genetic algorithms in search, optimization and machine learning.* Addison-Wesley (1989).
8. J. Lin: *IEEE-IT* **37**, 145 (1991).
9. J. Grefenstette: *IEEE-SMC* **16**, 122 (1986).
10. T. Caudell, C. Dolan: *Proc. 3rd ICGA*, 370 (1989).

VLSI OPTIMAL NEURAL NETWORK LEARNING ALGORITHM *

Valeriu Beiu†,‡ and John G. Taylor†

† King's College London, Centre for Neural Networks, Department of Mathematics, Strand, London WC2R 2LS, UK
‡ on leave of absence from "Politehnica" University, Computer Science, Spl. Independentei 313, RO-77206 Bucharest, România

Abstract **– In this paper we consider binary neurons having a threshold nonlinear transfer function and detail a novel direct design algorithm as an alternative to the classical learning algorithms which determines the number of layers, the number of neurons in each layer and the synaptic *weights* of a particular neural network. While the feedforward neural network is described by *m* examples of *n* bits each, the optimisation criteria are changed. Beside the classical *size*-and-*depth* we also use the *A* and the AT^2 complexity measures of VLSI circuits (*A* being the *area* of the chip, and *T* the *delay* for propagating the inputs to the outputs). We considering the maximum *fan-in* of one neuron as a parameter and proceed to show its influence on the *area*, obtaining a full class of solutions. Results are compared with another constructive algorithm. Further directions for research are pointed out in the conclusions, together with some open questions.**

I. INTRODUCTION

In this paper we shall consider only feedforward neural networks (**NNs**) made of threshold gates (**TGs**):

$$Z_k = (z_0, \dots, z_{n-1}) \in \{0,1\}^n,\ k = 1, \dots, m, \text{ and}$$

$$f(Z_k) = sgn\left(\textstyle\sum_{i=0}^{n-1} w_i z_i + \theta\right),$$

with w_i called the synaptic *weights*, θ known as the *threshold*, and *sgn* the sign. A feedforward NN will be a feedforward TG circuit having as cost functions: (i) *depth* (i.e. number of layers), and (ii) *size* (i.e. number of TGs). These are linked to *T* (*depth*) and *A* (*size*) of a VLSI chip. Still, TGs do not closely follow these proportionalities as: (i) the *area* of the connections counts; (ii) the *area* of one TG is related to its associated *weights*. That is why several authors have taken into account the *fan-in* [1], the total number of connections, the total number of bits needed to represent the *weights* or even more precise approximations like the *sum of all the weights and the thresholds* [2]:

$$area \propto \sum_{all\ TGs} \left(\sum_{i=0}^{n-1} |w_i^{TG}| + |\theta^{TG}| \right). \qquad (1)$$

Equation (1) has also been used to define the minimum-integer TG realisation of a function. The *depth* will be assumed to be a good approximation of the *delay* (we neglect the *delay* introduced by long wires [3]). Other sharp limitations for VLSI are: (i) the maximal *fan-in* cannot grow over a certain limit; (ii) the maximal ratio between the largest and the smallest *weight*.

Two natural questions arise when using NNs: (i) *"How to determine the synaptic weights?"*; and (ii) *"How many neurons should one use in each hidden layer?"*. These questions can either be solved by learning techniques or by constructive algorithms. *Learning algorithms* [4] suffer from the inherent error correction function which does not guarantee that the global minimum will ever be reached and the very long time needed for solving the problem. Other aspects, like the precision of the *weights* [5, 6] or the high *fan-in* of neurons, are completely neglected. These preclude their VLSI implementation. *Constructive algorithms* [7, 8] are error prone, so they will find the correct set of *weights*. They can be divided into *algebraic* [9], *network-based* [10], or *geometric* [11, 12]. Some of them are for particular functions, or use queries and/or hints [13]. An excellent overview of seven constructive algorithms can be found in [10]. There exist *alternate solutions* developed for overcoming some of the above mentioned deficiencies: *learning algorithms with quantified weights* [14–19], and *polynomial time direct design* [20].

In section 2 we shall shortly present some results concerning the *depth*, the *size* and the *area* of $F_{n,m}$. This is the class of Boolean Functions (**BFs**) of *n* bits having *m* groups of ones in their truth table. Based on already established *depth*, *size* and *area* values for implementing the COMPARISON of two *n*-bit numbers [2, 21], we review a constructive theorem [2] for implementing any function belonging to $F_{n,m}$.

In section 3 we detail the mathematical basis of the new constructive algorithm. The *m* given examples are represented on *n* bits each (either all the examples are Boolean, or they are reals which have been quantified on at most *n* bits). The output will be represented on *k* bits. The learning problem is thus equivalent to simultaneously implementing *k* BFs defined on (at most) *m* points. Extending the results from the previous section to this case, we determine the *depth*, the *size* and the *area* of such an implementation for a maximum given *fan-in*. We present the same measures for a *distributed normal form* (**DNF**) implementation (suggested in [7]). The two alternatives can be combined by using COMPARISONs only when they give a smaller *area* than the AND-equivalent gates needed to implement all the examples bounded by the COMPARISONs.

In support of our claim of VLSI-optimality we compare it with another constructive algorithm in section 4.

Conclusions and open problems are pointed in the last section.

* This research was carried out as part of an Individual Research Training Fellowship ERB4001GT941815 —Human Capital and Mobility programme— of the European Community and has been financed by the Commission of the European Communities under contract ERBCHBICT941741. The scientific responsibility is assumed by the authors.

II. PREVIOUS RESULTS

A BF is symmetric if and only if it remains unchanged by any permutation of its input variables [22]. Muroga [23] has shown that any symmetric function of n inputs can be realised by a *depth*-2 TG circuit of $size = n+1$. This result has been improved to $size = 2\sqrt{n} + O(1)$ in *depth*-3 [24]. Both constructions have polynomial *fan-in*, which can be reduced by decomposition. Symmetric functions can be build by using only MAJORITY functions (having only ± 1 *weights*). Any symmetric function can be decomposed in Δ-ary trees [2] having $size_{MAJ}(n,\Delta) = \lceil \frac{n-1}{\Delta-1} \rceil$, $depth_{MAJ}(n,\Delta) = \lceil \frac{\lg n}{\lg \Delta} \rceil$ and *area* $A_{MAJ}(n,\Delta) = \Delta \lceil \frac{n-1}{\Delta-1} \rceil$. In this paper Δ is the maximum *fan-in* [25], $\lceil x \rceil$ is the smallest integer greater or equal than x (or "ceiling"), and $\lg x$ is the base two logarithm of x ($\log_2 x$).

Another known class of BFs is $F_{n,m}$ which has been defined by Red'kin [26] as *"the class of Boolean functions that have exactly m groups of ones."* For constructing a TG circuit for any $f \in F_{n,m}$, Red'kin [26] uses $\varphi_i(\tilde{\sigma})$ which are COMPARISONs. It achieves an excellent *size* $2\sqrt{m} + 3$ in *depth*-3 (the existence lower bound being $\Omega(\sqrt{m})$ [7, 24]), but with: (i) exponential *weights* and *thresholds*, and (ii) polynomial *fan-in*. From the VLSI point of view these rule out his solution for $n > 8$.

We propose a solution which should use $2m$ COMPARISONs in a hidden layer to determine the beginning and the ending of each of the m groups, and one MAJORITY gate as output which, by taking alternate signs for the *weights*: +1, −1, +1, −1, ... (see Fig.1), will "ADD" the results from this hidden layer of COMPARISONs. We have proven [2, 21] that there are NNs for computing a 2 n-bit COMPARISON having *size* $O(n/\Delta)$ and *depth* $O(\lg n/\lg \Delta)$ for all the Δ in the range 3 to $O(\lg n)$. These have improved on the previous known *size* $O(n)$ [24]), while still having polynomial bounded *weights* (*weights* $\le \sqrt{n}$ for $\Delta \le \lg n$,). We have also shown [2] that by alternating the signs of the *weights* one can achieved the lowest possible *threshold* (−1) for all TGs.

By linking the results obtained for COMPARISON [21] with the ones for implementing MAJORITY functions, a systematic solution for decomposing $F_{n,m}$ functions has been developed [2]:

Theorem 1: Any function $f \in F_{n,m}$ can be computed by a NN with polynomially bounded integer *weights* in $depth_f(n,m,\Delta) = O(\lg(mn)/\lg \Delta)$ and

$$size_f(n,m,\Delta) = \begin{cases} O\left(\frac{mn}{\Delta}\right) & \text{if } 2m \le 2^\Delta \\ O\left(\frac{mn}{\Delta^2}\right) & \text{if } 2m > 2^\Delta \end{cases}$$

and occupying $A_f(n,m,\Delta) = O(mn \cdot 2^\Delta/\Delta)$ if $2m \le 2^\Delta$ for all the *fan-in* values in the range 3 to $O(\lg n)$.

The proof is given in [2].

Remark 1: For $2m > 2^\Delta$ we can determine the *area* by subtracting the *area* of the first layer and adding the *area* of all the 2^Δ different COMPARISONs:

$$A_f(n,m,\Delta) < \frac{8mn \cdot 2^\Delta}{\Delta^2} \cdot \left[\Delta + \frac{5}{4 \cdot 2^{\Delta/2}} - 2n\left(1 - \frac{2^\Delta}{2m}\right)\right].$$

Varying Δ from 3 to $c\lg n$, the full set of solutions with *area* ranging from $O(mn)$ up to $O\left(\frac{mn^{1+c}}{\lg n}\right)$, and *delays* ranging from $O(\lg(mn))$ to $O\left(\frac{\lg(mn)}{\lg n}\right)$ can be obtained. The AT^2 can also be computed: $O\left(mn \cdot \lg^2(mn) \cdot \frac{2^\Delta}{\Delta \lg^2 \Delta}\right)$ if $T \propto depth$.

Remark 2: These results also suggest that *the minimum area should be reached for* $\Delta_{opt\_A} = 3$, *while for minimising* AT^2 *one should use* $\Delta_{opt\_AT^2} = 6$. The values are not exact, but the optimum should be in very close vicinity (± 1 or ± 2), explained by the fact that the optimum Δ_{opt} have been obtained by estimating the derivatives (neglecting the ceilings) and numerically solving the resulting transcendental equations.

For a better understanding a small example is presented in Fig. 2. The function has 16 inputs and 3 groups of ones, while the *fan-in* has been limited to $\Delta = 4$. The overall structure has 4 levels: (i) the first 3 levels represent the decomposition of the "hidden layer" of COMPARISONs; (ii) the 4th level is the "output layer" (just one MAJORITY gate). The *size* is 105, but with a drastically limited *fan-in*: some TGs have 3 inputs while others have 4 inputs, leading to very small *weights* ($|weights| \in \{1, 2\}$) and extremely simple Boolean functions. This is an important feature when thinking to implement such structures in VLSI.

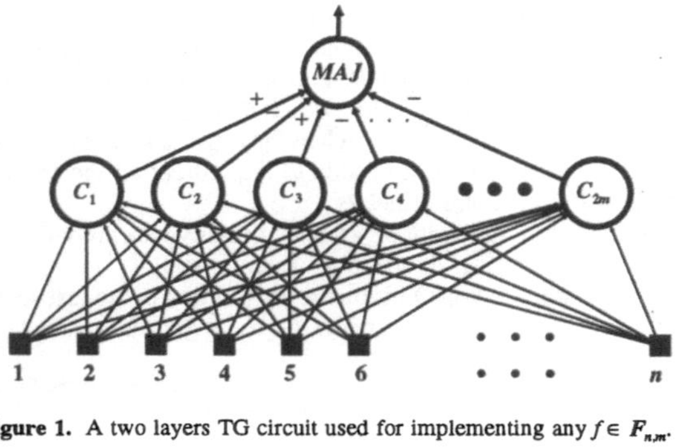

Figure 1. A two layers TG circuit used for implementing any $f \in F_{n,m}$.

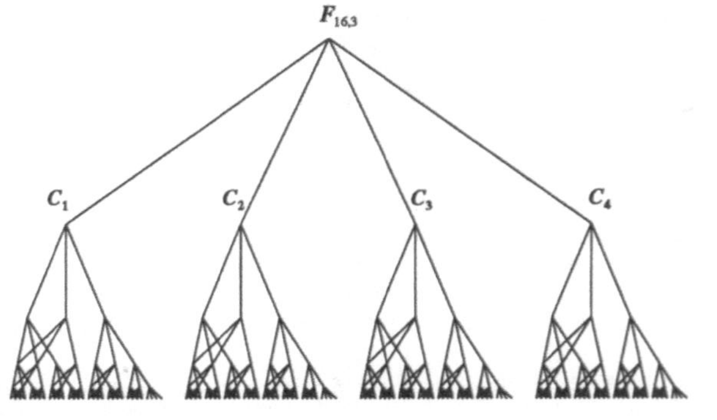

Figure 2. *Area*-efficient decomposition of $F_{16,3}$ ($n = 16$, $m = 3$) when $\Delta = 4$.

III. LEARNING FROM EXAMPLES

The following theorem shows that the decomposition method used for one $F_{n,m}$ function can be directly used to "learn" k functions from m examples (k $F_{n,i}$ functions, $i \le m$).

Theorem 2: Any set of k functions $f \in F_{n,m}$, $i = 1,2,\ldots,m$, with $i \le m \le 2^{\Delta-1}$, can be computed by a NN with polynomially bounded integer *weights* having *size* $O\left(m(2n+k)/\Delta\right)$ and *depth* $O\left(\lg(mn)/\lg\Delta\right)$ and occupying an *area* of $O\left(mn \cdot 2^{\Delta}/\Delta + mk\right)$ for all the values of the *fan-in* in the range 3 to $O(\lg n)$.

Proof: These results extend on the TG circuit used to implement one $F_{n,m}$ function (see Fig. 1 and [2]).

In the worst case the m given examples will be non-adjacent binary numbers, so at most $2m$ COMPARISONs are needed in the hidden layer. All the other k–1 functions are defined on the same m examples, so they can be realised by adding just k–1 TGs (one for each additional function) in the output layer. The circuit will have the same *depth* as the one implementing just one function:

$$depth_{kf}(n,m,\Delta) = depth_f(n,m,\Delta) = depth^{const}_{\mathrm{COMP}} + depth_{\mathrm{MAJ}} = \\ = \left\lceil \frac{\lg n - 1}{\lg\Delta - 1} \right\rceil + \left\lceil \frac{\lg m + 1}{\lg\Delta} \right\rceil = O\left(\frac{\lg(mn)}{\lg\Delta}\right) \quad (2)$$

but the *size* (for implementing one function) will be increased with k–1 MAJORITY gates:

$$size_{kf}(n,m,\Delta) = size_f(n,m,\Delta) + (k-1)\cdot size_{\mathrm{MAJ}}(m,\Delta) = \\ = 2m\left(\left\lceil \frac{2(n-2)}{\Delta-2} \right\rceil - \left\lceil \frac{\lg n - 1}{\lg\Delta - 1} \right\rceil\right) + k\left\lceil \frac{2m-1}{\Delta-1} \right\rceil = \\ = O\left(m(2n+k)/\Delta\right). \quad (3)$$

The *area* can now be computed (see [2)]:

$$A_{kf}(n,m,\Delta) = 2m\left(A^{const}_{\mathrm{C}}(n,\Delta) + A^{const}_{\mathrm{F}}(n,\Delta)\right) + k\,A_{\mathrm{MAJ}}(m,\Delta) < \\ < 2m\cdot\left(\frac{4n\cdot 2^{\Delta}}{\Delta} + \frac{5(n-\Delta)\cdot 2^{\Delta/2}}{\Delta(\Delta-2)}\right) + k\cdot\Delta\left\lceil \frac{2m-1}{\Delta-1} \right\rceil \cong \\ \cong 2mk + \frac{2m\cdot 2^{\Delta/2}}{\Delta(\Delta-2)}\cdot\left(4n\Delta\cdot 2^{\Delta/2} - 8n\cdot 2^{\Delta/2} + 5n - 5\Delta\right) = \\ = O\left(mn\cdot 2^{\Delta}/\Delta + mk\right) \quad (4)$$

the proof being concluded. ❑

Remark 3: For $2m > 2^{\Delta}$, many TGs needed in the first level of the $2m$ trees for decomposing COMPARISONs are redundant.

This result could still be improved as has been suggested by Baum [7]: implement directly the DNF. Instead of two COMPARISONs we might use just one TG, thus reducing the *size* by a factor of 2. The TG is an AND-equivalent gate having as inputs the direct value of the variables corresponding to ones and the complemented values for the variables corresponding to zeros. As the minterms are common to all the k functions, the worst case will require m AND-equivalent TGs in the hidden layer instead of $2m$ COMPARISONs. An AND-equivalent gate is simpler than a COMPARISON gate both with respect to *area*, and to the decomposition into a limited *fan-in* tree.

Theorem 3: Any set of k functions $f \in F_{n,m}$, $i = 1,2,\ldots,m$, can be computed by a NN with ± 1 *weights* having $O\left(m(n+k)/\Delta\right)$ *size* in $O\left(\lg(mn)/\lg\Delta\right)$ *depth* and $O(m(n+k))$ *area* for all the values of the *fan-in* in the range 2 to n.

Proof: The construction is the DNF:

- at most m AND gates of n inputs in a first layer;
- k TGs of m variables in the second layer which are OR gates of (at most) m inputs.

We can compute:

$$depth^{\mathrm{DNF}}_{kf}(n,m,\Delta) = \left\lceil \lg n/\lg\Delta \right\rceil + \left\lceil \lg m/\lg\Delta \right\rceil \\ = O\left(\lg(mn)/\lg\Delta\right) \quad (5)$$

$$size^{\mathrm{DNF}}_{kf}(n,m,\Delta) = m\left\lceil \frac{n-1}{\Delta-1} \right\rceil + k\left\lceil \frac{m-1}{\Delta-1} \right\rceil \\ = O\left(m(n+k)/\Delta\right) \quad (6)$$

and occupying:

$$A^{\mathrm{DNF}}_{kf}(n,m,\Delta) = m\Delta\left\lceil \frac{n-1}{\Delta-1} \right\rceil + k\Delta\left\lceil \frac{m-1}{\Delta-1} \right\rceil \\ = O(m(n+k)) \quad (7)$$

concluding the proof. ❑

Remark 4: The *area* estimate does not depend on the *fan-in*.

The improvements given by *Theorem 3* are: (i) a possible reduction with respect to *depth*; (ii) a reduction by a factor of almost 4 in *size*; (iii) a large reduction of the *area* ($8\cdot 2^{\Delta}/\Delta$). We do not know how to constructively reach the $\sqrt{m}$ existence *size* complexity for implementing any BFs with TGs [7, 24]. Still, we can combine together the idea of using COMPARISONs with the one of using AND gates such as to take the best advantage of both solutions and reduce as much as possible the *area*.

Theorem 4: It becomes more *area*-efficient to use two COMPARISONs instead of α AND gates only if the group of ones has a length of:

$$\alpha \ge \frac{6\cdot 2^{\Delta}}{\Delta} + \frac{10\cdot 2^{\Delta/2}}{\Delta(\Delta-2)}.$$

Proof: Computing the *area* we have:

- for two COMPARISONs (see [2])

$$A^{const}_{2\,\mathrm{COMP}} \cong 2\cdot\left[\frac{3n\cdot 2^{\Delta}}{\Delta} + \frac{5(n-\Delta)\cdot 2^{\Delta/2}}{\Delta(\Delta-2)}\right], \quad (8)$$

  where, instead the maximum *threshold* (2^{Δ}), we have taken *the average value for the threshold*: $2^{\Delta-1}$;

- for α MAJORITY gates

$$A_{\alpha\,\mathrm{MAJ}} = \alpha\,\Delta\left\lceil \frac{n-1}{\Delta-1} \right\rceil \cong \alpha n. \quad (9)$$

By solving $A_{2\,\mathrm{COMP}} = A_{\alpha\,\mathrm{MAJ}}$ we find the crossing point:

$$\alpha^{*} \approx \frac{6\cdot 2^{\Delta}}{\Delta} + \frac{10\cdot 2^{\Delta/2}}{\Delta(\Delta-2)} - \frac{10\cdot 2^{\Delta/2}}{n(\Delta-2)} \quad (10)$$

which concludes the proof. ❑

Remark 5: The length of the group of ones –starting from which it is more *area*-efficient to use COMPARISONs– grows exponentially with the *fan-in*. Still, for minimising the AT^2, small constant values of the *fan-in* ($\Delta_{opt} \cong 6 \ldots 8$) have to be used (cf. *Remark 2*).

Theorem 4 is properly valid only for the one dimensional case. The most simple way to extend it to the *d*-dimensional case is to work independently on each of the *d* axes. This is equivalent to *finding hyperplanes which are parallel to the axes* (these COMPARISONs with certain values are the first hidden layer). All the results from this first hidden layer are binaries and the function can be synthesised by a classical AND-OR structure. The NN has thus two more layers: (i) a second hidden layer of $2d$–inputs AND gates (each gate corresponds to a d–dimensional hypercube, and at most 2 COMPARISONs are needed for each of the d dimensions); and (ii) a third output layer of k OR gates.

For optimising the *area* some COMPARISONs from the first hidden layer should be replaced by AND gates (as proven in *Theorem 4*). This reduces the *area* of the chip but the *generalisation* capabilities of the network are also degraded: using AND gates the network will "recognise" only the training examples, while using COMPARISONs the network will *"generalise"* to hypercubes around the given examples.

The algorithm can now be described:

```
1. Find the max and the min on each axis.
2. Find the steps required on each axis (the
   initial steps are max-min and are reduced
   subsequently, until the classification of the
   m examples becomes possible).
3. Determine the constants for COMPARISONs (by
   scanning the space with the steps previously
   determined).
4. Reduce the number of COMPARISONs (by grouping
   together as many hypercubes as possible).
5. Use a 'generalisation' parameter to decide if
   either COMPARISONs or AND gates should be used
   in the final circuit.
```

IV. Comparison of Results

It is not very easy to compare our algorithm with others as our main aim was to minimise the *area*, and no other algorithms for doing that are known to the authors. Most of the direct design algorithms are based on a "learning" phase; as a result of which the *weights* are updated or: (i) a new neuron is added; (ii) a new layer is added. That is why it is not possible to bound the *size* of the network they "learn." Still, there is one algorithm we can compare with: that of Tan and Vandewalle [20]. While the *size* of the NN they build is of the same order $O(mn)$ as ours for constant *fan-in* ($O(mn/\Delta)$), the *depth* is reduced from linear $O(mn)$ to logarithmic $O(\lg(mn)/\lg\Delta)$. Moreover, the maximum *fan-in* is reduced from linear n – in their case – to constant Δ.

V. Conclusions

In this paper we have shown how to extend the applicability of a decomposition algorithm for functions belonging to $F_{n,m}$ to the problem of "learning from examples." While doing that we have proven that both the *size* and the *area* of a VLSI implementation of the NN grow linearly with the "problem" (i.e. n, m, k) for the two solutions discussed (with COMPARISONs or implementing the DNF). We have also discussed the influence of the *fan-in* on *size*, *depth* and *area*, and presented a better alternative by suggesting how to combine the two solutions proposed such as to minimise *area* even more.

Further research should be pursued to try to lower the *area* of COMPARISON and $F_{n,m}$ functions. Closer estimates could be achieved if the *area* of interconnections would be included in the cost function. For even better estimates, the additional *delay* introduced by long wires should not be neglected.

An interesting aspect which clearly deserves more attention is the balance between the *generalisation* capabilities and the minimum *area* of the chip. A software program is under development and the *generalisation–area* tradeoff is one of the topics currently under investigation.

References

1. Hammerstrom, D.: "The Connectivity Analysis of Simple Association –or– How Many Connections Do You Need," *Proc. NIPS'87* (Denver, November), Amer. Inst. Phys., 338-347 (1987).
2. Beiu, V., Peperstraete, J.A., Vandewalle, J., Lauwereins, R.: "Area-Time Performances of Some Neural Computations," in P. Borne, T. Fukuda and S.G. Tzafestas (eds.): *Proc. SPRANN'94* (Lille, April), GERF EC, Lille, 664-668 (1994).
3. Mead, C.A., Conway, L.: *Introduction to VLSI Systems*, Addison-Wesley, Reading (1980).
4. Rumelhart, D.E., McClelland, J.L., The PDP Research Group (eds.): *Parallel Distributed Processing: Explorations in the Microstructure of Cognition*, A Bradford Book, The MIT Press, Cambridge (1986).
5. Holt, J.L., Hwang, J.-N.: "Finite Precision Error Analysis of Neural Network Hardware Implementations," *IEEE Trans. on Comp.*, **C-42**(3), 281-290 (1993).
6. Wray, J., Green, G.G.R.: "Neural Networks, Approximation Theory, and Finite Precision Computation," *Neural Networks*, **8**(1), 31-37 (1995).
7. Baum, E.B.: "On the Capabilities of Multilayer Perceptrons," *J. Compl.*, **4**, 193-215 (1988).
8. Minsky, M.L., Papert, S.A.: *Perceptron: An Introduction to Computational Geometry*, The MIT Press, Cambridge (1969).
9. Hopcroft, J.E., Mattson, R.L.: "Synthesis of Minimal Threshold Logic Networks," *IEEE Trans. on Electr. Comp.*, **EC-6**, 552-560 (1965).
10. Śmieja, F.J.: "Neural Network Constructive Algorithm: Trading Generalisation for Learning Efficiency?" *Circuits, System, Signal Processing*, **12**(2), 331-374, 1993.
11. Bose, N.K., Garga, A.K.: "Neural Network Design Using Voronoi Diagrams," *IEEE Trans. on Neural Networks*, **NN-4**(5), 778-787, 1993.
12. Ramacher, U., Wesseling, M.: "A Geometrical Approach to Neural Network Design," *Proc. IJCNN'89* (Washington, January), IEEE Press, vol. **2**, 147-153 (1989).
13. Abu-Mostafa, Y.S.: "Learning from Hints in Neural Networks," *J. Compl.*, **6**, 192-198 (1990).
14. Armstrong, W.W., Gecsei, J.: "Adaption Algorithms for Binary Tree Networks," *IEEE Trans. on Systems, Man and Cybernetics*, **SMC-9**, 276-285 (1979).
15. Diederich, S., Opper, M.: "Learning of Correlated Patterns in Spin-Glass Networks by Local Learning Rules," *Phys. Rev. Lett.*, **58**(9), 949-952 (1987).
16. Fiesler, E., Choudry, A., Caulfield, H.J.: "A Universal Weight Discretization Method for Backpropagation Neural Networks," *IEEE Trans. on System, Man, and Cybernetics* (to appear).
17. Fontanari, J.F., Meier, R.: "Evolving a Learning Algorithm for the Binary Perceptron," *Network*, **2**, 353-359 (1991).
18. Gallant, S.I.: "Perceptron-Based Learning Algorithms," *IEEE Trans. on Neural Networks*, **NN-1**(2), 179-191 (1990).
19. Mézard, M., Nadal, J.-P.: "Learning in Feedforward Layered Networks: the Tiling Algorithm," *J. Phys. A: Math. Gen.*, **22**, 2191-2203 (1989).
20. Tan, S., Vandewalle, J.: "Efficient Algorithm for the Design of Multilayer Feedforward Neural Networks," *Proc. IJCNN'92* (Baltimore, January), IEEE Press, vol. **2**, 190-195 (1992).
21. Beiu, V., Peperstraete, J.A., Vandewalle, J., Lauwereins, R.: "Efficient Decomposition of COMPARISON and Its Applications," in M. Verleysen (ed.): *Proc. ESANN'93* (Brussels, April), Dfacto, Brussels, 45-50 (1993).
22. Bruck, J., Smolensky, R.: "Polynomial Threshold Functions, AC^0 Functions and Spectral Norms," *SIAM J. Comput.*, **21**(1), 33-42 (1992).
23. Muroga, S.: "The Principle of Majority Decision Logic Elements and the Complexity of Their Circuits," *Proc. Intl. Conf. Inform. Processing*, Paris (1959).
24. Siu, K.-Y., Roychowdhury, V., Kailath, T.: "Depth-Size Tradeoffs for Neural Computations," *IEEE Trans. on Comp.*, **C-40**(12), 1402-1412 (1991).
25. Shawe-Taylor, J.S., Anthony, M.H.G., Kern, W.: "Classes of Feedforward Neural Nets and Their Circuit Complexity," *Neural Networks*, **5**(6), 971-977 (1992).
26. Red'kin, N.P.: "Synthesis of Threshold Circuits for Certain Classes of Boolean Functions," *Kibernetika*, **6**(5), 6-9 (1970).

USING OF NEURAL-NETWORK AND EXPERT SYSTEM FOR IMISSIONS PREDICTION

E.Theocharidou , Z.Burianec

University of Chemical Technology ,Technicka 5 , 166 28 Prague, Czech Republic

The result of our analysis is oriented to the formation of a warning system by the prediction of the concentration of suphure dioxide (SO2) on the basis of a hydrometeorological forecast. The analysis carried out in this study is based on the application of neural nets (NN) and expert system (ES) methodology The prediction is made by putting to use both NN and ES in the form of an integrated system .

Introduction

The study of air pollution due to sulphur dioxide (SO2) emissions is very important, because SO2 has serious effects on living things, including human beings. The Czech Republic belongs to the regions with highest concentrations of pollutants in Europe. Also in the capital city Prague there is a critical quality of the atmosphere. The situation in Prague can be interpreted as a typical example of the complex interaction of emissions from different sources in the urban conditions.The concentrations of SO2 measured by the monitoring system, represent only a part of total SO2 emmissions. SO2 oxidation which occurs in the atmosphere can be distinguish into homogenous and heterogenous pathways. [1]

The state of knowledge about reaction mechanisms is unsufficient. The uncertainty in reaction conditions, the measurement of SO2 only, without the knowledge of sulphate concentrations and the interaction of many physical parameters create an extremely complex system,which belongs to the category of badly defined systems.

Data Sets

Data set representing half-hour data of SO2 concentration and hydrometeorological data measured at 7 a.m., 2 and 9 p.m. between 1st January and 31st December 1993 were obtained from Czech Hydrometeorological Institute in Prague. Half-hour missing hydrometeorological datas were generated by spline function. As a result the matrix 17520 x 10 was formed. The individual column vectors represent pressure, temperature, vertical temperature gradient, relative humidity, direction and velocity of the wind, fog, snow, rain and season .

Preliminary Analysis of the Data Set by the Time Series Analysis

The data were analyzed from the point of view of

autocorrelation functions and of spectra. The result shows, that the individual relations between SO2 and the most important explanatory hydrometeorological variables (temperature, pressure, velocity and direction of wind, humidity) can be represented satisfactory by the second-order autoregressive process models (AR models).

The results were applied below for the selection of regressors in NN.

Environmental Application of Neural Nets and Expert Systems

Neural nets could help by discovering patterns in the data,making recommendations e.g. for decisions about hazardous waste sites, about geophysical exploration for groundwater ect. But the activity in the field of atmosphere pollution is relatively low [2,3].

While expert systems technology has now existed for more than 20 years, enviromental expert systems are only almost 10 years old.It is possible to find only few papers concerning air pollution problems.[2,3,4]

Neural Nets

From the analyzed system which can be represented as system of partial differential equations including the concentration of various chemical - reactive imissions under unstationary conditions in the three-dimensional space of a concrete region, we measure the time-series of data in one point of this space.

Generally, this problem can be represented by the typical time-series data set.

The neural net structure gives several possibilities to construct different types of models [5] :

NNFIR,NNARX,NNOE,NNARMAX,NNBS.

Every type of NN-model has its own disadvantages. It shows that for our purposes the type NNFIR and NNARX are most convenient.

In the time of low wind velocity and temperature inverse situation SO2 concentration background changes very slowly and from this point of view were taken into calculation the day-average concentration delayed for one day ,which form the first regressor in NN. The other most important regressors which change relatively very fast are delayed for one and two time intervals in accordance with time series analysis.

Let us see the list of these regressors:

u1 : temperature　　u4 : wind velocity

u2 : pressure　　u5 : wind direction

u3 : humidity

regressors not delayed are:

u6 : vertical temperature gradient (the temperature difference between 2 and 1500 m altitude).

u7 : season: spring/automne, summer, winter

the other ones represent the presence or absence of

u8 : snow　　u9 : fog　　u10 : rain

The number of nodes in the input layer is equal to the number of regressors (20 regressors, in unusual situation 21 ones), the output layer has only one node corresponding to the concentration of SO2.

The number of hidden layers, number of nodes in the hidden layers and different transfer functions (linear,sigmoid and hyperbolic tangent) were checked for various combinations of the number of nodes and

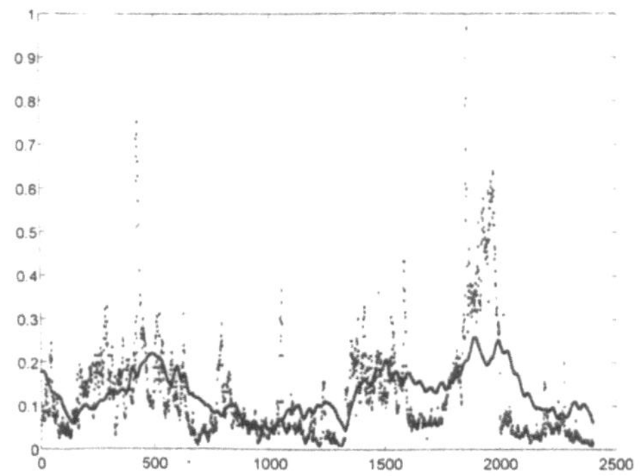

Fig. 1. Testing of the NN in the interval October - November (20-40-40-1), sigmoid function,
maximal value of SO2 concentration is 500 μg m^{-3}

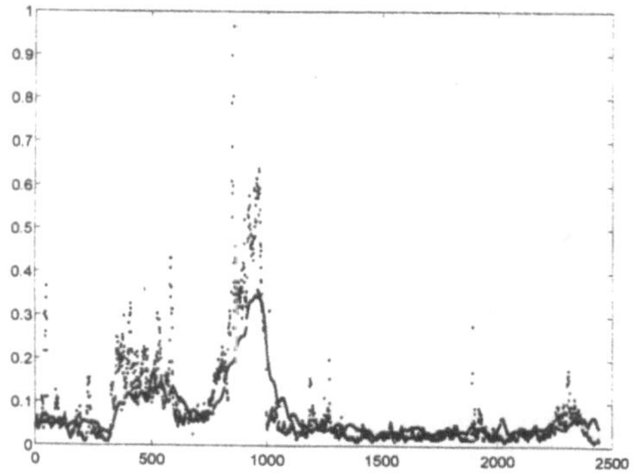

Fig. 2. Testing of the NN in the interval November - December (20-40-1), sigmoid function,
maximal value of SO2 concentration is 500 μg m^{-3}

hidden layers. (20-20-1, 20-40-1, 20-60-1, 20-20-20-1, 20-40-40-1,20-40-40-40-1).

The input data sets to the NN are nondimensionalized and these inputs should be within <0,1>.

The back-propagation training algorithm and delta-rule is used to update the weights of the network.

The network has been trained for 50000 to 100000 cycles using the training data of 8000 measurement patterns. The gain term and the momentum term were stepwise diminished, beginning with the values 0.35 and 0.25 . The training data set corresponds to the first half-year of 1993 and the test data set corresponds to the second half-year of 1993 in order to predict the imission concentrations on the basis of the former half-year.

The program NeuralWorks by NeuralWare Inc. and Matlab by Math.Works Inc. have been used.

The illustrations Fig. 1 and Fig. 2 show the testing.

Expert System (ES)

Two goals were observed:

- The construction of the expert system for the SO2 imission concentrations prediction on the basis of hydrometeorological data .
- The evolution of the integrated system, in which the NN output is combined with the ES intput in order to estimate the error of the NN- prediction.

Both goals were solved by utilization of an ES shell, Expert System Toolkit FLEX (Logic Programming Co.Ltd.,England) and the program KNOWLEDGE SEEKER, (Angoss Software Internat. Ltd.,Canada), which is able to generate production rules.

FLEX is a hybrid expert system toolkit, provided with Prolog, a multi-paradigm programming system, which includes frames and inheritance, forward and backward chaining and data driven programming.

KNOWLEDGE SEEKER is an automatic data analysis package that enables to analyze and understand the relationships between variables in data set and presents the results visually in the form of a statistical decision tree and production rules.

ES for the prediction of SO2-imission concentrations.

The training data set was the same as used in NN training but without the normalization. The result is formulated in 80 as many as 200 rules depending on the number of classes of individual variables.

The example of one rule is listed below:

RULE_28: IF DIRECTION[0,70) AND TEMPERATURE[8.5,15) AND HUMIDITY[60,85) AND TEMP.GRAD.[-3.5,-1.) AND . . . . THEN SO2 = 128.5; std = 48.7

ES for the estimation of NN predicted error value. The difference between measured and simulated values of SO2 were used as a dependent variable and the indepedent variables were all basic regressors used in NN. The goal was to compute the error of NN at concrete meteorological situation. But as the training data set only that part of data were used, which represent unusual meteorological situations. i.e. low or zero wind velocity and positive or small temperature gradient and/or fog and/or winter. In this case is very probable, that imission concetrations will be much higher than predicted by NN.

Discussion

The error in the SO2 concentration prediction by NN in standard hydrometeorological situations corresponds to the level of stochastic fluctuations (20 - 40 $\mu g\ m^{-3}$ SO2), in unusual situations (temperature inversions) is relatively high (150 -300 $\mu g\ m^{-3}$) caused probably also by the chemical reactions of SO2 and extremely complex interactions of physical and chemical parameters, which cannot be exactly measured.

ES give not only the value of the SO2 concentration, but also its standard deviation. In dependence on the concrete meteorological situation the standard deviation changes from 0.5 to 30 μgm^{-3}.

The other application of the ES is oriented to the error estimation in the SO2 concentration prediction made by NN during unusual meteorological conditions. In this case only the possibility of the existence of this error is computed. The probability cannot be estimated before long time data analysis.

Conclusion

The methodology for the creation of an integrated system NN - ES has been developed. This system allows to predict the short-time SO2 imission concentrations on the basis of the hydrometeorological forcast.

References

1. De Wispelaere C. (eds): Air Pollution Modeling and Its Application II. New York and London: Plenum Press 1983.
2. Huston Judith M. (eds): Expert Systems for Environmental Applications. Washington, American Chemical Society 1990.
3. Kohonen T. and coll.: Artificial Neural networks Elsevier Sci.Publ. ,North-Holland 1991, p.1371-74.
4. Kuzmin S.M.,Solovyov V.D.: Int.J.Appl.Exp.Syst. 1, 1, 1993.
5. Sjöberg J. and coll.: SYSID'94 2, 49-72 1994, Copenhagen Denmark.

THE COSTING OF PROCESS VESSELS USING NEURAL NETWORKS

LECK, M., BROMLEY, P., PEEL, D., GERRARD, A.M.

UNIVERSITY OF TEESSIDE, ENGLAND

Abstract

A set of commercial data relating physical dimensions and materials to the purchase cost of simple process vessels has been analysed using multilinear regression and artificial neural networks. The ability of each of these approaches to estimate purchase cost is compared with a third method which combines elements of the first two.

Introduction

The cost of process vessels required for the construction of new chemical and biochemical process plant often makes a significant contribution towards the total installation costs of that process plant. Indeed for the production of speciality chemicals, the cost of expensive plant items such as vessels may be a very large factor in the total plant cost. Consequently, during the competitive tendering stage, contractors find it desirable to have a quick, accurate method for estimating the cost of process vessels.

Commonly, the costs of process vessels are approximated using power law relationships based on the physical dimensions, and, occasionally, the operating characteristics of vessel. These relationships have been empirically derived over a number of years and produce accuracies of the order of 30% (on a mean absolute error basis).

This paper readdresses the problem of trying to improve the accuracy of the costing process by comparing the performance of existing empirical methods with that of linear relationships derived using least squares regression techniques, non-linear relationships attained using artificial neural network frameworks; and a hybrid method combining both linear and non-linear representations.

Data

The analyses were based on a set of data collected during 1979 (1) and divided into three distinct categories. These categories were: simple horizontal carbon steel drums; simple vertical carbon steel drums; and simple vertical carbon steel columns. The available descriptive data for each of the vessels is presented in Table 1.

Table 1

Maximum Inner Shell Diameter	m
Tangent to Tangent Length	m
Internal Design Pressure	bar
Design Temperature	C
Fabricated Mass	Kg
Maximum Shell Thickness	m
Number of Nozzles	--
'Delivered to Site' Cost	Rand

Basis for Comparison

In order to make the comparison a number of measures have been used. These measures are as follows :

Integral absolute error (IAE) :

$$\text{IAE} = \Sigma\, |(\text{cost - predicted cost})|$$

Integral square of the errors (ISE) :

$$\text{ISE} = \Sigma\, (\text{cost - predicted cost})^2$$

Mean normalised absolute percentage error :

$$\text{MNAE} = \frac{\Sigma\, (|(\text{cost - predicted cost})| / \text{cost} * 100\%)}{\text{number of data points}}$$

where each summation is carried out across the whole of the data set.

Method of Analysis

The PC based statistical software package 'Minitab' was used to implement least squares regression methods to investigate single variable and multivariable correlations against cost. These analyses were carried out on the raw data and also on the logarithm of the values (from which the power laws are obtained). An in-house neural network package was also used to analyse the data.

Using a representative subset of the original complete data set, the coefficients of the general power law relationships which give the most accurate fit can be found.

The artificial neural network analysis was carried out using 'back-error propagation' to determine the weights of the feed forward neural networks. However, due to the low number of data points available, only small networks have been considered. Each data set consisted of around thirty records which were then halved. The first half was used for the purposes of training and the second half used for validation purposes. The small number of training points meant that, statistically, it would be unsound for more than three or four parameters to be estimated at any one time.

The statistical restrictions placed upon us by the size of the data set proved to be convenient. In most cases, the best results were obtained by using only three or four variables to predict the cost of a vessel.

Results

Linear regression using 'Minitab' demonstrated a highly linear correlation between mass and cost. This is consistent with the known information in that the vessels are described as 'simple' which implies that they are relatively easy to manufacture and their cost will be dominated by material costs and not the actual fabrication costs of the vessel. However, as well as being expected, this relationship is also inconvenient. This is because the fabricated mass of a vessel is rarely available before its construction and so a linear relationship between mass and cost, however attractive, is almost useless to use at the early stages of vessel design.

Excluding mass from the list of variables available as inputs to the estimator, 'Minitab' gives reasonable results by utilising the vessel dimensions and the design pressure. The graphical representation is shown in Fig. 1. The ISE of this set of predictions is 2.67×10^8, with a MNAE of 16%. This prediction is acceptable and demonstrates the current 'state of the art'.

Figure 1

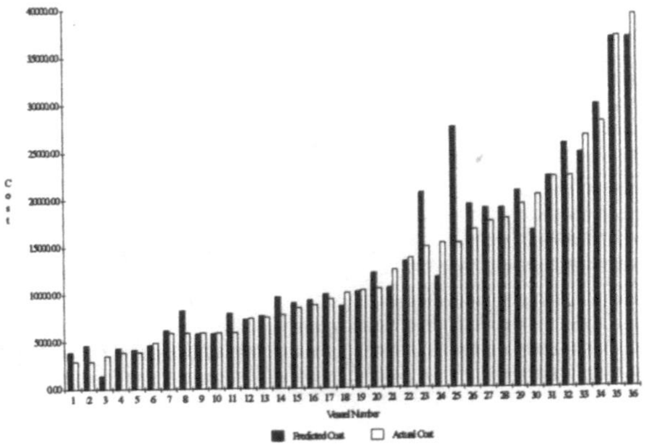

The neural network analysis yielded similar results. As discussed earlier, the data indicated a high degree of correlation between mass and cost. However, due to the inherently non-linear properties of a standard feed-forward neural network, the neural solution was unable to provide a correlation as good as the linear estimator derived from 'Minitab'.

Again, excluding mass from the list of predictors, the neural network yielded its best predictions when using the vessel dimensions and design pressure to predict the cost. The graphical representation of this prediction is shown in Fig. 2. The ISE for this prediction was 1.85×10^8, with a MNAE of 17%. This is an improvement on the linear regression result in terms of ISE.

Figure 2

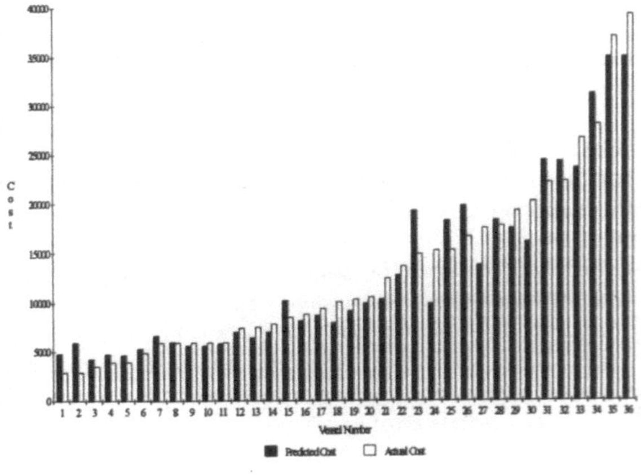

Note that the ISE in each case is quite different but that the mean absolute percentage error in each case is very similar. All the measures of error were investigated before making a decision as to the merits of each method.

These results indicate that using a neural network to replace a linear estimator in predicting the cost of process vessels would yield no improvement on the accuracy of predictions. Indeed, there is evidence to support the fact that such a replacement would be detrimental given the extra time needed to train a neural network.

Despite this conclusion there is still some merit in exploring the relationship between mass and cost. The lack of information before a vessels construction is a common problem. If the mass could be estimated from the known variables, then perhaps the estimated mass could be used to more accurately predict the cost.

The problem can therefore be separated into two stages. The first involves the estimation of the mass of the vessel and the second relates the estimated mass (or mass proxy) to the cost.

The Hybrid Model

Each stage described above can be accomplished in one of two ways, using 'Minitab' to find the coefficients of a linear predictor or using a neural network as a non-linear predictor. Both methods were applied to each stage and the results compared.

The neural network approach to the prediction of mass is more accurate than 'Minitab' (an ISE of 2.82×10^7 compared with 4.42×10^7). Each prediction assumes that only very basic vessel dimension data and design pressures are available. This result is instinctively attractive because the vessel dimensions are non-linearly related to its mass and based on this a neural network should assimilate the data with greater ease.

The second stage involves taking the best prediction of mass and using it along with the other known data in order to estimate the cost of the vessel. The prediction given by the neural network in this instance has an error of 3.05×10^8 where 'Minitab' predicted with a greater degree of accuracy, having an ISE of 1.11×10^8.

As stated earlier, there has been a high degree of linearity already demonstrated between mass and cost and so it should not be surprising that a linear predictor yields better results. The graphical representation of the prediction of cost given by the hybrid method is shown in Fig. 3.

Figure 3

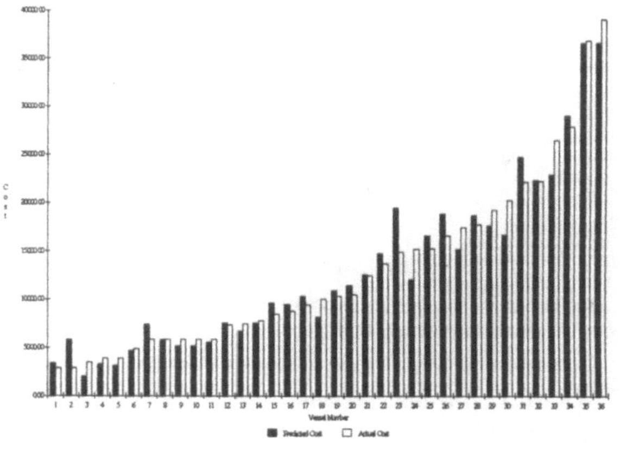

We are left with a model which comprises a non-linear element and a linear element and gives a more accurate description of the cost that either of the methods used singly.

Conclusions

Table 2 below shows the errors obtained from all three methods of estimating the cost. The errors are quoted in all three possible formats for comparison. The main criteria used for comparison in the investigation was ISE.

Table 2

	ISE	IAE	MNAE
Linear Regression	2.76E+08	6.14E+04	16%
Neural Network	1.85E+08	6.43E+04	17%
Hybrid Model	1.11E+08	4.83E+04	14%

The investigation shows that using a neural network to predict the cost of process vessels instead of a linear estimator does not lead to any significant improvement in accuracy. In fact, the additional time required to train a neural network would imply that neural networks are more cumbersome.

However, there is a notable improvement in prediction accuracy when a combination of prediction methods is used. A non-linear prediction of mass from physical dimensions using a neural network and then a linear prediction of cost using the predicted mass using regression yields significantly better results than either method used separately.

References

1. Philp, E.A. : Unpublished MSc Thesis : University of Witwatersrand. Republic of South Africa. 1988.

2. Wasserman, P.D. : Neural Computing - Theory and Practice. New York. Van Nostrand Reinhold. 1989.

3. Ott, L., Mendenhall, W. : Understanding Statistics. : 5th edition. Boston. PWS-KENT. 1990.

NEURAL NETWORK MODELLING OF FERMENTATION TAKING INTO ACCOUNT CULTURE MEMORY

P. Koprinkova, M. Petrova, T. Patarinska

Institute of Control and System Research, Bulgarian Academy of Sciences, Acad.G.Bontchev Str., Bl.2, P.O.Box 79, 1113 Sofia, Bulgaria

ABSTRACT: In the present paper the problem of chemostat modelling using the neural networks is considered. Two kinds of neural network models taking into account the culture memory are proposed. The first one is a feedforward neural network with time delayed feedback connections from and to the output neurons. The second one is a feedforward neural network with time delayed feedback connections from the input neurons describing only the specific growth rate. The obtained neural network model is adopted in the classical nonlinear model. Simulation investigations and comparison of the both models are carried out. The interpretation of the parameters according to their biological means is made.

INTRODUCTION

Nonlinear dynamic behaviour of chemostats is manifested by existence of multiple steady states and self-sustained oscillations [1]. Although these oscillations have been known for many years and many biochemical and physiological studies have been done on the oscillating cultures, the mechanism that originates this instability is not clear and there is no model able to describe the phenomenon in a satisfactory way. It has been recognized for a long time, that the observed response of a cell population at a certain time instant is the composite result of various biological processes that had been initiated at different time instants in the past as a response to the instantaneous environmental condition prevailing at each particular time. Because of that the kinetics of the fermentation processes (in particular of the chemostat) should be considered as a function not only of the current process state, but also of a weighted average of the states at past time instants. In the classical unstructured models the influence of the culture memory on the process dynamics is described by a time delay parameter or memory function [1,7].

The neural networks appear as a promising alternative for non-linear systems modelling [2]. Their advantages over the classical mathematical models are the simultaneous structure and parameter identification and the ability to learn by examples. Although there are a number of papers reporting about usage of neural networks as fermentation processes' models [3,4,5,9], investigations of the possibilities of neural network application as a modelling tool for fermentation processes is topical.

In the paper two kinds of neural network models taking into account the culture memory are developed. The first one is a feedforward network. Its input is the system state vector for the current time instant and the output is the same vector for the next time instant. The presence of the time delayed feedback connections from and to output neurons is the major peculiarity of the model. The second model is a classical unstructured model in which a feedforward neural network for the specific growth rate calculation is adopted. The both models are identified using backpropagation method [6]. Simulation investigations are carried out and the models are compared. The interpretation of the models' pa-rameters is made according to their biological means.

NEURAL NETWORK MODELS DESCRIPTION

Neural networks, like brain, consist of connected processing elements called neurons. For modelling of physical systems feedforward networks are usually used. They consist of a layer of input neurons, a layer of output neurons and one or several layers of hidden (intermediate) neurons. Information in such a kind of network is processed in forward order, from input to output. The considered neural network models are feedforward neural networks with three layers. Their specificity is the presence of the time delayed connections from some of the neurons, which allow to has memory for their past values. As output functions of all the neurons are the commonly used sigmoidal functions. The training algorithm is backpropagation through time methods [6], which realized by authors software product. On the example of biomass productivity of a strain of Saccaromyces cerevisiae on a glucose limited medium the proposed neural network models are investigated.

Chemostat neural network model (C-NNM)

The topology of the first neural network model called below chemostat neural network model (C-NNM) is shown on Fig. 1. Its input is the system state vector (biomass and substrate concentrations) and the control variable (dilution rate) for the current time instant. The output of the model is the state vector of the next time instant. The presence of the time delayed feedback connections from and to output neurons is the major peculiarity of the model. The values of the weights of all the possible time delayed connections from output neurons are investigated. Networks with time delayed connections for different number of time instants backward are trained and compared.

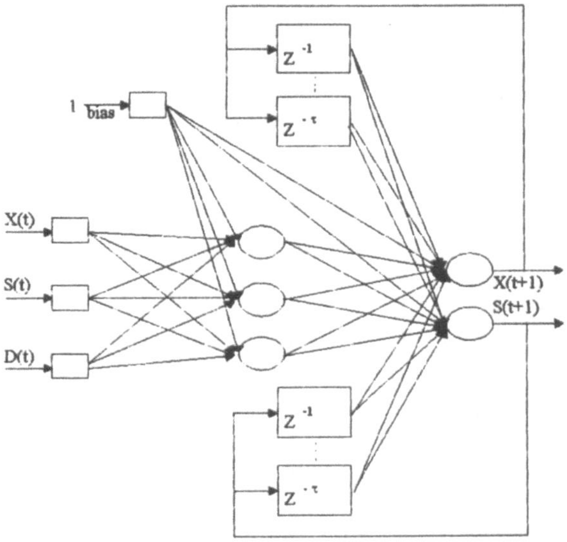

Fig.1 Topology of C-NNM

Specific growth rate neural network model (SGR-NNM)

The second neural network model describing the dependence of the specific growth rate on the process state variables, called SGR-NNM, is shown on Fig. 2.

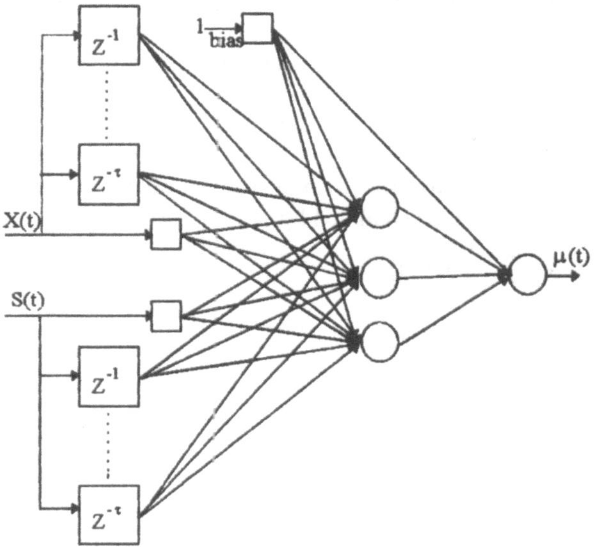

Fig. 2 Topology of SGR-NNM

Its inputs are the biomass and substrate concentrations at the current time instant and the output is the specific growth rate μ. There are time delayed connections from input neurons, so that the model has memory for the past biomass and substrate concentrations. This network is adopted in the classical unstructured model, described by the following equations:

$$X(t+1) = X(t) + \mu(t)X(t) - X(t)D(t) \qquad (1)$$
$$S(t+1) = S(t) - \frac{\mu(t)}{Y_s}X(t) + D(t)(S^0 - S(t))$$

where X(t) is the biomass concentration; S(t) is the substrate concentration; D(t) is the dilution rate; $\mu(t)$ is the specific growth rate, calculated from neural network; S^0 is the limiting substrate concentration in input flow; Y_s is the yield factor.

Networks with delayed connections for the both or one of the state variables and for different number of time instants backward are trained and compared.

SIMULATION RESULTS AND DISCUSSION

Continuous culture for the growth of a strain of Saccharomyces cerevisiae on a glucose limited medium as an example is considered. The data obtained by two types of experiments: steady-state

and dynamic step response [7] are used for the simulations investigations.
The training data set of the C-NNM consists of the normalized experimental data for the biomass and substrate concentrations. The topology of SGR-NNM imposes a change of the training data set. Using the equation, describing the biomass balance in a continuous process [8] the values of the specific growth rate are calculated by:

$$\mu(t) = \frac{dX}{dt}\frac{1}{X} + D(t) \tag{2}$$

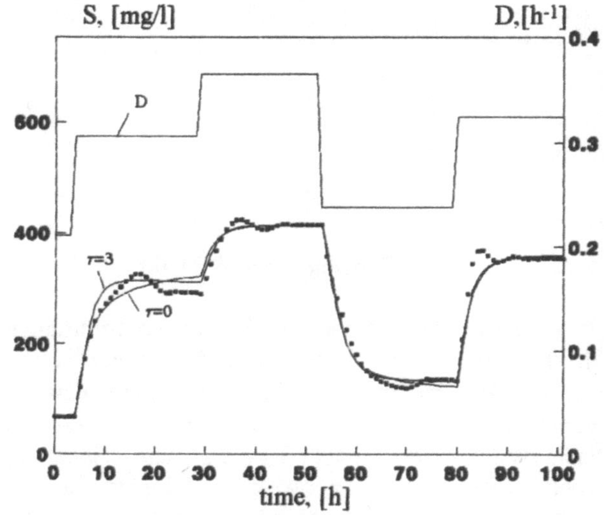

Fig.3a Substrate simulation by C-NNM

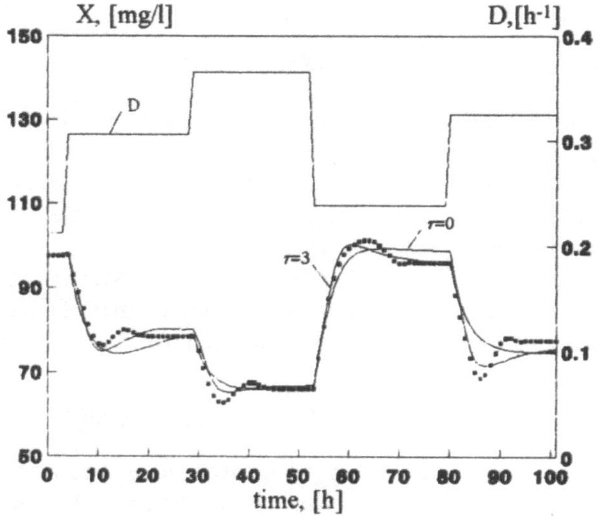

Fig.3b Biomass simulation by C-NNM

Simulation investigations of the both models with time delayed connections for two, three and four time instants backward (τ=2, τ=3 and τ=4 [h]) are carried out. For comparison simulations for the both models are made also by the usual feedforward network (network without time delayed connections, τ=0). The experimental data (pointed lines) and simulation results (solid lines) of the both models are showed on Fig. 3 and Fig.4. The C-NNM, obtained with time delayed connections with respect to substrate concentration for three time instants backwards, is shown on Figs. 3. The best SGR-NNM, obtained with time delayed connections with respect to biomass and substrate concentration for two time instants backward, is shown on the Figs. 4.
The following conclusions are based on the simulation results:
-as can be easily seen (Figs.3 and Figs.4) the networks without time delayed connections can not be learned to fit the specific dynamic responses of the cell population to the step changes of the dilution rate - the existence of oscillations or overshoots before reaching the steady state.
-the models with three and four time instants delay are almost identical.
The weights values of the time delayed connections in SGR-NNMs allow to derive the following conclusions:
-the time delayed connections weights for the fourth time instant backward have very small values in comparison with the other time delayed connections weights and could be neglected;
-the time delayed connections weights for the second time instant backward have the biggest values among all the time delayed connections weights; therefore, their influence on the output function of neurons is significant;
-the time delayed connections of the substrate concentration for the second time instant backward are with the biggest by modulus weight values; this confirms the significant influence of the limiting substrate concentration at past time instant on the fermentation process state.
-the time delayed connections of the biomass concentration for the first time instant backward have almost the same weights as connections for the present moment, which is in accordance with the biological point of view means that biomass concentration at the current moment is in a direct connection with its concentration one time instant backward.

The SGR-NNM with time delayed connections for both state variables shows better results in comparison with the other SGR-NNMs, which means that specific growth rate depends not only on substrate concentration several time instants backward but also on biomass concentration one time instant backward. For the C-NNM the time delayed connections weights can not be interpreted as the SGR-NNM ones. This is due to its structure. which don't allows to distinguish specific growth rate form the other part of the model.

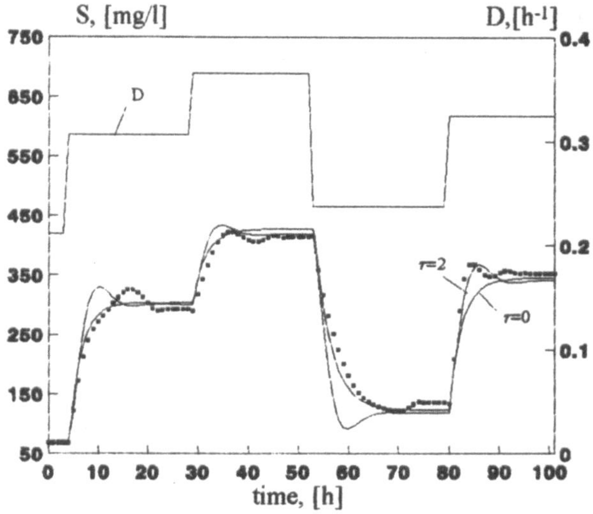

Fig.4a Substrate simulation by SGR-NNM

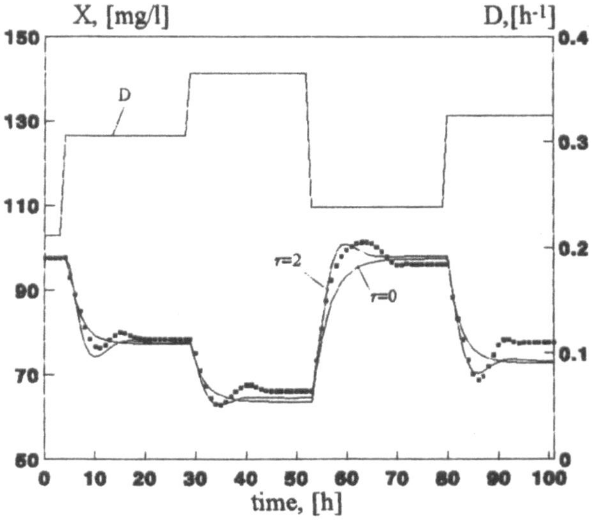

Fig.4b Biomass simulation by SGR-NNM.

The structure of the SGR-NNM has defined in advance and only the part of it (calculation of specific growth rate) is identified as neural network. This kind of model could be useful to find biological meaning of the connection weights because of clear distinguishing of specific growth rate function from the other part of the model.

CONCLUSION

In the paper two kinds of neural network models of fermentation process that taking into account the culture memory are proposed and investigated by simulations. The results of the both models show a good ability to fit the experimental data with satisfying accuracy. The interpretation of the values of the time delay connections weights is made. Further investigations on this direction are in progress.

ACKNOWLEDGMENTS

The partial support by the Bulgarian National Science Research Foundation under Grant SRTS 428/94 "Modeling and Control of Fermentation Processes Taking the Memory Effect into Account" is gratefully acknowledged.

REFERENCES

1. Wang, N.S., Stephanopoulos, G.: Biotechnology and Bioengineering Symp., 14, 635 (1984)
2. Thibault, J., Grandjean, B.: Proc. of the IFAC Int. Symp. Advanced Control of Chemical Processes ADCHEM'91, Toulouse, France, 295 (1991)
3. Thibault, J., Cheruy, A.: IMACS Annals on Computing and Applied Mathematics Proc. MIMS2'90, Brusels (1990)
4. Thibault, J., Van Breusegem, V., Cheruy, A.: B&B, 36, 1041 (1990)
5. Thibault, J., Van Breusegem, V.: Bioautomation, 12, 79 (1992)
6. Werbos. P.J.: Proc. of the IEEE, 78, 1050 (1990)
7. O'Neil, D.G., Liberatos, G.: B&B, 36, 437 (1990)
8. Grueger, W., Grueger, A.,, Brock, T.D. (eds): Biotechnology: A Textbook of Industrial Microbiology. Suderland, MA, USA: Sinauer Associates, Inc 1984.
9. Petrova, M., Koprinkova, P., Patarinska T.: Biotechnology and Biotechnological Equipment, 1, 88, (1994)

BRAIN ELECTROGRAPHIC STATE DETECTION USING COMBINED UNSUPERVISED AND SUPERVISED NEURAL NETWORKS

A.J.F. Coimbra[1], J. Marino-Neto[2], F.M. de Azevedo[3], C.G. Freitas[2], J.M. Barreto[3]

[1]Lab. of Neurophysiology, Université Catholique de Louvain, Belgium.
[2]Lab. de Neurofisiologia I, CFS-CCB,
[3]GPEB, Dept. Electrical Eng.
Universidade Federal de Santa Catarina, Brazil

Abstract

This work describes the development of a new approach to NN processing of sleep-related brain electrographic signals, using two sequentially combined unsupervised (Kohonen layer, KNN) and supervised (Widrow-Hoff layer, WHNN) algorithms. Twelve parameters extracted from physiological data (EEG, EMG and EOG, obtained from unrestrained rats through several sleep-waking periods), were first processed by a KNN, that detected different signal patterns. These patterns were further examined by an EEG expert, who identified them as belonging to one of the known sleep-waking stages, or to transitional and/or unknown signal combinations. Selected outputs of the KNN, classified in this way, formed the input vectors to a WHNN, that allowed fast and reliable tracking of changes in these states (both known and newly detected) during prolonged periods of time. Such an approach can represent an important aid for simultaneous exploration, detection and also temporal following of electrographic events along the sleep-waking cycle.

Introduction

In recent years, neural networks (NNs) have been proposed as useful tools for tackling analytical and classification problems in an increasingly high number of biological areas (anesthesiology, EEG classification and wave-form detection, regional blood flow estimation, diagnosis, see, [1] [6] [7] [11] [16]). The general advantages of the use of neural networks for biological signal processing include their ability to extract and recognize patterns and intrinsic characters from the typically complex and interactive biological processes, with high tolerance in dealing with noisy, incomplete and/or contradictory data. By far, the most promising biological field for NNs is neurobiology, particularly in processing brain generated electrical signals [1] [8] [9] [11] [13] [14]. In fact, the benefits of such an interaction are not confined to biology. The development of NNs methods for complex, "real word" data (like electroencephalographic ones) are supposed to bring improvements to NNs performances with other kind of data [8].

Electrographic sleep signals analysis takes into account a number of interacting phenomena, including electroencephalographic (EEG, electrical oscillations recorded at brain or scalp surface), electromiographic (EMG, electrical oscillations due to to the muscle activity), electrooculographic (EOG, electrical oscillations due to eye movements), and systemic and/or vegetative data, as arterial pressure, cardiorespiratory activity, metabolism and temperature. Sets and oscillations of all these attributes define a number of sleep-waking stages (or states), and can be used for research and diagnostic purposes for some sleep and mood disorders [2] [12] [15].

Most of recent NN approaches to sleep segmentation and analysis use a variety of topologies, with either supervised or unsupervised training algorithms. The approach of the later case, the approach of using unsupervised learning algorithms [13] [5] [4] has advantages for exploratory research activities, because it can extract and recognize unusual or formerly unnoticed signal patterns. In the second case (supervised algorithm)[8] [9] [11] [14], it can bring a number of problems in EEG analysis, including the existence of criteria disagreements among experts and different classifying rules for different abnormal sleep patterns. However, only supervised algorithms can bring the clinician/researcher expertise and objectivity to the NN output, while allowing for easily tracking temporally eventual changes in these states. The present work describes the development of a new approach to NN processing of sleep-related brain electrographic signals, using two sequentially combined unsupervised (Kohonen layer, KNN) and supervised (Widrow-Hoff layer, WHNN) algorithms. This approach is expected to combine the different capabilities of the KNN and WHNN paradigms for sleep-waking state analysis.

Methodology

The method developed here is based on the analysis of the dynamics of parameters extracted from bioelectrical potentials (EEG, EMG, EOG) related to the sleep-waking states in laboratory albino rats, chronically implanted with recording electrodes placed at frontal and occipital cortical surfaces (for EEG), neck muscles (for EMG) and periorbital bones (for EOG). These signals were continuously recorded from these unrestrained rats for as long as 8 hours, during the light phase of the day, so to record samples from most of the sleep-waking continuum. The signals were first amplified and filtered on a conventional polygraph (2nd order Butterworth low pass linear phase filter, 30 Hz cut-off frequency, 0.3 s time constant), digitized (12 bit AD converter), sampled at 64 Hz, and stored on an IBM compatible microcomputer hard disk. A software was specifically developed to control the calibration, acquisition and storage processes [3]. Recordings were sequentially partitioned in 4-s epochs (time frames), from which 11 numerical parameters were extracted: powers of 0.5 - 30, 0.5 - 4, 6.5 - 9 and 12 - 30 Hz EEG frequency bands, centroid frequency of each channel, and EMG integral. A 12th parameter was calculated as a linear combination of the normalized other 11, so to form 12 element vectors with norm 1, that formed the input vectors for a self-organizing Kohonen (KNN) array (10x10) of fully interconnected elements.

Training was performed through the standard Kohonen algorithm [10], usually with a set composed of the 2 first hours of the record. The initial connection weights were randomly assigned. Afterwards, each input vector was randomly picked up from the 1800 vectors pool and fed into the KNN for the updating of the connections weights; this training procedure was repeatedly carried out. After the training phase some "patches" of variable durations were chosen from records to evaluate the network behavior. After training, KNN was tested with signals from several recording periods other than those used for training. The KNN tests outputs (Fig. 3) and the corresponding raw data (tracings) were analyzed by an EEG expert to identify these periods as belonging to one of the known sleep-waking states, or as transitional periods between these states. Then, the KNN outputs for one recording period was connected to a supervised WHNN. This layer was further trained to recognize, e.g., three states (slow wave sleep, SWS; waking, W; SWS-W transition), as detected by the EEG expert on crude polygraphic data.

Results

Fig. 1 depicts moving averages of only 3 of the 12 parameters measured along an 8-hour record: the EMG module integral, and the powers of 0.5 - 4.0 and 6.5 - 9.0 Hz of the frontal EEG. Fig. 2 is a time amplification of Fig. 1 to show with more details the time evolution of the signals.

Fig. 3 shows a sample of the former, containing initially a slow wave sleep (SWS) period, followed by a transitional state and by a waking epoch at the end of the period. These, and the other 9 parameters, occurring at the time interval defined between the vertical dotted lines in A (Fig. 1), were

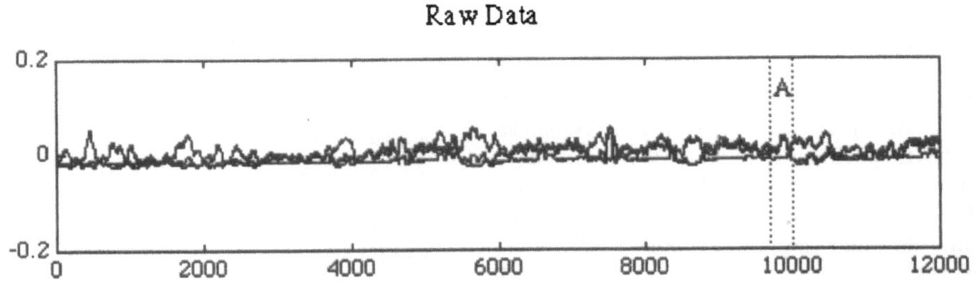

Figure 1: Moving average of some parameters of the record

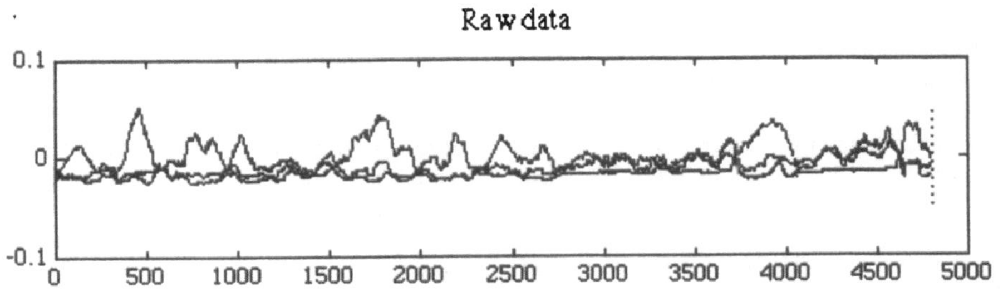

Figure 2: Time amplification of the signals represented in Fig. 1

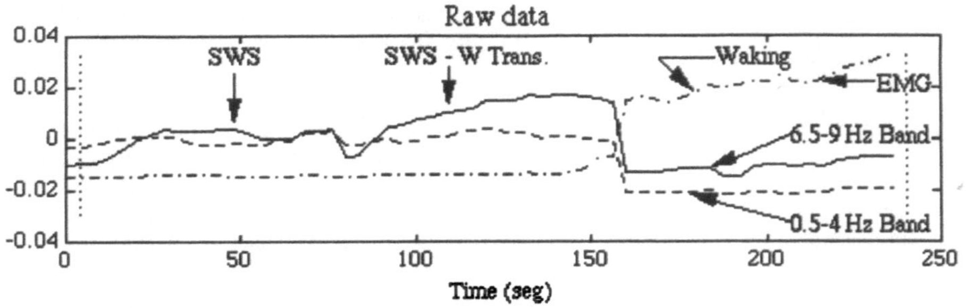

Figure 3: Detail of Fig. 1 containing a SWS, a Transitional and a Waking epoch

used in this example as input to the trained KNN, so to produce a 2-dimensional activation matrix (Fig. 4) illustrating the clusters (regions) of the output space of the KNN activated in that interval. Each of the mesh vertexes represents a processing unit, and the amplitude of each peak represents the frequency of unit activation during the analyzed period.

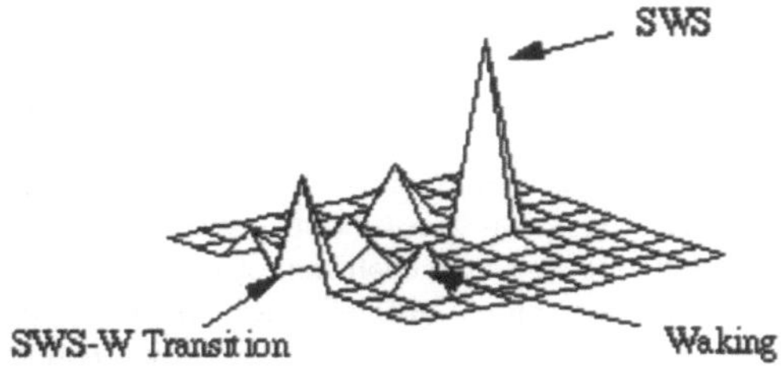

Figure 4: Kohonen Output Space

It is worth to remark that our EEG expert identified, from the crude recordings of this period, only a SWS period (from 0 to 155 s in the Fig. 3) followed by the final waking period. The analysis of the KNN activation clusters and the corresponding raw data revealed that the SWS period included two periods characterized by different parameter combinations. The first period was provisionally designed here as a "true" SWS epoch, and the final SWS period as a peculiar, transitional phases between SWS and waking, discovered with the aid of the KNN processing. These results indicate the suitability and sensibility of the KNN topology, the unsupervised algorithm and the parameters used for training and classification tasks in discriminating among different sleep-waking states, and also for uncovering distinctive periods within EEG epochs visually diagnosed as homogeneous by an EEG expert.

The output of a 3-neuron WHNN, each neuron trained to recognize one of the 3 "states" identified by the KNN in the epoch depicted in Fig. 3, is shown in the Fig. 5.

This figure shows the temporal evolution of each of the states detected by the KNN. It should be noted that there is a high level of interpolation between the SWS and the transitional state activations in the first third of the period. This could be interpreted as indicative of the presence of shared and unshared attributes between SWS and the transitional period, and that the SWS period is not a "pure" one, being contaminated by some parameters typical of the well defined transitional epoch. These results highlight the usefulness of a supervised network as final analytical step, in providing fast and reliable quantitative descriptions of the temporal architecture of electrographically recognized sleep-waking states, as well as of unknown or formerly uncovered aspects of these states.

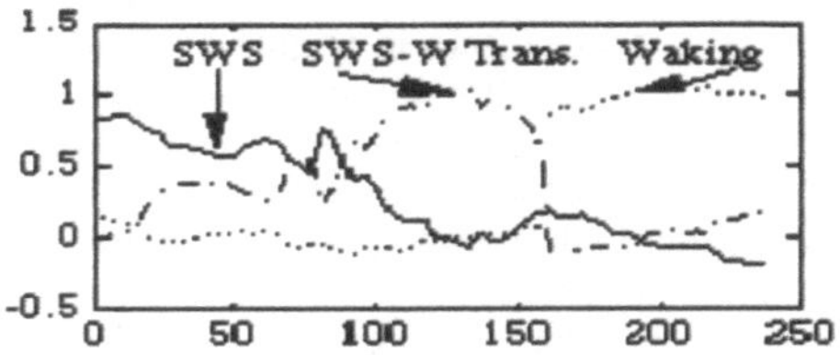

Figure 5: Widrow-Hoff Activations

Discussion

Our data indicate the suitability and sensitivity of this approach for training and automatic classification tasks in different brain electrographic states. Additional experiments, using the present methodology, are being conducted on a number of other sleep-waking states. To the moment, it can be suggested that such an architecture allows for clinicians/researchers the possibility to adapt or set new neurons in the supervised step, after detection and qualification of transitional, unusual (e.g., drug-related) or unknown pathological signal combinations by the unsupervised KNN step of the process. These neurons can be designed to automatically identify these states in long lasting routine recordings.

It is worth noting that the unsupervised nature of the training algorithm frees the method from any a priori (and thus possibly biased) information on these complex parameters sets, so it can also detect transitional and/or unusual association patterns. By handling these data in the sense of analyzing the KNN's feature map, one can also search for feature combinations associated to output events of this kind. Used in this way, such an approach can be employed not only as an automated state detector, but also as a tool for neurological and neurophysiological investigation, supporting the exploration and identification of novel dynamic characteristics of sleep-waking cycle, and of other behavioral/physiological states where electrographic examination holds.

Additional experiments, using the present methodology, are being carried out at our labs, assessing the behavior and suitability of different parameter sets for sleep-waking stage detection and in analysis of pharmacologically-induced changes in ongoing EEG. At present, however, it is reasonable to believe that further developments of self-organizing NN approaches would represent major improvements to the diagnosis and investigation in biological aspects of clinical relevance.

Acknowledgements

The first and the last authors acknowledge the CNPq, Brazil, for sponsoring partially this research.

References

[1] I. N. Bankman, V. G. Sigillito, R. A. Wise, and P. L. Smith. Feature-based detection of the k-complex wave in the human electroencephalogram using neural networks. *IEEE Transactions on Biomedical Engineering*, 39(12):1305–1310, 1992.

[2] M. A. Carskadon and W. C. Dement. Normal human sleep: an overview. In M.H. Kryger, editor, *Principles and Practice of Sleep Medicine*, pages 3–13. Saunders, New York, 1988.

[3] A. J. F. Coimbra. Análise computadorizada de sinais bioelétricos. Master's thesis, Center of Technology, Federal University of Santa Catarina, Brazil, 1994.

[4] A. J. F. Coimbra. Automatic detection of sleep-waking states using Kohonen neural networks. In *1 Congresso Brasileiro de Redes Neurais*, pages 327–331, Itajubá, Minas Gerais, Brasil, 1994.

[5] A. J. F. Coimbra. Electrographic analysis of brain states using neural networks. In *World Congress on Medical Physics and Biomedial Engineering*, page 463, Rio de Janeiro, Brasil, 1994.

[6] F. M. de Azevedo. *Contribution to the study of neural networks in dynamical expert systems.* PhD thesis, Institut d'Informatique, FUNDP, Belgium, 1993.

[7] R. O. Garcia. *Técnicas de inteligência artificial aplicadas ao apoio à decisão médica na especialidade de anestesiologia.* PhD thesis, Center of Technology, Federal University of Santa Catarina, Brazil, 1992.

[8] B. Klöppel. Application of neural networks for eeg analysis. *Neuropsychobiology*, 29:39–46, 1994.

[9] B. Klöppel. Classification by neural networks of evoked potentials. *Neuropsychobiology*, 29:47–52, 1994.

[10] T. Kohonen. *Self-Organization and Associative Memory.* Springer-Verlag, Berlin, 1984.

[11] A. N. Mamelak, J. J. Quattrochi, and A. Hobson. Automated staging of sleep in cats using neural networks. *Electroencephalography and Clinical Neurophysiology*, 79:52–61, 1991.

[12] R. Reimão. *Sono: Aspectos atuais.* Neurológica, Psiquiatria. Atheneu, São Paulo, 1990.

[13] S. Roberts and L. Tarassenko. New method of automated sleep quantification. *Medical & Biological Engineering & Computing*, 30:509–517, 1992.

[14] N. Schaltenbrand, R. Lengelle, and J.-P. Macher. Neural networks model: Application to automatic analysis of human sleep. *Computers and Biomedical Research*, 26:157–171, 1993.

[15] M. Timsit-Berthier. Approche neurophysiologique des états dépressifs. *Psychologie Medicale*, 22(8):757–763, 1990.

[16] F. Y. Wu and J. D. Slater. Regional cerebral blood flow estimation by neural network-based parametric regression analysis. *Int. Journal of Biomedical Computation*, 33:119–128, 1993.

FROM THE CHROMOSOME TO THE NEURAL NETWORK

Olivier MICHEL Joëlle BIONDI

om@alto.unice.fr jb@mimosa.unice.fr

Laboratory I3S — CNRS-UNSA, Bât. 4

Abstract

A proposal for a model of morphogenesis process taking inspiration from biology is presented in this paper. This process uses a chromosome as a production system to create an artificial neural network. It starts with a single cell containing the chromosome. Cells can divide and establish connections among them. Both structure and weights of the neural network are defined by the morphogenesis process. An application to a neural network driving an autonomous mobile robot is presented which exhibits encouraging first results.

1 Introduction

A promising direction for the search for optimal neural networks could be the use of genetic algorithms in order to evolve the structure and the weights of neural networks. The second section will point out the coding problem and give an overview of the state of the art in this research area. Then, in section three, it will be explained how biological inspiration could help to design a model for a morphogenesis process able to generate a neural network using a linear chromosome. The principles of such a process will be presented in section four, as well as the evolutionary approach. Section five gives results of the application of this process to a neural network driving an autonomous mobile robot.

2 State of the art

When genetically evolving neural networks, one has first to study the problem of the coding of the network on a chromosome structure that will be used by the genetic algorithm. Many solutions were investigated including linear chromosome [4, 7], matrix chromosome [11] or tree chromosome [6]. The use of fixed-length chromosomes implies an a priori fixed-dimensionality of the search space. This handicap can be avoided by using variable-length chromosomes [8, 3]. The first codings used direct encoding schemes. This too simple approach has one major drawback: the reuse of many identical sub-networks inside a global network is impracticable. Specialists of neural networks know that powerful networks are built of many similar substructures. This feature is explicit in topological maps [9] and can be found in a higher level in more specific applications [10]. New schemes have been proposed where loops or recursivity allow the development of modular structures [1, 6]. Recent investigations remain, for the moment, restricted to modular feedforward networks [6] or non-modular recurrent networks [3]. We try to go further in this paper by proposing a model, inspired from biology, allowing recurrent modular networks. Moreover, genetic operators used in the evolutionary process allow gradual complexification of the structures.

3 Inspiration from biology

The process described here was inspired from the biological principle of the proteins synthesis regulation. There are many ways to regulate this process, but we focused on two interesting points. Of course, we assumed many simplifications in order to be clearer, but the key ideas remain. If a particular protein called repressor is present, it can sit on the start codon of a specified gene on the chromosome, so that RNA polymerase cannot read it, and thus the corresponding proteins cannot be synthesized. This system can be recurrent when a synthesized protein can be a repressor for another gene. The other mechanism can be seen as the positive version of the first one. A molecule called activator is necessary to initiate the process of transcription of the DNA into mRNA at a specific locus, and thus to initiate the proteins synthesis process. Recurrence remains possible when a synthesized protein causes the synthesis of other ones. These processes can be seen as a kind of production system, where repressors and activators are conditions to the production of proteins.

4 Morphogenesis process

The cell growing mechanism starts on a single cell enclosing a chromosome and a message list. The message list corresponds to the set of proteins available in the biological cell. At the beginning, a single initial message is present in the message list.

4.1 Cell production system

A chromosome is a variable length list of rules. A rule is made of three binary strings, two condition parts and one action part: A set of messages whose *absence* in the cell message list is necessary for the rule to be fired (corresponding to the biological repressors), a set of messages whose *presence* in the cell message list is necessary for the rule to be fired (corresponding to the biological activators) and a set of produced messages which are added to the cell message list when the rule is fired. (corresponding to the biological synthesized proteins). There is no competition between rules: any rule that can be fired is actually fired. So the interpretation of the rules does not depend on their position in the chromosome (see Fig. 1).

A cell can perform operations on itself or on its neighborhood, such as setting a transfer function, giving birth to a new cell or establishing a connection to an other cell. The machineries performing these operations use messages as orders or parameters, they also may produce messages. For example, the *split machinery* will check if a message ordering split is present in the cell. If it finds one, then it will consume it and produce a cell division. The production system can be seen as such a machinery, consuming and producing messages. It could be shown that a cell has the properties of an autopoietic system [14].

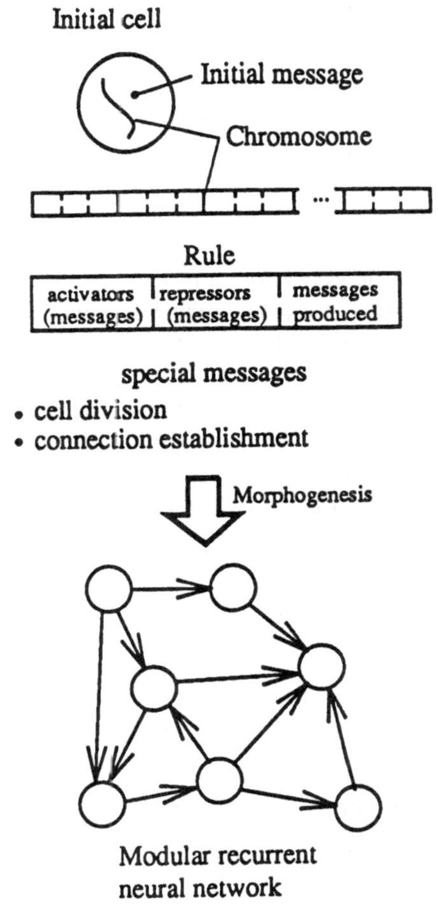

Fig. 1: Morphogenesis process: messages are encoded as binary strings

4.2 Resulting networks

This morphogenesis process is general: it is able to create any kind of neural network. It allows recurrent connections, different kinds of neurons with different transfer functions and different kinds of links with different learning rules. Thus, the space of neural networks explored is theoretically unlimited. Note that many different chromosomes may give birth to the same network if the same rules are organized in a different order on the chromosome, or use different internal messages to communicate between them. Rules which are never fired, or whose action don't change the morphogenesis operations can be added without modification on the resulting network.

4.3 Genetic Algorithm

We used a genetic approach [5] with a binary fitness (see next section) using classical random mutations and a genetic operator (a kind of mutation) which can add or remove a random element in one of the three parts of a rule, or add or remove whole rule at the end of the chromosome. This allows variable length chromosomes which are necessary for an increase in complexity during evolution. Crossover may occur only between two rules and never inside a rule: for example, the crossover of the chromosomes $R1a$, $R2a$, $R3a$ | $R4a$, $R5a$, $R6a$ and $R1b$, $R2b$, $R3b$ | $R4b$, $R5b$ at the designated breakpoints would give birth to the two offsprings $R1a$, $R2a$, $R3a$, $R4b$, $R5b$ and $R1b$, $R2b$, $R3b$, $R4a$, $R5a$, $R6a$. This operator tends to preserve subsets of rules in relatively close physical proximity to one another in the chromosome structures. Investigations for making the genetic algorithm more accurate for evolving this kind of chromosome are currently studied.

5 Application to an autonomous mobile robot

5.1 Description of the robot

In order to test the validity of our method, we evolved populations of chromosomes generating networks able to control a Khepera-like simulated mobile robot, keeping in mind a final application to the real robot Khepera [12, 13]. This robot resembles Braitenberg vehicles [2]. It has two motors, each controlling one wheel, and 8 infra-red proximity sensors able to detect obstacles at a distance of a few centimeters.

5.2 Evolutionary process

Natural selection uses a binary fitness function (*life* or *death*). We chose, as a first approach, to use this simple binary fitness function. According to this, if we want to evolve simple systems in order to get more complex and powerful systems, we need to evolve the environment in the same way: At the beginning, the environment must be so simple that very primitive behaviors are efficient for survival. Once a large enough population could survive, the environment would get a bit more hostile, so that the elite of the population could survive, and so on, until the individuals develop elaborated behaviors.

A simple goal, which is *go towards food*, was chosen. A kind of robot metabolism was introduced: A variable represents the robot's internal energy. This variable is increased each time the robot successfully eats food. It is decreased when the robot moves, and also slowly decreased even if the robot doesn't move, just like an electric battery. If this variable reaches zero, the robot "dies" and is eliminated from the population.

About 200 generations were necessary to obtain valid neural networks (a valid neural network has a number of inputs and outputs matching with the number of inputs and output of the robot model). At the beginning of the tests of valid networks, some individuals move following different trajectories (lines, circles), apparently ignoring food. After around 300 generations, the genetic algorithm has converged, robots become efficient: they are attracted by food. Populations of chromosomes, neural networks and behaviors seem to become homogeneous.

6 Conclusion

This first experiment using this morphogenesis process was successful. The behaviors discovered are yet very simple and can be compared to Braitenberg's simplest vehicles. To go forward, we need a more sophisticated robot equipped with more actuators and more sensors. This would allow complex behaviors in the perspective of ontogenetic learning. Moreover the use of a continuous valued fitness function as well as more accurate genetic operators may induce significant improvements allowing the exploration of more complex neural networks. The final step will be the transfer from the simulator to the real robot.

References

[1] E.J.W. Boers and H. Kuiper. *Biological metaphors and the design of modular artificial neural networks.* PhD thesis, Departments of Computer Science and Experimental Psychology at Leiden University, The Netherlands, 1992.

[2] V. Braitenberg. *Vehicles: Experiments in Synthetic Psychology.* MIT Press, Cambridge, 1984.

[3] Dave Cliff, Inman Harvey, and Phil Husband. Explorations in evolutionary robotics. *Adaptive Behavior*, 2(1):73 – 110, 1993.

[4] B. Fullmer and R. Miikkulainen. Using marker-based genetic encoding of neural networks to evolve finite-state behaviour. In F. Varela and P. Bourgine, editors, *Towards a Practice of Autonomous Systems, Proceedings of the First International Conference on Artificial Life, Paris.* MIT Press, 1992.

[5] D.E. Goldberg. *Genetic Algorithms in Search, Optimisation and Machine Learning.* Addison Wesley, Massachussets, 1989.

[6] F. Gruau and D. Whitley. The cellular developmental of neural networks : the interaction of learning and evolution. Technical report, Ecole Normale Superieure de Lyon, 46, Allee d'Italie, 69364 Lyon Cedex 07, France, January 1993.

[7] S. Harp and T. Samad. Genetic synthesis of neural network architecture. In Lawrence Davis, editor, *Handbook of Genetic Algorithms*, chapter 15, pages 202 - 221. Van Nostrand Reinhold, 1991.

[8] Inman Harvey. Species adaptation genetic algorithms: A basis for a continuing saga. In F. Varela and P. Bourgine, editors, *Towards a Practice of Autonomous Systems, Proceedings of the First International Conference on Artificial Life, Paris*, pages 346 - 354. MIT Press, 1992.

[9] T. Kohonen. *Self-Organization and Associative Memory.* Springer-Verlag, 1989. (3rd ed.).

[10] Y. Le Cun, B. Boser, J.S. Denker, D. Henderson, R.E. Howard, W. Hubbard, and L.D. Jackel. Handwritten digit recognition with a back-propagation network. In D.S. Touretzky, editor, *Advances in Neural Information Processing Systems II.* Morgan Kaufmann, 1990.

[11] G.F. Miller, P.M. Todd, and S.U. Hegde. Designing neural networks using genetic algorithms. In Schaffer J.D., editor, *Proceeding of the Third International Conference on Genetic Algorithms*, pages 379 - 384. Morgan Kaufmann, 1989.

[12] F. Mondada, E. Franzi, and P. Ienne. Mobile robot miniaturisation: A tool for investigation in control algorithms. In *Third International Symposium on Experimental Robotics*, Kyoto, Japan, October 1993.

[13] F. Mondada and C. Touzet. Quelques comportements adaptatifs pour le robot miniature khepera. In *Annales du Groupe CARNAC*, EPFL, Lausanne Suisse, 1993.

[14] J.F. Varela. *Autonomie et Connaissance.* Editions du Seuil, Paris, Janvier 1989.

CAM-BRAIN

The Evolutionary Engineering of a Billion Neuron Artificial Brain by 2001 which Grows/Evolves at Electronic Speeds inside a Cellular Automata Machine (CAM)

Hugo de Garis

Brain Builder Group, Evolutionary Systems Department,
ATR Human Information Processing Research Laboratories,
2-2 Hikaridai, Seika-cho, Soraku-gun, Kansai Science City, Kyoto, 619-02, Japan.
tel. + 81 7749 5 1079, fax. + 81 7749 5 1008, degaris@hip.atr.co.jp

Abstract

A "Darwin Machine" is defined to be a piece of hardware that evolves its own architecture at electronic speeds. A Darwin Machine is an example of "Evolutionary Engineering", which in turn is defined to be the building of complex structures and dynamics using evolutionary methods. The author believes that evolutionary techniques will dominate 21st century engineering, especially molecular scale (nanotech) engineering, with its gargantuan number of components and complexities. This paper reports on the second year of an ambitious 8 year research project which aims to implement a type of Darwin Machine in the form of a cellular automata based artificial brain with a billion neurons by 2001, which grows/evolves at (nano-)electronic speeds inside a Cellular Automata Machine - ATR's so-called "CAM-Brain Project". The basic idea is to use cellular automata based neural networks which grow under evolutionary control at (nano-)electronic speeds. The states of the cellular automata (CA) cells and the CA state transition rules can be stored cheaply in gigabytes of RAM. By using state of the art cellular automata machines, e.g. MIT's "CAM8" machine ($40,000), which can update 200 million CA cells a second, it may be technically feasible within a year or so to evolve artificial nervous systems containing a thousand neurons, and within 5 years, a million neurons. By the end of the current research project, i.e. 2001, it should be possible using nano-scale electronics to grow/evolve artificial brains containing a billion neurons and upwards. This is our aim.

Keywords

Evolutionary Engineering, Neural Networks, Genetic Algorithms, Cellular Automata, Cellular Automata Machines (CAMs), Nano-Electronics, Darwin Machines.

1. Introduction

Following on from the abstract above, to understand how CAs can be used to grow/evolve neural networks, imagine a 2D CA trail which is 3 cells wide (e.g. Fig. 2). Down the middle of the trail, send growth signals. When a growth signal hits the end of the trail, it makes the trail extend, or turn left, or right, or split etc., depending upon the nature of the signal. It is the sequence of these signals (fed continuously over time into an initial short trail) that is evolved. This sequence of growth signals is the "chromosome" of a genetic algorithm, and it is this sequence that maps to a cellular automata network. When trails collide, they form "synapses". Once the CA network has been formed in the initial "growth phase", it is later used in a second "neural signaling phase". Neural signals move along CA-based axons and dendrites, and across synapses etc. The CA network is made to behave like a conventional artificial neural network. The outputs of some of the neurons of the complex recurrent networks which result can be used to control complex time dependent behaviors whose fitnesses can be measured. These fitness values can be used to drive the evolution. By growing/evolving thousands of neural net modules and their interconnections in an incremental evolutionary way, it will be possible to build artificial brains. We expect that within a few years, a new field will be established, based partly on this work, called simply "Brain Building". According to the CAM developers at MIT, it is likely that the next generation of CAMs will achieve an increase in performance of the order of thousands, within 5 years. However, to be able to evolve a billion neuron artificial brain by 2001, a "nano-CAM" machine will need to be developed. To this end, we are collaborating with an NTT researcher who has developed a nanoscale electronics device, who wants to combine them to behave like molecular scale cellular automata machines.

In the summer of 1994, a two dimensional CAM-Brain simulation was completed which required 11,000 hand crafted state transition rules. It was successfully applied to the evolution of maximizing the number of synapses, outputting an arbitrary constant neural signal value, outputting a sine wave of a desired arbitrary period and amplitude and to the evolution of a simple artificial retina which could output the vector velocity of a "white line" which "moved" across an array of "detector" neurons. Next, will be to port the 2D simulation to a CM5 supercomputer to pursue more interesting and ambitious evolutionary experiments. Work on the 3D simulation should be completed early in 1995, and is expected to take about 60,000 rules. The Brain Builder Group of ATR took possession of one of MIT's CAM8 machines in the fall of 1994. An attempt will be made to port the rules of the 3D simulation to this machine. If this is not possible, then a CAM9 machine will be designed specifically for CAM-Brain.

The following paragraphs now present the CAM-Brain Project in more detail. As mentioned in the extended abstract, the "CAM-Brain Project" is an 8 year research project at ATR labs in Kyoto, Japan, which intends to build an artificial brain containing billions of artificial neurons. The complexity of such an artificial brain will make it largely undesignable, so a (directed) evolutionary approach called "evolutionary engineering" is being used. Neural networks based on cellular automata [Codd 1968], can be grown and evolved at electronic speeds inside state of the art cellular automata machines, e.g. MIT's "CAM8" machine, which can update 200 million cells per second [Toffoli & Margolus 1990]. Since RAM is cheap, gigabytes of RAM can be used to store the states of the CA cells used to grow the neural networks. CA based neural net modules are evolved in a two phase process. Three cell wide CA trails are grown by sending a sequence of growth signals (extend, turn left, turn right, fork left, fork right, T fork) down the middle of the trail. When an instruction hits the end of the trail it executes its function. This sequence of growth instructions is treated as a chromosome in a Genetic Algorithm [Goldberg 1989] and is evolved. This sequence maps to a CA network. When trails collide, they form "synapses". Once the CA network is grown, it is used as a neural network in a second neural signaling phase. Some of the neural signals can be tapped to control some process and the fitness of the control can be measured. This fitness is used to drive the evolution. Once gigabytes of RAM and electronic evolutionary speeds can be used, genuine brain building, involving millions and later billions of artificial neurons, becomes realistic, and should become concrete within a year or two. The CAM-Brain Project should revolutionize the fields of neural networks and artificial life, and in time help create a new specialty called "Brain Building", with its own conferences and journals.

This paper consists of the following sections. Section 2 describes briefly the idea of "Evolutionary Engineering", of which the CAM-Brain Project is an example. Section 3 describes how neural networks can be based on cellular automata [Codd 1968], and evolved at electronic speeds. Section 4 presents some of the details of CAM-Brain's implementation. Section 5 shows how using cellular automata machines will enable millions of artificial neural circuits to be evolved to form an artificial brain.

2. Evolutionary Engineering

Evolutionary Engineering is defined to be *"the art of using evolutionary algorithms (such as genetic algorithms [Goldberg 1989]) to build complex systems."* This paper reports on the idea of evolving cellular automata based neural networks at electronic speeds inside cellular automata machines. This idea is a clear example of evolutionary engineering. Evolutionary engineering will be increasingly needed in the future as the number of components in systems grows to gargantuan levels. Today's nano-electronics for example, is researching single electron transistors (SETs) and quantum dots. Probably within a decade or so, humanity will have full blown nanotechnology (molecular scale engineering), which will produce systems with a trillion trillion components [Drexler 1992]. The potential complexities of such systems will be so huge, that designing them will become increasingly impossible. However what is too complex to be humanly designable, might still be buildable, as this paper will show. By using evolutionary techniques (i.e. evolutionary engineering), it is often still possible to *build* a complex system, even though one does not understand how it functions. This arises from the notion of the "complexity independence" of evolutionary algorithms, i.e. so long as the (scalar) fitness values which drive the evolution keep increasing, the internal complexity of the evolving system is irrelevant. This means that it is possible to successfully evolve systems which function as desired, but which are to complex to be designable. The author believes that this simple idea (i.e. the complexity independence of evolutionary algorithms) will form the basis of most 21st century technologies (dominated by nanotechnology [Drexler 1992]). Thus, evolutionary engineering can "extend the barrier of the buildable", but may not be good science, because its products tend to be black boxes. However, confronted with the complexity of trillion trillion component systems, evolutionary engineering may be the only viable method to build them.

3. Cellular Automata Based Neural Networks

Building an artificial brain containing billions of artificial neurons is probably too complex a task to be humanly designable. The author felt that brain building would be a suitable task for the application of evolutionary engineering techniques. The key ideas are the following. Use evolutionary techniques to evolve neural circuits in some electronic medium, so as to take advantage of electronic speeds. The medium chosen by the author was that of cellular automata (CA) [Codd 1968], using special machines, called "Cellular Automata Machines (CAMs)", which can update hundreds of millions of CA cells a second [Toffoli & Margolus 1990]. CAMs can be used to evolve the CA based neural networks at electronic speeds. The states of the cellular automata cells can be stored in RAM, which is cheap, so one can have gigabytes of RAM to store the states of millions of CA cells. This space is large enough to contain an artificial brain. MIT's Information Mechanics Group (Toffoli and Margolus) believe that within a few years it will be technically possible to update a trillion CA cells in about 0.1 nanoseconds [p221, Toffoli & Margolus 1990]. Thus, if CA state transition rules can be found to make CA behave like neural networks, and if such CA based networks prove to be readily evolvable, then a potentially revolutionary new technology becomes possible. The CAM-Brain Project is based on the above ideas and fully intends to build artificial brains before the completion of the project in 2001. The potential is felt to be so great that it is likely that a new specialty will be formed, called "Brain Building".

For the first 18 months of the CAM-Brain Project, the author simulated a two dimensional version of CAM-Brain on a Sparc 10 workstation, hand coding over 11,000 (eleven thousand) CA state transition rules to get the CA-based neural nets to evolve. This work was completed in the summer of 1994. The 2D version was used briefly (before work on the 3D version was started) to undertake some evolutionary tests, whose results will be presented in the next section. The 2D version served only as a feasibility and educational device. Since trails are obliged to collide in 2D, the 2D version was not taken very seriously. Work was begun rather quickly on the more interesting 3D version almost immediately after the 2D version was ready. Proper evolutionary tests will be undertaken once the 3D version is ready, which should be by early 1995. To begin to understand how cellular automata [Codd 1968] can be used as the basis for the growth and evolution of neural networks, consider Fig. 1 which shows an example of a 2D CA state transition rule, and Fig. 2 which shows a 2D CA trail, 3 cells wide. All cells in a CA system update the state of their cells synchronously. The new state of a given cell depends upon its present state and the states of its nearest neighbors. Down the middle of the 3 cell wide CA trail, move "signal or growth cells" as shown in Fig. 2

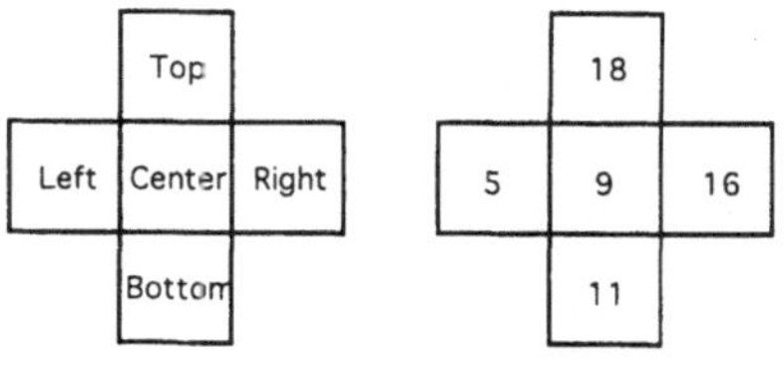

CTRBL -> Cnext 9.18.16.11.5 -> 4

Fig. 1 A 2D CA State Transition Rule

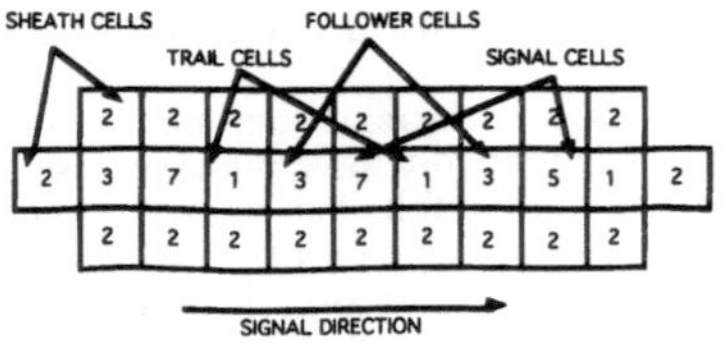

Fig. 2 Signal Cells Move Along a Cellular Automata Trail

As an example of a state transition rule which makes a signal cell move to the right one square, consider the right hand most signal cell in Fig. 2, which has a state of 5. The cell immediately to its right has a state of 1, which we want to become a 5. Therefore the 2D state transition rule to turn the 1 into a 5 is 1.2.2.2.5-->5. These signal or growth cells are used to generate the CA trails, by causing them to extend, turn left or right, split left or right, and Tsplit. When trails collide, they can form synapses. It is the sequence of these signal cells which determines the configuration of the CA trails, thus forming a CA network. It is these CA trails which later are used as neural network trails of axons and dendrites. Neural signals are sent down the middle of these CA trails. Thus there are two major phases in this process. Firstly, the CA trails are grown, using the sequence of signal cells. Secondly, the resulting CA trail network is used as a neural network, whose fitness at controlling some system can be measured and used to evolve the original growth sequence. To make this more explicit, it is the sequence of growth cells which is evolved. By modifying the sequence, one alters the CA network configuration, and hence the fitness of the configuration when it functions as a neural net in the second phase. From a genetic algorithm (GA) point of view, the format of the GA "chromosome" is the sequence of integers which code for the signaling or growth instructions. By mutating and crossing over these integers, one obtains new CA networks, and hence new neural networks. By performing this growth at electronic speeds in CAMs, and in parallel, with one CAM per GA chromosome, and attaching a conventional programmable microprocessor to each CAM to measure the user defined fitness of the CA based neural circuit, one has a means to evolve large numbers of neural modules very quickly. Using CAMs to evolve neural circuits, is an example of a type of machine that the author labels a "Darwin Machine", i.e. one which evolves its own structure or architecture. A related idea of the author concerns the concept of "Evolvable Hardware (EHW)" [de Garis 1993] where the software instructions used to configure programmable logic devices (PLDs) are treated as chromosomes in a Genetic Algorithm [Goldberg 1989]. One then rewrites the circuit for each chromosome.

4. Further Details

This section provides further details on the implementation of the CA based neural networks. There were three kinds of CA trails in CAM-Brain, labeled dendrites, excitatory axons and inhibitory axons, each with their own states. Whenever an axon collided with a dendrite or vice versa, a synapse was formed. When a dendrite hit an excitatory/inhibitory axon or vice versa, an excitatory/inhibitory synapse was formed. An inhibitory synapse reversed the sign of the neural signal value passing through it. An excitatory synapse left the sign unchanged. Neural signal values ranged between -240 and +240 (or their equivalent CA states, ranging from 100 to 580). The value of a neural signal remained unchanged when it was in an axon, but as soon as it crossed a synapse into a dendrite, the signal value (i.e. signal strength) began to drop off linearly with the distance it had to travel to its receiving neuron. Hence the signal strength was proportional to the distance between the synapse and the receiving neuron. Thus the reduction in signal strength acted like a weighting of the signal by the time it reached the neuron. But, this distance is evolvable, hence indirectly, the weighting is evolvable. CAM-Brain is therefore equivalent to a conventional artificial neural network, with its weighted sums of neural signal strengths. However, in CAM-Brain there are time delays as signals flow through the network. When two or three dendrite signals collide, they sum their signal strengths (within saturated upper or lower bounds).

Considering the fact that the 2D version took 11,000 rules, it is impossible in this short paper to discuss all the many tricks and strategies that were used to get CAM-Brain to work. That would require a book (something the author is thinking seriously about writing). However, some of the tricks will be mentioned here. One was the frequent use of "gating cells", i.e. cells which indicated the direction that dendrite signals should turn at junctions to head towards the receiving neuron. To give these gating cells a directionality, i.e. a "leftness" or a "rightness", special marker cells had to be circulated at the last minute, after the circuit had stabilized. Since some trails were longer than others, a sequence of delay cells were sent through the network after the growth cells and before the marker cells. Without the delay cells, it was possible that the marker cells would pass before synapses were formed.

Once the 2D simulation was completed (before the CAM8 was delivered) several brief evolutionary experiments using the 2D version were undertaken. The first, was to see if it would be possible to evolve the number of synapses. Fig. 4 shows the result of an elite chromosome evolved to give a large number of synapses. In this experiment, the number of synapses increased steadily. It evolved successfully. The next experiment was to use the neural signaling to see if an output signal (tapped from the output of one of the neurons) could evolve to give a desired constant value. This evolved perfectly. Next, was to evolve an oscillator of a given arbitrary frequency and amplitude, which did evolve, but slowly (it took a full day). Finally, a simple retina was evolved which output the two component directional velocity of a moving "line" which passed (in various directions) over a grid of 16 "retinal neurons". This also evolved but even more slowly. The need for greater speed is obvious.

The above experiments are only the beginning. The author has already evolved (not using CAs) the weights of recurrent neural networks as controllers of an artificial nervous system for a simulated quadruped artificial creature. Neural modules called "GenNets" [de Garis 1990, 1991, 1994] were evolved to make the creature walk straight, turn left or right, peck at food, and mate. GenNets were also evolved to detect signal frequencies, to generate signal frequencies, to detect signal strengths, and signal strength differences. By using the output of the detector GenNets, it was possible to switch motion behaviors. Each behavior had its own separately evolved GenNet. By switching between a library of GenNets (i.e. their corresponding evolved weights) it was possible to get the artificial creature to behave in interesting ways. It could detect the presence and location of prey, predators and mates and take appropriate action, e.g. orientate, approach, and eat or mate, or turn away and flee. However, every time the author added another GenNet, the motion of the simulated creature slowed on the screen. The author's dream of being able to give a robot kitten some thousand different behaviors using GenNets, could not be realized on a standard monoprocessor workstation. Something more radical would be needed. Hence the motivation behind the CAM-Brain Project.

5. A Billion Neurons in aTrillion Cell CAM by 2001

Fig. 3 shows some estimated evolution times for 10 chromosomes over 100 generations for a Sparc 10 workstation, a CAM8, and a CAM2001 (i.e. a CAM using the anticipated electronics of the year 2001) for a given application. These estimates raise some tantalizing questions. For example, if it is possible to evolve the connections between a billion artificial neurons in a CAM2001, then what would one want to do with such an artificial nervous system (or artificial brain)? Even evolving a thousand neurons raises the same question.

Sparc10	CAM8	CAM8	CAM8	CAM8	CAM2001	CAM2001
10000 CA cells	10000 CA cells	1 million CA cells	25 million CA cells	25 billion CA cells	25 billion CA cells	25 trillion CA cells
4 neurons	4 neurons	40 neurons	1000 neurons	1 million neurons	1 million neurons	1 billion neurons
1 Sparc10	1 CAM8	10 CAM8s	10 CAM8s	10 CAM8s	10 CAM2001s	100 CAM2001s
4.6 days	50 seconds	8 minutes	3.5 hours	5 months	13.6 hours	2 months

Fig. 3 Evolution Times for Different Machines & CA Cell, Neuron & Machine Numbers

One of the aims of the CAM-Brain research project is to build an artificial brain which can control 1000 behaviors of a "robot kitten" (i.e. a robot of size and capacities comparable to a kitten) or to control a household "cleaner robot". Presumably it will not be practical to evolve all these behaviors at once. Most likely they will have to be evolved incrementally, i.e., starting off with a very basic behavioral repertoire and then adding (stepwise) new behaviors. In brain circuitry terms, this means that the new neural modules will have to connect up to already established neural circuits. In practice, one can imagine placing neural bodies (somas) external to the established nervous system and then evolving new axonal and dendral connections to it. The CAM-Brain Project hopes to create a new tool to enable serious investigation of the new field of "incremental evolution." This field is still rather virgin territory at the time of writing. This incremental evolution could benefit from using embryological ideas. For example, single seeder cells could be positioned in the 3D CA space under evolutionary control. Using handcrafted CA "developmental or embryological" rules, these seeder cells could grow into neurons ready to emit dendrites and axons [de Garis 1992]. The CAM-Brain Project, if successful, should also have a major impact on both the field of neural networks and the electronics industry. The traditional preoccupation of most research papers on neural networks is on analysis, but the complexities of CAM-Brain neural circuits, will make such analysis impractical. However, using Evolutionary Engineering, one can at least build/evolve functional systems. The electronics industry will be given a new paradigm, i.e. evolving/growing circuits, rather than designing them. The long term impact of this idea should be significant, both conceptually and financially.

References

[Codd 1968] E.F. Codd, *Cellular Automata,* Academic Press, NY, 1968.

[de Garis 1990] Hugo de Garis, "Genetic Programming: Modular Evolution for Darwin Machines,"*ICNN-90WASH-DC,* (Int. Joint Conf. on Neural Networks), January 1990, Washington DC, USA.

[de Garis 1991] Hugo de Garis, "Genetic Programming", Ch.8 in book *Neural and Intelligent Systems Integration*, ed. Branko Soucek, Wiley, NY, 1991.

[de Garis 1992] Hugo de Garis, "Artificial Embryology : The Genetic Programming of an Artificial Embryo", Ch.14 in book *Dynamic, Genetic, and Chaotic Programming*, ed. Branko Soucek and the IRIS Group, Wiley, NY, 1992.

[de Garis 1993] Hugo de Garis, "Evolvable Hardware : Genetic Programming of a Darwin Machine", in *Artificial Neural Nets and Genetic Algorithms,* R.F. Albrecht, C.R. Reeves, N.C. Steele (eds.), Springer Verlag, NY, 1993.

[de Garis 1995] Hugo de Garis, *Brain Building: The Evolutionary Engineering of Artificial Nervous Systems,* Wiley, in prep., 1995.

[Drexler 1992] K.E. Drexler, *Nanosystems : Molecular Machinery, Manufacturing and Computation,* Wiley, NY, 1992.

[Goldberg 1989] D.E. Goldberg, *Genetic Algorithms in Search, Optimization, and Machine Learning,* Addison-Wesley, Reading, MA, 1989.

[Toffoli & Margolus 1987, 1990] T. Toffoli & N. Margolus, *Cellular Automata Machines,* MIT Press, Cambridge, MA, 1987; and *Cellular Automata Machines,* in Lattice Gas Methods for Partial Differential Equations, SFISISOC, eds. Doolen et al, Addison-Wesley, 1990.

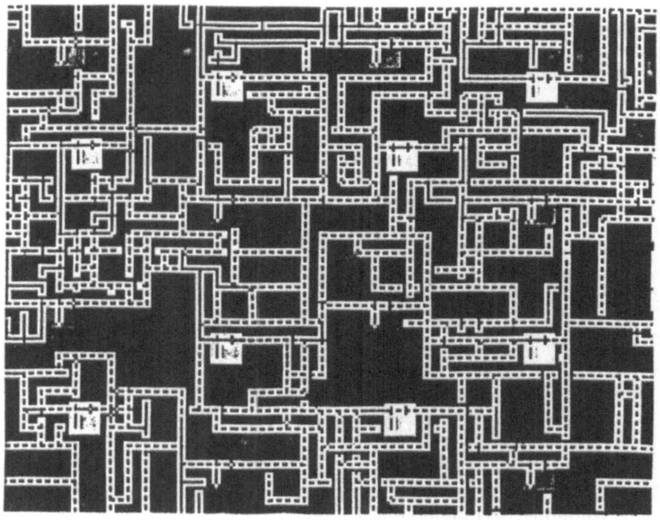

Fig. 4 2D CAM-Brain CA Network

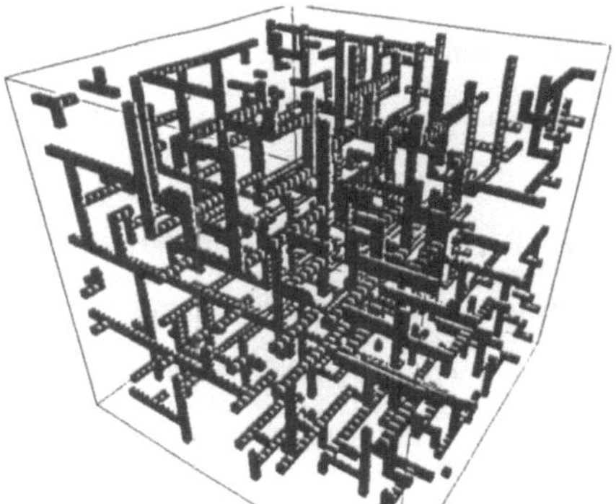

Fig. 5 3D CAM-Brain CA Network

A PERFECT INTEGRATION OF NEURAL NETWORKS AND EVOLUTIONARY ALGORITHMS

Percy P. C. Yip and Yoh-Han Pao

Case Western Reserve University and AI Ware, Inc.

Abstract. Evolutionary computation techniques need to interact with a fitness function or an objective function for the selection to be made properly. In cases objective functions are not well defined, evolutionary algorithms may not be able to perform properly. In this paper, we propose to use a well-trained neural network model to provide estimates of objective values at points that we have not experienced to support the selection process of the evolutionary algorithm. An example of gasoline blending task is used to demonstrate that such an integrated system functions as a powerful tool for product design.

INTRODUCTION

Evolutionary computation techniques are powerful search techniques that resulted from simulating natural evolutionary processes on a computer. In this paper, we call these techniques evolutionary algorithms, including Genetic Algorithms [1], Evolutionary Strategies [2] and Evolutionary Programming [3]. Applications of these techniques to many difficult problems have resulted in reports of considerable success. All these techniques require a common mechanism - "survival of fitness". That is, those that are better fitted to the environment have higher chances of being selected to survive and reproduce. Generally speaking, these evolutionary algorithms need to interact with a fitness function or an objective function for the selection to be made properly. In cases objective functions are not well defined, evolutionary algorithms may not be able to perform properly. An example of these is when properties of a mixture are known only for some instances of input ingredients. In the optimization process performed by an evolutionary algorithm, inputs that belong to the set of search space are randomly mutated. Points of inputs with unknown properties may be generated, and so the selection process of an evolutionary algorithm cannot be performed. In this paper, we will demonstrate how an integration of evolution algorithms and neural networks can perform such a task.

Artificial neural networks are computational paradigms developed in part because of inspiration by consideration of processes in biological brains. Artificial neural networks can perform many different functionalities, such as unsupervised learning, supervised learning and optimization [4]. Of these functionalities, a particular useful one is that of supervised learning. With supervised learning method, a neural network can perform a task of function approximation. Many applications of supervised learning have been reported. These include system identifications, and process modeling and control. During training of a model, instances of input-output pairs are passed onto a neural network, and the neural network learns the input-output behavior. After training, a neural network can be used to give reasonable good estimates of the outputs even for inputs that it has never experienced before. It is very useful for analysis purpose, but not for synthesis purpose. The following scenario may arise. "The market has changed. We need to produce a product with different properties. What should be the combination of the inputs?" This general question that product designers are often faced with is a search problem. A perfect solution for providing an answer to this question is the use of a powerful search method.

In this paper, we will describe a perfect integration of evolutionary algorithms and neural networks. The system that we propose consists of a neural network as a function approximator, and an evolutionary algorithm as an optimizer. The proposed system can answer the above questions. With use of a well-trained neural network model, reasonably good estimates of objective values at points that we have not experienced can be provided in support of the selection process of the optimizer. The evolutionary algorithm that we used is Guided Evolutionary Simulated Annealing [5], that combines genetic algorithms, evolutionary programming, and simulated annealing [6] in a novel way. An example of a gasoline blending problem is used to demonstrate that such an integration indeed can be used as a powerful computer-aided design tool.

SYSTEM DESCRIPTION

The system that we describe consists of six major modules, as indicated in Fig. 1. They are human interface module, cost evaluation module, constraint evaluation module, neural network module, optimizer module and a module that combines the goals of the specification. We will briefly describe the operation of the system.

In our proposed system, instead of using the neural network for evaluating the objective values, we suggest to use the neural network for learning a mapping of the input ingredients and output properties. An advantage of learning such a mapping is that users can specify the desired properties with ingredient constraints, and cost constraint. A module that combines the goals together returns the objective values so the evolution optimizer can operate properly.

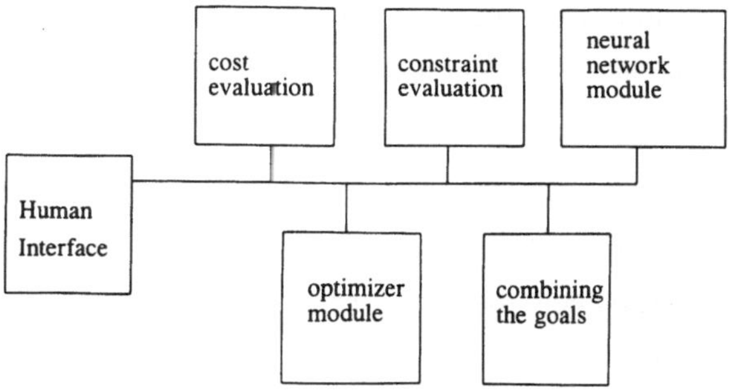

Fig. 1 Computer-Aided Decision System

Once the network has been trained, it is capable of estimating outputs for given inputs. A neural network is used as an expert and the system is ready for use. When the system is being used for an optimal selection of ingredients, an initial product specification is presented to the system. That means, the desired properties of the new product requirement and constraints of the input ingredients are specified. The system has to find out a combination of the inputs to achieve the goal. It is a search process. The evolution algorithm starts the search with a set of possible inputs. For each combination of inputs, the neural network module, the constraint evaluation module and the cost module give the estimates of the output properties, the constraint evaluation values, and the estimated cost. These values are combined to form an objective value, that returns to the optimizer. After the optimization process, an optimal combination of inputs is found. Sometimes, it is not possible to find the optimal combination of ingredients that meets all the desired specifications because one may have conflicting desired properties, or some of the specifications are not achievable under given constraints. However the system is capable of providing an optimal compromise between all the specified goals. The user can now use the found ingredients to build a prototype, or he can modify the specification and repeat the optimization process again.

In new product design problems, the mapping between inputs and output properties is so complex that analytical relationship is usually unavailable. The use of a well-trained neural network can provide an accurate estimate of the properties, thus supporting the functionality of the evolution optimizer.

RESULTS

To illustrate the use of this technique, we built a neural network model that maps the blending components and process conditions to different categories of emission levels and fuel octane level. The inputs of this mapping are Aromatics, methyl t-butyl ether (MTBE), Olefins, 90% distillation temperature (T_{90}), Reid vapor pressure (RVP), and Benzene. The outputs of this mapping are Hydrocarbon (HC), Non-Methane Hydrocarbon (NMHC), Carbon Monoxide (CO), Oxides of Nitrogen (NO_x), and Octane Rating. Property data of HC, NMHC, CO and NO_x were obtained from Hochhauser et. al. [7]. Property data of Octane Rating were generated using data from Snee [8]. Once a model has been trained, the system is ready for use in giving the optimal inputs to meet the design objective. The purpose of this experiment is to search for an optimal blending formulation that will minimize exhaust emissions (HC, NMHC, CO, NO_x) and still achieve a desired level of Octane Rating subject.

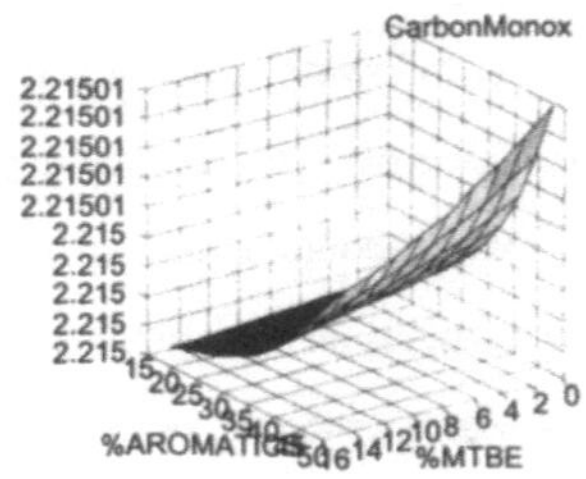

Fig. 2 Illustration of change of CO with respect to change of percentage of Aromatics and percentage of MTBE

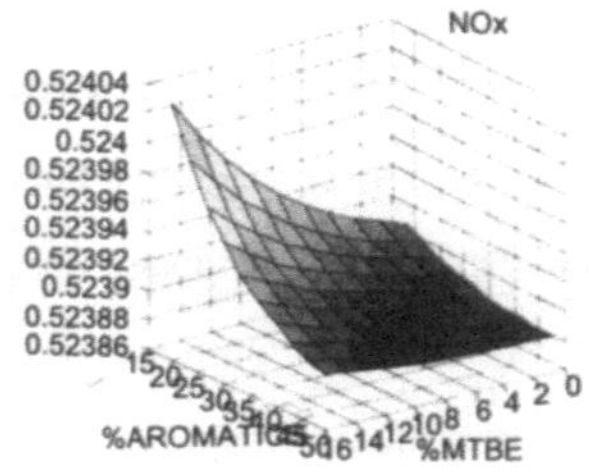

Fig. 3 Illustration of change of NOx with respect to change of percentage of Aromatics and percentage of MTBE

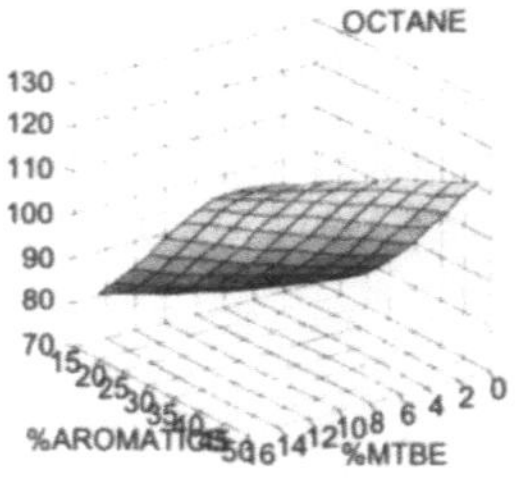

Fig. 4 Illustration of change of Octane with respect to change of percentage of Aromatics and percentage of MTBE

We illustrate the complexity of the task with Figures 2-4. Fig. 2 shows how the amount of Carbon Monoxide changes with respect to percentage of Aromatics and percentage of MTBE. Fig. 3 shows how the amount of NOx changes with respect to percentage of Aromatics and percentage of MTBE. Fig. 4 shows how the amount of Octane changes with respect to percentage of Aromatics and percentage of MTBE. These three figures indicate how difficult the problem is. CO increases with the percentage of Aromatics, whereas Nox decreases with the percentage of Aromatics. We have to make a compromise between desired properties. The system uses a weighted method to combine all the desired goals into one objective function.

Suppose we want to find a combination of input ingredients to minimize emission levels while maintaining an octane level of 100. If we used the available experimental data (i.e., the patterns in the training set) to provide a solution to achieve the criteria, the properties that we got were (0.201, 0.164, 0.2688, 0.615, 100). Fig. 5 shows the result of using the proposed system for providing such a solution. The properties that we got with use of the proposed system were (0.187572, 0.152999, 2.43732, 0.541199, 100). Clearly, the result obtained using the proposed system is far superior than the best training pattern.

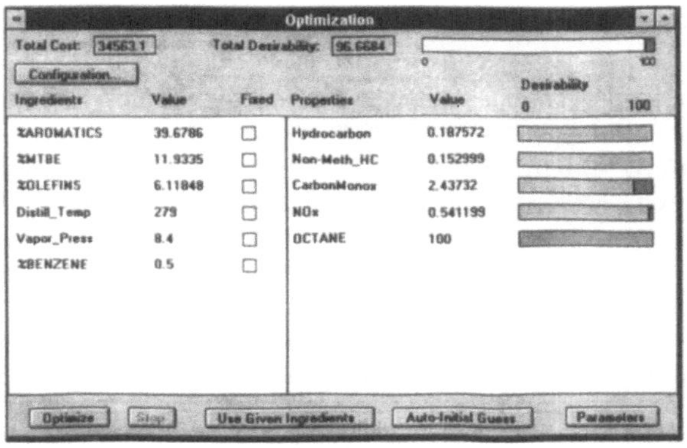

Fig. 5 Result of optimization process

CONCLUSION

An integration of neural networks and evolutionary algorithms has been proposed. Such a system can be used as an intelligent computer-aided design tool, that provides answers to "what if" as well as "if then" questions. An example of gasoline blending task has been used to demonstrate the capability of the proposed system.

REFERENCES

[1] Goldberg, D.E.: Genetic Algorithms in Search, Optimization & Machine Learning. Addison-Wesley, 1989.

[2] Schwefel, H.P.: Numerical Optimization of Computer Models. Chichester:John Wily, 1981.

[3] D. B. Fogel, D.B.: System Identification Through Simulated Evolution: A Machine Learning Approach to Modeling. Ginn Press, 1991.

[4] Pao, Y.H.: Adaptive Pattern Recognition and Neural Networks. Reading, MA: Addison-Wesley, 1989.

[5] Yip, P. P. C. and Pao, Y.H.: IEEE Trans. SMC 9, 1383 (1994).

[6] Aarts, E. and Korst, J.: Simulated Annealing and Boltzmann Machine, John Wiley & Sons, 1989.

[7] Hochhauser, A.M., Benson, J.D., Burns, V., Gorse, R.A., Koehl, W.J., Painter, L.J., Rppon, B.H., Reuter, R.M., and Rutherford, J.A.: SAE Transactions 100, 748 (1991).

[8] Snee, R.D.: Technometrics 23, 119 (1981).

G-LVQ, a combination of genetic algorithms and LVQ

J. J. Merelo, A. Prieto
Department of Electronics and Computer Technology, Facultad de Ciencias
Campus Fuentenueva, s/n, 18071 Granada (Spain)
e-mail: jmerelo@kal-el.ugr.es

Abstract

One of the ultimate goal in neural network research is the complete optimization of a neural network: topology, learning algorithm, initial weights and number of neurons. Up to now, only partial solutions have been found. This optimization should look at two conditions if the task assigned to the NN is going to be classification: accuracy, and obtention of a good representation of the sample. LVQ *neural nets are a supervised classification algorithms created by Kohonen. In this work,* LVQ *NNs will be optimized using a GA according to three criteria: accuracy, net size and distortion. These three criteria are considered a* vector fitness *with three components that must be optimized separately. In order to carry this out, variable-length genomes are used to represent the neural network; each neuron is codified together with its label. Results in synthetic and real-world problems show that* G-LVQ *is able to find an optimal size of the network, as well as combinations of weights that maximize classification accuracy.*

1 Main objectives of G-LVQ

G-LVQ (Genetic Learning Vector Quantization) is an algorithm that combines features taken from both genetic search and neural learning. In this way, it is faster and more accurate than a genetic algorithm by itself, because of the additional neural hill-climbing, and samples more efficiently the search space than a neural network alone.

Genetic algorithms (GAs) alone are usually slow in classification problems, since they are too coarse-grained to obtain a good solution quickly. On the other hand, neural algorithms are usually fast (in terms of processor cycles), if they are close to the solution, but tend to get stuck in local minima. Combining both kinds of algorithms manages to avoid local minima, and finds solutions accurately.

The utilization of variable length genome structure for neural network genetic coding is a problem that has been tackled recently, mainly from the field of robotic applications ([1],[2]) but also at classical classification problems ([3]).

In order to code genetically a NN of variable size and structure, several approaches are taken:

1. *Developmental*: in this approach, parameters for generating a NN ([4]) or a set of rules whose application would generate a connection array ([5]) or a NN ([3]) are coded into a genome.
2. *Node addressing*: genes that represent NN nodes refer by indirect or direct addressing to other parts of the genome that also codify nodes ([1]), [6]).
3. *Incremental/Decremental*, which are not purely GA approaches, but rather rely on heuristic rules to add nodes to a NN or prune them ([7]).

The approach presented in this paper is similar to the latest, but it does not apply increment and decrement of NN/genome blindly or heuristically, but following the spirit of GAs.

The organization of the paper is as follows: section 2 is devoted to an introduction to Kohonen's LVQ, to establish terminology, and point out its advantages and disadvantages, section 3 describes the algorithm proposed in this paper, G-LVQ, while section 4 is devoted to results and discussion.

2 Kohonen's LVQ

G-LVQ tries to optimize LVQ [8] neural nets. LVQ is a supervised classification algorithm. An LVQ network consists of an array of labeled vectors, which are conventionally called *weight vectors*. Each label corresponds to a class. Watching LVQ algorithm from another point of view, it is a *dictionnary design* algorithm, since it tries to compute an optimal group of labeled vectors that represent a sample and can later be used to recognize not previously seen members of that sample. A dictionnary is composed of l *reference vectors* or *codevectors*, being l called *number of levels* of the dictionnary. In the case of LVQ, a dictionnary is termed a neural net (NN), and a codevector a *neuron* or *weight vector*.

When training, each labeled vector $\vec{x}$ in the training set is submitted to the net. The closest weight vector (usually called the *winning* vector or neuron) is computed,

$$\vec{w}_i \quad / \quad d(\vec{x}, \vec{w}_i) \; < \; d(\vec{x}, \; \vec{w}_j) \; \forall j \neq i \qquad (1)$$

and this vector is updated, in such a way that it gets closer to the input vector if both belong to the same class, and is moved away from it if they belong to different classes

$$\Delta\vec{w}_i = \alpha(\vec{x} - \vec{w}_i) \quad (2)$$

if they belong to the same class, and

$$\Delta\vec{w}_i = -\alpha(\vec{x} - \vec{w}_i) \quad (3)$$

if they belong to different classes.

LVQ usually performs efficiently if the initial weight vectors are close to their final values. That is why some heuristic procedures are used for weight initialization, like Kohonen's SOM [8] or any other vector quantization procedure, like *k-means* [9]. One author [10] has even used a genetic algorithm to set the initial weights of a LVQ neural net.

Usually, the number of weight vectors remains fixed during training, and is usually set prior to training to be equal to the number of classes, or to a fixed number of weight vectors per class. Several sets of initial weight vectors are tested, and the best solution is chosen.

3 G-LVQ

G-LVQ algorithm is as follows:

1. Generate initial population of variable-length chromosomes, within a range.
2. Decode genomes into a neural network, and train them using LVQ algorithm and the training set.
3. Stop training, and use the same file as validation set. Evaluate accuracy and distortion.
4. Compare each net with the nets in its neighborhood. If it is the best one, do nothing. If it is not, apply genetic operators, including reproduction of the best of the neighborhood with another random one.
5. Back to step 2.

G-LVQ tries to overcome the limitations of LVQ by using genetic algorithms. But traditional GAs have also got some limitations, so that some improvements had to be added to the traditional GA paradigm, but remaining at the same time within the GA spirit.

The LVQ network should be optimized globally. This means optimizing the initial weight vectors, the learning algorithm, and the number of weight vectors as well. Nevertheless, the learning algorithm and its parameters seem to be optimal enough, so that they will not be taken into account.

In order to use a GA on a LVQ net the initial weight vectors must, then, be coded into a chromosome. All the components of each weight vector, along with its label, are coded together. As is usual in GAs, a population of chromosomes is used, each one is evaluated by decoding a net from the chromosome, and training and testing it on the training set, and a new generation is created from the fittest.

Usual GA operators, like mutation and uniform crossover, are used. But, besides, bearing in mind the length of the LVQ network must be also optimized, there must be some operators for altering the length of the chromosome. These operators add or eliminate the representation of a whole weight vector (including label), and are, namely:

- *duplicate*: this operator duplicates, with mutation, the representation of one of the weight vectors (a *gene*). The duplicated vector is the one that *wins* more in the previous training phase.
- *kill* deletes the representation of the weight vector that wins the least, or does not win at all, in the previous training/test phase, thus eliminating useless weight vectors from the neural net. In this way, the previous operator and this one make selection operate *inside* each neural net, cloning with mutation the best ones and eliminating the worst.
- *randomIncrement* adds a random weight vector (with random label) to the chromosome.

LVQ is a vector quantization algorithm usable for classification tasks. Two conditions should be met by an optimal classification NN (or, for that matter, any algorithmic classifier):

1. Accuracy in classification, measured by the number of misclassified samples.
2. A good representation of the training sample should be extracted from the NN, measured as the mean distance from each sample to the closest vector in the neural net, that is

$$D = \sum_{\vec{x} \in S} d(\vec{x}, \vec{w}_i) \quad (4)$$

where

$$\vec{w}_i \; / \; d(\vec{x}, \vec{w}_i) < d(\vec{x}, \vec{w}_j) \forall j \neq i \quad (5)$$

This quantity D is denominated *distortion*

With respect to the fitness criterium, it is usual in classification problems that adding more vectors to the classifier improves classification accuracy, but at a cost (the cost of computing the class raises with the number of vectors, since the input vector must be compared with each of the reference vectors). Besides, it is desirable that the classifier itself is close (in the euclidean sense) to the samples, in such a way that it becomes a good representation of the training sample. Thus, there are three quantities to optimize: classification accuracy, length of the dictionnary (neural net size) and distortion (or proximity from the neural net to the classified samples).

Usual scalar fitness cannot cope with this, since it is difficult to join all these criteria in a single fitness formula. Thus, G-LVQ uses a *vector fitness*, in such a way that two members of the population are compared for each of those quantities in turn. Accuracy is usually considered the most important criterium, being distortion and neural net size less important. However, the order of evaluation can be altered by the user.

Different evaluation orders give different results: if distortion is optimized with greater priority than size, usually very good results are obtained in the training phase, but the NNs obtained are large and generalize badly, giving bad results in the test phase. On the other hand, if accuracy in classification is considered the first to optimize, followed by net size, optimal nets are usually found.

On the other hand, G-LVQ algorithm is intended to be executed in parallel processor mesh machines, like PVM or a Connection Machine, that is why it has been designed to perform all GA operations locally, in such a way that the chromosomes in the population are considered to be in a grid. All fitness comparison and other operations, like reproduction, take place locally. For instance, each member of the population is only compared with those in the neighboring nodes of the grid, and mates also with one of the chromosomes in those nodes. Time spent in routing is reduced, and besides, this modification keeps the original spirit of GAs, distilled on the phrase "Distrust central authority" [14].

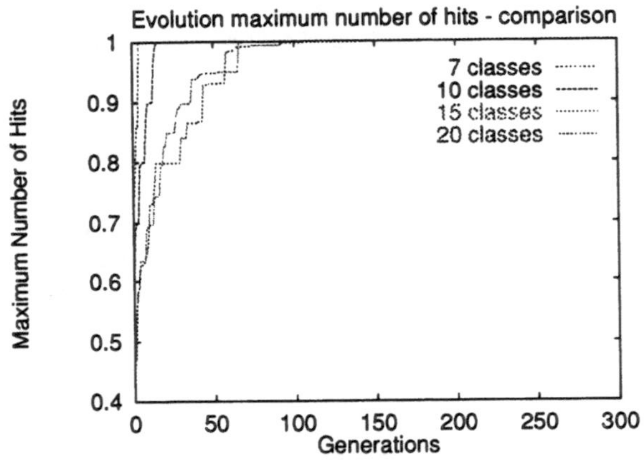

Figure 1: *Evolution of the classification accuracy of the best NN generated by* G-LVQ*in the population, for a simple gaussian cluster classification problem. Several number of clusters have been tested: 7, 10, 15 and 20 classes. The problem increases in difficulty with the number of classes. Each line represents the results of a single typical run.*

With respect to behavior, usually G-LVQ optimizes quantities in the same order that is set previously. As it can be seen in the previous figures, which show, respectively, evolution of classification accuracy, mean length and minimum distortion in a simple gaussian non-overlapping cluster classification problem, in each case first accuracy is achieved (in generation 70, in the case of 20-cluster classification), then a proper length (which, in the case of 20 clusters, is not achieved in mean, although the best one has size 20), and

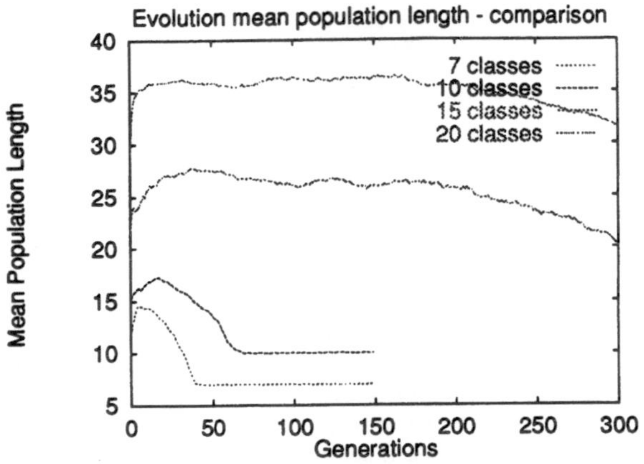

Figure 2: *Evolution of the average length for all the population, in the same problem than before, obtained by the* G-LVQ *algorithm*

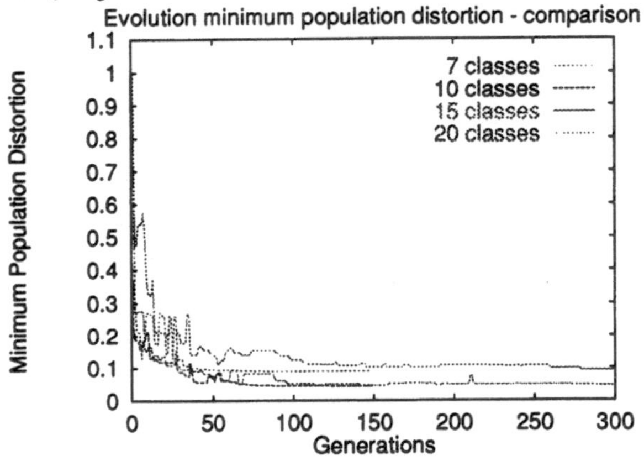

Figure 3: *Evolution of the distortion of the NN with the greatest accuracy, obtained by* G-LVQ *as shown in Fig. 1. The problem is the same than in the two previous figures.*

then a minimun distortion (which is achieved close to the end of training).

It must be enphasized that G-LVQ does not know in advance the proper size of the network, neither a "separation from optimal length" is optimized. G-LVQ tries to optimize the size, taking into account the constraint of keeping accuracy maximized.

4 Results

G-LVQ has been tested on easy synthetic problems (gaussian cluster classification), hard synthetic problems (Hart's problem) and real problems (macromolecule image classification and sonar image classification [13]). It usually obtains dictionnaries that are smallers and more accurate than other evolutionary or static (like simple LVQ) approaches.

Tests have been made on a problem taken from the PROBEN1 ([15]) benchmark, cancer, which is a problem of breast cancer diagnosis with 9 inputs and two possible outputs: sick or healthy. There are 699 examples, from which 3 partitions have been made, **cancer1**, 2 and 3. This data set was originally obtained from the University of Wisconsin Hospitals, Madison, from Dr. W. H. Wolberg [16].

In the following figures, HLD means that the quantities are optimized in hits (accuracy in classification), length (NN size) and distortion order; HDL puts distortion before length. Running parameters are as follows: population: 400-625 networks; generations: between 200 and 500; uniform crossover rate:0.1-0.2; mutation rate: 0.01; duplication rate: 0.05, kill rate: 0.03 and random increment rate: 0.01. The running time for each training is several hours in a Silicon Graphics Indigo R4000.

The following table shows a comparison of size and test set error with RPROP algorithm ([15]), using a multilayer perceptron with two hidden layers; the two numbers represented are the number of neurons in the first and second layer. Error and standard deviation are computed over the several optimal NNs obtained in a typical run (sometimes over 30), for a test set not previously used; network size is the typical net size obtained.

Algorithm	Error + Std. Dev	Network size
G-LVQ HDL	3.1 ± 0.7	8 neurons
G-LVQ HLD	3.1 ± 0.5	5 neurons
RPROP	1.14	4 + 2 neurons

Table I: Results for benchmark problem cancer1

Algorithm	Error + Std. Dev	Network size
G-LVQ HDL	4.2 ± 0.3	16 neurons
G-LVQ HLD	4.6 ± 0.1	3 neurons
RPROP	5.747	8 + 4 neurons

Table II: Results for benchmark problem cancer2

Algorithm	Error + Std. Dev	Network size
G-LVQ HDL	4.9 ± 0.3	33 neurons
G-LVQ HLD	3.45 ± 0.06	4 neurons
RPROP	2.299	4 + 4 neurons

Table III: Results for benchmark problem cancer3

As it can be seen, sometimes RPROP reaches better results than G-LVQ, although G-LVQ is consistently better in the cancer2 test set. However, usually nets obtained by G-LVQ are more compact (if the HLD ordering is used), the number of connections is far less (over all in cancer3 and 1 test set. Besides, G-LVQalgorithm and has got less free parameters.

References

[1] Cliff, D.; Harvey, I.; Husbands, P.; *Explorations in Evolutionary Robotics*, Adaptive Behavior, vol. 2, no. 1 1993, 32 pps.

[2] Nolfi, S.; Parisi, D.; *Growing Neural Networks*, Tech. Rep. PCIA-91-15, Department of Cognitive Processes and Artificial Intelligence, Institute of Psychology, CNR, Rome.

[3] Gruau, F.; *Genetic Synthesis of Modular Neural Networks*, Procs. of the 5th Int. Conf. on Genetic Algorithms, S. Forrest, Ed., pp. 318 ff; 1993.

[4] Harp, S.; Samad, T., Guha, A., *Towards the genetic synthesis of neural networks*, D.J. Schaffer (ed.), 3rd Int. Conf on Genetic Algorithms, pp. 360-369, 1989.

[5] Kitano, H.; *Designing neural network using genetic algorithm with graph generation system*, Complex Systems, 4:461-476, 1992.

[6] Polani, D.; Uthmann, T., *Training Kohonen Feature Maps in different Topologies: an Analysis using Genetic Algorithms*, Procs. of the 5th Int. Conf. on Genetic Algorithms, S. Forrest, Ed., pp. 326-333, 1993.

[7] Fritzke, B.; *Growing Cell Structures: a Self-organizing Network in k Dimensions*, Artificial Neural Networks 2, Aleksander, Taylor, Eds, 1993, pp 1051-1056.

[8] Kohonen, T.; *The Self-Organizing Map*, Procs. IEEE, 78, 1464 ss, 1990.

[9] Duda, R. O., Hart, P. E., *Pattern classification and scene analysis*, New York: Wiley & Sons, 1973.

[10] Monte, E.; Hidalgo, D.; Mariño, J.; Hernáez, I., *A vector quantization algorithm based on genetic algorithms and* LVQ, NATO-ASI Bubión 93, pp 231 ss, 1993.

[11] Yao, X.; *A Review of Evolutionary Artificial Neural Networks*, Technical Report, 1992

[12] Bengio, Y.; Bengio, S.; *Learning a synaptic learning rule*, Technical Report 751, Department d'Informatique et de Recherche Operationelle, Universitè de Montreal, Canada, November 1990.

[13] J. J. Merelo, *Study, applications and optimization of neural vector quantization algorithms using genetic algorithms*, PhD Thesis, University of Granada, Spain (in Spanish).

[14] Goldberg, D.; *Zen and the art of Genetic Algorithms*, Procs ICGA89, 3rd Int. Conf. on Genetic Algorithms, 1989, J. D. Schaffer, Ed. pp 80-85,.

[15] Lutz Pretchelt, PROBEN1, *A Set of Neural Network Benchmark Problems and Benchmarking Rules*, Technical Report 21/94, Univ. Karlsruhe, Germany.

[16] William H. Wolberg and O.L. Mangasarian, *Multisurface method of pattern separation for medical diagnosis applied to breast cytology* Proceedings of the National Academy of Sciences, U.S.A., Volume 87, December 1990, pp 9193-9196.

EVOLVING NEURAL NETWORK STRUCTURES: AN EVALUATION OF ENCODING TECHNIQUES

Stephen G. Roberts and Mike Turega

Department of Computation.
UMIST, Manchester, UK.
roberts@mp.co.umist.ac.uk

Abstract - *Feed-forward neural network structures, trained with back-propagation, are adapted by the use of the genetic algorithm (GA). Through this search technique problem specific topologies are found. The method used to represent network structure, in a form suitable for the GA, is investigated. Comparisons are made between three such encoding methods. Details are given of how these representational schemes can influence the performance of both the genetic algorithm and the resultant neural networks found by the GA.*

Introduction

In theory, a multilayer, feedforward, neural network can be used as a universal function approximator [1]. Given sufficient hidden nodes, virtually any non-random function may be approximated. Therefore, at a glance, the design of a multilayer feedforward architecture appears to be simple: provide a single layer of 'N' hidden nodes, each of which is fully connected to every input node and to every output node, as shown in figure 1. This is the most commonly used network design, and will therefore be referred to as the standard or traditional architecture.

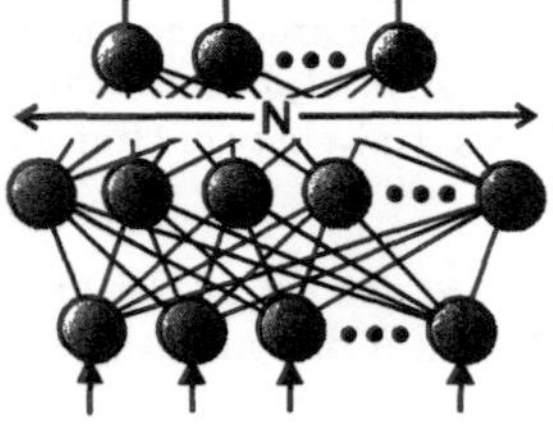

Figure 1: A Standard Feed-forward Architecture.

Unfortunately, designing a good network is not quite as easy as it may first appear. The problem in this case lies with the requirement that a "sufficient" number of hidden nodes be provided. A network given too few nodes will have a reduced learning ability, to the extreme of being unable to memorise any training patterns. Whereas, a topology displaying too many nodes will exhibit a symptom known as *overtraining*; the network will be less capable of forming generalised outputs for previously unseen inputs. Network design is further complicated if problem specific topologies are required. In performance critical areas, such as medical, military or financial situations, a network optimised for training speed, generalising ability or noise resistance may be of the utmost importance. Presently no well defined method exists for specifying such a network.

The location of the best network architecture is complicated by the size and nature of the search space of possible topologies. Even with a limited set of nodes, and the restriction of only allowing feed-forward connections, there are still a vast number of possible structures. In addition, the search space is *deceptive* [2]; two networks containing similar patterns of connectivity can show vastly different levels of performance. At the same time, two networks with completely different connectivities can have very similar performances; in this case the search space is said to be *multimodal* [2]. Finding the optimal architecture is therefore a most difficult task. In an attempt to overcome this most challenging problem, to separate the wheat from the chaff in neural network terms and to locate a network structure with a performance close to optimal, researchers have turned to the genetic algorithm search technique [3,4].

Evolving Neural Network Structures

A genetic algorithm (GA) employs a "survival of the fittest" mechanism on a population of strings or *chromosomes*. In the case of applying GAs to neural network design [5], each of these chromosomes will encode a network structure or *phenotype*. The encoded network structure is known as the *genotype*. Each one of these genotypes contains the information required for the creation of a network, subsequent evaluation of this network provides a measure of the fitness for its particular genotype. The fitness is then used to decide which members of the population shall live, die and mate. The phenotype-to-genotype mapping is the central focus of this investigation; more specifically, how the network structure is represented in the genotype, how this encoding scheme affects the convergence of the genetic algorithm, and how it influences the performances of the networks evolved by the GA.

Encoding Techniques

Miller et al.[2] have classified encoding methods into two categories: strong and weak specification schemes. They define the mapping from phenotype to genotype in terms of the level at which the encoding takes place. A weak network specification is defined as one which only considers structural elements at the highest level, namely the organisation of layers or nodes, not individual connections. A strong specification scheme is a low level form of encoding in which every gene (a parameter on the string encoding the network) represents a connection. We propose that there exists a third, intermediate, representational scheme, not covered by the original classification. This intermediate scheme can capture the modular nature of a network, while at the same time allowing for the specification of individual connections. Such a scheme is known as a *grammar encoding* [6]. Therefore, the three encoding schemes used in this investigation are as follows:

Strong Specification Schemes

In a strong specification scheme [2,7,8,9,10] (also known as *direct encoding*) the network structure is encoded as a "blueprint" of its topology. Each connection is specified individually within the genotype. In the classical form of the strong specification scheme, as employed by Miller et al.[2], a binary representation is used to denote connections, being set to '1' if that connection is present and '0' otherwise. Figure 2 shows how the XOR problem would be represented with this form of encoding.

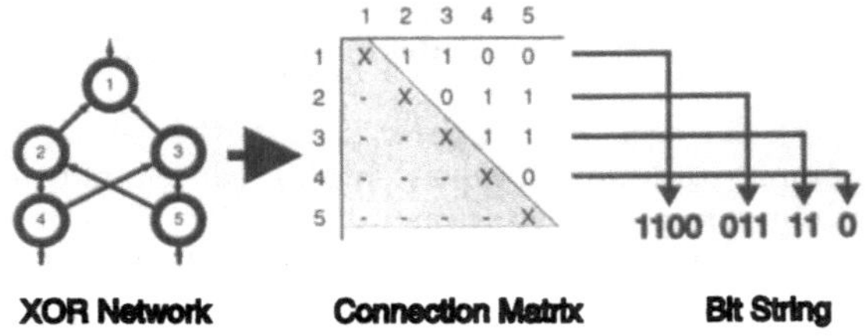

Figure 2: Phenotype to Genotype Mapping of an XOR Network for a Strong Encoding Scheme.

Weak Specification Schemes

When a network structure is represented by a weak specification scheme (*indirect encoding*) [11,1213,14] individual connections are not considered. Instead the parameters encoding the network define how specific modules are to be composed and connected. The total network topology is specified in terms of its layers and inter-layer connections.

This investigation employs a generic weak specification method based on that of Harp et al. [11,12,13]. This scheme employs a bit string to represent modules and the outward connections from these modules; a module being defined as a rectangular group of nodes that form a sub-network within the main structure. All layers within a module are fully-connected in a feed-forward manner, and each module is given an (not necessarily individual) identification number. Outgoing projections from a module are defined in terms of the identification number of the target module, the dimensions of the connecting footprint and the right offset of this connection block. Only the nodes in the highest layer of a module can have external connections, but these connections may be made to any number of external modules. Figure 3 shows the ordering of a module's parameters on a chromosome.

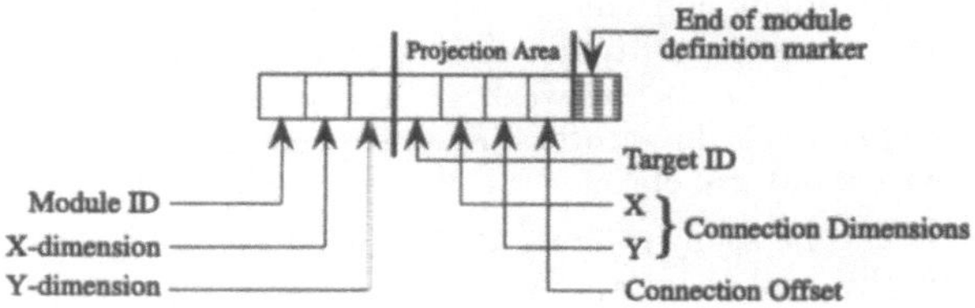

Figure 3: Parameter-ordering on a weak encoding scheme chromosome.

Consider the module encoding shown in figure 4 below.

Figure 4: An example of a weak coding chromosome.

In this example a module is defined consisting of a grid of nodes with dimensions 4x3. Two projection areas are shown, having connection footprints of 2x2 and 2x3 to modules 2 and 3 respectively. If module 2 is formed from 4 layers, each containing 5 nodes, and module 3 is a single layer of 4 nodes, then the sub-network created in this example would be as shown in figure 5. Note how the connection footprint to module 2 has been moved one node to the right due to the offset value for this projection. The parameters for the connection to module 3 specify a block of connections of size 2x3 with an offset value of 4. Since module 3 consists only of a single layer of nodes this connection is not possible. The vertical extent of the footprint is therefore truncated and the block is shifted right as many nodes as possible.

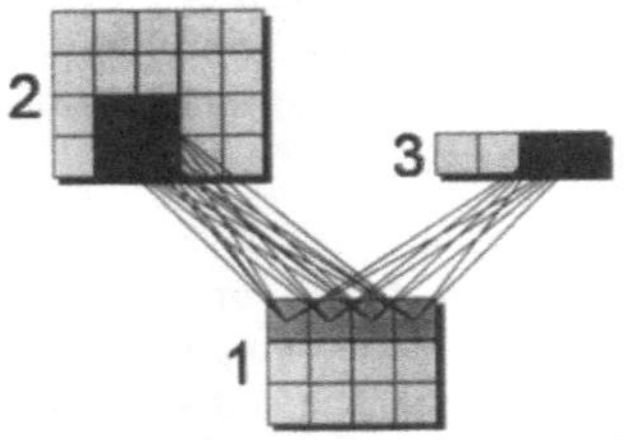

Figure 5: An example of a sub-network formed from a weak encoding scheme.

Structures are formed by creating such subnetworks between pre-defined input and output layers.

Intermediate Specification Schemes

Intermediate specification schemes, better known as *grammar encoding methods* [6,15,16] make use of a fractal technique known as *L-Systems.* Originally developed by Lindenmeyer [17] as a theory for the growth of filamentous organisms and branching filaments, the theory was extended by Prusinkiewicz [18] to form a string rewriting system or L-System. In this scheme an initial string, the *axiom*, is expanded by a set of production rules, through a given number of steps, to give a final output string. Since the development of the network architecture takes place through time, the process may be compared to the growth of natural objects [18].

As can be seen in figure 6, an initial axiom, formed from the alphabet {a..p,A..P}*, is expanded by given production rules. The genotype representing the network contains the axiom and variable rules A to P. The binary rules, *a* to *p*, remain fixed throughout the experiment. Expansion begins in the top left corner of the matrix and proceeds from left to right. Any characters being pushed out of the matrix will be lost.

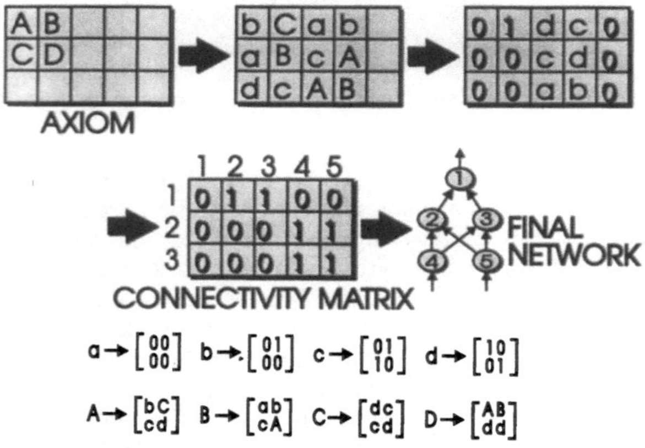

Figure 6: From genotype to XOR network for a grammar encoding method.

Empirical Method

To test the power of the three encoding schemes described, we have carried out two main experiments:

1. The Parity Problem: The standard test problem, as employed by Rumelhart et al. [19]. If an odd number of the network inputs are set to one then the network output should also be set to one. Increasing the number of inputs to the network increases the complexity of the parity problem.

To compare the results obtained for network structures produced by the genetic algorithm, 100 trials of a standard, fully connected, network have been carried out. As in Rumelhart and McClelland's [19] parity experiment networks have been given N hidden nodes for an N input problem. The parameters chosen for network training are a variation of those conventionally used for the parity problem (e.g. [19]). The momentum rate was lowered and the number of training cycles was raised as the problem size increased.

The genetic algorithm had its parameters set as follows;

Generations = 50	Population Size = 100
Crossover Rate = 0.900	Mutation Rate = 0.001

This experiment tests the ability of the network encoding scheme to generate networks with low training errors. The ability of the network to learn the problem is being evaluated. Each encoding scheme has been tested five times for every problem size. Figure 7 shows the average mean square error for the final generation of the run.

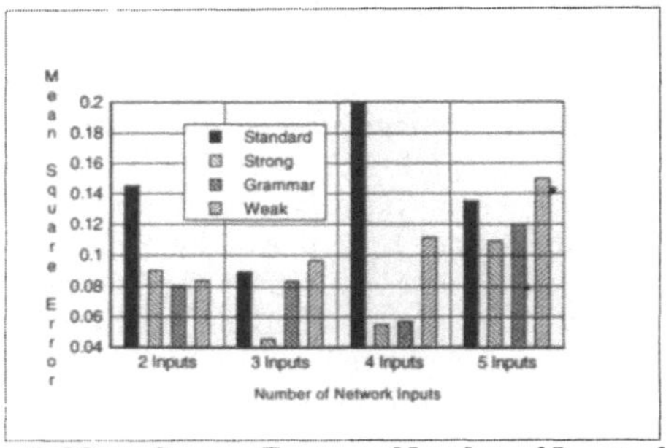

Figure 7: Mean Square Error vs. Number of Inputs for the Parity Problem: A Comparison of Encoding Techniques.

As can be seen from figure 7, both the strong and grammar techniques consistently produced networks with performances better than those shown by standard network structures. The weak scheme didn't perform quite as well.

2. The Two-or-More Clumps Predicate: A classical problem, introduced by Denker et al. [20] to evaluate the generalisation properties of a network. In this

* In this example only the subset {a..d,A..D} is being used.

experiment, the inputs to the network are arranged as a one-dimensional chain. A *clump* is defined as a set of adjacent nodes, each having a value of one, which are bounded on both sides by nodes of value zero. If there are two or more such clumps present on the chain then the output of the network should be set to one, otherwise it should be zero.

The network was trained using a subset of the training set used for the standard network evaluation. The remainder of this training set was then used to evaluate the generalisation properties of the structure, a network's fitness value being composed of both its training error and error on this unseen data. Both standard and evolved networks were then tested using a validation set. The results obtained are as shown in figure 8. This is for a 25 input problem with 25 hidden nodes.

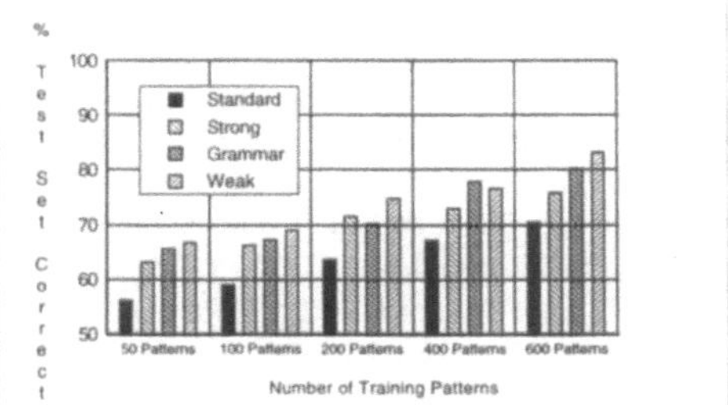

Figure 8: Percentage of test set correct vs. Number of training patterns for the two-or-more clumps predicate.

Figure 8 shows the results obtained during this experiment. It can be seen how all three encoding schemes produce networks more capable of generalising on the validation set than the traditional network. In four out of the five tests the weak encoding method produced the best networks. In contrast, in four out of the five tests the strong technique was the worst of the three schemes.

Conclusions

Our results indicate that the encoding method does indeed influence the quality of networks produced. Also, there appears to be a degradation in the performance of the GA as chromosome size is increased. This is accompanied by a fall in the resulting network's performance. On small problems, containing only a few nodes, the strong encoding scheme was found to outperform the other two techniques. As the number of network connections was increased the strong method produced fewer good networks. Whereas, as the problem size increased the weak encoding scheme started to perform well. The grammar encoding scheme appears to be relatively unaffected by problem size, consistently evolving networks better suited to the problem domain than standard architectures.

References

1 *Hornik,K., Stinchcombe,M., White,H.;* Multi-layer feedforward networks are universal approximators.; *Neural Networks, 2, pp359-68; (1989)*

2 *Miller,G.F., Todd,P.M., Hegde, S.U.;* Designing neural networks using genetic algorithms.; *In 3rd Int. Conf. Genetic Algorithms, Morgan Kaufmann, (1989)*

3 *Holland,J.H.;* Adaptation in natural and artificial systems.; *Ann Arbor (MI): University of Michigan Press, (1975)*

4 *Goldberg,D.;* Genetic algorithms in search, optimisation and machine learning; *Addison-Wesley (1988)*

5 *Jones,AJ.;* Genetic Algorithms and their Application to the Design of Neural Networks; *Neural Computing & Applications, 1, pp32-45;(1993)*

6 *Kitano,H.;* Designing neural networks using genetic algorithms with graph generation system.; *Complex Systems, (1990)*

7 *Whitley,D., Starkweather,T., Bogart,C.;* Genetic algorithms and neural networks: Optimising connections and connectivity.; *Parallel Computing, Vol.14,pp.347-361, (1990)*

8 *Maniezzo,V.;* Genetic evolution of the topology and weight distribution of neural networks.; *IEEE Transactions on Neural Networks, Vol.5, No.1, January (1994)*

9 *Robbins,G.E., Plumbley,M.D., Hughes,J.C., Fallside,F., Prager,R.;* Generation and adaptation of neural networks by evolutionary techniques; *Neural Computing and Applications, Vol.1, pp23-31,(1993)*

10 *Braun,H., Weisbrod,J.;* Evolving Neural Feedforward Networks; *Int. Conf. on Neural Networks & Genetic Algorithms, GA-ANN'93, Heidelberg: Springer-Verlag, pp25-32; (1993)*

11 *Harp,S.A., Samad,T.;* The Genetic Synthesis of Neural Network Architecture.; *Ch.15, in Davis, L.;* Genetic Algorithms and Simulated Annealing. *Pitman, London, (1987)*

12 *Harp,S.A., Samad,T., Guha,A.;* Designing application specific neural networks using the genetic algorithm.; *In Neural Information Processing Systems - Neural and Synthetic, pp.447-454. Morgan Kaufmann, (1989)*

13 *Harp,S.A., Samad,T., Guha,A.;* Towards the Genetic Synthesis of Neural Networks; *in 3rd Int. Conf. on Genetic Algorithms, Morgan Kaufmann, (1989)*

14 *Schaffer,J.D., Caruana,R.A., Eshelman,L.J.;* Using genetic search to exploit the emergent behaviour of neural networks. *Physica D, Vol.42, pp.244-248, (1990)*

15 *Boers,E.J.W., Kuiper,H.;* Biological metaphors and the design of modular artificial neural networks.; *Masters Thesis, Dept. of Computer Science & Experimental & Theoretical Psychology, Leiden University, Holland, (1992)*

16 *Merrill,J.W.L., Port,R.F.;* Fractally configured neural networks; *Neural Networks, 4, pp53-60; (1991)*

17 *Lindenmayer,A.;* Mathematical models for cellular interaction in development, parts I and II.; *in Journal of theoretical biology, 18,pp.280-315, (1968)*

18 *Prusinkiewicz,P., Lindenmayer,A.;* The algorithmic beauty of plants.; *Springer-Verlag, New York, (1990)*

19 *Rumelhart,D., McClelland,J.;* Parallel distributed processing: Explorations in the microstructure of cognition; *Vol.1, MIT Press, Cambridge (MA) (1986)*

20 *Denker,J., Schwartz, D.; Wittner,B.; Solla,S.; Howard, R.; Jackel,L.; Hopfield,J.;* Large Automatic Learning, Rule Extraction, and Generalsation.; *Complex Systems 1, pp877-922; (1987)*

APPLICATION OF RADIAL BASIS FUNCTION NETWORKS TO FAULT DIAGNOSIS FOR A HYDRAULIC SYSTEM

Dingli Yu[†], D.N.Shields[†], S.Daley[‡]

[†] *Control Theory and Applications Centre, Coventry University, Coventry CV1 5FB, U.K.*

[‡] *Mechanical Engineering Centre, European Gas Turbines Ltd, Leicester LE8 3HL, U.K.*

Abstract. Faults in a hydraulic test rig are detected and isolated by using two radial basis function networks (RBF). One RBF is used to model the test rig according to its nonlinear structure. The output prediction error, generated from the real responses and the model responses, is used as a residual to indicate the occurence of any fault. A second RBF is used to enhance the effect of an individual fault while reducing the effects of the other faults, in such a way that the fault is isolated. Simulation using real data collected from the rig demonstrates the effectiveness of this method.

Key words: Radial basis function, fault diagnosis, nonlinear system, robustness, hydraulic system.

1. INTRODUCTION

In recent years, fault detection and isolation (FDI) for non-linear systems have received increasing attention in both research and applications. However, the most of existing methods are based on linear models. Because most practical systems display a degree of nonlinearity, these methods generally only work well in small regions of the operating space. Although some authors have proposed nonlinear observer methods (Seliger and Frank, 1991), these methods are only effective for certain classes of systems. The main difficulty in developing a suitable nonlinear observer is in determining, to a reasonable degree of accuracy, the parameters of a norminal system model. Recently, robust observer schemes have been developed whereby the linearisation error is represented by unknown inputs (Frank and Wunnenberg, 1989, Patton and Chen, 1991, Yu and Shields, 1994a). However, for highly nonlinear systems and a large operating region the representating the modelling error is of a high dimension. Decoupling of residuals from high dimension unknown inputs implies a large number of measurements which are often unavailable for real application.

An alternative approach is one employing neural networks. It can be proved that a neural network with one hidden layer of a large enough number of neurons can model a nonlinearity to any accuracy. This feature of neural network makes it suitable for use in application to nonlinear system identification and control. The vast majority of applications make use of the multi-layer perceptron (MLP) network in conjunction with the back-propagation algorithm for training. However, the convergence of the weights for this sort of network often implies a long training time and the output error often converges to a local minimum. The RBF network, on the other hand, needs a smaller training time due to the linearity of the error function with respect to the weights (Chen and Billings, 1992, Leonard and Kramer, 1991).

RBF networks are used in this research. In the fault detection stage, a detection RBF is used to model the test rig. The output prediction error is used as a residual to indicate the occurence of faults. In the fault isolation stage, the residual vector is fed to an isolation RBF for which each specific output corresponds to each specific fault. This RBF is trained such that each specific output is only sensitive with respect to a specific fault, and is robust with respect to the other faults and the unknown inputs.

2. HYDRAULIC TEST RIG

The fault diagnosis method is applied to a hydraulic test rig which is considered to be more representative of a nonlinear system due to the factors of the nonlinear relation between flow rate and pressure, of the variations in oil viscosity and of the load-dependent characteristics, which all make linear methods less effective to apply (for more details see Daley, 1987, Yu and Shields, 1994b). The plant dynamics is presented as follows.

he spool valve displacement, $X_s(t)$, is

$$T_1T_2\ddot{X}_s(t) + (T_1+T_2)\dot{X}_s(t) + X_s(t) = K_s v(t) \tag{1}$$

where $v(t)$ is voltage input to the valve, T_1 and T_2 are electromagnetic and electro-mechanic time constants of the servo-valve respectively and K_s is valve gain. The flowrate, $Q_v(t)$, through the valve can be approximated by the square root relation of the orifice

$$Q_v(t) = K_\theta X_s(t)(P_s(t) - P_m(t))^{\frac{1}{2}} \tag{2}$$

where $P_s(t)$ is the supply pressure, $P_m(t)$ the pressure differencial across the motor and K_θ is the valve flow coefficient. For continuity of flow

$$Q_v(t) = C_r S_s(t) + \frac{V_t}{2\beta}\dot{P}_m(t) + K_l P_m(t) \tag{3}$$

where $S_s(t)$ is the shaft angular velocity, C_r is the motor displacement, V_t is the total trapped volume, β is the oil bulk modulus and K_l is a leakage coefficient. The motor torque is

$$T_m(t) = C_r\eta_m P_m(t) \tag{4}$$

where η_m is the efficiency of the motor. Neglecting static and coulomb friction,

$$T_m(t) = I\dot{S}_s(t) + DS_s(t) + T_p(t) \tag{5}$$

where I is the total inertia of the pump, motor and the shaft, D is the viscous friction coefficient and

$$T_p(t) = C_r \frac{1}{\eta_p} P_p(t) \tag{6}$$

where $P_p(t)$ is the pressure differential across the pump and η_p is the efficiency of the pump.

3. FDI USING RBF NETWORKS

Two outputs are chosen consisting of the shaft speed, $S_s(t)$, and the motor pressure, $P_m(t)$, and three inputs are chosen consisting of the setpoint of the speed, $v(t)$, the load pressure, $P_p(t)$, and the supply pressure, $P_s(t)$. The hydraulic test rig can then be approximated by a discrete, non-linear system as

$$y_{k+1} = f(y_k, u_k) + \omega_k \tag{7}$$

where u_k and y_k are the input and output vectors of the form

$$u_k = \begin{bmatrix} v_k \\ P_{pk} \\ P_{sk} \end{bmatrix}, \quad y_k = \begin{bmatrix} S_{sk} \\ P_{mk} \end{bmatrix} \tag{8}$$

where $f(*)$ represents a nonlinear function and ω a zero-mean white noise series. Here u_k, for example, represents the sampled value of $u(t)$ at $t = kT$.

3.1 RBF network

The nonlinear function in (7) can be modelled by a RBF network. The RBF network is a two layer processing structure. The hidden layer consists of an array of nodes. Each node contains a parameter vector called a centre. The node calculates the euclidean distance betreen the centre and the network input vector, and passes the result through a nonlinear function. The output layer is essentially a set of linear combiners. The overall input-output response of the RBF network is a mapping $\hat{f}$: $R^{n_I} \rightarrow R^{n_O}$, that is

$$\hat{f}_i(x) = \sum_{j=1}^{n_H} w_{ji}\phi(\| x - c_j \|, \rho_j), \quad i = 1, \cdots, n_O \tag{9}$$

where n_I, n_H and n_O are the number of inputs, the number of the nodes in the hidden layer and the number of outputs, respectively; w_{ji} are the weights of the linear combiners; $\| \cdot \|$ denotes the euclidean norm; ρ_j are some positive scalers called widths; $\phi(\cdot, \rho)$ is a nonlinear function from $R^+ \rightarrow R$; and c_j are known as the RBF centres.

It has been proved (Cybenko 1989, Park and Sanberg 1991) that under very mild assumptions on the nonlinear function, $\phi(\cdot)$, any continuous function f can be uniformly approximated to within an arbitrary accuracy by a RBF network, $\hat{f}$, provided that there are a sufficient number of hidden nodes. An assumption in the proof is that $\phi(\cdot)$ is continuous and bounded. According to this, a gaussian function

$$\phi(z, \rho) = exp(-z^2/\rho^2) \tag{10}$$

is selected in this work. In general, the network input data can only exist in some regions of the input space. It is resonable to allocate the RBF centres in these regions and to reflect the data patterns by the positions of the centres. A widely used clustering method is the k-means clustering technique. Two versions of the k-means method have been used in off-line and on-line modes. However, for the purpose of FDI all the parameters used in the identification for a healthy system must be stored and used on-line in the monitoring of the same system. Because of this, the k-means method can not be used in its usual form. The off-line method can not be used for FDI because FDI requires on-line monitoring. The k-means method adjusts the centres from samples to samples during the training. These changing centres can not repeat when the RBF is used on-line. Consequently, in this research, the centre of the system operating region is chosen as a constant centre vector for the whole process. Although the width in each node can have a different value, the same width for every node is sufficient for universal approximation (Park and Sandberg 1991). This means that all the widths, ρ_j, can be fixed at a value, $\rho_j = \rho$, giving a sufficient accuracy of approximation. In this research, the width ρ is selected as

$$\rho = (\frac{1}{n}\sum_{i=1}^{n} \| u_i - c \|^2)^{\frac{1}{2}} \tag{11}$$

where u_i lies in a subset of the input data space.

Having chosen the nonlinear function $\phi(\cdot)$ and formed the fixed centre c and the fixed width ρ, the RBF network can be used to model a multi-input multi-output nonlinear system

$$y_i = \sum_{j=1}^{n_H} \phi_j w_{ji} + e_i, \quad i = 1, \cdots, n_O \tag{12}$$

Here y_i is the ith output of the system to be modelled, e_i is the ith modelling error, ϕ_j is the nonlinear function output corresponding to the input $u(t)$ and the c_j is the jth centre of C. These quantities are defined as

$$\begin{aligned} y_i &= [y_i(1) \cdots y_i(N)]^T, \quad i = 1, \cdots, n_O & (13) \\ e_i &= [e_i(1) \cdots e_i(N)]^T, \quad i = 1, \cdots, n_O & (14) \\ \phi_j &= [\phi_j(1) \cdots \phi_j(N)]^T, \quad j = 1, \cdots, n_H & (15) \\ c &= [c_1 \cdots c_{n_H}] & (16) \end{aligned}$$

where N is the number of samples in a subset of the data space. Then, for $t = 1, \cdots, N$, equation (12) can be collectively arranged as

$$\begin{aligned} [y_1 \cdots y_{n_O}] &= [\phi_1 \cdots \phi_{n_H}] \begin{bmatrix} w_{11} & \cdots & w_{1n_O} \\ \vdots & & \vdots \\ w_{n_H 1} & \cdots & w_{n_H N_O} \end{bmatrix} \\ &+ [e_1 \cdots e_{n_O}] \end{aligned} \tag{17}$$

or, more concisely, in the following matrix form

$$Y = \Phi W + E. \tag{18}$$

The weight matrix W can be solved such that the error matrix E is minimized in the sense of least squares using both batch and recursive algorithms. Only the latter is suitable for FDI purposes. The recursive least square algorithm is chosen of the following form (Chen and Billings, 1992)

$$\begin{aligned} \hat{W}_k &= \hat{W}_{k-1} + \Psi_k \chi_k [\, y_k - \chi_k^T \hat{W}_{k-1} \,] & (19) \\ \Psi_k &= \Psi_{k-1} - \frac{\Psi_{k-1}\chi_k\chi_k^T\Psi_{k-1}}{1 + \chi_k^T \Psi_{k-1}\chi_k} & (20) \end{aligned}$$

Details of the algorithm can be found in (Chen and Billings, 1992).

3.2 Fault detection using RBF

A RBF network is used to model the nonlinear plant so that the modelling error is made very small for the non-fault condition of the system but made significantly greater than zero for the system when faults are present. The fault detection signal ϵ_D is defined as

$$\epsilon_D(k) = \| e_y(k) \|_2 \tag{21}$$

where $e_y(k)$ is the output estimation error defined as

$$e_y(k) = y_k - \hat{y}_k.$$

A fault situation is deemed to hold if

$$\epsilon_D(k) > \delta \tag{22}$$

where δ is a threshold which is pre-specified according to experience and the type of the system involved.

3.3 Fault isolation using RBF

From the analysis in (Yu and Shields, 1995) a reasonable assumption for many real systems is that faults with different directions (in the fault space) give rise to different directions for the vector $e_y(k)$. This assumption is made here to isolate faults. A fault isolation signal, ϵ_I, is defined as

$$\epsilon_I(k) = W e_y(k) \tag{23}$$

where $W \in \Re^{q \times p}$ is called the fault isolation matrix with q being the number of the possible faults and p being the number of outputs. When faults are present, the estimated error is assumed to be expressed as

$$e_y(k) = \sum_{i=1}^{q} v_i(k) f_i(k) + \omega, \tag{24}$$

where $v_i(k)$ is a distribution vector of the ith fault on $e_y(k)$, q is the number of faults to be considered and ω is the output estimation error vector in the no fault condition caused by system uncertainty and noise. The matrix W is constructed as

$$W = \begin{bmatrix} w_1 \\ \vdots \\ w_q \end{bmatrix} \tag{25}$$

where each row vector w_i is designed such that the following conditions are satisfied

$$w_i v_j = 0, \quad for\ i = j, \quad i = 1, \cdots, q \tag{26}$$
$$w_i v_j \neq 0, \quad for\ i \neq j, \quad j = 1, \cdots, q \tag{27}$$

It can be clearly seen from an algebric analysis that only the ith element of the fault isolation signal, $\epsilon_{I(i)}$, is zero when the ith fault occurs, provided that no two fault distribution vectors are in the same direction.

In fact, the left hand side in (24) can not be exactly zero because of system uncertainty and noise. Therefore, a RBF network is used to implement matrix W to optimally isolate faults. The RBF network is designed with p inputs and q outputs, with every output corresponding to a particular fault. The network is trained off-line such that, for the input data contaminated by the ith fault, only the ith output, $\epsilon_I(i)$, is zero. Thus, when the network is used on-line and when the detection network claims the occurence of a fault, one and only one output is zero and this particular output indicates uniquely which fault occurs.

4. REAL DATA SIMULATION

The simulation was performed using real data collected from the hydraulic test rig. The system was subjected to square wave inputs for both v_k and P_{pk} and 2000 samples of the system input and output were collected with a sample interval, $T = 10\ ms$.

The fault detection RBF network had three inputs, $v(k)$, $P_p(k)$ and $P_s(k)$, and two outputs, $S_s(k)$ and $P_m(k)$. A number, $n_H = 30$, for the hidden layer was found to be enough for this work. The constant centre and width were chosen as described in the last section. 1000 samples were used to train the network for only one cycle. The network output is displayed in Fig.1 together with the real output.

The fault conditions detected and isolated in the simulation were those of two sensor faults and one component fault – these were errors superimposed on S_{sk}, P_{mk} and D, respectively. The three faults are represented by the following equations.

$$\tilde{S}_s = (1 + \Delta_{S_s}) S_s \tag{28}$$
$$\tilde{P}_m = (1 + \Delta_{P_m}) P_m \tag{29}$$
$$\tilde{D} = (1 + \Delta_D) D \tag{30}$$

with

$$\Delta_{S_s} = \begin{cases} 0 & k \leq 100\ or\ k > 300 \\ 0.05 & 300 < k \leq 300 \end{cases}$$

$$\Delta_{P_m} = \begin{cases} 0 & k \leq 400\ or\ k > 600 \\ 0.05 & 400 < k \leq 600 \end{cases}$$

$$\Delta_D = \begin{cases} 0 & k \leq 700\ or\ k > 900 \\ 0.05 & 700 < k \leq 900 \end{cases}$$

where $\tilde{S}_s$, $\tilde{P}_m$ and $\tilde{D}$ are the faulty values of S_s, P_m and D, respectively.

The fault isolation network had two inputs, three outputs with 20 nodes in the hidden layer. The centre and width were chosen in the same way as that for the detection network. The training data for this network was simulated by superposing a 10% change on the real outputs for the two sensor faults and collected from a simulated nonlinear model when D had a 10% change for the component fault. After off-line training, the fault conditions described in (28)-(30) on the other set of real data were tested. Fault detection and isolation signals are displayed in Fig.2.

In order to display the three fault isolation signals, $\epsilon_{I(1)}$, $\epsilon_{I(2)}$ and $\epsilon_{I(3)}$, in one graph, the horizontal axis in Fig.2 is moved to values of 2 and 4 on the vertical axis for $\epsilon_{I(2)}$ and $\epsilon_{I(3)}$, respectively. It can be seen in Fig.2 that when the first fault occurs, only $\epsilon_{I(1)}$, which corresponds to fault 1, is zero, the other two signals, $\epsilon_{I(2)}$ and $\epsilon_{I(3)}$, are not zero. The same results repeat for the case of second and third faults occuring. From Fig.2 we note that the simulated faults have been clearly detected and isolated. Each fault diagnosis output is sensitive to one specific fault and is robust with respect to the modelling error, noise and any other fault.

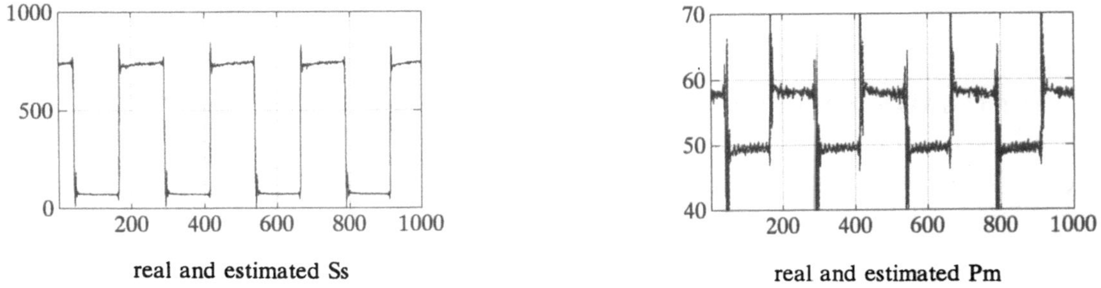

Fig.1 RBF output and real output

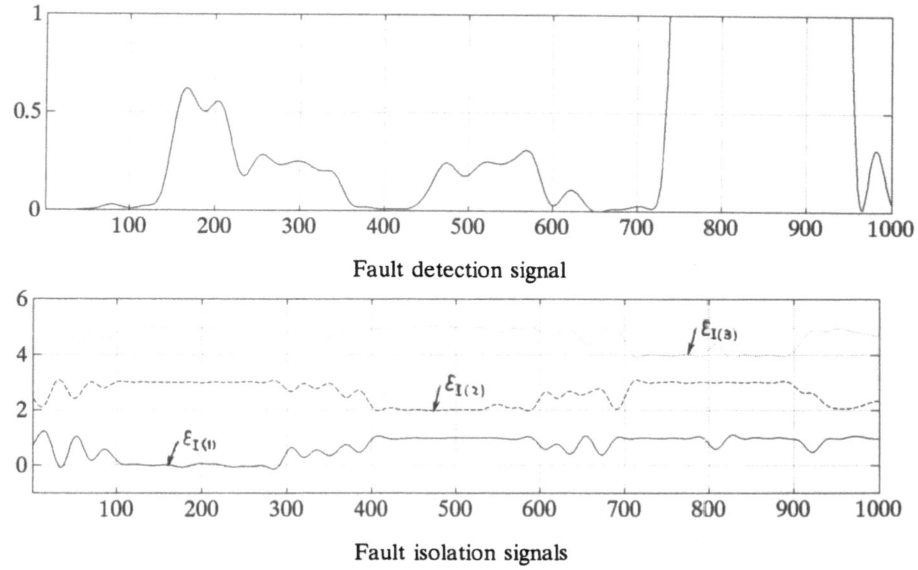

Fig.2 Fault detection signal and fault isolation signals

5. CONCLUSIONS

Fault detection and isolation for non-linear systems is realized by using RBF networks. One RBF network is used to model the non-linear system and generate the residuals and one to isolate faults. The method does not need an analytical model of the system which may be difficult to obtain. The simulation using real data indicates the effectiveness of the method for a real industrial application.

REFERENCES

Chen,S. and Billings,S.A., 1992, Neural networks for modelling and identifications, *International Journal of Control*, Vol.56, No.2, pp319-346.

Cybenko,G, 1989, Approximation by superpositions of a sigmoidal function, *Math. Control Signals Syst.*, Vol.2 pp303-314.

Daley,S., 1987, Application of a fast self-tuning control algorithm to a hydraulic test rig.*Proc Instn Mech Engrs.* Vol.201 No.C4, pp285-294.

Frank, P.M. Wunnenberg, J. 1989, Robust fault diagnosis using unknown input schemes. Chapter 3, in *Fault diagnosis in dynamic systems.* Patton *et al*, Prentice Hall.

Leonard,J., and Kramer,M., 1991, Rdial basis function networks for classifying process faults, *IEEE Control Systems Magazine*, No.4 pp31-38.

Park,J. and Sandberg,I., 1991, Universal approximation using radial basis function networks, *Neural Computing*, No.3 pp236-257.

Patton, R. J. and Chen, J. 1991, Optimal selection of unknown distribution matrix in the design of robust observers for fault diagnosis. *Proc.IFAC symposium on SAFEPROCESS'91*, Sept.10-13, Baden-Baden, Vol.2, pp1666-1675.

Seliger, R. and Frank, P.M. 1991, Robust component fault detection and isolation in nonlinear dynamic systems using nonlinear unknown input observers. *Preprints of SAFEPROCESS'91, Sept. 10-13, Baden-Baden, FRG.* **1**, 313-318.

Yu,D.L., Shields,D.N., Mahtani,J.L., 1994a, A nonlinear fault detection method for a hydraulic system, *Proc. 4th IEE Inter. Conf. on CONTROL'94*, University of Warwick, U.K., 21-24 Mar. Vol.2, pp 1318-1322.

Yu, D.L., Shields, D.N., 1995. A new fault isolation approach to linear and bilinear systems. Submitted to *European Control Conference ECC'95*, Sept. 5-8, Roma, Italy.

OPTIMALLY ROBUST FAULT DIAGNOSIS USING GENETIC ALGORITHMS

Dingli Yu, D.N.Shields

Control Theory and Applications Centre, Coventry University, Coventry CV1 5FB, U.K.
Tel. 0044 203 838972 Email: mty021@cov.ac.uk

Abstract. A new fault isolation method is proposed using state space parity equations and genetic algorithms and is deemed superior to a previous method proposed by the authors. Faults are isolated by making a residual vector zero for a particular individual fault and significantly non-zero for other faults. This method uses genetic algorithms to minimise the effect of a particular fault on the residuals in the presence of the unknown inputs and to maximise the effect for other faults. A real data simulation on a hydraulic test rig is performed to demonstrate the effectiveness of the method.

Key Words. Fault detection and isolation, genetic algorithms, parity equation, optimization, hydraulic system.

1. INTRODUCTION

The dedicated observer scheme proposed by Clark (1989) and the unknown input observer scheme proposed by Frank (1990) isolate faults by using a bank of observers. The fault detection filter (Beard, 1971, Jones, 1973) isolates faults in a simillar way but does not decouple the residual from the unknown input. Chow and Willskey (1984) proposed a systematic mechanism to generate parity equations and formulated a minimax framework for a robust design. Lou *et al* (1986) developed a design methodology that provides the most robust parity equations, given a finite set of uncertain plant models. Gertler and Singer (1990) proposed a structured parity equation for an input-output model. In this method a transformation matrix is used with parity checks to assess if a single fault residual lies in a prespecified direction. The authors (Yu and Shields, 1995) have previously proposed a method in which fault isolation was obtained more directly by using the idea of decoupling the effects of the faults on the residuals. This method is applied to the residuals generated by both the parity space method and the observer-based method and *High-threshold fault isolation*, definded by Gertler and Singer (1990), can be obtained by only one observer in the observer-based method. For the real application, however, due to modelling error and measurement noise, any decoupling of the fault isolation signal from the other faults may not be realistically feasible. In this paper, an optimization method is, therefore, employed to enhance the signal to noise ratio. A genetic algorithm is used to optimally choose the fault isolation matrix such that the effects of faults are isolated.

2. FAULT ISOLATION METHOD

Part of the method is a development of previous work of the authors (Yu and Shields, 1995). Several new features, however, are added, including the use of genetic algorithms and real data simulation. Consider a linear discrete-time model

$$x_{k+1} = Ax_k + Bu_k + Ed_k + Gf_{a(k)} \quad (1)$$
$$y_k = Cx_k + Du_k + Qf_{s(k)} \quad (2)$$

where $x_k \in \Re^n$, $u_k \in \Re^m$ and $y_k \in \Re^p$ are state, input and output vectors, respectively, and where $d_k \in \Re^l$, $f_{a(k)} \in \Re^r$ and $f_{s(k)} \in \Re^h$ are unknown input, component and actuator fault and sensor fault vectors, respectively. It is assumed that the distribution matrices of the unknown input, component faults and sensor faults, E, G and Q, are known and $rank\{E\} = l, rank\{G\} = r$ and $rank\{Q\} = h$. For linear systems only the state space parity space method is considered here. From the output values of system (1)-(2) over $s+1$ sampling intervals, the following equation is obtained,

$$\begin{bmatrix} y_{k-s} \\ \vdots \\ y_k \end{bmatrix} = Mx_{k-s} + H_1 \begin{bmatrix} u_{k-s} \\ \vdots \\ u_k \end{bmatrix} + H_2 \begin{bmatrix} d_{k-s} \\ \vdots \\ d_k \end{bmatrix} + H_3 \begin{bmatrix} f_{a(k-s)} \\ \vdots \\ f_{a(k)} \end{bmatrix} + H_4 \begin{bmatrix} f_{s(k-s)} \\ \vdots \\ f_{s(k)} \end{bmatrix} \quad (3)$$

where s is called the parity space order, and

$$M = \begin{bmatrix} C \\ CA \\ \vdots \\ CA^s \end{bmatrix}; \; H_1 = \begin{bmatrix} D & & & \mathbf{0} \\ CB & D & & \\ CAB & CB & D & \\ \vdots & & & \ddots \\ CA^{s-1}B & \cdots & \cdots \; CB & D \end{bmatrix};$$

where H_2 and H_3 are same as H_1 except for matrices D and B being replaced by 0 and E for H_2 and by 0 and G for H_3, respectively. Also, $H_4 = diag(\underbrace{Q \cdots Q}_{s+1})$. The parity check, ϵ_k, is defined as

$$\epsilon_k \triangleq \Omega(\begin{bmatrix} y_{k-s} \\ \vdots \\ y_k \end{bmatrix} - H_1 \begin{bmatrix} u_{k-s} \\ \vdots \\ u_k \end{bmatrix}) \quad (4)$$

where $\Omega \in P$ and P is a parity space defined as

$$P \triangleq \{\Omega \mid \Omega[M, \; H_2] = 0\} \quad (5)$$

From (3), (4) and (5), we obtain

$$\epsilon_k = \Omega H_3 \begin{bmatrix} f_{a(k-s)} \\ \vdots \\ f_{a(k)} \end{bmatrix} + \Omega H_4 \begin{bmatrix} f_{s(k-s)} \\ \vdots \\ f_{s(k)} \end{bmatrix} \quad (6)$$

Define

$$\Omega H_3 = Z \ \ and \ \ \Omega H_4 = J$$

Then equation (6) can be rearranged in a special form such that the values of a fault at various sampling intervals are grouped together. We define modified fault vectors ($\bar{f}_a(i)_k$, $\bar{f}_s(i)_k$) and matrices (Z_i, J_i) such that

$$\epsilon_k = [\ Z_1,\ \cdots,\ Z_r,\ J_1,\ \cdots,\ J_h\ ] \begin{bmatrix} \bar{f}_a(1)_k \\ \vdots \\ \bar{f}_a(r)_k \\ \bar{f}_s(1)_k \\ \vdots \\ \bar{f}_s(h)_k \end{bmatrix} \quad (7)$$

where

$$Z_i = [\ z_i \ \ z_{r+i} \ \ z_{2r+i} \ \ \cdots \ \ z_{sr+i}\ ],\ i = 1, \cdots, r$$
$$J_i = [\ j_i \ \ j_{h+i} \ \ j_{2h+i} \ \ \cdots \ \ z_{sh+i}\ ],\ i = 1, \cdots, h$$

Here, z_i is the ith column of matrix Z and j_i is the ith column of matrix J. Also, $\bar{f}_a(i)$ and $\bar{f}_s(i)$ are the ith modified component fault vector and ith modified sensor fault vector whose elements are such that

$$\bar{f}_a(i)_k = \begin{bmatrix} f_a(i)_{k-s} \\ \vdots \\ f_a(i)_k \end{bmatrix},\ i = 1, \cdots, r$$
$$\bar{f}_s(i)_k = \begin{bmatrix} f_s(i)_{k-s} \\ \vdots \\ f_s(i)_k \end{bmatrix},\ i = 1, \cdots, h$$

The fault detection signal ε_D is defined as

$$\varepsilon_D(k) = \| \epsilon_k \|_2 \quad (8)$$

A fault situation is deemed to hold if

$$\varepsilon_D(k) > \delta \quad (9)$$

where δ is a threshold which is pre-specified according to experience and the type of the system involved.

A fault isolation vector, $\varepsilon_I \in \Re^{r+h}$, is defined as

$$\varepsilon_I(k) = \Theta \epsilon_k \quad (10)$$

where Θ is called the fault isolation matrix. From equation (7) it is clear that if a matrix Θ can be designed such that

$$\Theta [\ Z_1,\ \cdots,\ Z_r,\ J_1,\ \cdots,\ J_h\ ] = \begin{bmatrix} \mathbf{0} & & \times \\ & \mathbf{0} & \\ & & \ddots & \\ \times & & & \mathbf{0} \end{bmatrix} \quad (11)$$

where $\times$ represents non-zero matrices, that is, that only the element matrices in the diagnal of the matrix on the right hand side of (11) are zero matrices. Then, when only one fault occurs, only the corresponding component of the fault isolation vector is zero. Other components are nonzero. In this way, when the fault detection signal is fired (i.e. (9) is true), we can decide which fault occurs.

Analysis. From (5) it can be shown using the singular value decomposition (SVD) of $[M,\ H_2]$, $[M, H_2] = [U_1, U_2]\Sigma V$, that the number of independent rows of Ω, β, is

$$\beta = p(s+1) - rank\{[M,\ H_2]\} \quad (12)$$

where $[M,\ H_2] \in \Re^{p(s+1)\times(n+sl)}$ and Ω is given as

$$\Omega = W U_2^T \quad (13)$$

where $W \in \Re^{\beta\times\beta}$ is an arbitrary matrix. Since the rank of a matrix is always less than or equal to the number of columns, equation (12) implies that

$$\beta \geq s(p-l) - (n-p) \quad (14)$$

Defining the fault isolation matrix Θ as

$$\Theta = [\theta_1^T, \cdots, \theta_{r+h}^T]^T,$$

where θ_i is the ith row vector of Θ. Satisfaction of equation (11) is equivalent to satisfaction of the following equations,

$$\theta_i Z_j \begin{cases} = 0 & \text{if } i = j \\ \neq 0 & \text{if } i \neq j \end{cases},\ \ i,j = 1, \cdots, r \quad (15)$$

$$\theta_{r+i} J_j \begin{cases} = 0 & \text{if } i = j \\ \neq 0 & \text{if } i \neq j \end{cases},\ \ i,j = 1, \cdots, h \quad (16)$$

and

$$\theta_i J_j \neq 0,\ \ i = 1, \cdots, r,\ \ j = 1, \cdots, h \quad (17)$$
$$\theta_{r+j} Z_i \neq 0,\ \ i = 1, \cdots, r,\ \ j = 1, \cdots, h \quad (18)$$

Since G and Q are of full column rank, if Q is not a span of $g_1, \cdots, g_r$, where g_i is the ith column of G, conditions (15)-(18) can be satisfied by designing θ_i as

$$\theta_i = w_i U_{Z2}^T,\ \ i = 1, \cdots, r \quad (19)$$
$$\theta_{r+i} = w_{r+i} U_{J2}^T,\ \ i = 1, \cdots, h \quad (20)$$

where U_{Z2} and U_{J2} are produced from the SVD of Z_i and J_i, respectively, as

$$Z_i = [U_{Z1}, U_{Z2}]\Sigma_Z V_Z^T,\ J_i = [U_{J1}, U_{J2}]\Sigma_J V_J^T,$$

and properly choosing w_i, $i = 1, \cdots, r+h$. Here, w_i is a row vector with the dimension of $(\beta - rank\{Z_i\})$ for $i = 1, \cdots, r$ and $(\beta - rank\{J_i\})$ for $i = r+1, \cdots, r+h$.

The existence conditions for Θ satisfying (19) and (20) are respectively,

$$\beta > max\{rank\{Z_i\},\ i = 1, \cdots, r\}, \quad (21)$$
$$\beta > max\{rank\{J_i\},\ i = 1, \cdots, h\} \quad (22)$$

Thus, considering (12) and conditions (21)-(22) together, the necessary and sufficient condition for the fault isolability of system (1)-(2) is, therefore,

$$p(s+1) > rank\{[M,\ H_2]\} + max\{rank\{Z_i\},\ rank\{J_{\bar{i}}\}\ ,\quad i = 1,\cdots,r; \bar{i} = 1,\cdots,h\} \quad (23)$$

From equation (7) it can be seen that $Z_i \in \Re^{\beta\times(s+1)}$, $i = 1,\cdots,r$ and $J_i \in \Re^{\beta\times(s+1)}$, $i = 1,\cdots,h$. Similar to (14), an explicit sufficient condition for the fault isolation of system(1)-(2) is

$$s(p-l-1) > n-p+1 \quad (24)$$

From equations (19), (20) we know that the row vectors of the fault isolation matrix are chosen from the left null space of the corresponding matrix Z or J, The effect of a individual fault on the corresponding element of ϵ_I must be zero. However, this element will never be zero due to the effects of system uncertainty and noise. Therefore an optimization method is prescribed for real applications. The parameter matrices w in (19)-(20) can be optimaly chosen such that the following objective functions are minimized,

$$\begin{aligned} \bar{J}_i &= \sum_{j=1}^{i-1} \|\frac{w_i U_{Z2}^T Z_i}{w_i U_{Z2}^T Z_j}\| + \sum_{j=i+1}^{r} \|\frac{w_i U_{Z2}^T Z_i}{w_i U_{Z2}^T Z_j}\| \\ &+ \sum_{j=r+1}^{r+h} \|\frac{w_i U_{Z2}^T Z_i}{w_i U_{Z2}^T J_j}\| \quad i = 1,\cdots,r \end{aligned} \quad (25)$$

$$\begin{aligned} \bar{J}_{r+i} &= +\sum_{j=1}^{r} \|\frac{w_{r+i} U_{J2}^T J_i}{w_{r+i} U_{J2}^T Z_j}\| + \sum_{j=r+1}^{r+i-1} \|\frac{w_{r+i} U_{J2}^T J_i}{w_{r+i} U_{J2}^T J_j}\| \\ &+ \sum_{j=r+i+1}^{r+h} \|\frac{w_{r+i} U_{J2}^T J_i}{w_{r+i} U_{J2}^T J_j}\|, \quad i = 1,\cdots,h \end{aligned} \quad (26)$$

These optimizations can be performed by using a genetic algorithm and an application example to a hydraulic test rig is presented in the next section.

3. An example

3.1 Genetic algorithm

The genetic algorithm used in the simulation is of a general form. The parameters to be optimised are ordered as a string of binary form. The length of the string depends on the number of the parameters and the accuracy required. This string is termed a chromosome. A set of chromosomes are randomly selected initially to compose the population of first generation. The typical operations, such as crossover and mutation, are then applied to the chromosomes in the existing population. The 'fitness value' is calculated for every chromosome by using the objective function to evaluate it and then the reproduction of the population of the next generation is done through a selection on a fitness basis, such that the better the fitness value, the higher the chance of it being chosen. This procedure is repeated until an acceptable fitness is achieved and the chromosome with best fitness gives rise to the optimal parameters for the objective function.

3.2 Hydraulic test rig model

An application of this optimal fault isolation method to a hydraulic test rig (Daley, 1987) is investigated by a simulation to demonstrate the application procedure. The test rig model used in this work is as follows (for the details, see Yu and Shields, 1994) This gives a third order continuous linear model of the system as follows

$$\dot{x}(t) = A_c x(t) + B_c u(t) + E_c d(t) \quad (27)$$

$$y(t) = Cx(t) \quad (28)$$

where the input, $u(t)$, output, $y(t)$, and the state, $x(t)$, are vectors selected as

$$\begin{aligned} u(t) &= [v(t) + v_0, P_p(t)]^T \\ y(t) &= [X_s(t), P_m(t), S_s(t)]^T \\ x(t) &= [X_s(t), P_m(t), S_s(t)]^T \end{aligned}$$

and where

$$A_c = \begin{bmatrix} -\frac{1}{T_1} & 0 & 0 \\ \frac{2\beta K_v}{V_t} & -\frac{2\beta(K_p+K_l)}{V_t} & -\frac{2\beta C_r}{V_t} \\ 0 & \frac{C_r \eta_m}{I} & -\frac{D}{I} \end{bmatrix}$$

$$B_c = \begin{bmatrix} \frac{K_s}{T_1} & 0 \\ 0 & 0 \\ 0 & -\frac{C_r}{I\eta_p} \end{bmatrix}, \quad C = I_3$$

where E_c, the distribution matrix of the unknown input, is unknown and will be specified later.

3.3 Real data simulation

Real data sampled from the test rid includes two inputs v_k and $P_{p(k)}$ and two outputs $P_{m(k)}$ and $S_{s(k)}$. Another output $X_{s(k)}$ which was not available was taken from a non-linear simulation of the rig because no possible fault was likely to occur with this signal. The distribution matrix of the unknown input, E, in the discretized form of model (35)-(36) was estimated using real data as $E = [0.0010,\ 11.7855,\ 57.8901]^T$. Two faults were simulated in which one was a component fault, a 10% change in the fiscous friction coefficient, D; and one a sensor fault, a 10% change in motor pressure sensor, P_m. The directions of the two faults are those of the vector $G = [-0.0001,\ -0.5725,\ 29.9558]^T$ and $Q = [0,\ 1,\ 0]^T$, respectively. In this example, $n = p = 3$, $l = 1$. The parity space order was selected as $s = 3$ so that the condition (24) was satisfied by $s(p-l-1) = 3$, which was greater than $n-p+1 = 1$. This left at least two parameters to be optimally chosen.

Calculation of the matrices Z and J in equation (7) gave $r = h = 1$. These matrices were represented with orthogonal columns, three for Z and two for J. Hence, w in both cases is a row vector with three elements for the first fault and with two elements for the second fault.

The genetic algorithm used in this example was with a population of 100 chromosomes. The word length of a parameter to be optimised was selected as 24 (in binary system) and the bounds to the parameters were set as -1, 1. Fig.1 displays the convergence of the fitness values for w_1 and w_2 after 20 generations. The solid line in Fig.1 represents the average fitness value of the chromosomes in every generation, whilst the doted line represents the minimum value in every generation.

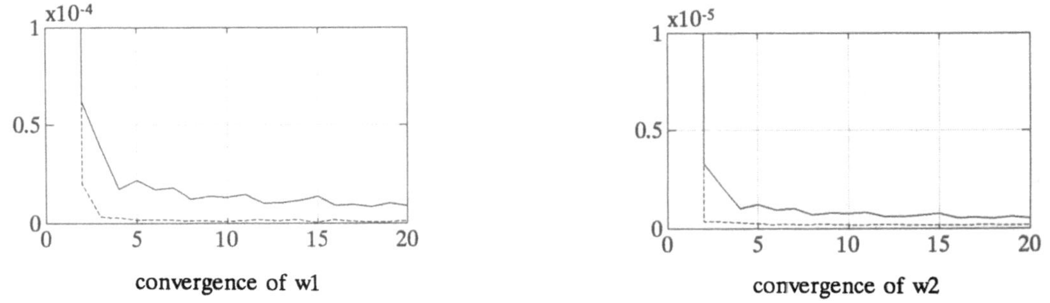

Fig.1 convergence of objective functions

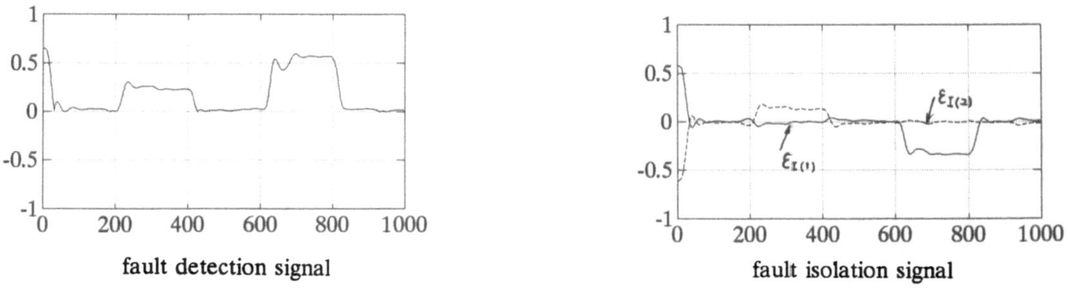

Fig.2 Fault detection and isolation signals

Using the optimal Θ, the fault isolation signal and the fault detection signal are displayed in Fig.2. It can be seen from Fig.2 that the fault detection signal diverges from zero when the first fault occurs, over the sample range $201 \leq k \leq 400$, and when the second fault occurs, over the sample range $601 \leq k \leq 800$. The first element of the fault isolation signal remains at zero value over the sample range $201 \leq k \leq 400$ when the first fault occurs while the second element is not zero over this range. The situation is reversed for the second fault over the range $601 \leq k \leq 800$. It is noted that, although the real data is subject to noise, the simulated faults are still isolated clearly. Note also, that the directions of the faults in this example are distinct.

4. CONCLUSIONS

The proposed new method for fault isolation does give a guarantee that an optimal fault isolation matrix can be chosen, where optimal means that faults are isolated as distinct as possible. The method uses a genetic algorithm and is shown to be an efficient approach to the problem of fault isolation. For simplicity, only two faults were simulated in the example. When more faults are taken account of, the use of a genetic algorithm for this task should prove more advantageous.

REFERENCES

Beard, R.V., (1971). Failure accommodation in linear systems through self-reorganization. Ph.D Thesis, MIT.

Chow, E.Y., Willskey, A.S. (1984). Analytical redundancy and the design of robust detection systems. *IEEE Tarns. Aut. Control*, Vol.29, No.7, pp 603-614.

Clark, R.N., (1989). State estimation schemes for instrument fault detection. *Chapter 2 in Fault diagnosis in dynamic systems, theory and application, Patton et al*, Prentice Hall.

Daley,S. (1987). Application of a fast self-tuning control algorithm to a hydraulic test rig.*Proc Instn Mech Engrs.* Vol.201 No.C4, pp285-294, 1987.

Frank, P.M. 1990, Fault diagnosis in dynamic systems using analytical and knowledge-based redundancy-A survey and some new results. *Automatica.* **26**, 459-474.

Gertler, J., Singer, D. (1990) A new structural framework for paroty equation-based failure detection and isolation. *Automatica*, Vol.26, No.2, pp 381-388.

Jones, H.L., (1973). Failure detection in linear systems. Ph.D. Thesis, MIT.

Lou, X., Willskey, A.S. and Verghese, G.C. (1986). Optimally robust redundancy relations for failure detection in uncertain systems. *Automatica*, Vol.22, No.3, pp 333-344.

Yu, D.L., Shields, D.N. (1994). A fault detection method for a nonlinear system and its application on a hydraulic test rig. *Preprints of IFAC Symposium on SAFEPROCESS'94*, Espoo, Finland, 13-16 June. Vol.2 pp 305-310.

Yu, D.L., Shields, D.N., 1995d. A new fault isolation approach to linear and bilinear systems. Submit to *European Control Conference ECC'95*, Sept. 5-8, Roma, Italy.

Development of a Neural-based Diagnostic System to Control the Ropes of Mining Shifts*

Ortega,F, Ordieres,J.B Menéndez,C González Nicieza,C.

Project Engineering Area Universidad de Oviedo
Applied Mathematics Area Universidad de Oviedo
Mining Area Universidad de Oviedo

Summary

Although the application of neural networks in quality control and maintenance is growing quickly from last years, they are just an incipient technology in the Mining Industry. At the same time, maintenance of the shift is probably the most important matter in Mining, considering that the shift could be the only way out for people and material in a colliery.

The Area of Project Engineering of the University of Oviedo has designed a system to control the state of the wire ropes for extraction in coal shifts, based on the information supplied by three groups of electromagnetic sensors (Inductive and Hall-Effect) placed in a head around the rope when inspection is carried out.

The system involves the use of three parallel neural subnetworks which output is introduced in common final layers to be definitely classified. This system allows to detect internal broken wires and to prevent more serious defects before they occur, in such a way that the rope can be maintained in service during a longer period of time, with the necessary equilibrium between security, reliability and economy. If this system is placed permanently on the shift, the risk of unexpected failure of the wire rope should be decreased to the minimum.

1.- Steel Ropes.

Steel ropes are basic elements in a lot of transport installations, not only at mines but also at de industry and public service installations. So, we used them in: elevation and support equipment (overhead cranes, tower cranes, travelling cranes), ground movement equipment (dredgers, diggers), transport services (cable railways, chair lifts, funiculars), buildings (lifts, hoists), civil structures (suspension bridges), and so on.

Steel ropes are composed by a metallic or textile heart involved in spirals of wires (see Figure I). When the time goes by, the rope losses some of its characteristics, specially due to the existence of broken wires or just wear or corrosion, so that it must be controlled carefully. As the number of broken wires grows, the cross-section to sustain the load is smaller and the risk of failure is bigger. The number of maximum allowed broken wires in this environment basically depend on rope kind, which is defined by structure (Seale, Warrington, Filler), cords number, wires per cord and kind of heart so as heart and wires mechanical characteristics. Usual ropes could have thirty or forty broken wires before it should be change.

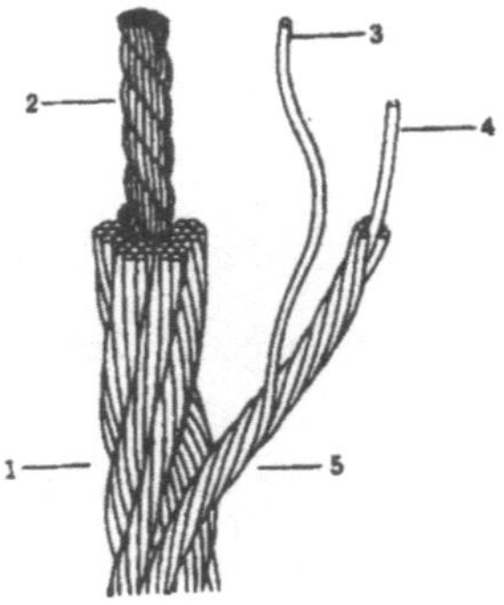

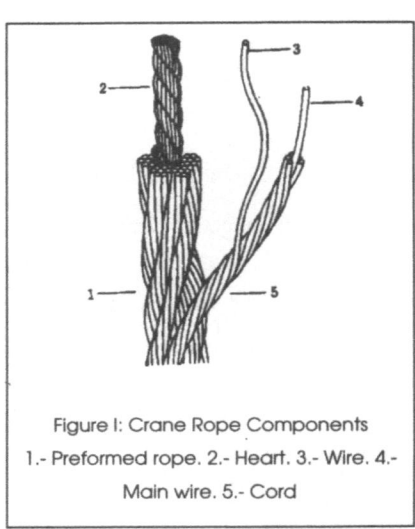

Figure I: Crane Rope Components

1.- Preformed rope. 2.- Heart. 3.- Wire. 4.- Main wire. 5.- Cord

Crane ropes that bear cages of the shaft, transport services or ladles at iron and steel industry must be more cautiously controlled because human lives could be involved. Usually, crane ropes inspection is done by visual examination; an human inspector realised a external and internal visual examination in order to detect defects such as broken wires, corrosion or wear. This method suffers from human inaccuracies such as fatigue, dangerous surrounding place, human subjectivity due to the responsibility if a break occurs, etc. Moreover, rope speed should not exceed of 0,3m/s and maximum visual inspection recommended time is 10 min.

The use of sensors has represented an important advance in rope control. So, some sensors are located surrounding the rope, and signals are continuous plotting. After that, inspection is done over the plot. The method suffers again from human subjectivity and fatigue.

2.- Measure Equipment.

The defects are detected in a magnetic-based way. When the metallic rope is fully magnetised with a permanent magnet o by means of a electromagnetic coil, it produces a magnetic field which should be uniform all through the rope length. If there is a defect in the rope, a change on the magnetic field flow is detected. Considering this idea, the acquisition system is based on a magnetic head manufactured in Poland by Mera-Ster. It is a cylinder, closed around the ropes diameter, with a big permanent magnet, strong enough to magnetise the rope up to saturation.

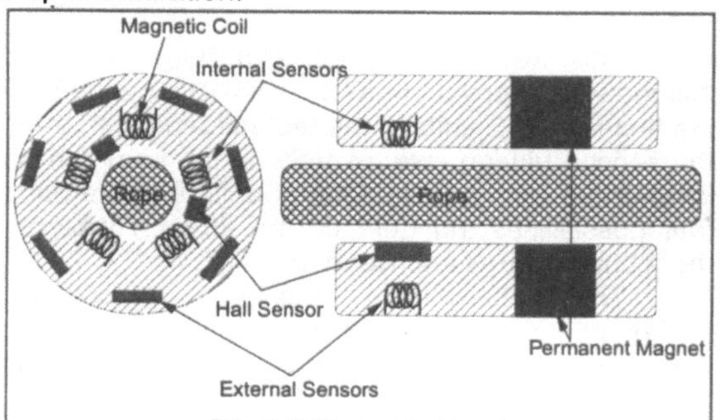

Figure II: Magnetic Head

System contains a group of five magnetic coils equi-spaced around a cross section of the rope which will be influenced mainly by surface alterations(see Figure II). Other group of seven coils, located further to the last ones, supplies information about inner and outer defects, complementing it. Besides these inductive sensors, there is an optical sensor used to measure the rope diameter. Lastly, there are two Hall-effect sensors, very suitable for mass-loss defects.

Both set of inductive sensors are serial connected. Internal sensors generate one signal and another signal are generated by external ones. Likewise, halls and optical sensors generate two analogic signals.

These three signals are sending by cable to an A/D converter card. Digital output signals are codified to eight bits, it give us enough good resolution.

The installation of an encoder on the main axis of crane pulley allows us to know rope speed.

Obviously, the signals are not clear at all. The irregular surface of the rope, due to the different wires, the hard environment full of small particles in the shift, the water, and even the noise in the electric supply to the equipment convert the interpretation of data in a very hard work, full of mistake possibilities. In order to substitute the visual inspection, the automatic systems should be independent of subjective criteria of human inspectors.

3.- Preparation of the data.

As explained before, the noise introduced in the capture of data due to very different reasons is considered impossible to be completely avoided. There are several techniques to be used in the "intelligent" classification of data and detection of pattern[1].

As the experience with this sensors is not long, expert systems doesn't fit for this proposal. Moreover the range of situations they can afford is limited by capacity reasons. After considering other methods, neural networks appear as the best solution for this classification task, specially operating with noise data.

The inspection is made with a constant speed of the rope (1m/s), sampling every 2 mm. Considering the effect of the defects on the signal, it was decided to divide the signals in pieces of 32 points, representing 64 mm on the rope with a 3x32 matrix. This length is enough to include the defect completely and the probability to find several errors in the same sample is small. Diameter is sampled only once every 64 millimetres, so every sample has only a diameter value.

We have about 100 samples per defect signal arising from previous test, use as learning data. Moreover, rope could be getting up or getting down. So, these samples plus their symmetric ones correspond with both movement possibilities. The amount of data is huge and must be reduced in some way. Tests developed in both Departments[2,3,4], have shown the importance of the preparation of the data before introduce it in the network. The 3x64+1 points could be introduced, but the number of links and weights should be enormous and the results are worse than those produced when the input is pre-processed.

The best results are obtained extracting some representative characteristics from the data. After combining some of them, it was decided to represent the magnetic signals with 6 characteristics: media, deviation, skewness, kurtosis, number of peaks and maximum absolute value. Diameter is introduced as the difference between the measured and the initial values. Before this extraction is done, magnetic signals are normalised and filtered, keeping the highest value.

There are several kinds of neural networks[5,6,7]. The choice between supervised or unsupervised learning systems is clear in this case: we can produce enough trials to train the net with cases we know perfectly, so this information should not be lost. The most common algorithms of supervised feedforward networks are based on the well-known backpropagation[8]. Its use is widely extended and only discussed when applied to cases including real-time training. In our case, this is not necessary, since the neural net is trained off-line and then introduced in the system. The algorithm is reduced to a number of sums and products, quick enough for real-time inspection of the rope.

Other algorithms where proved too, as Radial Basis Function[9], ART[10], DLVQ[11,12], SOM[13], etc, presenting the same or worse behaviour than the selected one.

Once the backpropagation-style networks were selected, several algorithms were tested, both batch and real-time learning, with different topologies. Quickprop[14] and Resilient Propagation[15] behaves unstable for different trials. The best percentages of success were obtained with Back Propagation and an added Momentum factor. Results exceeds the 95% of success, validating the technique and making possible the automatic classification of the data.

4.- Neural Network Topology.

Each group of sensors produce an output which provides information about any defect but with specific interest in one group. So, inner coils are more adequate to detect internal broken wires or heart problems, external coils allow the differentiation of external broken wires and Hall-effect sensors mark corrosion and wear.

The diameter sensor is placed to accomplish the normative about diameter reduction, but its output can supply more information about the nature of the defect (stretching or lengthening), so it is also introduced.

The data raised from the magnetic sensors is similar, so it is introduced into three subnetworks with the same topology but different weights. This subnetworks have a hidden layer with 5 neurones and 3 output units in the last layer (see Figure III). This number has been selected after several tests as a compromise between generalisation and possibility of classification, according to the number of patterns of training.

This subnetworks should produce a interpretation of the data, distinguishing if there are defects or not, but only all the information could determine the type of defect accurately.

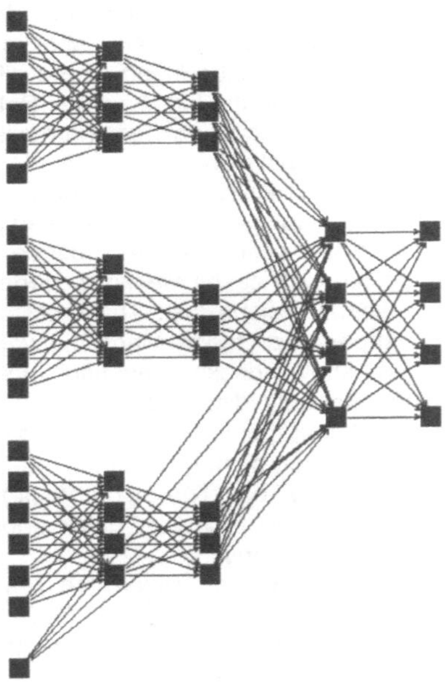

Figure III: Network Topology

From this point of view, the outputs of the subnetworks as well as the diameter signal feed other layers to produce the final identification. Every output unit defines a type of defect with an assigned importance. Other parts of the system (data base) valorise the rope status according to the defects detected in the proximity.

In order to train the whole network efficiently, all the subnetworks are connected and the error is back-propagated up to the inputs supplied by the sensors. This systems has proven a best behaviour than separated networks if enough number of test cases are provided.

This way, the network can be seen as a five layered Feed-Forward network, with 19 input neurones in the first layer, 12 units in the first hidden layer, 9 units in the second hidden layer, 4 units in the common hidden layer and 4 output units. Output units indicate currently, four possibilities: no defects, one broken wire, two or more broken wires and wearing defect.

The assigned training algorithm is backpropagation with a 0.15 momentum coefficient. Experiments has been carried out by randomly selection of two thirds as training data and the remainder as test data. We repeat this selection five times in order to avoid special cases which could introduce wrong behaviours, so results are an average over the five sets. Results are considered excellents with percentages always over 95% and usually close to 99%. The total data set manage about 1000 samples.

Currently the whole system is in the last stage of installation after the final tests.

5.- Conclusion.

A system to control the status of wire ropes in mines based on magnetic sensors is provided with automatic interpretation. After considering other possibilities, a neural network composed by 3 subnetworks is presented as the best solution. With this system, the automatic inspection of the ropes is possible without the collaboration of human inspectors.

As results are hopeful currently, we are evaluating the possibility to increase the number of detected defects: more broken wires, wires with mass loss, etc.

Likewise, another research way would be its application in coal and steel industry. In such a case, ladle speed compensation should be incorporated. Now, system is working with a fixed speed of 1m/s when rope is controlled, but it isn't the usual ladle velocity. Now, we are researching to raise a continuous inspection system, so we should compensate the speed.

References

[1] Schalkoff,R.J. (1992). Pattern Recognition: Statistical, Structural and Neural Approaches. Jhon Wiley and Sons, Inc.

[2] Menéndez,C., Ortega,F., González,C., Alvarez,A. (1993) "Aplicación de las redes neuronales al reconocimiento de patrones". IX Congreso Nacional de Ingeniería de Proyectos.Valencia.

[3] Ortega,F., Menéndez,C., Ariznavarreta,F., Taboada,J. (1993) "Proceso de Señal Procedente del Control On-

line de Cables de Grúas". IX Congreso Nacional de Ingeniería de Proyectos.Valencia.

[4] Ordieres,J.B., Ortega,F., Menéndez,C. & Alonso,E.. (1994) "Aplicación de las redes neuronales al mantenimiento de cables de grúa". XI Congreso Nacional de Ingeniería Mecánica.Valencia.

[5] Rumelhart,D.E., McClelland,J.L. & PDP Res. Group (1986). Parallel Distributed Processing: Explorations in the Microstructure of Cognition. Bradford Book. MIT Press. Cambridge.

[6] Kung,S.Y. (1993) Digital Neural Networks. PTR Prentice Hall. Englewood Cliff, New Jersey.

[7] Freeman,J.A. & Skapura,D.M. (1991). Neural Networks. Algorithms, Applications and Programming Techniques. Addison Wesley Pub. Co. Massachusetts. U.S.A.

[8] Rumelharht,D.E., McClelland,J.L. & P.D.P Res. Group. (1986). Parallel Distributed Processing. Vol. I. MIT Press

[9] Powell,M.J.D. (1987). "Radial Basis Functions for Multivariable Interpolation: a review". In Algorithms for Approximation. J.C.Mason and M.G.Cox (Eds.). Oxford. 143-167.

[10] Carpenter,G.A. & Grossberg,S (1987). "A masively Parallel Architecture for Selforganizing Neural Pattern Recognition Machine". Computer Vision, Graphics and Image Processing. 37, 54-115.

[11] Duda,R. & Hart,P.(1973) Pattern Classification and Scene Analysis. Wiley & Sons. 1973

[12] Schürmann,J. & Krebel,U. (1989) "Mustererkennung mit statistischen Benutzeroberfläche für einen Simulator konnektionistischer Netzwerke" Studienarbeit 746. IPVR. Universität Stuttgart.

[13] Kohonen,T. (1988) Self-Organisation and Associative Memory. Springer Verlag. 1988.

[14] Falhman,S.E. (1988) "Faster learning variations on back-propagation: an empirical study". In Sejnowski,T.J., Hinton,G.E. & Touretzky,D.S. (Eds.). Connectionist Model Summer School, San Mateo,CA. Morgan Kaufmann.

[15] Braun,H. & Riedmiller,M. (1992) "Rprop: a fast adaptive learning algorithm". In Proc. of the Int. Symposium on Computer and Information Science VII.

LIME KILN FAULT DETECTION AND DIAGNOSIS BY NEURAL NETWORKS

B Ribeiro*, E Costa and A Dourado
CISUC - Centro de Informática e Sistemas da Universidade de Coimbra
Phone:+ 351 39 7000040, Fax:+351 39 701266, e-mail:bribeiro@mercurio.uc.pt
Universidade de Coimbra, P-3030 Coimbra, Portugal

Abstract

Artificial neural networks have recently been used successfully for fault detection and diagnosis in chemical processes. In this paper, we present a study on fault detection and diagnosis of an industrial lime kiln which is a complex highly nonlinear process within the pulp and paper industry. We show the capability of neural networks to learn faults which can occur during steady state kiln operation, their adaptation to different input distributions involving nonlinear mappings and their capability to spontaneously generalize. We compare the performance of two arquitectures, nameley BPNN (Back Propagation Neural Network) and RBFNN (Radial Basis Function Neural Network), and investigate several topologies. Through this study, it can be concluded that the RBFNN arquitecture learns faster, with less error and performs better at classifying kiln malfunctions.

1 Introduction

Artificial neural networks (ANNs) have been shown to be very practical and useful in process engineering applications. They have learning abilities, can adapt to input distributions which involve arbitrary nonlinear mappings and do not rely on *a priori* process models. Besides they have the capability to store information about state process deviations caused by disturbances and, therefore, a diagnostic ability with respect to faults that occur in a process. In the area of chemical processes, numerous examples of fault detection and diagnosis have been successfully reported [1, 2, 3, 4]. These applications have used BPNN while others comprise RBFNN arquitectures [5, 6].

This paper describes the work done on using ANNs for fault detection and diagnosis of an industrial process - a lime kiln within the pulp and paper industry. The lime kiln is often subjected to several perturbations which difficult its control, increase operation costs and decrease the production level. The process is a multivariable one, and the responses to kiln input variables (manipulated variables) are difficult to measure and are not rigorous due to the harsh kiln environmental conditions (dust, heat, etc.). Additionally the possibility to measure the product quality on-line does not exist. These factors allow us to recognize the need for kiln automatic control as well as for a system of kiln fault detection and diagnosis as it will be shown herein.

2 Lime Kiln Process

The lime kiln is an essential component of the kraft recovery in the the causticizing plant of pulp and paper industry. Its main purpose is to recover CaO from the $CaCO_3$ according to the following chemical reaction: $CaCO_3 \rightarrow CaO + CO_2$. Ideally, the kiln should be able to produce sufficient, high-quality lime to meet the needs of the causticizing operation while minimizing its energy consumption and decreasing environmental pollution (i.e. total reduced sulphur (TRS) emissions). As shown in Fig. 1 the lime kiln is a very long direct-contact heat exchanger. Its function is to dry the incoming calcium carbonate ($CaCO_3$) mud and to heat it to a sufficient high temperature in order $CaCO_3$ to be calcinated into lime (CaO) and carbon dioxide (CO_2).

*Partially supported by Junta Nacional de Investigação Científica (JNICT) and PORTUCEL.

During lime kiln operation, heat energy is supplied by the combustion of fuel, which enters with the primary air at the hot kiln extremity. This energy is transferred to the $CaCO_3$ slurry which enters at the cold end extremity and flows down countercurrently to the gases, due to kiln inclination and rotation speed. The lime

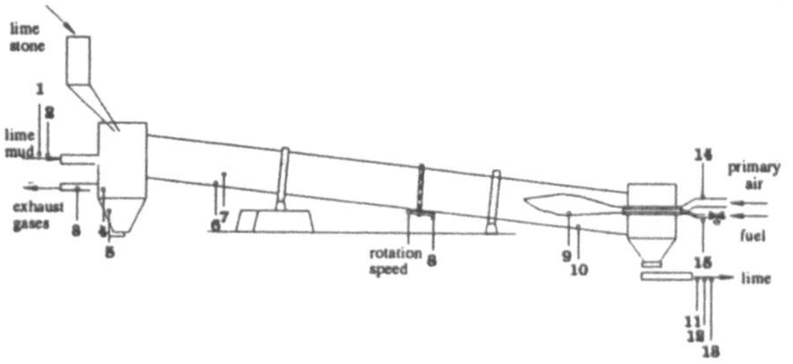

Fig. 1: Schematic representation of a lime kiln.

kiln is equipped with metallic chains which increase the heat transfer from combustion gases to solids. Table 1 presents the meaning of the measurements indicated in Fig. 1. Disturbances such as changes in lime mud

Measurement	Nr.	Symbol/Units
Lime mud feed flow	1	L (kg/s)
Mud feed humidity	2	Hum_L (%)
Draft flow rate	3	D (kg/s)
Cold end temperature	4	CET C
Excess O_2	5	exc O_2 (%)
Chain humidity	6	$Hum_{C}h$ (%)
Chain solids temperature	7	T_sCh (C)
Kiln rotation speed	8	RPM (r.p.m.)
Burning gases temperature	9	TBZ (C)
Hot end temperature	10	HET (C)
Product temperature	11	T_sProd (C)
Residual $CaCO_3$	12	$CACO_{3R}$ (%)
Production rate	13	CaO/d (ton/d)
Primary air	14	N (kg/s)
Fuel flow rate	15	F (kg/s)

Table 1: Lime kiln measurements.

feed composition and humidity, random process disturbances, long time constants and delays cause kiln operation problems. In a previous work [7] a control system based on an IMC structure which incorporates neural network models was proved to perform well with regards to robustness against disturbances. However, additionally to the kiln automatic control, a kiln monitoring system with learning abilities should be able to detect kiln malfunctions during its operation and identify their causes. Moreover, it should support kiln operator in ensuring smooth kiln operation which is essential to attain a high quality product. This system is based on neural networks which seem appropriate to tackle this problem as it will be explained in next section.

3 Kiln State Diagnosis

Usually kiln monitoring and diagnosis tasks have been left to the operator who should be able to analyse all possibles symptoms of kiln malfunctions and identify their causes. The multivariable and complex relationships involved, hampered them to take the correct decisisons in such critical situations. An automatic system can 'observe' the equipment behaviour in a more versatile and flexible way. Some of the kiln faulty process

Process state	Cause	
	Symbol	Description
'Underburned' lime	F^-	low
'Overburned' lime	F^+	high
High TRS (pollution)	D^-	low
Energy waste	D^+	high
Formation of lime balls and rings	RMP^-	low
Too diluted lime mud	RMP^+	high
'undercharged' kiln	L^-	low
'overcharged' kiln	L^+	high

Table 2: Kiln faulty process states.

states are illustrated in Table 2 and the corresponding possible causes of process upsets are listed in its second column as follows: lime mud feed flow rate, fuel flow rate, draft flow rate and kiln rotation speed, which are critical inputs to the operator running kiln conditions. The fault detection problem prompts us a pattern recognition task which can be resolved by ANNs.

3.1 Neural Networks on Fault Detection

Due to their structural and theoretical properties, ANNs can solve nonlinear problems by learning. They have proved to be very successful to solve technological problems and to have powerful capabilities in classification tasks. Clearly the neural networks are able to satisfy the requisites of monitoring, fault detection and diagnosis as they can 'observe' kiln variable deviations from their steady state values and, thus, detect the existence of faults and identify their causes. Figure 2

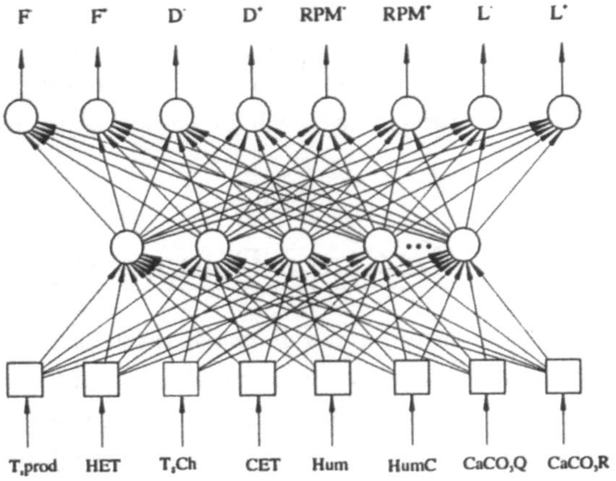

Fig. 2: Kiln diagnosis by ANNs.

shows a schematic representation of the neural network used to kiln monitoring and diagnosis. The network inputs are detailed in Table 3. The network outputs are variations (low, high) of kiln input conditions. In

Symbol	Measured variable
$T_s Prod$	Product temperature
HET	Hot end temperature
CET	Cold end temperature
$T_s Ch$	Chain temperature
HumL	Feed mud humidity
HumC	Chain solids humidity
$CaCO_3 Q$	$CaCO_3$ in burning zone
$CaCO_3 R$	Residual $CaCO_3$

Table 3: Neural network inputs.

Paragraph 3.3 the training algorithms used herein are outlined. Knowledge of the faults to be learned by the neural network are obtained from simulated data under normal and faulty conditions. This information is used together with the causes of faults (of the process to be diagnosed) as the teaching patterns. Following a training phase, through which the connection weights are adjusted, a new vector of sensor measurements can be sent to the input neurons and classified. It corresponds to the testing phase which will take very little computation time. This is an issue which plays an important role in the context of critical industrial situations.

3.2 Neural Network Training Patterns

The process input-output data (i.e., faults and causes) are pairs $(\mathbf{u}, \mathbf{y}_d)_i$, $i = 1, 2, \ldots, N_p$ which are presented to the network in order to allow it to learn the correct correspondence among kiln malfunctions and their causes. In order to obtain the neural network training data a distributed parameter kiln model was simulated using parameter correlations taken from literature as well as mill's kiln measurements. This was validated against real factory data. To obtain the relevant fault data for this process several simulations were carried out by introducing deviations ($-20\%; +20\%$) in all of the critical input variables listed in (Table 2). Table 4 shows the target patterns used to train the neural network in which a '1' is associated with a faulty condition and a '0' with a normal condition. We had considered a multicausal effect on the observed deviations from steady state and the uncertainty in data caused by white noise of zero average and standard deviation of 1%. However, it should be emphasized that the normal state was not used for training.

Symbol	Target output							
F^-	1	0	0	0	0	0	0	0
F^+	0	1	0	0	0	0	0	0
D^-	0	0	1	0	0	0	0	0
D^+	0	0	0	1	0	0	0	0
RPM^-	0	0	0	0	1	0	0	0
RPM^+	0	0	0	0	0	1	0	0
L^-	0	0	0	0	0	0	1	0
L^+	0	0	0	0	0	0	0	1
normal	0	0	0	0	0	0	0	0

Table 4: Neural network target output.

3.3 Training Algorithms

Two types of neural networks were used in this work: BPNN and RBFNN. This means that different algorithms to train the networks had to be considered. In the former, a backpropagation learning algorithm was used using a technique of improving convergence (adapted version); in the latter, a two step procedure was adopted consisting of a first phase unsupervised algorithm followed by a least mean square optimization algorithm.

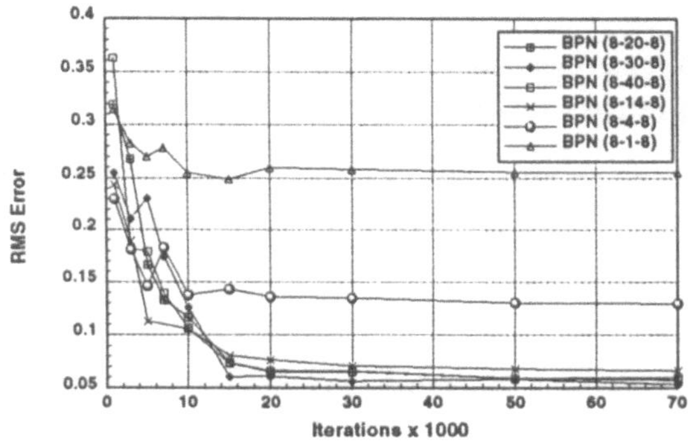

Fig. 3: Number of time steps for one hidden layers.

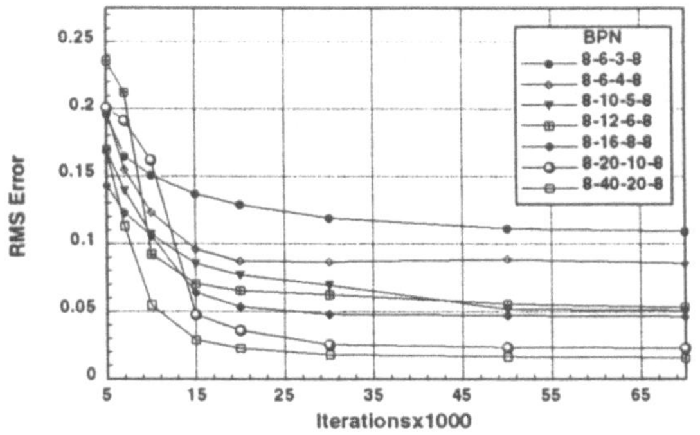

Fig. 4: Number of time steps for two hidden layers.

4 Simulation Results

In the BPNN arquitecture we had carried out simulations accounting for various topologies, i.e., a different number of hidden layers and neurons per layer. Figs. 3 and 4 show the results obtained after running a few simulation experiments. The networks which could recognize the deviation patterns were those trained with an acceptable error (≤ 0.05). Hence they could identify correctly the faulty conditions. It was possible some degree of generalization for never-seen test data, for instance, they could classify the normal state which has not been used in the training data. The results obtained with the RBFNN arquitecture with several number of RBFs are illustrated in Fig. 5. In this case, it was observed rapid convergence on the training task particularly due to the local 'tunning' of the RBF which is more suited for fast learning.

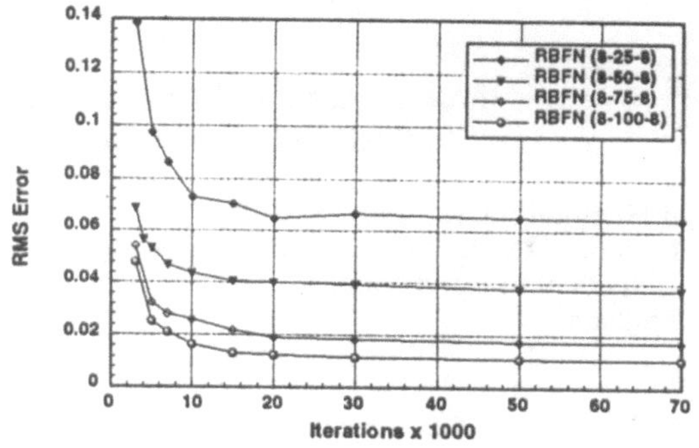

Fig. 5: Number of time steps for RBFNN approach.

5 Concluding Remarks

We have used and compared two arquitectures (BPNN and RBFNN) for fault detection and diagnosis of an industrial lime kiln. The neural networks used are able to accomplish these goals in 'observing' kiln state variable deviations from the steady state values and, thus, to detect the existence of process upsets and identify their causes. The main advantages of their use herein rely on their ability to classify faults which have not been used in the training set and generalize without the development of models (either qualitative or quantitative) to represent process knowledge. Besides, the in-built parallel processing allows fast computation which is essential in industrial cases. It can be concluded that the RBFNN arquitecture learns faster and performs better at classifying kiln malfunctions during its operation.

References

[1] Hoskins, J. C., Kaliyur, K. M., Himmelblau, D. M.: AIChE Journal, 37 1, 137 (1991).

[2] Watanabe, K., Matsuura, I., Abe, M., Kubota, M. Himmelblau, D. M.: AIChE Journal, 35 11, 1803 (1989).

[3] Ungar, L. H., Powell, B. A., Kamens, S. N.: Computers and Chemical Engineering, 14 4, 561 (1990).

[4] Venkatasubramanian, V., Vaidyanathan, R., Yamammoto, Y.: Computers and Chemical Engineering, 14 7, 699 (1990).

[5] Leonard, J. A., Kramer, M. A.: IEEE Control Systems, April, 32 (1991).

[6] Leonard, J. A., Kramer, M. A.: IEEE Expert, April, 44 (1993).

[7] Ribeiro, B., Dourado, A., Costa, E.: IJCNN'93, Vol II, 2037, (1993).

INVESTIGATIONS INTO THE USE OF WAVELET TRANSFORMATIONS AS INPUT INTO NEURAL NETWORKS FOR CONDITION MONITORING

J MacIntyre, BSc, IEng, AMBCS
School of Computing and Information Systems,
University of Sunderland
England

J C O'Brien, BSc, MPhil
School of Mathematical and Information Sciences,
Coventry University
England

ABSTRACT: Condition monitoring has become widely accepted in industry, with vibration providing the most useful condition parameter. Analysis of these vibration signals usually involves Fourier transforms and, occasionally, neural networks. However, a problem when using Fourier transforms as input into a neural network is the size of the input data set. Wavelet transforms provide an alternative which allows for a dramatic reduction in the size of the data set. This paper will explore the feasibility of using wavelet transforms as input into neural networks for condition monitoring.

1 Introduction

It has become widely accepted in industry that maintenance of process critical or high capital value plant can no longer be cost-effectively carried out without the use of condition monitoring. Furthermore, for most rotating machinery, vibration provides the most useful condition parameter, since changes in the condition of the machine will produce changes in how it vibrates.

However, the majority of vibration signals are too complicated to be utilized in their raw state and need to be processed in some way. The usual way of doing this is to use Fourier analysis to produce a frequency spectrum from which an experienced engineer can deduce a lot about the condition of the machine. There are many examples of attempts to improve upon the success of this technique by processing the resultant spectrum via some form of neural network. However, a major problem is determination of data to use as input - the size of the complete spectrum is usually prohibitive and selection of sub-sets is very subjective.

However, a new method of processing vibration data, known as wavelet transformation, has previously been shown by the authors to be an effective means of distinguishing machines in good and poor condition (4). Wavelet transforms can be considered as the localised equivalent of Fourier transforms and work on the principle that all signals can be constructed from sets of local signals of varying scale and amplitude, but constant shape.

The overall effect of the transformation is to transfer the data from one domain to another; whereas the Fourier transform moves from a time domain to a frequency domain with sines and cosines as the basis functions, the wavelet transform moves data from a space domain to a scale domain where wavelets are the basis function. One advantage of wavelet transformations is the ability to threshold the transformed data, resulting in a significant reduction in the size of the data set. This should overcome the problem of neural network input selection, since, in theory, the entire non-zero element of the threshholded wavelet transform is serviceable.

This paper will describe attempts to determine exactly how wavelet transformations can be employed to train neural networks to assess the condition of a given set of machinery.

2 An Introduction to Wavelets

2.1 What are Wavelets?

A family of wavelets consists of signals of a specified shape each of which is a scaled and translated version of a unique model, W(x), and can be identified by one label for position, k, and one for scale, j, (1). In general each wavelet within a family is of the form

$$W_{jk}(x) = W(2^j x - k) \quad \text{for } 0 \leq j \quad \text{and } 0 \leq k \leq 2^j - 1$$

The fundamental basis of wavelet theory is that any random signal, f(x), is of the form

$$\begin{aligned} f(x) &= a_0 + a_1 W(x) + a_2 W(2x) + a_3 W(2x-1) + a_4 W(4x) + \\ &\quad a_5 W(4x-1) + a_6 W(4x-2) + a_7 W(4x-3) + a_8 W(8x) + \ldots \\ &= a_0 + \sum_{j=0}^{\infty} \sum_{k=0}^{2^j-1} a_{2^j+k} W(2^j x - k) \end{aligned}$$

where a_0 is the overall mean of the signal and the remaining components are scaled members of the chosen wavelet family.

There are many different possible families of wavelets, each specified by their own set of wavelet coefficients. However, a commonly used one is that defined by Daubechies, where each member of the wavelet family is mutually orthogonal to every other member. The exact shape of these wavelets is defined by the 'order' of the family and the hierarchy, or level, within the family to which each individual wavelet belongs. In general, the higher the order the smoother the shape of the wavelets in the family, and the greater the hierarchy the narrower the wavelet.

The width of the wavelet determines the hierarchy to which the wavelet belongs. The data in the space domain must be an integer power of 2, i.e. of length 2^n. The wavelets are organised into a sequence of hierarchies of length 2^j where $1 \leq j \leq n-1$. The first hierarchy ($j=1$) is of length 2 and consists of wavelet numbers 3 and 4, the second hierarchy ($j=2$) is of length 4 and consists of wavelet numbers 5-8, etc. So if the wavelet is one-quarter the width of the space interval the wavelet belongs to the hierarchy containing four wavelets, i.e. wavelet numbers 5-8. Similarly if the wavelet is 1/512th the width of the time domain it belongs to the hierarchy containing wavelet numbers 513-1024, which contains 512 wavelets. Hence, it can be seen that the narrower the wavelet, the larger the size of the hierarchy to which it belongs. The position of the wavelet determines the wavelet number. Consider again the wavelet of width one-quarter of the space domain. If this wavelet starts on the left of the space domain, it is wavelet number 5; if the wavelet starts a quarter of the way through the space domain, it is wavelet number 6; if the wavelet starts half way along the space domain, it is wavelet number 7; and if the wavelet starts three-quarters of the way through the space domain, it is wavelet number 8.

Examples of Daubechies wavelets are shown in Figure 1. Full details on the derivation of wavelets and their coefficients is given by Daubechies [1], Mallat [2] and Newland [3].

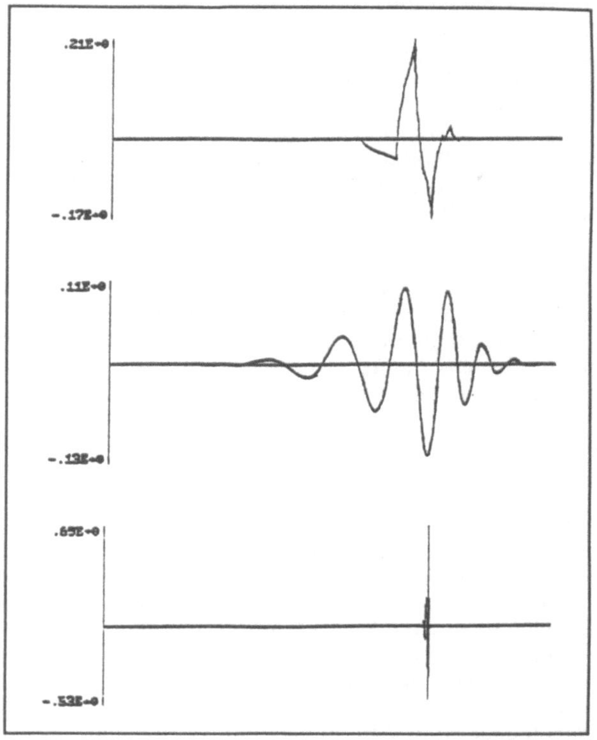

***Figure 1** - (top) wavelet number 24, order 4; (middle) wavelet number 24, order 20; (bottom) wavelet number 800, order 20 respectively.*

2.2 What are Wavelet Transforms?

Wavelet transforms are linear operators which operate on a 1-dimensional data vector or a 2-dimensional matrix to produce a new vector/matrix of the same size. The resulting vector/matrix of wavelet coefficients indicates the amplitude of each wavelet in the family, so that the original signal can be reconstructed by combining wavelets of these amplitudes together. Because wavelets are local functions many of the components of the transformed data will be zero, hence the original signal can be coded using less data points. It should be noted that is not correct to simply use the first N points in the transformation, as might happen with a Fourier transform; wavelet transforms should be truncated strictly according to the amplitude of their components.

The intricacies of the wavelet transform are detailed by Daubechies [1], Mallat [2], and Newland [3].

2.3 Wavelets for Condition Monitoring

The use of wavelets for condition monitoring is not as straight-forward as the use of Fourier analysis. Wavelets can be used to compare several signals only if each of the signals start at the same position in time or space. Consider two sets of vibration data taken from identical machines in identical condition, but each starting at a different point in the machine cycle. Theoretically, the resultant time signals will be identical except for a shift in location. The effect of this shift is that any transformed wavelet component remains in the same hierarchy because it is the same width, but the exact wavelet number alters because of the shifted position of the wavelet. Hence, simple wavelets cannot be used for comparison [3] because it is impossible to ascertain whether any change in the wavelet domain is due to a change in the signal being observed or simply due to a shift in the location of the signal. Therefore, for the comparison of data, such as in the case of condition monitoring, each data set must be matched so that it starts at the same position in time or space before it is transformed to the wavelet domain. In the data used here an external trigger, in the form of a small optical tachometer, ensured that the recorded signals started at the same point in the machine cycle.

3 Analysis of Wavelets using Neural Networks

The wavelet data used in this work comprised 1024 individual wavelets, and it was felt that an input vector of 1024 dimensions would create difficulties in training neural networks. Therefore, four separate approaches were adopted to pre-processing the wavelet data to reduce the dimensionality. The values extracted in the four approaches were:

i) The wavelet number of the 20 largest wavelet amplitudes ('Wavelet No.');

ii) The amplitude of the 20 largest wavelet amplitudes ('Max. Amp.');

iii) The mean amplitude (above a threshold) in each hierarchy ('Mean Amp.');

iv) The proportion of wavelets in each hierarchy above a threshold ('Proportion').

This then produced four data sets, with 20, 20, nine and nine inputs respectively. In each case, two outputs were defined, simply 'good' or 'bad' condition. The data contained approximately even numbers of 'good' or 'bad' condition records.

Network development for this work was carried out using NeuralWorks Professional II Plus, a neural networks CASE tool which allows rapid prototyping of various neural architectures. The networks used for this evaluation were standard multi-layer perceptrons using back propagation, as described by Hinton [5]. The architectures were 20:10:2 for the first two networks, and 9:5:2 for the last two. The RMS error threshold was set to 0.001 in each case, and the networks set to train for 100,000 iterations initially. Cross-validation techniques were used to prevent over-training.

Table 1 below outlines the performance of each of the networks in classifying the 'good', 'bad', and total data.

Data Type	% Good	% Bad	% Total
Wavelet No.	100	23	59
Max. Amp.	40	79	58
Mean Amp.	0	100	52
Proportion	0	100	52

Table 1 *- Percentage of 'good', 'bad', and total data correctly classified by each of the four networks.*

Both networks based on mean amplitudes and proportion of wavelets over a threshold did not converge successfully to an acceptable RMS error value, despite the use of a number of techniques to avoid local minima and optimise training. Consequently, the performance of these networks is poor. The networks based on wavelet number and maximum amplitudes did both converge; however, their generalisation properties still leave something to be desired. In overall performance, use of the wavelet number showed the best results; however, the authors believe that the network based on maximum amplitudes shows the greatest promise in generalisation. Further work is continuing to try to optimise the architecture of these networks, and to examine the potential for using other architectures, such as Radial Basis Function networks, to improve performance.

Figure 2 below shows a screenshot from NeuralWorks of the Wavelet Number network, while Figure 3 shows the poorly-performing Proportion network.

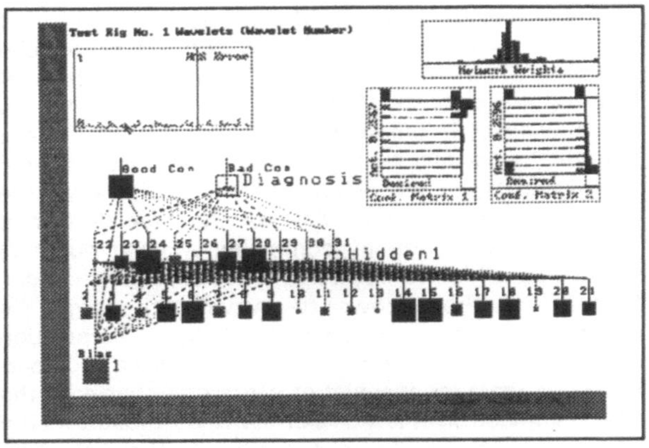

Figure 2 *- multi-layer perceptron network based on wavelet numbers data set.*

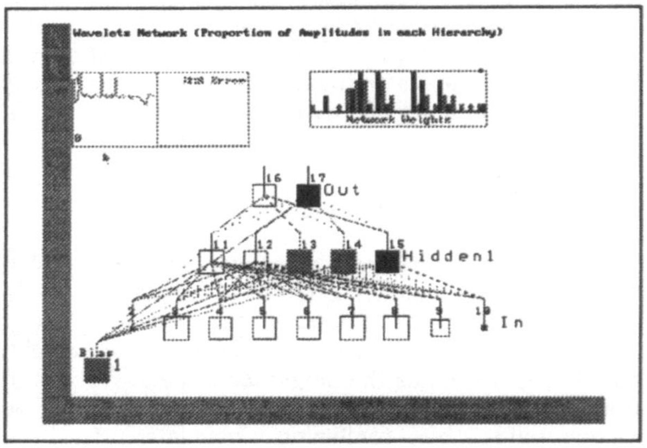

Figure 3 *- multi-layer perceptron network based on proportion data set.*

4 Conclusion

The work shows that wavelet transforms can be used as an alternative to Fourier transforms for the processing of condition monitoring data, and have an advantage where the element of the signal which is descriptive of the fault is not globally present in the signal. The data can be efficiently compressed to produce a signal indicative of a developing fault.

However, wavelet transforms cannot be interpreted as easily as Fourier transforms, and a reliable way of analyzing this data is required. Neural networks represent a possible solution to this problem, and the work shows some potential in indicating how to pre-process wavelet data for use as inputs to a neural network, and the difficulties of achieving adequate classification performance. Further work will build on the results shown here using different neural architectures.

References

[1] Daubechies I: The Wavelet Transform, Time-Frequency Localization and Signal Analysis. IEEE Transactions on Information Theory, Vol 6, No 5, Pg 961, Sept 1990.

[2] Mallat S G: A Theory for Multiresolution Signal Decomposition: The Wavelet Representation. IEEE Transactions on Pattern Analysis and Machine Intelligence, Vol 11, No 7, Pg 674, July 1989.

[3] Newland D E: An Introduction to Random Vibrations, Spectral and Wavelet Analysis (3rd Edition), Longman Scientific and Technical Publishers, 1993.

[4] O'Brien J C and MacIntyre J, Wavelets: An Alternative to Fourier Analysis. Seminar: Vibrations in the Power Industry, IMechE, 1994.

[5] Hinton G, How Neural Networks Learn from Experience. Scientific American, September 1992.

GENETIC ALGORITHM FOR NEUROCOMPUTER IMAGE RECOGNITION

Ernst M.Kussul, Tatyana N.Baidyk

Department of Neural Information Processing Systems,
V.M. Glushkov Institute of Cybernetics, Ukrainian Acad. Sci.,
pr. Acad. Glushkova 40, Kiev, 252207, Ukraine

Abstract

Genetic algorithms for optimization of feature set and internal structure of neural networks are considered. Results of experimental investigation of genetic algorithms are given. Experiments show that performance of neural networks after such optimization substantially increases.

Introduction

Many authors who used neural networks for pattern recognition consider that the main advantage of such networks is the automatic formation of feature set due to network learning [1,2]. The difficulties of this approach are connected with the fact that usually learning algorithms represent different variants of gradient descent method which gives rather good results for smooth cost functions. Real problems dealt with recognition of noisy objects have unsmooth cost functions. In such case it is better to use stochastic algorithms and especially genetic algorithms. Genetic algorithms are not sensible to the local properties of the cost function and algorithm is very proper to find the global extremum [3].

In the middle of the 60-s it was suggested to use genetic algorithm for optimization of the input structure of 3-layered perceptron [4,5]. In fact it's equivalent to the optimization of a feature set. Genetic algorithm implementation demands very high performance hardware but authors of perceptrons have modelled them on usual computers. That is why the potencial of genetic algorithms has not been realized.

Genetic algorithm for feature set optimization

We used genetic algorithms for feature set optimization and for optimization of perceptron input structure.

Feature set optimization.

The first problem was to optimize the feature set to recognize ten handwritten words which were written by different authors.

The initial set consisted of 40 features which were lines and arcs with different orientation. Apriori we didn't know what feature combinations would give us the best recognition results.

The task of genetic algorithm is to search for such combinations.

We shall define the "being" as 40-bit binary vector ($E(e_1,e_2,...,e_{40})$) in which every bit corresponds to one feature. Feature i is included into the set for recognition if $e_i = 1$ and feature i is not include to the set if $e_i = 0$. We shall define mutation m_i as changing of the i-th component of vector E (zero to one or one to zero). We explored the "single-sex" genetic algorithm in which every offspring was born from one parent. Hereat stochastic mutations take place. The mutation probability of every feature $p(m_i)$ depends only upon the cost function Q of parent:

$$(\forall i)\ p(m_i) = p(Q). \qquad (1)$$

The cost function Q was calculated as the 4-th power of the error number:

$$Q = c\ N^4, \qquad (2)$$

where N is the error number; c is a constant. Error number was counted during test recognition of ten handwritten words. To calculate the cost function of some being the neural network was trained to recognize words using feature set of that being. Words for training were written by 8 persons. For test we have used new words which were written by another 4 persons.

To model the natural selection we used the algorithm in which the probability of offspring generation was proportional to 1/Q, so that a being with a smaller cost function value usually generated more offsprings than a being with larger cost function.

Beings of initial generation were made by random number generator.

We made two series of experiments with our genetic algorithm. The results are given in table 1.

Table 1

k	1	2
	s=10	s=32
g	Q	Q
1	1056	834
2	659	704
3	540	792
4	662	576
5	651	630
6	893	692
7	813	538
8	475	624
9	555	673
10	397	614
11	469	631

In this table k is the experiment number; s is the being number in every generation; g is the generation number; Q is the average value of the cost function for one generation.

In these experiments high performance neurocomputer was used. The whole cycle of learning and recognition demanded several minutes per one being.

Optimization of input structure of perceptron.

We used perceptron with 4 layers [6,7] for optical character recognition. Connections between the 3-rd and the 4-th layer of perceptron were modifiable (synaptic weights of these connections were changed by training). The perceptron had 256 neurons in the third layer. Connections between other layers were not modifiable. We denote such structure as innate structure. Genetic algorithm was used for optimization of this structure. The characters which often are confused by recognition (e.g.I and l, b and h) were selected for training and examination of perceptron. Both training set and examination set included 120 character samples. The cost function was calculated as the sum of recognition errors during 5 examination trials. So the maximum number of errors could be 600.

Perceptron structure was optimized by the following parameters of the genetic algorithm.

In the first case, the parent number was PN=4, mutation number was MN=25. Each mutation consisted in elimination of one neuron of 3-th layer and addition of one new neuron which had random connections with the neurons of the first and the second layers. Every parent had 5 offsprings (the whole size of generation was GSZ=PNx5=20) and after reproduction it died. Mean cost function of parents for different generations is presented in Table 2 (case 1).

Table 2

case	1	2	3	
	PN=4 MN=25	PN=6 MN=25	PN=6 MN=7	PN=9 MN=7
g	GSZ=20	GSZ=21	GSZ=21	GSZ=45
0	353	360	360	355
20	323	275	291	281
40	301	272	262	277
60	299	270	258	273
80	302	268	254	272
100	280	264	254	271
120	281	264	250	262
140	284	261	250	253

In the second case the parent number was 6. The best parent had 6 offsprings, the second best parent had 5 offsprings, and so on. The worst parent had only 1 offspring. After reproduction each parent survived and could compete with offsprings of the next generation. The number of mutations was the same as in the first case. Mean cost function for this case is presented in Table 2 (case 2).

In the third case the number of mutations varied. The best parent had 7 mutations by generation of each offspring. The second parent had 14 mutations and so on. The worst parent had 42 mutations (Table 2, case 3).

In the forth case the number of parents was increased till 9 and the total number of springs was 45 (Table 2, case 4). In all the cases we can see the progress and decreasing of the error number.

Discussion

The obtained results show that evolution model gives the opportunity to select more optimal sets of features and innate structure of perceptron-like neural network. The neurocomputer performance was not high enough for detailed investigation of genetic algorithm properties. For such perpose it is necessary to increase the performance of neurocomputer so that the whole cycle of training and test recognition equals 1 s (or less). At present there are technical possibilities to create such neurocomputer [8].

Perceptron structure is more convenient for genetic algorithms and permits to implement them on personal computer, but hardware implementation could give better results.

Acknowledgements

We want to thank Dr. Dmitrij Rachkovskij for neurocomputer software that was used by us in implementation of genetic algorithm and Japanese company "WACOM" for providing with neurocomputer which was developed by this company within joint project in which this paper authors took part.

The research described in this publication was made possible in part by Grant number U4M000 from the International Science Foundation.

References

1. Fukushima, K. Applied Optics, 26(23), 4985, (1987).

2. Hecht-Nielsen, R. Neurocomputing. Addison-Wesley, Reading, MA, 1990.

3. Kussul, E.M., Luk, A.N. Soviet science review. Scientific developments in the USSR. 3 (3), 168 (1972).

4. Clopf, A.H., Gose, E.E. IEEE Trans. Syst. Sci. and Cybernet., 5 (3), 247 (1969).

5. Mucciardy, A.N., Gose, E.E. IEEE Trans. Electron. Computers, EC-15 (N2), 257 (1966).

6. Kussul, E.M., Baidyk, T.N., Lukovitch, V.V.,Rachkovskij, D.A. Adaptive high performance classifier based on random threshold neurons. Proc. of Twelfth European Meeting on Cybernetics and Systems Research (EMCSR-94), Austria, Vienna, April 5-8, 1994 in R.Trappl (ed.): Cybernetics and Systems'94, World Scientific Publishing Co.Pte.Ltd, Singapore. - P.1687-1695.

7. Kussul, E.M., Baidyk, T.N. Neural Random Threshold Classifier in OCR Application. Proc. of the Second All-Ukrainian Intern. Conf."UkrOBRAZ'94", Kyjiv, Ukraine, December 20-24, 1994. - P.154-157.

8. Kussul, E.M., Rachkovskij, D.A., Baidyk, T.N. Associative-projective neural networks: architecture, implementation, applications. Proc. of Fourth Intern. Conf. "Neural Networks & their Applications", Nimes, France, Nov.4-8 1991. - P. 463-476.

FEATURE MAP ARCHITECTURES FOR PATTERN RECOGNITION: TECHNIQUES FOR AUTOMATIC REGION SELECTION

B. Martín-del-Brío

Tecnología Electrónica
E.U.I.T.I., Universidad de Zaragoza
50009 Zaragoza, Spain

N. Medrano-Marqués, J. Blasco-Alberto

Dpto. de Ingeniería Eléctrica e Informática
Facultad de Ciencias, Universidad de Zaragoza
50009 Zaragoza, Spain

Abstract

In the present paper we address several issues about data processing with Kohonen maps. First, this neural system and multidimensional scaling are compared. Next, several unsupervised techniques for delimiting domains on a Kohonen map are described. The different procedures are illustrated by using real-world data.

1. Introduction.

The Self-Organizing Feature Map (SOFM) or Kohonen network [1] is a well known competitive unsupervised neural system used for pattern recognition, vector quantization, process monitoring [2] or as a tool for exploratory data analysis [3]. In this paper we address several issues about computing with SOFM architectures. First, we compare SOFM with some conventional statistical methods, as multidimensional scaling [4]. Next, we present several techniques for the unsupervised labeling of the SOFM neurons, in such a way that different regions or domains are delimited on the surface of the Kohonen map. Thus, those domains indicate the main features existing on the map, without requiring any prior (or little) knowledge of the problem.

The paper is organized as follows. In Sect. 2 we give a brief introduction to the SOFM model, which is compared to some statistical methods. In Sect. 3 the techniques for delimiting regions, quoted above, are described. Finally, In Sect. 4 the conclusions of our work are presented.

2. SOFM and Statistical Techniques.

The SOFM projects a high dimensional input space onto an usually one or two dimensional output space, represented by a discrete lattice of neurons (or units), usually arranged in a rectangular layout. Thus, this model transforms similarity relations between input patterns in neighbourhood relations on the map, in such a way that similar input patterns are mapped onto near neurons, preserving the topology of the input space. The Kohonen network is being applied to, for instance, speech recognition [1], robot control [5] or recognition of financial patterns [3].

Let us suppose a rectangular Kohonen network composed by $n_x \times n_y$ neurons, every one labeled with a pair of indices $\mathbf{i} \equiv (i,j)$ which give their location (x,y) on the map. In the recall phase, every neuron (i,j) computes the similarity (usually Euclidean distance) between the input vector $\mathbf{x}$, $\{x_k \,/\, 1 \le k \le n\}$, and its synaptic weight vector $\mathbf{w}_{ij}$ (or reference vector); the neuron $\mathbf{m}$ whose $\mathbf{w}_{ij}$ is more similar to $\mathbf{x}$ is called 'the winner'. In the learning phase, the weights of the neurons belonging to a neighbourhood of the winner are adjusted following a simple learning rule.

For illustration purpose, we present the result of processing by means of SOFM data from a real world problem: The Spanish banking crisis of 1977-85, when no fewer than 58 out of 108 Spanish banks were in crisis (this critical situation has been compared to the crash of 1929 in USA) [3]. The database used contains nine financial ratios from 66 banks (29 of then bankrupt). After the training of a 14x14 Kohonen network, we present the 66 patterns to the map, and label the winning neuron with the bank pattern number (Fig. 1). We can see how the SOFM has distinguished the solvent and crisis situations, projecting healthy and bankrupt patterns in different regions on the map.

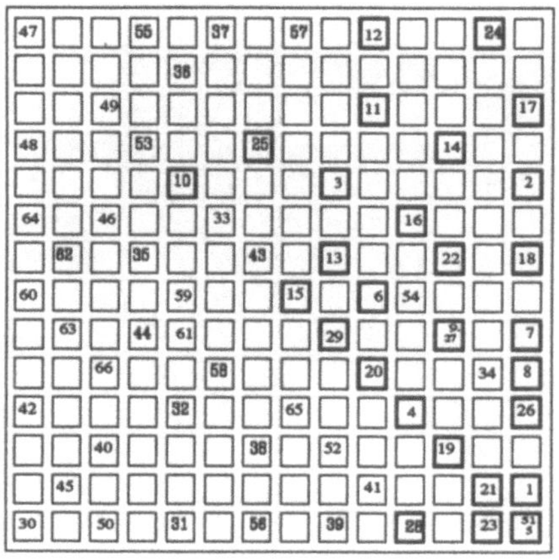

Figure 1. Patterns from the Spanish banking crisis on the SOFM. 1 to 29 are bankrupt banks, 30 to 66 are solvent.

SOFM and several conventional statistical methods frequently yield similar results. For instance, when the number of neurons is lesser than the number of patterns, these ones are clustered by the Kohonen units. Thus, SOFM with a few neurons works as a clustering algorithm [1].

However, if the map has a great number of neurons, the patterns are mapped onto different units, standing distributed on the map surface (Fig. 1). We say that SOFM has projected the (high-dimensional) input space onto a two-dimensional space (the map). In that sense, SOFM and multidimensional scaling (MDS), a conventional statistical technique [4], lead to surprisingly similar results (compare Fig. 1 and Fig. 2). MDS constructs a map of the locations of a set of patterns from a data matrix that specify how different they are (dissimilarities). Thus, SOFM and MDS lead to a map as their final result, but while in a MDS map distances between patterns are proportional to their similarity, in a Kohonen map patterns are uniformly distributed on its surface. Consequently, in a MDS map there are frequently areas where patterns are too close to be correctly distinguished (Fig. 2).

It is reasonable that conventional techniques and neural models arrive at similar solutions, because both try to solve similar problems. Establishing relationships between neural networks and traditional methods is very useful (and necessary) for both fields; much work is being carried out in this direction [6, 7].

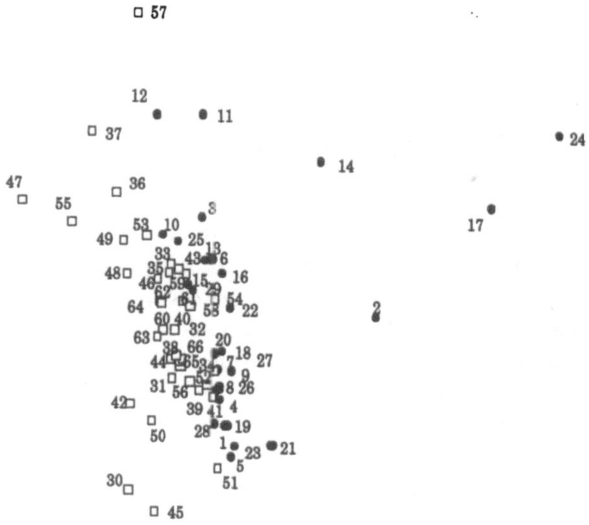

Figure 2. MDS of the patterns from the banking crisis.

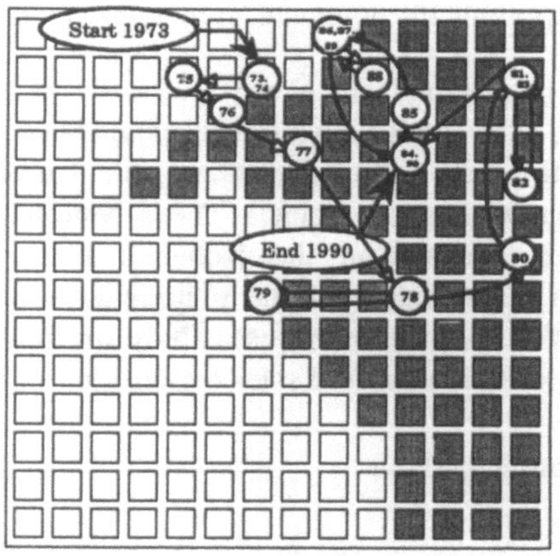

Figure 3. Time evolution of the 'Banco de Descuento' from 1973 to 1990. It went bankrupt in 1980, later reorganized, and went bankrupt again in 1990 (bankrupt area in grey).

3. Delimiting Regions on Self-Organizing Maps.

Due to its dimensionality reduction capabilities, the SOFM is a very powerful tool for nonlinear complex process monitoring [2]. For this task, the SOFM has several advantages, for instance, it does not require any specific (or only a limited) prior knowledge of the physical phenomenon, and it is a tool of easy use.

In [3] there is a good example of system monitoring with SOFM. If we present to the Kohonen map of Fig. 1 data for several consecutive years of a particular bank, we can see the time evolution of its financial state by tracing its trajectory on the map (Fig. 3). The same method can be applied to system state monitoring in many different problems [1, 2]

We can trace with more accuracy the evolution of a system state by delimiting regions (domains) on the Kohonen map. One way is to label every neuron in a supervised way, by making use of the available knowledge of the problem. For instance, in Fig. 3, two areas have been delimited by observing the kind of input pattern each neuron has tuned to (Fig. 1): A solvent (white neurons) and a critical area (in grey).

However, if no prior knowledge is available, or if we want the system to discover for itself the main features in the input data, unsupervised methods for labeling the neurons on the map (delimiting domains) are required. In the following points, some of such unsupervised techniques are presented.

3.1 Determining the dominant weight on each region.

The first method which we propose consists of studying the weights with more absolute value in every area of the map, and then clustering the neurons with the same dominant weights. These neuron clusters represent domains on the Kohonen map with a common main feature of the multidimensional input space. For instance, in our case study we find a large liquidity area, high cost of sales region, etc. (Fig. 4).

3.2 U-matrix method.

This method (introduced in [8]) basically consists of plotting perpendicular to the map the distances between the weight vectors of neighbouring neurons. Thus, a landscape of mountains and valleys is obtained; the mountains (representing neighbouring neurons with very different features) separate regions with similar weight vectors (valleys), delimiting areas of similar features. This procedure has not lead to satisfactory results in our problem, because only has marked out areas on the corners of the map, where the outlier patterns are located. The rest of the landscape is a great flat area, without significant regions.

3.3 Clustering of the reference vectors.

The following set of techniques that we propose are different clustering algorithms that, applied to the reference vectors of the SOFM, cluster together neurons with similar features, delimiting a domain on the map. This is related to reference [9].

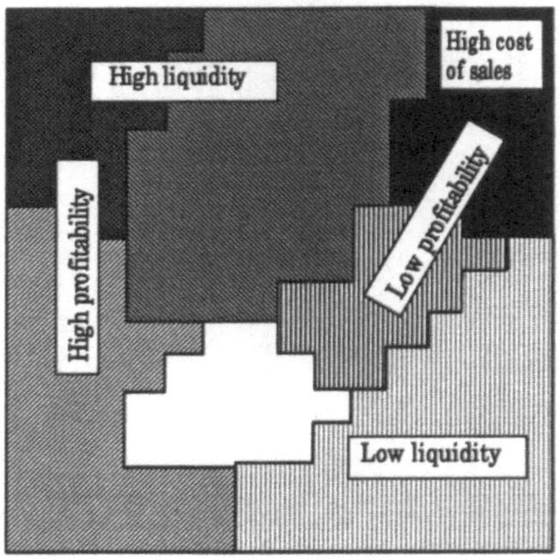

***Figure 4.** Regions by the dominant weight method (Sect. 3.1).*

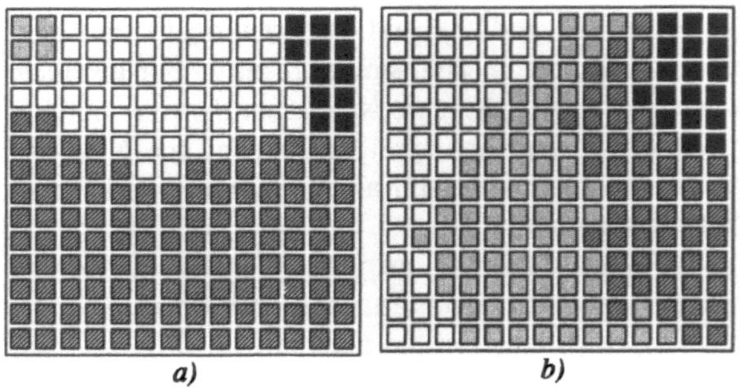

***Figure 5.** a) Regions delimited by k-means clustering of the vectors (Sect. 3.3). b) Regions by a cascaded map (Sect. 3.4).*

The first method uses the *k*-means algorithm. The results of the clustering of the $\mathbf{w}_{ij}$ with k=4 are in Fig. 5a. We can see four domains on the map: The worst (upper right) and the best (upper left) financial regions, a high liquidity (upper half, in white) and a low liquidity zone (lower half). Notice that, in this case, there is not a bankrupt and a solvent area; critical and solvent banks (Fig. 1) will be mixed in the upper (white) and lower (grey) areas.

3.4 Clustering with cascaded Kohonen maps.

In this case the clustering of the vectors is carried out by a second Kohonen layer (of only a few neurons) cascaded to the original SOFM map. This second net is fed with the reference vectors of the first one.

The results of this procedure applied to our case study are in Fig. 5b. A second Kohonen network of four neurons has delimited four zones in the map from Fig. 1: A solvent area (left), a bankrupt region (right), and two transition areas. Thus, this procedure do has find bankrupt and solvent features in the input data.

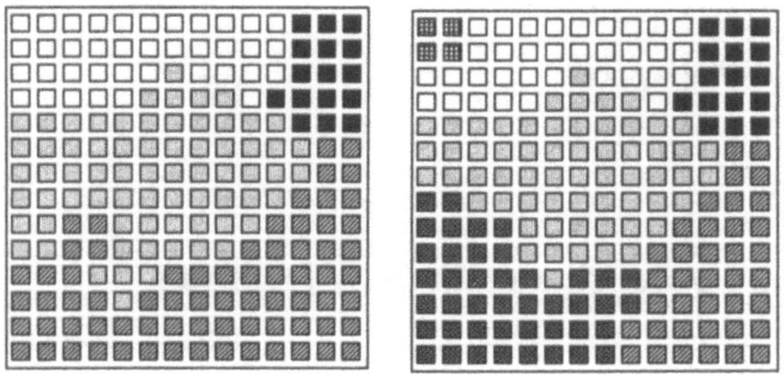

Figure 6. *Regions delimited by RPCL (Sect. 3.5).*

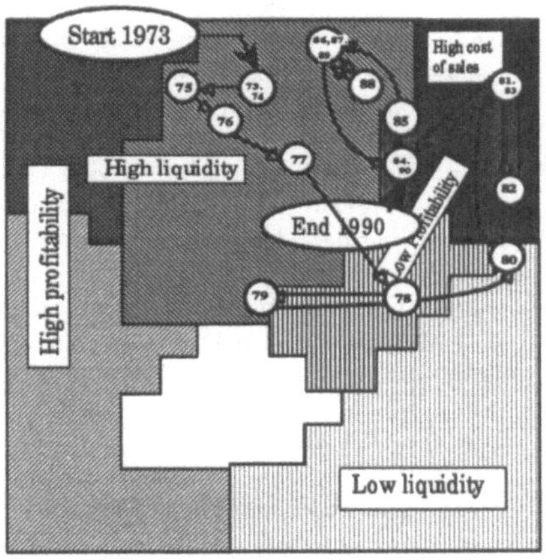

Figure 7. *Evolution of the 'Banco de Descuento' put on the top of the map from Fig. 4. Its financial features are identified here more accurately than in Fig. 3.*

3.5 Clustering with RPCL network.

RPCL (Rival Penalized Competitive Learning) is a new competitive network [10], with conscience mechanism and penalty of the rival neuron (the second more activated, after the winner). RPCL works discovering the proper number of clusters.

The results achieved by means of this method in our case study are depicted in Fig. 6. In a first run, RPCL has found four regions (Fig. 6a), while changing its learning parameters has found six (Fig. 6b), clearly related to those of Fig. 6a. Bankrupt and solvent regions are not accurately delimited with this method.

4- Discussion and Conclusions.

First, we have compared SOFM and MDS. It is necessary to notice that the kind of MDS that we have used in our work (with Euclidean distances) is equivalent to principal component analysis [4], thus the result of Fig. 2 is comparable to that obtained in [11]. Nevertheless, we think it is necessary a deep comparison between both algorithms, experimental and theoretically, involving the rest of types of MDS.

Afterwards, we have presented several unsupervised techniques for delimiting areas on a Kohonen map, which is very interesting for clustering, pattern recognition or process monitoring (Fig. 7). We have shown the results of the application of these procedures to a real case study, the Spanish banking crisis. From the viewpoint of the solvent and bankrupt situations, procedure 3.4 (cascaded Kohonen nets) has provided the best results. Nevertheless, all the discussed methods lead to significant results.

We have recently used these procedures for another very different real-world problem, the classification of electric hourly-load patterns [12]. In that case, best results are provided by RPCL (Sect. 3.5). Thus, we think that all the procedures should be applied to a given problem, and selected the best for it.

5. References.

1. Kohonen, T.: Proc. of the IEEE 78, 9, 1464 (1990)
2. Kangas, J.: PhD Thesis, Helsinki Inst. of Tech., May 1994.
3. Martín-del-Brío, B., Serrano, C.: Neural Computing and Applic., 1, 193 (1993)
4. Mar-Molinero, C. Research paper, Dept. of Accounting and Management Sci., Univ. of Southampton, UK, (1990)
5. Martinetz, T., Schulten, K.: IEEE 1993 Int. Conf. on Neural Networks, San Francisco, 820 (1993)
6. White, H.: AI Expert, 48, Dec. (1989)
7. Cherkasski, V., Friedman, J.H., Wechsler, H. (eds.): From Statistics to Neural Networks. Springer-Verlag 1994.
8. Ultsch, A., Siemon, H.P.: Proc. INNC90, Paris, 305 (1990)
9. Varfis, A., Versino, C. Neural Netw. World, 2, 813 (1992)
10. Xu, L., Krzyzak, A., Oja, E.: IEEE TNN, 4, 636 (1993)
11. Blayo, F., Demartines, P. Proc. Int. Work. on Artificial Neural Networks IWANN91, 469, Granada, Spain, (1991)
12. Martín-del-Brío *et al.* Proc of the Int. Conf. on Modelling, Identification and Control, Austria, Feb. (1995)

Adaptive Genetic Algorithms for Multi-Point Path Finding in Artificial Potential Fields

R.M. Rylatt, C.A. Czarnecki and T.W. Routen
Department of Computer Science
De Montfort University
Leicester LE1 9BH, UK

Abstract

We present research work in progress into the use of adaptive genetic algorithms (AGAs) to search for collision-free paths in an artificial potential field (APF) representation of a cluttered robotic work-cell. We argue that the AGA approach promises to avoid the drawback of other APF approaches which are vulnerable to entrapment by local minima.

1 Introduction

The programming of mechanical manipulators which operate in cluttered environments remains a problem in which research interest is keen. Zelinsky (1994) provides a good review and analysis of the different approaches tried to date. One approach to the problem relies on a special representation of the cluttered work-cell in terms of the manipulator's configuration space. This is known as an ***artificial potential field***. Essentially, zones of high potential represent obstacles and a point of lowest potential corresponds to the operating point which is the manipulator's goal. A suitable representation of the manipulator is subjected to the influence of these field effects, which causes repulsion from high potential in the region of the obstacles and attraction to the areas of lower potential between them. The collision avoidance problem is thus reduced to the minimization of a cost function. Typically, a gradient descent method is employed for its solution, but this method is prone to entrapment in local minima. Clearly, to be acceptable industrially, any risk that a manipulator will fail to reach an operating point must be nugatory. Genetic algorithms (GAs), as they do not depend on single point search, can avoid local minima by searching globally. Even so, the simple genetic algorithm (SGA) (Goldberg, 1989) is still vulnerable to premature convergence and so might not satisfy the particularly testing constraints of the proposed application. The research work presented here has therefore concentrated on the use and development of adaptive genetic algorithms (AGAs) which, across a range of problem types, have displayed a very low tendency to converge on local optima. AGAs also promise greater computational efficiency and, by eliminating the hand-tuning of genetic operator parameters which can make the SGA unwieldy, may overcome resistance to routine use.

2 Adaptive Genetic Algorithms

The GA is a search technique which roughly mimics evolutionary processes in natural systems. A population of individual potential problem solutions, whose representation is based on a broad interpretation of genetic structure, is randomly generated and subjected to the stochastic effects of genetic operators over successive generations. An objective function determines the fitness of each individual and its chance of survival. In the SGA, probabilistic genetic operator parameters are controlled externally by the user and so remain fixed for the duration of each run. Evidence suggests that these settings critically affect the performance of GAs thus implying that considerable expertise is necessary if GAs are to be successfully applied to real world problems.

More significantly for the problem at hand, Srinivas and Patnaik (1994) demonstrate that an Adaptive Genetic Algorithm (AGA), can respond to evolving population characteristics and so reliably avoid premature convergence on problems where the SGA shows a tendency to stick at local optima. In particular, they show that the AGA is markedly superior in solving problems where the decision space is both highly multi-modal and epistatic, which, according to Davidor (1991), are characteristic traits of manipulator path specification problems. Their results were achieved with crossover and mutation operator parameters which adapted to the current level of convergence as measured by an expression of the difference between the maximum fitness and the average fitness in the population. In effect, the probabilities of crossover and mutation increase whenever the characteristic signs of local optima entrapment are emerging in the population of candidate solutions, and decrease as such characteristics disappear. In work (Rylatt, 1994) related to the present research an AGA was implemented following these principles and was successfully applied to an inverse kinematics problem on which an SGA frequently became stuck at suboptimal solutions.

3 Artificial Potential Fields

The original Artificial Potential Field (APF) approach to obstacle avoidance was based on an operational space formulation to achieve low-level control for real-time obstacle avoidance (Khatib, 1986). High-level path planning on the other hand is expressed in term of manipulator arm configurations. Lozano-Perez (1987) defines the configuration of a moving object as any set of parameters that completely specify the the position of every point on the object. For example, a mechanical manipulator with n revolute joints has a configuration which is the set of its joint angles, and its motion can be described in terms of the characteristic n-dimensional space it occupies, which is known as the Configuration Space (C-space). Obstacles which a manipulator is required to avoid, whose position in relation to it are known and fixed, can be mapped into its C-Space.

Warren *et al.* (1989) describe an efficient method of determining the collision points on which they base an APF approach at the higher, path-planning level which aims to achieve a collision free trajectory through the cluttered space. The method of determining collision points relies on a geometrical description of the obstacles in terms of singular and regular points which correspond respectively to the vertices and quantised edges of convex polygons located in Euclidean space (E-space). Specifically, such points are identified where the end-effector would contact edges, and where link frames contact vertices (for the purposes of global planning, the mechanical arm can be modelled simplistically using line segments to represent each link frame in the open kinematic chain) and a point to represent the end-effector. Thus, all necessary points are identified in the subsets of points formed by the intersection of the set of obstacle edge points by the end-effector and the subsets of regular points along the manipulator line segments contactable by obstacle vertices. Each point is mapped into C-space using inverse kinematic transformations thus delineating so-called forbidden regions representing each obstacle. The centroids of these regions are then used as the reference points for calculating zones of high potential which tend to repulse the path representation, whilst outside the boundaries of the forbidden region low potential will result from computations based on the inverse of the distance to the closest forbidden region boundary. The path planning algorithm searches a space of paths, represented by linked nodes in C-Space, the first of which is located at the starting position, and the last at the goal position. Each node represents an arm configuration and all the nodes in the path are simultaneously subjected to a potential field, an improvement over the previous artificial potential field approach in which a single point representing an arm configuration in C-space moves away from obstacles (which have a repulsive field) towards the goal point (which has an attractive field). Algorithms based on this approach tended to become trapped at equilibrium points in the total potential field formed by the superimposition of the field potentials calculated at each discretized point in C-Space, but even the new approach does not completely overcome the problem of local minima. This way of representing possible solutions has clear similarities with the path specification GA approach of Davidor (1991) in that a sequence of arm configurations is used, although in this case the sequence is of fixed cardinality.

The underlying requirement for collision-free path finding without trial-and-error manipulation is that free space must be characterised explicitly. Global planning designed to move the end-effector from one operating point to the next requires less precision than local planning which requires numerous tight positional adjustments. It is therefore expedient to trade exactitude in the characterization of free space for computational efficiency by adopting a scheme which quantizes the former at a level of detail which is likely comfortably to satisfy practical operating requirements.

4 Solution Representation

Because via points must be visited in a sensible sequence, Davidor (1989) argues that path planning for mechanical manipulators falls into the category of order-based problems which require non-standard genetic operators or solution representations. He describes how a robot trajectory in C-space can be represented by strings of arm configurations whose orderliness is preserved by the application of a reproduction operator known as ***analogous crossover***. This matches cross-sites for disruption and recombination according to their phenotypic function - in this case the similarity of end-effector positions in the parent paths generated by a simulator in E-space. However, intermediate simulation is not suited to the problem at hand. Ahuactzin *et al.* (1993) use a Manhattan Motion representation of the solution individuals . In this scheme, a gene value represents a joint movement relative to the last position rather than a joint position in itself; a chromosome is an ordered sequence of such movements containing the series of arm configuration movements which define a fixed length path. Conventional crossover and mutation operators are used and the solutions are evaluated by testing whether each joint motion in turn leads to a collision and determining whether a simple collision-free motion exists from the last configuration to the pre-determined goal, penalising the solution according to the distance from the goal .

For the APF approach a somewhat different scheme is proposed. Its solution genotypes will consist of the configurations between the predetermined start and goal configurations. At the 'decoding' stage for fitness testing the first and last configurations in the solution will be 'connected' to the start and goal configurations in order to construct the phenotype - a virtual path of line segments between nodes in C-space, straddling a grid representation of the APF.

Standard GA operators require a problem representation with a fixed length structure which may impose unnatural constraints on possible solutions. Davidor (1989) reports encouraging results using a naturalistic representation for optimized robot path planning without collision avoidance in which the number of configurations is subject to mutation by a specialised operator. In GA terminology, this representation is a varying-length structure or a chromosome in which the number of genes is not fixed. The significance of this for the present problem is that, as Craig (1989) routinely observes, the number of via points in a conventionally planned path rises when obstacles have to be avoided. In spite of this, Warren *et al.* (1989) report good results using a non-genetic representation in which the initial structure consists of an arbitrary but fixed number of points, permitting only the distance between each point to vary in order to achieve optimization. As with all such methods, the question arises as to what heuristics should inform the intitial guess. Taking these observations in conjunction a good case can be made for choosing a varying-length genetic structure to represent the collision avoidance problem with the expectation that reliable collision avoidance can be achieved with optimized performance in terms of the shortest path. However, initial experiments designed to test the efficacy of the AGA in this domain conform to the fixed length regime typical of most GA work to date. An advantage of the relatively coarse discretization of the control space which satisfies the global requirements of the problem is that the solution structures can be kept reasonably short. A parameter p consists of n bits, where $n = 8$ would give the required resolution for each joint motion. A path length of m nodes where $m = 12$ and $p = 2$ gives a total string length of 192 bits.

5 Problem Representation

As collision avoidance is a constrained problem, the objective function for the AGA will have the form of a penalty or cost function. The problem is not highly constrained and there will be numerous feasible solutions. The fitness of a solution would therefore be related to the minimisation of the cost function. Warren *et al.* (1989) use a multi-objective function which combines and sums expressions for the maximum potential exerted on a line segment and the lengths of the path segments in a possible solution. A weighting multiplier constant is applied to the first expression to fix the relative influence of each expression on the value returned by the function. Problems with this trade-off approach in the GA context were encountered in work on the inverse kinematics problem mentioned earlier. Choosing an appropriate value can be difficult and time-consuming but the necessity to avoid infeasible solutions makes it a critical task. It is interesting to speculate whether the principle of adaptivity could be brought to bear on this problem just as it has addressed the hand-tuning drawback of genetic operators but at present the highly domain specific nature of the trade-offs makes it difficult to see how a sufficiently general criterion could be obtained to make robust central control of this aspect feasible.

Guided by the example of the non-genetic approach, the proposed GA implementation seems relatively straightforward. Decision parameters in the candidate solutions represent points in C-space. For n such points there will be $n+1$ line segments forming a 'path' from the predetermined start and end points. The lengths of the line segements can be calculated in the virtual space of configurations using standard methods of co-ordinate geometry. Their location with respect to the grid of discretized APF values can then be determined and the highest potential along each segment can be found and summed by the penalty function.

6 Conclusion

This paper has presented arguments for the complementary virtues of APFs and AGAs with respect to the collision avoidance problem for mechanical manipulators at the path planning level. Recognizing that the APF approach is a heuristic method designed to reduce the computational complexity faced by a planning algorithm, but one with a well-documented drawback which has inhibited its real-world adoption, the main conclusions arising from work to date are these:

- using APFs to characterize the problem space is a good way of reducing the naturally high computational demands on the objective function of a GA.
- the AGAs ability to avoid convergence on local optima identifies it as a highly promising approach to the outstanding problem of local minima entrapment in APFs.

Suitable problem and solution representations have been proposed which correspond closely to a successful non-genetic APF scheme. A promising variation on the scheme will be to explore the possibilities which solution structures of varying length appear to hold for the collision avoidance problem, and the problem domain itself is a fascinating test bed for further work on AGAs aimed at extending the principle of central control to other genetic parameters.

7 References

AHUACTZIN, J-M., TALBI E-G., BESSIERE, P and MAZER, E . (1993) Using Genetic Algorithms for Robot Motion Planning. IEEE-IROS'93 . Yokohama, Japan..

CRAIG, J.J. (1989) Introduction to Robotics: Mechanics and Control. Reading, MA: Addison Wesley.

DAVIDOR, Y. (1989) Analogous crossover. Proceedings of the Third Conference on Genetic Algorithms. San Mateo, CA: Morgan Kaufmann, 42-50.

DAVIDOR, Y. (1991) Genetic Algorithms: A Heuristic Strategy for Optimization. Singapore: World Scientific.

GOLDBERG, D.E. (1989) Genetic Algorithms in Search, Optimization and Machine Learning. Reading, MA: Addison Wesley.

KHATIB, O. (1986) Real-Time Obstacle Avoidance for Manipulators and Mobile Robots. The International Journal of Robotics Research. 5:1, 90-98.

LOZANO-PEREZ, T (1987) A Simple Motion-Planning Algorithm for General Robot Manipulators. IEEE Journal of Robotics and Automation, RA-3:3, 225-237.

RYLATT, R.M. (1994) M.Sc.dissertation. De Montfort University, Leicester, U.K.

SRINIVAS, M. and PATNAIK, L.M. (1994) Adaptive Probabilities of Crossover and Mutation in Genetic Algorithms. IEEE Transactions on Systems, Man and Cybernetics. 24:4, 656-667.

WARREN, C.W., DANOS, J.C. and MOORING, B.W. (1989) An Approach to Manipulator Path Planning. The International Journal of Robotics Research. 8:5, 87-95.

ZELINSKY, A. (1994) Using Path Transforms to Guide the Search for Findpath in 2D. International Journal of Robotics Research, 13:4, 315-325.

SPATIO-TEMPORAL MASK LEARNING: APPLICATION TO SPEECH RECOGNITION

Stéphane Durand & Frédéric Alexandre

CRIN-CNRS INRIA Lorraine
BP 239, F-54506 Vandoeuvre-lès-Nancy

Abstract

In this paper, we describe the "spatio-temporal" map which is an original algorithm to learn and recognize dynamic patterns represented by sequences. This work is slanted toward an internal and explicit representation of time which seems to be neuro-biologically relevant. The map involves units with different kinds of links: feed-forward connections, intra-map connections and inter-map connections. This architecture is able to learn sequences robust to noise from an input stream. The learning process is self-organized for the feed-forward links and "pseudo" self-organized for the intra-map links. An application to French spoken digits recognition is presented.

1. Introduction

Most real-world problems involve the temporal dimension. For instance, speech recognition, planning, or reasoning include the time parameter in an intrinsic manner. In speech recognition, the characteristic features of a phoneme are not the same if this phoneme is uttered either in a given context or in another: phonemes are context dependent. Phoneme discrimination is more accurate with contextual information which successively occurs in time. Thus, it is very important to introduce contextual or temporal aspects into neural networks if we want to use such a technique.
Current connectionist models are often static systems. They are powerful for patterns with no evolution in time, like in character recognition, but present some weaknesses if patterns involve a temporal component like in phoneme recognition.
However, some networks can also take time into account. In [6], we defined a classification of connectionist temporal models in order to situate our work. We can describe these networks according to the way time is physically represented. Two main classes of systems can be outlined: systems with an *external* representation, where the time is considered as an extra dimension in the input layer (the TDNN is an example [12]), and systems with an *internal* representation of time where the time is encoded in the activation of the network. More precisely, the latter class is divided into two sub-classes. One is devoted to the *implicit* representation, the other one is devoted to the *explicit* representation of time. The word *implicit* is chosen because time is given by the series of steady states of the network (recurrent neural networks are among this representation [9, 7]). *Explicit* refers to the existence of sequences in the network [3, 11].
Our work takes place in an internal and explicit representation of time. This representation seems to be more neuro-biologically plausible. For instance, a biological substrate of such a representation is the cortical column which has been observed in the cortex by neuro-biologists, is supposed to carry out higher functional computation and can especially differentiate its inputs.
The model that we propose in this paper follows the architecture presented in [6], and is inspired by the cortical column model [2]. It is a pseudo self-organized map which can learn spatial masks and temporal sequences. This map is able to learn statistically relevant sequences from the input stream. The connectivity involved in the map is composed of three type of links: feed-forward links corresponding to the spatial stimulus, intra-map links in order to build sequences and inter-map links for the transmission of the results toward higher level map. In the next section, we describe the spatio-temporal map designed to learn sequences, then we present in section 3 the results we obtained for spoken digit recognition. Perspectives and conclusions are given in the last section.

2. The spatio-temporal self-organized map

Although the mechanisms used in the cerebral cortex to treat complex dynamic tasks like speech recognition are not well-known, there are some hypothesis concerning the functional organization. For instance, it seems that cortical granularity refers rather to neural assemblies than to individual neuron itself. The cortical column [8, 1, 2] corresponds to this neural assembly and can be viewed like an "atomic" functional unit. The interest of a such representation is the decreasing of the combinatory concerning the number of neurons in a network and more especially the possibility to involve higher functionality levels. For example, such a unit has different kinds of input and output stream (feed-forward, intra-cortical, inter-cortical input/output in real cortical columns). We use this principle of differentiation of inputs to design a new basic functional unit able to take the time or the contextual information into account (fig. 1).

2.1 Three kinds of links

First, feed-forward weight matrix correspond

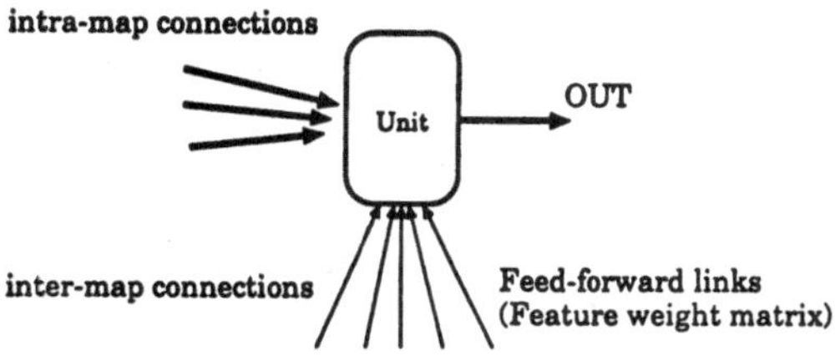

Figure 1: The basic unit

to spatial masks, that is to say, to information vectors extracted from the input space at time *t*. The learning process for these weights is self-organized and is discussed below. Second, the set of intra-map connections defines the "temporal" receptive field of the unit. Indeed, the unit cannot be activated if there isn't at least one of the units in the receptive field whose activation is greater than a threshold. So, to summarize, the unit reacts for a given space feature and for a given temporal information encoded with intra-map connections. The unit activation follows a law which is depicted in the fig. 2. High discharge (equal to 1) is obtained when both stimulus and intra-map receptive field are present. Then, the activation slightly decreases along the time down until it is re-activated. Third, inter-map links are only used for a simple labeling of units. These is a need of improvement for these links. We will only describe the two kinds of links below and their integration in a spatio-temporal map.

2.2 The spatio-temporal map

Like feature Kohonen maps [10] can extract spatial masks or classes, our goal is to design

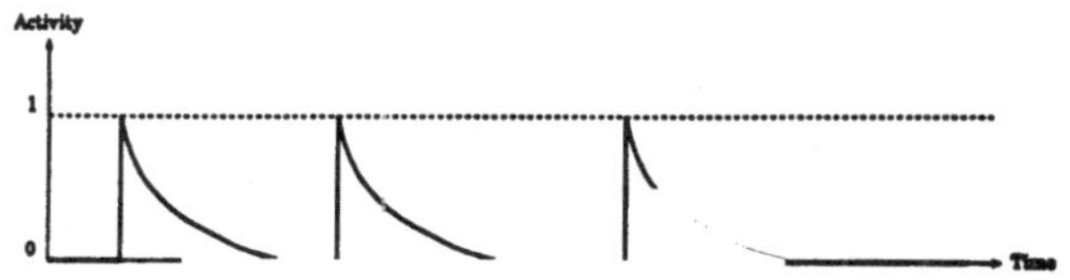

Figure 2: The activation law of a unit

a "spatio-temporal" map which is able to learn spatial and temporal masks. Temporal masks are built as a succession of events (spatial masks) represented by intra-map connections described above. So, patterns to learn on the map are represented by inter-connected units thanks to the intra-map connections. The propagation of activity through the network of intra-connections corresponds to a particular pattern. In figure 3, we can see an example of connectivity between units (small circles)

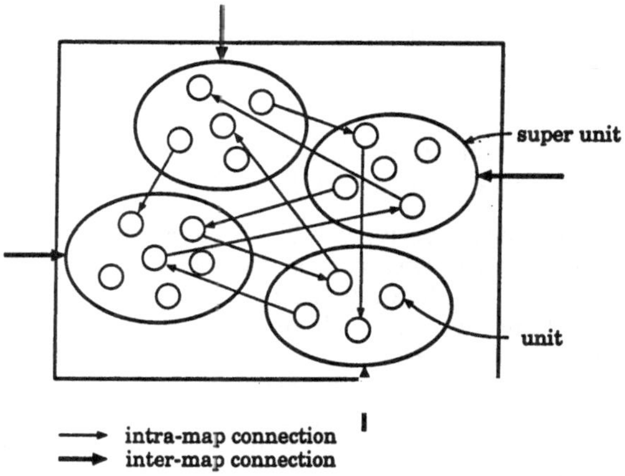

Figure 3: A spatio-temporal map

We want to include such units in a map in order to learn and recall sequences extracted from the input space. The propagation of activities thanks to the intra-links is characteristic of the dynamic pattern to recognize. The sequence of activities matches with the sequence of events in the pattern. But, if we directly involve the units as described above, a unit will be shared by several sequences. This phenomenon could lead to the activation of irrelevant sequences. So, in order to avoid this effect, we designed super-units whose aim is to gather units that differently react to the same stimuli thanks to their feed-forward links. These super-units include several units which share the same feed-forward links but each of them has its own intra-link receptive field. This process allows to enhance distinct sequences which can sometimes share the same stimulus at time *t*. A super-unit includes different units corresponding to different contexts and allows to avoid loops in chains of events. Intra-map links between super-units propagate information through time (see fig. 3). With this architecture, we can learn some sequences from the input data, and maps can be designed at different levels of integration. For instance, for spoken word recognition, we can use a first map which is designed to learn acoustic events during a short time (brief sequences), and a second map corresponding to higher acoustic events (such as phonemes), etc.

2.3 Functioning

The relaxation process is very simple and has been functionally described above. The activation ($A_{U_iSU_k}(t)$) of the unit U_i in a super-unit SU_k is equal to the following formulation:

$$\begin{cases} \overline{A_{U_iSU_k}(t)} & \text{if } \overline{A_{U_iSU_k}(t)} \geq ACT \\ 0.90 A_{U_iSU_k}(t-1) & \text{otherwise} \end{cases}$$

where:

$$\overline{A_{U_iSU_k}} = \frac{1}{\alpha+\beta}(\alpha SA + \beta TA)$$

and ACT, a threshold close to 1.

The spatial activity (SA) designs spatial contribution. In fact, this is the activation of the super-unit which codes the presence of the spatial event (this activity is 0 or 1, and is computed from the feed-forward weight matrix connected to the super-unit). The temporal activity (TA) represents the contextual effect. TA is equal to one if and only if a previous unit linked with an intra-link is activated. Coefficients α and β can modulate the participation of each parts (feed-forward and intra-link). They are set to 1 in our simulations.

The learning process uses a feature Kohonen map [10] to extract the spatial masks and to self-organize the super-units on the map. This process is carried out before the intra-link learning process. The intra-connections are dynamically learned. When a super-unit is activated by a feature in the input, two cases can happen. First, if any other unit is activated in the super-unit, a new unit is created with an intra-link coming from units whose activation is still superior to a threshold. Second, if a unit is activated in the super-unit, we add an intra-link between this unit and units whose the activation is still superior to a threshold. The learning process is "pseudo" self-organized because for each pattern to learn, we connect the last unit of the sequence to a higher level super-unit corresponding to the pattern.

3. Application to speech recognition

We have applied this architecture to French spoken digits recognition. We used one spatio-temporal map in order to extract sequences and one higher level map involving 10 super-units connected to the end of the sequences from the spatio-temporal map. Each super-unit in the higher map corresponds to a given digit (fig. 4). The recognition process uses one property of the Kohonen map: the spatial topology. Each super-unit on the map is connected to other super-units. Several kind of connectivity (which we call the local links) can be used, especially a 2D connections, but in our simulation we used a linear feature map. So, a super-unit is locally connected to two neighbors. Thus, when a vector from the pattern to learn comes into the map at one time, one super-unit is activated. In order to be robust for the spatial masks, the neighboring super-units are activated too, thanks to the local links. After that, the activation process described above is carried out and search different close sequences to activate.
For the moment, in our simulations, the intra-map links are not valuated. This can be embarrassing because a created link can be irrelevant for the remaining input stream. So, with the current algorithm, a new sequence is built for a new input sequence.
The database is speaker dependant. At the lowest level, 12 coefficients at each time serve as input vector to the input map. Input speech sampled at 16 kHz was hamming windowed and a 256-points MFCC (Mel Frequency Cepstrum Coefficients) [5] computed every 4 ms. The training set of patterns contains 5 occurrences for each digit (0 to 9). Less than 10 cycles are necessary to train the feed-forward links of the map which contains 50 super-units. The testing patterns correspond to 5 occurrences of each digit and the recognition rate is 90%. A remarkable behavior of the network can be viewed for the French digit "cinq". This word can be divided into three phoneme components: /s/, /in/ and /k/ which successively appear in time. The winning super-unit is the one corresponding to the digit 5 but we can see small activations for the super-units corresponding to the digits "un" (/in/) and "sept" (/s/). So, the map is able to statistically extract relevant sequences from the input space. First tests with a speaker-independent database TIDIGIT give a recognition rate of 70%. For this base, the number of occurences is bigger and the variability more important cause of the multi-speaker.

4. Conclusion

In this paper, we have described a new basic self-organized architecture, the self-organized spatio-temporal map. This map involves large neural assemblies, the super-units, which correspond to a general spatial stimulus extracted at each time

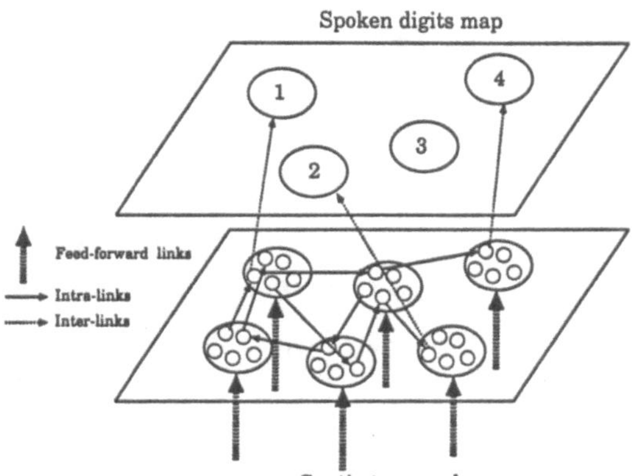

Figure 4: The speech recognition architecture

from the input stream. A first type of link, the feed-forward link, encodes the stimuli. These assemblies involve units which share a common stimulus. These units are interconnected with intra-map links, allowing the propagation through time of activities in the map. These propagations characterize the patterns to learn and recognize. So, the spatio-temporal map allows to learn (i) spatial features from the input space, the mask being coded thanks to feed-forward links, (ii) and relevant sequences thanks to intra-links. The system is robust to noise because the activations of the units decreasing according to a decay parameter allows an insertion of one or some events. The deletion of events is allowed by the fact that a unit is not only connected with its predecessors but with the predecessors of the predecessors up to a bound fixed by a threshold. Beyond the direct application to speech recognition, we can imagine to use such a technique to compute an automatic speech segmentation. For instance, if we learn a sequence corresponding to the silence, we can isolate words from the sentence. Moreover, our model is generic and can be used for other problems including the temporal dimension, especially in robot movements planning or in signal processing [4].

References

[1] F. Alexandre. *Une modélisation fonctionnelle du cortex : la colonne corticale. Aspects visuels et moteurs.* PhD thesis, Université Nancy I, 1990.

[2] F. Alexandre, F. Guyot, J P. Haton, and Y. Burnod. The cortical column : a new processing unit for multilayered networks. *Neural networks*, 4:15–25, 1991.

[3] B. Ans. Modèle neuromimétique du stockage et du rappel de séquences temporelles. t311, série iii, C. R. Acad. Sci. Paris, 1990.

[4] B. Colnet and S. Durand. Application of temporal neural networks to source localisation. In *ICANNGA, second international conference on artificial neural networks and genetic algorithms*, Alès, France, 1995.

[5] S. B. Davis and P. Mermelstein. Comparison of parametric representation for monosyllabic word recognition in continuously spoken sentences. *IEEE Transactions on acoustics, speech, and signal processing*, ASSP-28(4):357–366, 1980.

[6] S. Durand and F. Alexandre. A neural network based on sequence learning: Application to spoken digits recognition. In *7th international conference on Neural Networks and Their Applications*, pages 290–298, Marseille, 1994.

[7] J L. Elman. Finding structure in time. *Cognitive Science*, 14:179–211, 1990.

[8] D H. Hubel and T N. Wiesel. Functional architecture of macaque monkey visual cortex. *Ferrier Lecture Proc. Roy. Soc. Lond.B*, pages 1–59, 1977.

[9] M I. Jordan. Attractor dynamics and parallelism in a connectionist sequential machine. In Hillsdale, editor, *Proceedings of the Eighth Annual Conference of the Cognitive Science Society*. Erlbaum, 1986.

[10] T. Kohonen. *Self-Organization and Associative Memory.* Springer Series in Information Sciences. Springer-Verlag, third edition, 1989.

[11] V.I. Nenov and M.G. Dyer. Perceptually grounded language learning: Part1–a neural network architecture for robust sequence association. *Connection Science*, 5(2):115–138, 1993.

[12] A. Waibel, T. Hanazawa, G. Hinton, K. Shikano, and K J. Lang. Phoneme recognition using time-delay neural networks. *IEEE Transaction on Acoustics, Speech and Signal Processing*, 37(3):328–339, 1989.

Artificial Neural Networks for Motion Estimation

Olaf Schnelting, Bernd Michaelis and Rüdiger Mecke

Institute of Process Measurement Technology and Electronics
Otto-von-Guericke University of Magdeburg (Germany)
39106 Magdeburg, Universitätsplatz 2

Abstract

This paper deals with the using of Artificial Neural Networks (ANN) for motion estimation. By means of simple neural structures it is possible to improve the reliability and accuracy of block matching algorithms (BMA) by a postprocessing of the similarity criterion. The ANN dimensions the appropriate structures. The fundamental idea and some first results will be described. The performance capability of the proposed method is shown for selected synthetic one- and real world two-dimensional measuring situations which are not solvable by means of conventional BMA.

1 Introduction

In recent years interest in image processing has been directed towards applications in the field of image sequence analysis. Particularly, methods of motion estimation are increasing in their practical importance. Global object displacements as well as whole displacement vector fields are interesting. Well known application fields are e. g. image data compression for the digital video telephone, automatic traffic controlling, interpretation of satellite image data, automatic manufacturing, ecology and medical diagnostics.

Nevertheless, there are some unsolved problems which require innovative solutions. One example is the determination of complex object displacements in distorted industrial environments, e. g. bad image acquisition conditions as extreme illumination changes, overlapped movements or noise. The employment of neural structures to solve these problems is such an innovative way. There are only a few researches on this field [1, 2].

2 Problem Description and System Model

Because of the performance capability of biological systems in movement perception it seems to be useful to apply existing neurophysiological knowledge about natural information processing to the "technical vision". The aim is not the modeling of biological systems, rather it is planned to solve technical problems following such prototypes. Essential results in the investigation of the biological vision come from researches on insects, of which neural signal processing is easy to grasp.

Compared with such biological prototypes [3, 4] some simplifications are done. The starting point of investigation is the system model shown in Fig. 1. In the first step it realizes the well-known block matching method by comparison of consecutive image pairs.

Additionally to usual methods of motion estimation, the a-priori information about the image scene is taken into consideration. The neural network (Fig. 1, subsystem 3) diminishes errors in the displacement vector field.

Subsystem 2 corresponds to the widely known block matching algorithm [6]. The Mean Absolute Difference (MAD) is used as similarity criterion. Traditional matching algorithms exclusively perform a MAD minimum search for displacement detection.

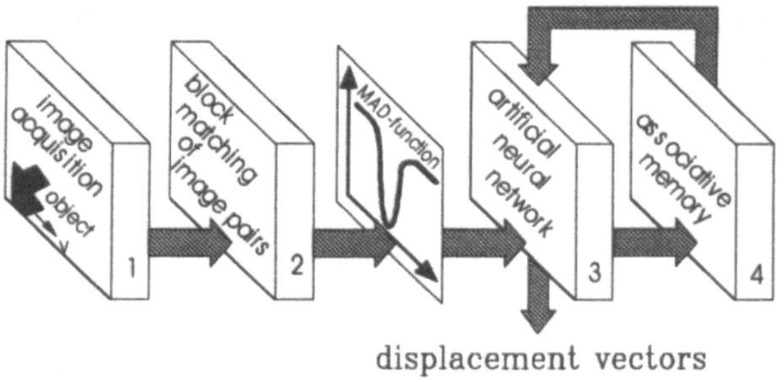

Fig. 1: System model for motion analysis with a neurobiological based structure

However, there are some problems of motion determination in case of image acquisition in distorted environments (e. g. illumination changes, overlapped movements in one block, textured foreground or background and noise) and high spatial resolution (block dimension 8x8 pixel or smaller) is required. One reason for this is the deformation of the similarity criterion. Particularly in the region of the MAD minimum it can cause wrong measuring results. Besides, the emergence of multiple similar minima can prevent the unique correspondence between the object movement and the coordinates of location.

The idea based on the system model (shown in Fig. 1) is the postprocessing of the MAD function in subsystem 3. This is realized by use of object-specific a-priori information. The associative memory shall reload the weights in subsystem 3 in dependence on the image contents. The main aim is the suppression of the above mentioned disturbances.

3 Improvement of Motion Estimation

Because most industrial applications of motion estimation have been directed towards certain classes of measuring objects, it can be assumed that information about object-specific parameters are known in advance. Investigations have shown, that essential object parameters are reflected in a characteristic form of the similarity criterion. Particularly in case of small block dimensions it is possible to reduce the variety of object-specific forms of similarity criteria to a number of significant basic patterns.

The special employment of neural structures for the improvement of motion estimation is not much investigated up to now. The ANN (Fig. 1, subsystem 3) has such a task and it uses trained knowledge about concrete objects for the determination of the minimum position of the MAD function.

In first fundamental researches on application of ANN for the solution of the above mentioned problems simple one-layer backpropagation networks were used. They were trained for the recognition of certain MAD forms using supervised learning. It is possible to restrict the consideration to one output neuron because the weights of this neuron are locally independent. This essentially simplifies the network structure and reduces the necessary calculations. The training vectors are created according to the MAD function (Fig. 2) for motionless objects in identical images in sense of the auto-correlation function.

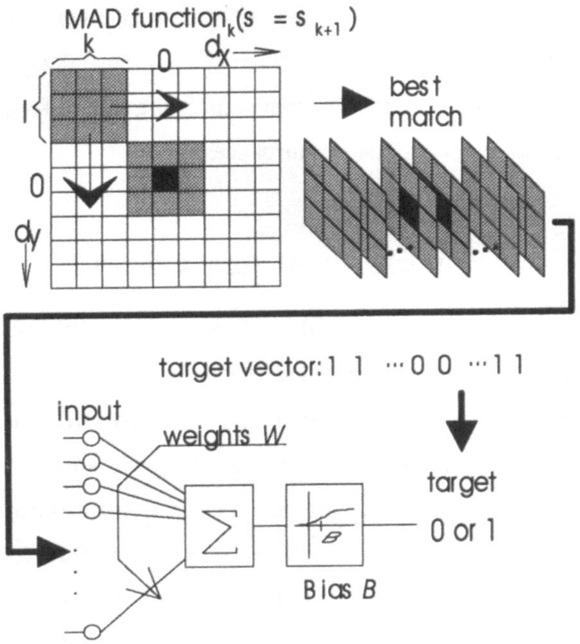

Fig. 2: Structure of the training data set and the simplified architecture of the neural network

The use of the MAD function in this manner based on different aspects. One of these is the fact, that the image contents is reflected in the form of the MAD function. Another one is the feature of local invariance regarding the object position. That means the form of the MAD minimum is independent of the object displacement in the case of undisturbed images.

The training phase is done off-line. The result of the training procedure is the matrix of weights ***W*** and the bias ***B***. The postprocessing of the distorted MAD function can be done on-line according to the transfer properties of the neural network. The postprocessed MAD function originate from the distorted similarity criterion. Ideally, this has the value zero at the detected position.

It seems efficient, starting from the similarity on the MAD level, to take a division of the objects respectively the corresponding MAD functions into determined classes. Among an empirical procedure for the division into classes some researches by using of self-organizing Kohonen Maps were done [2, 5]. This neural structures have the feature to group objects according to their similarity into local zones in a feature map. Thus it represents the corrected displacement vector.

For the described employment a classification of object-specific MAD functions is done. The trained Kohonen Map is suited to address the weight data sets (weight matrices), stored in an associative memory (Fig. 1, subsystem 4). The number of weight data sets corresponds to the number of classified feature groups.

4 Selected Example

The procedure explained in section 3 shall be illustrated by a selected example. Fig. 3 shows a used real image scene. The corresponding MAD function is depicted in Fig. 4. The clear correspondence of the existing minima to the actual object movement is not possible by conventional means. The in Fig. 3 (left side) visible overlapping between the foreground texture and the moving measuring object leads to the creation of two minima (see Fig. 4). The minimum that corresponds to the fixed foreground texture is dominant. In dependence of the technological problem this overlapping is mostly disturbing and a separation of both movements is desired. The resulting correspondence problem can be solved by the utilization of a-priori information about the measuring situation. The a-priori information is represented by the MAD function of the non-distorted motionless measuring object (Fig. 3, right side). This function is calculated in a previous off-line process.

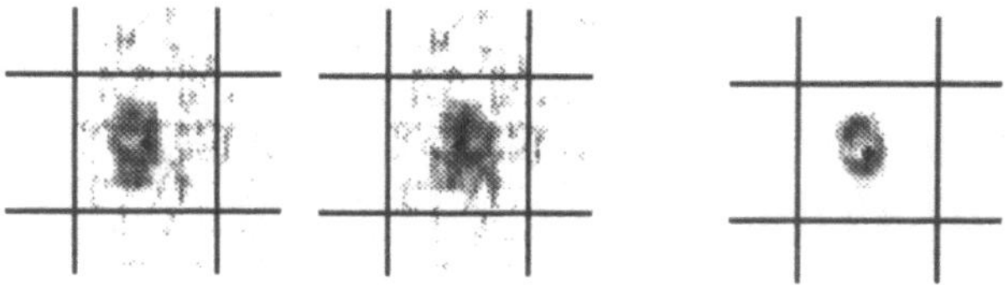

Fig. 3: Probing sequence (left side) and non-distorted measuring object (right side)

The used neural network were trained to the form of the MAD function according to the procedure, shown in Fig. 2. Fig. 4 (left side) depict the resulting set of weights ***W***. The postprocessed MAD function, calculated according to the network transfer properties, showes, that the minimum, corresponding to the foreground texture (position dx=dy=0), is completely suppressed. However, the interesting minimum is emphasized.

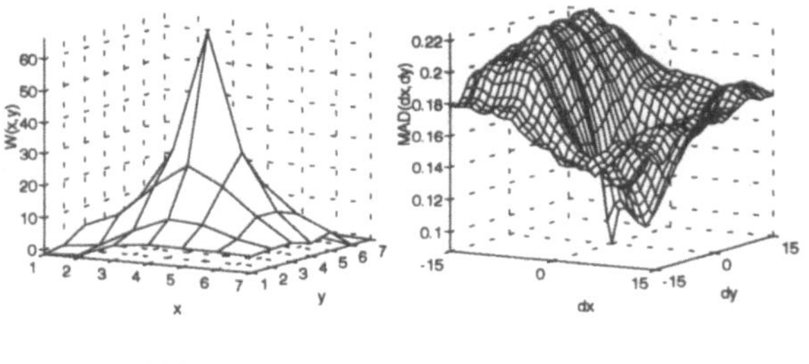

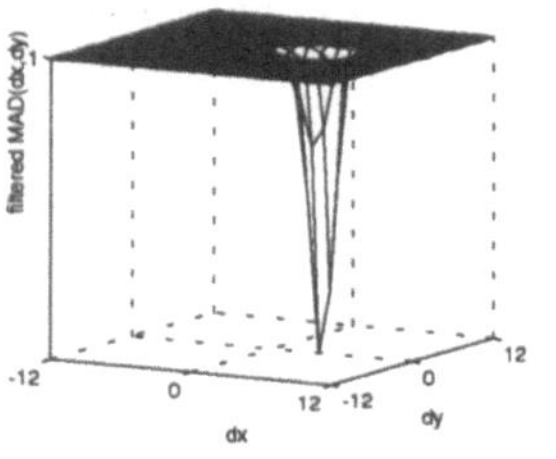

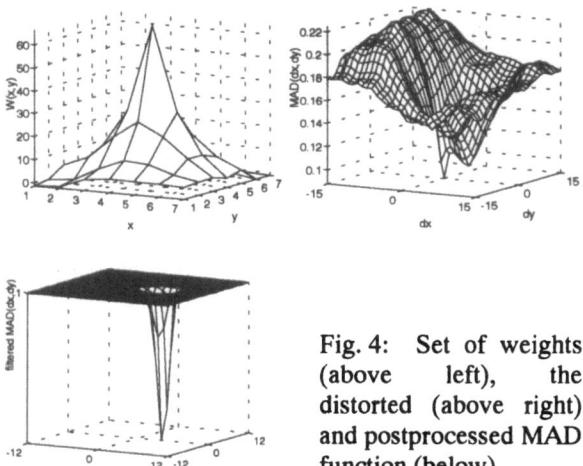

Fig. 4: Set of weights (above left), the distorted (above right) and postprocessed MAD function (below)

5 Conclusions

The described results show, that the structure according to Fig. 1 makes possible a considerable improvement of the motion estimation in the case of the above mentioned distorted measuring situations. This improvement is achieved by the utilization of object specific a-priori information. By using multiple layers of neurons respectively efficient classificators a further improvement of the efficiency of the procedure can be expected.

It is not possible to assign to each object a specific set of weights of the ANN in practical realizations. Therefore a creation of classes referring to object specific MAD functions is necessary. The filtering effect is scarcely influenced by rough object classes. It is efficient to analyze the creation of classes by means of self-organizing Kohonen Maps [2, 5]. A combination with the described backpropagation networks seems usefully to find new approaches for the object adaptive motion estimation in technical processes.

References

[1] Convertino, G.; et. al.: Hopfield: Neural Network for Motion Estimation and Interpretation. Proc. ICANN' 94, Sorrento, Italy, 26-29 May 1994, Vol. 1, pp. 78-81.

[2] Schnelting, O.; Seiffert, U.; Michaelis,B.: Bewegungsschätzung mit künstlichen neuronalen Netzen. 4. Dortunder Fuzzy-Tage '94 Dortmund, Germany.

[3] Zaagman, W.H.; et. al.: On the Correlation Model: Performance of a Movement Detecting Neural Element in the Fly Visual System. Biological Cybernetics 31 (1978), pp. 163-178.

[4] Poggio, T.; Reichart, W.: Considerations on Models of Movement Detection. Kybernetik 13 (1973), pp. 223-227.

[5] Seiffert, U; Michaelis, B.: Estimating Motion Parameters from Image Sequences by Self-Organizing Maps. Proc. 39. IWK Ilmenau, 1994.

[6] Musmann, H.-G.; Pirsch, P.; Grallert, H.-J.: Advances in Picture Coding. Proc. IEEE 73 (1985) No.4, pp. 523-530.

[7] Wiener, N.: The Extrapolation, Interpolation and Smoothing of Stationary Time Series. Wiley & Sons, New York, 1949.

ADVANCED NEURAL NETWORKS METHODS FOR RECOGNITION OF HANDWRITTEN CHARACTERS

Sławomir Skoneczny and Jarosław Szostakowski

Institute of Control & Industrial Electronics (ISEP),
Warsaw University of Technology,
00–662 Warszawa, ul. Koszykowa 75, Poland

ABSTRACT

In this paper, we present the efficient voting classifier for the recognition of handwritten characters. This system consists of three voting nonlinear classifiers: two of them base on the multilayer perceptron, and one uses the moments method. The combination of these kinds of systems showed superiority of neural techniques applied with classical against exclusive traditional approach and resulted in high percentage of correctly recognized characters. Also, we present a comparison of the recognition results.

1 Introduction

One the most often studied problem in practical pattern recognition is the classification of handwritten characters. Handwritten character recognition, especially for text that is mixed with graphics, would find many applications in office automation and computer aided design. Classical methods although still being developed and improved are far from such efficiency that would enable to be applied in practical using. Even when applying efficient neural network based classifiers the rate of correctly recognized characters is not satisfied for wide commercial applications. In our paper we describe the novel, efficient voting system bases on the different neural network models: multilayer perceptron with backpropagation learning algorithm and moments method.

2 Multilayer Perceptron with Backpropagation

One of the most useful method for the pattern recognition, especially for recognition of numerals and characters, is the multilayer perceptron with the backpropagation learning algorithm [1,2].

When an input pattern p is applied to the network, the activation of each unit is dynamically determined using the function :

$$o_{pj} = \frac{1}{1+e^{-(\sum_i w_{ji}o_{pi}+\theta_j)}} \tag{1}$$

where o_{pj} is the activation of unit j as a result of the application of pattern p, w_{ji} is the weight from unit i to unit j, and θ_j is the bias for unit j. Backpropagation is then invoked to update all of the weights in the network according to the following rule:

$$\Delta w_{ji}(n+1) = \eta \cdot \delta_{pj} \cdot o_{pi} + \alpha \cdot \Delta w_{ji}(n), \tag{2}$$

where n is the presentation number i.e., the number of times the system has been presented a pattern, η is the learning rate, δ_{pj} is the error signal for unit j, and α is the momentum factor. The error signal δ_{pj} for and *output* unit j is calculated from the difference between the target value and the actual value for that unit:

$$\delta_{pj} = (t_{pj} - o_{pj}) \cdot o_{pj} \cdot (1 - o_{pj}). \tag{3}$$

The error signal δ_{pj} for a *hidden* unit j is a function of the error signals of those units in the next higher layer connected to unit j and the weights of those connections:

$$\delta_{pj} = o_{pj} \cdot (1 - o_{pj}) \cdot \sum_k \delta_{pk} w_{kj}. \tag{4}$$

3 Moments Method

3.1 Legendre Moments

Legendre Moment rank (m, n) of two variable function is defined as:

$$\lambda_{m,n} = \frac{(2m+1)(2n+1)}{4} \int_{-\infty}^{\infty} \int_{-\infty}^{\infty} P_m(x)P_n(y)f(x,y)dxdy \tag{5}$$

where $P_i(x)$ is the Legendre polynomial:

$$P_i(x) = \frac{1}{2^i i!}\frac{d^i}{dx^i}(x^2-1)^i \tag{6}$$

These polynomials are the orthogonal basis if the double integral is the scalar product $< -1, 1 > \times < -1, 1 >$. In addition every function can be described by its moments:

$$f(x,y) = \sum_{m=0}^{\infty}\sum_{n=0}^{\infty} P_m(x)P_n(x) \tag{7}$$

But the domain of this function must be defined as $< -1, 1 > \times < -1, 1 >$. Therefore, if the Legendre moments are enough small, the function can be described by finite set of its moments. That is why the Legendre moments are used for characters recognition. In this case the function $f(x,y)$ is the image function and suitable integrals are replaced by the sums.

3.2 The feature vector creation

Each character is given by an array size of 48×48 pixels. The image of character is divided into 9 squared parts, which size is 16×16 pixels. Next for each part the moments rank $\{0,1,2\} \times \{0,1,2\}$ are calculated:

$$\lambda^p_{m,n} = \frac{(2m+1)(2n+1)}{4}\sum_{i=-8}^{8}\sum_{i=-8}^{8} P_m(x^p_i)P_n(y^p_i)f(i,j)dxdy \tag{8}$$

where $\lambda^p_{m,n}$ is the moment rank (m,n) for pth part:

$$x^p_i = i/8 \quad ; y^p_i = y/8 \tag{9}$$

Because the character image contains 9 parts, which are described by 9 moments, the whole character is described by 81 elements feature vector.

3.3 Classifier

For recognition task the modified stochastic classifier is used. The average feature vector is calculated for each character:

$$\bar{x}_i = \frac{1}{N}\sum_{j=0}^{N} x_{ij} \quad x \in \mathcal{R}^n \tag{10}$$

where $n = 81$, a x_{ij}–feature vector of ith character, and a covariance matrix is

$$R_i = \frac{1}{N}\sum_{j=0}^{N}(x_i - \bar{x}_i)(x_i - \bar{x}_i)^T \tag{11}$$

In modified stochastic classifier, the parameters g_i are calculated for each class of character:

$$\begin{aligned} g_i(x) &= (x-\bar{x}_i)^T(x-\bar{x}_i) - \\ &-\sum_{j=1}^{k}\frac{\lambda^i_j}{\lambda_j + h^2}(x-\bar{x}_i)^T R_i \Phi^T_i (x-\bar{x}_i)/h^2 + \\ &+ln[h^{2(n-k)}\Pi^k_{j=1}(\lambda^i_j + h^2)] \end{aligned} \tag{12}$$

where λ^i_j is the eigenvalue of covariance matrix and Φ_i is matrix of its eigenvectors. The parameters k, called the number of freedom degree, is equal the number of predominant covariance matrix eigenvalues (for this implementation is equal 12), and $h = \frac{k}{n}$.

During recognition the $g_i(x)$ is normalized:

$$w_i(x) = \frac{-g_i(x)}{|\max_k(g_k(x))| + |\min_k(g_k(x))|} \tag{13}$$

The class, for which $w_i(x)$ is the greatest, are chosen as recognized.

4 Mixed classifier

The mixed neural classifier was used for the characters recognition task (Fig. 1).

The words extracted from the text are segmented in the preprocessing stage. Because in some kind of fonts the characters are connected this process is under supervillence of the classifier. It means that the given word is segmented in such a way in order to obtain the greatest degree of reliability in character recognition stage.

As the first classifier (*Neuron1*) we used multilayer perceptron with two hidden layer (150 neurons in the first, and 130 in the second one). Characters extracted from the given word are next thinned and scaled. The feature vector which consist of 64 integer elements is create on the basis of the 16 nonoverlaping segments of character's field. In each the such segment we count number of character's lines segments in each of four directions. The neural net was learn using modified backpropagation algorithm.

As the second classifier, similar as in the first case, we use multilayer perceptron (*Neuron2*). Characters extracted from the word are first scaled and approximated by segments of line. Each such segment of line is described by the location of center, orientation and length [3]. Obtained in such way feature space was next divided on a small regions.

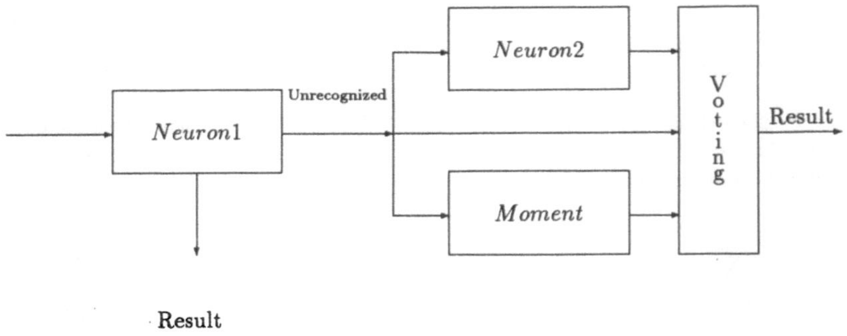

Figure 1: A Voting Classifier for the Recognition of Handwritten Characters

The feature vector consists of elements equal 0 or 1 in dependence of there is or not a segment of line in given region of feature space. The feature vector was next completed by some topological features. The net was learn as in the first case.

As the last classifier, the moments method (*Moment*), described in section 3, is used.

Given letter in each of the three classifiers can be classified to one or more class. It can also happen that a classifier is not able to indicate any class. If the first method (*Neuron1*) do not recognize the character, the second (*Neuron2*) and the third (*Moment*) methods are used. The finally decision is obtain during the voting.

Character	Rate of classified characters		
	GOOD	BAD	UNRECOGNIZED
A	90.0	4.0	6.0
a	86.0	4.0	10.0
B	88.0	8.0	4.0
b	80.0	4.0	16.0
K	88.0	2.0	10.0
k	80.0	2.0	18.0
P	90.0	0.0	10.0
p	88.0	4.0	8.0
Z	86.0	0.0	14.0
z	82.0	4.0	14.0
All characters	**85.1**	**4.7**	**10.2**

a)

Character	Rate of classified characters		
	GOOD	BAD	UNRECOGNIZED
A	56.0	18.0	26.0
a	68.0	19.0	13.0
B	54.0	36.0	10.0
b	44.0	18.0	38.0
K	40.0	18.0	42.0
k	54.0	14.0	32,0
P	44.0	22.0	34.0
p	56.0	8.0	36.0
Z	68.0	10.0	22.0
z	26.0	18.0	56.0
All characters	**56.0**	**15.3**	**28.7**

b)

Character	Rate of classified characters		
	GOOD	BAD	UNRECOGNIZED
A	0.0	0.0	100.0
a	0.0	0.0	100.0
B	0.0	0.0	100.0
b	0.0	0.0	100.0
K	0.0	2.0	98.0
k	0.0	0.0	100.0
P	0.0	0.0	100.0
p	0.0	0.0	100.0
Z	0.0	0.0	100.0
z	0.0	0.0	100.0
All characters	**0.4**	**21.0**	**78.6**

c)

Figure 2: Recognition results: a) multilayer perceptron I (*Neuron1*), b) multilayer perceptron II (*Neuron2*), c) moments method (*Moment*)

Character	Rate of classified characters		
	GOOD	BAD	UNRECOGNIZED
A	90.0	4.0	6.0
a	86.0	4.0	10.0
B	88.0	10.0	2.0
b	82.0	2.0	16.0
K	90.0	2.0	8.0
k	82.0	2.0	16.0
P	92.0	4.0	4.0
p	88.0	6.0	6.0
Z	88.0	2.0	10.0
z	82.0	4.0	14.0
All characters	**85.5**	**5.8**	**8.7**

Figure 3: Recognition results: Mixed Expert.

5 Experimental results

The performance was measured both on the training set of characters as well as on the test set. All simulations were performed on IBM PC 386/486. A special software package was built in C language (Watcom 9.0 C) in order to perform experiments with different network structures.

Our data base consisted of some characters which were written by about 200 persons. This set was divided on two subset: the training set and the test set. The training set to the test ratio was 2:1. The training set was used for classifiers creating. The methods were searched on the test set. Each character is given by an array size of 48x48 pixels.

The results rates of recognition for each method are given in Fig. 2. The expert system results are given in Fig. 3.

Especially the results for moments methods are interesting. The very small rate of recognition do not let use this method independently. But it can improve the rate of recognition for the expert system.

The mixed neuron expert seems to be an useful tool for the character recognition task. The recognition rate can by improved by:

- using the better feature vector for all classifiers learning,
- add the next recognition methods (classical or neural) to the expert system.

6 REFERENCES

[1] Y. Le Cun, B. Boser, J. S. Denker, D. Henderson, R. E. Howard, W. Hubbard, and L. D. Jackel, "Handwritten Digit Recognition with a Back-Propagation Network," in *Advances in Neural Information Processing Systems 2* (D. S. Touretzky, ed.), pp. 396–404, San Mateo, CA: Kaufmann, 1990.

[2] S. Skoneczny, R. Foltyniewicz, and J. Szostąkowski, "Neural Network Based Classifiers as Useful Tools in Zip Code Recognition Task," in *Proc. NOLTA'93*, vol. 3, (Hawaii, USA), pp. 941–944, December 1993.

[3] S. Kahan, T. Pavlidis, and H. S. Baird, "On the Recognition of Printed Characters of Any Font and Size," *IEEE Trans. on Pattern Analysis and Machine Intelligence*, vol. PAMI-9, pp. 274–288, March 1987.

THE USE OF A VARIABLE LENGTH CHROMOSOME FOR PERMUTATION MANIPULATION IN GENETIC ALGORITHMS

Phil Robbins

School of Computing and Information Systems, University of Greenwich, Wellington Street, London, SE18 6PF, U.K.

Abstract

Permutations are difficult to represent and manipulate as chromosomes in genetic algorithms. Simple crossover often yields illegal solutions, and repair mechanisms appear to be very disruptive. A new chromosome structure, the Variable Length Sequence (VLS), and associated operators have been developed. The rationale is that the crossover should guarantee that the resulting solution is legal, and that the effect of disruption should be reduced by the retention of genetic memory within the chromosome. VLS is applied to the travelling salesman problem (TSP), and the results compared with those obtained using the PMX and C1 operators. VLS out-performs the other operators over a wide range of parameters.

1. Introduction

Genetic algorithms have been applied to problems of sequencing, or permutation, with only limited success. Better results have been reported when using hybrid systems in which problem specific knowledge or heuristics are incorporated into the genetic algorithm. The work described here confines itself to the question of how to improve the performance of the genetic algorithm without introducing problem specific knowledge or heuristics.

If a straightforward representation of a sequence is used, then a standard crossover operator will often produce illegal solutions, as some alleles will appear more than once in the child chromosome, whilst others will not be represented at all. The responses to this problem have fallen mainly into two main categories, namely to either define a new crossover operator which is guaranteed to produce only legal solutions, or to allow illegal solutions to be created, and then to invoke a repair procedure to make the solution legal again. In either case, the disruption introduced may have a detrimental effect on the performance of the genetic algorithm.

A new chromosome representation, the Variable Length Sequence (VLS), and associated operators, are described in this paper. The rationale is that the crossover guarantees that the solutions produced will be legal, and that disruption is lessened by making use of genetic memory that is stored in the chromosome.

The *blind* travelling salesman problem is a prime example of the type of permutation problem that we are addressing. The performance of VLS has been tested on two TSP problems, and the results compared against those obtained from two previously established operators, namely the partially mapped crossover (PMX) [1], and the C1 crossover [2]. A range of combinations of parameters, affecting the mutation rate, degree of selective pressure and the absence/presence of elitism, were tested. VLS out-performs the other two algorithms in the most of the tests.

2. Biological Precedent[1]

The chromosomes of eukaryotes in the natural world carry redundant genetic material (redundant in the sense that is not expressed by coding for protein). Sequences of such material are referred to as introns (if the material is located between base pairs that contribute to a single gene) or spacing DNA (if it falls between genes). The benefit (if any) of such genetic material is not yet determined. Theories suggest that this redundant DNA may be a form of genetic memory, and/or may affect the linkage between coding base pairs. Levenick [8] has shown that the insertion of introns[2], which affect linkage, into bit strings processed by a genetic algorithm can lead to significant improvements in performance. The VLS representation uses introns as a form of genetic memory.

Genotype length in the VLS representation varies between individuals. At first this may appear unnatural. However, the genomic theory suggests that the vast increase in biodiversity in the Cambrian era ("the Cambrian explosion") was due, at least in part, to the fact that genomes did not at that time exercise such a high degree of self-regulation as is the case in modern times, thereby granting evolution and natural selection a greater range of new and radical variations to asses.

3. Description of VLS

Two main objectives drove the development of the VLS chromosome structure and associated crossover. Firstly, no illegal solutions should be generated. Secondly, the effects of disruption should be kept to a minimum.

3.1 VLS Chromosome Structure

A principle feature of the VLS system is the explicit separation of the genotype and the phenotype (problem solution) that the genotype decodes into. Repetition of allele values in the chromosome is allowed, but we insist that each allele is represented at least once in the chromosome. A chromosome (genotype) is decoded into the problem solution (phenotype) by reading each allele in the chromosome from left to right. If the

---

[1] Information for this section was drawn from [3], [4], [5], [6] and [7].

[2] Levenick's "introns" were actually analagous to spacing DNA, as they were located between genes. In the remainder of this paper, we will use Levenick's terminology.

allele does not yet appear in the phenotype, then it is added it onto the end of phenotype, otherwise it is ignored.

3.2 VLS Crossover

Firstly, two parent chromosomes are selected (in any of the standard ways). As an example, let us imagine that the two parents shown in Fig. 1 have been selected.

Parent 1

Genotype	a	d	a	c	e	a	b
Phenotype	a	d		c	e		b

Parent 2

Genotype	d	d	c	d	e	c	b	a
Phenotype	d		c		e		b	a

FIGURE 1 - TWO PARENT CHROMOSOMES

A random number between 1 and the number of different alleles (in a TSP, the number of cities) is generated. In our example, let us suppose that the number 3 was generated.

Next, we copy from the first parent into the offspring's genotype and phenotype until there are that number of expressed alleles in the offspring's phenotype. In the example, this brings us to the situation shown in Fig. 2.

Child

Genotype	a	d	a	c
Phenotype	a	d		c

FIGURE 2 - PARTLY-CONSTRUCTED OFFSPRING

The next step is to locate the locus on the genotype of parent 2 which results in the expression, in the parent's phenotype, of the last allele copied into the offspring's phenotype. In the example, we locate the allele 'c' at position 3 in parent 2's genotype.

Starting from the locus after that just identified, we copy from parent 2's genotype and phenotype into the offspring until all of the alleles are expressed in the child's phenotype. Figure 3 shows the result of doing this in the example. If the end of the parent's genotype is reached before all alleles are expressed in the offspring's phenotype, we continue from the left-hand end of the parent's genotype.

Child

Genotype	a	d	a	c	d	c	c	b
Phenotype	a	d		c		e		b

FIGURE 3 - THE COMPLETED OFFSPRING

3.3 VLS Mutation

The mutation operator is a type of transposition. Two random positions on the phenotype are selected, and the alleles in the genotype which give rise to the values in the phenotype at those positions are swapped. The phenotype is then reconstructed from the genotype.

4. Test Runs

The VLS system was tested against two established operators, namely PMX and C1. Each of the three algorithms was run, with 15 combinations of parameters, on two TSPs.

4.1 Test Run Parameters

The algorithms were each tested using all 30 combinations of three parameters, so that no one algorithm would gain an advantage in that the tests were carried out with parameters that particularly favour that algorithm. The parameters that were varied are:-

Probability of Mutation The probability of a chromosome being mutated was set to 0.2, 0.075, 0.05, 0.02 and 0.005.

Selection Pressure Parents were chosen to reproduce on the basis of their rank (as opposed to their fitness or scaled fitness value). Reeves [2] reports the use following probability function for parent selection:-

$$p([k]) = 2k / [\, M(M + 1)\,]$$

where k is the potential parent's ranked position, M is the population size and p([k]) is the probability of that chromosome being selected. Two related probability functions,

$$p_1([k]) = \sqrt{p([k])}$$

and

$$p_2([k]) = \sqrt{p_1([k])}$$

were also used. p([k]) produces a higher selective pressure than does $p_1([k])$, which in turn produces a higher selective pressure than does $p_2([k])$.

Elitism Elitism [9], which is the practice of copying the most fit member of one generation unaltered into the next generation (also known as cloning), could be enabled or disabled.

4.2 The Travelling Salesman Problems Used

Two related sets of test data were used. Each contained data on 29 cities in Bavaria. One set of data relates to the street distances between the cities, the other set to the geographic distances between the cities. This data was obtained via ftp. Details are provided in appendix A.

4.3 Test Run Results

The programs were run on the two sets of data. The parameters used for the runs are listed in Table 1. The results obtained using the street distance data are shown in Figs. 4 and 5.

TABLE 1 - PARAMETER SETTINGS USED IN TEST RUNS

Run Number	Parent Selection Probability Function	Probability of Mutation
1	p([k])	0.200
2	p([k])	0.075
3	p([k])	0.050
4	p([k])	0.020
5	p([k])	0.005
6	$p_1([k])$	0.200
7	$p_1([k])$	0.075
8	$p_1([k])$	0.050
9	$p_1([k])$	0.020
10	$p_1([k])$	0.005
11	$p_2([k])$	0.200
12	$p_2([k])$	0.075
13	$p_2([k])$	0.050
14	$p_2([k])$	0.020
15	$p_2([k])$	0.005

There was a very strong agreement between the results obtained using the street distance data and those obtained using the geographic distance data.

VLS performs well, in comparison with the other two operators tested, on both data sets in the majority of the tests.

VLS appears to be the most robust representation with respect to changes in parameters, particularly in the runs with low selection pressure (runs 11 to 15, which use $P_2([k])$) and without elitism. This could prove to be an important attribute of the VLS representation, as it is seldom known, without prior experimentation, which combination of parameters will be most suitable for a particular genetic algorithm representation operating against a particular problem data set.

It is interesting to observe the strong beneficial effect that elitism has, especially in the tests in which selection pressure was low.

It must be stated, however, that VLS is the most expensive representation in terms of processing time and memory requirements.

5. Conclusions and Further Work

In the experiments conducted to date, the VLS has performed well and appears to be tolerant of its parameter settings. Tests on other data sets are yet to be performed. The question of how well the VLS performs on large data sets is important for practical applications.

A detailed investigation into the internal workings of the VLS representation is also required. It may also be enlightening to investigate the operation of the VLS representation theoretically. In particular, it will be informative to investigate the cycles of dormancy and expression that the alleles undergo. It is expected that the performance of VLS may be due to the genetic memory.

Also of interest is the chromosome length. Initial investigation shows the average chromosome length rises sharply at the start of a run, peaks, and then falls sharply, settling somewhere just above the minimal genome length.

Acknowledgements

I would like to thank Dr. Alan Soper of the University of Greenwich for his constructive criticism of this work. I also thank SERC, the UK Science and Engineering Research Council, for funding me during this research.

References

1. Goldberg, D.E. and Lingle, R.: Alleles, Loci, and the Travelling Salesman Problem. International Conference on Genetic Algorithms and Their Applications, pp154-159. Hillsdale, NJ: Lawrence Erlbaum Associates 1985.
2. Reeves, C. Modern Heuristic Techniques for Combinatorial Problems, Blackwell Scientific Press 1993.
3. Dawkins, R.: The Blind Watchmaker. London: The Penguin Group 1991.
4. Dulbecco, R.: The Design of Life. Yale University 1987.
5. Jones, R.N. and Karp, A.: Introducing Genetics. London: John Murray (Publishers) Ltd. 1986
6. Lloyd, J.R.: Genes and Chromosomes. Basingstoke: Macmillan Education Ltd. 1986.
7. McLannahan, H.: Heredity and Variation. In Skelton, P. (ed.): Evolution : A Biological and Palaeontological Approach. Wokingham: Addison-Wesley Publishing Company 1993.
8. Levenick, J. R.: Inserting Introns Improves Genetic Algorithm Success Rate: Taking a Cue from Biology. Proceedings of the Fourth International Conference on Genetic Algorithms, pp123-127. Los Altos, CA: Morgan Kaufmann.
9. De Jong, K.A. Doctoral Dissertation. An Analysis of the Behavior of a Class of Genetic Adaptive Systems. University of Michigan 1975.

Appendix A - Details of TSP Data Source

The test data used in this paper is part of a larger collection of TSP data which has been compiled and made publically available by Gerhard Reinelt of the Institut für Mathematik, Universität Augsburg.

The data can be obtained by ftp as follows:-

>ftp 128.42.1.30
LOGIN USERID : **anonymous**
Password : **anonymous**
>cd public
>type binary
>get tsplib.tar

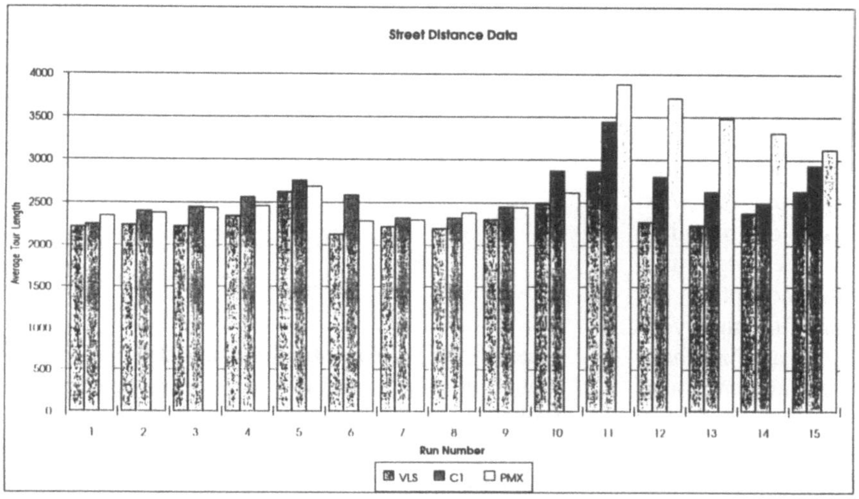

FIGURE 4 - AVERAGE BEST TOUR LENGTHS FOR STREET DISTANCE DATA

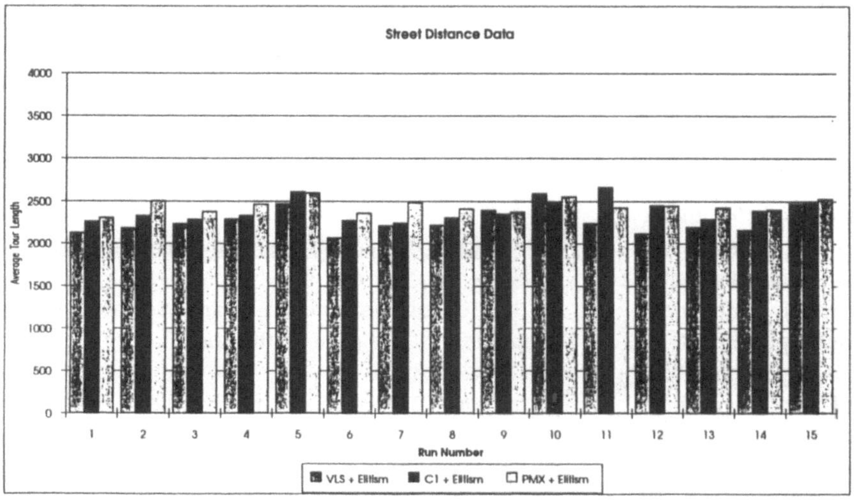

FIGURE 5 - AVERAGE BEST TOUR LENGTHS FOR STREET DISTANCE DATA (WITH ELITISM)

THEORETICAL BOUNDS FOR GENETIC ALGORITHMS

David Reynolds and Jagannathan Gomatam

Department of Mathematics, Glasgow Caledonian University
Cowcaddens Road, Glasgow G4 0BA U.K.

Abstract : We investigate various properties of Genetic Algorithms. We present 1) a result which highlights the potential existence of long-run crossover and mutation bias in a GA, together with a partial avoidance strategy based on mutation operators, 2) an analysis of GA premature convergence, which indicates the advantage of larger populations, 3) an application of the theory of metric spaces which describes an intrinsic smoothness property of GA populations and highlights the safety in numbers principle and 4) a lower bound on the convergence rate of a mutation-crossover GA. The models presented are generally independent of solution encoding and are thus applicable to a wide range of Genetic Algorithms.

1. Introduction

This paper is concerned with presenting analyses of various properties of Genetic Algorithms, such that observed phenomena of GA-based function optimization can be described theoretically. In this section we briefly review the work of [5]; there, GAs are classified in terms of the sampling strategies used on populations, and stochastic models formulated for GAs which implement solution sampling with and without replacement. In this paper, the class of GA/WI algorithms [5], which are a generalisation of the algorithms presented in [6] are analysed. The stochastic model for these algorithms, presented in [5] is

$$P_{\underline{XY}} = \frac{N!}{Y_1!Y_2!\ldots Y_k!}\prod_{i=1}^{k}(p_{xi})^{Y_i} \qquad (1.1)$$

$$p_{xi} = \sum_{j=1}^{k}\frac{X_j f_j}{\underline{X}\cdot\underline{f}}\sum_{d=1}^{k}\frac{X_d f_d}{\underline{X}\cdot\underline{f}}\sum_{b=1}^{k}\eta_{jb}\sum_{e=1}^{k}\eta_{de}C_{eb,i}^{*} \qquad (1.2)$$

$$C_{eb,i}^{*} = \begin{cases} C_{eb,i} & e \le b \\ C_{be,i} & e > b \end{cases} \qquad (1.3)$$

where $\eta > 0$ is a row-stochastic mutation probability matrix, and $C_{ij,m}$, $j \ge i$ is a row-stochastic crossover probability matrix (where the rows of the latter are indexed by ordered pairs of parent solutions). The population state-space of this algorithm is $S^{[k]} = \{\underline{X} = (X_1,\ldots,X_k): X_1+\ldots+X_k = N, X_i \text{ a positive integer}\}$, where X_i counts the number of solution i in population $\underline{X}$, and where $s = \{s_1,\ldots,s_k\}$ is the finite set of solutions to a given discrete maximisation problem; $\underline{f} = (f_1,\ldots,f_k)$ is the vector of associated costs (see [5] for an explicit formula for the size of $S^{[k]}$). Various properties of the model given in equations (1.1-1.3) are investigated in [5]; further properties are presented here.

2. Long-Run Crossover and Mutation Bias

In this section we proceed to analyse the long-run properties of GAs through this model;

Definition 2.1: Let e be a matrix whose rows store the population vectors in the same order in which they index **P** so that e is of size $|S^{[k]}| \times k$. Then $\underline{e}_t$ is a vector containing X_t of solution t across the populations.

Definition 2.2: Let $F(\underline{X})$ represent the population cost at stationarity, (i.e., $F(\underline{X}) = \underline{X}.\underline{f}$, where $\underline{X}$ is a random vector with probability distribution equal to the stationary distribution corresponding to equations (1.1-1.3)) and $\overline{F(\underline{X})}$ its expected value.

(The stationary distribution $q_{\underline{X}}$ is shown to exist in [5], when $\eta > 0$). An important result for GAs is

Theorem 2.1 : If $\eta > 0$ then

$$N\sum_{b=1}^{k} f_b \theta_b \leq \overline{F(\underline{X})} \leq N\sum_{b=1}^{k} f_b \varepsilon_b \tag{2.1}$$

where $\theta_b = \min C_{td,b}$, $\varepsilon_b = \max C_{td,b}$ $\forall$ $1 \leq t \leq d \leq k$ and $1 \leq b \leq k$; when $C_{ii,i}=1$ $\forall 1 \leq i \leq k$,

$$\overline{F(\underline{X})} \geq N\sum_{b=1}^{k} f_b \tau^2_b \tag{2.2}$$

where $\tau_b = \min \eta_{ib}$ $1 \leq b \leq k$, $\forall 1 \leq i \leq k$.

Proof: $\overline{F(\underline{X})} = \sum_{\underline{X}\in S^{[k]}} q_{\underline{X}} . F(\underline{X})$

$= \sum_{\underline{X}\in S^{[k]}} q_{\underline{X}} . (\underline{X}.\underline{f})$

$= \sum_{b=1}^{k} (\underline{q}.\underline{e}_b) f_b = \sum_{b=1}^{k} (\underline{q}.(P\underline{e}_b)) f_b = \sum_{b=1}^{k} (\underline{q} \cdot \underline{a}(b)) f_b$

[here $a_{\underline{X}}(b) = \sum_{\underline{Y}\in S^{[k]}} \binom{N}{\underline{Y}} \prod_{i=1}^{k} (p_{xi})^{Y_i} Y_b$]

$= N\sum_{b=1}^{k} [f_b \sum_{\underline{X}\in S^{[k]}} q_{\underline{X}} \{$

$$\sum_{j=1}^{k} \frac{X_j f_j}{\underline{X}\cdot\underline{f}} \sum_{d=1}^{k} \frac{X_d f_d}{\underline{X}\cdot\underline{f}} \sum_{g=1}^{k} \eta_{jg} \sum_{e=1}^{k} \eta_{de} C_{eg,b}^{*} \}] \tag{2.3}$$

so that (2.1) and (2.2) follow by direct substitution of the given restrictions into (2.3). (Note that (2.3) follows by the standard formula for the expected value of the multinomial distribution [2]).

Theorem 2.1 shows that the expected population cost in this GA may be biased in the long run by the crossover and mutation operators. As practical guidance, the introduction of a symmetric mutation matrix (achievable with a fixed length binary encoding discussed in [5]) means that τ_b is the same for all solutions.

3. Premature Convergence of GAs

A problem faced when implementing a GA is that of premature convergence [1]; clearly, it is important to analyse conditions under which the algorithm can be expected to converge prematurely.

Definition 3.1: Let $\alpha(\underline{X})$ denote the number of time-steps in which population $\underline{X}$ is occupied by GA/WI, once entered, and before exiting.

Clearly, by viewing the transitions of the algorithm as $\underline{X} \rightarrow \text{not}(\underline{X})$, the probability distribution of $\alpha(\underline{X})$ is geometric, so that

$$\Pr[\alpha(\underline{X}) = n] = (P_{\underline{X}\underline{X}})^n (1 - P_{\underline{X}\underline{X}}) \tag{3.1}$$

Of interest in measuring premature convergence are the statistics of (3.1), (occasionally referred to in the theory of Markov Chains as occupation time statistics [3]) in particular, its first moment; this is given by

$$E[\alpha(\underline{X})] = \frac{P_{\underline{X}\underline{X}}}{1 - P_{\underline{X}\underline{X}}} \tag{3.2}$$

From these definitions follows,

Theorem 3.1: Let $\underline{X} \in S^{[k]}$, where $X_i=N$ for some $1 \leq i \leq k$ (i.e., $\underline{X}$ is a convergent population). Then, if $C_{ii,i} = 1$, the following hold;

1) $E[\alpha(\underline{X})] > \beta > 0$ if $\eta_{ii} > (\frac{\beta}{\beta+1})^{1/2N}$ (N fixed)

2) $\beta > E[\alpha(\underline{X})] > 0$ if $N > \ln_{\gamma_i} \frac{\beta}{\beta+1}$ (η fixed)

where $\gamma_i = (\sum_{b=1}^{k} \eta_{ib} \sum_{e=1}^{k} \eta_{ie} C_{eb,i}^{*})$

Proof : By substitution in (1.1),

$$1)\ E[\alpha(\underline{X})] = \frac{P_{\underline{X}\underline{X}}}{1 - P_{\underline{X}\underline{X}}} = \frac{(\sum_{b=1}^{k} \eta_{ib} \sum_{e=1}^{k} \eta_{ie} C_{eb,i}^{*})^N}{1 - (\sum_{b=1}^{k} \eta_{ib} \sum_{e=1}^{k} \eta_{ie} C_{eb,i}^{*})^N}$$

$$\geq \frac{\eta_{ii}^{2N}}{1 - \eta_{ii}^{2N}} > \beta \Leftrightarrow \eta_{ii} > (\frac{\beta}{\beta+1})^{1/2N}$$

2) Setting $\gamma_i = (\sum_{b=1}^{k} \eta_{ib} \sum_{e=1}^{k} \eta_{ie} C_{eb,i}^{*})$ (fixed)

$$E[\alpha(\underline{X})] = \frac{\gamma_i^N}{1-\gamma_i^N} < \beta$$

$$\Leftrightarrow N > \ln_{\gamma_i} \frac{\beta}{\beta+1}$$

(It can be shown that $0< \gamma_i <1$ $\forall$i, whenever $\eta>0$). That the occupation time for the sub-optimal populations does not depend on their cost relative to the optimal population(s) is an important result, since it shows that premature convergence is an unavoidable feature of population based algorithms.

4. Metric Space Analysis for GAs

In this section the collection of population states of GA/WI is formulated as a metric space; an operator independent property of this space is described, which serves to highlight those conditions under which the 'safety in numbers' principle of GAs operates [1]. We define as a notion of population distance;

Definition 4.1: Let

$$\rho(\underline{X},\underline{Y}) = \sum_{j=1}^{k} (X_j - Y_j)\big|_{>0} \qquad \underline{X},\underline{Y} \in S^{[k]} \qquad (4.1)$$

Thus, $\rho(\underline{X},\underline{Y})$ represents the sum of the positive differences (in components) of $\underline{Y}$ from $\underline{X}$; it corresponds to the minimal number of solution changes to $\underline{X}$ in order to produce $\underline{Y}$ and gives a reasonably practical notion of population distances in a GA. $(S^{[k]}, \rho)$ has the following useful property;

Theorem 4.1: $(S^{[k]}, \rho)$ is a discrete metric space [4].

Proof : We check that the metric space axioms hold;
(i) Clearly, $\rho \geq 0$ by definition.
(ii) It can be shown that

$$\sum_{j=1}^{k} (X_j - Y_j)\big|_{>0} + \sum_{j=1}^{k} (X_j - Y_j)\big|_{<0} = \sum_{j=1}^{k} (X_j - Y_j) = 0$$

implies $\rho(\underline{X},\underline{Y}) = \rho(\underline{Y},\underline{X})$
(iii) It can also be shown that

$$\rho(\underline{X},\underline{Y}) = \frac{1}{2}\sum_{j=1}^{k} |X_j - Y_j|$$

$$\Rightarrow \rho(\underline{X},\underline{Y}) + \rho(\underline{Y},\underline{Z}) = \frac{1}{2}\left(\sum_{j=1}^{k} |X_j - Y_j| + \sum_{j=1}^{k} |Y_j - Z_j|\right)$$

$$\geq \rho(\underline{X},\underline{Z})$$

by application of the triangle inequality, component-wise.

Note that to each $\underline{X}$ in $S^{[k]}$ is associated a population cost, computed by $\underline{X}.\underline{f}$; we define a function which provides the mean relative change in cost about $\underline{X}$. First, we define;

Definition 4.2: Let $n_1(\underline{X})$ be defined as

$$n_1(\underline{X}) = \{\underline{Y} \in S^{[k]} : \rho(\underline{X},\underline{Y}) = 1\} \qquad (4.2)$$

It is a simple matter to show that

$$n_1(\underline{X}) \neq \emptyset \qquad \forall \underline{X} \in S^{[k]}$$

Thus $n_1(\underline{X})$ is the set of 'nearest-neighbours' of $\underline{X}$.

Definition 4.3: Let $\overline{\Delta F_r}(\underline{X})$ be defined by

$$\overline{\Delta F_r}(\underline{X}) = \frac{1}{|n_1(\underline{X})|} \sum_{\underline{Y} \in n_1(\underline{X})} \frac{\underline{Y}.\underline{f}}{\underline{X}.\underline{f}} \qquad \underline{X} \in S^{[k]} \qquad (4.3)$$

Thus, $\overline{\Delta F_r}(\underline{X})$ computes the mean relative change in cost about a population $\underline{X}$; we assume an ordering on the costs (since we do not assume any particular encoding for solutions), so that $f_1 \leq f_2 \leq ... \leq f_k$. Then the following holds;

Theorem 4.2:

$$1 - \frac{f_k - f_1}{Nf_1} \leq \overline{\Delta F_r}(\underline{X}) \leq 1 + \frac{f_k - f_1}{Nf_1} \qquad \underline{X} \in S^{[k]} \qquad (4.4)$$

Proof:

$$\overline{\Delta F_r}(\underline{X}) = \frac{1}{|n_1(\underline{X})|} \sum_{\underline{Y} \in n_1(\underline{X})} \frac{\underline{Y}.\underline{f}}{\underline{X}.\underline{f}} \qquad \underline{X} \in S^{[k]}$$

$$\leq \frac{1}{|n_1(\underline{X})|} \sum_{\underline{Y} \in n_1(\underline{X})} \frac{\underline{X}.\underline{f} + f_k - f_1}{\underline{X}.\underline{f}}$$

$$\leq \frac{1}{|n_1(\underline{X})|} \left(1 + \frac{f_k - f_1}{Nf_1}\right) \sum_{\underline{Y} \in n_1(\underline{X})} .$$

$$= 1 + \frac{f_k - f_1}{Nf_1}$$

Similarly,

$$\overline{\Delta F_r}(\underline{X}) \ge \frac{1}{|n_1(\underline{X})|}\sum_{\underline{Y}\in n_1(\underline{X})} \frac{\underline{X}.f-(f_k-f_1)}{\underline{X}.f}$$

$$\ge 1-\frac{f_k-f_1}{Nf_1} \qquad \underline{X}\in S^{[k]}$$

Thus, when specifying the population size, it can be taken large enough so that the resultant metric space contains populations which are very close in cost to their near-neighbours.

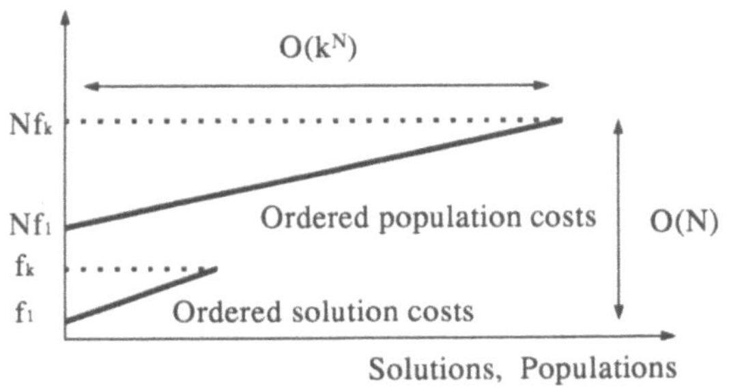

Figure 1: Transformations of Scale in GA

The transformations of scale inherent in using a GA are best summed up in Fig. 1 above.

5. Rate of Convergence of Simplified GA

In this section the rate of convergence of a mutation/crossover GA, under a fixed length (L) binary encoding for solutions is examined; to remove the selection operator we set the population size, N equal to 1. It can be shown that this type of convergence is dominated by the magnitude of the second eigenvalue, λ_2, of the transition matrix **P** given by (1.1) [3]. Since N=1, we can relabel the populations by the solution index to obtain from **P** a transition matrix **P'**, indexed by solutions; note that **P'** is identical to **P**. For this GA, with transition matrix **P'**, with eigenvalues $1=\lambda_1 > |\lambda_2| \ge \ldots |\lambda_k|$,

Theorem 5.1:

If $\eta_{ij} = p_m^{H_{ij}}(1-p_m)^{L-H_{ij}}$, and $C_{ii,i} = 1\ \forall i$

then $$p_m < 1-\left(\frac{\varepsilon-1}{2^L}+1\right)^{-2L} \Rightarrow |\lambda_2| > \varepsilon > 0 \qquad (5.1)$$

Proof: Since, $tr(P') = \sum_{i=1}^{k} P'_{ii} = \sum_{i=1}^{k}\lambda_i$, (given in [3])

if follows that (since $\lambda_1=1$)

$$1+\sum_{i=2}^{k}|\lambda_i| \ge \sum_{i=1}^{k} P'_{ii} \ge \sum_{i=1}^{k}\eta^2_{ii} = 2^L(1-p_m)^{2L}$$

$$\Rightarrow |\lambda_2| \ge 2^L(1-p_m)^{2L} - (1+\sum_{i=3}^{k}|\lambda_i|)$$

$$> 2^L(1-p_m)^{2L} - (1+2^L-2)$$

(since $0<|\lambda_i|<1$ for all $i\neq 1$)

$$= 2^L((1-p_m)^{2L}-1)+1$$

and so the result (5.1) follows by substitution. Thus, it is shown that the mutation/crossover GA can be parameterized so that convergence occurs only very slowly; the given mutation and crossover operators are commonly used, as are low mutation rates p_m.

6. Conclusion

Future work involves the combination of these results, together with those reported in [5], [6] and other works, to produce a theoretical framework for analysing GAs which is reasonably general and provides guidance towards important aspects of GA implementation and use.

References

1.Goldberg D.: Genetic Algorithms in Search, Optimization ,and Machine Learning, Reading, MA: Addison Wesley 1989.

2.Mendenhall, W., Wackerly, D.D., Scheaffer, R.L.: Mathematical Statistics with Applications. Boston : PWS-Kent.

3.Narayan Bhat, U.:Elements of Applied Stochastic Processes (2nd. Edn.), New York: Wiley 1984.

4.Pitts C.G.C.: Introduction to Metric Spaces. Edinburgh : Oliver and Boyd 1972

5.Reynolds, D., Gomatam, J.: Stochastic Modelling of Genetic Algorithms, to appear in *Artificial Intelligence* Journal.

6.Suzuki, J.: A Markov Chain Analysis on A Genetic Algorithm, in Proceedings of the Fifth International Conference on Genetic Algorithms, Forrest, S. (ed.), Urbana Illinois. Morgan Kaufmann 1993

AN ARGUMENT AGAINST THE PRINCIPLE OF MINIMAL ALPHABET

Gary M. GIBSON,

School of CIS, University of South Australia,
The Levels, Pooraka, S.A. 5095, AUSTRALIA.

For many years the field of Genetic Algorithms (GAs) has been dominated by bit-string based GAs. The argument as to why bit-strings are the best method of encoding parameters in GA strings is known as the principle of minimal alphabet.

In this paper, an alternative theory as to the activity that GAs carry out inside parameters encoded in bit-strings is presented. The results support what many researchers have known for a long time; that bit-strings have no special significance for GAs. The most interesting findings are: GAs do not process schemata within encoded qualitative parameters; and only process them within encoded quantitative parameters to a limited extent (simulating a binary search).

INTRODUCTION

A well established practice in the construction of GAs is to use bit-strings to represent parameters. There is ample evidence that approaches that do not employ bit-strings are just as effective. Genetic Programming (GP) [9] contains a full description of this field), for example, is an entire branch of GAs that generally does not employ bit-strings. Most GP applications instead use strings composed of parse trees.

Close examination of the activity of GAs inside qualitative and quantitative parameters encoded in bit-strings raises doubts as the wisdom of using bit-strings for all GA applications. First, a recap of the relevant theory on bit-string GAs (See [5] and [8] for in-depth theoretical discussion). Consider the problem of designing some device, with l design choices, where each design choice is between two possible alternatives. The search space of all possible designs therefore contains 2^l points. In this illustration, let l be arbitrarily set to 8 which means that there are $2^8 = 256$ possible designs. The problem can be represented as a binary string of length l. If, through trials of different designs, it is discovered that option '0' for design choice 'two' and option '1' for design choice 'five' seem to often be in the most successful designs then it would appear logical that designs should be tried which use both together.

Such an observation can be represented succinctly using *schemata* notation. In this notation one individual *schema* is represented as a similarity template of length l with '#'s replacing the bits which are not involved in the observation (i.e. the alphabet is {0,1,#}). For example, the observation described above could be represented by the schema

H = #0##1###

In the space of 2^l possible designs there are 3^l possible schema and hence 3^l possible similarities. This would appear to introduce more problems than it solves since the search space for similarities is larger than the original search space for designs (in the illustration there are $3^8 = 6561$ possible schema). However, this is not the case because schemata are not mutually exclusive. In the example above, the schema H will be present in all strings which have a '0' in position two and a '1' in position five. This is one full quarter of all possible strings. In fact a GA implicitly examines more than $O(n^3)$ schema in a population of n strings(See [5] pp. 40-41 for the formal proof.). This effect is called *implicit parallelism* and is very important since the GA carries out n fitness function evaluations to do this.

The reasoning as to why bit-strings are the best form of representation in GAs is known as the principle of *minimal alphabet* ([5], pp. 80-82.); a binary alphabet being the smallest. The argument is essentially that, the smaller the alphabet used to encode parameters, the longer the string containing those parameters will be. Since an alphabet of cardinality k can represent k^l different values, where l is the length of the string, it follows that, the smaller that k is, the larger l must be in order for the encoding to represent the same range of values. For a string composed entirely of elements of cardinality k there will be $(k+1)^l$ schemata present within the string (at each element, the similarity template may contain a value representable in the alphabet or, alternatively, a "#"). A larger l means more schemata when the degree of trade-off against k is considered. According to traditional theory, the more schemata present in a string, the more leverage the GA has in developing useful building blocks since each string is testing all the schemata present in it simultaneously.

Viewed from another perspective, however, this is not as ideal as it first appears. For the classic binary choice design problem mentioned above, Holland's [7] schemata theory holds true but what if there were more than two alternatives for a given choice?

QUALITATIVE PARAMETERS

Consider a design choice with three alternatives (A, B and C) instead of two. The binary representation would occupy two bits with three of the four possible bit patterns that these bits could produce in use. These are arbitrarily assigned to represent the alternatives (e.g. "00" could represent choice A, "01" could represent B, and "10" could represent C). Note that there is immediately a mapping problem. The unallocated pattern "11" must be either made invalid or mapped to another legal value; biasing the GA. It is usually is mapped, since an invalid string is **guaranteed** to be of no use.

Consider the two-armed bandit problem [7]. The argument is that the GA is attempting to find the 'best' allele at each bit position from the two competing alleles "0" and "1". Holland calculated that the optimal combination (in order to minimise expected loss) was to allocate slightly more than exponentially increasing trails to the observed best allele (bandit arm) (See [8] pp. 78-83 for the formal proof.). A GA assigns at least an exponentially increasing number of trials to the observed best building blocks (composed of particular allele), as shown by the *Fundamental Theorem of Genetic Algorithms* (See [5], pp. 28-33 for the full derivation.). Therefore the GA is utilising a near-optimal strategy **automatically** without any book-keeping or explicit statistics gathering.

In the binary design choice problem this is logical but what of problems possessing parameters with 3 (or more) alternatives? Since the choice parameter is qualitative, it follows that all states are effectively numerically independent; i.e. they are unrelated (recall that they are assigned to bit patterns entirely arbitrarily). For example, consider a 3 state parameter Q encoded into a set of bit positions (sub-string) s_q in the strings of a population. If a particular string has state B ("01") encoded for Q, then, according to traditional theory, all schemata with s_q set to "01","0#","#1" or "##" are simultaneously tested by that string as part of the implicit average fitness calculations taking place in the population for all the schemata represented in it.

If B encodes a state that is associated with strings of above average fitness then theoretically any other state that has the same schemata (ignoring the 'all' ("01") and 'nothing' ("##") cases) should also tend to be associated with strings of better fitness, although perhaps not to the same extent. For the example above, this implies that strings containing option *A* (with the pattern "00" that contains the schemata "0#") and whatever state that the pattern "11" is remapped to should have a higher average fitness than expected by chance. But there are no grounds for this supposition since there is no relationship between the patterns and the states they represent. It is analogous to stating that, "*if yellow cars are associated with fewer vehicle accidents, black and blue cars are also likely to be safer because they are encoded with a similar bit pattern to yellow*". This would bias a GA's search towards black and blue at the expense of other colours (such as white) on the basis of the performance of yellow.

The binary sub-string that represents an *m*-state parameter is totally epistatic (Epistatic problems are those with a high level of interdependence between parameters, or in this case, bits). Therefore, no useful average fitness information can be gathered for **any** of the bits in the sub-string since the fitness of a particular allele at one of the bits of *Q* is 100% dependent on the values of **all** the other bits in the sub-string. This means that while the string's fitness can be used in the calculation of the average fitness of schemata containing "01" or "##" (the 'all' and 'nothing' cases) at s_q, it cannot be used for those containing "0#" or "#1".

Grefenstette and Baker [6] touched on this issue when they showed that the estimate of fitness for the "0" and "1" states could be wrong when a highly epistatic sub-string is involved. They showed that, where a given allele was part of a highly fit larger pattern, the GA would rapidly fill the population with instances of that pattern, regardless of the average fitness of that allele compared to the other allele at that position. Their conclusion was that the two-armed bandit analogy did not hold under these circumstances. They state that "*The analogy is strongest when the hyperplanes* [i.e. schemata] *involved exhibit little variance in their fitness, and is weakest when the hyperplanes have high fitness variance...*". The variation of the fitness of the schemata sharing the same allele at a given bit is extremely high (the same as for completely random fitnesses on average) in encoded qualitative parameters since the patterns are unrelated, as show above. Therefore, **schemata processing is meaningless within a sub-string representing a qualitative parameter**.

The implication of this finding is that, although using a binary representation for representing qualitative parameters in strings will increase the number of schemata that the GA is processing, the GA is incapable of processing them effectively. In a string composed of *P* qualitative parameters, each of length *z*, there are only 2^P schemata effectively being tested in a string since, for each parameter, the choice is between the two states of 'known' and 'unknown' ("01" and "##" in the example above). This contrasts with the $2^{P \cdot z}$ ($P \cdot z = l$ where l is the length of the string in bits) suggested by traditional schemata theory. Consequently, any alteration of the sub-string representing a qualitative parameter by operators such as crossover will act to disrupt the value (with the mapping bias mentioned above) with no significant probability of improving it. It is therefore acting as a very high probability mutation operator. This explains why inserting introns (bits that do nothing) between encoded qualitative parameters [10] causes a performance improvement. Their presence reduces the probability of crossover disrupting parameter values since crossover will often cross between introns.

The probability p_d that a given crossover point will disrupt one of the parameters is given by the formula

$$p_d = \frac{P \cdot (z-1)}{P \cdot z - 1}$$

For example, the probability of a cross point disrupting a string with 5 qualitative parameters, each holding up to 8 states (3 bits), is 0.71, i.e. very likely. Mutation of such sub-strings is also problematic because bit-string mutation is biased. Regardless of the encoding scheme used (typically 2's complement or Gray-coding; see below), only a subset *x* of the possible states of the parameter are reachable from one particular state with a single bit mutation. For parameters encoded by *z* bits, *x* has *z* states. For a two or three state parameter this is not an issue since all other states are reachable in one mutation. However, for an 8 state parameter for example, this is only 3 states from 8. All other states (excluding the current one) are considerably more difficult to reach since they require two or more **simultaneous** mutations. If the probability of mutation is given by p_m, then those states reachable by two mutations, for example, have a probability of selection $p_m^2 \cdot (1-p_m)^{l-2} \cong p_m^2$ (for $0 < p_m << 1$) but those reachable by one mutation have a probability of selection $p_m \cdot (1-p_m)^{l-1} \cong p_m$, which is considerably greater.

The mutation operator will therefore be biased towards alternatives whose bit string encodings are similar to the bit pattern that is currently set. This bias can only be detrimental since there is no relationship between qualitative states encoded in similar patterns. Note that the mapping bias mentioned above will also affect mutation. None of these problems occur if the parameter is treated as an atomic unit of alphabet size *m*, where *m* is the number of alternatives in the qualitative variable, since mutation will be unbiased and crossover cannot disrupt the value. The mutation operator would then become the act of randomly choosing between the *m*-1 states not currently encoded in order to guarantee change (for two states this degenerates to flipping a bit to its alternate state). This operator has the disadvantages of needing a more complicated string structure, and requiring range information to be stored for each encoded parameter.

A note on 2's complement *vs.* Gray-coding; Caruana and Schaffer [1] found that GAs perform better with Gray-coded parameters than with 2's complement encoded parameters. In Gray-coding only one bit changes for every change in phenotypic value by 1 (see Caruana and Schaffer for an exact description). This avoids the problem found with 2's complement encoding known as *Hamming Cliffs* which is mentioned below.

QUANTITATIVE PARAMETERS

If bit-string based GAs are so destructive to simple qualitative parameters then why are they so successful in parameter optimisation problems generally? The answer may be that GAs have traditionally been used primarily to optimise quantitative parameters - a task they do well. But, even with parameters of this type, GAs do not perform true schemata processing within the parameter. As will be shown below, there are some useful effects of GAs operating on bit-strings encoding quantitative parameters but in the main these are **coincidental**. Further, it will be argued that there are only *z* potentially useful building blocks present in the sub-string representing a quantitative parameter (where *z* is the length of the sub-string) to a given problem.

Consider crossover and mutation applied to encoded numeric parameters. As shown in the section above, mutation of any sub-string representing a parameter will be biased towards bit patterns similar to the original sub-string. In 2's complement encoded (the most common) or Gray-coded strings, similar patterns are generally close in phenotypic value after decoding since the high order bits will generally be the same. Although every bit controls access to 50% of the search space, the highest order bit controls 50% of **contiguous** phenotypic search space. It is impossible for a sub-string to encode a number in the top half of the search space (ignoring double mapping) unless the top bit is a '1'. This contrasts with the lowest order bit which controls access to every second phenotypic value. Mutation will tend to, therefore, be biased towards the phenotypic local area - a useful effect in quantitative parameters since the chance that the optimum value for a given parameter is close to a known good value would appear to be better than chance in search spaces containing hills. Note that this is purely coincidental and is due to the way 2's complement and Gray-coded bit strings encode numbers - an encoding scheme that arbitrarily assigned values to bit patterns would not experience the same effect.

Crossover of bit-strings (2's complement or Gray-coded) reveals another significant coincident. Because bits are encoded adjacent to each other in descending order of impact on phenotypic value, traditional crossover is likely to cross bits of similar order of a quantitative parameter sub-string into a child string simultaneously. This is of particular importance when the bits in question are the highest order bits because it means that the GA will very often cross good sets of high order bits (controlling a contiguous section of the phenotypic search space) with different combinations of low order bits from other strings. The low order bits are meaningless when not placed in context with the high order bits but they serve as a pool of different bits which the GA is then able to combine with good sets of high order bits in order to find the optimum.

This general effect is already well-known [5] but what about its implications for schemata processing? According to traditional argument this is an example of the GA in action carrying out full schemata processing. However, this is not necessarily the case. Consider, for example, the pattern "100" as the last three bits of an eight bit 2's complement number. Under all circumstances it will add the value 4 to whatever the five high order bits represent. This pattern could therefore be part of a larger pattern representing a quantity anywhere from 4 to 252 (in a range of 0 to 255) in increments of 8. This effectivelv makes the calculation of average fitness for schemata that have fixed values (i.e. contain non-"#"s) for any of these three bits worthless.

More generally, schemata with bits set in a sub-string representing a quantitative variable can only be usefully processed if the set bits are contiguous and include the highest order bit. Even a schema which has a "#" in only its highest order bit position will not effectively be processed because the value of the top bit will move the decoded value of the whole pattern by 50% of the entire range representable by the encoding scheme. This makes it impossible for a useful average fitness value to be calculated for that schema.

As shown in the example above, all the higher order bits of a nominated bit must be known before even an estimate of the bit's impact can be determined since it is those higher order bits that control the section of the phenotypic search space that the fully decoded parameter value will be in. This is of fundamental importance to schemata processing because a GA cannot process schemata unless it can make a reasonable estimate of the fitness of each individual schema (this is done implicitly rather than explicitly but must still be possible).

The only schemata that can be usefully processed in a string containing a 2's complement encoded parameter of length z are either; those containing all "#"s, or those containing a contiguous series of fixed bits starting at the highest-order bit with "#"s for the other, lower order bits (there will be z such patterns). For example, in a 4 bit sub-string containing the pattern "1111", the 5 alternative schemata sections for which a reasonable estimate of fitness could be derived would be "####","1###", "11##", "111#" and "1111".

The number of schemata represented by a bit-string is 2^l ($l = P \cdot z$ in this example where P is the number of parameters encoded in the string) since there are two possible values (the allele at that position or a "#") at each of the l bit positions in the string. If each of the P numeric parameters is represented by a single atomic unit of the string, then this expression becomes $(z+1)^P$ since there are z+1 possible values for a schemata element at each position and P atomic units. These are the intermediate states between fully known and fully unknown where all known bits are contiguous and start at the highest order bit. For example, the number of schemata that can be usefully processed in a string containing 2 numeric parameters, each composed of 3 bits, is $(3+1)^2 = 16$. For the string "101 010" these are shown below. The other 48 schemata present in the string cannot have meaningful fitnesses calculated for them.

### ###	**### 0##**	**### 01#**	**### 010**
1## ###	**1## 0##**	**1## 01#**	**1## 010**
10# ###	**10# 0##**	**10# 01#**	**10# 010**
101 ###	**101 0##**	**101 01#**	**101 010**

Figure 1: Schemata processed usefully by a two parameter numeric string.

This is considerably less than the $2^{P \cdot z}$ (2^l) schemata suggested by traditional schemata theory. Note that this is still far more than would be present if crossover was prevented from operating within encoded quantitative parameters at all since, in that case, each parameter would be an atomic unit with only 2 possible states (known and unknown) resulting in only 2^P schemata being processed (they would then be treated essentially as qualitative variables). The GA is schemata processing but not to the same degree as suggested by traditional schemata theory. The only reason that it is schemata processing at all is because of the particular encoding schemes used.

As a consequence, a sub-string representing a quantitative parameter can only contain a set of 0 to l overlapping potential building blocks which all start at the highest order bit extending downwards. No other meaningful building blocks can exist inside that sub-string (i.e. it is impossible to have two adjacent building blocks). Gray-coded bit-strings experience the same effect since high order bits still dominate low order bits in defining the value of an encoded parameter.

Generally, given these potential building blocks, the development and refinement of an encoded quantitative parameter could be expected to proceed as follows. First, in an early generation, a string will generally emerge which has a set of good contiguous highest order bits for the parameter in question (how many will depend on the random initialisation). These bits form the meaningful building block B.

As generations progress, B can be expected to slowly grow in size one bit at a time (i.e. be replaced by a higher order building block that is 1 bit larger) as the value of the next highest order bit b not present in the building block is determined. This process will occur because, once B exists, the GA is able to gain leverage on b since it will dominate the other lower order bits in terms of effect on the phenotypic value. No matter what the values of the lower order bits, b will still control whether the phenotypic value is in the top or bottom half (with no overlap) of the area that it controls in the context of B.

This means that, at this time, useful average fitnesses can be determined for the competing alleles of this bit. If useful average fitnesses can be calculated then the two-armed bandit trade-off is valid. Schemata processing can now occur for this particular bit at this particular time. The succeeding allele was not a building block before this time because it could not have a meaningful average fitness calculated for it without all the higher order bits of it being known - it was dominated by the bit that was in turn higher order to it.

It is not only the GA's building block manipulation activity that is significant to this growing process. The probability that the next lower order bit from b will be correct by chance with no other contiguous bits of lower order set by chance is 0.25 (0.5 chance that the next bit has the correct value and 0.5 chance that the bit after it has the **incorrect** value), and the chance that the next two lower order bits will be correct by chance with the next bit not set correctly is 0.125. The cumulative total of expected extra gains in correct bit values e through chance is given by the formula

$$e = \sum_{i=1}^{x} \frac{1}{2^{i+1}} \cdot i$$

and so approaches 1.0 for large x. For 8 bit parameters, x is 4 on average ($e = 0.813$), and for 16 bit parameters x is 8 on average ($e = 0.980$). This implies that the building block B will grow in size, on average, at nearly double the rate at which the GA can determine the correct allele for a nominated bit position. A GA can take advantage of this because it is not explicitly 'at' a particular bit position - it only acts as if it is as far as the implicit process of determining the best allele at b is concerned. If more contiguous bits directly below b are set correctly then the GA will take advantage of them through its blind crossover operations. Note that this rate is not as high as suggested by traditional theory

As can be seen from the argument given above, the GAs schemata processing activity inside quantitative parameters is similar to a form of binary search; as the correct allele for each bit is determined, the search space left to be searched is halved. The effect of chance nearly doubles this rate, on average, but note that the rate at which the GA can progress in this manner is constrained by the rate at which cross points occur in the critical range of bit positions (any position higher than b assuming that the population has converged on a set of high order bits) in crossover operations (ignoring mutation).

While this process is quite efficient, it is a poor relation to true schemata processing where the 'best' alleles for all bits in the encoded parameter are determined in parallel. Crossover is acting as a local search or permutation operator within a quantitative operator - **not** as a recombination operator.

Note that if the optimum value for a parameter of a given string changes slightly over generations (as it almost certainly will since it is highly improbably that the GA will determine the optimum values for all encoded parameters at the same instant) then it will be difficult for the GA to alter the pattern to the new optimum since this will often involve changing many bits simultaneously, i.e. the encoding is brittle. For example, moving from the number 120 (encoded as "01111000") to the number 130 (encoded as "10000010") requires 5 bits to change simultaneously. Much of this is due to the phenomena of *Hamming Cliffs*. A Hamming Cliff is formed in 2's complement encoding when a contiguous set of lowest order bits of a bit string are all "1"s (e.g. the string "101111" is a Hamming Cliff). In order for the GA to increase the value by one (common in trying to locate the precise optimum) all of the contiguous '1's must change to '0's with the next highest bit set to a "1" (e.g. "101111" would change to "110000"). That means for a Hamming Cliff of length *b*, *b*+1 contiguous bits must invert.

However, of particular interest is that often high order bits will change as well as low order bits in order to move an encoded parameter a small distance phenotypically. High order bits may have to change for a small phenotypic movement due to the fact the a number must lay in one half of the phenotypic search spaced controlled by a given bit (in context of its higher order bits) or the other - a movement across the half-way point will require the bit to change. If only the highest order bit necessary changed in the above example then the value would change from 120 directly to 248, skipping completely past the phenotypic area of the desired value.

Since higher order bits are the first to have the best alleles determined for them, it is highly likely that alternative alleles at these bits will have been driven to extinction (or will only exist in lethal strings) by the time it is necessary for the GA to fine-tune parameter values. This means that mutation alone must often take responsibility for enabling the GA to move a parameter value a small distance. Unfortunately, movement of a high order bit will move the phenotypic value of a parameter by a large degree (by 2^{c-1} where c is the number of the bit in a numbering scheme that has the lowest order bit numbered as 1) so multiple simultaneous mutations of the same string may be necessary for an encoded parameter to change even slightly. The binary search yields initial search speed at the cost of flexibility in response to later change.

It could be argued that, if the only effect of schemata processing inside quantitative parameters is a local binary search, then any similar phenotypic operator would be just as valid. For example, a possible alternative operator could be to adjust the phenotypic value of a given parameter of a given string by a random amount in a random direction. To give the operator characteristics similar to the binary search of crossover, the maximum amount that the parameter value could be allowed to vary by could be gradually reduced over generations - making the operator almost identical to simulated annealing (within the parameter at least). Such an approach would not have the brittleness of the bit string crossover approach. An appropriate non-bit-string mutation operator could be one based on a normal distribution with the distribution centred at the current phenotypic value of the parameter. This would ensure that closer areas are searched more than further areas.

Alternatively, a hybrid scheme could be used. GA's manipulating 2's complement encoded numeric parameters already have a convenient, locally biased mutation operator and employ a fairly efficient binary search through the action of crossover. The addition of a small-distance phenotypic random movement operator (e.g. a creep operator as proposed by Davis [2]) would keep these desirable characteristics while overcoming encoding brittleness, thus enabling parameter values to be fine-tuned as other parameters change. Phenotypic movement operators would also overcome the Hamming Cliff problem. Gray-coded parameters would be a better choice than 2's complement encoded parameters because of Gray-coding's better mutation properties. The advantage, however, is marginal since it is the phenotypic movement operator that will be carrying out most of the local search operation (This approach is used in the research project described in [3] and [4]).

CONCLUSIONS

In summary, GAs do not manipulate schemata inside bit-string encoded qualitative parameters. In encoded quantitative parameters of length z they can only make use of z potential overlapping building blocks (varying between length 1 to z). Their activity in manipulating these building blocks is essentially a form of binary search.

Within bit-string encoded qualitative parameters, the effects of GA activities are disruptive, implying that the GAs should not represent qualitative parameters as bit strings but rather as atomic units of an *m*-element alphabet where *m* is the number of states for that parameter; i.e. *m*-state qualitative parameters should be treated in the same way as 2-state qualitative parameters.

Within bit-string encoded quantitative parameters, the problems are restricted to a bias in mapping and a difficulty in changing a parameter's value by small amounts in response to movement in other parameters during the optimisation process. Crossover actually behaves as a local search permutation operator when acting within quantitative parameters, as opposed to its role as a recombination operator when manipulating a string as a whole.

Since qualitative and quantitative variables are the basis of virtually all other data-structures manipulated by GAs, the view that strings should be always encoded in binary form is thrown into question. It could be argued that true schemata processing only happens when acting on good building blocks, be they groups of encoded parameters, single encoded parameters or parts of encoded parameters that can be decoded with a reasonable level of independence from other elements of the encoding.

This result is of particular concern when dealing with non-traditional data-structures. The guiding rule would appear to be that recombination and mutation within parameters should be of a form that makes sense for that particular parameter type. Mutation should search the entire space with a bias towards parts of the space close to the present string (There is evidence that this occurs in nature. See [11]). Crossover should logically combine the identifiable building blocks present in a data-structure. To be a building block, a section of the data-structure must have some utility (measurable effect on fitness) independent of other elements of the data-structure (An example of such an approach in use can be found in [3]).

BIBLIOGRAPHY

[1] Caruana R.A. and Schaffer J.D., Representation and Hidden Bias: Gray vs. Binary Coding for Genetic Algorithms, *Proc. Fifth International Conference on Machine Learning*, 153 (1988).

[2] Davis L., Adapting Operator Probabilities in Genetic Algorithms, *Proc. Third International Conference on Genetic Algorithms*, 61 (1989).

[3] Gibson G.M., Application of Genetic Algorithms to Mixed Schedule and Parameter Optimisation for a Visual Interactive Modeller, *Proc. Fifth Workshop on Neural Networks*, 119 (1993a).

[4] Gibson G.M., A Genetic Algorithm for Optimising Problems with Embedded Design Decisions, *Proc. 6th Australian Joint Conference on Artificial Intelligence*, 99 (1993b).

[5] Goldberg D.E., *Genetic Algorithms in Search, Optimisation & Machine Learning*, Addison-Wesley, New York NY, (1989).

[6] Grefenstette J.J. and Baker J.E., How Genetic Algorithms Work: A Critical Look at Implicit Parallelism, *Proc. Third International Conference on Genetic Algorithms*, 20 (1989).

[7] Holland J.H., *Adaptation in Natural and Artificial Systems*, Ann Arbor: The University of Michigan Press, (1975).

[8] Holland J.H., *Adaptation in Natural and Artificial Systems*, MIT Press, (1993).

[9] Koza J.R. 1992, Genetic Programming: On the Programming of Computers by Means of Natural Selection, MIT Press.

[10] Levenick J.R., Inserting Introns Improves Genetic Algorithm Success Rate: Taking a Cue from Biology, *Proc. Fourth International Conference on Genetic Algorithms*, 123 (1991).

[11] Wills C., *The Wisdom of the Genes*, Basic Books, New York, (1989).

Heterogeneous Co-evolving Parasites

J. Holmes, T.W. Routen and C.A. Czarnecki
Department of Computer Science
De Montfort University
Leicester LE1 9BH, UK

Abstract

This paper investigates a development of Hillis' co-evolving parasites scheme and looks at how it might be used for job-shop scheduling problems. It suggests that a problem defined by orthogonal constraints may be solved using multiple types of co-evolving parasite.

Areas of specialisation permit the development of solutions specialised for certain constraints; overlaps of these areas create hybrids which combat multiple kinds of parasite and therefore provide solutions.

The main advantage of such a technique is that it permits the use of an evolutionary approach without requiring the specification of a fitness function, which is problematic for many real-world problems.

1 Limited applicability of GA

The conventional GA [5] is a powerful general purpose problem solving technique but is difficult to apply to many real-world problems. It requires a fitness function which accurately reflects the usefulness of any proposed solution and can reduce its relative merits to a single numerical value. A problem under pressure from various forces, perhaps in conflicting directions, is hard to code into such a function. The problem is that by their nature, fitness functions apply all forces simultaneously. Furthermore, the conventional GA is frustrated by the possibility that the encoding method chosen may allow the evolution of invalid solutions. Each response to this problem has its drawbacks.

Hillis [4] points the way to an alternative scheme in which the emphasis is on adaptation rather than function optimisation. An initial experiment in evolving sorting networks with a conventional GA produced local optima in the form of networks which could sort the test cases, but were not general sorts.

Hillis' intriguing response was to allow the test cases to evolve independently of the sorting networks, creating a host-parasite relationship. When a certain sorting network becomes successful and therefore popular, it will be an attractive target for the evolving test cases. When a test case evolves that causes the popular network to have difficulties, the wealth of vulnerable "hosts" means that it will become popular itself. The strings present at early stages of the run largely inhabit the space of invalid solutions, but the parasites gradually shepherd them into the valid areas. Parasites are able to do this because each of their genes is an example of the problem, and effectively defines a small subspace of the invalid area where the candidate solutions found there fail to solve that example problem.

Co-evolving parasites encourage diversity because solutions are soon killed off if they share the same partial deficiencies as their neighbours because the area becomes attractive to parasites that can exploit that deficiency. Diversity means that the whole of the space defined by the chromosome is likely to be sampled, and that is likely to include all valid regions, however remote.

This approach replaces the task of writing a fitness function with that of defining parasites that work on reducing the number of invalid solutions in the population. The task of encoding the solutions into a chromosome still exists, and the specifications of the parasites depend on the encoding technique. Encoding is made easier by the fact that we need not try to conform to the assumptions made by the schema sampling theory, since we are no longer relying on it. Easier still, there is no need to avoid the invalid solutions that would slow down or mislead a conventional GA. In fact, invalid solutions are required for the parasites to work, so we must ensure that they are represented.

2 Multiple Parasites Types

Sorting networks are not fully representative of the kind of multiple constraint problem which we are interested in, and Hillis' scheme as it stands is of limited use in attacking such problems, since it is difficult to define a single parasite which can prey on all the relevant inadequacies in the required way. Suppose we have a problem defined by two constraints. In this case, we can generalise Hillis' scheme and use co-evolving parasites of two different types, where each type finds fault according to one of the constraints. It is not enough to have two separate populations of candidate solutions, each co-evolving with one of the parasite types in a 2D toroidal space, as in Hillis' work. We need to produce valid solutions that conform to both constraints. We need a single population of candidate solutions in a single space, where both types of parasite can influence the evolution of that population.

One way to do this would be to have each location in space occupied by one solution and two parasites, one of each type. Each solution is then under pressure from both constraints, and must adapt to the changing demands of both parasites if it is to survive. Two against one is not a fair fight. Each of the three organisms in a location has a chance to advance its chromosome only once per generation. From random beginnings, the solution is liable to succumb to one or other of the parasites.

An alternative is suggested by the work of Davidor [1] who describes the interesting effects of using locally acting genetic operators on a population defined on a 2D torus. His ECO GA defines a locality as a square block of 9 neighbouring locations, and in each generation, replaces poorer strings with the product of crossover between strings of better fitness found in the same locality. The effect of this localised breeding is that nearby locations hold strings that tackle the problem in a similar way. Locally defined operators mean that moderately good solutions that emerge early in a run will not immediately swamp the population, since progress across the space takes time. At early stages in a run, many islands of moderately good solutions are formed. As execution progresses, these small islands are replaced by larger areas containing better solutions. Finally, the global optimum takes over the entire space.

We can change the emphasis of the solutions likely to form in various parts of such a space by causing different areas to be concerned with different parts of the problem. Solutions specialising in conformance to one particular constraint will emerge from co-evolution with the corresponding type of parasite. If we divide the space into two areas of equal size, and assign a parasite to each area, then we have solutions specialising in satisfying the constraint ruling their area.

This arrangement biases the kinds of solutions that form on each side of the border, but do we get solutions satisfying both constraints in the centre? We have interaction between the specialised solutions that reside either side of the central border. This interaction is in the form of crossover, but is unlikely to result in the lasting hybridisation we desire, where islands of solutions conforming to both constraints are produced. Although islands forming close to the border are likely to be influenced by genes from both sides, when a solution passes a portion of its chromosome across the border, the substring that has been bred with respect to one of the constraints is now exposed entirely to the forces of the other constraint, which it is unlikely to conform to.

If we ***overlap*** the territories governed by the different parasites, we have a strip where both parasite types are represented and can avoid this out-of-the-Sahara-into-the-Arctic effect. Solutions in the overlap area each have two parasites to contend with, will suffer from the unfair fight described above, and at early stages are likely to be killed off. What makes the overlap approach different is that the neighbouring specialised areas can evolve at early stages, and will soon supply the overlaps with a variety of genes that have proved useful against one parasite type or another. The overlaps can be seen as a ground where hybrids are tested. Veterans of the specialised areas are combined there, and those combinations that survive are viable solutions in terms of both constraints.

3 Job Shop Scheduling

In this section we consider how a real-world problem could be handled within this scheme, and for this we choose the example of job-shop scheduling [2][3]. The first thing to do is to describe the constraints which define such a problem. There are several ways in which a schedule can be wrong. It may suffer from one or more of: ***resource contention*** (where a machine is scheduled to work on more than one task at a time); ***job contention*** (where a job is scheduled to be worked on by more than one machine at a time); ***precedence*** (where job tasks are executed in the wrong order); and ***incompleteness*** (where not all tasks required for a job are scheduled). A valid schedule is one that does not suffer from any of these problems. The path towards complete validity now consists of small steps along each of these dimensions, and even a randomly generated schedule is likely to achieve partial validity in one or more of them.

We can generalise the scheme devised above to handle multiple constraint problems. For example, in the case of three constraints, we can arrange each zone so that it borders on the other two, giving areas for testing hybrids of all combinations of two constraints. We also need an area where all three constraints are enforced, where the completely valid solutions are formed. Solutions in this area are under heavy bombardment, and it would be unreasonable to expect the specialised zones to provide anything very useful by crossover into there. Instead, the solutions that satisfy all three constraints can take promising material from solutions satisfying two of them. In essence, we hybridise the hybrids we get from the overlaps between the specialised areas. We can do this with a central area bordered by all overlapping areas.

Representing a schedule is relatively unproblematic when the existence of invalid codes is not detrimental to the performance of the system. Given that the candidate schedule chromosome is a set of specifications of tasks, then different sizes of space can be defined by different degrees of control on the values that specify a task. These values are the job number, machine number, start time and duration. To stay within the realms of the description of the problem, the job and machine numbers must identify valid jobs and machines. Values taken by start time and duration can only be determined with ***a priori*** knowledge of the tasks to be scheduled. We could allow them to take any value defined by the machine's integer implementation, but increasing the space size to this degree does not help solve the problem. With such a huge representable space, all the valid regions would be tucked down into one corner, and it just increases the time required to reduce the diversity of the solution population down to that area. Given that the parasites are of finite capacity, and there are a finite number of them, it may be impossible to stop solutions straying into this void.

So there is a reason to keep the space small, but the point of parasites is that they enable the use of the diversity available from a large space, so we must derive the happy medium. In the case of job shop scheduling, we do have knowledge of the tasks, and therefore we can determine the values taken by the start time and duration in valid solutions. The problem is such that invalid solutions will arise from invalid combinations of values, even when each value is within its valid set, so we are assured of a supply of invalid solutions for the parasites to attack.

Now that we have a coding technique for the solutions, we can design parasites to enforce constraints on encoded solutions. We saw that we can change the size of the space defined by a solution by defining the range of values it can take in each of its defining fields. In a similar way, we can control the size of the invalid region of space covered by a parasite gene. We could search the schedule for any occurrence of machine contention, but this would define a large proportion of the invalid space, and would kill most schedules at the beginning of a run and give nothing a chance to get a foothold. We introduce gradualness into this dimension by making the parasite more specific in what it looks for. If a machine is specified, then the size of the space is reduced, but will still kill very many solutions very easily. By specifying a machine and a time offset from the start of the schedule, we reduce the space to a reasonable level that will give solutions a chance. In the same way, the parasite for job contention can specify a job and a time offset.

Taking cues from Hillis, each parasite consists of several genes embodying specific tests that the solutions undergo. It is these tests that determine the fitness of a solution and the likelihood of its demise. The power of a parasite is determined by the amount of space it defines in the regions of invalid solution space. This is the size defined by one of its genes multiplied by the number of genes a parasite has, assuming they are all of equal size. Given that we do not want to make the parasites too strong or too weak, which would stop evolution, we can use the number of genes in a parasite as a convenient lever on its power. Since this lever effectively operates in units whose size depend on the size of space defined by one gene, then it is a good rule of thumb to have parasite genes defining as small an area of space as possible. We can increase the number of genes to a high level in order to raise the power of a parasite, but we cannot realistically lower it below 2 because crossover will be ineffective, preventing evolution.

The parasite that enforces the constraint that a schedule is complete must check that all tasks are included. At first glance, it seems that the smallest possible space could be defined by specifying a machine and job, and checking whether the schedule includes the job being scheduled on the machine for the required time. Early runs on a prototype implementation showed that this was not small enough. Even with only 2 genes, these parasites were just not dying. To address this, each gene was assigned a degree of enforcement (1, 2 or 3). A degree 1 gene checks only that there is a task referring to a particular job included in the schedule; a degree 2 checks that there is a task that refers to a particular job and specific machine; a degree 3 ensures that there is a task of a specific job, machine and duration on that machine. To enforce task precedence, a gene specifies a job and two consecutively used resources. The parasite checks that if the first is present in the schedule, the second occurs later on.

The scheme as presented is not exactly Mendelian. Complete adherence to one constraint will not survive the crossover process and so we do not firstly produce thoroughbreds in terms of each constraint, and then try to produce a superbeing by mating them. Instead, solutions under various pressures interact continually through all stages of development. When a partial solution to one constraint is mated with a partial solution to another, the result is less likely to be a complete cancelling-out of each other's fitness. We then have a viable solution that is beginning to satisfy both constraints, and will put up a good fight against both parasite types, and co-evolutionary cycles occur in the overlap.

It is important to note that Davidor-style islands are continually broken up by parasites, since they can attack any configuration that gets too popular within a small area. Rather than forming permanent contiguous islands, we can say that the local operators will cause a general increase in the sampling rate of those strings within the local area. This process causes continual movement through space by all genes, which increases the number and variety of viable specialised solutions that will be offered up to the overlap for evaluation. The artificial biological arms race means that any population becomes a churning mass whose genes are increasingly disposed towards high fitness as time progresses.

4 Experiments

A prototype was run using 3 of the 4 parasites defining the job shop scheduling problem. The runs were not expected to produce valid schedules, but at best to provide empirical data showing whether multiple parasites can be made to enter into co-evolutionary cycles with solutions. The data for these runs was taken from Beasley's OR library, and is based on the Muth and Thompson standard set of job shop scheduling problems.

Early runs of the embryonic system produced statistics on the death rates among the various types of organism suggesting that co-evolution was not progressing satisfactorily for all parasite types. In particular, the type 3 parasite (completion) was regularly killing most of its solutions, even when weakened by having only 2 genes. The concept of degree of effectiveness of a gene was introduced, as an additional lever to weaken it. This shows that it is an important task to calibrate the strengths of the parasites with the solution space so that a particular parasite is not too strong and kills virtually every solution it finds. The rules of battle can be amended to alleviate this problem. Each parasite gene is tested in turn and if it finds a fault with the solution, the solution is marked as dead and added to the hit list. Only when every gene fails to find fault with the solution is the parasite killed. This algorithm missed the fact that the culling needs to be probabilistic, biased by the number of genes that beat or are beaten by the solution. An improved version determines the fitness of a parasite by the number of genes that beat the solution. A random number is generated between 1 and the number of genes in the parasite, and if it comes out below the fitness, the solution is killed, otherwise the parasite is killed.

This change in the rule makes the relative strengths less critical because apocalypse is not a regular occurrence. Although co-evolution can at least now build on small gains made in each generation, the process will be slower because more evolutionary effort is required to respond to successful attacks by foe. Although these later runs do not show that co-evolution is not taking place, we cannot realistically conclude from them that it is. Nevertheless, there does seem to be distinct interaction between the death rates of the solutions and the most powerful (type 3) parasites, with the other parasites having a lesser relationship.

5 Conclusion

We have described reasons why an evolutionary scheme using heterogeneous parasites might be effective. There are difficulties in establishing a positive rather than non-negative conclusion. The scheme is very demanding of computational resource, although the fact that all interactions between organisms occur either within a location or with direct neighbours makes the scheme highly amenable to parallel implementation. Also it is difficult to know when the scheme has produced the optimal result. Some kind of stasis is usually the sign, but under this scheme it will only arise when every solution has found its way into the valid areas. The central overlap area is where the optimum is most likely to arise, so we need to watch the activity in this area. Finally, the use of parasites makes this type of GA hard to compare with others.

To use a fitness function solely for the purpose of comparison would not give a true reflection of progress, since traditional fitness functions derate invalid codes, but this scheme has them as a vital source of diversity.

There are no guarantees as to how well valid areas are searched, but most valid areas will be found. Therefore, we can say only that it is suitable for problems where the valid solutions are small islands distributed across the ocean of representable invalid solutions. The less contiguous are the valid regions, the better. This restricts the areas of application but it is not to brand this GA useless. Many real world problems do have these characteristics. Where almost any change to a valid solution turns it into an invalid one, we have this kind of problem. Examples previously discussed are the sorting networks and job shop schedules, but others are the travelling salesman problem and network topology optimisation.

6 References

[1] DAVIDOR, Y. (1991) A naturally occurring niche and species phenomenon: the model and first results. in R.K. Belew and L.B. Booker (eds.), Proceedings of the Fourth International Conference on Genetic Algorithms. Morgan Kaufman.

[2] DAVIS, L. (1985) Job shop scheduling with genetic algorithms. In J.J. Grefenstette (ed.), Proceedings of the First International Conference on Genetic Algorithms and their Applications, Lawrence Erlbaum Associates.

[3] FANG, H.L., ROSS, P., and CORNE, D. (1993) A promising genetic algorithm approach to job-shop scheduling, rescheduling, and open-shop scheduling problems. In S. Forrest (ed.), Proceedings of the Fifth International Conference on Genetic Algorithms. Morgan Kaufmann.

[4] HILLIS, W. D. (1990) Co-evolving parasites improve simulated evolution as an optimisation procedure. In S. Forrest (ed.), Emergent Computation: Self-Organising, Collective, and Cooperative Phenomena in Natural and Artificial Computing Networks, 228-234, MIT Press.

[5] HOLLAND, J. (1992) Adaptation in Natural and Artificial Systems, MIT Press.

TYPOLOGY OF BOOLEAN FUNCTIONS USING WALSH ANALYSIS

Cathy Escazut Philippe Collard

Laboratory I3S — CNRS-UNSA, Bât. 4
250, av. Albert Einstein, Sophia Antipolis, 06560 Valbonne, FRANCE

Much previous works deal with the functions that cause a genetic algorithm (GA) to diverge from the global optimum. It is now a fact: Walsh analysis allows the identification of GA-hard problems. But what about genetics-based machine learning? Do we know what makes a problem hard for a classifier system (CS)? In order to try to answer these questions, we describe the relation between CS performance and the structure of a given boolean function when it is expressed as a Walsh polynomial. The analysis of the relative magnitude of Walsh coefficients allows us to set up a typology of boolean functions according to their hardness for a CS. Thus, each function can be placed in a specific class of difficulty. Using the converse process, we can start from well chosen Walsh coefficients in order to build boolean functions hard for a CS to learn.

1 Introduction

Boolean functions have been deeply studied and belong now to a well known area in which it is easy to evaluate and to compare different approaches (neural networks, genetic algorithms, classifier systems...). In CSs, they are very attractive because they are binary and thus can be directly expressed by classifiers over the alphabet {0, 1, #} (0 means FALSE and 1 is for TRUE). In this paper, each function is named, following the numbering scheme proposed by Wolfram and used by Koza [5]. Let consider the binary values of a function F as the bits of a binary number, reading from the bottom up of the truth table of F, the decimal equivalent of this binary number is the name of the function F. For instance take the OR function with two arguments which is 1 (TRUE) if at least one of its argument is 1. The binary number corresponding is 1110, so the OR function is named 14 (decimal equivalent of 1110). Since there are $2^{2^2} = 16$ different boolean functions with two arguments and one output, they range from rule 00 (whose truth table consists of four 0's) to rule 15 (whose truth table consists of four 1's).

In the following section we review how Walsh analysis has been used to characterize the difficulty of function for a GA and describe how these results can be applied to the learning of boolean functions with a CS. We then propose a typology for boolean functions with two arguments: we show that the 16 different functions can be distributed among four classes according to the hardness they present. Finally, as D. Goldberg uses Walsh coefficients to construct deceptive functions for GAs [3], we generalize the classification to functions with more than two arguments in order to construct high-order hard functions for a CS.

2 Walsh analysis and boolean functions

A lot of researchers have pointed out the close connection existing between Walsh polynomials and GAs [2]. In particular, Bethke developped the *Walsh-Schema transform* in which particular basis functions are used to calculate schema average fitness efficiently [1]. He then used this transform to characterize functions as easy or hard for the GA to optimize. In this section we briefly describe the Walsh-Schema transform and then propose some modifications in order to use this tool in CS's analysis.

Before going further in our study, let us define the terminology: x is a λ-bit string, F is the fitness function, and $\psi_j(x)$ a set of basis functions, called *Walsh functions* and taking a value in {-1, +1}.

Since the Walsh functions form a basis set, any function defined on $\{0,1\}^\lambda$ can be written as a linear combination of Walsh functions:

$$F(x) = \sum_{j=0}^{2^\lambda-1} \omega_j \psi_j(x) \qquad (1)$$

where ω_j are real-valued coefficients called *Walsh coefficients*. Each ω_j is associated with the partition having the index j. The representation of a function F as a weighted sum of Walsh coefficient is called F's *Walsh polynomial*. Applying the converse analysis, Walsh coefficients can be obtained as follows:

$$\omega_j = \frac{1}{2^\lambda} \sum_{x=0}^{2^\lambda-1} F(x)\psi_j(x) \qquad (2)$$

The Walsh-Schema transform can be expressed quite similarly to the equation 1. The difference is that for $F(x)$, 2^λ coefficients must be summed. But to calculate the fitness of a schema s only coefficients corresponding to the partitions to which s belongs are needed. In this way, the schema average may be calculated as a partial, signed sum of Walsh coefficients:

$$F(s) = \sum_{j \in J_s} \omega_j \Psi_j(s) \quad (3)$$

J_s is the set of all the partitions that contain some schema s' such that $s' \supseteq s$, and $\Psi_j(s)$ is the value of $\psi_j(x)$ (i.e. +1 or -1) for any $x \in s$. From equation 3 we see that the Walsh-Schema transform is a sum of progressively higher-order Walsh coefficients. This illustrates how the GA actually estimates a schema s: first the algorithm bases its estimate on information about the low-order schemas containing s, and gradually it refines this estimate from information about higher and higher order schemas containing s. Thus, if the most important Walsh coefficients of a function F, are associated with short, low-order partitions, then F will be easy for the GA to optimize. In such cases, the location of the global optimum can be determined from the estimated average fitness of low-order, low-defining-length schemas. On the contrary, a function whose Walsh decomposition involves high-order j's with significant coefficients should be harder for the GA to optimize, since the algorithm will have a harder time constructing good estimates of the higher-order schemas belonging to the higher-order partitions j.

There is an obvious similarity between the chromosomes of a GA and the classifiers of a CS. Thus, why don't we apply after some modifications the results of Walsh analysis to CSs, in order to characterized the hardness of the learning of a boolean function? Learning a boolean function means acquiring the ability to give the correct output, for any input string. Each classifier consists of a single condition and an action. For λ-bit input string, the condition has length λ. The action is simply a single bit, +1 or -1. The action part of a classifier C is, in one way, the value of a function F whose arguments are the bits of the condition part of C. For instance, let us assume that the CS has to learn the boolean function AND. The classifier 01:-1 can be translated as $F(0,1) = -1$. In this way, it is possible to compute Walsh coefficients for a CS. But the term $F(x)$ in the equation 2 must be replaced by the action part of the classifier.

The value of the ω_j can give indications on the solution set: the partitions with the highest Walsh coefficients belong to the solution set, and as in GAs, the function is all the easier to learn as the Walsh coefficients with great value are associated with low-order partitions. Moreover, if a partition j, is associated with a Walsh coefficient, ω_j, equal to ± 1, then the others coefficients are null, and the solution set is only composed of all the classifiers of the partition j. In all other cases, the classifiers of the solution set are picked from different partitions. For instance let consider the boolean function F with three arguments such as the ouputs are all equal to -1, except for $F(0,1,1)$ and $F(1,0,0)$ whose value is +1. Figure 1 represents the distribution of the Walsh coefficients. The three coefficients ω_{011}, ω_{101} and ω_{110} are non null. Thus the solution set is composed of classifers picked from different partitions: ###:-1 from partition $j = 000$, and 011:+1, 100:+1 from partition of index $j = 111$. As the non zero Walsh coefficients are associated with 2-order partitions we can say that the function F is not very easy for a CS to learn.

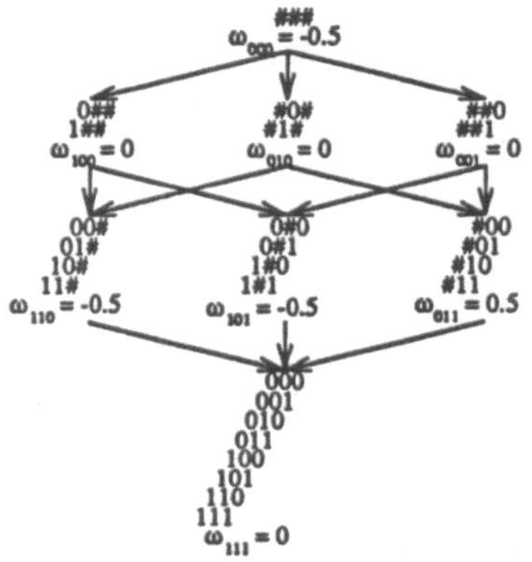

Fig. 1: Distribution of the Walsh coefficients.

In the next section, we propose a typology for boolean functions with two arguments according to their hardness.

3 Typology for minimal boolean functions

We consider here, only boolean functions with two arguments. All the 2^4 boolean functions with two arguments can be classified following four difficulty levels. We illustrate each level with one of its representative function.

Level I (Function 15):

All the Walsh coefficients of the function 15, ie. $F(p,q)$ =TRUE, are equal to 0, except the one associated with the partition 00: $\omega_{00} = +1$. The solution set is so composed of the unique classifier ##:+1. (Cf. figure 2)

The functions belonging to this first level are easy to learn because the highest Walsh coefficient is associated with the lowest-order partition.

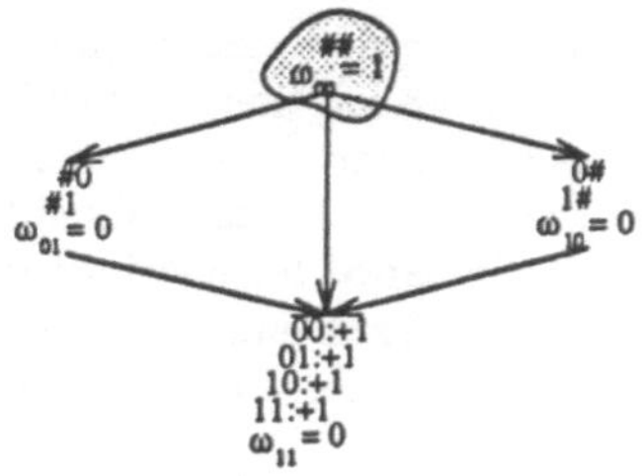

Fig. 2: Level I: Function 15

Level II (Function 03) :

Once again, only one Walsh coefficient of the function $F(p,q) =$ NOT p is non zero and equal to $+1$. But in this case, the partition containing the classifiers of the solution set, has the index $j = 10$. The solution is thus composed of the two one-order classifiers: 0#:+1, 1#:-1. (See figure 3)

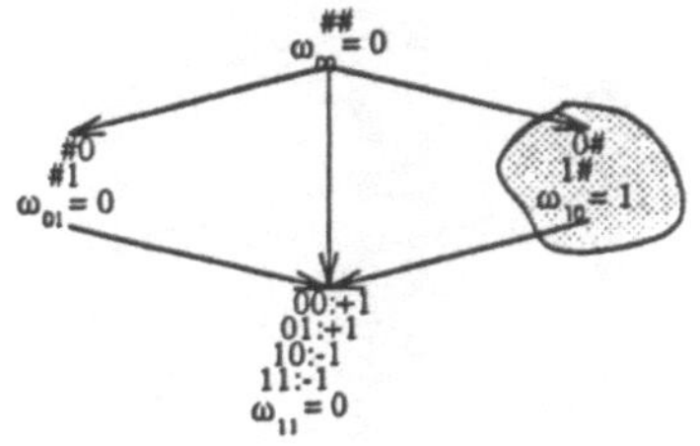

Fig. 3: Level II: Function 03

All the functions whose single Walsh coefficient non null is associated with a one-order partition (i.e. partitions whose $j = 01$ or $j = 10$), belong to the level II.

Level III (Function 08):

In the case of the function 08, i.e. $F(p,q) = p$ AND q, there is no Walsh coefficient equal to ± 1. Thus, the solution set is composed of classifiers picked in different partitions: the partitions with index $j = 00$ and $j = 11$. The rules of the solution form what Holland has called a *default hierarchy* [4]: the general rule is ##:-1, and the exception one is 11:+1. (Cf. figure 4)

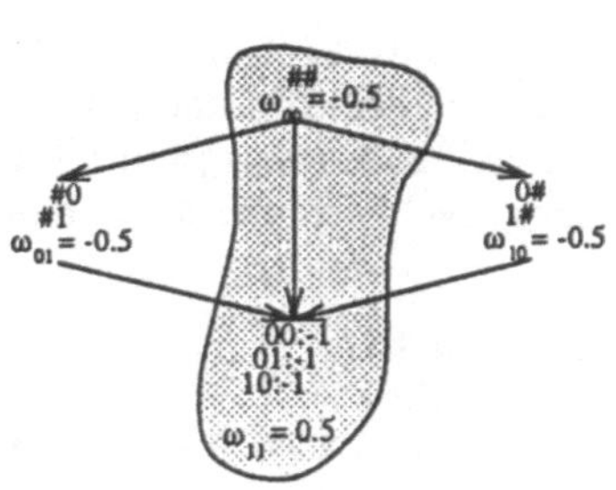

Fig. 4: Level III: Function 08

The functions whose solution set brings into play different partitions are put in the level III.

Level IV (Function 06):

The function 06 is the well known XOR (or *parity*) function. The only non zero Walsh coefficient is associated to the highest-order partition. The solution set contains the four totally instanciated classifiers: 00:-1, 01:+1, 10:+1 and 11:-1. The system has no other choice than learning the solution by heart. (See figure 5)

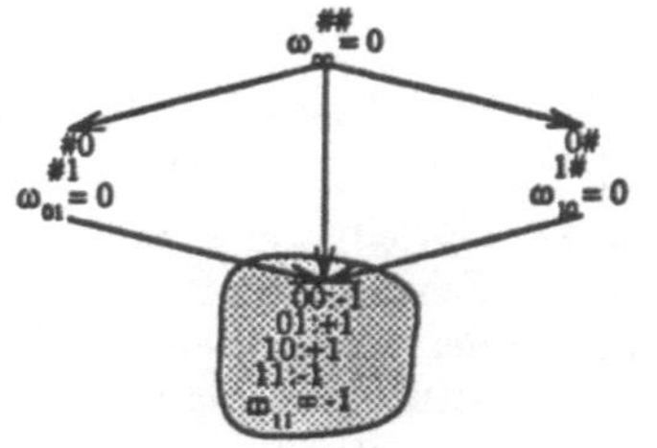

Fig. 5: Level IV: Function 06

The functions belonging to this level are the hardest ones: the one and unique Walsh coefficient non null is associated with the highest-order partition ($j = 11$).

All the 16 boolean functions with two arguments belong to one of these 4 levels. The table 1 represents this distribution. We treated here only boolean functions with two arguments but higher-order functions can be classed in a similar way. Of course the number of hardness levels increases with the number of arguments in the functions. But in all cases, functions whose $\omega_{0...0} = \pm 1$ are the easiest one, and functions for which $\omega_{1...1} = \pm 1$ are the hardest one.

Table 1: Distribution of minimal boolean functions.

	Functions	$F(p,q)$
Level I	00	FALSE
	15	TRUE
Level II	03	NOT p
	05	NOT q
	10	q
	12	p
Level III	01	NOT (p OR q)
	02	NOT ($q \Rightarrow p$)
	04	NOT ($p \Rightarrow q$)
	07	NOT (p AND q)
	08	p AND q
	11	($p \Rightarrow q$)
	13	($q \Rightarrow p$)
	14	p OR q
Level IV	06	(p XOR q)
	09	NOT (p XOR q)

4 Construction of high-order hard functions

As we can assess the hardness of a given function, we are able, by the inverse process, to construct a function more or less difficult for a CS. This is very useful to find appropriate test sets in order to evaluate and to compare the efficiency of different sytems.

Walsh coefficients, as we have defined them for boolean functions, present three main characteristics:

- Walsh coefficients have the form $(\pm 2k)/2^{\lambda}$, where λ is the length of the condition part of the classifier, and k is an integer such as $0 \leq k \leq 2^{\lambda-1}$.
- The sum of all Walsh coefficients is equal to $F(0,\ldots,0)$ in GAs; that is in our context, the action part of the classifier whose condition is $0\ldots0$. Thus: $\sum_{j=0}^{2^{\lambda}-1} \omega_j = \pm 1$.
- When one Walsh coefficient is equal to ± 1, all the others are null.

These three conditions are necessary but not sufficient to characterize a set of Walsh coefficients. Thus, using this constraints, it is practically impossible to find a set of accurate coefficients describing a boolean function. Indeed, this work becomes quite a blind random search as the number of arguments of the function increases. So we propose a solution which consists in mixing two functions easy to construct in order to obtain a harder one.

Before going further let us define F_i as a boolean function whose solution is a *local* one: its single Walsh coefficient non zero (i.e. equal to ± 1) is associated with the ith partition. In this case, the solution set is composed of all the classifiers of the single partition i. This class of function can be easily construct since only one of its Walsh coefficients is non null and has the value ± 1.

Let F_i and F_j be two local functions such as the partitions i and j have the same order. The solution of the function $F_k = F_i$ XOR F_j is also a local one: the solution set contains all the classifiers of the kth partition such as $k = i$ XOR j. This property allows to associate a single non null Walsh coefficient (so equal to ± 1) to a high-order partition. It has thus, no practical interest. But if we now define the function F_k as F_i OR F_j. The solution of F_k is not a local one: the Walsh coefficients associated to the four partitions of index i, j, i XOR j and 0 are equal to ± 0.5. In this way, we can construct high-order hard boolean functions from two functions owning a local solution. For instance, the function illustrated in the figure 1 can be obtained from the local functions F_{011} and F_{110}. Indeed, 011 XOR 110 = 101, so the partitions of index $j = 000$, $j = 011$, $j = 101$ and $j = 110$ are the only ones associated with Walsh coefficients non zero: $\omega_{000} = -0.5$, $\omega_{011} = 0.5$, $\omega_{101} = -0.5$ and $\omega_{110} = -0.5$.

The method proposed here is only a way of combining local functions. Other combinaisons can be imaginated in order to construct boolean functions with different degrees of hardness.

5 Conclusion

This paper deals with classifier systems learning boolean functions similar to ones used in Wilson's BOOLE [6]: the condition part is composed of one bit for each argument of the function, and the action part of the rule is a single bit whose value is +1 if the output of the function is TRUE and -1 otherwise. The likeness to GAs allow us to apply Walsh analysis to classifier systems. Indeed, the computation of Walsh coefficients is very useful to determine if a problem is hard or easy for classifier systems: the more coefficients with high value are associated with high-order partitions, the harder the problem is.

We have thus propose a typology for boolean functions with two arguments. Koza realized measures of the hardness of boolean functions with two and three arguments in genetic programming. His results are quite close to ours. In particular, in both approaches the hardest one is the parity function.

Finally we showed that the ability to characterize the hardness of a function allows us to synthesize high-order functions which can be hard for a CS to learn.

References

[1] A. D. Bethke. *Genetic algorithms as function optimizers.* PhD thesis, University of Michigan, 1980.

[2] D. E. Goldberg. Genetic algorithms and walsh functions. *Complex Systems*, 3:pp 129–171, 1989.

[3] D. E. Goldberg. Construction of high-order deceptive functions using low-order walsh coefficients. *Annals of Mathematics and Artificial Intelligence*, 5:pp 35–48, 1992.

[4] J. H. Holland. *Adaptation in natural and artificial systems.* Ann Arbor : University of Michigan Press, 1975.

[5] J. R. Koza. *Genetic programming: On the programming of computers by means of natural selection.* MIT Press, Cambridge, MA, 1993.

[6] S. W. Wilson. Classifier systems and the animat problem. *Machine Learning*, 2(3):199–218, 1987.

ARTIFICIAL NEURAL NETWORKS FOR NONLINEAR PROJECTION AND EXPLORATORY DATA ANALYSIS

Denis HAMAD and Mohamed BETROUNI

USTL, Centre d'Automatique de Lille, Bâtiment P2,
59655 Villeneuve d'Ascq, Cedex, France
e-mail : denis.hamad@univ-lille1.fr

Abstract. Mapping scheme consists in projecting data samples represented as points in high-dimensional data space onto a subspace of few dimensions, generally two dimensions. Mapping methods are used in order to eliminate statistical redundancies in the original data set and to facilitate the visual inspection of the data by the analyst which discover clusters between the data samples. A feedforward neural network trained by means of an unsupervised backpropagation algorithm is used for the nonlinear mapping. The Sammon's stress is used as an error function for the learning algorithm. The number of hidden units is related to the complexity of the nonlinear functions that can be generated by the network and is selected by means of an informational criterion. To provide some insight into the behavior of the interactive system, and to presents its main facilities, some results are reported.

1. Introduction

An important tool in interactive pattern classification consists to provide to the analyst a planar display of the high-dimensional data set. If it can be assumed that clusters on the plane are images of multidimensional clusters in the original data space, a human observer can usefully analyze graphic displays without conscious use of any analytical model of clusters, or any mathematical decision rule. It is hoped that any information lost in the dimensionality reduction process is fully compensated by the benefits associated with the integration of the human operator for organizing the data [1], [2].

Projection methods used in unsupervised pattern classification can be linear such as the principal component analysis PCA or nonlinear such as the Sammon's algorithm [3], [4]. Principal component analysis method attempts to preserve the variance of th: data in projection whereas Sammon's projection tries to preserve the interpoint distances. Unlike to PCA method, Sammon's algorithm detect nonlinear correlations between variables. However, the principal weakness of Sammon's algorithm is its non-analytical propriety i.e. it does not provide an explicit function governing the relationship between patterns in the original data space and in the projected space. Therefore it is impossible to place a new data in the plane created by Sammon's algorithm.

In the neural network context, the nonlinear projection can be obtained by the well known multilayer neural network trained by the backpropagation algorithm to perform the identity mapping. The network inputs are reproduced at the output layer through three hidden layers. The outputs of the internal hidden layer containing m units ($m < d$), are used to visualize the nonlinear projection of the data samples. The network can be splitted into two parts : the first one implements the mapping function and the second one realizes the demapping function [5].

The number of units in the hidden layer is related to the complexity of the topological relationships in the input data which can be captured by the network. The generalization ability of the network depends upon this number.

Recently, an alternative solution based on a feedforward neural network to overcome the non-analytical limitation of Sammon's algorithm has been proposed [6]. The neural network is made of an input layer which receives the information available from observations, one hidden layer and one output layer constituted of two units which provides the planar representation of the high-dimensional data set. The training phase, during which the network learns the structure of the data, is of upmost importance. The learning rule extends the backpropagation algorithm to unsupervised learning thanks to the implementation of Sammon's projection.

Our contribution lies in the automatic selection of the number of the hidden units by means of Akaike's information criterion applied to the multilayer feedforward neural network trained by the backpropagation algorithm to perform the identity mapping. The principle of our approach consists in varying the number of hidden units and in computing this criterion at the end of each learning phase. The optimal number of units corresponds to the minimum value of this criterion.

At the end of the learning step the mapping part of this network is then used to initialize the Sammon's neural network for an interactive pattern classification.
To provide some insight into the behavior of the interactive system, and to presents its main facilities, some results are reported.

2. Multilayer networks for nonlinear projection

The nonlinear mapping of Sammon is one of the first widely known nonlinear projection technique [4]. Let $\{X_1,\cdots,X_u,\cdots,X_v,\cdots,X_n\}$ denotes the Q unlabelled samples scattered in the n-dimensional data space R^n, where $X_u=(x_{u1},\cdots,x_{ui},\cdots,x_{ud})^T$ is the u-th sample and x_{ui} denotes the i-th component of the u-th sample X_u. This method creates a set of points on the plane, images of the original data points and denoted by $\{Y_1,\cdots,Y_u,\cdots,Y_v,\cdots,Y_n\}$, where each sample is represented by $Y_u=(y_{u1},\cdots,y_{ui},\cdots,y_{um})^T$. Since our emphasis is on exploratory data analysis, we will be mostly concerned with situations where m = 2. The points on the plane are placed in such a way that the distances between them approximate the distances between their respective original points. This is achieved by minimizing a mapping error E, called Sammon's stress :

$$E=\frac{1}{\lambda}\sum_{u=1}^{n-1}\sum_{v=u+1}^{n}E_{uv} \qquad (1)$$

where,

$$-\ \lambda=\sum_{u=1}^{n-1}\sum_{v=u+1}^{n}d^{*}_{uv} \qquad (2)$$

$$-\ E_{uv}=\frac{[d^{*}_{uv}-d_{uv}]^2}{d^{*}_{uv}} \qquad (3)$$

$$-\ d^{*}_{uv}=\sqrt{\sum_{i=1}^{d}(x_{ui}-x_{vi})^2}\ \ ;\ \ d_{uv}=\sqrt{\sum_{i=1}^{m}(y_{ui}-y_{vi})^2} \qquad (4)$$

d_{uv} and d^{*}_{uv} are the Euclidean distances between patterns u and v in the input space and in the projected space, receptively.
The set of projected points $\{Y_1,\cdots,Y_u,\cdots,Y_v,\cdots,Y_n\}$, where $Y_u=(y_{u1},\cdots,y_{ui},\cdots,y_{um})^T$ constitutes the $m.n$ optimization variables of mapping error E. Sammon used a gradient descent algorithm to find an n-points configuration mapping which minimizes E.
The main drawback of Sammon's mapping is that, it does not provide an explicit function governing the relationship between original data samples. This problem is known as generalization capability. In other word, when we would like to project new data sample, we have to run the algorithm again on the whole data samples (old data samples and the new data sample). Moreover, the computation time of Sammon's algorithm depends on the square of the number of points in the data space.
Jain and Mao proposed a multilayer feedforward network for Sammon's projection [6].The network architecture is composed of three layers : an input layer, a hidden layer and an output one (Fig. 1). The number of input units is set to the dimensionality of the input space. The number of output units is specified as the dimensionality of the projected space. There is no definitive method, however, for deciding a priori the number of units in the hidden layer. In section 3, we present an approach for selecting the number of hidden units h.

Fig. 1. Multilayer feedforward neural network

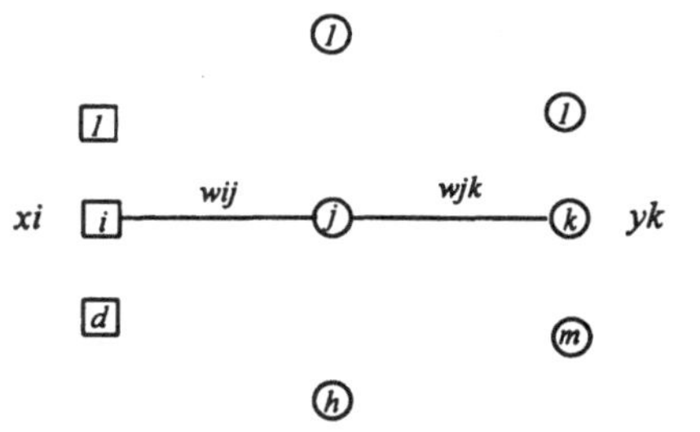

Let d be the number of units in the input layer and h the number of units in the hidden layer. The number of units in the output layer is equal to m (m is limited to 2 for a planar display). Let w_{ij} and w_{jk} be the weights on the interlayer connections from the i-th unit in the input layer to the j-th unit in the hidden layer and the j-th unit in the hidden layer to the k-th unit in the output layer respectively.
The output y_{ju} of the j-th unit in the hidden layer corresponding the input vector X_u is given by :

$$y_{ju}=g(p_{ju})=g(\sum_{i=1}^{d}w_{ij}x_{ui})\ \ ;\ \ j=1,\cdots,h \qquad (5)$$

Where $g(p_{ju})$ is the tanh function and x_{ui} is the *i-th* component of the input vector X_u :

$$g(p_{ju}) = \frac{1-e^{-p_{ju}}}{1+e^{-p_{ju}}} \quad (6)$$

Similarly, the kth output y_{ku} of the output layer is :

$$y_{ku} = g(\sum_{j=1}^{h} w_{jk} y_{ju}) \quad k = 1, \cdots, m \quad (7)$$

To minimize the Sammon's stress function E of equation (1) by the gradient descent technique, it is necessary to compute the sensitivity of E according to each connection weight in the network.

In a first time we present a pair of input vectors (X_u, X_v) at the input layer, and by applying equations (5), (6) and (7), we compute the outputs of hidden units and the output of output units respectively. This step, called the forward pass, is finished by computing E_{uv}.

The second pass, the backpropagation, propagates derivatives from the output layer to the hidden layer and from the hidden layer to the input one. The updating rules are:

$$\Delta w_{jk} = -\eta \frac{\partial E_{uv}}{\partial w_{jk}} \quad (8)$$

$$\Delta w_{ij} = -\eta \frac{\partial E_{uv}}{\partial w_{ij}} \quad (9)$$

where η is a constant that determines the rate of learning., the sensitivity factors in (8) and (9) are computed according to the chain rule [6].

3. Selection of the number of hidden units

The Sammon's network architecture is nothing else the mapping part (*d-h-m*) of a 5-layers (*d-h-m-h-d*) neural network trained to learn the identity mapping. The 5-layers network architecture is as follows : the input layer contains *d* units, the first hidden layer contains *h* units, the second hidden layer contains *m* units, the third hidden layer contains *h* units and the output layer contains *d* units. The backpropagation algorithm is then applied to estimate the weights of this network so that the reconstructed outputs $\hat{X}$ much the inputs X as closely as possible. Training consists to minimize the sum of squared errors given in Equation (10) :

$$J = \sum_{u=1}^{n} \sum_{i=1}^{d} (x_{ui} - \hat{x}_{ui})^2 \quad (10)$$

The number of units in the hidden layers is related to the complexity of the topological relationships in the input data which can be captured by the network. If too few mapping units are provided, accuracy might be low because the representational capacity of the network is limited. However, if there are too many mapping units, the network will be prone to "overfitting".

Since the input and the hidden layers of the neural network are biased, the total number of free parameters in the network can be written as :

$$N_f = [(d+1)h] + [(h+1)m] + [(m+1)h) + [h+1)d] \quad (11)$$

which can be rewritten as :

$$N_f = 2h(d+m+1) + d + m \quad (12)$$

Akaike proposed the use of functions that express the tradeoff between the fitting accuracy and the number of adjustable parameters in explicit terms. The Akaike's Information Criterion (AIC) is given by [5], [7] :

$$AIC = Ln(\bar{J}) + \frac{2N_f}{N} \quad (13)$$

where, $N = nd$ is the number of entries in the data matrix and $\bar{J} = J/2N$ is an average sum of squares error J.

The principle of our approach consists to variate the number of units in the hidden layer and to compute the AIC criterion for each number by using the Equ. 13. The adequate number of hidden units is given by the minimum value of this criterion.

At the end of the selection procedure of the number of hidden units of the 5-layers neural network, the mapping part of this network is used to initialize the Sammon's network.

4. Nonlinear mapping simulations

Fisher's Iris data has been used in many papers to illustrate various clustering techniques. These data consist of four variables on 50 sample vectors belonging to each of three groups of Iris : setosa, versicolor and virginica. The four variables are sepal length and width, and petal length and width. Therefore, the data set is constituted of 150 samples of 4-dimensional patterns (3 classes of 50

samples each). Two of the three classes are slightly overlapped in the data space.
Data samples have been normalized between [-1,1] and used to train a feedforward network in autoassociative mode. The number of units in each layer of the network are *4-h-2-h-4* respectively. The following Table shows the values of the AIC criterion and the squares error J as functions of the number of hidden units h which varies from 10 to 60. The number of hidden units corresponding to the minimum value of the AIC is equal to is equal to 19, while the error function J is equal to.
The mapping part constituted of *4-19-2* units of the above network is then trained by the Sammon's stress. The Fig. 2 shows the nonlinear map obtained by the neural network outputs. The Sammon's stress of the network is $6.162\ 10^{-3}$.

Table 1. AIC criterion and the squares error J as a function of number of units in hidden layers.

h	AIC	J
10	7.328	5.108
19	7.014	3.733
20	7.178	4.399
25	8.407	5.528
30	8.837	8.502
40	8.960	9.614
50	10.535	17.086
60	11.266	35.486

Fig. 2. A 2-space nonlinear projection of the IRIS data

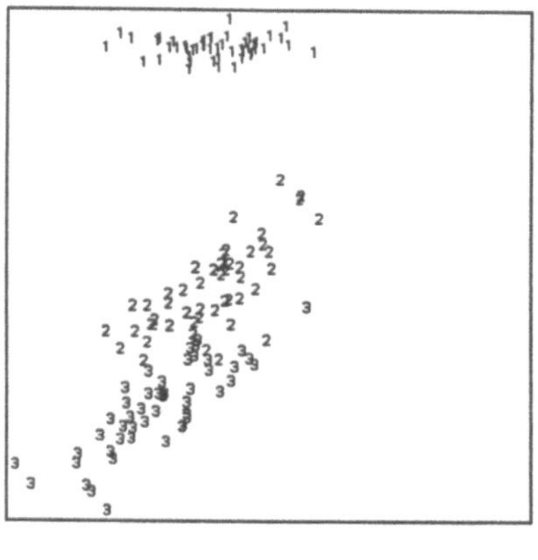

In Fig. 2, we visualize two clusters : the first one is constituted of 50 samples and the other is in fact constituted of two overlapped classes.
The user discover on the sceen the number of clusters and indicates by the mouse their centers. The K-means algorithm is then applied in the data space for the labelling procedure.

5. Conclusion

We have presented a feedforward neural network, trained by the Sammon's stress for unsupervised pattern classification. The architecture of this network has been initialized as the mapping part of a 5-layer feedforward neural network trained in autoassociation mode to perform the identity mapping. An informational criterion has been used for the automatic selection of the number of units in the hidden layers. The network outputs are used for visual inspection of the data by the analyst which discover the intrinsic structure of data samples. The results reported in this paper show that even someone who does not have any software system background can use the interactive system to classify multidimensional observations in an unsupervised context.

References

[1] Siedlecki W. Siedlecka K. and Sklansky J., "An Overview of Mapping Techniques for Exploratory Pattern Analysis," Pattern Recognition, Vol. 21, No. 5, pp. 411-429, 1988.
[2] Daoudi M., Hamad D. and Postaire J.-G., "A Display Oriented Technique for Interactive Pattern Recognition by Multilayer Neural Network". In Proc. of IEEE ICNN, Vol. III, pp. 1633-1637, San Francisco, CA, March 1993.
[3] Biswas G., Jain A. K. and Dubes R. C., "Evaluation of Projection Algorithms". IEEE Trans. PAMI., Vol. 3, NO. 6, pp. 701-708, Nov. 1981.
[4] Sammon Jr., J. W., "A Non-linear Mapping for Data Structure Analysis," IEEE Trans. on Computers, Vol. C-18, pp. 401-409, May 1969.
[5] Kramer M.A., "Nonlinear Principal Component Analysis Using Autoassociative Neural Networks", AIChE Journal, Vol. 37, N°. 2, pp. 233-243, February 1991.
[6] Jain A. K., and Mao J., "Artificial Neural Network for Non-linear Projection of Multivariate Data," IJCNN, Vol. III, pp. 335-340, Baltimore, Maryland, June 7-11, 1992.
[7] Akaike H., "A New Look at the Statistical Model Identification", IEEE Trans. Appl. Comp. vol. AC-19, pp. 716-723, 1974.

Generic Back-Propagation in Arbitrary FeedForward Neural Networks

Cédric Gégout[•◇⋆] Bernard Girau[•] Fabrice Rossi[◇⋆]

◇ École Normale Supérieure de Paris, 45 rue d'Ulm, 75005 PARIS France

• École Normale Supérieure de Lyon, L.I.P., 46 avenue d'Italie, 69007 LYON France

⋆ THOMSON-CSF SDC/DPR/R4, 7 rue des Mathurins, 92223 BAGNEUX France

Abstract

In this paper, we describe a general mathematical model for feedforward neural networks. The final form of the network is a vectorial function f of two variables, x (the input of the network) and w (the weight vector). We show that the differential of f can be computed with an extended back-propagation algorithm or with a direct method. By evaluating the time needed to compute the differential with the help of both methods, we show how to chose the best one. We introduce also input sharing and output function which allow us to implement efficiently a multilayer perceptron with our model.

1 Introduction

Léon Bottou and Patrick Gallinari have proposed in [1] a general framework for the cooperation of neural modules. In this paper we extend this model in order to derive a general model for feedforward neural networks. The model presented in section 2 can simulate multilayer perceptrons and related models, with the help of an extension proposed in section 4. We show in section 3 that an extended back-propagation algorithm allows us to compute the differentials of the function computed by the net and therefore to use gradient based neural training, as explained in section 6. We give in sections 5 and 6 complexity results allowing us to choose between direct calculation and retro-propagation in order to compute efficiently the gradient for a supervised learning.

Due to space limitation, the proofs are ommited in this paper. A technical report gives all the necessary details (see [2]).

2 The general model

2.1 The neuron

In our model, a neuron is a vectorial function of several variables. More formally, we have :

Definition 2.1 *Let n be a positive integer and let $I_1, \dots, I_n$, W and O be $n+2$ vectorial spaces on $\mathbb{R}$ of finite dimensions. A **n-input neuron** is a continuously differentiable function from $I_1 \times I_2 \times \dots \times I_n \times W$ to O.*

If N is such a neuron, we write dN_w its partial differential with respect to its $(n+1)$-th variable and dN_{i_k} its partial differential with respect to its k-th variable.

In this definition, W, I_k and O are respectively the weight space of the neuron, its$k$-th input space and its output space.

2.2 The neural net

In order to define a neural net, we need to introduce ordered graphs :

Definition 2.2 *An **ordered graph** is a triple $(\mathcal{N}, \mathcal{E}, <)$ fulfilling the following conditions :*

- *$\mathcal{N}$ is a **finite totally ordered set of nodes**. It is in fact a sequence of nodes and we will write its elements as $N^1, N^2, \dots$;*
- *$\mathcal{E}$ is a subset of $\mathcal{N}^2$. $e = (x, y) \in \mathcal{E}$ is an **edge** of the graph and it represents a connection from x to y ;*
- *$<$ is a function from $\mathcal{N}$ to $\mathcal{N}^2$. It associate to each node N^i in $\mathcal{N}$ a total order $<_{N^i}$ on its predecessor set, $Pred(N^i)$. This set is therefore a sequence and we can refer to $Pred(N^i)_k$ for its k-th element.*

The main difference between an ordered graph and a graph is of course the ordered structure of the former.

In order to simplify the rest of the paper we introduce here some notations : $\mathcal{G} = (\mathcal{N}, \mathcal{E}, <)$ is an ordered graph with exactly n nodes ; $N^1, \dots, N^n$ is the sequence of the graph nodes ; $P(N^k) = P(k)$ is the set of the predecessors of N^k ; $S(N^k) = S(k)$ is the set of the successors of N^k. We assume that node N^k has exactly p^k predecessors and s^k successors, and that we call $P(k)_1, \dots, P(k)_{p^k}$ the sequence of the predecessors of N^k and $S(k)_1, \dots, S(k)_{s^k}$ the sequence of the successors of N^k. In general, superscripts correspond to node numbers and subscripts correspond to input or output numbers. Furthermore, we won't make any distinction between a node N^k and its rank k.

We need also to generalize the notion of predecessors : for an arbitrary node N, we define $P^0(N)$ as $\{N\}$ and recursively $P^k(N)$ as $\{M \in \mathcal{N} \mid \exists Q \in P^{k-1}(N)$ so that $(M,Q) \in \mathcal{E}\}$. We have therefore $P^1(N) = P(N)$. We call also $P^+(N)$ the set $\cup_{k=1}^{\infty} P^k(N)$ and $P^*(N) = P^0(N) \cap P^+(N)$. Similar sets can be defined in order to generalize the successor notion.

We can now define a neural network :

Definition 2.3 *A* ***feedforward neural network*** *is an ordered graph $\mathcal{G} = (\mathcal{N}, \mathcal{E}, <)$ fulfilling the following conditions :*

1. *The graph has* ***no-cycle****.*
2. *The elements of $\mathcal{N}$ are* ***k-input neurons*** *(k depends of course on the neuron). The output space of N^k is O^k and its weight space is W^k.*
3. *If $p^k > 0$ then neuron N^k is a p^k-input neuron. That means it has exactly one input for each of its predecessors in the graph (but only if it has predecessors !). The input spaces of N^k are $I_1^k, \ldots, I_{p^k}^k$.*
4. *If $p^k = 0$ then neuron N^k is a 1-input neuron with input space I^k.*
5. *If $p^k > 0$ then for each input i of the neuron N^k the following condition holds : $\dim I_i^k = \dim O^{P(k)_i}$.*

Let us now introduce some additionnal definition related to the neural network.

In is the **subset of $\mathcal{N}$ which elements have no predecessor**, i.e., $In = \{N \in \mathcal{N} \mid P(N) = \emptyset\}$. *In* has *in* elements. As a subset of $\mathcal{N}$, *In* is totally ordered and $In_1, \ldots, In_{in}$ is the ordered sequence of its elements. The elements of *In* are connected to the "outside" by means of their inputs. As we make no disctinction between a node and its rank, we can call for instance O^{In_k} the output space of the node $N^j = In_k$, which is in fact O^j.

Out is the **subset of $\mathcal{N}$ which elements have no successor**, i.e., $Out = \{N \in \mathcal{N} \mid S(N) = \emptyset\}$. *Out* has *out* elements and is totally ordered. $Out_1, \ldots, Out_{out}$ is the ordered sequence of its elements.The elements of *Out* are connected to the "outside" by means of their outputs.

The vectorial space $I = \prod_{k=1}^{in} I^{In_k}$ is **the input space** of the neural network, the vectorial space $O = \prod_{k=1}^{out} O^{Out_k}$ **is the output space** of the neural network and the vectorial space $W = \prod_{k=1}^{n} W^k$ is **the weight space** of the neural network.

2.3 Computing the output

We define the output of the network like this :

Definition 2.4 *Let $\mathcal{G}$ be a feedforward neural network. Let $x = (x_1, \ldots, x_{in}) \in I$ be an input vector and let $w = (w^1, \ldots, w^n) \in W$ be a weight vector. For each l, $1 \le l \le n$, $o^l(x,w)$, the output of the neuron N^l, is computed with the help of the following recursive construction :*

- *If $N^l \in In$, we have $N^l = In_k$. Then $o^l(x,w) = N^l(x_k, w^l)$;*
- *If $N^l \notin In$, the output of N^l is obtained by $o^l(x,w) = N^l(o^{P(l)_1}(x,w), \ldots, o^{P(l)_{p^l}}(x,w), w^l)$.*

Finally, $G(x,w)$, the output of the network, is obtained by $G(x,w) = (o^{Out_1}(x,w), \ldots, o^{Out_{out}}(x,w))$.

Of course this definition is correct only because the underliying graph is non cyclic.

In the rest of the paper, we will call $I^l(x,w)$ the generalized input of node N^l, i.e :

$$(o^{P(l)_1}(x,w), \ldots, o^{P(l)_{p^l}}(x,w), w^l)$$

3 Computing the differential

3.1 Direct method

As a composed function, G is continuously differentiable. The first method to compute its differential is to use the classical derivation rule for composed function. This is the *direct method.*

Theorem 3.1 *Let $\mathcal{G}$ be a feedforward neural network. The o^l functions are continuously differentiable and we have :*

- *if $N^l \neq N^k$:*
  - *if $N^l \notin In$:*

$$\frac{\partial o^l}{\partial w^j}(x,w) = \sum_{k=1}^{p^l} dN_{i_k}^l \left(I^l(x,w)\right) \frac{\partial o^{P(l)_k}}{\partial w^j}(x,w) \qquad (1)$$

  - *if $N^l \in In$:*

$$\frac{\partial o^l}{\partial w^j}(x,w) = 0 \qquad (2)$$

- *if $N^l = N^k$:*

$$\frac{\partial o^l}{\partial w^l}(x,w) = dN_w^l \left(I^l(x,w)\right) \qquad (3)$$

We have a similar property if we consider $\frac{\partial o^l}{\partial x_i}$, where x_i is the input of the i-th input neuron.

3.2 Back-propagation

The key idea of the back-propagation algorithm is to consider $o^k(x,w)$, the output of neuron N^k, as a function of $o^l(x,w)$, the output of another neuron N^l. In fact, we define a new function $o^{k \to l}(x,w,f^l)$ and we have of course $o^{k\to l}(x,w,o^l(x,w)) = o^k(x,w)$. We have the following theorem :

Theorem 3.2 *Let $\mathcal{G}$ be a feedforward neural network. Let N^k, N^l, x and w be respectively two neurons of $\mathcal{G}$, an arbitrary input vector and an arbitrary weight vector. Then we have :*

$$\frac{\partial o^k}{\partial w^l}(x,w) = \frac{\partial o^{k\to l}}{\partial o^l}\left(x,w,o^l(x,w)\right) dN_w^l \left(I^l(x,w)\right) \qquad (4)$$

A similar equation is fullfilled for $\frac{\partial o^l}{\partial x_i}$.

The back-propagation algorithm gives a recursive method to compute $\frac{\partial o^{k\to l}}{\partial o^l}$. We need first a additionnal definition :

Definition 3.1 *Let $\mathcal{G}$ be an ordered graph and let N^k and N^l be two nodes of $\mathcal{G}$. $r(k,l)$ is the rank of N^k in the predecessor set of N^l, i.e., N^k is the $r(k,l)$-th predecessor of N^l.*

Theorem 3.3 *Let $\mathcal{G}$ be a feedforward neural network. Let N^k, N^l, x and w be respectively two neurons of $\mathcal{G}$, an arbitrary input vector and an arbitrary weight vector. Then we have :*

- *if $N^l = N^k$ then :*

$$\frac{\partial o^{k \to l}}{\partial o^l}(x, w, o^l(x,w)) = Id_{O^k}, \tag{5}$$

where Id_{O^k} is the identity function of O^k the output space of N^k.

- *if $N^l \notin P^*(N^k)$ then :*

$$\frac{\partial o^{k \to l}}{\partial o^l}(x, w, o^l(x,w)) = 0_{O^l,O^k}, \tag{6}$$

where $0_{A,B}$ is the null function from A to B, two vectorial spaces.

- *if $N^l \in P^+(N^k)$ then :*

$$\frac{\partial o^{k \to l}}{\partial o^l}(x, w, o^l(x,w)) = \tag{7}$$

$$\sum_{N^j \in S(l)} \frac{\partial o^{k \to j}}{\partial o^j}(x, w, o^j(x,w))\, dN^j_{i_{r(l,j)}}(I^j(x,w))$$

4 Input sharing

4.1 The model

The obtained model is rather powerfull but cannot imitate simple models such as multilayer perceptron (MLP). The main problem is that input nodes have different inputs whereas in a MLP they share the same input. In order to avoid this problem, we introduce an input sharing function :

Definition 4.1 *Let $\mathcal{G} = (\mathcal{N}, \mathcal{E}, <)$ be a feedforward neural network. Let NI be a vectorial space on $\mathbb{R}$ of finite dimension and let SF be a function from NI to I, the input space of the neural network. SF is called a* ***sharing function*** *for the neural network $\mathcal{G}$. It defines a new function G_{SF} from $NI \times W$ to O like this :*

$$G_{SF}(y,w) = G(SF(y), w) \tag{8}$$

By using a replicating function which maps one vector x to a n-tuple which elements are all equal to x, we can give the same input vector to all input neurons and therefore simulate a MLP. In fact, with this last definition, we can implement with our model all classic feedforward neural networks such as MLP, Radial Basis Function networks, wavelet networks, etc. The model is therefore a general representation of feedforward neural networks.

4.2 Differentials

The input sharing method does not change the differential of G with respect to its weight vector. In fact, we have $dG_{SF\,w}(y,w) = dG_w(SF(y), w)$. Therefore, we don't have to take care of a sharing function when evaluating the time needed to compute the differential of G with respect to its weight vector.

5 Complexity

In this section, we give the time needed to compute the matricial operations involved in both calculation methods. These times do not include the time needed to compute the local differential (e.g., dN^l_w) for individual nodes. They take only into account the matricial products and sums needed to compute the differential of the output of the network with respect to its weight vector. In the sequel, $|E|$ is the dimension of the vectorial space E.

5.1 Direct method

Theorem 5.1 *Let $\mathcal{G}$ be a neural network. The time needed to compute the matricial operations needed for calculating dG_w with the direct method is proportionnal to :*

$$\sum_{N^j \notin In} |O^j| \sum_{N^l \in P^+(j)} |W^l| \left(-1 + \sum_{N^k \in P(j) \cap S^*(l)} 2|O^k|\right) \tag{9}$$

5.2 Back-propagation

Theorem 5.2 *Let $\mathcal{G}$ be a neural network. The time needed to compute the matricial operations needed for calculating dG_w with the back-propagation method is proportionnal to :*

$$\sum_{k=1}^{out} |O^{Out_k}| \left[\sum_{N^l \in P^+(Out_k)} |W^l| \left(2|O^l| - 1\right) \right. \tag{10}$$

$$\left. + \sum_{N^l \in P^+(Out_k)} |O^l| \left(-1 + \sum_{N^j \in S(l) \cap P^+(Out_k)} 2|O^j|\right)\right]$$

5.3 Comparison

The formulae are not directly comparable. In fact, even for a classic MLP architecture, the number of neurons can be chosen so that the direct method is faster than the back-propagation. Therefore, in order to compute efficiently the differential of the output of a particular network with respect to its weight vector, the theoretical costs have to be compared.

6 Output function

6.1 Model

In order to train a neural network to approximate a function, we use an error function which is in fact a distance between the output of the network and the desired output. The error made by the network is a function of its weight vector and gradient based training methods need only the gradient of this function. In order to compute this gradient, we can handle the error E as a composed function of a distance d and the neural output $G(x, w)$. We compute dG_w with an arbitrary method and then use the classic chain rule to obtain ∇E as the result of a matricial product. Let us note that we need to introduce the desired output into the formula.

We can also see the couple (d, y), where d is a distance fonction and y a desired output, as a particular neuron with no weight vector and with *out* inputs (one input for each output neuron of $\mathcal{G}$). We can therefore apply the back-propagation to this model.

6.2 Back-propagation

We obtain the following theorem :

Theorem 6.1 *Let $\mathcal{G}$ be a feedforward neural network and let F be a continuously differentiable vectorial function with out variables from $S^1 \times \ldots \times S^{out}$ to O^F, with $S^i \simeq O^{Out_i}$. Let $F_{\mathcal{G}}(x, w)$ be :*

$$F\left(o^{Out_1}(x, w), \ldots, o^{Out_{out}}(x, w)\right) = F(G(x, w))$$

We define in a similar way $F^l_{\mathcal{G}}$, where N^l is an arbitrary node, with the help of $o^{Out_j \to l}$.

Let dF_{i_k} be the partial differential of F with respect to its k-th variable. We have :

$$\frac{\partial F_{\mathcal{G}}}{\partial w^l}(x, w) = \frac{\partial F^l_{\mathcal{G}}}{\partial o^l}(x, w, o^l(x, w)) dN^l_w(I^l(x, w)) \quad (11)$$

If $N^l = Out_k$:

$$\frac{\partial F^l_{\mathcal{G}}}{\partial o^l}(x, w, o^l(x, w)) = dF_{i_k}(G(x, w)) \quad (12)$$

If $N^l \notin Out$:

$$\frac{\partial F^l_{\mathcal{G}}}{\partial o^l}(x, w, o^l(x, w)) = \quad (13)$$

$$\sum_{N^j \in S(l)} \frac{\partial F^j_{\mathcal{G}}}{\partial o^j}(x, w, o^j(x, w)) dN^j_{r(l,j)}(I^j(x, w))$$

6.3 Complexity

Theorem 6.2 *With the hypothesis of theorem 6.1, we obtain that the time needed to compute the matricial operations needed for calculating dF_G with the back-propagation method is proportionnal to :*

$$|O^F| \left[\sum_{N^l} |W^l| (2|O^l| - 1) \quad (14) \right.$$
$$\left. + \sum_{N^l \notin Out} |O^l| \left(2 \sum_{N^j \in S(l)} |O^j| - 1\right) \right]$$

The complexity of the direct algorithm applied to this modified model is simply the sum of the complexity obtained in theorem 5.1 with the time needed to compute the product between dG_w and dF (i.e.,a time proportional to $|O^F||W|(2|O| - 1)$).

6.4 Comparison

Once again, it is not possible to compare directly the computation times for both methods. But we can prove that for a multilayer perceptron, if F is an error function (i.e., $|O^F| = 1$), the back-propagation is faster than the direct method (i.e., we rediscover this classic result). Another important result is that computing with our generalized back-propagation the gradient of the error mode by a MLP represented with the mathematical language of our model is as fast as doing it with the classic back-propagation algorithm. In fact, the formulae are identical for both methods.

7 Conclusion

In this paper we have presented a general model for feedforward neural networks. On one hand, this model is powerfull enough to describe classic architectures such as multilayer perceptrons or wavelet networks. On this other hand, the model is simple enough to allow the derivation of an extended back-propagation algorithm. Unfortunaltly, this algorithm is no longer faster than the direct method, except for particular cases (such as the MLP). The obtained differentials can be used for gradient based neural learning.

The general model gives a formal framework and allows us to study overall properties of feedforward neural networks, such as diffenrential computation. It has been used also to construct a simulator kernel which can handle arbitrary feedforward neural network in a very efficient way (see [3]).

References

[1] Léon Bottou and Patrick Gallinari. A Framework for the Cooperation of Learning Algorithms. In R.P. Lippmann, J.E. Moody, and D.S. Touretzky, editors, *Neural Information Processing Systems*, volume 3, pages 781–788. Morgan Kauffman, 1991.

[2] Cédric Gégout, Bernard Girau, and Fabrice Rossi. Les réseaux d'opérateurs. Technical report, Thomson-CSF/SDC, Juillet 1993.

[3] Cédric Gégout, Bernard Girau, and Fabrice Rossi. NSK, an Object-Oriented Simulator Kernel for Arbitrary Feedforward Neural Networks. In *Int. Conf. on Tools with Artificial Intelligence*, pages 93–104, New Orleans (Louisiana), November 1994. IEEE.

FROM PRIME IMPLICANTS TO MODULAR FEEDFORWARD NETWORKS

Uwe Hartmann

RAG, ZK 5.21, Shamrockring 1, D-44623 Herne, FRG

Abstract

The paper utilises prime implicants and minimal polynomials in order to reduce the size of the training set of a neural feedforward network. We propose a heuristic in order to compute reduced polynomials which are often able to reduce the training set since the computation of minimal polynomials is intractable. Further abstractions lead to modular feedforward sub-architectures of neural networks for special training patterns. Finally, we introduce overlapping modular sub-architectures for distinct training patterns.

Introduction

Back-propagation is a widely used, but time-consuming, learning algorithm for feedforward networks [24]. Several variants of the basic learning algorithm are proposed in order to improve its learning performance [3,7,9]. In practice back-propagation is often used to train feedforward networks with Boolean functions. We utilise ideas of the theory of Boolean functions in order to achieve an improvement of the learning performance and to study the architecture of neural networks.

Deriving optimal network architectures is proven to be intractable [14]. Our aim is to understand how the network architecture effects the performance of learning algorithms. Usually, architectures of neural networks are constructed according to the experience of the researcher and they are rated by the later performance of the learning algorithm. If architectures are constructed by certain algorithms (e.g. genetic algorithms) the quality of the architecture is often rated by measures of the learning performance [17,10,1]. Such measures, as the square error, depend on initial or learned weights as well as on the actual architecture [25].

Prime Implicants and Minimal Polynomials

In the theory of Boolean functions minimal polynomials and prime implicants are well-studied topics which are utilised to construct circuits, namely PLA's [any textbook on Boolean circuits].

A *monomial* is a conjunction of literals, where *literals* are either Boolean variables (e.g. a) or negated Boolean variables (e.g. $\bar{a}$). A *polynomial* is a disjunct of terms and each term is a monomial.

A monomial m is an *implicant* of a Boolean function f iff $m(x)=1$ implies $f(x)=1$. An implicant m of f is a *prime implicant* iff m includes no distinct monomial which also implies f.

Given a Boolean function as a function table the algorithm of Quine and McCluskey is a quite efficient algorithm to compute all the prime implicants of this function [15,22,23]. However, the construction of minimal polynomials is expensive. Given the prime implicants of the Boolean function we have to compute a minimum cover. This problem is proven to be NP-complete [8], i.e. not tractable in practice.

Modifying Training Data

We utilise minimal polynomials in order to modify the training data of a neural feedforward network. A back-propagation algorithm uses ternary input values: one, zero, and don't care (indicated by #). Instead of using the whole set of training examples we train the feedforward network using the monomials of the minimal polynomial.

In order to illustrate our approach we use the following simple example function:

$$f = abcd \vee ab\bar{c}\bar{d} \vee \bar{a}b\bar{c}d \vee \bar{a}\bar{b}\bar{c}d \vee \bar{a}\bar{b}\bar{c}\bar{d} \vee a\bar{b}\bar{c}\bar{d}$$

Description of network architectures	
Symbol	Interpretation
◯ (bold)	The activation and the error of this unit of the network architecture has to be computed for the current pattern.
━	This connection propagates the activation and the error for the current pattern.
○	This unit of the network architecture is not used to train the current pattern.
─	This connection is not used to train the current pattern.

Box 1

prime implicant	prime pattern	matching pattern	corresp. monomial
$abcd$	1111	1111	$abcd$
$\bar{a}\bar{b}\bar{c}$	000#	0000	$\bar{a}\bar{b}\bar{c}\bar{d}$
		0001	$\bar{a}\bar{b}\bar{c}d$
$a\bar{c}\bar{d}$	1#00	1000	$a\bar{b}\bar{c}\bar{d}$
		1100	$ab\bar{c}\bar{d}$
$\bar{a}\bar{c}d$	0#01	0001	$\bar{a}\bar{b}\bar{c}d$
		0101	$\bar{a}b\bar{c}d$

Table 1

Table 1 demonstrates that the corresponding pattern of the first prime implicant *abcd* matches only one training pattern while the other corresponding patterns match two training patterns. Consequently, the first pattern is trained with twice the learning rate as the other prime implicant compared with the original training patterns. Introducing the number of implied training patterns as a weight for each prime implicant is not sufficient.

prime pattern	number of matches	matching pattern	matched by primes
1111	1	1111	1
000#	2	0000	1
		0001	2
1#00	2	1000	1
		1100	1
0#01	2	0001	2
		0101	1

Table 2

As you see in table 2 pattern 0001 is matched by 000# as well as 0#01. Hence one pattern can also be matched by two prime patterns. In practice, however, we achieved quite good results using the number of # to the power of two of the prime pattern as a weight for the learning rate.

As mentioned before, we cannot expect to compute efficiently the minimal polynomial of a given Boolean function. Therefore we build polynomials consisting of prime implicants only, so-called reduced polynomials. These polynomials are smaller or equal to the size of the initial training set. That is to say we reduce the size of the initial training set. The reduced polynomial is constructed using a heuristic:

```
input:
  t = {1111,1100,0101,0001,0000,1100,1000}
     /* training patterns
  p = {1111,000#,1#00,0#01,#000}
     /* prime patterns
output:
  s = ∅
     /* solution set
while t ≠ ∅ do
  select an arbitrary member of p
  delete all the members of t which are
      matched by the selected member of p
  if there was a match then
    add the selected member of p to s
  fi
od
```

The empirical results show that reduced polynomials improve the learning performance of back-propagation. On average, however, these results are worse than the results for minimal polynomials.

Modular Feedforward Networks

Generalising the idea how to use prime implicants we derive a method to modularise neural feedforward networks. We can think of using training data including "don't care"-symbols either as presenting only the defined input values to the network or as training only connections of input units with defined input values. These different perspectives are illustrated by figure 2 and figure 3 (notice the description of network architectures in the box):

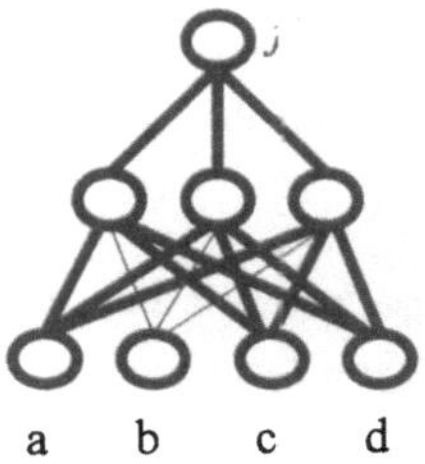

Figure 2

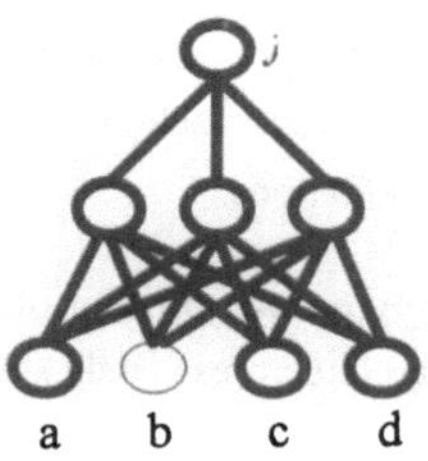

Figure 3

The alternative of this very fine distinction leads to our idea of modular network architectures. Only a part of the network architecture is used in order to train a certain subset of the training examples, other connections are ignored training this subset but might be included in order to train other examples. Therefore there are connections or paths in the network's architecture which are only used to train one or some examples while other connections might be used by all the training examples.

Figures 4 and 5 show the sub-architecture of a strictly modular network. Both figures refer to the same basic architecture (figure 1), however, figure 4 refers to a possible sub-architecture for prime pattern 000# while figure 5 refers to a possible sub-architecture for prime pattern #000.

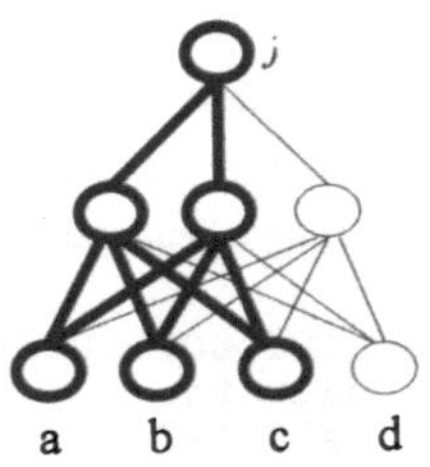

Figure 4

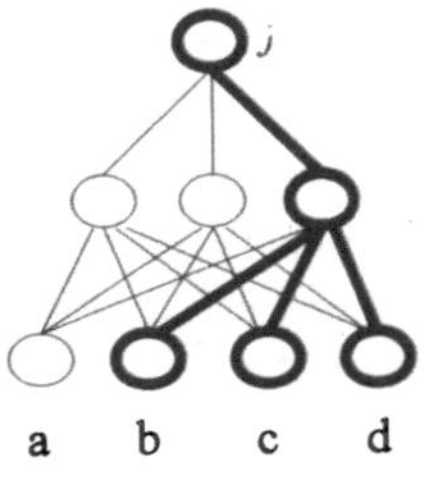

Figure 5

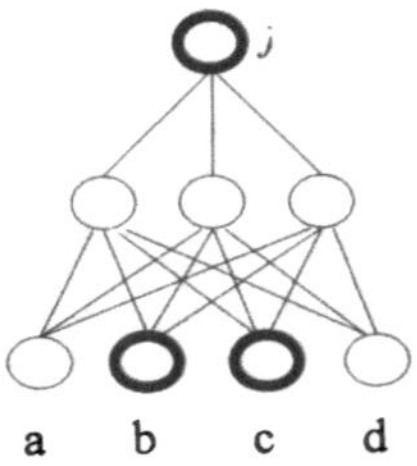

Figure 6

It can be seen (figure 6) that both architectures have got only input and output units in common.

Given minimal polynomials or at least reduced polynomials we are able to construct neural feedforward networks. Input units represent the input literals or arguments of the Boolean function, output units represent the output of the Boolean function, and each monomial of the polynomial is represented by a hidden unit. An input unit is connected to a hidden unit if the represented input literal belongs to the monomial represented by the hidden unit. Each hidden unit is connected to the output units in question. The weights are computed in the obvious way.

Overlapping Modular Architectures

However, we are not interested in constructing neural networks which are actually Boolean circuits based on threshold functions. We are interested in studying neural network architectures which provide distributed knowledge representations.

An alternative partition of the basic network architecture into sub-architectures is indicated by Figure 7 and 8.

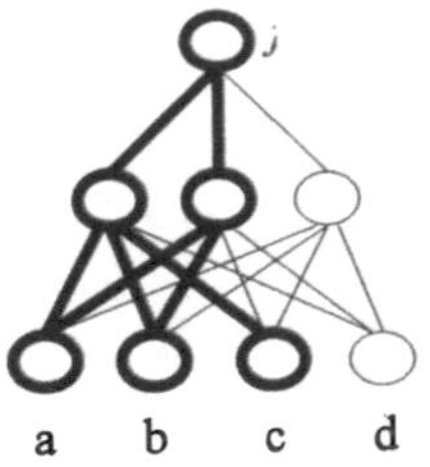

Figure 7

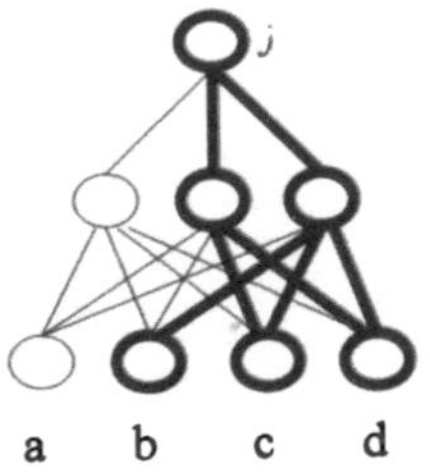

Figure 8

This partition into modules leads to common hidden units and common connections of these hidden units to the output units as well as common input units.

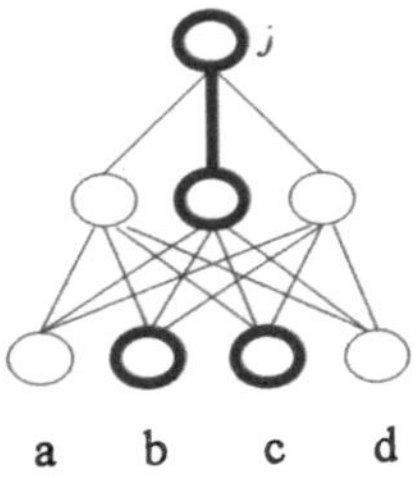

Figure 9

Consequently, the sub-architectures of figure 7 and 8 cannot be trained completely independent. After an initial independent training phase these sub-architectures have to be consolidated by training the whole (basic) architecture. It is, however, still possible to use our prime patterns instead of the original task.

Figure 10 and figure 11 show overlapping network sub-architectures. The overlapping part of the basic architecture comprise a complete sub-tree (figure 12).

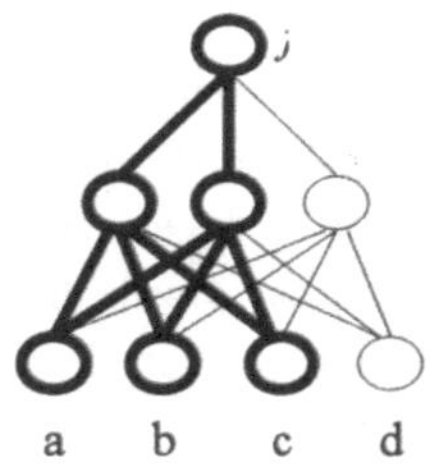

Figure 10

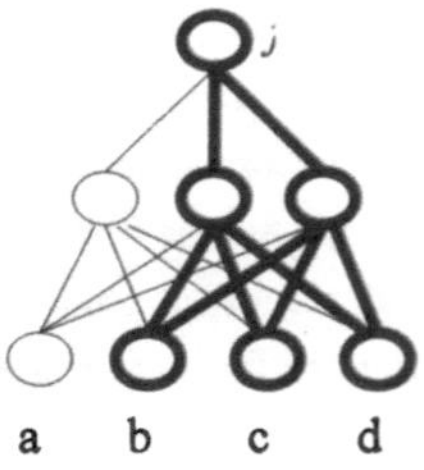

Figure 11

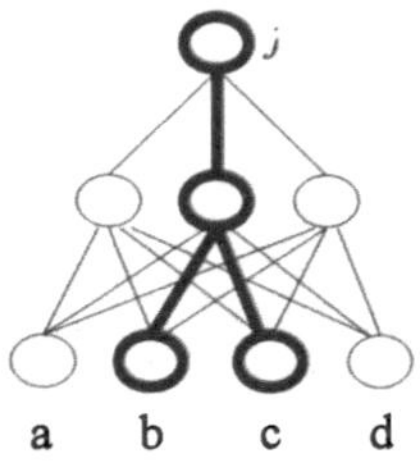

Figure 12

Again after the separate initial training phase of the sub-architectures we have to consolidate both architectures.

The developed method is a heuristic, i.e. it is supposed to improve back-propagation learning in many case. However, it does not guarantee to improve the performance.

Related Results

Our results are related to papers on

- improving the learning performance of back-propagation [27,16],
- designing feedforward network architectures [28,20,18,2,4],
- modularising network architectures [18,13,19],
- initialising neural networks [5,6,13], and
- modifying input data of neural networks [20,12,26].

Further Research

This paper explores the relation of learning tasks and network architectures using methods of a well-studied area. The next step is to study thoroughly the outlined modular and overlapping modular sub-architectures. Our long-term aim is to develop heuristic measures to rate the quality of a network's architecture independent of the current weights. Obviously, the complexity of the network's architecture has to represent the complexity of the learning task.

References

1. R. F. Albrecht, C. R. Reeves, N. C. Steele (eds.). ANNGA93, Springer, 1993.
2. G. Barna, K. Kaski. Choosing Optimal Network Structure. In: INNC-90, pp. 890 - 893, Kluwer, 1990.
3. S. Becker, Y. Cun. Improving The Convergence of Back-Propagation Learning With Second Order Methods. In: [29], pp. 29 - 37.
4. K. J. Cios, N. Liu. A Comparative Study of Machine Learning Algorithms for Generation of Neural Networks. In: [11], pp. I-189 - I-194.
5. T. Denoeux, R. Lengellé, S. Canu. Initialization of Weights in a Feedforward Neural Network Using Prototypes. In: [11], pp. I-623 - I-628.
6. G. P. Drago, S. Ridella. An Optimum Weights Initialization for Improving Scaling Relationships in BP-Learning. In: [11], pp. II-1519 - II-1524.
7. S. E. Fahlman. Faster Learning Variations on Back-Propagation: An Empirical Study. In: [29], pp. 38 - 51.
8. M. R. Garey, D. S. Johnson. Computers and Intractability. Freeman and Company, 1979.
9. H. Haario, P. Jokinen. Increasing the Learning Speed of a Backpropagation Algorithm by Linerization. In: [11], pp. I-629 - I-634.
10. S. A. Harp, T. Samad, A. Guha. Designing Application-specific Neural Networks Using the Genetic Algorithm. In D. S. Touretzky (ed.). IEEE CNIPS90, vol. 2. pp. 447 - 454, Morgan Kaufmann, 1990.
11. T. Kohnen, K. Mäkisara, O. Simula, J. Kangas (eds.). Artificial Neural Networks, North-Holland, 1991.
12. M. A. Kraaijveld, R. P. W. Duin. On Backpropagation Learning of Edited Data Sets. In: INNC-90, pp. 741 - 744, Kluwer, 1990.
13. Y. Lee, S. Oh, M. Kim. The Effect of Initial Weights on Premature Saturation in Back-Propagation Learning. In: IJCNN91. pp. I-765 - I-770., 1991.
14. J. Lin, J. S. Vitter. Complexity Issues in Learning by Neural Nets. In Ronald Rivest, David Haussler, Manfred K. Warmuth (eds.). COLT89. pp. 118 - 132, Morgan Kaufmann 1989.
15. E. J. McCluskey. Minimization of Boolean Functions. Bell Systems Tech. J. 35, pp. 1417 - 1444, 1956.
16. S. Makram-Ebeid, J.-A. Sirat, J.-R. Viala. A Rationalized Error Back-Propagation Learning Algorithm. In: IJCNN89. pp. II-373 - I-380, 1989.
17. G. F. Miller, P. M. Todd, S. U. Hedge. Designing Neural Networks Using Genetic Algorithms. In J. D. Schaffer (ed.). ICGA89. pp. 379 - 384, Morgan Kaufmann, 1989.
18. K. Möller, S. Thrun. Task Modularization by Network Modulation. In Neuro Nimes '90. pp. 419 - 432, 1990.
19. F. Nadi. Topological Design of Modular Neural Networks. In: [11], pp. I-213 - I-218.
20. N. K. Perugini, W E. Engeler. Neural Network Learning Time: Effects of Network and Training Set Size. In: IJCNN89. pp. II-395 - II-402, 1989.
21. D. Polani, T. Uthmann. Training Kohnen Feature Maps in Different Topologies: an Analysis Using Genetic Algorithms. In S. Forrest (ed.). 5th ICGA93, Morgan Kaufmann, pp. 326 - 333, 1993.
22. W. V. Quine. Two Theorems about Truth Functions. Bol. Soc. Math. Mex. 10, pp. 64 - 70, 1953.
23. W. V. Quine. A Way to Simplify Truth Functions. American Math. Soc. 62, pp. 627 - 631, 1955.
24. D. E. Rumelhart, J. L. McClelland (eds.). Parallel Distributed Processing, volume 1. The MIT-Press, 1986.
25. M. Schmitt, F. Vallet. Network Configuration and Initialization Using Mathematical Morphology: Theoretical Study of Measurement Functions. In: [11], pp. II-1045 - II-1048.
26. F. M. Silva, L. B. Almeida. Speeding-Up Backpropagation by Data-Orthonormalization. In: [11], pp. I-213 - I-218.
27. W. S. Stornetta, B. A. Huberman. An Improved Three-Layer, Back Propagation Algorithm. In: Maureen Caudill, Chales Butler. IEEE 1st ICNN87. pp. II-637 - II-644, San Diego, 1987.
28. G. A. Tagliarini, E. W. Page. Learning in Systematically Designed Networks. In: JCNN89. pp. I-497 - I-502, Washington, 1989.
29. D. Tourtzky, G. Hinton, T. Sejnowski. Proceedings of the 1988 Connectionist Models Summer School, Carnegie Mellon University, Morgan Kaufmann Publishers, 1988.

HIERARCHICAL BACKWARD SEARCH METHOD: A NEW CLASSIFICATION TREE USING PREPROCESSING BY MULTILAYER NEURAL NETWORK

Hiroto Yoshii

hiroto@cfp.canon.co.jp, Media Technology Laboratory, Canon Inc.
890-12 Kashimada Saiwai-ku Kawasaki-shi Kanagawa 211, Japan

Abstract

This paper proposes a novel pattern recognition algorithm. The new algorithm consists of two stages: the pyramid making stage and the classification making stage. The classification tree generated by the algorithm is called "PACT" (Pyramid Architecture Classification Tree), which is just a kind of classification tree but has a great ability to deal with huge dimensional real scale problem.

PACT does not only shows good performance but also induces a new idea: a fractal characteristic which a category set shows in huge dimensional feature space. The new idea accounts for how awkward conventional algorithms, including neural networks and classification trees, are when applied to huge dimensional pattern classification problems. The new idea gives a new positive reason for hierarchical pattern processing.

1. Introduction

The final goal of pattern recognition is the perfect--100%--recognition ability as well as human brain. Theoretically a recognition rate can be increased to as much as 100% for learning patterns. For test patterns, however, a recognition rate can't be increased simply; at some point of learning stages, it starts decreasing. This is called as overtraining/overfitting problem.

This problem is one of the most crucial problems lying on the way to the final goal. Neural networks and classification trees--two popular pattern recognition algorithms--avoid this problem by pruning algorithms: to prune hidden layers in neural networks and to prune subtrees in classification trees for adapting system's complexity to data's complexity.(1,2)

When overtraining problems are due to learning pattern's noises, for example overfitting problems in function fitting, these algorithms can be efficient; a system turns out to absorb learning pattern's noises and show the highest ability as possible. However, if the overtraining problem is due to feature extractions, these algorithms would be vain. The reason is the following.

Usually before making up a pattern recognition system, several appropriate features are extracted from raw learning patterns. It is as a matter of course that features--dimensions--decrease compared with full features which are available in raw learning patterns. This results in piles, folds and overlaps of learning patterns--the loss and confusion of information. Thus feature extractions constrain and limit system's recognition capacity and the pruning algorithms can neither deal with nor improve it. Consequently feature extractions turn out to be an essential cause of overtraining problems.

If one, on the other hand, dealt with raw patterns directly, one would have to give up learning because of considerable computational costs. Thus conventional pattern recognition algorithms, except for primitive pattern matching algorithms such as nearest neighbor matching algorithms, are forced to extract features which results in an imperfect recognition. This dilemma has been a severe bottleneck of pattern recognition problems for many yeas.

This paper proposes a novel algorithm which can directly deal with huge dimensions of real scale problems with low computational costs, overcome the dilemma mentioned above, and have an ability to realize the perfect recognition. The algorithm is a kind of classification tree but needs pyramid architecture neural network for dynamic feature extraction. Feature extractions in the algorithm are dynamic, automatic and systematic, though conventional feature extractions are static, manual and ad hoc. Also the algorithm induces a new idea: a fractal characteristic which a category set shows in huge dimensional feature space. The new idea will give a new positive reason for hierarchical computation.

2. Description of the algorithm

The algorithm consists of two stages: the pyramid making stage and the classification tree making stage. Notice that hierarchical structures appear in both stages and they are confusing, but they are different ones and have different meanings respectively. To make the situation clear, constituents of a pyramid are called as "cells", while constituents of a classification tree are called as "nodes" in the following explanation.

the pyramid making stage

All training patterns are preprocessed through a pyramid architecture neural network. For the sake of simplicity, a non-overlapped quad-pyramid structure is introduced as shown in

Fig. 1.[3] All cells in the pyramid have binary states--on/off; "black" is identical to "on" and "white" is identical to "off".

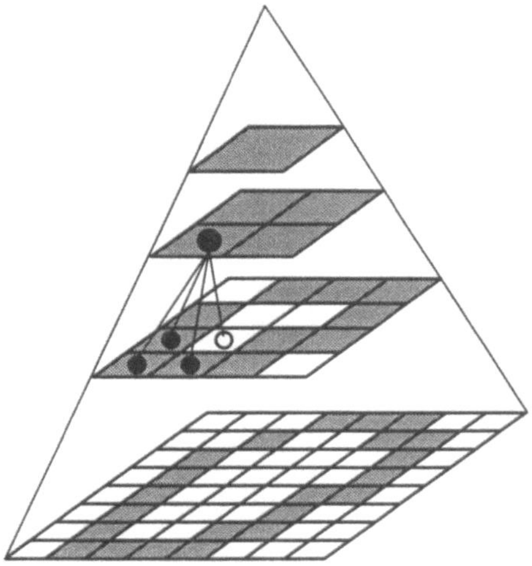

Fig. 1: the pyramid to preprocess training patterns

First, all training patterns--bit maps--are fed to the bottom layer of the pyramid. Next, informations are propagated hierarchically toward the apex cell of the pyramid. Information propagation rule is supposed to be very simple; when all of four children cells are "off", a parent cell is "off", otherwise "on". Consequently, fine_to_coarse representation of training patterns are realized. Notice that by this rule, if even one cell/bit is on in a training pattern, the apex cell will be "on". (We supposed this rule only because it is so easy to explain and understand. In fact the algorithm doesn't necessarily restrict the information propagation rule.)

The bottom layer consists of 8x8 cells in Fig. 1, but the algorithm does not restrict the size of a pyramid. It means that one can make the size of the bottom layer as bigger as possible, and the increase of layers does not influence the final size of a generated classification tree. (The reason will be mentioned at the end of the next subsection)

the classification tree making stage

In the classification tree making stage, a classification tree is made by using features generated in the pyramid making stage. However, the choice of feature subsets to be used at each node is a very crucial problem.[4] Conventional classification trees, applied to huge dimensional problems in special, make use of only one feature among all original features at each node, because it is impractical to use linear combinations of full features for computational costs. The strategy of the algorithm to deal with huge dimensions is the following; features used at each node are chosen from the apex--the most coarse feature--to the bottom layer--the original fine features--in the pyramid of the previous subsection.

Fig. 2 describes the classification making stage in detail. A classification tree starts from the root node and expands many descendent nodes. Every node except for the root node in a

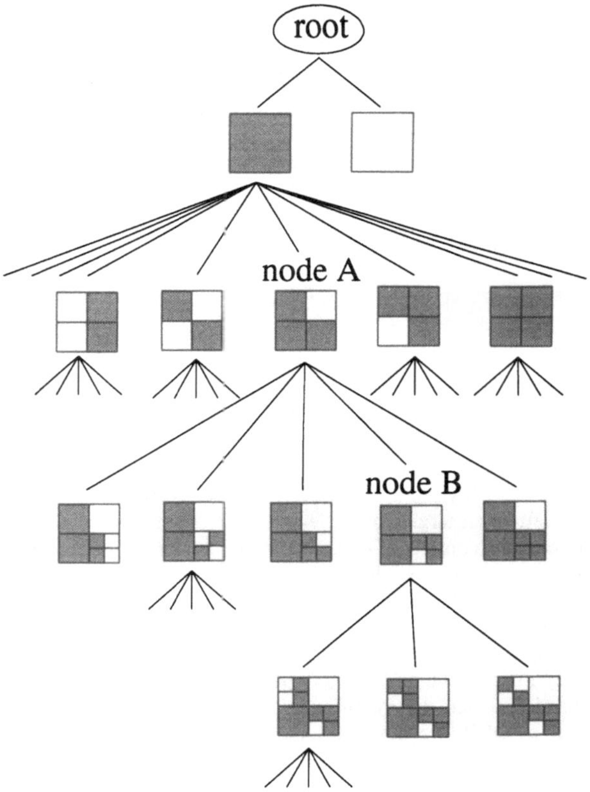

Fig. 2: a generated tree structure

classification tree is described by a subset of cells in the pyramid. Activities of a subset of cells in the pyramid are projected into heterogeneous meshes presented in Fig. 2.

At each internal node, one on-cell or a set of on-cells (number is n) is selected and only 4xn (n=1,2,...) cells below that/them are used in generating descendent nodes; $(2^{4xn} - 1)$ nodes are generated. For example, at node A in Fig. 2, the bottom-right on-cell is selected among three on-cells, and descendent nodes are generated with states of four cells below it, while at node B the top-left on-cell is selected among five on-cells.

In fact, there are three on-cells in node A and five on-cells in node B; there are three and five options about which on-cell should be selected. Which on-cell or set of on-cells is appropriate to select is decided in terms of "Classification Efficiency" which is Mutual Information / log(branch number). This measure ensures the optimal feature extraction at each internal node.

As long as there exist internal nodes, the process to generate descendent nodes will continue recursively, and a tree continues growing until there remains no internal node--there remain only null nodes and leaves.

It means that as a classification tree grows deeply, new and finer features are extracted dynamically, and when no more feature extraction is needed for classification, both tree growing and feature extractions stop. Therefore redundant features/cells in the pyramid are not used, and no matter how huge pyramid one made in the pyramid making stage, the size of a classification tree has nothing to do with it.

Fig. 3: examples of training patterns

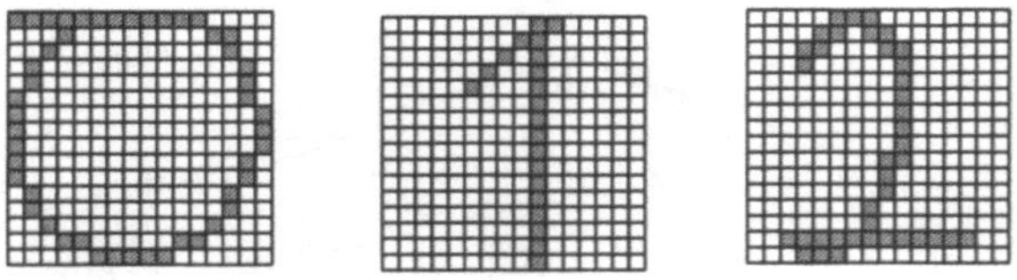

3. Experiment

To evaluate the ability of the proposed algorithm, a preliminary experiment was done. In the following explanation, the classification tree which is made by the algorithm is called "PACT" (Pyramid Architecture Classification Tree). Classification targets of the experiment are 10 numerals. The training set consists of 800 patterns per numeral sampled from one person--me--, while the test set consists of 50 patterns per numeral sampled from 50 people; each person wrote one set of the numerals 0-9. All patterns are sampled by pen-based computers which can digitize temporal series of pen strokes, in the experiment, however, temporal informations are abandoned and only configuration informations are used as well as OCR. To normalize the original patterns by size, a minimal square bounding box was fit to the set of nonzero pixels in each pattern, and the containing area was downsampled to a 16x16 pixel bitmap. No attempt was made to normalize for aspect ratio, shear, rotational variations. A subset of the training patterns is shown in Fig. 3.

To investigate a learning curve, different PACT's were made by different training sets in size. In the graph of Fig. 4, averages of recognition rates for 10 numerals are plotted as a function of training set size; every time a different PACT was made by 40,80,120 .. ,800 training patterns, recognition rates for 10 numerals were tested on 50 test patterns, and they were averaged. In the graph of Fig. 5, aggregations of the number of all leaves in the PACT's are plotted. Notice that recognition rates for the training set are perfect--100%--, because a generated PACT contains full informations of the training set.

According to Fig. 4, the recognition rate appears to saturate at approximately 90%, however, Fig. 5 shows that a learning stage is still continuing. In fact, the learning curve of PACT is thought to be identical to that of the nearest neighbor matching algorithm.(5) This means that more than 10,000 training patterns are required to realize 100% recognition rate. Of course it is true that the PACT explained in this paper is the most primitive one, and more sophisticated one could improve the ability--realize 100% recognition rate by fewer training patterns. However, even the most primitive PACT shows the overwhelming characteristics: no effort of skillful feature extractions, no need for decision about cessation of a learning (the recognition rate never decrease), and far smaller dictionary size and far faster classification speed compared with nearest neighbor matching algorithms.

Fig. 4: learning curve

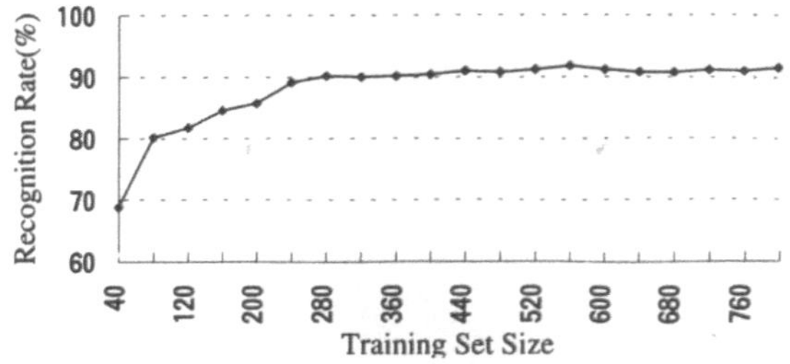

Fig. 5: aggregation of all leaves

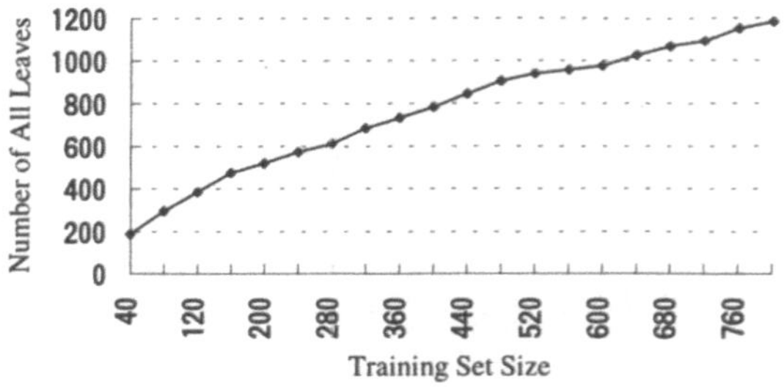

4. Discussions

The algorithm to generate PACT uses both the pyramid--a kind of neural network--and a classification tree. Thus PACT seems to resemble hybrid classification trees such as Neural Tree Networks whose nodes are composed by small neural networks.(6,7,8,9) However the hybrid classification trees never solve huge dimensional classification problems because they don't give the answer about which feature set is appropriate in each node. On the contrary, PACT finds the best features at each node and never extract. redundant fine features.

PACT was motivated from an idea to combine hierarchical pattern matching and dynamic feature extractions as mentioned in Section 1. This idea is not a new one; recently a few groups have proposed similar ideas(10), however, PACT suggests a much deeper idea: a fractal structure in a huge dimensional feature space.

Fig. 6: examples of leaves of category "0"

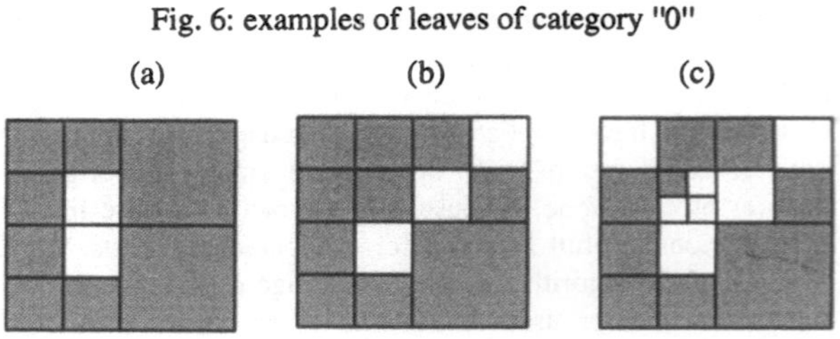

Fig. 6 shows examples of leaf shapes of category "0" which were produced by 800 training patterns. These patterns consist of different meshes in size. It means that these patterns contain various numbers of raw patterns or in these states various numbers of raw patterns degenerate. For example, 2^{224} bitmaps degenerate in leaf (a) in Fig. 6, 2^{208} bitmaps degenerate in leaf (b), and 2^{180} bitmaps degenerate in leaf (c); leaf (a) has as 2^{16} times raw patterns as leaf (b) has and as 2^{44} times raw patterns as leaf (c) has. Thus Fig. 6 suggests that whole set of leaf patterns--a dictionary--of one category should construct a very complicated domain in huge dimensional feature space, which may have fractal characteristic.

Fig. 7 illustrates the concept of PACT. In the middle of Fig. 7 the Kock curve is shown, inside which there exist various triangles of different size. Conceptually speaking, to make PACT turns to be the following. In the pyramid making stage, one prepares many hypercubes in huge dimensional feature space like various size triangles inside Kock curve. Next, in the classification making stage, one puts them into pieces and makes them up to construct the perfect dictionary shape in the huge feature space, whose surface has a fractal characteristic.

Fig. 7

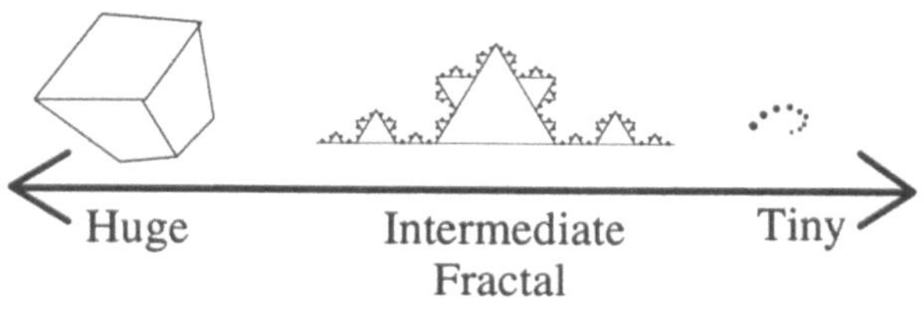

(Exactly, shapes of categories in huge dimensions may be not like the inside of Kock curve whose surface has a fractal characteristic, but like Cantor set which is itself a fractal set. This discussion requires more detail analysis of shapes in huge dimensions, and will appear in the next paper.) The surface's shape is so complicated that it can't be covered by several hyperplanes and inside of the surface is so large that it can't be filled by individual patterns.

A fractal in huge dimensions can not be discussed as well as a fractal in low dimension, thus this fractal characteristic is not confirmed one, but the result suggests that the fractal characteristic may be the cause of awkwardness which conventional pattern recognition systems have shown when applied to huge dimensional real scale problems. Covering by several hyperplane mentioned above corresponds to conventional perceptrons and classification trees, and filling by individual patterns corresponds to nearest neighbor matching algorithms. Even the hybrid classification trees such as Neural Tree Networks use pseudo-hyperplanes to separate a category domain, and correspond to the left hand illustration in Fig. 7.

Furthermore, the fractal characteristic gives a new positive reason for hierarchical matching. So far hierarchical classification methods have been discussed mostly in the context of computational costs. In this context, one need not introduce hierarchical classification methods positively; for example, in the experiment mentioned in Section 3, even a conventional perceptron could have the perfect classification ability by preparing huge ($2^{16x16}=2^{256}$) hidden units, and there seems to be no need of using hierarchical matching method. However it is empirically impractical and the reason is explained by only the word--"generalization".

If the shape of each category in huge dimensional feature space were fractal, hierarchical preprocessing would be needed essentially, because it is impossible to represent a category shape by a few parameters such as eigenvectors which conventional pattern recognition system have introduced. What is essential in huge dimensional pattern recognition problems is not to extract the best features once but recursive, successive and nested search of the best features, from global features to fine features.

Consequently, PACT gives a new positive reason for hierarchical representation and classification that they may be the most crucial for dealing with a real world, in which shapes of categories may have fractal characteristics.

5. Extensions and Conclusions

Extensions

Various extensions of PACT can be thought; the most simple one is variations of propagation rules in the preprocessing pyramid. Instead of the rule explained in Section 2, one can introduce the rule that if more than or equal to two child cells are "on", a parent cell is "on", otherwise "off". One can also change propagation rules in different positions in the pyramid. Pyramid structure can vary from irregular one to overlapped one.

Conclusions

This paper disclosed a novel classification method: PACT--Pyramid Architecture Classification Tree. By PACT, it becomes possible to deal with huge dimensional real scale problems. PACT also revealed a fractal characteristic of a category domain in huge dimensional feature space, which suggests that recursive and dynamic feature extractions are essential for real scale pattern recognition.

References

1. Russel Reed: IEEE Trans. NN 4(5), 740 (1993)
2. L. Breiman, J. H. Friedman, R. A. Olshen, and C. J. Stone: Classification and Regression Trees. Belmont CA: Wadsworth 1984.
3. J. M. Jolion and A. Rosenfeld: A Pyramid Framework for Early Vison. Netherlands: Kluwer Academic Publishers 1994.
4. S. R. Safavian and D. Landgrebe: IEEE Trans. SMAC 21(3) 660 (1991)
5. S. J. Smith, M. O. Bourgoin, K. Sims, and H. L. Voorhees: IEEE Trans. PAMI 16(9) 915 (1994)
6. J. A. Sirat and J. P. Nadal: Network 1 423 (1990)
7. A. Sankar and R. J. Mammone: Neural Networks: Theory and Applications. 281, New York: Academic Press 1991.
8. H. Guo and S. B. Gelfand: IEEE Trans. NN 3(6) 923 (1992)
9. I. K. Sethi and J. H. Yoo: Pattern Recognition 27(7) 939 (1994)
10. J. Zhou and T. Pavlidis: Pattern Recognition 27(11) 1539 (1994)

ADAPTATION ALGORITHMS FOR 2-D FEEDFORWARD NEURAL NETWORKS

Tadeusz Kaczorek

Warsaw University of Technology, Faculty of Electrical Engineering
00-662 Warszawa, ul. Koszykowa 75, POLAND

Abstract. The generalized weight adaptation algorithms presented in [6, 11] are extended for 2-D madaline and 2-D two-layer FNN's.

1. Introduction

A feedforward neural network (FNN) is a special type of nonrecurrent artificial neural network (ANN). It consists of layers of neurons with weighted links connecting the outputs of neurons in one layer to the inputs of neurons in the next layer [2, 6, 8-11]. Applications of FNN's to dynamical system indentification and control have been considered in [6].

In this paper the well-known [6, 11] generalized weight adaptation algorithm for the madaline (many adaptive linear elements) will be extended for the two-dimensional (2-D) case. The generalized weight adaptation algorithm will be also extended for the 2-D two-layer FNN. A need of this extension has been arisen in solving one practical problem.

2. 2-D madaline weight adaptation algorithm

A block diagram of the 2-D madaline weight adaptation algorithm is shown in Fig.1.

Weight matrix $W_{ij} = \begin{bmatrix} w_{ij}^1 & w_{ij}^2 & \dots & w_{ij}^n \end{bmatrix}$

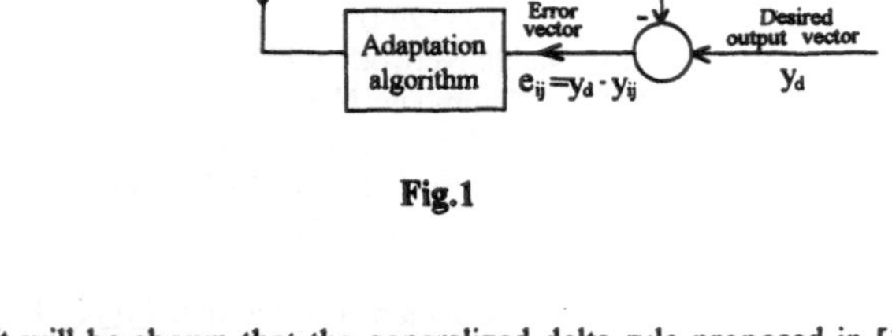

Fig.1

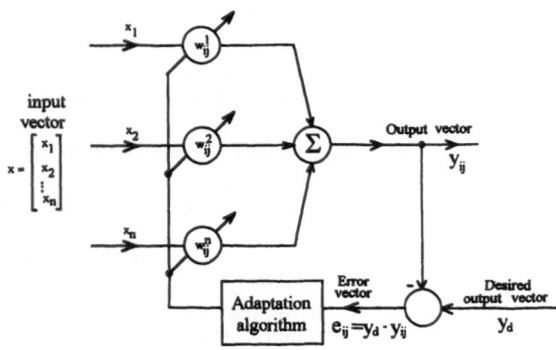

It will be shown that the generalized delta rule proposed in [11, 6] can be extended for the 2-D madaline.

Let Θ: $R^n \rightarrow R^n$ be an operator, for example

$$\Theta(x) := \begin{bmatrix} \operatorname{sgn}(x_1) \\ \vdots \\ \operatorname{sgn}(x_n) \end{bmatrix}, \quad \operatorname{sgn}(x_k) = \begin{cases} +1 & if \ x_k > 0 \\ -1 & if \ x_k < 0 \end{cases}, \quad k = 1,\dots,n$$

W_{ij} is a weight matrix with real entries depending on i and j and $i,j \in Z_+$, Z_+ is the set of nonnegative integers.

Let

$$e_{ij} := y_d - y_{ij} \tag{2.1}$$

be the error vector at the point (i,j), where y_d is the desired output vector, y_{ij} is the output vector at the point (i,j).

Theorem 1

If the weight matrix W_{ij} of the 2-D madaline is adapted according to the rule

$$W_{i+1,j+1} = W_{i+1,j} + W_{i,j+1} - W_{ij} + \Delta W_{ij}$$

where (2.2)

$$\Delta W_{ij} := \begin{cases} \left(Ae_{i+1,j} + Be_{i,j+1} + Ce_{ij}\right)\dfrac{\Theta^T(x)}{\Theta^T(x)x} & \text{if } \Theta^T(x)x \neq 0 \\ 0 & \text{if } \Theta^T(x)x = 0 \end{cases}$$

and the matrices A, B, C are chosen so that all roots (zero-manifolds) of the equation

$$\det\left[Iz_1z_2 + (A-I)z_1 + (B-I)z_2 + I + C\right] = 0 \tag{2.3}$$

are located in the unit bidisk $U:=\{(z_1, z_2) \in \mathbf{C}\times\mathbf{C}: |z_1|<1, |z_2|<1\}$ ($\mathbf{C}$ is the field of complex numbers) then the error vector e_{ij} converges asymptotically to the zero vector for i and j tending to infinity.

Proof

Using (2.1), (2.2) and $y_{ij}=W_{ij}x$ we have

$$e_{i+1,j+1} - e_{i+1,j} - e_{i,j+1} + e_{ij} = -y_{i+1,j+1} + y_{i+1,j} + y_{i,j+1} - y_{ij}$$
$$= -\left(W_{i+1,j+1} - W_{i+1,j} - W_{i,j+1} + W_{ij}\right)x$$
$$= -\left(Ae_{i+1,j} + Be_{i,j+1} + Ce_{ij}\right) \quad if \ \Theta^T(x)x \neq 0$$

and

$$e_{i+1,j+1} = (I-A)e_{i+1,j} + (I-B)e_{i,j+1} - (I+C)e_{ij} \tag{2.4}$$

The characteristic equation of (2.4) has the form (2.3). If all roots of (2.3) are located in the unit bidisk then the vector e_{ij} converges asymptotically to 0 for i and j tending to infinity [1,3-5]. □

If $A = B = -C = I$ then from (2.3) we have $z_1 = z_2 = 0$.

A 2-D madaline adaptation algorithm with $z_1 = z_2 = 0$ will be called the deadbeat one.

Let $A = aI$, $B = bI$ and $C = cI$ then

$$\begin{aligned}&\det[Iz_1z_2 + (A-I)z_1 + (B-I)z_2 + I + C] = \\ &= [z_1z_2 + (a-1)z_1 + (b-1)z_2 + 1 + c]^n = 0\end{aligned} \tag{2.3a}$$

It is easy to see that all roots of (2.3a) are located in the unit bidisk if for example $a = b = 1$ and $|c+1| < 1$.

3. 2-D two-layer FNN weight adaptation algorithm

A block diagram of the 2-D two-layer FNN weight adaptation algorithm is shown in Fig.2.

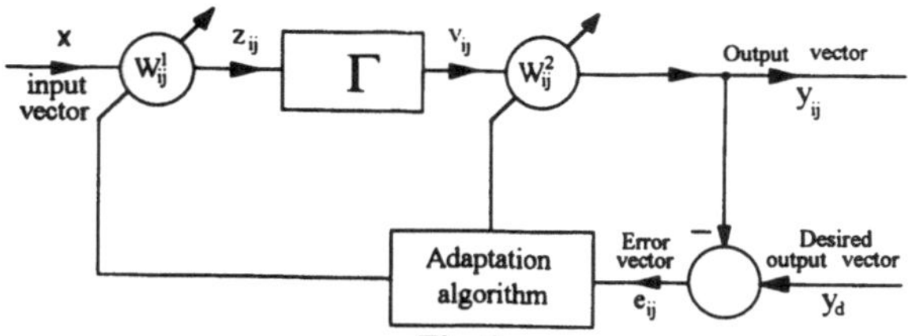

Fig. 2

The input vector $x \in R^n$ is constant (independent of i and j), $W^1_{ij} \in R^{n_1 \times n}$ and $W^2_{ij} \in R^{n_2 \times n_1}$ are weight matrices with real entries depending on i and j (i,j∈Z+), $z_{ij} \in R^{n_1}, v_{ij} \in R^{n_1}$ are real vectors, $\Gamma: R^{n_1} \to R^{n_1}$ is an activation operator satisfying the condition $\Gamma(-z_{ij}) = -\Gamma(z_{ij})$ for all i,j∈Z+ and $y_d \in R^{n_2}, y_{ij} \in R^{n_2}$ are the desired output vector and the output vector at the point (i,j).

Let

$$e_{ij} := y_d - y_{ij} \tag{3.1}$$

be the error at the point (i,j) and Θ_1 and Θ_2 be some nonlinear operators.

Theorem 2

If the weight matrices W^1_{ij}, W^2_{ij} of 2-D two-layer FNN are adapted according to the rule

$$W^k_{i+1,j+1} = W^k_{i+1,j} + W^k_{i,j+1} + W^k_{ij} + \Delta W^k_{i+1,j} + \Delta W^k_{i,j+1} + \Delta W^k_{ij}, \quad k = 1,2 \tag{3.2}$$

where

$$W^1_{i+1,j} = W^1_{i,j+1} = -W^1_{ij}$$

$$\Delta W^1_{i+1,j} = \Delta W^1_{i,j+1} = -\Delta W^1_{ij} = \begin{cases} \dfrac{2z_{ij}\Theta^T_1(x)}{\Theta^T_1(x)x} & if\ \Theta^T_1(x)x \neq 0 \\ 0 & if\ \Theta^T_1(x)x = 0 \end{cases} \tag{3.3}$$

$$\Delta W^2_{i+1,j} = \begin{cases} -2W^2_{i+1,j} - \dfrac{Ae_{i+1,j}\Theta^T_2(v_{ij})}{\Theta^T_2(v_{ij})v_{ij}} & if\ \Theta^T_2(v_{ij})v_{ij} \neq 0 \\ 0 & if\ \Theta^T_2(v_{ij})v_{ij} = 0 \end{cases}$$

$$\Delta W^2_{i,j+1} = \begin{cases} -2W^2_{i,j+1} - \dfrac{Be_{i,j+1}\Theta^T_2(v_{ij})}{\Theta^T_2(v_{ij})v_{ij}} & if\ \Theta^T_2(v_{ij})v_{ij} \neq 0 \\ 0 & if\ \Theta^T_2(v_{ij})v_{ij} = 0 \end{cases} \tag{3.4}$$

$$\Delta W^2_{ij} = \begin{cases} -2W^2_{ij} - \dfrac{Ce_{ij}\Theta^T_2(v_{ij})}{\Theta^T_2(v_{ij})v_{ij}} & if\ \Theta^T_2(v_{ij})v_{ij} \neq 0 \\ 0 & if\ \Theta^T_2(v_{ij})v_{ij} = 0 \end{cases}$$

and the matrices A,B,C are chosen so that all roots (zero manifolds) of the equation

$$\det[Iz_1z_2 - (I+A)z_1 - (I+B)z_2 + I - C] = 0 \tag{3.5}$$

are located in the unit bidisk then the error vector e_{ij} converges asymptotically to the zero vector for i and j tending to infinity.

Proof

Using (3.1), (3.2) and $z_{ij} = W^1_{ij}x,\ v_{ij} = \Gamma(z_{ij}),\ y_{ij} = W^2_{ij}v_{ij}$ we obtain

$$\begin{aligned}&e_{i+1,j+1} - e_{i+1,j} - e_{i,j+1} + e_{ij} = -y_{i+1,j+1} + y_{i+1,j} + y_{i,j+1} - y_{ij} \\ &= -W^2_{i+1,j+1}v_{i+1,j+1} - W^2_{i+1,j}v_{i+1,j} + W^2_{i,j+1}v_{i,j+1} - W^2_{ij}v_{ij} \\ &= -W^2_{i+1,j}[\Gamma(z_{i+1,j+1}) - \Gamma(z_{i+1,j})] - W^2_{i,j+1}[\Gamma(z_{i+1,j+1}) - \Gamma(z_{i,j+1})] \\ &-W^2_{ij}[\Gamma(z_{i+1,j+1}) + \Gamma(z_{ij})] - \Delta W^2_{i+1,j}\Gamma(z_{i+1,j+1}) - \Delta W^2_{i,j+1}\Gamma(z_{i+1,j+1}) \\ &-\Delta W^2_{ij}\Gamma(z_{i+1,j+1})\end{aligned}$$

From (3.3) it follows that

$$z_{i+1,j} = z_{i,j+1} = -z_{ij}$$

and by (3.2) we have

$$\begin{aligned}&z_{i+1,j+1} = W^1_{i+1,j+1}x = z_{i+1,j} + z_{i,j+1} + z_{ij} + \Delta W^1_{i+1,j}x + \Delta W^1_{i,j+1}x + \Delta W^1_{ij}x \\ &= z_{i+1,j} + \Delta W^1_{i+1,j}x = z_{i+1,j} - \frac{2z_{i+1,j}\Theta^T_1(x)x}{\Theta^T_1(x)x} = -z_{i+1,j}\end{aligned} \tag{3.6}$$

Taking into account (3.6), the assumption $\Gamma(-z_{ij}) = -\Gamma(z_{ij})$, and (3.4) we obtain

$$\begin{aligned}&e_{i+1,j+1} - e_{i+1,j} - e_{i,j+1} + e_{ij} \\ &= 2(W^2_{i+1,j}\Gamma(z_{i+1,j}) + W^2_{i,j+1}\Gamma(z_{i,j+1}) - W^2_{ij}\Gamma(z_{ij})) - \\ &(\Delta W^2_{i+1,j} + \Delta W^2_{i,j+1} + \Delta W^2_{ij})v_{ij} = Ae_{i+1,j} + Be_{i,j+1} + Ce_{ij}\end{aligned}$$

and

$$e_{i+1,j+1} = (I+A)e_{i+1,j} + (I+B)e_{i,j+1} + (C-I)e_{ij} \tag{3.7}$$

If all roots of (3.5) are located in the unit bidisk then the vector e_{ij} converges asymptotically to 0 for i and j tending to infinity [1,3-5]. □

4. Concluding remarks

The generalized weight adaptation algorithms have been extended for 2-D madaline and 2-D two-layers FNN's.

Theorem 2 can be extended for 2-D three-layer FNN and generally for 2-D n-layer FNN's (n>2).

Note that for the adaptation algorithm (3.2) another choice of $\Delta W^k_{i+1,j}, \Delta W^k_{i,j+1}$ and ΔW^k_{ij} for k=1,2 can also guarantee the convergence of e_{ij} for $i,j \to \infty$.

With slight modifications the above considerations can be extended for n-D, n>2 madaline and n-D many-layers FNN's.

References

1. A. R. E. Ahmed, *On the Stability of Two-Dimensional Discrete Systems*, IEEE Transactions on Automatic Control, vol. AC-25, No 3, June 1980, pp. 551-552
2. F. Chen, *Back-Propagation Neural Networks for Nonlinear Self-Tuning Adaptive Control*, IEEE Contr. Syst. Mag., vol. 10, Apr. 1990, pp. 44-48.
3. K. V. Fernando, H. Nicholson, *Stability Assessment of Two-Dimensional State-Space Systems*, Transactions on Automatic Control, vol. CAS-32, No 5, May 1985, pp. 484-487
4. E. Fornasini, G. Marchesini, *Stability of 2-D systems*, IEEE Transactions on Circuits System, vol. CAS-127, No 12, December 1980, pp. 1210-1217
5. T. Kaczorek, *Two-Dimensional Linear Systems*, Springer-Verlag, 1985
6. J. G. Kuschewski, S. Hui, S. H. Żak, *Application of Feedforward Neural Networks to Dynamical System Identification and Control*, IEEE Transactions on Control Systems Technology, vol. 1, March 1993, pp. 37-49
7. L. Pandolfi, *Exponential stability of 2-D systems*, Systems & Control Letters, No 4, September 1984, pp. 381-385
8. D. Psaltis, A. Sideris, A. A. Yamamura, *A Multilayered Neural Network Controller*, IEEE Contr. Syst. Mag., vol. 8, Apr. 1988, pp. 17-21
9. Special issue on intelligent control, Int. J. Contr., vol. 56, Aug. 1992
10. B. Widrow, M. A. Lehr, *30 Years of Adaptive Neural Networks: Perceptron, Madaline, and Back-Propagation*, Proc. IEEE, vol. 78, Sept. 1990, pp. 1415-1442
11. S. H. Żak, H. J. Sira-Ramirez, *On the Adaptation Algorithms for Feedforward Neural Networks*, in Theoretical Aspects of Industrial Design, D. A. Field, V. Komkov, Eds. New York: SIAM, 1990, pp. 116-133

SELECTING THE BEST SIGNIFICANT FRAGMENT TO THE INCREMENTAL HETEROASSOCIATIVE NEURAL NETWORK (RHI)

García Chamizo, J.M.;Satorre Cuerda, R.;Ibarra Picó, F.;Cuenca Asensi, S.

Departamento de Tecnología Informática y Computación.
Universidad de Alicante
Apdo. 99. Alicante. Spain
e-mail:juanma@dtic.ua.es

Abstract. The generality of the artificial neural networks models infers the requests based in the totality of the characteristics of the patterns. The RHI model infers just with a limited set of this characteristics, the significant fragment. This reason make RHI really appropriated by resolution of control and active vision problem. Although RHI model present high sensibility to distortion. In this paper it is developed the formalism to obtain the significant fragment in such a way it improve the noise tolerance.

Introduction

In general, an associative memory infers by using every component of input patterns used in recalling process. This is the case when hebbian correlations between the components of the learning patterns are used build the weight matrix [6,12,15]. The classification in this model is difficult because the linear dependencies between patterns bring up noise. Due to it, the recalling process consists sometimes of the orthogonalisation [9,10] of the learning patterns.

The RHI considers the components of each pattern which allow us to distinguish one pattern from the other and through these components, the model can infer, at least partially, the associated pattern to a given one. New components of this given pattern are then inferred in new iterations.

Total recall is not always possible, however, the model can indicate which components can not be obtained in an incremental manner.

Other artificial neural network models [1,4,7,8,9,11,13,14,16] use all the components of the patterns and, as a consequence, they can not used to extract the changing components in a set of patterns.

There could exist several significant fragments (a set of relevant components that are necessaries to infer a response). The components of the significant fragment particularised to the learning patterns represent a projection of them in a hyper-space.

In this work, an algorithm is proposed that finds a hyper-space to take out the projection of the learning patterns in such a manner that the relevant components corresponding every learning pattern form the maximum angles between them. That is the way to obtain the best distortion tolerance.

The RHI model

The RHI model is resumed:

There could be a way to obtain a weight matrix m_{kj} by solving equation systems so that, in the recalling process, the response were satisfactory.

To that end, let a_{ik} be the matrix whose row vectors A_i are the input to a table with q associated pairs, the dimension of which is n; let b_{ij} be the matrix whose rows are the p-dimensioned B_i vectors associated to A_i and let m_{kj} be the matrix resulting from the correlations between A_i and B_i pairs of vectors in the following way:

$$b_{ij} = \sum_{k=1}^{n} a_{ik} m_{kj} \quad \forall i = 1,\dots,q; \quad j = 1,\dots,p$$

Since we want to obtain m_{kj} to validate the expression, we have a system with $q \cdot p$ linear equations and $n \cdot p$ variables.

Alternatively, the equation system can be rewritten as a set of p simpler subsystems, one for each column of the second member of the matrix expression:

$$b_{ij'} = \sum_{k=1}^{n} a_{ik} m_{kj'} \quad \forall i = 1,\dots,q$$

A system with q equations is obtained for each j value..Note that j can change between 1 and p. That is, an equation system can be obtained for each column of the matrix m_{kj} of the variables; in other words, for each component of the B_i vectors.

It can be the case that some of the proposed equation subsystems could be solved whereas the rest could be incompatibles.

Let p_i' be the number of solvable columns and p_i'' the number of unsolvable columns. Then: $p_i' + p_i'' = p$.

It follows that the m'_{kj} matrix with p_i' columns and n rows could be obtained. As a consequence, the corresponding p' components of the vector B_i associated to any A_i could be calculated by making the product between A_i and m'_{kj}. They will be called subvectors B_i', the dimension of which will be

p_i'. All the other p_i'' components of vectors B_i can not be calculated. They will be called subvectors B_i'', the dimension of which will be p_i''. It is as, if the system can "remember" a part of the associated pattern from the variable pattern.

At this point, it is the time to go back to the that part the system has not been able to recall and treat to do it from the updated information; that is: the input pattern variable in addition to the associated pattern the system has been able to actually recall. What is then requested is that a new set of p_i'' subsystem be built having q linear equations and $n + p_i'$ variables.

However, the coefficients for the added up p_i' variables can not be the corresponding components for the column vectors already recalled. These have already been calculated by linear combination of the input vectors and as the result of the initial equations systems. The new p_i'' equations systems will remain incompatible, the way they previously were in the above argumentation.

To eliminate this limitation of algebraic kind and be able to obtain additional information from the already recalled associated pattern, it will be necessary to build de new p_i'' equation systems with the n initial variables in addition to p_i' variables so that the coefficients of the later ones be non-linear transformations of the recalled column vector. Discarded then the linear dependence, the possibility of actually recalling the missing information of the associated pattern can be expected.

By applying this technique repeatedly, the system could either definitively solve the problem or fail to do it because no one new component of the B_i vectors can be calculated in some iteration. In the latter case, the system can be told either to mark them with special symbols or to solve the problem by means of some heuristics.

Selecting the best significant fragment

In the precedent paragraphs not one hypothesis is made about the selection of the significant fragment, even of its possible repercussion about the quality of inference during recall. In this section we discuss about the existence of at least one optimal significant fragment, in order to get minimal distortion during the recalling of a noisy pattern.

Given a set of n learning patterns the significant fragment determine n vectors. These constitute the basis of one hyper-space the dimension of which is the cardinal of the significant fragment. If the angles among these vectors are too smalls, then any distortion could transform one of them in other vector of the base. The RHI answer will be wrong. We impose that the vectors of the significant fragment get a hyper-poliedra with maximum volume to obtain the best significant fragment; that is, the angles between two of theses vectors will be as big as possible.

NOTATION

$$A_i \in R^n; \quad i = 1..q; \quad A = \begin{bmatrix} A_1 \\ \cdot \\ \cdot \\ \cdot \\ A_q \end{bmatrix}$$ set of q learning patterns.

PREVIOUS: Eliminate null columns and repeated columns from A, because they generate null determinants.

$$A' = \begin{bmatrix} A_1' \\ \cdot \\ \cdot \\ A_q' \end{bmatrix}; \quad A_i' \in R^p; \quad i = 1..q$$

ALGORITHM

n = relevants_components_numbers (RHI, A)

INPUT: A'

OUTPUT: $FR \in R^{qxn}$ submatrix of optimal relevants components set.

Submatrix of optimal relevants

METHOD:

```
comb = C_{p,n} (number of combinations)
dmax = 0
FOR i=1.. comb
   generate C_i= matrix of columns of i-esima combination
   NC_i = Normalised (C_i)
   d = determinant (NC_i)
   IF (d > dmax )
      THEN dmax = d
           FR = C_i
   END_IF
END_FOR
END_METHOD
```

Example

Let the following pairs of patterns be used to teach a RHI:

$$A = \begin{bmatrix} A_1 = [010111010] \\ A_2 = [000111000] \\ A_3 = [101010101] \end{bmatrix} \rightarrow A' = \begin{bmatrix} 0111 \\ 0011 \\ 1001 \end{bmatrix}$$

In this case, only three not null different determinants could be built with the directions

$$\begin{vmatrix} 0 & \sqrt{2}/2 & \sqrt{2}/2 \\ 0 & 0 & 1 \\ 1 & 0 & 0 \end{vmatrix} = \sqrt{2}/2; \quad \begin{vmatrix} 0 & \sqrt{2}/2 & \sqrt{2}/2 \\ 0 & 0 & 1 \\ \sqrt{2}/2 & 0 & \sqrt{2}/2 \end{vmatrix} = 1/2; \quad \begin{vmatrix} \sqrt{3}/3 & \sqrt{3}/3 & \sqrt{3}/3 \\ 0 & \sqrt{2}/2 & \sqrt{2}/2 \\ 0 & 0 & 1 \end{vmatrix} = \sqrt{6}/6$$

As $\frac{\sqrt{2}}{2} > \frac{1}{2} > \frac{\sqrt{6}}{6}$, the first determinant bring up the best significant fragment.

In fact, the angles are: 90°, 90° and 45° and we have proved that the best distortion tolerance was obtained.

The second example consist of two sections: 1) the RHI is trained with a optimal fragment; 2) the RHI is trained with a randomly selected fragment. The RHI was trained with the same learningset in both cases and several noisy patterns were used to recalling.
The folowing results were obtained:

Low Distortion (10%)		
	optimal fragment	random fragment
% Recognition error	14%	22%

High Distortion (40%)		
	optimal fragment	random fragment
% Recognition error	49%	60%

It has been stablished that the optimal significant fragment could be founded by mean of the maximum determinant. Some heuristic is neccesary due to the number of determinants will be a combinational function of the data size. For instance: to extract the greatest one of a set randomly selected determinants.

Conclusions

This paper represent a improvement about the weakest aspect in the RHI model; that is, noise sensibility. It is a try to establish a formal basis to futures works. We are working in other ways about noise tolerance [5] with practise orientation.

References

[1] García-Chamizo,J.M.;Crespo Lorente,A. "Redes Neuronales Heteroasociativas Incrementales". IFIP Congress'92, Septiembre 1992

[2] García-Chamizo, J.M.; Crespo-Lorente, A.; Rizo Aldeguer,R.. "Extracción de fragmentos significativos de patrones para su posterior reconocimiento mediante Redes Neuronales RHI". Jornadas sobre Redes Neuronales 26-30 Octubre 1992. Centro Nacional de Microelectrónica (CSIC). Instituto de Neurociencias (UA).

[3] García-Chamizo, J.M.; Crespo-Lorente, A.; Rizo-Aldeguer, R. "Output Pattern Recalling Aided By Themselves: Incremental Heteroassociative Networks". Proceedings of the IJCNN, Beijing, November 1992

[4] García-Chamizo. "Semicoberturas heterogéneas de regiones bidimensionales morfológicamente no restringidas. Modelado conexionista aplicado". Doctoral Dissertation, February, 1994

[5] García Chamizo, J.M.; Mora Pascual, J.; Rizo Aldeguer, R.; Ledesma Latorre, B. "An incidence angle detection system for automatic assembly tools using the RHI network model". IEEE IAS International Conference on Industrial Automation and Control, Hyderabad (India), 1995.

[6] Grossberg, S. (editor). "The Adaptive Brain". Elsevier Science Publishing Co., Inc, 1987

[7] Ibarra Picó, F; García Chamizo, J.M. "A Generalized Bidirectional Associative Memory with a Hidden Orthogonal Layer". ICANN'94, Sorrento, May 1994

[8] Kandel, E. R.; Schwart, J. H. "Principles of Neural Science". Elsevier Science Publishing Co., Inc., 1985.

[9] Kohomen, T. "Self-Organization and Associative Memory". Springer-Verlag, 2nd. edition, 1988

[10] Kosko, B. "Adaptive Bidirectional Associative Memories". Applied Optics, vol 26, n 23, Dec. 1987

[11] Kosko, B. "Bidirectional Associative Memories ". IEEE Transactions on Systems, Man & Cybernetics. vol 18, nl, Jan./Feb. 1988

[12] Lancaster, P. & Tismenetsky, M. "The Theory of Matrices". Academic Press, 1985

[13] Pao, Y. "Adaptive Pattern Recognition and Neural Networks". Addison-Wesley Publishing Company, Inc, 1989

[14] Torregrosa, V. "Modelo Neuronal de Memoria Asociativa Ortonomalizada Adaptativa". P.F.C. U.P.V., 1993

[15] W.Thomas Miller,III;Richard S.Sutton; Paul I. Werbos. "Neural Networks For Control". Institute of Technology. Massachusetts, 1990

[16] Zurada, J.M. "Introduction to Artificial Neural Systems". West Publishing Company, 1992

AN ORTHOGONAL NEURAL NETWORK WITH GUARANTED RECALL BY ITERATIVE FILTERS AND ITS APPLICATION TO TEXTURE DISCRIMINATION

Ibarra-Picó, F.; García-Chamizo J.M.; Satorre-Cuerda, Rosana

Departamento de Tecnología Informática y Computación.
Universidad de Alicante.
Apdo. 99. Alicante. Spain.
email: ibarra@dtic.ua.es

Abstract. It is presented an orthogonal neural network which uses parallel iterative filters in its hidden layer. Orthogonality gives a high storage capacity of the network and its iterative filters get an accurate response. Also, this filters are implemented by a parallel process for getting a quick convergence time. As a simple example of application, we use it to learn several natural textures and its classification.

Introduction

The associative memories based on the conectionist paradigm [2]-[3], [7]-[8] have very limited capacity ($\sqrt{min(n,m)}$) for the BAM model [11], being n and m the size of the patterns; and only the recalling of a limited number of patterns could be guaranteed [12]-[13].
The general model of Bidirectional Associative OrthoGonalized Memory (BAOGM) [4]-[6] have a higher capacity that the others.
In this work a parallel filtering algorithm is proposed that guaranteed correct recalling being the energy of the system minimized. The algorithm is based in the fact that only a processing element in the intermediate layer is the winner corresponding a each learning pattern. Searching this processing element and its activation could be done in parallel by the interaction between the intermediate processing elements.

Learning of Patterns

Let a set of q patterns pairs (a_i,b_i) for i=1,..,q belonging to the vector spaces R^n and R^m, where the patterns a_i are normalized in some way so $|a_i| = k$ where $k \in R$ and i=1..q. We will build two learning matrices in the following way :

$$A = \begin{bmatrix} a_{11} & a_{21} & \cdots & a_{q1} \\ a_{12} & a_{22} & \cdots & a_{q2} \\ & \cdots\cdots & & \\ a_{1n} & a_{2n} & \cdots & a_{qn} \end{bmatrix}, \quad B = \begin{bmatrix} b_{11} & b_{21} & \cdots & b_{q1} \\ b_{12} & b_{22} & \cdots & b_{q2} \\ & \cdots\cdots & & \\ b_{1m} & b_{2m} & \cdots & b_{qm} \end{bmatrix} \qquad (1)$$

The memory is built as a neural network (see fig. 1) with two synaptic matrices (containing hebbian correlations) W and V, which are computed $W=AQ^t$ y $V=QB^t$. Where Q is an intermediate orthogonal matrix (Walsh, Householder, and so on) of dimensions qxq. The q_i vectors of Q are an orthogonal base of the vector space R^q. This characteristic of the qi vectors is very important to make accurate associations including noise patterns [10].

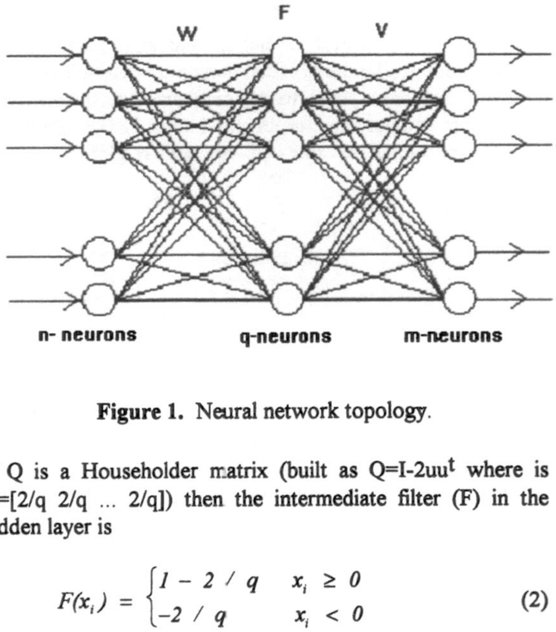

Figure 1. Neural network topology.

If Q is a Householder matrix (built as $Q=I-2uu^t$ where is $u^t=[2/q\ 2/q\ \ldots\ 2/q]$) then the intermediate filter (F) in the hidden layer is

$$F(x_i) = \begin{cases} 1 - 2/q & x_i \geq 0 \\ -2/q & x_i < 0 \end{cases} \qquad (2)$$

Where only the winner neuron x_i which is associated to the a_i pattern rises to 1-2/q.

Hidden Layer Topology

Let each processing element in the intermediate layer be connected to the others by a vicinity function that defines the topology of the interconnection.

Definition I. The vicinity (V_i) of a processing element i is the set of the accessible processing element in the same layer by a direct sinapsis between them.

Definition II. A conection path between two processing elements i and j in the same layer is a order set of procesing elements of that layer $\{p_1, p_2, \ldots, p_H\}$ that $i=p_1$, $j=p_H$ and $p_k \in V_{k+1}$ for k=1,2,...,k-1.

Definition III. A layer has a completely vicinity if every processing element has a conection path to the others.

Definition IV. The diameter of a layer (D) is the largest between the minimum paths, if exist, that connect every pair of processing neurons.

The diameter depends of the topology of the layer. For example for a 2-dimensional mesh $D = \sqrt{q}$, while for a n-dimensional hipercube D=log(q).

If in the intermediate layer a function of complete vicinity is established then a parallel filtering method could be defined. Now, We can define the asyncronous and syncronous parallel filtering algorithms to get the correct association in the network by looking for the winner neuron in a layer with complete vicinity..

Syncronous Filter

Let $X = \{x_j\}\ j = 1..q$ be the activation vector of the processing elements inside the hidden layer when the input pattern a_i is used to ask the system : $X = a_i^t W$. Only the i neuron must be activated and its value would be 1-2/q. Suppose that each processing element has an internal estate (E), where $E \in \{t1, t2\}$ and t1 is the winner. Then each neuron in a syncronous way can to search for the winner processing unit in the hidden layer as show the next algorithm.

In parallel for the neurons j=1...q
  E_j=t1
  For k=1..D
    If $x_j < max(V_j)$
      Then $x_j = max(V_j)$ $E_j = t2$
  IF E_j=t1
    Then $x_j = 1 - 2/q$
    Else $x_j = -2/q$

In this method, the filter always ends in a fixed number of D steps. This gives to us a total control of the system, but we could do more work that the necesary. Because, in last steps we can reach the minimum of the net energy before finishing.

Asyncronous Filter

Also, we can use an asyncronous method to get the correct association. In this case, each neuron converges in an asyncronous way to the final state when a new estimulus is applied to the net. A neuron uses local information into its vecinity V_i and makes a number of transitions to get its final state. The maximum number of iterations to get and stable state is limited by the topolgy, time of the response of a neuron and others constants of the system, but it will be limited by a constant K. In many cases, the neuron state will be stable in 2D steps. The general algorithm is

In parallel for the neurons j=1...q
  E_j=t1
  i=0
  c=0
  While (i<K and c<D)
    If $x_j < max(V_j)$
      Then $x_j = max(V_j)$ $E_j = t2$ c=0
      Else c=c + 1
    y = y +1
  IF E_j=t1
    Then $x_j = 1 - 2/q$
    Else $x_j = -2/q$

This algoritm have several advantages. It don't need a centralized syncronous control or global information access. So it's more fault-tolerant and easy to built for a high number of neurons. A disadvantage is that the exact point of convergence is without control, but we know that it's limited by K.

Guaranteed Recall

Each pair of learning patterns corresponds to a minimum of the energy function :

$$E = -(a_i^t \cdot W \cdot q_i + q^t \cdot V \cdot b) = -(n+m) \tag{3}$$

If the a_i pattern is used to ask the neural network, it can be demonstrated that the precedent algorithm produces an orthonormalized vector belonging to the base that the Q matrix defines. Besides, this vector has a positive component in the i position, the value of which is 1-2/q. The other components are negatives -2/q valued (theorems 1 and 2). In this way a minimum of the energy is reached.

Theorem I. (Consistency theorem) If the input pattern is a_i , then $x_i > x_k$ k=1..q, $k \neq i$.

Proof : Let be W the weight matrix which is built as

$$W = AQ^t = \sum_{j=1}^{q} a_j q_j^t \quad \text{where } a_j \in R^n\ q_j \in R^q \tag{4}$$

When, we ask for the a_i pattern the system makes the operation

$$a_i^t W = a_i^t \sum_{j=1}^{q} a_j q_j^t = a_i^t a_i q_i^t + a_i^t \sum_{\substack{j=1 \\ j \neq i}}^{q} a_j q_j^t \qquad (5)$$

The product $a_i^t a_i q_i^t$ will be expanded as

$$a_i^t a_i q_i^t = a_i^t \begin{bmatrix} a_{i1} \\ a_{i2} \\ \cdot\cdot \\ a_{ii} \\ \cdot\cdot \\ a_{in} \end{bmatrix} \begin{bmatrix} -\frac{2}{q} & \cdot\cdot & 1-\frac{2}{q} & \cdot\cdot & -\frac{2}{q} \end{bmatrix}$$

$$= a_i^t \begin{bmatrix} -\frac{2}{q}a_{i1} & \cdot\cdot & 1-\frac{2}{q}a_{i1} & \cdot\cdot & -\frac{2}{q}a_{i1} \\ -\frac{2}{q}a_{i2} & \cdot\cdot & 1-\frac{2}{q}a_{i2} & \cdot\cdot & -\frac{2}{q}a_{i2} \\ \cdot\cdot & & & & \\ -\frac{2}{q}a_{ii} & \cdot\cdot & 1-\frac{2}{q}a_{ii} & \cdot\cdot & -\frac{2}{q}a_{ii} \\ \cdot\cdot & & & & \\ -\frac{2}{q}a_{in} & \cdot\cdot & 1-\frac{2}{q}a_{in} & \cdot\cdot & -\frac{2}{q}a_{in} \end{bmatrix}$$

If we do the last product and put it in a more compact form, we have

$$= \begin{bmatrix} -\frac{2}{q}a_i^t a_i \\ -\frac{2}{q}a_i^t a_i \\ \cdot\cdot \\ 1-\frac{2}{q}a_i^t a_i \\ \cdot\cdot \\ -\frac{2}{q}a_i^t a_i \end{bmatrix}^t = |a_i^t||a_i| cos(a_i^t, a_i) q_i^t = k^2 q_i^t$$

The product $a_i^t \sum_{\substack{j=1 \\ j \neq i}}^{q} a_j q_j^t$ will be expanded as

$$a_i^t \sum_{\substack{j=1 \\ j \neq i}}^{q} a_j q_j^t = \sum_{\substack{j=1 \\ j \neq i}}^{q} a_i^t a_j q_j^t$$

$$= \sum_{\substack{j=1 \\ j \neq i}}^{q} |a_i^t||a_j| cos(a_i^t a_j) q_j^t$$

$$= K^2 \sum_{\substack{j=1 \\ j \neq i}}^{q} cos(a_i^t a_j) q_j^t$$

So the recall response in the hidden layer is

$$a_i^t W = a_i^t a_i q_i^t + a_i^t \sum_{\substack{j=1 \\ j \neq i}}^{q} a_j q_j^t$$

$$= K^2 C_{ii} q_i^t + K^2 \sum_{\substack{j=1 \\ j \neq i}}^{q} C_{ij} q_j^t$$

$$= K^2 \left(C_{ii} q_i^t + \sum_{\substack{j=1 \\ j \neq i}}^{q} C_{ij} q_j^t \right)$$

Where C_{ij} denotes $cos(a_i^t a_j)$. Then the response in the i component is

$$X_i = K^2 \left(1 - \frac{2}{q} - \frac{2}{q} \sum_{\substack{j=1 \\ j \neq i}}^{q} C_{ij} \right) \qquad (6)$$

And for other component $k \neq i$, we have

$$X_k = K^2 \left(-\frac{2}{q} + \left(1 - \frac{2}{q}\right) C_{ik} - \frac{2}{q} \sum_{\substack{j=1 \\ j \neq i \\ j \neq k}}^{q} C_{ij} \right) \qquad (7)$$

Now, we have to prove that $n_i > n_k \ \forall k, k \neq i$, so we have

$$n_i - n_k = 1 - C_{ik} \qquad (8)$$

As we Know $-1 < C_{ik} < 1$, because $k \neq i$, so $n_i - n_k > 0$. And i is the winner neuron in the hidden layer.

Corolary 1. The filtering algorithms leaves the winner processing element in the state t1.

Corolary 2. The distance between two neurons in the hidden layer $k, z \in \{1,2,..,q\}$ is $n_k - n_z = C_{ik} - C_{iz}$.

Computer Simulations

A simulation system has been developed and artificial neural network has been teached to learning several textures [1], [9] (see fig. 2). The neurons inside the hidden layer were connected in a bidimensional mesh topology which has a diameter of $D = \sqrt{q} - 1$. The recalling was correct in all the cases and the needed iteration number was $\leq D \approx 2$.

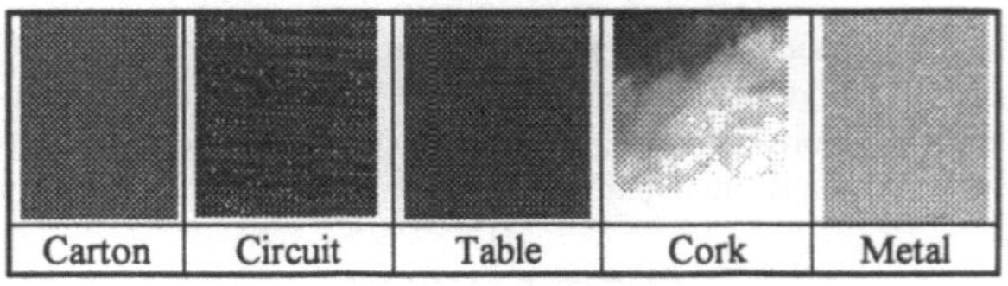

Figure 2. Textures for learning

The application of the network that was presented expands this simple example and goes from real time applications where a good classification is needed to memory associations and general pattern recognition.

References

[1] Chung-Ming Wu and Yung-Chang Chen: Multi-Threshold vector for texture analysis and its application to liver tissue classification. Pattern recognition Vol. 26 pg. 137-144, 1993.

[2] Hopfield J.J.:Neural Networks and physical systems with emergent collective computational abilities. Proceedings of the National Academy of Science, vol 79, pg. 2554-2558. 1984.

[3] Hopfield J.J.:Neural networks with graded response have collective Computational properties like those of two-state Neurons. Proceedings of the National Academy of Science, vol 81, pg. 3088-3092 .1984.

[4] Ibarra Picó, F. & García Chamizo, J. M.: Memorias Asociativas Bidireccionales Orthonormalizadas. V Conferencia de la Asociación Española Para la Inteligencia Artificial. Actas AEPIA. *Noviembre 1993.*

[5] Ibarra Picó, F. & García Chamizo, J. M.:Memorias Asociativas Bidireccionales Orthonormalizadas. Revista Novática. 1994..

[6] Ibarra Picó, F. & García Chamizo, J. M.: *A Generalized Bidirectional Associative Memory with a Hidden Orthogonal Layer.* ICANN'94, Sorrento, May 1994.

[7] Kosko, B.: Bidirectional Associative Memories. IEEE Tans. on Systems, Man & Cybernetics, vol 18, n 1 .1988.

[8] Kosko, B. : Competitive adaptative bidirectional associative memories. Procedings of the IEEE first International Conference on Neural Networks, eds M. Cardill and C. Butter vol 2. pp 759-66 .1988.

[9] McLean G.F. : Vector Quantization for Texture Clasification. IEEE Transactions on Systems man and Cybernetics vol. 23 No. 3. 1993.

[10] Pao You-Han : Adaptative Pattern Recognition and Neural Networks. Addison-Wesley Publishing Company, Inc. pg 144-148 .1989.

[11] Wang , Cruz F.J., Mulligan: "On Multiple Training for Bidirectional Associative Memory ". IEEE Tans. on Neural Networks, 1(5) pg 275-276.1990.

[12] Wang , Cruz F.J., Mulligan: Two Coding Strategies for Bidirectional Associative Memory. IEEE Tans. on Neural Networks, pg 81-92. 1990.

[13] Wang , Cruz F.J., Mulligan: "Guaranted Recall of All training Pairs for Bidirectional Associative Memory". IEEE Tans. on Neural Networks, 2(6) pg 559-567. 1991.

AUTOMATIC RADAR TARGET IDENTIFICATION USING NEURAL NETWORKS

CRIN-CNRS INRIA Lorraine
BP 239, 54506 Vandœuvre

LCTAR
BP 86, 78143 Vélizy

Abstract

Automatic radar target identification is a very difficult task because data are very noisy and very few is known to decide which part of the signal is important. Identification techniques could help to build more efficient radars but also to better understand how this signal could be interpreted. We present here some connectionist techniques that have been tested on this task. They offer rather good identification rate but do not yet permit to understand the hidden structure of the signal.

1. Introduction

Neural Networks are nowadays widely used for signal recognition. Most of the time, there also exists classical tools to perform similar processing and experts of the corresponding domain can also bring a very interesting knowledge. The work that we propose here is original concerning its domain of application. Automatic radar target identification has been rarely tackled by Artificial Neural Networks. This kind of work is all the more difficult as no expert is available to explain the meaning of the signal and direct the work of radar signal interpretation. In this paper, we will first introduce the radar domain and more precisely the task that we try to perform together with the classical methods that are used at the moment. Then we will describe the connectionist approaches that have been tried dealing with unsupervised and supervised models. We will conclude on these encouraging results while stating that improvements could come from an hybridization between such connectionist models and other knowledge based models.

2. The radar domain

One of the main new problems in the radar domain is automatic target identification, especially with stationary targets. The principle of a radar [5] [1], is to illuminate a target by transmitting properly designed signals, and to receive and process the returned echoes. Range localisation is obtained by measuring the time delay between the transmitted and received signals. For a radar transmitting pulses of duration t, targets closer to each other than $\delta r = ct/2$ cannot be distinguished, where δr is called the range resolution of the radar. The observation space is divided in range cells of length δr. A promising way to identify radar targets is to use a radar transmitting short pulses in order to obtain radar resolution cells thinner than the target spatial extent. The target is no more seen as a point but composed of scattering centers. The range profile [3] is the scattering distribution of a target along the radial distance. It is highly dependent on frequency and aspect angle. These fluctuations are all the more important as the radar cross-range resolution is larger with regard to the target spatial extent.

2.1. Description of the task

For the moment, as an initial work, our approach consists in testing neuronal techniques on automatic stationary target identification for high range resolution radars (the classical identification methods are described below). This study was realised on real targets, namely three kinds of terrestrial vehicle named target C, D, E. The data consist in range profiles of each target for different aspect angles. In high range resolution radar case, the database is collected by placing the target on a rotary turntable and slowly turning it over 360 degrees azimuth while the radar signatures are collected. Signatures of these vehicles were available at 0.2 degree azimuth intervals.

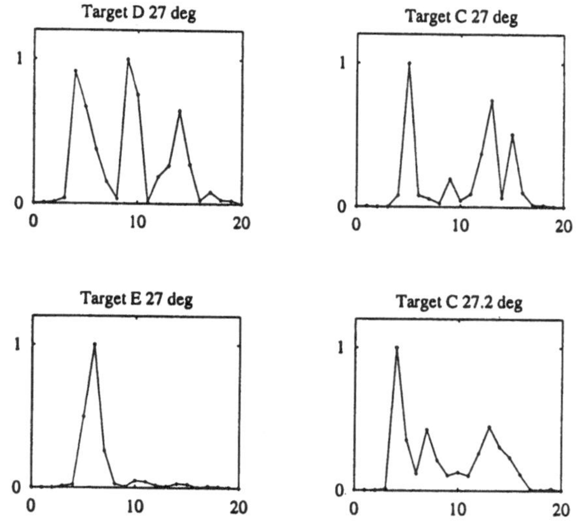

Figure 1: Examples of range profile

Fig. 1 gives an example of such signatures for different kinds of target. The y axis stands for the energy; the x axis represents the sampling along the radial distance. It must be underlined that such signals are very noisy and that two close signatures can be very different, because scattering centers can disappear or loose a part of their energy. Unfortunately such recordings are difficult to perform; therefore our corpus is very small-sized (for a 20 degrees sector angle, for each vehicle, each of the learning and test corpus gathers 100 examples), which is a problem for neuronal models.

The global goal of this task is to identify a target, given one signature for one unknown angle. A short term subgoal is to identify a target when the angle is known with one signature or even with a set of signature from close angles that could eliminate a part of noise. Radar experts would be also very interested in learning from the hidden structure of the radar signal.

2.2. Classical methods

We evoke here the methods that are today employed for this task. For this problem, classical identification methods are generally difficult to use and often lead to suboptimal algorithms. Deterministic methods would require a model of target (number of scattering centers, position and scattering strength,...) and lead to a non-linear optimization problem. Probabilistic methods require the knowledge of the probability density for the distribution of the scattering centers for each target. These probability densities are generally very hard to estimate.

A classical preprocessing is the normalised auto-correlation. Given a complex-valued range profile, $X(i,\theta)$ $i \in [0, N-1]$, θ the aspect angle, the normalized auto-correlation coefficient is

$$\frac{C_{XX}(i,\theta)}{C_{XX}(0,\theta)} = \frac{1}{C_{XX}(0,\theta)} \cdot \sum_{n=k}^{N-1} X(i,\theta)X^*(n-k,\theta)$$

This preprocessing is independent to translation and hence does not require target localisation. Nevertheless, it highly modifies and distort the signal. A classical classification method is a Bayesian method that consists in comparing examples to target prototypes. We will report results of these approaches on our task as we will compare them to those obtained with supervised connectionist models whose function is very close.

3. Neuronal techniques

We do not know which aspect of the signal is important and cannot design preprocessing functions that would explicitly give that knowledge to a neural network. As we had absolutely no information on this task, we chose to try several classical connectionist models to see if it was possible to extract any hints about this task or, at least, to get good results in classification.

3.1 Unsupervised approach

The first approach consisted in using an unsupervised model to automatically extract pertinent information. An unsupervised method is interesting here because of the lack of knowledge on the domain. We thus hope that a self organised model could extract pertinent structures by itself. On the other hand, such a model can also be considered as creating some kinds of prototype for each kind of target. Later, these prototypes can be used as references and examples can be matched with these prototypes to select their class. We chose the Neural Gas model [4] [2], a Kohonen-like model whose topology is created a posteriori and gives access to the topological aspect of the information. The principle of this algorithm permits to present all the occurrences of the target range profile for a given range of aspect angles and to get a vector quantization of the data while creating a model graph reflecting the topological structure of the data. At this stage, we obtained a rather dense graph, from which it was rather difficult to extract knowledge. The only hints that we could check was the importance of the onset of the peaks. We then chose to use these graphs for identification between the obtained models and target range profiles and to assign, for a given distance, a model to each target example.

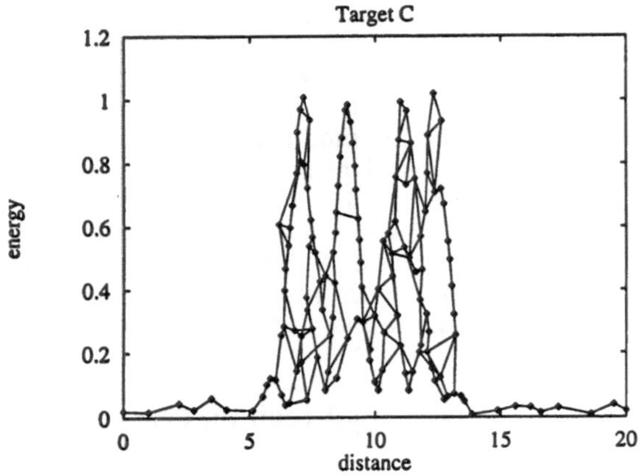

Figure 2: A graph obtained with Neural gas as a prototype of a target (here target C)

Fig. 2 represents a graph that is obtained after auto-organisation of a set of neurons on a set of examples of target C. Then, in the recognition phase, unknown examples are compared with the graphs representing all the targets and matched with the closer. The performances were homogeneous with an average rate of 80% for localised targets. Here, the problem is to make an efficient automatic localisation of the target.

3.2 Supervised approach

The second approach was clearly oriented toward the final goal of this task which is the classification of such signals along a set of known targets. For the moment, every trials that have been made to recognize a target on every angles at the same time have revealed unsuccessful. We thus report here trials for a sector angle of 20 degrees. All the supervised experiments have been carried out with Multi Layer Perceptrons (MLPs) with the back-propagation learning algorithm, a momentum and a learning coefficient per weight. Due to the limited learning corpus, we chose, for all the experiments reported below, an architecture with 20 neurons on the input layer (representing the 20 points of sampling on the radial distance), 7 for the hidden layer and 3 for the output layer (representing the 3 different possible targets).

The first problem that we faced was the one of preprocessing. One solution was to choose the one that is used with classical methods. In this case, the input of the MLP is the auto-correlation of the input profile. With the architecture and the corpus described above, and after 10,000 learning epochs, we obtain the results reported in Fig. 3.

In this figure below, the y axis represents the recognition rate and the x axis the number of averaged measures for recognition. As auto-correlation techniques deeply modify the shape of the signal, we also decided to test this connectionist supervised method on the rough signal, without auto-correlation preprocessing. In this case, we had to design an efficient localisation of the beginning of the signal. Here this localisation the obtained through a correlation between the signal and a step.

When the beginning of the target is localised, we then present the following 20 points of the target range profile as input to a MLP specialised for a small range of aspect angle.

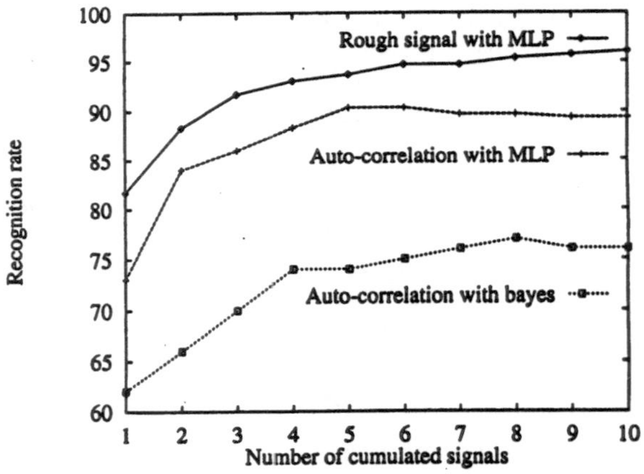

Figure 3: Recognition results

This figure indicates good results with neuronal techniques. Moreover, auto-correlation thus seems not appropriate as a preprocessing.

4. Conclusion

In conclusion this application is very formative, because it consisted, until now, in the blind utilisation of classical neural networks on a completely unknown and unexpertised problem. Despite this lack of expertise we obtained rather encouraging results for recognition. It is not the same with the extraction of pertinent hints to direct our framework of work (even the auto-correlation technique does not seem appropriate). Consequently, the main direction of the ongoing work consists in working with radar experts to try to get information on the structure of the target. This knowledge should lead to ameliorate the difficult aspects of our work : which pre-processing should be used for targets with unknown position and aspect angle? Moreover, it could also lead to hybridization between such connectionist models and other knowledge based or even probabilistic models.

The authors wish to thank the Direction des Recherches Etudes et Techniques (DRET) which gave them the opportunity to treat this subject and allowed them to publish this text.

References

[1] F. Le Chevalier. *Principes de traitement des signaux radar et sonar.* Masson, 1989.

[2] T. Martinetz J.A. Walter and K. Schulten. Industrial robot learns visuo-motor coordination by means of "neural gas" network. In T.Kohonen, K. Mäkisara, and J. Kangas, editors, *Proc. Int. Conf. Artificial Neural Networks*, pages 357–364. Elsevier Science Publishers B.V., 1991.

[3] H.J. Li and S.H Yang. Using range profile as feature vectors to identify aerospace objects. In *IEEE Transactions on antennas and propagation*, pages 261–268, 1993.

[4] T. Martinetz and K. Schulten. A "neural-gas" network learns topologies. In T.Kohonen, K. Mäkisara, and J. Kangas, editors, *Proc. Int. Conf. Artificial Neural Networks*, pages 397–402. Elsevier Science Publishers B.V., 1991.

[5] M.I Skolnik. *Introduction to radar systems.* Mc Graw-Hill, New-York, 1980.

KEYSTROKE DYNAMICS BASED USER AUTHENTICATION USING NEURAL NETWORKS

Roman Roštár, Jaroslav Olejár

Deptartment of Computer Science, Slovak Technical University, Ilkovicova 3, 812 19 Bratislava, Slovakia
e-mail: rostar@elf.stuba.sk

Abstract

We present a possible use of neural networks in the area of computer security. The back-propagation neural network is used to implement a prototype user authentication procedure using, capable to determine whether the user typing on the keyboard is valid. The obtained results are presented.

1. Introduction

Neural networks afford a flexible pattern recognition capability that can be adapted for many computer security problems, e.g. for user authentication [1],[2],[8] or intrusion detection [3],[4]. To start a session, user has to present his identity to the computer system, usually via user name or personal identification number. User authentication refers to the confirmation of the correctness of the alleged user's identity. Authentication may be based on combination of different user's characteristics (knowledge - password, possession - magnetic card, attribute - fingerprint). One possible solution is to analyse user's characteristics of typing on the keyboard, which is usually called keystroke dynamics. The idea of using keyboard characteristics in user authentication appears in many research studies [1],[2],[5],[6],[8].

The following notation is used to describe keystroke dynamics upon the user's typing. Let us denote Tp as a sequence of time samples, when keys were pressed, and Tr as a sequence, when they were released. From these two sequences we can compute time differences between two subsequent keystrokes $D_i = Tp_{i+1} - Tp_i$, periods during which a key was pressed $P_i = Tr_i - Tp_i$ or the time duration between release and adjacent keystroke $L_i = Tp_{i+1} - Tr_i$ (Fig.1). Pausing, hesitations and looking at the monitor contributes to the randomness of the measured data. The measurement vector D (Differences, sometimes called digraph latency times [6]) represents the real-time pairwise durations between the subsequent keystrokes and shows the speed of typing. For example, if the word WORD is the phrase, typed by the user, then the measurement vector D is created from the time durations between the letter pairs W-O, O-R, R-D. On the other hand, vector P (Pressure) consists from periods during which each particular key (W,O,R,D) was pressed and reflects the pressure of the typing. The L vector is complementary to the P vector, because $P_i + L_i = D_i$ .We will use the terms D-characteristic, L-characteristic and P-characteristic to denote user's characteristics obtained from D, L and P vector, respectively. In our experiments the D, L and P vectors are formed out of the predefined phrase typed by every monitored user in the login process.

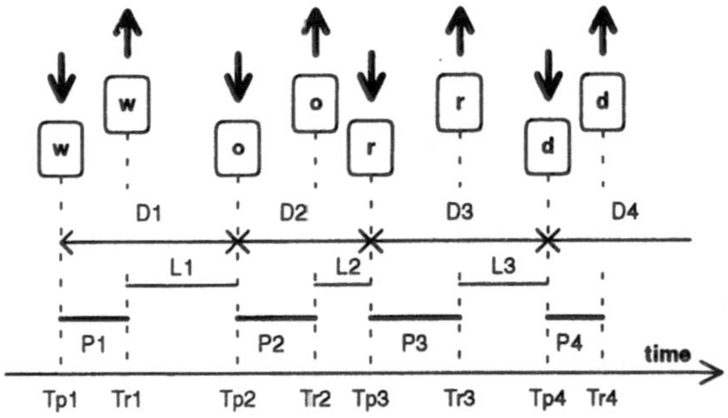

Fig. 1. Characteristics of typing on the keyboard.

The neural network analysis of the user keystroke dynamics was already applied in [1], [2]. Previous approaches were based on the analysis made upon D or conjunction of L and P vectors. Small data sets were used, or the data were gathered in only two user sessions (one for training and one for testing). We find that an analysis under previously described circumstances is insufficient. In our approach we consider D, P and L vectors and their combinations for every user. User data were gathered in continuous process of many user sessions during the certain period. The way of data collection should assure that the user characteristic reflects different states of user's mood and health and possible mutations through the time.

2. Experimental environment

To explore the user typing behaviour in real conditions we implemented experimental environment in the 386BSD UNIX system. We adapted typical UNIX login procedure to measure the Tp_i and Tr_i. time sequences. Every user had to type his login name and password as usually. After successful login special procedure was invoked and user was asked to type twice the phrase used for keystroke dynamics authentication.

One fixed 37 characters long phrase was used. The phrase was the same for all users. With intention to eliminate starting and closing periods we used only 18 characters long part from the middle of the phrase for the analysis.

Experiments were conducted with group of 11 users over a period of two months. A large amount of experimental data (1364 entry vectors, 124 vectors per user in 62 sessions) was collected.

3. Method

The back-propagation neural network is used as the pattern classification system. The keystroke dynamics vectors are presented on the input. Neural network's task is to confirm or to reject the measured vector for the given user.

For every user a unique neural network was created for each keystroke dynamics characteristic (D,L,P). In the training and testing process, the valid data were collected from the vectors made from the given user samples. Acquisition of the false intruder's data may be a problem in the real world. There is a question how to teach the network the false data. Because we have no real intruder's data, we have to model invalid entries from the vectors typed by other monitored users (all users typed the same phrase).

The continuous valued output of the net is the number in the range <0,1>. If close to zero, the vector is marked as invalid, if close to 1, it is marked as valid. A suitable threshold value must be chosen to distinguish the valid and invalid data sets.

Two measures are used for comparison of effectiveness of the authentication procedure. False alarm rate (FAR) is the number of authentic users rejected by the authentication method. Imposter pass rate (IPR) is the number of imposter samples incorrectly authenticated as valid data. Because these measures represent two contradictory properties of the authentication mechanism, secrecy and availability, decrease of IPR will usually cause the increase of FAR and vice versa.

4. Results

4.1 Data representation

We use two kinds of data. The first possibility is to analyse the raw data. The D and P vector elements are positive real numbers. On the other hand, the L vector elements can be frequently negative, since a key pressed with one finger was not released before another finger pressed another key. We examined simple preprocessing method - first successive difference. This method will cause the loss of the vector element absolute value, but represents the rhythm of typing. Achieved results show, that networks operating on these data were able to learn faster, but provided little bit worse results (especially for the P vectors, in Fig. 2). In light of this we used the raw data for analysis.

4.2 Hidden layer size

The network architecture M-N-1 was chosen. The number of input nodes M was usually 18. We evaluated how the number of neurons in the hidden layer influences the network performance.

One important property was the learning ability. The networks with larger number of hidden layer nodes were usually trained in smaller number of training epochs. The dependence of the network performance on the number of hidden layer neurons is showed in the Fig.2. The FAR values are comparable for all network architectures. These values slightly decreased only with the increase of the number of neurons. However, the IPR values decreased more significantly with larger hidden layer. Due to this we continued with 18-18-1 network architectures in further experiments.

4.3 Keystroke dynamics characteristics comparison

Values for the IPR and FAR for D, L and P characteristics of all users are in the Fig 3. According to the results, the suitability of D, L or P characteristics vary among the users.

We examined different threshold values for valid and invalid data. Results are given in the Table 1. It is obvious, that if we choose stronger threshold value, the resulting FAR will increase and IPR will decrease.

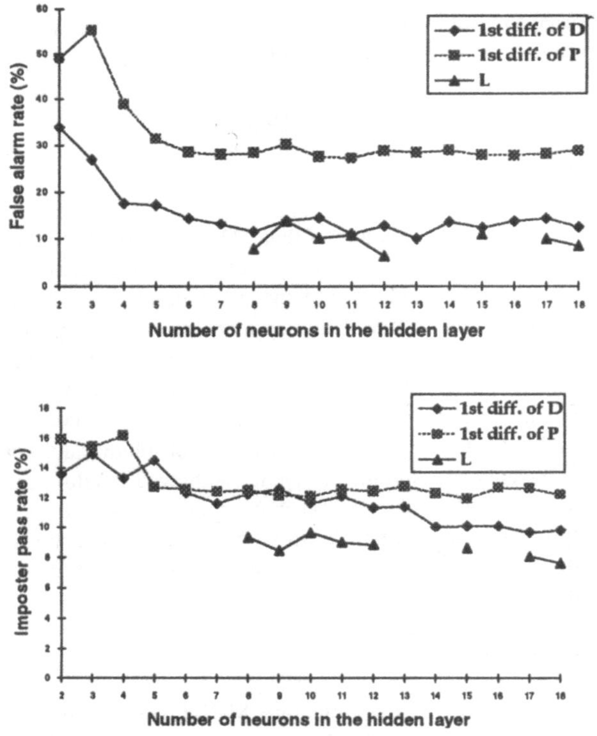

Fig. 2. Network performance versus the number of neurons in the hidden layer

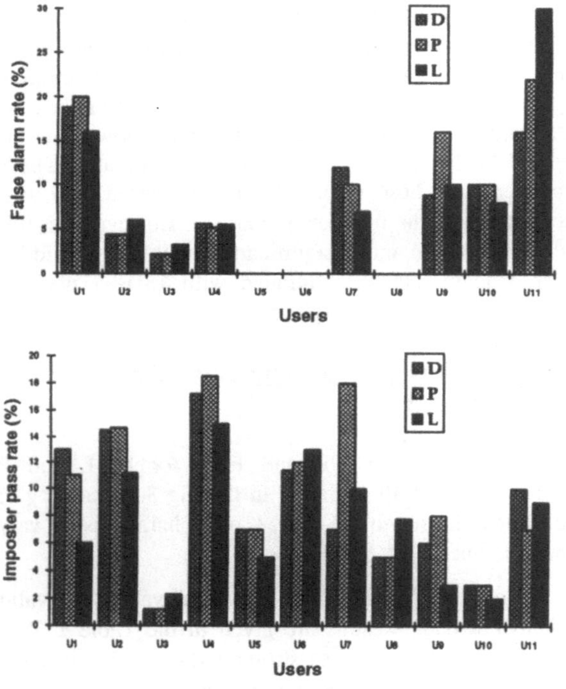

Fig. 3. Comparison of user IPR and FAR, thres. val.= 0.5.

However, for every user and further for every of his characteristics the value of the suitable "optimal" threshold may be different. It is very difficult, if not unrealistic, to find one universal threshold value, that could serve reliable for many users.

	D		L		P	
TV	FAR	IPR	FAR	IPR	FAR	IPR
0.5	9.50	9.83	8.49	7.61	8.99	8.79
0.7	12.63	8.36	9.70	6.66	10.91	7.38
0.9	15.86	6.28	11.11	5.25	14.44	5.82

Table 1. Comparison of average IPR and FAR for single characteristics and for different threshold values (TV).

In the experiments we are able to test results for different threshold values and to determine the optimal one. However in real situation this approach is not possible. We suggest the following modification of the authentication procedure algorithm: 1. training phase, 2. classification of a small testing set and the determination of the threshold value, 3. usage of the authentication procedure. Another possibility would be the periodical adaptation of the threshold value after every valid data classification.

4.4 Combination of the characteristics

The Table 2 shows the results for combined DP, DL and DLP characteristics. When combining single characteristics together we are increasing the number of input nodes of the network. Consequently, the calculation complexity increases, nevertheless, their learning ability grows. As it can be seen from the table 2., results obtained for DP, DL and DLP characteristics were better then results for separate D, L and P characteristics.

	DP		DL		DLP	
TV	FAR	IPR	FAR	IPR	FAR	IPR
0.5	9.80	7.45	9.90	7.41	7.37	8.56
0.7	11.11	6.52	10.40	6.33	8.79	7.44
0.9	13.23	5.35	14.34	4.77	11.21	5.90

Table 2. Comparison of average IPR and FAR for combined characteristics and for different threshold values (TV).

Another strategy would be to use a general resolver function to combine the outputs produced by networks analysing single characteristic. Such resolver function could use Boolean or fuzzy logic to provide the final result, as suggested in [9].

4.5 Training set size

There is another real world requirement, for the security reasons the training set should be small. Because of this we started with smaller training sets and we compared obtained results with the results for larger training sets. Achieved results indicated that the best balance was 34 vectors for the correct user and 9 vectors for every invalid user. When trained with the ratio of 24 valid user's vectors and 9 incorrect vectors for every invalid user the network's ability to recognize a valid user was insufficient (FAR was bigger). On the other hand when trained with 34 valid vectors and 6 incorrect vectors for every invalid user the network was showing worse IPR. The best training ratio can't be defined in general, it has to be determined upon experimental results.

5. Conclusions and further plans

It is quite possible, that user characteristics will change in time. Especially if small training sets were used some users were often rejected as invalid. To solve this larger training sets should be used (if possible), otherwise it is necessary to analyse recent history of vectors and adapt the network to changes in user's typing style.

If larger number of user is to be authenticated, we assume that only some of the possible invalid user's data vectors should be included in the training set. Suitable set of incorrect user's data should represent a large number of invalid users. If in the training period wide variety of different invalid data will be presented, the network will generalize with good accuracy which data should be considered invalid. Currently we are collecting data from large set of users to examine this problem.

Obtained results based on our experiments show that neural networks are suitable for keystroke dynamics based user authentication. If a proper threshold value is determined, it is possible to identify computer users upon D, L and P characteristics with good success rates. The best results can be achieved for combination of these characteristics.

Acknowledgements

This work was supported by the Slovak grant foundation under grant no. 1342/94

References

1. Bleha S.A., Obaidat M.S.: An Intelligent Neural System For Identifying Computer Users. C.Dagli et al. (eds.): Intelligent Engineering Systems through Artificial Neural Networks. New York, ASME Press, 953-959 (1991).
2. Brown, M., Rogers, S.J.: User identification via keystroke chara-cteristics of typed names using neural networks. Int. J. Man-Machine Studies **39**, 999-1024 (1993).
3. Debar H., Becker M., Siboni D.: A Neural Network Component for an Intrusion Detection Systems. Proc. 1992 IEEE Computer Society Symposium on Research in Security and Privacy, IEEE Computer Society Press 240-250 (1992).
4. Fox K.L., Henning R.R., Reed J.H., Simonian R.P.: A Neural Network Approach Towards Intrusion Detection. Proc. 13th National Computer Security Conference, vol.1, 125-134 (1990).
5. Joyce R., Gupta G. : Identity Authentication Based on Keystroke Latencies. Communications of the ACM 33 No. 2 168-176 (1990)
6. Legget J., Williams G., Usnick M., Longnecker M. : Dynamic identity veri-fication via keystroke characteristics. Int. J. Man-Machine Studies **35**, 859-870 (1991).
7. Lippmann R.P.: An introduction to computing with neural nets. IEEE ASSP Magazine, 4-22 (1987).
8. Roštár R., Durkovic M.: Two Methods for Computer Users Identification. Proc. 12th Int. Conference Applied Informatics, IASTED 180-183 (1994).
9. Roštár R., Olejar J., Safarik J.: Structured Approach to User Identification. Submitted to 10th IEEE Conf. on Computer Assurance COMPASS 1995.

APPLICATION OF LEARNING TO LEARN TO REAL-WORLD PATTERN RECOGNITION

Steve G. Romaniuk
Universal Problem Solvers
610 South Duncan Avenue
Clearwater, FL 34616, USA

Abstract

Pattern recognition has for decades played an important role in the development of intelligent systems and numerous algorithms have been proposed to deal with this important task. Though much interest and research has been invested into the construction of advanced pattern recognition systems, almost all of these algorithms are incapable of *meta-learning*. These *one-shot* learning systems do not address the problem of automatic adaptation of learning bias and hence lack one of the most fundamental requirements inherent to any truly autonomous pattern recognition system: the ability to *learn to learn*. The purpose of this paper is to report the findings of an extensive empirical study (i.e., including 26 real-world data sets) involving a system capable of meta-learning. Statistically significant reductions in both network units and training time are observed when utilizing a multi-shot learning environment.

1. Introduction

Pattern recognition algorithms can be classified into two groups: *one-shot* and *multi-shot* systems [13]. The former refers to systems which lack memory of problems encountered in the past, that is, once a pattern recognition problem has been learned and a new one is introduced, all knowledge gained from learning the first task is discarded. The vast majority of algorithms proposed by the learning community (i.e., machine learning, neural networks [1,4,6] and evolutionary computation [2,3,5,7,8,10]) fall into this class. This observation comes somewhat as a surprise, since it is a well known fact, that no *one-shot* system can equally well predict all functions belonging to the universal set. Instead, for every function for which an algorithm can obtain α percent prediction, there has to exist another function for which prediction is only $1-\alpha$. Consequently, any algorithm will perform well on some subset of functions, but fail on others. To overcome this pressing limitation, it is imperative that a learning system automatically adapt its built-in learning bias and allow it to be modified as new problems are encountered over time. In this context any system that supports adaptation of learning bias, is referred to as a *multi-shot* learner. In [11,13] an example of such a program is outlined. TDL (i.e., Trans-Dimensional Learner) has been found to display emergent behavior as found in complex systems. In a study consisting of 29 boolean functions, it was found that learning a subset of these functions was computationally more expensive than learning all functions at once [13]! This ability, where learning the whole can be simpler than learning some part of it, clearly suggests that emergent behavior is responsible for TDL's learning performance. Of course, it would be interesting to see, if the earlier obtained results can also be observed when utilizing real-world problems.

The next section provides a brief introduction to TDL, which is followed by the results of an extensive empirical study.

2. Trans-Dimensional Learning

TDL is introduced in [11,13] and simply refers to any pattern recognition system capable of tackling the subsequently listed task (i.e., multi-shot learning):

Given: A set of $P=\{p_1^{n1}, p_2^{n2}, \ldots, p_k^{nk}\}$, where p_i^{ni} in C^{ni}. The p_i^{ni} belong to the class of classificatory problems C^{ni} (i.e., continuous inputs/binary outputs) of variable input dimension.

Goal: Develop a representation R_k that can correctly classify all elements in P. No specific order is imposed on the problems in P when presented to the meta-learning system.

The actual TDL system as outlined in [13] is built upon a connectionist representation. The network is automatically constructed and dynamically adapted as new problems are encountered over time. All in all, there exist a total of 4 components:

- Network construction.

- Feature pruning.
- Feature selection.
- Knowledge forgetting.

We will briefly outline the first three components. The last component is excluded, since it was not utilized in this study. Those readers, who would like to find out more about knowledge forgetting within the TDL framework are referred to [15].

It was pointed out earlier, that TDL represents knowledge in the form of a self-configuring neural network. Hidden units are dynamically recruited and added to the current network, with one unit occupying a single layer initially. Newly added units are trained using the well known perceptron rule. A population of network units is trained concurrently on different subsets of the original training examples. The actual training subsets are evolved using evolutionary methods. In fact, the algorithm responsible for evolving the different training subsets and building the network is EGP (i.e., Evolutionary Growth Perceptrons) [10]. After a fixed number of generations have passed, the unit which displays the highest degree of recognition on **all** training patterns is permanently installed and its incoming weights are frozen.
An extension has been added to the basic EGP algorithm, that allows the generation of semi-weighted networks [16]. Instead of relying on connection weights and their adaptation by the perceptron rule, n-level transfer functions are combined with weightless neurons, which are trained by an evolutionary process. Both weighted as well as weightless neurons compete during the network construction phase.

As the connectionist network grows vertically, newly installed units receive inputs from previously added units. Consequently, the fan-in to recruited units increases rather quickly. To curb a potentially hazardous increase in connections, a feature pruning mechanism is included for trimming off any excess connections. The pruning algorithm utilized in TDL is non-destructive, meaning that the pruned unit recognizes all those training patterns correctly which were also recognized prior to pruning. In fact, it is possible that miss-classified patterns prior to pruning are afterwards correctly identified. For more information on the pruning algorithm itself consult [12].

Finally, feature selection is needed to decide which units in the already constructed network can best serve in recognizing a new set of training patterns. A measure is invoked to assess the overall quality of hidden units to act as good feature detectors. Only those units which display a quality measure exceeding some pre-defined threshold are connected to the currently constructed unit. For additional detail see [13].

This concludes the brief overview of the TDL system components.

3. Empirical Analysis

3.1 Data Sets

In order to assess TDL's ability to meta-learn, the following 26 data sets are considered:

- **Thyroid**: The thyroid disease diagnosis domain consists of 6 different data sets: **allrep** (4 outputs), **allbp** (3 outputs), **allhyper** (5 outputs), **allhypo** (5 outputs), **sick** (1 output), and **dis** (1 output). All data sets consist of 29 inputs and 972 examples.

- **Heart**: This diagnosis domain is composed of 4 data sets: **c-heart** (Cleveland version, 303 examples), **h-heart** (Hungarian version, 294 examples), **s-heart** (Switzerland version, 123 examples), and **va-heart** (V.A. Medical Center, Long Beach, CA, 200 examples). All data sets are described by 1 output and 13 inputs.

- **Monks**: The monks data sets come in three versions. They are described by 6 nominal inputs and 1 output. The monks problem was used in a first international machine learning algorithm competition.

- **Iris**: Plant recognition data set consists of 4 continuous inputs, 3 outputs, and 150 examples.

- **LED**: This data set is defined by 7 inputs representing the LED display and an additional 17 irrelevant attributes, which have been added to the example space. The values for the irrelevant attributes have been chosen randomly. There are a total of 24 binary inputs, 10 outputs, and 200 examples.

- **Harris**: Semiconductor waifer fault diagnosis data set is composed of 175 binary inputs, 28 outputs, and 828 examples. Outputs are not mutually exclusive.

- **Shuttle**: Space shuttle maneuvering data consists of 20 binary inputs, 1 output, and 15 examples.

- **Mushroom**: Poisonous mushroom classification is defined by 22 inputs, 1 output, and 1000 examples. The original data set consists of 8124 examples.

- **Tanks**: Two 16 x 16 images of tanks are described by 256 binary inputs, 2 outputs, and 157 examples.

- **Soybean**: Soybean disease diagnosis data set is composed of 108 binary inputs, 17 outputs, and 289 examples.

- **Voting**: Includes votes for each of the U.S. House of Representatives Congressman on 16 key votes identified by the CQA. A Total of 16 binary inputs, 1 output, and 435 examples are given.

- **2-spirals**: The task requires the separation of 2 intertwined spirals around the origin. This problem is highly non-separable, and has been extensively studied by the neural network community as a difficult to learn benchmark problem. There are 2 inputs, 1 output, and 192 examples.

- **Breast**: University of Wisconsin Hospitals, Madison breast cancer data set consists of 10 inputs, 1 output, and 699 examples.

- **Lung**: Lung cancer diagnosis is composed of 56 inputs, 3 outputs, and 33 examples.

- **Hepatitis**: Diagnosis of the hepatitis disease consists of 19 inputs, 1 output, and 155 examples.

3.2 Experiments

For this empirical study 2 experiments were conducted. The first experiment involved learning all 26 data sets by using TDL as a *one-shot* learner. For the second experiment TDL was trained in a *multi-shot* environment. For both experiments we report the number of generations (i.e., GE) of training time required and the final number of hidden units (i.e., HU) recruited. For all experiments average results over 10 randomly (i.e., uniform distribution) conducted trials are reported.

Tables 1 and 2 catalogue the number of hidden units and the number of generations for learning the 26 real-world pattern classification domains within a one-shot learning environment. Tables 3 and 4 report TDL's performance when the same data is multi-shot trained.
Clearly, for almost all domains multi-shot learning results in a decrease in the number of required hidden units and a substantial reduction in total training time (i.e., number of generations).
To determine the relevance of these experiments a statistical t-test has been conducted with a significance value of α=0.005! The percentage of domains that either significantly improve within the multi-shot environment, when compared against simple one-shot learning, are 70.6% when considering hidden units and 70% when examining the number of generations. Note, there is not a single domain which experiences a significant deterioration in the two performance measures when multi-shot trained.

A few more results: For all experiments the population size was set to 64 and the maximum number of epochs each perceptron was trained was 30. To learn all 26 data sets required on average about 350 generations, circa 100 combined units (i.e., hidden and output units), and around 5500 connections.

Finally, to gain access to a far more elaborate study (includes 38 real-world data sets) on this subject consult [14].

4. Concluding Remarks

The ability to automatically modify learning bias is essential to the construction of any truly autonomous pattern recognition system. Its main purpose is to allow the pattern recognizer to adapt to changes in the environment, as examplified, by an ever changing stream of pattern recognition tasks. In order to achieve this goal, learning has to commence in an *multi-shot* environment, akin to the one developed for Trans-Dimensional Learning. As the results of the herein detailed study have indicated, multi-shot learning as opposed to its one-shot counterpart, can help realize substantial savings both in knowledge representation (i.e., number of network units) as

Table 1. Part I: Number hidden units (HU) and number generations (GE) for one-shot learning of data sets.

	allrep	allbp	allhyper	allhypo	sick	dis	c-heart	h-heart	s-heart	va-hear	monk-1	monk-2
HU	0	1	0	0.6	1.7	0	3.7	1.3	0	2.5	2.8	6.3
GE	4	21.5	5	46.3	45	1	76.6	30.2	1	52.3	54.2	101.7

Table 2. Part II: Number hidden units (HU) and number generations (GE) for one-shot learning of data sets.

	monk-3	2-spirals	breast	lung	hep	iris	led	harris	shuttle	mush	tank-1	tank-2	soy	vote
HU	2.9	0.4	0	0	2.1	1	3.8	0	0.2	0	0.9	1.6	0	5.1
GE	46.1	12.4	5.8	3.4	47	15	67	20.1	5.7	1	23.5	31.4	17	88

Table 3. Part I: Number hidden units (HU) and number generations (GE) for multi-shot learning 26-domain data sets.

	allrep	allbp	allhyper	allhypo	sick	dis	c-heart	h-heart	s-heart	va-hear	monk-1	monk-2
HU	0	0	-0.1	-1	0	0	0	0	0	0	2.2	6.7
GE	4	3	4.9	4	1	1	1	1.1	1	1	41.5	113.6

Table 4. Part II: Number hidden units (HU) and number generations (GE) for multi-shot learning 26-domain data sets.

	monk-3	2-spirals	breast	lung	hep	iris	led	harris	shuttle	mush	tank-1	tank-2	soy	vote
HU	0.7	1.7	0	0	0	0	2.1	-9	0	0	0	1.4	0	0.9
GE	12.8	33.5	1.2	3	1	3	45	19	1	1	2	25.5	17	14

well as training time (i.e., number of generations). Additional experiments - not reported in this paper - have identified similar findings for the number of connections in a network and a networks predictive accuracy [14]. Additional studies are planned to further investigate the potential gains of multi-shot learning towards realizing the development of a truly adaptive real-world pattern recognition system.

References

1. Baffes, P.T. and Zelle, J.: Growing Layers of Perceptrons: Introducing the Extentron Algorithm, Proceedings of the 1992 International Joint Conference on Neural Networks (pp. II-392- II-397), Baltimore, MD., June, 1992.

2. Belew, R.K., McInerney, J., and Schraudolph, N.N.: Evolving Network: using the genetic algorithm with connectionist learning, In C.G. Langton, C. Taylor, J.D. Farmer and S. Rasmussen (Eds.) Artificial Life II, Redwood City, CA:Addison-Wesley, 1991.

3. Bellgard, M. I., Tsang, C. P.: Some Experiments on the Use of Genetic Algorithms in a Boltzmann Machine. International Joint Conference on Neural Networks, pp. 2645-2652, Singapore, 1991.

4. Fahlman, S.E. and Lebiere, C.: The Cascade-Correlation Learning Architecture, In D. Touretzky (Ed.), Advances in Neural Information Processing Systems 2 (pp. 524-532). San Mateo, CA.: Morgan Kaufmann, 1990.

5. Fogel, D.B., Fogel, L.J., and Porto, V.W.: Evolving Neural Networks, Biological Cybernetics, 63, 487-493, 1990.

6. Frean, M.: The Upstart Algorithm: A Method for Constructing and Training FeedForward Neural Networks, Neural Computation, 2, 198-209, 1991.

8. Marti, L.: Genetically Generated Neural Networks II: Search for an Optimal Representation, IJCNN'92, Vol. II, Baltimore, June, 1992.

9. Miller, G.F., Todd, P.M., and Hedge, S.U.: Designing neural networks using genetic algorithms, In J.D. Schaffer (Ed.) 3rd ICGA, San Mateo, CA:Morgan Kaufmann, 1989.

10. Romaniuk, S.G.: Evolutionary Growth Perceptrons, In S. Forrest (Ed.) Genetic Algorithms: Proceedings of the 5th International Conference, Morgan Kaufmann, 1993.

11. Romaniuk, S.G.: Trans-Dimensional Learning, International Journal of Neural Systems, Vol. 4, No. 2 (June), 171-185, 1993.

12. Romaniuk, S.G.: Pruning Divide & Conquer Networks, Network: Computation in Neural Systems, 4, 481-494, 1993.

13. Romaniuk, S.G.: Learning to Learn: Automatic Adaptation of Learning Bias, AAAI-94, Seattle, WA, USA, 1994.

14. Romaniuk, S.G.: Learning to Learn Real-World Pattern Recognition Problems, Unpublished report, Available from author, 1995.

15. Romaniuk, S.G.: Learning to Forget, Unpublished report, Available from author, 1995.

16. Romaniuk, S.G.: Semi-Weighted Neural Networks, Unpublished report, Available from author, 1995.

QUARRY AGGREGATES: A FLEXIBLE INSPECTION METHOD UTILISING ARTIFICIAL NEURAL NETWORKS.

DW Calkin RM Parkin

Mechatronics Research Group
University Of Technology, Loughborough, UK

ABSTRACT

Close monitoring of particle size and shape is essential if today's demanding material requirements for crushed rock aggregates are to be met.

An intelligent mechatronic inspection system is described here. On-line product sampling directs aggregate through an inspection chamber where a novel imaging system digitises the particle geometry. Data processing algorithms extract dimensional data and image features, allowing an artificial neural network classifier to assign qualitative shape descriptors to each particle, thus providing each sampled batch with a breakdown of constituent particle size and shape distributions.

INTRODUCTION

Escalating product costs within the aggregates industry have increased the need for new technology to improve product quality and reduce site operating costs. This paper describes current research concerning high speed inspection of crushed rock aggregates, to meet the ever increasing quality requirements for the production of high grade asphalt and precast concrete. Traditional quality control for aggregates production rely on the manual extraction of product samples from the crushing plant. These samples are then sent to a remote laboratory for analysis, resulting in poor plant performance and inconsistent product quality.

The addition of low cost automated inspection stations, distributed throughout the site will improve plant monitoring, allowing faster response to process changes and better control over each stage of the production process. This will help maximise the production of saleable aggregate, by ensuring undesirable product grades are minimised and that the finished product conforms to the customers quality requirements. The on-line aggregate inspection system described here compliments a new generation of Knowledge Based System (KBS) supervisory cone crusher controller[1] being developed by Pegson Ltd, a leading UK manufacturer of quarry equipment.

PLANT CONTROL

The British quarry industry produces 240 MTonnes of aggregate per annum, consuming 3% of all UK generated electricity[2]. Aggregate sources contain defects and impurities which must be corrected. Unwanted vegetation and clay, often found in sand and gravel deposits are removed by washing plants, while suitable product size and shape is developed through multiple crushing and screening stages. Cone crushers are employed mainly in crushing medium to hard abrasive rock types, typically processing 40-300 tonnes of material per hour.

Cone crusher performance depends on the hardness, size distribution and moisture content of the feed, discharge setting, power draw and liner wear. Inefficient crusher operation results in over crushed product which reduces throughput due to overloaded classifying screens and increased recirculating material within the plant. This leads to lost sales revenue from poor quality product, increased wear costs and wasted energy. Thus the scope for energy saving through the application of improved plant monitoring and control is enormous.

The Pegson Autocone series of cone crushing plant have hydraulic control over the product discharge setting making them ideally suited for automation. A supervisory control scheme utilising a KBS together with distributed control and condition monitoring, shown in Fig. 1, implements a control algorithm that predicts crusher power consumption and product grading by considering rock characteristics together with various parameters associated with crusher performance. The controller also protects the plant from an expensive breakdown by allowing a rapid discharge of material in the event of a machine stall.

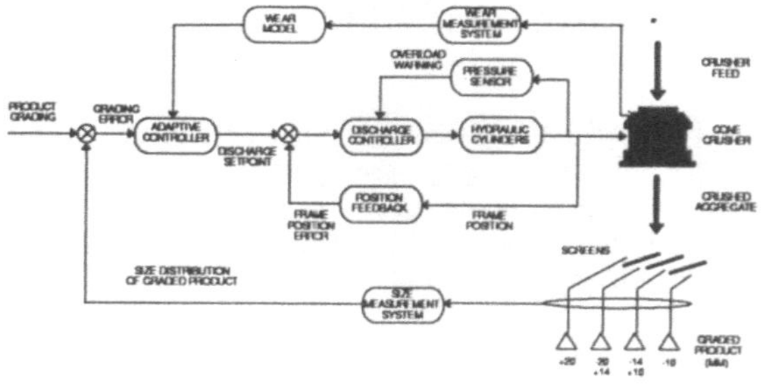

Fig. 1 Crusher control scheme

AGGREGATE INSPECTION SYSTEM

Many industrial processes utilise computer vision systems for particle shape analysis. However high installation costs and a large computational overhead still cause problems when required to analyse a large number of particles within a harsh, real time environment. The mechatronic aggregate inspection system illustrated in Fig. 2, provides an alternative solution to rapid, low cost optical sensing of large granular solids. Crushed aggregate is sampled from a stockpile or conveyor, then further subdivided into manageable quantities. Each subsample is dispersed into an optical inspection chamber, allowing the aggregate to fall freely through a sensing plane, where it is digitised by a novel scanning laser arrangement. The resulting data is compressed and stored, ready to be analysed once the current sample has been scanned.

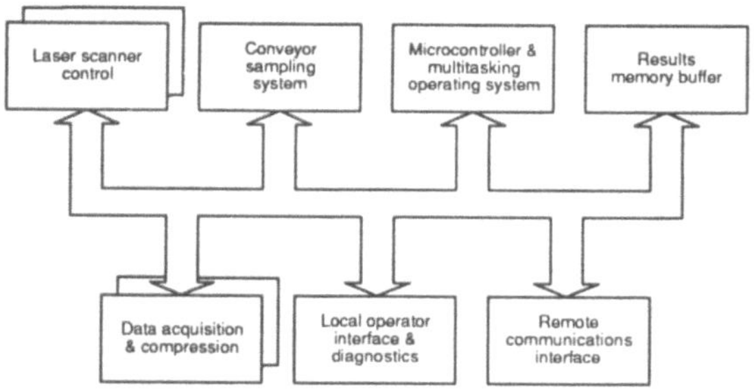

Fig. 2 Block diagram of inspection system

The acquisition of data, pre-processing, and subsequent analysis are co-ordinated by an embedded Motorola 68000 microprocessor based system, executing program control within a real time multitasking environment. Data processing activities are illustrated in Fig. 3. The pre-processing stage concentrates on improving the quality of the raw data and labelling of individual particles. Further analysis then extracts the physical dimensions of the particles and creates suitable pattern vectors, allowing a robust pattern recognition algorithm based around a neural network, to represent each particle by a suitable shape descriptor. Finally the computed error trends in the size and shape distributions of the crushed product provide a suitable machine feedback mechanism, allowing the KBS plant controller to compensate against unwanted process perturbations.

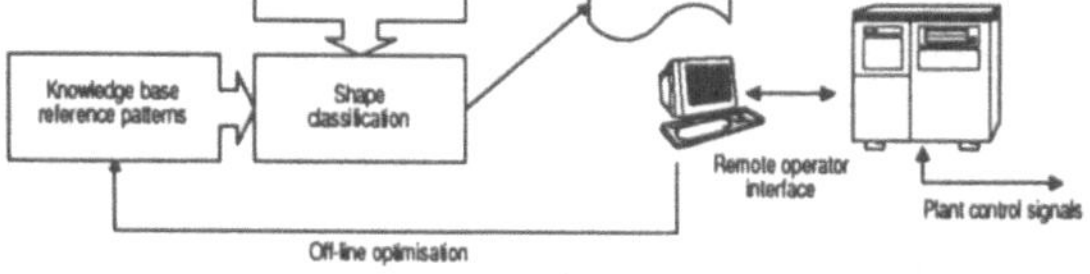

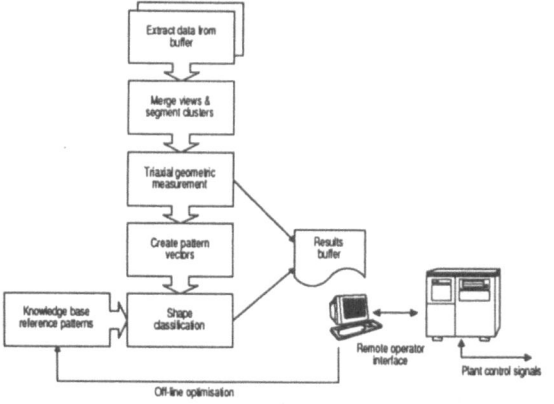

Fig. 3 Data processing

CHARACTERISATION OF PARTICLE SHAPE

Particle size and shape distributions within granular solids is important due to its influence on packing density and flow behaviour. Correct grading can improve some desirable feature or eliminate unwanted properties, allowing the supplier to sell the product for a premium price.

A collection of single size spherical particles can be described by a single dimension, diameter. A number of spherical particles of varying size can be described by an average particle diameter together with a statistical distribution. However, when the particles are non spherical it becomes necessary to take more care when describing size and shape. The principle methods available for describing particle shape include single number descriptors, geometric shape factors, characterisation of particle geometry by fractal dimensions, and by signature waveforms.[3,4] To allow the use of meaningful shape descriptions the particles are classified according to BS812[5], summarised in Table 1.

Table 1. Qualitative shape descriptors

Rounded	Completely shaped by attrition.
Irregular	Partly shaped by attrition having rounded edges.
Angular	Defined edges formed at the intersection of roughly planar faces.
Flaky	One dimension significantly smaller than the other two.
Elongated	One dimension significantly larger than the other two.
Flaky + Elongated	Three significantly different dimensions.

The aggregate inspection system described here aims to elicit subtle information about particle morphology. Single number descriptors are unsuitable for this task because a given value may be assigned to more than one shape. Instead, a hierarchical approach is used to characterise each particle. Triaxial measurements provide the physical dimensions of length, width, and breadth, allowing the particle to be broadly described as blocky, elongated or flaky. Then if required, the profile of the particle is analysed in detail to determine whether its surface is rounded, angular or irregular.

CLASSIFICATION

Pattern recognition attempts to match an unknown pattern vector to its closest equivalent from a set of reference models. Traditional classifiers utilise statistical methods, linear discriminant functions, or nearest neighbour techniques to divide the pattern space into a number of decision regions.

An alternative approach which does not require detailed knowledge of the products underlying statistical distributions is under evaluation. An artificial neural network is used to learn the classification rules which map example pattern vectors to predetermined class types. When properly trained the network is able to generalise, producing an output response to previously unseen input patterns.

The classification process is illustrated in Fig. 4. Harmonic series analysis[6], is utilised to create the pattern vectors. Once particles have been labelled, a rotating vector located within the original particle profile is used to obtain a normalised profile signature. This characterises the particle independently of translation and scale. Invariance to rotation is achieved by evaluating the fast fourier transform of this signature, this also provides significant reduction in the quantity of data that requires storage during the intermediate data processing stages. Re-entrant profile features may be overcome by enclosing the original profile with its convex hull approximation, alternatively the original profile may be described by arc length and change of slope[8]. The fourier coefficients are truncated and scaled to form a suitable pattern vector.

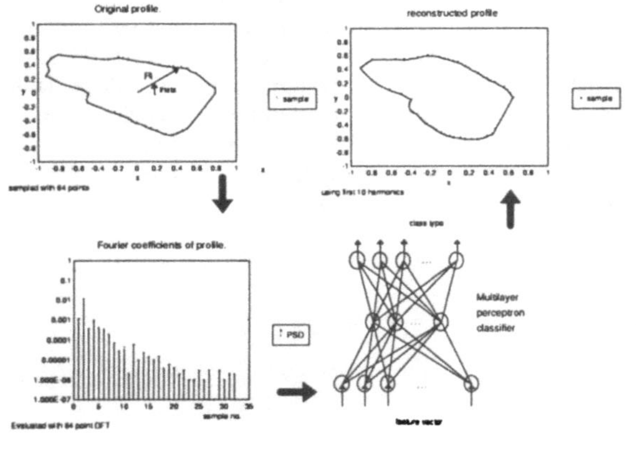

Fig. 4 Classification process

Although many types of neural network structure exist[9,10], the chosen pattern vectors limit the selection to those which can process continuous rather than binary input data. The aim of this work was not to research sophisticated network architecture's, but to apply and evaluate the performance of neural network methods within low cost embedded microprocessor applications.

The multilayer perceptron was selected because of its supervised learning procedure and ease of implementation. Each input node corresponds to a feature in the input pattern vector, while each output node represents a desired class type, e.g. irregular, rounded etc. Once the network structure has been defined the interconnecting weights are set to small random values. For each pattern in the training set a forward pass through the network is made to generate an output response. The output error, based on a mean square difference between the actual output and that specified by the training data, is used to modify the network weights. An iterative algorithm calculates the mean square error between the actual output and that specified by the training data. This error modifies the connection weights until the network error is reduced to an acceptable level.

The neural network currently exists as a software routine which is called from the main control program in the same way as any other software function/subroutine. Network connection weights can be stored within a disk file or as a look up table in memory, allowing the network to be replicated. This approach allows the operator to optimise the performance of the classifier off-line. Updated network weights can then be downloaded into the knowledge base of the on-line system when required. Future work will examine a possible migration of the network topology into hardware.

Each new pattern recognition problem to which the classifier is applied does not require a new program to be developed. The program code defining network structure and learning procedure remain unchanged, the only difference between applications lies in the values of the connection weights, which are modified by the iterative training algorithm for each new problem.

GENERIC APPLICATIONS

Although this work has been aimed at characterising granular materials within the quarry industry, the modular design allows the data acquisition and analysis techniques to be integrated within alternative applications. For example, noise emanating from an instrument binnacle is of significant concern throughout the automotive industry. This noise is frequently traced to the speedometer, which then requires costly replacement. Noisy speedometers result from several failure modes, including lack of grease, dirt ingress, incorrect assembly, and poor part tolerances. Existing quality procedures require an operator to run the speedometer up to a designated test speed, and then make a subjective pass/fail assessment based on its acoustic emission.

A new system is under development to automate this process. Acoustic signatures are captured via a microphone and their frequency content analysed. A neural network classifies the signature into categories of failure, thus permitting faulty items to be reworked in a realistic time frame at minimum cost.

CONCLUSIONS

The development of automated aggregate inspection compliments a new generation of rock processing plant. The system described here overcomes the limitations of existing solutions by providing each sampled batch with a breakdown of particle size distribution, along with qualitative descriptions of particle shape. This approach to direct product inspection offers tremendous benefits for plant monitoring and quality control throughout the aggregates industry.

ACKNOWLEDGEMENTS

The work outlined in this paper is supported by the Engineering Physical Sciences Research Council in collaboration with Pegson Limited.

REFERENCES

1. Bayliss, D C, Parkin, R M, Bearman, R A, 1994. Artificial intelligence in the control of cone crushers, *Mechatronics The Basis For New Industrial Development*, pp669-704. (Computational Mechanics Publications. ISBN 1 85312 367 6)

2. Bearman, R A, 1991. Energy utilisation in crushing, *Mine & Quarry*, Vol 20 No 10, pp35-37.

3. Barrett, P J, 1980. The shape of rock particles a critical review, *Sedimentology*, Vol 27, pp291-303.

4. Lloyd, P J, 1988. The characterisation of particle shape, *Particle Size Analysis*, pp347-355. (J Wiley & Sons Ltd. ISBN 0 7837 7654 3)

5. British Standards Institution, 1989. Testing aggregates methods for sampling, *BS812 Part 102*.

6. Holt, C B, 1981. Characterisation of the shape of particles produced by crushing using harmonic series analysis, *Institution Of Chemical Engineers Symposium Series No 63*, pp D2/H/1-D2/H/13.

7. Fong, S T, Beddow, J K, Vetter, A F, 1979. A refined method of particle shape representation, *Powder Technology*, Vol 22, pp17-21.
8. Carpenter, G A, 1989. Neural network models for pattern recognition & associative memory, *Neural Networks*, Vol 2, pp243-257.
9. Lippman, R P, 1987. An introduction to computing with neural nets, *IEEE Acoustics, Speech & Signal Processing Magazine*, Vol 4 No 2, pp4-22.

AN EVOLUTION MODEL FOR INTEGRATION PROBLEMS

Brigitta Lange

Fraunhofer Institute for Computer Graphics
Wilhelminenstr. 7
D-64283 Darmstadt
F.R.G.

Abstract

In this paper a general evolution model for integration problems is introduced and applied to the solution of the Rendering Equation which is a multidimensional integral equation modelling radiant light transfer. Often it is not possible to solve integral equations in a closed analytical form. Then an estimate for the integral has to be determined by numerical or stochastic approximation. Unfortunately stochastic integration techniques are extremely slow to converge. Numerical techniques are usually faster in convergence but their success strongly depends on function requirements. The evolution model presented here overcomes the disadvantages of these methods, i.e. loss of direction and missing innovation. It is far more flexible and produces reliable estimates even in the presence of multimodal integrands. Furthermore it provides a general framework for the description of a wide variety of adaptive sampling techniques.

Introduction

In science and engineering frequently integration problems occur where the function being integrated is of high complexity and cannot be represented or solved in a closed analytical way. In order to handle these problems a simulation model of less complexity has to be developed. The model should serve as a reasonably close approximation to the real function (integrand) and incorporate the most important aspects of the system.

All that is required to achieve good results is that there be a high correlation between the prediction by the model and what would actually happen with the real system [1]. But in most cases the verification of this correlation is difficult if not impossible, since the behavior of the real function is not known in advance. To ascertain wether the model satisfies the requirements, it is important to establish a control during the simulation process that is not restricted to a priori assumptions, but allows a flexible adaptation of the model to the real system.

In nature, evolution can be taken as an example for a very efficient adaptation process of living organisms to their environment. Thus it is very likely to develop nature analogous problem solving strategies in order to achieve an optimal adaptation of the simulation model and hence a better convergence than Monte Carlo methods or numerical integration techniques.

Monte Carlo integration is a stochastic process, where the integration domain is sampled by a finite set of random variables distributed according to a probability density function (PDF). The major problem is to determine an optimal location and density of the samples in order to guarantee some bound on the variance of the estimate. This is equivalent to finding a PDF that is a good primary estimator for the function being integrated. In general it is impossible to find a PDF that closely resembles the shape of the integrand, especially if this is a multimodal function. Here even variance reduction techniques (e.g. importance sampling, stratified sampling) fail to yield good approximations since they are mostly non adaptive and have to rely on a priori assumptions.

Unfortunately the only way to gain information about the shape of the function being integrated is by actually evaluating it. The goal, therefore, is to optimize the simulation process in such a way that this information is exploited efficiently, giving the sampling process a direction and allowing the system to adapt itself to the actual function.

Adaptive numerical integration techniques are much faster in convergence since they exploit information about gradients to direct the adaptation process effectively. The problem here is that these techniques are only applicable to a very small range of functions that fulfill all the requirements needed. In the case of multimodal functions these techniques tend to converge very fast to suboptimal solutions, since they have no innovation component. Their success in finding all significant function peaks depends very much on the initial spacing of knots.

The evolution model provided here contributes towards a solution to these problems. It has the capability to adapt even to extremely noisy and discontinuous integrands without needing any predefined model of the function and provides mechanisms for a flexible self-adaptation. The model is of general applicability and combines the advantages of both stochastic and numerical integration techniques: It is directed but also has a global search component.

First, we will formulate the integration process as an optimization problem, then suitable representations for this optimization problem as evolutionary algorithm are investigated. From this discussion a new evolution model for integration problems is derived. The concept here is to maximize the confidence in the approximation result. The model differs from classical evolutionary algorithms because it implements a population growth. Finally we verify our concept by applying this evolution model to the Rendering Equation [2] and show that in comparison to Monte Carlo Methods a better convergence of the estimate towards the solution of the Rendering Equation is achieved, which in turn results in an improvement of image quality.

Integration as Optimization Problem

The domain of integration problems remains largely unexplored, since the present effort in applying evolutionary algorithms has concentrated solely on parameter optimization problems where the absolute minimum or maximum of some function is searched for [3]. But here we are faced with a somewhat different kind of problem.

Let us consider an integral equation as objective function, then our concern is not the minimization or maximization of the integral value. Instead, we are looking for the best approximation.

Mathematically an approximation problem can be stated as follows: Given a function $f \in I$, then a function $\hat{g} \in G \subseteq I$, is searched for, such that

$$\|f - \hat{g}\| < \varepsilon, (\varepsilon > 0) \qquad (1)$$

Thus we want to minimize the approximation error, i.e replace an analytically non-calculable function f by a calculable function $\hat{g}$ and achieve maximum adaptation. In this case not the integral but the approximation error is the objective function that has to be minimized.

Now the question arises, what a suitable representation for this problem as evolutionary process will look like.

In the classical design of an evolutionary algorithm an arbitrarily initialized start population of individuals, each representing a point in the search space of potential solutions with an assigned quality measure related to the objective function, adapts to its environment by means of probabilistic selection and genetic operators.

In our approximation problem the potential solution, i.e. the estimated integral, can be represented by a weighted sum of variables $X \in D$:

$$\int_{x \in D} f(x) d\mu(x) = E\left[\frac{f(X)}{p_X(X)}\right] \approx \frac{1}{N} \sum_{i=1}^{N} \frac{f(X_i)}{p_X(X_i)}, \qquad (2)$$

where the X_i have been distributed probabilistically or deterministically (e. g. equidistant) over the integration domain by some distribution function $p_X(x)$.

One approach to minimize the approximation error is to form a population of estimates each represented by another randomly generated sample distribution. Then these estimates have to compete for survival in the sense of optimal adaptation. But if each individual represents an estimate of the actual integral, the computational cost for the evaluation is very high. Not only that for each individual N function evaluations at the sample locations (knots) have to be made, but also that there is a loss in information and therefore waste of computation time. Combining the samples of all individuals would certainly result in a much better approximation of the integral than given by a single individual consisting of only N samples. Furthermore it is very difficult to define a quality measure for the ranking of the individuals since the actual approximation error of each individual cannot be determined exactly. In a population of integrands with a fixed number of samples the design of appropriate mutation and crossover operators seems to be difficult. A random exchange of sample locations makes no sense in terms of achieving better individuals (i.e. better approximation results). If the number of samples does not increase no convergence can be achieved and also the mean quality of the population will vary randomly, making this representation rather impractical, not only from the computational efficiency point of view.

Therefore we have developed an alternative evolution model, that can be characterized as a model for population growth by cell-splitting, where each individual corresponds to a weighted sample, and the whole population forms the estimated integral.

An Evolutionary Integration Model

The concept is to maximize the confidence in our estimate of the integrand by an evolutionary stratification of the integration domain. Thus the integration domain represents a population of confidence intervals, where each individual is only a part of the estimate solution. Initially the domain is subdivided by samples into a few intervals (see Fig. 1). The fitness of each interval is given by an associated confidence value. The confidence is measured locally and globally; that means the confidence value is determined by the size of the individual interval (i.e. the proportion of interval height to width in the case of a one-dimensional function) as well as the variance to neighboring individuals. Each individual interval has a chance of being selected for reproduction by a probability proportional to its confidence. A selected individual is mutated through subdivision by new samples, hence it produces offsprings and then dies. From generation to generation the population size increases (see Fig. 1).

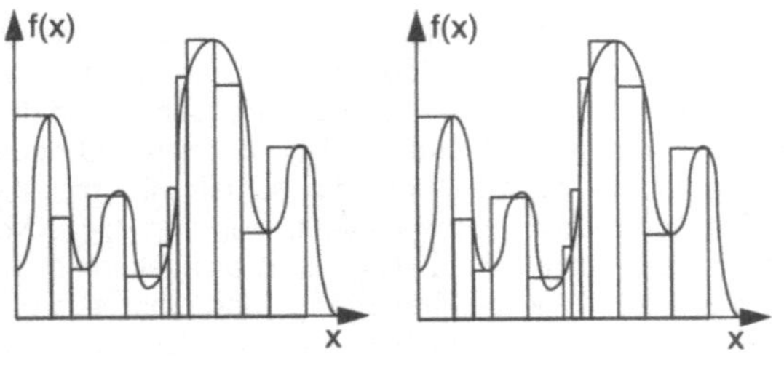

Fig. 1: Left: initial stratification by four individuals. Right: stratification after some generations.

The objective of the evolution process is to adapt the population to the integrand in such a way that the estimated value of the integral in each interval is equal. If the size of the intervals does not change within certain limits, the confidence is maximized and the optimal approximation result is reached. Thus the termination criterion can be determined by the generation differential, given as:

$$\delta G(i) = O(i) - O(i-1),\ i = 1, \ldots, n; \qquad (3)$$

where $O(i)$ is the sum of the evolved confidence intervals of one population at generation i, i.e. the estimated value of the integrand for one generation. In a converging evolutionary process the generation differential will approach zero with increasing number of generations:

$$\lim_{i \to \infty} \delta G(i) = 0 \qquad (4)$$

Hence we can control the progress of the adaptation process by analyzing the generation differential for a number of successive generations. If there is no significant change in the estimates $O(i)$ the process terminates.
This evolutionary integration process is superior to stochastic integration techniques in that it is directed and reduces variance significantly. Furthermore it is computationally more efficient because the sample set size needed to achieve a certain quality is much smaller. It even may outperform numerical techniques since it has an innovation component. Thus the result does not depend on the initial spacing and the algorithm has a chance to escape from local optima. It appears that a wide variety of adaptive integration processes can be viewed in a unified context provided by this evolutionary integration model. Dependent on the actual implementation Monte Carlo methods as well as deterministic algorithms can be interpreted as special cases of this evolution model.

Application of the Evolutionary Integration Model

In computer graphics realistic looking images are generated by simulating the propagation of light in an environment and calculating the amount of light reflected from every visible surface point.
The Rendering Equation is an integral equation modelling radiant light transfer. It determines the radiance of light reflected by a point on a surface in a certain direction (θ_r,φ_r). In order to calculate the reflected radiance L_o the incoming radiation (irradiance) L_i from the entire half-space has to be accounted for. This is achieved by integrating the reflected irradiance $\rho \cdot L_i$ over all directions (θ,φ) of the total hemisphere Ω above a surface point:

$$L_o(\lambda, \theta_r,\varphi_r) = \iint_{\theta\in\Omega\ \varphi\in\Omega} \rho(\lambda, \theta_r, \varphi_r, \theta, \varphi)\ L_i(\lambda,\theta,\varphi)\cos\theta\, d\omega \qquad (5)$$

Here λ is the wavelength of the light and ρ is the bidirectional reflectance function. Since the incoming irradiance is again a reflected radiance of another surface in the environment it is generally is not possible to find a closed analytical solution. Thus this complex multidimensional integral equation is solved by Monte Carlo methods, where light paths are followed from the eye all the way back to the light sources. These light paths correspond to Markov Chains ([1], [2]). Thus for every visible object point the hemisphere of incident radiation is sampled by a finite set of randomly selected rays [4].
Unfortunately this technique is extremely slow to converge and has an inherent limitation: sampling noise (see Fig 4). A specific cause of this noise is the highly non-uniform irradiance distribution. Since we only know the reflection function ρ and have no previous knowledge about the multimodal irradiance function it is hard to determine a PDF being a good primary estimator.
We have applied an adaptive algorithm, i.e. the evolutionary integration model in order to minimize the variance of the estimate. The individual confidence intervals correspond to irradiated solid angles, since the integration domain is a hemisphere. For computational efficiency each individual is defined by a spherical triangle with an associated mean irradiance value obtained by the sample rays at its vertices (see Fig. 2).
Initially the hemisphere is subdivided into a few spherical triangles of equal size. Each individual is evaluated by determining its local and regional fitness value. The local fitness is obtained by calculating the mean irradiance value. The regional fitness of an individual is determined by the maximum deviation (in RGB-primaries) from the

mean irradiance values of its direct neighbors weighted by its area size. Thus the regional fitness is a measure for confidence in the adaptation quality.
The goal now is to reduce the variance of the estimate by a stratification of the hemisphere into solid angles of equal confidence. In each generation the integral is evaluated and the size of the solid angles is adjusted. By applying the mutation operator, the selected spherical triangles are split up into four new triangles and the population size increases. The parents are selected by roulette wheel, where the chance of being selected is proportional to the triangles regional fitness. Thus the integration domain is subdivided primarily in those regions with a high variance in irradiance (see Fig. 3). The process terminates if all individuals have an almost equal confidence value.

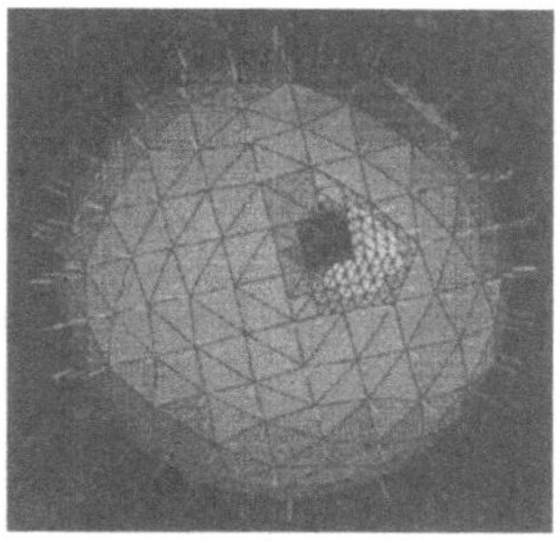

Fig. 2: Hemisphere as population of triangles.

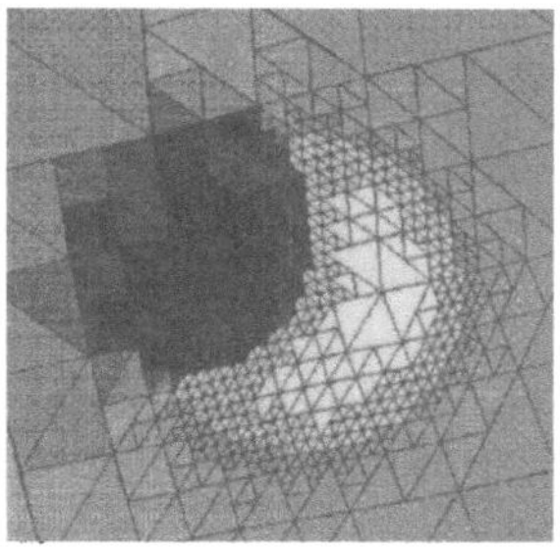

Fig. 3: Evolved stratification in detail.

The advantages of this method are its fast convergence and computational efficiency, because the evolution process and the evaluation of the integral can be effectively combined. It produces successively better approximations to the integral equation by exploring the hemisphere, searching for those solid angles that contribute significantly to the total irradiance. Thus increasing the average information value of the samples and converging towards the best approximation, which in turn results in an improvement of image quality (compare Fig. 4 and Fig. 5).

Fig. 4: Image generated by Monte Carlo integration

Fig. 5: Improved image generated by evolutionary integration

References

1. Rubinstein, R. Y.: Simulation and the Monte Carlo Method. New York: Wiley & Sons 1981.
2. Kajiya, J. T.: The Rendering Equation. Computer Graphics 20(4), August 1986, pp 143-150
3. Baeck, T.; Schwefel: H. P.: An Overview of Evolutionary Algorithms for Parameter Optimization. Evolutionary Computation1(1). Cambridge: The MIT Press, 1993.
4. Lange, B.: The Simulation of Radiant Light Transfer with Stochastic Ray-Tracing. Photorealistic Rendering in Computer Graphics. Berlin: Springer 1994, pp 30 - 44

CONVERGENCE OF ALGORITHM AND THE SCHEMA THEOREM IN GENETIC ALGORITHMS

Yoshinori Uesaka

Department of Information Sciences, Faculty of Science and Technology,
Science University of Tokyo

2641 Yamazaki, Noda-City, Chiba 278, Japan
Tel +81 − 471 − 24 − 1501; Fax +81 − 471 − 23 − 9764; e-mail uesaka@is.noda.sut.ac.jp

Abstract In this article two aspects of GA are commented from a mathematical point of view. One is concerned with the convergence of GA, and the other is a probabilistic interpretation of the schema theorem. GA produces a stochastic process (that is, Markov chain) of populations. A week sufficient condition which guarantees the convergence to a unique stable distribution will be given which is satisfied by a wide range of current GA. On the other hand, the inequality appeared in the Schema Theorem is not possible to be interpreted from a probabilistic point of view without replacing those random variables by their expectations since the theorem includes random variables. This replacement unfortunately prevents the theorem from being true, and hence it is forced to be revised. Two types of the correct versions of the schema theorem will be given after a rigorous derivation based on a probability theory.

1. Introduction

Genetic algorithms (GA) are known to be one of methods for solving combinatorial optimization problems[1]. Though a considerable amount of discussion on the effectiveness of GA is reported from an experimental point of view[say, 2], few of theoretical research are found in the literature[3-11]. Some of them have analyzed from a point of view of Markov chain, and the other have given interesting insights about the Schema Theorem.

In this article two aspects of GA are commented from a mathematical point of view. One is concerned with the convergence of GA, and the other is a probabilistic interpretation of the schema theorem.

GA produces a stochastic process (that is, Markov chain) of populations. Under what kind of condition does the process converge? A week sufficient condition which guarantees the convergence to a unique stable distribution will be given which is satisfied by a wide range of current GA[12].

On the other hand the schema theorem is known as one of the theoretical results in GA. The inequality appeared in the theorem is, however, not possible to be interpreted from a probabilistic point of view without replacing those random variables by their expectations since the schema theorem includes random variables. This replacement unfortunately prevents the theorem from being true, and hence the theorem is forced to be revised. Two types of the correct versions of the schema theorem will be given after a rigorous derivation based on a probability theory.

One is obtained by replacing random variables included in the theorem with their expectation, but the resultant is essentially different from the original version. This type holds for a finite number of individuals in a population. The other is obtained by limiting the number of individuals to infinity. That is, if each individual of the initial population is selected with an identical and independent probability distribution, then the original version of the inequality in the schema theorem becomes true with probability one.

2. Genetic Algorithms

In this section we shall make clear the genetic algorithms discussed in the present paper. As stated later, the genetic algorithms defined here are general frameworks including various types of genetic algorithms as special cases. Hence the results obtained in this article may be applied to almost all types of current genetic algorithms[12].

Let Σ denote the set of all strings consisting of l symbols 0 or 1: $\Sigma \equiv \{0\cdots 0, 0\cdots 01, \ldots, 1\cdots 1\}$. An element of Σ , that is, a string is also called an *individual*. An ordered n-tuple of individuals $x_1, \ldots, x_n \in \Sigma$ is called a population and denoted as $x = (x_1, \ldots, x_n)$. The number of individuals n in a population is assumed to be even for the convenience of crossover operation mentioned later. Furthermore, let S denote the set of all populations : $S \equiv \{x \mid x = (x_1, \ldots, x_n),\ x_i \in \Sigma\}$.

Suppose that a positive real-valued function defined on Σ , that is, $f : \Sigma \to \Re^+$ is given, where $\Re^+$ denotes the set of all positive real numbers. This function is called a *fitness function*. When we regard GA as an algorithm for finding the individuals which give the maximum value of the function, i.e., the maximum points of f, this function is also called an *objective* one.

The outline of the flow for GA discussed here is given as follows :

1. Select randomly an initial population $x(0) \in S$ according to a probability distribution π_0 defined on S ;
2. for $t := 0$ to ∞ do begin
3. *Reproduction* : Select randomly a population $y(t) \in S$ according to a probability distribution $g_r(x(t), \cdot)$ (of which detail will be given later) which depends on the population $x(t)$;
4. *Crossover* : Select randomly a population $z(t) \in S$ according to a probability distribution $g_c(y(t), \cdot)$ (of which detail will be given later) which depends on the population $y(t)$;
5. *Mutation* : Select randomly a population $u(t) \in S$ according to a probability distribution $g_m(z(t), \cdot)$ (of which detail will be given later) which depends on the population $z(t)$;
6. $x(t+1) := u(t)$;
7. end;

This algorithm produces a sequence of populations : $x(0), y(0), z(0), u(0) = x(1), y(1), z(1), \ldots, u(t-1) = x(t), y(t), z(t), \ldots$, which varies randomly. Thus these populations should be considered as random variables defined on a probability space (Ω, A, P) and denoted with a traditional fashion in the probability theory as $X(0), Y(0), Z(0), U(0) = X(1), Y(1), Z(1), \ldots, U(t-1) = X(t), Y(t), Z(t), \ldots$, where Ω, A and P are a sample space, a set of events, i.e., a σ-field and a probability measure, respectively.

Now the probability distributions $g_r(x(t), \cdot)$, $g_c(y(t), \cdot)$ and $g_m(z(t), \cdot)$ on S are given in detail. In the typical GA, reproduction selects the individuals with large fitness from a current population with a high probability. That is, the probability that a population $y = (y_1, \ldots, y_n)$ is selected by reproduction under the condition of the current population being $x = (x_1, \ldots, x_n)$ is usually given[1] by

$$g_r(x, y) = \frac{\sum_{i=1}^{n} f(x_i)\lambda(x_i = y_1)}{\sum_{i=1}^{n} f(x_i)} \times \cdots \times \frac{\sum_{i=1}^{n} f(x_i)\lambda(x_i = y_n)}{\sum_{i=1}^{n} f(x_i)}, \quad (1)$$

where $\lambda \equiv 1$ if a proposition A is true; $\equiv 0$ otherwise. As seen above the probability of a population y obtained from a population x by reproduction is often given by a utilization of the fitness function, and several kinds of the alternatives might be taken into consideration. The common feature of the probability is that any individual in the population y always belongs to the population x. Thus the probability of reproduction is assumed to satisfy the following condition : $g_r(x, y) > 0$ if $\varphi(y) \subseteq \varphi(x)$; $\equiv 0$ otherwise, where $\varphi(x)$ denotes the set of individuals in a population x, that is, $\varphi(x) \equiv \{x_1, \ldots, x_n\}$ for $x = (x_1, \ldots, x_n)$.

Second we check a condition which crossover should meet. As well known, one-point crossover, for instance, makes random mates (pairs) of individuals $y_i = y_{i1} \cdots y_{il}$ and $y_j = y_{j1} \cdots y_{jl}$ in a current population $y = (y_1, \ldots, y_n)$ and produces a pair of individuals such that $z_i = y_{i1} \cdots y_{ik} y_{j,k+1} \cdots y_{jl}$ and $z_j = y_{j1} \cdots y_{jk} y_{i,k+1} \cdots y_{il}$ with a random cut k of position. The obtained individuals by the crossover have a remarkable property such that for any j the number of "1's" of j-th bits of all individuals in the population is invariant before and after the crossover. Thus the probability of our crossover is assumed to satisfy the following condition : $g_c(y, z) \geq 0$ if $\forall j : \mathbf{1}_j(z) = \mathbf{1}_j(y)$; $= 0$ otherwise, where $\mathbf{1}_j(y)$ denotes the number of "1's" of j-th bits of all individuals in a population $y = (y_1, \ldots, y_n)$ ($y_i = y_{i1} \cdots y_{il}$) : $\mathbf{1}_j(y) \equiv \sum_{i=1}^{n} y_{ij}$.

Finally we consider a condition which mutation should satisfy. Because mutation occurs accidentally, at least one bit through the whole individuals in a population is considered to be reversed with a small probability and more than one bit mutation is not specified. That is, the probability of mutation is assumed to satisfy the following condition : $g_m(z, u) > 0$ if $H(z, u) \leq 1$; ≥ 0 otherwise, where $H(x, y)$ denotes the Hamming distance between populations $x = (x_1, \ldots, x_n)$ $(x_i = x_{i1} \cdots x_{il})$ and $y = (y_1, \ldots, y_n)$ $(y_i = y_{i1} \cdots y_{il})$: $H(xy) \equiv \sum_{i=1}^{n} \sum_{j=1}^{l} (x_{ij} \oplus y_{ij})$.

In the sequel we discuss the genetic algorithms which satisfies the week condition mentioned above. Hence the results obtained here might be applicable to a wide range of current GA's.

3. Sufficient Condition for Convergence

In this section it will be shown that the random process produced by GA is a finite Markov chain with the set of populations S as a state space, and this Markov chain is irreducible and aperiodic, and hence has a unique stable distribution being equal to the limiting distribution if week conditions are added to the crossover probability distributions $g_c(y, \cdot)$. These results will promote the various analyses of GA from a probabilistic point of view.

First note that the set of populations S consists of $N \equiv 2^{nl}$ elements. So let G_r, G_c and G_m be matrices of size N of which (x, y) component is $g_r(x, y)$, $g_c(x, y)$ and $g_m(x, y)$, respectively. Moreover let $G \equiv G_r G_c G_m$. Then, we have without difficulty :

Theorem 1 *The stochastic process $X(0), X(1), \ldots, X(t), \ldots$ of populations produced by our GA is a Markov chain on the state space S with the state transition matrix G .*

Since GA is regarded as a searching method, it is desired to get an optimum solution without depending on the initial state distribution. One of the first step to guarantee this is that the Markov chain of GA has a unique limiting distribution of states. On the other hand it is well known that a sufficient condition for a Markov chain having the unique limiting distribution is that it is irreducible and aperiodic. The Markov chain generated by our GA is unfortunately neither irreducible nor aperiodic as far as it is. So we are forced to add to the transition probability some conditions as week as possible.

Theorem 2 *If for any population $x \in S$ there is a positive integer s such that*

$$\sum_{y_1, \ldots, y_{s-1} \in S} g_c(x, y_1) g_c(y_1, y_2) \cdots g_c(y_{s-1}, x) > 0, \quad (2)$$

and there exists a population $x \in S$ such that $g_c(x, x) > 0$, then the Markov chain generated by our GA is irreducible and aperiodic.

As well known in the standard theory of finite Markov chains, any irreducible and aperiodic Markov chain has the unique stable distribution which is also the limiting distribution of states. Thus, we may conclude that *our GA has a unique stable and limitting distribution :*

$$\pi_\infty = \pi_\infty G \quad \text{and} \quad \forall \pi_0 : \pi_\infty = \lim_{t \to \infty} m\, \pi_0 G^t. \quad (3)$$

The constraints added to the crossover in Theorem 2 mean that any population may return back to itself within a finite number of crossover operations, and hence is said to require a sort of transition balance. It might be said that this condition is not so strong that almost all types of crossover, including one-point crossover, are expected to satisfy it. Thus, the convergence of almost all current GA's is said to be guaranteed.

4. Schema Theorem

The schema theorem is said to be one of most significant results from a theoretical point of view in GA. The inequality appeared in the theorem is not, however, possible to be mathematically interpreted as far as it is because it includes random variables. In this section the correct version of the schema theorem will be given after a rigorous derivation of the inequality from a probabilistic point of view.

Though we may find a considerable amount of alternatives of GA's[12], we here follow the version in [1], what is called, a simple genetic algorithm. That is, the probability of reproduction is assumed to be (1). Second the following one-point crossover is assumed. From a current population $y = (y_1, \ldots, y_n)$ we make $n/2$ pairs (mates) $(y_1, y_2), \ldots, (y_{n-1}, y_n)$ and apply a crossover operation to each pair with a probability p_c. For a *ing to a schema H has the following lower bound for* $i = 1, \ldots, n$ *:*

$$P(U_i(t) \in H) \geq (1 - p_m)^{o(H)} P(Z_i(t) \in H). \quad (11)$$

Comparing the inequality appearing in (4) with the one in Theorem 3, we know that, if the following

$$E\left(m(H,t) \cdot \frac{f(H)}{\bar{f}}\right) \geq E\left(m(H,t)\right) \frac{E(f(H))}{E(\bar{f})} \quad (12)$$

holds, the original version would be said to be also valid. This inequality is however not guaranteed in general as far as the number of individuals in a population is finite. In the next section the case of an infinitely many number of individuals will be discussed.

6. The Case of an Infinite Number of Individuals

In the previous section we have mentioned that there are two kinds of interpretation of the original version of the schema theorem : one is to replace the random variables appearing in the inequality with their expectations, and the other is to consider its validity in the sense of probability one (i.e., almost everywhere). In this section the second interpretation is discussed and the following result will be shown.

Theorem 4 *Suppose that the number of individuals n in a population is sufficiently large. Let* $X(0) = (X_1(0), \ldots, X_n(0))$ *be an initial population. If the set of random variables* $\{X_1(0), \ldots, X_n(0)\}$ *is independent and each individual* $X_i(0)$ *is selected according to an identical distribution, then the following schema theorem approximately holds :*

$$m(H, t+1) \geq m(H,t) \frac{f(H)}{\bar{f}} \left[1 - p_c \frac{\delta(H)}{l-1} - o(H) p_m\right] \; a.e., \quad (13)$$

where "a.e." means "almost everywhere", that is "with probability one".

The proof for this theorem requires many steps in which the following lemmata: First a simple type of the strong law of large number is presented which is easily found in the standard text of probability theory.

Lemma 7 *Let* $W_1, W_2, \ldots$ *be a sequence of independent real-valued random variables following an identical distribution. Suppose that for every n* $E(|W_n|) < \infty$. *Then,*

$$\lim_{n \to \infty} m \frac{1}{n} \sum_{i=1}^{n} W_i = E(W_1) \; a.e.. \quad (14)$$

Lemma 8 *Suppose that the number of individuals n in a population is sufficiently large. Let a population* $Y = (Y_1, \ldots, Y_n)$ *be produced from a population* $X = (X_1, \ldots, X_n)$ *by reproduction. If the set of pairs of the random variables* $\{(X_1, X_2), \ldots, (X_{n-1}, X_n)\}$, *is independent and each pair* (X_{2i-1}, X_{2i}) *has an identical distribution, then the set of pairs of the random variables* $\{(Y_1, Y_2), \ldots, (Y_{n-1}, Y_n)\}$, *is also approximately independent and each pair* (Y_{2i-1}, Y_{2i}) *has an identical distribution.*

We here prepare a simple lemma which is useful for proving the succeeding lemmata.

Lemma 9 *Let* X, Y, Z *be discrete-valued random variables. Suppose that the conditional probability* $P(Z = z \mid X = x \cap Y = y)$ *does not depend on y, that is, there exists a function, not depending on y, of two variables such that* $\forall x, y, z : Q(x,y) = P(Z = z | X = x \cap Y = y)$. *Then, the following holds :*

$$\forall x, y, z : P(Z = z | X = x \cap Y = y) = P(Z = z | X = x). \quad (15)$$

Lemma 10 *Let a population* $Z = (Z_1, \ldots, Z_n)$ *be produced from a population* $Y = (Y_1, \ldots, Y_n)$ *by crossover. If the set of pairs of the random variables* $\{(Y_1, Y_2), \ldots, (Y_{n-1}, Y_n)\}$ *is independent and each pair* (Y_{2i-1}, Y_{2i}) *has an identical distribution, then the set of pairs of the random variables* $\{(Z_1, Z_2), \ldots, (Z_{n-1}, Z_n)\}$ *is also approximately independent and each pair* (Z_{2i-1}, Z_{2i}) *has an identical distribution.*

Lemma 11 *Let a population* $U = (U_1, \ldots, U_n)$ *be produced from a population* $Z = (Z_1, \ldots, Z_n)$ *by mutation. If the set of pairs of the random variables* $\{(Z_1, Z_2), \ldots, (Z_{n-1}, Z_n)\}$ *is independent and each pair* (Z_{2i-1}, Z_{2i}) *has an identical distribution, then the set of pairs of the random variables* $\{(U_1, U_2), \ldots, (U_{n-1}, U_n)\}$ *is also approximately independent and each pair* (U_{2i-1}, U_{2i}) *has an identical distribution.*

Lemma 12 *Suppose that the number of individuals is sufficiently large, the set* $\{X_1(0), \ldots, X_n(0)\}$ *of random variables in an initial population* $X(0) = (X_1(0), \ldots, X_n(0))$, *and each individual* $X_i(0)$ *has an identical distribution. Let* $X(t) = (X_1(t), \ldots, X_n(t))$ *be a population of the t-th generation. Then, for any time t, the set of pairs of random variables* $\{(X_1(t), X_2(t)), \ldots, (X_{n-1}(t), X_n(t))\}$ *is independent and each pair of individuals* $(X_{2i-1}(t), X_{2i}(t))$ *has an identical distribution.*

References

[1] Goldberg. D.E. (1989) : Genetic algorithms in search, optimization, and machine learning, Addison-Wesley Publishing Company.

[2] Davis, L. (ed.) (1991) : Handbook of genetic algorithms, Van Nostrand Reinhold.

[3] Goldberg, D.E. and Segrest, J.C. (1987) : Finite Markov chain analysis of genetic algorithms, Proc. 2nd Int. Conf. Genetic Algorithm, 1-8.

[4] De Jong, K.A. and Spears, W.M. (1990) : An analysis of the interacting roles of population size and crossover in genetic algorithms, Proc. 1-st Int. Workshop on Parallel Problem Solving from Nature, Dortmund.

[5] Davis, T.E. and Principe, J.C. (1991) : A simulated annealing like convergence theory for the simple genetic algorithm. Proc. 4th Int. Conf. Genetic Algorithm. 174-181.

[6] Eiben, A.E.. Aarts, E.H.L. and Van Hee, K.M. (1991) : Global convergence of genetic algorithms : a Markov chain analysis, Lect. Notes Computer Science, Vol 496. 4-12.

[7] Vose, M.D. (1991) : Generalizing the notion of schema in genetic algorithms, Artificial Intelligence, 50, 385-396.

[8] Nix, A.E. and Vose, M.D. (1992) : Modeling genetic algorithms with Markov chains, Annals of Mathematics and Artificial Intelligence, 5, 79-88.

[9] Yanagiya. H. (1992) : Markov analysis of Genetic Algorithms, Technical Report of IEICE, NLP92-26, 19-26 (in Japanese).

[10] Vose, M.D. (1993) : Modeling simple genetic algorithms, in Whitley, L.D. (ed.) : Foundation of genetic algorithms 2, Morgan Kaufmann.

[11] Suzuki, J. (1995) : A Markov chain analysis on simple genetic algorithms, IEEE Trans. on Systems, Man, and Cybernetics, 25, 03.

[12] Kitano, H. (ed.) (1993) : Genetic algorithms, Sangyou-Tosho (in Japanese).

pair of (y_{2i-1}, y_{2i}), where $y_{2i-1} = a_1 \cdots a_j a_{j+1} \cdots a_l$ and $y_{2i} = b_1 \cdots b_j b_{j+1} \cdots b_l$ $(a_k, b_k = 0, 1)$, we select randomly a number j according to the uniform distribution on the set $\{1, \ldots, l-1\}$ and obtain a pair of individuals (z_{2i-1}, z_{2i}), where $z_{2i-1} = a_1 \cdots a_j b_{j+1} \cdots b_l$ and $z_{2i} = b_1 \cdots b_j a_{j+1} \cdots a_l$.

Finally the mutation operation is assumed to reverse independently each bit of each individual $z_i = z_{i1} \cdots z_{il}$ of a population $z = (z_1, \ldots, z_n)$ from 0 to 1 or vice versa with a small probability p_m.

Now we explain about the schema theorem. Consider a symbol $*$ in addition to symbols 0 and 1. A string $H = h_1 \cdots h_l$ consisting of the symbols $*$, 0 and 1 is called a *schema*. When we regard the symbol $*$ as *don't care* one, a schema might be identified with the set of individuals which is obtained by replacing $*$ into 0 or 1. For example the schema $H = 01*11*0$ is considered as a set $\{0101100, 0111100, 0101110, 0111110\}$. With this identification an individual $a = a_1 \cdots a_l \in \Sigma$ $(a_i = 0, 1)$ is said to *belong* to a schema H and denoted as a $a \in H$ if a coincides with H except $*$ positions.

For a schema H the difference of positions between the leftest symbol not being $*$ and the rightest symbol not being $*$ is called the *length* of H and denoted as $\delta(H)$. The number of symbols not being $*$ in a schema H is called the *order* of H and denoted as $o(H)$.

In order to state the schema theorem we prepare three kinds of random variables. Let H be a schema and $X(t) = (X_1(t), \ldots, X_n(t))$ be a population generated by our GA at a time t. Consider the number of individuals in $X_1(t), \ldots, X_n(t)$ which belong to the schema H and denote it as: $m(H,t) \equiv \sum_{i=1}^{n} \lambda(X_i(t) \in H)$, $\lambda(A)$ being 1 if a proposition A is true and 0 otherwise. Let $f(H)$ be the arithmetic average of fitness for the individuals included in H : $f(H) \equiv \sum_{i=1}^{n} f(X_i(t))\lambda(X_i(t) \in H) \big/ m(H,t)$, and $\bar{f}$ be the arithmetic average of fitness for all individuals in the population $X(t) = (X_1(t), \ldots, X_n(t))$: $\bar{f} \equiv \frac{1}{n} \sum_{i=1}^{n} f(X_i(t))$. Note that these three quantities are all random variables.

Now the schema theorem in [1] is stated as

Schema Theorem [1] *For our GA mentioned above, the following inequality holds : for* $t = 0, 1, \ldots$

$$m(H, t+1) \geq m(H,t) \frac{f(H)}{\bar{f}} \left[1 - p_c \frac{\delta(H)}{l-1} - o(H) p_m \right], \qquad (4)$$

where H *is a fixed schema and* p_c *and* p_m *are the probabilities of crossover and mutation, respectively.*

As mentioned above three quantities $m(H,t)$, $f(H)$ and $\bar{f}$ are all random variables which are measurable functions defined on a sample space Ω. So the inequality has no meaning without specifying elements in Ω, and we are forced to interpret it by replacing these random variables with their expectations or by considering its validity in the sense of probability one (i.e., almost everywhere). In the next section the first interpretation is discussed and the second in Section 6.

5. The Case of a Finite Number of Individuals

If we make the replacement above, we have that $E(m(H, t+1)) \geq E(m(H,t)) \frac{E(f(H))}{E(\bar{f})} \left[1 - p_c \frac{\delta(H)}{l-1} - o(H) p_m \right]$, but this inequality does not unfortunately hold. We will show that the correct version of the schema theorem should be stated as

Theorem 3 *For our GA mentioned above, the following inequality holds : for* $t = 0, 1, \ldots$

$$E(m(H, t+1)) \geq E\left(m(H,t) \frac{f(H)}{\bar{f}} \right) \left[1 - p_c \frac{\delta(H)}{l-1} - o(H) p_m \right], \qquad (5)$$

where H *is a fixed schema and* p_c *and* p_m *are the probabilities of crossover and mutation, respectively.*

This theorem can be obtained through a slightly long course in which the following lemmata are utilized :

Lemma 1 *Let a population* $X(t)$ *transit to a population* $Y(t) = (Y_1(t), \ldots, Y_n(t))$ *by reproduction when* $X(t) = x(t) = (x_1(t), \ldots, x_n(t))$. *Then the conditional probability of an individual* $Y_i(t)$ *belonging to a schema* H *is given by*

$$P(Y_i(t) \in H | X(t) = x(t)) = \frac{\sum_{j=1}^{n} f(x_j(t)) \lambda(x_j(t) \in H)}{\sum_{j=1}^{n} f(x_j(t))}. \qquad (6)$$

Lemma 2 *The sum of probabilities that each individual in the population* $Y(t) = (Y_1(t), \ldots, Y_n(t))$ *belongs to a schema* H *is given by*

$$\sum_{i=1}^{n} P(Y_i(t) \in H) = E\left(m(H,t) \cdot \frac{f(H)}{\bar{f}} \right). \qquad (7)$$

Lemma 3 *Let* l_H *and* r_H *be the numbers of the leftest and rightest bits not being* $*$ *in a schema* H. *Suppose that a population* $Z(t) = (Z_1(t), \ldots, Z_n(t))$ *is obtained from population* $Y(t) = (Y_1(t), \ldots, Y_n(t))$ *through crossover. Then, the probability of any odd-indexed individual has a lower bound for* $i = 1, \ldots, n/2$ *:*

$$\begin{aligned} P(Z_{2i-1}(t) \in H) &\geq P(Y_{2i-1}(t) \in H) \cap P(Y_{2i}(t) \in H) \\ &+ p_c \frac{l_H - 1}{l-1} P(Y_{2i-1}(t) \notin H) \cap P(Y_{2i}(t) \in H) \\ &+ \left[(1 - p_c) + p_c \frac{l - r_H}{l-1} \right] P(Y_{2i-1}(t) \in H) \cap P(Y_{2i}(t) \notin H). \quad (8) \end{aligned}$$

Lemma 4 *Let* l_H *and* r_H *be the numbers of the leftest and rightest bits not being* $*$ *in a schema* H. *Suppose that a population* $Z(t) = (Z_1(t), \ldots, Z_n(t))$ *is obtained from population* $Y(t) = (Y_1(t), \ldots, Y_n(t))$ *through crossover. Then, the probability of any even-indexed individual has a lower bound for* $i = 1, \ldots, n/2$ *:*

$$\begin{aligned} P(Z_{2i}(t) \in H) &\geq P(Y_{2i-1}(t) \in H) \cap P(Y_{2i}(t) \in H) \\ &+ \left[(1 - p_c) + p_c \frac{l - r_H}{l-1} \right] P(Y_{2i-1}(t) \notin H) \cap P(Y_{2i}(t) \in H) \\ &+ p_c \frac{l_H - 1}{l-1} P(Y_{2i-1}(t) \in H) \cap P(Y_{2i}(t) \notin H). \quad (9) \end{aligned}$$

Lemma 5 *The sum of probahilities that each individual in the population* $Z(t) = (Z_1(t), \ldots, Z_n(t))$ *which is obtained from the population* $Y(t) = (Y_1(t), \ldots, Y_n(t))$ *by crossover, belongs to a schema* H *has a lower bound :*

$$\sum_{i=1}^{n} P(Z_i(t) \in H) \geq \left[1 - p_c \frac{\delta(H)}{l-1} \right] \sum_{i=1}^{n} P(Y_i(t) \in H). \qquad (10)$$

Lemma 6 *Let* $U(t) = (U_1(t), \ldots, U_n(t))$ *be the population obtained from the population* $Z(t) = (Z_1(t), \ldots, Z_n(t))$. *Then, the probability of each individual in* $U(t) = (U_1(t), \ldots, U_n(t))$ *belong-*

INCORPORATING NEIGHBOURHOOD SEARCH OPERATORS INTO GENETIC ALGORITHMS

Christian Höhn
Institut für Grundlagen Elektrotechnik/Elektronik
Technische Universität Dresden, Germany *

Colin Reeves
School of Mathematical and Information Sciences
Coventry University, UK

Abstract

In this paper an abstract genetic algorithm (GA) is proposed which effectively merges local hill-climbing with recombination and population based selection in a general manner. This extension is possible because the traditional crossover can be resolved into two functions: one function is as a particular class of operator, which is actually distinct from the other which is the recombination itself. Thus traditional GAs can be classified as a special case of a more general approach in which recombination is applied along with other operators. In the work reported here, using the framework of an abstract GA, the performance of several operators as well as the effects of recombination are studied in the context of the graph bipartitioning problem.

The Abstract Genetic Algorithm

Traditional genetic algorithms have three main characteristics by which they are commonly distinguished from classical and modern local search techniques.

Population-based selection : In point-based local search methods selection is restricted to a choice between the parent and one or more offspring, but with GAs the presence of a population gives a spatial dimension to selection.

Recombination : Recombining two parent solutions the offspring usually inherits those variables for whose both parents share the same value. Thus recombination essentially results in a search space reduction.

Crossover : Traditionally crossover involves two (usually binary) strings. Having defined a set of crossover points, the intermediate bits are swapped.

*This work was undertaken when the first author was with the Control Theory and Appplications Centre, Coventry University, UK

While selection and recombination represent general search techniques crossover in essence stands for a special class of operators and thus can be replaced in the following by an abstract type of operator. In the context of local search an operator ω defines a neighbourhood $N(x,\omega) \subseteq S$ for all points x in the search space S. This general definition holds for a 1-bit Hamming hill climber as well as for many more sophisticated heuristics. However, crossover as introduced by genetic algorithms seems to be an exception in so far as two points, say x and y, are involved in the operation such that $N = N(x,y,\omega), x,y \in S$. Some implications of this are explored in [1], but it is obvious that defining a neighbourhood by more than one point complicates an analytical approach. Furthermore, in this way the crossover operation is inevitable linked to recombination leading to a superposition of both effects. As a consequence the terms are often used synonymously.

However, having 2 points effectively reduces the neighbourhood size by the total of all elements the two points have in common. This is an effect of recombination which occurs before any operators are applied. Following the ideas of Culberson [2] the crossover effect can then be implemented by a unary operator which always pairs a point $x \in S$ with its complementary point $\bar{x}$ in the reduced neighbourhood. This operator, which we term *complementary crossover*, thus fits into the general definition above. As a consequence, the traditional crossover may be decomposed into recombination and complementary crossover.

In traditional GAs the search space defined by both parents is usually explored by evaluating only one or two randomly selected offspring. Within the framework of an abstract GA local hill climbing can be applied to find the best offspring. In this manner the abstract GA also covers many hybrid forms. In this paper the following scheme will be used:

1. Initialize a population of size N.
2. Apply the local hill climber to each individual until a local optimum is reached or another stopping criterion is met.

3. Choose $2N$ individuals from the population according to fitness and pair them (randomly) so that each pair defines a reduced neighbourhood. This is passed to the next generation together with the complete bitstring and objective value from one parent.

4. All the offspring define the population for the next generation.

5. If not TERMINATED go to 2, otherwise stop.

The scheme is not restricted to a particular hill-climbing technique and thus provides a high degree of flexibility. Furthermore, for a given application several local hill climbing techniques may be evaluated individually before being extended by recombination and selection. In the rest of the paper such an approach is illustrated for the graph bipartitioning problem.

Bipartitioning and Graph Transformation

The uniform bipartitioning of a graph is a special case (i.e. $k = 2$) of the k-way graph partitioning problem. Given a graph $G = (V, E)$ with a cost function $c : E \subseteq V \times V \mapsto \mathbb{R}^1$ and a size $S(u)$ attached to each node, partition the nodes of G into k subsets of equal size such that the total cost of the edges cut is minimized. The uniform k-way partitioning problem is a constrained optimization problem, i.e. the search space usually contains an nonempty set of illegal solutions. By means of a graph transformation [3] the problem can be transformed into a free optimization problem. Given a graph $G = (V, E)$ the transformed graph $G^* = (V, E^*)$ is fully connected with the new cost function

$$c^*(e) = \begin{cases} S(u)S(v)R - c(e) & \text{if } e = \{u,v\} \in E; \\ S(u)S(v) & \text{if } e = \{u,v\} \in E^* - E; \end{cases}$$

where R is a positive real number, termed the *augmenting factor*. As shown in [3] if $R > \sum_{e \in E} c(e)$, any partition $\pi = (A, B)$ that maximizes

$$C^*(\pi) = \sum_{\substack{e = (u,v) \in E^* \\ \pi(u) \neq \pi(v)}} c^*(e)$$

will minimize

$$C(\pi) = \sum_{\substack{e = (u,v) \in E \\ \pi(u) \neq \pi(v)}} c(e)$$

under the constraint that $S(A) * S(B)$ is maximized. In the case $k = 2$, a solution can easily be encoded as a binary string of length $l = |V|$, each bit representing the partition to which the corresponding node belongs. For the following experiments all results are given for the objective function $C^*(\pi)$ and an augmenting factor $R = 20$ has been applied.

Evaluation Of The Operators

In the first step 1-point and 2-point complementary crossover (denoted by 1CX and 2CX respectively), a 1-bit Hamming hill climber (HC) and mutation (MUT) are used as universal operators and compared against the LPK-heuristic—a traditional method for solving the k-way graph partitioning problem as described by Lee *et al.* [3]. Table 1 shows some characteristics of these operators. A detailed discussion is given in [4]. Table 2 display the results for a single graph which underline the efficiency of LPK in terms of computational time as well as performance. Furthermore, the average number of moves gives a rough idea of the distribution of local optima in the search space of each operator.

Recombination

Recombining two binary strings of length l results in a search space $S_d = \{0,1\}^d$, $0 \leq d \leq l$ where d is the Hamming distance between both strings. Since the time required for performing an operation on a bit string is usually a (polynomial) function of l a reduced search space will speed up the operator and thus save computational time. This effect has been described elsewhere [1, 5].

However, for some operators recombination exhibits an even more interesting phenomenon which is now considered in more detail. Consider two bit strings, say x, y, both of length l. Applying a 1-bit Hamming hill climber the neighbourhood of both points contains l neighbors. Assuming that the Hamming distance between x and y is $0 < d < l$, recombination of x and y generates a neighbourhood $N(x, y, HC)$ containing d points. Obviously, $N(x, y, HC) \subseteq N(x, HC)$ and $N(x, y, HC) \subseteq N(y, HC)$. In contrast, as an easy example reveals, the last relation does not necessarily apply for 1-point complementary crossover. Here recombination is able to introduce new points into the reduced neighbourhood and thus it may help out of a local optimum. The same observation may be made for the 2-point complementary crossover or the LPK-heuristic.

Table 3 displays the results for a typical run of five generations. In particular the following observations may be made.

Op	N'hood	Time	Strategy	Termination
LPK	n	$O(n^2)$	SA	local optimum
1CX	n	$O(n^2)$	SA	local optimum
2CK	n^2	$O(n^3)$	NA	$< n$ trials/pass
HC	n	$O(n)$	SA	local optimum
MUT	2^n	$O(n^3)$	NA	$< n$ trials/pass

Table 1: Certain characteristic of the operators considered. SA and NA stand for steepest and next ascent.

Op	max	min	aver	dev	moves	time
LPK	572	562	569	3	2.280	2.0 s
1CX	532	488	511	8	1.720	1.0 s
2CX	530	498	515	8	6.340	8.0 s
MUT	552	498	531	13	13.060	7.0 s
HC	544	482	515	13	2.920	0.0 s

Table 2: Results for the random graph R.100.5 (100 vertices, mean degree 5, 252 edges). All the vertices of the graph are of the same size and all the edges have uniform costs. The average performance and the standard deviation are based on 50 runs.

- Owing to neighbourhood size reduction all the operators are speeded up. Thus five generations require less than double the time of the first generation.
- As predicted, recombination does not have any effect on the 1-bit Hamming hill climber. The increasing average performance is merely due to selection.
- Clearly, for 1CX and LPK, recombination leads the search away from the local optima found in the first generation. The increasing average number of moves indicates the progress.
- For mutation and 2CX the local search stops after n trials have been failed. However, in the next generation the selected points will be evaluated again which essentially exceeds the stopping criterion and accounts for the progress of one move per generation. Obviously, this effect is an artefact of the selection scheme and cannot be attributed to recombination. However, both operators benefit from the speed up.

However, since the LPK-heuristic found the best value within the first generation, it is not clear whether recombination is more effective than simply trying different initial points. To obtain a preliminary answer on this question a hill climber was compared to the GA when applied to a more ambitious random graph with 500 vertices and 1224 edges corresponding to a mean degree of 5. In 30 runs the CPU time has been fixed for both algorithms and the best values found per run are compared. The results are shown in Table 4. It appears that LPK coupled with recombination performs better than LPK alone. In particular, the worst value found in 30 runs with recombination exceeds the average performance of LPK on its own. The last column display the effective population size, i.e. 80 restarts require the same amount of time as 50 individuals evolved for 6 generations.

Conclusion

In this paper an abstract genetic algorithn has been presented and tested on a special case of graph partitioning. The proposed GA couples a local hill climbing techniques with recombination and selection in a general way, so that any local search technique may be accommodated within the proposed scheme. In this paper the investigation has focused on several operators and the effects of recombination. Initially the operators have been studied individually when implemented in the framework of a simple local hill climber. Such preliminary investigations may form a first step in the design of more sophisticated algorithms.

Furthermore, recombination has been proved to save computational time as well as offering a mechanism to cirumvent local optima for certain operators. Both characteristics may be used to guide the design of new operators. However, the efficiency of recombination certainly depends on the population size as well as the selection scheme applied. Both problems deserve further investigations. In future work the performance of the GA will be investigated when more sophisticated hill climbers are incorporated.

References

[1] Reeves, C.R. (1994) Genetic Algorithms and Neighbourhood Search. *Proc. of AISB Workshop on Evolutionary Computing*, Springer Verlag, Berlin, 115-130.

[2] Culberson, J.C. (1993) Mutation-crossover isomorphisms and the construction of discriminating functions. *Evolutionary Computation* (to appear)

[3] Lee, C.H. Park, C.I. Kim, M. (1989) Efficient algorithm for graph partitioning problem using a problem transformation method. *Computer Aided Design*, **21**, 611-618.

[4] Hoehn, C. (1995) Embedding local search operators in a genetic algorithm for graph bipartitioning. In preparation.

[5] Hoehn, C. (1994) *Heuristic Neighbourhood Search Operators for Graph Partitioning Tasks* Proc. 10th International Conference on Systems Engineering, Coventry, UK, 469 - 476

Operator	max	min	average	deviation	average moves	average size	time in s
LPK	1142	1124	1137	5	2.500	100.00	7.0
LPK	1142	1132	1140	2	2.920	34.96	8.0
LPK	1142	1140	1141	1	3.120	29.56	8.0
LPK	1142	1142	1142	0	3.120	31.16	9.0
LPK	1142	1142	1142	0	3.120	26.48	10.0
1CX	1058	972	1019	14	1.820	100.00	3.0
1CX	1064	1014	1035	14	2.700	46.00	3.0
1CX	1064	1030	1052	9	3.300	45.64	4.0
1CX	1070	1048	1059	5	3.900	30.92	7.0
1CX	1070	1056	1064	5	4.240	26.44	8.0
2CX	1060	1002	1029	12	7.540	100.00	59.0
2CX	1064	1020	1042	11	9.120	46.32	72.0
2CX	1074	1030	1054	10	10.680	43.16	86.0
2CX	1078	1040	1061	9	11.320	43.12	98.0
2CX	1092	1050	1067	7	12.320	35.60	112.0
MUT	1116	1040	1081	16	32.460	100.00	38.0
MUT	1116	1068	1092	11	34.700	44.52	41.0
MUT	1118	1082	1100	8	35.980	41.84	43.0
MUT	1120	1090	1106	8	37.200	38.40	45.0
MUT	1122	1094	1110	7	38.380	31.96	47.0
HC	1082	972	1030	26	4.500	100.00	1.0
HC	1082	994	1041	24	4.500	45.40	1.0
HC	1082	998	1053	20	4.500	44.60	2.0
HC	1082	1024	1058	18	4.500	39.36	3.0
HC	1082	1036	1066	15	4.500	41.08	3.0

Table 3: Typical run of the GA when applied to R.100.5

Operator	max	min	average	deviation	popsize
LPK	5778	5754	5762	6	80
GA + LPK	5788	5764	5774	5	50

Table 4: The LPK heuristic applied alone and coupled with recombination.

AUTOMATIC CHANGE OF REPRESENTATION IN GENETIC ALGORITHMS

Franz Oppacher and Dwight Deugo

e-mail: oppacher@scs.carleton.ca, dwightdeugo@scs.carleton.ca

Intelligent Systems Group, School of Computer Science
Carleton University, Ottawa, Canada, K1S 5B6.

Abstract: In the areas of Genetic Algorithms, Artificial Life and Animats, genetic material is often represented as a fixed size sequence of genes with alleles of 0 and 1. This is in accord with the 'principle of meaningful building blocks'. The principle suggests that epistatically related genes should be positioned very close to one another. However, in situations in which gene dependency information cannot be determined a priori, a Genetic Algorithm that uses a static, list chromosome structure will often not work. The problem of determining gene dependencies is itself a search problem, and seems well suited for the application of a Genetic Algorithm. In this paper, we propose a self-organizing Genetic Algorithm, and, after describing four different chromosome representations, show that the best one for a Genetic Algorithm to use to coevolve the organization and contents (gene dependencies and values) of a chromosome is a hierarchy.

1 The Evolution of Hierarchies

In the areas of Genetic Algorithms (GAs), Artificial Life and Animats, genetic material is often represented as a chromosome [1], [2] containing a fixed number of genes or features which can assume certain values, called alleles. The collection of alleles provides the genetic information required for constructing a phenotype or for determining a possible solution to a given problem. Under this representation, the chromosome is a static structure: every gene's locus is fixed. To simplify the encoding further, GAs often use alleles of 0 and 1 [1], [3], [4].

The simplification of a fixed size sequence of genes is in accord with the 'principle of meaningful building blocks'[1] which suggests that epistatically related genes should be positioned very close to one another so that crossover will be less likely to disrupt the building blocks. However, this is only possible if one knows a priori what the gene dependencies (building blocks) are, and if one knows a priori how many genes are required to encode the problem. It is, however, all but impossible to know the gene dependencies a priori without detailed knowledge of the fitness calculation. Therefore, a coding that does not violate the above principle is unlikely.

In situations in which gene dependency information cannot be determined a priori, a GA that uses a static, list chromosome structure will often not work. For example, a GA will not work if it uses a fitness function that requires two specific genes to be located beside one another for optimal fitness, and it does not provide an operator for this to occur in the chromosome, such as inversion. One of the problems with a unary operator such as inversion is that the new chromosome it produces does not gain from the experiences of other chromosomes. There are situations, however, where GAs work even though their gene dependencies cannot be determined a priori [5], [6]. These successes may be due to a lack of dependencies or to the fact that the encoding happened to reflect them.

The problem of determining gene dependencies is itself a search problem, and seems well suited for the application of a GA. In this paper, we propose a self-organizing GA, and, after describing four different chromosome representations, show that the best one for it to use to coevolve the organization and contents (gene dependencies and values) of a chromosome is a hierarchy.

2 Hierarchical Chromosomes

Is there any benefit in evolving the organization of the chromosome? As stated in the first section, in all but the simplest domains it is an impossible task to determine a problem's gene dependencies a priori. Therefore, if the fitness of the chromosome depends on the positioning of its genes, then to achieve the best results one must allow the genes to move within the chromosome so that their adjacency relations reflect their dependencies, or else be satisfied with suboptimal results. The benefits are obvious. Allowing the chromosome to determine its own gene dependencies without relying on static, hand-coded ones, makes it a much more flexible and adaptable structure.

Can one coevolve the organization of a chromosome, its components, and its contents? Exactly what is meant by a part of a chromosome? How does one measure a part's fitness? Are a chromosome and its parts processed separately or simultaneously? What should the representation of the chromosome be?

To help answer these questions, we construct and discuss four different GAs. Each one uses a different chromosome representation, but all share a common fitness function. The first GA uses a chromosome constructed as a linked list, and is ordered with the correct gene dependencies. The second GA uses a chromosome constructed as a linked list, but uses a random gene ordering. In these two organizations, the gene at the head of the list is known as the root gene. The third GA uses a chromosome constructed as a random, rooted tree, with genes represented by the nodes of the tree and their dependencies represented by its links. The tree is considered complete: every one of the chromosome's genes is present in the tree. The fourth GA uses a chromosome similar to the third's organization but, because the GA does not require that all genes be present in the tree, it is considered a subtree. In each representation, a gene's value can be either 0 or 1.

The fitness function used by all organizations depends on the ordering of a chromosome's genes and their values, and is described as follows:

Fitness

```
"self refers to the chromosome"
^ fitness := (self rootGene GeneFitness) / (2 *
numberOfGenesInChromosome).
```

<u>GeneFitness</u>

```
| sum |
"self refers to the gene"
sum := (self geneNumber) - (self parentGene geneNumber).
"Only reward a gene if its geneNumber is greater than its parent.
The maximum fitness contribution of a gene is 2, 1 for being in
the correct order - with respect to the parent gene - and 1 for a
gene value of 1"
(sum >= 0)
    ifTrue: [sum := 1 + self GeneAllele]
    ifFalse: [sum := self GeneAllele].
self linkedGenes do: [:attachedGene|
    sum := sum + attachedGene GeneFitness].
^ sum
```

The fitness of a chromosome is calculated as the fitness of its genes divided by two times the number of genes in the chromosome. The reason for the division is to establish a fitness value between zero and one, since the maximum fitness value of the genes is two times the number of genes in the chromosome.

The recursive function GeneFitness works through a chromosome's linked genes (lists or trees) starting at the root gene. If the gene number of a gene's predecessor, or parent (in the case of a tree) is less than it* , a value of 1 is added to the fitness total. Next, the gene's allele value is added to the total. Therefore, the contribution of each gene to the overall fitness can be either 0, 1, or 2.

Correct Order	Gene Value	Fitness Contribution
yes	0	1
yes	1	2
no	0	0
no	1	1

The fittest chromosome will have all alleles set to 1 and its genes ordered from the smallest to the largest gene number. The fitness function imposes a dependency on the chromosome: to receive the maximum fitness, a chromosome's genes must be in the correct order. Since this dependency is unknown to the chromosome, the chromosome must discover it and the correct allele settings, to achieve maximum fitness.

As an example of calculating a chromosome's fitness, consider the following randomly ordered, linked list, four gene example. In this example, the positions of the genes are randomly assigned in the linked list. In this case, the function GeneFitness would return 5: 1 + 2 + 0 + 2. Therefore, the fitness of the chromosome is 0.625 (5/8).

Gene Number:	3 <-	4 <-	1 <-	2
Gene Allele:	0	1	0	1
Fitness Contribution:	1	2	0	2
GeneFitness Total: 5				

In this example, gene number 1 is penalized since its parent gene's gene number - 4 - is greater than its own.

As another example, consider the following tree, four gene example. Gene 2 is the root gene, with allele 1, and has two children: gene 1 with allele 0, and gene 3 with allele 0. Gene 1 has one child: gene 4 with allele 1.

* The root gene's parent's gene number is the same as the root.

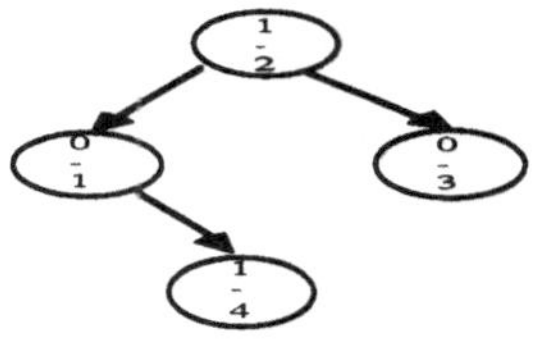

Figure 1. Example Tree Chromosome

In this case, the function GeneFitness would return 5: 0 + 2 + 1 + 2. Therefore, the fitness of the chromosome is 0.625 (5/8).

Gene Number:	4 ->	1 ->	2 <-	3
Gene Allele:	1	0	1	0
Fitness Contribution:	2	0	2	1
GeneFitness Total: 5				

In this example, gene number 1 is penalized since its parent gene's gene number - 2 - is greater than its own.

The initial fitness of a subtree is computed like that of a tree. However, because a subtree does not have all the genes of the complete chromosome, its fitness value must be made to reflect this fact. This fitness calculation is discussed in section 2.4.

In the following subsections, we describe the results of using each of the above representations in a GA, and describe the new operators required to support them. For each representation, the GA uses the same gene dependent fitness function described above.

2.1 Sequential Genetic Algorithm

The sequential GA's (SGA) chromosome is organized as a linked list of genes. Gene number one is the first element in the list, gene number two is the second element in the list, and is linked to the first one, and so on. As the genes are already in the correct order, e.g. each gene's preceding gene number is smaller than it own, all the SGA must do is determine the correct alleles for the genes - 0 or 1.

The SGA uses the two-point crossover operator [7] for combining chromosomes. One important feature of this operator, especially for the SGA, is that its application does not cause any change in the gene ordering of the resulting chromosome. The operator's application results in only an exchange of gene values, while maintaining gene ordering.

The fact that the SGA must only determine the correct gene values reduces the fitness function to that of the linear function $f(x) = c / n$. The value of the function is simply the number of 1 bits, c, in the input chromosome x divided by the total number of bits, n. This function presents no difficulty, since one hill-climb from anywhere in the space will find the maximum value [8].

The results in figure 4 prove this to be true. Figure 4 shows the median test run (out of 11) for the SGA to evolve the correct values for chromosomes of sizes of 5, 10, 20, and 40. The population sizes used by the SGA are a linear function of the size of the chromosome, defined as follows: PopulationSize(x) = 2 * size of x. As the graph indicates, the SGA has no problem in finding the best gene setting and quickly converges to the correct solution for varying chromosome sizes. This result is to be expected and establishes a baseline for the remaining GA test runs using different chromosome representations. However, remember that the SGA did not have to establish the gene ordering, because that was done a priori.

2.2 Random Sequential Genetic Algorithm

The random GA's (RGA) chromosome is organized, like the SGA, as a linked list of genes. However, genes are randomly assigned to a location in the linked list. Therefore, gene number one could be the last element in the list, gene number two could be the middle element in the list, and so on. As the genes are not in the correct order, e.g. each gene's preceding gene (or parent gene) number may not be smaller than its own, the RGA must determine the correct gene ordering and values.

Like the SGA, the RGA uses the two-point crossover operator. However, this causes a problem for the RGA. For example, given the following two chromosomes, A and B (with genes encoded as tuples: [value:gene-number], e.g. 0:3 means gene 3 has value 0), and crossover points 2 and 5, the application of the two-point crossover produces two chromosomes A' and B' as follows:

A:	1:1	1:3	1:5	1:7	1:9	1:2
	1:4	1:6	1:8	1:10		
B:	0:7	0:6	0:5	0:1	0:9	0:3
	0:10	0: 4	0:8	0:2		
A':	1:1	0:6	0:5	0:1	0:9	1:2
	1:4	1:6	1:8	1:10		
B':	0:7	1:3	1:5	1:7	1:9	0:3
	0:10	0: 4	0:8	0:2		

A good feature of this operator for the RGA is that its application causes a change in the gene orderings of the resulting chromosomes. Since the chromosome's gene ordering is randomly initialized, and only 1 in $n!$ gene orderings is correct, the RGA requires an operator that will alter the gene ordering. However, the operator's problematic feature is that its application results in both a duplication and loss of genes in the resulting chromosomes. For example, chromosome A' has genes 1 and 6 present twice, and for both values of 0 and 1. Also, genes 3 and 7 are missing from it. Without some other feature of the RGA to correct these problems, the RGA seems doomed to fail. This was in fact the end result. None of our test runs ever converged to the correct answer; therefore, these results have been omitted from figure 4. The main problem was that after a few generations, the population was filled with chromosomes with missing genes. Since mutation only changed the genes values and not the gene numbers, the missing genes could not return to the population. After this point, there was no way that the RGA could solve the problem.

Since it was next to impossible for the RGA to produce the correct chromosome using two-point crossover and mutation - there was always a remote chance that the RGA could do it, but it never happened in our experiments - the only other way for the RGA to produce it was to stumble upon it in the very first generation. The probability of this happening is $1 / 2^n * n!$, because there are $n!$ different gene orderings, each with 2^n different allele settings. We conclude that the RGA is not a successful approach for coevolving the organization and values of a chromosome.

2.3 Tree Genetic Algorithm

The tree GA's (TGA) chromosome is organized as a rooted tree of genes. Like the RGA, genes are randomly placed in the tree, organized initially as a binary tree. Therefore, gene number one could be a leaf node, gene number two could be an internal node, the last gene could be the root node, and so on. As the genes are not in the correct order, e.g. each gene's parent gene number may not be smaller than its own, the TGA must determine the correct gene ordering and values.

Although initially constructed as binary tree, the tree will become a n-ary tree as soon as crossover is applied. The TGA uses a new crossover operator called the Tree Crossover Operator (TreCoP). The TreCoP creates one new tree from two chromosomes. First a node (gene) is selected in the original tree. The selection process is weighted to select its root node with probability 0.5, and all other nodes with equal probability. Next, a node in the mate's tree is selected. This selection process is weighted to select a child node of its root with probability 0.5, and all other nodes with equal probability. Next, the original's subtree is replaced by the mate's subtree, unless the original subtree selected is the entire tree, in which case the mate's subtree becomes the new rooted tree. Then, because crossover may cause genes to be duplicated or missing from the resulting tree, the duplicate genes are removed from it and the missing ones added.

The reason for using the weighted selection functions is simple. They result in a constant pulling of nodes towards the root of the tree, which is desirable for reorganizing the nodes in the tree. Since the original tree's selected subtree is often the root (at least 50% of the time), the mate tree's selected subtree will replace it. Since the mate's subtree is often a child of the root (at least 50% of the time), and since the trees tend to be similar as the population converges to a solution, this results in elevating a child to the status of the root of the tree.

What is also desirable about this operator is that the tree it produces closely resembles the two parent trees. To do this, care must be taken when removing duplicate genes and restoring missing ones. The procedure of removing duplicate genes is done in a depth first manner. Once a duplicate gene is detected, it is removed from the tree and its children added as the children of the removed gene's parent. For example, removing gene 2 from the left tree of figure 2 results in the right tree of the figure.

If the gene being removed is at the root of the tree, one of its children is selected at random to become the new root. The remaining children of the old root are then added as the children of the new root. For example, removing gene 1 from the left tree of figure 3 results in the right tree of the figure.

Missing genes are added into the new tree in the same location as they were located in the original tree*. For example, if gene 2 was a child of gene 3 in the original tree, and it is missing from the resulting tree, it is added as a child of gene 3 in the resulting tree. If this is not possible because the parent gene - gene 3 - is not present in the resulting tree, the gene is randomly added to one of the leaf nodes. By ordering the missing genes in the way they are visited by a breadth first search of the original tree, this problem is mostly avoided. The reason is that if two dependent genes are missing from the resulting tree, the parent gene is always added first, ensuring that the child can always be attached to it and henceforth in its proper location.

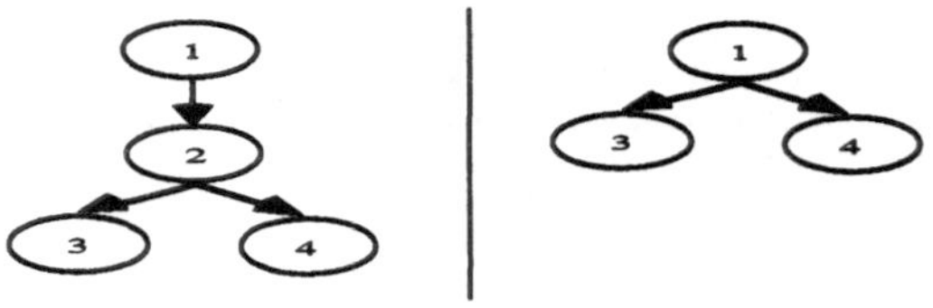

Figure 2 Removing a Gene

* Only those genes missing from the original tree, not the mate's tree, are added.

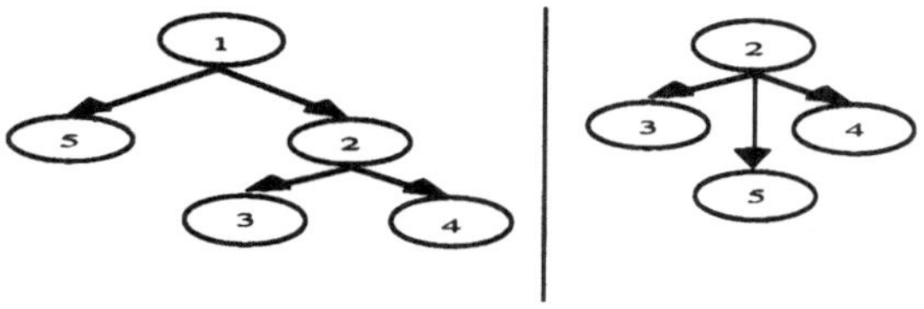

Figure 3 Removing the Root Gene

The TreCoP and its two supporting procedures for adding missing genes and removing duplicate ones attempt to maintain the same spirit of organization in the resulting chromosome as was found in the parent chromosomes, while also managing to change and reorganize the genes into a combination of the two. Since a TGA's chromosome ordering is randomly initialized, it is important that the TreCoP search for good gene orderings. However, it must not mix the existing chromosomes so much as to lose any information that has been gained from past explorations. The operator must simultaneously exploit old gene orderings while exploring new ones - the hallmark of GAs.

Does the TGA work? Does it locate the best gene ordering and values? The answer is yes. Figure 4 again shows the median test run (out of 11) for the TGA to coevolve the correct gene ordering and values for chromosomes of size 5, 10, 20, and 40. Although not as fast as the SGA, the TGA has no problem converging to a solution for varying chromosome sizes. The extra time required for the TGA to converge to a result, compared to the SGA, was to be expected. It must coevolve the organization and values of the genes. The SGA only evolves the values.

To get a rough estimate on the number of different trees in the TGA search space, consider the case of oriented trees. Oriented trees are not treated as different when they differ only in the respective ordering of subtree nodes, which is the case in the TGA's chromosome. For oriented trees with n labeled vertices (genes), there are n^{n-1} distinct oriented trees [9]. In addition to this space, the TGA searches the space of 2^n different gene values. Therefore, the TGA searches a space of size approximately: $2^n * n^{n-1}$.

Even with this enormous search space, the TGA works well, and is our first viable approach to coevolving a chromosome's organization and genes values. Can we do better?

2.4 Subtree Genetic Algorithm

The subtree GA's (STGA) chromosome is organized, similar to the TGA, as a rooted tree of genes. Unlike the TGA, the STGA's chromosome does not require the presence of all genes, and is usually missing some of them. Such a chromosome is known as a chromosomal subtree. However, a STGA chromosome ensures that no gene is duplicated in a subtree.

The chromosome is subjected to a new crossover operator called the SubTree Crossover Operator (STreCoP). The code for the STreCoP is identical to that of the TreCoP, but its behavior is slightly different in one respect. As in the TreCoP, missing genes are added to the resulting tree in the same location they were located in the original tree. However, this procedure adds only those genes that are in the original tree, not the mate's tree, and missing from the resulting tree. This feature makes little difference in the TreCoP, since the original tree has all of its genes, and so must its offspring. However, since now the original tree may not have all of its genes, those missing in it will also be missing in resulting tree, unless some of them were crossed over into the resulting tree from the mate's tree.

A minor difference between the TGA and the STGA is in the way the STGA measures fitness. Since STGA will have both complete and incomplete chromosomes of different sizes, a method is required to establish the fitness of each organization relative to the others.

The fitness of a chromosome is initially calculated as described at the beginning of section 2: the fitness of its genes divided by two times the current number of genes* in the chromosome. This method guarantees that all chromosomes have a fitness value between 0 and 1. Next, the fittest chromosome is selected from the STGA's population and its size - the number of nodes - is used to rank the remaining chromosomes relative to it. It is not the case that the best chromosome is always the largest. For example, a two node chromosome with both genes set to 1 and arranged in the correct order will have a fitness of 1. A ten node chromosome with all genes set to zero and in the wrong order will have a fitness of 0. Using the fittest chromosome's size ('relativeSize' below), we calculate the fitness of all other chromosomes as follows:

```
RelativeFitness: relativeSize
    "self refers to the chromosome "
    (self size > relativeSize)
    ifTrue: [
        "If my chromosome size is greater than the best chromo
    some's size, then decrease my fitness by an amount proportional
    to the ratio of the sizes"
        ^ fitness := self fitness * relativeSize / self size]
    ifFalse: [
            "If my chromosome size is less than the best chromo
    some's size, then decrease my fitness by an amount proportional
    to the ratio of the sizes"
            ^ fitness := self fitness * self size / relativeSize]
```

The fitness of a chromosome - referred to as 'self' - is changed proportionally to the ratio of its size and the size of the fittest chromosome. Chromosomes that have more or less nodes than the best chromosome will see their fitness values decreased. Chromosomes that have the same number of nodes will see no change in their fitness. The effect of this fitness function is to keep the STGA working at one level - size of chromosome - until good subtrees - building blocks - are produced. These subtrees are then the material for the construction of the next level. Once a level is established, i.e. the chromosomes at that level have superior fitness, the STGA's effort is concentrated there, developing the subtrees for the next level, and so on.

Does the STGA work? Does it locate the best gene ordering and values? The answer is obviously yes. The surprising result is that it works faster than the TGA. Figure 4 again shows the median test run (out of 11) for the STGA to coevolve the correct gene ordering and values for chromosomes of sizes 5, 10, 20, and 40. Although not as fast the SGA, the STGA has no problem in finding the best gene setting and quickly converges to a solution for varying chromosome sizes. The extra time required for the STGA to converge to a result is to be expected. It must coevolve both the organization and values of the genes. The SGA only evolves the values.

* In the TGA the number of genes in the chromosome is always the same. However, in the STGA this number may vary.

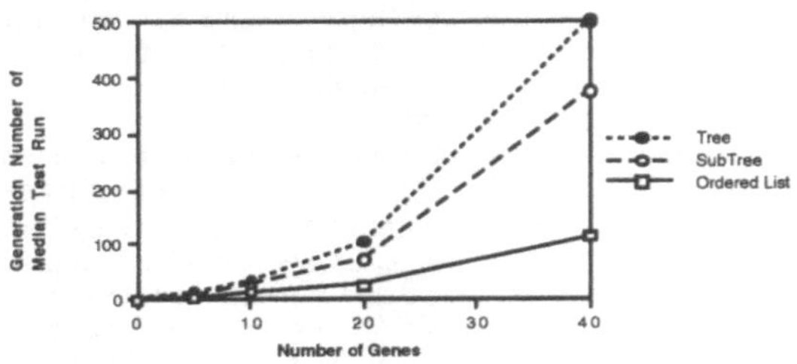

Figure 4 Gene Order Dependent GAs

3 Conclusion

Can one coevolve the organization of a chromosome, its components, and their contents? The answer is yes. Both the TGA and the STGA manage to do both. These approaches are not as fast as the SGA, but then the SGA already has the correct gene ordering which the TGA and the SGA must discover for themselves.

Exactly what is meant by a part of a chromosome? Both the TGA and the STGA use a chromosome structured as a rooted tree. Therefore, a part of their chromosome is nothing more than subtree.

A subtree is analogous to a GA's building block: a compact way to express similarities among chromosomes. However, rather than imposing a strict ordering on all of the genes in the chromosome, as a schema does, a subtree only imposes an ordering on those genes for which the values are known. For this reason, the number of complete chromosomes represented by a subtree is greater than that represented by a simple schema. Take for example the simple case of a schema of order n-1, with only one unknown gene value, e.g. 10*. This schema represents only two strings: 101, and 100. When the genes are positioned as nodes in an n-ary tree, the missing gene can be attached to any one of the existing n-1 nodes. Therefore, if we are using binary genes, the one subtree represents $2(n-1)$ different possible trees.

The difference between the number of actual strings represented by a schema and by a subtree increases as the number of missing genes increases. Therefore, if the power of a GA's search is related to the number of schemata in a population, then the use of subtrees, or even trees, in a strongly epistatic environment, should enhance a GA's performance.

How does one measure the fitness of a subtree? As with a tree, a subtree's fitness is calculated as a function of the gene values. If this function is dependent on the ordering of the genes, and unknown a priori, the TGA and STGA are two approaches that will discover the ordering.

How are a chromosome and its parts processed simultaneously? The answer is to first establish a method of determining the fitness of both complete and incomplete chromosomes, and then to rank each (sub)tree's fitness relative to the size of the fittest known chromosome, be it a whole or a part. When two chromosomes have the same fitness, but different sizes, the fittest chromosome is defined as the larger of the two. In this manner, parts are assembled into larger parts, and the process is repeated until the desired chromosome is constructed. By ranking all chromosomes and their parts relative to one another, they coevolve in the same population.

What should the structure of the chromosome be? When the gene dependencies are known a priori, a simple list provides the best structure, as demonstrated by the SGA. However, when the gene dependencies are unknown and a list structure is used, as in the RGA, the approach fails. The structure to use in the latter case is a tree, as demonstrated by the TGA; a tree can be used to coevolve both the organization and content of the chromosome. However, if fitness values can be determined for the parts of a tree, the best structure for a chromosome is a subtree that grows into complete trees, as demonstrated by the STGA.

Is there any benefit to evolving the organization of the chromosome? The answer is yes. Since in most complex systems it is an impossible task to determine all gene dependencies a priori, the only way genetic methods will succeed is if these dependencies can be determined by the GA itself. Since GAs are good at searching, it seems an obvious choice for them to evolve their own organization!

4 References

[1] Goldberg, D.E., Genetic Algorithms in Search, Optimization, and Machine Learning, Addison-Wesley, 1989.
[2] Rizki, M.M. & Conrad, M., Computing The Theory of Evolution, Physica D, 22, 83-89, 1986.
[3] Calloway, D.L., Using a Genetic Algorithm to Design Binary Phase-Only Filters for Pattern Recognition, Proceedings of the fourth International Conference on Genetic Algorithms, Morgan Kaufmann, 422-427, 1991.
[4] Nakano, R., Conventional Genetic Algorithm for Job Shop Problem, Proceedings of the fourth International Conference on Genetic Algorithms, Morgan Kaufmann, 474-479, 1991.
[5] Grefenstette, J.J., A System for Learning Control Strategies with Genetic Algorithms, Proceedings of the third International Conference on Genetic Algorithms, Morgan Kaufmann, 183-190, 1989.
[6] Gabbert, P.S., Brown, D.E., Huntley, C.L., Markowicz, B.P., and Sappington, D.E., A System for Learning Routes and Schedules with Genetic Algorithms, Proceedings of the fourth International Conference on Genetic Algorithms, Morgan Kaufmann, 430-436, 1991.
[7] Syswerda, G., Uniform Crossover in Genetic Algorithms, Proceedings of the Third International Conference on Genetic Algorithms, Morgan Kaufmann, 2-9, 1989.
[8] Ackley, D.H., An Empirical Study of Bit Vector Function Optimization. In Lawrence Davis (Ed.), Genetic Algorithms and Simulated Annealing, Morgan Kaufmann, 170-204, 1987.
[9] Knuth, D.E., The Art of Computer Programming: Fundamental Algorithms. V:1, Addison-Wesley Publishing Company, 391, 1973.

DOMINANT TAKEOVER REGIMES FOR GENETIC ALGORITHMS

David Noever
Biophysics Branch, ES76
National Aeronautics and Space Administration
George C. Marshall Space Flight Center
Huntsville, AL 35812 USA

Subbiah Baskaran
Institut fuer Molekulare Biotechnologie, e.V.
Beutenbergerstr. 11, DO-7745, Jena, Germany
Biophysics Branch, ES76, National Aeronautics and Space Administration, George C. Marshall Space Flight Center
Huntsville, AL 35812 USA

The genetic algorithm (GA) is a machine-based optimization routine which connects evolutionary learning to natural genetic laws [1–2]. The present work addresses the problem of obtaining the dominant takeover regimes in the GA dynamics. Estimated GA run times are computed for slow and fast convergence in the limits of high and low fitness ratios. Using Euler's device for obtaining partial sums in closed forms, the result relaxes the previously held requirements for long time limits. Analytical solutions reveal that appropriately accelerated regimes can mark the ascendancy of the most fit solution. In virtually all cases, the weak (logarithmic) dependence of convergence time on problem size demonstrates the potential for the GA to solve large N–P complete problems.

A central issue in how the GA processes strings (solution encoded genomes) is takeover times for the most fit genome [2]. Takeover times here refers to how the computational time or complexity scales with larger problem sizes. Short takeover times correspond to rapid convergence onto the most fit individual. Conversely, long takeover times correspond to slow convergence and sluggish dynamics. A historical discussion of computational complexity appears in reference 3.

Here we solve the standard GA models for generational reproduction [4], evaluate the takeover times, and compare the results using various approximation techniques. The approximations take place only in time and fitness space, such that the exact results can be compared directly. The aim is to identify and understand why the GA sorts through some regions of the fitness landscape so efficiently but gets bogged down in other regions.

For discrete time steps between generations, $t=\{0,1,2...\}$, let $P_{i,t}$ correspond to the proportion of alleles (or bit string values) set to the value 1 for a particular allele position i at generation t. Let $P_{i,o}$ represent P for the founder generation, $t=0$. For simplicity, all subsequent work will treat binary genetic algorithms which have alleles possessing either of two values, 0 or 1; the multivalue case is a trivial generalization at this level of proof. Let f_1 correspond to the organism's fitness (some survival probability) sampled with allele value 1 in a particular position j. Likewise, take f_o to represent the fitness of all organisms sampled with allele value 0 in position j. For any defined fitness ratio, $r=f_1/f_o$, the value will be considered time-independent and constant across generations.

Thus to begin, assume generational reproduction on a binary fitness landscape (f_o,f_1) with fitness ratio, $r=f_1/f_o$. Previous results have established an equivalence condition to match generational and steady-state reproduction, so we consider the following derivations to develop with some parallel applicability for steady-state GAs (e.g. Genitor, etc.). In the binary GA [5], generational reproduction implies that

$$P_{t+1} = (f_1 / S_t) P_t \tag{1}$$

where S_t defines the average population fitness, $S_t=f_1P_t+(1-P_t)f_o$.

An iterated recursion relation is complicated by the appearance of P_t in the average population fitness, S_t. Here $S_t=P_tf_1+(1-P_t)f_o$ is the total average population fitness. Physically the last term represents the copying of one individual with a reproductive rate, f_1/S_t.

The recursion relation governing inverse population growth is $(X=1/P)$[5]:

$$X_t = 1-\frac{1}{r}+\frac{X_{t-1}}{r} = A+BX_{t-1} \ . \tag{2}$$

Iterating and solving as a function of the fitness ratio and time gives:

$$X_t = A\sum_{n=0}^{t} B^n + B^t X_o = \frac{(r-1)}{r}\sum_{n=0}^{t}\left(\frac{1}{r}\right)^n + \frac{X_o}{r^t} \ . \tag{3}$$

Equation (3) can be inverted to solve for population dynamics, P_t:

$$P_t = \frac{P_o r^t}{1+P_o\Gamma_t r^{t-1}(r-1)} \tag{4}$$

where the term Γ_t is the binomial series

$$\Gamma_t = \sum_{n=0}^{t}\left(\frac{1}{r}\right)^n \ . \tag{5}$$

The aim of this analysis is to solve for closed form results in the limits of fitness space for long and short times. The approximate limits will parameterize the GA convergence for arbitrary fitness values.

The Case of Long Times ($t\to\infty$) and Fitness Ratios $r>1$

For an infinite series summed with large fitness ratios, then for $(1/r)<1$, the series (5) converges according to

$$\sum_{n=0}^{\infty} x^n = 1+x+x^2+....=(1-x)^{-1} \tag{6}$$

or

$$\Gamma_t = (1-\frac{1}{r})^{-1} = \frac{r}{r-1} \ . \tag{7}$$

Thus, in the limit of long times and fitness ratios greater than unity, the population changes according to the dynamical equation:

$$P_t = \frac{P_o r^t}{1+P_o r^t} \ . \tag{8}$$

Equation (8) can be solved for how the computational complexity varies with the population size, n. Two cases are examined, the worst and average complexities which in turn depend on the expected frequency of the final solution appearing in the initial population, P_o. In the average case, the initial population has a random frequency for the best solution, thus $P_o=0.5$. Alternatively in the worst case, the initial population has a minimum frequency for the best solution, thus $P_o=1/n$. Near convergence the final population approaches unity for all cases, thus, $P_f=1-(1/n)=(n-1)/n$. With this dependence on population size, it is possible to solve directly for the takeover time (or computational complexity) of the most fit member

$$t_c = \frac{\ln\left[P_f / P_o(1-P_f)\right]}{\ln(r)} \ . \tag{9}$$

For worst case convergence times, then

$$t_c = \frac{\ln[n(n-1)]}{\ln(r)} \sim O[\ln(n)] \ . \quad (10)$$

Similarly for average convergence times, then

$$t_c = \frac{\ln[2(n-1)]}{\ln(r)} \sim O[\ln(n)] \ . \quad (11)$$

Both cases for takeover times (10–11) demonstrate the same weak logarithmic dependence on population size.

The Case of Both Arbitrary Times and Fitness Ratios

For a finite series summed with arbitrary fitness ratios, then the series (5) converges according to

$$\sum_{n=0}^{t} x^n = 1 + x + x^2 + ... = \frac{(1-x^t)}{(1-x)} \quad (12)$$

or

$$\Gamma_t = \frac{1-(1/r)^t}{1-(1/r)} = \frac{1}{r^{t-1}}\left[\frac{r^t-1}{r-1}\right] \ . \quad (13)$$

Thus, for finite times and arbitrary fitness ratios, the population changes according to the dynamical equation:

$$P_t = \frac{P_o r^t}{1+P_o r^{t-1}(r-1)\Gamma_t} = \frac{P_o r^t}{(1-P_o)+P_o r^t} \ . \quad (14)$$

This agrees with the result derived independently by Deb and Goldberg [6] and Ankenbrandt [4].

Equation (5) can be solved for the computational complexity with the population size, n and fitness ratio, r.

$$t_c = \frac{\ln\left[\frac{P_f(1-P_o)}{P_o(1-P_f)}\right]}{\ln(r)} \ . \quad (15)$$

Again, two cases are examined, the worst and average complexities. For worst case convergence times, then

$$t_c = \frac{2\ln(n-1)}{\ln(r)} \sim O[\ln(n)] \ . \quad (16)$$

Similarly for average convergence times, then

$$t_c = \frac{\ln(n-1)}{\ln(r)} \sim O[\ln(n)] \ . \quad (17)$$

Both cases for takeover times demonstrate the same weak logarithmic dependence on population size.

The Case of Long Times (t-->∞) and Low Fitness Ratios (r<1)

For an infinite series summed with low fitness ratios, then the series (10) converges according to

$$\Gamma_t = \frac{1-(1/r)^t}{1-(1/r)} = \frac{1}{r^{t-1}}\left[\frac{r^t-1}{r-1}\right] = \lim_{t\to\infty}\frac{r^t-1}{r-1}$$
$$= (1-r)^{-1} \text{for } r<1 \ . \quad (18)$$

Thus, for infinite times and low fitness ratios, the population changes according to the dynamical equation:

$$P_t = \frac{P_o r^t}{1+P_o r^{t-1}(r-1)\Gamma_t} = \frac{P_o r^t}{1-P_o} \ . \quad (19)$$

Equation (5) can be solved for the computational complexity with the population size, n.

$$t_c = \frac{\ln\left[P_f(1-P_o)/P_o\right]}{\ln(r)} \ . \quad (20)$$

Again, two cases are examined, the worst and average complexities. For worst case convergence times, then

$$t_c = \frac{\ln\left[(n-1)^2/n\right]}{\ln(r)} \ . \quad (21)$$

For large populations, n-->∞, then the logarithmic numerator goes to the limit, $\ln[n-2+(1/n)] = \ln(n-2)$. For small populations, n-->0, then the logarithmic numerator goes to the limit, $\ln[n-2+(1/n)] = \ln[(1/n)-2]$.

Similarly for average convergence times, then

$$t_c = \frac{\ln\left[(n-1)/n\right]}{\ln(r)} \ . \quad (22)$$

For large populations, n-->∞, then the logarithmic numerator goes to the limit, $\ln[1+(1/n)] = 0$.

The Case of Finite Times and Arbitrary Fitness Ratios r

For a finite series summed with arbitrary fitness ratios, then the series (5) converges according to an ingenious tool sometimes called Euler's device [7]. For this case, the series is rewritten for $z=(-1/r)$ and the geometric series can be summed for any intermediate time (in powers of $p=2^q$ $q=1,2,3...$)

$$\sum_{n=0}^{t} z^n = 1 - z + z^2 - z^3 + ... \quad (23)$$

$$\sum_{n=0}^{t} z^n = 1 - z + z^2 - z^3 + ... \ . \quad (24)$$

To evaluate any intermediate sum, Γ_{2p-1}, Euler's device gives:

$$\Gamma_1 = (1-z) = 1+1/r; \Gamma_{2p-1} = \left(1+z^p\right)\Gamma_{p-1} = \left[1+(1/r)^p\right]\Gamma_{p-1} \ . \quad (25)$$

This case represents the main exposition of this paper. It enables one to inspect the GA dynamics in intermediate time intervals without a lack of analytical generality. For any intermediate generation, the series can be summed in terms of previously known terms. As an example, consider the series up to generation 15:

$$\Gamma_3 = \left(1+1/r^2\right)\Gamma_1; \Gamma_7 = \left(1+1/r^4\right)\Gamma_3$$
$$= 1+1/r+1/r^2+...+1/r^7; \Gamma_{15} = \left(1+1/r^8\right)\Gamma_7 \ . \quad (26)$$

Thus, in the limit of any finite time and arbitrary fitness ratios, the population changes according to the dynamical equation:

$$P_{2p-1} = \frac{P_o r^{2p-1}}{1+P_o\left(1+1/r^p\right)\Gamma_{p-1} r^{2p-2}(r-1)} \ . \quad (27)$$

Fig. 1 shows the stepped amplification for discrete iterations using Euler's device on the series sum and the population.

The principal value of the partial summation formalism is to allow intermediate stages of the growth cycle to be expressed in closed form without resorting to any assumptions of long times or constraints on the fitness ratio. The appeal is that programmed changes in population (injection or withdrawal of strings) can be undertaken without loss of analytic versatility. Additionally, the time complexity for regions of rapid convergence (following such injection or withdrawal) can be monitored without resetting the population balance with a new fitness ratio. Finally, the appeal of a 2^q representation space for population changes automatically suggests a simple mapping onto the hypercube of available search space. The generation steps thus naturally can be fitted to vertices of an ever enlarging (hypercube) search space.

Not only does (27) give the partial summation in a closed (polynomial) form, but also allows for the immediate production of interval summations, e.g. the growth that occurred between generations 7 and 15 (or $p-1$ to $2p-1$). Rewriting Euler's device in terms of q for powers of $p=2^q$ gives:

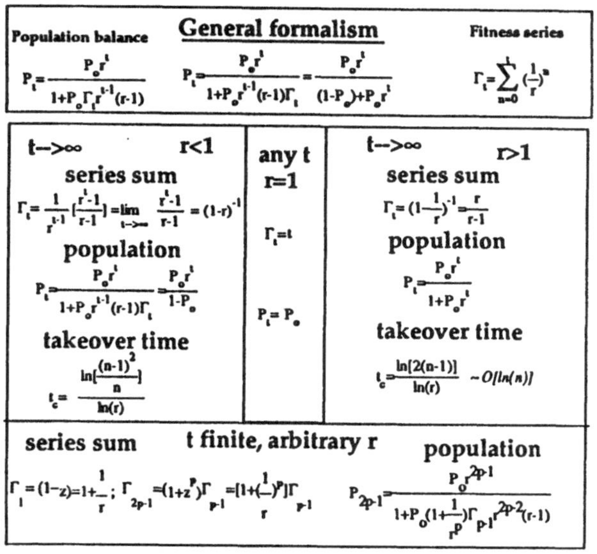

Fig. 1. Stepped population growth and series sum using Euler's device to obtain partial summations without constraining the fitness space or simulation time. The bars highlight the points of partial summation on a binary landscape (t=2p–1 where p=2^q).

$$q:\Gamma_{2^{q+1}-1}=\left(1+z^{2^q}\right)\Gamma_{2^q-1} \tag{28}$$

$$q-1:\Gamma_{2^{q-1}-1}=\left(1+z^{2^{q-1}}\right)\Gamma_{2^{q-1}-1} \ . \tag{29}$$

By subtracting (29) from (28) and rewriting in terms of p gives the interval summation formula:

$$\Gamma_{2p-1}-\Gamma_{p-1}=\Gamma_{2p-1}-\left(1+z^{p/2}\right)\Gamma_{p/2-1} \ . \tag{30}$$

For example, the population growth between generations 7 and 15 can be written in exact closed form as:

$$\Gamma_{15}-\Gamma_7=\Gamma_{15}-\left(1+1/r^4\right)\Gamma_3 \ . \tag{31}$$

As shown schematically in Fig. 2, a complete interval analysis becomes possible for subsets of the partial summation. By breaking the population growth curve into discrete intervals, a useful formalism evolves for addressing intermediate changes in GA dynamics

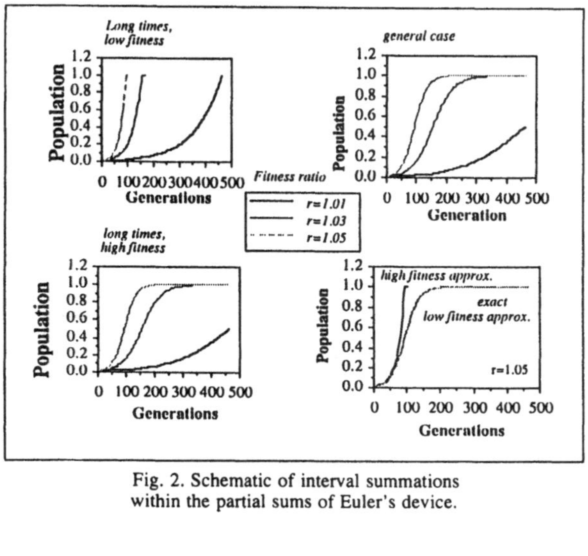

Fig. 2. Schematic of interval summations within the partial sums of Euler's device.

(e.g. withdrawal or addition of strings, changes in fitness). Thus the objective of tracking regions of accelerated convergence is simplified without loss of mathematical generality.

The Case of Arbitrary Times (t-->∞) and Fitness Ratios r=1

For completeness sake, consider the trivial case of equal fitness, r=1. When the two solutions have indistinguishable performance, the population should remain fixed on average at the initial value, P_o. As a check on the previous formalism, such a result does indeed follow when the infinite series sums for equal fitness, then for $(1/r)$=1 the series (5) converges according to

$$\sum_{n=0}^{t} x^n = 1+x+x^2+\ldots=t \tag{32}$$

or

$$\Gamma_t = 1+1+1+\ldots+1=t \ . \tag{33}$$

Thus, in the limit of arbitrary times and equal fitness ratios, the population remains unchanged according to the dynamical equation:

$$P_t=\frac{P_o r^t}{1+P_o t\, r^{t-1}(r-1)}=P_o \ . \tag{34}$$

A somewhat shorter derivation follows from observing that $(r-1)$=0 in this case, thus for any finite number of terms in the summation, the dependence on time should vanish entirely from the denominator of (30).

Summary of Results

A summary of the various approximations is shown in Fig. 3. The graphical difference between exact and approximate population dynamics is highlighted in Fig. 4. The time complexities calculated in the various approximate limits are compared in Fig. 5. In general the weak (logarthmic) dependence of the GA processing on problem size gives it great potential compared to other sorting and search routines (Fig. 6).

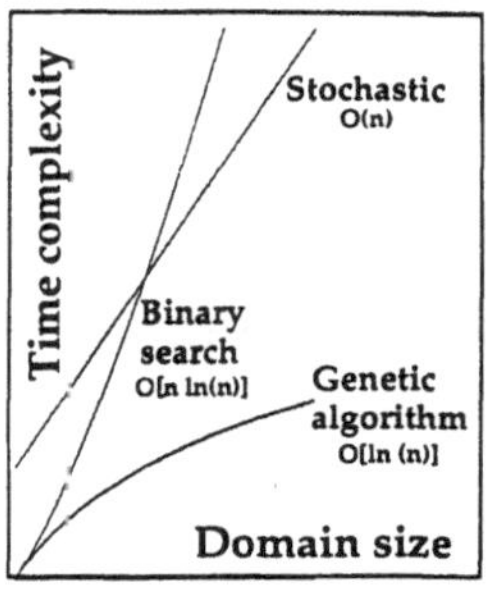

Fig. 3. Phase diagram for series summation in approximate limits of short and long times and arbitrary fitness values.

Much work has been published on the compromises struck between rapid GA convergence and processing parameters (e.g. reference 6). A preliminary discussion here presents the fundamental issue in terms of accelerated regions of GA convergence. In general, GA populations move toward optimum following an S-curve or logistic growth. The central steep ascendency of the S-curve corresponds to highly efficient GA processing. In this regime, genetic mixing improves overall performance, such that neither good strings get heavily disrupted (strong crossover) nor do bad strings get unnecessarily preserved (low selection). Thus GA efficiency can match with the genetic terms of a balanced tradeoff struck between diversity and selectivity. Over time, this tradeoff can be imagined schematically to combine two composite operators which evolve according to Fig. 7. For the same operations mapped into time and fitness space, neighborhoods of high diversity correspond to high fitness ratios, while neighborhoods of strong selectivity correspond to high fitness ratios. In this translation, regimes of accelerated convergence can be identified directly with their approximate time complexities (Fig. 3).

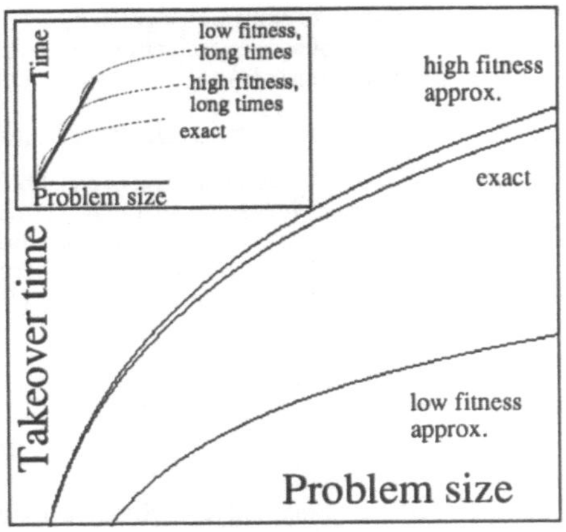

Fig. 4. Summary of low and high fitness approximations and their effects on population growth. Lower right shows the comparison between exact and approximate solutions. Note that the exact and high fitness regimes essentially overlap.

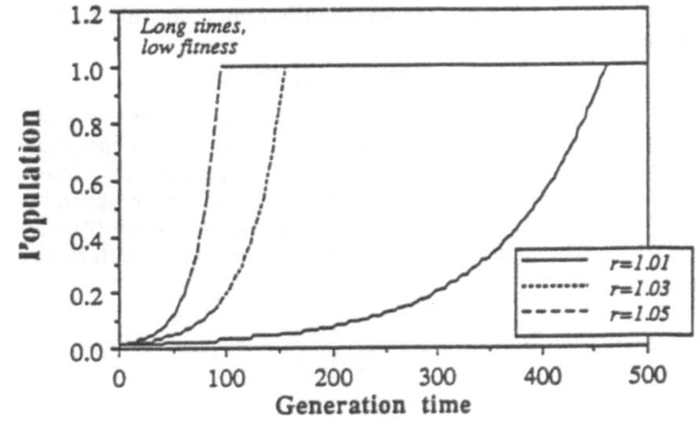

Fig. 5. Comparative time complexities (takeover times) for the approximate limits. Short times correspond to regions of accelerated convergence.

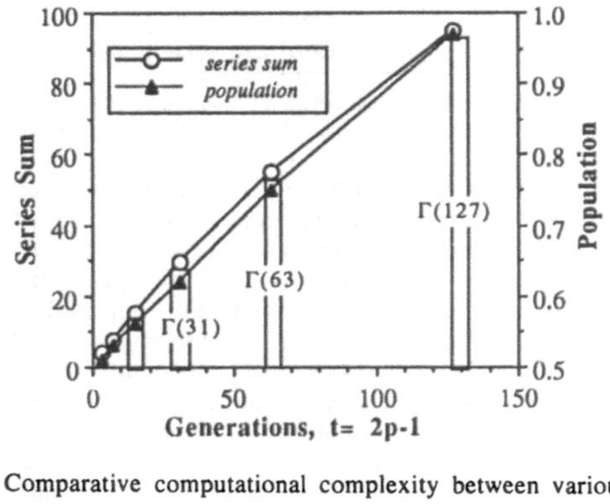

Fig. 6. Comparative computational complexity between various sort routines.

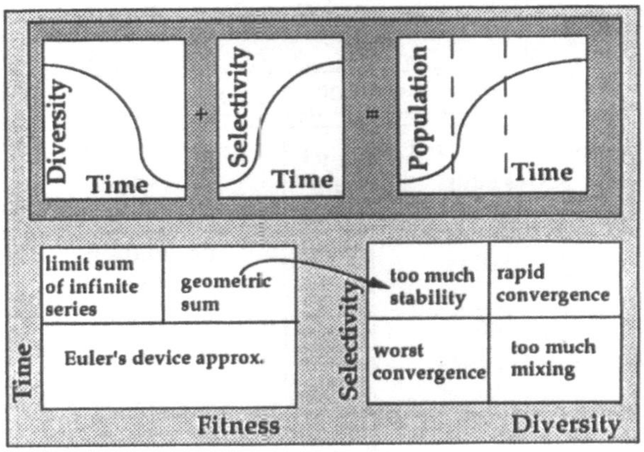

Fig. 7. Diversity-selectivity profiles for GA convergence.

To summarize, the work has analyzed the population dynamics of a general GA in fitness space. The iterated population yields recursion relations and a convenient formalism is developed for the approximate cases of long and short times as well as high and low fitness values. The validity of these closed forms is compared to the exact result. By using the elegant tool of partial summation (Euler's device) then the analytical basis of the GA is prepared to handle discontinuous changes in parameters without loss of mathematical generality. For example, the introduction or removal of strings can be safely accounted for by using the discrete summation formalism in a particularly transparent way. An additional advantage in using the partial series summation (Fig. 1) is to relax the requirements for long (infinite) time series. In practice, most GAs run no longer than a few decades of generational time steps. Future work will examine the potential for abrupt population changes or more dynamic fitness landscapes (time dependent optimizations) in a focused effort to understanding GA processing and convergence.

References

1. Holland, J.H.: Adaptation in Natural and Artificial Systems. Ann Arbor, Michigan: University of Michigan Press 1975.
2. Goldberg, D.E.: Genetic Algorithms in Search, Optimization and Machine Learning. Reading, Massachusetts: Addison Wesley 1989.
3. Cook, S.A.: Comm. of ACM 26, 401 (1983).
4. Ankenbrandt, C.: (1991) "Time Complexity and Convergence of Genetic Algorithms".Rawlins, G.J.E. (ed.): Foundations of Genetic Algorithms. San Mateo, California: Morgan Kaufmann Publishers 1991.
5. Noever D. and Baskaran, S.: "Steady State vs. Generational Genetic Algorithms: A Comparison of Time Complexity and Convergence Properties" Santa Fe Institute preprint series, 92-07-032 (submitted to Machine Learning) (1992).
6. Goldberg, D.E. and Deb, K.: "A Comparative Analysis of Selection Schemes Used in Genetic Algorithms," Rawlins, G.J.E. (ed.): Foundations of Genetic Algorithms. San Mateo, California: Morgan Kaufmann Publishers 1991.
7. Broucke, R.A.: Comm. of the ACM 14, 34 (1971).

INTERACTVE EVOLUTIONARY ALGORITHMS IN DESIGN

Jeanine Graf

Dortmund University, Computer Science Department
44221 Dortmund, Germany

Abstract

Interactive Evolutionary Algortihms use the judgment of a human user as the fitness or objective function in a genetic search. Most approaches to interactive evolution take advantage of the human capability to instantly grasp the value, usefulness or beauty of images. The human fantasy and creativity is supported by the genetic operators of recombination and mutation. Artificial evolution can thus serve as a useful tool for achieving flexibility and complexity in constructive synthesis applications, with a moderate amount of user-input and detailed knowledge. In this paper we investigate the novel idea of using operators related to warping and morphing techniques as evolutionary operators in an image search space. We also present the IDEA system, an interactive X-window tool, a tool for interactive evolution of bitmap images. Our main interest is in the evolution of design pictures, which we examplify with, the design of cars and houses.

1 Introduction

Evolution and design both aims at creating new forms, and we can view biological evolution as a prototype for an approach to design. In the book the Blind Watchmaker, Dawkins [4] demonstrated convincingly the potential of Darwinian variation and selection in graphics. He evolved 2-D graphic objects termed *biomorphs* which were produced from a collection of genetic parameters in interaction with the user. Recently, some research has been directed into the application of genetic algorithms to image and graphics problems, such as the segmentation of range images [12] or pattern identification [8]. Sims [15] used genetic algorithms for interactive generation of color art; Todd and Latham [16] have considered similar ideas to reproduce computer sculptures through structural geometric techniques.

The novel idea offered in this article is to provide a user with new technique to evolve 2-D (bitmap) design images that can be applied to many fields of interest like design of cars, houses, airplanes. Methods that transform one two-dimensional image into another have gain wide acceptance recently. Image processing techniques, have reached common use in the entertainment industry.

2 Simulated Evolution

The best-known representatives of this class of algorithms are genetic algorithms (GAs), developed in the U.S. by J.H. Holland [10], evolution strategies (ESs), developed in Germany by I. Rechenberg [13] and H.-P. Schwefel [14] and evolutionary programming (EP), developed in the U.S. by L.J. Fogel [5]. In all of these cases, the selection criteria are fixed and held constant traditionally from the start of the simulation, therefore these criteria must be detailed explicitly beforehand. This constitutes a significant problem in many realistic applications (apart from optimization), because an explicit fitness function may not be available in closed form. Recently, various work-arounds have been tried, one of the most prominent being *co-evolution*. In this method, rather than use one population to search for the best solution, two or more antagonistic populations are run which compete against each other. The realized fitness in this case is in part determined by the relationship of one population to the other, and does not have to be defined explicitly beforehand [9].

This paper makes use of the alternative method to generate fitness by involving the computer-user into the selection process of artificial evolution. The computer graphics domain is ideal for interactive evolution, because of the unparallelled human visual capacity.

3 Interactive Algorithms

In *interactive evolutionary algorithms*, the user selects one or more favorite model(s) which survive(s) and reproduce(s) (with variation) to constitute a new generation. These techniques can be applied to the production of computer graphics in order to create forms, textures, and motion pictures [1, 6]. Potential applications of interactive evolution include different design problems, e.g. houses, cars, airplanes and consumer products.

3.1 Phenotypes and Genotypes

We shall need to discern between genotypes and phenotypes in interactive evolution. Both of these terms are also basic concepts for biological evolution. The biological genotype is the information that codes for the development of an individual. In interactive evolution, genotypes are represented as numerical data and real values, collections of procedural parameters, symbolic expressions or compound data structures (e.g., trees). The phenotype is the realized behavior of the individual itself, i.e., the product of an interpretation of the underlying genotypic representation. In our case, the phenotype is the resulting graphical image.

In our system the genotype is a bitmap and an underlying structure defining 2-D points in the images which are used by the recombination and mutation operators.

3.2 Fitness and Selection

The term fitness in interactive evolution is the capability of an individual or model to survive into the next generation, and therefore is tied directly to selection. Usually, fitness is not defined explicitly but is instead a relative measure of success following from the selection activity of a human user. Here, it is even based on non-quantifiable measures like visual preference or personal taste. Hybrid systems, however, are reasonable as well. Certain predefined criteria (for example: the physical constraints of the construction) help to sort candidates for survival from the set of all variants, among which a human user finally selects the next generation.

3.3 Variation

In interactive evolution, the true realm of the computer is the automatic generation of variants. Variation is accomplished by defining problem-specific mutation and recombination operators that constantly propose new variants to the presently existing population of graphic models on the screen. A certain amount of knowledge has to be invested in order to find appropriate operators for an application domain. In the next section we shall provide appropriate operators for manipulation of bitmap graphics.

To our knowledge no effort has been made to apply evolutionary algorithms to the evolution of *bitmap* design images directly. Our variation operators are thus partly applied to bitmaps. In the following the emphasis should be on this possibility.

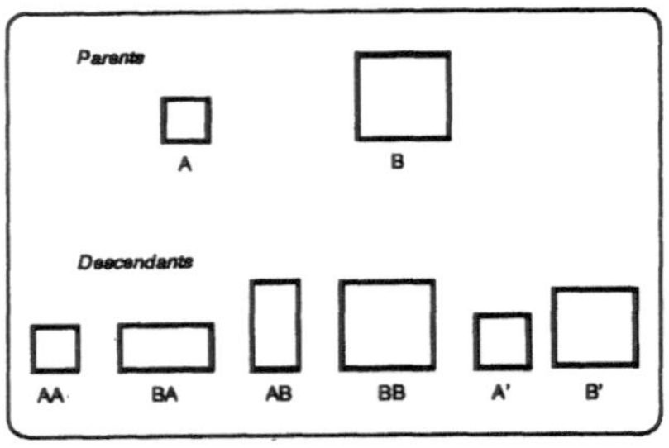

Figure 1: Generation of variants by recombination and mutation of graphical models. AA and BB are the unchanged parents, BA and AB are generated by recombination and A' and B' are generated by mutation

3.4 Bitmap Images

Bitmaps and other forms of direct encoding of images have found an excellent niche in computer graphics, video composition and image rendering [6, 11]. Any 2-D shape can be represented as a sequence of points or vertices, with each vertex consisting of an ordered pair of numbers (x, y), its coordinates. The array of pixel values of a 2-D image, however, has nothing to do with the structure being represented in the image. This constitutes the challenge to finding appropriate operators for the generation of new variants of an existing image, because structural or functional conservation of the image content is of utmost importance in application.

We solve the problem for realizing evolutionary operations by using *warping* and *morphing* to create variations.

Warping is a method which, by using tiepoints in two images it allows for the creation of intermediate images [17]. Basically, these intermediate images are interpolations along an abstract axis from the two pictures. The tiepoints are constraints of the interpolation. Morphing require the transformation of of one 2D image into another 2D image. Variations of these effects have been used to create astonishing special techniques for commercials, music videos, and movies.

We adopt this novel approach for the artificial evolution of design images. By specifying tiepoints, sufficient control can be exerted about structure in 2-D images as to provide useful variants to the images being varied. Whereas in conventional computer graphics morphing is used for the purpose of transforming images in a dynamic animation series, we use the generated intermediate images as variants of the original images in the process of evolution. Mutation is commonly considered to be a local operation that does not radically change the resulting phenotype. In order to provide this feature in bitmap image evolution, we propose to use a very small number of tiepoints. For the variation process a one-point mutation selects one tiepoint in an image. The corresponding tiepoint in a second image would then be used as a source for novelty, by providing information into which direction to evolve the original image. Structure is conserved because tiepoints in both images correspond to each other. A parameter would be used to quantify the degree of substitution in the image.

The first generation images in a way help to form equivalence classes between structures expressed as tiepoints. Some domain knowledge must be used in the process of tiepoint selection for the first generation.

Recombination is implemented as a more global operation by which two images exchange information. We propose to use as many tiepoints as necessary to conserve the underlying structure in two images . A recombination would then be quantified in the image space by a certain parameter indicating the degree of "intermediateness" of a variant. Figures 2 and 3 demonstrate a recombination. Different variants between the original cars and houses are shown.

4 IDEA: A System for Interactive Evolution

IDEA (Interactive Design with Evolutionary Algorithms) is a digital image warping and morphing program, with evolutionary operators, that allows the evolution of designs, and runs under the X Window System [3]. IDEA loads and saves image populations. It provides facilities to store tiepoints in images, to warp images and to apply the evolutionary process. Tiepoints are inherited from generation to generation, with the first generation provided by the user. With a very small population, between 5 and 20 graphical models per generation and over a short time, a human user, for instance an architect, can select new generations of images. This process will be repeated until an acceptable individual in the population has been generated.

5 Summary

We demonstrate how interactive evolution can be applied to 2-D bitmap images and a generalization to 3-D representation is outlined. The main idea is to combine the concepts from interactive evolutionary algorithms with the concepts of warping and morphing from computer graphics. Structure within images is substituted by a collection of tiepoints. By providing a first generation of images, where a structure in the images can manually be defined by the user. Evolution then proceeds along the paths constrained by the set of these tiepoints in all of the images. In our version of *interactive evolution*, the user selects his favourite individual which then is reproduced to constitute the next generation. These techniques can be applied to the design of cars, planes, engineering components and construction projects. Our interactive simulations have shown that interesting results can be achieved even with low population sizes and few generations. This makes the system applicable to quick design and prototyping, in a large variety of application areas. One of the main benefits from this technique is the ability to help the human user to get new ideas and increase his creativity.

Acknowledgement

Funding from the German Bundesministerium fuer Forschung und Technologie (BMFT) under project EVOALG is gratefully acknowledged. Discussions with Peter Nordin have been very helpful.

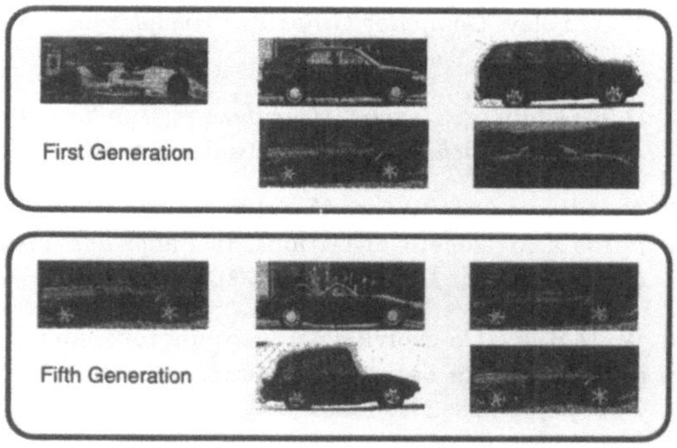

Figure 2: This example shows the evolution of bitmap images. After fourth generation some models are found which are closer to the users taste and to his target model

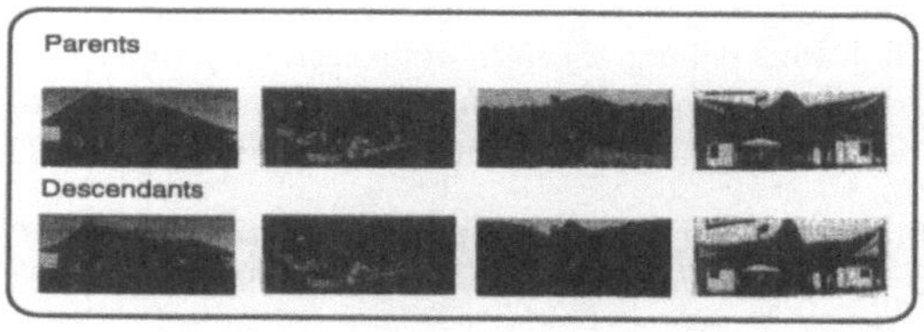

Figure 3: This example shows the evolution of architectural design, between parents and descendants

References

[1] J. Arvo. *Graphics Gems.* Academic Press, Inc, California, 1991.

[2] C. Caldwell and V. Johnston. Tracking a criminal suspect through face-space with a genetic algorithm, 1991.

[3] E. Cutler, D. Gilly, and T. O'Reilly. *The X Window System in a Nutshell.* The Definitive Guides to the X Window System. O'Reilly & Associates, Inc., Sebastopol, CA, sep 1993.

[4] R. Dawkins. *The Blind Watchmaker.* Longman, Harlow, 1986.

[5] L. J. Fogel, A. J. Owens, and M. J. Walsh. *Artificial Intelligence through Simulated Evolution.* Wiley, New York, 1966.

[6] T. A. Foley. *Computer Graphics Principles and Practice.* Addison-Wesley, 1992.

[7] D. E Goldberg. *Genetic Algorithms in search, optimization and machine learning.* Addison-Wesley, 1989.

[8] A. Hill and C.J. Taylor. Model based image interpretation using genetic algorithms. In *Image and Vision Computing, vol.10,* pages 295–300, June 1992.

[9] W. D Hillis. Co-evolving parasites improve simulated evolution as an optimization procedure. In *Physica D42,* pages 228–234, 1990.

[10] J. Holland. *Adaption in Natural and Artificial Systems.* Ann Arbor, The University of Michigan Press, 1975.

[11] D. Kirik. *Graphics Gems III.* Academic Press, Inc, California, 1992.

[12] A. Meygret, D. Levine, and G. Roth. Robust primitive extraction in a range image. In *Conf. on Pattern Recognition, Vol. III,* pages 193–196, 1992.

[13] I. Rechenberg. *Evolutionsstrategie: Optimierung technischer Systeme nach Prinzipien der biologischen Evolution.* Frommann–Holzboog, Stuttgart, 1973.

[14] Hans-Paul Schwefel. *Numerical Optimization of Computer Models.* Wiley, Chichester, 1981.

[15] K. Sims. Artificial evolution for computer graphics. In *Computer Graphics, Vol.25,* pages 319–328, Jul 1991.

[16] S.P. Todd and W. Latham. *Mutator, a Subjective Human Interface for Evolution of Computer Sculptures.* IBM United Kingdom Scientific Center Report, 1991.

[17] G. Woldberg. *Digital Image Warping.* IEEE Computer Society Press, 1990.

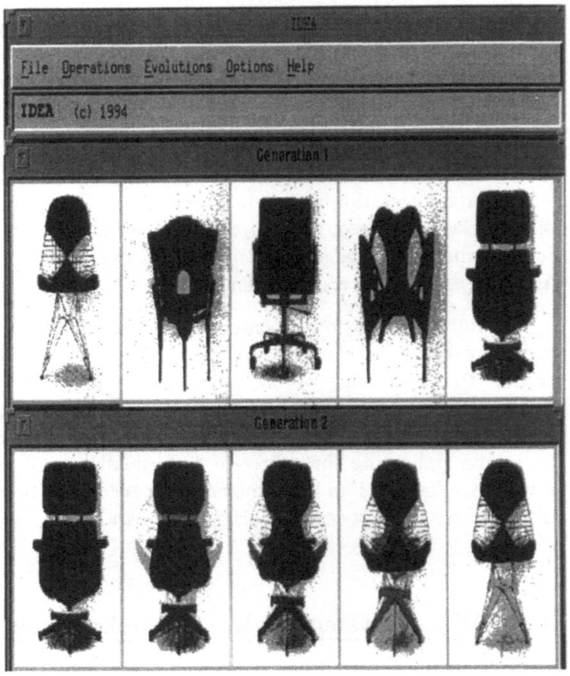

Figure 4: Screen dump of the IDEA system.

DECRYPTING NEURAL NETWORK DATA: A GIS CASE STUDY

R.A. Bustos and **T.D. Gedeon**

School of Computer Science Engineering
University of New South Wales, Sydney NSW 2052, Australia

ABSTRACT

The problem of data encoding for training back-propagation neural networks is well known. The basic principle is to avoid encrypting the underlying structure of the data. This is not easy in the real world, where we often receive data which has been processed by at least one previous user. The data may contain too many instances of some class, and too few instances of other classes. Then topology, and parameters settings designed. Finally, the network produces some results which need to be explained, or decrypted.

We present our experience and results on some satellite data augmented by a terrain model. The task was to predict the forest supra-type based on the available information. In this process we were forced to invent some methods to deal with very large amounts of erratically reliable data, and to produce meaningful predictions at the end.

INTRODUCTION

The choice of a suitable pattern representation is often crucial for good learning and generalisation in machine learning in general, and also for neural networks. In this case study we focus initially on the Engineering aspects of this area and then on the Scientific. Thus, we first present our experience on a set of raw data as provided to us by a Geographer, and the decisions we made based on analysis of the data. This is essentially an engineering approach. Then we present the new method we needed to discover to improve the performance of the neural network on the data.

The raw data for this study comes from a forest in New South Wales, Australia. The available information is from a rectangular grid of 244,494 points, and is a vector of 16 values, 7 from satellite images, augmented by 9 values from a terrain model derived from soil maps and aerial photography and so on. The last 5 are the forest supra-type.

The data as provided by the Geographer had already been somehow encoded. There was some available information on the encoding, as shown below:

- ◊ Aspect: 0: flat, 10: North, 20: NE, 30: East, ..., 70: West, 80: NE.
- ◊ Sin & Cos Aspect: some unknown encoding has been applied.
- ◊ Altitude: metres above sea level.
- ◊ Topographic position: 32: gully, 48: lower slope, 64: mid-slope, 80: upper slope, 96: ridge.
- ◊ Slope: 10: < 1%, 20: < 2.15%, 30: < 4.64%, 40: < 10%, 50: < 21.5%, 60: < 46.4%, 70: < 100%, 80: > 100%
- ◊ Geology descriptor: unknown encoding.
- ◊ Rainfall: (mm - 801)/5.
- ◊ Temperature: (degrees - 11)*30.
- ◊ Landsat tm bands 1 to 7: values in range 0 to 256, rescaled to 0 to 99.

DATA ANALYSIS (EXAMPLES)

Aspect:

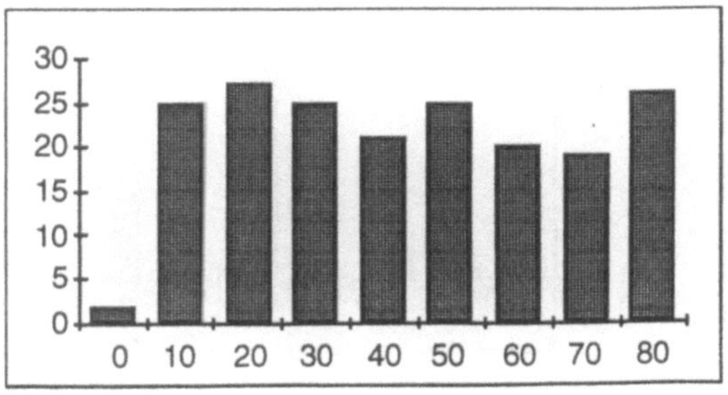

◊ This is a 'circular' value, so AS=80, AS=10 may cause a net to generalise to 45!

◊ The 0 value for aspect is redundant since the slope degree also encodes this information.

◊ The sin and cos of the aspect are also redundant.

Altitude

Statistics

Maximum Value	71
Minimum Value	7
Average Value	34
Standard Deviation	13
Average Absolute Deviation	11

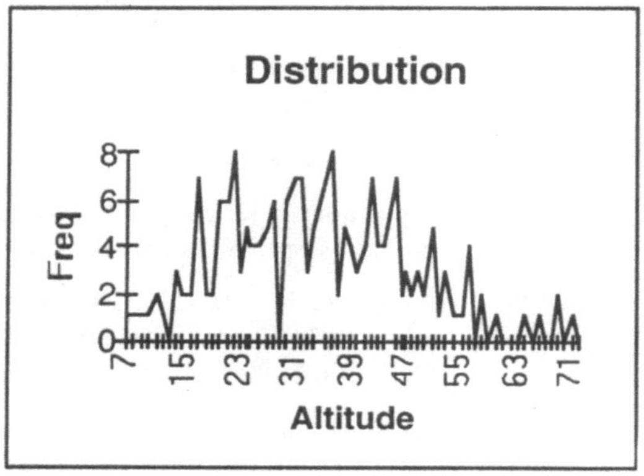

◊ Need to normalise data over the range 0 - 1 for the network, for the logistic function.

◊ Consider using statistical Z function to remove bias of lowest and highest values.

Geology descriptor:

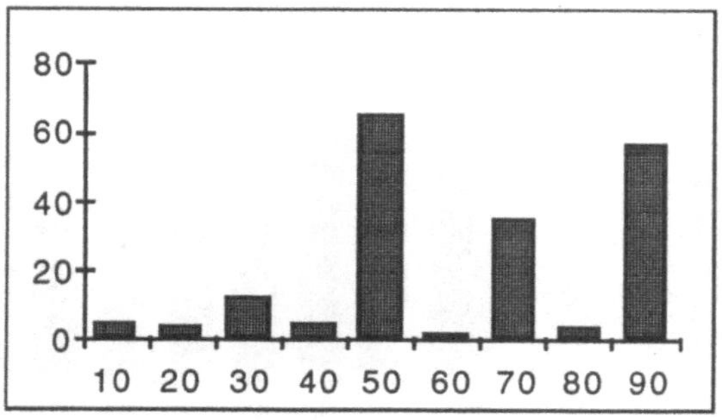

◊ 3 significant groups, not distributed normally.

◊ Patterns with '50' output similar to those with '70' output may cause spurious '60' results.

◊ No information was available on the encoding used by the geographer on this field.

Slope degree

Encoding format:

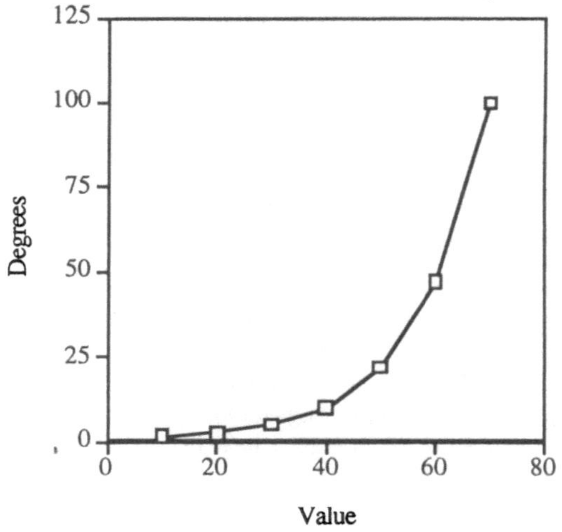

◊ The degrees increase exponentially with the values. This logarithmic encoding is unlikely to be coincidental, and we assume incorporates some domain knowledge of the Geographer.

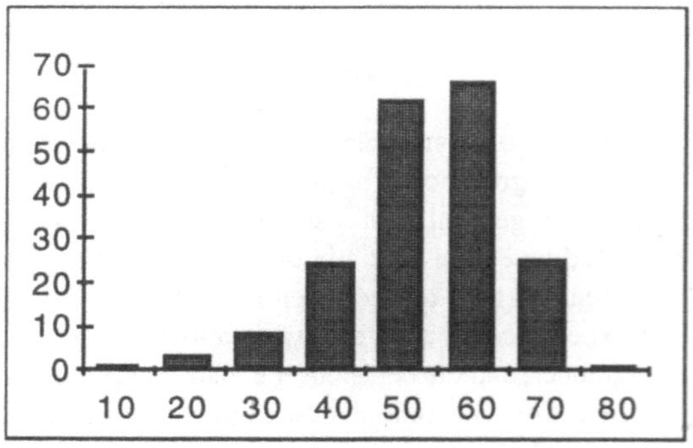

◊ Note that the slope degree categories are not linear in their variation with the percentage of slope.

◊ Consider using a single continuously valued unit or a number of category units with soft boundaries.

ENCODING DECISIONS (EXAMPLES)

The encoding decisions described are the result of trialing a number of alternative representations.

Where there were no significant differences, the simplest is presented, otherwise the best result is described.

Aspect

◊ The best input coding for generalisation would be a category for each direction, with a redundant gaussian activating the input and the adjacent inputs to a lesser amount. This would require 8 inputs (too many).

◊ The raw encoding of *aspect* does not reflect the circular nature of the information.

◊ As a compromise, code the *aspect* into 4 units, using one input to represent each point of the compass.

As	Dir.	A1	A2	A3	A4	Σ activ
0	Flat	0	0	0	0	0
10	N	1	0.5	0	0.5	2
20	NE	1	1	0	0	2
30	E	0.5	1	0.5	0	2
40	SE	0	1	1	0	2
50	S	0	0.5	1	0.5	2
60	SW	0	0	1	1	2
70	W	0.5	0	0.5	1	2
80	NW	1	0	0	1	2

◊ Total activation for valid aspects is 2. The activation is apportioned in a way that should aid generalisation as there is a gradual change between directions.

Altitude:

◊ Distribution of values in the training set is fairly normal, so use a simple linear squashing function.

Slope Degree:

◊ Retain existing logarithmic encoding. Normalise to range 0 to 1 using a single continuous valued output.

Geology descriptor

◊ From the data it appears to be a nominal value. There is no particular distribution, three of the types are quite common and the others are rare.

◊ A single continuously value input is not appropriate. The categories will be represented by 4 inputs, distinguishing between the popular types and the rare ones, thus losing some information. In any case, the input vector over these inputs would be quite sparse.

VALUE	G1	G2	G3	G4
10	0.9			
20	0.9			
30	0.9			
40	0.9			
50		0.9		
60	0.9			
70			0.9	
80	0.9			
90				0.9

◊ All unmarked activations are 0.1. Cases activating G1 are similar in total to the other three categories.

OUTPUT ENCODING

The network is a classifier, and the output vectors are very sparse. This can lead to difficult learning, since the tss can improve by pushing output vectors to all 0's.

This can be avoided by equilateral coding [1], so the 5 possibilities are represented by 4 units. All units have some activation on each pattern, and the maximum distance between vectors is maintained.

Category	Unit 1	Unit 2	Unit 3	Unit 4
Scrub	0.1838	0.3174	0.3709	0.4
Dry scler.	0.8162	0.3174	0.3709	0.4
Wet-dry scler.	0.5	0.8651	0.3709	0.4
Wet scler.	0.5	0.5	0.8872	0.4
Rain Forest	0.5	0.5	0.5	0.9

◊ To retrieve the category, calculate Euclidean distance between the output vector and the above values:

Output category = MIN(distance (U_i))

where

$$\text{distance}(U_i) = \sqrt{\sum_{j=1}^{4} (y_j - u_{j_i})^2}$$

where

$$U_i = (u_{j_S}, u_{j_{DS}}, u_{j_{WDS}}, u_{j_{WS}}, u_{j_{RF}})$$

PATTERN REDUCTION

We have done some work in speeding up learning by removing outliers [2, 3]. On very large sets of patterns, such methods can be too slow to be useful.

We have also done some work on a simple heuristic method of reducing training sets to half or quarter the size and retaining and sometimes improving performance [4]. Such reduction in size of a training set speeds up training by a commensurate ratio.

RESULTS

◊ 12 hidden unit network

epochs	tss train	tss test	actual train	perform. test
750	12.9	3.9	77%	58%
1050	11.4	4.4	82%	55%
1350	9.7	4.8	84%	55%
1650	9.1	5.0	84%	51%

Twelve hidden units provided the best performance.

Note that generalisation suffers as more epochs are run.

Adding Random noise

◊ Trained with normal, and randomly distorted data in 50 epoch alternation.

Note the oscillation of the test set results as the training continues. This effect was more pronounced with higher levels of added noise (or *randomness*), this observation lead to our decreasing amplitude noisy training.

epoch	tss train	tss test	actual train	actual test	random-ness
200	19.2	3.8	58%	55%	0.4
300	17.9	3.5	69%	58%	0.3
400	16.7	3.1	70%	62%	0.2
500	16.3	3.6	67%	44%	0.15
600	18.9	3.6	72%	58%	0.1
700	16.9	3.4	71%	55%	0.05
800	13.2	3.3	77%	68%	0.0

CONCLUSION

We have shown using our example of Geographical Information System (GIS) data pre-encoded in a fashion appropriate to the use of the data in Geography.

To produce a neural network which performed as well as the standard statistical results of 57% accuracy on the test set, some effort was required on encoding of the data suitably for the network, as well as using pattern reduction techniques to reduce the large number of points without losing significant information, and hence again degrading network performance.

To outperform the statistical result, we developed a method for alternating noisy training with normal training.

REFERENCES

1. Masters, T, *Practical Neural Network Recipes in C*, Academic Press, Boston, 1993.

2. Slade, P and Gedeon, TD "Bimodal Distribution Removal," in Mira, J, Cabestany, J and Prieto, A, *New Trends in Neural Computation*, pp. 249-254, Springer Verlag, Lecture Notes in Computer Science, vol. 686, 1993.

3. Wong, PW and Gedeon, TD "The Error Sign Testing Method in Pattern Reduction," *International Journal of Systems Research and Information Science*, (in press), 1994.

4. Gedeon, TD and Bowden, TG, "Heuristic Pattern Reduction II," *Proceedings ICCS*, Invited Position Paper, pp. 3.43-3.45, Beijing, 1993.

A Framework for Creating Societies of Agents

Jean-François Arcand and Sophie-Julie Pelletier

Centre for Information Technology Innovation (CITI), Industry Canada
1575 Chomedey Boulevard, Laval (Quebec), Canada H7V 2X2
e-mail: jarcand@citi.doc.ca, spelletier@citi.doc.ca

Abstract

The ADN project constitutes a generic tool for the analysis and development of distributed intelligent agents. The introduction presents the impetus driving ADN as well as a brief history of the project. Next, the parallel between cognitive psychology and the knowledge modeling architecture are laid open. A brief description of ADN's functionalities is also presented.

1. Introduction

Since we work primarily in the field of Performance Support Systems [7], our research is largely conducted in the domains of distributed artificial intelligence, collaborative work and cognitive psychology. The concepts of encapsulation of knowledge, distribution of knowledge, sharing of knowledge, knowledge hierarchy abstraction and communication between different knowledge structures were used in developing ADN. Research is directed towards creating a tool that will allow the development of intelligent agent societies [2] [3].

In ADN, we use the term society to represent a collection of agents capable of communicating, exchanging and learning. The term agent signifies: "All cognitive processes or parts of cognitive processes which possess their own behaviour, their own learning style and their own knowledge representation model and are able to interact with other agents regardless of their type (i.e. neural agent, genetic agent, etc.)" [3]. Because several knowledge modeling techniques exist (artificial intelligence, genetic algorithms, artificial neural networks, etc.), we decided to develop a platform that would provide some of these techniques and that would allow the integration of other techniques and their variants.

The paragraphs that follow describe the psychological basis of our knowledge modeling approach as well as the functionalities associated with artificial neural networks.

2. Psychological Foundation

Since the beginning of the 60s, humans have attempted to reproduce their behaviour using computers. Some of this work led to systems able to reproduce parts of the human experience. The problem with this type of modeling is that humans do not fully understand how their own brains work. Several models have tried to explain human thought —Rumelhart and Normand [12], Minsky [8], Rasmussen [10], Rouse [11], Wickens [16] and Moray [9]— some better than others. The modeling of ADN's knowledge is based on a mixture of the six aforementioned models (described below). Due to space limitations for this article, we will develop only three of the six models cited above. This type of modeling allows ADN to reproduce and combine those models.

2.1 The Rasmussen Model

The knowledge representation mode implanted in ADN provides a structure which allows the three types of performance proposed by Rasmussen [10] to be represented: "Skill-based, rule-based and knowledge-based levels." Basically, a collection of societies allows the modeling of each performance type with a more or less elevated level of complexity. For example, the "skill-based" level can be represented by a society composed of neural agents (that allows pattern matching); the "rule-based" level can be represented by a society composed of artificial intelligence-based agents; the "knowledge-based" level can be represented by a society composed of neural agents and genetic agents. The relationship between the three performance levels will be modeled in a hierarchical fashion, that is, the stimulus will first pass through the "skill" level. If the society recognizes the stimulus, an answer is proposed, if not, the stimulus moves to the "rules" society. Again, if the stimulus is recognized, an answer is proposed, if not, the stimulus moves to the "knowledge" society. The "knowledge" society, which represents the mental knowledge model, can be decomposed into the three parts of the human thought process proposed by Atkinson and Schiffrin [4]: The sensorial information register, short-term memory and long-term memory. We feel that, modeled this way, the structure proposed by ADN allows a degree of reproduction of the human brain. This hypothesis will be validated during phase two of the ADN project.

2.2 The Wickens Model

According to Wickens [16], it is more difficult to discern a stimulus which affects various objects than it is to discern several stimuli affecting a same object. As a way of overcoming this, Wickens proposes representing interrelated variables in an integrated fashion in order to guarantee coherence in the representation, if the representation is separate. To ensure this, we used an object-oriented approach for application development, as well as a number of ergonomic principles for interface development which, in our opinion, will allow effective use of the Wickens model.

Wickens states that parallel information processing used by humans while simultaneously participating in more than one activity is characterized by four factors:

1. Spatial proximity: for visual stimuli, parallel processing is favoured, though not guaranteed, by coexistence in the visual field. For auditory stimuli, coexistence is detrimental. ADN's mode of knowledge representation allows this conflict to be avoided because operations are effected in two different societies and in two different spaces (the software being composed of collaborative tools allowing the division of a society across a computer network).

2. Common object: integration of information distinct to an object favours parallel processing. Since ADN is developed using an object-oriented methodology, the encapsulation of information is effected at both the society and agent levels.

3. Separate resources: mental processes do not all use the same resources [15]. Two tasks requiring different resources can be done in parallel without inference. These resources are defined by three dimensions:

- Processing steps: encoding, central processing and response;
- Input modes: visual versus auditory;
- Processing type: spatial versus verbal.

The Wickens model does not predict perfect resource sharing but rather an improvement in tasks requiring different resources. For ADN, societies could represent the dimensions proposed by Wickens through the addition of a society that directs stimuli according to its input mode or the type of processing the information requires.

4. Task difficulty: a task's degree of difficulty not only affects performance but also the type of resources needed to accomplish it. In the future, ADN will offer artificial intelligence possibilities as much as it will offer artificial neural network-based properties. One technique could then be favoured for task implantation.

2.3 The Rumelhart and Normand Model

The schema [12] modeling approach particularly attracted our attention. Rumelhart and Normand define a schema in the following manner: "In general, a schema consists of a network of interrelation among its constituent parts, which themselves are other schemata. Schemata is the primary meaning and processing unit of the human information-processing systems." In ADN, the term society corresponds to schema and the term agent corresponds to schemata. Following this approach, ADN provides a structure that allows modeling in accordance with this formalism.

3. ADN's Object-Oriented Modeling

To date, ADN's object-oriented analysis has followed two main axes: long-term vision, that is, task modeling by society hierarchy, and short-term vision, the modeling of knowledge by neural agents, which is in fact a specialization of a class of societies.

3.1 Society Hierarchies: Modularity and Common Language

The major attraction of modeling by society hierarchies, which can be associated with modeling by agents [13], resides in information encapsulation and its processes, and therefore in the modularity of implantation. Society hierarchy is established naturally according to the decomposition of the task to be modeled: agents (neural, genetic or other) are at the heart of societies, which are in turn grouped in a recursive fashion into levels of societies. We believe that this approach facilitates the elaboration and acquisition of knowledge because it is similar to the hierarchical task analysis model proposed by Annet and Duncan [1]. In effect, if knowledge acquisition is done by hierarchical analysis, either a society level or agent level could represent each task (simple or complex).

In general, every society that contributes to completion of a task follows three steps: it obtains data elements (knowledge) from the mother society, it internally processes the data according to its knowledge representation model, and then it provides new data elements which help progress task execution. This uniformity facilitates the coming together of societies, the elaboration of society hierarchies (for fluid task division), the direct manipulation of societies (i.e. copy, save), as well as the reutilization of societies according to various granularities associated with the different hierarchy levels of the model created. Once designed, societies from different levels can be uniformly manipulated, regardless of the knowledge modeling technique behind them and regardless of the level of constituent society and agent overlap.

Encapsulation requires the development of a language common to all societies, and the passage of information from one level to another within the hierarchy. By common language, we mean the posting of general interest knowledge, in a generic format, so that it may be understood by all societies. Using a common language, two societies, one implanted by neural agents and the other implanted by society hierarchy, could use the same knowledge elements. All societies used in ADN must be able to understand and talk this common language.

As for the passage of information from one level to another in the hierarchy, we advocate the use of hierarchical blackboards, as presented by Dai [6] and Carver [5], with each society having an associated blackboard. The society updates information it receives from its component societies or agents and posts the information received from its mother society (or the inaugural knowledge for the first society in the hierarchy). Cognitive filtering [14] is both ascendant and descendant, each society knowing the needs and interests of its mother society and, its own sub-elements.

Though we are rapidly progressing in our efforts to refine the objects needed to create society hierarchies, more research must be effected before a model offering all the desired flexibility can be put in place.

3.2 Neural Agents

3.2.1 Hierarchy and Internal Language: Basically, a neural agent encapsulates a collection of interlinked artificial neural networks. In the same way as societies, neural agents can be hierarchically assembled in order to allow implantation by successive refinement of the knowledge base. Knowledge is distributed in the hierarchy's last level of agents, each of which encapsulates a specialized neural network.

Communication between neural networks around and across the hierarchy is effected via the agents' input and output synapses. In the neural network, the input synapses provide a data stimulation vector while the output synapses communicate the results of neural network activation. The output synapse of an agent must be connected to the input synapse of another agent in order to effect information passage

The concept of neural agent communication by synapses constitutes a parallel to the blackboard method used at the level of societies. The use of blackboards is not considered optimal for neural agents, the flood of information being too voluminous, risking congestion. Communication between neural agents participating in the same sub-task does not need to follow the societies' common language, since communication is part of the neural agents' internal knowledge representation. A translation protocol must be provided inside the neural agent so that neural agents directly linked to a parent society can communicate outside.

Even though, from a conceptual point-of-view, blackboards and synapses are clearly distinct, we believe that a unique presentation metaphor can group the two concepts at the user-interface level.

3.2.2 Knowledge Modeling by Neurons: ADN has a few particularities, identified through our experience using neural networks, when compared to its peers. Among the particularities is the possibility of creating mixed architecture neural networks, that is, networks in which different training algorithms can be associated. In order to allow this type of architecture, the training algorithm is not encapsulated at the network level. Often, many neurons are associated with the same trio of algorithms (input, activation and training). To increase modularity, ADN encapsulates the trio of algorithms in a group object. Neurons are then associated with the groups which represent them. Algorithms are also implanted in object form, in order to allow variation in associated parameter values, from one group to another.

3.2.3 The Shared Neuron: To allow effective and optimal modeling is to propose a "non-duplication" of knowledge across a society. Conceptually, this leads us to talk of shared neurons [3], which represent another particularity in ADN's neural network implantation. A shared neuron is inspired by the structure of the biological neuron, but can simultaneously belong to many neural networks. Communication by synapses allows this type of neuron to be implanted. Though, physically, the neuron can be situated in only one neural network, it is virtually present inside several networks via the input/output synapses. Synapses permit an activation state to travel to a neuron of a hidden or output layer of a second neural network. This is equivalent to considering the shared neuron as part of the input layer of the second neural network.

4. Conclusion

We have presented part of the research linked to the ADN project, a generic tool for the analysis and development of distributed intelligent agent systems. The knowledge modeling proposed by ADN—by society hierarchies and agent hierarchies—is inspired by numerous models from the domain of cognitive psychology. The goal is to reproduce the effectiveness of the human brain's knowledge representation.

Over the next few months, three principal activities will be completed within the framework of the ADN project: object persistence, inter site communication and concept validation. This consists of continuing analysis and programming work regarding the persistence of objects manipulated by the software. We will also work at developing utility objects for inter site communication in order to allow the distribution of societies and agents on a computer network. Finally, ADN will be used within the framework of other research projects and will be made available to various user types.

5. Acknowledgements

We would like to thanks Mr. Simon Legault and Mr. Martin Deveault of CITI for there contributions to the object-oriented model of ADN.

6. References

[1] Annet, J.; Duncan, K.D. Task Analysis and Training Design. Occupational Psychology, Vol. 41, pp. 211-221 (1967).

[2] Arcand, J.-F.; Pelletier, S.-J. ADN : Analysis and Development of a Distributed Neural Network for Intelligent Applications. In IEEE World Conference on Neural Networks, Orlando, Florida. pp. 1516-1522 (1994).

[3] Arcand, J.-F.; Pelletier, S.-J. A Framework for Distributed Neural Networks. In 1994 International Symposium on Artificial Neural Networks, Tainan, Taiwan. pp. 204-214 (1994).

[4] Atkinson, R.C.; Shiffrin, R.M. Human Memory: a Proposed System and Its Control Processes. In The Psychology of Learning and Motivation, Vol 2, eds. K.W. Spence and J.T. Spence. New-York : Academic Press, 1968.

[5] Carver, N.; Lesser, V. Evolution of Blackboard Control Architectures. Expert Systems With Applications, Vol. 7, pp. 1-30 (1994).

[6] Dai, H.; Hughes, J. G.; Bell, D. A. A Distributed Real-Time Knowledge-Based System and Its Implementation Using Object-Oriented Techniques. pp. 23-30. IEEE, 1993.

[7] Dalkir, K.; Trevail, R. Canadian Research and Development in Performance Support Systems. Learntec Conference, Stuttgart (October 1993).

[8] Minsky, M. The Society of Mind. New-York : Simon and Schuster, 1986.

[9] Moray, N. Intelligent Aids, Mental Models, and the Theory of Machines. International Journal of Man-Machine Studies, No. 27, pp. 619-625 (1987).

[10] Rasmussen, J. Skill, Rules, and Knowledge; Signals, Signs, and Symbols, and Other Distinction in Human Performance Models. IEEE Transactions on Systems, Man, Cybernetics, Vol. SMC-13, May/June , pp. 257-266 (1983).

[11] Rouse, W.B. Models of Human Problem Solving: Detection, Diagnosis, and Compensation for Systems Failures. Automatica, Vol 19, No 6, pp. 613-625 (1983).

[12] Rumelhart, D.E.; Norman, D.A. Accretion, tuning and restucturing : Three Modes of Learning. In Semantic Factor in Cognition. ed. Cotton, J.W.; Klatsky, R.L. pp. 37-60. New Jersey : Hillsdale, 1978.

[13] Shoham, Y. Agent-Oriented Programming, AI, Vol. 60, pp. 51-62 (1993).

[14] Stary, C.; Stumptner, M. Representing Organizational Changes in Distributed Problem Solving Environments. IEEE Transactions on Systems, Man, and Cybernetics, Vol. 22, No. 5, September/October, pp. 1168-1177 (1992).

[15] Wickens, C.D. Engineering Psychology and Human Performance. Ohio : Charles E. Merrill, 1984.

[16] Wickens, C.D. Information Processing, Decision-Making, and Cognition. In Handbook of Human Factors, ed. Slavendy, G. pp. 72-107. New-York : Wiley., 1987.

ADAPTIVE SCALING OF CODEBOOK VECTORS

S. Haring J.N. Kok

Department of Computer Science, Utrecht University,
P.O.Box 80.089, 3508 TB Utrecht, The Netherlands
E-mail: bas@cs.ruu.nl

Abstract

In this paper we introduce a vector quantization algorithm in which the codebook vectors are extended with a *scale* parameter to let them represent Gaussian functions. The means of these functions are determined by a standard vector quantization algorithm; and for their scales we have derived a learning rule. Our algorithm estimates probability densities efficiently. The main application is pattern classification.

Introduction

Pattern classification is trivial if a function is available which describes the probability distribution of the classes in the pattern space. In that case we can use the Bayesian classifier; classify a pattern according to the class with the largest probability at the respective sample position in the pattern space. The obvious problem is then to find a function which describes this probability distribution. A standard method is *Parzen window estimation* [1]. In this method each pattern is seen as a Gaussian distribution whose mean equals the pattern position and whose standard deviation (or *scale*) has some arbitrary value. The estimated probability function is the average of all Gaussians. The major advantage of this method is that any probability function will be estimated correctly if the number of samples reaches infinity. The main disadvantages are: (i) a large number of samples is necessary to obtain a reasonable estimate; (ii) the probability function is in terms of a large number of Gaussians and evaluation of the function is time and memory consuming, and (iii) the choice of the standard deviation is arbitrary. In this paper we propose a new method which reduces the mentioned disadvantages but keeps the nice properties of Parzen window estimation.

Related Work

Our work is related to that in the field of *radial basis function* networks [2]; these feedforward networks have one hidden layer in which the units have a Gaussian activation function. Regularly only the means of these Gaussian activation functions are adaptively tuned; the scales are set after training of the means using some heuristic. Most of these heuristics make use of distance from the mean to K nearest means [2, 3]. But also more advanced algorithms are known to tune the scales [4, 5, 6]. We do not use radial basis function networks, but we propose variants of vector quantization algorithms.

Vector Quantization

Assume given training patterns $\vec{x} \in R^D$ and a much smaller number of codebook vectors $\vec{m}_i \in R^D$. The codebook vectors are given random initial values. Repeatedly training patterns are drawn randomly from the pattern distribution. The codebook vectors are modified according some form of the vector quantization update rule:

$$\vec{m}_i(t+1) = \begin{cases} \vec{m}_i(t) + \alpha(\vec{x}(t) - \vec{m}_i(t)) & \text{if some condition is satisfied} \\ \vec{m}_i(t) & \text{otherwise,} \end{cases} \tag{1}$$

with $0 < \alpha \ll 1$. The condition varies with the specific vector quantization algorithm. After having labeled the codebook vectors, classification of patterns may straightforwardly be done by "K-nearest codebook classification". This classification scheme has the disadvantages that (i) it is sensitive to the correct labeling of the codebook vectors and (ii) for classification of a pattern only the nearest K codebook vectors are taken into account. In our experiments we use the Self Organizing Map (SOM) [7] algorithm. In the SOM algorithm the condition is twofold: either $\|\vec{x}(t) - \vec{m}_i(t)\|$ is smallest for $\vec{m}_i$—the so-called "winner"—, or $\vec{m}_i$ is in the neighborhood (according a topology imposed on the codebook vectors) of the winner. The SOM algorithm provides a codebook such that few codebook vectors are placed in sparse areas of the pattern space and *vice versa*, that is, the codebook vectors preserve the probability distribution of the input space. There is no proof that the probability preservation is exact [8], but by adding some additional features, like a "conscience" mechanism [9], the approximation is very good. Hence we can use the codebook vectors as a basis for Parzen window estimation, thereby limiting the number of Gaussians that are needed.

Scale Tuning

In our algorithm each codebook is a set of Gaussians (elements of which we call codebook Gaussians):

$$G_i(\vec{x}) = \frac{1}{\sqrt{2\pi s_i^2}^D} \exp - \frac{(\vec{x} - \vec{m}_i)^T(\vec{x} - \vec{m}_i)}{2s_i^2}.$$

We want to tune the scales s_i of the codebook Gaussians in a manner similar to equation (1). We look for a learning rule such that the correct estimate of the probability distribution is an equilibrium point of the algorithm (we refer to this constraint as the "equilibrium constraint"). Hence if there is only one codebook Gaussian, if the input patterns are drawn from a D-dimensional Gaussian distribution with mean $\vec{\mu}$ and standard deviation σ and if the codebook Gaussian already has the correct values $\vec{m} = \vec{\mu}$ and $s = \sigma$, then, after application of one more step of the learning rule, $E(\vec{m}) = \vec{\mu}$ and $E(s) = \sigma$ (where $E(\cdot)$ denotes the expected value).

We propose the following format for the update rule for the scales of the codebook Gaussians:

$$s_i(t+1) = \begin{cases} (1-\beta)s_i(t) + \gamma\|\vec{x}(t) - \vec{m}_i(t)\| & \text{if } \vec{m}_i \text{ was modified} \\ s_i(t) & \text{otherwise,} \end{cases} \quad (2)$$

with $0 < \beta \ll 1$.

By choosing the parameters β and γ we can obtain the equilibrium constraint. For example, if patterns $\vec{x}$ drawn from a D-dimensional Gaussian distribution with mean $\vec{\mu}$ and standard deviation σ. Then the equilibrium constraint gives

$$E(s(t+1)) = (1-\beta)s(t) + \gamma E(\|\vec{x}(t) - \vec{\mu}(t)\|).$$

Hence we can take

$$\frac{\beta}{\gamma} = \frac{E(\|\vec{x} - \vec{\mu}\|)}{\sigma}.$$

The expected value $E(\|\vec{x}-\vec{\mu}\|)$ is easily calculated if we change to polar coordinates.

Simulations show that one codebook Gaussian represents any D-dimensional Gaussian pattern distribution correctly, using the learning rules (1) and (2). Further research is needed for situations with more than one codebook Gaussian. In experiments we already found good values for $\frac{\beta}{\gamma}$, but it would be nice to derive them analytically. Figure 1 shows a simulation of the algorithm in which samples from a uniform two-dimensional pattern distribution are input to 16 codebook Gaussians.

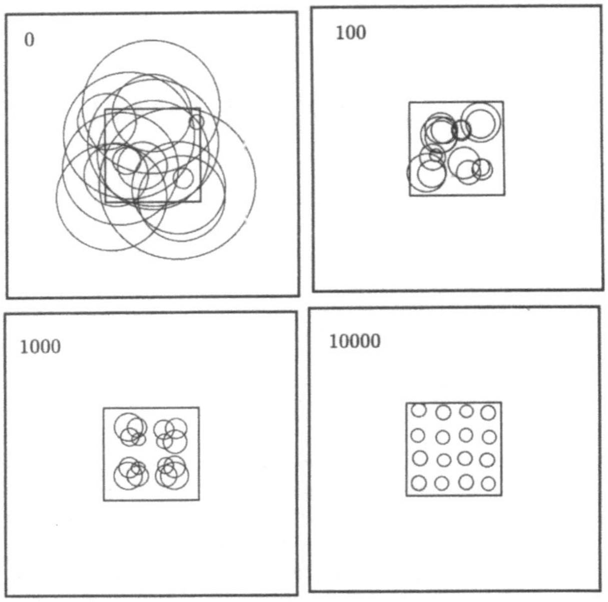

Figure 1: Each circle corresponds to a codebook Gaussian; the center of a circle represents the mean, and the radius of a circle equals the standard deviation. Samples are uniformly drawn within the plotted square area. The number of iterations is shown in each picture. The means and standard deviations are initialized in a random fashion, but as processing proceeds the codebook Gaussians start to produce a proper representation of the input pattern distribution, and finally the process stabilizes.

Classification

To apply the algorithm for classification we assign each class a separate set of codebook Gaussians. Each set is only trained with patterns from a single class. After having trained the codebook Gaussians, each set provides an estimate of the probability function of one class; just as with Parzen window estimation, we take as the estimate of the pattern distribution the average of all Gaussians in the set. Classification of a pattern may now be done by calculating the probability of each class at the respective sample point, and assigning to the pattern the class with the highest probability. Hence the whole codebook plays a role in the classification of patterns. This is *not* the case with regular classification schemes using codebooks.

GCC	0.893 ± 0.005
SOM	0.840 ± 0.005
LVQ	0.883 ± 0.005
PWE	0.896
NN	0.849
NM	0.532

Table 1: The results of various algorithms on the described artificial classification problem. The number denotes the fraction of correct classified patterns. With GCC we mean our algorithm; it stands for "Gaussian Codebook Classification". In the test we applied 100 codebook Gaussians. Applying SOM, we also used 100 codebook vectors. PWE denotes Parzen window estimation, in the test we applied standard deviation of 0.1. NN means nearest neighbor classification, and with NM we mean nearest mean classification.

We have tested the classification scheme on several classification tasks including the two spiral problem 2. We compared our algorithm to various other classification algorithms and it came out second 1; the best algorithm for the applications is the Parzen window estimation. However, the computing time and memory for Parzen window estimation are excessive when compared to our algorithm, and hence, in practical situations, our algorithm is to be preferred.

Conclusion

We have developed a fast algorithm which combines attractive properties of both Parzen window estimation and vector quantization. The scale parameter is tuned adaptively and, therefore, is not set in an *ad hoc* manner. It allows a classification strategy in which *all* the codebook vectors are taken into account. This yields better results than the standard vector quantization techniques. An interesting topic for further research is to use radially non-symmetric Gaussians.

References

[1] E. Parzen. On estimation of a probability density function and mode. *Annual Mathematical Statistics*, 33:1065–1076, 1962.

[2] J. Moody and C.J. Darken. Fast learning in networks of locally-tuned processing units. *Neural Computation*,

[4] M.T Musavi, W. Ahmed, K.H. Chan, K.B. Faris, and D.M. Hummels. On the trainig of radial basis function classifiers. *Neural Networks*, 5:595–603, 1992.

[5] P.D. Wasserman. *Advanced Methods in Neural Computing.* Van Nostrand Reinhold, 1993.

[6] B. Kosko. *Neural Networks and Fuzzy Systems.* Prentice-Hall International Editions, 1992.

[7] T. Kohonen. *Self-Organization and Associative Memory.* Springer-Verlag, 1984.

[8] H. Ritter, T. Martinetz, and Klaus Schulten. *Neural Computation and Self-Organizing Maps.* Addison Wesley, 1992.

[9] D. DeSieno. Adding a conscience to competitative learning. In *Proceedings of the International Joint Conference on Neural Networks*, pages 117–124. IEEE, 1988.

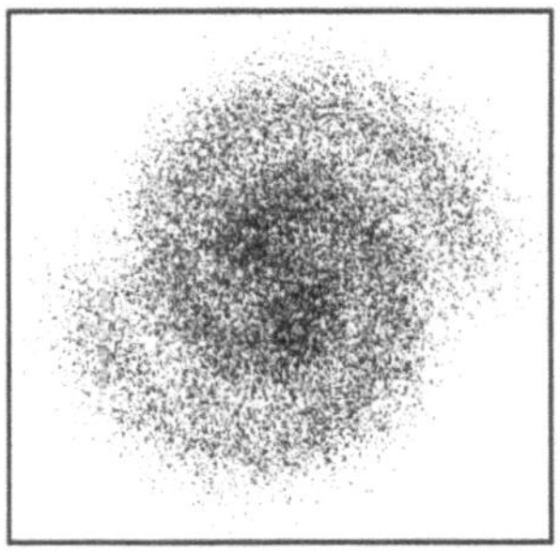

Figure 2: Two intermingled two-dimensional spirals each comprising 10000 patterns. One has parametrization $x(t) = (t+1)\cos t + \varepsilon$, $y(t) = (t+1)\sin t + \varepsilon$, the other $x(t) = -(t+1)\cos t + \varepsilon$, $y(t) = -(t+1)\sin t + \varepsilon$, with $0 \leq t < 2\pi$, and ε is Gaussian noise with standard deviation 1.

BIAS ESTIMATION FOR NEURAL NETWORK PREDICTIONS

Colin R Reeves
School of Mathematical and Information Sciences
Coventry University, UK

Abstract

This paper looks at the problem of performance assessment in the use of neural networks for classification tasks. It is well-known that the prediction obtained from a trained neural network is subject to errors, both in terms of bias and variance in the estimated error rate.

In order to estimate these measures, it is customary to reserve some data as a test set. This is reasonable if data are plentiful, but when the data set is small in size, this is likely to reduce the accuracy of the network estimates, simply because there are not enough data left for adequate training. An alternative approach will allows the use of all the data, is to employ the bootstrap method.

Here we give a brief introduction to the bootstrap, and then report on some computational experiments on artificial data sets in order to investigate the potential of this approach for the estimation of error bias.

Introduction

In many applications of artificial neural networks (ANNs), it is necessary to evaluate the performance of the proposed ANN by means of statistical procedures. A common approach is to partition the available data into two sets, one of which is used to train the net, while the other is used as a 'test set' to measure the generalization capability of the trained net. This is particularly important in the case of classification problems where the most important criterion is usually the *mis-classification rate* (MCR) of the ANN model. It is obvious that the MCR for the training set—sometimes called the *apparent* mis-classification rate—is likely to be biased downwards, since we are in some sense minimising the errors, whereas the MCR for the test set will be unbiased. The difference between the two provides an estimate of the bias.

This works well where there are sufficient data, but when data are scarce it can cause difficulties. In [1] we reported some experimental results using ANNs to classify medical diagnostic data, in which we found that the test set MCR was liable to vary quite markedly depending on the particular partition of the data into training and test sets. This highlights the fact that ideally we would like to give not only an estimate of the MCR, but also an estimate of its variance.

This is reminiscent of the familiar 'bias/variance dilemma' (Geman *et al.* [2]). In principle, if the size of the training set is increased, we can expect the 'model variance' to become smaller (i.e. models generated by different partitions will become more similar), and in the limit, if all the data are used, we will have just one model[1], but in order to get as precise an estimate of the bias as possible, we want the test set to be as large as possible.

Perhaps the most common and simplest alternative to training/test set partition is the 'leaving-one-out' method, where the model is fitted using all but one of the data points. The resulting model is then used to predict the remaining case. This is repeated N times (where N is the number of cases in the training set), and the overall MCR is an unbiased estimate of the true MCR. This method has also been called *cross-validation*, although the term is often used (see e.g. [3]) in the more general sense of splitting a data set into two portions for the purpose of estimation and assessment of a statistical parameter.

In principle, this clearly involves considerable computation, although with conventional statistical methods it is often possible to calculate this estimate with a relatively small addition to computational requirements. Unfortunately, no short-cuts are possible with ANNs, and to repeat the process of training a network N times might often need a prohibitive amount of computation. This can be modified to give k-fold cross-validation, whereby k subsets of the data are chosen, and the model fitted using $k-1$ of them, the kth being used for testing.

By partitioning the data, information is sacrificed which might actually help to estimate a better model. As remarked above, ideally we would like to use all the information available, especially when the total amount is small, but if we use all the data, how can we estimate the bias in the apparent MCR?

The Bootstrap

The past decade has seen an explosion of interest in statistical circles in the use of Monte Carlo resampling tech-

[1] In practice, this may not be true: different models can be generated from the same data in the case of multi-layer perceptrons for example, because it may not be possible to find a single global optimum of the error function.

niques for estimating 'difficult' quantities, where the necessary simplifying assumptions about sampling distributions cannot be made. The concepts are now fairly well known in the statistical community, but they have not been widely used in ANNs.

The problem can be summarized as follows: we wish to estimate a parameter $\theta(X;\mathcal{F})$ using a sample $X = \{\mathbf{x}_1,\ldots,\mathbf{x}_n\}$, where $\mathbf{x}_i$ is a realisation of a random variable $\mathbf{X}_i$, the probability distributions being assumed to be *iid*. We denote this distribution by $\mathcal{F}$. (It is important to realize that not only is θ a function of the data$X$, but it is also a functional with respect to $\mathcal{F}$.

In order to make statistical statements about $\theta(X;\mathcal{F})$ (for example to estimate its bias or variance), the conventional approach would be to assume some particular parametric form of $\mathcal{F}$ from which exact or asymptotic results could be derived.

Efron [4] suggested that, rather than parameterize $\mathcal{F}$, we could use the empirical distribution $\hat{\mathcal{F}}$, and repeatedly sample from it (with replacement), thus generating a sequence of B samples $\hat{X}_b$. Suppose $\hat{\theta}_b = \theta(\hat{X}_b;\hat{\mathcal{F}})$, then

$$\lim_{B\to\infty}\frac{\sum_{b=1}^{B}\hat{\theta}_b}{B} = \theta(X;\hat{\mathcal{F}}),$$

where $\theta(X;\hat{\mathcal{F}})$ is the non-parametric maximum likelihood estimator of θ, under certain conditions. Roughly speaking, provided $\hat{\mathcal{F}}$ is sufficiently 'close' to $\mathcal{F}$, the estimate of θ thus obtained should be a good one. The general approach has become known as a *bootstrap* procedure. Unfortunately, the empirical distribution may not be close enough, especially for small samples, and in practice B is necessarily finite, but there are ways of improving the simple bootstrap to try to account for this. Further details are given by Efron [4], while a comprehensive account of the current state of the art is given by Hall [5].

Bootstrapping ANNs

Previously Reeves and O'Brien [6] carried out some computational experiments on classification problems in the context of applying bootstrap methods to ANNs, while Paass [7] has also investigated their application in some small-scale simulation experiments.

In the context of using an ANN as a classifier, it means we first train the network, using all the data, and estimate the apparent MCR. In order to estimate the bias in the apparent MCR, we now take B samples $\{\hat{X}_b; b = 1\ldots,B\}$ each of size N with replacement from the data X. For each sample,

1. train the network using $\hat{X}_b$ and calculate the apparent MCR $e_{\hat{X}_b}$;
2. calculate e_X, the MCR for the *full* data set X;
3. estimate the bias by

$$w_b = (e_{\hat{X}_b} - e_X)$$

By repeating B times, the mean bias can be estimated by

$$\bar{w} = \frac{\sum_b w_b}{B}$$

and its variance (see Efron [8]) by

$$\sum_b\left(\frac{(w_b-\bar{w})^2}{B-1}\right)$$

Efron used this approach to estimating the bias in MCR estimates obtained from linear discriminant functions, where he found that it generally *under-estimated* the bias. A modification which often performed better was the '632' bootstrap: this uses the fact that only about 63.2% of the data X can be expected to appear in $\hat{X}_b$. Intuitively, therefore, the points of X that are *not* in $\hat{X}_b$ are less likely to be classified correctly by the network trained on $\hat{X}_b$. Efron suggests this can be exploited to obtain the '632' bootstrap estimate of MCR

$$e^{632}_{\hat{X}_b} = 0.368e_{\hat{X}_b} + 0.632e_{X\setminus\hat{X}_b}.$$

Empirical evidence [8] suggested that this was superior to the simple bootstrap.

Apart from the use of a linear rather than a non-linear classifier, the context for Efron's experiments is almost identical to that of many ANN applications. However, the implications of using rather general non-linear models of the type implemented by ANNs are not totally clear, and is clearly an area which needs theoretical study. In this work, we report on some simple simulation experiments where we investigate the performance of bootstrap techniques on a problem where we could evaluate the bias exactly.

Experimental Results

We used a simple 2-class problem (the 'circle in a square' problem) with 2 inputs $\{(x_1,x_2) : -1 \le x_i \le 1; i = 1,2\}$, defined as follows:

$$x_1^2 + x_2^2 < 2/\pi \to C_1$$

$$x_1^2 + x_2^2 \ge 2/\pi \to C_2.$$

Because the correct decision boundary is known, the errors in the model generated (and hence the bias), could be evaluated exactly.

A set of 200 data points was generated, and fitted using a radial basis function (RBF) network. (Using RBFs has the advantage that speed of learning is not an issue, so that the intensive computation required by bootstrap methods is eminently feasible.) Repeated experiments established that reasonably good fits could be obtained using between 4 and 7 centres, chosen by the k-means clustering algorithm, and in the subsequent work all nets used 6 centres. It is not clear how many bootstrap samples are necessary: in Efron's experiments some hundreds were sometimes needed, although as few as 30 were often fairly informative. We decided to proceed sequentially, adding samples as they were needed, and assessing convergence of $\bar{w}$ empirically.

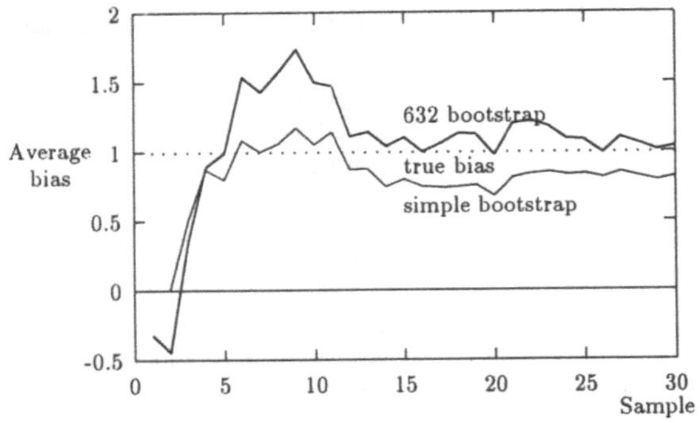

Figure 1: Convergence of average bias estimates: case 1

Bootstrap samples were thus taken and estimates of ω using both the simple and 632 bootstraps were made. The graph in Fig. 1 shows the results of one such run.

These results appear quite promising. However, it is necessary to repeat the complete procedure several times for increased confidence, and on doing so it was found that the bootstrap procedure did not always converge to the correct value. Usually the 632 estimate was closer than the simple one, and both under-estimated the bias, but in some cases, as in Fig. 2, they both *over*-estimated it.

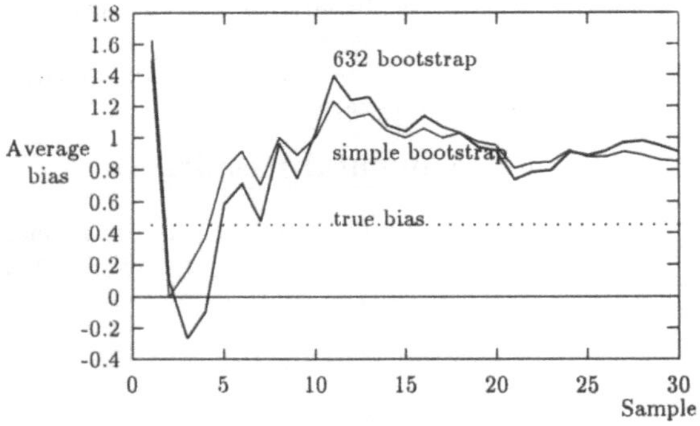

Figure 2: Convergence of average bias estimates: case 2

Conclusions

These experiments suggest that bootstrap methods can give a clearer picture of the likely bias in the apparent MCR. Further experiments are needed with different problems in order to study the performance of the bootstrap procedure. There is also clearly a need for more theoretical analysis of the methodology in the context of non-linear regression models such as those used by ANNs.

References

[1] C.R.Reeves and N.C.Steele (1993) Neural networks for multivariate analysis: results of some cross-validation studies. *Proc. of 6th International Symposium on Applied Stochastic Models and Data Analysis*, World Scientific Publishing, Singapore, Vol II, 780-791.

[2] S.Geman, E.Bienenstock and R.Doursat (1992) Neural networks and the bias/variance dilemma. *Neural Computation*, **4**, 1-58.

[3] M.Stone (1974) Cross-validatory choice and assessment of statistical predictions (with Discussion). J.R.Statist.Soc. B, **36**, 111-147.

[4] B.Efron (1979) Bootstrap methods: another look at the jackknife. *Annals of Statistics*, **7**, 1-26.

[5] P.Hall (1992) *The Bootstrap and Edgeworth Expansion*, Springer-Verlag, New York.

[6] C.R.Reeves and J.O'Brien (1993) Estimation of misclassification rates in neural network applications. *British Neural Net Society Symposium on Neural Networks*, Birmingham, UK, January 29, 1993.

[7] G.Paass (1993) Assessing and improving neural network predictions by the bootstrap algorithm. In S.J.Hanson, J.D.Cowan and C.L.Giles (1993) *Advances in Neural Information Processing Systems 5*, 196-203.

[8] B.Efron (1983) Estimating the error rate of a prediction rule: improvement on cross-validation. *J.Amer.Statist.Ass.*, **78**, 316-331.

ANN REALIZATIONS OF LOCAL APPROXIMATIONS SCHEMES

A.A.Pervozvanskii

Dept.of Mechanics and Contr.Processes,SPb State Technical University
29,Polytechnicheskaja str.,St.Petersburg,I9525I,Russia
E mail - root fmf.stu.spb.su

Abstract. Some schemes of the local approximation (LA) of functions defined on multidimensional space are described. For the simplest variants,namely,for the space filtration scheme and Nadraja - Watson scheme,the approximation precision is estimated. It is shown that ANN based on LA can solve the approximation problem with a good precision if the space dimension is moderate and some special elements are used.

I. Introduction

There are two principal types of ANN. The Ist type is based on local (potential) elements and the 2nd one uses sigmoidal elements. As it was shown in [I] the sigmoidal ANN of perception structure allows to achieve a good precision of approximation if their parameteres are adjusted optimally,but the optimization problem is unsolved.

The ANN based on local elements are connected with ideas of non-parametric approximation [2] . Their adjustment is much simpler. However the approximation precision of the local schemes in multidimensional case is not investigated in detail.

The main goal of this paper is a presentation of some local approximations (LA) schemes and an estimation of their precision. The estimates show that LA may be efficient tool for the approximation of smooth functions if the local functions have some appropriate properties.

2. Methods of local approximation

This section gives a brief overview of some basic schemes of LA approach (a more systematic presentation can be found in [2]).

Let $L=\{x^{(k)}, y^{(k)}, k=I,\dots,N\}$ be a given sequence. It is supposed that

$$y^{(k)}=y(x^{(k)}),\ k=I,\dots,N$$

but the form of the function $y(x)$ is unknown.

One looks for an estimate $\hat{y}(x) \in Y$ where Y is a class of functions defined on a set $X \subset R^d$, with $x^{(k)} \in X$.

The parametric approach uses a class of estimates in the form $\hat{y}(x)=F(C,x)$, where $F(\cdot,\cdot)$ is a known function and $C \in R^n$ is a parameter vector to be adjusted for a proximity of $\hat{y}(x)$ and $y(x)$.

As a rule,a quadratic norm of errors is used and one looks for a C_N which minimizes

$$J_N(C)=\sum_{k=I}^{N} [y^{(k)}-F(C,x^{(k)})]^2 \mu(x^{(k)})$$

where $\mu(x)$ is a weighting function.

For two-layer perception,

$$\hat{y}(x)=F(C,x)=\sum_{i=I}^{n} D_i \sigma(A_i x+b_i) \qquad (I)$$

where $\sigma(u), u\in R'$ is a sygmoidal function,i.e.

$$\sigma'(u) \geq 0;\ \lim_{u\to\infty}\sigma(u)=I;\ \lim_{u\to-\infty}\sigma(u)=0 \qquad (2)$$

and $C=(A_i,b_i,D_i)$ is the parameter vector.

The LSE problem is nonconvex in A_i, b_i and there do not exist efficient algorithms to solve it. The most popular backpropagation algorithm and its modifications may pretend on a local minimization in the parameter space.

All schemes of ANN based on potential function have a similar structure

$$\hat{y}(C,x)=\sum_{i=I}^{n} D_i y^{(i)} h\left(\frac{x-x^{(i)}}{l}\right), \qquad (3)$$
$$(x^{(i)},y^{(i)})\in L_n \subset L$$

but here $h(u), u\in R^d$ is a function having other properties

$$\lim_{|u|\to\infty} h(u)=0; \qquad (4)$$
$$h(0)>0; \int_{R^d} h(u)du=I$$

$C=\{l,D_i\}$ and l is a small positive scalar value which is called the locality parameter. It defines a vicinity of the point $x^{(i)}$ where the i-th term of the sum (3) has an essential influence on the behaviour of F(C,x). The locality parameter is only one which is included nonlinearly in the estimate (3) and it facilitates the adjustment problem. Moreover there are some reasonable ways how to choose the parameteres D_i under a fixed l.

The simplest one is a scheme of space filtration (SF) where

$$\hat{y}(C,x)=\hat{y}_n(l,x)=(nl^d)^{-I}\sum_{i=I}^{n} y^{(i)} h\left(\frac{x-x^{(i)}}{l}\right) \quad (5)$$

The main idea of SF based on the observation that (5) is a quadrature formula for the integral presentation

$$\hat{y}(l,x)=l^{-d}\int_{R^d} y(u)\, h\left(\frac{x-u}{I}\right)du \qquad (6)$$

and, for l small enough, y(l,x) is similar to y(x).

It can be shown that the transformation (6) is a space filtration of y(x) having the frequency response $H(j\omega l)$ if $H(j\omega)$ is the Fourier transformation of h(u) and

$$H(0)=I$$

For small l, we have a broad space frequency band and y(l,x) looses high-frequency components of y(x) only but the quadrature formula is less exact for a quickly changed kernel h(u,l). A receipt of a reasonable choice of the locality parameter will be presented in the following section.

Another way is used in the non-parametric approach [2] . The simplest non-parametric estimate is given by Nadaraja-Watson formula where

$$\hat{y}_n(l,x)=D(l,x)\sum_{i=I}^{n} y^{(i)} h\left(\frac{x-x^{(i)}}{I}\right) \qquad (7)$$

and

$$D^{-I}(l,x)=\sum_{i=I}^{n} h\left(\frac{x-x^{(i)}}{I}\right)$$

3. Precision of some LA estimates

The precision of LA depends on a smoothness of the approximated function, on properties of the locality function and, of course, on a repreesentability of the learning sequence.

Definition I. Let $f(x), x=(x_j)\in X\subset R^d$ be such that

$$|D^s f(x)| \le L_f, s=I,\dots,p,$$

$$|D^p f(x)-D^p f(x+u)| \le L_f \sum_{j=I}^{d} |u_j|$$

where D^s is any partial derivative of order s. Then

$$f(x)\in H(p), \quad \|f(x)\|_{H(p)}=L_f$$

Definition 2. A function $h(u), u\in R^d$ is called the potential having the astatism of order a, $h(u)\in S_a$, if

$$\int_{R^d} h(u)du=I; \int_{R^d} p^s(u)h(u)du=0, s=I,\dots,a$$

where $p^s(u)$ is an arbitrary polynom of power s.

Theorem. Let $\hat{y}_n(l,x)$ be given by the formulas (5) or (7). Let $x^{(k)}$ C^d where C^d is a unite hypercube, then, for any l, small enough,

$$I_n(l) = \sup_{x \in C_l^d} |y(x) - \hat{y}_n(l,x)| \leq \qquad (8)$$

$$\leq A_0 l^{r_0} + A_I l^{-r_1} \; g(n)$$

where

$$C_l^d = x / \; l^{I-\varepsilon} \leq x_j \leq I - l^{I-\varepsilon} \;, \; \varepsilon > 0 \;.$$

If $y(x) \in H(p), h(u) \in H(p), h(u) \in S_a$ and

$$|h(u)| \leq A \exp\{-\sum_j |u_j|\} \;, u \in C^d \qquad (9)$$

then $r_0 = p+I, p = \min(p,a), r_I = p$.

The values A_0, A_I do not depend on l and n. The function $g(n)$ is defined by a degree of uniformity of L_n on C^d. There exists a mesh $x^{(i)}, i = I, \ldots, n$, such that

$$g(n) = O(n^{-\frac{p+I}{d}}) \qquad (I0)$$

For any fixed L_n, there exists $l = l^*$,

$$l^* = O\left(g^{-\frac{r_0}{r_0 + r_1}}\right) \qquad (II)$$

such that

$$I_n^* = I_n(l^*) = O(n^{-\bar{n}}), \qquad (I2)$$

$$\bar{n} = \frac{(\bar{p}+I)(p+I)}{(p+\bar{p}+I)d}$$

<u>Proof</u> of the Theorem is based on the following lemmas.

<u>Lemma I</u>. Let $y(x) \in H(p), x \in R^d, h(u) \in S_a$, $u \in R^d$. Then

$$|\hat{y}(l,x) - y(x)| \leq C l^{\bar{p}+I}$$

<u>Lemma 2</u>. Let $\|y(x)\| \leq L_y, x \in C^d, h(u)$ be satisfying (9), then

$$\hat{y}(l,x) = l^{-n} \int_{C^d} y(u) h(\tfrac{x-u}{I}) du + O(\exp\{-l^{\varepsilon}\}) \,,$$

if $x \in C_l^d$.

<u>Lemma 3</u>. Let $f(x) \in H(p), x \in C^d$, then

$$\left| \int_{C^d} f(x)dx - \frac{I}{n} \sum_{k=I}^{n} f(x^{(k)}) \right| \leq L_f g(n)$$

and a mesh $x^{(k)}$ can be shown such that

$$g(n) = O(n^{-\frac{p+I}{d}}).$$

The Lemma I is a simple modification of the analogous result in [2]. The Lemma2 is proved by the direct estimation. The Lemma 3 is due to N.Bakhvalov [3]. The results of the Theorem can be obtained now if one takes into account the inequality

$$|y(x) - \hat{y}_n(l,x)| \leq |y(x) - \hat{y}(l,x)| + |\hat{y}(l,x) - \hat{y}_n(l,x)|$$

and the following evident properties

$$\|h(\tfrac{x-u}{l})\|_{H(p)} \leq l^{-p} \|h(u)\|_{H(p)}$$

$$\|y(u)h(\tfrac{x-u}{I})\|_{H(p)} \leq l^{-p} \|y(u)\|_{H(p)} \|h(u)\|_{H(p)}$$

for a l small enough.

The Theorem shows that the estimates (5), (7) have bad precision if the approximated function is not differentiable (p=0) and the space dimension d is high. However, for functions which are smooth enough, one can have good result for multidimensional case too if one uses the potentials of appropriate order of astatism.

The results can be compared with the Barron's estimate for the sigmoidal perceptron. It was shown in [I] that, for any fixed measure $\mu(x)$ there exists $\hat{C}(\mu)$ that

$$\int_{B^d} |\hat{y}(x) - y(x)|^2 d\mu = O(n^{-I}),$$

where $\hat{y}(x)$ is given by (I), $C = \hat{C}(\mu)$, B^d is the unit ball, and $y(x)$ is differentiable. The precision does not depend on d but, unfortunately, there are no ways how to find that $\hat{C}(\mu)$.

4. Conclusion

The local approximations schemes give reasonable results for the smooth functions and a moderate space dimension. The ANN realization can be based on 6π - elements [4] or special elements having the properties of the potential functions of some order of astatism. One can see that the classical Gaussian elements have the order a=I and it is rational to use such elements for functions $y(x) \in H(I)$. The results presented above can be extended on other classes of functions. In particular, it is rather simple to obtain precision estimates for functions having a bounded spectrum or some other classes of nondifferentiable functions.

References

I. Barron,A.: IEEE Tr.,v.IT-39,930(I993)
2. Katkovnik,V.: Non-parametric identification and data smoothing: method of local approximation,M.: Nauka I985
3. Bakhvalov,N.: Vestnik MGU,Mat.ser., N$^{\underline{o}}$5,3,(I959)
4. Gurney,K.: Neural Networks,v.5,289 (I992)

CHANGING NETWORK WEIGHTS BY LIE GROUPS

David W. Pearson

Laboratoire de Génie Informatique et d'Ingénierie de Production
EMA-EERIE
Parc Scientifique G. Besse
30000 Nîmes, France.
Email : pearson@eerie.fr

Abstract

In this paper we investigate how stability properties of a continuous recursive neural network within neighbourhoods of certain equilibrium points can be changed whilst keeping the equilibrium points fixed. This can be achieved, under certain conditions, by the use of Lie group transformations operating on the synaptic weight matrix.

Introduction

Assuming that a synaptic weight matrix has been calculated for a continuous recursive neural network, directly or by training, and certain equilibrium points exist. We investigate a method of modifying the synaptic weight matrix in order to change the structure of the vector field, defined by the network equations, in neighbourhoods around certain equilibrium points, whilst ensuring that these particular equilibrium points stay the same.

The fundamental tools that we make use of are Lie groups and Lie algebras. These methods form the basis of many studies of nonlinear dynamical systems. There are of course many references within the literature, but the much cited paper by Brockett [1] contains most of the background material used in this paper.

Preliminaries

The evolution of the network considered is governed by a differential equation that can be written in the form

$$\dot{x} = -x + \theta(Wx) \tag{1}$$

where $x \in R^n$, $W: R^n \to R^n$ is the synaptic weight matrix, $\dot{x} := \dfrac{dx}{dt}$ and $\theta: R^n \to R^n$ is a suitable C^∞ sigmoid function which we consider to act on each element of its argument. So, for example, the ith row of equation (1) would be

$$\dot{x}^i = -x^i + \frac{1 - e^{-w_{ij}x^j}}{1 + e^{-w_{ij}x^j}} \tag{2}$$

where x^i is the ith element of x , w_{ij} are the elements of W and we use the summation convention on "dummy" indices so that $w_{ij}x^j := \sum_{j=1}^{n} w_{ij}x^j$.

Suppose that a Lie transformation group G acting on R^n is given, then a subgroup of G called the isotropy group can be defined as [1]

$$H := \{h \in G : hx = x, x \in R^n\}$$

where x is arbitrary but fixed. Now if $gx_1 = x_2$ for $g \in G$ and H is the isotropy group at the point x_1 , then it can easily be proved that gHg^{-1} is the isotropy group at x_2 .

A (matrix) Lie algebra is defined by the usual operations from matrix algebra (here we consider an algebra over the reals) plus an extra operation defined by the Lie bracket [1] (note that we use the reverse to that in [1] but this is of no consequence in this paper)

$$[A, B] := BA - AB$$

for two $n \times n$ matrices A and B . Let L denote a Lie algebra, then a subalgebra $L_1 \subset L$ is defined simply by

$$\{A_i, A_j \in L_1 : [A_i, A_j] \subset L_1\}$$

which is written symbolically as $[L_1, L_1] \subset L_1$. To obtain a group element, we exponentiate an element from the algebra, ie we solve a differential equation in the form

$$\begin{aligned} \dot{g} &= Ag \\ g(0) &= 1 \end{aligned} \tag{3}$$

where 1 denotes the unit element of the group (the $n \times n$ identity matrix) and A is an element of the algebra. The solution to (3) is the absolutely convergent series

$$g(t) = \exp(tA) = 1 + tA + \frac{t^2}{2!}A^2 + \cdots$$

Also we have $g(t)g(-t) = 1 \Rightarrow g(t)^{-1} = g(-t)$.

It can be proved [1] that if K and L are subalgebras and $K \subset L$ then the corresponding subgroups $\exp(K)$ and $\exp(L)$ satisfy $\exp(K) \subset \exp(L)$.

Method

Assume that the set $\{x_1,\ldots,x_r\}$ is a subset of equilibrium points for the network (1) for a given synaptic weight matrix W. We want to be able to change the synaptic weight matrix by multiplying it by a group element that preserves the specific subset of equilibrium points, whilst changing some features of the vector field defined by (1) at the equilibrium points.

If H is the isotropy group at the point x_1 then we have

$$-x_1+\theta(Whx_1)=-x_1+\theta(Wx_1)=0, h\in H$$

and so multiplication on the right by an isotropy group element preserves the equilibrium point. Now, as shown above, we can transport the isotropy group structure to other points via further group transformations. Imagine now that it is possible to calculate a sequence of subgroups such that

$$G_k:=\{g\in G: gx_{k-1}=x_k\}, k=2,\ldots,r \tag{4}$$

$$G_r\cdots G_2HG_2^{-1}\cdots G_r^{-1}\subset\cdots\subset G_2HG_2^{-1}\subset H \tag{5}$$

then clearly

$$G_r\cdots G_2HG_2^{-1}\cdots G_r^{-1}x_i=x_i, i=1,\ldots,r \tag{6}$$

Hence the problem reduces to two steps. First of all, the calculation of a sequence of subgroups that satisfies (4) and (5), and secondly the choice of a particular element from the final subgroup to satisfy some criteria such as stability.

From [1] we see that a sequence of subgroups can be calculated satisfying (4) and (5) if we can compute a set of matrices to span the associated subalgebras which satisfies the following.

Symbolically $G=\exp(L)$ so L is the overall algebra and G the group. Now find $A\subset L$ such that $[A,A]\subset A$ and $\exp(t_kA_k)x_1=x_1, A_k\in A$ so A is a subalgebra corresponding to an isotropy group at x_1. Then calculate $K^{(1)}\subset L$ such that $[K^{(1)},A]=A^{(1)}\subset A$ and $\exp(s_1K_j^{(1)})x_1=x_2, K_j^{(1)}\in K^{(1)}$, followed by $K^{(2)}\subset L$ such that $[K^{(2)},A^{(1)}]=A^{(2)}\subset A^{(1)}$ and $\exp(s_2K_j^{(2)})x_2=x_3, K_j^{(2)}\in K^{(2)}$ and so on. This process has to stop eventually of course due to the finite dimensions involved. The dimension of the final subalgebra will depend upon the two parameters n,r. The synaptic weight matrix can then be modified by

$$\begin{aligned}
&W\mapsto Wh\\
&h=g_{r-1}(s_{r-1})\cdots g_1(s_1)h_p(t_p)\cdots h_1(t_1)g_1(-s_1)\cdots g_{r-1}(-s_{r-1})\\
&g_i(s_i):=\exp(s_iK_i^{(i)}), K_i^{(i)}\in K^{(i)}\\
&h_k(t_k):=\exp(t_kA_k), A_k\in A^{(r-1)}
\end{aligned} \tag{7}$$

The differential of the vector field defined by (1),(2) evaluated at an equilibrium point x_k is

$$-1+\theta'(Whx_k)Wh=-1+\theta'(Wx_k)Wh$$

by virtue of (6), where $\theta'(Wx_k):=\left[\dfrac{\partial\theta^i(Wx_k)}{\partial x^j}\right]$ and θ^i denotes the ith component of the function θ. Hence it is clear that if μ is an eigenvalue of the matrix $\theta'(Wx_k)Wh$ then $\lambda=-1+\mu$ is an eigenvalue of the network Jacobian matrix. Let

$$\theta'(Wx_k)Wz_i=\mu_iz_i$$
$$y_i^*\theta'(Wx_k)W=\mu_iy_i^*$$

be the left and right eigenvectors corresponding to a particular eigenvalue of the unmodified system, then it can be shown that for the modified system resulting from (7) the eigenvalues can be considered as functions of the t_j and near the origin [2]

$$\left.\frac{\partial\mu_i}{\partial t_j}\right|_{t_j=0}=\frac{y_i^*\theta'(Wx_k)WA_jz_i}{y_i^*z_i} \tag{8}$$

A solution of the set of equations (8) will then tell us how the eigenvalues will be modified by choices of the t_j, provided that the t_j are sufficiently small. Thus

$$\mu_i\approx\mu_i(0)+\frac{y_i^*\theta'(Wx_k)WA_jz_i}{y_i^*z_i}t_j$$

where $\mu_i(0)$ is the eigenvalue corresponding to the unmodified system.

Example

In the first instance we have investigated a very simple case where $G=GL(n)$, the general linear group of dimension n with an algebra spanned by the matrices E_{ij} which have zeros everywhere except in the ith row and jth column where there is a 1. We then find a set of matrices A_k whose exponentials generate the isotropy group at the point x_1. For example in the three dimensional case if $x_1^*=[\alpha,\beta,\gamma]$ then an isotropy group can be generated by exponentiating terms from the subalgebra spanned by the six matrices

$$A_1=\begin{bmatrix}-\beta&\alpha&0\\0&0&0\\0&0&0\end{bmatrix}, A_2=\begin{bmatrix}0&-\gamma&\beta\\0&0&0\\0&0&0\end{bmatrix}, A_3=\begin{bmatrix}0&0&0\\-\gamma&0&\alpha\\0&0&0\end{bmatrix}$$

$$A_4=\begin{bmatrix}0&0&0\\-\beta&\alpha&0\\0&0&0\end{bmatrix}, A_5=\begin{bmatrix}0&0&0\\0&0&0\\-\gamma&0&\alpha\end{bmatrix}, A_6=\begin{bmatrix}0&0&0\\0&0&0\\-\beta&\alpha&0\end{bmatrix}$$

We need to find a subalgebra, $A^{(1)}$, of this subalgebra and a set of E_{ij} such that

$$[A_k, A_p] = c^{v}_{kp} A_v$$
$$[E_{ij}, A_k] = c^{v}_{ij,k} A_v \qquad (9)$$
$$A_k, A_p \in A^{(1)}$$

for a set of real constants $c^{v}_{kp}, c^{v}_{ij,k}$.

It is easily verified that if we choose $A^{(1)} = \{A_1, A_8\}$ then (9) is satisfied by selecting $E_{ij} = E_{31} = \begin{bmatrix} 0 & 0 & 0 \\ 0 & 0 & 0 \\ 1 & 0 & 0 \end{bmatrix}$.

Hence the second equilibrium point is parametrised by s , ie

$$x_2 = \exp(sE_{31})x_1 = \begin{bmatrix} 1 & 0 & 0 \\ 0 & 1 & 0 \\ s & 0 & 1 \end{bmatrix} \begin{bmatrix} \alpha \\ \beta \\ \gamma \end{bmatrix} = \begin{bmatrix} \alpha \\ \beta \\ \gamma + s\alpha \end{bmatrix}$$

A rapid calculation corresponding to (7) then yields

$$\exp(sE_{31})\exp(t_2A_8)\exp(t_1A_1)\exp(-sE_{31}) =$$
$$\begin{bmatrix} e^{-\beta t_1} & \frac{\alpha}{\beta}(1-e^{-\beta t_1}) & 0 \\ 0 & 1 & 0 \\ s(e^{-\beta t_1}-1)-\beta t_2 e^{-\beta t_1} & \frac{s\alpha}{\beta}(1-e^{-\beta t_1})+\alpha t_2 e^{-\beta t_1} & 1 \end{bmatrix}$$

We chose

$$x_1 = \begin{bmatrix} 0.1 \\ 0.2 \\ 0.3 \end{bmatrix}, x_2 = \begin{bmatrix} 0.1 \\ 0.2 \\ -0.3 \end{bmatrix} = \exp(-6E_{31})x_1, x_3 = \begin{bmatrix} 0.5 \\ 0.4 \\ 0.1 \end{bmatrix}$$

where the points x_1, x_2 are the equilibrium points of interest to us and x_3 was chosen simply to make the set linearly independent. We calculated a synaptic weight matrix to fix these as equilibrium points by a simple gradient method. The eigenvalues at each of the two equilibrium points of interest were practically the same, $\{1.2526621e-1, -3.0747089e-2, -6.1123867e-2\}$, indicating unstable equilibria. In Fig. 1 we plot the vector field of (1) in the (x^1, x^2) plane whilst keeping $x^3 = 0.3$. Clearly $(0.1, 0.2)$ is an unstable equilibrium point and we can see the effect of this by solving the equation (1) starting at the initial point $x(0)^* = [0.3 \quad 0.3 \quad 0.3]$, the results for x^1, x^2 are shown in Fig. 2 where $* = x^1, + = x^2$.

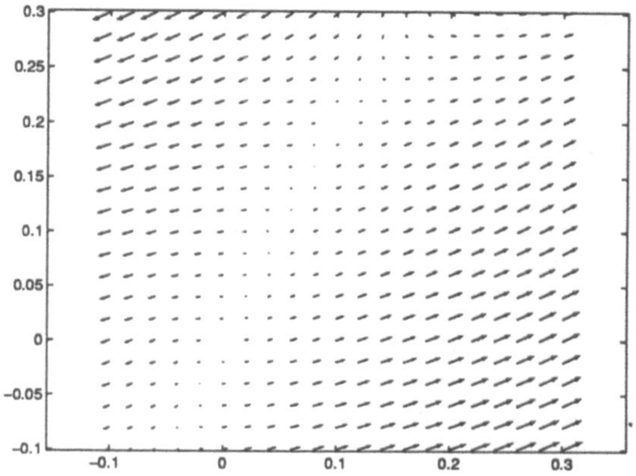

Figure 1

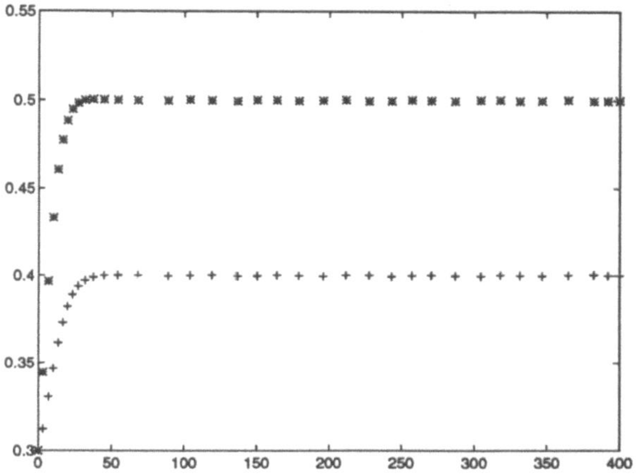

Figure 2

We then calculated for the point x_1 from (8)

$$\left.\frac{\partial\lambda_1}{\partial t_1}\right|_{t_1=0} = -2.1948026e-1, \left.\frac{\partial\lambda_1}{\partial t_2}\right|_{t_2=0} -1.3110475e-17$$
$$\left.\frac{\partial\lambda_2}{\partial t_1}\right|_{t_1=0} = -4.8003108e-3, \left.\frac{\partial\lambda_2}{\partial t_2}\right|_{t_2=0} = -4.1348659e-17$$
$$\left.\frac{\partial\lambda_3}{\partial t_1}\right|_{t_1=0} = 2.6544821e-18, \left.\frac{\partial\lambda_3}{\partial t_2}\right|_{t_2=0} = 5.0991620e-17$$

and the corresponding values calculated at the point x_2 were almost identical. Hence, these equilibrium points can be stabilised by a group transformation corresponding to $\exp(t_1A_1)$, the transformation $\exp(t_2A_8)$ having practically no effect. We chose $t_1 = 1, t_2 = 0$ and so calculated

$$\exp(-6E_{31})\exp(A_1)\exp(6E_{31}) =$$

$$\begin{bmatrix} 8.1873075e-1 & 9.0634623e-2 & 0 \\ 0 & 1 & 0 \\ 1.0876155 & -5.4380774e-1 & 1 \end{bmatrix}$$

The eigenvalues at the two equilibrium points of the modified system were again almost identical, $\{-1.9204353e-2, -8.9552375e-2, -6.1123867e-2\}$, indicating as predicted that the equilibria have been stabilised. In Fig. 3 we plot the vector field again and we can see how the transformation has affected this. Finally, in Fig. 4 we plot the convergence of x^1, x^2 with the transformed synaptic weight matrix and we see that they converge to $(0.1, 0.2)$ as we wanted them to.

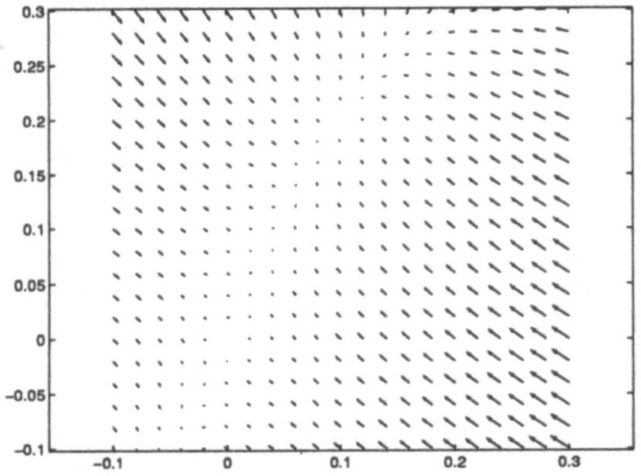

Figure 3

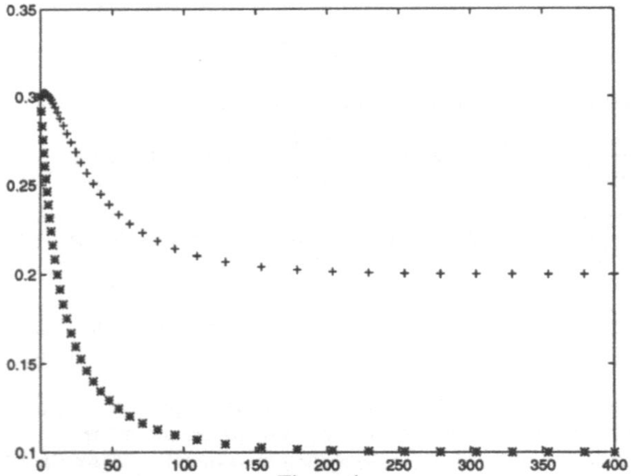

Figure 4

Conclusions

In this paper we have shown one way in which Lie groups may be used to change certain properties of a trained neural network, whilst keeping certain other desired properties fixed.

The results, whilst promising, certainly need more clarification. There are also many more theoretical issues to be researched, such as the choice of group transformations involved (we could replace $GL(n)$ by $O(n)$ for example). Our work is continuing in the application of these methods to the field of neural computing.

References

[1] Brockett, R.W.: SIAM J. Control, Vol. 10, No. 2, 265, (1972).

[2] Golub, G.H. and Van Loan, C.F.: Matrix Computatons, London, North Oxford Academic, 1986.

UNSUPERVISED LEARNING OF TEMPORAL SEQUENCES BY NEURAL NETWORKS

B. GAS

ENSEA-ETIS, 6 av. du Ponceau,
95014 Cergy-Pontoise cedex, FRANCE
(33) 1 30 73 62 85 ; email : gas@ensea.fr

R.NATOWICZ

ESIEE, Cité Descartes BP 99,
93162 Noisy-Le-Grand, FRANCE
(33) 1 45.92.67.14; email : natowicr@esiee.fr

Abstract

We propose to define a new model of formal neural network. This model extends existing Hopfield networks to process temporal data and achieve a non-supervised learning of them. We propose a learning law to adress in this context the sensitivity to input changes. A spatial representation of network's temporal activity is given by which learnt sequences can be identified. An example of such a network is given and the results of the simulation are presented.

1 Introduction

The work presented here aims at defining and simulating a formal neural network for non supervised learning and recognition of temporal sequences of symbols.
A symbol sequence is coded by spike trains presented on network input cells. Figure 1 is an example of such a sequence, $e_1, e_2, e_3, e_4, e_1, e_2, \ldots$, that can be learnt by a network having two input cells. Symbol e_1 is encoded by simultaneous presentation of a spike burst of same length on both input cells, symbol e_2 is coded by presentation of spike burst on input cell 1 and silence of same length on input cell 2, etc. Such a network will react to a learnt sequence by synchronizing its "activity" with it. The "average activity" of the network takes the shape of a regular motif, characterizing the sequence (figure 3). Such a network will not react to an unknown sequence and one will observe along the time a centered noise around an average activity (figure 4). To achieve such a network global behaviour, one associate two computation functions to every cell:

1. an "evoked" computation function : binary cell output is deterministic and depends only on its inputs (output of pre-synaptic cells and input signal for network input cells, output presynaptic cells for non input ones).
2. a "spontaneous" computation function : binary cell output does not depend on its past or present inputs. A random or regular spike generators are instance of such functions.

At every discrete time, every cell selects its computation function according to its past and present inputs. The network "average activity" is, at every time, the ratio of the number of cells computing in the evoked mode to the number of cells computing in the spontaneous mode. Defining two modes of computation for every cell can also find a biological justification in the fact that "spontaneous discharges" can be measured on axons of isolated "real neurons" [3]. By extension, one assumes that "spontaneous discharges" could be measured on axons of connected "real neurons".
In our model, synaptic efficacy of a connection is an integration time of signals going through it and the learning process is exclusively an adjustment of these integration times.
In the following, one formally defines the discrete sequences processed, the network and learning model linked to the recognition criterion and the spatial representation of resonance between input signal and cell average activity. One also gives examples of simulations.

2 Temporal sequences

One considers a network of fully connected cells and distinguishes a small subset $\mathcal{IC}$ of "input" cells on each of which one can force temporal sequences as inputs. Such inputs are sequences built on three specific trains of constant time length τ (time is discrete). Denoting $pr(s_{t-u} = 1)$ the probability to have a spike at time $t-u$, the three specific trains are:

- "firing train" $st_{fire}(t, \tau)$:
  $\forall u \in [0, \tau], pr(s_{t-u} = 1) = 1;$
- "void train" $st_{void}(t, \tau)$:
  $\forall u \in [0, \tau], pr(s_{t-u} = 1) = 0;$
- "random train" $st_{rand}(t, \tau)$:
  $\forall u \in [0, \tau], pr(s_{t-u} = 1) = 1/2.$

Let $\mathcal{I} = \{st_{fire}, st_{void}, st_{rand}\}$ be the set of input trains. At time $t = 0$, one imposes a spike train on every input cell. This input is a vector of $\mathcal{I}^{|\mathcal{IC}|}$. The set $\mathcal{E} = \{e_1, e_2, ..., e_{|\mathcal{E}|}\}$ of input vectors is the alphabet of symbols on which one can build any input sequence for the network. Thus, any sequence $\sigma = \sigma_1\sigma_2...$ with $\sigma_i \in \mathcal{E}$ is a possible input sequence. One will focus attention on the learning of non random regular sequences $\sigma = \sigma_1...\sigma_k\sigma_1...\sigma_k... = (\sigma_1...\sigma_k)^+$ where k is the length of the shortest repeated pattern of the sequence and, by abusive extension, the length of the sequence. Figure 1 illustrates the definition of input sequences for a network with two input cells. One has $|\mathcal{IC}| = 2, \overline{\mathcal{E}} = \{(st_{fire}, st_{fire}), (st_{void}, st_{fire}), (st_{fire}, st_{void}), (st_{void}, st_{void})\} \subset \mathcal{E}, |\overline{\mathcal{E}}| = 2^{|\mathcal{IC}|} = 4$ and $\sigma = (e_1e_2e_3e_4)^+$.

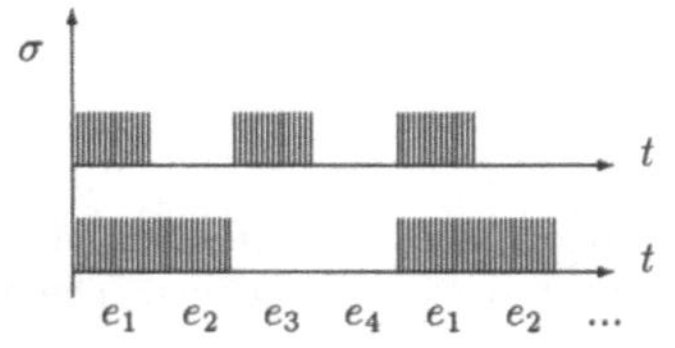

Figure 1 : A temporal sequence example

3 Extending discrete Hopfield networks for unsupervised learning of temporal sequences

In Hopfield discrete time neural networks [4], whether synchronous or asynchronous, any cell of the network computes its output as a threshold of its membrane potential, this membrane potential being the sum of its inputs, weighted by the strengthes of its presynaptic connections. One proposes to extend this computation into two directions in order to process temporal sequences and unsupervised learning of regular sequences. The first extension consists in defining the strength of a synapse as the width of a temporal interval for summation of the signal going through it. One stresses that the connection strengthes are the only parameters subjet to learning in our model and that in the learning process, cells modify the strengthes of their presynaptic connections according to a local criterion.
The second extension is motivated by non supervision, which means in our context that no other signal than the input sequence is sent to the network (reward/punishment, error signal or any other). Qualitatively one wants the network to have an "activity" in resonance with a recognized input sequence and conversely, a random "activity" if not. To achieve this kind of behavior one proposes to let every cell of the network choose at any time one function out of two, by which it will compute its output at that time. The first function is the classical threshold of membrane potential and one calls it the evoked activity law. Through the second function a cell can decorelate its output from past and present output of its presynaptic cells. One calls such a function the spontaneous activity law. In the present model we decided to define this spontaneous activity law as a probabilistic one.

4 Formal definition of the model

Every cell's connection computes a temporal integration of the spikes going through it. This temporal integration is called "contribution of the connection to cell's membrane potential".
Let us denote

- ω_{ij} the jth connection of cell C_i,
- $pre(\omega_{ij})$ be presynaptic cell of connection ω_{ij} and $pre(C_i)$ the set of all presynaptic cells of cell C_i,
- $sgn(\omega_{ij}) \in \{-1, 1\}$ the class, excitatory or inhibitory, of presynaptic cell $pre(\omega_{ij})$,
- $\Omega(C_i)$ the set of all presynaptic connections of cell C_i,
- $T_{ij}(t) \in \mathbf{N}$ the strength of the connection at time t,

and let $\nu_{ij}(t)$ be the summation of spikes input to connection ω_{ij} during the last $T_{ij}(t)$ units of times :

$$\nu_{ij}(t) = \sum_{u=1}^{T_{ij}(t)} s_j(t-u)$$

were $s_j(t-u) \in \{0, 1\}$ is the output of cell $C_j \in pre(C_i)$ at time $t-u$.
The contribution of connection ω_{ij} to membrane potential of cell C_i is such that firing and void incoming trains have a high contribution and random trains a low one; it is obtained by centering summation $\nu_{ij}(t)$ at 0 :

$$v_{ij}(t) = \nu_{ij}(t) - \tfrac{1}{2}T_{ij}(t)$$

Let $\mathcal{V}_i(t)$ denote the potential membrane of cell C_i at time t, $\mathcal{V}_i(t)$ is the sum of input connection contributions:

$$\mathcal{V}_i(t) = \sum_{\omega_{ij} \in \Omega(C_i)} sgn(\omega_{ij})v_{ij}(t)$$

Since all contributions to membrane potential are centred on 0, one has :

$$-\mathcal{V}_i^{max}(t) \leq \mathcal{V}_i(t) \leq \mathcal{V}_i^{max}(t)$$

where $\mathcal{V}_i^{max}$ denotes the maximum value of membrane potential $\mathcal{V}_i(t)$, i.e. the total integration time :

$$\mathcal{V}_i^{max}(t) = \tfrac{1}{2} \sum_{\omega_{ij} \in \Omega(C_i)} T_{ij}(t)$$

The evoked and spontaneous activity laws f_{ev} and f_{sp} are, $\forall i, \forall t$:

$$\begin{cases} f_{ev}(\mathcal{V}_i(t)) = \begin{cases} 1 \text{ if } \mathcal{V}_i(t) \geq 0 \\ 0 \text{ if } \mathcal{V}_i(t) < 0 \end{cases} ; \\ f_{sp} \quad : \quad pr(s_i(t) = 1) = 1/2. \end{cases}$$

The "evocation criterion" P_{ev} is the probability the evoked activity to be chosen by cell $\mathcal{C}_i$ at time t :

$$pr(f_i(t)=1) = P_{ev}(\mathcal{V}_i(t), \mathcal{V}_i^{max}(t))$$

were :

$$\begin{cases} f_i(t)=1 & \Leftrightarrow \quad \mathcal{C}_i \text{ is evoked at time } t \\ f_i(t)=0 & \Leftrightarrow \quad \mathcal{C}_i \text{ is spontaneous at time } t \end{cases}$$

Qualitatively one wants any cell to choose the evoked activity law if it "perceives" its presynaptic neighbourhood as evoked. This situation occurs when the membrane potential is very high ("strong exitation") or very low ("strong inhibition"). More precisely, evocation criterion P_{ev} is defined to fulfill the three following requirements :

1. Every network cell $\mathcal{C}_i$, when evoked, sends on its output a firing train or a void train. (Because of function f_{ev} shape, it leads $\mathcal{C}_i$ to be evoked only if the membrane potential $\mathcal{V}_i(t)$ approaches its border values $\pm\mathcal{V}_i^{max}(t)$).
2. For a network cell to be evoked it is necessary that a subset of its presynaptic cells are evoked.
3. A subset of evoked presynaptic cells can force a cell to be spontaneous (depending on the ratio of exitatory to inhibitory cells).

In this paper, one proposes the following symetric evocation criterion (figure 2):

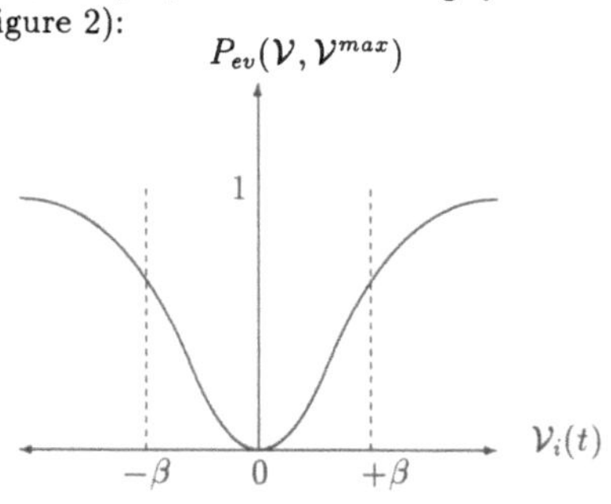

Figure 2 : evocation criterion

where β is a parameter of the model (cf. [2] for more details). One can notice that recurrent boolean networks are limit cases of the model for $\beta = 0$ and $T(\omega_{ij}(t) = 1, \forall\, i, j, t$.

5 Learning criterion

In a previous paper [1] one presented such a model to learn and recognize binary temporal sequences. A very simple "interest function" was proposed as the goal to the adaptive process : *any cell of the network "tries", when evoked, to increase its evocation.* This law leads the membrane potential of every network cell $\mathcal{C}_i$ to increase up to its border values ($\pm\mathcal{V}_i^{max}(t)$). As a consequence, cells became more and more insensitive to small variations of their input signals and the network stability took shape of insensitivity to new sequences. To overcome this problem one proposes to define the following goal to the learning process :

any cell of the network "tries", when reaching the evoked mode of processing, to increase it.

Our translation of this local criterion of learning is the following one : denoting $\Delta T_{ij}(t)$ the integration time modification of connection ω_{ij} at time t and $a_i(t) \in \{0,1\}$ be a function such that cell $\mathcal{C}_i$ is learning at time t iff $a_i(t) = 1$,

$$T_{ij}(t+1) = T_{ij}(t) + a_i(t)\Delta T_{ij}(t)$$

with

$$a_{ij}(t) = \begin{cases} 1 \text{ if } f_i(t-1) = 0 \text{ and } f_i(t) = 1 \\ 0 \text{ else} \end{cases}$$

One proposes to increase the current cell evocation $P_{ev}(\mathcal{C}_i, t)$ by decreasing the ratio $N_{soll}^*(\mathcal{C}_i, t)/N_{soll}(\mathcal{C}_i, t)$ where $N_{soll}(\mathcal{C}_i, t)$ $= |\Omega_{soll}(\mathcal{C}_i, t)|$ is the number of connections that currently support firing or void trains and $N_{soll}^*(\mathcal{C}_i, t) = N(\mathcal{C}_i, t) - N_{soll}(\mathcal{C}_i, t)$ is the number of the other ones:

$$\Delta T_{ij}(t) = \begin{cases} +1 & \text{if} \quad \omega_{ij} \in \Omega_{soll}(\mathcal{C}_i) \\ -1 & \text{if} \quad \omega_{ij} \in \Omega_{soll}^*(\mathcal{C}_i) \end{cases}$$

6 Recognition criterion and simulations

A measure of the network evocation, the "average activity", was given in [1] :

$$\Upsilon(t) = \frac{1}{N(\mathcal{R})} \sum_{i=1}^{N(\mathcal{R})} f_i(t),$$

$N(\mathcal{R})$ denoting the network's cells number. This average activity brings the following recognition criterion: *a temporal sequence is considered as recognized by the network if its presentation brings a totally evoked state of the network, that is* $\Upsilon(t) = 1$. This could not fit with the task of recognizing several sequences but only with the task of recognizing or not a single one;

We propose in this paper to give a spatial representation of the network's average activity by which learnt sequences can be identified. Qualitatively, the network remains sensitive if its evocation $\Upsilon(t)$ is synchronized on the current input sequence. Because of the input signal periodicity (length $l(\sigma)$ of the input sequence σ), one must have $\Upsilon(t + k.l(\sigma)) = \Upsilon(t)$ for some $k > 0$. It is then possible to build a spatial representation of the network activity in order to observe its sensitivity to input sequences. We propose the following polar coordinates one :

$$\begin{cases} x(t) = \Upsilon(t)\cos(2\pi\frac{t}{l(\sigma)}) \\ y(t) = \Upsilon(t)\sin(2\pi\frac{t}{l(\sigma)}) \end{cases}$$

Figures 3,4,5 and 6 are examples of this representation. Each figure required 10 presentations of the input sequence which approximatively corresponds to 1500 measures of the average network evocation $\Upsilon(t)$.

One describes now the simulation of a neural network, fully connected with self-connections. It has $N(\mathcal{R}) = 32$ cells (16 exitatory and 16 inhibitory) with 2 input cells.

1. Sequence σ_A learning. In this experimentation, an average of 20 presentations of sequence $\sigma_A = (e_2, e_3)^+$ (see figure 1 for sequence components e_i) is enough to obtain a stable regular pattern of activity. Figure 3 represents the network activity during 10 presentations of sequence σ_A after that the network stabilized its connection strengthes.

2. Unknown sequence σ_B as input. In this experiment, the sequence $\sigma_B = (e_1, e_4)^+$ is presented after sequence σ_A learning. The network cells remain spontaneous and this leads the pattern of activity to be a zero-centered noise (figure 4). Sequence σ_B is not recognized.

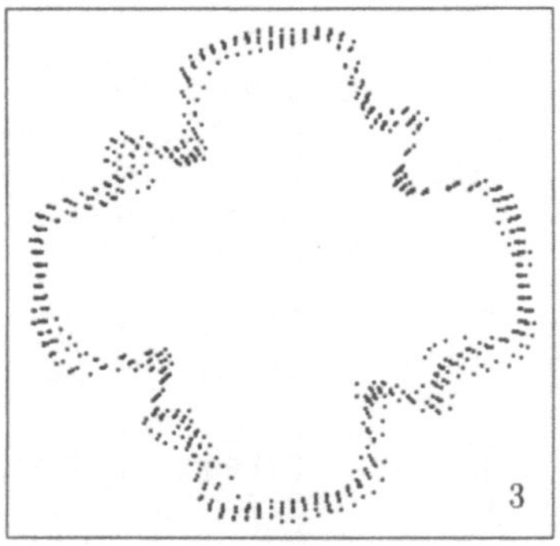

3

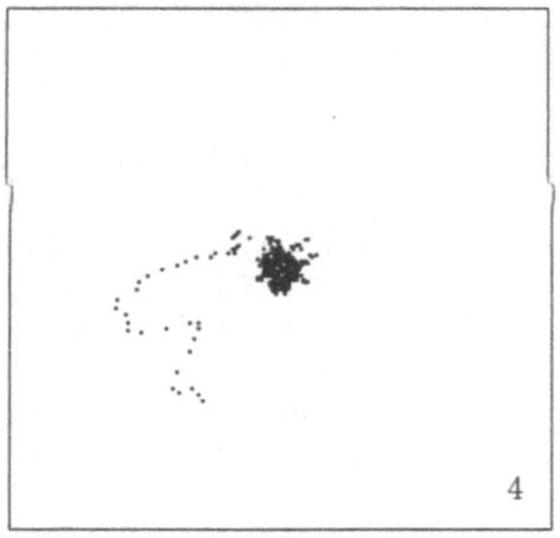

4

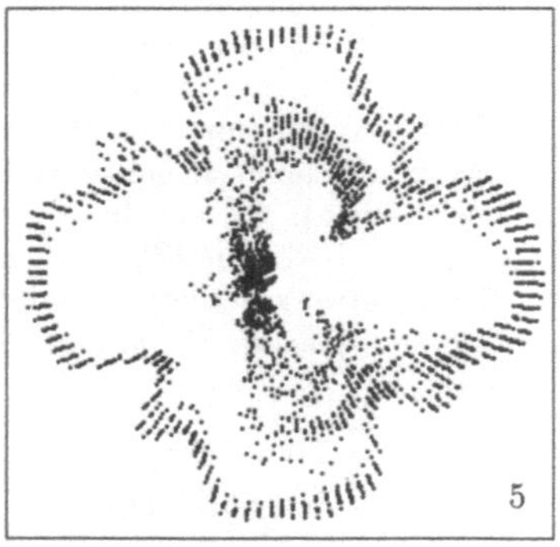

5

Figure 5 shows the pattern of activity when sequence $\sigma_C = (\sigma_A\sigma_B)^+$ is input to the network. Sub-sequence σ_B drives network's activity to zero. Network recovers the activity pattern characteristic of sub-sequence σ_A at every time it appears on the input.

3. Sequence σ_B learning. Same results hold when sequence σ_B is learnt and sequence σ_A is input. Figure 6 shows that the regular pattern for σ_B can be easely distinguished from the pattern of σ_A. It also shows the low activity (inside the σ_B motif) obtained when presenting the unknown sequence σ_A to the network, depicting the fact that sequence σ_A is not recognized.

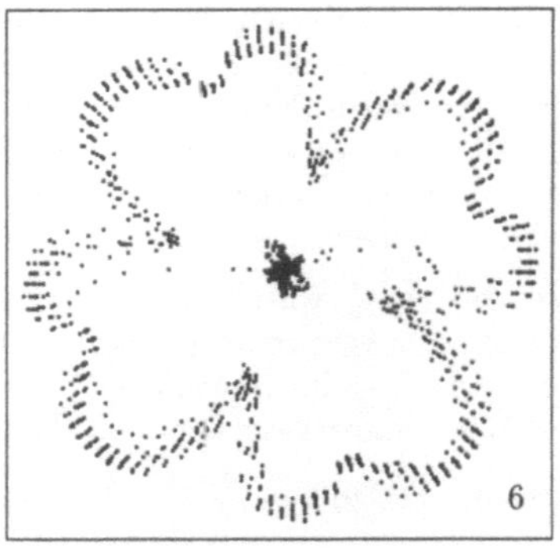

6

7 Conclusion

One defined a connectionist model which extends Hopfield model to process temporal data and implement a non-supervised learning of them.

The non supervised learning was achieved under the hypothesis that any cell of the network can have two distinct modes of processing, spontaneous and evoked, depending upon the activity of its presynaptic cell neighbourhood.

A spatial representation of the average network activity was given in order to observe the network activity and sequence recognition. Examples of simulations were described.

Further developments will be devoted to explicit classification of sequences by the network itself.

References

[1] B. Gas, R. Natowicz. *A model of formal neural networks for unsupervised learning of binary temporal sequences.* IJCNN, Baltimore (1992).

[2] B. Gas. *Un modèle connexionniste non supervisé pour l'apprentissage et la reconnaissance de séquences temporelles.* PhD thesis, Université de Paris XI (1994).

[3] B. Changeux. *L'Homme Neuronal.* Fayard 106,110 (1983).

[4] J.J. Hopfield. *Neural networks and physical systems with emergent collective computational abilities* Proc. Nat. Acad. Sci. USA, vol.79, pp.2554-2558 (1982).

MULTIVARIATE TIME SERIES MODELLING OF FINANCIAL MARKETS WITH ARTIFICIAL NEURAL NETWORKS

Thomas Ankenbrand
Institut d'Informatique
Faculté des Sciences
University of Lausanne
Switzerland

Marco Tomassini
Swiss Centre of Scientific Computing
Manno
and
Laboratoire de Systèmes Logiques
Departement d'Informatique
Ecole Polytechnique Fédérale de Lausanne
Switzerland

Abstract

This work presents an integrated approach for modelling the behaviour of financial markets with Artificial Neural Networks (ANNs). The model allows to forecast financial time series. Its originality lies in the fact that it is based on statistics and macroeconomics principles and it integrates fundamental economic knowledge in a multivariate nonlinear time series ANN model. The model is applied to real-life case studies and the results are discussed.

1. Introduction

ANNs are successfully applied on supporting technical analysis of financial markets. But the univariate technical approach could fail for several reasons. The market could be efficient and only driven from outside indicators. Or the available time series are too short for a significant technical analysis with the chosen forecasting horizon. This modelling method integrates the multivariate fundamental forces with ANNs. It allows to forecast difficult financial markets under difficult circumstances.

We firsts present a method for the identification of the relevant market indicators. Traditional multivariate statistics and a simple ANN test identify linear and nonlinear relations between predictor and possible indicators. These results allow to design an ANN for the forecast. The implicit knowledge of the ANN is a stable quantitative fundamental multivariate model. The described method is applied on different financial markets. This paper presents an application for monthly forecasting of the Swiss Performance Index (SPI). The difficulties are the undersampling and the trend over the whole training period.

Section two contains a description of the system and the feasibility analysis. The feasibility analysis contains the identification process of indicators with linear and nonlinear, univariate and multivariate analysis. In section three the ANN is designed based on the results of the feasibility analysis. Section three discuss also the results of the SPI application. Section four gives our conclusion and a future outlook of the work.

2. System Description and Feasibility Analysis

The SPI is a stock index of Switzerland. It shows the development of a broad diversified portfolio. It contains all titles of the three Swiss bourses. The SPI is member of the SwissIndex-family. The calculation is based on the Laspeyres formula with the weight of capitalisation. The data are available monthly from January 1987 to December 1994. Possible indicators for modelling are the S&P 500, the DEM/USD exchange rate and the average bond interest rate of CHF and USD. The values are the close of the latest trading day of the month. They are available for the same period as for the SPI.

2.1. Data Preprocessing

Plotting the observation against time is the first step in any time series analysis. Fig. 1 shows the values of the SPI index over the 1987-1994 period and gives a first idea of the long-term behaviour.

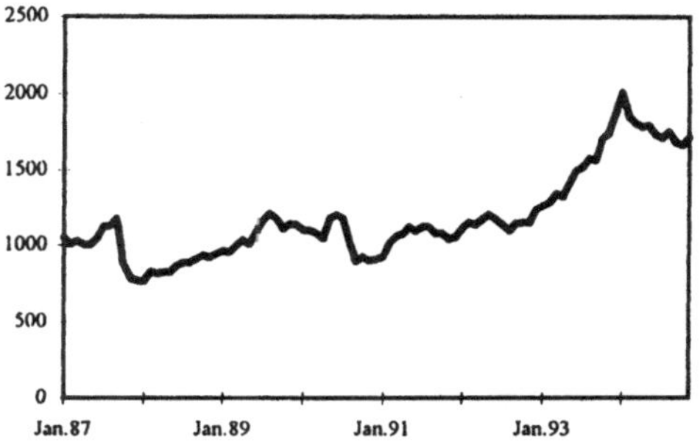

Figure 1

The use of the monthly differences of the SPI detrends the predictor. Another possibility is the adjusting of the SPI with the inflation rate [1].

Since the network has a sigmoid output node and the indicators are of different orders and magnitude, there is a need for scaling for the predictor and the indicators. The predictor must lie between 0 and 1, because the sigmoid function produces a value in the range [0...1], but it is hard to reach the extremes of the range. For that reason and to be aware of exceeding the limit by future values the difference of the SPI is scaled among 0.2 and 0.8. The indicators are scaled in the range [-1...1]. The logarithms of the two stock indices are used, because they have some outliners. The scaling formula for the predictor and the indicators is [2]:

$$y = \frac{x(T_{max} - T_{min})}{x_{max} - x_{min}} + T_{min} - \frac{x_{min}(T_{max} - T_{min})}{x_{max} - x_{min}}.$$

where

y is the rescaled value.
x is the raw or unscaled data value.
T_{min} = target minimum (0.2 for predictor, -1 for indicators)
T_{max} = target maximum (0.8 for predictor, 1 for indicators)
x_{min} = raw minimum
x_{max} = raw maximum.

The monthly differences of the SPI are scaled with the following scaling and offset factors:

$$y = \frac{0.2x * \log(|x|)}{|x|} + 0.5$$

The scaling factors for the indicators are:

DEM/USD $y = 3.39x - 5.71$

difference S&P 500 $y = \frac{x * \log(|x|)}{|x|}$

difference bond CHF $y = 1.83x + 0.05$

2.2. Univariate Analysis

The monthly differences of the SPI do not show any significant linear autocorrelations at the 95 per cent significance level. The results come from the autocorrelation and the partial autocorrelation function.

According to the cumulated periodogramm test, SPI appears to be a random walk. The cumulated periodogramm is a nonlinear parametric test. A nonlinear nonparametric test is the calculation of the Hurst exponent (H). It is a measure of bias in a random walk process which is based on the growth of the standard deviation by the number of observations. A normal distributed random process grows with the square root of the number of observations. The square root means a Hurst exponent of 0.5. A Hurst exponent greater than 0.5 ($0.5 < H < 1$) means a trend-reinforcing behaviour of the time series. With H lesser than 0.5 ($0 < H < 0.5$) the series is antipersistent [3]. The SPI is a trend-reinforcing process with a Hurst exponent of 0.78. Such a process has a local randomness but a global structure. It is not a random walk and therefore it should be predictable to some extent.

2.3. Multivariate Analysis

The linear dependencies between the indicators and the predictor are measured with the cross correlation function. Both the raw indicators and the monthly differences are tested. The 95 per cent significance level is about +/- 0.22. The following Fig. 2 shows the results of the cross correlation of the raw indicators.

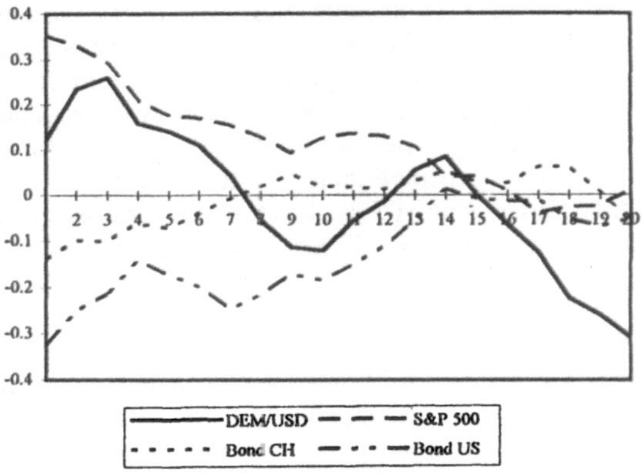

Figure 2

Fig. 3 shows the results of the cross correlation's of the monthly differences.

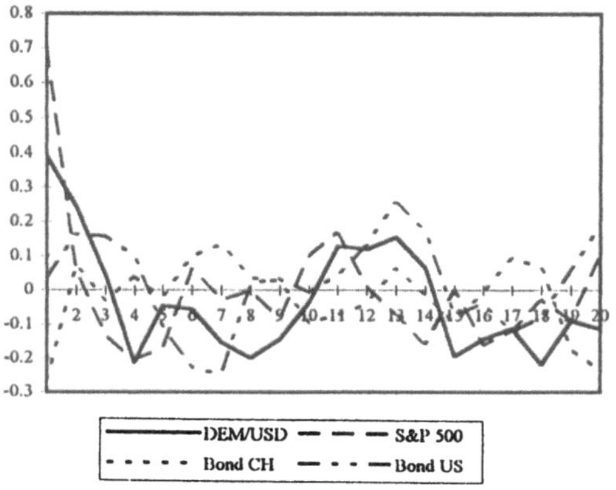

Figure 3

The nonlinear nonparametric multivariate dependencies are measured with an ANN [4]. In this case the ANN has 3 input, 1 hidden and 1 output unit. The input is the indicator time series and the output the predictor series. The test counts the learning cycles. The learning stops if the error function increases. The learning rule is backpropagation with a learning rate and momentum of 0.5. A visual inspection of the error curve allows the ranking of the indicators. The strengths of the nonlinear influence are:

- difference interest rate CHF
- interest rate CHF
- DEM/USD
- differences S&P 500
- differences DEM/USD
- differences interest rate USD

The other indicators do not have any significance since the ANN is not able the learn a relation between those indicators and the predictor. The error curve does not decrease and does not converge to a minimum.

2.4. Economic Review

The results of the feasibility analysis are not surprising. The influence of the interest rate CHF on the stock market is well known. If the interest rates are high it is more interesting to invest money in bonds than in stocks. Consequently the price of stocks decreases. The influence of the exchange rate DEM/USD is strong for the Swiss market, because the whole industry of Switzerland is very sensitive to exports. The relation of American and the Swiss market manifests itself in the relation of the S&P 500 and SPI and also in the relation of the interest rates of USD and SPI.

The advantage of our ANN model is the combination of the influences from the indicators onto the SPI. It is a quantitative fundamental model without any technical influence.

3. Neural Network Design and Results

There are hardly any systematically investigations for the optimal design. The degrees of freedom of the ANN are determined by the number of connections. An important rule of thumb is the relation of degrees of freedom of the ANN and the number of datasets.

$$c = \frac{1}{10} n$$

where

c is the number of connections.
n is the number of datasets.

The time series have only 96 datapoints. An expansion of the time series is impossible, because the calculation of the SPI starts in 1987. The maximum size for ANN is 3 input, 2 hidden and 1 output units with 10 connections or degrees of freedom. More degrees of freedom allow the ANN to adapt to noise. This overlearning decreases the generalisation capabilities. The ANN does not recognise the general structure of the market.

The results of the feasibility analysis determine the kind of input. The DEM/USD exchange rate, the monthly differences of S&P 500 and the monthly differences of the CHF interest rate have the strongest nonlinear influence. The lagged value of the SPI has also a strong influence. With a Hurst exponent of 0.78 it is a strong trend reinforcing process. This means that the chance of the same direction of change is good for the SPI. The ANN would learn this relation with a lagged SPI value as input. The result would be a trend following system. But in a trend market the traditional linear models work very well. The nonlinear possibilities of ANN can be used for forecasting trend breaks. In this area the linear trend following models fails. The ANN forecasting model is:

$$y(t+1)-y(t) = f[e(t),\ s(t)-s(t-1),\ i(t)-i(t-1)]$$

where

y is the SPI.
e is the DEM/USD exchange rate.
i is the interest rate CHF.
s is the S&P 500.

The learning rule is backpropagation. The training set consists of the learning and validation/test set. The validation and test sets are not separated, because the data base is too small. The validation/test set is built by taking data every ninth data sample. The size of the learning set is 84 and that of the validation set is 10. The learning stops if the error over the validation set increases. This prevents the network from overfitting. The learning rate and the momentum are 0.8 for the first 400 cycles, then 0.5 for 200 cycles and for the last 200 cycles 0.2. Starting with a high learning rate and momentum prevents the ANN from being trapped in a local minimum.

A formal performance indicator is the normalised mean square error, but for practitioners in the financial field the trend forecasting is a more useful indicator. The chosen ANN correctly forecasts the trend in 78 per cent of the cases in-sample. The out-of-sample trend forecasting performance over the validation/test set is 70 per cent, which is quite good. The estimation of the trading profit is difficult, because the test set does not consist of successive data samples.

4. Conclusions

The performance of the model seems to be very good. A definitive judgement could be made after more and following test samples with the estimation of the trading profit. The main difficulty of forecasting the SPI on a monthly basis is the short time series. The model is based only on fundamental indicators. It does not use any lagged SPI values. This seems to be the reason for the good middle term accuracy of the prediction.

Future work will include a deeper analysis of the results and comparisaison with univariate ANN, a technical ANN model with technical indicators like volume, volatilities etc. and traditional statistical linear forecasting models.

5. References

1. Peters, E.E.: Chaos and Order in the Capital Markets. John Wiley & Sons, 1991, 165.

2. Klimasauskas, C.C.: Neural Network Techniques. In Deboeck, G.J. (Ed.): Trading on the Edge. John Wiley & Sons, 1994, 13f..

3. Peters, E.E.: Fractal Market Analysis. John Wiley & Sons, 1994, 56ff..

4. Lee, T.-H., White, H., Granger, W.J.: Testing for neglected nonlinearity in time series models. In Journal of Econometrics. Elsevier Science Publishers, 1993, 269ff..

NEURAL NETWORK APPROACH TO PISARENKO'S FREQUENCY ESTIMATION

Fa-Long Luo and Rolf Unbehauen (Fellow, IEEE)

Lehrstuhl für Allgemeine und Theoretische Elektrotechnik, Universität Erlangen-Nürnberg, Cauerstraße 7, 91058, Erlangen, Germany

Abstract

This paper proposes a neural network approach for computing in real-time the required eigenvector in Pisarenko's frequency estimation method. We will show both analytically and by simulations that this neural network approach has some advantages over the available neural network methods and is more suitable for real-time applications of Pisarenko's frequency estimation method.

1. Introduction

The estimation of the frequencies of multiple sinusoids in the presence of noise is a significant problem in signal processing. Many high-resolution frequency estimation techniques have been developed and extensively analysed. Pisarenko's method is an example, which utilises the orthogonal relationship between the signal vector and the eigenvector corresponding to the minimum eigenvalue of the covariance matrix of the underlying data. However, it is not easy to implement Pisarenko's method in real-time mainly because of the costly computational complexity of the eigendecomposition. Although some effective sequential methods have been reported [1-2], it is still difficult to provide the desired performance.

Recently, some neural network approaches have been reported for estimating the eigenvector corresponding to the minimum eigenvalue of the covariance matrix of the underlying data. One is the anti-Hebbian learning algorithm proposed by Xu, Oja and Suen in Reference [3], one is the normalized version of the anti-Hebbian learning algorithm reported in the same reference and further studied in Reference [4]. The convergence of the anti-Hebbian algorithm can not be guaranteed because the magnitude of the weights of the neural network could tend to infinity as pointed out in [4]. The normalized version of this anti-Hebbian algorithm does not encounter this problem, however, this improved algorithm requires the division computation in the learning phase. Although the algorithm developed by Oja in Reference [5] eliminates the division computation, it requires the assumption that the minimum eigenvalue of the covariance matrix of the data is less than unity. Mathew and Reddy have also developed a neural network approach for computing the eigenvector required in Pisarenko's frequency estimation by constructing an appropriate energy function and using the penalty function approach [6]. They have shown that the eigenvector corresponding to the minimum eigenvalue is a minimizer of the energy function and all its minimizers are global minimizers. However, the norm of the eigenvector provided by this neural network approach is not unity so that additional division computations are needed for the desired unit eigenvector in Pisarenko's frequency estimation method. In addition, it is necessary in this neural network approach to estimate the trace of the covariance matrix for selecting appropriate penalty factors.

For the purpose of a more practical use, this paper proposes an alternative neural network approach for computing the required eigenvector in Pisarenko's frequency estimation method. The key features of the proposed approach are as follows: 1. This approach is based on an analog circuit architecture which has continuous-time dynamics, as a result, the convergence time of the network (the time to reach closely the stable state) is during an elapsed time of only a few characteristic time constants of the circuit, which means that this approach can provide in real-time the desired eigenvector. 2. This approach can still provide the desired eigenvector even if the minimum eigenvalue of the covariance matrix is not single, that is, we have no assumption for the minimum eigenvalue of the covariance matrix. 3. The norm of the output vector of the network is invariant and equal to the norm of the initial vector during the time evolution, so that we can provide the desired unit eigenvector by letting the norm of the initial vector be unity. This property is also useful in designing a hardware implementation of the proposed approach. 4. The covariance matrix is directly taken as the network parameters without any computation, so we do not encounter the "programming complexity" problem pointed out in [7]. 5. Not any division computation is needed in this proposed approach.

All the above demonstrates that the proposed approach is suitable for real-time applications of Pisarenko's frequency estimation method.

This paper is organized as follows: In Section 2, Pisarenko's frequency estimation method will be recalled. Section 3 will deal with the architecture and theoretical analysis of the proposed neural network model. In Section 4 and 5, we will show how to use this network to compute the eigenvector required in Pisarenko's frequency estimation method and simulation results.

2. Pisarenko's Frequency Estimation Method

Let us consider a received signal consisting of P sinusoids corrupted by white noise, i.e.

$$x(n) = \sum_{i=1}^{P} \alpha_i cos(\omega_i n + \theta_i) + w(n) \tag{1}$$

$$(\text{for } n = 1, 2, ..., L)$$

where $x(n)$ and $w(n)$ are the measured samples and noise samples, respectively. α_i, ω_i and θ_i denote the amplitude, frequency(normalized) and initial phase (uniform in $[0, \quad 2\pi]$) of the $i'th$ sinusoid. $w(n)$ is zero mean and has variance σ^2.

In order to estimate the frequencies of the P sinusoids from the available data samples, Pisarenko's frequency estimation method mainly includes the following steps:
1) To estimate the covariance matrix $\boldsymbol{R}$ of the size $N \times N$ from the measured samples. The accurate covariance matrix $\boldsymbol{R}$ is given by

$$R(i,j) = \sum_{l=1}^{P} \frac{\alpha_l^2}{2} cos(\omega_l k) + \sigma^2 \delta_{ij} \tag{2}$$

$$(\text{for } i, j = 1, 2, ..., N)$$

where $k = i - j$.
2) To compute the eigenvector corresponding to the smallest eigenvalue of the estimated covariance matrix $\boldsymbol{R}$.
3) To compute the roots of the polynomial formed by the elements of the above eigenvector. This polynomial will have roots located at $exp(\pm j\omega_i)$ $(i = 1, 2, ..., P)$.

Note that the second step involves the matrix eigendecomposition and therein lies the major computational burden of Pisarenko's frequency estimation method, and this makes it very difficult to implement the method in real-time. To tackle this problem, we present in the next section a new computational model which will be shown to be very powerful for computing in real-time the eigenvector corresponding to the smallest eigenvalue of the estimated covariance matrix $\boldsymbol{R}$.

3. The Architecture and Theoretical Analysis of a Neural Network

The neural network proposed in this paper is shown in Figure which can be represented by the continuous-time differential equations

$$C_i \frac{dU_i(t)}{dt} = \phi(t) V_i(t) - \psi(t) \sum_{j=1}^{n} D_{ij} V_j(t) \tag{3}$$

$$(\text{for } i = 1, 2, ..., n),$$

where the various quantities are described as follows: $U_i(t)$ and $V_i(t)$ stand for the input and output voltages of the $i'th$ neuron, respectively. $V_i(t) = g_i(U_i(t))$, $g_i()$ is selected to be a monotonically increasing function. C_i is the input capacitor of the $i'th$ element. Furthermore

$$\psi(t) = \sum_{i=1}^{n} (V_i(t))^2, \tag{4}$$

$$\phi(t) = \sum_{i=1}^{n} \sum_{j=1}^{n} V_i(t) D_{ij} V_j(t), \tag{5}$$

$\boldsymbol{D} = \{D_{ij}\}$ (for $i = 1, 2, ..., n; j = 1, 2, ..., n$) is called connection conductance matrix and is selected to be positive definite.
In order to guarantee that this neural network is stable, we define an energy function as:

$$E(t) = \frac{\phi(t)}{\psi(t)}. \tag{6}$$

It follows that

$$\begin{aligned} \frac{dE(t)}{dt} &= \frac{2}{\psi(t)^2} \{ \psi(t) \sum_{i=1}^{n} \frac{dV_i(t)}{dt} \sum_{j=1}^{n} D_{ij} V_j(t) \\ &- \phi(t) \sum_{i=1}^{n} V_i(t) \frac{dV_i(t)}{dt} \} \end{aligned} \tag{7}$$

$$\frac{dE(t)}{dt} = \frac{2}{\psi(t)^2} \{ \sum_{i=1}^{n} \frac{dV_i(t)}{dt} (\psi(t) \sum_{j=1}^{n} D_{ij} V_j(t) - \phi(t) V_i(t)) \}. \tag{8}$$

Comparing (3) with (8) gives

$$\begin{aligned} \frac{dE(t)}{dt} &= -\frac{2}{\psi(t)^2} \sum_{i=1}^{n} C_i \frac{dU_i(t)}{dt} \frac{dV_i(t)}{dt} \\ &= -\frac{2}{\psi(t)^2} \sum_{i=1}^{n} C_i (g_i^{-1}(V_i(t)))' (\frac{dV_i(t)}{dt})^2. \end{aligned} \tag{9}$$

Since C_i is positive and $g_i^{-1}(V_i(t))$ is a monotonically increasing function, each term on the right side of (9) is nonnegative and

$$\frac{dE(t)}{dt} \leq 0, \qquad \frac{dE(t)}{dt} = 0 \quad \rightarrow \quad \frac{dV_i(t)}{dt} = 0 \tag{10}$$

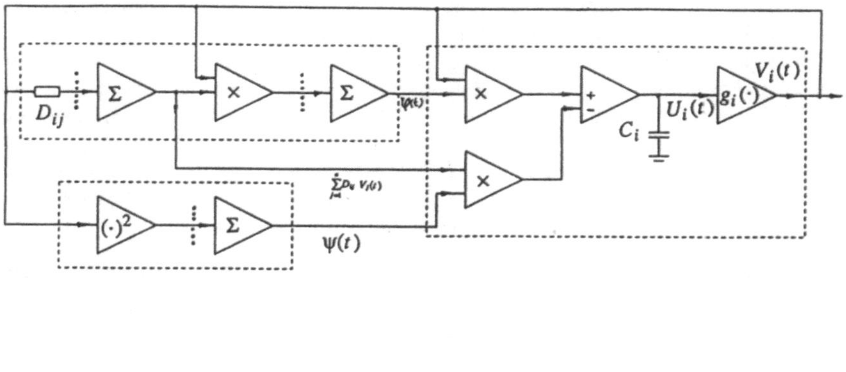

Together with the fact that $E(t)$ is also non-negative (hence, $E(t)$ is bounded from below), (10) shows that the time evolution of this neural network is a motion in the state space that seeks out the minimum to $E(t)$ and comes to a stop at such a point. Thus, any optimization problem with an objective function that may be mapped to (6) can be solved by this proposed neural network. One of the minimum points of $E(t)$ is provided by

$$\frac{dU_i(t)}{dt} = 0, \quad or \quad \frac{dV_i(t)}{dt} = 0 \tag{11}$$

$$(\text{for } i = 1, 2, \ldots\ldots, n).$$

In the next section, we will show how to use this network to compute the desired eigenvector in Pisarenko's frequency estimation method.

4. Real-time Computation of the Desired Eigenvector by Use of the Proposed Neural Network

In order to use the proposed neural network to compute the desired eigenvector corresponding to the smallest eigenvalue of the estimated covariance matrix $\boldsymbol{R}$ in Pisarenko's frequency estimation method, we choose

(1) $\boldsymbol{R} = \boldsymbol{D}$, that is, the covariance matrix $\boldsymbol{R}$ is directly taken as the connection conductance matrix of the network,

(2) $C_i = C$ (for $i = 1, 2, \cdots, N$),

(3) $g_i(u) = K_1 u$, which means that the $i'th$ neuron is modeled as an amplifier with voltage gains K_1.

According to the above relationships, we can write (3) in matrix form

$$\frac{d\boldsymbol{V}(t)}{dt} = K\{\phi(t)\boldsymbol{V}(t) - \psi(t)\boldsymbol{R}\boldsymbol{V}(t)\} \tag{12}$$

where $\boldsymbol{V}(t) = [V_1(t), V_2(t), \ldots.., V_N(t)]^T$ and $K = \frac{K_1}{C}$ is a positive constant. We also have the following relationships in matrix form:

$$\psi(t) = \boldsymbol{V}^T(t)\boldsymbol{V}(t), \tag{13}$$

$$\phi(t) = \boldsymbol{V}^T(t)\boldsymbol{R}\boldsymbol{V}(t). \tag{14}$$

About the dynamics of this network for Pisarenko's frequency estimation method, we have the following theorems:

Theorem 1.

The norm of the output matrix $\boldsymbol{V}(t)$ of the network is invariant during the time evolution and is equal to the norm of the initial state $\boldsymbol{V}(0)$, that is

$$\| \boldsymbol{V}(t) \|^2 = \| \boldsymbol{V}(0) \|^2, \quad t \geq 0, \tag{15}$$

$$\psi(t) = \psi(0), \quad t \geq 0 \tag{16}$$

The proof of Theorem 1 reads:

Multiplying (12) by $\boldsymbol{V}^T(t)$ on the left yields

$$\frac{d \| \boldsymbol{V}(t) \|^2}{dt} = -2K\{\psi(t)\boldsymbol{V}^T(t)\boldsymbol{R}\boldsymbol{V}(t) - \phi(t)\boldsymbol{V}^T(t)\boldsymbol{V}(t)\}. \tag{17}$$

Substituting (13) and (14) into (17) gives:

$$\frac{d \| \boldsymbol{V}(t) \|^2}{dt} = -2K\{\psi(t)\phi(t) - \phi(t)\psi(t)\} = 0 \tag{18}$$

which shows that (15) and (16) hold and concludes the proof of Theorem 1.

Because the norm of the desired eigenvector in Pisarenko's frequency estimation method is unity, we select $\psi(0) = \| \boldsymbol{V}(0) \|^2 = 1$. If we set

$$\boldsymbol{V}(t) = \sum_{i=1}^{N} y_i(t)\boldsymbol{S}_i \tag{19}$$

Theorem 1 shows

$$0 \leq y_i^2(t) \leq 1, \quad t \geq 0. \tag{20}$$

$$\psi(t) = 1, \quad t \geq 0 \tag{21}$$

where $0 < \lambda_1 \leq \lambda_2 \leq \cdots \leq \lambda_N$ and $\boldsymbol{S}_1, \boldsymbol{S}_2, \cdots, \boldsymbol{S}_N$ are the eigenvalues and the corresponding orthonormal eigenvectors of the matrix $\boldsymbol{R}$.

Theorem 2.

If the initial state of the network satisfies $\boldsymbol{V}^T(0)\boldsymbol{S}_1 \neq 0$, the network is asymptotically stable and

$$\lim_{t\to\infty} \boldsymbol{V}(t) = \boldsymbol{S}_1, \tag{22}$$

$$\lim_{t\to\infty} \phi(t) = \lambda_1. \tag{23}$$

The stability of this proposed network has been proved in the previous section, in the following we prove that (22) and (23) hold.

We substitute (19) into (12) and obtain

$$\frac{dy_i(t)}{dt} = K(-\lambda_i y_i(t) + \phi(t)y_i(t)) \tag{24}$$

According to the assumption of the initial state $\boldsymbol{V}(0)$ of the network , we may define

$$z_i(t) = \frac{y_i(t)}{y_1(t)} \tag{25}$$

$$(\text{for } i = 2, 3, \cdots, N).$$

Moreover

$$\frac{dz_i(t)}{dt} = \frac{y_1(t)\frac{dy_i(t)}{dt} - y_i(t)\frac{dy_1(t)}{dt}}{(y_1(t))^2} \tag{26}$$

Combining (24) and (26) gives the differential equation

$$\frac{dz_i(t)}{dt} = K(\lambda_1 - \lambda_i)z_i(t) \tag{27}$$

with the solution

$$z_i(t) = K_{i1}exp(K(\lambda_1 - \lambda_i)t) \tag{28}$$

where K_{i1} is a constant depending on K , the initial value and the eigenvalues of the matrix $\boldsymbol{R}$. Using (25) yields

$$y_i(t) = K_{i1}y_1(t)exp(K(\lambda_1 - \lambda_i)t) \tag{29}$$

$$(\text{for } i = 2, 3, \cdots, N).$$

Because of $y_1^2(t) \leq 1$, we have

$$y_i^2(t) \leq M_{i1} exp(2K(\lambda_1 - \lambda_i)t) \qquad (30)$$

where $M_{i1} = K_{i1}^2$. If $\lambda_i > \lambda_1$, we know

$$\lim_{t\to\infty} y_i(t) = 0 \qquad (31)$$

(for $i = 2, 3, ..., N$).

Together with $\| V_1(t) \|^2 = 1$, (19) and (31) show

$$\lim_{t\to\infty} y_1(t) = \pm 1 \qquad (32)$$

and

$$\lim_{t\to\infty} V(t) = \pm S_1. \qquad (33)$$

Since $-S_j$ is also a unit eigenvector corresponding to the eigenvalue λ_1 of the matrix R, (33) may be written in the unified form

$$\lim_{t\to\infty} V(t) = S_1. \qquad (34)$$

For the case in which the smallest eigenvalue λ_1 is not single but multiple, that is, $\lambda_1 = \lambda_2 ...... = \lambda_M = \lambda$, $M \leq N$, we have

$$\lim_{t\to\infty} V(t) = \sum_{i=1}^{M} \lim_{t\to\infty} y_i(t) S_i.$$

Multiplying the above equation by R on the left yields

$$\begin{aligned}\lim_{t\to\infty} RV(t) &= \sum_{i=1}^{M} \lim_{t\to\infty} y_i(t) RS_i \qquad (35)\\ &= \sum_{i=1}^{M} \lim_{t\to\infty} \lambda_i y_i(t) S_i\\ &= \lambda \sum_{i=1}^{M} \lim_{t\to\infty} y_i(t) S_i\\ &= \lambda \lim_{t\to\infty} V(t) \qquad (36)\end{aligned}$$

which means that $\lim_{t\to\infty} V(t)$ is still the eigenvector corresponding to the multiple eigenvalue λ of the matrix R.

This may conclude the proof of Theorem 2.

Theorem 1 and 2 guarantee that the output of the network in the steady state provides the desired eigenvector for Pisarenko's frequency estimation method.

5 Simulation Results

We have simulated the proposed network for computing the desired eigenvector in Pisarenko's frequency estimation method. We used $K_1 = 1$, $C = 10^4 pF$. Two examples are given in this paper. In the first example, two sinusoids are considered with the frequencies 0.1 and 0.11, respectively, and their associate signal-to-noise ratios being $10dB$. In the second example, three sinusoids are considered with the frequencies 0.1, 0.11 and 0.12, respectively, and their associate signal-to-noise ratios being $10dB$, $10dB$, and $0dB$. In these simulations, R is the covariance matrix with 4×4 elements; $V(f)$ and $V(0)$ are the outputs of the network in the steady state and the initial state, respectively. $\phi(f)$ is the eigenvalue provided by the output $\phi(t)$ in the steady state, $\lambda(a)$ is the accurate eigenvalue of the matrix R. Obviously, $\phi(f)=\lambda(a)$.

Example 1:

$$R = \begin{pmatrix} 12.6000 & 11.9338 & 11.7358 & 11.4083 \\ 11.9338 & 12.6000 & 11.9338 & 11.7358 \\ 11.7358 & 11.9338 & 12.6000 & 11.9338 \\ 11.4083 & 11.7358 & 11.9338 & 12.6000 \end{pmatrix}$$

$$V(f) = \begin{pmatrix} 0.5675 \\ -0.7242 \\ -0.1864 \\ 0.3533 \end{pmatrix}, \quad V(0) = \begin{pmatrix} 1.0000 \\ 0.0000 \\ 0.0000 \\ 0.0000 \end{pmatrix}$$

$$\phi(f) = 0.6004, \quad \lambda(a) = 0.6000$$

Example 2:

$$R = \begin{pmatrix} 22.0000 & 20.8784 & 20.5150 & 19.9142 \\ 20.8784 & 22.0000 & 20.8784 & 20.5150 \\ 20.5150 & 20.8784 & 22.0000 & 20.8784 \\ 19.9142 & 20.5150 & 20.8784 & 22.0000 \end{pmatrix}$$

$$V(f) = \begin{pmatrix} 0.5543 \\ -0.7311 \\ -0.1816 \\ 0.3691 \end{pmatrix}, \quad V(0) = \begin{pmatrix} 1.0000 \\ 0.0000 \\ 0.0000 \\ 0.0000 \end{pmatrix}$$

$$\phi(f) = 1.0001, \quad \lambda(a) = 1.0000$$

References

[1] V.U.Reddy, B.Egardt and T.Kailath, "Least squares type algorithm for adaptive implementation of Pisarenko's harmonic retrieval method," IEEE Trans. On ASSP, Vol.30, 1982, pp.399-405.

[2] Z.Banjanian, J.R.Cruz and D.S.Zrnic, "Eigendecomposition methods for frequency estimation: A unified approach," Proc. of ICASSP'90, pp.2595 -2598.

[3] L.Xu, E. Oja and C.Y.Suen, "Modified Hebbian learning for curve and surface fitting," Neural Networks, Vol.5, 1992, pp.441-457.

[4] L.Xu, A.Krzyzak and E. Oja, "Neural nets for dual subspace pattern recognition method," International Journal of Neural Systems," Vol.2, 1991, pp.169-184.

[5] E.Oja,"Principal components, minor components and linear neural networks," Neural Networks, Vol.5,1992, pp.927-935.

[6] G.Mathew, V.U.Reddy, "Development and analysis of a neural network approach to Pisarenko's harmonic retrieval method," IEEE Trans. on SP, Vol.42, No.3, 1994. pp.663-667.

[7] M.Takeda and J.W.Goodman,"Neural networks for computation: Numerical representation and programming complexity," Applied Optics, Vol.25,1986, pp.3033-3052.

ROUNDING FIR FILTER COEFFICIENTS USING GENETIC ALGORITHMS

D.H. Horrocks**, N. Karaboğa* and M. Alçı*

*Erciyes Üniversitesi, Mühendislik Fakültesi, Elektronik Bölümü, Kayseri, Turkey.
**University of Wales College of Cardiff, Signal Processing Laboratory, Cardiff,UK.

Abstract

The quantisation process of filter coefficients required due to the finite length of the register provided to store the coefficients is an important step in the realization of a given filter. This process must be optimally carried out to minimise the error in the filter response caused by the quantisation of coefficients. Genetic algorithm is a heuristic optimisation procedure which is able to optimise both discrete and continuos functions efficiently. This paper explains how to use a genetic algorithm to round the coefficients of an FIR (finite-impulse-response) filter represented using fixed-point numbers and presents the simulation results.

1. Introduction

In the design of FIR and IIR filters in hardware or software, the filter coefficients found by an approximation algorithm are rounded since the length of the register that is provided to store the coefficients is limited. The poles and the zeros of the system function will be different from the desired poles and zeros because the coefficient used are not exact. It means that we obtain a filter having a frequency response that is different from the desired frequency response of the filter [1]. The shortest coefficient word-length which satisfies the desired frequency specifications must be determined to save in cost of hardware and to increase the processing time. So, the rounding process of the coefficients must be carried out optimally, but it is usually done by hand which produces sub-optimal solution rather than optimal for rounding problem. Genetic algorithm (GA) found by J.H. Holland [2] is a robust and efficient optimisation procedure which is able to find global optimal solution for a difficult problem to solve.

A filter implementation can assume many forms depending on the type of machine arithmetic used. The arithmetic can be of the fixed-point or floating-point type. Floating-point arithmetic leads to increased dynamic range and improved precision of processing. But it also leads to increased cost of hardware and to reduced speed of processing. For software real-time implementations fixed-point arithmetic is usually preferred since the speed of processing is a significant factor [3]. This type of arithmetic, depending on the representation of negative numbers, can assume different forms: signed magnitude, one's complement, two's complement.

This work presents an approach based on genetic algorithm which optimally rounds the FIR filter coefficients represented using signed-magnitude numbers. Section 2 describes the optimisation problem.

Section 3 explains how to use a genetic algorithm for rounding the coefficients. In Section 4, the simulation results obtained for a bandpass filter are given to show the performance of the proposed approach.

2. Description of the problem

Consider a digital filter characterized by $H(z)$ and let $M(w) = |H(\exp(jwt))|$ where $M(w)$ is the amplitude response of the filter with unquantised coefficients. The effect of coefficient quantisation is to introduce an error ΔM in $M(w)$ given by

$$\Delta M = | M(w) - M_Q(w) | \qquad (1)$$

where $M_Q(w)$ is the amplitude response of the filter with the rounded coefficients. When the maximum permissible value of ΔM is denoted by $\Delta M_{max}(w)$, the desired specifications are met if

$$\Delta M \leq \Delta M_{max}(w) \qquad (2)$$

for $0 <= w <= w_p$ and $w_a <= w <= w_{s/2}$. Here w_p, w_a and w_s are passband edge, stopband edge and sampling frequencies, respectively [4].

The optimum word-length can be determined by rounding the coefficients optimally and this requires evaluating ΔM as a function of frequency for successively larger values of the word length until equation (2) is satisfied.

Let q be a vector of rounded coefficients of the filter. The space of possible vectors is potentially very large. The rounding process can be regarded as the selection of an optimal vector q' that will minimise ΔM subject to the constraint that q' belongs to Q.

$$\min_{q \in Q} \Delta M \qquad (3)$$

In general, the constrained design space Q is a complex multidimensional space . To find the optimal vector q' requires a robust search algorithm capable of efficiently exploring such a space. This work adapted genetic algorithm, stochastic search procedure that has been used successfully for many difficult optimisation problem [5,6].

3. Application of GA to the problem

After all original coefficients are scaled between certain limits defined according to the length of the register provided, the scaled coefficients are optimally rounded using genetic algorithm.

In the application of a genetic algorithm to any problem, the first stage is to decide a way to represent the parameters to be optimised in a string form since genetic algorithm works with a set of strings called population. For our problem, each string in the population is in the following form:

$$\boxed{\Delta_1}\boxed{\Delta_2}\boxed{\Delta_3}\cdots\boxed{\Delta_i}\cdots\boxed{\Delta_k}$$

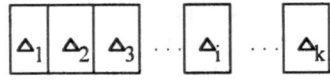

Here, Δ_i which is an integer number to be added to the integer value of ith coefficient obtained after the scaling and truncating process is itself a binary string. The total length of a string is related to the number of the coefficients, n , and the number of the bits used to represent Δ_i.

The second stage is the design of a function called fitness function to assess the fitnesses of solutions to the problem automatically. In this work, the following function is employed to calculate the fitness value of a possible solution.

$$fit(i)=A-\max_{k=1}^{l}[|(M_I(W_k)-M_Q(W_k))|\xi(k)] \qquad (4)$$

where $M_I(w)$ is the ideal amplitude response of the filter, A is a positive constant, l is the number of frequencies at which the difference (error) between the ideal and the actual amplitude responses are calculated, $\xi(.)$ is the weight assigned to each error and fit(i) is the fitness of the ith string (solution).

4. Simulation results

The simulation results have been obtained for several linear-phase filters designed using HYPER SIGNAL software package [7]. Just as an example, the results belonging to a banddpass filter of which the length is 42 are given below. The magnitude responses of the filter obtained using the coefficients unquantised, rounded by hand and rounded by the genetic algorithm are given in Figures 1, 2 and 3, respectively. The rounded coefficients by hand and the genetic algorithm are presented in Table 1.

Since the filter designed is a linear-phase filter the coefficients are symmetric. So, the number of the parameters represented in a string is 21 (i.e. k=21). The total length of a string in bits is equal to 42 because Δ_i is represented with 2 bits. The value of $\xi(.)$ is 1, 4, and 0 for the bandpass, bandstop and transition regions, respectively.

From Figures 2 and 3, it is seen that the filter with the coefficients rounded by genetic algorithm shows a better performance than the filter with the coefficients rounded by hand.

5. Conclusion

The results obtained demonstrates that the coefficients of FIR filter represented using fixed-point (signed-magnitude) numbers can be efficiently rounded by a genetic algorithm. The filter with the coefficients rounded by genetic algorithm shows better performance than the filter with the coefficients rounded by hand.

6. References

1. Jenkins, W.K. and Leon, B.J. An analysis of quantisation error in digital filters based on interval algebra's, IEEE Trans. Circuits Syst., vol. Cas-22, 223, (1975).

2. Holland, J.H. Adaptation in natural and artificial systems, Ann Arbor: University of Michigan Press, USA, 1975.

3. Proakis, J.G. Digital signal processing (Principles, algorithms, and applications) (Sec. ed.), Macmillan Publishing Company, New York, 1992.

4. Antoniou, A. Digital Filters (Analysis, design, and applications), (Sec.ed.), McGraw-Hill, Inc., USA, (1993).

5. Davis, L.(ed.) Handbook of genetic algorithms, Van Nostrand Reinhold, NY, New York, 1991.

6. Karaboğa N., Wani A. and Karaboğa D. Optimisation of multivariable functions, Proc. of the Third. Conf. on Information Technology and its Applications, ITA'93, Markfield Conf. Centre, Leichester, April, 14-16, UK, 1994.

7. Hyper signal DSP software, (Users manual, Reference guide, Hardware and interface source code), Hyperception, Inc. 9550 Skillman LB125, Dallas, 1991.

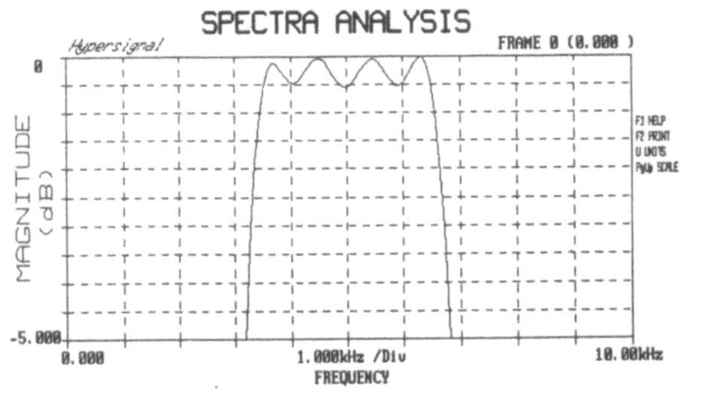

a)

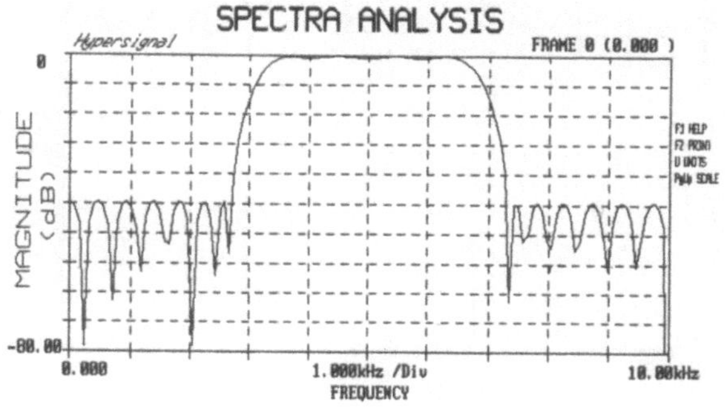

b)

Figure 1. Response of the filter with the unquantised coefficients
a) Passband region
b) Stopband region

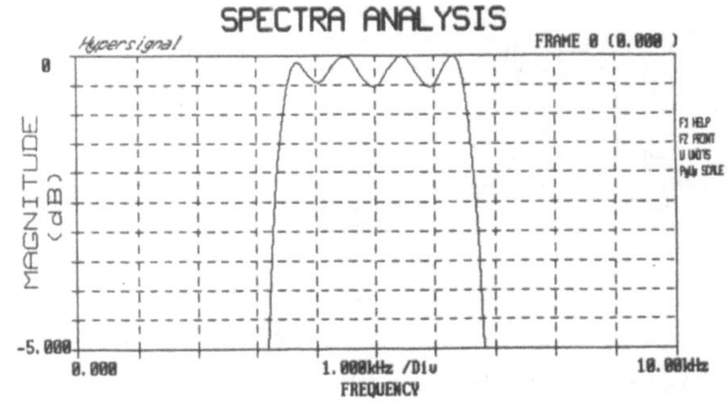

a)

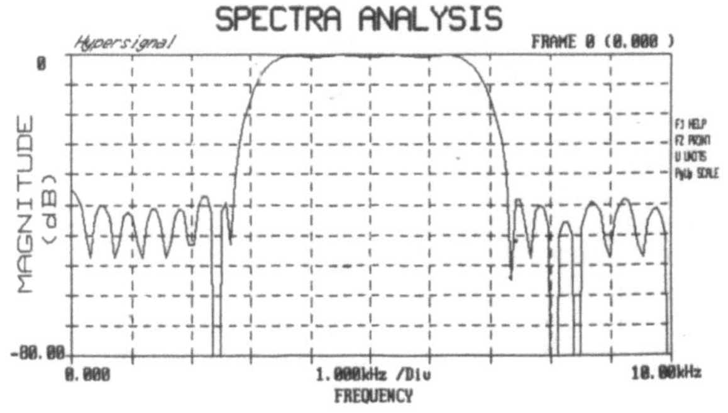

b)

Figure 2. Response of the filter with the coefficients rounded by hand
a) Passband region
b) Stopband region

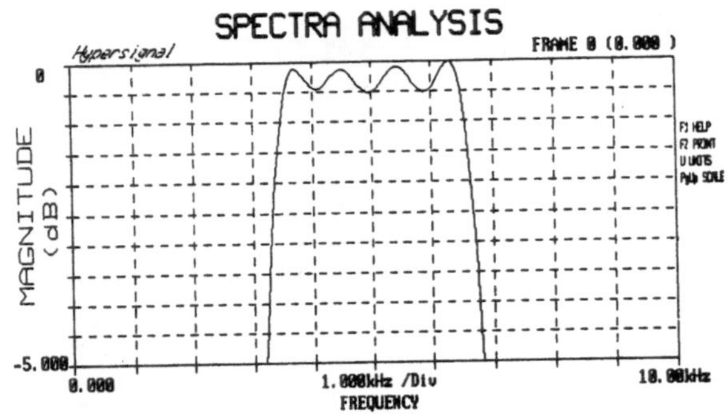

a)

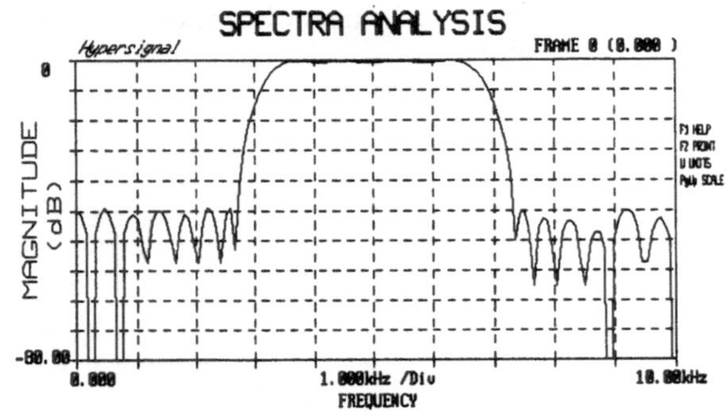

b)

Figure 3. Response of the filter with the coefficients rounded by genetic algorithm
a) Passband region
b) Stopband region

Table 1. Filter coefficients rounded by hand and genetic algorithm

Filter coefficients	Rounded by hand	Rounded by GA
a(0) = a(41)	-7	-7
a(1) = a(40)	-11	-10
a(2) = a(39)	13	12
a(3) = a(38)	9	8
a(4) = a(37)	-4	-3
a(5) = a(36)	5	6
a(6) = a(35)	-15	-15
a(7) = a(34)	-22	-22
a(8) = a(33)	21	21
a(9) = a(32)	12	12
a10) =a(31)	4	5
a(11)=a(30)	26	26
a(12)=a(29)	-43	-43
a(13) = (28)	-50	-50
a(14)=a(27)	38	38
a(15)=a(26)	6	6
a(16)=a(25)	45	45
a(17)= (24)	109	110
a(18) = (23)	-173	-173
a(19)=a(22)	-226	-226
a(20)=a(21)	255	255

EVOLUTIONARY ADAPTIVE FILTERING

D. Ait-Boudaoud, R. Cemes, S. Holloway
School of Electronics,Bournemouth University
Poole House, Fern Barrow
Poole, Dorset, BH12 5BB,UK.

Abstract

Evolutionary algorithms have seen an ever increasing use in a variety of applications owing to their robustness and ease of implementation. This paper considers the performance of these algorithms when applied to adaptive filtering, with particular emphasis on the direct system modelling problem. The proposed approach also introduces a two - layer genetic algorithm that performs a coarse grain search at the word level followed by a fine grain optimisation of the filter weights at the bit level.

1. Introduction

Adaptive filters have the ability to adjust their filtering coefficients when operating in non - stationary, and time varying environments. Their scope includes: noise cancellation, system identification, adaptive equalisation, periodic interference cancellation, etc. The most common structure used in adaptive filtering is the Finite Impulse Response (FIR) structure whose coefficients are adjusted by some adaptive algorithm.

Several algorithms have been suggested for the filter coefficient adaptation procedure varying from the optimal Wiener filter to the more practical Least Mean Square (LMS) algorithm [1]. These adaptive algorithms have different performance characteristics in terms of convergence rate, computational requirement, knowledge of environment, input signal statistics etc. In this paper, we provide an analysis of the behaviour of adaptive filters that use a different class of algorithms namely, evolutionary algorithms. These algorithms are mainly based on Genetic Algorithms (GA), Simulated Annealing (SA), and their hybrids. We also introduce a two layer genetic algorithm that performs a coarse grain search at the binary word level on the randomly created initial population, followed by a fine grain optimisation performed at the bit level. A system identification problem is used as a vehicle to assess the proposed procedures. This will provide the basis for further extensions to solve other adaptive filtering problems.

2. Direct Systems Modelling

The direct systems modelling problem is concerned with identifying an unknown system using some adjustable filter, as illustrated in Fig. 1. The technique is based on minimising the error signal $e(n)$ to make $y(n)$ and $\hat{y}(n)$ equal so that the adaptive filter mimics the response of the unknown system. The adaptive filter consists of two sections, namely the filter structure, and the adaptive algorithm.

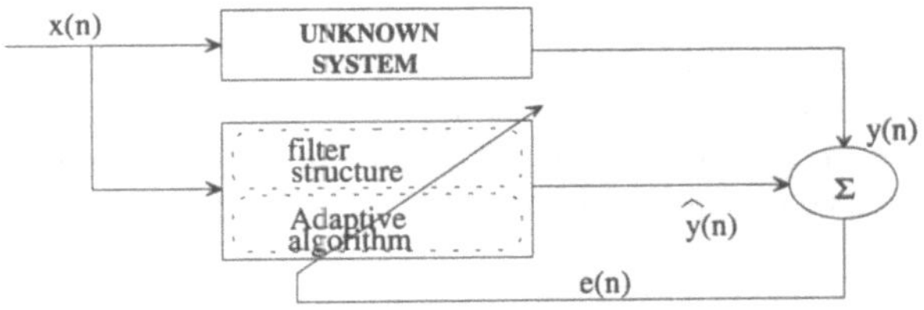

Fig. 1: Systems Identification

In our analysis, the chosen filter structure is that of an FIR filter, and the adaptive algorithm is based on evolutionary algorithms - namely genetic algorithms, simulated annealing and the combination of them.

3. Genetic based adaptive algorithm

Genetic algorithms are search algorithms based on the process of natural selection and natural genetic manipulation [2]. They form part of the set of evolutionary algorithms, since the concept relies upon the survival of the fittest through the creation of offspring from an initial randomly generated population. The simple genetic algorithm performs three basic operations:

1. Reproduction

2. Fitness Evaluation

3. Genetic Manipulation (Crossover, Mutation, . . .)

The most challenging aspect of GAs is the identification of the cost (fitness) function which can determine the outcome of the algorithm. This function basically undertakes the process of assessing the appropriateness of new offspring to form a new generation.

Fortunately, in the case of the adaptive filter, the conventional adaptive algorithms also use some form of cost function which seems to be appropriate for GAs. The most widely used cost function is based on the mean square of the error signal (MSE). The inverse of the MSE has been adopted as the fitness function for our GA, so that the problem is now that of obtaining a maximum fitness solution. Also, in order to assign considerably less fitness to larger errors, the cost function used here is given by:

$$fitness = \frac{1}{(\frac{1}{N}\sum_n e_n)^6}$$

A two layer genetic algorithm has been used in this problem. The first layer performs genetic operations at the word level on the coefficients randomly created in the initial population. This procedure is fast and allows convergence to the most appropriate generation at this level. The second layer takes the generation from the first layer as a starting population. The second layer applies genetic operators at the bit level. The top level Two - Layered Genetic Algorithm is shown below in Fig. 2.

Experiments have indicated that the initial population plays an important role in the convergence of the algorithm. As a result, this two - layered approach provides an added benefit to one that relies upon only one form of representation.

4. Simulated Annealing

Simulated Annealing (SA) is a naturally motivated robust optimisation technique, analogous to the physical process of annealing. SA has the inherent ability to "climb hills" under the control of a probabilistic criterion thus allowing the algorithm to escape from local minima. SA is essentially an iterative random search technique with adaptive moves along the coordinate directions.

```
Two-Layer Genetic Algorithm
$Word_level _Layer
Initialise
        Pop:            Randomly created population
        pcross:         Crossover Rate
        pmutation:      Mutation rate
        max_gen:        Maximum number of generations

procedure
        compute_fitness (Pop)
        While(gen<=max_gen) | ~Converged)
                New_Pop=Select (Pop)
                Pop=Crossover(New_Pop,pcross,Word_level,1-point)
                Pop=Mutate(Pop,pmutate,Word_level)
                compute_fitness(Pop)
        EndWhile

$Bit_Level_Layer
        bPop=Convert(Pop)
        While (~Converged)
                bNew_Pop=Select(bPop)
                bPop=Crossover(bNew_Pop,pcross,Bit_level,2-point)
                bPop=Mutate(Pop,pmutate,Bit_level)
                compute_fitness(bPop)
        EndWhile
EndProcedure
```

Fig. 2: Two layer genetic algorithm

The SA algorithm is based on laws of statistical and thermal mechanics, which state that for a system in a heat bath, the probability that it is in a state with energy E is proportional to the Boltzmann distribution, $e^{-E/T}$, where T is the temperature of the bath [3].

For low temperatures, a system in thermal equilibrium is, with a high probability, in a state of low energy. The process of annealing proceeds iteratively: first, the system is heated up to a sufficiently high temperature (melting point). Then, the temperature T is gradually decreased and, at each temperature, a series of random moves in a searching space is performed generating thus new points X_{new}. A new point X_{new} is accepted according to the Metropolis criterion [4]:

$$f(X_{new}) < f(X_{current}) : \text{ P\_accept}(X_{new}) = 1;$$

$$f(X_{new}) > f(X_{current}) : \text{ P\_accept}(X_{new}) = e^{\frac{-(X\_new - X\_current)}{T}}$$

The high level pseudo code of the Simulated Annealing algorithm is shown in Fig. 3. The initial temperature was set at 3000°C.

Simulated Annealing Algorithm

```
$ main SA loop
initialise;
while ~frozen
        while ~equilibrium {
                search move for a new point;
                metropolis criterion;
                decide on the new point;
                check equilibrium;
        EndWhile
        get new temperature;
EndWhile
// quench frozen configuration
while ~equilibrium
                search move for a new point;
                metropolis criterion;
                decide on the new point;
                check equilibrium;
EndWhile
```

Fig. 3: Simulated Annealing Algorithm

5. Experiments and results

Several experiments have been carried out for the analysis of the performance of the evolutionary algorithms. In general, these algorithms have been compared with the performance of the well known Least Mean Square algorithm. It could be argued that this comparison is biased, since the techniques differ in their approaches. However, it is worth highlighting that the objective of this research is to assess the effectiveness of evolutionary algorithms for the direct systems modelling problem. Accordingly, the Least Mean Square based adaptive algorithm will only be used as a point of reference to provide us with some indications on their performance.

Experiment 1

The two layered GA was employed to identify a black box (unknown) system. This system performs the operation of a three tap FIR digital filter. The input signal *x(n)* was obtained by passing a white noise sequence *w(n)* through a noise shaping filter in order to adjust the Eigenvalue Ratio (EVR) of the signal correlation matrix of *x(n)* [5]. With a large EVR the convergence of the LMS can be slow, and the steady - state error can be larger than the possible minimum depending on the step size used in the LMS algorithm. The results clearly indicate that there is some scope for evolutionary algorithms in adaptive filtering, since the EVR of the input signal has no effect on the performance of the GA. Fig. 4 illustrates the mean square error of the best chromosome of the GA at each generation. The mean squared error of the LMS algorithm for several EVR levels is plotted on the same axis. The lms step size parameter was 0.02 for all cases.

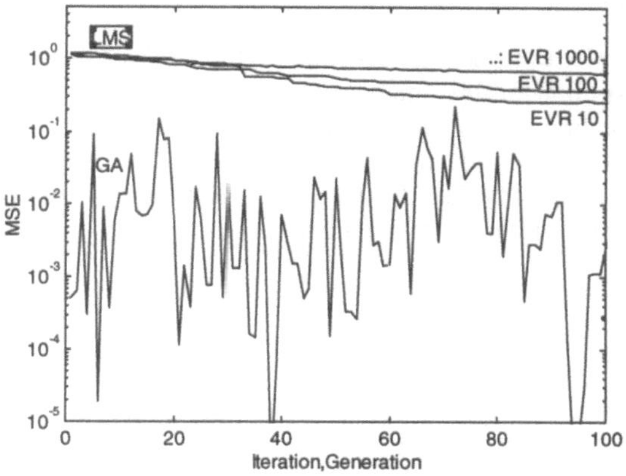

Fig. 4 : Performance comparison of GA with LMS

Fig. 5 shows the coefficient trajectories for the LMS and GA algorithms with EVR = 100.

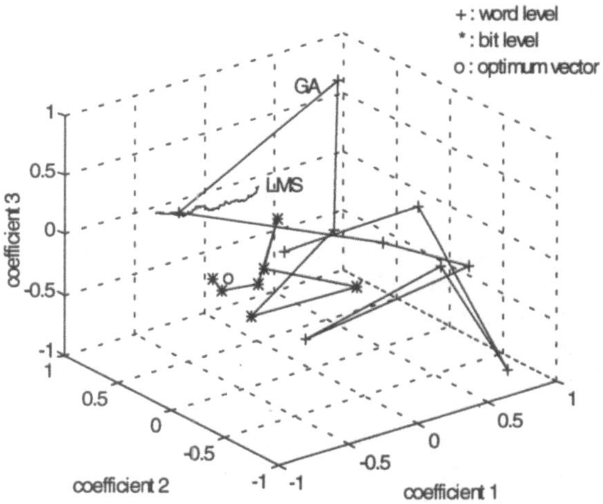

Fig. 5 : coefficient vector trajectories, LMS and GA

Experiment 2

The initial population was constructed from the first few iterations of the LMS algorithm. This technique is aimed at localising the region of search thus increasing the convergence rate of the GA. So far, the results have indicated that there is no striking benefit from the experiments conducted. Nevertheless, we believe the principle is valid and that further refinements and investigations need to be carried out to confirm this claim.

Experiment 3

In order to further enhance the performance of the genetic algorithm, a fine tuning procedure based on simulated annealing was incorporated. This has clearly improved the performance of the algorithm, and the results are highlighted in Fig. 6. This hybridisation has enormous scope for future consideration.

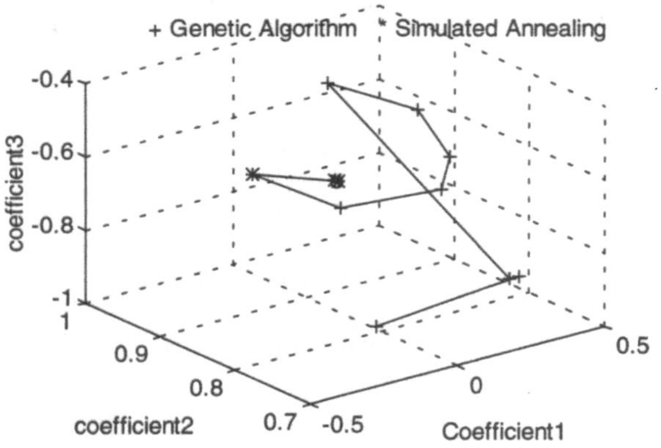

Fig. 6: Experiment 3- Hybrid Genetic-Simulated Annealing algorithm

6. Conclusions

From this initial investigation, it has been shown that in some cases, evolutionary algorithms could outperform existing classical methods for the adaptive filtering problem. The main conclusions can be summarised as follows:

1. Evolutionary algorithms are less sensitive to the environment and the characteristics of the input sequence.
2. The incorporation of knowledge could improve the convergence rate of the GA.

3. Hybridisation of GA with other search techniques could lead to more robust algorithms.

It should also be emphasised that the LMS algorithm is still the preferred option for its practicality and ease of implementation. This research was not aimed at ousting the LMS algorithm but its objective was to analyse the behaviour and performance of evolutionary algorithms to implement adaptive filtering. The research is being extended to non - linear systems identification problems.

REFERENCES

1. Widrow,B., Stearns, S.D.: Adaptive Signal Processing, Prentice - Hall, 1984

2. Goldberg, D.E.: Genetic Algorithms in Search, Optimization and Machine Learning. Addison Wesley, Reading, MA, 1989

3. Reif, F.: 'Fundamentals of Statistical and Thermal Physics, McGraw-Hill, New York, 1965

4. Corana, A., Marchesi, M., Martini, C. and Ridella, S.: 'Minimising multimodal functions of continuous variables with the Simulated Annealing algorithm', ACM Trans. Mathematical Software, Vol.13, pp.262-280, Sept.1987

6. Haykin, S.: Adaptive Filter Theory. Prentice - Hall, 1991

A GENETIC ALGORITHM FOR SOME DISCRETE-CONTINUOUS SCHEDULING PROBLEMS

Joanna Józefowska, Rafał Różycki and Jan Węglarz

Poznań University of Technology, Institute of Computing Science,
Piotrowo 3A, 60-965 POZNAŃ, POLAND

Abstract

In this paper a Genetic Algorithm (GA) for solving some discrete-continuous scheduling problems is proposed. In this algorithm, each genotype represents a solution which is derived directly from the idea of a feasible sequence rather than being coded as a binary string. Applying suitable genetic operators allows to conduct the searching process in a feasible region only.

1. Problem formulation.

Discrete-continuous scheduling problems occur when jobs simultaneously require for their processing discrete (i.e. discretely - divisible) and continuous (i.e. continuously - divisible) resources.

In the simplest situation of this type each job requires a machine from a set of identical parallel machines and an amount, unknown in advance, of a continuous renewable resource (e.g. power). Assume that jobs are nonpreemptable and independent and all are available at the start of the process. In [1] the formulation of the problem is discussed in which processing rate of job i is described by the equation:

$$\dot{x}_i(t) = \frac{dx_i(t)}{dt} = f_i[u_i(t)],\ x_i(0) = 0,\ x_i(C_i) = \tilde{x}_i \quad (1)$$

where

$x_i(t)$ is the state of job i at time t,

$u_i(t)$ is the amount of the resource allotted to job i at time t,

f_i is a continuous, nondecreasing function, $f_i(0)=0$,

C_i is (unknown in advance) completion time of job i,

$\tilde{x}_i$ is the final state of job i.

Assume that $0 \le u_i(t) \le 1$, $\sum_{i=1}^{n} u_i(t) \le 1$ for every t.

The problem is to find a sequence of jobs on machines, and, simultaneously, a continuous resource allocation, which minimize the schedule length.

Of course the defined problem is NP-hard since scheduling nonpreemptable jobs on parallel machines to minimize the schedule length is NP-hard.

It has been proved in [1] that for concave f_i, $i=1,2,\ldots,n$ each feasible schedule can be divided into $p \le n-m+1$ intervals Δ_k such that the number of jobs performed in Δ_k is $\le m$ and the resource allocation among jobs remains constant in Δ_k. Let Z_k denote the combination of jobs corresponding to the interval Δ_k. Thus, with each feasible schedule, a feasible sequence S of combinations Z_k is associated, where the number of elements in each combination does not exceed m, each job appears in at least one combination, and if a job appears in more than one combination then these combinations are consecutive ones (nonpreemptibility).

Example. Consider the following feasible schedule of 5 jobs on 3 machines in which it is assumed for simplicity $u_i(t)=1/3$ for every t (Fig. 1).

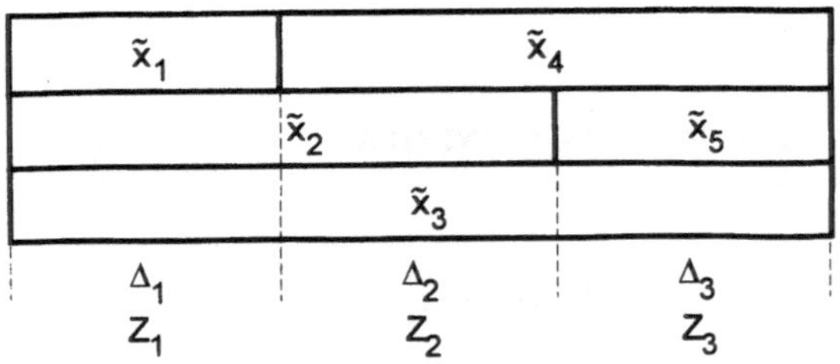

Fig. 1 The idea of discrete - continuous scheduling. An example of a feasible schedule.

The corresponding feasible sequence of combinations is of the form

$$S=\{1,2,3\},\{4,2,3\},\{4,5,3\}.$$

For a given feasible sequence S, an optimal division of final states of jobs, $\tilde{x}_i$, $i=1,2,\ldots,n$ among combinations in S can be found. However, the number of all feasible sequences grows exponentially with the number of jobs. Thus it is justified to apply local search metaheuristics operating on a space of feasible sequences. In the following sections we present one of them, namely a genetic algorithm (GA). Although its idea is a general one, a computational experiment will be performed for functions f_i, $i=1,2,\ldots,n$ of the form

$$f_i=c_i u_i^{1/\alpha},\ \alpha>1,\ i=1,2,\ldots,n \qquad (2)$$

This assumption significantly simplifies the solution of the "continuous part" of the problem.

2. An application of GA

Genetic algorithms operate on a population of genotypes each one representing a solution of the considered problem. In the presented implementation a genotype is defined as a feasible sequence S of combinations Z_k. Random nature of genetic algorithms, assuming accidental changes of a randomly selected genotype from the current population, results in the difficulty in maintaining the feasibility of genotypes during the evolution process. Under these circumstances it was impossible to define an efficient (maintaining feasibility and easy to implement) crossover operator. However, genetic algorithms are flexible enough to reach satisfactory solutions by using other genetic operators and proper parameters.

2.1. Genotypes

A genotype representing a feasible sequence has to be implemented as a complex structure, e.g. array of arrays or list of arrays. In our implementation the first possibility has been chosen. The size of this structure is constant and depends on the number of jobs and machines. It is worth mentioning that the above representation is redundant assuming the class of functions (2), however the aim was to develop a general approach.

2.2. Creating an initial population

The simplest method of creating an initial population - the random creating - can not be adapted to the discrete-continuous problems directly. Thus, in our implementation, firstly a set of m feasible sequences is generated using the algorithm proposed in [1]. New feasible sequences are obtained by a random exchange of a job index in a sequence randomly selected from this set. The genotype representing this new sequence is appended to the created population. The aforementioned process is repeated until the size of the initial population achieves the desired size (pop_size).

2.3. Fitness of a genotype

The performance of a genotype is defined as the minimum schedule length for a relevant feasible sequence S. For the considered class of functions f_i it is given by the following formula

$$M_s^* = \sqrt[\alpha]{\left(\sum_{i\in P_1} \frac{\tilde{x}_i}{c_i}\right)^{\alpha} + \ldots + \left(\sum_{i\in P_m} \frac{\tilde{x}_i}{c_i}\right)^{\alpha}} \qquad (3)$$

where P_i ($i=1,\ldots,m$) denotes the set of indices of jobs assigned to machine i.

Formula (3) follows from the fact (see [1]) that in this case the minimum schedule length depends on how the jobs are assigned to the machines but is independent of jobs order on a particular machine.

As for all minimization problems the fitness of a genotype differs slightly from the performance measure. In the presented implementation a simple ranking of genotypes has been applied, as described in Section 2.4.

2.4. Genetic operators

The main genetic operator is the selection operator. Usually the selection operator uses genotype fitness value directly. It means that the probability that a genotype will survive to the next population is proportional to its fitness. Of course, for the considered discrete-continuous problems genotype fitness is inversely proportional to schedule length. Moreover, for the assumed class of functions f_i the minimum schedule lengths of various genotypes do not differ much. This fact would cause a slow convergence to the required solution. For these reasons the ranking method was used in the selection procedure. The genotypes were ordered in nondecreasing order of their fitness. Probability p_i of the selection of genotype i (i=1,2,...,pop_size) to the next population depends on its position a_i in the ranking and is equal to

$$p_i = \frac{(pop\_size - a_i + 1)}{pop\_size * (pop\_size + 1)/2} \quad (4)$$

It is easy to notice that $\sum_{i=1}^{pop\_size} p_i = 1$.

There are two recombination operators especially suited for the discrete-continuous scheduling problems. Both of them, the mutation and the renumeration operators work on a single genotype.

The mutation operator randomly selects a genotype, an interval Δ_k and a job from combination Z_k in this genotype. A new genotype is created in the way described below. Assume that k is a randomly chosen index of a combination and i is a randomly chosen index of a job from combination Z_k, which is going to be changed. A job i in Z_k is replaced by either

(i) job j from Z_{k-1} if (j is not in Z_k) and (i is in Z_{k-1}) or

(ii) job l from Z_{k+1} if (l is not in Z_k) and (i is in Z_{k+1})

For example if n=6, m=3, and k=2, i=2 and genotype S={1,2,3},{**2**,3,4},{4,5,6} are selected by the mutation operator, then the offspring genotype will be the following one: S'={1,2,3},{**1**,3,4},{3,4,5}.

The mutation operator fails in two cases:

(i) if nonpreemptibility of jobs would be violated; (e.g. for k=2, i=3 and S={1,2,3},{2,**3**,4},{3,4,5}) or

(ii) if the selected job occurs only in one combination in S;(e.g. for k=1, i=1 and S={**1**,2,3},{2,3,4},{3,4,5})

In such cases the mutation operator is called again.

The second recombination operator is the renumeration operator. This operator exchanges a number of randomly generated job numbers in a randomly selected genotype. Assume that S = {1,2,3},{1,2,4},{1,2,5},{1,5,6} and job numbers 2,4,6 are selected by the renumeration operator. Then swapping 2→4, 4→6, and 6→2 will be produce the following offspring genotype :

S'={1,4,3},{1,4,6},{1,4,5},{1,5,2}

We have not defined any crossover operator due to the strong feasibility conditions for genotypes.

The computational experiment shows, however, that applying only mutation and renumeration operators gives satisfactory results.

2.5. Other parameters of the genetic algorithm

The population size was set at 50 genotypes, the size recommended by many researchers (e.g. [2]). The given number of evolution cycles without improvement was the termination criterion and was fixed at 10 cycles. Other parameters (mutation rate p_m and renumeration rate p_r) were examined during the computational experiment.

3. The computational experiment

The GA has been tested to study its sensitivity to the values of the parameters. For every of 1000 instances of various problem sizes 9 combinations of the mutation rate p_m and the renumeration rate p_r have been tested. The obtained results were compared with the lower bound calculated from (3) under the assumption that nonpreemptibility is violated. The minimum schedule length is reached if the particular components of (3) equal to each other.

The algorithm has been implemented in C++ and the experiment has been carried out on Silicon Power Challenge computer with four RISC TFP 75 MHz processors in the Poznań Supercomputing & Networking Center .

The average relative deviation (r) from the lower bound, the number of instances (l) for which the schedule length obtained using the combination of parameters p_m/p_r is not greater than the best schedule length found for this instance, and the computational time (t) are presented in Table 1.

4. Conclusions

In this paper, we present an application of a genetic algorithm to a class of discrete-continuous scheduling problems. The aim of the computational experiment was to examine the influence of the mutation rate and the renumeration rate on the performance of the algorithm. The results show that, in general, the algorithm performs better for larger values of the mutation rate. The influence of the renumeration rate depends on the problem size. For large problems (50x3), small renumeration rate is advisable, while for small problems (up to 20x3), it is quite different. Moreover, the average distance from the lower bound does not depend on the problem size.

The computational time grows with the increase of each of the following: the mutation rate, the renumeration rate and the problem size (assuming the stopping criterion defined in Section 2.5.).

In the further research it is planned to define a crossover operator and/or develop a new solution representation. Parallelization of some procedures of the algorithm can make the searching process much more efficient.

References

[1] Józefowska, J., Węglarz, J., Report of the Inst. of Computing Science, Poznań University of Technology, RA-94/003

[2] Liepnis, G.E., Hilliard, M.R., AOR 21, 31 (1989)

[3] Chuen-Lung, C., Venkateswara, S., Nasser, A.: EJOR 80, 389 (1995)

[4] Michalewicz, Z.: Genetic Algorithms + Data Structures = Evolution Programs. Berlin Heidelberg: Springer - Verlag 1992 .

Table 1. The results of the computational experiment.

n x m	20 x 2			20 x 3			50 x 2			50 x 3			100 x 10		
p_m/p_r	r	l	t [s]	r	l	t [s]	r	l	t [s]	r	l	t [s]	r	l	t [s]
0.1/0.1	1.000333	135	0.244	1.002695	69	0.406	1.000337	150	0.998	1.000374	91	1.466	1.002737	77	26.059
0.1/0.25	1.000329	204	0.335	1.001903	101	0.584	1.000336	225	1.322	1.000359	108	2.004	1.002726	65	37.065
0.1/0.4	1.000343	215	0.443	1.001916	100	0.756	1.000336	281	1.678	1.000365	62	2.669	1.003419	16	47.231
0.25/0.1	1.000330	157	0.257	1.001966	118	0.443	1.000336	224	1.060	1.000354	161	1.586	1.001706	189	29.482
0.25/0.25	1.000318	212	0.285	1.001801	124	0.658	1.000336	267	1.489	1.000353	121	2.288	1.001799	162	46.256
0.25/0.4	1.000313	231	0.631	1.001811	103	1.026	1.000336	241	2.405	1.000358	73	3.902	1.002467	31	71.213
0.4/0.1	1.000315	201	0.294	1.000337	140	0.486	1.000336	258	1.183	1.000350	185	1.744	1.001596	238	32.112
0.4/0.25	1.000314	261	0.520	1.000336	164	0.858	1.000336	275	1.999	1.000351	119	3.054	1.001646	186	59.732
0.4/0.4	1.000315	278	1.121	1.000336	133	1.885	1.000336	289	4.773	1.000359	83	7.427	1.002325	36	155.43

HYBRIDIZING GENETIC ALGORITHMS WITH BRANCH AND BOUND TECHNIQUES FOR THE RESOLUTION OF THE TSP

C. Cotta, J.F. Aldana, A.J. Nebro, J.M. Troya

Dept. of Lenguajes y Ciencias de la Computación, Univ. of Málaga

Plaza de El Ejido s/n, E-29013 - MÁLAGA (ESPAÑA)

aldana@tecma1.ctima.uma.es

Abstract - In this paper some mixed techniques are outlined in order to combine the advantages of two very different methods for the resolution of combinatorial optimization problems (*Genetic Algorithms* and *Branch and Bound Techniques*), simultaneously avoiding their drawbacks. Due to the disparity between the basic techniques it is not suitable that they work at the same level so two models have been developed in which each technique assumes the role of being a tool of the other one. Parallelism is an important issue in these techniques.

1. INTRODUCTION

Both *Genetic Algorithms* (GAs) and *Branch and Bound Techniques* (B&B) are heuristic procedures for the resolution of optimization problems [1] but they are very different in their philosophy and in the way they work. A first-look comparison between them shows the typical computational dilemma space vs. time. GAs realize a continued treatment on a finite (and small) subset of the whole work-area. This process may require a long time before acceptable solutions are founded (obtaining the optimal is not guaranteed). On the contrary, B&B techniques perform an intelligent exploration of the work-area. This exploration is relatively fast (when compared with GAs) but, as the size of this area grows, requires great amounts of memory. However, these techniques do guarantee achieving the optimal solution.

A question arises from what above stands: is it possible to develop a hybrid technique which combines the advantages of both approaches and, simultaneously, avoids or minimizes their drawbacks, thus constituting a synergyc system?. The existence of such a technique is the subject of this paper, in which some possibilities to achieve that integration are studied on the TSP problem.

2. COMBINATION OF GENETIC ALGORITHMS AND BRANCH & BOUND

2.1. Direct collaboration. Issues

A first possibility for GAs and B&B techniques to work together is executing both procedures independently and in parallel, that is, at the same level. Both processes will enclose the solution but coming from different directions. There are two ways of obtaining a benefit of this parallel execution (Fig 1):

- Use the upper bounds provided by the GA to purge the problem queue of the B&B algorithm, deleting those subproblems whose lower bound is bigger than the one obtained from the GA.
- Inject into the GA information about those subtours the B&B algorithm considers more promising. They would be the *building blocks* of the TSP representation.

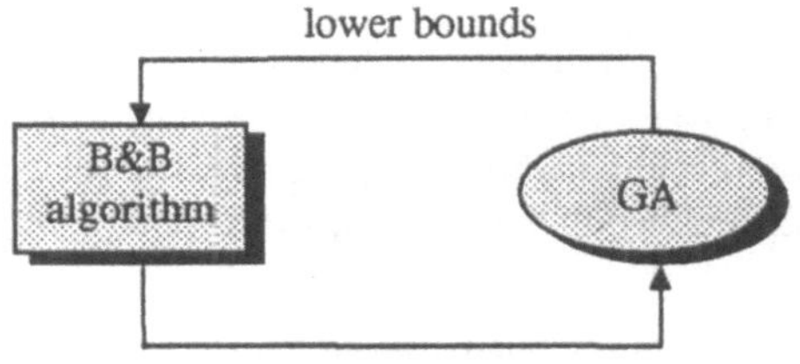

Fig 1. Direct collaboration scenario

These options are, however, difficult to carry out for many reasons:

- Execution times are quite different, so establishing an effective information exchange is very difficult.

For example, injecting some high-valued building blocks in early stages of the GA could yield the apparition of superindividuals [2], affecting diversity and polarizing further evolution.

- There is a big gap between the accuracies of both methods. Specifically for the TSP, B&B algorithms start from an initial approximation that is usually within a 2% of the optimum [3], whilst GAs do not usually achieve a high accuracy until late stages of evolution, so it is too late for a B&B algorithm to get a benefit by purging the problem queue.

Clearly, it is necessary to find an alternative approach in order to establish a productive collaboration between both techniques. According to the functioning features mentioned, there exist at least two ways of harmonizing their performances: on the one hand, it is possible to reduce the combinatorial explosion which limits the effectiveness of a B&B algorithm by imposing some restrictions to the problem. On the other hand, parallel execution of many GAs may heighten their power.

2.2. Branch and Bound working for a Genetic Algorithm

In the model described next, the GA plays the role of a master and the B&B algorithm is incorporated as a tool of it.

2.2.1. Hybridizing a Genetic Algorithm. The key feature of a GA is its robustness: binary representation and simple binary crossover and mutation operators provide problem-domain independence. However, such an algorithm will be normally outperformed by a specialized algorithm.

It is possible, however, to improve a GA performance applying some techniques and knowledge used by other algorithms, although this implies a loss of robustness. In [2] some guidelines in order to incorporate those items are described.

2.2.2. A Hybrid Genetic Algorithm to solve the TSP. A promising way of hybridizing a GA incorporating B&B techniques and knowledge is to build a hybrid operator with the philosophy of B&B. The idea underlying this option is shown next (see Fig 2).

All information related to the goodness of a certain tour is contained in the edges between nodes. The classical crossover operators usually introduce a great amount of new information since they produce offsprings including edges not appearing in any ancestor. Some advanced operators like *Edge-Recombination* [4] guarantee that an offspring will inherit a 95% of its genetic information from its ancestors. The operator described next assures that an unique offspring will be created using only edges from the ancestors (i.e. without any implicit mutation).

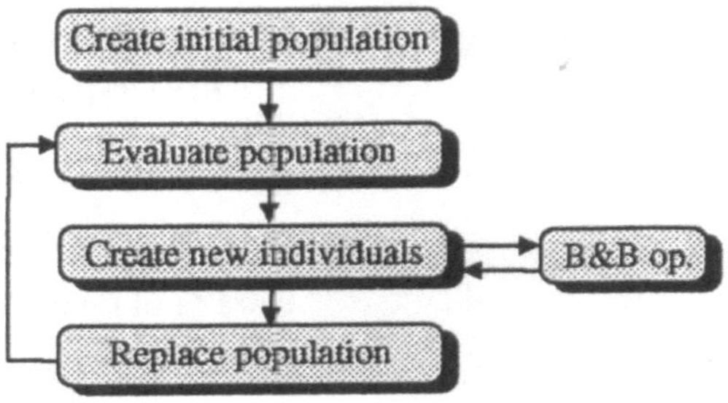

Fig 2. Outline of the Hybrid Genetic Algorithm.

The functioning of this operator is as follows: the B&B algorithm starts from a list of all edges in the graph with information related with associated weights. This operator is a restricted version of the B&B algorithm in which only the edges that appear in the ancestors are taken into account. The offspring obtained will not only 100% composed by its ancestors' edges but it will also be the best solution that can be build using those edges.

The net effect of this operator is to link low-cost subtours existing in the ancestors, or their construction. In fact, considering these elements as the building blocks of the problem the B&B algorithm would build them and glue together (injecting them into the GA as mentioned in 2.1).

2.2.3. Issues using the hybrid operator. It is clear that the restricted TSP solution carried out by the described operator is faster and does not have the memory limitations of the base case. However, this operator is going to be used a high number of times and its computational complexity is much higher than other operators such as edge-recombination, so execution times of a GA using it intensively might reach untenable values.

A first preventive measure in order to decrease the computational effort of the algorithm is a combined use of many crossover operators. Since the hybrid-operator complexity is determined by the number of edges taken into account, its use could be restricted to those occasions in which the ancestors share a good amount of edges, using a standard operator in any other situation.

The described model has not only the advantage of reducing execution time but it also mitigate another drawback of the operator, the greediness of its heuristic functioning. This model resembles in some way a thermodynamic system: in early stages temperature (disparity between individuals) is high and the standard operator is often used, thus grouping building blocks. As the system freezes the heuristic operator becomes more used. In those late stages the probability of an important block not being carried due to the greediness of the operator is lower.

Another less desirable side effect of this operator is the tendency to a fast convergence that induces in the GA, due to apparition of superindividuals. This effect can be relieved introducing a higher mutation rate, with the risks that come with it.

2.2.4. Improving the model. Parallelism. In spite of the use of the above described model, execution times (although decreased) are still high so another strategy must be used to optimize the process. One of the most suitable options is parallelizing it. There exists some alternatives in order to carry out that parallelization. The following ones can be mentioned:

- Parallelizing the hybrid operator [5]: this option would only reduce the execution time of the GA, remaining unaltered its behaviour.
- Overlapping the hybrid operator with the rest of the GA (Fig 3): this option consists of the parallel launching of many hybrid crossover operations. The GA continues its functioning without waiting for these operations to be concluded. As they finish the generated offsprings are inserted into the population. This option do introduce qualitative behaviour differences in relation to the sequential version. The main one is the absolute disappearance of the *generation* concept since a great gapping [6] occurs (individuals resulting from crossover operations launched many iterations before are inserted into the population).

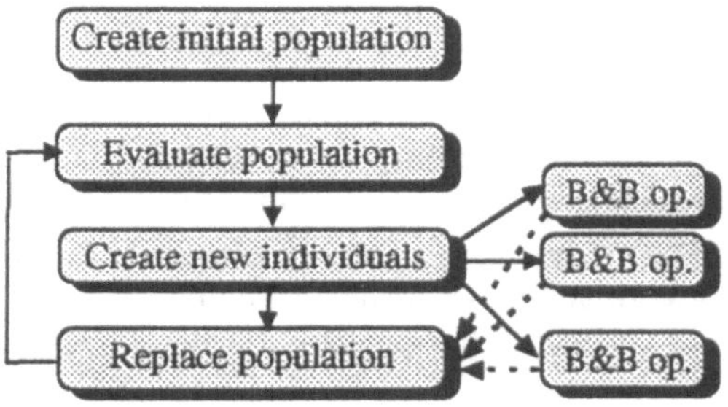

Fig 3. Outline of the Asynchronous Hybrid Genetic Algorithm (AHGA).

It is possible that this overlapping provides a positive side effect related with the problems caused by superindividuals mentioned in 2.2.3: inserting them in the population many evolutionary cycles later may allow average fitness to increase enough not to suffer (or to suffer less) because of the presence of those individuals.

2.3. Branch and bound and parallel evolution

The other way to close up characteristics of B&B techniques and GAs is to heighten the power of the later ones. This heightening can be achieved by means of parallelly executing many GAs. These GAs can independently execute or establish some kind of collaboration between them, periodically interchanging a part of their populations [7].

2.3.1. Describing the model. In the simplest scenario, a B&B algorithm executes in parallel with a certain number of GAs (Fig 4). Many GAs executing independently will generally produce solutions of similar magnitude but different structure, so the problem queue will seldom be purged.

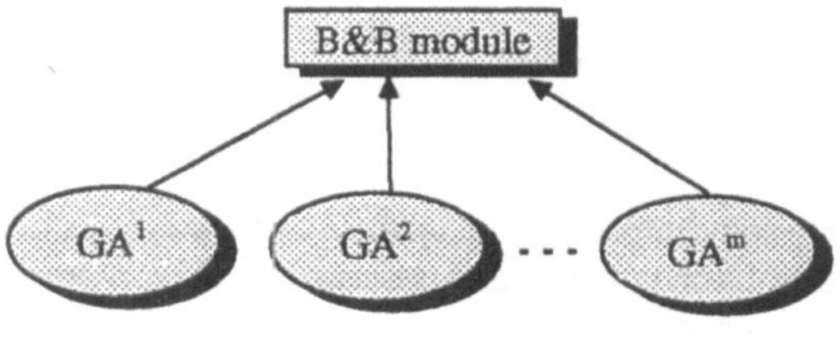

Fig 4. Outline of the parallel evolution model

Notice one of the main problems of the B&B algorithm when applied to the TSP: a high number of edges must be considered. The complexity of the algorithm can be decreased by reducing this number of edges. In order to ignore "*bad*" edges, information from the GAs can be used.

The model outlined above can be described as follows. There is a generalized hybrid operator that recombines n individuals. Such operator is placed in a master process that controls m GAs, where $n = k{\cdot}m$. Periodically, every GA send to the operator k individuals (e.g. the best k). The probable diversity in the structure of individuals provided by independent GAs will contribute to make that almost every edge suitable to be part of the optimal solution be included in any individual, and many non-suitable edges will not be taken into account. Achieving the optimum is anyway not guaranteed.

2.3.2. Variations and improvements. There exist many variations that take as reference the strategy above. two fairly significant ones are:

- Perform an information exchange between GAs as stated in the migratory models described in [7].

- Take $n = k \cdot m+1$, that is, send to the hybrid operator not only the individuals provided by the GAs but also the best obtained so far. This is the B&B counterpart of GA elitism.
- Inject into the GAs the individual created by the hybrid operator. This is a dangerous action since it may cause convergence problems.

3. RESULTS AND CONCLUSIONS

3.1 Results

The set of experiments realized so far has not been very large so only a first impression of the behaviour of each technique has been obtained. All experiments have been carried out with randomly generated euclidean problems of size 75 and 100 running over a network of SUN workstations.

As a first stage, every problem was solved using *pure* GAs and B&B algorithms. The so obtained results will be used as reference for further experiments.

Next, some probes were realized with the Synchronous Hybrid Genetic Algorithm described in 2.2.2. Although a small accuracy improvement takes place (about 10% reduction in the mean error), execution times are too large (two magnitude orders larger) so only a few probes were finished.

Subsequently, a model based on the Asynchronous Hybrid Genetic Algorithm was tested. At this point, the results were a little bit shocking as no apparent regular behaviour was observed. In fact and depending upon the test-problem, variable accuracy improvements were achieved (including worse results as those of the *pure* GA, mainly with a 1GA+1B&B model) using the same parameter configuration.

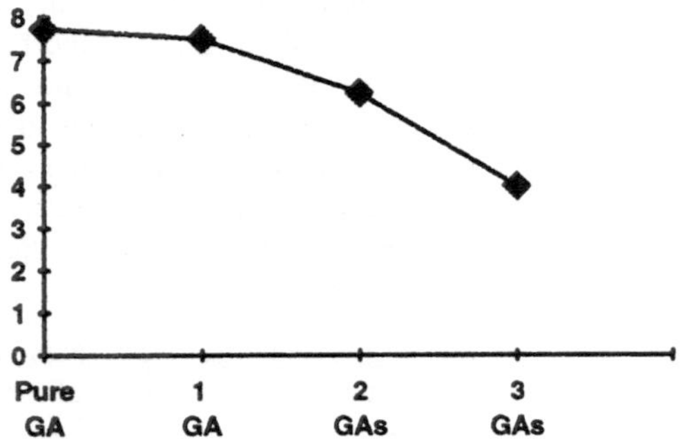

Fig 5. Mean error in some executions of the Parallel Evolution model on a 75-city problem.

Finally, the experiments focused of the model outlined in 2.3.2 including the variations described in 2.3.3 (except inter-GA communication). This model shows the best results. Moreover, its behaviour appears to be more predictable as shown in Fig 5, where it can be seen that the mean error decreases as the number of parallel GAs increases. Specifically, the best improvements achieved were 48% for a 75-city problem (that is, halving the mean error) and 61% in a 100-city problem.

3.2 Conclusions

The results above should be carefully studied since they are only preliminary results. For example, it is possible that an AHGA exhibits a more predictable behaviour with a high number of B&B modules running in parallel. Experiments must also be done with larger problem instances in order to confirm tendencies. Anyway, if a conclusion is to be extracted of obtained results, we can say that hybridization of GAs with B&B techniques will normally improve the quality of results when compared with more classical approaches.

REFERENCES

1. Reeves C.R., “Modern Heuristic Techniques for Combinatorial Problems”, Blackwell Scientific Publications, 1993
2. Davis L., “Handbook of genetic algorithms”, VNR Computer Library, 1991
3. Volgenant T., Jonker R. "A branch and bound algorithm for the symmetric travelling salesman problem based on the 1-tree relaxation.", Univ. of Amsterdam, 1981
4. Michalewicz Z., “Genetic Algorithms + Data Structures = Evolution Programs”, Springer Verlag, 1992
5. Nebro Urbaneja A.J. "Implementation and Evaluation of Parallel Schemes for Branch and Bound Techniques" MCS Tesis, Univ. of Málaga, 1991
6. Alba Torres E., Aldana Montes J.F. "Genetic Algorithms as Heuristic in Optimization Problems. An Overview" (in spanish), Technical Report, Dept. of Lenguajes y Ciencias de la Computación (Univ. of Málaga), 1992
7. Suárez del Rey V. "Parallel Genetic Algorithms for Combinatorial Problems over Shared-Memory Mutiprocessor Systems" (in spanish), MCS Tesis, Univ. of Málaga, 1994

PERFORMANCE OF GENETIC ALGORITHM USED FOR ANALYSIS OF CALL AND SERVICE PROCESSING IN TELECOMMUNICATIONS

Vjekoslav Sinković, Ignac Lovrek

University of Zagreb, Faculty of Electrical Engineering and Computing
Department of Telecommunications, 41000 Zagreb, Unska 3, Croatia

Abstract - A genetic algorithm applied for finishing time determination in the simulation method used for analysis of call and service processing in telecommunications has been described. The evaluation of a performance of the genetic algorithm including influences of the initial string population size on the convergence, differences coming from crossover and mutation probabilities variation, relations between the number of processors and the convergence as well as termination criterion have been discussed.

Introduction

A genetic algorithm is applied for finishing time determination in a simulation method used for the analysis of parallel processing in telecommunications. This parallelism can be expressed in two ways - as a simultaneous handling of different calls and services and as an inherent parallelism within call and service processes. The hierarchy call - call phase - elementary task describes call decomposition; a call phase is initiated with a processing request, an elementary task represents a call grain. A stochastic nature of processes is defined with probability functions for processing requests arrival characteristics and number of elementary tasks per request. A granularity of processes determines their computational properties [1].

The simulation method includes three steps - request flow generation, finishing time determination and response time calculation which are repeated as many times as simulation samples is defined. The simulation is based on a random generation of processing request flow with random structures with respect to the number of elementary tasks and their precedence relations. To obtain response time, T_q, two parameters have to be evaluated - finishing time, T_s and waiting time, T_w. The problem is similar to the problem of multiprocessor scheduling: T_s corresponds to the optimum finishing time for a given set of elementary tasks and T_w to the time needed to complete the previous set of elementary tasks.

The simulation method is described briefly in the first section, the genetic algorithm itself in the second. The third section includes performance evaluation results.

1. Analysis of call and service processing

The simulation procedure is shown in Fig. 1: requests enter the request queue limited to N places. If the number of requests is greater than N , a loss occurs. Requests are waiting until the processing of the previous group of requests has been completed. After that all requests from the queue are moved to the task pattern formatting stage where a set of elementary tasks (ET) with their precedence relations is associated to every request. The next step, the elementary task scheduling is performed by using the genetic algorithm (GA). Finally, elementary tasks enter processor queues.

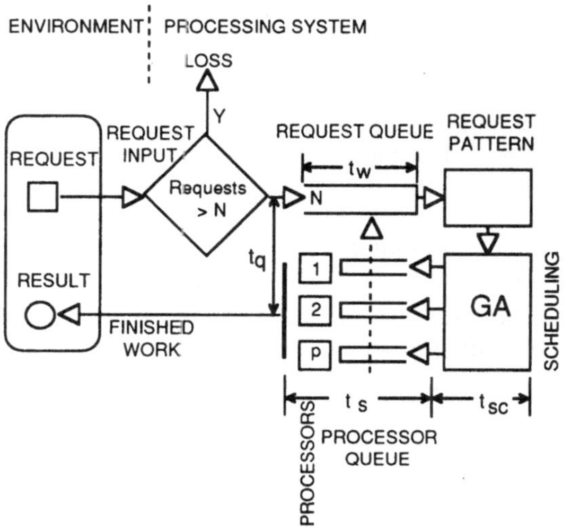

Fig. 1 - Simulation procedure

As an example, Fig. 2 shows the case with 3 requests in the queue, their task patterns, total number of ETs and resulting number of the ETs per time slot, Δt. All ETs are assumed to have the same duration (1 Δt) in this example. The genetic algorithm gives scheduling for 9 processors with the obtained finishing time of 14 Δt.

The structure of a task pattern which corresponds to a specific request r_i is represented as follows:

$$r_i \rightarrow \{s_1, s_2, \ldots, s_j, \ldots, s_{mi}\}$$

where s_j is the number of ETs i.e. the level of parallelism in the j-th time slot. All tasks from the slot j-1 precede all tasks from the j-th slot.

The index j has the meaning of the height for a given task partition.

Requests - 3
Request patterns
{{1, 9, 1, 9, 1}
{0, 1, 6, 1, 1, 1, 8, 1, 4},
{1, 10, 1, 4}}
Σ ETs - 60
ETs per time slot
{2, 20, 8, 14, 2, 1, 8, 1,4}
Height - 9

Processors - 9
Optimal finishing time - 12
GA scheduling
{{0, 2, 1, 2, 0, 0, 1, 0, 0},
{0, 2, 1, 2, 0, 0, 1, 0, 0},
{0, 4, 0, 1, 1, 0, 1, 0, 1},
{0, 3, 1, 1, 0, 0, 1, 0, 0},
{0, 2, 1, 2, 0, 0, 1, 0, 0},
{0, 1, 1, 2, 0, 0, 1, 0, 1},
{0, 1, 1, 2, 0, 0, 0, 0, 1},
{0, 2, 1, 2, 0, 0, 1, 0, 1},
{2, 3, 1, 0, 1, 1, 1, 1, 0}}
Obtained finishing time - 14

Fig. 2 - Simulation example

A stochastic nature of the call and service request flow is defined by using two probability functions: exponential for the request interarrival time and Erlangian for the number of ETs within the request. The similar approach is used in [2].

Two additional options are included: the control of ET parallelism and task pattern shifting [3].

Task patterns with a two-stage parallelism can be generated - low parallelism (LP) is followed by high parallelism (HP) then again LP, HP and so on. The number of ETs in parallel is a stochastic value between LP-minimum and HP-maximum which have to be defined.

Because requests are independent, all even requests can be shifted for one time slot:

$$r_i' \rightarrow \{s_1' = 0, s_2' = s_1, \ldots, s_{j+1}' = s_j, \ldots, s_{mi+1}' = s_{mi}\}$$

to obtain concordance of LP of odd requests with HP of even requests.

That is the way to obtain more uniform distribution of ETs per time slots.

2. Genetic algorithm

The genetic algorithm includes the following steps:

(1) Definition

(1.1) The definition of processor scheme

The number of processors, P, must be defined. Bounds for the number of processor are evaluated using methods from [4, 5].

(1.2) The definition of fitness function

The finishing time is used for evaluating different strings in the population.

(1.3) The determination of the probabilities for controlling genetic operations mutation and crossover.

(2) Initial population generation

ETs are allocated randomly to each processor with precedence relations preserved. The allocation is represented with lists of ETs, one for each processor. Single allocation corresponds to a string as a member of an initial population. The population size must be defined.

(3) Fitness function evaluation

The evaluation is based on the difference between the highest calculated finishing time and the finishing time for the string under consideration. The number of generations is also controlled.

(4) Genetic operations reproduction, crossover and mutation, based on the approach formulated in [6]

(4.1) Reproduction

A selection of strings in the current generation is done. Selection criteria are based on fitness function, i.e. strings with greater fitness value have a greater possibility to enter the new generation. A roulette wheel where each string occupies a slot size proportional to its fitness value is used. Additionally, the best string in the current generation is included in the next one.

(4.2) Crossover

An exchange of portions of two strings to obtain the new two is done with the predefined probability.

(4.3) Mutation

An exchange of ETs of the same height (randomly generated) in the best string to obtain a new one is done with the predefined probability.

A modification is done in our model for ETs of the same duration: after the height in the best string has been generated randomly, positions (processors) are searched for minimum and maximum values of the number of allocated ETs; ETs are redistributed between two processors in a way that half of them are allocated to each one.

All genetic operations preserve precedence relations.

(5) Convergence testing

If the algorithm is not converging repeat steps (2) - (4) and perform genetic operations on a new generation.

Otherwise, the resulting strings and finishing time are obtained.

Two versions of the genetic algorithm are developed, the first dealing with ETs of the same duration only, and the second one for a general case.

3. Performance evaluation

A series of experiments is done to evaluate a performance of the genetic algorithm itself. Different situations, from regular to a heavy request flow, including request bursts are analyzed. The finishing time is taken as a comparison criterion.

3.1. Task pattern shape

Task patterns with the different number of ETs, their duration and precedence relations as well as different interpretation of the same pattern are analyzed:

- Case A: task patterns with ETs differing in duration $d_i\ \Delta t\ (1 \leq d_i \leq d)$.
- Case B: task patterns with the maximum number of ETs having the same duration (1 Δt).
- Case C: task patterns with the minimum number of ETs having the longest duration $(1 \leq d_i \leq d_{max})$.

The same processing request can be represented by using task patterns of different shapes. If the starting pattern is one defined in Case A, then Case B can be derived by splitting all tasks with $d_i > 1$ on d_i ETs of 1 Δt; Case C by combining serially connected tasks in a new one with he duration $\Sigma\ d_i$.

The analysis is done for the optimum number of processors, P, when all inherent parallelism can be exploited and for P - 1 processors.

An example with 35/47/20 tasks having the maximum duration 2/1/5 in cases A/B/C is shown in Fig. 3. The resulting finishing time, FT, is the worst one when all ETs are of the same duration. This has been the reason for the redefinition of the mutation and the development of two GA versions.

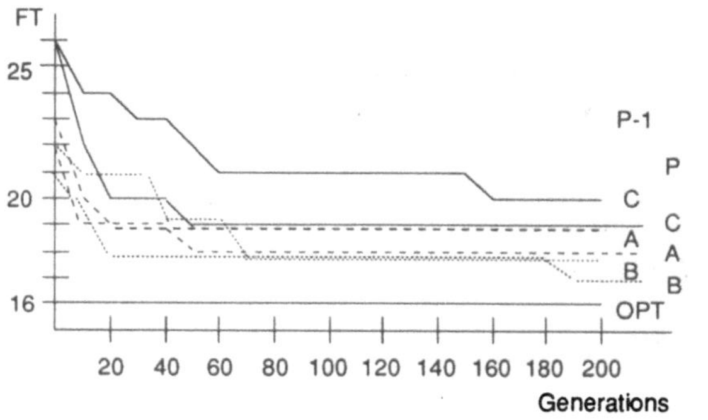

Fig. 3 - Examination of different task pattern shapes

3.2. Initial population size and number of generations

The initial population with respect to generated strings (schedules) and size have been evaluated. The comparison criterion was the best string in the population (the string with the lowest finishing time).

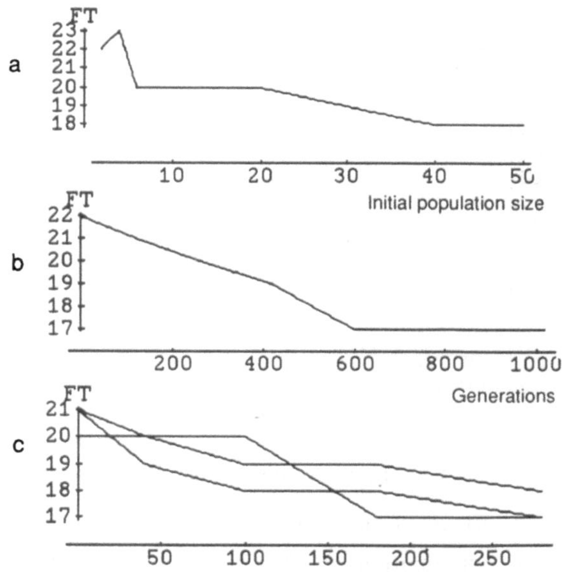

Fig. 4 - Initial population testing

Experiments show the following: small (< 20) to modest (20 - 40) initial population size would be sufficient for the purpose under consideration (Fig. 4a), the generation number has to be relatively high (Fig. 4b), optimization trends being the same applying GA more time on just one initial population or producing each time a new initial population of the same size (Fig. 4c). In this example 96 ETs are scheduled on 9 processors.

3.3. Crossover and mutation probabilities

The crossover probability is proposed to be held in the range of 0.6. - 0.9, the mutation probability depending on the task pattern shape: very low(< 0.1) for task patterns differing in duration, modest (0.2 - 0.3) for equal elementary tasks. An example for equal ETs is shown in Fig. 5. A comparison between GA and optimal scheduling is included.

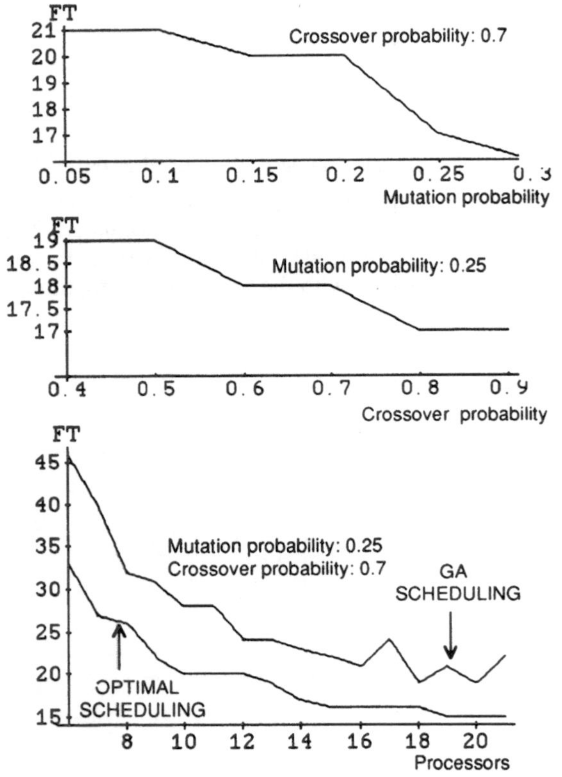

Fig. 5 - Genetic operations probabilities testing

The effects of mutation are twofold:

- the redistribution of tasks between two processors may improve the population and
- the migration of some tasks from one processor to another may help in establishing load balance.

Some points are specific to the situation when tasks having equal duration are analyzed. In that case a part of tasks having the same height may be migrated from one processor to another, so that mutation is performed not only "one for one" but also "one for many" (modified mutation).

3.4. Termination criterion

The termination criterion includes two components: the finishing time value "stabilization" and the number of generations.

4. Conclusions

An application of genetic algorithms to the analysis of parallel processing of calls and services in telecomunications is presented. Stochastics in the problem is expanded with stochastics in the method for solving the problem. In the next step adaptive [7] and parallel genetic algorithms will be evaluated.

References

1. Sinkovic, V., I. Lovrek: Proceedings MPCS'94 Conference on Massively Parallel Computing Systems (MPCS): the Challenges of General Purpose and Special-Purpose Computing, Part II (Draft), May 2-6, 1994, Ischia, Italy, 373(1994).
2. Fleischmann, G., M. Gente, F. Hofmann, G. Bolsch: Parallel Computing, Vol. 20, 1583(1994).
3. Huang, J-H., L. Kleinrock: IEEE Transactions on Parallel and Distributed Systems, Vol.4, No.3, 306(1993).
4. Fernandez, E.B., B. Bussell: IEEE Transactions on Computers, Vol. C-22, No. 8, 745(1973).
5. Kleinrock, L., J-H. Huang: IEEE Transactions on Software Engineering, Vol. 18, No. 5, 434(1992).
6. Hou, E.S.H., N. Ansari, H. Ren: IEEE Transactions on Parallel and Distributed Systems, Vol. 5, No. 2, 113(1994).
7. Srinivas, M., Patniak: IEEE Transactions on Systems, Man and Cybernetics, Vol. 24, No. 4, 656(1994).

EVOLUTIONARY HEURISTICS FOR THE BIN PACKING PROBLEM

Sami Khuri
Dept. of Math. & Computer Science
San José State University
One Washington Square
San José, CA 95192-0103, U.S.A.

Martin Schütz
Universität Dortmund
Fachbereich Informatik, LS XI
D–44221 Dortmund, Germany

Jörg Heitkötter
EUnet Deutschland GmbH
Research & Development
Emil-Figge-Str. 80
D–44227 Dortmund, Germany

Abstract

In this paper we investigate the use of two evolutionary based heuristic to the bin packing problem. The intractability of this problem is a motivation for the pursuit of heuristics that produce approximate solutions. Unlike other evolutionary based heuristics used with optimization problems, ours do not use domain-specific knowledge and has no specialized genetic operators. It uses a straightforward fitness function to which a graded penalty term is added to penalize infeasible strings. The encoding of the problem makes use of strings that are of integer value. Strings do not represent permutations of the objects as is the case in most approaches to this problem. We use a different representation and give justifications for our choice. Several problem instances are used with a greedy heuristic and the evolutionary based algorithms. We compare the results and conclude with some observations, and suggestions on the use of evolutionary heuristics for combinatorial optimization problems.

Introduction

The bin packing problem (bpp) consists in placing n objects, each of which has a weight ($w_i > 0$), in the minimum number of bins, such that the total weight of the objects in each bin does not exceed the bin's capacity. All bins have the same capacity ($c > 0$). The bpp can be viewed as a multiple subset problem. The goal is to partition n positive numbers (the weights) into the smallest possible number of subsets B_j (the bins), such that the sum of the integers in each subset does not exceed a positive number c.

The bpp belongs to the class of NP-hard problems. The tripartite matching problem can be reduced to it [10]. Thus, unless $P = NP$, the search for exact algorithms is replaced by the design of approximate algorithms. Incidently, only a handful of exact algorithms do exist. Martello and Toth's branch-and-bound based "first-fit decreasing" heuristic is the most adequate algorithm for large problem instances [9].

In this work we describe the results of applying a genetic algorithm (GA) [6, 7] and an Evolution Strategy (ES) [12] to the bpp. The approach used here differs in more than one way to the traditional evolutionary algorithms. First, unlike most encodings that are binary, our population in the GA consists of n-ary strings $x_1x_2\ldots x_n$, where each x_i, $1 \le x_i \le n$, is a real number. The ES makes use of integer valued parameter vectors rather than the standard real-valued parameters. Second, unlike many traditional approaches that use domain-specific knowledge and specialized genetic operators, the work presented in this paper makes use of a graded penalty term incorporated in the fitness function of the GA as well as the ES.

In the next section, we give the formal definition of the bin packing problem including three greedy based heuristics. The paper then shifts to the evolutionary based heuristics. The different ingredients with justifications, for our GA implementation are then presented and encoded in LibGA [3], the GA package we use in this work. For the ES, we make use of the package EVOS$^+$ [11], which was successfully used as an optimization tool for mixed integer problems in the field of optics [1]. In the third section, several problem instances are used with both evolutionary based heuristics and the greedy heuristic and the results are compared. Our work concludes with some observations from our findings, and some suggestions on the use of evolutionary heuristics for other combinatorial optimization problems.

The Bin Packing Problem

The following is a formal definition of the bpp in which we make use of Stinson's terminology for combinatorial optimization problems [13] and in which we introduce concepts and notations that we use in subsequent sections of the study.

Problem Instance:

Objects:	1	2	...	n
Weights:	w_1	w_2	...	w_n

n bins, each of capacity c. The weights and capacity are positive real numbers and $w_i < c$ for $i = 1, 2, \ldots, n$.

Feasible Solution:

A vector $\vec{x} = x_1x_2\ldots x_n$ where $x_i = j$ means that the ith object is placed in bin j, $1 \le j \le n$; and for every j we have

$$\sum_{i \in B_j} w_i \le c$$

where $B_j = \{i \mid x_i = j\}$; i.e., B_j represents the objects that are in bin j.

Objective Function:

A function $P(\vec{x})$ = number of different values of the components of $\vec{x}$, where $\vec{x} = x_1x_2\ldots x_n$ is a feasible vector. In other words, $P(\vec{x})$ denotes the number of bins required for the solution represented by $\vec{x}$.

Optimal Solution:

A feasible vector $\vec{x}$ that gives the smallest possible $P(\vec{x})$.

The following are straightforward, simple, greedy based heuristics for the bpp. In the following three algorithms we consider placing an object in previously used bins (non-empty) if such a bin can be found, otherwise we place the object in a new one. In the First Fit Algorithm each element is placed in the first bin that has enough room for it, while in the Best Fit, the object is placed in the best bin that can still contain it. The best bin means the most filled one that still has enough room for the object under consideration.

We can do better than both, the First and the Best Fit Algorithms by saving the objects with small weights for the last steps of the procedure, thus preventing them from taking up space that would be better filled with heavy objects first. More formally, the First Fit Decreasing Algorithm first sorts the weights in decreasing order and then uses the First Fit Algorithm. Note that with the three algorithms, we assume that bins are considered in increasing order of their indices.

We define $W = \sum_{i=1}^{n} w_i$ and by borrowing Garey and Johnson's notation [5], we denote the number of non-empty bins obtained when First Fit Algorithm is applied to problem instance I by FF(I); and by FFD(I) the number of non-empty bins when the First Fit Decreasing Algorithm is used with problem instance I. Similarly, let OPT(I) denote the number of bins obtained by the using an exact algorithm (thus OPT(I) is the optimum packing). It can be shown that [5]:

$$FF(I) \leq \left\lceil \frac{2W}{c} \right\rceil \tag{1}$$

$$FFD(I) = \frac{11}{9} OPT(I) + 4 \tag{2}$$

Since Result 2 shows that the First Fit Decreasing Algorithm always yields solutions that are within 22% of the optimal for large problem instances, we chose to implement it so as to compare its performance with our evolutionary based heuristic.

First Fit Decreasing Algorithm

As mentioned above, the heavy objects are considered before the lighter ones. The actual filling of the bins is thus preceded by the reinitialization of the objects in decreasing order of their weights.

Algorithm First Fit Decreasing
  begin
    reinitialize objects in decreasing order of weights
    for $i = 1, n$ **do**
      for $i = 1, m$ **do**
        if $B[j] + w_i \leq c$ **then**
          $B[j] := B[j] + w_i$
  end

Evolutionary Based Heuristics

First, recall that GAs work with encodings of the problem rather than the problem itself. We use the encoding described in "Feasible Solution", i.e., our population consists of strings of length n, $x_1x_2\ldots x_n$, except that we can sometimes get a better range than $1 \leq x_i \leq n$ for the values of x_i. Inequality 1 gives an upper bound on FF(I). Since we are interested in designing a heuristic that outperforms the First Fit Algorithm on the average, it does make sense to restrict the values of x_i to be no greater that m, where $m = \min\{n, \lceil \frac{2W}{c} \rceil\}$.

Our initial population is thus generated by producing random strings of length n, $x_1x_2\ldots x_n$, where $1 \leq x_i \leq m$.

Next we need to define a fitness function.

The bpp is a highly constrained combinatorial optimization problem. Nevertheless, infeasible strings are not tossed out; they remain in the population but their fitness is weakened by adding a penalty term to the fitness function. Moreover, the penalty is graded since all infeasible strings are not equal. Recall that the number of different values of the components of $\vec{x}$ determines the number of non-empty bins used by $\vec{x}$, which we denoted by $P(\vec{x})$ in "Objective Function".

More precisely, our implementations for the GA and for the ES make use of the following fitness function to be minimized:

$$f(\vec{x}) = P(\vec{x}) + s \cdot \left(m + \sum_{j=1}^{m} \max\{w(B_j) - c,\ 0\} \right)$$

where $s = 0$ when $\vec{x}$ is feasible, $s = 1$ when $\vec{x}$ is infeasible, and $w(B_j)$ is the sum of the weights of the objects in B_j, i.e., $w(B_j) = \sum_{l \in B_j} w_l$

The first term of the fitness function yields the number of bins given by $\vec{x}$. The second term is the penalty and is activated only when $\vec{x}$ is infeasible. The first term, m, of the penalty is an offset term whose sole purpose is to guarantee that infeasible vectors will always yield fitness values that are inferior to the ones obtained from feasible vectors. Moreover, the penalty is highly graded; not only does it take into consideration the number of overfilled bins ($\sum_{j=1}^{m}$ in the second term of $f(\vec{x})$), it also takes into account the distance from feasibility for every bin ($w(B_j) - c$).

We next consider a simple problem instance in which we illustrate the concepts and notations we have introduced.

Example

Consider the following problem instance of the bin packing problem:

Objects:	1	2	3	4	5	6	7	8
Weights:	6	7	5	9	3	4	5	4

Suppose that the capacity of each bin is 12.5.

We first compute $m = \min\{8, \lceil \frac{2(W)}{12.5} \rceil\}$, where $W = 43$ is the sum of the weights of the objects. Thus, $m = 7$. In other words we consider only 7 (rather than $n = 8$) bins.

Consider the string $\vec{a} = 4\ 5\ 6\ 3\ 5\ 1\ 6\ 1$.

The contents of the 5 bins are $B_1 = \{6, 8\}$ since $x_6 = x_8 = 1$, $B_3 = \{4\}$ since $x_4 = 3$, $B_4 = \{1\}$ since $x_1 = 4$, $B_5 = \{2, 5\}$ since $x_2 = x_5 = 5$, and $B_6 = \{3, 7\}$ since $x_3 = x_4 = 6$. Note that $B_2 = B_7 = \emptyset$.

$\vec{a}$ represents a feasible solution since $w(B_i) \leq c$ for $1 \leq i \leq n$. Consequently, $f(\vec{a}) = P(\vec{a}) = 5$.

On the other hand, $\vec{b} = 6\ 5\ 1\ 3\ 1\ 6\ 3\ 1$ represents an infeasible solution. The contents of the 4 bins are $B_1 = \{3, 5, 8\}$, $B_3 = \{4, 7\}$, $B_5 = \{2\}$ and $B_6 = \{1, 6\}$. The total weight of the objects in B_3 is 14 which exceeds the bin's capacity. Since $\vec{b}$ is infeasible, $f(\vec{b}) = P(\vec{b}) + 7 + (14 - 12.5) = 12.5$.

The First Fit Algorithm applied to this problem instance yields 5 bins: $B_1 = \{1, 3\}$, $B_2 = \{2, 5\}$, $B_3 = \{4\}$, $B_4 = \{6, 7\}$, and $B_4 = \{8\}$.

The Best Fit Algorithm applied to this problem instance yields 4 bins: $B_1 = \{1, 6\}$, $B_2 = \{2, 3\}$, $B_3 = \{4, 5\}$, and $B_4 = \{7, 8\}$.

The First Fit Decreasing Algorithm considers the objects according to their weights (in decreasing order) and then uses 4 bins: $B_1 = \{4, 5\}$, $B_2 = \{2, 3\}$, $B_3 = \{7, 1\}$, and $B_4 = \{6, 8\}$. Note that we might get different fillings of the 4 bins since some of the objects have equal weights.

Experimental Results

For our experimental runs, we make use of the 40 bin packing problem instances found in Beasley's OR-library [2]. These are divided into two groups. The first (binpack4 from the OR-library) has 20 problems, each consisting of 1000 objects and the bin capacity is 150. The second (binpack8) also has 20 problem instances, each consisting of 501 objects with bin capacity 100.

For the ES, we use a (15, 100)-ES, i.e. 15 parents and 100 offspring (see [12] for more details) with standard parameter settings. As for the GA, each run consists of 1000 generations with 500 strings in each. We note that only a very small part (500×10^3) of the search space (e.g. m^{1000} where $m \approx 800$ for binpack4) is explored by the GA. No attempt was made to fine-tune any of the evolutionary heuristics' parameters.

binpack4				*binpack8*			
best	**FFD**	**GA**	**ES**	**best**	**FFD**	**GA**	**ES**
399	403	512	509	167	190	202	211
406	411	530	519	167	191	203	213
411	416	525	526	167	190	204	210
411	416	525	532	167	190	206	216
397	402	513	498	167	191	206	213
399	404	514	506	167	190	203	207
395	399	506	497	167	190	201	209
404	408	514	521	167	189	202	208
399	404	511	512	167	191	202	209
397	404	512	513	167	190	206	210
400	404	516	504	167	190	204	212
401	405	515	508	167	190	202	213
393	398	499	494	167	190	206	209
396	401	509	511	167	190	203	205
394	400	509	504	167	189	207	214
402	408	516	516	167	190	202	209
404	407	525	515	167	189	206	209
404	409	515	524	167	191	203	210
399	403	510	514	167	189	203	206
400	406	512	509	167	191	204	209

Table 1: Runs performed by the greedy heuristic (FFD), and the evolutionary heuristics, a Genetic Algortihm (GA), and an Evolution Strategy (ES) on binpack4 and binpack8 from OR-Lib.

Conclusion

In this work, two evolutionary based heuristic are used as approximation algorithms with a highly constrained combinatorial optimization problem. The solutions are compared to one of the best greedy based procedures, namely the First Fit Decreasing Algorithm. As can be seen from Table 1, the results obtained by the First Fit Decreasing Algorithm are better than the ones obtained by the evolutionary based heuristics. Although it might not be very fair to compare a specialized greedy procedure to general purpose evolutionary heuristics, we do offer these two observations in an attempt to shed some light on our results. First, it is very possible that evolutionary heuristics' representation, and more importantly, stochastic genetic operators are not appropriate for problems such as bin packing that exhibit a "grouping property". In other words, our work would tend to agree with Falkenauer's findings about GAs [4]. He believes that neither standard nor ordering operators are suitable for such problems and offers a hybrid Group Genetic Algorithm as a remedy. Second, we believe that our findings are not conclusive; more work has to be done. After all, similar GA approaches have worked quite well with other combinatorial optimization problems [8].

References

[1] Th. Bäck and M. Schütz. Evolution strategies for mixed-integer optimization of optical multilayer systems. In *Proceedings of the 4th Annual Conference on Evolutionary Programming*, 1995. (accepted for publication).

[2] J. E. Beasley. OR-Library: Distributing test problems by electronic mail. *Journal of the Operational Research Society*, 41(11):1069–1072, 1990.

[3] A. L. Corcoran and R. L. Wainwright. Libga: A user-friendly workbench for order-based genetic algorithm research. In *Proceedings of the 1993 ACM/SIGAPP Symposium on Applied Computing*, pages 111–117. ACM, ACM Press, February 1993.

[4] E. Falkenaur. A new representation and operators for genetic algorithms applied to grouping problems. *Evolutionary Computation*, 2(2):123–144, 1994.

[5] M. R. Garey and D. S. Johnson. *Computers and Intractability — A Guide to the Theory of NP–Completeness.* Freemann & Co., San Francisco, CA, 1979.

[6] D. E. Goldberg. *Genetic algorithms in search, optimization and machine learning.* Addison Wesley, Reading, MA, 1989.

[7] J. H. Holland. *Adaptation in natural and artificial systems.* The University of Michigan Press, Ann Arbor, MI, 1975.

[8] S. Khuri, T. Bäck, and J. Heitkötter. An evolutionary approach to combinatorial optimization problems. In D. Cizmar, editor, *Proceedings of the 22nd Annual ACM Computer Science Conference*, pages 66–73. ACM, ACM Press, 1994.

[9] S. Martello and P. Toth. *Knapsack Problems: Algorithms and Computer Implementations.* Wiley, Chichester, West Sussex, England, 1990.

[10] C. H. Papadimitriou. *Computational Complexity.* Addison Wesley, Reading, MA, 1994.

[11] M. Schütz. Eine Evolutionsstrategie für gemischt-ganzzahlige Optimierungsprobleme mit variabler Dimension. Diplomarbeit, Universität Dortmund, Fachbereich Informatik, 1994.

[12] H.-P. Schwefel. *Evolution and Optimum Seeking.* Wiley, New York, 1995.

[13] D. R. Stinson. *An Introduction to the Design and Analysis of Algorithms.* The Charles Babbage Research Center, Winnipeg, Manitoba, Canada, 2nd edition, 1987.

A CLAUSAL GENETIC REPRESENTATION AND ITS EVOLUTIONARY PROCEDURES FOR SATISFIABILITY PROBLEMS

Jin-Kao Hao

LGI2P
EMA-EERIE
Parc Scientifique Georges Besse
F-30000 Nîmes
France
email: hao@eerie.fr

Abstract

This paper presents a clausal genetic representation for the satisfiability problem (SAT). This representation, CR for short, aims to conserve the intrinsic relations between variables for a given SAT instance. Based on CR, a set of evolutionary algorithms (EAs) are defined. In particular, a class of hybrid EAs integrating local search into evolutionary operators are detailed. Various fitness functions for measuring clausal individuals are identified and their relative merits analyzed. Some preliminary results are reported.

1. Introduction

The satisfiability problem or SAT [4] is of great importance in computer science, both in theory and in practice. The statement of the problem is very simple. Given a well-formed boolean formula F in its conjunctive normal form (CNF)[1], is there a truth assignment that satisfies it? In theory, SAT is at the very core of NP-complete problems. In practice, many applications can be formulated with it. One important topic related to SAT is to find such a satisfiable truth assignment for a satisfiable formula F. In this paper, we are interested in this model-finding problem for SAT.

Although SAT is NP-complete, many methods have been devised for practical needs. These methods can be roughly classified into two large categories: complete and incomplete methods [1]. As examples of the first category, we can mention the logic-based approach (Davis & Putnam algorithm, Binary Decision Diagrams), the constraint-network-based approach (CSP, local propagation), and 0/1 linear programming. The second category includes such methods as simulated annealing [14][15], local and greedy search [6][9][13] and evolutionary algorithms (EAs) [2][3][7][16].

Evolutionary algorithms are essentially based on three elements: an internal representation, an external evaluation function and an evolution mechanism. It is commonly recognized that the internal representation greatly conditions the success or failure of an EA. The integration of specific knowledge at different levels in an EA may also greatly improve its performance.

1. A CNF formula is a set of clauses linked by logic **and**, a clause being a set of literals linked by logic **or**, a literal being a boolean variable or its negation.

SAT has a natural binary representation (what we call variable-coding): an individual represents a potential solution to the boolean formula F, each gene coding a variable of F. Genetic algorithms can easily be developed using the variable-coding and the traditional genetic operator set. This representation is also the one used by all the reported work mentioned above.

Surprisingly, however, no efficient EA has been found yet using this variable-based representation. In fact, the reported EAs for SAT are far from being comparable with GSAT [13][14], a simple local search based procedure using the same variable-coding. Why this happens is not clear, but we have enough reasons to think that we can devise EAs outperforming GSAT. First, the evolutionary framework allows GSAT to be simulated with a singleton population and a controlled mutation operator. Therefore, we have already a special EA which can do as well as GSAT does. Second, we can easily integrate the basic idea of GSAT into EAs, which should lead to better performance. We are working on these issues, results of that will be reported elsewhere.

Another aspect for designing efficient EAs consists of looking for more intelligent codings, which is the topic of the current paper. The proposed coding is based on the following observation. For a given CNF formula, there are some intrinsic relations among the clauses of the formula. These relations are defined by shared variables and important for problem-solving (model-finding here). However, they are lost in the variable-coding in which only the values of variables are manipulated. In this paper, we explore an alternative coding called clausal representation (CR) which takes into account these relations between clauses.

The paper is organized as follows. In §2, the clausal representation is defined. In the next two sections, various fitness functions and CR-based hybrid EAs are presented. In particular, the integration of local search within the evolution mechanism are detailed. In the last two sections, preliminary experimental results and future work are discussed.

2. CR-A Genetic Clausal Representation

In order to simplify our presentation, without losing the generality, we will consider the 3-SAT problem. Recall that a 3-SAT instance F is defined by a set of L clauses of length 3 connected by logic ***and*** ($\wedge$), L being called the length of F. Each clause is composed of 3 different literals linked by logic ***or*** ($\vee$). A literal is either a boolean variable (called a positive literal) or the negation of a boolean variable (called a negative literal). Note that 3-SAT is the preferred form of many studies, and most methods mentioned above deal with 3-SAT instances.

Given a 3-SAT instance $F \equiv C1 \wedge C2 \wedge ... \wedge CL$ containing N different variables. We encode clauses instead of variables. More precisely, for any clause Ci ($1 \leq i \leq L$), there are in total $2^3=8$ possible assignments {000, 001, 010, 011, 100, 101, 110, 111}. Among them, there are exactly 7 satisfying assignments and 1 non-satisfying one. For example, {100} is the only non-satisfying assignment for the clause ($\neg a \vee b \vee c$). Since this non-satisfying assignment will never participate in any solution for SAT instances containing this clause, it will be advantageous to eliminate this assignment from chromosomes at the beginning. This analysis leads us to the following genetic clausal representation (CR).

For any N-variable SAT instance $F \equiv C1 \wedge C2 \wedge ... \wedge CL$, each individual (chromosome) is composed of a string having a length 3*L divided into L parts (genes) <G1, G2,...,GL>, each gene Gi ($1 \leq i \leq L$) taking as its values a *satisfying* assignment of the corresponding clause Ci. In other words, we will forbidden the non-satisfying assignments to be part of any chromosome. For example, {010 101} is one valid chromosome for the formula $F \equiv (\neg a \vee b \vee c) \wedge (a \vee \neg c \vee b)$ while {100 101} is not since 100 is not a satisfying assignment for the first clause.

One interesting property of CR is that each gene of such a chromosome is locally *consistent* with its clause, which is a necessary condition for any solution to the SAT instance[1]. Moreover, the elimination of non-satisfying values from chromosomes implies the reduction of the search space from the beginning. Also the relations between clauses become more explicit in CR and may be efficiently exploited.

With CR, a variable may have the two values (0/1) in a chromosome. Such a variable is said to be *inconsistent*. It is easy to see that a chromosome is a solution iff each of the N variables of the formula F has a unique 0 or 1 value in the chromosome. Consequently, the number of inconsistent variables can be used as a good indicator to measure the fitness of individuals.

3. Fitness Evaluation

The aim of this evaluation is to measure the "fitness" of each individual (chromosome) with respect to the given formula. The difficulty concerning this evaluation is that no exact measuring rule is available and only approximate rules can be tried. Several possibilities exist to define such a fitness function Eval with our clausal representation. The basic idea is to use the number of inconsistent variables to define these evaluation functions since we want to reduce this number to 0. In the following, we give some such functions. Each evaluation function **Eval(CH, F)** takes as its input a chromosome **CH** and a CNF formula **F**, and gives an integer or a real number as the fitness of **CH** w.r.t. the given formula **F**.

1. Note that this is not case with the variable-coding.

a) Eval_1=ΣW(Vi) ($1 \leq i \leq N$) where Vi are variables
W(V)=1 if variable V is inconsistent
W(V)=0 otherwise

This function simply counts the number of inconsistent variables. It is easy to see that the value returned by this function is between 0 (which means a solution) and N (which means all variables in the chromosome are inconsistent). The smaller the score, the better the chromosome. However, this function cannot differentiate two chromosomes which have the same number of inconsistent variables.

b) Eval_2=ΣC(Vi) ($1 \leq i \leq N$) where
C(V) = # of clauses containing V or $\neg$V
if variable V is inconsistent
C(V) = 0 otherwise

This function, which is also defined with inconsistent variables, takes into account the importance of variables[2] in order to differentiate chromosomes containing the same number of inconsistent variables.

c) Eval_4=Σ|1−2*N1(Vi)/N01(Vi)| ($1 \leq i \leq N$) where
N1(Vi) = # of 1s taken by Vi in CH.
N01(Vi) = # of 0s and 1s taken by Vi in CH,
which is the total number of literals of Vi.

This function returns a real number between 0 and N. The value 0 means that all the variables have the same number of 0s as 1s while the value N corresponds to a solution.

d) Eval_3=Σ|N0(Vi)-N1(Vi)| ($1 \leq i \leq N$) where
N0(Vi) = # of 0s taken by Vi in CH.
N1(Vi) = # of 1s taken by Vi in CH.

The value returned by this function is an integer between 0 (all variables have the same number of 0s and 1s)[3] and ΣN01(Vi) (all variables have an unique value 0 or 1). A CH having this biggest value is a solution while a CH having the value 0 can be considered to be less fitted.

Moreover, it is possible to combine some of them to define multiple-step evaluation. Finally, instead of defining these functions statically, they may be made self-adapting during the evolution of the population.

4. CR-based Evolutionary Procedures

Using the clausal representation, various evolutionary algorithms can be directly built with a population of clausal chromosomes. Operators such as random crossover and mutation can be directly applied to this population with some minor constraints. For instance, in order for crossover to be meaningful, a single point crossover will take place at a point 3*i ($1<i<L-1$, L being the length of the formula F). Mutation will change a 3-bit valid value of a selected gene to another 3-bit valid value. We will not detail this kind of algorithm since it is straightforward to build them. In the

2. The importance of a variable is defined by the number of its appearances in the formula.
3. If there are odd appearances of variables, the minimum value becomes the sum of such variables.

following, we first investigate a special population for clausal individuals. Then we investigate a class of hybrid algorithms based on this population.

In order to build an evolutionary procedure, an initial population must be generated in some way. Although this population may contain any number of chromosomes, a more intelligent organization can be devised with the clausal representation. In fact, since each gene in CR corresponds to a clause which has exactly 7 satisfying assignments, one natural way of realizing the initialization is to generate a population of 7 chromosomes such that for each gene we make sure that the gene has all of its 7 satisfying assignments (alleles). Here is an example.

$$F \equiv (\neg a \vee b \vee c) \wedge (a \vee \neg c \vee b) \wedge (c \vee \neg a \vee \neg b) \wedge (\neg a \vee \neg c \vee \neg b)$$

<u>Population</u>

Clauses	C1	C2	C3	C4
Genes	G1	G2	G3	G4
Ind.1	0 0 0	0 0 0	0 0 0	0 0 0
Ind.2	0 0 1	**0 0 1**	**0 0 1**	**0 0 1**
Ind.3	**0 1 0**	0 1 1	0 1 0	0 1 0
Ind.4	0 1 1	1 0 0	1 0 0	0 1 1
Ind.5	1 0 1	1 0 1	1 0 1	1 0 0
Ind.6	1 1 0	1 1 0	1 1 0	1 0 1
Ind.7	1 1 1	1 1 1	1 1 1	1 1 0
Forbidden-values	1 0 0	0 1 0	0 1 1	1 1 1

The 7 possible values for each gene are put in increasing order in this example; the order is not important in the general case.

One nice property of this population P is that all the solutions for a given formula are covered by it. In fact, any solution may be "constructed" from P by concatenating good values of the genes. This is because a solution will necessarily be a combination of all the L genes, each taking a value from the 7 satisfying ones for a clause. Therefore, what an EA must do is to make these chromosomes evolve towards solutions by using some evolution mechanisms. In the following, we will see how hybrid EAs can be built using such a population.

There are many good reasons and different ways to build hybrid EAs. We will not discuss this general issue here, see for example [10] for more detail. We will instead consider the hybridization of the evolutionary framework with local search using our clausal coding.

Local search groups a class of heuristics based on random search and neighborhood relations. They are known to be very efficient for finding local optima for certain classes of problems. A typical local search procedure begins with a starting point $s0$, often generated randomly. Then an iterative process follows to produce a series of points $s1, s2, \ldots, sn$. The last point sn will be the best solution found. Each si ($1 \leq i \leq n$) is chosen from a set of neighbors associated to $si\text{-}1$ which are defined by a neighborhood function. In essence, a local search procedure is characterized by this neighborhood function and the neighbor selection strategy (that decide which the neighbor to take). Various hillclimbers (steepest ascent, next ascent), simulated annealing [8] and Tabu search [5] are typical examples of local search. Integrating local search into EAs makes it possible to combine the diversity offered by EAs and the efficiency of local search.

There are essentially two possibilities to realize this hybridization. First, we can use the local search to improve all the individuals or a subset of a population at each generation. This integration which is easy to realize can be qualified as a loosely coupled approach. Second, local search can be directly incorporated into genetic operators. More precisely, applying local search at the level of genetic operators consists in using *controlled* crossover or mutation to locally improve the quality of some chromosomes. For example, crossover can produce, according to the local search techniques used, the best children or the next better ones. Similarly, mutation can look for the best or next better offspring. If a local search technique like simulated annealing is used, degenerate offsprings might also be accepted with certain fixed or variable probability.

The following pseudo code gives an example in which we decide to improve a single (the first) chromosome by local search.. By convention, we use P[I,J] to indicate the Jth gene of the Ith chromosome ($1 \leq I \leq 7$, 7 being the population size, $1 \leq J \leq L$, L being the length of the 3-SAT instance F), and we use CH[I] to represent the Ith chromosome of the population P.

Procedure Local_Genetic

```
{
1. generate (P);                    /* as described above */
2. while not stop_condition do
3. for J=1 to L                     /* L=length of F */
4.    for I=2 to population_size    /* equal to 7 */
5.       if improve(P[1,J], P[I,J])
6.       then swap(P[1,J], P[I,J]);
}
```

The "improve" operation tests if swapping the two alleles for a given gene improves the fitness of the given chromosome. This test can be carried out by using the evaluation functions described in §3. Remember that a chromosome is a solution iff all its variables have a unique 0/1 value. Thus improving an individual means to reduce the number of its inconsistent variables. The "swap" operation takes the corresponding genes of two chromosomes and effectively exchanges their alleles. It is interesting to note that "swap" can be considered as either mutation or a special two-point crossover operation. Note also that selection is realized implicitly; the two children produced by "swap" replace their parents. The "stop_condition" in the **while** loop may be defined according to different criteria. One possible way is to fix the number of iterations. Another way is to check if the chromosome which is being locally improved is effectively improved during the last iteration. The complexity analysis of this algorithm is straightforward and thus omitted.

From this simple algorithm, many variants can be easily deduced. Here are some examples, First, this algorithm systematically tries to improve all of the L genes of a given chromosome. A test can be added between the lines 3 and 4 to select only the genes containing at least an inconsistent variable. Second, for a given gene, the algorithm may swap at most 6 alleles due to the **for** loop at the line 4. This loop can be replaced by a procedure which may look for the next better allele or the best one, and then swap only once. Third, this algorithm looks for improving a single chromosome. It can be easily modified to allow all the chromosomes of the population to be improved at each **while** iteration. Finally, classic crossover can be similarly adapted to include local search.

5. Experimental Results

Experiments are being carried out. To test the different methods described above, we are using SAT instances such as those of the second Dimacs challenge and random instances as described in [11]. We observe that the methods based on local search are very efficient to reach local optima (which may or may not be solutions). Very often, local optima can be found after a small number of generations for formulae containing 50, 100, 200 variables. However, it is too early to draw any final conclusion yet given the fact that only partial experiments have been realized. We hope more results will be available in the near future.

6. Discussion & Conclusions

We presented an original genetic representation for SAT based on clause assignments and its associated evolutionary procedures. In particular, a class of procedures integrating the local search within the evolution mechanism are introduced. Several evaluation functions are identified.

Experiments are being carried out to study the effectiveness of these ideas. The limited experimental results prevent us from giving any conclusive remarks, but show that these hybrid procedures need only a very small number of generations to find (local or global) optima for the set of SAT instances tested. Note that the population size (only 7 individuals) used is dramatically small compared with the search space visited.

One possible disadvantage of the clausal-coding may be its bigger search space compared with that of the variable-coding given the fact that there are often more clauses than variables in hard SAT instances. Also there are more alleles for each gene with the clausal-coding than with the variable-coding. However, with the clausal coding which conserves relations among variables, there may exist more room for us to devise more accurate fitness functions and some intelligent genetic operators which permit to efficiently exploit these relations.

Another interesting point worthy of further study is the landscapes related to the two different codings. For example, the distributions and depths of local optima of these two codings are *a priori* very different. A better understanding of these issues will help to design more appropriate fitness functions and genetic operators.

We continue to experiment with various EAs for SAT using both the variable-coding and clausal-coding. We wish to be able to compare the two codings and their related algorithms. Finally, the ultimate goal of our study on SAT is to develop hybrid EAs (which may use either of the two codings) more efficient than the simple local search procedures like GSAT.

Acknowledgments

The author would like to thank R. Dorne and S. Vautier for the many useful discussions. Thanks are also due to Ph. Bogy and W. Lallart for their implementation effort. The work is partially supported by the Bahia project (PRC-IA the French national coordinated research programme for Artificial Intelligence).

References

[1] Collective paper of Bahia project. "Comparative study of three formalisms in propositional logic" (in French). 5ème Journées Nationales PRC-GDR en Intelligence Artificielle, Nancy, February 1995.

[2] De Jong K.A. and Spears W.M. "Using genetic algorithms to solve NP-complete problems". Intl. Conf. on Genetic Algorithms (ICGA'89), Fairfax, Virginia, June 1989, pp124-132.

[3] Dorne R. and Hao J.K. "New genetic operators based on a stratified population for SAT" (in French). Evolution Artificielle, Toulouse, France, September 1994.

[4] Garey M.R. and Johnson D.S. "Computers and intractability: a guide to the theory of NP-completeness". Freeman, San Francisco, CA, 1979.

[5] Glover F. and Laguna M. "Tabu search". In [12].

[6] Gu J. "Efficient local search for very large-scale satisfiability problems". SIGART Bulletin, Vol. 3, No. 1, January 1992.

[7] Hao J.K. and Dorne R. "A new population-based method for satisfiability problems". Proc. of 11th European Conf. on Artificial Intelligence (ECAI'94), Amsterdam, Aug. 94, pp135-139.

[8] Kirkpatric S., Gelatt C.D. and Vecchi M.P. "Optimization by simulated annealing". Science, Vol. 220, May 1983, pp671-680.

[9] Koutsoupias E. and Papadimitriou C.H. "On the greedy algorithm for satisfiability". Information Processing Letters, Vol. 43 1992, pp53-55.

[10] Lawrence D. "Handbook of genetic algorithms". Van Nostrand Reinhold, New York, 1991.

[11] Mitchell M., Selman B. and Levesque H.. "Hard and easy distributions of SAT problem". Proc. of the AAAI'92, San Jose, CA, 1992, pp459-465.

[12] Reeves C.R. (Ed.) "Modern heuristic techniques for combinatorial problems". Blackwell Scientifique Publications, Oxford, 1993.

[13] Selman B., Levesque H. and Mitchell M. "A new method for solving hard satisfiability problems". Proc. of the AAAI'92, San Jose, CA, 1992, pp440-446.

[14] Selman B., Kautz H.A. and Bram C. "Noise strategies for improving local search". Proc. of the AAAI'94, Seattle, WA, 1994.

[15] Spears W.M. "Simulated annealing for hard satisfiability problems". NCARAI Technical Report #AIC-93-015, Washington D.C. July 1993.

[16] Young R.A. and Reel A. "A hybrid genetic algorithm for a logic problem". Proc. of the 9th European Conf. on Artificial Intelligence, Stockholm, Sweden, Aug. 1990, pp744-746.

MASSIVELY PARALLEL TRAINING OF MULTI LAYER PERCEPTRONS WITH IRREGULAR TOPOLOGIES

D. Koll, M. Riedmiller, H. Braun

Institut für Logik, Komplexität und Deduktionssysteme
Universität Karlsruhe[1]

***Abstract*—In this paper we present an approach to the training of feed forward neural networks on massively parallel SIMD-architectures. In order to cover a wide field of applications we focus our attention on the flexibility of the load balancing routines. Our approach is characterized by three important properties: 1. All four types of parallelism inherent in the training phase are used. 2. In a preprocessing step neural networks are transformed into equivalent topologies, more suited for parallel computation. 3. Each learning task can be parallelized in a number of different ways, the best of which is chosen according to estimations of the computing efficiency.**

Following these concepts we developed PINK[2], a massively parallel simulator kernel for the MasPar MP1216. In contrast to most known approaches, efficient only for special topologies, it achieves good computing performance on a broad range of differing benchmark problems.

I. Introduction

The vast requirements of computing power for the training of neural networks and their inherent parallelism made neural computation to an interesting and thoroughly investigated field of application for parallel and massively parallel machines. Apart from dedicated hardware solutions for the training and recall phase of feed forward neural networks, the most promising results have been shown by SIMD-architectures with often some thousand processing elements (e.g. [1], [2], [3], [4], [5]).

However, while attaining good peak values on special neural networks their average performance on other network topologies is mostly comparatively poor. It depends heavily on the structure of the given learning problem, i.e. the number of networks to be trained, the number of training examples, the number and width of the network layers and the structure of interconnection between them.

To put it another way: each parallelization is tacitly or explicitly tailored to a special structure of learning problems; their load balancing routines are designed for those special cases. As a result the scope of most massively parallel neural simulators is not only bounded by algorithmic restrictions (as in sequential implementations), but also by the structure of the learning problem.

Subject of our work was the development of a fairly problem structure independent, massively parallel simulator kernel for feed forward neural networks (trained with backpropagation or its improved variants, e.g. Quickprop [6] or the adaptive learning rule RPROP [7]). Instead of concentrating on fully connected, regularly layered networks only, we aimed at a solution capable of training all sorts of large[3] neural learning problems.

Besides usual large problems like training big networks or handling large pattern sets another (quite nasty) example for such a 'large' problem is the training of a genetic algorithm's population consisting of a number of small, sparsely connected neural networks with different, typically irregular network topologies ([8], [9], [10]). This class of learning problems has not found much interest in the context of massively parallel computation, because each single net (a population member) may be far too small to be computed on a SIMD-computer with thousands of processing elements efficiently. The overall task (the training of the whole population) usually is by far big enough.

The concepts described in section II were applied in the development of PINK, a simulator kernel on a MasPar MP1216, a SIMD-computer with up to 2^{14} processing elements (PEs) arranged in a two-dimensional mesh of 128×128 processors ('PE-array') [11]. Since we left out as many machine specific considerations as possible, the following should be interesting for most massively parallel SIMD-architectures.

II. Intelligent Load Balancing

A parallelization of the neural training phase has to use its intrinsic parallelism to divide a learning task into a given number of pieces each of which has essentially the same processing requirements. We distinguish four levels of parallelism[4] inherent in the class of neural learning problems considered, namely:

---

[1]ILKD, Universität Karlsruhe, D-76128 Karlsruhe;
e-mail: koll@ira.uka.de, riedml@ira.uka.de, braun@ira.uka.de

[2]Parallel Intelligent Neural Network Simulator Karlsruhe

[3]The problem size is defined as the product of the number of networks to be trained with the number of network connections and training patterns (batch update).

[4]Some authors count another two types: layer-parallelism (horizontal slicing), and learning phase-parallelism. We consider them to be variants of pattern set parallelism.

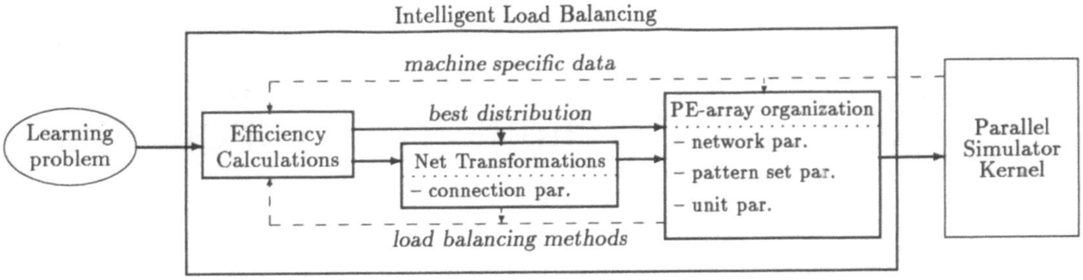

Figure 1: Intelligent load balancing routines: the distribution of a learning problem is selected out of a set of parallelization variants according to efficiency estimations.

1. network parallelism(training several independent neural nets of different topologies in parallel)
2. pattern set parallelism (batch- and epoch-learning only)
3. unit (node) parallelism
4. connection (synapse) parallelism

Since we do not want to make any restrictions to the task's structure apart from the (necessary) claim for a minimum size of the problem we will have to exhaust all four levels of parallelism: Big neural networks require to be distributed at least on node level. For load balancing reasons and in order to be able to train very large networks it will be useful to distribute networks on connection level too, parallelizing the computation of the nodes' net input. In the genetic algorithm learning task mentioned above the number of processing elements typically exceeds the number of connections per network. In order to make efficient use of the PE-array it is therefore necessary to use the other two types of parallelism additionally.

Unfortunately there exists no simple combination of these levels that is equally suitable for all classes of learning tasks.

The conventional solution would be to define one strict distribution function that performs well on a special class of problems. Instead of this, our load balancing routines do not determine one fixed parallel representation for a given learning problem, but a range of possible ones. The actually used distribution can be chosen out of this set of variants arbitrarily.

In doing this we have to address three problems: 1. We have to ensure that the set of possible distributions is big enough to comprise good variants for a wide range of problems. 2. Each parallel representation should be as simple to compute as possible. 3. For a given learning problem we have to identify the best out of the set of distributions.

Fig. 1 sketches the structure of the load balancing routines: for a given learning problem the efficiencies of all parallel representations are calculated (II.3.). The best distribution variant is chosen and controls the further processing of the problem: first the neural networks are transformed into better distributable topologies. As we will see in II.2. in this step we profit from the connection parallelism. After that we generate the parallel structures, using the other three types of parallelism (II.1.).

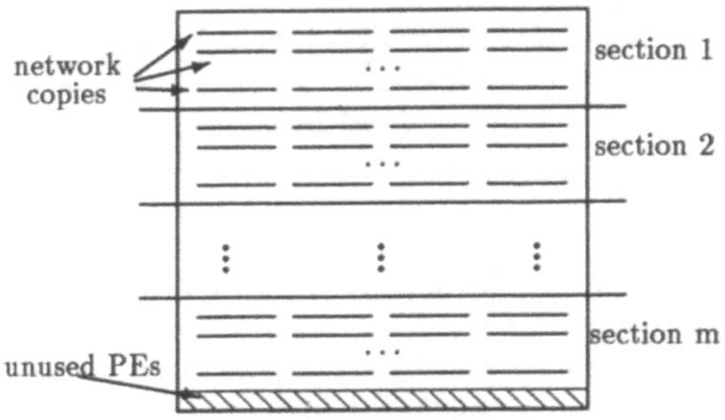

Figure 2: the processor array is partitioned into sections, each of which contains multiple copies of the neural network computed in it.

In the following we will describe the single steps in more detail.

II.1. PE-array organization

*1.1.*Network Parallelism: Given a set of independent neural networks to be trained, the PE-array can be divided into same sized sections, each of which holds one network of the set. Because the section's structure is determined by the biggest networks it is in principle possible to train neural nets of totally different topologies simultaneously. However this is neither efficient nor necessary: most applications require to train networks of the same topology (for example with different learning parameters or initial weights) or with similar ones (e.g. genetic algorithms) only.

*1.2.*Training Set Parallelism: In order to benefit from the pattern set parallelism of epoch- or batch learning several identical copies of the parallel representation of a neural net are stored on the processor array ('network replication'). So each section subdivides into subsections, containing each one copy of the neural network trained in the corresponding section (fig. 2) and a subset of the training examples.

All copies compute the error gradient $\frac{\partial E}{\partial w}$ on their subset of the training patterns simultaneously. During update phase these terms has to be summed up over all network copies for each weight w in order to compute the weights update.

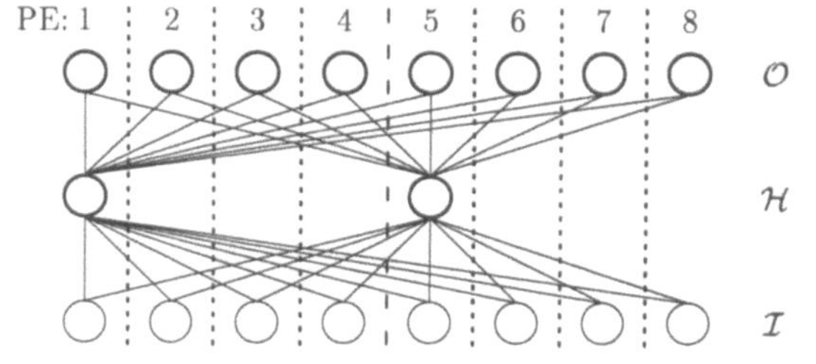

*1.3.*Unit Parallelism: The distribution of a single neural net is based on the vertical slicing approach [2]. While being a very simple and efficient method of computing regular neural networks, there is one serious drawback in 1:1–vertical slicing: only one characteristic feature of a neural net, namely the number of units in the broadest network layer, determines the PE-array-representation of the whole problem (fig. 3 shows the distribution of an 8-2-8–network). Because other structure parameters can not be taken into account this form of distributing a network is rather inflexible and not favourable for irregularly layered and/or connected networks (e.g. the 8-2-8–net). Therefore we resolved the strictness of this method by allowing to assign more than one neuron slice per PE. In the example of figure 3 we could increase the training efficiency significantly by putting two or four neuron slices (dashed line) to each PE.

By choosing the slices which are assigned to the same processor in a sensible way we are also able to balance varying in-degrees in irregularly connected networks.

II.2. Net Transformations

In 1.3. we increased the efficiency simply by choosing a more compact representation of a network, i.e. by reducing the number of processing elements involved. This is a feasible solution only if it is possible to use the PE-array to its capacity by exhausting the other types of parallelism.

Another way to improve the flexibility of vertical slicing is to split the neural slices further, assigning only parts of a slice to each processing element. In order to keep the parallel code and communication operations simple we do this by net transformations: neural networks are converted into better distributable topologies with the same learning- and recall-behaviour before being transferred to the PE-array.

In most MLP-learning rules this can be done by inserting so called 'pseudo-units'. These units are special in two ways:

1. activation function: $f_{act} \equiv \text{id}$
2. the bias is set to 0, it cannot be changed during training (it is 'frozen' to 0)

A pseudo-unit can be connected to an existing regular unit or a previously added pseudo-unit by a link, whose weight is set and frozen to 1 ('pseudo-links', as you might have guessed). All incoming or outgoing connections of the changed unit, depending on whether we have inserted the pseudo-unit in front of or behind it, can be redirected to point to, respectively to go out from the new pseudo-unit.

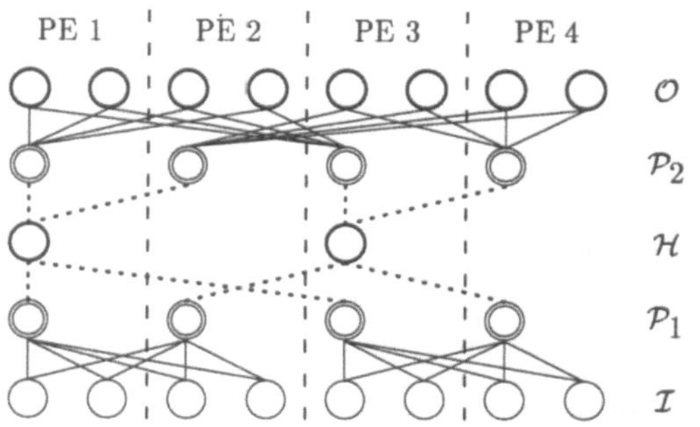

Figure 4: applying net transformations to increase the computing efficiency of an 8-2-8–network divided on 4 PEs

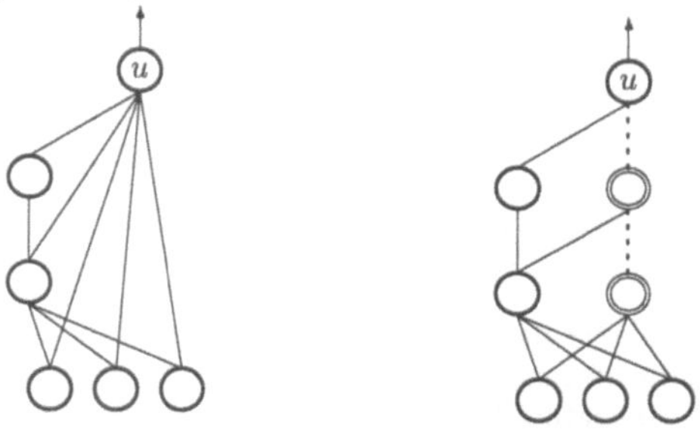

Figure 5: eliminating the shortcut connections of unit u by inserting two pseudo-units

An example shall clarify this idea: fig. 4 shows an extended version of the 8-2-8–network. We have inserted two layers of pseudo-units ($\mathcal{P}_1$ and $\mathcal{P}_2$) and distributed the links formerly connected to the hidden layer $\mathcal{H}$ evenly on the corresponding pseudo-units. When splitting up this modified network as usual in vertical unit slices (in fig. 4 on 4 PEs) we actually distribute the network on connection level although the parallel kernel does divide networks on unit level only.

In the example net transformations were used to balance the processor load of an irregularly layered network. This method allows us also to level out strongly varying in-degrees in irregularly connected networks by distributing the connections of units with high fan-in on several pseudo-units. Further the parallel simulator kernel is thereby able of handling neural topologies with nodes of almost arbitrary fan-in.

Last but not least net transformations are a very simple and efficient way to handle shortcut connections: shortcuts are broken up into connections between adjacent layers by inserting an appropriate number of pseudo-links (fig. 5).

The main advantage of net transformations: the parallel code doesn't have to provide any means to distribute networks on connection level or to handle shortcuts. The computational overhead caused by the additional units and links is small due to their very special quality (no bias, $f_{act} \equiv \text{id}$, connection weights 1).

II.3. Efficiency Calculations

By now we have said nothing about how we actually distribute a given learning problem. The distribution methods described so far left the following questions open: how many

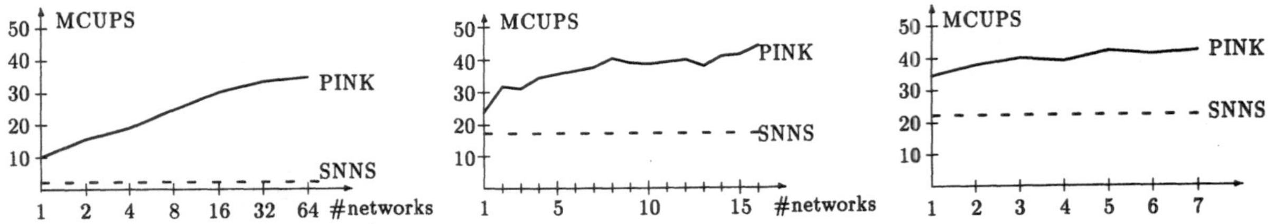

Figure 7: training performance of PINK and SNNS on a) 256-8-256–Encoder b) 14-28-1–Parity c) NETTalk networks

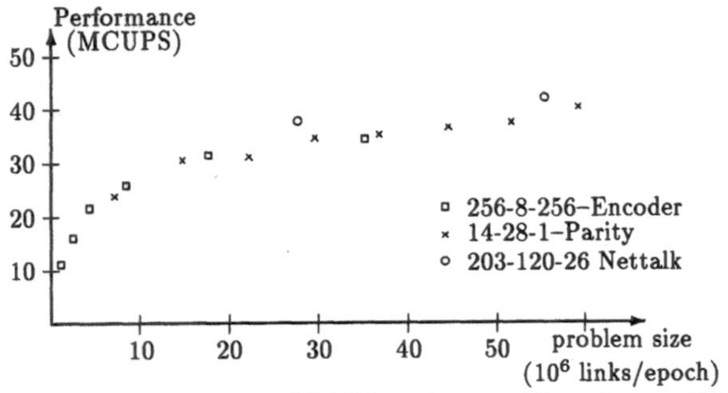

Figure 6: Performance of PINK against problem size on different network topologies. The performance depends on the problem size rather than on its structure.

sections and network copies per section are allocated, in how many slices a neural network is cut and what kind of net transformations are applied.

The PE-array-representation of a given learning problem is no longer uniquely defined (in contrast to other approaches). Each learning problem possesses a set of different distributions, one of which can be chosen arbitrarily. However due to the number of (conflicting) influencing factors there seems to be no simple heuristic that allows to detect in any case the best variant.

Therefore PINK selects the actual distribution of a problem according to efficiency estimations for each variant (cf. fig. 1): the load balancing methods determine the set of possible distributions for a given learning problem, constrained by machine characteristics, e.g. the number of processing elements and their memory capacity. For each member of this set the computing efficiency is estimated, taking into account the homogeneity of load balancing (all processing elements should compute approximately the same number of connections) and machine specific considerations (communication structure of the PE-array, relative cost of inter-processor-communication and floating point operations). The variant with highest efficiency is chosen automatically to compute the considered problem.

This way we evade the dilemma of load balancing: the overall distribution function is very flexible and quite complex, but each possible PE-array-presentation of a given problem is comparatively simple and thereby fast to compute.

III. The Results

In section II we have outlined the basic load balancing methods of PINK. The performance of this simulator is far less dependent on the structure of a learning problem (fig. 6), than the other approaches mentioned above. As expected its peak performance of about 68 MCUPS[5] is lower than those of more specialized and thereby less complex simulator kernels (using the 1:1–vertical slicing approach with network replication the MasPar MP1216-kernel of the SNNS [4] attains 129 MCUPS). But for typical neural network applications the benefits of this structure independence does not only compensate the cost of the more complex distributions: our system surpasses previous approaches on them by far. Figure 7 a)–c) summarize the results on some common network topologies: in spite of its lower peak performance the PINK outperforms the 1:1–vertical slicing approach of the SNNS on a single 256-8-256–Encoder network by a factor of 10. By training several Encoder networks simultaneously this factor goes up to about 30. The quite regular NETTalk problem is computed approximately twice as fast as by the MasPar-kernel of the SNNS.

III.1. Conclusion

We have presented an adaptive load balancing method for massively parallel training of feed forward neural networks. The described preprocessing steps (net transformations) and the optimization of the parallel representation by a distribution arbiter turned out to be very efficient on SIMD-architectures. Using this approach, PINK's training performance shows little dependencies of the learning task's structure, still compares to those of more specialized parallel kernels on their domain and clearly outperforms them on other problems.

References

[1] A. Singer: Parallel Computing 14, 305 (1990)
[2] X. Zhang, M. Mckenna, J.P. Mesirov, D.L. Waltz: NIPS 2, 801 (1990)
[3] Wei-Ming Lin, V.K. Prasanna, K.W. Przytula: IEEE Transactions on Computers, Vol. 40, No. 12, 1390 (Dec 1991)
[4] N. Mache, Master thesis, University of Stuttgart, 1992
[5] I. Pitas (ed), A. Petrowski, H. Paugam-Moisy: *Parallel Algorithms* pp 259-328 Wiley 1993
[6] S.E. Fahlmann: CMU–CS-88-162 (1988)
[7] M. Riedmiller, H. Braun: Proc. ICNN '93, 379 (1993)
[8] H. Braun, J. Weisbrod: Proc. ICANNGA '93, 25 (1993)
[9] J. Schaefer, H. Braun: Proc. ICANNGA '95
[10] H. Braun, P. Zagorski: Proc. 3rd PPSN, 444 (1994)
[11] W. Butscher: Supercomputer '91, 187 (1991)
[12] A. Zell: *Simulation Neuronaler Netze*, Addison Wesley 1994

[5]Millions of Connection Updates per Second [12]

Parallel Boltzmann Machine Topologies for Simulated Annealing Realisation of Combinatorial Problems

I. Ashman, T. Vladimirova, C. Jesshope and R. Peel

Electronic Engineering, University of Surrey, England.

Abstract

In this paper we investigate the convergence of parallel Boltzmann Machine topologies for solving the maximum matching problem. The simulations of the Boltzmann Machine topologies are done in Occam on transputers. The results of tests carried out by varying some implementation parameters are presented. A description is given of a Boltzmann Machine in terms of an asynchronous iterative algorithm allowing a new approach to the proof of theoretical convergence.

1 Introduction

The use of combinatorial optimisation is increasing. A wide variety of problems have emerged from such diverse areas as management science, computer science, engineering, VLSI design etc. The problems encountered keep increasing in size so that exact optimisation algorithms have become too computationally expensive. This increase in size and complexity creates the need for general approximation or heuristic algorithms. Even with these algorithms the size of the problems means that there is a need for the algorithms to be run in a parallel form.

Simulated annealing is a nature-based stochastic computational technique which comes from condensed matter physics, where annealing is a thermal process for obtaining low energy states of a solid in a heat bath. In combinatorial optimisation it uses the theory of Markov chains in conjunction with a decreasing control parameter, temperature, to control the minimisation search. A simulated annealing algorithm has been used in a sequential form to demonstrate that the algorithm converges to the global optima or within a small value of the global optima in solving the maximum matching problem.

Simulated annealing is basically a sequential algorithm which does not lend itself that well to parallelisation. The Boltzmann machine (BM) is a neural network model, that can be viewed as a parallel implementation of simulated annealing, consisting of a large network of connected simple computing elements. The units have two states either 'on' or 'off' and the connections have real valued strengths that impose local constraints on the states of the individual units. A consensus functions gives a quantitative measure for the 'goodness' of a global configuration of the BM, determined by the state of the individual units.

The BM is conceptually simple, but allows massive parallelism and can easily be mapped onto VLSI chips improving the parallel performance even further. The BM has been used initially in a sequential form as there is a theoretical proof of asymptotic convergence to a global optimum in this form. A number of different approaches are considered to realise parallel state transitions in BM's: limited synchronous parallelism, unlimited synchronous parallelism, limited asynchronous parallelism and unlimited asynchronous parallelism.

In synchronous parallelism sets of state transitions are scheduled in successive trials, each trial consisting of a number of individual state transitions. After each trial the accepted state transitions are communicated through the network so that all units have up to date information about the states of their neighbours before the next trial is initiated. During each trial a unit is allowed to propose a state transition exactly once. If the BM is fully connected it requires a global clocking scheme, however, if it is more sparsely connected it can use local interaction or a loosely asynchronous clocking scheme to control synchronisation. In limited synchronous parallelism, units may change their state in parallel only if they are not adjacent. Whereas in unlimited synchronous parallelism, units may change their states in parallel whether or not they are adjacent, introducing the possibility of using out of date information in the calculation of the consensus function. In asynchronous parallelism state transitions are evaluated simultaneously and independently. Units continuously generate state transitions and accept or reject them on the basis of information that is not necessarily up to date, since the state of its neighbours may have changed in the meantime. Clearly, asynchronous parallelism does not require a global clocking scheme. Units may change their states simultaneously, but state transitions are no longer combined in successive cycles of the clocking scheme. It is again possible to distinguish between limited and unlimited parallelism. Limited parallelism can be realised by introducing a blocking scheme which prohibits adjacent units carrying out state transitions simultaneously, therefore, no erroneously calculated state transitions will occur. In the case of unlimited parallelism, no blocking scheme is used and erroneously calculated state transitions will occur, through carrying out the calculations using out of date information.

The limited synchronous parallelism form has a theoretical proof of global convergence using the theory of Markov chains, the proof does not extend to unlimited synchronous parallelism due to the possibility of incorrect state calculations. The asynchronously parallel form has no theoretical proof of global convergence, only an intuitive argument. This is due to the fact that the usual description method and proof techniques use the theory of Markov chains which are no longer valid in this mode as the asynchronous nature of the units means that they cannot be modeled as sequential trials.

This paper looks at the use of the BM in two different parallel topologies, unlimited synchronously parallel and unlimited asynchronously parallel, for solving the maximum matching problem. The main focus is on the convergence to final solutions that do not include illegal states. The BM topologies are implemented in Occam on T800 Transputers and experiments are carried out to confirm the convergence. Also a description is given of a BM in terms of an iterative algorithm. This leads to the use of a form of chaotic relaxation to prove the theoretical convergence for the unlimited synchronously parallel mode.

2 Boltzmann Machine Topologies for Maximum Matching

Maximum matching is a fundamental problem in combinatorial optimisation. An instance of the maximum matching problem is a simple graph G = (V,E), where V denotes the set of nodes of G and E denotes the set of (undirected) edges of G [5]. A matching M in G is a subset of E such that no two edges in M share a node. The maximum matching problem for instance G is to find a matching in G with maximum cardinality.

This problem has initially been mapped onto a sequential BM. The tests showed the need for a parallel implementation to increase the speed of the computationally expensive sequential algorithms.

BM's can be used to solve combinatorial optimisation problems [1],[2]. The approach is based on the observation that the structure of many combinatorial optimisation problems can be directly mapped onto the structure of a Boltzmann machine by choosing the right connection pattern and strengths. In this way minimising the consensus in a BM is equivalent to finding the optimal solutions of the corresponding optimisation problem.

If the inherent parallelism of a BM can be efficiently used for solving combinatorial optimisation problems, then this approach can be viewed as a massively parallel implementation of simulated annealing.

General strategy for optimisation.
Consider an instance of a combinatorial optimisation problem as a tuple (S, S', f), where S denotes the finite set of solutions, $S' \subseteq S$ the set of feasible solutions, ie, the set of solutions satisfying the constraints that go with the problem, and $f : S \rightarrow \Re$ the cost function that assigns a real number to each solution. The problem is to find a feasible solution for which the cost function f is optimal.

A solution of a combinatorial optimisation problem can be characterised by a finite set of discrete variables $X = (x_1, ...x_n)$. If the domain of each variable $x_i \in X$ is given by a finite set X_i, then the set of solutions S is given by $S = X_1 \times X_2 \times ... \times X_n$.

To use the BM for solving combinatorial optimisation problems, a bijective function m : $R \rightarrow S$ is defined which maps the set of configurations R onto the set of solutions S, by applying the following general strategy [1].

- Formulate the optimisation problem as a 0-1 programming problem, ie, formulate the problem such that $X_i = (0,1)$, for i = 1 ...n.
- define a BM B = (U,C), such that the state of each unit $u_i \in U = (u_1, ...u_n)$ determines the value of the variable u_i. If unit u_i is on then $x_i = 1$; if u_i is off then $x_i = 0$. Thus, $x_i = k(u_i)$. Clearly, this defines a bijective function m which maps the set of configurations R onto the set of solutions S.
- Define the set of connections C and the corresponding strengths such that the consensus function is feasible and order preserving.

This general strategy shows that the BM is tailored to the particular problem and the algorithm is designed so that no conditions can occur that break the constraints. This is true for all BM's where neighbouring units cannot change their states simultaneously, but if neighbouring units can change their states simultaneously illegal states will occur which break the constraints of the problem.

The limited synchronous parallelism form has a theoretical proof of global convergence using the theory of Markov chains. The proof does not extend to unlimited synchronous parallelism or unlimited asynchronous parallelism as it is possible for incorrect state calculations and that state changes can no longer be modeled sequentially.

The incorrect or illegal states that the unlimited BMs can produce can be described as asynchronous distortions. In asynchronous algorithms asynchronous distortions occur and cause the algorithm to enter a state that could not be reached by successive applications of the iteration operator. The causes of the conditions are identical, that is the different speeds of calculating elements and the communication time of updating the changed state of the element.

The simulations were carried out on transputers using the programming language Occam. The BM was tested in two different topologies; unlimited asynchronous parallelism and unlimited synchronous parallelism. The intuitive proof of convergence uses the idea that as the temperature tends to zero fewer state transitions occur and therefore the chance of having an illegal state tends to zero. Therefore the cooling schedule needs to be long, ie, go to a very low temperature value to allow for asynchronous distortions. This can undo some of the speed advantages that the use of asynchronous models aims to provide. Therefore in these tests a relatively short cooling schedule was selected and not varied throughout the tests. The tests were designed to look at the effect of the calculation time of the consensus function of each unit in comparison to the time to communicate the state changes to other units. This was varied by adding an artificial delay in the calculation of the consensus value and varying the size of this delay. The results are summarised, for a 50% connected, weighted, undirected graph of 16 nodes requiring a BM with 32 units, in Table 1. The rows of the table with varying delay relate to the number of clock cycles that each iteration is forced to halt for in the asynchronous model, the last row relates to the synchronous model. The columns are: Res is the integer value of the weights in the final matching, the average result 136 being the global optimum for the graph used; Loop is the average loop time; Comm is the worst case communication time; Av Comm is the average case communication time; Optimum % is the percentage of times that the test found the optimum; % Illegal is the percentage of tests that gave an illegal result; Total Time is the normalised average total time for the complete test. Where there is an N/A entry is due to illegal states making the results invalid, for example, you could have an average result of 140 greater than the global optimum.

Table 1 shows that in the case of 32 units in a BM all of the results that do not give illegal states achieve the global optimum every time, although the important issue is the number of illegal states and not the magnitude of the result. Also as can be seen from Table 1 the loop time increases as the communication time decreases. The reason for this is that as the delay increases the throughput demand on the communication channel decreases allowing it to run more efficiently. This phenomena also explains the decrease in program run time with an increase in the delay around the 2250 Delay, as the increase in communication efficiency outweighs the increased delay. The main result of interest is the reduction of illegal states with the increased delay. This time is directly related to how out of date the information used in the update is. In the unlimited synchronous parallel mode the BM always uses the most recent value in the update procedure as these are synchronised. In this mode the simulation always gave the global optimum with no illegal states. In the asynchronous case, if the delay was large making the communication time small in comparison with the consensus calculation time again generally the most recent value would be used in the update and there was convergence with no illegal values. As the communication time becomes significant the occurrence of illegal states increases, this is directly related to using more and more out of date values in the update. The transition from having illegal states to no illegal states in the results is quite sharp and appears to coincide with the use of the most recent value for the update.

3 Boltzmann Machine as an Iterative Asynchronous Algorithm

The first publication on the convergence of asynchronous algorithms was [4] which introduced the technique of chaotic relaxation. Its main purpose was to account for the parallel implementation of iterative methods on a multiprocessor system so as to reduce the communication and synchronisation between the cooperating processes. This reduction is obtained by not forcing the processes to follow a predetermined sequence of computations, but simply by allowing a process, when starting the evaluation of a new iterate, to choose dynamically not only the components to be evaluated but also the values of the previous iterates used in the evaluation [3]. In [4] the antecedent values used in the evaluation of an iterate was bounded by an integer value. This restriction was relaxed by [3] to the antecedent values used only needing to be finite and not necessarily bounded. In [6] as in [3] no assumptions are made on the computational model, except for the weakest fairness assumptions. For this general model asynchronous contraction defines the largest possible class of asynchronous iterative algorithms that are correct, at least for finite data domains [6]. In [6] the convergence is proved for a more restricted form of asynchrony that does not require asynchronous contraction.

The definition given below formulates the general model of asynchronous iterative algorithms [6]. It is defined in terms of a schedule ($\mathcal{S}$), an iteration operator and initial data.

Definition 1. An asynchronous iteration with respect to the schedule $\mathcal{S} = \{(\alpha(k)), (\beta(k))\}$, corresponding to **F** and starting with $x(0) \in$ S, is a sequence $\{\mathbf{x}(k)\}$ such that

$$x_i(k) = \begin{cases} x_i(k-1) & if \quad i \notin \alpha(k) \\ F_i(x_1(\beta_1(k)), .., x_n(\beta_n(k))) & if \quad i \in \alpha(k) \end{cases}$$

for all $k \in N^+$ and $i \in$ **I**, and is denoted by (**F**, **x**(0), $\mathcal{S}$), where $\{\alpha(k)\}$ is a sequence of subsets of **I** and $\{\beta(k)\}$ is a sequence of elements of N^n. Where S is the cartesian set of n sets from which the data take values; **F**: S $\rightarrow$ S is a mapping function; N is the set of nonnegative integers; N^+ is the set of positive integers and $\mathbf{I} = (1, 2, ..., n)$ is the index set.

$\{\alpha(k)\}$ is the update set at k, where the ith component is updated at k if $i \in \alpha(k)$; $u_j(k) = x_j(\beta_j(k))$ is called the input of the kth update. A schedule ($\mathcal{S} = (\{\alpha(k)\}, \{\beta(k)\})$) is assumed to satisfy the following conditions [6]:

A1 $\beta_i(k) < k$ for all $k \in N^+$ and $i \in$ **I**; that is , The value of an element can only be dependent on its past values.

A2 Each $i \in$ **I** occurs in an infinite number of $(\alpha(k))$'s, that is, Each element is updated infinitely often, ie, nonstarvation.

A3 For each $l \in N$ and each $i \in$ **I**, there can only exist a finite number of k's such that $\beta_i(k) = l$, that is, After a value is generated it can only be used in a finite number of updates.

These conditions describe the most general form of asynchronous iterative algorithms and are very natural and weak conditions.

Most proofs use the idea of contracting operators in the convergence conditions. After the application of the update operator the result is a smaller or equal subset of the input, for all possible inputs. This condition clearly is not the case in a BM implementation. In [6] however, a proposition is given that does not require that the operator is contracting. It does impose a condition on the update of a component that the update is carried out using the most recent value of that component as the update.

Proposition 6 [6]: Let D(0) be a finite set. Then, an asynchronous iteration (**F**, **x**(0), $\mathcal{S}$) converges to a fixed point of **F** in D(0) if the following conditions hold:

A4 For all $k \in N^+$, $\mathcal{S}$ satisfies

$$i \in \alpha(k) \Rightarrow \beta_i(k) = k - 1$$

A5 **F** is closed in D(0)

A6 **F** is nonexpansive in D(0), that is, for all $a \in D(0)$, **F(a)** $\preceq$ **a**

Where $\preceq$ is an ordering relation.

This makes (**x**(k)) monotonically decreasing, and since D(0) is finite, it converges to some ξ (fixed point). Therefore, there exists an M such that $\mathbf{x}(k) = \xi$ for all $k \geq \psi(M)$ [6].

$\psi(k)$ is an increasing integer sequence, where between $\psi(k)$ and $\psi(k+1)$ every component is updated and components used as the input for an update must have been updated on or after $\psi(k-1)$.

The update strategy of a BM only becomes monotonically decreasing when the temperature of the cooling schedule no longer accepts deterioration in the consensus, that is, a small value. Until this time A6 of the above proposition does not hold.

Traditionally the BM is thought of as a graph with units as nodes and connections as edges, here we will consider a BM as a vector. Each unit is an element of the vector and the bias and con-

Test	Res	Loop	Comm	Av Comm	Optimum %	% Illegal	Total Time
no Delay	N/A	1846	14105	7052.5	N/A	83	1
500 Delay	N/A	1910	8888	4444	N/A	76	1.11
1500 Delay	N/A	2259	6875	3437.5	N/A	67	1.31
2000 Delay	N/A	2283	5712	2856	N/A	24	1.41
2250 Delay	N/A	2434	3421	1710.5	N/A	9	1.35
2500 Delay	136	2733	1560	780	100	0	1.38
3000 Delay	136	3152	528	264	100	0	1.49
5000 Delay	136	5079	370	185	100	0	2.05
15000 Delay	136	15044	208	104	100	0	4.93
Synchronous	136	4066	2676	1338	100	0	1.73

Table 1: Comparison Results

nection strengths are components stored for use in the iteration step of calculating the new value of the element. The number of iterations is controlled by two parameters, called temperature which decreases slowly and count giving the iterations within the temperature loop. The state is decided in the usual way for a BM with the temperature being used in the acceptance criteria. With this in mind we will look at how convergence can be guaranteed for the unlimited synchronously parallel topology.

An unlimited synchronously parallel BM model where the ith unit has an update that can be described in the following form:

$$\mathrm{U_i}(\mathbf{x}) = f(C_i + \sum_{i \neq j} C_{ij} x_j)$$

where C_i is the bias, C_{ij} is the connection strength, x_j is the most recent state stored of the unit U_j and f includes the acceptance criteria. If we assume that $\mathbf{x}$ can only be zero or one, that is 'on' or 'off', then the values will be in the domain D of the form $D = \{0,1\}^n$.

Then adapting proposition 8 of [6], we can say that all asynchronous iterations ($\mathbf{U}$, $\mathbf{x}(0)$, $\mathcal{S}$) with $\mathbf{x}(0) \in$ D such that D is in the form, $D = \{0,1\}^n$, converge for all schedules satisfying A4 of proposition 6.

4 Conclusion

This paper presents the results of the theoretical and experimental work to tackle the global convergence of asynchronous Boltzmann Machines. It is suggested that there may be a direct correlation between the use of the most recent value for the update and convergence to a desired result in an unlimited synchronously parallel BM. Tests on different sized graphs are carried out to reinforce the proposition of convergence to a desired state when the most recent value is used in the update procedure. The next step is to look at how out of date the information used in the update could be whilst guaranteeing convergence.

References

[1] Aarts, E. and Korst, J. "*Simulated Annealing and Boltzmann Machines*" John Wiley and Sons, 1989

[2] Aarts, E. and Korst, J. "*Boltzmann Machines as a Model for Parallel Annealing*" Algorithmica, Springer Verlag, 1991, No. 6, pp. 437-465.

[3] Baudet, G. M. "*Asynchronous Iterative Methods for Multiprocessors*" J. ACM 25, 2 (APR 78), pp. 226-244.

[4] Chazan, D. and Miranker, W. "*Chaotic Relaxation*" Linear Algebra and its Applications, 1969, pp. 199-222.

[5] Lovasz, L. and Plummer, M. D. "*Matching Theory*" Elsevier Science, 1986.

[6] Uresin, A. and Dubois, M. "*Parallel Asynchronous Algorithms for Discrete Data*" J. ACM 37, (July 90), pp. 588-606.

RADIAL BASIS FUNCTION NEURAL NETWORKS IN CREDIT APPLICATION VETTING SYSTEMS

Williamson A.G. and Munson P.

N219 Division of Computing and Software Engineering, Coventry University, U.K.

ABSTRACT

This paper describes an investigation carried out at Coventry into the suitability of Radial Basis Function(RBF) Neural Networks for use in credit vetting systems. The RBF used is an All Classes in One configuration with unweighted Euclidean distance measure. The paper examines the performance of the RBF network in classifying good and bad loan cases over a data set supplied by a substantial finance organisation. The effects of changing the number of centres and the training regime are examined. The network prediction performance is compared over the same data set with both a manually configured, and a genetic algorithm designed Back Propagation Trained Multi-Layer-Perceptron.

The performance of the RBF network in classifying cases is found to be comparable with these two solutions, providing similar prediction rates. The relative simplicity of the RBF solution gives greatly reduced computing time for comparable performance, and potentially easier routes to providing information about the decisions reached. Ideas for enhancing the future performance of the system are discussed.

INTRODUCTION

The use of Artificial Neural Networks(ANNs) for business forecasting is now established, and recently networks have been applied to areas such as the currency markets by Refenes[1][2] with some success. In the area of credit vetting however the take up has been slower. In spite of some applications [3][4], many problems still persist even for substantial finance companies, as described by Bazley[5].

The problems for the traditional Back Propagation trained Multi-Layer-Perceptron centre around the training process and the network's subsequent ability to generalise, leading to difficulties in selecting training sets and network parameters. Work by Williamson[6] showed that an effective optimum might be found using Genetic Algorithms but the ultimate restriction appeared to be the characteristics of the network and the Back Propagation training. This has led to an investigation using RBF neural networks for credit application vetting, in order to compare their performance with these existing network configurations.

The basic principle of credit vetting is to use knowledge of the application for credit to make a decision whether to make the loan or not. It is also a well established principle that the decision will be aided by considering the results of previous loans made to similar applicants. These principles can be applied to the use of a neural network for assisting in this credit vetting process, as a network operates in a similar fashion, as a classifier. The network inputs are coded information from loan application forms and the outputs the decision to lend or not.

THE RADIAL BASIS FUNCTION NETWORK

The Radial Basis Function network forms complex decision boundaries by using overlapping radial kernel functions. Kernel nodes compute radial symmetric functions, which give the highest output when an input is close to its centre. These networks are trained with supervision and use matrix algorithms such as singular value decomposition (SVD) to calculate the weights connecting the kernel nodes to the output nodes. The number of kernel nodes, their centres and widths are adjusted to give the best results on the training data. The centres of nodes can be calculated by using unsupervised clustering techniques, random selection of training examples from the training data or by a supervised training method.

The basis function in an RBF classifier is a form of distance measure and the Euclidean distance function is the most commonly used. The basis function can also be defined by a weighted Euclidean distance measure, or as an inner product. The weighted Euclidean distance has been shown to improve classification results in some cases, although the same improvements can be obtained by using a supervised learning method to calculate the centre locations[7].

RBF Network Structure

The RBF network structure used was an All Classes in One Network (ACON) structure. Some problems arise with this network structure in that the hidden layer nodes have to satisfy more than one class at a time. The One-Class-in-One-Network is a variation of this presented by S.Y. Kung[8]and another variation of an OCON is the hidden-node structure[8]. The OCON network has a faster convergence rate than an ACON structure, as the hidden layer nodes do not get influenced by conflicting signals during training. It is however, considerably more complex.

During the training of an RBF classifier, irrespective of the network structure used, there are two unknown parameters which have to be found for each of the nodes in the hidden layer, these are the node's centre (mean of Gaussian) and centre width (variance of Gaussian). When an ACON or a hidden node OCON structure is used a third parameter must be learnt, this is the weight connecting each of the nodes in the hidden layer to the output layer which is done by using a supervised method.

The RBF classifier using an unsupervised k-means algorithm for clustering and an ACON net structure and SVD to determine the connecting weights was finally chosen for this investigation.

RBF Network Usage

In this application, an RBF network is first configured with sample input data using a "K-means" clustering algorithm. This algorithm clusters irrespective of case outcome in an unsupervised process. The resulting network is then trained with either the same data or a new training set in a supervised configuration. The trained network will then be used to classify the suitability of issuing credit on new loan applications not previously presented to the network.

THE LOAN DATA

The test data was supplied by a commercial company. It consisted of information on 1283 completed 'good' mortgage applications and 1120 completed 'bad' applications. Each case initially had 63 input data fields associated with it, but many of these were clearly not useful and were eliminated during preliminary investigations. Some of the fields used are listed below:

Freehold/leasehold property	Number of applicants
Amount of loan approved	Payment Protection Plan
Completion month and year	Pension case
Purchase price of property	Remortgage case
Property valuation	Area of country
Applicant's income	Local Authority

Before the data was presented to the network, it was transformed into all numeric values. The data was then normalised to fall within the range -1 to +1. The alternative to this is to use input scaling, which is now available on most commercial artificial neural network packages.

Network Clustering and Training Set

101 instances of each type of case were used for clustering. In one set of experiments these same cases were used for training. In the second experiment a disparate group of 642 cases was used for training. The outputs used for the cases in training were zero (0) for a bad case and one (1) for a good case.

RESULTS

In the first experiment, the clustering set and training set were the same 202 cases. The number of centres was varied and the complete data set (including the 202 cases) was classified. The results are shown in table 1.

Table 1: RBF network with common cluster and training data

No of Centres	Training Accuracy	All Cases Accuracy	Good Cases Accuracy	Bad Cases Accuracy
1	55%	54%	84%	20%
2	86%	70%	96%	40%
3	100%	81%	90%	72%
4	100%	83%	89%	75%
5	100%	82%	90%	72%
6	100%	82%	90%	72%
7	100%	83%	90%	74%
8	100%	83%	90%	74%
9	100%	83%	89%	76%
10	100%	83%	89%	76%
12	100%	83%	89%	77%
20	100%	82%	90%	73%
30	100%	83%	89%	76%
40	100%	81%	91%	71%
50	100%	83%	90%	75%
60	100%	83%	89%	76%
70	100%	84%	89%	78%
80	100%	84%	88%	80%
100	100%	83%	90%	76%

In the second experiment the same clustering set and an extended training set were used, the results are shown in table 2.

Table 2: RBF network with different training data

No of Centres	Training Accuracy	All Cases Accuracy	Good Cases Accuracy	Bad Cases Accuracy
1	31%	52%	78%	22%
2	64%	75%	94%	52%
3	76%	86%	85%	86%
4	79%	87%	83%	91%
5	81%	87%	81%	95%
6	81%	87%	80%	95%
7	82%	87%	81%	95%
8	82%	87%	81%	95%
9	83%	87%	79%	97%
10	83%	87%	79%	97%
12	83%	87%	78%	97%
20	86%	86%	76%	98%
30	86%	86%	76%	98%
40	86%	87%	78%	98%
50	87%	87%	78%	97%
60	87%	86%	77%	97%
70	87%	86%	77%	97%
80	87%	85%	76%	97%
100	87%	84%	76%	94%

In the second experiment 642 cases were used for training, and the performance of the network was evaluated over the remaining 1752 cases. In this experiment the training cases were removed from the final classification test. The number of centres was varied and the data set without the training set classified as shown in table 2. As the number of cases available is fixed the results with respect to generalisation are considerably less significant.

In a parallel experiment a manual operator was set the task of configuring a Back Propagation trained Multi-Layer-Perceptron with the cluster set as a training set. In order to speed up this experiment Principal Components Analysis was used to reduce the input dimensionality of the data set from 27 to 20, losing 7% of the available information. This was judged acceptable for a comparative study. The result was a 20-20-1 network. Additionally a similar Back Propagation trained Multi-Layer-Perceptron network was designed using a Genetic Algorithm. The result of this was a 27-10-2-1 network.

Table 3 shows the comparative performance of the best RBF network trained on the same data set used for clustering, the manual network and the Genetic Algorithm designed network. The networks were trained on 101 good and 101 bad cases, which were included in the all cases evaluation.

Table 3: Comparative Network performance

Network	Training Accuracy	All Cases Accuracy	Good Cases Accuracy	Bad Cases Accuracy
RBF	100%	84%	89%	78%
Manual	100%	77%	86%	66%
GA	100%	81%	86%	76%

CONCLUSIONS

From these experiments it can be seen that in this context the RBF network has a similar classification performance to those using back-propagation. However the training times are greatly reduced, at the expense of having a greater memory requirement. This is generated by the need for more nodes and therefore connection weights. The matrix based approaches used in calculating the weights between the kernel nodes and the output layer makes the memory requirements during training greater still, due to the need for simultaneous access to all the training examples.

The results of the second experiment show an exceptional performance on the detection of bad unseen cases. This could be a major financial benefit to the loan company if it could be scaled to realistic sized data sets.

In operating a vetting system it is not usual to have a clear cut binary classification, and the introduction of a 'referred' class into the network would be a requirement of most institutions. The capacity to post process difficult cases by hand will almost certainly offer cost-benefit advantages.

It is noted that the three compared network designs have comparable performance with 12 nodes over the supplied data set. It appears that greater complexity yields minimal improvements in accuracy. This will be driven by statistical features of the supplied data.

Traditional and network credit vetting systems operate under evolving procedures. These will include near 100% accuracy on the training set and good performance on known cases. For credit vetting it is apparent that the RBF network could be substituted in existing network based system, and show improvements in update times. There may be some outright improvements in classification, but more data would have to be processed to re-enforce this view. The question raised here is what procedures should be used in association with the RBF network to ensure satisfactory performance ?

Unlike Back Propagation training the network performance on the training set need not be 100% for effective overall performance. It is therefore necessary to seek a replacement measure for terminating the network configuration evaluation. Techniques for establishing the number of centres will be beneficial in terminating the search, or an iterative experiment could be used.

Once again the need to optimise such a system suggests that a Genetic Algorithm might be applied to automate this task.

Secondly the identification of training sets which appropriately represent the overall population is a crucial issue. Substantial finance houses may have 2-3 million customers on their books. The availability of data is therefore not an issue, but selecting data for use in network classification still needs effective procedures. These may be borrowed from scorecards in the first instance until the characteristics of networks can be accommodated.

Currently solutions produced by a back-propagation trained network are difficult to understand apart from trivial cases, and the networks frequently require many nodes and connection weights. As loan companies endeavour to present a caring customer image, using a system which does not yet clearly identify the reasons for refusing credit is likely to be unpopular. RBF classifiers offer potential improvements in this area through their improved operational simplicity.

Finally it should be noted that a credit vetting system would not normally use raw data from the customer application, but would usually combine factors into known effective measures, as is common with scorecards.

FUTURE WORK

A number of considerations have arisen during the investigation which warrant further work. Considering the RBF classifier itself, an investigation into the modification of the node variance both globally and on individual nodes might reap improvements. An alternative to the SVD technique for computing the connecting weights is to use the Gaussian Elimination technique. This was started during the project but problems occurred with the accuracy of the floating point numbers in the computer. When two equations were near parallel the computer's high accuracy became a problem. Wong[9] identified that the Gaussian function might be too smooth and that the use of a box function such as that used in the Cerebellar Model Articulation Controller might yield better results. Wong also notes that this function might not be smooth enough and that a triangle function could be better. The use of a fuzzy decision neural network(FDNN) could be incorporated as this would smooth decision boundaries and allow for the best compromise in terms of error. Kung[8] presents a suitable algorithm as a base for this idea.

Supervised learning of centre locations had been shown to improve the classification results obtained [7], and this might show improvements for the system. Further research into unsupervised clustering techniques such as the VQ[8] and the ART introduced by Carpenter and Grossberg[10] could be carried out. These techniques have the added benefit of determining how many kernel nodes are required to adequately cluster the training data. Neither the VQ or ART clustering algorithms take into account the class of the training data. Musavi et.al.[11] noted that clustering in relation to class could be an important factor which affected the classification results achieved. An algorithm is presented, although step four is rather vague.

REFERENCES

1. Refenes A.N. *Neural Network Applications in Investment and Finance Services*. Chapter 27, E. Turban, R Trippi. USA. Probus Publishing. 1992.
2. Refenes A.N.. Managing Exchange Rate Prediction Strategies with Neural Networks. In *Techniques and Applications of Neural Networks*, Chapter 7, Taylor and Lisboa (eds). Chichester. Ellis Horwood. 1993.
3. Gardner R.M. Neural Networks: Hardware silicon for 'wetware' algorithms. In *Proceedings of the Symposium on VLSI Technology*. Digest of Technical Papers. p5-6. 1991.
4. Roy J. Suret J.M. A clever screening system for commercial loan applications. In *Proceedings of the Conference on Expert Systems in Economics, Banking and Management*. 1989.
5. Bazley G. Report of the Practical Application of a Neural Network in Financial Service Decision Making. In *Proceedings of the International Conference on Artificial Neural Nets and Genetic Algorithms*. Innsbruck. Vienna. Springer Verlag. 1993.
6. Williamson A.G. Fitness Criteria for the Genetic Algorithm Optimization of a Neural Network Credit Application Vetting System. In *Proceedings of the Conference on Financial Information Systems*. Sheffield Hallam University. Sheffield. 1994.
7. Wettschereck D. Dietterich T. Improving the performance of Radial Basis Function Neural Networks by Learning Centre Locations. In *Proceedings of the Conference on Neural Information Processing Systems-Natural & Synthetic*. 1991.
8. Kung S.Y. *Digital Neural Networks*. Prentice Hall. 1993.
9. Wong Y. How Gaussian Radial Basis Functions Work. In *Proceedings of the International Joint Conference on Neural Networks*. Seattle.1991.
10. Carpenter G.A Grossberg S. Self-organization of stable category recognition code for analog input patterns. In *Proceedings of the IEEE International Conference on Neural Networks*. San Diego. IEEE. 1987.
11. Musavi M.T. Faris K.B. Chan K.H. Ahmed W. On the Implementation of Neural Network Applications. In *Proceedings of the Conference on Analysis of Neural Networks Applications*. 1991.

A DYNAMICAL ARCHITECTURE FOR A RADIAL BASIS FUNCTION NETWORK

LEMARIÉ Bernard, DEBROISE Anne-Gaelle.

Service de recherche Technique de la Poste, (SRTP)10, Rue de l'Ile Mabon
F-44063, NANTES CEDEX. e-mail:lemarie@srtp.srt-poste.fr

Abstract : .We present a dynamical architecture for a Radial Basis Function Network. The scheme is based on the Simulated Annealing procedure for learning. Increase of performances with respect to classical methods and opportunity to vary the size of the network are reported.

1. Introduction.

Radial Basis Function (RBF) Networks have been extensively studied in the past few years. Briefly, the advantages of such network models can be expressed in term of functional approximation, thanks to the regularization theory support [8], of statistical tool through non parametric estimation methods like Parzen window's, of simplicity with only a linear output weights layer and experimentally good performances[5][6][7]. RBF networks are not only gaussian networks, and for example wavelets networks can also be considered, yet with specific properties[10].

However and even with this kind of network, faced to a classification problem some challenges remain opened such as: determination of an optimal number of cells, specialisation of the network after a first phase of learning, risk of local minima with the use of the Back-Propagation (BP) algorithm learning.

In this paper we present the formalism of a dynamical RBF-network architecture based on a Simulated Annealing (SA) procedure [4]. This should help in avoiding the risk of local minima. A promising aspect of the proposed architecture is the opportunity to include it in a incremental or dynamical building of the network. Also, as a consequence of this architecture, output are normalized.

The organization of the paper is as follows: In the next section we develop the needs for improvements of the RBF network model. The third section is dedicated to the presentation of the new architecture. Then, the section 4 describes the first experiments realized with this architecture. Conclusion and perspectives are drawn in section 5.

2. RBF models.

As many others, our current RBF-network is built in a static way. First of all, a clustering is done on the reference set by a supervised method [5]. The method is different of other methods like classically K-means in the sense that the number of cells is found automatically on the basis of a probabilistic estimation of the separability of the classes. After initializing the network by this technique, we minimize the mean square classification error by a back propagation procedure applied to the network parameters. These are the ray around each center and the output weights.

In spite of the good results yet encountered with this architecture, for example for a handwritten digit recognition task [5], some improvements for this model and generally for RBF-networks were identified as follows: First of all, the back propagation algorithm is sensible to the presence of local minima. As a consequence the good performance of the learning phase strongly depends on the initial state of the network. Next, the learning process is separated in two parts: clustering and learning that is parameter adjustment. Consequently, if the learning set is modified in anyway, for example in term of number of examples or label values, the learning phase but also the clustering phase have to be done again. Due to the generally important computing cost of those algorithms and especially backpropagation, the use of the RBF-network turns out to be impracticable in this situation. A last drawback of the architecture lies in the fact the clustering method does not use the same convergence criteria than the back-propagation learning phase. This last is most often the minimization of the Mean Square Error on the network outputs (MSE), a well-suited criteria to the classification task since it has been proved that this criteria leads to asymptotically approximate the bayesian probability [3]. So even if the clustering method produces a number of centers according to its own criteria we can't assure that this number is optimal for the classification problem. A cyclic and for example incremental learning could also be considered [10] but no proof exist for the convergence of such an iterative procedure. A last improvement is needed which concerns the stochastic nature of the network outputs. In many applications it is preferable to be able to consider the networks output as probabilities so as to delay the decision to the subsequent process. Thus including the stochastic constraint in the learning phase is fruitful.

3. The dynamic model.

Let us consider a C class classification problem in the representation space R^d. The aim is to approximate the bayesian probability $p(c_k|x)$ of a pattern x to be of class c_k, k=1, 2, ..., C . Given a learning set of N elements $L = \{x_1, x_2, ..., x_M\}$ and the corresponding label function $\delta(x, k) \in \{0, 1\}$, k=1, 2, ..., C, we wish to build a network with M centers $\{u_1, u_2, ..., u_M\}$ of R^d. The k-th output approximates $p(c_k|x)$ and is given by:

$$S(c_k, x) = \sum_{j=1}^{M} \lambda(c_k|u_j) \cdot f(u_j|x) \qquad (1)$$

The two terms in the right side of (1) are computed from the output value of a center u activated by a pattern X:

$$g(x, u) = \exp\left(-\frac{d^2(x, u)}{2r^2(u)}\right)$$

where $d(.,.)$ is the euclidean distance in R^d and $r(u)$ the ray of the center u. $\lambda(c_k|u_j)$ is the weight function and is computed with the learning examples:

$$\lambda\left(c_k|u_j\right) = \frac{\sum_{x \in L,\, \delta(X, k) = 1} g(x, u_j)}{\sum_{x \in L} g(x, u_j)}$$

$f(u_j|x)$ is called the membership function in reference to the fuzzy clustering approach. It is given by :

$$f\left(u_j|x\right) = \frac{g\left(x, u_j\right)}{\sum_{l=1}^{M} g(x, u_l)}$$

Thus, unlike classical RBF models where output weights are free parameter, the network is here completely specified by giving the centers and the associated rays. The question is now to find a method so as to select the number an the position of the centers.

Part of the solution can be found in recent studies on clustering methods [2][9]. For example in [2], the centers are first determined by a simulated annealing based clustering algorithm. All rays are equal and relied with a temperature parameter. After this clustering process, the clustering set is considered as a RBF-network center sets and this network is used for a recognition task with a expression similar to the formula (1).

The method is appealing because it can be used so as to build the cluster set in a dynamical way. But in this clustering phase, the criteria minimized is the weighted distance*:

$$E = \left(\frac{1}{N}\right) \sum_{x \in L} \sum_{j=1}^{M} f(u|x) \times d^2(x, u)$$

As yet pointed out his error criteria is not appropriate for the bayesian probability approximation and our idea is to try to directly compute the mean square error for each cluster set:

$$MSE = \frac{1}{N} \sum_{x \in L} \sum_{k=1}^{C} \left(S\left(c_k, x\right) - (\delta(x, k)) \right)^2$$

The scheme is the following. We define a state space consisting of all the possible networks from the learning set:

$$S = \{ \{u_1, u_2, \ldots\ldots, u_M\} \times \{ r\left(u_1\right), r\left(u_2\right), \ldots\ldots, r\left(u_M\right) \} \}$$

$$\text{with} : 1<M<N$$

and $u_i \in L, r(u_i) \in R^+, \forall i, 1<i<M$

For each state we can then compute the mean square error (6). Completed with an appropriate neighbourhood definition in the space state, the framework will allow to minimise the MSE through a simulated annealing procedure. Starting from an initial set of centers an adequate SA procedure will select a global minimum of the MSE. Note that centers position is restricted to the learning set elements position so as to limit the computing time since distance can be stocked rather computed after each transition.

Nevertheless, this approach is still costly since it requires a M*N computing time for each transition, that is the same time than for only one backpropagation learning cycle. This obstacle can be favourably overcome if the transitions are selected in a fashionable way. Let us consider a first kind of transition at time t by the modification $\Delta(r)$ of the ray of the center u_l. The activation between this center and all the pattern of L will change. The weight function values will change only for the center u_l and the membership function value will change for every pattern and every center. Fortunately, thank's to the stochastic characteristics of the weight and membership functions we can show that:

$$S^{t+1}\left(c_k, x\right) = a\left(x, u_l\right)S^t\left(c_k, x\right) + b\left(x, u_l\right)\left(c\left(x, u_l\right)\lambda^{t+1}\left(c_k|u\right) - \lambda^t\left(c_k|u\right)\right)$$

with the function a, b and c depending only on the transition center and the current pattern. The computing time for a transition is therefore in N rather than in M*N and the simulated annealing optimisation becomes feasible. It is easy to show that the same simplification occurs for the other kind of transitions we have selected: mutation, that is moving a center toward another point of the learning set L, creating a new center from the learning set L and suppression of a center.

4. Experiments.

Numerical experiments with real data from french handwritten postal code digits are now reported. We use a learning set of 1000 patterns from the ten different classes.The generalisation aspect are not reported here and thus, we do not consider any validation set.

In a first experiment the number and the position of centers remain fixed and the learning phase adjust the ray values (SA-net(I)). The 71 elements set of centers is achieved with our standard clustering method [5]. The performance is compared with the BP-net's one starting from the same network.

Results are presented in the table 1. In both algorithms, learning is stopped when the error decrease becomes slow. In term of the MSE, the SA learning is clearly much better than the BP one. Note that since the back-propagation method adjusts the rays and the output while the simulated annealing works only on the rays the optimization space for the new architecture is thus embedded in the BP one. Therefore, nothing but a local minima prevents the BP algorithm to reach the better solu-

* In [2], other terms are introduced in the clustering error like a "supervision complexity cost". They do not change the scope of this work and therefore are not considered here.

tion proposed by the SA algorithm.

net	MSE	# centers
BP-net	0.16	71
SA-net (I)	0.092	71
SA-net (II)	0.021	71
SA-net (III)	0.00032	997
SA-net (IV)	0.0091	150

Table 1 : results on the learning set.

The mutation process is then introduced in a second experience (SA-net(II)). It still reduces the MSE error by selecting the best centers in the reference set.

The complete learning dynamic includes the generation and the suppression of centers. This is illustrated by SA-net (III) in table 1. Still starting with an initial center set of 71 elements, the final center set is almost equal to the reference set and the MSE is now very low. This proves the ability of the SA procedure to run through the optimisation space even from a far initial state.

Of course this solution presents nothing but a numerical interest. Indeed, we can anticipate that, besides the high computing cost, the generalisation ability of this network should not be very good. In practice, some criteria must be introduced so as to limit the network size in relation with the generalization capacity of the model such as cross validation, introduction of a complexity or regularization term in the error. Almost all kind of error which can be first computed for each state and next cheaply estimated for each transition will be accepted by the simulated annealing procedure.

These topics are currently under experimentation and will be reported later, but another experience, illustrated by the SA-net (IV) in the table 1, can be yet pointed out. In this experience we have intentionally promote the suppression transitions. The final network does not correspond to the global optimum as before but presents still a low MSE with a intermediate number of centers.

The theoretical explanation of this result is the lost of symmetry of the state generation procedure in the SA-numerical scheme. In this case we have no guarantee of the convergence of the SA procedure towards the global optimum [1]. however, this method could be applied for practical situations so as to limit the number of centers and we believe that the theoretical implications of this learning dynamic should be explored more deeply.

5. Conclusion

In this paper, we have presented a new model of RBF-network. It is a dynamic model in the sense that it integrates in an unique learning phase the selection of the number and position of centers and the adjustment of the network parameters.The model is based on a statistical physics approach and the use of the simulated annealing algorithm.

We showed that the numerical scheme remains practicable for learning. Experiments validate the model and reveal its superiority on classical back-propagation learning. They also integrate dynamical modifications of the size of the network.

The next step is to proposed a scheme for building the network based on the generalisation capacity.

6. References.

1. AArts E., Korst J.: Simulated annealing and Boltzmann Machine. John Wiley & Sons, (1989).
2.Buhman J, Kuhnel H.: Neural Comp., 5, 75, (1993).
3. Hampshire J. B., Pearlmutter B.: Proceedings of the 1990 connectionnist models summer school, Morgan Kaufmann, 159, (1990).
4. Kirkpatrick S., Gelatt C., Vecchi M.: Science 220, 671, (1983).
5. Lemarié B.:IJCNN'93, 331, (1993).
6. Moody J., Darken C. J.: Neural Comp., 1, 281, (1989).
7. Ng, Lipmann R.P.: NIPS 3, (1991).
8. Poggio T., Girosi F.: Proceedings of the IEEE, 78, 9, (1990).
9. Rose K., Gurewitz E., Fox G. C.: Physic. Rev. Let., 65, 945, (1990)
10. Zang Q.: IRISA, internal report No 709, France, (1993).

Centre Selection for Radial Basis Function Networks

K. Warwick, J.D. Mason and E.L. Sutanto

Department of Cybernetics, University of Reading
Reading, UK

Abstract

This paper is concerned with radial basis function neural networks. A new method is described by which means radial basis function centres can be selected. The method is based on a mean tracking clustering algorithm and it is shown how the approach provides a good procedure for radial basis function centre selection. Results are given which indicate how the method realises a network with excellent approximation properties.

Introduction

Recently Radial Basis Functions (RBFs) have been used to form alternative neural network solutions to the more commonly encountered Multi-Layer Perceptrons (MLPs). They have the ability to model, in a fairly straightforward way, any arbitrary non-linear mapping [1], even though the output layer of an RBF network is merely a linear combination of hidden layer signals, there being only one hidden layer. RBF networks therefore offer considerable promise in terms of network stability and robustness proofs, largely because the network can be easily defined by means of an equation set [2].

As is the case with modelling methods in general, and particularly with neural networks, such as MLPs, a problem occurs in terms of making a suitable selection of RBF network structure. The key problem is deciding both the quantity and position of basis function centres, this is discussed further in the section that follows. The choice of centres directly affects the goodness of fit of the overall network approximation capabilities achieved, with a requirement that they must certainly be sufficient in number to span the entire input domain [3]. However use of a large number of basis function centres obviously raises questions in terms of computational effort, particularly insofar as the number of basis function centres increases exponentially with regard to input space dimension [4]. In this way it can be considered that RBF networks become unviable in the presence of a high dimensional input space or, looked at in another way, RBF networks are best suited to less complex problems.

In this paper a method of basis function centre selection based on cluster analysis is put forward [5,6]. Essentially a batch of data is collected from the input variables and an n-dimensional plot in phase space is obtained where *n* is the number of input variables. The complete plot of the data points then indicates the possible existence of cluster points, as shown in Figure 1 .

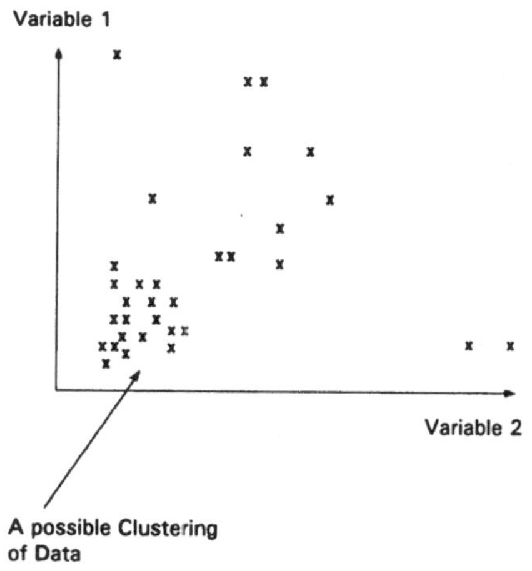

Fig. 1: Clustering by eye

In the section that follows, a description is given of radial basis function networks, with emphasis being placed on basis function centres. Following this the mean clustering algorithm is described, including the allocation of basis function centres from the clusters found. Finally some experimental results are obtained based on data from a flexible robot arm.

Radial Basis Function Networks

Multi-layer perceptron networks have been found to be problematic in a number of ways, not the least of which is the poor convergence of network weights under training due to the inherent non-linear nature of the weight updating relationships [4]. Recently, attention has moved towards RBF networks, which have much better adaptive features. This is mainly because of the straightforward linear relationship between the network weights and the output, such that a linear parameter estimation technique, for example linear least squares [7] can be employed for weight updating.

If the *n*-dimensional input vector is given by *u*, then the RBF network output *y*, as a single variable, is

$$y = \sum_{i=1}^{N} \omega_i R_i(u) + \omega_o \qquad (1)$$

in which ω_i (I = 1,2, ...N) are the network weights, ω_o is a bias term and R_i are activation functions such that

$$R_i(u) = \phi\left[\| u - c_i \|\right] \qquad (2)$$

where c_i(i = 1,2,.......,N) are the basis function centres and ϕ(.) is the radial basis function, where typically, and certainly here, the same function ϕ is applied for all N summations. It is of course quite possible to employ a different ϕ for each i. Further a vector of network outputs can easily be formed just by taking multiples of the individual network in (1).

The radial basis function type ϕ can be selected in one of a number of ways[2], a popular form being Gaussian, where

$$\phi(u) = \exp\left[-(\|u - c_i\|)^2 / \sigma^2\right] \qquad (3)$$

in which σ is a scaling parameter.

However the authors' experience indicates that the thin plate spline works very well in practice, this being defined by:

$$\phi(r) = r^2 \log(r) \qquad (4)$$

where

$$r = \|u - c_i\| \qquad (5)$$

In standard RBF network operation, as shown in Fig. 2, the centres c_i are fixed and function learning is carried out by adjustment of the weights ω_i. When dynamic sampled systems are being considered in a discrete-time situation, the input vector will most likely take on a form

$$u = (y_{t-1} y_{t-2}, \ldots\ldots : u_{t-1}, u_{t-2}, \ldots.) \qquad (6)$$

where y_{t-1} and u_{t-1} respectively represent the values of output signal and input signal one sample period previous.

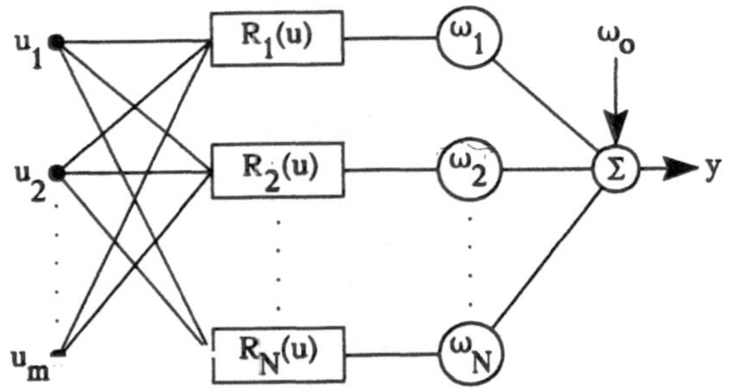

Fig. 2 : A radial basis function network

As has been mentioned earlier however, a critical structural point occurs in terms of how many centres should be selected and where should they be positioned in the *n*-dimensional space.

Mean Tracking Clustering Algorithm

Analysis of clusters has been used for a wide range of applications [8,9], however it has only recently been connected with radial basis function centre selection [3]. The mean-tracking algorithm is employed here, as an RBF centre selection tool, for the first time. The algorithm is quite capable of dealing readily with large quantities of data.

Data is assumed to be multivariable of n-dimensions of real numbers. Also, if M data samples are collected, an n-dimensional search window is placed on the plotted data space, the lower and upper limits on window size being given by:

$$\left[(c_j - a_j), (c_j + a_j)\right] \quad j = 1, \ldots.., n \qquad (7)$$

where c_j is the search window centre on the j-axis
a_j is the predefined window radius on the j-axis
n is the number of dimensions.

If the set $\{x_j : j = 1, \ldots, n\}$ denotes each data point, giving $M\{x_j\}$ data vectors in total, the centre of gravity (c.o.g.) for a search window is found in terms of the mean value of all the vector points which lie in the window, i.e.

$$c.o.g._{kj} \frac{\sum_{r=1}^{P_k} x_{rj}}{P_k} \quad j = 1, \ldots , n \tag{8}$$

where $c.o.g._{kj}$ is the value of the c.o.g. of the jth dimension at the kth iteration and P_k is the number of data vectors present in the window boundary of (7) at the kth iteration.

The window is recentered, thus

$$c_{k+1,j} \;=\; c.o.g._{kj} = 1, \ldots, n \tag{9}$$

thereby generating the next window iteration for (7), a procedure which is repeated until

$$c_{k+1,j} \;=\; c_{kj} \qquad j = 1, \ldots, n \tag{10}$$

and the window is stationary until an iteration limit is reached or until the change in centre is below a previously defined minimum. A cluster has therefore been found, with the c.o.g. being taken as the centre of the cluster, the procedure being depicted in Fig. 3, for the two-dimensional (n = 2) case.

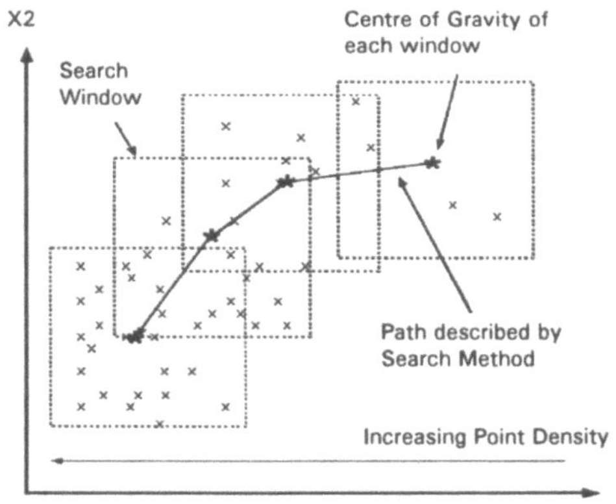

Figure 3: The Mean-Tracking Cluster Search Method.

Multiple clustering is obtained by placing a number of search windows in the data space in order to find possible clusters. Duplicate or closely linked clusters can then be eliminated either by killing off one cluster in favour of the other or by merging two clusters [10].

All points in the data space are checked for their membership of a cluster simply by assessing whether or not the Euclidean distance of that point from each cluster centre is below the preset bound a_j: j = 1, ...n.

It is reasoned here that a useful measure is the standard deviation, γ, of each variable which can then be used to determine the closed interval of the axis which that variable represents. So the offset a_j in (7) is given by

$$a_j \;=\; \lambda_c \; \gamma_j / 2 \quad j \;=\; 1, \ldots n \tag{11}$$

In which

$$\gamma_j \;=\; sqrt\left(\sum_{l=1}^{M} \left(x_{lj} \;-\; \bar{x}_j\right)^2 / M \right) : \; j = 1, \ldots, n \tag{12}$$

and λ_c is a positive definite scalar.

The measure $\lambda_c \, \gamma$ indicates a good initial window size for cluster location, because γ is a measure of the density of concentration of data, the smaller the value of γ, so the more concentrated is the data.

In fact major clusters are likely to be located by most (bounded) window sizes, and it is really for the location of small clusters that the size of the window is important. In this sense $\lambda_c \gamma$ represents a compromise solution.

Experimental Results

The mean-tracking algorithm has been tested on a set of flexible manipulator arm data generated by the Department of Mechanical Engineering, Katholic University of Leuven (KUL) and Department ELEC, Vrije Universiteit Brussels (VUB). This data is commonly available in MATLAB.

The input to the flexible arm was a variable torque applied to the vertical axis at one end of the arm, and the resulting tangential acceleration was measured at the opposite end. The input was a multisine with frequency components in the range 0.125 Hz to 25Hz in steps of 0.125 Hz. Input and output samples were taken at 500 Hz, with trials consisting of 4096 samples.

The mean-tracking algorithm was used to find RBF network centres with the RBF network modelling the flexible arm as a system identification exercise, as shown in Fig. 4. Two sets of trials were conducted: the first using a 7-input RBF network employing 3-past arm input values, 3 past output values and the most recent input value. The second trial meanwhile employed a 9-input RBF network employing 4-past arm input values, 4-past output values and the present input value.

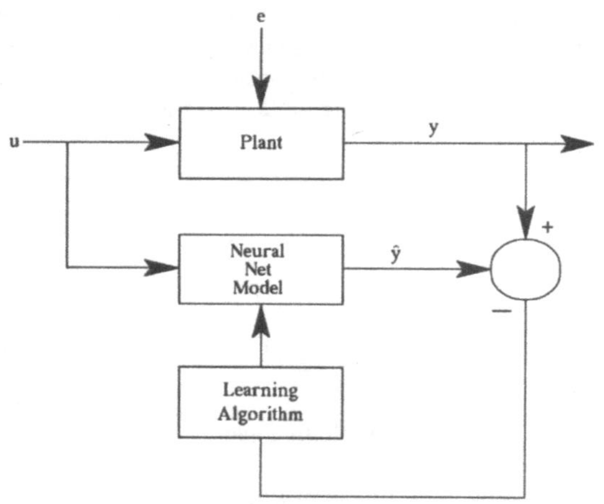

Fig. 4: RBF Neural network for system identification

The results given indicate the mean square error as a percentage of the mean square output signal amplitude, the role of the network being as a one-step ahead predictor. The size of the search window used in the mean-tracking algorithm is given as the W/size multiple of the standard deviation in each axis direction and W/No is the number of search windows in the data space. An example plot of flexible arm output and one-step ahead prediction error is shown in Fig. 5.

Table 1: Flexible arm test results

Centres	Network Inputs	W/size	W/No	Train Error %	Test Error %
14	7	1.036	455	16	15
23	7	1.036	1365	2.2	5.8
40	7	0.995	1365	1.4	1.2
43	7	0.954	1365	1.2	1.9
72	7	0.878	1365	0.7	0.5
39	9	1.036	455	1.4	1.0
59	9	1.036	1365	1.0	0.7
76	9	0.995	1365	0.9	0.7
103	9	0.954	1365	0.8	0.6

In Table 1 the mean square error over 4096 points has been taken, and the column "train error" indicates the percentage error taken over the data points (4096) on which the RBF network was trained. The "test error" column meanwhile indicates the percentage error taken over a subsequent set of test data (4096 points again) from the same flexible arm.

Conclusions

The results in general show, as would be expected, that the more centres that are used, so a better function approximation can be obtained. However, as can be seen when looking at the error difference with a

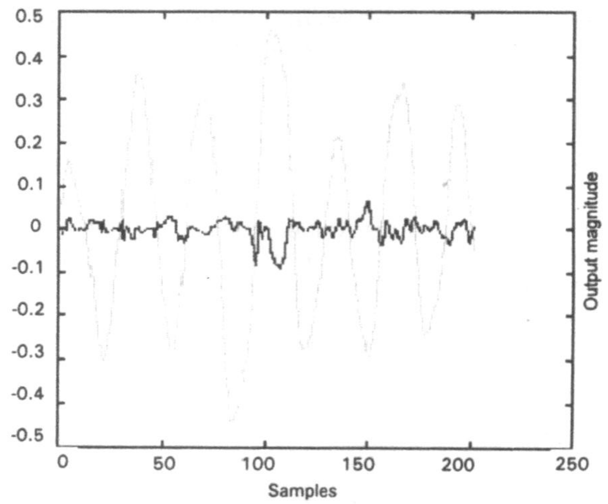

Fig. 5: Plant output (larger signal) and prediction error

network based on 59 centres and one employing 103 centres, there is a very little difference in the modelling accuracy with respect to the computational increase.

References

[1] Girosi, F., Poggio,T.:Biol. Cybernetics, 63, 169, (1990).

[2] Cichocki, A., Unbehauen, R.: Neural Networks for Optimization and Signal Processing. John Wiley and Sons, New York 1993.

[3] Chen, S.,, Billings, S.A., Cowan, C.F., Grant, P.M.: Int. Journal of Control, 52, 1327 (1990).

[4] Narendra, K.S.: Adaptive control of Dynamica Systems using Neural Networks, in Handbook o Intelligent Control, van Nostrand 1994.

[5] Sutanto, E., Warwick, K.: Proc. Cybernetics and Systems, 1, ed. Robert Trappl, 327 (1994).

[6] Craddock, R.J., Mason, J.D., Warwick, K: Proc IMACS Int. Symposium on Signal Processing Robotics and Neural Networks, 302 (1994).

[7] Soderstrom, T., Stoica, P.: Systen Identification. Prentice-Hall (1989).

[8] Vogt, W., Nagel, D.: Clinical Chemistry, 38, 18 (1992).

[9] Huth, R., Nemesova, I., Klimperova, N.: Inte national Journal of Climatology, 13, 817 (1993)

[10] Sutanto, E.L.: The Design and Development c Multivariable Cluster Analysis for High-Spee Machinery. PhD Thesis, University of Readir (1995).

FLEXIBLE USER-GUIDANCE IN MULTIMEDIA CBT-APPLICATIONS USING ARTIFICIAL NEURAL NETWORKS

Klaus Langer, Prof. Dr. Freimut Bodendorf

University of Erlangen-Nuernberg, Department of Information Systems
Lange Gasse 20, 90403 Nuernberg, Germany, langer @ wiso.uni-erlangen.de

Abstract

The paper decribes a system architecture for CBT-applications that includes artificial neural networks (ANNs) for dynamically creating user-specific paths through a database of multimedia objects. The organisation of these objects is based on principles of hypersystems. Additionally a semantic data model is involved to describe the objects´ characteristics. The ANNs represent experiences and decisions of former users, based on the object´s characteristics. During a CBT-session, the ANNs are used by a run-time controller in order to retrieve objects, which are appropriate candidates for continuing the lesson, from the semantic data model .

Keywords: computer-based training, artificial neural network, hypermedia, semantic network.

1 Introduction

One objective of the system architecture described in this paper is to increase the *reusability* of multimedia resources. For modern CBT applications, designing and creating multimedia resources such as text, graphics, pictures, animation, audio, and video sequences require a considerable amount of time and effort. During the project, a number of authoring methods and tools have been developed. These tools enable authors to re-configure existing modules for different educational settings, e.g., teaching basics, brush-up, training etc., and various user classes. Besides reusability, increasing *flexibility* to adapt to each user's preferences, motivation, and experiences is another objective. For each user, a specific course is dynamically created at run-time, using a run-time controller developed within the project. This controller contains a fuzzy logic component for representing pedagogical knowledge by fuzzy rules and an artificial neural network (ANN) component. The latter one is included to represent experiences and decisions of former users (other learners or even authors) for their successors.

2 Semantic data model

Learning paths are created by sequentially activating multimedia objects, using the run-time controller mentioned above (see Fig. 2). For a dynamic, user-individual design of learning paths, the run-time controller must have access to "knowledge" about the objects' contents, media design, pedagogical functions, etc.

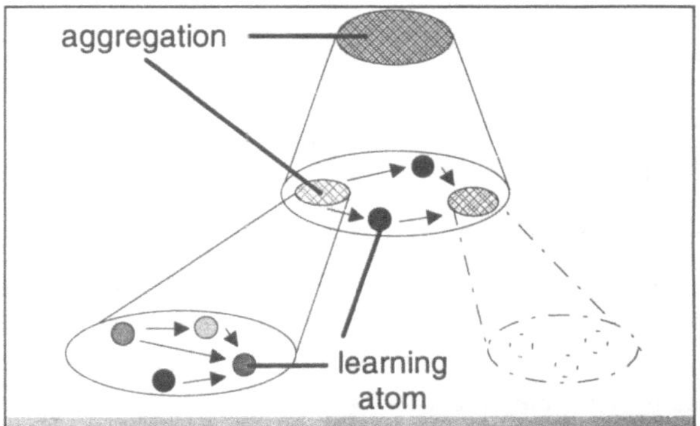

Fig. 1: Aggregation & learning atom

These semantics of the multimedia objects can only be described within the context of each course. For this purpose, a level of declarative elements of knowledge exists within the architecture. This level contains information about four different types of elements [4],[5]:

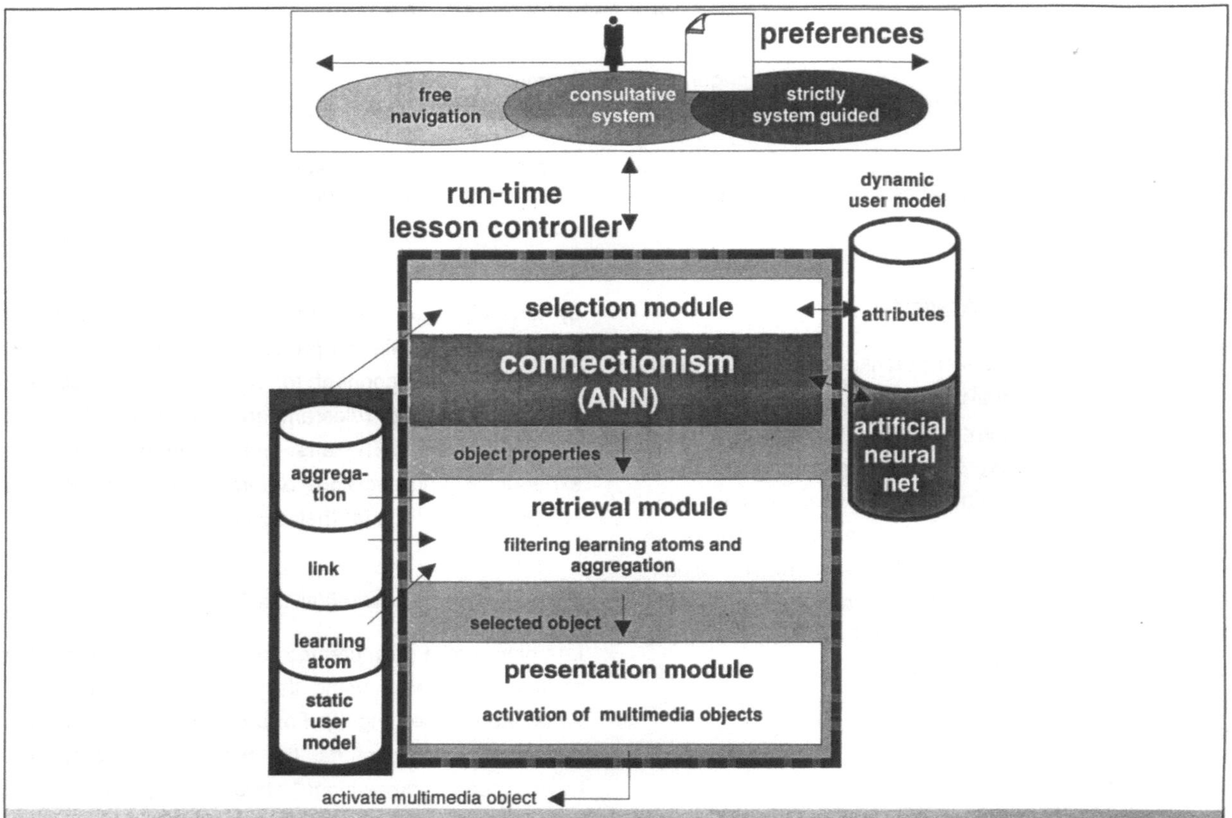

Fig. 2: Run-time controller & artificial neural network

- ❒ learning atoms,
- ❒ aggregations,
- ❒ links, and
- ❒ user-classes.

From a pedagogical point of view, minimal elements of knowledge are called *learning atoms.* They should be as small as possible, in order to ensure high flexibility. At the same time they must be large enough to form a complete, but minimal step for the student's learning process. Learning atoms are results of the conceptual design and refinement of CBT applications. *Aggregations* are (associative) collections of learning atoms or other aggregations. They are used to represent the hierarchical structures that are present in most CBT lessons (see Fig. 1).

An important part of the system architecture is a model for representing learning atoms, aggregations and their relationships. This model combines characteristics of hypersystems and semantic networks.

3 Artificial neural network

The knowledge represented by the semantic data model is used as input for an ANN with a backpropagation topology [1],[2] (see Fig. 3). More precisely, all attribute values of instances of the four element types mentioned above may be selected as components of the input vector. This selection is part of the design stage of the ANN. It is accomplished by using a network editor specially developed for this project, because CBT authors with little experience in assigning and transforming values of attribute-oriented data models into domains of ANNs should be able to define input and output of the ANN [2],[3].

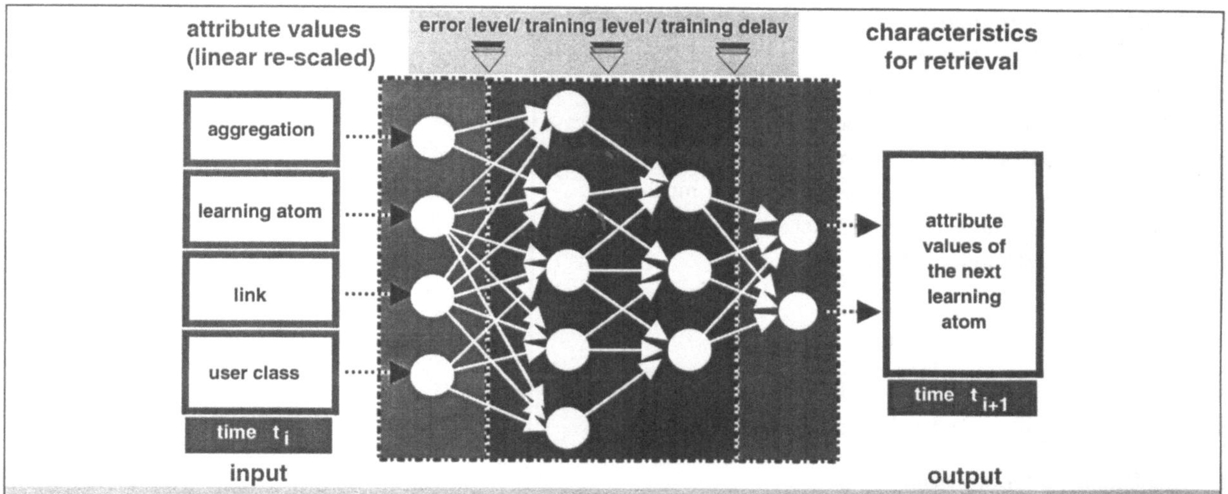

Fig. 3: Integrating the artificial neural network and semantic data model

The editor provides a graphical-oriented interface and supports the user during the definition process, e. g., all available attributes of each element type are presented in list boxes. If an attribute is selected, the number of necessary input/output neurones will be automatically calculated depending on the attribute's domain type. Moreover, transformations of input/output scales (e. g., Likert-scales for 'level of interest') are also done automatically. This conversion feature is possible since all available attributes for a network's input/output are described in detail on a meta level of the knowledge base (data dictionary).

A database that contains learning paths of former users is used during *training* the ANN. For successful training, these users have to be experienced learners or authors following a well-defined strategy that can be "discovered" by the ANN. While *processing*, the attribute values of the currently activated learning atom, aggregation, link, and user-class are fed into the network. The ANN's resulting output vector is transformed into an attribute-based description of all the properties the next learning atom should have. Subsequently, a retrieval module sets up a list of appropriate learning atoms from which the user or the system may select.

4 Experiences

Testing the ANN's functionality within the architecture has started with small, "artificial" courses. Nevertheless, it reveals interesting results. If experienced users (e.g., authors themselves who simulate being a "learner" belonging to a specific user-class) have created the learning paths for training the ANN, the ANN provides useful recommendations for subsequent users: The learning atoms selected fit in the "strategy" of the former user. The ANN works like an adaptive, auto-learning filter, though in all cases with some more complex input/output vectors the training has not shown convergence. However, the ANN approach shows some weakness if strict system guidance is desired. The selection of the next "best" learning atom tends to nominate one global "champion" that is chosen all the time. Here at least a hybrid combination with rule-based mechanisms (e.g., fuzzy logic) is required.

Summarising, these findings are encouraging, because involving an ANN enables an author to configure user-specific CBT courses by just walking along an appropriate path of learning atoms. The strategy is automatically learned by the ANN and can be applied to subsequent users.

5 Implementation

The architecture has been implemented on top of Microsoft (MS) Windows™ as GUI. The run-time controller and fuzzy logic system have been implemented using MS C 7.0/SDK 3.1™. The ANN and multimedia object management are based on MS Visual C++ 1.0™. The editor for maintaining the knowledge-base (SemEd) has been built with Turbo-Pascal-for-Windows™. The multimedia database system that contains elements for presentation is based on Gupta's SQL Server™. The multimedia object management makes extensive use of Window's Media Control Interface (MCI).

Bibliography

1. Kosko, B.: Neural Networks and Fuzzy Systems. Englewood Cliffs: Prentice-Hall 1992.
2. Hofmann, O.: Entwicklung einer lernfähigen konnektionistischen Komponente zur Ergänzung einer Ablaufsteuerung für CBT-Applikationen. Thesis, Department of Information Systems at the University of Erlangen-Nuernberg, Nuernberg (Germany): 1994.
3. Hinton, G. E.: Connectionist Learning Procedures. In: Kodratoff, Y., Michalski, R. (eds.): Machine Learning, Vol. III. San Mateo 1990.
4. Merrill, M. D.: Component Display Theory. In: Reigeluth, C. M. (ed.): Instructional Design - Theories and Models: An Overview of their Current Status. pp. 279. Hillsdale: Lawrence Erlbaum Associates 1983.
5. Mühlhäuser, M.: Requirements and Concepts for Networked Multimedia Courseware Engineering. In: Maurer, H. (ed.): Computer Assisted Learning. Lecture Notes in Computer Science. No. 360, pp. 400. Berlin: Springer 1989.
6. Halasz, F., Schwartz, M.: The Dexter Hypertext Reference Model. Communications of the ACM 2, 30 (1994).
7. Marchionini, G.: Hypermedia and Learning: Freedom and Chaos. Educational Technology 11, 8 (1988).
8. Wenger, E.: Artificial Intelligence and Tutoring Systems. Los Altos (CA): Kaufmann 1987.

USE OF GENETIC AND NEURAL TECHNOLOGIES IN OIL EQUIPMENT COMPUTER-AIDED DESIGN

R.M.Vahidov, M.A.Vahidov, Z.E.Eyvazova

Azerbaijan State Oil Academy
20 Azadlyg
Baku 370601, Azerbaijan

Abstract

Oil pumping equipment designers have to solve different types of optimization problems. Use of strong mathematical means is frequently very difficult, or, even impossible because of complexity of those problems. This paper suggests using genetic algorithms as an alternate facility to find optimal parameters of pumping unit under given particular conditions. Neural networks are employed to approximate the best solution using statistics on already found solutions for a set of conditions. Experimental results are discussed.

1. Introduction

Oil pumping equipment designers inevitably have to face with different types of optimization problems, i.e., they have to choose such values of equipment parameters that fit best for given conditions. There are strong mathematical means to solve optimization problems, e.g., linear programming, gradient ascent/descent etc. Unfortunately, it is frequently difficult or, sometimes, impossible to apply them directly to real problems because of the complexity involved.

Genetic algorithms (GA) [7 et al] are evolution-based means for solving optimization problems of wide class. GAs are not so "strong" but much easier to apply. They operate on population of individuals. Each individual (or chromosome) represents a possible solution to a problem. Therefore, a "population" is a set of possible solutions. Evolution forces the population to undergo selection, reproduction, mutation and crossover processes. Selection provides "good" chromosomes, i.e. chromosomes with high fitness with high likelihood to survive and produce offspring. In terms of a problem to be solved the fitness is the estimate of the particular solution. Mutation and crossover are genetic operators which involve randomness in solution search process and provide information exchange between different directions of search.

This paper suggests application of GA to oil pumping unit design. Fitness function for this problem incorporates different particular criteria for solution estimating. Those criteria are weighted and summed to give global estimation. The weights put are dependent on current situation and they determine conditions for a problem.

A neural network (NN) is employed to provide mapping conditions-solution. Neural networks have been proven to be universal approximators, i.e. they are capable of approximating any function [6]. NNs are simulators of brain and they possess such valuable characteristics as learning and generalization. Having statistics on conditions-solution pairs, one can train a NN that builds internal representation of knowledge on the area. When new set of conditions is entered to the trained network, it will produce an approximation to the sought solution. This generated output can be then processed genetically to search for a possible better solution.

2 Oil Pumping Unit Design Problem

Pumping unit is a complicated mechanical device that enables oil extraction. It is of great importance to choose its parameters in optimal way. This necessitates description of the problem mathematically. Finding a unique universal criterion is not a trivial procedure, because there are different particular criteria, such as cost, power consumption, etc. Which criterion is to be focused on and which is not so important depends on the current situation. For example, if cost of electrical energy increases, then --large focus should be placed on the corresponding particular criterion.

To build the unique global criterion a number of particular criteria are multiplied by weight coefficients and summed:

$$C = \sum_{i=1}^{N} w_i C_i \qquad (1)$$

where N is the number of criteria, w_i are weight coefficients, f_i are particular criteria, and C is the global criterion. Particular criteria include cost, power consumption, reduced stress in sucker-rods of a pump, charge on reducer and pressure loss in suction valve. They are built in such a manner that the higher is their value the better is solution. For example, the higher is the value of the criterion associated with the cost, the lower is the real cost. There are complicated procedures for criteria calculation. Weight coefficients are the means of adjusting the problem to current situation. They are estimated by experts and all sum up to one.

Thus, we have an objective function to optimize. In our case optimization means maximization of the global criterion.

To this end we have to choose parameters of pumping unit in a due way. Those parameters include ratios of lengths of links (ratios of the radius of crank to the length of back arm of rocker and to the length of connecting-rod, ratio of the length of front arm of rocker to the length of its back arm), desaxial angle, stroke, etc. Values of the parameters are chosen from predetermined intervals. Combinations of those values represent a single solution that can be estimated by (1).

To solve the above problem one might like to use strong mathematical techniques to find ideally optimal solution. Unfortunately, the complexity of formulas involved in calculations does not allow us to consider this opportunity seriously.

Now the Nature offers an assistance for those who are baffled by tremendous mathematical expressions. Recently, genetic algorithms had appeared and have been conquering more and more space in research and developments.

3. Genetic Algorithms

Genetic algorithms devised and developed by Holland [5] and other researchers [1], [2], [3], [4], are becoming very popular in modern research and developments. The idea behind GAs is borrowed from the evolution of living organisms. These algorithms consider the development of some population of individuals (chromosomes). Each individual has a certain degree of "fitness", i.e. possibility to survive in the current environment to produce offspring. Mutations provide the chromosomes with possibility to modify, thus, implementing random search. Most of mutations are negative, however, if positive mutation occurs it contributes to a chromosome's fitness. Crossover also takes place in the population with some probability, causing two chromosomes to exchange their parts. Due to selection, mutation and crossover evolution leads the whole population to a higher degree of fitness.

For solving an optimization problem with the use of GAs it must be represented in terms of the process described above. Each solution is represented by a chromosome which contains genes. There are two versions of GAs: binary and floating point. In binary representation each gene is represented either by 0 or 1. In this case potential solutions to a problem must be encoded in binary form. The floating point representation does not require such a transformation. In both cases a chromosome is represented by a vector.

Selection is the process that involves randomness to decide whether a particular chromosome is going to survive or die. This simulates the aggressive nature of the environment. Chromosomes with higher fitness have better chances to survive than those having lower fitness. In our case selection is implemented by "roulette" [7] and the fitness is measured by the objective function. Each individual has a corresponding sector on the roulette. The higher is fitness of an individual, the broader is its sector. The roulette is rotated as many times as many individuals are present in the population. Each time one individual is selected for survival. Those chromosomes which are not selected at all should die.

Mutation adds variability to the population during evolution, in fact implementing random search. As the result of mutation an individual acquires new features. Most of the time those features turn out to be negative, i.e. fitness-decreasing, however, some of them may be positive, in which case they are reinforced by selection. The intensity of mutation is managed by its probability. In binary representation case mutation just switches a selected gene's value to the opposite one. In floating point representation a value of a gene is chosen randomly from the predetermined interval.

Crossover is also applied in a random manner to pairs of chromosomes in the population. It causes chromosomes to exchange parts of genetic codes that they carry. To this end a partition point is chosen (randomly) in a chromosome and two chromosomes recombine with each other. This process is managed by the probability of crossover. In fact, crossover provides information interchange between different directions of search.

The overall process, progressing step by step in simulated environment is believed to bring the population close to the optimal point. As one can see, GAs do not involve complex mathematical transformations and place minimum requirements to a problem. Moreover, the risk of being trapped by local optima on the objective function surface is little, because the search is performed in several directions.

4. Experiment Design and Results

We have chosen 22 variables describing a pumping unit design. They include different ratios, stroke, desaxial angle and other parameters. The criteria included into objective function estimate a particular solution (chromosome) in respect to the basic design. The basic design is chosen beforehand from the predetermined intervals of values of variables. In our case the form of objective function coincide with that of the fitness function, since the former meets the requirements placed by GAs: it takes non-negative values and requires maximization.

We experimented both with floating point and binary representations of individuals. In first case the length of chromosomes was equal to the number of variables. In second case the length of binary chromosomes was equal to 120. In the latter case the variables were encoded into binary format with the consideration of accuracy requirements. We used the fixed population size of 50 in our experiments.

The probabilities of mutation and crossover were set at 0.001 and 0.25 respectively. For floating point representation we utilized non-uniform mutation with probabilty 0.01 and conventional crossover along with arithmetical crossover [7].

We performed over sixty experiments with software combining different values of weight coefficients in the objective function, thus, simulating different external conditions. The average number of generations was about 1000. The difference between solutions found by floating point and binary versions was calculated using the following formula:

$$d=\frac{1}{M}\sum_{i=1}^{M}\frac{|b_i-f_i|}{r_i}\cdot 100\% \qquad (2)$$

Here, d is the difference, b_i is the value of the i-th variable found by binary version, f_i is that value found by floating-point version, r_i is the range of i-th variable, and M is the length of vector of variables. This difference on the average was about 1.8%. The floating point representation version in most cases slightly outperformed the binary version. This is due to the sacrifice of representation precision in binary version. Small differences in values of variables, however, have little significance for the problem considered here. On the average, GA was able to find solution which were 2.7 times better than the basic ones.

5. Neural Networks

Artificial neural networks are proven to be universal approximators [6]. Devised in early Forties [8], criticized in late Sixties [9] they are becoming of great interest after developing efficient learning techniques, such as Error Backpropagation algorithm [10], [11]. NNs are built to simulate the principles of human brain. The most of the research work today is done with the use of feed-forward backpropagation type neural networks. They consist of input and output layers of neurons and may include one or more hidden layers to reach the required computational power. Each neuron (except for the input neurons which do no computations, but simply distribute input signals to other neurons in fan-out manner) receives weighted input from other neurons and applies transference function to define its output signal. The output signal is then sent to other neurons or to the external environment.

NNs possess a number of useful features, among which we are interested in learning and generalization capabilities. Learning enables a NN to perform required mapping given a set of samples by manipulating weights of connections in NN. Generalization allows a NN to respond to new unknown inputs. Given a sufficient number of hidden neurons and samples a NN is capable of approximating any function with the desired level of accuracy.

For our experiments we had designed three-layer feed-forward NN with five input nodes, twelve hidden nodes and twenty two output neurons. Input neurons receive a vector of conditions from the objective function. The corresponding optimal solution found by GA is connected to the output layer of the network. We made an assumption that there is a complex functional dependency between the set of conditions and the optimal solution. The NN was trained using sixty samples to detect that dependency, i.e. to find the corresponding set of weights.

We used the backpropagation algorithm [10], based on steepest gradient descent to train the network. After 25,000 iterations he learning was stopped and tested on twenty new inputs. Then the values of variables obtained through the network were compared to those found by GA. We used the same formula (2) to calculate the difference, using NN predictions instead of solutions found by binary version of GA. On the average the difference was found to be about 4.7 %.

The reason for using NN is that it allows to find immediately a rough approximation to the solution sought as the new conditions are entered. This approximation can be then processed genetically to find a possible better solution. This allows to reduce significantly the search time. The solution found by GA is then entered to the training set to improve NN's performance(Fig.1).

6 Conclusion

We have described an approach for designing oil pumping units with utilization of genetic algorithms and neural networks. Experimental results indicating closeness of solutions found by different versions of GAs demonstrated usefulness of the suggested approach. The outcomes of experiments had been used to train a neural network for approximating optimal solutions. The network was tested on the set of new examples and it demonstrated the capability to find rough approximation to sought solutions. Those approximate solutions help to speed up genetic search of better solutions.

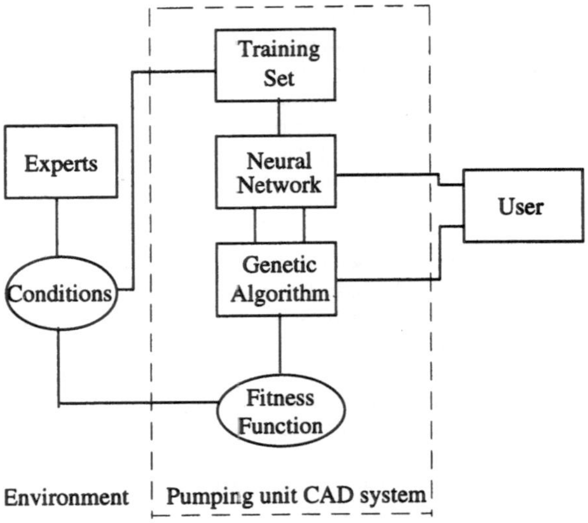

Figure 1. Pumping unit design process

References

[1] A.D. Bethke, "Genetic algorithms as function optimizers", Ph.D. thesis, Dept. Computer and Communication Sciences, Univ. of Michigan, 1981.

[2] L.B. Booker, "Intelligent behavior as an adaptation to the task of environment", Ph.D. thesis, Dept. Computer and Communication Sciences, Univ. of Michigan, 1982.

[3] D.J. Cavicchio, "Adaptive search using simulated evolution", Ph.D. thesis, Dept. Computer and Communication Sciences, Univ. of Michigan, 1970.

[4] K.A.DeJong, "Analysis of the behavior of a class of genetic adaptive systems", Ph.D. thesis, Dept. Computer and Communication Sciences, Univ. of Michigan, 1975.

[5] J.H.Holland, Adaptation in Natural and Artificial Systems, Univ. of Michigan, Ann Arbor, MI, 1975.

[6] K.Hornik, M.Stinchcombe and H.White, "Multilayer feedforward networks are universal approximators", Neural Networks, 2(5), 1989, p. 359

[7] Z.Michalewicz, Genetic Algorithms + Data Structures = Evolution Programs, Berlin; New York: Springer-Verlag, 1992.

[8] W.S.McCulloch, W.Pitts, "A logical calculus of the ideas immanent in nervous activity", Bulletin of Mathematical Biophysics, Vol.9, 1943, p. 127.

[9] M.Minsky, S.Papert, Perceptrons, Cambridge, Massachusetts: The MIT Press, 1969.

[10] D.E.Rummelhart, G.E.Hinton, R.J.Williams, "Learning representations by back-propagating errors", Nature 323, 1986, p. 533.

[11] D.E.Rummelhart, G.E.Hinton, R.J.Williams, "Learning internal representations by back-propagating errors" In PDP Vol.1: Foundations, Ed D.E.Rummelhart, J.L.McClelland and The PDP Research Group, Cambridge, Massachusetts: The MIT Press, 1986, p. 318.

GENESIS: AN OBJECT-ORIENTED FRAMEWORK FOR SIMULATION OF NEURAL NETWORK MODELS.

L. Fuentes, J.F. Aldana, J.M. Troya

Dept. of Lenguajes y Ciencias de la Computación, Univ. of Málaga

Plaza de El Ejido s/n, E-29013 - MÁLAGA (SPAIN)

e-mail: lidiaf@tecma1.ctima.uma.es

Abstract - In this paper we present GENESIS as a highly flexible and graphical environment for simulating ANN based on the object-oriented paradigm. This includes a hierarchy of classes that models ANNs being characterized by a high degree of modularity. The use of object-oriented techniques provides the data abstraction facilities necessary to support modification and extension of the simulation system at a high level of abstraction, from a graphic user interface.. Furthermore, ANN specification is suitable for joining networks in a neural macro-structure.

1. INTRODUCTION

Neural networks have many things in common with object-oriented methodology, such as construction, concurrence and message passing between the units of the system[1]. We consider the development of simulation capabilities that support flexible graphic interface, as well as automatic generation of ANNs code. Our approach allows ease of use, extensive model design with minimal usage of training, resulting in an effective simulation tool. The aim of this work is the building of an object-oriented system that can be easily modified according to user requirements[2].

Another issue that increases the flexibility of the system is the construction of neural macro-structures. The object-oriented design of the system allows to ensemble several ANNs into a new one. Each network accomplishes a specific task, and all together achieve the resolution of a unique problem. Hybrid neural models may also be implemented by the ensemble of different types of ANNs[2].

2. NEURAL NETWORK HIERARCHY

The resurgence in neural network research has resulted in the development of a wide variety of neural network models. The issues that differentiate the various models are: network topologies, computation inside neurons, learning and recall algorithms[3]. All these characteristics are already defined in the abstract class *Network*. Starting from this *Network* class, the new ones redefine some methods according to the specific models. The details of the particular architecture or learning algorithm can be incorporated into the system through the use of inheritance. In this process, only the data and procedures that are different from the original object or contain additional functionality or information, need to be described in the new object.

The hierarchy of classes created by inheritance is shown in figure 1. The root is the base class *Network* and the terminal nodes are specific implementations of neural network models such as Hopfield, Backpropagation, etc. The second level encapsulates specific characteristics that define neural families[4].

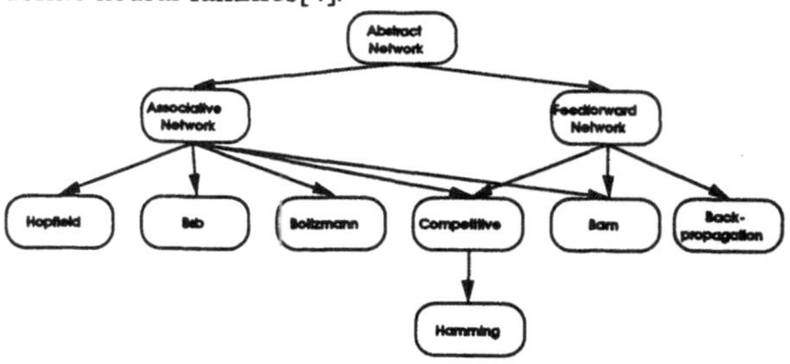

Fig.1 Hierarchy of neural networks GENESIS.

The design choices were based in two different points: recall and learning algorithms. The recall algorithm may be classified in two types: associative and feedforward. We find that associative networks usually use unsupervised learning and that feedforward networks use supervised one. Therefore, we mapped into separated classes: *Associative Network(NetOsc)*, and *Feedforward Network(NetFf)*. The hierarchy is based on this criterion.

The *Feedforward Network* class implements a feedforward recall algorithm with supervised learning. These are the main characteristics of neural models such as Perceptron, Adaline and Madaline, therefore they were all simulated by the same class with specialized functional parameters. To implement the Backpropagation model we build a new network class in which we add specific fun ctions. The Backpropagation model defines a specific learning algorithm that involves an error back propagation and the delta learning rule.

These additions were accomplished in a straight forward manner redefining some methods provided by the *Feedforward Network* class.

The *Competitive* network model constitutes a hybrid network with associative and feedforward characteristics. Therefore it inherits two intermediate classes: *Associative Network* and *Feedforward Network* by using the multiple inheritance technique. An important aspect of this approach is that we can simulate hybrid neural network models in a more abstract fashion. That is, a new ANN model doesn't need to inherit the particular implementation details of an existing network class. The classes shown in figure 1 represent a small portion of ANNs that can be simulated by this system.

Moreover, to complete the ANN object-oriented design, there is a Layer and a Learning Controller hierarchy of classes. The Layer hierarchy fits the Neural Network one. The Learning Controller hierarchy was made according to unsupervised and supervised learning clasification.

3. INTERACTIVE CONSTRUCTION OF A NEURAL NETWORK

The system has a graphical user interface that allows the use of the set of simulation classes already defined. It is based on the XWindows graphics environment. It was developed to aid the user of the simulator in the testing and evaluation of new ANN models and the existing ones. With this tool the user can create new network structures using a mouse. Simulate a neural prototype implies the instantiation of the *Network* class desired. The simulator has the pointer to the object created. The user might now send, through the graphic environment, the appropiate information to this object about the desired prototype[5]. Menus provide a large ensemble of functions to store, load, modify and run simulations.

The network architecture, along with the initial weight values and patterns was incorporated into a file structure that was read by a simulation program to construct the network and supply computational parameters, recall and learning algorithms that users choose from a menu.

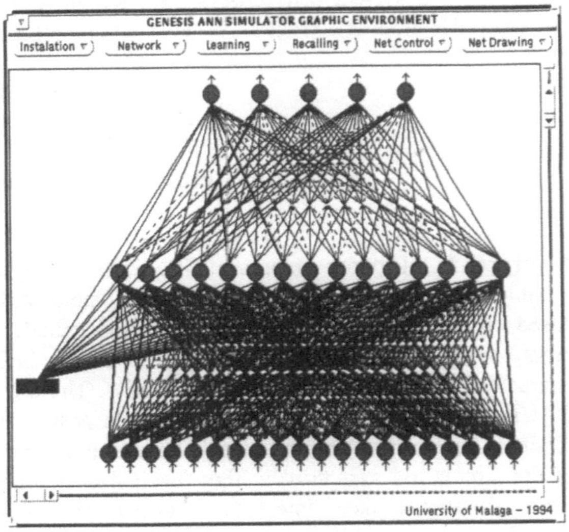

Fig.2 Example Screen of GENESIS.

Several common network algorithms along with several processing element activation functions, ports and links synaptic functions are provided. These algorithms and functions can be modified to create completely new networks. The entire object-oriented facilities are available to the user. An example screen is shown in figure 2. This figure illustrates the graphic nature of the program. Using the menu, the researcher can develop network models with new activation functions quickly and easily. Also, the user may modify code that controls the simulation environment.

4. MODIFICATION OF NEURAL NETWORK MODELS.

The user may alter or extend existing ANN models by writing new code and by menu manipulations. The creation of a new ANN model is performed by inheriting code from an existing model. The user simply selects a ANN model from a menu of available models and provides a new name for the novel model to be created. The system modification may be accomplished by redefining or extending a *Network, Layer* and/or *Learning Controller classes.* The system then creates a new class using code inherited from the selected class. This process involves creating new declaration, definition and the incorporation of the new class to the system. When created, a new ANN model is functionally identical to its ancestor. In order to develop a proper new model the user must modify methods that describe the ANN's, layer's or learning controller's behavior.

The user can also create new functions or parameters if needed to describe the model in a better way. The ANN model, layer or learning controller to be modified is selected by setting the current ANN model to the appropriate model. The user then uses an edit menu to modify existing functions and create new functions. For instance, a new activation function may be defined according to the users' requirements. For a new method this involves creating a C++ function [6] and allowing the user to define it implementation. The new function might keep the original definition. After the user has made any modifications, it has to execute a makefile. If errors occur, the process can be stopped and corrections made in the code. If the rebuild process is successful then the newly modified model can be tested and evaluated using the bar menu.

4.1. Implementation of the Hamming network.

In this section we explain how to build a new neural network from an existing one. We will extend the Competitive network and at last we will obtain a Hamming network. A competitive network has an input, an output and one hidden layer. The three layers are connected in a feedforward manner to allow the input, proccesing and the output of patterns. Also the neurons in the hidden layer have lateral connections to compete in a winner-take-all way. This competition is implemented in a competitive layer by a method call *Oscillate.*

The Hamming network operates by calculating a Hamming distance between the input pattern and each of the exemplar patterns. The input pattern is transformed to the output pattern that has the closest Hamming distance to it. The learning phase implies the initialization of the feedforward links with fixed weights. During the recall phase the neurons compete through lateral inhibit connections. Neurons belonging to the category (hidden) layer are connected through fixed weighted links. There are some differences between a generic competitive network and the Hamming network. The Hamming network has two layers, the weights for the neuron interconnections are fixed and have other characteristics that are incorporated to the system in a new class derived from the *Competitive Network* class.

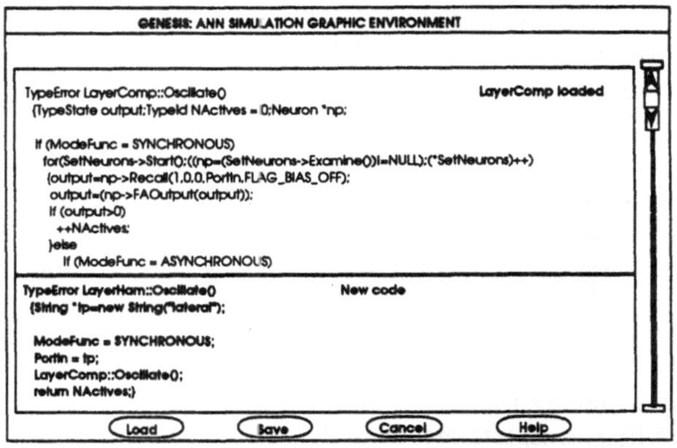

Fig.3 Example screen of the file editor window. New code for Hamming neural network model.

The Hamming network was implemented using the GENESIS graphical environment. It has been stated that it is a particular competitive network, therefore, the necessary classes are built deriving from the *Competitive* classes explained above. Furthermore, this model has particular characteristics that we have to add. By using the graphical user interface we were able to incorporate these specific characteristics into a new class *Hamming Network*. Also we define a new *Hamming Layer* class and the corresponding *Learning Controller* class. Figure 3 shows how the user redefines the method *Oscillate* in the new class *LayerHam*. The new function *Oscillate* initialize the lateral port with inhibits connections, the operate mode as sychronous and then call the competitive *Oscillate* function.

5. NEURAL NETWORK MACRO-STRUCTURES.

The partition of a problem is some times useful and necessary to achieve the solution. To carry out complex tasks, more than one network is needed. In GENESIS, a neural network model is built from connected objects. This approach allows the construction of multi-networks which have independent processing and interact across links. These are the neural macro-structures[7]. The networks are arranged in macro-layers with the outer layers corresponding to input and output, and they communicate with each other by passing real-valued activation across the inter-network links. We can arrange neural networks with different topologies. These networks could cooperate in the resolution of the problem. One network will usually only do one task or make one type of classification[7]. Often, complex problems require multiple stages of processing. This also requires the definition of multiple networks.

One of the most common fields that applies this kind of neural system is the image recognition[8]. One common approach is to create a hierarchy of networks of a similar type (fig.4.).

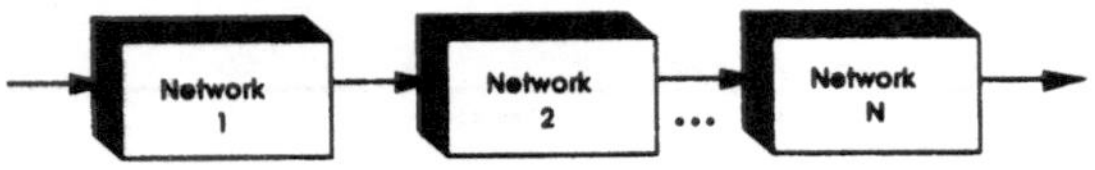

Fig.4 Neural macro-structure. Hierarchical type.

Another approach is to partition a problem so that different aspects are worked upon in parallel. This is useful when several different analyses must be performed on the same incoming data. Also the user may be interested in building a multiple parallel network to match input data against several different models (fig.5.).Yet another approach to the designing system of neural networks is to build a structure involving different types of networks (fig.6.).

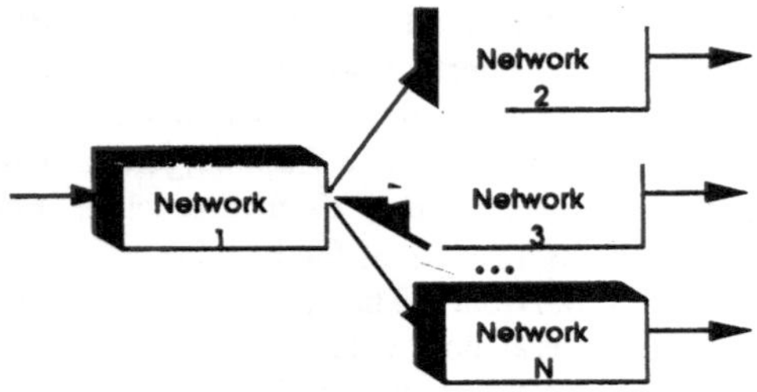

Fig.5 Neural macro-structure. Parallel type.

For example we build a two-layer structure composed of a Hopfield network (nethop) and a backpropagation (netbp). The autoassociative network cleans the input patterns, and the backpropagation classifies the M output patterns in N sets.

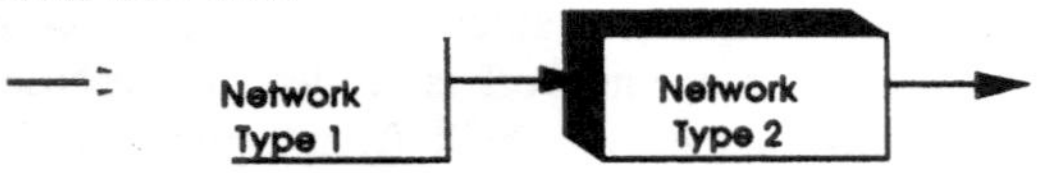

Fig.6 Neural macro-structure. Heterogeneous type.

Finally, hybrid networks can be built using multiple inheritance or by means of the specification of a macro-structure. We have already talked about one hybrid network: the Hamming network that GENESIS implements by the use of multiple inheritance.

6. CONCLUSION.

A general environment for simulation of ANN models has been presented that take advantage of object oriented programming techniques. The basic components of the simulation system are generic base classes that capture the functionality of general neural network models. The information is distributed among objects that cooperate in neural processing. Specific neural network models are created through the inheritance of these classes. The object-oriented design of the system is suitable for joining networks by weighted links of communication. This allows an easy and fast implementation of macro-structures whose elements cooperate in the resolution of a partitioned problem.

Examples were given of the use of the system, including the user modification of the code for a selected model. The final software product is implemented using Open Windows libraries and the object-oriented language C++ at a SUN4 workstation.

7. REFERENCES.

1. Gael de la Croix, Catherine Molinoux, Benol Derot. "The N programming Language", NATO ASI Series, Vol. F68 Neurocomputing. pp. 89-92, Springer Verlag Berlin Heidelberg 1990.
2. L. Fuentes, J.F. Aldana, J.M. Troya. "URANO: An Object-oriented Artificial Neural Network Simulation Tool". IWANN'93. LNCS no. 686, pp. 364-369. Barcelona:Springer Verlag 1993.
3. Gregory L. Heilemen, Michael Georgiopoulos, y William D. Roome. "A General Framework for Concurrent Simulation of Neural Network Models". IEEE Transactions on Sotfware Engineering 1992. Vol 18. No. 7, pp. 551-562.
4. L. Fuentes, J.F. Aldana, J.M. Troya. "Modelado Redes Neuronales mediante un lenguaje lógico concurrente orientado a objeto" Technical Report.. Dpto de Lenguajes y Ciencias de la Computación. Universidad de Málaga. 1992.
5. Darkui Shouren Hu. "An Object Oriented Neural Network Language", CH 3065 IEEE pp.1606-1611, 1991.
6. Scott Robert Ladd. "C++ Techniques & Applications", Prentice Hall.
7. A. Maren, C. Harston, R. Pap. "Handbook of Neural Computing Applications". Academic Press, Inc. 1990.
8. L. Fuentes, J.F. Aldana, J.M. Troya. "Modelado de Redes Neuronales con URANO". Technical Report.. Dpto. de Lenguajes y Ciencias de la Computación. Universidad de Málaga. 1992.

NNDT – A NEURAL NETWORK DEVELOPMENT TOOL

Björn Saxén and Henrik Saxén

Heat Engineering Laboratory, Department of Chemical Engineering
Åbo Akademi University, Biskopsgatan 8, FIN 20500 Åbo, Finland
Phone: 358-21-2654 311, Fax: 358-21-2654 792, E-mail: bjorn/henrik.saxen@abo.fi

Abstract

A tool for analysis, modelling, simulation and prediction with feedforward and recurrent neural networks is presented. The *Neural Network Development Tool* (*NNDT*) is implemented in Visual Basic and C and runs under MS Windows on personal computers. Network training is carried out by the Levenberg-Marquardt method and the user interface facilitates interactive analysis and modification of parameters as well as illustration of results. The features offered by the tool are especially useful in the difficult process of training recurrent networks, but are also of value, e.g., in scanning the performance of feedforward networks for analysing if and when overfitting occurs.

1. Introduction

The use of neural network models normally involves specification and analysis of a large number of parameters as well as large data sets. The development of such models is facilitated by software with an interactive graphical user interface [1]. Moreover, the recent interest in modelling, simulation and analysis of data with nonlinear and temporal dependence [2,3] has resulted in a need for software implementing recurrent networks. This paper presents a tool for modelling feedforward and (discrete time) recurrent neural networks; network architectures and training are discussed and the user interface is illustrated with two test problems.

2. Network Models

In NNDT, standard feedforward multilayer perceptron (MLP) networks [4] and a class of recurrent networks are implemented. Up to three hidden layers are possible, and there are several alternatives for the node activation functions in the hidden and output layers.

The recurrent networks are obtained by modifying the feedforward architecture as follows: The output from some units in the output layer are fed back with unit time lag to context units (fictitious input nodes) in the input layer. Additional feedback nodes (fictitious output nodes) may be included in the output layer. Each fictitious input node has a trainable initial state which corresponds to the node's activation for the first pattern in the data set; several initial states can be estimated to utilise training data composed of separate periods. In contrast to feedforward networks, the recurrent ones do not necessarily require input signals, since they can implement autonomous systems (cf. Figure 1). The feedback nodes, and, especially, the fictitious output nodes, may here act as *internal states* of the model [5].

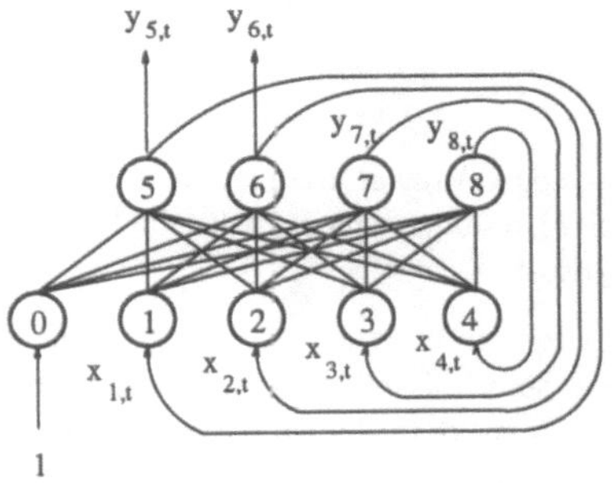

$x_{1,t} = y_{5,t-1}$

$x_{2,t} = y_{6,t-1}$

$x_{3,t} = y_{7,t-1}$

$x_{4,t} = y_{8,t-1}$

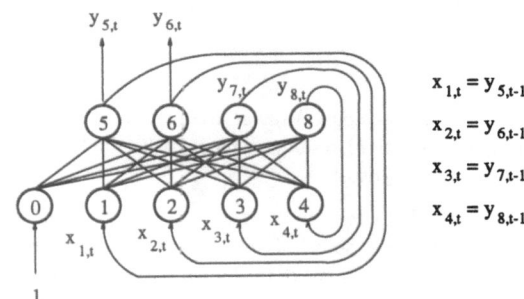

Fig. 1 A fully recurrent 4-node network.

3. Network Training

The Levenberg-Marquardt method [6] is used to adjust the unknown network parameters in order to minimise the sum of squares (SSQ) of residuals, calculated as the differences between network outputs and target outputs. This training method is significantly faster than standard backpropagation [1].

The optimisation problem can be formulated as

$$\min_{\mathbf{w}}\left\{E = \mathbf{r}^T\mathbf{r}\right\} \tag{1}$$

where $\mathbf{r}$ is the residual vector and the network unknowns, i.e. the weights and the initial states, are collected into a vector, $\mathbf{w}$. At each iteration, i, the optimisation method adjusts the unknowns according to

$$\mathbf{w}^{(i+1)} = \mathbf{w}^{(i)} + \boldsymbol{\delta}^{(i)} \tag{2}$$

where the correction, $\boldsymbol{\delta}^{(i)}$, is obtained from

$$\left(\mathbf{J}^{(i)}\mathbf{J}^{(i)T} + \lambda^{(i)}\mathbf{I}\right)\boldsymbol{\delta}^{(i)} = -\mathbf{J}^{(i)}\mathbf{r}^{(i)} \tag{3}$$

In the equation above, $\mathbf{J}$ is the Jacobian matrix with first derivatives of the residuals with respect to the unknowns and $\mathbf{I}$ is the identity matrix. The Marquardt parameter λ is automatically adjusted during the training [7]; as limiting cases, the method approaches Gauss-Newton if $\lambda \to 0$ and steepest descent with a small step length if $\lambda \to \infty$. Analytical expressions are derived for calculation of the Jacobian [8].

For certain problems, the training method may abort at a point far from optimum due to saturation of nodes caused e.g. by single large weight values. To overcome this difficulty, several methods for limiting the weight values are implemented. Box constraints for the weight values can be specified, and an upper limit for the relative weight change, $|\delta_j / w_j|$, can be given. In both methods, the step suggested by the search method is restricted if the change in any parameter would exceed the limits. Finally, large weight values may also be avoided by introducing a penalty term as an additional residual to be minimised in the training. The penalty residual is calculated as

$$r_{M+1} = \gamma \sum_{j=1}^{n} |w_j|, \tag{4}$$

where γ is a (non-negative) factor specified by the user, M is the original number of residuals and n is the total number of unknowns in the network.

There is a possibility to reduce the number of parameters in the network by keeping some of the network weights constant during training, or keeping the values of some weights equal. For a weight which is specified as constant, its initial value is maintained throughout training.

For recurrent networks with feedback connections from true output nodes, the convergence in training may be enhanced if the target values, instead of network outputs, are fed back occasionally. The user specifies the interval between such *teacher forcing* actions [9].

4. Additional Features

The data in the pattern file can be filtered by feeding it through a sliding window where arithmetic mean values are calculated. The size of the training set can be reduced by specifying that only every *n*:th (filtered) observation be used for training. Also, the user can specify the number of lines to be ignored in the beginning of the pattern file and give a limit for the number of patterns to be used in training. In addition to the pattern file with training data, the user can specify a separate test file. The test file option enables evaluation of the generalisation properties of the network throughout training by means of observations which are not included in the training data. The test data is fed through the network after each iteration and the root mean square (rms) error, $\sqrt{E/M}$, where M is the number of residuals, for the test patterns is presented.

5. User Interface

The user interface, developed with MS Visual Basic 3.0, was built to be both flexible and easy to use. All parameters defining the network and its training are presented in setup windows and stored in a file which can subsequently be read by the program. On-line help is available for all windows in the program; the help page is shown by pressing F1. The training can be interrupted at any time retaining all network information. The result from a training or an evaluation, i.e. input signals, node activations and target values for all patterns, can be saved in a file for later use. Although the user interface is mainly developed for interactive use, non-interactive runs may be carried out by starting the program with optional command line arguments. Also, several instances of the program may run simultaneously.

During training, the network and its performance can be analysed in several ways. Network outputs and desired outputs are presented in a graph which is updated after each iteration. Alternatively, the residuals (i.e. differences between network outputs and desired outputs), the weights or the activations of internal network nodes can be plotted. The sum of squared errors (SSQ), the rms error and the Marquardt parameter (λ) are shown after each iteration and written to a log table, which can be analysed later. Optionally, the contents of the log table may also be saved in a file, especially suitable for non-interactive runs. In a window showing the training progress, the rms error is plotted vs. iteration index. If a test file is used, the rms error for the test patterns is also plotted. The graphs can be copied to the clipboard as Windows metafiles. Network weights, initial states for fictitious input nodes and node activations are easily examined in the network state window.

5.1 A Feed-Forward Network

This example illustrates the user interface and shows how node saturation is avoided by restricting the weight changes during training. The training data consists of 100 patterns describing

$$f(x) = \sin(10x) + \sqrt{x}, \quad x = 0 \ldots 1 \tag{5}$$

We use a network with three hidden nodes in one layer; the standard sigmoid is chosen as activation function for the hidden nodes whereas the output node has a linear activation function. The main window of the program, which holds a picture of the network specified, is shown in Figure 2.

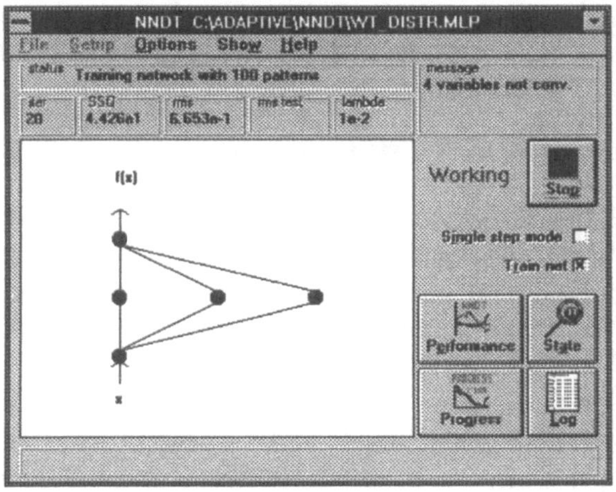

Fig. 2 The main window in NNDT

For the problem studied, the algorithm fails to find an acceptable solution with most initial weight guesses. This happens, e.g., when the network parameters are initialised in the range [-0.1, 0.1] using 50 as seed for the random number generator. After 100 iterations, the rms error is 0.67 and a plot of the performance shows that only the first few patterns are fitted by the network, see Figure 3.

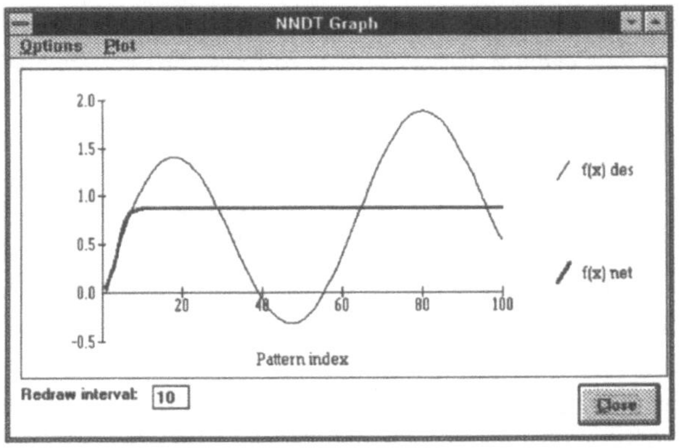

Fig. 3 Target values and network outputs after 100 iterations without limitation of the relative weight changes.

An analysis of the network shows that some weight values are several orders of magnitudes higher than the other ones. This has lead to saturation, i.e. paralysis, of two hidden nodes throughout the training set. The problem can be avoided, e.g., by specifying a limit for the relative weight changes (see Chapter 3). With this limit set to 100%, a training started from the same initial weights leads to an rms error of 0.011 after 50 iterations. As can be seen in Figure 4, all the hidden nodes now operate in their active ranges. It may be noted that even if two of the hidden nodes have almost similar outputs at this solution, three hidden nodes are required to describe the relation (5), since the oscillating output is created by magnifying the small differences between the signals from nodes 2 and 3.

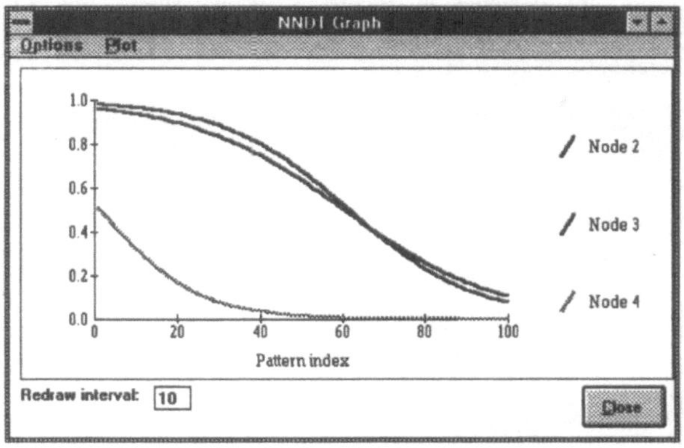

Fig. 4 The outputs from the hidden nodes after 50 iterations with a limit of 100% for the relative weight changes.

5.2 A Recurrent Network

Neural networks have been applied by several investigators to approximate and predict non-linear time series. It is easy to show that autoregressive processes can be modelled by feedforward neural networks, while for moving average or mixed (NARMA) processes recurrent networks are required [10]. An important problem in applying neural networks for approximation/prediction is how to avoid overfitting of the training patterns. This problem can partly be tackled by making an appropriate choice of network size [11]. Another approach, which is well-known in the field of statistical model building, is to select the complexity of the model based on its performance on an independent *test set*; a model of "optimal size" shows minimum test set error. An innovation from the field of neural networks is to scan the performance on an independent test set of the network model during training, and interrupt the search when the test set errors are at their minimum [12-14].

A simple non-linear time series is used to illustrate the features of NNDT. We study 2×300 observations of the series

$$x_{t+1} = 0.5 - 0.5x_t + 3x_t\varepsilon_t + \varepsilon_{t+1} \tag{6}$$

where $\varepsilon \sim N(0, 0.01)$ is white noise. The training set consists of 300 observations while the remaining 300 observations are used in the test set. Because of the bilinear (BL) term [15], the network must include a feedback connection (in order to reconstruct ε_t [10]) and a hidden layer (in order to "multiply").

Here we use a small recurrent network, with three hidden units and a feedback connection from the single output unit. Since saturation of hidden nodes is a common problem in training recurrent networks, we make use of the weight penalty (γ=0.05) during the first 15 iterations, and the rms error decreases to approximately the level of the best linear model (rms ≈ 0.15). After this, the training is continued with relaxed penalty (γ=0): The evolution of the training and test set rms errors are depicted in Figure 5, which shows that the performance on the test set deteriorates strongly after 6-7 iterations.

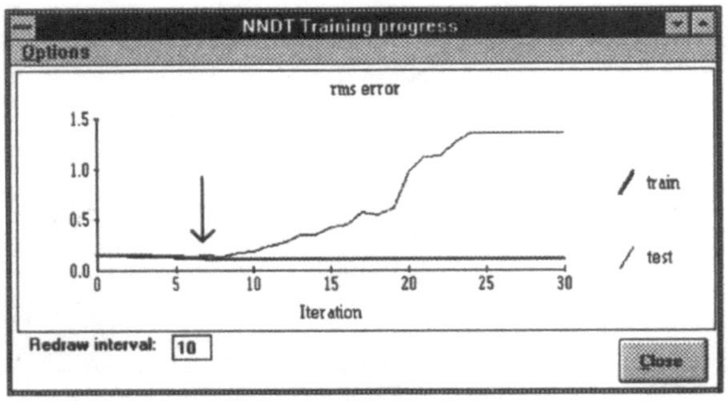

Fig. 5 Rms errors on training and test sets after relaxing the penalty.

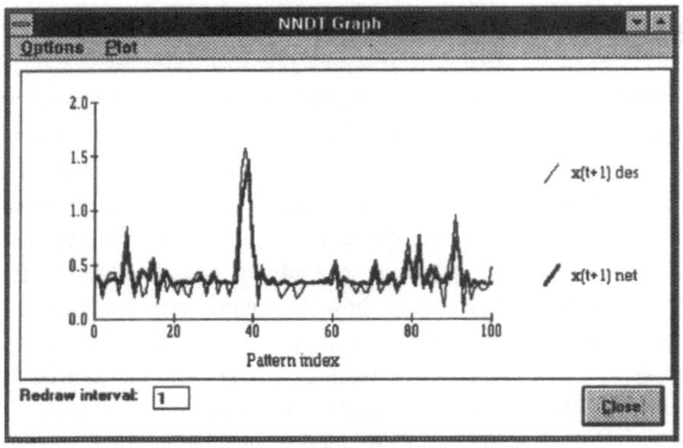

Fig. 6 The first 100 observations of the training set and the fit at the point indicated by the arrow in Fig. 5.

At this point the rms of the model is close to $\sigma_\varepsilon = 0.1$. (Figure 6 shows the 100 first observations of the training set and the fit at this point.) The overfitting which follows may be attributed to the fact that the nodes of the network after this moment tend to adapt to apparent relations between x_t and x_{t+1} in the training set. This led to saturation of some of the nodes on several occasions in the test set.

6. Concluding Remarks

A neural network development tool, NNDT, has been described and illustrated by two examples. The program, which implements feedforward and recurrent networks trained by the Levenberg-Marquardt method, includes a user interface facilitating interactive analysis and modification of parameters as well as illustration of results. The current version of the program, which runs under MS Windows 3.1 on personal computers, can be copied from /pub/pc/win3/demo/nndt110.zip by anonymous ftp connection to ftp.cica.indiana.edu.

References

[1] Demuth, H. and M. Beale, *Neural Network ToolBox for use with MATLAB*, The MathWorks Inc., 1993.

[2] Tsung, F-S., *Modeling Dynamical Systems with Recurrent Networks*, Ph.D. Dissertation, University of California, San Diego 1994.

[3] *IEEE Transactions on Neural Networks* (Special Issue on Dynamic Recurrent Neural Networks) **5**, No. 4, 1994.

[4] Rumelhart, D. E. and J. McClelland (Eds.), *Parallel Distributed Processing*, MIT Press, 1986.

[5] Bulsari, A. B. and H. Saxén, *Neurocomputing* **7**, 29-40, 1995.

[6] Press, W. H., B. P. Flannery, S. A. Teukolsky, and W. T. Vetterling, *Numerical Recipes*, Cambridge University Press, Cambridge 1986.

[7] Marquardt, D. W., *J. SIAM* **11**, 431-441, 1963.

[8] Williams, R. J. and D. Zipser, *Neural Computation* **1**, 270-280, 1989.

[9] Jordan, M. I., Proceedings of the *Eight Annual Conference of the Cognitive Science Society*, Amherst, 531-546, Hillsdale: Erlbaum 1986.

[10] Bulsari, A. B. and H. Saxén, Proceedings of *International Conference on Neural Networks and Genetic Algorithms (ICANNGA'93)*, Innsbruck, Austria, (Eds. R. F. Albrecht, et al.), 285-291, Springer-Verlag, Wien 1993.

[11] Bulsari, A. and H. Saxén, Proceedings of the *International Joint Conference on Neural Networks (IJCNN'93-Nagoya)*, Nagoya, Japan, October 1993, Vol. 1, 995-998.

[12] Weigend, A. S., B. A. Huberman and D. E. Rumelhart, *International Journal of Neural Systems* **1**, 193-209, 1990.

[13] Finnoff, W., F. Hegert and H. G. Zimmermann, *Neural Networks* **6**, 771-783, 1993.

[14] Ljung, L. and J. Sjöberg, *Neural Networks for Signal Processing II* (eds. S. Y. Kuhn, et al.), 423-435, IEEE 1992.

[15] Tong, H, *Non-linear Time Series*, Oxford University Press, Oxford, 1990.

GENETIC SYNTHESIS OF TASK-ORIENTED NEURAL NETWORKS

Andrej Dobnikar
Faculty of Electrical Engineering and Computer Science
University of Ljubljana
Trzaska 25, 61000 Ljubljana, SLOVENIA

Abstract
A stochastic search technique based on genetic algorithms for design of task-oriented neural networks is described in the paper. Although the theory of the algorithms is clear, its implementation in the design of neural structures is not yet well investigated. With the help of two case studies, we want to outline the new design approach.

I. INTRODUCTION

The problem of optimising neural network structure for a given application is complicated. There are many variables, both discrete and continuous, and they interact in a complex manner. The evolution of a given design is a noisy affair, since the efficacy of training depends on initial conditions that are typically random. In short, the problem is a logical application for a genetic algorithm.

There have already been some attempts to apply a genetic algorithm to the problem of neural-network design (Montana et al., 1989, Whitley et al., 1989, Harp et al., 1991, Cliff et al., 1992). Our approach is similar to that of Cliff et al., 1992 , although we made some important modifications. We also tested our approach on a time-series prediction problem.

II. PROBLEM DESCRIPTION

Our basic goal was to avoid some empirical choice of network structure and conventional learning procedure. Instead, we want to use some stochastic search processes based on genetic algorithms to approximate the optimal neural network structure for some predefined application.

Basically, there are three steps that form the big loop of the evolution procedure:

1. Random generation of the set (initial generation) of neural nets;
2. Performance evaluation (application-oriented) of every net in the generation;
3. Ordering of all nets in the generation (based on the results of step 2), selection of the "mates" for genetic operations and new generation formation. If no terminating criterion is met, the procedure repeats step 2.

For practical reasons (computing time), we added some constraints to the choice of parameters allowed to be modified in the evolution. This significantly decreased the potential number of different designs, and thereby the evolution time. We limited each neuron in the net so that they have exactly 5 excitatory and 5 inhibitory inputs which can be connected to the output of every neuron (including itself) or can be simply disconnected. This limitation is the result of our experimental work related to our problems, and therefore can not be treated as universal. Neurons which communicate with the world via input signals have an additional (sensor) input. All input signals are in the closed interval [0,1]. Output signals are calculated using the expression:

$$N = \tanh\left(\sum_{j=1}^{5} Ei_j\right) * H_{El}\left(\sum_{j=1}^{5} (sgn(Ei_j) * Ew_j)\right) * \left(1 - H_{Il}\left(\sum_{j=1}^{5} Ii_j\right)\right) \qquad 1)$$

where:

- Ei_j is the value of the j-th excitatory input;
- Ii_j is the value of the j-th inhibitory input;
- Ew_j is the weight of the j-th excitatory input;
- El is the excitatory threshold;
- Il is the inhibitory threshold;
- H_x is the Heaviside function with step at x.

The layout of a neuron, together with an explanation of Eq.1, can be found in Fig.1. If an external (sensor) input exists, it is treated like other excitatory inputs.

All the parameters from Eq.1 (Ew_j, El, Il) are included in the genetic plan of every neuron. Ew_j is initially set to 0.5 and can be changed inside the interval [0,1]. The weights of external (sensory) inputs are fixed at value 2 and don't change. Value El is initialised to 1 and can be changed within [0,2]. Il is first set to 1 and is allowed to be modified inside [0,3].

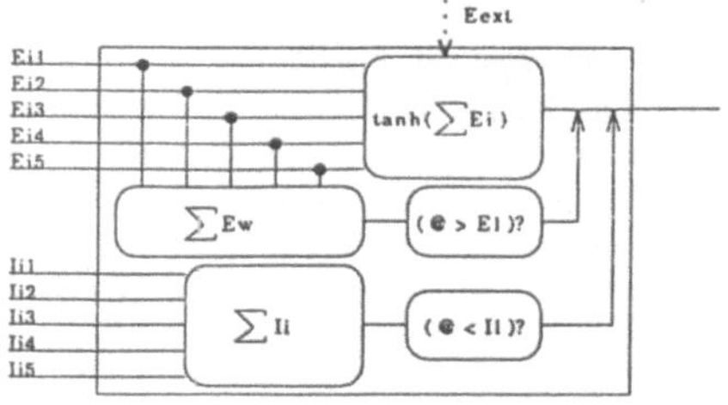

Fig.1 The layout of neuron (see also Eq.1)

The genetic plan of the neuron also includes the paths of all inputs (excitatory and inhibitory), which means that the origin of every input is defined. At the start, all the connections are set randomly. The list of all genetic plans of neurons in the net constitutes the chromosome or "blueprint", which describes the whole neural network. In the evolution phase, its contents change according to the genetic algorithms.

Within the big loop of the evolution procedure, only genetic operators are application-free. They will therefore be described first. Basically, there are three operators for reproduction: crossing-over, prolongation and mutation function.

a) Crossing-over: is perhaps the most important operator, because it significantly distinguishes the genetic approach from other stochastic search techniques. It needs two mates or their blueprints for reproduction. It operates in such a way that for two randomly defined neurons, one in each mate's chromosome, their contents from blueprints are exchanged.
b) Prolongation: extends each blueprint by one neuron with some small probability. Its parameters and paths are determined as with neurons of 0-generation.
c) Mutation: exposes the parameters of all neurons to stochastic changes inside the interval of their possible values. The probability (rate) of changes is below 0.01. The amount of change is also random, and shouldn't exceed 0.5 of the interval size. It decreases with the time of evolution.

The evolution procedure enables the formation of new generations of neural nets. From one generation we first randomly select the two mates from the upper half of the nets. Then we utilise the genetic operators: first mutation then prolongation and finally crossing-over. In this way, we get two offspring. The procedure is repeated until we reach the same number of nets as in the original generation. Before the procedure is repeated, we combine this set with the original generation by selecting the best nets from both sets according to performance evaluation so that the number of nets in the new generation is the same as before.

III. TEST RESULTS

We conducted two experiments in order to prove the efficiency of the proposed design technique. The first deals with a robot-planning problem. A special task is planned for the robot and its performance is evaluated. We were looking for the optimal neural net that would best accomplish the given task. With the introduction of the new evaluation function, we were able to achieve even better results than those published in (Cliff et al., 1992). In the second experiment, we tried to find a net that would be able to predict the future value of the randomly chosen analogue function from some large set of likelihood functions, based on its last few values.

2.1 Robot planning

We want first to design the robot for some special task, implemented by the neural network. We consider the arena where the robot is placed randomly. Its task is to move towards the centre of the arena; when it gets there, it must stay there. There are four input sensors that help the robot to orient itself, and there are also two motors controlled by the robot that enable it to move inside the arena.

Given the constraints of neural nets that we introduced above, all input and output data are normalised. We therefore suggest an arena with a radius of unity, with black walls and white ceiling and floor. The robot is half the height of the arena. There are four sensors, two eyes, one sensor of the smallest distance to the wall (radar) and one sensor for touch. The two parameters that define the characteristics of the eyes are modified in the evolution procedure. These are the visual angle (ZK) and the distance angle (KO), shown in Fig.2.

Visual data represents the percentage of black space in the scene. For example, if black covers all visual angles, visual data is 1. The robot sensor "radar" measures the smallest distance to the wall. The corresponding input data is the ratio of the smallest distance and the radius of the arena. The touch sensor is Boolean: it gives 1 if the robot is pushing the wall, and 0 otherwise.

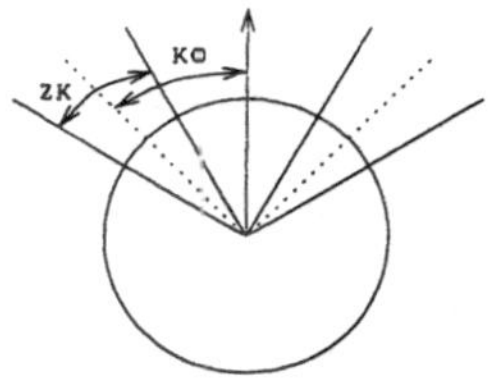

Fig.2. The layout of the visual sensors of the robot

There are also two motors, controlled by the outputs of the two neurons. The first enables forward and backward motion regarding the orientation of the robot. Maximum forward speed is twice as high as maximum backward speed. The latter turns the robot around its axis in left or right direction.

We started evolution with 0-generation, composed of 100 neural nets, each with 10 neurons: 6 external (4 inputs and 2 outputs) and 4 hidden. All initial connections were randomly selected.

Each net was tested four times, and we took the worst result as the score. In this way, we wanted to avoid some stochastic coincidences. For the testing time, we chose the travelling time of the robot along the diameter of the arena at full speed. We also multiplied the output values of the robot during the testing time by some random value from the interval [0.8;1.2] in order to include noise. The two mates for the genetic operations were always chosen from the upper (better) half of the generation.

We made two running tests. In the first, we estimated the behaviour of the robot using the function:

$$=\sum_{t}\{\exp(-r(t)^2)*(1-H_{0.7}(r(t)) \, , \qquad 2)$$

where r(t) is the distance of the robot from the centre of the arena at time t. Fig.3 shows the simulated behaviour of the best robot from the 180-th generation.. The number of neurons in this net increased to 22.

In the second run, we estimated the behaviour using the function:

$$=\sum_t \{(1-2*r(t))-C_1*v(t)-C_2*w(t)\}, \qquad 3)$$

where:

v(t) means absolute value of the speed at time t;
w(t) is absolute value of the angular speed at time t;
C_1, C_2 are positive constants.

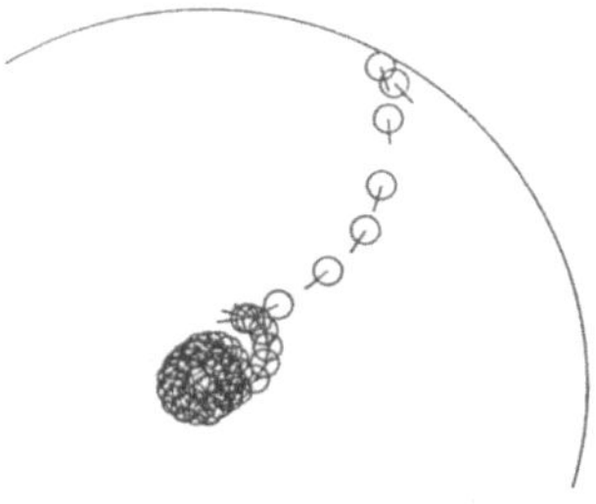

Fig.3 The behaviour of the robot from the 180-th generation of the first run

This time, the estimation was more rigorous. For $r > 0.5$, the robot got a negative score. The same was also true for moving and rotating. We wanted to punish the redundant waste of energy. After 160 generations, the best robot finally matched our expectations. Fig.4 shows the behaviour. The number of neurons this time increased only to 13.

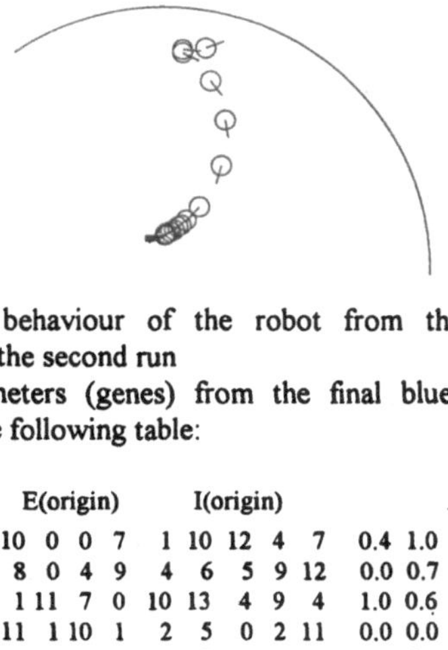

Fig.4 The behaviour of the robot from the 160-th generation of the second run

The parameters (genes) from the final blueprint are outlined in the following table:

n	El	Il	E(origin)					I(origin)					Ew_j				
1	0.1	2.7	8	10	0	0	7	1	10	12	4	7	0.4	1.0	1.0	1.0	1.0
2	1.4	2.2	9	8	0	4	9	4	6	5	9	12	0.0	0.7	0.0	0.9	0.0
3	0.3	2.2	10	1	11	7	0	10	13	4	9	4	1.0	0.6	1.0	0.0	1.0
4	1.0	0.9	7	11	1	10	1	2	5	0	2	11	0.0	0.0	0.0	0.8	0.8
5	0.1	1.0	13	9	6	4	0	1	8	6	4	2	0.5	0.6	0.7	0.5	0.6
6	0.1	1.1	2	1	9	10	4	11	8	8	4	0	0.6	1.0	0.8	0.5	0.0
7	1.0	1.0	2	3	10	9	5	3	2	9	0	1	0.8	0.9	0.9	0.8	0.0
8	1.3	3.0	7	10	10	6	7	10	1	11	8	11	1.0	0.5	0.7	0.5	0.6
9	1.6	0.7	10	5	4	6	3	0	9	3	10	8	1.0	0.4	0.8	0.9	1.0
10	2.0	2.0	4	4	3	9	6	12	7	8	5	2	0.6	0.4	0.1	0.3	1.0
11	1.3	2.9	1	7	10	11	4	4	8	3	11	12	1.0	0.1	0.0	0.3	0.0
12	0.5	1.7	11	0	7	5	7	2	8	10	7	11	1.0	0.5	0.7	0.6	0.6
13	0.7	0.6	0	10	0	0	10	0	0	0	0	13	0.5	0.9	0.5	0.6	0.7

Table1: The final structure of the net from the 160-th generation

The visual angle and distance angle of the same blueprint were set to 0.5468 and 0.2180 radians. In this table, the neurons with input sensors are numbered 1-4, and neurons with external outputs with 5-6. The remaining neurons belong to hidden layers.

2.1 Predicting time-series

The problem here was to design a neural network that is able to predict the future value of some smooth curve based on the last five function values. There were no major limitations in the selection of the curves, as long as they are sufficiently smooth and their values lay within the interval [0,1].

For this experiment, we made some small modifications to the constraints of the parameters for the evolution. We deleted the thresholds El and Il, since they might enter some singularities. We also allowed all weights (of excitatory and inhibitory inputs) to be modified within the interval [0,2], with initial values set to 1.

We tested the nets with the help of the function:

$$f(t) = \sin\left|\bar{a} + \bar{b} * t\right|, \qquad 4)$$

where $\bar{a}, \bar{b}$ are two-dimensional vectors, chosen randomly for each test. For each net, we conducted five test runs and selected the worst. The trajectory chosen was the result of the intersection of the vertical plane $\bar{a} + \bar{b} * t$ with the function :

$$z(x,y) = (1 + \sin\sqrt{(x^2 + y^2)})/2 \qquad 5)$$

In this way, a great variety of different trajectories can be generated.

In order to speed up the evolution, we first generated 1500 nets for the numbers of neurons between 6 and 20. We evaluated them in such a way that 500 of the best were chosen for the 0-generation. Then we made 50000 mutations in the following way:

- we randomly selected a net from the 0-generation and made a copy of it;
- the copy was mutated, with the rate and size dependent on the fitness (score) of the original;
- the resulting net was tested and if good enough, it was included on the list of current generation, while the worst net was deleted from the list.

After this procedure, we obtained a list of nets all with the same number of neurons - 9. This was somehow a surprising result, because we didn't stimulate this number in any way. One possible explanation is that this number, given the constraints, offers the best structure for the required task. Then the real evolution with mutations and crossing-over operations (another 50000 steps) followed. The results were further improved.. We tried this example with a standard big loop as well, but convergence proved much slower then with the above modifications. Unfortunately, there is no straightforward way to combine genetic operators most effectively for some dedicated task. At the end of the above evolution, the net with the highest score had the following structure:

n	$E_j;Ew_j$					$I_j;Iw_j$				
1	2;0.6	9;0.0	0;1.2	0;0.5	5;1.1	2;0.1	0;1.3	8;0.3	3;1.2	2;0.5
2	4;0.0	0;0.8	0;0.4	8;0.0	4;0.3	4;0.6	9;0.7	1;0.7	9;0.6	1;0.7
3	9;0.0	5;0.5	0;1.0	6;0.5	0;0.4	6;0.4	0;0.3	0;0.0	6;0.0	4;0.5
4	7;0.9	4;0.0	0;0.5	9;0.4	2;0.0	0;0.2	8;0.9	3;0.4	6;0.7	3;0.0
5	5;0.2	3;0.6	0;0.1	0;0.7	5;0.4	4;0.5	3;0.9	1;0.3	0;0.5	3;0.7
6	4;0.5	4;0.9	5;0.4	5;1.1	2;0.3	9;0.3	0;1.0	3;0.3	1;0.1	7;0.0
7	5;0.6	0;0.8	0;0.2	0;0.0	0;1.0	1;0.0	0;0.0	2;0.4	0;0.7	4;0.0
8	0;0.0	0;0.2	6;0.2	0;1.0	7;0.0	9;0.0	4;0.3	9;0.3	3;0.0	4;0.6
9	5;0.0	7;1.0	8;0.6	0;1.9	9;0.0	5;1.3	0;0.4	3;0.1	7;0.6	0;0.4

Table 2: The best neural network for time-series prediction

In the above table of connections and weights, 0 means no connection. Again, the first five neurons have external inputs, while neuron number 6 has external output.

Fig.5 outlines the results of tracking the trajectories obtained in the same way as in the evolution phase. Fig.6 shows the results of tracking when the trajectory is defined by $z = x^4$ (only a part, within the interval [0,1], is considered) and when the trajectory is non-derivable. Finally, trajectories with smaller amplitudes than in the learning phase are tested. The results are given in Fig.7. In all the tests "o" means true value and "x" predicted.

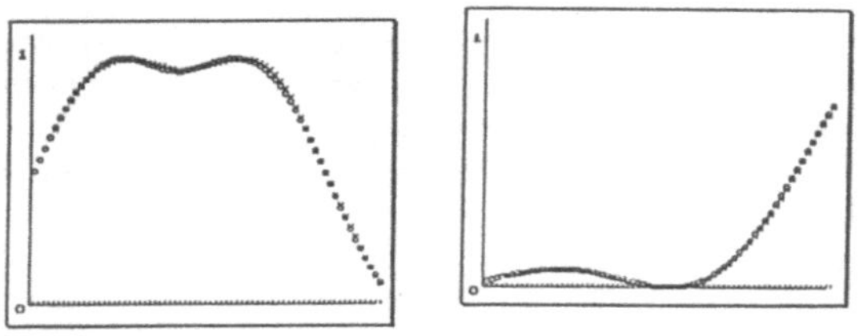

Fig. 5 Tracking with optional trajectories from the training set

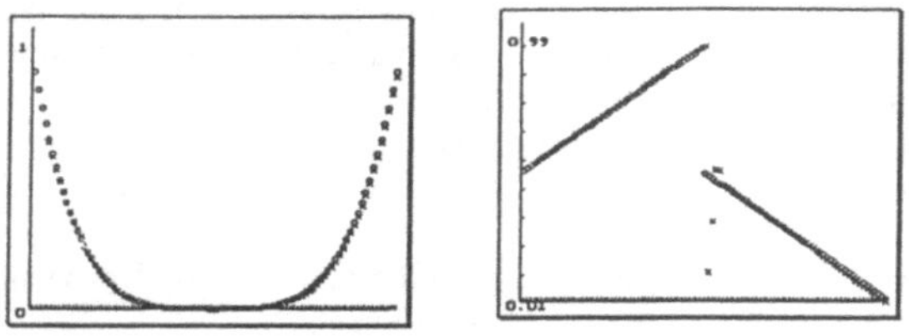

Fig.6 Tracking behaviour with x^4 and non-derivable trajectories

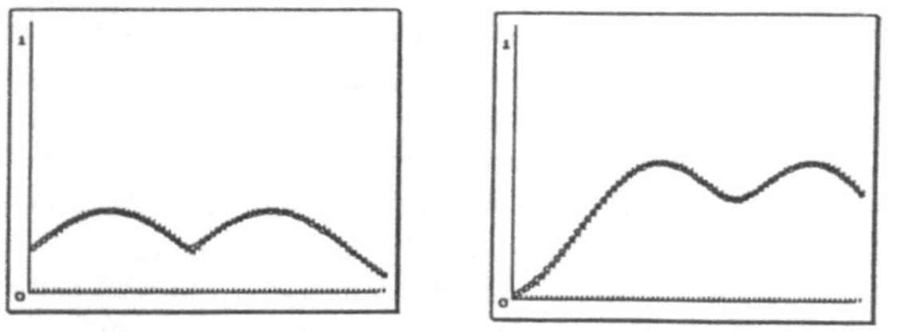

Fig.7 The result of tracking the damped trajectory regarding training set

With the help of a design based on genetic operators, we obviously succeeded in generating neural nets capable of tracking a very large set of likelihood functions. A strong generalisation feature is also present in the set. In all the above experiments, we had to multiply all function values by 0.91. The reason for this lies in the constraints requiring only five excitatory and five inhibitory inputs for each neuron.

V. CONCLUSION

A new approach to the design of application-oriented neural-network structures is discussed in this paper. It is based on well-known genetic algorithms. Although the theory of the algorithms is clear, its implementation in the design of neural structures is not yet well investigated. The experiments that we used in this paper are easy to understand, and yet they represent quite heavy problems for many classical design techniques. We hope that with their outlines we will create a better understanding of the new design approach.

REFERENCES

1. Montana,D.J., Davis,L.: Training feedforward neural networks using genetic algorithms. Proc. of the 11-th Int. Joint Conf. on AI, pp762-766, 1989

2. Whitley,D., Hanson,T.: Optimising neural networks using faster more accurate genetic search. In J.D.Schaffer (ed.), Proceedings of the Third Int. Conf. on Genetic Algorithms, 1989

3. Harp,S.A., Samad,T.: Genetic Synthesis of Neural Network Architecture. In L.Davis(ed.), Handbook of genetic algorithms, 1991

4. Cliff,D., Harvey,I., Husbands,P.: Visual Sensory-Motor Networks Without Design: Evolving Visually Guided Robots. Int workshop on Mechatronic Computer Systems for Perception and Action, Halmstad, Sweden, 1992

5. Dobnikar,A, Likar,A., Trebar,M.: Real time tracking with Stochastic Reinforcement Learning Algorithm, Int. workshop on Mechatronic Computer Systems for Perception and Action, Halmstad, Sweden, 1992

6. Tadel,M., Grabensek L., Dobnikar A. : Neural Networks without design - evolution with genetic algorithms and genotypes of variable length, int. report (in slovenian language), LASPP-FER, Ljubljana, 1993.

EVOLVING NEURAL NETWORKS USING THE "BALDWIN EFFECT"

Egbert J.W. Boers, Marko V. Borst and Ida G. Sprinkhuizen-Kuyper

Department of Computer Science, Leiden University,
Niels Bohrweg 1, 2333 CA Leiden, The Netherlands

Abstract—*This paper describes how through simple means a genetic search towards optimal neural network architectures can be improved, both in the convergence speed as in the quality of the final result. This result can be theoretically explained with the Baldwin effect, which is implemented here not just by the learning process of the network alone, but also by changing the network architecture as part of the learning procedure. This can be seen as a combination of two different techniques, both helping and improving on simple genetic search.*

Introduction

Recently, several papers appeared which describe the optimization of artificial neural networks using evolutionary computation, e.g. [13, 15]. There are many approaches to this mixture of biologically inspired methods. Various aspects of artificial neural networks can be optimized, and several varieties of evolutionary computation exist.

There are many ways to represent the different aspects of artificial neural networks in the 'genetic material' of the evolutionary computation algorithms. These ways range from 'blueprints' to codings which make use of a kind of 'recipe' describing only the generation of the network [3, 4]. This last method is more scalable towards real-life problems: one recipe can be used to generate the appropriate networks for a whole class of problems. Furthermore, very large architectures can be generated with just a small recipe, reducing the search space of the evolutionary computation algorithm.

A known problem with evolutionary computation is the *fine tuning* of parameters. When the weights of a fixed architecture are optimized with e.g. a genetic algorithm, this fine tuning problem can be solved by using the genetic algorithm to find good starting points for a training process.

This fine tuning problem also arises when optimizing artificial neural network *architectures*. What is needed here are algorithms that can optimize the architecture after the evolutionary computation has found an approximate solution. Several pruning methods exist, but no existing algorithm is able to *add* connections or nodes in an *existing* architecture.

This paper will introduce a new algorithm that, using heuristics, is able to determine the position in a modular network architecture where more nodes are needed, and in this way dynamically changes the network architecture as part of the training process.

This local search, applied at each fitness evaluation of the evolutionary computation, decreases the time needed to find an approximation. This is called the 'Baldwin effect', named after the one that first observed this in biology [1].

Evolutionary Computation

This section will outline the three mainstreams in simulated evolution used for optimization, see e.g. [8]. All three kinds of evolutionary computation are based on the same principle: they work on a population of individuals, each representing a possible solution to the problem to be optimized. Each member is awarded a fitness measure, corresponding with the quality of the proposed solution; selection for reproduction is either based on the (scaled) fitness or on the *rank* of the member in the population sorted on fitness. The main differences between the methods are: the way in which the individuals are represented in the population, the methods for reproduction and the ways in which new populations are generated.

Genetic algorithms
This method operates on binary strings, containing a coding of the parameters, each of which is called a *gene*. Usually a complete new generation is created by repeatedly selecting two parents and applying a cross-over operator to merge substrings of both parents into the new individual, which is placed in the new generation.

Evolutionary strategies
This method generally works with real numbers coding the parameters of the problem. The genetic operator that changes these parameters usually adds a standard Gaussian random variable to each parameter. New generations are created by combining the newly generated individuals with the old population, and selecting the best.

Evolutionary programming
This paradigm differs from the previous two in that the individuals do not consist of just parameters, but of actual functions, often written in LISP. These functions, coded as trees, are recombined by exchanging subtrees of two parents. The successive populations are generated in the same way as in evolutionary strategies.

Coding schemes for neural networks

The different coding schemes for representing artificial neural networks in the 'genes' of a population are strongly related to the type of evolutionary computation and the neural network training paradigm.

Blueprint representations
Here, a complete one-to-one relation exists between the network's weight and/or architecture and its genetic representation, e.g. [14].

Recipes
In this approach, not the complete network is coded, but just an algorithmic description of how to create the network. Boers et al. [3, 4] used L-systems, coded in the chromosomes of a genetic algorithm, to grow the architecture of feedforward networks. The fitness was calculated by looking at the generalization of the resulting networks after training.

Gruau [11] proposed a similar approach, using *cellular encoding*. His method uses the tree representation of genetic programming to store grammar trees, containing instructions which describe the architecture as well as the weights of the network. This has the consequence that when recursion is used, all weights conform to the same layout.

The philosophy behind these and other [16, 19] rewriting systems is the *scalability* of the process, which can not be achieved using blueprint methods.

On-line adaptation of architecture

An other way to find the 'correct' network architecture is to incorporate on-line architectural change in the learning algorithm. Most of the existing methods for on-line architecture adaptation can be classified into two categories:

- *constructive* algorithms, which add complexity to the network starting from a very simple architecture until the network is able to learn the task. When the remaining error is sufficiently low, this process is halted [7, 9, 17, 18].
- *destructive* algorithms, which start with large architectures and remove complexity, usually to improve generalization, by decreasing the number of free variables of the network. Nodes or edges are removed until the network is no longer able to perform its task. Then the last removal is undone [6, 20, 21].

Destructive algorithms leave us with the problem of finding an initial architecture. Existing constructive algorithms produce architectures that, with respect to their shape, are problem independent. Only the *size* of the produced architecture varies. Since the architecture of a network greatly affects its performance, this is a serious restriction.

Here, we propose a constructive method that is able to work on *modular* architectures [3, 4, 13], the initial architecture of which is found using evolutionary computation. This algorithm determines during training *where* to perform the adaptation [5]. We considered the following possibilities:

- adding nodes to existing modules,
- adding connections between modules and
- adding modules.

Adding a module, in most architectures, is not a 'small' adaptation; i.e. it influences the predefined modular structure in a major way. Adding a connection between previously unconnected modules has the same problem. We implemented the first possibility, because it is the simplest method in the sense that it only needs local information.

A module, when seen as feature detector, should be able to detect whether it is powerful enough for the number of features it is presented with. The method presented here looks at the weight changes of nodes in a module during training. When, after a number of training cycli these changes remain relatively large, a *computational deficiency* is detected. Our current approach is to add one node at a time to the module with the highest computational deficiency. This, of course, can be done in each module independently, but this requires an extra threshold and can lead to adding more nodes than is really needed.

Further, experiments were done with different ways to calculate the computational deficiency, e.g. looking at incoming and outgoing weights or looking at changes in the sign of the weights. This rarely led to differences in the path followed by the algorithm, indicating the robusteness of the method. An other important issue is the initialization of the weights of the added node, and the possible ways to treat the existing weights. Ideas taken from cascaded-correlation [7], and growing cell-structures [10] were tried, which increased the learning speed compared with random initializations [5].

To give an impression of the results of this relatively simple algorithm we show some experiments we did with the 'What/Where' problem [22], where 9 different 3x3 patterns are presented on all 9 possible positions in a 5x5 grid, giving a total of 81 different input/output patterns. The network has to learn *which* pattern is presented *where*. Strictly speaking, this problem can be learned *without* hidden layer. When however a hidden module is present, the problem becomes more difficult to solve (due to *interference* of the two tasks). Then, better results are obtained when the hidden module is divided among the two separate tasks giving a separate hidden module for each task. The original experiment was neurological

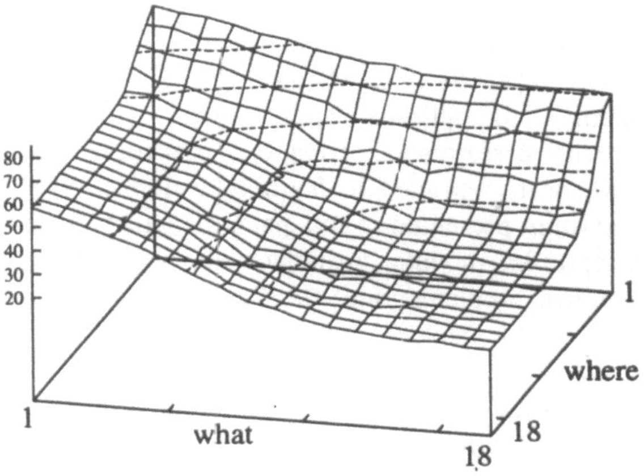

Figure 1: The remaining error plotted as function of the sizes of the two hidden modules.

in nature: it tried to explain why there are separate neurological pathways in the brain concerned with the detection of objects and the position of those objects. Fig. 1 gives the average sum-squared error of ten repetitions of training 300 cycli using backpropagation with momentum for several sizes of the two hidden modules. Table 1 gives the consecutive module sizes of our algorithm for the two subnetworks. It is easy to see that our algorithm follows the optimal path, and learns the task, which demonstrates that it is correctly determining the module where the next node should be added. More experiments are described in [5].

Table 1: Results of the what/where experiment.

Step	0	1	2	3	4	5	6	7	...	19
What	1	1	1	1	1	2	2	3	...	15
Where	1	2	3	4	5	5	6	6	...	6

Initiating the "Baldwin effect"

Baldwin was the first to recognize the impact of adaptive behavior of individuals on evolution [1]. He showed that 'Lamarckism' was not necessary to explain that *learned behaviour* seems to propagate through the genes of successive generations, but that instead, the inherited character was the *ability to learn*, with a profitable effect on the fitness of the individual.

The same effect can be used in evolutionary computation applied to neural networks. When learning is part of the fitness evaluation when searching for a good set of weights for a given architecture, a significant speed-up and final quality of solution can be achieved [2, 15]. Also when using evolutionary computation to optimize architectures, learning can increase performance [3,12], but sofar these attempts have been restricted to learning weights. With the algorithm presented in this paper, it will be possible to optimize modular artificial neural network architectures, 'implementing' the Baldwin effect not just by learning weights, but by adapting the modular structure itself as well.

However, as already observed by Whitley et al. [23], the Baldwin effect usually results in better solutions than the Lamarckian approach, but it also takes more time. Therefore, we are currently implementing this scheme on a parallel supercomputer (CM5), to try some real world problems.

Conclusions and further work

Looking at several recent studies in the area of neural network architecture optimization, our conclusion is that current computers are just beginning to be fast enough to test existing methods on really large problems. Recognizing the need for more scalable methods has resulted in several approaches using grammars. Moreover, effects known from biology that can increase the speed of artificial evolution are getting more and more attention. In order to cope with the increasing complexity of our world, fully understood methods with deterministic operations will in the future no longer have the needed strength. What might be called a *messy* approach will be better suited. Evolutionary computation will perhaps continue to be the only search strategy able to generate the desired complexity.

References

1. J.M. Baldwin; 'A new factor in evolution.' In: *American Naturalist*, **30**, 441–451, 1896.
2. R.K. Belew; 'When both individuals and populations search: adding simple learning to the genetic algorithm'. In: J.D. Schaffer (Ed.); *Proceedings of the third International Conference on Genetic Algorithms*, 34–41, Kaufmann, San Mateo, CA, 1989.
3. E.J.W. Boers and H. Kuiper; *Biological Metaphors and the Design of Modular Artificial Neural Networks*. MSc. Thesis, Leiden University, 1992.
4. E.J.W. Boers, H. Kuiper, B.L.M. Happel and I.G. Sprinkhuizen-Kuyper; 'Designing modular artificial neural networks'. In: H.A. Wijshoff; *Computing Science in The Netherlands: Proceedings (CSN'93)*, Ed.: H.A. Wijshoff, 87–96, Stichting Mathematisch Centrum, Amsterdam, 1993.
5. M.V. Borst; *Local Structure Optimization in Evolutionairy Generated Neural Network Architectures*. MSc. Thesis, Leiden University, 1994.
6. Y.L. Cun, J. Denker and S. Solla; 'Optimal brain damage'. In: *Advances in Neural Information Processing Systems*, **2**, 598–605, 1990.
7. S.E. Fahlman and C. Lebiere; 'The Cascaded-Correlation Learning Architecture'. In: *Advances in Neural Information Processing Systems*, **2**, 524-532, 1990.
8. D.B. Fogel; 'An introduction to simulated evolutionary optimization'. In: *IEEE Transactions on Neural Networks*, **5**, 3–14, 1994.
9. M. Fréan; 'The Upstart algorithm: a method for constructing and training feedforward neural networks'. In: *Neural Computations*, **2**, 198–209, 1990.
10. B. Fritzke; 'Growing cell structures - A self-organizing network for unsupervised and supervised Learning. TR-93-026, 1993.
11. F. Gruau; *Neural Network Synthesis Using Cellular Encoding and the Genetic Algorithm*. PhD. Thesis, l'Ecole Normale Supérieure de Lyon, 1994.
12. F. Gruau and D. Whitley; 'Adding learning to the cellular development of neural networks: evolution and the Baldwin effect'. In: *Evolutionary Computation*, **1**, 213–233, 1993.
13. B.L.M. Happel and J.M.J. Murre; 'Design and evolution of modular neural network architectures'. In: Neural Networks, **7**, 985–1004, 1994.
14. S.A. Harp, T. Samad and A. Guha; 'Towards the genetic synthesis of neural networks'. In: J.D. Schaffer (Ed.); *Proceedings of the third International Conference on Genetic Algorithms (ICGA)*, 360–369, Kaufmann, San Mateo, CA, 1989..
15. G.E. Hinton and S.J. Nowlan; 'How learning can guide evolution'. In: *Complex Systems*, **1**, 495–502, 1987
16. H. Kitano; 'Designing neural network using genetic algorithm with graph generation system'. *Complex Systems*, **4**, 461–476, 1990.
17. M. Marchand, M. Golea and P. Ruján; 'A convergence theorem for sequential learning in two-layer perceptrons'. In: *Europhysics Letters*, **11**, 487–492, 1990.
18. M. Mezard and J.-P. Nadal; 'Learning in feedforward layered networks: the Tiling algorithm'. In: *Journal of Physics A*, **22**, 2191–2204, 1989.
19. E. Mjolsness; 'Bayesian interference on visual grammars by neural nets that optimize'. Technical Report YALEU-DCS-TR-854, Yale University, 1990.
20. M. Mozer and P. Smolensky; 'Skeletonization: a technique for trimming the fat from a network via relevance assessment'. In: *Advances in Neural Information Processing Systems*, **1**, 107–115, 1989.
21. C.W. Omlin and C.L. Giles; *Pruning recurrent neural networks for improved generalization performance*. Revised Technical Report No. 93-6, Computer Science Department, Rensselaer Polytechnic Institute, Troy, N.Y., 1993.
22. J.G. Rueckl, K.R. Cave and S.M. Kosslyn; 'Why are "what" and "where" processed by separate cortical visual systems? A computational investigation'. In: *Journal of Cognitive Neuroscience*, **1**, 171–186, 1989.
23. D. Whitley, V.S. Gordon and K. Mathias; 'Lamarckian evolution, the Baldwin effect and function optimization'. In: Y Davidor, H.-P. Schwefel and R. Männer (Eds.); *Lecture Notes in Computer Science*, **866**, 6–15, Springer-Verlag, 1994.

MODIFICATION OF HOLLAND'S REPRODUCTIVE PLAN FOR DIPLOID POPULATIONS

V.B. Klepikov, K.V. Mahotilo and S.A. Sergeev

Kharkov State Polytechnical University, Kharkov, 310002, Ukraine

Abstract

A modification of Genetic Algorithm capable of dealing with a diploid population is described. Each individual of such a population should be represented as a double bitstring chromosome. Both a procedure of creation of the initial population using genomic mutation and a version of gene dominancy evolution are suggested. Results of numerical experiments proving high search ability of the algorithm are presented.

Introduction

Nearly 20 years have passed since the first book by J. Holland [1] was published having broadened the horizons of evolutionary modelling and initiated studies on development and applications of Genetic Algorithms (GA). Within this period several original interpretations and versions of GA were suggested which substantially improved its search ability. The majority of the publications, however, kept to Holland's reproductive plan ignoring its internal contradiction between the fashion of describing individuals and the list of the permitted genetic operators. In particular, according to the plan, each individual is represented as a single binary string, i.e. like a haploid organism. At the same time, crossover operator which is inherent only in a diploid organism is applied to a pair of the haploid individuals.

Only a few papers describe the attempts to approach modelling of the evolution of a truly diploid population. Specific attention should be paid to the research conducted by J. Born and K. Bellmann [2]. They implemented in their algorithm the conception of genetic load, the phenomenological explanation of the reasons for higher adaptability of a diploid population than that of a haploid one. A diploid population is known to possess a hidden reserve of variability which is contained in the recessive genes of its individuals. To enhance the similary of a modelled population to a diploid one the authors of [2] brought into population some haploid individuals which were stated unchangeable and which simulated the genetic load.

We have chosen another way and modified Holland's reproductive plan so that it can deal with individuals described as diploid organisms. While doing it we faced two problems. The first was the reasoned way of creation of the initial population. It was not clear enough how to map the phenotype of an individual, which we considered as a vector or real-value coordinates of a point in the search space, in a bitstring genotype. The other problem was the evolution of dominancy between the contrasting genes located in identical loci on homologous chromosomes.

Here we present the possible solutions to these problems.

Description of the algorithm

The initial diploid population is created in two stages. First, we form a temporaryhaploid population of randomly generated chromosomes. Then to make it diploid we perform a genomic mutation over all the individuals, thus doubling each individual's chromosome. No matter what dominancy relations are initially set. They do not influence an individual phenotype, because both its chromosomes are identical. To be more certain, we specify all the genes of the primary chromosome as dominant (active) and the genes of its copy - as recessive (passive).

Our reproductive plan comprises the following steps :

(i) random selection of mates;
(ii) extraction of a gamete from each parent;
(iii) gene mutation in the gametes;
(iv) merger of the gametes into zygote;
(v) reformation of the dominancy relations within zygote;
(vi) replacement of the least fit individuals by new offsprings;
(vii) ranging of the population.

The steps (i) and (vi)-(vii) are traditional. As for the others, they contain the following features.

While producing the gametes in step (ii) we perform both the ordinary one point crossover and polycrossover. Every time only one gamete of the two available is used. It is extremely important that the gamete inherits information not only about gene alleles (figures 0 or 1), but also about the state (active or passive) of these genes in the parental chromosomes.

Fig. 1 illustrates this procedure. The dark squares denote the dominant genes and the light squares - the recessive ones. Here P1 and P2 are parents from the initial diploid population, O1 and O2 - hybrids.

The probability of gene mutation within step (iii) is the same for all the genes of the gamete. Whether the gene is active or passive is not taken into account. Besides, mutation does not change the gene's state. The mutated genes are pointed out by black triangles.

In step (iv) the pair of the gametes merge into zygote. So, each couple produces one offspring. For further operation with it we have to reform the dominancy relations.

Solving this problem within step (v) we proceed from the assumption that the homologous genes of the zygote may not simultaneously be active or passive. That is why in situations similar to that in section B (Fig. 1) we switch the state of the gene, which belongs to the gamete of the less fit parent, to the opposite one. Due to this we strengthen influence of the more successful parent's genes on the offspring.

In step (vi) we decode the zygote into phenotype. Only dominant genes are used to compile a single bitstring, certain segments of which denote the coordinates of a vector of variables represented in binary code. We reckon the best choice is Gray's code, in which a change of only one bit in a segment results in a discretization step shift in the phenotype space.

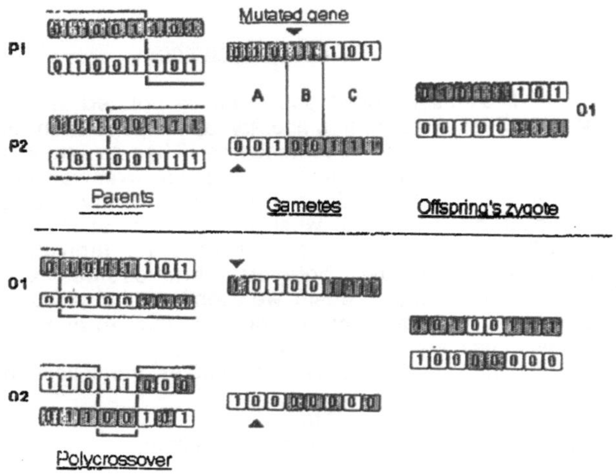

Fig. 1 Generation of the offspring in the first and one of the next iterations of the algorithm.

Results of numerical experiments

In order to probe the developed GA in more severe conditions and understand how diploid exerts an influence on its ability to avoid stagnation, we gave it trials with excluded mutation operator. We took the well-known test function f(x) numerously reffered to in papers on the topic, to evaluate the fitness of the individuals

$$f(x) = -\sum_{i=1}^{10} x_i \sin\sqrt{|x_i|}, \quad -500 \le x_i \le 500$$

Its global minimum is located in the point with coordinates

$$\bar{x}_i^0 \cong 420.968 \quad i = 1,\ldots,10$$

and is equal to

$$f(\bar{x}^0) = -4189.828$$

The results of computations performed over the population of 100 individuals and the distance from it to the global extremum.

The results confirmed our expectations that a diploid population has higher search ability than haploid one. It may be explained by slower exhaustion of a diploid population under the selective pressure. Though such a population has lower search rate, which is estimated as the average increase of an absolute value of the objective function through the genrations, it exploits the combination reserve of the gene pool more carefully and copes with the complex relief of the function much better than a haploid one.

Table 1 The comparaison of the different population seach efficiency

	Objective function/Distance to the optimum		
Number of function evaluations	Results from [2] (with mutation)	Haploid population	Diploid population
0	-1805/2910	-1133.9/ 1242.6	-133.9/1242
2000	- 4142.92/19, 3	- 3923.4/737. 9	- 3517.6/125 4
3000	-	- 3945.5/737. 8	-3876.1/719
5000	- 4189.83/0,0 28	- 3945.5/737. 8	- 4136.4/20.7

The only shortcoming of the developed algorithm is a comparatively large number of generations required to find the global extremum. We believe, however, this imperfection is not fatal in designing problems, taking into account the high search ability of the method.

Conclusion

Describing individuals by means of double chromosomes permits to utterly solve the problem of losing genetic diversity during evolution. Like a haploid population, a diploid one can destroy diversity in phenotypes but never in genotypes. Even lacking in the mutation component of the viariability a diploid population of moderate size proves to be capable of finding the global extremum.

Acknowledgement

We would like to thank Dr. V.F. Cheshko from Biology Department of Kharkov State University for the valuable discussions on the evolution of a diploid population.

References

1. Holland, JH : Adaptation in natural and artificial systems. Ann Arbor, MI : University of Michigan Press 1975

2. Born, J, Bellmann, K : Numerical adaptation of parameters in simulation models using evolution strategis / Molecular Genetic Information Systems : Modelling and Simulation (Ed K. Bellmann) : Berlin : Akademie-Verlag. 1983, pp 291-320.

SIMULATED ANNEALING ARTIFICIAL NEURAL NETWORKS

FOR THE SATISFIABILITY (SAT) PROBLEM

T.Tambouratzis
Institute of Informatics & Telecommunications, NRCPS "Demokritos", Aghia Paraskevi 153 10, Greece.

Address for correspondence: Department of Mathematics, Agricultural University of Athens, Iera Odos 75, Athens 118 55, Greece.

Abstract

A simulated annealing artificial neural network, which is based on the principles of Harmony Theory [1], is proposed for the solution of the satisfiability (SAT) problem of propositional calculus. The structure of the employed network is such that not only is a valid solution to any given SAT problem automatically provided, but - furthermore - this solution contains the values of the variables which render true the greatest possible number of clauses.

The number of clauses which are satisfied is automatically revealed by the quantitative measure of harmony [1] once the simulated annealing settling procedure has been completed. Additionally, both the identity of the satisfied clauses and the actual values of all variables are determined by the settled-upon activation values of the network nodes.

1. Definitions

In order to be able to understand and fully appreciate the satisfiability (SAT) problem as well as the rationale of the proposed parallel implementation and solution, a number of definitions and salient properties of propositional calculus are given (also see [2] for more details):

- Provided that a **logical expression** E of propositional calculus involves n **logical variables** $x_1, x_2, \ldots, x_n$ these can take the binary values {0,1} specifying the **"true"** or **"false"**, respectively, **state** of the variable.

- The logical variables of E can appear inside the expression as themselves or as their negations/complement forms ($x_1', x_2', \ldots, x_n'$), both of which are collectively known as the **literals** of E.

- A (sub)set of literals of E can be connected via:

  (a) **disjunctions** (V, OR, + relations) whereas **sum terms** are created,

  (b) **conjunctions** (Λ, AND, . relations) whereas **product terms** are created.

  Both sum and product terms are known as **clauses.**

- For the sake of uniformity, expression E must contain clauses of only one kind, i.e. either only sum terms, or only product terms.

  (a) If the expression contains sum terms $C_1, C_2, \ldots, C_m$ these are connected via conjunctions (Λ, AND, . relations). Then E is expressed as a product of sums, i.e. $E = C_1 \wedge C_2 \wedge \ldots \wedge C_m = C_1 C_2 \ldots C_m$ (the dots are omitted as in numeric multiplication) and E appears in **conjunctive form** (sometimes also called **conjunctive normal form**).

  (b) If the expression contains product terms $D_1, D_2, \ldots, D_l$ these are connected via disjunctions (V, OR, + relations). Then E is expressed as a sum of products, i.e. $E = D_1 \vee D_2 \vee \ldots \vee D_l = D_1 + D_2 + \ldots + D_l$ and E appears in **disjunctive form** (sometimes also called **disjunctive normal form**).

- (a) If a sum term C of the conjunctive form of E involves all the logical variables $x_1, x_2, \ldots, x_n$, C is called **standard sum** or **maxterm.** There are 2^n maxterms at most, combining the possible states of the n logical variables. If each sum term C_i of the conjunctive form of E is a maxterm, E constitutes a **product of maxterms** and appears to be in the **conjunctive canonical form**. Any expression E which appears in the conjunctive form can be transformed in the conjunctive canonical form.

(b) If a product term D of the disjunctive form of E involves all the logical variables $x_1, x_2, ..., x_n$, D is called **standard product** or **minterm**. There are 2^n minterms at most, combining the possible states of the n logical variables. If each product term D_j of the disjunctive form of E is a minterm, E constitutes a **sum of minterms** and appears to be in the **disjunctive canonical form**. Any expression E which appears in the disjunctive form can be transformed in the disjunctive canonical form.

Any expression which appears in the disjunctive canonical form can be transformed into the conjunctive canonical form and vice versa. By extension, any expression which appears in disjunctive (normal) form can be transformed into conjunctive (normal) form and vice versa, through repeated transformation into and from the corresponding canonical forms. In short, **the two canonical forms are equivalent**. **The two normal forms are** also **equivalent**.

2. Existing Solutions to the SAT Problem

The satisfiability (SAT) problem of an expression in propositional calculus can be expressed as follows:

Given an expression E involving n logical variables $x_1, x_2, ..., x_n$, the task is to determine whether there exists a truth assignment $\{0,1\}^n$ for all variables such that E can be satisfied.

Subsequently, the SAT problem aims at:

(i) **establishing whether there exist appropriate values** for the logical variables x_i which render expression E true and - if so -

(ii) **determining these values** for all variables.

The SAT problem has been characterised as an NP-complete problem. A number of solutions to the SAT problem have already been proposed, involving serial as well as parallel implementations [3-12]. The existing research has mainly concentrated on the solution of the SAT problem for expressions of the propositional calculus which appear in the conjunctive (normal) form, i.e. as conjunctions of sum terms. In this case, the SAT problem is solved by **finding the greatest set of clauses which can be made true simultaneously**. If this set coincides with the total set of clauses in the logical expression, satisfiability is established. Failing this, unsatisfiability is suggested.

3. Proposed Solution For The SAT Problem

In the present piece of research, a simulated annealing artificial neural network, which is based on Harmony Theory [1], is employed for the solution of the SAT problem. For this implementation, the SAT problem is solved for logical expressions of the propositional calculus which appear into disjunctive (normal) form, i.e. as disjunctions of product terms. Due to the last point of section 1 - the conjunctive (normal) form and the disjunctive (normal) form of a logical expression are equivalent and can be transformed from and into each other [2] -, this manner of representation of the logical expression does not put any limits on the types of expressions that can be tested for satisfiability.

Solutions to the SAT problem for an expression E which involves n logical variables $x_1, x_2, ..., x_n$ and which appears in the disjunctive (normal) form focuses on determining whether there exists a truth assignment $\{0,1\}^n$ for all variables such that E can be satisfied. This entails **finding at least one clause which can be satisfied** for appropriate values of the variables. If this is true, the satisfiability of the logical expression is established. Else, unsatisfiability is deduced.

Harmony Theory networks [1] comprise two layers of nodes with binary activation values and a sigmoid activation function. Network connections are symmetric and only possible between nodes of different layers. The weight of each connection depends upon the excitatory/inhibitory nature of the connection as well as upon the connectivity of the linked network nodes (see [1] for a more detailed explanation).

The **inbuilt quantitative measure of harmony** gives an evaluation of the internal consistency of the network state. This is calculated over the activation values of all network nodes and the weights of all connections. The **settling procedure** is based on the **simulated annealing** process of Harmony Theory. The procedure begins from a random initial state of the network and, subsequently, states with higher harmony acquire greater probabilities of being preferred over states with lower harmony. The difference between state-probabilities is magnified as simulated annealing progresses. As a result, at the completion of the settling procedure, the state of greatest internal consistency is settled upon with an infinitely greater probability than all the other network states. Simulated annealing is, thus, chiefly responsible for the progressive settling of the Harmony Theory network upon this "**mostly harmonious**" network state, which coincides with the most valid solution to the SAT problem.

4. Network Construction For The SAT Problem

Given an expression of propositional calculus - appearing in the disjunctive (normal) form - to be tested for satisfiability, the Harmony Theory simulated annealing artificial neural network is constructed as follows:

- Each **variable** is expressed in a **node of the lower layer**.

- Each **clause (product term)** is represented by a **node of the upper layer**.

- The **connectivity** between each node of the upper layer to the nodes of the lower layer acquires:

(a) a **positive** connection, if the clause concerned contains the corresponding variable as a literal,

(b) a **negative** connection, if the clause contains the negation of the variable as a literal,

(c) **no connection**, if the variable does not appear in the clause.

Simulated annealing is applied to the Harmony Theory network via an appropriately chosen settling procedure. The network gradually becomes configured into the **most correct decision** concerning the satisfiability of an expression of propositional calculus (i.e. the solution of maximum satisfiability), where:

- The **greatest number of clauses (product terms) that can be simultaneously satisfied** is output by the activated nodes of the **upper layer**.

- The **desired values of the variables** are produced by the activation values of the nodes of the **lower layer**.

- The **number of clauses** involved in the most valid solution of the SAT problem is directly provided by the **harmony of the end-state**.

If the harmony of the end-state equals 0, no appropriate values of the variables can be found to render any clause true. In this case the expression is characterised as unsatisfiable. It is in this manner that the proposed artificial neural network implementation of the SAT problem can determine satisfiability as well as unsatisfiability.

5. Experimental Results

The simulated annealing artificial neural network which is proposed for the solution of the SAT problem has been implemented and tested on a considerable number of logical expressions involving five or more logical variables (and hence ten or more literals) and up to fifty clauses. The satisfiability of these expressions has always been correctly determined, resulting in:

- The **correct and most valid solution** in the case of satisfiable logical expressions. This is encoded by the active nodes of the lower layer (selected values for the logical variables) and the upper layer (clauses that are satisfied). Furthermore, this solution involves the **greatest set of clauses which can be simultaneously satisfied** and, thus, **maximises the harmony** of the annealed Harmony Theory network.

The harmony value expresses the greatest possible number of simultaneously satisfied clauses.

- A **harmony value of zero** in the case of unsatisfiable logical expressions. No nodes of the upper layer become active, signifying the inability of the proposed parallel solution to satisfy any single clause of the expression.

6. References

1. Smolensky, P., (1986). *Information Processing In Dynamical Systems: Foundations Of Harmony Theory,* pp. 194-281, in D.E.Rumelhart, J.L.McClelland (eds), **"Parallel Distributed Processing: Foundations (Vol. 1)"**. MIT Press, Cambridge.
2. Mano, M., (1984). **"Digital Design"**. Prentice-Hall, Inc. Englewood Cliffs, New Jersey.
3. Davis, M., Putnam, H., (1960). *A Computing Procedure For Quantification Theory,* **Journal of the Association for Computing Machinery** 7, pp. 201-215.
4. Dowling, W.F., Gallier, J.H., (1984). *Linear-Time Algorithms For Testing The Satisfiability Of Propositional Horn Formulae,* **Journal of Logic Programming** 3, pp. 267-284.
5. Purdom, P.W., (1984). *Solving Satisfiability With Less Searching,* **IEEE Transactions on Pattern Analysis and Machine Intelligence** 6, pp. 510-513.
6. Monier, B., Speckenmeyer, E., (1985). *Solving Satisfiability In Less Than* 2^n *Steps,* **Discrete Applied Mathematics** 10, pp. 287-295.
7. Gallo, G., Urbani, G., (1989). *Algorithms For Testing The Satisfiability Of Propositional Formulae,* **Journal of Logic Programming** 7, pp. 45-61.
8. van Gelder, A., (1988). *A Satisfiability Tester For Nonclausal Propositional Calculus,* **Information and Computation** 44, pp. 1-21.
9. Hansen, P., Jaumard, B., (1990). *Algorithms For The Maximum Satisfiability Problem,* **Computing** 44, pp. 279-303.
10. Pinkas, G., (1990). *Energy Minimisation And The Satisfiability Of Propositional Logic,* in D.S.Touretzky, J.L.Elman, T.J.Sejnowski, G.E.Hinton (eds) Proceedings of the **"1990 Connectionist Summer School"**, San Mateo, California.
11. d'Anjou, A., Grana, M., Torrealdea, F.J., Hernandez, M.C. (1993). *Solving Satisfiability Via Boltzmann Machines,* **IEEE Transaction on Pattern Analysis and Machine Intelligence** 15, pp. 514-521.
12. Gu, J, (1993). *Local Search For Satisfiability (SAT) Problem,* **IEEE Transactions on Systems Man and Cybernetics** 23, pp. 1108-1129.

POLICY OPTIMIZATION BY NEURAL NETWORK AND ITS APPLICATION TO QUEUEING ALLOCATION PROBLEM

Hisashi Sato*, Yutaka Matsumoto** and Norio Okino*
* Division of Applied Systems Science
Faculty of Engineering, Kyoto University
** I.T.S., Inc., Japan

Abstract
The problem of allocating an arriving customer to one of parallel servers has been actively studied in queueing theory for load balancing in computer networks or in multi-processor systems. To theoretically derive the optimal allocation policy, the assumption of identical servers is usually required. However this assumption is unrealistic in many applications. This paper considers the queueing allocation problem with non-identical servers and multi-class customers. The goal is to optimize the allocation policy with respect to the mean delay of an arbitrary customer. To this end, we represent the allocation policy by a neural network; namely, we allocate an arriving customer according to the output of the neural network, where the inputs to the neural-net are the numbers of queueing customers at each server and the class of the arrival. By using the simulated annealing method, we search the optimal allocation policy in the weight space. Numerical results show that the present procedure significantly reduces the mean delay in comparison to an empirical policy.

1 Introduction

The queueing allocation problem treats how to assign an arriving customer to one of parallel servers upon arrival. Depending on the availability of information, allocation policies can be classified into two types: *static* or *dynamic*. Policies are called static if customers are allocated in a predetermined manner (e.g., in round robin) or dynamic if the system state (e.g., length of each queue) is observed and this information is effectively utilized for allocation.

The goal of the queueing allocation problem is to derive such a policy that optimizes the allocation in terms of a certain objective function. The objective function usually considered can be cost, throughput or mean delay of customers. Allocation policies that optimize such an overall performance measure are called *socially optimal*. On the other hand, policies, according to which each individual customer benefits most, are called *individually optimal*. For example, the SSQ (Send-to-the Shortest Queue) policy that allocates an arriving customer to the server with the shortest queue is obviously individually optimal in the sense of minimizing delay if all servers have the same service speed. The difficulty of the queueing allocation problem is in that individual optimality does not always imply social optimality as it is often the case that individuals have to accept sacrifices for the benefit of society at large [1].

The queueing allocation problem arises in many practical situations such as routing in computer communication networks, load balancing in multi-processor systems and job scheduling in manufacturing systems. For importance in applications, its theoretical analysis based on Markov decision theory has been extensively studied [2]. Provided that 1) all servers are stochastically identical, 2) the service time is i.i.d. (independent and identically distributed) exponential, and 3) the arrival process is Poisson, Winston proved that the SSQ policy maximizes, with respect to stochastic order, the discounted number of customers to complete their service in any time t [3]. Later Weber extended this result to the case where the service time distribution has an increasing failure rate and the arrival process is a generic point process [4]. These works are, however, based on the assumption of stochastically identical servers, which is too restrictive in practical applications, so that the result is not directly applicable to real situations.

Recently, Takinami *et. al.* have applied neural networks (Back-Propagation and LVQ3) to the queueing allocation problem with non-identical servers [5]. They presented a simple on-line training procedure which allocates an arriving customer to the server that maximizes the probability of starting his service in the earliest time. Here "on-line" means that the neural network is trained and therefore the policy is updated whenever service to a customer is finished. Thus the allocation policy is inherently adaptive to the change of traffic situations. We note, however, that the allocation policy in [5] is not socially optimal but individually optimal, so that the mean delay of an arbitrary customer cannot be minimized by the training procedure.

This study aims at social optimization of a dynamic allocation policy in the queueing allocation problem with non-identical servers and multi-class customers. We apply the optimization procedure discussed in [6] to the present problem, where the LVQ (Learning Vector Quantization) neural network [7] is adopted for policy representation. This study is different from [5] in the following points: 1) in compensation for giving up the on-line training, we obtain a socially optimal policy, and 2) we take multi-class customers into account. The objective function to be minimized is the mean delay of an arbitrary customer, where the delay is defined as the time elapsed from the arrival epoch of a customer to the end of his service. We note that the present procedure is not restricted to this performance measure.

This paper is organized as follows. Section 2 describes the mathematical model of the queueing system and the LVQ neural network. Section 3 summarizes the optimization procedure of the neural allocation policy. Simulation results are shown in Section 4 and we conclude this paper in Section 5.

2 The Model

2.1 Queueing System

We consider a queueing system with N parallel servers and C classes of customers (see Fig. 1). Each server has in general a distinct service time distribution for each class of customers. The service time of a class-j-customer by server i is assumed to be an independent random variable, whose first moment is denoted by $1/\mu_{ij}$ ($i = 1, \cdots, N; j = 1, \cdots, C$).

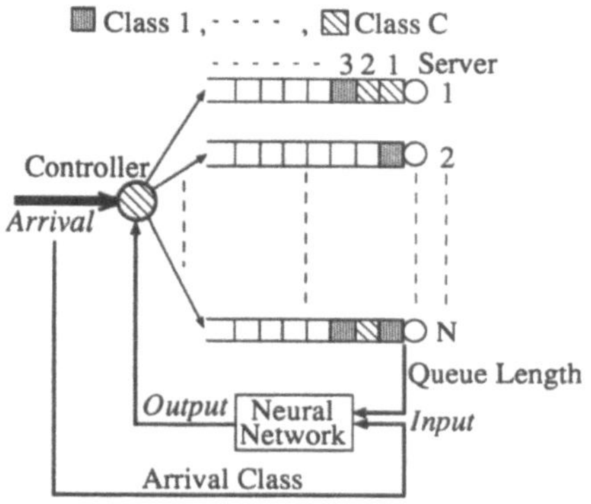

Figure 1: Queueing Model

Customers arrive at the system according to an independent general process and we denote the arrival rate of class-j-customers by λ_j ($j = 1, \cdots, C$). We assume that the complete information on 1) the class of the arrival and 2) the number of customers in each class at each server is instantaneously available for allocation decision. As soon as a customer arrives at the system, the controller sends him to one of the parallel servers according to an allocation policy and if there is any customer at the server, he forms a line, irrespective of class, to receive service. Once a customer is allocated to a queue, no jockeying among queues is allowed and each server serves his customers according to the first-in first-out (FIFO) discipline, irrespective of class.

2.2 Neural Network

There are many types of neural networks which can be used for policy representation. This study adopts the LVQ [7] because of its structural simplicity. The LVQ neural network consists of two layers: input layer and output layer. We suit the LVQ to the present problem as follows. The input layer has $(N+1)C$ neurons, where the first $N \times C$ neurons receive the information on the number of queueing customers in each class at each server, while the rest of the neurons are used to identify the class of the arrival. More concretely, suppose that a class-c-customer ($c = 1, \cdots, C$) arrives at the system at time t. Let $q_{ij}(t)$ denote the number of class-j-customers at server i (including the one being served, if any) at time t ($i = 1, \cdots, N; j = 1, \cdots, C$). Let $\boldsymbol{x}$ denote the input vector to the LVQ for allocation. Then $\boldsymbol{x}$ is given by

$$\boldsymbol{x} = [q_{11}(t), \cdots, q_{1C}(t), q_{21}(t), \cdots, q_{NC}(t), c_0\delta_{1c}, \cdots, c_0\delta_{Cc}], \tag{1}$$

where δ_{ij} is Dirac's delta and c_0 is a constant.

We note that $\boldsymbol{x}$ should be classified into one of N categories, namely, one of N servers that serves the new arrival in class c. Therefore the output layer should have N categories, where category i corresponds to server i ($i = 1, \cdots, N$). Let us assume that there are K codebook vectors in each category. Then the output layer contains $N \times K$ codebook vectors in total.

Let $\boldsymbol{m}_k^{(n)}$ and $m_{kl}^{(n)}$ ($k = 1, \cdots, K; l = 1, \cdots, (N+1)C; n = 1, \cdots, N$) denote the k-th codebook vector for category n and its weight from neuron l in the input layer, respectively. The LVQ classifies each input into the category with the closest codebook vector to the input [7]. In our context, for the given input $\boldsymbol{x}$, the controller sends the new arrival to server s, such that

$$s = \arg_n \min_{k,n} \|\boldsymbol{x} - \boldsymbol{m}_k^{(n)}\|, \tag{2}$$

where we use Euclidean distance for the norm.

3 Optimization Procedure

The goal of this study is to obtain an optimal allocation policy in the sense of minimizing the mean delay for a given traffic condition. The problem is that we cannot explicitly know to which queue a customer should be sent for optimality. In other words there is no teacher to provide the correct answer. The only way to cope with this problem is to use a random search technique in the weight space (i.e., weight perturbation [8]).

In this study we apply the optimization procedure discussed in [6], where the simulated annealing method [9] is adopted for stochastic search. Since the simulated annealing is an optimization algorithm for a finite discrete space, we modify the LVQ learning algorithm in such a way that all weights have upper & lower bounds $[M_{min}, M_{max}]$ and they may take only discrete equidistant steps. Namely, for quantization level L, $m_{kl}^{(n)}$ ($k = 1, \cdots, K$; $l = 1, \cdots, (N+1)C$; $n = 1, \cdots, N$) may equal one of the following $(L+1)$ discrete values: M_{min}, $M_{min} + \Delta$, $M_{min} + 2\Delta$, $\cdots$, M_{max}, where $\Delta \stackrel{\text{def}}{=} \frac{M_{max} - M_{min}}{L}$. In the following the above LVQ is abbreviated to wq-LVQ. The training algorithm of the wq-LVQ can be easily obtained by a simple modification to the LVQ algorithm in [7] (for detail, refer to [6]).

We note that the wq-LVQ has $(L+1)^{(N+1)CNK}$ distinct states, each of which may represent a distinct allocation policy. We denote this finite discrete space by S and we attempt to search the optimal policy in S by the simulated annealing method. Since S is vast even for small L and N, it is expected that the convergence is very slow. To reduce computation time, it is very important to start with an appropriate initial policy. As for the initial policy we train the SSED (Send-to-the-queue-with Shortest Expected Delay) policy to the wq-LVQ because it is individually optimal and in case of identical servers, it is in agreement with the SSQ policy that is proved socially optimal under some conditions [3], [4]. The detail of the SSED policy will be described in Section 4.

Let $\pi(t)$ ($t = 0, 1, \cdots$) and π^* denote the allocation policy represented by the wq-LVQ at the t-th iteration and the current solution, respectively. In addition, let $\bar{D}(t)$ and $\bar{D}^*$ denote the mean delays under $\pi(t)$ and π^*, respectively. We update the neural allocation policy as follows [6]:

Step 1) Initialization: Teach the SSED policy to the wq-LVQ to obtain $\pi(0)$. Evaluate $\bar{D}(0)$ under $\pi(0)$. Let

$t \Leftarrow 0$, $\pi^* \Leftarrow \pi(0)$, and $\bar{D}^* \Leftarrow \bar{D}(0)$.

Step 2) Perturbation: Let $t \Leftarrow t+1$. Choose a weight $m_{kl}^{(n)}$ randomly out of the $(N+1)CNK$ weights and add ξ to it to obtain a new policy $\pi(t)$, where ξ is a random number which takes one of the discrete steps between $\max(M_{min}, m_{kl}^{(n)} - \eta\Delta)$ and $\min(M_{max}, m_{kl}^{(n)} + \eta\Delta)$ with equal probability, and η $(1 \leq \eta \leq L)$ is an integer which controls the scope of perturbation. Evaluate $\bar{D}(t)$ under $\pi(t)$.

Step 3) Acceptance: Accept $\pi(t)$ as the current solution with probability P_{ac}, where P_{ac} is given by

$$P_{ac} = \begin{cases} 1, & \text{if } \bar{D}(t) \leq \bar{D}^*, \\ \exp\left(-\frac{\bar{D}(t)-\bar{D}^*}{T(t)}\right), & \text{if } \bar{D}(t) > \bar{D}^*, \end{cases} \quad (3)$$

where $T(t)$ is the *temperature* at the t-th iteration. If $\pi(t)$ is accepted, then let $\pi^* \Leftarrow \pi(t)$ and $\bar{D}^* \Leftarrow \bar{D}(t)$. Otherwise, no change.

Step 4) Termination: If $t > \nu$, then stop, where ν is a stopping criterion. Otherwise, go to Step 2.

4 Simulation Result

In simulation experiments we consider an asymmetric queueing system of $N = 3$ different servers and $C = 2$ classes of customers. We assume that for $j = 1$ and 2, the arrival processes of class-j-customers are mutually independent Poisson with rate $\lambda_j = 1.0$. The service times of class-j-customers by server i are assumed to be exponentially distributed with the service rates given in Table 1.

Table 1: Service Rate for Each Class of Customers

μ_{ij}	Class 1	Class 2
Server 1	1	1/2
Server 2	1/2	1
Server 3	2/3	2/3

As for the wq-LVQ, the number of codebook vectors in each category is $K = 5$ and the bounds are $[M_{min}, M_{max}]$=[-50,50], whose interval is quantized into $L = 200$ levels, so that the step size is $\Delta = 0.5$. The perturbation scope is set at $\eta = 25$ and the constant in the input vector is $c_0 = 10.0$.

As explained in Section 3, we teach the SSED policy to the wq-LVQ as the initial policy. We note that since the service time distribution is assumed to be exponential in our example, we do not need to take account of the elapsed service time to calculate the expected delay (i.e., memoryless property [1]). Suppose that at time t, a class-c-customer ($c = 1$ or 2) arrives at the system whose state is in $\{q_{ij}(t) | i = 1,2,3; j = 1,2\}$. Then, under the SSED policy, this customer is sent to server s, such that

$$s = \arg\min_i \sum_{j=1}^{2} \frac{q_{ij}(t)}{\mu_{ij}} + \frac{1}{\mu_{ic}}. \quad (4)$$

To teach the SSED policy, we first create training inputs by selecting one of the two classes randomly and generating random integers between 0 and 50 for q_{ij} $(i = 1,2; j = 1,2,3)$. Then their target categories are determined by Eq.(4), where if there is a tie, we assume that one of them is selected with equal probability. We train the wq-LVQ for 200,000 training samples and we check if more than 90 % are correctly classified in accordance with the SSED policy at a test bed, where a test set of 1,000,000 new samples is randomly generated in the same manner as the training set. In our example the percentage of correct classification becomes 96.0%. We note that the correct classification rate cannot be 100 % due to the possible tie. We adopt the resultant policy as $\pi(0)$.

In the optimization procedure, we empirically set the parameters of the simulated annealing at $T(t) = 0.15/\ln(10t)$ and $\nu = 1,000$. Since it is impossible to derive the mean delay analytically, we evaluate $\bar{D}(t)$ by simulation, where the waiting room at each server has a capacity of 100 customers (excluding the one being served, if any), irrespective of class, and if a customer overflows this capacity, we discard the policy with probability one at Step 3. To take account of the stationary state in simulation, we do not start collecting data until 10,000 customers are served and we evaluate the delay for the succeeding 100,000 customers during iteration. For $\pi(0)$ and π^*, we carry out 25 runs with different initial numbers to evaluate the 95% confidence interval, where the mean delay for 1,000,000 customers (instead of 100,000 customers) is evaluated for preciseness.

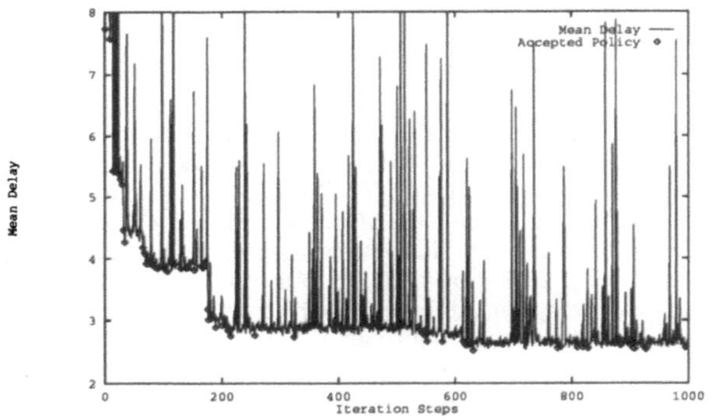

Figure 2: Optimization Process of Allocation Policy

Fig. 2 shows the optimization process of the allocation policy. In the simulation new policies are accepted 78 times, which are plotted by ◇. As we can see, the mean delay exponentially decreases with iteration steps.

Table 2: Mean & Variance of Delay

	Neural Net	SSED
Mean	2.648	5.258
95% Confi.	±0.007	±0.006
Variance	6.042	24.96

Table 2 compares the mean and variance of delay under π^* with those under the SSED policy. The 95% confidence interval of the mean delay is also shown. We observe that the mean delay achieved by the neural network is half as much as the mean delay under the SSED policy. The SSED has also much larger variance of delay in comparison to the neural network.

Table 3: Mean & Variance of Delay for Each Class

		Neural Net			SSED		
		Allo.	M_{ij}	V_{ij}	Allo.	M_{ij}	V_{ij}
Class 1	Serv 1	74.4%	2.626 ±0.007	5.809	51.7%	5.036 ±0.006	24.65
	Serv 2	1.5%	3.009 ±0.015	7.814	18.7%	5.578 ±0.006	24.74
	Serv 3	24.0%	2.563 ±0.007	5.786	29.6%	5.443 ±0.006	25.47
Class 2	Serv 1	0.6%	3.117 ±0.022	7.804	18.7%	5.578 ±0.006	24.75
	Serv 2	73.0%	2.718 ±0.007	6.368	51.7%	5.036 ±0.006	24.64
	Serv 3	26.4%	2.559 ±0.006	5.850	29.6%	5.443 ±0.006	25.48

The result of Table 2 is detailed in Table 3, where M_{ij} and V_{ij} denote the mean and the variance of delay for class-j-customers at server i, respectively. In the column *Allo.*, we also show how many percent of class-j-customers are sent to server i in simulation. It is interesting to observe that under the SSED policy, the allocation percentage, the mean and variance of delay are symmetric between class 1 and class 2 as the classes have the same arrival rate and the symmetric service rates (cf., Table 1).

The reason why π^* has much shorter mean delay can be explained in the following. According to Table 1, server i ($i = 1, 2$) serves class-i-customers at the fastest speed among the three servers. Therefore, from the viewpoint of work loaded by a customer, it would be most efficient to send all class-i-customers to server i. For the system to be stable, however, server i cannot be the sole server to class-i-customers as $\lambda_i/\mu_{ii} \geq 1$. In consequence a clever strategy is that only two servers with higher service rates 1 and 2/3 are mostly used for each class of customers. As can be seen in Table 3, the neural network carries out this strategy.

Table 4: Allocation Table by Neural Net ($q_{21} = q_{22} = 0$)

			Server 3					
	Exp. Delay		1.5	3.0		4.5		
		q_{i1}, q_{i2}	0,0	1,0	0,1	2,0	1,1	0,2
Server 1	1.0	0,0	1	1	1	1	1	1
	2.0	1,0	3	1	1	1	1	1
	3.0	2,0	3	1	1	1	1	1
		1,1	3	1	1	1	1	1
	4.0	3,0	3	3	3	1	1	1
		1,1	3	3	3	1	1	1
	5.0	4,0	3	3	3	2	2	2
		2,1	3	3	3	2	2	2
		0,2	3	3	3	2	2	2

Table 4 shows an example of allocation by the neural network. Suppose that a class-1-customer arrives at the system when server 2 has an empty queue (i.e., $q_{21} = q_{22} = 0$). If this customer is sent to server 2, he can receive service immediately, so that his expected delay is $1/\mu_{21} = 2.0$. We notice, however, that according to the neural allocation policy, this customer is not sent to server 2 unless servers 1 and 3 have so many waiting customers that he incurs expected delay of 5.0 at server 1 and that of 4.5 at server 3. On the other hand, the SSED policy is more *myopic* in the sense that this customer is sent to server 2 once his expected delay exceeds 2.0 at servers 1 and 3, which results in heavier work load to the system. We also observe that if the expected delay at server 1 is the same as that at server 3, the neural policy always selects server 1 as this choice causes lighter load to the system.

5 Conclusion

In this paper we applied the optimization procedure in [6] to the queueing allocation problem with non-identical servers and multi-class customers. The procedure was based on the policy representation by the wq-LVQ neural network and its optimization by the simulated annealing. The numerical example showed that the present procedure significantly reduced the mean delay in comparison with the SSED policy.

The extention of the service time to an independent general process is straightforward if we take the elapsed service time into account in the input vector. Rule extraction from optimal weights remains for future research. The application of the present approach to dynamic routing in telecommunication networks will be also an interesting research topic.

Reference

[1] J.Walrand, *An introduction to queueing networks,* Prentice Hall, NJ, 1988.

[2] M.Ohnishi, "On two control problems of queues," *Journal of the Operations Research Society of Japan*, no.4, pp.179-184, 1991 (in Japanese).

[3] W.Winston, "Optimality of the shortest line discipline," *J. Appl. Prob.*, vol.14, pp.181-189, 1977.

[4] R.R.Weber, "On the optimal assignment of customers to parallel servers," *J. Appl. Prob.*, vol.15, pp.406-413, 1978.

[5] J.Takinami, Y.Matsumoto, N.Okino, "Performance evaluation of neural networks applied to queueing allocation problem," *Artificial Neural Nets and Genetic Algorithms*, R.F.Albrecht *et. al.* (eds.), Springer-Verlag, pp.316-323, 1993.

[6] Y.Matsumoto, "On optimization of polling policy represented by neural network," *Computer Communication Review*, vol.24, no.4, pp.181-190, October 1994.

[7] T.Kohonen, "The self-organizing map," *Proceedings of the IEEE*, vol.78, no.9, pp.1464-1480, September 1990.

[8] S.Markon, H.Kita and Y.Nishikawa, "Reinforcement learning for stochastic system control by using a feature extraction with BP neural networks," *Tech. Rep. of IEICE,* NC91-126, pp.209-214, 1991.

[9] E.Aarts and J.Korst, *Simulated annealing and Boltzmann machines,* John Wiley & Sons, 1989.

Comparison of Genetic and Other Unconstrained Optimization Methods

Antti Autere
Helsinki University of Technology, Department of Computer Science
Otakaari 1 A, SF-02150 Espoo, Finland. email: aau@cs.hut.fi

Abstract

This paper presents and compares genetic and non-genetic algorithms for unconstrained optimization of differentiable functions $f : R^n \mapsto R$ with many local optima. The topics discussed here are: population size, the role of crossover, including gradient line search into GA and memorizing the calculated gradients.

One thousand 10 dimensional test functions with 10-20 randomly located optima have been minimized by the tested algorithms. The population sizes of the algorithms were bigger than the number of the local optima.

A rank comparison test, the Friedman test, have been used to find the statistically significant differences between the convergence of the algorithms.

1 Test Functions

The test functions are based of the modified 10-dimensional Shekel's function [1]:

$$f(\mathbf{x}) = \sum_{i=1}^{m}(s_i/((\mathbf{x}-\mathbf{a}_i)^T\mathbf{H}_i(\mathbf{x}-\mathbf{a}_i)+c_i). \quad (1)$$

The parameter m represents the number of local optima and is uniformly distributed $(10, 20)$ The vectors $\mathbf{a}_i \in (0, 10)^{10}$ represent the places of the local optima and their components are uniformly distributed. The signs $s_i = +1, -1$ are selected at random and tell if there is a local minimum or maximum near $\mathbf{a}_i$. The scalars c_i scale the values of local optima and are uniformly distributed $(0.1, 1.0)$. $\mathbf{H}_i$s are diagonal matrices, $\mathbf{H}_i = diag(h_{1,i}, ..., h_{10,i})$ that make the contours elliptical near the local optima. $h_{j,i}$s are uniformly distributed $(0.01, 1.0)$.

To ensure that local optima are at $\mathbf{a}_i$ the previous function is modified as:

$$f(\mathbf{x}) = s_k/((\mathbf{x}-\mathbf{a}_k)^T\mathbf{H}_k(\mathbf{x}-\mathbf{a}_k)+c_k) \quad (2)$$

in a set $0 \le x_j \le 10$ $(j = 1, ..., 10)$, where k is the index of the nearest $\mathbf{a}_i$ to $\mathbf{x}$. These functions possess derivatives almost everywhere. The gradients are calculated from the first equation.

2 Testing Method

One thousand 10-dimensional test functions were randomly selected from the set described above. All the algorithms tried to minimize every function once. The percentages of the global minima that a particular algorithm reached after these 1000 minimization tasks were recorded. The length of one run was limited by a predefined number of the function evaluations. The cost of one gradient evaluation was 10 function evaluations.

A rank comparison test for randomized complete block designs - the Friedman test - was used to find the statistically significant differences between the convergence of the algorithms [4] pp. 299-305. In the usual terminology of this type of test one test function represents a *block* and one algorithm represents a *treatment*.

The data is put into a matrix whose one row corresponds the results of the algorithms for one test function. Hence the matrix has 1000 rows. The algorithm that is best in one row receives the highest rank number (credit), the second best receives the second highest rank and so on. At the end of each column the sum of the ranks the algorithm possesses is calculated. Let us call these numbers *rank sums*.

The null hypothesis for this test is that no algorithm is superior to any of the others. If the null hypothesis can be rejected below some risk level (e.g. 1 percent) then the differences between each two algorithm can be tested at the same level. The test statistics for both these tests are based on the rank sums.

The null hypothesis is rejected at the level α if the test statistic exceeds the $1-\alpha$ quantile of the F-distribution. The test statistic for pairwise comparisons utilizes the $1-\alpha/2$ quantile of the t-distribution. Any two algorithms are regarded as unequal if the absolute difference of their rank sums exceeds it.

3 Tested Algorithms

3.1 Random Search: RA

RA generates uniformly distributed solution candidates within the interesting set $S \subset R^{10}$ one after another and remembers the best one. In GA terminology this is the *mutation* operator.

3.2 Genetic Algorithm: GA

GA has the features typical of the genetic algorithms: population and selection mechanism, mutation and crossover operators.

The algorithm uses the ranking selection scheme. A new member is accepted only if its fitness is less than the fitness of the last individual in the list (population).

Crossover produces a child $\mathbf{c} \in S \subset R^{10}$ from two parents $\mathbf{p}_1, \mathbf{p}_2 \in S$ according to a formula:

$$\mathbf{c} = \mathbf{R} * \mathbf{p}_1 + (\mathbf{I} - \mathbf{R}) * \mathbf{p}_2, \quad (3)$$

where $\mathbf{R}$ is a diagonal matrix, $\mathbf{R} = diag(r_1, ..., r_{10})$, whose elements are uniformly distributed (0,1). $\mathbf{I}$ is the identity matrix. We may think that the child is selected from the (hyper)rectangle whose opposite corners the parents are.

Mutation generates an individual $\mathbf{c} \in S$ at random.

3.3 GA with Gradient Search: GAG

The algorithm is the same as GAG but in addition to mutation and crossover there is a third operator called *gradient line search* along the direction of the negative gradient [1]:

$$\mathbf{c} = \mathbf{p1} - t * (\nabla f(\mathbf{p1}))/(\| \nabla f(\mathbf{p1}) \|), \tag{4}$$

where **p1** is the parent and t is a step size. Golden section line search is fully described e.g. in [3] p. 259.

The step size $t = uniform(min, \| \mathbf{p1} - \mathbf{p2} \|)$ or $t = uniform(min, max)$, where $min = max/200$ and $max = s_{ave} = 10$. s_ave is the average diameter of the set S where the minimization is done. **p2** is another vector of the population.

3.4 Memorize Gradients: GAGM

GAGM is an algorithm, described in detail in [2]. It is the same as GAG except it has *two populations*: one sorted by the function values and the other by the lengths of the gradients. The calculated gradient vectors are included in the latter population. Every time one of these populations is selected first before applying the crossover or the gradient search operation.

When crossover has produced a child it calculates its gradient with probability 0.05. When golden section search has produced a new candidate it calculates its gradient with $p2$ [1]. All the individuals do not have their gradients calculated. If an individual is in the gradient population then it is in the both populations but not the other way round. It follows that the intersection of the two populations is not necessarily empty.

4 Results

4.1 Convergence and Population Size

There are 1000 test runs for every algorithm variants. Simulation runs were done with four population sizes: 20, 50, 100 and 200. The lengths of the runs varied from 1000 to 70000 criteria function evaluations.

Figure 1. shows these results. The solid lines show the average percentages of the global optima GAG and GA reached. The dashed lines show the percentages of the 1000 runs that found the value of 90 percent or more of the global optima (for GAG only).

For GAG, smaller population sizes seem to yield quicker convergence than bigger ones. However, bigger populations give better results once they have converged.

GA has only one curve (solid line) that shows the average percentages of the global optima GA reached. The curve combines the results with different population sizes: 20, 50, 100 and 200. This shows that GAG is clearly superior to GA.

The curves with population sizes 20 - 100 show that sometimes shorter runs give better results that longer ones. The reason for this is that a shorter run is not part of the longer one but is a completely independent trial.

The lengths of all the next test runs were selected as 100*pop.size i.e. 2000, 5000, 10000 and 20000 criteria function evaluations, respectively. These lengths were chosen not to represent the fully convergent runs but just "good" ones, see fig. 1.

[1] $p2$ is calculated according to some rules described in [2]

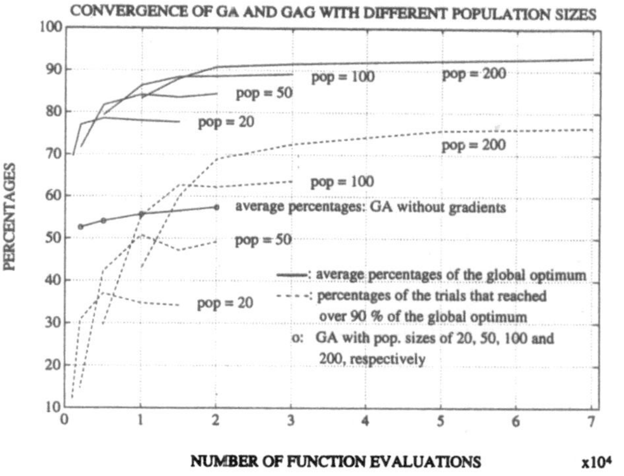

4.2 Including Gradients

The effect of including gradient search into the GA is studied here more deeply. Three algorithms are selected for these tests: random search, RA, genetic algorithm without gradients, GA, and genetic algorithm with gradient line search, GAG.

There are three variants of GAG. GAG1 approximates the gradient by calculating only one randomly chosen component (dimension) of the gradient vector. Similarly, GAG5 approximates the gradient by 5-dimensional vector with randomly chosen dimensions. Finally, GAG uses a true gradient vector.

Table 1. shows the results with population size 20. The first row shows the algorithms studied. The second row shows the average percentages of the global optima an algorithm reached after minimizing the 1000 test functions. The third row shows the percentages of the 1000 minimization trials that reached over 90 percent of the global optima. The fourth row shows the rank sums for each algorithm. The structures of all the next tables are similar.

The test statistic of the Friedman test is 3411.3. This implies that the null hypothesis, that all the algorithms are similar, is rejected at the level 0.001. The 0.999 quantile of the F-distribution with 4 and 1000 degrees of freedom was found to be about 3.75 by quadratic extrapolation of the 0.95, 0.975 and 0.99 quantiles.

Any two algorithms whose rank sums are more than 110.8 units apart may be regarded as unequal. These comparisons yield a complete ordering between the algorithms. Let us mark it as:

$$RA < GA < GAG1 < GAG5 < GAG. \tag{5}$$

The notation "$A1 < A2$" means that "the algorithm A2 found better solutions than A1" and the difference between them is statistically significant (at the level of 0.001). If A1 and A2 cannot be regarded as unequal then this is marked as: $A1 = A2$.

Table 2. shows the results with population size 50. The test statistic is 3951.1 and the null hypothesis is rejected at the level 0.001. The limit for pairwise comparisons is 104.6. The same complete ordering between the algorithms is found as in table 1.

Algorithm	RA	GA	GAG1	GAG5	GAG
Average	51.9	52.6	70.0	75.2	77.0
P90	0.0	0.0	12.9	28.4	30.7
Rank Sum	1411	1614	3531	4121	4323

Table 1: Including gradients: pop. size 20

Algorithm	RA	GA	GAG1	GAG5	GAG
Average	51.8	54.1	73.0	80.0	81.7
P90	0.0	0.5	21.8	39.0	42.2
Rank Sum	1307	1720	3409	4175	4389

Table 2: Including gradients: pop. size 50

Table 3. shows the results with population size 200. The test statistic is 5380.2 and the limit for pairwise comparisons is 92.1. The conclusions are the same as before (tables 1., 2.).

The same conclusion holds for all the test results of the different population sizes. Random search was worse than GA and including gradient line search into GA clearly improves the performance.

4.3 Crossover and Weighting

The previous test showed that including the crossover made the algorithm better than pure random search. This holds when no gradient information was utilized. These tests studies further what is the role of the crossover when gradients are included.

Also the effect of changing the probability of applying the gradient search during one run is studied. In the beginning of the run this probability is 0.1 and after a while it is increased to 0.9.

Two classes of algorithms were selected for these tests: genetic algorithm with gradient line search, GAG, and gradient line search without crossover, MG. There are two variants of MG and three variants of GAG.

GAG has the probability of applying the mutation operator $p(mut) = 0.05$, the crossover $p(box) = 0.45$ and gradient line search $p(grad.) = 0.50$ all the time. GAG1/4 has $p(mut) = 0.01$, $p(mut) = 0.89$ and $p(grad.) = 0.10$ before it has reached 1/4 of the total run length. After that these probabilities are changed to $p(mut) = 0.01$, $p(mut) = 0.09$ and $p(grad.) = 0.90$. GAG3/4 is the same as GAG1/4 except the time of changing the probabilities is now 3/4 of the total run length.

MG has no crossover. MG/0.50 has the probabilities of applying both the mutation and gradient line search 0.50. MG/0.95 has $p(mut) = 0.05$ and $p(grad.) = 0.95$.

Table 4. shows the results with population size 20. The test statistic is 21.7 and the null hypothesis is rejected at the level 0.001. The limit for pairwise comparisons is 230.3. No statistically significant difference between GAG, MG/0.50 and GAG1/4 exists. However, these three algorithms are better than MG/0.95. No difference between GA and GAG3/4 exists but these two are again better than MG/0.95. Finally, GAG3/4 is better than both MG/0.50 and GAG1/4 that are better than MG/0.95. So, this yields only partial ordering between the algorithms, represented not as a single chain or list but a set of lists. The following describes this with the notation used here:

Algorithm	RA	GA	GAG1	GAG5	GAG
Average	51.5	57.4	76.1	87.4	90.8
P90	0.0	1.4	30.1	57.9	68.9
Rank Sum	1245	1803	3170	4285	4497

Table 3: The effect of including gradients: pop. size 200

Algorithm	GAG	MG/0.95	MG/0.50	GAG1/4	GAG3/4
Average	77.0	74.9	77.2	74.6	75.1
P90	30.7	24.0	30.6	26.0	28.2
Rank Sum	3135	2654	3020	2928	3263

Table 4: Crossover and weighting: pop. size 20

Algorithm	GAG	MG/0.95	MG/0.50	GAG1/4	GAG3/4
Average	81.7	81.0	83.7	76.4	76.5
P90	42.2	39.3	45.8	30.2	30.1
Rank Sum	3245	2882	3226	2677	2970

Table 5: Crossover and weighting: pop. size 50

$$\{(GAG = MG/0.50 = GAG1/4) > MG/0.95,$$

$$(GAG = GAG3/4) > MG/0.95,$$

$$GAG3/4 > (MG/0.5 = GAG1/4) > MG/0.95\}.$$

Table 5. shows the results with population size 50. The test statistic is 23.5 and the null hypothesis is rejected at the level 0.001. The limit for pairwise comparisons is 230.1. This data suggests a complete ordering between the algorithms if GAG and MG/0.50 are put into one equivalence class and MG/0.95 and GAG3/3 into another by using the "=" -relation, i.e.:

$$(GAG = MG/0.50) > (MG/0.95 = GAG3/4) > GAG1/4. \quad (6)$$

Table 6. shows the results with population size 100. The test statistic is 54.4 and the null hypothesis is rejected at the level 0.001. The limit for pairwise comparisons is 226.7. The conclusions are the same as before (table 5.)

Table 7. shows the results with population size 200. The test statistic is 98.0 and the null hypothesis is rejected at the level 0.001. The limit for pairwise comparisons is 222.2. This data suggests a complete ordering if GAG and MG/0.50 are put into one equivalence class, i.e.:

$$(GAG = MG/0.50) > MG/0.95 > GAG3/4 > GAG1/4. \quad (7)$$

All the test results of the different populations sizes show that there is not a statistically significant difference between the algorithms GAG with crossover and MG/0.50 without it.

Algorithm	GAG	MG/0.95	MG/0.50	GAG1/4	GAG3/4
Average	86.4	85.8	88.7	79.2	78.6
P90	55.3	50.6	60.2	36.5	35.2
Rank Sum	3381	2967	3323	2521	2808

Table 6: Crossover and weighting: pop. size 100

Algorithm	GAG	MG/0.95	MG/0.50	GAG1/4	GAG3/4
Average	90.8	90.5	92.1	81.4	80.8
P90	68.9	69.0	74.5	42.6	39.6
Rank Sum	3520	3141	3302	2363	2674

Table 7: Crossover and weighting: pop. size 200

Algorithm	GAGM	MGM/0.5	GAG	MG/0.5
Average	79.5	81.8	81.8	83.0
P90	38.6	42.5	42.9	44.5
Rank Sum	2375	2163	2651	2511

Table 8: GAGM: pop. sizes 25, GAG pop. size 38

GAG3/4 is always better than GAG1/4. This would encourage to give the gradient line search operator more resources only at the end of the run. However, GAG3/4 and GAG1/4 are worse than GAG and MG/0.50 except when the population size is 20.

MG/0.95 become relatively better when the population size increases. In three of the four tests it is better or as good as GAG3/4, surprisingly.

4.4 Memorizing Gradients

The effect of utilizing the calculated gradients are studied here. four algorithms are selected for these tests: GAGM, MGM/0.5, GAG and MG/0.5

GAGM and MGM/0.5 have two populations the intersection of which may not be empty. That is why the "effective" sizes of the populations of GAGM and MGM/0.5 are difficult to estimate. The following rule is used. Both the populations of GAGM and MGM/0.5 have equal sizes, say x. Then the sizes of the population of GAG and MG/0.5 are $1.5 * x$

In other aspects GAGM and MGM/0.5 are similar to GAG and MG/0.5, respectively.

Table 8. shows the results of GAGM and MGM/0.5 with population and gradient population sizes 25 and GA and MG/0.5 with population sizes 38. The test statistic is 8.0 and the null hypothesis is rejected at the level 0.001. The 0.999 quantile of the F-distribution with 3 and 1000 degrees of freedom was found to be about 4.3 by quadratic extrapolation of the 0.95, 0.975 and 0.99 quantiles. The limit for pairwise comparisons is 189.3. The only statistically significant difference is found between GAG and GAGM: $GAG > GAGM$, although the second and third rows of the table 8. may suggest an other conclusion.

Table 9. shows the results of GAGM and MGM/0:5 with pop. and grad. pop. sizes 50 and GA and MG/0.5 with pop. sizes 75. The test statistic is 4.6 and the null hypothesis is rejected at the level 0.001. The limit for pairwise comparisons is 189.7. The conclusions are the same as before (table 8.).

Table 10. shows the results of GAGM and MGM/0.5 with pop.

Algorithm	GAGM	MGM/0.5	GAG	MG/0.5
Average	84.9	87.2	86.1	88.7
P90	52.0	57.6	55.1	62.6
Rank Sum	2393	2527	2602	2478

Table 9: GAGM: pop. sizes 50, GAG pop. size 75

Algorithm	GAGM	MGM/0.5	GAG	MG/0.5
Average	89.3	91.7	90.5	92.4
P90	64.3	73.0	68.3	74.9
Rank Sum	2571	2680	2500	2249

Table 10: GAGM: pop. sizes 100, GAG pop. size 175

and grad. pop. sizes 100 and GA and MG/0.5 with pop. sizes 175. The test statistic is 20.5 and the null hypothesis is rejected at the level 0.001. The limit for pairwise comparisons is 188.2.

This time MG/0.5 was the worst. No statistically significant differences between the other three algorithms was found. This implies that

$$(GAGM = MGM/0.5 = GAG) > MG/0.5. \tag{8}$$

These tests suggest that memorizing the gradients in the form of another population does not have any effect on the convergence (disappointingly !). Again, the crossover has no effect.

5 Conclusions and Future

Based on the tests of this paper the following conclusions can be drawn. The algorithms with large populations converged more slowly but produced better results than the ones with small populations. This holds for both the gradient and non-gradient based algorithms, see figure 1.

The algorithms with gradient line search were clearly superior to the ones without gradients, see tables 1. - 3.). For the algorithms without gradients the crossover improved the convergence (GA and RA in tables 1. - 3.). However, when the gradients were utilized no statistically significant difference existed between the algorithms with and without the crossover (GAG and MG/0.5 in tables 4. - 7.) (GAGM and MGM/0.5 in tables 8. - 10.).

Increasing the probability of applying the gradient search during one run did not improve the convergence (GAG, GAG1/4 and GAG3/4 in tables 4. - 7.). Also, memorizing and utilizing the calculated gradient information did not improve the convergence (GAGM and MGM/0.5 in tables 8. - 10.).

Finally, one can argue that the only difference that would matter in practise is the one found between the non-gradient and gradient based algorithms, see the two last lines of every table.

All these tests were done with the population sizes bigger than the number of local optima of the test functions. So, it would be interesting to carry out similar tests when the population sizes are substantially smaller than the number of local optima.

References

[1] Antti Autere. Combining gradient search and genetic algorithm for optimizing differentiable functions. In Jarmo T. Alander, editor, *The Second Finnish Workshop on Genetic Algorithms*, University of Vaasa, Palosaari, Wolffintie 36, Vaasa, Finland, March 1994.

[2] Antti Autere. Employing gradient information in a genetic algorithm. In *Second European Congress on Intelligent Techniques and Soft Computing, EUFIT '94*, Aachen, Germany, September 1994.

[3] M.S. Bazaraa and C.M Shetty. *Nonlinear Programming: Theory and Algorithms*. Wiley, 1979.

[4] W.J. Conover. *Practical Nonparametric Statistics*. Wiley, 1980.

NEURAL NETWORKS FOR DYNAMICAL CROP GROWTH MODEL REDUCTION AND OPTIMIZATION

Ilya Ioslovich, Ido Seginer

Department of Agricultural Engineering, Technion, Haifa, 32000, Israel

Abstract.

A dynamic crop growth model of undeterminate tomato variety (TOMGRO) consists of a large number (69) of difference equations that describe age classes of different organs. In order to find a fast working and low-dimensional equivalent with dynamic Neural Networks (NN), several model reduction approaches has been tried, including aggregation, PCA transformation, and bottleneck NN compression. Reduced data were used to train dynamic NN. Simulations with the NN model produced trajectories which agreed well with the original trajectories of TOMGRO.

1. Introduction.

Modern agricultural technologies are dealing with complicated environmental control in greenhouses to operate optimally with crop growth processes. The aim is to increase quantity and quality of production under a set of technological, economical, physiological and climatological constraints. A very essential point is the use of sophisticated but consistent and validated models of plant growth. For tomato such a model (TOMGRO) [1], contains a set of difference equations, describing time trajectories for characteristics of age classes of different organs. All of them, (7x10+1), are the number and weight of leaves, the number and weight of stems, leaf area index (LAI), the number and weight of fruits, and total number of nodes (plastochrone).

Weather data give additional inputs, including solar radiation, length of day, maximum and minimum temperature of particular day. From these data a set of hourly inputs are produced, so the model operates as a set of difference equations with time-dependent right hand side and hourly step. In addition to this a greenhouse model can be optionally added, to consider different balances in the greenhouse and to determine steady state points for special variables that describe the working regime of the greenhouse and use of control equipment (fan, heater and CO_2 injector). The growing season for tomato is about nine months, so for dynamical optimization dimension of dynamic system is too high . Thus a conflict exists between the requirements of optimal control policy determination (that deals with rather low-dimensional models), and adequate crop model description (69 variables). To solve this problem we are considering different approaches for model reduction by the application of NN . The idea is to find a transformation (and if possible also an inverse transformation) from the original space of 69 state variables (that presumably are partially correlated) to some low-dimensional reduced space, in which the original trajectories could be mapped and a dynamic NN can be designed to fit one-day steps dynamics (instead of the hourly steps of the original TOMGRO) of reduced TOMGRO trajectories. Thus both "horizontal and vertical", (time and state) compression of data would be achieved. If successful, this method could be applied to mimic a daily periods of the trajectories with a local optimal control of greenhouse equipment. This gives the possibility to make quick iterative seasonal runs for the realization of an hierarchical scheme of a seasonal search of an optimal trajectory. This will not be considered here.

The Neural Networks MATLAB Toolbox [2] was used for all NN applications of this study.

2. Description of the methodology.

The proposed methodology generally consists of four stages:
a) Static model reduction and data compression,
b) Mapping data to a reduced space and training of dynamic NN, c) Simulation of dynamic trajectories in the reduced space, d) Inverse transformation of the results in original space of TOMGRO variables and comparison with the original TOMGRO trajectories.
Three different methods were used for stage a) that also leads to correspondent differences in the other stages. In all cases the same original data were used that were obtained by running the TOMGRO model with six different weather sequences from Gainesville (Florida) according to six different pairs of consecutive years. Another three pairs of years also from Gainesville were used for verification. The three different reduction methods are now described.

2.1. Aggregation.

This is the simplest and most traditional approach to model reduction. All age classes of selected variables and sometimes several types of variables are aggregated (lumped), such that the reduced variables can be easy measured. Mostly the reduced variables are some totals. The age classes of number and weight of fruits are divided on 9 classes of green fruits, that are state variables, and 10th class represents mature fruits. This age class doesn't affects the growth of plant and doesn't enter into the right hand part of difference equations of TOMGRO. It only represents some accumulated value, that is calculated along the trajectory. Two variants of aggregation (AGV6 and AGV4) were compared. First variant (AGV6) has six aggregated state variables as part of input, that are: 1) total number of nodes, 2) total number of living leaves, 3) total number of green fruits, 4) leaf area index, 5) total dry weight of shoot (green fruits, stems and leaves altogether), 6) dry weight of green fruits.
In the second variant (AGV4) the variables 2 and 3 were omitted. The aggregated data were obtained

from the data of original TOMGRO trajectories. The transformation is completely deterministic and independent of any methodological parameters and preliminary operations. On the other hand there is no definite inverse transformation, therefore the only comparison of results can be done in reduced space after dynamical simulation. The main parameters for comparison are related to the dry weight of mature fruits, that are not state variables of TOMGRO. These values are used in the dynamical training as additional outputs, therefore its original data can be compared with NN output also for these variants.

2.2. Principal Component Analysis.

This method is widely used for compression of statistical information and its further use in a rather compact form by replacing a wide set of correlated original data by a reduced number of uncorrelated transformed data (see [3]). A set of TOMGRO trajectories were sampled and the sampled data of 324 vectors were centered to obtain zero means. Then the covariance matrix of these data was obtained and diagonalized, that gives direct and inverse transformations to uncorrelated variables. The uncorrelated variables which corresponds with the less significant eigenvalues were erased so that the only predefined number of principal components were left. An inverse linear transformation was done to reconstruct the data in the original space which were then compared with the original data. To evaluate deviations of PCA from TOMGRO, a correlation index (see [4]) for each variable was computed. The results of this transformation are dependant of the procedure of sampling, that should uniformly cover all the region of the possible data for the TOMGRO trajectories. It is assumed that each of the variables has approximately the same mean value on all the trajectories. This means that the trajectories are on average independent of the weather sequences. It is of course an approximation. The numbers of principal components were chosen from 2 to 16. The mean over all variables value of correlation index for PCA transformation from were changed from 0.7457 to 0.997. These data are the means for the training set sampled of six trajectories from Gainesville (Florida) with planting date at 1 of August of each year.

2.3. NN bottleneck compression (NNB).

A training set of complete state variable vectors (69 elements in each) is produced by simulating with original TOMGRO model over a range of weather conditions. The same set was used to determine PCA transformation. The training is performed with 5-layer NN, presented with the training vectors at both the input and output layers. The middle layer serves as a bottleneck through which the information has to pass. The number of nodes in this layer is reduced as much as possible within the permissible error margin. This NN can be presented as a sequence of two different NN, each one with 3 layers, where output of the first NN serves as input to the second NN. For this presentation the 3rd layer of 5-layers NN has to be considered as virtually divided on two identical layers with identical transformation between them. This layer represents a reduced vector which contains most of the original information. If we shall consider 5-layer NN as a sequence of two 3-layers NN, then first of them will perform direct transformation from original to reduced space, that gives compression of input information, while second NN performs inverse transformation from reduced to original space or decompression of information. For number of nodes in second layer (and the same in fourth layer) the experimental formula was used

$$n = 2\sqrt{(I+O)}, \qquad (1)$$

where I is number of inputs and O is number of outputs. This time this formula is applied to the couple of intermediate NN: one from layer 1 to layer 3, that is considered as intermediate output, and second from layer 3 to layer 5, where layer 3 is considered as intermediate input. We should note that there is some similarity with technique that was used by Sejnowski [5], for the problem of pattern recognition. Sejnowski has used simple 3-layer NN with identical input and output to compress visual information of pictures and images and has used a compression network to preprocess images into a feature vector before they are recognized. Transformation from 1st to 2nd layer gives compression, supposing that 2nd layer has less nodes then 1st. On the next stage this transformation was used to compress images and these compressed versions were recognized instead of actual images. In his case two separate 3-layers NN were built: one for compression and decompression and one for recognition of compressed information. We are using 5-layers NN for compression and decompression and then dynamical NN that is trained on compressed information. For training of 5-layer NN an adaptive algorithm based on backpropagation procedure of MATLAB (learnbp) was coded. Nonlinear tansigmoidal functions were used in this NN. The original information was normalized to range from -1 to 1. A rather small initial step of about 2.e-4 was used for the learning process. For each variable a correlation index was computed to evaluate the agreement between TOMGRO variables and corresponding NN outputs. For the training set the mean correlation value over all variables was increased from 0.907 to 0.982, when the number of nodes in the middle layer was changed from 2 to 16. Comparison of original and retransformed information for training set is shown on Figure 1 for variable Weight of Green Fruits, Age Class 6.

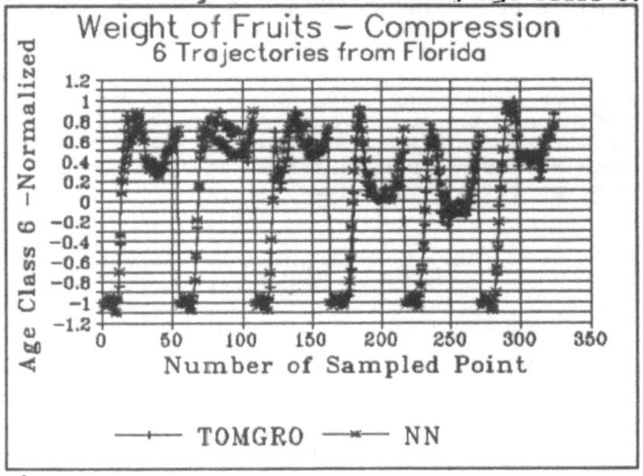

Figure 1

NNB compression of sampled set. TOMGRO and NN data for Weight of Fruits, age class 6. Number of NNB nodes 4.

we should note that technological parameters related to the training process could affect effectiveness of NNB transformation.

2.4. Dynamic NN.
Next step for all methods of data compression is to create a dynamic NN (DNN), that would work like a system of difference equations in reduced space. It's input consists of weather summary and transformed (reduced) state variables, and its output consists of daily increments of reduced state variables and increments of yield (daily harvest-number and dry weight of mature fruits). For this purpose each of the TOMGRO trajectories that were included in the training set was transformed by corresponding transformation for model reduction. The compressed data together with additional information of weather summary and daily harvest were sampled to receive data for training set and daily increments for target set. Difference system of such type can be written in the form

$$[dP^k,\ dF^k] = DNN\{P^k,\ W^k\}, \qquad (2)$$
$$[P^{k+1}, F^{k+1}] = [P^k, F^k] + [dP^k, dF^k],$$

where k counts daily time step, P is vector of reduced state variables, F is two-dimensional vector of yield (number and dry weight of mature fruits), W is a weather summary vector. DNN is a 3-layers NN, For its training a special MATLAB procedure was written, which realized the method of conjugated gradients (see [6]). To find a current gradient the backpropagation routine was used.

2.5. Dynamic simulation.
Trained DNN was used for dynamic simulation. For this purpose six TOMGRO trajectories were selected, three from which were used also to produce training set and three were used for verification only. All these trajectories were transformed to reduce state trajectories. Then daily increments and weather summary were normalized to make inputs to DNN. Outputs from DNN were denormalized and reduced vector along trajectory was computed using equation (2). This trajectory could be compared with reduced trajectory of TOMGRO. We should stress, that this comparison includes not only artificial data of reduced variables, but also original data of yield, that were included in the output of DNN without reduction. In Figure 2 plots of both TOMGRO and DNN trajectories, variable Dry Weight of Mature Fruits, are presented for the weather sequence from Florida, 1958. This trajectory was part of the verification set.

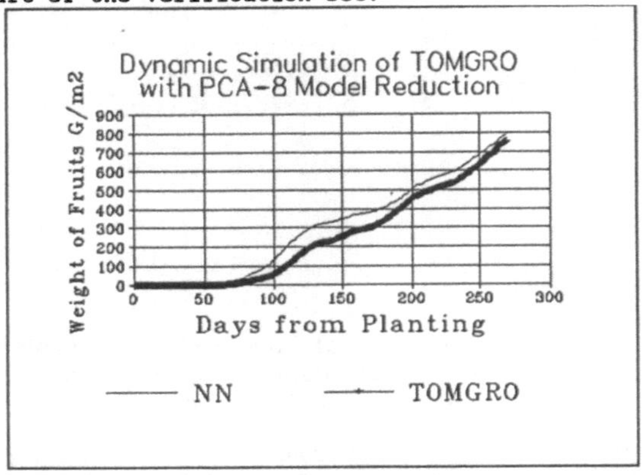

Figure 2. Dynamic simulation of TOMGRO trajectory for Florida (1958), Dry Weight of Mature Fruits, Model Reduction by PCA-8.

2.6 Inverse transformation and comparison with TOMGRO.
After the simulation, the DNN output was transformed back by inverse transformation into original space of TOMGRO variables. This transformation gives the mapping of DNN trajectory, which should be similar to corresponding TOMGRO trajectory. Figure 3 presents comparison of the TOMGRO trajectory for the variable Weight of leaves, Age class 4, with retransformed DNN output. Weather sequence Florida, 1958 was used. For numerical evaluation of the differences the correlation index for each variable was computed.

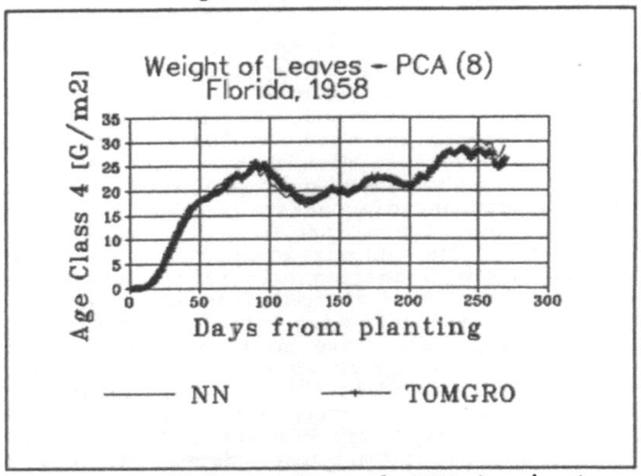

Figure 3. Retransformed DNN trajectory and the TOMGRO trajectory of the variable Weight of Leaves, Age class 4. Weather from Florida (1958). Model Reduction by PCA-8.

3. Results and discussion.

The main interest of investigation was concerned with dynamic simulation using reduced TOMGRO variables. However also the comparisons of static transformations are of interest. Figure 4 presents data for comparison of static transformation by NNB and PCA with different dimensionality of reduced space (different number of principal components and number of nodes in the middle layer, respectively). Data are computed for the training set.

A wide range of simulations was done and different characteristics, including the correlation indexes, R^2 were calculated. We present here only a small part of these data, that will form following Table 1.

Table 1.
Mean R^2 over all variables for dynamical simulation of trajectories from verification set, Florida.

n	PCA	NNB	AGV
4	0.77	0.86	0.92
6			0.94
8	0.85	0.92	--

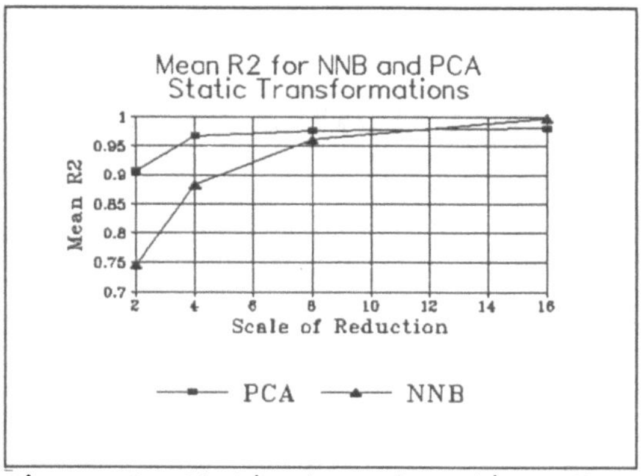

Figure 4. Static transformation. Mean R2 over all variables for different numbers of principal components and NNB nodes. Data for training set.

Here n denotes number of PCA variables or NNB nodes, and also aggregated variables, so this number is characteristic of reduction. This table shows rather better values for NNB then for PCA and still better values for AGV6. To show an example of the work of DNN based on AGV6, we present here Figure 5, that contains both TOMGRO and AGV6 plots of trajectories with 1958 weather from Florida, for aggregated variable Dry Weight of Green Fruits.

It seems rather natural that some advantage exists of the nonlinear transformation of NNB comparing to the linear transformation of PCA. The transformation obtained by aggregation is based on completely deterministic procedure, so probably more exact tuning of some parameters, used in PCA and NNB could improve their effectiveness. Generally speaking, our conclusion is that all three methods of dynamical model reduction could be successfully used and should help to decrease the amount of calculations to an acceptable level. The artificial NN seems to be, a powerful tool for this purpose.

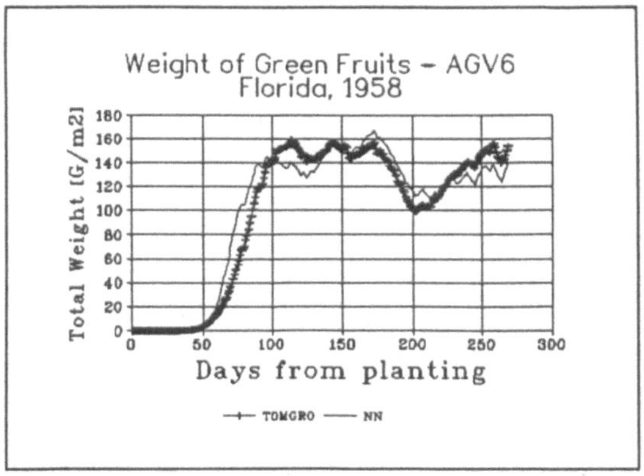

Figure 5. Dry Weight of Green Fruits, Florida (1958). Dynamic simulation, DNN is trained on aggregated data AGV6.

References.

1. Dayan, E., van Keulen, H., Jones, J.W., Shmuel, D., Challa, H.: Agric. Systems, 43, 165-183, Development, calibration and validation of a greenhouse tomato growth model: I. Description of the model. Agric. Systems, 43,145-163.

II.Field calibration and validation. (1993).

2. Demuth, H., Beale, M.: Neural Network Toolbox for Use with Matlab. Users Guide, Natick, MA 01760, USA: MathWorks, Inc., 1992.

3. Kendall, M.G., Stuart, A., Ort, J.K.: The Advanced Theory of Statistics. Vol 3., Design and Analysis, and Time Series, 4th edition, London: Griffin 1983.

4. Arkin, H., Colton, G.: Statistical Methods, New York, USA: Barnes and Noble, Inc. 1953.

5. Sejnowski, T., Lawrence, D. and Golomb, B.: Sexnet: A Neural Network Identifies Sex from Human faces, in Advances in Neural Information processing systems 3. Touretzky, D., and Lippman, R., (eds.), San Mateo, CA: Morgan Kaufmann 1991.

6. Polyak, B.T.: Introduction to Optimization, New York: Optimization Software 1987.

MODIFIED KOHONEN'S LEARNING LAWS FOR RBF NETWORK

Tommi Ojala and Petri Vuorimaa

Tampere University of Technology, Signal Processing Laboratory
P.O. Box 553, FIN-33101 Tampere, Finland
tel. +358-31-316 1835, fax +358-31-316 1857
email: to88407@cs.tut.fi, pv@cs.tut.fi

Abstract - **A hybrid training method for the radial basis function (RBF) network is presented. The method applies the Kohonen's self-organizing map (SOM) and a modified learning vector quantization (LVQ) algorithms. Learning algorithms are derived for two alternative RBF network structures exploiting local Gaussian basis functions. The potential of the proposed methods is demonstrated by a function approximation example.**

I. INTRODUCTION

Kohonen's self-organizing map (SOM) [5] has special properties among neural networks. First, it creates spatially organized internal representations of features of input signals. The main idea, in contrast to the other competitive algorithms, is that not only the best-matching cell is updated, but also its neighbourhood which is defined as a neuron lattice where adjacent cells have interconnections. Second, due to topological preservation, the SOM is capable of visualizing high-dimensional data in a lower-dimensional discrete space. For pattern classification problems, the accuracy of SOM can be increased by using learning vector quantization (LVQ) algorithms [5,6], which tune the map to be specific for pattern classes. The basic map as such is not intended for function approximation, but still some researchers have proposed extensions of the SOM to improve its capability to approximate continuous functions [2,9,10]. The main application area of the SOM is pattern recognition, especially speech recognition and process state monitoring [13].

Radial basis function (RBF) networks are commonly considered as a viable alternatives to multilayer perceptron networks which are highly nonlinear in the parameters and suffer from the lack of intuitive understandability. Among the first, Moody and Darken [8] employed RBF technique as an artificial neural network model. They also emphasized the relation to the biological neurons which seems to be selective for some range of input signals. This kind of neurons can be found in many parts of nervous systems arranged as cortical maps.

RBF networks are used as pattern classifiers in many applications, but they are also commonly employed in function approximation. In this paper, we are mainly interested in the latter case.

In this paper, modified Kohonen's learning laws are exploited to the RBF model. The idea is to first find the initial centers by using the SOM algorithm and then fine-tune the parameters of the network by using the modified LVQ algorithm. Recently, one of the authors [11] has used a similar technique to tune a neuro-fuzzy model called fuzzy self-organizing map.

The organization of this paper is as follows: in Sect. II., the basic structure of RBF networks is briefly presented. After that, the SOM learning method is presented. In Sect. III., we show how the SOM and modified LVQ algorithms are employed in the tuning of the RBF network. In Sect. IV., the proposed method is verified with a function approximation task. Finally, the potential of the method and future work are discussed.

II. NETWORK STRUCTURES AND TRAINING

A. RBF NETWORK

The RBF network is a three layer feedforward network, as illustrated in Fig. 1. The input layer consists of neurons which only function is to distribute the incoming signals to the neurons of the hidden layer. In the hidden layer, neurons perform nonlinear mapping from input space to a space which dimension is equal to the number of the neurons in the hidden layer. In our work, the neurons are defined as local receptive fields using Gaussian exponent function. The responses R_i of the neurons are computed according to

$$R_i(\boldsymbol{x}) = e^{-\|\boldsymbol{x} - \boldsymbol{c}_i\|^2/\sigma_i^2}, \qquad (1)$$

where $\boldsymbol{c}_i$ is the center of the receptive field unit and $\boldsymbol{x}$ is the input vector. Parameter σ_i defines the width of the receptive field. The neurons of the output layer combine linearly the responses of the hidden layer units. Two alterative structures are introduced [8]. In the first network, the output of the radial basis function network is computed as

$$f(\boldsymbol{x}) = \sum_{i=1}^{n} w_i R_i(\boldsymbol{x}), \qquad (2)$$

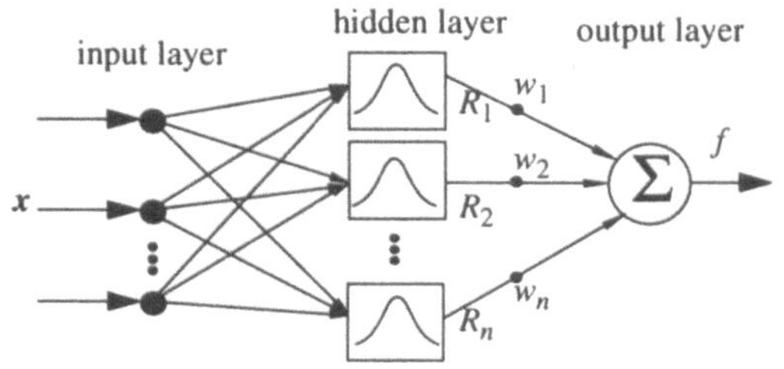

Fig. 1. A general structure of a single output Gaussian RBF network.

where w_i are the output weights of the neurons. Variable n is the number of neurons in the hidden layer. In the second approach, lateral connections between the processing elements are assumed. The network produces the normalized response of neurons according to

$$f(\boldsymbol{x}) = \frac{\sum_{i=1}^{n} w_i R_i(\boldsymbol{x})}{\sum_{i=1}^{n} R_i(\boldsymbol{x})}. \tag{3}$$

The training of the RBF network can be done in several ways. Moody and Darken [8] proposed a hybrid method exploiting k-means clustering, P-nearest neighbour width selection and LMS algorithm.

Currently, perhaps, the most effective way to estimate the parameters of the RBF networks is the orthogonal least squares (OLS) method introduced by Chen [1] and Wang [12]. The method is based on systematic selection of the most significant basis functions by using a batch procedure. The centers and the weights are optimally chosen one by one from the training set. After that, the network can be further tuned by using the back-propagation algorithm. On the other hand, Lee and Kil [7] suggested a dynamic construction approach where they combined a heuristic adding of the units and the backpropagation of errors.

Methods presented above are very efficient. However, authors' intention is to explore iterative and biologically more relevant learning laws than the back-propagation and the OLS.

B. SELF-ORGANIZING MAP

Kohonen's SOM [4,5] is a vector quantizer which clusters the input data. The SOM network consists of neuron cells which are interconnected to each other according to a predefined structure or topology. Each neuron cell contains a codebook vector (i.e., set of input weights).

The self-organizing map algorithm to adapt iteratively the network consists of two steps:

1) *Similarity matching.* The nearest codebook vector $\boldsymbol{m}_c$ of the input vector is searched according to

$$\|\boldsymbol{x}(k) - \boldsymbol{m}_c(k-1)\| = \min_i \{\|\boldsymbol{x}(k) - \boldsymbol{m}_i(k-1)\|\}. \tag{4}$$

2) *Updating.* The best matching cells and its neighbourhood cells are shifted towards the input sample:

$$\begin{aligned} \boldsymbol{m}_i(k) &= \ldots \\ &\boldsymbol{m}_i(k-1) + \alpha(k) h(k) [\boldsymbol{x}(k) - \boldsymbol{m}_i(k-1)], \end{aligned} \tag{5}$$

where $\alpha(k)$ is a decaying function of time so that $0 \le \alpha(k) \le 1$ and $h(k)$ is a factor determining the net topology so that it gives the largest values for the best matching cell and the cells which are close to it in the network topology, $0 \le h(k) \le 1$.

Learning Vector Quantization LVQ [5] is a supervised learning algorithm for pattern classification. In the Kohonen's LVQ2.1 algorithm [6], the two closest neurons of the input vector are tuned when the input sample falls into a "window" between them. The window is defined by relative distances from the input sample. The consequence of learning law is that the window's position is shifted, when the two closest neurons are updated. The basic idea is to shift the window so that, in the long run, the window superimposes the decision border between two different output classes.

III. HYBRID LEARNING METHOD

We suggest a three phase learning method for both of the RBF structures presented in previous section. In the first phase, the self-organizing algorithm (SOM) is used to find the initial centers of the neurons in the hidden layer. After that, the receptive fields are formed around the centers and the output weights are initialized. Finally, a modified LVQ2.1 algorithm is employed to tune all the parameters of the network.

A. SELF-ORGANIZATION

The centers of the basis functions are self-organized according to standard Kohonen's SOM learning laws presented in previous section (cf. Eq. (4) and (5)).

B. FORMATION OF RECEPTIVE FIELD UNITS

Next, the receptive fields are formed around the centers and the output weight are initialized. The widths of the units are initialized so that they sufficiently overlap each other. The P-nearest neighbour heuristics is used:

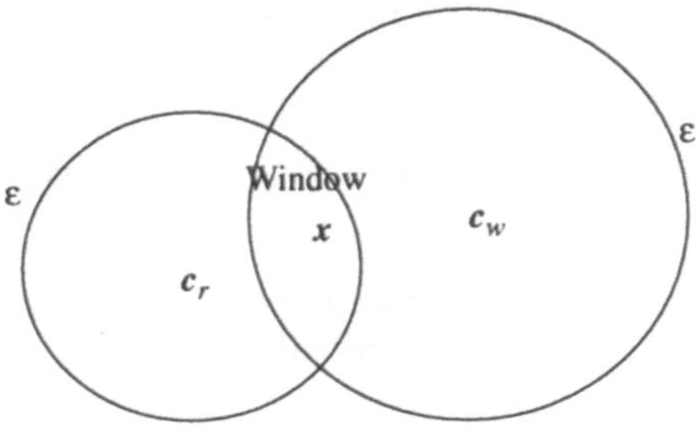

Fig. 2. The "window" is defined as intersection of the most firing "winner" receptive field unit w and the "first runner-up" unit r.

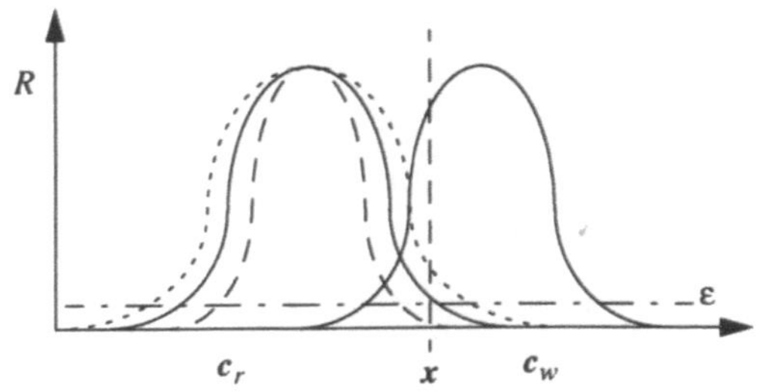

Fig. 3. The width of the first runner-up receptive field unit is either decreased or increased according to the modified LVQ2.1 learning rule.

$$\sigma_j = \sqrt{\frac{1}{P}\sum_{i=1}^{P}\|c_j - c_i\|}, \tag{6}$$

where c_i are the P closest centers of the center c_j. In addition, the output weights w_i are set to zero.

C. FINAL TUNING

Modified LVQ2.1 learning laws are employed in the final tuning. Now, the window is simply defined as the overlapping area of two most firing receptive field units where the firing strengths are greater than a threshold value ε, as shown in Fig. 2. The unit w is "winner" firing most and the unit r is the "first runner-up". When at least two receptive field units fire simultaneously, a learning law for the updating of the width is used. The actual updating is done by either decreasing or increasing the width of the first runner-up unit σ_r. The updating law for the width is:

$$\begin{aligned} \sigma_r(k+1) &= \sigma_r(k) - \operatorname{sign}(w_r)\, g_r(k)\, \sigma_r(k), \\ &\quad \text{if } (f-y) > 0, \\ \sigma_r(k+1) &= \sigma_r(k) + \operatorname{sign}(w_r)\, g_r(k)\, \sigma_r(k), \\ &\quad \text{otherwise}, \end{aligned} \tag{7}$$

where f is the output of the RBF network and y is the desired value of the output. Parameter $g_r(t)$ is the learning rate of the receptive field unit so that $0 \le g_r(t) < 1$. Fig. 3 illustrates the width updating.

For the *normalized response* structure, the corresponding updating laws are

$$\begin{aligned} \sigma_r(k+1) &= \sigma_r(k) - g_r(k)\, \sigma_r(k), \\ &\quad \text{if } \operatorname{sign}(f-y) = \operatorname{sign}(w_r - w_w), \\ \sigma_r(k+1) &= \sigma_r(k) + g_r(k)\, \sigma_r(k), \\ &\quad \text{otherwise}. \end{aligned} \tag{8}$$

The center of the winner receptive field unit is updated according to

$$c_w(k+1) = c_w(k) + g_r(k) R_w [x(k) - c_w(k)], \tag{9}$$

and the output weight is updated according to

$$w_w(k+1) = w_w(k) + g_w(k) R_w [y(k) - f(k)]. \tag{10}$$

where $(0 \le g_w(t) < 1)$ is the learning rate of the output weights.

IV. SIMULATION EXAMPLE

The proposed method was verified by a simple function approximation problem [3,11]. The function to be approximated is a two dimensional nonlinear *sinc* function:

$$z = \operatorname{sinc}(x, y) = \frac{\sin(x)}{x} \cdot \frac{\sin(y)}{y}. \tag{11}$$

A training data set of 225 samples was taken uniformly from an area of inputs (x, y) [-10,10], [-10, 10] and the output z ranged between [-0.21, 1]. A validation data set of 625 was generated from the same ranges of the input. In the beginning of the training, the centers of the receptive field units were given random values from the training data and the topology of the map was a 5×5 matrix. The self-organization took about 45 epochs (i.e., runs through training data set). The widths of the receptive field units were set according to 2-nearest neighbour heuristic ($P = 2$). The final tuning was done for both network structures. The learning rates were decreased monotonously during the learning procedure. Fig. 4 shows how the training error of structure (2) converged in 10 separate training sessions. The corresponding results for structure (3) are presented in Fig. 5. We also compared our method to a hybrid method similar to [8] which exploited SOM, P-nearest neighbour heuristics, and LMS algorithm. The RMS of this method ($P = 2$) was 0.0425 on training data set and 0.0415 on validation data set.

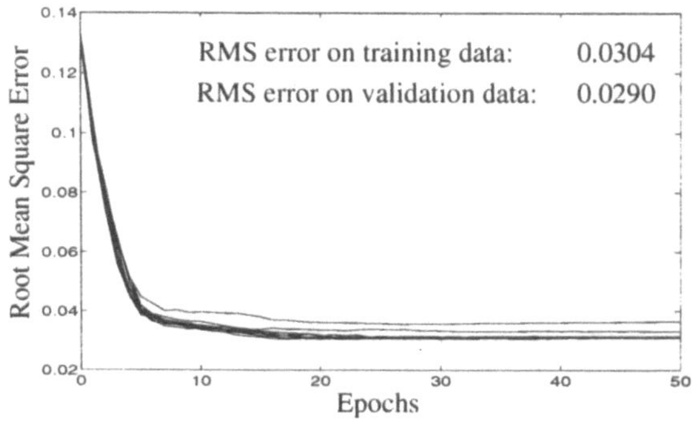

Fig. 4. RMS error convergence of RBF structure (2). Initial learning rates were $g_r(t) = 0.015$ and $g_w(t) = 0.33$.

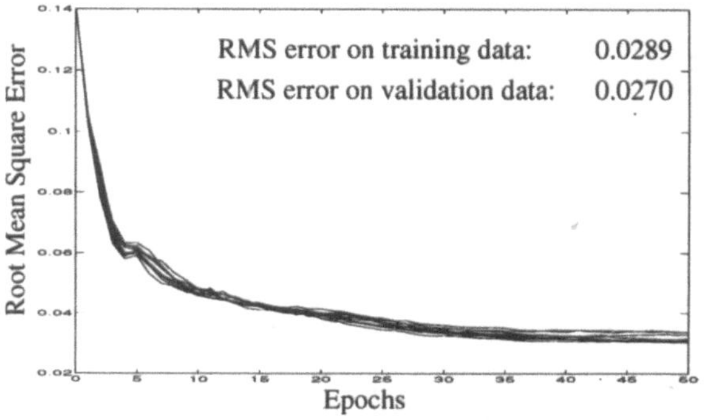

Fig. 5. RMS error convergence of RBF structure (3). Initial learning rates were $g_r(t) = 0.02$ and $g_w(t) = 0.3$.

V. CONCLUSIONS

In this paper, we have introduced a hybrid learning method for the RBF network. This method uses learning algorithms based on Kohonen's SOM and LVQ. The proposed learning algorithms converge fast and seem to be robust. In addition, they are very simple and use local information. The results given in Sect. IV are promising. In our future work, we plan to study the use of the proposed method in real-world problems.

REFERENCES

1. S. Chen, C.F.N. Cowan and P.M. Cowan: IEEE Trans Neural Networks 2, 302 (1991)
2. D. Fox, V. Heinze, K. Möller, S. Thrun, and G. Veenker, Artificial Neural Networks, T. Kohonen, K. Mäkisara, O. Simula, and J. Kangas (eds.) 201, 1991.
3. J.-S. Jang: IEEE Trans. Neural Networks, 23, 665 (1993)
4. T. Kohonen: Self-organization and Associative Memory, 2nd ed., Berlin: Springer Verlag 1988.
5. T. Kohonen: Proc. IEEE, 1464 (1990)
6. T. Kohonen, Proc. Int. Joint Conf. Neural Networks, TIJCNN-90 I, 545 (1990)
7. S. Lee and R. M. Kil: Neural Networks 4, 207 (1991)
8. J. Moody and C. J. Darken: Neural Computation 1, 281 (1989)
9. H. Ritter, T. Martinez, and K. Schulten: IEEE Trans. Neural Networks 1, 131 (1990)
10. H. Ritter, Artificial Neural Networks, T. Kohonen, K. Mäkisara, O. Simula and J. Kangas (eds.) 379, 1991.
11. P. Vuorimaa: Fuzzy Sets and Systems 66, 223 (1994)
12. L.-X. Wang and J. M. Mendel: IEEE Trans. Neural Networks 3, 807 (1992)
13. 1XXX studies of the self-organizing map (SOM) and learning vector quantization (LVQ). Helsinki University of Technology, Laboratory of Computer and Information Science, 1995. Available at the internet address 130.233.168.48 via anonymous FTP.

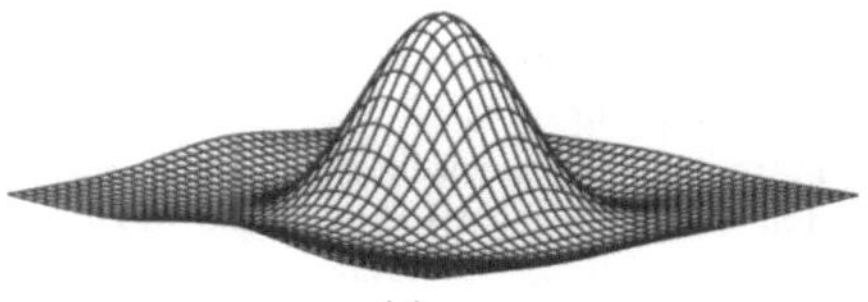

(a)

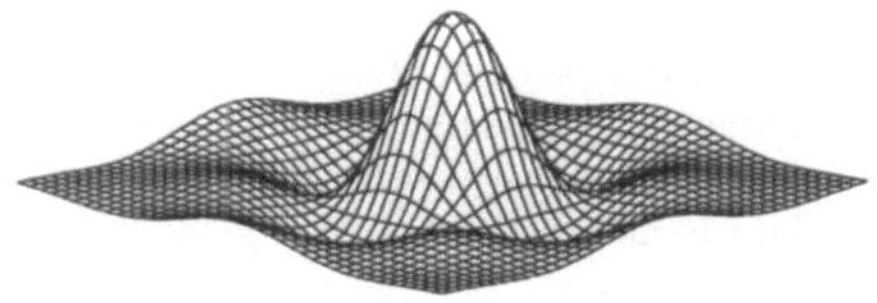

(b)

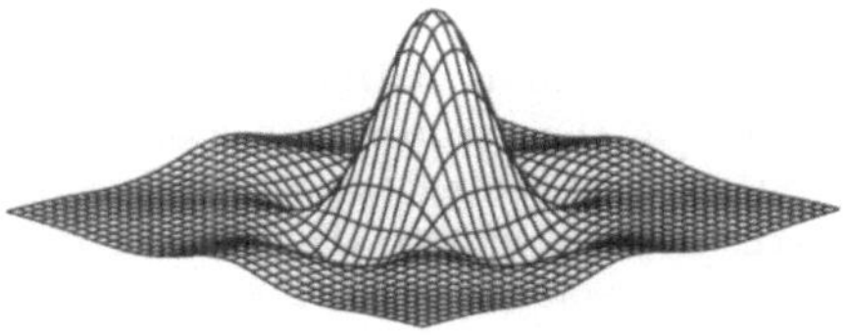

(c)

Fig. 6. Approximation of sinc function after (a) one epoch, (b) 5 epochs and (c) 40 training epochs.

KOHONEN'S MAPS FOR CONTOUR AND "REGION-LIKE" SEGMENTATION OF GRAY LEVEL AND COLOR IMAGES

René Natowicz (1) Lothar Bergen (1) Bruno Gas (2)

(1) **E.S.I.E.E.**—Cité Descartes–BP 99–93162 Noisy le Grand cedex France—email: natowicz@esiee.fr

(2) **E.N.S.E.A.**—6, av. du Ponceau, 95014 Cergy-Pontoise cedex France—email: gas@ensea.fr

Abstract

One propounds two methods of image contour segmentation based on the tonotopy property of Kohonen's self-organizing maps (SOM). The first method consists in quantizing the set of gray levels of the image by making use of a one dimensional SOM, then detecting the spatial discontinuities of the involved mapping of quantized gray levels onto map's cells. The same method brings color image segmentation when quantizing the color values of the image by making use of a three dimensional SOM. This first method is multiresolution acording to gray level or color accuracy. Examples of gray level and color image segmentations illustrate its outcome.

The second method presented consists in quantizing the set of spatial and gray level pixel coordinates by a three dimensional SOM. This quantization brings a "region-like" pixel clustering out of which one computes contour segmentation of the image. This method is monoresolution and yelds contour segmentations less noisy than the one obtained by the first method. This second method is illultrated by an example of gray level image segmentation.

1 Introduction

Self-organizing feature maps (SOM) [1,2] allow to compute vector quantizations through which any two points of the quantized set, close according to the distance defined on the set, are mapped onto two cells close according the distance defined on the map ("tonotopy" property). In what follows one will be interested in the quantization mapping K, $K : V \longrightarrow M, v \mapsto c$ which assigns any vector v of the quantized set its representing cell c on the map. Let V be the quantized vector set, $dim(V) = dim(\{\alpha \cdot u + v; u \in V, v \in V, \alpha \in R\})$ be the dimension of the set of linear combinations of its elements, and let M be the quantizing map. Tonotopy property holds when map's cells are at vertices of a grid of same dimension $dim(V)$. The four following cases are of interest in the following:

- $V = G$. The image's gray level values are the quantized set. One has $dim(V) = 1$ and a one dimensional quantizing map (1D-SOM) composed of n linearly connected cells
- $V = C$. The image's color values are the quantized set. One has $dim(V) = 3$ and a 3D-SOM composed of n^3 cells at vertices of a cubic grid
- $V = P$ and gray level image. The spatial and gray level components of a gray level image's pixels are the quantized set. One has $dim(V) = 3$ and a quantizing 3D-SOM composed of n^3 cells at vertices of a cubic grid
- $V = P$ and color image. The spatial and color components of a color image's pixels are the quantized set. One has $dim(V) = 5$ and a quantizing 5D-SOM (composed of n^5 cells at vertices of a hypergrid).

Situations in which dimension of the map is lower than the dimension $dim(V)$ of the quantized set bring mappings for which tonotopy property does not hold. In [1,2] is an instance of such a situation: a subset of the 2D-plane with a uniform probability density is quantized by a 1D-SOM which organizes as a "peano-like" curve. One can notice that two close points of the 2D quantized set can be mapped onto two distant map's cells.

2 Contour segmentation by detecting quantization mapping's spatial discontinuities

The first method of gray level image contour segmentation set out in this paper consists in quantizing the set of gray levels present on an image by making use of a 1D-SOM. Thanks to tonotopy property, any two close gray levels are mapped onto two close map's cells (possibly the same). Contour segmentation consists in detecting spatially close pixels mapped onto distant cells [3,4]. Formally, let G be the set of gray levels, M a 1D-SOM composed of linearly connected cells and $K : G \longrightarrow M, g \mapsto K(g) = c$ be the quantization mapping involved by the quantization. Mapping K's domain is extended to set P of image's pixels, $K : P \longrightarrow M, p = (x, y, g) \in G \mapsto K(g) = c$; any pixel p is assigned a fixed spatial neighborhood $N_P(p) \subseteq P$ on the image (classical 8-nearest neighbors of central pixel p); any cell c_i is assigned a spatial neighborhood $N_M(c_i, r) \subseteq M$, defined as the set of cells whose path length to cell c_i is less than integer r: $N_M(c_i, r) = \{c_k \in M, |i - k| \leq r\}$. Integer parameter r is the radius of the neighborhood centered on cell c_i and acts as a gray level resolution parameter. The set $S(r)$ of segmentation pixels computed at resolution r is

$$S(r) = \{p \in P; \exists \pi \in N_P(p), K(\pi) \notin N_M(K(p), r)\}.$$

The segmentation is multiresolution according to gray level resolution parameter r as the following property holds:

$$r < r' \Longrightarrow S(r) \supseteq S(r').$$

Letting $\delta(M)$ be the diameter of map M, i.e. the maximum shortest path length between two map's cells, one has the two following limit cases:

- $r > \delta(M)$. As $r > \delta(M) \Longrightarrow S(r) = \emptyset$, it is the lowest segmentation resolution
- $r < 0$. As $r < 0 \Longrightarrow S(r) = P$, it is the highest segmentation resolution.

Between these limit cases, segmentation sets $S(r)$ computed at high gray level resolutions (i.e. with "small" radius $r's$ values) are noisy because even irrelevant small local variations of gray levels can bring spatial discontinuities of the quantization mapping. These noise points of the segmentation progressively vanish when lowering the resolution (i.e. increasing radius $r's$ value) but at the detriment of real contour points. Figure 1. is an instance of gray level segmentation at various resolutions.

Similar segmentations of color images are achieved with the same method of spatial discontinuity detection: the set of colors is quantized with a 3D-SOM for tonotopy to be a property of the quantization mapping and the definition of segmentation sets $S(r)$ remains unchanged. One decided map $M's$

Figure 1: **Contour segmentation of a gray level image.** Six contour segmentations at decreasing resolutions (highest resolution is for map's neighborhood radius $r = 1$). Quantizing map is composed of 50 linearly chained cells. The original image has 256 gray levels.

cells to be at vertices of a cubic grid. The distance $d_M(c, c')$ between cells c and c' is defined as the shortest path length on the grid between the two cells and $N_M(c, r) = \{\sigma \in M, d_M(c, \sigma) \leq r\}$ is the neighborhood of radius r centered on cell c. Figure 2. is an instance of such a color segmentation at various resolutions. The same remarks hold about noise vanishing at the detriment of real contour points when lowering color resolution (i.e. increasing 3D-SOM's neighborhood radius r). One can notice that some contours are present on the color image segmentation which were not on the gray level corresponding one and that segmentations are more noisy.

The noise present on image segmentations computed by this method which consists in detecting spatial

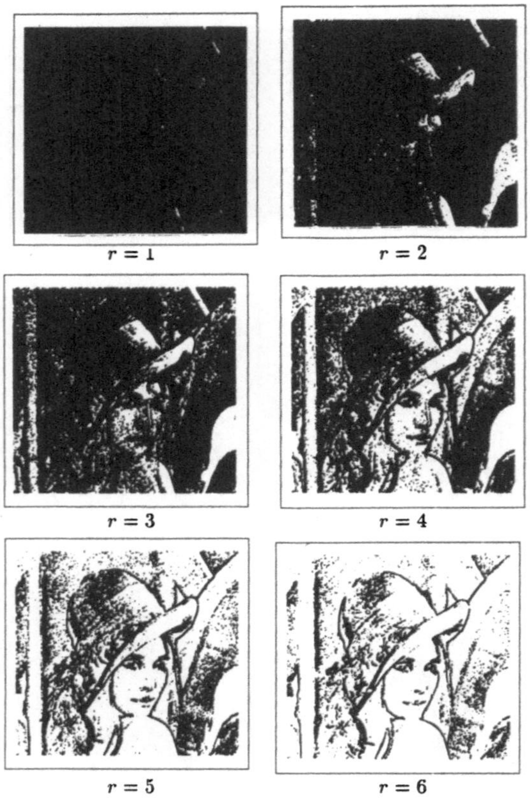

Figure 2: **Contour segmentation of a color image.** Six contour segmentations at decreasing resolutions (highest resolution is for map's neighborhood radius $r = 1$). Quantizing map composed of $10 \times 10 \times 10$ cells connected in a cubic grid. The original image has 16,000 colors.

discontinuities of the quantization mapping motivates the second method of contour segmentation through "region-like" mappings which is defined in what follows.

3 Contour segmentation out of "region-like" mappings

The segmentation method set out in section 2. was based on a vector quantization of gray levels or colors which does not take into account pixel spatial closeness. In order to compute less noisy segmentations preserving real contours of gray level images, one will consider the pixels as a set of points in the 3D-space of spatial and gray level coordinates and quantize it by a 3D-SOM for tonotopy property to hold. Thus, the distance between pixels in the space of spatial and gray level coordinates will be evenly represented by the distance between map's cells. One expects

1. any cell to represent a set of spatially close pixels with close gray level values
2. any couple of connex cells to represent two sets of spatially close pixels with distant gray level values
3. any couple of distant cells to represent two sets of spatially distant pixels.

The inverse mapping image $K^{-1}(c)$ of any cell c will be a subset of pixels close in the 3D-space of spatial and gray level coordinates but it is important to notice that this subset is not necessary connex –even though generally connex–; e.g. think of a spatial 5-neighborhood on the image of a chessboard: the inverse mapping $K^{-1}(c)$ of cell c can include two different white squares. Because of this connectivity lack, one will call this pixel clustering a "region-like" mapping.

On figure 3. a–d are four different "region-like" segmentations defined as the borders of any two "region-like" pixel sets mapped onto a couple of connex cells. These four "region-like" segmentations only differ by the codebooks initially input to the process of map's self-organization (random vectors initially assigned to map's cells). The quantized image has 256 gray levels and the quantizing maps have $6 \times 6 \times 6$ cells at vertices of a cubic grid.

On figures 3. a–d one can notice

- "tesselation-like" segmentation lines in which real contour points are present
- "tesselation-artefacts" corresponding to almost regular tesselation of homogeneous gray level areas
- apart from "tesselation-artefacts", the segmentations are far less noisy that the ones obtained through the first method of segmentation
- "tesselation-artefacts" are sensitive to initial codebooks while real contours points are less.

This last point imposes a simple (and perhaps "tricky") way to obtain contour segmentation out of these "region-like" segmentations by simply computing the image intersecting the "region-like" segmentations. Figure 3. e is the resulting intersection of figures 3. a–d. One can observe that little noise is present and real contour points are preserved.

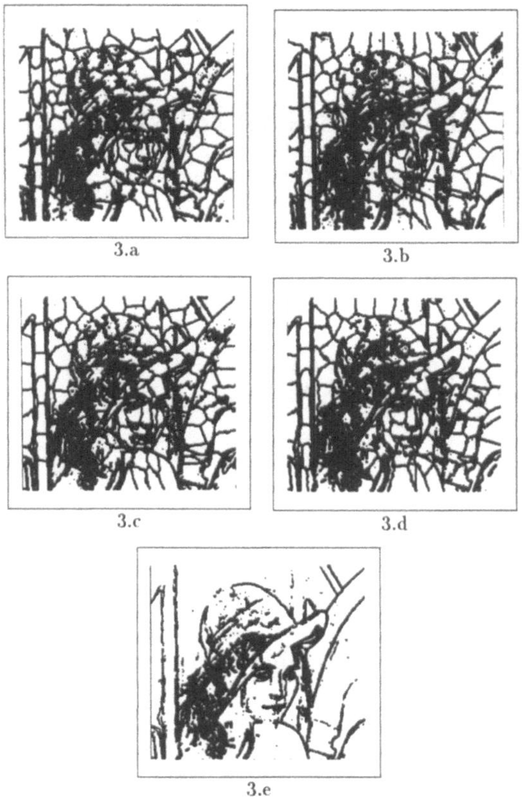

Figure 3: **Region-like and contour segmentations.** Figures a–d: four region-like segmentations each obtained from a separate map composed of $6 \times 6 \times 6$ cells connected in a cubic grid. Figure e: Contour segmentation is the intersection of the four region-like segmentations.

4 Discussion

Two different methods of contour segmentation were set out, both relying on the tonotopy property of Kohonen's SOMs which imposes the dimension of the quantizing map. In both methods, quality of the segmentation depends upon the number of cells which, at the present time, is not automatically determined.

The first method consists in detecting spatial discontinuites of the quantization mapping of either gray level or color value set of an image. This method is multiresolution according to a gray level or color accuracy parameter. One noticed that noise segmentation points can only be discarded at price of real contour points.

The second method is monoresolution and aims at computing noiseless contour segmentations preserving real contour points. It consists in a clustering of the pixels (spatial and gray level coordinates) which, thanks to tonotopy property of SOMs, brings a "region-like" segmentation computed as the border points of the inverse mapping image of all couples of connex cells. It appears that this method brings less noisy segmentations while preserving real contour points.

The extension of this second method towards color image segmentation was not set out in the present paper. It must be thought as a straightforward extension consisting in clustering the pixels in the five dimensional space of spatial and color coordinates by making use of a 5D-SOM.

The proposed approach for processing monoresolution contour segmentation of gray level images by computing the intersection of several "region-like" segmentations asks for several quantizations of the same image. A more satisfying approach for computing multiresolution contour segmentations out of a single "region-like" clustering is under study. This approach combines "region-like" clustering and spatial discontinuity detection in order to progressively discard segmentation points corresponding to small gray level differences between connex regions (which comprises "tesselation-like" artefacts). To achieve such a segmentation of gray level images, one quantizes the set of gray levels using a 1D-SOM and, in parallel, one computes a "region-like" segmentation using a 3D-SOM. At gray level resolution r on the 1D-SOM, solely are kept in the segmentation point set $S(r)$ the couples of border pixels of the "region-like" segmentation whose gray level values are coded by cells at distance greater than resolution r. This approach under study is multiresolution as it holds that segmentation sets are included according to gray level resolution r.

References

1. T. Kohonen, "Self-organization and associative memory" , Springer-Verlag Berlin, 1984.
2. T. Kohonen, "The self-organizing feature map"; proceedings of the I.E.E.E., vol. 78, n. 9, September 1990.
3. R. Natowicz, R. Sokol, "Self-organizing feature maps for image segmentation", I.W.A.N.N.'93, lecture notes in computer science, vol. 686, Springer-Verlag, 1993.
4. R. Natowicz, "Segmentation d'images par cartes de Kohonen", colloque Gretsi, Juan les Pins, 1993.

USING GENETIC ALGORITHM IN SELF-ORGANIZING MAP DESIGN

Ari Hämäläinen

Rolf Nevanlinna Institute, P.O.Box 26 (Teollisuuskatu 23), 00014 University of Helsinki, Finland

Abstract A new method for self-organizing map design is proposed. The method is based on a genetic algorithm. Some simulations are also reported.

1 Introduction

The self-organizing map (SOM) [4] has two interesting features: The density of the reference vectors reflects the density of the input data and the reference vectors of neighboring units are near each other in the input space (topological ordering). The ordering feature has some applications in robotics [2, 8] and it can be useful in displaying the distances between different classes (e.g., semantic maps) [7]. Usually the topology is fixed beforehand and typical choices are a chain, a two or three dimensional rectangular array and a hexagonal array [4, 8].

In some cases, e.g., when the dimension of the training data is high and the shape of the density is unknown, it might be useful to have an automatic method for SOM topology design. An example is given in Fig. 1. If the SOM with the lattice topology is trained by the data from the annulus, some of the reference vectors will converge to the center of the annulus. On the other hand the ring topology is not the best possible for the two dimensional uniform density.

The so called neural-gas algorithm of Martinetz and Schulten [5] has the capability to "learn" the topology. Martinetz has proposed also another learning rule to form topologically ordered maps [6]. In this paper a new method for SOM design is proposed. The method is based on a genetic algorithm (GA) [1].

2 SOM Training

A SOM training step is

$$m_i(t+1) = m_i(t) + \alpha(t)h_{ic}(x(t) - m_i(t)),$$

where m_i is the reference vector at node i, α is the gain, h_{ic} is the neighborhood function

$$h_{ic} = \begin{cases} 1 & \text{if } i \in N_c \\ 0 & \text{if } i \notin N_c \end{cases}$$

and N_c is the neighborhood of the winner node c. The winner node is the one whose reference vector is nearest to the training vector x. The neighborhood of the node c is the set of nodes which are within a certain radius r from node c (e.g., there is a connection between the c and the other node ($r = 1$) or there are at most two connections between c and the other node ($r = 2$)). Also some other neighborhood functions can be used instead of the one given above [8].

3 Genetic Operations

For the GA, the connections were encoded using a connection matrix C. If, e.g., a SOM has six nodes (two rows three columns), C is as given in Fig. 2, where 1 means that there is a connection between the nodes (every node is connected to itself). The upper triangle of the matrix was used as the chromosome.

The genetic operations were crossover and mutation. In the crossover operation a substring was chosen randomly from a chromosome and then the corresponding substrings between two randomly chosen parents were swapped. In mutation every bit was changed with probability p. The parents were ordered using their fitness and then the best 25 % of the population was used for reproduction (the parents were also included in the next generation). The reference vectors of the next generation were initialized randomly.

The fitness of the individuals were tested using

$$f = \left(\sum_{j=1}^{n} \sum_{x_i \in F_j} \| x_i - m_j \|^2 \right) * (1 + M)$$

as the fitness function for the trained SOMs. The measure M is the measure of disorder proposed [3]. Also other measures given in literature can be used (e.g., [9]) or one may set $M = 0$. F_j is the Voronoi region of m_j, i.e., for every $x \in F_j$, m_j is the nearest of all m_i's. Those maps that broke into several parts were also penalized. This fitness function is not the only possible, but it seemed to work well in the simulations.

4 Simulations

The method was tested in simple examples. In the first example the training data (random sample of size 1000) were from the one dimensional uniform density. The connection matrices C for each individual were initialized so that there was a connection between nodes with probability 0.2 (the diagonal elements are 1). Each individual was trained using 100*(# nodes) iterations, α decreased from 0.3 to 0.01 and the neighborhood radius r decreased to 1 (i.e., the final neighborhood of node i includes only those nodes which are connected to it). The mutation probability was $p = 0.01$ and the population size was 48. A typical result is shown in Fig. 3. The left figure is the result given by GA. The smallest reference vector is at node 6 and this node is connected to the node 7, which has the second smallest reference vector etc. The result is clearer if we renumber the nodes so that the smallest one is 1 etc. (right)

$$[67582109134] \rightarrow [12345678910].$$

The connections are depicted as lines between nodes. The resulting topology is a chain as may be expected.

In the second example the data were from the uniform annulus density. Again, the connection matrices C were initialized so that there was a connection between two nodes with probability 0.2 and the mutation probability was $p = 0.01$. The population size was 48. The result is shown in Fig. 4.

5 Conclusions

A new method for SOM design was proposed. In the simple test cases the method worked well. It is to be seen how well the method works in more complex tasks (i.e., in higher dimensions, with more complicated training data distributions). Computational problems may arise because the training time for large SOMs must be longer and also the population size must be larger. One may try to avoid these problems by choosing the initial population carefully, e.g., by including the chain and the lattice topology or some other standard topology in the initial population.

In these tests the training sample was fixed. The SOM algorithm is an on-line method and the training data do not have to be fixed beforehand. It is possible to use the GA -based method for online training if the fitness is tested, e.g., using the next 1000 data points coming from the source. Note that during testing the training of SOMs must be stopped, otherwise the fitness function will change during testing. The initialization of the next generation can be done using the parent population, e.g., if the mutation operation was used the reference vectors of the offspring are the reference vectors of the parent with some connections removed and some added and the training of SOM goes on. These ideas are not tested in the present paper, but they will be a topic for further research.

References

[1] Lawrence Davis, editor. *Genetic Algorithms and Simulated Annealing.* Morgan Kaufman Publishers, Inc., 1987.

[2] Daryl H. Graf and Wilf R. LaLonde. A neural controller for collision-free movement of general robot manipulators. In *IEEE, ICNN-88 San Diego, California, 24.-27.7.1988*, volume 1, pages 77–84, 1988.

[3] Ari Hämäläinen. A measure of disorder for the self-organizing map. In *Proceedings of the 1994 IEEE International Conference on Neural Networks. Walt Disney World Dolphin Hotel, Orlando, Florida. June 28 - July 2*, pages 659–664, 1994.

[4] T. Kohonen. *Self-Organization and Associative Memory.* Springer-Verlag, third edition, 1989.

[5] T. Martinetz and K. Schulten. A "neural-gas" network learns topologies. In T. Kohonen, K. Mäkisara, O. Simula, and J. Kangas, editors, *Artificial Neural Networks*, volume 1, pages 397–402. North-Holland, 1991.

[6] Thomas Martinetz. Competitive Hebbian learning rule forms perfectly topology preserving maps. In Stan Gielen and Bert Kappen, editors, *ICANN '93. Proceedings of the International Conference on Artificial Neural Networks*, pages 427–434. Springer-Verlag, 1993.

[7] H. Ritter and T.Kohonen. Self-organizing semantic maps. *Biol. Cybern.*, 61:241–254, 1989.

[8] Helge J. Ritter, Thomas M. Martinetz, and Klaus J. Schulten. Topology-conserving maps for learning visuo-motor-coordination. *Neural Networks*, 2:159–168, 1989.

[9] Hans Ulrich Bauer and Klaus R. Pawelzik. Quantifying the neighborhood preservation of self-organizing feature maps. *IEEE Trans. on Neural Networks*, NN-3(4):570–579, July 1992.

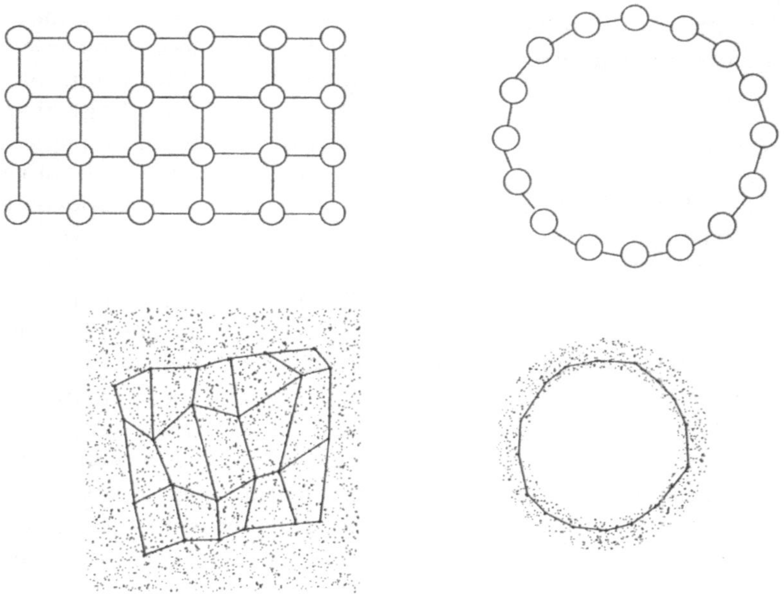

Figure 1: Self-organizing maps. The topologies of the maps (above) and the reference vectors in the two dimensional space (below).

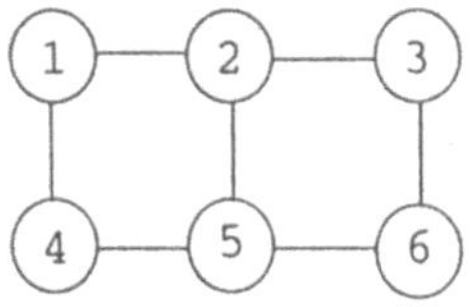

$\Rightarrow C$ is

	1	2	3	4	5	6
1	1	1	0	1	0	0
2	1	1	1	0	1	0
3	0	1	1	0	0	1
4	1	0	0	1	1	0
5	0	1	0	1	1	1
6	0	0	1	0	1	1

$\Rightarrow$ 101001010001101

Figure 2: The connection matrix C.

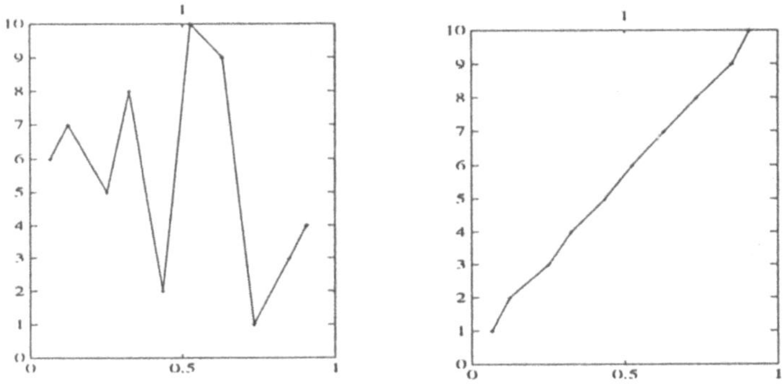

Figure 3: The result in a one dimensional test case (left) and the re-numbered SOM (right). The values of the reference vectors are on the x-axis and the number of the node is on the y-axis. The SOM is ordered.

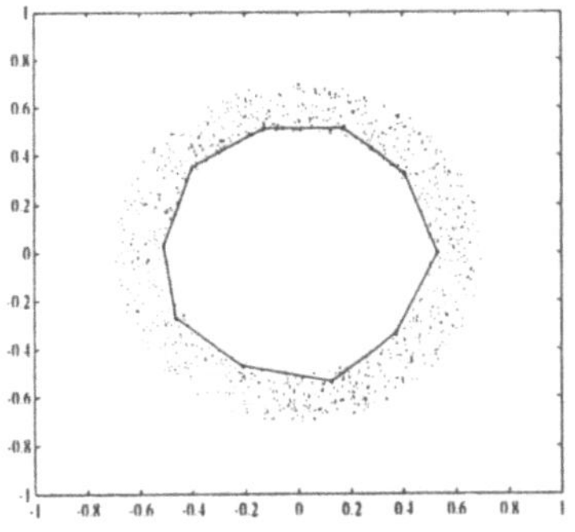

Figure 4: The result in a two dimensional test case.

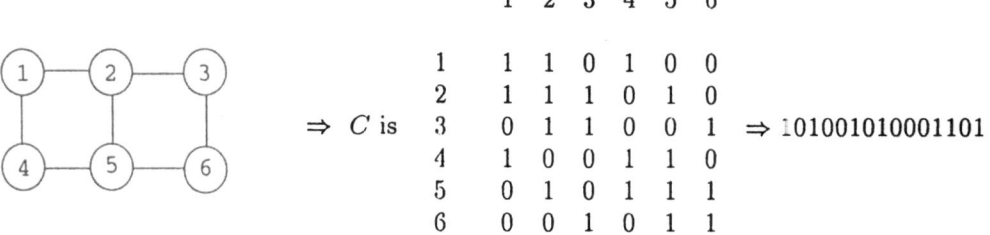

MODULAR LABELING RAAM

Alessandro Sperduti, Antonina Starita

Dipartimento di Informatica, Università di Pisa
Corso Italia, 40 - 56125 PISA, Italy

Abstract

The Labeling RAAM model is a neural network able to encode labeled graphs in fixed size representations. In order to speed up the training procedure and for reducing in size the developed compressed representations, we propose a modular Labeling RAAM. In order to develop the modular system, we face two main problems: the mapping problem, i.e., how to map components of the structures into modules; and the membership problem, i.e., discovering which module must be used for decoding a compressed representation. The mapping between components and modules can be decided on the basis of the strongly connected components of the structures. The membership problem is solved by resorting to the BAMs derived from each LRAAM module. Preliminary results on the modular system are encouraging.

INTRODUCTION

In the last few years, several researchers have tried to demonstrate how symbolic structures can be represented and manipulated in a connectionist system, while still preserving all the computational characteristics of connectionism (and extending them to the symbolic representations) [5, 8, 9, 10, 13]. The goal is to highlight the potential of the connectionist approach in handling domains of structured tasks. The common background of their ideas is an attempt to achieve distal access and, consequently, compositionality.

The Labeling RAAM (LRAAM) is one of this models. It can encode labeled graphs with cycles by representing pointers explicitly into an encoder network. Data encoded in an LRAAM can be accessed by a pointer as well as by content. Specifically, the LRAAM network can be transformed into an analog Bidirectional Associative Memory (BAM) [6] by connecting the output units with the input units. The BAM is, then, used to access by content the structures encoded into the LRAAM. Different access procedures can be defined depending on the access key [11]. In [12] we gave sufficient conditions for the stability of equilibria arising in the BAM.

Recent works have established that LRAAM can be exploited in at least three different domains: *i)* in knowledge representation, by encoding *Conceptual Graphs* [3, 4]; *ii)* in natural language processing, by representing *syntactical trees* [1]; *iii)* in image coding, by storing *Quadtrees* [7];

However, even if in some cases an LRAAM can store structures of the order of hundreds, the dimension of the hidden layer do not scale logarithmically with the size of the training set. As a consequence of that, when dealing with a large set of structures, due to the considerable size of the network, the learning time is usually long.

In general, when the problem at hand can be decomposed, modularization is often used to improve the performances of the network. In this paper, we present preliminary results on the modularization of the LRAAM. Modularity allows us to have a good scaling of the capacity of encoding as well as a reduction in the learning time. In the next sections we will discuss our approach to the efficient synthesis of modular LRAAMs.

THE LABELING RAAM

In Fig. 1 we have summarized some of the concepts underpinning an LRAAM. At the top of the picture, an LRAAM network with one label and two reduced descriptor

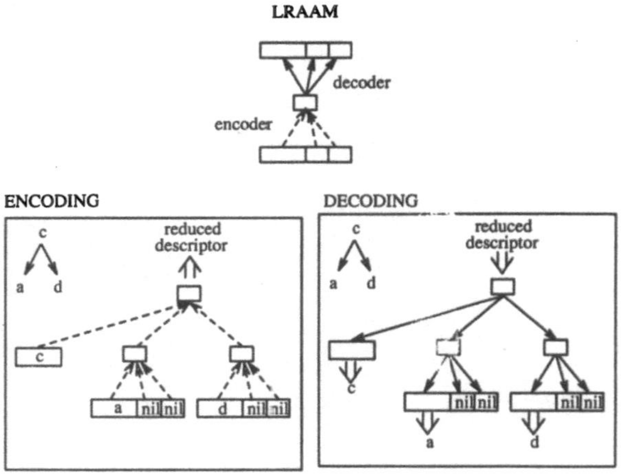

Figure 1: Encoding and decoding of a small labeled † using an LRAAM (top). Encoding connections are repr. sented by dashed lines, while decoding connections are represented by solid lines.

fields is shown. The set of connections from the input to the hidden layer of the network implements an encoder which is able to devise reduced descriptors of structures presented in input to the network. Since the size of a reduced descriptor is equal to the size of the reduced descriptor fields in input, reduced descriptors can be used recursively to encode complex structures. An example of encoding is given at the bottom left of the picture, where we have explicited the encoding process for a small tree. The information contained into a reduced descriptor can be accessed by using the decoder function implemented by the set of weights from the hidden to the output layer of the LRAAM. At the bottom right of the picture, the decoding process of the reduced descriptor devised by the encoding of the small tree is shown. Access by content can be obtained by connecting the output of the LRAAM with the input, obtaining in this way an analog Bidirectional Associative Memory (BAM).

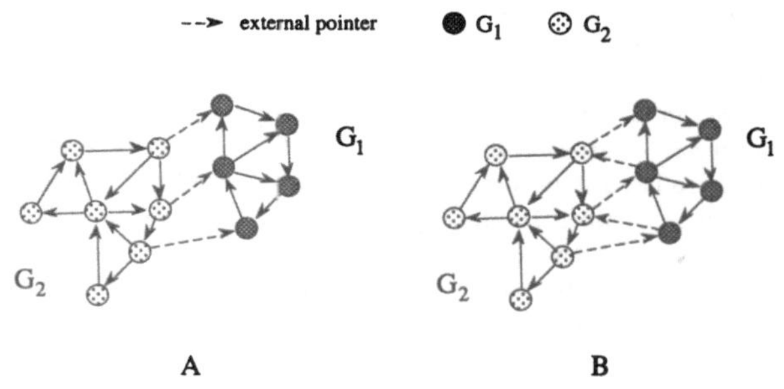

Figure 2: Moving target among modules.

MODULAR LRAAM

The synthesis of a modular LRAAM must solve two main problems: the *mapping problem* and the *membership problem*. The mapping problem consists in the choice of which components of the structure are encoded by a module, i.e., we must define how to map components to modules. At first glance this problem seems trivial, since any balanced partition of the structure seems to be a good solution to the problem. However, if we want to reduce the moving target problem at the minimum, care must be taken in defining such mapping. We will see that, even if in the general case there is no way to avoid the moving target problem to propagate among modules, when the structure we want to codify has *local cycles*, then it is possible to avoid the moving target problem among modules at the cost of a reduced degree of parallelism in the training.

The membership problem, instead, arises in the decoding phase and consists in to discover which module must be used to decode a given pointer, i.e., to which module a pointer belongs. The easiest approach to solve this problem is to build a reference table for the pointers containing the membership information. This reference table may be implemented by a standard classifier, however we will see that this information can be retrieved without extra neural structures and learning time by using the BAM defined for each module.

The Mapping Problem

The problem is to find a partition $PART_G=\{G_1,\cdots,G_k\}$ of the nodes of the graph G we want to encode* , such that each part G_i is encoded by the module $LRAAM_i$ and the moving target problem among modules is reduced at minimum. Of course, each module must codify a sufficient number of nodes to justify its existence. The moving target among modules happens when changes in the training set of one module propagates to other modules. For example, consider the partition of the graph shown in Fig. 2A. The set of nodes G_1 is encoded by $LRAAM_1$ and G_2 by $LRAAM_2$. Since there are pointers from G_2 to G_1, the training set for $LRAAM_2$ contains pointers (we will name them *external pointers*) whose values are established by the training of $LRAAM_1$. If the modules are trained altogether, $LRAAM_2$ will experience a moving target effect for the external pointers due to the training of $LRAAM_1$. This is not the case for $LRAAM_1$, whose training set does not depend on $LRAAM_2$. In this situation, the moving target among $LRAAM_1$ and $LRAAM_2$ can be removed if the training of $LRAAM_2$ is started only at the end of the training of $LRAAM_1$, when the external pointers have reached a stable representation. Even if this technique reduces the parallelism of the training, it produces a benefit on the efficiency of the training, i.e., the total number of operations employed by the learning is generally reduced.

In the general case, however, this technique cannot be applied. In fact, for example, if pointers from nodes in G_1 to nodes in G_2 are present (see Fig. 2B), also the training set of $LRAAM_1$ depends on the training of $LRAAM_2$, and consequently the modules *must* be trained altogether.

Thus, the proposed technique can be applied only if a cycle is codified inside a module. More formally, we must consider the *strongly connected components* of the graph. A direct graph G is said to be *strongly connected* if for every pair of distinct nodes v_i, v_j in NODE(G) there is a directed path from v_i to v_j and also from v_j to v_i. A *strongly connected component* (SCC) is a maximal subgraph that is strongly connected. According to the previous definitions we are interested in the *abstract graph* AG of the graph G, defined as the graph with: *i)* one node n_i for each strongly connected component SCC_i of G; *ii)* one arc from n_i to n_j if there exists at least one arc in G from a node in SCC_i to a node in SCC_j. Strongly connected components can be computed very efficiently with standard sequential techniques (see, for example, [2]). Note that, by definition, an abstract graph is acyclic. The basic idea is that each module can codify one or more strongly connected components according to the maximum number of elements it is able to encode. This

*A training set containing several graphs can be considered as containing a single disconnected graph.

combination of SCCs must be performed carefully, since the union of SCCs of an abstract graph may give rise to a cyclic graph. In order to avoid this event, a *topological sort* of the nodes in the abstract graph can be used to drive the merging process: a subset of nodes X in NODE(AG), can be merged if there is no other remaining node in AG whose topological order is between the topological order of two nodes in X.

Once a satisfactory graph FG (the *final graph*) is found, it can be used to define the mapping of the nodes of G, i.e., the nodes of G belonging to the set of SCCs represented by a node in FG are mapped in the same module. Moreover, the topology of FG can be used to define the modules' dependencies which must be respected during training.

Note that, this method do not require a sequential learning of the modules. Actually, its aim is to avoid the moving target among modules while preserving the maximal degree of parallelism in the training of the modules. In fact, according to the topology of FG, multiple modules may be *trained in parallel and in a totally independent manner.*

An example which clarify this statement is enlightened in Fig. 3, where it is shown how a tree can be mapped in six modules. The resulting abstract graph (which in this case is also the FG) is again a tree. The training of the modules will proceed from the bottom to the top of this tree, allowing modules on the same level to be trained in parallel and in an independent manner. Moreover, the training of a module can start as soon as the modules which are children of it have terminated their training. Thus, for example, if the module $LRAAM_4$ terminates before modules $LRAAM_5$ and $LRAAM_6$, the training of $LRAAM_2$ can start in parallel with $LRAAM_5$ and $LRAAM_6$.

The Membership Problem

Once the mapping problem is solved and the graph encoded in the modular system, the problem is how to keep trace of where a component of the structure is codified. In fact, when a pointer must be decoded, the right module must be chosen.

The key idea is to use the BAM associated to an LRAAM to discover if a given pointer belongs or not to it, i.e., if the LRAAM has generated it. The *membership test* of a given pointer $\vec{p}$ can be performed as follows:

1. consider all the BAMs, $BAM_1, \cdots, BAM_k$, derived by the LRAAMs, $LRAAM_1, \cdots, LRAAM_k$, in the modular system;
2. initialize the hidden layer output of the BAMs to $\vec{p}$ and leave the BAMs to relax; the activation vector at time t of the hidden layer in BAM_i is named $\vec{h}_i(t)$;
3. the pointer $\vec{p}$ belongs to $LRAAM_j$, where $j = arg\ \min_i \| \vec{h}_i(t_{test}) - \vec{p} \|$ and t_{test} must be experimentally defined.

This procedure relies on the observation that it is unlikely that a compressed representation devised by an LRAAM is the same as the one devised by another LRAAM trained on different patterns. Note that, in the hypothesis of perfect learning, t_{test} can be taken equal to one, since the pointer $\vec{p}$ will be distorted, at the first iteration, by all the BAMs derived by LRAAMs which did not devise it. However, in practice, perfect learning is never obtained and a validation of the membership method must be performed. This is the reason for which we did not link t_{test} to a particular value. We will see in the next subsection how to choose a value for t_{test}.

An example of modular LRAAM

Consider the simple tree in Fig. 4. We use two modules to encode the tree. The nodes of the tree are mapped on the modules as shown in Fig. 4. To avoid the moving target between modules, $LRAAM_1$ was trained first and then the obtained pointer to the root of the subtree was enclosed in the training set of $LRAAM_2$. A 12-3-12 encoder was used for both of them and the learning parameters were set as in the training of another LRAAM (18-6-18) encoding the whole tree and used for comparison. $LRAAM_1$ employed about 865 epochs to encode the subtree and $LRAAM_2$ 1123 epochs. The LRAAM encoding the whole tree took 1719

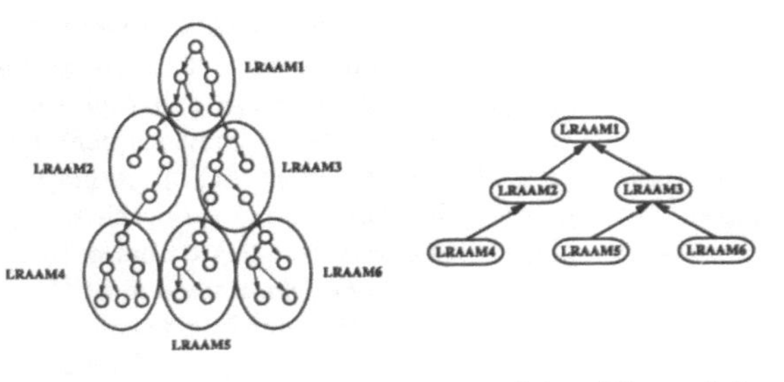

Figure 3: Mapping of a tree to six modules. The training of the modules proceeds in a bottom-up fashion. A module can start its training as soon as its children modules have terminated their training.

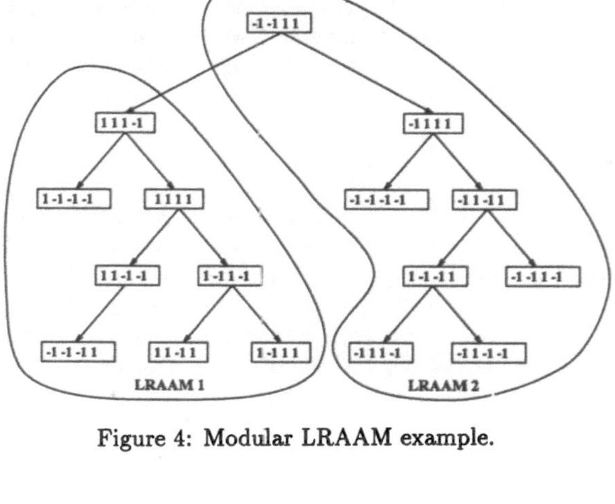

Figure 4: Modular LRAAM example.

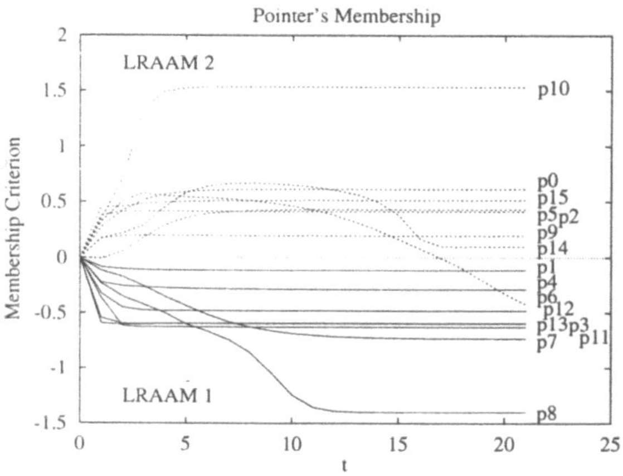

Figure 5: Plots of the membership criterion (see text) for the pointers in the tree. A positive value for the criterion implies the membership of the pointer to $LRAAM_2$, and a negative value the membership of the pointer to $LRAAM_1$.

epochs to converge. The total number of parameters in this LRAAM was 240 (216 weights + 24 biases) and each epoch computed 240×16 changes in the parameters (16 patterns in the training set). Thus the computational cost of the learning was around $240 \times 16 \times 1719$. The total cost for the modular system was instead given by $87 \times 8 \times (865 + 1123)$, with an improving in the efficiency of a factor 4.77. This factor was confirmed by other simulations. Moreover, if it is considered that one epoch of training in the monolithic system took the double of the time employed by one epoch in a module, then the modular system results to be faster than the monolithic system even considering a parallel implementation of them.

The value for t_{test} was computed on the basis of the following membership criterion:

$$\| \vec{h}_1(t) - \vec{p} \| - \| \vec{h}_2(t) - \vec{p} \|, \quad (1)$$

i.e., t_{test} can take any value such that for any pointer $\vec{p}$ belonging to $LRAAM_1$ the criterion is negative and for any pointer $\vec{p}$ belonging to $LRAAM_2$ the criterion is positive. Fig. 5 shows the plots of the criterion for all the pointers in the tree. From these plots it is clear that any value of t_{test} in the range $[2, 16]$ leads to a correct membership test.

CONCLUSIONS

In this paper, we proposed a modularization of the Labeling RAAM. The main motivations for developing a modular LRAAM were the speed up in learning and the reduction in size of the compressed representations. In order to develop the modular system, we faced two main problems: the mapping problem, i.e., how to map components of the structures into modules by avoiding the moving target among modules; and the membership problem, i.e., discovering which module must be used for decoding a compressed representation.

Regarding the former problem, we concluded that the moving target problem can be solved only if the structures have local cycles. In this case, the mapping between components and modules can be decided on the basis of the strongly connected components of the structures represented in the training set. The latter problem was, instead, solved by resorting to the BAMs derived from each LRAAM module. Preliminary results on the modular system are encouraging. However, the validity of the system must be checked on larger training sets, and this is one of the tasks we shall consider for future work.

REFERENCES

[1] V. Cadoret. Encoding syntactical trees with labelling recursive auto-associative memory. In *Proceedings ECAI*, 555 (1994). Amsterdam.

[2] T. H. Cormen, C. E. Leiserson, and R. L. Rivest. *Introduction to Algorithms*. The MIT Press, 1990.

[3] M. deGerlache, A. Sperduti, and A. Starita. Encoding conceptual graphs by Labeling RAAM. In *International Conference on Artificial Neural Networks*, 272 (1994). Sorrento.

[4] M. deGerlache, A. Sperduti, and A. Starita. Using Labeling RAAM to encode medical conceptual graphs. In *International Conference on Neural Networks and Expert Systems in Medicine and Healthcare*, 1994. Plymouth.

[5] G. E. Hinton. Mapping part-whole hierarchies into connectionist networks. *Artificial Intelligence*, Vol. 46, 47 (1990).

[6] B. Kosko. *Neural Networks and Fuzzy Systems*. Prentice Hall, 1992.

[7] S. Lonardi, A. Sperduti, and A. Starita. Encoding Pyramids by Labeling RAAM. In *IEEE Workshop on Neural Networks for Signal Processing*, 651 (1994). Ermioni, Greece.

[8] T. Plate. Holographic reduced representations. Technical Report CRG-TR-91-1, Department of Computer Science, University of Toronto, 1991.

[9] J. B. Pollack. Recursive distributed representations. *Artificial Intelligence*, Vol. 46, 77 (1990).

[10] P. Smolensky. Tensor product variable binding and the representation of symbolic structures in connectionist systems. *Artificial Intelligence*, Vol. 46, 159 (1990).

[11] A. Sperduti. Labeling RAAM. Technical Report 93-029, International Computer Science Institute, 1993. A revised version will appear on Connection Science.

[12] A. Sperduti. On some stability properties of the LRAAM model. Technical Report 93-031, International Computer Science Institute, 1993. A revised version will appear on IEEE Transactions on Neural Networks.

[13] D. S. Touretzky. Boltzcons: Dynamic symbol structures in a connectionist network. *Artificial Intellicence*, Vol. 46, 5 (1990).

SELF-ORGANIZING MAPS FOR SUPERVISION IN ROBOT PICK-AND-PLACE OPERATIONS

Enrique Cervera and Angel P. del Pobil

Computer Science Dept., Jaume I University
Campus Penyeta Roja, E-12071 Castelló, Spain
{ecervera, pobil}@inf.uji.es

ABSTRACT

A scheme for supervised learning based on multiple self-organizing maps (SOMs) is presented, and its application to robotic tasks, namely pick-and-place operations, is outlined. The advantage of this multiple organization is that the learning method is simplified because the problem is divided into several SOMs, which are trained in the standard unsupervised way. The resulting network preserves the SOM properties like dimensionality reduction and cluster formation, and its classification performance is comparable to other supervised methods like backpropagation networks.

1. INTRODUCTION

In this paper, we present a supervised learning method using Self-Organizing Maps, an unsupervised neural network introduced by Kohonen [1]. We have applied successfully this method to robotic pick-and-place tasks using force sensing. These tasks are very error-sensitive, and neural networks are well suited for obtaining information from the real state of the system with the help of force sensors.

Self-Organizing Maps can be used in an unsupervised way for process monitoring [2]. However, if multiple SOMs are used, and provided that the training samples are labeled with their corresponding class information, the supervised learning method presented in this paper allows to classify properly new data. Figure 1 depicts several contact states in a simulated insertion example. Weight and contact forces are shown.

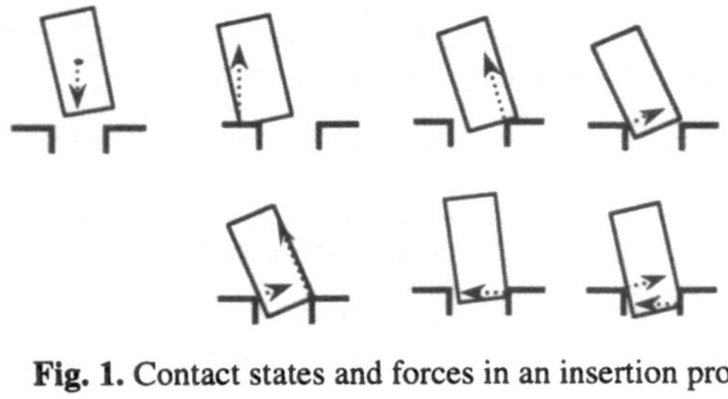

Fig. 1. Contact states and forces in an insertion problem.

In order to perform a correct insertion, the system should detect the current state and perform adequate actions. Force signals can help to find this state. However, in a real application, there are six temporal sequences of force and torque measurements, as shown in figure 2. In addition, noise is usually present, thus a robust classification method is desired.

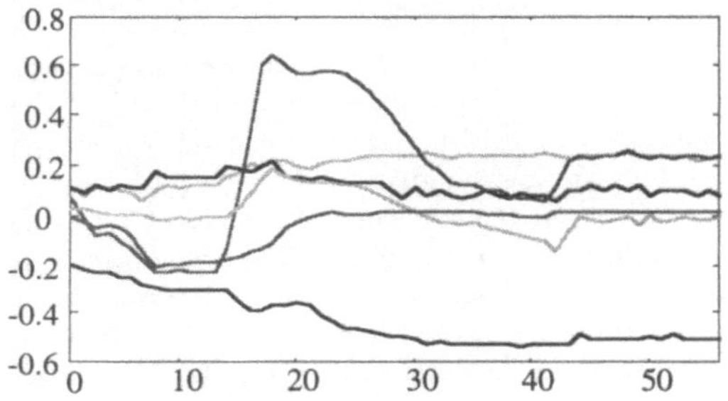

Fig. 2. Force and torque measurements in a real insertion task.

The main advantage of our method is the addition of classification properties to Self-Organizing Maps, while preserving all their nice properties like topological mapping and dimensionality reduction. Our scheme shows how to combine supervised and unsupervised learning, and how to simplify the learning process by means of the division of the problem among several SOMs. However, we recognize that the classification accuracy of our method is not better than pure supervised neural networks like back-propagation. But we have got promising results in real robotic problems.

Due to space limitation, we will focus only on the neural network approach. The description of the robotic problem and the force sensing method along with simulation results can be found in [2]. In the rest of this paper, the scheme for supervised learning based on multiple self-organizing maps is presented and its classification performance is compared in a pattern classification benchmark.

2. SUPERVISED LEARNING WITH MULTIPLE SELF-ORGANIZING MAPS

The process in which a SOM is formed is an unsupervised learning process. There are N continuous-valued inputs, ξ_1 to

ξ_N, defining a point ξ in an N-dimensional real space. The output units O_i are arranged in an array (generally one- or two-dimensional), and are fully connected via w_{ij} to the inputs. A competitive learning rule is used, choosing the winner c as the output unit with weight vector closest to the current input ξ:

$$\|w_c - \xi\| \leq \|w_i - \xi\| \quad \text{(for all } i\text{)}. \tag{1}$$

And the learning rule is:

$$\Delta w_{ij} = \Lambda(c,i)\,(\xi_j - w_{ij}) \tag{2}$$

for *all i* and *j*. A typical choice for $\Lambda(c,i)$ is

$$\Lambda(c,i) = \eta \exp(-\|r_c - r_i\|^2/2\sigma^2), \tag{3}$$

where $r_c \in \mathbf{R}^2$ and $r_i \in \mathbf{R}^2$ are the radius vectors of nodes c and i, respectively, in the array, and both η and σ are some monotonically decreasing values.

Rule (2) drags the weight vector $\mathbf{w}_c$ belonging to the winner towards ξ. But it also drags the $\mathbf{w}_i$'s of the closest units along with it. Therefore we can think of a sort of elastic net in input space that wants to come as close as possible to the inputs; the net has the topology of the output array (i.e., a line or a plane) and the points of the net have the weights as coordinates.

This training algorithm is unsupervised, since only input values are used. The resulting map, however, can be used in classification tasks if its units are *labeled*. This can be done by introducing a number of known examples into the map, and looking where the winners lie. The map units are then labeled according to the majority of examples of a given class 'hitting' a particular map unit. Then, an unknown example can be classified by computing the winner unit, and its label will give us the class of that example.

Though classification accuracies obtained by this method are not very high, it is very useful for monitoring and visualization of high-dimensional data. We have used it successfully in sensory mapping in robotic applications [2].

2.1 Probability Density Approximation and Bayesian Classification

Our approach is an extension of this scheme. In order to increase the accuracy, we might use the output information of the training data during the learning process, thus performing *supervised* learning.

The key is that a SOM approximates the probability density of the training set. If the probability densities of each class (denoted by $\mathbf{p}(\xi|\mathbf{C}_i)$ for class i) were known, we could use Bayesian theory to minimize the average rate of misclassifications, assuming that the a priori probabilities of each class (denoted by $\mathbf{P}(\mathbf{C}_i)$) are also known [3]. Then a sample ξ would be assigned to class C_i iff:

$$p(\xi|C_i)\,P(C_i) > p(\xi|C_j)\,P(C_j) \qquad \text{for all } j \neq i \tag{4}$$

Although the distribution of the SOM units is not exactly the same as the distribution of the underlying training density (for the one-dimensional case the density of output units is proportional to $P(\xi)^{2/3}$ around point ξ [4]), this does not matter in Bayesian classification where only ratios of class densities are important. Furthermore, dimensionality of the SOM is less than that of the training set, thus a SOM is a "nonlinear projection" of the probability density function of the high-dimensional input data onto the one- or two-dimensional array of units. For the sake of simplicity, we will use the distance of the input vector ξ to the best matching unit as a measure of the probability density at that point.

2.2 Supervised Learning Procedure

Our simple yet powerful idea is based on *training a different SOM for each class*. Thus, output data from the training set is used to divide all data in several subsets, each one containing only data from one class. This is the supervised part of the procedure. After this, the SOMs will be trained as usual, in an unsupervised way. A similar idea was used by Kurimo in [5], where the system is initialized with little SOMs, one for each phoneme class, which are combined and applied to other methods. Another similar system was explained in [6] where one SOM is trained for each speaker. In our approach, however, classification is performed with only one vector of input data. In the experiment reported in [6], a whole sequence of inputs were tested on each SOM and the SOM which gave the smallest quantization error was the 'winner'. Our approach is a generalization of this scheme which should be able to classify properly not only sequences but single data samples.

The complete learning process has three phases:

1) Division of the training set into several subsets, as many as different classes.

2) Training of different SOMs, each with only a subset of data of the same class, following the unsupervised learning procedure defined in previous section.

3) Labeling of units of all the SOMs. This can be done in two ways. Each SOM is separately labeled, thus all its units will be labeled as belonging to the class it was trained with. Or all SOMs are labeled at once with all training data. We have found the latter method more accurate when there is overlapping between classes.

Now, classification of unknown data is done with all the SOMs together. When an example is presented, the best matching unit of **all** maps will give the correspondent class.

3. A CLASSIFICATION EXAMPLE

This is a problem introduced by Kohonen [7] that consists of two normally distributed classes in a d-dimensional Euclidean space with equal class a priori probabilities. In the "easy" test the first class has an $N(0,I_d)$ distribution and the second class an $N(\mu,4I_d)$ distribution with $\mu=(2.32,0,...,0)\in\mathbf{R}^d$. the "hard" test has the two classes distributed as $N(0,I_d)$ and $N(0,4I_d)$, respectively.

The dimensions considered are d=2,4,6. The SOM sizes are 6x6, 8x8, and 10x10, thus the number of units increases approximately linearly with the number of dimensions. Two independent sets for training and testing were used. Each set consisted of 2000 randomly chosen samples from each class. Data and the trained SOMs for each class in the two-dimensional "easy" case are shown in figure 3.

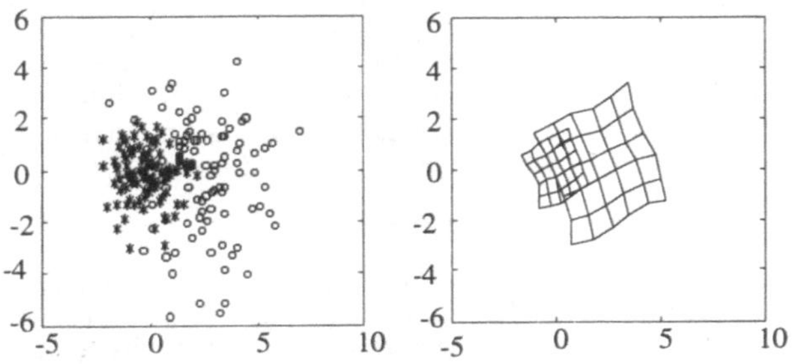

Fig. 3. Sample data and SOMs trained in the two-dimensional "easy" problem.

A summary of the results is given in table I. The second column gives the theoretically optimal classification errors. The third and fourth columns give the results for our method with separate labeling -Msom(1)- and common labeling -Msom(2)- respectively. The next two columns give the results for backpropagation and LVQ as reported in [7].

Table 1. Test results for the two class problem.

Easy problem

D	*theor.*	*Msom(1)*	*Msom(2)*	*BP*	*LVQ*
2	16.4	21.7	18.4	16.4	17.0
4	11.6	19.3	16.3	12.5	13.1
6	8.4	16.4	15.9	10.8	10.7

Hard problem

D	*theor.*	*Msom(1)*	*Msom(2)*	*BP*	*LVQ*
2	26.4	30.3	26.6	26.3	26.5
4	17.6	24.3	23.7	19.4	18.8
6	12.4	20.9	20.8	20.7	15.3

Classification results are not better than those obtained by purely supervised methods like backpropagation and LVQ, but keep rather close, specially in the two-dimensional case. This is due to the fact that we are using a two-dimensional SOM, with which is hard to approximate a normal distribution of more dimensions, since it is quite homogeneous. In addition, good results are obtained in the hard problem, thus indicating that our scheme is suitable in highly overlapping problems.

Furthermore, we are using more units than in the other methods, but, on the other hand, algorithm complexity is somewhat smaller, specially than backpropagation. Moreover, training a SOM in this experiment only takes 25 epochs (presentations of the complete training set). However, we recognize that training a SOM usually requires fixing the values of some parameters in order to get good convergence, and maybe a large number of units would be necessary in order to get a good approximation of a tricky class, but these are not major drawbacks since the algorithm is computationally cheap. Moreover, our scheme can be easily parallelised.

Theoretically, one should get near-optimal results with a great number of units in each SOM, but SOMs are best suited to data which is distributed in a subspace of the high dimensional original space. Ideally, the SOM number of dimensions should be equal to the dimension of this subspace.

The main advantage of our method is that it simplifies the original problem. Training smaller networks with subsets of data is easier and faster than training a bigger network with all the data. Furthermore each independent SOM might be reused in other problems of classification of its class with other different classes. Thus we could get a collection of SOMs and pick the relevant ones depending on the classes involved in a given problem.

Another advantage is that this approach gives a reply to a criticism frequently made of neural networks, regarding the lack of significance of the weights of the units, and *how* the network solves the problem. In our method, each SOM is an approximation of the probability density of a class, and classification is done by Bayesian theory.

4. CONCLUSIONS

A method for supervised learning using multiple Self-Organizing Maps has been presented. The learning procedure is straightforward, since it only requires the division of the training set in separate subsets, one for each data class. Then a different SOM is trained in a unsupervised way with the subset of a data class. Classification is based on the SOM approximation of the probability density of each class and Bayesian decision. We know not only *what* computes the network but also *how* it does.

Training is simplified since input data is simpler than the complete set, thus a complex problem is split in several easier problems, and the significance of the units is shown as approximators of the densities of data. Furthermore, SOM properties like dimensionality reduction and cluster formation are preserved and allow the integration of this method with other AI techniques.

Test results demonstrate that this method is comparable to other supervised methods widely used -like backpropagation networks- on hard problems, and should be tested on other tasks. Our aim is to use this method in robotic operations, with complex and noisy data. However, further studies are required in order to get better approximations of probability densities with SOMs.

ACKNOWLEDGMENTS

This work has been funded by the CYCIT under project TAP92-0391, and by a grant of the FPI Program of the Spanish Department of Science and Education. We wish to thank Prof. Kohonen and Prof. Oja for making possible a stay of one of the authors at the Laboratory of Computer and Information Science at Helsinki University of Technology, and specially Dr. Kangas, from the mentioned laboratory, and Prof. Holmström, from Rolf Nevanlinna Institute at University of Helsinki, for helpful discussions.

REFERENCES

1. Kohonen, T.: Biol. Cybern. 43, 59, 1982.

2. Cervera, E., del Pobil, A.P., Marta, E., Serna, M.A.: Unsupervised Learning for Error Detection in Task Planning, Computer Science Dept. Tech Report, Jaume I University, 1994.

3. Devijver, P.A., Kittler, J.: Pattern recognition: A statistical approach. London: Prentice Hall 1982.

4. Hertz, J., Krogh, A., Palmer, R.G.: Introduction to the Theory of Neural Computation. Addison-Wesley 1991.

5. Kurimo, K.: Proc. IEEE Workshop on Neural Networks for Signal Processing, 362, 1994.

6. Naylor, J., Higgins, A, Li, K.P., Schmoldt, D.: Neural Networks, 1, 311, 1988.

7. Kohonen, T., Barna, G., Chrisley, R.: Proceedings of the IEEE International Conference on Neural Networks, San Diego, 1, 61, 1988.

Q-LEARNING AND PARALLELISM IN EVOLUTIONARY RULE BASED SYSTEMS

Antonella Giani, Fabrizio Baiardi, and Antonina Starita

Dipartimento di Informatica, Università di Pisa
Corso Italia, 40 - 56125 PISA, Italy

Abstract

We present PANIC (Parallelism And Neural networks In Classifier systems), a parallel learning system which uses a genetic algorithm to evolve behavioral strategies codified by sets of rules. The fitness of an individual is evaluated through a learning mechanism, QCA (Q-Credit Assignment), to assign credit to rules. QCA evaluates a rule depending on the context where it is applied. This new mechanism, based on Q-learning and implemented through a multi-layer feed-forward neural network, has been devised to solve the rule sharing problem posed by traditional credit assignment methods. To overcome the heavy computational cost of this approach, we propose a decentralized and asynchronous parallel model of the genetic algorithm.

INTRODUCTION

A large amount of Machine Learning research has been recently dedicated to the evolution of populations of agents to solve a given task. Genetic Connectionism [1, 2, 3] uses Genetic Algorithms (GA) to evolve neural networks, while in Genetic Programming [9] each individual in the population is a program in a high level language. This paper concentrates on Evolutionary Rule Based Systems (ERBS) [13, 6, 7]. In ERBSs, a GA is used to evolve a population where each individual is a set of rules, codifying the behavior of an agent. The fitness of a set of rules is related to its ability to obtain good performance in a little known environment, according to a reinforcement learning paradigm. Learning consists in assigning a credit to each rule, reflecting the environmental reward expected by its use.

PANIC (Parallelism And Neural networks In Classifier systems) is a parallel ERBS, which introduces a new mechanism, QCA (Q-Credit Assignment), to allocate credit to rules. QCA is based on Q-learning [17], a temporal difference (TD) architecture for reinforcement learning, and it is realised through a neural network (NN). To overcome the shared rule problem of traditional credit assignment strategies, QCA evaluates each rule according to the context where the rule is applied.

To overcome the large computational load, due both to the cost of QCA and to the evaluation of alternative sets of rules, a decentralized and asynchronous parallel model of the GA has been devised. Experimental results show that the model allows the system to evolve a population of genotypes with high fitness in an acceptable time, as well as a good scalability of the model, with respect to both the evaluation and the time to form genetic groups [5]. The paper is focused on the learning at the individual's level.

Q-CREDIT ASSIGNMENT

Credit assignment and the rule sharing problem

Credit assignment is fundamental in any learning system, expecially in delayed reinforcement learning tasks. Evolutionary Rule Based Systems usually assign credit to rules through mechanisms inspired by those of Holland's classifier systems [8, 11], namely the Bucket Brigade Algorithm (BB) [8] and epochal algorithms [6, 7]. Both mechanisms assign to each rule a scalar value, called strength, which predicts the reinforcement expected by the use of the rule. While the BB back-propagates the strength along an activation sequence incrementally, epochal methods distribute every external reinforcement among the rules activated since the last reinforcement. As Sutton points out [14], in reinforcement learning tasks as those tackled by ERBSs and classifier systems, subgoal-reward schemes, like the BB, are faster and more accurate than goal-reward schemes, like epochal methods. However, both mechanisms *fail* if some rules are shared among activation chains leading to different outcomes.

This common weakness of both the BB and epochal methods id due to the single credit paired with a rule, independently of the situation where it fires. If a *single* credit value is assigned to a rule that can fire in different contexts and *with different outcomes*, then this value must be wrong in at least one case. Rule sharing cannot be avoided, because ERBSs and classifier systems have been devised to operates in complex environments, with a huge number of states. To use a utility measure of rules based on the context, while maintaining most advantages of traditional methods,we propose the reinforcement learning architecture Q-learning [17] as the credit assignment mechanism in ERBSs.

Q-learning

Q-learning is a Temporal Difference (TD) [14] reinforcement learning method designed to acquire an optimal policy in Markovian domains by experience, without requiring an exact model of the environment. TD methods have been widely used in the reinforcement learning field, mainly as an on-line learning scheme in neural networks [10, 15, 16]. The task domain is modeled as a Markov process, where the agent acts as controller. At each time step, the agent knows the state of the process and executes an action according to an estimate of its consequences in terms of return, i.e. a long-term discounted reward.

More formally, given an action a and a state s, Q-learning predicts the expected return of the execution of a in s. The prediction is based upon an evaluation function Q of pairs (state, action) and a state evaluation function V, defined as follows:

$$Q(s,a) = r + \gamma V(s')$$
$$V(s') = \max_{k} Q(s',k)$$

where s' is the state reached by executing a in s, γ is the discount factor and r is the reward in s'. Thus, the evaluation $V(s)$ of a state s is the best possible outcome from s. The TD error, used to update Q, is the difference between the current prediction $Q(s,a)$ and the value $r + \gamma V(s')$, i.e. the prediction which is currently considered as the correct one, according to the current evaluation function V. The best policy is the greedy policy with respect to $Q(s,a)$.

How to adapt Q-learning to Classifier Systems

To use Q-learning in a rule based system we have to define (a) the meaning of "state" and "action", i.e. to identify the Markov process to be controlled, and (b) how the return prediction has to be computed and updated according to the TD error. The module of PANIC which evaluates a set of rules is very similar to Holland's classifier systems: rules can fire in parallel and interact through message passing, using a message list as a short term memory. To satisfy the path independence property required by Q-learning, the process to be controlled by the mechanism has been identified with the *internal state* of the system, i.e. the message list, while an "action" is the posting of any message. As a matter of fact, at each time step, messages are posted according to the message list contents, and they modify that contents (process state) because (a) at the next step these messages will form the new list, and (b) they can produce external actions which modify the external environment, thereby causing the posting of new messages.

In PANIC, the function $Q(s,a)$ is implemented by a multi-layer, fully connected feed-forward neural network, which takes as input the message list contents together with a message produced by a satisfied rule, and produces as output the evaluation function of the input pair. The network has a hidden layer with symmetric sigmoidal activation function and one linear output unit. The learning mechanism is implemented by a combination [10, 14] of a TD method and of the backpropagation algorithm [12]: the TD error between two successive evaluations is back propagated in the network to update the weights of the network and improve the evaluation function. At each iteration, the patterns corresponding to the current state and the enabled actions, are presented to the network twice: at first with old weights, to evaluate the TD error on previous iteration, then with updated weights, to compute the Q values for the current competition. While Q-learning assumes that only one action can be executed at each iteration, QCA allows several rules to fire in parallel: the TD error is computed for each action and the network learning is applied to the corresponding pattern.

The main motivation of the adoption of a neural network is the complexity of the task domain, i.e. the huge number of possible states and actions to be evaluated. A look-up table representation does not seem feasible, while a NN can store the function in a distributed manner on connection weights. Furthermore, the generalisation capability of a NN provides an inductive mechanism to exploit experiences in evaluating states and actions never seen before, but "similar" to already explored ones. This could increase the ability of the system to discover regularities in the input and to absorb new rules without losing capabilities previously acquired. See [5] for experiments about generalization and changing environments.

Like the BB, QCA is domain independent and incremental, since graceful learning is a feature of TD methods. However, it is not parallel, since a single network evaluates every (s,a) pair. A first advantage of QCA lies in its ability to evaluate the convenience of activating a rule according to the situation. Hence, as shown by the experimental results, QCA is more appropriate when a rule is shared among several activation sequences. Furthermore, while QCA does not require structural links among rules in order to assign credit to a temporal activation sequence, BB requires rule coupling to propagate the strength through rule chains back to stage setters.

A drawback of QCA is a large computational cost. At each iteration, it requires $2n$ evaluations, where n is the number of matches, while the number of patterns to be learnt at iteration t is equal to the number of messages produced by the winning rules at iteration $t-1$. Furthermore, the network size grows with the message list and with the message length. Instead, in the BBA the strength of each rule is immediately available and the updating is very simple.

EXPERIMENTAL RESULTS

The task domain chosen to test the system is the FSW (Finite State World) [11], where each world is modelled as a finite Markov process. Each world is defined by a set of states and by a transition probability matrix. A state is identified by a number and may be associated with a non null payoff. The probability matrix P is function of e, a parameter specified by messages directed to the output interface. This domain allows the modelling of worlds of desired complexity, where rules have to be organised in special structures, like chains or default hierarchies, to obtain the

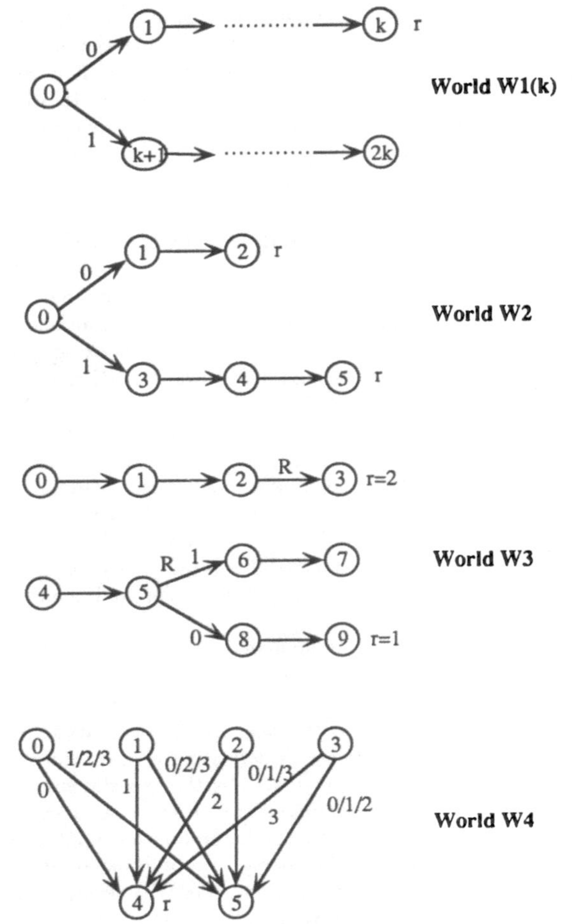

Figure 1: Some environments where the QCA has been tested.

maximum payoff. These structures have been advocated as the basic building blocks to construct more complex knowledge structures [11] and it is essential that the system can both create them and assign them the correct credit.

The experiments reported here concern the credit assignment mechanism only, i.e. an execution corresponds to the evaluation of a set of rules by the Problem Solving module. For a more complete set of tests of PANIC, see [4, 5]. The system is initialised with a set of rules ad hoc for each analysed task. The set does not change during the execution and includes both rules leading to a reward and wrong ones. The system has to allocate correct credit to the rules and learn how to obtain the maximum payoff. The performance measure reported in all graphs is the ratio of the obtained payoff to the possible maximum, versus the number of iterations. We consider an on-line performance Pn, defined as the average on the last n iterations. At iteration t:

$$Pn(t) = \frac{\sum_{\tau=t-n}^{t} r_\tau}{n \cdot r_{\max}}$$

QCA has been tested in different environments, all shown in Fig. 1, to test its behavior with respect to particular structures of rules. The performances obtained in these environments are reported in Fig. 2. All the performances are averaged on 10 executions, starting with a different random seed.

The parametric world $W1(k)$ has been used to test how QCA allocates credit to stage-setters at the beginning of an activation sequence. In $W1(k)$ two actions are allowed, corresponding to the effector settings $e = 0$ and $e = 1$. From the initial state 0, the agent can go either to 1, by setting $e = 0$, or to $k + 1$, by setting $e = 1$, while from k and $2k$ it always goes to 0. A reward r is associated with k. The agent has been initialised with a set of rules allowing all transition represented by arrows. The graph shows the performance of the agent in $W1(k)$ for $k = 4$ and $k = 8$, using the QCA mechanism to assign credit. The world W2 presents two paths of different lengths, leading to the same reward. As shown in the graph, QCA learned the shortest path. In the world W3 the rule R is shared between two chains leading to different outcomes. This is a typical situation where the BB fails, thus we have compared the BB and QCA in this environment with the same set of rules. As shown in the graph, the BB performance sharply falls down to the minimum, $P = 0.66$, while the QCA succeeds in learning the correct path.

Finally, the system has been tested in the world W4, with a set of rules implementing a default hierarchy. As shown in the graph, the QCA assign a correct credit to the rules in the hierarchy.

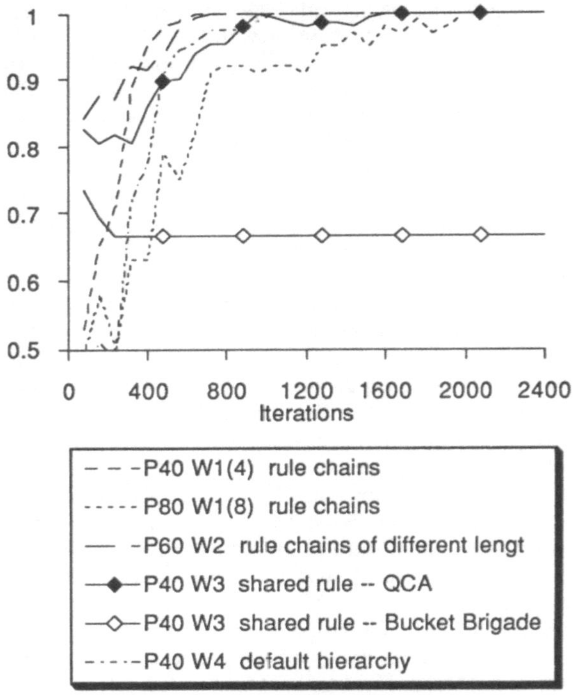

Figure 2: Performances of QCA in different environments. In the world W3, QCA is tested against the Bucket Brigade algorithm.

CONCLUSIONS

This paper has presented PANIC, a parallel Evolutionary Rule Based System, whose main features are asynchronous and decentralised genetic algorithm and QCA, a new credit assignment method to rules. QCA is implemented by a neural network and it tries to solve the rule sharing problem by evaluating rules depending on the context where they are applied. As our results show, QCA is able to allocate correct credit to rules and to learn how to obtain the maximum payoff in situations where traditional algorithms fail. The main disadvantage of QCA is a heavier computational load, which is balanced in PANIC by the parallel model of the GA. We argue that PANIC shows the feasibility of the integration of different paradigms, such as rule based systems, neural networks and evolutionary algorithms, to define powerful and robust learning systems.

REFERENCES

[1] P. J. Angeline, G. M. Saunders, and J. B. Pollack. An evolutionary algorithm that constructs recurrent neural networks. *IEEE Transaction on Neural Networks*, 1994.

[2] R. K. Belew, J. McInerney, and N. N. Schraudolph. Evolving networks: using the genetic algorithm with connectionist learning. Technical Report CSE CS90-174, Comp. Sci. and Engr., Univ. California, 1990.

[3] D. J. Chalmers. The evolution of learning: An experiment in genetic connectionism. In *Proceedings of the 1990 Connectionist Models Summer School*, 1990.

[4] A. Giani. Un nuovo approccio alla definizione e all'implementazione di sistemi a classificatori. Master's thesis, Dip. di Informatica, University of Pisa, Italy, 1992.

[5] A. Giani, F. Baiardi, and A. Starita. Panic: A parallel evolutionary rule-based system. In *Proceedings of the Fourth Annual Conference on Evolutionary Programming*, 1995.

[6] J. J. Grefenstette. Credit assignment in rule discovery systems based on genetic algorithms. *Machine Learning*, 3(2-3), 1988.

[7] J. J. Grefenstette. A system for learning control strategies with genetic algorithms. In *Proceedings of the Third International Conference on Genetic Algorithms and Their Applications*, 1989.

[8] J. H. Holland. *Escaping brittleness: The possibilities of general-purpose learning algorithm applied to parallel rule-based systems*, volume 2 of *Machine learning: An artificial inteligence approach.* Morgan Kaufmann, 1986.

[9] J. Koza. *Genetic programming: On the programming of computers by the means of natural selection.* MIT Press, 1992.

[10] L. Lin. Self-improving reactive agents: Case studies of reinforcement learning frameworks. In *From Animals to Animats: Proceedings of the First International Conference on Simulation of Adaptive Behaviour*, 1990.

[11] R. L. Riolo. *Empirical studies of default hierarchies and sequences of rules in learning classifier systems.* PhD thesis, University of Michigan, 1988.

[12] D. E. Rumelhart, G. E. Hinton, and R. J. Williams. *Learning internal representation by error propagation*, volume 1 of *Parallel Distributed Processing.* MIT Press, 1986.

[13] S. F. Smith. *A learning system based on genetic adaptive algorithms.* PhD thesis, University of Pittsburgh, 1980.

[14] R. S. Sutton. Learning to predict by the methods of temporal differences. *Machine Learning*, 3, 1988.

[15] R. S. Sutton. Reinforcement learning architectures for animats. In *From Animals to Animats: Proceedings of the First International Conference on Simulation of Adaptive Behaviour*, 1990.

[16] G. J. Tesauro. Practical issues in temporal difference learning. Technical Report RC 17223, IMB T. J. Watson Research Center, Yorktown Heights, NY, 1991.

[17] C. J. C. H. Watkins. *Learning with delayed rewards.* PhD thesis, University of Cambridge, England, 1989.

COMPARING PARALLEL TABU SEARCH AND PARALLEL GENETIC ALGORITHMS ON THE TASK ALLOCATION PROBLEM

I. De Falco, R. Del Balio, E. Tarantino

Istituto per la Ricerca sui Sistemi Informatici Paralleli (IRSIP)-CNR
Via P.Castellino, 111, 80131 Naples (ITALY)

Abstract

There exist many sequential heuristic combinatorial techniques to efficiently solve a wide set of problems. Researchers are proposing new parallel versions for them so as to take advantage of the power offered by parallel computers. In this paper new locally-linked parallel versions for two such techniques, Tabu Search and Genetic Algorithms, are proposed and tested on different-sized items of the Task Allocation Problem, one of the most challenging optimisation problems.

1. Introduction

With the wide diffusion of parallel computers, aiming at the minimisation of the execution time of time-consuming highly computational tasks, new challenges have to be faced in order to efficiently exploit the capabilities offered by such machines. Among them, one of the most intriguing and, at the same time, important is related to the problem of optimally allocating the processes composing the parallel program on an MIMD parallel architecture. The choice of an optimal or suboptimal allocation, in fact, results in a lower execution time for the program at hand. Several evaluation models have been proposed to assess the effectiveness of a given allocation: among them at least the summed cost model [1] and the minimax model [1] should be mentioned; we will make reference, in our paper, to the minimax model. The Task Allocation Problem (TAP) is clearly an optimisation problem, like many others people must deal with when working in very different fields ranging from the scientific (Operational Research, dynamic load balancing for fluid dynamic applications, weather forecasting) to the industrial (VLSI module placement, gas pipeline control, just to mention some) to the financial one (stock market forecasting, foreign currency exchange, credit evaluation). Many possible approaches have been designed, resulting in a wide set of techniques, in order to solve optimisation problems. Among them, we should mention at least the Genetic Algorithms (GAs)[2] and the Simulated Annealing (SA)[3]. Quite recently, a new promising technique has been introduced, the Tabu Search (TS)[4,5]. In very recent years, the wide availability on the market of powerful parallel machines has made it possible to design and implement parallel versions for such techniques, resulting in solutions of higher quality found in a time lower with respect to the sequential behaviour. In this paper we propose a novel approach for the parallelisation of these techniques; specifically, we design and implement parallel versions for Tabu Search and for Genetic Algorithms. In order to test their effectiveness for the proposed problem, we compare their results experimenting with mapping problems of different sizes. In Section 2 a general description of the Tabu Search strategy is given. Section 3 contains a brief introduction to Genetic Algorithms. In Section 4 issues on the introduction of parallelism in these optimisation techniques are reported, together with our proposed approach. In section 5 a more formal description of the mapping problem is provided, and a testbed of randomly-generated mapping problems is considered to compare the two above techniques; experimental results are shown. Finally, Section 6 reports our conclusions and future work in this field.

2. The Tabu Search technique

Tabu Search is quite a recent heuristic technique based on the idea of randomly choosing an initial solution, and then considering the set of its neighbouring solutions, i.e. the solutions that can be directly reached from it by means of an elementary operation called *move*; all these solutions are evaluated by means of a suitable fitness function *fc* and the best among them will constitute the new current best solution to be used in the next iteration, even if the move is a non-improving one, i.e. it does not lead to a solution better than the current best solution.

This mechanism needs that a solution to the problem may be represented in some form easy to handle, like for instance an array of components or a matrix; the elements may be, depending on the problem, integers or binary elements. The most suitable among all the possible moves for the problem at hand has to be determined. In some cases, it may consist in the random alteration of a component of the solution (*mutation*), or in the *swap* between the contents of two locations of the current best solution. Of course, as the whole set of neighbours is too large to be examined in one iteration, only a subset, whose size, let us say it SIZE, has to be defined and is a parameter for the search process, will be taken into account at each iteration. Better and better solutions are very likely to be found as the number of iterations increases.

The other key-point of this technique is the *tabu list*: it is a FIFO list, whose length is called TENURE, which contains the most recent moves leading to the current best solution. Each time a move is chosen as it leads to the best solution in the considered neighbourhood, it is inserted in the tabu list, i.e. it receives a tenure value equal to TENURE. The reverse of all moves in the tabu list are forbidden, to prevent the search process from being stuck in local optima. At each iteration, the TENURE value of each move in the tabu list is decreased by one, so that after TENURE iterations a move (better, its reverse) originally tabu can again be considered feasible. Because of this, this mechanism represents a kind of short-term memory. Both SIZE and TENURE are crucial parameters for the TS execution. Several variants are possible for the TS; the most interesting one is based on the "aspiration criterion". According to this criterion, the reverse of a tabu move can be anyway executed, provided that it leads to a new solution better than any already visited so far; in fact, it is evident that in this case it cannot originate an already-existing cycle (which would result in the search becoming prisoner of a local optimum). We will make use of this aspiration criterion within this paper. The scheme of a TS program is the following:

```
program tabu_search;
begin
  select randomly a starting solution s^start;
  set s^best=s^start;
  while not(terminated) do
    evaluate a subset S(s^best) of neighbours of s^best;
    evaluate fc for each solution in S(s^best);
    choose the best neighbour s^next in S(s^best);
    record the related move m^next;
    if (fc(s^next)<fc(s^best)) or (m^next not in tabu list)
      set s^best=s^next;
    update the tabu list by inserting the move m^next;
    fi
    decrement by one the tenure for all moves in tabu list;
    check the termination conditions;
  od
end.
```

3. Genetic Algorithms

Genetic algorithms are based on the idea of mimicking natural behaviour in terms of biological evolution in order to reach the best possible solution to a given problem.

Basically, to solve a problem we randomly generate an initial population of items (hereinafter referred to as *chromosomes*), each one consisting of a vector of components whose type depends on the particular problem to be solved. Each of these chromosomes is a potential solution to the given problem, and is associated to its fitness value, expressing how well this solution solves the problem. This population is let free to evolve, i.e. its components may vary in time according to predefined laws. More specifically, the population undergoes a *reproduction* mechanism to generate a new one in such a way that fitter elements will spread their genes with higher probability; then couples of chromosomes may mate and generate offspring based on genetic operators. The classical operators proposed in literature are the *crossover*, which consists in an exchange of sections of the parents' chromosomes starting from a randomly chosen point, and the *mutation*, consisting in the random alteration of an element of the chromosome. These operators are applied with a given, fixed probability. Other operators, too numerous to be listed here, have been proposed by researchers; among them at least the *inversion* (random choice of two points and reinsertion of the contained part of the chromosome in reverse order) and the *swap* (exchange of the contents of two components of the chromosome) should be mentioned. As the number of generations increases, the population seen as a whole evolves, and the new chromosomes are most likely to represent good solutions to the problem. Within this frame, the role of some parameters like for instance the population size (*POP*), the crossover probability (p_{cross}) and the mutation probability (p_{mut}), becomes crucial.

The GA structure is given in the following:

```
program genetic_algorithm;
begin
  randomly generate an initial population with POP elements;
  evaluate the fitness function fc for each individual;
  save the best s^best;
  while not (termination) do
    perform reproduction to obtain a new population;
    for i=1 to POP/2 do
      select randomly two elements of the population;
      perform crossover with probability p_cross;
      perform mutation with probability p_mut;
      substitute these new elements to their parents;
      evaluate the fitness of these new elements;
    od
    if exists some s^el such that fc(s^el)<fc(s^best)
      s^best=s^el;
    fi
    check the termination conditions;
  od
end.
```

4. Introducing parallelism in optimisation techniques

Starting from the basic structures of these optimisation techniques, many researchers are designing and implementing parallel versions for them in order to substantially reduce the response time provided by the sequential versions. A particular remark must be made for the GAs, as for them some authors, like for instance Manderick and Spiessens [6], have considered the possibility of parallelising the control structure of the GA itself, e.g. the genetic operators. Apart from this idea, for both the techniques many of the proposed approaches are very similar, so that we can list them independently of the peculiar technique to be parallelised. Some researchers [7,8] have implemented parallel versions consisting in a set of independent searches, and a master simply collects the best results from all the search processes. Another idea consists in the implementation of a broadcasting mechanism among all the searchers, so that all the search trajectories are linked very closely. As an example, Malek [3] has proposed a parallel TS in which all the processors perform independent searches, and at given times a master halts all the slaves and determines the global best solution among all the local best solutions; then a broadcast takes place of such a solution to all the processes which are then restarted with this received solution as the new initial one. In our opinion this approach has, as a drawback, the probability of getting stuck in local optima, since there is a high loss of diversification within the set of candidate solutions. Furthermore, the broadcasting overhead is dependent on the number of processors.

With the aim to reduce these problems, we have designed [9,10] a parallel version based on the replication of the same search process on all the processors of a torus; each of these searches performs a normal, sequential search process, and the search trajectories are locally linked, in the sense that communication of the current local best takes place in each processor only with the other N_Neighbours directly connected processes within the network, and, if some of the received solutions is better than the current local best, the former replaces the latter. Therefore, a diffusion process of good local information takes place in the network, saving in the meanwhile the diversification among searches, differently from other parallel approaches. These are concepts borrowed directly from parallel evolutionary and genetic algorithms, as they refer to the idea of socialisation among different subpopulations. A master process, needed only as interface towards the user, receives at each iteration the best values from all the slave processes, and then selects the best among all the received values and considers it as the global best for that iteration. Note that in our approach, differently from Malek's, the overhead due to information exchange is little and independent of the number of processors.

The general scheme proposed is detailed in the following, where the box SOT represents the sequence of actions typical of the chosen Sequential Optimisation Technique (tabu search or genetic algorithm), while s^{vec} is a vector of N_neighbours components used to contain the best elements sent by the neighbours.

```
procedure optimiser;
begin
  initialisation phase (as in the sequential version);
  while not (terminated) do
    send s^best to the master process;
    SOT and, if necessary, update s^best;
    perform the exchange of s^best with neighbours;
    for i=1 to N_neighbours do
      if (fc(s^vec[i]<fc(s^best))
        s^best=s^vec[i];
        update local information (tabu list) for TS;
      fi
    od
    check termination conditions;
  od
end;

program parallel_optimiser;
pardo
  master;
  for i=1 to N_searchers do optimiser;
  od
parend
end.
```

By using this approach, we have implemented a Parallel Tabu Search (PTS) and a Parallel Genetic Algorithm (PGA).

5. Experimental results

5.1 Formalisation of the TAP

Given a parallel program P, consisting of N processes, to be mapped on a parallel machine PM with M processors (with N>M), to each of its composing processes P_i we associate W_i, the computational weight of the process, and C_{ij}, the communication weight (the number of communications) between P_i and every P_j (i<>j). The system is assumed to be homogeneous, meaning that all the processors are equally powerful and all the communication links are capable of the same communication rate. The processors are assumed to either execute a computation or perform a communication at any given time, but not simultaneously do both. The cost of a communication is assumed to be proportional to the size of the message and to the distance d_{ij} between the sender and the receiver (i.e. the minimum number of links to be traversed). We define Task Set of a processor k (TS_k) as the set of processes mapped onto it, and Communication Set of a processor k (CS_k) as the set of the communications to be performed by the processes of TS_k with processes not allocated on k (i.e. not belonging to TS_k). With these premises, the total workload L_k for a processor k will be:

$$L_k = \sum_{i \in TS_k} w_i + \sum_{(i,j) \in CS_k} c_{ij} d_{ij}$$

and, according to the chosen minimax model, the fitness function for a given mapping will therefore be:

$$L = max\{L_k\} \quad k = 1, \ldots, M$$

and this quantity has to be minimised.

5.2 The testbed

To perform all our experiments we have used a Meiko Computing Surface with 65 INMOS T800 transputers, so our programs have been written in OCCAM2 language. As a testbed we have considered a set of four randomly-generated problems consisting, respectively, in the mapping of 30 processes on 9 processors, of 50 processes on 16 processors, of 100 processes on 30 processors, and of 150 processes on 45 processors. The processors have been supposed to be connected as a torus. A solution to an N-sized TAP has been represented as a string *stringsol* of N integers in the range 1..N, so that *stringsol*[i]=j means that the i-th process is allocated on the j-th processor. In a previous paper [10] we have shown that, when performing a PTS, the most suitable operator for the task allocation problem is represented by the sequence of mutation and swap, which has overcome the mutation applied on its own, and other operators like for instance the inversion, so we have made use of this operator in our experiments, for both the PGA and the PTS. Due to the hardware available, i.e transputers with four bidirectional communication links, N_neighbours has been fixed equal to four. Some preliminary runs have been made to tune the parameters for both the techniques as a function of the different problem size. After this phase, each algorithm has been run 10 times on the same problem, the stopping condition being the completion of the same predefined amount of time for both techniques. In Table 1 the best final value (*best_val*), its convergence time (*best_time*), the average final value over the 10 runs (*aver_val*) together with the relative average time (*aver_time*) are reported for the PGA for each problem.

Table 1: The performance of the PGA.

Size	*best_val*	*best_time*	*aver_val*	*aver_time*
30	540	1' 55"	552.1	3' 24"
50	839	5' 25"	858.6	7' 51"
100	697	19' 28"	712.3	19' 34"
150	1300	31' 49"	1342.5	30' 55"

Table 2 contains the same values relative to the PTS. The results show that in all cases the performance of the PTS is better than that reached by the PGA as regards the solution quality.

Table 2: The performance of the PTS.

Size	*best_val*	*best_time*	*aver_val*	*aver_time*
30	519	1' 54"	536.6	2' 24"
50	799	8' 01"	807.9	8' 23"
100	661	21' 09"	677.7	20' 13"
150	1260	31' 40"	1287.1	31' 08"

It is worth noting from the comparison of the two above Tables that, for all the problems taken into account, the average values obtained by the PTS technique are better than the best values reached by the PGA technique. Table 3 shows the difference in percentage between the PGA and the PTS in terms of relative distance for both the best values found (*best_val*) and the average values (*aver_val*). It can be seen from this Table that the distance of the PGA performance from the PTS performance in terms of quality of solutions found is within the range of about 3 to 6 per cent, and this independently of the problem size.

Table 3: Distance in percentage between PGA and PTS in terms of solution quality.

Size	*30*	*50*	*100*	*150*
best_val	3.88	4.76	5.16	3.07
aver_val	2.80	5.90	4.85	4.12

Furthermore, these experiments have revealed that the time needed by the PTS to reach the same best values found by the PGA is, in average, about 50 per cent of the time required by the PGA, so that the remaining 50 per cent of the time is profitably exploited by PTS to further improve the solution. So, the superiority of the PTS is not only in its better solution quality, but also in its lower convergence time to a same solution. This feature has been confirmed also by experiments concerning other optimisation problems (like Travelling Salesman Problem and Quadratic Assignment Problem [11]) and may be very important for real problems when the response time becomes a crucial factor.

6. Conclusions

In this paper the Task Allocation Problem for MIMD machines has been taken into account. Two combinatorial optisation techniques, the Genetic Algorithms and the Tabu Search, have been considered for the solution of four different-sized items of such problem. Parallel versions have been implemented for these techniques, allowing the local exchange of good solutions within the searchers. Experimental results have revealed that, for the problem at hand, the Parallel Tabu Search outperforms the Parallel Genetic Algorithm, in terms of both absolute solution quality and convergence time to a same value. This is, in our opinion, probably due to the lack of any kind of memory in classical versions of GAs, so that the search cannot avoid re-examining already visited parts of the solution space. This problem is avoided by the Tabu Search, which is provided with a short-term memory represented by the tabu list. Because of this, we plan to experiment in our future with the introduction of some form of memory in Genetic Algorithms, both trying to somehow mimic the memory shown by Tabu Search and considering other forms of memory proposed in literature, like for instance the diploidity for chromosomes with the related concepts of dominance and recessivity.

References

[1] F. Ercal, J. Ramanujam and P. Sadayappan, "Task Allocation onto a Hypercube by Recursive Mincut Bipartitioning", in *Proc. of the Third Conference on Hypercube Concurrent Computers and Applications. Volume I: Architecture, Software, Computer Systems and General Issues*, Pasadena, CA, pp. 210-221, G. Fox ed., ACM, New York, NY, 1988.

[2] D.E. Goldberg, *Genetic Algorithms in Search, optimization and Machine Learning*, Addison-Wesley Publishing, 1989.

[3] M. Malek, M. Guruswamy, M. Pandya and H. Owens, "Serial and Parallel Simulated Annealing and Tabu Search Algorithms for the Travelling Salesman Problem", *Annals of Ops. Res.*, vol. 21, pp. 59-84, 1989.

[4] F. Glover, "Tabu Search - Part I", *ORSA J. on Computing*, vol. 1 no. 3, pp. 190-206, 1989.

[5] F. Glover, "Tabu Search - Part II", *ORSA J. on Computing*, vol. 2 no. 1, pp. 4-32, 1990.

[6] B. Manderick and P. Spiessens, "Fine-grained Parallel Genetic Algorithms", in *Proc. of the Third Int. Conf. on Genetic Algorithms*, Schaffer J.D. ed., Morgan Kauffmann Pub., pp. 428-433, 1989.

[7] C. N. Fiechter, A. Rogger and D. de Werra, "Basic Ideas of Tabu Search with an Application to Traveling Salesman and Quadratic Assignment", *Ricerca Operativa*, no. 62, pp. 5-27, 1992.

[8] R. Battiti and G. Tecchiolli, "Parallel Biased Search for Combinatorial Optimization: Genetic Algorithms and TABU", *Microprocessors and Microsystems*, vol. 16, no. 7, pp. 351-367, 1992.

[9] I. De Falco, R. Del Balio, E. Tarantino and R. Vaccaro, "Simulation of Genetic Algorithms on MIMD Multicomputers", *Parallel Processing Letters*, vol. 2 no. 4, pp. 381-389, 1992.

[10] I. De Falco, R. Del Balio and E. Tarantino, "Solving the Mapping Problem by Parallel Tabu Search", in *Proc. of the twelfth IASTED International Symposium on Applied Informatics*, Annecy, France, pp. 264-267, May 18-20, 1994.

[11] I. De Falco, R. Del Balio, E. Tarantino and R. Vaccaro, "Parallel Tabu Search versus Parallel Evolution Strategies", in *Proc. of the First International Conference on Massively Parallel Computing Systems - MPCS'94*, Ischia, Italy, pp. 564-569, May 2-6, 1994.

DISTRIBUTED GENETIC ALGORITHMS WITH AN APPLICATION TO PORTFOLIO SELECTION PROBLEMS

A. Loraschi

SIGE Consulenza S.P.A., Milano, Italy

Marco Tomassini

Centro Svizzero di Calcolo Scientifico, Manno
and Laboratoire de Systèmes Logiques
Ecole Polytechnique Fédérale de Lausanne
Switzerland

A. Tettamanzi

Universita' degli Studi di Milano
Dip.to di Scienze dell'Informazione, Milano, Italy

Paolo Verda

Université de Fribourg, Institut d'Informatique
Fribourg, Switzerland

Abstract

This paper presents a PVM-based coarse-grained distributed genetic algorithm implemented on workstation clusters. After successfully evaluating the algorithm with standard test functions, we apply it to a hard real-world portfolio selection problem. The distributed version easily outperforms sequential genetic algorithms and shows promise for difficult management applications.

Distributed Genetic Algorithms

Genetic Algorithms (GAs) have enjoyed an increasing popularity as reliable stochastic optimization, search and rule-discovering methods in the last few years. The original formulation of GAs by Holland and others in the seventies was a sequential one. This approach made it easier to reason about mathematical properties of the algorithms and was justified at the time by the lack of adequate software and hardware. However, it is clear that GAs offer many natural opportunities for parallel implementation [1]. There are several possible parallel GA models, the most popular being the fine-grained or *grid* models [2], and the coarse-grain or *island* models. In the grid models, large populations of individuals are spatially distributed on a low-dimensional grid and individuals interact locally within a small neighborhood. Massively parallel SIMD machines are especially suited for the grid model.

In the island model the population is subdivided into smaller subpopulations which evolve independently and simultaneously according to a standard GA. Periodic migrations of some selected individuals between islands allow to inject new diversity into converging subpopulations. Microprocessor-based multicomputers and workstation clusters are well suited for the implementation of this kind of parallel GA. Since we are targeting real-world industrial problems, concerns of availability, cost and portability in non-academic environments prompted us to implement the island model using workstation clusters and the PVM environment [3]. Being coarse-grained, the island model is less demanding in terms of communication speed and bandwidth, which makes it a good candidate for a cluster implementation. We employed an SPMD (Single Program Multiple Data) model in which different subpopulations exchange groups of individual in a loosely synchronous manner. The following is an outline of the basic algorithm as it has been proposed for example in [4,5,8]:

```
1 - initialize P subpopulations
    of size N each
2 - for each subpopulation do in
    parallel for M generations
      evaluate and select
      individuals by fitness.
      recombine individuals.
      mutate individuals.
      send K<N best individuals to
      a neighbouring subpopulation.
      receive K individuals from a
      neighbouring population.
      replace K individuals in
      the subpopulation.
3 - if termination criterion met
    then stop, otherwise go to 2
```

We started from two well-known public domain GA packages: GENEsYs [6] and Genitor [7]. After the code parallelization, the first step is to establish suitable values for a number of parameters of the distributed algorithm. At the very least, one has to define the topology of the communicating processes, the size N of the subpopulations, the migration frequency M, the number of individuals K involved in the migration and the individual replacement policy. One problem with the distributed/parallel GAs is that there are no rigorous prescriptions for setting those parameters, the only practical way being experimental. We found the following values by trial and error, running the distributed algorithm on a few standard test functions. The size of the subpopulations was between 100 and 200, the migration frequency was 10 generations and the number of individuals exchanged 5 to 10% of the subpopulation size. Not surprisingly, these values turned out to be comparable to those used in other studies [4, 5]. Besides, our experiments showed that small variations affect the results in a negligible manner. We experimented with only one process topology: the ring, and we tested two exchange policies: simply passing individuals to the next process in the ring, alternating directions at each swap, and passing individuals "modulo" the swap number (see [8]). The "modulo" swap gave the best results and we retained it for the subsequent tests. The replacement policy was extremely simple: the new K individuals displaced the worst K individuals of the receiving population. Here again other solutions are possible.

The goal of the next step was to test the distributed approach in order to see if it offered real advantages with respect to the standard sequential algorithm. To this end, we selected the following four hard standard test functions: Schwefel's function, low autocorrelation binary sequences, fully deceptive function and Griewank's function. They are described in detail in [6]. The task is to find the absolute minimum of each function in a given domain. In order for the results to be comparable, we experimentally fixed the number of function evaluations for each problem and imposed that both the sequential and the distributed algorithm made this same number of function evaluations. For each problem we ran both the sequential and distributed GA 30 times and took the average and best results over the whole number of experiments.

We found that, although the distributed algorithm was not optimized in any significant way, it consistently outperformed the sequential version, both on the average and best values. In all cases the distributed algorithm found the global optimum much more frequentely than the sequential one within the given number of function evaluations and its speed of convergence was always greater. Our workstation cluster was under normal usage by other people during all of our tests, giving rise to strong fluctuations in computing time. Furthermore, PVM processes are given low priority. But this is exactly the way in which these clusters are best used, since what one wants to maximize is total system throughput and PVM nicely exploits idle time on the machines. Under these conditions one can hardly talk about speedups in the usual way. We did some time measurements nevertheless and found that with eight Sun Sparc workstations the average speedup was about 4.5. Had we run the tests during nighttime, the results would have certainly been better.

These results confirm the findings of other authors [4,5,8] and show that distributed GAs are not only advantageous in terms of computing time, they also often solve problems more easily than the sequential algorithm and we remarked that they can better help track multiple optima. Qualitatively, this can be attributed to a reduction of premature convergence due to the exchange of fresh genetic material. This effect could perhaps be enhanced by using different GA parameter settings for different island but we did not experiment with this approach. In the next section we will put the distributed algorithm to a more stringent test: the portfolio selection problem, a difficult real-world optimisation problem.

Distributed Portfolio Selection

This section deals with an application of GAs to some issues raised by recent research in Portfolio Theory (PT) concerning the use of non-trivial indices of risk. The central problem of PT has to do with selecting the weights of assets in a portfolio that minimize a certain measure of risk for any given level of expected return. According to mainstream PT [9], a meaningful measure of risk is given by the variance of the distribution of portfolio returns. However, while leading to elegant analytical results and easy practical implementation, this approach fails to capture what an investor perceives as the essence of risk, that is the chance of incurring a loss. In this respect, the economic literature has recently proposed alternative measures of risk; an interesting family of measures, defined as lower partial moments [10], refers to the down-side part of the distribution of returns and has therefore become known under the name of downside risk.

A number of difficulties are encountered when trying to apply this new approach even to simple problem instances. First of all, the nature of these risk measures prevents one from devising general analytical solutions and quadratic optimization techniques cannot be applied because the shape of the objective surface is in general non-convex. Moreover, the typical sizes of real-world portfolio selection problems are of the order of several tens or even hudrends of assets; expected return and risk are calculated by using historical series made up of hundreds of historical returns. Even the software packages that apply the conventional mean-variance approach through quadratic optimization suffer limitations as the size of the problem grows beyond a certain threshold. Things become more involved as it is usual practice among fund managers to impose various constraints on their optimal portfolios.

We have carried out experiments applying genetic algorithms to the portfolio selection problem using the downside risk approach on a sequential machine [11], and obtained satisfactory results. However, the structure itself of the portfolio selection problem suggests a distributed architecture for a suitable evolutionary algorithm. The problem can be viewed as a convex combination parametric programming problem [12], in that an investor wants to minimize risk while maximising expected return. As the level of risk increases, the expected return attached to optimal portfolios draws a convex non-decreasing curve, which is called efficient frontier, which is the set of Pareto-optimal, i.e. non-dominated, portfolios. In other words, on the efficient frontier, a larger expected return corresponds to a greater risk. This can be expressed as a two-objective optimization problem:

$$\min_{\mathbf{w}}\{\text{Risk}(\mathbf{w})\}$$

$$\max_{\mathbf{w}}\{\text{Return}(\mathbf{w})\}$$

subject to

$$\sum_{w_i \geq 0} w_i = 1$$

These two objectives can be parametrized to yield a parametric objective

$$\min_{\mathbf{w}}\{\lambda\,\text{Risk}(\mathbf{w}) - (1-\lambda)\,\text{Return}(\mathbf{w})\},$$

where parameter λ is a trade-off coefficient ranging between 0 and 1. When $\lambda = 0$ the investor disregards risk and only seeks to maximize expected return; when $\lambda = 1$ the risk alone is minimized, whatever the expected return. Since there is no general way to tell which particular trade off between risk and return is to be considered the best, optimizing a portfolio means finding a whole range of optimal

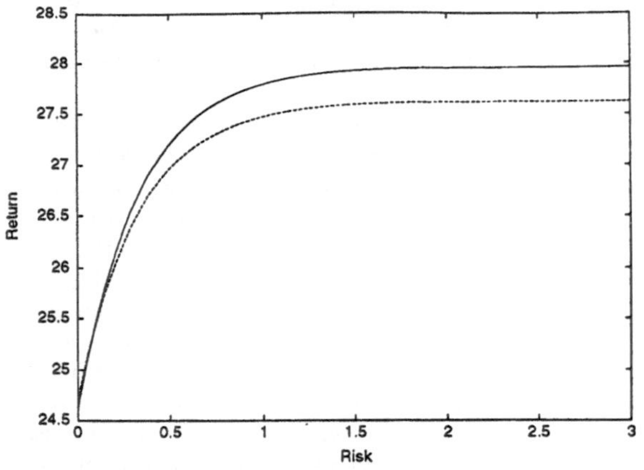

Figure 1: Efficient frontier for a 53 assets portfolio. Sequential (dashed line) and distributed (thin line).

portfolios for all the possible values of the trade off coefficient; the investors will thus be able to choose the one they believe appropriate for their requirements. A natural way to achieve that in an evolutionary setting is to have several distinct populations evolve for a number of trade-off coefficient values. The greater this number, the finer the resolution with which the investor will be able to explore the efficient frontier. Because it is likely that slight variations in the trade-off coefficient do not significantly worsen a good solution, a natural way to sustain the evolutionary process is to allow migration or cross-breeding between individuals belonging to populations corresponding to values of the trade-off coefficient that are close together. This suggests a distributed implementation where populations are linearly arranged according to their relevant trade-off value. The distributed genetic algorithm discussed in the previous section has been slightly adapted to the present problem. The only difference being the subpopulations topology: a two-way line in which exchanges take place only between nearest neighbours i.e., those with similar trade-off coefficients. The results have been very encouraging. Fig. 1 shows a fully developed efficient frontier for a portfolio with 53 assets.

It is clearly seen that the distributed version can find portfolio weightings that offer a significantly better return for a given risk level over nearly all the risk/return region of interest. We also experimented with very large portfolios (more than 100 assets). In all cases, the distributed GA outperformed the sequential algorithm. In a selection problem with 140 assets the distributed algorithm running on a normally loaded Sun Sparc cluster with 10 machines has been able to develop acceptable solutions in a few hours time, while we had to stop the sequential version after two days, before any reasonable curve had formed.

Conclusions

PVM-based distributed genetic algorithms proved to be a useful tool for solving difficult optimization problems. Both standard hard test functions and a difficult financial problem were optimized faster and more reliably by the distributed GAs than by their sequential counterparts. This opens up a whole new range of possibilities for hard financial applications, whereby larger and more realistic problem instances can be dealt with in reasonable time. The distributed GA is only slightly more complicated than the sequential version and the PVM implementation can make use of existing workstations, thus efficiently exploiting available CPU time. This makes this kind of widely available architecture more affordable than dedicated parallel computers for non-academic applications.

In the future we will pursue the application of the distributed GA to other important financial and industrial problems. The algorithm presented here is of the loosely synchronous type. We think that an asynchronous approach would be more efficient since a given subpopulation would not exchange individuals at fixed times irrespective of its internal degree of uniformity, but only when some convergence level has been attained. Work is planned for experimenting with this model.

Acknowledgement

We are grateful to Professor Gianni Degli Antoni of the Computer Science Department of the University of Milano for useful discussions and his continuing support.

References

1. H. Mühlenbeinm : *Parallel Genetic Algorithms, Population Genetics and Combinatorial Optimization*, in Proc. of the Third Int. Conf. on Genetic Algorithms, J. D. Schaffer (Editor), Morgan Kauffman, 416, 1989.

2. B. Manderick and P. Spiessens: *Fine-Grained Parallel Genetic Algorithms*, in Proc. of the Third Int. Conf. on Genetic Algorithms, J. D. Schaffer (Editor), Morgan Kauffman, 428, 1989.

3. PVM3 User's Guide and Reference Manual, Oak Ridge National Laboratory, Tennessee, 1993.

4. J.P. Cohoon, S.U. Hegde, W.N. Martin and D. Richards: *Punctuated Equilibria: a Parallel genetic Algorithm*, in Proc. of the Second Int. Conf. on Genetic Algorithms, J.J Grefenstette (Editor), Lawrence Erlbaum Associates, 148, 1987.

5. R. Tanese: *Parallel genetic Algorithm for a Hypercube*, in Proc. of the Second Int. Conf. on Genetic Algorithms, J.J Grefenstette (Editor), Lawrence Erlbaum Associates, 177, 1987.

6. T. Bäck: *A User's Guide to GENEsYs 1.0*, University of Dortmund, Department of Computer Science, 1992.

7. D.L. Whitley: *Getting Started with Genitor*, Computer Science Department, Colorado State University, 1990.

8. T. Starkweather, D. Whitley and K. Mathias: *Optimization Using Distributed Genetic Algorithms*, in Parallel Problem Solving from Nature, Lecture Notes in Computer Science Vol. 496, H.-P. Schwefel and R. Männer (Editors), Springer-Verlag, 176, 1991.

9. H. M. Markowitz: *Portfolio Selection*, New York, John Wiley & Sons, 1959.

10. H. V. Harlow: *Asset Allocation in a Downside-Risk Framework*, Financial Analysts Journal, Sep/Oct, 30, 1991.

11. S. Arnone, A. Loraschi and A. Tettamanzi: *A Genetic Approach to Portfolio Selection*, Neural Network World 3(6), 597, 1993.

12. R.E. Steuer: *Multiple Criteria Optimization: Theory, Computation and Application*, John Wiley & Sons, New York, 1986.

Interval arithmetic and genetic algorithms in global optimization

Jarmo T. Alander
Department of Information Technology and Industrial Management
University of Vaasa, P. O. Box 700, FIN-65101 Vaasa, Finland, e-mail: jal@uwasa.fi

Abstract

In this work we have combined two promising global optimization methods, interval arithmetic methods and genetic algorithms, into a set of hybrid algorithms that share the good properties of both methods while avoiding some drawbacks of the constituent methods. Interval aritmetic gives us some simple arithmetic methods to estimate the range of rational functions to be optimized, while genetic algorithms are able to adaptively guide the search.

Keywords: genetic algorithms, global optimization, hybrid methods, interval aritmetic, multimodal optimization

1 Introduction

In this paper we discuss the following simple type of global optimization problem:

$$\min f(x) = f(x_1, x_2, \ldots, x_n)$$

$$l_i \le x_i \le u_i, \quad i = 1, \ldots, n$$

where $f : R_n \to R$. We denote by $\mathbf{X}$ the box $\{x = (x_1, \ldots, x_n) | l_i \le x_i \le u_i, i = 1, \ldots, n\}$. There are many well-known local optimization methods that can be used to solve this problem when the function f is unimodal. In the case that f is multimodal the situation is totally different and the problem can be extremely complex. The book by Horst and Tuy [1] contains a good review of the known global optimization methods.

1.1 Interval aritmetic

Interval arithmetic operations [2] are defined for ordered pairs of real numbers,[1] which represent closed intervals on the real line.

$$A = [a_1, a_2] = \{x \in R | a_1 \le x \le a_2\}.$$

The first element of the pair is the minimum value of the interval, often denoted by $\inf(A)$, while the second element is the maximum denoted by $\sup(A)$. Real numbers x can be identified with degenerate intervals $[x, x]$.

[1]It is also possible to define interval aritmetic on complex numbers. Then intervals are circles on the complex plane [3, 4].

The interval arithmetic operators are usually defined as follows:

$$A + B = [a_1, a_2] + [b_1, b_2] = [a_1 + b_1, a_2 + b_2] ,$$

$$A - B = [a_1, a_2] - [b_1, b_2] = [a_1 - b_2, a_2 - b_1] ,$$

$$A \times B = [a_1, a_2] \times [b_1, b_2]$$

$$= [\min(a_1 b_1, a_1 b_2, a_2 b_1, a_2 b_2),$$

$$\max(a_1 b_1, a_1 b_2, a_2 b_1, a_2 b_2)] ,$$

$$A / B = [a_1, a_2]/[b_1, b_2] = [a_1, a_2] \times [1/b_2, 1/b_1] ,$$

$$\text{if } 0 \notin [b_1, b_2] .$$

The most valuable property of interval aritmetic in numerical analysis and especially in optimization is its inclusion monotonicity. Let A, B, X, and Y be intervals such that $A \subseteq X$ and $B \subseteq Y$, then

$$A + B \subseteq X + Y ,$$

$$A - B \subseteq X - Y ,$$

$$A \times B \subseteq X \times Y ,$$

$$A/B \subseteq X/Y \text{ ,if } 0 \notin Y.$$

In general, it is true for any rational function $F(A_1, \ldots, A_n)$: If $A_1 \subseteq X_1, \ldots, A_n \subseteq X_n$, then

$$F(A_1, \ldots, A_n) \subseteq F(X_1, \ldots, X_n)$$

for all values $X_1, \ldots, X_n$, for which F is defined [2]. If we replace the interval values $A_1, \ldots, A_n$ by degenerate intervals (scalars) $x_1, \ldots, x_n$, we have for all i: $x_i \in X_i$

$$f(x_1, \ldots, x_n) \in F(X_1, \ldots, X_n) .$$

In other words this means that

$$f^*(X_1, \ldots, X_n) \subseteq F(X_1, \ldots, X_n) ,$$

i.e. the interval extension F of a scalar function f is a more or less good upper bound of the range of f (denoted by f^*) over the given hyperbox $\mathbf{X} = \mathbf{X_1 X_2} \cdots \mathbf{X_n}$.

1.2 Related work

Interval methods have not been used much with genetic algorithms. According to the bibliography [5] compiled by the author there are only the following papers, that are related to intervals [6, 7, 8, 9, 10]. As there are so few interval GA papers, one objective of this paper is to make interval aritmetic and its potential application in genetic algorithms more familiar to the GA community and *vice versa* the author also hopes that this paper makes GAs and their potential in optimization more familiar to the interval aritmetic and global optimization community. According to [5] only four papers dealing with GA have appeared in the prominent optization journals. This is quite a small number for an inherently optimization method such as GAs!

2 Interval aritmetic

The goodness of the interval range approximation depends on the form of the function to be evaluated. In the special case when every variable x_i occurs textually exactly once and to the first power in f that the interval generalization gives the exact range f^* [12]. Chuba and Miller have shown that the error is between $O(w)$ and $O(w^2)$, where w is the width of the argument intervals [13]. A comparison of different interval methods can be found e.g. in [14]. The so called centered forms seems to be rather efficient in approximating f^*.

2.1 Centered forms

H. Ratschek gives a definition of the centered form based on the Taylor expansion of the object function [15]. Let $f(x) = p(x)/q(x)$ be a rational function of one real variable $x \in X$ and let n be the maximum of the degrees of the polynomials p and q. The centered form of f is, whenever $G(X)$ is defined

$$F(X) = f(m) + G(X - m) ,$$

where m is the mid point of X,

$$G(Y) = \frac{\sum_{i=1}^{n} a_i Y^i/i!}{\sum_{i=0}^{n} q^{(i)}(m)Y^i/i!}$$

and $a_i = p^{(i)}(m) - f(m)q^{(i)}(m)$. This expansion can be derived by substituting $p(x)$ and $q(x)$ by their Taylor series at m.

The above definition of centered forms can be generalized to higher orders by expanding f to its Taylor series and taking the first k terms:

$$\begin{aligned} F_k(Y) &= f(m) + f'(m)Y/1! \\ &+ \cdots + f^{(k-1)}(m)Y^{k-1}/(k-1)! \\ &+ \frac{a_k^{(k)}Y^k/k! + \cdots + a_{k+n-1}^{(k)}Y^{k+n-1}/(k+n-1)!}{q(m) + q'(m)Y + \cdots + q^{(n)}(m)Y^n/n!} , \end{aligned}$$

where

$$a_i^{(k)} = p^{(i)}(m) - \sum_{j=0}^{k-1} \binom{i}{j} f^{(j)}(m)q^{(i-j)}(m) .$$

The obvious drawback of cenetered forms is the need of derivatives of f.

2.2 Intervals in optimization

Interval methods for optimization are based on the inclusion monotonicity of interval generalizations of scalar functions. In global optimization, the branch and bound principle is usually used to find the exact global optimum. The argument intervals are subdivided until the optimum of given tolerance is found [16, 17]. Ratschek and Rokne have proposed the following simple interval optimization method [18, Algorithm 2 of Chapter 5]:

1. put all uninspected boxes into a list L. At the beginning, L contains only the search box $\mathbf{X}$.
2. subdivide the first box of L into boxes $\mathbf{X}_1$ and $\mathbf{X}_2$.
3. put box $\mathbf{X}_i$ at the end of L unless it can not contain a global solution
4. if a stopping criterion is met, terminate, else go to step 2.

During the search we can eliminate all those boxes $\mathbf{X}_j$ for which applies:

$$\inf F(\mathbf{X}_j) > \sup F(\mathbf{X}_i).$$

since these can not contain the global minimum of f. Let us call the set of boxes for which this relation holds the passive set and the set of all other boxes be called the active set. Any one of the active boxes can contain the global minimum. In order to be sure that we will finally find the global optimum, we must have all elements of the active set available. In practise this can lead to a need to save quite large box sets, which may cause some troubles with memory utilization. The number of active boxes also inevitably correlates with the processing time needed. Therefore all possibilities to reduce the active set as much as possible would be beneficial in practical applications of the interval global optimization approach.

3 Hybrid optimization methods

There are two main principal ways to use GAs together with interval aritmetic in global optimization. The first approach is to apply interval aritmetic in GA based optimization by replacing at least partly the function to be optimized by some of its interval extensions. The other is approach is to apply GAs in some internal problems of the interval arithmetic global optimization methods.

3.1 FIGA

The first approach is quite straightforward. One way of realizing it is to evaluate the fitness function in intervals that are getting smaller and smaller while more generations are evaluated. We call this as the focusing interval GA method (FIGA). This method has some similarity with the well known simulated annealing type global optimization, where the lowering temperature parameter is similar to our shortening interval lengths.

The obvious drawback of this approach is that the resulting method is inherently stochastic and may easily lose the global optimum. We can try to make FIGE less stochastic e.g. by setting the starting population to cover the search space by equally spaced intervals. In fig. 1 you can see an example where the modified FIGA tries to find the global optimum of the function

$$f_2(x_1, x_2) = (x_1 - 1)^2(x_1 + 1)^2(x_2 - 1)^2(x_2 + 1)^2,$$

which actually has four equal global optima at points $x_{1,2} = \pm 1$. Interval package called IP++ written in C++ was used in interval calculations [19] and F_2 was evaluated using the Ratschek's centered form [15]. The GA used was a slightly modified version from our earlier project [20]. The intervals were contracted by a factor of 0.85 for each generation. The population size was the traditional 50 and 30 generations were evaluated. 50% of the best specimens were always selected to the next generation.

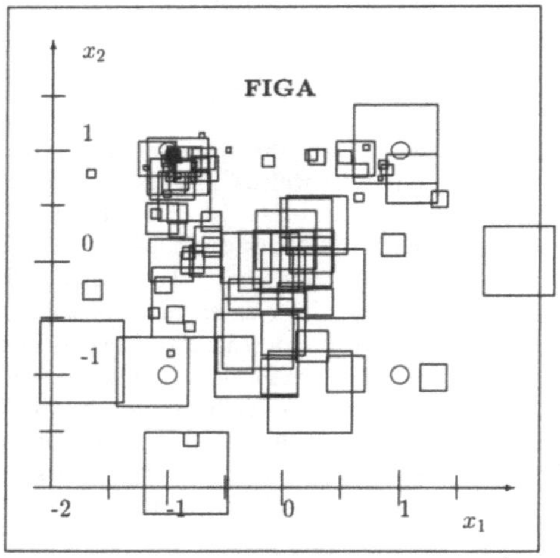

Figure 1: Focusing interval GA method solving f_2. The global optima are marked with o. (Only every 5th search interval from the second generation and on is shown.)

3.2 IAGA

In repeated runs with different random number seeds FIGA indeed found the different optima. However, in order to get a more robust optimizer it seems inevitable to exploit some more tools and methods provided by interval aritmetic. We can also see if we could use GAs in interval arithmetic global optimization methods that are quarantined to find the global optimum.

In the first case we could use the information from the evaluated function values and intervals as some kind of active and passive areas maps. If search seems to stuck to a local optimum, new areas could be searched by guiding mutations to the most promising alternative areas.

In the latter case there are several possible ways to exploit GAs in interval global optimization methods. These include

- selection of boxes to be subdivided
- selection of variables to be subdivided
- selection of interval extension of f

while keeping the active list as short as possible and thus minimizing the function evaluations and processing time.

Our first step towards a more rigorous interval arithmetic GA optimizer (IAGA) is the following one: We save the selected intervals of the first generation of FIGA as a kind of active set or a set which gives quite good estimation of the possible location of the optima[2].

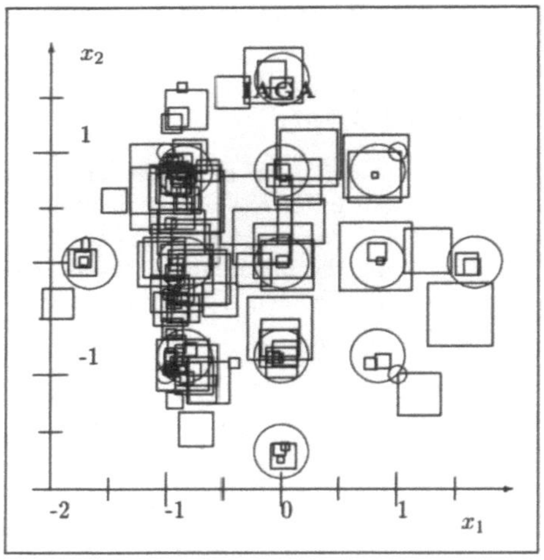

Figure 2: Interval arithmetic GA method (IAGA) solving f_2. The global optima are marked with o, while the bigger circles denote the simple active list elements. (Only every 5th search interval from the third generation and on is shown.)

The ability of IAGA to find more than one optimum for f_2 is clearly demonstrated in fig. 2. However, the performance could be better and so our search for a better interval aritmetic GA hybrid optimizer continues.

[2]Remember that the starting population covered the whole search space.

3.3 CIAGA

The last hybrid interval aritmetic GA optimization methods class is to use both interval methods and GA methods concurrently (CIAGA) (fig. 3): while the interval method is carefully searching every possible location for an optimum, a GA method can freely do random search. Every now and then the methods exchange information: GA might have found values that help interval method to reduce the active list while the GA may get rid of a local optimum by taking new start points from the active list of the interval method.

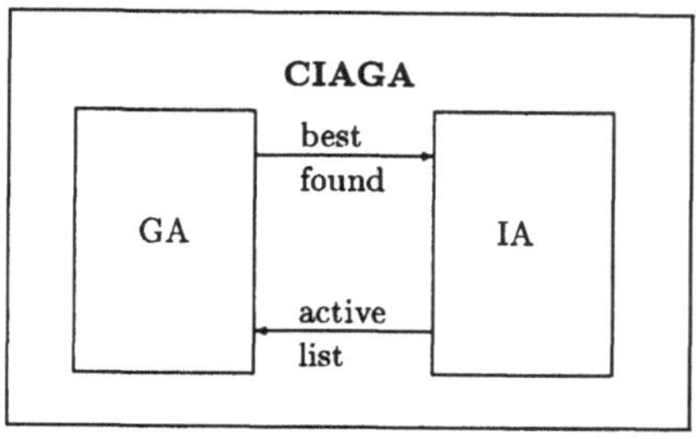

Figure 3: Concurrent interval arithmetic GA method (CIAGA).

We could also include other robust optimization methods like tabu search, simulated annealing, or factor analysis into a concurrent hybrid method, in which the constituent methods could exchange information during search. The extra benefit of this kind of concurrent arrangement is that the resulting system would be more reliable than any of the constituents alone. In some applications this could be a really valuable feature.

The description of this hybrid class is unfortunately beyond the scope of this paper and is reported elsewhere [21].

4 Conclusions and future

Some hybrids of genetic algorithms and interval aritmetic global optimization methods have been presented in this paper. While the results so far are still preliminary, the hybrids of these global optimization methods seems to be beneficial and deserve more research. We are still working on testing and analysing a set of GA and interval aritmetic methods and the detailed results will be reported elsewhere [21]. We hope that this paper has drawn the attention of the GA research community to the potential applications of interval aritmetic in some GA based global optimization methods.

5 Acknowledgements

The author wants to acknowledge all those who have helped him during this work. Thanks also to William F. Nice for his kind help with the proofreading of the manuscript.

References

[1] Reiner Horst and Hoang Tuy. *Global Optimization*. Springer-Verlag, Berlin, 1990.

[2] R. E. Moore. *Interval Analysis*. Prentice-Hall, Inc., Englewood Cliffs, 1966.

[3] Jon Rokne and T. Wu. The circular complex centered form. *Computing*, 28:17–30, 1982.

[4] Jon Rokne and T. Wu. A note on the circular complex centered form. *Computing*, 30:201–211, 1983.

[5] Jarmo T. Alander. *An indexed bibliography of genetic algorithms: Years 1957-1993*. Art of CAD Ltd., Vaasa (Finland), 1994. (over 3000 GA references).

[6] T. M. Murdock, W. E. Schmitendorf, and Stephanie Forrest. Use of genetic algorithm to analyze robust stability problems. In *Proceedings of the American Automatic Control Conference*, pages 886–889, Boston, MA, 26.-28. June 1991. American Automatic Control Council.

[7] Marco Muselli and Sandro Ridella. Global optimization of functions with the interval genetic algorithm. *Complex Systems*, 6(3):193–212, June 1992.

[8] Larry J. Eshelman and J. David Schaffer. Real-coded genetic algorithms and interval-schemata. In Whitley [22], pages 187–202.

[9] Kwong Sak Leung, Yee Leung, Leo So, and Kim Fai Yam. Rule learning in expert systems using genetic algorithm: 1, concepts. In *Proceedings of the 2nd International Conference on Fuzzy Logic and Neural Networks (IIZUKA'92)*, volume 1, pages 201–204, Iizuka (Japan), 17.-22. July 1992. Fuzzy Logic Systems Institute.

[10] Kwong Sak Leung, Yee Leung, Leo So, and Kim Fai Yam. Rule learning in expert systems using genetic algorithm: 1, empirical studies. In *Proceedings of the 2nd International Conference on Fuzzy Logic and Neural Networks (IIZUKA'92)*, volume 1, pages 205–208, Iizuka (Japan), 17.-22. July 1992. Fuzzy Logic Systems Institute.

[11] Peng Tian and Zihou Yang. An improved simulated annealing algorithm with genetic characteristics and the traveling salesman problem. *J. Inf. Optimization Sci. (India)*, 14(3):241–255, September 1993.

[12] R. E. Moore. *Methods and Applications of Interval Analysis*. SIAM, Philadelphia, 1979.

[13] W. Chuba and W. Miller. Quadratic convergence in interval arithmetic. Part I. *BIT*, 12:284–290, 1972.

[14] Jarmo T. Alander. On interval arithmetic range approximation methods of polynomials and rational functions. *Computers&Graphics*, 9(4):365–372, 1985.

[15] H. Ratschek. Centered forms. *SIAM Journal of Numerical Analysis*, 17:656–662, 1980.

[16] E. R. Hansen. Global optimization using interval analysis — the one-dimensional case. *J. Optim. Theory and Appl.*, 29:331–344, 1979.

[17] E. R. Hansen. Global optimization using interval analysis — the multi-dimensional case. *Numerische Matematik*, 34:247–270, 1980.

[18] Helmut Ratschek and Jon Rokne. *New computer methods for global optimization*. Ellis Horwood, New York, 1988.

[19] Jarmo T. Alander. Programmers manual of interval package IP++, Version 0.0 and comparison to IP. Technical Report HTKK-TKO-B62, Helsinki University of Technology, Laboratory of Information Processing Science, 1987.

[20] Jarmo T. Alander. On robot navigation using a genetic algorithm. In R. F. Albrecht, C. R. Reeves, and N. C. Steele, editors, *Artificial Neural Nets and Genetic Algorithms*, pages 471–478, Innsbruck, Austria, 13.-16. April 1993. Springer-Verlag, Wien.

[21] Jarmo T. Alander. Some new global optimization algorithms using genetic algorithms and interval arithmetics. Technical Report 95-1, University of Vaasa, Department of Information Technology and Industrial Management, Vaasa (Finland), 1995. (to appear; anonymous ftp site `ftp.uwasa.fi` directory `cs` file `report95-1.ps.Z`).

[22] Darrell Whitley, editor. *Foundations of Genetic Algorithms — 2 (FOGA-92)*, Vail, CO, 24.-29. July 1992 1993. Morgan Kaufmann: San Mateo, CA.

We now consider the following problem, that will be solved using GAs:

How can we determine a polarity for a given Boolean function f such that the number of terms in the corresponding FPRM is minimized?

3 Genetic Algorithm

In this section we describe the *Genetic Algorithm* (GA) that is applied to the problem given above.

3.1 Representation

For each variable a polarity must be chosen. Thus, each element of the population corresponds to an n-dimensional binary vector. A population is a set of vectors. Using this binary encoding each string represents a valid solution.

3.2 Objective Function and Selection

As an *objective function* that measures the *fitness* of each element we use the number of terms in the FPRM corresponding to the chosen polarity. This function has to be minimized to find a small two-level representation of the function. The number of terms is determined using OFDDs (see Section 2). An OFDD with a given polarity p is transformed to an OFDD with polarity p' by performing EXOR-operations. Each EXOR-operation on an OFDD G of size $|G|$ has runtime $O(|G|^2)$, and thus can be performed efficiently.

The selection is performed by *roulette wheel selection*. Additionally, we also make use of *elitarism* [2]: Some of the best elements of the old population are included in the new one anyway. This strategy guarantees that the best element never gets lost and that a faster convergency is obtained. GA practice has shown that this method is usually advantageous. The size of the population is adapted dynamically using problem specific measures. (The choice of the population size will be explained later.)

3.3 Initialization

At the beginning of each GA-run an initial population is randomly generated and the fitness is assigned to each element.

Often it is helpful to combine GAs with problem specific heuristics [2]. The resulting GAs are called *Hybrid GAs* (HGAs). In our application we use the following *greedy heuristic* from [5]:

Greedy Heuristic (GRE): Start with an OFDD and change the polarity of variable x_1. Choose the best polarity for x_1 and go to the next variable (if it exists).

In our applications we focus on the question, whether it is ingenious to consider HGAs with respect to runtime and quality of the result. The initial population is further optimized by applying the greedy algorithm GRE to i elements randomly chosen ($i \in \{0, .., pop\}$), where pop denotes the size of the population. (If $i = 0$ we have a *pure GA*, i.e. a GA without application of heuristics.) A frequent application of GRE to the initial population guarantees that the starting points are not too bad, thus the convergency is speeded up. The application of GRE itself is time consuming, but forces a faster convergency. Therefore it is only applied at the beginning and at the end to the best element for a final optimization to avoid local minima.

3.4 Genetic Operators

In the following we describe the genetic operators used in our application. All operators are directly applied to binary strings of finite length that represent elements in the population. The parent(s) for each operation is (are) determined by the mechanisms described in Subsection 3.2.

Reproduction: Copying strings according to their fitness.

Crossover: Construction of two new elements x_1 and x_2 from two parents y_1 and y_2, where the first (second) part of x_1 (x_2) up to a randomly chosen cut position i is taken from y_1 and the second (first) part is taken from y_2.

2-time Crossover: Construction of two new elements x_1 and x_2 from two parents y_1 and y_2, where the first (second) part of x_1 (x_2) up to a cut position i is taken from y_1 (y_2), the second part up to a second cut position $j > i$ is taken from y_2 (y_1) and the last part is again taken from y_1 (y_2).

Mutation: Construction of a new element from a parent by copying the whole element and randomly changing its value at mutation position i.

2-time Mutation: Construction of a new element from a parent by copying the whole element and randomly changing its value at mutation positions i and j ($i \neq j$).

Mutation with neighbour: Construction of a new element from a parent by copying the whole element and randomly changing its value at mutation positions i and $i + 1$.

Obviously, all genetic operators only generate valid solutions, if they are applied to the binary strings.

A GENETIC ALGORITHM FOR MINIMIZATION OF FIXED POLARITY REED-MULLER EXPRESSIONS *

Rolf Drechsler   Bernd Becker   Nicole Göckel

Comp. Sc. Dept., J. W. Goethe-University
60054 Frankfurt am Main, Germany
email: <name>@kea.informatik.uni-frankfurt.de

Abstract

A Genetic Algorithm (GA) is developed to find small or minimal Fixed Polarity Reed-Muller expressions (FPRMs) for large functions. We combine the GA with greedy heuristics, i.e. we use Hybrid GAs (HGAs). We show by experiments that results superior to all other approaches for large functions can be obtained using GAs.

1 Introduction

Genetic Algorithms (GAs) are often used in optimization and machine learning [6, 2]. In many applications they are superior to the classical optimization techniques, e.g. gradient-descent. Recently, GAs have succesfully been applied to several hard problems in CAD [3].

The high complexity of modern VLSI circuitry has shown an increasing demand for synthesis tools. In the last few years synthesis based on AND/EXOR realizations has gained more and more interest, because AND/EXOR realizations are very efficient for large classes of circuits [9, 8].

In the following we consider a restricted class of *EXOR Sum-Of-Products expressions* (ESOPs), called *Fixed Polarity Reed-Muller expressions* (FPRMs), and present methods to minimize them by GAs. First preliminary ideas on how to model FPRM-minimization in a genetic environment have been presented in [1]. In this paper we concentrate on more sophisticated genetic algorithms (including hybrid approaches) and show that results superior to all previously published methods can be obtained.

The fitness function of our algorithm can be evaluated quickly by using a problem dedicated multi-level data structure, called *Ordered Functional Decision Diagrams* (OFDDs). The correspondence between OFDDs and FPRMs was already used in [5] and is also used here in combination with genetic operators. In contrast to other methods [7] this allows the minimization of large functions with several outputs in parallel. Thus, our algorithm matches more closely the minimization problem for PLAs. Experimental results on large functions that improve on the results in [5] are given to show the efficiency of our approach. Finally we discuss the quality-time trade-off that results from different population sizes and from different applications of the greedy heuristic.

*This work was supported in part by DFG grant Be 1176/4-2

2 Problem Domain

An *FPRM* is an exclusive-OR of AND product terms, where each variable only appears in complemented or uncomplemented form, but not both. FPRMs are a canonical representation of Boolean functions $f : \mathbf{B}^n \rightarrow \mathbf{B}$, if the polarity for each variable is fixed. The choice of the polarity largely influences the size of the resulting FPRM, as is shown by the following example:

Example 1 Let $f : \mathbf{B}^n \rightarrow \mathbf{B}$, given by $f = \overline{x}_1\overline{x}_2..\overline{x}_n$. Thus, only one term is needed, if all variables are complemented. If all variables are uncomplemented the resulting expression consists of 2^n terms, i.e. $f = 1 \oplus x_1 \oplus x_2 \oplus ... \oplus x_1x_2..x_n$.

For the representation of Boolean functions we use a multi-level data structure, called *OFDDs*, as defined in [5]. An OFDD for a function f is a rooted directed acyclic graph giving a recursive AND/EXOR based decomposition of f into simpler subfunctions, constant 0 and 1 being the simplest subfunctions. An OFDD for f directly corresponds to an FPRM of f with fixed polarity. Furthermore, OFDDs allow efficient synthesis operations, at least with respect to our application [4, 5]. Using the OFDD it is possible to determine very fast the number of terms of the corresponding FPRM and to change the polarity of the FPRM by manipulation of the OFDD.

```
genetic_algorithm (benchmark, i) {
 generate_random_population () ;
 calculate_fitness () ;
 if(HGA) optimize_elements_with_GRE (i) ;
 {
   apply_operators_with_corresponding_probabilities () ;
   calculate_fitness () ; update_population () ;
 } while (not terminal case) ;
 if(HGA) apply_GRE_to_best_element () ;
 return ;
}
```

Figure 1: Sketch of basic algorithm

3.5 Algorithm

Using the operators introduced above our genetic algorithm works as follows:

1. Initially a random population of binary finite strings is generated and i elements are optimized by the greedy heuristic as described in Subsection 3.3.

2. The better half of the population is copied in each iteration without modification. Then the genetic operators *reproduction* and *crossover* are applied to another $\frac{pop}{2}$ elements. The elements are chosen according to their fitness as described in Subsection 3.2. The newly created elements are then mutated by one of the three mutation operators with a given probability.

3. The algorithm stops if no improvement is obtained for $50 \cdot log(best\_fitness)$ iterations, where $best\_fitness$ denotes the fitness of the best element in the population. Finally, if $i > 0$, the greedy algorithm is applied to the best element.

A sketch of the algorithm is given in Figure 1.

3.6 Parameter Settings

The size of the population is adapted dynamically. The initial population has size half of the number of variables of the considered Boolean function. After initialization the population size depends on the average fitness of the population. If the average fitness is smaller than 100 the size is decreased to constant 5. Experimental results have shown that convergency and runtime are speeded up without loss of quality.

The genetic operators are iteratively applied corresponding to their probabilities:

1. *Reproduction* is performed with a probability of 20%.

2. *Crossover* and *2-time crossover* are performed with a probability of 80%.

3. *Mutation*, *2-time mutation* and *mutation with neighbour* are carried out on the newly generated elements with a probability of 15%.

name	*in*	*out*	exact		HGA	
			terms	time	terms	time
add6	12	7	132	408.2	132	26.3
gary	15	11	349	31093.8	349	72.5
m181	15	9	67	1577.6	67	13.2
tial	14	8	3683	14012.0	3683	286.4

Figure 2: Exactly minimized FPRMs for small circuits

4 Experimental Results

In this section we present experimental results that were carried out on the package presented in [4] on a *Sun Sparc 10*. All times given are measured in CPU seconds.

We applied our algorithm to several benchmarks from LGSynth91. In [5] OFDD based heuristics have been compared to several other approaches. There it has been shown that with this method the best results were obtained. Thus, in the following we restrict ourselves to a comparison with [5] with respect to runtime and quality of the results.

In a first series of experiments the HGA is applied to *small* benchmarks with up to 15 variables. For these functions the optimal solution with respect to the considered objective function has been determined in [5]. The results are given in Figure 2. *in* (*out*) denotes the number of inputs (outputs). Column *exact* gives the results of the exact method of [5] and HGA denotes the approach presented in this paper. *time* gives the runtime of the corresponding algorithm. *terms* denotes the number of terms in the resulting FPRM. For all benchmarks we used the same parameter set, i.e. we applied the greedy heuristic to $\frac{pop}{4}$ elements. The HGA *always* determined the minimal solution and additionally is much faster than the exact method, e.g. the HGA is more than 420 times faster for benchmark *gary*.

In a next series of experiments we applied the HGA to larger benchmarks for which the optimal solution is not known. We compare our results with the most powerful (and most time consuming) heuristic from [5]. The results are presented in Figure 3. The HGA always determines the same or better results than the heuristic approach in [5]. Especially, for very large

name	*in*	*out*	heuristic		HGA	
			terms	time	terms	time
apex7	48	37	604	7.7	515	648.1
bc0	26	11	1118	10.8	1117	334.1
bcb	26	39	2241	30.4	1892	1086.7
chkn	29	7	900	5.4	900	377.2
cps	24	109	293	10.1	291	368.2
ibm	48	17	1220	6.8	1181	736.6
in2	19	10	381	2.1	355	57.9
s1196	32	32	7951	16.6	7012	2320.1
vg2	25	8	5899	5.3	5290	661.6
x6dn	39	5	593	5.9	536	213.6

Figure 3: Minimized FPRMs for large circuits

functions the results of the HGA are much better than in the heuristic approach (see e.g. *apex7*), but the HGA is more time consuming.

Finally, we discuss the influence of the frequency of the application of the greedy heuristic and the size of the population to runtime and quality.

For this we consider benchmark *vg2* as an example. The results are given in Figure 4. i in the top row denotes the number of applications of the greedy heuristic after initialization. *pop* denotes the size of the population. The upper row for each population size gives the resulting number of terms and the lower row gives the runtime measures in CPU seconds.

As can easily be seen good results are not obtained, if the population size is chosen too small. In contrast a too large population size wastes resources without any gain. The pure GA, i.e. the GA without application of the greedy heuristic, performs not very good, since the starting points are too bad. This avoids a fast convergency. Our experimental results have shown, that our parameter settings lead to good results with respect to runtime and quality on average.

5 Conclusions

An OFDD based method to minimize FPRMs using hybrid GAs was presented. We focused on the minimization problem of large functions and showed that our results improve significantly on the best results known before [5].

It is focus of current work to generalize this approach and to apply it to more general classes of AND/EXOR networks.

vg2	$i = 0$	$i = \frac{pop}{4}$	$i = \frac{pop}{2}$	$i = pop$
pop=1	36481	8682	8682	8682
	2.8	24.1	24.1	24.1
pop=5	6520	6062	6062	6062
	201.3	257.7	258.2	257.6
pop=12	5290	5290	5290	5290
	1062.6	661.6	661.3	661.7
pop=25	5290	5290	5290	5290
	1069.0	896.2	898.0	899.4

Figure 4: Results for benchmark *vg2*

References

[1] B. Becker and R. Drechsler. OFDD based minimization of fixed polarity reed-muller expressions using hybrid genetic algorithms. In *Int'l Conf. on Comp. Design*, pages 106–110, 1994.

[2] L. Davis. *Handbook of Genetic Algorithms*. van Nostrand Reinhold, New York, 1991.

[3] R. Drechsler, H. Esbensen, and B. Becker. Genetic algorithms in computer aided design of integrated circuits. Technical report, Universität Frankfurt, Fachbereich Informatik, 1994.

[4] R. Drechsler, A. Sarabi, M. Theobald, B. Becker, and M.A. Perkowski. Efficient representation and manipulation of switching functions based on ordered kronecker functional decision diagrams. In *Design Automation Conf.*, pages 415–419, 1994.

[5] R. Drechsler, M. Theobald, and B. Becker. Fast OFDD based minimization of fixed polarity reed-muller expressions. In *European Design Automation Conf.*, pages 2–7, 1994.

[6] D.E. Goldberg. *Genetic Algortithms in Search, Optimization & Machine Learning*. Addision-Wesley Publisher Company, Inc., 1989.

[7] A. Sarabi and M.A. Perkowski. Fast exact and quasi-minimal minimization of highly testable fixed-polarity and/xor canonical networks. In *Design Automation Conf.*, pages 30–35, 1992.

[8] T. Sasao. *Logic Synthesis and Optimization*. Kluwer Academic Publisher, 1993.

[9] T. Sasao and Ph. Besslich. On the complexity of mod-2 sum PLAs. *IEEE Trans. on Comp.*, 39:262–266, 1990.

IDENTIFICATION AND ADAPTIVE CONTROL OF NONLINEAR PROCESSES USING COMBINED NEURAL NETWORKS AND GENETIC ALGORITHMS

A. H. Abu-Alola and N. E. Gough

School of Computing and Information Technology, University of Wolverhampton, Wulfruna Street, Wolverhampton WV1 1SB, UK

Abstract

This paper presents a new combined neural-genetic method for identifying processes comprising static nonlinearities and linear dynamics, and for adapting the model to linear and nonlinear parameter changes. The approach is based on multilayer and recurrent neural networks. It is shown that the powerful properties of genetic algorithms may be used to advantage in a learning neural network. We solve a number of problems and compare the results and the method with those previously reported by Narendra & Parthasarathy.

1 Introduction

In the last decade, neural networks (NNWs) have been increasingly applied to the problem of identification and control of dynamical systems. Considerable research effort has gone into investigating various algorithms to determine the values of weights, thresholds and reduction factors. However, there are still drawbacks to applying these algorithms generally and this raises the question if there are any new tools that can be applied to overcome these limitations. One area of identification which has made slow progress is that involving nonlinear (NL) models. In a key work, Narendra & Parthasarathy [1] demonstrated that different classes of NNWs are able to identify a complicated NL dynamical system. Their results showed considerable promise but nevertheless suffered from the following disadvantages:

- a fairly good knowledge of the system structure is needed to decide how many delayed outputs should be fed back
- some mathematical analysis is usually required to arrive at a method for updating the NNW parameters
- the learning time is invariably long and hence problematical for potential real-time applications
- the degree of the accuracy in the light of the excessive learning time is poor.

The present paper re-examines the basic neural approach adopted by Narendra & Parthasarathy with a view to overcoming these problems by using a genetic algorithm (GA). §2 reviews some limitations of GAs in NNWs and §3 gives the proposed new method. Its application to NL system identification is then described in §4, including results for two plants, and the effect of model delay and network topology. §5 then extends the method to cover adaptation and simulated results are presented. Finally, §6 discusses the significance of the results for real-time control.

2 Genetic Algorithms in NNWs

GAs use populations of individuals, encoded as binary strings and initially chosen at random. They carry out a stochastic global search to minimise a fitness function. A reproductive stage selects the more fit individuals (chromosomes), the genes of which are then modified by genetic operators to give the next generation. Recombination operators give pairs or larger groups exchanging information and mutation operators produce occasional random changes. After one stage, the chromosomes are decoded, the fitness function is computed, and the next stage begins. The GA finds its major application in the field of optimisation. GA's are less likely to converge to local optima, do not require the fitness function to be well-behaved, tolerate discontinuities and noisy signals, and converge to a global optimum, even for difficult constrained problems [2].

These advantages lead us to use a GA to train a NNW. An extended direct encoding scheme is used, where each network connection is represented by a binary code. Fig. 1 shows the simple topology used initially. We studied this form and found that direct GA techniques can be used successfully using only 30 bits, with each weight and neuron encoded with 3 bits.

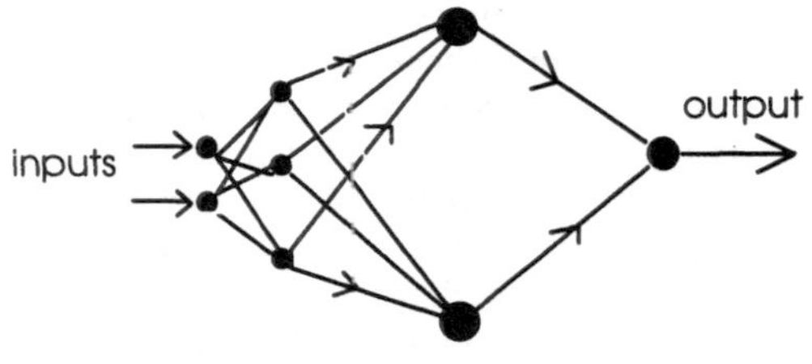

Figure 1. Basic NNW topology used

However, more complex NNW design involves a optimization search at two levels of network architecture: evolution of network connectivity and adaptation of connection weights. The latter has attracted considerable research effort. Previous work confirms that difficulties are encountered when GAs are applied to NNW design problems: a NNW is very sensitive to the weights and a basic GA is limited in that it has only three operators and one string. Consequently, we modified the basic method to use a diploid (chromosome pair) and exploit the concept of dominance (genotype-to-phenotype mapping). We have found that the GA's major contribution to searching is to provide a mechanism for remembering alleles that were previously useful and to protect these alleles in the future.

3 Proposed Modified Neural-Genetic Method

If a NNW has excessive connectivity and sensitivity, huge genes are needed, and in addition the limits of weight values are not known. Thus the search for a point is very difficult. In large NNWs such as those suggested by Narendra & Parthasarathy [1] we found that a large string is needed and the basic GA does not perform well when identifying a NL system. To overcome this limitation and provide the mechanism for memorising the alleles, we developed the GA to be more effective in learning NNWs. In the adopted paradigm, we retain one string (haploid) as the initial guess. In the second run, instead of discarding the old and trying another, the string is used to suggest a route from the old solution and to modify it at each sample. In this way the GA does not search directly for the solution but looks for a way to guide the solution. This method bases the change of each weight on the current value and the error between the desired output and actual one and provides the required memory.

To implement the new algorithm, we use GAs largely in the standard way but introduce the following variations: a) the thresholds and reduction factors are chosen randomly and b) the strings are chosen for each element to decide whether to increase of decrease the values depending on i) the value of the string of that element; ii) the error between the desired output and actual one; iii) the value of the element itself. At each instant in time, the weights and reduction factors are adjusted using the same string values that have been chosen for that particular trial of population for 100 iterations. After trying each one in the population, the GA then selects the new strings - including the number of delayed outputs and inputs for each trial of the population - and again runs for a further 100 iterations.

4 Application to Nonlinear Dynamic System Identification

In control systems, it is an important requirement to identify plant dynamics, especially for the purposes of adaptation. Although a GA has the potential to deal with direct adaptation as will be seen in §5, the ability to provide a specific identification model such as that required in indirect adaptation is important because a GA makes a number of trials which can be applied to the model produced by an identification.

To demonstrate clearly the advantages of the modified algorithm, we used a model similar to that of Narenda & Parthasarathy, as shown in Fig. 2. However, unlike Narendra, we made no assumptions about the structure of the system. The GA thus chooses the delayed inputs and outputs automatically.

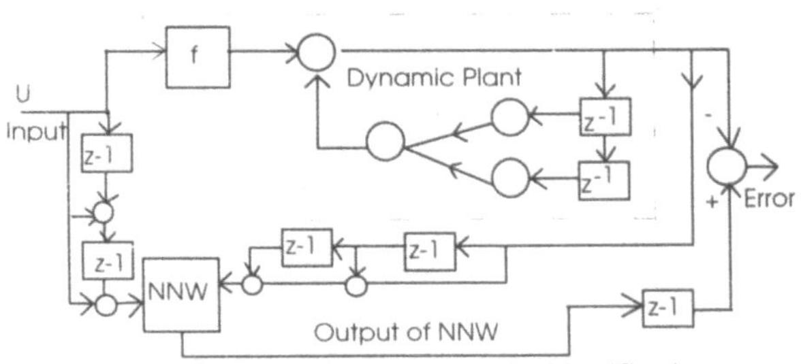

Figure 2 Series-parallel model for identification

Four examples are introduced to illustrate various features of the modified NNW-GA approach based on two plants:

Plant A $y_p(k+1) = 0.3y_p(k) + 0.6y_p(k-1) + f[u(k)]$ (1)
where $f[u(k)] = u(k)^3 + 0.3u(k)^2 - 0.4u(k)$ (2)

Plant B $y_p(k+1) = 0.8y_p(k) + f[u(k)]$ (3)
where $f[u(k)] = (u-0.8)u(u+0.5) = u(k)^3 - 0.3u(k)^2 - 0.4u(k)$ (4)

In both cases, we first identify the static NL function f[u(k)] only for input $u=\sin(2\pi k/250)$. Then we identify the whole plant including the NL and linear dynamic parts. The NNW topology is similar to Narendra & Parthasarathy's i.e. 2 input nodes, one for the test signal and the other a feed back from the plant output (Fig. 2); 20 nodes in the first hidden layer, 10 nodes in the second and 1 node in the output layer. The string length was 510 bits, 500 to encode the weights and 10 to encode the reduction factor and determine the number of input delays. The procedure was programmed in Pascal based on Goldberg [2].

4.1 Results of Identification Simulations

Example 1: Plant A, static nonlinearity only

In this example, there is no linear part to the model. The NL function (2) is assumed to be unknown and the aim was to identify it. Although this function has no feedback from its output, it was found that the GA must use the delayed output and feed it back to the input of the NNW: When no delayed values of the function are fed back to the NNW input, the results are poor. Conversely, when the delayed outputs are fed back, excellent results were obtained as shown in Fig. 3.

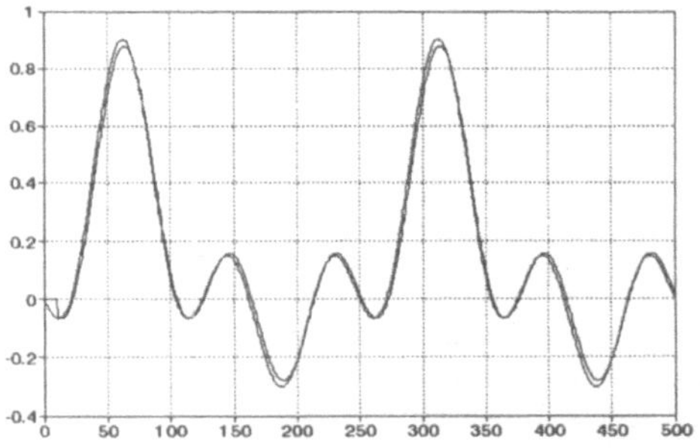

Figure 3 Actual and estimated outputs for static NL in plant A

Example 2: Plant B, static nonlinearity only

Here, the NL was chosen to be (4) with no dynamic part to the plant. Again good convergence was achieved, see Fig. 4.

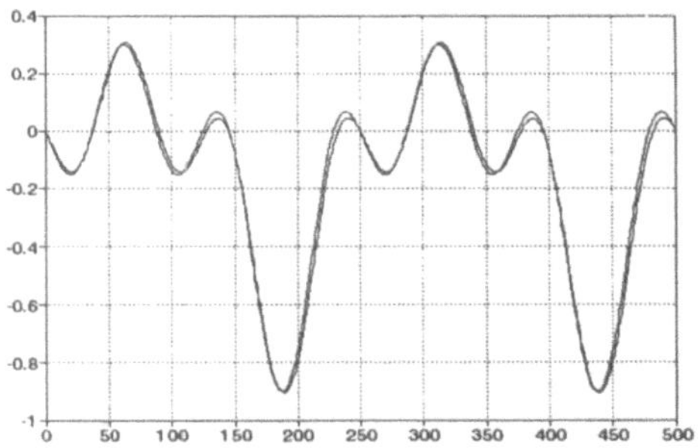

Figure 4 Actual and estimated outputs for static NL in plant B

Example 3: Plant A, combined dynamic and NL model
In this example we assume the dynamic NL model in which the overall system is described by (1) and (2). It is also assumed that the internal feedbacks are unknown. Fig. 5 shows the result.

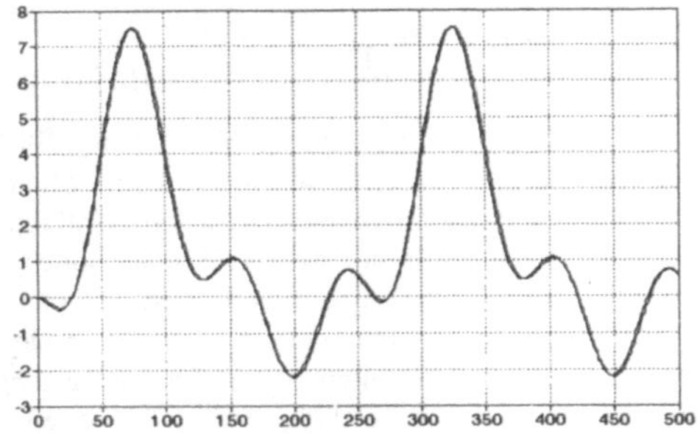

Figure 5 Outputs of dynamic NL plant A and estimated model

Example 4: Plant B, combined dynamic and NL model
In this example we assume the dynamic NL model in which the overall system is described by (3) and (4). The internal feedbacks are unknown. Figure 6 shows the result.

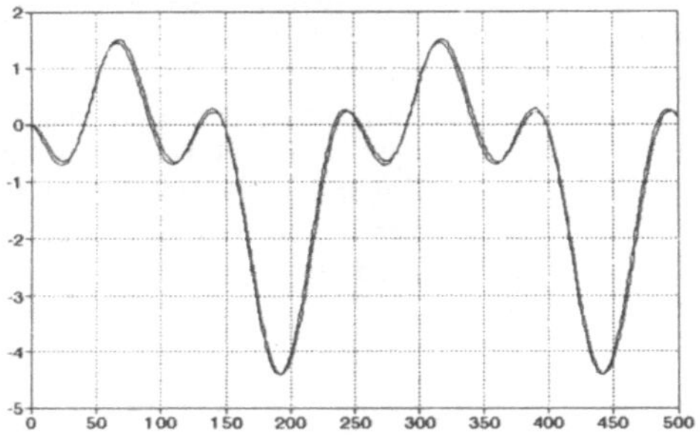

Figure 6 Outputs of dynamic NL plant B and estimated model

4.2 Effect of Delay in the Model

The previous results demonstrate the capability of the combined NNW and GA approach to identify certain classes of bilinear models. However the model shown in Fig. 2 was used for which the output of neural network is delayed and then compared to the present plant output. We have carried out simulations in which the delay has been omitted. A typical result for plant A in Fig. 7 shows that the identified model approaches, but does not fully converge to the desired solution.

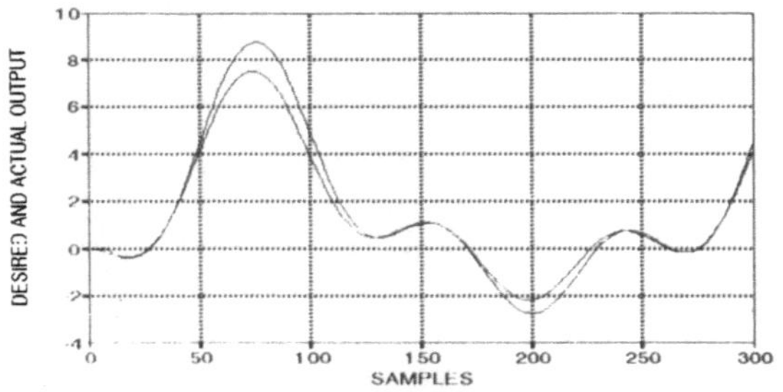

Figure 7 Results for identification of plant A, for no model delay

4.3 Effect of NNW Topology

For the previous examples, the NNW topology was the same as that used by Narendra & Parthasarathy. We now adopt a different topology in which the number of nodes in both hidden layers are reduced, which in turn reduces the string size. Several examples were tried. Fig. 8 shows the results when the number of nodes in the first hidden layer was reduced from 20 to 5, reducing the string size to 120. The model in Fig. 2 is used for Plant B. It can be seen that the GA is still able to learn with a simpler NNW topology and produce excellent results as before. Furthermore, this is achieved is about the same time as before.

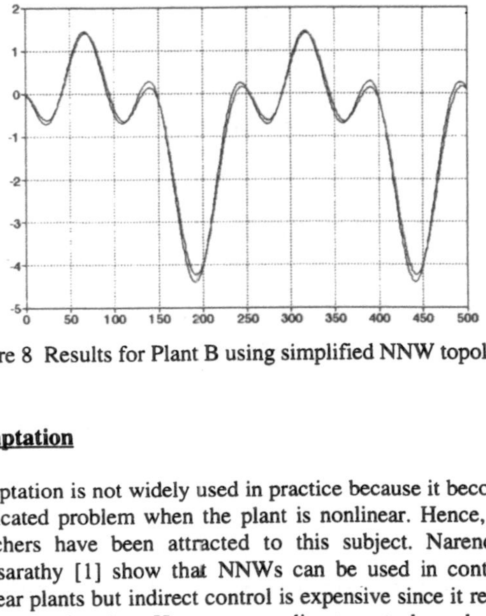

Figure 8 Results for Plant B using simplified NNW topology

5 Adaptation

Adaptation is not widely used in practice because it becomes a complicated problem when the plant is nonlinear. Hence, many researchers have been attracted to this subject. Narendra & Parthasarathy [1] show that NNWs can be used in control of nonlinear plants but indirect control is expensive since it requires parameter estimation. Here we use direct control as shown in Fig. 9. Two NNWs are used. The first has 3 inputs, 2 hidden layers, 12 neurons in the first hidden layer and 10 in the second. The output of this network is the input to the plant u_{ad}. The plant output, y_{ad} is compared with the output of a reference model y_r, and the error passes to a second NNW. This has 2 inputs, 2 hidden layers each with 8 neurons. The output of this network is fed as one of the inputs to the first network. Two examples are demonstrated for plant A.

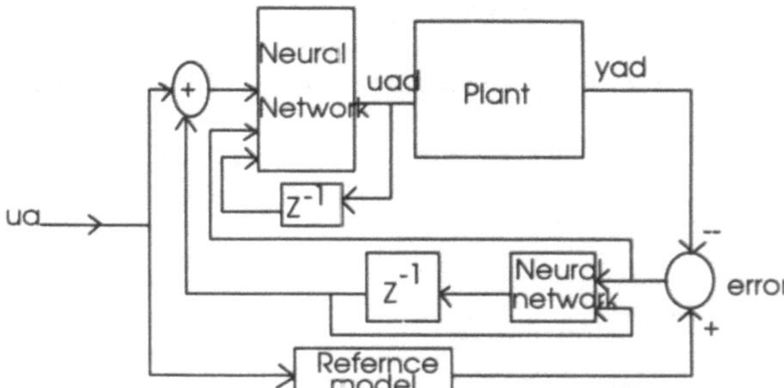

Figure 9 Model used for adaptation.

Example 5: Adapting to linear model parameter changes

In this example we assumed that the one of the linear parameters of the plant has been changed to give

$$y_p(k+1) = 0.2y_p(k) + 0.6y_p(k-1) + f[u_a(k)] \qquad (1a)$$

where u_a is the reference model input ($u=u_a=\sin[2\pi/250]$). Figure 10 shows the desired output and the actual output before adaptation and Fig. 11 shown the results after adaptation.

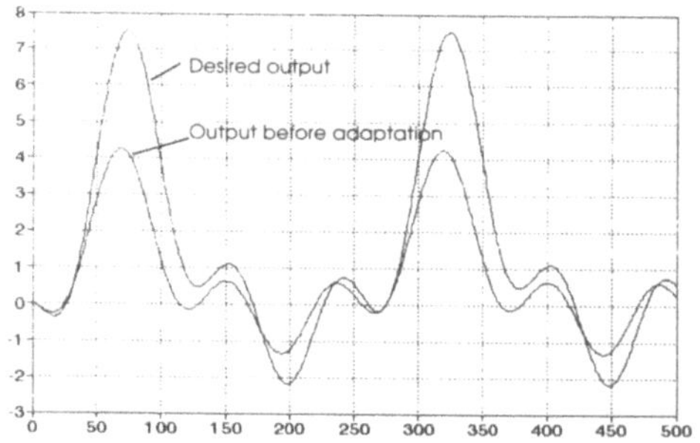

Figure 10 Desired and actual output before adaptation

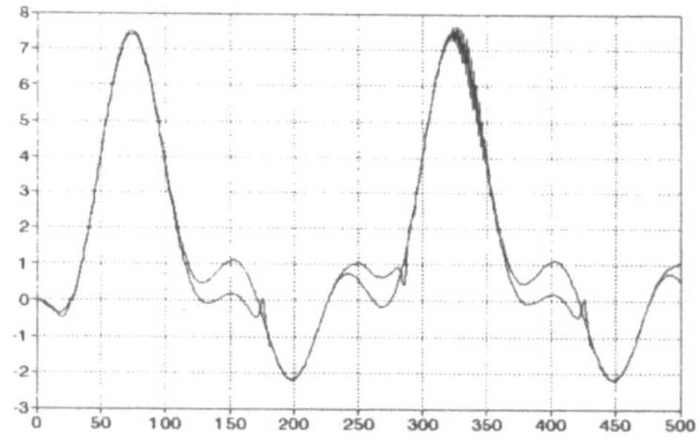

Figure 11 Desired and actual output after adaptation

Example 6: Adapting to dynamic and static NL changes

In this example we assumed that the plant had changes in the linear parameter as before and in addition the NL part in (2) has changed to $f = u^3 + 0.2u^2 - 0.4u$. Figures 12 and 13 show the reference output compared with the non-adapted output and adapted output, and Fig. 14 shows the difference between the input to the reference model and the input to the plant. It can be seen that in spite of the differences between the inputs, the neural network has adapted well to changes in both linear and nonlinear model parameters.

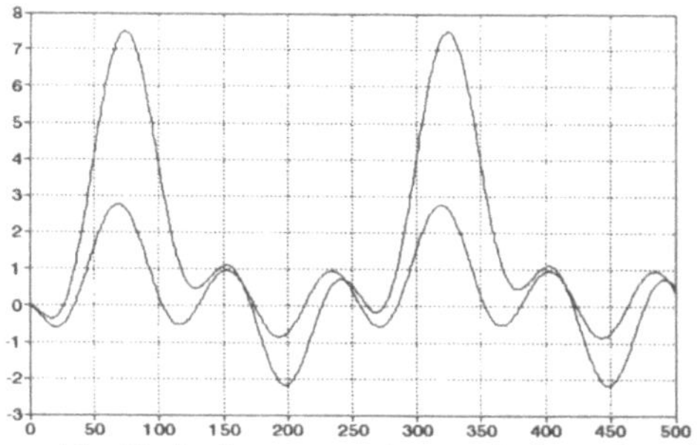

Figure 12 Desired output and output before adaptation following both linear and non-linear parameter changes

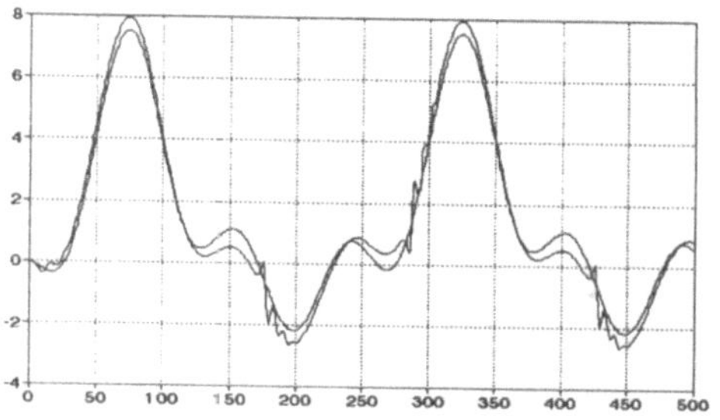

Figure 13 Desired output and output after adaptation

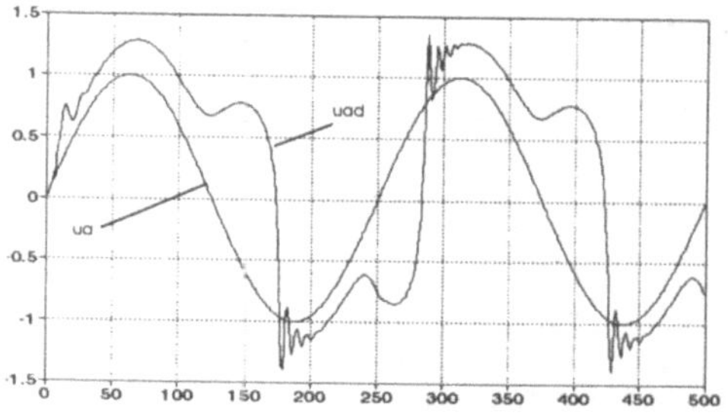

Figure 14 Input to reference model u_a and input to plant u_{ad}

6 Conclusions

This paper has proposed a new combined NNW-GA method that may be used for identification and adaptation in NL dynamic control systems. An essential part of the method is that the desired output of the NNW must be fed to the input. Using two examples previously given by Narendra & Parthasarathy, we have demonstrated that a NNW may be trained with a modified GA to identify both nonlinear static and dynamic plants. Furthermore, it was shown that the new method makes it possible to dramatically reduce the number of neurons in the hidden layers, thereby reducing the string length. In the second part of the paper, we extended these ideas to tackle successfully the problem of adaptation. The identification never required more that 180 samples to converge and the adaptation always converged in less than 210. Our method thus offers potential for

- very fast convergence to the optimal values of the weights, which makes the method suitable for on-line control
- a simple but powerful adaptation method that needs no knowledge of the underlying model structure or theory
- A method that adapts from information about outputs only, thus obviating the need for state estimation.

Current work is taking place on model parameter estimation, reduction and linearization using the NNW-GA approach.

Acknowledgements The authors wish to acknowledge the support of the Saudi Royal Navy in this research.

References

[1] Narendra, K.S. & Parthasarathy, K. Identification and control of dynamical systems using neural networks, *IEEE Trans. on Neural Networks*, **1**, 1, 4-27 (1990).

[2] Goldberg, D.E. *Genetic Algorithms in Search, Optimization and Machine Learning*, Addison-Wesley Publishing Co. (1989).

Elevator group control using distributed genetic algorithm

Jarmo T. Alander, Jari Ylinen
University of Vaasa, Department of Computer Science and Production Economics
P. O. Box 700 FIN-65101 Vaasa, Finland, e-mail jal@uwasa.fi
Tapio Tyni
Kone Elevators Research Center
P. O. Box 6 FIN 05800 Hyvinkaa, Finland, e-mail hattty@hat-fi.kone.com

Abstract

In this work we have studied the applicability of genetic algorithms to the optimization of elevator group control parameters. The goal was to minimize the average passenger waiting time in a simulated office building. The elevator group control consists of a set of elevators and controllers. At the highest level of the system hierarchy the elevator group controller decides which elevator serves which call. The decision is based on the actual calls, state of the elevators (location, direction, load) and on the estimation of the traffic situation like: incoming, interfloor, outgoing, which gives some idea how the calls will be distributed within a few minutes time period. Especially in office buildings the traffic type depends on the time of the day. This allocation problem, which seems to be quite simple, turns out to be a very difficult control problem in practise and is one of the features how elevator companies can competed with one another.

Keywords: control, C++, distributed processing, elevators, genetic algorithms, local area networks, optimization, simulation.

1 Introduction

Usually the goal of the elevator group controller is to allocate hall calls so that average passenger waiting time is minimized. At the same time the waiting time distribution should be as uniform as possible. Especially occasional long waiting times are not acceptable. The passenger trips by elevators form the passenger traffic flow. Typically in office buildings the traffic type depends on the time of the day [1]. When traffic flows from the building's entrance floor to the upper parts of the building the type is incoming traffic, see fig. 1. Outgoing traffic is contrary to the incoming traffic; all traffic goes from upper floors to the entrance floor. The third type is the interfloor traffic; passengers travel between floors above the entrance floor.

How the elevator group controller algorithm behaves and handles the traffic affects greatly the passenger waiting times and group transportation capacity [2, 3]. Because it is difficult to gather data of each individual passenger in real elevator installation, the evaluation of elevator group performance is usually done with simulators [4, 5]. These simulate the building itself, elevators, elevator movements, and passenger behaviour.

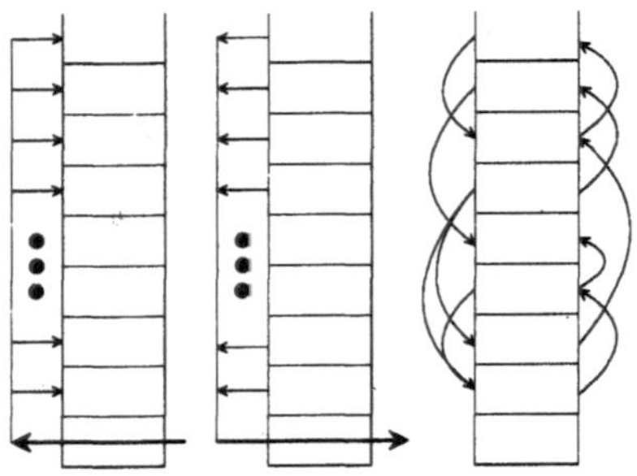

Figure 1: Basic passenger traffic types: incoming, outgoing, and interfloor.

In our work we have optimized some of the elevator group controller parameters with a GA. The goal was to search parameter values that minimize the average passenger waiting times. Although there was a comprehensive set of passenger data (waiting times, journey times etc.) available from the ALTS (Advanced Lift Traffic Simulator [4]) elevator simulator program, it was decided at the beginning to concentrate on one cost objective only.

1.1 Related work

As far as the authors know there is only one article published dealing with GAs and elevators [6]. That is the work by Ricardo R. Gudwin and Fernando A. C. Gomide, who also used GAs in elevator group control design [7]. Their system is based on modeling the controller by an automaton [8].

2 Parameters of the group controller

The group controller has a set of parameters related to the hall calls. Each floor has two hall call buttons, one for up call and the other for down call. Each call has an integer weight coefficient having values ranging from 1 to 8. A call weight matrix W is

$$W = [w_{ij}] ,$$

where $i = 1, 2, 3, \ldots,$ number of floors, $j = 1, 2$. The bigger the weight the more the associated call will be priorized within the group controller algorithm, thus giving better service to that call. The cost or fitness function to be minimized is the average passenger waiting time over the whole simulation run:

$$F = 1/N \sum_{i=1}^{N} T_i(W) = \min ,$$

where T_i is the waiting time of the ith passenger and N is the total number of passengers. It is obvious that each traffic type requires its own weight matrix W. Figure 2 shows the principle of the optimization concept. Three separate programs run together: a GA based optimizer, ALTS simulator and a statistics program to evaluate the fitness value from the simulator output data. The simulator reveives new parameters for each run from the optimizer.

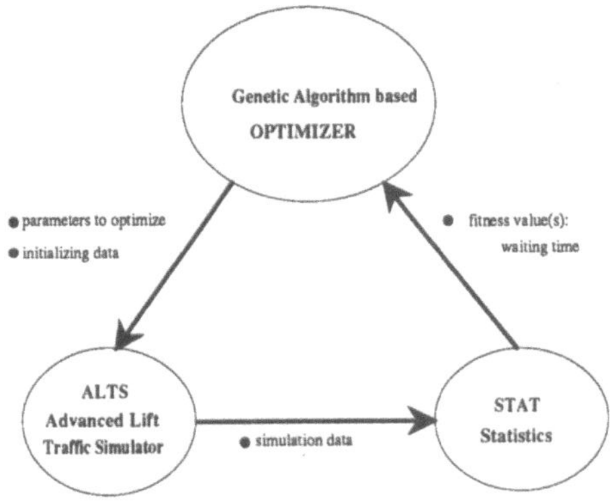

Figure 2: Elevator group controller parameter optimization.

3 Computing environment

The distributed computing environment of the optimization system consisted of several PCs ranging from i386 to i486DX2 machines connected by a local area network. One computer was acting as a master, controlling the other machines and running the GA, while the others we busy evaluating the fitness function by running the elevator simulator program (fig. 3).

3.1 Programs

The GA and the master program was written in C++ while the simulator program was mostly written in Fortran 77. Only a few modifications were necessary to integrate the interactive elevator group control simulator program to the distributed optimization system.

The master program consists of two logically separate parts: the genetic algorithm part that optimizes the parameters and the job manager part, which delivers the fitness function calculations to the PCs, which happen to be idle. The job manager stores the state of the slave machines and the states of the jobs completed and the jobs in the task queue. The job manager allows new machines to enter and leave the system on the fly. The distribution is based on a network file system common to all the PCs involved. All information between different nodes is transferred via files which are accessible to all nodes and located in fixed locations within the file system.

The slave machines run three separate programs: PC job executive, **ALTS**, and **Stat** the statistics program. The PC job executive starts elevator traffic simulations when the master gives the order and passes information to the master when the slave is ready for another job.

3.2 Distribution

Data transmission is solved in a very straightforward manner. Every slave has its own directory in the server disk via which all data between the master and the slave is transferred. Messages between the master and slaves are kept as few as possible. To configure the system is simply to read in the list of directories of the machines involved in the processing. The number of machines is in principle unlimited, however, the population size of the GA sets a practical limit on how many machines is feasible to use in processing. Due to the network load the maximum setup of computers is lower than the population size.

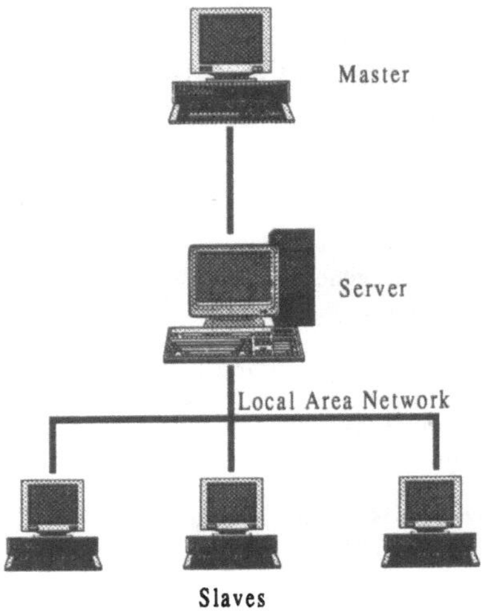

Figure 3: Overview of the test computing environment.

4 Results

The test case was an office building with 16 floors and a group of three elevators. The controller simulated was for high rise buildings and it was modified only slightly for the optimization. The modifications were mainly for data transfer between the different programs involved.

The traffic type simulated was pure outgoing with different intensities. The hall call weights were optimized at 40, 60, ..., and 120 % intensities. Two reference controllers were used: a group controller intended for mid rise buildings and the original, unmodified high rise controller.

Figure 5 shows the improvement of the average passenger waiting times when compared to reference controllers at different traffic intensities. The result of the optimization was unexpectedly good. The improvement is significant even from the original high rise controller to the optimized one. Traditionally the performance improvement of order 2–5% has been considered as a very tough task and has required much effort by skilled elevator control engineers.

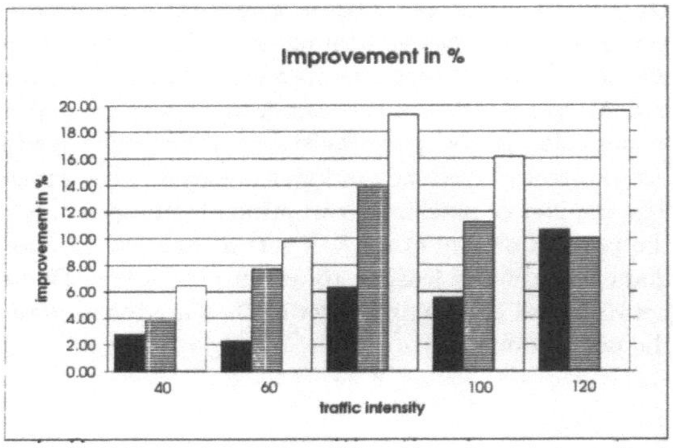

Figure 4: Passenger average waiting time improvements. ■ = middle range controller to high range controller, ▤ = high range controller to GA optimized high range controller, □ = middle range controller to GA optimized high range controller.

The new parameter values shortened the average passenger waiting times about twice as much as the difference between controller generations. Benefits of the GA based control parameter tuning include:

- problem non-specificity and flexibility
- easy to make parallel for more efficient distributed processing
- optimization is in any case generally beneficial in product design
- does not need mathematical modeling of the optimized system

while the obvious drawbacks include:

- GAs techniques are not well known
- GAs are not currently suited for real time control tuning
- GAs need quite much processing power
- the method is based on a very limited number of test cases
- need expertise with GAs

5 Conclusions and future

Although the results of the work can not be directly implemented to the end product, the test case has shown that there is still hidden capacity in the current group controller algorithm which may be released with proper arrangement and optimization. The distributed optimization environment will be used in future in other optimization problems at Kone Elevators. The GA itself has proven to be robust and reliable method and its capability to distributed computation shortens optimization times to acceptable levels.

The overall positive result motivates us to study further the use of GAs in elevator group control optimization and indeed further studies have already been started. The potential application of GA in elevator control includes:

- fine tuning of several parameter in several traffic situations
- optimization of certain functions of the controller (genetic programming)
- optimization of the traffic type estimator
- generation of demanding test cases for the simulator
- control program quality testing by generating difficult test cases

In general the work done shows the potential benefits of using GA optimization in product design and control rather than tedious manual parameter fine tuning and simulator running.

6 Acknowledgement

The work was supported by the Finnish Technology Development Center (Tekes) and was a part of the GATE (Geneettisten Algoritmien TEolliset sovellutukset[1])-project. Thanks also to William F. Nice for his kind help with proofreading of the manuscript.

References

[1] G. C. Barney and S. M. dos Santos. *Elevator Traffic Analysis Design and Control.* Peter Peregrinus Ltd, UK, 1985.

[2] N.-R. Roschier and M. J. Kaakinen. New formulae for elevator round trip calculation. *Supplement to Elevator World for ACUST members*, pages 189–197, 1979.

[1] Industrial applications of GAs

[3] M. Kaakinen and N.-R. Roeschier. Integrated elevator planning system. *Elevator World*, pages 73–76, March 1991.

[4] M.-L. Siikonen. Computer modelling of elevator traffic and control. Licentiate thesis, Helsinki University of Technology, 1989. (in Finnish).

[5] M.-L. Siikonen. Elevator traffic simulation. *Simulation*, pages 257–267, October 1993.

[6] Jarmo T. Alander. *An indexed bibliography of genetic algorithms: Years 1957-1993*. Art of CAD Ltd., Vaasa (Finland), 1994. (over 3000 GA references).

[7] Ricardo R. Gudwin and Fernando A. C. Gomide. Genetic algorithms and discrete event systems: an application. In *Proceedings of the First IEEE Conference on Evolutionary Computation*, volume 2, pages 742–745, Orlando, FL, 27.-29. June 1994. IEEE, IEEE.

[8] P. J. Ramadge and W. M. Wonham. The control of discrete event systems. *Proceedings of the IEEE*, 77(1) , January 1989.

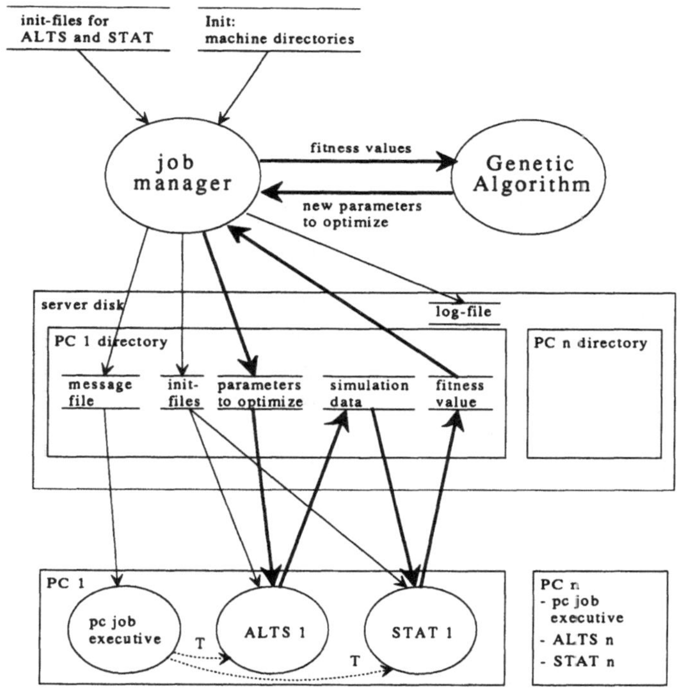

Figure 5: System architecture.

NEURAL CONTROL OF NONLINEAR NON-MINIMUM PHASE DYNAMICAL SYSTEMS

Abdelmoumène TOUDEFT

CEMAGREF
BP 5095
34033 Montpellier cedex 1, France
e-mail: toudeft@dir.cemag-montp.fr

LAFORIA
Université Paris 6
France

Abstract

This paper argues that the 'optimization along trajectories formulation of the distal learning approach [1] doesn't work on non-minimum phase dynamical plants [2] and the 'local optimization' one doesn't work on plants with variable or badly-estimated time-delay. It then proposes a modification of the procedure and illustrates improvements on the control of a simulated irrigation canal.

1. Introduction

Recently, neural networks have found many applications in the domain of system control. Adaptive and learning neural controllers have been implemented [3-6]. One of the most important problem met is to estimate the controller's output error from plant's output one. Following this purpose, Jordan and Rumelhart [1] have proposed a control architecture to deal with distal in space errors and Jordan has shown how to include generic constraints in their model [7]. Although they have presented their approach in the general context of supervised learning, the method found lot of applications in the control context. For the control problem, the approach allows to use a feedforward model of the plant to estimate desired outputs for the controller from plant's outputs errors.

In this paper, we apply the distal learning approach to control a non-minimum phase plant with variable time-delay. Non-minimum phase plants are frequently met in practice since even a minimum phase continuous plant can loose this characteristic after discretization [8]. Also, in classical and adaptive linear control theories, methods developped for minimum phase systems are different from those developped for non-minimum phase ones [9]. We show that the 'optimization along trajectories' formulation of the distal learning approach [1] doesn't work on non-minimum phase dynamical plants because inversion of the linear approximation of the plant model around actual trajectory leads to instability. The 'local optimization' one doesn't work on plants with variable or badly-estimated time-delay. Irrigation canals are examples of plants with variable time-delay. The time-delay is due to the time needed for the water to travel from the upstream to the downstream of the canal. On such systems, dynamic backpropagation leads to instability and static backpropagation cannot learn the controller.

In section 2, we present static and dynamic backpropagation in the control context. We investigate the use of this algorithm to generate learning errors for a neural controller. In section 3, static, dynamic backpropagation and three other schemes are compared on the basis of their capacity to reconstruct a perturbed command signal from plant output error. The most useful of these schemes is incorporated in a controller learning algorithm in section 4. Because the controller command obtained is highly oscillating, a smoothness constraint is added to the algorithm. Improvements are presented in section 5.

2. Static and dynamic back-propagation in learning an inverse model

Suppose a differentiable model M of the plant P to be controlled is already identified and described by the following equation:

$$y_m(t) = M[y_m(t-1), .., y_m(t-n), u(t-d-1), .., u(t-d-p)] \quad (1)$$

where u and y_m are model input and output, respectively.

There are basically two ways to derive estimation of the controller output error from the plant's output error using the model M: local optimization- which uses static backpropagation [10], and optimization along trajectories- which uses dynamic backpropagation [2]. In the former, plant's output error at time t ($e_y(t)=y(t)-y^*(t)$) is statically backpropagated through M and an error vector Δx is derived such that:

$$\Delta x = J^T.e_y(t) \quad (2)$$

where J^T is the jacobian transpose of M evaluated at its actual inputs.

The component Δx_{n+1} of Δx corresponding to u(t-d-1) is then used as a learning error:

$$\Delta w(t) = - \alpha.\Delta x_{n+1}.\partial u(t-d-1)/\partial w(t) \quad (3)$$

where w(t) is the controller weight vector at time t.

In the later, temporal dimension is taken into account and the error on the command u(t-d-1) is obtained by using equation (4) where the partial derivatives are obtained by backpropagation. $A(z^{-1})/B(z^{-1})$ is the linear approximation of M around actual trajectory.

$$e_u(t-d-1) = \frac{1-\sum_{i=1}^{n}\frac{\partial y_m(t)}{\partial y_m(t-i)}.z^{-i}}{\sum_{j=1}^{p}\frac{\partial y_m(t)}{\partial u(t-d-j)}.z^{-j+1}}.e_y(t) \quad (4)$$

$$= \frac{B(z^{-1})}{A(z^{-1})}.e_y(t)$$

Unfortunately, in some cases- for example, when the time-delay d is variable or underestimated- Δx_{n+1} is zero even if $e_y(t)$ is not and the weights, and hence the command, are not corrected. In such cases, and also when the model is not in minimum phase, dynamic backpropagation cannot be used, since the equation (4) leads to instability and e_u grows in an unbounded fashion. To avoid this problem, someone can use static gain:

$$e_u(t-d-1) = \frac{B(1)}{A(1)}.e_y(t) \quad (5)$$

or

$$e_u(t-d-1) = \frac{B(z^{-1})}{A(1)}.e_y(t) \quad (6)$$

but these schemes have to be evaluated.

We propose the following method to deal with non-minimum phase plants with variable or under estimated time-delay: static backpropagation is first used to compute Δx as in equation (2). Then, rather to use only one component, we use all those corresponding to u(t-d-1),...,u(t-d-p):

$$\Delta w(t) = -\alpha.\sum_{i=1}^{p}\Delta x_{n+i}\frac{\partial u(t-d-i)}{\partial w(t)} \quad (7)$$

$$= -\alpha.\sum_{i=1}^{p}\frac{\partial y_m(t)}{\partial u(t-d-i)}.e_y(t).\frac{\partial u(t-d-i)}{\partial w(t)}$$

to minimize:

$$E = \sum_{i=1}^{p}\Delta x_{n+i}^2$$

In this manner, it suffices to have

$$\sum_{i=1}^{p}\left(\frac{\partial y_m(t)}{\partial u(t-d-i)}\right)^2 < 2$$

to garantee that weights will be modified in the desired direction. This condition is necessary to be useful to use J^T instead of J^{-1}. Otherwise, J has to be partially normalized.

In the following section, we compare the schemes described by equations (3-7) on the basis of their capacity to correct the command signal. The best one is then used to learn a neural controller in section 4.

3. Approximating command error

A two-layered neural network model M of a plant P corresponding to a simulated irrigation canal is obtained by backpropagation. M is described by equation (1) with n=2, p=27. u and y_m correspond to upstream and downstream flows, respectively. The time-delay d is varying between 30 and 37 with respect to the average flow u_m in the canal:

$$d = \alpha.u_m^{\beta}.l$$

with α=0.55, β=-0.288. l=109 (kms) is the canal length.
and

$$u_m(t) = \frac{1}{2}\left(\frac{u(t-1)+..+u(t-50)}{50} + \frac{y(t-1)+y(t-2)}{2}\right)$$

Schemes described by equations (3-7) are used to reconstruct a perturbed command signal as follows:
Two command signals u and u_p are separately applied as inputs to the model M and two separate outputs y and y_p are obtained, respectively. The signal u is a rectangular signal with randomly selected amplitudes and lengths. u_p is the signal u perturbed as follows:

$$u_p(t) = u(t) + s(t).\delta$$

where Prob(s(t)=1) = p; Prob(s(t)=-1) = 1-p; δ is positive constant.

Let e_y=y-y_p and e_u=u-u_p.

Table 1. Comparison of different estimation schemes.

	$\delta = 0.1$		$\delta = 0.5$	
	p = 0.9	p = 0.5	p = 0.9	p = 0.5
$\|u-u_p\|^2$	0.5	0.5	12.5	12.5
$\|u-u_1\|^2$	0.5018	0.5028	12.4386	12.4648
$\|u-u_2\|^2$	1.02x10^68	8.95x10^69	3.24x10^72	3.90x10^72
$\|u-u_3\|^2$	0.2226	0.7102	7.8011	14.6757
$\|u-u_4\|^2$	0.2418	0.6972	7.4685	14.3387
$\|u-u_5\|^2$	0.2018	0.3505	7.8035	9.2349

Equations (3-7) are used to find from e_y different estimations $e_{u1},..,e_{u5}$ of e_u, respectively. Five corrected signals $u_1,..,u_5$ are obtained: $u_i = u_p+e_{ui}$, $i=1,..,5$.
The schemes are compared on the basis of the distance: $||u-u_i||^2$. Results are shown on Table 1.

For a scheme i to be useful, the distance $||u-u_i||^2$ has to be smaller than $||u-u_p||^2$. Otherwise, the correction is done in a bad direction. Results reveal that the first scheme (static backpropagation) isn't useful. This was expected since most of the time, the time-delay is greater than d and hence u(t-d-1) has no effect on the output y(t). We can also see that the second scheme (dynamic backpropagation) leads to instability and the error e_{u2} grows in an unbounded fashion. The 3rd and 4th schemes are not useful when the command error is highly oscillating (p=0.5). Finally, only the 5th scheme corrects the error in all cases.

4. Learning a controller

The fifth scheme, described by equation (7), is used to generate learning errors for a multilayered neural network controller. The trajectory to be learned is shown on Fig. 1. This reference trajectory is submitted to the controller iteratively. Controller weights are adapted at each time-step by the backpropagation algorithm. Sum-squared plant's output error is shown on Fig. 2.

Although this error decreases, commands generated by the controller are highly oscillating making physical implementation difficult, if not impossible (Fig. 3).
In the following section, smoothness contrainst is added to the learning error to obtain less oscillating command.

5. Learning with smoothness constrain

The previous learning error

$$E = \sum_{i=1}^{p} \Delta x_{n+i}^2$$

is changed to

$$E = \sum_{i=1}^{p} \Delta x_{n+i}^2 + \lambda . \sum_{j=1}^{p} (u(t-d-j) - u(t-d-j-1))^2$$

Because a small positive value of λ cannot avoid command oscillations and a big value leads to a poor command, λ has to be choosen judiciously. Rather than to choose λ by trial and error tests, which are not compatible with real time learning, we propose the following adaptive law:

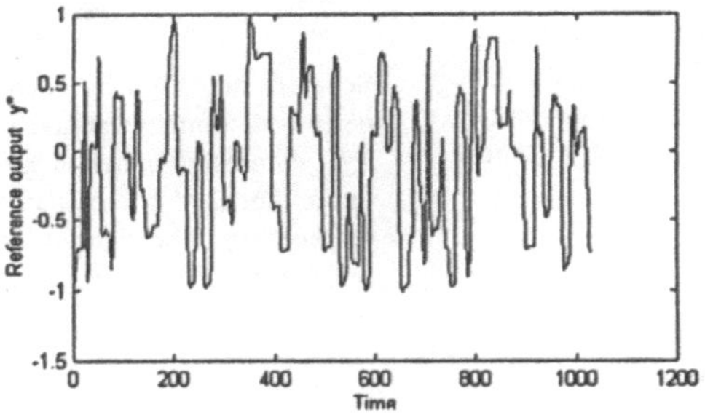

Fig. 1. Reference Trajectory.

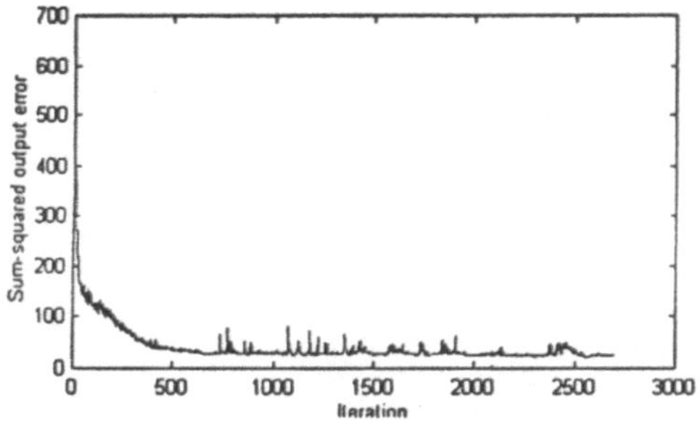

Fig. 2. Sum-squared output error.

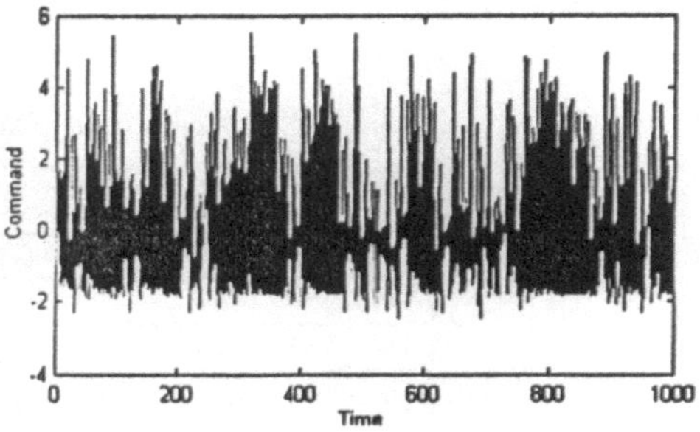

Fig. 3. Command signal after learning.

```
Initialize λ.
For each iteration k do:
  if (Ey(k)-Ey(k-1)).(Ey(k-1)-Ey(k-2))<0
  then (increase λ): λ = iλ.λ
  else
      if (Ey(k))>θ.(Ey(k-1))
      then (decrease λ): λ = dλ.λ
      end
  end
end
```

where $E_y(k)$ is the sum-squared plant's output error at iteration k. $i\lambda>1$, $0<d\lambda<1$ and $0<\theta<1$ are constants. We have used $i\lambda=1.05$, $d\lambda=0.95$ and $\theta=0.95$. Initially $\lambda=0.5$.
The learning process is applied with this adaptation law. Sum-squared plant's output error is shown on Fig. 4. Fig. 5 shows the improvements: command oscillations are largely reduced. However, much work is needed to find the best adaptation law.

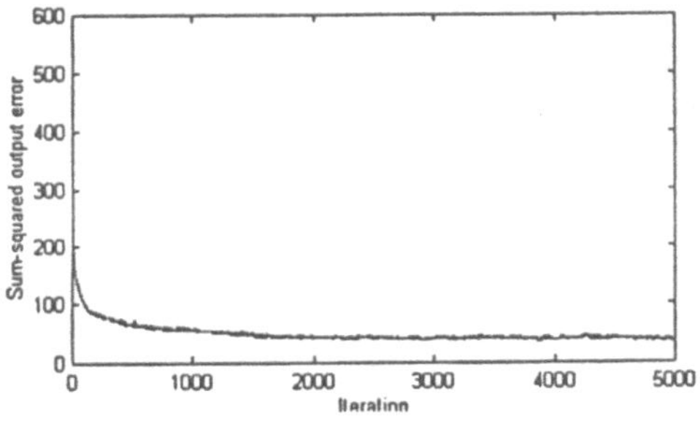

Fig.4. Sum-squared output error. With adaptive λ.

6. Conclusion

A modified version of the distal learning approach is proposed. It shows better characteristics when applied to a non-minimum phase dynamical system with variable or badly estimated time delay. Several schemes are compared on the basis of their capacity to reconstruct a perturbed command signal. A modified version of static backpropagation is shown to be the most useful. This scheme is used in learning a controller. Since the plant is not in minimum phase, this controller is unstable and the command signal is highly oscillating. To solve this problem, a λ-weighted smoothness constraint is added to the performance measure. Because a small value of λ has no effect and a big value limits command signal amplitude, an adaptation law of λ is proposed. Strong improvements are obtained but further investigations are needed to find the best adaptation law.

References

1. Jordan, M.I., Rumelhart, D.E.: Cognitive Science, 16, 307 (1992).

2. Narendra, K.S., Parthasarathy, K.: IEEE Trans. on Neural Networks, 1-1, 4 (1990).

3. Toudeft, A., Kosuth, P., Gallinari, P.: ICANN, 2, 1211 (1994).

4. Fukuda, T., Shibata, T., Tokita, M., Mitsuoka, T.: IEEE Trans. on Industrial Electronics, 39-6, 497 (1992).

5. White, D.A., Sofge, D.A., (Eds): Handbook of Intelligent Control: Neural, Fuzzy, and adaptive approaches. Van Nostrand Reinhold, 1992.

6. Yabuta, T., Yamada, T.: IEEE Trans. on Systems, Man, and Cybernetics. 22-1, 170 (1992).

7. Jordan, M.I.: IJCNN. I, 217 (1989).

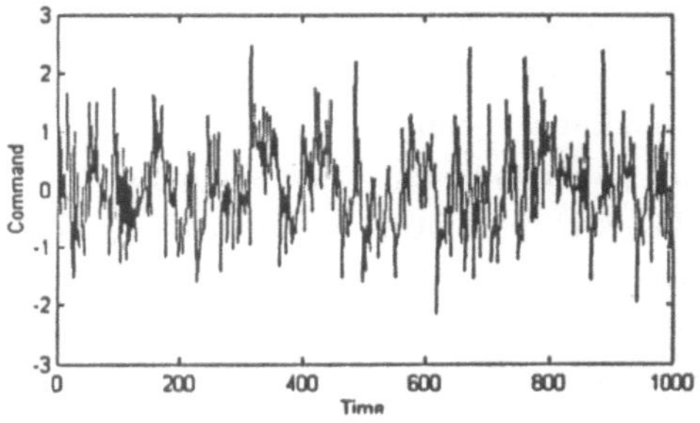

Fig. 5. Command signal after learning. With adaptive λ.

8. Landau, I.D., Dugard, L. (eds.): Commande Adaptative: aspects pratiques et théoriques. Masson 1986.

9. Astrom, K.J., Wittenmark, B. (eds.): Adaptive Control. MA, Addison-Wesley 1989.

10. Rumelhart, D., McClelland, J. (eds.): Parallel Distributed Processing. 1, Cambridge, MA: MIT Press 1986.

INDUSTRIAL KILN MULTIVARIABLE CONTROL: MNN AND RBFNN APPROACHES

B Ribeiro*, A Dourado and E Costa
CISUC - Centro de Informática e Sistemas da Universidade de Coimbra
Phone:+ 351 39 7000040, Fax:+351 39 701266, e-mail:bribeiro@mercurio.uc.pt
Universidade de Coimbra, P-3030 Coimbra, Portugal

Abstract

Artificial neural networks have been recognized as a valuable framework for nonlinear identification and control. In this paper we discuss and compare the use of two types of neural network arquitectures (1) MNN (Multilayer Neural Network) and (2) RBFNN (Radial Basis Function Neural Network) for modelling a second order nonlinear chemical process - a lime kiln in the pulp and paper industry. The simulation results showed that MNN performs better in this practical case. Therefore, it was used in an IMC (Internal Model Control) strategy. The neurocontroller was analysed with regards to performance and robustness against disturbances.

1 Introduction

In recent years there has been an extensive research on the development of applications of artificial neural network (ANNs) in many fields of science and engineering. Based on the fact that neural networks can model nonlinear mappings with an arbitrary accuracy, through a learning process driven by the minimization of an error signal, researchers have thoroughly exploited their use for nonlinear identification and control [1, 2, 3, 4].

This paper reports some aspects of the work done on the field of controlling a multivariable nonlinear dynamic system - a lime kiln in the pulp and paper technology. As it will be explained in the following section, the lime kiln has several caracteristics, such as long time delays, long and short term process disturbances and high nonlinearity, which make it appropriate to be tackled with neural network approaches. In this work we have chosen two different feedforward network types: MNN and RBFNN for lime kiln plant modelling. The results obtained herein have shown that the high accuracy achieved with the former approach made it possible to synthesize the kiln neurocontroller through inverse plant modelling. Both neural network models were incorporated into an IMC (Internal Model Control) structure that ensured the required robustness and free off-set control.

*Partially supported by Junta Nacional de Investigação Científica (JNICT) and PORTUCEL.

2 Lime Kiln Control

The process in a rotary lime kiln is highly nonlinear, multivariable and often subjected to disturbances during operation. The objectives for its control are mainly to obtain a high quality lime and to reduce the fuel costs and environmental pollution. To meet these requirements, a control strategy stabilizing the kiln temperature profile has proved to be adequate. As shown in Fig. 1 the manipulated variables are the fuel flow rate u_1 and the draft flow rate u_2; the controlled variables are the burning end temperature y_1 (hot kiln end) and the temperature of the outlet gases y_2 (cold kiln end). In order to obtain the simulated training data, a distributed parameter model of the process (constituted of first-order hyperbolic PDEs) was solved by numerical integration. It was developed previously and validated with real factory data and can be found elsewhere [5]. Selecting a nonlinear black-box approach, the dynamic system representation at discrete time $k+1$ is a nonlinear function of past inputs, past outputs and noise and is form of the NARX (Nonlinear Auto Regressive

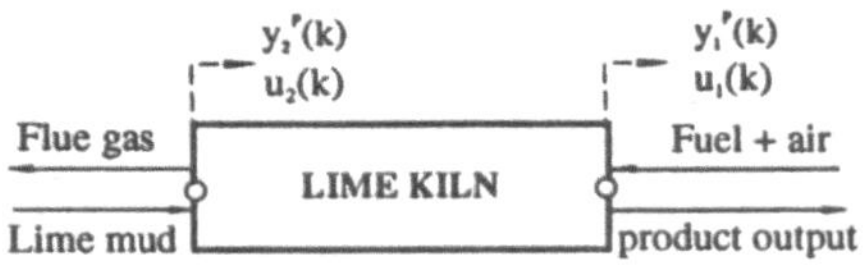

Fig. 1: Schematic lime kiln representation.

with Exogenous Inputs) model [6] given by (1):

$$\mathbf{y}^P(k+1) = \mathbf{f}\,[\mathbf{y}^P(k),\ldots,\mathbf{y}^P(k-n_y+1), \mathbf{u}(k),\ldots,\mathbf{u}(k-n_u+1)] + \mathbf{e}(k). \quad (1)$$

In this work $n_y = n_u = 2$ are the corresponding time lags in the inputs and the outputs; the vectors in (1) are defined by: $\mathbf{y}(k) = [y_1(k), y_2(k)]^T$, $\mathbf{u}(k) = [u_1(k), u_2(k)]^T$, $\mathbf{e}(k) = [e_1(k), e_2(k)]^T$. Clearly, the nonlinear characteristics of neural networks encourages their use for direct and inverse plant modelling.

2.1 Direct Model

The neural network response that approximates the dynamic process is (2):

$$\mathbf{y}^M(k+1) = \widehat{\mathbf{f}}\,[\mathbf{y}^P(k),\ldots,\mathbf{y}^P(k-n_y+1), \mathbf{u}(k),\ldots,\mathbf{u}(k-n_u+1)] \quad (2)$$

2.2 Inverse Model

The inverse model response is given by (3):

$$\mathbf{u}(k) = \widehat{\mathbf{f}^{-1}}\,[\mathbf{y}^M(k),\ldots,\mathbf{y}^M(k-n_y+1), \mathbf{y}_d(k+1),\mathbf{u}(k-1)\ldots,\mathbf{u}(k-n_u+1)]. \quad (3)$$

The neural networks that model the process (direct model) and its inverse (inverse model) are used in an IMC (Internal Model Control) structure, which has been shown to be a suitable approach for nonlinear control [7]. In the scheme, the main idea is to compensate the nonlinear relation among the lime kiln plant variables, by introducing the nonlinear inverse model of the system in the closed loop, as shown in Fig. 2. The filter is a first order one with unit gain and is essential to project the error signal $\mathbf{e}$ into the controller input space.

2.3 Using MNN and RBFNN

The approximation capabilities of two different types of network arquitectures, namely, (1) MNN (2) RBFNN

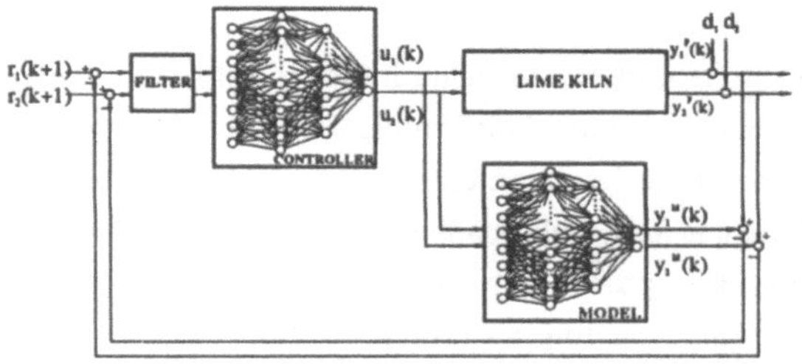

Fig. 2: IMC (Internal Model Control) strategy.

have been investigated in the literature and it has been proved that they can approximate nonlinear mappings within arbitrarily desired accuracy. The main characteristics of any network are defined by the the way the neurons are interconnected, their activation functions and the algorithm used to adjust their parameters (or weights). Following these features we briefly describe both arquitectures below.

(1) Multilayer Neural Networks (MNNs)
In the MNNs the neurons are organized in layers (usually completely connected) being the network specification parameters, the number of layers ($l = 1,\ldots,L$) and the number of neurons per layer n_l. The neural network response can be represented by: $y = f[\mathrm{u}] = \Phi\left(\sum_{i=1}^{n_{L-1}} w_i^L y_i^{L-1} + w_0^L\right)$ where Φ is a nonlinear function (usually a sigmoid), w_i^L are the connection weights between the output layer (L) and the previous layer, y_i^{L-1} are the outputs from neurons in the layer $L-1$ and w_0^L the bias. To fully specify the neural network the connection weights w_i have to be chosen in order to achieve a desired input-output relationship. This can be done by a learning algorithm, the backpropagation (BP) algorithm, through the minimization of a quadratic error function between the actual output $\mathbf{y}$ and the desired output $\mathbf{y}_d$ according to (4):

$$\mathbf{w}(k+1) = \mathbf{w}(k) + \eta\left(-\nabla[\mathbf{e}(k)^T\mathbf{e}(k)]\right) + \alpha\Delta\mathbf{w}(k-1). \quad (4)$$

where $\mathbf{w}$ is the vector of network weights, η is the learning rate and α is the *momentum* term. In the BP algoritm, small randomly generated weight values are chosen as the starting point for the learning iterations. The weights are then updated iteratively in a steepest descent manner. The shortcomings associated with this method are, at one hand, the uncertainty associated with finding a global minimum of the error function and the long training times to obtain acceptable errors and, at the other hand, the slow convergence to a local or global minimum. However, here we have used a 'robust' version of BP which follows established techniques for accelerating convergence through the use of adaptive parameters (learning rate and *momentum* term) and rescaling variables.

(2) Radial Basis Function Networks (RBFNNs)
The RBFNN may be considered as a two layer network in which the hidden layer makes a fixed nonlinear transformation (without adjustable parameters) in such a way that the input space is mapped in a new space. The output layer combines linearly the outputs from the new input space. The RBFNN response can be given by: $y = f[\mathbf{u}] = \sum_{i=1}^{n_L} w_i \Phi_i(r) + w_0$ where w_i are the network weights. The activation functions $\Phi_i : \mathbb{R}^n \rightarrow \mathbb{R}$ can be written as $\Phi_i(r) = \Phi_i(\| \mathbf{u} - c_i \|)$, with $c_i \in \mathbb{R}^n$. One of the most common RBF choices is the Gaussian function defined by $\Phi_i = \exp(-\frac{r^2}{\beta})$. The RBFs are characterized by their centers c_i and width β which is usually proportional to the mean distance between all the centers. In order to obtain these parameters an unsupervised learning algorithm suggested in [8] has been used. The input vectors are then clustered in order to determine a set of points representative of all input-output pairs $(\mathbf{u}, \mathbf{y_d})_i$. After the centers and widths of RBFs have been specified, the weights w_i have been adjusted by a simple least square method. The advantages of RBFNN rely the use of this method, the form of RBFs, the rapid training and the ease of analysis.

3 Simulation Results

A two step procedure to incorporate the neural network models directly in the IMC strategy is used. In the first step the direct model is trained by an error signal which is the difference between the plant output and the network output. In the second one, the inverse plant model is trained by an algorithm which uses the difference between the model output (obtained in the previous learning step) and the reference signal as the error training signal. Following the standard IMC structure the inverse of the plant model is used as the controller.

3.1 Lime Kiln Identification

A training set of input output pairs ($N_p = 4000$) is used in order to provide sufficient information to model the nonlinear plant. The input signals were chosen to excite all the dynamic modes of the system and cover the whole amplitude range of interest. The results shown in Fig. 3 respect network configurations described as follows (we use here conventional notation): (8- 20-10-2) for the MNN arquitecture and (8-60-2) for the RBFNN one. The learning algorithms used have been outlined in Paragraph 2.3. The results illustrate the RMS (root mean square) error of 0.4% for the MNN arquitecture (simulation (b)) and of 1.6% for the RBFNN arquitecture (simulation (c)) when both were subjected to PBRS inputs plotted in (a). The representation properties of RBFNNs depend on the number of RBFs used in the network. A result in [9] provide a theoretical justification to the fact that the size of a Gaussian network with center in a regular mesh required to achieve some fixed accuracy ϵ grows exponentially with the dimension (m) of the input space for large m. As in our practical case the dimension of the input space is high, the accuracy of the plant model obtained with RBFNN arquitecture (results in (c)) was restricted. So far it has not been feasible to synthesize the inverse model due to the complexity of the problem.

3.2 Kiln Internal Model Control (IMC)

The MNN was able to copy the lime kiln plant dynamics. With respect to the inverse model, the error obtained with training data derived from simulated plant was 1% over the complete space of training patterns. Enough accuracy was obtained to allow it to be used in the closed loop as the lime kiln controller. Several simulations were then carried out in order to test the performance of the controller. In Fig 4 two practical situations are illustrated: reference tracking (a) and process regulation (b) and (c) where ξ_1, ξ_2 and ξ_3 are process disturbances. The control results show high control precision which may be explained by (1) the use of an inverse model in the closed loop and (2) the accuracy of the plant model and its inverse. According to the properties of the IMC structure, if a good model of the plant is available, the closed loop will achieve ex-

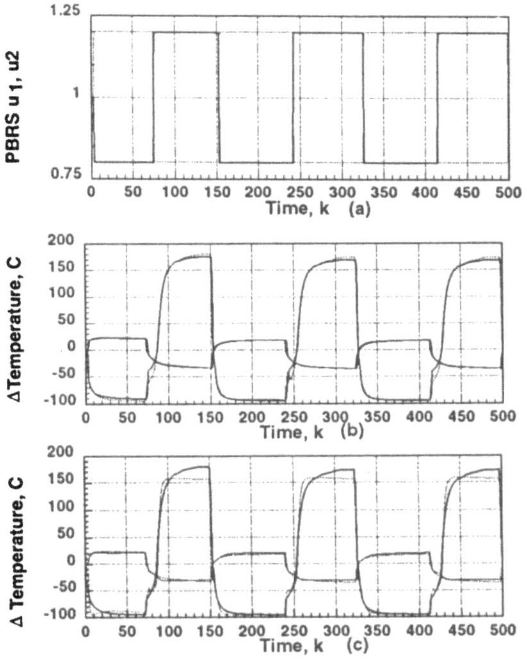

Fig. 3: Identification results. Step changes (a) in PBRS input signals. MNN (b); RBFNN (c). Thick line - y_1; thin line -y_2; dotted line - modelled output.

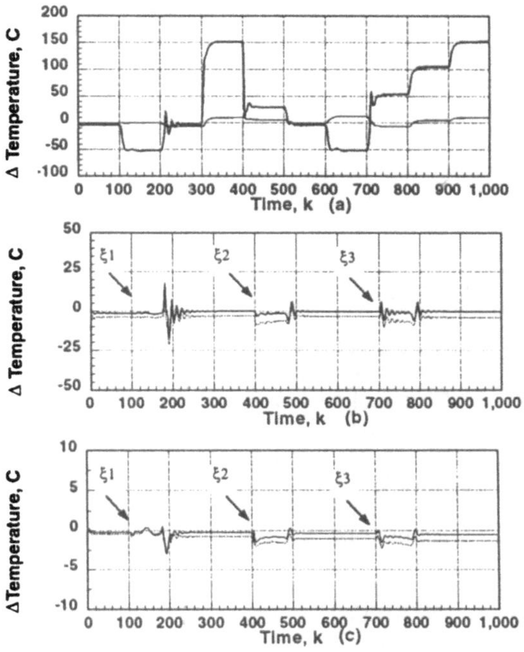

Fig. 4: Process control. Reference Tracking (a); Regulation (b) (c). Thick line - y_1; thin line -y_2; dotted line - modelled output.

act setpoint tracking despite unmeasured disturbances acting on the lime kiln. The filter introduced the required robustness into the structure.

4 Concluding Remarks

Based on some theoretical established considerations and the practical results obtained in a industrial lime kiln, it can be concluded that the MNN approach is more suitable for controlling the plant. In the presented IMC, two MNNs have been used for different tasks, so different learning concepts had to be applied. The MNN appears to offer distinctive advantages over the RBFNN approach despite of, in the former, the parameters occur nonlinearly in the error function leading to the use of gradient methods. From this results a slower and less efficient learning than in the RBFNN approach. However, in our practical case, simulation results have shown that the number of RBFs for accurate kiln modelling is prohibitive, mainly due the expo nential growth in complexity with the high dimension of the input space we have to deal with.

References

[1] Barto, A. G. in Neural Networks for Control. Miller III, W. T.,Sutton, R. S., Werbos, P. J. (eds.). Cambridge MA 02142: The MIT Press 1990.

[2] Hunt, K. J., Sbarbaro, D., Żbikowski, R., Gawthrop, P: J.: Automatica, 28 6, 1083 (1992).

[3] White, D. A., Sofge, D. A. (eds.): Handbook of Intelligent Control: Neural, Fuzzy, and Adaptive Approaches. New York: Van Nostrand Reinhold, 1992.

[4] Trentelman, H.L., Willems, J. C. (eds.): Essays on Control: Perspectives in the Theory and its Applications. Boston: Birkhäuser, 1993.

[5] Ribeiro, B., Correia in Energy Efficiency in Process Technology. A. D., Pilavachi, P. A., (eds.). London: Elsevier Science Publishers LTD, 1993.

[6] Chen, S., Billings, S. A.: International Journal of Control, 49, 1013 (1989).

[7] Economou, G. G., Morari, M., Palsson, B. O, Ind. Eng. Chem. Process Des. Dev.: 25, 403 (1986).

[8] Moody, J., Darken, C., Neural Computation: 1 2, 281 (1989).

[9] Liu, B., Si, J., IEEE Transactions on Neural Networks: 5 5, 845 (1994).

GENETIC TUNING OF NEURAL NON-LINEAR PID CONTROLLERS

A H Jones

Intelligent Machinery Division
Research Institute for Design, Manufacture, and Marketing
University of Salford

ABSTRACT

The techniques of genetic algorithms are proposed as a means of tuning neural non-linear PID control systems. It is shown that the use of genetic algorithms for this purpose results in highly effective genetic neural non-linear PID control systems. These results are illustrated by designing a genetic neural non-linear PID control systems and contrasting the results with an optimally tuned linear PID controllers.

1. INTRODUCTION

The potential of computers to address the problem of intelligent control was realised as early as 1958, when *Kalman [1]* examined the problem of building a machine which adjusts itself automatically to control an arbitrary dynamical system. But, whilst it must be acknowledged that, in the intervening years, computer technology has made the implementation easier and more efficient, it must also be acknowledged that this technology has not made the practice of control any easier. Recently, considerable effort has been deployed in the field of intelligent control systems in order to provide the techniques for designing expert controllers *[2]*, fuzzy controllers *[3]*, and neural controllers *[4]*. With the technology having reached a fairly advanced state, some manufacturers have now integrated such tuning methods into their commercially available products *[5]*. However, none of these technologies has achieved universal acceptance and each has its own limitations.

The recently developed genetic tuners for digital PID controllers developed by *Porter and Jones [6]* offer an alternative to the tuning problem which can overcome many of the limitations of other controller tuning techniques. Indeed, if controller tuning is viewed as a parameter optimisation problem subject to plant-dependent constraints, the genetic algorithm approach can be seen as a natural evolutionary tuning mechanism which has all the features of an ideal adaptive controller.

This genetic approach to controller tuning has been presented *[6][7]* for both single-input/single-output plants (for which genetic algorithms were used to tune digital PID controllers so as to minimise the integral square error (ISE) in response to a unit-step command), and linear multi-input/multi-output plants for which genetic algorithms were used to tune a multivariable PID controller. In the latter case, a generalised integral square error function was minimised, the effects of cross-channel interaction were included and the technique can be properly considered as a form of auto-tuning. However, it is well-known that if non-linear gains are used in the PID controller then superior performance can be achieved. The major problem with such a technique is how to define and tune the non-linear functions. The problem of defining the function can be avoided by defining the function explicitly via a neural network. Once the problem has been posed as a neural network problem, the emphasis of the problem then shifts to obtain the optimum weight of the neural network. Such problems have been addressed *[8]* using a modified form of this back-propagation algorithm. However, such techniques do require explicit information of the plant transfer function. Clearly, this information is not always available and this motivates the possibility of using genetic algorithms for the design of such neural control systems. Indeed, it is shown that the case of genetic algorithms greatly facilitates the design of such neural control systems such that a time domain cost function is minimised. These results are illustrated by designing a neural non-linear PID controller and contrasting the results with an optimally tuned linear PID controller when controlling the same plant.

2. ANALYSIS

The controller structure proposed in this paper for the neural non-linear PID control systems involves three separate one-input/one-output neural networks representing the non-linear gains for the proportional, integral, and derivative gains. In this structure, the single-input/single-output plant is assumed to be governed by linear time-invariant state and output equations of the respective forms

$$x(t) = Ax() + Bu(t) \quad \ldots . (1)$$

and

$$y(t) = Cx(t) \quad \ldots . (2)$$

where the state vector $x \in R^n$, the input $u \in R$, the output $y \in R$, the plant matrix $A \in R^{nxn}$, the input matrix $B \in R^{nx1}$, the output matrix $C \in R^{1xn}$.

The network topology chosen for the neural controllers is a fragmented multilayer neural network. It consists of three independent multilayer networks representing proportional, integral, and derivative gains. The signals from these three networks are connected to a linear output neuron to generate the output of the neural controller. The inputs to the neural network are error, sum of error, and derivative of error, so that the neural controller can be viewed as a neural non-linear PID controller, where the non-linear PID gains are represented by neural networks. This feature enables both linear and non-linear gain mappings to be learnt by the neural networks. Hence, the neural controller can be represented as

$$u(k) = \{nnP\}e(k) + \{nnI\}z(k) + \{nnD\}de(k) \quad \ldots . (3)$$

where the scalar $e(k) \in R, e(k) = v - y(k)$ is the error, $v \in R$ is the set-point, $u(k) \in R$ is the controller output, $z(k) \in R$, where $z(k+1) = z(k) + Te(k)$ is the digital integrator, $T \in R$ is the sampling period, $de(k) = (e(k) - e(k-1))/T$ is the derivative of error, and the non-linear PID controller gains are given by the neural networks $\{nnP\}, \{nnI\}$, and $\{nnD\}$, respectively.

In such genetic neural control systems, the objective is to change the weights in the neural networks representing the non-linear PID controller gains such that a performance cost function, such as ISE or ITAE, is minimised. This is achieved by encoding the weights of the neural network's parameters in each set $(\{WnnP\}, \{WnnI\}, \{WnnD\})$ involved in the design equation (3) in accordance with a system of concatenated, mutli-parameter, fixed-point coding *[3]*. Thus, each set $(\{WnnP\}, \{WnnI\}, WnnD\})$ of neural network weights is represented by a string of binary digits. Then, following a random initial choice, entire generations of such strings can be readily processed in accordance with the basic genetic operations of selection, crossover, and mutation *[7]*. In particular, the selection process ensures that the successive generations of neural PID controllers designed by genetic algorithms exhibit progressively improving behaviour in respect of the defined fitness measure.

3. ILLUSTRATIVE EXAMPLE

This procedure for the tuning of genetic neural control systems can be conveniently illustrated by designing a neural non-linear PID control (GNNC) system for the open-loop single-input/ single-output plant with transfer function

$$g(z) = \frac{0.8}{z^4(z-0.9)}$$

where the sampling period is 0.25 *sec*. The neural network controller comprises three networks each with one input neuron, four neurons in the hidden layer and one output neuron. The inputs to these three networks are error, integral of error, and derivative of error, respectively. Each of the output neurons of these networks is connected to a linear output neuron, which provides the output from the neural controller.

The results obtained by implementing the genetic neural control system incorporating a neural non-linear PID controller such that an integral square of error to step input is minimised is considered.

The results of solving this problem by means of a genetic algorithm with a population size $N = 100$, a crossover probability, $p_c = 0.6$, and a mutation probability, $p_m = 0.001$ are shown after *1000* generations. The closed-loop step response of the neural control system is shown in Figure 1, and the plots of the non-linear gains, K_p, K_i, and K_d, are shown in Figure 2. In this case, the genetic neural controller has steadily reduced the ISE to a value of *1001* Finally, to contrast the results, the same plant was controlled by a linear PID controller and the gain genetically optimised in the manner of *Porter and Jones [6]*. The results of solving this problem by means of a genetic algorithm with a population size $N = 100$, a crossover probability of $p_c = 0.6$, and a mutation probability of $p_m = 0.001$ are shown after *100* generations in Figure 3. Here, the closed-loop step response of the linear PID control system is shown and it is interesting to note that the ISE of the linear PID controller is *1092.5*. These results clearly demonstrate the added performance that can be attained if non-linear gains are used in preference to linear gains.

4. CONCLUSION

The techniques of genetic algorithms have been proposed as a means of tuning neural non-linear PID control systems. It has been shown that the use of genetic algorithms for this purpose results in highly effective genetic neural non-linear PID control systems such that the integral square of error is minimised. These results have been illustrated by designing genetic neural non-linear PID control systems and contrasting these results with an optimised PID controller.

REFERENCES

[1] R E Kalman, 1958, "Design of self-optimising control systems", Trans ASME, Vol 80, pp 468-478.

[2] A H Jones and B Porter, 1985, "Expert tuners for PID controllers", Proc IASTED Conference on Computer-Aided Design and Applications, Paris.

[3] C C Lee, 1990, Fuzzy logic in control systems", IEEE Trans on Systems, Man, and Cybernetics, SMC, Vol 20, pp 404-435.

[4] W Godenthal and J Farrell, 1990, "Applications of neural network to automatic control", Proc AIAA Guidance, Navigation, and Control Conference, pp 1108-1112, Portland, USA/

[5] D G Schwartz, 1992, "Fuzzy logic flowers in Japan", IEEE Spectrum, July, pp 32-35.

[6] B Porter and A H Jones, 1992, "Genetic tuning of digital PID controllrs", Electron Lett, Vol 28, pp 843-844.

[7] B Porter, S S Mohamed, and A H Jones, 1993, "Genetic tuning of multivariable PID controllers", Proc ECC, Groningen, Netherlands.

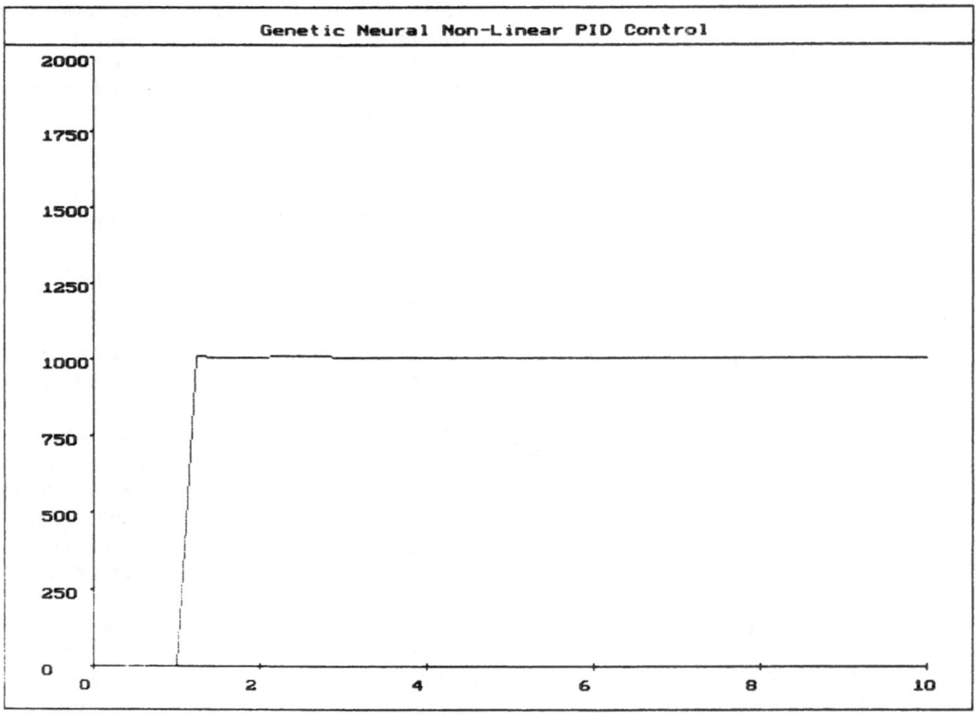

Figure 1: Closed-loop step response of the genetically-tuned neural PID non-linear controller

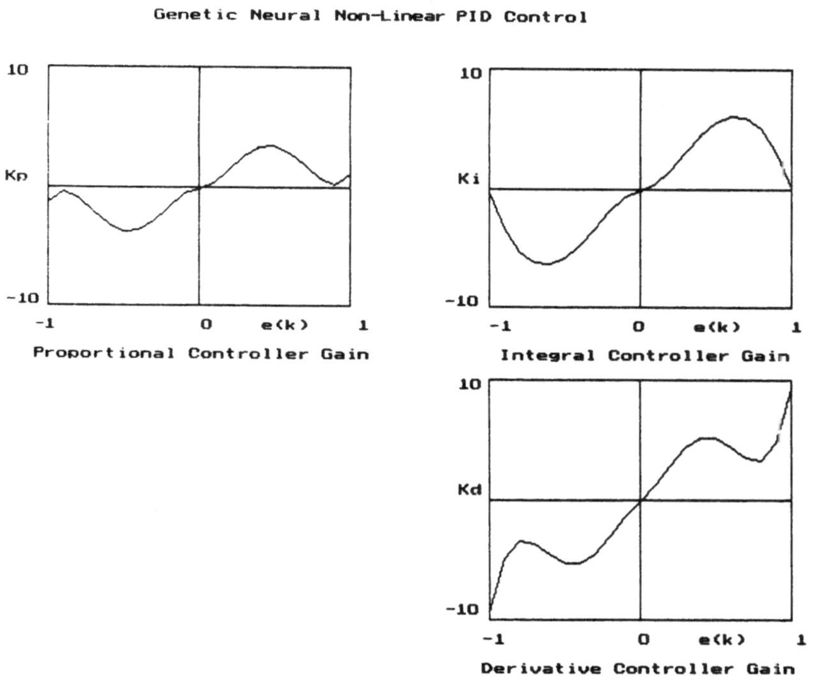

Figure 2: Non-linear PID gains of the neural networks

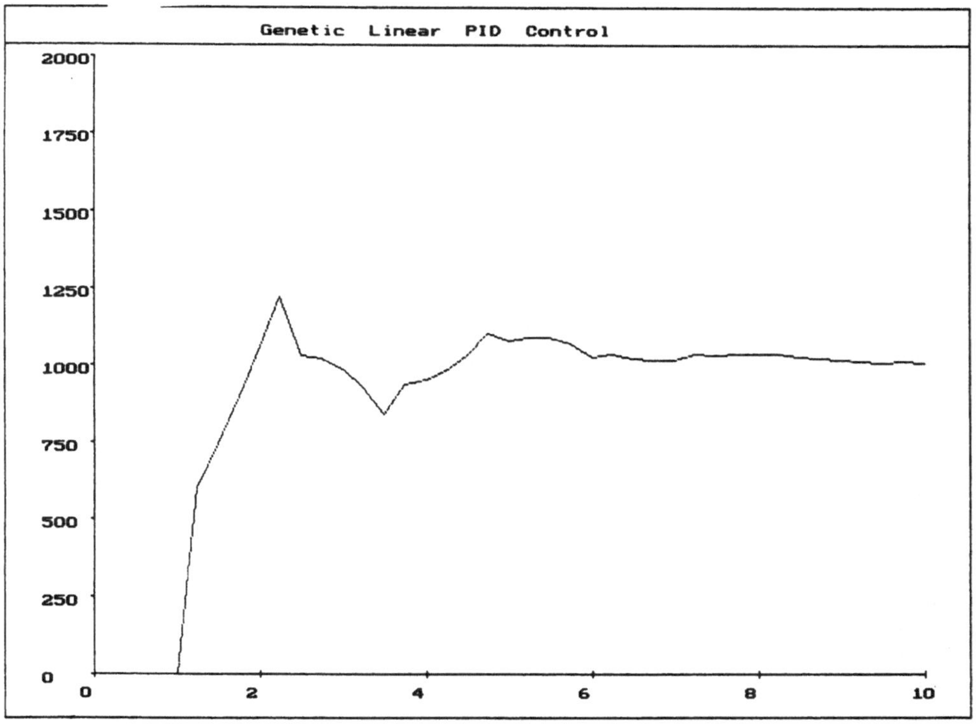

Figure 3: Closed-loop step response of the genetically-tuned PID controller

CONNECTIONNIST ALGORITHM FOR A 3D DENSE IMAGE BUILDING FROM STEREOSCOPY

Marc-Noël FAUVEL, Pascal AUBRY

CEA/ DTA/ CEREM/ DPSA/ STR
CEA- CEN FAR - BP n°6

ABSTRACT

In this paper, we present a neuron-like network able to build 3D dense maps from stereoscopic image pair. The process of stereopsis is encoded by an energy function, which controls the evolving of the network. This one has the same structure than original images, and evolutes on a simple gradient steep. Thus, the system is fully parallel and could be hardware implemented for a real time use.

I. INTRODUCTION

Since many years, the STR works on intervention systems such as autonomous and teleoperated robots, that could solve critical problems in nuclear plant environments. Following that way, the teleoperation of mobile systems need dense perceptual informations such as vision, which can help the operator to command the system.

Intervention robotics is confronted with two important problems: The perception, which gives a representation of the environment, and the navigation, which must be able to find a available way for the robot progression. These two problems are not independents. On the contrary, navigation needs the best an complete representation of the environment to compute the right way. If the above representation is uncompleted or less accurate, the navigation couldn't process correctly. Perception tools suited to robot navigation are of two kinds: Active or passive perception captors. The first ones emits physical signals and use the reflected signal to determine objects localisation. The seconds don't have to generate any signal, and simply use natural characteristics of the objects, or ambient illumination.

Because of its polyvalence and its accuracy, passive stereoscopic vision [1, 2, 3] has been chosen. It is certainly the perception method the best suited to our environment, and our buildings. However, this powerful tool need a complex processing to extract 3D informations from a pair of stereoscopic images. But if a system is able to extract as many 3D informations as the number of pixels on the images, it gives a precious dense 3D map for the navigation level. So our goal is to build a 3D map of the environment as dense as possible, i.e. with one 3D information for each pixel of the images.

II. PRINCIPLE

Because we choose to build a 3D map with one disparity for each pixel, we will have to process a very large set of datas. The only way to limit the processing time is to build a massively parallel system, containing as many nodes as the number of pixels in the original image. Then we use a connectionist evolving algorithm which is designed to process such a set of datas. From the example of Hopfield's [6, 7] networks, we build a network of nodes, one by image pixel, each node activity is the disparity d of its associated pixel in the image.

To get the very first 3D information, we use the correlation method. The problem with this method, is that it gives wrong disparities at areas of homogenous intensity. However, this difficulty can be reduced by closely adapting the size of the correlation window to the image content. So we following the next very simple algorithm, to determine the dimension L_f of the squared correlation window:

1. For each pixel *(i,j)*
   1.1 $L_f = 25$
   1.2 While $\sigma(i,j) > \sigma_f$
      1.2.1 $L_f \leftarrow L_f - 2$

where $\sigma(i,j)$ is the variance of disparities over the window of dimension L_f around the pixel *(i,j)*, and σ_f a variance threshold. This simple algorithm ensures that the dimension of the correlation window is large enough for homogenous intensity areas, and small enough around edges, to preserve accuracy of edge reconstruction. However, an important percentage of the disparities computed above are wrong disparities. We must introduce a confirmation process between the disparities computed on right image, and those computed on the left one. This process, first proposed by Fua [4], improve disparity fiability but produce lakes of disparities. Because left and right computed disparity are different, we are enable to give the concerned pixel a disparity. So we have to interpolate it with neighbourhood ones. Moreover, we build an energy function $E(d)$ that contains all desired behaviour. So that, each node (pixel) will evolving and changing its disparity d by following the gradient steep:

$$\Delta d = -\eta . \frac{\delta E}{\delta d} \tag{1}$$

III. THE ENERGY FUNCTION

We then build the energy function $E_i(d)$ for each node (pixel) i. First, we introduce a term that proposes the disparity given by the correlation method as the most probable disparity. We call it *disparity term* and note it E_i^{disp}. In order to write a continuous energy function, we build E_i^{disp} as a Gaussian G_1 centred on the disparity given by the correlation method. If the disparity of the pixel i was not confirmed by the Fua's algorithm, the disparity term for this pixel is taken as the mean between the two found disparities. Then, we compute the characteristics of G_1 by viewing the neighbourhood disparities. Indeed, we adjust the standard deviation to be high when the pixel i is in a area of homogenous intensity, and small if the pixel belong to an edge area. By this, we permit the disparity of a pixel from homogenous area to fluctuate in large proportions, and that of edge pixels to fluctuate lower, because the correlation from edges is reliable, and not that for homogenous areas. Thus, to determine if the processed pixel is in edge or homogenous area, we compute the number of pixel which have the same intensity (in a ratio ρ) than the processed pixel. We consider that two pixels have a comparable intensity when:

$$\frac{\min(I_1, I_2)}{\max(I_1, I_2)} \geq \rho \tag{2}$$

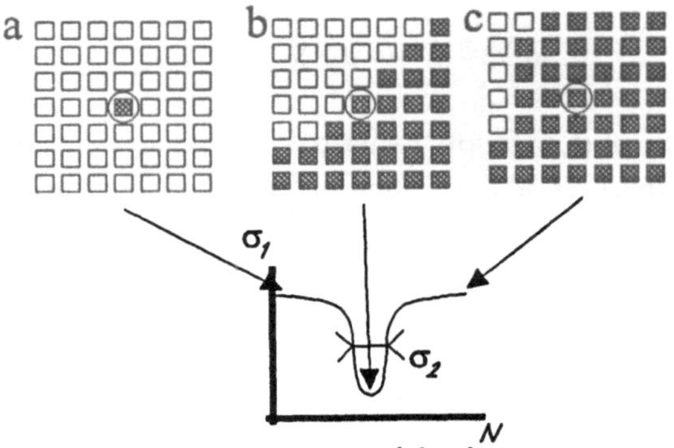

Fig. 1: Standard deviation of the disparity term

As we can see in figure 1, the standard deviation σ_1 of G_1 can also be approximated by a Gaussian G_2 if we represent $\sigma_1(N)$ where N is the number of pixels of same intensity over a domain D_p. So we write the disparity term:

$$E_i^{disp}(d) = -\exp\left(-\frac{\left(d - d_i^{m}\right)^2}{2\sigma_1^2}\right) \tag{3}$$

where d_i^{m} is the disparity given by the correlation method for the pixel i, and σ_1 has the general form:

$$\sigma_1 = \sigma_1^0 + A.\left(1 - \exp\left[-\frac{(N-B)^2}{2\sigma_2^2}\right]\right) \tag{4}$$

where A and B are two parameters fixed by experience. So our energy function has the form:

$$E_i(d) = E_i^{disp}(d) \qquad (5)$$

Of course this term alone is not able to interpolate 3D disparities on a relatively large area. We must introduce one supplement term we call *coherence term,* which does this work. Following one of the Marr's principles [5], we say that a set of pixels of comparable intensities belong to the same object. From this principle, we define a domain D_z around the processed pixel, where disparities of the pixels of same intensity than the first are used in a plane regression. Then, the disparity d_i^{coh} of the processed pixel, placed at the center of the domain, is given by the best plane:

$$d_i^{coh} = \overline{d} - \alpha.\overline{y} - \beta.\overline{x} \qquad (6)$$

where $\overline{d}, \overline{x}, \overline{y}$ are respectively the means of the disparities, abcsissis and ordinates of the pixels of same intensity than i in the domain D_z. The others parameters are given by:

$$\begin{cases} \alpha = \dfrac{\vartheta(X).co\vartheta(Y,D) - co\vartheta(X,Y).co\vartheta(X,D)}{\vartheta(X).\vartheta(Y) - co\vartheta^2(X,Y)} \\ \beta = \dfrac{co\vartheta(Y,D) - \alpha.\vartheta(Y)}{co\vartheta(X,Y)} \end{cases} \qquad (7)$$

X, Y and D are the random variables respectively associated to x, y and d. To integrate that disparity in the energy function, we build a parabolic hole around d_i^{coh}. So our energy function has the form:

$$E_i(d) = E_i^{disp}(d) + E_i^{coh}(d) \qquad (8)$$
$$= -\exp\left(-\frac{(d - d_i^{m})^2}{2\sigma_1^2}\right) + \frac{A_z}{2}.(d_i^{coh} - d)^2$$

with A_z a parameter adjusting influence between disparity and coherence term.

Finally, we have to add a last term, which ensures that homogenous intensity areas will be tied to edges. Moreover, we introduce an *interaction term,* that uses disparities of pixels of different intensity than processed pixel. We compute the d_i^{int} disparity by taking the mean of the precedent pixels, between the height immediate neighbourhoods. Thus, a parabolic hole is built around this disparity and the final energy function form is:

$$E_i(d) = E_i^{disp}(d) + E_i^{coh}(d) + E_i^{int}(d) \qquad (9)$$
$$= -\exp\left(-\frac{(d - d_i^{m})^2}{2\sigma_1^2}\right) + \frac{A_z}{2}.(d_i^{coh} - d)^2 + \frac{A_h}{2}.(d_i^{int} - d)^2$$

where A_h is a new influence parameter.

IV. EXPERIMENTAL RESULTS

We present now some experimental results on indoor and outdoor scenes. On each figure, we find on upper line the stereo original pair; below on left, the correlation 3D map with mutual confirmation of disparities, and at right the final 3D map resulting of network evolution. On 3D maps, dark pixels are nearest.

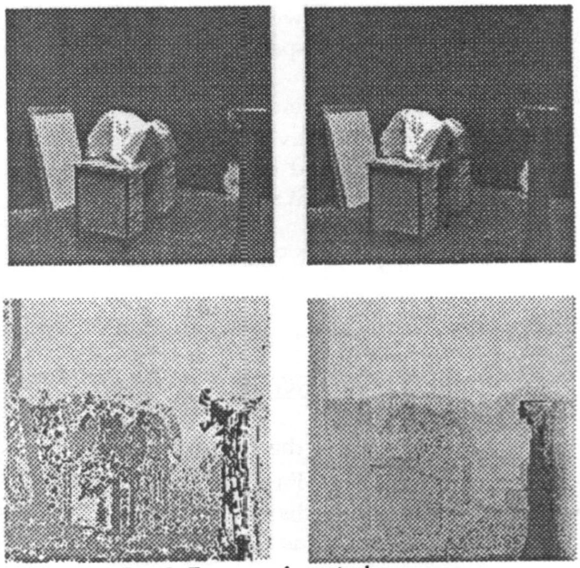

Fig. 2: First result on indoor scene.

On figure 2, we can see an indoor desk scene. The final 3D map clearly shows space occupation for the desk, and the object at first plane is accurately reconstructed, with clean edges.

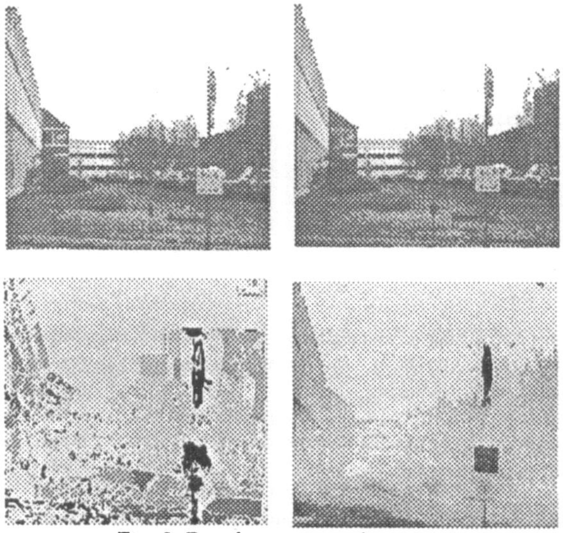

Fig. 3: Result on an outdoor scene

On figure 3, the network has reconstructed the left side of the map, even though there was no disparity information at this location. Also here, the panel at first plane is clean. The trees at right show a disparity progression, from a progressive removing in the scene.

In the figure 4 a scene of stairs is presented. The disparities of different elements represent clearly the disposition of the objects in the scene.

Fig. 4: Arranged indoor stairs scene.

V. CONCLUSIONS

In this paper, we have presented a network that build 3D dense maps, following a massively parallel algorithm. The problem of stereopsis has been written in terms of an energy function, which can be easily modified, completed and accurate. One can also add terms for integrate others methods of stereovision and even more active vision methods.

We seen that the network was able to interpolate the 3D informations from correlation method by very simple neighbourhood considerations; particularly in medium-textured scenes. Furthermore, the parallel aspect of the network suggests a future hardware implementation and perhaps a real-time 3D map generator.

REFERENCES

[1] N. Ayache, B. Faverjon: Un algorithme rapide de stéréoscopie passive utilisant la prédiction et vérification récursive d'hypothèse. 5è congrès de reconnaissance des formes et intelligence artificielle: Grenoble 1985.

[2] H.H. Bake: Depth from edge & intensity based stereo. Technical repport AIM-347, Stanford University: California, september 1982.

[3] S.D. Cochran: Surface Description from binocular stereo. DARPA F33615-87-C-1436: Los Angeles, California, november 1990.

[4] P. Fua: A parallel stereo algorithm that produces dense depth maps and preserves image features. INRIA rapport de recherche 1369: janvier 1991.

[5] D. Marr, T.Poggio:, Cooperative computation of stereo disparity. Science 194:283-287, 1976.

[6] J.J. Hopfield: Neural networks and physical systems with emergent collective computational abilities. Proceedings of the national academy of sciences 79:2554-2558, 1982.

[7] J.J. Hopfield: Neurons with graded response have collective computational properties like those of 2-state. Proceedings of the national academy of sciences 81:3088-3092, 1984.

A Unified Neural Network Model for the Self-organization of Topographic Receptive Fields and Lateral Interaction

Joseph Sirosh and Risto Miikkulainen
Department of Computer Sciences
The University of Texas at Austin, Austin, TX 78712
email: sirosh,risto@cs.utexas.edu

Abstract

A self-organizing neural network model for the simultaneous development of topographic receptive fields and lateral interactions in cortical maps is presented. Both afferent and lateral connections adapt by the same Hebbian mechanism in a purely local and unsupervised learning process. Afferent input weights of each neuron self-organize into hill-shaped profiles, receptive fields organize topographically across the network, and unique lateral interaction profiles develop for each neuron. The resulting self-organized structure remains in a dynamic and continuously-adapting equilibrium with the input. The model can be seen as a generalization of previous self-organizing models of the visual cortex, and provides a general computational framework for experiments on receptive field development and cortical plasticity. The model also serves to point out general limits on activity-dependent self-organization: when multiple inputs are presented simultaneously, the receptive field centers need to be initially ordered for stable self-organization to occur.

1 Introduction

The response properties of neurons in many sensory cortical areas are ordered topographically, i.e. nearby neurons respond to nearby areas of the receptor surface. Such topographic maps form by the self-organization of afferent connections to the cortex, driven by external input (Hubel and Wiesel 1965; Miller et al. 1989; von der Malsburg 1973). Several neural network models (Amari 1980; Kohonen 1982, 1993; Miikkulainen 1991; Willshaw and von der Malsburg 1976) have demonstrated how the global topographic order can emerge from local cooperative and competitive lateral interactions within the cortex. Such models are based on predetermined lateral interaction and focus on explaining how the afferent connections become ordered. The self-organized nature of the lateral interactions themselves has not been addressed in these models, perhaps because it was not discovered until very recently.

A number of recent neurobiological experiments indicate that lateral connections self-organize like the afferent connections: (1) The lateral connectivity is not uniform or genetically predetermined, but forms during the early development based on external input (Katz and Callaway 1992; Löwel and Singer 1992). (2) In the primary visual cortex, lateral connections are initially widespread, but develop into clustered patches at the same time as the orientation and ocular dominance columns form (Burkhalter et al. 1993; Katz and Callaway 1992; Löwel 1993). (3) Lateral connections primarily connect areas with similar response properties, such as columns with the same orientation or eye preference (Gilbert 1992; Löwel and Singer 1992). To fully account for cortical self-organization, a cortical map model must demonstrate that both afferent and lateral connections can organize simultaneously, from the same external input, in a mutually supportive manner.

A new model of cortical self-organization called LISSOM (Laterally Interconnected Synergetically Self-Organizing Map: Sirosh and Miikkulainen 1993, 1994) was developed with this goal in mind. Lateral and afferent connections self-organize cooperatively in LISSOM, and with suitably abstracted visual inputs, LISSOM can model self-organization of visual cortical structures. For example, the development of ocular dominance columns can be modeled with 3-D input (similar to that of Obermayer et al. 1992), where two components stand for the retinotopic coordinates and the third represents an ocular dominance value. In the resulting map, the afferent weights exhibit rough retinotopy and alternating patches of eye preference similar to ocular dominance columns in the visual cortex. The lateral connections predominantly connect neurons with similar ocular dominance, as has been observed in the primary visual cortex of strabismic cats (Löwel and Singer 1992).

This paper advances previous work on LISSOM in two ways: it presents a unified model for the simultaneous development of receptive fields and smooth lateral interaction profiles, and it shows how LISSOM can work with realistic, high-dimensional retinal input. It turns out that with such input, the LISSOM computations are actually simpler and biologically more realistic: it is not necessary to normalize the input, and the same multiplicatively normalized Hebbian rule can be used for the adaptation of both afferent and lateral connections. The new model unifies previous self-organizing models (Kohonen 1982; Willshaw and von der Malsburg 1976) under a common framework, and establishes general limits on activity-dependent self-organization: when multiple inputs are presented simultaneously, self-organization can occur only when the receptive field centers are initially ordered.

2 The LISSOM Model of Receptive Field Development

The LISSOM network is a sheet of interconnected neurons (figure 1). Through the afferent connections, neurons receive input from a receptive surface or "retina". In addition, each neuron has reciprocal excitatory and inhibitory lateral connections with other neurons. Lateral excitatory connections are short-range, connecting only close neighbors. Lateral inhibitory connections run for long distances, and may even implement full connectivity between neurons in the network. Each connection has a characteristic strength (or weight), which may be any positive value.

Neurons receive afferent connections from overlapping

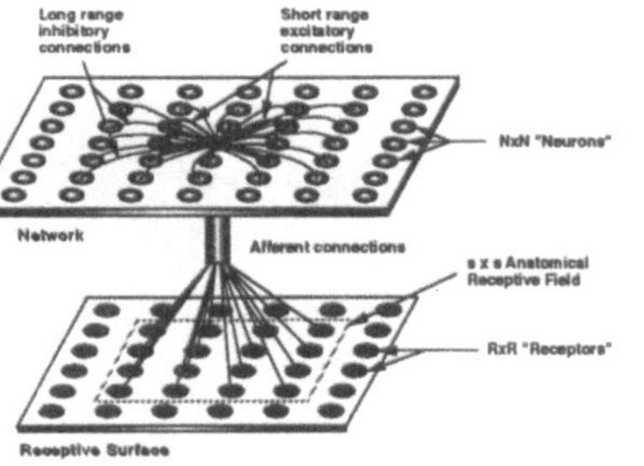

Figure 1: **The LISSOM architecture.**

patches on the retina called anatomical receptive fields, or RFs. The patches are distributed with a given degree of randomness. The $N \times N$ network is projected on the retina of $R \times R$ receptors, and each neuron is assigned a receptive field center (c_1, c_2) randomly within a radius $\rho * R$ of the neuron's projection. Afferent connections are set up so that the neuron receives input from receptors in a square area around the center, with side s. Depending on its location, the number of afferents to a neuron could vary from $\frac{1}{2}s \times \frac{1}{2}s$ (at the corners) to $s \times s$ (at the center).

The external and lateral weights are organized through a purely unsupervised learning process. At each training step, neurons start out with zero activity. The initial response η_{ij} of neuron (i, j) is based on the scalar product

$$\eta_{ij} = \sigma\left(\sum_{r_1,r_2} \xi_{r_1,r_2}\mu_{ij,r_1r_2}\right), \qquad (1)$$

where ξ_{r_1,r_2} is the activation of a retinal receptor (r_1, r_2) within the receptive field of the neuron, μ_{ij,r_1r_2} is the corresponding afferent weight, and σ is a piecewise linear approximation of the familiar sigmoid activation function. The response evolves over time through lateral interaction. At each time step, the neuron combines retinal activation with lateral excitation and inhibition:

$$\begin{aligned}\eta_{ij}(t) = \sigma\big(& \textstyle\sum_{r_1,r_2} \xi_{r_1,r_2}\mu_{ij,r_1r_2} + \\ & \gamma_e \textstyle\sum_{k,l} E_{ij,kl}\eta_{kl}(t-1) - \\ & \gamma_i \textstyle\sum_{k,l} I_{ij,kl}\eta_{kl}(t-1)\big),\end{aligned} \qquad (2)$$

where $E_{ij,kl}$ is the excitatory lateral connection weight on the connection from neuron (k, l) to neuron (i, j), $I_{ij,kl}$ is the inhibitory connection weight, and $\eta_{kl}(t-1)$ is the activity of neuron (k, l) during the previous time step. The constants γ_e and γ_i are scaling factors on the excitatory and inhibitory weights and determine the strength of the lateral interactions. The activity pattern starts out diffuse and spread over a substantial part of the map, and converges iteratively into a stable focused patch of activity, or activity bubble. The convergence is rapid, and the activity settles typically within a few iterations of equation 2. After settling, the connection weights of each neuron are modified. Both afferent and lateral connection weights adapt according to the same mechanism—the Hebb rule, normalized so that

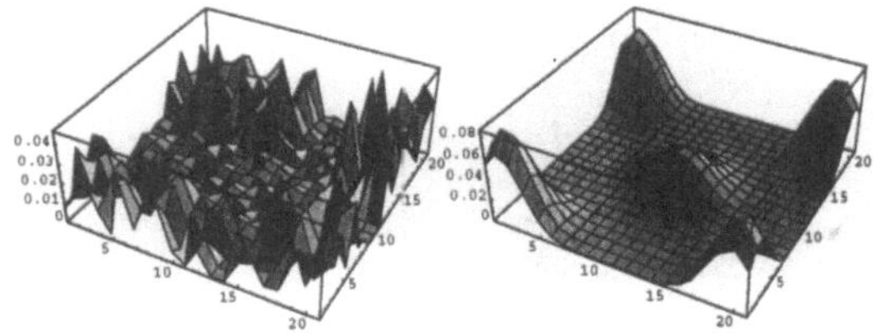

(a) Initial random weights (b) Final Organized RFs

Figure 2: **Self-organization of the afferent input weights into receptive fields.** The afferent weights of five neurons (located at the center and at the four corners of the network) are superimposed on the retinal surface in this figure. The RF centers were randomly placed around their optimal locations, and the weights were initialized randomly (a). There are four concentrated areas of weights slightly displaced from the corners, and one larger one in the middle. At the corners, the profiles are taller because the normalization concentrates the total afferent weight in fewer connections. As the self-organization progresses, the weights organize into smooth hill-shaped profiles (b).

the sum of the weights is constant:

$$w_{ij,mn}(t + \delta t) = \frac{w_{ij,mn}(t) + \alpha\eta_{ij}X_{mn}}{\sum_{mn}\left[w_{ij,mn}(t) + \alpha\eta_{ij}X_{mn}\right]}, \qquad (3)$$

where η_{ij} stands for the activity of the neuron (i, j) in the settled activity bubble, $w_{ij,mn}$ is the afferent or lateral connection weight (μ_{ij,r_1r_2}, $E_{ij,kl}$ or $I_{ij,kl}$), α is the learning rate for each type of connection (α_a for afferent weights, α_E for excitatory, and α_I for inhibitory) and X_{mn} is the presynaptic activity (ξ_{r_1,r_2} for afferent, η_{kl} for lateral).

In contrast to an earlier LISSOM model of ocular dominance, which required two different normalization rules (Sirosh and Miikkulainen 1994), in the receptive field model not only the lateral connections, but also the afferent connections can adapt by the same biologically appealing (Kohonen 1982; Miller and MacKay 1992) sum-of-weights normalization mechanism. It is interesting to analyze how such simplification is possible. The normalization constrains afferent weights of each neuron (i, j) to the hyperplane defined by $\sum_{r_1,r_2} \mu_{ij,r_1r_2} = 1$. When inputs sparsely populate a similar high-dimensional space, as in the case of simple inputs on a retina, the projection of the inputs on to a hyperplane does not drastically alter the input density. Therefore, such normalization is sufficient for good maps to form. However, if the input distribution is dense, the constraint surface for afferent weights must match the input manifold for a correct map to develop. One way to achieve this is to transform the m dimensional inputs to points on a unit $m+1$ dimensional hypersphere, establish $m+1$ corresponding afferent weights, and use the hypersphere defined by $\sqrt{\sum_{r=1}^{m+1} \mu_{ij,r}^2} = 1$ as the matching constraint surface (Miikkulainen 1991). Such a normalization was necessary in LISSOM models of low-dimensional input but interestingly is not required in the biologically more realistic model.

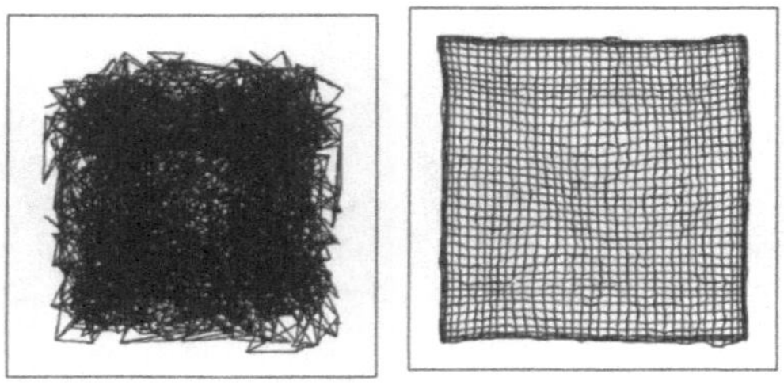

(*a*) Initially random map (*b*) Final ordered map

Figure 3: **Self-organization of scattered receptive fields into a topographic map.** The center of gravity of the afferent weight vector of each neuron in the 42×42 network is projected onto the receptive surface (represented by the square). Each center of gravity point is connected to those of the four immediately neighboring neurons by a line. The resulting dark grid depicts the topographical organization of the map. The RF centers were randomly scattered within a radius of 4 from their ideal position, so that any two neighboring neurons may be centered up to 8 neurons apart. The afferent weights were initially random (*a*). As the self-organization progresses, the network unfolds and the weight vectors spread out to form a regular topographic map of the receptive surface (*b*).

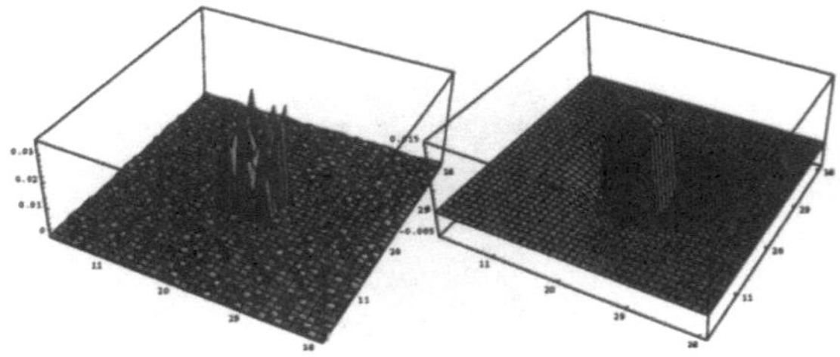

(*a*) Initial rough profile (*b*) Final smooth interaction

Figure 4: **Self-organization of the lateral interaction.** The lateral interaction profile for a neuron at position (20, 20) in the 42×42 network is plotted. The excitation and inhibition weights are initially randomly distributed within a radius of 3, and a radius of 18, respectively. The combined interaction is the sum of the excitatory and inhibitory weights and illustrates the total effect of the lateral connections. The total weights of excitatory and inhibitory connections were chosen to be equal, but because the excitatory connections had a shorter range, the interaction has the shape of a rough plateau (*a*). During self-organization, smooth patterns of excitatory and inhibitory weights evolved, resulting in a smooth "Mexican hat" shaped lateral interaction profile (*b*).

3 Simulation results

The LISSOM receptive field model was simulated with single and multiple gaussian spots of "light" on the retina as input. At each presentation, the activation ξ_{r_1,r_2} at the receptor (r_1, r_2) was given by:

$$\xi_{r_1,r_2} = \sum_{i=1}^{n} exp(-\frac{(x-x_i)^2 + (y-y_i)^2}{a^2}) \qquad (4)$$

where n is the number of spots, a^2 is a constant, and the spot centers (x_i, y_1): $0 \leq x_i, y_i < R$, were chosen randomly. It turns out that the self-organizing abilities of the network with single and multiple inputs depend on the degree of initial ordering of the afferent connections.

The extent of order and spread of afferents to the undeveloped cortex is not well known. It is possible that the afferent connections are scattered randomly across the retina. It is also possible that the afferent connections are ordered in overlapping patches, but the synaptic weights are random. A third (but remote) possibility is that the afferents overlap completely over large regions of the cortex, so that there is no initial topographic order at all. Corresponding to these alternatives, three network structures were simulated. Each network had initially random synaptic weights and either

1. randomly scattered receptive field centers, or
2. topographically ordered RF centers, or
3. each neuron was connected to every retinal receptor.

In each case, the afferent and lateral connections self-organized robustly when the light spots were presented one at a time. Figures 2, 3 and 4 illustrate the self-organization in the first network. The initial rough pattern of afferent weights of each neuron evolved into a hill-shaped profile (figure 2). The afferent weights of different neurons peaked over different parts of the retina, and their center of gravities (calculated in retinal coordinates) formed a topographical map (figure 3). Simultaneously, the lateral interaction evolved into a smooth "Mexican hat" shaped profile around each neuron (figure 4).

When multiple spots were presented ($n > 1$), the first network did not self-organize into a topographic map, whereas the second and the third did. In these cases, the activity bubbles on the network matched the topographic order of the input spots. This correlation in activity leads to the development of topographic order by the Hebbian weight adaptation mechanisms. When the receptive fields are scattered as in the case of the first network, the activity bubbles do not have topographical correlations with the input spots. Therefore, no topographic order is produced. Further, even if self-organized order were to be established with single spots initially, the resulting structure is easily perturbed, and loses its order when multiple spots are presented.

4 Discussion

The simulations serve to point out general limits on activity-dependent self-organizing processes. Unless the afferent connections are ordered so that the network activity patterns are broadly correlated at some scale, dynamic and stable self-organized structures cannot be formed or maintained. Therefore, in cortical maps, an initial ordering of afferent connections must be established by other means. Perhaps the receptive field centers are genetically preordered. If not, other processes such as selective attention must exist for selecting the matching activity patterns and inputs.

LISSOM is in many ways a generalization of several self-organizing models of cortical development. The second network discussed above is similar to that of Willshaw and von der Malsburg (1976), and Miller et al. (1989). However, these networks modeled lateral interactions by simple and fixed neighborhood functions. In the LISSOM process, the shape of the lateral interaction is automatically extracted from the statistical properties of the external input. The self-organization of afferent weights "bootstraps" itself by using the afferent input to develop the appropriate lateral interaction profiles. The resulting maps are in a dynamic and continuously adapting equilibrium with the input, and can reorganize in response to changes in the input distribution, such as those caused by lesions in the receptor surface.

The third LISSOM network is similar to Kohonen's (1982) Self-Organizing Map (SOM) in structure and self-organization of afferent connections, but constitutes a local, neural implementation of the process. In SOM, inputs are presented one at a time, and an external supervisor finds the maximally responding neuron and determines a neighborhood around it for weight adaptation. In LISSOM, adaptation neighborhoods are defined by the activity bubbles themselves, formed through the lateral interactions. Several inputs can be presented at the same time, and the weights can adapt in several neighborhoods simultaneously. Another main difference is that SOM aims at modeling a small patch of cortex where the afferents overlap completely, whereas LISSOM networks define anatomical receptive fields for each neuron and self-organize with varying numbers of connections from different parts of the receptive surface. A LISSOM network can be extended in a simple fashion by adding more neurons and afferents, making it possible to construct large scale simulations of cortical maps as well.

5 Conclusion and Future Work

The LISSOM model of cortical development demonstrates that the only major requirements for topographical organization are short range excitation, longer range inhibition and a Hebbian mechanism for weight modification. Starting from random-strength connections, neurons develop localized receptive fields and lateral interaction profiles cooperatively and simultaneously. Self-organization takes place with multiple retinal inputs, and with multiple anatomical receptive fields of varying sizes, but requires ordered receptive field centers. The resulting topographic maps are in a dynamic and continuously adapting equilibrium with the external input. Recent neurobiological results (Gilbert 1992; Kaas 1991; Merzenich et al. 1990) show that cortical structures are in a similar dynamic state and remain plastic throughout adult life. Preliminary experiments show that LISSOM can account for several observations on cortical plasticity, such as the expansion of receptive fields following retinal lesions (Gilbert 1992; Pettet and Gilbert 1992). An interesting possibility for future research is to develop a comprehensive model of activity-dependent cortical plasticity based on LISSOM.

Acknowledgments

This research was supported in part by National Science Foundation under grant #IRI-9309273. Computer time for the simulations was provided by the Pittsburgh Supercomputing Center under grants IRI930005P and TRA940029P.

References

Amari, S.-I. (1980). Topographic organization of nerve fields. *Bulletin of Mathematical Biology*, 42:339–364.

Burkhalter, A., Bernardo, K. L., and Charles, V. (1993). Development of local circuits in human visual cortex. *Journal of Neuroscience*, 13:1916–1931.

Gilbert, C. D. (1992). Horizontal integration and cortical dynamics. *Neuron*, 9:1–13.

Hubel, D. H., and Wiesel, T. N. (1965). Receptive fields and functional architecture in two nonstriate visual areas (18 and 19) of the cat. *Journal of Neurophysiology*, 28:229–289.

Kaas, J. H. (1991). Plasticity of sensory and motor maps in adult animals. *Annual Review of Neuroscience*, 14:137–167.

Katz, L. C., and Callaway, E. M. (1992). Development of local circuits in mammalian visual cortex. *Annual Review of Neuroscience*, 15:31–56.

Kohonen, T. (1982). Self-organized formation of topologically correct feature maps. *Biological Cybernetics*, 43:59–69.

Kohonen, T. (1993). Physiological interpretation of the self-organizing map algorithm. *Neural Networks*, 6:895–905.

Löwel, S. (1993). Private Communication

Löwel, S., and Singer, W. (1992). Selection of intrinsic horizontal connections in the visual cortex by correlated neuronal activity. *Science*, 255:209–212.

Merzenich, M. M., Recanzone, G. H., Jenkins, W. M., and Grajski, K. A. (1990). Adaptive mechanisms in cortical networks underlying cortical contributions to learning and nondeclarative memory. In *Cold Spring Harbor Symposia on Quantitative Biology, Vol. LV*, 873–887. Cold Spring Harbor, NY: Cold Spring Harbor Laboratory.

Miikkulainen, R. (1991). Self-organizing process based on lateral inhibition and synaptic resource redistribution. In *Proceedings of the International Conference on Artificial Neural Networks* (Espoo, Finland), 415–420. Amsterdam; New York: North-Holland.

Miller, K. D., Keller, J. B., and Stryker, M. P. (1989). Ocular dominance column development: Analysis and simulation. *Science*, 245:605–615.

Miller, K. D., and MacKay, D. J. C. (1992). The role of constrains in Hebbian learning. CNS Memo 19, Computation and Neural Systems Program, California Institute of Technology, Pasadena, CA.

Obermayer, K., Blasdel, G. G., and Schulten, K. J. (1992). Statistical-mechanical analysis of self-organization and pattern formation during the development of visual maps. *Physical Review A*, 45:7568–7589.

Pettet, M. W., and Gilbert, C. D. (1992). Dynamic changes in receptive-field size in cat primary visual cortex. *Proceedings of the National Academy of Sciences, USA*, 89:8366–8370.

Sirosh, J., and Miikkulainen, R. (1993). How lateral interaction develops in a self-organizing feature map. In *Proceedings of the IEEE International Conference on Neural Networks (San Francisco, CA)*, 1360–1365. Piscataway, NJ: IEEE.

Sirosh, J., and Miikkulainen, R. (1994). Cooperative self-organization of afferent and lateral connections in cortical maps. *Biological Cybernetics*, 71(1):66–78.

von der Malsburg, C. (1973). Self-organization of orientation-sensitive cells in the striate cortex. *Kybernetik*, 15:85–100.

Willshaw, D. J., and von der Malsburg, C. (1976). How patterned neural connections can be set up by self-organization. *Proceedings of the Royal Society of London B*, 194:431–445.

A NEURAL NETWORK IMPLEMENTATION FOR A ELECTRONIC NOSE

Philip B James-Roxby

UMIST, PO Box 88, Manchester, England, M60 1QD

A new implementation of a multi-layer perceptron neural network is presented where activation levels within the network are encoded using Sigma-Delta modulation. Large, hardware networks can be constructed, which can be trained using the standard back-propagation algorithm. The network has been used to form a stand-alone electronic nose system capable of distinguishing between four odours.

Introduction

The human sense of smell is the least understood of the senses, and is widely regarded as the least useful. However, in many industry sectors, a keen nose is invaluable; for example, food production.

The sense of smell is influenced by many criteria such as age, gender and smoking. An instrument to perform odour classification and intensity without being influenced by such diverse criteria would be highly desirable.

This paper describes the design of such an instrument, with particular regard to the classification of odours.

An electronic nose

Advances in odour sensor technology and general circuit miniaturisation have made the construction of an instrument capable of odour classification possible. The instrument is normally called an electronic nose, a phrase attributed to Gardner [1], who defines an electronic nose as "*an instrument which comprises of an array of electronic chemical sensors with partial specificity and an appropriate pattern-recognition system, capable of recognising simple or complex odours*"

There are several types of odour sensors; tin-oxide Taguchi gas sensors [2], Langmuir-Blodgett films [3], and conducting gas polymers [4] have all been successfully used. In the system under consideration, the sensors are conducting gas polymers deposited on a ceramic substrate with controlling circuitry.

The resistance of the polymers change when gas molecules are adsorbed on the surface. Each class of polymers change resistance by a different amount when different gases are present. By measuring the amount the base resistance of each sensor changes, that is the $\Delta R/R$, a unique gas fingerprint is generated for each gas. Examples of sensor responses are shown in figure 1.

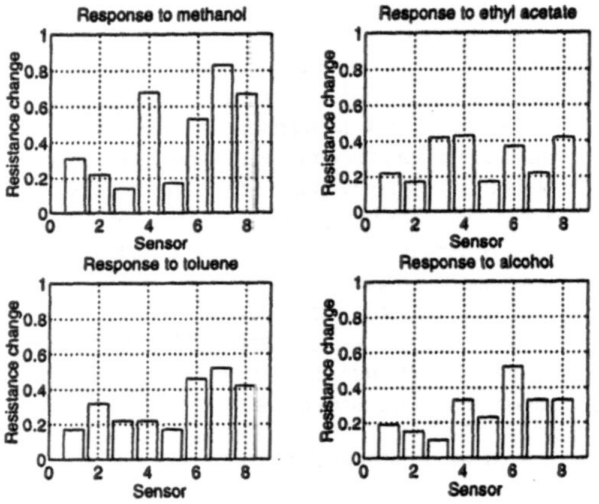

Figure 1
Example of sensor responses

Classification

The classification task is the matching of an unknown pattern of resistance changes against library responses. There are many suitable algorithms, but artificial neural networks (ANN) have several advantages, the tolerance to noise being the most important.

Noise is a particular problem in the electronic nose since the sensors respond to temperature, humidity and light. In addition, the sensors age in a non-deterministic manner. Therefore, the patterns to be classified rarely resemble the library patterns.

Integration

The electronic nose currently uses a host computer to perform a software implementation of a neural network. The aim is to develop a hand-held instrument which could be used for example for the monitoring of food quality at any point in production.

A hardware implementation of a neural network would remove the need for a host computer, making the electronic nose a truly portable instrument.

Hardware implementations

Hardware implementations of ANNs are still in their infancy. A number of successful implementations do exist, and can be categorised into analog and digital systems. Analog implementations have the advantage of small elements, but suffer due to the lack of an effective analog memory cell to store the weight values.
The problem with digital implementations is that the basic elements required for an ANN tend to be large if a coding mechanism such as pulse-code modulation is used. Multipliers in particular are large, and since a great number are required in ANNs, this would lead to an extremely large system implementation.
In addition, due to the tremendous inter-connections required in ANNs, if a multi-bit coding mechanism such as PCM is used, these paths will dominate the available area, leaving little room for the arithmetic elements.
If a single-bit coding mechanism is used, the basic elements and communication paths would be smaller. However, a bit-serial approach has a number of disadvantages; most importantly, the susceptibility to bit-wise errors.
A mechanism where every bit has equal importance would therefore be ideal for use in an ANN. Such a coding mechanism is Sigma-Delta modulation (Σ–Δ). Σ–Δ modulation is normally used in the implementation of high-accuracy A-to-D converters [5].

Arithmetic using Σ–Δ bitstreams

A Σ–Δ bitstream representing the analog value 0.75 would consist of the repeating binary pattern 111011101110... The analog value can be reconstructed from the bitstream by taking the average value of the bitstream over time; in general, if the bitstream contains a repeating pattern with N logical 1's, and M logical 0's, the bitstream represents the analog value N/(N+M).
The linear operations of multiplication, addition and subtraction can be performed on Σ–Δ bitstreams by simple logical structures. The multiplication of two Σ–Δ bitstreams can be performed by an AND gate. Addition and subtraction can be performed by larger structures.

A Σ–Δ based Artificial Neural Network

For the electronic nose, the well known Multi-Layer Perceptron (MLP) model was chosen. This is a relatively well understood topology, with many training algorithms available.

The basic elements required for the MLP are signed multiplication by a programmable weight, addition of the weighted values, and a non-linear operation on this sum to provide the neuron output, as shown in figure 2.

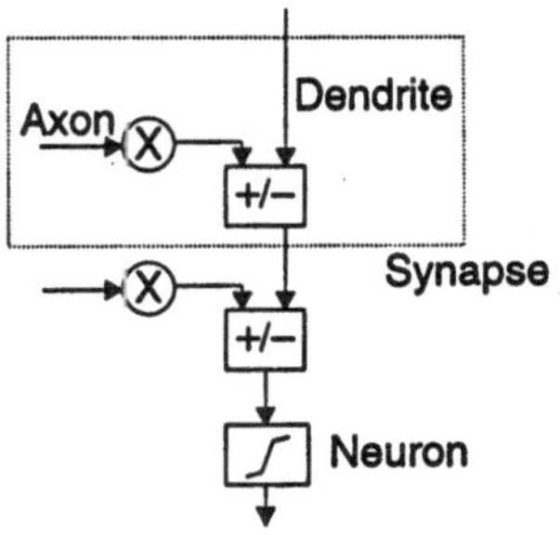

Figure 2
Basic elements required in an MLP

Synapse design

The synapse is required to weight the incoming activity on the axon, and add this weighted value to the activity level on the dendrite, as shown in figure 2. The weight must be programmable to allow training.
The synapse structure is shown in figure 3. The weight is held in Σ–Δ form in the circulating shift register; the sign bit determines whether the axon input is excitory or inhibitory.

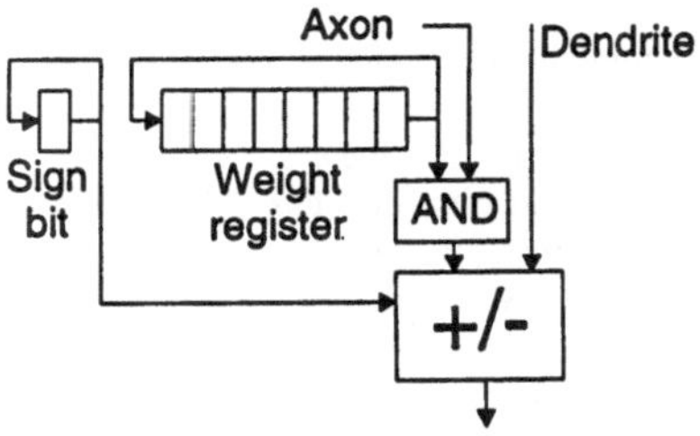

Figure 3
Σ–Δ based synapse structure

There are two modes of operation determined by a MODE input. When MODE is high, the weight and sign bits are serially shifted in; when MODE is low, the weight register circulates, presenting the weight in Σ–Δ form to the AND gate performing the multiplication. The sign bit determines whether the weighted value is added or subtracted to the incoming value on the dendrite.

Neuron structure

In a similar implementation in [6], a step function is implemented for a pulse-density based system. ANNs with step functions are prone to noise, and are difficult to train, since no gradient information is available.
A differentiable sigmoid function allows a gradient descent training algorithm such as back-propagation to be used. Therefore, an element is required to map an input Σ–Δ bitstream to its sigmoided output.

It was decided to use a small finite state machine (FSM) as the mapping element; the FSM has one input, one output and can be in one of 16 different states at a time, with the state held in a four-bit shift register.
The value of the output bit depends on the input bit and the current state, and the next state the machine goes to is determined by the input bit. The machine moves from one arbitrary state to another in a complex way with the average value of the output bit forming the sigmoid of the value represented by the input bitstream. The machine is impossible to design by hand.
The machine can be defined by a table containing the next state the machine will move to given the current state and the input, with the corresponding state output as shown in figure 4

Input	State	Next state	State output
0	S0	S3	0
1	S0	S7	0
0	S1	S1	1
1	S1	S9	1
.	.	.	.
0	S15	S2	0
1	S15	S4	1

Figure 4
Table representing the FSM

Effectively, the FSM is defined by a 32x2 matrix, with the first column containing the next state, and the second containing the state output.
For each row of the matrix, there are 16 possible next states, and 2 possible output bits; therefore 32 possible row entries. Given there are 32 rows, the number of FSMs available is 32^{32}, or 1.46×10^{48}. Clearly, an exhaustive search is impossible.
The Genetic Algorithm (GA) [7] is a suitable searching algorithm for a number of reasons. Firstly, the problem is easily encoded into a digital chromosome; the next states and state outputs can be encoded directly onto a string. Secondly, fitness measures are directly available through measuring how closely a test machine implements an ideal sigmoid function.

Design of an FSM by the Genetic Algorithm

The GA finds excellent solutions to problems by the promotion of encoded partial solutions called schemata. From the Schema Theorem in [7], the promotion of a schema depends on two factors; firstly, the degree to which the schema is above average fitness, and secondly whether the schema has a short defining length. Short, fit schemata will be promoted at the expense of longer, weaker ones.
Considering the implications of the Schema Theorem on encoding shows that the standard GA will only work on a problem where closely related features of the problem are encoded close together on the string.
This is not the case with the design of the FSM. The related features are the sequences of state transitions and corresponding outputs for a sequence of input bits. Since the states are not necessarily close together on the string, the standard crossover operation will split useful state transitions.
The standard GA works on linear chromosomes; for this design, linear chromosomes are not suitable. Koza in [8] shows how crossover and mutation can be extended to a tree-shaped chromosome. For the design of an FSM based mapping element, the most suitable shaped chromosome is a cyclic graph structure shown in figure 5.

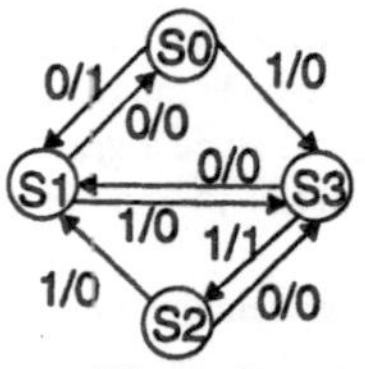

Figure 5
Cyclic graph representing an FSM

The notation used is the standard for illustrating transitions in FSMs. The M/N refers to the input pattern M and output pattern N. Thus when the machine is in state 0 with an input 0, the machine outputs a 1 and moves to state 1.
Where in [8] crossover became the exchanging of sub-trees, crossover in this case becomes the exchange of sub-graphs, representing the response of the FSM to a sequence of input bits. Schemata are now useful state transitions with their corresponding state outputs, and will be promoted in the normal way based on their fitness.
Mutation remains the random changing of string entries; though mostly destructive, when beneficial improves the output bit pattern by either altering state outputs, or next state values. Table 1 shows the GA parameters used in the design.

P(Reproduction)	0.5
P(Crossover)	0.5
P(Mutation)	0.1
Population size	200
Max. Generations	500

Table 1
Genetic Algorithm parameters used

The fitness function was a measure of how closely the decimated output of the machine represented the sigmoid of the input for 1024 input values. The GA consistently found excellent solutions in less than the maximum number of generations permitted. Figure 6 shows the performance of a generated mapping element.

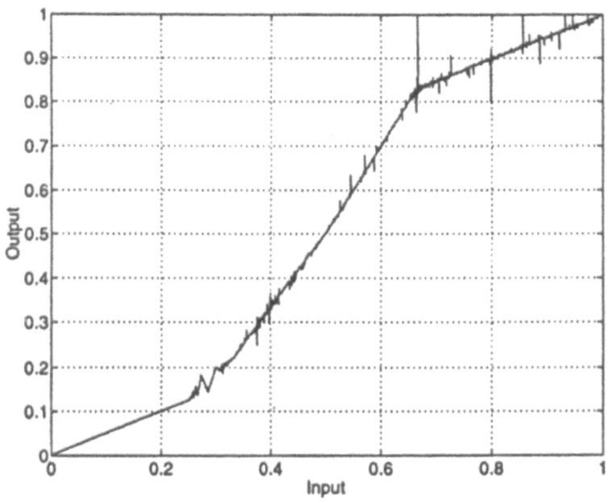

Figure 6
Performance of generated mapping element

Implementation

Each element was implemented separately on reprogrammable gate arrays, before a one-time programmable ACTEL gate array was designed. This device was programmed to contain 3 neurons each with 4 synapses. By using a modular design, it is possible to cascade similar devices together to form realistically sized networks. Because all inter-chip communication is one-bit wide, there is no need to multiplex output pins to provide the required connectivity.
The network is currently being tested with the conducting gas polymers. Initial findings show that a two-layer network with 8 input nodes, 16 hidden nodes and 4 output nodes can discriminate between 4 odours with 95% accuracy, though temperature must be very carefully controlled.

Conclusion

A new hardware implementation of a multi-layer perceptron neural network has been presented. Such an implementation is useful where classification is required without a host computer. The basic elements of the network are small, and since all communication paths are 1-bit wide, realistically sized hardware networks are possible.

References

[1] Gardner, J.W., Bartlett, P.N.: Sensors and Actuators B, vol 18, no 1-3, 211 (1993)
[2] Shurmer, H.V.: IEE Proceedings, vol 37, pt G, no 3, 197 (1990)
[3] Moriizumi, T.: Thin Solid Films, 160, 413 (1988)
[4] Bartlett, P.N., Archer, P.B.M, Ling-Chung, S.K : Sensors and Actuators, 19, 125 (1989)
[5] Uchimura, Hayashi, Kimura, Iwata : IEEE Trans. Acoustics, Speech and Sig.Proc, 36, 1899 (1986)
[6] Tomberg, J.E., Kaski, K. K. K.: IEEE J. of Solid-State Circuits, vol. 25, no. 5, 1277 (1990)
[7] Goldberg, D.E. : Genetic Algorithms in search, optimization and machine learning. Addison Wesley 1989
[8] Koza, J.R. : Genetic Programming. MIT Press 1992

GENETIC ALGORITHMS (GAs) IN THE ROLE OF INTELLIGENT REGIONAL ADAPTATION AGENTS FOR AGRICULTURAL DECISION SUPPORT SYSTEMS

Gianni Jacucci, Mark Foy, and Carl Uhrik

Laboratorio di Ingegneria Informatica (LII)
Università degli Studi di Trento
Via Fortunato Zeni 8; I-38068 Rovereto (TN) Italy

ABSTRACT

An AI adaptation methodology designed to assist in transporting agricultural models between regions is presented. A methodology to perform model adaptation (viz. localization) is frequently necessary when models are transported because models developed in one region often do not produce valid results when used in a different region. In this methodology, a GA plays the role of the adaptation agent. By linking a GA to an agricultural model, the model become more robust because it is able to adapt to the region in which the model is being used. This methodology has been implemented within a decision support system (DSS) developed within an EC project called SYBIL. The DSS helps farmers predict when diseases and funguses will attack their plants, so they can make intelligent decisions on preventing these attacks. Preliminary testing within this environment indicates this adaptation methodology has the ability to allow agricultural models developed in one area to be effectively utilized in other regions.

INTRODUCTION

In the domain of agriculture, the utilization of already developed models in a broad area is often hindered by one or more factors. One frequent factor which impedes transportation is model inaccuracy. For example, when models that perform well in one region, are transported to be used in a different region, they often do not give accurate output (such as, recommendations, results, and/or indicators) in their new environment (i.e., when they are run in a new region).

This is one of the major difficulties of *model technology transfer*. To address this difficulty, an artificial intelligence (AI) methodology is proposed. In this methodology, a GA plays the role of an intelligent adaptation agent by adjusting agricultural risk assessment models to particular regions. The general component created by this combinational methodology is called an 'Agricultural Model-GA' or an AGMOD-GA.

A decision support system (DSS) (which addresses grape and apple management) developed under an EC project called SYBIL has utilized this AI adaptation methodology. The models included within this DSS are designed to help growers determine when it is necessary to spray against certain pests. One of the models in this DSS is called the P.R.O. model, and the methodology has been applied to this model to create a PRO-GA component.

The following sections will give an overview of the adaptation methodology, discuss under what conditions this methodology can be applied, describe the methodology's elements, and then give details about linking a GA to a model under this scenario. Finally, conclusions will be given.

OVERVIEW OF THE AI ADAPTATION METHODOLOGY

The theory of this AI adaptation methodology is that by utilizing historical data from a particular region, a model's parameter settings can be adapted so that the new parameters allow the model to work well in the particular region. This adaptation is done by trying to match the model parameter settings (which should be easy to change) to the particular region. That is, new model parameter settings are found which allow this model to give accurate output values (such as, recommendations, results, and/or indicators) when run in the particular region. To find matching model parameter settings, intelligent search is performed which utilizes historical outcome data as part of the objective function (i.e., the search attempts to fit the model parameter settings such that when used in the model, the model gives output near to the given historical outcome data).

For example, suppose a model has been developed in **RegionA** that gives an assessment of the risk that a certain plant will be damaged by a certain fungus. This model uses an instantiation of model parameter setting values (such as values for thresholds and coefficients, etc.) called **ModelParmsA**. The model has proved quite useful in **RegionA**, and now agriculturalists in **RegionB** are interested in using this model. Upon first running this model in **RegionB** (with the original model parameter settings, i.e., **ModelParmsA**), the model gives output values (recommendations, results, and/or indicators) which are inaccurate. As an attempt to make this model work accurately in **RegionB**, new model parameter settings (which are different from **ModelParmsA**) could be derived. By applying the methodology discussed here (which requires historical data and an intelligent search method to be described below), it is possible to locate new model parameter settings, and once this is done, this new instantiation of settings can be called **ModelParmsB**. After applying this methodology, the model should produce accurate output values in **RegionB** when using **ModelParmsB**.

Overall, by following this methodology, a component will be created which can search for good model parameter settings such that when the given model is applied and run at the location in question, the output values given will be consistent with the historical outcome data. Moreover it is hoped that this will also allow the model to be generally used in this region, producing accurate output values on data which it has not seen. This component that performs this search/adaptation can be called an expert system component since it modifies and adjusts a model to work in a new location in the same way an expert would modify and adjust a model.

Additionally, it should be emphasized that this methodology is particularly appealing because it is not a strictly empirical or analytical, but both. That is, this methodology does not perform a search to fit the historical data from a particular location into an empirical algorithm; rather it performs the search in a larger context, fitting the model parameter settings to a particular location. Therefore, the resulting instantiation of the adapted/localized agricultural model (with the new parameter settings inside) is as good (or as bad) as the original model. Consequently, if the model is biologically significant (e.g., if it simulates biological events), then this feature is not lost by this adaptation methodology since the model is used in the same form (i.e., the structure of the model is left intact), only the model parameter settings are changed.

Before describing the elements of this methodology, a discussion of the limitations of the methodology will be given

APPLICABILITY OF THE METHODOLOGY

The difficulty of transporting models from one location to another is not limited to being explain in one way (i.e., one or more different reasons could explain model inaccuracy in a new location). Likewise, the methodology described here is not applicable to all models that have difficulty being transported; therefore, not all models that have difficulty being transported will necessary benefit by following this methodology.

Distinguishing models that can be treated by this methodology from those that can not, agricultural models can be divided into two cases:

(Case1) models that are not transportable because they are too location-specific and not robust enough to allow different conditions to enter into the model and

(Case2) models that are transportable, but the model parameter settings need to be altered to allow the model to give accurate output values for the new region.

In **Case1**, the model is just too specific to operate outside one area. This situation is difficult to address and solve if there is a strong desire to transfer the model. This case will not be addressed here. On the other hand, **Case2** can be addressed by the adaptation methodology described here which determines location-specific model parameter settings so that the model functions accurately in a new location.

ELEMENTS OF THE METHODOLOGY

Generally stated, this methodology prescribes the utilization of:
(i) historical situation data,
(ii) historical outcome data,
(iii) the model, and
(iv) an intelligent search method (in this case, a GA).

Historical situation data is the basic data required by the model in question. In the domain of agricultural models, this often includes meteorological data since this is frequently an important input to the model. In this methodology, the more historical situation data that is available, the better.

The other type of historical data needed is **historical outcome data**. The methodology assumes that for every piece of historical situation data, there is a corresponding outcome. Both the situation and the outcome are observable in the real world. The model accepts situations and computes outcomes. For risk assessment models, historical outcome data regards the occurrence of fungus or pest problems in past years (epidemiological data); or for crop growth models, historical outcome data regards crop yield in past years. Thus, historical outcome data is a particularly notable part of this methodology because it is used to *fit* the model parameter settings to the new region in question. Therefore, when constructing a component using the methodology described here, it must be possible to match model outputs to some combination of outcomes and/or events in the real-world. For example, with a risk assessment model, if the only historical epidemiological outcome data available is a set of historical primary infection dates, then the model should be able to produce a primary infection date (albeit a guess of the primary infection date) as one of it's outputs. In this way, it will be possible for the model to be adapted such that the primary infection date produced by the model nearly matches a set of actual primary infection dates (i.e., dates specified by the historical outcome data).

In this methodology, the agricultural **model** is fundamental because it will be used to evaluate how well particular model parameter settings work in the given region (i.e., with the given historical data). In particular, the intelligent search method will repeatedly call upon the model as it constructs new model parameter settings that need to be evaluated.

This methodology has the capability to address many types of agricultural models: risk assessment models, damage prediction models, crop growth simulation models, and other models that use meteorological data.

The **intelligent search method** is an important part of this component because, in this particular domain of agricultural models, knowledge of the domain is often hard to codify (i.e., 'rules of thumb' are vague and difficult to construct), and the selection of an intelligent search method can help to alleviate this difficulty. This is due to the fact that intelligent search methods do not rely on 'rules of thumb'.

The selection of the actual intelligent search method to be employed was made among the following possible methods: hill-climbing, simulated annealing, and GAs. In the end, GAs were selected because of their unbiased, effective search capabilities and their applicability to real-world problems [1] [2] [3] [4].

UTILIZING THE AI ADAPTATION METHODOLOGY - THE AGRICULTURAL MODEL-GA (AGMOD-GA)

Linking an Agricultural Model to a GA

To allow a GA to search the space of an agricultural model's parameters, the GA uses the model as the evaluation function. Furthermore, the model uses the historical situation data (as this is necessary to run the model in the given historical years), and the GA additionally uses the historical outcome data (discussed earlier) in combination with the output of the model. Whenever the GA wants to evaluate one instance of model parameter settings, the agricultural model is called, and the final outcome is returned through an objective function to the GA so that a "fitness" can be computed. This resulting general component is called an 'Agricultural Model-GA' or an AGMOD-GA.

Figure 1 illustrates an agricultural model and a GA linked to form an AGMOD-GA.

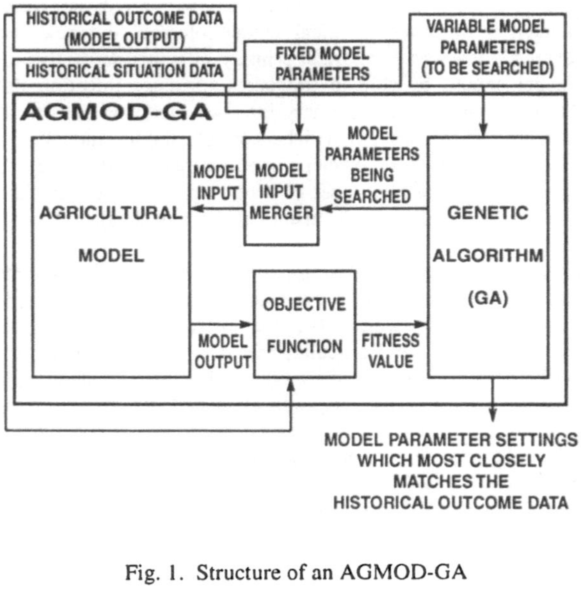

Fig. 1. Structure of an AGMOD-GA

The function of the AGMOD-GA is to find near-optimal model parameter settings for the given desired behavior (i.e., matching the given historical outcome data). There are three main steps involved in the execution of a typical AGMOD-GA. First, the agricultural model and the GA are initialized. The second step in a typical AGMOD-GA is the fitness computation. This involves taking each GA population member and executing one or more model simulations using the model parameter settings represented by this member. This fitness evaluation step is executed many times because new population members are continually being generated by the GA. Fitness evaluation is usually continued until the GA has converged on a suitable optimal or quasi-optimal solution. The last step is the evolution of the GA population. This involves applying operations to the population members. The three operators used in a typical GAs are reproduction, crossover, and mutation. They act by treating the GA bit strings (which represent model parameters) in a way analogous to the evolution of chromosomes in genetics [1].

AGMOD-GA Performance

When particular instantiations of this adaptation methodology are employed, AGMOD-GA search runs provide the user with model parameter settings that the operator can use to ameliorate the original model. Instantiations of the AGMOD-GA perform similarly to most GAs, in that the average fitness of a generation, over time, will increase. That is, the members of later populations will converge on maximum fitness members of the space (in this case, on optimal sets of model parameter settings). This performance of a typical AGMOD-GA is shown in Figure 2. The two curves ("POPULATION MAXIMUM FITNESS" and "POPULATION AVERAGE FITNESS") illustrate the GA converging on high fitness members. This type of performance has in fact been achieved on an instantiation of the AGMOD-GA called the PRO-GA. This PRO-GA will be discussed in the next section.

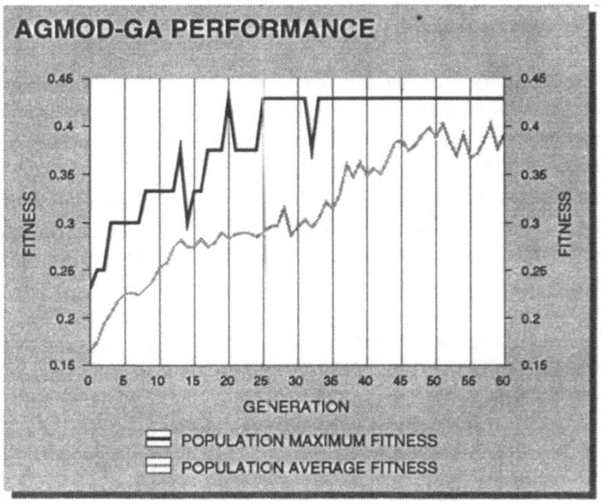

Fig. 2. Typical AGMOD-GA Performance
(Note: The values plotted are from a PRO-GA run)

An Example: The PRO-GA

The AI adaptation methodology described has been employed within a DSS developed under an EC project called SYBIL. In particular, this methodology has been utilized within one of the developed DSSes to adapt the P.R.O. model (i.e., find localized model parameter settings for the P.R.O. model). The P.R.O. model is a biological life cycle model that simulates the growth of downy mildew (viz. Plasmopara viticola, also called peronospora) on grape vines. This model is designed to help growers determine when it is necessary to spray against downy mildew. The model has been used in Germany to help grape growers reduce the amount of fungicide needed to control this fungus [5] [6] [7] [8] [9] [10]. The AI component intelligently searches the space of P.R.O. model parameter settings and locates settings that match local conditions. Preliminary testing with the P.R.O. model indicates that this adaptation methodology has the capacity to allow the parameter values of regional models to be effectively adapted to regions other than where they were developed [11].

CONCLUSION

This AI methodology addresses *model technology transfer* (i.e., the moving of agricultural models that are developed in one location to a new location so they can be used in this new location). In particular, it addresses one of the major difficulties within this area, namely, model accuracy; i.e., it addresses the instance when a useful model is transported from a region where it is functioning accurately (viz. producing accurate recommendations, results, and/or indicators) to a new region where it subsequently does not function accurately.

The methodology employs four main elements, with a GA at the center. By employing this AI component in conjunction with the engine of an agricultural model and historical data, model parameter settings can be adapted to new locations, allowing the model to give accurate results when run in the new location. Specifically, the module created by this methodology can be applied to localize models by deriving new model parameter settings that can be employed in the particular location to give good suggestions/decision support.

With the assumption that model technology transfer is an advantageous action (refer to [11] for an elaboration of advantages and disadvantages in transporting models between regions), this AI methodology has been found to efficiently addresses this issue and improves the current state-of-the-art in model technology transfer.

This is briefly discussed here through an example which describes the utilization of this methodology with the P.R.O. model which is contained within one of the decision support systems (DSSes) developed under an EC project called SYBIL. The resulting PRO-GA component has been used to produce new model parameter settings, and testing with these settings has shows that this methodology has great potential to localize model parameter settings, and this should assist in achieving the goal of making sound models more widely available.

REFERENCES

[1] Goldberg, D.E. 1989. Genetic Algorithms in Search, Optimization, and Machine Learning. Addison-Wesley, Reading, MA, USA.

[2] Grefenstette, J.J. 1985, ed. Proc. of an International Conference on Genetic Algorithms and Their Applications. Carnegie-Mellon University, Pittsburgh, PA, USA, July 24-26, 1985. U.S. Navy Center for Applied Research in Artificial Intelligence, USA.

[3] Grefenstette, J.J. 1987, ed. Proc. of the Second International Conference on Genetic Algorithms. Massachusetts Instituted of Technology, Cambridge, MA, USA, July 28-31, 1987. L. Erlbaum Associates, USA.

[4] Schaffer, J.D. 1989, ed. Proceedings of the Third International Conference on Genetic Algorithms. George Mason University, Fairfax, VA, USA, June 4-7, 1989. Morgan Kaufmann Publishers, Inc., USA.

[5] Hill, G.K. 1989. Mit P.R.O. System Spritzungen Sparen. Die Weinwirtschaft, 125, pp. 28-31.

[6] Hill, G.K. 1990. Das Peronospora-Riskoprognosemodell Oppenheim (P.R.O.) im Praxistest. Der Deutsche Weinbau, 44 (12/27 April 1990), pp. 514-517.

[7] Hill, G.K. 1990. Studies on Plasmopara Viticola Epidemics in the Vineyard. Proc. Atti del Convengo su Modelli Euristici ed Operativi in Agricoltora, Caserta, Italy, 27-29 Sept. 1990, Societa Italiana de Fitoiatria ed., pp. 266-273.

[8] Hill, G.K. 1990. Plasmopara Risk Oppenheim - A Deterministic Computer Model for the Viticultural Extension. Proc. Atti del Convengo su Modelli Euristici ed Operativi in Agricoltora, Caserta, Italy, 27-29 Sept. 1990, Societa Italiana de Fitoiatria ed., pp. 182-194.

[9] Hill, G.K. 1993. Severe Primary Infections of Plasmopara Viticola at bloom-Causal Climatic Conditions and Threshold of Damage on Bunches. Proc. OILB Meeting on Integrated Control in Viticulture, Bordeaux, France, 2-5 March, 1993.

[10] Hill, G.K., Breth, K., Spies, S. 1993. The Application of the P.R.O. - Simulator for Minimizing of Plasmopara Sprays in the Frame of an Integrated Control Project in Rheinhessen/Germany. Vitic. Enol. Sci., 48, pp. 176-183

[11] Jacucci, G., Foy, M., and Uhrik, C. 1994. SYBIL DSS:Localization of Agricultural Risk Assessment Models. Proceedings of Decision Support 2001 (DSS2001) (The 17th Annual Geographic Information Seminar & The Resource Technology '94 Symposium), Toronto, Canada, 12-16 Sept. 1994, American Society of Photogrammetry and Remote Sensing.

GENETIC ALGORITHM APPLIED TO RADIOTHERAPY TREATMENT PLANNING

O.C.L. Haas*, K.J. Burnham*, M.H. Fisher*, J.A. Mills+

* Control Theory and Applications Centre, Coventry University, U.K.
+ Department of Clinical Physics and Bioengineering, Walsgrave Hospital, U.K.

Conformal therapy attempts to produce accurate medical treatment that will produce a uniform dose over cancerous regions whilst at the same time sparing healthy tissues, especially the organs at risk. The constrained optimisation problem, that consists of working back from a given dose specification to elemental beam weight intensities is referred as the so called inverse problem. Due to the nature of the problem and in particular to its conflicting objectives, it is believed that heuristic techniques may have advantages over direct methods to solve for the beam intensities.

This paper formulates the problem, which may be regarded as a multiple extremum problem, within a genetic algorithm framework. It describes the relative merits arising from the use of genetic algorithms over existing techniques, such as methods based on iterative/adaptive least squares.

Introduction

External radiation therapy use x-ray photon or electron beams to treat many different forms of cancer. The computation and planning process to determine an acceptable treatment plan, that produces a high uniform dose in the treatment region and minimises the dose delivered to the other healthy tissues and in particular in the organs at risk, is known as radiotherapy treatment planning (RTP). Traditional RTP attempts to produce an acceptable treatment plan using a forward approach, that is basically a trial and error procedure. This procedure is dependent on the experience of the treatment planner, and does not allow the use of a wide variety of complex beam modifiers due largely to time constraints. Conformational RTP, is a relatively new technique which is concerned with solving the *inverse problem* [1, 2, 3, 4] that is to work back from the prescribed dose distribution to determine a set of elemental beam intensities that will achieve a dose distribution that is considered to be optimal in some sense.

The choice of beam configuration, that is the number and positioning of the beams, is vital to produce an acceptable plan. If the beam configuration is not well chosen, or if the treatment plan is very complex, the solution to the inverse problem may lead to physically unrealisable beam intensities. In order to prevent the possibility of an unrealisable solution, constraints have to be assigned to the elemental beam intensities. The basic level of constraint is that the elemental beams be positive or null, as it is not possible to extract the dose received by previous irradiation, i.e. a negative beam intensity is meaningless. A further level of constraint depends on the techniques used to produce intensity modulated beams. These techniques include individually shaped compensators and multileaf collimators (MLC); the latter approach is presently regarded as the best method to produce arbitrary and irregular shaped fields [1, 2, 4]. However, the robustness offered by the use of individual compensators is appealing and much effort is currently being invested in research into the development of automated RTP systems in which the design, specification and manufacture of patient specific compensators form an integral part of the overall plan. In the case of compensators, a constraining factor is that the rate of change of adjacent elemental beam intensities has to be kept within realistic limits.

In this paper the inverse optimisation problem is addressed and an approach derived from iterative least squares (ILS) is compared to heuristic techniques that make use of genetic algorithms (GAs). Whilst the ILS approach attempts to solve the inverse problem directly, the heuristic approaches attempt to solve the problem in a forward manner which could be viewed as an intelligent trial-and-error procedure. Having demonstrated the feasibility of the two approaches on standard test cases and shown that they give comparable results, the problem is then formulated in terms of a multiobjective cost function, with each objective being treated independently. The result is a set of solutions rather than a unique solution; namely a pareto-optimal set [5].

In the case of the RTP problem, it is shown that the pareto-optimum set may be represented as a surface in three dimensional space; the axes of which correspond to errors in the three main regions of interest, namely the treatment area, the organs at risk and the other healthy tissues. Finally, methods of decision making that are required in order to select the most appropriate cost function weightings are discussed.

Problem formulation

The problem is formulated on the plane (x,y) as shown in Figure 1, within the Cartesian frame of co-ordinates (x,y,z), where z is the patient head to toe axis and the plane (x,y) is a transversal view of the patient that could be obtained using computed tomograpy (CT) or magnetic resonance imaging (MRI). In the plane of interest the clinician outlines the treatment regions and the critical organs, and assigns maximum and minimum dosage levels.

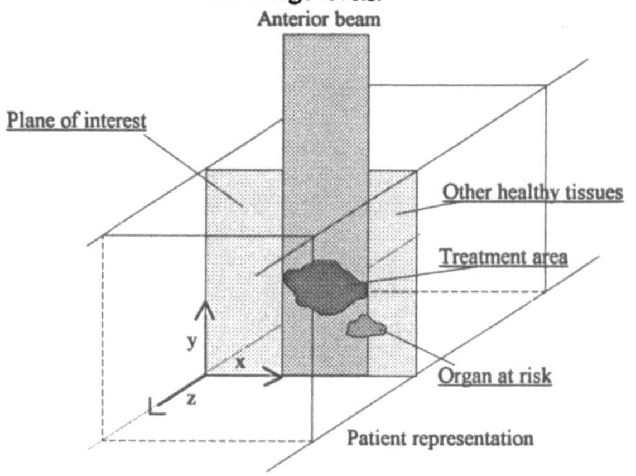

Figure 1: Treatment plan

In order to proceed, it is necessary to express the clinical requirements in terms of an objective function (or functions) which, for a given beam configuration, can be manipulated in order to obtain the optimum elemental beam intensities. Factors affecting the acceptance of a given RTP include compliance with clinical requirements in terms of dose constraints and statistical relationships between the control of a tumour and the likelihood of complication involving normal tissues. In this paper consideration is restricted to that of satisfying the dose constraints. The following criteria form the basis of the optimisation:

(i) The dose delivered to the treatment area (TA) should be uniform and as close as possible to the prescribed dose, that is normalised to 100%.
(ii) The dose delivered to the organs at risk (OAR) should be as low as possible and less than the maximum tolerable value, which is chosen as 20%.
(iii) The dose delivered to the other healthy tissues (OHT) surrounding the treatment area and the organs at risk should be kept low, around 30%.

The greater the number of beams, the smaller is the contribution from each of the beams, and the lower the dose received by the healthy tissues on each beam path.

To solve the optimisation problem it is necessary to make use of beam models that include beam intensity modulation facilities. A number of approaches to solve the inverse problem have been studied by researchers in the field. In [1, 3, 4, 6, 7] use is made of deterministic techniques to formulate the problem so that the number of elemental beam intensities is equal to the number of dose samples; the samples being taken at random points of interest. Other similar approaches exist, but common to all is the need to develop a beam model that can be expressed in matrix form, where a dose vector is given by a matrix product between the dose calculation matrix and the elemental beam intensity vector:

$$\mathbf{d} = \mathbf{\Phi}\, \mathbf{b} \tag{1}$$

where $\mathbf{d} \in \Re^n$ is the dose vector, $\mathbf{b} \in \Re^m$ is the elemental beam intensity vector and $\mathbf{\Phi}$ is an n by m dose calculation matrix.

Then, knowing the dose specification vector $\delta \in \Re^n$, given by the clinician, and given the dose calculation matrix $\mathbf{\Phi}$, the corresponding beam intensities could ideally be obtained from:

$$\mathbf{b} = \mathbf{\Phi}^{-1}\, \delta. \tag{2}$$

In previous work [8, 9, 10] a beam model of the form (1) has been developed, where each beam involved in the treatment plan is modelled by a matrix product between a dose calculation matrix, including both primary and scattered contributions, and the elemental beam intensity vector. For each beam, the initial elemental beam intensity vector is modified by both a flattening filter and penumbra.

In contrast to direct approaches to a solution, such as ILS, when use is made of heuristic techniques, such as simulated annealing or GAs, inversion of a dose calculation matrix is not required. As such the optimisation algorithm is less environment dependent, and any beam model in which elemental beam intensity modulated facilities are available can be employed.

Brief review of existing RTP optimisation schemes

The only stochastic techniques that have been used to solve the RTP problem are simulated annealing and fast simulated annealing [2]. Whilst these techniques have provided some promising results, most of the methods used are based on deterministic strategies. In many of the deterministic algorithms [1, 3, 4, 5, 6, 7], it is not possible to obtain a direct inversion of the dose calculation matrix. In [1, 3, 4, 6, 7] use is made of Newton's method of multiobjective minimisation in conjunction with an approximation to the inverse of the dose calculation matrix to iteratively determine the elemental beam intensities.

In [11] techniques used by tomographic imaging give good results, in particular Bayesian optimisation has been combined with algebraic reconstruction techniques, and maximum entropy optimisation has been used with multiplicative algebraic reconstruction techniques.

In [8, 9, 10] an approximate beam model is used to calculate the dose delivered to elemental cells belonging to a regular grid and an ILS procedure is used to calculate the elemental beam intensities. However, since the dose calculation matrix in this approach is non-square, use is made of the pseudo inverse such that the beam intensity vector is given by:

$$\mathbf{b} = [\mathbf{\Phi}^T\mathbf{\Phi}]^{-1}\mathbf{\Phi}^T\, \delta \tag{3}$$

If only one beam were used, this would give a unique least squares solution to the problem. However, as more than one beam is used, an iterative approach is required to optimise the dose delivered by each of the beams involved.

In this paper a new application of GAs to RTP is described and an extension to multiple objective genetic algorithms (MOGA) is discussed.

Whilst most of the deterministic approaches use quadratic objective functions, methods that are based on GAs enable the use of a wider variety of objective criteria, such as, for example the maximum dose over a region criterion.

Formulation of objectives

The traditional approach to solve this problem is to combine these competing objectives into a single objective function. In order to do so weightings are associated to each of the objectives using empirical knowledge. The weightings are assigned such that more emphasis is focused on criteria (i) and (ii) with less attention being given to criteria (iii).
In previous work [8, 9, 10] the following objective function has been used. It takes the form:

$$J=E\{(\alpha(\mathbf{d}_{TA}-\delta_{TA})^2+\beta(\mathbf{d}_{OAR}-\delta_{OAR})^2+\gamma(\mathbf{d}_{OHT}-\delta_{OHT})^2)\} \quad (4)$$

where $E\{.\}$ is the expectation operator, $\mathbf{d}_{TA}$, $\mathbf{d}_{OAR}$, $\mathbf{d}_{OHT}$ and δ_{TA}, δ_{OAR}, δ_{OHT} are, respectively, the calculated dose and the prescribed dose vectors representing the treatment area, the organ at risk and the other healthy tissues. α, β and γ are scalar weighting factors for each of the three regions and provide 'tuning knobs' that can be tailored for any specific optimisation problem. Usually the weighting factors are chosen such that $\alpha \cong \beta >> \gamma$. This corresponds to the goal of a small error for the objectives (i) and (ii) and a larger tolerable error for objective (iii). The weight associated with the third objective depends on the number of beams used in a treatment plan. In the case of the OAR and the OHT regions, only the dose above the prescriptions are penalised.

Using this empirical technique of combining the objectives it is possible to determine the 'best' solutions for a given set of weights, but it does not necessarily guarantee to achieve the overall 'best' solution; implying that the choice of the weighting factors may itself be considered as an optimisation problem. To illustrate the importance of the choice of weighting factors in cost function (4) a standard test case is selected and the weighting factors are varied over a range of values. An ILS algorithm is run for 50 iterations to optimise the dose distribution given by a three field brick composed of two parallel opposed beams and one anterior beam. The beam configuration and the desired dose distribution is shown in Figure 2. The resulting solutions are ranked to obtain a pareto-optimal surface as shown in Figure 3.

Using the definition of inferiority as given in [5], the points marked 'x' in Figure 3 constitute the non-inferior solutions to the multiobjective problem and they form a surface in three-dimensional space. Note that the lowest errors in each of the three axes corresponds to the highest cost weightings.

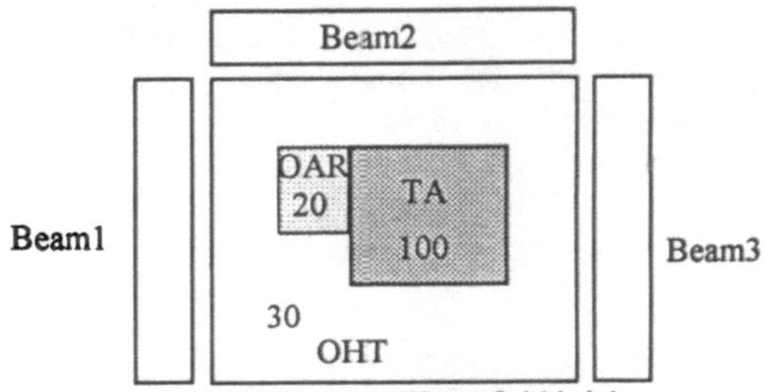

Figure 2: Three field brick.

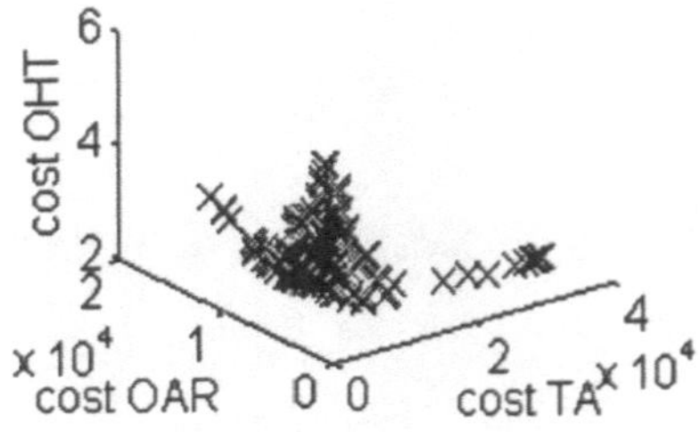

Figure 3: Pareto-optimal set

Genetic algorithm applied to RTP

Until recently most of the optimisation procedures used in RTP were derived from the deterministic approaches that are used in image processing, with the objective functions being represented in the frequency domain [1, 3, 6]. This duality can be explained by historical reasoning, as RTP interacts closely with computed tomography and thus image reconstruction techniques.

In this section the performances of a GA are compared with the ILS scheme studied in [8, 9, 10] when applied to standard test case problems. Two standard test cases are used: a hinged pair composed of one anterior and a left hand beam and; a three field brick, composed of two parallel opposed beams and one anterior beam. In the case of the three field brick the desired treatment plan is as given in Figure 2 and in the case of the hinged pair the desired dose distribution is shown in Figure 4. The influence of different GA parameters are studied, as well as the introduction of *a priori* knowledge into the optimisation procedure. The results given in Table 1 represent the performances of the various schemes in terms of mean dose, integral of absolute error (IAE) and root mean square value (RMS) of the error in the three areas of interest: TA, OAR and OHT. The results show that given the same initial conditions, a GA can perform at least as well as ILS. In each case the GAs are used with a population of 50 chromosomes. Each elemental beam intensity is constrained between the values 0.1 and 1.5 and is coded over ten bits using binary code. Each chromosome is 200 bits long for the hinged pair and 300 bits long for the three field brick (i.e. ten bits per elemental beam). The results for the hinged pair are given in columns 1 and 2 and for the three field brick in columns 3 and 4. The GA requires 10000 and 12000 iterations to achieve comparable results to those obtained from ILS in 50 and 150 iterations respectively. However it is believed that better results may be possible by combining the GA and ILS approaches.

The best results have been obtained with a high cross-over probability (90%) and low mutation probability, typically (1%). Since the number of parameters to be estimated is considered to be reasonably large (i.e. 20 and 30 for the standard test cases chosen), multiple point cross-over and uniform cross-over have been found superior to single point cross-over. Also tournament selection [12], with a tournament size of three has been shown in this particular case to lead quickly to better solutions than ranking selection mechanisms.

The ability of a GA to use empirical knowledge and in this case clinical knowledge on the overall beam intensities has been found to be quite useful. In particular, seedings in the form of approximations to solutions enable the introduction of *a priori* knowledge on likely solutions which has been found to speed up the GA. (It is noted however that if the seedings are too close to a solution the GA may become trapped in a local optima.)

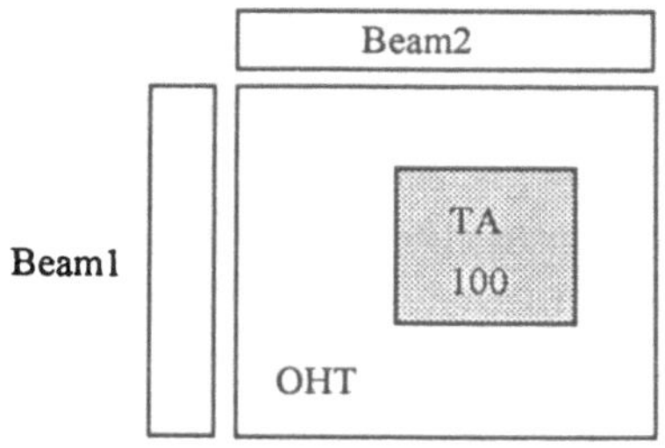

Figure 4: Hinged pair

$\alpha\,\beta\,\gamma$		3.5,0,0	3.5,0,0	3.5,3.5,0	3.5,3.5,0
No of iterations		50	10000	150	12000
No of beams		2	2	3	3
Scheme		ILS	GA	ILS	GA
	Mean dose	99.99	100.04	99.96	99.8
TA	IAE	0.35	0.33	0.86	0.6
	RMS	0.49	0.42	1.08	0.82
	Mean dose	NA	NA	20.14	22.19
OAR	IAE	NA	NA	2.24	2.98
	RMS	NA	NA	2.65	3.29
	Mean dose	36.47	32.81	37.35	32.39
OHT	IAE	23.28	22.8	30.39	27.8
	RMS	24.12	24.17	33.75	32.65

Table 1: Comparison of results

Formulation of the multiobjective problem.

When consideration is given to each of the objectives independently, the solutions can be represented as a pareto optimal set. Traditionally in RTP the clinicians are only interested in obtaining a small number of acceptable treatment plans. In order to determine these acceptable plans a decision maker needs to be used.

A decision maker linked to a more classical optimisation procedure allows decisions to be taken according to a wide range of criteria that needs to be previously defined by clinicians. It is necessary to identify the minimum requirements as well as the relative importance of a set of criteria relating to the patient condition and/or the type of disease to be treated. The decision maker must be able to have some degree of adaptability to accommodate different beam configurations, different patients and different diseases.

One approach would be to make use of MOGAs and to consider each objective independently. The selection of the population can be based on a ranking mechanism as described in [5] but with interaction from the decision maker. The decision maker can take many forms including goal programming, fuzzy logic or even another GA.

Conclusions

The paper has shown that genetic algorithms can be employed to solve standard test cases within the radiotherapy treatment planning problem. The use of genetic algorithms effectively negates the requirement for a mathematical model of the beam since no direct inversion is required, the problem being solved in a forward manner. In addition a wider choice of cost function criteria can be readily incorporated, thus making the approach a realistic option in radiotheapy treatment planning. Whilst the use of genetic algorithms has been demonstrated and shown to provide comparable results to the more traditional iterative least squares procedures, it is believed that further improvements may be possible by combining the two approaches into a single scheme. The latter observation forms the basis of on-going research in the field.

References

[1] Brahme A., Radiotherapy and Oncology, 12, 129, 1988
[2] Webb S., Phys. Med. Biol., 1689, 1991
[3] Kallman P., Lind B., Ecklof A., Brahme A., Phys. Med., Biol., 11, 1291, 1988
[4] Bortfeld T., Boyer A.L., Schlegel W., Kahler D.L., Waldron T.J., Proc. XIth ICCR, Manchester, UK, 180, 1994
[5] Fonseca C.M., Fleming P.J., Proc. Int. Conf. on Genetic Algorithms 416, (ed) Stephanie Forrest, 1993
[6] Holmes T. Mackie R.T., Phys. Med. Biol. 39, 91, 1994
[7] Svenson R., Kallman P., Brahme A., Proc. XIth ICCR, Manchester, UK, 56, 1994
[8] Haas O.C.L., Beam profile determination for conformation therapy, MSc dissertation, Coventry University, 1993
[9] Haas O.C.L., Burnham K.J., Fisher M.H., Mills J.A., Radiology & Oncology 94, British Journal of Radiology 67 (Congress Supplement), Work in Progress May 1994
[10] Haas O.C.L., Burnham K.J., Fisher M.H., Mills J.A., Proc. Int. Conf. Systems Engineering, ICSE' 94 Coventry, U.K., Vol. 1, 434, 1994
[11] Yuan Y., Sandham W.A., Durrani T.S., Mills J.A., Deehan C., Applied Sig. Process, 20, 1994
[12] Reeves C.R. (ed.), Modern heuristic techniques for combinatorial problems. Blackwell Scientific publication , 1993

AN EXPERIENCE BASED COMPETITIVE LEARNING NEURAL MODEL FOR DATA COMPRESSION

Dr Jianmin Jiang, *MIEEE, MIEE, CEng*

School of Engineering, Bolton Institute, Bolton BL3 5AB, United Kingdom

ABSTRACT: This paper presents a novel design of a multi-layer neural model which implements an experience based learning algorithm to train the neural network and achieve comparable data compression. This design takes 16 bits each cycle as the input to the network. A competitive learning based operation is then applied to the input to find the winner which exactly matches the 16 bits. Another hidden layer is further designed to produce various outputs corresponding to the different outcomes of the competitive learning. The output then controls the coder to complete the encoding. Finally, an experience based learning algorithm is developed to train the network to make the best use of the statistical information from input to achieve the highest possible compression.

1. INTRODUCTION

Lossless data compression requires statistical information to establish the best strategy or minimum length codes to reduce the redundancy for its input data. When the statistical information is gathered in the form of frequency counts, probability based model is often adopted to compress input files. Typical algorithms developed along this direction are Huffman coding, arithmetic coding and their variations. Another alternative for compression is to make use of the statistical information in the form of matching strings. In other words, a limited number of model strings are used to match input strings. The index of the model strings is then used to address the matching process for coding and decoding. This model is often referred to as dictionary based. Since Ziv-Lempel pioneered the idea of dictionary based compression, a whole family of various algorithms and numerous papers have been published. The most representative, however, is limited to LZW, LZFG algorithms. All of the algorithms in this nature has the feature which involves searching a dictionary updating the dictionary and encoding the index of matching.

Four sections are included in this paper. Except this section is presented as an introduction, section 2 describes the structure of the neural network. In section 3, the neural algorithm is discussed to outline its operation for data compression. The section 4, finally, gives the experimental results and conclusions.

2. NEURAL NETWORK STRUCTURE

The neural network structure can be illustrated in Figure 1. It consists of four layers. One input layer, one output layer and two hidden layers. The input layer will accept

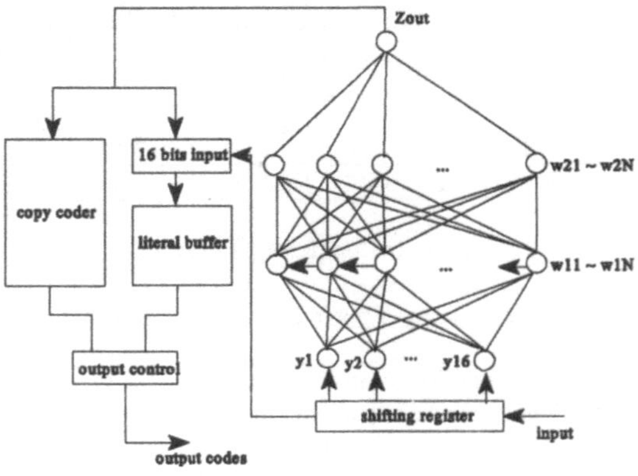

Figure 1

16 bits each basic cycle, and forward them to the first hidden layer where a competitive learning is operated. For lossless data compression, the comparison has to be performed on the basis that whatever compared should be exactly the same. In other words, the weight of the winning neuron has to be exactly the same as the input. Otherwise, no winning neuron is selected. In this multi-layer model, it is designed to allow four different competitive outcomes, which will be described specifically in the learning algorithm. The second hidden layer is designed to process the four different outcomes and finally the output neuron controls the encoding of each input 16 bits. The learning rule is designed to train the network in a way that the weights stored in the first hidden layer could produce winners as many times as possible. Under this scheme, therefore, the network can be viewed as learning from the statistics of input data.

On the left hand side of Figure 1, the whole block describes the operation of encoding the outcomes of neural network. Basically, it consists of two major parts. One is copy coder, and the other literal coder. Similar to Ziv-Lempel style, the copy coder uses fixed length codes depending on the number of neurons to encode those winning neurons. When 128 neurons are designed in the network, for example, the fixed length copy codes will be 7 bits. In literal coder, only unmatched inputs are encoded which is triggered by the copy codes. This means that all unmatched input words will be kept in the literal buffer until a winning neuron is found. The literal words are coded in the way similar to Ziv-Lempel algorithms which have two fields. One identifies the literal codes and the other identifies the length of literal words. By words, it means the whole input bits appearing at the input layer of the network at each cycle. In the design presented in this paper, 16 bits are involved altogether.

3. THE ALGORITHM

Step 1. Given a set of input bytes: $\mathbf{x} \in \mathbf{R}^8$, two bytes are taken into the shifting register, $\mathbf{y} = \{\mathbf{x}_1, \mathbf{x}_2\} = \{\mathbf{y}_1, \mathbf{y}_2, \ldots \mathbf{y}_{16}\}$. Thus the input layer will have 16 bits ready to be forwarded to the processing layers.

Step 2. For the neuron i at the first hidden layer, the PE function is:

$$f_i = \begin{cases} 0 & w_{1,i} \neq y \\ 1 & 8\ \mathit{MSB\ bits\ of}\ w_{1,i} = 8\ \mathit{MSB\ bits\ of}\ y \\ 2 & 8\ \mathit{LSB\ bits\ of}\ w_{1,i} = 8\ \mathit{LSB\ bits\ of}\ y \\ 3 & w_{1,i} = y \end{cases}$$

where: $1 \leq i \leq N$ and $N = 2^M$.

Step 3. At the second hidden layer, all of the neuron weights are N dimensional, which are also fixed and can be specified as:

$$\begin{bmatrix} w_{2,1} \\ w_{2,2} \\ \vdots \\ w_{2,N} \end{bmatrix} = \begin{bmatrix} 1 & 2 & 1 & 2 & \ldots & 1 & 2 \\ 0 & 1 & 2 & 1 & \ldots & 2 & 1 \\ 0 & 0 & 1 & 2 & \ldots & 1 & 2 \\ & & \ldots & & & & \\ 0 & 0 & 0 & \ldots & 0 & \ldots & \end{bmatrix}$$

The PE function:

$$z_i = \begin{cases} 2 & \mathit{any}\ 3\ \mathit{is\ in\ the\ input\ vector} \\ 1 & \mathit{any}\ \{1\ 2\}\ \mathit{is\ in\ the\ input\ vector} \\ 0 & \mathit{else} \end{cases}$$

Step 4. At the output layer, the PE function is:

$$z_{out} = \begin{cases} 2 & \mathit{any\ element\ of\ the\ input\ vector\ is}\ 2 \\ 1 & \mathit{any\ element\ of\ the\ input\ vector\ is}\ 1 \\ 0 & \mathit{else} \end{cases}$$

Step 5. Learning rule:

The learning rule is designed as experience based, which means the learning procedure is dependent on the output. When the output value is '0', this means no match of the input 16 bits can be found from the first processing layer. Therefore the network is required to learn to take this into effect. The simplest design for this outcome is to shift the neuron weights left at the first hidden layer to allow the input 16 bits to be shifted into the right most neuron and meanwhile, the weight at the far left neuron is discarded. This mechanism is designed in the principle that every weight of 16 bits has N chances to be the winner. If it fails throughout the cycle, it has to be discarded on the ground that it is less likely to occur frequently in subsequent inputs until further justified. When the output value is greater than 0, which means one of the neurons is a winner by matching the input either fully or partly, therefore more chances should be given to allow the neuron stay longer. Various schemes are developed by a whole family of dictionary based data compression algorithms, which will be detailed by the full paper. At this stage, we simply hold the neuron without left shifting to give the neuron one more chance to stay in the network. In addition, more advanced rules can be developed in this regard to train the network to catch the best possible information and learn the statistics of input data. This could well be an important direction for further research.

Step 6. Output codes:

Three status are identified by the output values, 0, 1 or 2, which corresponds to no match, partly match and full match of the input word. Overhead codes, 10, 11, and 0, are designed to address the three different status respectively. For no match, the word is sent to the buffer to be output as literal codes. For partly match, the number of the first neuron is coded by M bits, and for full match, M bits are also used to address the corresponding neuron in the network.

4. EXPERIMENTS AND CONCLUSIONS

Extensive experiments are carried out for N=128 and 256. The value of M, therefore, is 7 and 8 respectively. The compression ratio, defined as the compressed output divided by the original input, varies from 30% to 70% depending on the type of input files.

Specifically, about six files are designed to test this algorithm. Its performance in terms of compression ratio are summarized in Table I. In comparison with other traditional algorithms it can be seen that the neural network algorithm is not the best. This is because the learning algorithm designed is based on simple comeptitive learning and experience based updating scheme, which is essentially similar to the dictionary based traditional algorithms. On the other hand, small redundancy within the process of addressing the dictionary or neurons is not reduced in the same way as LZFG in introducing PATRICIA tree structure to address copy words by tree nodes rather than fixed length coding. The results in Table I, however, does show improvement over traditional statistical algorithms

Table I. Compression Performance

File	Size/bytes	Neural Net	LZW	LZFG	Huffman	Arithmetic Coding
Graph	1445	0.59	0.65	0.51	0.58	0.64
Binary	24576	0.32	0.36	0.2	0.3	0.24
Image	10051	0.76	0.75	0.78	0.72	0.72
Text 1	46386	0.53	0.49	0.41	0.56	0.56
Text 2	4081	0.67	0.66	0.55	0.59	0.61
Source Program	16709	0.49	0.68	0.32	0.65	0.67

such as Huffman and arithmetic coding with 0-order Markov models.

The traditional algorithms compared in Table I include two typical dictionary based algorithms, LZW and LZFG, as well as two typical statistical compression algorithms which are Huffman coding and arithmetic coding. For arithmetic coding, only the 0-order Markov model is implemented [3]. Arithmetic coding with multiple Markov models is not considered due to its huge amount of computation and long time taken for the computation. The sample files for experiments are designed under the principle that they represent typical variations of practical inputs in computer networks.

From the discussions and experiments, the following conclusions can be drawn to summarize the research work presented in this paper.

- The learning rules and coding methods adopted in the design of neural network is mainly developed from existing dictionary based algorithms such as LZ77 and LZFG. The scheme actually splits the input message up into each word with 16 bits, and the word is used to train the network independently without considering the correlation between words, and the overall effect upon the network learning. Further research is required to improve the situation along this line.

- One of the main features in the neural model is that compression of input words can only be achieved by those winning neurons, and the winner is picked up on the basis that the comparison made is exactly the same as the input word. This nature makes it difficult for the algorithm to outperform the existing algorithms. One possible option is to consider the outcome of those unsuccessful comparisons and encode the difference as well to further reduce its redundancy.

- The novelty of this design of the multi-layer neural model is that it indicates a new direction for the application of neural networks to lossless data compression. In this paper, the simplest learning rule is adopted. On one hand, more advanced learning schemes developed by the family of dictionary based algorithms can be implemented and on the other, better and new algorithms can also be developed both in the framework of neural network technology and other new research areas, such as fuzzy and generic learning etc. In addition, the highly parallel processing feature of neural networks offers tremendous potentials for high speed real time data compression in both telecommunications and computer science.

REFERENCES

[1] Welch T.A. 'A technique for high-eperformance data compression', IEEE Computer, 17(6), June 1984, pp 8-19.

[2] Fiala, E.R. and Greene, D.H. 'Data compression with finite windows', Commun. of the ACM, 32 (4), April 1989, pp 490-505.

[3] Jiang, J. and Jones S. 'Parallel design of arithmetic coding', IEE Proceedings: Computers and Digital Techniques, Vol 141, No 6, November 1994, pp 327-333.

[4] Knuth D.E. 'Dynamic Huffman Coding' J. Algorithms 6, 163-180, 1985.

USING NEURAL NETWORKS FOR GENERIC STRATEGIC PLANNING

Ray Wyatt

The University of Melbourne, Parkville 3052, Australia

Abstract

This paper argues the scarcity of strategic planning software is due to Western philosophical traditions which see strategy hypothesising as a mysterious, intuitive process that resists analysis and computerisation. But progress is possible if one extracts, from the tactical planning literature, eight key, generic, strategy-evaluation criteria. An experiment then tests whether scores on such criteria can be used to power machine learning of overall strategy desirabilies, and whether such learning is better achieved using multiple regression analysis or a simulated neural network. Both methods were successful, but the neural network was clearly the most accurate. It therefore constitutes a promising basis for self-improving, strategic planning software.

Introduction

There is a plethora of available software which helps users to become better tactical planners. There is a very much smaller amount of available software which helps users to become better *strategic* planners. The latter usually works by asking a user to input strategy alternatives and to then score such alternatives on various strategy-evaluation criteria. It then somehow combines such criterion scores to advise the user what seems to be the best strategy [1].

But however much this a sorting, remembering and clarifying process might help the user, the software generates nothing other than what the user contributed. Hence this paper aims at having the computer generate something new of its own.

More specifically, it seeks to have software learn, from its previous users, how a strategy's criterion scores can be most validly combined to generate that strategy's overall desirability score. Of course, previous users would have to act as "trainers" of the software. They would need to observe which strategy the software recommends and correct this if it does not "seem" right. Such user input can be prompted using suitably-written interactive code. Hence, over time, the software will gradually learn to replicate the strategy-assessment processes of its user community, and it will make better and better strategy recommendations the more it is used.

Accordingly, this paper focuses on the mechanisms whereby a computer can actually learn to become a better and better recommender of strategies. To achieve such an ambition two things are needed - a good understanding of the strategic planning problem and a careful experiment to see whether computerised learning of the strategy-evaluation process is actually possible. What follows will therefore be divided into two major sections - "the strategic planning problem" and "the experiment".

The Strategic Planning Problem

Planning is here defined as deciding what to do in the future. Some psychologists refer to this as "the formulation of behavioural projects" [2]. But our focus is not on planning when there is just one, or a few goals, such as in company planning where there is usually one goal - profit, or production planning where there are usually two goals - fast completion time and low resources usage. The focus here is on planning where there are multiple, simultaneous goals [3].

Such situations are often socially sensitive and politically delicate because of all the actors involved, along with all their hidden agendas [4]. Hence planning within such contexts can be extremely problematic, and so it has been described as a "wicked" problem [5]. Not only are there multiple, conflicting, simultaneous and hidden goals, but there might also be incomplete knowledge about how to achieve such goals. Moreover, ignorance about possible alternatives to the identified goals might be profound, and the goals themselves might change as planning proceeds.

It is therefore little wonder that social science has come up with very few tools to help humans perform such a daunting task. But if it could be performed just a little better, we might make some headway towards solving the massive problems of our time. These include inequality, justice, hunger, industrial relations, energy policy, war and peace. How well we solve such problems will determines our future quality of life.

But "hypothesising" what ought to be done is usually seen as a mysterious, cerebral activity which is resistant to improvement by technology. It appears to be a uniquely human activity powered by those ill-defined beliefs and desires which reside at the bottom of people's psyches. Therefore, we have only developed techniques to help us decide *how* to do things. Hypothesising *what* to do in the first place has long been regarded as non-technical.

For example, Aristotle divided human activity into two types : thinking and doing. Thinking was concerned with contemplating our place in the cosmos. Doing was concerned with practical things, like tactical plan-formulation. Hypothesising what to do, so that it could later be formulated into a tactical plan of action, was part of thinking, not doing.

Such attitudes extended into the Renaissance and so presented philosophers of the day with a dilemma. If the "scientific method" comprised observing, hypothesising and verifying, how could the middle part, hypothesising, be mechanised into standard steps like observing and verifying can? Clearly, hypothesising is resistant to standardisation.

More recently, Popper has maintained that hypothesising is beyond the realms of rational analysis. Moreover, the German philosopher Habermas believes that human activity is based on three types of attitudes : the psychological, the political and the dramaturgical. That is, people's planning stems partly from how they see themselves, partly from their ideal socio-political system(s), and partly from their dramaturgical attitudes : their deepest beliefs and desires. It is these latter which drive hypothesising. It is they which are so resistant to investigation.

Hence in practice, comparisons between alternative, hypothetical strategies are usually ignored, or at least done only cursorily. It often seems more "practical" to planners to get started straight away with detailed, tactical plan formulation, and this is a root causes of why poor planning pollutes our planet [6].

An example should clarify this. Suppose there are two competing companies and one decides that the best strategic direction is to increase turnover. The second, by contrast, decides that the best overall strategy is to reduce costs. The first, turnover-boosting company will formulate tactical plans for product discounting, aggressive advertising and other promotional activities. By contrast, the second, cost-cutting company will plan for workshop re-tooling, office re-organisation and other cost-reduction measures.

The winner will probably be the company who has chosen the most appropriate initial direction - increased turnover or reduced costs. That is, it usually matters little whose tactical planning techniques are the best; high-quality strategizing is more important. Although the first company had a notion that increased turnover was best, the second company had an equally strong feeling about the wisdom of costs reduction. One of them was wrong [7].

Hence it seems wise to try to improve performance in the initial, strategic phase of the planning process - the phase where general strategic directions of intent are first chosen [8], [9]. We shall refer to this phase as "strategizing". With greater understanding of it, practitioners can be advised how consistently and sensibly they are performing it, Hence the overall quality of planning should be improved.

But strategizing software invariably finds it difficult to penetrate very deeply into the process [10], [11]. Hence any planner-assisting software tends to be just that - a mere assistant to planners. It is not a substitute for human thinking. It simply helps the human to better evaluate alternative directional thrusts.

But despite such apparent shallowness, such software might still improve strategic planning, for two reasons. The first is that humans have never managed to plan socially-sensitive situations very well on their own. The second is that none of the tactical planning techniques focuses back on the initial, strategizing task.

Now, if software is serious about comparing alternative strategies, it must know their relative importances. That is, it should know each strategy's "intensity" or "desirability", which is sometimes called its "valance". Valance is in fact approximated, in some unique way, by all tactical planning techniques. Hence it would seem advisable to look across all the tactical planning literature to see what valance criteria have been used.

Strategy-evaluation Criteria

For instance, those planners who always conceive of planning as a scheduling task will invariably focus on the "critical path" of required actions. They seek to speed up the completion times of such actions. Hence it would seem logical that "speed", or time in which a strategy can be achieved, might be an important consideration in (sub-conscious) strategizing.

Secondly, those planners who conceive of planning as a decision tree process always focus on "expected worth", which is the product of "probability" and "utility". Again, these two criteria, which we shall respectively call "likelihood" and "effectiveness" (at achieving the overall aim), will be adopted as plausible criteria for better evaluating broad strategies.

In the same way one can look at still other Operations Research techniques to extract still other concepts which intuitively seem useful for strategizing. For example, hill-climbing, optimisation techniques yield the concept of "marginal utility", or "returns for effort" or "marginal return". Moreover, those who plan by simulating the likely outcomes of proposed plans - the simulation modellers, are always very conscious of how different goals interact. They therefore learn about the "autonomy" of goals. Goals which depend for their success on the success of many simultaneous goals are not very autonomous; goals which require few simultaneous goals to be achieved are highly autonomous. Again, autonomy is probably another criterion which can be used in strategizing.

There are some planners who reject Operations Research on the grounds that it is too mechanistic to reflect wicked reality. For example, those who "satisfice" [12] believe that real world planners are frequently satisfied by the first plan they find that improves the current situation. But the odds of finding such a satisficing plan are proportional to the potential for improvement, or the "improvability" of each goal. Hence improvability will be adopted here as another strategy-evaluation criterion.

Finally, still other planners reject Operations Research in favour of "workshops" - collections of people who have some knowledge of the planning situation. Although such an approach is intuitive, non-rigorous and subject to dominance by dominant individuals, the workshops culture has still generated some useful strategizing concepts. These include the degree of difficulty of a goal, or its "easiness", and the degree to which achieving a goal does *not* hampers achievement of other goals - its "harmlessness". The latter concept actually has a rich pedigree, being known as "flexibility" in budget planning, "resilience" in ecological planning, "hedging" in financial planning and "robustness' in public policy planning.

Thus our all-purpose group of generic, strategy-evaluation criteria comprises -

speed, likelihood, effectiveness, marginal return, improvability, autonomy, easiness and harmlessness.

Although one would expect strategies which score highly on such criteria to be highly desirable, this might not always be the case. For example, planners who wish to appear "brave" might seek difficult rather than easy strategies and planners who wish to look like "winners" might seek already-satisfied rather than improvable strategies. That is, strategizing can be very complex, and so any strategizing software needs to "learn" about such complexity if it is to ever contribute something new of its own. Accordingly, it is to such learning that we now turn.

The Experiment

An experiment was set up to see whether indeed there is a "learnable" relationship between people's overall scores for strategies and their criterion scores for the same strategies. More exactly, about sixty people were given questionnaires which asked them to rate, on a one to five scale, the overall desirability of three strategies. The questionnaire also asked for strategy scores, again on a one to five scale, for speed, likelihood, effectiveness, marginal return, improvability, autonomy, easiness and harmlessness.

The relationship between strategies' overall scores and their criterion scores was then "learned" using both conventional statistical analysis and neural network software. From this it could be seen whether the best learning mechanism was conventional statistics or neural network technology.

More specifically, the author obtained three strategies, from an environmental expert. They all aimed at reducing the amount of car travel, and consequent air pollution, within Melbourne, Australia. The first was "dearer petrol". If fuel is made more expensive then, in theory, people will be discouraged from driving their cars. The second strategy was "more public transport" which, in theory, will entice people out of their cars and into trains, buses and trams. The third strategy was "denser suburbs", or limited sprawl which, in theory, reduces the need to travel and so reduces automobile use.
A randomly chosen first half of the subjects were asked to assign overall scores to strategies and then to assign criterion scores.

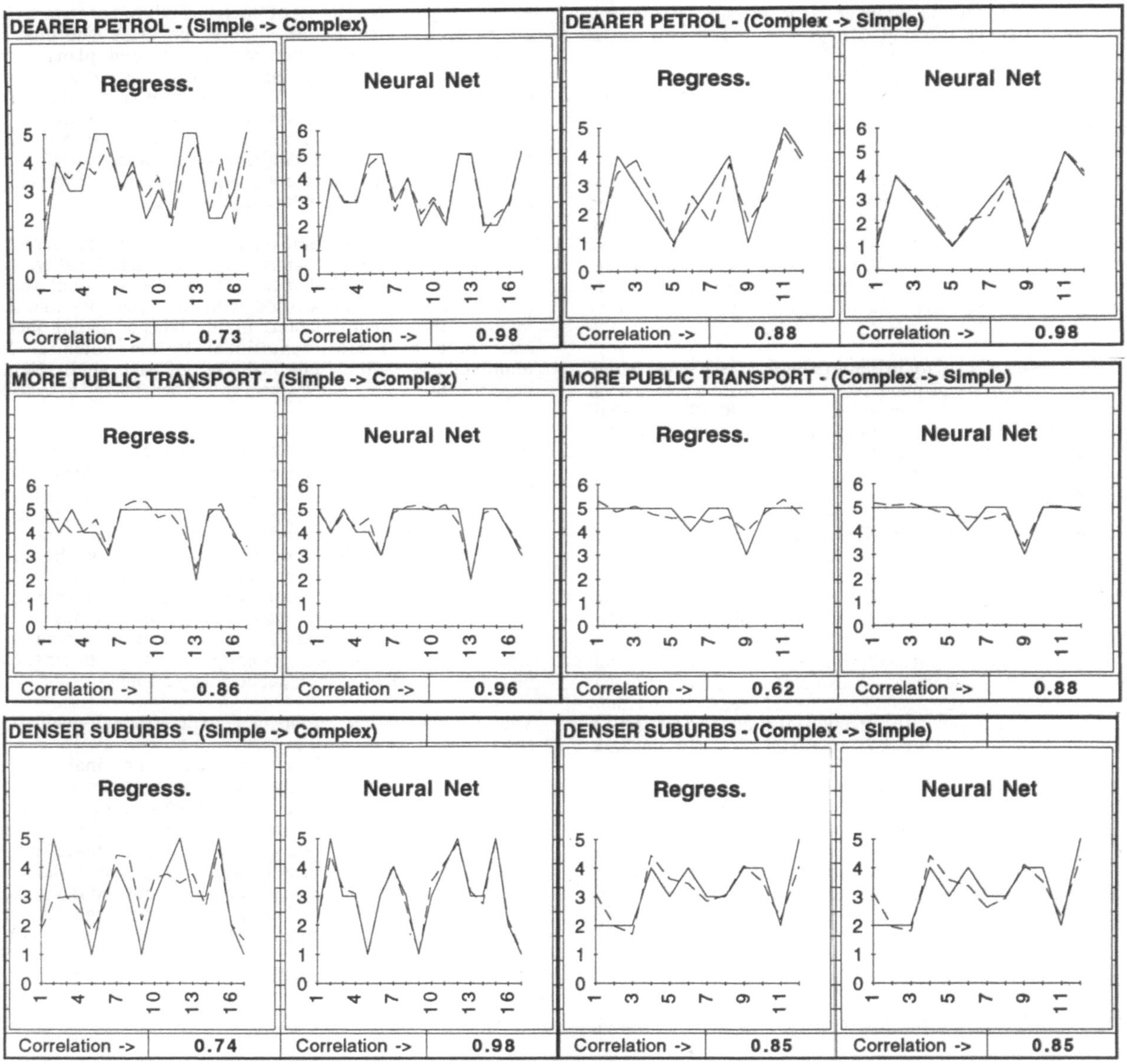

Figure 1 - Regression-based and Neural Network-based Learning

The other half were asked to assign criterion scores and then overall scores. Thus half of the respondents were asked to think of simple, holistic scores followed by detailed concepts - the "simple -> complex" group; the others were asked to think about complex criteria and then simple, overall scores - the "complex -> simple" group. At the time of writing, group sizes were seventeen respondents for the "simple -> complex" group, and twelve respondents for the "complex -> simple" group.

Tests were then performed to see whether the computer could "learn" to predict a strategy's overall score from its performance on the eight, strategy-evaluation criteria, and the first learning method used was multiple regression analysis. More specifically, the SYSTAT package was used, running on a 486 PC computer The dependent variable was overall score; the independent variables were criterion scores and one set of responses about one strategy, from one subject, represented one data point. Responses covering all three strategies were *not* lumped together because this would have meant that each person contributed three pieces of data, thereby violating regression's underlying assumption of independent data inputs.

Once the multiple regression equation of "best fit" was found for each strategy, each respondent's dependent variable (overall score) was written to a file along with its "residual". The latter was the distance, from actual value, of the regression line's prediction. These two values were then imported into an EXCEL spreadsheet, within which two columns were produced to respectively show "target" scores and "estimated" (regression-predicted) scores. The correlation between such target and estimated scores was then measured to see how well the multiple regression process had "learned" to predict overall score.

The second "learning" method was an artificial neural network extracted from a disk at the back of McClelland and Rummelhardt's book [14]. The back-propagation network option, incorporating a Sigmoid type transformation function, was chosen. There were, of course, eight input nodes - one for each of the eight criterion scores, and one output node - for overall score. One set of such scores, from each respondent, constituted one element of the "training set".

Several hidden node structures, involving both one and two layers, were experimented with. Satisfactory network convergence, to the 0.01 "total sum of squares" error level, was eventually obtained using just one internal layer containing four hidden nodes. The smallest number of training epochs required to achieve this was 887, and the largest was 2242. After convergence, each person's (overall) score, along with the neural network's output score was written to an ASCII ".log" file and later imported into an EXCEL spreadsheet. Here the "target" and "output" values could be directly compared to regression analysis' "target" and "estimated" values. Both sets of values were therefore charted.

Results

Such charts and correlations are shown above in Fig. 1. Both methods "learned" how to predict fairly well. Correlations between targets and regression-estimated scores ranged from 0.62 to 0.88. For the neural network they ranged from 0.85 to 0.98. Hence it is obvious that the neural network software was invariably a better predictor of overall strategy scores. Thus if any strategizing software is to learn how to predict overall strategy score, based on what a current user believes to be that strategy's criterion scores, a neural network trained on past users' responses is likely to be more promising than multiple regression analysis.

Such a result conforms to expectations. It is extremely unlikely that the relationship between criterion scores and overall score is so simple that it is "understandable" by multiple regression and its restrictive assumptions about linearity and data continuity. It is far more likely that the mysterious, sub-conscious, strategizing process is plagued by non-linearities, discontinuities and threshold phenomena, all of which are better accommodated using neural networks rather than conventional statistics. Hence neural networks seem to have more potential for use in self-improving strategizing software.

Conclusion

This paper attempted to explain how Western philosophical traditions have discouraged the mechanisation, and hence the software-guided improvement of strategizing. It then argued that some progress in this area might be possible by extracting, from the tactical planning literature, eight key, generic, strategy-evaluation criteria.

An experiment was then conducted to find the best way that a computer might learn how to predict overall desirability from strategies' scores on such criteria. The neural network approach was shown to be clearly superior for this purpose. Thus we have improved the prospect of developing some strategizing software which keeps improving by "learning" from more and more of its past users.

References

1. Friend, J. :"New Directions in Software for Strategic Choice", Eur. J. of OR 61, 154 1992.
2. Nuttin, J. : Motivation, Planning and Action. Lawrence Erlbaum : Hillsdale, New Jersey 1984.
3. Wilensky, R. : Planning and Understanding. Addison-Wesley : London 1983.
4. Sillince, J. : A Theory of Planning. Gower : Aldershot 1986.
5. Rosenhead, J. (ed) : Rational Analysis for a Problematic World. John Wiley : Chichester 1989.
6. Wyatt, R. : Intelligent Planning. Unwin : London 1989.
7. Reger, R.K., Huff, A.S. : "Strategic Groups : A Cognitive Perspective", Strategic Management J. 14 1, 103 1993.
8. Fabian, J. : Creative Thinking and Problem Solving. Mich Lewis : Chelsea 1990.
9. Primozic, K.I., Primozic, E.A., Lefen, J. : Strategic Choices. McGraw Hill : New York 1991.
10. Fetzer, J.H. : Artificial Intelligence : Its Scope and Limits. Kluwer : Dordrecht 1990.
11. Collins, H. : "Will Machines Ever Think?, New Scientist June 30, 36 1992.
12. Simon, H.A. : Reason in Human Affairs. Stanford University Press : Stanford 1983
13. Eden, C. (1992), "Strategy Development as a Social Process", J. of Management Sce. 29 6, 799 1992
14. McClelland, J.L., Rumelhart, D.E. : Explorations in Parallel Distributed Processing. MIT Press : Cambridge, Mass. 1988.

NEURAL NETWORK TRAINING COMPARED FOR BACKPROP QUICKPROP AND CASCOR IN ENERGY CONTROL PROBLEMS

Alexei N. Skurikhin

Mathematics Department, Institute of Physics
and Power Engineering 249020 Obninsk, RUSSIA
ippe.asku@rbv.rosmail.msk.su Fax: 7 095 230 23 26

Alvin J. Surkan

Department of Computer Science and Engineering,
University of Nebraska Lincoln, NE 68588-0115 USA
surkan@cse.unl.edu Fax: 402-272-7767

Abstract

Gradient-based neural network learning algorithms, specifically the backpropagation (BACKPROP), quickpropagation (QUICKPROP), and cascade correlation (CASCOR), have been re-implemented, tested, and compared for making predictions of time series for daily energy (gas) consumption. Daily records of energy consumption form a time series with information that can predict energy demand for a few days. Initially, the basic prediction is made from the past series without taking external influences into consideration. Later, other independent predictive data series can be included to improve prediction. Repeated presentation of hundreds of sample patterns systematically trains a network which converges upon connection configurations that predict expected demand. It has been demonstrated that the learning algorithms can consistently train networks to predict, one-day-ahead over the succeeding year, with the average error reduced to below 20%.

Key words: neural network, training, prediction, cascade correlation, energy

Introduction

Industrial decision makers often depend on energy predictions based on the time series consisting of sequences of daily values. Important sources of time series include electrical power and natural gas utilization. Such energy series result from interactions in complex systems of commercial and residential energy consumption over large geographic areas and of inter-area exchanges. BACKPROP, QUICKPROP, CASCOR and other feed-forward architectures provide no short-term memory mechanism in their networks.

At any instant the outputs are assumed to depend only of the current inputs and the network's weights. This assumption was made for simplicity and to improve control in the testing and comparing the applicability of these three gradient-based, learning algorithms for predicting energy consumption. Additional experiments developed recurrent gradient-based architectures and simulated annealing algorithms.

Basics of BACKPROP, QUICKPROP and CASCOR

Currently, many systems of neural networks use some form of the backpropagation algorithm [Rumelhart, 1986] for learning. Most learning algorithms use variants of the gradient-descent method. During training the network weights are iteratively modified to minimize, the overall mean square error E between desired and actual output values for all output units over all input patterns. The errors and weight updates are calculated from:

$$E = \frac{1}{2} \sum_{t=1}^{T} \sum_{j=1}^{N} \left(D_j(t) - A_j(t)\right)^2 \tag{1}$$

$$\Delta W(t) = -\epsilon \frac{\partial E}{\partial W(t)} + \alpha \ \Delta W(t-1) \tag{2}$$

where $\partial E/\partial w(t)$ is the error derivative for that weight. Usually, back-propagation learning is too slow. The QUICKPROP algorithm [Fahlman, 1988] improves the basic back-propagation algorithm. It manages the problem of step-size for back-propagation types of learning algorithms. QUICKPROP computes the partial derivative $\partial E/\partial W(t)$ just as in standard BACKPROP, but then instead of using simple gradient descent, it updates the weights using second-order differences in a variant of Newton's method. The BACKPROP method assumes: (1) the error function differentiated with respect to each weight is locally quadratic, and (2) small changes in one weight have little effect on the error gradient observed at other weights. The computation uses only the information local to the weight being updated:

$$\Delta W(t) = \frac{\frac{\partial E}{\partial W(t)}}{\frac{\partial E}{\partial W(t-1)} - \frac{\partial E}{\partial W(t)}} \tag{3}$$

The CASCOR algorithm [Fahlman, 1990] has several advantages over BACKPROP and QUICKPROP. CASCOR automatically determines the size and topology of the network. This method also preserves the structure as it builds even after subsequent changes in the training set. It does not depend on error back-propagation for training. CASCOR is a supervised learning method that, during training, tends to build a network with a minimal topology. Initially, the network has only inputs and output units with the connections linking them.

The single layer of connections is trained using the QUICKPROP algorithm through error minimization. When the error level stops decreasing, the network's performance is evaluated. Training halts when the error level is acceptable. Otherwise, a new hidden unit is added to the network in an attempt to reduce the residual error. Prior to introducing a new hidden unit, the algorithm begins with a pool of candidate units. Each candidate receives inputs through weighted network connections from the input patterns and from any hidden units already present. At this stage, the outputs of candidate units still remain to be connected into the active network. Multiple passes through the training set adjust incoming weights for each candidate unit. Adjustment, is aimed at maximizing the correlation between the network's output and the residual error in the active net.

$$\text{Correlation} = \left| \sum_{I} \left(A(t) - \bar{A} \right) \bullet \left(E(t) - \bar{E} \right) \right| \quad (4)$$

where: $\bar{A}$ and $\bar{E}$ are the values **A** and **E** averaged over all patterns. To maximize this correlation, QUICKPROP computes partial derivatives with respect to each of the candidate unit's incoming weights. When the correlation stops improving, the best candidate unit is selected and the input weights are frozen. That unit is added to the network as one of the weights leading to the output units, along with the weights leading to it. This process of adding a new hidden unit and re-training the output layer continues until the error is negligible or the training is halted by some other criterion.

Preprocessing Observed Input Data

To scale the data, a simple linear mapping of the variable's observed extremes onto the network's practical operating limits was used. Data values are mapped either to the range (0.1 to 0.9), or (-0.5 to 0.5), or (-0.9 to 0.9), depending on which activation function was being used. The learning algorithm used the errors accumulated over batches of input training patterns. Before beginning the training of a network, the data records were partitioned into three non-overlapping sets. The first set was used for training, the second in selecting a better trained network, and the third set for testing of selected better performing network. The first data were daily values of energy consumption over the three year period from 1987 to 1989. The second data were daily energy consumption values from 1990, and the third were the same series of values from 1991.

Results of Simulation Experiments

The effects of different activation functions were tested before proceeding with an extensive series of simulation experiments. Preliminary tests showed that the preferred combination of activation functions included a hyperbolic tangent as the hidden node nonlinearity. This nonlinearity was used in conjunction with linearity at the output nodes. This combination of nonlinear and linear activation functions produced the fastest convergence. Table 1 compares the average number of iterations necessary for the three algorithms to attain nearly the same asymptotic accuracy of 19.5% over the training set.

TABLE 1
Results giving the average number of iterations (for 10 trials) for the network when using a hyperbolic nonlinearity at the hidden layer and a linear function at output layer.

For networks with a hidden layer of two nodes single linear nodes at both the input and output		
BACKPROP	QUICKPROP	CASCOR
5200	3500	250

The training process always converged in fewer than 600υ iterations. As expected, the QUICKPROP and CASCOR algorithms learn significantly faster than did BACKPROP. Of the three algorithms, CASCOR was the fastest at learning. Moreover, CASCOR eliminated the need for advance guessing of the network's size, depth, and topology. The CASCOR algorithm automatically builds a small network.

Experimentally it was found that networks trained by different algorithms achieved comparable accuracy when provided the same number of hidden units. Increasing the number of hidden units never results in any additional improvement. Also, artificial input signals such as cos(x) introduced as season indicators were found not to be useful as supplementary inputs. Also, no improvement resulted when the first differences were introduced as supplementary inputs. Figure 1 provides a graph which compares the predicted values and those values known for the test data.

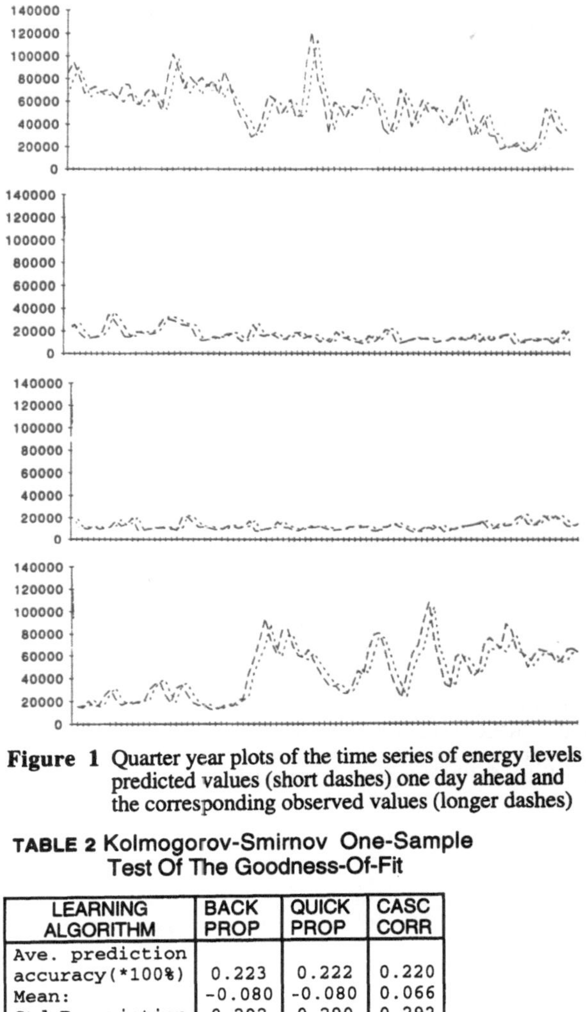

Figure 1 Quarter year plots of the time series of energy levels predicted values (short dashes) one day ahead and the corresponding observed values (longer dashes)

TABLE 2 Kolmogorov-Smirnov One-Sample Test Of The Goodness-Of-Fit

LEARNING ALGORITHM		BACK PROP	QUICK PROP	CASC CORR
Ave. prediction accuracy(*100%)		0.223	0.222	0.220
Mean:		-0.080	-0.080	0.066
Std Deveviation		0.292	0.290	0.292
ESTIMATED	D+	0.049	0.049	0.055
KOLMOGOROV	D_	0.091	0.089	0.091
STATISTICS	Dn	0.091	0.089	0.091
Significance		0.005	0.006	0.005

Statistical Study of the Prediction Results

Statistical calculations on the errors to compare the distribution of errors in residuals or prediction residuals with those expected for a normal distribution. The integrated periodogram verifies that error values from the time series prediction are random because the cumulative sum of the periodogram ordinates are within the expected 75 and 95% Kolmogorov-Smirnov bounds for the ordinates from a uniform distribution.

Greater refinement in the examination of residuals and their histograms is obtained with the use of the chi-square and Kolmogorov-Smirnov one-sample tests. A study of the figures 2, 3, 4 and table 2 reveals that the histograms of residuals are not consistent with those of normal distribution. This is true even after training was continued to the point at which improvements can no longer can be obtained.

The difference of the error residuals from those expected to correspond with the ideal normal is explained largely by the absence of memory in the algorithms being evaluated. This is confirmed by the results when the three algorithms were applied to the same data. The classical ARIMA model was also run to get some benchmark results which had an accuracy of 19.6%. This was almost the same (insignificantly better) as that achieved with the three compared neural network algorithms which gave a best accuracy of 21%. It was also found that the histogram or error residuals from ARIMA was more normal in shape than the best of the others. Similar results were obtained by the authors [6][7] when the results of a memory neuron network model. This gave an error level as low as 17% which was better than the best of the three neural network models of this study.

BACKPROP STATISTICAL RESULTS

Lower Limit	Upper Limit	Obs. Freq	Exp. Freq	CHISQR Value	FREQUENCY Obs = W Exp = B
-inf	-0.7	9	6	1.3580	
-0.7	-0.6	4	7	1.6219	
-0.6	-0.5	16	14	0.3862	
-0.5	-0.4	25	22	0.3203	
-0.4	-0.3	14	32	0.4421	
-0.3	-0.2	25	42	6.7762	
-0.2	-0.1	63	48	4.6091	
-0.1	0.0	60	49	2.3444	
0.0	-0.1	46	45	0.0275	
0.1	0.2	49	36	4.3434	
0.2	0.3	33	26	1.7008	
0.3	0.4	10	17	2.8305	
0.4	0.5	9	10	0.0488	
0.5	+inf	1	9	6.6695	

Chisquare = 43.48 for 11 deg. fr Sig. = 0.00000896

Figure 2 Histogram of the residuals and their Chi-square values for counts of prediction errors in gas sendout predicted by a BACKPROP trained network with a (1:2:1) structure. The histogram shows the occurence frequencies observed and predicted for the categories given at left.

QUICKPROP STATISTICAL RESULTS

Lower Limit	Upper Limit	Obs. Freq	Exp. Freq	CHISQR Value	FREQUENCY Obs = W Exp = B
-inf	-0.7	9	6	1.5931	
-0.7	-0.6	4	7	1.5332	
-0.6	-0.5	15	14	0.1515	
-0.5	-0.4	23	22	0.0262	
-0.4	-0.3	17	32	7.3223	
-0.3	-0.2	27	42	5.3472	
-0.2	-0.1	60	48	2.8008	
-0.1	-0.2	61	50	2.6545	
.0	0.1	46	45	0.0178	
0.1	0.2	49	37	4.2683	
0.2	0.3	33	26	1.7158	
0.3	0.4	10	17	2.7663	
0.4	0.5	9	10	0.0341	
0.5	inf	1	8	6.4540	

Chisquare = 36.69 with 11 degees f. Sig. = 0.00013

Figure 3. Histogram of residuals and their Chi-square values for counts of errors in gas sendout predictions by a QUICKPROP trained network with a (1:2:1) structure. The histogram shows the occurence frequencies observed and predicted for the categories given at left.

CASCOR STATISTICAL RESULTS

Lower Limit	Upper Limit	Obs. Freq	Exp. Freq	CHISQR Value	FREQUENCY Obs = W Exp = B
-inf	-0.7	10	5	3.845	
-0.7	-0.6	2	7	3.411	
-0.6	-0.5	15	13	0.417	
-0.5	-0.4	23	21	0.183	
-0.4	-0.3	15	31	8.271	
-0.3	-0.2	29	41	3.368	
-0.2	-0.1	52	48	0.414	
-0.1	-0.2	56	49	0.864	
.0	0.1	53	46	1.137	
0.1	0.2	47	38	2.278	
0.2	0.3	43	28	8.488	
0.3	0.4	9	18	4.551	
0.4	0.5	9	11	0.215	
0.5	+inf	1	10	7.607	

Chisquare = 45.05 with 11 degrees f Sig = 0.0000048

Figure 4 Histogram of the residuals and their Chi-square values for counts of errors in ranges of gas sendout predicted by a CASCOR-trained network with eight hidden nodes. The histogram shows the occurence frequencies observed and predicted for the categories at left.

Distributions of Residuals Predictions Errors from BACKPROP, QUICKPROP & CASCOR Training

The residual prediction of one-day-ahead value errors from networks trained by BACKPROP, QUICKPROP and CASCOR have histograms which are qualititatively and quantitatively very similar. For all three, the smallest (middle) and most negative (leftmost) error categories in the predictions have almost the same frequencies. These were generally all higher than would be expected with residuals distributed normally. The frequency of occurrence of the errors resulting from prediction was lower than expected in the moderate-sized categories and also for those errors whic fell in the largest positive categories.

Comparison of Results with the ARIMA Benchmark

For comparison purposes the ARIMA model was applied to the same data. The results show that the accuracy achieved by this benchmarking model is effectively the same (at 19.65%) as the highest accuracy obtained from BACPROP, QUICKPROP, or CASCOR(the best being 21%) methods. Also, the histogram of error residuals is closer to a normal distribution than those obtained by the other three algorithms in the comparison. Similar results were obtained when the same three methods were compared with a memory neuron network [6,7] which gave a best accuracy of 17%

Observations and Conclusions

An error reduction of 10-fold has been realized after training the network with fewer than 6000, 4000, and 400 iterations of the BACKPROP, QUICKPROP, and CASCOR algorithms, respectively. The most important finding was that one input signal is sufficient only for achieving an asymptotic accuracy represented by an error that does not exceed 21%. Further improvements did not result from any other modifications of the three training methods. To improve accuracy, it is necessary to provide some other independent and predictive input parameters to the neural network.

CASCOR learns much faster than does BACKPROP and QUICKPROP while at the same time it automatically derives the network's structure. Statistical calculations made on the residual errors produced histograms that did not approximate normal distributions even at the point when further improvements of trained networks by these three selected learning algorithms had ceased. The final accuracy became fixed and asymptotic for each combination of training algorithm and input data sample.

Simulations showed CASCOR was the preferred algorithm, It took less training time to achieve the same accuracy than did the two other algorithms. Also, there is no need to define the network structure before starting. Also, statistical tests showed that the frequency distributions of the residuals after CASCOR was superior while after learning by BACKPROP and QUICKPROP they were inferior. CASCOR is thus recognized as a fast and approximate network training method.

Suggested Directions For Future Research

Extended studies of all three of these training methods are indicated. Experiments are planned for exploring network performance by adding new independent input parameters to the networks. It should be informative to test the performance of the BACKPROP, QUICKPROP, and CASCOR algorithms relative to that of the simulated annealing-based algorithms such as Alopex which was recently reported by Unnikrishnan et al [4][5].

This can be explained by the fact that the algorithms considered had no memory, (i.e. did not have any mechanism for representing time-delay effects. This indication of further research is consistent with the results from the application of algorithms with memory to the same data. Future research is planned on the development of hybrid techniques that combine the strengths of genetic algorithms or fuzzy logic or conventional approaches with artificial neural networks with and without the incorporation of memory mechanisms in the neural networks structure. Also, this system can be extended to applications with multi-variate time series.

Acknowledgments

The authors acknowledge Scott Fahlman's papers for the descriptions of the QUICKPROP and CASCOR algorithms.

References

[1] Speed in Backpropagation Networks", June 1988, Technical Report, CMU-CS-88-162. Computer Science Department, Carnegie Mellon University, Pittsburgh, PA.

[2] Fahlman, S.E., Lebiere, C. (1990) "The Cascade-Correlation Learning Architecture", February 1990, Technical Report, CMU-CS-90-100. Computer Science Department, Carnegie Mellon University, Pittsburgh, PA.

[3] Rumelhart, D.E., Hinton, G.E., and Williams, R.J. (1986) "Learning Internal Representations by Error Propagation", In: Parallel Distributed Processing, MIT Press, Vol.1, pp.318-362.

[4] Unnikrishnan, K. P. and Venugopal, K. P. "ALOPEX: A correlation-based learning algorithm for feed-forward and recurrent neural networks" R & D Publication GMR-7919, GM Technical Center, MI. (1993)

[5] Unnikrishnan, K. P. and Venugopal, K. P. "Learning in connectionist networks using the ALOPEX algorithm" Proceedings of the International Joint Conference on Neural Networks IJCNN Part I, pp. 926-931. (1992)

[6] Surkan, A.J., Skurikhin, A.N. "Memory neural networks applied to the prediction of daily energy usage," Proceedings of the World Congress on Neural Networks - San Diego, June 5-9, Volume 2, pages 254-259, (1994) Publisher: Lawrence Erlbaum Associates, Hillsdale, NJ

[7] Sastry, P.S., Santharam, G. and Unnikrishnan, K.P. "Memory neuron networks for identification and control of dynamical systems", R&D Publication GMR-7916, GM Technical Center, MI (1993)

APPLICATION OF NEURAL NETWORK FOR HOME SECURITY SYSTEM

Turgay IBRIKCI, and V. KRISHNAMURTHI

CUKUROVA UNIVERSITY, Faculty of Engineering. & Archicture.,
Department. of Electrical & Electronics Engineering, 01330 Adana, TURKEY.
01330 Adana-TURKEY

Abstract

In this study, the application of Artificial Neural Network (ANN) for home security system (HSS) is investigated. The HSS consists of a neural network with input, hidden and output layers. The HSS detects the presence of unidentified person in the house and initiates an alarm. It has its own fault-diagnostic algorithm. In case any system component fails, it gives an alarm which is different from the other, to indicate the system failure. The minimum number of nodes, the optimum number of hidden layers, the speed, the accuracy and learning time for back propagation (BP) algorithm [1] are investigated in the present paper.

Introduction

The past two decades have witnessed the exponential growth of research and development works in the field of Artificial Neural Network. The ANNs are capable of performing many human-like tasks and exhibit intelligent abilities such as learning, forgetting, parallel processing, pattern recognition, self organization; generalization and abstraction [2-3]. The neural networks function as mathematical models of biological brain for analysis and comparison of its activities. There are many neural network applications that can control costs, increase accuracy and consistency, and use resources efficiently[2].

Neural Network Architecture

Neural Networks are large networks of simple processing elements or nodes. Nodes are simplified models of neurons. They process the information dynamically in response to external stimuli. The knowledge in neural network is distributed throughout the network in the form of internode connections and weighted links which form the inputs to the nodes. Fig. 1 shows a generalized multilayer ANN.

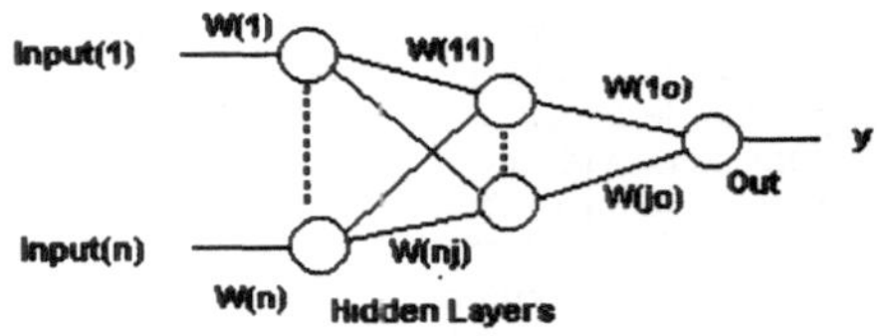

Fig. 1 Generalized Multilayer ANN

Home Security System(HSS)

The Neural Network is proving to be useful tools in problem solving. Fig. 2 shows the home security system (HSS). It is composed of two subsystem SS1 and SS2. The subsystem SS1 consists of a computer(A1) which has the room scene embedded in its memory, relay (B1), alarm (C1) and a battery(D1) that activates the relay and alarm. The second subsystem SS2 consists of a relay (B2), battery(D2) and alarm(C2). V is the video camera which transmits the room scene to the computer.The computer(A1) is vital element of the HSS.

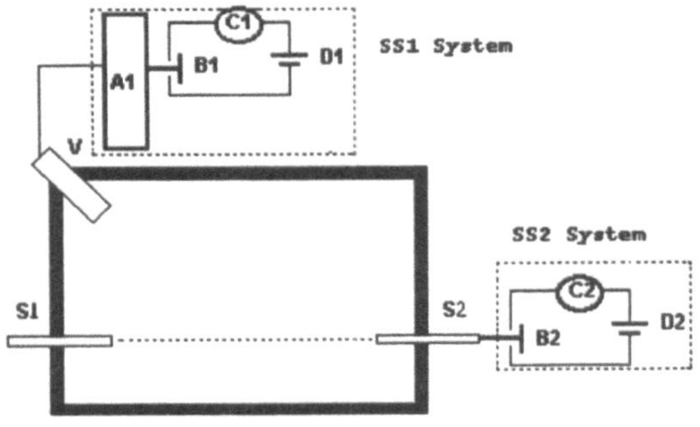

Fig. 2 Model of Home Security System

Logic Structure

The goal of logic structure is to create a functioning neural network for HSS which provides the most consistent and robust security mechanism. The logic diagram of HSS is shown in Fig 3. We can formulate the equation for a correct operation as

$$OUT = (A1 * B1 * C1 * D1) + (A2 * B2 * C2) \quad (1)$$

Then the failure equation may be written as

$$\overline{OUT} = \overline{(A1*B1*C1*D1) + (B2*C2*D2)} \quad (2)$$

$$= \overline{(A1*B1*C1*D1)} * \overline{(B2*C2*D2)} \quad (3)$$

Failure means an unidentified visitor in the room without sounding an alarm. Failure of circuit elements V, S2, D1 and D2 would cause the alarm to sound, if the rest of the circuit work. Thus, the anamoly will be immediately known as soon as it takes place.

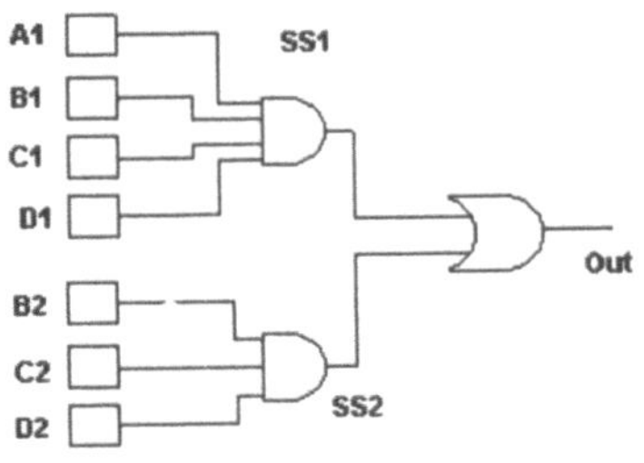

Fig. 3 Logic Diagram

The Neural Network Implementation

The architecture of NN for the HSS is shown in Fig. 4. It has one input layer, two hidden layers and an output layer . There are seven inputs that are A1,B1,C1,D1,B2,C2 and D2.

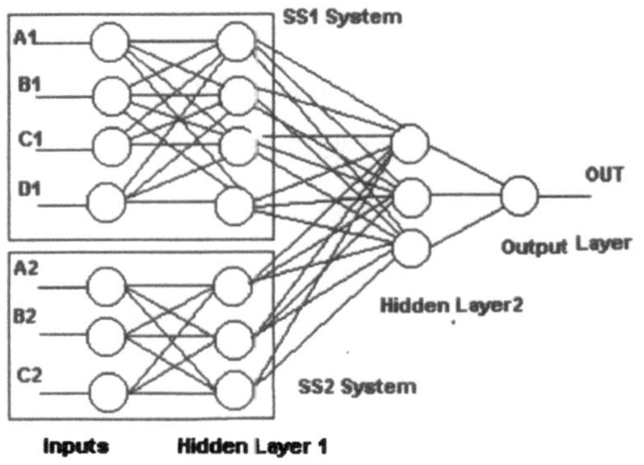

Fig. 4 The Neural Network structure of The HSS

They are grouped into two, one having four and the other having three nodes. The first hidden layer has seven nodes and it is also grouped into two as the input layer. All the four nodes in the first group of the input layer are connected to the four nodes of hidden layer 1 and similarly for the second group, so that the input layer partly connects with the first hidden layer. The first hidden layer fully connects with second hidden layer that has three nodes. The second hidden layer is connected to the output layer. The information from ANN is available to outside world through the output node.

The Stages of Learning Algorithms

For this HSS, we applied BP proposed by Rumelhart, Hinton, and Williams using generalized delta rules [1]. The rule is based on the concept of continuously modifying the strengths of the connections to reduce the difference between the desired and current output value of processing element. The main purpose is to find correct weight values from the initial random weights assigned. The algorithm adapts the beginning weight values to correct weights.

There are two stages of passes called forward pass and backward pass.

Forward Pass

Step 1 : Initialize all the 49 connection weights by randomly choosing them between 0 and 1.

Step 2 : Apply all possibilities of the input patterns at the input layer. For each input pattern,the output is obtained by

$$NET = \Sigma \text{ weight}[n] * \text{input}[n] \quad (4)$$

$$OUT = 1 / (1 + e^{-NET}) \quad (5)$$

where out is the actual output.

Backward Pass

Step 3 : Calculate the error δ and the error derivative $D\delta$ at the output according to the equations

$$\delta = OUT*(1-OUT) *(TARGET - OUT) \quad (6)$$

$$D\delta = OUT * (1 - OUT) * \Sigma \delta * W(n)) \quad (7)$$

where TARGET is the desired output.

Step 4 : Change each weight according to following equations:

$$\Delta W(n) = \eta * D\delta * OUT + \alpha * \Delta W(n-1) \quad (8)$$

$$W(n+1) = W(n) + \Delta W(n) . \quad (9)$$

$\Delta W(n)$ is the increment of weight rate

η is training-rate coefficient value.

α is smoothing coefficient value.

Step 5 : Repeat this operation until the error is within the permissible error which is 0.01.

Results And Conclusion

We have investigated the HSS with two and three hidden layers. In the case of two hidden layers, the first hidden layer had 7 nodes and the second one 3. In case of three hidden layers, the number of nodes were 7 , 3 and 2 respectively.We found that the maximum suceessful results were obtained with two hidden layers . With three hidden layers, complexity had increased, and the learning time had increased too.

Determining learning rate is an important task in neural network and operation must be convergent. In this study,we invetigated with several combination values of η and α . We found that smoothing and training coefficient values for efficient training are $\eta = 0.8$, $\alpha = 0.7$ for our application.

Acknowledgment

The authors wish to thank Dr. A.Hamit Serbest for his encouragement and support.

References

1. D.E. Rumelthart, Hinton, and R.Willaiams : Parallel Distributed Processing Vol I. Chap. 8. (1990)
2. David L.Bailey and Donna Thompson: Developing Neural Network Applications AI Expert 34 (1990)
3. David L.Bailey and Donna Thompson : How to Develop Neural Network , AI Expert, June 1990, pp 36 - 47.

APPLICATION OF GENETIC ALGORITHMS IN SLIDING MODE CONTROL DESIGN

N.H. Moin, A.S.I. Zinober and P.J. Harley
University of Sheffield, England

ABSTRACT

A Genetic Algorithm (GA) is a stochastic adaptive algorithm whose search method is based on simulation of natural genetic inheritance and Darwinian striving for survival. The GA has been adapted to study the problem of designing a stable sliding mode which yields robust performance in variable structure control systems. For various cases, we show that GA is viable and has great potential in the design of sliding mode control systems.

1. INTRODUCTION

Mathematical models of actual systems contain uncertainty terms which reflect the designer's lack of knowledge of parameter values and disturbances. Uncertainties arising from imperfect knowledge of system inputs and inaccuracies in the mathematical modelling itself, contribute to performance degradation of the feedback control system. In self-tuning and other stochastic adaptive control systems, the parameter values and disturbances are constantly monitored using on-line identification algorithms, and appropriate adaptive globally stable controllers are implemented. These schemes, however, are costly and result in additional complexity. In direct contrast to these adaptive controllers, Sliding Mode Control systems have *fixed nonlinear* feedback control functions incorporating a switching function, which operate effectively over a specified magnitude range of system parameter variations and disturbances, without any on-line identification of the system parameters.

2. SLIDING MODE

The essential feature of a variable structure control (VSC) system is that the generally non-linear feedback control function switches between distinct values when the state crosses certain manifolds in the state space. This implies that the feedback structure is altered (switched) on the discontinuity manifolds. The central feature of the VSC system is sliding motion. This occurs when the system state repeatedly crosses and immediately re-crosses a switching manifold, because all motion in the neighbourhood of the manifold is directed inwards (i.e. towards the manifold). When sliding occurs on all the discontinuity surfaces simultaneously, the system is said to be in sliding motion. The motion of the system is then effectively constrained to lie within a certain subspace of the full state space, and the system is thus formally equivalent to a lower order system called the equivalent system.

The design of a VSC system consists of the choice of the switching surfaces, the specification of discontinuous control functions and the determination of the switching logic associated with the discontinuity surfaces. The aim of the design is to drive the system from an arbitrary initial condition to the intersection of the switching surfaces. This is achieved by a suitable choice of control functions and is known as the reachability problem.

The first stage of the VSC design process involves selecting a suitable switching hyperplane which will ensure that the state approaches the state space origin within the sliding mode. The second stage of the design process is to select the control which will ensure that the chosen sliding mode is attained. This includes the problem of determining a control structure and associated gains which ensure the reaching or hitting of the sliding mode.

In this paper we have adapted GA to study the design of a variable structure control systems. GA is used to study the design of a static gain matrix for use in the reaching phase of a variable structure control systems. The results obtained using GA are compared with those found using the techniques employed in Heck et al. (1993). The new MATLAB Genetic Algorithm Toolbox (Chipperfield et al. 1993) has been used.

3. GENETIC ALGORITHM (GA)

A Genetic Algorithm is an adaptive search technique that mimics the process of evolution based on Darwin's survival-of-the-fittest strategy. When applied to a problem, GA uses a genetics-based mechanism to iteratively generate new solutions from currently available solutions. It then replaces some or all of the existing members of the current solution pool with the newly created members. The motivation behind the approach is that the quality of the solution pool should improve with the passage of time.

The underlying basis of a GA is represented by a string, more commonly known as a chromosome. Normally chromosomes are represented in binary form, however some researchers have used other representations such as integer and floating point to suit the needs of their problems. These chromosomes undergo three basic genetic operations; reproduction, crossover and mutation.

Many researchers (Dorling and Zinober 1988, Heck et al. 1993, Oh et al. 1993) have resorted to conventional optimisation techniques in the design of variable structure control systems. The process can involve certain transformation of the variables into a form suitable for the use of the optimisation techniques. In this paper we will highlight the fact that GA works with the objective value only, without the need for a complex computational algorithm.

4. THE SYSTEM

Consider the system

$$\begin{aligned} \dot{x}(t) &= Ax + Bu \\ y &= Cx \end{aligned} \qquad (1)$$

where the state $x \epsilon R^n$, the control $u \epsilon R^m$, the output $y \epsilon R^p$ and CB has full rank. Sliding Mode Control based on *output* feedback, has the ith component

$$u_i(y) = \begin{cases} u_i^+(y) & \text{if } s_i(y) \geq 0 \\ u_i^-(y) & \text{if } s_i(y) < 0 \end{cases}$$

The switching function has the form $s = Gy$ where $s = 0$ denotes the sliding surface. Setting $\dot{s} = 0$ and solving for the equivalent control, yields the equation of motion on the sliding surface

$$\dot{x} = (A - B(GCB)^{-1}GCA)x \qquad (2)$$

where it is assumed that $(GCB)^{-1}$ exists.

The two step design process is to choose matrix G to give desired behaviour on the sliding mode surface, and then to select the control law to ensure that the sliding mode exists and is reachable.

Heck et al. (1993) studied the design of a static gain matrix for use in the reaching phase of a variable structure control system, and the methods are applicable for both static and dynamic output feedback. The design techniques yield the control which satisfies the reaching condition. The control law has the form

$$u = -(GCB)^{-1}Py - \alpha(GCB)^{-1}\mathrm{SGN}(s) \qquad (3)$$

where s = Gy and SGN(s) is a vector with components $\mathrm{sgn}(s_i) = 1$ if $s_i \geq 0$ and $\mathrm{sgn}(s_i) = -1$ if $s_i < 0$. The gain matrix P is chosen to satisfy the reaching condition

$$s^T \dot{s} = x^T C^T G^T (GCA - PC)x - \alpha \|s\|_1 < 0$$

where $\|s\|_1 = \sum_{i=1}^m |s_i|$.

Let $Q(N)$ represent the symmetric part of $C^T G^T (GCA - PC)$,

$$Q(P) \triangleq \left\{C^T G^T (GCA - PC) + (GCA - PC)^T GC\right\}/2$$

Note that for $\alpha > 0$, the reaching condition can be restated as

$$\lambda_{\max}\{Q(P)\} \leq 0$$

Even though the matrix P can be found by minimizing the maximum eigenvalue of $Q(P)$, a straightforward minimization might yield too large a gain. Therefore, the weighted average of the norm of P and the maximum eigenvalue of $Q(P)$ is minimized instead. The objective function is thus defined as

$$\min \phi(P) = \lambda_{\max}\{Q(P)\} + \rho\|P\|^r \qquad (4)$$

where $\rho \geq 0$ is a constant and $r \epsilon \{1, 2\}$.

Since the maximum rank of GC is m, which gives the maximum rank of $Q(P)$ as $2m$ which is usually less than dimension of n, there will be multiple values of the zero eigenvalue. The determination of P requires an algorithm that can handle nonsmooth optimisation. Heck et al. (1993) reformulated the objective function into a form suitable for using a cutting plane method and the interior point method.

The Genetic Algorithm (GA) developed here is based on the multi-population GA with the objective function (4). The population consists of q subpopulations with l number of individuals per subpopulation. Each individual (string) consists of mp decision variables since the gain matrix P is an $m \times p$ matrix.

The chromosomes are represented as real values and created using the function CRTRP. This function creates a random valued matrix of size $(mp \times q) \times l$ ((no.of individuals × no. of subpopulation)×no. of variables). The values of q and l are chosen such that they are dependent on the number of decision variables. The function crtrp also ensures that all the values generated for each decision variable in each individual are within the defined boundaries. The objective value of each individual is then calculated and is ranked separately for each subpopulation. In order to prevent premature convergence, it is important that no individuals generate an excessive number of offspring (Chipperfield et al. 1994). Therefore, we have employed Baker's method (Baker 1989) which assigns a fitness value to each individual according to its rank in the population rather than raw performance. A variable, selective pressure, is used to determine the bias towards the most fit individuals and the fitness of the others is determined by the following rules:

(a) MIN = 2.0 − MAX
(b) INC = 2.0 × (MAX − 1.0)/NIND
(c) LOW = INC/2.0

where MIN is the lower bound, INC is the difference between the fitness of adjacent individuals, NIND is the number of individuals and LOW is the expected number of trials (number of times selected) of the least fit individual. MAX is normally chosen in the interval [1.1,2.0]. Here the selective pressure is chosen as 2.0.

Within each subpopulation, individuals are selected for breeding independently using Stochastic Universal Sampling which has been shown to have a minimum spread and zero bias (Baker 1989). A generation gap, the fractional difference between the new and old population sizes, of 0.8 has been selected and the pairs are recombined using discrete recombination. This method creates a crossover mask randomly and the parity of the bits in the mask indicates which parent will supply the offspring with which bit. Since real-valued representation is used, mutation is achieved by either perturbing the gene values or a random selection of new values within the allowed range. In this algorithm the MUT-BGA operator (breeder GA proposed by Mühlenbein) (Mühlenbein et al. 1993) is used to mutate the offspring. This method of mutation uses a non-linear term for the distribution of the range of mutation applied to gene values. This is done by producing an internal mask table that determines which variable to mutate and the sign for adding δ, the normalised mutation stepsize. This mutation operator is able to generate most points in the hypercube defined by the variables of the individual and the range of the mutation. However, it tests more often near the variable than otherwise. This means that the probability of small stepsizes is greater than large stepsizes. The offsprings are then reinserted into the populations. Migration of individuals between subpopulations takes place at every 20 generations and 20 % of the most fit individuals of each subpopulations are selected for migration. Nearest neighbour subpopulations then exchange these individuals and uniformly reinserting the immigrant individuals.

We have used the example in Heck et al. (1993) to demonstrate the application of GA (see the Appendix).

We applied GA and the program was run for $\rho = 1 \times 10^{-5}$, $r = 1$ and $\rho = 2 \times 10^{-4}$, $r = 2$ so as to compare with the results obtained by Heck.

We have chosen the same boundary for each of the decision variables (i.e. each element in the gain matrix N). In order to select a suitable boundary for these variables, multiple runs were carried out with different boundary settings. If a boundary is too large the programs would take considerable additional time to run. On the other hand, if the interval is too small, the objective value might not converge to the best objective value as the path to the optimum objective value might require values outside the boundaries as intermediate values. In this case a boundary of [−60, 60] was found to be sufficient.

We first studied the effect of mutation rate on the best objective value and the results are summarised in Table 1.

Mutation Rates	Best Objective Values
0.1250	$9.79010E-04$
0.1875	$8.6410E-04$
0.2500	$8.0469E-04$
0.3125	$7.2793E-04$
0.5000	$6.8277E-04$
0.6250	$7.5266E-04$
0.7500	$7.6158e-04$
0.8750	$6.8600E-04$
1.0000	$6.8314E-04$

Table 1: Results for various mutation rates

In general, the results seem to suggest that higher mutation rates give better objective values. This is because a higher mutation rate encourages greater exploration of the solution space. Note that the mutation rate of 0.5 gives the best result.

The best objective value is an improvement on the result obtained in Heck et al. (1993) by the cutting plane method. However, this is at the expense of a slight increase in the norm of the gain matrix P. The best objective value using GA has

$$P = \begin{bmatrix} -0.7961 & 0.0001 & 2.5225 & 8.9308 \\ 16.1753 & -33.9957 & -1.3876 & 25.0443 \end{bmatrix}$$

If the values of the elements of the gain matrix P are restricted to the interval $[-7, 5]$ (as in Heck et al. 1993) the results obtained are comparable to that obtained by the cutting plane method.

In order to compare our results with those obtained by the interior point method, the program was run with $\rho = 2 \times 10^{-4}$ (see Table 2). The best boundary for this problem is enclosed in the interval $[-10, 10]$. In contrast to the first problem, the results obtained seem to indicate that a higher mutation rate is not effective for a boundary with small intervals. For this problem, GA produced results that are comparable to the interior point method in Heck et al. It is interesting to note the best objective value obtained agrees up to six decimal places and the values of the feedback matrix P are also comparable with Heck et al.

Constants	Max. Eigenvalue		Best Objec. Val.	
	Heck	GA	Heck	GA
$r = 1$, $\rho = 1E-05$	0.00156	0.0003429	0.0016	0.0006828
$r = 2$, $\rho = 2E-04$	0.00168	0.00168	0.003322	0.003322

Table 2

5. CONCLUSIONS

We have successfully used GA in the design of a variable structure control system. The results obtained, are slightly better than those found using the conventional optimisation techniques.

The results also suggest that GA promises a very bright future in the process of sliding mode and other control designs. The design problem can be solved without having to reformulate the problems into a suitable form as GA can be applied directly to the objective functions. The results also show that GA is very *straightforward* to implement and the GA MATLAB Toolbox can easily be employed to solve a wide range of problems.

We have used real representation instead of binary/Gray coding because as pointed out by Wright (1991), real-valued genes offer a number of advantages in numerical optimisation. It is claimed that efficiency of the GA is increased as there is no need to convert chromosomes to phenotypes (decision variables) before each function evaluation; less memory is required as efficient floating point internal computer representations can be used directly; there is no loss in precision by discretisation to binary or other values and there is greater freedom to use different genetic operators. Furthermore, Wright (1991) and Janikow and Michalewicz (1991) demonstrated how real-coded GA can take advantage of higher mutation rates than binary coding, increasing the level of possible exploration of the search space without adversely affecting the convergence characteristics. The results obtained from the test problem have confirmed these factors.

6. REFERENCES

Baker, J.E., 1985, Adaptive Selection Methods for Genetic Algorithm, *Proceeding of ICGA 1*, 101–111.

Chipperfield, A., Fleming, P.J., Pohlheim, H. and Fonseca, C.M., 1993, Genetic Algorithm Toolbox for use with MATLAB

Dorling, C.M. and Zinober, A.S.I., 1988, Robust hyperplane design in multivariable variable structure control systems, *Int. J. Control*, 48, 2043–2054.

Fleming, P.J. and Fonseca, C.M., 1993, Genetic Algorithms in control system engineering, *Research report no. 470*, Dept of Automatic Control and Systems Engineering, The University of Sheffield.

Goldberg, D.E., 1989, *Genetic Algorithms in search, optimization and machine learning*, Addison Wesley, Reading, MA

Heck, B.S., Yallapragada, S.V. and Fan, M.K.H., 1993, Numerical methods to design the reaching phase of output feedback variable structure control, (submitted to Automatica)

Michalewicz, Z., 1992, *Genetic Algorithm + Data Structures = Evolution Programmes*, Springer-Verlag

Mühlenbein, H., Gorges-Schleuter, M. and Kramer, O., 1988, Evolution algorithms in Combinatorial optimization, *Parallel Computing*, 7, 65–85

Mühlenbein, H. and Schlierkamp-Voosen, D., 1993, Predictive models for the breeder genetic algorithm: I. Continuous parameter optimization, *Evolutionary Computation*, 1, 25–49

Oh, M., Gu, D. and Spurgeon, S.K., 1993, Robust pole assignment in a specified region using output feedback, *Optimal Control Applications & Methods*, 14, 57–66

Wright, A.H., 1991, Genetic Algorithms for Real Parameter Optimization in *Foundations of Genetic Algorithms*, Rawlins, J.E.(Ed), Morgan Kaufmann, 205–218

Zinober, A.S.I. (editor), 1990, *Deterministic control of uncertain systems*, Peter Peregrinus Press, London

Appendix:

For

$$A = \begin{bmatrix} 0 & 0 & 1 & 0 & 0 & 0 & 0 \\ 0 & -.15 & -.004 & 1.54 & 0 & -.74 & -.03 \\ 0 & .25 & -1 & -5.2 & 0 & .34 & -1.12 \\ .039 & -.996 & -.0003 & -2.12 & 0 & .02 & 0 \\ 0 & 0.5 & 0 & 0 & -4 & 0 & 0 \\ 0 & 0 & 0 & 0 & 0 & -20 & 0 \\ 0 & 0 & 0 & 0 & 0 & 0 & -25 \end{bmatrix}$$

$$B = \begin{bmatrix} 0 & 0 \\ 0 & 0 \\ 0 & 0 \\ 0 & 0 \\ 0 & 0 \\ 20 & 0 \\ 0 & 25 \end{bmatrix}$$

$$C = \begin{bmatrix} 0 & -0.154 & -0.0042 & 1.54 & 0 & -0.744 & -0.032 \\ 0 & 0.249 & -1 & -5.2 & 0 & 0.337 & -1.12 \\ 1 & 0 & 0 & 0 & 0 & 0 & 0 \\ 0 & 0 & 0 & 0 & 1 & 0 & 0 \end{bmatrix}$$

the output feedback matrix

$$G = \begin{bmatrix} -0.0067 & 0.0167 & 0.0033 & 0 \\ 0.0167 & -0.0333 & 0 & 0.0333 \end{bmatrix}$$

(Heck et al. 1993) yields the desired sliding mode dynamics.

APPLICATION OF TEMPORAL NEURAL NETWORKS TO SOURCE LOCALISATION

Brigitte Colnet, Stéphane Durand

CRIN-CNRS INRIA Lorraine BP 239, 54506 Vandœuvre-lès-Nancy, France

Abstract

In this paper, we present several neural networks approaches to deal with the problem of source localisation in signal processing. To locate a far field source, we have to retrieve the propagation delays between sensors. They are the only relevant temporal information we can handle on the signal.

As we have to process temporal data, we study different time representations in neural networks. First, time may be represented explicitly and externally with multilayer perceptrons. Then, we use a time-delay neural network that appears to be more suitable to represent explicitly the temporal nature of the signal. Last, we study recurrent neural networks that represent time in their internal organisation.

1. Introduction

Neural Networks (NNs) have been successfully applied in many domains. In some cases, they must take temporal information into account like in speech recognition or signal processing. But it is not always easy to represent temporal data, especially when they simultaneously come from several sensors.

NNs which integrate time can be classified into different categories according to the way time is encoded. NNs with an *external* time representation encode this parameter as an additional dimension of the input space. Conversely, if the time is not represented in the input, it can be taken into account in the internal organisation of the network: it is an *internal* representation. In this latter class, we can distinguish two cases: an *implicit* representation where the succession of the steady states of the network intrinsically encodes the time, and an *explicit* representation where the succession of events is encoded by specific neurons [6].

These different types of temporal NNs perform different kinds of temporal processing. They also lead to different interpretations of temporal information. We will illustrate this phenomenom here and explain how the different interpretations correspond to different a priori knowledge integration in neural models. This will be done by applying different kinds of NNs to the same problem: source localisation. First, we will introduce the source localisation problem. Second, we will use an external representation of time with a multilayer perceptron (MLP); this architecture is adapted to represent knowledge we get in the data. Then, we will present a time-delay neural network (TDNN), which represents more information, integrating acoustic events and the relationships between these events. Finally, we will describe a recurrent neural network (RNN) with an *internal* representation that implicitly encodes time keeping track of past events in a synthetic way.

2. Problem definition

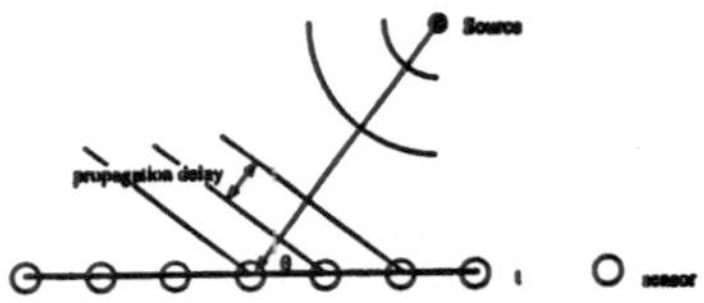

Figure 1: Propagation model

In the far field source detection and localisation problem, relevant information is provided by the propagation time delays between the array sensors. It is the time for a wave impinging on a given sensor to reach other sensors. We consider that waves emitted by a far field source are plane when they impinge on an array of N_s equally spaced sensors. According to the propagation model [3] depicted in figure 1, equation of signal on sensor number i at time t when the direction of arrival of signal is θ is

$$s_i(t) = s_j(t - \tau_{ij}) \text{ with } \tau_{ij} = \frac{X_{ij} \sin\theta}{c} \qquad (1)$$

where X_{ij} is the distance separating sensors i and j and c the speed of sound. With the knowledge of the set of delays τ_{ij} we can retrieve the direction of arrival of the signal.

3. External representation of time with a MLP

In general, a MLP can represent temporal data by encoding them on the input layer. Time is represented spatially. In our case, available relevant information is provided by temporal data and the way data may be related to each other. The approach we propose consists in dividing the whole horizon from -90 to $+90$ degrees in N_a angular sectors.

A specific MLP is attached to each angular sector and dedicated to detect and enhance the signal coming from this area of the space [4].

3.1. MLP architecture

Let Γ_j be the network dedicated to the angular sector $[\theta_{j-1}, \theta_j]$ with $\theta_j = \frac{180}{N_a - 1} - 90$ degrees. The accuracy of localisation depends on the number of MLPs used. Γ_j is a three-layered network.

The input layer consists of a temporal observation window of the sampled signal on the N_s sensors. The window size is chosen to take into account the maximum propagation delay between the two end sensors.

3.2. Connections

Units in the input layer are connected to hidden units in such a way to enhance the signal coming from the direction associated with the network. The input units corresponding theoretically to the same wave front reaching sensors at different time create a *delay line* and are connected to the same group of hidden units (see figure 2). With this specific way of connecting neurons from input to hidden layer, Γ_j learns that units belonging to the same delay line have theoretically the same value. Thus, the network does not learn the signal form itself, which is highly dependent on parameters such as amplitude and frequency.

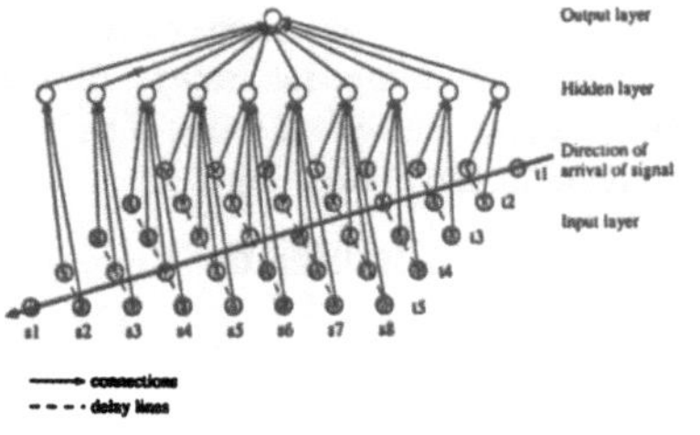

Figure 2: MLP architecture

3.3. Global approach using MLP

We build N_a MLPs with the architecture described above. Each network is trained to identify the signal coming from the angular sector it is dedicated to. The training step is performed with a simulated signal while recognition is performed with real data. Figure 3 compares the results obtained with MLP to those obtained with beamforming. The specificity of the connection topology and the representation of time as a second dimension on the input layer allow us to deal with the temporal nature of signal. With the MLP approach, available information is represented both in the input layer and by the connections' organisation. Nevertheless, the size of the temporal observation window depends on the frequency of the signal. Time is externally and explicitly represented: it does not

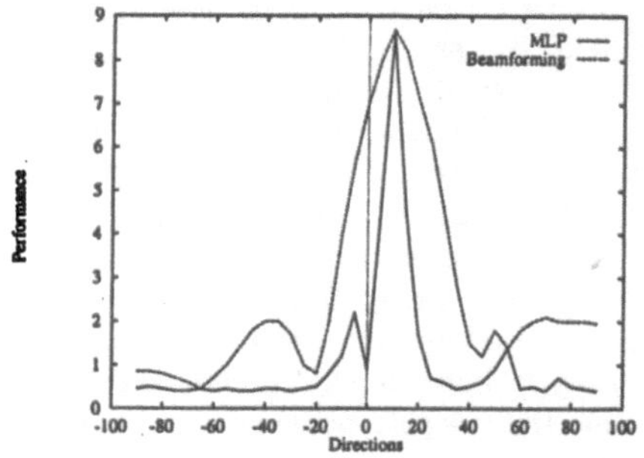

Figure 3: Results obtained on simulated noisy data ($\theta = 10$ degrees)

4. External representation of time with TDNNs

Waibel shown in [15] that TDNNs are well suited to deal with the dynamic nature of a signal. They take into account temporal relationships among acoustic events whatever time they occur. Constraints encountered in the source localisation issue are the same as those encountered by Waibel in phoneme recognition: relationships between spatio-temporal events are time-invariant. In our case, delays between sensors do not depend on time (see equation 1) and do not change over time because we assume that the array geometry is fixed and the emitting source does not move.

4.1. TDNN architecture

The basic unit of a TDNN is a delayed unit. We construct a set of TDNNs each focusing on some angular sector of the whole horizon. A TDNN attached to an angular sector indicates in which part of this sector the emitting source is located. It has four layers.

Input layer. The input layer of the TDNN we use consists of N_s units corresponding to the N_s sensors with their delays. The input layer is supplied with the sampled signal received by the array sensors. Because coefficients are the samples of the signal on sensors at a given time, introducing delays is like considering the signal through a temporal observation window sliding over the data.

Hidden layers. The first hidden layer has $nh1$ units expanded out temporally. The second hidden layer is made up of $nh2$ delayed units. The number of delays for the hidden units depends on the number of input delays and on the size of the window frame sliding on the previous layer.

Output layer. It is obtained by integrating over time the $nh2$ hidden units. The number of output units is equal to $nh2$ and $nh2$ is also the number of parts into which we divide the angular sector focused on by the TDNN.

Connections. The originality of TDNNs lies in delayed units and their feed-forward connections with shared weights. A frame window scans the input layer. Each unit in this window is connected to one frame of $nh1$ units in the first hidden layer. The frame window size is chosen to integrate low level temporal events of the signal arriving on sensors: it depends on the propagation delay between two adjacent sensors. In the second hidden layer, each frame of $nh2$ units receives activation from a larger sliding window on the first hidden layer. This allows the network to process temporal events on a larger scale. Additionally, since we need to link temporal events whatever time they occur, connection weights between units corresponding to the same time shift are forced to have the same value. They are updated with the average of all corresponding time-delayed weight changes computed using the standard back-propagation algorithm [13]. Shared weights reflect a priori knowledge that learned acoustic events could happen on any sensor.

To locate sources, TDNNs are encompassed in a global successive-refinement approach [5].

4.2. Performance of the TDNN approach

This approach takes advantage of the TDNNs specificity: they are able to handle the dynamic nature of signal and parameters such as frequency, phase and amplitude are not taken into account. As with MLPs, they are supplied with raw temporal data.

It appears from results obtained with real data that the performance is not as good as that obtained with MLP (see figure 4). One explanation is that TDNN are not robust to temporal distortion. Indeed, we know that TDNN are well suited to relate time-invariant events but the signal we have to process may suffer from temporal distortion. For instance, when the signal frequency changes, signal periodicity is changed too, and the TDNN seems to be sensitive to such distortion. This problem comes from the externally explicit representation of time as delayed units with shared weights.

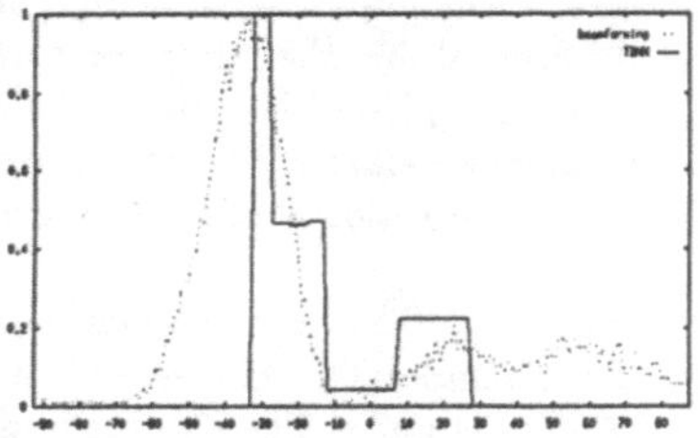

Figure 4: Comparison of results between TDNN and beamforming. Tests are performed on real data with a signal whose direction of arrival is -34 degrees.

5. Internal and implicit representation with RNNs

To avoid drawbacks that can arise from the external representation of time, our current work is to represent time internally with RNNs. They encode time in their internal structure in a relevant way without the need for other information. We build an Elman network [8] to process temporal signal provided by sensors array. Time is not represented as a second dimension of the input but encoded in the network structure: the hidden units are duplicated on the input layer as contextual units, the input layer receiving the signal coming from the sensors. The aim of the hidden layer is to integrate the temporal acoustic events successively presented to the network. This layer contains the largest part of the network knowledge. Investigations in this layer are necessary to understand how knowledge is represented.

Architecture The RNN architecture is described in [8].

Input layer It consists of N_s units receiving the signal from external world and nh contextual units whose inputs are the activation states of the hidden units at the previous step.

Hidden layer It is made up of nh units fully connected to all input units.

Output layer It is made up of units that indicate the direction of arrival of the input signal. They are fully connected to the hidden units.

Time coding Conversely to previous neural networks we have described, the temporal nature of the signal is encoded intrinsically by contextual units. The network keeps track of past events in the hidden layer.

6. Discussion

What can be our choice faced with a dynamic problem such as the source localisation problem described above? Unfortunately, the response is not easy and strongly depends on the problem nature and more especially on the input coding. Two kinds of representations can be used as we saw in the introduction: the external representation of time and the internal representation. The first seems to be easier to implement and test from an engineering point of view but is inevitably limited by the length of the temporal window which is a priori fixed. The second, which seems to be more plausible from a neuro-mimetic or cognitive point of view, is less used and corresponds to ungoing research. RNNs are among the systems with an internal representation of time. Although popular, they are not very efficient in real world problems.

Concerning the external representation of time, we have used two kind of models to solve the source localisation problem: an MLP [13] and a TDNN [15]. In these two cases, time is considered as an additionnal dimension of the input, so time is spatialised. The results obtained with a multi-layer perceptron seems to be better than those with the TDNN.

Moreover, the learning process with the TDNN is longer and less effective. Studies have shown that the hidden layers of the TDNN extract the derivative vectors (or the variation vectors). Using the vectors and their derivatives in the input of an MLP can improve the network efficiency.

Concerning the internal representation of time, two major modes exist: the implicit and the explicit modes. Recurrent neural networks [8, 9] are among the systems with an internal and implicit representation of time. In this case, the sequence of information is coded in the hidden layer. The succession of the steady states of the network matches with the temporal sequence. Few recurrent neural networks have been applied to real world problems [11]. The last type of representation (internal and explicit) is in the research domain. Time is explicitely represented, *i.e.* the succession can physically appear on the neural net. This involves the definition of a new functional basic unit to deal with time either in the unit level or in the connection level or in both levels. Some models exist and are generally inspired from the neurobiology, for instance, the short term memory column [2], the Katamic memory [12] or models inspired by the cortical column [1]. Some architectures have been applied to real world problems like speech recognition [7]. These methods appear very promising for applications like the source localisation problem.

7. Conclusion

The results obtained with these temporal NNs are comparable to those obtained with classical localisation methods such as beamforming. Moreover, we take advantage of NNs properties: no pre-processing (data are directly obtained from the sensors), ability to generalise (networks are trained with simulated data while the recognition phase is performed with real noisy data), and robustness to noise. It is interesting to use temporal NNs as described above because they integrate the temporal dimension that we need to solve our source localisation problem. We saw that MLPs with an appropriate organisation are suitable for processing explicit temporal data with additional information. In addition TDNNs take relationships between invariant temporal events into account while recurrent networks integrate all the knowledge implicitly in their internal structure.

The approaches presented here correspond to external representation and internal implicit interpretation of time. In further work, it would be interesting to test biologically-inspired models such as delay tuned neurons [14] or interaural intensity difference tuned-neurons [10]. These models use an internal and explicit representation of time.

References

[1] F. Alexandre, F. Guyot, J P. Haton, and Y. Burnod. The cortical column: a new processing unit for multilayered networks. *Neural networks*, 4:15–25, 1991.

[2] B. Ans. Modèle neuromimétique du stockage et du rappel de séquences temporelles. *Académie des Sciences Paris*, 3:7–12, 1990.

[3] W.S. Burdic. *Underwater acoustic system analysis.* Prentice-Hall signal processing series, Englewood Cliffs, New-Jersey, 1984.

[4] B. Colnet and J.P. Haton. Far field array processing with neural networks. In *Proceedings of the IEEE International Conference on Acoustics, Speech, and Signal Processing*, pages 281–284, Adelaid, Australia, April 1994.

[5] B. Colnet and J.C. Di Martino. Bearing estimation with time-delay neural networks. In *Proceedings of the IEEE International Conference on Acoustics, Speech, and Signal Processing*, Detroit, Michigan, May 1995.

[6] S. Durand and F. Alexandre. A neural network based on sequence learning: Application to spoken digits recognition. In *7th international conference on Neural Networks and Their Applications*, pages 290–298, Marseille, 1994.

[7] S. Durand and F. Alexandre. Spatio-Temporal Mask Learning: Application to Speech Recognition. In *International Conference on Artificial Neural Networks and Genetic Algorithms*, Alès, April 1995.

[8] J.L. Elman. Finding structure in time. *Cognitive Science*, 14:179–211, 1990.

[9] M I. Jordan. Attractor dynamics and parallelism in a connectionist sequential machine. In Hillsdale, editor, *Proceedings of the Eighth Annual Conference of the Cognitive Science Society.* Erlbaum, 1986.

[10] E.I. Knudsen, S. du Lac, and S.D. Esterly. Computational maps in the brain. *Annual Review of Neuroscience*, 10:41–65, 1987.

[11] R.P. Lippmann. An introduction to computing with neural nets. *IEEE Acoustics Speech and Signal Processing Society*, 4(2):4–22, April 1987.

[12] V.I. Nenov and M.G. Dyer. Perceptually grounded language learning: Part1–a neural network architecture for robust sequence association. *Connection Science*, 5(2):115–138, 1993.

[13] D.E. Rumelhart and J.L. McClelland. *Parallel distributed processing: explorations in the microstructure of cognition*, volume 1. MIT press, 1986.

[14] N. Suga. Cortical computational maps for auditory imaging. *Neural Networks*, 3:3–21, 1990.

[15] A. Waibel, T. Hanazawa, K. Shikano, and K.J Lang. Phoneme recognition using time-delay neural networks. *IEEE Transactions on Acoustics, Speech, and Signal Processing*, 37(3):328–339, March 1989.

USING GENETIC ALGORITHMS FOR OPTIMAL DESIGN OF AXIALLY LOADED NON-PRISMATIC COLUMNS

Carlos A. Coello & Alan D. Christiansen

Department of Computer Science, Tulane University, New Orleans, LA 70118, USA

Abstract: *This paper presents a method for optimizing the design of axially loaded non-prismatic columns using genetic algorithms (GAs). The design problem was formulated as an optimization problem in which the objective function is to minimize the volume of a column under a given load by changing its shape, subject to both buckling and strength constraints. Both floating point representation and binary representation (with and without Gray coding) were used and compared against a mathematical programming method based on the generalized reduced gradient method. Our results show that the floating point representation scheme provides the best solutions, both in terms of precision and in terms of computing time. This problem is of great interest in engineering, since considerable savings can be achieved due to the effective use of material through the optimal shape of a column, mainly for mass production.*

Introduction

The optimization of structural members subject to compression forces has been of great interest for a long time. Leonhard Euler derived first the formula for the critical buckling load of an ideal slender column [6], and was also the first to solve the problem of inextensible elastica (i.e., for constant modulus of elasticity and modulus of inertia, he found the curve of prescribed length with prescribed terminal displacements and slopes and minimum stored energy). Euler solved the case of a column that has the lower end fixed and the upper end free. Later he extended [5] his work on columns, and even today it has a great influence in every strength of materials textbook. In fact there were very few contributions to his work until Lamarle [10] noticed that Euler's formula should be used only for slenderness ratios over a certain limit, and that the experimental data should be applied only to small ratios. In 1889 the French engineer Considère [2] performed a set of 32 tests on columns, establishing the so-called theory of the reduced module. During that same year, the German engineer F. Engesser [4] independently suggested the theory of tangential module. From these two theories, the first one dominated until 1946, when American professor F. R. Shanley indicated the logical paradoxes of both theories. In a remarkable paper of only one page [14] he explained not only what was wrong with both theories, but he also proposed his own theory that solved the paradoxes.

On the other hand, the problem of non-prismatic columns (i.e., those with a variable cross-section area) has been studied more recently. A. N. Dinnik [3] discussed the design of columns in which the moment of inertia of the cross-section areas varies according to a power of the distance along the member axis. Keller [9] and later Tadjbakhsh and Keller [17] derived optimal solutions to the strongest-column problem, which was characterized in the following way: "For a column of given length and volume of material, determine the column shape for which the Euler buckling load is maximum". In their analysis, Tadjbakhsh and Keller established the necessary conditions for a maximum by performing variations on the differential equations of equilibrium and associated boundary conditions, and the constraint of constant volume.

J. Taylor [18] studied the same problem using an energy approach, and presented a method to calculate a lower bound to the maximum eigenvalue. Spillers and Levy extended Keller's solution for the optimal design of columns to the case of plates [15] and later, for axisymmetric cylindrical shells [16]. However, in all these works, only the constraint of constant volume was considered and, as Fu and Ren [7] point out, in a practical design, material strength constraints are equally important. With that in mind, these two last authors added such constraint to the optimization problem and used an algorithm called the generalized reduced gradient method [13] to select the design variables at nodal points. The results that they obtained are very reasonable and verifiable. The generalized reduced gradient method linearizes the non-linear constraints of this problem, and uses the convex simplex method to select the best direction of search from all the candidate directions which are both feasible and descent. Then the search for the optimum is started from the feasible initial point. Newton iteration is employed to adjust the basic variable to maintain feasibility. The convergence is accelerated by incorporating conjugate direction or quasi-Newton constructions. Naturally, the existence of continuous differentiability of the problem functions is a fundamental requirement for using the method.

Our work consisted followed this last approach, and applied the Genetic Algorithm (GA) instead of the generalized reduced gradient method. Some problems had to be faced, though, namely the representation scheme and the parameters to be optimized. However, our results are practically as good as those found by Fu and Ren [7], and in at least one case we found a better solution than them. Binary and floating point representation schemes were tried, since this problem has a continuous search space. Nevertheless, the search space can be easily discretized, since in real designs there is always a lower and an upper bound on the dimensions of the column.

Statement of the Problem

Given a column subject to axial load along the horizontal direction, the governing differential equation is

$$EIy'' + Py = 0 \tag{1}$$

where E is the modulus of elasticity, I is the moment of inertia, P is the axial load, and y is the function that represents the slenderness of the column. Let's assume that the column that we are going to study is divided into 6 equal-sized segments throughout its length. Then, Equation 1 may be expressed in a finite difference form, as [7]:

$$\frac{E}{h^2}\begin{bmatrix} -2I_2 & I_2 & 0 \\ I_3 & -2I_3 & I_3 \\ 0 & 2I_4 & -2I_4 \end{bmatrix}\begin{bmatrix} y_2 \\ y_3 \\ y_4 \end{bmatrix} + P\begin{bmatrix} y_2 \\ y_3 \\ y_4 \end{bmatrix} = \begin{bmatrix} 0 \\ 0 \\ 0 \end{bmatrix} \tag{2}$$

For non-trivial solution, the determinant must vanish, namely

$$\begin{vmatrix} \left(-2+\frac{Ph^2}{EI_2}\right) & 1 & 0 \\ 1 & \left(-2+\frac{Ph^2}{EI_3}\right) & 1 \\ 0 & 2 & \left(-2+\frac{Ph^2}{EI_4}\right) \end{vmatrix} = 0 \tag{3}$$

or, in linear form:

$$\frac{P^3h^6}{E^3I_2I_3I_4} - 2\frac{P^2h^4}{E^2}\left(\frac{1}{I_2I_3}+\frac{1}{I_3I_4}+\frac{1}{I_2I_4}\right) + \frac{Ph^2}{E}\left(\frac{2}{I_2}+\frac{4}{I_3}+\frac{3}{I_4}\right) - 2 = 0 \tag{4}$$

where, for round and regular polygonal sections, the moments of inertia will be given by

$$I_i = \alpha D_i^4 \tag{5}$$

D_i is the diameter for round sections, or the side length for regular polygonal sections. Table 1 shows the values of α for the most commonly used cross-sections.

In general, for an n-sided regular polygon, α may be derived as

$$\alpha = \frac{n}{192}Cot\frac{\pi}{n}\left(3Cot^2\frac{\pi}{n}+1\right) \tag{6}$$

For rectangular sections where width b is assumed to be constant throughout the length of the column,

$$I_i = \frac{bD_i^3}{12}, i = 2,3,4. \tag{7}$$

Round Section	Square Section	Triangular Section
$\frac{\pi}{64}$	$\frac{1}{12}$	$\frac{\sqrt{3}}{96}$

Table 1: Value of α for the most common sections. Notice that when we refer to triangular sections, we assume an equilateral triangle.

In a column design, Equation 4 represents a buckling constraint. Furthermore, a compressive strength constraint must also be satisfied, that is

$$\frac{P}{A_1} \leq \sigma_y, \tag{8}$$

where A_1 is a function of D_1. Since P and σ_y are always given values, we can compute the minimum D_1 or A_1 for each particular problem. This means that we may express the compressive strength constraint only in terms of D_1 or A_1.

Now, we have all the necessary elements to express the column design problem as an optimization problem. If we assume that P, h and σ_y are given, the objective is to minimize the volume of the column. Therefore, we'll consider 2 cases:

(1) Square or Round Columns: The objective function may be stated as [7]

Minimize:

$$V_s = K(D_1^2 + 2D_2^2 + 2D_3^2 + D_4^2 + D_1D_2 + D_2D_3 + D_3D_4) \tag{9}$$

where K is a constant defined according to Table 2, and V_c is the volume of the round or square column.

The objective function is subjected to the equality constraint defined by Equation 4, and the following additional inequality constraints:

$$C_1 < D_i < C_\mu, i = 1,2,3,4 \tag{10}$$

where D_i are the design variables; C_1 and C_μ are, respectively, the lower and upper bounds of the design variables.

(2) Rectangular Columns: The objective function will be [7]:

Minimize:

$$V_r = \frac{bl}{9}(D_1 + 2D_2 + 2D_3 + D_4 + \sqrt{D_1D_2} + \sqrt{D_2D_3} + \sqrt{D_3D_4}) \tag{11}$$

where V_r is the volume of the rectangular column.

The objective function is subjected to the equality constraint defined by Equation 4, and the following similar equation that is derived on the basis of buckling in the orthogonal direction:

$$\frac{P^3h^6}{E^3I_2'I_3'I_4'} - 2\frac{P^2h^4}{E^2}\left(\frac{1}{I_2'I_3'}+\frac{1}{I_3'I_4'}+\frac{1}{I_2'I_4'}\right) + \frac{Ph^2}{E}\left(\frac{2}{I_2'}+\frac{4}{I_3'}+\frac{3}{I_4'}\right) - 2 = 0 \tag{12}$$

where

Round Section	Square Section	Triangular Section
$\frac{\pi l}{36}$	$\frac{1}{9}$	$\frac{l\sqrt{3}}{36}$

Table 2: Value of K for the most common sections. Notice that when we refer to triangular sections, we assume an equilateral triangle.

$$I_i' = \frac{D_i b^3}{12} i = 1, 2, 3, 4 \quad (13)$$

Furthermore, an additional set of inequality constraints have to be satisfied

$$\left.\begin{array}{c} b \times D_1 > A_1 \\ C_l < D_i < C_\mu \\ C_l < b < C_\mu \end{array}\right\} i = 1, 2, 3, 4 \quad (14)$$

where $A_1 = P/\sigma_y$.

Use of the Genetic Algorithm

To solve this problem, we used the Simple Genetic Algorithm (SGA) proposed by Goldberg [8]. An issue in this application is the representation scheme, because we are dealing with real-valued parameters, and therefore it is necessary to use some kind of discretization, so that we can apply a binary representation scheme. We used a linear discretization that uses the upper and lower bounds given by the user to compute the decoded value with a 3 decimal precision.

We also tried to use Gray codes as suggested by Goldberg [8]. The Gray code representation has the property that any two points next to each other in the problem space differ by only one bit [11]. In other words, an increase of one step in the parameter value corresponds to a change of a single bit in the code. This is a well known technique used to reduce the distance of two points in the problem space, and it is argued to bring some benefit because of their adjacency property, and the small perturbation caused by many single mutations. However, the use of Gray codes didn't help much in this particular application, as we'll see in the next section.

Finally, we used a floating point representation, since it is conceptually closest to the problem space [11], and allows the easy and efficient implementation of closed and dynamic operators. We'll see how this last approach provided the best results, both in terms of the precision obtained and in terms of the computation time needed.

The fitness function that we used is illustrated by the following algorithm:

check1 = Error in Equation 4

If $P/(A_1 \times \sigma_y) - 1.0 > 0.0$ then *check2* = 1.0

else *check2* = 0.0

fitness = $1.0/(vol \times (check1 + check2) + 1.0)$

As we can see, if our answer violates the constraint imposed by Equation 4 then the fitness function will be penalized by the error produced. On the other hand, if it violates the stress constraint (i.e. $P/A_1 \leq \sigma_y$), then the penalty is 1.0. For rectangular columns, the constraint is that $b \times D_1 > P/\sigma_y$. In this last case, we must also check the orthogonal direction, so that we have three penalty values instead of two. These values are added and the result is multiplied by 1000—we magnify the error—so that we "punish" our result. Note that when there are no violations to any of the constraints the fitness function returns the inverse of the volume.

We used a simple genetic algorithm (SGA) as described by Goldberg [8], but with some modifications: we used two-point crossover and binary tournament selection. The four diameters that we want to find were represented by consecutive binary strings of the same length. The halting criteria was through a maximum number of generations. The GA was implemented in Turbo Pascal 7.0 using the technique proposed by Porter [12] for dynamic memory management. We found experimentally that the following parameters seem to give the best results:

Population size ⇒ 400
Crossover probability ⇒ 0.80
Mutation probability ⇒ 0.01

Method	Ex. 1	Ex. 2	Ex. 3	Ex. 4
Fu and Ren	1642.400	1608.300	1507.000	1617.952
GA (FP)	1642.660	1608.790	1509.729	1213.857
GA (B)	1642.713	1609.027	1509.864	1626.735
GA (G)	1647.990	1641.357	1516.663	1722.564

Table 3: Comparison of final volumes for the four examples. The (FP) indicates floating point representation. The (B) indicates binary representation without Gray coding, and the (G) indicates the use of it.

Examples

The following examples were taken from Fu and Ren [7]:

(1) Example 1: Select the best diameters at nodal points for a steel round column of 10' length which is subjected to an axial load of 400 kips. The modulus of elasticity is $E = 30 \times 10^6$ psi and the yield strength, σ_y is 60,000 psi. Thus, the minimum diameter may be computed as 2.914". The design variables are the diameters at nodal point, D_1, D_2, D_3 and D_4. The lower and upper bounds, C_l and C_μ are 2.914" and 20", respectively. The size of the search space for this problem is $(20000 - 2914)^4 \cong 8.52 \times 10^{16}$.

(2) Example 2: Select the best side-widths at nodal points for a steel square column of 10' length which is subjected to an axial load of 400 kips. The modulus of elasticity is $E = 30 \times 10^6$ psi and the yield strength, σ_y is 60,000 psi. Thus, the minimum diameter may be computed as 2.582". The design variables are the diameters at nodal point, D_1, D_2, D_3 and D_4. The lower and upper bounds, C_l and C_μ are 2.582" and 20", respectively. The size of the search space for this problem is $(20000 - 2582)^4 \cong 9.20 \times 10^{16}$.

(3) Example 3: Select the best side-widths at nodal points for a steel equilateral triangle column of 10' length which is subjected to an axial load of 400 kips. The modulus of elasticity is $E = 30 \times 10^6$ psi and the yield strength, σ_y is 60,000 psi. Thus, the minimum diameter may be computed as 3.924". The design variables are the diameters at nodal point, D_1, D_2, D_3 and D_4. The lower and upper bounds, C_l and C_μ are 3.924" and 20", respectively. The size of the search space for this problem is $(20000 - 3924)^4 \cong 6.70 \times$

10^{16}.

(4) Example 4: Select the best side-widths at nodal points for a steel rectangular column of 10' length which is subjected to an axial load of 400 kips. The modulus of elasticity is $E = 30 \times 10^6$ psi and the yield strength, σ_y is 60,000 psi. Thus, the minimum diameter may be computed as 1.500". The design variables are the diameters at nodal point, D_1, D_2, D_3 and D_4. The lower and upper bounds, C_l and C_μ are 1.500" and 20", respectively. The size of the search space for this problem is $(20000 - 1500)^4 \cong 1.20 \times 10^{17}$. In this problem we had to use a larger population than in the others (500 chromosomes as compared to the 400 used in the others), and we ran the algorithm for 100 generations, instead of the 50 generations used before. The reason for this change of parameters was the extra length added to our chromosomic strings in this case (we have 15 extra bits) because of the extra parameter needed (the width b of the column).

Table 3 shows the final results of our experiments. We can see how, in general, the floating point representation gave the better results, and the binary representation using Gray coding produced the worst. This wasn't only in terms of the volume, but also in terms of the convergence time and the maximum error produced. Fu and Ren [7] don't report the times that their approach took, but only mention the number of cycles required. In our case, we can say that in all tests, the computing time didn't exceed one minute, in an IBM PC 486/DX running at 66 MHz. For details, refer to our technical report [1].

Future Work

We are now working on a generalization of this problem, so that columns of any material (probably even of composite materials) can be considered. This will complicate the analysis a bit more. Also, we are considering the problem of using more than one objective function at a time (multi-objective optimization), in which we could have conflicting objectives. This will introduce some changes in the way the GA has to be applied, but it will drive us towards a more realistic view of structural design.

Conclusions

We have shown another successful application of the genetic algorithm to an engineering optimization problem. In this case we saw how a floating point representation worked better when dealing with a continuous search space. This representation not only generates superior solutions than the binary representation scheme (with or without Gray encoding) in terms of the precision obtained, and also in terms of the speed. Because we can use the same genetic operators with only slight modifications, the convergence will be faster since the chromosomes are of shorter length. This problem is an interesting one because, even when its analysis is very simple, it normally has fairly large search spaces and several constraints. The penalty technique that we used has proved to be useful in incorporating the constraints into the fitness function for this particular application, as can be seen from our results.

The problem treated here is one that still challenges mathematicians and engineers all over the world, because it refers to the design of the optimal geometrical shape of a column. Even if the precise mathematical answer of our method may not be suitable for practical civil engineering applications, because of the practical limitations of current building procedures, it may be very useful in industry, mainly for mass-production of structural elements.

References

[1] Carlos A. Coello and Alan D. Christiansen. Using genetic algorithms for optimal design of axially loaded non-prismatic columns. Technical Report TUTR-CS-95-101, Tulane University, January 1995.

[2] A. Considère. Résistance des pièces comprimées. In *Congrès International des Procédés de Construction*, volume 3, page 371. Libraire Polytechnique, Paris, 1891.

[3] A. N. Dinnik. Design of columns of varying cross-section. *Transactions ASME*, 54, 1932.

[4] F. Engesser. Uber die knickfestigkeit gerader Stäbe. *Zeischrift für Architektur und Ingenieurwesen*, 35(4):455–562, 1889.

[5] Leonhard Euler. Euler's calculation of buckling loads for columns of non uniform section. In *The Rational Mechanics of Flexible or Elastic Bodies 1638-1788*, pages 345–7. Orell Füssli Turici, Societatis Scientiarum Naturalium Helveticae, 1960. Originally published in 1757.

[6] Leonhard Euler. Euler's treatise on elastic curves. In *The Rational Mechanics of Flexible or Elastic Bodies 1638-1788*, pages 199–219. Orell Füssli Turici, Societatis Scientiarum Naturalium Helveticae, 1960. Originally published in 1743.

[7] K. C. Fu and D. Ren. Optimization of axially loaded non-prismatic column. *Computers and Structures*, 43(1):159–62, 1992.

[8] David E. Goldberg. *Genetic Algorithms in Search, Optimization and Machine Learning.* Reading, Mass. : Addison-Wesley Publishing Co., 1989.

[9] Joseph B. Keller. The shape of the strongest column. *Archive for Rational Mechanics and Analysis*, 5:275–85, 1960.

[10] A. H. E. Lamarle. Mémoire sur la flexion du bois, 1845.

[11] Zbigniew Michalewicz. *Genetic Algorithms + Data Structures = Evolution Programs.* Springer-Verlag, second edition, 1992.

[12] Kent Porter. Handling huge arrays. *Dr. Dobb's Journal of Software Tools for the Professional Programmer*, 13(3):60–3, March 1988.

[13] G. V. Reklaitis, A. Ravindran, and K. M. Ragsdell. *Engineering Optimization. Methods and Applications.* John Wiley and Sons, 1983.

[14] F. R. Shanley. The column paradox. *Journal of the Aeronautical Sciences*, 13(12):261, December 1946.

[15] W. R. Spillers and Robert Levy. Optimal design for plate buckling. *Journal of Structural Engineering*, 116(3):850–8, March 1990.

[16] W. R. Spillers and Robert Levy. Optimal design for axisymmetric cylindrical shell buckling. *Journal of Engng Mech. ASCE*, 115:1683–90, 1991.

[17] I. Tadjbakhsh and J. B. Keller. Strongest columns and isoperimetric inequalities for eigenvalues. *Journal of Applied Mechanics*, 29(1):159–64, March 1962. Transactions ASME Series E.

[18] J. E. Taylor. The strongest column: an energy approach. *Journal of Applied Mechanics*, 34:486–7, June 1967. Transactions of the ASME.

BEHAVIOUR LEARNING BY A REWARD-PENALTY ALGORITHM : FROM GAIT LEARNING TO OBSTACLE AVOIDANCE BY NEURAL NETWORKS.

Isabelle SARDA and Anne JOHANNET

LGI2P, EMA - EERIE ; 6, av. de Clavières, 30319 Alès cedex, FRANCE
Tel : (33) 66.38.70.21- Email : sarda@eerie.fr and johannet@eerie.fr.

ABSTRACT

Designing a mobile robot capable to evolve on various floors is a hard, and not yet solved task. Among the numerous ways, the biological inspired researches are attractive. Taking inspiration from such devices, a study on walk learning and obstacle avoiding by neural networks is presented. After a general presentation of the learning method, and its adaptation to neural networks, results on gait learning are described and first simulations on obstacle avoidance are also presented. Finally, a global network performing the two tasks with success is given.

I. INTRODUCTION

The realisation of an autonomous mobile robot is still a difficult task. Generally two types of problems are studied : the first one deals with locomotion (stability, efficiency) and the second one deals with guidance in order to reach a goal and to avoid the obstacles. The major difficulty encountered in such a work is the extreme variability of the environment in interaction with the robot. Obviously nobody tries to realise a robot able to evolve in all types of environment but the intrinsic variability of each situation is sufficient to lead the classical method of modelling to a relative failure [1] [2]. In this context, this study focuses on the capacity of the neural networks to learn an appropriate behaviour in an unpredictable environment. Two types of tasks have been investigated : the obstacle avoidance and the gait learning by a legged robot. The particular case of legged robots has been extensively studied, for it appears to be more appropriate for uneven ground than the wheeled robots [3]. The applications are numerous, for instance the manipulations in hostile and unpredictable surroundings or the automation of some processes in agriculture, with the advantage of a better preservation of the natural environment.
Based on the studies centred on stochastic automata, and on supervised pattern classification, an algorithm proposed by Barto *et al* has been applied to the learning of neural networks. According to the aim of realising the simplest device, in order to interpret easily its behaviour [4], it has been chosen to design two separate neural networks, each one specialised in one task : the first one in locomotion and the second one in obstacle avoidance. The simultaneous working of both networks has been satisfactorily tested in simulation, while the locomotion network has been implemented on a real six legged robot.

II. LEARNING THE TASK

II.1. A.R.P. Algorithm

The learning algorithm is the associative reward-penalty (A_{rp}) algorithm. This algorithm had first been proposed by Barto et al [5], the authors took inspiration from two previous algorithms : first the *stochastic learning automaton algorithm*, where the automata's behaviour is governed by a probabilistic evaluative feedback of the environment, and secondly from the *supervised learning pattern classification* algorithm which minimises a cost function via a supervised gradient descend method. The framework in which the A_{rp} algorithm had been described consists in an automaton and an environment connected in a feedback loop. The automaton produces an action, and the environment responds to this action by an evaluative signal which is usually interpreted as a success or a failure. The way of an automaton for producing an action is inspired from the *supervised learning pattern classification* algorithm and depends on a vector of parameters termed C, and on its input vector termed I. In the A_{rp} algorithm, the decision of the automata is probabilistic and depends on a noisy threshold : t. At the step k, the automaton decision s^k can be computed as :

$$s^k = 1 \text{ if } C^{kT}.I^k + t^k \geq 0 \text{ ; and } s^k = -1 \text{ otherwise.}$$

The parameter vector is updated depending on the response of the environment (termed b^k). In the case of success ($b^k = +1$) the vector is updated according to :

$$C^{k+1} = C^k - \mu^k.[E\{s^k|C^k,I^k\}-b^k.s^k].I^k.$$

In the case of failure, the vector is updated according to :

$$C^{k+1} = C^k - \alpha.\mu^k.[E\{s^k|C^k,I^k\}-b^k.s^k].I^k.$$

where $\mu > 0$ and $0 < \alpha < 1$. The scalar α is used for privilege reward or penalty.
As the description above lead us to think, the A_{rp} algorithm is more understandable in the case of two automata actions ($s^k = \{+1,-1\}$).The parameter vector adaptation is then more explainable : if the action at the step k conduces to a success response, the same action is encouraged by the term $b^k.s^k$, in the opposite case, in the case of failure, the opposite action is then encouraged by the same term $b^k.s^k$. One theorem of convergence has been proved by authors in the case of input vectors linearly independent and in the case of a two-action-automaton. Extension of this theorem to several automata is not trivial and has not, to our knowledge, yet been done. Nevertheless the neural network presented in the next section is one example of these stochastic automata sets.

II.2. Application to one layer network

II.2.a. Architecture of the network

The application of the A_{rp} algorithm to a neural network is straightforward, and suggested by the authors. We have used the simplest network : a one layer network with binary stochastic neurons. Binary stochastic neurons are necessary for the respect of the original algorithm. The architecture with only one layer consists on an assembly of elementary "stochastic automata" as defined by Barto.
A posteriori it appears that the better network in both cases of gait learning and obstacle avoidance is similar, its synthetic version is described hereafter.
The network is composed of six neurons arranged in one layer like a perceptron. Each neuron is affected to one particular leg.

Globally, the neuron's output codes the movement whereas the inputs receive the position of the legs (Fig. 1).

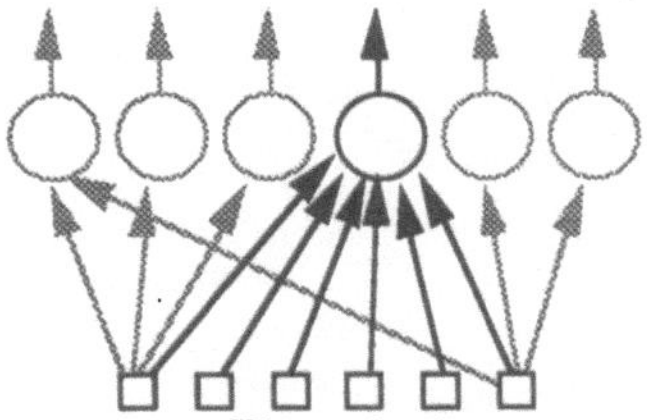

Outputs (S) : movement of the leggs (s_j, i=1 to 6)

Inputs (I) : position of the leggs (i_j, j=1 to 6)

Figure 1 : One layer neural networks. Neurons are shown like circles, their inputs are visualised as squares and the coefficients figure as links. Each neuron is connected to all inputs. The connections of one neuron are fully shown in black links, others connections are not totally visualised (some of them appear in grey links).

The iterations of the network are parallel. Each iteration is decomposed such that :

-> For each neuron (i = 1 to 6), compute the weighted sum :

$v_i : \Sigma_{j=1,6}\, c_{ij}.i_j$.

The parameters c_{ij} are the synaptic coefficients (the parameters of the automaton for Barto *et al*), i_j is the input producing the position of the leg j, and v_i is usually termed the "potential", in analogy with the biologic post-synaptic potential).

-> For each neuron, compute its output (or its action, for Barto et al). The output is evaluated applying a noisy threshold to the potential. One has :

$s_i = \mathrm{sign}(v_i + b_i)$,

where sign(.) is the sign function and b_i is a noisy biais. The noise is chosen in order to obtain the binary output (si) depending on the potential value :

$s_i = +1$ with the probability : $P(+1) = \dfrac{1}{1+\exp(-2ßv_i)}$.

$s_i = -1$ with the probability : $P(-1) = \dfrac{1}{1+\exp(+2ßv_i)}$.

ß is a parameter which has to be fixed before learning.

II.2.b. Learning

The learning procedure of a neural network consists in the computation of the synaptic coefficients (c_{ij}, i and j = 1,6).

At the beginning, the coefficients are initialised so as to have equiprobabilty of movement for each neuron (Prob=0.5). At each iteration the environment response (r) is taken into account and modifies the coefficients in order to increase the probability of "good" movements. When the procedure works well, at the end of learning, the behaviour is deterministic.

The coefficient's modification proposed by Barto *et al* uses the expectation of the output s_i, this value can be computed (*and not estimated by statistics*) in this way :

$E(s_i|c_{ij}, i_j) = (+1).P(+1)+(-1).P(-1) = \mathrm{th}(ß.v_i)$.

Then the learning rule can be written as :

If r = +1 (reward), $\Delta c_{ij} = \mu^{+}(r.s_i - E(s_i|c_{ij}, i_j))i_j$.

If r = -1 (penalty), $\Delta c_{ij} = \mu^{-}(r.s_i - E(s_i|c_{ij}, i_j))i_j$.

μ^{+} and μ^{-} are the two parameters controlling the step of the increment. As ß, μ+ and μ- have to be chosen before learning, their role will be described in a future section.

III. Hexapod Locomotion

III.1. Formulation of the problem

III.1.a. Coordination problem

The problem of locomotion of a legged robot is to perform the leg coordination in order to adapt itself to rough ground. Among the possible approaches, the one which takes inspiration in biological systems is interesting. Usually, the modelling of insect gait is based on precise knowledge of dynamics and physiological processes. Previous researches about gait learning have shown that the leg coordination can be obtained without the requirement for a central supervisor [6]. Our realisation of a legged robot aims to validate these studies. The starting point is that coordination and therefore walk are not inborn, but arise from a learning. The use of the A_{rp} algorithm leads to a partially supervised learning : with the only aim to go forwards without falling down, and without an exact knowledge of how to move, the robot adopts a behaviour in reaction to external stimuli (*i.e* the information of fall or advance from sensors). At the beginning of procedure, the robot moves its legs randomly. The movements which lead to an advance are encouraged and the probability of their occurrence increases. The movements which lead to a fall are progressively eliminated.

III.1.b. Why use six legs ?

The choice of an hexapod robot leads in the first stage to a simplification of the real device since the walk can be implemented without taking into account the dynamical model. The gait is stable if the centre of gravity is always inside the basis of support. Restriction to a walk (opposite to a trot or a gallop), limits essentially the speed of the robot.

III.1.c. Neural network definition

The neural network is the one presented in section II.2. As defined previously, each neuron output controls the movement of one leg : put forward or put backward. When at least one leg on each side of the robot is put backward, the robot takes a step.

All the steps have the same length then if the output of a neuron is identical two times successively, the movement can't be redone, then it is inhibited. Because of parallel iterations, movements are synchronous.

We have chosen to code the leg's movement rather than their position in order to decrease the dimension of the output space. Consequently, the time of convergence will be decreased.

The learning procedure is the one described in section II.2.b. It can be explained like this : after one movement, if the robot fails (fall down or immobility), the previous movement was wrong, then the robot is encouraged to do the opposite movement. If the robot takes a step, it is encouraged to do the same movement.

III.2. Result of the simulation

III.2.a. Interaction with exterior world

Two studies have been done concurrently : the first one is the electromechanical conception of the robot and the second one is the realisation of the software, to simulate learning in a first step, and after to implement it into the robot.

In order to minimise the differences between the two aspects of virtual and real world we took into account some interactions :

- Initial position of the robot.
- Position after a fall.
- Static model of the robot.
- Forbidden movements : these movements, avoided by "muscular" filters had been distinguished in two categories : those which break the robot, and those which are physically impossible to do (move the robot forward with only one leg...).

III.2.b. Tripod Walk

When the parameters are well adjusted, and if the robot is initialised in a stable position, the learning converges towards a solution, in less than 100 parallel iterations, with a probability of 0.7.

The most frequently found walk is a "tripod" walk (fig.2.a). This walk is biologically plausible and used by insects [7]. Depending on the initial position, another walk (Fig. 2.b) can occur.

III.2.c. Learning convergence

The convergence towards an hexapod gait depends on three parameters : the slope of the probability (ß), and the two learning parameters μ^{+} and μ^{-}. The role of these parameters and of the initialisation is described hereafter.

III.2.d. Influence of $\mu^{\pm}$

The relative magnitude between μ^+ and μ^- has to reflect the ratio between the number of states contributing positively (μ^+) and the number of states contributing negatively (μ^-) in the state space during the learning. In the case of gait learning, there are fewer "good" than "wrong" actions, therefore we have to choose $\mu^+ >> \mu^-$.

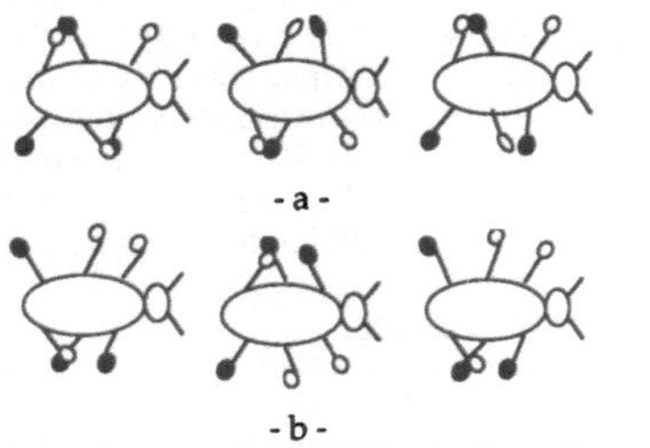

Figure 2: Walks found by the learning.
The foot coloured in black represents a leg on the ground (push backward), and the foot coloured in white, a leg moving forward.

When the learning works well, the convergence of the coefficients evolve according to two behaviours : the first one corresponds to a movement leading to a penalty, this case is the more frequent and induces the coefficients to evolve softly (small slope on the figures); the second case corresponds to a "good" movement, this case is rare and leads generally (depending on the term $(r.s_i - E(s_i|c_{ij}, i_j))$ to a great increment (large step on the figures : $\mu^+ >> \mu^-$). The convergence occurs always following a reward increment.
Depending on the initialisation and stochastic process, the learning search into the output state space is never identical : an example of two searches is shown (Fig. 3.a., 3.b.).
If the parameters μ^+, μ^- do not have a good ratio the learning procedure may not converge (Fig.3.c).

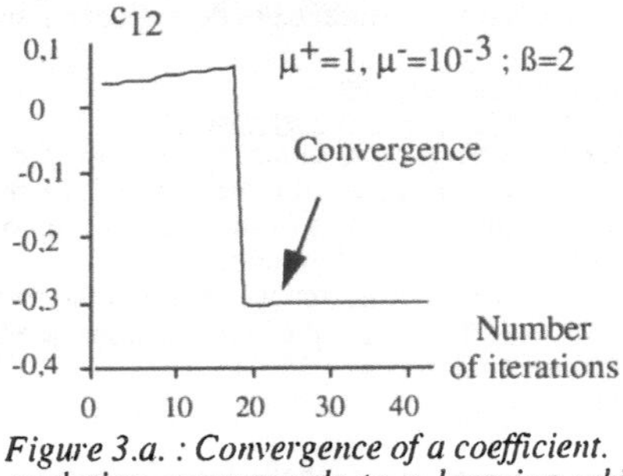

Figure 3.a. : Convergence of a coefficient.
This evolution corresponds to a learning whith a good choice of parameters : $\mu^+ >> \mu^-$ ($\frac{\mu^+}{\mu^-} = 10^3$). The convergence occurs after about 20 iterations.

III.2.e. Influence of the noise

The higher the slope of the probability function, the lower the effect of the noise. The noise is crucial because it allows the exploration of the output state space in order to find an original solution. If the slope is too high, the network is deterministic and doesn't learn (cycles). If the slope is too shallow, the learning process doesn't converge (Fig. 4).
The magnitude of the slope is estimated in relation with the amplitude of the parameters μ^+ and μ^-.

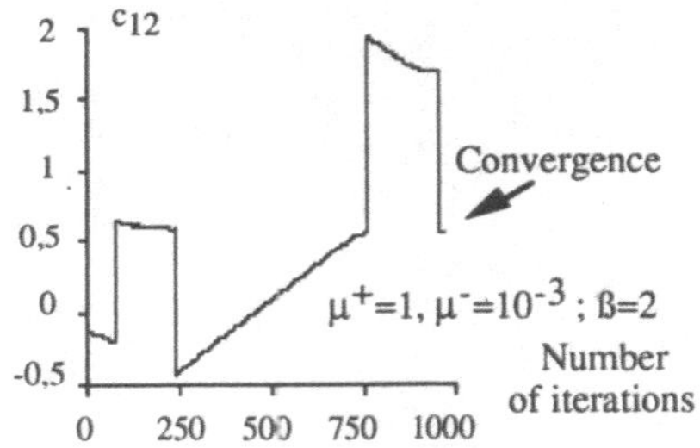

Figure 3.b. : Convergence of a coefficient.
This example is taken with the same good choice of parameters ; the learning stochastic evolution leads in this case to a slow convergence.This case is relatively rare.

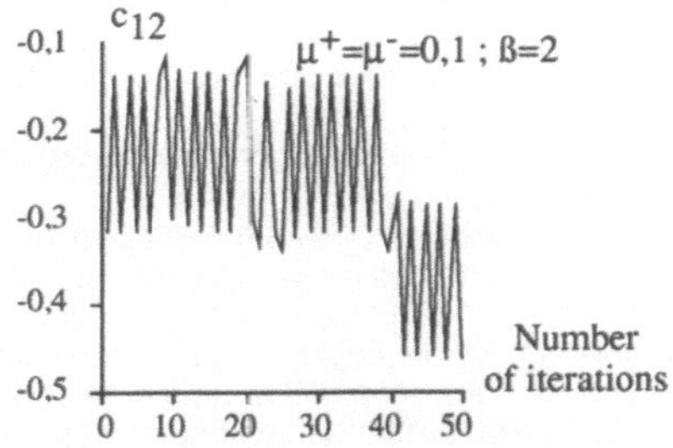

Figure 3.c. Evolution of a coefficient.
This figure shows an example of bad evolution of "learning" because of the inappropriate choice of parameters. In this case we have chosen $\mu^+ = \mu^-$. The "learning" can't converge because the penalties are preponderant relatively to the rewards. The learning doesn't work, there is no convergence.

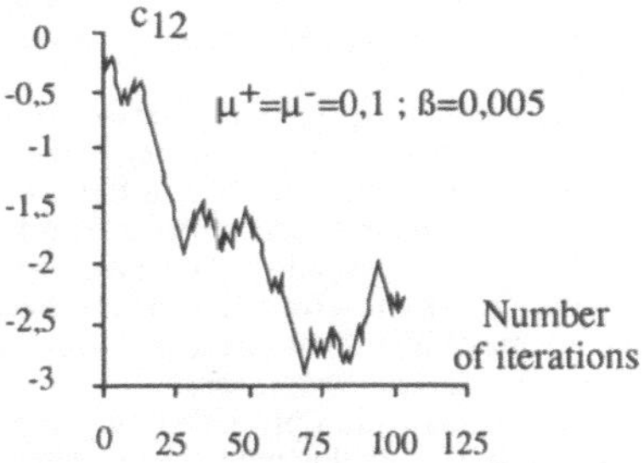

Figure 4. : Evolution of a coefficient.
The slope ß is very low, then the noise is greater than the potential of the neuron : the behaviour of the learning procedure is essentially stochastic ; there is no convergence.

III.3. Comparison with the real robot behaviour.

The electromechanical design of the hexapod has been constrained in consideration of dimensions, weight, architecture, cost, mechanical and electrical power. Sensors provide information to the robot about its leg's positions and a possible fall. The robot is 27 cm long and 15 cm wide. All the sensors, signal processing and motors control electronics is on board. Only the power source is external and comes to the robot via a cable [8].
The high potential capabilities of the robot allow first the control of amplitude and speed of the movement, and secondly the management of exteroceptive sensors. They should give us the tool to investigate some problems as the following of curved

trajectories, the crossing of obstacles and the changes of gait. Unhappily, the mechanical realisation of the robot doesn't allow us to test as many configurations as we would like to. Nevertheless the behaviour of the robot, taking into account its limitations is in good accordance with what we could wish. An other version of the robot, with a new mechanical conception, is in progress.

IV. OBSTACLE AVOIDANCE

The network devoted to control the obstacle avoidance is similar to the first one, used for gait learning. It is composed of six neurons, one for each leg. The inputs are also the leg's position. The difference between the two networks is the response of the environment. In the case of gait learning the robot has to advance and the response depends on the success or failure of this task. In the case of obstacle avoidance, the robot has a more sophisticated task to perform which includes the previous task (advance) adding the function of "no obstacle". This function is defined using a fictitious half of a sphere at the head of the robot in which no obstacle can stay (as it should be performed by an ultrasonic sensor). An evaluation of this task is done by the "environment" which responds by a binary signal. This signal is used during learning of obstacle avoidance.
Only simulations have been done on this problem and the results are satisfactory. When the simulated robot meets an obstacle, it tries first to go left and to go right randomly. After this first stage the simulated robot takes a "strategy" and chooses between left and right. This choice led it to shunt the obstacle. The time taken for evasive action is small : it can be 40 parallel iterations.
The behaviour of the obstacle avoidance is more difficult to appreciate than the one related to gait learning because of the increased complexity of the task. We can observe that the simulated robot avoids obstacles but we are searching a way of measuring its learning. In other words we can't say now if the robot learned the task, or if it is always adapting itself to a new configuration. This interesting question is actually one of our themes of research in order to pursue this study.

V. GLOBAL DESCRIPTION

The proposed global architecture includes the two networks as separated networks. The solution using only one network computing the two tasks of gait learning and obstacle avoidance leads to a more complicated problem for two reasons :
- The "alone" network needs to be more complicated than the two simpler networks, thus the search in the space of the coefficients would be more difficult.
- The binary information coming from the environment is different for the two tasks, thus there is no sense in applying it to the same coefficients.
One global simulation has been done including the two networks and the two responses coming from the environment (Fig.5). The switch between one network to the other is triggered by the sensing or the vanishing of an obstacle. In this case, the behaviours of both networks are strictly identical to the behaviours obtained separately.

VII. CONCLUSION

In the field of the very hard problem of the mobile robot conception, following the idea stating that a complicated behaviour could be obtained from simple devices, and inspired from the biological example, this paper has presented neural methods able to learn walking and to avoid obstacles. The simplicity of these methods is surprising and leads us to think that the spirit of this study could be generalised to other problems. The real experimentation of the algorithms is not forgotten and for validating the results a new version of the actual little hexapod robot built at the LG2IP Laboratory is in progress.

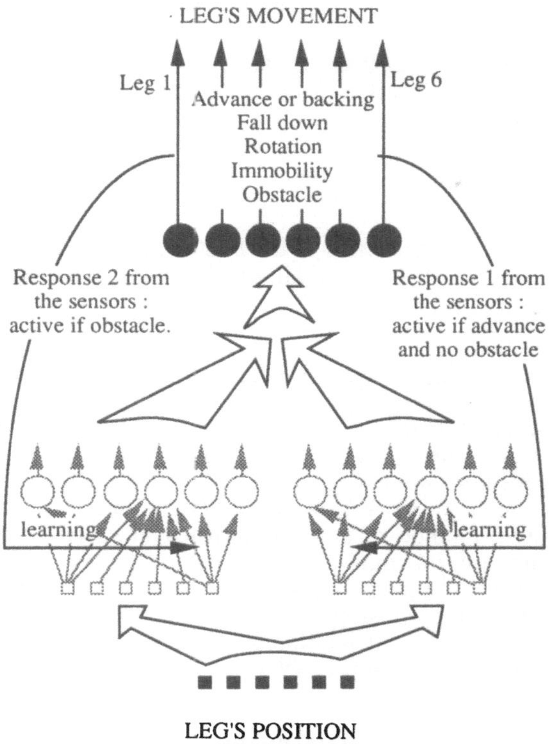

Figure 5 : The global network architecture. The global architecture includes the both processes of gait and obstacle avoidance learning. The two tasks are implemented by two networks. Only one network is active at time, the choice of the active network is triggered by the informations incoming from the environment (obstacle or not). The environment has two responses depending on the active function.

ACKNOWLEDGEMENTS

The authors want to thank warmly M. ARTIGUE, M. MEIMOUNI and M. SILVAIN for their contribution to the realisation of the robot.
M. CANAL and M. LANG, students from the Alès School of Mines, are also greatly thanked for their enthusiasm and their work on the difficult problem of obstacle avoidance.

REFERENCES

1. R.A. Brooks, "Intelligence Without Representation". Artificial Intelligence 47, 139 (1991).
2. R.A. Brooks, "A Robot that Walks:Emergent behaviors from carefully evolved network,". Neur. Comput. 1(2):253-(1989).
3. A. Frank, "Walkers". "Encyclopaedia of robotics: application and automation" DGRF R.C. ed. Wiley Interscience, (1988).
4. "Vehicles". V. Braitenberg; MIT Press 1984.
5. A. G. Barto et P. Anandan,."Pattern Recognition Stochastic Learning Automata". IEEE Trans. S.M.C. 15 : 360 (1985).
6. C.Touzet, O.Sarzeaud. "Application d'un algorithme d'apprentissage à la génération de formes locomotrices hexapodes". Journées de Rochebrune AFCET (1992).
7. D. M. Wilson. "Insect Walking". Annual Rewiew of Entomology 11 : 103-122. (1966).
8. A. Johannet, I. Sarda. "Gait Learning of an Hexapod Robot : from Simulation to Realization". Proceedings of the 2nd Japan-France Congress on Mechatronics. Takamatsu, Japan,(1994).

USING EVOLUTIONARY COMPUTATION TO GENERATE TRAINING SET DATA FOR NEURAL NETWORKS†

Dan Ventura
Tim Andersen
Tony R. Martinez

Provo, Utah 84602Computer Science Department, Brigham Young University
e-mail: dan@axon.cs.byu.edu, tim@axon.cs.byu, martinez@cs.byu.edu

Most neural networks require a set of training examples in order to attempt to approximate a problem function. For many real-world problems, however, such a set of examples is unavailable. Such a problem involving feedback optimization of a computer network routing system has motivated a general method of generating artificial training sets using evolutionary computation. This paper describes the method and demonstrates its utility by presenting promising results from applying it to an artificial problem similar to a real-world network routing optimization problem.

Introduction

Many inductive learning algorithms based on neural networks, machine learning, and other approaches have been developed and have been shown to perform reasonably well on a variety of problems [2][4]. Typically, neural networks (NN) perform inductive learning through the presentation of preclassified examples; however, one of the largest obstacles faced in applying these algorithms to real-world problems is the lack of such a set of training examples. Many times collecting data for a training set is the most difficult facet of a problem.

This paper presents such a real-world problem -- one for which no training data exists and for which gathering such data is at best extremely expensive both in time and in resources. To remedy the lack of training set data, a method using evolutionary computation (EC) [3][8] is described in which the survivors of the evolution become the training examples for a neural network. The synthesis of EC with NN provides both initial unsupervised random exploration of the solution space as well as supervised generalization on those initial solutions. Work involving a combination of EC and NN is becoming more prevalent; the reader is referred to [1][5][6][7] for examples.

† This research was funded in part by a grant from Novell, Inc.

Problem Description

Although we describe here a specific problem, the approach described in the next section is general enough to apply to a large class of optimization/feedback problems. We are specifically concerned, however, with a computer networking application and the constant setting/adjusting of certain control variables depending on the values of certain status variables. More formally, given a network, Θ, the state of Θ may be described at time t by a vector of status variables, s^t. Control of the network is effected by the setting of variables in a control vector, c. That is, given a network Θ at time t described by vector s^t, the setting of the values of the vector c will result in a different network Θ' at time $t+\delta$ described by the vector $s^{t+\delta}$. The problem is, given a status vector, s^t, what modifications should be made to the control vector c such that $s^{t+\delta}$ describes a better network, if possible, than s^t? Here, the best network is defined as the network that maximizes some function of its status variables, $f(s)$. The operation of Θ is continuous and feedback from Θ in the form of s is continuous as well. A neural network is expected to monitor the values of s either continuously or periodically and to update the values of c, the goal being to maximize the performance of Θ over time. In this particular problem 21 status variables (e.g. AverageRoundTripTime, TotalPacketsSent, TotalPacketsReceived, etc.) and 11 control variables (e.g. SendWindowSize, RetransmitTimeInterval, MaxRetransmitCount, etc.) have been identified.

Generating the Training Set

Assuming that the status variables are defined over even a modest range, it is obvious that Θ may be described by any of an extremely large number of unique state vectors. In other words, the search space defined by the number of unique networks is enormous. Even if we limit ourselves for a moment to considering a single network by "freezing" the values of s, we are still faced with a similar problem in choosing values for c. With the intractable search spaces involved, evolutionary computation suggests itself for the exploration of those spaces. Actually, we limit ourselves to exploring, via evolutionary computation, the solution space for c only.

From the space defined by s that describes Θ we choose a representative set of network states by choosing n initial status vectors. We denote these s_i, $0<i\leq n$ and refer to the network state described by s_i as Θ_i, $0<i\leq n$. These choices could of course be biased by any a priori heuristics as to what constitutes a realistic network. In choosing this set of status vectors, s_i, $0<i\leq n$, we have chosen the left hand sides of our training instances. We now use evolutionary computation to discover "good" right hand sides yielding training instances of the form $s_i^{t=0} \rightarrow c_k$.

Assume a fitness function f that takes as input a status vector s and returns a real-valued fitness measure. Now for each $s_i^{t=0}$, randomly initialize a population of m control vectors, denoted c_k, $0<k\leq m$. Evaluate the initial population by simulating the workings of $\Theta_i^{t=0}$ for each c_k for δ time steps, where δ time steps are sufficient for Θ_i to stabilize, and then applying fitness function f to $s_i^{t=\delta}$. Next choose parents and use genetic operators to produce m offspring. Now evaluate the children and select m survivors from amongst the parents and children. The algorithm is sketched in figure 1.

```
evolve()
    generate s_i
    for each s_i^{t=0}
        initialize population c_k, 0<k≤m.
        evaluate(c_k,Θ_i)
        until(done)
            select parents from c_k
            apply genetic operators to parents
            evaluate(children,Θ_i)
            c_k ← choose m survivors from c_k and
children

evaluate(c,Θ_i)
    run Θ_i^{t=0} for δ time steps with control vector c
    return f(s_i^{t=δ})
```

Figure 1. Sketch of algorithm for evolving training set

Finally, choose j individuals from each of the n populations and build a set of jn training examples of the form $s_i^{t=0} \rightarrow c_k$. (To avoid ambiguity in the training set we could set $j=1$.)

We now return to the infinite space defined by s. Since we have only chosen a finite number of seed points from this space, our evolutionary computation has found approximate solutions for only these n points in the space and can say nothing about any other points, many of which we are likely to encounter during normal execution of Θ. Therefore it becomes necessary to generalize on this relatively small set of approximate solutions. Using this set of approximate solutions as training examples, an NN model can be trained to develop a general hypothesis over the entire space defined by s.

Simulation

In order to establish proof of concept, simulations using artificially generated problems were run. The simulation process includes the following steps:

1. Generate a problem definition
2. Create a training set using the algorithm
3. Train an NN with the training set
4. Create a test set
5. Test the NN network on the training set

An artificial problem generation/simulation program was used for two reasons. First, it is much easier to work with in terms of analysis, reproduction of results, etc. Second, it is possible to create a test set which can be used to show how well the NN is performing in relation to optimum, and thus to establish (to some extent) the quality of the EC-generated training set.

Generating a problem definition

In the networking problem, s is a vector of p status variables and c is a vector of q control variables. At any time t, $s^t = g(s^{t-\delta}, c^{t-\delta})$, where g represents the operation of the network for time δ. Unfortunately, because of the network's asynchronous nature, it is impossible to determine the length of δ necessary for the network to stabilize or to accurately define g.

Therefore, we develop an artificial problem that is a simple analog of the network routing problem that preserves the relationship among s^t, s^{t-1}, and c^{t-1} but that operates in definite time steps and employs a simpler function H. Given a vector s of p integer variables (this vector is the analog of the status vector so we abuse notation and call it s) and a vector c of q integer variables (the analog of the control vector), define the relationship $s^t = H(s^{t-1}, c^{t-1})$ where H is in turn defined as a vector of functions $h_1, h_2, ..., h_p$. Now, to generate a problem, generate a $p \times q$ matrix G whose elements are randomly selected from {+,-}. The function h_i calculates the value of s_i^t from s_i^{t-1} and c^{t-1} as follows

$$s_i^t = h_i\left(s_i^{t-1}, c^{t-1}\right) = \sum_{k=1}^{q} G_{ik}\left(s_i^{t-1} - c_k^{t-1}\right).$$

Thus, a problem is completely defined by H and G, which together represent a simple analog for the operation of a network. Once H and G have been defined, the combination of EC (to create a training set) and NN (to generalize on that training set) can attempt to learn the problem. To generate a new and different problem, randomly regenerate the matrix G; this allows for the generation of $\binom{2q}{p}$ unique problems. Note that s_i^t (c_i^t) has been used to indicate the *ith* vector at time t, while s_i^t (c_i^t) is now used to indicate the *ith element* of a vector s (c) at time t, a slight abuse of notation.

As an example, suppose p=3 and q=3, that is, there are 3 "status" variables and 3 "control" variables. In order to generate a problem, randomly generate a 3×3 matrix G. One possibility is

$$G = \begin{bmatrix} + & - & - \\ + & + & + \\ - & + & - \end{bmatrix}$$

The system of equations that define the problem is then:

$$s_1^t = +\left(s_1^{t-1} - c_1^{t-1}\right) - \left(s_1^{t-1} - c_2^{t-1}\right) - \left(s_1^{t-1} - c_3^{t-1}\right)$$
$$s_2^t = +\left(s_2^{t-1} - c_1^{t-1}\right) + \left(s_2^{t-1} - c_2^{t-1}\right) + \left(s_2^{t-1} - c_3^{t-1}\right)$$
$$s_3^t = -\left(s_3^{t-1} - c_1^{t-1}\right) + \left(s_3^{t-1} - c_2^{t-1}\right) - \left(s_3^{t-1} - c_3^{t-1}\right).$$

Creating a training set

Once the problem has been defined, the evolutionary algorithm is employed to generate a training set. The function H and matrix G are substituted for the step "run $\Theta_i^{t=0}$ for δ time steps with control vector c" in the *evaluate* procedure. The fitness function may be arbitrary, and the one used in these simulations was simply:

$$f(s) = \sum_{i=0}^{p} s_i$$

All simulations used training sets of size 1000.

Training the NN

The PDP group's software implementation of the backpropagation algorithm[4] was used for all simulation results. For the simulations we defined 5 integer "control" variables and 10 integer "status" variables on the range [0, 100). Note that even though we limit the problem to integer values, the search spaces are extremely large: 100^5 for "control" vectors and 100^{10} for "status" vectors. The inputs to the backprop (the "status" variables) were binary encoded, while the outputs were simple localist nodes, one for each control variable. Fifteen hidden units were used so that the entire network consisted of 70 input nodes, 15 hidden units, and 5 output nodes. The output nodes produce values in the range [0, 1) and these are simply multiplied by 100 to produce a "control" vector in the desired range. All simulations were run with the default settings and training was only allowed to proceed for 500 epochs.

Creating a test set

The main advantage of having an explicitly defined function H representing our problem is that given a status vector s^t, the optimal control vector c (the one that optimizes s^{t+1}) can be calculated. The simplest method is by brute force, though this can be significantly improved in the case of this particular function H. To create a test set, randomly generate a vector s and then determine the optimal vector c. Add the instance $s \rightarrow c$ to the training set and repeat for as many instances as desired. All simulations used a test set of size 50.

Testing the NN network

Typically, when evaluating how well the NN performs on a given training set we determine the number of correctly classified instances from the test set. However, because this is an optimization problem and because the output nodes are localist and expected to produce outputs in the range [0, 1) (rather that just being active or not), the NN is not expected to produce the optimum vector c_{opt} but rather a good approximation of it, c_{approx}. Thus, given a status vector s the NN generates c_{approx}, while c_{opt} and c_{worst} (needed for normalization) may be found by brute force (or some more efficient method). A measure of the correctness (normalized % of optimum) of c_{approx} is

$$\frac{f(s_{approx}) - f(s_{worst})}{f(s_{opt}) - f(s_{worst})}$$

where $s_{approx}=H(s,c_{approx})$, $s_{worst}=H(s,c_{worst})$, and $s_{opt}=H(s,c_{opt})$.

Empirical Results

The simulation process as described in the previous section was performed for genetic population sizes (the upper bound m in the *evolve* algorithm) of 20, 100, and 200. Ten different simulations were run for each value of *m* and the results are reported in table 1 The % of optimum is the correctness measure discussed above where c_{approx} is produced by either the GA (training set) or the NN (test set). The average pss is the average sum-squared error per pattern.

Table 1. Simulation results for different population sizes

	Population Size (*m*)					
	20		100		200	
	mean	sd	mean	sd	mean	sd
% opt. (train)	73.8	8.1	87.6	5.5	89.9	4.6
% opt. (test)	70.8	10.3	80.9	6.8	85.1	8.6
avg pss (train)	.237	.037	.128	.023	.103	.023
avg pss (test)	.667	.145	.364	.102	.327	.102

A number of interesting observations may be made from the table. First, the values for percent of optimum for the training set indicate that the GA is finding artificial instances that seem reasonable in the sense that they are a significant percentage of optimum. Second, the values for percent of optimum for the test set indicate that the training set produced by the GA enabled the NN to learn the problem fairly well, since it performs almost as well as the GA. Even more significant is the fact that as we increase the population size, both the GA and the NN perform closer to optimum with smaller standard deviations. Also, as the population size is increased, the sum-squared error on both the training set and test set decreases indicating that the training set more closely represents the true function.

Conclusion

Empirical simulation results suggest that at least for some problems, evolutionary computation is indeed capable of generating a training set that closely represents the actual underlying function. The key to this process is the assumption that an appropriate fitness function *f* can be defined. We conjecture that in many cases this will indeed be the case (e.g. in the networking problem *f* would attempt to maximize throughput while minimizing resource usage). This paper has developed proof-of-concept on a relatively simple problem. Current research involves application of this method to the real network optimization problem and other difficult real-world applications. Unfortunately, ascertaining what percentage of optimum the approximate solutions achieve will not be possible (since the function *g* will be unknown). A measure of performance will have to be obtained by comparing this approach with current methods for solving these problems. As a final note, although generating the training set and training the NN is potentially time consuming, this need be done only once (or very infrequently), and it is conceivable that further training could be ongoing in the background using results obtained during execution. The actual execution would be extremely fast.

References

[1] Caudell, T. P. and Dolan, C. P., "Parametric Connectivity: Training of Constrained Networks using Genetic Algorithms", *Proceedings of the Third International Conference on Genetic Algorithms*, 1989.

[2] Falhman, S. E. and Lebiere, F., "The Cascade-Correlation Learning Architecture", Advances in Neural Information Processing 2, D. S. Touretzky (ed.), Morgan Kaufman, 1990.

[3] Goldberg, D. E., Genetic Algorithms in Search, Optimization, and Machine Learning, Addison-Wesley Publishing, 1989.

[4] McClelland, James L. and Rumelhart, David E., Explorations in Parallel Distributed Processing, MIT Press, Cambridge, Massachusetts, 1988.

[5] Harp, S. A., Samad, T., and Guha, A., "Designing Application-Specific Neural Networks Using the Genetic Algorithm", *NIPS-89 Proceedings*, 1990.

[6] Montana, D. J. and Davis, L., "Training Feedforward Neural Networks Using Genetic Algorithms", *Proceedings of the Third International Conference on Genetic Algorithms*, 1989.

[7] Romaniuk, Steve G., "Evolutionary Growth Perceptrons", *Genetic Algorithms: 5th International Conference (ICGA-93)*, S. Forrest (ed.), Morgan Kaufman, 1993.

[8] Spears, W. M., Dejong, K. A., Baeck, T., Fogel, D., de Garis, H., "An Overview of Evolutionary Computation", *European Conference on Machine Learning (ECML-93)*, 1993.

OPTIMAL TRAINING PATTERN SELECTION

USING A CLUSTER-GENERATING ARTIFICIAL NEURAL NETWORK

T.Tambouratzis,

Institute of Informatics & Telecommunications,
NRCPS "Demokritos",
Aghia Paraskevi 153 10, Greece.

D.G.Tambouratzis,

Department of Mathematics,
Agricultural University of Athens,
Iera Odos 75, Athens 118 55, Greece.

Address for correspondence: Department of Mathematics, Agricultural University of Athens, Iera Odos 75, Athens 118 55, Greece.

Abstract:

In this piece of research, the problem of optimal pattern selection for artificial neural network training is investigated. Given an initial set of training patterns, the objective is to extract the minimal subset which accurately represents the initial set.

The training pattern selection strategy which is presented here is based on the clustering capability of the Harmony Theory simulated-annealing harmony-maximisation artificial neural network [1]. Correction and reduction of the training set is achieved by rectifying the repetition and misclassification errors as well as by organising and unifying the training patterns in terms of their similarities.

1. Introduction

The **extraction of a number of appropriate patterns which accurately represent a particular domain** is a crucial issue for the purposes of curve-fitting through a number of observations. It is worth mentioning that curve-fitting may be performed in many ways, e.g. by interpolation, by extrapolation or by artificial neural network construction and training. Pattern extraction aims at minimising the classification error, in other words, at accomplishing successful separation of the classes which exist within the domain.

Another equally important issue - especially for practical applications - is the **ability to select the minimal <u>and</u> most representative subset of a selected set of patterns**. Efficient selection of such a subset not only lessens the computational load required for class separation, but also keeps the classification error sufficiently low. The two issues of pattern extraction and pattern selection are, indeed, related: the **quality of the selected optimal subset of patterns** is **directly limited by** the **quality of the extracted original set of patterns**.

This piece of research is concerned with the selection of a minimal and optimal subset of patterns and does not - thus - refer to the degree to which the initial set of patterns accurately represents the particular domain. The issue under examination is known as **optimal (training) pattern selection**, the term "training" being employed especially when optimal pattern selection is performed for the purposes of artificial neural network construction and training. The **clustering capability of the Harmony Theory simulated-annealing harmony-maximisation artificial neural network** is utilised here in order to select the minimal and optimal training pattern set from a given set of patterns.

2. Artificial Neural Network Trainin. and Optimal Training Pattern Selection

The efficacy of artificial neural network training is affected [2] by such factors as:

- network architecture,
- activation function,
- training parameters,

- initial connection weights,
- training pattern selection strategy,
- number of training epochs,
- order and frequency of presentation.

As far as efficient artificial neural network training is concerned, the issue of optimal training pattern selection has not yet received much attention. Under most artificial neural network training methods, all training patterns are presented equally often, in random order for each training epoch and for a generally large number of epochs. **Pedagogical training** [3-4], that can be employed instead, increases the presentation frequency of those training patterns which produce high error values relative to those training patterns which produce low error values. Additionally, a number of highly specialised strategies has been presented for the construction of optimal training pattern sets (e.g. [5]). These strategies, however, are only applied during the fine-tuning of partially pre-trained artificial neural networks.

In the unlikely but **ideal case** that the initial set of patterns perfectly represents the problem domain, the issue of optimal training pattern selection refers solely to the **reduction of the redundancy and**, consequently, the **cardinality of the initial set**, without allowing a rise in the classification error. Here, both the initial set and the selected subset of patterns must produce zero classification error.

Choosing the optimal and minimal subset of patterns becomes, however, particularly critical when the initial set does not perfectly represent the problem domain, but some of its training patterns are noised, erroneous or if a number of salient, atypical patterns of the domain are missing. Such phenomena frequently occur in **practical applications**, where the training patterns constitute the result of a feature-extraction or some other pre-processing stage. Consequently, serious **classification problems** appear, the most obvious and more easily rectifiable of which are:

- **The problem of repetition**, i.e. the existence of identical training patterns belonging to the same class.

- **The problem of misclassification**, i.e. the existence of identical training patterns belonging to different classes.

For non-perfect initial sets of patterns, the best possible use of the available training patterns must be made in order to:

- **Overcome the** two kinds of **classification errors** (repetition and misclassification).

- **Reduce the cardinality** of the training set by organising and unifying the training patterns in terms of their similarities and, thus **minimise** the time spent on artificial neural network **training and - subsequently - testing**.

- **Maximise correct classification** of the training set and - by extension - of the entire problem-domain during artificial neural network training and, subsequently, of the test set during artificial neural network testing.

3. A Clustering Artificial Neural Network for Optimal Pattern Selection Based on Harmony Theory

The proposed technique introduces a **novel (training) pattern selection strategy** which is based on a Harmony Theory simulated-annealing harmony-maximisation artificial neural network [1].

Harmony Theory constitutes *"a theoretical framework for studying a class of dynamical systems that perform cognitive tasks according to the account of the subsymbolic paradigm."* (see [1]). For a brief overview of the basic construction characteristics of Harmony Theory the reader is referred to [6] and for a more detailed treatise - especially concerning its theoretical and underlying mathematical aspects - to [1]. Harmony Theory artificial neural networks have been constructed, implemented and tested on **inference** (e.g. the electricity problem [1], the scene analysis problem of AI [7]) and **combinatorial optimisation** (e.g. the map-colouring problem [8]) tasks. Here, the **clustering capability** of Harmony Theory is introduced for optimal (training) pattern selection. The problem is treated as a **completion** task, and - more specifically - as a **pattern reconstruction and noise elimination task**.

The clustering capability of the Harmony Theory artificial neural network derives from the parameter *k* [1] of Harmony Theory. This parameter always lies in the interval (-1,1) and acts as a **criterion of compatibility** between the activation values of connected nodes **within the artificial neural network**:

- Theoretically (as stated in [1]) it has been proposed that, as the value of *k* approaches 1 (**perfect matching limit criterion** [1]), the activation values of all the nodes in the artificial neural network assume their desired values.

- In practice (for inference and constraint optimisation tasks), an **optimal interval** around 0.5 has been supported for *k* [7-8]. Outside this interval the parameter acts as an **inverse threshold**. At lower values, **partial activation** of the artificial neural network occurs, with only the most crucial nodes (the ones chiefly responsible for the activation flow and the simulated annealing process of the artificial neural network) becoming active. At higher values, **excessive and erroneous activation** occurs, with more nodes becoming active than is desirable for the accurate settling for the artificial neural network.

4. The Proposed Strategy

The Harmony Theory artificial neural network which accomplishes optimal pattern selection is constructed as follows:

- The **lower layer** represents the **feature-vector** of the patterns of the domain.

- The **upper layer** encodes the **set of patterns** from which the selected subset is to be derived.

Each pattern of the initial set constitutes a candidate for selection in the optimal subset and is entered, in turn, in the lower layer of the artificial neural network. The preliminary aim is to detect and rectify the repetition and misclassification problems, whereas a **first reduction and improvement of the initial set of patterns** is accomplished. This is not required if the initial set perfectly represents the domain. Subsequently, the **selection of the minimal subset of patterns** which accurately represent the initial set is performed.

The first reduction and improvement of the given set of patterns is executed by **assigning a very high *k* value** (above the perfect matching limit) to the Harmony Theory artificial neural network. At such a value, only "perfect matches" become activated, i.e. exactly those nodes of the upper layer become active whose encoded patterns coincide with the pattern which has been input in the lower layer.

Each candidate pattern is input, in turn, to the lower layer. If no nodes of the upper layer - except from the node representing the input pattern - become active, the pattern is retained for further processing. If, however, a **cluster of nodes** becomes active, **reduction** can ensue. In the case that the cluster represents patterns of exactly one class, a **repetition error** is detected and **all the other patterns coinciding with the input pattern are deleted from the initial set**. Alternatively, in the case that the cluster represents patterns of more than one class, a **misclassification error** is detected. Again, **all the other patterns coinciding with the input pattern are deleted** from the initial set but, additionally, **all classes represented by the cluster of nodes become attached to the retained pattern**.

The final selection of the minimal subset of patterns proceeds once the patterns which have been deleted from the initial set, as well as the patterns retained from misclassification errors, are also pruned from the Harmony Theory artificial neural network.

Each candidate pattern of the reduced set is input into the reduced artificial neural network and, for each, the **most appropriate value of the *k* parameter** is found. This value is determined by applying the **inverse of the specialised clustering technique** described in [9]. Each *k* is chosen so that the cluster of nodes contains as many patterns of the same class as the candidate pattern, but no patterns from other classes. Beginning from a high value of *k* at which only the node corresponding to the input pattern becomes active (e.g. a value equalling the perfect matching limit), k is progressively lowered until the cluster of nodes represents more than one class. The lowest value of k for which **the cluster of nodes contains patterns of only one class** is the value sought for the current input pattern. The cluster of nodes of the upper layer as well as the class they represent are retained for further processing.

Once all candidate patterns have been tested, **the optimal subset of patterns can be selected**. The **clusters of nodes are grouped** together so that each group of clusters is disjoint from the other groups, while each cluster of a group has a non-empty intersection with at least one cluster of the same group. More than one group of clusters may be needed to describe a single class, but only one class is represented by each group (due to the appropriate selection of the value of k). The **representative of each group** is determined by taking a **"majority" vote** between the patterns of all the clusters in it. **The selected representatives - one for each group - together with the patterns arising from misclassification errors constitute the optimal training set** devised by the proposed strategy. The patterns resulting from misclassification errors are not tested during the final selection of the minimal subset of patterns, since at no k value can patterns from a single class belong to their cluster of nodes. They are, however, added to the optimal subset of patterns as they constitute critical patterns for successful class separation.

The cardinality of the selected optimal set of training patterns depends upon the configuration of the domain. Well separated classes - which are topologically far apart from each other and whose patterns are closely grouped within each class - require few prototypes. Less easily separable classes demand more prototypes, especially at irregular borders between them.

5. Conclusions

A novel training pattern selection strategy has been presented. The strategy is based on the clustering capability of the k parameter of the Harmony Theory artificial neural network. The most appropriate values of this parameter are found. They accomplish rectification of the two kinds of classification errors (repetition and misclassification) and, additionally, assemble the patterns in terms of their similarities and the relative locations of the groups of clusters within as well as between classes.

7. References

1. Smolensky, P., (1986). *Information Processing In Dynamical Systems: Foundations Of Harmony Theory*, pp. 194-281, in D.E.Rumelhart, J.L.McClelland (eds), **"Parallel Distributed Processing: Foundations (Vol. 1)"**. MIT Press, Cambridge.
2. Rumelhart, D.E., McClelland, J.L., (1986). **"Parallel Distributed Processing: Explorations in the Microstructure of Cognition (vol. 1)"**. MIT Press, Cambridge, MA.
3. Munro, P.W., (1992). *Repeat Until Bored: A Pattern Selection Strategy*, in J.E.Moody, S.J.Hanson, R.P.Lippman (eds) **"Advances in Neural Information Processing"** 4, pp. 1001-1008. Morgan Kaufmann Publishers, San Mateo, CA.
4. Cachin, C., (1994). *Pedagogical Pattern Selection Strategies*, **Neural Networks** 7, pp. 175-181.
5. Hwang, J.N., Choi, J.J., Seho, O.C., Marks, R.J., (1991). *Query-based Learning Applied To Partially Trained Multilayered Perceptrons*, **IEEE Transactions on Neural Networks** 2, pp. 131-136.
6. Tambouratzis, T., (1995). *Simulated Annealing Artificial Neural Networks For The Satisfiability (SAT) Problem*, in Proceedings of the **"International Conference on Artificial Neural Networks and Genetic Algorithms - 1995"**, Mines d' Ales, France.
7. Tambouratzis, T., (1991). *An Implementation Of A Harmony Theory Network For Interpreting Line Drawings*, **Network: Computation in Neural Systems** 2, pp. 443-454.
8. Tambouratzis, T., (1993). *A Harmony Theory Network Solution To The Map-Colouring Problem*, in Proceedings of the **"International Joint Conference on Neural Networks 1993"**, Nagoya, Japan, pp. 1545-1548.
9. Sabourin, M., Mitiche, A., (1993). *Modelling And Classification Of Shape Using A Kohonen Associative Memory With Selective Multiresolution*, **Neural Networks** 6, pp. 275-284.

TRAINING SET SELECTION IN NEURAL NETWORK APPLICATIONS

Colin R Reeves
School of Mathematical and Information Sciences
Coventry University, UK

Abstract

In some applications of ANNs to classification problems, training can be inefficient simply because of the high volume of data available for training purposes. The question arises whether it is necessary to use all the data in order adequately to approximate the decision boundaries.

It is intuitively obvious that points near to the boundaries are likely to have more influence over the network weights than those which are far away. Of course, as the boundaries are initially unknown, the status of any particular data point is also unknown. Here we describe an approach which uses an initial crude estimate of the decision boundaries to select appropriate training data in the case of the Multi-Layer Perceptron, followed by a phased addition of points to the training set. We compare this approach with the standard method on both artificial and real data sets, and report results which demonstrates the potential for improved performance in terms of both efficiency and reliability.

Introduction

In many applications of artificial neural networks (ANNs), it is customary to partition the available data into (at least) two sets, one of which is used to train the net, while the other is used as a 'test set' to measure the generalization capability of the trained net.

It is often difficult to achieve the twin goals of adequate training and proper evaluation because of a lack of data, and some consequences of this are considered elsewhere [1]. However, it is not commonly recognized that it may also be a problem to have too much data, and it is this that is the subject of this paper. In what follows, we shall assume that the ANN which is being used is a multi-layer perceptron (MLP) which is being trained by back-propagation, but it should be realized that the problem is not limited to ANNs of this type, although it is particularly acute in such cases.

Motivation

The studies reported on here were originally motivated by an application of ANNs to the problem of DNA sequence prediction [2]. DNA is an extremely long string of base-pairs; some regions of this string are coding regions (or *exons*) which contain the 'instructions' for the cell to manufacture amino-acids, while other regions (*introns*) are not. Identifying whether an arbitrary region is an exon or an intron is a two-class discrimination problem. Various statistical indicators can be calculated for gene sequences for which the coding status is known, and these can then be used as input to train a neural net to distinguish between exons and introns.

The problem is that this leads to huge data sets—typically something like 10^6 cases can be generated even for relatively short genes. Training would thus inevitably be a long process, even for 'fast' ANN paradigms and/or training methods. The problem is not confined to this application—optical character recognition problems form another class of problems where vast amounts of data are generated, as reported recently [3]. It therefore seemed appropriate to investigate this problem in more detail. This paper will report on some preliminary trials of a method for reducing the size of training sets in the context of some rather smaller problems.

Training set selection

The 'ideal' training set would be a faithful reflection of the underlying probability distribution over the input space, but most actual training sets are far from ideal. Even in the 'simple' case of a random sample from a uniform distribution over a 2-dimensional input space, 'holes' and 'clumps' readily appear unless the sample is large. It would thus appear that a large training set is a good thing.

If the goal is to build up a complete picture of the underlying class probability distributions, a large data set is important. For the lesser task of producing a model that discriminates between the classes, it can be argued that not all the data are equally useful in training the net. For example, in a discrimination problem, it is intuitively plausible that data points which fall near the decision 'boundary' between two classes are likely to be more influential on the weight-changing algorithm than points which are well 'inside'. Similarly, if several points from the same class are very close to each other, the information they convey is virtually the same, so are they all necessary?

A random sample of the complete data set is clearly one possibility, but the above discussion suggests the need for some more systematic selection procedure. Work on the importance of boundary points has been reported before [4], but this considered only a simple problem, and the emphasis was on the effect on generalization performance of the exclusion of certain points, rather than on actively trying to find a good set. Since this work commenced, Plutowski has described a very comprehensive study [5] into the general question of training set selection, but in the context of function approximation rather than classification problems, as in the work reported here. The same is true of a similar approach reported by Röbel [6].

In order to investigate the systematic selection route more fully, it was decided that initially a fairly simple artificial 2-dimensional problem would be experimentally studied. This was a 2-class problem ('the circle in a square problem') with 2 inputs $\{(x_1, x_2) : -1 \leq x_i \leq 1; i = 1, 2\}$, defined as follows:

$$x_1^2 + x_2^2 < 2/\pi \rightarrow C_1$$

$$x_1^2 + x_2^2 \geq 2/\pi \rightarrow C_2.$$

The advantage of using such a test problem is that the errors in the model generated could be evaluated exactly, without the additional concern as to how these were to be estimated. Initially, there were 210 points in the complete data set. (This is much less than would be available in really large problems but it was necessary, in order to obtain results in a reasonable amount of time, to restrict the amount of computation to be carried out.)

There are clearly many ways of defining a selection strategy. The procedure we adopted was to construct a training set in *phases*, as is shown in Fig. 1.

This procedure requires the specification of the parameter θ and a test for the potential boundary status of a point at step 2, and at step 3 a definition of convergence. It is also necessary to decide on the size of the initial training set in step 1.

Boundary status

The definition of a PBP adopted was a simple intuitive one: a point is a PBP if there exist no other points of the *same* class which are closer than the closest point of *another* class. (Closeness here was measured in terms of Euclidean distance.)

Investigation of parameters

Experiments were carried out by generating many instances of the 'circle-in-a-square' problem. Initially the learning parameters were explored for the full data set, in order to find a reliable architecture, learning rate, stopping criterion etc. These settings were then used throughout the subsequent tests with the method of phases. In the latter case, subsidiary experiments were also carried out in which the parameter θ and the size $|T|$ of the initial sample were varied.

1. Take a random sample $\mathcal{T}$ from the complete data set $\mathcal{S}$.
2. If the proportion of potential boundary points (PBPs) in $\mathcal{T}$ is less than θ, draw further points from $\mathcal{S}\backslash\mathcal{T}$; if they are PBPs, update $\mathcal{T}$ by exchanging them with non-PBPs; stop when the criterion is satisfied.
3. Train the net using $\mathcal{T}$ until convergence.
4. Test the trained net on $\mathcal{S}\backslash\mathcal{T}$; add any misclassified points to $\mathcal{T}$.
5. Repeat steps 3 and 4 until no further improvement.

Figure 1: Proposed selection strategy

In each case the results (in terms of *true* error) were compared with those obtained by training on the full data set.

It was found that $|T| = 25$ and $\theta = 20\%$ gave satisfactory results for the method of phases, but provided they were not set too small, these choices were not critically important.

Computational results

In most of the experiments, the method of phases gave slightly better generalization performance than using the full set, as is shown in Fig. 2 which displays the performance of a typical run. The growth in sample size is shown in Fig. 3.

The method of phases generally appeared to need more training cycles than the approach of training with the full data set. However, as the amount of computation per cycle is reduced, the overall average amount of computation (measured by the number of back-propagation steps) was also usually less for the experiments reported above.

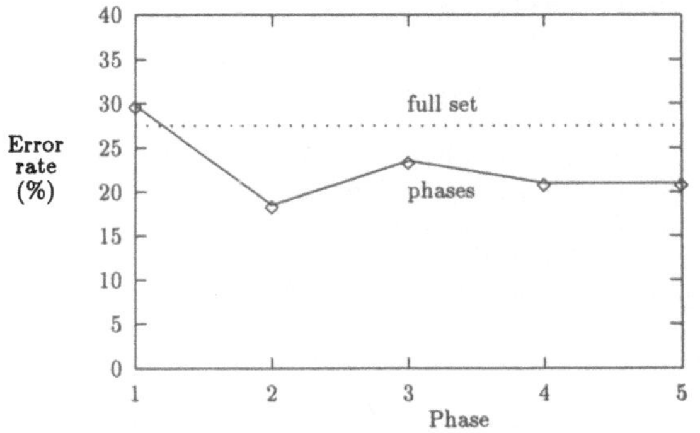

Figure 2: Error rate for method of phases

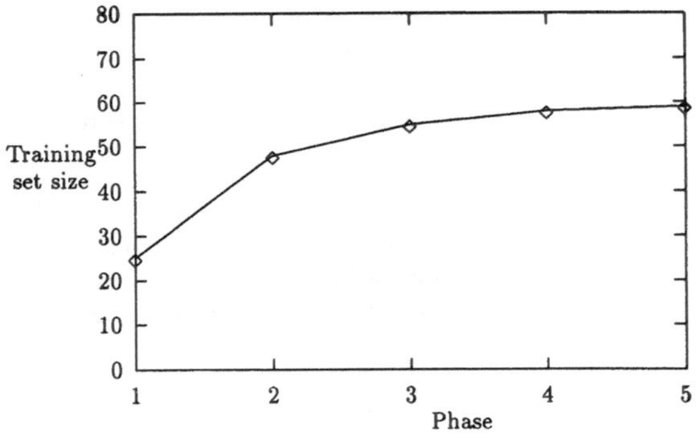

Figure 3: Training set size for method of phases

Further experiments were made using a larger, real-world problem. This was the Wisconsin breast-cancer data set [7] which contains measurements of 9 variables on 535 suspected breast cancer patients, where the classification relates to the malignancy or otherwise of the tumour. The full set was randomly split into 210 cases used for training and 325 reserved as a test set (the true error rate is unknown here). Suitable architecture and learning parameters were known from previous work [8], and these were used with the full training set of 210 cases to train the network; then the method of phases was used with the same parameters, and with $|T| = 50$ and $\theta = 20\%$. The whole procedure was repeated 9 times.

On 2 of the 9 occasions, the 'full set' approach failed to converge even after more than 10^6 back-propagation steps, but the method of phases produced acceptable results every time. On average, for the 7 cases where convergence was achieved by both methods, the generalization error was about 0.25% better using phases. This is not statistically significant, but it shows that at least using phases was not worse than using the full set. In terms of computational speed-up, the method of phases was significantly better by a factor of 3 on average (i.e. it used 1/3 of the computer time needed by the full set approach).

Conclusions

More research is needed to investigate training set selection methods more comprehensively, but this work has demonstrated their potential. The approach used here was less sophisticated than that advocated by Plutowski [5] or Röbel [6], but still on average it tended to produce better generalization at a lower computational cost. This work concentrated on some relatively small problems in order to keep the computational costs manageable, but it is intended now to apply these and other ideas to the DNA sequence prediction problem which originally motivated this work.

Acknowledgment

The author would like to acknowledge the assistance of Danny Wan, who wrote some of the code and carried out the experiments reported above.

References

[1] C.R.Reeves (1995) Bias estimation for neural network predictions. *This volume.*

[2] E.C.Uberbacher and R.J.Mural (1991) Locating protein-coding regions in human DNA sequences by a multiple sensor neural network approach. *Proc.Nat.Acad.Sci.USA*, **88**, 11261-11265.

[3] P.Herdman(1994) Report in *Networks: the Official Newsletter of the Neural Computing Applications Forum*, November 1994.

[4] M.Wann, T.Hediger and N.N.Greenbaum (1990) The influence of training sets on generalization in feed-forward neural networks. *Proc. International Joint Conference on Neural Networks*, Vol III, 137-142.

[5] M.Plutowski (1994) *Selecting Training Exemplars for Neural Network Learning.* PhD Dissertation, University of California, San Diego.

[6] A.Röbel (1994) *The Dynamic Pattern Selection Algorithm: Effective Training and Controlled Generalization of Backpropagation Neural Networks.* Technical Report, Technical University of Berlin.

[7] W.H.Wolberg and O.L.Mangasarian (1990) Multisurface method of pattern separation for medical diagnosis applied to breast cytology. *Proc.Nat.Acad.Sci.USA*, **87**, 9193-9196.

[8] C.R.Reeves and N.C.Steele (1993) Neural networks for multivariate analysis: results of some cross-validation studies. *Proc. of 6th International Symposium on Applied Stochastic Models and Data Analysis*, World Scientific Publishing, Singapore, Vol II, 780-791.

A GENETIC ALGORITHM FOR LOAD BALANCING IN PARALLEL QUERY EVALUATION FOR DEDUCTIVE RELATIONAL DATA BASES

E. Alba, J. F. Aldana, J. M. Troya

Dpto. Lenguajes y Ciencias de la Computación. Universidad de Málaga.
Plaza El Ejido s/n. 29013 Málaga. (Spain)
{alba, aldana, troya}@tecma1.ctima.uma.es

Abstract. *Datalog is a query language for deductive databases. This language allows to evaluate a query incrementally on a network of processes. The tuples flow among them in parallel in order to compute the solution to the query. A major issue in this evaluation is the problem of assigning the processes to processors in a multiprocessors system. Not only load balancing is wanted but lowering the communication costs among the processors is crucial. We have tackled this problem by using a GA that works on a specific fitness function that allows to meet these goals. The techniques and results are presented here.*

1.- PARALLEL DATAFLOW EVAL. OF DATALOG

In recent years, research on deductive databases has become increasingly important [1]. It is commonly assumed that deductive databases, defined by Datalog programs are usually evaluated bottom-up. Datalog is a set oriented declarative query language for deductive databases that avoids all the undesirable Prolog features. Substantial research has been focused on optimization techniques for the sequential evaluation of Datalog programs [2], [3]. These have been proposed in several approaches and vary considerably. More recently research has shifted to parallel evaluation as a way of improving performance, which is particularly important in deductive databases as they require very expensive computational processes. Datalog is a logic programming language which was conceived as a database query language.

A Datalog query can be evaluated by means of a dataflow parallel computation [4]. The Datalog program is described by a network of parallel processes. The relations of the extensional database send tuples across the channels. Every process receives one or two streams of tuples (unary or binary operations) and generates one stream of new tuples. The streams of tuples make several loops on this network of processes until a special set of tuples called the *Incremental Relations* becomes empty (fixed point iteration).

Different processors can work on different set of tuples simultaneously and the tuples belonging to the same relation can be spread along the set of processes. The streams of tuples are continuously feeding tuples from one process to another one(s).

We have developed a cost model in order to know the computational cost of every process and the load of the communications on every data link channel. This model uses several important parameters. One of the most important is K, the *hashing value* of a Join process. From the set of possible processes (Join, Project, Select, Union, Difference, Extensional Relation Feeders and Incremental processes) the Join of two relations is the most heavier. Internally we have implemented it as a set of K cooperating Join subprocesses. The best value of K has to be selected in order to ease the load balancing and keep at a low cost the communications overhead, since the subJoins of a given Join can be placed on different processors.

We have applied a GA in order to decide how to allocate the processes of the network to processors. In the next sections we present the characteristics of this GA and the analysis of the results. For the balancing we have tried a ring of a varying number of processors. The tests are valid for any specific physical system like a network of workstations or a Transputer multiprocessors system. The Datalog network we have used (see appendix A) is a representative example of query having the main characteristics that may be present in a typical query.

2.- A GA FOR LOAD AND COMM. BALANCING

In order to solve the twofold balancing problem we have encoded in the strings an assignment of processes to processors [5]. The other important aspect to devise before applying the GA to this specific problem is to define some kind of *fitness function* in order to compute how good any given string is. The following subsections explain the main characteristics of the used GA.

2.1.- Parameters

We have used a steady state GA [6]. The GA initiates by generating a random population of strings. Every string is divided into GENNUMBER *genes* encoded as GENLENGTH binary digits. Every gene in the string is the binary representation of the number of a processor in which to place the associated process.

The process associated with any gene (encoded processor) is known because of the position of the gene in the string: the value of the *i*-th gene in the string is the processor in which to place the process number *i*. Of course, processes are been numbered.

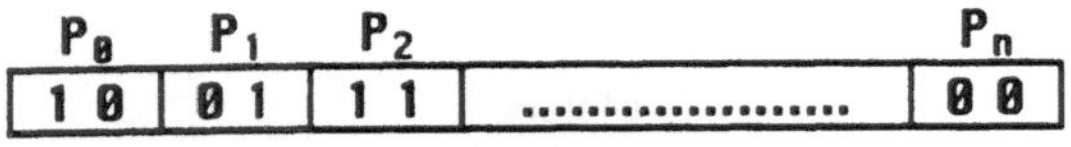

Fig 1. *String encoding the mapping for assigning processes to processors. We are assuming 4 processors (2 bits needed for encoding this number) and* P_{n+1} *processes.*

The number of genes is *Processes_Number* + *Number_of_Joins**(**K**-1) where **K** is the hashing factor in a join (number of internal joins per join in the implementation).

The *selection* consists in a random choice of a pair of individuals (strings) according to their fitness values (Roulette Wheel). Reproduction applies a two-points *crossover* to the pair of strings giving one new string and then mutation takes place on this offspring.

Before inserting the offspring in the population one individual must leave the population if we want to keep constant the population size. The selection of the individual to be replaced is made from among a set of randomly selected individuals. The worst of them is replaced. The whole scheme is elitist because the best string present in one iteration of the algorithm is present in the next iteration.

Strings are longer as K increments but population sizes of 1000 individuals have shown good results. Also for the tests we have used 10^4 recombinations (steps), Pc=1 and Pm=0.1 (somewhat high but good).

2.2.- Fitness

FITNESS = MEPV - PAYOFF

where the MEPV is a constant selected as the max estimated payoff value. The expression of the genotype in phenotype is a simple binary-to-decimal decoding. Thus while maximizing the fitness value in the GA we are actually minimizing the payoff function.

PAYOFF = *Cw* * **Comm** + *Uw* * **Unb**

Cw and *Uw* are two values for weighting the relative importance of the communication and unbalancing factors, i.e., the costs of making a wrong communication load on the dataflow channels and the cost of having a bad assignment of the processes to processors. The total cost is optimized when both parameters are 1.0.

$$\textbf{Comm} = \sum_{\forall \text{Channel } i \dashrightarrow j} \text{distance}(P_{P_i}^{a}, P_{P_j}^{b}) * c(P_i, P_j)$$

$$\textbf{distance}(P_{P_i}^{a}, P_{P_j}^{b}) = \textbf{if } P^a < P^b \textbf{ then } P^b - P^a \textbf{ otherwise } \textit{Processors\_Number} - (P^a - P^b)$$

$c(P_i,P_j)$ is the accumulated number of tuples that process *i* sent to process *j* in the simulation of the Datalog network made before the GA runs.

where $P_{P_i}^{a}$ stands for the process number *i* (*i*: 0..*E_Processes_Number*) that is contained in the processor number *a*: 0..*Processors_Number*-1.

$$\textbf{Unb} = \sum_{\forall \text{Processor}_i} \text{abs}(1 - (\textit{Processor\_Load}_i / \textit{Mean\_Processor\_Load}))$$

where *Processor_Load*$_i$ is the sum of the cost of all the processes assigned to processor *i*, and *Mean_Processor_Load* is the mean load for the computed assignment.

The term **Unb** expresses the cost of making an unbalanced allocation. Only when all the processors have the same computational load does this term decays to zero. The *Comm* and *Unb* operands are normalized to make a proper analysis of the results. Besides the normalization a *penalty function* is used in the GA for ensuring that the whole pool of processors is being used in the solution string.

3.- PERFORMANCE ANALYSIS

We have made tests on the cost of the whole system taking account of the communication and balancing of the processes. We have changed the number of processors (2,4,8,16,32) and for every value we have tried different values for K of 1/2, 1 and 2 times the number of processors (PN=Pn+1).

There is a trend in the GA of working out results progressively better as K becomes larger. Clearly this is because every join is splittered in more processes and the whole set of processes is bigger. This allows a better balancing (otherwise a join could take a processor in property).

This splittering can cause more communication but this overhead is overcome by the savings of a best load balancing.

An exception has to be made when using two processors because there exist more than four big joins in the network and their placement is definitive for the quality of the results. In this case when using K=1 the overall cost is smaller than K=2 or K=4, but anyway the balancing between the 2 processors is being made correctly by the GA.

The balancing achieved by the GA is better when the unbalancing and communication factors are equally important, yet better results have been got when the communication cost is enforced. This is because when using low values for the communication weight the numerical apportation to the payoff value is minor (since the values of communication are a few order of magnitude smaller than the computational load).

We are conducting some more experiments intended to found the best assignment of weights (although it seems to be at the middle of the graph). Also we have supposed to use a weights that is the complementary value of the other, and this is not a requirement of the system.

4.- CONCLUSIONS AND FUTURE WORKS

We have constructed a GA and applied it to a task allocation problem. The steady state approach is well suited for this purpose. By tuning the population size and the probability of applying mutation we have avoided local optima. The GA has solved the problem in a good manner where other techniques could have required a deeper management of the model. We have designed an elaborated fitness function that eases the work of the GA.

The splittering of processes in subprocesses decrements the overall cost in terms of processor load and communication. The communication overhead of separating processes among the processors is highly compensated by the saving met by the better load balancing the GA can achieve.

In the tests we have tried a single point crossover versus a double point crossover and concluded that the last one is the best. The balancing achieved by a single point crossover operator is sensibly worse than the balancing got when using two points crossover.

For real time application the GA can be used for a network of about 8 processors. If a larger number of processors is wanted the GA becomes slow. A solution can be to use a RISC architecture that (we have tested for genetic neural design) improves notably the execution of the GA.

5.- REFERENCES

[1] Ceri, Gottlob and Tanca "Logic Programming and Databases". Surveys in Computer Science. Springer Verlag. 1990.

[2] Bancilhon, "Magic Sets and Other Strange Ways to Implement Logic Programs". In Proceedings of the 5th ACM Symposium on PODS. 1986.

[3] Ullman, "Principles of Database and Knowledge-Base Systems". Vol. 2. Computer Science Press, New York. 1989.

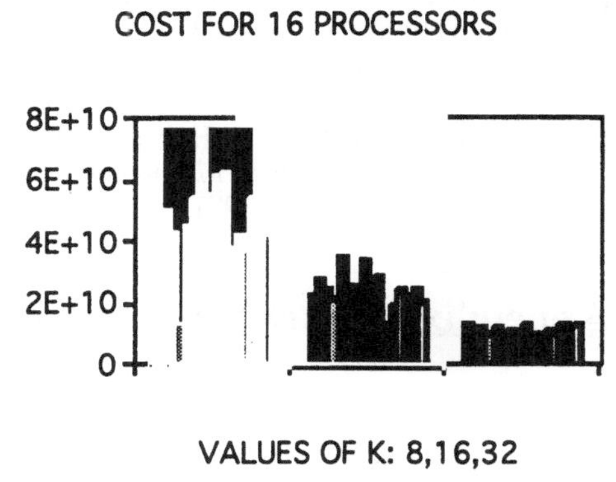

(a) *16 processors*

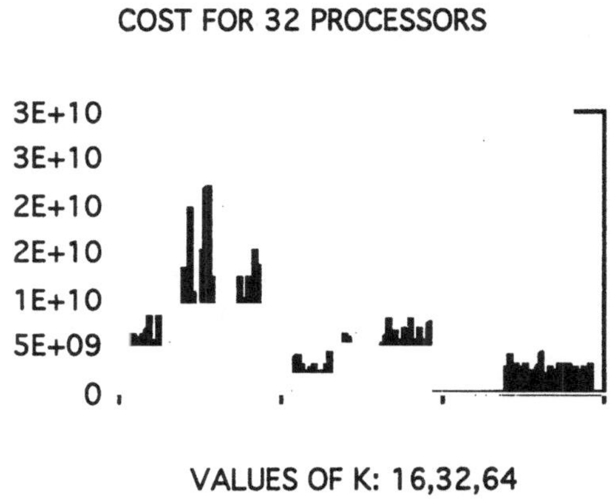

(b) *32 processors*

Fig 2. *Heuristic choice of the K value depending on the number of processors in the ring. In every graph K is set to 1/2, 1 and 2 times the number of processors.*

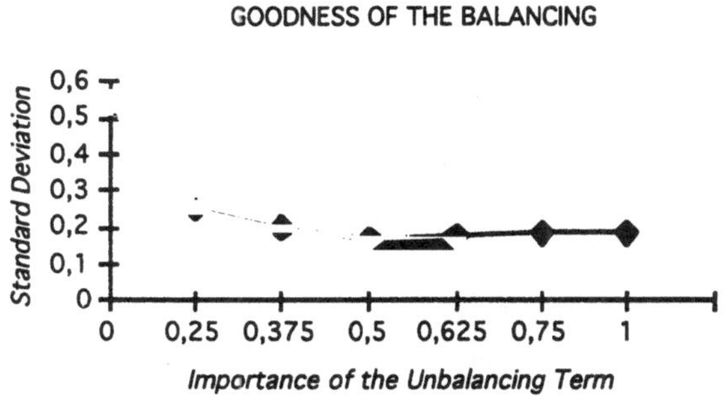

Fig 3. *The balancing equally spreads the processes on the processors after 0.5. A slight worsening appears as communication is forgotten (also we are reducing the numerical importance of the communication factor).*

[4] Aldana "A Dataflow Model for Datalog Parallel Evaluation" Tech. Rep. Dpt Lenguajes y Ciencias de la Computación. Univ. of Málaga, 1993.

[5] Mansour and Fox, "A Hybrid Genetic Algorithm for Task Allocation in Multicomputers". ICGA-91, Morgan Kaufmann, 1991.

[6] Whitley "The GENITOR Algorithm and Selection Pressure: Why rank based allocation of reproductive trials is best", ICGA, Springer-Verlag, 1989.

[7] Bennet, Ferris and Ioannidis, "A Genetic Algorithm for Database Query Optimization". ICGA-91, Morgan Kaufmann, 1991.

Appendix A: *Intensional DataBase, Datalog Equations and Graph of Processes*

IDB:

(1) ***p(X,Y) :- r(Z,Y) & X=a.***
(2) ***p(X,Y) :- s(X,Z1) & t(Z1,A,B,Z2) & r(Z2,Y).***
(3) ***q(X,Y) :- p(X,b) & X=Y.***
(4) ***q(X,Y) :- p(X,Z) & s(Z,Y).***
(5) ***t(X,Y,Z,T) :- r(X,Z1) & s(Z1,Y) & q(X,Z) & p(Z,T).***
(6) ***u(X,Y) :- r(B,Z) & s(A,Z) & q(Y,Z) & p(B,X).***

Datalog Equations:

$$P(X,Y) \supseteq \Pi_{X,Y}(R(Z,Y) \times [a](X))$$

$$P(X,Y) \supseteq \Pi_{X,Y}(S(X,Z1) \bowtie \Pi_{Z1,Y}(T(Z1,A,B,Z2) \bowtie R(Z2,Y)))$$

$$Q(X,Y) \supseteq \sigma_{X=Y}(\Pi_X(\sigma_{Z=b} P(X,Z)) \times \Pi_Y P(Y,W))$$

$$Q(X,Y) \supseteq \Pi_{X,Y}(P(X,Z) \bowtie S(Z,Y))$$

$$T(X,Y,Z,T) \supseteq \Pi_{X,Y,Z,T}(R(X,Z1) \bowtie (S(Z1,Y) \bowtie (Q(X,Z) \bowtie P(Z,T))))$$

$$U(X,Y) \supseteq \Pi_{X,Y}(R(B,Z) \bowtie (S(A,Z) \bowtie (Q(Y,Z) \bowtie P(B,X))))$$

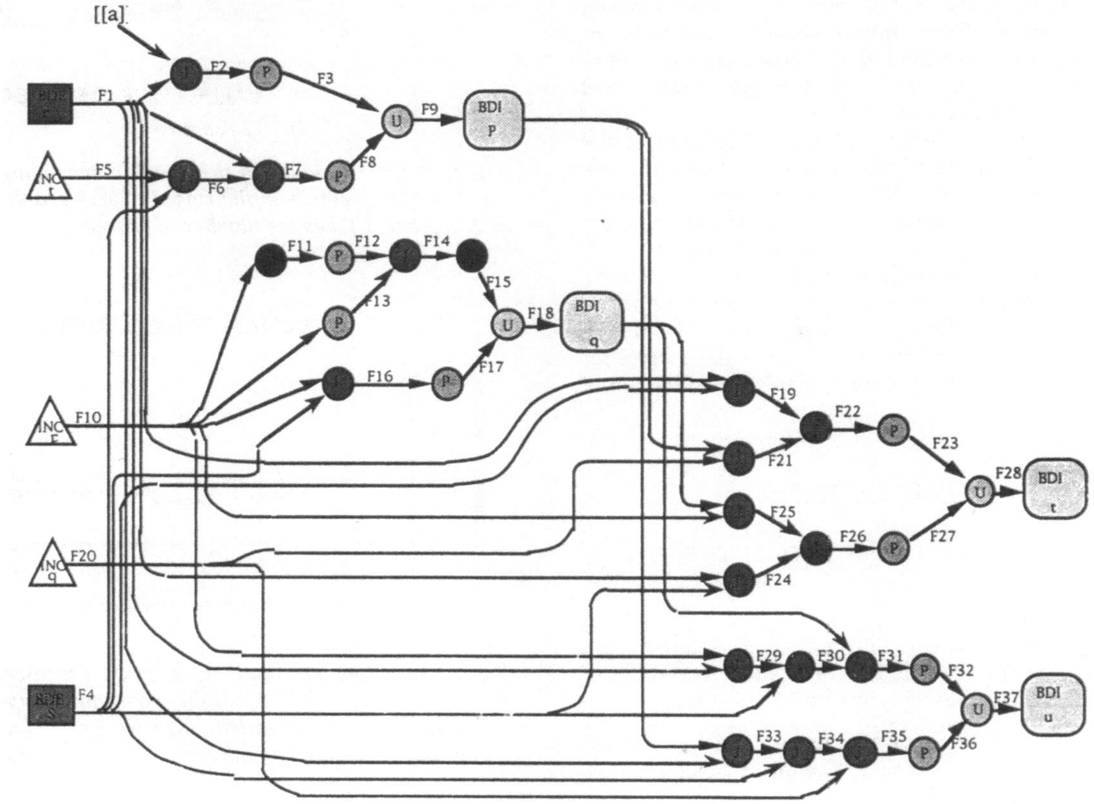

COMBINING DISTRIBUTED POPULATIONS AND PERIODIC CENTRALIZED SELECTIONS IN COARSE-GRAIN PARALLEL GENETIC ALGORITHMS

R. Bianchini, C. M. Brown, M. Cierniak, and W. Meira

{ricardo,brown,cierniak,meira}@cs.rochester.edu

Department of Computer Science
University of Rochester
Rochester, New York 14627

Abstract

In this paper we demonstrate that parallel genetic algorithms can profit from performing periodic centralized selections of distributed populations. With this combination, implementations can benefit from the variety of environments provided by distributed approaches, while periodically being able to consider the population as a whole and disregard very unfit individuals. We study four different parallel genetic algorithm implementation strategies; each of them striking a different balance between centralization and distribution. These strategies are applied to several well-known benchmark problems. Our results show that the implementations using periodic centralized selections exhibit remarkable robustness in terms of their search quality, while keeping running time overheads under control. Our main conclusion is that performing centralized selections represents an improvement to the traditional population distribution approaches, which are susceptible to somewhat inefficient search.

1 Introduction

Genetic Algorithms (GAs) have proven to be very effective approaches to solving various hard problems in state space search and optimization. However, some of these algorithms may require a large amount of time in order to find good problem solutions. A common approach to reducing this excessive execution time is to resort to parallel GAs (e.g. [1, 3, 7, 10, 12]).

Many search quality issues must be considered when designing a parallel GA for a certain application. Although centralization of the population (in the form of centralized selections) can cause premature convergence, it allows for global knowledge of the best individuals at each moment in time. In contrast, in a distributed organization, processors replicate individuals in the context of local smaller subpopulations. This isolation can cause some loss of computing cycles, whenever individual processors spend time working on very unfit individuals or get stuck on local minima (maxima). Nevertheless, distributed algorithms have the great advantage that a larger number of different environments composing the population (each of which corresponding to a separate subpopulation) may improve the search.

Many researchers have studied the effects of the population distribution on the search quality of parallel genetic algorithms. Neuhaus [10] mentions that a centralized solution to the process mapping problem delivered better search quality than a distributed solution to the same problem. Cohoon *et al.* [3] present results in which distributed algorithms with migration found better solutions than a sequential GA for optimization problems. Kommu and Pomeranz [9] present the same type of results for the Set Covering Problem. In [12], Tanese concluded that a distributed algorithm without migration searched at least as effectively as a distributed implementation with periodic migrations for several function optimization problems. In these latter experiments, both distributed implementations found fitter individuals than the sequential GA. Other work (e.g. [4]) considers the local selection of distributed populations as leading to a faster, more diverse, and more robust type of search than its global counterpart. Collectively these results show that, although distributed populations generally search better and faster, centralized ones can also deliver good results, depending on the problem at hand.

With respect to execution performance, centralization of the population entails potentially large amounts of communication and creates execution bottlenecks, causing serialization and reduced performance. Distributed implementations, on the other hand, lack central bottlenecks and have reduced communication needs.

In this paper we investigate the profitability of using distributed populations with periodic centralized selections. With such a hybrid approach, implementations can benefit from the variety of environments provided by distributed approaches, while periodically being able to consider the population as a whole and to discard very unfit individuals. In order to assess the search quality and execution time performance of parallel GAs with periodic centralized selections, we compare a set of implementation alternatives [2]. Our results show that hybrid strategies are extremely robust and do not degrade running time performance significantly. Our main conclusion is that periodically performing centralized selections represents an improvement to traditional distributed GAs.

The remainder of this paper is organized as follows. Section 2 describes the different parallel GA strategies we study. In section 3 we describe our benchmark suite, the parameters used in our experiments, and our main search quality and execution time results. Section 4 concludes the paper.

Strategy	*Topology*	*Comm Style*	*Centralized Selections?*
Cent	star	master-slave	yes
Semi-Dist	clusters of stars	master-slave + near-neighbor	yes
Dist	torus	near-neighbor	no
Tot Dist	torus	–	no

Table 1: Main Implementation Characteristics

2 Implementation Strategies

Table 1 summarizes the main characteristics of the different parallel GA implementation strategies we study. Our first strategy, called *Centralized (Cent)*, is organized in a master-slave fashion, interconnected logically as a star, with the master in the center. The computation proceeds with the master and each slave processor performing reproduction, crossovers, and mutations on different groups of individuals for a certain number of generations. After computing independently for a while, the slaves communicate with the master so that it can perform a completely centralized reproduction. After each centralized reproduction phase, the population is redistributed for one more round of independent computations.

Our second implementation strategy, *Semi-Distributed (Semi-Dist)*, provides a greater degree of data distribution than the centralized algorithm, while alleviating the performance bottleneck caused by having a single master processor in that algorithm. The idea is to have clusters of processors that work as in the centralized implementation. Clusters can be made sufficiently small so that contention is relieved. The clusters' masters are connected to form a (logical) torus topology. Communication between masters happens with some desired frequency, exchanging some of the best individuals between clusters of processors.

In contrast with the aforementioned algorithms, our *Distributed (Dist)* implementation does not perform centralized selections; it is based on the traditional organization of parallel GAs for distributed-memory architectures (e.g. [12]). There are no shared populations; each processor holds a piece of the total population and runs a complete sequential GA on it. Communication happens periodically, when processors send some of their best individuals to their nearest neighbors. Each processor includes the individuals it receives in its population. The processor interconnection topology we use in this implementation is again (logically) a torus.

Finally, our *Totally Distributed (Tot Dist)* strategy is a simple variation of the distributed implementation; one in which individuals are never transferred between processors. Processors only communicate in the initialization and termination phases of the algorithm.

3 Experimental Results

Our evaluation is centered on a number of well-known benchmark problems:

- DeJong's five function minimization problems (*f1*–*f5*) [8].

$$f1(x_i) = \sum_1^3 x_i^2$$
$$f2(x_i) = 100(x_1^2 - x_2)^2 + (1 - x_1)^2$$
$$f3(x_i) = \sum_1^5 integer(x_i)$$
$$f4(x_i) = \sum_1^{30} i x_i^4 + Gauss(0,1)$$
$$f5(x_i) = 0.002 + \sum_{j=1}^{25} \frac{1}{j + \sum_{i=1}^{2}(x_i - 1)^6}$$

- Traveling Salesman Problem (TSP). TSP consists of finding the shortest route through n cities, such that each city is visited only once. Our instance of TSP involves 58 cities.

- 0-1 Integer Linear Programming problem (ILP). ILP [11] is an *NP*-complete optimization problem where one wants to minimize a linear function of the form $f(x_1, \ldots, x_n)$, subject to a set of linear constraints. The variables $(x_1, \ldots, x_n)$ can only take the values 0 or 1. The problem can be stated as the minimization of $f = \sum_{j=1}^{n} c_j x_j$, subject to the constraints: $\sum_{j=1}^{n} a_{i,j} x_j \geq b_i$, where $a_{i,j}$, b_i and c_j are real constants; $i = 1, \ldots, m$ and $j = 1, \ldots, n$. Our instance of this problem assumes $n = 50$ and $m = 10$.

Our implementations use binary encoding and perform simple (one-point) crossover and mutation on randomly-picked individuals. Evolution is generational with an elitist selection strategy based on the *stochastic remainder without replacement* method [6].

Our experiments were run on an 8-processor SGI Challenge using the Mercury runtime library [5] for parallelism. We present a summary of the results of 15 runs for each of the example functions; each run corresponds to a different random seed controlling the search. Each experiment was run for 2000 generations, except for the ones optimizing *f4* and TSP which were run for 4000 generations. Crossover and mutation probabilities were .8 and .01, respectively. Population sizes roughly correspond to the complexity of the landscape being searched (72 for *f1*, 96 for *f2*, 168 for *f3*, and

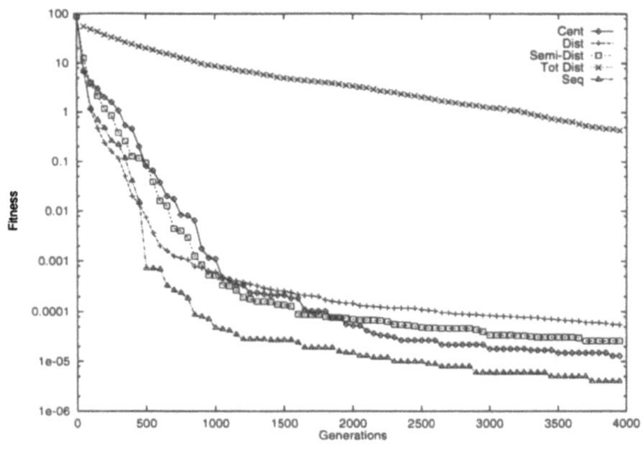

Figure 1: Search Quality on *F4*.

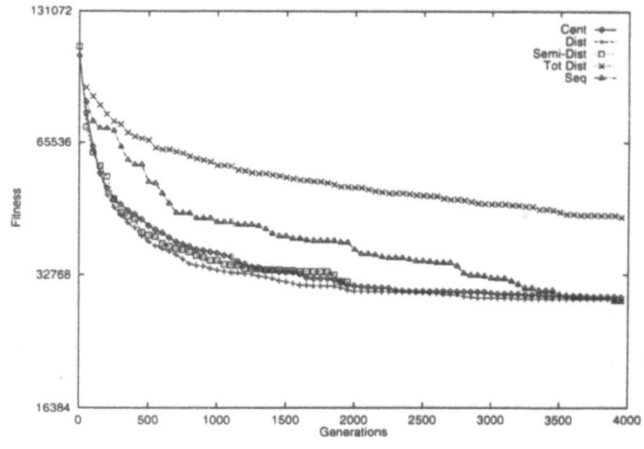

Figure 2: Search Quality on TSP.

216 for *f4*, *f5*, ILP, and TSP). We studied several frequencies of communication; from 5 to 40 generations between centralized replications (Cent and Semi-Dist) or near-neighbor communication (Dist). The frequency of communication between the two master processors in Semi-Dist was kept constant, occurring once every 50 generations. In Dist, each processor sends its best individual to each neighbor during the migration phase.

Our results show that the strategies with periodic centralized selections exhibit remarkable robustness in terms of their search quality. In cases where the sequential GA achieves good performance (*f2* and *f4*), Cent and Semi-Dist approach or surpass that performance. When the more distributed strategies entail the best performance, i.e. the sequential GA exhibits poor performance (*f1*, *f3*, *f5*, ILP, and TSP), the implementations with centralized selections closely approach those results.

As representative examples of these two scenarios, figures 1 and 2 present the overall best fitness at each generation averaged over 15 different runs for each of our implementation strategies running on 8 processors. Figure 1 shows results achieved for *f4*. In these experiments Cent and Semi-Dist perform master-slave communication every 5 generations, while near-neighbor migrations are performed every 10 generations in Dist.

Figure 2 shows the best results achieved for TSP. For this program near-neighbor migrations are again performed every 10 generations. Best performance for Cent and Semi-Dist was also achieved with this frequency of communication between master(s) and slaves.

As an overview of our results, table 2 presents the implementation and period between communication that achieved the best search quality performance (in terms of the average of the best fitnesses for 15 runs) for each of our benchmarks. The results in the table show that, besides being robust, implementations with periodic centralized selections frequently deliver the best overall performance for our benchmarks. Dist also exhibits good performance for several benchmarks, while Tot Dist invariably delivers significantly worse search quality than all the other implementations.

As representative examples of these two scenarios, figures 1 and 2 present the overall best fitness at each generation averaged over 15 different runs for each of our implementation strategies running on 8 processors. Figure 1 shows results achieved for *f4*. In these experiments Cent and Semi-Dist perform master-slave communication every 5 generations, while near-neighbor migrations are performed every 10 generations in Dist.

Figure 2 shows the best results achieved for TSP. For this program near-neighbor migrations are again performed every 10 generations. Best performance for Cent and Semi-Dist was also achieved with this frequency of communication between master(s) and slaves.

As an overview of our results, table 2 presents the implementation and period between communication that achieved the best search quality performance (in terms of the average of the best fitnesses for 15 runs) for each of our benchmarks. The results in the table show that, besides being robust, implementations with periodic centralized selections frequently deliver the best overall solutions for our benchmarks. Dist also exhibits good performance for several benchmarks, while Tot Dist invariably delivers significantly worse search quality than all the other implementations.

In addition to good search quality, parallel GAs must also deliver good running time performance. Table 3 presents the execution times of each implementation optimizing *f3* on 8 processors. The results in this table show that, at least for the small machine configurations we studied, centralized selections do not entail significant performance degradation. The greatest performance difference between Dist and the implementations with centralized selections is only about 12%. Note also that Semi-Dist performed slightly worse than Cent, even though the former strategy is supposed to address a bottleneck of the latter. The reason for this result is that Semi-Dist has a more complicated organization than Cent and our multiprocessor is based on a shared bus,

Benchmark	*Strategy/Interval Between Comms*
f1	Dist/40
f2	Cent/5
f3	Semi-Dist/40 Dist/5,10,40
f4	Seq
f5	Dist/5,10,40
ILP	Cent/5,10,40 Semi-Dist/5,10,40 Dist/5,10,40
TSP	Dist/5

Table 2: Average of Best Fitnesses

which renders Semi-Dist's clusters ineffective. On multiprocessors with scalable interconnection media, we expect that Semi-Dist shall easily surpass Cent's execution time performance.

4 Conclusions

Our main conclusion is that performing centralized selections represents an improvement to the traditional population distribution approaches. Strategies performing periodic centralized selections are extremely robust and frequently achieve superior search results, while keeping execution time overheads under control.

Strategy	*Comm Freq*	*Run Time*	*Speedup*
Seq	–	17.1	1.0
Cent	5	8.1	2.1
	10	7.5	2.3
	40	7.0	2.4
Semi-Dist	5	8.1	2.1
	10	7.6	2.3
	40	7.4	2.3
Dist	5	7.5	2.3
	10	7.2	2.4
	40	7.2	2.4
Tot Dist	–	6.7	2.6

Table 3: Running Times (in Seconds) and Speedups for *f3*

Acknowledgements

We would like to thank Leonidas Kontothanassis for his help with the Mercury runtime library. This work was supported by Brazilian CAPES, CNPq, and NUTES/UFRJ fellowships, the National Science Foundation under Grants numbered IRI-8920771 and CDA-8822724, and Darpa contract MDA972-92-J-1012. The Government has certain rights in this material.

References

[1] S. Baluja. Structure and performance of fine-grain parallelism in genetic search. In *Proc. of the 5th Int. Conf. on Genetic Algorithms and their Applications*, pages 155–162, 1993.

[2] R. Bianchini and C. Brown. Parallel genetic algorithms on distributed-memory architectures. In *Transputer: Research and Applications*, May 1993.

[3] J. P. Cohoon, W. N. Martin, and D. S. Richards. A multi-population genetic algorithm for solving the k-partition problem on hyper-cubes. In *Proc. of the 3rd Int. Conf. on Genetic Algorithms and their Applications*, 1989.

[4] R. J. Collins and D. R. Jefferson. Selection in massively parallel genetic algorithms. In *Proc. of the 4th Int. Conf. on Genetic Algorithms and their Applications*, pages 249–256, 1991.

[5] R. J. Fowler and L. I. Kontothanassis. Mercury: Object-affinity scheduling and continuation passing on multiprocessors. In *Parallel Architectures and Languages Europe (PARLE) '94*, June 1994.

[6] D. E. Goldberg. *Genetic Algorithms in search, optimization and machine learning*. Addison-Wesley, 1989.

[7] J. J. Grefenstette. Parallel adaptive algorithms for function optimization. Technical Report CS-81-19, Nashville: Vanderbilt University. Computer Science Dept., 1981.

[8] K. A. De Jong. *Analysis of the Behavior of a Class of Genetic Algorithms*. PhD thesis, University of Michigan, 1975.

[9] V. Kommu and I. Pomeranz. Effect of communication in a parallel genetic algorithm. In *Proc. of the 1992 Int. Conf. on Parallel Processing*, pages III:310–317, 1992.

[10] P. Neuhaus. Solving the mapping-problem – experiences with a genetic algorithm. In *Parallel Problem Solving from Nature*, pages 170–175. Springer-Verlag, 1991.

[11] H. A. Taha. *Integer Programming — Theory, Applications and Computations*. Academic Press, 1975.

[12] R. Tanese. Distributed genetic algorithms. In *Proc. of the 3rd Int. Conf. on Genetic Algorithms and their Applications*, 1989.

MODIFIED GENETIC ALGORITHMS BY EFFICIENT UNIFICATION WITH SIMULATED ANNEALING

S. Ghoshray, K. K. Yen, and J. Andrian

Department of Electrical and Computer Engineering
Florida International University
Miami, Florida 33199
E-mail: Sghosh01 @servax.fiu.edu

Abstract

Genetic algorithms (GA) are stochastic search techniques based on the mechanics of natural selection and natural genetics. By using genetic operators and cumulative information, genetic algorithms prune the search space and generate a set of plausible solutions. This paper describes a Modified Genetic Algorithm (MGA) that is developed by making a marriage between the Simple Genetic Algorithm (SGA) and the Simulated Annealing (SA). In this proposed algorithm, all the conventional genetic operators, such as, selection, reproduction, crossover, mutation, have been used, but they have been modified by a set of new functions such as, a evaluation function, a selection function, a mutation function, etc., which utilizes the concept of successive descent as seen in simulated annealing. In this way, MGA can be implemented to solve combinatorial optimization problems more accurately and quickly.

KEY WORDS : modified genetic algorithm, simulated annealimg, combinatorial optimization, fitness evaluation function, neighborhood acceptance, probabilistic selection

I. Introduction

John Holland laid the foundation of GAs with the goal to create computer algorithms by simulating the characteristics of a natural system [1]. It was intended that, if nature can produce from a random population, a population with individuals that are better fit to the environment, it is possible to develop an algorithm to solve complex problems by utilizing the concept from nature. In this respect, a wide but diverse range of applications [2, 3, 4, 5] bear testimony to the fact that GAs are very useful in solving complicated problems by mimicking some facets of natural evolution. Similarly, Simulated Annealing (SA) is a stochastic based computational technique that owes its origin to statistical mechanics. Kirkpatrick et al [8] were the first to propose and demonstrate the application of simulation techniques borrowed from statistical physics to the large optimization problems.

A combinatorial optimization problem is either a minimization problem or a maximization problem and consists of : (1) a set of instances, (2) a finite set of candidate solutions for each instance and (3) a function that assigns to each instance and each candidate solution a positive solution value. This combinatorial optimization problem is solved by finding an optimal solution for each instance of the optimization problem. In this research, we propose a modified genetic algorithm by combining the powers of both, a simple genetic algorithm and simulated annealing. The solution we try to obtain here is found near to the optimum, but with more accuracy and in a shorter run-time. We have organized the paper as follows: section II provides an overview of GAs, section III gives a probabilistic view of different genetic operators, section IV introduces the concept of SA by presenting some pseudo-codes. The next section brings us to the proposed Modified Genetic Algorithm (MGA) with different new definitions and algorithms. Finally, we conclude by summarizing and focusing on future work.

II. Overview of Genetic Algorithms

Genetic algorithms represent a highly idealized model of a natural process and as such can be legitimately viewed as a very high level of abstraction. Genetic algorithms are loosely based upon the Darwinian principles of biological evolution. Biological strategies of behavior adaptation and synthesis are used to enhance the probability of survival and propagation during their evolution. Environmental pressures requiring these strategies have affected profound changes in biological organisms. These changes are manifested in structural and functional organization, and internal knowledge representations [6]. A GA randomly generates a set of possible solutions to the problem which is being investigated. This is termed the initial generation. Each successive incremental improvement in a solution structure becomes the basis for the next generation. The process continues until the desired number of generations has been completed.

In a natural genetic system, the chromosomes consist of genes. Each gene has a value and position. The combination of chromosomes form the total genetic prescription for the construction and operation of some organism [7]. While developing a GA, strings and characters are used in order to simulate chromosomes and genes corresponding to the natural genetic system. When using a genetic algorithm to solve a problem, the problem is represented by a string, and an evaluation function is defined. The evaluation function uses the value of the string as a parameter to evaluate the results of the problem Similar to natural systems, in an artificial genetic system we have populations and generations. A set of strings is used to represent the populations. Each string is evaluated through the evaluation function and the new generation is formed by using the specific genetic operators.

A genetic algorithm is an interactive procedure that maintains a population of strings which constitute the set of candidate solutions to the specific problem. During each generation, the strings in the current population are rated for their effectiveness as solutions. On the basis of these evaluations, a new population of candidate solutions is formed using genetic operators such as reproduction, crossover, and mutation. There are four major preparatory steps required to use the conventional genetic algorithm to solve a problem [8], namely determining

(1) the representation scheme,
(2) the fitness measure,
(3) the parameters and variables for controlling the algorithm, and
(4) a way of designating the result and a criterion for terminating a run

III. A Probabilistic View of The Genetic Operators

In this section, we are going to briefly describe the genetic operations mentioned earlier. The genetic operation of *reproduction* is based on the Darwinian principle of reproduction and survival of the fittest. Probability plays a very important role in the overall implementation of genetic algorithm. In the reproduction operation, an individual is probabilistically selected from the population on the basis of its fitness and the selected individual is then copied into the next generation of the population without any change. In this selection process, the better an individual's fitness is the more is the likelihood of its selection.

The genetic operation of *crossover* allows the creation of new individuals which accounts for the generation of new points in the search space to be tested. Crossover starts with two parents independently selected probabilistically from the population on the basis of their fitness.

Individuals from the population can be selected and, in general, are selected more than once during a generation to participate in the operations of reproduction and crossover. Indeed, the differential rates of survival, reproduction, and participation in genetic operations by more fit individuals is an essential part of the genetic algorithm.

The operation of *mutation* begins with the probabilistic selection of an individual from the population on the basis of its fitness. A mutation point along the string is chosen at random, and the single character at that point is randomly changed. The altered individual is then copied into the next generation of the population. The usefulness of mutation comes in situations requiring the restoration of genetic diversity that may have been lost in a population because of premature convergence. Mutation is used very sparingly in most genetic algorithm work.

The Darwinian selection of individuals to participate in the operation of reproduction, crossover, and mutation on the basis of their fitness is an essential aspect of the genetic algorithm. When an individual is selected on the basis of its fitness to be copied (with or without mutation) into the next generation of the population, the effect is that the new generation contains the characteristics it embodies. These characteristics consists of certain values at certain positions of the character string and, more importantly, certain combinations of values situated at two or more positions of the string. When two individuals are selected on the basis of their fitness to be recombined, the new generation contains the characteristics of both of these parents. The probabilistic selection used in the genetic algorithm is an essential aspect of the algorithm. The genetic algorithm allocates every individual, however poor its fitness may be, some chance of being selected so that it can participate in the operations of reproduction, crossover, and mutation. In that respect, the genetic algorithm is not merely a greedy hill climbing algorithm, rather, the genetic algorithm contains the characteristics of simulated annealing [9, 10].

IV. Simulated Annealing Algorithms

Simulated Annealing (SA) is a stochastic computational technique derived from statistical mechanics for finding near globally minimum-cost solutions to large optimization problems. Kirkpatrick et al [9]. First proposed and demonstrated the application of simulation techniques from statistical physics to problems of combinatorial optimization. Finding the global minimum value of an objective function with many degrees of freedom subject to conflicting constraints is an NP-complete problem. Therefore, the objective function will tend to have many local minima. A procedure for solving optimization problems of the above nature can be implemented by following the Successive Descent algorithm by kirkpatrick et al [8] which follows the evolution of a solid thermodynamic equilibrium with a decreasing succession of temperature values. The pseudo code for this Descent algorithm goes as follows:

```
begin Successive Descent
        Select an initial state i in the search space of E
        such that i ∈ E
        Repeat

                Generate a random state j in the
                 search space of E such that j is
                        a neighbor of i

                if Cj - Ci < 0

                then i: = j
```

until Cj > Ci for all j in the
neighborhood of i

end Successive Descent

In this algorithm, E is a finite set of possible configurations that constitute a finite search space. Ci is the cost function associated with configuration i. In SA, the algorithm attempts to accept a neighboring configuration that increases the cost function C. The probability of accepting such a neighborhood move is determined by Pi = exp(-s/T), where s = an increase in cost function = Cj - Ci, T = the control parameter in the optimization process analogous to temperature in SA. With this in mind, therefore, the minimization algorithm can be written in pseudo-code as follows:

```
begin minimization

    select an initial state i ∈ E
    select an initial control parameter T > 0
    select parameter change counter t = 0

    Repeat

    set repetition counter r = 0
        Repeat

        Generate state j : j is a neighbor of i, i ∈ E
        compute S = Cj - Ci

        If S < 0 then i : = j
            else if rand (0,1) < exp(-s/T)

                then i: = j

        r : = r+1
        until r = R (t)
        t : = t +1
        T : = T (t)
    until stopping Rule: = True
end minimization
```

V. Modified Genetic Algorithm (MGA)

Let us now present a modification of GA introduced above that would converge both quickly and accurately. This proposed Modified Genetic Algorithm (MGA) is developed by unifying Genetic Algorithm (GA) and Simulated Annealing (SA) The following pseudo code presents an insight into proposed MGA

```
begin  MGA

    input system specification
    evaluate fitness function
    generate genetic codes at random
    assign randomly generated genetic codes as
    the initial population
    while TERMINATION CRITERION is True
    and initial population <> NIL do

        begin
        input initial population to the fitness
          evaluation function
          while FITNESS CRITERION is True do

            begin
                assign initial population to parents
                mate
                add the children to the population

                while pm < 0.005 do
                  begin
                    mutate
                  end
            end
    reduce population
    end
    output the result

end    MGA
```

In the above algorithm, p_m is the probability of mutation. The success of the MGA depends on how effective we are in defining the genetic operators used in this modified algorithm. Let us now define the genetic operators as follows:

A. Size of Initial Population: i ; this represents the randomly generated genetic codes which are to be given as input to the fitness evaluation function. That means, i is the initial population which depends on problem specification and, also i ∈ E, E is the search space for the specific problem.

B. Fitness Evaluation Function : f_{eval} ; this represents the function that is defined in accordance with the problem specification. It takes as the input the initial population (set of randomly generated genetic codes), tests for truth condition for fitness and returns a set of codes as the parents for the next generation.

C. Set of Possible parents: S_i ; this represents all the possible populations which are to be used for mating.

D. Set of Extended Populations: S_{i+} ; this represents all the populations which is generated by adding the children into the original populations of parents

E. A Selection Function Z: $f_s(p, x)$; it is a function of two variables. The variable p is the parameter that randomizes the selection function f_s and x is any member of S; $x \in S_i$ such that $f_s(p,x)$ selects the set of parents that are chosen for reproduction at the onset of the algorithm.
$f_s(p, x) = y \subseteq x , p \in P$

F. A Reproduction Function: f_{RP} (q, y) ; this function is randomized by q and it works over populations $y \subseteq S$; to produce offspring z.
$f_{RP}(q,y) = z, z \in Si , q \in Q.$

G. A Mutation Function: f_M (r, z) ; this function is randomized by parameter r, $r \in R$. The goal here is not to be entrapped in the local minima and accept the neighborhood move by invoking the following function call:
f_M (r, z) = Z' , Z' $\in$ neighborhood of Z such that C (z') > C (z) , that is, cost function assigned with configuration Z' is greater than that assigned with Z.

H. A Reduction Function: f_{Rd} (r, t_+) = t ; here the extended populations are reduced to regular populations of original size, signaling the end of the genetic programming loop.

I. A Boolean Function: f_{contrl} (x) ; this is a Boolean function that controls the termination of the algorithm.
Based on the above notations and definitions, the MGA is presented as follows :

```
begin  MGA (Modified Genetic Algorithm)

generate random genetic code i
f_eval (i) = x ∈ S_i

while f_contrl (x) = True do

begin

      get p, q, r, s
      y = f_s (p, x)
      z = f_RP (v, y)
      z' = f_m (r, z)
      t = x ∪ z'
      x = f_Rd (s, t)

      end
      output all x

end MGA
```

VI. Conclusion

This paper has presented a Modified Genetic Algorithm by efficiently combining simple genetic algorithms with Simulated Annealing. The empirical results show that Modified Genetic Algorithms provide efficient search heuristics for solving various combinatorial optimization problems. The model proposed has a short run time compared to that of SGA and the results produced have been qualitatively correct. However, the main difficulty remaining is in the efficient modeling of evaluation function that selects the parent population from the initial population. We hope to address this issue in more detail in a future publication. Our forthcoming research would also focus on developing more efficient parallel selection algorithms by using fractal methods. This will enhance the capability to change genetic algorithm parameters dynamically.

BIBLIOGRAPHY

1. Holland, J. H. *Adaptation in Natural and Artificial Systems.* University of Michigan Press, Ann Arbor, 1975.

2. Pettey, C. B., Leuze, M. R., Grefenstette, J. J. *A Parallel Genetic Algorithm.* In Proceedings of the Second International Conference on Genetic Algorithms, 1987.

3. Grefenaatette, J. J. & Baker, J. E. *How Genetic Algorithms Works: A Critical Look at Implicit Parallelism.* In Proceedings of the Third International Conference on Genetic Algorithms, 1989.

4. Schaffer, J. D., Caruana, R. A., Eshelman, L. J. & Das, R. *A Study of Control Parameters Affecting Online Performance of Genetic Algorithm for Function Optimization.* In Proceedings of the Third International Conference on Genetic Algorithms, 1989.
5. Lucasins, C. B.,. & Kateman, G. *Application of Genetic Algorithms in Chemometric.* In Proceedings of the Third International Conference on Genetic Algorithms, 1989.

6. Austin, S. *An Introduction to Genetic Algorithms.* AI Expert, March 1990.

7. Goldberg, D. E. *Genetic Algorithms in Search, Optimization, and Machine Learning.* Addison-Wesley, 1989.

8. Koza, J. R. *Genetic Programming II Automatic Discovery of Reusable Programs.* The MIT Press, 1994.

9. Kirkpatrick, S., Gelatt, C. D. & Vechi, M. P. *Optimization by Simulated Annealing.* Science 220, 1983. pgs. 671-680.

10. Aarts, E. & Korst, J. *Simulated Annealing and Boltzman Machines.* Wiley 1989.

VLSI STANDARD CELL PLACEMENT BY PARALLEL HYBRID SIMULATED-ANNEALING AND GENETIC ALGORITHM

Karl Kurbel#, Bernd Schneider#, Kirti Singh$

#Institute of Business Informatics, University of Muenster, Grevener Strasse 91, D-48159 Muenster, Germany

$Institute of Computer Science, Electronics and Instrumentation, Devi Ahilya University, Khandwa Road, Indore, M.P. 452 001, India

Abstract

Placement of standard cells is a part of physical VLSI chip design. In order to achieve high performance, area of the chip and lengths of wires connecting cells have to be minimized. In the placement step, the goal is to place cells in such a way that total wire-length is as short as possible. Since this problem is NP-hard, heuristic techniques have to be applied. Modern approaches include simulated annealing and genetic algorithms. In this paper, we discuss those methods and show that they can be improved by combination. A heuristic technique called *parallel recombinative simulated annealing (PRSA)* is described. It integrates features of both simulated annealing and genetic algorithms. Behavior of PRSA is studied with respect to different parameter settings.

1 Standard Cell Placement

Since integration in VLSI (Very Large Scale Integration) design is increasing fast, managing layout complexity has become an important issue. In this paper, we address the problem of placing so-called standard cells (i.e. rectangular logic modules with predesigned internal layouts) on a chip. The problem occurs during the stage of physical chip design. Major steps of this stage are breaking down the circuit into groups of components (partitioning), determining positions of the components (placement), connecting the cells by wires (routing), and compressing the layout of the chip (compaction). In the placement step, one tries to find optimal positions for a given number of cells in a given layout pattern. "Optimal" are positions that will minimize the total length of interconnecting wires.

Standard cell layout techniques are very common because they provide a semi-regular layout which can be accomplished in a relatively short time. Other techniques not considered here are gate-array and macro cell placement. Standard cells are rectangular blocks of the same height but varying widths depending on their functionality. Figure 1 shows a chip with standard cells. Cells are placed in rows and interconnecting wires run through channels between rows. Bondpads are placed around the chip. Some macro blocks are also included. Macro blocks are logic modules that do not have the standard cell format. They are preassigned by the designer and thus given restrictions for standard cell placement.

Interconnections between cells are pre-defined by so-called netlists. A netlist consists of nets which in turn are made up of those cells that are to be connected together. For a given number of cells with their dimensions and a given netlist, the task is to place the cells on the given layout in such a way that overall interconnection wire-length is minimal.

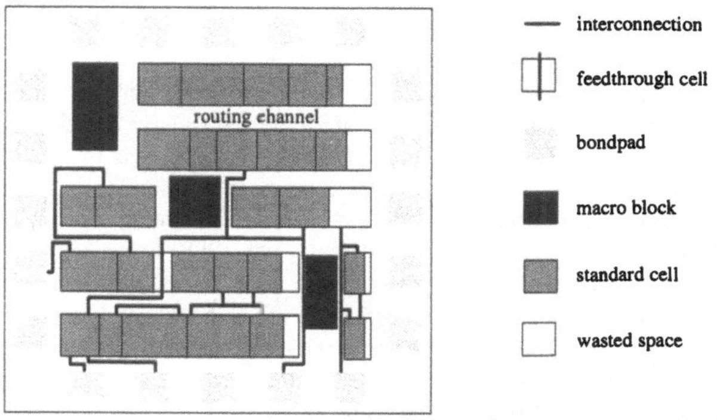

Fig. 1: Chip layout using standard cells

Wire-length has to be estimated in the placement step because actual routing will only be done later [3, 18]. Good estimates are provided by the bounding-rectangle method. A bounding rectangle is the smallest (imaginary) rectangle surrounding all cells of a net. The wire-length of such a net is calculated as half the perimeter of its bounding rectangle. The optimization goal is then to minimize half the sum of perimeters of all rectangles.

Many attempts have been made to solve the standard cell placement. Stochastic techniques like simulated annealing [15, 16], genetic algorithms [17], and evolution based method [10] have been tried as well as deterministic techniques like min-cut placement, force-directed methods [18], and Hopfield neural networks [20]. General findings were that deterministic techniques can solve only simplified versions of the problem, and even then they are not very efficient [19]. Performance of stochastic techniques was found better in many cases [19, 21].

2 Features of Simulated Annealing and Genetic Algorithms

Simulated annealing is a stochastic heuristic optimization technique derived from the physical process of crystallization [9]. It has proved to be effective in generating good standard cell placements [18]. The stochastic component is controlled by an external parameter called temperature. The algorithm starts with an initial feasible solution and a very high temperature, then a flip operator interchanges locations of two cells selected randomly. The new solution is accepted if its wire-length is less than the previous one; otherwise it is accepted with a probability that depends on the temperature. Then the temperature is decreased and the next pair of cells is selected. The process continues until the temperature reaches a value near zero or until some other termination criterion is satisfied. A very significant feature of simulated annealing is that it converges eventually, although not necessarily to the global optimum; a formal proof of convergence exists [1].

The main drawback is excessive computing time which quickly grows with the problem size [15, 21]. Several approaches have been proposed to speed up computation (e.g. [22]). Improvements in simulated annealing applied to standard cell placement were reported elsewhere [14, 15, 16].

Recently, *genetic algorithms* were also tried for standard cell placement [17]. Genetic algorithms start with a set of solutions (individuals) called a population. New individuals are generated by cross-over and mutation. The *cross-over* operator takes two individuals (parents) and recombines them in such a way that the resulting individual (child) inherits features from both parents. Generally, children are expected to have better "fitness" than their parents. The *mutation* operator selects two cells of an individual and interchanges their positions randomly. The concept of mutation safeguards that if some knowledge (gene) is lost during the optimization process, it may still be regained later. After creating new individuals, a new population is formed by selecting a set of individuals from the old and the newly generated individuals according to some selection strategy [5].

It has been found that genetic algorithms can produce as good solutions as those obtained by simulated annealing [17], but they are very slow for larger numbers of cells.

Several approaches have been tried to *couple* simulated annealing and genetic algorithms. The simplest form is to use both methods one after the other, either in such a way that the initial population is pre-optimized by simulated annealing, or that parts of the final population are handed over to simulated annealing and post-optimized there [12]. Another combination is to employ simulated annealing as a special kind of a "genetic operator" to change individuals in some way. Simulated annealing would thus act like other (genetic) operators and run under control of the genetic algorithm [2].

Approaches to *integrate* simulated annealing and genetic algorithms have two origins, either extending the sequential method of simulated annealing in such a way that it can be parallelized, or enriching genetic algorithms so that they benefit from the convergence property of simulated annealing. Integration-oriented approaches often use the Boltzmann criterion to accept or reject new individuals after application of genetic operators [7, 13].

3 Parallel Recombinative Simulated Annealing

The approach outlined in this section inherits implicit and explicit parallelism from genetic algorithms and the convergence property from simulated annealing [13]. PRSA starts with a population and then generates new individuals by applying genetic recombination and flip operators. It starts with a very high temperature which is then lowered at every step by a very small fraction. New populations are constructed by selecting from old and new individuals by Boltzmann trials [1]. In contrast to normal genetic algorithms, the selection process is controlled by the temperature. Like in simulated annealing, selections are mostly random when the temperature is high. As the temperature goes down, the degree of randomness decreases and selection of an individual depends increasingly on its fitness just as in a genetic algorithm.

For problem representation, a rather simple format could be used. If macro blocks are left out for simplification, then just the order of cells and their positions on the chip (indicated by coordinates) have to be mapped. This information is represented by a list of triplets as shown in the upper part of figure 2. The order of the list defines the order of cells on the layout. The bottom part of the figure illustrates the abstract layout equivalent to the example list. For this specific representation, an appropriate cross-over operator had to be constructed. We used *cyclic cross-over* [7] in order to safeguard that new strings correspond to *feasible* solutions only.

Cell	A	B	C	D	E	F	G	H	I	J
X coordinate	0	20	50	75	90	0	30	55	70	95
Y coordinate	0	0	0	0	0	50	50	50	50	50

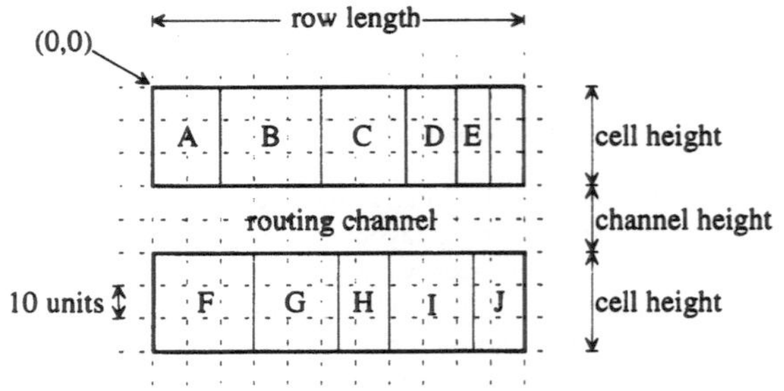

Fig. 2: Representation of standard cell placement as a list (example)

The *objective function* we used for the placement problem consists of two terms W and P_R. The term W represents the total weighted wire-length, using the bounding-rectangle measurement:

$$W = \sum_{n=1}^{N} S_x(n)H(n) + S_y(n)V(n)$$

with N = total number of nets
n = index of net ($n = 1, ..., N$)
$S_x(n)$ = horizontal length of bounding rectangle for net n
$S_y(n)$ = vertical length of bounding rectangle for net n
$H(n)$ = horizontal weighting factor for net n
$V(n)$ = vertical weighting factor for net n

Heights and widths of the rectangles are controlled by weighting factors H and V given by the designer. Those factors influence the strategy of positioning cells of a netlist. If $H(n)$ is constant and $V(n)$ is increased, the height of the bounding rectangle will decrease but the width will increase. So the tendency would be to place cells of one net in a rather "linear" form, side by side, instead of in a "rectangular" form.

Available row lengths are given by the designer. Generally, they are not equal to the physical lengths. This desired length of row r is denoted by $L_d(r)$. The part of row r that will actually be used by cells is $L(r)$. Obviously, $L(r)$ should be as close to $L_d(r)$ as possible. To control row lengths, a penalty term P_R (with R = total number of rows) is introduced as

$$P_R = \sum_{r=1}^{R} |L(r) - L_d(r)|$$

Since minimum total wire-length and minimum wasted space on the chip are both desirable features of a layout, the objective function C is composed of the two terms W and P_R. If influence of the penalty term is controlled by a weighting factor c the objective function becomes:

$$C = W + c\,P_R$$

4 Implementation of PRSA on Transputer-based Systems

Parallelization of simulated annealing has been tried by others before. Usually some sort of coordinated search is applied [4, 7, 8, 11, 12, 13, 22]. Coordination often follows a master-slave strategy [1, 6]. When PRSA is the basis for parallelization, the parallelism property inherited from genetic algorithms can be exploited.

The hardware platforms used for our implementations were two transputer systems: one Parsytec MultiCluster/2 and one Parsytec SuperCluster 320. The first one has 32 processor nodes, Inmos 805/25, each one equipped with 4 MB of RAM. The second one consists of 320 nodes of the same type. Both systems have Parix 1.2 as operating system. All nodes are connected via links as a high speed bus system. An efficient way to use this type of parallel machine is to maximize the number of tasks which can be executed in parallel and to minimize communication overhead between these tasks.

In the parallel version of PRSA, one PRSA algorithm similar to the one outlined in section 3 runs on each transputer node. Individuals are exchanged between nodes at certain intervals (migration). Those individuals are selected randomly. Although it might seem intuitively favorable to take the best individuals, this strategy would not support the concept of merging sub-populations well. Global convergence is better if selection is random.

In migration, individuals from one processing node are sent to those nodes directly connected by links. To keep sub-population sizes constant during the program run, the number of migrating individuals is the same for all nodes. When individuals arrive at some node, they replace the individuals sent by that node. Figure 3 summarizes the parallel algorithm for PRSA.

Communication among the PRSA algorithms running in parallel means sending and receiving selected individuals. Since communication follows the links between nodes, the virtual arrangement of nodes (i.e. the *communication topology*) has some impact on performance. Several topologies are common for transputer systems, e.g. master-slave topology, ring topology, and different grid topologies. We used a structure where all processors are arranged in form of a grid and linked with their nearest neighbors.

```
FOR all nodes DO in parallel:
  Set temperature T to a sufficiently high value.
  Build a population of size N by
    generating N individuals which represent consistent place-
        ments,
    calculating x and y coordinates for each cell of all indi-
        viduals,
    evaluating the fitness of each individual.
  REPEAT
    Mark all individuals as "unused".
    DO N/2 times:
      Select two "unused" individuals (parents) and apply cy-
          cle cross-over operator to create two new indi-
          viduals (children).
      Apply mutation operator to each child once.
      Calculate x and y coordinates for each cell of the chil-
          dren in order to evaluate the fitness of the children
          with regard to the objective function.
      Select two individuals from the two parents and the two
          children by Boltzmann trial. Mark winning indi-
          viduals as "used" and replace the two parents by
          those individuals.
    Lower the temperature T.
    Select individuals for migration and send them to other
        processors.
    Receive new individuals from other processors. Replace
        the individuals sent before by received individuals.
  UNTIL termination criterion is satisfied.
```

Fig. 3: Algorithm for parallel PRSA

5 Exploration of Parameters and Test Results

Implementation of parallel PRSA was tested with regard to different parameter settings. On the one hand, there are "ordinary" genetic parameters like the strategy of selecting individuals for the next generation. On the other hand, various parameters relate to parallelization of the method and to parallel processing. In our tests, we experimented with combinations of the following parameters:

(1) Selection strategy: a) Each child competes against one of his two immediate parents, b) "good" children compete against "bad" parents and vice versa, c) each child competes against one randomly selected parent.
(2) Migration strategy: a) Individuals to be sent to other nodes are the ones with best fitness, or b) individuals are randomly selected ones.
(3) Sub-population size (i.e. number of individuals to form a local population on one node): 4, 8, 16, and 32.
(4) Migration rate (i.e. number of individuals sent from one node to other nodes): 0 to 25% of sub-population size.
(5) Number of nodes: 1, 4, and 8 transputer nodes working in parallel.

One noticeable result of the test runs was that *combinations* of certain parameter settings have stronger effect on the resulting wire-length than variations of any single parameter. In particular, different combinations of settings of parameters (1) and (2) lead to significantly different results. The best selection strategy proved to be having each child compete against one of his immediate parents (selection strategy a)). As to migration, strategy a), i.e. choosing the "best" individuals to migrate, mostly performed better than strategy b). Nevertheless, the very best solution average was obtained by strategy b), but here the difference between the two strategies was very small (only 0.01 % difference between average minimal wire-lengths). With regard to computing time, migration strategy a) proved to be better than strategy b). Variations of the parameters sub-population size, migration rate, and number of nodes had less significant effects on solution quality.

In comparison with simulated annealing and genetic algorithms, PRSA performed slightly better (0.8 and 0.04 %, resp.), but computing times were 12 times and 4 times longer, resp.

6 Outlook

In future work we will attempt to improve performance of parallel PRSA by exploiting other communication topologies. Since merging of sub-populations and thus global convergence will be the better the more individuals migrate, the number of immediate neighbors to a node (i.e. nodes connected directly by hardware links) has to be increased. In our grid topology, all inner nodes are fully connected but outer nodes have at least one unused link. This drawback can be overcome if the grid is extended to a *torus* where outer nodes are also connected with their respective counterparts on the opposite side of the grid.

Migration will be adapted to the torus and new strategies (e.g. selecting individuals according to fitness distribution) will be explored. Experiments will also be made regarding the procedure of building new sub-populations (individuals received from other nodes will have to pass Boltzmann trials to be accepted). Test runs will be performed with different parameter settings to find out favorable constellations.

References

[1] Aarts, E.H.L., Korst, J.H.M.: Simulated Annealing and Boltzmann machines: a stochastic approach to combinatorial optimization and neural computing; New York 1989.

[2] Brown, D.E., Huntley, C.L., Spillane, A.R.: A parallel genetic heuristic for the quadratic assignment problem; in: Schaffer, J.D. (eds.): Proceedings of the Third International Conference on Genetic Algorithms; San Mateo, CA 1989, pp. 406-415.

[3] Chang, S.: The generation of minimal spanning trees with a Steiner topology; in: Journal of the ACM 19 (1972) 4, pp. 699-711.

[4] Chen, H., Flenn, N.S.: Parallel Simulated Annealing and Genetic Algorithms: A Space of Hybrid Models; in: Davidor, Y., Schwefel, H.-P., Männer, R. (eds.): Parallel Problem Solving from Nature - PPSN III; Berlin et al. 1994, pp. 428-439.

[5] Davis, L.: Handbook of Genetic Algorithms; New York 1991.

[6] Diekmann, R., Lüling, R., Simon, J.: Problem independent simulated annealing and its application; Technical Report, University of Paderborn, Germany 1992.

[7] Goldberg, D.E.: A Note on Boltzmann Tournament Selection for Genetic Algorithms and Population-Oriented Simulated Annealing; in: Complex Systems 4 (1990) 4, pp. 445-460.

[8] Hoffmann, K.H., Sibani, P., Pedersen, J.M., Salamon, P.: Optimal Ensemble Size for Parallel Implementations of Simulated Annealing; in: Applied Mathematics Letters (1990) 3, pp. 53-61.[9] Kirkpatrick, S., Gellat, S.D., Vecchi, M.P.: Optimization by Simulated Annealing; in: Science 220 (1983) 5, pp. 671-680.

[10] Kling, R.M., Banerjee, P.: ESP - A new standard cell placement package using simulated evolution; in: Proceedings of 24th ACM/IEEE Design Automation Conference; New York 1987, pp. 60-66.

[11] Laursen, P.S.: Problem-Independent Parallel Simulated Annealing Using Selection and Migration; in: Davidor, Y., Schwefel, H.-P., Männer, R. (eds.): Parallel Problem Solving from Nature - PPSN III; Berlin et al. 1994, pp. 408-417.

[12] Lin, F.-T., Kao, C.-Y., Hsu, C.-C.: Incorporating genetic algorithms into simulated annealing; in: Cantu-Ortiz, F. J., Terashima-Marin, H. (eds.): Proceedings of International Symposium on Artificial Intelligence; Cancun, Mexico 1991, pp. 290-297.

[13] Mahfoud, W.S., Goldberg, D.E.: Parallel recombinative simulating annealing: a genetic algorithm; Technical Report No. 92002, University of Illinois, 1992.

[14] Mallela, S., Grover, L.K.: Clustering Based Simulated Annealing for Standard Cell Placement; in: Proceedings of 25th ACM/IEEE Design Automation Conference; New York 1988, pp. 312-317.

[15] Sechen, C., Lee, K.W.: An improved simulating annealing for row based placement; in: Proceedings of IEEE International Conference on Computer Aided Design; New York 1987, pp. 478-481.

[16] Sechen, C., Sangiovanni-Vencentelli, A.: TimberWolf 3.2: A new standard cell placement and global routing package; in: Proceedings of 23rd Design Automation Conference; New York 1986, pp. 432-439.

[17] Shahookar, K., Mazumder, P.: A genetic approach to standard cell placement using meta-genetic parameter optimization; in: IEEE Transactions on Computer-aided Design 9 (1990) 5, pp. 500-511.

[18] Shahookar, K., Mazumder, P.: VLSI placement techniques; in: ACM Computing Surveys 23 (1991) 2, pp. 143-220.

[19] Sherwani, N.: Algorithms for VLSI physical design automation; Boston et al. 1993.

[20] Singh, K., George, S.M., Rambabu, P.: Recurrent Networks for Standard Cell Placement; in: Balagurusamy, R., Sushila, B. (eds.): Artificial Intelligence Technology - Applications and Management; New Delhi, New York et al. 1993, pp. 141-148.

[21] Singh, K., Schneider, B., Kurbel, K.: Solving Sequencing Problems using Recurrent Neural Networks and Simulated Annealing - A Structural and Computational Comparison; to appear in: Proceedings of the International Conference on Operations Research (OR '94), Berlin 1994.

[22] Azencott, R.: Simulated Annealing: Parallelization Techniques; New York 1992.

PERFORMANCE OF GENETIC ALGORITHMS IN THE SOLUTION OF PERMUTATION FLOWSHOP PROBLEM

Nicoletta Sangalli, Quirico Semeraro, Tullio Tolio

Dipartimento di Meccanica -Politecnico di Milano
Via Bonardi 9 - 20133 Milano - E-mail: sangalli@itiasrvr.itia.mi.cnr.it

Abstract

Scope of this paper is to investigate the applicability of Genetic Algorithms to the solution of Static Permutation Flowshop problem and to compare their performance with those obtained with Taboo Search Algorithms. Since in order to obtain good results with Genetic Algorithms (Ga_s) it is necessary to tune a set of parameters, then a second purpose of the study is to devise a methodology to perform this tuning. The methodology used is the "Response Surface Methodology".

1 INTRODUCTION

The Static Permutation Flowshop problem consists of finding the best schedule for N jobs on M machines for a given objective function knowing that:

- all the jobs visit all the machines in the same order.
- all the machine process all the jobs in the same order.

In the study carried out the Makespan (completion time of the last job on the last machine) was adopted as the objective function.

Some different Ga_s were devised which differ in the crossover operator (one point crossover or PMX) and in the composition of the initial population (random or with some "good" solution found with fast "Single Shot" algorithms[1]). Regarding the selection strategy and the reproduction strategy, the Tournament Selection and the Steady State Reproduction respectively were applied to all algorithms. In order to make the comparison between the implemented Ga_s and Taboo Search it was necessary to tune the parameters, crossover rate, mutation rate and population size. The Response Surface Methodology was applied to each genetic algorithm in order to find the optimal parameter values.

Finally a comparison with Taboo Search heuristics [2],as other author made[3], was carried out, by using the MANOVA approach (Multivariate Analysis Of Variance) [1].

2 GENETIC ALGORITHM CHARACTERISTICS

Three different genetic algorithms were studied. The "tournament selection "strategy [4] was applied to all algorithms. The applied strategy is little different from the tournament procedure suggested by Wetzel. Here two individuals are randomly selected, they are compared and only the best individual becomes the potential parent. This procedure is repeated for another copy. In this way two parents are selected and crossover can be applied on them.

The "steady state reproduction" [5] was used to replace individuals. With this strategy only one individual is replaced every generation.. The new individual takes the place of the worst individual contained in the population.

2.1 Genetic Algorithm with one point crossover (GA_1)

This algorithm uses the "numerical coding" as representation, and the "one point crossover" [6] as crossover.

Every individual is represented as a gene sequence, which is a sequence of integer or floating point [7]. A decoding algorithm decodes the gene sequence in a job sequence in order to obtain a feasible solution (i.e. jobs must not be repeated inside the sequence). The job sequence is a string of number in which each number represents a job. The job sequence is used to evaluate the fitness of the individual. The following figure (Fig. 1) shows the relation between gene and job sequences and the rule of decoding algorithm.

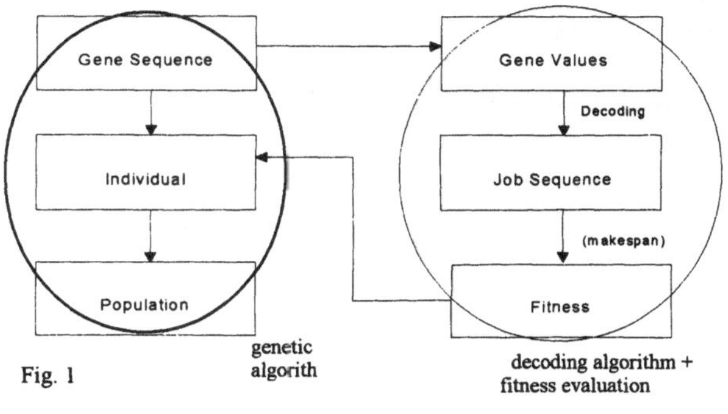

Fig. 1

The decoding algorithm can be formalized in the following way:

Decoding algorithm:

```
for i=1 to job number
 {  j=argmax(gene_sequence[h])
    job_sequence[i]=j
 }
```

Decoding algorithm steps:

- Find the maximum number in the gene sequence;
- The next job in the job sequence becomes the corresponding position of the maximum number found in the first step;
- Set the maximum number of the gene sequence found to "0".
- Go to the first step until the whole job sequence was completed.

Example

The figure shows two gene sequences and their corresponding job sequences:

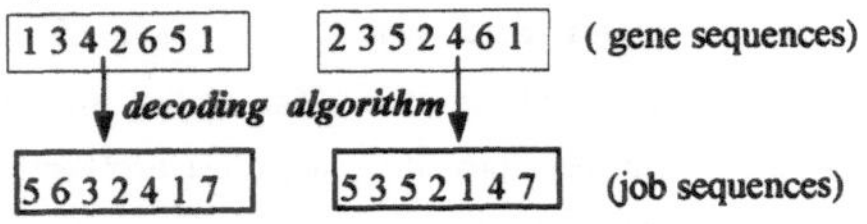

The one point crossover is applied to the gene sequences always obtaining, after the decoding, a feasible solution.

Example:

The figure shows the crossover between the two gene sequences in the 3rd position, the offspring generated and the decoding algorithm:

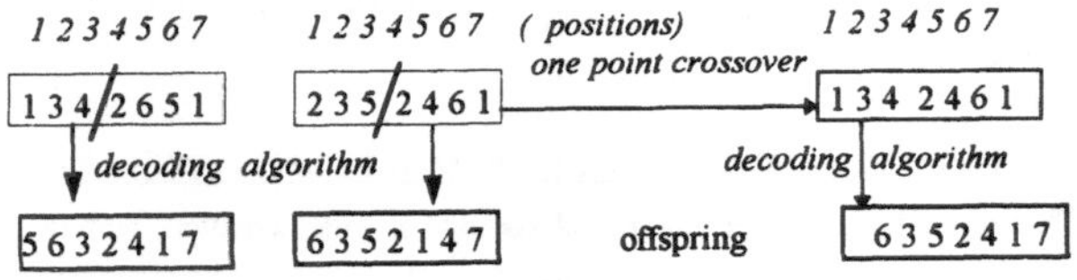

The mutation works on each bit that constitutes the gene value.

The initial population is generated randomly.

2.2 Genetic Algorithms with PMX (GA_2)

This algorithm is different to the first one in the crossover, that is here the "pmx"[8]. The gene sequences represent the job sequences because with the crossover "pmx" the solutions are always feasible. This is the most common genetic algorithm used in permutation flowshop or in travelling salesman problems. The mutation operator changes the position of two genes. The initial population is generated randomly.

2.3 Genetic Algorithm with good initial solution (GA_3)

This algorithm differs from the first one (GA_1) in the way in which the initial population is generated. A lot of individuals are created randomly and few others are generated by some Single Shot algorithms [1].

3 PARAMETER TUNING

In order to set the optimal values for the population size, crossover probability and mutation probability, the Response Surface Methodology (RSM) [9] was applied. It would be interesting to know how important these parameters are on the genetic algorithm performance and if the parameter setting changes for every algorithm, for every problem class (job number, machine number, processing times associated with the jobs) or it is a general result.

3.1 Response Surface Methodology

The response surface methodology (RSM) [9] is a technique set that encompasses:

1. setting up a series of experiments (design a set of experiments) that will yield adequate and reliable measurements of the response of interest
2. determining a mathematical model that best fits the data collected from the experiments chosen (1), by conducting appropriate tests of hypotheses concerning the model's parameters
3. Determining the optimal settings of the experimental factors that produce the maximum (or minimum) value of response .

The RSM includes the application of regression as well as other techniques in an attempt to gain a better understanding of the characteristics of the response system under study (in our case the genetic algorithm performance). In order to apply the RSM the following steps must be performed:

1. An initial range of parameters must be defined (in our case it is a cube, see Fig. 2).
2. For each extreme point of the range considered the response of the system must be performed (a certain number of repetitions must be carried out). A Taylor polynomial, that approximates the real curve in the considered range, is evaluated on the collected data. The order of the polynomial is chosen as the minimum that allows the rejection of the hypothesis of lack of fit. (see the table 2).
3. It is possible to visualize the shape of the approximating response surface by using the contour plot (Fig. 3). Each contour represents a constant value of the function in the examined range. In this way it is possible to know the direction where the function improves and what value must be given to the parameters in the next experiments.
4. The response function must be calculated in points outside the parameter range, in the direction suggested by the contour plot (for each point a certain number of repetitions must be carried out). The comparison between the mean value obtained in two consecutive points is realized by using the t_student test. The search is carried out until the function becomes worse (Fig. 3).
5. The previous best point becomes the central point of a new range and another approximating function must be evaluated. The analysis is repeated starting from step 3, until a parameter range (as in our case) or a point, in which the response is optimal, are reached (see the F-

Fisher probability of the linear component in table 2). The corresponding parameter values are the optimal parameters.

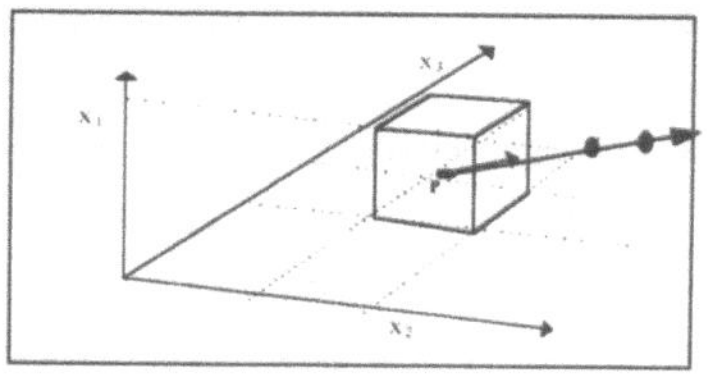

Fig. 2

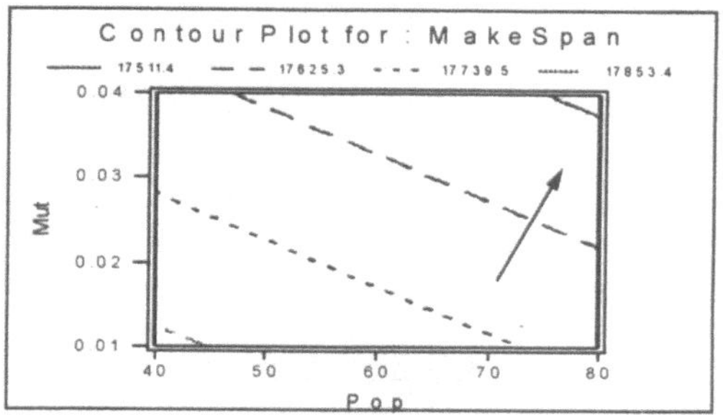

Fig. 3

4 Experimentation

In order to answer the validity of Ga, implemented and of the methodology used to tune the parameters an experimentation phase was carried out.

4.1 Experiment Setting

A flowshop problem is defined by the number of jobs. the number of machines and he processing times. In our experimentation we considered problems of the following class: 15 jobs. 15 machines and processing times drawn from an uniform distribution [0-999]. The RSM methodology was applied for each Ga on one problem of class.

During the search in every point (see section 3.1) each Ga was repeated for 60 times. in order to have the power of F_test of 85%. At the end we found three optimal parameter ranges. one for each Ga (see result section).

After the application of RSM and after the demonstration of its validity. the comparison of the optimal Ga, (GA_1. GA_2. GA_3) with a Taboo Search was carried out on 300 test problems of the class described above. The setting of Taboo Search parameters was: Taboo list is set to 12 (12 forbidden moves) and the consecutive cycles without improvement are set to 15 [2]. In order to make a fair comparison the Ga, were stopped when the total number of alternative solutions considered was equal to the number of solution tested by the Taboo Search.

5 Results

The optimal parameter ranges for the three algorithms considered are shown in the following table 1:

OPTIMAL RANGES

for GA_1

population size	crossover rate	mutation rate (on bit)
90-110	0.6-0.7	0.04-0.07

for GA_2

population size	crossover rate	mutation rate (on gene)
85-95	0.35-0.45	0.12-0.14

for GA_3

population size	crossover rate	mutation rate (on bit)
90-110	0.6-0.7	0.04-0.07

table 1

The parameter range obtained for each Ga represents an optimum solution for the single problem analysed (it can be a local optimum however). The table 2 shows the output of RSM methodology. The p_value of the linear and interaction components show that these terms do not explain the function. so only the constant term is significant. This means that these parameter ranges identify a flat region for the response function.

table2

Flat Zone for GA_1

Estimated Regression Coefficients for Makespan

Term	Coef	Stdev	t-ratio	p
Constant	18225	540.46	33.721	0.000
Pop	-5	5.12	-0.949	0.343
Mut	-11094	5249.46	-2.113	0.035
Cros	-240	87.7	-0.349	0.727
Pop*Mut	66	42.86	1.530	0.127
Pop*Cros	-0	6.43	-0.038	0.970
Mut*Cros	5439	4286.16	1.269	0.205

s = 140.9 R-sq = 2.5% R-sq(adj) = 1.2%

Analysis of Variance for Makespan

Source	DF	SeqSS	AdjSS	AdjMS	F	P
Linear	3	160814	91735	30578	1.5	0.203
Interaction	3	78430	78430	26143	1.3	0.268
Residual	473	9384738	9384738	19841		
-Lack-of-Fit	1	9991	9991	9991	0.5	0.479
-Pure Error	472	9374746	9374746	19862		
Total	479	9623982				

Flat Zone for AG_2

Estimated Regression Coefficients for Makespan

Term	Coef	Stdev	t-ratio	p
Constant	17536.1	128.500	136.468	0.000
Pop	-0.8	1.088	-0.692	0.489
Mut	-455.2	543.984	-0.837	0.403
Cros	149.2	108.797	1.371	0.171

s = 119.2 R-sq = 0.6% R-sq(adj) = 0.0%

Analysis of Variance for Makespan

Source	DF	Seq SS	Adj	AdjM	F	P
Linear	3	43465	43465	14488	1.0	0.38
Residual Error	476	676114	67611	14204		
-Lack-of-Fit	4	41739	41739	10435	0.7	0.57
-Pure Error	472	671940	67194	14236		
Total	479	680461				

Flat Zone for AG_3

Estimated Regression Coefficients for Makespan

Term	Coef	Stdev	t-ratio	p
Constant	17519	359.17	48.775	0.000
Pop	-0	3.40	-0.005	0.996
Mut	-6064	3488.59	-1.738	0.083
Cros	215	457.08	0.469	0.639
Pop*Mut	47	28.48	1.647	0.100
Pop*Cros	-3	4.27	-0.676	0.499
Mut*Cros	3504	2848.42	1.230	0.219

s = 93.61 R-sq = 5.6% R-sq(adj) = 4.5%

Analysis of Variance for Makespan

Source	DF	Seq	Adj SS	AdjMS	F	P
Linear	3	2070	32463	10821.	1.2	0.296
Interaction	3	4102	41028	13676.	1.5	0.198
Residual Error	473	4144	41447	8762.6		
-Lack-of-Fit	1	579	579	578.8	0.0	0.797

The results obtained by the application of the RSM methodology show that the parameter tuning depends on the type of genetic algorithm (they could also depend on the class of problem). It is important to verify if the range found is actually a good range for all the problems of the class considered. In order to test the validity of these results we compared the performance, obtained by using genetic algorithm with the optimal parameter values (GA_opt), with those obtained by setting the parameters in a arbitrary way (GA_arb). The results of a GA_opt and of a GA_arb, tested for 300 problems, were compared by means of a sign test. For all the three algorithm considered (GA_1, GA_2, GA_3) this validation step was carried out. As you can see in the table 3 it is possible to reject the hypotheses of equality between the performance of GA_opt and GA_arb with one sided sign test at 1% significance.

table 3

```
MTB >Let C17='GA_opt'-'GA_arb'(for genetic algorithm GA1)
MTB > STest 0.0 C17;
SUBC>    Alternative 0.
SIGN TEST OF MEDIAN = 0.00000 VERSUS  N.E.   0.00000
       N    BELOW   EQUAL  ABOVE    P-VALUE      MEDIAN
C17  300     211       9     80     0.0000       -120.5

MTB>Let C3='GA_opt'-'GA_arb'(for genetic algorithm GA2)
MTB > STest 0.0 C3;
SUBC>    Alternative 0.
SIGN TEST OF MEDIAN = 0.00000 VERSUS  N.E.   0.00000
           N  BELOW  EQUAL  ABOVE    P-VALUE      MEDIAN
C3       300  267      6       27     0.0000      -259.0

MTB> Let C3 ='GA_opt'-'GA_arb'(for genetic algorithm GA3)
MTB > STest 0.0 C3;
SUBC> Alternative 0.
SIGN TEST OF MEDIAN = 0.00000 VERSUS  N.E.   0.00000
           N  BELOW    EQUAL  ABOVE    P-VALUE      MEDIAN
C3       300  219        8     73      0.0000       -125.0
```

In the next step of the analysis was to compare the performance of the Ga, with those obtained with Taboo Search. The multivariate analysis (MANOVA) applied to the results obtained with the three Ga, and the Taboo Search showed that there is for some algorithms (see the table 4) no statistical evidence to reject the hypothesis of equality of means. Therefore we can say nothing about these heuristic algorithms (the symbol "=" in the table 4). The best between of each compared couple of algorithms is called with its name. It seems for this class of problem that the worst algorithm is AG_1.

These results are referred only to the class of problems considered.

	TABOO	GA_1	GA_2	GA_3
TABOO	/	TABOO	=	=
GA_1	TABOO	/	GA_2	=
GA_2	=	GA_2	/	=
GA_3	=	=	=	/

table 4

6 FUTURE WORK

It would be interesting characterize the shape of the response function in order to discover if there are local optima. It would be interesting to considered other classes of problems (with processing times drawn from different distributions of probability or a bigger number of jobs) in order to generalize the results obtained with this study.

References

1 . M.Bolognini,R. Borelli, T. Tolio, Q.Semeraro:Influence of the structure of Static Permutation Flowshop problem on the performance of Single Shot Heuristics. Computers ind.Engng. Vol.26,N°3,pp437-450,(1994).

2 .M.Bolognini,R. Borelli, T. Tolio, Q.Semeraro:.Analysis and Evaluation of Taboo Search Heuristics for Scheduling of a Static Permutation Flowshop Proceeding Computer-Aided Production Engineering. Edinburgh(1992).

3 Colin Reeves. Genetic Algorithm for Flow Shop Sequencing. Computers &Opns.Res

4 A.Wetzel. Evaluation of the Effectiveness of genetic algorithms in combinatorial optimization. Unpublished manuscript, University of Pittsburgh (1983).

5 D.Whitley. GENITOR:a different genetic algorithm. In proceedings of the Rocky Mountain Conference on Artificial Intelligence. Denver, Colo (1988).

6 D.E. Goldberg. Genetic Algorithms in Search Optimization and Machine Learning. Addison Wesley. (1989)

7 James C. Bean. Genetic and Random Keys for Sequencing and Optimization. Technical Report, Department of Industrial & Operations Engineering University Michigan(1992.).

8 I.M. Oliver, D.J.Smith and J.R.C. Holland. International Conference on Genetic Algorithms and their Application..224-230. 1987.A Study of Permutation Crossover Operators on the Travelling Salesman problem. Proc. 2nd.

9 Andrè I. Khuri, John A. Cornell. Response Surfaces Designs and Analyses .Marcel Dekker,Inc.ASQC Quality Press. New York.

A GENETIC ALGORITHM FOR THE MAXIMAL CLIQUE PROBLEM

Cristina Bazgan Henri Luchian
e-mail: hluchian@uaic.ro

Faculty of Computer Science, "Al.I.Cuza" University of Iasi, Romania

Abstract. The maximal clique problem (MaxClique) is known to be difficult (in terms of complexity) both in its exact and in the approximate form. Therefore, probabilistic algorithms for this problem are worthwhile to study. In this paper, we present an evolutionary (genetic) approach to solving the maximal clique problem (we are not directly concerned here with complexity). As opposed to previous work, our solution is inspired by a theoretical result [1] which improves the genetic algorithm. Experimental results for graphs with various properties are given; the convergence towards a good solution (in almost all of the runs, the global optimum) turns out to be quick.

1. Introduction

We consider the problem of finding a maximal clique (its cardinality is called "clique number") in a given graph. This is a special problem w.r.t. approximation. Although the decision problems associated to MaxClique and MinVertexCover are equivalent (they are NP-complete [5]), their approximated versions are not: MinVertexCover is in APX (the class of problems for which an approximate solution can be found in polynomial time), but MaxClique is not in APX [8], [9].

Since MaxClique is NP-complete, the search for an optimal solution should be replaced by the search for a good heuristic / polynomial time algorithm which can find solutions close to the optimum (approximation algorithms). The quality of a deterministic approximation algorithm A is measured by its performance (solution found) against the optimal solution:

$$\mathbf{q} = \sup(OPT(I)/A(I)), \qquad (1)$$

(for a given instance I of MaxClique, OPT(I) is the optimal solution and A(I) is the solution given by A). For MaxClique, the best known **q** is achieved by an algorithm due to Boppana and Haldorsson ([6]); **q** is there of the order $O(n/\log^2 n)$. It is known that no determinist polynomial time approximation algorithm for MaxClique can achieve a constant **q** (unless P=NP); furthermore, there exists a constant c such that no determinist polynomial time approximation algorithm for this problem can have $\mathbf{q}=n^c$, unless P=NP ([7]). Since (as shown by the two negative results quoted above) the approximate version of MaxClique ("find a near-optimal solution") is also difficult, genetic algorithms may be an important tool for solving it. Some work has already been done in this respect ([3]). The algorithm given there is based on a simple approach, where the fitness function is a direct -though elegant- expression of how good a clique is (no theoretical properties of the problem are used). Some related graph problems have also been undertaken recently in an

H.Luchian and C.Bazgan are joint first authors of this paper.

evolutionary setting ([4]). We propose a new genetic approach to MaxClique. The fitness function is based on a theoretical result of Croitoru e.a. [1]; the quick convergence in almost all runs to the optimal solution is probably mainly due to the strength of this result. In section 2 we give details of the algorithm; section 3 presents the experimental results and in the last section we draw our conclusions.

2. The genetic algorithm

An instance of MaxClique is a graph G=(V,E) where V={1,...,n} is the set of vertices and E is the set of edges. A feasible solution is a set V' of vertices s.t.: $\forall$ i,j $\in$ V', (i,j) $\in$ E. The objective function is the size of the set V', |V'|. An optimal solution is a clique V' which maximizes |V'|.

Our algorithm is characterized by: classical mutation and one-point crossover operators; constant-size population; roulette-wheel selection; elitism. We chose the following representation: a chromosome is a binary string (x[1], x[2], ... ,x[n]), where a 0 or a 1 in the i-th gene position indicates the absence or presence of the i-th vertex of the graph in the subgraph which represents the candidate solution encoded by the chromosome. Note that a particular bitstring may represent a solution which is not feasible. This is an obvious and handy representation for problems involving subgraphs (see [3] and [4], where the same representation is used).

The fitness function to be maximized by the genetic algorithm adds, for a given chromosome x, the products x[i]*x[j], $\forall$(i,j)$\in$E and penalizes with n times the sum of the products x[i]*x[j], $\forall$(i,j)$\notin$E, i $\neq$j. This function enforces the algorithm to leave the regions of the search space that represent unfeasible solutions and to orient the search towards more promising regions.

$$f(x) = \sum_{(i,j)\in E} x[i] * x[j] - n * \sum_{(i,j)\notin E} x[i] * x[j] \qquad (2)$$

The maximal value of this function is w(G)*(w(G)-1)/2, where w(G) is the size of a maximal clique of the graph G; the corresponding chromosome represents a maximal clique of the graph (this is inspired from the theoretical result of C.Croitoru e.a.[1]).

3. Experimental results

Experiments with a genetic algorithm are somewhat difficult to design. Besides the probabilistic and tunable nature of the algorithm, one has to face the problem of choosing adequate instances of the problem at hand (there are also other design objectives, which are not considered in this paper, e.g. comparisons between different algorithms). In designing our experiments, we confined ourselves to: a) finding "good" values for the parameters of the algorithm (mutation / crossover rates, pop_size, maximum number of generations); b) for the values found in the first stage, testing the reliability of the genetic algorithm (percentage of runs in which the global-best is found). This is consistent with the definition of **q** for deterministic approximation algorithms, since, in (1), A(I) can be considered to represent, in the case of probabilistic approximation algorithms, the average, over a given number of runs, of the objective function values for the solutions found by the algorithm. This seems to be a good basis for a comparative study of deterministic and probabilistic approximation algorithms, with respect to their accuracy. The study could be further developed by adding complexity concerns (as in [6], [7]).

In our experiments, rather than using random graphs (as in [3]) or one scaleable graph (as in [4]), we have chosen five instances of MaxClique which are particularly interesting / difficult.

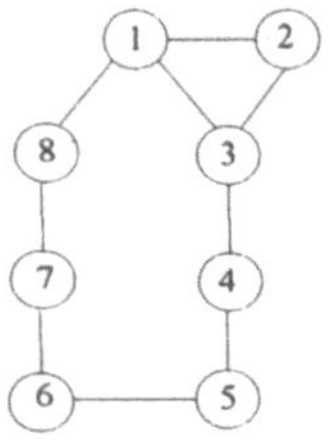

Fig. 1

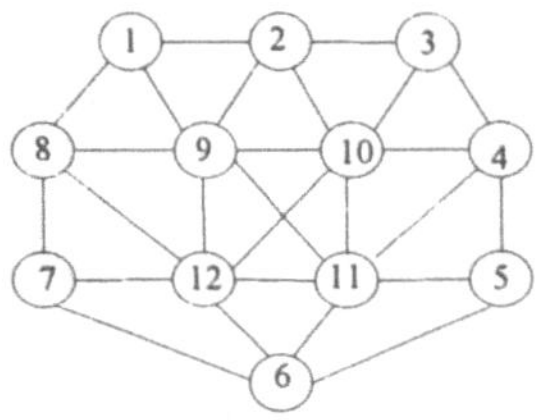

Fig. 2

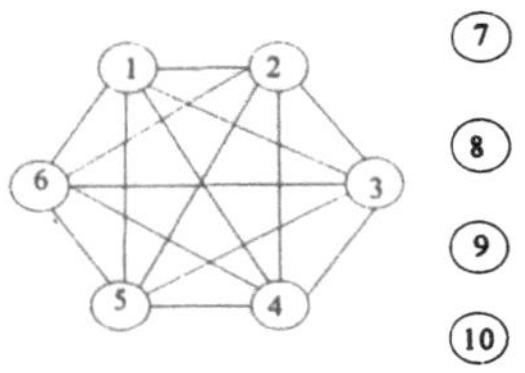

Fig. 3

The first graph (Fig. 1) has only one maximal clique of cardinality 3, whose size is close to the size of nine non-optimal cliques. The second graph (Fig. 2) has many edges, but only one maximal clique of cardinality 4. The third graph has n=14 vertices which are split into four classes: {1,2,3}, {4,5,6}, {7,8,9,10}, {11,12,13,14}; each vertex in each class is connected with all the vertices in the other classes; for this graph, w=4. This graph: i) has the maximum number of edges among all the graphs with n=14 and w=4; ii) has the maximum number of maximal cliques among all these graphs. The fourth graph (Fig. 3) has n=10 and only one clique of maximum cardinality, 6; this graph has a minimal number of edges among all the graphs with n=10 and w=6. The last graph has n=10; it is the complement of a bipartite graph and it has one clique of cardinality 4 and one of cardinality 6.

For these instantiations of MaxClique, we performed two stages of the experiments: i) in search for good values of the parameters, we have run 20 times the algorithm for each graph and for each combination of parameter values (8 values for Pc - the crossover rate, and 4 values for Pm - the mutation rate, a total of 32 combinations). ii) for two of the best combinations (for all graphs) found above, the genetic algorithm was run N=100 times, using different values for the population size and the maximum number of generations.

Part of the results (average fitness of solutions over 20 runs) are summarized in table 1 (first stage for the graph in Fig.2) and table 2 (second stage; we have counted the number of runs which gave the optimal clique vs. the number of fails). In a quick summary, it seems that a few combinations of crossover and mutation rates are very good for all examples, though there is some variability.

Pm\Pc	0,2	0,4	0,6	0,8
0,03	13,75	13,25	14,25	14,25
0,06	14	13,75	13	14,25
0,1	14,25	14	15	14,5
0,2	14,75	14,5	14,75	14,75
0,3	14,75	14,25	14,5	15
0,4	15	14,5	14,75	14,75
0,5	15	14,75	15	15

Table 1

Graph G	1		2		3		4		5	
W(G)	3		4		4		6		6	
	fitness	N	itness	N	fitness	N	fitness	N	fitness	N
Pm = 0.4 Pc = 0.6 pop_s = 20 nr_g = 100	3	100	3 6	44 56	6	100	15	100	10 15	15 85
Pm =0.2 Pc = 0.6 pop_s =20 nr_g = 100	3	100	3 6	38 62	6	100	15	100	10 15	13 87
Pm =0.2 Pc = 0.6 pop_s =40 nr_g = 200	3	100	3 6	12 88	6	100	15	100	15	100
Pm =0.1 Pc = 0.6 pop_s =40 nr_g = 200	3	100	3 6	9 91	6	100	15	100	10 15	2 98

Table 2

More important, the algorithm has a stable behavior ("acts almost like a deterministic algorithm"): if properly tuned, it finds the optimal clique in at least 90% of the runs, for all the examples we considered.

4. Conclusions and further work

It is our opinion that, while being widely applicable, the genetic approach should not be hastily implemented to many problems following the paradigm: "write one GA for one problem, then switch to the next problem". On the contrary, we think that various genetic algorithms should be considered for the same problem (with different representations, fitness functions, operators, selection strategies, thus incorporating various problem-specific elements) in order to obtain what should be "an optimal genetic algorithm for a given problem" (whatever "optimal" could mean for a user).

We have presented a new genetic algorithm for MaxClique. Our algorithm seems to be quicker and more reliable than the genetic algorithms previously proposed by other authors ([3], [4]). This is because more problem-specific knowledge was included by means of the fitness function. We have also suggested a way to directly compare deterministic and probabilistic approximation algorithms. The experiments give evidence that: - there are certain combinations of the GA-parameters, each of which is suitable for different instances of MaxClique; - once tuned by means of such a combination, the algorithm finds the maximal clique in more than 90% of the runs (for some graphs, in all of the runs).

Our algorithm can be further improved in several ways: the parameters of the genetic algorithm should be optimized by means of a meta-GA; other, problem-oriented operators (such as the "partial copy crossover" suggested in [3]) should be taken into account. We are currently implementing the other existing algorithms for a fair and direct comparison; this comparison will be concerned with both the accuracy of solutions found by each algorithm and the complexity of these algorithms (in terms of CPU time; to this end, larger examples are considered).

References

[1] - Croitoru,C., Radu,C.: "Quadratic programs on graphs", The Scientific Annals of the "Al.I.Cuza" University, section Ia, Mathematics and Computer Science, Iasi, Romania, vol. xxxiv, 1988.

[2] -Michalewicz,Z.: "Genetic Algorithms+Data Structures = Evolutionary Programs", Springer Verlag, 1992.

[3] - Murthy,A.S., Parthasarathy,G., Sastry,V.U.K.: "Clique Finding - A Genetic Approach", Proc. of the First IEEE ICEC Conference, Orlando, USA, 1994, pp. 18-21.

[4] - Back, T., Khuri, S. "An Evolutionary Heuristic for the Maximal Independent Set Problem", Proc. of the first IEEE ICEC Conference, Orlando, USA, 1994, pp. 531-535.

[5] - Garey, M.R., Johnson, D.S.:" Computers and Intractability" Freeman, San Francisco, 1979.

[6]-Boppana,R., Halldorsson,M.M. :" Approximating maximal independent set by excluding subgraphs", BIT 32, 1992, pp.180-196.

[7] - Hougardy,S., Promel,H.J., Steger,A.: "Probabilistically checkable proofs and their consequences for approximation algorithms", research report, Forschungsinstitut fur Diskrete Mathematik, Univ. of Bonn, 1993.

[8] - Arora, S., Safa, S.: " Approximating MaxClique is NP-Complete", manuscript, 1992.

[9] - Kann, V.: " On the Approximability of NP-Complete Optimisation Problems", Phd Thesis, Royal Institute of Technology Stockolm, 1992.

MINIMAL ERROR RATE CLASSIFICATION IN A NON-STATIONARY ENVIRONMENT VIA A MODIFIED FUZZY ARTMAP NETWORK

Chee Peng Lim and Robert F. Harrison
Department of Automatic Control and Systems Engineering
The University of Sheffield, Sheffield, United Kingdom

Abstract

This paper investigates the feasibility of the fuzzy ARTMAP neural network for statistical classification and learning tasks in an on-line setting. The inability of fuzzy ARTMAP in implementing a one-to-many mapping is explained. Thus, we propose a modification and a frequency measure scheme which tend to minimise the misclassification rates. The performance of the modified network is assessed with noisy pattern sets in both stationary and non-stationary environments. Simulation results demonstrate that modified fuzzy ARTMAP is capable of learning in a changing environment and, at the same time, of producing classification results which asymptotically approach the Bayes optimal limits. The implications of taking time averages, rather than ensemble averages, when calculating performance statistics are also studied.

1 Introduction

Feedforward neural networks such as the Multilayered Perceptron (MLP) networks and the Radial Basis Function (RBF) networks possess some attractive properties when the objective is to develop a classifier to operate in a probabilistic environment. These network structures have been proven to be able to represent any smooth enough function to an arbitrary degree of accuracy [1,2]. Thus, it is likely that feedforward networks can offer a direct solution to the problem of developing a one-from-many classifier. However, such an approach is only viable when there is good reason to believe that the data environment is stationary and that the data sample used in training is sufficiently representative. In cases where learning takes place in a non-stationary environment, it is either necessary to allow the feedforward networks to carry on learning or to re-train them off-line. Nevertheless, it is well-documented that networks of the Adaptive Resonance family offer a way out of this problem—the so-called stability-plasticity dilemma [3,4]. They are able to learn continuously in a changing data environment and acquire knowledge *in situ* whilst simultaneously providing useful classification results. This paper investigates the classification ability of fuzzy ARTMAP (FAM) [5], a variant of the supervised Adaptive Resonance Theory (ART) networks, in purely statistical learning tasks and compares the results with Bayesian decision theorem.

2 Fuzzy ARTMAP (FAM)

FAM is an extension of the ARTMAP network [6] which makes use of the operation of fuzzy set theory instead of the classical set theory that governs the dynamics of ARTMAP. A FAM network consists of two fuzzy ART [7] modules, ART_a and ART_b, connected by a map field as shown in Fig. 1. During supervised learning, an input pattern vector ***a*** is fed to ART_a with its target vector ***b*** to ART_b. ART_a and ART_b cluster their input vectors independently. An intervening map field (F_{ab}) adaptively associates predictive antecedents in ART_a with their consequents in ART_b.

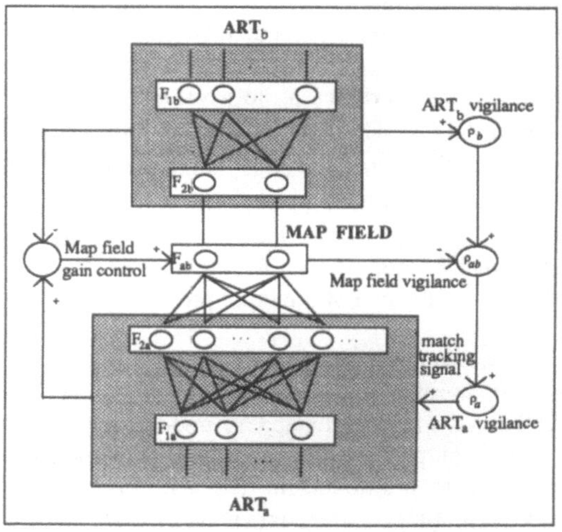

Fig. 1 A schematic diagram of the fuzzy ARTMAP network

In ART_a (as well as in ART_b), an input vector ***a*** first registers itself at the F_{1a} layer in complement-coded format, *i.e.* $A=(a, 1-a)$, to avoid the category proliferation problem [7,8]. This pattern vector is faned-out to all the nodes in the F_{2a} layer via a set of Long Term Memories (LTMs) or weights. The response of each F_{2a} node is based upon a fuzzy *choice function*

$$|A \wedge w_{a-j}| / (\alpha_a + |w_{a-j}|)$$

where w_{a-j} is the weight vector of the jth F_{2a} node, α_a is the choice parameter [5,7] of ART_a and the fuzzy "and" operator ($\wedge$) and the norm $|.|$ are defined as: $(x \wedge y)_i \equiv \min(x_i, y_i)$ and $|x| = \sum_i |x_i|$ [9].

The maximally activated node is selected as the winner and all other nodes are suppressed in accordance with the winner-take-all competitive structure. The winning F_{2a} node then feeds back its weight vector to F_{1a}. This weight vector represents the category prototype of the winning node and is used for comparison with the input vector against a vigilance threshold. Resonance is said to occur if the vigilance test is satisfied, *i.e.*

$$|A \wedge w_{a-J}| / |A| \geq \rho_a$$

where ρ_a is the vigilance parameter of ART_a and w_{a-J} is the winning Jth node in F_{2a}. Otherwise, a mismatch signal is sent to F_{2a} to reset the winning node for the rest of the pattern matching cycle. The input vector A is now re-transmitted to F_{2a} to select a new winner. This search cycle ends when the current category prototype is able to meet the vigilance test or a new node is recruited in F_{2a} with the input pattern coded as the prototypical weight vector.

After resonance has occurred in ART_a and ART_b, a predictive signal is sent from the winning F_{2a} node to the map field. If this prediction is disconfirmed by the winning node in F_{2b}, *i.e.* the map field vigilance test fails, a control strategy called match-tracking is initiated. Match-tracking increases ρ_a to a value which triggers a search in ART_a. Thus, ρ_a is made slightly greater than $|A \wedge w_{a-J}| / |A|$ to cause the ART_a vigilance test to fail. In such a way, match-tracking provides a means to select a node in F_{2a} which fulfils both the ART_a and the map field vigilance tests. If such a node does not exist, F_{2a} is *shut down* for the rest of the input presentation [5,6].

2.1 One-to-many Mapping

One-to-many mapping is defined as the formation of an association from an F_{2a} category node to more than one F_{2b} target output via the map field. Obviously, this association is prohibited in the ARTMAP (both the binary and fuzzy versions) networks to avoid any confusion during recall, hence prediction, from an F_{2a} category node to its F_{2b} predictive answer. However, in statistical pattern classification, overlapping regions can occur in the input space where the same cluster may belong to more than one target output, subject to different probabilities of class membership. It is therefore useful if this one-to-many mapping can be established.

2.2 Modified Fuzzy ARTMAP

One way to implement the one-to-many mapping has been proposed in [10] and is explained below. When the input vector is a fuzzy subset of an F_{2a} category node, it will match perfectly with the category prototype and the ART_a vigilance test produces $|A \wedge w_{a-J}| / |A| = 1$. If the winning category has previously been associated with a different target output, the prediction will be disconfirmed and match-tracking is triggered. So, ρ_a has to be increased to a value slightly greater than unity. This implies that no other nodes can satisfy the ART_a vigilance test and the current input will be ignored.

In view of the above scenario, we propose that during match-tracking, the ART_a vigilance parameter is constrained by

$$0 \leq \rho_a \leq \min\left(1,\ |A \wedge w_{a-J}| / |A|\right)$$

where A is the current input vector to F_{2a} in complement-coded format and w_{a-J} is the winning Jth node in F_{2a}. Thus, if no other F_{2a} node is able to meet the vigilance test, a new node can be recruited to code the input vector. Now it seems that it is possible to have two similar category nodes to map to different target outputs. This modification is applicable to both the binary and fuzzy ARTMAP networks. Indeed, in the simulations described later which involve only binary data, the fuzzy and binary realisations of ARTMAP are identical.

In FAM, if a tie occurs in the choice function, the winning F_{2a} node is selected in the sequence of 1,2,...[5]. To ensure that similar category nodes have a fair chance to be selected as the winner, we introduce a frequency measure scheme. This frequency measure records the number of correct predictions an F_{2a} category node has accomplished. This information not only facilitates the selection of the winner but also reflects the prior probabilities represented by each F_{2a} category prototype. There are two variants of the frequency measure scheme [10]: INC and INC/DEC based on the reward and reward-penalty rationale. The INC method INCreases the frequency count of each F_{2a} node for correct predictions and no penalty is imposed for incorrect predictions. Conversely, the INC/DEC method INCreases or DECreases the frequency count of each F_{2a} node for correct or incorrect predictions accordingly.

3 Simulation Studies

We assess the performance of FAM and modified FAM in classifying *noisy* data set into two classes in stationary and non-stationary environments. The class distributions are fully governed by the prior class probabilities ($P(c_1)$ and $P(c_2)$) and the conditional probabilities or likelihoods ($P(x|c_1)$ and $P(x|c_2)$). By applying Bayes' theorem to these parameters, a data set with a specific posterior probability distribution can be generated. Appendix A shows the parameters used in the following simulations. All the experiments below employ the single-epoch, on-line learning strategy with fast-learning [5,6] and the INC method is adopted for the frequency count. The on-line operational cycle proceeds as follows: an input vector is first presented to ART_a and a prediction is sent to ART_b. The predicted output is compared with the actual output and the outcome gives a classification result (prediction). Learning then ensues to cluster the input and target vectors (learning).

3.1 Stationary On-line Learning & Classification

Two classes of noisy data samples for the two-bit parity (commonly known as XOR) and four-bit parity problems were generated. In each case, 5000 samples were used and a 1000-sample window was applied to calculate the on-line classification results, *e.g.*, the accuracy at sample 2000 was the number of correct predictions from samples 1001-2000 expressed as a percentage. Although the data statistics are time-invariant, tackling the task on-line is in fact a non-stationary process owing to the build-up of templates—the so-called finite-operating-time problem.

Table 1 shows the average accuracy (Acc.) of 5 runs at the end of the experiment. Although the standard deviation (Std. Dev.) is estimated from a small sample size (5 runs), it does serve as an indication of how the results disperse across the averages. Fig. 2 depicts some typical accuracy plots against increasing number of samples. From the results, it is clear that modified FAM outperforms FAM in all cases with closer proximity to the Bayes limits and smaller standard deviations. Note that some of the results exceed the theoretical Bayes limits because of the use of the window method in calculating the accuracy.

Bayes Limits	XOR				Four-bit Parity			
	FAM		Mod. FAM		FAM		Mod. FAM	
Acc. (%)	Acc. (%)	Std. Dev.	Acc. (%)	Std. Dev.	Acc. (%)	Std. Dev.	Acc. (%)	Std. Dev.
55	50.1	1.4	55.6	1.3	49.8	2.3	53.6	0.3
65	51.2	8.3	64.6	1.0	56.8	8.2	64.9	0.3
75	57.3	11.2	75.7	1.0	67.4	5.2	75.2	0.9
85	70.5	15.7	85.2	1.8	77.1	3.6	85.4	0.6
95	85.9	12.3	95.0	0.3	93.1	1.8	95.0	0.8

Table 1 Results for the XOR and four-bit parity problem

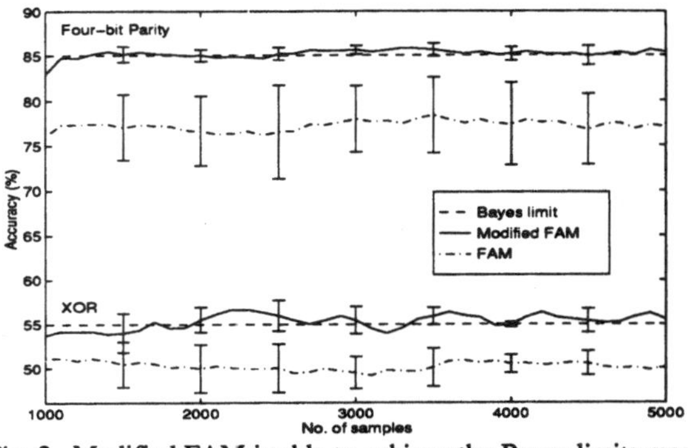

Fig. 2 Modified FAM is able to achieve the Bayes limits more closely than FAM with smaller standard deviations in a stationary environment.

3.2 Non-stationary On-line Learning & Classification

In this simulation, two classes of 25000 data samples were generated for the four-bit parity problem. The distribution parameters were subject to step changes every 5000 samples. This simulates a severe non-stationary scenario since in many applications, one might expect to experience a gradual change in the environment rather than a step change. As can be seen in Fig. 3, modified FAM is able to approach the Bayes limit more closely than FAM.

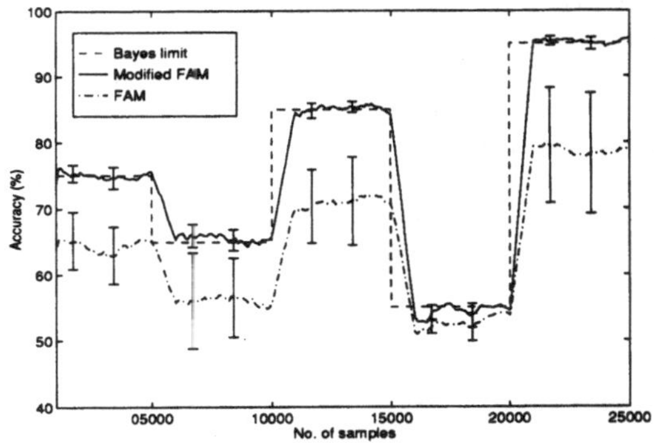

Fig. 3 Modified FAM tracks non-stationarity in the data environment and simultaneously achieves the Bayes limit. The average result of modified FAM also shows smaller standard deviations than FAM.

3.3 Ensemble Average and Time Average

So far we have only used the time average to calculate the on-line results. However, in a stochastic process, each realisation represents only one of the many possible outcomes. Thus, in order to measure the general performance of modified FAM in a non-stationary environment, an ensemble of 1000 networks has been used. The results were calculated across 1000 realisations and compared with the time average of single realisation with a 1000-sample window as in Fig. 4.

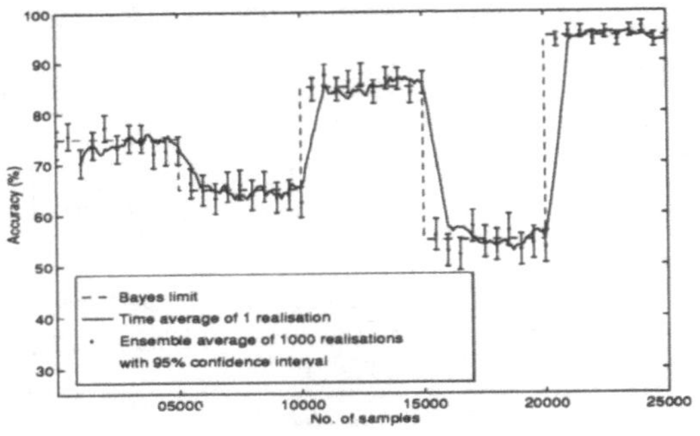

Fig. 4 A comparison between the ensemble and time averages. The Bayes limit and the time average of single realisation fall mostly within the 95% confidence interval of the ensemble average.

4 Summary

From the above simulations, modified FAM is able to achieve classification results which closely approximate the Bayes limits in both stationary and non-stationary environments. As might be expected, FAM only creates 4 nodes for the noisy XOR data set and 16 nodes for the four-bit parity data set whereas modified FAM creates 8 and 32 nodes respectively. The formation of category prototypes depends merely on the orderings of input presentation. Once an association has been established for a particular input pattern, FAM will ignore the same pattern when it appears to be a member of a different class. If spurious prototypes (owing to noise) have been developed at the early stage, most of the patterns will be incorrectly classified which directly leads to a degradation in performance. However, owing to the proposed modification, modified FAM is able to create two category nodes in ART_a to map to different target outputs for a specific input pattern. Two identical category nodes are set up with one set serving as the frequently excited prototypes while the other set acts as the spurious prototypes. The frequency measure scheme then ensures that the most probable prototypes are selected to predict an output and thus minimises the overall misclassification rates.

In the non-stationary experiment, modified FAM is competent to recover from drastic changes in data statistics. It tracks the non-stationarity very well and at the same time achieves the Bayes limit. The suitability of using the windowed time average in calculating the results has also been validated with an ensemble of networks. From the comparison, it implies that the time average adequately indicates the ensemble result except close to severe changes in statistics. Note that the time average of single realisation consistently approaches the ensemble average and falls mostly within the 95% confidence interval.

In conclusion, we demonstrate that modified FAM is capable of classifying binary-valued patterns optimally in stationary and non-stationary environments. It creates more category prototypes than the original network in order to implement a one-to-many mapping. Based on the frequency count information, the most probable prototypes are selected to make a prediction which in turn tends to minimise the misclassification rates and thus asymptotically approaches the Bayes limits.

References

[1] Cybenko, G.: *Mathematics of Control, Signals and Systems*, **2**, pp 303. (1989).

[2] Girosi, F., Poggio, T.: *Biological Cybernetics*, **63**, pp 169. (1990).

[3] Carpenter, G.A., Grossberg, S.: *Computer Vision, Graphics and Image Processing*, **37**, pp 54. (1987).

[4] Carpenter, G.A., Grossberg, S.: *IEEE Computer*, **21**, pp 77. (1988).

[5] Carpenter, G.A., Grossberg, S., Markuzon, N., Reynolds, J.H., Rosen, D.B.: *IEEE Trans. on Neural Networks*, **3**(5), pp 698. (1992).

[6] Carpenter, G.A., Grossberg, S., Reynolds, J.H.: *Neural Networks*, **4**, pp 565. (1991).

[7] Carpenter, G.A., Grossberg, S., Rosen, D.B.: *Neural Networks*, **4**, pp 759. (1991).

[8] Moore, B.: *Proc. 1988 Connectionist Models Summer School*, pp 174. (1988).

[9] Zadeh, L.: *Information and Control*, **8**, pp 338. (1965).

[10] Lim, C.P., Harrison, R.F.: To appear in *Neural Networks*.

Appendix A

Bayes	x = Even-parity Samples			x = Odd-parity Samples		
Limits	$P(c_1)$	$P(x\mid c_1)$	$P(x\mid c_2)$	$P(c_2)$	$P(x\mid c_2)$	$P(x\mid c_1)$
55%	0.5	0.275	0.225	0.5	0.275	0.225
65%	0.5	0.325	0.175	0.5	0.325	0.175
75%	0.5	0.375	0.125	0.5	0.375	0.125
85%	0.5	0.425	0.075	0.5	0.425	0.075
95%	0.5	0.475	0.025	0.5	0.475	0.025

Table A1 Parameters used for the XOR data set

Bayes	x = Even-parity Samples			x = Odd-parity Samples		
Limits	$P(c_1)$	$P(x\mid c_1)$	$P(x\mid c_2)$	$P(c_2)$	$P(x\mid c_2)$	$P(x\mid c_1)$
55%	0.5	0.06875	0.05625	0.5	0.06875	0.05625
65%	0.6	0.08125	0.04375	0.4	0.08125	0.04375
75%	0.4	0.09375	0.03125	0.6	0.09375	0.03125
85%	0.3	0.10625	0.01875	0.7	0.10625	0.01875
95%	0.1	0.11875	0.00625	0.9	0.11875	0.00625

Table A2 Parameters used for the four-bit parity data set

DECISION MAKING IN UNCERTAIN ENVIRONMENT WITH GA

Christiaan Perneel* and Marc Acheroy†

* Dpt of Applied Mathematics, Royal Military Academy, 30 Renaissance av, B-1040 Brussels, Belgium. (perneel@elec.rma.ac.be)

† Electrical Engineering Dpt, Royal Military Academy, 30 Renaissance av, B-1040 Brussels, Belgium.

Abstract

In this work, we develop a particular approach based on Genetic Algorithms (GA) for solving the decision-making problem in an uncertain environment, expressed as a graph-search problem. We show how Genetic Algorithms can be used as a new kind of heuristic search technique. The algorithm is illustrated in a particular decision-making application, namely an expert system which performs the automatic target recognition of armoured vehicules from short-range infra-red images. To illustrate the flexibility of the Genetic Algorithms approach for decision making, two implementations (single and multi-stage GA) are build and compared.

Introduction

Decision-making systems are usually designed to solve complex combinatorial problems with numerous candidate solutions. The brute approach of evaluating all the possible decisions is generally impracticable. When decision making in an uncertain environment is considered as a graph-search problem, several heuristic search techniques are well known to furnish good solutions in an efficient way, provided that specific conditions of applicability are fulfilled. Most of these classical methods result in a unique final solution, which it is hoped will be an optimum one. However, in an uncertain environment, the human decision-maker usually prefers to have available a set of alternative solutions, since the final decision provided by these methods could well turn out to be unsatisfactory.

We consider the increasingly popular techniques of Genetic Algorithms as a new kind of heuristic search technique for decision making in an uncertain environment. Indeed, Genetic Algorithms appear to overcome the limitations inherent the previous methods (specific applicability conditions, availability of good estimates, uniqueness of the solution), or at least to have different and less stringent requirements.

The aim of this paper is to investigate the feasibility of the Genetic Algorithms approach for decision-making problems and to compare it with more classical techniques.

Problem statement

We consider the following decision-making problem: to find a decision consisting of a sequence of decision elements (or hypotheses) optimizing some criteria in an uncertain environment. Let $\mathbf{D}$ be a candidate decision consisting of n decision elements d_i; each decision element d_i belongs to a finite, discrete set D^i:

$$\mathbf{D} = (d_1, d_2, ..., d_n) \qquad d_i \in D^i \qquad |D| = \prod_{i=1}^{n} n_i$$

$$\mathbf{D} \in D = D^1 \times D^2 \times ... \times D^n \qquad n_i = |D^i|$$

As a link between the decision and the uncertain environment, a number M of measurements (or observations) are available:

$$m_i : D \rightarrow \mathcal{M} : \mathbf{D} \rightarrow m_i(\mathbf{D}) \qquad i = 1, ..., M$$

To each measurement m_i corresponds one heuristic h_i which describes how well a decision (and its associated measurement) fits in with the environment:

$$h_i : \mathcal{M} \rightarrow \mathcal{R} : m_i \rightarrow h_i(m_i) \qquad i = 1, ..., M$$

Each heuristic can be considered as a piece of knowledge usually coming from an expert. Heuristics are combined to form a global rating r, which is a measure of the quality of the decision:

$$r = O[h_1(m_1(\mathbf{D})), h_2(m_2(\mathbf{D})), ..., h_M(m_M(\mathbf{D}))]$$

where O is the combination operator across all heuristics.

Let

$$\mathbf{Q}(\mathbf{D}) = (q_1(\mathbf{D}), q_2(\mathbf{D}), ..., q_M(\mathbf{D})) \qquad q_i = h_i \circ m_i$$

$$L : D \rightarrow \mathcal{R} : \mathbf{D} \rightarrow r = L(\mathbf{D}) = O(\mathbf{Q}(\mathbf{D}))$$

The function L associates a rating r to each decision D, and results from the composition of the measurements, heuristics and combination operators.

The initial decision-making problem can now be formalized as an optimization problem: to find the maximizer of $L(\mathbf{D})$.

In the remainder of this paper, the decision-making

problem will be considered as a hierarchical graph-searching problem; the graph consists of several nodes grouped by levels, each node of level k representing a partial decision $(d_1, d_2, ..., d_k)$ for the problem ; **D** can be associated with a specific path in the decision graph.

At this stage, there is no means of evaluating the quality of a partial decision. In other words, only the terminal nodes (the nodes of the last level n) are given a rating, which can be used to guide the search for the best decision. Methods to solve this particular kind of problem with poor or unstructured information are scarce and relatively inefficient (systematic enumeration or random walk). Nevertheless, Genetic Algorithms with their ability to cope with little information about the search space, certainly constitute a promising alternative. This is one of the approaches developed in the present work.

Now, let us assume further that knowledge is revealed partially at each level in an incremental fashion; this means that **Q** can be decomposed in the following manner :

$$\mathbf{Q} = [q_{11}(d_1), ..., q_{1M_1}(d_1), q_{21}(d_1, d_2), ..., q_{2M_2}(d_1, d_2), ...$$

$$q_{n1}(d_1, d_2, ..., d_n), ..., q_{nM_n}(d_1, d_2, ..., d_n)]^t,$$

with $\sum_{i=1}^{n} M_i = M$. Each function q_{ij} is the individual rating of heuristic j at level i and depends only on the partial decision $(d_1, d_2, ..., d_i)$. At each level i, it is assumed that the individual heuristic ratings q_{ij} can be combined into a partial rating $L_i(d_1, d_2, ..., d_i)$ which represents the quality of the partial decision up to this level, given the partial knowledge of this level. The global rating can be expressed as a combination of the partial ratings $L_i : L[L_1(d_1), L_2(d_1, d_2), ..., L_n(d_1, d_2, ..., d_n)]$ The previous assumptions allow more efficient strategies to guide the search in a smarter way by exploiting the partial information available at each node. Branch-and-bound methods can be applied provided that, for every node of level k, it is possible to find an upper limit to the global rating of all the decisions containing the partial decision represented by the node considered. The efficiency of the branch-and-bound method relies on the basic assumption that the main contribution to the global rating function comes from the first levels of the graph. In other words, it is expected that the progression in the tree by addition of new decision elements to the current partial decision does not decrease greatly its performance. Of course, Genetic Algorithms can still be applied with this kind of ordered information. However, their fundamental underlying requirements are more general and much less restrictive than those of branch-and-bound methods. To be efficient, Genetic Algorithms do not require that the most significant decision elements be placed at the first levels, nor that an optimum solution can be found by sequentially refining a current partial decision with new good decision elements. Genetic algorithms only require that the optimum decision can be built from a combination of good partial decisions arbitrarily located in the sequence; moreover, the identification and combination of good partial decisions is performed in an automatic and implicit way.

Decision making in uncertain environment with GA

In this paper, we have investigated the usefulness of Genetic Algorithms (GA) to the resolution of the graph-search problem.

Genetic Algorithms need a roughly constant time before finding a satisfactory solution, whereas this time is variable for branch-and-bound methods (short if the efficiency requirements are satisfied; long otherwise). However, although Genetic Algorithms are known to be highly reliable, they are not absolutely guaranteed to find the global best solution, which will be certainly found by branch-and-bound methods.

To solve the graph-searching problem with Genetic Algorithms, a population of individuals is formed. Each individual consists of a particular sequence of decision elements $\mathbf{D}_i = (d_1^i, d_2^i, ..., d_n^i)$ grouped in the population $\mathbf{P} = \{\mathbf{D}_1, \mathbf{D}_2, ..., \mathbf{D}_p\}$. The chromosome length is thus n, each gene d_i^j taking n_i allele values. The fitness function of individual $\mathbf{D}_i$ is taken as the global measure of quality (rating) of the decision $L(\mathbf{D}_i)$. The fitness function therefore depends on the particular approach adopted to cope with the uncertainties (Bayesian approach, Fuzzy approach,...). When structured information is available, it is also possible to exploit it by designing more efficient strategies based on Genetic Algorithms which involve applying several times GA-based search in an incremental fashion (multi-stage GA). For example, a multi-stage GA can be designed which first solves the partial decision problem limited to the first k levels, and then solves the complete decision problem by adding grouped levels of decision gradually.

Description of the Application

The problem consists in identifying armoured vehicles based on short distance 2-D infrared images. The major difficulty lies in the lack of knowledge of the position and orientation of the vehicle with respect to the camera. Therefore, the problem is divided in two subproblems: the first subproblem is the determination of the orientation and position of the vehicle based on a crude model of the vehicle; and the second subproblem is the identification of the vehicle in a reference position and orientation. The first task, position and orientation detection, consists in putting a system of three axes, $\{X, Y, Z\}$, on the image of the vehicle according to predefined conventions. In our application, the conventions are the following:

- X **axis**: the line on the side of the vehicle between the wheel train and the ground.
- Y **axis**: the line on the front or on the rear of the vehicle between the front or rear wheels and the ground.
- Z **axis**: the vertical direction of the vehicle, normalized on the wheel train height if the image gives a side view of the vehicle or on the distance between the floor of the vehicle body and the ground if the image gives a front or rear view of the vehicle.

The position and orientation detection is then divided into 5 ($N = 5$) subtasks. These 5 subtasks correspond to 5 different levels of increasing knowledge of the position and orientation of the vehicle.

- **Level 1**: Determination of one principal direction out of the eight most important orientations found on the image. This principal orientation is either the direction of the X axis or the direction of the Y axis ($n_1 = 8$).
- **Level 2**: Determination of the line of the X axis if it is a side view or the line of the Y axis if it is a front or rear view ($n_2 = 30$).
- **Level 3**: Determination of the position of the origin if the image is a side view or the position of the engine if the image is a front/rear view ($n_3 = 2$).
- **Level 4**: Determination of the Z axis ($n_4 = 9$).
- **Level 5**: Determination of the third axis ($n_5 = 16$).

Once the first task of detecting the position and orientation of the vehicle is accomplished, it remains to identify the type of vehicle. Only the first part of this method, position and orientation detection, will be used as application in this paper.

Results

Description of the current Genetic Algorithms

The GA parameters used in our application are summarized in table 2 : The fitness value of each individual

Number of individuals	40
Chromosome length	5
Maximum number of generations	100
Mutation probability	0.1
Cross-over type	Uniform
Cross-over probability	0.7
Elitism	Yes
Selection pressure	1.7

Table 1: GA parameters used in our application

is the global rating of the decision represented by the individual. The fitness values are scaled to belong to the interval [0,500]. For computing fitness values, a Bayesian and a fuzzy logic approach (see de Mathelin *et al* [1]) was examined and compared.

single-phase GA

Fig.1 shows the learning curves (evolution of the fitness) for fuzzy logic approaches over 50 experiments with different initial populations. The diagrams are given for one particular image but similar results are observed for most of the available images in our database. Starting from top to bottom of the dia-

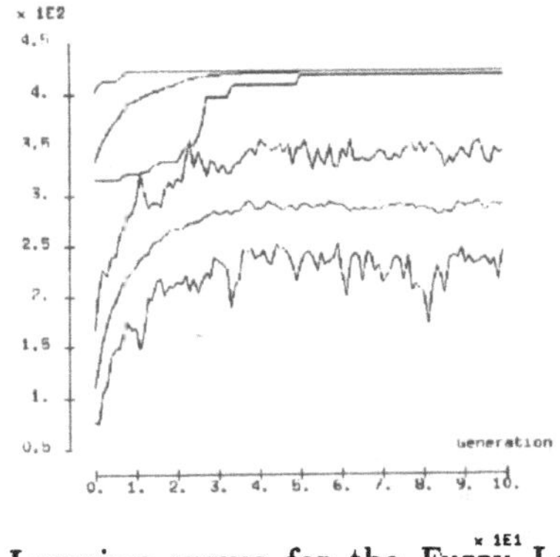

Figure 1: Learning curves for the Fuzzy Logic approach

grams the curves represent respectively (over 50 experiments) the highest value of the best fitness of each population, the mean value of the best fitness of each population, the lowest value of the best fitness of each population, the highest value of the mean fitness of each population, the mean value of the mean fitness of each population, the lowest value of the mean fitness of each population.

As expected, the reliability of the GA search is excellent since all experiments result in the optimum (or near-optimum) decision. Fig.1 shows that a satisfactory result can be expected after 40 generations, corresponding to at most 1600 nodes tested (well below the 69120 terminal nodes of the overall tree). The success rates of the Bayesian and fuzzy logic approaches are compared in Fig.2. Starting from bottom to top of the diagrams, the curves represent respectively the success rate using the Bayesian rating, the success rate using the fuzzy-derived rating. The third curve corresponds to the multi-stage GA and will be described later. Considering the first two curves, it turns out that the fuzzy logic approach converges more rapidly. This could be explained by the fact that the operators used to build the global rating are more adapted to the fundamental mechanisms and requirements of GA. The percentage of success (decisions identical or nearly identical to the true optimum one) was 96 % for all the 135 images of our database.

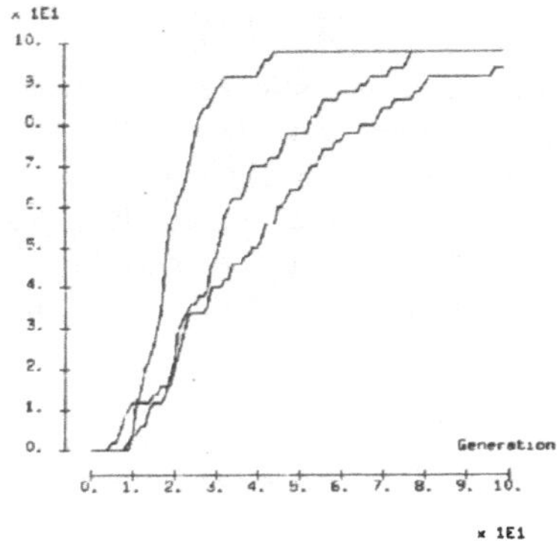

Figure 2: Comparaison of the success rates

Multi-stage GA

We chose to decompose the GA search into two stages. In the first stage, GA performs a search limited to the first two levels, which carry the most important part of the information. An individual therefore represents a partial decision (d_1, d_2). Only the level ratings L_1 and L_2 (respectively of level 1 and 2) are used and combined to compute the fitness of a partial decision. At the beginning of phase 2, a new population of complete decisions is formed from the last generation of phase 1 by adding randomly 3 decision elements. During phase 2, which involves a normal GA-search on the 5 levels, the first two genes (or levels) are still allowed to change. Phase 1 can thus be considered as a means to choose a biased starting population for usual single-phase GA. Fig.3 shows the learning curves for a multi-stage GA using the same conventions as Fig.1. Looking again at Fig.3, it is observed that two-stage GA outperforms the single-stage methods : a faster convergence is achieved.

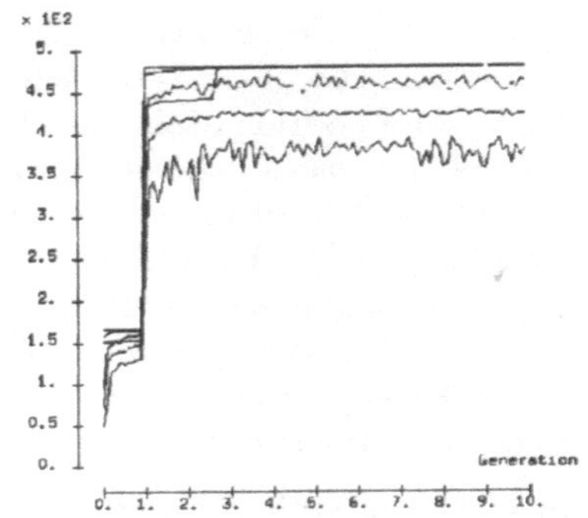

Figure 3: Learning curves for the multi-stage approach

Conclusions

All things considered, it appears that Genetic Algorithms are well adapted to the resolution of our graph-search problem, especially multi-stage Genetic Algorithms which fully exploit the information. On the other hand, for this application, branch-and-bound methods are much more efficient in most cases, because their efficiency requirements are satisfied here. However, Genetic Algorithms remain attractive since they provide the user with a family of interesting solutions. They can thus present a significant advantage especially when the choice of heuristics does not fulfill the basic requirement that the maximum value of L is really the optimum solution to the underlying problem of identification.

References

[1] M. de Mathelin, C. Perneel, and M. Acheroy, "Bayesian estimation vs fuzzy logics for heuristic search algorithms," in *Proceedings Second IEEE International Conference on Fuzzy Systems*, pp. 944–951, 1993.

THE USE OF FUZZY ARTMAP TO IDENTIFY LOW RISK CORONARY CARE PATIENTS

J. Downs, R.F. Harrison
Dept. of Automatic Control
and Systems Engineering
University of Sheffield, UK

R.L. Kennedy
Dept. of Medicine
University of Edinburgh, UK

K. Woods
Dept. of Pharmacology
University of Leicester, UK

Abstract The performance of fuzzy ARTMAP and modified fuzzy ARTMAP is compared using real-world data from a medical domain, the task being to predict the death or survival of patients admitted to a coronary care ward. Modified fuzzy ARTMAP is shown to perform consistently more accurately than fuzzy ARTMAP and is also much less prone to variations in performance with different orderings of training data. However, modified fuzzy ARTMAP does not show as large an improvement in performance as fuzzy ARTMAP when employed in the voting strategy. When unanimous voting decisions alone are considered, fuzzy ARTMAP is able to increase significantly accuracy in identifying survivors at the cost of decreased coverage of cases. This allows the identification of a subset of patients who have a low-risk of death from their condition and are thus potentially suitable for early discharge from hospital.

Introduction

Since the fifties, there has been a progressive reduction in the recommended length of hospital stay for patients admitted to coronary care units. This change from six weeks bed rest to the current 7 to 10 days has occurred without significant effects upon patient mortality rates, and with obvious economic benefits. It is hoped to continue this trend yet further. However, early hospital discharge requires very accurate identification of those patients at minimal risk of death from their condition. Furthermore, such identification must occur soon after a patient is admitted. We have adopted a neural network approach to the task. To this end, a database of approximately 5500 patient records has been collected from the coronary care unit of Leicester Royal Infirmary over a six year period. Once records with incomplete data were removed, 4200 data items remained. Each record consisted of 43 items of clinical or electrocardiographic data considered to be useful for patient prognosis, together with the outcome for the patient's stay in hospital—death or survival. One of the items had an integer value, all others were binary-valued. (The Appendix provides a full listing of the input features.)

The generation of accurate predictions from this set of patient records is a very difficult learning task for a neural network. Specifically, two features of the data make the problem hard to solve. Firstly, the data is "noisy"—it is gathered from a real-world medical domain and has no simple indicators delineating category boundaries. Secondly, the distribution of categories is skewed—only 7.1% of all patients admitted die while on the ward. This poses particular difficulties for learning algorithms such as backpropagation [1] where weights are refined by a process which effectively averages together similar cases. In such circumstances rare events like patient deaths become completely submerged by the greater numbers of surviving patients who show similar data features.

This work forms part of a programme to evaluate the utility of ART models for application in decision-support tools for a variety of medical classification tasks. Useful results have previously been demonstrated in the early diagnosis of heart attacks [2] and the diagnosis of breast cancer [3].

Fuzzy ARTMAP and Modified Fuzzy ARTMAP

Fuzzy ARTMAP [4] was selected for the application since it is claimed to possess a number of capabilities that are particularly suited to this and other medical domains. First, fuzzy ARTMAP does not perform optimization of an objective function and is not therefore prone to the problem of local minima. Instead it forms a structuring of the data for itself (self-organisation) into prototypical category clusters. Also, fuzzy ARTMAP has few user-changeable parameters, which allows the model to be tuned to a particular problem without undue effort. (The single most important fuzzy ARTMAP parameter being that of *vigilance*, which controls the size of the category clusters formed). Additionally, a modified version of fuzzy ARTMAP [5] has been demonstrated to show optimal data classification in the Bayesian sense.

Very importantly, the model is able to discriminate rare events from a "sea" of similar cases with different outcomes. This is because the family of adaptive resonance theory (ART) models to which fuzzy ARTMAP belongs all incorporate a feedback mechanism based on top-down matching of learned categories to input patterns [6]. Learning in fuzzy ARTMAP can also occur with only one pass through the data set (single-epoch training). Furthermore, the model is capable of incorporating new data items without degradation of performance on previous data, or the necessity of retraining on such past data. This solution to the so-called *stability-plasticity dilemma* is claimed to be a feature unique among neural networks to the ART models [7]. Collectively these features make fuzzy ARTMAP potentially suitable for on-line learning, even in non-stationary environments.

Modified Fuzzy ARTMAP [5] was developed from fuzzy ARTMAP for use with statistical data—the demonstration task being to separate two classes of Gaussian distributed random variables. The model was shown to provide superior performance to fuzzy ARTMAP on this problem, and can approach the Bayes optimal classification rates for the domain. To achieve this, modified Fuzzy ARTMAP records data on the usage of category cluster nodes, and resolves ties between nodes competing to encode an exemplar by selecting the node with the greatest usage.

Method

For the purposes of this application off-line learning was employed. The data were partitioned into a training set, comprising the first 3000 patient records, and a test set comprising the remaining 1200 records. Twenty different orderings of the training set were derived and served as input data to separate instances of fuzzy ARTMAP and modified fuzzy ARTMAP using single-epoch training. (The order of presentation of data items is known to have quite large effects on category formation in fuzzy ARTMAP [4]) The vigilance parameter was set low (0.3) to avoid excessive cluster formation—a notable problem for fuzzy ARTMAP [8]. Other parameters were set to their "standard" values [9]; the learning rate being set to its maximum value of one (so-called fast learning) and the category choice parameter being set close to zero (0.000001).

The voting strategy [4] was also employed on the test data. This works as follows: a number of ARTMAP networks are trained on different orderings of the input data. During testing, each individual network makes its prediction for a test item in the normal way. The number of predictions made for each category is then totalled and the one with the highest score (or the most "votes") is the final predicted category outcome. The voting strategy can provide improved ARTMAP performance in comparison with the individual networks. In addition it also provides an indication of the confidence of a particular prediction, since the larger the voting majority, the more certain is the prediction.

A range of 3 to 13 odd numbered voters was used (odd numbers ensuring no tied decisions occurred), choosing those fuzzy ARTMAP instances from the pool of 20 that had achieved the highest individual accuracy scores. The same voting procedure was then repeated using the modified fuzzy ARTMAP networks.

Results

Initial performance on the test set proved disappointing. Accuracy for the individual fuzzy ARTMAP networks ranged between 73.2% and 87.5% with a mean of 81.1%. For modified fuzzy ARTMAP, accuracy ranged between 87.3% and 89.6% with a mean of 88.3%. This compares with a default accuracy of 92.9% for the simple assumption that all patients will survive. The reason for this was that fuzzy ARTMAP in particular over-represents the rare cases of patient deaths in excess of their actual frequency within the data set. (This is probably because such cases were not tightly clustered together but widely spread throughout the feature space.) This effect was reduced, if not entirely overcome, with modified fuzzy ARTMAP. Thus fuzzy ARTMAP appears to suffer from the opposite problem to backpropagation—too much credence, rather than too little, is given to rare cases.

The general effect of the voting strategy was to increase accuracy for both fuzzy ARTMAP and modified fuzzy ARTMAP to around 89-91%, still slightly below baseline performance. However, the voting strategy with fuzzy ARTMAP did provide useful results for the important special case of high-confidence predictions of patient survival. (A high confidence prediction being one upon which all fuzzy ARTMAP voters agreed.) Such patients are the most suitable for early hospital discharge.

With 3 voters, a unanimous survival decision

accounted for 911 of the data items and was proved wrong 44 times. This translates to 95.2% accuracy covering 75.9% of the 1200 test items. With extra voters, accuracy steadily improved at the cost of decreased coverage (see table 1), until at the 13 voter case an accuracy of 99.3% covering 34.0% of the data was achieved.

Table 1: Voting Strategy Performance for Unanimous Survival Decisions

	Fuzzy ARTMAP		Modified Fuzzy ARTMAP	
Number of Voters	Accuracy (%)	Coverage of Cases (%)	Accuracy (%)	Coverage of Cases (%)
3	95.2	75.9	94.0	87.0
5	95.6	64.5	94.2	81.3
7	97.8	53.0	94.8	75.3
9	98.2	45.3	95.1	71.8
11	98.1	40.3	94.9	69.1
13	99.3	34.0	95.1	66.3

Discussion

The results for the individual networks show that modified fuzzy ARTMAP consistently performs better than fuzzy ARTMAP with this real-world data. Moreover, the modified fuzzy ARTMAP networks show a much smaller variation in accuracy for different orderings of the training data. However, modified fuzzy ARTMAP does not gain as much benefit from the voting strategy as fuzzy ARTMAP. This is particularly marked when unanimous votes alone are considered (see table 1). With fuzzy ARTMAP, an increase in the number of voters tends to increase accuracy while reducing the number of cases. This effect is much less pronounced in modified fuzzy ARTMAP

Collectively these findings seem to indicate that modified fuzzy ARTMAP is not prone to the ordering effects of training data that occur with fuzzy ARTMAP. Thus modified fuzzy ARTMAP tends to form similar category clusters regardless of the order in which exemplars are presented. A single modified Fuzzy ARTMAP network therefore provides more reliable pattern classification than a Fuzzy ARTMAP network, and hence seems better suited to on-line learning tasks for example.

However, modified Fuzzy ARTMAP's relative immunity to ordering effects is not an unmitigated advantage. If the classification performance of an individual network is relatively unaffected by the ordering of the input data, individual networks cannot usefully be pooled using the voting strategy. If each network has learned approximately the same mapping, performance will not be greatly improved by using the networks collectively.

This has the drawback that no estimate of the confidence of a particular prediction is available. Moreover, one cannot easily use the unanimous voting strategy to trade coverage for accuracy by increasing the number of voters, thus allowing a partitioning of the test data to identify those items on which the networks' have a high-certainty of making a correct prediction.

Indeed, this latter ability used with fuzzy ARTMAP yielded the only results of practical value for medical decision-support in this domain. The figures for 13 voter fuzzy ARTMAP (see table 1) are nearly identical to those achieved by Parsons et al. [10] using the statistical technique of logistic regression upon a different data set collected for the same purpose. (Parsons et al. achieved 99.2% accuracy in a third of all cases using a data set of 5746 training items and 1000 test items respectively.) However, the advantage of the fuzzy ARTMAP unanimous voting strategy is its ability to gain wider coverage of the data set with only a small decrease in accuracy by reducing the number of voters.

Future work might include applying symbolic rule extraction capabilities [11] to the fuzzy ARTMAP networks. This will allow a qualitative description of a typical low-risk patient to be constructed, thus providing a useful supplement to the network's predictive capabilities.

Another area of further work is in identifying high-risk patients. None of the methods employed in this study provided reliable predictions of patient mortality, probably due to the low frequency of occurrence of deaths in combination with the non-uniformity of the features shown by such patients. However, an expanded version of the data set should shortly become available which provides mortality figures for patients up to 12 months of discharge from hospital. It is hoped that more reliable prognoses may be yielded from this data.

Acknowledgement

This research was supported by the Science and

Engineering Research Council (SERC) of the UK, grant number GR/J/43233.

References

1. Rumelhart, D., Hinton, G., Williams, R.: Nature, 323, 533 (1986).

2. Harrison, R.F., Lim, C.P., Kennedy, R.L.: Proc. of the Int. Conf. on Neural Networks and Expert Systems in Medicine and Healthcare, 15 (1994).

3. Downs, J., Harrison, R.F., Cross, S.S.: Proc. of the 10th Conf. of the Society for the Study of A.I. and Simulation of Behaviour (In Press).

4. Carpenter, G.A., Grossberg, S., Markuzon, N., Reynolds, J.H., Rosen, D.B.: IEEE Transactions on Neural Networks, 3, 698 (1992).

5. Lim, C.P., Harrison, R.F.: Neural Networks (In Press).

6. Carpenter, G.A., Grossberg, S. (eds): Pattern Recognition by Self-Organizing Neural Networks. Cambridge, MA: MIT Press 1991.

7. Carpenter, G.A., Grossberg, S.: IEEE Computer, 21, 77 (1988).

8. Marriott, S., Harrison, R.F.: Neural Networks (In Press).

9. Kasuba, T.: AI Expert, 8, 18 (1993).

10. Parsons, R.W., Jamrozik, K.D., Hobbs, M.S.T., Thompson, D.L.: British Medical Journal, 308, 1006 (1994).

11. Carpenter, G.A., Tan, A.H.: Proc. of the World Congress on Neural Networks, Volume I, 501 (1993).

Appendix: Network Input Features

1. Atrial Fibrillation
2. Supraventricular Tachycardia
3. Ventricular Tachycardia/Fibrillation
4. Bundle Branch Block
5. ST Depression—Inferior
6. ST Depression—Anteroseptal
7. ST Depression—Anterolateral
8. ST Elevation—Inferior
9. ST Elevation—Anteroseptal
10. ST Elevation—Anterolateral
11. T Wave—Inferior
12. T Wave—Anteroseptal
13. T Wave—Anterolateral
14. Q Wave—Inferior
15. Q Wave—Anteroseptal
16. Q Wave—Anterolateral
17. Age<40
18. Age 40-50
19. Age 50-60
20. Age 60-70
21. Age>70
22. Time since Cardiac Event>24 Hours
23. Family History of Ischaemic Heart Disease
24. Smoker
25. Ex-Smoker
26. Glucose>7.9
27. Glucose>8.9
28. Glucose>10.9
29. Pulmonary Venous Engorgement
30. Pulmonary Oedema
31. Chest Pain
32. Short of Breath
33. Syncope
34. Nausea
35. Sweating
36. Palpitations
37. Cardiac Arrest while in Coronary Care Unit
38. Creatine Kinase>600
39. Creatine Kinase>1000
40. Creatine Kinase>3000
41. Sex
42. Number of Previous Cardiac Episodes
43. Previous Cardiac Episode(s)

NeuroGraph - A Simulation Environment for Neural Networks, Genetic Algorithms and Fuzzy Logic

P. Wilke J. Rehder G. Billing C. Mansfeld J. Nilson

c/o Lehrstuhl für Programmiersprachen der Universität Erlangen-Nürnberg
Martensstrasse 3, 91058 Erlangen, Germany
E-Mail: wilke@informatik.uni-erlangen.de

Abstract

NEUROGRAPH is a simulation environment for neural networks. It provides an easy to use graphical user interface to design, construct and execute neural networks. The most important design goals were easy extensibility, good performance and the ability to solve real world problems. Extensibility is ensured by an extremely flexible, hierarchical internal representation (data structure), which can be easily manipulated by graphical tools. Current available extensions to the neural network simulator are components for genetic algorithms and fuzzy logic. As examples the automated generation of a neural network by a genetic algorithm and a fuzzy controller are shown. This data structure can be interpreted for debugging purposes or can be executed directly for high performance computing. In the later case C or C++ code is generated from the internal representation of the neural, gentic or fuzzy functions, compiled and stored in an executable process (library). NEUROGRAPH has a file or process interface to interconnect which other processes. It is also possible to generate C or C++ code representing the neural network, genetic algorithm or fuzzy component which can be incorporated in other applications.

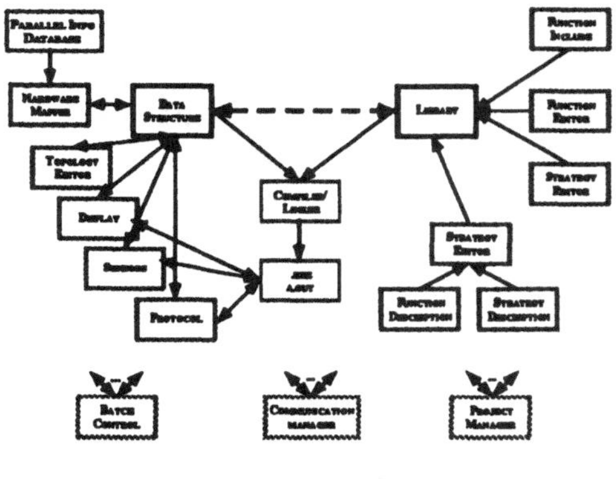

Figure 1: NEUROGRAPH's components

1 The Main Components of NeuroGraph

The newest release V3.0 of NEUROGRAPH is the first soft-computing simulation environment combining neural networks, fuzzy logic and genetic algorithms [4, 5]. The Figure on this page shows the main processes of the NEUROGRAPH system. The most important components are the data structure and the function library. The data structure stores all relevant information of a network concerning functionality or topology. Data is typically entered interactivly via the graphical user interface of the topology editor (see fig. 5) and send from there to the data structure using the communication interface. Alternatively it is possible to construct a neural network by an evolutionary strategy using genetic algorithms (see fig. 4). The data in the data structure share all its information with all other processes. Storage in files can be done in compressed or expanded format. The data structure supports parallel distributed computations on multiprocessors and/or workstation-clusters because of its capability to split the data among the available computers. This splitting is transparent to all other NEUROGRAPH processes, so they are not concerned with distributed communication. Consistency is ensured by the communication interface.

The function library contains all neural, genetic and fuzzy functions in compiled and source format, which can be directly executed or interpreted. Most standard models (e.g. for neural networks: Back Propagation, Pattern-Associator, Hopfield, Kohonen, Boltzmann Machine, etc. including their specific functions, for genetic algorithms: mutation, cross-over, ranking, for fuzzy components: fuzzyfication, de-fuzzyfication) come with NEUROGRAPH. Users can exchange their libraries to grant others access to their user defined models and function. To protect know-how, libraries can be exchanged in execute-only format.

Trained networks, genetic algorithms and fuzzy systems can run as a NEUROGRAPH component or as a stand-alone-process. The compiler generates ANSI-C or C++ source code from the network description stored in the data struc-

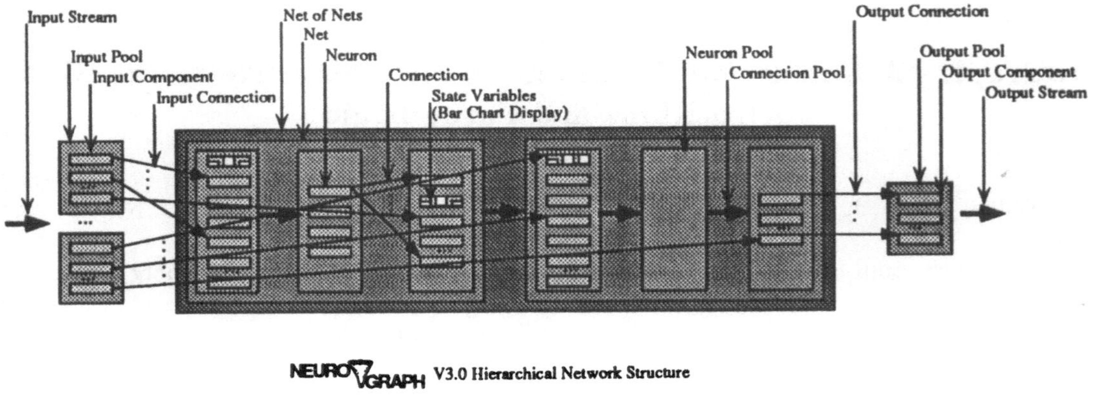

Figure 2: Hierarchical structure of a neural network: The network consists of two three-layer feed-forward network, connected with each other. The input stream consists of two, the output stream of one file or pipe.

ture. This source code can be compiled on every computer running a C or C++ compiler, therefore porting the code to non-unix-like operating systems (e.g. for PCs) is easy. All parameter files (e.g. connection weights etc.) are stored in ASCII format so that they can be used on all platforms.

Compiler options allow the integration of NEUROGRAPH visualization components or user-defined visualization routines. Also the automated generation of graphical user interfaces made up of OSF/Motif widgets is supported.

The topolopy editor serves as a graphical tool for the description of neural networks topologies, genetic algorithms and fuzzy systems as well as a visualization tool for the current state of the computation. For example the description of a neural network topology is done by drawing active elements (i.e. neurons and connections) on a virtual infinite drawing area. The constructed networks can be structured by assigning elements to pools, layers or networks. Mechanisms like convolution operators supplied by the data structure simplify the handling of very large networks. Among these mechanisms are multi-neurons, multi-connectors and networks-of-networks, i.e. trained networks can be connected for recalling. The visualization of the network's current state divides up into the visualization of the active elements and the topology, both displays are available in 2D or 3D. The visualization of active elements allows the representation of an arbitrary number of variables in compact form, e.g. active elements can be drawn as polyeders, where each surface represents a variable whose value determines the fill pattern or filling degree. A numerical exact representation is available, too.

Functions are defined using the function editor for graphical input of algorithms, e.g. neuron functions, network models and graphic functions. All atomic functions are predefined, a library of all important functions comes with NEUROGRAPH. In debug-mode functions can be executed in single step mode. The evaluation and upcoming results are animated so making debugging easier. Fig. 3 shows the delta learning rule for a connection in a back-propagation network. This function is stored in the library 'Backprop'.

2 Finding an Optimal Topology for Neural Networks using Genetic Algorithms

A separate process for the specification and execution of genetic algorithms using user-defined genetic operators allows the evolutionary construction of neural networks with optimized topology. Fig. 4 shows an example of a genetic algorithm for optimization of the neural network topology for the XOR-problem. The individuals enter the algorithm from the start object and are processed by the operators, e.g. mutation, crossover, etc. Specific operators control the path of the individuals, e.g. distribution operator. On the right-hand side a sensor shows the decreasing fitness (optimal when zero) of the generated nets. The results are shown in fig. 5, note the different number of neurons and connections. The genetic operators are described using the function editor and stored in a library. The genetic algorithms itself is constructed within the genetic algorithms editor, which supports the design and test.

3 Fuzzy Systems

In NEUROGRAPH, fuzzy systems are represented as there equivalent neuronal network. In this form of representation it is very easy to combine fuzzy systems and standart neuronal networks (or any other system written in the NEURO-

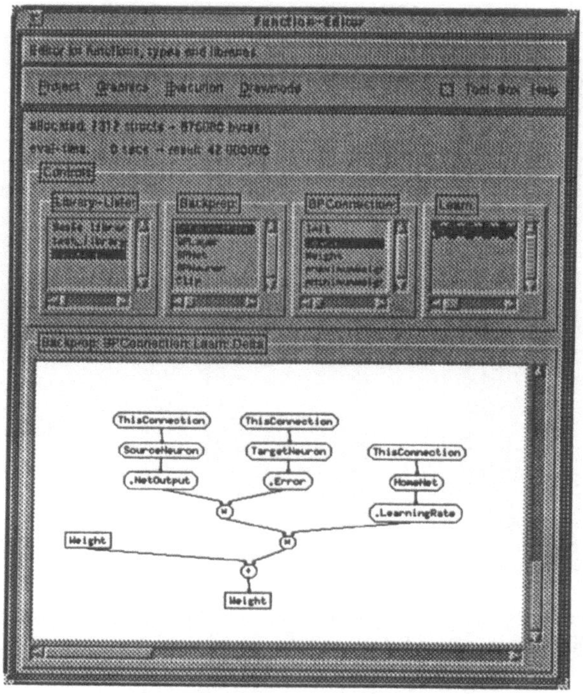

Figure 3: Representation of a learning rule for a connection in a back-propagation network

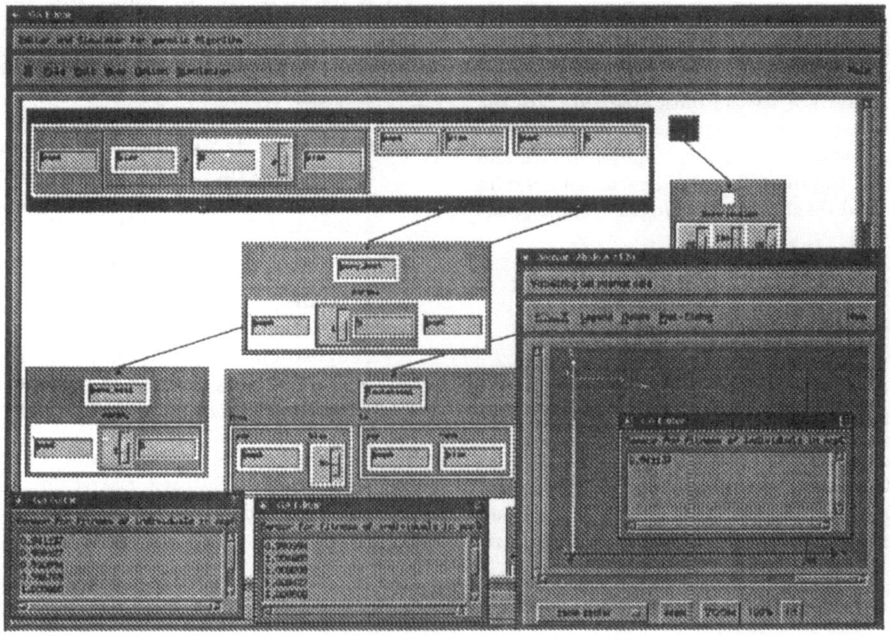

Figure 4: A genetic algorithm for the generation of neural networks for the XOR-problem

GRAPH-language) to a hybrid system. It is also possible that fuzzy systems control special parts of an other network (i.e. Parameters of an genetic algorithm)

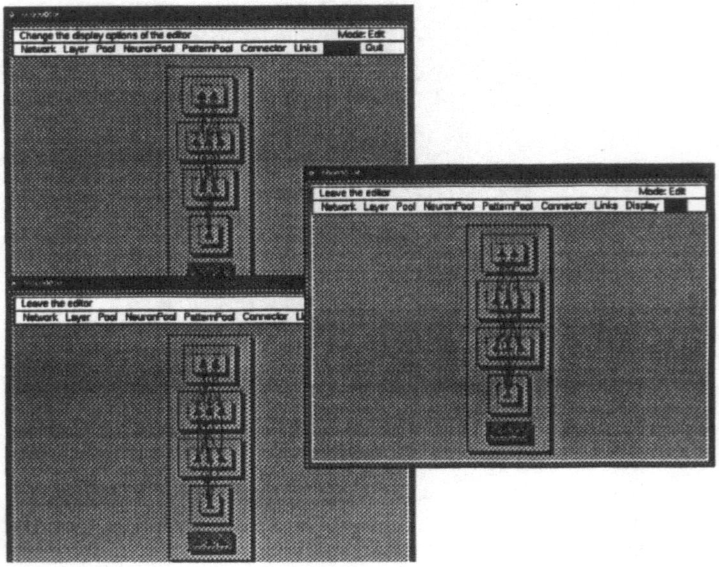

Figure 5: Three neural networks solving the XOR-problem generated by a genetic algorithm: the number of nodes and the way the neurons are connected differ from network to network.

4 Integration of new Components into the NeuroGraphSystem

Integration of new components like genetic algorithms or fuzzy logic is quite simple. the new processes just have to have the common interface to the data structure via the communication process. In this way e.g. the genetic algorithm editor has access to the stored genetic operators and can remotely execute the individuals as neural networks in the data structure and display them using the topology editor.

5 Further information

Acknowledgements

We would like to thank the NEUROGRAPH Team Members Anke Richter, Sandra Frings, Ralf Scholz, Rolf Beck, Bodo Bauer, Oliver Heggelbacher and Paul Cash for their contributions.

References

[1] Hertz, J., Krogh, A., Palmer, R.G. (1991): Introduction to the Theory of Neural Computation, Addison-Wesley

[2] Khanna, T. (1990): Foundations of Neural Networks, Addison-Wesley

[3] P. Wilke, C. Jacob,: The NeuroGraph Neural Network Simulator, Proceedings Int. Workshop on Modeling, Analysis and Simulation of Computer and Telecommunication Systems MASCOTS93, Simulation Series Vol.25,1, San Diego, 1993, p. 341pp

[4] P. Wilke, et. al.: NeuroGraph V3.0 Neural Network Simulator Product Overview, WCNN'94 World Congress on Neural Networks 1994, Ed.: International Neural Network Society, 5-9 June, San Diego, Programm Addendum, p. 26 - 30, 1994

[5] P. Wilke, et. al.: NeuroGraph V3.0 Dokumentation, Forschungsbericht des Lehrstuhls für Programmiersprachen, Universität Erlangen-Nürnberg, 1994, ca. 200 pages

COMPARISON OF IDENTIFICATION TECHNIQUES FOR NONLINEAR SYSTEMS

FRENCH, I. G., HO, S., ADGAR, A.

SUNDERLAND UNIVERSITY
CHESTER ROAD
SUNDERLAND, SR1 3SD

1: Introduction.

The problem of identifying a dynamic process from experimentally derived data has received much attention in the technical literature. Classically, a least squares based method is used, having first established a suitable structure, to estimate the parameters of a linear discrete time model. Whilst recognising that such linear models are not suitable for all applications, they have gained widespread acceptance, since, in the main they are used with controller design algorithms which require models of that form.

Recently, however, the advent of model based predictive controllers has provided a mechanism by which improved control can be readily achieved by the inclusion of information relating to plant nonlinearity. This usually takes the form constraints placed upon a linear model. However, there is considerable potential for still greater improvement if the model itself is able to capture any nonlinear operating characteristics present in the plant.

The problem with any technique used to model nonlinear behaviour is one of dimensionality. As an example consider a simple ARMA structure containing 3 numerator and 3 denominator terms. If expanded as a NARMAX model using 3rd order polynomials then the resulting structure may require the estimation of as many as 84 parameters. Further, this dramatic increase is not restricted to the NARMAX model since a neural network model, containing one hidden layer with 5 nodes, may also require the estimation of 40 plus parameters. Fortunately, experience (especially in the case of the NARMAX model) shows that only a restricted number of the parameters are significant and that the remainder may be neglected. The problem, therefore, becomes one of establishing the most appropriate model structure. The paper considers this problem and shows how genetic algorithms can provide an efficient means of establishing the structure. Results will be presented which compare the relative merits of different modelling approaches in terms of model accuracy, model complexity and the computational burden required in model generation.

2: ARMAX Models.

The term *system identification* generally brings to mind the application of a least squares type algorithm (GLS, IV, AML,..) [1] to estimate the parameters of a linear difference equation (or ARMA) model of the form:

$$y(t) = b_1 u_{(t-d)} + \ldots + b_q u_{(t-d-q+1)} - (a_1 y_{(t-1)} + \ldots + a_p y_{(t-p)}) \qquad [1]$$

The reasons for this are perhaps two fold. Firstly, from a historical perspective, such methods have been utilised for many years and an inspection of the technical literature reveals a large number of papers and several major texts [1,2,3] dedicated to the subject. Secondly, from a practical engineering view point, such methodologies are also simple and quick and therefore may be used as a starting point for many practical identification studies.

One of the prime requirements of any identification exercise is the estimation of an unbiased model. To achieve this, assuming data is presented in an appropriate form and that a suitable estimation algorithm is used, the key is to select the most appropriate model structure. In the case of the ARMA model the modern approach to structure selection is by direct search based on a statistical measure of model quality such as AIC, MDL [2] or YIC [3]. A typical algorithm may take the form:

1. Define the search space for structure parameters p, q and d.
2. For each combination of p, q and d evaluate the individual a and b parameters. Hence, evaluate the chosen quality measure and rank the model according to quality. This procedure would typically take the form of three loops where in this example $p = 1$ to 3 $q = 1$ to 3 and $d = 1$ to 5.

```
for p = 1 to 3
        for q = 1 to 3
                for d = 1 to 5
                        evaluate 'a' and 'b' values for given p, q and d
                        evaluate quality index
                        if quality index improves
                                save p, q and d values
                        end
                end
        end
end
```

The above approach has been described as an exhaustive search [4]. Using the above defaults, the algorithm would evaluate the *'a'* and *'b'* coefficients for a total of 45 different model structures.

3: NARMAX Models.

Compared with linear models, nonlinear representations can provide the complexity needed to obtain satisfactory solutions to a far wider variety of problems. Within the class of non-linear representations, the NARMAX model has proved popular because of its ability to characterise nonlinear behaviour with a relatively small number of terms. This is, of course, important in any control application because of stringent real time requirements.

The structure of a typical NARMAX model may take the form:

$$y(t)=\sum_{i=1}^{q} b_i u_{(t-d-i-1)} - \sum_{i=1}^{p} a_y y_{(t-i)}$$

Nonlinear Terms

$$+ \quad \sum_{i=1}^{p}\sum_{j=1}^{q}\sum_{k=1}^{l-1}\sum_{m=1}^{l-1} c_{ijkm} u_{(t-d-i+1)}{}^{k} y_{(t-j)}{}^{m} \Big|_{\max j+m=l}$$

$$+ \quad \sum_{i=1}^{q}\sum_{j=1}^{q}\sum_{k=1}^{l-1}\sum_{m=1}^{l-1} d_{ijkm} u_{(t-d-i+1)}{}^{k} u_{(t-d-j+1)}{}^{m} \Big|_{\max j+m=l}$$

$$+ \quad \sum_{i=1}^{p}\sum_{j=1}^{p}\sum_{k=1}^{l-1}\sum_{m=1}^{l-1} e_{ijkm} y_{(t-i)}{}^{k} y_{(t-j)}{}^{m} \Big|_{\max j+m=l} \qquad [2]$$

From above we can see that the NARMAX model is an extension to the ARMA model which incorporates nonlinearity by the use of nonlinear data elements within a linear-in -the-parameters structure.

As with ARMA models one of the key requirements for unbiased estimation is the selection of an appropriate model structure. In the case of the NARMAX model, the traditional approach to structure selection is to initially define a model containing all terms and then eliminate those which are not significant. Hence, defining the search space for the model as p = 1 to 3, q = 1 to 3, d = 1 and l (the power of the polynomial expansion) = 2 would result in a model with 28 terms. This gives a maximum of 2^{28} (268,435,456) possible model structure available to an exhaustive search.

With such a large search space, an exhaustive search is clearly inappropriate. However, several alternative search procedures exist, one of which is a search based on genetic algorithms. This is outlined by Li and Jeon [5]. The principal feature of the method of Li and Jeon is the coding of the model structure into a binary string. In this string a '1' represents the inclusion of a model term and '0' represents its absence. For example defining the data vector as:

[dc,u(t-1),u(t-2),y(t-1),y(t-2),u(t-1)^2,u(t-2)^2,...,]

The model: y(t) = dc + b1.u(t-1) + a1.y(t-1) + d1.u(t-1)^2

may be expressed by the string: struct = 1101010...0

However, the disappointing feature of the approach, as outlined by the authors, is the very weak termination condition (based on the first peak in fitness) which does not allow the algorithm to find a global minimum. Also, since model fit is the only cost there is no penalty for an overparameterised model.

4: Artificial Neural Network Models.

Recently, the artificial neural network has proved increasingly popular as a model for nonlinear systems. In this field, perhaps the most commonly encountered network structure is the multilayered perceptron containing one hidden layer and sigmoidal activation functions. The structure is illustrated in Figure 1.

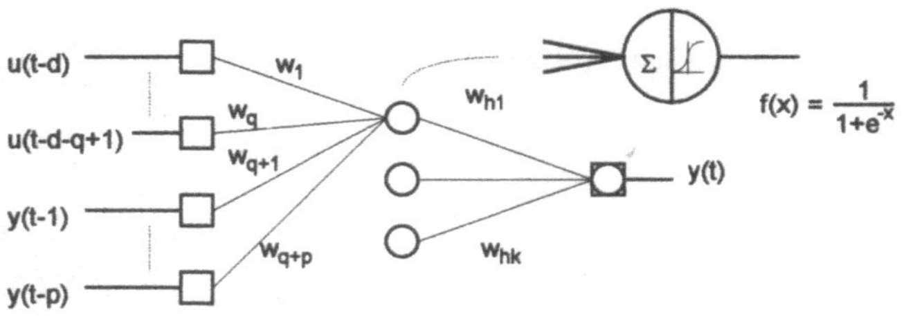

Figure 1: Multilayer Perceptron Model.

If we compare this model with that of a conventional ARMA structure (see Figure 2) it is clear that the ARMA model may be considered to be a special case of a neural network with zero nodes in the hidden layer.

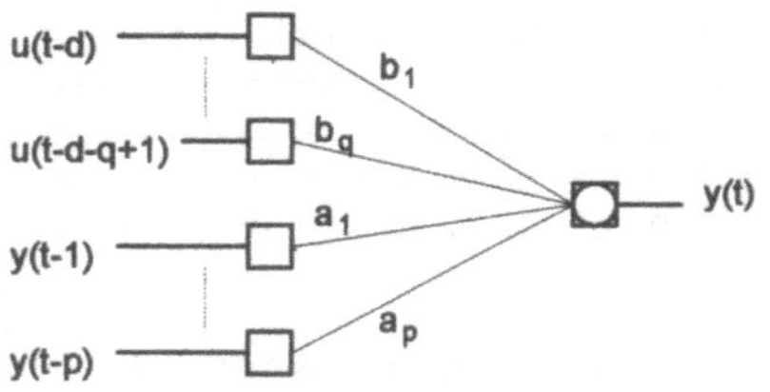

Figure 2: ARMA Network.

Once again, the key to unbiased estimation (and the avoidance of the overtraining problem) with the multilayer perceptron network is the selection of an appropriate network structure. In this case defining p = 1 to 3, q = 1 to 3, d = 0 to 5 and k (the number of hidden layer nodes = 1 to 10 gives, by analogy with section 1, a search space of 450 different network structures. This is not an unreasonable number when compared to the NARMAX structure of section 2, however, it must be remembered that training of each of the structures may take several hundred passes through the data. As a consequence on automated structure selection procedure, similar to that outlined for the conventional ARMA model, is likely to take an unreasonable period of time.

One point of note is that Riva et al [6] have shown an equivalence between the network of Figure 1 and the conventional NARMAX model. Therefore, given that automated structure selection methods are available for such models, these may prove more appropriate under certain circumstances.

5: Radial Basis Function Networks.

Radial basis function (RBF) networks may be considered as a variation of the three layer network shown in Figure 1. In the case of radial basis function networks, however, we consider the special case when the weights from the input layer to the hidden layer are all unity. Also, the radial basis function network uses an activation function which responds to a local field; that is f(x) is largest at the centre and decreases to zero either side of this [7].

$$f(x)=e^{-D} \qquad \text{where } D=(\underline{x}-\underline{c})^T(\underline{x}-\underline{c})/\sigma$$

The advantage of this approach lies in the fact that once the centres, $\underline{c}$, have been specified (this can be done by a simple clustering algorithm or by uniform distribution), the output from each neuron in the hidden layer can be calculated directly. Consequently, the weights Wh1 to Whk connecting the hidden layer to the output can be evaluated by a simple least squares algorithm, thus avoiding the problem of a lengthy training period. Because of this, it becomes possible to use an automatic structure selection algorithm when using RBF models. A genetic algorithm based method is outlined below.

Defining the structure of the network in terms of p, q, d, k (the number of nodes in the hidden layer) and n (the number of centres to be used to define the scaling factor σ [7]), the structure of the network may be represented by the string:

$$struct = p\ q\ d\ k\ n$$

1. Define the search range for each entry in the string.
2. Generate an initial population of say 20 randomly chosen individuals.
3. Having established the centres, $\underline{c}$, and scaling values σ, use least squares or alternative to estimate the weights from the hidden layer. Hence, rank each member of the population using:

$$score = 1 - NEV - (p+q)/100 - k/200$$

$$NEV = \text{Normalised Error Varience}$$

4. Terminate the algorithm if any model within the population contains only parameters that satisfy:

$$|W_{hi}| > 1.95\sqrt{Varience\_W_{hi}}$$

and has a fit of greater than 80%.

5. Based on the score define a population for mating.
6. Mate the population by randomly exchanging information between pairs of structures. Pairings are chosen at random. Also the number and location of exchange sites within the string are randomly selected.
7. Mutations are introduced by randomly selecting elements from within the population and altering these to other valid values from within the search range. A typical mutation rate is 1 every 200 information exchanges.
8. goto 3

6: Test Examples.

Three different sets of test data will be used to illustrate the effectiveness of the modelling approaches described in the previous sections. The systems from which this data is taken are:

1. a linear difference equation.
2. a NARMAX system.
3. a laboratory scale 'industrial' flow process.

In each case results will be given which contrast the complexity of the competing model structures, the relative accuracies of the model in terms of fit and the average number of candidate configurations evaluated in arriving at the final structure.

6.1 The Linear Difference Equation Model.

The model used for testing is given by:

$$y(t) = 0.5y(t-1) + 0.25u(t-2) + \varepsilon(t)$$

The input u(t) was chosen as a PRBS and the noise term $\varepsilon(t)$ as a uniformly distributed random sequence. The signal to noise ratio at the output was 20:1. 300 data pairs were used for the identification process.

The results of the three automated structure selection procedures are presented in the following table.

Type of Model	No. Terms / Nodes	Structure	Fit %	Av. Model Evaluations
ARMA	2	p=1, q=1, d=2	95.6	45
NARMAX	4	see App-T1	98.9	400
RBF	6	p=1, q=1, d=2, k=6, n=4.	98.6	80

Table 1

6.2 The NARMAX System.

The NARMAX system is defined by the expression:

$$y(t) = 0.5y(t-1) + u(t-2) + 0.2u(t-3) + 0.1u(t-1)^2 + \varepsilon(t)$$

The input sequence, u(t) was defined as a multi level random sequence. Once again 300 data pairs were used in the identification process and the signal to noise ratio at the output 20:1.

The findings of the automated structure selection tests are given in Table 2:

Type of Model	No. Terms / Nodes	Structure	Fit %	Av. Model Evaluations
ARMA	2	p=1, q=1, d=2	86.9	45
NARMAX	10	see App-T2	97.6	560
RBF	8	p=1, q=1, d=2, k=6, n=4.	97.7	80

Table 2

6.3 Flow Rig Data.

Flow control differs from other process control problems in that (for short pipe lengths) the lag in the process is negligible. Consequently, the response of the system depends mainly on the characteristics of the actuators, transducers, controllers and transmission lines, all of which have extensive non-linear operating regimes. Flow control is difficult because both steady state and dynamic response vary with set point. The task of control is not made easier by the fact that the measurement signals are usually entrenched in noise.

A randomly stepped PRBS test was used to gather data from across the operating profile of the test equipment. A total of 800 data pairs were obtained: this data is presented in Figure 3.

The results of the structure selection tests based on this data are presented below.

Type of Model	No Terms / Nodes	Structure	Fit %	Av. Model Evaluations
ARMA	3	p=1, q=2, d=1	95.6	45
NARMAX	6	see App-T3	98.8	520
RBF	6	p=1, q=2, d=1, k=6, n=4.	99.3	120

Table 3

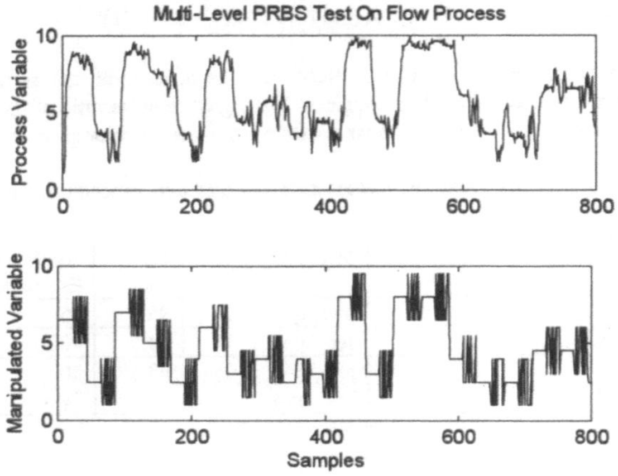

Figure 3: Flow Rig Test Data.

7: Conclusion.

The paper has contrasted the application of several system identification techniques on a range of test examples. From the results, certain observations can be made.

1. In the examples considered there is no significant improvement in accuracy to be gained by using a nonlinear representation.

2. The nonlinear forms contain 2-4 times the number of terms calculated in the linear representation.

3. The nonlinear forms require 4-10 times the computational effort, in terms of model evaluations, when compared with the linear representations.

It must be noted, however, that the range of test examples chosen was very limited and, therefore, the above comments should not be taken as general statements regarding the viability of the methods described. Indeed, if the relatively small nonlinear term present in the second, NARMAX, example (0.1 $u(t-1)^2$) is replaced by a more pronounced nonlinearity (say 1.0 $u(t-1)^2$) then the best linear model with structure p=1, q=2, d=1, k=16, n=15 can produce a 97% 'fit' and, of course, a NARMAX model would also produce an almost 100% 'fit'.

References.

[1] Goodwin, G.C. & Payne R.L., *Dynamic System Identification: Experiment Analysis and Design.* Academic Press, 1977.

[2] Johansson, R.,. *System Modelling & Identification.* Prentice Hall, 1993.

[3] Young, P.C., *Recursive Estimation and Time Series Analysis.* Springer Verlag, New York 1984.

[4] Haest, M. et al, *ESPION- An Expert System for Identification.* Automatica, V26, page 85-95.

[5] Li, C.J. & Jeon, Y.C., *Genetic Algorithms in Identifying Nonlinear Auto Regressive with Exogeneous Inputs Models for Nonlinear Systems.* Proc. Am Control Conf., IEEE, page 2305-2309, 1993

[6] Riva, A. et al, *Dynamic System Identification by means of Neural Networks.* Politecnico di Torino, Italy, 1993.

[7] Hofland, A.G. et al, *Radial Basis Function Networks Applied to Process Control.* University of Newcastle, UK, 1992.

Appendices.

App-T_1: Terms of NARMAX Model

$y(t-1)$, $u(t-2)$, $y(t-2)^2$, $u(t-1)^2$

App-T_2: Terms of NARMAX Model

$y(t-1)$, $u(t-2)$, $u(t-3)$, $u(t-1)^2$, $u(t-2)^2$, $u(t-3)^2$, $y(t-1)u(t-4)$, $y(t-2)u(t-5)$, $y(t-4)u(t-3)$, $u(t-1)u(t-4)$

App-T_3: Terms of NARMAX Model

$y(t-1)$, $u(t-1)$, $u(t-2)$, $y(t-1)^2$, $y(t-2)^2$, $u(t-2)^2$

Computing

Archives for Informatics and Numerical Computation
Archiv für Informatik und Numerik

Computing publishes original papers and short communications from all fields of scientific computing in English (preferably) or German. Contributions may be of theoretical or applied nature, the essential criterion is computational relevance. Subject areas include discrete mathematics, symbolic computation, parallel computation, computer arithmetic, architectural concepts for computers and networks, operating systems, programming languages, software engineering, performance and complexity evaluation, data bases, image processing, computer graphics, pattern recognition, artificial intelligence, optimization, numerical analysis, and numerical statistics.

Subscription Information:
1995. Vols. 54–55 (4 issues each):
DM 916,–, öS 6.412,–, plus carriage charges
approx. US $ 607.00 incl. postage and handling
ISSN 0010-485X   Title No. 607

Prices are subject to change without notice

Springer-Verlag Wien New York

Sachsenplatz 4–6, P.O.Box 89, A-1201 Wien · 175 Fifth Avenue, New York, NY 10010, USA
Heidelberger Platz 3, D-14197 Berlin · 3-13, Hongo 3-chome, Bunkyo-ku, Tokyo 113, Japan

R. Albrecht, G. Alefeld, H. J. Stetter (eds.)

Validation Numerics

Theory and Applications

1993. 23 figures. IX, 291 pages.
Soft cover DM 160,–, öS 1120,–
ISBN 3-211-82451-0
(Computing / Supplement 9)

The articles in this book give a comprehensive overview on the whole field of validated numerics. The problems covered include simultaneous systems of linear and nonlinear equations, differential and integral equations and certain applications from technical sciences. Furthermore some papers which improve the tools are included. The book is a must for scientists working in numerical analysis, computer science and in technical fields.

G. Farin, H. Hagen, H. Noltemeier (eds.)
in cooperation with W. Knödel

Geometric Modelling

1993. 175 figures and 6 plates. VIII, 316 pages.
Soft cover DM 160,–, öS 1120,–
ISBN 3-211-82399-9
(Computing / Supplement 8)

In this volume experts from university and industry are presenting new technologies for solving industrial problems as well as important and practicable impulses for new research.

The following topics are treated: solid modelling, geometry processing, feature modelling, product modelling, surfaces over arbitrary topologies, blending methods, scattered data algorithms, smooting and fairing algorithms, NURBS.

21 articles are giving a state-of-the-art survey of the relevant problems and issues in the rapidly growing area of geometric modelling.

G. Tinhofer, E. Mayr, H. Noltemeier, M. M. Syslo (eds.)
in cooperation with R. Albrecht

Computational Graph Theory

1990. 68 figures. VII, 282 pages. ISBN 3-211-82177-5
Soft cover DM 148,–, öS 1036,–
(Computing / Supplement 7)

U. Kulisch, H. J. Stetter (eds.)

Scientific Computation with Automatic Result Verification

1988. 22 figures. VIII, 244 pages. ISBN 3-211-82063-9
Soft cover DM 128,–, öS 900,–
(Computing / Supplement 6)

K. Böhmer, H. J. Stetter (eds.)

Defect Correction Methods

Theory and Applications
1984. 32 figures. IX, 243 pages. ISBN 3-211-81832-4
Soft cover DM 72,–, öS 504,–
(Computing / Supplement 5)

R. Albrecht, U. Kulisch (Hrsg.)

Grundlagen der Computer-Arithmetik

1977. 13 Abbildungen. IX, 150 Seiten.
ISBN 3-211-81410-8
Broschiert DM 69,–, öS 480,–
(Computing / Supplement 1)

* *10% price reduction for subscribers to "Computing"*

Also available:

B. Buchberger, G. E. Collins, R. Loos (eds.)
in cooperation with R. Albrecht

Computer Algebra

Symbolic and Algebraic Computation
2nd edition
1983. 5 figures. VII, 283 pages. ISBN 3-211-81776-X
Soft cover DM 70,–, öS 490,–

Prices are subject to change without notice

Springer-Verlag Wien New York

Sachsenplatz 4–6, P.O.Box 89, A-1201 Wien · 175 Fifth Avenue, New York, NY 10010, USA
Heidelberger Platz 3, D-14197 Berlin · 3-13, Hongo 3-chome, Bunkyo-ku, Tokyo 113, Japan

Springer-Verlag and the Environment

WE AT SPRINGER-VERLAG FIRMLY BELIEVE THAT AN international science publisher has a special obligation to the environment, and our corporate policies consistently reflect this conviction.

WE ALSO EXPECT OUR BUSINESS PARTNERS – PRINTERS, paper mills, packaging manufacturers, etc. – to commit themselves to using environmentally friendly materials and production processes.

THE PAPER IN THIS BOOK IS MADE FROM NO-CHLORINE pulp and is acid free, in conformance with international standards for paper permanency.

NATO ASI Series

Advanced Science Institutes Series

A series presenting the results of activities sponsored by the NATO Science Committee, which aims at the dissemination of advanced scientific and technological knowledge, with a view to strengthening links between scientific communities.

The series is published by an international board of publishers in conjunction with the NATO Scientific Affairs Division

A	**Life Sciences**	Plenum Publishing Corporation
B	**Physics**	New York and London
C	**Mathematical and Physical Sciences**	D. Reidel Publishing Company Dordrecht, Boston, and Lancaster
D	**Behavioral and Social Sciences**	Martinus Nijhoff Publishers
E	**Engineering and Materials Sciences**	The Hague, Boston, Dordrecht, and Lancaster
F	**Computer and Systems Sciences**	Springer-Verlag
G	**Ecological Sciences**	Berlin, Heidelberg, New York, London,
H	**Cell Biology**	Paris, and Tokyo

Recent Volumes in this Series

Volume 133—Membrane Receptors, Dynamics, and Energetics
edited by K.W.A. Wirtz

Volume 134—Plant Vacuoles: Their Importance in Solute Compartmentation in Cells and Their Applications in Plant Biotechnology
edited by B. Marin

Volume 135—Signal Transduction and Protein Phosphorylation
edited by L. M. G. Heilmeyer

Volume 136—The Molecular Basis of Viral Replication
edited by R. Perez Bercoff

Volume 137—DNA—Ligand Interactions: From Drugs to Proteins
edited by Wilhelm Guschlbauer and Wolfram Saenger

Volume 138—Chaos in Biological Systems
edited by H. Degn, A. V. Holden, and L. F. Olsen

Volume 139—Lipid Mediators in Immunology of Shock
edited by M. Paubert-Braquet

Volume 140—Plant Molecular Biology
edited by Diter von Wettstein and Nam-Hai Chua

Series A: Life Sciences

Plant Molecular Biology

Edited by

Diter von Wettstein

Carlsberg Laboratory
Copenhagen, Denmark

and

Nam-Hai Chua

Laboratory of Plant Molecular Biology
Rockefeller University
New York, New York

Plenum Press
New York and London
Published in cooperation with NATO Scientific Affairs Division

Proceedings of a NATO Advanced Study Institute on
Plant Molecular Biology,
held June 10-19, 1987,
at the Carlsberg Laboratory, Copenhagen, Denmark

ISBN 978-1-4615-7600-6 ISBN 978-1-4615-7598-6 (eBook)
DOI 10.1007/978-1-4615-7598-6

Library of Congress Cataloging in Publication Data

NATO Advanced Study Institute on Plant Molecular Biology (1987: Carlsberg Laboratory)
Plant molecular biology / edited by Diter von Wettstein and Nam-Hai Chua.
p. cm.—(NATO ASI series. Series A, Life sciences; vol. 140)
"Published in cooperation with NATO Scientific Affairs Division."
Includes bibliographies.
ISBN 978-1-4615-7600-6
1. Plant molecular genetics—Congresses. 2. Plant molecular biology—Congresses. I. Wettstein, Diter von. II. Chua, N.-H. (Nam—Hai) III. North Atlantic Treaty Organization. Scientific Affairs Division. IV. Title. V. Series: NATO ASI series. Series A, Life sciences; v. 140.
QK981.N367 1987
581.8′8—dc19 87-24389

Softcover reprint of the hardcover 1st edition 1987
A Division of Plenum Publishing Corporation
233 Spring Street, New York, N.Y. 10013

PREFACE

The present volume contains the invited lectures and abstracts of the posters presented at the NATO Advanced Study Institute on Plant Molecular Biology held at the Carlsberg Laboratory in Copenhagen, June 10 to 19.

The Directors of the Advanced Study Institute would like to express their most sincere gratitude for financial support to the following:

NATO
Danish Natural Science Research Council
Danish Agricultural and Veterinary Science Research Council
Carlsbergs Mindelegat for Brygger J.C. Jacobsen
Carlsberg and Tuborg Breweries
International Society for Plant Molecular Biology
Bayer AG, Leverkusen
Hoechst AG, Frankfurt-Main
Biotechnology, Ciba-Geigy, Research Triangle Park, North Carolina
Calgene, Davis, California
Danisco Biotechnology, Copenhagen
E.I. DuPont de Nemours & Co., Wilmington, Delaware
Plant Science Foundation
Holmegaards Glasværker, Fensmark
Imperial Chemical Industries, Bracknell
Martinus Nijhoff Publishers, Dordrecht
Monsanto & Co., St. Louis, Missouri
Novo, Copenhagen
Rhône-Poulenc Agrochimie, Lyon
Schering AG, Berlin
Shell Research Ltd., London
Unilever PLC, London
Zoecon Research Institute, Sandoz, Palo Alto, California

Without this generous support it would not have been possible to bring together so many outstanding plant molecular biologists for exchange of experience and knowledge and for animated as well as fruitful discussions on future experiments which can advance the understanding of plant biology. Also discussed were the possibilities to apply this knowledge to crop improvement.

Diter von Wettstein

Nam-Hai Chua

CONTENTS

MITOCHONDRIAL GENES

tRNA AND TRANSCRIPTION OF ORGANELLE DNA

TRANSPOSONS

BREEDING BY SOMATIC HYBRIDIZATION

INTERESTING PATHWAYS I

INTERESTING PATHWAYS II

HERBICIDE RESISTANCE

STRESS REACTIONS

VIRUS

VIROIDS

NITROGEN FIXATION

AGROBACTERIUM

TRANSGENIC PLANTS

FROM MOLECULAR BIOLOGY TO USEFUL PLANTS

THREE-DIMENSIONAL STRUCTURE OF RIBULOSE-1,5-BISPHOSPHATE CARBOXYLASE/ OXYGENASE FROM RHODOSPIRILLUM RUBRUM

Gunter Schneider, Ylva Lindqvist, and Carl-Ivar Brändén
Department of Molecular Biology, Swedish University of Agricultural Sciences, Biomedical Center, Box 590, S-75124 Uppsala, Sweden

George Lorimer
Central Research and Development Department
E.I. du Pont de Nemours and Company
Wilmington, Delaware 19898, USA

INTRODUCTION

Ribulose-1,5-bisphosphate carboxylase/oxygenase (Rubisco) is the key enzyme in photosynthetic carbon dioxide fixation. Due to its central role in the carbon metabolism of plants, Rubisco has been intensively studied (for a review see (1)). The enzyme catalyses the initial step in photosynthetic carbon dioxide fixation, the carboxylation of ribulose-1,5-bisphosphate, yielding two molecules of phosphoglycerate. This reaction of the enzyme results in the annual fixation of 10^{11} tons CO_2. The same enzyme also catalyses the oxygenation of ribulose-1,5-bisphosphate, the first step in photorespiration. This reaction substantially reduces the overall rate of photosynthesis and consequently plant productivity. The dual function of Rubisco makes it a challenging target in attempts to improve the efficiency of photosynthesis and thus increase crop productivity with the help of modern DNA-techniques.

Rubisco from higher plants, algae and most photosynthetic micro-organisms is a complex multisubunit protein. The enzyme consists of eight large (MW 56000 d) and eight small (MW 15000 d) subunits, building up a L_8S_8 type enzyme. The catalytic function resides on the large subunit; the function of the small subunits is still unknown. The primary structures for the large subunits of higher plant and algal carboxylases are very similar. The amino acid sequence homology amongst the carboxylases of L_8S_8 type is in the range of 70 - 80 % (1).

In contrast to these L_8S_8 type carboxylases, the enzyme from the photosynthetic bacterium Rhodospirillum rubrum differs considerably in amino acid and subunit composition. This carboxylase is only a dimer of large subunits and lacks the small subunits. The overall amino acid homology to higher plant type large subunits is 25 % (2,3). However, despite this low overall sequence homology, some peptide regions are highly conserved amongst all the carboxylases. Three of these peptide regions have been identified as parts of the active site (4,5,6). This indicates, that the topology of the active site and the overall tertiary structure for the large subunit are very similar in the Rh. rubrum and the higher plant enzymes.

The gene for the Rh. rubrum Rubisco has been cloned and expressed in E. coli. The gene product, obtained by Somerville and Somerville (7), is a fusion peptide containing an additional 24 amino acid peptide from β -galactosidase as the N-terminus. The kinetic parameters of the wild - type enzyme and this recombinant Rubisco are however indistinguishable. Recently, Larimer et al. (8) obtained an authentic, wild-type Rubisco without the 24 additional amino acids from β-galactosidase at the N-terminus.

A number of site-directed mutagenesis experiments have been carried out without the knowledge of the three-dimensional structure of the enzyme, probing the active site (9 - 13). The crystal structure of Rubisco from Rh. rubrum has recently been determined to 2.9 Å resolution (14). In the following, we summarize the structure determination of Rubisco and describe its three-dimensional structure.

CRYSTALLISATION

Rubisco from Rh. rubrum, both the authentic wild-type and the recombinant enzyme, has been crystallized under a number of different conditions (15 - 18). Four different crystal forms have been obtained so far (Table I). Under identical conditions, both the wild-type and the recombinant Rubisco, containing the 24 additional amino acide peptide, crystallize in the same spacegroup with identical cell dimensions (17).

Table 1. Crystal forms of Rubisco from Rhodospirillum rubrum

Source of Rubisco	pH	Spacegroup	Cellparameters (Å)	Reference
native	8.0	$P4_12_12$	82. 82. 290.9	(16)
native and recombinant	7.8-8.2	$P4_12_12$	82. 82. 290	(17)
recombinant	6.5-6.8	$P4_12_12$	82.4 82.4 324.	(15)
recombinant	5.6	$P2_1$	65.5 70.6 104.1 β =92.1°	(18)
recombinant	5.6	$P2_12_12_1$	70.9 100.1 131.	(18)

Structure Determination

The monoclinic crystal form (spacegroup $P2_1$) was used for the structure determination. Recombinant Rubisco from Rh. rubrum, containing the 24 additional amino acid peptide as the N-terminus, was used for all X-ray work. The monoclinic crystal form contains the whole dimer in the crystal asymmetric unit and has a local, non-crystallographic twofold rotation axis, relating the two subunits of the of the molecule. Intensity data for the native enzyme and two heavy metal derivatives were collected to 2.9 Å resolution. X-ray data for four additional heavy metal derivatives were collected to lower resolution. The local non-crystallographic axis was used to refine the initial MIR-phases, calculated from the heavy metal positions. An initial model was built from the averaged electron density map. A more detailed acccunt of the structure determination can be found in reference (14).

The crystallographic refinement of our model of Rubisco from Rh. rubrum is in progress. At present, the crystallographic R-factor is 26.5 % at 2.9 Å resolution.

The enzyme subunit

The enzyme subunit consists of two domains. The N-terminal domain comprises amino acid residues 1 - 137. The larger C-terminal domain consists of residues 138 - 466. Figure 1 shows a schematic view of the subunit.

The N-terminal domain is built up of a central, mixed five-stranded β-sheet. Two α-helices are found on one side and one α-helix on the other side of the β-sheet. The connection to the C-terminal domain is a short α-helix, followed by a piece of extended chain.

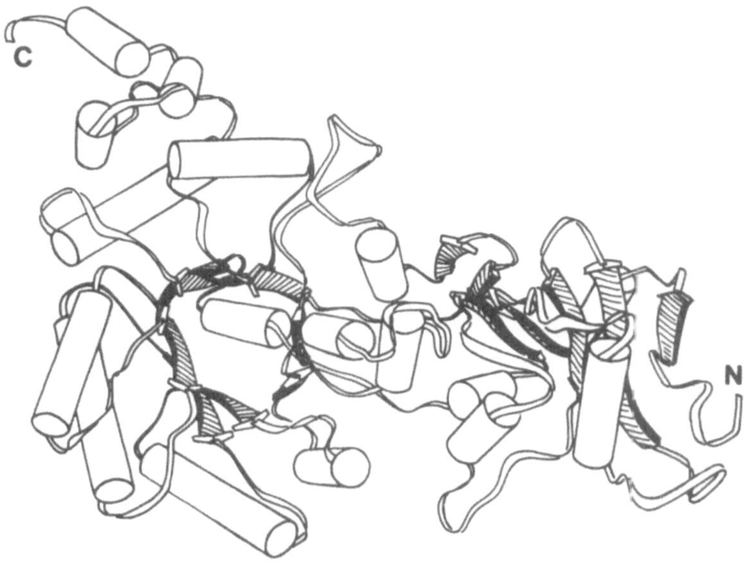

Figure 1: Schematic view of the subunit of Rubisco from Rhodospirillum rubrum. Cylinders represent α-helices and arrows represent β-strands.

The C-terminal domain has an α/β barrel structure, as found in triose phosphate isomerase (19) and other functionally non-related proteins. This fold consists of eight α/β units, which are joined adjacent to each other sequentially along the polypeptide chain. The eight parallel β-strands form the core of the barrel, with the eight helices on the outside.

This domain starts with an α-helix, which does not belong to the α/β barrel motif. This helix is located at the bottom of the α/β barrel and closes off the barrel from this side. After this helix, the chain enters strand no 1 of the α/β barrel. Additional secondary structural elements are found in this domain. After helix no 6., the polypeptide chain forms two additional antiparallel β-strands, before the chain enters strand no 7 of the barrel (figure 1). These two antiparallel β-strands are involved in domain-domain interactions. A small additional α-helix is found in the loop between strand no 8 and helix no 8 of the α/β barrel. At the C-terminal end of the polyopeptide chain, we find three consecutive α-helices.

At the present state of resolution and refinement, two loop regions of the subunit are not defined in our electron density maps. One such loop region is found in the N-terminal domain and comprises amino acid residues 54 - 63. The second loop region is found in the C-terminal domain and comprises residues 325 - 333. These residues are located in loop no 6 of the α/β barrel. We do not observe electron density for the additional 24 amino acid residues from β-galactosidase at the N-terminus. These residues are probably disordered in this crystal form.

The dimeric molecule

Figure 2 shows a picture of the C α backbone atoms for the Rubisco dimer from Rh. rubrum. The molecule has approximate dimensions 50 x 72 x 105 Å.

It can be seen from figure 2, that the subunit-subunit interactions are tight and extensive. Two main interface areas are found. One such area is between the C-terminal domains of the two subunits, which built up the core of the molecule. The second contact area is beween the C-terminal domain of one subunit and the N-terminal domain of the second subunit. Loop regions between the β-strands and the α-helices of the α/β barrel from the C-terminal domain are involved in subunit-subunit interactions at both interface regions. Amino acid residues from the loops no 1,2 and 3 interact with the N-terminal domain of the second subunit. These loop regions are at the carboxy end of the β-strands, which built up the α/β barrel. Loop no 1 and 2 exhibit extensive amino acid sequence homology between the bacterial and higher plant enzymes. Parts of these conserved peptide regions are involved in subunit-subunit interactions. This is strong evidence that some of these interactions, observed in the Rh. rubrum enzyme, are preserved in the higher plant enzymes.

Subunit contacts are also formed through homologous interactions of residues from loops no 3, 4 and 5 across the local twofold axis with corresponding residues of the C-terminal domain of the second subunit. No obvious amino acid sequence homology is found for these regions. As a consequence, the dimer interactions in the Rh. rubrum enzyme might not be identical with those in the plant enzymes.

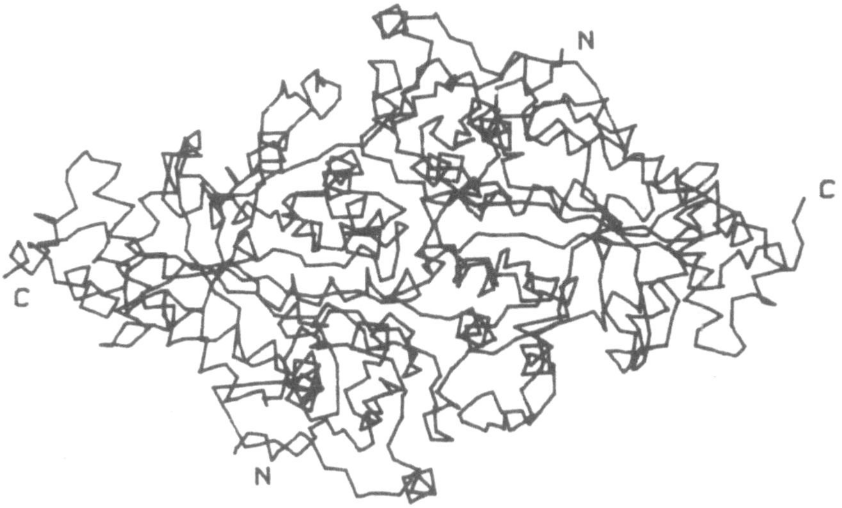

Figure 2: View of the Cα backbone of the dimer of Rubisco from Rhodospirillum rubrum. The N- and C-termini of the two subunits are indicated.

The active site

The active site of Rubisco is found at the carboxy end of the eight parallel β-strands of the C-terminal domain, at one end of the barrel. Active site residues are located at the carboxy end of the β-strands or in the loops between the strands and the helices of the barrel.

This side of the barrel is partly covered by the N-terminal domain of the second subunit. Amongst those residues of this domain which are close to the active site, is glu 48, which is part of a highly conserved region in the amino acid sequence. In our model, which represents the decarbamylated enzyme, these residues are too far away to participate directly in catalysis. However it is conceivable, that a conformational change upon activation or substrate binding might decrease the distance for some of these residues to the active site.

Three conserved lysine residues, 166, 191 and 329 have been suggested to be involved in either catalysis or the activation process (4,5,20). It has been shown (4), that lys 191 is the site of carbamylation. Recently, it has been suggested, that lys 166 is the base, which initiates catalysis by abstracting the C-3 proton from ribulose-1,5-bisphosphate (13,21): lys 329 has been identified as an active site residue by chemical modification. All three lysine residues are found at the carboxy end of β-strands or in loops between the carboxy end of β-strands and α-helices of the α/β barrel. Lys 166 is located in the loop after strand no 1, lys 191 is the last residue in strand no 2 and lys 329 is found in loop no 6.

The function of lys 191 is best understood. This residue is involved in the activation process of the enzyme, a process common to all carboxylases. During activation, a carbamate is formed between the ε-NH_2 group of lys 191 and an activator CO_2 molecule. This carbamate is then stabilized by an Mg ion. Figure 3 shows the surroundings of lys 191 in the deactivated enzyme. Two acidic residues, asp 193 and glu 194, are found in close proximity of lys 191. These two residues are conserved in all the carboxylases. It is very likely, that these two residues are part of the Mg^{2+} binding site in the activated ternary complex. In close proximity of this site is the sidechain of his 287. From our present model, it is possible that this sidechain is also involved in the binding of the metal ion in the ternary complex. This would be in agreement with EPR-measurements of the activated ternary complex, where Mg^{2+} had been replaced by Cu^{2+} (22,23). From these studies, a nitrogen ligand has been proposed to be part of the coordinationssphere of the metal ion. Upon formation of the quaternary complex, this nitrogen ligand is displaced, presumably by oxygen atoms from the substrate (22,23,24).

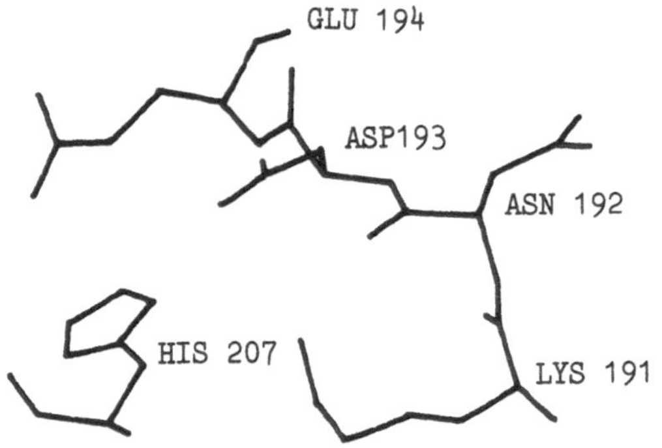

Figure 3: Picture of the Mg^{2+} binding site in Rubisco. Amino acid residues in the immediate surrounding of the activator lysine (lys 191) are shown.

Site-directed mutagenesis

Site-directed mutagenesis experiments have been carried out to probe the function of certain conserved amino acid residues (9 - 13). Two such mutatagenesis experiments have focussed on residues, involved in activation and assumed to participate in metal binding (9,10).

The activator lysine 191 has been replaced by a glutamic acid, such preserving the negative charge at this residue as in the carbamylated enzyme. The obtained mutant is inactive, probably due to the fact that the mutant enzyme is unable to bind Mg^{2+} (10). Modelbuilding experiments of the ternary complex, assuming coordination of the metal ion to asp 193, glu 194 and the carbamate show, that the replacement of the carbamylated lysine sidechain by the shorter glu sidechain destroys the metal binding site. The oxygen atoms of the carboxyl group of glu 191 are too far away from the metal for direct coordination. Consequently, the proper surrounding of the Mg binding site is not maintained and the ability of the enzyme can no longer bind Mg in its proper place.

The amino acid sequence in the proximity of lys 191 is highly conserved and contains a number of acidic residues (table II). Asp 188 has been replaced by glu via site-directed mutagenesis to probe the presumed function of this residue in metal binding (9). It is now obvious from the crystal structure, that this residue cannot be involved in the binding of the metal ion. Asp 188 is located at the bottom of the α/β barrel, more than 10 Å away from the metal binding site. The slight changes in the EPR-spectra upon replacement of asp 188 by glu probably reflect a local structural disturbance, introduced by the larger sidechain of the glutamic acid.

Table II: Active site loop 2 of Rubisco barrel domain. The amino acid sequence surrounding the activator lysine of Rubisco is shown. Conserved amino acid residues are enclosed in brackets. Corresponding secondary structural elements are indicated.

```
---------------------------------------------------------------------
┌────β 2────┐
188         191       194                       199
[ASP-PHE]-ILE-[LYS]-ASN-[ASP-GLU]-PRO-GLN-GLY-ASN-[GLN-PRO-PHE]-ALA-PRO-

   205
LEU-[ARG-ASP-]
   └─────────→ α 2
---------------------------------------------------------------------
```

Conclusions

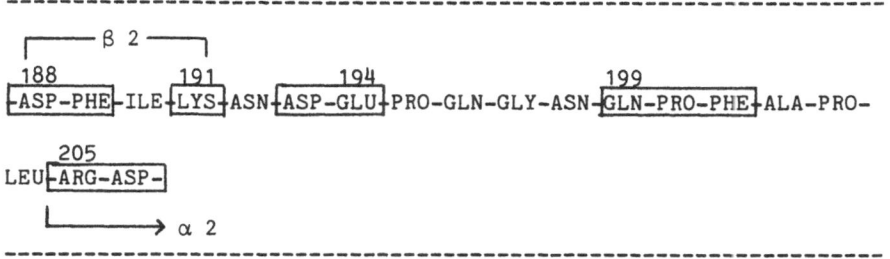

X-ray data for the monoclinic crystal form have been collected to 1.7 A resolution. These data are now subsequently included in the crystallographic refinement and in the near future, a refined model of Rubisco from Rh. rubrum to 1.7 Å will be available. Studies of complexes of Rubisco with inhibitors/substrates are in progress. Furthermore, the higher plant Rubisco will eventually yield to a high-resolution structure analysis. For two of these species, the enzyme from spinach (25) and Alcaligenes eutrophus (26) the crystals studied are of the activated quaternary complex, enzyme - Mg - HCO_3^- - CABP. The structure of this complex is of particular interest to the catalytic mechanism, since CABP is thought to mimic the reaction intermediate 3-keto-2 CABP.

All these crystallographic studies will provide a structural framework, to which the wealth of biochemical data on this enzyme can be related. The combination of the high-resolution structures with site-directed mutagenesis will help to elucidate the function of specific active-site residues and allow us to establish structure function relationships and to map catalysis in structural terms. This knowledge will then be the basis for rational site-directed mutagenesis experiments, with the aim of improving the efficiency of Rubisco as a carboxylase.

This work was supported by a grant from the Swedish Natural Science Research Council.

References

1. Miziorko, H.M. and Lorimer,G. (1983), Ann. Rev. Biochem. 52, 507 -535
2. Hartman,F.C., Stringer,C.D. and Lee, E.H. (1984), Arch. Biochem. Biophys. 232, 280 - 295
3. Nargang,F., McIntosh, L. and Somerville,C. (1984), Molec. gen. Genet. 193, 220 - 224
4. Lorimer,G. (1981), Biochemistry 20, 1236 - 1240
5. Herndon,C.S., Norton,I.C. and Hartman,F.C. (1982), Biochemistry 21, 1380 -1385
6. Fraij, B. and Hartman,F.C. (1982), J. Biol. Chem. 257, 3501 - 3505
7. Somerville,C.R. and Somerville,S. (1984), Mol. Gen. Genet. 193, 214 - 219
8. Larimer,F.W., Machanoff, R. and Hartman,F.C. (1986), Gene 41, 113 - 120
9. Gutteridge,S., Sigal,I., Thomas,B., Arentzen,R., Cordova,A. and Lorimer,G. (1984), EMBO J. 3, 2737 - 2742
10. Estelle,M., Hanks,J., McIntosh,L. and Somerville, C. (1985), J. Biol. Chem. 260, 9523 - 9526
11. Niyogi,S.K., Foote,R.S., Mural,R.J., Larimer,F.W., Mitra,S., Soper,T.S., Machanoff,R. and Hartman,F.C. (1986), J. Biol. Chem. 261, 10087- 10092
12. Terzaghi,B.E., Laing,W.A., Christeller,J.T., Petersen, G.B. and Hill, D.F. (1986), Biochem. J. 235, 839 - 846
13. Hartman, F.C., Soper,T.S., Niyogi, S.K., Mural,R.J., Foote,R.S., Mitra,S., Lee,E.H., Machanoff,R. and Larimer,W.F. (1987), J. Biol. Chem., in press
14. Schneider,G., Lindqvist,Y., Brändén, C.-I. and Lorimer,G. (1986b), EMBO J. 5, 3409 - 3415
15. Schneider,G., Brändén, C.-I. and Lorimer,G. (1984), J. Mol. Biol. 175, 99 -102
16. Janson,C.A., Smith,W.W., Eisenberg,D. and Hartman,F.C. (1984), J. Biol. Chem. 259, 11594 - 11596
17. Choe, H.-W., Jakob,R., Hahn,U. and Pal,G.P. (1985), J. Mol. Biol. 185, 781 -783
18. Schneider,G., Brändén, C.-I. and Lorimer,G. (1986), J. Mol. Biol. 187, 141 -143
19. Phillips,D.C., Sternberg, M.J.E., Thornton,,J.M. and Wilson,I.A. (1978), J. Mol. Biol. 119, 329 - 351
20. Herndon,C.S. and Hartman,F.C. (1984), J. Biol. Chem. 259, 3102 -3110
21. Hartman, F.C., Milanez, S. and Lee, E.H. (1985), J. Biol. Chem. 260, 13968 -13975
22. Brändén,R., Nilsson,T. and Styring,S. (1984), Biochemistry 23, 4373 - 4378
23. Styring,S. and Brändén,R. (1985), Biochemistry 24, 6011 - 6019
24. Miziorko,H.M. and Sealy, R. (1984), Biochemistry 23, 479 - 485
25. Andersson, I. and Brändén, C.-I. (1984), J. Mol. Biol. 172, 363 -366
26. Pal,G.P., Jakob, R., Hahn,U., Bowien,B. and Saenger,W. (1985), J. Biol. Chem. 260, 10768-10770.

FUNCTION OF ACTIVE-SITE RESIDUES OF RIBULOSE BISPHOSPHATE CARBOXYLASE/OXYGENASE

Fred C. Hartman, Robert S. Foote, Frank W. Larimer,
Eva H. Lee, Richard Machanoff, Sylvia Milanez, Sankar Mitra,
Richard J. Mural, Salil K. Niyogi, Harry B. Smith,
Thomas S. Soper, and Claude D. Stringer

Protein Engineering and Molecular Mutagenesis Program of the
Biology Division, Oak Ridge National Laboratory,
Oak Ridge, Tennessee 37831

INTRODUCTION

Ribulose 1,5-bisphosphate carboxylase/oxygenase (ribulose-P_2 carboxylase) (EC 4.1.1.39) is a bifunctional enzyme that catalyzes competing reactions of ribulose-P_2 with CO_2 (to form two molar equivalents of D-3-phosphoglycerate) and with O_2 (to form one molar equivalent each of D-3-phosphoglycerate and phosphoglycolate)[1,2]. Although multiple substrate specificities are not unusual, the bifunctionality of the carboxylase is perhaps unprecedented in that the two reactions catalyzed are the initial steps in competing metabolic pathways --- photosynthetic assimilation of CO_2 and energy-wasteful photorespiration. In recent years, the structure, mechanism, and regulation of ribulose-P_2 carboxylase have been under intense study in numerous laboratories, because of the enzyme's pivotal role in plant productivity and because modulation of its substrate specificity may provide a means to increase plant productivity[3,4].

As one facet of elucidating the mechanism of ribulose-P_2 carboxylase and of ultimately evaluating the feasibility of altering the carboxylase/oxygenase ratio, active-site characterization has received considerable attention. Chemical modification has revealed five disparate segments of primary structure that appear to constitute the active site. The amino acid sequence of each segment is highly conserved among evolutionarily diverse carboxylases. In this connection, the homodimeric carboxylase from the purple, non-sulfur photosynthetic bacterium Rhodospirillum rubrum provides a stringent test for sequence conservation, because it shows only ~30% invariance[5] in comparison to the catalytic subunit of the heterohexadecameric carboxylase as found in eucaryotic organisms.

One of the active-site segments includes the lysyl residue that functions in an unique, obligate activation process, entailing chemical reaction of its ε-amino group with CO_2 to form a carbamate[6]. Although stabilized by the essential divalent cation (Mg^{2+}), this structure is too labile to permit direct characterization. However, the carbamylated enzyme forms an exchange-inert complex with the transition-state analogue carboxyarabinitol-P_2, which is readily isolated by gel filtration[7]. Treatment of the complex with diazomethane results in esterification of the carbamate to form a very stable derivative[8]. Although the esterification is not specific, only one site is labeled with a radioactive marker provided that the complex has been prepared with $^{14}CO_2$. Subsequent characterization of labeled peptides in proteolytic digests reveals that the carbamate-forming lysyl residue occupies position 201 in the spinach enzyme[8,9] and position 191 in the _R. rubrum_ enzyme[5,10]:

(spinach)	D-F-T-$\overset{201}{\text{K}}$-D-D-E
(_R. rubrum_)	D-F-I-$\overset{191}{\text{K}}$-N-D-E

This lysine is the only active-site residue whose function has been unequivocally established.

Three other active-site segments of ribulose-P_2 carboxylase have been identified by direct chemical modification of amino acid side-chains with the affinity labels depicted in Fig. 1. Collective data[5] provided by several of these reagents have revealed the presence of two distinct active-site lysyl residues. These occupy positions 175 and 334 of the large subunit of the enzyme from spinach and positions 166 and 329 of the enzyme from _R. rubrum_:

(spinach)	T-I-K-P-$\overset{175}{\text{K}}$-L-G-L-S
(_R. rubrum_)	I-I-K-P-$\overset{166}{\text{K}}$-L-G-L-R
(spinach)	G-$\overset{334}{\text{K}}$-L-E-G-E
(_R. rubrum_)	G-$\overset{329}{\text{K}}$-M-E-G-E

One of the affinity labels shown in Fig. 1 suggests that the active-site domain also includes a segment of the polypeptide near the NH_2-terminus.

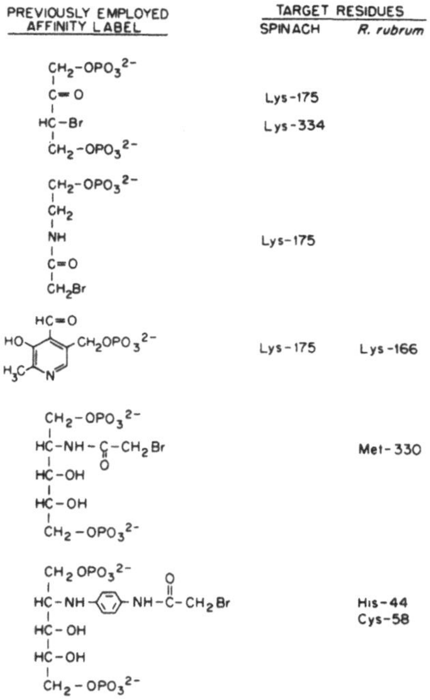

Fig. 1. Affinity labels for ribulose-P_2 carboxylase and the targets of modification. From top to bottom, the reagents are 3-bromo-1,4-dihydroxy-2-butanone 1,4-bisphosphate, N-bromoacetylethanolamide phosphate, pyridoxal phosphate, 2-bromoacetylaminopentitol 1,5-bisphosphate, and 2-(4-bromoacetamido)anilino-2-deoxypentitol 1,5-bisphosphate.

This reagent labels both His-44 and Cys-58 of the R. rubrum enzyme in a mutually exclusive fashion[11]; neither residue appears to be essential since they are not conserved:

(spinach)		G-A-A-V-A-A-E-S-S-T-G-T-W-T-T-V-W-T
	44	58
(R. rubrum)		A-A-H-F-A-A-E-S-S-T-G-T-N-V-E-V-C-T

However, the high degree of homology within this region indicates essentiality of at least one of the conserved residues.

Selective modification of His-298 of the spinach carboxylase by diethylpyrocarbonate provides the basis for attributing a fifth segment of polypeptide to the active site[12]:

	298	
(spinach)	H-R-A-M-H- A-V-I	
	291	
(R. rubrum)	H-R-A-G-H-G-A-V-T	

Although judicious application of chemical modification, particularly with affinity labels, can provide credible evidence of the identity of active-site residues, their function is rarely revealed. Ongoing efforts are designed to determine whether the residues assigned to the active site on the basis of chemical modification studies are indeed essential and, if so, the nature of their precise functions. Approaches include a determination of the pK_a values of the implicated residues based on the pH-dependencies of chemical modifications, a determination of their proximity based on covalent cross-linking with bifunctional reagents, and a determination of the consequences on catalytic functionality of their selective replacement with other amino acids by site-directed mutagenesis.

pK_a VALUES OF ACTIVE-SITE RESIDUES

Trinitrobenzenesulfonate arylates Lys-166 of the *R. rubrum* enzyme and Lys-334 of the spinach enzyme with a very high degree of specificity[13]. Hence, a single reagent, applied to the carboxylase from two different organisms, can provide the pK_a values and the intrinsic reactivities (k_o) for two different active-site lysines (Table 1). Especially noteworthy are the enhanced nucleophilicities of the two protein ε-amino groups compared to acetyllysine, despite their stronger acidities. These unusual properties are highly suggestive of catalytic functionality of the active-site lysines.

His-298 of the spinach carboxylase (which corresponds to His-291 in the *R. rubrum* enzyme) exhibits a pK_a of 6.8 on the basis of the pH dependency of inactivation by diethylpyrocarbonate[14]. This histidyl residue, which has been assigned to the active site[12], thus appears "normal" in terms of its acid/base properties. However, the lack of specificity of diethylpyrocarbonate complicates the correlation of an inflection in the pH-curve with the ionization of a specific residue. Furthermore, the possibility of a pH dependency of reagent specificity has not been evaluated.

Table 1. pH Dependencies of Modifications by Trinitrobenzenesulfonate

Sample	pK_a	k_o ($M^{-1}min^{-1}$) Observed	k_o ($M^{-1}min^{-1}$) Predicted (by Bronsted relationship)
N-α-Acetyllysine	10.8	1250	1200
Lys-166 (*R. rubrum* carboxylase)	7.9	670	120
Lys-334 (spinach carboxylase)	9.0	4500	280

His-291 and Lys-166 have both been proposed[13,14] as the base that initiates catalysis by promoting the enolization of ribulose-P_2. Irrespective of residue assignment, this essential base exhibits a pK_a of 7.5 (a value intermediate between those for His-291 and Lys-166) based on the pH dependency of the deuterium isotope effect with $[3-^2H]$ribulose-P_2 as substrate[15]. As judged by the insensitivity of its pK_a to the dielectric constant of the solvent, the base could be either a lysyl or histidyl residue[15]. As these latter two experimental observations do not allow a distinction to be made between His-291 or Lys-166 as the essential base, this question has been addressed by site-directed mutagenesis (see below).

ACTIVE-SITE MAPPING WITH CROSS-LINKING REAGENTS

If the suggestions, based on indirect evidence, of functionality of active-site residues are correct, these implicated residues must be rather close to each other, given the small size of ribulose-P_2 (in its fully-extended conformation, the distance between phosphorus atoms is only 11 Å). One approach to measure interresidue distances is through the use of bifunctional protein reagents. This approach seems practical with the carboxylase, because the enhanced nucleophilicities of active-site Lys-166 and Lys-329 increase the likelihood of achieving the required specificity of modification. Among the various species of ribulose-P_2 carboxylase that have been well characterized, the enzyme from *R. rubrum* appeared to be a logical choice because of its simple homodimeric structure.

Although the suitability of numerous bifunctional reagents has been examined, only two meet criteria for prompting in-depth studies: enzyme inactivation, protection by the competitive inhibitor carboxyribitol bisphosphate against inactivation, and a high degree of selectivity as evaluated by peptide mapping. These two reagents are 4,4'-diisothiocyano-2,2'-disulfonate stilbene, which spans 12 Å, and 4,4'-difluoro-3,3'-dinitrodiphenylsulfone, which spans 9 Å. The former reagent bridges Lys-166 with Lys-329 of the same subunit[16], providing direct evidence of their proximity as dictated by their purported catalytic functionality. The latter reagent also reacts with active-site Lys-166, but the cross-link is completed by involvement of Cys-58 (also a target of an affinity label, see Introduction) of the other subunit[17]. This finding of an intersubunit contact in the vicinity of a well-documented essential residue indicates that the active site is positioned at an interface between subunits. Elaboration of this postulate is found in the next section.

At the time cross-linking studies were initiated, the 3D structure of ribulose-P_2 carboxylase from any source had not been solved. Subsequently, the 3D structure of the R. rubrum carboxylase at 2.9 Å resolution has been reported[18]. Regrettably, interresidue distances are not yet revealed, because amino acid side-chains have not yet been fit to the electron density map. However, an interaction between the NH_2-terminal segment of one subunit and an active-site region of the other subunit is evident, consistent with intersubunit proximity of Lys-166 and Cys-58 as demonstrated chemically.

SITE-DIRECTED MUTAGENESIS

Site-directed mutagenesis has been used to clarify further the functions of Lys-166, Lys-329, His-291, and Glu-48 of ribulose-P_2 carboxylase from R. rubrum. These two lysines were selected for examination because of the wealth of data discussed earlier that had implicated their catalytic involvement. His-291 was scrutinized because of the suggestion[14] of its participation, rather than Lys-166, as the proton transfer group that enolizes ribulose-P_2. The reasons for inspecting Glu-48 were circumstantial. Within the homologous region flanked by His-44 and Cys-58, targets of an affinity label[11], Glu-48 is the only acid/base group and hence a logical candidate for functionality. Furthermore, crystallographic[18] and chemical cross-linking[17] studies had placed this NH_2-terminal region near the active site.

The cloned R. rubrum carboxylase (rbc) gene is experimentally convenient for mutagenesis work, in contrast to genes for the plant carboxylases, because the functional enzyme requires expression of a single gene product. The rbc gene was originally cloned, giving rise to a fusion protein[19]; a reconstruction[20] of the original clone expressed wild-type carboxylase and has been used to generate the mutant proteins described herein. Mutations were introduced into the rbc gene by single-primer extension[21-23] of an appropriate M13 vector or by a technique that utilizes an expression plasmid directly[24,25]. In all cases, the mutant genes were expressed in E. coli JM107[26], and the mutant proteins were purified by immunoaffinity chromatography[23].

Catalytic deficiency of a carboxylase mutant could be due to any one of several possibilities: (a) improper folding of the polypeptide, (b) failure of subunits to associate to form a dimer, (c) inability to undergo carbamylation as required for activation, (d) failure to bind substrates, or (e) loss of a group that participates directly in catalysis. Some of these possibilities are readily distinguished by direct experimentation. Semi-quantitative estimation of molecular weight reveals whether the mutant is a monomer or dimer. Because the transition-state analogue carboxyarabinitol-P_2 binds tightly only to the activated (i.e. carbamylated) form of the carboxylase, demonstration of quaternary complex formation (protein CO_2 Mg^{2+} analogue) verifies carbamylation and ligand (and by inference ribulose-P_2) binding. Decreased $\underline{k}_{cat}$ as a consequence of a small conformational change is difficult to exclude without extensive X-ray crystallographic data.

If all data are consistent with catalytic essentiality of the residue under consideration, it should then be possible to correlate the overall deficiency with the inability of the mutant protein to catalyze one of the discrete partial reactions. Procedures have been reported[27] for assaying the initial enolization of ribulose-P_2, the forward partitioning of the carboxylated intermediate to yield 3-phosphoglycerate, and the backward partitioning of the carboxylated intermediate to release CO_2.

A listing of mutant proteins generated and some of their properties are compiled in Table 2. One obvious conclusion is that His-291 of the carboxylase cannot be the essential base that enolizes ribulose-P_2, because the corresponding alanyl mutant retains 40% of wild-type activity[23]. This observation does not discount the presence of His-291 in the active-site region but does exclude any obligatory function of the residue.

All position-166 mutants are severely deficient in carboxylase activity[28]. An inspection of the properties of the Gly-166 mutant provides especially strong evidence for the direct participation of Lys-166 in catalysis. Despite the total absence of carboxylase activity, the Gly-166 mutant undergoes carbamylation by CO_2 and subsequently is able to form the quaternary complex with carboxyarabinitol-P_2. Furthermore, this mutant catalyzes formation of D-3-phosphoglycerate from the carboxylated reaction intermediate (3-keto-2-carboxyarabinitol 1,5-bisphosphate, which has been isolated and characterized[27]) but is unable to catalyze exchange of solvent protons with the C3 proton of ribulose-P_2, the reaction indicative of enolization[29]. The properties of the Gly-166 mutant are entirely consistent with the postulate that the Lys-166 ε-amino group is the base that abstracts the C3 proton of ribulose-P_2 and thereby initiates the catalytic pathway.

Among the position-166 mutant proteins, only the Asp-166 mutant fails to form a dimer. Given the intersubunit proximity of Lys-166 and Cys-58 as demonstrated by their covalent cross-linking with a 9-Å bifunctional reagent[17], this finding is most readily explained by charge-charge

Table 2. Characterization of Mutant Proteins

Amino Acid at:	Quaternary Structure	k_{cat} (% Wild-Type)	Binding of Carboxyarabinitol-P_2
Position-291			
His (wild-type)	dimer	100	++
Ala	dimer	40	+
Position-166			
Lys (wild-type)	dimer	100	++
Arg	dimer	0.02	-
His	dimer	< 0.01	-
Cys	dimer	< 0.01	+
Ala	dimer	0.1	+
Ser	dimer	0.2	+
Gln	dimer	< 0.01	-
Gly	dimer	< 0.001	+
Asp	monomer	< 0.001	-
Position-329			
Lys (wild-type)	dimer	100	++
Arg	dimer	< 0.01	-
Gln	dimer	< 0.01	-
Glu	dimer	< 0.01	-
Ser	dimer	< 0.01	-
Gly	dimer	< 0.01	-
Cys	dimer	< 0.01	-
Ala	dimer	< 0.01	-
Position-48			
Glu (wild-type)	dimer	100	++
Gln	dimer	0.05	+

repulsion between the cysteinyl-58 thiolate (or perhaps the glutamyl-48 γ-carboxylate) of one subunit and the introduced aspartyl-166 β-carboxylate of the other subunit.

The essentiality of Lys-329 is also underscored by the absence of carboxylase activity associated with a variety of amino acid substitutions. Although these substitutions do not preclude association of subunits, none of the position-329 mutants that have been examined form a stable complex with carboxyarabinitol-P_2. These results do not reveal whether Lys-329 is indirectly involved in the activation process or perhaps directly involved in stabilization of the transition state.

It is unlikely that the catalytic deficiencies of the position-329 mutants reflect gross distortions of the wild-type conformation, because the sulfhydryl group of the cysteinyl mutant exhibits unusual reactivity reminiscent of the corresponding ε-amino group of the native enzyme.

This unusual reactivity provides an avenue for converting the cysteinyl side-chain to a lysyl-like side-chain,

$$R\text{-}CH_2SH + BrCH_2CH_2NH_2 \longrightarrow R\text{-}CH_2SCH_2CH_2NH_2$$

thereby generating a novel carboxylase which differs from wild-type only in containing a sulfur atom in place of the γ-methylene of lysyl-329. Treatment of the Cys-329 mutant with bromoethylamine restores carboxylase activity to a level equal to ~25% that of wild-type[30]. The carboxylase/oxygenase ratio of the aminoethylated mutant protein is unchanged compared to wild-type enzyme.

Virtually all carboxylase activity is lost upon replacement of Glu-48 with glutamine[31], which abolishes the acid/base functionality of the glutamyl side-chain without introducing significant steric perturbation or change in hydropathy. As this replacement does not prevent dimerization of subunits, carbamylation by CO_2, or binding of the transition-state analogue (and by inference substrate), the drastic reduction in $\underline{k}_{cat}$ must reflect the presence of Glu-48 at the active site and indicates that this residue serves an important function. In conjunction with the known proximity of the NH_2-terminal region of one subunit with an active-site lysyl residue of the other subunit[11,13,18], the finding of essentiality of Glu-48 strongly suggests that the active site of the carboxylase is located at an interface between subunits. Although direct binding studies have shown that both dimeric and hexadecameric forms of ribulose-P_2 carboxylase contain the same number of catalytic sites as large subunits[7,10,32,33], the issue of whether a completely functional catalytic site comprises a single subunit or requires polypeptide regions from more than one subunit had not been previously addressed.

Current investigations include hybridization of position-166 and position-48 mutants to document rigorously the intersubunit location of the active site and thorough characterization of all the mutant proteins with respect to their potential functionality in the discrete partial reactions that have been described for wild-type carboxylase.

ACKNOWLEDGEMENT

Research sponsored by the Office of Health and Environmental Research, U.S. Department of Energy, under contract DE-AC05-84OR21400 with the Martin Marietta Energy Systems, Inc.

REFERENCES

1. G. Bowes and W. L. Ogren, Oxygen inhibition and other properties of soybean ribulose 1,5-diphosphate carboxylase, J. Biol. Chem. 247:2171 (1972).
2. G. H. Lorimer, T. J. Andrews, and N. E. Tolbert, Ribulose diphosphate oxygenase. II. Further proof of reaction products and mechanism of action, Biochemistry 12:18 (1973).
3. H. M. Miziorko and G. H. Lorimer, Ribulose-1,5-bisphosphate carboxylase/oxygenase, Ann. Rev. Biochem. 52:507 (1983).
4. R. J. Ellis and J. C. Gray, Eds., Ribulose bisphosphate carboxylase-oxygenase, Phil. Trans. R. Soc. Lond. B313:303 (1986).
5. F. C. Hartman, C. D. Stringer, and E. H. Lee, Complete primary structure of ribulosebisphosphate carboxylase/oxygenase from Rhodospirillum rubrum, Arch. Biochem. Biophys. 232:280 (1984).
6. G. H. Lorimer and H. M. Miziorko, Carbamate formation of the ε-amino group of a lysyl residue as the basis for the activation of ribulosebisphosphate carboxylase by CO_2 and Mg^{2+}, Biochemistry 19:5321 (1980).
7. H. M. Miziorko and R. C. Sealy, Characterization of the ribulosebisphosphate carboxylase-carbon dioxide-divalent cation-carboxypentitol bisphosphate complex, Biochemistry 19:1167 (1980).
8. G. H. Lorimer, Ribulosebisphosphate carboxylase — amino acid sequence of a peptide bearing the activator carbon dioxide, Biochemistry 20:1236 (1981).
9. G. Zurawski, B. Perrot, W. Bottomley, and P. R. Whitfeld, The structure of the gene for the large subunit of ribulose 1,5-bisphosphate carboxylase from spinach chloroplast DNA, Nucleic Acids Res. 9:3251 (1981).
10. M. I. Donnelly, C. D. Stringer, and F. C. Hartman, Characterization of the activator site of Rhodospirillum rubrum ribulosebisphosphate carboxylase/oxygenase, Biochemistry 22:4346 (1983).
11. C. S. Herndon and F. C. Hartman, 2-(4-bromoacetamido)anilino-2-deoxypentitol 1,5-bisphosphate, a new affinity label for ribulose bisphosphate carboxylase/oxygenase from Rhodospirillum rubrum, J. Biol. Chem. 259:3102 (1984).
12. Y. Igarashi, B. A. McFadden and T. El-Gul, Active site histidine in spinach ribulosebisphosphate carboxylase/oxygenase modified by diethyl pyrocarbonate, Biochemistry 24:3957 (1985).
13. F. C. Hartman, S. Milanez and E. H. Lee, Ionization constants of two active-site lysyl ε-amino groups of ribulosebisphosphate carboxylase/oxygenase, J. Biol. Chem. 260:13968 (1985).

14. C. Paech, Further characterization of an essential histidine residue of ribulose-1,5-bisphosphate carboxylase/oxygenase, Biochemistry 24:3194 (1985).

15. D. E. Van Dyk and J. V. Schloss, Deuterium isotope effects in the carboxylase reaction of ribulose-1,5-bisphosphate carboxylase/oxygenase, Biochemistry 25:5145 (1986).

16. E. H. Lee, C. D. Stringer, and F. C. Hartman, Distance between two active-site lysines of ribulose bisphosphate carboxylase from Rhodospirillum rubrum, Proc. Natl. Acad. Sci. USA 83:9383 (1986).

17. E. H. Lee, T. S. Soper, R. J. Mural, C. D. Stringer, and F. C. Hartman, An intersubunit interaction at the active site of ribulose bisphosphate carboxylase/oxygenase revealed by cross-linking and site-directed mutagenesis, Biochemistry, in press (1987).

18. G. Schneider, Y. Lindqvist, C. I. Branden, and G. Lorimer, Three-dimensional structure of ribulose-1,5-bisphosphate carboxylase/oxygenase from Rhodospirillum rubrum at 2.9 Å resolution, EMBO Journal 5:3409 (1986).

19. C. R. Somerville and S. C. Somerville, Cloning and expression of the R. rubrum ribulosebisphosphate carboxylase gene in E. coli, Mol. Gen. Genet. 193:214 (1984).

20. F. W. Larimer, R. Machanoff and F. C. Hartman, A reconstruction of the gene for ribulose bisphosphate carboxylase from Rhodospirillum rubrum that expresses the authentic enzyme in Escherichia coli, Gene 41:113 (1986).

21. M. J. Zoller and M. Smith, Oligonucleotide-directed mutagenesis of DNA fragments cloned into M13 vectors, Methods Enzymol. 100:468 (1983).

22. I. T. Nisbet and M. W. Beilharz, Simplified DNA manipulations based on in vitro mutagenesis, Gene Anal. Techn. 2:23 (1985).

23. S. K. Niyogi, R. S. Foote, R. J. Mural, F. W. Larimer, S. Mitra, T. S. Soper, R. Machanoff, and F. C. Hartman, Nonessentiality of histidine 291 of Rhodospirillum rubrum ribulose-bisphosphate carboxylase/oxygenase as determined by site-directed mutagenesis, J. Biol. Chem. 261:10087 (1986).

24. R. J. Mural and R. S. Foote, "Bandaid" mutagenesis: A novel technique for oligonucleotide-directed site-specific mutagenesis, DNA 5:84 (1986).

25. J. Childs, K. Villanueba, D. Barrick, T. D. Schneider, G. D. Stormo, L. Gold, M. Leitner, and M. Caruthers, Ribosome binding site

sequences and function, in: "Sequence Specificity in Transcription and Translation," R. Calendar and L. Gold, eds., Alan R. Liss, New York (1985), pp. 341-350.

26. C. Yanisch-Perron, J. Vieira, and J. Messing, Improved M13 phage cloning vectors and host strains: nucleotide sequences of the M13ml8 and pUC19 vectors, Gene 33:103 (1985).

27. J. Pierce, T. J. Andrews, and G. H. Lorimer, Reaction intermediate partitioning by ribulose-bisphosphate carboxylases with differing substrate specificities, J. Biol. Chem. 261:10248 (1986).

28. F. C. Hartman, T. S. Soper, S. K. Niyogi, R. J. Mural, R. S. Foote, S. Mitra, E. H. Lee, R. Machanoff, and F. W. Larimer, Function of Lys-166 of Rhodospirillum rubrum ribulosebisphosphate carboxylase/ oxygenase as examined by site-directed mutagenesis. J. Biol. Chem. 262:3496 (1987).

29. G. H. Lorimer and F. C. Hartman, unpublished observations.

30. H. B. Smith and F. C. Hartman, Restoration of enzyme activity to the Cys-329 mutant of ribulose bisphosphate carboxylase/oxygenase by aminoethylation, Fed. Proc. 46: 1987.

31. F. C. Hartman, F. W. Larimer, R. J. Mural, R. Machanoff, and T. S. Soper, Essentiality of Glu-48 of ribulose bisphosphate carboxylase/oxygenase as demonstrated by site-directed mutagenesis, Biochem. Biophys. Res. Commun., in press (1987).

32. M. R. Badger and G. H. Lorimer, Interaction of sugar phosphates with the catalytic site of ribulose-1,5-bisphosphate carboxylase, Biochemistry 20:2219 (1981).

33. D. B. Jordan and R. Chollet, Inhibition of ribulose bisphosphate carboxylase by substrate ribulose 1,5-bisphosphate, J. Biol. Chem. 258:13752 (1983).

PARTIAL REACTIONS OF RIBULOSE BISPHOSPHATE CARBOXYLASE: THEIR UTILITY IN THE STUDY OF MUTANT ENZYMES

George H. Lorimer, Steven Gutteridge and Michael W. Madden

Central R.& D. Department, E.I. du Pont de Nemours & Co.

Experimental Station, Wilmington, Delaware 19898, USA.

ABSTRACT

Four partial reactions, independent of overall catalysis, have been exploited to analyse the properties of site-directed mutants within the active-site of the dimeric, recombinant Rubisco from Rhodospirillum rubrum. (1) Formation of the activator carbamate-Mg^{2+} complex; (2) Enolization of ribulose bisphosphate; (3) Decarboxylation of the 6-carbon reaction intermediate, 2'carboxy-3-keto-D-arabinitol 1,5-bisphosphate and (4) Hydrolysis of 2'-carboxy-3-keto-D-arabinitol 1,5-bisphosphate.

Crystallographic studies [Schneider et al., these proceedings] suggest that the carbamate of lys191, together with asp193, glu194 and his287 provide ligands to the divalent metal ion. Site-directed mutations at each of these sites have been created to examine the role of the metal ion in catalysis. Excepting asn287 (1% activity) and lys287 (0.1% activity), all mutants were unable to catalyse the overall carboxylation reaction. However, glu191, asn193, glu194, val194, asn287 catalysed the decarboxylation of 2'-carboxy-3-keto-D-arabinitol 1,5-bisphosphate in the absence of Mg^{2+}. In the presence of 100mM Mg^{2+} asn193 catalyzed the hydrolysis of 2'-carboxy-3-keto-D-arabinitol 1,5-bisphosphate but did not catalyze the enolization of ribulose bisphosphate. This suggests that the coordination to the metal ion necessary to achieve enolization of ribulose bisphosphate is different from that required to bring about the hydrolysis of the 6-carbon intermediate.

It is a biologically bizarre fact that a single enzyme, Rubisco, catalyses the initial reactions of disparate metabolic processes, photosynthetic CO_2 fixation and photorespiration. This abundant, but fickle protein, catalyses either the carboxylation or the oxygenation of RuBP at a common active site. In an attempt to understand the underlying chemical reason for such enzymatic dichotomy, a series of mechanistic and structural studies have been undertaken [see Pierce, 1986 and Andrews and Lorimer, 1987 for recent reviews]. Much progress has been made towards an understanding of the carboxylation reaction. With some confidence, we may dissect the overall catalytic reaction into five discrete steps (Scheme 1).

(1) *Enolization*, the abstraction of the C-3 proton, creates a nucleophilic center at C-2 in the form of the 2,3-enediol(ate).

(2) *Carboxylation* of the 2,3-enediol(ate) generates the 6-carbon intermediate, 2'-carboxy-3-keto-D-arabinitol-1,5-bisphosphate (6-CI) in the C-3 keto form.

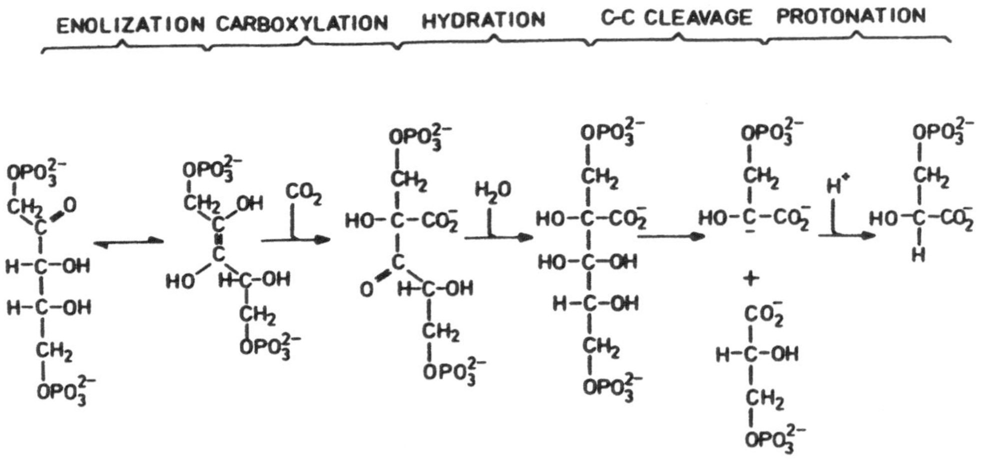

Scheme 1 : The pathway of carboxylation of ribulose 1,5-bisphosphate

(3) *Hydration* of the C-3 carbonyl yields the gem-diol form of 6-CI.
(4) Deprotonation of the hydrated 6-CI leads to *carbon bond cleavage*, giving a molecule of *lower* 3-P-glycerate and the C-2 carbanion form of the *upper* 3-P-glycerate.
(5) Stereospecific *protonation* of the C-2 carbanion generates the second molecule of product.

When stereochemical considerations (Lorimer et al., 1986) are superimposed upon the reactions illustrated in Scheme 1, some important mechanistic constraints are introduced. However, such descriptions of enzymatic mechanisms are inherently unsatisfying for they are disembodied from the reagent responsible for bringing them about, the enzyme. For this reason, studies aimed at defining structures within the active site are especially important. With respect to Rubisco, the affinity labeling studies of Hartman's group are exemplary [for recent examples, see Herndon and Hartman, 1984; Fraij and Hartman, 1982, 1983 and Hartman et al., 1985]. Consequently, lys 166, lys 329 and a conserved octapeptide ala46-thr53 are known to be within the domain of the active site. Cross-linking experiments (Lee et al., 1986) indicate that lys166 and lys329 can be juxtaposed to within 12A. Additionally, spectroscopic studies [reviewed by Pierce, 1986] have demonstrated that the activator carbamate-divalent metal-ion complex, centered on the ϵ-amino group of lys191, is also within the active site. The three-dimensional relationship of these active-site residues, one to another, has recently been determined by x-ray crystallography to 2.9A resolution (Schneider et al. 1986). This structural information has proven valuable in selecting residues within the catalytic site for site-directed mutational analysis. Although these structural studies severely limit the number of residues which may directly participate in catalysis, they do not define the function of these components. In order to relate structure to function, we have exploited four partial reactions of Rubisco in conjunction with a series of site-directed mutations within the catalytic site.

PARTIAL REACTIONS OF RUBISCO

Each of the four partial reactions described below can be monitored independently of the overall carboxylation reaction, a property which is especially useful when analyzing mutants which to unable to perform the overall reaction. This permits conclusions to be drawn concerning the functional role of various residues in the catalytic process.

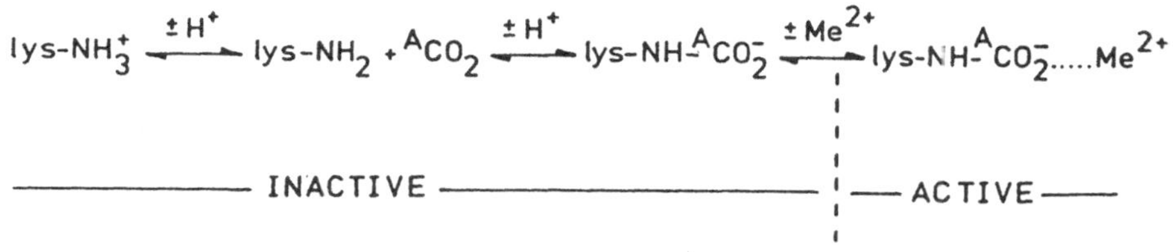

Scheme 2 : The activation of Rubisco via carbamylation of lys191

Carbamylation of Lys191

The activation of Rubisco via carbamylation of lys191 is most conveniently followed by measuring catalytic activity. However, this method is precluded with mutants that are catalytically inactive. Even with mutants displaying partial activity, it is not possible to quantitatively assess their ability to undergo carbamylation by measuring catalytic activity. Two other methods may be used to assess the ability of mutants to undergo carbamylation. (1) *Quaternary complex formation*: This method takes advantage of the fact that when the transition-state analog, 2'carboxy-D-arabinitol 1,5-bisphosphate (CABP) is added to the carbamylated ternary complex, an exceptionally stable quaternary complex is formed. The ligands

$$\text{enzyme}\cdot{}^{A}CO_2\cdot\text{Mg} + \text{CABP} \longrightarrow \text{enzyme}\cdot{}^{A}CO_2\cdot\text{Mg}\cdot\text{CABP}$$

of the quaternary complex do not readily exchange with unbound ligand. Thus, enzyme is first incubated with ^{14}C-labeled ligand (CO_2 or CABP) and subsequently challenged with a vast molar excess of unlabeled ligand. As Table 1 indicates, wild type enzyme and the mutant asn287 are capable of forming stable quaternary complexes. The mutant glu191 is obviously unable to form a quaternary complex containing CO_2 as a carbamate nor is it able to bind CABP tightly (Estelle et al., 1985). The properties of the mutant asn193, however, point to a shortcoming of this method. It is not an entirely independent method of assessing carbamylation for it depends upon the ability of the enzyme to bind CABP. Thus, a mutant such as asn193 which binds CABP much less tightly, also appears to bind the carbamate CO_2 less strongly. (2) A completely independent, but more demanding, method of assessing the formation of the carbamate-divalent metal-ion complex has been demonstrated by O'Leary et al. (1979), namely *^{13}C NMR*. A resonance attributable to the carbamate was clearly evident in solutions of the carbamylated ternary complex of wild type *Rhodospirillum rubrum* enzyme. We plan to examine some of our mutant enzymes by this method which requires that the clones be cultured on a much larger scale than usual.

Enolization

The abstraction of the C-3 proton of ribulose bisphosphate by an essential base in the active site represents the first step in catalysis.

Table 1. The Stoichiometry of Quaternary Complex Formation by Wild Type and Mutant Rubiscos

Enzyme	Overall Activity (%)	$^{14}CO_2$/protomer	^{14}CABP/protomer
wild type	100	1.01	0.99
asn287	1	0.97	1.08
glu191[a]	0	0	<0.01
asn193	0	0.24	0.08

[a] see also Estelle et al. (1985)

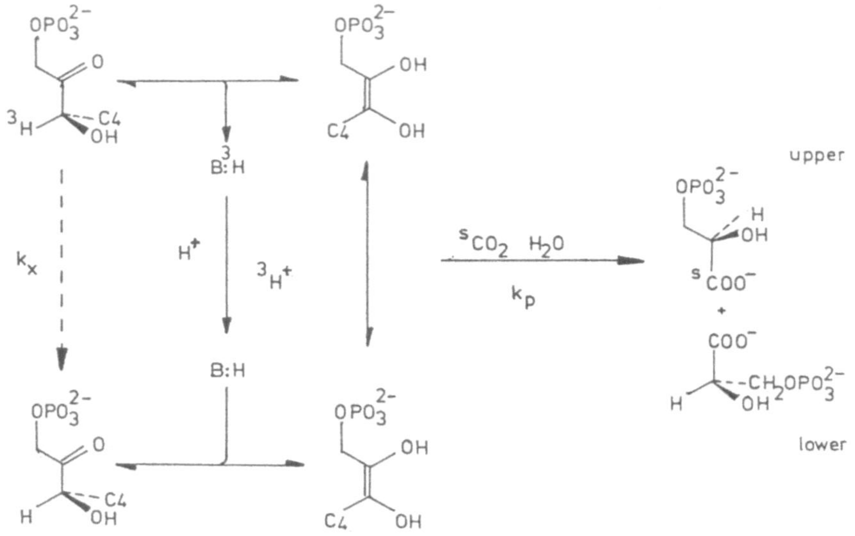

Scheme 3: Partial reactions of Rubisco II - the enolization step can be assayed by the exchange of the C-3 proton of ribulose bisphosphate with the medium.

Enolization has been shown to proceed independently of the subsequent steps in catalysis (Saver and Knowles, 1982; Gutteridge et al., 1984 and Pierce et al., 1986). This partial reaction may be conveniently monitored by following the loss of ^{3}H from [3-^{3}H] ribulose bisphosphate (Scheme 3). This is a sensitive assay and rates of ^{3}H loss equivalent to ~0.2% of the wild type rate have been measured. However, as Fig. 1 illustrates, the enolization reaction is dependent upon the enzyme being in the carbamylated state.

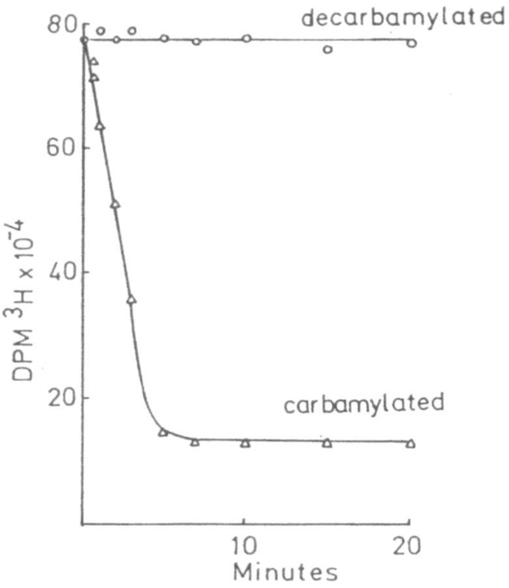

Fig. 1: The loss of ^{3}H from [3-^{3}H] ribulose bisphosphate by exchange with the medium is catalyzed by carbamylated Rubisco but not by enzyme in the decarbamylated state.

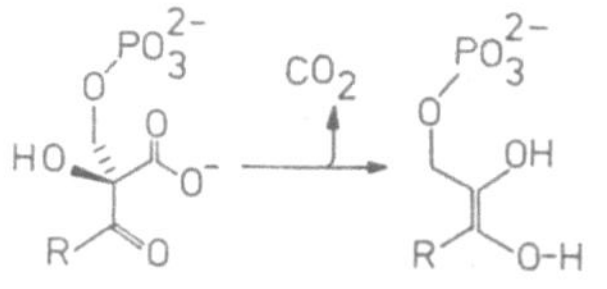

Scheme 4: Partial reactions of Rubisco III - the decarboxylation of the 6-carbon intermediate, 2'-carboxy-3-keto-D-arabinitol-1,5-bisphosphate yielding the 2,3-enediol(ate).

The six-carbon reaction-intermediate

The last two partial reactions of Rubisco depend upon the relatively stable nature of the six-carbon reaction intermediate, 2'carboxy-3-keto-D-arabinitol 1,5-bisphosphate (Lorimer et al., 1986). When Rubisco, activated with $^{14}CO_2$, is reacted with a slight molar excess of ribulose bisphosphate for a brief period (~12ms) before being quenched with acid, 14C-labeled reaction intermediate is released into the medium. This intermediate is sufficiently stable to be partly purified (freed of unreacted ^{14}C, buffer salts, denatured enzyme, etc). When stored at -80° it appears indefinitely stable. However, at 25° in free solution it undergoes decarboxylation with a half-time of ~1h. In hot acid it undergoes rapid decarboxylation and can thus be distinguished from the acid-stable product, 3-P-glycerate. The radioactivity in the reaction intermediate can be stabilized by reduction with borohydride. Thus, methods have been developed for synthesizing and assaying the six-carbon reaction-intermediate. Therefore, it can be used as a substrate to analyze the terminal stages of the reaction, independently of the preceding steps.

Decarboxylation of the six-carbon reaction intermediate

When decarbamylated wild-type Rubisco is offered the six-carbon reaction-intermediate, it catalyzes its decarboxylation (Scheme 4 and Fig. 2). This reaction is inhibited by the analog CABP, so it seems specific for the active-site (Pierce et al., 1986). Since catalysts alter the rates of both forward and reverse reactions, the decarboxylation of the six-carbon reaction-intermediate catalyzed by the decarbamylated enzyme most probably represents a simple reversal of the carboxylation of the 2,3-enediol(ate) that is catalyzed by the carbamylated enzyme.

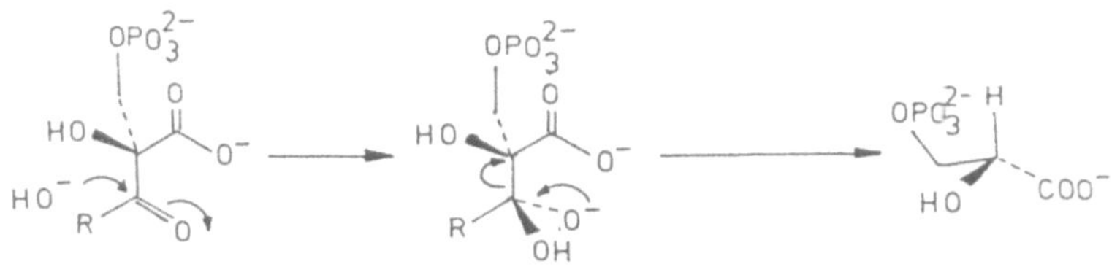

Scheme 5: Partial reactions of Rubisco IV - the hydrolysis of the six-carbon intermediate, 2'-carboxy-3-keto-D-arabinitol 1,5-bisphosphate to yield the product, 3-P-glycerate. Only the upper molecule of product is shown.

Hydrolysis of the six-carbon reaction-intermediate

When carbamylated wild-type Rubisco is given the six-carbon reaction intermediate, it catalyzes its hydrolysis to the product, 3-P-glycerate (Scheme 5 and Fig. 3).

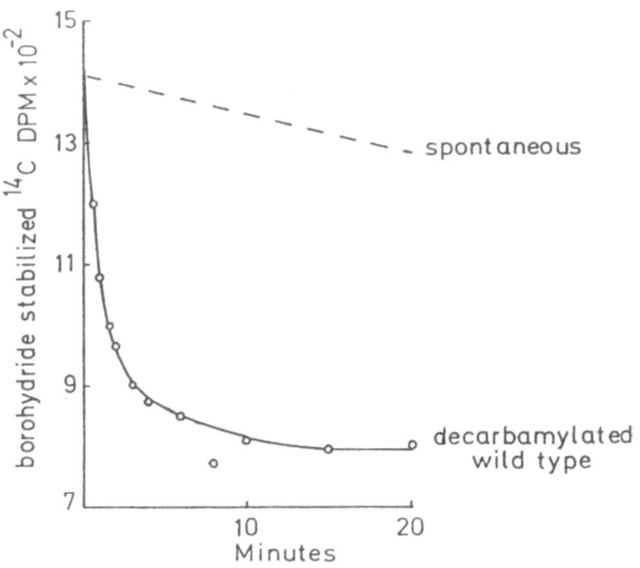

Fig. 2: The decarboxylation of the six-carbon reaction-intermediate, [2'-^{14}C] 2'-carboxy-3-keto-D-arabinitol 1,5-bisphosphate, followed by the loss of borohydride-stabilizable ^{14}C radioactivity, proceeds spontaneously ($t_{1/2}$ ~ 1h at 25°) but is catalyzed by decarbamylated wild-type Rubisco.

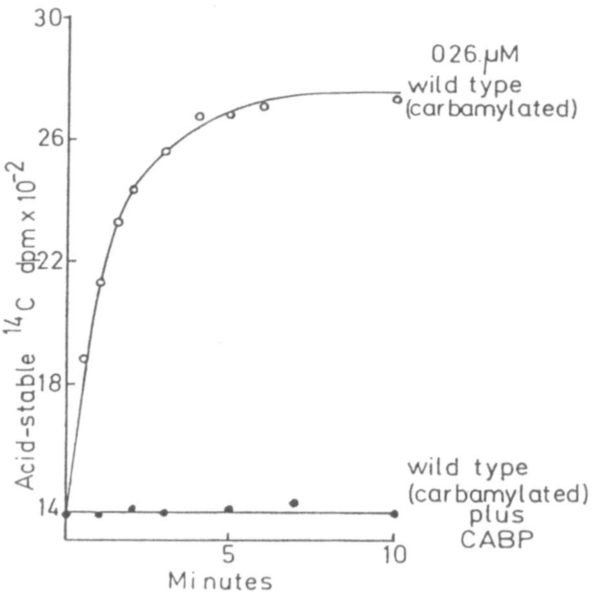

Fig. 3: The hydrolysis of the six-carbon reaction-intermediate, [2'-^{14}C] 2'-carboxy-3-keto-D-arabinitol 1,5-bisphosphate, followed by the increase in acid-stable ^{14}C radioactivity, is catalyzed by carbamylated wild-type Rubisco. Catalysis is inhibited by the analog, CABP.

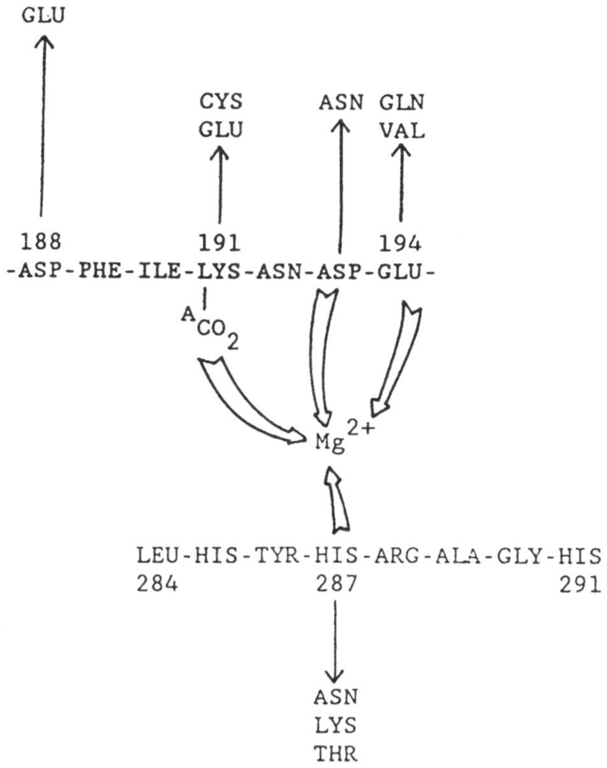

Scheme 6: Site-directed mutations in the putative metal-ion binding ligands

SITE-DIRECTED MUTAGENESIS OF RUBISCO

Crystallographic Analysis

The three-dimensional structure of *Rhodospirillum rubrum* Rubisco has been solved to 2.9A (Schneider et al., 1986). The crystals are of the decarbamylated enzyme. While the positions of many of the active-site components have been experimentally defined with some confidence, the ligands responsible for coordinating the divalent metal ion cannot be determined with these crystals. Nevertheless, model building suggests that the metal ion is most probably coordinated by the carbamate on the ϵ-amino group of lys191, by the carboxyl groups of asp193 and glu194 and by the imidazole ring of his287. Accordingly, we have selected these residues as targets for site-directed mutational analysis (Scheme 6).

Mutagenesis

The genetic manipulations are being performed on a derivative of pRR2119, , the expression plasmid containing the gene for the dimeric Rubisco from *Rhodospirillum rubrum* (Somerville and Somerville, 1984; Narang et al., 1984). The regions of the gene encoding the active-site constitute a small portion of the total gene. As indicated in Scheme 7, in addition to the naturally occuring unique restriction sites, we have engineered additional unique restriction sites so as to create "cassettes" which can be replaced with synthetic, double-stranded, complementary oligonucleotides, containing the desired mutation (Wells et al., 1985). Additionally, a silent mutation which destroys a restriction site within the wild-type cassette is

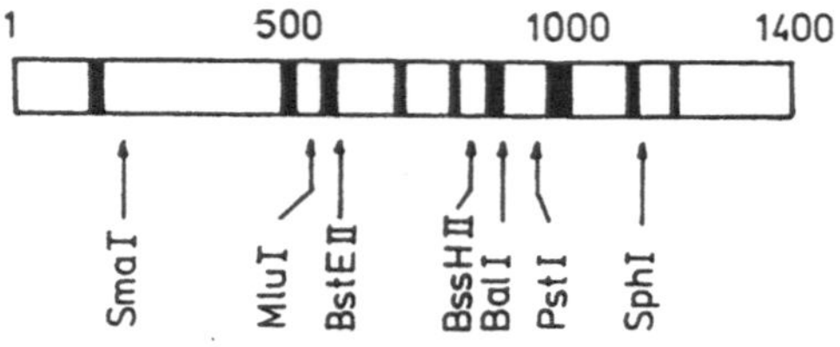

Scheme 7: The gene for *Rhodospirillum rubrum* Rubisco contains a limited number of regions (shaded) encoding the amino acid residues of the active site. These are bracketed by the indicated naturally occurring and engineered unique restriction sites. These sites were used to excise the wild-type sequence and to reinsert the mutant sequences. Thus, the *MluI* - *BstEII* cassette was used to create the mutations at lys191, asp193 and glu194, while the *BssHII-BalI* cassette was used to create mutations at his287.

incorporated into the replacement cassette so as to distinguish the mutant from the wild-type. Each mutant is subsequently sequenced in both directions through the restriction sites constituting the cassette.

Mutations at lysine 191

Two mutations have been made at this position, glu191 and cys191. The former was first prepared by Estelle et al., (1985). In agreement with these authors, no overall carboxylation could be detected with glu191 (Table 2). Glu191 did not form a stable quaternary complex (Table 1). The mutant cys191 also lacked any detectable overall activity. Neither mutant catalyzed the exchange of the C-3 proton (Table 2). However, glu191 catalyzed the decarboxylation of the six-carbon intermediate both in the absence and the presence of 20mM Mg^{2+} (data not shown). The ability of glu191 to catalyze the decarboxylation reaction in a similar manner to the decarbamylated wild-type suggests that the structure of the active-site has not been grossly distorted by the replacement of a lysyl residue with a glutamyl residue. Certainly these mutations establish that lys191 is not obligatorily involved in catalyzing the decarboxylation of the six-carbon intermediate. The properties of glu191 are consistent with (but do not prove) the carbamate on lys191 being an essential ligand to the metal ion and with the postulate that the carbamate-metal ion plays an essential role in enolization and hydrolysis.

Table 2: The relative rates of overall carboxylation and enolization by wild-type and mutant Rubiscos

Enzyme	Overall Carboxylation[a] (%)	Enolization[b] (%)
wild type	100	100
glu191	~0	~0
cys191	~0	~0
asn193	~0	~0
gln194	~0	~0
asn287	1	1
lys287	0.2	0.2
thr287	~0	~0

a measured as incorporation of $^{14}CO_2$ into acid-stable material
b measured as 3H loss from [3-3H] ribulose bisphosphate

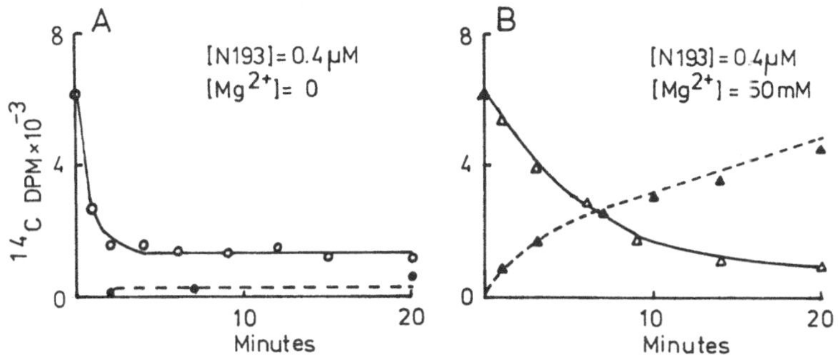

Fig. 4: (A) The decarboxylation of the six-carbon reaction-intermediate catalysed by the mutant asn193 in the absence of Mg^{2+}.
(B) The hydrolysis of the six-carbon reaction-intermediate catalysed by the mutant asn193 in the presence of 50mM Mg^{2+}. The open symbols represent borohydride stabilizable ^{14}C (i.e. six-carbon reaction-intermediate) while the closed symbols represent acid-stable ^{14}C (i.e., 3-P-glycerate).

A mutation at aspartate 193

Thus far, only one mutation has been made at this position, asparagine replacing wild-type aspartate. This mutant did not catalyze any detectable overall carboxylation nor did it catalyze the exchange of the C-3 proton with the medium, even when the [Mg^{2+}] was raised to 100mM (Table 2). In the absence of Mg^{2+}, asn193 catalyzed the decarboxylation of the six-carbon intermediate as expected (Fig. 4A). In the presence of Mg^{2+}, asn193 also catalyzed the forward hydrolysis of the six-carbon intermediate (Fig. 4B). However, compared with the wild-type, asn193 required a 10-fold greater [Mg^{2+}] to achieve the forward hydrolysis of the six-carbon intermediate (Fig. 5). Paradoxically, asn193 did not form a very stable quaternary complex (Table 1). Under conditions of elevated [Mg^{2+}] asn193 catalyzed the hydrolysis reaction but not the enolization reaction. This suggests that the coordination to the metal ion necessary to catalyze the hydrolysis of the six-carbon reaction intermediate is different from that required to bring about the enolization of ribulose bisphosphate. The results with asn193 are consistent with the notion that asp193 provides a ligand to the metal ion that is essential for enolization to occur.

Mutations at glutamate 194

The two mutations at position 194, gln194 and val194 did not catalyse any detectable overall carboxylation nor did they catalyze the exchange of the C-3 proton with the medium (Table 2). In the absence of Mg^{2+}, both mutants catalyzed the decarboxylation of the six-carbon intermediate (Fig. 6). In the presence of elevated [Mg^{2+}] (50mM) the decarboxylation reaction was markedly inhibited (Fig. 6). However, unlike the situation with asn193, little or no increase in acid stable ^{14}C radioactivity occurred in the presence of elevated [Mg^{2+}], so that these mutants have either a very limited or no capacity whatsoever to catalyze the forward hydrolysis of the six-carbon reaction intermediate. This uncertainty reflects the rather limited dynamic range of the assays involving the six-carbon intermediate.

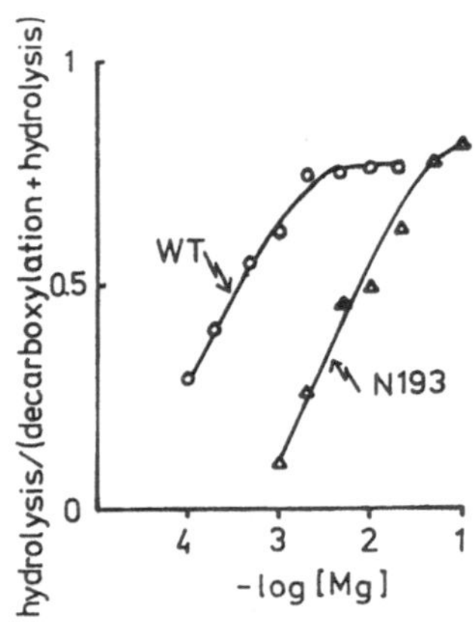

Fig. 5: The partitioning of the six-carbon reaction-intermediate between decarboxylation and hydrolysis as a function of $[Mg^{2+}]$ for the wild-type and asn193 mutant Rubiscos.

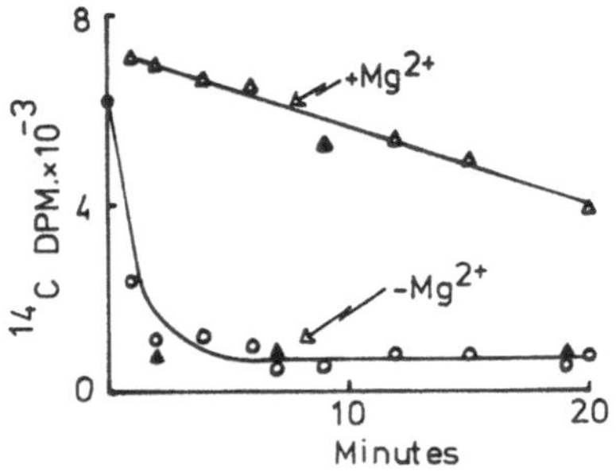

Fig.6: The decarboxylation of the six-carbon reaction-intermediate catalyzed by the mutant gln194 in the absence of Mg^{2+} (open circles) and in the presence of 50mM Mg^{2+} (open triangles). Note that there was little or no increase in the quantity of acid-stable ^{14}C (closed triangles) in presence of Mg^{2+} indicating that decarboxylation predominated over hydrolysis.

Mutations at histidine 287

Of the three mutations at this position, two (lys287 and asn287) showed detectable but limited overall carboxylase activity, while the third, thr287, did not (Table 2). Asn287 formed a relatively stable quaternary complex (Table 1). Both asn287 and lys287 catalyzed the loss of the C-3 proton to the medium at rates that were proportional to their ability to catalyze the overall reaction (Table 2). In the absence of Mg^{2+}, asn287 catalyzed the decarboxylation of the six-carbon intermediate at rates that were 10-20%

of that of the wild-type. Even in the presence of elevated [Mg^{2+}], however, the decarboxylation reaction predominated over hydrolysis so that little or no increase in acid-stable ^{14}C was observed (data not shown), despite the fact that in order for any overall carboxylation to have occurred, some hydrolysis must have taken place. This paradoxical result again reflects the limited sensitivity of the assay.

REFERENCES

Andrews, T.J. and Lorimer, G.H., 1987, in "Plant Biochemistry", M.D. Hatch, ed., Academic Press, N.Y.

Estelle, M., Hanks, J., McIntosh, L. and Somerville, C.R., 1985, *J. Biol. Chem.* 260; 9523.

Fraij, B. and Hartman, F.C., 1982, *J. Biol. Chem.* 257; 3501.

Fraij, B. and Hartman, F.C., 1983, *Biochemistry* 22; 1515.

Gutteridge, S., Parry, M.A.J., Schmidt, C.N.G. and Feeney, J., 1984, *FEBS Letters* 170;355.

Hartman, F.C., Milanez, S and Lee, E.H., 1985, *J. Biol. Chem.* 260; 13968.

Herndon, C.S. and Hartman, F.C., 1984, *J. Biol. Chem.* 259; 3102.

Lee, E.H., Stringer, C.D. and Hartman, F.C., 1986, *Proc. Nat. Acad. Sci., USA*, 83; 9383.

Lorimer, G.H., Andrews, T,J., Pierce, J. and Schloss, J.V., 1986, *Phil. Trans. R. Soc. Lond. B* 313; 397.

Narang, F., McIntosh, L. and Somerville, C.R., 1984, *Mol. Gen. Genet.* 193; 220.

O'Leary, M.H., Jaworski, R.J. and Hartman, F.C.,1979, *Proc. Nat. Acad. Sci., USA* 76; 673.

Pierce, J., 1986, *Plant Physiol.* 81;

Pierce, J., Andrews, T.J. and Lorimer, G.H., 1986, *J. Biol. Chem.* 261; 10248.

Pierce, J., Lorimer., G.H. and Reddy, G.S., *Biochemistry* 25; 1636.

Saver, B.G. and Knowles, J.R., 1982, *Biochemistry* 21; 5398.

Schneider, G., Lindqvist, Y., Branden, C.I. and Lorimer, G.H., 1986, *EMBO J.* 5; 3409.

Somerville, C.R. and Somerville, S.C, 1984, *Mol. Gen. Genet.* 193; 214.

Wells, J.A., Vasser, M. and Powers, D.B. (1983) *Gene* 34; 315.

THE RUBISCO LARGE SUBUNIT BINDING PROTEIN - A MOLECULAR CHAPERONE?

R.John Ellis and Saskia M.van der Vies

Department of Biological Sciences
University of Warwick
Coventry CV4 7AL, U.K.

Sean M.Hemmingsen

Plant Biotechnology Institute
110 Gymnasium Road
Saskatoon, Saskachewan
Canada S7N OW9.

INTRODUCTION

Evidence is accumulating for the existence of a class of proteins whose function it is to ensure that the folding and assembly of other proteins into oligomeric structures occurs correctly (Pelham, 1986). We propose to term such proteins 'molecular chaperones', because their role is to prevent improper interactions, which may occur when hydrophobic or charged surfaces normally involved in protein-protein interactions are exposed during synthesis, and to disassemble insoluble aggregated structures which form during stresses such as heat shock.

The term 'molecular chaperone' was first used by Laskey et al (1978) to describe nucleoplasmin, a soluble nuclear protein which is required for the correct assembly of nucleosomes from DNA and histones in Xenopus extracts. Nucleoplasmin is required only for assembly and does not itself form part of the nucleosome, nor does it carry any information specific for nucleosome assembly; this information resides in the histones. It is part of our definition of molecular chaperones that they do not form part of the final structure nor do they necessarily possess information specifying assembly.

Table 1 presents a list of proteins which we suggest can be regarded as molecular chaperones. In this report we summarize recent data on the Rubisco large subunit binding protein, which we have suggested is a molecular chaperone that ensures the correct assembly of large subunits with the small subunits of Rubisco in higher plants (Musgrove and Ellis, 1986).We regard the evidence for this proposal as strong but not conclusive. It is important to pursue this hypothesis because attempts to produce mutant forms of higher plant Rubisco by expressing cloned genes in E.coli have so far been hampered by the failure of the large subunits to assemble with small subunits to produce active enzyme (Gatenby et al, in press; van der Vies et al, 1986). The emphasis of our research is on the characterisation of the binding protein so that cDNA sequences encoding this protein can be expressed in the same bacterial cells as the Rubisco large and small subunit sequences; in this way we hope to rescue the assembly of Rubisco in E.coli.

Table 1. Examples of Molecular Chaperones

Example	Proposed Function	Authors
Nucleoplasmin	Nucleosome assembly	Laskey et al, 1978
Scaffolding protein	T4 phage assembly	Laemmli & Favre 1980
Immunoglobulin heavy chain binding protein ('BiP')	Immunoglobulin assembly	Munro & Pelham, 1986
Heat shock protein (hsp-70)	Disassembly of insoluble nuclear aggregates	Lewis & Pelham, 1985
Rubisco large subunit binding protein	Rubisco assembly	Barraclough & Ellis, 1980.

DISCOVERY OF THE RUBISCO LARGE SUBUNIT BINDING PROTEIN

Barraclough and Ellis (1980) observed that the majority of the Rubisco large subunits synthesized by isolated intact chloroplasts of *Pisum sativum* do not comigrate with the Rubisco holoenzyme on non-denaturing polyacrylamide gels, but comigrate instead with another abundant stromal protein of higher Mr than Rubisco. Comigration of the labelled large subunits with the unlabelled stained protein is exact, indicating that the two are bound together, but this binding is non-covalent since it can be disrupted by detergent. A recent repetition of these observations is shown in Figure 1. It is important to note that it is only newly-synthesized large subunits that are attached to the binding protein. These newly-synthesized large subunits can be visualized by autoradiography but not by staining, whereas the reverse is true for the binding protein which is not synthesized in isolated chloroplasts. Thus the molar ratio of binding protein to attached large subunits is very high and it is this aspect of phenomenon that prompts us to add a question mark to our title.

PURIFICATION AND PROPERTIES

The Rubisco large subunit binding protein has been purified from *Pisum sativum* (Hemmingsen and Ellis, 1986) and *Hordeum vulgare* (Musgrove et al, 1987). The binding protein occurs as an oligomer of apparent Mr about 720 000, and is composed of equal amounts of two different subunits, termed alpha and beta. These subunits have apparent Mr values of 61 000 (alpha) and 60 000 (beta), suggesting an oligomer composition of $\alpha_6\beta_6$. The oligomer from *Hordeum vulgare* and *Triticum aestivum* has slightly large subunits than that from *Pisum sativum*. The two subunits from *Pisum sativum* are immunologically distinct, show different partial protease digestion patterns, and have different aminoterminal sequences (Figure 2). However we suspect that the two subunits are related because some sequence homology can be detected if gaps are introduced (Figure 2). Neither subunit is synthesized by isolated chloroplasts and higher molecular weight precursors are synthesized when cytoplasmic leaf polysomes are translated in a

RUBISCO LARGE SUBUNITS NEWLY–SYNTHESIZED BY INTACT ISOLATED PEA CHLOROPLASTS ARE BOUND TO ANOTHER PROTEIN

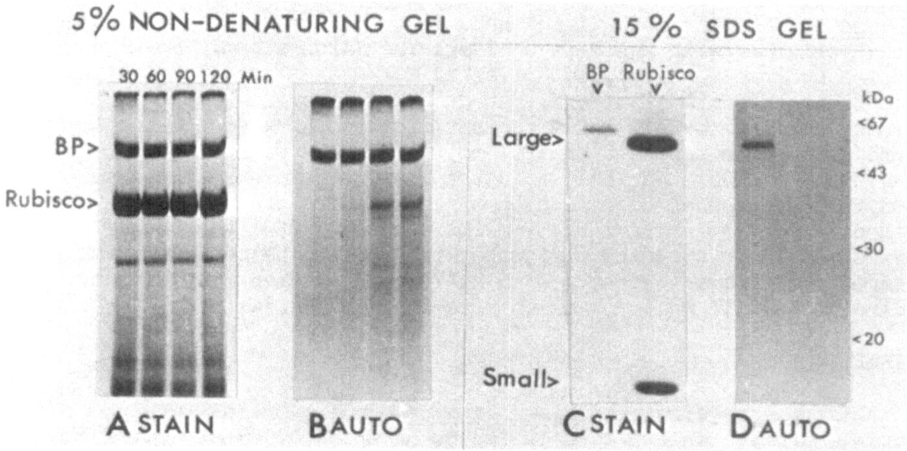

Fig.1. Binding of newly-synthesized Rubisco large subunits to another stromal protein. Isolated intact chloroplasts of Pisum sativum were illuminated at 20^{0}C in sorbitol resuspension medium with S35-methionine. Samples were removed at intervals, the chloroplasts lysed in hypotonic buffer, and the stromal fractions electrophoresed on a non-denaturing polyacrylamide gel. The gel was stained in Coomassie Blue (A) and autoradiographed (B). The staining bands marked BP and Rubisco were excised from the 30 min track and analyzed separately on an SDS polyacrylamide gel. The SDS gel was stained (C) and autoradiographed (D). Symbols: BP, binding protein; Rubisco, holoenzyme of ribulose bisphosphate carboxylase-oxygenase; large and small, large and small subunits of Rubisco. This experiment was performed by R.A.Johnson.

alpha	AAKDIAFDQHSRSAMQAGID??ADAVGL
beta	AKELHFNKDGSAIRKLQNGVNKLADLVG
alpha	AK F SA--- Q G AD VG
beta	-AK F -SA Q G AD VG

Fig.2. Aminoterminal sequences of the alpha and beta subunits of the binding protein from Pisum sativum (Musgrove et al, 1987). The lower pair of sequences show the homology that appears if gaps, indicated by the dashes, are introduced (Hemmingsen et al, 1987).

cell-free system. We suggest that the two subunits are encoded by at least two nuclear genes which may be evolutionarily related by duplication.

The binding protein oligomer is reversibly dissociated by MgATP but not by other nucleotides (Bloom et al, 1983; Hemmingsen et al,1986; Musgrove et al, 1987). Dissociation produces both monomers, but attempts to detect the phosphorylation of either subunit have been unsuccessful (Hemmingsen et al, 1987). The binding protein subunits can be distinguished from the phosphorylated 67 kDa stromal protein (J.Bennett, personal communication).

The fate of the bound large subunit on treatment of the binding protein oligomer with MgATP is unclear. The large subunit moves near the top of a sucrose density gradient after MgATP treatment, suggesting that it occrs in the form of dimers, but on non-denaturing gradient polyacrylamide gels this large subunit migrates as a continuous smear extending over an apparent Mr of 2 to 4 x 10^5. We do not understand this behaviour and we are not able to determine whether these 'released' large subunits are attached to subunits of the binding protein. We have not managed to confirm the report by Milos and Roy (1984) that newly-synthesized large subunits comigrate with Rubisco holoenzyme after treatment of stromal extracts with MgATP. The precise role, if any, of the binding protein in Rubisco assembly remains to be clarified.

DISTRIBUTION

Antibodies against the Pisum sativum binding protein detect cross-reacting material in extracts of leaves of spinach, wheat, barley and tobacco, as well as in extracts of leucoplasts from the endosperm of Ricinus communis. Analyses of extracts prepared from leaves of the C4 plant Zea mais show that the binding protein occurs in the bundle sheath cells but not in the mesphyll cells(Figure 3). There is thus a correlation between the occurrence of the binding protein and the occurrence of Rubisco in higher plants which is consistent with our hypothesis that the binding protein is involved in Rubisco assembly.

Cross-reacting material has also been detected in extracts of a range of bacterial species (Hemmingsen et al, 1987).

CLONING

A cDNA sequence encoding the alpha subunit of the binding protein from Ricinus communis has been isolated from a lambda gt 11 cDNA expression library (Hemmingsen et al ,1987). The recombinant plasmid (pG3BP2) contains 90% of the mature polypeptide nucleotide sequence and its aminoterminal region shows homology with that of the alpha subunit from Pisum sativum. There is no apparent sequence homology with either the large or the small subunit of Rubisco, the immunoglobulin heavy chain binding protein or the hsp 70 protein. However there is good homology both to a protein from E.coli and to a different protein from mycobacteria (Hemmingsen et al, 1987); no function has been attributed to either protein.

We have screened alambda gt 11 cDNA expression library of Triticum aestivum with antibodies raised against the binding protein from Pisum sativum. We have isolated a cDNA fragment of 1.85 kbp which shows strong homology with the alpha subunit of the binding protein from Ricinus communis, as can be demonstrated by DNA/DNA hybridization (Figure 4). The size of the cDNA fragment suggests that it might encode the complete nucleotide sequence for the alpha subunit precursor polypeptide of the binding protein. We are currently attempting to isolate full-length cDNA for the beta subunit from Triticum aestivum, so that both subunits can be expressed in the same bacterial cells that are synthesizing Rubisco subunits from wheat (van der Vies et al, 1986; Gatenby et al, in press).

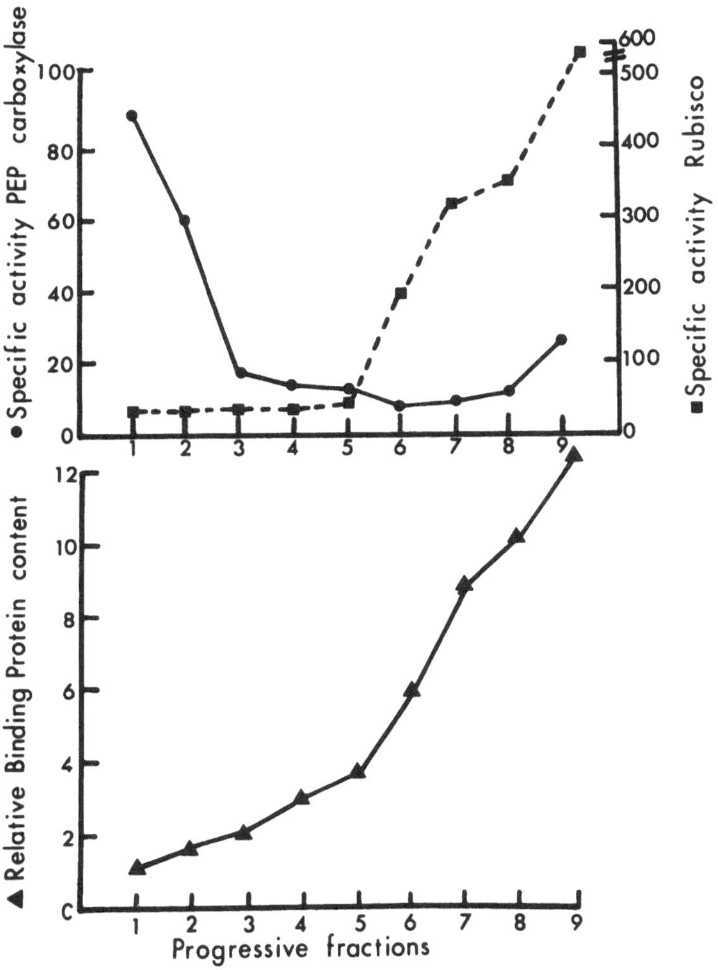

Fig.3. The occurrence of binding protein in extracts of Zea mais leaves containing increasing amounts of bundle sheath cell protein. Extracts of maize leaves were prepared by sequential grinding (Kruger et al, 1986) so that successive extracts were enriched in the contents of the bundle sheath cells. The extracts were assayed for Rubisco, PEP carboxylase and the binding protein. The binding protein was measured by means of an ELISA technique, using total maize leaf extract to construct a calibration curve. The value for the content of binding protein in the first extract was normalised to 1.0. We are grateful to N.J.Kruger for help with this experiment.

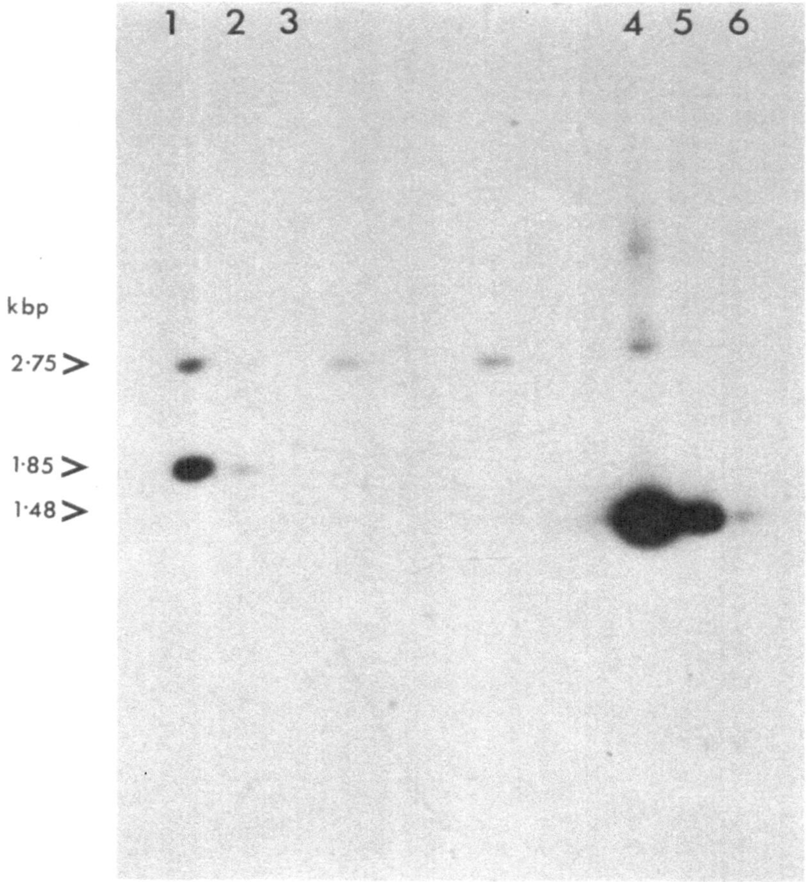

Fig.4. cDNA sequence homology of the alpha subunit of the binding protein from Triticum aestivum and Ricinus communis. A wheat leaf lambda gt 11 cDNA expression library, kindly supplied by T.A.Dyer and C.Raines, was screened with antibody raised against binding protein from Pisum sativum. Phage DNA was isolated from a plaque-purified recombinant and digested with Eco R1. The 1.85 kbp fragment was ligated into pGEM4 to give pSV8. The Figure shows an autoradiograph of a Southern blot in which pSV8 (lanes 1-3) and pG3BP2 (lanes 4-6) DNA was digested with Eco R1 and analyzed on a 1% agarose gel. A P32-labelled 1.48 kbp DNA fragment, isolated from pG3BP2 and containing the sequence for the alpha subunit from Ricinus communis (Hemmingsen et al, 1987) was used for hybridization. Lanes 1 and 4, 0.5 ug DNA; lanes 2 and 5, 0.25 ug DNA; lanes 3 and 6, 0.1 ug DNA. The filter was washed in 0.1 SSC containing 0.1 % SDS at 65^0C for 2 h. The 2.75 kbp fragment represents contaminating pGEM4 vector DNA.

SPECULATIONS

We suggest that we have stumbled on a representative of the class of proteins that Pelham (1986) has proposed is involved in the assembly and disassembly of oligomeric proteins. Such molecular chaperones would be expected to occur in all cells which synthesize oligomeric proteins whose subunits have the potential to interact incorrectly. The Rubisco large subunit binding protein may have evolved from this general class of proteins to mediate Rubisco assembly specifically. Such an origin would account for the immunological and sequence similarities that we have found between the large subunit binding protein from higher plants and certain bacterial proteins.

If this general argument is correct, there may exist other molecular chaperones involved transiently in the assembly of oligomeric proteins especially where the constituent subunits originate in different subcellular compartments, where problems of folding, stability and transport are more acute. We recommend the use of non-denaturing gradient polyacrylamide gels to detect the possible involvement of binding proteins in the assembly of other oligomeric proteins, because this technique permits the detection of non-covalently linked protein complexes with high resolution.

REFERENCES

Barraclough,R. and Ellis,R.J., 1980, Protein synthesis in chloroplasts IX. Assembly of newly-synthesized large subunits into ribulose bisphosphate carboxylase in isolated intact chloroplasts, Biochim. Biophys.Acta, 608:19.

Bloom,M.V., Milos,P. and Roy,H., 1983, Light-dependent assembly of ribulose bisphosphate carboxylase, Proc.Natl.Acad.Sci.USA, 80:1013.

Gatenby,A.A., van der Vies,S.M. and Rothstein,S.J. (in press), Expression of both the maize large and wheat small subunit genes of ribulose bisphosphate carboxylase in Escherichia coli, Eur.J.Biochem.

Hemmingsen,S.M., Denis,D.T. and Ellis,R.J., 1987, The Rubisco large subunit binding protein, Proc.Conf,'Rubisco 87', Tucson, Arizona (eds.Bohnert,H.J. and Jensen,R.).

Hemmingsen,S.M. and Ellis,R.J., 1986, Purification and properties of the ribulose bisphosphate carboxylase large subunit binding protein, Plant Physiol. 80:269.

Kruger,N.J., Hemmingsen,S.M. and Denis,D.T., 1986, Distribution of pyrophosphate:fructose 6-phosphate phosphotransferase in maize leaves, Plant Physiol.81:12.

Laemmli,V.K. and Favre,M., 1980, Maturation of the head of bacteriophage T4, J.Mol.Biol. 80:575.

Laskey,R.A., Honda,B.M., Mills,A.D. and Finch,J.T., 1978, Nucleosomes are assembled by an acidic protein which binds histones and transfers them to DNA, Nature Lond.275:416.

Lewis,M.J. and Pelham,H.R.B., 1985, Involvement of ATP in the nuclear and nucleolar functions of the 70 kd heat shock protein, EMBO.J.4:3137.

Milos,P. and Roy,H., 1984, ATP-released large subunits participate in the assembly of RuBP carboxylase, J.Cell.Biochem.24:153.

Munro,S. and Pelham,H.R.B., 1986, An hsp 70-like protein in the ER: identity with the 78 kd glucose-regulated protein and the immunoglobulin heavy chain binding protein, Cell 46:291.

Musgrove,J.E. and Ellis,R.J., 1986, The Rubisco large subunit binding protein, Phil.Trans.R.Soc.Lond.B313:419.

Musgrove,J.E., Johnson,R.A. and Ellis,R.J., 1987, Dissociation of the ribulose bisphosphate carboxylase large subunit binding protein into dissimilar subunits, Eur.J.Biochem. 163:529.

Pelham,H.R.B., 1986, Speculations on the functions of the major heat-shock and glucose-regulated proteins, Cell, 46: 959.
van der Vies,S.M., Bradley,D. and Gatenby,A.A., 1986, Assembly of cyanobacterial and higher plant ribulose bisphosphate carboxylase subunits into functional homologous and heterologous enzyme molecules in E.coli., EMBO.J. 5:2439.

THE MOLECULAR PHOTOPHYSIOLOGY OF GREENING IN ETIOLATED PEA SEEDLINGS

Winslow R. Briggs

Department of Plant Biology
Carnegie Institution of Washington
290 Panama St.
Stanford, CA 94305 U. S. A.

INTRODUCTION

The past decade has seen a rapid increase in the pace of research probing molecular aspects of chloroplast development and regulation. Prominent in this increase are studies of changes occurring during the greening of etiolated seedlings, and changes in green plants placed in darkness and subsequently returned to light. Testimony to this activity is to be found in the several recent reviews devoted partially or wholly to such studies (Thompson et al., 1985; Tobin and Silverthorne, 1985; Sharma, 1985; Schäfer and Briggs, 1986; Kuhlemeier et al., 1987; Silverthorne and Tobin, 1987). Since major changes during greening involve photosynthetic proteins encoded in nuclear genes, synthesized in the cytoplasm and transported into the chloroplasts, it is not surprising that workers have focused on these proteins, their mRNAs, and their genes. Included prominently among these are the small subunit of ribulose bisphosphate carboxylase/oxygenase (SS), the light-harvesting, chlorophyll *a/b*-binding protein (LHCP); NADH-proto-chlorophyllide oxidoreductase (Pchl-reductase), and ferredoxin (FD).

The purpose of the present paper is to review studies on the molecular photophysiology of greening of etiolated pea seedlings with emphasis on steps beyond transcription. The reader is referred to the above-cited reviews for more comprehensive treatments, and in particular to the review by Kuhlemeier et al. (1987) which emphasizes the progress made in under-standing photoregulation at the gene level through the use of transgenic plants. Although there is increasing evidence for involvement of a blue light photoreceptor in regulation of levels of some of the chloroplast proteins (Kaufman et al., 1985b; Fluhr and Chua, 1986), emphasis here will be on regulation through phytochrome.

STUDIES OF mRNA ABUNDANCE

Thompson et al. (1983) reported results of screening a cDNA library for clones corresponding to mRNAs which were under some form of light regulation in etiolated pea and mung bean seedlings (We'll restrict our remarks here to results with pea). They tested by dot blot hybridization the effect of various light treatments on the abundance of the mRNAs corresponding to 23 different clones. These treatments included a saturating pulse of red light (or red followed by far red) on days four, five, and six after imbibition with or without additional white light 24 hours after the terminal red treatment; or white light alone applied to seven-day-old dark-grown seedlings for various durations.

Several patterns of response were apparent from inspection of the dot blots. Certain mRNAs were fully induced by the red light treatment alone, for example, that for the LHCP, while others showed significant additional induction by subsequent white light, for example that for the SS. All cases which showed induction by red light also showed far red reversibility. One mRNA, recognized by a probe designated pEA 207, showed clear negative regulation by red light, a change that was also far-red reversible. Not unexpectedly several mRNAs showed no phytochrome regulation at all though one showed clear induction by white light nonetheless. These mRNAs provided the basis for the detailed photobiological studies to be reviewed below. Results with five clones are selected here for detailed consideration as they illustrate all of the major response patterns found: clones for SS mRNA and LHCP mRNA; a clone designated pEA 46, subsequently identified by Dobres et al. (1987) as ferredoxin (FD); and two clones designated pEA 170 and 207 respectively. Brief mention will be made of four others: pEA 13, pEA 25, pEA 215, and pEA 277.

Fluence-response studies

It is well known that the range of fluences of red light over which etiolated plants may respond physiologically can extend over eight orders of magnitude. In many cases fluence-response studies reveal two distinct ranges: A very low fluence (VLF) range where saturation may occur with only one percent or less of the phytochrome in its far red-absorbing form, Pfr, and a low fluence (LF) range where saturation requires the bulk of the phytochrome to be in the Pfr form. Since far red light alone may induce phototransformation of as much as 3 percent of the phytochrome to Pfr at photostationary equilibrium (depending upon the spectral quality of the far red light source used), it is obvious that only LF responses should show far red reversibility, and that VLF responses should themselves be inducible by far red light (See Briggs et al., 1985). A first question, therefore, was whether both LF and VLF responses could be detected in the phytochrome regulation of mRNA abundance.

Kaufman et al. (1984, 1985a) investigated fluence-response relationships for red light induction of a number of the pea mRNAs previously identified by Thompson et al. (1983) as being under phytochrome control. The protocol was simplified to a single red light pulse of varying fluence on day 6, 24 hours of darkness, and then harvest and RNA extraction. Under these conditions, these workers obtained all of four possible responses: First, the SS (Kaufman et al., 1985) and FD mRNAs (Kaufman et al., 1986a) increased only after pulses in the LF range, and showed full far red reversibility - classic phytochrome responses. Second, by contrast, the LHCP mRNA, showed both an LF and a VLF component; as one would predict, only the LF component showed far-red reversibility. Third, mRNA recognized by the clone designated pEA 170 showed only a VLF response, and a complete lack of far-red reversibility. Finally, the mRNA recognized by clone pEA 207, which showed down regulation in the protocol used by Thompson et al. (1983) showed no light regulation whatsoever following a single pulse.

Kaufman et al. (1985a) also measured induction by 24 hours of white light alone on day 8, and obtained fluence-response curves for plants given a red light pulse as before, on day 6, 24 hours of darkness, and an additional 24 hours of white light prior to harvest. The results, summarized in Table 1, provide additional evidence that the situation is an extremely complicated one. The LHCP mRNA response is saturated by a single red treatment, and 24 hours of white light alone give no further increase. However, after 24 hours of white light, a VLF component is no longer detectable. By contrast, neither the SS nor the ferredoxin mRNAs are fully induced by a single pulse of red light. Both show significant increases in abundance in white light, with the FD increase being the largest. Their response remains in the LF range, but with significantly decreased sensitivity.

The mRNAs recognized by clones pEA170 and pEA 215 behave in an entirely different manner (Table 1). Although both are up-regulated by a pulse of red light, both are strongly down-regulated by continuous white light alone. A red light-induced increase is still detectable after the

Table 1. Comparison of induction ratios and response types for mRNAs from pea plants given various light treatments or darkness only

Clone	Induction Ratio		Response Type	
	Red/Dark	White/Dark	Red alone	Red plus White
LHCP	5.0	5.6	VLF plus LF	LF only
SS	5.2	9.7	LF only	LF only[a]
FD	1.8	10.0	LF only	LF only[a]
pEA 170	3.2	0.5	VLF only	LF only
pEA 215	1.6	0.2	LF only	VLF plus LF
pEA 207	1.1	0.5	No response	No response

[a]Threshold increased by one order of magnitude, relative to red alone.

white light treatment, indicating that a photoreceptor other than phytochrome may be involved in this down regulation. Note, however, that the nature of the phytochrome responses for these mRNAs changes: pEA 170, no longer shows sensitivity in the VLF range, responding only to LF fluences, and pEA 215, showing only a LF response to red alone reveals a VLF component if the RNA is extracted following white light treatment. Finally, pEA 207 mRNA shows no sensitivity to a single pulse of red light whether or not measured after subsequent white light treatment, but shows down regulation in white light alone - again indicative of participation by a photoreceptor other than phytochrome.

Time-course studies

The fluence-response studies discussed above were all done at a single time point, either 24 hours following red light treatment or after an additional 24 hours of white light. Kaufman et al. (1986) also investigated the time course for the changes in mRNA abundance following a single red light pulse sufficient to saturate the LF responses. Here, four widely different responses were obtained. First, a number of mRNAs began increasing gradually without a discernible lag period and continued to increase over most of the following 24 hour dark period. In this group were both SS and LHCP mRNAs. Another group including pEA 170 only increased after a lag of about 16 hours; and a third group including FD increased rapidly, reaching their maximum values within 2 hours and remaining there for the rest of the 24 hour dark period. Finally, pEA 13 was unique among the mRNAs in that it increased in the darkness unlike the others which remained unchaged. Red light accelerated the increase, followed by a plateau, so that by 24 hours of darkness the dark control level had caught up with it

Escape from far red reversibility

Another parameter investigated for the various phytochrome-regulated mRNAs was their escape from far red reversibility with increasing time in darkness following an inductive red pulse (Kaufman et al., 1986). These workers investigated escape over a period of 7 hours following red treatment sufficient to saturate the LF responses. Here again several widely different patterns were obtained. A number of the mRNAs, among them that for the LHCP, began to escape immediately following red light treatment, without showing any lag. In the case of the LHCP, Horwitz et al. (1987) subsequently showed that this gradual escape continued over most of the ensuing 24 hour dark period. By contrast, the SS mRNA showed no escape for about 2 hours and then escaped completely by the end of 7 hours. Surprisingly, the FD mRNA showed no escape whatsoever over the 7 hour period investigated, suggesting that the continued presence of Pfr is required to maintain it in the elevated state it reaches within 2 hours of red treatment.

Independence of photoregulatory properties for different mRNAs

It is clear from the above discussion that fluence-response relationships, induction kinetics, and escape kinetics must be independently regulated for these photoregulated mRNAs. Note that while the SS and FD mRNAs show identical fluence-response relationships, they show very different time courses; and while SS and LHCP mRNAs show similar time courses, they have very different fluence-response relationships. Finally, the three mRNAs show widely divergent kinetics for escape from far red reversibility.

Obviously regulation of mRNA abundance can be either at the level of transcription or at the level of mRNA stability or both. There are now several reports of regulation of transcription of SS and LHCP genes in peas, as detected by run-on transcription measurements with isolated nuclei (Gallagher and Ellis, 1982; Ellis et al., 1984; Gallagher et al., 1985). Hence, although detailed photobiological investigations are still lacking, it is reasonable that the SS and LHCP mRNA changes seen here are at least partially transcriptionally regulated. If such regulation is indeed transcriptional, then it should be possible to determine what structural elements of the genes are responsible for these various photobiological properties and how they differ from gene to gene. Ingenious use of artificial gene constructs in transgenic plants holds considerable promise as a way to address these questions. Information on the effect of Pfr on the stability of these mRNAs will also be of great assistance.

Partitioning between nucleus and cytoplasm

The above studies by Kaufman et al. (1984, 1985a, 1985b, 1986) all utilized total RNA and a slot blot technique for quantification. This procedure does not distinguish between processed mRNA in the cytoplasm, which is translatable into polypeptides, and unprocessed mRNA still in the nucleus, which is not. It is conceivable that some phytochrome regulation of gene expression might occur at the level of processing and transport out of the nucleus. One could detect such light regulation in one of two ways: First, one could compare light-induced changes as determined by slot blot hybridization with those determined by in vitro translation. A six-fold increase in translatable mRNA following red light treatment, for example, accompanied by a very much smaller increase in total hybridizable mRNA, might suggest some sort of regulation at the processing-transport level (although other interpretations are possible). In practice, where the two techniques have been compared, for example for the down regulation of the mRNA for phytochrome itself (Colbert et al., 1983, 1985), no such discrepancies have been detected. Such negative cases do not eliminate the possibility that such a mechanism might function for some mRNA species, however.

The second approach to detecting regulation at the transport-processing level is to measure separately nuclear and cytoplasmic mRNA fractions. A light-induced change in the ratio of the two could be interpreted as photoregulation at this step in the transduction chain. Sagar et al. (1987a) have examined the effects of various light treatments on this ratio for the mRNAs recognized by six different cDNA clones. Light treatments were selected to induce mRNA increases to several different levels prior to cell fractionation, RNA extraction, and slot blot assay. Dark control mRNA levels were set at unity for both nucleus and cytoplasm and normalized values for the increments in cytoplasmic and nuclear mRNA plotted against each other to determine whether light treatment had any effect of the nuclear/cytoplasmic ratio. Not unexpectedly this ratio was unaffected by light treatment for pEA 207 mRNA, eliminating the possibility that phytochrome could regulate the expression of the gene coding for pEA 207 mRNA at the level of processing and transport without any effect on its overall abundance. For the other mRNAs, showing light-induced increases, the plots were all straight lines. In no case did a any particular light treatment alter the ratio. The slopes of these lines are shown in Table 2.

There are three points to be stressed from these experiments. First, there is no evidence for an effect of light on the partitioning of any of these mRNAs between nucleus and cytoplasm. Phytochrome is presumably not regulating gene expression for these mRNAs at the processing-transport level. Second, there are large intrinsic differences between the various transcripts with respect to the nuclear/cytoplasmic ratio. The slopes range from 2.44 for clone pEA 25 all the way to 17.5 for FD. The structural basis for these differences is unknown.

The third point to be stressed, and not shown in Table 2, is that it is the ratio in the *increment* which remains constant rather than the total ratio. Expressed differently, not all of the straight lines extrapolate

Table 2. Ratio of nuclear to cytoplasmic fraction of the mRNA increment inducible by various light treatments for 5 phytochrome-regulated mRNAs (after Sagar et al., 1987a)

Clone	Ratio of cytoplasmic to nuclear increment
pEA 25	2.44
LHCP	3.50
SS	4.39
pEA 277	10.3
FD	17.5

through zero. It is known that some of these genes, for example those for SS and LHCP mRNA in pea (Coruzzi et al., 1983; J. C. Watson and W. F. Thompson, unpublished observations) and other plants (e. g. petunia, Dunsmuir et al., 1983a, 1983b) are members of multigene families. If some members of the family for a particular gene were unaffected by light while others were being induced, then one would not expect the slopes to pass through zero.

THE REGULATION OF GREENING

There is considerable evidence that for chlorophyll to remain stable in vivo it must be associated with LHCP polypeptide (Apel and Kloppstech, 1980; Bennett, 1981). Hence it is conceivable that regulation of greening might under certain circumstances be limited by synthesis of LHCP polypeptides, and that LHCP synthesis might in turn be limited by the abundance of LHCP mRNA. That this need not be the case is illustrated by studies with chlorophyll-deficient mutants of maize. These completely lack LHCP proteins although the mRNA for these proteins is present at normal levels (Harpster et al., 1984). On the other hand, mutants lacking carotenoids (and which therefore lose through photodestruction any chlorophyll synthesized) lack both the LHCP protein and its mRNA (Mayfield et al., 1984). Likewise treatment of seedlings with the herbicide norflurazon, which blocks carotenoid synthesis, and therefore produces seedlings functionally identical to those genetically deficient in carotenoid synthesis, also lack either the LHCP or its corresponding mRNA (Mayfield and Taylor, 1984; Oelmüller and Mohr, 1986; ,Oelmüller et al., 1986).

Chlorophyll accumulation versus LHCP mRNA: comparative photobiology

It has been known for some time that a brief pulse of red light followed by a suitable dark period will eliminate a lag period for chlorophyll accumulation during subsequent exposure to white light (See Virgin, 1972). Indeed, this phenomenon is well documented for pea seedlings (Henshall and Goodwin, 1964; Raven and Spruit, 1972a, 1972b; Raven and Shropshire, 1975; Thompson et al., 1983). The action spectrum in pea matches the absorption spectrum of Pr (Raven and Spruit, 1972a), clearly implicating phytochrome as the photoreceptor, and far-red reversibility has been reported to be poor (Raven and Spruit, 1972b) suggesting the presence both of VLF and LF components (Henshall and Goodwin reported full far-red reversibility, but neglected to present any data).

Horwitz et al. (1987) recently completed comparative photobiological studies both on chlorophyll accumulation and on the abundance of the mRNA for the LHCP in dark-grown peas given various irradiation treatments. They first verified that a red light pulse would reduce the lag in subsequent chlorophyll accumulation in white light, showing that VLF pulses were active but less effective than LF pulses. They next compared the time courses for increases in fresh weight, elimination of the lag in chlorophyll synthesis, and increase in LHCP mRNA synthesis following both VLF and LF pulses. In all three cases the kinetics were similar with the change following the VLF pulse smaller than that following the LF pulse. The increases following LF pulses persisted for almost the entire 24 hour dark period while those for chlorophyll accumulation and mRNA increase following LF pulses leveled off at about 8 hours (fresh weight continued to increase). Only the LF component for chlorophyll accumulation showed repeated red - far red photoreversibility, in keeping with the partial photoreversibility reported earlier for accumulation of the LHCP mRNA (Kaufman et al., 1984). Finally, the kinetics of escape from photoreversibility both for chlorophyll accumulation and for mRNA accumulation were virtually identical: both began without apparent lag and continued steadily over the entire 24 hour dark period. Comparison of these escape data with those for mRNA accumulation suggests that removal of Pfr at any time may simply prevent any further increase in mRNA. The continual presence of Pfr is required for the increase to take place.

Thus far the data of Horwitz et al. (1987) are consistent with the hypothesis that the level of LHCP mRNA is limiting for chlorophyll accumulation. However, two kinds of experiments indicate that this hypothesis is incorrect. First, both the fluence-response curve for reduction in the lag in chlorophyll accumulation and that for the increased level of LHCP mRNA at the start of the white light treatment both show VLF and LF components. However, the curve for chlorophyll accumulation is roughly an order of magnitude above that for LHCP mRNA accumulation - it requires ten times as much light to saturate chlorophyll accumulation as it does to saturate the mRNA increase (Horwitz et al., 1987). The similarity in induction and escape kinetics for the changes therefore indicates only coordinate change rather than any causal relationship.

The second type of experiment involved growing the pea seedlings under various fluence rates of continuous red light (log fluence rate from -2.7 to -0.6 μmol m^{-2} s^{-1}). The concentration of chlorophyll increased roughly ten fold from the lowest to the highest fluence rate, but the LHCP mRNA level remained absolutely constant. As with the fluence-response experiments, it was possible to uncouple chlorophyll accumulation completely from steady-state levels of LHCP mRNA. It seems probable, therefore, that chlorophyll accumulation is limited by the synthesis of chlorophyll itself in greening pea seedlings. The nature of this limitation is unknown, although it oould be at the level of the synthesis of the chlorophyll precursor δ-aminolevulinic acid, which is known to limit chlorophyll synthesis (Beale, 1978; see Briggs et al., 1987a). Jenkins and Smith (1985) did a slightly different experiment, growing pea seedlings under constant white light, and then providing supplemental far red to achieve a range of red to far-red ratios between 0.1 and 1.87. They failed to find any effect on the abundance either of the LHCP or the SS mRNAs. Hence under continuous illumination neither relative fluence rate nor the ratio of Pr to Pfr affect the levels of these transcripts.

Studies with the herbicide norflurazon

As mentioned above, seedlings treated with the herbicide norflurazon, an inhibitor of carotenoid synthesis, lack both the LHCP protein and its mRNA. Sagar et al. (1987b) have recently used this herbicide to study its effect on greening processes in pea seedlings grown under a low fluence rate of continuous red light (fluence rate 0.5 μmol m^{-2} s^{-1}). The fluence is permissive for considerable chlorophyll accumulation. Sager et al. (1987b) measured changes both in the levels of chlorophyll over 24 hours in white light and in various mRNAs in the presence or absence of the herbicide. A summary of the results appears in Table 3.

There are several interesting aspects of these experiments. First, the loss of chlorophyll parallels the loss of LHCP mRNA. Kinetic studies for these losses (not shown in Table 3) show that the loss of the mRNA lags at least a couple of hours behind the chlorophyll loss. Second, some mRNAs are more drastically effect than others. Loss of LHCP mRNA is more severe than loss of SS mRNA, and FD mRNA declines only slightly. Robert Elliott (personal communication) has recently found that in peas treated with norflurazon and transferred to white light, transcriptional activity by nuclei isolated after 24 hours is severely decreased for the SS and LHCP mRNAs, but not strongly affected for FD. These results are consistent with those in Table 3 for the abundance of these three mRNAs. Finally, the loss of a chloroplast-encoded mRNA, that for the carboxylase large subunit (LS) parallels that for its nuclear counterpart, that for the small subunit, although their rates of increase in white light differ significantly from each other. These results again emphasize the divergent behavioral patterns of different mRNAs sharing at least in part a common photoregulatory system.

Table 3. Relative changes in several chloroplast components for plant grown under dim red light for 6 days and then transferred to white light (data after Sagar et al., 1987b)[a]

Component measured	minus norflurazon 0 hours	minus norflurazon 24 hours	plus norflurazon 0 hours	plus norflurazon 24 hours
Chlorophyll	100	246	100	17
LHCP mRNA	100	104	100	14
FD mRNA	100	114	100	84
SS mRNA	100	246	100	57
LS mRNA[b]	100	127	100	56

[a]Results normalized to values at the beginning of white light treatment.
[b]LS: large subunit of ribulose bisphosphate carboxylase/oxygenase, a chloroplast-encoded mRNA.

CONCLUSIONS

The evidence discussed above, obtained from pea seedlings indicates the extraordinary complexity of light regulation of the greening process. The fluence requirements for induction, induction kinetics, escape kinetics, and stability in the presence of norflurazon for several nucleus-encoded mRNAs all appear to vary independently despite photoregulation by a common photoreceptor, phytochrome; they partition differently between nucleus and cytoplasm; and at least one of them, that for the LHCP, does not limit the abundance of its translation polypeptide: the abundance of the polypeptide is regulated by Pfr through some other mechanism. possibly a step in the synthesis of chlorophyll itself. These studies complement related molecular physiological studies of phytochrome-regulated greening in barley (Briggs et al., 1987a, 1987b; Mösinger et al., 1987) which also highlight the complexity of these light responses at the molecular level.

Probably the most promising approach to begin disentanglement of this intricate regulatory system will be careful structural studies both of the properties of the 5' and 3' regulatory regions and of the coding regions of the genes involved. The use of transgenic plants in combination with use of the available technologies for precise manipulation of sequences in the regulatory and coding regions of genes to be inserted provide extraordinarily powerful tools for kind of research. Indeed, studies using these tools is already under way in several laboratories (see Kuhlemeier et al., 1987).

ACKNOWLEDGEMENTS

The author is deeply grateful to Drs. Anurag Sagar and James Shinkle for their careful review of this paper. This is Carnegie Institution of Washington Department of Plant Biology Publication No. 985.

REFERENCES

Apel, K., and Kloppstech, K., 1980, The effect of light on the biosynthesis of the light-harvesting chlorophyll a/b protein. Evidence for the requirement for chlorophyll a for the stabilization of the protein, Planta, 150:426.

Beale, S. I., 1978, Δ-aminolevulinic acid in plants: its biosynthesis, regulation, and role in plastid development. Annu. Rev. Plant Physiol., 29:95.

Bennett, J., 1981, Biosynthesis of the light-harvesting chlorophyll a/b protein. Polypeptide turnover in darkness. Eur. J. Biochem., 118:61.

Briggs, W. R., Mandoli, D., Shinkle, J. R., Kaufman, L. S., Watson, J. C., and Thompson, W. F., 1985, phytochrome regulation at the whole plant, physiological, and molecular levels, in "Sensory perception and transduction in aneural organisms," G. Columbetti and P.-S. Song, eds., Plenum, New York, p. 265.

Briggs, W. R., Mösinger, E., Batschauer, A., Apel, K., and Schäfer, E., 1987a, Molecular events in photoregulated greening in barley leaves, in "Molecular Biology of Plant Growth Control," J. E. Fox and M. Jacobs, eds., Alan R. Liss, New York, p. 413.

Briggs, W. R., Mösinger, E. and Schäfer, E., 1987, Phytochrome regulation of greening in barley. Effects on chlorophyll accumulation. Submitted to Plant Physiol.

Colbert, J. T., Hershey, H. P., and Quail, P. H., 1983, Autoregulatory control of translatable phytochrome mRNA levels, Proc. Natl. Acad. Sci. USA, 80:2248.

Colbert, J. T., Hershey, H. P., and Quail, P. H., 1985, Phytochrome regulation of phytochrome mRNA abundance. Plant Mol. Biol., 5:91.

Coruzzi, G., Broglie, R., Cashmore, A., and Chua, N.-H., 1983, Nucleotide sequences of two pea cDNA clones encoding the small subunit of ribulose bisphosphate carboxylase and the major chlorophyll a/b-binding thylakoid polypeptide, J. Biol. Chem., 258:1399.

Dobres, M. S., Elliott, R. C., Watson, J. C., and Thompson, W. F., 1987, A phytochrome regulated pea transcript encodes ferredoxin, Plant Mol. Biol., in press.

Dunsmuir, P., Smith, S. M., and Bedbrook, J., 1983a, The major chlorophyll *a/b*-binding protein of petunia is composed of several polypeptides encoded by a number of distinct nuclear genes, J. Mol. Appl. Genet., 2:285.

Dunsmuir, P., Smith, S. M., and Bedbrook, J., 1983b, A number of different nuclear genes for the small subunit of RuBPCase are transcribed in petunia. Nucleic Acids Res., 11:4177.

Ellis, R. J., Gallagher, T. F., Jenkins, G. I., and Lennox, C. R., 1984, Photoregulation of the biosynthesis of ribulose bisphosphate carboxylase. J. Embryol. Ext. Morph., 83(Suppl.):163.

Fluhr, R., and Chua, N.-H., 1986, Developmental regulation of two genes encoding ribulose-bisphosphate carboxylase small subunit in pea and transgenic petunia plants: phytochrome response and blue-light induction, Proc. Natl. Acad. Sci. USA, 83:2358.

Gallagher, T. F., and Ellis, R. J., 1982, Light-stimulated transcription of genes for two chloroplast polypeptides in isolated pea leaf nuclei. EMBO J., 1:1493.

Gallagher, T. F., Jenkins, G. I., and Ellis, R. J., 1985, Rapid modulation of transcription of nuclear genes encoding chloroplast proteins by light. FEBS Letters, 186:241.

Harpster, M. H., Mayfield, S. P., and Taylor, W. C., 1984, Effects of pigment-deficient mutants on the accumulation of photosynthetic proteins in maize, Plant Mol. Biol., 3:59.

Henshall, J. D., and Goodwin, T. W., 1964, The effect of red and far red light on carotenoid and chlorophyll formation in pea seedlings, Photochem. Photobiol., 3:243.

Horwitz, B. A., Thompson, W. F., and Briggs, W. R., 1987, Phytochrome regulation of greening in *Pisum*: chlorophyll accumulation and abundance of mRNA for the light-harvesting chlorophyll *a/b*-binding proteins. Plant Physiol., in press.

Jenkins, G. I., and Smith, H., 1985, Red:far-red ratio does not modulate the abundance of transcripts for two major chloroplast polypeptides in light-grown *Pisum sativum* terminal shoots. Photochem. Photobiol., 42:679.

Kaufman, L. S., Briggs, W. R., and Thompson, W. F., 1985a, Phytochrome control of specific mRNA levels in developing pea buds. The presence of both very low fluence and low fluence responses. Plant Physiol., 78:388.

Kaufman, L. S., Roberts, L. L., Briggs, W. R., and Thompson, W. F., 1986, Phytochrome control of specific mRNA levels in developing pea buds. Kinetics of accumulation, reciprocity, and escape kinetics of the low fluence response, Plant Physiol., 81:1033.

Kaufman, L. S., Thompson, W. F., and W. R. Briggs, 1984, Different red light requirements for phytochrome-induced accumulation of *cab* and *rbc*S RNA, Science, 226:1447.

Kaufman, L. S., Watson, J. C., Briggs, W. R., and Thompson, W. F., 1985, Photoregulation of nucleus-encoded transcripts: blue-light regulation of specific transcript abundance in "Molecular biology of the photosynthetic apparatus," K, Steinback, C. J. Arntzen, L. Bogorad, and S. Bonitz, eds., Cold Spring Harbor Press, N. Y., p. 367.

Kuhlemeier, C., Green, P. J., and Chua, N.-H., 1987, Regulation of gene expression in higher plants, Annu. Rev. Plant Physiol., 38, in press.

Mayfield, S. P., and Taylor, W. C., 1984, Carotenoid-deficient maize seedlings fail to accumulate light-harvesting chlorophyll a/b binding protein (LHCP) mRNA, Eur. J. Biochem., 144:79.

Mösinger, E., Batschauer, A., Apel, K., Schäfer, E., and Briggs, W. R., 1987, Phytochrome regulation of greening in barley. Effects on mRNA abundance and on transcriptional activity of isolated nuclei. Submitted to Plant Physiol.

Oelmüller, R., Levitan, I., Bergfeld, R., Rajasekhar, V. K., and Mohr, H., 1986a, Expression of nuclear genes as affected by treatments acting on the plastids, Planta, 168:482.

Oelmüller, R., and Mohr, H., 1986b, Photooxidative destruction of chloroplasts and its consequence for expression of nuclear genes, Planta, 167:106.

Raven, C. W., and Spruit, C. J. P., 1972a, Induction of rapid chlorophyll accumulation in dark grown seedlings. I. Action spectrum for pea. Acta Bot. Néerl., 21:219.

Raven, C. W., and Spruit, C. J. P., 1972b, Induction of rapid chlorophyll accumulation in dark grown seedlings. II. Photoreversibility. Acta Bot. Néerl., 21:640.

Raven, C. W., and Shropshire, W., Jr., 1975, Photoregulation of logarithmic fluence-response curves for phytochrome control of chlorophyll formation in *Pisum sativum* L. Photochem. Photobiol., 21:423.

Sagar, A. D., Briggs, W. R., and Thompson, W. F., 1987, Post-transcriptional regulation of nuclear-cytoplasmic partitioning of phytochrome-sensitive transcripts in *Pisum sativum*. In preparation.
Sagar, A., Horwitz, B. A., Elliott, R. C., and Briggs, W. R., 1987, Regulation of several nucleus-encoded mRNAs and a chloroplast-encoded mRNA during photobleaching in *Pisum sativum* L.
Schäfer, E., and Briggs, W. R., 1986, Photomorphogenesis from signal perception to gene expression, Photobiochem. Photobiophys., 12:305.
Sharma, R., 1985, Phytochrome regulation of enzyme activity in higher plants, Photochem. Photobiol., 41:747.
Silverthorne, J., and Tobin, E., 1987, Phytochrome regulation of nuclear gene expression, BioEssays, in press.
Thompson, W. F., Everett, M, Polans, N. O., Jorgensen, R. A., and Palmer, J. D., 1983, Phytochrome control of RNA levels in developing pea and mung-bean leaves, Planta, 158:487.
Thompson, W. F., Kaufman, L. S., and Watson, J. C., Induction of plant gene expression by light, BioEssays, 3:153.
Tobin, E. M., and Silverthorne, J., 1985, Light regulation of gene expression in higher plants, Annu. Rev. Plant Physiol., 36:569.
Virgin, H. I., 1972, Chlorophyll biosynthesis and phytochrome action, in "Phytochrome," K. Mitrakos and W, Shropshire, Jr., eds., Academic Press, N. Y., p. 371.

LIGHT-DEPENDENT REGULATION OF GENE EXPRESSION IN BARLEY

Klaus Apel[a], Egon Mösinger[b,c], Klaus Kreuz[a,d], Katayoon Dehesh[a,e], Alfred Batschauer[a,f], and Eberhard Schäfer[b]

[a]Botanisches Institut der Christian-Albrechts-Universität Kiel, F.R.G.
[b]Biologisches Institut II der Albert-Ludwigs-Universität Freiburg, F.R.G.

INTRODUCTION

Angiosperms are unable to form chloroplasts in the absence of light. When seedlings are grown in the dark the proplastids of the developing leaves differentiate into another plastid type, the etioplast. Upon illumination of dark-grown higher plants a rapid transformation of etioplasts into chloroplasts is induced. Protochlorophyllide is photoreduced to chlorophyllide, the prolamellar body disintegrates and an activation of the porphyrin pathway leads to the accumulation of chlorophyll. Parallel to chlorophyll accumulation a rapid formation of the thylakoid membranes occurs, leading to the assembly of the photosynthetic apparatus of the mature chloroplast[1]. These light-induced changes are based on an intimate interaction of the two genetic systems within the nucleus and the plastid[2]. Light does not only exert its control on processes which occur within the plastid but it simultaneously triggers reactions in the cytoplasm which contribute to chloroplast formation. It is not surprising, therefore, that not only protochlorophyllide within the plastid but also at least one additional photoreceptor, phytochrome, is involved in the light-dependent control of the complex interaction between plastids and the nucleus[3, 4]. We have tried to analyze the cooperation of the two photoreceptors and to understand the molecular basis of their functions during chloroplast development in barley.

[c-f]present address:

[c]Pflanzenphysiologisches Institut der Universität Bern, Switzerland

[d]CIBA-GEIGY, Basel, Switzerland

[e]Department of Botany, University of Wisconsin, Madison, U.S.A.

[f]Biologisches Institut II der Universität Freiburg, F.R.G.

RESULTS AND DISCUSSION

The complex interaction of different photoreceptors during the light-dependent transformation of etioplasts into chloroplasts in barley seedlings is demonstrated in Fig. 1. Seedlings were kept in the dark for 4 days and were either exposed directly to continuous white light or were treated for 12 h with farred light prior to the exposure to white light. During the farred light treatment sufficient amounts of phytochrome (P_r) are transferred into the active form P_{fr} to allow the maximum response of phytochrome-dependent processes to occur in the seedling. In both plant samples chlorophyll had not been accumulated to detectable levels at the beginning of the white light treatment. Thus, the presence of P_{fr} alone had no effect on the amount of chlorophyll within the plastid. However, during the subsequent illumination with white light in farred light pretreated seedlings chlorophyll began to accumulate instantaneously at the maximum rate while in dark-grown control plants the maximum rate of chlorophyll accumulation was only reached after a lag period of 3 to 4 h (Fig. 1). The maximum rate of chlorophyll accumulation does not depend only on the light-induced reduction of protochlorophyllide to chlorophyllide but also on the presence of active phytochrome P_{fr}. In the following work we have tried to untangle the complex interaction of the two photoreceptors, phytochrome and NADPH-protochlorophyllide-oxidoreductase (PCR), and to understand the molecular basis of their function during the light-dependent control of plastid development.

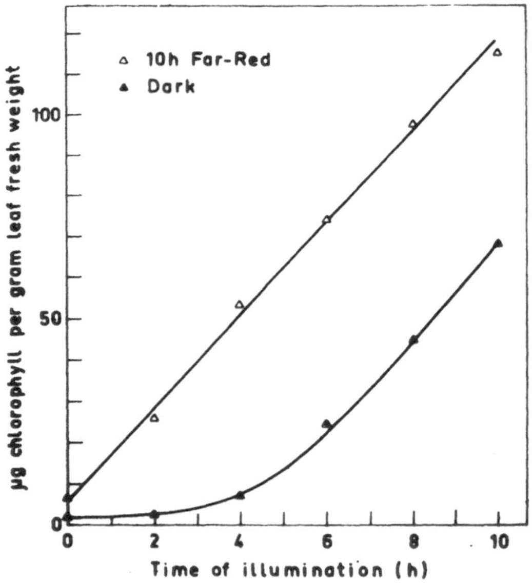

Fig. 1. The effect of phytochrome (P_{fr}) on the light-dependent chlorophyll accumulation in barley. 5 d old etiolated seedlings were directly transferred to continuous white light or were exposed for 10 h to farred light prior to the white light treatment.

During the light-dependent transformation of etioplasts to chloroplasts two of the most prominent plastid membrane proteins, whose concentration is affected dramatically by light are the light-harvesting chlorophyll a/b protein (LHCP) and the NADPH-protochlorophyllide-oxidoreductase (PCR). While the appearance of the LHCP within the plastid membrane fraction occurs only during illumination, the PCR is affected inversely by the white light treatment, its concentration rapidly declines during illumination[5,6]. Both these proteins are encoded by nuclear genes and in both cases the biosynthesis of the proteins has been shown by measurements of nuclear run-off-transcripts to be regulated by light at the level of gene transcription[7].

In subsequent work we have analyzed more extensively the effect of phytochrome (P_{fr}) on the rate of transcription of the two genes in isolated nuclei[8]. The concentrations of P_{fr} were manipulated experimentally either by illuminating intact plants or macerated plant material prior to the isolation of nuclei or by adding purified undegraded phytochrome (P_{fr}) to the isolated nuclei. Our results with nuclei isolated from preilluminated plants indicate that phytochrome controls the light-induced changes in the transcription of the two specific genes[7,8]. There are at least two different steps involved in this control which can be distinguished experimentally.

Firstly, in plants grown for 5 days in the dark an inductive light pulse leads to a rapid and transient change in transcription rates of LHCP mRNAs. During its initial phase this step is saturated by even very low amounts of P_{fr}. During the first 90 min following a R- as well as a FR-pulse treatment the kinetics of increase in LHCP mRNA transcription were indistinguishable[8] (Fig. 2). Thus, the rate-limiting step in the signal transduction chain connecting the phytochrome system and the control loci of LHCP was saturated at even very low amounts of P_{fr} as established by a saturating FR-pulse, less than 0.1 % of the total cellular phytochrome[9].

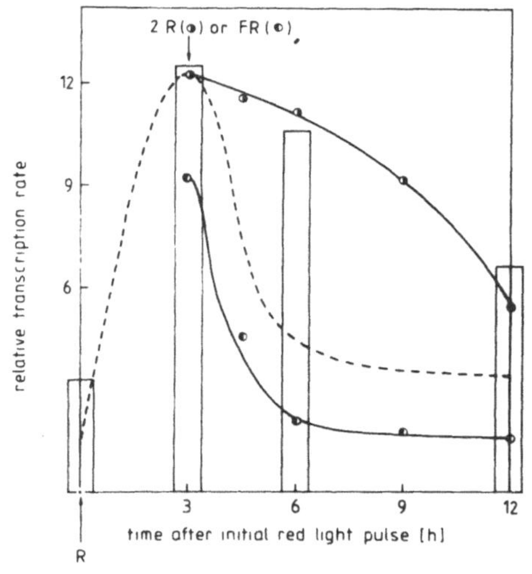

Fig. 2. Relative LHCP-mRNA transcription rates of isolated nuclei of barley leaves after a second irradiation or P_{fr} addition. Assay conditions were as described in (8). Dotted lines: Transcription after a single R pulse taken as reference. - ● -: Transcription rates of nuclei isolated at the times indicated from plants which were irradiated with 1 min R at t=3 h after the initial R. - ● -: Same except R at t=3 h replaced by 5 min FR. Columns: Nuclei isolated 0, 3, 6 and 12 h after an initial R pulse and treated with isolated P_{fr} in vitro.

Secondly, the duration and the maximum range of the transient change in transcription rates is controlled by the actual P_{fr} concentration. This second step is not saturated by the very low amounts of active phytochrome produced by an initial FR pulse or the remaining P_{fr} 6 h after an initial R pulse. Therefore, the second step reacts very sensitively and rapidly to changes in the P_{fr} concentration as induced by secondary irradiations. To be effective the first irradiations must be applied to the intact plant indicating that cytoplasmic components other than phytochrome and intact structures are required for the initial signal transduction. Second irradiation, FR as well as R, can be applied effectively also during or at the end of the maceration period when cells are heavily plasmolysed by a high mannitol concentration, implying a different mode of action compared with that required for first irradiations[8].

In the case of the negatively controlled regulation of the PCR gene by phytochrome there are also clear differences in the sensitivity of P_{fr} concentrations between the control of gene expression during the initial phase of response to R and the responsiveness to second R treatments[8].

In the second part of this work we used a reconstituted nuclear system to which varying amounts of purified phytochrome were added in order to test some of our conclusions derived from the two-step model of phytochrome action described above. Recently, Ernst and Oesterhelt[10] have reported a stimulation of the overall transcription rate if 124-kDa phytochrome from rye was added to nuclei isolated from rye seedlings. In contrast to those results we were unable to demonstrate a stimulation of the overall transcription if purified phytochrome was added to nuclei of dark-grown plants. A stimulatory effect of added phytochrome could be observed only if nuclei were used that had been isolated from preirradiated plants.

Since we knew from our in-vivo irradiations that nuclei isolated from plants at the maximum for the primary induction response had not responded very well to a secondary R pulse we used nuclei from plants which had already passed that peak, namely t=6 h for R induction. Nuclei isolated from dark-grown plants which had been prepared rapidly under dimmed green safelight showed only a very weak response to the added P_{fr}. Nuclei isolated from R-pulsed plants, by contrast, showed a dramatic response, a virtual doubling of the transcription rate[8].

Results on transcriptional changes in a reconstituted nuclear system should reflect at least some of the control mechanisms which operate also in the intact plants. Thus, we have made special efforts to minimize the possible interference of artefacts. Rather than measuring changes in the overall rate of transcription we have monitored changes in the rate of transcription of two specific genes which are known to be regulated inversely by phytochrome in the intact plant. While the rate of transcription of LHCP genes was stimulated after the R treatment of dark-grown barley seedlings, the rate of transcription for the reductase decreased rapidly. The preparation of purified oat phytochrome affected the rate of transcription of these two genes in the reconstituted nuclear system in a manner similar to that found in the intact plant[8]. These results seem to indicate that the addition of purified phytochrome to isolated nuclei may induce some of the effects which are also important during the light-dependent regulation of gene expression in the intact plant. Our confidence in the fidelity of results obtained with the reconstituted nuclear system was further strengthened by the striking similarity between the kinetics of changes in the rate of transcription which were induced either by the application of a second irradiation to the intact plant given 3 h after the initial R pulse or by the addition of purified oat phytochrome to isolated nuclei (see Fig. 2). Phytochrome (P_{fr}) added to nuclei isolated immediately after the R pulse or at t=3 h produced little or no stimulation of LHCP transcription. On the other hand P_{fr} added to nuclei harvested 6 or 12 h after an initial R pulse produced a stimulation in transcription of LHCP mRNA which was nearly equal to the effect of a second R pulse applied at t=3 h in vivo (compare columns and the upper solid curve in Fig. 2).

A similar result was observed in the same experiments for reductase mRNA transcription. At t=0 and t=3 h after the initial R pulse the P_{fr}-treated samples showed a transcriptional activity slightly but insignificantly higher than the untreated controls. At t=6 and t=12 h however, the measured transcription of reductase mRNA was even more repressed on P_{fr} addition than was the case in samples from plants treated with a second R pulse at t=3 h after induction in vivo[8].

Again, these similarities are consistent with our assumption that the addition of phytochrome to the reconstituted nuclear system mimics primarily the second phytochrome-dependent step in the signal-transduction chain. Furthermore, these results indicate that it should be possible to study further the molecular basis of phytochrome function in such a reconstituted nuclear system.

So far we have concentrated only on the light-dependent control of nuclear genes. However, chloroplast formation also depends on the activity of plastid-specific genes. Whereas light-dependent control at the level of transcription has been studied intensively for nuclear genes, much less is known about the light-dependent control of plastid gene expression.

A phytochrome-dependent change in the rate of transcription similar to that of nuclear genes has been proposed as a major control by which light may regulate the expression of plastid-specific genes [11,12]. However, recent studies of several plastid-specific genes including those encoding the 32 kDa herbicide-binding polypeptide of photosystem II and the large subunit of ribulose 1,5-bisphosphate carboxylase have indicated that the expression of these genes during the light-dependent formation of chloroplasts may not be regulated exclusively at the level of transcription but also at one of the subsequent steps of protein biosynthesis[13,14].

The P700 reaction centre chlorophyll a protein of photosystem I is a major plastid membrane protein whose accumulation is induced by light[15]. The genes encoding polypeptides of the P700 chlorophyll a protein complex have been located on the plastid DNA of different higher plants[16,17].

We have determined the relative level of transcripts encoding the P700 chlorophyll a protein of photosystem I reaction centre. Equal amounts of total RNA isolated from leaves of dark-grown and 12 h illuminated barley plants were separated electrophoretically and blotted onto nitrocellulose. The blot was hybridized with the spinach gene probe for the P700 chlorophyll a apoprotein. Only one mRNA species of approximately 4.7 kb was detectable, which hybridized to the radioactively labelled DNA probe[18].

The amount of this transcript relative to the total cellular RNA, which is mainly ribosomal RNA, did not change drastically upon illumination of dark-grown seedlings. This observation was further substantiated by quantitative dot-blot hybridization assays. The data show that the steady-state level of hybridizable mRNA complementary to the probe for the P700 chlorophyll a apoprotein remained fairly constant throughout an extended illumination period except for a slight transient increase after 3 h of illumination. In contrast, the mRNA sequence level for the 32 kDa herbicide-binding protein increased drastically in response to light. Also in this case there is a readily detectable transcript pool already present in dark-grown plants[18].

Based on the results presented above it can be inferred that the accumulation of the P700 chlorophyll a protein during light-dependent chloroplast formation in barley is not coupled to its transcript level.

Further attempts were made to address the question of whether or not translational control may be involved in the light-dependent accumulation of the P700 chlorophyll a protein. The levels of hybridizable mRNA in membrane-bound and free polysomes from etioplasts and chloroplasts isolated during the early phase of greening when a maximum rate of protein accumulation occurred, were estimated by the quantitative dot-blot hybridization. A high concentration of mRNA for the P700 chl a protein was detected in the fraction of membrane-bound polysomes of etioplasts, the concentration of mRNA within this polysomal fraction remained almost constant during greening (Fig. 3).

The uptake of a given mRNA into the membrane-bound polysomes might be regarded as indicative of the active translation of this mRNA. In such a case the synthesis of the P700 chl a protein of photosystem I would occur in dark-grown plants but since there is no chlorophyll available the protein would be rapidly destroyed. We have tested this possibility by measuring the rate of synthesis of the P700 apoprotein in dark-grown and illuminated barley seedlings. ^{35}S-methionin was fed to cut leaves for brief periods of time and the radioactively labelled apoprotein was immunoprecipitated, separated electrophoretically and visualized by autoradiography. In addition to wild type barley plants also the chlorophyll-deficient mutant xantha-l[81] was used. In this mutant the biosynthetic pathway of chlorophyll is blocked completely after Mg-protoporphyrin[19]. In the mutant the relative concentration of transcripts coding for the P700 apoprotein was the same as in the wild type seedlings and apparently was not affected by light (Fig. 4). However, synthesis of the apoprotein was detectable only in wild type plants during illumination but not in the xantha mutant (Fig. 4). Even though we cannot rule out that our experimental approach was not sensitive enough to

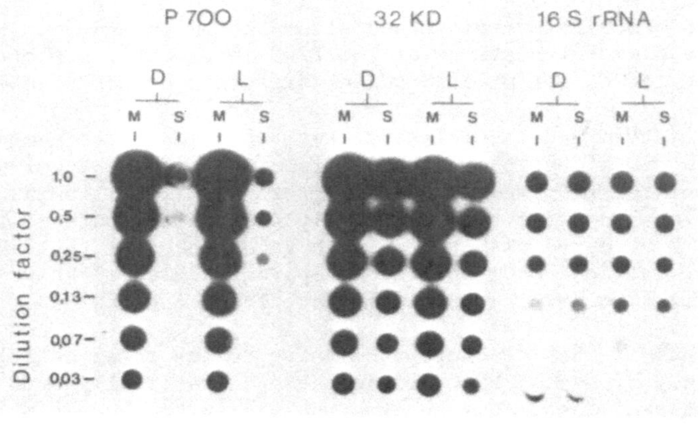

Fig. 3. Relative amounts of mRNA encoding the P700 chlorophyll a apoprotein (P700) and the 32000-M_r herbicide binding protein (32 kD) and the 16S rRNA in membrane-bound (M) and soluble (S) plastid polysomes of dark-grown (D) and 6 h illuminated (L) barley seedlings. RNA levels of polysomal fractions were determined by dot-blot hybridization using radioactively labelled gene-specific probes[18]. For measurements of the 16S rRNA only 5 % the amount of RNA used in the other experiments was dotted onto nitrocellulose.

detect low levels of freshly synthesized apoprotein which may be rapidly degraded in the absence of chlorophyll we don't think it very likely that etiolated seedlings would use their limited amounts of storage material for such a wasteful synthetic step. Thus, at present it appears to be more likely that the polysomal mRNA for the P700 apoprotein is not translated in the dark and that its translation is controlled by a light-dependent step within the plastid. The only known photoreceptor within the plastid compartment of higher plants that might be responsible for a direct light-dependent control of plastid gene expression at a translational or posttranslational level is the NADPH-protochlorophyllide oxidoreductase (PCR).

The enzyme protein and its regulation by light has been subject of numerous studies (e.g. 4). It has been suggested that the enzyme is responsible only for overall chlorophyll synthesis during light-induced chloroplast formation and in fully green plants (e.g. 20).

Quite surprisingly, however, a rapid light-dependent drop in enzyme activity has been found in a number of different higher plants[6,21,22]. In several instances it could be demonstrated that the decrease in enzyme activity is accompanied by a decrease in the concentration of enzyme protein within the plastid[6,23,24] and mRNA sequences encoding the PCR[25].

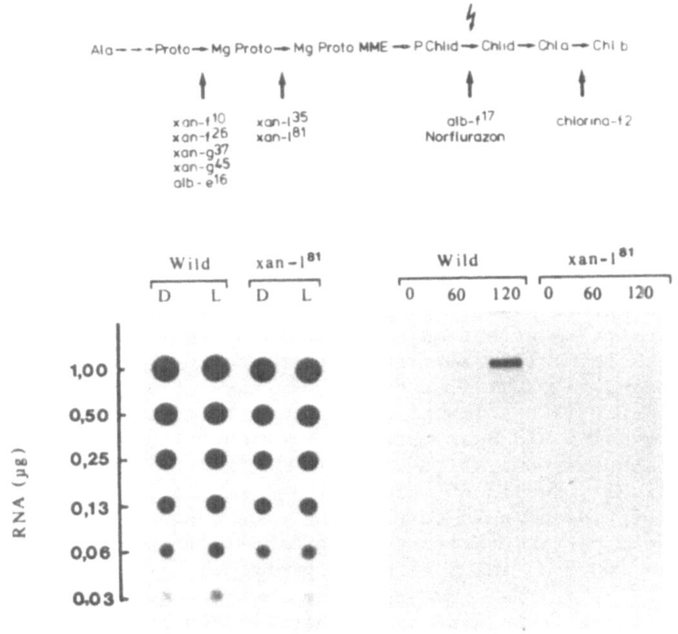

Fig. 4. Light-dependent changes in the amounts of mRNA and of the freshly synthesized P700 chlorophyll a apoprotein in wild type and the xantha-l[81] mutant of barley. RNA levels in dark-grown (D) and illuminated (L) plants were determined by dot-blot hybridization as described previously[18]. The synthesis of the apoprotein was determined by feeding S^{35}-methionine for 4 h to dark-grown leaves. During incubation leaves were kept in the dark (0) or exposed for the last 60 (60) or 120 (120) min to continuous white light. The labelled apoprotein was immunoprecipitated from total leaf extracts, separated electrophoretically and visualized by autoradiography. The upper part of the figure illustrates the step of chlorophyll biosynthesis which is blocked in the xantha mutant.

It has been suggested that the massive accumulation of the enzyme protein and its mRNA in etiolated plants may be the consequence of the extended dark period to which these plants had been exposed and thus may reflect an experimental artefact[20]. Similar light- induced drastic changes in the concentration of the enzyme protein and its mRNA could also be detected in 3-d-old germinating seeds from which the coleoptile and the primary leaf had just begun to emerge. These results indicate that the function of the PCR may be more complex than previously assumed. This suggestion received additional support by the results of recent work on the distribution of the 36000 M_r polypeptide of the PCR within different cellular compartments of barley[26,27]. In addition to the PCR which is present in the plastid a second, closely related polypeptide is present outside the plastid in the area of the plasmalemma. The antigenic properties and the apparent molecular weight of the PCR and the 36000-M_r polypeptide outside the plastid were so similar as to indicate that the two proteins may be of common origin. The two protein populations were differently affected by light: while the plastid-specific protein rapidly declined during illumination the extraplastidic protein appeared to be relatively stable[26,27]. Some of the experimental evidence on which these conclusions are based upon is shown in Fig. 5.

In the barley mutants xantha-l[81] and xantha-f[10] the biosynthetic pathway of chlorophyll is blocked completely after Mg-protoporphyrin or protoporphyrin IX, respectively[19]. These mutant plants lack protochlorophyllide and they are unable to form a prolamellar body. In total leaf extracts of dark-grown mutant plants, however, the immunoreactive 36000-M_r, polypeptide was present even though its concentration was lower than in protein extracts of dark-grown wild-type plants (Fig. 5). In xantha-l[81] mutant plants which had been exposed to dim white light (80 lx) for 12h the concentration of the 36000-M_r polypeptide remained unchanged. However, when the light intensity was increased to 8000 lx, the concentration of this polypeptide in leaf extracts declined and the protein was barely detectable after 12 h of illumination[27]. Identical results were obtained for the xantha-f[10] mutant. The cellular distribution of the 36000 M_r polypeptide of the PCR was determined in ultrathin sections of leaves by the method of immunogold labelling. In dark-grown mutant plants the immunoreactive polypeptide was not detected in the plastid compartment but only in the surrounding cytoplasma in the area of the plasmalemma (Fig. 5). An almost identical distribution of gold particles was found in ultrathin sections of leaves from mutant plants which had been exposed for 12 h to continuous dim white light[27]. However, in ultrathin sections which had been obtained from mutant plants treated for 12 h with continuous strong white light (8000 lx) the concentration of gold particles drastically declined[27]. These results clearly demonstrate that a nonspecific adsorption of gold to the leaf tissue is negligible and that the protein A-gold particles reacted specifically with antigen-antibody complexes of the 36000-M_r immunoreactive protein.

At present there is no hint as to what kind of purpose the extraplastidic protein might serve. Perhaps two findings are relevant in this respect. In the barley mutant xantha-l[81] the concentration of the extraplastidic protein was altered under the influence of different light intensities. At higher light intensities the concentration of this protein was drastically reduced. Other proteins like the ATPase and the ribulose-1,5-bisphosphate carboxylase were not visibly affected in the mutant plant under these conditions. This could mean that the extraplastidic protein absorbs light in the visible range. It will be important to determine whether or not a pigment is associated with this protein, and if so, to identify its nature. Secondly, the extraplastidic protein is not distributed randomly in the cytoplasm but appears highly concentrated in the area of the plasmalemma. In this context it is interesting that protochlorophyllide has been shown to interfere with the transport of Ca^{2+} in a microsomal membrane fraction isolated from various plant tissues[28]. Even though, as suggested by the authors, this effect of protochlorophyllide might be a purely artificial one, it may also offer a first hint from which one might start to unravel the cryptic function of the extraplastidic protein.

Collectively these results indicate that the function of the PCR may not be restricted to that of an 'ordinary' enzyme equivalent to others which take part only in the biosynthesis of chlorophyll. In addition to phytochrome the PCR could play an important role as photoreceptor during the regulation of photomorphogenesis in higher plants.

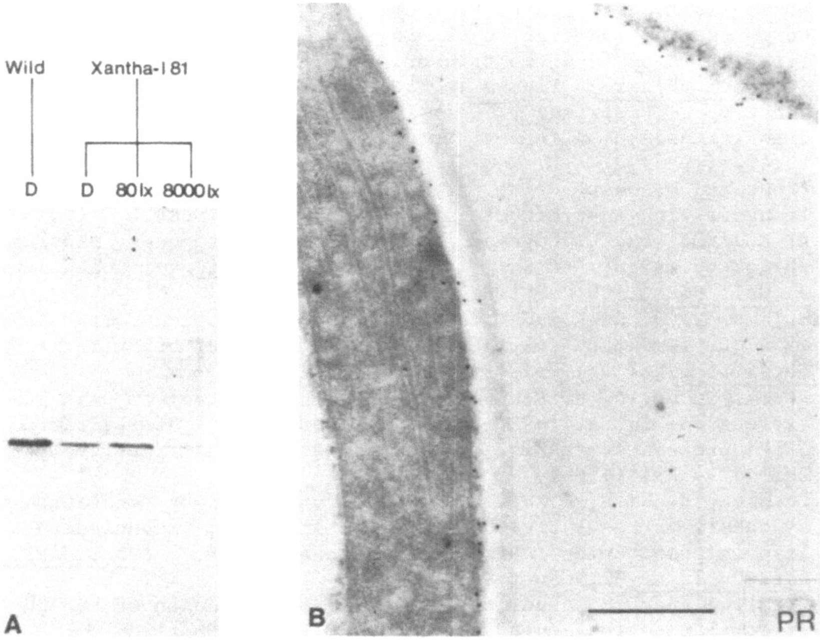

Fig. 5. Immunological detection of the 36000 M_r polypeptide of NADPH-protochlorophyllide oxidoreductase (PCR) in wild-type plants and in xantha-l^{81} mutants of barley. A) Plants had been grown in the dark for 6 days (D) or were exposed afterwards to continuous dim (80 lx) or strong (8000 lx) white light for 12 h. Equal amounts of total leaf protein were separated electrophoretically; the polypeptides were blotted onto nitrocellulose and immunostained as described in (27). B) Localization of the 36000 M_r immunoreactive polypeptide in leaves of dark-grown barley mutant. Sections of Lowicryl-embedded tissue slices were reacted with an antiserum against PCR, followed by 10 nm protein A-gold. Bar = 0.5 µm.

REFERENCES

1. P. A. Castelfranco and S. I. Beale, Chlorophyll biosynthesis: Recent advances and areas of current interest, Annu. Rev. Plant Physiol. 34:241 (1983).
2. R. J. Ellis, Chloroplast proteins: Synthesis, transport and assembly, Annu. Rev. Plant Physiol. 32:111 (1981).
3. E. M. Tobin and J. Silverthorne, Light regulation of gene expression in higher plants, Annu. Rev. Plant Physiol. 36:569 (1985).
4. M. Harpster and K. Apel, The light-dependent regulation of gene expression during plastid development in higher plants, Physiol. Plant. 64:147 (1985).
5. K. Apel and K. Kloppstech, The plastid membranes of barley (Hordeum vulgare). Light-induced appearance of mRNA coding for the apoprotein of the light-harvesting chlorophyll a/b protein, Eur. J. Biochem. 85:581 (1978).
6. H. J. Santel and K. Apel, The protochlorophyllide holochrome of barley (Hordeum vulgare L.). The effect of light on the NADPH-protochlorophyllide oxidoreductase, Eur. J. Biochem. 120:95 (1981).
7. E. Mösinger, A. Batschauer, E. Schäfer and K. Apel, Phytochrome-control of in-vitro transcription of specific genes in isolated nuclei from barley (Hordeum vulgare), Eur. J. Biochem. 147:137 (1985).
8. E. Mösinger, A. Batschauer, R. Vierstra, K. Apel and E. Schäfer, Comparison of the effects of exogenous native phytochrome and in-vivo irradiation on in-vitro transcription in isolated nuclei from barley (Hordeum vulgare), Planta 170:505 (1987).
9. E. Schäfer, G. Löser and B. Heim, Formalphysiologische Analyse der Signaltransduktion in der Photomorphogenese, Ber. Dtsch. Bot. Ges. 96:497 (1983).
10. D. Ernst and D. Oesterhelt, Purified phytochrome influences in vitro transcription in rye nuclei, EMBO J. 3:3075 (1984).
11. S. R. Rodermel and L. Bogorad, Maize plastid photogenes: mapping and photoregulation of transcript levels during light-induced development, J. Cell Biol. 100:463 (1985).
12. Y. S. Zhu, S. D. Kung and L. Bogorad, Phytochrome control of levels of mRNA complementary to plastid and nuclear genes of maize. Plant Physiol. 79:371 (1985).
13. H. Fromm, M. Devic, R. Fluhr and M. Edelman, Control of psbA gene expression: in mature Spirodela chloroplasts light regulation of 32-kd protein synthesis is independent of transcript level, EMBO J. 4:291 (1985).
14. G. Inamine, B. Nash, H. Weissbach and N. Brot, Light regulation of the synthesis of the large subunit of ribulose-1,5-bisphosphate carboxylase in peas: evidence for translational control, Proc. Natl. Acad. Sci. U. S. A. 82:5690 (1985).
15. E. Vierling and R.S. Alberte, Regulation of synthesis of the photosystem I reaction center, J. Cell Biol. 97:1806 (1983).
16. P. Westhoff, J. Alt, N. Nelson, W. Bottomley, H. Bünnemann and R. G. Herrmann, Genes and transcripts for the P700 chlorophyll a apoprotein and subunit 2 of the photosystem I reaction center complex from spinach thylakoid membranes, Plant Mol. Biol. 2:95 (1983).
17. L. E. Fish, U. Kück and L. Bogorad, Two partially homologous adjacent light-inducible maize chloroplast genes encoding polypeptides of the P700 chlorophyll a protein complex of photosystem I, J. Biol. Chem. 260:1413 (1985).
18. K. Kreuz, K. Dehesh and K. Apel, The light-dependent accumulation of the P700 chlorophyll a protein of photosystem I reaction center in barley. Evidence for translational control, Eur. J. Biochem. 159:459 (1986).

19. D. v.Wettstein, K. W. Henningsen, J. E. Boynton, G. C. Kannangara and O. F. Nielsen, The genetic control of chloroplast development in barley, in: "Autonomy and biogenesis of mitochondria and chloroplasts" N. K. Boardman, A. W. Liannane and R. M. Smillie. eds., North-Holland Publishing Company, Amsterdam (1971).
20. W. T. Griffiths, A. S. Kay and R. P. Oliver, The presence of photoregulation of protochlorophyllide reductase in green tissues, Plant Mol. Biol. 4:13 (1985).
21. E. R. Mapleston and W. T. Griffiths, Light modulation of the activity of protochlorophyllide reductase, Biochem. J. 189:125 (1980).
22. M. Ikeuchi and S. Murakami, Changes in the properties of protochlorophyllide reductase during early greening of etiolated squash cotyledons, in: "Protochlorophyllide and greening" C. Sironval and M. Brouers, eds., Martinus Nijhoff/Dr. W. Junk publishers, The Hague, Boston, Lanaster (1984).
23. S. A. Kay and W. T. Griffiths, Light-induced breakdown of NADPH-protochlorophyllide oxidoreductase in vitro, Plant Physiol. 72:229
24. I. Häuser, K. Dehesh and K. Apel, The proteolytic degradation in vitro of the NADPH-protochlorophyllide oxidoreductase of barley (Hordeum vulgare), Arch. Biochem. Biophys. 228:577 (1984).
25. K. Apel, The protochlorophyllide holochrome of barley (Hordeum vulgare L.). Phytochrome-induced decrease of translatable mRNA coding for the NADPH-protochlorophyllide oxidoreductase, Eur. J. Biochem. 120:89 (1981).
26. K. Dehesh, M. Klaas, I. Häuser and K. Apel, Light-induced changes in the distribution of the 36000-M_r polypeptide of NADPH-protochlorophyllide oxidoreductase within different cellular compartments of barley (Hordeum vulgare L.) I. Localization by immunoblotting in isolated plastids and total leaf extracts. Planta 169:162 (1986).
27. K. Dehesh, B. v. Cleve, M. Ryberg and K. Apel, Light-induced changes in the distribution of the 36000-M_r polypeptide of NADPH-protochlorophyllide oxidoreductase within different cellular compartments of barley (Hordeum vulgare L.) II. Localization by immunogold labelling in ultrathin sections of barley leaves, Planta 169:172 (1986).
28. I. Gross, A. Ayadi and D. Marmé, Protochlorophyll(ide) photosensitizes active Ca^{2+} accumulation in microsomal and mitochondrial fractions isolated from plants, Photochem. Photobiol. 30:615 (1979).

SPLIT GENES AND *CIS/TRANS* SPLICING IN TOBACCO CHLOROPLASTS

Masahiro Sugiura, Kazuo Shinozaki, Minoru Tanaka, Nobuaki Hayashida, Tatsuya Wakasugi, Tohru Matsubayashi, Chikara Ohto, Keita Torazawa, Bing Yuan Meng, Tadashi Hidaka and Norihiro Zaita*

Center for Gene Research, Nagoya University, Chikusa, Nagoya 464, Japan

INTRODUCTION

Chloroplast DNAs of higher plants are circular molecules with a size of 120-217 kbp. One of the outstanding features of chloroplast DNAs of most higher plants is the presence of a large inverted repeat. These sequences are separated by a large and a small single-copy region. Chloroplast DNAs are known to contain all the chloroplast rRNA genes (four genes in higher plants) and tRNA genes (ca. 30 genes) and probably all the genes for proteins synthesized in the chloroplast (ca. 100 genes) (Dyer, 1984; Gray et al., 1984; Palmer, 1985; Shinozaki and Sugiura, 1986).

Introns in chloroplast genes were first reported for the 23S rRNA gene of *Chlamydomonas reinhardii* (Rochaix and Malone, 1978) and then isoleucine-tRNA and alanine-tRNA genes in the 16S-23S rRNA gene spacer of maize (Koch et al., 1981). Since then many split genes have been reported in a variety of chloroplast genomes (i.e. Hallick et al., 1985; Shinozaki et al., 1986a). Split genes so far analyzed in higher plant chloroplast genomes contain single introns while *Euglena* and *Chlamydomonas* genes for proteins contain multiple introns (i.e. Montandon and Stutz, 1983; Koller et al., 1984; Erickson et al., 1984; Hallick et al., 1985). Recently an unique split gene which requires *trans* splicing has been identified (Zaita et al., 1987; Koller et al., 1987).

In this article, we will describe the structure of split genes so far found in the tobacco chloroplast DNA and *cis/trans* splicing of their precursor RNAs.

SPLIT GENES IN TOBACCO CHLOROPLAST DNA

Tobacco (*Nicotiana tabacum* var. Bright Yellow 4) chloroplast DNA is a circular molecule of 155,844 bp (Sugiura et al., 1986; Shinozaki et al.,

*Present address: Hoechst-Japan Research and Development Laboratories, Kawagoe 350, Japan

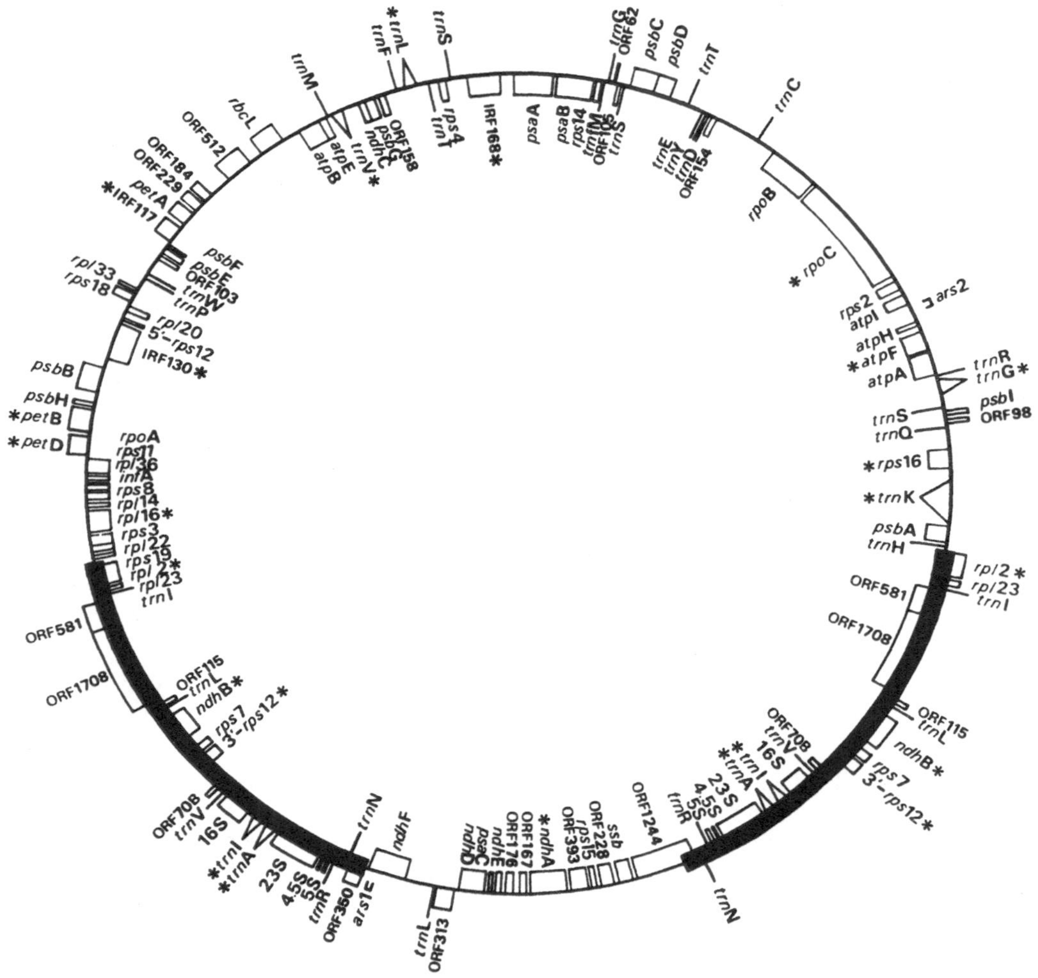

Fig. 1. Circular gene map of the tobacco chloroplast genome. Genes shown outside the circle are on the A strand (the DNA strand which codes for rbcL) and transcribed counterclockwise. Genes shown inside the circle are on the B strand and transcribed clockwise. Asterisks indicate split genes. Major ORFs and IRFs are included. The nomenclature for genes follows the proposals of Hallick and Bottomley (1983). ars is an autonomously replicating sequence in yeast. Bold lines show inverted repeats. The entire DNA sequence has been deposited with the EMBL database and presented in Shinozaki et al. (1986d).

1986c). It contains two copies of an identical 25,339 bp inverted repeat, which is separated by a 86,684 bp and a 18,482 bp single-copy region. The genes for 4 different rRNAs, 30 different tRNAs, 44 different proteins and 9 other predicted protein-coding genes have been located (Fig. 1). Among them, 15 genes contain introns. Blot hybridization revealed that all rRNA, tRNA and protein genes are transcribed in the chloroplast. L36 is a new ribosomal protein species recently found associated with the 50S subunit of E. coli (Wada and Sako, 1987) and the corresponding chloroplast gene is located between infA and rps11 (rpl36, formerly designated as secX) (Fig. 2). The gene (psaC) for the 9 kd polypeptide (a possible apoprotein for the iron-sulfur centers A and B) of photosystem I has been located between ndhE and ndhD (Hayashida et al., 1987). This psaC gene was identified based on the N-terminal amino acid sequence of the spinach 9 kd polypeptide which was determined by Dr. T. Hiyama's group in Saitama University. ORF52 located between trnS and trnQ (Deno and Sugiura, 1983) corresponds to the gene for the 6 kd component of photosystem II recently reported in the maize genome (Kato et al., 1987).

The presence of possible introns in chloroplast tRNA genes was first demonstrated in maize trnI and trnA (Koch et al., 1981). Six different tRNA genes have been shown to contain single introns (503-2526 bp) among 30 different tRNA genes in the tobacco chloroplast genome (Deno et al., 1982; Takaiwa and Sugiura, 1982; Deno and Sugiura, 1984; Sugita et al., 1985; Sugiura et al., 1985; Yamada et al., 1986; Wakasugi et al., 1986) (see Fig. 1). The primary transcripts and spliced tRNAs from the six genes have been detected by Northern blot hybridization. Their intron sites have been determined by S1 mapping (Deno and Sugiura, 1984) or deduced through their homology with the corresponding chloroplast tRNAs (Osorio-Almeida et al., 1980; Guillemaut and Weil, 1982) and E. coli tRNAs. Fig. 3 shows the structure of these six tRNA genes.

The trnK intron contains an open reading frame (ORF) of 509 codons and its predicted amino acid sequence has a local homology with that of the ORF present in the yeast mitochondrial oxi3 intron 2 (Sugita et al., 1985; Shinozaki et al., 1986c). The 888 bp intron in Chlamydomonas chloroplast 23S rDNA contains a 489 bp ORF encoding a potential polypeptide that is related to mitochondrial maturases (Rochaix et al., 1985). The trnK intron (2574 bp) of mustard chloroplasts has also an ORF of 524 codons whose derived polypeptide appears to be structurally related to maturases (Neuhaus and Link, 1987). It is interesting whether a maturase-like activity is in fact involved in chloroplast splicing.

Introns within a chloroplast protein gene were first identified in the Euglena gracilis rbcL (Stiegler et al., 1982). In the tobacco chloroplast genome nine different protein genes contain single introns (536-1148 bp) as shown in Fig. 3 (Zurawski et al., 1984; Shinozaki et al.,

```
Tobacco CL36   MKIRASVRKICEKCRLIRRRGRIIVICS-NPRHKQRQG
               ** **** * *  *    * * * ****  * ******
E.coli    L36  MKVRASVKKLCRNCKIVKRDGVIRVICSAEPKHKQRQG
```

Fig. 2. Comparison of the amino acid sequences deduced from the gene for ribosomal protein CL36 of tobacco chloroplasts (rpl36) and from the gene for ribosomal protein L36 of E. coli (rpmJ).

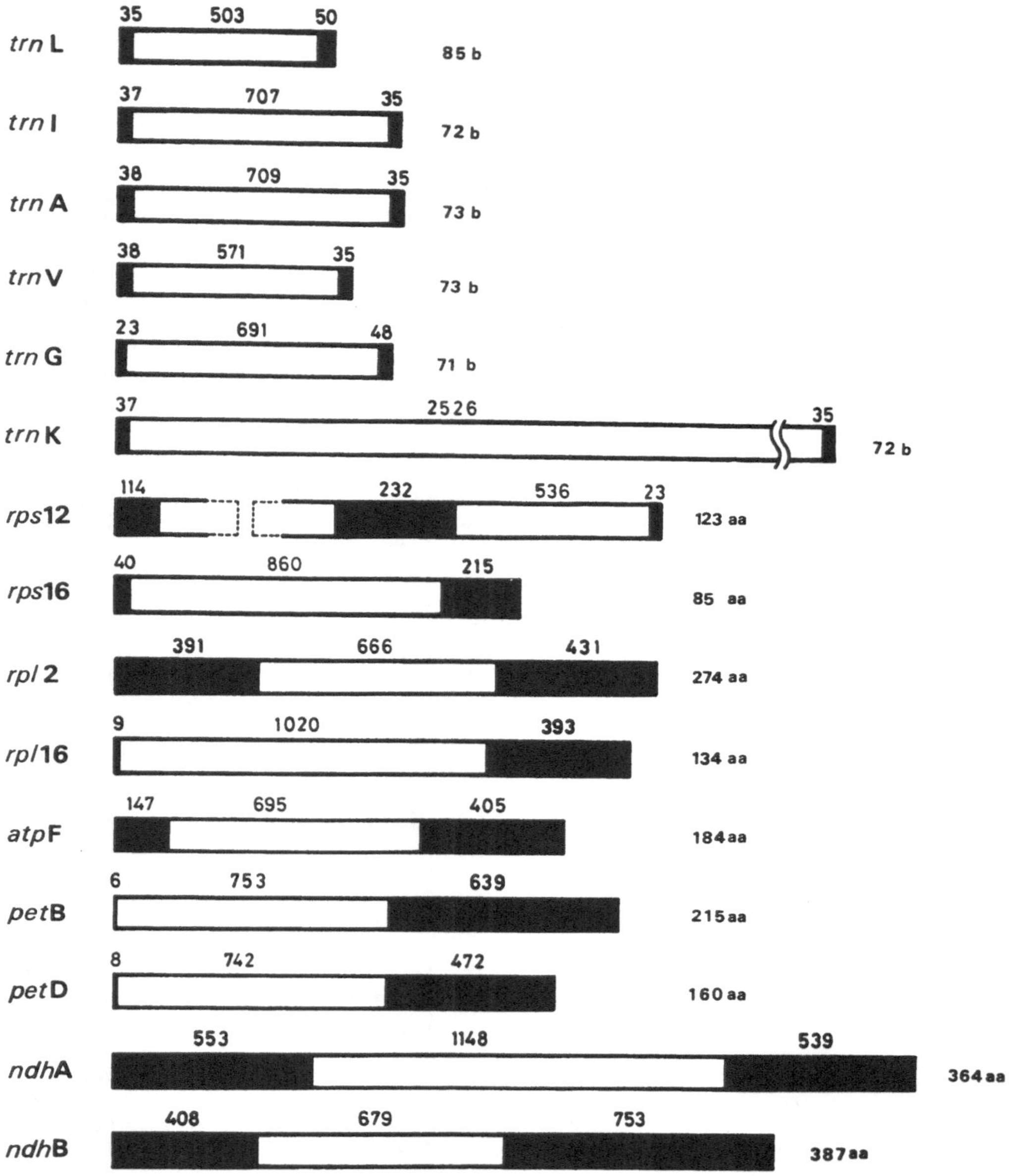

Fig. 3. Schematic presentation of the structures of 15 split genes found in tobacco chloroplast DNA. Filled boxes indicate exons and open boxes intron. Numerals above boxes are bp. Dotted boxes in rps12 show transons.

1986a; Shinozaki et al., 1986b; Tanaka et al., 1986; Fromm et al., 1986; Torazawa et al., 1986; Tanaka et al., 1987; Matsubayashi et al., 1987). Six tobacco chloroplast DNA sequences whose predicted amino acid sequences resemble those of components (ND1, 2, 3, 4, 4L and 5) of the respiratory-chain NADH dehydrogenase from human mitochondria have been found (Shinozaki et al., 1986c; Meng et al., 1986). As these sequences are highly expressed in tobacco chloroplasts, they are most likely to be the genes for components of a chloroplast NADH dehydrogenase (ndhA, B, C, D, E and F) (Matsubayashi et al., 1987). Two of them, ndhA and ndhB, contain introns.

The primary transcripts and/or spliced RNAs from these genes (except for atpF) have been detected by Northern blot hybridization. The intron sites for the ribosomal protein genes are deduced through their homology with the corresponding E. coli genes. The intron sites of petB, petD, ndhA and ndhB have been determined experimentally (see the next section). The atpF intron was assigned through its homology with the wheat atpF gene (Bird et al., 1985). A region similar to the E. coli rpoC sequence has been found next to rpoB in tobacco chloroplast DNA (Shinozaki et al., 1986c) and its expression is currently being studied.

Introns found in chloroplast genes can be classified into three groups on the basis of the intron boundary sequences and of the possible secondary structures (Shinozaki et al., 1986a). Fig. 4 shows the intron boundary sequences of tobacco split genes so far found.

1) Group I introns. The intron of trnL can be folded into a secondary structure which is similar to the postulated structure of the intron of autosplicable rRNA precursor of Tetrahymena (Michel and Dujon, 1983; Bonnard et al., 1984; Steinmetz et al., 1982). The intron of 23S rDNA from Chlamydomonas belongs to this group. Chloroplast group I intron sequences may be removed by an autocatalytic reaction as in the case of the Tetrahymena rRNA precursor or by a maturase as in the case of yeast cytochrome b and cytochrome oxidase mRNA precursors.

2) Group II introns. Introns of trnA and trnI can be folded into a secondary structure which is similar to the postulated structure of introns of the genes for maize mitochondrial cytochrome oxidase and yeast cytochrome oxidase and cytochrome b (Michel and Dujon, 1983). However their intron boundary sequences do not fit into the conserved sequences of the group III introns (see below).

3) Group III introns. Introns of trnV, trnG, trnK, rps12, rps16, rpl2, rpl16, atpF, petB, petD, ndhA and ndhB have conserved sequences at their boundaries (Fig. 4). The conserved boundary sequences of the group III introns are GTGYGRYYYR at the 5' ends and RYCNAYY(Y)YNAY at the 3' ends. These conserved boundary sequences resemble those of Euglena gracilis chloroplast introns (Hallick et al., 1985); they also resemble partly the conserved boundary sequences of nuclear gene introns. The group III intron sequences may be removed by a mechanism similar to that which worked for nuclear mRNA precursors.

INTRON-CONTAINING READING FRAMES

At least 50 ORFs are found in the tobacco chloroplast genome. Some of them contain sequences homologous to the group III intron boundary sequences. These ORFs have been subjected to Northern blot analysis to detect transcripts from the ORFs and possible spliced products. The sequences from which transcripts have been detected are likely to be protein-coding genes and we have designated intron-containing reading frames (IRF). Fig. 5 shows the three examples; IRF168 located between trnS and psaA, IRF117 between petA and psbF, and IRF130 between psbB and 5'-rps12. The intron sites of these IRFs have not been determined experimentally and hence the length of exons and introns are tentative.

		5'EXON	INTRON		3'EXON	INTRON SIZE (bp)
Group I	trnL-UAA	ACGGACTT	AATTGGATTGAGCCTTG---------	GAAATTTATCGTAAGAGG	AAAATCCG	503
Group II	trnI-GAU	CCTGATAA	TTGCGTCGTTGTGCCTG---------	TGCGATGATTTACTTCAC	GGGGCGAG	707
	trnA-UGC	TCTTGCAA	TTGGGTCGTTGCGATTA---------	GGCGGTGGTTTACCCTGC	GGCGGATG	709
Group III	trnV-UAC	GTTTACAC	GCGCGCCAATGTTTTTC---------	AGTTTGACCTGTT TTAC	CGAGAAGG	571
	trnG-UCC	TGGTAAAA	GTGTGATTCGTTCTATT---------	ATCGTCGTCGACTATAAC	CCCTAGCC	691
	trnK-UUU	GCTTTTAA	GTGCGGCTAGTCTCTTT---------	GAACTTATCTACT CCAT	CCGACTAG	2526
	rps12	GCGTTCTA	GTGCGTTGTAGATTCTT---------	TTCGAGATCCACC CTAC	AATATGGG	536
	rps16	AAAGCAAC	GTGCGACTTGAAGGACA---------	ATCTACATCAATCCCAAT	GAGCCGTC	860
	rpl2	ACCTTTGA	GTGCGGTTTGAACTATT---------	AGAAGAATCTACT TCAA	CCGATATG	666
	rpl16	TGCTTAGT	GTGTGACTCCGTTGGTT---------	ACAACCATCAACTATAAC	CCCAAAAG	1020
	atpF	GGGAGTGT	GTGCGAGTTGTTTATTT---------	AAAATAATCTACTTTCAT	TAAGTGAT	695
	petB	GTATGAGT	GTGTGACTTGTTATAAT---------	AGTTCCGCCTATCTCAAT	AAAGTATA	759
	petD	ATGGGAGT	GTGTGACTTGAACTATT---------	GAATTCACCTATCCCAAT	AACAAAAA	742
	ndhA	ATCTCTAC	GTGTGATTCGGTGAGAC---------	GTTATCATCGACTATGAT	TATCTAAC	1148
	ndhB	ACGAAGGA	GTGCGGTTCGTTCGAGA---------	CCTCTTTTCGACTCTGAC	TCTCCCAC	679
Conserved sequence	Group III		GTGYGRYYYR	RYCNAYY(Y)YNAY		
	Euglena		GTGYG	TARTTNTAY		
	Nuclear	AAG	GTGRAGT	YYYYYYNCAG	GG	

Fig. 4. Comparison of the exon-intron boundary sequences of 15 split genes from tobacco chloroplasts (see Fig. 3). The conserved boundary sequences of *Euglena* introns (Hallick et al., 1985) and of introns of nuclear protein genes (Cech, 1983) are shown below.

At present no known sequences similar to these IRFs have been found.

CIS SPLICING

The presence of introns can be suggested based on sequence homologies with the corresponding E. coli genes (e.g. tRNA genes, ATP synthase genes and ribosomal protein genes) and on the conserved intron boundary sequences. It is not easy to deduce the 3' ends of potential introns because of the low conservation (see Fig. 4). Both primary transcripts and spliced RNA products should be detected to identify given sequences as split genes. To confirm the existence of introns and to determine the splice-sites of pre-mRNAs from petB, petD, ndhA and ndhB, reverse transcription analysis was carried out (Tanaka et al., 1987; Matsubayashi et al., 1987). Primers were prepared from the coding strands of the putative second exons (either cut from cloned DNA fragments or chemically synthesized) and hybridized to total tobacco chloroplast RNA. cDNAs were synthesized from the primers using AMV reverse transcriptase in the presence of dideoxyribonucleoside triphosphates. The sequence ladders obtained in this way indicate exact splice-sites of the pre-mRNAs as shown in Fig. 4. This method was successfully applied for the above gene products because their primary transcripts seem to be spliced very rapidly. The primary transcripts from these tobacco genes were hardly dectable in our growth conditions.

TRANS SPLICING

Tobacco rps12 is divided into one copy of 5'-rps12 and two copies of 3'-rps12 (Trazawa et al., 1986; Fromm et al., 1986). The 5'-rps12 contains exon 1 of 38 codons and 3'-rps12 consists of exon 2 of 78 codons, a 536 bp intron and exon 3 of 7 codons (Fig. 5). We have designated this gene structure as a "divided" gene (Shinozaki et al., 1986c). Northern blot hybridization revealed that both 5'-rps12 and 3'-rps12 are transcribed in the chloroplasts. A possible primary transcript from 5'-rps12 is likely to contain both IRF130 and rpl20 sequences and that from 3'-rps12 contains the rps7 sequence (see Fig. 5c and d). Interestingly the Euglena rps12 is not split and is co-transcribed with the downstream rps7 (Montandon and Stutz, 1984).

Reverse transcription analysis indicated that trans splicing between the 5'-rps12 and 3'-rps12 transcripts occurs in vivo (Zaita et al., 1987). Trans-spliced CS12 mRNA has also been observed under an electron microscope using an artificially combined exons 1, 2 and 3 DNAs (Koller et al., 1987).

The 3'-flanking sequence of exon 1 has been designated as "transon" 1 and the 5'-flanking sequence of exon 2 as "transon" 2 (Zaita et al., 1987). The sequence GTGCGAC at the 5'-end of transon 1 and the sequence GTCAACTTTTCC at the 3'-end of transon 2 fit the conserved boundary sequences of the chloroplast group III introns. A long complementary structure (about 70 bp) can be constructed between transon 1 and transon 2. This structure, if any, may be necessary for or at least enhance trans splicing as has been reported in trans splicing in vitro (Solnick, 1985; Konarska et al., 1985). Trans splicing has also been reported in the case of "mini-exon" sequences from the trypanosome genome (Murphy et al., 1986; Sutton and Boothroyd, 1986). However none of the trypanosome genes described to date contain introns in their coding sequences and their transcripts do not undergo cis splicing (Van der Ploeg, 1986). It is noteworthy that the tobacco rps12 requires both cis and trans splicing for a mature mRNA.

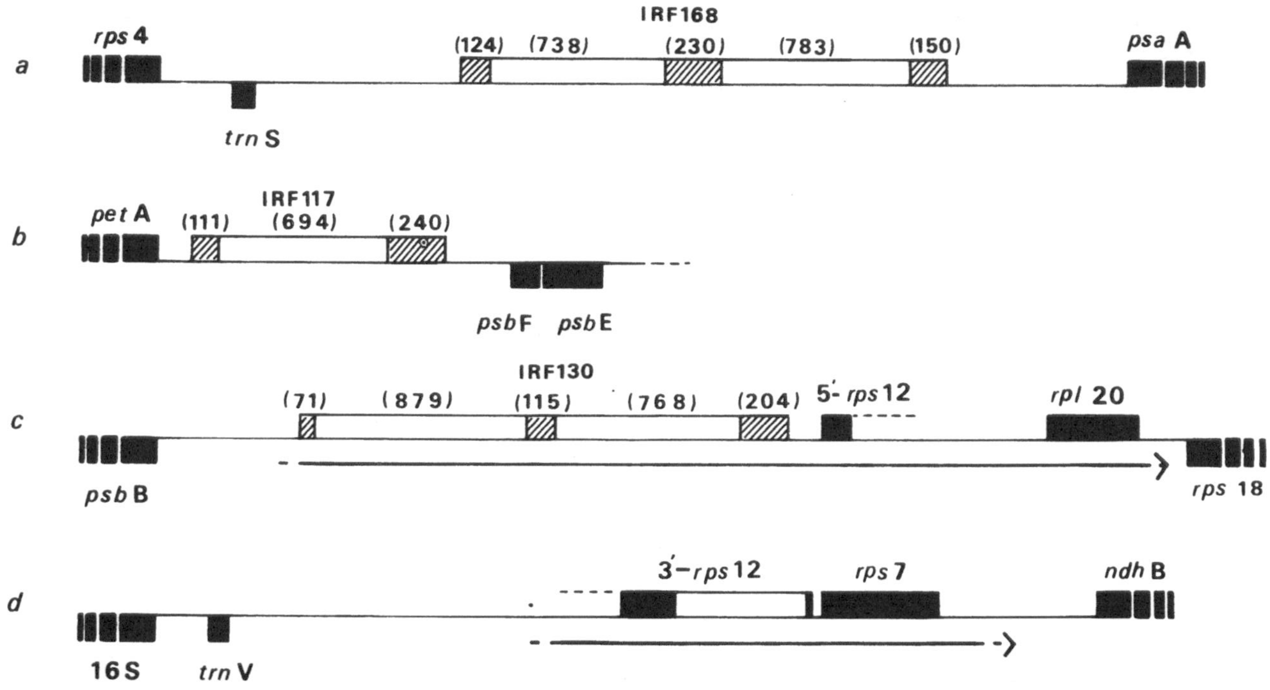

Fig. 5. Schematic presentation of the structure of IRF168(a), IRF117(b), IRF130 + 5'-rps12(c) and 3'-rps12(d) and their surrounding regions. Filled boxes indicate coding regions and exons, open boxes introns, and shaded boxes possible exons in IRFs. Numerals in parentheses are tentative bp. Dotted lines flanking 5' and 3'-rps12s are transons. Arrows in c and d show possible primary transcripts.

REFERENCES

Bird, C.R., Koller, B., Auffret, A.D., Huttly, A.K., Howe, C.J., Dyer, T.A. and Gray, J.C., 1985, The wheat chloroplast gene for CFO subunit of ATP synthase contains a long inton, EMBO J., 4:1381-1388.

Bonnard, G., Michel, F., Weil, J.H. and Steinmetz, A., 1984, Nucleotide sequence of the split tRNA-Leu(UAA) gene from Vicia faba chloroplasts: evidence for structural homologies of the chloroplast tRNA-Leu intron with the intron from the autosplicable Tetrahymena ribosomal RNA precursor, Mol. Gen. Genet., 194:330-336.

Cech, T., 1983, RNA splicing: three themes with variations, Cell, 34:713-716.

Deno, H., Kato, A., Shinozaki, K. and Sugiura, M., 1982, Nucleotide sequences of tobacco chloroplast genes for elongator tRNA-Met and tRNA-Val(UAC): the tRNA-Val(UAC) gene contains a long intron, Nucl. Acids Res., 10:7511-7520.

Deno, H. and Sugiura, M., 1983, The nucleotide sequences of tRNA-Ser(GCU) and tRNA-Gln(UUG) genes from tobacco chloroplasts, Nucl. Acids Res., 11:8407-8414.

Deno, H. and Sugiura, M., 1984, Chloroplast tRNA-Gly gene contains a long intron in the D stem: Nucleotide sequences of tobacco chloroplast genes for tRNA-Gly(UCC) and tRNA-Arg(UCU), Proc. Natl. Acad. Sci. USA, 81:405-408.

Dyer, T.A., 1984, The chloroplast genome: Its nature and role in development, in "Chloroplast Biogenesis", N.R. Baker and J. Barber, eds., Elsevier, Amsterdam.

Erickson, J.M., Rahire, M. and Rochaix, J.D., 1984, Chlamydomonas reinhardii gene for the 32,000 mol. wt. protein of photosystem II contains four large introns and is located entirely within the chloroplast inverted repeat, EMBO J., 3:2753-2762.

Fromm, H., Edelman, M., Koller, B., Goloubinoff, P. and Galun, E., 1986, The enigma of the gene coding for ribosomal protein S12 in the chloroplasts of Nicotiana, Nucl. Acids Res., 14:883-898.

Gray, J.C., Phillips, A.L. and Smith, A.G., 1984, Protein synthesis by chloroplasts, in "Chloroplast Biogenesis", R.J. Ellis, ed., Cambridge University Press, Cambridge.

Guillemaut, P. and Weil, J.H., 1982, The nucleotide sequence of the maize and spinach chloroplast isoleucine transfer RNA encoded in the 16S to 23S rDNA spacer, Nucl. Acids Res., 10:1653-1659.

Hallick, R.B. and Bottomley, W., 1983, Proposals for the naming of chloroplast genes, Plant Mol. Biol. Rep., 1:38-43.

Hallick, R.B., Gingrich, J.C., Johanningmeier, U. and Passavant, C.W., 1985, Introns in Euglena and Nicotiana chloroplast protein genes, in "Molecular Form and Function of the Plant Genôme", L. van Vloten-Doting, G.S.P. Groot and T.C. Hall, eds., Plenum Press, New York, pp.211-231.

Hayashida, N., Matsubayashi, T., Shinozaki, K., Sugiura, M., Inoue, K. and Hiyama, T., 1987, The gene for the 9 kd polypeptide, a possible apoprotein for the iron-sulfur centers A and B of the photosystem I complex, in tobacco chloroplast DNA, Curr. Genet., in press.

Kato, K., Sayer, R.T. and Bogorad, L., 1987, Expression of the PSII-I gene in maize chloroplasts (in Japanese), Proc. Ann. Meeting Jpn. Soc. Plant Physiol., Urawa.

Koch, W., Edwards, K. and Kössel, H., 1981, Sequencing of the 16S-23S spacer in a ribosomal RNA operon of Zea mays chloroplast DNA reveals two split tRNA genes, Cell, 25:203-213.

Koller, B., Fromm, H., Galun, E. and Edelman, M., 1987, Evidence for in vivo trans splicing of pre-mRNAs in tobacco chloroplasts, Cell, 48:111-119.

Koller, B., Gingrich, J.C., Stiegler, G.L., Farley, M.A., Delius, H. and Hallick, R.B., 1984, Nine introns with conserved boundary sequences in the Euglena gracilis chloroplast ribulose-1,5-bisphosphate carboxylase gene, Cell, 36:545-553.

Konarska, M.M., Padgett, R.A. and Sharp, P.A., 1985, Trans splicing of mRNA precursors in vitro, Cell, 42:165-171.

Matsubayashi, T., Wakasugi, T., Shinozaki, K., Yamaguchi-Shinozaki, K., Zaita, N., Hidaka, T., Meng, B.Y., Ohto, C., Tanaka, M., Kato, A., Maruyama, T. and Sugiura M., 1987, Six chloroplast genes (ndhA to F) homologous to human mitochondrial genes encoding components of the respiratory-chain NADH dehydrogenase are actively expressed: Determination of the splice-sites in ndhA and ndhB pre-mRNAs, Mol. Gen. Genet., in press.

Meng, B.Y., Matsubayashi, T., Wakasugi, T., Shinozaki, K., Sugiura, M., Hirai, A., Mikami, T., Kishima, Y. and Kinoshita, T., 1986, Ubiquity of the genes for components of a NADH dehydrogenase in higher plant chloroplast genomes, Plant Sci., 47:181-184.

Michel, F. and Dujon, B., 1983, Conservation of RNA secondary structures in two intron families including mitochondrial-, chloroplast- and nuclear-encoded members, EMBO J., 2:33-38.

Montandon, P.E. and Stutz, E., 1983, Nucleotide sequence of a Euglena glacilis chloroplast genome region coding for the elongation factors Tu; evidence for a spliced mRNA, Nucl. Acids Res., 11:5877-5892.

Montandon, P.E. and Stutz, E., 1984, The gene for the ribosomal proteins S12 and S7 are clustered with the gene for the EF-Tu protein on the chloroplast genome of Euglena gracilis, Nucl. Acids Res., 12:2851-2859.

Murphy, W.J., Watkins, K.P. and Agabian, N., 1986, Identification of a novel Y branch structure as an intermediate in trypanosome mRNA processing: evidence for trans splicing, Cell, 47:517-525.

Neuhaus, H. and Link, G., 1987, The chloroplast tRNA-Lys(UUU) gene from mustard (Sinapis alba) contains a class II intron potentially coding for a maturase-related polypeptide, Curr. Genet., 11:251-257.

Osorio-Almeida, M.L., Guillemaut, P., Keith, G., Canaday, J. and Weil, J.H., 1980, Primary structure of three leucine transfer RNAs from bean chloroplast, Biochem. Biophys. Res. Commun., 92:102-108.

Palmer, J.D., 1985, Comparative organization of chloroplast genomes, Ann. Rev. Genet., 19:325-354.

Rochaix, J.D. and Malnoe, P., 1978, Anatomy of the chloroplast ribosomal DNA of Chlamydomonas reinhardii, Cell, 15:661-670.

Rochaix, J.D., Rahire, M. and Michel, M., 1985, The chloroplast ribosomal intron of Chlamydomonas reinhardii for a polypeptide related to mitochondrial maturases, Nucl. Acids Res., 13:975-984.

Shinozaki, K., Deno, H., Sugita, M., Kuramitsu, S. and Sugiura, M., 1986a, Intron in the gene for the ribosomal proteins S16 of tobacco chloroplast and its conserved boundary sequences., Mol. Gen. Genet., 202:1-5.

Shinozaki, K., Deno, H., Wakasugi, T. and Sugiura, M., 1986b, Tobacco chloroplast gene coding for subunit of proton-translocating ATPase: comparison with the wheat subunit I and E. coli subunit b, Curr. Genet., 10:421-423.

Shinozaki, K., Ohme, M., Tanaka, M., Wakasugi, T., Hayashida, N., Matsubayashi, T., Zaita, N., Chunwongse, J., Obokata, J., Yamaguchi-Shinozaki, K., Ohto, C., Torazawa, K., Meng, B.Y., Sugita, M., Deno, H., Kamogashira, T., Yamada, K., Kusuda, J., Takaiwa, F., Kato, A., Tohdoh, N., Shimada, H. and Sugiura, M., 1986c, The complete nucleotide sequence of tobacco chloroplast genome: its gene organization and expression, EMBO J., 5:2043-2049.

Shinozaki, K., Ohme, M., Tanaka, M., Wakasugi, T., Hayashida, N., Matsubayashi, T., Zaita, N., Chunwongse, J., Obokata, J., Yamaguchi-Shinozaki, K., Ohto, C., Torazawa, K., Meng, B.Y., Sugita, M., Deno, H., Kamogashira, T., Yamada, K., Kusuda, J., Takaiwa, F., Kato, A., Tohdoh, N., Shimada, H. and Sugiura, M., 1986d, The complete nucleotide sequence of tobacco chloroplast genome, Plant Mol. Biol. Rep., 4:110-147.

Shinozaki, K. and Sugiura, M., 1986, Organization of chloroplast genomes, Adv. Biophys., 21:57-58.

Solnick, D., 1985, Trans splicing of mRNA precursors, Cell, 42:157-164.

Steinmetz, A., Gubbing, E.J. and Bogorad, L., 1982, The anticodon of the maize chloroplast gene for tRNA-Leu(UAA) is split by a large intron, Nucl. Acids Res., 10:3027-3037.

Stiegler, G.L., Matthews, H.M., Bingham, S.E. and Hallick, R.B., 1982, The gene for the large subunit of ribulose-1,5-bisphosphate carboxylase in Euglena gracilis chloroplast DNA: location, polarity, cloning, and evidence for an intervening sequence, Nucl. Acids Res., 10:3427-3444.

Sugita, M., Shinozaki, K. and Sugiura, M., 1985, Tobacco chloroplast tRNA-Lys(UUU) gene contains a 2.5-kilobase-pair intron: An open reading frame and a conserved foundary sequence in the intron, Proc. Natl. Acad. Sci. USA, 82:3557-3561.

Sugiura, M., Shinozaki, K. and Ohme, M., 1985, Tobacco chloroplast genes for transfer RNA, in "Molecular Form and Function of the Plant Genome", L. van Vloten-Doting, G.S.P. Groot and T.C. Hall, eds., Plenum Press, New York, pp.325-334.

Sugiura, M., Shinozaki, K., Zaita, N., Kusuda, M. and Kumano, M., 1986, Clone bank of the tobacco (Nicotiana tabacum) chloroplast genome as a set of overlapping restriction endonuclease fragments: Mapping of eleven ribosomal protein genes, Plant Sci., 44:211-216.

Sutton, R.E. and Boothroyd, J.C., 1986, Evidence for trans splicing in trypanosomes, Cell, 47:527-535.

Takaiwa, F. and Sugiura, M., 1982, Nucleotide sequence of the 16S-23S spacer region in an rRNA gene cluster from tobacco chloroplast DNA, Nucl. Acids Res., 10:2665-2676.

Tanaka, M., Obokata, J., Chunwongse, J., Shinozaki, K. and Sugiura, M., 1987, Rapid splicing and stepwise processing of a transcript from the psbB operon in tobacco chloroplasts: Determination of the intron sites in petB and petD, Mol. Gen. Genet., in press.

Tanaka, M., Wakasugi, T., Sugita, M., Shinozaki, K. and Sugiura, M., 1986, Genes for the eight ribosomal proteins are clustered on the chloroplast genome of tobacco (Nicotiana tabacum): Similarity to the S10 and spc operons of Escherichia coli, Proc. Natl. Acad. Sci. USA, 86:6030-6034.

Torazawa, K., Hayashida, N., Obokata, J., Shinozaki, K. and Sugiura, M., 1986, The 5' part of the gene for ribosomal protein S12 is located 30 kbp downstream from its 3' part in tobacco chloroplast genome, Nucl. Acids Res., 14:3143.

Van der Ploeg, L.H.T., 1986, Discontinuous transcription and splicing in trypanosomes, Cell, 47:479-480.

Wada, A. and Sako, T., 1987, Primary structure of and genes for new ribosomal proteins A and B in Escherichia coli, J. Biochem., 101:817-820.

Wakasugi, T., Ohme, M., Shinozaki, K. and Sugiura, M., 1986, Structures of tobacco chloroplast genes for tRNA-Ile(CAU), tRNA-Leu(CAA), tRNA-Cys(GCA), tRNA-Ser(UGA) and tRNA-Thr(GGU): a compilation of tRNA genes from tobacco chloroplasts, Plant Mol. Biol., 7:385-392.

Yamada, K., Shinozaki, K. and Sugiura, M., 1986, DNA sequences of tobacco chloroplast genes for tRNA-Ser(GGA), tRNA-Thr(UGU), tRNA-Leu(UAA), tRNA-Phe(GAA): the tRNA-Leu gene contains a 503 bp intron, Plant Mol. Biol., 6:193-199.

Zaita, N., Torazawa, K., Shinozaki, K. and Sugiura, M., 1987, Trans splicing in vivo: joining of transcripts from the 'divided' gene for ribosomal protein S12 in the chloroplasts of tobacco, FEBS Lett., 210:153-156.

Zurawski, G., Bottomley, W. and Whitfeld, P.R., 1984, Junctions of the large single copy region and the inverted repeats in *Spinacia oleracea* and *Nicotiana debneyi* chloroplast DNA: sequence of the genes for tRNA-His and the ribosomal proteins S19 and L2, Nucl. Acids Res., 12:6547-6558.

CHLOROPLAST-SPECIFIC IMPORT AND ROUTING OF PROTEINS

Peter Weisbeek, Johan Hageman, Sjef Smeekens, Douwe de Boer and Fons Cremers

Department of Molecular Cell Biology and Institute of Molecular Biology, University of Utrecht, Utrecht, The Netherlands

INTRODUCTION

The genetic information of the chloroplast and therefore its protein-coding capacity is much smaller than what is needed for the many photosynthetic and metabolic processes that occur in this organelle. The remaining proteins are coded for by the nuclear DNA, synthesized in the cytosol and transported into the chloroplast post-translationally. The population of proteins present in chloroplasts is therefore of mixed genetic origin; one part is encoded by organellar DNA and the other much larger part by the nuclear DNA. The plant cell therefore has developed a mechanism for transport of proteins across the chloroplast envelope membranes. Once inside the chloroplast routing towards the correct compartment is necessary; either towards stroma, inner membrane, intermembrane space, thylakoid membrane or thylakoid lumen. Similarly the proteins that are made inside the chloroplast have to be sorted properly.

Proteins can not traverse membranes unless they are equipped with specific information for the interaction with the membrane-associated translocation mechanisms. In eukaryotic cells three different types of transport-signals are known for the primary targeting of proteins made in the cytosol. These are i) signalpeptides which in general are used in the cotranslational transport of these proteins into or across the membranes of the endoplasmatic reticulum, ii) nuclear targeting signals for, presumably posttranslational, import into the nucleus, and iii) transitpeptides and presequences and internal targeting signals for posttranslational import into chloroplasts, mitochondria and endosomes. Signalpeptides, transitpeptides and presequences are mostly present as amino-terminal extensions of the mature protein that are cleaved off during or shortly after translocation; the precursor than has a larger size than the functional mature protein. An increasing number of proteins however has no cleavable peptide and also the targeting information need not necessarily be present at the amino-terminus of the protein.

A transitpeptide contains the information necessary for binding to and translocation over the chloroplast membrane as shown by the transit-peptide dependent import into chloroplasts of a bacterial enzyme neomycin phosphotransferase (Van den Broeck et al., 1985; Schreier et al., 1985), a cytosolic soybean shock protein (Lubben and Keegstra, 1986) and the yeast mitochondrial manganese superoxide dismutase (Smeekens et al., 1986b). The transitpeptide most likely also determines the organelle-specificity; the precursor has to go to the chloroplast and not to the mitochondrion.

The process of chloroplast protein import can be distinguished in the following steps : i) binding of the precursor to the chloroplast, ii) translocation across the envelope membranes, iii) routing to one of the intra-organelle compartments and iv) removal of the transitpeptide by a specific protease. How the transitpeptide functions in each of these steps is at present unclear and only very recently has it been shown that envelope transport and routing inside the chloroplast can be separate processes.

If one assumes that the chloroplast developed from a free-living photosynthesizing unicellular organism (with a set of approx. 2000 genes), than a major portion of this DNA has been removed from the organelle during evolution. Probably most of this DNA did become integrated in the nuclear DNA. In all higher plants the same set of chloroplast genes are found. The reason for this conservation is unknown; some of these proteins have been shown to be imported quite efficiently in vitro. It may well be that the main reason for the movement of the DNA lies at the level of control of gene-expression; all nuclear encoded proteins analyzed so far are regulated at the level of transcription whereas in the chloroplast this regulation is mainly post-transcriptionally.

TRANSITPEPTIDE SEQUENCES AND COMPARISON

The nucleotide sequences of the cDNA's for ferredoxin (FD), plastocyanin (PC), the small subunit (SS) and the chlorophyll a/b binding protein (CAB) from Silene pratensis, the genomic DNA coding for plastocyanin from Arabidopsis thaliana and the PC gene of the cyanobacterium Anabaena variabilis ATCC29413 were determined (Smeekens et al., 1985a, 1985b, 1986a; Vorst et al., submitted; Van der Plas, manuscript in preparation) and the parts of the precursor that are involved in transport were compared. The plant proteins are targeted for three different compartments of the chloroplast; ferredoxin and the small subunit are stromal proteins, the CAB protein is integrated in the thylakoid membrane and plastocyanin is located in the thylakoid lumen. Therefore, transitpeptide sequences are available for three different location inside the chloroplast from the same species.

Comparison of the four Silene transitpeptides sequences shows that there is very little homology in length or in amino acid sequence between them. For the transitpeptides of CAB, SS, FD and PC three regions can be found that appear to have some homology; this is shown in Fig.1A. The amino terminal part of these transitpeptides contain mostly hydrophobic and hydroxy amino acids (serine and threonine), whereas near the processing site at the carboxy terminus the sequence GRV or a functionally similar sequence is frequently found (helix destabilizer/positive charge/hydrophobic amino acid). The latter region is in general able to form a structure with a beta turn and this may be important for the interaction of the transitpeptide with the membrane and for processing. These three regions are separated by non-homologous amino acid sequences of varying length. The plastocyanin transit sequence is different from the other three. It has an amino terminal part of comparable composition but its carboxy terminal halve is completely different.

Comparison of the plastocyanine sequence from different plant species and from cyanobacteria showed that the divergence in the transit peptide can also be very significant for the same protein. This is shown in Fig. 1B where also the sequence of the spinach protein (Rother et al., 1986) is included. All plant PC's have a stretch of about

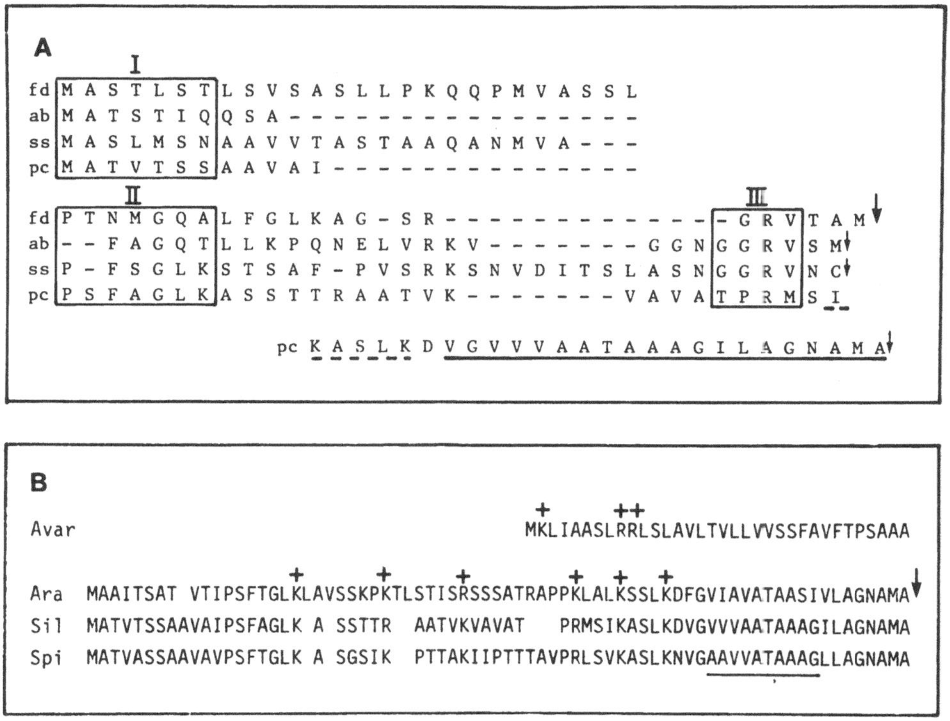

Fig. 1. A. Alignment of the Silene transitpeptides. Homology regions are indicated by roman numerals. The arrows indicate the processing sites. For plastocyanin the intermediate processing region is dashed and the thylakoid transfer domain is underlined. B. Alignment of the transitpeptides of plastocyanin from Silene pratenis, Arabidopsis thaliana, Spinacia oleracea and Anabaena variabilis. Positively charged amino acids are indicated, the arrow indicates the processing site and the hydrophobic region is underlined.

20 hydrophobic though variabel amino acids just in front of the mature protein that is completely absent from the other precursors. It is therefore tempting to assume that this region is involved in the transfer of the thylakoid membrane. This is strengthened by the comparison with a cyanobacterial plastocyanin. In cyanobacteria no import of proteins is necessary but plastocyanin still has to be translocated across the thylakoid membrane. In this case the precursor PC has an N-terminal extension of 34 amino acids as compared to the ca. 60 amino acids found in the plant PC's but the major property of this part of the cyanobacterial precursor is that is very hydrophobic. Is resembles in primary structure a signalpeptide as used in protein transport in prokaryotes and it resembles the carboxy terminal part of the plant PC's. This stresses the involvement of this part in thylakoid passage and it suggests also that this transport process may well be of prokaryotic nature.

An interesting question in the light of this conclusion is whether thylakoid transfer of an imported protein makes use of the same mechanism as thylakoid transfer of a chloroplast-encoded protein (e.g.

cytochrome f) or whether these are two independent processes. Such a post-translational membrane transfer involving a basically cotranslational type of signal sequences has been described for the bacterial export system (Randall and Hardy, 1984) and, more recently, also in the animal cotranslational transport systems (Perara et al., 1986). A consequence of such a situation is that a thylakoid-located signalpeptidase also functions in the processing of the plastocyanin precursor to its mature size.

IMPORT OF PRECURSOR AND FUSION PROTEINS

Protein import into chloroplasts has been studied by incubation of in vitro synthesized proteins with intact chloroplasts followed by fractionation of the chloroplasts to determine the intra-organellar location of the imported proteins. The synthesis of the proteins was done by in vitro transcription of the isolated DNA sequences in an SP6 RNA polymerase dependent transcription system and subsequent translation of this RNA in a wheat germ translation system (Smeekens et al., 1986c). Fusion proteins used are FD-PC and FD-SOD where the ferredoxin transitpeptide is linked to the mature proteins of plastocyanin and yeast mitochondrial superoxide dismutase (SOD) respectivily, and PC-FD and PC-SOD in which the PC transit is coupled to the mature proteins of FD and SOD respectively. The primary structure around the site of fusion in these proteins is given in Fig. 2. The results of incubation of intact chloroplasts with the precursors for ferredoxin and plastocyanin and the fusion proteins and the subsequent fractionation is given in Fig. 3 and 4. It shows that precursor and fusion proteins all bind efficiently to the chloroplast and that they are all internalized and processed. Ferredoxin and the fusion proteins carrying the FD transitpeptide (Fig. 3) are found inside the chloroplast as the mature protein and they are exclusively located in the stroma. This means that the

	-4	-3	-2	-1 ▼	1	2	3	4	5	6	7			
PC	asn	ala	met	ala	ala	glu	val	leu	leu	gly	ser			
FD	val	thr	ala	met	ala	thr	tyr	lys	val	thr	leu			
	-4	-3	-2	5	6	7								
PCFD	asn	ala	met	val	thr	leu								
	-4	-3	-2	-1	1									
PCSOD	asn	ala	met	ala	ala	leu	gln	val	asp					
	-4	-3	-2	-1	1	2	3	4	5	6	-2	-1	1	2
FDPC	val	thr	ala	met	ala	thr	tyr	lys	val	thr	met	ala	ala	glu
	-4	-3	-2	-1	1	2	3	4	5	6				
FDSOD	val	thr	ala	met	ala	thr	tyr	lys	val	thr	met	ala	ala	

Fig. 2. The effect of the fusion procedures on the amino acid sequences near the putative processing sites.

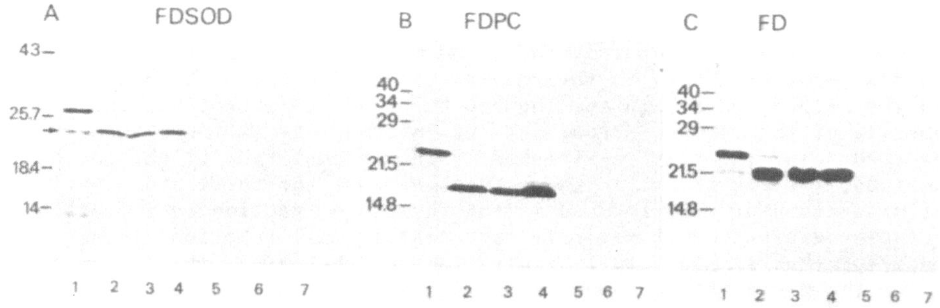

Fig. 3. Import into and distribution inside intact chloroplasts of the precursor for ferredoxin and the FD-SOD and FD-PC fusion proteins. Lanes 1: the in vitro translation mix, lanes 2: import into chloroplasts, lanes 3: import followed by protease treatment, lanes 4: stromal fraction, lanes 5: envelope fraction, lanes 6: thylakoid fraction, lanes 7: protease treated thylakoid fraction.

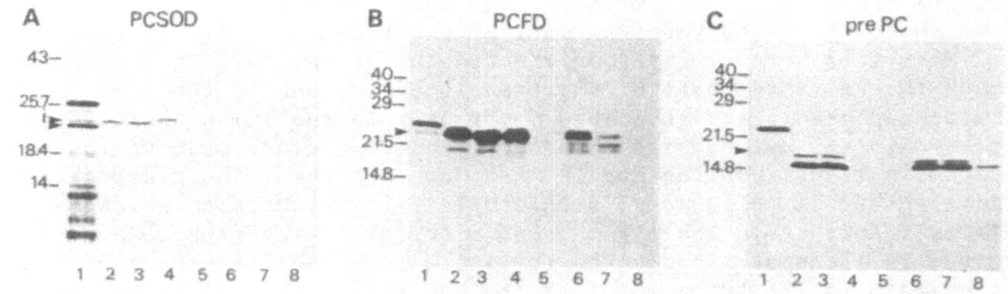

Fig. 4. Import into and distribution inside intact chloroplasts of the precursor for plastocyanin and the PC-SOD and PC-FD fusion proteins. Lanes 1: the in vitro translation mix, lanes 2: import into chloroplasts, lanes 3: import followed by protease treatment, lanes 4: stromal fraction, lanes 5: envelope fraction, lanes 6: thylakoid fraction, lanes 7: protease treated thylakoid fraction and lanes 8: sonicated and protease treated thylakoids.

FD transitpeptide can equally well induce the import of other proteins as it effects the import of the mature FD. This is in line with results obtained for the SSU transitpeptide (Van den Broeck et al., 1985; Lubben et al., 1986). It also means that the processing of the fusion proteins appears not to be influenced seriously by the change in the amino acid sequence after the first six amino acids of the FD mature sequence (Fig. 2).

The import of plastocyanin and the fusion proteins carrying the PC transit is more complex (Fig. 4). The mature form of plastocyanin (Fig. 4C) is only found in the thylakoid fraction in a protease-protected manner as might be expected from the normal location of this protein. Next to the mature plastocyanin however also an intermediate-sized protein is observed that is found in the stroma and/or attached to the stromal face of the thylakoid membrane. This intermediate is a real intermediate because in time-course studies it is converted into the mature protein and this is found in the lumen.

The PC-FD fusion protein (Fig. 4B) is equally well imported and processed. It also gives two bands; the major portion of the imported protein (more than 90%) is intermediate in size between the precursor and the mature ferredoxin and the second, minor component has the mobility of the mature ferredoxin. The intermediate is found in part in the stroma and the rest is attached to the stromal face of the thylakoid membrane, i.e. it is all on the stromal side of the thylakoid membrane. The minor component is located in the thylakoid fraction in a protease protected manner and it may well represent a small fraction of the precursor that is succesfully routed towards the lumen. The fusion of the two parts has been imperfect; one amino acid of the PC transit is lacking and this may have caused the almost complete absence of lumen transport.

In the PC-SOD fusion the complete plastocyanin transitpeptide was present. In this case (Fig. 4A) however a single imported protein is found with a molecular weight that is in between the precursor and the SOD mature protein. More important even this molecule was found only in the stroma and was not associated at all with the thylakoid membrane.

On the basis of the experiments with the PC precursor and the fusion proteins a model for the routing of plastocyanin towards the lumen was developed as shown in Fig. 5. It involves two distinct steps; first pre-plastocyanin is translocated over the envelope into the stroma where the amino-terminal part of the transitpeptide is cleaved off. Next this stromal intermediate recognizes the thylakoid membrane through the remaining part of the transitpeptide and is then translocated across the thylakoid membrane into the lumen where it is processed to the mature size. Based on these and other observations a "two domain transitpeptide model" has been proposed. The plastocyanin transitpeptide is composed of a chloroplast import domain, which mediates transport to the stroma, and a thylakoid transfer domain, involved in transport across the thylakoid membrane. Transport of plastocyanin to the lumen involves the successive removal of the two domains in two different processing steps. A partially purified stromal processing protease (Robinson and Ellis, 1984) was capable of removing the first part of the plastocyanin transit peptide but it does not process the precursor to its mature size. A second and enzymatically different protease could subsequently be identified in the thylakoid fraction (Hageman et al., 1986) and this protease can remove the last part of the PC-transit. This new protease is able to process the intermediate to mature-sized protein but it does not recognises the precursor as a substrate. Complete processing of plastocyanin precursor is therefore only accomplished by successive processing events; the first cleavage at the intermediate processing site mediated by the stromal protease and the second cleavage at the mature processing site mediated by the thylakoidal protease. The results obtained with the two fusion proteins (PC-FD and PC-SOD) make it evident that transport across the thylakoid membrane is a much more sensitive process than translocation across the chloroplast envelope and that it critically depends on the nature of the passenger protein. In both cases the plastocyanin transitpeptide could direct import of the passenger proteins into chloroplasts but was not capable of transporting the proteins across the thylakoid membrane. This ability to block one process while allowing the other to occur, suggests that the two transport events involve significantly different mechanisms. Similar conclusions can be drawn from the results obtained with mutants that have deletions in their transit peptides.

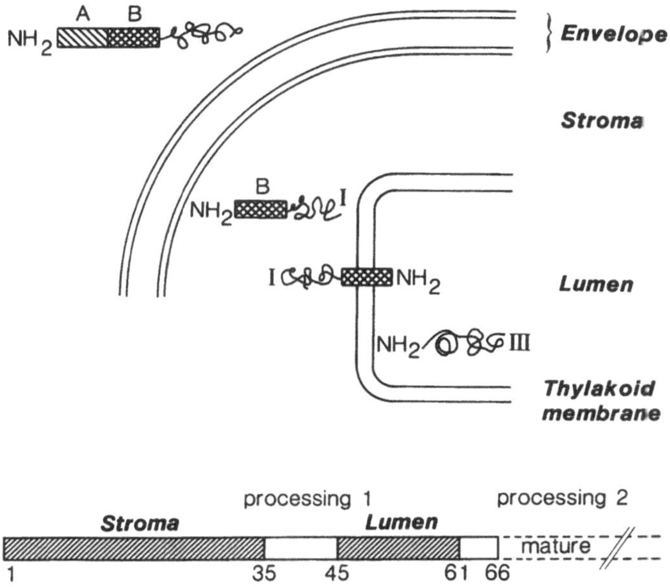

Fig. 5. The two-step transport of plastocyanin to the thylakoid lumen (upper part) and the two transport-domains present in the transitpeptide of the precursor (lower part).

MUTANTS IN THE FD TRANSITPEPTIDE

A set of overlapping deletions, starting from the carboxy-terminus of the transitpeptide towards its amino-terminus were made for the ferredoxin precursor. The resulting mutant precursors with deletions ranging from two to thirty-six amino acids and the result of the import-experiment are given in Fig. 6. Deletion of two amino acids at the junction between transit and mature protein does not affect binding and import and this mutant is also processed to a protein similar in size to mature ferredoxin and therefore reacts indistinguishably from the wildtype ferredoxin precursor. In addition, the efficiency of mutant t2 in binding and import is comparable to the wildtype precursor. Removal of the first seven or more amino acids (t7-t36) completely abolishes binding and translocation. The sequence between -2 and -7 appears therefore to be critical in the binding of the precursor to the envelope membrane.
At present transitpeptide sequences are available for four distinct classes of chloroplast specific precursor proteins: pSS and pCAB sequences from many different species (for a recent compilation see Karlin-Neumann and Tobin, 1986), preferredoxin (Smeekens et al, 1985a) and preplastocyanin (Smeekens et al, 1985b). The -2/-7 regions all

	Ferredoxin -48 … -1	1 5---98	Binding	Import	Processing
wt	MASTLSTLSVSASLLPKQQPMVASSLPTNMGQALFGLKAGSRGRVTAM	A TYKVTL	+	+	+
t2	MASTLSTLSVSASLLPKQQPMVASSLPTNMGQALFGLKAGSRGRVT	A TYKVTL	+	+	+
t7	MASTLSTLSVSASLLPKQQPMVASSLPTNMGQALFGLKAGS	P TYKVTL	−	−	−
t10	MASTLSTLSVSASLLPKQQPMVASSLPTNMGQALFGLK	A TYKVTL	−	−	−
t13	MASTLSTLSVSASLLPKQQPMVASSLPTNMGQALF	A TYKVTL	−	−	−
t15	MASTLSTLSVSASLLPKQQPMVASSLPTNMGQA	S TYKVTL	−	−	−
t16	MASTLSTLSVSASLLPKQQPMVASSLPTNMGQ	A TYKVTL	−	−	−
t17	MASTLSTLSVSASLLPKQQPMVASSLPTNMG	P TYKVTL	−	−	−
t18	MASTLSTLSVSASLLPKQQPMVASSLPTNM	A TYKVTL	−	−	−
t20	MASTLSTLSVSASLLPKQQPMVASSLPT	T TYKVTL	−	−	−
t36	MASTLSTLSVSA	S TYKVTL	−	−	−

Fig. 6. Amino acid sequence of the transitpeptides of mutantferredoxin precursors and their properties in binding, import and processing.

contain the sequence 'GRV' or a functionally very similar sequence (helix destabilizer / positive charge / hydrophobic amino acid). The helix destabilizing and charged amino acids in this region could be involved in surface exposure of this region.

The main conclusion that can be drawn from this deletion analysis is that the C-terminus of the transitpeptide contains essential information for binding of the precursor to the envelope.

MUTANTS IN THE PC TRANSITPEPTIDE

A set of overlapping deletions, starting from the carboxy-terminus of the transitpeptide towards its amino-terminus were also made for the plastocyanin precursor. The amino acid sequence of the resulting mutant precursors with deletions ranging from two to twenty-five amino acids and the result of their import is given in Fig. 7 and 8. The deletions were made only in the thylakoid domain of the transitpeptide.
The mutant precursors all bind to the envelope, are translocated into the chloroplast and become processed to the intermediate form only. Deletion of two or three amino acids near the junction between transit and mature protein slows down import considerably and the imported processed precursor binds to the thylakoid membranes without being translocated (they are protease-sensitive). Removal of seven or more

	Plastocyanin transit -66 … -1		Import	Routing
wt	MATVTSSAAVAIPSFAGLKASSTTRAATVKVAVATPRMSIKASLKDVGVVVAATAAAGILAGNAMA		+	lumen
t2	MATVTSSAAVAIPSFAGLKASSTTRAATVKVAVATPRMSIKASLKDVGVVVAATAAAGILAG	MA	±	thylakoid
t3	MATVTSSAAVAIPSFAGLKASSTTRAATVKVAVATPRMSIKASLKDVGVVVAATAAAGILA	MA	±	thylakoid
t7	MATVTSSAAVAIPSFAGLKASSTTRAATVKVAVATPRMSIKASLKDVGVVVAATAAA	MA	+	stroma
t11	MATVTSSAAVAIPSFAGLKASSTTRAATVKVAVATPRMSIKASLKDVGVVVAA	MA	+	stroma
t21	MATVTSSAAVAIPSFAGLKASSTTRAATVKVAVATPRMSIKAS	MA	+	stroma
t25	MATVTSSAAVAIPSFAGLKASSTTRAATVKVAVATPRMS	MA	±	stroma

Fig. 7. Amino acid sequence of the transitpeptides of mutantplastocyanin precursors and their in vitro routing in chloroplasts.

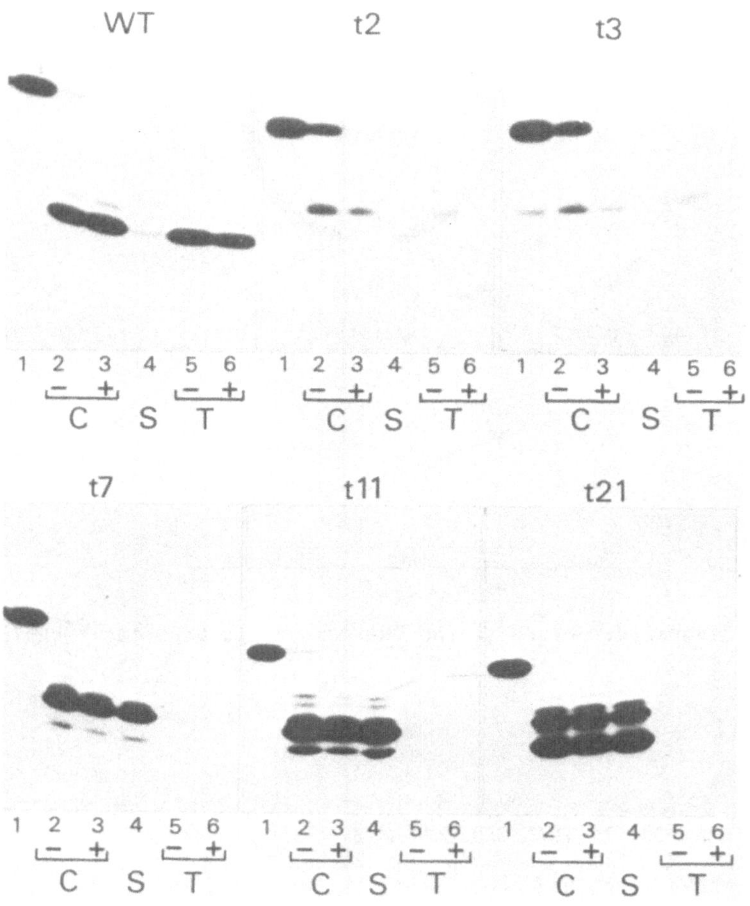

Fig. 8. Import into and distribution inside intact chloroplasts of the mutant-plastocyanin precursors. Lanes 1: the in vitro translation mix, lanes 2: import into chloroplasts, lanes 3: import followed by protease treatment, lanes 4: stromal fraction, lanes 5: thylakoid fraction, lanes 6: protease treated thylakoid fraction.

amino acids restores the wild type efficiency of import but no affinity for the thylakoid membrane is left; the imported protein fractionates completely in the stroma.

The conclusions from this analysis are that i) the thylakoid domain of the transitpeptide does influence the transport over the envelope, ii) the thylakoid transfer is more sensitive to disturbance of the regular transitpeptide sequence than the envelope transport and iii) binding to and translocation over the thylakoid membrane can be separated whereas this has not been observed with the envelope transport mutants.

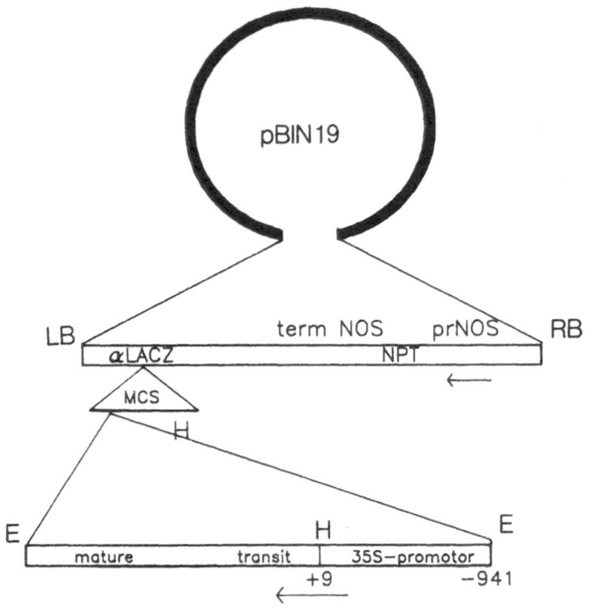

Fig. 9. General structure of the DNA constructs used for transformation.

IN VIVO TRANSPORT IN TRANSGENIC PLANTS

The above described experiments reflect upon the properties of the proteins when tested in an in vitro or semi in vitro system. Research in many other areas of molecular biology has shown that results obtained in in vitro experiments in general are valid for the in vivo situation but that in details there can be quite some variation . Therefore the DNA of the precursor proteins and fusion proteins was transformed into the nuclear DNA of tomato plants and the transformed plants were subsequently tested for integration of the new DNA, for the expression of these genes and for the transport properties of these added proteins. Details of this analysis will be published elsewhere.

DNA constructs for preFD and PC-FD were linked to the 35S CMV promotor (Morelli et al., 1985) and integrated in the pBIN19 vector (Bevan, 1984). The structure of the resulting recombinant plasmids is given in Fig. 9 and they were introduced in tomato plant cells with the Agrobacterium tumefaciens leaf disc transformation assay. The selected (kanamycin-resistant) plants were analyzed by Southern hybridization for the chromosomal situation of the integrated plasmid DNA. All selected plants had a single insert of the right size (data not shown). For a number of these plants total RNA was isolated and screened with probes specific for the integrated DNA (Northern analysis). This is shown in

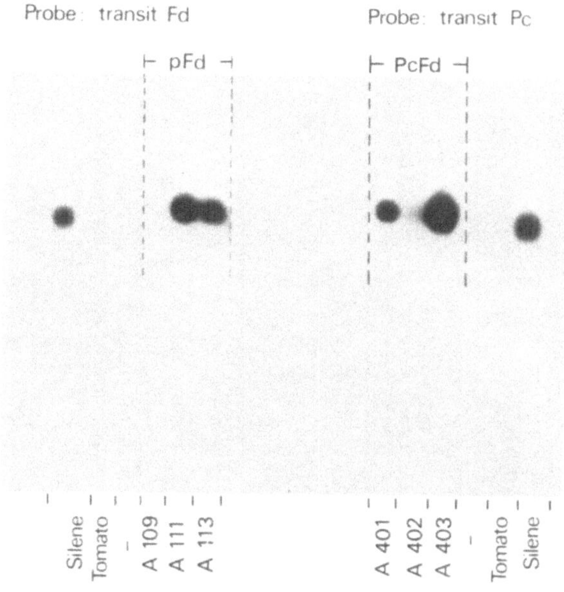

Fig. 10. Northern blot of total leaf-RNA of transformed tomato plants. A109, 111 and 113 contain the preFD gene; A401, 402 and 403 contain the PC-FD gene. The probe used was transit FD DNA (left part) or transit PC DNA (right part).

Fig. 10 and it demonstrates that both the preFD and the PC-FD DNA are expressed at the level of transcription and that the transcripts have the expected size when compared to the Silene RNA of known size. The analysis at the level of protein synthesis is complicated by the similarities in size between the added Silene ferredoxin and the endogenous tomato ferredoxin which means that in tissue synthesizing tomato ferredoxin expression of the new proteins is difficult to monitor. In petals and roots normally no ferredoxin is synthesized; this was checked by immunoblotting of protein-gels and by electron microscopical analysis of cryo-coupes of petals and roots treated with anti-ferredoxin serum and immunogold. The immunoblots gave only very weak bands for ferredoxin as compared to leaf tissue and in the EM analysis no FD-specific labelling could be detected. Therefore the expression of the Silene DNA in transformed plants could be followed in these plant parts.

Western blots of petal and root-proteins probed with anti-ferredoxin serum showed a cross-reacting band of the size of mature ferredoxin (data not shown). The fact that the proteins are found in the mature size indicates that processing has occurred; this was investigated

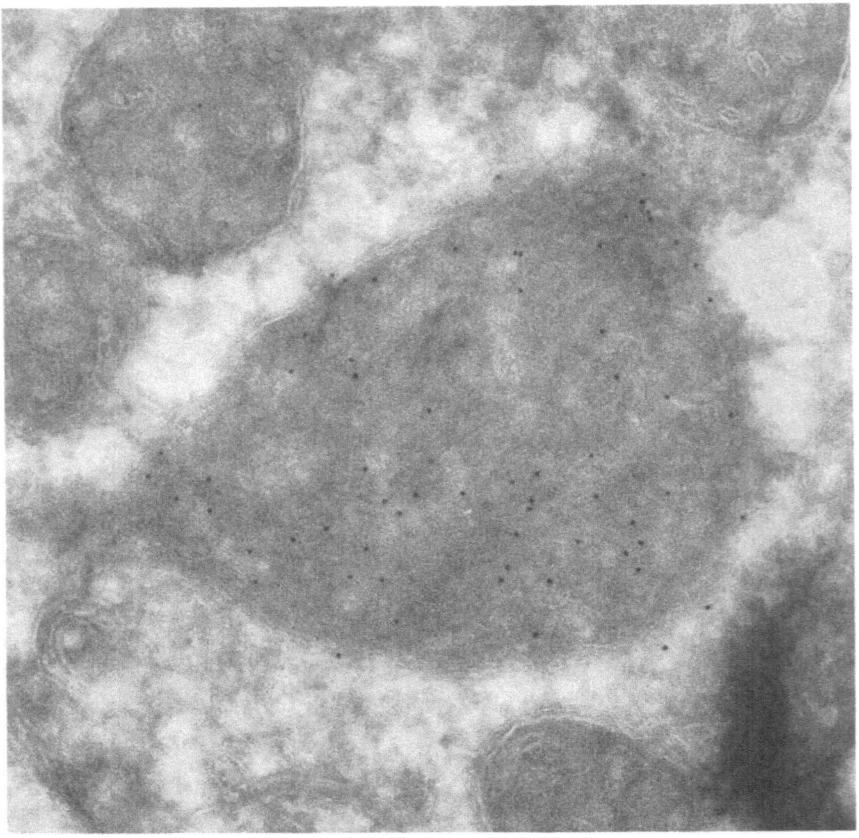

Fig. 11. Electron microscopical analysis of petals of tomato plants transformed with Silene preFD DNA. The cryocoupes were incubated with anti-ferredoxin serum and the antibody-ferredoxin interaction was visualized with proteinA-gold spheres.

further by EM immuno-gold analysis of this tissue. Coupes of roots and petals of transformed plants were found to react efficiently with anti-ferredoxin serum whereas the untransformed plants did not. The goldlabel is found in the plastids (chromoplasts for petals and leukoplasts for roots) whereas the remaining parts of the cells do not react significantly. This is shown for petals in Fig.11.

Although this analysis has not been completed yet, it does indicate strongly that in petals and roots of the transformed plants the precursors are transported to the chloroplast-related organelle and are imported and processed. One important conclusion that can be drawn from these experiments is that the protein-import mechanisms of different plastids do recognize and interact with precursors that are normally not made in that specific tissue. It appears therefore that the translocation systems of different plastids are very similar or even identical. Another conclusion can be that the number of different plastid-envelope receptors is very small (or even just one) and that they are identical between the different plastids.

CONCLUSIONS

The major aspects of the mechanisms for protein transport and sorting in plant cells still have to become elucidated. It may however be expected that the strategies that are employed in other eukaryotic cells will also be valid here. The routing towards the chloroplast and the sorting within this organel are however unique mechanisms for the plant cell and the first insights in how cytoplasmically synthesized proteins reach the chloroplast have been obtained.

The results obtained with the plastocyanin precursor and fusion proteins have made visible the basic concept of how proteins reach the thylakoid lumen. Here two different transport steps occur, one for passage of the envelope membranes and the other to pass the thylakoid membrane; whether they happen indepently or are in some way linked together has been undetermined sofar. Further research has to focus on the topology of the transitpeptide to determine what amino acid sequence or other property determines the targeting-specificity for this organel, whether the information for targeting and translocation can be separated, and how the transitpeptide functions in the intra-organellar routing. The analysis about the transport mechanisms in the envelope membranes and the thylakoid membrane still has to be started. There are indications for the involvement of a proteinacious receptor but direct interactions between the transit peptide and the membrane is another interesting possibility for the first step in translocation that has not been worked out yet. The available information with regard to the thylakoid membrane is even more limited. So far thylakoid translocation appears to be a very sensitive process; knowledge of the real mechanism is necessary to learn how to overcome the inability to introduce foreign proteins into this compartment.

The targeting and sorting of proteins of the inner membrane and also of the thylakoid membrane itself, with the exception of the chlorophyll a/b protein, has not been studied yet. The first experiments with the chlorophyll a/b protein (Cline, 1986) indicate that it behaves very differently compared to plastocyanin.

The results obtained sofar with the stroma-targeted proteins show that we are able to use these transitpeptides for import into the stroma of a variety of foreign proteins without knowledge about the translocation mechanism in which these peptides function. It enables us to study and interfere with many metabolic processes that occur in the chloroplast. Similar possibilities will soon be available for the other compartments of the plant cell.

The first experiments on the transport of proteins in different tissues indicate that the different plastids have very similar ways for import; it may well be that during the development of the plant no major changes occur in the mechanism and specificity of this process. It seems unattractive for a plastid to have specific receptors for proteins that have never to be imported into that organelle; for that reason we favour the presence of one or a few receptors with affinity for all the proteins that are imported by the different plastids together.

REFERENCES

Bevan, M., 1984. Binary Agrobacterium vectors for plant transformation. Nucleic Acids Res. 12, 8711-8721.

Cline, K., 1986. Import of proteins in chloroplasts. Membrane integration of a thylakoid precursor protein reconstituted in chloroplast lysates. J. Biol. Chem. 261, 14804-14810.

Hageman, J., Robinson, C., Smeekens, S., Weisbeek, P., 1986. A thylakoid located processing protease is required for complete maturation of the lumen protein plastocyanin. Nature 324, 567-569.

Karlin-Neumann, G.A., Tobin, E.M., 1986. Transit peptides of nuclear encoded chloroplast proteins share a common amino acid framework. EMBO J. 5, 9-13.

Lubben, T., Keegstra, K., 1986. Efficient in vitro import of a cytosolic heatshock protein into pea chloroplasts. Proc. Natl. Acad. Sci. USA. 83, 5502-5506.

Morelli, G., Nagy, F., Fraley, R.T., Rogers, S.G., Chua, N-H., 1985. A short conserved sequence is involved in the light-inducibility of a gene encoding ribulose 1,5-biphosphate carboxylase small subunit of pea. Nature 315, 200-204.

Perara, E., Rothman, R.E., Lingappa, V.R., 1986. Uncoupling of translocation from translation: implications for transport of proteins across membranes. Science 232, 348-352.

Randall, L.L., Hardy, S.J.S., 1984. Export of proteins in bacteria. Microbiological Reviews 48, 290-298.

Robinson, C., Ellis, R.J., 1984. Transport of proteins into chloroplasts. Partial purification of a chloroplast protease involved in the processing of imported precursor polypeptides. Eur. J. Biochem. 142, 337-342.

Rother, C., Jansen, T., Tyagi, ., Tittgen, J. and Herrmann, .G., 1986, Plastocyanin is encoded by an uninterrupted nuclear gene in spinach, Curr. Genet., 11:171-176.

Schreier, P.H., Seftor, E.A., Schell, J., Bohnert, H.J., 1985. The use of nuclear encoded sequences to direct the light regulated synthesis and transport of a foreign protein into plant chloroplasts. EMBO J. 4, 25-32.

Smeekens, S., Van Binsbergen, J., Weisbeek,P., 1985a. The plant ferredoxin precursor: nucleotide sequence of a full length cDNA clone. Nucl. Acids Res. 13, 3179-3194.

Smeekens, S., De Groot, M., Van Binsbergen, J., Weisbeek,P., 1985b. The sequence of the precursor of the chloroplast thylakoid lumen protein plastocyanin. Nature 317, 456-458.

Smeekens, S., Van Oosten, J., De Groot, M., Weisbeek, P. 1986a. Silene cDNA clones for a divergent chlorophyll-A/B-binding protein and a small subunit of ribulosebiphosphate carboxylase. Plant Mol. Biol. 7, 433-441.

Smeekens, S., 1986b, Transport of nuclear-encoded chloroplast proteins. Thesis, University of Utrecht, Utrecht, The Netherlands.

Smeekens, S., Bauerle, C., Hageman, J., Keegstra, K., Weisbeek, P., 1986c. The role of the transit peptide in the routing of precursors towards different chloroplast compartments. Cell 46, 365-375.

Van den Broeck, G., Timko, M.P., Kausch, A.P., Cashmore, A.R., Montagu, M. van, Herrera-Estrella, L. 1985. Targeting of foreign protein to chloroplasts by fusion to the transit peptide from the small subunit of ribulpse-1,5-biphosphate carboxylase. Nature 313, 358-363.

Vorst, O., Smeekens, S., Weisbeek, P., 1987. Molecular cloning, sequence analysis and transport of the Arabidopsis thaliana plastocyanin precursor. Submitted.

CELL-SPECIFIC EXPRESSION OF LHCII AND THE ORGANISATION OF THE PHOTOSYNTHETIC REACTION CENTRES IN CHLOROPLAST THYLAKOIDS

D.J. Simpson, R. Bassi[1] and U.G. Hinz[2]

Department of Physiology, Carlsberg Laboratory,
Gamle Carlsberg Vej 10, DK-2500 Copenhagen, Valby. Denmark
[1] Dipartimento di Biologia, Universita' di Padova, PD 3513 Italy
[2] Friedrich Miescher Institut, CH-4002 Basel, Switzerland

INTRODUCTION

The photosynthetic membranes of plant chloroplasts capture light energy and convert it into high energy chemical compounds (ATP and NADPH), releasing molecular oxygen from water in the process. The absorption of light is mediated by pigments (xanthophylls, carotenoids, chlorophyll b and chlorophyll a) bound non-covalently to protein. Most of the pigments are associated with chlorophyll-proteins which function as light-harvesting antennae and transfer the captured excitation energy to reaction centres. Each type of reaction centre (photosystem I and photosystem II) has its own set of antennae, but the major one in thylakoids is the light-harvesting chlorophyll a/b-protein complex of photosystem II, or LHCII. The complexity of the chlorophyll-protein composition of thylakoids can be seen in Figure 1a, which is an unstained non-denaturing polyacrylamide gel. The intensity of the bands reflects the amount of chlorophyll in each one, and shows how much is present in LHCII, especially in grana membranes. The chlorophyll-proteins and chlorophyll-protein complexes (those with an asterisk) of grana membranes (Figure 1b) belong to photosystem II.

THE FUNCTION OF LHCII

Isolated LHCII has a chlorophyll a/b ratio of 1.37, with between 11-13 molecules of chlorophyll per 25 kDa polypeptide chain (Mullet 1983, Hinz and Welinder 1987). It contains large amounts of neoxanthin and violaxanthin, plus β-carotene (Rawyler et al. 1980). The most pure LHCII is that isolated as the oligomer by non-denaturing sodium dodecyl sulphate polyacrylamide gel electrophoresis (SDS-PAGE). This can be resolved into several bands in the 25-29 kDa range, depending on the gel system and the plant material.

Most of the LHCII in thylakoids is located in the appressed grana regions. This is clear from the chlorophyll-protein composition of grana (Figure 1b) and is demonstrated by immunogold labelling of isolated thylakoids using an antibody to LHCII (Figure 2). Its function is to transfer excitation energy to the photosystem II reaction centres in appressed grana lamellae, although under certain circumstances (see below), LHCII can transfer energy to photosystem I. When LHCII is absent, such as in the chloroplasts of the chlorophyll b-less mutant of barley (chlorina-f2), the

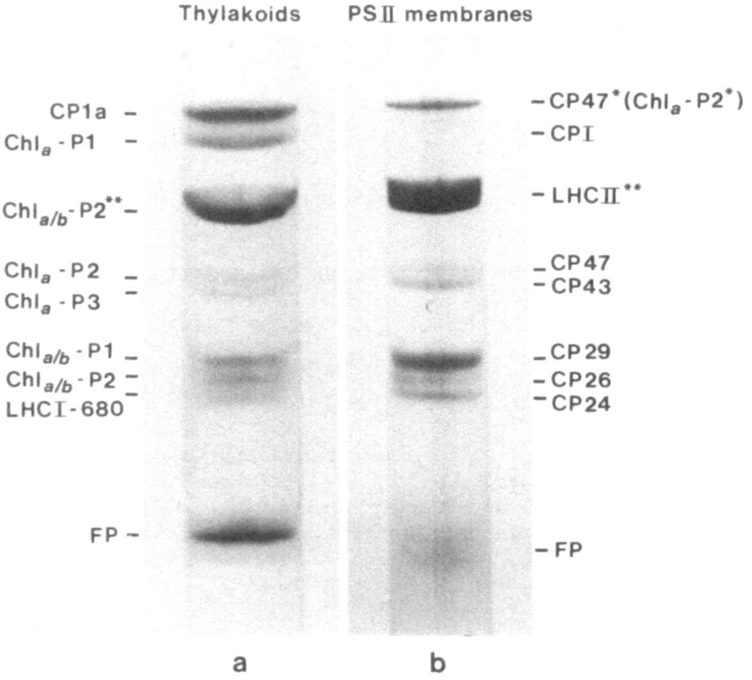

Figure 1. Non-denaturing gel electrophoretic separation of the chlorophyll-proteins of thylakoids and isolated grana appressed membranes. Alternative nomenclature is shown for some chlorophyll-containing bands.

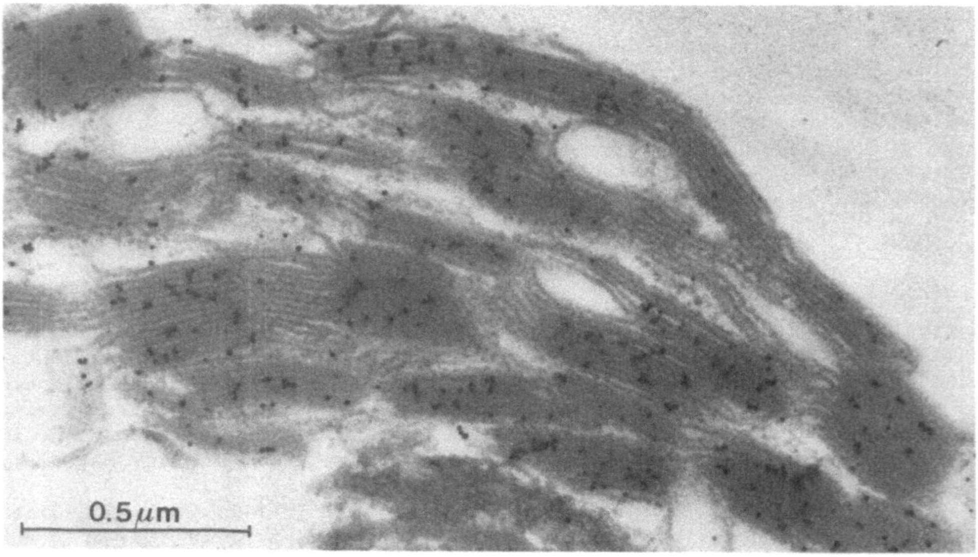

Figure 2. Immunogold labelling of isolated thylakoids, showing localisation of LHCII. Thin sections are incubated with an antibody to LHCII and visualised by reacting with protein A labelled with colloidal gold. Each dark sphere thus corresponds to the location of an LHCII antibody, and hence LHCII, which is mostly in the appressed grana regions.

thylakoids are still able to form grana (Simpson 1979), but the Mg^{++} concentration required to maintain appression in vitro is raised from 2 mM to about 25 mM. In this mutant, other components of photosystem II maintain thylakoid appression in vivo (Bassi et al. 1985), but the photosystem II reaction centres cannot transfer excitation energy between themselves, and because of their smaller antenna size, are less efficient in capturing light energy.

LHCII thus has many functions in thylakoids. It transfers excitation energy to photosystem II, and by transferring energy from one LHCII assembly to the next, it facilitates interconnection of photosystem II reaction centres. It can also transfer excitation energy to photosystem I (Bassi and Simpson 1987) and has an important role in maintaining thylakoid appression. LHCII consists of at least 5 polypeptide bands, and molecular biological evidence has shown that an even greater number of different LHCII genes are expressed. This article examines the regulation of LHCII expression at the protein level in relation to its function in different cell types and in different regions of the thylakoid membrane.

SYNTHESIS AND GENETICS OF LHCII

LHCII contains about 230 amino acids, as deduced from the nucleotide sequences of cDNA derived from LHCII mRNA, corresponding to a molecular weight between 25 and 29 kDa (Dunsmuir 1985, Lamppa et al. 1985, Karlin-Neumann et al., 1985, Pichersky et al. 1986). The mRNA for LHCII is abundant during greening (Apel and Kloppstech 1978) and specific for leaf tissue. Its accumulation in response to light is mediated by phytochrome (Apel 1979), although transcription may also be regulated by products of plastid ribosomal synthesis (Mayfield and Taylor 1984) or precursors of chlorophyll biosynthesis, such as Mg protoporphyrin methyl ester (Johanningmeier and Howell 1984). LHCII is synthesised on cytoplasmic ribosomes as a larger precursor (pre-LHCII), which is water-soluble and contains an N-terminal transit sequence of about 34 amino acids. Pre-LHCII binds to the outer chloroplast envelope and is post-translationally transported into the plastid if ATP is present in the medium (Cline et al. 1985). ATP is further required for the integration of pre-LHCII into the thylakoid membrane, where it is processed into the mature, water-insoluble form by cleavage of the transit sequence (Cline 1986, Chitnis et al. 1986). This is in contrast to proteins transported across the thylakoid membrane into the thylakoid lumen, where the envelope transit sequence is cleaved by a stromal protease before the intermediate precursor crosses the thylakoid and is cleaved to the mature form (Hageman et al. 1986).

Although transcription and translation of LHCII mRNA does not require chlorophyll _a_ (Apel 1979) or chlorophyll _b_ (Bellemare and Chua 1982), both are required for stable integration of LHCII into the membrane, although photosystem II reaction centres are not, since mutants lacking photosystem II, contain LHCII (Simpson et al. 1977). In the absence of chlorophyll _b_, as for example in the barley mutant _chlorina-f2_, LHCII turns over rapidly and only low levels are present in the thylakoid (Bellemare and Chua 1982). If this mutant is grown under continuous white light, then a significant amount of LHCII can be detected by non-denaturing SDS-PAGE (Bassi et al. 1985). But it contains no chlorophyll _b_ and is too unstable to be eluted from the gel.

Sequencing of different cDNA clones derived from the major LHCII mRNA, revealed that in petunia there were 5 multigene families, each with several members, with a minimum of 16 genes in all (Dunsmuir 1985). This family codes for the larger, more abundant LHCII polypeptides and is designated type I. Type II LHCII mRNA has been cloned from _Lemna_ (Karlin-Neumann et al. 1985) and petunia (Stayton et al. 1986). These differ from type I LHCII genes in having a small intron located between the codons for amino acids 13 and 14 of the mature LHCII polypeptide in both species. They have a lower amino acid sequence homology to type I LHCII (86 vs 96-98%), especially at the transit sequence (16 vs 88-91%), and code for a polypeptide with a higher electrophoretic mobility (Stayton et al. 1986). A similar genetic complexity for the LHCII genes has been found for Lemna (Karlin-Neumann et al. 1985) and tomato (Pichersky et al. 1985), and analysis of tomato genomic clones has revealed 2 loci, Cab-1 and Cab-3, located on chromosomes 2 and 3, respectively. At the Cab-1 locus, 4 LHCII genes are found in tandem within a 10 kb stretch of DNA, and because of their similarity to one another, are postulated to have arisen by gene duplication events. The deduced amino acid sequences of the different type I LHCII polypeptides differ mainly at the N-terminus of the mature polypeptide, and in the transit sequences (Dunsmuir 1985, Pichersky et al. 1985). It is not known why so many different LHCII genes are expressed in these plants. Part of the reason may lie in slightly different functions, or differential expression in response to changes in development or the environmental conditions, or different types of tissue. In Arabidopsis, a flowering plant with a very small genome, there are, however, only 3 copies of the LHCII gene (Meyerowitz and Pruitt 1985).

THE STRUCTURE OF LHCII

LHCII can be isolated in relatively pure form by solubilisation of thylakoids in Triton X-100, followed by sucrose gradient ultracentrifugation and Mg^{++} precipitation of the fluorescent LHCII band (Burke et al. 1978). It has been recently shown, however, that such preparations are contaminated with another chlorophyll a/b-protein of photosystem II designated CP29 (Høyer-Hansen et al. 1987). Although this is not a problem in chemical and spectroscopic analysis of LHCII, CP29 is a good antigen, and antibodies raised against such an LHCII preparation will react with CP29, and cross-react with the light-harvesting chlorophyll a/b-proteins of photosystem I (LHCI). Thus, although there are likely to be primary sequence homologies between the different chlorophyll a/b-proteins of plant thylakoids, some of the work with antibodies raised against LHCII prepared by the method of Burke et al. (1978) may need to be re-evaluated.

Even when LHCII is isolated as the oligomer by SDS-PAGE, it is still not known how many different polypeptides are present, or even if they all bind chlorophyll. Thus, the number of chlorophylls bound per polypeptide chain may have been underestimated. Of the different spectroscopic techniques available for monitoring the integrity of isolated LHCII, the most sensitive is circular dichroism (CD) spectroscopy, which in the red region, measures the interactions between chlorophyll molecules, and in the UV region, reflects the organisation of the polypeptide backbone. Addition of detergents, such as Triton X-100 or SDS, caused a progressive loss of the negative CD peak at 688 nm with time, indicating changes in the organisation of chlorophyll a (Hinz and Welinder 1987). This was not due to the dissociation of LHCII oligomers into monomers, since detergent-treated LHCII eluted from gel filtration columns with an apparent molecular weight of 275-300 kDa. Molecules of chlorophyll b were much more resistant to the action of detergents, indicating that they are better protected by the folding of the LHCII polypeptide chain.

Extensive trypsin-treatment of isolated LHCII caused relatively small changes in the CD spectrum (Figure 3), with reductions in the peaks arising from chlorophyll a and xanthophylls. This was accompanied by an increase in the electrophoretic mobility of denatured LHCII, corresponding to an apparent molecular weight of 16 kDa. The chlorophyll a/b ratio decreased from 1.37 to 1.25, with no change in the molar ratio of chlorophyll b : protein, indicating the loss of only 1 molecule of chlorophyll a per poly-
195, 546-557 (1979)

was identical to a stretch of 14 amino acids in the wheat sequence, starting with serine 52 (Lamppa et al. 1985). Based on the molecular weight of the tryptic LHCII fragment, and the deduced amino acid sequence, the probable site of C-terminal cleavage is at lysine 203 (Hinz and Welinder 1987)

It has been established by Andersson et al. (1982), using inside-out and right-side out thylakoid vesicles, and antibodies to LHCII, that the LHCII polypeptide spans the membrane, while Mullet (1983) has shown that the N-terminus is located at the stromal side. The three-dimensional structure of LHCII in 2-dimensional crystals has been reported at 16 Å resolution, and indicates the presence of 3 or possibly 4 transmembrane α-helices (Kühlbrandt 1984). Based on hydropathy plots, Karlin-Neumann et al. (1985) have proposed a model for the folding of LHCII with 3 such membrane-spanning helices. This places the N- and C-termini on opposite sides of the membrane. Anderson and Goodchild (1987) have used antibodies raised against a synthetic peptide based on the C-terminal sequence to show that the C-terminus is located on the stromal side. The hypothetical model in Figure 4 is based partly on hydropathy values, and partly on algorithms for the prediction of β-turns in membranes. The folding pattern differs from that of Karlin-Neumann et al. (1985) mainly in having an extra transmembrane α-helix from valine 90 to leucine 113, placing the N- and C-termini on the same side of the membrane. In this model, most of the trypsin-susceptible lysine and arginine residues are buried in the hydrophobic α-helices, where most of them are potentially stabilised by salt bridges to glutamate residues, and are thus resistant to trypsin attack. There are, furthermore, no proline residues located in the transmembrane α-helices of this model.

The pigment analysis shows that, with the exception of 1 molecule of chlorophyll a, all of the chlorophyll associated with LHCII remains bound to the 16 kDa tryptic fragment. They are presumably bound to the 4 transmembrane helices, but the chemical nature of the association is not known. Initial experiments with photoaffinity chemical cross-linkers in vitro has

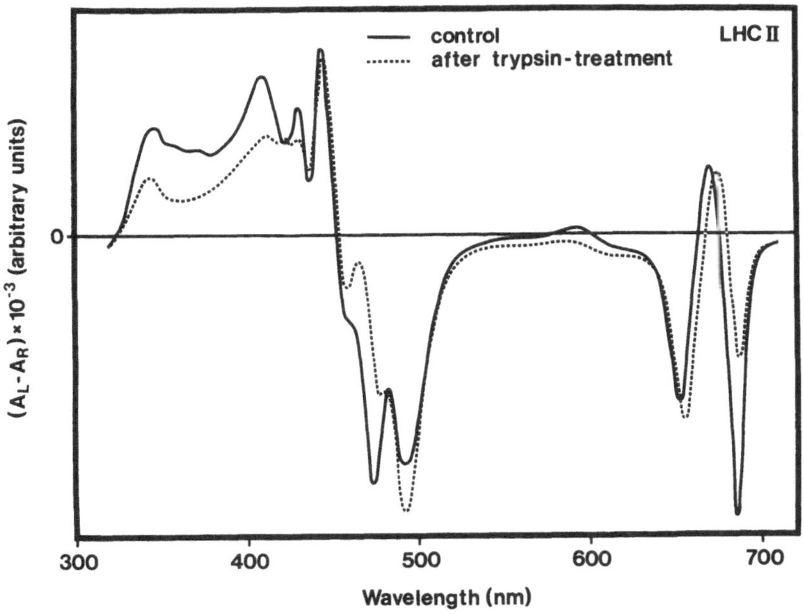

Figure 3. CD spectrum of purified LHCII before and after trypsin treatment. Note the changes in the regions due to chlorophyll a (688 nm) and xanthophylls (476 and 493 nm).

been promising (Hinz, unpublished results), and may enable the elucidation of the binding sites of chlorophyll in LHCII when used in combination with the reconstitution techniques of Plumley and Schmidt (1987). The recent isolation of three-dimensional crystals of LHCII (Kühlbrandt 1987), which produced X-ray reflections to a resolution of 3.7 Å may ultimately reveal the orientation and association between chlorophyll molecules, and between chlorophylls and the side chains of the LHCII polypeptide backbone.

CELL-SPECIFIC EXPRESSION OF LHCII

The leaves of C_4 plants contain two distinct types of chloroplasts located in different types of cells. The mesophyll cell chloroplasts contain stromal and grana lamellae as well as photosystem I and photosystem II. They fix CO_2 by phosphoenolpyruvate carboxylase (PEPC) and transport it as malate to the bundle sheath cells. In species which contain high levels of the NADPH-malic enzyme, such as sorghum and maize, the bundle sheath chloroplasts of fully developed leaves are agranal and contain no functional photosystem II activity (Schuster et al. 1985, Bassi and Simpson 1986). In bundle sheath cells the malate is decarboxylated to yield CO_2 and NADPH, and the ATP required for CO_2 fixation by ribulose 1,5 bisphosphate carboxylase (Rubisco) is generated by cyclic electron flow around photosystem I.

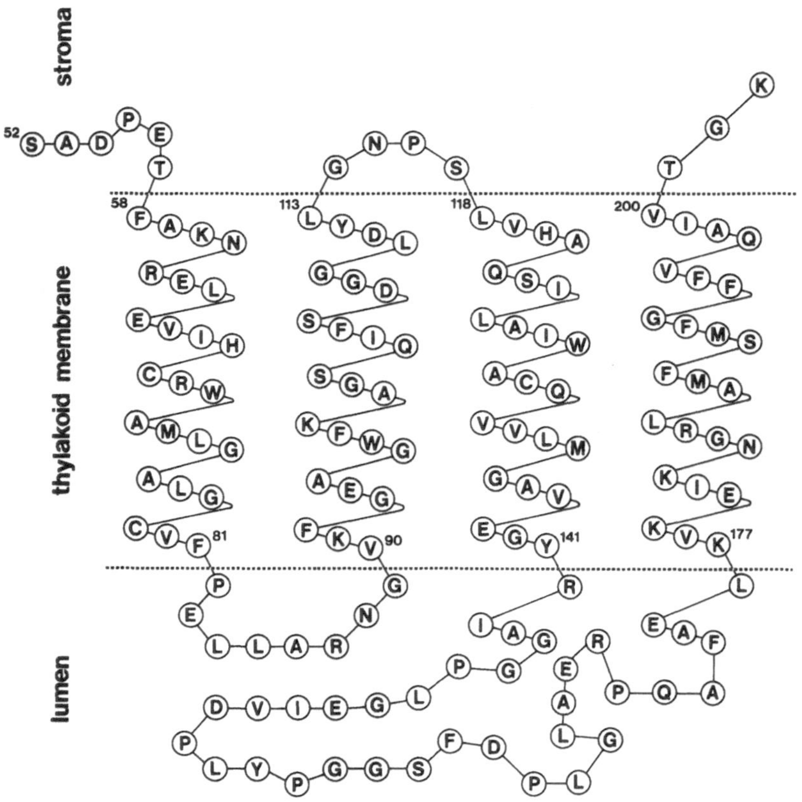

Figure 4. Hypothetical model for the folding of the 16 kDa tryptic fragment of LHCII, based of the sequence deduced from wheat. There are four transmembrane helices, which probably bind all of the chlorophylls. The presence of charged amino acids (lysine and arginine) in these helices can be neutralised by the formation of salt bridges with glutamate residues, and are protected from trypsin attack.

Some of the soluble enzymes, such as Rubisco and PEPC, which are present in one type of chloroplast and not the other, have been shown to be regulated at the mRNA level (Link et al. 1978). Although some of the intrinsic polypeptides of the photosystem II reaction centre have been reported to be present in bundle sheath chloroplasts (Broglie et al. 1985), they do not bind [^{3}H]DCMU (Schuster et al. 1986) and there are very low levels of the mRNA for the DCMU-binding protein (Schuster et al. 1985). In view of the absence of functional photosystem II reaction centres, the presence of LHCII in bundle sheath chloroplasts is interesting. We wanted to know its function, and whether its polypeptide composition differed from that of LHCII from mesophyll chloroplasts.

Large quantities of purified bundle sheath thylakoids were isolated by using a homogeniser with replaceable razor blades to cut maize seedling leaves, and removing the mesophyll cells from the debris by digestion with cellulase (Bassi and Simpson 1986). Because these thylakoids are relatively deficient in chlorophyll a/b-proteins, their chlorophyll a/b ratio is a guide to the degree of purity. Generally this should be >5.5, with higher values obtained from bundle sheath thylakoids isolated from plants grown under higher light intensities. Measurement of photosystem II activity (from 1,5-diphenyl carbazide to 2,6-dichlorophenol indophenol) indicated a low level of contamination with mesophyll thylakoids (between 2-3%), and this was confirmed by low temperature, solid dilution fluorescence emission spectroscopy (Figure 5A). The absence of a 682 nm fluorescence peak due to bundle sheath LHCII shows that LHCII transfers excitation energy to photosystem I. The chlorophyll-protein composition of bundle sheath and mesophyll thylakoids revealed the presence of Chl_a-P1, LHCI and LHCII (Figure 5B).

Because such large amounts of purified bundle sheath thylakoids could be prepared, we were able to isolate LHCII by the method of Burke et al. (1978). The polypeptide composition of isolated mesophyll LHCII is shown in Figure 6A, and consists of four bands at 30, 29.5, 28.8 and 28.5 kDa, with a fifth at 26 kDa. Of these, only the 26 kDa band could not be

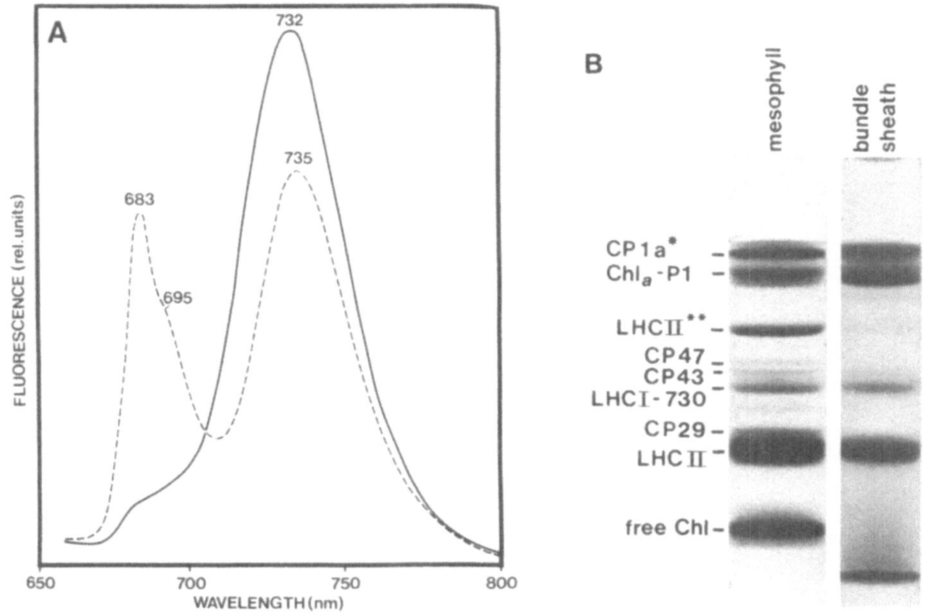

Figure 5. (A) Low temperature, solid dilution, fluorescence emission spectra of purified mesophyll (dotted line) and bundle sheath thylakoids (solid line). (B) Chlorophyll-protein composition of mesophyll and bundle sheath thylakoids resolved by non-denaturing SDS-PAGE.

phosphorylated with [^{32}P] γ-ATP (Bassi and Simpson 1987). The LHCII from bundle sheath thylakoids consists of two polypeptide bands at 28 and 27.5 kDa, neither of which could be phosphorylated in vitro (Bassi and Simpson 1986). It is not known whether this is because the kinase could not be activated, or if bundle sheath LHCII is permanently phosphorylated.

A further difference between LHCII from mesophyll and bundle sheath chloroplasts was their ultrastructure when examined by freeze-fracture electron microscopy. Both were characterised by the typical, densely-packed 80 Å circular particles, but only in LHCII from mesophyll thylakoids were they aligned in regular arrays (Figure 7a). Furthermore, although bundle sheath LHCII was organised into large sheets lying parallel to each other, they were not in close contact, whereas mesophyll LHCII membranes were tightly appressed (Figure 7). This difference in organisation must reflect differences in LHCII-LHCII interactions, which may relate to functional differences in vivo. Appression of thylakoids in grana regions is strengthened by the presence of LHCII, although its presence is not absolutely required (Simpson 1979, Bassi et al. 1985). Removal of the N-terminal 8 amino acids by trypsinisation abolishes this appression at low Mg^{++} concentrations (Steinback et al. 1979, Mullet 1983). So, differences in the N-terminus of bundle sheath LHCII polypeptides may account for its appearance in vitro. Similarly, the lateral interactions by LHCII in grana, or between LHCII and photosystem II reactions centres, may be reflected in their ability to form hexagonal arrays in vitro. Since bundle sheath thylakoids lack both functional photosystem II reaction centres and appressed thylakoids, bundle sheath LHCII may not interact and be unable to form arrays. Removal of the N-terminal amino acids by mild trypsin treatment, does not affect the ability of bundle sheath LHCII to transfer energy to photosystem I, since there is no change in the low temperature fluorescence emission spectrum (Figure 6B).

Sheen and Bogorad (1986) have demonstrated that the LHCII polypeptide differences correspond to differential expression of the LHCII genes in mesophyll and bundle sheath cells of maize. They isolated 6 different cDNA clones from LHCII mRNA, which accounted for 95% of the total LHCII produced from in vitro translation assays. Of these, three hybridised predominantly with mRNA from mesophyll cells, and one almost exclusively with bundle sheath

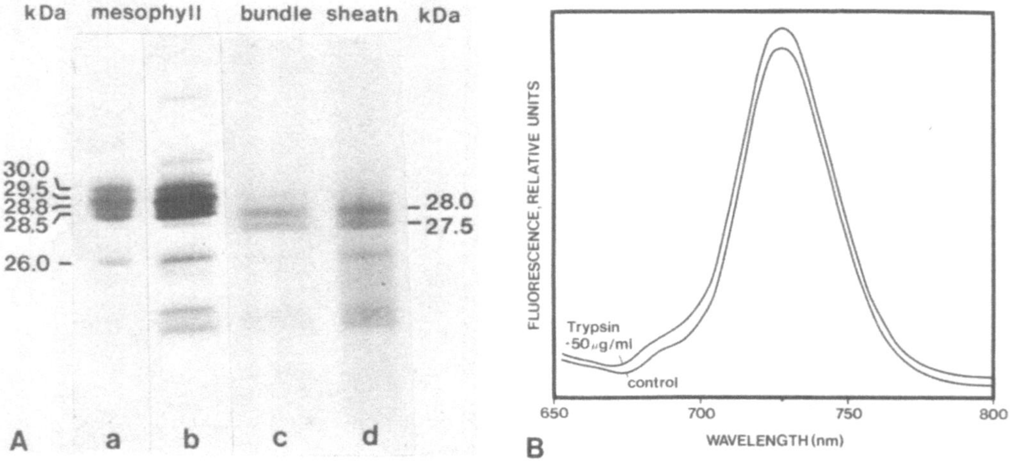

Figure 6. (A) Polypeptide composition of LHCII and isolated thylakoids from mesophyll (lanes a,b) and bundle sheath cells (lanes c,d). Note the differences in polypeptide numbers and molecular weights.
(B) Low temperature fluorescence emission spectra of control and trypsin-treated bundle sheath thylakoids.

cell mRNA. Two others coded for LHCII polypeptides which were present in equal, but low levels in mesophyll and bundle sheath cells (Sheen and Bogorad 1986). Thus the different LHCII polypeptides from mesophyll and bundle sheath cells are likely to be the products of different LHCII genes, rather than different post-translational products of the same genes.

LHCII PHOSPHORYLATION AND REGULATION OF EXCITATION ENERGY DISTRIBUTION

Plants adapt to their environment to maximise their photosynthetic efficiency. An individual leaf can be exposed to full sunlight or deep shade within a short time. Shaded leaves receive light enriched in its far-red component, which is preferentially absorbed by photosystem I (PSI), while in full sunlight more of the excitation energy is absorbed by photosystem II (PSII). Short term changes in the extent of shading would therefore unbalance the energy distribution between the two photosystems, with a reduction in the electron transport rate and overall efficiency, unless a mechanism existed for the dynamic adaptation of the antenna size of PSI and PSII reaction centres.

When leaves are exposed to the dark, or to light which preferentially excites PSI, they undergo a transition to state 1 in which the light energy absorbed by LHCII is preferentially transferred to PSII. This state is characterised by a high fluorescence yield. Under conditions in which light is preferentially absorbed by PSII, a transition to a low fluorescence state 2 is induced over a time scale of several minutes, leading to a redistribution of excitation energy allowing more light to be absorbed by PSI.

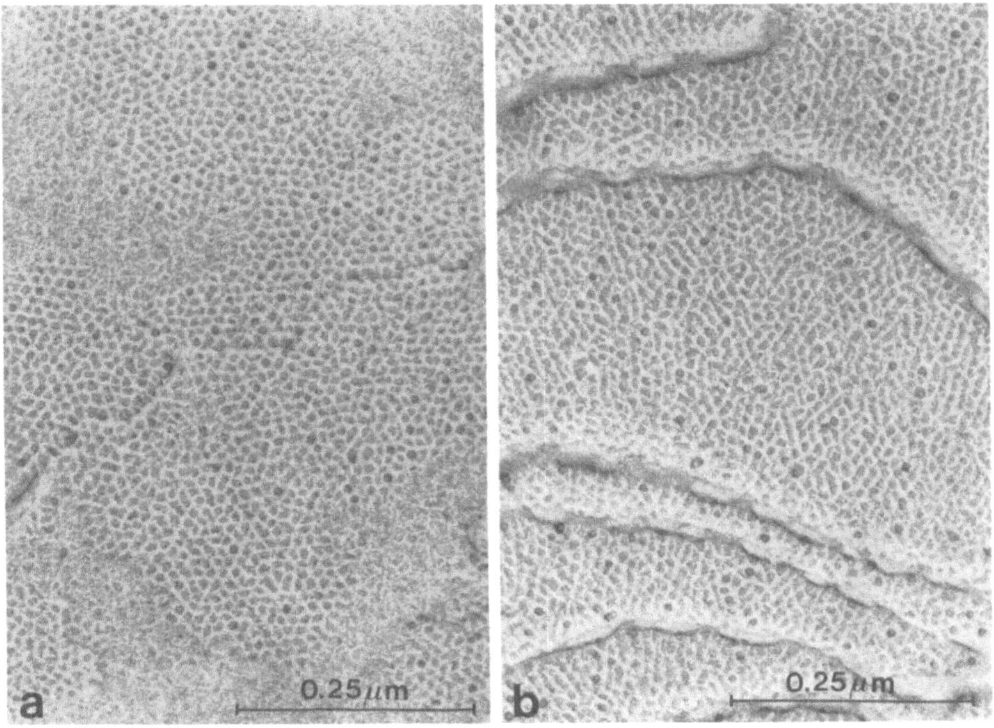

Figure 7. Freeze-fracture appearance of LHCII isolated from (a) mesophyll and (b) bundle sheath thylakoids. LHCII from bundle sheath thylakoids does not form regular arrays, and membranes sheets are not as tightly appressed as those from mesophyll LHCII.

Although changes in the cation concentration in vitro have been shown to induce changes resembling state transitions, it is the phosphorylation of LHCII which ultimately regulates energy distribution between the two photosystems as a result of state transitions. This phosphorylation is thought to be mediated by a thylakoid bound, Mg^{++}-dependent kinase, whose activity is regulated by the redox state of the plastoquinone pool, which in turn reflects the excitation energy distribution between PSII and PSI. The addition of negatively charged phosphate groups to the LHCII complex causes it to dissociate from PSII, and favours its migration out of the appressed regions of the grana and into the stroma lamellae, where PSI centres are located.

However usually only a small percentage of the granal LHCII moves into the stroma lamellae (about 17%, corresponding to a 15% increase in PSI antenna size). The maximum increase in PSI antenna size as a result of LHCII migration in vivo was 50% for a photosystem II mutant of Chlamydomonas (Delepelaire and Wollman 1985), which in higher plants would correspond to approximately two-thirds of the LHCII. The question then arises whether the LHCII which is phosphorylated and migrates out of the appressed regions has the same properties and composition as the LHCII which remains.

Stroma lamellae were isolated from mesophyll thylakoids of maize plants exposed to dark (state 1) and high intensity white light (state 2). Such preparations consisted of small, spherical, right side-out vesicles, and were contamined by less than 0.1% grana appressed lamellae. The state 1 - state 2 transition resulted in a decrease in the chlorophyll a/b ratio of the isolated stroma lamellae due to lateral migration of LHCII as a result of its phosphorylation in the grana. Fluorescence spectroscopy showed that this "mobile LHCII" became functionally attached to the photosystem I reaction centre, with a resulting 15% increase in antenna size, as determined by measurement of the chlorophyll : P700 ratio and kinetics of P700 photo-oxidation under light-limiting conditions. Analysis of the chlorophyll-protein composition of isolated stroma lamellae confirmed the absence of chlorophyll-proteins associated with the photosystem II core, and showed that only the level of LHCII changed in response to the state 1 - state 2 transition.

This mobile LHCII could be resolved by gel electrophoresis into 4 polypeptides with an approximate 1:1:1:1 stoichiometry. This differed significantly from the polypeptide composition of grana LHCII, which had a different stoichiometry, and an additional polypeptide of lower apparent molecular weight which did not become phosphorylated. Furthermore, the phosphorylated LHCII in stroma lamellae would only form oligomers during SDS-PAGE if Mg^{++} was present during electrophoresis, unlike LHCII in appressed grana lamellae (Bassi et al., unpublished data). This could be a result of the negatively charged phosphate groups, or an intrinsic property of the LHCII polypeptides.

So it is clear that LHCII is able to associate functionally with photosystem I and transfer excitation energy to the reaction centre. This can occur in bundle sheath thylakoids, or as the result of phosphorylation, or even as an isolation artefact (Bassi and Simpson 1987). In the first two cases, the LHCII can be distinguished from that located in the appressed grana regions by virtue of an altered polypeptide composition. It remains to be determined if the LHCII which remains in the grana after maximum lateral migration, is able to be phosphorylated. In tomato, LHCII polypeptides coded at the Cab-1 locus, lack threonine residues (the site of phosphorylation) in the first eight N-terminal amino acids (Pichersky et al. 1985) and may therefore never become phosphorylated. This diversity of LHCII function may go some way in explaining the complexity of the gene family coding for LHCII.

REFERENCES

Anderson, J.M. and D.J. Goodchild: Lateral heterogeneity of thylakoid complexes and the transverse arrangement of the light-harvesting chlorophyll a/b-protein of photosystem II of thylakoid membranes. Chemica Scripta, in press (1987)

Andersson, B., J.M. Anderson and I.J. Ryrie: Transbilayer organization of the chlorophyll-proteins in spinach thylakoids. Eur. J. Biochem. 123, 465-472 (1982)

Apel, K.: Phytochrome-induced appearance of mRNA activity for the apoprotein of the light-harvesting chlorophyll a/b-protein of barley (Hordeum vulgare). Eur. J. Biochem. 97, 183-188 (1979)

Apel, K. and K. Kloppstech: The plastid membranes of barley (Hordeum vulgare). Light-induced appearance of mRNA coding for the apoprotein of the light-harvesting chlorophyll a/b-protein. Eur. J. Biochem. 85, 581-588 (1978)

Bassi, R., U.G. Hinz and R. Barbato: The role of the light harvesting complex and photosystem II in thylakoid stacking in the chlorina-f2 barley mutant. Carlsberg Res. Commun. 50, 347-367 (1985)

Bassi, R. and D.J. Simpson: Differential expression of LHCII genes in mesophyll and bundle sheath cells of maize. Carlsberg Res. Commun. 51, 363-370 (1986)

Bassi, R. and D.J. Simpson: Chlorophyll-protein complexes of barley photosystem I. Eur. J. Biochem. 163, 221-230, (1987)

Bassi, R. and D.J. Simpson: The organisation of photosystem II chlorophyll-proteins. Prog. Photosyn. Res. Vol. 2, pp. 81-88 (Ed. J. Biggins, publ. Martinus Nijhoff, Dordrecht) (1987)

Bellemare, G., S.G. Bartlett and N.-H. Chua: Biosynthesis of chlorophyll a/b binding polypeptides in wild type and the chlorina f2 mutant of barley. J. Biol. Chem. 257, 7762-7767 (1982).

Broglie, R., G. Coruzzi, B. Keith and N.-H. Chua: Molecular biology of C_4 photosynthesis in Zea mays: differential localization of proteins and mRNAs in the two leaf cell types. Plant Mol. Biol. 3, 431-444 (1984)

Burke, J.J., C.L. Ditto and C.J. Arntzen: Involvement of the light-harvesting complex in cation regulation of excitation energy distribution in chloroplasts. Arch. Biochem. Biophys. 187, 252-263 (1979)

Chitnis, P.R., E. Harel, B.D. Kohorn, E.M. Tobin and J.P. Thornber: Assembly of the precursor and processed light-harvesting chlorophyll a/b protein of Lemna into the light-harvesting complex II of barley etiochloroplasts. J. Cell Biol. 102, 982-988 (1986)

Cline, K.: Import of proteins into chloroplasts. Membrane integration of a thylakoid precursor protein reconstituted in chloroplast lysates. J. Biol. Chem. 261, 14804-14810 (1986)

Cline, K., M. Werner-Washbourne, T.H. Lubben and K. Keegstra: Precursors to two nuclear-coded chloroplast proteins bind to the outer envelope membrane before being imported into chloroplasts. J. Biol. Chem. 260, 3691-3696 (1985)

Delepelaire, P. and F.-A. Wollman: Correlations between fluorescence and phosphorylation changes in thylakoid membranes of Chlamydomonas reinhardtii in vivo: a kinetic analysis. Biochim. Biophys. Acta 809, 277-283 (1985)

Dunsmuir, P.: The petunia chlorophyll a/b binding protein genes: a comparison of Cab genes from different gene families. Nuc. Acid Res. 13, 2503-2518 (1985)

Hageman, J., C. Robinson, S. Smeekens and P. Weisbeek: A thylakoid processing protease is required for the complete maturation of the lumen protein plastocyanin. Nature 324, 567-569 (1986)

Hinz, U.G. and K.G. Welinder: The light-harvesting complex of photosystem II in barley. Carlsberg Res. Commun. 52, 39-54 (1987)

Høyer-Hansen, G., R. Bassi, L.S. Hønberg and D.J. Simpson: Immunological characterization of chlorophyll a/b-binding proteins of barley thylakoids. Planta, in press (1987)

Johanningmeier, U. and S.H. Howell: Regulation of light-harvesting chlorophyll-binding protein mRNA accumulation in Chlamydomonas reinhardi. Possible involvement of chlorophyll synthesis precursors. J. Biol. Chem. 259, 13541-13549 (1984)
Karlin-Neumann, G.A., B.D. Kohorn, J.P. Thornber and E.A. Tobin: A chlorophyll a/b-protein encoded by a gene containing an intron with the characteristics of a transposable element. J. Mol. Appl. Gen. 3, 45-61 (1985)
Kühlbrandt, W.: Three-dimensional structure of the light-harvesting chlorophyll a/b-protein complex. Nature 307, 478-480 (1984)
Kühlbrandt, W.: Three-dimensional crystals of the light-harvesting chlorophyll a/b-protein complex from pea chloroplasts. J. Mol. Biol. 194, 757-762 (1987)
Lamppa, G.K., G. Morelli and N.-H. Chua: Structure and developmental regulation of a wheat gene encoding the major chlorophyll a/b-binding polypeptide. Mol. Cell. Biol. 5, 1370-1378 (1985)
Link, G., D.M. Coen and L. Bogorad: Differential expression of the gene for the large subunit of ribulose bisphosphate carboxylase in maize leaf cell types. Cell 15, 725-731 (1978)
Mayfield, S.P. and W.C. Taylor: Carotenoid-deficient maize seedlings fail to accumulate light-harvesting chlorophyll a/b-binding protein (LHCP) mRNA. Eur. J. Biochem. 144, 79-84 (1984)
Meyerowitz, E.M. and R.E. Pruit: Arabidopsis thaliana and plant molecular genetics. Science 229, 1214-1218 (1985)
Mullet, J.E.: The amino acid sequence of the polypeptide segment which regulates membrane adhesion (grana stacking) in chloroplasts. J. Biol. Chem. 258, 9941-9948 (1983)
Pichersky, E., R. Bernatzky, S.D. Tanksley, R.B. Breidenbach, A.P. Kausch and A.R. Cashmore: Molecular characterization and genetic mapping of two clusters of genes encoding chlorophyll a/b-binding proteins in Lycopersicon esculentum (tomato). Gene 40, 247-258 (1985)
Plumley, G.F. and G.W. Schmidt: Reconstitution of chlorophyll a/b light-harvesting complexes: xanthophyll-dependent assembly and energy transfer. Proc. Nat. Acad. Sci. USA 84, 146-150 (1987)
Rawyler, A., L.E.A. Henry and P.A. Siegenthaler: Acyl and pigment lipid composition of two chlorophyll-proteins. Carlsberg Res. Commun. 45, 443-451 (1980)
Schuster, G., I. Pecker, J. Hirschberg, K. Kloppstech and I. Ohad: Transcription control of the 32 kDa-Q_B protein of photosystem II in differentiated bundle sheath and mesophyll chloroplasts of maize. FEBS Lett. 198, 56-60 (1986)
Schuster, G., I. Ohad, B. Martineau and W.C. Taylor: Differentiation and development of bundle sheath and mesophyll thylakoids in maize. Thylakoid polypeptide composition, phosphorylation, and organisation of photosystem II. J. Biol. Chem. 260, 11866-11873 (1985)
Sheen, J.-Y. and L. Bogorad: Differential expression of six light-harvesting chlorophyll a/b binding proteins genes in maize leaf cell types. Proc. Nat. Acad. Sci. USA 83, 7811-7815 (1986)
Simpson, D.J.: Freeze-fracture studies on barley plastid membranes. III. Location of the light-harvesting chlorophyll-protein. Carlsberg Res. Commun. 44, 305-336 (1979)
Simpson, D.J., G. Høyer-Hansen, N.-H. Chua and D. von Wettstein: The use of gene mutants in barley to correlate thylakoid polypeptide composition with the structure of the photosynthetic membrane. Proc. Fourth Internat. Photosyn. Cong., Reading pp. 537-548 (1977)
Stayton, M., M. Black, J. Bedbrook and P. Dunsmuir: A novel chlorophyll a/b binding (Cab) protein gene from petunia which encodes the lower molecular weight Cab precursor protein. Nuc. Acid Res. 14, 9781-9796 (1986)
Steinback, K.E., J.J. Burke and C.J. Arntzen: Evidence for the role of surface-exposed segments of the light-harvesting complex in cation-mediated control of chloroplast structure and function. Arch. Biochem. Biophys. 195, 546-557 (1979)

SYNTHESIS OF ELECTRON TRANSFER COMPONENTS OF THE PHOTOSYNTHETIC APPARATUS

John C. Gray, Paul P.J. Dunn, Christopher J. Eccles, Sean M. Hird, David I. Last, Barbara J. Newman and David L. Willey

Botany School, University of Cambridge
Downing Street, Cambridge CB2 3EA, U.K.

INTRODUCTION

The photosynthetic electron transfer chain of higher plant chloroplasts transfers electrons from water to $NADP^+$ using the three supramolecular complexes of photosystem II, the cytochrome *b-f* complex and photosystem I and a number of electron carriers operating between the membrane complexes. Plastoquinone operates as a lipid-soluble shuttle between photosystem II and the cytochrome complex whereas the water-soluble copper protein plastocyanin transfers electrons between the cytochrome complex and photosystem I. Ferredoxin and ferredoxin-$NADP^+$ reductase are responsible for the reduction of $NADP^+$ by photosystem I, and one, or both, are involved in transferring electrons back to the cytochrome complex in cyclic electron flow.

Our aim is to understand how the components of the electron transfer chain are synthesised, and assembled into the appropriate membrane complexes, to produce a functional photosynthetic electron transfer chain. It has become clear over the past few years that the thylakoid membrane and its associated proteins are the product of a complex interaction between the nucleo-cytoplasmic and chloroplast genetic systems, with each system providing approximately half of the structural genes for photosynthetic protein components of the membrane complexes. We have located and characterised all the known chloroplast genes for components of photosystems I and II and the cytochrome *b-f* complex in chloroplast DNA from pea or wheat, and are currently characterising cDNA and genomic clones for nuclear-encoded electron transfer proteins.

PHOTOSYSTEM II

Photosystem II carries out the light-driven transfer of electrons from water to the plastoquinone pool. Light energy is absorbed by the light-harvesting complex (LHCII) composed of chlorophyll *a*/*b*-binding polypeptides of approximately 25kDa. The light energy is transferred via two antenna chlorophyll proteins of 47kDa and 44kDa which make up part of the core complex to the reaction centre P680. P680 is associated with two polypeptides of approximately 33kDa, called D1 and D2, which also bind chlorophyll a, phaeophytin and the primary and secondary quinone acceptors Q_A and Q_B (Nanba & Satoh, 1987). The Q_B-binding site on D1 is the site of action of a large number of herbicides, such as diuron and atrazine, which prevent electron transfer from the photosystem II core to the plastoquinone pool. The oxidation of water with the release of O_2 is carried out by a complex of polypeptides located on the lumenal side of the thylakoid membrane. Extrinsic polypeptides of 33kDa, 23kDa, 16kDa and 10kDa appear to be associated with intrinsic membrane proteins of 24kDa and 22kDa (Ljungberg et al, 1984). The 33kDa polypeptide has been implicated in Mn-binding and shows sequence similarities to bacterial Mn-superoxide dismutases in the region of the proposed Mn ligands (Oh-oka et al, 1986). Cytochrome b-559, composed of 9kDa and 4kDa polypeptides, is associated with the core complex but its role in electron transfer is not clear. Other small polypeptides of 7kDa, 6.5kDa, 5.5kDa and 5kDa have been shown to be associated with photosystem II preparations but their function is unknown. A 10kDa polypeptide which is reversibly phosphorylated similar to LHCII polypeptides is also associated with photosystem II preparations and may be involved in regulating excitation energy transfer.

The genes for seven identified polypeptides of photosystem II have been located in pea and wheat chloroplast DNA where they are organised in four separate transcriptional units. The gene (psbA) for the 32kDa D1 or Q_B-protein encodes a polypeptide of 353 amino acid residues with five putative membrane-spanning regions (Oishi et al, 1984; MDC Barros, unpublished). This gene is transcribed separately whereas the other photosystem II genes are arranged in clusters. The genes (psbD and psbC) for the D2 and 44kDa chlorophyll a-binding polypeptides are cotranscribed and overlap by 50bp (Rasmussen et al, 1984; SM Hird, unpublished). The D2 polypeptide is homologous to D1, and to the L and M subunits of the Rhodopseudomonas viridis reaction centre complex (Michel et al, 1986). By analogy with the L and M subunits, the D1 and D2 polypeptides each form five membrane-spanning regions and provide binding sites for P680, phaeophytin, Q_A, Q_B and an associated Fe ion. The 44kDa polypeptide is composed of 473 amino acid residues and is homologous to the 47kDa chlorophyll a-binding protein. Both polypeptides are predicted to form seven membrane spans and contain membrane-located histidine residues which may act as ligands for chlorophyll binding. An open reading frame of 62 codons is also included in this transcriptional unit (Quigley & Weil, 1985) but it has not been established if this putative protein is a photosystem II component. This open reading frame is conserved in a similar position with

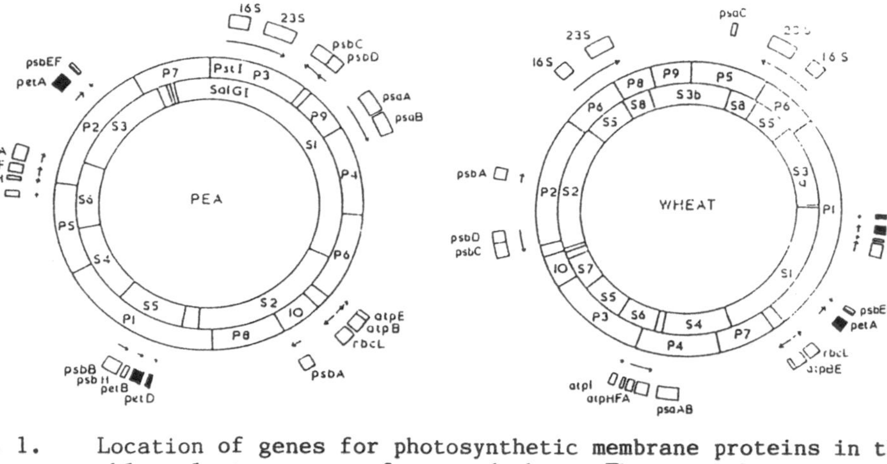

Fig. 1. Location of genes for photosynthetic membrane proteins in the chloroplast genomes of pea and wheat. The maps show restriction sites for PstI(P) and SalI(S). The direction of transcription of genes is indicated by arrows.

respect to the psbD and psbC genes in chloroplast DNA from all plants examined. The deduced protein sequence is highly conserved and predicts two membrane-spanning regions. The genes (psbB and psbH) for the 47kDa chlorophyll a-binding polypeptide and the 10kDa phosphoprotein are part of a complex transcriptional unit including two genes (petB and petD) for components of the cytochrome b-f complex. The psbH gene was identified because of similarities between the protein encoded by an open reading frame of 73 codons and the 10kDa phosphoprotein purified from spinach (Farchaus & Dilley, 1986; Hird et al, 1986). The 10kDa phosphoprotein is predicted to be oriented in the thylakoid membrane with its N-terminus in the stroma and its C-terminus in the thylakoid lumen, separated by a single hydrophobic membrane-spanning region. The protein shows weak homology with the N-terminal region of LHCII polypeptides (Hird et al, 1986). An open reading frame potentially coding for a hydrophobic protein of 38 amino acid residues is located between the psbB and psbH genes. This open reading frame is conserved in liverwort and tobacco chloroplast DNA (Ohyama et al, 1986; Shinozaki et al, 1986) and predicts a basic protein with a single membrane-spanning region.

The genes (psbE and psbF) for the 9kDa and 4kDa polypeptides of cytochrome b-559 have been characterised from both pea and wheat (Hird et al, 1985; DL Willey, unpublished). Open reading frames of 83 and 39 codons encode polypeptides which are similar at their N-termini where they each contain a histidine residue in a putative membrane-spanning region. Two histidines are required as the ligands for the haem group of cytochrome b-559 (Babcock et al, 1985) indicating that two or more polypeptides are necessary for the structure of cytochrome b-559. In both pea and wheat the genes are co-transcribed into a prominent transcript of 1100 nucleotides which also contains two additional open reading frames of 38 and 40 codons. These open reading frames each potentially code for a hydrophobic polypeptide with one putative membrane span. The function of these polypeptides and their possible association with photosystem II have not been established.

The location of the genes for photosystem II polypeptides in chloroplast DNA from pea and wheat is shown in Fig. 1. More detailed arrangement of the genes is shown in Fig. 2.

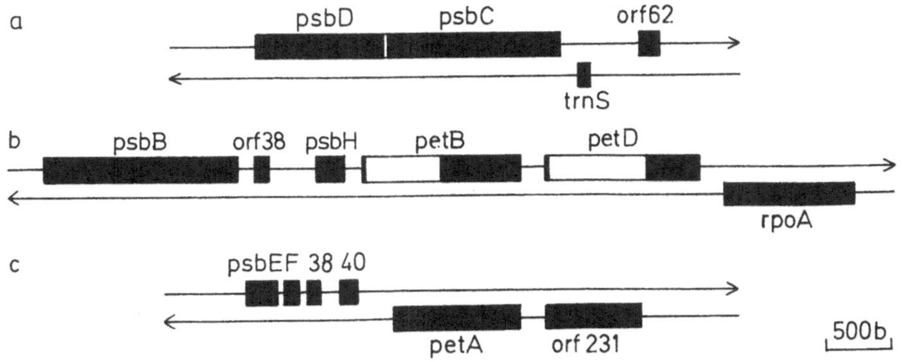

Fig. 2. Organisation of genes for components of photosystem II and the cytochrome b-f complex in chloroplast DNA. (a) Organisation of psbD and psbC genes in wheat; (b) organisation of psbB, psbH, petB and petD genes in wheat. Introns in petB and petD are shown as open boxes; rpoA is the putative gene for the α subunit of RNA polymerase. (c) organisation of the psbE, psbF and petA genes in pea.

The three major extrinsic polypeptides of the oxygen-evolving complex of photosystem II are synthesised as larger precursors on translation of polyA-RNA in vitro, and are imported and processed to their mature form by isolated chloroplasts (Westhoff et al, 1985). cDNA clones for the spinach 33kDa, 23kDa and 16kDa polypeptides have been isolated (Tittgen et al, 1986). We have isolated cDNA clones for the pea 33kDa and 23kDa polypeptides from a λgt11 library produced from pea leaf polyA-RNA (Gantt & Key, 1986). These clones are currently being characterised.

CYTOCHROME b-f COMPLEX

The cytochrome b-f complex transfers electrons from plastoquinol to plastocyanin in the linear photosynthetic electron transfer chain. The cytochrome complex is composed of four intrinsic membrane components, three of which, Rieske Fe-S protein, cytochrome f and cytochrome b-563, contain redox centres, whereas the 15kDa polypeptide probably contributes to the plastoquinol-binding site on the lumenal side of the complex. Electron flow through the complex is not fully understood although a linear flow from plastoquinol via Rieske Fe-S and cytochrome f to plastocyanin is well documented. However it seems likely that cytochrome b-563 is involved in a Q-cycle type mechanism responsible for the translocation of additional protons across the thylakoid membrane. The cytochrome complex is also involved in cyclic electron flow around photosystem I, but the route of electrons from photosystem I to the cytochrome complex is not clear. A direct route from ferredoxin, perhaps involving an as yet uncharacterised ferredoxin-plastoquinol reductase, seems most likely, but ferredoxin-$NADP^+$ reductase has been implicated in the route of cyclic electron flow (Shahak et al, 1981) and may be associated with the cytochrome b-f complex (Clark et al, 1984).

The genes for three components of the cytochrome b-f complex have been located and characterised in chloroplast DNA from pea and wheat. The gene (petA) for cytochrome f encodes a polypeptide of 320 amino acid residues, of which 285 residues constitute the mature polypeptide and 35 residues comprise an N-terminal presequence (Willey et al, 1984 a,b). The mature polypeptide is organised with the N-terminal portion, containing the haem-binding site, in the lumenal space and the C-terminal 15 amino acid residues accessible to the stroma. A single hydrophobic membrane-spanning region near the C-terminus anchors the polypeptide in the thylakoid membrane (Willey et al, 1984a). The 35 amino acid residue presequence is proposed to act as a signal sequence directing the polypeptide to the membrane, and has been shown, in gene-fusion experiments, to direct β-galactosidase to the inner membrane in Escherichia coli (Rothstein et al, 1985). The petA gene is co-transcribed with an open reading frame potentially coding for a protein of 231 amino acid residues (Willey et al, 1984a). The function of this protein is unknown but it appears to be synthesised with a putative cleavable signal sequence (DL Willey, unpublished).

The genes (petB and petD) for cytochrome b-563 and the 15kDa polypeptide are located close together and co-transcribed with the psbB and psbH genes. Both genes contain introns that were unsuspected in earlier publications (Heinemeyer et al, 1984; Phillips & Gray, 1984). S1 nuclease analysis and comparison with concensus intron boundary sequences indicates that the first exon is composed of 6bp in petB and 8bp in petD. The cytochrome b-563 polypeptide is composed of 215 amino acid residues and is predicted to form five membrane-spanning segments. Four membrane-located histidine residues have been identified as ligands for the two haem groups which are located on opposite sides of the membrane (Widger et al, 1984). The 15kDa polypeptide is composed of 160 amino acid residues and is predicted to form three membrane-spanning regions. The cytochrome *b*-563 and 15kDa polypeptides are homologous to *N*-terminal and *C*-terminal sequences of mitochondrial cytochrome *b*. The 15kDa polypeptide shows high homology with a region of mitochondrial cytochrome *b* that has been implicated in ubiquinol binding at a mucidin-susceptible site (Subik & Takacsova, 1978). This would correspond to a plastoquinol-binding site on the lumenal side of the thylakoid membrane.

The Rieske Fe-S protein has been shown to be synthesised as a larger precursor from polyA-RNA *in vitro* (Alt et al, 1983). We have isolated cDNA clones for the pea Rieske Fe-S protein from a λgt11 library and these clones are currently being characterised. Tittgen et al (1986) have reported the isolation of cDNA clones for the spinach Rieske Fe-S protein.

Plastocyanin has also been shown to be synthesised as a larger precursor prior to import into chloroplasts (Grossman et al, 1982). cDNA clones for plastocyanin have been isolated from *Silene* and spinach libraries (Smeekens et al, 1985a; Tittgen et al, 1986). We have isolated cDNA clones for pea plastocyanin from a library, prepared from pea leaf polyA-RNA, in M13 K8.2 (Waye et al, 1985) following identification by hybridisation with an oligonucleotide predicted from the known amino acid sequence of pea plastocyanin. Nucleotide sequence analysis of the cDNA clones suggested that all were derived from a unique mRNA sequence. Hybridisation to a Southern blot of pea genomic DNA indicated the presence of a single copy of the plastocyanin gene in the pea haploid genome. Nucleotide sequences of the cDNA and genomic clones indicate that the pea plastocyanin gene does not contain any introns, similar to the spinach plastocyanin gene (Rother et al, 1986). Nucleotide sequence indicates that the protein is synthesised with an *N*-terminal extension of 69 amino acid residues. The presequence is similar to those for the *Silene* and spinach proteins in the *N*-terminal and *C*-terminal regions, but differs considerably in the central part of the sequence (Gray et al, 1987). The *C*-terminal residues are mainly hydrophobic and may constitute a signal sequence for transfer across the thylakoid membrane (Smeekens et al, 1986).

A 3.3kbp *Eco*RI fragment of pea nuclear DNA containing the plastocyanin gene has been inserted in the *Agrobacterium* binary vector Bin19 (Bevan, 1984) and transferred to tobacco by the leaf disc method. Kanamycin-resistant transgenic plants were shown to contain pea plastocyanin, by electrophoresis of leaf extracts on non-denaturing acrylamide gels (DI Last, unpublished). This method is able to resolve pea and tobacco plastocyanins (Fig. 3).

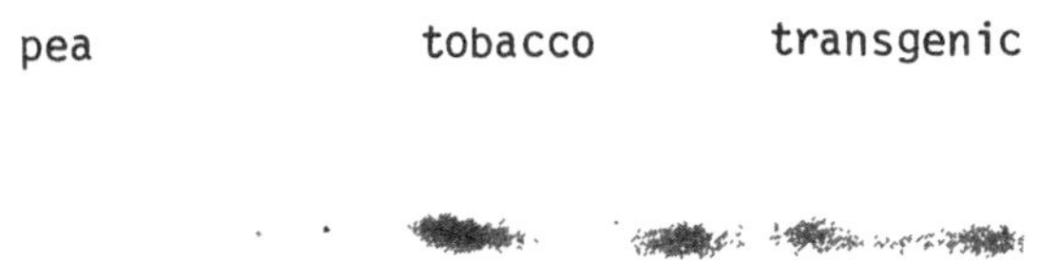

Fig. 3. Polyacrylamide gel electrophoresis of plastocyanin extracted from pea, tobacco and a transgenic tobacco plant containing the pea plastocyanin gene.

Southern blots of nuclear DNA from the transgenic tobacco plants demonstrate the presence of the pea 3.3kbp EcoRI fragment, indicating that this fragment contains the necessary sequences for high-level expression of plastocyanin in tobacco leaves. These plants are currently being examined for possible gene dosage effects on the amounts of plastocyanin and other electron transfer components.

PHOTOSYSTEM I

Photosystem I carries out the light-driven transfer of electrons from plastocyanin, in the thylakoid lumen, to ferredoxin, in the stroma. Reduced ferredoxin is the source of electrons for a wide variety of biochemical reactions in the chloroplast stroma, including $NADP^+$ reduction, nitrite reduction, glutamate synthesis, sulphate reduction, fatty acid desaturation and thioredoxin reduction. Light energy is absorbed by the light-harvesting complex (LHCI) composed of chlorophyll a/b-binding polypeptides of 22-25kDa (Haworth et al, 1983). The light energy is transferred to the reaction centre P700 which is believed to be associated with two polypeptides of 66kDa. These polypeptides are also believed to bind electron acceptors A_o, a specialised chlorophyll molecule, A_1, a phylloquinone molecule and the Fe-S centre X. The electrons then move via Fe-S centres A and B to ferredoxin. A polypeptide of 8kDa has been suggested to be the location of centres A and B on the basis of its high cysteine content (Lagoutte et al, 1984). Photosystem I preparations also contain a number of polypeptides of unknown function. A polypeptide of 20kDa in the spinach complex has been implicated in electron transfer from plastocyanin to P700 (Bengis & Nelson, 1977) but it is not clear if the polypeptide provides a binding site for plastocyanin or has some other role.

Three genes for photosystem I polypeptides have been located in chloroplast DNA from pea or wheat. Two genes (psaA and psaB) for the 66kDa reaction centre polypeptides have been characterised from pea (Lehmbeck et al, 1986). Nucleotide sequence analysis revealed the presence of two large open reading frames of 761 and 734 codons separated by only 25bp. The deduced amino acid sequences of the polypeptides indicate that they are homologous, with about 45% identical amino acid sequences. The two genes have been shown to code for the two polypeptides of the photosystem I core (Fish et al, 1985). The two polypeptides are each predicted to form 12 membrane-spanning segments and each contains two conserved cysteine residues which have been suggested to be the ligands to the iron atoms in the Fe-S centre X (Hoj & Moller, 1986).

The gene (psaC) for the 8kDa polypeptide binding the Fe-S centres A and B has recently been identified. An open reading frame of 81 codons in the small single copy region of wheat chloroplast DNA encodes a protein with a similar amino acid composition to the spinach 8kDa protein of Lagoutte et al (1984). The deduced amino acid sequence contains 9 cysteine residues and shows high homology to bacterial ferredoxins containing two Fe_4S_4 centres (Fig. 4.). An homologous open reading frame has been located in liverwort chloroplast DNA, where it has been designated frxA (Ohyama et al, 1986), and the insertion of a single nucleotide into the tobacco sequence generates a similar open reading frame (Shinozaki et al, 1986).

The presence of another gene for a photosystem I component in chloroplast DNA is suggested by the synthesis of a polypeptide on chloroplast ribosomes. In *Spirodela*, pea and wheat a polypeptide of variously 12kDa, 15kDa or 17kDa has been shown to be synthesised on chloroplast ribosomes *in vivo* or *in vitro* (Nechushtai et al, 1981; Mullet et al, 1981; Smith and Gray, 1984; Obokata, 1986).

```
                              10             20               30
Wheat          M S H S V K I Y D T|C|I G|C|T Q|C|V R A|C P T|D V L E M I P W D G C K A K
P. elsdenii          M H V I S D E|C|V K|C|G A|C|A S T|C P T|G A I E E G E T
                                      10              20
                   40                 50              60            70              80
Wheat          Q I A S A P R T E D|C|V G|C|K R|C|E S A|C P T|D F L S V R V Y L G P E T T R S M A L S Y
P. elsdenii          K Y V V T D S|C|I D|C|G A|C|E A V|C P T|G A I S A E
                     30                 40              50
```

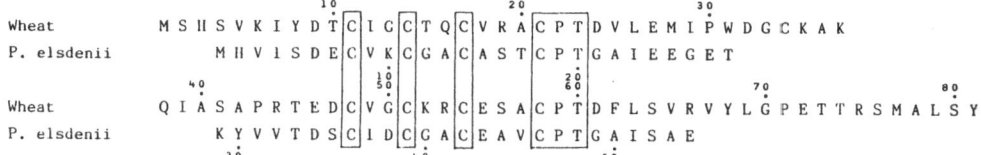

Fig. 4. Comparison of the amino acid sequence of the wheat psaC gene product and the Peptostreptococcus elsdenii ferredoxin (Yasunobu & Tanaka, 1980). The sequences have been aligned at the cysteine residues involved in Fe-S centre ligation.

cDNA clones for several of the smaller polypeptides of the spinach photosystem I core have been isolated (Tittgen et al, 1986). We have isolated cDNA clones for the pea 21kDa polypeptide from a λgt11 library. This polypeptide may correspond to the 20kDa polypeptide implicated in the interaction of plastocyanin and P700 (Bengis & Nelson, 1977). cDNA clones for ferredoxin have been isolated from Silene (Smeekens et al, 1985b), pea (Dobres et al, 1987) and spinach (Tittgen et al, 1986). The Silene clone indicates that ferredoxin is synthesised initially with a presequence of 44 amino acid residues, which is responsible for targetting the protein to the chloroplast stroma.

We have isolated cDNA clones for pea ferredoxin-$NADP^+$ reductase from a λgt11 library using monospecific antibodies raised to the purified pea stromal enzyme. The cDNA encodes a polypeptide of 360 amino acid residues, of which 308 residues comprise the mature protein and 52 residues constitute an N-terminal presequence. The 40kDa precursor protein, synthesised in vitro from an SP6 transcript, is imported and processed to the mature 34kDa protein by isolated pea chloroplasts (BJ Newman, unpublished). Hybridisation to a Southern blot of pea genomic DNA indicates the presence of more than one (probably two) copies of the ferredoxin-$NADP^+$ reductase gene in the pea haploid genome. Pea genomic clones for the protein have been isolated and are currently being characterised.

ACKNOWLEDGEMENTS

We are grateful to Dr J.S. Gantt for the pea leaf λgt11 cDNA library. D.L.W. is The Royal Society Rosenheim Research Fellow. This work was supported by grants from SERC and AFRC.

REFERENCES

Alt, J., Westhoff, P., Sears, B.B., Nelson, N., Hurt, E., Hauska, G. and Herrmann, R.G. (1983) EMBO J. 2, 979-986.

Babcock, G.T., Widger, W.R., Cramer, W.A. Oertling, W.A. and Metz, J.G. (1985) Biochemistry 24, 3638-3645.

Bengis, C. and Nelson, N. (1977) J.Biol.Chem. 252, 4564-4569.

Bevan, M. (1984) Nucleic Acids Res.12, 8711-8721.

Clark, R.D., Hawkesford, M.J., Coughlan, S.J., Bennett,J. and Hind, G. (1984) FEBS Lett. 174, 137-142.

Dobres, M.S., Elliott, R.C., Watson, J.C. and Thompson, W.F. (1987) Plant Mol. Biol. 8, 53-59.

Farchaus, J. and Dilley, R.A. (1986) Arch. Biochem. Biophys. 244, 94-101.
Fish, L.E., Kuck, U. and Bogorad, L. (1985) J. Biol. Chem. 260, 1413-1421.
Gantt, J.S. and Key, J. (1986) Mol. Gen. Genet. 202, 186-193.
Gray, J.C., Dunn, P.P.J., Eccles, C.J., Hird, S.M., Hoglund, A.S., Last, D.I., Newman, B.J. and Willey, D.L. (1987) in Plant Membranes: structure, function, biogenesis. Leaver, C.J. and Sze, H. eds. A.R. Liss, New York.
Grossman, A.R., Bartlett, S.G., Schmidt, G.W., Mullet, J.E. and Chua, N.H. (1982) J.Biol.Chem. 257, 1558-1563.
Haworth, P., Watson, J.L. and Arntzen, C.J. (1983) Biochim.Biophys.Acta 724, 151-158.
Heinemeyer, W., Alt, J. and Herrmann, R.G. (1984) Curr. Genet. 8, 543-549.
Hird, S.M., Dyer, T.A. and Gray, J.C. (1986) FEBS Lett. 209, 181-185.
Hird, S.M., Willey, D.L., Dyer, T.A. and Gray, J.C. (1985) Mol. Gen. Genet. 203, 95-100.
Hoj, P.B. and Moller, B.L. (1986) J.Biol.Chem. 261, 14292-14300.
Lagoutte, B., Setif, P. and Duranton, J. (1984) FEBS Lett. 174, 24-29.
Lehmbeck, J., Rasmussen, O.F., Bookjans, G.B., Jepsen, B.R., Stummann, B. M. and Henningsen, K.W. (1986). Plant Mol. Biol. 7, 3-10.
Ljungberg, U., Akerlund, H.E., Larsson, C. and Andersson, B. (1984) Biochim.Biophys.Acta 767, 145-152.
Michel, H., Epp, O. and Deisenhofer, J. (1986) EMBO J. 5, 2445-2451.
Mullet, J.E., Grossman, A.R. and Chua, N.H. (1981) Cold Spring Harbor Symp. Quant. Biol. 46, 979-984.
Nanba, O. and Satoh, K. (1987) Proc. Natl. Acad. Sci. USA 84, 109-112.
Nechushtai, R., Nelson, N., Mattoo, A.K. and Edelman, M. (1981). FEBS Lett. 125, 115-119.
Obokata, J. (1986) Plant Physiol. 81, 705-707.
Oh-oka, H., Tanaka, S., Wada, K., Kuwabara, T. and Murata, N. (1986) FEBS Lett.197, 63-66.
Ohyama, K., Fukuzawa, H., Kohchi, T., Shirai, H., Sano, T., Sano, S., Umesono, K., Shiki, Y., Takeuchi, M., Chang, Z., Aota, S., Inokuchi, H. and Ozeki, H. (1986) Nature 322, 572-574.
Oishi, K.K. Shapiro, D.R. and Tewari, K.K. (1984) Mol.Cell.Biol. 4, 2556-2563.
Phillips, A.L. and Gray, J.C. (1984) Mol.Gen.Genet. 194, 477-484.
Quigley, F., and Weil, J.H. (1985) Curr.Genet.9, 495-503.
Rother, C., Jansen, T., Tyagi, A., Tittgen, J. and Herrmann, R.G. (1986) Curr.Genet.11, 171-176.
Rothstein, S.J., Gatenby, A.A., D.L. Willey and J.C. Gray (1985) Proc. Natl. Acad. Sci. USA 82, 7955-7959.
Shahak, Y., Crowther, D. and Hind, G. (1981) Biochim.Biophys.Acta 636, 234-243.
Shinozaki, K., Ohme, M., Tanaka, M., Wakasugi, T., Hayashida, N., Matsubayashi, T., Zaita, N., Chungwonse, J., Obokata, J., Yamaguchi-Shinozaki, K., Ohto, C., Torazawa, K., Meng, B.Y., Sugita, M., Deno, H., Kamogashira, T., Yamada, K., Kusuda, J., Takaiwa, F., Kato, A., Tohdoh, N., Shimada, H. and Sugiura, M. (1986) EMBO J. 5, 2043-2049.
Smeekens, S., de Groot, M., van Binsbergen J. and Weisbeek, P (1985a) Nature 317, 456-458.
Smeekens, S., van Binsbergen, J. and Weisbeek, P. (1985b) Nucleic Acids Res.13, 3179-3194.
Smeekens, S., Bauerle, C., Hageman, J., Keegstra, K. and Weisbeek P (1986) Cell 46, 365-375.
Smith, A.G. and Gray, J.C. (1984) Adv. Photosynth. Res. 4, 513-516.
Tittgen, J., Hermans J., Steppuhn, J., Jansen, T., Jansson, C., Andersson, B., Nechushtai, R., Nelson, N., and Herrmann, R.G. (1986) Mol. Gen. Genet. 204, 258-265.
Waye, M.M.Y., Verhoeyen, M.E., Jones, P.T. and Winter, G. (1985) Nucleic

Acids Res. 13, 8561-8571.
Widger, W.R., Cramer, W.A., Herrmann, R.G. and Trebst, A. (1984) Proc. Natl. Acad. Sci. USA 81, 674-678.
Willey, D.L., Auffret, A.D. and Gray, J.C. (1984a)Cell 36, 555-562.
Willey, D.L., Howe, C.J., Auffret, A.D., Bowman, C.M., Dyer, T.A. and Gray, J.C. (1984b)Mol.Gen.Genet. 194, 416-422.
Yasunobu, K.T. and Tanaka, M. (1980) Methods Enzymol. 69, 228-238.
Rasmussen, O.F., Bookjans, G., Stummann, B.M. and Henningsen, K.W. (1984) Plant Mol. Biol. 3, 191-199.
Subik, J. and Takacsova, G. (1978). Mol. Gen. Genet. 161, 99-108.
Westhoff, P., Jansson, C., Klein-Hitpass, L., Berzborn, R., Larsson, C. and Bartlett, S.G. (1985). Plant Mol. Biol. 4, 137-146.

STRUCTURE AND TRANSCRIPTION OF THE <u>OENOTHERA</u> MITOCHONDRIAL GENOME

Wolfgang Schuster, Rudolf Hiesel, Bernd Wissinger, Werner Schobel and Axel Brennicke
Lehrstuhl für Spezielle Botanik der Universität, Auf der Morgenstelle 1, D-74 Tübingen, FRG

INTRODUCTION

The mitochondrial genome encodes only a small portion of the information necessary for assembly of these organelles, the majority of structural and enzymatic functions are synthesized from nuclear genes. Mitochondrial biogenesis thus requires coordinated expression of nuclear and mitochondrial genetic information. The sites of transcription initiation, termination and processing in mitochondria could potentially be used in such control and coordination signalling from the nucleus. Mediators in these regulatory could be the enzymatic activities and accessory factors involved in transcription (and replication) of the mitochondrial DNA, which are coded for in the nucleus (Clayton 1984). Thus control foci most likely involved in regulatory functions might be found in nucleic acid sequence regions around the sites of transcript initiation and/or termination and processing.

As part of the program to elucidate potential mechanisms of coordinate control of the different cell compartments we have investigated transcription signals of the plant mitochondrial genome in <u>Oenothera</u>. Another way of controlling plant mitochondrial transcription activity in higher plants could involve regulation of gene copy number by varying the pool of actively transcribed genes through genomic sequence rearrangements, which could transform active cistrons into silent pseudogenes and/or activate a given gene by introduction of functional promoter regions upstream of the coding region.

RESULTS AND DISCUSSION

We have determined the regions of potential transcript initiation and termination by identifying the precise transcript termini of a number of genes in the mitochondrial genome of Oenothera. Sequence comparison of these regions reveals conserved sequence and structure elements that can now be analysed for functional significance.

Transcript sizes

Most genes identified to date in Oenothera mitochondria are transcribed into monocistronic RNAs, for examples the reading frames encoding subunits 6 and 9 of the ATPase (ATP 6 and ATP 9; Schuster and Brennicke in preparation) and subunit III of the cytochrome oxidase (COX III; Hiesel et al. 1987).

An example for a bicistronic transcript has been found with the genes for the two small ribosomal RNAs (18S and 5S), that are closely linked and can be cotranscribed, as nuclease protection anlyses have shown (Schuster et al. 1987a).

Table I: Transcripts of Oenothera mitochondrial genes

	5'leader	Coding region	3'trailer
5S rRNA	0-5	118	ND
18S rRNA	125-140	1898	ND
26S rRNA	29-39	3244	ND
ATPA	250	1533	262-268
ATP 6	222	948	118-120
COX I	337	1584	47-52
COX II	207	774	421-422
COX III	1473	795	315

Other loci are possibly transcribed into multicistronic transcripts, like the putative small ribosomal protein S13 (S13) - NADH dehydrogenase subunit I (NDH 1) linkage group (Schuster and Brennicke 1987). Multiple transcripts are detected of this locus up to a length of 5 kb, their individual structure being as yet unclear however.

Precursor molecules for the ribosomal RNAs and individual mRNAs for the different protein coding genes contain leader sequences varying from 30 nucleotides in the 26S rRNA precursor to more than 1400 transcribed nucleotides preceding the COX III reading frame (Table I).

5'termini of transcripts

The genomic sequences surrounding transcription initiation sites are often important in promoter function and recognition (Rosenberg and Court 1979; Breathnach and Chambon 1981; Dynan and Tjian 1985). Prokaryotic and eukaryotic nuclear transcription is dependent on conserved sequences 5' to the actual transcription initiation sites. The conserved sequences considered the core-promoter region are important for the binding of RNA polymerase usually under the guidance of additional proteins. Auxiliary polypeptides often recognize additional sequence elements in the vicinity of the core promoter.

As a first step towards the definition of plant mitochondrial promoters we have determined the 5'termini of transcripts from a number of Oenothera mitochondrial genes and analysed the surrounding sequences in comparison with corresponding data from other plant mitochondria for homologies (Schuster et al. 1987b). Statistical analysis of these regions shows a conserved stretch of 16 nucleotides where transcription appears to be initiated (Fig.1).

The 5' termini of most transcripts are located in the highly conserved tetranucleotide 5'-TAAG-3'. The most frequent terminal nucleotide is the first A in this tetranucleotide. This sequence element forms the central part of a region with a conserved biased nucleotide distribution. Especially noteworthy is the low content of cytosine, which is favored in only one position 3 nucleotides upstream of the most frequent transcript termini.

The sequence region conserved at the 5'transcript termini in Oenothera appears to be a general phenomenon in higher plant mitochondria, as alignment of transcript termini determined in mitochondria of other plant species suggests (Fig.1).

Oe 26S rRNA	TAAGGTTAGTAAGGTTAGGG
Oe 18S rRNA	GAAATGTCATAAGTGATGTT
Oe 5S rRNA	ATCGGAAATCAAGACAAACC
Oe COX I	ACAATTGCGTAAGTGAGGGT
Oe COX II	AAAATCTCGTATGAGAATCA
Oe COX III	ACAATTGCGTAAGTGAGGGT
Oe ATP 6	ATATTATCATAAGTGAGGAG
Oe ATPA	GATAAATCATAAGAGAAGCA
Sugarbeet 1.9 kb	CTAAAATCATAAGTGATATC
Pea COX II site 1	TAAATTTACTAAGAGAAGAA
site 2	GAAATCACGTAAGTGATAGA
Sorghum COX I site 1	GGCTGCGCTCAAGAACTAGT
Petunia ATP 9	GAAATTTCATAAGATAAGAG
Maize S-2	AATCTACATAAAGATACCAA
Maize COX I	AGAAACTCATAAGTAATCCA
Consensus	AAATNTC$^{A}_{G}$TAAG$^{A}_{T}$GA
Yeast	ATATAAGTA

Fig.1 Alignment of nucleotide sequences surrounding the 5'-termini of transcripts in Oenothera mitochondria with the RNA termini determined in other higher plant mitochondria. The most frequent 5' RNA termini are found at the first A in the highly conserved tetranucleotide TAAG. The consensus sequence obtained from probability anaysis is given underneath the aligned experimetally determined genomic sequences (Schuster et al. 1987b). Statistical analysis gives no indication for other significant consensus motifs further upstream at the -10 or -35 regions. Data for the sugarbeet 1.9 kb minicircle are taken from Munk Hansen and Marcker (1984); for pea COX II from Moon et al. (1985); maize COX I from Isaac et al. (1985); Petunia ATP 9 from Young et al. (1986); sorghum COX I from Bailey-Serres et al. (1987) and maize S-2 from Traynor and Levings (1986). In this summary only transcripts have been included with termini clearly identified by S-1 mapping and in some cases supporting primer extension data.

The core element of four nucleotides (5'-TAAG-3') in the conserved region has also been found in the yeast mitochondrial promoter sequence (Fig.1; Isaac et al. 1985), where the guanosine has a crucial function in determining promoter strength (Osinga and Tabak 1982; Christianson and Rabinowitz 1983). This homology with the yeast promoter is suggestive of a potential significance of the conserved sequence in the transcription initiation process in plant mitochondria. Details of the recognition/initiation process must differ though between plant and yeast mitochondria, as transcription in yeast mitochondria is initiated at the terminal nucleotide of the entire promoter sequence two nucleotides behind this tetranucleotide. Initiation within the conserved nucleotide region appears to be a common theme of mitochondrial promoters, since mammalian mitochondrial transcripts are likewise initiated within a region of conserved sequence blocks (CBS; Chang and Clayton 1984; Clayton 1984).

The precise nucleotide assignment of 5'transcript termini in higher plants now allows the development of functional test systems to evaluate the significance and requirements in the described conserved sequence region for initiation of transcription in higher plant mitochondria.

3'termini of transcripts

Identification of 3'termini in Oenothera mitochondrial transcripts has revealed a conserved structural motif in the terminal transcribed nucleotides, that might be involved in the termination of transcription in plant mitochondria (Schuster et al. 1986). This functional role is suggested by analogy to known unaided bacterial terminators, that show a similar potential for two consecutive hairpins just before the site of pausing and termination of the RNA polymerase activity. Such structures are found at some of the Oenothera mitochondrial transcript termini (ATPA, COX II, ATP 6) determined by nuclease protection and cDNA cloning.

It is as yet unclear, whether different factors might be involved in generating the observed termini of other transcripts, where these structures cannot be identified, and whether these other termini are genuine transcription termination products or result from processing events. This latter explanation seems possible since the next genes downstream from these termini are at least physically closely linked with only a few hundred nucleotides intervening spacer sequence in these loci without potential hairpin structures at the 3'transcript termini.

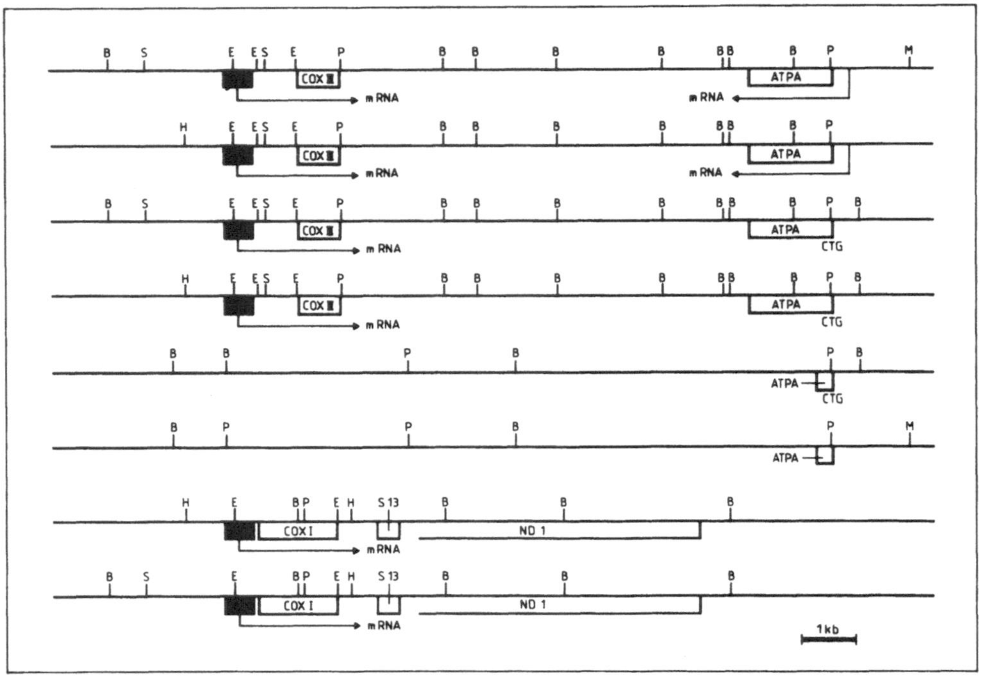

Fig.2: Intragenomic sequence rearrangements involving the promoter region (black bar) of the cytochrome oxidase subunit I and III genes (COX I and COX III) and the alpha subunit gene of the ATPase (ATPA) in the mitochondrial genome of Oenothera. The genes encoding subunit I of the NADH dehydrogenase (ND 1) and ribosomal protein S13 are closely linked to the COX I gene. Such rearrangements lead to multiple linkage groups and presumably molecules in the mitochondrial DNA of this species.

Genomic sequence rearrangements

Genomic sequence rearrangements have been found in the genomes of some organisms to be a means of regulating transcription (Borst and Greaves 1987). Such examples are found in the immunoglobulin genes of humans, in the "sex" determination locus of yeast and the variable surface antigenes of trypanosomes. Selection of active genes in the latter occurs apparently randomly by casette transposition of one gene from the silent pool to a location behind an active promoter sequence.

A similar event seems to have occurred in Oenothera mitochondria, where a promoter region has been duplicated and been placed upstream of two genes, that are now transcribed from identical promoter sequences (Hiesel et al. 1987). Interestingly these two genes encode two different subunits of one holoenzyme, subunits I and III of the cytochrome oxidase complex (Fig.2).

Other examples of duplicated and rearranged sequences in the mitochondrial genome of Oenothera involve duplications of entire genes and cistrons in different linkage groups and on different molecules. The 5S rRNA gene for example is present once in tandem behind the 18S rRNA gene and once in a different linkage group due to a rearrangement in the transcribed intergenic spacer (Fig.3; Schuster et al. 1987a). The entire COX II coding region is contained in the population of large mitochondrial DNA molecules and also in one of the smaller circular molecules (Hiesel and Brennicke 1983).

Some genes and coding regions are however interrupted by such sequence rearrangements and found as truncated copies or pseudogenes in the mitochondrial genome.

One example for the generation of pseudogenes by sequence rearrangements has been found in sequences derived from the coding region of the alpha subunit of the ATPase (ATPA), where two rearrangement events within the open reading frame cause the generation of three pseudoalleles with upstream sequences, downstream regions and with an internal portion of the gene (Fig.2; Schuster and Brennicke 1986).

A rearrangement event within the first intron of the gene encoding subunit 5 of the NADH dehydrogenase (ND 5) leads to the presence of a pseudogene containing only part of the first intron and the last two exons (Fig.3; B.Wissinger, R.Hiesel, W.Schuster and A.Brennicke, in preparation).

Another rearrangement event circularizes the 3'portion of the large ribosomal RNA gene and downstream sequences into a small circular molecule (Fig.4; Manna and Brennicke 1986). The 5' portion of this gene is either selectively destroyed or simply not amplified and maintained as an independent replication unit.

These rearrangements might be involved in quantitative and qualitative transcriptional control by adjusting the amount of template available for active transcription. Coordination of the gene copy number could be achieved by potential nuclear control of the recombination activity and specificity. This specificity might be obtained through individual additional factors needed for each rearrangement site. The rearrangement sites analysed so far in the Oenothera mitochondrial genome show very little homology with each other between the different

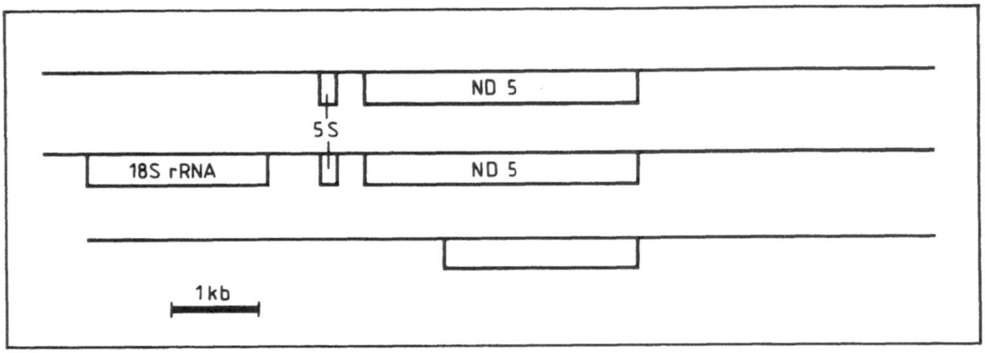

Fig.3: One rearrangement site occurs within the spacer separating the two small rRNA genes (5S and 18S rRNA), duplicating the 5S rDNA and the downstream sequences encoding subunit 5 of the NADH dehydrogenase (ND 5). A second rearrangement event in the first exon of the ND 5 gene creates a pseudoallele with only part of the first intron and the last two exons present (B.Wissinger, R.Hiesel, W.Schuster and A.Brennicke, in preparation).

genes involved. The only homology of bordering sequences between two different events intriguingly involves related genes, the above mentioned events at the 26S rRNA and the 18S-5S rRNA genes. Secondary structural elements might also be involved in recognition, since the 26S rRNA recombination event occurs in the same region of the molecule only 21 nucleotides apart from the integration event of plastid sequences with the 3' part of the plastid 23S rRNA (Schuster and Brennicke 1987a).

All other rearrangement sites must be controlled by their own specificities which appear to very precise, since pseudogene sequences and intact active gene sequences are identical in all the examples analysed.

These pseudogenes are found consistently in different preparations in the apparently unchanging mitochondrial genome of tissue culture cells. The steady state relations between intact alleles and pseudogenes thus remain constant in these cells. Gene specific alterations might only be detectable if individual mitochondrial genomes could be followed during the ontogenesis of individual cells or during the development and differentiation of plant tissues.

Genome rearrangement in plant mitochondria also exerts a perceptable influence on transcription in an evolutionary scale during the

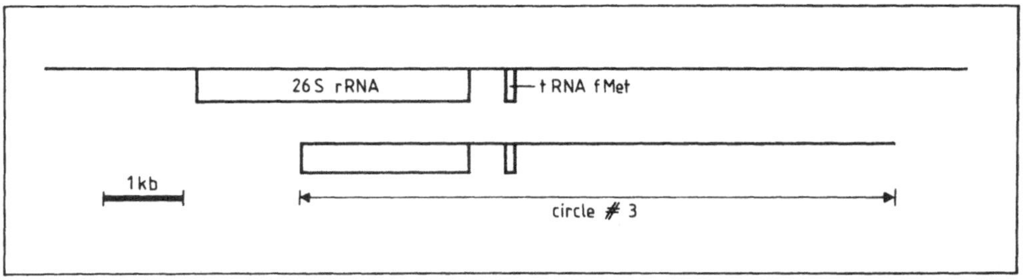

Fig.4: A small circular molecule is excised from the main mitochondrial genome of Oenothera at a ten nucleotide repeat occurring once within the 26S rRNA gene and once 7.5 kb downstream. This new molecule (circle # 3; Manna and Brennicke 1986) contains an intact copy of the gene for tRNA fMet.

genesis of new genes by sequence casette rearrangements. One of the byproducts of these events seems to have been found in the mosaic genes causing cytoplasmic male sterility, that arose from rearrangements linking transcription signals and pieces of different genes (See also articles by Drs.C.J.Leaver et al. and D.Lonsdale et al., this volume).

Changes in major control functions of copy numbers and relative stochiometries of different genome arrangements appear to be involved in plasma type and potentially in species evolution (Small et al. 1987).

Transcription of foreign sequences in the mitochondrion

The Oenothera mitochondrial genome contains sequences imported from both plastids and the nucleus. Plastid sequences involve part of the ribosomal cistron containing the 3'portion of the plastid 23S rRNA sequence, the intergenic spacer towards the 4.5S rRNA, the entire 4.5S rRNA and a large part of the adjacent intergenic spacer towards the 5S rRNA sequence (Schuster and Brennicke 1987a). Another event has transferred the 3' portion of the small ribosomal protein S4, the adjacent intergenic nucleotides and the $tRNA^{Ser}$ sequence (Schuster and Brennicke 1987b).

Nuclear sequences in the mitochondrial genome of Oenothera contain part of the 18S rRNA sequence and an open reading frame with high homology to reverse transcriptase from transposable elements (Schuster and Brennicke 1987b).

It is unclear as yet whether any of the entire genes in the sequences transferred from the plastid are transcribed in the mitochondrion, this would be especially interesting with respect to the 4.5S rRNA and $tRNA^{Ser}$ coding regions.

So far only the open reading frame potentially encoding a reverse transcriptase has been investigated for the presence of transcripts in the mitochondrial RNA population. This reading frame seems to be transcribed at a very low level, since homologous transcripts can be detected in RNA blot hybridizations.

This reading frame, if encoding a functional reverse transcriptase polypeptide, could be involved in the integration of RNAs imported from either the cytosol or the plastid into the mitochondrial genome. A function of this enzyme in controlling the plant mitochondrial RNA pool accessible for active transcription would require a number of auxiliary functions in guiding the reverse transcribing activity. Transcripts would need to be specifically protected at their 3'termini from the reverse transcriptase activity, that could inactivate the transcript by syntheziseing an RNA-DNA hybrid after specific release of the terminal block. This would be an energy and resource expensive way of regulation, far exceeding the efficiency of for example transcript stability control by access control of RNAses.

CONCLUSION

Specific nucleotide sequence signals have been found for transcription initiation and termination in Oenothera mitochondria that are highly conserved in mitochondria in other higher plants.

Sequence elements found at the 5'termini of transcripts show analogy to mitochondrial promoters from other species and appear to be distinct from prokaryotic and eukaryotic promoter recognition signals.

Secondary sequence structures at the 3'termini of transcripts show analogy with simple prokaryotic terminators suggesting a common mechanism for transcript termination in bacteria and plant mitochondria.

ACKNOWLEDGEMENTS

We are grateful to Drs.C.J.Leaver F.R.S. and P.G.Isaac for many discussions. V.Uhle-Schneider did the artwork, and C.Specht photography. This work was supported by grants from the Deutsche Forschungsgemeinschaft and a Heisenberg-Fellowship (A.B.)

REFERENCES

Auble, D.T., Allen, T.L. and deHaseth, P.L. (1986) Promotor recognition by Escherichia coli RNA polymerase. J.Biol.Chem. 261, 11202-11206

Bailey-Serres, J., Hanson, D.K., Fox, T.D. and Leaver, C.J. (1987) Mitochondrial genome rearrangement leads to extension and relocation of the cytochrome c oxidase subunit I gene in sorghum. Cell 47, 567-576

Borst, P. and Greaves, D.R. (1987) Programmed gene rearrangements altering gene expression. Science 235, 658-667

Breathnach, R. and Chambon, P.A. (1981) Organization and expression of eucaryotic split genes coding for proteins. Annu.Rev.Biochem. 50, 349-383

Chang, D.D. and Clayton, D.A. (1984) Precise identification of individual promotors for transcription of each strand of human mitochondrial DNA. Cell 36, 635-643

Christianson, T. and Rabinowitz, M. (1983) Identification of multiple transcriptional initiation sites on the yeast mitochondrial genome by in vitro capping with guanylyltransferase. J.Biol. Chem. 258, 14025-14033

Clayton, D.A. (1984) Transcription of the mammalian mitochondrial genome. Ann.Rev.Biochem. 53, 573-594

Dynan, W.S. and Tjian, R. (1985) Control of eukaryotic messenger RNA synthesis by sequence specific DNA-binding proteins. Nature 316, 774-778

Hiesel, R. and Brennicke, A. (1983) Cytochrome oxidase subunit II gene in mitochondria of Oenothera has no intron. EMBO J. 2, 2173-2178

Hiesel, R., Schobel, W., Schuster, W. and Brennicke, A. (1987) Cytochrome oxidase subunit I and III genes are transcribed from identical promotor sequences. EMBO J. 6, 29-34

Isaac, P.G., Jones, V. and Leaver, C.J. (1985) The maize cytochrome c oxidase subunit I gene: sequence, expression, and rearrangement in cytoplasmic male sterile plants. EMBO J. 4, 1617-1623

Manna, E. and Brennicke, A. (1986) Site-specific circularisation in Oenothera mitochondria. Mol.Gen.Genet. 203, 377-381

Moon, E., Kao, T. and Wu, R. (1985) Pea cytochrome oxidase subunit II gene has no intron and generates two mRNA transcripts with different 5'termini. Nucleic Acids Res. 13, 3195-3212

Munk Hansen, B. and Marcker, K.A. (1984) DNA sequence and transcription of a DNA minicircle isolated from male-fertile sugar beet mitochondria. Nucleic Acids Res. 12, 4747-4756

Osinga, K. and Tabak, H.F. (1982) Initiation of transcription for mitochondrial ribosomal RNA in yeast: comparison of the nucleotide sequence around the 5'ends of both genes reveals a homologous stretch of 17 nucleotides. Nucleic Acids Res. 10, 3617-3626

Rosenberg, M. and Court, D. (1979) Regulatory sequences involved in the promotion and termination of RNA transcription. Ann.Rev.Genet. 13, 19-53

Schuster, W. and Brennicke, A. (1986) Pseudocopies of the ATPase a-subunit gene in Oenothera mitochondria are present on different circular molecules. Mol.Gen.Genet. 204, 29-35

Schuster, W., Hiesel, R., Isaac, P.G., Leaver, C.J., and Brennicke, A. (1986): Transcript termini in messenger RNAs in higher plant mitochondria. Nucleic Acids Res. 14, 5943-5954

Schuster, W. and Brennicke, A. (1987a) Plastid, nuclear and reverse transcriptase sequences in the mitochondrial genome of Oenothera: Is genetic information transferred between organelles via RNA ? Manuscript submitted

Schuster, W. and Brennicke, A. (1987b) Plastid DNA in the mitochondrial genome of Oenothera: inter- and intraorganellar rearrangements involving part of the plastid ribosomal cistron. Manuscript submitted

Schuster, W., Hiesel, R., Manna, E., Schobel, W., Wissinger, B. and Brennicke, A. (1987a) Molecular analysis of the mitochondrial genome of Oenothera. Plant Phys.Biochem., in press

Schuster, W, Hiesel, R. and Brennicke, A. (1987b) Common sequence motifs at the 5' termini of plant mitochondrial transcripts. Submitted

Small, I.D., Isaac, P.G. and Leaver, C.J. (1987) Stochiometric differences in DNA molecules containing the atpA gene suggest mechanisms for the generation of mitochondrial genome diversity in maize. EMBO J. 6, 865-869

Traynor, P.L. and Levings, C.S.III (1986) Transcription of the S-2 maize mitochondrial plasmid. Plant Mol.Biol. 7, 255-263

Young, E.G., Hanson, M.R. and Dierks, P.M. (1986) Sequence and transcription analysis of the Petunia mitochondrial gene for the ATP synthase proteolipid subunit. Nucleic Acids Res. 14, 7995-8006

THE BETA SUBUNIT OF A PLANT MITOCHONDRIAL ATP SYNTHASE HAS A PRESEQUENCE INVOLVED IN MITOCHONDRIAL TARGETING

Marc Boutry[1,2], Ferenc Nagy[2] and Nam-Hai Chua[2]

[1]Laboratoire d'Enzymologie
Université de Louvain
Place Croix du Sud, 1
1348 Louvain-la-Neuve
Belgium

[2]Laboratory of Plant Molecular Biology
The Rockefeller University
New York, N.Y. 10021-6399

INTRODUCTION

Plant mitochondria are the major source of ATP in non photosynthetic tissues and in green tissues during the dark period. A key enzyme of the mitochondrial energy machinery is the ATP synthase complex which converts ADP into ATP from the energy released in the respiratory chain. The mitochondrial ATP synthase is well concerved among different organisms and is very similar in structure and function to the bacterial and chloroplast ATP synthases. Therefore all these complexes seem to have evolved from a common ancestor. However particular properties characterize complexes from different sources. For instance, the gene organization of the various subunits is quite distinct. *E. coli* genes for the ATP synthase are organized in an operon of nine genes (1). In eukaryotes on the contrary, the genes are dispersed on two genomes : the nuclear DNA and the chloroplast or mitochondrial DNA. Moreover the partition between both genomes is not unique. Most of the chloroplast ATP synthase subunits are encoded in the chloroplast genome (2) while in animals and yeast, a majority of the mitochondrial ATP synthase subunits are encoded in the nucleus. Few genes for the plant mitochondrial ATP synthase have been localized so far. Like in yeast and mammals, subunits 6 and 9 are encoded by mitochondrial genes (3-5). Also dependent on the mitochondrial genome is the alpha subunit which, in yeast and in mammals, is encoded in the nucleus (6,7).

TWO GENES FOR THE MITOCHONDRIAL ATP SYNTHASE BETA SUBUNIT

We have undertaken a gene analysis of the alpha and beta subunits of the mitochondrial ATP synthase in Nicotiana plumbaginifolia. As in maize (6,7), the gene for the alpha subunit is located on the mitochondrial genome. It is a single gene which has been sequenced (F. Chaumont and A. Vassarotti, unpublished results). The nuclear gene for the beta subunit was obtained after screening cDNA and genomic libraries with the corresponding gene from yeast (8). Southern blot analysis indicated that N. plumbaginifolia contains two genes for the beta subunit. They are both expressed, at least in leaf, since we isolated cDNA clones corresponding to each of them (9). One gene (atp2-1) has been analyzed in detail (9). The sequence of the second gene (atp2-2) has been almost completed. Figure 1 compares the structure of both genes. Their coding sequences are interrupted at the sames positions by 8 intervening sequences. However their lengths and compositions are not identical. In average the introns show less than 50 % of homology. On the other hand, the coding sequence is well conserved (more than 90 % of homology at the nucleotide level). Most changes are silent since only 4 amino acid residues distinguish both genes (results not shown).

We observed in intron VIII of atp2-1 the presence of an insertion element (Inp) which is present in multiple copies in the N. plumbaginifolia genome (9). However Inp is not present in the second gene. Southern blot analyses of 6 different Nicotiana species have shown that they all contain two atp2 genes but that in each case, only one gene contains an Inp sequence (not shown). These results suggest that both genes diverged from a common ancestor before the emergence of current species belonging to the genus Nicotiana. Whether both genes are expressed at a similar level in different tissues is unknown. The availability of DNA probes specific to each of them should allow us to elucidate that question.

A PRESEQUENCE FOR MITOCHONDRIAL TARGETING

Figure 2 depicts the N-terminal amino acid sequences of the beta subunits from various organisms. The sequence homology among the bacterial, yeast, human, tobacco chloroplast and N. plumbaginifolia mitochondrial beta subunits is very high all along the protein (over 65 % in any combination, not shown). Only the N-terminal region appears to be organism specific. The yeast, human and N. plumbaginifolia mitochondrial beta polypeptides are encoded in the nucleus and consequently have to be translocated from the cytosol to the organelle. The yeast and human beta polypeptides have N-terminal extensions (presequences) which are involved in mitochondrial import and removed once inside the organelle. The yeast beta precursor is matured at lys19-glu20 (10) while it has been suggested that the human beta presequence is 59 amino acid residues long (11). A longer

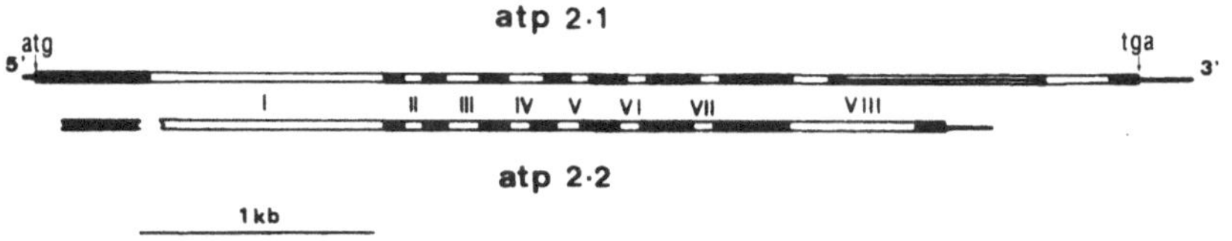

Figure 1. Structure and organization of atp2-1 and atp2-2 genes. Exons are represented in closed boxes and introns (numbered I-VIII) in open boxes. The line inserted in intron VIII of atp2-1 represents the insertion element.

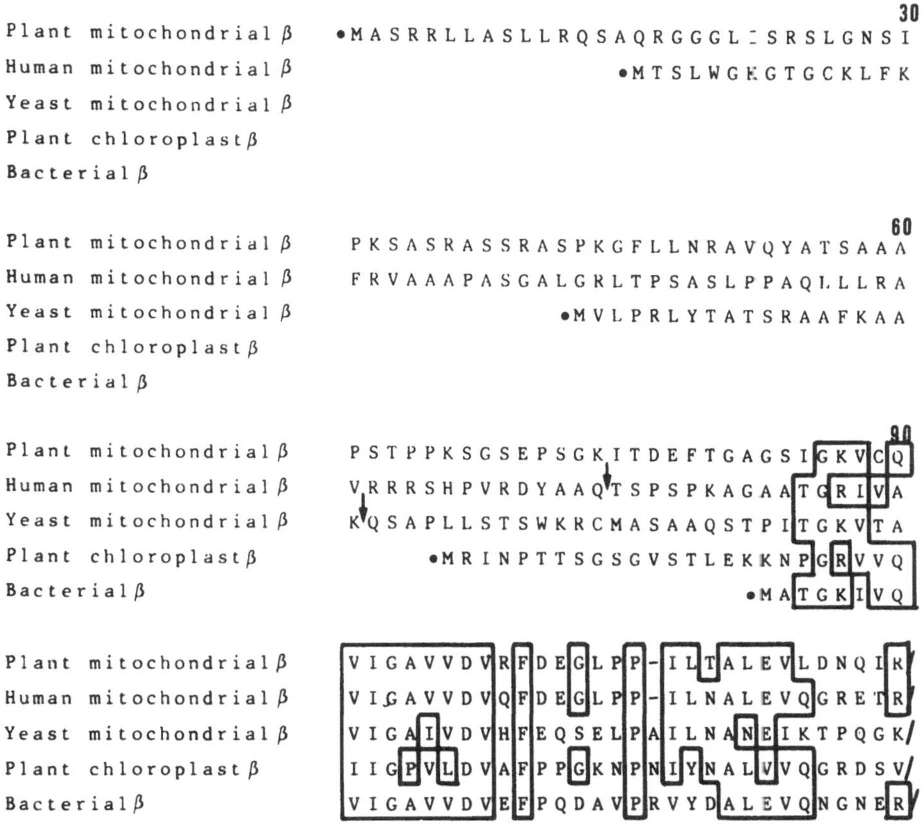

Figure 2. Comparison of the N-terminal amino acid sequences of the ATP synthase beta subunits from various organisms. The amino acid sequences were predicted from the nucleotide sequences for the N. plumbaginifolia mitochondrial (9), the human mitochondrial (11), the yeast mitochondrial (10), the tobacco chloroplast (23) and the bacterial (22) beta polypeptides. Amino acid residues are boxed if a least three residues in a row are homologous. Arrows indicate the position of cleavage site. Note that only 120 residues are shown.

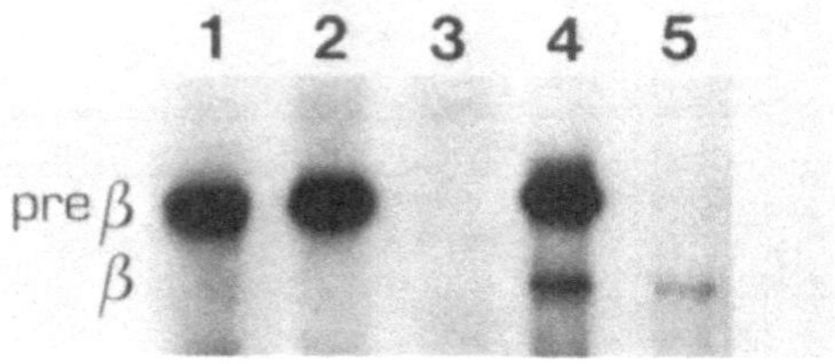

Figure 3. In vitro uptake and processing of the beta polypeptide precursor. An atp2-1 cDNA was subcloned in the expression vector pDS5 (13) and then submitted to an in vitro transcription-translation system as described (13). Lane 1, in vitro synthesized product; lanes 2-5, in vitro synthesized products were mixed to freshly isolated tobacco mitochondria (24) in 30 ul of 250 mM mannitol, 25 mM Hepes (pH 7.5), 50 mM K-acetate, 2 mM Mg acetate, 1 mM ADP and 5 mM K_2HPO_4, in the absence (lanes 2-3) or presence of (lanes 4-5) of glycine. After incubation at 20°C for 30 min, the mitochondria were pelleted in a microfuge through a cushion of 0.5 ml of 1 M sucrose, 10 mM K_2HPO_4 (pH 7.2), 1 mM PMSF. In lanes 3 and 5, mitochondria were treated with 1 ug of proteinase K for 10 min on ice before pelleting.

presequence is predicted for the beta polypeptide from N. plumbaginifolia. A comparison between the size of the precursor (Mr = 59 866) and that of the mature subunit (Mr = 51 000) suggests a presequence of about 9 000 daltons. We have conducted in vitro uptake experiments to confirm the existence of the beta presequence. A complete atp2-1 cDNA clone was placed in the expression vector pDS5 (13) under the control of a T5 coliphage promoter. In vitro synthesized RNA was then translated in a reticulocyte lysate to give a precursor to the beta polypeptide with the expected size of Mr = 60 000 (Figure 3, lane 1). We incubated the precursor with purified tobacco mitochondria and then reisolated the organelles through a sucrose cushion. In these conditions, the beta precursor remained bound to mitochondria (lane 2) and was degraded after addition of proteinase K (lane 3). In the presence of a respiratory substrate, part of the precursor was converted to a smaller size corresponding to that of the mature beta polypeptide (lane 4). After treatment with proteinase K, the precursor form completely disappeared while the processed form remained unaffected (lane 5). This indicated the intramitochondrial location of the processed form. In a similar experiment, we performed an in vitro import of the N. plumbaginifolia beta precursor in broad bean mitochondria (not shown). The precursor was partly converted to a mature form, indicating that the uptake machinery is conserved among those species.

The variation in the presequence length among different organisms (19, 59 and over 75 amino acid residues for the yeast, human and plant mitochondrial beta subunits, respectively) possibly reflects a distinct evolution of the machinery responsible for mitochondrial targeting. However it has been suggested that the secondary structure of the presequence plays a role in the import process. In this respect, it is interesting to note that the three beta presequences are able to form amphiphilic helices like other mitochondrial presequences (12).

IN VIVO TARGETING OF A FOREIGN PROTEIN TO MITOCHONDRIA IN TRANSGENIC PLANTS

It is well documented that yeast, *Neurospora* or mammalian mitochondrial presequences are able to drive *in vitro* as well as *in vivo* mitochondrial import of foreign proteins fused to them (14-17). It has also been shown that the chloroplast presequence of the ribulose-1,5-bisphosphate carboxylase small subunit is able to target a bacterial neomycin phosphotransferase to chloroplasts (18,19).

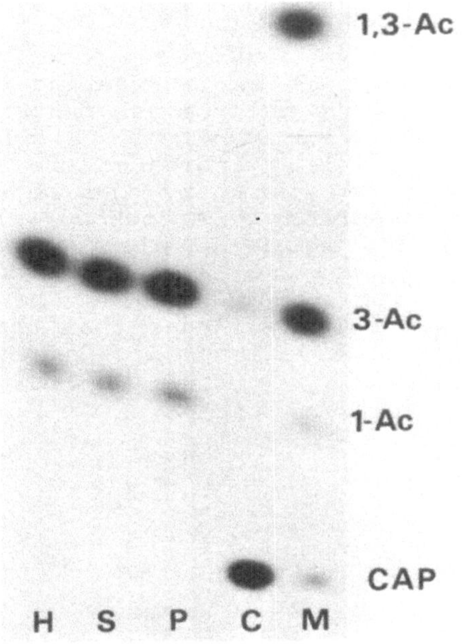

Figure 4. Distribution of CAT activity in different subcellular fractions of a plant transformed with a prebeta-CAT gene. The coding sequence of an *E. coli* CAT gene was fused to the *atp2-1* sequence corresponding to the first 90 amino acid residues. The hybrid gene placed under the control of the CaMV 35S promoter was introduced into tobacco by *A. tumefaciens* mediated transformation (21). An homogenate (H) prepared from leaf material was centrifuged to give a crude cytosolic supernatant (S) and a crude organellar pellet (P). The latter was further purified onto Percoll gradients to give a chloroplast (C) and a mitochondrial (M) fractions. The construction of the hybrid gene and the subcellular fractionation will be detailled elsewhere. CAP, chloramphenicol; 1-Ac, 1-acetyl-chloramphenicol; 3-Ac, 3-acetyl-chloramphenicol; 1,3-Ac, 1,3-diacetyl-chloramphenicol.

We have undertaken *in vivo* mitochondrial targeting experiments to bring a response to two questions : 1/ Can the *atp2-1* presequence target a foreing protein to plant mitochondria ? 2/ How specific is the targeting process ?

We have fused the *E. coli* chloramphenicol acetyltransferase (CAT) to the sequence of *atp2-1* corresponding to the first 90 amino acid residues, which is predicted to encompass the entire presequence as well as a few residues of the mature beta polypeptide. This hybrid gene placed under the control of the CaMV 35S promoter (20) was introduced into tobacco via *A. tumefaciens* transformation (21). As a control we used the CAT gene without presequence. Transformants were regenerated in whole plants and shown to express CAT activity in a crude cellular extract. Figure 4 reports the CAT activity detected in different subcellular fractions of a transgenic plant with the prebeta-CAT construct. The homogenate (H) was centrifuged to give a crude cytosolic supernatant (S) and a crude organellar pellet (P). The latter was further purified on Percoll gradients to give a mitochondrial (M) and a chloroplast (C) fraction. Clearly CAT activity was associated with the crude organellar fraction and was enriched in the mitochondrial fraction. The CAT activity found in the cytosolic fraction can be attributed to enzyme released from mitochondria broken during homogenization. It is interesting to note that little activity was found in the chloroplast fraction. Probably most, if not all, of that activity can be due to mitochondrial contamination. This result indicates the high specificity of the import process mediated by the beta presequence. We analyzed by immunodection a mitochondrial soluble fraction after separation by SDS-PAGE electrophoresis (Figure 5). Antibodies directed against *E. coli* CAT detected in the mitochondrial fraction a CAT protein, the apparent molecular weight of which was slighly larger than that of the native *E. coli* protein. This observation indicates that the

Figure 5 Immunodetection of CAT in a mitochondrial fraction. The mitochondrial matrix from a transgenic plant with the prebeta-CAT gene (lane 1) and 1 ug of purified *E. coli* CAT (lane 2) were separated by SDS-PAGE and transferred to a nitrocellulose paper. The filter was immunodecorated with an anti-CAT serum followed by ^{125}I-protein A incubation. Radioactive bands were visualized by autoradiography.

prebeta-CAT hybrid precursor was processed to the mature form. The small difference (1 500 daltons) probably corresponds to beta polypeptide residues located dowstream to the processing site.

We have shown that the mitochondrial ATP synthase beta subunit from Nicotiana plumbaginifolia is synthesized as a larger precursor with an N-terminal presequence which is removed during or soon after mitochondrial import. We used a chimaeric gene (prebeta-CAT) to direct the transport of the CAT protein to plant mitochondria. The results indicated that the beta subunit presequence is sufficient to specifically direct that transport.

ACKNOWLEDGMENTS

We thank M.-Y. Hsu, M. Wong and A.M. Faber-Hubermont for excellent technical help. M.B. is Research Associate at the F.N.R.S. (Belgium). M.B. wishes to thank Drs A. Goffeau and M. Briquet (University of Louvain) for their support and interest. The work done at The Rockefeller University was supported by grants from Monsanto, GM 30726, and a Winston Foundation Fellowship to M.B.

REFERENCES

1. M. Futai and H. Kanazawa, Structure and function of proton-translocating adenosine triphosphatase: biochemical and molecular biological approaches, Microbiol. Reviews 47: 285-312.
2. J. Henning and R.G. Herrmann, Chloroplast ATP synthase of spinach contains nine nonidentical subunit species, six of which are encoded by plastid chromosomes in two operons in a phylogenetically conserved arrangement, Mol. Gen. Genet. 203: 117-128 (1986).
3. R.E. Dewey, C.S. Levings III and D.H. Timothy, Nucleotide sequence of ATPase subunit 6 gene of maize mitochondria, Plant Physiol. 79: 914-919 (1985).
4. E. Hack and C.J. Leaver, Synthesis of a dicyclohexylcarbodiimide-binding proteolipid by cucumber mitochondria, Current Genetics 8: 537-542 (1984).
5. R.E. Dewey, A.M. Schuster, C.S. Levings III and D.H. Timothy, Nucleotide sequence of Fo-ATPase proteolipid (subbunit 9) gene of maize mitochondria, Proc. Natl. Acad. Sci. USA 82: 1015-1019 (1985).
6. P.G. Isaac, A., Brennicke, S.M. Dunbar and C. Leaver, The mitochondrial genome of fertile maize (Zea mays L.) contains two copies of the gene encoding the alpha-unit of the F1-ATPase, Curr. Genet. 10: 321-328 (1985).
7. C.J. Braun and C.S. Levings III, Nucleotide sequence of the F1-ATPase alpha subunit gene from maize mitochondria, Plant Physiol. 79 : 571-577
8. M. Takeda, A. Vassarotti and M.G. Douglas, Nuclear genes coding the yeast mitochondria adenosine triphosphatase complex, J. Biol. Chem. 260: 15458-15465 (1985).

9. M. Boutry and N.-H. Chua, A nuclear gene encoding the beta subunit of the mitochondrial ATP synthase in Nicotiana plumbaginifolia, EMBO J. 4: 2159-2165 (1985).
10. A. Vassarotti, W.-J. Chen, C. Smagula and M.G. Douglas, Sequences distal to the mitochondrial targeting sequences are necessary for the maturation of the F1-ATPase beta-subunit precursor in mitochondria, J. Biol. Chem. 262: 411-418 (1987).
11. S. Ohta and Y. Kagawa, Human F1-ATPase: molecular cloning of cDNA for the beta-subunit, J. Biochem. 99: 135-141 (1986).
12. G. von Heijne, Mitochondrial targeting sequences may form amphiphilic helices, EMBO J. 5: 1335-1342 (1986).
13. D. Stveber, I. Ibrahimi, D. Cutler, B. Dobberstein and H. Bujard, A novel in vitro transcription-translation system: accurate and efficient synthesis of single proteins from cloned DNA sequences, EMBO J. 3: 3143-3148.
14. S. Emr, A. Vassarotti, J. Garret, M. Geller, M. Takeda and M. Douglas, The amino terminus of the yeast F1-ATPase beta-subunit precursor functions as a mitochondrial import signal, J. Cell Biol. 102: 523-533 (1986).
15. E. Hurt, U. Miller and G. Schatz, The first twelve amino acids of a yeast mitochondrial outer membrane protein can direct a nuclear encoded cytochrome oxidase subunit to the mitochondrial inner membrane, EMBO J. 4: 3509-3518 (1985).
16. A. Horwich, F. Kalousek, I. Mellman and I. Rosenberg, A leader peptide is sufficient to direct mitochondrial import of a chimeric protein, EMBO J. 4: 1129-1135 (1985).
17. A.P. van Loon, A.W. Brändli and G. Schatz, The presequences of two imported mitochondrial proteins contain information for intracellular and intramitochondrial sorting, Cell 44: 801-812 (1986).
18. P. Schreier, E.A. Seftor, J. Schell and H.J. Bohnert, The use of nuclear-encoded sequences to direct the light-regulated synthesis and transport of a foreign protein into plant chloroplasts, EMBO J. 5: 1335-1342 (1985).
19. G. Van den Broeck, M. Timko, A. Kausch, A.R. Cashmore, M. Van Montagu and L. Herrera-Estrella, Targeting of a foreign protein to chloroplasts by fusion to the transit peptide from the small subunit of ribulose 1,5-biphosphate carboxylase, Nature 313: 358-363 (1985).
20. J.T. Odell, F. Nagy and N.-H. Chua, Identification of DNA sequences required for activity of the cauliflower mosaic virus 35S promoter, Nature 313: 810-812 (1985).
21. S.G. Rogers, R.B. Horsch and R.T. Fraley, Gene transfer in plants: production of transformed plants using Ti plasmid vectors, Methods in Enzymol. 118: 627-640 (1986).
22. M. Saraste, N.J. Gay, A. Eberle, M.J. Runswick and J.E. Walker, The atp operon: nucleotide sequence of the gene for the gamma, beta and epsilon subunits of Escherichia coli ATP synthase, Nucleic Acids Res. 9:5287-5296 (1981)
23. K. Shinozaki, H. Deno, A. Kato and M. Sugiura, Overlap and cotranscription of the genes for the beta and epsilon subunits of tobacco chloroplast ATPase, Gene 24: 147-155 (1983).
24. M. Boutry, A.-M. Faber, M. Charbonnier and M. Briquet, Microanalysis of plant mitochondrial protein synthesis products, Plant Mol. Biol. 3: 445-452 (1984)

TRANSCRIPTIONAL AND POST-TRANSCRIPTIONAL REGULATION OF CHLOROPLAST GENE EXPRESSION

Wilhelm Gruissem, Xing-Wang Deng, Helen Jones, David Stern, John Tonkyn and Gerard Zurawski*

Department of Botany, University of California, Berkeley, California 94720, and *DNAX Research Institute of Molecular and Cellular Biology, 901 California Avenue, Palo Alto, California 94304

INTRODUCTION

The development of photosynthetically active chloroplasts from proplastids and etioplasts is one of the most important morphogenetic changes during leaf formation in higher plants. In other non-photosynthetic plant organs or cells, proplastids usually differentiate into highly specialized plastid types, e.g. amyloplasts in roots and proteinoplasts in parenchymatic tissues. Photosynthetically active chloroplasts can also differentiate into plastid types of other distinct functions, examples being the chromoplasts in flower petals and tomato fruits (1,2). The development and differentiation of the various plastid types requires the coordinate interaction of the nuclear and plastid genomes to achieve the controlled expression of genes for the components of the genetic and photosynthetic apparatus. Previous studies have shown that mRNA levels for several photosynthetic proteins synthesized in the cytoplasm and in the organelle undergo significant changes during chloroplast development and chromoplast differentiation (3-10). After illumination of etiolated seedlings, mRNAs from several plastid and nuclear genes accumulate during chloroplast development, but the extent to which these mRNAs increase in the light is variable. Similarly, cytoplasmic and chloroplast mRNA levels for photosynthetic proteins decrease during chromoplast differentiation in tomato fruit, but the decline appears to be specific and characteristic for each mRNA. The application of nuclear run-on transcription has provided information that such changes for nuclear encoded mRNAs (e.g. the small subunit of ribulose-1,5-bisphosphate carboxylase) is at least in part controlled at the transcriptional level (11,12). Based on competitive hybridization studies with several maize plastid genes it was concluded that the control of changes in their mRNA levels is also fundamentally transcriptional (5). In addition to the observed changes in mRNA levels, more recent work has shown that the expression of chloroplast genes for photosystem I and II proteins, as well as the large subunit of ribulose-1,5-bisphosphate carboxylase, can also be adjusted rapidly at the translational level (13,14). However, the translational control is often superimposed on the long-term changes at the RNA level, again indicating that transcriptional regulation may serve as a pivotal control of their expression.

Our current understanding of transcriptional control of genes on the chloroplast genome is mostly derived from the molecular analysis of chloroplast promoter regions. Prokaryote-like promoter sequences have been identified upstream of several chloroplast genes. The structure and function of such sequences as promoter elements has been demonstrated for tRNA and protein-coding genes in spinach, maize, mustard and pea (15-20), but the presence of such sequences is not a strict requirement for transcription initiation of all plastid genes (21). Most of the work in this field has been recently reviewed by Hanley-Bowdoin and Chua (22). Transcription of spinach chloroplast genes in vitro has indicated that the different strengths of chloroplast promoter regions may be primarily responsible for their relative transcription activities, but these experiments do not exclude the modulation of promoter activities by trans-acting factors during chloroplast development and differentiation (16). The analysis of transcriptional regulation of chloroplast genes is also complicated by the multicopy nature of the chloroplast genome. The DNA copy number in individual plastids can vary by as much as five- to ten-fold, depending on the developmental stage of chloroplasts in leaves or the specialized plastids in different tissues (23,24). It is therefore possible that the availability of transcriptionally competent DNA molecules may play an important role in the control of gene expression.

The differential changes of RNA levels and stabilities of individual plastid mRNAs during chloroplast development and chromoplast differentiation in consequence could determine the translation efficiency of the respective proteins, although much less is known about such regulation in plastids. Earlier work has shown that synthesis of two prominent proteins, the large subunit of ribulose-1,5-bisphosphate carboxylase and the Q_B-binding protein of photosystem II, can be rapidly depressed in the dark without changes in their mRNA levels in Spirodela and Amaranth (13, 25), but light is not a strict requirement for the translation of several chloroplast membrane and soluble proteins in barley (14). Thus, the differences in translational control may reflect the contrast in developmental programs for these plants, as well as switches in the regulation of chloroplast gene expression at specific stages of leaf development.

The regulation of chloroplast gene expression thus includes several control steps, each of which may be of specific importance for certain classes of plastid genes, critical gene products or particular stages of chloroplast development and plastid differentiation. In an effort to understand the regulatory mechanisms that may operate at different levels and their relative contribution to the control of plastid gene expression, we have examined the transcriptional and post-transcriptional regulation of several plastid genes during chloroplast development and chromoplast differentiation (6,7,26,27). This review summarizes our current view of regulatory mechanisms that operate at the transcriptional and post-transcriptional level in higher plant chloroplasts.

CHLOROPLAST PROMOTER REGIONS: THE BASIC CONTROL STEP

The DNA sequences flanking the 5' ends of chloroplast genes are highly conserved for each gene in different higher plants. For example, the 5' regions of rbcL genes are more than 80% homologous between dicotyledons (spinach, tobacco, tomato and pea) and monocotyledons (maize, wheat and barley), and similar homologies have also been observed for psbA (for review see 22). These regions contain DNA sequences that resemble the -10 (TATAAT) and -35 (TTGACA) consensus prokaryotic promoter elements. Their function as chloroplast promoter regions was first inferred from DNA sequence comparisons, their location relative to respective transcription initiation sites, and binding studies with

```
                                                     ▪
psbA     TTGGTTGACACGGGCATATAAGGCATGTTATACTGTTGAATAA
                 <......18 bp.....>

trnM2    ATTCTTGCTTATATATAATATTTGATTTATAATCAATCTATGG
                 <......17 bp....>
                                                     ▪
rbcL     TGGGTTGCGCCATATATATGAAAGAGTATACAATAATGATGGA
                 <......18 bp.....>
                                                     ▪
atpB     AGTCTTGACAGTGGTATATGTTGTATATGTATATCCTAGATGT

               CGC                     C C
             TTGACA.........?........TATAAT
                TT
```

Figure 1 Comparison of spinach chloroplast promoter regions. The 5' regions which flank the transcription initiation sites (▪) are compared for conserved sequence elements. The function of the underlined DNA sequences as promoter elements has been verified in chloroplast transcription extracts (15,16). The bottom line shows the prokaryotic consensus -35 and -10 promoter elements with modifications in the chloroplast DNA sequences.

E.coli RNA polymerase. The development of chloroplast in vitro transcription systems from various plants has allowed the functional assay of these DNA sequences for transcription initiation. We have shown by 5' deletions in the spinach <u>trn</u>M2 upstream region that the sequence 5'TTGCTT is critical for transcription initiation. The functional significance of the <u>trn</u>M2 upstream region has been tested further by analyzing the effects of deletions, insertions and base substitutions in this region on transcription of <u>trn</u>M2 in vitro. This work has confined the chloroplast DNA sequences critical to <u>trn</u>M2 promoter function in vitro to the underlined sequences shown in Figure 1 (15). The spinach chloroplast transcription extract has also been used to locate the promoter regions for <u>psb</u>A, <u>rbc</u>L and <u>atp</u>B. Templates constructed using synthetic oligonucleotides spanning the 5' upstream regions of these genes and a promoter-deficient <u>trn</u>M2 mutant were assayed for their ability to support the transcription of <u>trn</u>M2 in vitro. These experiments have allowed us to define the <u>psb</u>A, <u>rbc</u>L and <u>atp</u>B promoter regions to the DNA sequences shown in Figure 1 (16). Therefore, the identified regions are likely to contain the essential cis-acting elements required for transcription initiation of these genes in the chloroplast genome, but we do not exclude that 5' DNA sequences eliminated in this assay are required for the efficient transcription of <u>psb</u>A, <u>trn</u>M2, <u>rbc</u>L and <u>atp</u>B in vivo.

Since tRNA processing is efficient and the mature tRNA transcripts are relatively stable in the spinach chloroplast extract (28), the promoter-<u>trn</u>M2 deletion mutant constructs have allowed us to determine the relative strengths of several promoter regions by quantitation of the $tRNA^{Met}$ product (16). In Figure 1 the promoter regions from <u>psb</u>A, <u>trn</u>M2, <u>rbc</u>L and <u>atp</u>B are listed in order of their relative strengths, with <u>psb</u>A being the strongest promoter that we have analyzed in vitro. Thus, differential promoter strengths may represent the primary control step in the regulation of chloroplast gene expression. Such control mechanism could be directly related to the interaction of RNA polymerase(s) with the promoter regions shown in Figure 1, and thus changes in the DNA sequence of the promoter elements may influence the relative strengths of different promoter regions. A detailed mutational analysis of the spinach <u>psb</u>A promoter (22,29), as well as specific base substitutions in the trnM2 and rbcL promoter regions (15,16), support this notion. Base substitutions in the -10/-35 chloroplast promoter elements almost invariably result in down mutations, but mutations that modify the spacing between these two elements can also increase the relative strength of individual promoters. Alternatively, the relative promoter strengths of chloroplast genes may be determined by the interaction of trans-acting factors with the DNA sequences and RNA polymerase(s). To test for this hypothesis, we have used the promoter regions shown in Figure 1 in binding assays with chloroplast extracts. The specific binding of multiple protein components to the spinach psbA promoter region is shown as an example in Figure 2. When the DNA fragment is reacted with chloroplast proteins, we can identify at least two prominent binding components which interact

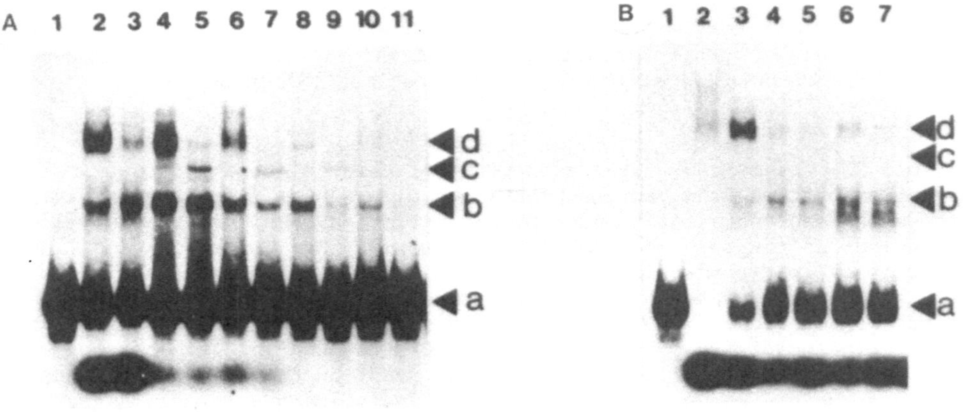

Figure 2 DNA sequence-specific binding of protein(s) to the spinach chloroplast psbA promoter region. The labeled 50 bp psbA promoter fragment was incubated in the chloroplast transcription extract, and the DNA-protein complexes were separated on on a 5% non-denaturing polyacrylamide gel. (A) Specific binding was tested in competition reactions using a 100-fold excess of unlabeled psbA promoter fragment and a 6,000-fold excess of the non-specific competitor poly d(IC). Lane 1: psbA promoter fragment, no poly d(IC), no extract. Lanes 2-11 represent reactions with different concentrations of extract. Protein concentrations were 100 µg (2 and 3), 30 µg (4 and 5), 10 µg (6 and 7), 3 µg (8 and 9) and 1 µg (10 and 11). (B) Competition of the psbA promoter with homologous and heterologous 5' regions in the presence of a 6,000-fold excess of non-specific competitor. Lane 1: Labeled psbA promoter, no poly d(IC), no extract. Lane 2: 100 µg protein, no poly d(IC). Lanes 3-5: competition with poly d(IC) and 20- (4) and 60-fold (5) excess of unlabeled psbA promoter fragment. Lanes 6 and 7: Competition with 50- (6) and 250-fold (7) excess of unlabeled trnS1 5' DNA fragment, which has no promoter function in vitro. a: 50 bp psbA promoter fragment. b,c and d: mobility shifts of the psbA promoter fragment after incubation in the chloroplast extract and in competition with unlabeled DNAs. The labeled low molecular weight band most likely results from a phosphatase activity in the extract.

with the psbA promoter region. The significant retardation of the psbA probe is observed over a wide protein concentration. When unlabeled psbA DNA or DNA from the 5' upstream region of trnS1, which does not contain promoter function in vitro, is used to determine if the protein:DNA complex is specific for the psbA promoter, only the unlabeled psbA promoter eliminates effectively the protein:DNA complex formation in band d (Figure 2). The complex in band b is not significantly affected, indicating the binding of a protein which may be abundant in the chloroplast extract. Together, these data show that chloroplast proteins form a specific complex with the psbA promoter, and that this complex is not formed effectively with chloroplast DNA sequences that have no promoter function. DNAase I footprint experiments are currently underway to determine the DNA sequences that are protected in the protein:psbA promoter complex.

Although the concept of 5' prokaryote-like promoter regions may apply to most chloroplast genes in higher plants, it is not a strict requirement for all genes. When several spinach chloroplast tRNA genes were analyzed for their promoter regions, we found that a subpopulation of tRNA genes does not require upstream promoter elements for transcription (21). The absence of functional promoter elements in the 5' regions of at least two tRNA genes, trnR1 and trnS1, is supported by several criteria. First, comparison of 5' upstream sequences with the defined chloroplast promoter regions shown in Figure 1 does not reveal any regions of significant homology. Second, deletion of 5' DNA sequences to or beyond the limit of minimal DNA sequence required for promoter function (i.e. 45 bp) has little or no effect on the transcription of these genes in vitro. Third, DNA sequences from the 5' upstream regions of these genes do not support the transcription of trnM2 promoter deletion mutants. At present we can only speculate about the location of the promoter regions for trnS1 and trnR1, but it appears that a prokaryote-type promoter structure can not be generalized for higher plant chloroplast genes.

THE ROLE OF TRANSCRIPTIONAL REGULATION DURING CHLOROPLAST DEVELOPMENT AND PLASTID DIFFERENTIATION

The changes in the levels of different plastid RNAs during chloroplast development and plastid differentiation, as well as the different strengths of plastid promoter regions and alterations in plastid DNA levels, have prompted us to examine the role of transcriptional regulation. The morphogenetic changes associated with the formation of photosynthetically competent chloroplasts in cotyledons and leaves of higher plants are characterized by significant changes in the mRNA levels for components of the photosynthetic apparatus. For example, the light-induced development of chloroplasts in spinach cotyledons is associated with a 5- to 10-fold increase in the levels of psbA and atpB/E mRNAs, but the rbcL mRNA levels is not significantly affected (Figure 3). Similarly, RNA levels for most (e.g. rbcL, psaA, psbB, psbC/D), but not all (psbA, rrn) chloroplast genes decrease rapidly during chromoplast differentiation in tomato fruit (6). It is most likely, therefore that those stages in chloroplast development and differentiation represent periods of highly coordinated and differential gene activity. The regulatory mechanism(s) controlling plastid gene activity during these developmental changes may be established through the modulation of relative transcription activities at the promoter level by stage-specific, trans-acting factors. Such regulation could operate in conjuction with changes in plastid RNA polymerase levels and/or the restricted availability of transcriptionally active chloroplast templates. Alternatively, differential changes in plastid RNA levels could be a consequence of posttranscriptional changes in the stability of individual RNA species. To dis-

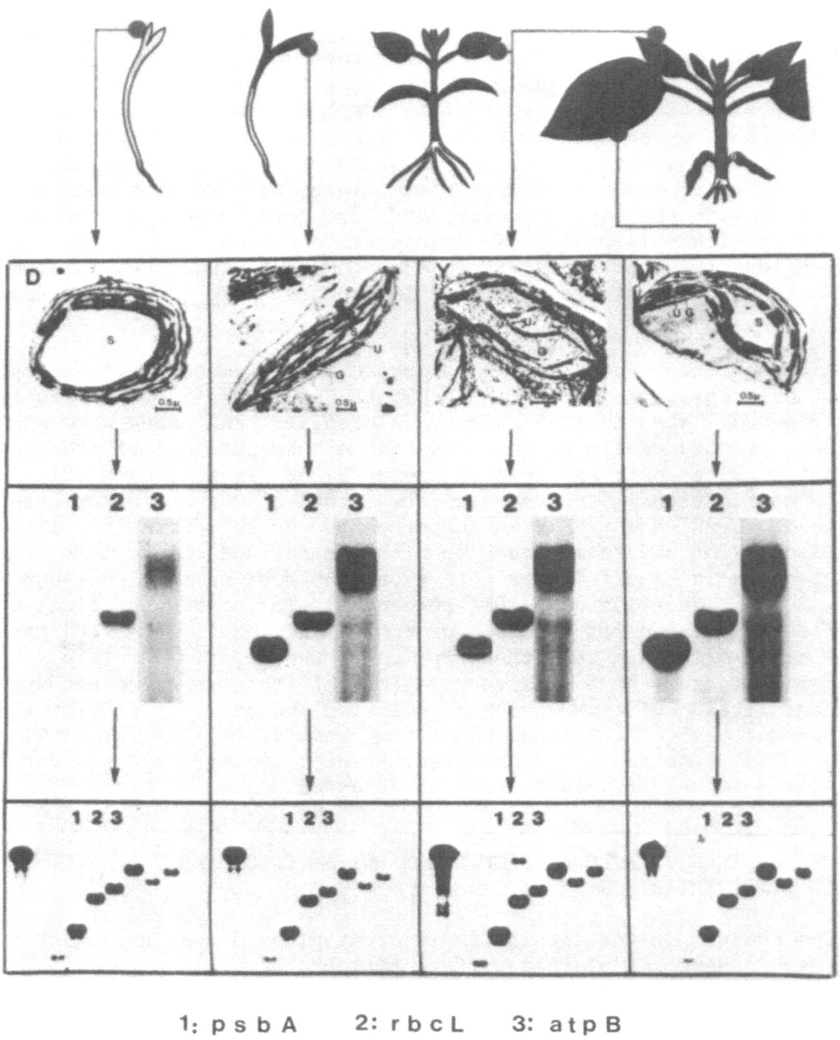

Figure 3 Transcriptional and posttranscriptional regulation of chloroplast gene expression. Plastids were isolated from etiolated spinach seedlings grown in the dark for 3 days (D) and etiolated seedlings transferred to light for 24 hours (24), and chloroplasts were isolated from young (Y) and mature (M) leaves. The upper panel shows the morphogenetic changes that are observed at the ultrastructural level during chloroplast development in cotyledons and leaves. The Northern analysis of mRNA levels for different genes in the middle panel shows psbA (1), rbcL (2) and atpB (3) as examples. The steady-state mRNA levels were determined using synthetic 49 base oligonucleotides homologous to the protein-coding regions of the respective genes. The lower panel summarizes the relative transcription activities of ten spinach chloroplast genes determined by hybridization of labeled run-on transcripts to gene-specific probes. Lanes labeled 1, 2 and 3 contain probes for psbA, rbcL and atpB, respectively. A complete presentation of the results has been published elsewhere (26).

tinguish between these possibilities, we have developed a plastid run-on transcription system that permits us to examine the overall transcriptional activity of plastid DNA and the relative transcriptional activities of individual plastid genes during chloroplast development and plastid differentiation (27).

When plastids at different stages of chloroplast development in spinach or chromoplast differentiation in tomato were tested for their overall transcriptional activities, we found no correlation of transcriptional activities with changes in the plastid DNA levels at the respective stages. Most strikingly, while the chloroplast:nuclear DNA ratio increased approximately 30% during spinach leaf maturation, the transcriptional activity in the chloroplasts decreased more than 5-fold (26). In contrast, the decrease in overall transcriptional activity during chromoplast differentiation in tomato fruit is not associated with significant changes in the level of plastid DNA (30). Together, these results suggest that the differences in overall transcriptional activities and RNA accumulation or decrease cannot simply be consequences of changes in plastid DNA levels. They do not exclude, however that template availability is limiting for transcriptional activity in plastids. If such a model would be considered, we would expect only small changes in the relative transcriptional activity of plastid genes. Furthermore, this model would predict that the different strengths of promoter regions constitute the primary determinant for the relative transcriptional activities of individual genes at all developmental changes. It is interesting to note that the different in vitro promoter strengths for psbA, rbcL and atpB (in this order) are also reflected by their relative transcriptional activities in the plastid run-on system (Figure 2). To address this problem, studies are required to determine if all or only a fraction of the genomes in each plastid are engaged in transcription at all times, and in which molecular conformation the supercoiled plastid DNA exists at different developmental stages. Also possible, although not supported by the experimental evidence, are changes in limiting RNA polymerase levels. If such a model of overall transcriptional regulation is considered, we would expect stronger promoter regions, in the absence of regulatory, trans-acting factors, to compete more effectively for RNA polymerase, thus resulting in significant changes in relative transcriptional activities of individual genes.

To determine the extent of transcriptional activity at the genome and gene level, we have hybridized run-on transcripts to restriction enzyme-digested chloroplast DNA and gene-specific probes. At a genomic resolution, we find only few differences in the hybridization pattern of run-on transcripts from plastids of etiolated and greening spinach cotyledons, young and mature spinach leaves, or chloroplasts and differentiating chromoplasts from leaves and fruit of tomato, respectively. This strongly suggests that the significant changes in the steady-state mRNA levels of individual genes are not merely consequences of their preferential activation or inactivation at different developmental stages. At the gene level, the hybridization of run-on transcripts from plastids of etiolated and light-exposed spinach cotyledons, as well as young and mature spinach leaves, to ten different chloroplast genes shows very little difference in the relative transcriptional activities of these genes, although their RNA levels can vary significantly at all four developmental stages. This is most striking for psbA, rbcL and atpB, which are shown as examples in Figure 3. Both psbA and atpB accumulate little mRNA in the dark, but the mRNA level rises rapidly after transfer of the seedling to light. In contrast, the change in the rbcL mRNA is less then two-fold, and relatively high levels are already visible in the dark. However, the differences in mRNA accumulation are not associated with changes in the relative transcriptional activities of these genes. Together, these results provide evidence that the mRNA levels are not established through a preferential activation (or inactivation) of the individual genes, but argue that adjustments of plastid RNA stability during light-induced chloroplast development in spinach cotyledons and during the synthesis of photosynthetic membranes in young and mature leaves is a major control step in the regulation of chloroplast gene expression and the accumulation of their RNAs. Although we have not completed a similar analysis for individual genes in tomato, we would expect that a similar regulatory mechanism is also operational during chromoplast differentiation (30).

For most chloroplast genes it appears, therefore that their transcriptional control is limited and based on two principal mechanisms: the strengths of promoter regions for respective plastid genes and the modulation of overall transcription activity at different developmental stages. This control is not sufficient to explain the differential accumulation of plastid mRNAs during development, and thus the above results imply that gene expression in plastids can be effectively controlled at the posttranscriptional level. Although this model may be applicable to most chloroplast genes, certain genes appear to be under transcriptional rather than posttranscriptional control. The spinach chloroplast gene which shows the most prominent difference in relative transcriptional activity is trnK, which is located upstream of psbA. In all higher plants that have been analyzed the trnK gene is interrupted by an approximately 2.5 Kb intron (31,32). Part of this intron contains an open reading frame that has homology to the yeast maturase gene, and it is also interesting to note that trnK is the only gene on the chloroplast genome that codes for a lysine tRNA. The relative transcriptional activity of trnK decreases 2- to 3-fold during leaf maturation, but shows no significant change during light induced chloroplast development. At present we can only speculate about the significance of transcriptional control for this locus, until both the function of the intron-encoded open reading frame and a possible regulatory role of $tRNA^{Lys}$ in translation have been established.

PROTEIN SYNTHESIS DURING CHLOROPLAST DEVELOPMENT: THE ROLE OF TRANSLATIONAL REGULATION

Several laboratories have recently provided evidence that synthesis of proteins for photosynthetic complexes is controlled at the translational level during light/ dark transitions (13,14,25,33-36), although only few studies have excluded the rapid turnover of the polypeptides in the dark (13,14). The polypeptides that appear to be primarily affected by this regulatory mechanism include the photosystem I chlorophyll apoproteins and the 32-kDa protein of photosystem II (14). The large subunit of ribulose-1,5-bisphosphate carboxylase is another example of a chloroplast protein that is regulated at the translational level, although the control of its synthesis appears to be strikingly different in monocotyledons and dicotyledons. Whereas the synthesis of the large subunit is most active in plastids of 4.5-day old dark-grown barley seedlings (14), synthesis rapidly declines after 4 days in dark-grown Amaranth cotyledons (13). In both plants, however, the short-term translational control is superimposed on long-term changes in the rbcL mRNA level. We have analyzed the levels of selected proteins in dark-grown and illuminated spinach cotyledons, and in young and mature spinach leaves (37). The results shown in Figure 4 for the large subunit of ribulose-1,5-bisphosphate carboxylase and the 32 kDa protein of photosystem II as examples correspond to the developmental stages of plastids in cotyledons and chloroplasts in leaves shown in Figure 3. Steady-state levels of the large subunit protein can be detected at all four stages, and the changes

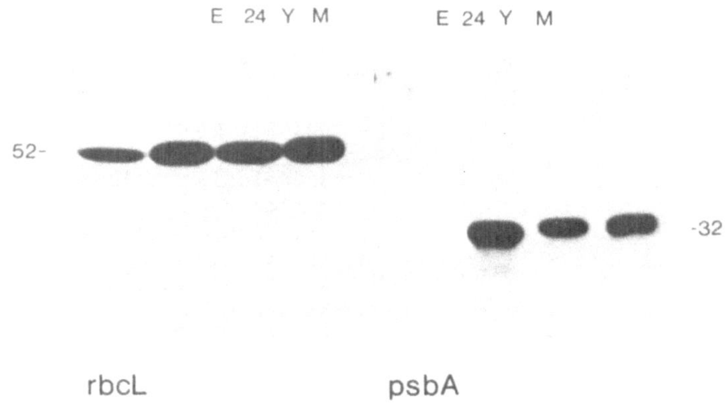

Figure 4 Protein levels during chloroplast development and maturation in spinach cotyledons and leaves. Total proteins from isolated plastids and chloroplasts were separated on 12% PAGE-SDS, transferred to nitrocellulose, and reacted with antibodies specific for the large subunit of ribulose-1,5-bisphosphate carboxylase (A) and the 32 kDa protein of photosystem II (B). Plastids were isolated from etiolated (E) and 24 hr illuminated (24) cotyledons, and chloroplasts were isolated from young (Y) and mature (M) leaves.

in the protein level appear to correlate with the changes of the steady-state rbcL mRNA levels. In contrast, no 32 kDa protein is detectable in the dark, although a low level of psbA mRNA is detectable. The protein level rises significantly after 24 hr illumination, which is concomitant with the increase of the mRNA level. However, although the psbA mRNA level continues to increase during leaf maturation (Figure 3), the steady-state protein level remains unchanged. This result supports previous studies that have shown the lack of correlation between synthesis and steady-state levels of of the 32 kDa protein and its mRNA in a light/dark cycle. Although we have not determined changes in the synthesis of photosynthetic polypeptides, the discordant mRNA and protein levels at different developmental stages of spinach cotyledons and leaves, and during chromoplast differentiation in tomato fruit (38), clearly indicate that translational control has an important function for the regulation of plastid gene expression.

DIFFERENTIAL mRNA STABILITY: A ROLE FOR THE 3' END?

The differential accumulation of plastid mRNAs during chloroplast development, together with the limited transcriptional regulation of the respective chloroplast genes, have raised our interest in the mechanism(s) responsible for the adjustments of plastid mRNA stability in the chloroplast. Such posttranscriptional regulatory mechanism(s) are apparently specific for different plastid loci. For example, the transcript from psbA accumulates approximately 10-fold after illumination of dark-grown spinach cotyledons, whereas the rbcL mRNA accumulates only 2-fold during the 24 hr illumination period (Figure 3). The differential accumulation of mRNAs is not unique to psbA and rbcL, but quantitative and, in the case of polycistronic mRNA and processing products, qualitative changes are also found for other spinach chloroplast genes (39). Similar

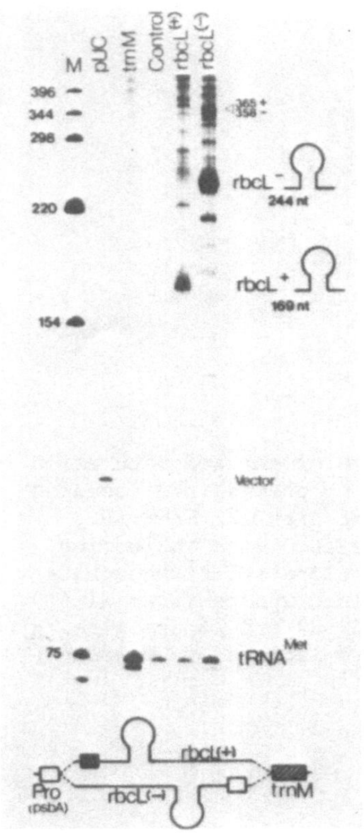

Figure 5
Transcription termination versus mRNA 3' end processing in vitro. The inverted repeat DNA sequence 3' of the rbcL coding region was inserted between the psbA promoter region and the promoter deletion trnM2 mutant gene. The inverted repeat DNA sequences was tested for its ability to terminate transcription in vitro. The lane labeled pUC indicates transcription from the vector DNA alone, trnM shows transcription of the wild type trnM2 gene, and control represents the trnM2 promoter deletion mutant fused to the psbA promoter which is separated from trnM2 by a polylinker region. Lanes labeled rbcL(+) and rbcL(-) represent constructs in which the rbcL 3' inverted repeated DNA sequence was inserted into the polylinker in both orientations (see schematic illustration). Indicated are the $tRNA^{Met}$ transcripts and 3' processing products which are obtained in the in vitro transcription reactions.

quantitative and qualitative changes in plastid mRNA accumulation during light-induced chloroplast development have recently been reported for plants such as barley (4), mustard (3) and maize (5), indicating that adjustments in mRNA stability may be a general regulatory mechanism in higher plant chloroplasts. We therefore decided to investigate the molecular mechanism(s) that could result in the stabilization of different spinach chloroplast mRNAs. Upon examination of chloroplast genes we have noted that most genes and polycistronic loci are flanked by inverted repeats located 3' of the coding regions, and most often are part of the 3' non-translated mRNA sequences (40). Several investigators have proposed that these sequences, which potentially can fold into stem/loop structures, may serve as transcription termination signals for the chloroplast RNA polymerase(s), but this conclusion was based exclusively on DNA sequence and nuclease S1 analysis of the respective transcription units (for review see 41). It is possible, however that the 3' end of chloroplast mRNAs also have a fundamental function in the protection of the transcripts against nucleolytic degradation, and as such may serve as control steps for their stabilization at different developmental stages. To determine the role of chloroplast mRNA 3' termini, and to distinguish between their possible function in transcription termination and transcript stabilization, we have first constructed templates that would allow us to test the 3' inverted repeat DNA sequences for their ability to terminate transcription in vitro. In these experiments we took advantage of the available promoter-deficient trnM2 gene, and fused this to the psbA promoter and region containing the psbA transcription initiation site, to allow transcription of $tRNA^{Met}$. The psbA promoter and trnM2 were separated by a polylinker region, in which different 3' ends of spinach chloroplast genes could be inserted. Figure 5 shows the example of an experiment, in which a DNA fragment containing the 3' end of the

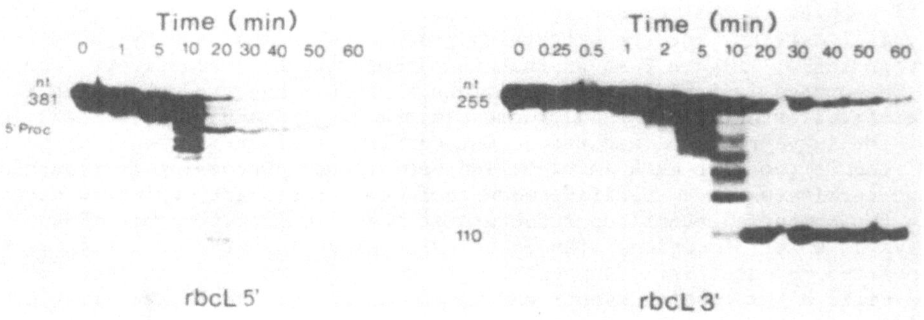

Figure 6 Processing of chloroplast mRNA 3' ends in vitro. Synthetic RNAs from the 5' and 3' ends of rbcL were incubated in the chloroplast extract, and the processing products were analyzed by gel electrophoresis. (A) The 5' end RNA was synthesized from and 381 bp EcoRI/DraI fragment, which includes the rbcL transcription start site and 39 bp of the coding region. The lower molecular weight RNA species most likely represents the 5' processed end of the rbcL mRNA (42). (B) The 3' end RNA was synthesized from a 255 bp EcoRI/BamHI fragment, which includes 24 bp of the rbcL 3' coding region. The lower molecular weight processing product (110 nt) has been analyzed by S1 nuclease mapping, and corresponds to the 3' end of the rbcL mRNA.

rbcL locus, including a portion of the coding region, was inserted into the polylinker region in both orientations. Transcription of these templates in the chloroplast transcription extract reveals no substantial difference in the amount of $tRNA^{Met}$ transcribed in the presence or absence of the rbcL 3' inverted repeat DNA sequence, or with the rbcL 3' DNA fragment inserted into the vector construct in + or - orientation (43). This strongly suggest that the 3' DNA region of the rbcL does not function efficiently in transcription termination in vitro. Experiments with other 3' DNA fragments from different chloroplast genes show essentially similar results, indicating that the inverted repeats may have functions other than transcription termination in the chloroplast. It is also possible, however that the chloroplast transcription system is deficient of transcription termination factors which are required for efficient termination in vitro. This appears to be unlikely, since we have already demonstrated for trnS1 that efficient transcription termination occurs in the chloroplast extract (21), and trnH1 and the T7e terminator are also 85% and 75% efficient in termination under the same experimental conditions.

Figure 5 demonstrates that the transcription of template DNA also results in the appearance of higher molecular weight RNA products, which represent processing products from larger transcripts that include the rbcL 3' sequences and $tRNA^{Met}$. When the rbcL 3' DNA fragment is inserted into the vector in the + orientation, the higher molecular weight processed rbcL mRNA is shorter, however, than would be expected from a single RNAase P-like cleavage event that would separate the $tRNA^{Met}$ from the larger transcript. We therefore considered the possibility that specific processing events occur at the 3' end of the rbcL mRNA in the chloroplast. To test this hypothesis, we constructed templates that include the 3' end of the rbcL gene in SP6/T7 transcription vectors to synthesize

^{32}P-labeled rbcL 3' transcripts. These RNAs were incubated in the chloroplast extract, and the processing products were analyzed by gel electrophoresis. Figure 6 shows that the primary rbcL 3' transcript from the + DNA strand is rapidly processed into a shorter RNA product, which is relatively stable in the chloroplast extract. The analysis of the 3' end of the in vitro processed transcript confirms that it is identical to the 3' end of the rbcL mRNA in vivo, indicating that processing is accurate and terminates at a specific sequence in the transcript, which is distal to the potential stem/loop structure at the 3' end of the rbcL mRNA. Several other 3' regions of spinach chloroplast loci that are flanked by inverted repeats, including psbA, petD and rpoA, are also subject to specific 3' processing events and the formation of stable RNAs in vitro (43). It could be argued that termination of nucleolytic processing is a general feature of inverted repeat sequences in the chloroplast extract, but this argument may not be relevant since most often chloroplast loci on opposite DNA strands share the same inverted repeat in their 3' regions. The rapid degradation of transcripts from the coding region of petB, or the 5' end of rbcL (and psbA, not shown) without formation of stable processing products supports the notion that the formation of stable mRNA 3' ends may be a highly specific process in the chloroplast of higher plants (Figure 6). It appears therefore that the 3' inverted repeat sequences flanking chloroplast genes are primarily involved in the maturation of the respective mRNAs, but not in the transcription termination of the primary transcript. Preliminary evidence also indicates that the 3' processing events result in stable mRNAs in the chloroplast, but fail to produce stable processing products in extracts of plastids isolated from etiolated cotyledons (43). This is interesting to note, since we and others have shown that several chloroplast mRNAs do not accumulate (or to a lower extent) in etiolated seedlings, indicating that most mRNAs, although transcribed in the dark as shown in Figure 3, are not protected against nucleolytic degradation in the dark. Again, it is possible that the 3' ends are involved in the stabilization of chloroplast mRNAs in the light, and preliminary results indicate that such stabilization might be achieved through the formation of RNA:protein complexes that involve the inverted repeat (43).

CONTROL OF CHLOROPLAST GENE EXPRESSION: AN ARRAY OF COMPLEX REGULATORY EVENTS

Taken together, a scheme for the control of chloroplast gene expression emerges that emphasizes the importance of posttranscriptional and translational processes in the formation of stable mRNA and protein products during chloroplast development. Our work has more clearly defined the role of transcriptional regulation in the control of plastid gene expression. The experimental evidence suggests that transcriptional control is based on two principal mechanisms: the strengths of promoter regions for respective plastid genes and the modulation of overall transcriptional activity at different developmental stages. Although several details of these regulatory steps still need to be defined more accurately, it is clear that their role is limited, but essential, for the control of chloroplast gene expression. Exclusive transcriptional regulation may be operational for the control of only few specific plastid genes, and their role in regulatory mechanism(s) remains to be established. Transcription termination, if functional for the transcription of mRNAs in the chloroplast, is at best inefficient and does not provide a significant regulatory step in the control scheme. The discordant transcriptional changes and the differential mRNA accumulations, however, provide strong evidence that gene expression in plastids of higher plants can be effectively controlled at the posttranscriptional level. This step clearly involves changes in mRNA stabilities, which could be achieved by nucleolytic processing events and/or RNA:protein interactions that effect the 3' ends of the respective mRNAs. Most recent results obtained in our laboratory indicate that differential stabilities of the 5' ends of chloroplast mRNAs may also contribute to the posttranscriptional control step. Superimposed on these regulatory events are changes in the translatability of mRNAs at different developmental stages, which may serve as important short-term control steps in the assembly and maintenance of functional photosynthetic complexes. Although the relative importance of these different control steps has to be defined more clearly for individual genes or classes of mRNAs and proteins, it is interesting to speculate why, in contrast to the importance of transcriptional control of nuclear genes, posttranscriptional and translational control steps may have a more significant role for the regulation of chloroplast gene expression. Considering the photosynthetic function of the chloroplast, as well as the maintenance of the genetic apparatus of this organelle in specialized plastids, posttranscriptional and translational control mechanisms may have evolved as rapid and more efficient circuits to regulate chloroplast gene expression during plant development

H.J. and D.S. are supported by postdoctoral fellowships from the National Science Foundation, X.W.D. holds a graduate student fellowship from the University of California, and J.T. acknowledges the partial support by a fellowship from the McKnight Foundation. This work was supported by a grant from the U.S. Department of Energy to W.G.

REFERENCES

1. Kirk JTO and Tilney-Basset RAE (1978) The Plastids, Elsevier/North-Holland Biomedical Press, Amsterdam, New York
2. Thomson WW and Whatley JM (1980) Ann Rev Plant Physiol 31: 375-394
3. Link G (1984) Plant Mol Biol 3: 243-248
4. Poulsen C (1983) Carlsberg Res Commun 48: 57-80
5. Rodermel SR and Bogorad L (1985) J Cell Biol 100: 463-476
6. Piechulla B, Chonoles-Imlay K and Gruissem W (1985) Plant Mol Biol 5: 373-384
7. Piechulla B, Pichersky E, Cashmore AR and Gruissem W (1986) Plant Mol Biol 7: 367-376
8. Sugita M and Gruissem W (1987) Proc Natl Acad Sci USA: in press
9. Harpster M and Apel K (1985) Physiol Plant 64: 147-152
10. Tobin EM and Silverthorne J (1985) Ann Rev Plant Physiol 36: 569-593
11. Gallagher TF and Ellis RJ (1982) EMBO J 1: 1493-1498
12. Silverthorne J and Tobin EM (1984) Proc Natl Acad Sci USA 81: 1112-1116
13. Berry JO, Nikolau BJ, Carr JP and Klessig DF (1986) Mol Cell Biol 6: 2347-2353
14. Klein RR and Mullet JE (1987) J Biol Chem 262: 4341-4348
15. Gruissem W and Zurawski G (1985) EMBO J 4: 1637-1644
16. Gruissem W and Zurawski G (1985) EMBO J 4: 3375-3383
17. Hanley-Bowdoin L, Orozco EM Jr and Chua NH (1985) Mol Cell Biol 5: 2733-2745
18. Link G (1984) EMBO J 3: 1697-1704
19. Bradley D and Gatenby A (1985) EMBO J 4: 3641-3648
20. Boyer SK and Mullet JE (1986) Plant Mol Biol 6: 229-244
21. Gruissem W, Elsner-Menzel C, Latshaw S, Schaffer M and Zurawski G (1986) Nucleic Acids Res 18: 7541-7556
22. Hanley-Bowdoin L and Chua NH (1987) Trends Biochem Sci 12: 67-70
23. Scott NS and Possingham JV (1983) J Exp Bot 34: 1756-1767
24. Thompson WF, Everett M, Polans NO, Jorgensen RA and Palmer JC (1983) Planta 158: 487-500

25. Fromm H, Devic M, Fluhr R and Edelman M (1985) EMBO J 4: 291-295
26. Deng XW and Gruissem W (1987) Cell 49: 379-387
27. Deng XW, Stern DB, Tonkyn JC and Gruissem W (1987) J Biol Chem: in press
28. Greenberg BM, Gruissem W and Hallick RB (1984) Plant Mol Biol 4: 97-109
29. Zurawski G and Gruissem W (1987) Manuscript in preparation
30. Gruissem W, Callan K, Lynch J, Manzara T, Meighan M, Narita J, Piechulla B, Sugita M, Thelander M and Wanner L (1987) In Tomato Biotechnology, Nevins D (ed), Alan Liss, New York, in press
31. Sugita M, Shinozaki K and Sugiura M (1985) Proc Natl Acad Sci USA 82: 3557-3561
32. Neuhaus H and Link G (1987) Curr Genet 11: 251-257
33. Slovin JP and Tobin EM (1982) Planta 154: 465-472
34. Nelson T, Harpster MH, Mayfield SP and Taylor WC (1984) J Cell BIol 98: 558-564
35. Bennett J, Jenkins GI and Hartley MR (1984) J Cell Biochem 25: 1-13
36. Berry JO, Nikolau BJ, Carr JP and Klesig DF (1985) Mol Cell Biol 5: 2238-2246
37. Deng XW, Jones H and Gruissem W, unpublished results
38. Piechulla B, Glick RE, Bahl H, Melis A and Gruissem W (1987) Plant Physiol: in press
39. Deng XW and Gruissem W, unpublished results
40. Shinozaki K, Ohme M, Tanaka M, Wakasugi T, Hayashida N, Matsubayashi T, Zaita N, Chunwongse J, Obokata J, Yamaguchi-Shinozaki K, Ohto C, Torazawa K, Meng BY, Sugita M, Deno H, Kamogashira T, Yamada K, Kusuda J, Takaiwa F, Kato A, Tohdo N, Shimada H and Sugiura M (1986) EMBO J 5: 2043-2050
41. Crouse EJ, Bohnert Hj and Schmitt JM (1984) In Chloroplast Biogenesis, Ellis RJ (ed), University Press, Cambridge, pp 83-136
42. Mullet JE, Orozco EM Jr and Chua NH (1985) Plant Mol Biol 4: 39-54
43. Stern D and Gruissem W, unpublished results

LOCALIZATION OF tRNA GENES ON MAIZE AND WHEAT MITOCHONDRIAL GENOMES

Jean-Michel Grienenberger, Henri Wintz, Pia Runeberg-Roos, Laurence Marechal, Genevieve Jeannin and J.H. Weil

Institut de Biologie Molécularie et Cellulaire
Université Louis Pasteur, 15 Rue Descartes
F-67084 Strasbourg, France

I. INTRODUCTION

Plant mitochondrial (mt) genomes have been shown to consist, in most species studied so far, of a number of circular molecules deriving from a master chromosome by recombination events, plus a few small plasmid-like molecules (1,2). Together, these DNA molecules constitute large genomes, the size of which differ, depending on the plant considered, from about 200 to 2500 kb (1,3,4), and are much more complex than animal or fungal mt genomes (5,6).

Up to now only a few restriction maps of plant mt genomes have been established. The map of the maize mt genome was obtained by Lonsdale et al (1), that of the wheat mt genome by Quetier et al (7). As very few genes have been identified and localized so far on the plant mt genome, we have decided to study the localization of tRNA genes on the mt DNA of maize and wheat. Mitochondria are known to have their own translation system, different from that found in the cytoplasm and in the chloroplasts, but we don't know the number of plant mt tRNAs and whether they are all encoded in the mt genomes.

It has been shown that plant mt genomes contain chloroplast (cp) DNA insertions, some of them including tRNA genes (8,9). It was therefore interesting to study whether this is a general phenomenon and whether cp tRNA genes inserted into mt DNA are transcribed in the mitochondria.

II. LOCALIZATION OF tRNA GENES ON THE MITOCHONDRIAL GENOME, BY HYBRIDIZATION OF MITOCHONDRIAL DNA RESTRICTION FRAGMENTS

1) Hybridization to total mitochondrial tRNAs

The cosmid bank of maize mt DNA was obtained from D. Lonsdale (Plant Breeding Institute, Cambridge, England). The restriction map of the maize mt master chromosome, established using Sma I, Sst II and Xho I (1), is shown in linearized from in fig. 1. The 53 Sal I cloned fragments covering the whole genome of wheat mt DNA were obtained from F. Quetier and B. Lejeune (Biologie Moléculaire Végétale, Orsay, France). The restriction map of the wheat mt master chromosome (7) is shown on fig. 2.

Total maize or wheat mt tRNAs were prepared as described by Marechal et al (10). They were specifically labeled at their 3' end, using α-^{32}P-ATP and yeast tRNA nucleotidyl-transferase as previously described (11). The labeled tRNAs were hybridized to maize or wheat mt DNA restriction fragments as already described (12).

The results of these hybridization experiments in the case of maize mt DNA are shown on fig. 1. Total maize mt tRNAs were hybridized to 26 cosmids digested either by Sma I or Xho I and the positive fragments are indicated as dotted areas. There are about 20 positive fragments, indicating that there are at least 20 mt tRNA genes (but we know that some fragments contain more than one tRNA gene). It should be noticed that, in two instances, positive fragments are part of a 12 kb direct repeat or of a 14 kb inverted repeat (see fig. 1)

The results of the hybridization performed with wheat mt Sal I DNA fragments are shown on fig. 2. There are 14 positive fragments, but here also we already know (see below) that some fragments contain several tRNA genes. Again, it should be pointed out that some of these positive fragments coincide with repeated sequences as for instance in the case of repeat sequence n°1, which contains a tRNA^{Met} gene, and is present 3 times on the map (13).

The number of tRNA genes present in each fragment, and the nature of the aminoacid corresponding to the tRNA gene, are determined by cloning and sequencing sub-fragments (see below).

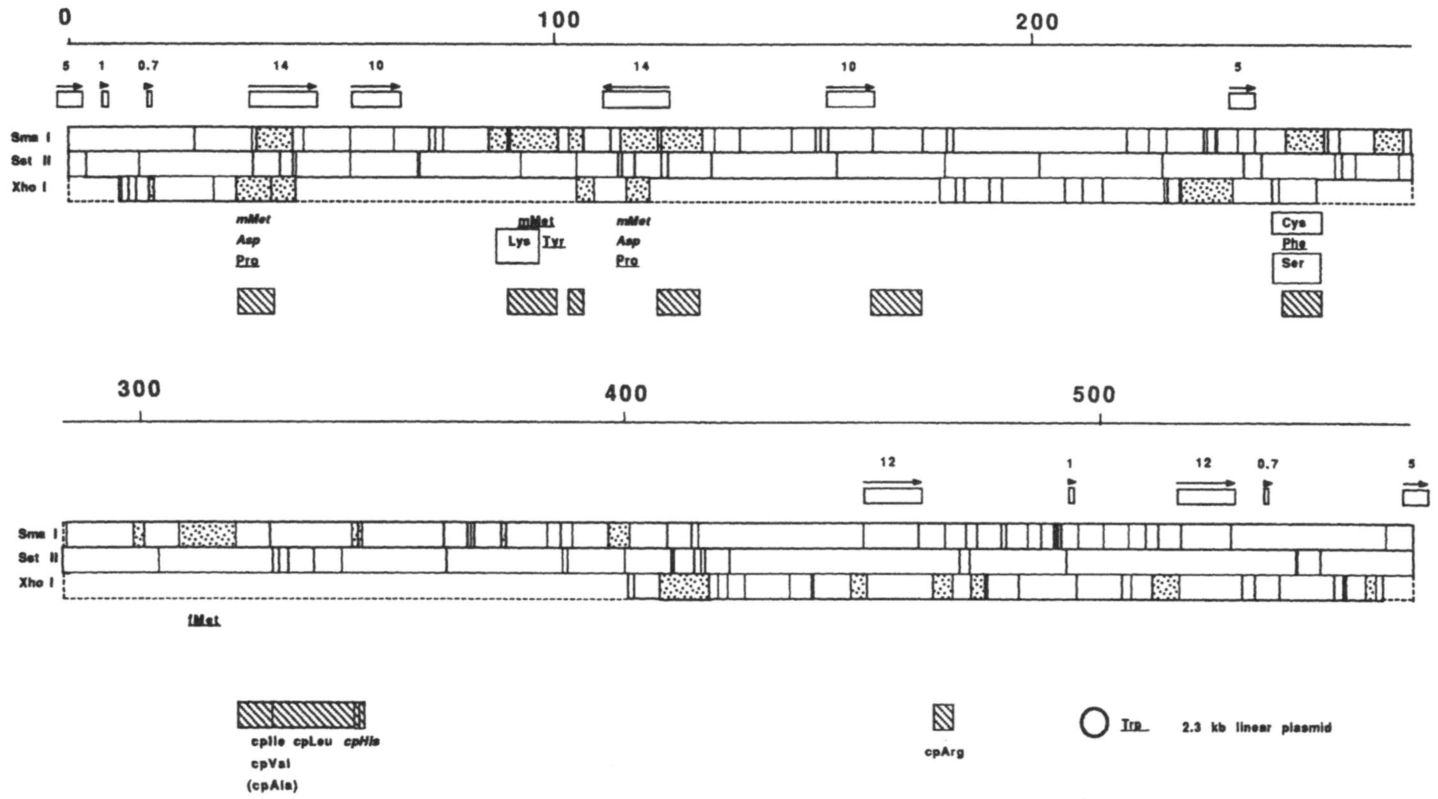

Figure 1. Localization of tRNA genes on the mtDNA of maize. The map of the maize mt DNA is represented in a linear form, upon digestion by Sma I, Sst II and Xho I. Restriction fragments, either on the Sma I or on the Xho I map, hybridizing to total mt tRNAs are shown as dotted boxes and those hybridizing to total cp tRNAs as hatched boxes. The localization of identified tRNA genes is indicated by the name of the corresponding aminoacid, with the following code : to indicate the identification method : 1. hybridization with pure bean mt tRNAs : underlined ; 2. hybridization with oligonucleotides : italics ; 3. sequenced tRNA genes : boxes.

2) Hybridization to total chloroplast tRNAs

As plant mt DNA is known to contain cp DNA insertions, some of them including tRNA genes (8,9), it was important to check whether the positive hybridization results obtained with mtDNA restriction fragments (see above) could be due to contaminations of mt tRNAs by cp tRNAs, because of the high degree of homology observed between some mt tRNAs and their cp counterparts. For instance, there is 97% homology between bean mt and cp $tRNAs^{Trp}$ (14).

Total maize or wheat cp tRNAs were prepared as previously described (15), labeled at their 3' end (11) and hybridized to maize or wheat mt DNA restriction fragments as already described (12).

The results of these hybridization experiments are shown in fig. 1 (maize) and fig. 2 (wheat). In the case of maize, there are about 10 positive fragments, and 3 in the case of wheat. Each of these fragments may contain in fact more than one tRNA gene, as shown for instance in the case of wheat mt DNA fragment D1 which can be further digested into 3 distinct sub-fragments which are still hybridizing to total cp tRNAs, and in the case of a maize mt DNA region, located on the map between positions 325-337, which is known to contain a 12 kb cp DNA insertion (8). This cp DNA insertion is part of the inverted repeat of maize cp DNA and contains the 3' half of the $tRNA^{Ala}$ gene, the complete $tRNA^{Ile}$ split gene, the 16S rRNA gene, a $tRNA^{Val}$ gene and a $tRNA^{Leu}$ gene (16).

A more detailed analysis of this region has shown that Sma I fragments 7.0 and 16.9 hybridize to total maize cp tRNAs, but not to total maize mt tRNAs (fig. 3). This indicates that these cp tRNA genes (and especially the 3 complete ones) are not expressed in the mitochondrion, and that our preparations of maize mt tRNAs are not contaminated by cp tRNAs (otherwise we should have obtained positive hybridization of these fragments to mt tRNAs).

3) Hybridization to purified bean mt tRNA isoacceptors

Several isoaccepting tRNAs from bean (**Phaseolus vulgaris**) mitochondria have been obtained in pure form and sequenced, namely $tRNA^{Phe}$ (12), $tRNA^{Trp}$ (14), $tRNA^{Tyr}$ (17), $tRNA^{Met}$ (18) and $tRNA^{Pro}$ (19). These tRNAs were labeled at their 3' end with [^{32}P] pCp using T_4 RNA ligase (20), and hybridized to mt DNA restriction fragments from maize and wheat. The results obtained are shown in fig. 1 and 2.

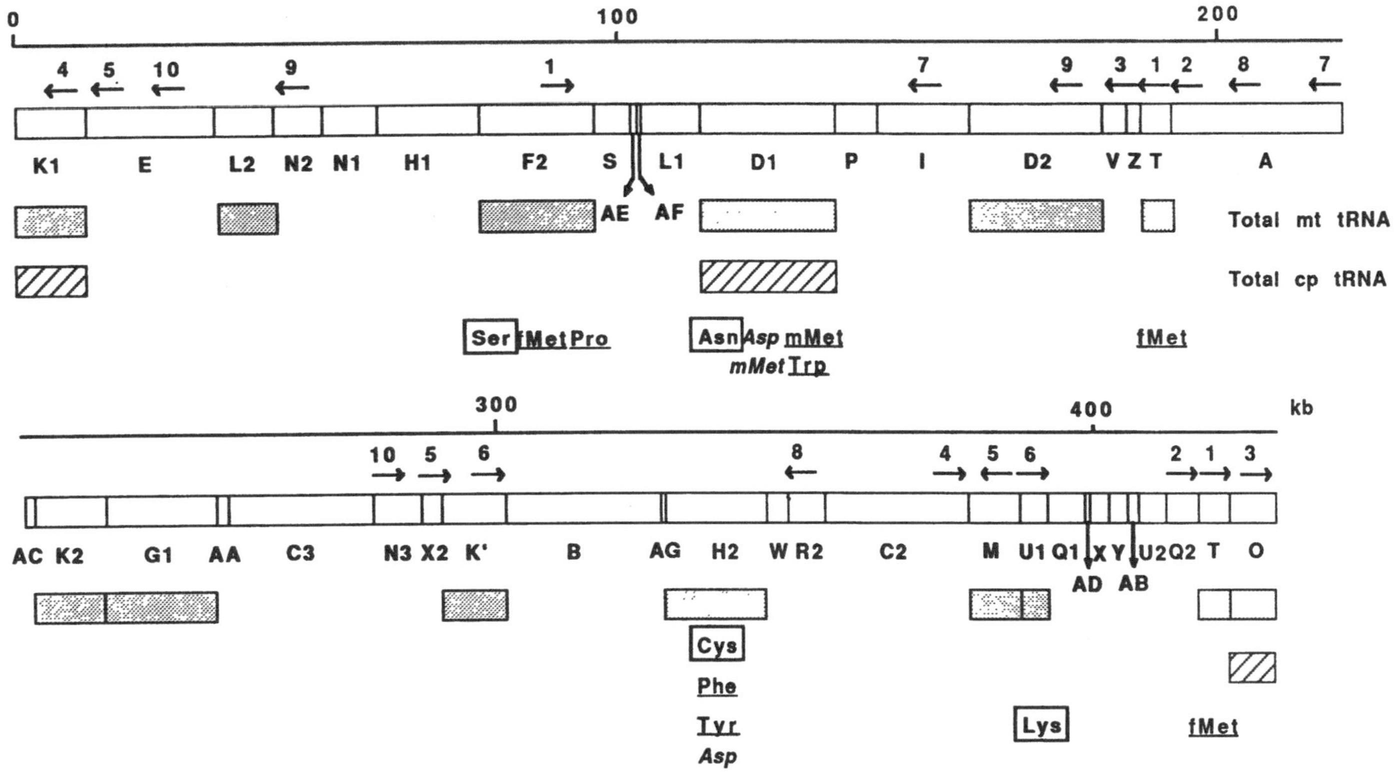

Figure 2. Localization of tRNA genes on the mt DNA of wheat. The map of the wheat mt DNA is represented in a linear form, upon digestion by Sal I. Restriction fragments hybridizing to total mt tRNA are shown as dotted boxes and those hybridizing to total cp tRNAs as hatched boxes. The localization of identified tRNA genes is indicated by the name of the corresponding aminoacid, with the following code to indicate the identification method : 1. hybridization with pure bean mt tRNAs : underlined ; 2. hybridization with oligonucleotides : italics ; 3. sequenced tRNA genes : boxes.

Whereas only one $tRNA^{Pro}$ gene was detected in the case of wheat, two $tRNA^{Pro}$ genes were found on maize mt DNA, as this gene is located on a 14 kb inverted repeat. In wheat, there are 3 $tRNA^{Met}$ genes on the master chromosome, located on repeat sequence n°1 next to the 18S rRNA gene (21). Whereas a $tRNA^{Trp}$ gene was found on fragment D1 of the wheat mt DNA (for its location see fig. 2, and for its sequence see below) labeled mt $tRNA^{Trp}$ did not hybridize to any of the cosmids of the maize master chromosome (although it hybridized to maize mt genomic DNA) ; the $tRNA^{Trp}$ gene was found to be encoded in a 2.3kb linear plasmid present in maize mitochondria (22).

4) Hybridization to synthetic oligodeoxyribonucleotides

Oligodeoxyribonucleotides (25-mer) were synthesized according to sequence data published for $tRNA^{Met}$, $tRNA^{Asp}$ (23,24) and for $tRNA^{His}$ (9) in the case of maize mt DNA ; these oligonucleotides were complementary to the 5' end of these 3 tRNA genes. They were labeled at their 5' end, using α^{32}P-ATP and polynucleotide kinase, and hybridized to maize and wheat mt restriction fragments.

It had been previously shown (24) that the $tRNA^{Asp}$ and $tRNA^{Met}$ genes were located on the same DNA fragment. We have shown that this fragment is part of the 14 kb inverted repeat. These two tRNA genes are also located on the same fragment (D1) in the wheat mt genome (see fig. 2). There is an additional $tRNA^{Asp}$ gene, revealed by the same oligonucleotide probe, on fragment H_2 of wheat mt DNA. There is also an additional $tRNA^{Met}$, both on the wheat (fragment D1) and the maize mt genomes, but this gene is revealed by hybridization to the bean mt $tRNA^{Met}$, not when using the oligonucleotide probe. The $tRNA^{Met}$ gene revealed by the oligonucleotide and that revealed using bean mt $tRNA^{Met}$ have little homology ; the fact that the latter has a high enough degree of homology to hybridize to a $tRNA^{Met}$ extracted from mitochondria indicates that it is transcribed.

5) Localization of mitochondrial tRNA genes upon sequencing

When neither the pure mt tRNA, nor sequence data for the corresponding mt tRNA gene (to synthesize an oligonucleotide) are available, localization can be achieved through sequencing of fragments hybridizing to total mt tRNAs, and identification of the tRNA gene then results from the determination of the anticodon. This approach has allowed the localization of $tRNA^{Lys}$, $tRNA^{Cys}$, $tRNA^{Ser}$ genes on the maize mt genome and of the $tRNA^{Asn}$ on the wheat mt genome.

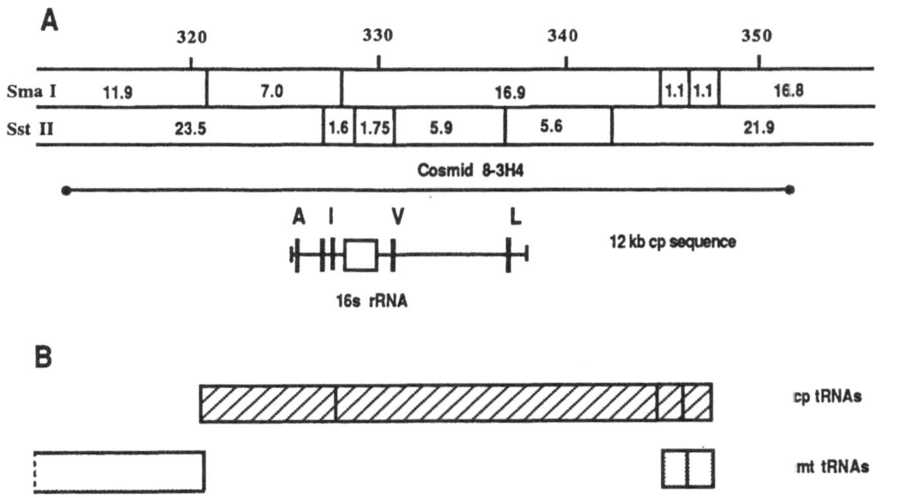

Figure 3. A. Restriction map of cosmid 8-3H4 with the position of the 12 kb cp sequence containing the genes for 16S rRNA, the first exon of $tRNA^{Ala}$ (A), the complete $tRNA^{Ile}$, (I), $tRNA^{Val}$ (V) and $tRNA^{Leu}$ (L). B. Patterns obtained by hybridization of a Sma I digest of cosmid 8-3H4 with total maize cp tRNAs (hatched boxes) and total maize mt tRNAs (dotted boxes) specifically labeled at their 3' end.

III. EXPRESSION OF tRNA GENES PRESENT AS CHLOROPLAST DNA INSERTIONS IN THE MITOCHONDRIAL GENOME

As there is a very high degree of homology (97%) between the mt and cp $tRNAs^{Trp}$ of the same organism (bean), and as bean mt $tRNA^{Trp}$ was used to localize the corresponding gene on the wheat mt genome (see above), we decided to analyze the sequence of the cp and mt $tRNA^{Trp}$ genes and of their flanking regions, to see if the sequence homology extends beyond the $tRNA^{Trp}$ gene itself (as might be expected in the case of a cp DNA insertion into mt DNA).

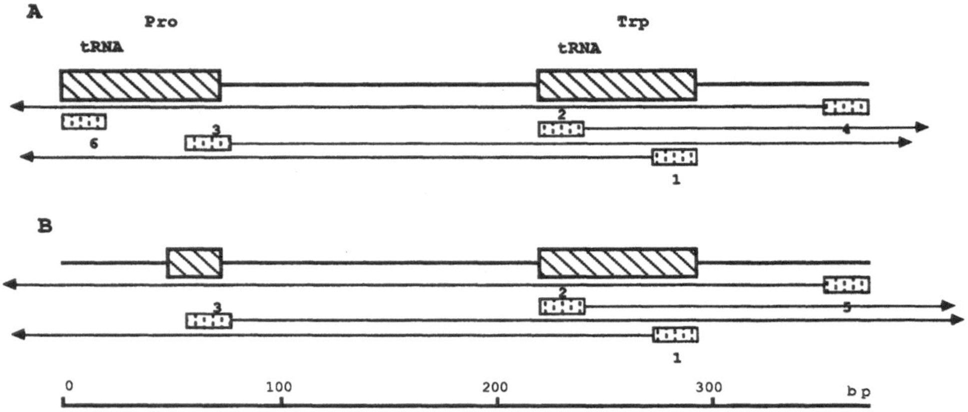

Figure 4. Sequencing strategy for : A. The region of the wheat cp DNA subfragment including the cp $tRNA^{Trp}$ and $tRNA^{Pro}$ genes (hatched boxes). B. The region of the wheat mt DNA subfragment including the mt $tRNA^{Trp}$ and the sequence homologous to the 3' end of the cp $tRNA^{Pro}$ gene (hatched boxes). Dotted boxes indicate the location of the sequences homologous to the oligonucleotides used as primers for sequencing (1,2,3,4,5) or for hybridization studies (6). Arrows indicate the direction and extent of DNA sequencing.

The sequences of both wheat mt and cp $tRNA^{Trp}$ genes and of their flanking regions were established using the following strategy (fig. 4) : Two oligonucleotides (1 and 2) were synthetised, respectively homologous to the 3' end and to the sequence complementary to the 5' end of bean mt $tRNA^{Trp}$. These oligonucleotides were annealed to the wheat mt or cp recombinant phage DNAs containing the mt or cp $tRNA^{Trp}$ gene and the sequencing reactions were performed using the dideoxynucleotide method. The resulting sequences were used to define the sequences of three other oligonucleotides. These oligonucleotides (3,4 and 5) were used as primers for the sequence determination of about 300 bp on both sides of the $tRNA^{Trp}$ genes (fig. 4). Because of the homologies between the mt and the cp sequences, oligonucleotides 1, 2 and 3 have been used for the sequencing of both mt and cp subfragments. The number and location of the oligonucleotides used during these sequencing experiments are indicated in fig. 4.

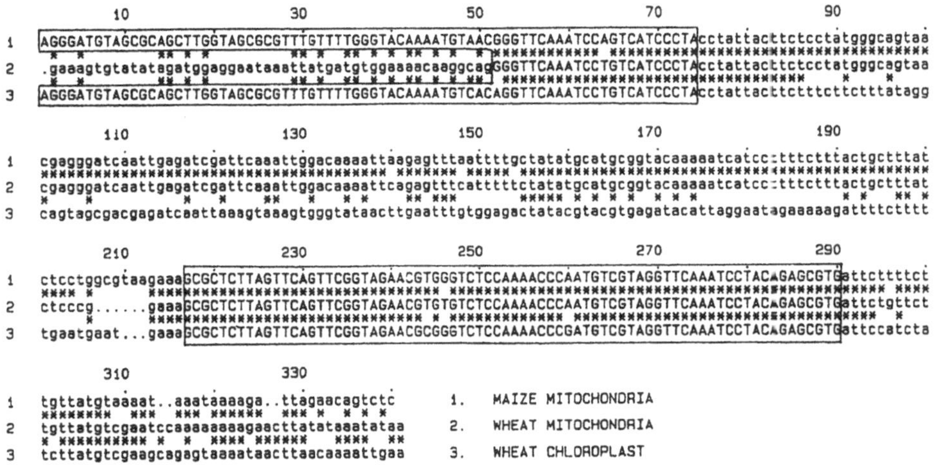

Figure 5. Nucleotide sequences of the regions containing the $tRNA^{Trp}$ gene from : 1. maize mitochondria (P. Leon and P. Bedinger, personal communication), 2. Wheat mitochondria, 3. Wheat chloroplast. These sequences are aligned on the $tRNA^{Trp}$ genes. The $tRNA^{Trp}$ and (total or partial) $tRNA^{Pro}$ genes are boxed. Sequence homologies are shown by stars.

Figure 5 shows part of the sequences which were determined. As expected, these two sequences were shown to contain a gene coding for a $tRNA^{Trp}$, 74 nucleotide long, with a CCA anticodon. To enable a better comparison, the two sequences were aligned on the $tRNA^{Trp}$ gene. The wheat cp and mt $tRNA^{Trp}$ genes show a high sequence homology (96%) with only three changes.

Upstream of the wheat cp $tRNA^{Trp}$ gene (fig. 5), another tRNA gene, 74 nucleotide long and with a UGG anticodon corresponding to proline, was found. The localization of a $tRNA^{Pro}$ gene on the same fragment of wheat cp genome was already shown by Mubumbila et al. (25) and the close linkage of these two tRNA genes is in fact a common feature of all higher plant cp genomes (16,26,27). The 3' end of this cp $tRNA^{Pro}$ gene is located 139 bp upstream of the 5' end of the cp $tRNA^{Trp}$ gene, on the same strand.

When the wheat mitochondrial sequence was compared to the chloroplast one, a sequence, 23 nucleotide long, homologous to the 3' end of the cp $tRNA^{Pro}$ gene was shown to be present 136 bp upstream of the mt $tRNA^{Trp}$ gene on the same strand. This partial mt $tRNA^{Pro}$ sequence is 100% homologous to the corresponding sequence of the chloroplast DNA. This 100% homology extends up to 12 nucleotides in the 3' flanking region of the $tRNA^{Pro}$ genes (fig. 4).

In order to look for the 5' part of the $tRNA^{Pro}$ gene in the wheat mitochondrial genome, an oligonucleotide (n°6 in fig. 4) was synthetized and used as a heterologous probe in hybridization experiments with all the 54 Sal I fragments which constitute the whole wheat mitochondrial genomic library (7). These experiments were negative, but a complete $tRNA^{Pro}$ (UGG) gene is not necessary in this region of the wheat mt genome, as another $tRNA^{Pro}$ (UGG) gene is located on fragment F2 as shown on fig. 2 (19).

P. Leon and P. Bedinger (unpublished results) have sequenced the corresponding region in maize mt DNA (fig. 5). There is a very high degree of homology between maize and wheat mt genomes in this region, namely 99% in the $tRNA^{Trp}$ genes, 93% in the intergenic regions and 96% in the parts of the $tRNA^{Pro}$ gene present in both genomes. The fact that the general organization of the two tRNA genes and the intergenic region is very similar in cp and mt genomes suggests that these sequences are part of a cp insertion into the mitochondrial genome. That the $tRNA^{Trp}$ gene, present in this cp DNA insertion, is transcribed in mitochondria, is suggested by the fact that the product of this gene is found in mitochondria (at least in bean mitochondria) and does not hybridize to any other mt DNA fragment.

Another mt tRNA gene apparently present as a cp DNA insertion and which is very homologous (99%) to its chloroplast counterpart, namely the $tRNA^{Cys}$ gene located in maize mt DNA (fig. 1) and sequenced (unpublished results), also appears to be transcribed in the mitochondria, as there is no other sequence to which it hybridizes on the maize mt genome.

Using these different approaches, it has been possible to localize 12 tRNA genes on the mt genomes of wheat and maize, corresponding to 10 aminoacids (see fig. 1 and 2). It is not possible to know, at this time, how many isoaccepting tRNAs (and therefore how many tRNA genes) exist for each aminoacid. The total number of tRNA genes present and expressed in the mitochondria is thus not known and it is impossible to say whether plant mitochondrial DNA encodes a complete set of tRNAs, functioning as postulated by the wobble hypothesis (28) or by a "two out of three" mechanism (29), or whether some mitochondrial tRNAs are imported into plant mitochondria as postulated in the case of **Tetrahymena** mitochondria (30).

REFERENCES

1. D.M. Lonsdale, T.P. Hodge and C.M.R. Fauron
The physical map and organization of the mitochondrial genome from the fertile cytoplasm of maize
Nucl. Acids Res., 12:9249-9261 (1984).
2. R.J. Kemble and J.R. Bedbrook
Low molecular weight circular and linear DNA in mitochondria from normal and male-sterile **Zea mays** cytoplasm
Nature, 284:565-566 (1980).
3. J.D. Palmer and C.R. Shields
Tripartite structure of **Brassica campestris** mitochondrial genome
Nature, 307:436-440 (1984).
4. B.L. Ward, R.S. Anderson and A.J. Bendich
The mitochondrial genome is large and variable in a family of plants (Cucurbaceae)
Cell, 25:793-803 (1981).
5. S. Anderson, A.T. Bankier, B.G. Barrell, M.H.L. de Bruijn, A.R. Coulson, J. Drouin, I.C. Eperon, D.P. Nierlich, B.A. Roe, F. Sanger, P.H. Schreier, A.J.H. Smith, R. Staden and I.G. Young
Sequence and organization of the human mitochondrial genome
Nature, 290:457-465 (1981).
6. B. Dujon
Mitochondrial genes, mutants and maps : a review
In : Schweyen R.J., Wolf K., Kaudewitz F. (Eds) Mitochondria 1983. Walter de Gruyter, Berlin, New York, pp 1-24, (1983).
7. F. Quetier, B. Lejeune, S. Delorme, D. Falconet and M.F. Jubier
in : "Molecular form and function of the plant genomes", van Vloten-Doting L., Groot G.S.P. and Hall T.C. eds., NATO ASI series A : life sciences 83, pp 413-420 (1985).
8. D.B. Stern and D.M. Lonsdale
Mitochondrial and chloroplast genomes of maize have a 12-kilobase DNA sequence in common
Nature, 299:698-702 (1982).
9. K.P. Iams, J.E. Heckman and J.H. Sinclair
Sequence of histidyl tRNA present as a chloroplast insert in mtDNA of **Zea mays**
Plant Mol. Biol. 4:225-232 (1985).
10. L. Marechal, H. Wintz, J.M. Grienenberger, P. Guillemaut, G. Jeannin, J.H. Weil and D.M. Lonsdale
Transfer RNAs of higher plant mitochondria gene localization and sequence determination
In "Achievements and perspectives of mitochondrial research" Quagliarello E., Slater E.C., Palmieri F., Saccone C. and Kroon A.M. eds., Elsevier science publishers, Amsterdam, pp 123-132 (1986).

11. M.N. Silberklang, A.M. Gillum and U.L. RajBhandary
Use of **in vitro** ^{32}P labelling in the sequence analysis of non radioactive tRNAs
In "Methods in Enzymology" Moldave K., Grossman L. Eds, Academic Press, New York, vol LIX, pp. 58-109 (1979).
12. L. Marechal, P. Guillemaut, J.M. Grienenberger, G. Jeannin and J.H. Weil
Structure of bean mitochondrial $tRNA^{Phe}$ and localization of the $tRNA^{Phe}$ gene on the mitochondrial genomes of maize and wheat
FEBS Lett. 184:289-293 (1985).
13. M.W. Gray and D.F. Spencer
Wheat mitochondrial DNA encodes a eubacterial-like initiator methionine transfer RNA
FEBS Lett., 161:323-327 (1983).
14. L. Marechal, P. Guillemaut, J.M. Grienenberger, G. Jeannin and J.H. Weil
Sequence and codon recognition of bean mitochondria and chloroplast $tRNAs^{Trp}$: evidence for a high degree of homology
Nucl. Acids Res., 13:4411-4416 (1985).
15. F. Quigley, J.M. Grienenberger and J.H. WEIL
Localization and nucleotide sequences of the $tRNA^{Gly}$ (GCC), $tRNA^{Asp}$ (GUC) and $tRNA^{Cys}$ (GCA) genes from wheat chloroplast
Plant Mol. Biol., 4:305-310, (1985).
16. R.F. Selden, A. Steinmetz, L. McIntosh, L. Bogorad, G. Burkard, M. Mubumbila, M. Kuntz, E.J. Crouse and J.H. Weil
Transfer RNA genes in **Zea mays** chloroplast DNA
Plant Mol. Biol., 2:141-153 (1983).
17. L. Marechal, P. Guillemaut and J.H. Weil
Sequence of two bean mitochondria $tRNAs^{Tyr}$ which differ in the level of post-transcriptional modification and have a prokaryotic-like large extra-loop
Plant Mol. Biol., 5:347-351 (1985).
18. L. Marechal, P. Guillemaut, J.M. Grienenberger, G. Jeannin and J.H. Weil
Sequences of initiator and elongator methionine tRNAs in bean mitochondria
Plant Mol. Biol., 7:245-253 (1986).
19. P. Runeberg-Roos, J.M. Grienenberger, P. Guillemaut, L. Marechal, V. Gruber and J.H. Weil
Localization, sequence and expression of the gene coding for $tRNA^{Pro}$ (UGG) in plant mitochondria
Plant Mol. Biol., in press (1987).
20. T.E. England and O.C. Uhlenbeck
3'-terminal labelling of RNA with T4 RNA ligase
Nature, 275:560-561 (1978).
21. D. Falconet, B. Lejeune, F. Quetier and M.W. Gray
Evidence for homologous recombination between repeated sequences containing 18S and 5S ribosomal RNA genes in wheat mitochondrial DNA
EMBO J., 3:297-302 (1984).
22. L. Marechal, P. Runeberg-Roos, J.M. Grienenberger, J. Colin, J.H. Weil, B. Lejeune, F. Quetier and D.M. Lonsdale
Homology in the region containing a $tRNA^{Trp}$ gene and a (complete or partial) $tRNA^{Pro}$ gene in wheat mitochondrial and chloroplast genomes
Curr. Genet., in press (1987).
23. T.D. Parks, W.G. Dougherty, C.S. III Levings and D.H. Timothy
Identification of two methionine transfer RNA genes in the maize mitochondrial genome
Plant Physiol., 76:1079-1082 (1984).

24. T.D. Parks, W.G. Dougherty, C.S. III Levings and D.H. Timothy
Identification of an aspartate transfer RNA gene in maize mitochondrial DNA
Curr. Genet., 9:517-519 (1985).
25. M. Mubumbila, C.M. Bowman, F. Droog, T. Dyer, M. Kuntz and J.H. Weil
Chloroplast transfer RNAs and tRNA genes of wheat
Plant Mol. Biol., 4:315-320 (1985).
26. M. Ohme, T. Kamogashira, K. Shinozaki and M. Sugiura
Locations and sequences of tobacco chloroplast genes for $tRNA^{Pro}$ (UGG), $tRNA^{Trp}$, $tRNA^{fMet}$ and $tRNA^{Gly}$ (GCC) : the $tRNA^{Gly}$ contains only two base-pairs in the D stem
Nucl. Acids Res., 12:6741-6749 (1984).
27. P. Bergman, P. Seyer, G. Burkard and J.H. Weil
Mapping of transfer RNA genes on tobacco chloroplast DNA
Plant Mol. Biol., 3:29-36 (1984).
28. F.H.C. Crick
Codon-anticodon pairing : the wobble hypothesis
J. Mol. Biol., 19:548-555 (1966).
29. U. Lagerkvist
"Two out of three" : an alternative method for codon reading
Proc. Natl. Acad. Sci. USA, 75:1759-1762 (1978).
30. Y. Suyama
Two dimensional polyacrylamide gel electrophoresis analysis of Tetrahymena mitochondrial tRNA
Curr. Genet. 10, 411-420 (1986).

TRANSPOSABLE ELEMENTS AND THEIR ROLE IN PLANT EVOLUTION

Heinz Saedler and Zsusanna Schwarz-Sommer

Max-Planck-Institut für Züchtungsforschung
Egelspfad
5000 Köln 50, FRG

Transposable elements are observed in many plant species (Nevers et al. 1986). Their presence in the genome is often seen by their ability to transpose into genes affecting an easily recognizable phenotype, like pigmentation or flower morphology. Most of such TE induced mutations are somatically unstable resulting in a variegated phenotype. Molecularly TEs have been isolated mostly from Antirrhinum majus and Zea mays (Nevers et al. 1986).

Whatever the structural organization of these elements may be, they apparently all produce a short sequence duplication upon integration.

This duplicate is now flanking the integrated element. The number of basepairs duplicated is usually characteristic for a given element with the exception of Cin4, which apparently can produce duplication of different sizes (see Schwarz-Sommer this volume).

This property of duplicating target sequences during the integration process is shared by most elements in all organismal kingdoms.

However, plant TEs have another unique property not seen with TEs from non-plant (bacteria, fungi, flies and other animals) organisms, i.e. their mode of transposition, more precisely the excision products formed. This might be of particular evolutionary relevance.

a) Evolution of proteins

Genetic evidence suggests that transposition of En occurs via excision and re-integration (Peterson 1970, Nowick and Peterson 1981). DNA sequence analysis of many excision products reveals that the excision process is rather sloppy but still follows certain rules. These have been described previously (Saedler and Nevers 1985). To generate the end-products of excision only one element-encoded function, namely transposase, is needed. All other enzymatic activities required are recruited from the DNA-repair machinery of plant cells. Transposase is assumed to produce staggered nicks with protruding 5' fringes thus generating substrates for DNA polymerase and 5' exonuclease. The resultant products are ligated to yield the final excision product. This product by chance can be sequence-identical with the original wildtype gene. In most of the cases, however, altered sequences, socalled footprints, are left behind (Schwarz-Sommer et al. 1985b).

If the insertion of the TE occured within a gene encoding for a protein its footprint upon excision will alter the primary structure of that protein in the resulting revertant (Sachs et al. 1983, Pohlman et al. 1985) Schwarz-Sommer et al. 1985b). These alterations may also affect the function of the affected protein as revealed by analysis of several excision products (Chen et al. 1986, Wessler et al. 1986). In most of the sequenced cases the alteration is not of a simple base substitution type, generating a single aminoacid exchange, but they are typical in-frame TE footprints leading to additional aminoacids within the protein. These observations indicate that TEs can serve as generators for protein sequence diversity. In fact, footprints also can be found comparing alleles. They most frequently occur within introns, where apparently all sorts of DNA alterations are tolerated (Schwarz-Sommer et al. 1985, Werr et al. 1985, Zac et al. 1986). But footprints also occur within exons resulting in allelic variants of wildtype proteins (Doyle et al. 1986, Schwarz-Sommer 1987b). Therefore, the visitation of a gene by a TE transiently produces a mutation due to the integration of the element. However, during somatic development of the plant, the element might have already moved away from the locus but leaving a footprint behind. If this cell or progeny end(s) up as a germinal cell such footprints become manifest in the plants progeny and hence selection can operate on this variant gene product. In short, visitations of a locus by TEs generate the sequence diversity needed in evolution (Schwarz-Sommer et al. 1985).

Aside from producing mutations which could affect the structure and function of proteins, TEs also can serve as moduls in generating new regulatory units.

b) Evolution of a regulatory unit

Whenever TEs transpose into a gene they have various effects on gene expression even in leaving the locus again. They either restore the activity of the locus or the locus becomes defective by leaving a mutation or else by leaving parts of the elements themselves behind. It is the latter possibility which concerns us here, the others have already been discussed in the previous section.

An autonomous element like En (Spm) has integrated into for example the *a1*-gene of *Zea mays*, one of the many genes involved in anthocyanin-biosynthesis (Coe and Neuffer 1977). This locus encodes an NADPH-dependent dihydroquercitin reductase (Rohde et al. 1986, Schwarz-Sommer et al. 1987) needed for color expression in various parts of the plant. The presence of En at *A1* results in a variegated phenotype, i.e. colored spots (as the result of the insertion mutation prohibiting gene expression) in the developing kernels. Among many variegated kernels rare stable colorless or pale-colored exceptions are seen.

Both of these stable exceptions, however, are only stable in the absence of activity of an autonomous En(Spm) element somewhere else in the genome. If En is crossed into the genome of these mutants different and adverse phenotypes are observed. The pale-colored mutant which has been termed by McClintock *a1-m1* (1965) becomes colorless with many colored spots and the colorless mutant *a1-m2* (McClintock 1962, Reddy and Peterson 1985) becomes colored with a few darker spots. This shows that both of the mutants still respond to En functions and hence these alleles are said to be responsive or receptive. This can be understood if a defective element is still present at the locus, though either at different positions or with a different constitution in the two mutants.

Molecular cloning of the two mutations clearly showed that the residual defective elements responsible for the mutations are at different positions within the al-gene and also are of slightly different constitution (Schwarz-Sommer et al. 1987). Both defective elements are deletion derivatives of an autonomous Spm element in which internal sequences are deleted. The ends of the defective TEs are still intact.

As mentioned already, the integrity of the termini is required in the excision/transposition process mediated by the En(Spm)-encoded transposase. In this response the two mutable al alleles are similar as revealed by spotting patterns in the presence of an active En(Spm). But the two Al-alleles differ substantially in their response to other En-encoded functions. In al-ml the homogeneous pigmentation seen in the kernel in the absence of En(Spm) is totally suppressed in the presence of an active autonomous En(Spm) somewhere else in the genome. This essentially means that the Al locus has come under the control of En-functions. This system is negatively controlled because an En-function interferes with the expression of this modified al-gene.

Converse is the situation with al-m2. Here the modified al-gene cannot express its function in the absence of an autonomous En. The independently located active En(Spm) turns-on al-gene expression in al-m2. Therefore, expression of al in al-m2 is positively dependent on an En-encoded product.

Provided either of the remnant inserts (defective TE) at the two al alleles could be stabilized such that they could not respond to En-transposase by excision (revealed by the absence of spots) and provided also En became stable with respect to its location, i.e. transposition defective, then a new regulatory unit had evolved.

A cis-acting modul, the defective element at the locus (here the al-gene), responds to transacting signals emitted by the regulator gene (here the En element) either negatively or positively by turning off or turning on the expression of the adjacent gene.

Mutations stabilizing TEs have recently been described. Either they affect the transposase (Pohlman et al. 1984, Pereira et al. 1986) or the substrates of the enzyme, the termini of the element (Hehl et al. 1986). Thus stabilized neither the cis-acting module at the locus, nor the independently located autonomous element could move any more, but the former still would respond to the latter functions other than transposase. Mutational drift except for the essential signals within the cis and the trans elements would diverge these sequences to the extent that their common origine might go undetected once cloned and sequenced. Genetically the locus affected by an insertion would be another wildtype.

In principle this shows that TEs like En(Spm) have the capacity to serve as moduls in the generation of new regulatory units. Whether nature took advantage of this particular property is hard to prove, because of the drift mentioned above.

REFERENCES

Chen, C.H., Freeling, M. and Merckelbach, A. (1986) Maydica, XXXI, 93-108
Coe, E.H. and Neuffer, M.G. (1977) In Sprague G.F. (ed.) Corn and Corn improvement, American Society of Agronomy, Inc., Madison, WI Usa, pp.11-223
Doyle, J.J., Schuler, M.A., Godette, W.D., Zenger, V. and Beachy, R.N. (1986) J. Biol. Chem., 261, 9118-9238
Hehl, R., Sommer, H. and Saedler, H., Mol Gen Gent (1987) 207:47-53

McClintock, B. (1962) Carnegie Inst. Wash. Year Book, 61, 448-461
McClintock, B. (1965) Brookhaven Symp. Biol., 18, 162-184
Nevers, P., Shepherd, N. and Saedler, H. (1986) Advances in Botanical research, 12, 103-203
Nowick, E.A. and Peterson, P.A. (1981) Mol Gen Genet, 183:367-377
Pereira, A., Cuypers, H., Gierl, A., Schwarz-Sommer, Zs., and Saedler, H. (1986) EMBO J., 5, 835-841
Peterson, P.A. (1970) Theor. Appl. Gen, 40, 367-377
Pohlman, R.F., Fedoroff, N. and Messing, J. (1984) Cell, 37, 635-643
Reddy, L.V. and Peterson, P.A. (1985) Mol Gen Genet, 200:211-219
Rohde, W., Barzen, E. Marocco, A., Saedler, H. and Salamini, F. (1986) Barley Genetics, V, in press
Sachs, M.M., Dennis, E.S., Gerlach, W.L. and Peacock, W.J. (1986) Genetics, 113, 449-467
Saedler, H. and Nevers, P. (1985), EMBO J., 4, 585-590
Schwarz-Sommer, Zs., Gierl, A., Berndtgen, R. and Saedler, H. (1985a) EMBO J., 4, 2439-2443
Schwarz-Sommer, Zs., Gierl., A., Cuypers, H., Peterson, P.A. and Saedler, H. (1985b) EMBO J., 4, 591-597
Schwarz-Sommer, Zs., Shepherd, N., Tacke, E., Gierl, A. Rohde, W., Leclercq, L., Mattes, M., Berndtgen, R., Peterson, P.A. and Saedler, H. (1987) EMBO J., 6, 287-294
Schwarz-Sommer, Zs. (1987) in Structure and Function of the Eukaryotic Chromosomes (ed. W. Hennig), Springer Verlag Heidelberg, in press
Werr, W., Frommer, W.B., Maas, C. and Starlinger, P. (1985) EMBO J., 4, 1373-1380
Wessler, S.R., Baran, G., Varagona, M. and Dellaporta, S.L. (1986) EMBO J., 5, 2427-2432

EFFECTS OF TRANSPOSABLE ELEMENTS ON SPATIAL PATTERNS OF GENE EXPRESSION IN *Antirrhinum majus*

Enrico S. Coen, Tim P. Robbins, Andrew Hudson, Jorge Almeida, Cathie Martin and Rosemary Carpenter

AFRC Institute of Plant Science Research
John Innes Institute
Colney Lane
Norwich NR4 7UH, UK.

I. INTRODUCTION

One of the most interesting features of transposable elements is that they can change the level and pattern of gene expression in a variety of different ways. Analysis of the mechanisms which underly this phenomenon can provide an understanding of the genetic factors which determine pattern and can also shed light on the genetic behaviour of transposable elements. The most intensive studies of this kind in plants have been on patterns of coloration since these are readily observed and do not significantly affect organism viability in the laboratory.

Variation in pigmentation patterns may be classified into two types: clonal and non-clonal. Clonal patterns arise from genetic changes in cells that occur during the development of the organism and which are subsequently passed on to daughter cells through mitosis. The resulting patterns consist of clonal sectors of altered pigmentation which reflect cell lineage during ontogeny. For example, variegated flowers of *A. majus* which carry the *pallida*^{recurrens}*-2* (*pal*^{rec}-2) allele have spots and sectors of red pigment on an ivory background (Fig. 1b). Close inspection of these sectors reveals their clonal origin since the number of cells in a sector is normally 1,2,4,8..., as expected if they originated 0,1,2,3... divisions before complete flower development. The molecular basis of this clonal pattern is now understood (Martin et al., 1985). The *pal*^{rec}*-2* allele has a transposable element (Tam3) inserted in the *pal* gene, blocking its expression. Since the *pal* gene encodes an enzyme required for pigment biosynthesis, the presence of the element prevents the production of colour. However, during the development of the flower, Tam3 may excise from the *pal* locus to restore a functional *Pal*$^{+}$ gene which is then faithfully replicated in subsequent cell divisions to give red spots and sectors in the flower. Other mechanisms can also give rise to clonal pigmentation patterns such as position effect variegation at the white locus of D. melanogaster (Demerec, 1940) or random X-chromosome inactivation in mammals (Lyon, 1961). The analysis of clonal patterns provides important information on the cell lineage of a structure and hence is an essential background to the study of non-clonal patterns.

Non-clonal patterns do not have a simple explanation in terms of cell lineage and are usually the more common type of pattern. They include cases in which distinct tissues or organs are differentially pigmented and also cases in which an apparently uniform structure contains differentially pigmented regions. The study of these types of pattern can therefore provide a general model for understanding differential gene expression in development. Although inter-cellular signalling must be involved at some stage in setting up non-clonal patterns, most mutations which alter these patterns of pigmentation are in genes whose action is cell-autonomous. These mutations therefore affect the realisation or execution of pre-existing patterns or pre-patterns. The term pre-pattern was first introduced by Stern (see Stern, 1968) and will be used here to indicate specific spatial distributions of molecules which regulate gene expression. The nature of pre-patterns and the diverse ways in which they can be revealed is now open to molecular analysis in a few genetically well-characterized systems and in this review we consider some of the recent findings in the study of spatial patterns in A. majus.

Three types of mutation may alter the spatial pattern or intensity of pigmentation. The first type consists of cis-acting mutations which affect genes encoding enzymes involved in pigment biosynthesis. In A. majus two such loci have been most intensively studied; nivea (niv) which encodes the enzyme chalcone synthase (Spiribille & Forkmann, 1982) and pallida (pal) which encodes dihydroflavonol-4-reductase (E. Coen, J. Firmin & R. Carpenter, unpublished). The second type of mutation affects genes such as delila (del) which encode trans-acting factors which regulate the biosynthetic pathway. The third type are mutations which occur in biosynthetic genes such as niv but act in trans as well as having an effect in cis. These three types of mutation will now be considered separately.

II. CIS-ACTING MUTATIONS

The numerous alleles at the niv and pal loci of A. majus which have so far been studied may be divided into two classes: stable alleles, that confer reduced intensities or altered spatial patterns of flower colour, and alleles that show instability caused by the activity of transposable elements. Well characterized cases in the latter category are the pal^{rec}-2 allele described above and niv^{rec} alleles containing Tam1 (Bonas et al., 1984a), Tam2 (Upadhyaya et al., 1985) or Tam3 (Sommer et al., 1985). A feature common to alleles in both classes is that they generally carry sequence alterations which affect gene expression in cis. Several stable cis-acting mutations at the pal and niv loci have been characterised in detail and will be considered first.

A. Stable cis-acting mutations

Most alleles in this class have arisen in the progeny of pal^{rec}-2 or niv^{rec} lines. The majority of stable alleles produced from unstable lines confer full red colour on the flowers (revertants) and are due to excision of transposable elements in the germinal tissue (Bonas et al., 1984b; Sommer et al., 1985; Martin et al., 1985; Upadhyaya et al., 1985). Less frequently, a number of stable mutants conferring reduced intensities of coloration ranging from almost full red to very pale, or determining pigmentation confined to certain areas of the flower have been recovered in the progeny of pal^{rec}-2 and niv^{rec} (Fincham & Harrison, 1967; Carpenter et al., 1984; Harrison & Carpenter, 1973; Carpenter et al., 1987).

One of the best characterized series of alleles which alter the intensity or distribution of pigment are those derived from pal^{rec}-2. This allele contains Tam3 located 70bp upstream of the start of pal transcription (Coen et al., 1986). The transposon is flanked by a 5 bp direct duplication of host sequence, the target duplication. Precise excision of a transposon will restore the wild type sequence of the locus at which it was inserted, a single copy of the target being retained. However, analysis of the sequences around the excision sites in full red revertants at pal and niv has shown that excision is not precise in general since, in the alleles examined the target duplication is retained in a modified form (Bonas et al., 1984b; Sommer et al., 1985; Coen et al., 1986; Hudson et al., 1987; Hehl et al., 1987). For example, alleles pal-501 and pal-520, which confer a full red phenotype, contain insertions, relative to wild type, of 1bp and 7bp respectively at the site of Tam3 excision. The 7bp insertion in pal-520 is the result of deletion of 1bp of each of the copies of the target duplication and an inverted duplication of one of the copies. This type of rearrangement has been postulated to arise either as a result of DNA polymerase strand switching (Nevers & Saedler, 1985) or following the resolution of intermediate hairpin structures generated on excision of Tam3 (Coen et al., 1986).

The structures of several other stable alleles derived from pal^{rec}-2 have also been determined. In pal-518, which confers reduced pigmentation correlating with decreased levels of pal transcript (14%) relative to full red flowers, 3bp of the target duplication and 10bp to its left have been deleted. It is very unlikely that the differences in transcript levels between pal-518 and pal-501 or pal-520 are due to differences in genetic background since the lines that carry these alleles have descended from a common pal^{rec}-2 progenitor which was inbred for many generations. Therefore, the sequence missing from pal-518 is necessary in cis for wild type levels of pal transcription. A reduction in transcription has also been shown in niv alleles carrying deletions generated by excision of Tam1 (Sommer, Bonas & Saedler, unpublished).

The remaining alleles analysed determine altered patterns of flower colour (Fig. 1) which may be divided into two classes. The first class includes alleles pal-33, pal-32 and pal-15, which cause pigmentation to be restricted mainly to the flower tubes. The second class contains pal-41 which gives a more complex pattern. In the alleles of the first class, excision of Tam3 has generated an inverted duplication of 4 or 5bp of the target sequence. However, in contrast to the origin of pal-520, this rearrangement has been accompanied by deletions with end points approximately 10, 20 and 100bp upstream of the target in pal-33, pal-32 and pal-15 respectively. The generation of these deletions can be explained by assuming that in the model of Coen et al. (1986) the endonuclease nicks that lead to resolution of the hairpins may occur not only within the target duplication, as in pal-520, but also at several sites beyond it. The second class (pal-41) shows the most radical alteration of the pal promoter: the entire region 5' to the Tam3 excision site has been replaced by a foreign sequence. This is thought to have arisen from an aberrant transposition of Tam3 which, from recombination analysis, appears to have inverted a chromosome segment extending at least 6 map units from the pal locus (Robbins, Carpenter and Coen, unpublished).

One explanation for the generation of specific patterns of pigmentation by these sequence alterations is that the wild type pal promoter contains a set of sequences that respond to diverse regulatory signals spatially arranged as a pre-pattern in the flower. Novel spatial patterns are produced by mutations which change the interpretation of the pre-pattern by modifying the affinity of the *pal* promoter for different regulatory molecules.

Such a model, involving the interaction between transcription factors and the *pal* promoter can be invoked to explain the generation of patterns that are specific to the corolla tube. Although there is no direct evidence for such interactions, it is worth emphasizing some sequence features that might provide clues as to how modifications of DNA-protein and/or protein-protein interactions could arise as a result of the mutations described. The lobes of full red flowers are more intensely pigmented and contain significantly higher amounts of *pal* transcript than the tubes. Therefore, an overall reduction of *pal* transcription would be expected to give flowers which still have darker lobes than tubes. However, removal of 10bp to the left of the target sequence in *pal*-518 results in a reduction of pigmentation that seems more accentuated in the lobes than in the tubes (Fig.1d). This is more extreme in *pal*-33 which carries a deletion similar to that in *pal*-518. Other stable alleles derived from *pal*^{rec}-2 whose structures have not been investigated in detail, show a gradation of colour pattern between those of *pal*-518 and *pal*- 33. In *pal*-32, a further deletion of about 10 bp to the left of the target sequence causes an almost complete inhibition of expression in the lobes. The region deleted in these alleles might therefore be a candidate for the binding of a hypothetical transcription factor, that differentially enhances *pal* expression in the lobes. Evidence for a trans-acting factor that plays a role in determining differential expression of *pal* in tubes and lobes is presented in section III. In *pal*-15, a deletion of approximately 100 bp to the left of Tam3 results in lack of pigmentation in the whole corolla except for a ring at the base of the tube (Fig.1). This ring of pigment is formed at a very early stage of flower development in both *pal* mutants and wild type, suggesting that the base of the tube represents an area where *pal* expression is subjected to different control mechanisms, presumably independently of sequences within 100bp to the left of Tam3. It appears, therefore, from the analysis of this set of alleles, that separable cis-acting regulatory elements might confer spatial-specific expression on *pal*. The involvement of complex sets of cis-acting sequences in conferring cell type-specific expression has also been postulated for the yellow gene of *Drosophila* (Chia et al., 1986).

The pattern produced by *pal*-41 is quite distinct from that of the tube specific alleles (Fig. 1k) and requires a different explanation. It is conceivable that this pattern partly reflects a pre-pattern different from the one that normally interacts with the *pal* promoter, since in this allele the entire region to the left of the Tam3 excision site has been replaced by foreign sequences. For example, in *Drosophila* an inversion between the *Antp* gene and a nearby gene results in the exchange of cis-acting sequences with the result that the *Antp* gene is now under the control of the promoter of the other gene (Schneuwly et al., 1987). However, the foreign sequence brought next to the *pal* gene in *pal*-41 does not promote transcription in flowers when in its original position (Robbins, Carpenter & Coen, unpublished). This indicates that the pattern observed in *pal*-41 may reflect a position effect of the adjacent foreign chromatin rather than the introduction of a new promoter. Such effects have been noted for particular transformants of the *white* gene in *Drosophila* resulting in novel distributions of the red eye pigment (Levis et al., 1985).

B. Unstable cis-acting mutations

Stable alleles showing altered gene expression can be produced by the excision of a transposable element from an unstable allele. An unstable allele can also give rise to new alleles with altered intensity or patterns of expression whilst retaining the transposable element at the locus. In these cases host gene expression can occur with the transposable element present, to produce a non-clonal background pigmentation with a superimposed clonal pattern of revertant cells. Such a mutation may owe its altered phenotype to changes within the transposable element, or to its reinsertion at a new position in the gene. These mutations are of interest for several reasons. Firstly, the transposable element alters the expression of its host gene and, like the mutations caused by excision, can reveal something of the wild-type function of the sequence into which it is inserted. Secondly, the frequency of reversion provides information about the behaviour of the transposable element after it has been altered or has transposed. Thirdly, the way in which the transposable element changes the expression of the host gene may be clarified by examining the effect of changing the element or moving it to a new position. Several unstable anthocyanin mutations containing the transposable elements Tam1, Tam2 and Tam3 have been studied in A. majus (see Coen and Carpenter, 1986 for review). The most important factor determining the effect of each transposable element on expression appears to be its location within the host gene. Alleles containing insertions in the promoter, in introns or exons, will each be considered in turn.

The pal^{rec}-2 allele of A. majus contains the 3.5 kb transposable element, Tam3 inserted into the promoter of the pal gene (Coen et al. 1986). The flowers show a high frequency of dark red sites, caused by somatic excision of Tam3, against a colourless background, indicating that no expression of the gene occurs with Tam3 in position (Fig 1b.). This null phenotype might be expected to be due to the interruption of the promoter sequence, part of it being displaced over 3.5 kb away from the gene. However, several pieces of evidence suggest that Tam3 may not simply act as a passive spacer sequence, but may actively inhibit expression of the host promoter. Firstly, Tam3 completely prevents expression of pal, whereas even the large deletions of promoter sequences upstream of the insertion site in the alleles pal-15 and pal-41 do not. Further evidence is provided by a consideration of alleles derived from pal^{rec}-2, and of other alleles containing transposable elements in the promoter.

The allele pal-510 was produced from pal^{rec}-2. It shows a much reduced frequency of somatic excision, against a light red background pigmentation (Fig 1i.). This suggests that Tam3 no longer prevents expression of the gene completely, when in position. DNA studies show that Tam3 is in the same position as in its progenitor, but has been altered slightly at the end nearer the pal gene (Hudson, Carpenter and Coen, unpublished results). This alteration is consistent with an excision of this end of Tam3, followed by immediate re-integration into its original position. That such a slight change can permit background pal expression suggests that the disrupting effect of the intact transposable element was not due entirely to interruption of the promoter, but was partly due to active inhibition by Tam3. Alternatively, the altered transposable element might still disrupt the promoter but now partially replace the missing function with its own internal sequences. Such expression of genes by promotion or enhancement from insertion sequences has been reported for the bacterial IS2, IS3 and IS5 (Glandsdorff et al., 1981; Jund and Loison, 1982), Ty1 of yeast (Roeder et al., 1985), copia in Drosophila (Scott et al., 1983) and many vertebrate retroviruses (eg. Luciw et al., 1983). In the case of pal-510, no transcript starting within Tam3 and running into the gene can be detected, and it is difficult to imagine how the altered transposable element could supply an enhancer or promoter function, when the intact Tam3 did not. This strongly suggests that the effect of the intact Tam3 is an active inhibition of pal expression.

The allele *pal*-42 was also derived from *pal*rec-2, and it shows certain similarities to *pal*-510 in exhibiting a reduced frequency of somatic excision against a background pigmentation. However, whereas the background pigmentation produced by *pal*-510 has the same distribution as in the wild type flower, so that it is most intense at the base of the tube and in the face and lobes of the corolla, that caused by *pal*-42 shows a novel pattern of concentration in the top of the tube between the lobes (Fig 1j). Molecular analysis of *pal*-42 reveals that Tam3 has undergone a small deletion on the side nearer the gene, similar to that in *pal*-510 but extending further into the promoter. It differs dramatically at the other end of the element, where the same sequence as in *pal*-41 is found, suggesting that *pal*-41 and *pal*-42 are derived from a common progenitor (Robbins, Carpenter and Coen, unpublished results). It seems probable that *pal*-42 possesses background pigmentation for the same reasons as *pal*-510. The pattern of pigmentation in this allele may be due to the presence of the new sequence at the upstream end of Tam3, as has been discussed in section II A.

The behaviour of Tam1 in the promoter of the *niv*-53 allele is similar to that of Tam3 in pal^{rec}-2 (Bonas et al., 1984). In this case Tam1 blocks expression when in position, but restores it on excision. The allele *niv*-46, is an apparently stable derivative of *niv*-53 showing a pale pigmentation and contains a 5 bp deletion in the end of Tam1 further from the gene (Hehl et al., 1987). This deletion therefore restores some gene expression whilst abolishing or greatly reducing the ability of Tam1 to excise. There are a number of ways in which a transposable element in a promoter region might actively inhibit expression. One possible mechanism is that it might prevent expression by disrupting the secondary structure of the chromatin in its environment. A *Drosophila* transposable element, *HMS Beagle*, is known to increase the DNA'ase I sensitivity of a neighbouring DNA sequence, suggesting that the element alters the chromatin structure (Eissenberg et al., 1985). However, such changes are usually associated with an increase in expression. Another possibility, proposed for inhibition of the *yellow* gene of *Drosophila* by *gypsy* is that the transposable element competes with the host promoter for transcription factors (Chia et al., 1985).

The similarity of the alleles *pal*-510, *pal*-42 and *niv*-46, which show a reduced frequency of excision and background pigmentation, suggests another way in which Tam1 and Tam3 may inhibit expression. The three alleles all show alterations at one end of the transposable element. The ends are the proposed recognition sites for transposase molecules (Saedler and Nevers, 1985), and so the reduced frequency of excision presumably reflects a reduced affinity of transposase for the altered ends. Thus the inhibitory effect of the intact tranposable elements may be due to transposase binding, possibly because both termini are brought together in a complex with transposase to produce a structure which physically restricts access of transcription factors to the host promoter. Alterations in the ends of the transposable elements may then reduce the affinity for transposase and the probability of forming the transposition complex, thus allowing some expression of the gene. A similar model has been put forward to explain the action of the maize transposable element *I* in the *a*1 gene (Schwarz-Sommer et al., 1985).

However, even an intact Tam3 does not always inhibit expression when located in the promoter. In the allele niv^{rec}-98, Tam3 both excises with a high frequency and allows background expression (Fig 1c.) (Sommer et al., 1985). It may be that the orientation of the element is also important in determining its phenotypic effects, since in niv^{rec}-98 Tam3 is in the opposite orientation to that found at *pal*. Alternatively the different structure of the promoters or sites of insertion of the element at *niv* and pal may affect the interaction between the element and the host gene.

In addition to insertions in the promoter, other transposon locations have been described in A. majus. The niv-44 allele contains Tam2 at exactly the first exon-intron boundary (Upadhyaya et al, 1985). In general, insertions in introns cause either a reduction or a complete abolition of expression of the host gene. No expression occurs when the transposable element contains poly-adenylation signals which terminate transcription. This is the case for the copia transposable element in the white^{apricot} allele of Drosophila, (Zachar et al., 1985) and the maize Spm/En transposable element in the wx gene (Gierl et al., 1985). When the transposable element contains no poly-adenylation signals a reduced level of expression usually results; probably because its inclusion in the pre-mRNA interferes with RNA processing. The maize Adh1-S5340 allele containing Mu in the first intron appears to fall into this category (Bennetzen et al., 1984). In the case of the niv-44 allele the insertion disrupts the splice-donor sequences. However, the absence of the original donor sequence may not be resposible for the phenotype since the new donor site is still consistent with the cónsensus sequence proposed by Mount (1982). The presence of Tam2 allows no expression, so it may either cause termination of transcription or provide a splice-acceptor site resulting in a mature transcript which contains Tam2 sequences. A reduced level of expression is observed in niv-567 which has been derived from niv-44, possibly by an internal deletion of Tam2. It shows a background pigmentation with both red and colourless sites superimposed. The background pigmentation suggests that whatever was preventing expression in niv-44 has now been deleted. A similar deletion in the I transposon of maize allows some gene expression by removing poly-adenylation signals. The sites seen on the corolla of niv-567 are probably caused by excision of the remainder of Tam2 to produce either mutant or wild-type splice-donor sequences.

The consequence of transposable element insertions into coding regions are the most straightforward to understand. The allele niv-540 was derived from niv-98, which contains Tam3 in the promoter. In contrast to its progenitor, niv-540 produces no background pigmentation and has a reduced frequency of red sites (Fig. 1h). The new allele contains two copies of Tam3. One copy, about 3 kb upstream of the promoter does not affect the phenotype of flowers, while the second is within the last exon of the niv gene and blocks niv expression completely. The reduced frequency of wild-type sites seen on niv-540 flowers is presumably a reflection of the location of Tam3 within a coding region, since almost all imprecise excisions will cause null mutations caused by amino acid substitutions and frame shifts. Amongst all these alleles only pal-42 shows a novel spatial distribution of pigmentation. This may reflect the change in relative position on the chromosome of the pal gene in this allele as described earlier for pal-41 (see section II A).

III. TRANS-ACTING MUTATIONS

A trans-regulatory gene may be defined as one which encodes a factor that interacts with one or more genes at other loci so as to produce altered transcriptional activity. The cumulative action of several such factors acting directly or indirectly on the structural genes of the anthocyanin pathway is thought to be responsible for the normal distribution of anthocyanin in the flower. Cis-alterations of structural genes may alter their interaction with particular trans-acting regulators to produce altered distributions of pigment that reveal underlying pre-patterns not visible in the wild type flower. It follows that mutations in the genes which encode trans-acting regulators might also be expected to reveal such pre-patterns.

In A. majus, the mutation of a single gene, delila (del), generates an altered spatial pattern of pigmentation by a trans-regulation of the anthocyanin pathway. The most dramatic change in plants homozygous for the recessive del allele is the loss of anthocyanin pigment from the corolla tube whilst apparently normal levels are maintained in the corolla lobe (Fig. 11). Several lines of evidence indicate that the del mutation alters the distribution of pigment synthesis, at least in the corolla, by acting in trans to change the expression of several structural genes required for anthocyanin biosynthesis and conceivably acts on the entire pathway. This has been most clearly demonstrated by the analysis of transcript levels for two structural genes of the pathway, niv and pal. By extracting RNA from dissected wild type (Del+) and mutant (del) flowers it was shown that, whilst apparently normal levels of expression were maintained in the lobes of del flowers, the tubes revealed a 5-10 fold reduction in niv gene expression and an undetectable level of pal expression. Additional evidence from the feeding of a leucocyanidin intermediate to flowers (Carpenter and Coen, unpublished data) and measuring levels of the late acting enzyme UDP-Glucose: flavonoid 3-0-glucosyltransferase (Martin et al., 1987) suggests that subsequent steps in the pathway may also be altered at the transcriptional level in del mutants.

An unstable del mutation, del^{rec}, which restores pigment synthesis to single cells in the corolla tube, provides evidence that the Del+ gene product can act in the final stages of flower development, in an intracellular or cell autonomous fashion. This argues against any role for the Del+ gene product in the establishment of pre-pattern but instead suggests a late acting intracellular role in the trans-activation of anthocyanin biosynthetic genes in the corolla tube. Because of the high levels of pigment synthesis in the lobes of del flowers it is not clear whether the Del+ gene product also activates expression in the lobes to a small degree. Whether the Del+ gene product itself might be differentially expressed between lobes and tubes is therefore uncertain. In addition to the corolla tube, a number of other normally pigmented structures lack colour in del plants, notably the lower stem and leaves and the stamens. Conversely a number of structures including the sepals and style, retain some level of pigment. In specifically reducing the synthesis of pigment in a subset of normally pigmented structures the action of del is closely analagous to that of certain regulatory loci in maize. These loci appear to regulate the distribution of anthocyanin in the various tissues of the maize plant in a cell autonomous manner (Coe and Neuffer, 1977). Two of these regulatory loci, R and C, alter the levels of at least two enzymes encoded by structural genes of the pathway (Dooner, 1983). However, whilst the action of these loci distinguishes between recognisable tissues or organs of the plant, the most dramatic result of del is to produce a spatial pattern resulting from differential expression within a single organ (the corolla). A pattern of pigmentation defined within the corolla of Petunia hybrida has also been described (Mol et al., 1983). Therefore it is possible for trans-regulatory factors to produce patterns of gene expression which need not correlate with the accepted divisions within a plant, of organ or tissue type. Gene expression which is specific to organs, tissues or regions within organs can therefore be considered to be particular cases of more general pre-patterns that are set up during development.

We have argued that the Del+ gene product confers a pre-pattern which confers corolla tube pigmentation on the flower. It is important to distinguish this pre-pattern from the underlying pre-pattern of lobe-specific expression revealed in the absence of the Del+ gene product. Presumably the latter pre-pattern is also present in the wild type flower but is not revealed until the Del+ gene product is removed. Clearly this pre-pattern has a strict morphological correlation with the lobe and tube structures. However the analysis of sectors of cells in pal^{rec} flowers indicates that these two structures are not derived from separate cell lineages since clonal sectors of pigment can span both tubes and lobes. This is consistent with the view that the pre-pattern revealed by the del mutation reflects a division within the corolla ultimately determined by some form of intercellular signalling. Evidence for a pre-pattern conferring expression in the lobes also comes from a particular distribution of the yellow pigments termed aurones. These pigments are derived from chalcones which form the first step of the anthocyanin biosynthetic pathway. Aurones can be most readily observed in flowers blocked in the later stages of anthocyanin biosynthesis (eg. pal-15). If such plants are also homozygous for the recessive sulfurea (sulf) mutation the spatial distribution of aurones in the corolla is similar to that of anthocyanins in a del plant. The yellow pigments are present throughout the lobes but are absent from the tube (cf. Figs. 1m and 1l). It is possible that this similarity arises from the response of both pigment pathways to the same lobe-specific pre-pattern. In contrast, the normal distribution of aurones in a dominant Sulf+ flower is a restriction to the "lip" of the lower lobe (eg. Fig. 1g). This bears a limited resemblance to the distribution of anthocyanin that occurs in plants carrying a dominant Eluta (El+) allele (not shown) or the pal-41 allele already described. This raises the possibility that both structural genes and trans-regulatory genes of pathways giving rise to two distinct floral pigments may be responding to the same pre-patterns within the corolla which can confine expression to the lobe or "lip" areas.

IV. MUTATIONS WHICH ACT BOTH IN CIS AND TRANS

The majority of mutations in biosynthetic genes such as niv and pal described above reduce the overall quantity of gene expression. These mutations are recessive to wild-type since a single dose of the wild-type allele encodes sufficient gene product to confer the full red phenotype. However, several niv alleles have been described which are semi-dominant to wild-type (Carpenter et al., 1987). Plants homozygous for the niv-525 allele have very pale flowers with pigment concentrated in the tubes and around the lower lip (Fig. 1n). When crossed to wild-type, the F1 plants (Niv^{+}/niv-525) have a similar spatial distribution of pigment as niv 525/niv-525 but of a slightly darker intensity. Self-pollination of the F1 plants gives F2 progeny containing niv-525 homozygotes, niv-525/Niv^{+} heterozygotes and Niv^{+} homozygotes in a ratio of 1:2:1. The niv-543 allele is also semi-dominant but, unlike niv-525, it is unstable and has spots and sectors of pigment on the flower (Fig. 1o).

The chalcone synthase protein encoded by the niv locus is thought to be a multimeric enzyme, so the semi-dominance of niv-525 might be explained if this allele produces an altered chalcone synthase monomer which interacts with the wild-type protein to give heteromers with reduced enzyme activity. However, analysis of chalcone synthase protein and mRNA levels in different genotypes containing the niv-525 allele shows that niv-525 reduces the quantity of niv gene expression rather than altering the type of chalcone synthase produced (Coen & Carpenter, unpublished results). The niv-525 mutation therefore acts both in cis and in trans to alter the spatial distribution and quantity of niv gene expression. What might be the mechanism underlying this phenomenon?

The semi-dominant alleles niv-525 and niv-543 were both derived from niv-98::Tam3, suggesting that they resulted from sequence changes induced by Tam3. This has been confirmed since an inverted duplication of about 200bp, including the untranslated leader and part of the first exon, occurs in the niv-525 allele at the site of Tam3 excision (Coen & Carpenter, unpublished results). The structure of niv-525 is readily explained by the

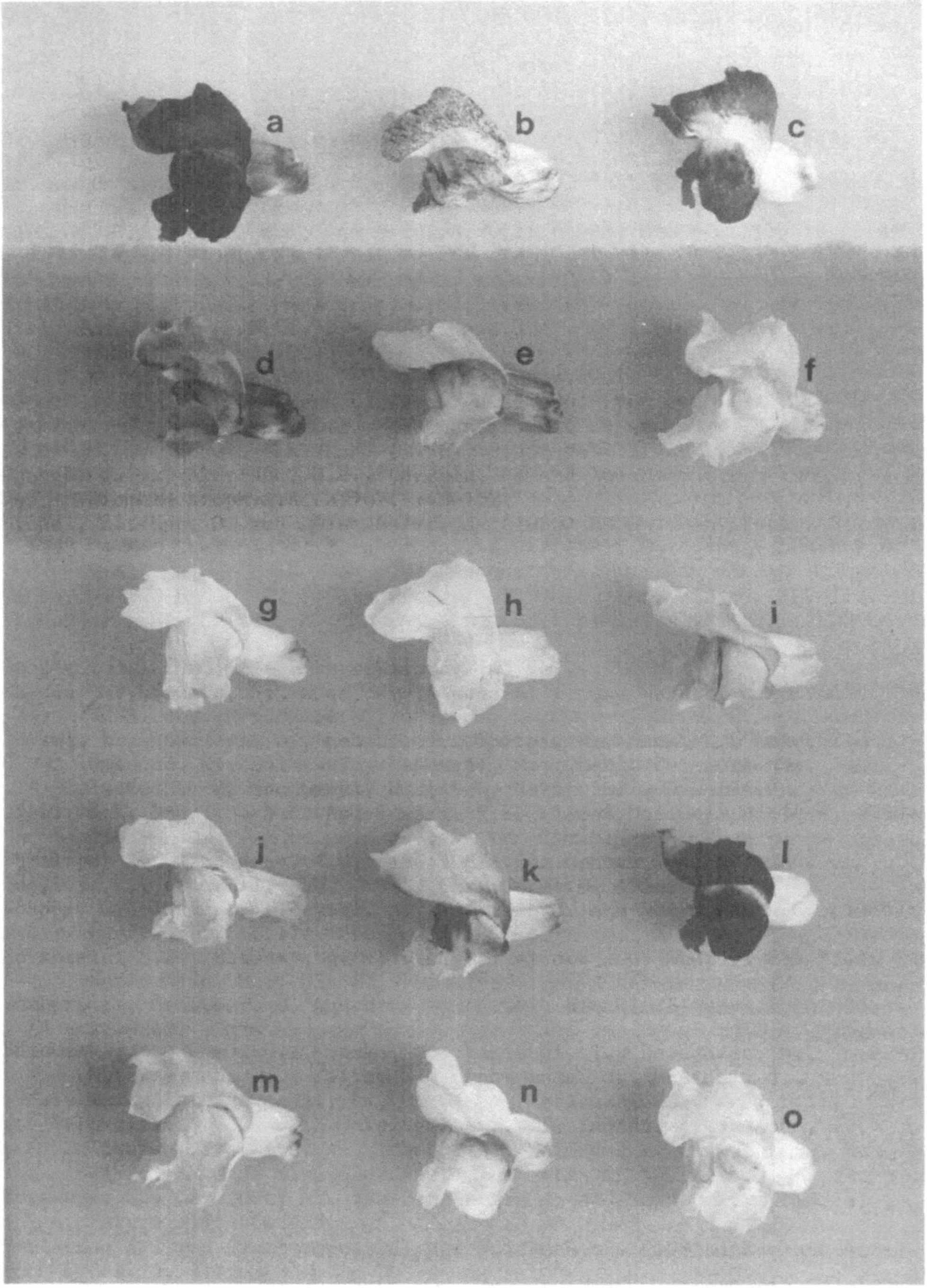

Fig. 1. Flowers illustrating various mutations in *Antirrhinum majus*. Flowers show: (a) full red wild-type (*pal*-501); (b) *pal*^{rec}-2; (c) *niv*^{rec}-98; (d) *pal*-518; (e) *pal*-33; (f) *pal*-32; (g) *pal*-15; (h) *niv*-540; (i) *pal*-510; (j) *pal*-42; (k) *pal*-41; (l) *delila*; (m) *sulfurea* which also carries the *pal*-15 mutation to reveal the yellow pigment; (n) *niv*-525; (o) *niv*-543.

hairpin model of transposon excision (Coen et al., 1986) in a similar manner to the explanation of the short inverted duplications produced by imprecise excision of Tam3 from the pal locus described above. The presence of an extra 200 bp in the niv promoter would be expected to have an effect in cis on the transcription of the niv gene. However, the effect of niv-525 on a Niv^+ allele in trans requires a different explanation.

A possible model for the semi-dominance of niv-525 is that it is due to the production of anti-sense RNA. Deletion analysis of the Niv+ promoter showed that a region of 1.2 kb, which lies upstream of the TATA box, is required for expression of the promoter (Kaulen et al., 1986). Most of this region is retained intact, upstream of the inverted duplication of niv-525, and would be expected to drive transcription to the right if provided with a suitable TATA box. The sequence TTATATAAA, identical to that of the normal TATA box, is located within the inverted duplication near the junction of the inverted leader and coding sequence. Transcription might therefore be driven from this new TATA box to produce an RNA with 40 bp at its 5' end of the anti-sense strand of leader sequence. We propose that in a niv-525/Niv^+ heterozygote the niv-525 allele produces an anti-sense RNA molecule which hybridizes to the first 40 bp of the leader RNA from the Niv^+ allele and that the resulting double-stranded RNA is degraded in the nucleus. The spatial pattern of pigmentation observed in heterozygotes could be explained if greater levels of expression of anti-sense RNA occurred in regions of the flower with less colour. The distribution of anti-sense RNA would therefore reflect a novel pre-pattern, distinct from those described in previous sections.

One problem with the anti-sense model is that previous studies using DNA transfection indicate that inhibition by anti-sense RNA requires a 10 to 100-fold excess of DNA expressing anti-sense over DNA expressing sense RNA (Izant & Weintraub, 1985). However, in these experiments the DNA encoding anti-sense RNA is introduced at random locations in the genome whereas the niv-525 allele is located at a chromosome position which is homologous to the position of the Niv^+ allele which it inhibits. A strong effect of chromosome position on allelic interactions has been described in D. melanogaster and is termed transvection. For example, a dominant allele at the white locus, wDZL, acts in trans to reduce expression of a w^+ allele only when the two alleles are at homologous chromosome positions (Bingham and Zachar, 1985). Such effects of chromosome position might be explained if the trans-effects are mediated by a labile RNA species (e.g. anti-sense RNA) which requires close proximity between interacting genes in order to maintain a high local concentration of the labile RNA (Jack & Judd, 1979).

V. CONCLUSION

The analysis of mutations which alter pigmentation has revealed many non-clonal pre-patterns in the flower. Cis-acting mutations in pigment biosynthetic genes can reveal several different patterns by changing the affinity of the promoters for spatially distributed trans-acting transcription factors, hence giving different interpretations to the pre-patterns. Analysis of one such trans-acting factor, encoded by the del locus, shows that its own spatial distribution is in turn determined by a pre-pattern since it behaves in a cell-autonomous fashion. This provides a molecular model for Stern's notion that "development may be conceived as a consecutive series of patterned events in which a specific pre-pattern becomes expressed in a subsequent pattern, the subsequent pattern serving as a pre-pattern for the next patterned process" (Stern, 1968). Most mutations which produce non-clonal patterns in other organisms, such as Z. mays or D. melanogaster, act in a cell-autonomous way and hence affect genes which interpret a pre-pattern. This raises the question as to what determines pre-patterns.

Pre-patterns must ultimately derive from genes whose products act in an inter-cellular or non-autonomous manner. Mutations in this type of gene might be expected to have consequences on general aspects of morphogenesis as well as on pigment patterns. For example, the *pal*-41 allele produces a complex pattern of pigmentation: the upper lobes of the flower have little colour, most pigment being in a diffuse area around the lip of the lower lobes (Fig. 1k). The morphological mutation *cycloidea*^{radialis} results in radially symmetrical flowers in which all lobes resemble the lower middle lobe. If such a flower also contained the *pal*-41 allele, a relatively simple pattern of pigment would be expected in which a symmetrical diffuse band of pigment would be seen around the lip of the entire flower. The complex pattern observed with *pal*-41 in normal flowers therefore partly reflects a transformation of a simpler pattern through the action of the *Cycloidea*^{radialis} gene product which changes the flower morphology. This leads to the conclusion that our understanding of pre-patterns which determine the spatial distribution of pigment may ultimately depend on how they interact with genes involved in morphogenesis.

REFERENCES

Bennetzen, J.L., Swanson, J., Taylor, W.C., Freeling, M., 1984: DNA insertion in the first intron of maize *Adh1* affects message levels: Cloning of progenitor and mutant *Adh1* alleles. *Proc. Natl. Acad. Sci. USA* 81:4125-4128

Bingham, P.M and Zachar, Z.,1985: Evidence that two mutations, *wDZL* and *z1*, affecting synapsis-dependent genetic behaviour of white are transcriptional regulatory mutations. *Cell* 40:819-825

Bonas, U., Sommer, H., Harrison, B. J. and Saedler, H., 1984a: The transposable element *Tam1* of *Antirrhinum majus* is 17 kb long. Mol. Gen. Genet. 194: 138-143.

Bonas, U., Sommer, H. and Saedler, H., 1984b: The 17kb *Tam1* element of *Antirrhinum majus* induces a 3bp duplication upon integration into the chalcone synthase gene. *EMBO J.* 3: 1015-1019.

Carpenter, R., Coen, E.S., Hudson, A.D. and Martin, C.R., 1984: Transposable genetic elements and genetic instability in *Antirrhinum.* Seventy-third Annual Report of the John Innes Institute. (Norwich: Crowe and Sons Ltd.), pp. 52-64.

Carpenter, R., Martin, C. and Coen, E.S., 1987: Comparison of genetic behaviour of the transposable element Tam3 at two unlinked pigment loci in *Antirrhinum majus. Mol. Gen. Genet.* 207: 82-89.

Chia. W., Howes, G., Martin, M., Meng, Y., Moses, K. and Tsubota, S., 1986: Molecular analysis of the yellow locus of *Drosophila. EMBO J.* 5: 3597-3605.

Coe, E.H. and Neuffer, M.G., 1977: The Genetics of Corn. In: Corn and corn improvement. Sprague, G.F. (ed), American Society of Agronomy, Madison, Wisconsin.

Coen, E.S. and Carpenter, R. 1986: Transposable elements in *Antirrhinum majus:* generators of genetic diversity. *Trends in Genet.* 2: 292-296.

Coen, E.S., Carpenter, R. and Martin, C., 1986: Transposable elements generate novel spatial patterns of gene expression in *Antirrhinum majus. Cell* 47:285-296

Demerec, M., 1940: Genetic behaviour of euchromatic segments inserted into heterochromatin. *Genetics* 25: 618-627.

Dooner, H.K., 1983: Coordinate genetic regulation of flavonoid biosynthetic enzymes in maize. *Mol. Gen. Genet.* 189: 136-141

Fincham, J.R.S. and Harrison, B., 1967: Instability at the *Pal* locus in *Antirrhinum majus.* II. Multiple alleles produced by mutation of one original unstable allele. *Heredity* 22: 211-227

Eissenberg, J.C., Kimbrell, D.A., Fristrom, J.W. and Elgin. S.C.R., 1984: Chromatin structure at the 44D larval cuticle gene in *Drosophila:* the effect of a transposable element insertion. *Nucl. Acids Res.*12:9025-9037

Gierl, A., Schwarz-Sommer, Z. and Saedler, H. 1985: Molecular interactions between the components of the En-I transposable element system of Zea mays. EMBO J. 4: 579-583.

Glandsdorff, N., Charlier, D., and Zafarullah, M., 1980: Activation of gene expression by IS2 andIS3. Cold Spring Harbor Symp. Quant. Biol. 45: 153 -156.

Harrison, B.J. and Carpenter, R. 1973: A comparison of the instabilities at the nivea and pallida loci in Antirrhinum majus. Heredity 31: 309-323.

Hehl, R., Sommer, H. and Saedler, H., 1987: Interaction between the Tam1 and Tam2 transposable elements of Antirrhinum majus. Mol. Gen. Genet. 207:47-53.

Hudson, A., Carpenter, R. and Coen, E.S. 1987: De novo activation of the transposable element Tam2 of Antirrhinum majus. Mol. Gen. Genet. 207: 54-59.

Izant, J.G. and Weintraub, H., 1985: Constitutive and conditional suppression of exogenous and endogenous genes by anti-sense RNA. Science 229: 345-352.

Jack, J.W. and Judd, B.H. 1979: Allelic pairing and gene regulation: A model for the zeste-white interaction in Drosophila melanogaster. Proc. Natl. Acad. Sci. USA 76: 1368-1372.

Jund, R. and Loison, G., 1982: Activation of transcription of a yeast gene in E. coli by an IS5 element. Nature 296: 680-681.

Kaulen, H., Schell, J. and Kreuzaler, F. 1986: Light induced expression of the chimeric chalcone synthase-NPTII gene in tobacco cells. EMBO J. 5, 1-8.

Levis, R., Hazelrigg, T. and Rubin, G.M. 1985: Effects of genomic position on the expression of transduced copies of the white gene of Drosophila. Science 229: 558-561.

Luciw, P.A., Bishop, J.M., Varmus, H.E. and Cappechi, M.R., 1983: Location and function of retroviral and SV40 sequences that enhance biochemical transformation after microinjection of DNA. Cell 33: 705-716.

Lyon, M.F., 1961: Gene action in the X-chromosome of the mouse (Mus musculus L.). Nature 190: 372-373.

Martin, C.R., Carpenter, R., Coen, E.S. and Gerats, A.G.M. 1986: The control of floral pigmentation in A. majus. In: Soc. Expt. Biol. Symp. 32: 19-52. Thomas and Grierson (eds.), Cambridge University Press.

Martin. C.R., Carpenter, R., Sommer, H., Saedler, H. and Coen, E.S., 1985: Molecular analysis of instability in flower pigmentation of Antirrhinum majus, following isolation of the pallida locus by transposon tagging. EMBO J. 4, 1625-1630.

Mol, J.N.M., Schram, A.W., de Vlaming, P., Gerats, A.G.M., Kreuzaler, F., Hahlbrock, K., Reif, H.J. and Veltkamp, E., 1983: Regulation of flavonoid gene expression in Petunia hybrida: Description and partial characterization of a conditional mutant in chalcone synthase gene expression. Mol. Gen. Genet. 192: 424-429.

Mount, S.M., 1982: A catalogue of splice-junction sequences. Nucl. Acids Res. 10:459-472

Roeder, G.S., Rose, A.B. and Perlman, R.E., 1985: Transposable element sequences involved in the enhancement of yeast gene expression. Proc. Natl. Acad. Sci. USA, 82:5428-5432.

Saedler, H. and Nevers, P., 1985: Transposition in plants: a molecular model. EMBO J. 4: 585-590.

Schwartz-Sommer, Zs., Gierl, A., Berntgen, R. and Saedler, H., 1985: Sequence comparison of "states" of a1m1 suggests a model for Spm (En) action. EMBO J. 4: 2439-2443.

Schneuwly, S., Kuroiwa, A. and Gehring, W.J., 1987: Molecular analysis of the dominant homoeotic Antennapedia phenotype. EMBO J. 6: 201-206.

Scott, M.P., Weiner, A.J., Hazelrigg, T.I., Polisky, B.A., Pirotta, V., Scalenghe, F. and Kaufman, T.C., 1983: The molecular organisation of the Antennapaedia locus of Drosophila. Cell 35: 763-776.

Sommer, H., Carpenter, R., Harrison, B.J. and Saedler, H., 1985: The transposable element Tam3 of Antirrhinum majus generates a novel type of sequence alteration upon excision. Mol. Gen. Genet. 199: 225-231
Spiribille, R. and Forkmann, G. 1982: Genetic control of chalcone synthase activity in flowers of Antirrhinum majus. Phytochemistry 21:2231-2234.
Stern, C. 1968: Developmental genetics of pattern. In Genetic mosaics and other essays. pp 130-173. Harvard press.
Upadhyaya, K.C., Sommer,H., Krebbers, E. and Saedler, H., 1985: The paramuatagenic line niv-44 has a 5kb insert, Tam2, in the chalcone synthase gene of Antirrhinum majus. Mol. Gen. Genet. 199: 201-207.
Zachar, Z., Davidson, D., Garza, D. and Bingham, P.M., 1985: A detailed developmental and structural study of the transcriptional effects of insertion of the copia transposon into the white locus of Drosophila melanogaster. Genetics 111: 495-515.

MOLECULAR GENETIC ANALYSIS OF SEQUENCES HOMOLOGOUS TO MUTATOR TRANSPOSABLE ELEMENTS IN NON-MUTATOR MAIZE STOCKS

Luther E. Talbert, Devon Turks, Catherine Faber, Laura Mann, Karen Sylvester, Fenella Raymond and Vicki L. Chandler

Institute of Molecular Biology
University of Oregon
Eugene, OR 97403

INTRODUCTION

Robertson (1978) described a heritable Mutator system in maize that increased mutation rates at many loci 20-50 fold. Approximately 90% of the outcross progeny of a Mutator parent retain the high mutation rate, which correlates with the transposition of a family of transposable elements. A 1.4 kb insertion, designated Mul, was cloned from Mutator-induced mutations at Adhl (Bennetzen et al., 1984) and Al (O'Reilly et al., 1985). Mul has the structure of a transposable element including long terminal inverted repeats flanked by 9 bp target site duplications (Barker et al., 1984). A second element known to transpose and cause mutations in Robertson's Mutator stocks is Mul.7 (Taylor et al., 1986), which differs from Mul primarily due to an extra 380 bp segment of internal sequence (Taylor and Walbot, 1987). Approximately 10-60 copies of these elements are found in Mutator lines, and most mutant alleles that have been examined with molecular probes contain insertions of Mul- or Mul.7-like elements (Freeling, 1984). Although Mul and Mul.7 appear to be the major cause of Mutator-induced mutations, a third type of Mu element has been cloned as a result of its insertion into the Adhl gene. This element, termed Mu3, has the terminal inverted repeats of a Mu element, but its internal sequence is not related to Mul or Mul.7 (K. Oishi and M. Freeling, pers. comm.).

The regulation of Mutator transposition is not understood. It is not known whether any of the identified Mu elements encode trans-acting factors required for transposition, or if other elements are required. One level of regulation of Mu elements may be DNA modification. There is a correlation between modification of certain methylation-sensitive restriction sites within Mu elements and loss of Mutator activity in previously active lines (Chandler and Walbot, 1986; Bennetzen, 1987). However, in some cases, loss of Mutator activity is not accompanied by modification of Mu elements (Robertson, 1986; V. Chandler, unpublished data) indicating other mechanisms of regulation must exist.

Standard non-Mutator stocks of maize contain sequences with homology to Mu elements (Chandler et al., 1986). As shown in Figure 1, a probe for the internal region of Mul hybridizes to 2 fragments in the non-Mutator plants,

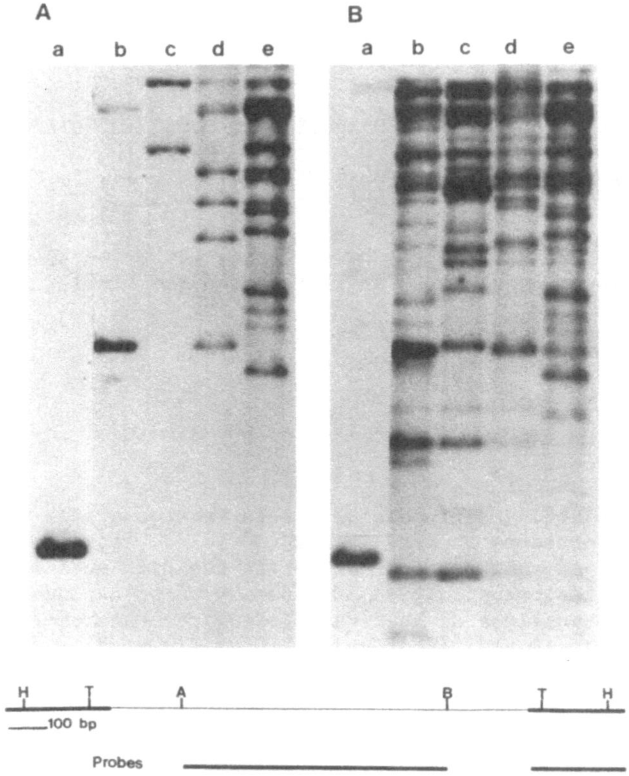

Figure 1. Presence of Mu-homologous sequences in non-Mutator maize lines. A. EcoRI digest of DNA from two non-Mutator and two Mutator maize plants, blotted onto Genetran and hybridized to the Mu1 internal probe. Lane a: three copy reconstruction with Mu1 plasmid DNA. b,c: non-Mutator plants. d,e: Two individuals derived by outcrossing a Mutator plant to the non-Mutator plant shown in lane b. The individual in lane d contained modified Mu elements and had no detectable Mutator activity. The individual in lane e was an active Mutator plant. B. Same blot hybridized to the Mu1 terminal probe. A partial restriction map of Mu1 and the DNA fragments used as probes are shown. The restriction sites indicated are H: HinfI, T: Tth111-I, A: AvaI, B: BstNI.

and 8-20 fragments in the Mutator lines. In contrast, a probe specific for the Mu termini hybridizes to a large number of fragments in both Mutator and non-Mutator lines. Most of the restriction fragments that hybridize to internal Mu probes also hybridize to the terminal repeats, but there are many additional fragments homologous to the termini only. We have cloned and characterized several types of Mu-homologous elements from non-Mutator lines. Each of these has unique features that may contribute to an understanding of the evolution and biology of the Robertson's Mutator system. In this paper, we describe their structure, hypothesize on their relationship to active Mu elements, and discuss the basis of their apparent inactivity in the genome.

SOME MAIZE LINES CONTAIN AN INTACT MU1 ELEMENT WHICH MAY BE INACTIVE DUE TO DNA MODIFICATION.

An intact Mu1-like element, termed MuE1, was cloned from the inbred line B37. The sequence of MuE1 (Chandler, Talbert and Raymond, in prep.) is virtually identical to that of Mu1 (Barker et al., 1984). Only a single nucleotide difference was detected and this base substitution is silent in regard to altering any protein which may be encoded by the major Mu1 open reading frames. MuE1 is not found in the same genomic location in other maize lines, and it is flanked by 9 bp direct repeats. Thus, it appears that MuE1 transposed into its current genomic location in B37.

The comparison of MuE1 and Mu1 suggests that they should be equally capable of transposition since their termini are identical and all open reading frames are conserved. However, MuE1 appears to be inactive in B37 and other inbreds since these lines show no increase in number of Mu1-like elements and have a normal spontaneous mutation rate. One reason for this may be DNA modification. In genomic DNA of non-Mutator plants the HinfI sites in the MuE1 termini are modified such that they are inaccessible to digestion (Chandler et al., 1986). This modification is probably 5-methyldeoxycytosine and is similar to that observed for Mu elements in Mutator lines that have lost activity (Chandler and Walbot, 1986). Given the strong correlation with DNA modification and lack of activity in Mutator stocks, a necessary prerequisite for the activation of MuE1 may be that it loses its DNA modification.

To test whether the modification of the HinfI sites in MuE1 is stable in the presence of Mutator activity, the MuE1 element was introduced into an active Mutator background by genetic crosses. DNA was isolated from F1 and F2 individuals and the accessibility of the MuE1 HinfI sites was examined using Southern blots. In active Mutator plants, the HinfI sites in the MuE1 element are accessible to digestion, demonstrating that the modification state of the MuE1 element can change depending on the genetic background of the stock (Chandler, Talbert, and Raymond, in prep.). Furthermore, in Mutator stocks that lose activity in subsequent generations, MuE1 again becomes modified as do all the Mu elements in the plant. Thus, the modification state of MuE1 reflects that of all the Mu elements in a Mutator line, suggesting that MuE1 is capable of responding to factors in a manner similar to other Mu elements.

ALL MAIZE LINES CONTAIN A SEQUENCE WITH HOMOLOGY TO THE INTERNAL REGION OF MU1.7.

A sequence, termed MuE2, that is homologous to the internal regions of Mu1 and Mu1.7 was cloned from a complex maize hybrid and sequenced (Talbert and Chandler, in press). Subsequent genomic restriction mapping indicates that it derived from the inbred line W23. Figure 2 shows a comparison of the structure of MuE2 with the transposable element Mu1.7. MuE2 does not have the structure of a transposable element in that it lacks Mu termini. MuE2 also has a large deletion (401 bp) relative to the internal region of Mu1.7, but the remaining 893 bp is 97% conserved (Talbert and Chandler, in prep.)

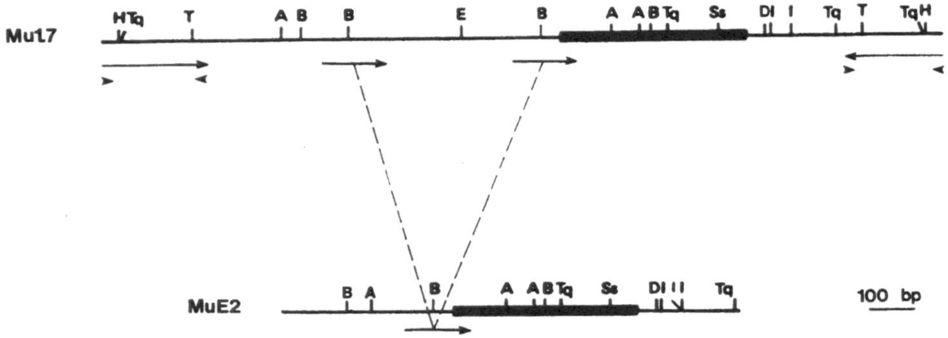

Figure 2. Comparison of MuE2 with the transposable element Mu1.7. The thickened bar represents Mu1.7-specific sequence relative to Mu1. The homology between MuE2 and Mu1.7 ceases 1 bp from the Mu termini. Within the region of shared homology, not all restriction sites are conserved due to slight sequence differences.

Explanations for the structure of MuE2 are suggested by comparison with the Mu1.7 element. One major structural difference is the 401 bp of DNA missing in MuE2 relative to Mu1.7. Mu1.7 has internal direct repeats of 138 bp each with 263 bp between them as diagrammed in Figure 2. The 401 bp of DNA missing in MuE2 corresponds to a deletion of one copy of the direct repeat and the intervening 263 bp sequence. Such a deletion could result from homologous recombination between the direct repeats. The second structural difference is the lack of Mu termini in MuE2. This may relate to the fact that the terminal inverted repeats of Mu1.7 terminate in small (8 bp) inverted repeats (Figure 2) which may allow for the transposition of the termini alone as observed for IS sequences in bacteria (Kleckner, 1983).

This is consistent with our sequencing data in that the homology between MuE2 and Mu1.7 ceases 1 bp from the most internal member of these small repeats on each side of the element. The existence of elements that contain Mu termini but no Mu1 or Mu1.7 internal sequences (see below) is also consistent with the idea that Mu termini may transpose independently.

Independent transposition of Mu termini suggests two hypotheses for the relationship of MuE2 to Mu1.7. One, MuE2 may be a rearranged, deleted version of a Mu1.7 element. Alternatively, a MuE2-like sequence may have been encompassed by Mu termini to generate an ancestral Mu1.7-like element. Subsequent deletion of an internal segment of Mu1.7 might then generate a Mu1 element as proposed by Taylor and Walbot (1987). If MuE2 was formerly a transposable element, we would expect it to be found in different locations in different lines much as we observe for MuE1. However, MuE2 is flanked by the same DNA in all maize stocks and maps to the same chromosome arm in two inbred lines (Talbert and Chandler, in prep.), suggesting it is a stable component of the maize genome. Furthermore, MuE2 is highly conserved; it is in all maize stocks and related Zea species examined. The stability of MuE2 in the genome suggests that it may perform some function. Consistent with this idea, an RNA transcript that hybridizes both to MuE2 and its flanking DNA is found in all Mutator and non-Mutator plants examined (Talbert and Chandler, in prep.). Our current hypothesis is that MuE2 is a stable, functional region of the maize genome and is related to a sequence that was encompassed by Mu termini to generate an ancestral Mu1.7-like element.

ALL MAIZE LINES CONTAIN ELEMENTS HOMOLOGOUS TO MU TERMINI ONLY.

Two elements homologous to the Mu termini, but not homologous to internal regions of Mu1 and Mu1.7, were cloned from the inbred line B37. Their structures, as determined by DNA sequencing, are shown in Fig. 3. Both elements, designated Mu4 and Mu5, have Mu terminal inverted repeats that are approximately 90% homologous to those of Mu1 (Chandler, Faber, Mann and Sylvester, in prep.). Mu4 is 2.0 Kb in length and is flanked by 9 bp direct repeats while Mu5 is 1.3 Kb and is flanked by 7 bp direct repeats. The internal portions of Mu4 and Mu5 bear no sequence homology to Mu1, Mu1.7 or Mu3. An unusual structural characteristic of Mu4 and Mu5 is that the terminal inverted repeats extend internally from the Mu termini approximately 290 and 150 bp, respectively. However, aside from the Mu termini Mu4 and Mu5 are not homologous to each other.

The presence of multiple copies of Mu termini (see Fig. 1) raises the question of how many of the termini are represented by Mu4 and Mu5 elements. As shown in Figure 4, internal probes for Mu4 and Mu5 hybridize to 2-4 and 8-12 fragments, respectively, in both Mutator and non-Mutator lines. This demonstrates that only a subset of the Mu termini are represented by these elements, suggesting additional related elements remain to be characterized. In both cases, some but not all of the fragments homologous to the internal probes are also homologous to the Mu terminal probe (data not shown). This indicates that the internal regions of Mu4 and Mu5 may exist independently of Mu termini, much as we have observed for Mu1 and Mu1.7.

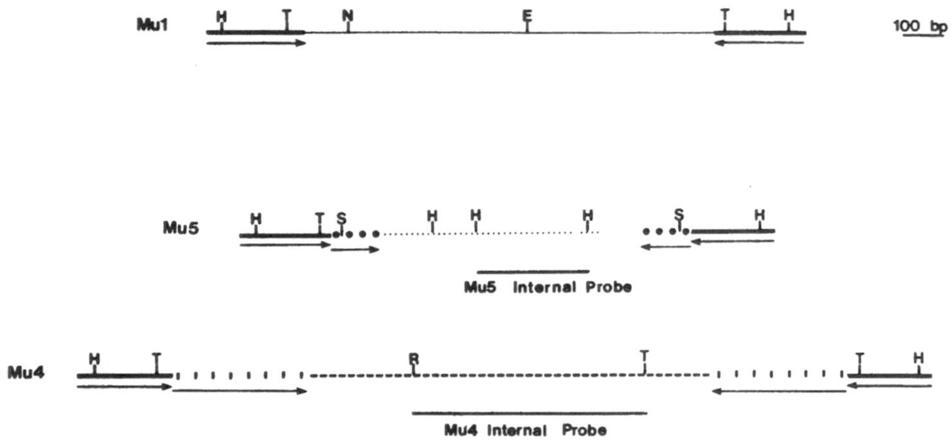

Figure 3. Comparison of Mu4 and Mu5 with the transposable element Mu1. The Mu terminal inverted repeats which are the only regions of homology are indicated by thickened bars. The restriction sites represented are H: HinfI, T: Tth111-I, N: NcoI, E: BstEII, S: SalI, R: EcoRI. There are additional HinfI sites in Mu4 which are not shown.

Our sequencing data suggest that Mu4 and Mu5 represent previously uncharacterized transposable elements in the Robertson's Mutator family. Several questions are yet to be answered regarding these elements. First, are they completely inactive in non-Mutator backgrounds, or might they contribute to the background spontaneous mutation rate due to a low level of transposition? Second, do these elements transpose in a Mutator background? Regarding the second question, an analogy might be drawn with the Ac-Ds transposable element family (reviewed by Courage et al., 1984). Ds elements are defined by their ability to transpose in the presence of an active Ac. This transposition is presumably dependent upon a transposase encoded by Ac. All characterized Ac and Ds elements share an 11 bp terminal inverted repeat sequence. However, the internal regions of Ds elements are quite variable. Some Ds elements appear to be deleted versions of Ac, while others share little homology with Ac except for the 11 bp terminal inverted repeats. However, all of these elements can respond to the Ac transposase. If a similar situation exists for the Mutator family, we might expect Mu4 and Mu5 to respond to the same transposase factors as do Mu1, Mu1.7, and Mu3. Genetic and molecular experiments are in progress to monitor transposition of Mu4 and Mu5 in a Mutator background.

CONCLUSIONS

A satisfying hypothesis for the origin of Robertson's Mutator activity is that it arose from endogenous components already existent in the maize genome. Activation might occur simply by the genetic construction of a stock which assembled all the neccessary components for Mu transposition into a single genome. An alternative, but not mutually exclusive, model is that all the components may exist in certain stocks and activation might occur as a result of a genomic shock which releases cryptic elements, possibly by demethylating them. Once activated, the elements might increase in number up to the level observed for Mu1 and Mu1.7 in Mutator stocks.

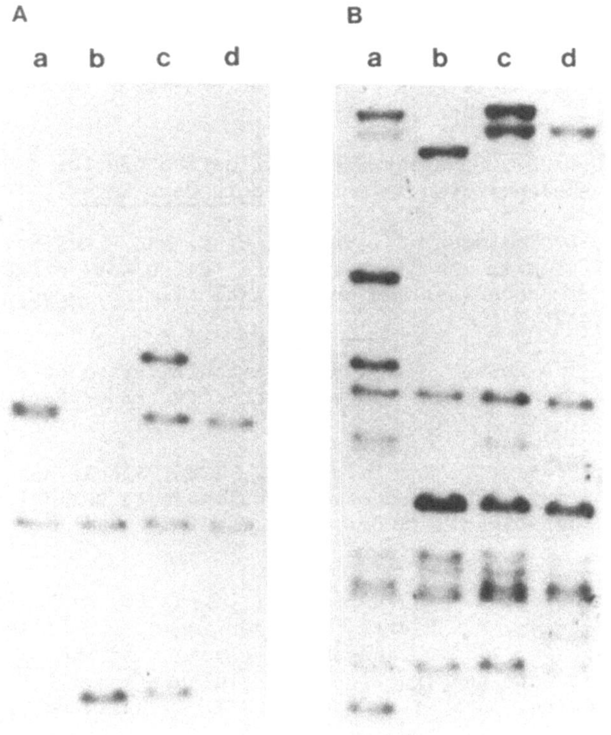

Figure 4. Number of Mu4 and Mu5 elements in maize lines. A. EcoRI/HindIII double digests of DNA from two non-Mutator (lanes a, b) and two Mutator (lanes c, d) maize plants, blotted onto Genetran and hybridized to the Mu4 internal probe (as shown in Figure 3). B. Same blot hybridized to the Mu5 internal probe (as shown in Figure 3).

ACKNOWLEDGEMENTS

We wish to thank Elizabeth Cooksey for her help in manuscript preparation. This research was supported by a National Science Foundation grant (DCE-8451656) and matching funds from Pioneer Hi-Bred International and CIBA-GEIGY.

REFERENCES

Barker, R. F., Thompson, D. V., Talbot, D. R., Swanson, J., and Bennetzen, J. L., 1984, Nucleotide sequence of the maize transposable element Mu1, Nucl. Acids Res., 12:5955.

Bennetzen, J., 1987, Covalent DNA modification and the regulation of Mutator element transposition in maize. Mol. Gen. Genet., in press.

Bennetzen, J. L., Swanson, J., Taylor, W. C. and Freeling, M., 1984, DNA insertion in the first intron of maize Adh1 effects message levels: cloning of progenitor and mutant Adh1 alleles, Proc. Natl. Acad. Sci. USA, 81:4125.

Chandler, V., and Walbot, V., 1986, DNA modification of a maize transposable element correlates with loss of activity, Proc. Natl. Acad. Sci. USA, 83:1767.

Chandler, V., Rivin, C., and Walbot, V., 1986, Stable non-Mutator stocks of maize have sequences homologous to the Mu1 transposable element, Genetics 114:1007.

Courage, U., Döring, H. -P., Kunze, R., Laird, A., Merkelbach, A., Müller-Neumann, M., Riegel, J., Starlinger, P., Tillman, E., Weck, E. and Yoder, J., 1984, Transposable elements Ac and Ds of Zea mays, in: "Molecular form and function of the plant genome," L. van Vloten-Doting, G. S. P. Groot, and T. C. Hall, eds., Plenum Press, New York.

Freeling, M., 1984, Plant transposable element and insertion sequences. Ann. Rev. Plant Physiol. 35:277.

Kleckner, N., 1983, Transposon Tn10, in: "Mobile Genetic Elements," J. A. Shapiro, ed., Academic Press, New York.

O'Reilly, C., Shepherd, N. S., Pereira, A., Schwarz-Sommer, Z., Bertram, I., Robertson, D. S., Peterson, P. A., and Saedler, H., 1985, Molecular cloning of the a1 locus of Zea mays using the transposable elements En and Mu1, EMBO J. 4:877.

Robertson, D. S., 1978, Characterization of a Mutator system in maize, Mutat. Res. 51:21.

Robertson, D. S., 1986, Genetic studies on the loss of Mu mutator activity in maize, Genetics 113:765.

Taylor, L. P., Chandler, V. L., and Walbot, V., 1986, Insertion of 1.4 Kb and 1.7 Kb *Mu* elements into the *bronze*-1 gene of *Zea mays* L., *Maydica* XXXI:31.

Taylor, L. P., and Walbot, V., 1987, Isolation and characterization of a 1.7 Kb transposable element from a Mutator line of maize, *Genetics*, in press.

Cin4, A RETROTRANSPOSON-LIKE ELEMENT IN *Zea màys*

Zsuzsanna Schwarz-Sommer, Lise Leclercq and Heinz Saedler

Max-Planck-Institut für Züchtungsforschung
5000 Köln 30 FRG

INTRODUCTION

DNA mediated transposition is a universal phenomenon in prokaryotes and in eukaryotes allowing rapid reorganisation of their genomes. In contrast, RNA-mediated mobility of genetic information seemed for a long time to be an attribute of higher eukaryotes since it was restricted mainly to mammals (Rogers,1985). The detection of reverse transcriptase activity or homology to reverse transcriptase-like proteins in so-called viral retrotransposons as *Ty* elements of yeast (Boeke et al.,1985), the *copia* element of *Drosophila* (Saigo et al.1984) and the *CaMV* virus in plants (Pfeiffer and Hohn, 1983) indicates that reverse transcriptase-governed transposition *via* an RNA intermediate may be a more general phenomenon. However, non-viral retrotransposons, like processed pseudogenes and other dispersed repetitive intronless sequences terminating in a polyA track have only been found until recently in mammals. One class of such sequences, termed LINEs (or L1 family), deserves special interest. These elements are highly reiterated in the genome of mammals (for review see Rogers,1985). DNA sequence analysis of L1 elements of human (Hattori et al.,1986), mouse (Loeb et al,1986), cat (Fanning and Singer,1987) and rabbit (Demers et al.,1986) indicates a general principle in the organisation of these sequences. The most striking feature of LINEs and related sequences is the conservation of a region within a long open reading frame (ORF) with homology to retroviral reverse transcriptases (Hattori et al.,1986). A similar class of elements has recently been identified in non-mammals, like the I-elements of *Drosophila* (Fawcett et al.,1986) and the Ingi-elements of *Trypanosoma* (Kimmel et al.,1987). L1, Ingi and I-elements show no other sequence homologies than their similarity in overall organisation and the presence of the highly conserved region within a long ORF mentioned above. These data suggest that such elements are functionally related and that they originate from a common progenitor during evolution of at least the animal kingdom.

Amplification and propagation of genetic information by reverse transcription does not seem to occur in lower eukaryotes. There is no evidence for genome-derived reverse transcription in plants and there is only circumctantial evidence for the occurance of pseudogenes in yeast (Fink,1987). In our present report we shall provide evidence that the Cin4 element in maize is homologous to L1, Ingi and I elements in all characteristic features. This homology defines Cin4 as the first non-viral retrotransposon of plants.

RESULTS AND DISCUSSION

Detection of Cin4 as an insert within the *A1* transcription unit

As reported by us previously, the wildtype *A1* gene of maize has two allelic forms termed type 1 and type 2 (Schwarz-Sommer et al.,1987). The type 2 allele differs from type 1 by insertion of the 1.1kb long Cin4-1 element into the 3' coding untranslated region of the type 1 allele (Figure 1). Termination of transcription in the type 2 allele occurs within the Cin4-1 sequence. The two *A1* alleles, therefore, contain unrelated sequences within the last 200bp of their 3' end. Due to a frameshift mutation occuring upstream of the insertion site of Cin4-1, the translation product of the type 2 allele contains 14 additional amino acids as compared to the product of type 1.

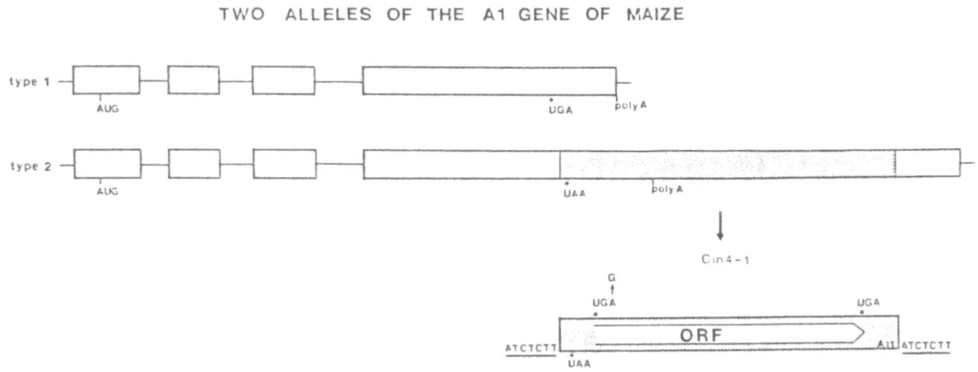

Figure 1 Structure of the type 1 and type 2 wild type alleles of the *A1* gene of *Zea mays*

The boxes represent exons and introns are shown by horizontal lines. The position of translation signals and the polyadenylation site of the *A1* gene is indicated below the scheme representing the two alleles. The 1.1kb long Cin4-1 element is shown by a shaded box. The structure of Cin4-1 is depicted at the lower part of the Figure. The flanking target site duplication is underlined. Translation start and stop for the ORF within Cin4 is shown above the structure.

An 0.6kb long segment of Cin4-1 is conserved in 80% of its 50-100 copies in the maize genome

Southern blot analysis was carried out to estimate the copy number of Cin4-1 in the maize genome. The copy-number of Cin4-1-related sequences is about 50-100 per diploid maize genome. This rough estimation is based on the number of differently sized fragments lighting up with the central portion of Cin4-1 as a probe (represented by the Hinf fragment in Figure 2) if genomic maize DNA is digested with an enzyme not restricting Cin4-1 (BamHI in right panel of Figure 2). The number of multiple copies within a single-sized band cannot be estimated with accuracy. We also have evidence from the analysis of genomic clones that duplication of Cin4-containing regions may occur involving flanking regions. As also documented in Figure 2, the reiteration pattern of BamHI fragments carrying Cin4-1-homologous sequences varies comparing different maize lines. This can indicate frequent changes in the chromosomal location of the element. Whether these are due to transposition of Cin4 or reflect restriction fragment length polymorphism of BamHI sites cannot be deduced from such type of experiments.

The most striking result from Southern blot experiments is revealed by the fact that an 0.6kb long portion of Cin4, produced by restriction with Hinf, is conserved in 80% of the genomic copies (Figure 2, left panel). This portion of Cin4-1 corresponds to the long ORF within the element (Figure 2, discussed in the following section). The data can be interpreted as if the ORF of Cin4-1 had some functional relevance for the element and not only occurs by chance. In addition, Cin4-homologous sequences only occur in plants related to the genus *Zea* and are not present even in closely related species of other genera (Figure 2). The only exception is the genus of *Tripsacum* in which the copy number of Cin4 is low, but homologous sequences can weakly be detected. The relevance of these observations for the origin and evolution of Cin4 is not clear yet.

Structural features of the 1.1kb long Cin4-1 element

DNA sequence analysis of Cin4-1 revealed several intriguing features: the insert is flanked in the type 2 *A1* allele by direct duplication of 7bp occuring only once in the type 1 allele (Figure 1). This indicates that the integration of Cin4-1 was preceeded by generation of a 7bp long staggered nick within the type 1 allele. Integration by generation of staggered nicks is characteristic for transposons which, in addition often show other common structural features revealed by terminal repeats (for review see Nevers et al.,1986, Schwarz-Sommer,1987). Cin4-1, however, does not possess structured termini neither in direct, nor in inverted form. We found instead 11 A residues at its end - a feature common to pseudogenes or non-viral retroposons (see Rogers,1985) which integrate *via* reverse transcription of an RNA intermediate. In fact, the polyA track of Cin4-1 is preceeded in a distance of 38bp by a potential polyadenylation signal AATAT (Proudfoot and Brownley,1976) suggesting its origin from a polyadenylated transcript.

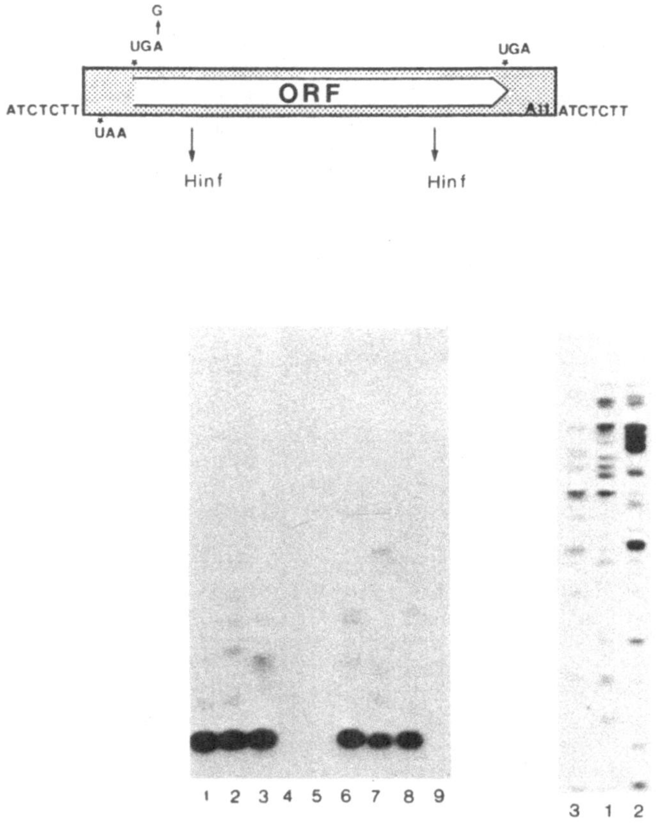

1 *Zea mays ssp. mays LC*
2 *Zea mays ssp. mays 5719*
3 *Zea mays ssp. parviglumis*
4 *Tripsacum dactyloides*
5 *Chionachne sp.*
6 *Zea mays ssp. mays NalTel*
7 *Zea diploperennis*
8 *Zea luxurians*
9 *Trioblachne sp.*

Figure 2 Conservation of Cin4-related sequences in the maize genome

Southern blot analysis was carried out with HinfI (left panel) and BamHI (right panel) digested genomic DNA from maize and related species. The slots contain 7ug of DNA. The internal HinfI fragment of Cin4 (as indicated below the scheme at the top) was used as a probe. All methods used are described elsewhere (Schwarz-Sommer et al.,1987).

These observations and also the fact that Cin4-1 contains an 0.8kb long ORF (Figure 1) prompted us to ask, whether Cin4-1 represents a family of elements in plants which move after reverse transcription of an RNA intermediate.

Indications for the RNA-mediated transposition of Cin4

DNA-mediated and RNA-mediated transposition of genetic information is readily distinguishable by the following two criteria: First, transposons have well defined termini, which do not undergo changes during transposition (except aberrant events). Pseudogenes and non-viral retrotransposons may be 5' truncated and also the length of the polyA track defining their 3' end may vary in length. Second, every class of transposon generates a characteristic and class-specific size of target site duplication upon integration, governed by element-specific "transposases". Retrotransposons, except viral ones possessing viral "integrases", integrate into preformed staggered nicks. The size of the flanking duplication at new locations are therefore variable even within a single class of elements.

We have cloned and analysed 30 genomic copies of Cin4 to be able to define it as a transposon or as a retrotransposon by the above mentioned criteria. The results can be briefly summarized as follows. Using a probe derived from the 5' portion of Cin4-1 one can identify other elements with further 5' extending homology. These elements are identical in structure to each other toward their 3' end as indicated by restriction (and also by DNA sequence) analysis. Cin4-1 therefore is a 5' truncated copy of a larger element. By deriving new "5-prime" probes from these longer copies still longer elements can be identified. This method does not allow to identify the longest existing copy if it is present in low copy number in the genome. It allows, however to state, that the majority of Cin4 copies occur in a 5' truncated form.

Indicative for the movement of large Cin4 copies to their location by transposition is the presence of direct duplications flanking the elements. The 5' end of the element in this case is defined by its homology to a still longer copy of Cin4. The 3' end of the element is defined by the polyA track preceeded by sequences identical in all copies. To determine the size of the target duplication and the size of the 3' terminal polyA track 5 independent copies of Cin4 were sequenced. As outlined above, the longest of these copies, the 3430bp long Cin4-198 element flanked by a duplication,

Table 1 Analysis of truncated forms of Cin4

Clone #	total length (bp)	length of the 3'polyA track(bp)	size of the flanking target duplication (bp)
Cin4-1	1081	11	7
Cin4-162	1239	8	6
Cin4-232	1506	6	16
Cin4-151	2947	8	6
Cin4-198	3423	7	7

represents the second longest copy of Cin4 elements. The 5' end of the longest copy analysed by us until now is not yet determined. The relevant data obtained are compiled in Table 1.

According to the criteria mentioned above these data indicate that Cin4 is an element which moves after reverse transcription because:

- independent copies of the element occur in 5' truncated forms
- the size of the 3' terminal polyA track varies comparing different truncated copies
- the size of the duplication flanking the element is variable among independent copies

Cin4 copies in the genome are pseudogenes with the potential to encode reverse transcriptase

The question, whether truncated copies of Cin4 represent processed pseudogens generated by reverse transcription of a mosaic gene and/or whether Cin4 represents a novel class of nonviral retrotransposons cannot yet be answered definitely and must await identification of the fullsize (non-truncated) Cin4 copy. Nevertheless, analysis of numerous truncated copies allows some insight into the nature of the Cin4 elements. Detailed description of this information is subject of a forthcoming report. Here we only want to summarize briefly our results obtained by sequence analysis of 5 truncated Cin4 copies.

Comparison of sequences determined within homologous regions of independent Cin4 copies indicates a high degree of conservation at the nucleic acid level. The common regions sequenced sofar are over 90% homologous to each other and differ only by base substitutions. The only exception is Cin4-232 which in addition contains insertions arising by duplication of Cin4 sequences. Restriction analysis of other truncated copies indicates that larger inversions and deletions may also occur. Furthermore, sequence analysis of Cin4 elements revealed that the translation stop UGA in Cin4-1 is due to a base substitution not present in the other copies. All other copies thus contain a further upstream extending ORF the total length of which in Cin4-198 (the longest copy analysed sofar) is 3198bp. The majority of the base substitutions mentioned above do not affect the ORF but may generate conservative or non-conservative amino acid exchanges. We therefore argue that truncated copies of Cin4 represent pseudogenes in analogy to the nature of truncated copies of L1 elements in mammals (Skowronski and Singer,1986) and the Ingi elements in *Trypanosoma* (Kimmel et al.,1987).

As mentioned in the Introduction, several non-viral retroposons detected in mammals, *Trypanosoma* and *Drosophila* reveal striking homology in conservation of a region resembling the conserved amino acid sequence within viral reverse transcriptases (compiled by Toh et al.,1983). This conserved region is also present within the long ORF of Cin4-198 (data not shown). The spacing between the homology blocks among non-viral retrotransposons is different from that of authentic reverse transcriptases (Fawcett et al.,1987) but show distinct homology to each other. Cin4 belongs to the non-viral group with respect to the spacing, but also with respect to the conservation of typical regions within these homology blocks.

The non-viral retrotransposons also contain conserved putative DNA-binding domains, located downstream of the region with homology to reverse transcriptases (Fanning and Singer,1987). This location is unusual, since DNA-binding "fingers" identified in retroviruses are present in the "gag" region, e.g. upstream of the location of the reverse transcriptase coding "pol" region (Covey,1986). This putative DNA-binding domain is present in Cin4-198 in a location similar to that in non-viral retrotransposons (Schwarz-Sommer et al., manuscript in preparation).

CONCLUSION

The presence, location and characteristics of two conserved domains with homology to reverse transcriptase and to DNA-binding proteins also conserved in viral and non-viral retrotransposons indicates that the Cin4 element of *Zea mays* is a non-viral retrotransposon. Cin4 copies occur in a 5'truncated form and perhaps represent pseudogenes derived by reverse transcription of a mRNA transcribed from an active copy. Detection and analysis of the structure and the function of this functional copy will shed light on the biological function of the element.

REFERENCES

Boeke, J.D., Garfinkel, D.J., Styles, C.A. and Fink, G.R., 1985, Ty elements transpose through an RNA intermediate, Cell, 40, 491-500

Covey, S., 1986, Amino acid sequence homology in *gag* region of reverse transcribing elements and the coat protein gene of cauliflower mosaic virus, Nucleic Acids Res., 14, 623-633

Fanning, T. and Singer, M., 1987, The LINE-1 DNA sequences in four mammalian orders predict proteins that conserve homologies to retrovirus proteins, Nucleic Acids Res., 15, 2251-2260

Demers, G.W., Brech, K. and Hardison, R.C., 1986, Long interspersed L1 repeats in rabbit DNA are homologous to L1 repeats of mouse and human in an open reading frame, Mol. Biol. Evol., 3, 179-

Fawcett, D.H., Lister, C.K., Kellett, E. and Finnegan, D.J., 1986, Transposable elements controlling I-R hybrid dysgenesis in *D. melanogaster* are similar to mammalian LINEs, Cell, 47, 1007-1015

Fink, G., 1987, Pseudogenes in yeast?, Cell, 49, 5-6

Hattori, M., Kuhara, S., Takenaka, O. and Sakaki, Y., 1986, L1 family of repetitive DNA sequences in primates may be derived from a sequence encoding a reverse transcriptase-related protein, Nature, 321, 625-628

Kimmel, B.E., Ole-Moiyoi, O.K. and Young, J.R., 1987, Ingi, a 5.2kb dispersed sequence element from *Trypanosoma brucei* that carries half of a smaller mobile element at either end and has homology with mammalian LINEs, Mol. Cell. Biol., 7, 1465-1475

Loeb, D.D., Padgett, R.W., Hardies, S.C. Shehee, W.R., Comer, M.B. Edgell, M.B. and Hutchison, C.A., 1986, The sequence of a large L1Md element reveals a tandemly repeated 5' end and several features found in retro-transposons, Mol. Cell. Biol., 6, 168-182

Nevers, P., Shepherd, N.S. and Saedler, H., 1986, Plant transposable elements, Adv. Bot. Res., 12, 103-203

Pfeiffer, P. and Hohn, T., 1983, Involvement of reverse transcription in the replication of cauliflower mosaic virus: a detailed model and test of some aspects, Cell, 33, 781-789

Rogers, J.H., 1985, The origin and evolution of retroposons, Int. Rev. Cytol., 93, 188-279

Saigo, K., Kigimiya, W., Matsuo, A., Inouye, S. and Yoshioka, K., 1984, Identification of the coding seuqence for a reverse transcriptase-like enzyme in a transposable genetic element in *Drosophila melanogaster*, Nature, 312, 659-661

Schwarz-Sommer, Zs. 1987, The significance of plant transposable elements in biological processes, in: "Results and problems in cell differentiation, 14, Structure and function of ekaryotic chromosomes" W. Hennig ed., Springer Verlag Berlin-Heidelberg, in press

Schwarz-Sommer, Zs., Shepherd, N., Tacke, E., Gierl, A., Rohde, W.,Leclercq, L., Mattes, M., Berndtgen, R., Peterson, P.A. and Saedler, H., 1987, Influence of transposable elements on the structure and function of the *A1* gene of *Zea mays*, EMBO J., 287-294

Skowronski, J. and Singer, M.F., 1986, The abundant LINE-1 family of repeated DNA sequences in mammals: Genes and pseudogenes, Cold Spring Harbor Symposia on Quant.Biol., 51, 457-464

Toh, H., Hayashida, H. and Miyata, T., 1983, Sequence homology between retroviral reverse transcriptase and putative polymerases of hepatitis B virus and cauliflower mosaic virus, Nature, 305, 827-829

CYBRIDS IN NICOTIANA, SOLANUM AND CITRUS: ISOLATION AND CHARACTERIZATION OF PLASTOME MUTANTS: PRE-FUSION TREATMENTS, SELECTION AND ANALYSIS OF CYBRIDS

Esra Galun[1], Dvora Aviv[1], Adina Breiman[2], Hillel Fromm[1], Avihai Perl[1], and Aliza Vardi[3]
[1]Department of Plant Genetics, Weizmann Institute of Science, Rehovot
[2]G.S.Wise Faculty of Life Science, Tel Aviv University, Tel-Aviv
[3]Division of Horticulture, ARO, Volcani Center, Bet-Dagan, Israel

INTRODUCTION

The fusion of two plant protoplasts, in one of which nuclear-division was arrested previous to fusion by X- or gamma irradiation, will result in a fusion product having one functional nucleus and a mix of organelles from the two fusion partners. Cell division of hetero-fused protoplasts results commonly in sorting out of organelles, leading ultimately to cybrid plants having novel organelle compositions. The protoplast in which nuclear division was arrested serves as donor of chloroplasts and mitochondria. We therefore termed this fusion method "donor-recipient protoplast-fusion". The utilization of this method was reported in several of our previous publications (e.g. Zelcer et al., 1978; Aviv and Galun, 1980; Galun et al., 1982; Fluhr et al., 1983; Aviv et al., 1984; Aviv and Galun, 1985; Aviv et al., 1986; Aviv and Galun, 1986; Vardi et al., 1987; see review: Galun and Aviv, 1986). This or similar methods were also utilized by others to produce cybrid plants (e.g. Menczel et al., 1982; Menczel et al., 1983; see review: Galun and Aviv, 1986).

In this presentation we shall briefly review our collection of plastome mutants in Nicotiana, indicating their usefulness in the establishment and characterization of cybrids and comment the employment of these mutants in the molecular study of the chloroplast. We shall also focus on two specific aspects in Nicotiana cybridization: the possibility to eliminate and re-establish cytoplasmic male sterility and the impact of pre-fusion treatments by iodoacetate and Rhodamine-6-G on the organelle constitutions of cybrids. Finally, examples will be provided for the utilization of the donor-recipient protoplast-fusion method in cybridization of potato and Citrus.

PLASTOME MUTANTS IN NICOTIANA

Using procedures described previously (Fluhr et al., 1985), which were based on mutagenesis of seeds by N-nitroso-N-methylurea, we assembled 71 plastome (chloroplast-genome) mutants and alloplasmic lines in Nicotiana. These may be grouped in the following categories:

- 26 single mutants with antibiotics resistance
- 9 single mutants with pigmentation deficiency.
- 6 double mutants with antibiotics resistance and pigmentation deficiency
- 13 double mutants with resistances to two antibiotics
- 5 triple mutants with resistances to three antibiotics
- 12 alloplasmic lines: mutated or wild type plastomes were transferred to alien species

Plastome mutants served as markers and for the selection of cybrids in our previous cybridization studies (Fluhr et al., 1983; Fluhr et al., 1984; Aviv and Galun, 1985; Aviv et al., 1986). Pigmentation deficient mutants are especially useful in recipients fused with donors having normal chloroplasts: all normal-green fusion derivatives should be cybrid plants (e.g. Aviv et al., 1984). Maliga and his associates at the Hungarian Academy of Sciences isolated antibiotic-resistant Nicotiana mutants, ascertained their maternal inheritance and used them in several of their protoplast-fusion studies (e.g. Maliga et al., 1975; Menczel et al., 1982; Cseplö and Maliga, 1982 and see review: Fluhr and Cseplö, 1986).

The pigmentation-deficient mutant Xa pig^{-1} (=XV1) served in our Department to study the assembly of ribulose-1,5-bisphosphate carboxylase subunits (Fluhr, 1985 and poster by Avni et al. in this meeting). Several streptomycin and spectinomycin resistant mutants were employed (H.Fromm,M. Edelman and E. Galun - unpublished) to locate change in base sequences of the ctDNA which are correlated with these resistances. In the search for possible base sequence changes correlated with streptomycin resistance of the mutants 92 str^R-6 and 92 str^R-7 we were led into a different research line: the structure of rps 12 which codes for one of the small subunit's ribosome protein. A gene with interupted exons was revealed and evidence for trans-splicing of the respective mRNA was provided (Fromm et al., 1986; Koller et al., 1987). The utilization of several other plastome mutants in donor-recipient protoplast fusions will be described below.

CYTOPLASMIC MALE STERILITY IN NICOTIANA

In previous studies we reported that the fusion mediated transfer of cytoplasmic male sterility (CMS) from a CMS donor to N.sylvestris is correlated with the transfer of the chondriome (the mitochondrial genome) or parts of this genome (Zelcer et al., 1978; Galun et al., 1982) and that male fertility of a CMS recipient can be restored by fusion with protoplasts of a fertile donor (Aviv and Galun, 1980). Recently we reported that the fusion of protoplasts from two different CMS cybrids may result in fertile plants (Aviv and Galun, 1986). The restriction patterns of mtDNAs in the fertile plants is rearranged relative to both the fusion partners. Such rearrangements are common in protoplast-fusion progenies but it should be noted that rearrangements of mtDNA per se is not necessarily leading to changes in male-ferility.

We asked whether in such a cybrid, with restored fertilty, the nuclear genome underwent a change (in the course of fusion, culture and regeneration), causing a nuclear-coded restoration of fertility. To answer this question we used a fertility restored derivative of such a cybrid, G-2-1-1-k/8, as a recipient and fused it's protoplasts with protoplasts of a CMS donor 92 str^R-7, spe^R-5. The recipient protoplasts were pre-fusion treated with either 0.3 or 0.5mM iodoacetate (IA) and the donor protoplasts were X-irradiated. The resulting calli were then regenerated in the presence or absence of streptomycin. The cybrids resulting from the fusion of 0.5mM IA treated recipient protoplasts with donor protoplasts were analysed for CMS. All cybrids which regenerated in the presence of streptomycin were CMS (12 plants). Of the cybrids which regenerated in the absence of streptomycin, 10 were CMS and one was fertile. Similar results were obtained from the fusion in which the recipient protoplasts were treated with 0.3mM IA but under these conditions even upon regeneration in the presence of streptomycin only 4 out of 6 cybrids were CMS and only half of the cybrids regenerated in the absence of streptomycin were CMS. Obviously IA had a detrimental effect on the organelles of the recipient's protoplasts. Most cybrids, whether or not regenerated in the presence of streptomycin, were found to be streptomycin (and spectinomycin) resistant. The message was rather clear: the nucleus of G-2-1-1-k/8 did not prevent the expression of CMS caused by the plasmone (most probably the chondriome) of 92 str^{R}-7, spe^{R}-5 (which harbors CMS caused by its N.undulata-type chondriome). It should be noted that one of the two CMS fusion partners from which G-2-1-1-k/8 was derived had N.undulata cytoplasm. Further verification for the lack of restorer genes in G-2-1-1-k/8 will be provided by it's use as pollen parent in sexual crosses to its two CMS fusion progenitors.

THE EFFECT OF PREFUSION TREATMENTS ON ORGANELLE COMPOSITION OF CYBRIDS

As indicated above IA treatment of recipient protoplasts clearly reduced or even eliminated the recipient's chloroplasts from the resulting cybrids. Furthermore the transfer of CMS, previously shown to be chondriome-controlled in Nicotiana, from the donor into the cybrid, was also facilitated by IA treatments of the recipient protoplasts.

Rhodamine-6-G (R6G) is a lipophilic dye which binds to mammalian mitochondrial membranes and impairs their functionality; mammalian cells which were treated with R6G and were subsequently fused with other cells did not contribute their mitochondria to the fusion-product heterocaryons. We recently investigated the effect of R6G in plant protoplasts which were used as donor-recipient fusion partners (Aviv et al., 1986). We found that cybrid Nicotiana plants can be produced efficiently when X-irradiated donor protoplasts are fused with R6G treated recipient protoplasts. The latter protoplastsusually did not divide unless hetero-fused with the donor protoplasts while the donor protoplasts did not produce colonies because of nuclear-division arrest. Moreover when X-irradiated protoplasts of N.rustica were fused with R6G protoplasts of the double mutant SR1 pig^{-}35 (streptomycin resistant, albino) chondriome analyses of the cybrids indicated that the cybrids had only either of two compositions. They had either N.tabacum mtDNA (as the recipient) or N.rustica mtDNA. We did not reveal novel, rearranged, mtDNA in the cybrids. In a previous fusion between N.rustica protoplasts (donors) and N.sylvestris protoplasts (recipient) none of the resulting (41) cybrid plants had N.rustica mtDNA, all analysed cybrids had either "pure" N.sylvestris mtDNA or their mtDNA restriction patterns were almost identical to the pattern of N.sylvestris mtDNA. But, the above mentioned effect could be specific to fusion of N.rustica because in another donor-recipient fusion in which SR1 pig^{-}35 protopalsts were pre-fusion treated with R6G but fused with 92 lin^{R}-17 donor protoplasts (having N.undulata mtDNA) rearranged mtDNA was found in the cybrid derivatives. Only one cybrid, out of 12 which were anlysed, contained the donor's (i.e. N.undulata) mtDNA. The other 11 cybrids contained novel type mtDNA. The effect of R6G on the mitochondria of treated recipient protoplast could be dose-dependent. We therefore performed a rather extensive fusion experiment in which the donor protoplasts were X-irradiated as well as treated with various doses of R6G. The recipient protoplasts were prefusion treated with 0.5mM IA. 92 lin^{R}-17 and SR1 pig^{-35} served as donor and recipient, respectively. The results are summarized in Table 1 and an example of Southern blot hybridization of the cybrids' mtDNA is shown in Figure 1. Of the 28 cybrids in which the mtDNA was analyzed 24 had novel or

Table 1. MtDNA analysis by Southern blot hybridization of cybrid plants derived from the fusion of X-irradiated and Rhodamine-6-G (R6G) treated donor protoplasts of 92 lin^R-17 (CMS) with iodoacetate treated recipient protopalsts of SR1 pig$^-$-35.

Plant designation	R6G conc. (μg/ml)	Anther type	Hybridization pattern with radioactive probes								
			pm	*Syl*	*Sa-1*	*pm*	*Syl*	*Sa-2*	*pm*	*Syl*	*Sa-8*
			T	U	Novel	T	U	Novel	T	U	Novel
a22	10	fertile	+++	+	-	+++	-	-	+++	-	-
a23	10	sterile	+++	+	-	NT	NT	NT	+++	-	-
a31	10	mixed	+++	+	-	+++	-	-	+++	-	-
a52	10	mixed	+++	+	-	+++	-	-	+++	-	-
a53	10	sterile	+++	+	-	+++	-	-	-	+++	++
b13	15	sterile	+++	+	-	NT	NT	NT	NT	NT	NT
b16	15	sterile	+++	+	-	NT	NT	NT	NT	NT	NT
b21	15	fertile	+++	+	-	NT	NT	NT	NT	NT	NT
b25	15	mixed	+++	+	-	NT	NT	NT	NT	NT	NT
b42	15	fertile	+++	+	-	+++	-	-	+++	-	-
b44	15	fertile	+++	+	-	+++	-	-	+++	-	++
b45	15	sterile	+++	+	-	NT	NT	NT	NT	NT	NT
b11(G)	15	sterile	+++	-	-	+++	-	-	+++	-	-
b12(G)	15	mixed	+++	-	-	+++	-	-	+++	-	-
c13	20	sterile	+++	+	-	-	+++	-	+++	-	-
c22	20	fertile	+++	+	-	NT	NT	NT	+++	-	-
c25	20	sterile	+++	+	-	+++	+++	novel	+++	-	-
c26	20	mixed	+++	+	-	NT	NT	NT	+++	-	-
c34	20	fertile	+++	+	-	+++	-	-	+++	-	-
c65	20	Mixed	+++	+	-	+++	-	-	+++	-	-
c75	20	fertile	+++	+	-	+++	+++	-	+++	+	-
c92	20	fertile	-	+	++	+++	-	-	+++	-	-
d12	25	sterile	+++	+	-	NT	NT	NT	+++	-	-
d26	25	sterile	+++	+	-	+++	-	-	+++	-	-
d12(G)	25	sterile	+++	+	-	+++	-	-	+++	-	-
d13(G)	25	sterile	+++	+	-	+++	-	-	+++	-	-
d34(G)	25	fertile	+++	-	-	+++	-	-	+++	-	-
d51(G)	25	fertile	+++	-	-	+++	-	-	+++	-	-

T= N.tabacum - pattern; U=N.undulata (=line 92) pattern; (-)=no bands; (+) = faint bands; (+++) = strong bands; (NT)=not tested.

rearranged mtDNA as indicated by the mtDNA probes. Two cybrids (b 11 G and b 12 G, resulting from treatment of donor protoplasts with 15μg/ml R6G) had apparently N.tabacum (i.e. recipients) mtDNA; but these plants were CMS or mixed thus some (undetected by Southern blot hybridization) mtDNA from the donor was probably transferred to these cybrids. Only 2 cybrids (d 34 G and d 51 G) had apparently pure recipient's mtDNA and were fertile. These plants were indeed cybrids because they had green chloroplasts (from the donor). These two cybrids were derived from a fusion

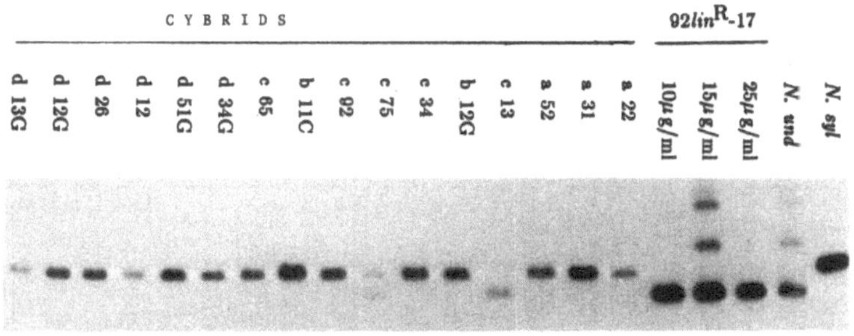

Figure 1. Southern blot hybridization of mtDNA from cybrid plants derived from the fusion of X-irradiated and R6G treated 92 lin^R-17 protoplasts (donor) with iodoacetate treated SR1 pig$^-$35 protoplasts (recipient). SalI digest; probe: pmSylSa-2

with donor protoplasts treated with the highest level of R6G (25µg/ml). The size of the sample is not sufficient to firmly establish a positive correlation of a high R6G dose with the elimination of treated-donors' mitochondria from the fusion derivatives. On the otherhand such a correlation seems plausible and further tests which are under way should provide an answer to the question whether or not R6G can be used to regulate the transfer of mtDNA in donor-recipient protoplast fusions.

CYBRIDIZATION WITH POTATO

Potato protoplasts served in the pioneering work of Melchers et al. (1978) to produce intergeneric somatic hybrids. Intergeneric and interspecific somatic hybrid plants were also obtained in subsequent studies (e.g. Shepard et al., 1983; Austin et al. 1986) and researchers at the Department of Plant Pathology of the University of Wisconsin, Madison - where A.J. Riker initiated the use of tissue culture for the study of plant pathology - produced tetraploid somatic hybrid potatoes by fusing protoplasts of two diploid potato clones (Austin et al., 1985).

We asked whether the donor-recipient protoplast-fusion procedure developed in Nicotiana, could be applied to produce potato cybrids. In one specific research line we attempted to transfer CMS into certain potato cultivars which are used as seed-parents for the production of true-seed F_1 hybrid potato cultivars. The international Potato Center (CIP) in Lima, Peru, and other potato breeding enterprises are pursuing the reproduction of potatoes through true-potato-seeds (TPS) for specific farming systems, especially in developing countries. There is a great advantage of having CMS in cultivars which are intended to serve as seed-parents for hybrid TPS. We were thus asked by CIP to transfer CMS into such seed-parent cultivars. Consequently we used protoplasts of a CMS potato clone, Y245.7, as donors and fused them to protoplasts of the cultivar Atzimba. The donor protoplasts were exposed, before fusion, to 10 kR gamma irradiation and the Atzimba protoplasts were prefusion treated with either R6G or IA. Table 2 provides data from two such fusion experiments. As flower pigmentation is nuclear coded fusion derivatives with white petals are assumed to have the recipients nuclear genotype. Fusion-derived plants with violet petals are probably escapes

Table 2. Analysis of potato plants derived from the fusion: Gamma-irradiated (10kr) Y245.7 protoplasts (CMS, donor) with pre-fusion treated (IA or R6G) protoplasts of the cultivar Atzimba

CALLUS SOURCE OR DESIGNATION	FLOWER PIGMENTATION	NUMBER OF ANALYSED PLANTS		MALFORMATION IN POLLEN OR ANTHERS
		WITH STERILE POLLEN	WITH FERTILE POLLEN	
Y245.7 (Donor)	VIOLET	MANY	NONE	[3]TETRADES
ATZIMBA (Recipient)	WHITE	NONE	MANY	
[1]I-B	WHITE	3	0	
I-C	WHITE	0	4	
I-E	WHITE	1	2	
I-F	WHITE	2	3	
I-G	VIOLET	1	1	TETRADES
I-H	VIOLET	5	0	TETRADES
I-P	WHITE	3	0	
I-S	WHITE	0	3	
I-X	WHITE	2	2	
I-8	WHITE	0	4	
[2]II-A	WHITE	1	3	
II-B	WHITE	1	1	
II-C	WHITE	2	0	SHRUNKEN POLLEN
II-D	WHITE	2	3	
II-E	WHITE	0	1	
II-F	WHITE	3	0	PETIOLATED ANTHERS
II-G	WHITE	1	4	
II-I	WHITE	0	2	
II-L	WHITE	0	2	
II-M	WHITE	1	3	SHRUNKEN POLLEN
II-N	WHITE	0	2	

[1]I-Prefusion treatment with R6G; [2]II- Prefusion treatment with IA. [3]-Pollen development arrested

Table 3. Update of donor-recipient protoplast-fusion experiments in which protoplasts of potato cultivars served as recipients (May, 1987).

FUSION PARTNERS		AIM OF FUSION	STAGE OF EXPERIMENT
DONOR[1]	RECIPIENT[2]		
Y245.7 (CMS)	S.tuberosum cv Atlantic	Transfer of CMS	Flowering cybrids
Y245.7 (CMS)	S.tuberosum cv Atzimba	Transfer of CMS	Flowering cybrids
S.commersonii	S.tuberosum cv Ticahousi	Nucl/Cyto compatibility	Calli-slagish regeneration
S.brevidens	S.tuberosum cv Desirée	Nucl/Cyto compatibility	Regenerated cybrids
S.demissum	S.tuberosum cv Desirée	Nucl/Cyto compatibility	Mature cybrid plants
S.chacoense	S.tuberosum cv Desirée	Nucl/Cyto compatibility	Regeneration of cybrids
S.berthaulti	S.tuberosum cv Desiree	Nucl/Cyto compatibility	Regenerated cybrids
S.etuberosum	S.tuberosum cv desirée	Nucl/Cyto compatibility	Regenerated cybrids

[1]10 KR gamma rays; [2] 0.25 mM iodoacetate.

from the gamma irradiation. The derivatives of I-G & I-H having violet petals also had tetrades rather than normal pollen which is the form of CMS manifestation in Y245.7. Among the fusion derivatives with white petals we found several plants which were male sterile. Most of these did produce pollen grains but the grain did not germinate; some plants produced only shrunken pollen while plants derived from one callus (II-F) conspicuously produced petiolated anthers which were almost devoid of pollen and none of the very few pollen grain could germinate. The organelle composition of the fusion-derived plants is presently being analysed. Since the two fusion partners differ in their isozyme patterns we shall also verify the recipients' nuclear genome in the male-sterile fusion-derivatives.

In addition to the fusion combination detailed above, we performed several other donor-recipient fusions as updated in Table 3.

CYBRIDIZATION IN CITRUS

Somatic hybridization and cybridization in Citrus and its relatives are of special interest for cultivar and rootstock breeding. Breeding and genetic studies of Citrus by sexual crosses is problematic because of several mechanisms such as nucellar embryogeny, heterozygosity, self-incompatibility and sterility. The regeneration of Citrus trees from isolated protoplasts was developed by us in previous studies (e.g. Vardi et al., 1975, 1982). These studies established the basis to obtain protoplast-fusion derived Citrus trees. Ohgawara et al. (1985) showed that somatic hybrid trees can be derived from the fusion of orange (Citrus sinensis) protoplasts with protoplasts of the Citrus relative Poncirus trifoliata.

Our intention was to obtain cybrids having the nuclear genome of a given Citrus cultivar but the organelles (either or both chloroplasts and mitochondria) of another Citrus type or of a related species. Consequently we used the donor-recipient protoplast-fusion method to obtain Citrus cybrids.

The following fusion combinations were performed:
(1) Poorman X Poncirus trifoliata (donor) with Villafranca lemon (recipient)
(2) Poorman X Poncirus trifoliata (donor) with Sour orange (recipient)
(3) Sour orange (donor) with Villafranca lemon (recipient)
(4) Microcitrus sp. (donor) with Rough lemon (recipient)
(5) Microcitrus sp. (donor) with Sour orange (recipient)

The donor protoplasts were exposed to 50-60 krad gamma irradiation and the recipient protoplasts were pre-fusion treated with iodoacetate (0.25mM).

The progeny of the three first fusion combinations were recently analyzed (Vardi et al., 1987). In these combinations the morphology of the fusion partners is rather different. All the fusion-derived trees had the morphology of the respective recipients (hence had probably the recipients nuclear genome). MtDNA analyses of the derivatives of these fusion combinations indicated that at least part of the chondriome from the donor was transferred to the fusion derivatives. It should be noted that the plastomes of the fusion partners in these three combinations do not differ (Green et al., 1986) thus transfer of chloroplasts could not be revealed by ctDNA analysis. In combinations (4) and (5) the plastomes of the donors do differ from the respective recipients. We are presently analysing the plastomes and the chondriomes of the fusion derivatives of these 2 combinations.

Acknowledgements. The research reported in this publication was supported by the following grantors: (1) The NCRD (Israel) and the GSF (F.R.Germany) Fund to D.A.; AID grant No. DFE 5542-G-SS-4030 to E.G and D.A.; (3) The L & F. Forschheimer Fund for Molecular Genetics, to E.G.; (4) The United States-Israel Binational Agricultural Research and Development Fund (BARD) to A.V. and E.G.. We are grateful to Pazia Arzee-Gonen, Shlomit Bleichman,

Ada Dantes and Ahuva Frydman-Shani for devoted technical assistance.

REFERENCES

Austin, S., Baer, M., Ehlenfeldt, M., Kazmierczak, P.J. and Helgeson, J.P., 1985, Intra-specific fusion in Solanum tuberosum, Theor. Appl., Genet. 71: 172.

Austin, S., Ehlenfeldt, M.K., Baer, M.A. and Helgeson, J.P., 1986, Somatic hybrids produced by protoplast-fusion between S.tuberosum and S.brevidens: phenotypic variation under field conditions, Theor.Appl. Genet. 71: 682.

Aviv, D., Arzee-Gonen, P., Bleichman,S. and Galun, E., 1984, Novel alloplasmic Nicotiana plants by "Donor-Recipient" protoplast fusion: cybrids having N.tabàcum or N. sylvestris nuclear genomes and either of both plastomes and chondriomes from alien species, Mol.Gen.Genet. 196: 244.

Aviv, D., Chen, R., and Galun, E., 1986, Does pretreatment by Rhodamine 6-G affect the mitochondrial composition of fusion-derived Nicotiana cybrids? Plant Cell Reports 5 : 227.

Aviv, D. and Galun, E., 1980, Restoration of fertility in cytoplasmic male sterile (CMS) Nicotiana sylvestris by fusion with x-irradiated N.tabacum protoplasts, Theor. Appl. Genet. 58: 121.

Aviv, D. and Galun, E., 1985, An in vitro procedure to assign pigment mutations in Nicotiana to either the chloroplast or the nucleus, J. Heredity, 76: 135.

Aviv, D., and Galun, E., 1986, Restoration of male fertile Nicotiana by fusion of protopalsts derived from two different cytoplasmic-male-sterile cybrids, Plant Mol. Biology 7: 411.

Cseplö, A., and Maliga, P., 1982, Lincomycin resistance, a new type of maternally inherited mutation in Nicotiana plumbaginifolia, Current Genetics 6: 105.

Fluhr, R., 1983, Molecular genetics of the Nicotiana tabacum chloroplast, Ph.D. Thesis, Feinberg Graduate School of the Weizmann Institute of Science, Rehovot, Israel.

Fluhr, R., Aviv, D., Edelman, M. and Galun, E., 1983, Cybrids containing mixed and sorted-out chloroplasts following interspecific somatic fusions in Nicotiana, Theor. Appl. Genet. 65: 289.

Fluhr, R., Aviv, D., Galun, E., and Edelman, M., 1985, Efficient induction and selection of chloroplast-coded antibiotic-resistant mutants in Nicotiana, Proc. Nat. Acad. sci. (Wash.) 82: 1485.

Fluhr, R., Aviv, D., Galun, E., and Edelman, M., 1984, Generation of heteroplastidic Nicotiana cybrids by protoplast fusion: analysis for plastid recombinant types, Theor. Appl. Genet. 67: 491.

Fluhr, R., and Cseplö, A., 1986, Induction and selection of chloroplast-coded mutations in Nicotiana. Methods in Enzymology 118: 611.

Fromm, H., Edelman, M., Koller, B., Goloubinoff, P., and Galun, E., 1986, The enigma of the gene coding for ribosomal protein S12 in the chloroplasts of Nicotiana, Nucleic Acids Research 14: 883.

Galun, E., Arzee-Gonen, P., Fluhr, R., Edelman, M., and Aviv, D., 1982, Cytoplasmic hybridization in Nicotiana mitochondrial analysis in progenies resulting from fusion between protoplasts having different organelle constitutions, Mol. Gen. Genet. 186: 50.

Galun, E., and Aviv, D., 1986, Organelle Transfer, Chapter in: Methods in Enzymology, 118: 595.

Green, R., Vardi, A., and Galun, E., 1986, The plastome of Citrus: Physical map, variation among Citrus cultivars and species and comparison with related genera, Theor. Appl. Genet. 72: 170.

Koller, B., Fromm, H., Galun, E., and Edelman, M., 1987, Evidence for in vivo trans splicing of pre mRNAs in tobacco chloroplasts. Cell, 48: 111.

Maliga, P., Breznovits, Sz., and Marton, L.,1975, Non Mendelian streptomycin resistant tobacco mutant with altered chloroplasts and mitochondria. Nature, 255:401.

Menczel, L., Galiba, G., Nagy, F., and Maliga, P., 1982, Effect of radiation dosage on efficiency of chloroplast transfer by protoplast fusion in Nicotiana, Genetics 100: 487.

Menczel, L., Nagy, F., Lazar, G., and Maliga,P., 1983, Transfer of cytoplasmic male sterility by selection for streptomycin resistance after protoplast fusion in Nicotiana, Mol. Gen. Genet. 189: 365.

Melchers, G., Sacristan, M.D., and Holder, A.A., 1978, Somatic hybrid plants of potato and tomato regenerated from fused protoplasts, Carlsberg Research Communication 43: 203.

Ohgawara, T., Kobayashi, S., Ohgawara, E., Uchimiya, H., and Ishii, S., 1985, Somatic hybrid plants obtained by protoplast fusion between Citrus sinensis and Poncirus trifoliata, Theor. Appl. Genet. 71: 1.

Shepard, J.F., Bidney, D., Barsby, T., and Kemble, R., 1983, Genetic transfer in plants through interspecific fusion. Science 219: 683.

Vardi, A., Breiman, A., and Galun, E., 1987, Citrus cybrids: production by donor-recipient protopalst-fusion and verification by mitochondiral-DNA restriction profiles, Theor. Appl. Biol. (in press).

Vardi, A., Spiegel-Roy, P., and Galun, E., 1975, Citrus cell culture: isolation of protoplasts, plating densities, effect of mutagens and regeneration of embryos, Plant Sci. Lett. 4: 231.

Vardi, A., Spiegel-Roy, P., and Galun, E., 1982, Plant regeneration from Citrus protoplasts: variability in methodological requirement among cultivars and species, Theor. Appl. Genet. 62: 171.

Zelcer, A., Aviv, D., and Galun, E., 1978, Interspecific transfer of cytoplasmic male sterility by fusion between protoplasts of normal Nicotiana sylvestris and x-ray irradiated protoplasts of male-sterile N. tabacum, Z. Pflanzenphysiol. 90: 397.

INTERSPECIFIC SOMATIC FUSIONS BETWEEN SOLANUM BREVIDENS AND S. TUBEROSUM

Sandra Austin and John P. Helgeson*

Department of Plant Pathology and *USDA, ARS
University of Wisconsin
Madison, Wisconsin

ABSTRACT

We have produced interspecific hybrids from several different combinations of Solanum brevidens and S. tuberosum by protoplast fusion. Plants were tested under field conditions and shown to have distinctive morphologies intermediate to those of parental lines. The degree of variation between individual hybrid combinations was assessed. Ploidy levels, for the most part, were those expected from the combination of parental lines used in the fusion. Hybrids expressed disease resistance genes of parental lines. Furthermore, some hybrids were fertile and were used in conventional crosses with potato cultivars. Chloroplast genomes of hybrids segregated between parental types and the assigned type was inherited maternally, as expected, when hybrids were used as females in sexual crosses. The potential use of somatic hybridization as part of a conventional crop improvement program is discussed.

INTRODUCTION

Wild Solanum species have proved valuable as sources of desirable agronomic features in the development of potato cultivars. Full utilization of related species can be limited, however, by sexual incompatibilities. An alternative hybridization pathway is the direct fusion and culture of isolated vegetative, somatic cells. Thus, agronomically important traits of related wild species can theoretically be incorporated into breeding programs for crop improvement by means of protoplast fusions between sexually incompatible species. Full expression of the traits in the original somatic hybrids is essential and more importantly the hybrids must be fertile to allow the subsequent sexual transfer of the desirable qualities.

Somatic hybrids theoretically have the nuclear and cytoplasmic genomes of both parents. In practice segregation of organelle genomes tends to occur and novel variation can arise due to gene recombination and somaclonal variation (Evans, 1983). The production of somatic hybrids between Solanum species using protoplast fusion has been undertaken by several researchers (e.g. Barsby et al., 1985; Binding et al., 1982; Butenko and Kuchko, 1979; Shepard et al., 1983; Gleddie et al., 1986; Butenko et al., 1982; Melchers et al., 1978; Puite et al.,

1986). Our work has centered primarily on an evaluation of protoplast fusion for the transfer of disease resistances from wild Solanum species into breeding lines of S. tuberosum Gp Tuberosum, the cultivated potato. Inter- and intraspecific hybrids have been produced (Austin et al., 1985a, 1985b, 1986) and hybrids have been shown to express resistance genes of fusion parents (Helgeson et al., 1986). Furthermore, some of the somatic hybrid plants possess sufficient fertility to be used in conventional sexual crosses (Ehlenfeldt and Helgeson, 1987). Thus we are able to examine the sexual transfer of germplasm derived from wild Solanum species. This report will summarize our protoplast fusion work and attempt to evaluate the contribution of somatic hybridization to crop improvement programs.

FUSION PROCEDURE AND SELECTION OF HYBRIDS

Production of somatic hybrid plants requires a reproducible method for isolation of viable protoplasts, an efficient fusion protocol and a method for selecting hybrid cells capable of morphogenetic regeneration. Protoplast isolation and regeneration of Solanum species is well documented (Shepard, 1980). We have developed protocols, modified from those used by Lam (1977), and Shepard and Totten (1977), that facilitate protoplast isolation and subsequent regeneration from several related Solanum species as well as cultivated potato lines (Haberlach et al., 1985). The cultural requirements for all the Solanum lines used in our fusion studies are investigated prior to hybridization attempts.

Cell fusions have been achieved using a variety of procedures and fusogenic chemicals. These include sodium nitrate (Carlson et al., 1972), high pH and calcium (Keller and Melchers, 1973), polyethylene glycol (Kao and Michayluk, 1974) and electrofusion (Zimmermann and Scheurich, 1981). The procedure found to be the most satisfactory in our studies is a modification of that used by Melchers et al., (1978) and is a combination of PEG agglutination coupled with high pH-calcium elution (Austin et al., 1985a). Fusion frequency using this method is typically 1-2%. Higher rates can be achieved by increasing PEG concentration and/or length of exposure to the fusigen, but in our experience this tends to give rise to multiple fusion events or to reduce protoplast survival.

After the cell fusion procedure a mixture of unfused protoplasts, homokaryons and heterokaryons is obtained. Since the desired hybrid cells are generally in a minority within the mixture, it is advantageous to select for them at the cellular level. The most widely used selection system is based on genetic complementation of nonallelic albino mutants (e.g. Melchers and Labib, 1974; Scheider, 1977; Evans et al., 1980). Other systems rely on complementation of auxotrophic mutants (Hein et al., 1983) or use selected cell lines with antibiotic resistance (Maliga et al., 1977). Such systems utilizing unique cell lines or mutants, though successful, are obviously of limited value for breeding purposes. Micro-isolation of hybrid cells based on morphological differences, (for example green mesophyll protoplasts fused with colorless cell suspension culture cells), is perhaps the most promising general approach. One major drawback, however, to the use of cultured cells is that the material may be aneuploid or polyploid (Bayliss, 1980).

Since the aim of our research was to determine if somatic hybridization could contribute to a potato improvement program it was essential to work with protoplasts derived from genetically stable, nonmutated Solanum species. Characteristics of Solanum materials used in protoplast fusions to generate hybrids in our research are listed in

Table 1. Characteristics of *Solanum* lines used in fusion studies

Species	Identifier	Code	Chromosome No.	Remarks
S. brevidens	PI 218228	6A	24	Res to PLRV,PVY and
	PI 245763	5B	24	Frost; non-tuber-bearing
S. tuberosum	PI 203900	R4	48	Late blight differential Res to RO, Sus to R4
	USW 13,114	5	24	cv Superior haploid
	77-1	13	24	GP Phureja/Stonotomum metribuzin sensitive
	77-19	19	24	Gp Phureja/Stonotomum metribuzin resistant
	USW 9545.99	3	24	Red tuber flesh
	USW. 9310.3	9	24	Yellow tuber flesh

Table 1. Care was taken to have a genetically uniform population of plants for the preparation of protoplasts. Clonal stock lines were obtained originally from a single meristem or a single seed. This was important both for cultural characteristics of a given line and also for the expression of disease resistance. For example lines of *S. brevidens* derived from individual seeds of the same accession behave differently in culture and have different responses to bacterial wilt infection, *Pseudomonas solanacearum* (Sequeira and Helgeson, unpublished data). We have found that the most efficient system was a selection of hybrid cells based on the cultural requirements of parental lines. This is feasible since the ability to regenerate from protoplasts behaves as a dominant trait in somatic hybrids (Cocking et al., 1977: Schenck and Robbelson, 1982) and hybrid cells generally have a more vigorous growth pattern than parental lines. An example of hybrid selection, in this case tetraploid *S. tuberosum* (PI 203900) fused with diploid *S. brevidens* (PI 218228), based on media selection is outlined in Table 2. Other hybrid combinations were derived in a similar manner. Selected shoot-producing callus thought to be derived from hybrid cells was transferred to proliferation medium to give multiple plantlet formation such that 10-50 plants could be derived from an individual callus (Austin et al., 1985a). This gave a large population of hybrids for assessment and selection. Wherever possible protoclones of parental lines were also obtained for comparative purposes.

Table 2. Example of selection using cultural requirements of parental lines.

R4 = *S. tuberosum* (PI 203900) 6A = *S. brevidens* (PI 218228)

Culture Step	Unfused Controls 1.6A	2.R4	Fusion Experiments 1.R4+R4 (self fusion)	2.6A+6A (self fusion)	3.R4+6A. (Mixed fusion)
Development of Microcalli	+	+	(1)–	+	+
Regeneration on J1	(2)–	+	–	–	+

(1)R4 protoplasts do not survive the fusion step

(2)6A protoplasts do not regenerate on J1

Only hybrid calli regenerate

FIELD TESTING OF SOMATIC HYBRIDS

Plants obtained from fusion experiments were assessed under field conditions and subjected to standard agricultural practices for growing commercial lines of potato. This is a rigorous assessment, but the value of material that is not fully competent under field conditions is limited. In all cases hybrid plants had a distinctive morphology intermediate to that of parental lines used in the fusion (Figure 1). Plants were assessed weekly for vigor and flowering. Overall, material was rated for forty-two phenotypic characteristics covering vegetative appearance, flowering incidence, flower form, tuber yield and tuber type. Clonal and protoclonal parental lines were rated similarly for comparative purposes. This data not only clearly showed the intermediate nature of hybrid morphology as compared to parental lines, but also gave a measure of phenotypic variation within groups of different hybrid combinations and protoclones of parental lines. The callus origin of plants was documented such that variation of plants generated from the same callus as compared to those from other calli could be assessed. Results from such a detailed analysis for one hybrid combination have been given elsewhere (Austin et al., 1986). A brief general summary of these results and those from four other fusion combinations is presented in Table 3.

Tubers obtained from individual hybrids were assessed in the field over two successive generations and compared with clonally maintained *in vitro* copies. Results showed that phenotypes were remarkably stable.

Several General Conclusions Can Be Drawn From This Work:

1. Somatic hybrid combinations had a distinct phenotype easily recognizable when plants were grown to maturity under field conditions. This could not be confused with somaclonal variation of parental lines. In most cases hybrids showed considerably increased vigor over parental lines.

2. Hybrid populations showed considerable variation, similar to that variation seen within protoplast derived plants of parental lines.

3. Severely abnormal plants for the most part either failed to root in culture or did not establish in the field. If they did survive under field conditions, their phenotype was so distinctive that these plants could easily be rogued out, as would be required in an applied breeding program. The same callus gave both phenotypically abnormal and normal plants.

4. Tuberization in hexaploid hybrids, (Fusion A, see Table 3), was variable but most plants did produce tubers. Tuberization in tetraploid interspecific hybrids, however, where one partner was non-tuber-bearing (fusions B-E) was poor and the "tubers" did not store well enough to be reliably used as propagules for further field testing (Figure 2). One contributing factor was the extremely late maturity of these plants. This trait appears to be derived from *S. brevidens* because in our growing conditions this species does not senesce and continues to grow until it is killed by frost.

5. Hybrids were stable through successive tuber generations.

Figure 1. Morphological characteristics of somatic hybrids as compared to parental lines. Top: Parental lines 77-1 (left), S. brevidens Pl 245763 (right) and somatic hybrid (center). Middle: Leaves of S. brevidens Pl 218228 (left) and S. tuberosum Pl 203900 (right) and somatic hybrid (center). Bottom: Flowers of S. brevidens Pl 218228 (left), cv. Superior haploid (right) and somatic hybrid (center).

Table 3. Summary of some field characteristics of somatic hybrids compared with protoclones and clonal copies of parental lines.

Type	No. of shoot-producing calli used	number planted in field	number established	number phenotypically normal	number flowered
Clonal R4	-	63	60(95%)	60(100%)	60(100%)
Protoclonal R4	8	89	72(81%)	70(97%)	18(25%)
Clonal 6A	-	45	43(95%)	43(100%)	43(100%)
Protoclonal 6A	6	26	24(92%)	(1)11(45%)	7(29%)
Fusion A (R4+6A)	21	174	156(90%)	138(88%)	123(79%)
Clonal 13	-	45	43(95%)	43(100%)	43(100%)
Protoclonal 13	7	48	32(66%)	27(84%)	21(65%)
Clonal 5B	-	45	44(97%)	44(100%)	44(100%)
Protoclonal 5B	12	81	75(92%)	(2)37(49%)	36(48%)
Fusion B (5B+13)	29	205	185(90%)	148(80%)	145(78%)
Clonal 19	-	45	42(93%)	42(100%)	41(97%)
Protoclonal 19	7	33	22(66%)	16(73%)	10(45%)
Fusion C (5B+19)	20	127	96(75%)	26(27%)	18(18%)
Fusion D (6A+19)	4	43	30(70%)	11(36%)	11(36%)
Clonal 5	-	27	27(100%)	27(100%)	27(100%)
Fusion E (6A+5)	3	37	37(100%)	35(94%)	35(94%)

(1)11 diploid, 10 tetraploid, 3 hexaploid, 0 grossly abnormal
(2)37 diploid, 33 tetraploid, 5 grossly abnormal

The range of somaclonal variation and the production of aberrant plants within hybrids groups was not too surprising and has been documented by several other researchers (see review by Evans, 1983). It is noteworthy, however, that phenotypically normal vigorous hybrids were found from each hybrid combination field tested. Combinations A,B and E were particularly successful.

CYTOLOGY AND FERTILITY

Interspecific hybrids are useful for plant breeding only if they possess fertility. Hybrids obtained from fusion combinations A through D shed viable pollen; plants obtained from fusion E however, though vigorous and free flowering, were male sterile. Ploidy levels were obtained from anther squashes of floral buds whenever possible. In general hybrid plants had ploidy levels which were the sum of parental lines used in the fusion. Plants which were grossly abnormal vegetatively were aneuploid and tended to have much higher chromosome numbers than expected.

Results of fertility assessments for two interspecific hybrid combinations of S. brevidens and S. tuberosum, hexaploid hybrids from Fusion A (2x + 4x) and tetraploid hybrids from Fusion B (2x + 2x), have been presented elsewhere (Ehlenfeldt and Helgeson, 1987). In summary, pollen stainability ranged from 0-23% in hexaploid hybrids and 0-83% in the tetraploid combinations. Tetraploids had more regular meiosis, lower levels of micropollen and fewer unassociated chromosomes than hexaploids. Although pollen from both sets of hybrids was ineffective in pollinations, both hybrid groups were effective as females. Thus actual pollinations rather than stainability of pollen should be used to evaluate fertility.

Hexaploid fusion hybrids crossed particularly well with two tetraploid potato cultivars, Katahdin and Norland. Progeny from these crosses had 5x

Figure 2. Tubers of somatic hybrids as compared to parental lines. Top: Representative hill of fusion parent S. tuberosum Pl 203900 (center) and 4 hills of somatic hybrid tubers from a fusion of S. brevidens Pl 218228 (non-tuber-bearing) and S. tuberosum Pl 203900. Note extreme variation in tuber production between different hybrid lines. Bottom: Tubers of superior haploid (center) compared to those produced by somatic hybrids of S. brevidens Pl 218228 and the cv. Superior haploid (right and left).

or nearly 5x ploidy levels. Preliminary field evaluations of these materials showed that most of the progeny were vigorous plants that shed viable pollen. Tuber production was much improved as compared to the original hybrid (Figure 3). Several of the pentaploid progeny were further crossed to the Katahdin and this second generation is currently under evaluation.

DISEASE RESISTANCE SCREENING

Phenotypically normal somatic hybrids with the expected complement of chromosomes were screened for resistance to PLRV and late blight. Representatives of both tetraploid and hexaploid somatic hybrids between S. brevidens and S. tuberosum expressed resistance to PLRV derived from S. brevidens (Austin et al., 1985a; Helgeson et al.,1986). A small scale field test, using PLRV-infected source plants and relying on natural aphid transmission, showed that hexaploid hybrids did not acquire PLRV. Hybrid material is currently undergoing further field evaluation for expression of resistance. Preliminary screening of twelve pentaploid progeny derived from a sexual cross between a resistant hexaploid somatic hybrid and the cultivar Katahdin (PLRV-susceptible), showed that 5 were resistant to PLRV and 7 were susceptible. Thus it appears that the resistance incorporated by protoplast fusion is sexually transferable.

The tetraploid S. tuberosum parent used in Fusion A is a late blight differential line possessing the R4 gene which confers resistance to race 0 of P. infestans, S. brevidens is susceptible to R0 of P. infestans. Hexaploid somatic hybrids derived from Fusion A were resistant to race 0 of P. infestans (Helgeson et al., 1986). Thus we have demonstrated the expression of disease resistance genes in somatic hybrids incorporated from both parental lines.

METRIBUZIN RESISTANCE

In our fusion studies two of the diploid S. tuberosum Gp Phureja-Stenotonum lines (supplied by De Jong) were chosen initially because they were susceptible (line 77-1, coded as 13) or resistant (line 77-19 coded as 19) to the herbicide metribuzin. It was suggested that this could serve as a marker gene system in somatic hybridization studies (De Jong, 1983). Sensitivity to the herbicide is determined by a single recessive gene. Resistance to the herbicide is related to the rate of metabolic detoxification (Hinks, 1977 and Stephenson et al., 1976). Preliminary experimentation to determine if resistance or susceptibility was expressed in disorganized or in shoot-regenerating callus derived from the above marker lines gave disappointing results. A differential response was seen, particularly in green callus tissue with shoot initials, but the results were inconsistent. Some calli, (3-8%), from the known susceptible line were unaffected by metribuzin included in the medium. Nodal cuttings of test lines exposed to metribuzin in vitro gave consistent and highly reproducible results. All other lines used in fusion experiments were also screened in vitro for their response to metribuzin using the response of the known susceptible and resistant lines as reference. The response of somatic hybrids combinations, (using at least 15 different representatives of each combination), was then compared to that of parental lines.

For the assay technical grade metribuzin (supplied by DuPont de Nemours and Company) was added to propagation medium (Haberlach et al., 1985) to give a range of concentrations between 0 and 20 ug / ml. The bioassay was performed using 1 cm nodal sections from in vitro propagated

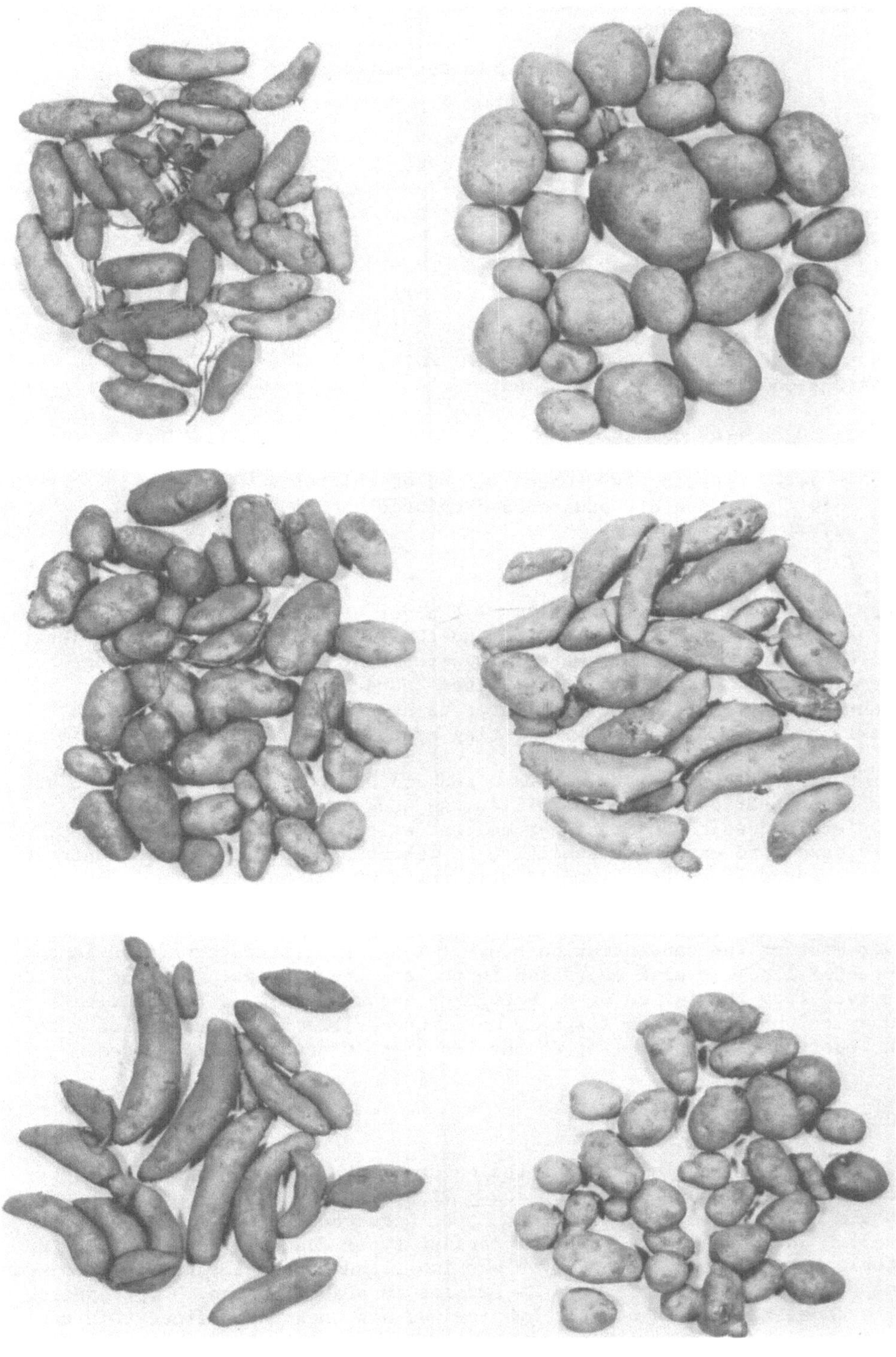

Figure 3. Representative tuber hills from a hexaploid somatic hybrid (top left) between S. brevidens Pl 218228 and S. tuberosum Pl 203900, the cultivar Katahdin (top right), and 4 progeny derived from a sexual cross of the somatic hybrid with Katahdin (middle and bottom).

Table 4. In-vitro metribuzin screening of test lines and interspecific somatic hybrids.

Test line	Metribuzin Concentration 0	0.5	1.0	2.5	5.0	10.0	20.0
6A	0	++	++	++++	++++	++++	++++
R4	0	0	0	0	0	+	++
R4+6A(Fusion A)	0	0	0	0	0	+	++
5B	0	0	++	+++	+++	++++	++++
13	0	+++	+++	++++	++++	++++	++++
5B+13(Fusion B)	0	+	+++	+++	++++	++++	++++
19	0	0	0	0	0	0	++
5B+19(Fusion C)	0	0	0	0	0	0	+
5	0	0	0	0	0	+	+
5+6A(Fusion E)	0	0	0	0	0	++	+++

0 = no damage
+ = slightly stunted
++ = obviously stunted, signs of chlorosis
+++ = severely stunted and chlorotic
++++ = death

test lines or hybrids. Generally 4-5 nodes of test material were taken for each metribuzin concentration and the bioassay performed on at least two different occasions. Materials were maintained under conditions normally used to maintain clonal lines (Haberlach et al., 1985). Plantlets were assessed visually for signs of metribuzin damage on a weekly basis for 8 weeks. The rating system used is given in Table 4.

Susceptible plantlets were stunted and had poor root development at low concentrations of the herbicide and were severely chlorotic or completely dessicated at higher concentrations. Both accessions of S. brevidens were extremely sensitive to metribuzin (Table 4). In contrast, S. tuberosum lines R4, 5 and 19 were resistant at most of the concentrations tested. Somatic hybrids derived from fusions between susceptible S. brevidens and resistant S. tuberosum were also resistant over most of the concentration range. Thus, resistance expressed in one parental line was also expressed in the somatic hybrids. Somatic hybrids derived from the fusion of S. brevidens and a haploid of the cultivar Superior, however, were slightly more susceptible to the herbicide than the resistant parental line at the two highest concentrations used.

CHLOROPLAST DNA

Preliminary evaluations of the nature of the chloroplast DNA of somatic hybrids between S. tuberosum PI 203900 and S. brevidens showed that a segregation of chloroplast types occurred (Kitzinger and Helgeson, 1987). In no case were both chloroplast types found in the same hybrid (Table 5). The chloroplast type was inherited maternally as expected when the somatic hybrids were used as females in sexual crosses. Segregation of organelle types in somatic hybrids was not unexpected since this has been found by several other researchers (for example Barsby et al., 1984). There were no obvious differences in hybrid phenotype or fertility which could be linked to a particular chloroplast genotype.

An overall summary of characteristics of interspecific hybrids between S. brevidens and S. tuberosum is given in Table 6.

Table 5. Determination of the chloroplast DNA of somatic fusion hybrids and progeny of the crosses of the hybrids with S. tuberosum cv. katahdin.

	Clone	Chloroplast type S. tuberosum (PI 2039000)	Chloroplast type S. brevidens (PI 218228)
A. hexaploid somatic hybrids			
	206		X
	209		X
	240		X
	242		X
	249		X
	252		X
	920		X
	903	X	
	937	X	
	1687	X	
	1861	X	
B. Sexual crosses			
206 X Katahdin	4711		X
	4714		X
937 X Katahdin	4676	X	
	4686	X	
	4806	X	
	4807	X	

CONCLUSIONS

At this point in our investigations any conclusions as to the potential contributions of somatic fusions to potato improvement must be considered preliminary. However, several of our findings have led us to be optimistic. Firstly some of the somatic hybrid plants were vigorous, phenotypically normal plants with ploidy levels equal to the sum of the parental lines used in the fusion. A substantial amount of somaclonal variation was generated and some combinations gave a large number of aberrant plants. Field testing of large numbers of hybrids allowed selection of competent individual hybrid lines for further evaluation. Overall it appears that the nuclear genomes of interspecific hybrids between S. brevidens and S. tuberosum are compatible and stable through successive tuber generations.

A second finding that gives us optimism is that disease and herbicide resistances of both of the donor species were expressed in somatic hybrids. Clonal copies of each individual somatic hybrid were maintained for screening purposes so that we could determine expression of parental genes using the same hybrid material. In almost every case individual hybrid lines expressed resistances from both parental lines (Helgeson et al., 1986).

A third, and even more important cause for optimism, is the fact that a number of plants derived from different combinations of S. brevidens and S. tuberosum were highly fertile. Thus the possibility for passing the traits "captured" from the wild species on to other potato lines can be tested. Indeed, the preliminary data indicate that sexual transfer of the traits does occur. Also, as crosses with good tuber-bearing plants progress into additional generations tuber quality is much improved.

Table 6. Summary of characteristics of interspecific hybrids between S. tuberosum and S. brevidens

A.	R4+6A (S. tuberosum PI 203900) +S. brevidens PI 218228)	Generally vigorous and free-flowering. Majority hexaploid, pollen shed 0-23%. Tuber production variable, tendency of tubers to elongate. Used as females in sexual crosses. Metribuzin resistant, PLRV resistant, resistant to race 0 of P. infestans. Chloroplast genomes segregated, hybrids had cpDNA of either parent.
B.	5B+13 (S. tuberosum 77-1) +S. brevidens PI 245763)	Vigorous and free-flowering, Majority tetraploid, pollen shed 0-83% Tuber production negligible, generally enlarged stolons. Used as females in sexual crosses. Metribuzin sensitive, PLRV resistant.
C.	5B+19 (S. tuberosum 77-19 +S. brevidens PI 245763)	Generally poor, majority were severely stunted and aberrant. Phenotypically normal hybrids examined were tetraploid and shed pollen with high viability. Tuber production as B. Metribuzin resistant.
D.	6A+19 (S. tuberosum 77-19 +S. brevidens PI218228)	As above, phenotypically normal plants were tetraploid and shed viable pollen. Tuber production as B. PLRV resistant.
E.	6A+5 (S. tuberosum USW-13,114 +S. brevidens PI 218228)	The most vigorous and free-flowering of all hybrid combinations All plants were tetraploid. Tuber production as B. No viable pollen shed. Metribuzin resistant, PLRV resistant.

Our approach has been to thoroughly evaluate the performance of protoplasts from our various test species prior to attempting somatic fusions. This has been very useful as it has permitted the selection of fusion products by manipulation of media. Hybrid cells appear to be capable of regeneration if one of the parent lines is responsive. Thus, lack of regeneration ability in one fusion parent is actually beneficial in the selection process. The morphological characteristics of the hybrids are so distinct, and so obvious early in the growth of the hybrid shoots from calli, that selection of hybrids is really rather easy and can be done quite early.

We have chosen to work with normal cell lines from wild-type plants rather than to devise selections schemes based on mutant cell lines. As mentioned above, we have obtained substantial variation and a number of highly abnormal plants have been produced even though we started with materials that had no obvious mutations. The fact that we have used this material may be the reason for the observed fertility of our hybrids. Not all of our hybrids were fertile; those of S. brevidens and a male sterile clone of a diploid extracted from the cultivar Superior appeared to be infertile. Fusions between male fertile and male sterile plants can produce fertile hybrids, however. A 2x+2x fusion of male sterile and male fertile S. tuberosum haploid lines yielded tetraploid hybrids which were highly self-fertile (Austin et al., 1985b).

Acknowledgments

Part of this work was supported by USDA-CRGO grants 84-CRCR-1-1512 and 86-CRCR-1-2142 to JPH. The authors thank Beth Howman-Combs for technical assistance to S. Austin throughout this work.

REFERENCES

Austin, S., Baer, M., and Helgeson, J. P., 1985a, Transfer of resistance to potato leaf roll virus from Solanum brevidens into Solanum tuberosum by somatic fusion. Plant Science, 39:75.

Austin, S., Baer, M.,Ehlenfeldt, M., Kazmierczak, P.J., and Helgeson, J. P., 1985b, Intraspecific fusions in Solanum tuberosum, Theor. Appl. Genet., 71:172

Austin, S., Ehlenfeldt, M., Baer, M., and Helgeson, J. P., 1986, Somatic hybrids produced by protoplast fusion between S. tuberosum and S. brevidens: phenotypic variation under field conditions, Theor. Appl. Genet., 71:682

Bayliss, M. W., 1980, Chromosomal variation in plant tissues in culture, in: International Review of Cytology, Suppl. 11A, G. H. Bourne, J. R. Danielli, and K. W. Jeon, eds., Academic Press, Florida

Barsby, T., Shepard, J., Kemble, R., and Wong, R., 1984, Somatic hybridization in the genus Solanum: S. tuberosum and S. brevidens, Plant Cell Reports, 3:165.

Butenko, R., and Kuchko, A., 1979, Somatic hybridization of Solanum tuberosum L. and Solanum chacoense Bitt. by protoplast fusion, in: Advances in Protoplast Research, Akad. Kiado, Budapest, 293.

Butenko, R., Kuchko, A., and Komarnitsky S., 1982, Some features of somatic hybrids between Solanum tuberosum and S. chacoense and its F_1 sexual progeny, p. 643, in A, Fujiwara (ed)., Proc. 5th Intern. Cong. Plant Tissue and Cell Culture, Japan. Assoc. Plant Tissue Culture, Tokyo.

Carlson, P., Smith, H., and Dearing, R., 1972, Parasexual interspecific plant hybridization, Proc. Nat. Acad. Sci., 69:2292.

Cocking, E. C., George, D. Price-Jones M. J., and Power, J. B., 1977, Selection procedures for the production of interspecific somatic hybrids of Petunia hybrida and Petunia parodii, Albino complementation selection, Plant Sci. Lett., 10:7.

De Jong H., 1983, Inheritance of sensitivity to the herbicide metribuzin in cultivated diploid potatoes, Euphytica 32:41.

Evans, D., Wetter, L., and Gamborg, O., 1980, Somatic hybrid plants of Nicotiana glauca and Nicotiana tabacum obtained by protoplast fusion, Physiol. Plant, 48:225.

Evans, D., 1983, Agricultural application of plant protoplast fusion, Biotechnology, 1:253.

Gleddie, S., Keller, W. A., and Setterfield, G., 1986, Production and characterization of somatic hybrids between Solanum melongena L. and S. sisymbriifolium Lam, Theor. Appl. Genet, 71:613

Haberlach G. T., Cohen B. A., Reichert, N. A., Baer, M. A., Towill L. E., and Helgeson, J. P., 1985, Isolation, culture and regeneration of protoplasts from potato and several related Solanum species, Plant Sci. Lett., 39:67.

Hein, T., Przewozny, T., and Schieder, O., 1983, Culture and selection of somatic hybrids using an auxotrophic cell line, Theor. Appl. Genet., 64:119.

Helgeson, J. P., Hunt, G. J. Haberlach, G. T., and Austin S., 1986, Somatic hybrids between Solanum brevidens and Solanum tuberosum: Expression of a late blight resistance gene and potato leaf roll resistance, Plant Cell Rep., 3:212.

Hinks, T. G., 1977, The nature of the tolerance of potatoes to metribuzin herbicide, M. Sc. Thesis. Univ, Guelph 69p.

Kao, K. N., and Michayluk, M. R., 1974, A method for high-frequency intergeneric fusion of plant protoplasts, Planta, 115:355.

Keller, W., and Melchers, G., 1973, The effect of high pH and calcium on tobacco leaf protoplast fusion, Z. Naturforsch, 28:737.

Kitzinger, C., and Helgeson, J. P., 1987, Segregation of cpDNA in somatic hybrids of Solanum and their sexual progeny, manuscript in preparation.

Lam, S. L., 1977, Regeneration of plantlets from single cells in potatoes, Amer. Potato Journ., 54:575.

Maliga, P., Lazar, G., Joo, F. Nagy, A. H. and Menczel, L., 1977, Restoration of morphogenic potential in Nicotiana by somatic hybridization, Molec. Gen. Genet., 157:291.

Melchers, G. and Labib, G., 1974, Somatic hybridization of plants by fusion of protoplasts. I. Selection of light resistant hybrids of "haploid" light sensitive varieties of tobacco, Molec. Gen. Genet., 135:277.

Melchers A., Sacristan, M. D., and Holder, S. A., 1978, Somatic hybrid plants of potato and tomato regenerated from fused protoplasts, Carlsberg Res. Comm., 43:203.

Puite K. J., Roest S., and Pijnacker, L. P., 1986, Somatic hybrid potato plants after electrofusion of diploid Solanum tuberosum and Solanum phureja, Plant Cell Reports, 5:262.

Schenck, H. R. and Robbelen, G., 1982, Somatic hybrids by fusion of protoplasts from Brassica oleracea and B. campestris, Z. Pflanzenzuch, 89:278.

Schieder, O., 1977, Hybridization experiments with protoplasts from chlorophyll--deficient mutants of some Solanaceous species, Planta, 137.

Shepard, J. F., and Totten, R. E., 1977, Mesophyll cell protoplasts of potato:isolation, proliferation and plant regeneration, Plant Physiol., 60:313.

Shepard, J. F., 1980, Mutant selection and plant regeneration from potato mesophyll protoplasts, in Genetic Improvement of Crops, I. Rubenstein, B. Gengenbach, R. L. Phillips, C. E. Green ed., 185.

Shepard, J., Bidney, D., Barsby, T., and Kemble, R., 1983, Genetic transfer in plants through interspecific protoplast fusion, Science, 219:683.

Stephenson, G. R., McLeod, J. E., and Phatak, S. C., 1976, Differential tolerance of tomato cultivars to metribuzin, Weed Sci., 24: 161.

Zimmermann, U., and Scheurich, P., 1981, High frequency fusion of plant protoplasts by electric fields, Planta, 151:26.

BRASSICA CYBRIDS AND THEIR UTILITY IN PLANT BREEDING

Tina L. Barsby, Roger J. Kemble and Stephen A. Yarrow

Agriculture Division, Allelix Inc., 6850 Goreway Drive, Mississauga, Ontario, L4V 1P1, Canada

ABSTRACT

Brassica napus represents most of the Canola acreage in Canada. The increased market potential of F1 hybrids has stimulated an interest in *in vitro* manipulation of cytoplasmic male sterile (cms) cytoplasms. We have shown that the laboratory technique of protoplast fusion can be a useful adjunct to a conventional Canola breeding program, either in providing material which cannot be produced sexually, or in speeding up the production of alloplasmic substitution lines. The production of cybrids in a crop plant for the purposes of utilising them in a breeding program is a quite different prospect to a purely scientific endeavour. The process must be cost-efficient and reproducible over a wide range of genotypes. In addition the regenerated plants must be stable with respect to important agronomic traits.

The opportunities which protoplast fusion affords for crop improvement by cytoplasmic manipulation have begun to be realised. But there are still many potential targets waiting to be explored.

THE CROP

Although rapeseed is the generic name for the crop, 'Canola' is the commodity name for *B. napus* or *B. campestris* which has less than 2% erucic acid in the extracted oil, and less than 50 μmoles of glucosinolates per gram of remaining meal. Canola-quality rapeseed was first developed in Canada (Steffansson and Kondra, 1975). Recently, the U.S. conferred GRAS (Generally Regarded As Safe for human consumption) status on Canola, which stimulated increased commercial interest worldwide. Canola oil is used in cooking and salad oils, and high erucic forms of rapeseed have industrial uses. In Canada 3.5 - 4 million acres are planted with spring *B. napus*, 2.5 - 3 million with spring *B. campestris* and about 30,000 acres with winter *B. napus*. This winter acreage is expected to expand and most of the potential in the U.S. is for winters. Most of the European acreage is winter planted genotypes.

Interest in manipulating the cytoplasm of rapeseed was stimulated when the potential for increasing yield through F1 hybrid production was demonstrated (Schuster and Michael, 1976; Sernyk and Steffansson, 1983; Grant and Beversdorf, 1985). In order to produce hybrids economically, some genetic means of preventing self fertilisation is desirable, and cytoplasmic male sterility is one of the few agronomically important traits known to be associated exclusively with a cytoplasmic organelle (the mitochondrion - Pelletier et al., 1983; Barsby et al., 1987a). Several forms of cms in *B. napus* are known (Shiga, 1980). Among the leading candidates for use in hybrid production are 'Ogura' (from *Raphanus*), 'Polima' and 'Nap'.

It seems logical that those interested in cms should turn to protoplast fusion. Effects which have been demonstrated in model species could be useful in improving cms if applicable to *B. napus*. Protoplast fusion must either provide something which cannot be achieved conventionally, or accelerate an existing process. One of the limitations of cms improvement by manipulation of the cytoplasm is that maintainer and restorer characteristics are controlled by nuclear genes. However, there are examples where altering the cytoplasm alone can be beneficial.

Cytoplasmic triazine tolerance is a chloroplast-encoded trait which is attractive to rapeseed breeders. The cytoplasm was transferred from a tolerant line of *B. campestris* by conventional crossing to *B. napus* (Beversdorf et al., 1980). This culminated in the release of the cultivar 'Triton' which was planted in approximately 250,000 acres in Western Canada in 1986, despite a yield penalty associated with the cytoplasm. *B. napus* could not otherwise be grown in these weed-infested areas. Beversdorf et al., (1985) also proposed a hybrid production scheme involving cytoplasmic triazine tolerance and cms. These two traits can be combined only by protoplast fusion, since the cytoplasm is maternally inherited in sexual crosses. In the hybrid production field, males and females are interplanted and after pollination, spraying with a triazine herbicide would eliminate the sensitive males whereas the resistant cms females would survive to produce hybrid seed.

THE TECHNOLOGY

The potential which protoplast fusion affords for the introgression of valuable nuclear traits into crop plants has been recognised for some time. One realisation of that potential has been suitably demonstrated in the preceding paper. In addition, protoplast fusion may be used to transfer cytoplasmic traits between genotypes, without the involvement of nuclear DNA. Most of the basic work concerning cybrid recovery was performed with *Nicotiana* (Zelcer et al., 1978; Aviv and Galun, 1980) or *Petunia* (Izhar and Power, 1979) where protoplast culture techniques have been available and reproducible longer than in crop species. The wealthy background of knowledge concerning the ability to manipulate cytoplasms paved the way for applications in crop plants. The limiting factor in all such applications is the ability to recover whole plants from single protoplasts. Rapeseed was among the first crops where this technology became available, and there are several publications pertaining to techniques (e.g. Kartha et al., 1974; Thomas et al., 1976; Glimelius, 1984).

Protoplast regeneration techniques are difficult to reproduce from laboratory to laboratory, and often are variety specific. The system used in our experiments is generally reproducible and applicable to a wide variety of both spring and winter genotypes (Barsby et al., 1986). The regeneration frequency varies between cultivars, but plants have been recovered from all the B. napus genotypes so far attempted. B. campestris does not respond to these conditions.

Having developed a protoplast regeneration protocol, a method for selecting fusion products must be chosen. PEG-mediated protoplast fusion (modified after Wallin et al., 1974, and Kao and Michayluk, 1974) usually results in less than 5% heterokaryons. If the process is to be reproducible over a range of cultivars, it is preferable that the selection methods not be restricted to certain genotypes. The use of albinos or other mutants is impractical, since the mutation must be available in all genotypes utilised. The isolation of individual heterokaryons is one method which can be applied. Following fusion, heterokaryons may be identified by morphological characteristics, if the parental protoplasts were derived from different tissue sources. Alternatively, the protoplasts may be stained with non-lethal fluorescent dyes which enable the visualisation of fusion products. Heterokaryons may then be isolated using a micropipette, and cultured either in microdrops (Gleba and Hoffman, 1980) or with a 'nurse' culture (Patniak et al., 1982).

Differences in growth requirements between the species to be fused may be exploited in the selection of fusion products (Carlson et al., 1972). Usually, differential response to media components allows only heterokaryons to survive and regenerate. However, this method requires a genetic difference between the fusion partners and is probably not applicable to fusions between close relatives.

The above methods do not select against somatic hybrids, and their usefulness in cybrid production relies on the random occurence of either a complete lack of nuclear fusion or specific elimination of one set of parental chromosomes during culture. The use of irradiation to prevent the involvement of one parental nucleus in a fusion product has been described (Zelcer et al., 1978), and is called the 'donor-recipient' mode of selection. The frequency of cybrid recovery is increased over that expected without such treatment. The effectiveness of the technique is enhanced by the use of IOA to prevent division in the other fusion partner (Sidorov et al., 1981).

As is the case with most tissue culture techniques, the best documented examples of cytoplasmic organelle transfers, segregation, and rearrangements following fusion are in tobacco. It has been shown that all possible combinations of parental mitochondria and chloroplasts are recoverable in the regenerant population. Recombination between the mitochondrial DNAs has been demonstrated (Nagy et al., 1981, 1983; Galun et al., 1982; Aviv et al., 1984; Belliard et al., 1979), and in one case chloroplast DNA clearly underwent similar rearrangements (Medgyesy et al., 1985). Mitochondrial

DNA recombination has also been reported in Petunia (Boeshore et al., 1983; Rothenberg et al., 1985), Solanum (Kemble et al., 1986a), Daucus (Matthews and Widholm, 1985) and Brassica (Chetrit et al., 1985). Our objective has not been to collect similar information for Brassica, but to apply the techniques available to the improvement of the crop. The aforementioned problems in directly adopting tissue culture, and especially protoplast culture techniques from other labs necessitated the development of our own methods, based on those already available. Since we have used protoplast fusion to create cybrids in Brassica for about three years, we have accumulated enough information to enable us to comment on the effectiveness of certain techniques and to report the occurrence of certain phenomena. More importantly we have been able to see the results of our fusions be incorporated into Allelix's Canola breeding programme. This has resulted in some changes in our approach to cybrid production. In early experiments we analysed most carefully all regenerants for cp and mt DNA composition (Yarrow et al., 1986; Barsby et al., 1987a; 1987b). The most recent material resulting from fusions is primarily subjected to the plant breeders requirements for degree of sterility, seed quality, uniformity of progeny and ultimately yield.

BRASSICA CYBRIDS: CHLOROPLAST REPLACEMENT

One of the problems with Brassica cms was tackled by Pelletier et al., (1983). The 'Ogura' cms mitochondrion is associated with 'defective' chloroplasts which confer a yellowing response at low temperatures. Using protoplast fusion the 'ogura' chloroplasts were replaced with the normal Brassica ('nap') chloroplasts. The resultant cybrid had no low temperature problems and so could potentially be useful in an 'ogura'-based hybridisation scheme. In the same paper the combination of the ogura cms mitochondrion with the triazine resistant chloroplast from B. campestris was described. This cytoplasm could be useful in the scheme proposed by Beversdorf et al., (1985).

With the support of the National Research Council, Canada, and in collaboration with Beversdorfs' group at the University of Guelph, we undertook to produce cybrids carrying the 'Polima' cms mitochondrion and the triazine tolerant chloroplasts. Our two groups used different methods to produce the cybrids (Barsby et al., 1987a). The Guelph group adopted a system which involved screening the total regenerant population for the desired characteristics. The Allelix group used the donor-recipient system. Gamma-irradiation and iodoacetic acid (IOA) treatment were combined to selectively eliminate unfused parental protoplasts, resulting in fewer regenerants. These were analysed for organelle composition by restriction fragment length polymorphisms (RFLPs) of cp and mt DNA (Kemble, 1987). Eco RI restriction digests will differentiate most Brassica cytoplasms and these are performed routinely on all potential cybrids. We also examined this material with Sal I, Pst I, Bcl I and Xba I. No evidence for mitochondrial recombination was found. Table 1 summarises cybrid recovery in this type of experiment.

The desired cybrids were evident amongst the regenerant population. Individual plants were increased by crossing with the 'Polima' cms maintainer genotype. After two sexual

TABLE 1. Construction of 'Polima' (POL)-mt / triazine tolerant 'campestris' (ttCAM)-cp *B. napus* cybrids by iodoacetic acid / gamma-irradiation selection.

Number of plants produced per organelle combination:

	MT/CP-DNA			
Expt.	POL/POL	POL/ttCAM*	ttCAM/POL	ttCAM/ttCAM
1.	1	0	0	0
2.	1	1	0	1
3.	1	0	0	1
4.	0	1	0	1
Total	3	2	0	3

* - desired cybrid combination.

Protoplasts were prepared from dark grown hypocotyls of *B. napus* 'Polima Regent' (cms) and from leaves of *B. napus* 'tt Regent' ('Regent' with triazine tolerant *B. campestris* cytoplasm). 'Polima Regent' protoplasts were treated with 2-4 mM iodoacetic acid, and fused with 'tt Regent' protoplasts, which were subjected to 30 krad gamma-irradiation.

generations sufficient seed was available to permit small field plots (6m x 6 row). Tolerance to field application of the triazine herbicide 'Bladex' was demonstrated, and normal seed set was observed. Despite low female fertility in the regenerant generation, sexual crosses resulted in the recovery of full female fertility.

Similar cybrids were recovered by exploiting differences in protoplast growth requirements between *B. napus* and *B. campestris*. Protoplasts of *B. campestris* do not divide in the media devised for *B. napus* (Barsby et al., 1986). *B. napus* protoplasts were treated with 3mM IOA prior to fusion. See Table 2.

The 'Nap' cms system may also be suitable for hybrid production. In this case the mitochondrion is that native to *B. napus* and the cms phenotype is expressed only in the absence of the nuclear restorers which normally exist. The 'Nap' mitochondrion and the triazine tolerant chloroplast were combined by protoplast fusion (Yarrow et al., 1986). In this case manual heterokaryon isolation was used to select fusion products, and cybrids were identified amongst the regenerants. See Table 3.

In all of the above examples there is no assurance that true cybrids were produced and not nuclear hybrids. The aneuploid condition of most of the regenerants suggests that they were unbalanced somatic hybrids. Alternatively, the aneuploidy may have been induced in one parental nucleus. These alternatives were not explored. Seed set was extremely difficult to achieve, but the progeny were usually diploid. Subsequent generations are being evaluated by Allelix's plant breeders.

TABLE 2. Construction of 'Polima' (POL)-mt / triazine tolerant 'campestris' (ttCAM)-cp *B. napus* cybrids by iodoacetic acid / differential growth selection.

Number of plants produced per organelle combination:

	MT/CP-DNA			
Expt.	POL/POL	POL/ttCAM*	ttCAM/POL	ttCAM/ttCAM
1.	5	0	0	1
2.	0	1	0	0
3.	1	1	0	0
Total	6	2	0	1

* - desired cybrid combination.

Protoplasts were prepared from dark grown hypocotyls of *B. napus* 'Polima Regent' (cms) and from leaves of *B. campestris* (triazine tolerant line). 'Polima Regent' protoplasts were treated with 2 mM iodoacetic acid, and fused with *B. campestris* protoplasts, which were untreated. No growth was observed in the controls.

BRASSICA CYBRIDS: ALLOPLASMIC SUBSTITUTION

The chloroplast replacement experiments described above were performed with spring genotypes. The greatest yield potential is exhibited by winter types as a result of their protracted growing season. In order to utilise the cms-based pollination control mechanism, the cms-conferring mitochondrion must be exchanged for that occuring naturally in the winter genotype. To effect this exchange by conventional means the cms spring genotype is pollinated with the winter. This process is repeated until the nuclear-encoded winter characteristics are recovered. Since the cytoplasm is maternally inherited, the cms mitochondrion will be retained in the winter nuclear background. The generation time of most winter lines is around six months, including ten weeks vernalisation to induce flowering. Consequently the whole backcross process takes about three years to accomplish. The result is an alloplasmic substitution since the nuclear component is essentially similar to that of the original winter genotype and only the cytoplasm is altered.

Our experience with fusion experiments in *B. napus* suggested that it would take approximately nine months to accomplish similar alloplasmic substitutions using protoplasts (i.e. from receiving the seed of the two parents, growing plants to a suitable size for protoplast isolation (5 weeks), performing the fusion and recovering shoots (12 weeks), rooting and potting regenerants (4 weeks), obtaining leaf material and performing DNA analysis (4 weeks), vernalisation (10 weeks), flowering (3 weeks), pollination and seed set (4 weeks), harvest and dry down seed (3 weeks). (These are conservative estimates allowing for unpredicted biological phenomena!). This obviously represents a considerable saving in time over the conventional process.

TABLE 3. Construction of 'Napus' (NAP)-mt / triazine tolerant 'campestris' (ttCAM)-cp B. napus cybrids by manual heterokayon isolation selection.

Number of plants produced per organelle combination:

Mesophylls untreated:

	MT/CP-DNA			
Expt.	NAP/NAP	NAP/ttCAM*	ttCAM/NAP	ttCAM/ttCAM
1.	2	2	0	0
2.	2	0	0	0
Total	4	2	0	0

Mesophylls irradiated:

	MT/CP-DNA			
Expt.	NAP/NAP	NAP/ttCAM*	ttCAM/NAP	ttCAM/ttCAM
1.	5	1	0	1
2.	0	1	0	0
Total	5	2	0	1

* - desired cybrid combination.

Protoplasts were prepared from dark grown hypocotyls of B. napus 'Regent' and from leaves of B. napus 'tt Regent' ('Regent' with triazine tolerant B. campestris cytoplasm). 'Regent' protoplasts were treated with fluorescein diacetate, and fused with 'tt Regent' protoplasts, which were either untreated or subjected to 20 krad gamma-irradiation. Heterokaryons were identified under U.V. microscopy and transferred to nurse culture.

In order for this procedure to be adopted, it must be demonstrated that the resultant plants are comparable to their sexual counterparts using criteria acceptable to the plant breeder. These criteria include: full female fertility; complete male sterility; the normal euploid chromosome number of B. napus; uniform progeny; and stable quality characteristics such as oil and meal composition. Since variation for agronomic traits may be expected among protoclones, such stability appears difficult to acheive. However, in the first successful experiment of this type, at least one satisfactory individual was recovered.

The donor-recipient mode of cybrid selection was again employed in the alloplasmic substitution experiments (Barsby et al., 1987b). Leaf mesophyll protoplasts of the desired cms donor (spring type) were fused with the recipient winter line. Regenerants were analysed for chloroplast and mitochondrial type. Following vernalisation, crossing to the maintainer was attempted and fertility recorded. Plants with high seed set were noted, and progeny of these taken forward as the cms line, eventually to become the female population in winter hybrid seed production fields. The earliest lines carrying 'Polima' cms have been increased through three generations (two in the field) and are currently being evaluated in seed production mini-plots (6m x 6 row).

To date, by protoplast fusion, 'Polima' cms has been introduced into 6 different winter lines from the Allelix germplasm collection. The procedures have been modified for cost-effectiveness. The first forty regenerants were examined with the five restriction enzymes listed above for possible rearrangements in mt and cp DNA. The next one hundred were examined only with Eco RI, which will distinguish the Brassica cytoplasms (Kemble, 1987). In future, DNA composition will only be analysed in the first sexual generation, so that expensive resources are not used on female steriles which are of no direct use to the breeder.

Table 4 summarises the mt/cp DNA complement of all Polima cms winter cybrids produced to date. In addition, two winter cybrids with Ogura mitochondria and normal Nap chloroplasts were also produced.

BRASSICA CYBRIDS: THE FUTURE

Hence the two primary goals of cybridization in Brassica viz., the creation of cms/triazine tolerant spring lines and accelerating the production of cms alloplasmic substitutions in winter B. napus were achieved. The process of alloplasmic substitution by protoplast fusion will continue since there are many other winter lines into which cms must be incorporated. Although we consider the techniques to be efficient compared to others available, improvements to the frequency of recovery of useful cybrids are desirable.

Recently we have recovered cybrids carrying mitochondria of uncultivated species of Brassica (Yarrow, in prep.). These plants will be assessed for their role in the development of new cms systems. Novel combinations of the mitochondria and chloroplasts already involved in existing systems are also being sought.

In addition there are other potential improvements to cms which could be approached through protoplast fusion. Perhaps rearrangements to the mitochondrial genome would improve the functional efficiency of the cms mitochondrion. Our experiments to date have not revealed any such rearrangements but they have been documented in Brassica by Chetrit et al., (1985). We assume that the phenomenon is not unique to any particular cytoplasm, and ideally in the future its occurence will be controlled through the techniques utilised.

Another interesting phenomenon associated with our protoplast fusion experiments is the behaviour of the 11.3 kb mitochondrial plasmid. Although it is not associated with the cms phenotype (Kemble et al., 1986b), other roles may be envisioned. The plasmid is mobilised independently of the mitochondrion by fusion. Loss or gain of the plasmid in the mitochondria of regenerated plants has been documented (Kemble, in prep.).

TABLE 4. Construction of 'Polima' (POL) cms winter *B. napus* cybrids by iodoacetic acid / gamma-irradiation selection.

Number of calli producing plants per organelle combination (calli often produced more than one shoot, sometimes with different organelle combinations):

Winter lines with 'napus' (NAP/NAP) cytoplasm:

Winter line (No. calli)	MT/CP-DNA POL/POL*	POL/NAP*	NAP/POL	NAP/NAP
AWR0019 (1)	1	0	0	0
AWR0027** (6)	2	0	0	5
AWR0028 (6)	0	1	0	5
AWR0033 (1)	1	0	0	0
AWR0052 (1)	1	0	0	0

Winter line with 'campestris' (CAM/CAM) cytoplasm:

Winter line (No. calli)	MT/CP-DNA POL/POL*	POL/CAM*	CAM/POL	CAM/CAM
AWR0009** (19)	9	3	0	6

* - desired cybrid combinations.
** - regenerants subjected to DNA restriction analysis using *Eco* RI plus five other restriction enzymes.

Protoplasts were prepared from dark grown hypocotyls of various winter genotypes of *B. napus* and from leaves of *B. napus* 'Polima Regent'. Protoplasts from the winter lines were treated with 2-4 mM iodoacetic acid, and fused with 'Polima Regent' protoplasts, which were subjected to 30 krad gamma-irradiation.

DEFINITION

The definition of the word cybrid has been under some dispute. We prefer the following; a cybrid is a derivative of a protoplast fusion product in which no nuclear fusion is detectable, possessing a novel nucleo-cytoplasmic combination. This definition includes all types of organellar hybrids from the mixing of one chloroplast with the other mitochondrion, to any new types of mitochondria or chloroplasts derived from recombination of organellar DNAs. The existence of the cytoplasmic organelles of one parent with the nucleus of the other parent also constitutes a cybrid if derived from protoplast fusion. See also Kumar and Cooper-Bland (1986).

PROCEDURES

Detailed descriptions of procedures may be found in the following publications: Barsby et al., (1986, 1987a., 1987b); Yarrow et al., (1986); Kemble (1987).

REFERENCES

Aviv, D., Arzee-Gonen, P., Bleichman, S. and Galun, E., 1984, Novel alloplasmic Nicotiana plants by "donor-recipient" protoplast fusion: cybrids having N. tabacum or N. sylvestris nuclear genomes and either or both plastomes and chondriomes from alien species, Mol. Gen. Genet., 196:244.

Aviv, D., and Galun, E., 1980, Restoration of fertility in cytoplasmic male sterile (CMS) Nicotiana sylvestris by fusion with X-irradiated N. tabacum protoplasts, Theor. Appl. Genet., 58:121.

Barsby, T. L., Choung, P. V., Yarrow, S. A., Wu, S-C., Coumans, M., Kemble, R. J., Powell, A. D., Beversdorf, W. D. and Pauls, K. P., 1987a, The combination of Polima cms and cytoplasmic triazine resistance in Brassica napus, Theor. Appl. Genet., 73:809.

Barsby, T. L., Yarrow, S. A., Kemble, R. J. and Grant, I., 1987b, The transfer of cytoplasmic male sterility to winter-type oilseed rape (Brassica napus) by protoplast fusion, Submitted to Plant Science.

Barsby, T. L., Yarrow, S. A., and Shepard, J. F., 1986, A rapid and efficient alternative procedure for the regeneration of plants from protoplasts of B. napus, Plant Cell Rep., 5:101.

Belliard, G., Vedel, F., and Pelletier, G., 1979, Mitochondrial recombination in cytoplasmic hybrids of Nicotiana tabacum by protoplast fusion, Nature, 281:401.

Beversdorf, W. D., Erickson, L. R., and Grant, I., 1985, Hybridization process utilizing a combination of cytoplasmic male sterility and herbicide tolerance, U.S. Patent No. 4517763.

Beversdorf, W. D., Weiss-Lerman, J., Erickson, L. R., and Souza-Machado, Z., 1980, Transfer of cytoplasmically inherited triazine resistance from bird's rape to cultivated oilseed rape (Brassica campestris and B. napus), Can. J. Genet. Cytol., 22:167.

Boeshore, M. L., Lifshitz, I., Hanson, M. R., and Izhar, S., 1983, Novel composition of mitochondrial genomes in Petunia somatic hybrids derived from cytoplasmic male sterile and fertile plants, Mol. Gen. Genet., 190:459.

Carlson, P. S., Smith, H. H., and Dearing, R. D., 1972, Parasexual interspecific plant hybridisation, Proc. Natl. Acad. Sci. U.S.A., 69:2292.

Chetrit, P., Mathieu, C., Vedel, F., Pelletier, G., and Primard, C., 1985, Mitochondrial DNA polymorphism induced by protoplast fusion in Cruciferae, Theor. Appl. Genet., 69:361.

Chuong, Phan V., Pauls K. P., and Beversdorf W. D., 1985, A simple culture method for Brassica hypocotyl protoplasts, Plant Cell Reports, 4:4.

Fan, Z., Stefansson, B. R., and Sernyk, J. L., 1985, Maintainers and restorers for three male sterility inducing cytoplasms in rape, Brassica napus, Can. J. Plant Sci., 66:229.

Galun, E., Arzee-Gonen, P., Fluhr, R., Edelman, M., and Aviv, D., 1982, Cytoplasmic hybridization in Nicotiana: mitochondrial DNA analysis in progenies resulting from fusion between protoplasts having different organelle constitutions, Mol. Gen. Genet., 186:50.

Gleba, Y. Y., and Hoffman, F., 1980, "Arabidobrassica": A novel plant obtained by protoplast fusion, Planta, 149:112.

Gleba, Y., Kolesnik, N. N., Meshkene, I. V., Cherep, N. N., and Parokonny, A. S., 1984, Transmission genetics of the somatic hybridization process in Nicotiana 1.Hybrids and cybrids among the regenerates from cloned fusion products, Theor. Appl. Genet., 69:121.
Glimelius, K., 1984, High growth rate and regeneration capacity of hypocotyl protoplasts in some Brassicaceae, Physiol. Plant., 61:38.
Grant, I., and Beversdorf, W. D., 1985, Agronomic performance of triazine-resistant single-cross hybrid oilseed rape (Brassica napus L.), Can. J. Plant Sci., 65:889.
Izhar, S., and Power, J. B., 1979, Somatic hybridization in Petunia: a male sterile cytoplasmic hybrid, Plant Sci. Lett., 14:49.
Kao, K. N., and Michayluk, M. R., 1974, A method for high frequency intergeneric fusion of plant protoplasts, Planta, 115:355.
Kartha, K. K., Michayluk, M. R., Kao, K. N., Gamborg, O. L., and Constabel, F., 1974, Callus formation and plant regeneration from mesophyll protoplasts of rape plants (Brassica napus L. cv. Zephyr), Plant Sci. Lett., 3:265.
Kemble, R. J., 1987, A rapid, single leaf, nucleic acid assay for determining the cytoplsmic organelle complement of rapeseed and related Brassic species, Theor. Appl. Genet., 73:364.
Kemble, R. J., Barsby, T. L., Wong, R. S. C. and Shepard, J. F., 1986a, Mitochondrial DNA rearrangements in somatic hybrids of Solanum tuberosum and Solanum brevidens, Theor. Appl. Genet., 72:787.
Kemble, R. J., Carlson, J. E., Erickson, L. R., and Sernyk, J. L. 1986b, The Brassica mitochondrial DNA plasmid and large RNAs are not exclusively associated with cytoplasmic male sterility, Mol. Gen. Genet., 205:183.
Kumar, A., and Cooper-Bland, S., 1986, Organelle genetics of somatic hybrid-cybrid progeny in higher plants, in: "The Chondriome," S. H. Mantell, G. P. Chapman, and P. F. S. Street, eds., Longman, England.
Matthews, B. F., and Widholm, J. M., 1985, Organelle DNA composition and isoenzyme expression in an interspecific somatic hybrid of Daucus, Mol. Gen. Genet., 198:371.
Medgyesy, P., Fejes, E., and Maliga, P., 1985, Interspecific chloroplast recombination in a Nicotiana somatic hybrid, Proc. Acad. Sci. USA., 82:6960.
Nagy, F., Torok, I., and Maliga, P., 1981, Extensive rearangements in the mitochondrial DNA in somatic hybrids of Nicotiana tabacum and N. knightiana, Mol. Gen. Genet., 183:437.
Nagy, F., Lazar, G., Menczel, L., and Maliga, P., 1983, A heteroplasmic state induced by protoplast fusion is a necessary condition for detecting rearrangements in Nicotiana mitochondrial DNA, Theor. Appl. Genet., 66:203.
Patniak, G., Cocking, E. .C., Hamill, J., and Pental, D., 1982, A simple procedure for the manual isolation and identification of plant heterokaryons, Plant Science Lett., 24:105.
Pelletier, G., Primard, C., Vedel, F., Chetrit, P., Remy, R., Rouselle, P., and Renard, M., 1983, Intergeneric cytoplasmic hybridization in Cruciferae by protoplast fusion, Mol. Gen. Genet., 191:244.

Rothenberg, M., Boeshore, M. L., Hanson, M. R., and Izhar, S., 1985, Intergenomic recombination of mitochondrial genomes in a somatic hybrid plant, Curr. Genet., 9:615.
Rouselle, P., 1981," Etude de systemes d'androsterilite chez le colza (Brassica napus)," These univ, Rennes 1.
Schuster, W., and Michael, J., 1976, Unterschungen uber Inzuchtdepressiones und Heterosiseffekte bei Raps (Brassica napus oleifera), Z. Pflanzenzucht., 77:55.
Sernyk, J. L., Stefansson, B. R., 1983, Heterosis in summer rape, Can. J. Plant Sci., 63:407.
Shiga, T., 1980, Male sterility and cytoplasmic differentiation, in: "Brassica Crops and Wild Allies". S. Tsunoda, K. Hinata, and C. Gomez-Campo, eds., Japan. Scientific Society Press, Japan.
Sidorov, V. A, Menczel, L., Nagy, F., and Maliga, P., 1981, Chloroplast transfer in Nicotiana based on metabolic complementation between irradiated and iodoacetate treated protoplasts, Planta 152:341.
Stefansson, B. R., and Kondra, Z. P., 1975, Tower summer rape, Can. J. Plant Science 55:343.
Thomas, E., Hoffmann F., Potrykus, I., and Wenzel, G., 1976, Protoplast regeneration and stem embryogenesis of haploid androgenetic rape, Mol. Gen. Genet., 145:245.
Wallin, A. K., Glimelius, K., and Erikson, T., 1974, The induction of aggregation and fusion of Daucus carota protoplasts by polyethylene glycol, Z. Pflanzenphysiol., 74:64.
Yarrow, S. A., Wu, S-C., Barsby, T. L., Kemble, R. J., and Shepard, J. F., 1986, The introduction of cms mitochondria to triazine tolerant Brassica napus L., var. "Regent", by micromanipulation of individual heterokaryons, Plant Cell Rep., 5:415.
Zelcer, A., Aviv, D., and Galun, E., 1978, Interspecific transfer of cytoplasmic male sterility by fusion between protoplasts of normal Nicotiana sylvestris and X-ray irradiated protoplast of male sterile N. tabacum, Z. Pflanzenphysiol., 90:397.

SELECTION AND CHARACTERIZATION OF CARROT SOMATIC HYBRIDS

Jack M. Widholm

University of Illinois
Department of Agronomy
1102 South Goodwin
Urbana, Ill. 61801 USA

Many protoplast fusion studies have been carried out with carrot since the protoplasts are easy to prepare and grow and there are numerous mutations available to use as markers. Carrot cells are usually capable of plant regeneration although in most of the studies described here plants were not regenerated. In this summary I would like to describe several carrot protoplast fusion studies which have demonstrated some fundamental basic findings, taken in large part from work done in this laboratory. A more complete review of carrot protoplast fusion was presented by Widholm et al. (1984).

Somatic Hybrid Selection Using Amino Acid Analog Resistance

In our initial carrot protoplast fusion work we fused one carrot strain (C123) selected as resistant to both a tryptophan analog, 5-methyltryptophan (5MT)* and to a proline analog, azetidine-2-carboxylate (A2C) (Harms et al., 1981). The respective resistances are due to increased levels of free tryptophan and proline and the tryptophan increase is due to a feedback altered form of the tryptophan biosynthetic control enzyme, anthranilate synthase. The other fusion partner was the carrot strain C81 which was resistant to aminoethylcysteine (AEC) due to a decreased rate of uptake (Matthews et al., 1980). Following polyethylene glycol fusion thousands of colonies were selected which grew on 5MT and AEC levels which would inhibit the growth of the parental protoplasts alone or when mixed together without fusion. Further testing showed that most doubly resistant clones also carried A2C resistance as expected since the C123 parents was also resistant to A2C. The resistance traits were

*Abbreviations: A2C, azetidine-2-carboxylate
AEC, aminoethylcysteine
EPSPS, enolpyruvylshikimic acid-3-phosphate synthase
Glp, glyphosate
mt, mitochondrial
5MT, 5-methyltryptophan

generally maintained for at least 110 cell doublings without selection pressure and the putative hybrid clones had near tetraploid chromosome numbers. Later studies (Harms et al., 1982) showed that the resistance traits were generally maintained for at least 250 cell doublings without selection pressure and the hybrids and C123 all contained a unique 94,000 kD polypeptide visualized by SDS-polyacrylamide gel electrophoresis.

These studies clearly show that certain resistance traits can be used to select for protoplast fusion hybrids due to dominant expression. This seems expected in the case of the 5MT resistance where expression of some amount of the feedback altered anthranilate synthase should cause resistance, however, in the case of the AEC resistance caused by decreased uptake it is not clear that dominant expression would be expected. Recently (ca. 6 years after the initial fusion experiments) we studied two of the fusion hybrids further (Widholm and Harms, 1987). Even though the hybrids H1B-2 and H2A2A had been grown without selection pressure, both retained the A2C resistance but the formeR had lost 5MT resistance. The 5MT resistance correlated with increased free tryptophan levels and the presence of the feedback altered form of anthranilate synthase. The A2C resistance correlated with increased free proline levels.

Both hybrids still retained the AEC resistance which was shown to be due to decreased AEC uptake. This dominant expression of a deficiency seems unlikely since the fusion partner should contribute an active uptake system which should be expressed, but examples from animal cell fusion, as discussed by Davidson (1974) show that actinomycin D resistance was expressed dominantly in hybrids (Sobel et al., 1971) and this resistance was caused by decreased uptake (Biedler and Riehm, 1970). One might postulate a multicomponent uptake system which requires all components to be active for normal uptake to occur. Another possibility would be a system to expel lysine at a high rate which could be expressed dominantly. Also the expression could be caused by a trans-acting inhibitor made in the C81 cells which acts to inhibit the lysine uptake system. We have no further evidence to help clarify the mechanism, however.

More in depth studies of the AEC uptake by C81 cells has shown that the decreased uptake is caused by either a shift in K_m of a single carrier system to a higher value or that the low K_m carrier (<300 µM AEC or lysine) is decreased in activity while the higher K_m carrier (>300 µM AEC or lysine) is unchanged (D.R. Duncan and J.M. Widholm, unpublished).

The amino acid analog resistance selection system was also used to select somatic hybrids between 5MT resistant N. tabacum and C81, the AEC resistant carrot line (Hauptmann et al., 1984). Comparison of polyethylene glycol, polyvinyl alcohol and dextran as fusigens showed them to be similar in their ability to form valid stable fusion hybrids. The doubly resistant selected clones initially contained complete sets of the carrot and tobacco chromosomes, but the double resistance was largely lost within 7 months when followed in 546 clones. The carrot chromosomes were preferentially lost with time but some were retained in clones which retained AEC resistance. These results show that intergeneric fusion hybrids can be selected using analog resistance and that the resistances can be lost as the chromosomes of one parent are lost.

Somatic Hybrid Selection Using Resistance and Plant Regeneration Ability

The carrot strain C123, which is incapable of regenerating plants, was also fused using dextran, with protoplasts from a Daucus capillifolius strain (DC) which could regenerate plants (Kameya et al., 1981). Colonies were selected for the 5MT resistance carried by the C123 cells and then for embryoid formation and green pigmentation, the capability carried by the DC cells. Shoots were regenerated which were intermedite in morphology between those of carrot and D. capillifolius. Suspension cultures initiated from shoots contained near the additive chromosome number, had intermediate 5MT resistance and A2C resistance near that of the C123 parent even though this resistance was not selected for. The hybrids also contained anthranilate synthase activity with feedback inhibition characteristics intermedite between those of the parents. Isozyme analysis also indicated that the selected cultures were hybrids. These studies show that 5MT and A2C resistance and plant regeneration are both dominant or semidominant traits in fusion hybrids.

mtDNA Recombination in Somatic Hybrids

Studies carried out about 4 years after the fusion with one of the fusion hybrids between C123 and DC described above, showed that both parental forms of the enzyme homoserine dehydrogenase were still being expressed and that the mitochondrial (mt) DNA of the hybrid had undergone recombination following fusion (Matthews and Widholm, 1985). Recombination was indicated by the presence of new restriction endonuclease fragments and the loss of some parental fragments in the mtDNA of the hybrid. The DNA changes should not have been due to random occurrences over time since several different strains, like C123 and C81 which were selected from a single parent carrot line C1 in 1977 (Widholm, 1977) and 1973 (characterized by Matthews et al., 1980), respectively, all had identical patterns even though the strain selections had occurred several years earlier. The C1 parental cell line was initiated in 1967. Carrot cell mtDNA stability was also demonstrated when six plants were regenerated from a carrot suspension culture and cultures reinitiated from these six plants were found to have mtDNA identical to that of the original culture when tested with five restriction endonucleases (Matthews and DeBonte, 1985).

Further evidence for specific mtDNA recombination in carrot fusion hybrids was obtained when strains carrying resistance to glyphosate(Glp, the strain denoted PR), 5MT, sodium selenate or selenocystine were fused with dextran in three combinations (Kothari et al., 1986). Doubly resistant clones were selected using the corresponding inhibitors where unfused mixture controls showed no growth. The clones were independently derived since the fused protoplasts were immobilized in agar 14 days after the fusion and the clones used came from different plates as well. Suspension cultures initiated from the selected clones retained the double resistance and contained near the additive chromosome number of the parental strains.

Restriction endonuclease analysis of mtDNA of the parental strains showed that three different patterns were found which matched the culture origins. The sodium selenate and selenocystine resistant strains had identical mtDNA and had originated from the haploid HA line

(Furner and Sung, 1982, 1983). The PR and 5MT resistant strains which had been selected from two different wild type lines had mtDNA different from each other and from the other two strains and also had been selected from different wild type lines. In two of the fusions of parental combinations with different mtDNAs (5MT resistant with selenocystine resistant and PR with sodium selenate resistant), the hybrids contained mtDNA identical to one of the parents (the former parent in the two cases). This indicates that mt sorting out appears to occur which would result in the loss of one parental type.

In the other combination, where the PR strain was fused with the 5MT resistant strain, the mtDNA restriction patterns of eight of the fusion hybrids examined were different from that of either parent but were identical to each other. Gel analysis of the parental mtDNAs following SalI digestion shows that the PR and 5MT resistant strains have 29 and 30 fragments larger than one kb, respectively. Sixteen of these fragments are common to each strain while the others are unique. The same analysis of the fusion hybrids produces 38 fragments with three being new fragments not found in the parents and with four parental fragments being lost. When the fragment sizes are all added together taking into account the stoichiometries of each fragment size estimated by densitometry, the hybrid mtDNA genome size is ca. 100 kb larger than the ca. 250-265 kb genome size of the parents.

This data would indicate that mtDNA recombination occurred only in one of the three fusion combinations utilized here and that identical recombinational events occurred in the eight hybrids examined. This apparently identical recombination pattern might result from recombination at specific sites such as the repeats found in maize (Lonsdale et al., 1984) and Brassica campestris (Palmer and Shields, 1984) which have been implicated in normal recombinational events which produce subgenomic size molecules. If this is true and the repeats are identical then each new unique fragment found in the hybrids should contain a repeat which would show homology to each other. However, when one of the fragments was isolated from a gel and labeled with ^{32}P and used as a probe, no hybridization to the other unique fragments was observed.

Somatic Hybrid Selection Using a Universal Hybridizer

We have also utilized a universal hybridizer system, i.e. a strain carrying both a dominant and recessive selectable marker. Such a strain is called a universal hybridizer since it can be fused with any wild type line and the hybrids be selected. We used the nitrate reductase deficient strain of Nicotiana plumbaginifolia, NX1, isolated by Marton et al. (1982) and selected for A2C resistance (Ye et al., 1987). Thus this strain, denoted NXA^r, carrying the recessive nitrate reductase deficiency and the dominant A2C resistance can be fused with any wild type line and the hybrids be selected for by their ability to grow on nitrate and A2C containing medium. The hybrids can grow since the wild type would supply nitrate reductase and the universal hybridizer the A2C resistance while the wild type alone would be inhibited by A2C and the universal hybridizer alone cannot grow on nitrate medium.

When PR, a Glp resistant carrot line (Nafziger et al., 1984), and NXA^r protoplasts were fused with dextran and were plated after a 14 day incubation in complete medium into A2C containing medium with nitrate as the nitrogen source, 528 colonies formed after 6 to 8 weeks from 2×10^6 fused protoplasts. No colonies formed when unfused mixtures of protoplasts were cultured in the same way. When these colonies were retested for growth in medium containing nitrate, A2C and in addition, Glp, which was not used in the original selection, 93% of the colonies grew. This indicates that most if not all of the selected colonies were indeed hybrids between the universal hybridizer and the PR line carrying Glp resistance. Further analysis of suspension cultures of several of the hybrids demonstrated near additive chromosome numbers, intermediate resistance to A2C and intermediate nitrate reductase levels. These hybrids also contained phosphoglucoisomerase, isocitrate dehydrogenase and malate dehydrogenase isozyme banding patterns largely matching those of both parents.

Expression of Amplified Genes in Somatic Hybrids

The Glp resistance carrot line used in this fusion experiment had previously been shown to contain 12-fold increased levels of the Glp target enzyme, enolpyruvylshikimic acid-3-phosphate synthase (EPSPS) with kinetic characteristics similar to the enzyme in the wild type line (Nafziger et al., 1984). Since a similar selection with *Petunia* cultures increased the EPSPS levels due to gene amplification (Shah et al., 1986) we used the *Petunia* EPSPS cDNA clone and antibodies to study the PR strain. These studies showed that gene amplification had occurred in the PR line and that this leads to the production of increased levels of EPSPS (R. M. Hauptmann, G. della-Cioppa, A. G. Smith and J. M. Widholm, unpublished). In higher plants, the only other case of selected gene amplification is with the nonselective herbicide L-phosphinothricin (L-glufosinate), which inhibits glutamine synthetase (Dunn et al., 1984). Alfalfa (*Medicago sativa*) suspension cultures were selected gradually on increasing L-phosphinothricin levels to produce strains with increased resistance (20 to 100-fold), increased glutamine synthetase activity (3 to 7-fold), increased glutamine synthetase mRNA levels (8-fold) and increased glutamine synthetase gene number (4 to 11-fold).

The *Petunia* EPSPS cDNA clone was also used to probe the hybrids formed between the PR and NXA^r strains and the amplified gene number was maintained in these hybrids (Ye et al., 1987). This probing with a clone showing homology to both parental strain genes, also showed that homologous sequences from both parents were indeed present in the hybrids, clearly confirming the hybrid nature.

The Pr strain with the EPSPS amplified gene number was also fused with dextran with protoplasts from two cloned lines of strain C123 which carried 5MT and A2C resistance (R.M. Hauptmann, G. della-Cioppa, A. G. Smith and J. M. Widholm, unpublished). Somatic hybrids were selected by plating into medium containing A2C and Glp concentrations which were inhibitory to the parental strains not selected for resistance to the respective inhibitor. The doubly resistant colonies which were recovered at a frequency near 0.1% were all (408 colonies tested) capable of growing on normally inhibitory levels of 5MT as well. Most of the selected putative hybrids contained additive chromosome numbers with no evidence of double minute chromosomes. Most of the selected fusion hybrids maintained the resistances for a three year period without selection pressure. The Glp resistance was again shown to be due to EPSPS gene amplification in the hybrids which caused increased EPSPS enzyme activity and EPSPS protein levels as measured by western hybridization analysis.

Conclusion

The studies discussed here on carrot protoplast fusion have demonstrated several phenomena including the very efficient selection of hybrids using amino acid analog and herbicide resistances, plant regeneration ability or a universal hybridizer cell strain. Even a decreased uptake resistance was used effectively as a selectable trait since it was expressed in somatic hybrids. Having unselected resistances carried by the fusion partners can be useful in proving the hybridity of the putative hybrids. Usually the selectable traits are expressed stably for long periods in the hybrids except in the intergeneric fusion example where carrot chromosomes and the resistance they carried were regularly lost in a somatic hybrid with *N. tabacum*. These studies have also shown that amplified genes are stably maintained and expressed in somatic hybrids indicating that the genes are stably integrated chromosomally. Southern hybridization analysis demonstrating gene polymorphism can also be used to confirm the parentage. Recombination of mtDNA can be demonstrated in specific somatic hybrid combinations and this recombination appears to be identical in independently derived hybrids. The three fusigens used, dextran, polyethylene glycol and polyvinyl alcohol, all produce similar numbers of viable fusion hybrids.

The systems demonstrated here need to be studied further especially at the molecular level to learn more about the basic mechanisms involved in phenomena such as gene amplification, dominant uptake resistance and mtDNA recombination.

References

Biedler, J. L. and Riehm, H., 1970, Cellular resistance to actinomycin D in Chinese hamster cells in vitro: cross-resistance, radioautographic, and cytogenetic studies, *Cancer Res.*, 30:1174.

Davidson, R. L., 1974, Gene expression in somatic cell hybrids, *Ann. Rev. Genet.* 8:195.

Donn, G., Tischer, E., Smith, J. A. and Goodman, H. M., 1984, Herbicide-resistant alfalfa cells: an example of gene amplification in plants, *J. Molec. Applied Genet*. 2:621.

Furner, I. J., and Sung, Z. R., 1982, Regulation of sulfate uptake in carrot cells: properties of a hyper controlled variant. *Proc. Natl. Acad. Sci. USA*, 79:219.

Furner, I. J. and Sung, Z. R., 1983, Characterization of a selenocystine resistance carrot cell line, *Plant Physiol*. 71:547.

Harms, C. T., Oertli, J. J., and Widholm, J. M., 1982, Characterization of amino acid analogue resistant somatic hybrid cell lines of *Daucus carota* L., *Z. Pflanzenphysiol.*, 106:239.

Harms, C. T., Potrykus, I., and Widholm, J.M., 1981, Complementation and dominant expression of amino acid analogue resistance markers in somatic hybrid clones from *Daucus carota* after protoplast fusion, *Z. Pflanzenphysiol.*, 101:377.

Hauptmann, R., Kumar, P., and Widholm, J., 1983, Carrot x tobacco somatic cell hybrids selected by amino acid analog resistance complementation, in: "6th International Protoplast Symposium, 1983" I. Potrykus, C. T. Harms, A. Hinnen, R. Hutter, P. J. King, and R. D. Shillito, eds., Birkhauser Verlag, Boston, pp. 92.

Kameya, T., Horn, M.E., and Widholm, J. M., 1981, Hybrid shoot formation from fused *Daucus carota* and *D. capillifolius* protoplasts, *Z. Pflanzenphysiol*. 104:459.

Kothari, S. L. Monte, D. C., and Widholm, J. M., 1986, Selection of Daucus carota somatic hybrids using drug resistance markers and characterization of their mitochondrial genomes, Theor. Appl. Genet., 72:494.

Lonsdale, D. M., Hodge, T. P., and Fauron, C. M. R., 1984, The physical map and organization of the mitochondrial genome from the fertile cytoplasm of maize, Nucleic Acids Res., 12:9249.

Marton, L., Dung, T. M., Mendel, R. R., and Maliga, P., 1982, Nitrate reductase deficient cell lines from haploid protoplast culture of Nicotiana plumbaginifolia. Mol. Gen. Genet. 186:301.

Matthews, B. F., and DeBonte, L. R., 1985, Chloroplast and mitochondrial DNAs of the carrot and its wild relatives, Plant Mol. Biol. Reporter, 3:12.

Matthews, B. F., Shye, S. C. H., and Widholm, J. M., 1980, Mechanism of resistance of a selected carrot cell suspension culture to S(2-aminoethyl)-L-cysteine, Z. Pflanzenphysiol., 94:453.

Matthews, B. F., and Widholm, J. M., 1985, Organelle DNA compositions and isoenzyme expression in an interspecific somatic hybrid of Daucus, Mol. Gen. Genet., 198:371.

Nafziger, E. D., Widholm, J. M., Steinrucken, H. C., and Killmer, J. L., 1984, Selection and characterization of a carrot cell line tolerant to glyphosate, Plant Physiol., 76:571.

Palmer, J. D., and Shields, C. R., 1984, Tripartite structure of the Brassica campestris mitochondrial genome, Nature 307:437.

Shah, D. M., Horsch, R. B., Klee, H. J., Kishore, G. M., Winter, J. A., Tumer, N. E., Hironaka, C. M., Sanders, P. R., Gasser, C. S., Aykent, S., Siegel, N. R., Rogers, S. G., and Fraley, R. T., 1986, Engineering herbicide tolerance in transgenic plants, Sci., 233:478.

Sobel, J. S., Albrecht, A. M., Riehm, H., and Biedler, J. L., 1971, Hybridization of actinomycin D- and amethopterin-resistant Chinese hamster cells in vitro, Cancer Res. 31:297.

Widholm, J. M., 1977, Relation between auxin autotrophy and tryptophan accumulation in cultured plant cells, Planta, 134:103.

Widholm, J. M., and Harms, C. T., 1987, Dominant expression of an amino acid uptake deficiency in carrot somatic fusion hybrids, Plant Cell Reports, 6:138.

Widholm, J. M., Hauptmann, R. M., and Dudits, D., 1984, Carrot protoplast fusion, Plant Mol. Biol. Reporter, 2:26.

Ye, J., Hauptmann, R. M., Smith, A. G., and Widholm, J. M., 1987, Selection of a Nicotiana plumbaginifolia universal hybridizer and its use in intergeneric somatic hybrid formation, Mol. Gen. Genet., in press.

CANDIDATE FOR A SEXUAL REPRODUCTION SPECIFIC PATHWAY IN PLANTS: PUTRESCINE TO GABA VIA HYDROXYCINNAMIC AMIDE INTERMEDIATES

Philip Filner

Sungene Technologies Corporation
3330 Hillview Avenue
Palo Alto, California 94304, U.S.A.

INTRODUCTION

In biochemical and molecular terms, the reproductive development of plants is still largely an enigma. We are quite familiar with the anatomical, morphological changes, but we know relatively little about reactions which are causal in reproductive development. Most of the reactions which are known, e. g. accumulation of storage proteins in embryonic cotyledons, are consequences of the process, not primary causal events. Progress is hampered by our limited knowledge of causal reactions in reproductive development. Out of the work discussed and presented in this paper, a picture is emerging of unique metabolism which appears likely to have a central role in reproductive development of all plants. The as yet uncharacterized enzymes involved in this metabolism, and the genes encoding them, merit more attention than they have so far received.

POLYAMINE CONJUGATES IN FLOWERS

Martin-Tanguy and coworkers at Dijon (Martin-Tanguy et al., 1978; Martin-Tanguy, 1985) discovered that the reproductive organs of apparently all plant species contain millimolar concentrations of amides which are conjugates of hydroxycinnamic acids (p-coumaric, caffeic, ferulic or sinapic acid) and aliphatic polyamines (putrescine, spermidine or spermine) or aromatic monoamines (tyramine or dopamine). Prior to the work of the Dijon group, there were a few isolated reports of occurrence of hydroxycinnamic amides (HCAs) in plants (Mbadiwe, 1973). HCAs of putrescine were first encountered in tobacco by Mizusaki et al. (1971), when they fed (C-14)putrescine to cultured tobacco cells in a study of nicotine biosynthesis. Shortly thereafter, Martin-Tanguy et al. (1973) found these compounds in TMV-infected leaves of Nicotiana tabacum cv. Xanthi-nc and showed that the compounds interfered with TMV infection. Further work done at Dijon, particularly the thesis work of F. Cabanne (Cabanne et al., 1977; Cabanne, 1980; Cabanne et al., 1981) revealed that the HCAs were absent from healthy, uninfected vegetative tobacco plants, but were present in the inflorescences of reproductive tobacco plants, and in the leaves immediately adjacent to the inflorescences.

Within flowers, the compounds were concentrated in the anthers and ovaries. Anthers contained predominantly neutral di-HCAs, and ovaries contained predominantly basic mono-HCAs. While elevated temperature could prevent flowering of tobacco, it did not prevent "ripening to flower", i.e. the switching from a vegetative to a pre-reproductive state in which the HCAs began to be appear in the top leaves as they would in a flowering plant, but they did not accumulate to the high concentrations characteristic of a flowering plant (Cabanne et al. 1981).

At about the same time In the U.S., Widholm (Palmer & Widholm, 1975; Berlin and Widholm, 1978) selected variants of the XD line of cultured tobacco cells originally developed by Filner (1965), coincidentally from N. tabacum cv. Xanthi-nc. The variants selected for resistance to p-fluorophenylalanine accumulated phenolics made from phenylalanine, rather than the expected phenylalanine. Subsequently, Berlin (1981) identified the phenolics as hydroxycinnamic amides of putrescine, particularly caffeoylputrescine.

Up to this point, HCAs appeared to be just another group of natural products made by a few plant species. Then the Dijon group did a survey of the plant kingdom, and found HCAs in the reproductive organs of every species examined (Martin-Tanguy et al., 1978). Furthermore, while most HCAs were found in more than one species, and more than one HCA was detected in almost all species, the mix of HCAs was species-specific. Three types of polyamines and four types of hydroxycinnamic acids can potentially be combined in 16 mono-HCAs of polyamines (because the two primary amines of spermidine are not equivalent), and 12 homodi-HCAs of polyamines , i.e. with the same hydroxycinnamic acid at both ends of the polyamine. Heterodi-HCAs of polyamines might also exist. There also are 8 possible HCAs of aromatic monoamines. Taking HCAs in sets of 2, 3, 4, etc, there is a very large number of possible mixes of HCAs, thus making possible species-specificity of the mix.

The possibility that the sex organs of plant species can be distinguished from each other by the mix of this one family of compounds is intriguing. The concentrations at which they occur-micromoles per g fresh weight-are indicative of a major metabolite rather than a hormone or other type of regulatory effector. The HCAs can contain 20% to 100% as much nitrogen as all protein in sex organs of the plant. It seems highly unlikely that all plant species would invest so much of their reproductive energy in HCAs unless they make an important contribution to successful sexual reproduction.

One can imagine a number of possible functions for these compounds. They absorb UV light and fluoresce, so one possible function may be to attract pollinating insects. These compounds at high concentrations have antibiotic activities, so they may protect the exposed sex organs from pathogens. However, these possible functions should not require different compounds in the male and female organs. Some complementary metabolic relationship is required for species-specific fertilization. Perhaps these compounds are involved in that as yet unknown complementary metabolism. One can readily imagine that in an interspecific incompatible mating, one or more HCA would be contributed by one gamete without the enzymes necessary for its essential metabolism, and the gamete from the other species could lack them or the signal to induce them.

The Dijon group asked the question: do HCAs have a function in the sex organ which is important to the reproductive mechanism? Martin-Tanguy et al (1982) examined the occurrence of the HCAs in cytoplasmically male sterile maize with and without a restorer gene present. In both C-type and T-type cms, the concentration of the major HCAs in anthers, feruloylputrescine and diferuloylputrescine, were 1/30 - 1/100 the levels in anthers of wild type fertile maize, or in anthers of fertile plants carrying both T cytoplasm and a nuclear restorer gene. Furthermore, the concentration of diferuloylspermidine, the major HCA in ovaries of maize, was not significantly affected by the cms mutations, which cause the pollen to abort. The HCAs in the anthers of male fertile plants are localized in the pollen (Martin-Tanguy, 1985).

In another system studied by the Dijon group, the inflorescences of Araceae (Ponchet,et al.,1982), male sterile and fertile flowers appear on the same inflorescence. Again, they found that male sterility was associated with a deficiency in HCAs of the male organ.

Additional evidence for a link between flowering and hydroxycinnamic amides comes from another piece of work from Dijon. A tobacco mutant, RMB7, has lost the ability to flower. Leaf discs of this mutant will not proliferate callus on Murashige & Skoog media, but leaf discs of normal tobacco will callus on the same medium. Addition of HCAs of putrescine, the ones which occur in inflorescences of normal tobacco, cause the RMB7 leaf discs to produce abundant callus (Martin, et al.,1985).

POLYAMINES IN REPRODUCTIVE DEVELOPMENT OF PLANTS

Polyamines are synthesized via either of two pathways (Tabor & Tabor, 1984). The first begins with decarboxylation of L-arginine, catalyzed by arginine decarboxylase, to yield agmatine. Agmatine is probably first deamidated to yield ureidoputrescine, and then the ureido group is lost, resulting in putrescine. The alternative path is decarboxylation of L-ornithine to yield putrescine directly. Alpha difluoromethylarginine (DFMA) specifically inhibits arginine decarboxylase and alpha difluoromethylornithine (DFMO) specifically inhibits ornithine decarboxylase. Spermidine is synthesized from putrescine by donation of an aminopropyl group from the methionine portion of S-adenosylmethionine (SAM) to one of the amino groups of putrescine. Methylglyoxal-bis(guanylhydrazone), known as MGBG, has been used by some researchers as an inhibitor of SAM decarboxylase, the first step in the making of the aminopropyl group for spermidine synthesis. However, MGBG inhibits at least one other enzyme in polyamine metabolism, diamine oxidase and induces others (Pegg et al, 1985), so interpretation of its effects is not straightforward.

The first evidence that putrescine was essential for reproductive development in plants came from Cohen et al (1982). They showed that the growth of the tomato ovary was inhibited by DFMO, and the inhibition was overcome by putrescine. Feirer et al (1984) showed that at sublethal concentrations, DFMA partially inhibited embryogenesis in carrot cell cultures, and spermidine overcame the inhibition.

Malmberg (Malmberg, et al., 1985) has selected tobacco cells for resistance to inhibitors of polyamine metabolism, such as MGBG, and regenerated mutant plants from such cells. Many of these mutant plants look like homeotic mutants-the sex organs, or parts of them, form in the wrong place in the flower. He has found altered levels of enzymes of polyamine metabolism in these mutants.

Taken together, these results indicate that normal polyamine metabolism is essential for normal reproductive development in plants. These results do not necessarily mean that the free polyamine is the biologically important compound. It could also be a metabolite derived from the polyamine. The abundance of conjugated polyamines in HCAs of reproductive tissue makes this a key point to clarify in future work. Slocum and Galston (1985) determined free and base-hydrolyzable conjugates of polyamines in developing tobacco flowers. Although they did not identify the chemical nature of the conjugates, it is reasonable to presume that they were HCAs. They found that the bulk of the polyamines were conjugated. They also found that total polyamines reached a peak early in reproductive development, then declined during seed maturation. Both free and conjugated pools of polyamines declined.

While the prevailing view among workers in the field seems to be that free polyamines are the biologically important species, and conjugates are some storage form of polyamines, there is no evidence to support this bias. On the contrary, the very close association of occurrence of HCAs and normal, functional reproductive development in plants, leads this writer to favor the importance of the HCAs over the free polyamines. The latter can occur anywhere in a plant, and have no special link to reproductive development.

OXIDATIVE METABOLISM OF PUTRESCINE

In bacteria, fungi, animals, and some plants, there exists a two-step pathway by which putrescine is oxidized to 4-aminobutryic acid (gamma aminobutyric acid, GABA). Animals also synthesize N-acetylputrescine, and a pathway of N-acetylated intermediates was hypothesized (Seiler et al, 1975), but not proven to exist in animals. More recently, genetic and biochemical evidence for an oxidative pathway of N-acetylated intermediates has come from studies of a yeast, Candida boidinii (Haywood and Large, 1986), when it grows on putrescine as sole nitrogen source. The usual path from putrescine to GABA begins with the diamine oxidase reaction, which yields 4-aminobutyraldehyde (GABaldehyde). This compound is in equilibrium with its dehydration product, pyrroline. The second enzyme in the pathway is pyrroline dehydrogenase, which catalyzes the synthesis of GABA from pyrroline. An alternative pathway begins with conversion of putrescine to spermidine, followed by the action of polyamine oxidase, which decomposes spermidine to yield GABaldehyde. Pea contains diamine oxidase and pyrroline dehydrogenase, while oat has in addition the polyamine oxidase route (Flores & Filner, 1985a).

The polyamine oxidases of animals actually are more active on N-acetyl spermidine and N-acetylspermine than on the free polyamines. They catalyze oxidation of the secondary amines, cleaving the molecule into aminopropionaldehyde and putrescine or spermidine (not the N-acetyl compounds), depending on the substrate (Tabor & Tabor, 1984). Interestingly, partially purified polyamine oxidase of oats also is active on both mono and diactylpolyamines, but cleaves on the other side of the secondary amine, releasing diaminopropane and pyrroline (Suzuki, et al.,1985). On the other hand, an amine oxidase from soybean oxidizes but does not cleave mono-N-acetylpolyamines (Suzuki, et al., 1985), so it is presumed to oxidize the primary amine at the end of molecule to an aldehyde.

PUTRESCINE UTILIZATION BY CULTURED TOBACCO CELLS

In a survey of the ability of the XD line of cultured tobacco cells to assimilate various organic nitrogen compounds, we found that the cells could use the nitrogen in GABA as well as the nitrogen in nitrate, both of which supported a doubling time of two days. However, the cells could not grow on putrescine as sole nitrogen source. Because the literature indicated that there was a two-step pathway between putrescine and GABA in some organisms, this looked like a potential opportunity to select for amplification of an enzyme. Selection for growth on putrescine as sole source of nitrogen was therefore undertaken.

When the cells were incubated in putrescine medium, they stayed alive, as indicated by cytoplasmic streaming, but they did not proliferate. However, they developed a blue color. After harvesting and resuspending the cells in fresh putrescine medium every few weeks for about half a year, a cell line which grew rapidly on putrescine emerged.

To determine if these cells were growing on putrescine nitrogen via conversion to GABA, the inhibitor Gabaculine was used. This compound has a Ki of 2 nM when inhibiting GABA:pyruvate transaminase extracted from the XD or Put cells. When added to culture media, it inhibited growth of the putrescine-utilizing cells (Put cells) on putrescine, or the growth of XD cells on GABA, at less than 10 nM. However, growth on nitrate or urea was unaffected until concentrations close to 10 uM were reached. The Put cells were found to be very efficient at converting (14-C)putrescine to (14-C)-GABA, but the XD cells could not perform that conversion (Flores & Filner, 1985b).

The above results convinced us that the Put cells were oxidizing putrescine to GABA. We were therefore very surprised when we could not find diamine oxidase, polyamine oxidase or pyrroline dehydrogenase in the XD or Put cells, although we were able to extract and assay these enzymes from pea and oat. Mixing experiments indicated that the enzymes should have been detected if they were in the tobacco cell extracts.

The apparent absence of the best known pathways from putrescine to GABA led us to look for alternatives. In the experiments in which (14C)-putrescine was fed to XD or Put cells, the only other major peak of radioactivity was in HCAs of putrescine, which was mostly caffeoylputrescine. We thought that perhaps hydroxycinnamoylputrescines might be used in plants in an oxidative pathway to GABA with N-acylated intermediates, similar to the pathway of N-acetylated intermediates from putrescine to GABA which which had been hypothesized, but never shown, to occur animal cells (Seiler et al., 1975).

Aminooxyphenylpropionic acid (AOPP) is a specific inhibitor of phenylalanine ammonia lyase, the enzyme which catalyzes the synthesis of trans-cinnamic acid from phenylalanine. There is no reason to expect AOPP to inhibit growth on one nitrogen source more than another, unless some derivative of trans-cinnamic acid is involved in the assimilation of the nitrogen source. We found that 5 uM AOPP inhibited accumulation of HCAs of putrescine in the XD and Put cells, regardless of nitrogen source. That concentration of AOPP did not inhibit growth on nitrate or urea, but did inhibit growth of Put cells on putrescine. This was the first positive indication that perhaps the hypothesized pathway of HCA intermediates actually existed (Flores & Filner, 1985b).

We set out to look for the one hypothetical intermediate which we thought might be stable: caffeoylGABA (Balint, et al, 1987). This compound had not been described in the literature, neither as a synthetic compound nor as a natural product. In order to detect it as a natural product, we needed a synthetic standard which could be used to set up analytical methods suitable for use with plant extracts. The compound turned out to be somewhat more difficult to synthesize directly in good yield than expected. The problem was the reactivity of the terminal carboxylate with the amide nitrogen. There is a marked tendency for the compound to cyclize, forming the pyrrolidone ring. As it turned out, diacetoxycaffeoylpyrrolidone (DACP) was relatively easy to synthesize and it behaved well in both gas chromatography and mass spectrometry. It could be readily converted to caffeoylGABA, and vice versa. The analytical procedure which we eventually adopted involved converting caffoylGABA partially purified by HPLC to DACP for identification and quantification.

Using this procedure, we were able to detect caffeoylGABA in extracts of Put cells, but not in extracts of XD cells. Thus a key prediction of the hypothesis of a pathway of HCA intermediates from putrescine to GABA turned out to be correct. The level of caffeoylGABA was quite low, however, compared to that of caffeoylputrescine, usually being 1/30 to 1/100 the concentration of caffeoylputrescine (Balint et al., 1987). This may mean that the rate-limiting step of the proposed pathway is between caffeoylputrescine and caffeoylGABA.

Knowing of the work of the Dijon group, we looked for caffeoylGABA in vegetative and reproductive tobacco. We found that the compound was absent from vegetative tobacco, but present in flowers and leaves adjacent to flowers of reproductive tobacco, but not in lower leaves. That is, the distribution closely paralleled that of caffeoylputrescine, but as in the cultured cells, the level of caffeoylGABA was only a few percent of the level of caffeoylputrescine (Balint et al., 1987). This pattern of occurrence of caffeoylGABA is what one would expect if the proposed pathway existed in reproductive organs of tobacco, but not in vegetative tobacco.

Finding caffoylGABA, and inhibiting growth on putrescine with AOPP, does not prove that the proposed pathway exists. We still must show that putrescine goes through the proposed intermediates, and detect the proposed enzymes. It also will be important to determine whether analogous reactions involving other HCAs occur in other plant species.

QUESTIONS ABOUT ASSIMILATION OF POLYAMINES VIA HCA INTERMEDIATES

We do not know if the HCAs are made in the reproductive tissue or are imported perhaps from nearby leaves. If they are imported, they have some attractive features. GABA provides nitrogen with minimal dilution by carbon, and the carbon can be conserved. Little of the carbon need be lost as carbon dioxide because GABA leads to the tricarboxylic acid cycle at succinate, and the three decarboxylation steps lie between pyruvate and succinate. Also, the nitrogen is readily available via transamination to pyruvate, yielding alanine, and then transamination from alanine to alpha-ketoglutarate, yielding glutamate, which is virtually a universal precursor for nitrogen compounds. Finally, the hydroxycinnamoyl moiety can be used as is in the synthesis of anthocyanins to pigment the flower. If they are imported, the HCAs are almost ideally structured to be used in a highly efficient manner by developing reproductive structures.

The XD cells readily make large amounts of caffeoylputrescine, but cannot further metabolize it. This compound is normally made in reproductive tobacco but not vegetative tobacco. It may be significant that the XD line was started from stem pith of a bolted tobacco plant, i.e. one in the early part of the reproductive state. The Put cells were selected from the XD cells. They not only make caffeoyl putrescine; they appear to oxidize it to caffeoylGABA and release GABA. Perhaps the metabolic pathway was turned on in the cultured cells in two steps corresponding to the "ripening to flower" and "flowering" stages which the Dijon group discerned in their studies of accumulation of HCAs in tobacco. It will be interesting to see in the future if there are two stages in the development of the capacity to oxidize HCAs of polyamines in developing flowers.

The existence of a different mix of HCAs in the male and female parts of flowers raises the question: if they are imported, how are the different HCAs sorted ? If they are made in the sex organ in which they are found, are there sex-organ specific enzymes for each type of HCA ? If the HCAs are oxidized in the sex organs in which they are found, are their sex-organ specific oxidative enzymes ? The transport model suggests functions for the different N-acyl groups: "ticketing" and "metering" the supplies of metabolites for building vital structures in the sex organs.

Future research in this area would be greatly facilitated if some of the HCAs of polyamines were commercially available. They are needed for analysis standards, enzyme assays, evaluation as physiological effectors, etc. Specific inhibitors of synthesis or further metabolism would also be very helpful. The agricultural chemical companies are a potential source of such compounds, since many of them have sought compounds which could be used to control reproductive development in plants. If the HCAs of polyamines play a central role in reproductive development as proposed, then some of the compounds already discovered may act by perturbing this metabolism. Conversely, chemicals designed to perturb this metabolism may do interesting and useful things to the reproductive process in plants.

ACKNOWLEDGEMENT

The research on putrescine utilization by cultured tobacco cells, and identification of caffeoylGABA, was done at what was then the ARCO Plant

Cell Research Institute, Dublin, California, in collaboration with Hector Flores, who is now at Louisiana State University; and Robert Balint, Geoffrey Cooper and Mark Staebell. The latter three remain on the staff of the reconstituted Plant Cell Research Institute, the major owner of which is the Montedison Company of Milan.

REFERENCES

R. Balint, G. Cooper, M. Staebell and P. Filner, N-caffeoyl-4-amino-n-butyric acid (caffeoylGABA) a new flower-specific metabolite in cultured tobacco cells and tobacco plants, J. Biol. Chem., in press (1987).

J. Berlin, Formation of putrescine and cinnamoyl putrescines in tobacco cell cultures, Phytochem. 20:53 (1981).

J. Berlin and J. M. Widholm, Metabolism of phenylalanine and tyrosine in tobacco cell lines resistant and sensitive to p-fluorophenylalanine, Phytochem. 17:65 (1978).

F. Cabanne, Etude de nouveaux dérivés hydroxycinnamiques au cours du développement de Nicotiana tabacum L. cv. Xanthi n.c., Thèse d'Etat, Dijon (1980).

F. Cabanne, M. A. Dalebroux, J. Martin-Tanguy and C. Martin, Hydroxycinnamic acid amides and ripening to flower of Nicotiana tabacum var. xanthi n.c., Physiol. Plantar. 53:399 (1981).

F. Cabanne, J. Martin-Tanguy & C. Martin, Phénolamines associés à l'induction florale et à l'état reproducteur du Nicotiana tabacum L. var. Xanthi n.c., Physiol. Vég. 15:429 (1977).

E. Cohen, S. Arad, Y. Heimer, Y. Mizrahi, Participation of ornitihine decarboxylase in early stages of tomato fruit development, Plant Physiol. 70:540 (1982).

R. P. Feirer, G. Mignon and J. D. Litvay, Arginine decarboxylase and polyamines required for embryogenesis in the wild carrot, Science 223:1433 (1984).

P. Filner, Semi-conservative replication of DNA in a higher plant cell, Exptl. Cell Res. 39:33 (1965).

H. Flores and P. Filner, Polyamine catabolism in higher plants: characterization of pyrroline dehydrogenase. Plant Growth Reg 3:277 (1985a).

H. Flores and P. Filner, in Primary and Secondary Metabolism in Plant Cell Cultures (K. H. Neumann, W. Barz & E. Reinhard, eds), Springer-Verlag, Berlin pp 174-185 (1985b).

G. W. Haywood and P. J. Large, 4-Acetamidobutyrate deacetylase in the yeast Candida boidinii grown on putrescine or spermidine as sole nitrogen source and its probable role in polyamine catabolism, J. Gen. MIcrobiol. 132:7 (1986).

R. Malmberg, J. McIndoo, A. C. Hiatt, B. A. Lowe, Genetics of polyamine synthesis in tobacco, developmental switches in the flower, Cold Spr. Harb. Symp. 50:475 (1985).

C. Martin, G. Kunesch, J. Martin-Tanguy, J. Negrel, M. Paynot and M. Carre, Effect of cinnamoyl putrescines on in vitro cell multiplication and differentiation of tobacco explants, Plant Cell Reports 4:158 (1985).

J. Martin-Tanguy, The occurrence and possible function of hydroxycinnamic acid amides in plants, Plant Growth Reg 3:381 (1985).

J. Martin-Tanguy, C. Martin & M. Gallet, Présence de composés aromatiques liés à las putrescine dans divers Nicotianas virosés, C. R. Acad. Sci. 276:1433 (1973).

J. Martin-Tanguy, E. Pedrizet, J. Prevost and C. Martin, Hydroxycinnamic acid amides in fertile and cytoplasmic male sterile lines of maize, Phytochem. 21:1939 (1982)

J. Martin-Tanguy, F. Cabanne, E. Pedrizet and C. Martin, The distribution of hydroxycinnamic acid amides in flowering plants, Phytochem. 17:1927 (1978).

E. I. Mbadiwe, Caffeoylputrescine from Pentalethra macrophyta, Phytochem. 12:2546 (1973).

S. Mizusaki, Y. Tanabe, M. Noguchi and E. Tamaki, p-Coumaroylputrescine, caffeoylputrescine and feruloylputrescine from callus tissue culture of Nicotiana tabacum, Phytochem. 10:1347 (1971).

J. E. Palmer and J. Widholm, Characterization of carrot and tobacco cell cultures resistant to p-fluorophenylalanine, Plant Physiol. 56:233 (1975).

A. E. Pegg, B. G. Erwin and L. Persson, Induction of spermidine/spermine N1-acetyltransferase by methylglyoxal-bis(guanylhydrazone), Biochim. Biophys. Acta 842:111 (1985).

M. Ponchet, J. Martin-Tanguy, A. Marais and C. Martin, Hydroxycinnamoyl acid amides and aromatic amines in the influorescences of some Araceae species, Phytochem. 21:2865 (1982).

R. D. Slocum and A. W. Galston, Changes in polyamine biosynthesis associated with post-fertilization growth and development in tobacco ovary tissues, Plant Physiol. 79:326 (1985).

N. Seiler, U. Lamberty and M. J. Al-Therib, Acetyl-coenzyme A: 1,4-diaminobutane N-acetyltransferase: activity in rat brain during development in experimental brain tumours and in brains of fish of different metabolic acitivity, J. Neurochem. 24:797 (1975).

O. Suzuki, T. Matsumoto and Y. Katsumata, Oxidation of acetylpolyamines by amine oxidase from soybean and oat seedlings, Biogenic Amines 2:37 (1985).

C. W. Tabor and H. Tabor, Polyamines, Ann. Rev. Biochem. 53:749 (1984).

HYDROXYCINNAMIC ACID AMIDES, HYPERSENSITIVITY, FLOWERING AND SEXUAL ORGANOGENESIS IN PLANTS

Josette Martin Tanguy, Jonathan Negrel
Michel Paynot and Claude Martin

INRA, Station de Physiopathologie végétale
BV 1540, F - 21034 Dijon cedex

INTRODUCTION

In recent years polyamines have been shown to participate in various aspects of plant growth processes [1,2,3]. In general, free putrescine spermidine and spermine were the only amines measured, and only little attention has been given to bound or conjugated polyamines.

The amine conjugates formed with hydroxycinnamic acids (called hydroxycinnamic acid amides) contain hydroxycinnamic acids and amines linked by an amide bound. There are two types : basic amides (water-soluble) having a primary amine function and neutral amides (water-insoluble) showing no strongly ionizable functions. The amines of basic amides include aliphatic, di-and polyamines such as putrescine, spermidine and spermine whereas those of neutral amides include both aliphatic and aromatic amines (tyramine, dopamine, serotonine, octopamine,tryptamine and so on). The hydroxycinnamic acid amides have now been found throughout the plant kingdom. In a survey of 20 species representing 13 families they occurred as the main phenolic constituents of the reproductive organs and seeds of all plants supplied [4]. In some cases of host-pathogen interactions, especially during the hypersensitive reaction towards fungi and viruses, leaves of vegetative plants synthesize some phenolic amides at the same time as defence mechanism limit pathogen expansion [5,6]. It is well known that resistance to viruses is often associated with flowering. Virus particles are nearly absent from meristems, sex organs and seeds where hydroxycinnamic acid amides accumulate in large amounts.

All of these results prompted us to examine the possible role of these amides in the regulation of plant development. In this review some of the recent data concerning the correlation between synthesis of large amounts of phenolic amides with virus resistance, initiation of floral development, and sexual organogenesis will be reported and discussed.

HYDROXYCINNAMIC ACID AMIDES AND VIRUS MULTIPLICATION IN NICOTIANA TABACUM PLANTS

Hydroxycinnamic acid amides. Tobacco mosaic virus (TMV), when inoculated into a plant, especially into a tobacco plant, can either produce a systemic infection (*Nicotiana tabacum* variety Samsun nn) or necrotic local lesions (*Nicotiana tabacum* variety Xanthi n.c.). This latter reaction is called a hypersensitive reaction. In inoculated *N*. Xanthi n.c. the virus remains confined to the necrotic lesions and to its immediate environs [7,8]. The behaviour of TMV in *N*. Xanthi n.c. also depends on the temperature at which infected plants are growing [7,8]. At temperatures below 29 °C the strain produces only necrotic lesions. At 32 °C, the type of reaction changes and the plants develop systemic symptoms [7,8]. It is now known that the localization of the virus is not caused by the necrosis but takes place in the living cells around the necrotic lesions [7,8]. The reason for the inability of virus to move through the host tissue is not understood, but there is evidence that a protective mechanism is developed by the infected plant. At 20 °C *N*. Xanthi n.c. accumulates large amounts of feruloylputrescine, diferuloylputrescine, feruloyltyramine and the β-glucoside of feruloyltyramine after infection with TMV [5,6,9]. They appear at the time of lesion formation and are accumulated mainly in the living cells surrounding the necrotic lesions. At the beginning of infection (48 hr after inoculation) feruloylputrescine reaches a concentration of about 0,5 μ mol/g fresh weight. All of these compounds are absent from the fully expanded vegetative leaves and from the control leaves inoculated with water. An increase in temperature which leads to the generalization of TMV, suppresses the accumulation of amides. No production of amides occurs when the infected plant is not hypersensitive to TMV (*N*. Samsun nn). Application of amides to excised leaf discs at ca. 1mM caused a significant reduction in the number of lesions on infection with TMV by comparison with an untreated control [6]. The formation of local necrotic lesions on *N*. Xanthi n.c. inoculated at 20 °C does not strongly affect free amine levels in comparison with the water inoculated control [10]. An interesting case is one of a *Nicotiana* hybrid (RMB 7) obtained from the seventh back cross on *N. rustica* by *N. tabacum* var. *Maryland Mammoth* [9]. This tobacco has lost its flowering capacity and remains vegetative under normal growth conditions. It never synthesizes detectable amount of amides at any stage of development. Inoculated with TMV it produces necrotic lesions like classical hypersensitive varieties, but the virus spreads and the necrogenic process develops over the entire plant. No production of amides occurs. Polyacrylic acid (PA) injected at 20 °C into a tobacco Xanthi n.c. induces complete resistance to infection with TMV [11]. The effect is maximal when the injection is made 2 to 3 days before inoculation. This induced-resistance disappears when plants are kept at 32 °C but the effect of temperature is reversible. Extracts from PA-injected leaves at 20 °C contain large levels of feruloylputrescine, diferuloylputrescine, feruloyltyramine and the β-glucoside of feruloyltyramine. The accumulation is transient ; the maximum occurs 3 days after injection. The resistance to virus infection disappears when the injected plants are kept for 2 days at 32 °C. Leaf extracts made after this treatment did not contain phenolic amides. Detailed studies relative to the biosynthesis of amides during hypersensitivity were achieved in our laboratory.

Ornithine decarboxylase activity. In tobacco, putrescine is known to be formed from ornithine and arginine. The activity of ornithine decarboxylase (ODC) is increased 20 fold in leaves of *Nicotiana tabacum* var Xanthi n.c. following infection with tobacco mosaic virus at 20 °C [10]. This activation is transient and the activity declines rapidly during the following days (Figure. 1). In contrast, we could not detect any activation of arginine decarboxylase (ADC). In mature leaves ADC activity remained low

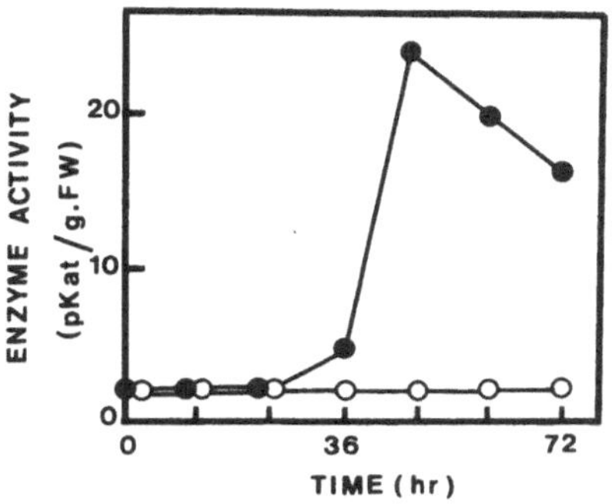

Fig. 1. Effect of TMV inoculation on ODC activity in Xanthi n.c. leaves at 20°C. TMV (●), water inoculated control (○). Necrotic lesions appeared 36-40 hr after inoculation.

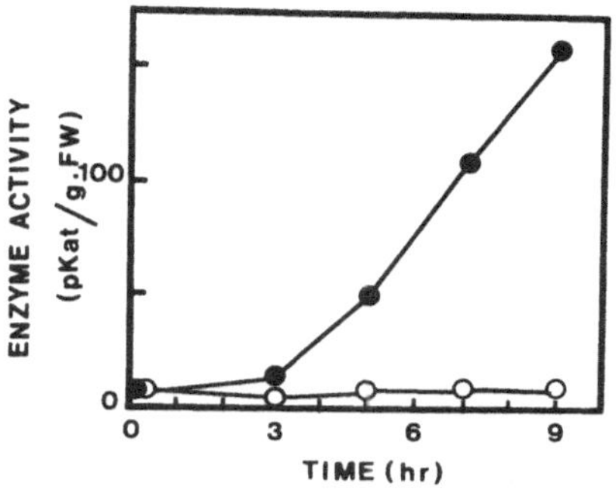

Fig. 2. ODC activity in the apical part of tobacco plants transferred from 32°to 20°three days after inoculation with TMV (●), water inoculated control (○).

and nearly undectable using the standard assay. There is little or no increase in ODC activity in plants kept at 32° C when infection is systemic. ODC activity remains unchanged and very low when the plant is healthy or when infected plant is not hypersensitive (N. Samsun n.n.). However, if the infected plants are transferred to 20°C, a marked and rapid increase in ODC activity occurs in the invaded leaves, which collapse 7 to 9 hr after the transfer (Figure 2) and die very quickly afterwards. The specific activity of ODC reaches 30 pkat/mg protein in the infected cells before their death. Tyrosine decarboxylase (TDC) was found to be activated under the same conditions [10]. At 20° the time course of changes was very similar to one observed for ODC, with a maximum 48 hr after inoculation. However TDC remained 3-5 fold less active than ODC. During the hypersensitive reaction, the increase of ODC or TDC parallels the activation of phenylalanine ammonia lyase (PAL) activity [10].

Tyramine feruloyl-CoA transferase activity (TFT). An enzyme which synthesizes feruloyltyramine from tyramine and feruloyl. CoA has been detected in crude extract of 72 hr infected leaves at 20°C [12]. This was the first report of the occurrence of tyramine feruloyl CoA transferase. At 20 °C in TMV inoculated leaves of N. Xanthi n.c. local necrotic lesions appeared 36 hr after inoculation. TFT activity reached a maximum 24 hr later i.e. (Figure 3) after the death of most of the infected cells. The fact that TFT is not enhanced in the living cells which are not going to die was shown by temperature shift experiments. One mature leaf was inoculated at 20° and the plant was then kept at 32° for 72 hr. At this temperature TMV multiplication is systemic and when the plant is transferred back to 20 the part of the plant in which the virus is multiplying, i.e., the apical part, and the inoculated leaf collapse and die within 8-9 hr. No increase of TFT activity could be detected in the apical leaves during such experiments. The activity remained at levels similar to those found in healthy leavec at 20°C. TFT activity during the hypersensitive reaction in tobacco is transitory. The enzyme remains very active only 1-2 days. This implies that the regulation of its synthesis (or activation) must be strictly controlled.[13]

Lastly *in vivo* feeding experiments using ^{14}C-tyramine and ^{14}C-feruloyltyramine and experiments based on microscopic evidence using the natural fluorescence of the amides indicate that feruloyltyramine synthetized during hypersensitivity has been found to leave the cells and bind covalently to the cell wall constituents (especially in the lignin).

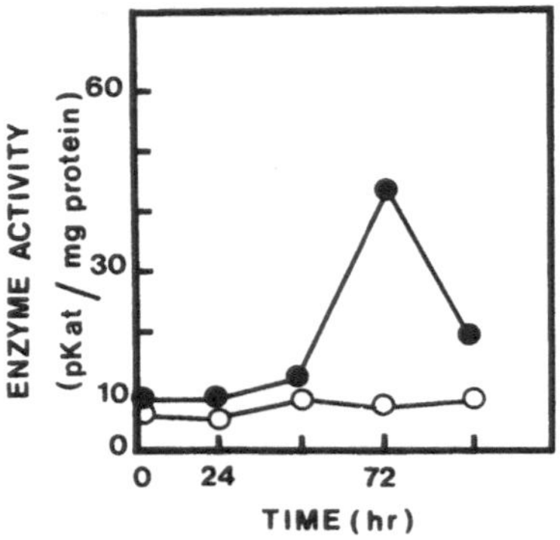

Fig. 3. TFT activity during the hypersensitive reaction to TMV in Xanthi n.c. mature leaves at 20°C. TMV (●), water inoculated control (○)

Microscopic studies are in agreement with the metabolic activation during the hypersensitive reaction. The mesophyll cells surrounding the necrotic local lesions exhibit structural features suggestive of metabolically active tissues. Endoplasmic reticulum with associated ribosomes is especially abundant.

HYDROXYCINNAMIC ACID AMIDES, FLOWERING AND RIPENESS TO FLOWER IN NICOTIANA TABACUM

When tobacco plants (N. variety Xanthi n.c.) are grown at 20°C, a complete development is observed, and three chronological reference points are utilized [14]. The first is situated at 65 days after sowing : 5-week-old seedlings placed at 32°C, never enter flowering but 65 days (or older) seedlings placed at 32°C do enter flowering. In other words, the high temperature inhibitory effect against flowering does not exist in the latter case. The other two points are situated at 78 and 85 days and correspond to initiation of the floral complex, and its emergence, respectively. This chronology characterizes the usual rate of development (a-chronology). A quicker development was exceptionnally noticed. In this case, emergence was observed earlier at 79 days (b-chronology).

Accumulation of amides during plant development. Between sowing and 47 days, it is not possible to detect any amide either in the stem or in the foliar system [14]. After 47 days and until emergence of the floral complex, only the last initiated leaves accumulate amides. These were water-soluble compounds, as p-coumaroyl-, caffeoyl-, feruloylputrescine, p-coumaroyl-, caffeoyl- and feruloylspermidine. The flowering plants contain little or no amides in leaves ; these compounds are found in large levels in the inflorescence, mainly in flowers, except for calyx and corolla. The amides that accumumate in anthers, for instance di-p-coumaroylputrescine, di-p-coumaroylspermidine and p-coumaroyltyramine are neutral compounds. Those which accumulate in scape and massively in ovaries are water-soluble compounds. Caffeoylputrescine and caffeoylspermidine are the main amides accumulated, with concentrations close to 10 µmol./g fresh weight for caffeoylputrescine and to 20 µmol./g fresh weight for caffeoylspermidine at anthesis. The concentration of these compounds quickly and drastically decreases after fertilization, while feruloyl tyramine and di-feryloyl-spermidine become predominant [15]. Feruloyl tyramine is accumulated massively in the grain [15]. A group of amides is characteristic of specific leaves whereas another would be typical of reproductive system.

The concentrations of free putrescine, spermidine and spermine remain relatively unchanged with age, although free tyramine increases 5-fold and is found especially in anthers [16]

Relationship between water-soluble amides and flowering

1. Amides are quantified in shoots of plants grown at 20 °C. The plants constitute 2 groups with different development rates, i.e., with different earlinesses of flowering (a and b chronologies). Statistical analysis shows that chronologies a and b, though both of exponential type, are significantly distinct from one another (Figure 4).

2. Topping induces accumulation of water-soluble amides only in the first leaves near the stem top of plants cultivated at 20°C, and only in plant that are at least 66 days old [14]. On regeneration of the axillary buds, this concentration declines. In another series of trials, the stems of plants from the same batch are cut at different levels ; on each plant, the first leaf under the stem cutting are taken 3 days later. Analysis of amides indicates that the induced concentration of amides depends upon the level at which the stem is cut ; the lower the cutting the lower the amide amount in the axillary shoots. Since conversion of axillary shoots of Nicotiana tabacum cv. Wisconsin 38, also a day-neutral plant, from vegetative development to flowering depends upon their position on the main axis [17], the later the flowering the lower the amount of amides in the axillary shoots [14]. It is clear that accumulation of amides is bound to the physiological development in tobacco plant.

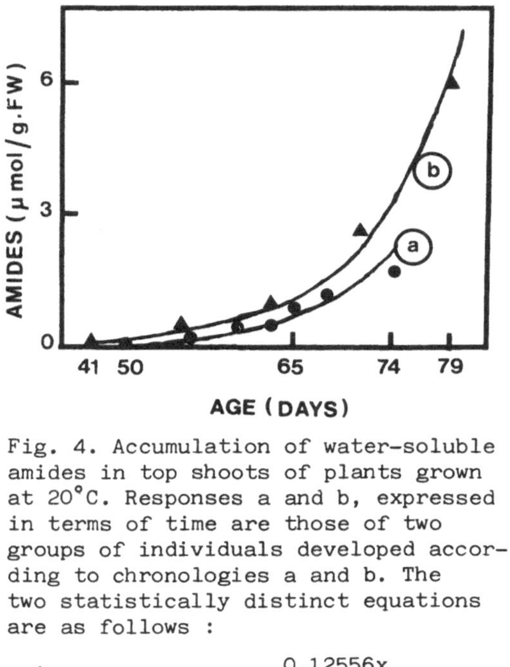

Fig. 4. Accumulation of water-soluble amides in top shoots of plants grown at 20°C. Responses a and b, expressed in terms of time are those of two groups of individuals developed according to chronologies a and b. The two statistically distinct equations are as follows :

$$a\ \hat{y}_{1,i} = 0.000229e^{0.12556x_{1,i}}$$

$$b\ \hat{y}_{2,i} = 0.000343e^{0.12556x_{2,i}}$$

A "shoot" corresponds to a sample of the plant apex including a 1 to 2 cm fragments of the foliar system.

Relationship between basic amides and ripeness to flower. The presence of amides is investigated in plants cultivated a 30°C, a temperature that inhibits flowering. The plants are placed at 30°C from about 35 days on ; sixteen day later, water-soluble amides are observed in shoots. However, concentrations are relatively low (0.1 µ mol caffeoyl putrescine p/g. fresh weight) and remain so when plants become older. Thus water soluble amides, which are never found in young plants, always appear during development regardless of the temperature. How can the presence of amides in plants grown at 32°C be explained in terms of the relationship between amide accumulation and flowering ? A study of leaf emergence in vegetative plants grown under temperature conditions compatible (20°C) and incompatible (30°C) with flowering furnishes some interesting informations.

Rate of leaf emergence. This character is studied in vegetative plants grown under compatible temperature conditions (20°C) and incompatible (30°C) with flowering.

Plants at 20°C. Statistical analysis shows that the responses can not be described by means of one equation only. Two curves of different types, namely, a straight line and an exponential curve are needed to account for the rate of leaf emergence in terms of time ; the change in shape of the responses occurs at about 41 days (Figure 5).

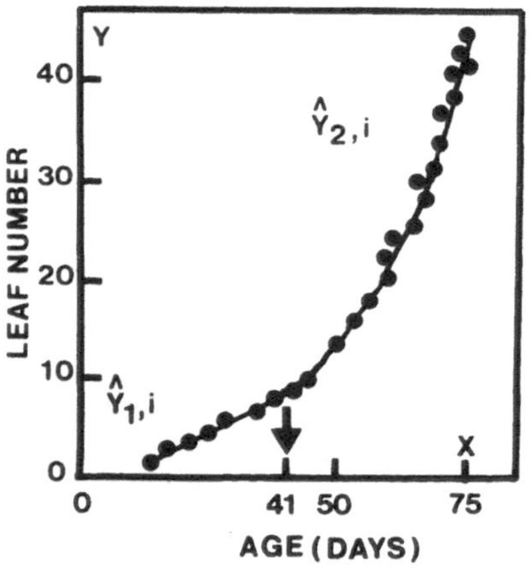

Fig. 5. Rate of leaf emergence at 20°C. From a homogeneous group of plants, 2 or 3 individuals are taken every second or third day and the total number of leaves counted. Statistical analysis led to split the response into.
Two parts :

$$\hat{y}_{1,i} = 1.53 + 0.249x_{1,i}$$

$$\hat{y}_{2,i} = 1.175e^{0.0488x_{2,i}}$$

Plants at 30°C. Here also, the rate of leaf emergence is irregular within the time-range studied (19-105 days). Again, two equations are necessary to describe the response ; unlike what happens at 20°C, two straight lines, intersecting at about 42 days, are statistically satisfactory (Figure 6).

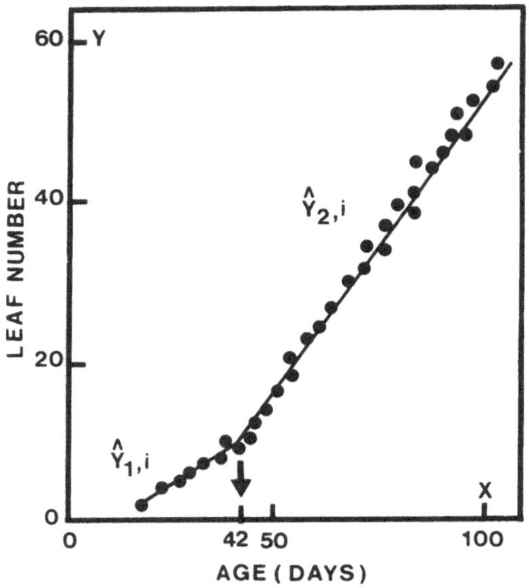

Fig. 6. Rate of leaf emergence at 30 °C. The leaves are counted as Fig.5. Statistical analysis led to split the response into two parts

$$\hat{y}_{1,i} = 3.21 + 0.295x_{1,i}$$

$$\hat{y}_{2,i} = 22.8 + 0.763x_{2,i}$$

The rate of leaf emergence in plants grown at 20°C does not lead to a new stable value but yields a constantly increasing rate (exponential curve). If an exponential increase in leaf emergence can be regarded as a criterion of proximity to flowering then linearity of leaf emergence at 32°C indicates that flowering is strongly delayed or even suppressed at this temperature. Consequently, high temperature is a favorable condition for acquiring an intermediate state which has been interpreted as ripeness to flower [18,19,20,21]. In other words, high temperature, which inhibits flowering, would not inhibit ripeness to flower which would start at about 41 days regardless of the temperature. Thus the appearance of amides in plants at 32 °C is not inevitably in disagreement with the amides flowering correlation, which has been established for plants grown at 20 °C. Is the amide concentration of some importance in this phenomenon ? Some data are in agreement with this hypothesis [13].

Thus vegetative development may be subdivided into two periods : a juvenile period apparently free of amides and a mature period with a measurable accumulation of amides. At 20°C the change in rate of leaf emergence and the appearance of amides both occur before the first observable transformation of the meristem into a floral structure. At 32 °C, the appearance of amides indicate ripening to flower. Finally, regardless of the temperature the appearance of basic amides indicate that the tobacco plant is ripening to flower.

HYDROXYCINNAMIC ACID AMIDES FLOWERING AND SEXUAL REPRODUCTION IN OTHER PLANTS

Studies of amides in different botanical species has revealed that amides begin to accumulate whenever a plant is ripening to flower [22,23,24]. These compounds are closely related to the flowering process and to sexual organogenesis [15,23,24,25,26]. Male and female sex organs can be separated by their different amide contents [14,23,25]. It must be noted that different associations of amides are observed in the foliar and reproductive systems. Phenolic amides are absent from sterile reproductive organs in several plants and amides appear to constitute biochemical markers of pollen and ovule fertility [23,25,27]. The close association of these amides with fertile reproductive organs in maize [25,27] and in other plants [23] suggests the possibility that their formation is functionally linked with reproduction. The fertile and cytoplasmic male sterile lines of maize chosen for this study were F7N, F7T and F7C ; they were isogenic and had normal, Texas and C cytoplasms respectively [25,27]. A two-locus system restores the normal fertility of plants with Texas cytoplasm. Line FC31, used here, had Texas cytoplasm and was homozygous for the two dominant genes Rf1 and Rf2 ; this was male fertile. Cytoplasmic male sterility is characterized both in F7T and F7C by pollen abortion and normally developed tassels.

Phenolic amides are found in large quantities in the anthers of male fertile maize (F7N). The amides accumulate are feruloylputrescine, di-feruloyl putrescine, di-feruloylspermidine and feruloyltyramine. Feruloylputrescine, di-feruloylspermidine and feruloyl tyramine are present in the largest quantities with concentrations close to 7 μmol/g. fresh weight for feruloyl putrescine, to 3 μmol g. fresh weight for di-feruloyl-spermidine and 1 μmol/g. fresh weight for feruloyltyramine. These compounds are absent from the anthers of the two male sterile lines one with cytoplasm T (F7T), and the other with cytoplasm C (F7C). Restoration of fertility (line TFC31) is associated with the production of these substances. These compounds are localized in the pollen. Female organs contain mainly neutral amides and there are no differences in the amide contents of female organs, irrespective of the presence or absence of T or C cytoplasm and the restorer genes.

Considerable variatious are observed in the distribution and concentration of amides at different stages of growth and grain maturation. Large amounts of amides are present within the cells at all stages of growth and grain maturation [15,27]. During germination, the content of these substances decreased drastically.

CONCLUSIONS

Hydroxycinnamic acid amides do not normally occur in leaves or other vegetative tissues, but they accumulate in shoot apices upon floral initiation, leading to a proposal that may be involved in the physiology of flowering. Male and female organs can be separated by their different amide contents. Phenolic amides are absent from sterile reproductive organs in several plants, and they appear to constitute biochemical markers of pollen and ovule fertility. In addition, hydroxycinnamic acid amides appear to be among the major phenolic compounds of most plant reproductive tissues examined to date ; which would suggest that they do indeed play functional role in reproductive physiology or development.

In some cases of host-pathogen interactions, specially during the hypersensitive reaction towards fungi or viruses, leaves of vegetative plants synthesize some hydroxycinnamic acid amides at the same time as defence mechanisms limit pathogen expansion. It is known that viruses are rarely seed-transmitted, and that resistance to viruses is often associated with flowering. In addition, since the rate of viral multiplication is associated with the formation of amides and appears to be inhibited in their presence, a role in virus resistance is implied. Hydroxycinnamic acid amides may confer virus resistance to the seed or other reproductive tissues.

The functions of hydroxycinnamic acid amides in various aspects of plant growth processes are still unclear. Are they a storage form of polyamines as such or are they active as conjugates ? It has been suggested [26] that there is only a limited exchange between free and conjugated polyamines, at least in tobacco, however the hypothesis that the conjugation to cinnamic acids may be a way of regulating the free polyamine pool in the cell has been suggested [28].

In vitro free putrescine does not promote the cell multiplication of tobacco, but when conjugated with a cinnamic acid it interferes with hormones in the control of cell multiplication and differentiation [29].

Recently we have shown [30] that in foliar explants of tobacco cultivated in vitro in a medium promoting bud formation, growth (increase in fresh weight) and bud formation (time of emergence, number of buds) are strongly influenced by K-nutrition . In K-deficient medium and in high K medium, growth and bud formation are markedly inhibited. No apparent relationship is found between free putrescine and growth and bud formation. In contrast, changes in hydroxycinnamoyl putrescine levels are shown to correlate well with growth and bud formation. The greatest stimulation of budding and growth is correlated with the greatest accumulation of hydroxycinnamoylputrescines. The latter results obtained in vitro with the use of specific inhibitors of putrescine biosynthesis in N. Xanthi n.c. leaf explants (J. Martin-Tanguy, unpublished data) indicate that : (1) the intracellular levels of hydroxycinnamoyl putrescines are markedly enhanced when the cultures are induced to proliferate the rapid and permanent depletion of hydroxycinnamoyl putrescines caused by specific inhibitors of putrescine biosynthesis in a medium promoting bud differentiation is correlated well with the depressive effect on budding activity (number of buds) and growth ; (2) high intracellular levels of hydroxycinnamic acid amides strongly inhibit cell multiplication and suppress bud formation ; (3) higher levels of hydroxycinnamic acid amides occur in cultures producing callus than those producing buds ; and (4) the formation of buds is associated with a significant decrease of these metabolites. In fact, the highest contents of these amides are found during the first days in culture when an intensive cell division takes place. Then, they decline sharply as the rate of cell division decreases and differentiation occurs. In a medium producing callus formation without differentiation, inhibition of putrescine activity (by a suicide-inhibitor) resulting in an amide decrease facilitates the expression of differentiated cell functions.

Recent results also show a relationship between the rate of tobacco cell growth in suspension culture and the formation, of cinnamoyl-putrescines. It has been observed [32] that in a variant tobacco XD cell line selected for growth on putrescine as the sole nitrogen source, cinnamoylputrescines may have a primary role in putrescine metabolism and serve as carrier for the oxidation of putrescine to γ-aminobutyric acid.

Whether these substances play a similar role in the entire plant is an open question. However, the occurrence in plants of amine conjugates of cinnamic acids, which exhibit a biological activity in vitro, may be an indication of their importance in plant metabolism.

REFERENCES

1. A.W. Galston and R. Kawr Sawhney. Polyamines and plant cells. What's new. Plant Physiol. 11:1 (1970).
2. T.A. Smith. The occurrence, metabolism and function of amines in plants. Biol. Rev. 46:201 (1971).

3. A. Altman and U. Bachrach. Involvement of polyamines in plant growth and senescence in Advances in Polyamine Research C.M. Caldarera, U. Zappia and U. Bachrach. ed., Raven Press, New York (1981).
4. J. Martin-Tanguy, F. Cabanne, E. Perdrizet and C. Martin. The distribution of hydroxycinnamic acid amines in flowering plants. Phytochemistry. 17:1927 (1978).
5. J. Martin-Tanguy, C. Martin and M. Gallet. Présence de composés aromatiques liés à la putrescine dans divers Nicotiana virosés. C.R. Acad. Sc. 276:1433 (1973).
6. J. Martin-Tanguy, C. Martin and M. Gallet. Sur de puissants inhibibiteurs de multiplication du virus de la mosaïque du tabac. C.R. Acad. Sc. 282:2231 (1976).
7. C. Martin and M. Gallet. Hypersensibilité aux virus, température et induction florale chez les végétaux. C.R. Acad. Sc. 262:997 (1966).
8. C. Martin and M. Gallet. Contribution à l'action de la température sur la réaction d'hypersensibilité de certains hôtes à l'égard du virus de la mosaïque du Tabac. C.R. Acad. Sc. 262:646 (1966).
9. C. Martin and J. Martin-Tanguy. Polyamines conjuguées et limitation de l'expansion virale chez les végétaux. C.R. Acad. Sc. 292:249 (1981).
10. J. Negrel, J.C. Vallée and C. Martin. Ornithine decarboxylase activity and the hypersensitive reaction to tobacco mosaic virus in Nicotiana tabacum . Phytochemistry 23:2747 (1984).
11. S. Gianinazzi and B. Kassanis. Virus resistance induced in plants by polyacrylic acid. J. Gen. Virol. 23:1 (1987).
12. J. Negrel and C. Martin. The biosynthesis of feruloyltyramine in Nicotiana tabacum. Phytochemistry 23:2797 (1984).
13. J. Negrel. Aspects du métabolisme de la putrescine et de la tyramine au cours de la réaction hypersensible au virus de la mosaïque de Tabac chez Nicotiana tabacum. Thèse de Doctorat Paris (1984).
14. F. Cabanne, M.A. Dalebroux, J. Martin-Tanguy and C. Martin. Hydroxycinnamic acid amides and ripening to flower of Nicotiana tabacum var. Xanthi n.c. Physiol. Plant. 53:399 (1981).
15. F. Cabanne, J. Martin-Tanguy and C. Martin. Phénolamines associées à l'induction florale et à l'état reproducteur du Nicotiana var. Xanthi n.c. Physiol. Veg. 15:429 (1977).
16. E. Perdrizet and J. Prévost. Aliphatic and aromatic amines during development of Nicotiana tabacum. Phytochemistry 20:2131 (1981).
17. C.N.Mc Daniel and F.C. Hsu. Position-dependant development of tobacco meristems. Nature 259:564 (1976).
18. A. Lance. Transformation du point végétatif d'Aster sinensis en méristème d'inflorescence. C.R. Acad. Sc. 238:2437(1954).
19. G. Bernier. Etude physiologique et histochimique de l'évolution du méristème apical de Sinapis alba L cultivé en milieu conditionné et en diverses durées de jours favorables ou défavorables à la mise à fleur. Acad. R Belg Men (Sci) TXU:71 (1964).
20. C. Besnard-Wibout. Réponses du méristème caulinaire à différents types d'induction florale. Ann. Biol. 16:385 (1977).
21. G. Bernier. The sequences of floral induction in La Physiologie de la floraison CNRS ed. Paris (1979).
22. J. Belliard, J. Pernes and M. Sandmaier. Les différentes phases du développement chez le Mil (Pennisetum typhoides Stapfand Hubbard) et la recherche de marqueurs. Physiol. Veg. 17:387 (1979).
23. M. Ponchet, J. Martin-Tanguy, A. Marais and C. Martin. Hydroxycinnamic acid amides and aromatic amines in the inflorescence of some Araceae species. Phytochemistry. 21:2865 (1982).

24. J. Martin-Tanguy, J. Margara and C. Martin. Phénolamides et induction florale de Cichorium intybus dans différentes conditions de culture en serre et in vitro. Physiol. Plant. 61:259 (1984).
25. J. Martin-Tanguy, A. Deshayes, E. Perdrizet and C. Martin. Hydroxycinnamic acid (HCA) in Zea mays : Distribution and changes with cytoplasmic male sterility. Febs Lett. 1:176 (1979).
26. R.D. Slocum and A.W. Galston. Changes in Polyamine biosynthesis associated with post-fertilization growth and development in tobacco ovary tissues. Plant Physiol. 79:336 (1985).
27. J. Martin-Tanguy, E. Perdrizet, J. Prévost and C. Martin. Hydroxycinnamic acid amides in fertile and cytoplasmic sterile lines of maize. Phytochemistry 21:1939 (1982).
28. H.E. Flores and P. Filner. Polyamine catabolism in higher plants : characterization of pyrroline dehydrogenase. Plant Growth Regiel 3:277 (1985).
29. C. Martin, G. Kunesch, J. Martin-Tanguy, J. Negrel, M. Paynot and M. Carré. Effect of cinnamoyl putrescine on in vitro cell multiplication and differentiation of tobacco explants. Plant Cell Reports 4:158 (1985).
30. S. Klinguer, J. Martin-Tanguy and C. Martin. K. Nutrition, growth bud formation and amine and hydroxycinnamic acid amide contents in leaf explants of Nicotiana tabacum variety Xanthi n.c. cultivated in vitro. Plant Physiol. 82:561 (1986).
31. O. Schiel, K. Jarchow-Redecker, G.W. Piehl, J. Lehman and J. Berlin. Increased formation of cinnamoyl putrescines by fed batch fermentation of cell suspension cultures of Nicotiana tabacum. Plant Cell Reports 3:18 (1984).
32. H.E. Flores, P. Filner. Metabolic relationships of putrescine GABA and alkaloids in cell and root cultures. In primary and secundary metabolism of plant cell cultures. K.H. Newman, W. Barz and E. Reinhard eds, Springer Verlag, New York (1985).

MOLECULAR STRUCTURE OF PLANT FATTY ACID SYNTHESIS ENZYMES

AR Slabas, A Hellyer, C Sidebottom, H Bambridge, IR Cottingham, R Kessell, CG Smith, P Sheldon,* RGO Kekwick,* J de Silva, C Lucas, J Windust, CM James, SG Hughes and R Safford

Biosciences Division, Unilever Research, Colworth House, Sharnbrook, Bedford MK44 1LQ, UK
*Department Biochemistry, University of Birmingham, PO Box 363, Birmingham, B1J 2TT UK

Introduction

Acyl lipids are a major constituent of plant tissues. They can be broadly considered to serve three functions:

1) Structural - ie. glycosyldiacylglycerols in the chloroplast[1] or cutins and waxes on the outside of the plant,[2]
2)Biosynthetic - ie. as acyl ACP's and acyl CoA's as intermediates in complex lipid biosynthesis,[3] and phosphatidylcholine as a substrate for fatty acid desaturation,[4]
3) Storage - as triacylglycerols in oil bodies of seeds and other lipid rich materials.[5] All of these types of molecules share a common hydrophobic acyl side chain - the greatest diversity in structure of which is present in triacylglycerols.[6] It is the enzymes involved in the biosynthesis of fatty acids and their regulation which will be considered in this article.

Enzymes of De Novo Fatty Acid Biosynthesis

Early cell fractionation studies have indicated that the de novo biosynthesis of fatty acids is localised to plastids, both in seeds[7,8] and in other parts of the plant.[9,10,11] The biosynthesis of fatty acids from acetyl CoA is catalysed by two enzymes 1) acetyl CoA carboxylase (ACC) and 2) fatty acid synthetase (FAS). A single report of a barley plant mutant has suggested that the FAS components are nuclear coded.[12]Both ACC and FAS have been studied, from a variety of plant sources, and dependent on the source of the material different biological questions present themselves - such as; light/dark regulation, tissue specificity, organellar localisation, differential gene expression during seed development and the regulation of lipid biosynthesis. It is such questions which make the study of these enzymes of particular importance. The information which is available on these enzymes will now be considered.

Acetyl CoA Carboxylase

Structure

ACC catalyzes the two step ATP dependent carboxylation of acetyl CoA:

$$ATP + CO_2 + BCCP \xrightarrow{\text{Biotin Carboxylase}} CO_2 \sim BCCP + ADP + Pi \quad (1)$$

$$CO_2 \sim BCCP + acetyl\ CoA \xrightarrow{\text{transcarboxlase}} malonyl\ CoA + BCCP \quad (2)$$

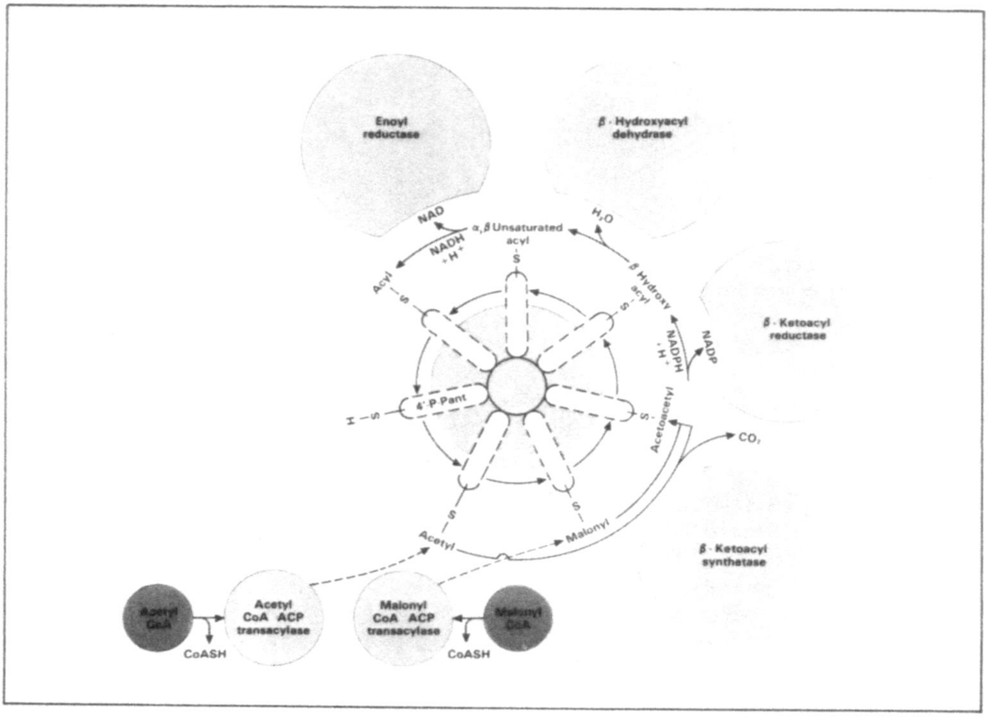

Figure 1
Schematic representation of the components of plant FAS and their relationship.

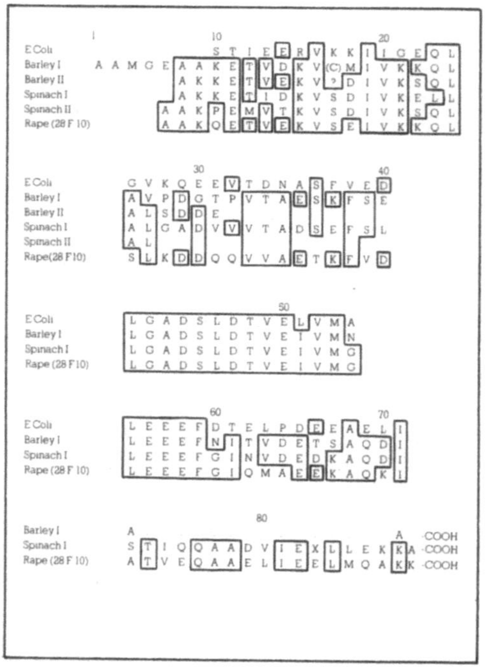

Figure 2
Amino acid sequences of ACP from E coli, spinach and barley leaf, and rape embryo [only N-terminal data available for spinach/barley ACP II]. The sequences are aligned via Ser 45, the phosphopantetheine binding site.

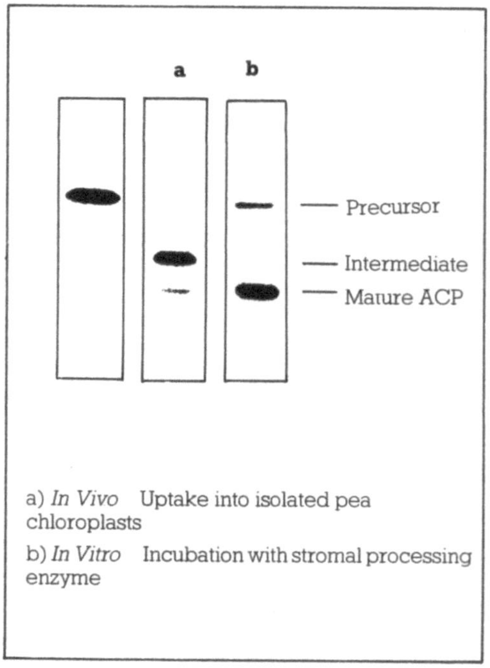

Figure 3
Processing of rape embryo ACP precursor. Precursor is synthesized in vitro via SP6 transcription/wheat germ translation. a) Uptake into isolated pea chloroplasts. b) Incubation with partially purified pea stromal processing peptidase.

In bacteria the enzyme is composed of 3 different polypeptide chains, biotin carboxylase, biotin carboxyl carrier protein (BCCP) and transcarboxylase.[13] In yeast and animals all these three functional domains are present on one polypeptide chain.[14,15]

Studies on the enzyme from numerous plant sources have yielded non uniform results on the polypeptide composition of the plant enzyme.[16] Experiments have suggested that the chloroplast enzyme is composed of several polypeptides like the bacterial enzyme. Technical difficulties are, however, encountered in plant acetyl CoA carboxylase isolation and its identification using western blotting.[16] Hence such results require careful evaluation.

The enzyme was first isolated in a high molecular weight sub-unit polypeptide form, in excess of 220 KDa, from wheat germ and parsley culture.[17] It has also been purified, in high molecular form, from both rape seed[18] and rape leaf;[19] indicating that, in this species at least, there is no fundamental difference between the sub- unit polypeptide molecular weight of the enzyme from leaf and seed. The biotin domain of the enzyme from spinach leaf has been localised to the thylakoid membrane[20] and the significance of this will be discussed later on.

Regulation

The activity of the enzyme has been studied in two different experimental systems;

1) The chloroplast - where it is light dark regulated [16] and

2) The developing seed - where it is induced prior to the onset of lipid deposition, and rapidly declines once full lipid deposition has been achieved [21]

Fatty Acid Synthetase

Structure

FAS catalyzes the sequence of consecutive reactions shown in figure 1 below. The enzyme contains 6 catalytic domains and one structural - acyl carrier protein (ACP) domain. In plants[22], like bacteria[23] all the domains are present on separate polypeptides. This situation is in marked contrast to the situation in mammals[24] and yeast,[25] where the domains are present on one and two polypeptides respectively. Purification and characterisation of the proteins from a number of different plant sources has been the subject of intensive research in several laboratories over the last few years.

Properties of the individual components

ACP

ACP has a dual metabolic role in the cell. It is a component of plant FAS and is also a cofactor/substrate for glycosyldiacylglycerol biosynthesis in the chloroplast.[3] The product of plant FAS synthesis in vitro are acyl ACP's.[26] There is, however, no proof, to date, that it is the physiological substrate for triacylglycerol biosynthesis. Acyl CoA's are probably the substrate for this reaction.

ACP has been the most intensely studied of all the components of plant FAS. It was first purified and sequenced from *E coli* by Vanaman *et al.*[27] A fully biologically active molecule has also been chemically synthesised.[28] The protein isolated from all sources is relatively small, approximately 10 KDa, and with the notable exception of seed ACP's in general, is acid and heat stable.[29] The protein is post-translationally pantethenelated. Comparison of the amino acid sequence of ACP from various sources (Figure 2) shows a high degree of sequence conservation, in particular the domain surrounding the pantethenyl site (Ser 39) is absolutely conserved.

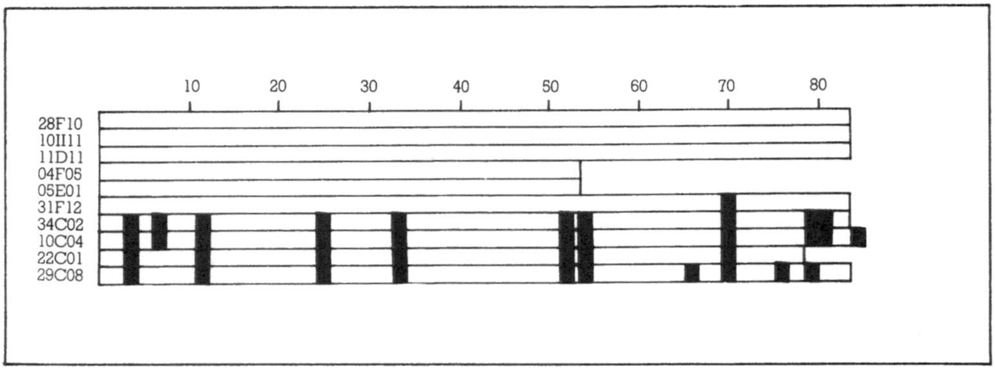

Figure 4a

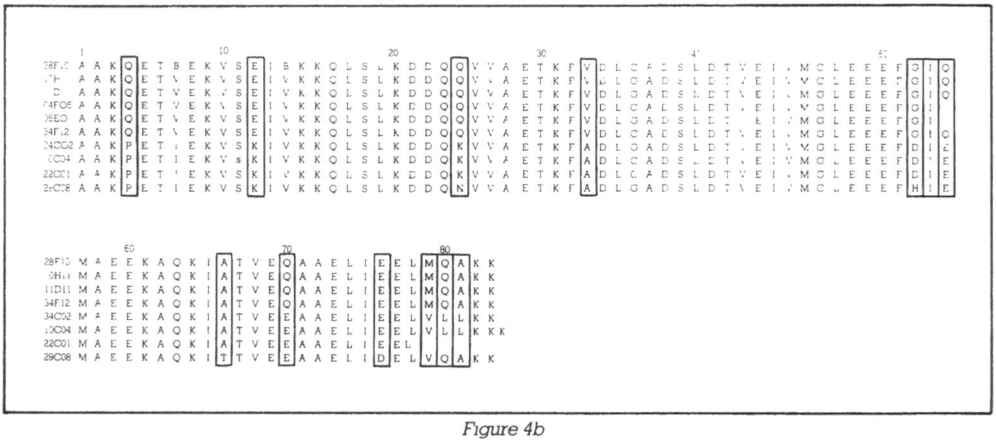

Figure 4b

Figure 4
Amino acid sequence of mature rape embryo ACP deduced from 10 cDNA clones. a) Schematic representation of sequence heterogeneity denotes variant residues. b) Single letter amino acid code sequences - variant residues are boxed. NB cDNAs 04F05/05E01 are incomplete at the c-terminus.

Two forms of the protein have been isolated from leaf material, which differ in electrophoretic mobility, and amino acid sequence.[30,31] Immunological studies, however, indicate that only one form is present in the seed, when analysed using Western blotting and anti- spinach leaf ACP antibodies.[31] Two laboratories have been involved in cloning plant ACP genes. At Unilever the cDNA encoding ACP has been cloned using mRNA isolated from, the developing seeds of *Brassica napus*,[32] whilst spinach leaf ACP cDNA has been cloned at the Calgene Laboratory.[33] Rape ACP is nuclear encoded, synthesised as a precursor with a transit sequence, which is processed into the mature form during import into the plastid (Figure 3).[34] This finding is consistent with immunocytochemical studies showing the plastid location of ACP in rape seed material as discussed below.

Cloning the rape embryo ACP gene has revealed heterogeneity both at the nucleotide and deduced amino acid level. Such heterogeneity is present in both the mature protein and the transit sequence. Figure 4 shows the heterogeneity present in the mature protein deduced from cDNA sequence analysis. Rape Seed ACP has been isolated, fortuitously, as a mixture of all its isoforms. Direct amino acid sequencing of such material (Figure 5) has allowed the ratio of the 3 isoforms of the protein to be estimated in rape seed material.[29]

Despite the high level of sequence homology of ACP's of the amino acid level (see Figure 2) this level of conservation does not appear to be present at the nucleotide level. We have found that rape embryo ACP cDNA fails to cross hybridise with rape leaf mRNA although the integrity of ACP mRNA from these sources was demonstrated using a mixed olegonucleotide probe to the conserved sequence in northern blot experiments (Figure 6). This implies that in rape, seed and leaf, ACP is encoded by distinct gene families. This observation is confirmed by failure of spinach leaf ACP cDNA to hybridise to spinach seed mRNA.[35] It is implicit from this observation that any studies performed on the expression and regulation of proteins involved in seed lipid biosynthesis must be conducted with the genes isolated from that tissue.

ACP was first localised to the chloroplast, following subcellular fractionation, by Ohlrogge *et al.*[11] We have recently directly localised it, using immunogold labelled antibodies, in a wide variety of tissues at the EM level.[36] It is almost exclusively localised in plastids and in the chloroplast is localised to the thylakoid membrane (Figure 7). Taken in conjunction with the reported localisation of BCCP of ACC to the same area (20) it would tend to indicate that the two proteins are in close spacial proximity, with concompitant implications of localised high substrate concentrations.

Acetyl CoA ACP transacylase

This enzyme has been purified 175 fold from spinach leaf, but is not available in homogenous form.[37] Its molecular weight is 48 KDa as judged by gel filtration on S300.[37] The enzyme is of low abundance, and has been postulated as the rate limiting enzyme in fatty acid biosynthesis.[37] It has also been implicated in medium chain (C8-C10) fatty acid biosynthesis[37] a view inconsistent with its elevated level in rape seed, which does not make medium chain fatty acids in vivo - but can be made to make most fatty acids *in vitro*.[26]

Malonyl CoA ACP transacylase

The enzyme has been purified to near homogeneity from both barley[38] and avocado,[39] SDS/PAGE data, on the barley enzyme, taken in conjunction with gel filtration studies, indicates that it is probably monomeric with an Mr of 34.5 KDa. Two forms of the enzyme

ACP Sequence Data from 3 Separate Runs

Residue No	Amino Acid	p moles	Average	Corrected
1	A	580, 446, 464	497	
2	A	530, 388, 483	467	
3	K	215, 236, 276	242	
4	P Q	120, 118, 116 93, 83, 88	118 88	140, 152, 142 93, 88, 83
5	E	86, 127, 116	110	
6	T	160, 97, 183	147	
7	V I	89, 76, 49 37, 39, 25	71 34	119, 95, 53 55, 54, 34
8	E	49, 55, 53	52	
9	K	78, 101, 94	91	
10	V	59, 65, 51	58	
11	S	32, 36, 30	33	
12	K E	12, 24, 19 5, 19, 4	18 9	12, 32, 25 5, 25, 4
13	121, 25, 18	21		

Figure 5a

Variant / Residue	1	2	3	4	5	6	7	8	9	10	11	12	13	% total ACP
ACP_{SI}	A	A	K	Q	E	T	V	E	K	V	S	E	I	42.7
ACP_{SII}	A	A	K	P	E	T	V	E	K	V	S	K	I	24.9
ACP_{SIII}	A	A	K	P	E	T	I	E	K	V	S	K	I	32.4

Figure 5b

Figure 5
N-terminal amino acid sequencing of purified rape seed ACP (a) heterogeneity is shown at position 4, 7 and 12 (b) deduced ratio of the major forms of ACP obtained from a consideration of the data in 5a and 4a.

have been reported in soybean leaf tissue, whilst one form is predominant in soybean seed.[40] This situation is analogous to the situation with ACP.

ß-Ketoacyl ACP Synthetase

Is commonly termed the condensing enzyme. Two forms of the enzyme with different substrate specifications have been identified in plants. Synthetase I is more active on short chain acyl ACP's and has been identified as the synthetic sub-unit for *de novo* synthesis up to C16:0.[41] Synthetase II, on the other hand, is the synthetic sub-unit for elongation of C16:0 to C18:0 ACP.[42]

Undoubtedly several other synthetases might well exist, involved in the biosynthesis of substituted, unsaturated or longer fatty acids. The enzyme has not been purified to homogeneity. Its molecular weight, estimated by gel filtration on S300 is spinach leaf I - 56 KDa,[41] spinach leaf II - 57.5 KDa,[42] rape seed II - 52.5 KDa.[43] Purification of the enzyme to homogeneity is currently being pursued in several laboratories.

ß-Ketoacyl ACP reductase

Two forms of the enzyme have been identified - one utilising NADH and the other NADPH.

In avocado the NADPH form has been clearly resolved from the NADH form.[39] The NADPH form has been purified to electrophoretic homogeneity from spinach leaf[44] and has a Mr of 24.2 KDa. Gel filtration indicates that the enzyme is tetrameric.[44]

The NADH enzyme has been purified to near homogeneity from Avocado.[39] The NADPH enzyme has recently been purified to homogeneity from avocado.[45] The enzyme has a Mr 28 KDa, on SDS/ PAGE, and is dimeric on gel filtration. Its amino acid composition is different from that reported for the spinach enzyme. The NADPH enzyme has also been purified from rape seed. The purified enzyme shows several bands, when analysed by SDS/PAGE, in the region of Mr 20-30 KDa. Clevland mapping has shown that all these bands are related.[45] Amino acid sequencing of peptides isolated from the NADPH enzyme, from rape seed and avocado, indicates that there are regions of conserved sequence homology between the two enzymes.[45]

ß-hydroxy ACP dehydratase

The enzyme has been purified to electrophoretic homogeneity from spinach[44] and the active species is reported to be tetrameric. The Mr is 19 KDa, by SDS/PAGE, and the amino acid composition has been determined.

Enoyl ACP reductase

Two forms of this enzyme have been detected in plants. Type I utilises Crotonyl ACP as substrate, and is NADH specific, and type II utilises 2-decenoyl-ACP as substrate, and requires NADPH, rather than NADH. Seed material has type I & type II, whilst only type I has been found in leaves.[46] The type I enzyme has been purified from both spinach[44] and rape seed.[47] The spinach enzyme has a Mr of 32.5 KDa when analysed by SDS/PAGE.[44] The purified rape seed enzyme has two components, when analysed by SDS/PAGE; of 34.8 and 33.6 KDa.[47] These two species are related by a six amino acid extension, which has been determined by amino acid sequencing[48] (Figure 8). The rape enzyme has an essential arginine residue[49] and its amino acid composition is remarkably similar to that of spinach. The genes encoding the rape enzyme is currently being cloned from seed material in our laboratory.

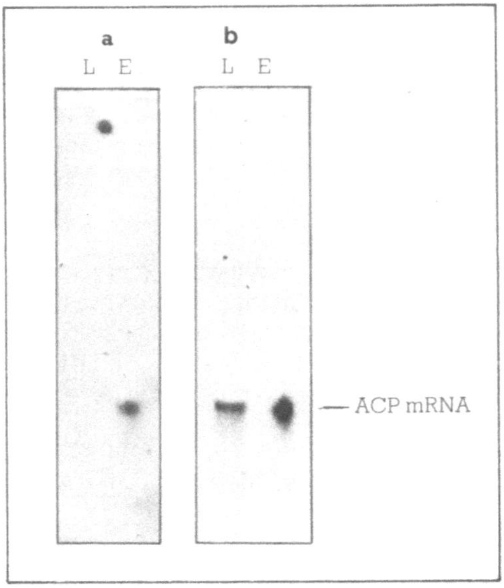

Figure 6
Northern blot of B napus leaf (L) and embryo (E) poly (A) + RNA hybridized with A) Nick-translated rape embryo Acp cDNA B) 14 mer-oligonucleotide derived from a conserved region of ACP (LEEEF). No hybridization of embryo cDNA to leaf poly (A) + RNA is observed. As a control, the mixed synthesis oligonucleotide probe, continuing all possible coding sequences, is seen to hybridize to ACP mRNA from both sources.

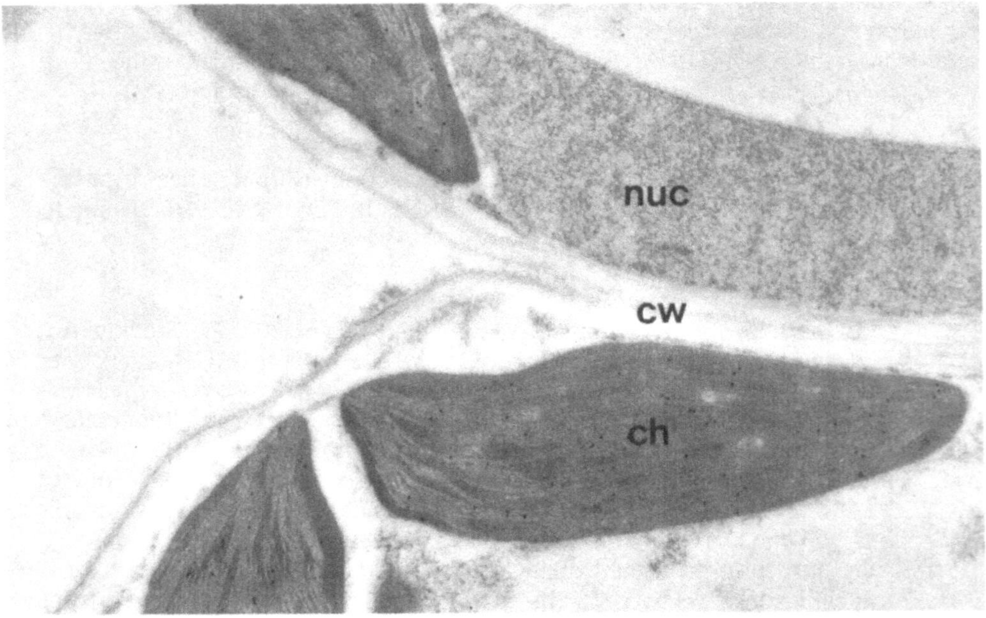

Figure 7a

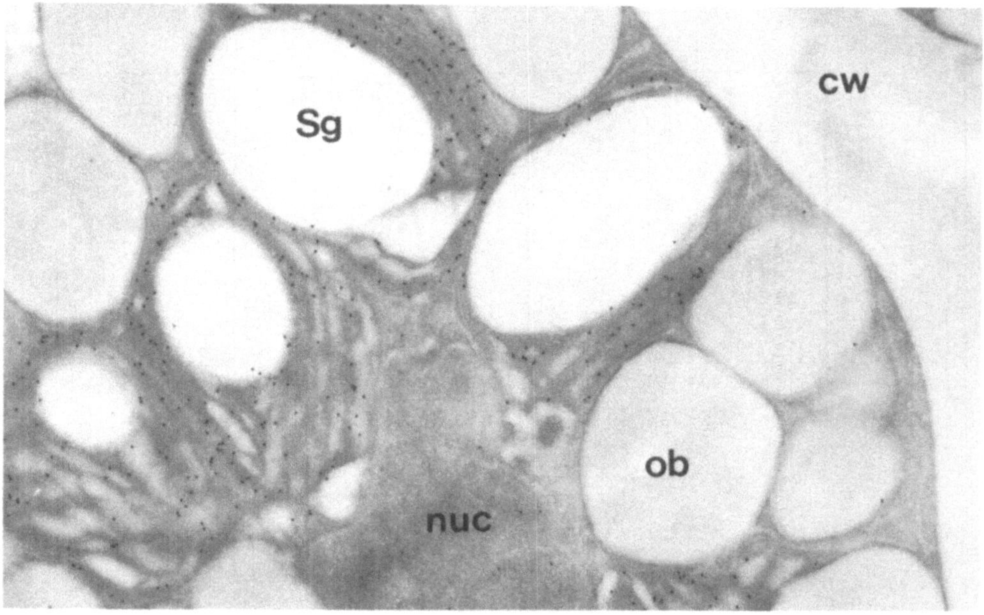

Figure 7b

Figure 7
Immunogold localization of ACP In (A) leaf and (B) seed material. Rabbit anti-spinach leaf ACP antibodies were used and these were localized with protein A - gold. cw = cell wall, sg = starch grain, ob = oil body, nuc = nucleus, Ch = chloroplast.

∝	S	E	S	S	E	S	K	A	S	S	G	L	P	I	D	L	R	G	K
ß							K	A	S	S	G	L	P	I	D	L	R	G	K

Figure 8
N-terminal amino acid sequence of the ∝ and ß form of rape seed enoyl ACP reductase.

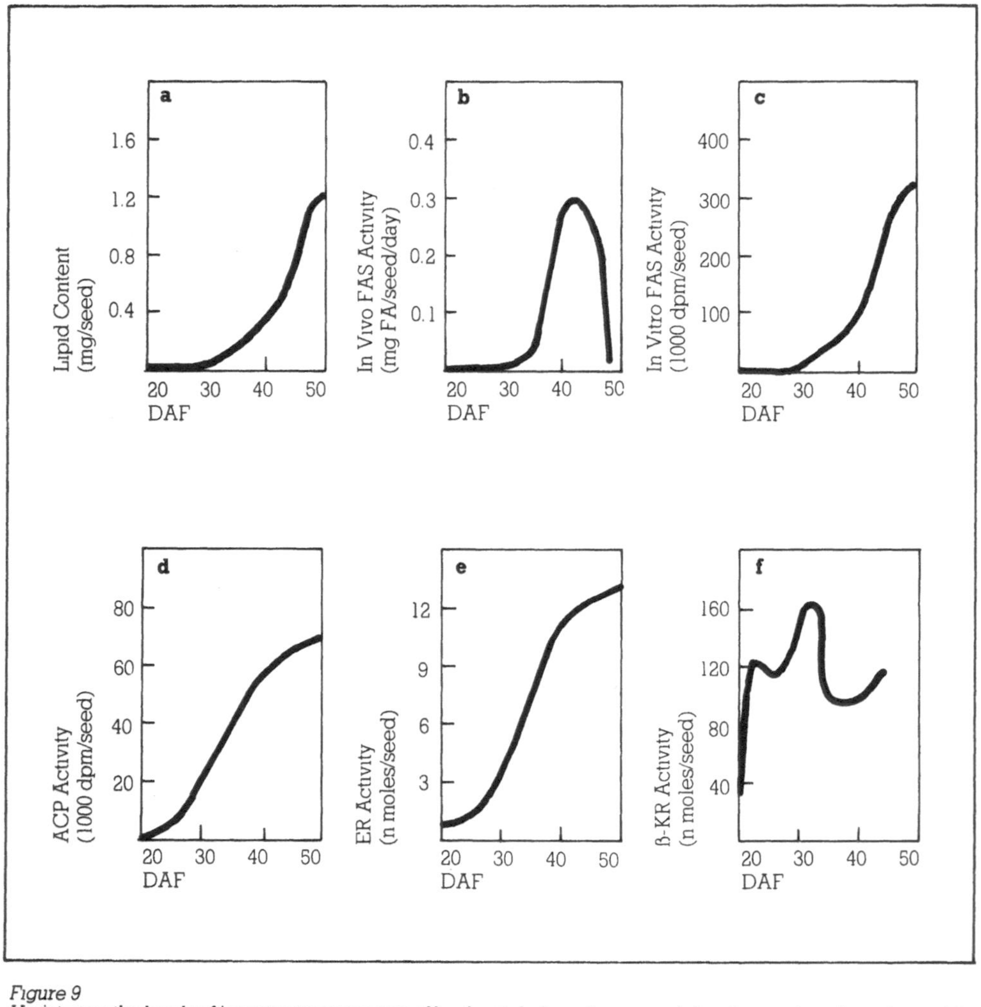

Figure 9
Variation in the levels of important components of lipid metabolism during seed development in oil seed rape (a) lipid content (b) in vivo FAS activity (c) in vitro FAS activity (c) ACP activity (d) enoyl ACP reductase activity (e) ß-keto acyl ACP reductase activity.

Summary

4 of the components of plant FAS have been purified to homegeneity - the rest require purification

Association of plant FAS components

It seemed intuiative that the components of plant FAS are associated *in vivo*. One would imagine that nature would have devised such an association to maximise efficiency of the enzyme reaction. Shimakata & Stumpf,[44] in a series of elegant experiments, tried to demonstrate such an association of purified, remixed, components of plant FAS following gel filtration. This did not prove possible. The demonstration of such an association will require re-evaluation as and when *all* the components have been purified. Even then, experimental conditions may be hard to define as subtleties may be present which we do not yet understand. Smith [50] for example has been able to demonstrate a function interaction between Rat FAS and acylthioesterase II but it has not been possible to show a direct physical interaction. Estimations of the molecular weight of plant FAS are premature at present.

Regulation of plant FAS components during seed development

We have measured the levels of several FAS components during rape seed development together with the in vivo rate of fatty acid biosynthesis. The levels of *in vitro* FAS activity, ß-ketoreductase, enoyl reductase and ACP activity rise prior to the onset of lipid deposition (Figure 9). Unlike ACC activity they remain high once lipid biosynthesis has ceased. This indicates that it is probably the level of substrate - malonyl CoA which is regulating total lipid synthesis.

Further metabolism of fatty acids

As mentioned earlier, the product of plant FAS is almost exclusively ACP. Further metabolism of the acyl ACP could occur by it being used directly as a substrate - as for example in glycosyldiacylglycerol biosynthesis in the chloroplast. Alternatively the acyl ACP could be hydrolysed by an acyl ACP thioesterase[51] and then be exported as the free fatty acid for further metabolism. The modified fatty acid could then be re-imported to the plastid.

Conclusion

In conclusion much is known now about the components of plant FAS, several of them have been purified to homogeneity. Amino acid sequence data is becoming available and gene cloning is in progress for several components. The molecular regulation of one of the components, ACP, is now under investigation using the cloned gene. Similar analyses will be possible on the other components when purified. These studies should advance our understanding of lipid metabolism and its regulation. In particular the key questions which have to be addressed are the regulation of components of fatty acid synthesis at the tissue, temporal and organellor level. The answers to such questions will be eventually obtained by a comparison of several of the components of plant FAS.

References

[1]Douce R, and Joyard J (1980). Biochemistry of plants, vol 4, 321-362, Academic Press New York. (Eds Stumpf PK and Conn EE).

[2]Kolattukudy PE (1980). Biochemistry of plants, vol 4, 571-646, Academic Press New York (Ed Stumpf PK, and Conn EE).

[3]Frentzen M, Hares W and Schiburr A (1984). Structure, function and metabolism of plant lipids. 105-110 Elsevier Scientific Publishers, Amsterdam. (Ed Siegenthaler P, and Eichenberger W).

[4]Browse J, Roughan G, and Slack CR (1981). Biochem J *196*, 347-354.

[5]Gurr MI, Blades J, Appleby RS, Smith CG, Robinson MP, and Nichols BW, (1974). Eur J Biochem *13*, 281-290.

[6] Gurr MI, and Jones A. (1971). Lipid Biochemistry on introduction. Chapman and Hall (London).

[7]Zilkey BF and Canvin DT (1971). Can J Bot *50*, 323-326.

[8]Yamada M and Usami Q, (1975). Plant Cell Physiol *16*, 879-884.

[9]Weaire PJ, and Kekwick RGO, (1973). Biochem J *146*, 425-437.

[10]Vick B and Beevers H, (1978). Plant Physiol *62*, 173-178.

[11]Ohlrogge JB, Kuhn DN, and Stumpf PK, (1979). Proc Natl Acad Sci USA *76*, 1194-1198.

[12]AJ, Corde JP, Joyard J, Horner T and Douce R, (1982). Plant Physiol *69*, 1467-1470.

[13]HC, and Barden RE, (1977). Ann Rev Biochem *46*, 385-413.

[14]PG, and Cohen P, (1978). FEBS Letts *91*, 1-7.

[15]Mishina M, Roggenkamp R, and Schweitzer F, (1980). Eur J Biochem *111*, 79-87.

[16]Hellyer A, Bambridge HE, and Slabas AR, (1986). Biochem Soc Trans *14*, 565-568.

[17] Egin Buhler B, Loyal R, and Ebel J, (1979). Arch Biochem Biophys *203*, 90-100.

[18]Slabas AR, and Hellyer A, (1985). Plant Sci *39*, 177-182.

[19]Slabas AR, Hellyer A, and Bambridge HE, (1986). Biochem Soc Trans *14*. 714-715.

[20]Kannangara CG, and Jensen CJ, (1975). Eur J Biochem *54*, 25-30.

[21]Turnham E, and Northcote DN, (1983). Biochem J *212*, 223-229.

[22]Ohlrogge JB, (1982). Trends in Biochemistry. 385-387.

[23]Vagelos PR, (1974). In "MTP International Review of Science, Biochem of Lipids" (TW Goodwin, ed) vol 4 pp100-140. Butterworths, London.

[24]McCarthy AD and Hardie DG, (1983). Eur J Biochem 130, 185-193.

[25]Stoops JK and Wakil SG (1981). J Biol Chem *256*, 8364-8370.

[26]Slabas AR, Roberts P and Ormester J (1982). Biochemistry and metabolism of plant lipids 251-256. Elsevier Biomedical Press - Amsterdam (Eds Winterman JFGM and Kuiper PJC).

[27]Vanaman T, Wakil SJ, and Hill RL 1968). J Biol Chem *243*, 6420-6431.

[28]Prescott DJ, and Vagelos PR, (1972). Adv Enzymol *36*, 269-311.

[29]Slabas AR, and Sidebottom CM Manuscript in preparation.

[30]Hoj PB, and Svendson IB, (1984). Carlsberg Res Commun *49*, 483-492.

[31]Ohlrogge JB, and Kuo TM, (1985). J Biol Chem *260*, 8032-8037.

[32]Safford R, Windust JHC, Lucas C, de Silva J, James CT, Hellyer A, Smith CG, Slabas AR, and Hughes SG - manuscription preparation.

[33]Scherer D, Shintani DK and Knauf VC, (1986). Abstract - International Symposium of the structure and function of plant lipids - Davis, California.

[34]Safford R, and de Silva J, manuscript in preparation.

[35]Scherer D (personal communications).

[36]Slabas AR, and Smith CG, Manuscript in preparation.

[37]Shimakata T, and Stumpf PK, (1983). J Biol Chem-*258*, 3592-3598.

[38]Hoj PB and Svendson I, (1983). Carl Res Commun *48*, 285-305.

[39]Caughey I and Kekwick RGO, (1982). Eur J Biochem *123*, 552-561.

[40]Guerra DJ, and Ohlrogge JB, (1986). Arch Biochem Biophys *246*, 274-285.

[41]Shimakata T, and Stumpf PK, (1983). Archives Biochem Biophys. *220*, 39-45.

[42]Shimakata T, and Stumpf PK, (1982). Proc Natl Acad Sci (US), *79*, 5808-5812.

[43]Slabas AR, Harding J, Hellyer A, Sidebottom C, Gwynne H, Kessell R, and Tombs MP, (1984). Structure, function and metabolism of plant lipids. Elsevier-Amsterdam 3-10 (Eds. Siegenthaler P and Eichenberger W).

[44]Shimakata T, and Stumpf PK, (1982). Arch Biochem Biophys *218*, 77-91.

[45]Shelden P, Kekwick RGO, and Slabas AR, Unpublished results.

46 Shimakata T, and Stumpf PK, (1982). Archives Biochem Biophys *217*, 144-154.

[47]Slabas AR, Sidebottom CM, Hellyer A, Kessell RMJ, and Tombs MP (1986). Biochemica Biophysica Acta *877*, 271-280.

[48]Slabas AR, and Cottingham I Unpublished data.

[49]Cottingham I and Slabas AR Unpublished data.

[50]Smith S, Mikkelson J, Witkowski A, and Libertini LJ, (1986). Biochem Soc Trans *14*, 583-584.

[51]Joyard J, Stumpf PK, (1981). Plant Physiol *67*, 250-256.

ASPECTS OF GLUTATHIONE FUNCTION AND METABOLISM IN PLANTS*

Heinz Rennenberg

Fraunhofer Institute for Atmospheric Environmental Research
Kreuzeckbahnstrasse 19
D-8100 Garmisch-Partenkirchen, F.R.G.

INTRODUCTION

Glutathione (γ-glutamyl-cysteinyl-glycine) is a product of the primary metabolism that is thought to be present in all living cells. Still, convincing evidence for its wide distribution in the Plant Kingdom has so far not been established (Rennenberg, 1982). This lack of knowledge is mainly a consequence of the observation that glutathione is the major low-molecular-weight thiol in several of the few plant species analyzed. As the free thiol content of plant cells can easily be determined by simple methods (Lay and Casida, 1976), but more fancy methods are needed for the quantification of glutathione (Newton et al., 1981; Law et al., 1983), many investigators measured thiol, but claimed that glutathione has been determined.

The assumption that the thiol content of plants represents the amount of glutathione present can be a considerable source of error. Both, glutathione and cysteine undergo seasonal fluctuations (Esterbauer and Grill, 1978). These fluctuations result in changes in the relative proportion of the composition of the cells' thiol from these compounds. Further, glutathione is not the major free, low-molecular-weight thiol in all plants investigated. In several legumes the thiol-containing tripeptide γ-glutamyl-cysteinyl-ß-alanine (homo-glutathione) was found instead of glutathione (Price, 1957; Carnegie, 1963). The advantage of the non-proteinogenic amino acid ß-alanine at the C-terminal site of this peptide may be that it cannot be attacked by unspecific carboxypeptidases. Therefore, the finding of other thiol containing γ-glutamyl tripeptides in the Plant Kingdom with this property of homo-glutathione would not be surprising.

The isolation of mutants of E. coli lacking the enzymes of glutathione synthesis (Apontoweil and Berends, 1975) shows that glutathione is not a compound essential for life. Though glutathione could not be detected in the cells, the mutants grew at the same rate observed with the parent strains. As these mutants were much more susceptible to a wide range of chemical agents than the parent cells, it has been concluded that detoxification of compounds unfavorable for

* Dedicated to Prof. Dr. Ludwig Bergmann on the occasion of his 60th birthday

growth is a major function of glutathione in microorganisms. In higher plant cells, it is obvious that glutathione plays a central role in the removal of toxins generated during cellular metabolism, or acting on plants in their environment. Glutathione is an important metabolite, e.g., in the detoxification of toxic oxygen species (Halliwell, 1984), of pesticides (Lamoureux and Rusness, 1981), and of heavy metal ions (Robinson and Jackson, 1986). The present report deals with the roles of glutathione in pesticide metabolism and the synthesis of heavy metal chelating compounds (phytochelatins) with special emphasis on recent advances in the molecular biology of these processes. As these processes are closely connected with the metabolism of glutathione, synthesis and degradation of this peptide are discussed.

SYNTHESIS AND DEGRADATION OF GLUTATHIONE

Biosynthesis of Glutathione

In animal and bacterial cells (Meister and Tate, 1976) glutathione is not translated directly on mRNA, but is synthetized enzymatically in a two-step procedure (Fig. 1A, B). In the initial step, the dipeptide γ-glutamylcysteine is produced in an ATP-dependent reaction from glutamate and cysteine. This reaction is catalyzed by a γ-glutamyl-cysteine synthetase (EC 6.3.2.2) in much the same way glutamine is synthesized by glutamine synthetase (Meister and Tate, 1976). In the second step of glutathione synthesis, glycine is added at the C-terminal site of γ-glutamylcysteine to yield glutathione. This ATP-dependent reaction is catalyzed by a glutathione synthetase (EC 6.3.2.3).

In higher plant cells glutathione is likely to be synthetized by the same two-step process. The glutathione metabolite γ-glutamylcysteine has been isolated from several plant species; glutathione is synthetized

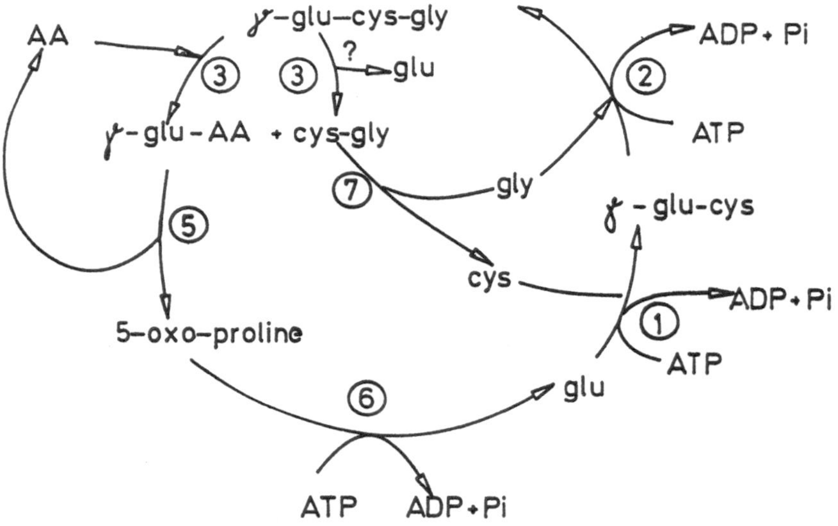

Figure 1 A Synthesis and degradation of glutathione
The γ-glutamyl-cycle

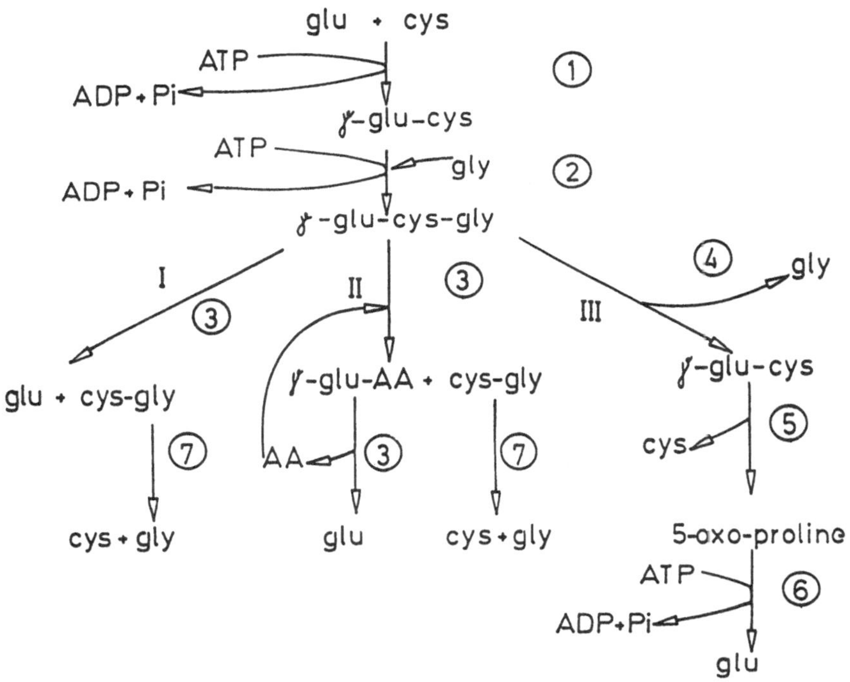

Figure 1B Synthesis and degradation of glutathione
Possible path of glutathione metabolism
in plant cells

1: γ-glutamylcysteine synthetase, 2: glutathione synthetase,
3: γ-glutamyltranspeptidase, 4: GSH-carboxypeptidase,
5: γ-glutamylcyclotransferase, 6: 5-oxo-prolinase,
7: dipeptidase.

in cell homogenates in vitro upon addition of glutamate, cysteine, glycine, ATP and Mg2+ (Rennenberg, 1982). Though earlier reports on the presence of γ-glutamylcysteine synthetase and glutathione synthetase in plant cell homogenates (Webster, 1953a,b; Webster and Varner, 1954), were found not to be reproducible in other laboratories, recent studies (Law and Halliwell, 1986; Klapheck et al., 1987) clearly show that glutathione synthetase is a constituent of plant cells. 100-fold purified preparations of the enzyme from spinach showed optimum activities at pH 8.5-9.0 in the presence of Mg2+ and substrate concentrations of 10-30 mM. Millimolar concentrations of heavy metal ions inhibited the enzyme, whereas H2O2 or ascorbate did not affect its activity. Product inhibition by ADP was observed, as previously reported for the animal enzyme (Meister and Tate, 1976); however, no inhibition by the product glutathione could be detected at mM concentrations (Law and Halliwell, 1986). Several pieces of evidence suggest that at least part of glutathione synthesis in green cells is localized in the chloroplasts. The concentration of glutathione in the chloroplast is high compared with its concentration in the cytoplasm (Rennenberg, 1982); glutathione synthesis and efflux is 12-14 times higher in green, photoheterotrophic than in heterotrophic tobacco cells (Bergmann and Rennenberg, 1978; Rennenberg and Bergmann, 1979). Glutathione synthetase activity is about equally distributed between chloroplasts and cytoplasm of pea leaf cells (Klapheck et al., 1987). Whether the enzyme is coded by the nuclear, or by the chloroplastic genome, or both, has so far not been elucidated. Though these findings strongly suggest that a significant portion of the synthesis of glutathione proceeds in the chloroplasts, data on the catalytic properties and the subcellular distribution of γ-glutamylcysteine synthetase are urgently needed to proof this assumption.

Enzymatic studies on the synthesis of homo-glutathione have so far not been reported. The observation that many legumes in which homo-glutathione is present do not contain glutathione (Klapheck, personal communication) suggests differences in the catalytic properties of glutathione synthetase and homo-glutathione syntetase to be responsible for the synthesis of the two different tripeptides. Studies on the substrate specificity of the enzymes will proof this hypothesis.

Degradation of Glutathione

In animal cells glutathione is degraded in the series of reactions shown in Fig. 1A. In this pathway the γ-glutamyl moiety of glutathione is transferred by a γ-glutamyl transpeptidase (EC 2.3.2.2) to an amino acid acceptor (Meister and Tate, 1976). At which extent the hydrolytic activity of the same enzyme, resulting in a direct release of glutamate from glutathione (Fig. 1A), is active in vivo still remains to be elucidated. The dipeptide cys-gly, produced by both the transfer and the hydrolytic activity of γ-glutamyl transpeptidase, is hydrolyzed into the constituent amino acids by a dipeptidase (EC 3.4.13.6). The γ-glutamyl-moiety of the second product of the transfer reaction, a γ-glutamyl-amino acid, is cyclized by a γ-glutamylcyclotransferase (EC 2.3.2.4) to 5-oxo-proline, the cyclic lactam of glutamic acid. The 5-oxo-proline produced this way is hydrolyzed in an ATP-dependent reaction to glutamate. This reaction, catalyzed by a 5-oxo prolinase, appears to be the rate limiting step in glutathione degradation in animal cells (cf. Rennenberg, 1982). From this pathway of glutathione degradation and the path of glutathione synthesis Meister and coworkers suggested the γ-glutamyl cycle and proposed that the uptake of amino acids via a membrane-bound γ-glutamyl transpeptidase is a major function of this cycle (cf. Rennenberg, 1982). Though this function of the γ-glutamyl cycle is still a controversial issue (cf. Rennenberg, 1982) the sequence of reactions outlined above can considered to be definitely established in mammalian cells.

The path of glutathione degradation in plant cells is less certain. Still glutathione degradation is an essential process for the sulfur nutrition of higher plants. Sulfur is available to plants predominantly in the form of sulfate to the roots, but is almost exclusively needed in its reduced form (oxidation state -2). As the major part of the plants sulfate reduction takes place in the chloroplasts of the leaves, reduced sulfur has to be translocated from the leaves to the roots to meet the nutritional needs of the roots for sulfur. Glutathione is the main long distance transport form of reduced sulfur in higher plants (Rennenberg et al., 1979; Bonas et al., 1982). This peptide, therefore, has to be degraded in significant amounts in the roots to make reduced sulfur available for protein synthesis. Several pieces of evidence suggest that in plant cells glutathione degradation proceeds via pathways different from the path established in animal cells.

Experiments with 14C-glu and 35S-cys labeled glutathione showed that γ-glu-cys and 5-oxo-proline are intermediates of glutathione degradation in higher plants (Rennenberg et al., 1980; Steinkamp and Rennenberg, 1985). In addition, the analysis of the degradation products of glutathione conjugates with pesticides in plants showed γ-glu-cys and cys-derivatives to be produced (Lamoureux and Rusness, 1981).
Cys-gly or cys-gly-derivatives, consistently observed in animal cells as metabolites of glutathione or glutathione conjugates, were not found in these studies. From these data it has been suggested that glutathione is degraded in plant cells via path III in Figure 1B. In this pathway first the glycine moiety of glutathione is split by a carboxypeptidase. The remaining dipeptide, γ-glutamylcysteine, may then be further degraded to cysteine and glutamate via 5-oxo-proline in the sequence of reactions suggested in the γ-glutamyl-cycle. Alternatively, γ-glu-cys may be used for de novo synthesis of glutathione. This pathway would not require γ-glutamyltranspeptidase activity, an enzyme not only widely distributed among animals and microorganisms (Meister and Tate, 1976) but also among plants (Kosai and Larson, 1980). Studies with tobacco γ-glutamyltranspeptidase showed, however, that glutathione is not acting as a donor, but as an acceptor of γ-glutamyl-moieties in the reaction catalyzed by this enzyme (Steinkamp and Rennenberg, 1985). This observation suggests that γ-glutamyltranspeptidase is not participating in glutathione degradation in these cells. Further, pathway III (Fig. 1B) requires a γ-glutamylcyclotransferase with a high specificity for S-containing γ-glutamyldipeptides. Such an enzyme has recently been demonstrated in tobacco cells (Steinkamp et al., 1987). 5-Oxo-prolinase activity, also needed in this pathway of glutathione degradation, has been observed in many higher plant species (Rennenberg, 1982). The enzyme was found to be regulated by the sulfur nutrition of the cells (Rennenberg et al., 1983), a pre-requisite for a role of this enzyme as the rate-limiting factor in glutathione degradation. Both γ-glutamylcyclotransferase and 5-oxo-prolinase were found to be soluble enzymes localized in the cytoplasm. Apparently, glutathione degradation in plant cells is restricted to the cytoplasm, and does not take place in the organelles (Steinkamp et al., 1987).

These observations strongly support the idea that degradation of glutathione in plant cells proceeds via a path different from that in animal cells. In higher plants the series of reactions shown in Figure 1B, III may indeed be responsible for glutathione degradation. Experiments with yeast (Jaspers et al., 1985) show that nutritional factors may highly influence the path of glutathione degradation. When ammonium was used as the nitrogen source, γ-glutamyl-transpeptidase activity was low, and glutathione degradation virtually did not exist. In the presence of urea γ-glutamyl-transpeptidase activity was high and glutathione degradation proceeded via pathway I, Figure 1B, because of the low content in free amino acids of the cells. When glutamate was supplied as the nitrogen source, the free amino acid content and the γ-glutamyl- transpeptidase activity were high; consequently, glutathione was degraded via pathway II, Figure 1B. Both γ-glutamyl-cyclotransferase and 5-oxo-prolinase were generally not found in the yeast cells. These data show that nutritional factors can control the path of glutathione degradation in plant cells. The mechanisms by which glutathione degradation is adjusted to the plants' nutritional needs are entirely unknown. An identification of these regulatory devices is, however, essential for an understanding of the role of glutathione in detoxification processes. Upon removal of many toxins, such as pesticides or heavy metals, glutathione or its precursors are channeled into new metabolic pathways. To prevent that detoxification is achieved at the risk of an insufficient supply of glutathione for other cellular processes, a mutual regulation of the paths of glutathione synthesis and degradation and the paths of detoxification has to be assumed.

THE ROLE OF GLUTATHIONE IN PESTICIDE METABOLISM AND TOXICITY

Many plants, including important agronomic, horticulture, and weed species form conjugates of glutathione with a broad range of intercellularly produced or extracellularly applied electrophilic substances. Because of the growing interest in the metabolic fate of pesticides in plants, most studies on the formation of glutathione-conjugates have been focused on herbicides, fungicides, insecticides, and antidotes. Though spontaneous conjugation of glutathione has been observed in vitro (Leavitt and Penner, 1979), in the living cell this process appears to be catalyzed by a group of enzymes named glutathione-S-transferases (EC 2.5.1.18). The reactions catalyzed by this group of enzymes can apparently prevent toxic effects of many pesticides in plants. The huge amount of data published on this subject has recently been reviewed in much detail by Lamoureux and Rusness (1987). For the present paper only a few important findings have been excerpted.

Synthesis and Metabolism of Glutathione Conjugates

Formation of glutathione conjugates with pesticides in plants can be the result of different types of reactions. These reactions include displacement of the chloro-group (e.g. symmetrical chloro-S-triazine herbicides like atrazine, chloroacetamides like propachlor) and nucleophilic displacement ether cleavage (diphenyl ether herbicides like fluorodifen) as the initial step in pesticide metabolism, but also conjugation following an activation reaction. The latter type of conjugation is best documented with the thiocarbamate herbicide EPTC that has to be oxidized to become a substrate for glutathione-S-transferase (cf. Lamoureux and Rusness, 1987).

The broad range of pesticides conjugated, and the different type of reactions involved easily explain the necessity for the multiple forms of isozymes of glutathione-S-transferase observed in plant cells. Intensive studies of the enzyme have mainly been performed with glutathione-S-transferase from corn utilizing atrazine and several analogs of this herbicide as substrates. This enzyme represents about 1% of the soluble protein in corn. Its activity is more than 2-fold higher in the roots than in the shoots and is doubled upon treatment with certain herbicide antidotes (see below). Like many other glutathione-S-transferases the corn enzyme is a dimeric protein of a cellular weight of approximately 50,000. It conjugates glutathione with atrazine at a rate of at least 1 nmole per g freshweight of tissue and hour (cf. Lamoureux and Rusness, 1987).

Glutathione conjugates are metabolized in living cells by numerous enzymatic and non-enzymatic reactions. Figure 2 shows a selection of pathways important for the degradation of glutathione conjugates. The initial sequence of reactions in the catabolism of glutathione conjugates leads to the synthesis of cysteine conjugates in both animal and plant cells. Still, the pathways are different in animal and plant

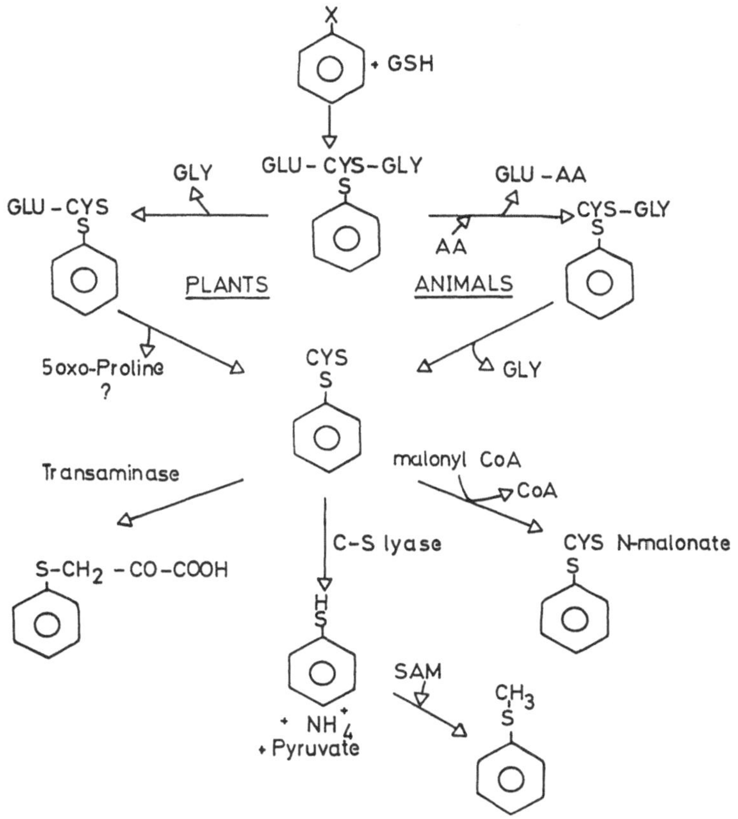

Figure 2 Pesticide metabolism in animal and plant cells

cells. In most higher plants, γ-glutamylcysteine derivatives are produced as intermediates, in animals cysteinylglycine derivatives are usually formed. The cysteine conjugates may be the final metabolites, but in many plants they are catabolized to other products (Figure 2). N-malonylcysteine derivatives have been identified as the most abundant end products of these reactions in plants; mercapturic acids are the most important end products in animal cells. Formation of malonyl derivatives is thought to be a common mechanism in plants to block further metabolism. Though it is unknown whether N-malonylcysteine conjugates are deposited in the vacuole, such a subcellular distribution is suggested by the finding that malonylglucosides of flavonoids are predominantly localized in this compartment. Among many other pathways, catabolism of glutathione conjugates to S-methyl derivatives has been reported in plants (cf. Lamoureux and Rusness, 1987). However, the C-S lyase activity needed in this path has only been observed in the brassica family, several legumes and in onion.

Glutathione Conjugation, a Detoxification Process

Conjugation of glutathione with atrazine in atrazine resistant in contrast to atrazine susceptible species is one of the best documented examples of herbicide selectivity in plants based on metabolic differences. Atrazine, an inhibitor of photosynthetic electron transport, inhibits photosynthesis in both resistant (e.g. corn, sorghum, sugarcane) and susceptible (e.g. barley, oats, pea, wheat) species, but photosynthesis is restored in resistant species upon conjugation of glutathione with atrazine. High atrazine metabolism occurs in resistant species containing high glutathione-S-transferase activity; little atrazine metabolism is observed in susceptible species containing low glutathione-S-transferase activity. In a similar way, atrazine resistance of inbred corn lines was connected with high glutathione-S-transferase activities and a high rate of atrazine metabolism by glutathione conjugation. Also the selectivity of several other herbicides was found to be due to rapid conjugation with glutathione (Lamoureux and Rusness, 1987).

Many herbicides can cause injury to a crop at dosages needed to sufficiently control weeds. This lack of selectivity can in some cases be overcome by the addition of compounds called antidotes, protectants, or safeners. These compounds do not reduce herbicide injury to weeds, while decreasing herbicide injury to crops. Thus, selectivity is transferred from the herbicide to the mode of action of an antidote. Analysis of the mechanisms responsible for the protection of plants from herbicide injury by antidotes showed many examples of an antidote elevated glutathione-S-transferase activity (Rennenberg, 1982; Lamoureux and Rusness, 1987; Zama and Hatzios, 1986). On the other hand, synergistic effects in pesticide toxicity could be attributed to an inhibition of glutathione conjugation at the level of glutathione-S-transferase (Lamoureux and Rusness, 1987).

The increase in glutathione-S-transferase activity upon antidote treatment in corn was found to be caused by both, the induction of a constitutive enzyme (GST I) and the induction of an enzyme absent in untreated corn (GST II). To determine whether antidotes act at the transcriptional level, cDNA clones encoding the major glutathione-S-transferase subunit have been isolated from maize (Shah et al., 1986; Wiegand et al., 1986; Moore et al., 1986). Using one of these clones as a probe for Northern analysis, a 3-4 fold increase in the steady state level of the corresponding mRNA was found in maize tissue grown from antidote treated seeds. These observations suggest an action of antidotes at the transscriptional level. Structural analysis of the GST I gene and its mRNA showed a 107 bp 5' untranslated region, a 642 bp coding region, and a 340 bp 3' untranslated region. The sequence of the 214 amino acids of the monomeric polypeptide was deduced from the nucleotide sequence. It did not contain apparent homology with the published sequences of animal glutathione-S-transferases (Shah et al., 1986). Expression of one of the clones isolated from a maize cDNA library in E. coli resulted in glutathione-S-transferase activity (Moore et al., 1986). For two reasons these results encourage the view that herbicide resistance to susceptible crops may be obtained by the use of methods of plant molecular biology. With the expression of maize glutathione-S-transferase in E. coli, it will be possible to analyze the structure/function relationship of the enzyme ex planta, in order to obtain an understanding of the mode of action of antidotes at the molecular level. In addition, the isolation of a glutathione-S-transferase gene from maize makes it possible to approach the transfer of this gene to susceptible species. However, efforts dealing with glutathione-S-transferase alone will not solve the problems of herbicide resistance to susceptible crops. In order to stimulate conjugation of glutathione with herbicides, enhanced glutathione synthesis is needed in addition to enhanced glutathione-S-transferase activity. As glutathione appears to be a feedback inhibitor of its own synthesis (Rennenberg, 1982) it might be assumed that its removal by conjugation will sufficiently enhance its production. Studies with several antidotes, however, did not only show an enhanced glutathione level of the antidote treated cells, but also an enhanced activity of glutathione synthesis <u>in vitro</u>. In corn, even an antidote increased uptake of sulfate and an enhanced sulfate activation for reduction has been reported (Lamoureux and Rusness, 1987). These observations show a whole range of additional metabolic processes to be involved in herbicide detoxification by conjugation with glutathione that have to be considered in genetic engineering of herbicide resistant plants.

SYNTHESIS AND FUNCTION OF PHYTOCHELATINS

In both animal and plant cells the presence of metal-binding proteins has unequivocally been proven (Robinson and Jackson, 1986). From the first group of metal-binding proteins discovered, these compounds have been named metallothioneins (MT) or "metallothionein-like" metal complexes (MTLs). MTs are characterized by their low molecular-weight, their heat-stability, an unusually high cysteine content, and a high affinity for certain metal ions. The cellular concentrations of metal ions were found to regulate the rate of transcription of MT-genes, either in a direct or in an indirect way. One important function of these metal-binding proteins seems to be the participation in metal ion homeostasis, including detoxification of heavy metal ions. Recently, a new class of MTLs, the so-called phytochelatins (PCs) has been discovered in plants. The structure of these compounds, i.e. $(\gamma\text{-glutamyl-cysteinyl})_n$-glycine, strongly suggests that these peptides are not encoded by structural genes, but are the product of a biosynthetic pathway in which glutathione and/or its metabolites are involved.

Cd-binding peptides with PC-like structure have first been analyzed in yeast grown in the presence of 1 mM CdCl2 (Murasugi et al., 1981; Kondo et al., 1983 and 1984). Since these initial reports, PCs have been isolated and characterized in many higher plant species (Grill et al., 1985 and 1986; Fujita and Kawanishi, 1986; Obata and Umebayashi, 1986; Steffens et al., 1986). In legumes, containing homo-glutathione, homo-PCs are produced; in legumes containing both, glutathione and homo-glutathione, PCs and homo-PCs have been found (Grill et al., 1986). These oberservations strongly support the idea that glutathione and homo-glutathione are precursors of PCs and homo-PCs in plants. The production of PCs or homo-PCs was found to be induced in the presence of Cd2+ , Zn2+, Cu2+, Hg2+, Pb2+, Ag2+, Sb3+, AsO42-. Induction of homo-PCs was not observed with Al3+, B3+, Ba2+, Ca2+, Ce3+, Cs+, K+ , Mg2+, Mn2+ (Grill et al., 1985 and 1986). Small amounts of PCs in the absence of excess metal ions have been reported (Steffens et al., 1986), but are not generally observed (Grill et al., 1985).

In cells of tomato suspension cultures, selected for Cd2+-resistance, PCs accumulated in the presence of Cd2+. Buthionine sulfoximine, a specific inhibitor of γ-glutamylcysteine synthetase, prevented accumulation of PCs and caused Cd2+ sensitivity of the

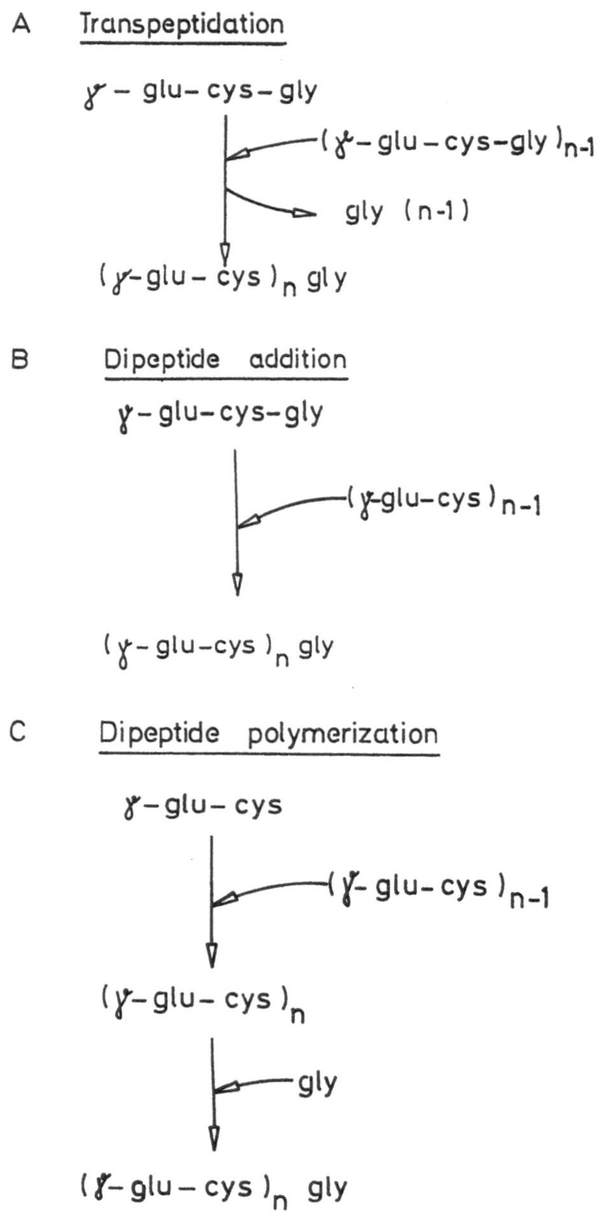

Figure 3 Possible paths of phytochelatin synthesis

resistant cell line (Steffens et al., 1986). These data clearly show that PCs can represent the primary mechanism for metal detoxification in plant cells. In addition they show, that glutathione or γ-glutamylcysteine are essential precursors of PCs. This is also indicated by the observation of a transient decline of the glutathione content during the initial synthesis of PCs upon addition of metal ions. During prolonged exposure to metal ions, the glutathione content recovered. One may assume from these findings that the γ-glutamylcysteine moities of the PCs partially originate from glutathione degradation. Still at least 3 different pathways of PC-synthesis have to be considered (Fig. 3). Glutathione may be a starter for the successive addition of γ-glutamylcysteine residues from glutathione by a transpeptidation reaction (Fig. 3A); alternatively, γ-glutamylcysteine may be added directly to a glutathione-starter (Fig. 3B). Polymerization of γ-glutamylcysteine with subsequent addition of glycine or glutathione (Fig. 3C) can be the path of PC synthesis as well. From the data available so far none of these pathways can be excluded.

The presence of small amounts of PCs in the absence of toxic amounts of heavy metals and the evolution of copious amounts of H2S upon acidification of PCs have led to the conclusion that PCs may represent the acceptor of activated sulfate in the reaction catalyzed by adenosine 5' phosphosulfate sulfotransferase during sulfate reduction (Steffens et al., 1986). Though this suggestion is intriguing, experimental evidence for such a function of PCs has so far not been reported.

What are the molecular changes occuring upon addition of heavy metal ions to plant cells leading to PC-synthesis? What are the differences at the molecular level between heavy metal resistant cells producing high amounts of PCs and sensitive cells producing small amounts of these peptides? Increased expression of the structural genes of MTs has been observed in animal cells. As the structure of PCs rules out the direct synthesis on ribosomes, modulation of the path of PC-biosynthesis appears to be responsible for enhanced PC production. Such a regulation may be obtained by one or more regulatory enzymes of the path of PC-biosynthesis. The observation of an elevated production of certain mRNA sequences during metal ion induced synthesis of MTL complexes supports the idea of a transcriptional control of this pathway (Jackson et al., 1984). Such a control of PC-synthesis would comprise the possibility to identify an environmentally regulated gene. Still, the path of PC-synthesis may be regulated by substrate or co-substrate availability, product inhibition, or feedback regulation of allosteric enzymes as well. In this context, the binding of metal ions to the PCs may represent a process of product removal. In other words, the regulation of PC-synthesis is so far not understood.

CONCLUSIONS

In vascular plants the γ-glutamyl tripeptide glutathione apparently plays a central role in the detoxification of pesticides and in the synthesis of phytochelatins, a group of peptides capable of heavy metal ion detoxification. At present, progress in these areas of research is limited mainly because of the lack of knowledge on the biochemistry of the paths of glutathione metabolism and the regulation of the processes involved. This lack of knowledge is at least partially due to the fact that analysis of the biochemistry and physiology of sulfur compounds in plants has for many years been considered rather to be an esoteric subject than a relevant area of plant research. Combined efforts of plant physiologists and plant molecular biologists are urgently needed to overcome this point of view that hinders progress in an important area of environmental research.

REFERENCES

Apontoweil, P., and Berends, W., 1975, Isolation and initial characterization of glutathione-deficient mutants of Escherichia coli K 12, Biochim. Biophys. Acta, 399:10.

Bergmann, L. and Rennenberg, H., 1978, Efflux and production of glutathione in suspension cultures of Nicotiana tabacum, Z. Pflanzenphysiol., 88:175.

Bonas, U., Schmitz, K., Rennenberg, H., Bergmann, L., 1982, Phloem transport of sulfur in Ricinus, Planta, 155:82.

Carnegie, P.R., 1963, Structure and properties of a homologue of glutathione, Biochem. J., 89:471.

Esterbauer, H., and Grill, D., 1978, Seasonal variation of glutathione and glutathione reductase in needles of Picea abies, PLANT PHYSIOL., 61:119.

Fujita, M., Kawanishi, T., 1986, Purification and characterization of a Cd-binding complex from the root tissue of water hyacinth cultivated in a Cd2+-containing medium, Plant Cell. Physiol., 71:1317.

Grill, E., Gekeler, W., Winnacker, E.-L., Zenk, M.H., 1986, Homo-phytochelatins are heavy metal-binding peptides of homo-glutathione containing Fabales, FEBS-Letters, 205:47.

Grill, E., Winnacker, E.-L., Zenk, M.H., 1985, Phytochelatins: The principal heavy-metal complexing peptides of higher plants, Science, 230:674.

Halliwell, B., 1984, Chloroplast metabolism. The structure and function of chloroplasts in green leaf cells, Clarendon Press, Oxford.

Jackson, P.J., Roth, E.J., McClure, P.R., Naranjo, C.M., 1984, Selection, isolation and characterization of cadmium resistant Datura innoxia suspension cultures, Plant Physiol., 75:914.

Jaspers, C.J., Gigot, D., Penninckx, M.J., 1985, Pathways of glutathione degradation in the yeast saccharomyces cerevisiae, Phytochemistry, 24:703.

Kosai, T., and Larson, P.O., 1980, Chemistry and biochemistry of γ-glutamyl derivatives from plants including mushrooms (basidiomycetes), in: Progress in the Chemistry of Natural Organic Products Products, 39:173.

Klapheck, S., Latus, C., Bergmann, L., 1987, Localization of glutathione synthetase and distribution of glutathione in leaf cells of Pisum sativum L., J. Plant Physiol., in press.

Kondo, N., Imai, K., Isobe, M., Goto, T., 1984, Cadystin A and B, major unit peptides comprising cadmium binding peptides induced in fission yeast - separation, revision of structure and synthesis, Tetrahedron Letters, 25:3869.

Kondo, N., Isobe, M., Imai, K., Goto, T., 1983, Structure of cadystin, the unit-peptide of cadmium-binding peptides induced in a fission yeast, Schizosaccharomyces pombe, Tetrahedron Letters, 24:925.

Lamoureux, G.L., and Rusness, D.G., 1981, Catabolism of glutathione conjugates of pesticides in higher plants, in: Sulfur and Pesticide Action and Metabolism, J.D. Rosen, P.S. Magee, and J.E.Casida, eds., ACS Symp. Ser., 158:133, Am. Chem. Soc., Washington, DC.

Lamoureux, G.L., and Rusness, D.G., 1987, The role of glutathione and glutathione-S-transferase in pesticide metabolism, selectivity, and mode of action in plants and insects, in: Coenzymes and Cofactors, J. Wiley, New York, in press.
Law, M.Y., Charles, S.A., Halliwell, B., 1983, Glutathione and ascorbic acid in spinach (Spinacea oleracea) chloroplasts. The effect of hydrogen peroxide and paraquat, Biochem. J., 210:899.
Law, M.Y., and Halliwell, B., 1986, Purification and properties of glutathione synthetase from spinach (SPINACEA OLERACEA) leaves, Plant Sci., 43:185.
Lay, M.M., and Casida, J., 1976, Dichloroacetamide antidotes enhance thiocarbamate sulfoxide detoxification by elevating corn root glutathione content and glutathione-S-transferase activity, Pestic. Biochem. Physiol., 6:442.
Leavitt, R., and Penner, D., 1979, In vitro conjugation of glutathione and other thiols with acetanilide herbicides and EPTC-sulfoxide and the action of the herbicide antidote R-25788, J. Agric. Food Chem., 27:533.
Meister, A., and Tate, S.S., 1976, Glutathione and related γ-glutamyl compounds: Biosynthesis and utilization. Ann. Rev. Biochem., 45:559.
Moore, R.E., Davies, M.S., O'Connell, K.M., Harding, E.I., Wiegand, Tiemeier, D.C., 1986, Cloning and expression of a cDNA encoding a maize glutathione-S- transferase in E. coli. Nucleic Acids Res., 14:7227.
Murasugi, A., Wada, C., Hayashi, Y., 1981, Cadmium-binding peptide induced in fission yeast, Schizosaccharomyces pombe. J. Biochem., 90:1561.
Newton, G.L., Dorian, R., Fahey, R.C., 1981, Analysis of biological thiols: derivatization with monobromobimane and separation by reverse-phase high-performance liquid chromatography, Anal. Biochem. 111:383.
Obata, H., and Umebayashi, M., 1986, Characterization of cadmium-binding complexes from the roots of cadmium-treated rice plant, Soil Sci. Plant Nutr., 32:461.
Price, C.A., 1957, A new thiol in legumes, Nature, 180:148.
Rennenberg, H., 1982, Glutathione metabolism and possible biological roles in higher plants. Phytochemistry, 21 2771.
Rennenberg, H., and Bergmann, L., 1979, Influences of ammonium and sulfate on the production of glutathione in suspension cultures of Nicotiana tabacum, Z. Pflanzenphysiol., 92:133.
Rennenberg, H., Polle, A., Grundel, I., 1983, 5-Oxo-prolinase activity in tobacco suspension cultures: regulation by sulfate nutrition. Physiol. Plant., 59:61.
Rennenberg, H., Schmitz, K., Bergmann, L., 1979, Long-distance transport of sulfur in Nicotiana tabacum, Planta, 147:57.
Rennenberg, H., Steinkamp, R., Polle, A., 1980, Evidence for the participation of a 5-oxo-prolinase in degradation of glutathione in Nicotiana tabacum, Z. Naturforsch., 33c:708.
Robinson, N.J., and Jackson, P.J., 1986, "Metallothionein-like" metal complexes in angiosperms; their structure and function, Physiol. Plant., 67:499.
Shah, D.M., Hironaka, C.M., Wiegand, R.C., Harding, E.I., Krivi, G.G., Tiemeier, D.C., 1986, Structural analysis of a maize gene coding for glutathione-S-transferase involved in herbicide detoxification. Plant Molec. Biol., 6:203.

Steffens, J.C., Hunt, D.F., Williams, B.G., 1986, Accumulation of non-protein metal-binding polypeptides (γ-glutamyl-cysteinyl)n- glycine in selected cadmium-resistant tomato cells. J. Biol. Chem., 261:13879.

Steinkamp, R., and Rennenberg, H., 1985, Degradation of glutathione in plant cells: evidence against the participation of a γ-glutamyl-transpepdidase. Z. Naturforsch., 40c:29.

Steinkamp, R., Schweihofen, B., Rennenberg, H., 1987, γ-glutamylcyclotransferase in tobacco suspension cultures: catalytic properties and subcellular localization. Physiol. Plantarum, 69:499.

Webster, G.C., 1953a, Enzymatic synthesis of gamma-glutamyl-cysteine in higher plants, Plant Physiol., 28:728.

Webster, G.C., 1953b, Peptide-bond synthesis in higher plants. I. The synthesis of glutathione. Arch. Biochem. Biophys., 47:241.

Webster, G.C., and Varner, J.E., 1954, Peptide-bond synthesis in higher plants. II. Studies on the mechanism of synthesis of γ-glutamylcysteine, Arch. Biochem. Biophys., 52:22

Wiegand, R.C., Shah, D.M., Mozer, T.J., Harding, E.I., Diaz-Collier, J., Saunders, C., Jaworski, E.G., Tiemeier, D.C., 1986, Messenger RNA encoding a glutathione-S-transferase responsible for herbicide tolerance in maize is induced in response to safener treatment. Plant Molec. Biol., 7:235.

Zama, P., and Hatzios, K.K., 1986, Effects of CGA-92194 on the chemical reactivity of metolachlor with glutathione and metabolism of metolachlor in grain sorghum (Sorghum bicolor), Weed Sci., 34:834.

RELATIONSHIP BETWEEN ANAEROBIC INDUCIBILITY AND TISSUE-SPECIFIC EXPRESSION FOR THE MAIZE ANAEROBIC GENES

Barbara Kloeckener-Gruissem and Michael Freeling

Department of Genetics
University of California, Berkeley
Berkeley, CA 94720

INTRODUCTION

If maize seedlings are temporarily placed under low oxygen conditions, as in flooded cornfields, a drastic metabolic change can be observed. Cells switch from aerobic respiration to fermentative glycolysis which results in a decrease in the intracellular pH and in the accumulation of ethanol. The physiological events involving glycolytic enzymes have been previously reviewed (Bennett and Freeling, 1987).

The anaerobic response in maize seedlings is characterized by the synthesis of a few (approximately 10 major and 10 minor) soluble proteins (the anaerobic proteins or ANPs) while total protein synthesis is decreased markedly. A detailed anaerobic protein profile has been described by Sachs et al. (1980). Primary roots from anaerobically treated 6 to 10 day old seedlings were used to incorporate radio-labeled amino acids into polypeptides. The results were visualized using autoradiography following 2-dimensional gel electrophoresis. During the first hour of anaerobiosis the aerobic protein synthesis pattern disappears; instead polypeptides of approximately 33 kilo dalton in molecular weight are synthesized. After 5 hours of anaerobicsis these become almost undetectable. Sachs et al. (1980) called this class of proteins "transition polypeptides". After 2 hours of anaercbiosis a simpler protein synthesis profile emerges: approximately 20 polypeptides with molecular weights between 30 and 90 kilo dalton were identified. They account for approximately 70% of the newly synthesized proteins. The quantitative protein pattern did not change over a period of 72 hours of anaerobic treatment. If the primary roots are returned to aerobic conditions after 12 hours of anaerobiosis, a nearly "aerobic" protein pattern can be seen after 5 hours. However the pattern is not fully restored, even after 30 hours. The primary root apical meristem will die in such recovered roots but the formation of adventitious roots allows the seedlings to grow into mature and fertile plants.

In an attempt to understand the physiological reactions elicited by anaerobic conditions the enzymatic activity of some of the 20 ANPs have been identified: Alcohol dehydrogenase ADH1 and ADH2 (Sachs and Freeling, 1978; Ferl et al. 1979), pyruvate decarboxylase (Wignarajah and Greenway, 1976; Laszlo and St.Lawrence, 1983), glucose phosphate isomerase (Kelley and Freeling, 1984), aldolase (Kelley and Freeling 1984b) and sucrose synthase-1 (Springer et al. 1986). All of these enzymes are involved in glycolysis. Only ADH1 and sucrose synthase-1 are have been proven to be essential enzymes for the survival of maize seedlings under anaerobic conditions. Adh1 deficient mutants die after a few hours of anaerobiosis (Schwartz, 1969) and Sh1 mutants (with deficiencies in sucrose synthase-1) germinate extremely poorly under hypoxic conditions (Freeling lab, unpublished). The other ANPs may be necessary for survival of hypoxia as well.

The induction of the ANPs results from increased steady-state mRNA levels. Gerlach et al. (1982) and Hake et al. (1985) were able to isolate cDNAs after anaerobic induction. These cDNAs were generated from transcripts of Adh1, Adh2, aldolase and sucrose synthase-1 (Sh1). Additional anaerobic cDNAs have been isolated but have yet to be correlated with known genes/proteins (Hake et al. 1985). Anaerobic induction results in a 50 fold increase in ADH1-mRNA levels. A maximum is reached within 5 hours. When Hake et al. (1985) analyzed the Northern blot hybridization data quantitatively it became obvious that the different anaerobic mRNAs do not accumulate simultaneously. The rate of induction, the lag period before anaerobic response, and the level of accumulated mRNAs is not the same for all 5 transcripts studied. However, ADH1 and sucrose synthase-1 have similar or identical induction kinetics. It is possible that they are regulated by the same trans-acting molecule. In contrast to the differences in response to anaerobiosis, all 5 mRNAs under investigation disappeared at the same rate when seedlings were returned to air.

ORGAN-SPECIFIC ANAEROBIC RESPONSE

The anaerobic response described above, with its typical 2-D anaerobic protein profile, exhibits a characteristic quantitative and qualitative pattern of polypeptide synthesis in primary roots. Is the anaerobic response restricted to roots? After all, the roots are intimately connected to a flooded cornfield. Okimoto et al. (1980) addressed this question and studied the protein synthesis profile in several organs of the maize plant and compared the polypeptide patterns obtained under aerobic and anaerobic conditions. From studies of the primary roots, the immature endosperm, the immature scutella and whole anthers (harvested 1 day before dehiscence), two major conclusions can be drawn: (1) The protein profile differs dramatically for major polypeptides synthesized in the presence of air. (2) Upon anaerobiosis the profiles become far less complex and very similar for all four organs. A closer examination indicates some diagnostic differences characteristic of each organ. These include the appearance of new polypeptides in the endosperm and scutellum and different rates of synthesis as well as qualitative differences when compared to those proteins synthesized in the anaerobic primary roots. Okimoto and coworkers also attempted to analyze leaf tissue for its ability to respond to low oxygen. It was possible to label aerobic leaf tissue but no incorporation could be found in anaerobic tissue; leaves actually died after seedlings were treated for 5 hours under argon.

The experiments reported by Okimoto et al. (1980) raise several questions: (1) Does the same regulatory mechanism determine the anaerobic response in roots, endosperm, scutellum and anthers/pollen? (2) What signals are responsible for the organ-specific anaerobic response? (3) Why do leaves fail to respond by synthesizing ANPs? Following Schwartz's (1969) suggestion that Adh1 was maintained over time by a selective advantage in suboptimal environments, Okimoto and coworkers offered some evolutionary-type answers to some of these questions. The authors reason that seed and seedling organs synthesize ANPs in order to survive low levels of oxygen. Similarly it is possible that anthers need to respond to short-term anaerobic conditions , caused by water accumulation in the leaf whorl of the plant after long periods of rain. The extended leaf blades, in contrast, are normally not required to survive short-term drowning. Such explanations are only suggestions; they do not explain why Adh1 deficient mutants survive in the field when exposed to a naturally rainy environment.

The studies described above did not address questions about cell-specific responses in the various organs tested. It seems quite possible that only specialized cells are equipped to respond to the low-oxygen stress signals. At this point we do not have proof of a cell-specific response only some circumstantial evidence. Induced and uninduced primary roots from Adh1 active and Adh1 deficient seedlings were subjected to in situ staining for ADH1 activity (Karoly, 1987). It was observed that cells in the root cap and some cells of the root steele (vascular system) express Adh1 during aerobic conditions. This activity may be considered "constitutive" in comparison with ADH activity under anaerobic conditions, which can be detected as an overall dark blue staining covering the complete cross-section surface exposed to the stain. In Adh1 deficient mutants neither the constitutive nor the induced activity can be observed. Staining fir ADH activity stain is an extremely sensitive method which allows a detailed analysis of individual cells types and their responses to anaerobic conditions.

GENETIC STUDIES

Genetic studies on the regulation of Adh1 were reviewed recently (Freeling and Bennett, 1985). Woodman and Freeling (1981) found that each naturally occurring Adh1 allele tested in the 20 diverse families from 6 inbred lines or races examined carried cis-acting information as to organ-specific, quantitative expression. A unifying rule, called the "reciprocal effect", emerged: an allele that is expressed strongly in the scutellum (of the kernel) is compensatorily weakly expressed in the primary root, and vice versa. Karoly (1987) imagined a simple, but untested, model that explains these data. She suggested that each allele carries an enhancer-like sequence(s) that binds a similar regulatory protein in both scutellum and root. If binding leads to enhanced expression in the scutellum, but down-regulates in the root, or vice versa, the reciprocal effect would be achieved. Birchler (1981) has used aneuploid studies to identify a locus that inversely affects Adh1 expression. This "inverse effect" locus is loosely linked to Adh1 on the long arm of chromosome 1.

MUTANTS

In general, the analysis of complex biological problems is aided by the isolation and characterization of interesting mutants. Many mutants have been identified which are deficient in Adh1 expression, but most of the lesions map to the coding sequence of the gene and abolish or alter Adh1 expression in all organs simultaneously (Freeling and Bennett, 1985). However, among our mutant collection there are several mutants that seem helpful in approaching the organ-specific problem. Mutant FkF3037 arose as a spontaneous mutation. Allozyme balance studies (Freeling et al. 1982) show that this mutant over-expresses ADH1 polypeptides in the scutellum but is normal in primary roots. In FkF3037 the reciprocal effect rule is violated. A more conclusive answer has to await cloning of the mutant allele and its nucleotide sequence analysis.

Another mutant, Adh1-3F1124, alters the organ-specific expression of Adh1. Using the conditional anaerobic lethal selection scheme described by Chen et al. (1986), three families were isolated as conditional anaerobic lethals (Chen et al. 1978). In two of the three families scutellar ADH1 activity is present at normal levels. Two possible explanations for this phenotype are: (1) The mutation maps to a gene other than Adh1, which encodes a product required for anaerobic metabolism. In a most interesting case, this gene product might regulate the anaerobic response. Alternatively, the gene might encode one of the ANPs, which would provide proof of the ANP's essential nature. (2) The mutation could map to the Adh1 locus and destroy specifically the region that is involved in the anaerobic response. The analyses of these two and similar mutants are in progress in the Freeling lab.

The third conditional anaerobic lethal mutant in our collection showed low ADH1 activity in the scutellum (Chen et al. 1987). Analysis of scutellum and primary anaerobic roots from heterozygotes Adh1-3F1124/1S allowed us to quantify the low activity, which we determined to be 6% of the wild type (progenitor Adh1-3F/1S) activity. The reciprocal effect is not distorted - low expression affects both primary roots and the scutellum. Analysis of mature pollen grains resulted in the surprising finding that ADH1 activity was indistinguishable between mutant and progenitor. The low activity in roots of Adh1-3F1124 seems a direct result of decreased synthesis of ADH1 polypeptides, which in turn seems a direct result of reduced accumulation of steady-state mRNA. This low level of mature ADH-mRNA is synthesized in response to anaerobic conditions (Fig. 1A). Northern blot analysis of anther mRNA (anthers were harvested that were 1-3 days before dehiscence) from progenitor and mutant showed indistinguishable ADH1 transcript levels (Fig. 1B).

The mutant phenotype is somatically unstable. When the aleurone layer of a 4 day old germinated seedling is peeled off the seed and stained for ADH activity, an overall light purple color can be seen, representing the 6% mutant enzyme activity. Single cells or small groups of cells which stain dark blue are present at a high frequency (10^{-3} to 10^{-4}), indicating there is a reversion event which leads to normal ADH activity. This observed instability of the mutant phenotype as well as the line's genetic background suggest the involvement of a transposon in the mutant phenotype. (The mutant comes from a line carrying Robertson's mutator (Robertson, 1978). This line has been shown to result in the occurrence of new mutants at a very high frequency, due to the activity of the transposable element Mu (for review see Lillis and Freeling, 1986). Cloning and sequencing of the mutant Adh1-3F1124 allele demonstrated the insertion of a 1.85 kb transposon, Mu3, into the 5' upstream region of the Adh1-3F coding sequence (Chen et al. 1987). Mu3 creates a duplication of 9 target host nucleotides upon insertion. In this case the "TATA" box was duplicated. This insertion and duplication generated the genomic configuration diagrammed in Figure 2. Traveling from the 5' end of the Adh1 cistron, the Adh1 5' upstream region is authentic until the first (distal) TATA box. his is followed by Mu3 sequences which end in the second (proximal) TATA box. Intact Adh1 sequence continues downstream of the second TATA. The insertion is the only detectable change that occurred at the DNA sequence level. The mutant phenotype and the insertion of this 1.85 kb transposable element (Mu3) are linked in segregation studies. Analysis of the revertants to be described prove that Mu3 insertion caused the organ-specific phenotype of Adh1-3F1124.

How can the insertion of Mu3 interfere with transcript accumulation in the seed and seedling? We know one example where another mutator

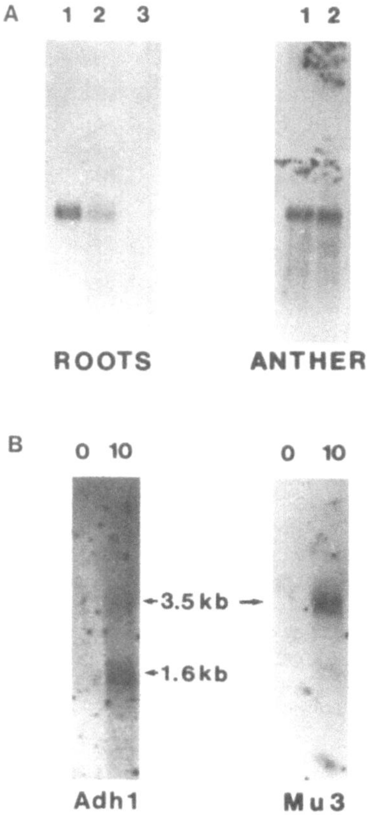

Figure 1: Northern blot analysis of steady-state mRNA from mutant and progenitor primary roots and anthers followed by gel autoradiography. Poly A+ RNA was electrophoretically separated through 1% agarose-formaldehyde, transferred to nitrocellulose and hybridized to specific DNA as indicated. Panel A roots: 2ug RNA from anaerobically treated progenitor roots (lane 1); 10ug RNA from anaerobically treated mutant seedlings (lane 2) and 20 ug RNA from mutant seeds grown in air (lane 3). Panel A anthers: 20 ug RNA from progenitor anthers (lane 1); 20 ug from mutant anthers (lane 2). All RNA is hybridized to Adh1 specific DNA. Panel B: mutant RNA from aerobically grown seedlings (0) and anaerobically treated seedlings (10) are hybridized to Adh1 and Mu3 specific DNA. The size of the transcripts is given in kilo bases (kb).

element (Mu1), inserted into the first intron of Adh1 interferes with transcriptional elongation and termination. The mutant, Adh1-S3034, expresses only 40% of the wild-type Adh1-S message levels. Vayda and Freeling (1987) showed by in vitro run-off transcription experiments that initiation of transcription was not affected in this mutant, but rather that the polymerase seems to be hindered when proceeding through Mu1 sequences. The authors suggested that either the high G/C content of the Mu1 sequences or secondary structures of the transposable element (or

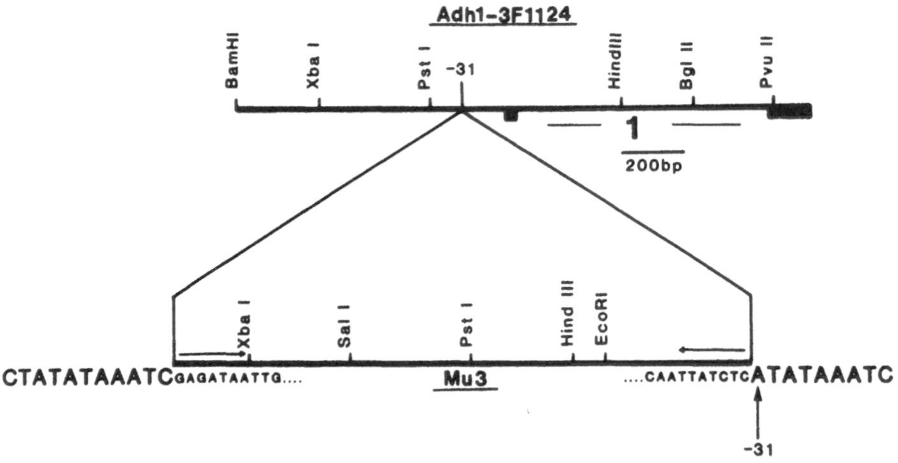

Figure 2: Genomic configuration of Adh1-3F1124 mutant allele (Chen et al. 1987). The 1.85 kb Mu3 element is inserted 31 nucleotides upstream of the transcriptional initiation site of Adh1-3F. The 9 base pair direct repeat duplication of Adh1-3F sequence is shown in enlarged capital letters. This duplication includes the consensus TATA box sequence. The small capitalized nucleotides are part of the Mu3 terminal inverted repeats, which are indicated by the long arrows. The number "1" between the two black boxes, the first two exons, denotes the first intron of Adh1-3F.

both) present steric hindrance to the movement of the RNA polymerase II complex, resulting in premature chain termination. Run-off transcription studies of 3F1124 have not yet been done.

The new genomic configuration in 3F1124 raises the question: are both TATA boxes utilized as transcriptional promoter elements? When mRNA of mutant roots is hybridized to an Adh1-specific DNA probe, two Adh1 transcripts are detected in an autoradiograph after a long exposure: the smaller mRNA corresponds in size to the mature 1.6 kb Adh1 transcript while the larger mRNA (3.5 kb) is equivalent in size to the sum of the Adh1 transcript and Mu3. If the same RNA is hybridized to Mu3 specific DNA, only the larger transcript hybridizes (Fig. 1B). These results suggest that transcription of the Mu3 sequence causes the low level of mature mRNA. There are at least two different ways this may happen: (1) Only the distal TATA box is utilized for initiation of transcription; this generates a hybrid transcript of 3.5 kb. Processing of this large mRNA would yield the mature mRNA. Since processing would require a new mutant-specific activity, it might not be very efficient, leading to the low levels of both mRNAs. The low levels of both transcripts could result from inefficient usage of the distal TATA box. (2) Both TATA boxes may be used, generating a large transcript (Mu3 +Adh1) from the distal promoter and a smaller wild-type transcript (Adh1) from the proximal promoter. If there is a factor limiting transcription under anaerobic conditions, both TATA boxes might compete for that factor, resulting in the accumulation of lower levels of Adh1 transcripts. We conducted one experiment which yielded results favoring the second explanation. We determined the location of the 5' end of the mRNA in mutant homozygotes which hybridizes to an Adh1 specific oligonucleotide (Fig. 3) and compared it to the wild type (progenitor) 5' end of the Adh1 transcript. An important finding can be summarized from these primer extension experiments: The 5' ends of the smaller mutant and wild type Adh1-mRNA are indistinguishable. If the mutant uses only the distal promoter and generates the smaller transcript from the precursor, the postulated processing activity would have to splice the pre-mRNA at precisely that site which leads to a "wild type" mature 5' end. We feel that the fortuitous existence of such a processing activity would be highly unlikely. Experiments to determine the 5'end of the Mu3 containing mRNA are still necessary.

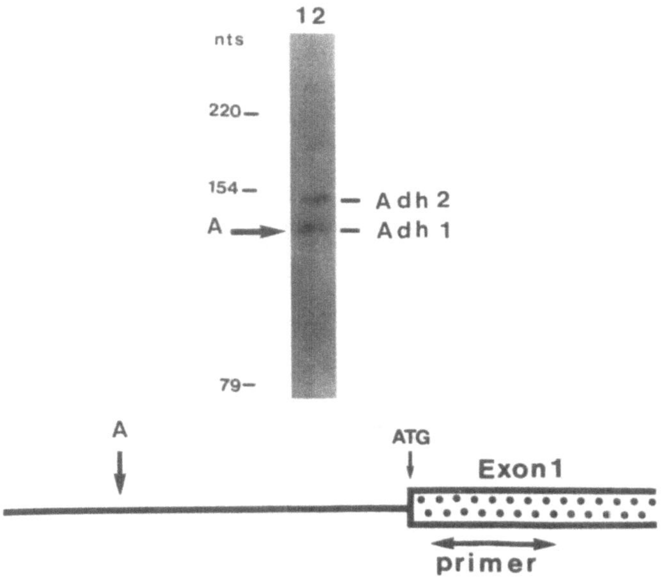

Figure 3: Primer extension of progenitor (3F) and mutant (3F-11124) Adh1-mRNA. The primer is an oligonucleotide of 21 nucleotides in length which starts at the second Adh1 codon. (The primer was synthesized as a gift by O. Anderson, USDA, Albany, California.) The primer was hybridized in solution to anaerobically induced poly A+ mRNA from progenitor 3F (lane 1) and mutant 3F1124 (lane 2) primary roots, and elongated with reverse transcriptase. Extension products were separated on 5% acrylamide gels. The nucleotide designations (nts) represent a size marker. The letter "A" indicates the 5' end location of the Adh1-mRNA.

The anaerobic responsiveness of the mutant 3F1124 is demonstrated by an increase of message levels during anaerobic induction (Fig.1). The Mu3-Adh1 hybrid transcript as well as the normal 1.6 kb Adh1 transcript are inducible under anaerobic conditions (Fig. 1). These findings indicate that those cis-acting signals that are responsible for transmitting the anaerobic response have not been destroyed by the insertion of Mu3. Ellis et al. (1987) reported studies on wild type Adh1 sequences that identified those regions which are involved in the anaerobic response. The authors tested sections of Adh1 upstream regions for anaerobic inducibility of a chloramphenicol acetyl transferase reporter gene in a transient expression system. The results indicate that sequences starting at the translational initiation codon, extending through the TATA box up to 247 bp upstream are sufficient to induce the anaerobic response. The fact that mutant 3F1124 transcribes both mature Adh1-mRNA and hybrid Mu3-Adh1-mRNA upon anaerobic treatment is consistent with Ellis et al.'s (1987) finding on the 5' location of the anaerobic response signal (see also Dennis E.S. this volume), as long as the anaerobic site(s) does not require fixed spatial relationship. That is, the cis-acting anaerobic sequence discovered by the peacock lab must be able to act over variable distances to the TATA box.

ISOLATION OF GERMINAL REVERTANTS

The Mu3 insertion did not destroy the mutant's ability to respond to anaerobiosis but it did separate those signals responsible for Adhl expression in roots and scutellum from those responsible for expression in pollen. We decided to take a genetic approach in order to identify the sequences involved in organ-specificity. We isolated germinal revertants of Adhl-3F1124.

Homozygous mutant 3F1124 seeds were germinated in a partially-aerated aqueous chamber for 7 days. Under these conditions, Adhl positive seeds will germinate and grow primary roots several mm in length while Adhl deficient seeds will not germinate at all. A total of 47,850 3F1124 homozygotes were subjected to this selection. Table 1 gives the data that lead to the recovery of 50 revertant seeds. A small sliver of the scutellum from each germinating seedling was tested for ADH1 activity. Only a small fraction of these seeds actually exhibited ADH1 activity (Table 1). The other germinating seeds may have resulted from either an alternative pathway for anaerobic metabolism or may have been revertants with ADH1 activity below our detection limits. The allozyme

Table 1: The number of homozygous seeds of the mutant Adhl-3F1124 and revertants is given for two families which were subjected to the revertant selection .

Family	#seeds tested	#seeds germinating (%)		#seeds with ADH1 activity (%)		Frequency
6629	40,770	573	(1.5)	48	(8.3)	1 : 850
6640	7,080	242	(11.7)	2	(0.8)	1 · 3500

ratios were analyzed in 5 of the revertants (Table 2). Revertants r-16 and r-17 gained only partial ADH1 activity in the scutellum and anaerobic roots. Revertant r-18 and r-21 express almost wild type levels of ADH1 activity in scutellum (91% and 98%) and anaerobic roots (100% and 96%). In contrast, revertant r-19 shows very different levels of activity in scutellum and roots; while the reversion resulted in a gain of full activity in the roots, ADH1-F activity is relatively underexpressed in the scutellum. This is not the only case in which we found an organ-specific reversion event. When pollen was subjected to an ADH activity staining assay, we found the following surprising results: in two independent revertants, r-17 and r-19, pollen expression was lowered while, simultaneously, roots and scutellum expression was raised.

Table 2: ADH1 activity in scutellum and anaerobically treated (10 hours) roots was measured as allozyme ratio from native starch gels as described (Woodman and Freeling 1981). Aleurone layers were peeled from 5 day old germinated seedlings and stained in situ for ADH activity. The ADH activity from freshly shed pollen grains was determined as described (Freeling and Birchler, 1981).

genotype	% ADH1 activity in: scutellum	roots	aleurone	pollen
3F (WT)	100	100	normal	normal
3F1124 (mutant)	6.4	6	low	normal
r-16	27	27	normal	normal
r-17	62	68	normal	low
r-18	91	100	normal	normal
r-19	75	100	normal	low
r-21	98	96	normal	normal

How can a reversion event affect the expression in two separate organs? This situation is quite reminiscent of the reciprocal effect. First, the insertion of Mu3 lowers ADH1 expression in seedling but not in pollen. Now, revertants restore restore expression in the seedling but simultaneously lower expression in pollen.

To come closer to an understanding of these effects we analyzed genomic DNA from the revertants. Some revertants were heterozygous and some were homozygous (Fig. 4). An Adh1 specific probe was used to determine whether the 1.85 kb Mu3 insert was still present, either fully or in part at its original insertion site and whether new insertions had occurred within the Adh1 cistron. Figure 4A. shows restriction fragment analysis of 4 different revertants. The position of the fragments expected for the various possible Adh1 alleles is indicated. Summarizing the data, all but one revertant (r-16) studied so far are indistinguishable from the progenitor 3F allele, indicating that the excision of Mu3 caused reversion. It becomes critical to study the nucleotide sequence around the insertion site to identify small changes that result in the various revertant phenotypes. Data for the one exception, in which partial reversion leads to 27% activity in roots and scutellum, is presented in Figure 4B. Three siblings were analyzed in this Southern hybridization and compared to DNA from the homozygous mutant. All three siblings carry an Adh1 fragment which has the same length as the corresponding mutant fragment.

Is was possible that the partial ADH activity in revertant r-16 is caused by a new suppressor mutation (unlinked to Adh1) rather than changes of the Mu3 insertion? Suppressors have been identified in Drosophila, in which they suppress - partially or fully - the phenotype of mutants caused by the insertion of transposable elements (for review see Kubli, 1986). We tested linkage between the revertant r-16 and the mutant Adh1-3F1124 by a backcross of the homozygote carrying the revertant r-16 allele and an electrophoretically distinguishable allele Adh1-S, to the homozygous mutant Adh1-3F1124. r-16 maps to the Adh1 locus.

We have been extraordinarily lucky that the 3F1124 revertants behave in such developmentally interesting ways. We have hopes that the "transposon-directed, site-specific mutants" will help us to understand mechanical relationships between *cis*-acting sequences involved in anaerobic inducibility, expression in organs (scutellum and root) that respond to anaerobiosis, and expression in pollen.

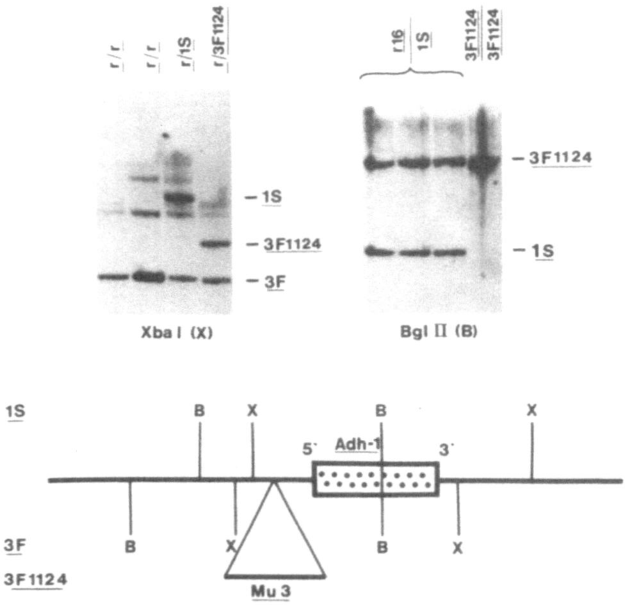

Figure 4: Autoradiography of southern blot hybridization of genomic 3F1124 revertant DNA. Genomic DNA was cut with XbaI (X) or BglII (B), electrophoretically separated through 0.7% agarose and transferred to nitrocellulose filter. DNA was hybridized to an *Adh* specific probe. 3F-1124, 3F and 1S indicate the positions of the allele specific restriction fragment. The first gel shows DNA, digested with the restriction enzyme XbaI (X), from 4 different revertants with the genotypes r/r/ (r=revertant); r/r/; r/1S and r/3F1124. The second gel shows in the first three lanes DNA, cut with the restriction enzyme BglII, from three siblings of revertant r-16. The last lane shows DNA from homozygous mutant 3F1124 leaves. The faintly hybridizing bands, indicated by the open circle, correspond to cross-hybridization to *Adh2*-specific restriction fragments. The restriction map below indicates the position of the XbaI (X) and BglII (B) sites at the *Adh1* locus for the 1S allele, the progenitor 3F allele and the mutant 3F1124 allele.

REFERENCES

Bennet DC and M Freeling. 1987. CRC Model Building in Plant Physiology / Biochemistry / Biotechnology. in press.

Birchler JA. 1981. Genetics 97, 625-637.

Chen C-H, Freeling M and A Merckelbach. 1986. Maydica XXXI 93-108.

Chen C-H, Oishi KK, Kloeckener-Gruissem B and M Freeling. 1987. Genetics, in press.

Ellis JG, Llewellyn DJ, Dennis ES and WJ Peacock. 1987. EMBO J. 6, 11-16.

Ferl RJ, Dlouhy SR and D Schwartz. 1979. Mol. Gen. Genet. 169, 7-12.

Freeling M and JA Birchler. 1981. In: Genetic engineering, principles and methods III. 223-264. Stelow JK and A Hollaender eds. Plenum Press, New York.

Freeling M, Chen DS-K and M Alleman. 1982. Genetics 3, 179-196.

Freeling M and DC Bennett. 1985. Ann. Rev. Genet. 19, 297-323.

Gerlach WL, Pryor AJ, Dennis ES, Ferl RJ, Sachs MM and WJ Peacock. 1982. Proc. Natl. Acad. Sci. 79, 2981-2985.

Hake S, Kelley PM, Taylor WC and M Freeling. 1985. J. Biol. Chem. 260, 5050-5054.

Karoly C. 1987 Ph. D. dissertation University of California, Berkeley.

Kelley PM and M Freeling. 1984. J. Biol. Chem. 259 673-677.

Kelley PM and M Freeling. 1984. J. Biol. Chem. 259, 14180-14183.

Kubli E. 1986. Trend in Genetics. August .

Laszlo A and P StLawrence. 1983. Mol. Gen. Genet. 192, 110-121

Lillis M and M Freeling. 1986. Trends in Genetics, July .

Okimoto R, Sachs MM, Porter EK and M Freeling. 1980. Planta 150 89-94.

Robertson D. 1978. Mutation Research 51, 21-28.

Sachs MM and M Freeling. 1978. Mol. Gen. Genet. 161, 111-115.

Sachs MM, Freeling M and R Okimoto. 1980. Cell 20, 761-767.

Schwartz D. 1969. Am. Nat. 103, 479-481.

Springer B, Werr W, Starlinger P, Bennett DC, Zokolica M and M Freeling. 1986. Mol. Gen. Genet. 205, 461-468.

Wignarajah K and H Greenway. 1976. New Phytol. 77, 575-584.

Woodman JC and M Freeling. 1981. Genetics 98, 357-378.

Vayda ME and M Freeling. 1987. Plant Mol. Biol. 6, 441-454.

BARLEY RAINCOATS: BIOSYNTHESIS AND GENETICS

Penny von Wettstein-Knowles

Institute of Genetics, University of Copenhagen, and
Department of Physiology, Carlsberg Laboratory,
Gamle Carlsberg Vej 10, DK-2500 Copenhagen, Denmark

INTRODUCTION

The outermost surface of a plant cuticle serves as its raincoat. The molecular structure of this epicuticular wax layer varies markedly, in part due to differences in composition[1]. A wide range of relatively non-polar lipid classes have been identified therein which have in common very long chain carbon skeletons. Their origin is attributed to elongases which are enzyme complexes condensing short carbon chains to a primer and preparing the growing chain for the next addition. Soluble plastid fatty acid synthetase (FAS) in higher plants is a special example in which the initial primer is acetyl-acyl carrier protein (ACP), the donor of C_2-units is malonyl-ACP and the 16 carbon product is palmityl-ACP. Addition of the next C_2-unit, however, is accomplished by a different complex, namely the soluble plastid palmityl elongase[2,3]. Other elongases are generally believed to be located in the epidermal cells where they are affiliated with or part of the microsomal membranes[1,4,5,6]. They carry out analogous C_2-unit additions until the carbon skeletons of the epicuticular waxes are completed. Coenzyme A (CoA) rather than ACP activated derivatives are thought to serve as their substrates[1,4,6,7]. Subsequent action of associated enzyme systems yields the diverse lipid classes which in barley include hydrocarbons, free and esterified alkan-1-ols, aldehydes, free acids, β-diketones, hydroxy-β-diketones, esterified alkan-2-ols plus esterified 7-oxoalkan-2-ols.

Three disparate types of raincoats occur on wild type barley plants: (i) leaf blades and the lower leaf sheaths have highly lobed plates (Fig. 1a), large amounts of primary alcohols (predominantly C_{26}) and an absence of β-diketone lipids; (ii) lemmas, glumes and uppermost leaf sheaths and internodes have long, thin tubes (Fig. 1c) and β-diketone lipids; and (iii) awns have thin plates, frequently horizontal to the surface, major amounts of hydrocarbons and esters and an absence of β-diketone lipids[8]. Mutations in genes potentially affecting wax synthesis and/or its deposition have been isolated in a number of species because of an alteration in the phenotype of the cuticle surface. While they are often referred to as waxless, glossy and non-glaucous, a collection in barley[11] (1548 mutants distributed among 84 complementation groups) and one in Arabidopsis[12] are called <u>eceriferum</u> (<u>cer</u>). Figures 1b, d and e illustrate the raincoat ultrastructure of three <u>cer</u> mutants. Whereas water drops are readily shed by uneven leaf blade and lemma surfaces (Figs. 1a, c and d) they are retained by smooth ones (Figs. 1b and e).

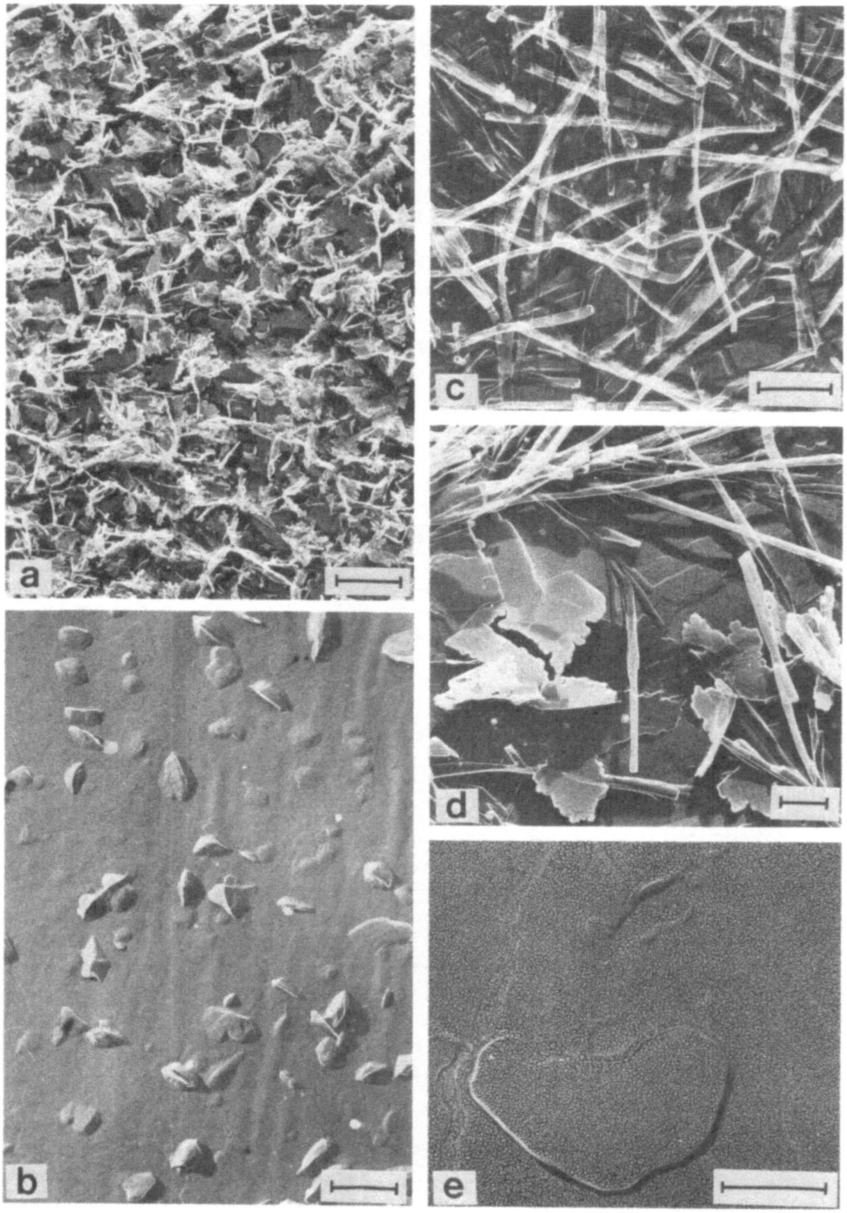

Fig. 1. Raincoats of barley leaf blades (a and b) and lemmas (c, d and e) are revealed by transmission electron micrographs of shadowed carbon replicas[9,10]. a) Small, lobed plates of the wild type Bonus. b) Small, moundlike structures scattered among thin plates horizontal to the cuticle of _cer-j_[71]. c) Long thin tubes of Bonus. d) Long thin tubes, very thin upright lobed plates and thin plates lying on the cuticle of _cer-u_[69]. e) A thin plate appressed to the cuticle of _cer-cu_[108]. Bar = 1 μm.

The focus of this paper is on the major components of the second type of raincoat listed above. I present a genetic analysis of the structural gene *cer-cqu*. This is followed by a description of the 3-oxoacyl biosynthetic pathway, and the functions of a β-ketoacyl elongase and *cer-cqu* therein. Finally, dominantly inherited *cer* mutants and their potential roles in the 3-oxoacyl pathway are considered.

GENETICS OF THE STRUCTURAL GENE *cer-cqu*

The most frequently induced *cer* mutations modifying the blue grey color of the wild type spikes and uppermost leaf sheaths plus internodes to yellow green are those of the *cer-cqu* locus (522/900)[11]. While 204, 157 and 148 of these have been assigned by genetic analyses to the *cer-c*, *-q* and *-u* complementation groups, another 13 belong to more than one of

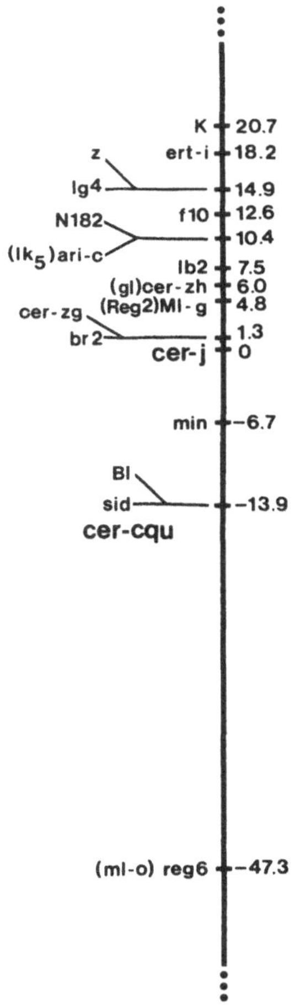

Fig. 2. Segment of barley chromosome 4 showing the relative location of *cer-cqu* and *cer-j*[11,15]. The statistically derived sequence and distances are in cM. Genes within 0.8 cM of one another are grouped together at their average value. The gene *cer-cqu* is placed in the vicinity of *sid* and *Bl* instead of at -14.5 as the accompanying S.D. for this estimate was 5.8.

these, that is, -cu, -cq, -qu or -cqu. Genetic testing with the latter revealed initially that the cer-c, -q and -u complementation groups were very tightly linked (on the order of 0.0012 cM), and secondly that they belonged to the same gene[13]. The latter conclusion was derived from the recovery of cer-cqu revertants to wild type after NaN_3 mutagenesis which would have been highly improbable if they had originated from concurrent mutational events[14].

Before carrying out the NaN_3 treatment the multiple mutants were crossed with another marker on chromosome 4 (Fig. 2), namely cer-j which alters only leaf blade raincoats (Fig. 1a vs b). The double mutants were selected, and the seed increased. The cer-j marker permitted discrimination between NaN_3 induced wild type revertants and those from cross pollinations by stray pollen or inmixtures etc during the several years long testing period which these experiments required. All 13 apparent multiple mutants have been reverted and hence are attributable to individual mutational events. The average reversion frequency of the cer-cqu mutants (5.6×10^{-5}, ranging from 2.6-15.1×10^{-5}) is very similar to that for cer-j^{59} (4.2×10^{-5})[14].

The 36 cer-cqu revertants recovered from the above experiments can have arisen from mutational events within the same or other loci. To clarify this question homozygous revertant cer-cquR,j^{59} strains are being identified and crossed to the wild type Bonus. Thus far, subsequent genetic analyses divulge that 12 of the revertants stem from mutations within the cer-cqu locus, and that one of them cer-cu^{947}R:2 is due to mutation of a non-linked gene to a dominant suppressor of the cer-cu^{947} allele[14]. Furthermore, the suppressor locus (Su-cer-cu^{947})[11] is not linked to cer-j and does not influence the expression of the cer-j^{59} allele. Genetic analyses of the 36 revertants are being continued. Combined with the original 522 cer-cqu mutants, they provide a large store house of tools for the molecular dissection of the cer-cqu locus when cloned sequences of this locus become available.

THE ROLES OF cer-cqu AND A β-KETOACYL ELONGASE IN 3-OXOACYL WAX LIPID SYNTHESIS

The summarized genetic analyses intimate that cer-cqu determines a polypeptide having at least three functional domains corresponding to the complementation groups cer-c, -q and -u. Furthermore, the polypeptide functions as a multimer with a minimum of two units. As a first step toward revealing the nature of these functions, the wax composition on various organs of the wild type Bonus and a cer-c, -q and -u mutant were determined. Analyses of the results led to two deductions[10,16]. Firstly, barley wax lipids result from two parallel elongation systems. The β-ketoacyl elongase gives rise to the 3-oxoacyl wax classes (Fig. 3) and the acyl elongase(s) to all the others. Secondly, the mutants create blocks at different sites: cer-q^{42} previous to a precursor of both β-diketones and esterified alkan-2-ols; cer-c^{36} beyond the common precursor in the β-diketone branch; and cer-u^{69} in the associated enzyme system inserting a hydroxyl group into the β-diketone carbon chain. Disclosure of the sequence of reactions in the 3-oxoacyl pathway as highlighted below was accomplished using radioactive tracers and/or inhibitor pretreatments with wild type and cer-cqu mutants serving as the plant material.

As shown at the top of Fig. 3, the condensing activity of the β-ketoacyl elongase yields 3-oxoacyl compounds which may have one of two fates. If the C_2-unit is added to a C_{12}- or C_{14}-CoA primer (Fig. 3 left-side)[7,16,17,18], the β-carbonyl is not immediately reduced. Instead an apparent decarboxylation to a methyl ketone (I) occurs before reduction of the oxo group to a hydroxyl group (II). The resulting C_{13} and C_{15} alkan-2-

ols are invariably esterified (III). Thereafter, an oxo group may be inserted into the ester on carbon 7 of the C_{15}-2-ol moieties. When C_{16}-CoA serves as primer the β-carbonyls are retained in two successive rounds of elongation (Fig. 3 rightside) although the subsequent ones are eliminated presum-

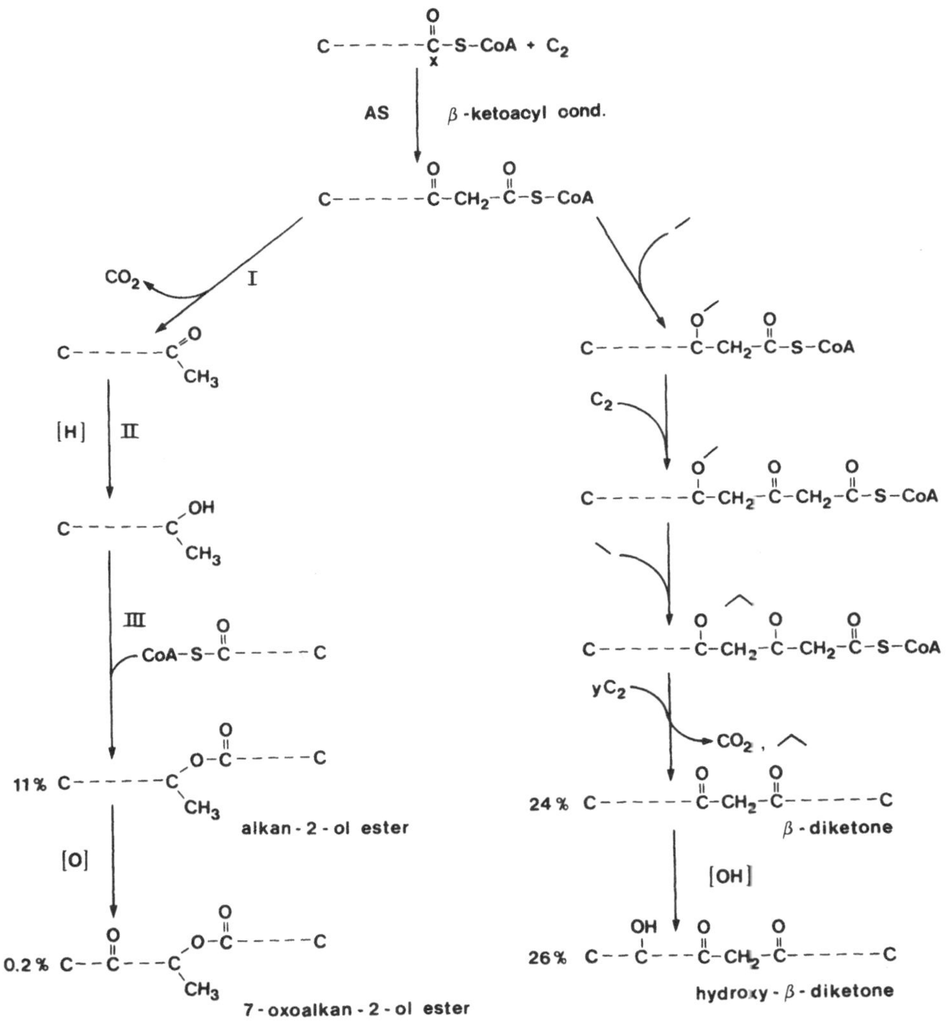

Fig. 3. Scheme for synthesis of barley 3-oxoacyl-CoA derived epicuticular lipids. The condensing enzyme of the β-ketoacyl elongase (β-ketoacyl cond.) is presumed to be the arsenite sensitive moiety of this enzyme complex (see text). In alkan-2-ol ester synthesis (left), x is primarily 12 and 14, rarely 10 and 16. The three tightly coupled reactions I, II and III are described in the text. In β-diketone synthesis (right) x is predominantly 16, rarely 14 and y is generally 6 although sometimes 5. Retention of 3-oxo groups in two successive rounds of elongation is shown by / and ∧. The amounts of the four lipid classes in Bonus spike waxes are given as weight percentages. Sites of action of cer-cqu mutants within this pathway are summarized in Fig. 4.

ably in an analogous fashion to that in saturated fatty acid synthesis. An apparent decarboxylation[19] yields a completed β-diketone (96% hentriacontan-14,16-dione) into which a hydroxyl group may be inserted on carbon 25. How protection of one or two β-carbonyl groups is accomplished by the elongase is unknown. One can envisage either interaction with a prosthetic group such as copper[20] and/or the absence or inability of the requisite enzyme activities to act on the specified chain lengths even though the condensing activity is unimpeded[19], as exemplified by chalcone synthase[21] and 6-methylsalicylic acid synthetase[22], respectively. This conservation of specific oxo groups is one of the unique features of the β-ketoacyl elongase which distinguishes it from the other elongases participating in raincoat biosynthesis and from plastid FAS and palmityl elongase.

Among the facets of the pathway illustrated in Fig. 3, two are of particular interest in the present context. Firstly, appropriately activated and radioactively labelled substrates for three successive steps in a cycle of elongation, that is C_{16} acyl-, 3-oxoacyl- and 3-hydroxyacyl-CoAs, were fed to tissue slices prepared from spikes minus awns of cer-c[36], -q[42] and -u[69] plants[7]. The acyl- and 3-oxoacyl-CoAs served as precursors for esterified alkan-2-ols in cer-c[36] tissue and for both esterified alkan-2-ols and β-diketones in cer-u[69] tissue. While neither of these yielded labelled β-diketones with cer-q[42] tissue, the 3-oxoacyl-CoA but not the acyl-CoA proved a good substrate for esterified alkan-2-ols. That the 3-oxoacyl-CoA failed as a β-diketone precursor in cer-q[42] tissue is presumably due to the fact that, in contrast to alkan-2-ol synthesis, two successive rounds of 3-oxo conservation are required in β-diketone synthesis[17] (Fig. 3). The 3-hydroxyacyl-CoA was ineffective as a precursor in all cases. These results not only supported the contention that a 3-oxoacyl compound is the key intermediate in β-diketone and esterified alkan-2-ol syntheses, but also unveiled one of the functions of the cer-cqu polypeptide; namely, the β-ketoacyl elongase condensing activity which is defective in cer-q[42] plants.

Secondly, synthesis of the 3-oxoacyl derived lipids is very sensitive to arsenite. A likely explanation is that the inhibition occurs in a biosynthetic step common to all of them. Confirmation of this supposition and identification of the step as that forming the 3-oxoacyl-CoA compounds themselves (Fig. 3) was obtained from the following sorts of observations. The C_{14} fatty acid myristate is a far better precursor in a cer-c[36] tissue slice system of C_{15}-2-ols than of the longer alkan-1-ols. While pretreatment with 0.1 mM arsenite has no effect on the latter, synthesis of the former is totally blocked[23]. Acetate fed to cer-u[69] spikes labels both the C_1 and C_{31} ends of the β-diketone carbon chain. After pretreatment with arsenite, however, all label is in the C_1 end demonstrating that only chains already containing two protected oxo groups were being used as β-diketone precursors[20]. Combining the arsenite sensitivity of the β-ketoacyl elongase condensing activity with the relative insensitivity of the synthesis of the acyl elongase derived wax lipids recalls an analogous well known disparity between palmityl elongase and FAS. In spinach this has recently been pinpointed to the presence in these two complexes of different condensing enzymes[2,3]. Another similarity between the palmityl and β-ketoacyl elongases is that C_{12}, C_{14} and C_{16} chains are the preferred primers although the former uses ACP and the latter CoA derivatives[7,23]. Structures of β-diketones and alkan-2-ols from other plants intimate, however, that β-ketoacyl elongases can use primers of very varied chain length[17,24]. While the noted parallels between palmityl and β-ketoacyl elongases may be serendipitous, they suggest that one possible approach for obtaining the cer-cqu gene whose product is part of the epidermal cells' plasmalemma and/or walls[5] would be to exploit the potential homology between coding sequences of the two condensing enzymes[17]. Thus, we have started to isolate the soluble plastid palmityl elongase condensing enzyme for antibody preparation.

Fig. 4 summarizes the sites of action of 39 cer-cqu mutants in the 3-oxoacyl lipid pathway (Fig. 3) as deduced from compositional analyses of spike waxes. The figure shows that mutants belonging to different complementation classes can have the same effect on wax composition. Interestingly, only the cer-u and -q classes functioning in the first and last steps of the pathway are involved. What is not revealed by Fig. 4 is that with the exception of the nine mutants totally blocked in 3-oxoacyl-CoA synthesis, the other four groups of mutants can be divided into two or three subgroups each on the basis of their effects on the esterified alkan-2-ol lipids[14]. For example, (i) of the eight blocked cer-c mutants three have no effect, three increase somewhat and two increase drastically their synthesis. This is not a simple rechanneling of C_{18} 3-oxoacyl-CoA intermediates at the branch point as C_{17}-2-ols are not increased proportionally. (ii) Of the 11 mutants unable to insert a hydroxyl group into the β-diketone carbon chain, three markedly reduce esterified alkan-2-ol formation while the others have no effect. Combined these observations illustrate the lack of a direct correlation between complementation groups and their phenotypes as well as the non-predictable effects of a mutant on distinct biosynthetic steps in the 3-oxoacyl pathway. Such behavior reflects the complex nature of the interactions among the various domains of the cer-cqu polypeptides. A parallel situation in lipid biosynthesis has been described for the FAS1 and FAS2 multifunctional genes in yeast[25,26].

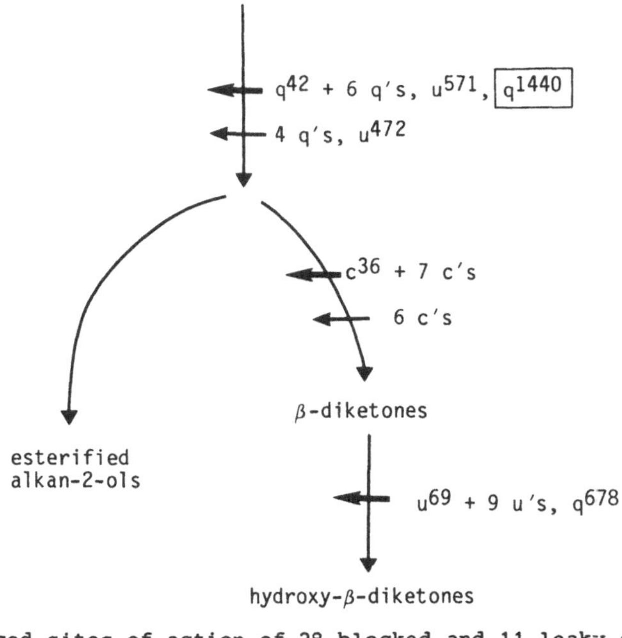

Fig. 4. Deduced sites of action of 28 blocked and 11 leaky cer-cqu mutants in the 3-oxoacyl pathway (Fig. 3) are denoted by heavy and light arrows, respectively. The term 6 q's includes mutants 35, 56, 82, 629, 829 and 1464; 4 q's mutants 50, 1368, 1400 and 1459; 7 c's mutants 29, 44, 61, 73, 89, 95 and 602a; and 9 u's mutants 21, 58, 76, 80, 119, 120, 189, 602b and 699. All mutants are recessive except the boxed one which is dominant. Data on which this figure are based are presented in reference 14.

DOMINANT MUTATIONS AT THE cer-cqu, -n, -t AND Cer-yy LOCI

The mutant cer-q[1440], which is totally blocked in synthesis of the 3-oxoacyl derived wax lipids (Fig. 4), is the only one of the cer-cqu mutants which is dominantly inherited[11]. The other 521 behave as recessives. This situation is not unique, but applies also to the cer-n and -t loci at which a total of 56 and 50 mutant alleles[11], respectively, are known. The gene cer-n, as cer-cqu, modifies the raincoats on both spikes and the uppermost leaf sheaths and internodes, whereas cer-t alters that of the spikes only. Although some information has accrued, the precise roles of cer-n and -t in raincoat formation are unknown at the moment[11]. The presence of dominant and recessive mutant alleles at a locus certainly intimates a minimum of two activities. Does this mean then that cer-cqu has in fact a fourth function? Also in maize a dominantly and many recessively inherited mutations of the c locus have been identified. This gene is thought to regulate several enzymes of the flavonoid pathway in the aleurone during seed development[27]. Recently parts of the wild type and the dominant C-I alleles have been cloned[28]. The identification of several transcripts from the wild type is certainly intriguing, and possibly an explanation of the allele inheritance patterns will be forthcoming. On the other hand the GAL4 locus in yeast determines a regulatory protein having at least two active sites. Mutations in an internal region representing 4% of the genetic length of this locus are dominantly inherited, those in the other regions recessively[29].

In marked contrast to the above genes, all 18 mutant alleles of Cer-yy are dominantly inherited[11]. This gene as cer-t affects only spike waxes. Compositional analyses[30] of such waxes from five mutants reveals an absence of all 3-oxoacyl lipids and a marked increase in the relative amount of the primary alcohols. The chain length distributions of the latter differ markedly in waxes from leaf blades and spikes. That is, 90% C_{26} vs 51% C_{26}, 16% C_{28}, 12% C_{30} and 9% C_{32}. The primary alcohols from Cer-yy spikes are 86% C_{26}. Alterations in chain length distributions toward those characteristic of leaf blades were found for the other acyl elongase derived wax lipids on Cer-yy spikes. To summarize, Cer-yy mutant spike wax is a phenocopy of that from leaf blades. This suggests that the Cer-yy gene determines a regulatory component that upon mutation activates in the spike the genes for the leaf blade acyl elongase system(s) and at the same time represses those for the spike β-ketoacyl and acyl elongase systems[30].

REFERENCES

1. P. von Wettstein-Knowles, Genetics and biosynthesis of plant epicuticular waxes, in: "Advances in the Biochemistry and Physiology of Plant Lipids," L.-Å. Appelqvist and C. Liljenberg, eds., Elsevier/North-Holland Biomedical Press, Amsterdam pp. 1-26 (1979).
2. T. Shimakata and P.K. Stumpf, Fatty acid synthetase of Spinacia oleracea leaves, Plant Physiol. 69:1257 (1982).
3. T. Shimakata and P.K. Stumpf, Purification and characterization of β-ketoacyl-ACP synthetase 1 from Spinacia oleracea leaves, Arch. Biochem. Biophys. 220:39 (1983).
4. C. Cassagne and R. Lessire, Biosynthesis of saturated very long chain fatty acids by purified membrane fractions from leek epidermal cells, Arch. Biochem. Biophys. 191:146 (1978).
5. J. D. Mikkelsen, Synthesis of lipids by epidermal and mesophyll protoplasts isolated from barley leaf sheaths, in: "Biogenesis and Function of Plant Lipids," P. Mazliak, P. Benveniste, C. Costes, and R. Douce, eds., Elsevier/North-Holland Biomedical Press, Amsterdam pp. 285-290 (1980).

6. V. P. Agrawal, R. Lessire, and P.K. Stumpf, Biosynthesis of very long chain fatty acids in microsomes from epidermal cells of Allium porrum L., Arch. Biochem. Biophys. 230:580 (1984).
7. J. D. Mikkelsen, Biosynthesis of esterified alkan-2-ols and β-diketones in barley spike epicuticular wax: synthesis of radio-active intermediates, Carlsberg Res. Commun. 49:391 (1984).
8. D. Simpson and P. von Wettstein-Knowles, Structure of epicuticular waxes on spikes and leaf sheaths of barley as revealed by a direct platinum replica technique, Carlsberg Res. Commun. 45:465 (1984).
9. P. von Wettstein-Knowles, The molecular phenotypes of the eceriferum mutants, in: Barley Genetics II, R.A. Nilan, ed., Washington State Univ. Press, Pullman, pp. 146-193 (1971).
10. P. von Wettstein-Knowles, Genetic control of β-diketone and hydroxy-β-diketone synthesis in epicuticular waxes of barley, Planta 106:113 (1972).
11. B. Søgaard and P. von Wettstein-Knowles, Barley: genes and chromosomes, Carlsberg Res. Commun. 52:123 (1987).
12. L. Dellaert, J. van Es and M. Koornneef, Eceriferum mutants in Arabidopsis thaliana (L.) Heynh: II. Phenotypic and genetic analysis, Arab. Inf. Serv. 16:10 (1979).
13. P. von Wettstein-Knowles and B. Søgaard, The cer-cqu region in barley: gene cluster or multifunctional gene, Carlsberg Res. Commun. 45:125 (1980).
14. B. Søgaard and P. von Wettstein-Knowles, Dissection of the cer-cqu locus, in: "Barley Genetics V," Proc. Fifth Intl. Barley Genet. Symp. 1986, Okayama (in press).
15. J. Jensen, Linkage map of barley chromosome 4, in: "Barley Genetics V," Proc. Fifth. Intl. Barley Genet. Symp. 1986, Okayama (in press).
16. P. von Wettstein-Knowles, Biosynthetic relationships between β-diketones and esterified alkan-2-ols deduced from epicuticular wax of barley mutants, Molec. Gen. Genet. 144:43 (1976).
17. P. von Wettstein-Knowles, Role of cer-cqu in epicuticular wax biosynthesis, Biochem. Soc. Tran. 14:576 (1986).
18. P. von Wettstein-Knowles and J.Ø. Madsen, 7-oxopentadecan-2-ol esters - a new epicuticular wax lipid class, Carlsberg Res. Commun. 49:57 (1984).
19. P. von Wettstein-Knowles, Genes, elongases and associated enzyme systems in epicuticular wax synthesis, in: "Biochemistry of Plant Lipids: Structure and Function," P.K. Stumpf, J.B. Mudd and W.D. Nes, eds., Plenum, New York (in press).
20. J. D. Mikkelsen and P. von Wettstein-Knowles, Biosynthesis of β-diketones and hydrocarbons in barley spike epicuticular wax, Arch. Biochem. Biophys. 188:172 (1978).
21. R. Schüz, W. Heller, and K. Hahlbrock, Substrate specificity of chalcone synthase from Petroselinum hortense, J. Biol. Chem. 258:6730 (1983).
22. F. Lynen, H. Engeser, J. Friedrich, W. Schindlbeck, R. Seyffert, and F. Wieland, Fatty acid synthetase of yeast and 6-methylsalicylate synthetase of Penicillium patulum - two multienzyme complexes, in: "Microenvironments and Metabolic Compartmentation," P.A. Srere and R.W. Estabrook, eds., Academic Press, New York pp. 283-303 (1978).
23. P. von Wettstein-Knowles, Effects of inhibitors on synthesis of esterified alkan-2-ols in barley spike epicuticular wax, Carlsberg Res. Commun. 50:239 (1985).
24. P. von Wettstein-Knowles, J.D. Mikkelsen, and J.Ø. Madsen, Nonan-2-ol esters in sorghum leaf epicuticular wax and their collection by preparative gas chromatography, Carlsberg Res. Commun. 49:611 (1984).
25. E. Schweizer, K. Werkmeister and M.K. Jain, Fatty acid biosynthesis in yeast, Mol. Cell. Biochem. 21:95 (1978).

26. K. Werkmeister, R. Johnston and E. Schweizer, Complementation in vitro between purified mutant fatty acid synthetase complexes of yeast, Eur. J. Biochem. 116:303 (1981).
27. H. K. Dooner, Coordinate genetic regulation of flavonoid biosynthetic enzymes in maize, Mol. Gen. Genet. 189:136 (1984).
28. J. Paz-Ares, U. Wienand, P.A. Peterson and H. Saedler, Molecular cloning of the c locus of Zea mays: a locus regulating the anthocyanin pathway, EMBO J. 5:829 (1986).
29. Y. Oshima, Regulatory circuits for gene expression: The metabolism of galactose and phosphate, in: "Molecular Biology of the Yeast Saccharomyces. Metabolism and Gene Expression," Cold Spring Harbor Laboratory pp. 159-180 (1982).
30. U. Lundqvist and P. von Wettstein-Knowles, Dominant mutations at Cer-yy change barley spike wax into leaf blade wax, Carlsberg Res. Commun. 47:29 (1982).

STIMULATION OF PHENYLPROPANOID PATHWAYS BY ENVIRONMENTAL FACTORS

Dierk Scheel, Jeffery L. Dangl, Carl Douglas, Karl Dietrich Hauffe, Anette Herrmann, Heidi Hoffmann, Edmundo Lozoya, Wolfgang Schulz and Klaus Hahlbrock

Max-Planck-Institut für Züchtungsforschung, Abteilung Biochemie, D-5000 Köln 30, FRG

INTRODUCTION

Higher plants accumulate and excrete many kinds of phenylpropanoid products. Lignin, suberin, flavonoids, stilbenoids or coumarins are only a few examples. Their functions as components of the cell wall, flower pigments, allelochemicals or as protective agents against various kinds of stress reflect their structural diversity.

Phenylpropanoid compounds are synthesized from phenylalanine *via* general phenylpropanoid metabolism, finally branching into different pathways (Fig. 1). Since the need for various phenylpropanoid products varies during plant development and with changing environmental conditions, the intensity and direction of metabolic flux through general phenylpropanoid and subsequent pathways is under strict control.

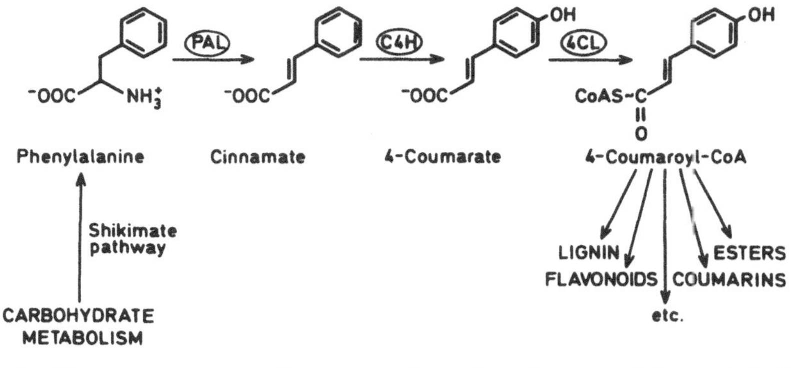

Fig. 1. The enzymes of general phenylpropanoid metabolism: Phenylalanine ammonia-lyase (PAL), cinnamic acid 4-hydroxylase (C4H) and 4-coumarate:CoA ligase (4CL). The first and the last enzyme catalyze the branch point reactions from primary metabolism (PAL) and - although not proven in all cases - to various specific pathways (4CL).

We study the differential response of cultured parsley cells to UV irradiation, a physical stress, and to treatment with fungal elicitor, simulating infection, as a model system for rapid regulation of phenylpropanoid pathways.

DIFFERENTIAL INDUCTION OF FLAVONOID AND FURANOCOUMARIN BIOSYNTHESIS

Cultured parsley cells respond to UV irradiation with vacuolar accumulation of UV absorbing flavonoids (Hahlbrock, 1981; Hahlbrock et al., 1982; Matern et al., 1983). Upon treatment with a hyphal cell wall preparation (elicitor) from the soybean pathogen *Phytophthora megasperma* f. sp. *glycinea* the cells excrete a complex mixture of coumarin derivatives with antifungal activity into the culture medium (Tietjen et al., 1983; Hauffe et al., 1986; Scheel et al., 1986). No flavonoids accumulate after elicitor treatment and no coumarins are synthesized upon UV irradiation.

Flavonoids and coumarins originate from general phenylpropanoid metabolism (Fig. 1) (Ebel and Hahlbrock, 1982; Murray et al., 1982), which is stimulated in parsley after both treatments. Following either UV irradiation or addition of elicitor, rapid, transient increases in transcription rates of the genes for phenylalanine ammonia-lyase (PAL) and 4-coumarate:CoA ligase (4CL) are induced (Chappell and Hahlbrock, 1984). This gene activation is followed by coordinate increases of respective mRNAs and enzymes (Schröder et al., 1979; Hahlbrock et al., 1981; Ragg et al., 1981; Tietjen and Matern, 1983; Kreuzaler et al., 1983; Kuhn et al., 1984; Hauffe et al., 1986).

All responses beyond the hypothetical branch point of the biosynthetic pathways are signal specific. For example, UV irradiation induces increases in transcription, translation and activity of chalcone synthase (CHS), and stimulates the catalytic activity of another approximately thirteen enzymes specifically involved in flavonoid synthesis (Ebel and Hahlbrock, 1982; Kreuzaler et al., 1983; Chappell and Hahlbrock, 1984; Hahlbrock et al., 1985). CHS is the first enzyme of this pathway. It catalyzes the stepwise condensation of three molecules of malonyl-CoA with 4-coumaroyl-CoA to form naringenin chalcone, the starting substrate for all classes of flavonoids (Heller and Hahlbrock, 1980).

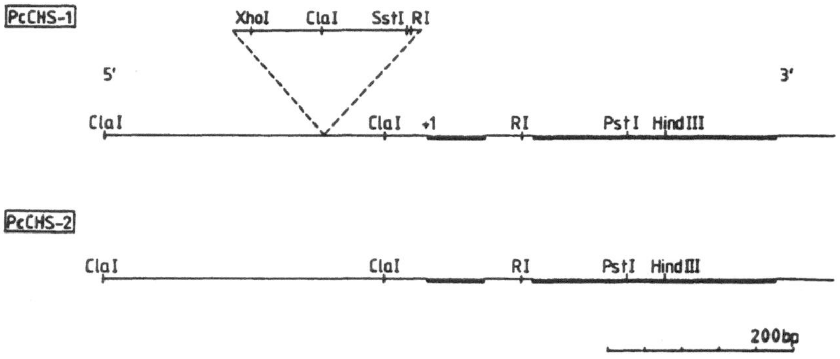

Fig. 2. Structure of PcCHS-1 and PcCHS-2, showing exons (filled-in boxes) and the 5' insertion in PcCHS-1.

Using a nearly full size CHS cDNA (Reimold et al., 1983), two different classes of CHS clones were isolated from a genomic library of cultured parsley cells (Herrmann, 1987). One representative clone of each group, containing the complete CHS gene, was analyzed in detail. Southern blot and restriction enzyme analysis indicated that additional copies of CHS are not present in the parsley genome. Quantitative Southern analysis of DNA from individual plants showed that some plants were homozygous for either one of the two types of CHS, identifying them as alleles rather than separate genes.

The restriction patterns of both alleles, PcCHS-1 and PcCHS-2, are identical except for a fragment of about 1 kbp, which is only present in the 5' noncoding region of PcCHS-1 (Fig. 2). This result was confirmed by analyses of heteroduplices of the two alleles among one another and with the cDNA. The complete nucleotide sequence of PcCHS-1 was determined,

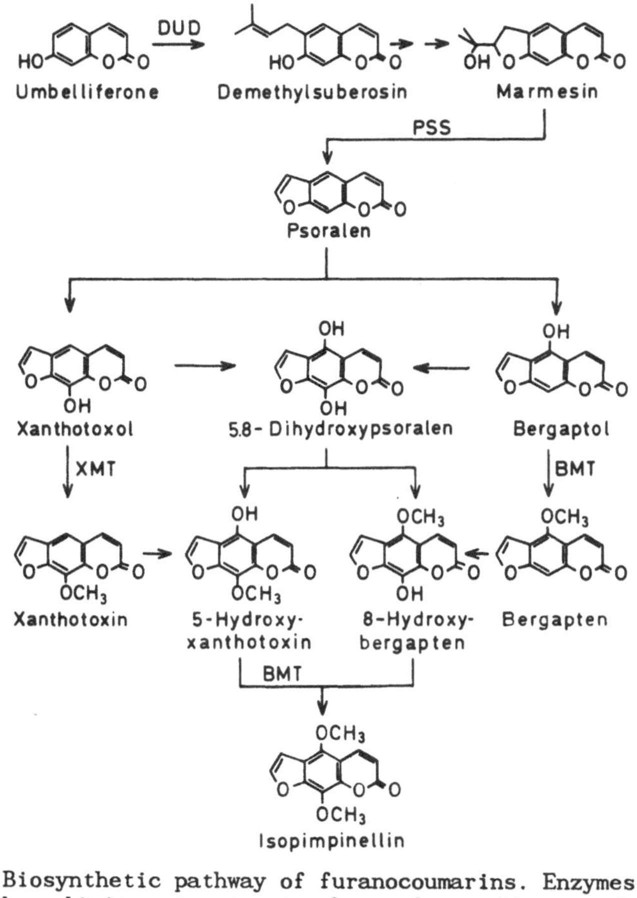

Fig. 3. Biosynthetic pathway of furanocoumarins. Enzymes stimulated by elicitor treatment of parsley cells are indicated by their abbreviated names (DUD = dimethylallyl diphosphate: umbelliferone dimethylallyltransferase; PSS = psoralen synthase; XMT = S-adenosyl-L-methionine:xanthotoxol O-methyltransferase; BMT = S-adenosyl-L-methionine: bergaptol O-methyltransferase).

whereas from PcCHS-2 only the region neighbouring this insertion was sequenced. The coding regions consist of two exons of 328 and 1225 bp in lengths interrupted by a 263 bp intron. Complete homology exists between the cDNA and the protein coding region of sequenced PcCHS-1. Two transcription start sites were identified by S1 nuclease protection and primer extension assays at approximately 135 and 90 bp upstream of the translation start codon.

The identification of parsley plants homozygous for either PcCHS-1 or PcCHS-2 enabled us to determine the expression and inducibility of both alleles in cell cultures initiated from these plants. After UV irradiation of both types of cultures, accumulation of flavonoids, increase in CHS enzyme activity, and accumulation of CHS mRNA were similar to the heterozygous parsley cell culture. This indicates that the two CHS alleles are expressed and induced by UV light, and that the 5' insertion of PcCHS-1 does not play a decisive role for UV induction.

Elicitor treatment of parsley cells stimulates the furanocoumarin pathway, which is less well understood than the flavonoid pathway (Fig. 3). Two membrane associated enzymes of this biosynthetic sequence increase transiently in activity upon elicitor addition to the cells (Tietjen and Matern, 1983; Wendorff and Matern, 1986). Dimethylallyl diphosphate:umbelliferone dimethylallyltransferase (DUD) catalyzes the prenylation of C-6 of umbelliferone to form demethylsuberosin which is converted to marmesin. Psoralen synthase (PSS) then uses marmesin as a substrate to form psoralen by an unknown mechanism. We purified and

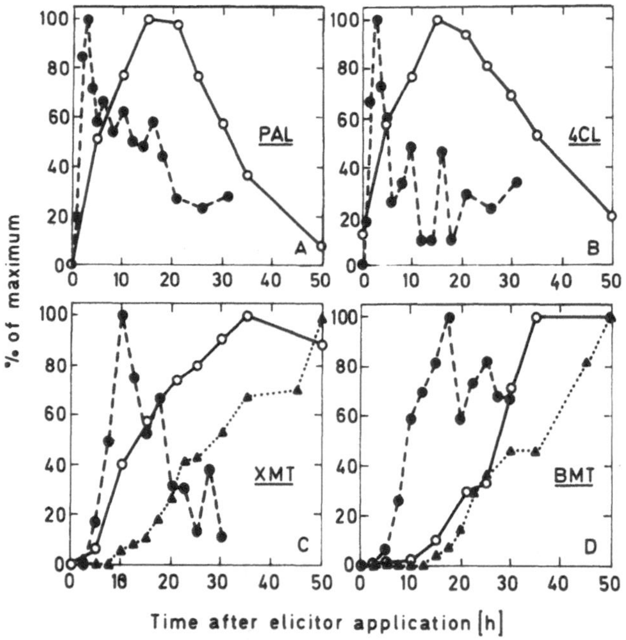

Fig. 4. Time courses of induction of enzyme and mRNA translational activities of PAL (A), 4CL (B), XMT (C) and BMT (D) in elicitor-treated parsley cells and subsequent accumulation of the products xanthotoxin (C) and bergapten (D) (o—o enzyme activities; •---• mRNA translational activities; ▲······▲ product accumulation).

characterized two O-methyltransferases, both stimulated in activity by elicitor treatment, and produced polyclonal antisera (Hauffe et al., 1986). S-adenosyl-L-methionine:xanthotoxol O-methyltransferase (XMT) and S-adenosyl-L-methionine:bergaptol O-methyltransferase (BMT) catalyze final methylation steps of the three major parsley phytoalexins xanthotoxin, bergapten and isopimpinellin. Enzyme activity and *in vitro* translational activity of XMT and BMT are transiently stimulated by elicitor treatment with a distinctive lag compared to PAL and 4CL (Fig. 4). Finally, the respective products are excreted into the culture medium. The accumulation of both proteins in elicitor-treated parsley cells closely resembles the time courses of their activity increases (Fig. 5). Poly(A)$^+$ RNA isolated from cultured parsley cells 10 and 20 h after elicitor application was used to construct cDNA libraries in the expression vector λgt11, which are now being screened with the antisera.

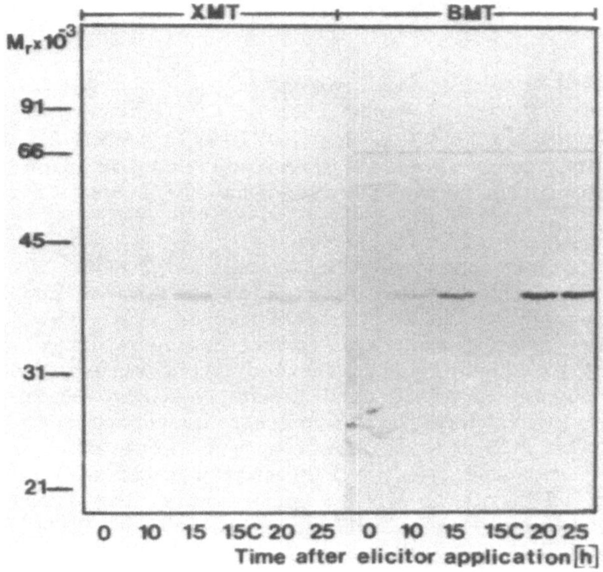

Fig. 5. Western blot analysis of XMT and BMT from elicitor-treated parsley cells (C = untreated control cells).

INDUCTION OF GENERAL PHENYLPROPANOID METABOLISM

In contrast to the flavonoid and furanocoumarin pathways, general phenylpropanoid metabolism is stimulated by both UV irradiation and elicitor treatment of parsley cells. This pathway consists of the three enzymes PAL, cinnamic acid 4-hydroxylase (C4H) and 4CL (Fig. 1). The first and the last of these enzymes are of special interest, because they catalyze the branch point reactions from primary metabolism (PAL) and - although not proven in all cases - to various specific pathways (4CL).

Two 4CL genes, Pc4CL-1 and Pc4CL-2, are present in parsley (Douglas et al., 1987a). The complete nucleotide sequences of both genes (Fig. 6) (Douglas et al., 1987a) and their corresponding nearly full size cDNAs were determined (Hoffmann, 1987). Restriction enzyme analysis and genomic blots of DNA from cultured cells and from leaves indicate that no additional copies of 4CL are existing in the parsley genome. The genes contain five exons separated by introns of varying lengths (Douglas et al., 1987a). Except for one major difference in the size of the second intron and a 190 bp deletion in the 5' noncoding region of Pc4CL-2, the two genes are highly homologous. Upstream of the transcription start sites, which were determined by S1 nuclease and primer extension assays, sequences homologous to CAT (-120 bp) and TATA (-20 bp) boxes and several elements with potential regulatory functions were found.

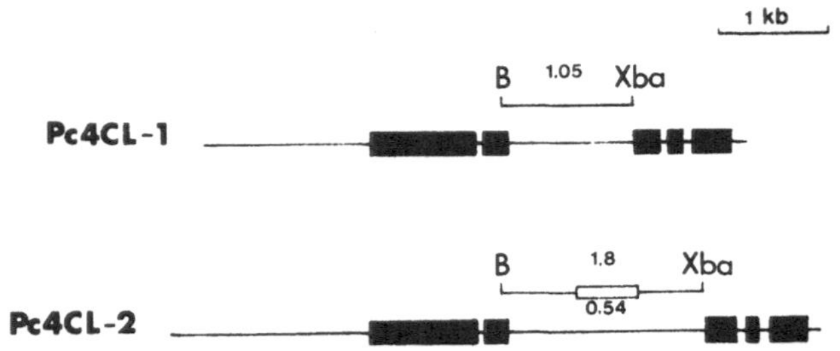

Fig. 6. Structure of Pc4CL-1 and Pc4CL-2 (5' ends to the left), showing exons (filled-in boxes) and location of runoff transcription probes derived from the second introns.

Since cDNA clones corresponding to either Pc4CL-1 or Pc4CL-2 were found in cDNA libraries derived from elicitor-treated and UV-irradiated cells, both genes appear to be transcriptionally activated after either treatment. This result was confirmed by hybridization of runoff transcripts from UV- and elicitor-treated cells with a Pc4CL-2 specific probe (540 bp fragment in the second intron, see Fig. 6) in comparison to a probe recognizing both genes. Transcription specific to the Pc4CL-2 probe was activated ~ 6-fold by elicitor treatment and ~ 2.5-fold by UV irradiation. The increase in signal strengths after either treatment was two to three times stronger with a probe measuring transcription from both 4CL genes (Douglas et al., 1987a).

The amino acid sequences deduced from the sequences of the cDNAs for Pc4CL-1 and Pc4CL-2 differ in only three amino acids (Gln/Leu; Asn/Asp; Arg/Lys) (Hoffmann, 1987). Since both genes are expressed after UV and elicitor treatment, we postulated the existence of two slightly different isoenzymes. RNA from elicitor-treated cells was translated *in vitro* and the translation products precipitated with 4CL antisera. Two polypeptides with a molecular weight around Mr 60,000 and slightly different isoelectric points (about pH 4.9 and 5.0) were separated by two-dimensional polyacrylamide electrophoresis. Using FPLC anion exchange chromatography as a final step we purified two 4CL isoenzymes from cultured parsley cells (Fig. 7A). Both isoenzymes are present in untreated, UV-irradiated and elicitor-treated cells with considerably different total amounts, but always at a ratio of 1:1.

In order to structurally compare these two isoenzymes with the translation products of Pc4CL-1 and Pc4CL-2, the corresponding cDNAs were cloned into Bluescribe vector (Vector Cloning Systems, San Diego, USA) and expressed in *E. coli*. A purification procedure identical to that used for parsley 4CL isoenzymes was applied to the enzymes produced in the *E. coli* strains. Both plasmid constructs produced active 4CL which each comigrate with only one of the two isoenzymes isolated from parsley cells (Fig. 7B and C). The protein synthesized from Pc4CL-2 binds more strongly to the anion exchange resin than that of Pc4CL-1. This is expected since one of the three amino acid differences (Asn in Pc4CL-1, Asp in Pc4CL-2) reduces the isoelectric point and increases the negative charge of the Pc4CL-2 product in comparison to that of Pc4CL-1. These results strongly suggest that the two 4CL isoenzymes isolated from parsley cells correspond to the two 4CL genes present in the parsley genome.

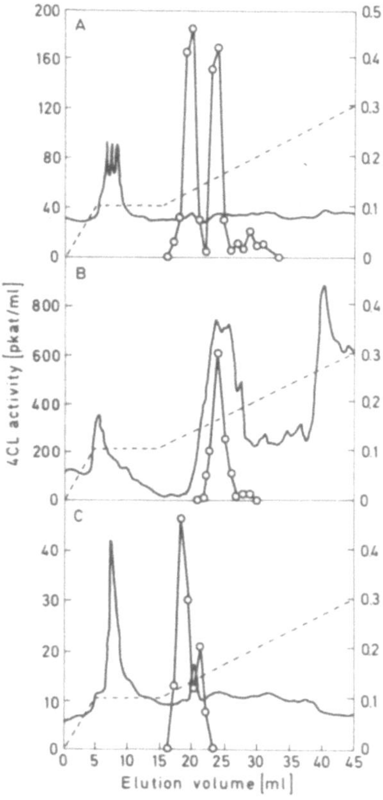

Fig. 7. Anion exchange chromatography of different 4CL preparations. Separation of two 4CL isoenzymes from elicitor - treated parsley cells (A). Products of Pc4CL-2 (B) and Pc4CL-1(C) expressed in *E. coli* (see text for details; —— absorption at 280 nm; --- KCl gradient; o—o 4CL enzyme activity).

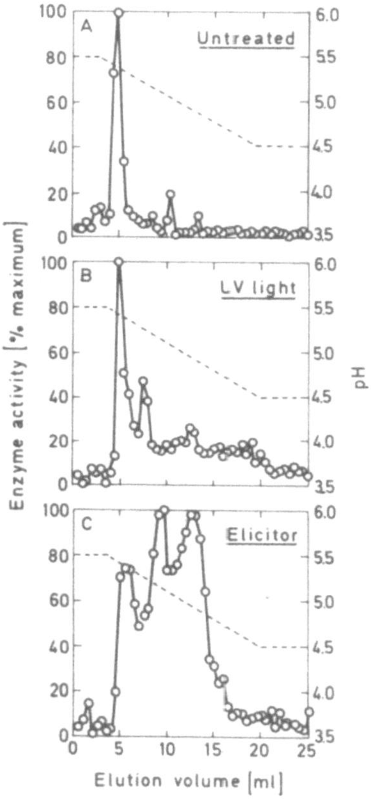

Fig. 8. Chromatofocusing of PAL from cultured parsley cells before (A) and after UV irradiation (B) or treatment with elicitor (C) (o—o PAL enzyme activity; ---- pH gradient).

The organization and expression of PAL genes appears to be much more complicated than that of 4CL. A number of different PAL isoforms can be separated by chromatofocusing of enzyme extracts from untreated, UV-irradiated or elicitor-treated cells (Fig. 8). Although the putative isoenzymes have not been purified and characterized, the different elution profiles indicate differential stimulation of PAL by different treatments. Preliminary results of the analysis of PAL at the genomic level support this interpretation. At least four to five different PAL genes are present in the parsley genome. These genes contain conserved regions as well as regions with little homology. Two cDNAs, originating from UV-irradiated and elicitor-treated cells, respectively, strongly differ in their hybridization efficiency in blots of RNA from cells triggered with the respective treatment. These results indicate the presence of PAL genes differentially activated by both stimuli.

GENE TRANSFER INTO PROTOPLASTS: AN ASSAY SYSTEM FOR INDUCIBLE GENE EXPRESSION?

Coordinate activation of genes by a particular stimulus, as described here for different genes of phenylpropanoid metabolism, may reflect the occurrence of common regulatory sequences in their promoters. In order to identify and analyze these regulatory elements, we developed a gene transfer system for parsley protoplasts.

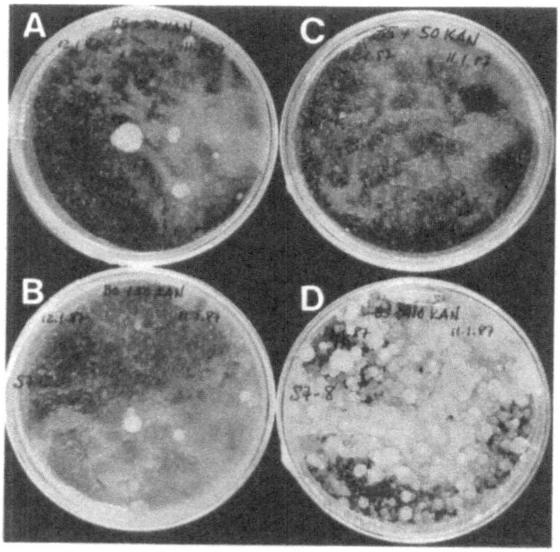

Fig. 9. Stably transformed parsley calli growing on media containing kanamycin. Parsley protoplasts were transformed with vectors containing the CaMV-nptII chimeric gene (see text). Each plate contained 2-5x10^5 protoplasts which were grown in liquid culture for 2 weeks before plating onto agar. Plates C and D are from the same experiment, but transformed only with pUC19. Plates A, B and C contain 50 µg/ml kanamycin, while D contains no kanamycin. Calli appearing in A and B were transferred to 100 µg/ml kanamycin and continued growing. This picture was taken 5 weeks after plating, and 7 weeks after transformation.

Parsley cell suspension cultures release protoplasts in high yield upon enzymatic treatment (Dangl et al., 1987). These protoplasts rapidly regenerate new calli, which under constant shaking disaggregate to form cell suspensions after transfer into liquid medium.

Stably transformed parsley cells were generated by transformation of protoplasts with plasmid DNA using a modification of published methods (Krens et al., 1982; Hain et al., 1985). In each case, the plasmids contained the selectable neomycin phosphotransferase II (nptII) gene, encoding resistance to kanamycin, linked to the cauliflower mosaic virus (CaMV) 35S promoter and 3' regions. This base plasmid (kindly provided by R. Töpfer, Max-Planck-Institut für Züchtungsforschung, Köln, FRG) also contains a pUC18 polylinker. We cloned a marked derivative of the Pc4CL-1 structural gene (Douglas et al., 1987a), containing various portions of 5' sequences, into this polylinker. Transformed protoplasts were grown for two to three weeks in non-selective media and then directly plated onto agar plates containing 100 µg/ml kanamycin. Calli were easily visible after eight weeks (Fig. 9) and were then assayed for nptII activity (Reiss et al., 1984; Schreier et al., 1985). All transformants expressed the nptII gene, but the amount of activity varied even between clones transformed with the same plasmid (Fig. 10).

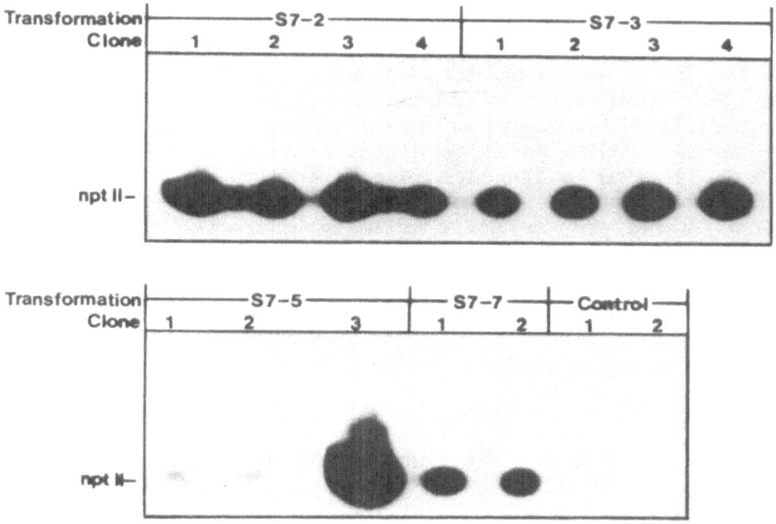

Fig. 10 Kanamycin resistant parsley calli express the nptII gene. Independent calli from transformations with one of four vectors (S7-2, -3, -5, -7) were tested for nptII activity. Control calli transformed with pUC19 and grown without kanamycin selection show no nptII activity. Exposure time was 30 min at -80°C. Clones S7-5.1 and S7-5.2 subsequently grew slowly on 100 µg/ml kanamycin.

We used the same plasmid and transformation systems to develop a transient expression assay in parsley protoplasts (Douglas et al., 1987b). In this case maximum expression of nptII was observed between three and five days after transfer and nptII activity was still detectable after seven days.

Analysis of induced expression of introduced genes relies, however, on availability of a system which has no background stimulation, yet responds to appropriate signals in a manner identical to the intact cell. Each of these conditions is met by the parsley protoplast system, where both common and specific features of the responses to UV light and elicitor are maintained (Dangl et al., 1987). In addition, cell suspension cultures initiated from stably transformed calli shown in Fig. 9 did not contain flavonoids or coumarins, indicating that there was no background stimulation. The cultured cells were either irradiated with UV light or treated with elicitor and accumulation of flavonoids and coumarins was determined. In each case, UV-irradiated cells accumulated typical amounts of flavonoids, but no coumarins, whereas elicitor-treated cells excreted coumarin derivatives, but did not synthesize flavonoids.

We therefore feel that both assay systems are suitable to study the inducible expression of constructs containing 5' sequences of the isolated 4CL and CHS genes. We thereby hope to define regulatory elements, which determine the inducibility of these genes by UV light and/or elicitor.

CONCLUSIONS

Flavonoids and coumarins are synthesized *via* branched phenylpropanpoid pathways (Fig. 1). The three enzymes common to the two biosynthetic sequences are induced by both treatments, whereas flavonoid and coumarin specific reactions are only stimulated by either one of the stimuli. The genes encoding 4CL, which forms the hypothetical branch point of this pathway, are likewise activated by both treatments. Therefore, direction and intensity of metabolic flux into subsequent products appears not to be under direct control of different 4CL genes. In contrast, several different genes encoding PAL, the first common enzyme of phenylpropanoid pathways, are present in parsley. In addition, these genes appear to be differentially activated (reflected in the synthesis of distinct isoenzymes) upon various treatments. A careful structural and functional analysis of these genes and isoenzymes is needed to decide whether differential PAL activation directly controls channeling into different phenylpropanoid pathways.

ACKNOWLEDGEMENTS

We thank R. Töpfer, Max-Planck-Institut für Züchtungsforschung, Köln, FRG, for providing the pRT100neo vector, and C. A. Corr, M. Jung, E. Logemann and R.-M. Voigt for skilled technical assistance. This work was supported by the Max-Planck-Society and fellowships from NSF, USA, to J. L. D. and from CONACYT/CINVESTAV-IPN U. Irapuato, Mexico, to E. L.

REFERENCES

Chappell, J., and Hahlbrock, K., 1984, Transcription of plant defence genes in response to UV light or fungal elicitor, *Nature*, 311:76.

Dangl, J. L., Hauffe, K. D., Lipphardt, S., Hahlbrock, K., and Scheel, D., 1987, Parsley protoplasts retain differential responsiveness to UV light and fungal elicitor, *EMBO J.*, in press.

Douglas, C., Hoffmann, H., Schulz, W., and Hahlbrock, K., 1987a, Structure and elicitor or u.v.-light-stimulated expression of two 4-coumarate:CoA ligase genes in parsley, *EMBO J.*, 6:1189.

Douglas, C. J., Dangl, J. L., Hoffmann, H., Lipphardt, S., and Hahlbrock, K., 1987, Analysis of fungal elicitor- and UV light-induced gene expression in parsley cells, *in*: "Plant Gene Systems and their Biology", L. McIntosh and J. Key, eds., Alan R. Liss, New York.

Ebel, J., and Hahlbrock, K., 1982, Biosynthesis, *in*: "Advances in Flavonoid Research 1975-1980", J. B. Harborne and T. J. Mabry, eds., Chapman and Hall, London.

Hahlbrock, K., 1981, Flavonoids, *in*: "The Biochemistry of Plants", Vol. 7, P. K. Stumpf and E. E. Conn, eds., Academic Press, New York.

Hahlbrock, K., Chappell, J., and Scheel, D., 1985, Genes involved in resistance reactions in higher plants: possible candidates for gene transfer?, *in*: "The Impact of Gene Transfer Techniques in Eukaryotic Cell Biology", Springer-Verlag, Berlin.

Hahlbrock, K., Lamb, C. J., Purwin, C., Ebel, J., Fautz, E., and Schäfer, E., 1981, Rapid response of suspension-cultured parsley cells to the elicitor from *Phytophthora megasperma* var. *sojae*, *Plant. Physiol.*, 67:768.

Hahlbrock, K., Kreuzaler, F., Ragg, H., Fautz, E., and Kuhn, D. N., 1982, Regulation of flavonoid and phytoalexin accumulation through mRNA and enzyme induction in cultured plant cells, *in*: "Biochemistry of Differentiation and Morphogenesis", L. Jaenicke, ed., Springer-Verlag, Berlin.

Hain, R., Stabel, P., Czernilofsky, A. P., Steinbiβ, H. H., Herrera-Estrella, L., and Schell, J., 1985, Uptake, integration, expression and genetic transmission of a selectable chimeric gene by plant protoplasts, *Mol. Gen. Genet.*, 199:161.

Hauffe, K. D., Hahlbrock, K., and Scheel, D., 1986, Elicitor-stimulated furanocoumarin biosynthesis in cultured parsley cells: S-adenosyl-L-methionine:bergaptol and S-adenosyl-L-methionine:xanthotoxol O-methyltransferases, Z. *Naturforsch.*, 41c:228.

Heller, W., and Hahlbrock, K., 1980, Highly purified "flavanone synthase" from parsley catalyzes the formation of naringenin chalcone, *Arch. Biochem. Biophys.*, 200:617.

Herrmann, A., 1987, "Charakterisierung der beiden Allele des Chalkon-Synthase-Gens aus Zellsuspensionskulturen von *Petroselinum crispum* L.", Ph. D. thesis, Köln.

Hoffmann, H., 1987, "Zwei Isoenzyme der 4-Cumarat:CoA Ligase in Petersilie-Zellkulturen: Aufklärung von Primärstruktur und Expression durch Bestimmung der Nukleotidsequenzen von cDNAs", Ph. D. thesis, Köln.

Krens, F. A., Molendijk, L., Wullems, G. J., and Schilperoort, R. A., 1982, *In vitro* transformation of plant protoplasts with Ti-plasmid DNA, *Nature*, 296:72.

Kreuzaler, F., Ragg, H., Fautz, E., Kuhn, D. N., and Hahlbrock, K., 1983, UV-induction of chalcone synthase mRNA in cell suspension cultures of *Petroselinum hortense*, *Proc. Natl. Acad. Sci. USA*, 80:2591.

Kuhn, D. N., Chappell, J., Boudet, A., and Hahlbrock, K., 1984, Induction of phenylalanine ammonia-lyase and 4-coumarate:CoA ligase mRNAs in cultured plant cells by UV light or fungal elicitor, *Proc. Natl. Acad. Sci. USA*, 81:1102.

Matern, U., Heller, W., and Himmelspach, K., 1983, Conformational changes of apigenin 7-O-(6-O-malonyl-glucoside), a vacuolar pigment from parsley, with solvent composition and proton concentration, *Eur. J. Biochem.*, 133:439.

Murray, R. D. H., Mendez, J., and Brown, S. A., 1982, "The Natural Coumarins. Occurrence, Chemistry and Biochemistry", Wiley and Sons, New York.

Ragg, H., Kuhn, D. N., and Hahlbrock, K., 1981, Coordinated regulation of 4-coumarate:CoA ligase and phenylalanine ammonia-lyase mRNAs in cultured plant cells, *J. Biol. Chem.*, 256:10061.

Reimold, U., Kröger, M., Kreuzaler, F., and Hahlbrock, K., 1983, Coding and 3' non-coding nucleotide sequence of chalcone synthase mRNA and assignement of amino acid sequence of enzyme, *EMBO J.*, 2:1801.

Reiss, B., Sprengel, R., Will, H., and Schaller, H., 1984, A new sensitive method for qualitative and quantitative analysis of NPT in crude cell extracts, *Gene*, 30:211.

Scheel, D., Hauffe, K. D., Jahnen, W., and Hahlbrock, K., 1986, Stimulation of phytoalexin formation in fungus-infected plants and elicitor-treated cell cultures of parsley, *in*: "Recognition in Microbe-Plant Symbiotic and Pathogenic Interactions", B. Lugtenberg, ed., Springer-Verlag, Berlin.

Schreier, P., Seftor, E. A., Schell, J., and Bohnert, H. J., 1985, The use of nuclear sequences to direct the light-regulated synthesis and transport of a foreign gene into plant chloroplasts, *EMBO J.*, 4:25.

Schröder, J., Kreuzaler, F., Schäfer, E., and Hahlbrock, K., 1979, Concomitant induction of phenylalanine ammonia-lyase and flavanone synthase mRNAs in irradiated plant cells, *J. Biol. Chem.*, 254:57.

Tietjen, K. G., and Matern, U., 1983, Differential response of cultured parsley cells to elicitors from two non-pathogenic strains of fungi. 2. Effects on enzyme activities, *Eur. J. Biochem.*, 131:409.

Tietjen, K. G., Hunkler, D., and Matern, U., 1983, Differential response of cultured parsley cells to elicitors from two non-pathogenic strains of fungi. 1. Identification of induced products as coumarin derivatives, *Eur. J. Biochem.*, 131:401.

Wendorff, H., and Matern, U., 1986, Differential response of cultured parsley cells to elicitors from two non-pathogenic strains of fungi. Microsomal conversion of (+)marmesin into psoralen, *Eur. J. Biochem.*, 161:391.

A PHOSPHO-OLIGOSACCHARIDE SYNTHASE

IN DEVELOPING MAIZE ENDOSPERMS*

David Pan and Oliver E. Nelson

Laboratory of Genetics
University of Wisconsin
Madison, Wisconsin 53706

KEYWORDS/ABSTRACT: enzyme complex, phospho-oligosaccharide synthase/maize endosperm/starch primer

Developing maize endosperms contain a phospho-oligosaccharide synthase as part of a multi-enzyme complex that may be responsible for synthesizing primer molecules to which chain-elongating enzymes add glucose units. The synthase activity is resistant to boiling for 7 minutes as a step in purification. It utilizes Glc-1-P as a substrate, but Glc-1,6-bisP, which can be synthesized slowly from Glc-1-P by the complex, markedly accelerates the incorporation of Glc-1-P and is itself incorporated. The phosphate groups in the products are largely esterified to C6 and are derived from the Glc-1,6-bisP incorporated.

INTRODUCTION

Starch, an abundant and economically significant natural polymer, is present in many tissues in most green plants. Starch is an important end product of carbon fixation during photosynthesis and is used for respiration in the dark. In seeds which accumulate starch during development, starch degradation in these tissues commences at time of germination when it is used as a source both of carbon skeletons and energy.

Most, if not all, starch synthesis is catalyzed by the nucleoside diphosphate glucose-starch glucosyl transferases, both starch granule-bound and soluble[1,2,3]. The soluble starch synthases require an oligosaccharide primer(s) as does starch phosphorylase[3,4,5]. There have been reports of de novo synthesis of oligosaccharides that might serve as primers for the enzymes that elongate alpha-1,4 glucan chains. Lavintman, et al.[6] have reported that potato tubers contain an enzyme capable of transferring glucose from UDPGlc to a protein acceptor. Nelson, et al.[2] have detected the presence in starch granules from waxy maize endosperms (which lack the major starch granule-bound glucosyl transferase) of a minor glucosyl transferase that has a low apparent K_m for its substrate, ADPGLc. MacDonald and Preiss[1] who have solubilized the starch granule-bound starch synthases have confirmed the existence of two starch synthases. Schilling[7]

* Research supported by the College of Agricultural and Life Sciences, University of Wisconsin-Madison and by Department of Energy Contract DE-AC02-82ER12031. This is paper No. 2719 from the Department of Genetics.

has recently identified a maltose synthase in spinach leaves which synthesizes maltotetraose and higher homologues with Glc-1-P as a substrate and has suggested that this enzyme functions to provide primer molecules for starch synthesis. The synthesis of primer molecules to which starch-synthesizing enzymes add glucose molecules is, therefore, an important problem in starch synthesis[3].

In this paper, we report a soluble enzyme which is capable of the de novo synthesis of short chain length phospho-oligosaccharides which can act as primer molecules to which the starch synthases could add glucose molecules. The designation of the enzyme as soluble indicates only that it is found in the supernatant fraction following a buffer extraction of developing maize endosperms and does not imply that it is cytosolic rather than organellar in location. This enzyme, which appears to be part of a multienzyme complex, was initially believed to utilize only Glc-1-P as a substrate. Subsequent experiments showed that Glc-1,6-bisP markedly accelerates the reaction and is an essential component. The phosphate groups linked to C6 of some glucose moieties of the phospho-oligosaccharides are apparently derived from the Glc-1,6-bisP molecules incorporated into the chain.

We have also detected the existence of a starch granule-bound enzyme which catalyzes the synthesis of similar products. The investigation of this second enzyme is to be reported in a separate paper.

MATERIALS AND METHODS

Enzyme Preparation

Developing kernels (22 days after self-pollination) of the W64A inbred line of maize were frozen in liquid nitrogen and stored at -20C. The pericarp and embryo were removed from kernels (15 g) prior to homogenization in a mortar with the addition of 7.5 ml of 50mM Tris-HCl buffer (pH 7.0). The resultant slurry was centrifuged at 30,000g for 20 min. Two ml of the supernatant fraction was put in a 15 ml Corex test tube which was placed in a boiling water bath for specified periods. The heat-treated crude extract was centrifuged to remove the pellet. The clear supernatant fraction, following dialysis to remove endogenous metabolites, was used for enzyme assays. For larger scale enzyme preparations, 350 grams of endosperms were homogenized with 175 ml of buffer in a blender. Following centrifugation, heat treatment, and a second centrifugation, the protein

TABLE 1 ENZYME PURIFICATION

STEPS	Vol. ml	Total Protein mg	Specific Activity nmol./mg/min	Purification
1. Crude Extract	338	12708.8	0.094	1
2. 90°C (7 min)	259	5568.0	12.3	130.8
3. Sephacel gel filtration	164	101.0	21.6	229.7
4. DEAE cellulose (52) column chromatography I	50	48	74	787.3
II	23	17	23	224.6
III	12	0.72	84	893.6
5. Sephacel gel filtration of DEAE - II enzyme	19	0.312	93	989.4

precipitating between 25 and 40% ammonium sulphate saturation was dissolved in the extraction buffer for dialysis overnight against the buffer. The dialyzate was applied to a Sephacel column which had been equilibrated with the same buffer. The enzyme was eluted with the same buffer and the fractions monitored for enzyme activity as described below. The fractions containing enzyme activity were pooled and precipitated with ammonium sulfate (60%) saturation), dissolved in 15 ml of 50mM Tris-HCl buffer containing 10% glycerol, and dialyzed overnight. The dialyzate was applied to a DEAE-cellulose column (16 x 1 cm) preequilibrated with the same buffer and eluted with that buffer. The increase in specific activity following the steps in purification is given in Table 1.

Reagents

The (^{14}C) Glc-1-P (250 mCi/mmol) was purchased from ICN Radiochemicals. 5'-(-^{32}P)ATP (3000 Ci/mmol) was obtained from Amersham. Glucose-6-phosphate dehydrogenase (torula yeast), alkaline phosphatase (calf intestine), amyloglucosidase (Rhizopus), phosphorylase b (rabbit muscle), phosphorylase a and b (rabbit heart), sugars, sugar phosphates, malto-oligosaccharides (a mixture of G_1 to G_{10}, and G_2, G_3, G_4, G_5, G_6, G_7), and acid phosphatase (potato) were obtained from Sigma. All other chemicals were analytical grade.

Enzyme Assays

Unless otherwise indicated, the reaction mixtures contained 5 ul of (^{14}C)Glc-1-P (0.1 umol., 500 cpm/nmol.), 100 ul of Sephacel gel-purified enzyme in a final volume of 135 ul of 0.125 M MES buffer, pH 6.0, and with or without the addition of phosphate sugar (0.1 umol). The reaction was started by adding enzyme and incubated at 37°C in a waterbath for various times. The reaction was stopped by quick freezing with dry ice. For treatment of the end products with potato acid phosphatase, 20 ul of the supernatant fraction was incubated with 20 ug (1 ul) of phosphatase at 25° for 100 min, after which an aliquot of the reaction mixture was taken for charcoal-celite column chromatography. In the experiments where the products were used as primers for phosphorylase b, the reaction mixture contained 100 ul of enzyme (ca 2-4 ug protein) and 10 ug of Glc-1-P (2 umol) in a final volume of 130 ul of 0.125 M MES buffer, pH 6.0. The reaction was incubated at 37° for 18 hours before using in the phosphorylase b enzyme assays. It should be noted that the commercial (^{14}C)Glc-1-P used was slightly contaminated with oligosaccharides. The treatment of Glc-1-P with Norite or passage through a charcoal-celite column removes the impurities[4]. For specific identification of the reaction products, the purified (^{14}C)Glc-1-P was used in enzyme assays.

Gel-filtration

Gel-filtration of proteins was accomplished on a Sephadex G-200 column (100 x 2.5 cm) that had been equilibrated with 50 mM Tris-HCl buffer, pH 7.0. The void volume was determined with blue dextran (Sigma). The molecular weight of the soluble enzyme was determined according to the procedure recommended by Pharmacia. Nonenzymatic protein molecular weight markers (Schwarz/Mann) were used.

Protein Determination

Protein was determined according to the method of Bradford[8] using rabbit gamma globulin as a standard or by measuring absorbance at 280 nm.

Synthesis of (^{32}P)Glc-1-P

($^{32}P)PO_4$ was obtained from (gamma-^{32}P)ATP by acid hydrolysis in 1 N HCL with boiling as described by Leloir and Cardini[9]. After neutralizing, (^{32}P) was separated from (^{32}P)ATP according to the method of Crane and Lipmann[10], and then desalted on a Sephadex G-25 column (1 x 150 cm). (^{32}P)Glc-1-P was synthesized from ($^{32}P)PO_4$ and glycogen as substrates with rabbit muscle phosphorylases a and b according to the procedure of Craig et al. [11]. The (^{32}P)Glc-1-P was again separated from ($^{32}P)PO_4$ by chromatography on Whatman No. 1 papers. After desalting and lyophyllizing, the (^{32}P)Glc-1-P was used as a substrate for the formation of phospho-oligosaccharides by the phospho-oligosaccharide synthases.

Synthesis of Glc-1,(6-^{32}P)-BISPHOSPHATE

Glc-1, (6-^{32}P)-bisP was synthesized according to the method of Eyer et al.[12] modified as follows: 2 mM $MgCl_2$, 5 mM Glc-1-P, 0.2 mM ATP (1 uCi of gamma-^{32}P-ATP), and 10 units of phosphofructokinase (rabbit muscle, Sigma) in a final volume of 100 ul of 50 mM Tris-HCl buffer, pH 7.5. The reaction mixture was incubated for 1 hr at 37°C. After incubation, the reaction product was applied, together with authentic Glc-1,6-bisP to an AG1-X2 anion exchange column, and eluted with 0.2 to 0.6 M NaCl linear gradient for separating Glc-1,(6-^{32}P)-bisP from (gamma-^{32}P)ATP. The identity of Glc-1,6bisP was verified according to the method of Eyer et al.[12]. After desalting on a Sephadex G-25 column, the synthesized Glc-1,(6-^{32}P)-bisP was used as a substrate in the synthase enzyme assay.

The Reaction Products as Primers

The reaction system modified from the method of Craig et al.[11] for testing whether reaction products of the soluble oligosaccharide synthase served as primer molecules for phosphorylase b was as follows. Five ul of Mg^{2+} (0.1 umol); 5 ul of 5'-AMP (0.25 umol), 5 ul of phosphorylase b (0.4 mg); and 5 ul of (^{14}C)Glc-1-P (ca. 4 x 10^5 cpm) were mixed with 130 ul of the products obtained from the enzyme reaiciton. The final volume of 150 ul contained 1.0 mM DTT. The reaction mixture was incubated at 25°C for 1 hr and the reaction stopped by adding 0.5 ml (5 mg) of a solution of sugary maize phytoglycogen and precipitated with 1 ml of 95% ethanol. The precipitate was collected and washed four times with a total of 4 ml of 95% ethanol before suspending in 0.5 ml of water.

RESULTS

Identification of End Products

When the heat-treated, dialyzed enzyme preparation is incubated for various time periods up to 30 minutes with (^{14}C) Glc-1-P (0.4 micromol, 10^5 cpm) as a substrate, only slight enzymatic activity is observed. If the substrate is Glc-1,6-bisP (0.4 micromol) plus 172 picomol (^{14}C) Glc-1-P (10^5 cpm), considerable incorporation of counts into phospho-oligosaccharides is observed (Table 2). Further heat treatment of the enzyme preparation reduces but does not eliminate the synthesis of phospho-oligosaccharides. The demonstration that the products of the reaction are phospho-oligosaccharides was made as described in Materials and Methods following treatment with acid phosphatase to convert Glc-1-P to Glc. Figure 1 provides a graphic demonstration of the reaction products following treatment with acid phosphatase and separation by paper chromatography. In this experiment, the acid phosphatase treatment has not completely converted the Glc-1-P to Glc. The bulk of the reaction product is resistant to acid phosphatase, however, and retains at least one phosphate group. It is interesting to note that phytic acid (inositol hexaphosphate) is not hydrolyzed by potato acid phosphate either (data not shown). If the compounds identified as phospho-oligosaccharides are eluted

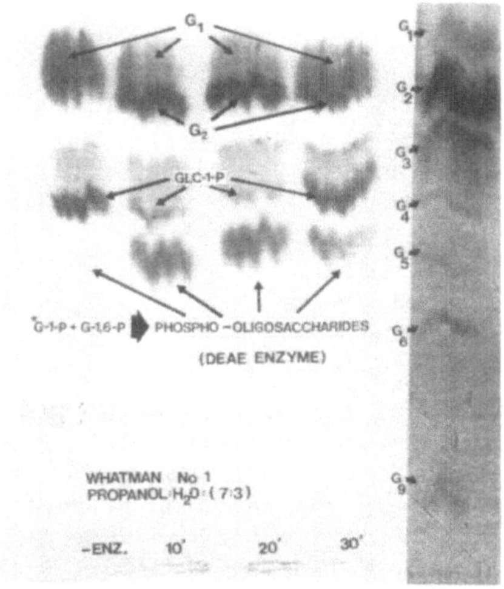

Fig. 1

Table 2

ASSAY SYSTEM		INCORPORATION OF (^{14}C) GLC-1-P INTO PHOSPHO-OLIGOSACCHARIDES nmole/mg/min	
STEP 2, 7' boiled enzyme.			
(^{14}C) GLC-1-P + GLC-1-P	0,4 µmole	0.05	
(^{14}C) GLC-1-P + Glc-1,6-bisPHOSPHATE	0,4 µmole	13.80	
STEP 2, 40' boiled enzyme.			
(^{14}C) GLC-1-P + Glc-1,6-bisPHOSPHATE	0,4 µmole	7.50	
STEP 4 FRACTION II DEAE-ENZYME		nmole/mg	
(^{14}C) GLC-1-P + Glc-1,6-bisPHOSPHATE	0,4 µmole	115.25	(5')
		211.12	(10')
		314.21	(15')

from the paper, treated with alkaline phosphatase, and rechromatographed with the same solvent system, there is a separation of the oligosaccharides formed in the reaction as shown in the right portion of the figure.

If the heat-treated enzyme preparation is not dialyzed before use, it is capable of catalyzing the formation of phospho-oligosaccharides with Glc-1-P (data not shown). We suggest that the presence of such low molecular metabolites as Glc-1,6-bisP in small quantities is responsible for the effect.

There are earlier experimental results obtained from enzyme preparations made before we developed the enzyme purification protocol described in Materials and Methods. These preparations which involved ammonium sulphate fractionation, gel filtration, and ion exchange chromatography resulted in less active enzyme preparations than our latest protocol, but the results of experiments with those preparations are still illuminating. Most of the results to be discussed from this point forward involve these earlier preparations.

The resistance of some of the linkages between glucose and PO_4 to hydrolysis when the phospho-oligosaccharide products are boiled in 1.0 N HCl also indicates that some phosphate groups are attached to carbon 6 of glucosyl residues of oligosaccharides[10]. These results are consistent with the report of the esterification of phosphate groups to carbon 6 of the glucosyl residues of starch [13, 14, 15, 16, 17, 18, 19, 20]. Furthermore, the radioactive peaks eluted by high salt indicate that some phospho-oligosaccharides have more than one phosphate group per molecule of oligosaccharide.

In view of the apparent linkage of phosphate groups to the C_6 of some glucose molecules, we investigated the possibility that Glc-6-P might be the substrate utilized by the phospho-oligosaccharide synthase and that the Glc-6-P was formed by the action of phosphoglucomutase present in the enzyme preparation. This possibility was tested by furnishing either (^{14}C)Glc-1-P or (^{14}C)Glc-6-P as a substrate (3.6mM) in reaction mixtures. When Glc-6-P was the substrate, the mean net incorporation was 1.39 nmoles Glc/mg protein/min. With Glc-1-P, the net incorporation as 6.0 nmoles Glc/mg/min indicating the preference for Glc-1-P as a substrate.

As shown in Table 2, the incorporation of the label from ^{14}C-Glc-1-P into phospho-oligosaccharides is markedly accelerated in the presence of Glc-1,6-bisP if the enzyme preparation has been heated in a boiling water bath for 7 min and dialyzed. This suggests that Glc-1,6-bisP is a substrate for the reaction forming phospho-oligosaccharides and further that heating has damaged the ability of the enzyme preparation to synthesize Glc-1,6-bisP from Glc-1-P. Other activities that appear to be associated with the enzyme complex are:

1) Glc-1-P + Glc-1-P <---> Glc-1,6-bisP + Glc
2) Glc-1-P + Glc-6-P <---> Glc-1,6-bisP + Glc
3) Glc-6-P + Glc-6-P <---> Glc-1,6-bisP + Glc
4) Glc-1,6-bisP + Amylopectin ---> Amylopectin-Glc-6-P

Reactions 1, 2, and 3 were demonstrated by showing the production of Glc (by glucose oxidase) but not Pi in reaction mixtures containing Glc-1-P alone, Glc-6-P alone, or both Glc-1-P and Glc-6-P (data not shown). Reaction 1 is also shown to occur in Figure 2 where the reaction products of an incubation of the heat-treated enzyme with Glc-1,6-bisP and Glc as substrates were separated by paper chromatography following treatment with acid phosphatase. The formation of phospho-oligosaccharides in this reaction is readily apparent.

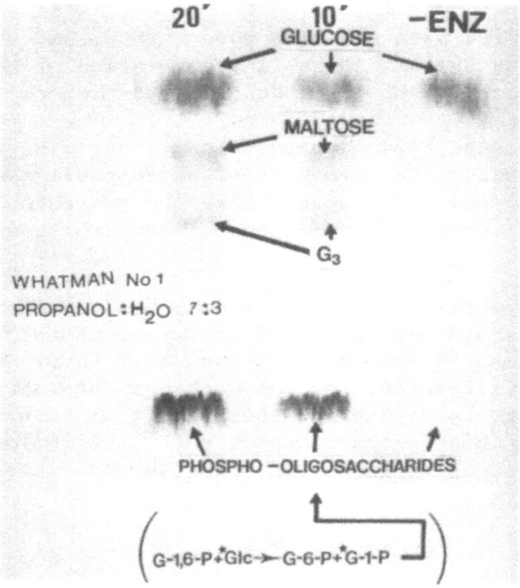

Fig. 2

The incorporation of Glc-6-P from Glc-1,6-bisP into amylopectin molecules was shown in reaction mixtures which contained either amylopectin or amylose as acceptors, (^{14}C)Glc-1,6-bisP as a substrate, and the enzyme preparation. No or very low incorporation was observed when amylose was the acceptor while incorporation into amylopectin molecules was readily observable (Fig. 3). The incorporation of Glc-6-P into amylopectin molecules may account for the phosphate groups which are located close to the branch points in potato amylopectin[20].

The formation of Glc-6-P from Glc-1,6-bisP, as monitored by Glc-6-P dehydrogenase, is completely dependent on the presence of Glc in the reaction. Leloir et al.[21] have demonstrated that extracts of E. coli can synthesize Glc-1,6-bisP from Glc-1-P. In maize, the ability of either Glc-1-P or Glc-6-P to serve as substrates for the formation of Glc-1,6-bisP and ultimately phospho-oligosaccharides implies the existence of

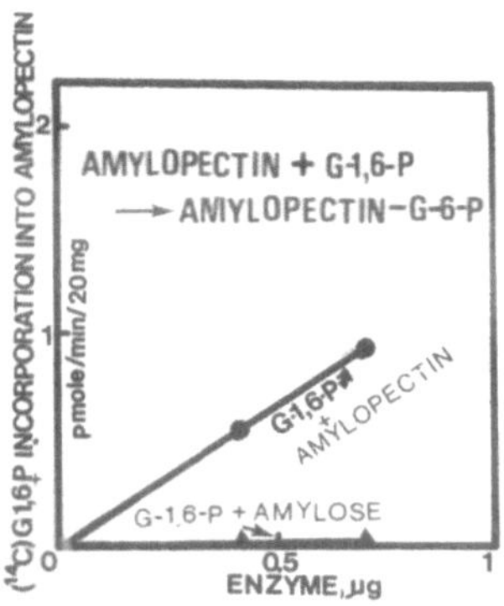

Fig. 3

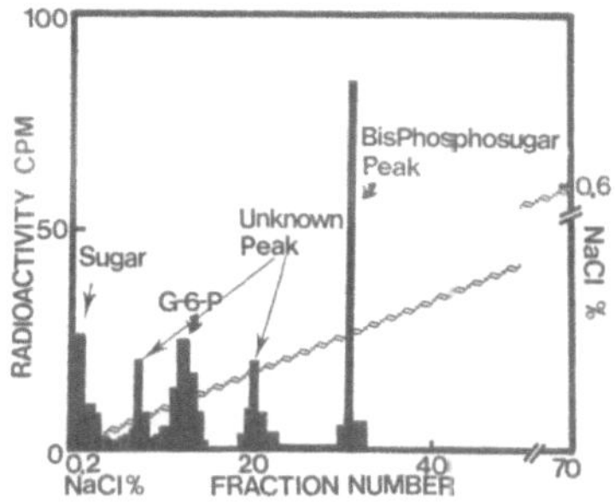

Fig. 4

phosphoglucomutase activity associated with the complex, and this activity can be demonstrated directly in the phospho-oligosaccharide synthase enzyme preparation before heat treatment but not afterwards (data not given).

These observations of activities other than phospho-oligosaccharide synthase associated with the enzyme complex aid in explaining the use of both Glc-1-P and Glc-6-P by the synthase and the acceleration of the reaction by Glc-1,6-bisP.

Utilization of Glc-1,(6-^{32}P)-Bisphosphate

Since the presence of Glc-1,6-bisP in the reaction mixture markedly enhances the incorporation of the label from ^{14}C-Glc-1-P into phospho-oligosaccharides and since many of the phosphate groups in the phospho-oligosaccharides are linked to C6 of the Glc moieties, it is a natural question as to whether the C6-linked phosphate groups of Glc-1,6-bisP are found in the phospho-oligosaccharide product of the reaction. To test this point, Glc-1,(6-^{32}P)-bisP synthesized as described in Materials and Methods, was added to the reaction mixtures together with various quantitites of Glc-1-P. The data in Table 3 show that there is radioactivity present in the phospho-oligosaccharide product and thus Glc-1,6-bisP is being utilized as a substrate in the synthesis. Less incorporation is measured in the reactions where labelled Glc-1,6-bisP was present than in those where labelled Glc-1-P was present since the separation of Glc-1,6-bisP from the phospho-oligosaccharides permitted only the measurement of phospho-oligosaccharides with 3 or more phosphate groups. The data also show that there is greater incorporation of label from Glc-1,(6-^{32}P)-bisP with increasing concentrations of Glc-1-P just as there is increased incorporation of label from (^{14}C)-Glc-1-P into phospho-oligosaccharides with increasing concentrations of Glc-1,6-bisP so both compounds are being used as substrates, and the presence of both may be necessary for the reaction to proceed. When the phospho-oligosaccharide reaction products are hydrolyzed by boiling in 1N HCl and the hydrolyzed mixture separated on an anion exchange column (AG1-X2) as shown in Fig. 4, it is observed that labeled Glc-6-P is one of the products of hydrolysis. There is also a peak of radioactivity eluting from the column in fractions whose position indicates that two phosphate groups are present. We believe that this peak results from a small phospho-oligosaccharide or several such compounds which are resistant to hydrolysis.

Molecular Weight Determination

The molecular weight of the purified phospho-oligosaccharide synthase was estimated by Sephadex G-200 gel filtration and by 10% polyacrylamide

Table 3

ASSAY SYSTEM	INCORPORATION OF Glc-1,(6-^{32}P)bisPHOSPHATE INTO PHOSPHO-OLIGOSACCHARIDES
PLUS GLC-1-P μmole	μmole/mg/min
.02	.11
.06	.32
.10	.47

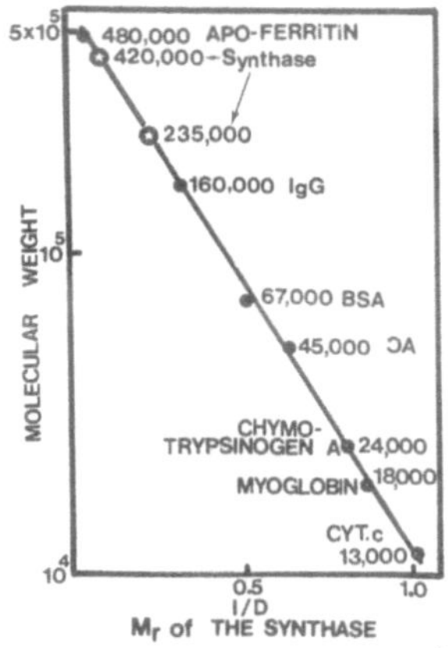

Fig. 5

gel (PAG) electrophoresis in the presence of SDS. Following gel filtration, two peaks of activity were found corresponding to proteins of 235 and 420 kDa suggesting that these forms may be related to each other as monomer and dimer (Fig. 5).

The Effect of Substrate Concentration

Figure 6A shows a double reciprocal plot relating the production of phospho-oligosaccharides to the concentration of Glc-1-P at various concentrations of Glc-1,6-bisP. The presence of Glc-1,6-bisP markedly accelerates the velocity of the reaction and at the lower concentrations of Glc-1,6-bisP leads to a pronounced departure from linearity. The same K_m for Glc-1-P (1.66mM) is estimated both in the absence and presence of Glc-1,6-bisP.

In figure 6B, the same set of data are used to estimate the K_m for Glc-1,6-bisP in the presence of varying concentrations of Glc-1-P. While increased quantities of Glc-1-P result in enhanced synthesis of phospho-oligosaccharides, the estimated K_m (0.2mM) for Glc-1,6-bisP is not altered.

Reaction Products as Primers

If the phospho-oligosaccharide synthases are responsible for the synthesis of the molecules to which glucose can be added by starch-

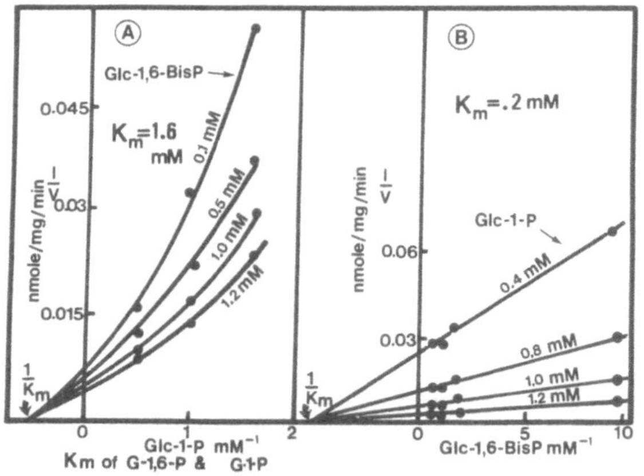

Fig.6

Table 4

ASSAY SYSTEM	INCORPORATION OF (^{14}C) GLC-1-P INTO THE END PRODUCTS (PRIMERS) OF THE SYNTHASE REACTION BY PHOSPHORYLASE b
	nmole/mg/min
a. Incubation of G-1-P with enzyme followed by the addition of (^{14}C) GLC-1-P & phospho rylase b.	161.50
b. Incubation of enzyme (no G-1-P) followed by the addition of (^{14}C) GLC-1-P & phospho rylase b.	0.20
c. Incubation of G-1-P with enzyme followed by the addition of (^{14}C) GLC-1-P only. The reaction was stopped immediately after the addition of phospho rylase b (0 time control).	0.00
d. Incubation of G-1-P with buffer only (no enzyme) followed by the addition of (^{14}C) GLC-1-P & phospho rylase b.	0.01

synthesizing enzymes, then one should be able to demonstrate that the products of the phospho-oligosaccharide synthase serve as primers for a chain-elongating enzyme; e.g., rabbit muscle phosphorylase b. The results of these experiments are shown in Table 4. It can be seen from these data that the activity of phosphorylase can be observed to any extent only in the presence of the reaction products obtained from the soluble enzyme suggesting the formation of primer molecules to which glucose molecules can be added by phosphorylase b. The minor activity of phosphorylase b in the absence of the products of oligosaccharide synthase reaction is probably due to the slight contamination of commercial Glc-1-P or phosphorylase b itself by oligosaccharides. The priming effect can be demonstrated only after the reaction mixture has been incubated with the synthase for a period of time. The lack of priming effect by synthase and phosphorylase b added simultaneously indicates that the synthase, itself, can not act as a primer for phosphorylase b, nor does the synthase contain a bound oligosaccharide which can act as a primer.

DISCUSSION

The enzyme system being reported in this paper together with a phospho-oligosaccharide synthase which is tightly bound to the starch granules (unpublished data) may play a role in the initiation of starch synthesis by providing short chain lengths of alpha-1,4 glucans which can serve as the primer molecules required by all starch synthases reported to date[3,5]. The initial products of the reaction are phospho-oligosaccharides, and many of the phosphate groups are linked to C6 of the glucose moiety as shown by examining the compounds formed by hydrolyzing the reaction products. The phosphate groups present in the primer molecules as well as the ability of this enzyme system to insert glucose molecules with a phosphate group at the nonreducing end of chains in amylopectin molecules may account for the phosphate content reported to be present in many starches[13,16,17,20] in the amylopectin fraction but infrequently in the amylose component[17]. The phosphate groups are esterified to the hydroxyl of C6 and limit the attack of both alpha-amylase and beta-amylase on starch so that a series of limit phospho-dextrins is produced[14,15,17,18,20].

The substrate for the enzyme was initially believed to be Glc-1-P since the presence of Glc-1-P in the reaction mixture led to the formation of phospho-oligosaccharides. The marked acceleration of the incorporation of the label from ^{14}C-Glc-1-P into the phospho-oligosaccharides by Glc-1,6-bisP and the linkage of many phosphate groups to C6 of glucosyl residues prompted a reexamination of the role of Glc-1,6-bisP in the reaction. The finding using Glc-1,6-bisP labelled within ^{32}P at C6 in the reaction mixture that the label was present in the phospho-oligosaccharide product made it clear that Glc-1,6-bisP is utilized together with Glc-1-P as a substrate by the enzyme preparation and that its presence may well be obligatory. Under this view, the formation of phospho-oligosaccharides with only Glc-1-P as a substrate is preceded by the formation of some Glc-1,6-bisP from Glc-1-P, a reaction which we have shown that the enzyme complex can catalyze. The presence of Glc-6-P together with Glc-1-P in the reaction mixture also stimulates the formation of phospho-oligosaccharides and, in this case, by facilitating the synthesis of Glc-1,6-bisP.

The enzyme complex, which is capable of carrying out several diverse reactions, is a multimer composed of two different subunits, at least. Although we observe subunits of 23 and 24 kDa following PAGE in the presence of SDS, we cannot yet be certain that there is only one type of subunit within each of these molecular weight classes.

The ability of the enzyme to incorporate glucosyl residues with a phosphate group at C6 from Glc-1,6-bisP at the nonreducing ends of amylopectin molecules but not amylose may account for the presence of phosphate groups, which have been shown to be present in the amylopectin fraction but infrequently in the amylose fraction of many starches[18].

A second phospho-oligosaccharide synthase has also been identified. This second enzyme is tightly bound to the starch granules in contrast to the soluble enzyme reported here. The relative roles of these enzymes in forming primer molecules has not been clarified at this time.

REFERENCES

1. F. D. MacDonald, and J. Preiss, Solubilization of the starch granule-bound starch synthetase of normal maize kernels. Plant Physiol. 73:175 (1983).
2. O. E. Nelson, P. S. Chourey, and M. T. Chang, Nucleoside diphosphate sugar-starch glucosyl transferase activity of wx starch granules. Plant Physiol. 62:383 (1978).
3. J. Preiss, and D. A. Walsh, The comparative biochemistry of glycogen and starch. In: "The Biology of Carbohydrates" Vol I, V. Ginsburg, Ed., John Wiley and Sons, Inc., N.Y. (1981).
4. A. Kamogawa, T. Fukui, and Z. Nikuni, Potato alpha-glucan phosphorylase: Crystallization, amino acid composition and enzymatic reaction in the absence of added primer. J. Biochem. Tokyo 63:361 (1968).
5. S. Schiefer, E. Y. C. Lee, andW. J. Whelan, The requirement for a primer in the in vitro synthesis of polysaccharide by sweet corn (1,4)-alpha-D-glucan synthase. Carbohydr. Res. 61:239 (1978).
6. N. Lavintman, J. Tandecarz, M. Carceller, and C. E. Cardini, Role of UDP-glucose in the biosynthesis of starch. Mechanism of formation of a glucoproteic acceptor. Eur. J. Biochem. 50:145 (1974).
7. N. Schilling, Characterization of maltose biosynthesis from alpha-D-glucose-1-phosphate in Spinacia oleracea L.. Planta 154:87 (1982).
8. M. Bradford, A rapid and sensitive method for the quantitation of microgram quantitites of protein utilizing the principle of protein-dye binding. Anal. Biochem. 72:248 (1976).
9. L. F. Leloir, and C. E. Cardini, Characterization of phosphorus compounds by acid lability. In: "Methods in Enzymology" Vol. III, S. P. Colowick and N. O. Kaplan, Eds., Academic Press, N.Y. (1957).
10. R. K. Crane, and F. Lipmann, The effect of arsenate on aerobic phosphorylation. J. Biol. Chem. 201:235 (1953).

11. H. D. Craig, L. H. Schliselfeld, and I. G. Krebs, Phosphorylase b and a Isozyme I from rabbit heart. In: "Methods in Enzymology" Vol. VIII, E. F. Neufeld, and V. Ginsburg, Eds., Academic Press, N.Y. (1966).
12. P. Eyer, W. Hofer, E. Krystek, and D. Pette, Synthesis of Glc-1,6-bisphosphate by the action of crystalline rabbit muscle phosphofructokinase. Eur. J. Biochem. 20:153 (1971).
13. S. Hizukuri, K. Shirasaka, and B. O. Juliano, Phosphorus and amylase in rice starch granules. Starch 35:348 (1983).
14. L. McBurney, and M. D. Smith, Effect of esterified phosphorus in potato starch on the action pattern of salivary alpha-amylase. Cereal Chem. 42:161 (1965).
15. F. Parrish, and W. J. Whelan, The structure of phosphomaltotetraose, an alpha-limit dextrin of starch. Starke 13:231 (1961).
16. T. Posternak, On the phosphorus of potato starch. J. Biol. Chem. 188: 317 (1951).
17. M. W. Radomsky, and M. B. Smith, Location and possible role of esterified phosphorus in starch fractions. Cereal Chem. 40:31 (1963).
18. S. Tabata, S. Hizukuri, and K. Nagata, Action of sweet potato beta-amylase on the phosphodextrin of potato starch. Carbohyd. Res. 67: 189 (1971).
19. Y. Takeda, and S. Hizukuri, Re-examination of the action of sweetpotato beta-amylase on phosphorylated (1,4)-alpha-D-glucan. Carbohydrate Res. 89:174 (1981).
20. Y. Takeda, and S. Hizukuri, Location of phosphate groups in potato amylopectin. Carbohydr. res. 102:321 (1982).
21. L. F. Leloir, R. E. Trucco, C. E. Cardini, A. C. Paldini, and R. Caputto, The Formation of Glucose diphosphate by E. coli. Arch. Biochem. 23:65 (1949).

ACETOLACTATE SYNTHASE, THE TARGET ENZYME OF THE SULFONYLUREA HERBICIDES

Barbara J. Mazur, S. Carl Falco, Susan Knowlton, and
Julie K. Smith
Agricultural Products Department
E. I. Du Pont de Nemours and Co., Inc.
Experimental Station 402
Wilmington, Delaware, 19898, USA

INTRODUCTION

Traditional approaches to herbicide development have relied on the synthesis and testing of chemicals which are toxic to weeds but not to crops, usually due to their selective metabolism to non-toxic compounds by crops. The discovery of compounds which have such chemical selectivity, has, however, often required the synthesis of thousands of compounds before a herbicide candidate can be identified. An alternative approach to obtaining specificity is to introduce genetic selectivity into crop plants. Such plants can be created in several ways. Selective tolerance can be achieved by introducing genes coding for enzymes that metabolize herbicides, or by engineering the overproduction of the herbicide target enzyme so that some enzyme remains free of herbicide and active. Selective resistance can be achieved by introducing genes coding for target proteins that are not inhibited by herbicides. We have primarily used the latter approach to produce crop plants that are resistant to field application rates of sulfonylurea herbicides.

BACKGROUND

The sulfonylurea herbicides are a relatively new class of chemicals, first discovered in 1975. A large number of herbicidally active structural variants of the sulfonylureas have been synthesized. The structure of a typical sulfonylurea compound is shown in Figure 1. The sulfonylureas are effective at doses as low as two grams/hectare, an application rate that is 1000 fold lower than the rates required for some older classes of herbicides. At the same time, the sulfonylureas are non-toxic to mammals; they have proven less toxic than sodium chloride in rat feeding studies. One reason for this non-toxicity is that the target for the sulfonylureas is the enzyme acetolactate synthase, or ALS, which is the first common enzyme in the biosynthesis of the branched chain amino acids isoleucine, leucine, and valine. These are essential amino acids, and must be ingested in the diet of mammals, which lack the metabolic pathways leading to their synthesis (reviewed by LaRossa et al., 1987a).

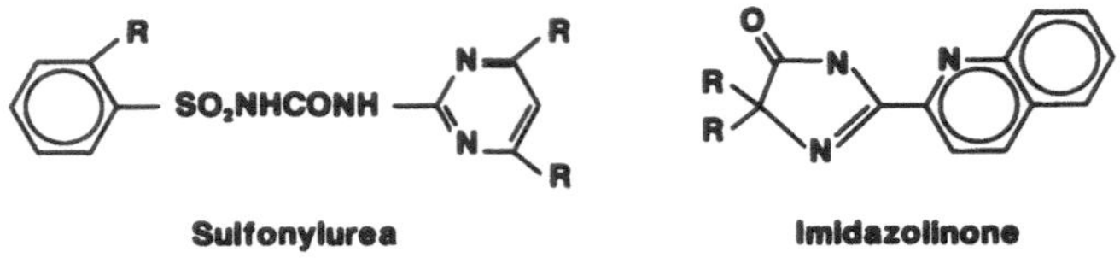

Fig. 1. Herbicide structures. The structure on the left is a sulfonylurea compound (marketed by Du Pont). The structure on the right is an imidazolinone compound (marketed by American Cyanamid).

Our interest in making plants resistant to the sulfonylurea herbicides began with the finding of Chaleff and Ray (1984) that sulfonylurea resistant tobacco lines could be selected *in vitro* using cell culture methods. Chaleff and Ray were able to select a number of mutants resistant to the sulfonylurea herbicides Glean® and Oust®. Genetic analyses of plants regenerated from these mutants indicated that the mutations segregated as single nuclear genes in two linkage groups, and were dominant or semi-dominant. This finding suggested that the genes coding for sulfonylurea herbicide resistance might be amenable to recombinant DNA manipulations. The primary molecular target of these herbicides remained elusive, however, until the discovery by LaRossa and Schloss (1984) that ALS was the target of the sulfonylureas in bacteria. This mode of action was deduced through physiological studies, in which some bacteria were shown to be sensitive to sulfonylureas when growing on minimal media, but tolerant when growing on rich media or on minimal media supplemented with branched chain amino acids. These physiological studies were complemented by biochemical studies on the enzyme, which indicated that ALS activity was highly sensitive to nanomolar concentrations of sulfonylureas, and by genetic studies, which indicated that herbicide resistant bacteria produced a herbicide insensitive form of ALS. This finding was quickly followed by work which demonstrated that ALS is also the target of the sulfonylureas in plants and in yeast (Chaleff and Mauvais, 1984; Ray, 1984; Falco and Dumas, 1985). In tobacco, biochemical and genetic studies were combined in order to show that a herbicide resistant form of the ALS enzyme co-segregated with the herbicide resistant whole plant phenotype (Chaleff and Mauvais, 1984). In yeast, an ALS gene was isolated from a genomic library by plating transformed yeast cells on media containing a sulfonylurea herbicide; transformants receiving a sensitive ALS gene on the multicopy plasmid vector produced increased amounts of enzyme which could overcome the lethality of the herbicide. Resistant yeast mutants were subsequently selected. Genetic and biochemical studies of these resistant mutants were used to confirm that ALS is the target of these herbicides in yeast (Falco and Dumas, 1985).

Studies to deduce the mode of action of a structurally unrelated class of compounds, the imidazolinones (Figure 1), indicated that ALS is also the primary target of these compounds (Shaner et al., 1984). Thus, ALS is the target of at least two unrelated classes of herbicides, and appears to be a particularly efficacious target for herbicides. Experiments of LaRossa et al. (1987b) have addressed the issue of why ALS is a preferred target for herbicides. This work has indicated that the activity of these compounds in bacteria is potentiated by the accumulation of an ALS substrate, α-ketobutyrate, which is toxic. LaRossa et al. have suggested that it is both the deficiency of branched chain amino acids, and the increase in concentration of the toxic intermediate, that combine to make ALS a particularly potent target for herbicides.

RESULTS AND DISCUSSION

Isolation and Characterization of Plant ALS Genes

Upon the isolation of a clone carrying the yeast ALS gene, several methods were used to localize the gene on the cloned genomic DNA fragment, including transposon mutagenesis (Van Dyk et al., 1986), deletion mapping (Falco and Dumas, 1985) and heterologous hybridization to a cloned *Salmonella typhimurium* ALS gene. The observed hybridization of the yeast and *Salmonella* genes indicated an unanticipated level of conservation between ALS enzymes, and lead to an attempt to detect homology among ALS genes from other species. A segment of the yeast ALS gene that spanned most of the coding region was used as a probe, and hybridizations were carried out under low stringency conditions. Homology was initially detected between a fragment of the yeast ALS gene and DNA from the prokaryotic cyanobacterium *Anabaena* 7120. Subsequently, homology to cloned DNA from the higher plants *Arabidopsis thaliana* and *Nicotiana tabacum*, or tobacco, was detected. Phage carrying putative ALS genes were isolated from all three species (Mazur et al., 1985).

In order to confirm that these isolated clones carried ALS genes, the clones were mapped and partially sequenced (Mazur et al., 1985). Comparison of the partial deduced amino acid sequences of the putative cyanobacterial and plant ALS genes with those of the yeast and three *Escherichia coli* ALS genes confirmed the identity of the clones. All three putative ALS clones shared conserved sequences with the yeast and bacterial enzymes, which had been shown in previous studies to share three domains of homology interspersed with four domains of non-homology (Falco et al., 1985).

Subsequent complete sequence analysis of the two plant ALS genes has indicated that they code for proteins of 667 and 670 amino acids, with predicted molecular weights of approximately 73000 daltons. Neither gene has introns. The two plant ALS sequences are conserved relative to each other throughout most of the length of the genes, including those domains which are not conserved even among the three *E. coli* isozymes. Approximately 75 percent of the nucleotides and 85 percent of the encoded amino acids are conserved between the two genes (Mazur et al., 1987). A comparison of the deduced amino acid sequences of the tobacco and *Arabidopsis* ALS genes is shown in Figure 2.

There is, however, one region which is not highly conserved between the two plant ALS sequences, at the 5′ end of the coding sequences. This region is assumed to encode chloroplast transit sequences, in that ALS is a nuclear encoded but chloroplast localized enzyme in plants (Miflin, 1974; Jones et al., 1984; Chaleff and Ray, 1984). The overall homology in this region is in fact greater at the nucleotide level than at the amino acid level, suggesting that there are few constraints upon mutations in this region. Despite this lack of nucleotide and amino acid sequence homology, however, hydrophobicity/hydrophilicity profiles of the two transit regions show a number of similar domains (Figure 3), suggesting that the polarity of the amino acids in this portion of the peptide is more important than their primary structure. In order to determine the precise positions at which the transit sequences are cleaved from the mature proteins, the tobacco and *Arabidopsis* ALS genes have been cloned into SP6 and T7 riboprobe vectors. *In vitro* transcribed RNA has been translated in a rabbit reticulocyte lysate in the presence of radioactive amino acids, and incubated with isolated chloroplasts. The chloroplast localized, processed products have been subjected to N-terminal amino acid analyses. Preliminary results indicate that the plant ALS transit peptides extend to the homologous regions of the sequences, and are approximately 100 amino acids in length, making them the longest transit sequences characterized to date (Bascomb et al., manuscript in preparation).

```
MAAAA..PSPSSSA.FSKTLSPSSSTSSTLLPRSTFPFPHHPHKTTPPPLHLTHTHIHIHSQRRRFTISNVISTNQKVSQTEKTETFVSRFAPDEPRKGS
MAAATTTTTTSSSISFSTKPSPSSSKSPLPISRFSLPFSLNPNKSSSSSRRRGIKSSSPSSISAVLNTTTNVTTTPSPTKPTKPETFISRFAPDQPRKGA

DVLVEALEREGVTDVFAYPGGASMEIHQALTRSSIIRNVLPRHEQGGVFAAEGYARATGFPGVCIATSGPGATNLVSGLADALLDSVPIVAITGQVPRRM
DILVEALERQGVETVFAYPGGASMEIHQALTRSSSIRNVLPRHEQGGVFAAEGYARSSGKPGICIATSGPGATNLVSGLADALLDSVPLVAITGQVPRRM

IGTDAFQETPIVEVTRSITKHNYLVMDVEDIPRVVREAFFLARSGRPGPILIDVPKDIQQQLVIPDWDQPMRLPGYMSRLPKLPNEMLLEQIVRLISESK
IGTDAFQETPIVEVTRSITKHNYLVMDVEDIPRIIEEAFFLATSGRPGPVLVDVPKDIQQQLAIPNWEQAMRLPGYMSRMPKPPEDSHLEQIVRLISESK

KPVLYVGGGCSQSSEDLRRFVELTGIPVASTLMGLGAFPTGDELSLSMLGMHGTVYANYAVDSSDLLLAFGVRFDDRVTGKLEAFASRAKIVHIDIDSAE
KPVLYVGGGCLNSSDELGRFVELTGIPVASTLMGLGSYPCDDELSLHMLGMHGTVYANYAVEHSDLLLAFGVRFDDRVTGKLEAFASRAKIVHIDIDSAE

IGKNKQPHVSICADIKLALQGLNSILESKEGKLKLDFSAWRQELTEQKVKHPLNFKTFGDAIPPQYAIQVLDELTNGNAIISTGVGQHQMWAAQYYKYRK
IGKNKTPHVSVCGDVKLALQGMNKVLENRAEELKLDFGVWRNELNVQKQKFPLSFKTFGEAIPPQYAIKVLDELTDGKAIISTGVGQHQMWAAQFYNYKK

PRQWLTSGGLGAMGFGLPAAIGAAVGRPDEVVVDIDGDGSFIMNVQELATIKVENLPVKIMLLNNQHLGMVVQWEDRFYKANRAHTYLGNPSNEAEIFPN
PRQWLSSGGLGAMGFGLPAAIGASVANPDAIVVDIDGDGSFIMNVQELATIRVENLPVKVLLLNNQHLGMVMQWEDRFYKANRAHTFLGDPAQEDEIFPN

MLKFAEACGVPAARVTHRDDLRAAIQKMLDTPGPYLLDVIVPHQEHVLPMIPSGGAFKDVITEGDGRSSY
MLLFAAACGIPAARVTKKADLREAIQTMLDTPGPYLLDVICPHQEHVLPMIPSGGTFNDVITEGDGRIKY
```

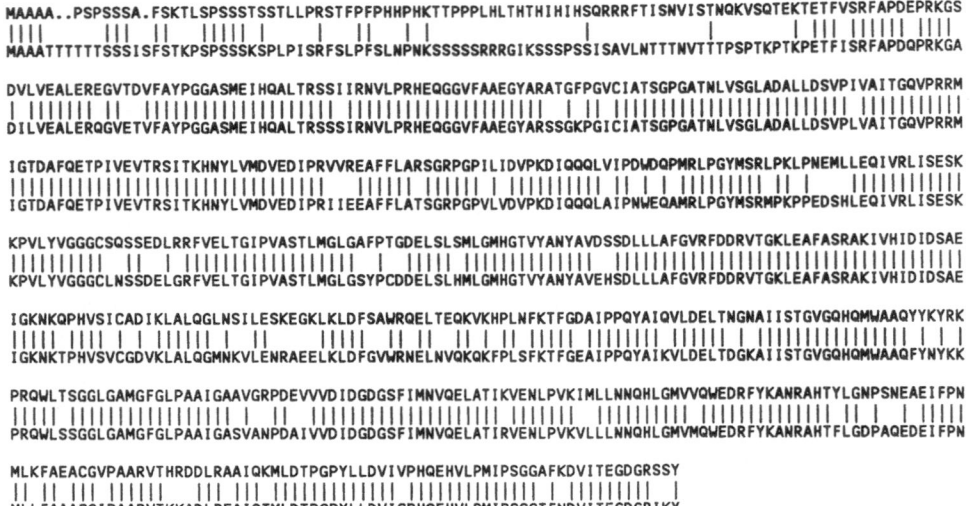

Fig. 2. Deduced amino acid sequences of ALS. The deduced amino acid sequence of a tobacco ALS gene is shown on the top line, and that of an *Arabidopsis* ALS gene is shown on the bottom line. Vertical lines highlight identical amino acids.

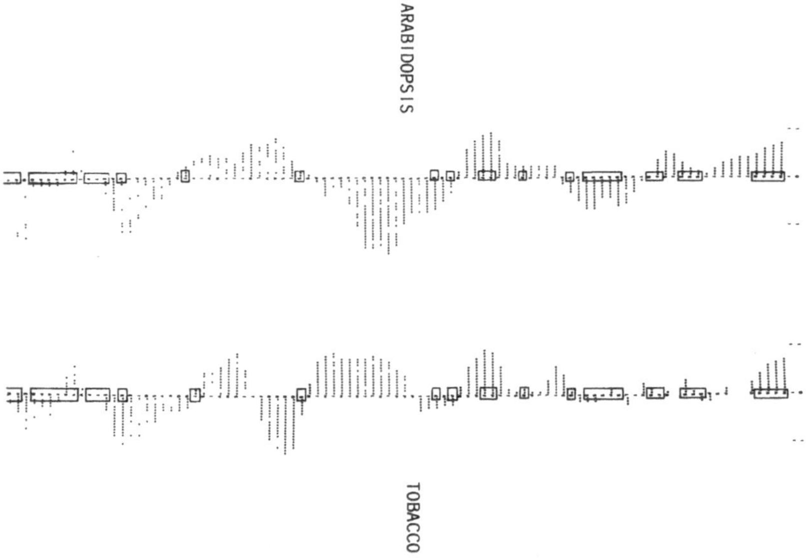

Fig. 3. Hydrophobicity plots. Hydrophobicity/hydrophilicity plots of the N-terminal deduced amino acids of *Arabidopsis* (left) and tobacco (right) ALS enzymes. Amino acids which are conserved in both sequences are boxed. Hydrophobic regions lie to the left of center, and hydrophilic regions to the right.

Retrospective comparisons of the DNA sequences of the yeast, tobacco, and *Arabidopsis* ALS genes have given some insight into the homologies which allowed these clones to be identified. Both the yeast and tobacco and yeast and *Arabidopsis* ALS gene pairs share scattered blocks of homology throughout the length of the genes. Several of these homology blocks range over 20 to 30 nucleotides with few mismatches, and have high GC contents. The most striking of these blocks has 36/40 conserved nucleotides between the yeast and tobacco genes, and 27/31 conserved nucleotides between the yeast and *Arabidopsis* genes. Both sets of gene pairs have several scattered sections of 10-20 identical nucleotides. We expect that these sequence homologies will also be conserved in other ALS genes.

The physical organization of the ALS genes in tobacco and *Arabidopsis* has been determined by Southern blot analyses. Homologous cloned ALS genes were used as probes. A single ALS gene hybridizes to the probe in *Arabidopsis*, while two hybridize in tobacco, which is an allotetraploid (Mazur et al., 1987). The presence of two ALS genes in tobacco is consistent with genetic data, which had indicated that tobacco mutations which confer herbicide resistance define two loci (Chaleff and Ray, 1984). Southern blot analyses of DNA from several other crop species have indicated that many carry multiple ALS genes.

The expression of the tobacco ALS genes has been measured in various plant tissues by Northern blot hybridizations and by RNase protection experiments. These analyses have revealed that both ALS genes are strongly expressed in flowers, are moderately expressed in leaves, and are weakly expressed in roots (Martin et al., manuscript in preparation). This finding may help to explain the sensitivity of roots to sulfonylurea herbicides which is sometimes observed.

Introduction of herbicide resistant ALS genes into plants

The cloned plant ALS genes have been used as probes to isolate genes carrying ALS mutations from herbicide resistant plants. In tobacco, the *Hra* line, which is mutated at the *SuRB* locus and which is 1000 fold more resistant to sulfonylureas than are wild type lines (Chaleff et al., 1987), has been used as one source of mutant ALS genes. A second tobacco line, C3, which carries a mutation at the *SuRA* locus (Chaleff and Ray, 1984), has also been used as a source of mutant ALS genes. Genes representing each ALS locus have been isolated from both lines and have been reintroduced into tobacco. The molecular characterization and reintroduction of mutant and wild type genes from each plant line has permitted the assignment of the genes to the appropriate genetic locus. These experiments have indicated that mutant genes from either locus can confer herbicide resistance in transformed cells. Plants regenerated from cells transformed with the *Hra* gene are resistant to field application rates of sulfonylurea herbicides (Lee et al., manuscript in preparation).

The tobacco *Hra* ALS gene has also been used to transform a number of heterologous species, and has conferred selectable levels of herbicide resistance in them. The sensitivity of ALS activity in transgenic tomato plants to a sulfonylurea herbicide is shown in Figure 4. ALS activity was measured in leaf extracts of regenerated plants. Three different herbicide inhibition profiles are apparent among the five transformants shown; two pairs of transformants probably originate from single transformation events. At a concentration of 10 parts per billion of herbicide, approximately 60% of the activity in plants 4 and 4c, 40% of the activity in plant 3, and 30% of the activity in plants 4b and 4d remains uninhibited by the sulfonylurea. These results demonstrate that the tobacco *Hra* gene can be

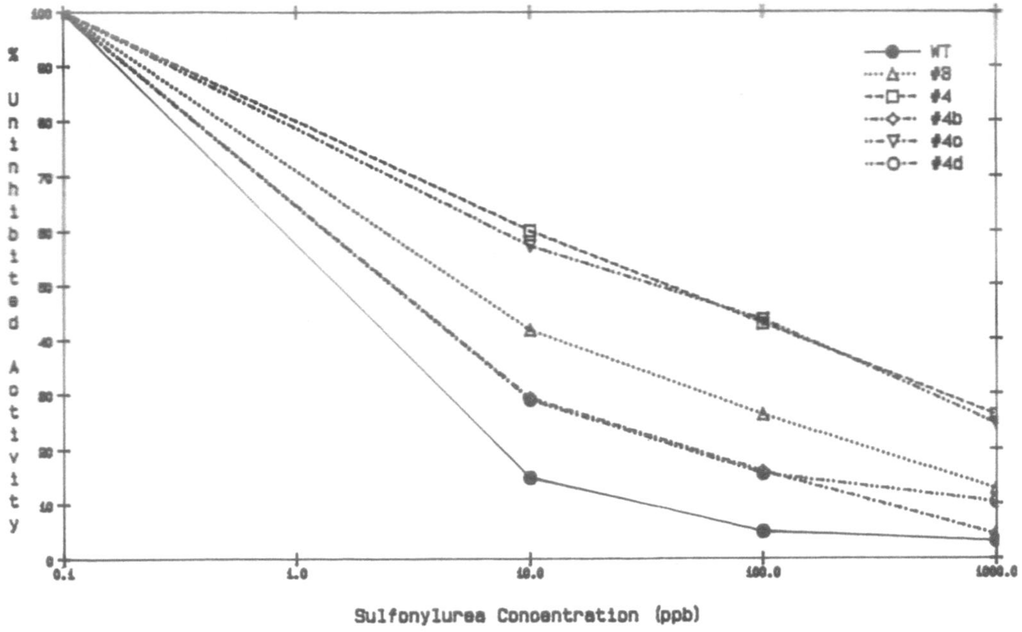

Fig. 4. ALS activity in transgenic tomatoes. Leaves of tomato plants, which had been transformed with the tobacco *Hra* ALS gene, were assayed for herbicide resistant ALS activity. The vertical axis shows ALS activity in the presence of herbicide, as a percentage of activity in the uninhibited control reaction. The horizontal axis shows the concentration of herbicide used in the assays, in parts per billion. The solid circles show activity for a wild type line of tomato; the open symbols show activity for transformants, as noted on the figure.

highly effective in conferring herbicide resistance in heterologous species. Similar but more extensive studies using commercially important cultivars of tobacco have also demonstrated the effectiveness of this gene, as described below.

The *Hra* gene has been introduced into a number of commercial lines of tobacco, and regenerated plants have been assayed for levels of sulfonylurea resistance. Resistance was measured by assaying leaf ALS activity in the presence of herbicide, in experiments similar to those shown in Figure 4, by measuring secondary callus growth in the presence of increasing concentrations of herbicide, by monitoring the ability of progeny seeds to germinate and grow in the presence of increasing concentrations of herbicide, and by monitoring plant phytotoxicity after foliar spray applications of herbicide. The results of each of these tests are consistent, yet indicate the need for monitoring resistance by several methods in order to identify those lines most suitable for crop breeding.

Figure 5 shows the percentage of leaf ALS activity in transgenic tobacco plants which is uninhibited by 10 parts per billion of a sulfonylurea herbicide (Classic®), and the ability of secondary callus initiated from these leaves to grow in the presence of the herbicide. Herbicide resistant ALS activity varies between 30 and 70 percent of the total ALS activity. The ability of the secondary callus to grow on particular levels

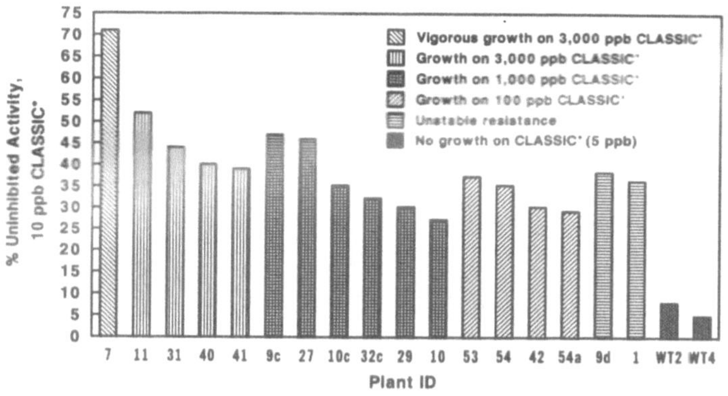

Fig. 5. Resistant ALS activity and secondary callus growth in transgenic tobacco lines. The figure shows herbicide resistant ALS activity in leaf extracts of individual tobacco transformants, assayed in the presence of 10 parts per billion of a sulfonylurea, as a percentage of the activity in uninhibited control reactions. Also shown in the figure is the growth on herbicide of secondary callus tissue initiated from leaves of the transformants.

of herbicide is generally, but not strictly, correlated with the amount of herbicide resistant ALS. The number of loci which produce these levels of herbicide resistance, as determined by segregation analyses of progeny plants produced from self-fertilization, is shown in Table 1. The highest level of resistant ALS activity is found in a plant in which ALS genes are integrated at four loci. Most plants have ALS genes integrated at only one or two sites. The number of loci at which resistant ALS genes are integrated is not the sole factor which affects the degree of resistance of the transformants, however; position effects on gene expression and/or the presence of tandem gene copies at a locus also appear to influence expression. For example, secondary callus from plants 40 and 41 grows equally well in the presence of herbicide, and the plants have equivalent levels of resistant enzyme. Yet, segregation analyses indicate that plant 40 has two resistant ALS loci, while plant 41 has one. It can therefore be inferred that either the ALS gene in plant 41 is integrated in a particularly favorable position for expression, or that tandem copies of the gene are present at the single locus. For plant breeding purposes, a high level of herbicide resistance originating from a single genetic locus is preferred.

As a measure of agronomically useful herbicide resistance, the tobacco transformants were sprayed with sulfonylurea herbicides and evaluated for phytotoxic symptoms. In the experiment shown in Figure 6, foliar sprays were applied at rates corresponding to 0, 4, 8, and 16 grams of herbicide/hectare; a typical field application rate is equivalent to 8 grams/hectare of herbicide. Transformed plants show no damage at application rates as high as 16 grams/hectare, a rate twice that of the field application rate. Wild type plants show damage at an application rate of 4 grams/hectare. Thus, transformation with the *Hra* gene provides a highly effective means of conferring sulfonylurea resistance in crop species.

A similar series of experiments has been carried out on a line of *Arabidopsis* which had been selected for sulfonylurea resistance following seed mutagenesis (Haughn and Somerville, 1986). The resistant ALS gene has been isolated using the wild type *Arabidopsis* ALS gene as a hybridization probe. The resistant gene has been subcloned and introduced into tobacco by *Agrobacterium* mediated transformation. Transformed tobacco plants are

Table 1. Progeny Analysis of Tobacco Transformants

Plant	Resistant/Sensitive[a]	Segregation Ratio[b]
27	129/45	3:1
31	106/35	3:1
54	099/38	3:1
41	231/75	3:1
40	218/18	15:1
07	990/04	255:1

[a]Transgenic herbicide resistant tobacco plants were self-fertilized, and the resulting seeds were germinated in the presence of herbicide. The number of seedlings that grew in the presence of 100 parts per billion of a sulfonylurea herbicide is shown, for representative lines.
[b]A 3:1 (resistant:sensitive) ratio indicates a single genetic locus for herbicide resistance, a 15:1 ratio indicates two loci, and a 255:1 ratio indicates 4 loci.

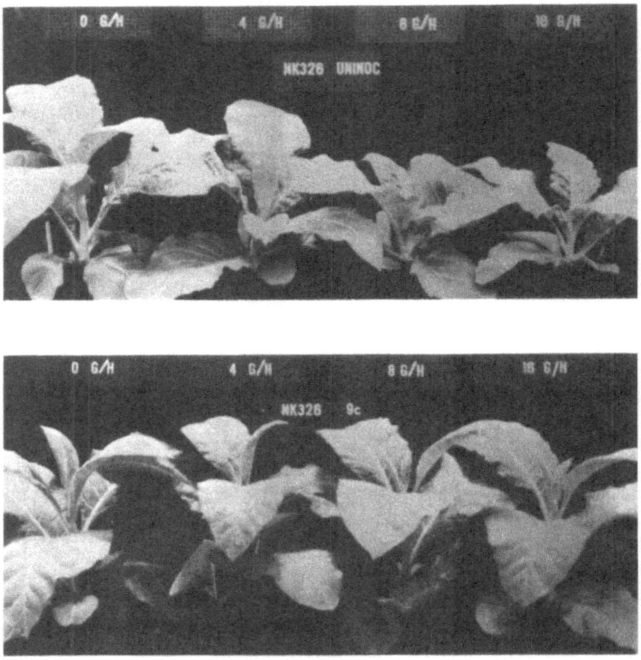

Fig. 6. Foliar applications of herbicide to transgenic and wild type tobacco lines. Commercial cultivars of tobacco, which had been transformed with the *Hra* resistant tobacco ALS gene, were sprayed with a sulfonylurea herbicide at rates corresponding to 0, 4, 8, or 16 grams/hectare. Typical field application rates correspond to the 8 grams/hectare rate. Top: wild type (non-transformed) plants Bottom: transgenic plants with *Hra* ALS gene.

resistant to sulfonylureas, demonstrating that the *Arabidopsis* gene is functionally expressed in a heterologous plant (Haughn et al., 1987).

Characterization of herbicide resistance mutations

The sequences of the mutant ALS genes isolated from the herbicide resistant tobacco and *Arabidopsis* lines have been determined. This analysis has shown that genes from both plants are mutated at a proline residue which had previously been shown to be the site of a herbicide resistance mutation in the yeast ALS gene (Yadav et al., 1986). This proline residue is in one of the conserved regions of ALS. The mutant *Hra* tobacco ALS gene has a substitution of alanine for this proline (Lee et al., manuscript in preparation), while the mutant Arabidopsis ALS gene has a substitution of serine at this position (Haughn et al., 1987). The yeast mutation also causes a serine substitution at this position. The tobacco *Hra* gene has a second mutation in it, in which a tryptophan is substituted with a leucine.

This coincidence of mutations at the proline site, as well as the discovery of a mutation in the yeast ALS gene analogous to the tobacco mutation at the tryptophan site, lead us to postulate that mutations to herbicide resistance might generally be in conserved regions of the ALS protein. We have therefore taken two approaches to exploit this observation. In one approach, mutations have been selected in the yeast ALS gene. In the second approach, herbicide resistance mutations in the *Arabidopsis* ALS gene have been studied in *E. coli*.

To determine the sites at which the yeast ALS gene can mutate to produce a functional yet herbicide resistant enzyme, random spontaneous mutants have been selected in the cloned yeast ALS gene and have been characterized by DNA sequencing. At sites at which spontaneous mutants have been found, all possible amino acid substitutions have been made using oligonucleotide directed mutagenesis. In order to assess the relative merits of these substitutions in conferring useful herbicide resistance, the resulting mutant ALS enzymes have been assayed to determine their activity. These analyses have indicated that many of the mutations have little adverse effect on the catalytic activity of the enzyme. The resistance characteristics of these mutant enzymes have been further elucidated using disc diffusion assays, in order to determine the utility of particular herbicide-mutant enzyme combinations (Falco et al., manuscript in preparation).

Similar studies have been carried out with the *Arabidopsis* ALS gene. In this case, the *Arabidopsis* gene has been functionally expressed in *E. coli*, such that the plant gene complements a branched chain amino acid auxotrophy in the bacteria (Smith et al., manuscript in preparation). This has allowed rapid screening of the herbicide sensitivity of the *Arabidopsis* ALS gene, either by *in vivo* herbicide disc diffusion assays, or by *in vitro* enzyme assays. For example, ALS assays of *E. coli* extracts containing the mutant *Arabidopsis* ALS described above have indicated that the mutant enzyme is completely resistant to the sulfonylurea herbicide Glean®, is less resistant to the sulfonylurea herbicide Classic®, and is sensitive to the imidazolinone herbicide Scepter®. The relative sensitivities of the enzyme to Glean® and Scepter®, as measured in *E. coli* extracts, are consistent with those measured in plant extracts (Haughn and Somerville, 1986). Similar data have been obtained by herbicide disc diffusion assays. These two types of assays thus provide a rapid and facile means of determining the sensitivity of mutant plant enzymes to a variety of herbicides. These assays are particularly useful for quickly assessing the phenotypes of *in vitro* generated ALS mutations, and for designing optimum combinations of mutated genes and herbicides.

CONCLUSIONS

This paper has described the isolation, molecular characterization, and reintroduction into plants of ALS genes. Tobacco and *Arabidopsis* ALS genes have been isolated using a yeast ALS gene as a heterologous hybridization probe. The genes have been sequenced and their primary structures compared; the locations of their chloroplast transit peptide cleavage sites have been determined. The chromosomal organization of the genes has been analysed, and the tissue specificity of their expression monitored. Mutated ALS genes have been isolated from herbicide resistant plants, and have proven to be highly effective in conferring herbicide resistance when reintroduced either into the original species or into heterologous crops. The molecular substitutions in many herbicide resistant mutants have been characterized, and studies are now in progress to determine the most effective gene-herbicide combinations for field usage.

ACKNOWLEDGEMENTS

We would like to acknowledge our many colleagues who have participated in these studies. Our co-workers at Du Pont have included Chok-Fun Chui, Sharon Martin, Ray McDevitt, Mary Hartnett, Tony Guida and Tim Ward on molecular analyses of the ALS genes, Jeff Mauvais, Todd Houser, and Chris Kostow on plant transformations, Gary Fader on greenhouse spray tests, and Newell Bascomb and Ken Leto on chloroplast uptake experiments. Naren Yadav, Joan Odell, Bob LaRossa, Roy Chaleff, John Schloss and Tom Ray have also contributed to this project in many ways, both directly and indirectly. Finally, we have enjoyed collaborations with George Haughn and Chris Somerville at Michigan State University, and with John Bedbrook, Kathy Lee, Jeff Townsend, and Pamela Dunsmuir at Advanced Genetic Sciences, Inc.

REFERENCES

Chaleff, R. S. and Mauvais, C. J., 1984, Acetolactate synthase is the site of action of two sulfonylurea herbicides in higher plants, Science, 224:1443.

Chaleff R. S., and Ray, T. B., 1984, Herbicide-resistant mutants from tobacco cell cultures, Science, 223:1148.

Chaleff, R. S., Sebastian, S. A., Creason, G. L., Mazur, B. J., Falco, S. C., Ray, T. B., Mauvais, C. J. and Yadav, N. S., 1987, Developing plant varieties resistant to sulfonylurea herbicides, in: "Molecular Strategies for Crop Protection," Alan R. Liss, Inc. New York.

Falco S. C., and Dumas, K. D., 1985, Genetic analysis of mutants of *Saccharomyces cerevisiae* resistant to the herbicide sulfometuron methyl, Genetics, 109:21.

Falco S. C., Dumas, K. D. and Livak, K. J., 1985, Nucleotide sequence of the yeast *ILV2* gene which encodes acetolactate synthase, Nucleic Acids Res., 13:4011.

Haughn, G., Smith, J. K., Mazur, B. J. and Somerville, C., 1987, An *Arabidopsis* acetolactate synthase gene in tobacco confers resistance to sulfonylurea herbicides, Mol. Gen. Genet., submitted.

Haughn G., and Somerville, C., 1986, Sulfonylurea-resistant mutants of *Arabidopsis thaliana*, Mol. Gen. Genet., 204:430.

Jones A. V., Young, R. M. and Leto, K., 1985, Subcellular localization and properties of acetolactate synthase, target site of the sulfonylurea herbicides, Plant Physiol., 77:S293.

LaRossa, R. A., Falco, S. C., Mazur, B. J., Livak, K. J., Schloss, J. V., Smulski, D. R., Van Dyk, T. K. and Yadav, N. S., 1987, Microbiological identification and characterization of an amino acid biosynthetic enzyme as the site of sulfonylurea herbicide action, in "Biotechnology in Agricultural Chemistry", LeBaron, H. M., Mumma, R. O., Honeycutt, R. C., and Duesing, J. H., eds., American Chemical Society, Washington, D. C..

LaRossa R. A., and Schloss, J. V., 1984, The sulfonylurea herbicide sulfometuron methyl is an extremely potent and selective inhibitor of acetolactate synthase in *Salmonella typhimurium*, J. Biol. Chem., 259:8753.

LaRossa, R. A., Van Dyk, T. K. and Smulski, D. R., 1987a, Toxic accumulation of α-ketobutyrate caused by inhibition of the branched-chain amino acid biosynthetic enzyme acetolactate synthase in *Salmonella typhimurium*, J. Bacteriol., 169:1372.

Mazur, B. J., Chui, C.-F., Falco, S. C., Mauvais, C. J. and Chaleff, R. S., 1985, Cloning herbicide resistance genes into and out of plants, in "The World Biotech Report 1985," Online International, New York.

Mazur, B. J., Chui, C.-F. and Smith, J. K., 1987, Isolation and characterization of plant genes coding for acetolactate synthase, the target enzyme for two classes of herbicides, Plant Physiol., submitted

Miflin B. J., 1974, The location of nitrite reductase and other enzymes related to amino acid biosynthesis in the plastids of root and leaves, Plant Physiol., 54:550.

Ray, T. B., 1984, Site of action of chlorsulfuron, Plant Physiol., 75:827.

Shaner D. L., Anderson, P. C. and Stidham, M. A., 1984, Imidazolinones (Potent inhibitors of acetohydroxyacid synthase), Plant Physiol., 76:545.

Van Dyk, T. K., Falco, S. C. and LaRossa, R. A., 1986, Rapid physical mapping by transposon Tn5 mutagenesis to localize the cloned yeast *ILV2* gene, Appl. Environ. Microbiol., 51:206.

Yadav N, McDevitt, R. E., Benard, S., and Falco, S. C., 1986, Single amino acid substitutions in the enzyme acetolactate synthase confer resistance to the herbicide sulfometuron methyl, Proc. Nat. Acad. Sci. U.S.A, 83:4418.

EXPRESSION OF COAT PROTEIN GENES IN TRANSGENIC PLANTS CONFERS PROTECTION AGAINST ALFALFA MOSAIC VIRUS, CUCUMBER MOSAIC VIRUS AND POTATO VIRUS X

Nilgun Tumer, Cynthia Hemenway, Keith O'Connell, Maria Cuozzo[1], Rong-Xiang Fang[1], Wojciech Kaniewski and Nam-Hai Chua[1]

Monsanto Company, St. Louis, MO and [1]The Rockefeller University, New York, N. Y.

Cross-protection, a phenomenon in which infection of a plant with one strain of virus protects against superinfection with a second related virus, has been observed with a number of viruses and viroids. Cross-protection has been successfully applied in agriculture and has been effective with a number of viral diseases. Some of the examples include protection of citrus trees from citrus tristeza virus, protection of glasshouse-grown tomatoes from tomato mosaic virus and protection of papaya plants from papaya ringspot virus (Fulton 1986). The use of cross-protection in agriculture, however, is not widespread due to its potential drawbacks. Some of these are that it requires 1) isolation and characterization of an appropriate mild strain, 2) a virus present in the plant could react synergistically with another virus or it might mutate to a more severe form, 3) mild protecting virus may spread to other hosts in which its effects may be more severe, 4) inoculating an entire crop is difficult and costly and 5) the mild strain itself may cause a reduction in yield.

The disadvantages of classical cross-protection can be overcome if cross-protection is achieved in plants by expression of specific viral gene(s) rather than inoculation by the intact virus. Several research groups have successfully introduced viral genes into plants and verified their expression (Bevan et al., 1985, Powell Abel et al., 1986, Baulcombe et al., 1986, Tumer et al., 1987).

Powell Abel et al. (1986) constructed an expression vector containing a cloned cDNA of the coat protein gene of U1 strain of tobacco mosaic virus, introduced it into tobacco cells and regenerated plants. Progeny of self-fertilized transgenic plants expressing high levels of coat protein (0.05 to 0.1% of total soluble protein) either did not develop an infection or developed disease symptoms slower than the control plants after inoculation with the U1 strain of TMV. The plants expressing the coat protein (+CP) were also delayed in symptom development after inoculation with a severe TMV strain, PV230, which is immunologically related to the U1 strain (Nelson et al., 1987).

We have extended the phenomenon of genetically engineered cross-protection to alfalfa mosaic virus (AlMV), potato virus X (PVX), and cucumber mosaic virus (CMV). These viruses differ from TMV in many respects. TMV is a rigid rod shaped virus that encapsidates a single RNA molecule. It encodes at least four proteins in three open reading frames (Hirth and Richards, 1981). In contrast AlMV is a bacilliform-shaped virus that has a tripartite genome consisting of RNA molecules 1, 2 and 3. RNA 4, a subgenomic RNA of RNA 3, encodes the coat protein (Jaspars, 1985). Unlike infection with TMV RNA , which does not require coat protein, infection by AlMV RNA requires addition of RNA 4 or coat protein (Bol et al., 1971; Houwing and Jaspars, 1978). AlMV is transmitted mechanically and by aphids; TMV is primarily mechanically transmitted. Thus, TMV and AlMV are clearly distinguished by their morphology, genome structure, strategies of viral gene expression, early steps in replication and modes of transmission.

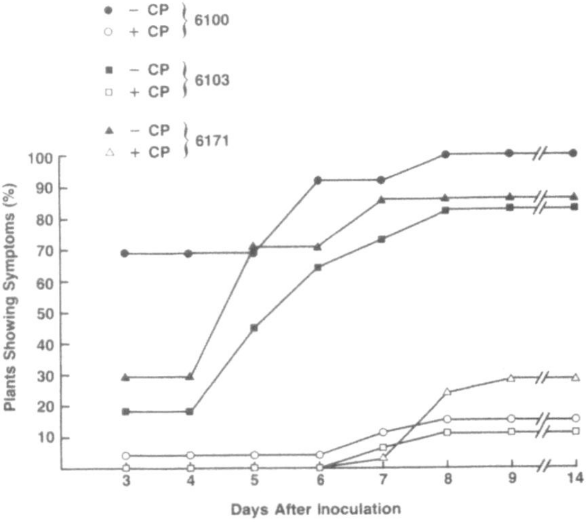

Figure 1: The percentage of plants showing symptoms on inoculated leaves succeeding days after inoculation with ~5 μg/ml of AlMV. The sample sizes were 13 -CP and 27 +CP seedlings from 6100, 11 -CP and 17 +CP seedlings from 6103, 7 -CP and 50 +CP seedlings from 6171. +CP: progeny from self-fertilized transgenic plants that express the coat protein gene. -CP: progeny from self-fertilized transgenic plants that do not express the coat protein gene.

We have cloned RNA 4 encoding the CP of AlMV and engineered it into an expression vector downstream from the CaMV 35S promoter. The cDNA was flanked at the 3' end by the nopaline synthase polyadenylation site. Leaf discs of tobacco and tomato were transformed with A. tumefaciens containing this construct. Transformed cells were selected for kanamycin resistance and regenerated into plants (Horsch et al., 1985; McCormick et al., 1986). The amount of AlMV CP expressed in different transgenic tobacco plants varied between 0.1 to 0.4% of total extractable leaf protein. The amount of of AlMV CP expressed in transgenic tomato plants was between 0.1 to 0.8% of the total extractable leaf protein. The segregation ratio of the CP gene was determined in the seedling progeny of the self-fertilized transgenic plants.

To study the disease development in the +CP and -CP seedlings, the progeny of three different transgenic tobacco plants (6100, 6103 and 6171) were inoculated with 5 μg/ml of AlMV. These plants were held in a growth chamber under conditions that allowed rapid appearance of disease symptoms. Under these conditions, typical symptoms of AlMV infection appeared in 3-4 days on the inoculated leaves of the control seedlings. In contrast, the

+CP seedlings were significantly delayed in symptom development. Of the seedlings that expressed the AlMV CP, 85% of 6100, 89% of 6103 and 72% of 6171 did not develop symptoms on the inoculated leaves 2 weeks after inoculation, while 82-100% of the controls developed symptoms by the end of 2 weeks (Figure 1). The small percentage of +CP plants that showed disease symptoms, did so 4-6 days later than the -CP plants on their inoculated leaves. In addition, the severity of systemic symptoms on +CP plants was attenuated compared with the symptoms observed in -CP plants.

In another experiment progeny from a single transgenic tobacco plant were inoculated with different concentrations of AlMV (20 µg/ml, 10 µg/ml and 5 µg/ml) and symptom production was monitored in a growth chamber. Of the seedlings that expressed the AlMV CP, 80% of the plants inoculated with 20 µg/ml of AlMV and 90% of the plants inoculated with 5 and 10 µg/ml of AlMV did not develop symptoms by the end of two weeks. Decreasing the concentration of AlMV in the inoculum, decreased the numbers of plants that became infected. It also reduced the numbers of lesions that developed and increased the delay in symptom development by one to four days.

Seedlings that were inoculated with 20 µg/ml and 10 µg/ml of crude AlMV suspension were observed over a 45 day period. At the end of 45 days, 81% of +CP progeny inoculated with 10 µg/ml of AlMV and 60% of the +CP progeny inoculated with 20 µg/ml of AlMV did not develop symptoms.

In a similar experiment, the progeny of self-fertilized transgenic tomato plants were inoculated with AlMV. One week after infection, plants that expressed the CP gene did not produce any lesions, while the plants that did not have the CP gene and the wild type plants contained substantially higher numbers of lesions on the inoculated leaves. Most of the plants that did not develop symptoms on the inoculated leaves did not accumulate the virus in the systemic leaves.

We have extended the phenomenon of genetically engineered cross-protection to two additional viruses, potato virus X (PVX) and cucumber mosaic virus (CMV). The filamentous PVX is a type member of the potexvirus group and possesses a single RNA of about 6.4 kb. The 5' end of PVX RNA is capped and the 3' end is polyadenylated (Dolja et al. 1987). A full length cDNA clone encoding the PVX coat protein was reconstructed from a partial PVX CP cDNA clone. The final construct contained 18bp of 5' noncoding region, 657 bp of coding region, 72 bp of 3' noncoding region and 40 bp of poly A tail in pEMBL12. The coat protein gene was then transferred to a plant expression vector downstream from the CaMV 35S promoter. The CP sequences were flanked by the pea small subunit gene E9 polyadenylation site at the 3' end. The CP cDNA was cloned into this vector in both orientations and introduced into tobacco plants. The presence of PVX CP transcripts in transgenic tobacco plants was examined by Northern analysis. A single RNA band of expected size (1.2 kb) was observed in different transgenic plants containing the PVX CP in sense or antisense orientation. The levels of the antisense RNA were lower than the levels of the sense RNA in the transgenic plants. The amount of PVX CP that accumulated in the transgenic plants was determined by immunoblot analysis. The level of the PVX CP in different transgenic plants was about 0.1% of the total soluble protein.

Progeny of self-fertilized transgenic plants 6856 and 6857 that contained the CP gene and progeny of the plants containing the vector sequences alone were inoculated with 5 and 20 µg/ml of PVX, and symptom development was monitored. Chlorotic lesions were observed on the inoculated leaves of the +CP and the vector control plants 6 to 7 days after inoculation; however, the numbers of these lesions were reduced in the +CP plants compared to the vector controls. With higher concentrations of virus (20 μg/ml) greater numbers of lesions were produced on the inoculated leaves and again the numbers of these lesions were reduced in the transgenic plants. Plants containing the vector sequences alone started showing systemic symptoms 8 to 9 days after inoculation, the +CP plants did not show any systemic symptoms up to thirteen days after inoculation. One month after inoculation 100% of the vector controls had systemic symptoms, while only 40 to 50 % of the transgenic plants showed systemic symptoms. In addition, the severity of systemic symptoms in these plants was attenuated compared with the symptoms observed in -CP plants.

Accumulation of PVX in the inoculated and the systemic leaves of the transgenic and the vector control plants was examined by ELISA analysis. Six days post inoculation, the inoculated leaves of the +CP plants contained one half the level of virus present in the inoculated leaves of the vector control plants. Thus the decrease in the accumulation of PVX in the inoculated leaves of the transgenic plants correlated well with the lower number of lesions observed on the inoculated leaves of these plants. The virus accumulation in the first and second systemic leaves of the transgenic plants was inhibited to an even greater extent than in the inoculated leaves. The first systemic leaves of the transgenic plants contained 10 to 25% of the virus present in the first systemic leaves of the vector control plants and the second systemic leaves of the transgenic plants contained 5 to 6% of the virus present in the second systemic leaves of the vector control plants. The decreased number of lesions on the inoculated leaves of the transgenic plants could partially be responsible for the lower levels of virus in the systemic leaves; however there may also be a block in cell to cell spread of the virus, systemic spread of the virus or in the establishment of infections in the upper leaves of these plants.

We have demonstrated that genetically engineered cross-protection can be extended to cucumber mosaic virus (CMV) which is a member of the cucumovirus group. CMV infects cereals, forages, woody plants, vegetables and fruit trees. It is transmitted by aphids in a nonpersistent manner, it can also be mechanically transmitted. The CMV genome consists of three messenger sense RNAs. A fourth RNA, RNA 4 is the subgenomic messenger RNA for the coat protein synthesis. Infection by CMV RNA requires only RNAs 1, 2 and 3. The two largest RNAs, RNA 1 and 2, are encapsidated separately, while RNAs 3 and 4 are packaged together in polyhedral particles (Peden and Symons, 1973, Lot et al., 1974).

A full length cDNA encoding the CP of the D strain of CMV was synthesized from CMV RNA which was enriched for RNA 4 and polyadenylated using *E. coli* poly A polymerase. The complete sequence of the CP was determined and the CP gene was inserted into an expression vector downstream from the CaMV 35S promoter in both sense and antisense orientations. The coat protein sequences were flanked at the 3' end by the pea small subunit E9 polyadenylation site.

The presence of the CMV CP specific transcripts in transgenic tobacco plants was examined by Northern analysis. A single transcript of expected size (1.4 kb) was observed in the transgenic plants. Equal levels of the CP specific transcripts were observed in the plants containing the sense the antisense constructs. The amount of CMV CP that accumulated in the different transgenic plants was examined by immunoblot analysis. There was some variation in the levels of the coat protein present in different transgenic plants and CP was not detected in the transgenic plants that contained the antisense construct.

Progeny of the self-fertilized plants 6631 and 6795 that contained the CMV coat protein gene in sense orientation, 6796 which contained the CMV CP gene in the antisense orientation, 6918 which contained the vector sequences alone and the wild type tobacco plants were inoculated with 1 µg/ml and 5 µg/ml of CMV and symptom development was monitored. The +CP plants that were inoculated with 1 µg/ml of CMV did not show any systemic symptoms three weeks after inoculation, while 80% of the vector control plants and the wild type tobacco plants showed symptoms by the end of three weeks. Only 40% of the plants that contained the CP gene in the antisense orientation showed symptoms three weeks after inoculation with 1 µg/ml of virus. When the virus concentration is increased to 5 µg/ml, 30 to 40% of the +CP plants showed systemic symptoms, while 100% of the controls showed symptoms by the end of three weeks. The plants containing the antisense construct exhibited one to two day delay in symptom production after inoculation with 5 µg/ml of CMV; however 100% of these plants showed systemic symptoms by the end of three weeks. In subsequent experiments, we have increased the inoculum concentration to 25 µg/ml of CMV. Even at this concentration, 40 to 90 % of the +CP progeny did not show symptoms three weeks after inoculation.

Our results with AlMV, PVX and CMV demonstrate that genetically engineered cross-protection is a generally applicable method for conferring disease resistance in plants. An important question which remains to be answered is if the protection observed in transgenic plants containing the viral CP is the same as that found in plants that are cross-protected against virus infection. Although the answer is not known, the two types of protection show similarities. Both are due to a delay in disease development. Some plants escape infection completely and if infection is established, spread of virus is slowed down in the transgenic plants. The protection observed in the transgenic plants is effective against different strains of AlMV, CMV and PVX. Cross-protection is also most effective against viral strains that are closely related and it has traditionally been used for virus identification.

It is still too soon to tell whether the protection observed in transgenic plants and in classical cross-protection results from similar mechanisms. The results described here, however, are consistent with the hypothesis that initial virus infection is prevented in the inoculated leaves of the transgenic plants. It appears that the presence of the free CP in the cell prevents virus uptake into the cell or virus uncoating, since viral RNA can partially overcome the protection observed in transgenic plants (Nelson et al., 1987). The decrease in the numbers of lesions and the delay observed in lesion development on the inoculated leaves of the transgenic plants suggest inhibitory effects at later stages of infection, either reduced rate of multiplication or a block in cell to cell transport of the virus. There could also be a block in systemic spread of the virus or in the establishment of the infection in the upper leaves of the transgenic plants, since the systemic leaves of the +CP plants contain much lower concentration of virus than the systemic leaves of the control plants. Additional experiments are in progress to determine the specific mechanism(s) involved in the protection observed in transgenic plants. The results described here also demonstrate that it is possible to confer protection in transgenic plants by expressing the antisense RNA to the viral CP gene. Protection observed in these plants is effective only against low inoculum concentrations (1 µg/ml CMV) and can be overcome by increasing the inoculum concentration to 5 µg/ml. Additional experiments are in progress to determine the mechanism of the antisense protection in transgenic plants.

References

Bevan, M. W., Mason, S. E., and Goelet, P., 1985, Expression of tobacco mosaic virus coat protein by a cauliflower mosaic virus promoter in plants transformed by Agrobacterium, EMBO Journal, 4:1921.

Bol, J. F., Van Vloten-Doting, L. and Jaspars, E. M. J., 1971, A functional equivalence of top component a RNA and coat protein in the initiation of infection by alfalfa mosaic virus, Virology, 46:73.
Dolja, V. V., Grama, D. P., Morozov, S. Y., and Atabekov, J. G., 1987, Potato virus X-related single and double stranded RNAs, FEBS Letters, 214:308.
Fulton, R. W., 1986, Practices and precautions in the use of cross-protection for plant virus disease, Annual Review of Phytopathology, 24:67.
Hirth, L. and Richards, K. E., 1981, Tobacco mosaic virus: Model for structure and function of a simple virus. Advances in Virus Research 26:145.
Horsch, R. B., Fry, J. B., Hoffman, N. L., Eicholtz, D., Rogers, S. G., and Fraley R. T., 1985, A simple and general method for transferring genes into plants, Science 227:1229.
Houwing, C. J. and Jaspars, E. M. J., 1978, Coat protein binds to 3' terminal part of RNA 4 of alfalfa mosaic virus, Biochemistry, 17:2927.
Jaspars, E. M. J., 1985, Interaction of alfalfa mosaic virus nucleic acid and protein, in: "Molecular Plant Virology," J. W. Davies, ed., Volume 1, p. 155, CRC Press, Boca Raton.
Lot, H., Marchoux, G., Marrou, J., Kaper, J. M., West, C. K., Van Volten-Doting, L., and Hull, R., 1974, Evidence for three functional RNA species in several strains of cucumber mosaic virus, 1974, J. Gen. Virol., 22:81.
McCormick, S., Niedermeyer, J., Fry, J., Barnason, A., Horsch, R., and Fraley, R., 1986, Leaf disc transformation of cultivated tomato (L. esculentum) using Agrobacterium tumefaciens, Plant Cell Reports, 5:81.
Nelson, R. S., Powell Abel, P. and Beachy, R., 1987, Fewer viral infection sites in transgenic tobacco plants expressing the coat protein gene of tobacco mosaic virus, Virology, 158:126.
Peden, K. W. C. and Symons, R. H., 1973, Cucumber Mosaic Virus Contains a Functionally Divided Genome, Virology, 53-487.
Powell Abel, P., Nelson, R. S., De, B., Hoffman, N., Rogers, S. G., Fraley, R. T. and Beachy, R. N., 1986, Delay of disease development in transgenic plants that express the tobacco mosaic virus coat protein gene, Science, 232:738.
Tumer, N. E., O'Connell, K. M., Nelson R. S., Sanders, P. R., Beachy, R. N., Fraley, R. T., and Shah, D. M., 1987, Expression of alfalfa mosaic virus coat protein gene confers cross-protection in transgenic tobacco and tomato plants, EMBO Journal, 6:1181.

MUTATIONS RESISTANT TO PHOTOSYSTEM II HERBICIDES

Joseph Hirschberg, Adi Ben Yehuda, Iris Pecker and Nir Ohad
Department of Genetics
The Hebrew University of Jerusalem
Jerusalem, Israel

Dedicated to the memory of professor Menashe Marcus

Herbicide inhibition of photosystem II electron transport

Photosynthetic electron transport takes place in the thylakoid membranes of the chloroplasts. Several classes of compounds such as triazines, triazinones, pyridazinones, ureas and nitrophenols, inhibit electron transport in the thylakoid-membrane-bound photosystem II (PS II) [1]. The light-driven electron transport in PS II from H_2O to plastoquinone (PQ) is illustrated in the following scheme.

$$H_2O \rightarrow Z \rightarrow p680 \rightarrow \text{pheophytine} \rightarrow Q_A \rightarrow Q_B \rightarrow PQ$$

Z - a transient intermediate (possibly a quinone); P680 - reaction-center chlorophyll a; Q_A and Q_B - the primary and secondary quinone acceptors.

These electron carriers are part of the PSII core complex which consists of five membrane-bound protein subunits: D1 - the 32KDa apoprotein of Q_B; D2 - the 34 KDa apoprotein of Q_A; CP43 and CP47 - the 43KDa and 47KDa chlorophyll-binding polypeptides, and the 10KDa-apoprotein of cytochrome b559 [for recent review see 2]

PSII-herbicides inhibit the electron flow from Q_A to Q_B [2] by binding to the thylakoid membrane [3] at a stoichiometry of one molecule per electron transport chain [4], which results in a quinone displacements [5,6]. Direct competition between herbicides and quinone and among different herbicides for the same binding site, has been demonstrated [7,8].

Since mild trypsinization of thylakoid membranes results in a concomitant loss of herbicide binding and inhibition of electron flow [9,10], it was postulated that the binding site is a protein component the of PSII complex. This protein was identified in the laboratory of C. Arntzen by photoaffinity labelling of thylakoid membranes with ^{14}C-azido atrazine [11]. After UV-irradiation, azidoatrazin, as well as other azido-substituted herbicides [12,13], were found to bind covalently to a 32KDa polypeptide identified as the D1-protein of PSII [14,15].

The D1 protein has been studied extensively in various plants [16]. It has been shown to have a very rapid turn-over in the light relative to other protein in the thylakoid membrane [17,18]. Its degradation and synthesis are light regulated and appear to be a consequence of light induced damage [19]. This feature of D1 probably reflects the invovlement of the protein in other functions of PSII in addition to binding Q_B.
The D1 polypeptide is encoded by the chloroplast gene psbA [20]. It is synthesized within the chloroplast as a 34KDa precursor polypeptide [17,21,22] which is assembled into PSII complexes following processing and postranslational acylation [23].
The psbA gene of several higher plants, algae and cyanobacteria has been cloned and sequenced. The deduced amino acid sequence indicates that D1 is a hydrophobic protein of 353 amino acid residues in higher plants, 352 in Chlamydomonas, 345 in Euglena and 360 in Cyanobacteria [reviewed in 24]. Comparative analysis revealed a remarkable conservation in the deduced amino acid sequence of D1 - 98% homology among higher platns and 85-90% among all species.

The molecular basis of herbicide resistance

The first triazine - resistant mutant was reported in 1970 in the weed common groundsel (Senecio vulgaris L.) [25]. Following an extensive use of herbicides, many more mutants have since been found among 38 weed species [26]. All of the mutants thus far analysed exhibit a similar phenotype: high resistance (over 100 fold) to s-triazines i.e. atrazine and terbutryne, and no resistance to diuron - a urea-type herbicide. Triazine resistance is maternally inherited [27].

Electron transport in isolated thylakoids of resistant plants is unaffected to added atrazine. Although D1 protein is present in chloroplasts of the resistant plants, its binding affinity to s-triazines is greatly reduced compared to the wild type [15]. Thus it was suggested that a subtle change in the herbicide-binding protein - D1, results in a selective loss of affinity to triazines without loss of function in electron flow.

In order to characterize this modification in D1 protein, the psbA gene was cloned from triazine-resistant and triazine - susceptible (wild type) biotypes of the weed Amaranthus hybridus [28]. DNA sequence analysis revealed a nucleotide change in the resistant mutant which resulted in a single amino acid substitution at position 264 in the D1 polypeptide from serine to glycine. The same substitution has also been detected in two atrazine-resistant biotypes of Solanum nigrum [29,30], in Brassica campestris (after transfer of the gene by sexual crosses to B. napus) [31] and recently in Phalaris paradoxa [32 and this report]

Similar mutations in psbA, that change the serine residue at 264 to alanine were found also in diuron-resistant mutants of the algae Chalmydomonas reinhardii [33] and Euglena gracilis [34] and the cyanobacterium Synechococcus R2 (Anacystis nidulans R2) [35,36].

Mutations in four other sites in psbA, which have been identified in Chlamydomonas reinhardii, were found to confer a different phenotype of herbicide cross-resistance. These mutations include the following amino acid changes: Valine 219 to isoleucine [33], alanine 251 to valine [37], phenylalanine 255 to thyrosine [33] and leucine 275 to phenylalanine [Erickson and Rochaix personal communication]. All of the above mutations and their phenotypes are summarized in Table 1.

Recently we have isolated and characterized a variety of herbicide-resistant mutants in the cyanobacterium Synechococcus R2 [38]. PSII structure and function in this organism is essentially similar to higher plants. Yet their prokaryotic genetic nature enables isolation of a large number of mutants and allows for genetic manipulations such as DNA - mediated genetic transformation [39]. Synechococcus R2 has three psbA genes [40]. The deduced amino acid sequences of the D1 proteins are 85 -90% homologous to higher plants [36,40]. One copy - psbA-"copyI" was found to account for 95% of psbA transcripts [40]. This is probably the reason why all the herbicide resistant mutations that have been characterized so far were found in psbA-copyI [35, 36].

Table 1. Mutations in D1 conferring herbicide resistance. For references see text.

		Relative resistance			
Species/mutant	Change in D1	Atrazine	Metribuzin	Diuron	Bromacil
C. reinhardii; Dr2	219 Val → Ile	2	15	1	1
C. reinhardii; MZ2	251 Ala → Val	25	1000	5	
C. reinhardii; Ar207	255 Phe → Tyr	15		0.5	1
A. hybridus	264 Ser → Gly	1000	200	1	
S. nigrum	264 Ser → Gly	1000		1	20
B. campestris	264 Ser → Gly	600	50	1	
P. paradoxa	264 Ser → Gly	>120	23	1.3	62
C. reinhardii; DCMU4	264 Ser → Ala	100		10	
E. gracilis	264 Ser → Ala				
Synechococcus R2; R2D2-X1, Di1	264 Ser → Ala	17	5000	150	300
C. reinharddi;	275 Leu → Phe			5	4
Synechococcus R2; D5	264 Ser → Ala 255 Phe → Tyr	360	200	300	330
Synechococcus R2;Di22	264 Ser → Ala 255 Phe → Leu	2.5	175	2650	35

In addition to the single mutation in serine 264, we have selected two double mutants -D5 and Di22 (table 1). In both of them a serine to alanine substitution at position 264 in the D1 protein is accompanied by an additional change at position 255. When the phenylalanine at 255 is changed to tyrosine in the mutant D5, the phenotype is high resistance to atrazine and diuron (table 1); but when changed to leucine in mutant Di22, resistance to atrazine is almost completely abolished and diuron resistance is increased dramatically (table 1). This result suggests that different herbicides interact with the D1 polypeptide at slightly different binding niches. Several amino acids are invovled in this interaction as can be seen in mutant Di22, where leucine at position 255 instead of phenylanine (or tyrosine), restores atrazine binding which was lost by the serine to alanine change at 264. A model of two receptor binding site has been proposed to explain herbicide cross resistance in other mutants[41].

The fine structure of the herbicide resistant site

Current models of the architecture of PSII complex are based on a wealth of new information about the structural organization of the photosynthetic reaction center of purple bacteria [42,43]. There are several analogies in the chemical nature between bacterial photosystem and PSII. Both contain a reaction center chlorophyll, pheophytin, Q_A and Q_B as electron carriers. Three integral membrane proteins - H, M and L, exist in the photosynthetic reaction center of purple bacteria. M is the apoprotein of Q_A while Q_B binds to L[44]. There is homology in the amino acid sequence between M and L and both

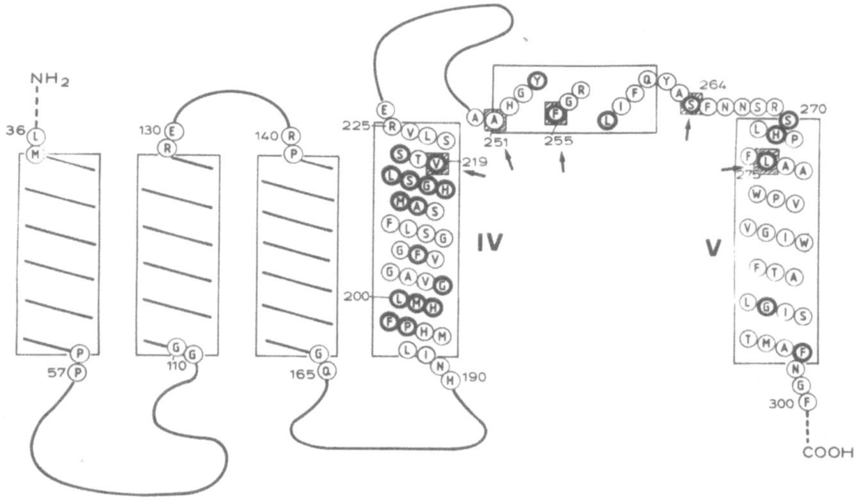

Fig. 1 A model for the topology of D1 in the thylakoid membrane. Hydrophobic regions are boxed. Amino acids are shown in transmembrane helices IV and V and in the parallel helix, and at the borders of the transmembrane regions. Thick circles idicate amino acid residues that are conserved between D1 and L subunit of R. capsulata. Arrows mark amino acids that are changed in herbicide resistant mutants. After A. Trebst [48,49]

share some sequence homology to the D1 and D2 of proteins of PSII [43,44]. In addition, triazines inhibit electron transport from Q_A to Q_B also in the purple bacteria, and the herbicide binding site was found to be the L subunit[45]. Recently, herbicide resistant mutations were found to change L subunit in a similar way to D1 [45,46,46]. These similarities in function of the L and M subunits to the D1 and D2 proteins have led A. Trebst to propose a model [48] for the topology of D1 and D2 in the thylakoid membrane, which is based on the structure of L and M that has been determined by X-ray analysis of crystallized bacterial photosystem [42]. According to the model, D1 has 5 membrane-spanning helices and one short helix parallel to the menbrane [49] (figure 2). Similarities in the folding of L and D1 and conserved amino acid residues suggest the possible involvement of histidin 215, phenylalanine 255 and serine 264 in the binding of Q_B. The amino acid residues in D1 that are involved in herbicide resistance are located in close vicinity near the parallel helix, thus indicating the herbicide binding niche. Iterestingly, all the triazine-resistance mutations in the purple bacteria occur in the homologous region in L subunit [45 - 47].

Yield penalties in herbicide resistant crops?

Introducing triazine resistance into crop plants is of great significant agronomical importance. This has been achieved in Canada where triazine resistance has been transferred by sexual crosses from the weed Brassica campestris into several Brassica crops of commercial importance: oil rapeseed, rutabga and Chineses cabbage, [50].

However, in all the triazine resistant biotypes of higher plants and also in mutant DCMU4 of C. reinhardii, the resistant phenotype was always accompanied by a slower electron transport from Q_A to Q_B [51] and lower quantum yield of CO_2 reduction [52]. The reduced photosynthetic capacity due to slower Q_A to Q_B electron transport was detected even after six generations of backcrossing of the resistant biotype to an isogenic nuclear background [53]. Other alterations in the photosynthetic apparatus, which have also been reported in herbicide-resistant mutants, include: changes in lipid composition of the chloroplast membranes [54-56], lower chlorophyll a/b ratio [57] and changes in chloroplast ultrastructure [58]. Reduced fitness, as measured by accumulation of dry matter, seed germination etc., has also been reported in triazine-resistant mutants [26,59].

Recently, a triazine-resistant biotype of hood cannarygrass (Phalaris paradoxa) has been identified in Israel [60]. The resistant biotype of this species was found to be equal or superior to the susceptible biotype for several photosynthetic and growth parameters [32,61] (table 2).

We have cloned and sequenced the psbA genes from resistant and susceptible biotypes of P. paradoxa. The deduced amino acid sequence of the D1 protein in this species is 98-99% homologous to other higher plants. However, it differs from all other plants and algae in having a lysine residue, instead of argenine, at position 238. Like other atrazine - resistant mutants, the only difference found in D1 protein between the two biotypes was a serine to glycine substitution at position 264. It is possible that lysine at position 238 reduces the detrimental effects of the serine to glycine change at position 264. However, slower electron flow from Q_A to Q_B in the resistant mutant was still apparent in low light intensities [61]. Recently, a triazine

Table 2. Photosynthetic and growth parameters ratios for triazine resistant (R) and triazine susceptible (S) biotypes of *Phalaris paradoxa* [32, 61].

Parameter	R/S
Electron transport at low light intensity (50 µE m^{-2} s^{-1})	0.84
electron transport at saturating light intensity (>1000 µEm^{-2} s^{-1})	1.21
Quantum yield for electron transport	0.70
CO_2 uptake at light saturation	1.05
Quantum yield for CO_2 uptake	0.83
Plant dry weight	1.06
Main spike dry weight	1.17
Seed germination	1.3
Seedlings emergence	1.83

- resistant biotype of *Chenopodium album* was also reported to be superior to the susceptible biotype in different parameters of growth performances [62] but the molecular basis of this mutation is yet uknown.

It is concluded that chloroplast-encoded triazine resistance does not necessarily lead to loss of vigor. The differences in photosynthetic and growth performances between biotypes could arise from a natural polymorphism in chloroplast genes other than psbA, which are inherited maternally along with the herbicide - resistant psbA.

The genetic nature of herbicide resistance

Triazine and urea-type of herbicide resistance is a chloroplastic trait. The frequency of occurance of chloroplst-encoded herbicide resistance in higher plants was estimated to be 1:10^9-10^{11} [63]. Such a low frequency does not allow the isolation of mutant plants by a classical mutagenesis and selection procedure. Chloroplast mutations for antibiotic resistance have been selected in *Nicotiana* following mutagenes and selection in tissue culture [64]. Because of the nature of herbicide action namely, inhibition of photosynthesis, herbicides are not suitable for selection under regular tissue culture conditions where cell proliferation is not photosynthesis dependent. However, triazine resistant mutants have been isolated in photomixotrophic cultures of *Nicotiana* [65].

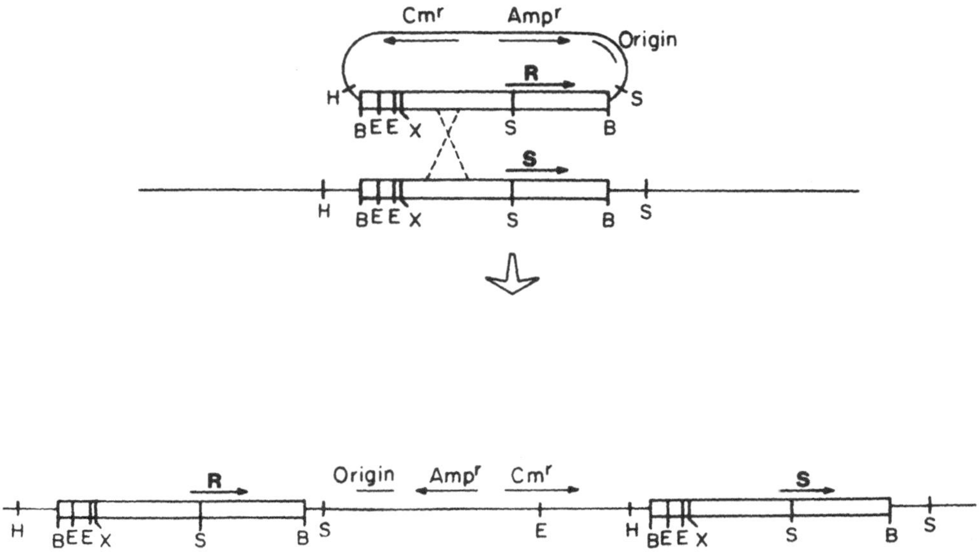

Fig. 2 Construction of a strain of Synechococcus R2 heterozygous for atrazine resistance. A pBR328 plasmid carrying the psbA gene from mutant D5 is used transformed into herbicide suceptible strain of Synechococcus R2. Following homologous recombination, the plasmid integrates into the bacterial chromosome. Boxed regions mark a 3.5 kb fragments that contains the psbA gene. R - resistant, S - sensitive.

Since triazine resistance is a single-gene trait it can be transferred to crop plants by genetic-engineering manipulation. Reliable procedures for stable transformation of plastids are not available yet but when such procedures are developed, it would be necessary to understand the phenotypic expression of the triazine resistant gene.

In order to determine the dominance - recessiveness nature of triazine resistance, we have constructed a strain of synechococcus R2 so that it became partially diploid to psbA-copyI and heterozygous to triazine resistance (figure 2). Several lines of evidence demonstrated that this strain contained two functional psbA genes of which only one was the herbicide-resistant allele [66]. Sensitivity to herbicide of this strain was similar to the susceptible wild-type, indicating that triazine resistance is a recessive trait in Cyanobacteria. The recessive nature of atrazin resistance in Chlamydomonas has been described by Metz et al. [67] and is discussed in ref. 68. A contradicting result namely, that diuron resistance is dominant in Synechococcus R2, has been reported by Brusslan et al. [69].

Acknowledgement

We thank Ana Rahat for excellent technical assistance. This Research was supported by a grant from NCRD Israel and GSF Munchen, Germany.

REFERENCES

1. A. Trebst, and W. Draber, in: "Advances in Pesticide Science" Vol. 2, Geissbuhler ed., Pergamon Press, N.Y.(1979) pp. 223-234.
2. P. Mathis, in: "Progress in Photosynthesis Research", J. Biggens ed., Martinus Nijhoff Publishers, Dordrecht, (1987) Vol.I pp. 151-160.
3. W. Tischer, and H. Strotmann, Biochim. Biophys. Acta 460, 113-125 (1977).
4. K. Pfister, S.R. Radosevich, and C.J. Arntzan, Plant Physiol. 64, 995-999 (1979).
5. B.R. Velthuys, FEBS Lett. 126, 277-281 (1981).
6. C.A. Wraight, Israel J. Chem. 21, 348-354 (1981).
7. J. Lavergne, J. Biochim Biophys Acta 679, 12-18 (1982).
8. W.F.G. Vermas, C.J. Arntzen, L. -Q. Gu. and C. A. Yu, Biochim. Biophys. Acta 723, 266-275 (1983)..
9. J.E. Mullet, and C.J. Artnzen, Biochim. Biophys. Acta 589, 100-117 (1980).
10. K.J. Leto, A. Kerensztes, and C.J. Arntzen. Plant Physiol. 69, 1450-1458 (1982).
11. K. Pfister, K.E. Steinback, G. Gardner, and C. J. Arntzen, Proc. Natl. Acad. Sci. USA 78, 981-985 (1981).
12. W. Oettmeier, K. Masson and U. Johanningmeier, Biochim, Biophys. Acta 679, 376-383 (1982).
13. A. Boschetti, M. Tallenbach, and A. Gerber, Biochim. Biophys. Acta 810, 12-19 (1985).
14. A.K. Matoo, U. Pick, H. Hoffman-Falk, and M. Edelman. Proc. Natl. Acad. Sci. USA 78, 1572-1576 (1981).
15. K. E. Steinback, L. McIntosh, L. Bogorad, and C. J. Arntzen, Proc. Natl. Acad. Sci. USA 78, 7463-7467 (1981).
16. D.J. Kyle, Photochem. Photobiol. 41, 107-116 (1985).
17. A. Reisfeld, A.K. Matoo, and M. Edelman, Eur. J. Biochem. 124, 125-129 (1982).
18. A.K. Matoo, H. Hoffman-Falk, J. B. Marder, and M. Edelman, Proc. Natl. Acad. Sci. USA 81, 1380-1384 (1984).
19. I. Ohad, D.J. Kyle, and J. Hirschberg, The EMBO J. 4, 1655-1654 (1985).
20. L. Bogorad, S.O. Jolly, G. Link, L. McIntosh, C. Poulsen, Z. Schwartz, and A. Steinmetz. in: "Biological Chemistry of Organelle Formation", T. Bucher, W. Sebald and H. Weiss eds., Springer, Berlin, (1980) pp. 87-96.
21. R.J. Ellis, Ann. Rev. Plant Physiol. 32: 111-137 (1981).
22. Grebanier, A.E., Coen, D.M., Rich, A. and Bogorad, L. J. Cell Biol. 78, 734-746 (1978).
23. F.E. Callahan, M. Edelman, and A. K. Matoo, in:"Progress in Photosynthesis Research", J. Biggens ed. Vol. III, Martinus Nijhoff Publishers,(1987) pp. 799-802.
24. J.M. Erickson, J.-D.Rochaix, and P. Delepelair, in:"Moelcular Biology of the Photosynthetic Apparatus", K.E. Steinback, S. Bonitz, C.J. Arntzen and L. Bogorad eds., Cold Spring Harbor Laboratory, (1985) pp. 53-65.
25. G.F. Ryan, Weed Sci. 18, 614-616 (1970).
26. J. Gressel, in: "Weed Physiology". S.E. Duke ed., CRC Press, Boca Raton,(1985) pp. 159-189.
27. V. Souza-Machado in: "Herbicide Resistance in Plants", H. M. LeBaron and J. Gressel eds., John Wiley & Sons, New York, (1982), pp. 275-273.
28. J. Hirschberg, and L. McIntosh, Science 222, 1346-1349 (1983).
29. J. Hirschberg, A. Bleecker, D. J. Kyle, L. McIntosh, and C. J. Arntzen, Z. Naturforsch 39c, 412-420 (1984).

30. P. Goloubinoff, M. Edelman and R.B. Hallick, Nuc. Acid. Res. 24, 9489-9496 (1984).
31. M. Reith, and N.A. Straus, Theor. Appl. Genet. 73, 357-363 (1987).
32. M. Shonfeld, T. Yaacoby, A. Ben Yehuda, B. Rubin, and J. Hirschberg, Z. Naturforsch. 42C, 88-91 (1987).
33. J.M. Erickson, M. Rahire, J.-D. Rochaix, and L. Metz, Science 228, 204-207 (1985).
34. U. Johaningmeier, and R.B. Hallick, unpublished (quoted in 30).
35. S.S. Golden, and R. Haselkorn, Science 229, 1104-1107 (1985).
36. N. Ohad, I. Pecker, and J. Hirschberg, in: "Progress: Photosynthesis Research" J. Biggens ed., Martinus Nijhoff Publishers, (1987) Vol.III pp. 807 - 810.
37. U. Johanningmeier, U. Bodner, and G.F. Wildner, FEBS Letters 211, 221-224 (1987).
38. J.Hirschberg, N., Ohad, I., Pecker, and A. Rahat Z. Naturforsch, 42C 109-112 (1987).
39. J.G.K. Williams, and A.A. Szaley, Gene 24: 37-51 (1983).
40. S.S. Golden, J. Brusslam, and R. Haelkorn. The EMBO J. 11, 2789-2798 (1986).
41. A. Thiel and P. Böger, Pesticide Biochem. Physiol. 22, 232-242 (1984).
42. J. Deisenhofer, O. Epp., K. Miki, R. Huber and H. Michel Nature 318, 618-624 (1985).
43. H. Michel, and J. Deisenhofer in: "Progress in Photosynthesis Research" J. Biggens ed., Martinus Nijhoff Publishers, (1987) Vol.I pp. 353-362.
44. D.C. Youvan, E.J. Bylina, M. Alberti, H. Begusch and J.H. Hearst, Cell, 37, 949-957 (1984).
45. C.W. Gilbert, J.G.K. Williams, K.A. Williams, and C.J. Arntzen, in: "Molecular Biology of the Photosynthetic Apparatus" K.E. Steinback, S. Bonitz, C.J. Arntzen and L. Bogorad eds., Cold Spring Harbor Laboratory, (1985) pp. 67-71.
46. I. Sinning, and H. Michel. in: "Progress of Photosynthesis Research" J. Biggens, ed.,Marthinus Nijhoff Publishers, (1987) Vol. III pp 771-773.
47. M.L. Paddock, J.C. Williams, S.H. Rongey, E.C. Abresch, G. Feher, and N.Y. Okamura ibid Vol III 775-778.
48. A. Trebst, Z. Naturforsch. 41C, 240-245 (1985).
49. A. Trebst and W. Draber, Photosynthesis Res. 10, 381-392 (1986).
50. W.D. Beverdorf, J. Weiss-Lerman, L.R. Erickson Crop Science, 20, 284 (1980).
51 C.J. Arntzen, K. Pfister, and K.E. Steinback, in: "Herbicide resistance in plants", H. LeBaron and J. Gressel, eds., John Wiley and Sons, new York, pp. 185-214 (1985).
52. D.R. Ort, H. Ahrens, B. Martin, and E.W. Stoller. Plant Physiol. 72, 925-930 (1983).
53. A. Ali, P.E. Fuerst, C.J. Arntzen and V. Souza-Machado, Plant Physiol. 80, 511-514 (1986).
54. P. Pillali and J.B. St. John, Plant Physiol 68, 585-587 (1986).
55. L. Lehoczki, E. Polos, G. Laskay and T. Farkas, Plant Science 42, 19-24 (1982).
56. D.J. Chapman, J. De-Felica and J. Barber, Planta 166, 280-285 (1985).
57. J.J. Burke, R.F. Wilson and J.R. Swalford, Plant Physiol. 70, 24-29 (1982).
58. K.C. Vaughn and S.O. Duke, Plant Physiol. 62, 510-520 (1984).
59. S.G. Conrad and S.R. Radosevich J. Appl. Ecol. 16, 171-177 (1979).
60. T. Yaacoby, M. Schonfeld, and B. Rubin Weed. Sci. 34, 181-184 (1986).

61. M. Schonfeld, T. Yaacoby, O. Michael and B. Rubin Plant Physiol. 83, 329-333 (1987).
62. M.A.K. Jansen, J.H. Hobe, J.C. Wesselius, and J.J.S. Van Rensen, Physiol. Veg. 24, 475-484 (1986).
63. C.J. Arntzen, and J.H. Deusing in: "Advances in Gene Technology: Molecular Genetics of Plants and Animals", Miami Winter Symposium Vol. 20., Academic Press, New York (1984). pp. 273-294.
64. R. Fluhr, D. Aviv. E. Galun, and M. Edelman. Proc. Natl. Acd. Sci. 82, 1485-1489,(1985).
65. A. Cseplo, P. Medgyesy, E. Hideg, S. Demeter, L. Marton and P. Maliga. Mol. Gen. Genet. 200, 508-510 (1985).
66. I. Pecker, N. Ohad and J. Hirschberg in: "Progress in Photosynthesis Research" J. Biggens ed, Martinus Nijhoff Publishers, (1987) Vol. III pp. 811-814.
67. L.J. Metz, R.E. Galloway, R.E. and J.E. Erickson in: "Biotechnolgoy in Plant Science" M. Zaitlin, P. Day and A. Hollaender eds., Academic Press, new York (1985) pp. 301-312.
68. D. Robertson, Plant Mol. Biol. Reporter 3, 99-106 (1985).
69. J. A. Brusslan, S.S. Golden and R. Haselkorn. in:"Progress in Photosynthesis Research" J. Biggens ed. Martinus Nijhoff Publishers, Dordrecht,(1987) Vol. IV pp. 821-824.

DIURON RESISTANCE IN THE *psbA* MULTIGENE FAMILY OF *Synechococcus* PCC7942

Judy Brusslan and Robert Haselkorn

Department of Molecular Genetics and Cell Biology

University of Chicago, Chicago, IL 60637

The cyanobacteria comprise a vast and diverse group of microorganisms, some of which are adapted to growth in the most hostile environments on earth. All cyanobacteria carry out photosynthesis like that of green plants, using two photosystems and evolving oxygen. We will focus on the use of one cyanobacterial strain, *Synechococcus* PCC7942 (*Anacystis nidulans* R2), to analyze the mode of action of a class of herbicide.

Modern herbicides act selectively on plants by inhibiting reactions uniquely found in chloroplasts: aromatic amino acid biosynthesis, branched chain aliphatic amino acid biosynthesis, carotenoid synthesis, or the electron transfer steps of photosynthesis itself (1). Green plant photosynthesis requires two separate chlorophyll-containing protein assemblies embedded in a membrane. The absorption of light energy by a special chlorophyll molecule initiates a series of electron transfers that results in the reduction of pyridine nucleotide, the oxidation of water, and the translocation of protons across the membrane. One of the key protein assemblies, termed the photosystem II reaction center, includes an integral membrane protein called D1 or the QB protein or the 32 kDa protein (2). One function of this protein is to provide a binding pocket for plastoquinone (Q_B), the ultimate acceptor of electrons from the special chlorophyll. Quinones exchange between this pocket and the lipid phase of the membrane, providing the mobile carrier that creates the proton gradient. The herbicides diuron (DCMU) and atrazine bind to this pocket, prevent the binding of plastoquinone to the D1 protein, and thereby inhibit electron flow from photosystem II (3,4).

Atrazine and diuron have been sprayed widely to control the growth of weeds. Inevitably, herbicide-resistant plants appeared. Some of these plants have been analyzed by cloning and sequencing the chloroplast DNA fragment containing the *psbA* gene, which encodes the D1 protein, from both herbicide-resistant and sensitive plants. In each case it was found that a change in the nucleotide sequence, corresponding to replacement of serine at position 264 of the wild type protein with alanine or glycine, is correlated with herbicide resistance (5,6,7). As far as plants are concerned, this observation remains a correlation, since transformation with chloroplast DNA has not yet been achieved.

Proof that the ser -> ala change at position 264 of the D1 protein is sufficient to confer diuron resistance was provided by Susan Golden's experiments with Synechococcus (8). This organism can be transformed with linear or circular DNA. Linear fragments can be as small as 700 bp, although the efficiency of transformation with such fragments is improved at least 1000-fold by incorporating them into circular molecules, e.g. by cloning into pBR328. The latter plasmid does not replicate in Synechococcus, but it survives long enough to donate the homologous fragment through a double exchange or to be incorporated into the chromosome via a single exchange (Figure 1).

Efficient homologous recombination between transforming DNA and resident chromosomes, as shown in Figure 1, makes it possible to replace, duplicate, or inactivate genes. The first example of these procedures is a simple replacement (8). Wild type Synechococcus cells were mutagenized with nitrosoguanidine and then clones resistant to diuron were selected. DNA from one resistant line was used to prepare a library of recombinant fragments in bacteriophage lambda. From this library, a clone was selected on the basis of homology with a DNA probe containing the psbA gene from spinach chloroplasts. DNA from this clone could transform wild type Synechococcus to diuron resistance. The cloned DNA was digested with restriction enzymes to produce a number of small fragments, one of which, the 700 bp TaqI fragment mentioned above, retained the biological activity of transforming Synechococcus to diuron resistance. When this piece of DNA was sequenced and compared to the corresponding DNA from wild type Synechococcus, we found a single base pair changed in the mutant DNA. This change resulted in the conversion of serine 264 in the wild type D1

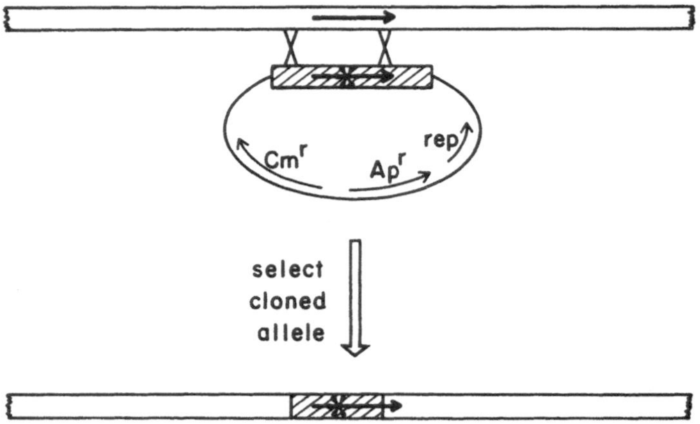

Fig. 1. Mechanism of gene replacement by transformation of Synechococcus with cloned DNA fragments. The open bar represents the chromosome of the recipient cell with the coding region of a gene (arrow). The circle represents pBR328 containing, in addition to antibiotic resistance genes, a cloned fragment of Synechococcus DNA (shaded) which includes the coding region of the same gene (arrow) cloned from a herbicide-resistant mutant. The mutation is represented by an asterisk. Transformation with plasmid DNA, selecting for herbicide resistance, yields transformants resulting from double exchanges of the type shown. Selection for one of the plasmid-borne antibiotic resistance genes yields transformants resulting from single exchanges, in which the entire plasmid is incorporated into the chromosome and the shaded region of Synechococcus DNA is duplicated.

protein to alanine in the mutant D1 protein. These experiments provide the genetic proof that the serine to alanine mutation in the D1 protein is sufficient to confer herbicide resistance in Synechococcus (8).

In the course of these experiments we found that Synechococcus contains three different DNA fragments each of which is related in sequence to the psbA gene of spinach (9). Earlier work from our laboratory had reported the existence of multiple copies of the psbA gene in Anabaena, a different cyanobacterium (10). In that work it was not possible to draw firm conclusions about the function of the multiple genes because the genes could not be mutated individually. In Synechococcus, however, the transformation system provided the opportunity to answer questions of function unequivocally. Each of the three copies of the psbA gene in Synechococcus was cloned, sequenced, and the start site of transcription determined by S1 nuclease protection (9).

Our current information about the three genes in wild type Synechococcus can be summarized as follows: each open reading frame contains 1080 nucleotides and each transcript starts about 50 nucleotides before the open reading frame. Nucleotide homologies among the three genes are: psbAI-psbAII, 91.2%; psbAI-psbAIII, 91.3%; psbAII-psbAIII,

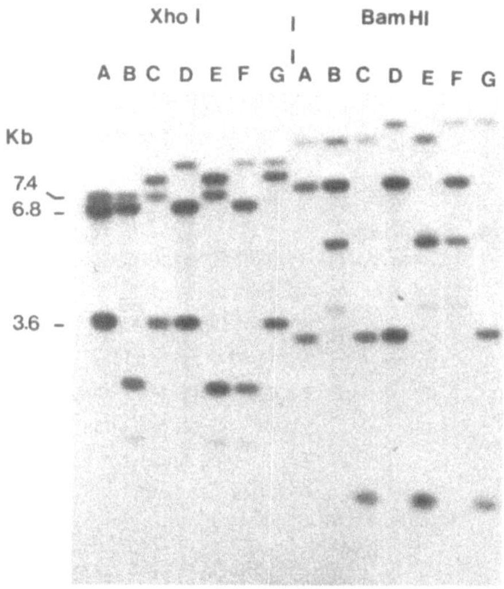

Fig. 2. Southern analysis of DNA from psbA-inactivated strains. Total DNA was isolated from wild-type Synechococcus cells and from strains in which one or two of the psbA genes had been inactivated. The DNA was cleaved with XhoI (left) or BamHI (right), electrophoresed, blotted, and hybridized with a probe which recognized all psbA sequences. Lanes contain DNA samples from the following Synechococcus strains: (A) wild type; (B) psbAI containing Tn5, called K1; (C) psbAII containing a spc^r cassette, called S2; (D) psbAIII containing a cam^r cassette, called C3; (E) the K1S2 double mutant; (F) the K1C3 double mutant and (G) the S2C3 double mutant. The three wild type XhoI fragments of 7.4, 6.8 and 3.6 kb contain psbAIII, psbAII and psbAI, respectively. The corresponding BamHI fragments are 12.0, 8.0 and 3.3 kb, respectively.

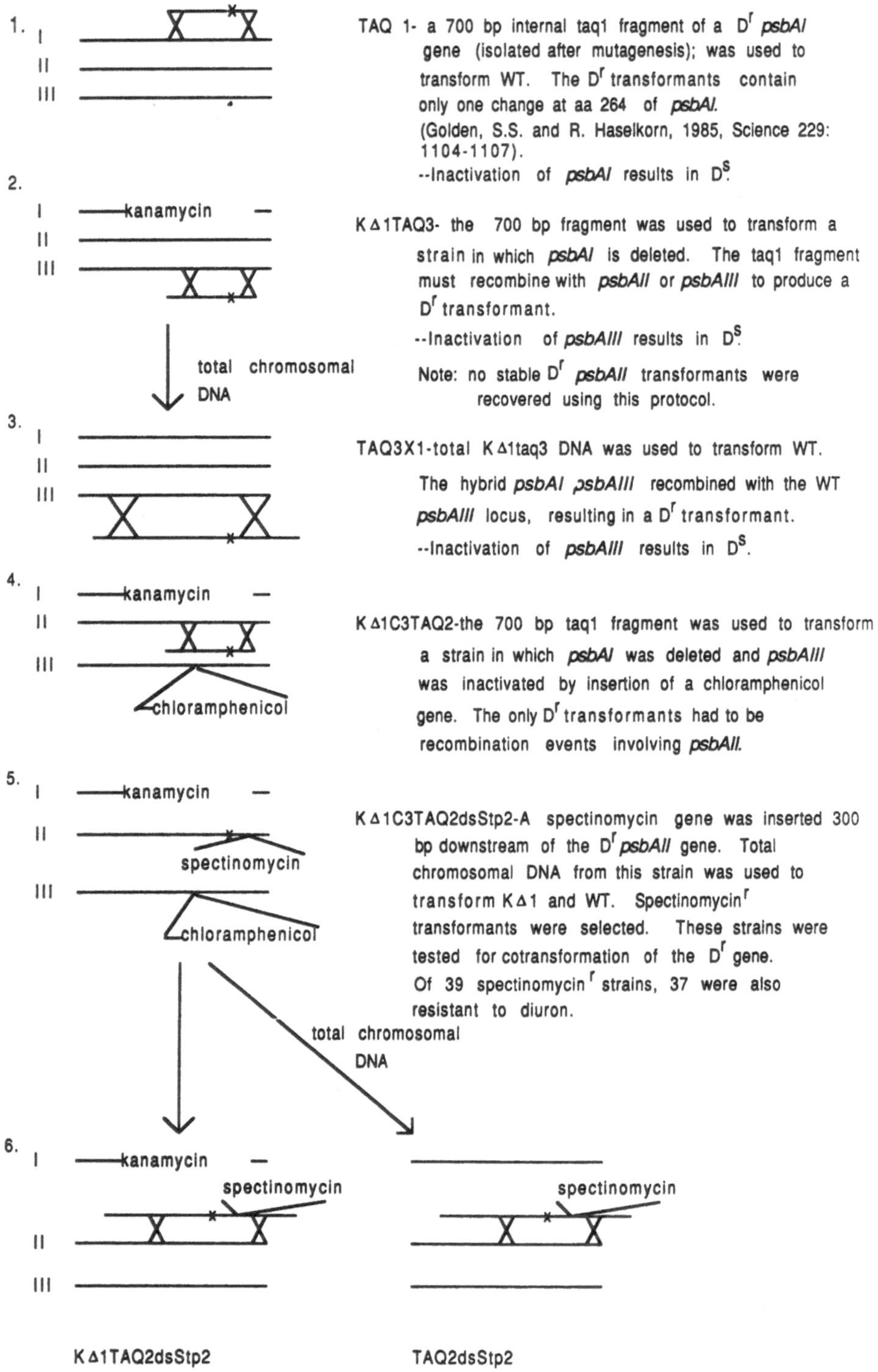
1.
I
II
III
TAQ 1- a 700 bp internal taq1 fragment of a D^r *psbAI* gene (isolated after mutagenesis); was used to transform WT. The D^r transformants contain only one change at aa 264 of *psbAI*.
(Golden, S.S. and R. Haselkorn, 1985, Science 229: 1104-1107).
--Inactivation of *psbAI* results in D^s.
2.
I
kanamycin
II
III
KΔ1TAQ3- the 700 bp fragment was used to transform a strain in which *psbAI* is deleted. The taq1 fragment must recombine with *psbAII* or *psbAIII* to produce a D^r transformant.
--Inactivation of *psbAIII* results in D^s.
Note: no stable D^r *psbAII* transformants were recovered using this protocol.
total chromosomal DNA
3.
I
II
III
TAQ3X1-total KΔ1taq3 DNA was used to transform WT.
The hybrid *psbAI* ,*psbAIII* recombined with the WT *psbAIII* locus, resulting in a D^r transformant.
--Inactivation of *psbAIII* results in D^s.
4.
I
kanamycin
II
III
chloramphenicol
KΔ1C3TAQ2-the 700 bp taq1 fragment was used to transform a strain in which *psbAI* was deleted and *psbAIII* was inactivated by insertion of a chloramphenicol gene. The only D^r transformants had to be recombination events involving *psbAII*.
5.
I
kanamycin
II
spectinomycin
III
chloramphenicol
KΔ1C3TAQ2dsStp2-A spectinomycin gene was inserted 300 bp downstream of the D^r *psbAII* gene. Total chromosomal DNA from this strain was used to transform KΔ1 and WT. Spectinomycin^r transformants were selected. These strains were tested for cotransformation of the D^r gene.
Of 39 spectinomycin^r strains, 37 were also resistant to diuron.
total chromosomal DNA
6.
I
kanamycin
spectinomycin
II
III
spectinomycin
KΔ1TAQ2dsStp2
TAQ2dsStp2

Fig. 3. Construction of Synchecococcus PCC7942 strains containing one diuron resistant allele in each of the three psbA loci.

98.6%. The predicted protein sequences from II and III are identical; they both differ from I at 25 positions, most of which are located near the amino-terminal end of the protein. Each gene gives rise to 1.2 kb messenger RNA (9). Any one gene can provide enough D1 protein for normal photosynthetic rates. This conclusion was reached by selective inactivation of each of the psbA genes, singly or in pairs. Inactivation was accomplished by transformation with cloned psbA genes, interrupted by insertion of DNA fragments encoding proteins that confer resistance to kanamycin, spectinomycin, or chloramphenicol. By successive transformation it was possible to inactivate each of the psbA genes singly or in pairs. In general, it was not possible to obtain strains in which all three genes are inactivated, an expected result because Synechococcus requires photosystem II to grow. Verification of the single and double inactivation events is illustrated in Figure 2 (6). All of the strains whose DNA is shown in Figure 2 grow photosynthetically and evolve oxygen at the same rate under saturating light.

With this information as background, we can return to the original question of herbicide resistance, re-analyze the simple transformation experiment presented initially, and then explore dominance relations among the various psbA alleles. The first question to answer is: what is the genotype of the strain made resistant to diuron by transformation with the 700 bp psbAI TaqI DNA fragment from the D^r mutant (Fig. 3,1.)? If recombination occurred exclusively at the psbAI locus of the recipient, and no gene conversion occurred, the transformants would be D^r at psbAI and D^s at psbAII and psbAIII. This is the case, proved by inactivation of the D^r psbAI gene by transformation with the inactivating kan^r-containing psbAI gene. Kan^r transformants are all diuron-sensitive (11).

How do we explain the D^r phenotype of the original transformant? 57% of the psbA transcripts are from psbAI (See table 1, below). If the translation yields from the three mRNA's are similar, the majority of photosystem II reaction centers will be D^r, which could account for the D^r phenotype. But with the tools at hand it is possible to study the question of dominance much more carefully. For this purpose, we began with a strain of Synechococcus in which the psbAI gene is deleted. (Deletion was accomplished by transformation with a deleted psbAI clone carrying a kan^r cassette.) The deletion strain (called KΔ1) was then transformed with the 700 bp TaqI D^r psbAI DNA, selecting for herbicide resistance (Fig. 3,2). Since psbAI was deleted from the recipient, only psbAII and psbAIII were available to provide homology for recombination. In fact, the majority of D^r transformants selected were at psbAIII. This was determined by the ability subsequently to inactivate the D^r phenotype by transformation with interrupted psbAIII DNA but not by transformation with interrupted psbAII DNA (11). One of these D^r transformants was named KΔ1TAQ3. To construct a strain with a D^r psbAIII gene and an intact psbAI gene, total chromosomal DNA from KΔ1TAQ3 was used to transform wild-type Synechococcus. D^r clones were selected, and in all cases subsequent inactivation of psbAIII resulted in D^s. Thus these strains contain one D^r gene(psbAIII) and two D^s genes(psbAI and psbAII). One of these clones was named TAQ3X1 (Fig. 3,3). The psbAIII gene contributes only 4% to the total psbA transcript pool in wild-type cells, yet the genotype of this allele determines cellular phenotype. In classical genetic terms, D^r is dominant to D^s.

Transformation of KΔ1 with the internal TaqI fragment of the D^r psbAI gene did not result in the production of any stable strains containing a D^r psbAII gene. It appeared that either the TaqI fragment could not recombine with the psbAII locus or that the genotype of psbAII was unable to determine phenotype in the presence of the D^s psbAIII gene. To

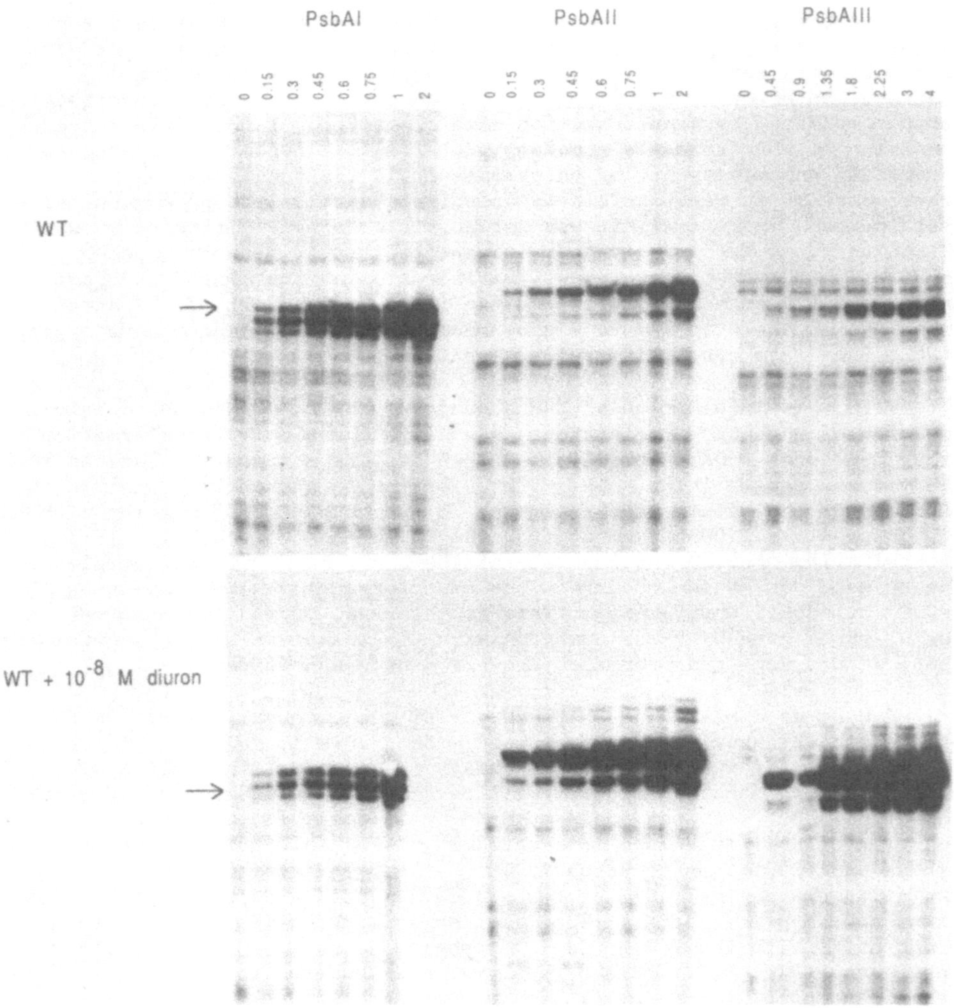

Fig. 4. Quantitative nuclease protection assays to determine the abundance of *psbA* gene transcripts. Wild type cells of *Synechococcus* were grown without (above) or with 10^{-8}M diuron (below). Total RNA prepared from each was annealed with copy-specific riboprobe (RNA complementary to mRNA, synthesized *in vitro* using T7 RNA polymerase), using increasing concentrations of total RNA (0.15, 0.3 μg etc.). Each annealed mixture was digested with RNAse A and T1 and then analyzed by electrophoresis on acrylamide gels. Autoradiograms of the gels are shown. The bands indicated by arrows were cut and counted. The slopes (cpm/μg RNA) were normalized to correct for the size and base compositions of each probe. The quantitative results are listed in Table 1. It is obvious from the figure that the level of *psbAII* and *psbAIII* transcripts increases dramatically in response to a sublethal concentration of herbicide.

distinguish between these two possibilities the Dr TaqI fragment was used to transform a strain containing a deleted psbAI gene and a disrupted psbAIII gene (KΔ1C3). D^r transformants were recovered. These had to contain a D^r psbAII gene because this is the only intact psbA gene (Fig. 3,4.). Thus recombination was not the problem in the formation of a strain with a D^r psbAII gene. If the presence of an intact D^s psbAIII gene were to mask the D^r psbAII gene, it would be necessary to mark the D^r psbAII gene in another way in order to select for its inclusion in a diploid Synechococcus strain. We decided to place a gene encoding spectinomycin resistance just downstream of the D^r psbAII open reading frame. This was accomplished by transformation of KΔ1C3TAQ2 with a plasmid containing the last 200 bp of a cloned psbAII gene (the D^r mutation is 300 bp upstream from the end of the ORF) plus 5 kb of downstream sequence with a spc^r gene inserted 300 bp from the end of the psbAII ORF (Fig. 3,5). The chromosomal DNA from Spc^r transformants was then used to transform KΔ1 and wild-type (Fig. 3,6). Spc^r transformants were selected in the expectation that in some cases the D^r allele would co-transform. Thus, if psbAII were able to manifest its genotype as cellular phenotype, them some of the Spc^r transformants should also be D^r. Indeed, 18 of 19 KΔ1 and 19 of 20 wild-type Spc^r transformants were D^r. Clones of each strain were named KΔ1TAQ2dsStp2 and TAQ2dsStp2, respectively (Fig. 3,6). Thus psbAII is able to express D^r at the cellular level even in the presence of intact psbAI and psbAIII genes. Again the D^r phenotype of this strain indicates the dominance of D^r.

The relative abundance of the three psbA transcripts was measured using quantitative S1-type solution hybridization titrations using copy-specific riboprobes. In these experiments varying amounts of Synechococcus RNA were hybridized to an excess of labelled riboprobe. After hybridization the single-stranded RNA was digested with RNAse A and RNAse T1. The undigested, protected RNA was analyzed by electrophoresis on a denaturing sequencing gel. The labelled bands were cut from the gel and counted. The slope, cpm/μg input Synechococcus RNA, is proportional to the amount of mRNA arising from each psbA gene. Slopes can be compared to provide the relative amounts of the three mRNAs.

In the wild-type strain the ratio of psbaI:psbAII:psbAIII mRNAs was found to be 13:9:1 (Table 1). When the wild-type strain was grown in a sublethal concentration of diuron, this ratio changed to 1:4:1. By comparison of slopes it can be seen that the psbAI transcript level remained constant, but that the levels of psbAII and psbAIII increased 7 fold and 16 fold, respectively. When the TAQ1 strain was grown in the presence of diuron the transcript levels also changed. In this case, psbAI decreased by 1/2 whereas psbAII and psbAIII increased 2 fold and 5 fold, respectively. This induction appears to be a general stress response in that it is independent of the presence of a D^r allele. It is not yet known if this induction leads to an increase in the number of reaction centers in the thylakoid membrane. Other stresses which elicit this response are currently under study.

The induction of the TAQ1 strain is curious. It appears that the relative contribution of the psbAI mRNA, the sole source of a D^r Q_B protein, decreases in the presence of diuron. Thus the final question remaining is whether both D^r and D^s transcripts are translated to provide functional D1 protein. To answer this question we obtained help from Dr. H. Robinson, who measured the decay of variable fluorescence, following a saturating flash of light, from photosystem II reaction centers in the various strains of Synechococcus described above (12). The fluorescence

Table 1. Relative Steady State Transcript Levels of the Three psbA Genes in Different Strains of *Synechococcus* PCC 7942 Grown in the Presence or Absence of Diuron

Strain	Diuron	Doubling Time	Slopes (cpm/μg RNA)		Ratios	Induction +/-diuron
WT	0	6h	I	3200	13	
			II	2300	9	
			III	230	1	
WT	10^{-8}M	15.5h	I	3800	1	1x
			II	18900	4	7x
			III	4500	1	16x
TAQ1	0	10h	I	1800	5	
			II	6200	16	
			III	390	1	
TAQ1	5×10^{-7}M	13h	I	1000	1	0.5x
			II	12300	13	2x
			III	2000	2	5x

measurement tells, at a given time, what proportion of the reaction centers, which were closed by the flash, have reopened. Reopening a center requires that an electron be transferred to Q_B. Addition of the herbicide diuron to a sensitive system blocks electron transfer, resulting in inhibition of the normal fluorescence decay.

Robinson determined the proportion of closed centers, as a function of diuron concentration, for the wild type strain, TAQ1, and KΔ1TAQ3 (12). For wild type *Synechococcus*, 50% of the centers are closed at 2×10^{-8} M diuron and nearly all are closed at 10^{-7} M. The strain TAQ1, which has the D^r allele in *psbAI* only, shows a similar titration curve displaced to 300-fold higher concentration of diuron. The critical experiment with the strain KΔ1TAQ3 clearly shows two components in the herbicide titration. About one-third of the centers are closed at 10^{-7} M diuron (i.e. sensitive) while 50% of the remainder remain open at 2×10^{-5} M. Clearly the transcript levels do not reflect the proportion of D^r and D^s reaction centers. In the TAQ1 strain the *psbAI* transcript is only 1/6 of the transcript pool, yet it encodes all of the Q_B proteins in reaction centers. The transcript levels in the KΔ1TAQ3 strain have not yet been determined, but once again it may be that the least abundant transcript, that which arises from *psbAIII*, provides Q_B protein for the majority of reaction centers. Clearly posttranscriptional events determine the composition of reaction centers. Binding assays, using radiolabelled diuron, will have to be used to determine the absolute amount of Q_B protein in the thylakoid membranes in each of our strains.

The studies reported here, using a prokaryotic organism to analyze dominance relations among alleles encoding photochemical reaction center components, have significant implications for plant breeding. The mutation to herbicide resistance, as might be expected, affects the normal function of the reaction center. Photosynthetic efficiency in plants homozygous for herbicide resistance is impaired (13). It seems possible that a heterozygote with respect to herbicide resistance would have the advantage of resistance to herbicide when it is applied but normal photosynthetic efficiency between applications.

REFERENCES

1. C. Fedtke, Biochemistry and Physiology of Herbicide Action. Springer-Verlag, Berlin (1982).
2. A. R. Crofts and C. A. Wraight. Biochim. Biophys. Acta 636, 218-233 (1983).
3. A. K. Mattoo, U. Pick, H. Hoffman-Folk and M. Edelman. Proc. Natl. Acad. Sci. USA 78, 1572-1576 (1981).
4. K. Pfister, K. E. Steinback, G. Gardner and C. J. Arntzen. Proc. Natl. Acad. Sci. USA 78, 981-985 (1981).
5. A. K. Mattoo, J. B. Marder, J. Gressel and M. Edelman. FEBS Lett. 140, 36-39 (1982).
6. J. Hirschberg and L. McIntosh. Science 222, 1346-1348 (1983).
7. P. Goloubinoff, M. Edelman and R. B. Hallick. Nucl. Acids Res. 24, 9489-9496 (1984).
8. S. S. Golden and R. Haselkorn. Science 229, 1104- 1107 (1985).
9. S. S. Golden, J. Brusslan, and R. Haselkorn. EMBO J., 5, 2789-2798 (1986).
10. S. E. Curtis and R. Haselkorn. Plant Mol. Biol. 3, 249-258 (1984).
11. J. A. Brusslan, S. S. Golden, and R. Haselkorn. In: Progress in Photosynthesis Research, Vol. 4, pp. 821-824. J. Biggins, ed. Martinus Nijhoff, Dordrecht (1987).
12. H. Robinson, S. Golden, J. Brusslan and R. Haselkorn. In: Progress in Photosynthesis Research, Vol. 4, pp. 825-829. J. Biggins, ed. Martinus Nijhoff, Dordrecht (1987).
13. D. R. Ort, W. A. Ahrens, B. Martin and E. W. Stoller. Plant Physiol. 72, 925-930 (1983).

A POSSIBLE ROLE FOR 3' SEQUENCES OF THE WOUND-INDUCIBLE POTATO PROTEINASE INHIBITOR IIK GENE IN REGULATING GENE EXPRESSION

Gynheung An, Robert W. Thornburg[2], Russell Johnson,
Gerry Hall and Clarence A. Ryan.
Institute of Biological Chemistry
Program in Biochemistry/Biophysics and
Program in Plant Physiology
Washington State University
Pullman, WA 99164-6340

INTRODUCTION

Proteinase Inhibitors are a diverse group of small proteins that are present in storage organs and vegetative tissues as part of the array of defensive chemicals that plants employ to discourage pest attacks (1,2). Whereas most known seeds or storage organs contain relatively high levels of these protein inhibitors of serine endopeptidases, leaves of certain Solanaceae species, including potato and tomato (3), and of some Leguminosae species such as alfalfa (4) and soybean (5), can accumulate considerable quantities of proteinase inhibitor proteins in response to insect attacks or other mechanical wounds. These responses are systemic and can be reproduced in excised plants by supplying oligouronide fragments derived from the plant cell wall back to the plants through their cut petioles (4,6). This response has provided one of the least complicated systems available for studying the expression of specific genes coding for well characterized proteins in response to an environmental stress i.e., wounding.

In leaves of potato and tomato plants, members of two small nonhomologous serine proteinase inhibitor families (7), called Inhibitor I (Mr 8100) and Inhibitor II (Mr 12,300), can be induced to accumulate in response to wounding (8). Members of these inhibitor families were originally found in potato tubers (9,10) where they cumulatively comprise over 10% of the soluble proteins of the cortical tissues (11). More recently another proteinase inhibitor, a member of the Bowman-Birk family of inhibitors, was found to be wound-inducible in alfalfa leaves (12). Thus among the six or so nonhomologous families of proteinase inhibitors (7), three are induced to accumulate in plant leaves.

Wound-inducible Inhibitor I and II genes have been isolated from tomato and potato genomes (13-16). These genes are under intensive studies in our laboratory as part of our efforts to elucidate the biochemical steps involved in the activation of the proteinase inhibitors in response to signals released from wounds. The identity of the wound-inducible, cis-acting sequences of the genes may facilitate the identification of transacting factors that regulate wound-inducible

[2] Department of Biochemistry, Iowa State University, Ames, IA 50011

proteinase inhibitor gene expression. The regulatory regions of one of the wound-inducible proteinase inhibitor genes from potato called the Inhibitor IIK gene (15), have been studied in detail by constructing chimeric genes containing the 5' and 3' regions of the inhibitor gene fused with the chloramphenicol acetyltransferase CAT structural gene (15), which provides a sensitive reporter enzyme to monitor the levels of wound-inducible gene expression.

RESULTS AND DISCUSSION

Wound-Inducible Expression of Fused Inhibitor IIK-CAT Gene Fusion. A 1000-bp restriction fragment (TaqI-ScaI) from the 5' flanking region of the potato Inhibitor IIK gene was fused to the open reading frame of the CAT gene (15). This fusion was terminated in two ways: (i) with a 1000bp (RsaI-TaqI) terminator sequence from the wound-inducible potato Inhibitor IIK gene (pRT45) and (ii) with a 730 bp (SmaI-NaeI) terminator sequence from the 6b gene of Ti plasmid pTiA6 (pRT50) (Figure 1).

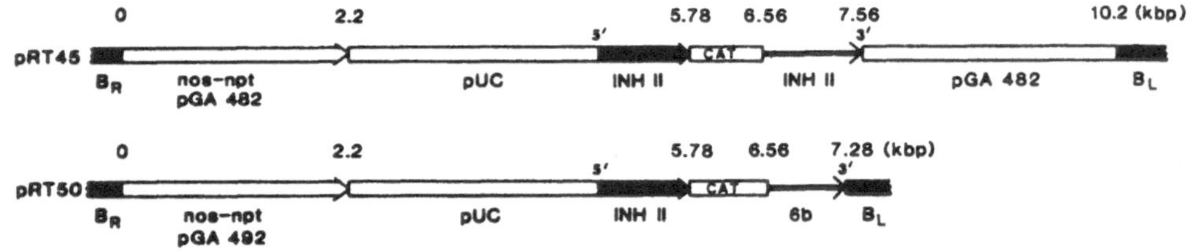

Figure 1. Constructs in the binary Ti vector containing 1000 bp of Inhibitor IIK 5' region, fused with the CAT gene. The fused gene is terminated with; Top, the inhibitor IIK terminator; and Bottom, the 6b terminator from the Ti plasmid A6.

These chimeric genes were transferred into tobacco cells via a binary Ti vector system (16), and transgenic plants were regenerated. The chimeric CAT gene was expressed in response to wounding in leaves of tobacco plants transformed with the pRT45 vector shown in Figure 1. The wound-inducible expression of CAT activity was systemic and was induced in tissues both proximal and distal to the wounded tissues. The time course of wound induction of CAT activity in transgenic tobacco leaves is similar to that found for wound-inducible inhibitor I and II mRNAs in tomato leaves (Figure 2). However, plants transformed with pRT50 which

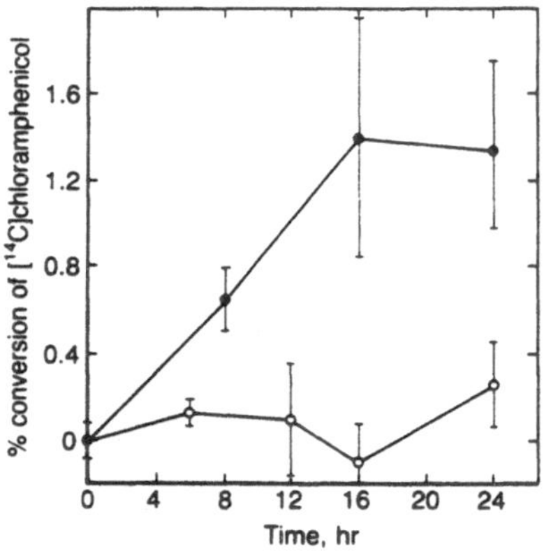

Figure 2. Time course of the wound-induced CAT activity in leaves of tobacco plants transformed with pRT45 (●) and pRT50 (○).

contains the 6b terminator did not express CAT in response to wounding (15). These results demonstrate that sequences necessary and sufficient for wound inducibility are present within about 1000 bp of the 3' and/or 5' regions of the Inhibitor IIK gene and that wound-inducible components of tobacco leaf cells can regulate these sequences. The absence of wound inducibility in the construct containing the 6b terminator suggests a possible role of a 3' element in regulating wound-induction.

The levels of the wound-induced CAT expression in intact leaves was only about 10% of that driven by a gene fusion in which the nos promoter (17) and the 6b terminator regulate CAT expression. However, wound-inducible CAT expression of the pRT45 transformants was elevated to that of the nos promoter (Figure 3) when leaf discs were excised from the leaves and floated on MS medium (18).

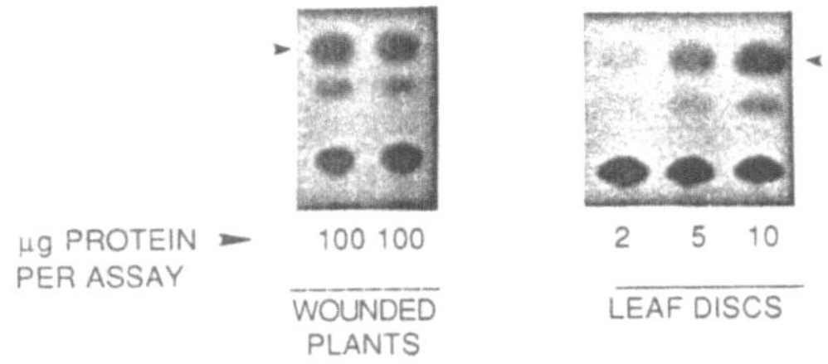

Figure 3. CAT activity in leaf discs of transformed tobacco plants (pRT45 transformants) driven with the Inhibitor IIK 5' and 3' regions. The discs were cut from tobacco leaves and floated on MS media (18) for 24 hr and assayed.

Thus, the potential expression of the CAT gene under wound-inducible regulation in tobacco leaves appears to be limited in intact leaves. The cause of this limitation is not understood, but in leaf discs the response is somehow enhanced. The causes of this enhancement are currently under investigation.

Expression of the Inhibitor IIK-CAT fusion gene in tobacco protoplasts via electroporation. The 15.7 kb pRT45 and pRT50 vectors that were used to transform tobacco plants were also used in attempts to express the Inhibitor IIK-CAT fusion genes in tobacco protoplasts following electroporation. In typical experiments, twenty μg plasmid/ml was electroporated at 400 V. As with the transformation experiments, expression of

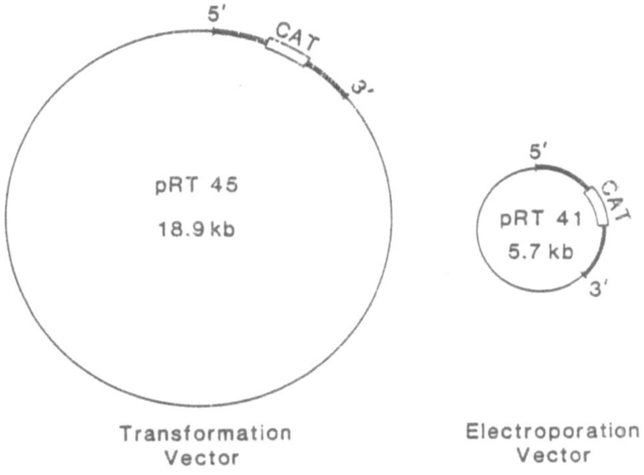

Figure 4. Comparison of the sizes of vectors employed for electroporation of the Inhibitor IIK CAT fusion gene.

CAT was achieved only with the pRT45 plasmid but not with pRT50. However, the level of expression using pRT45 was very low when compared with, for example, expression of the nos promotor in a smaller pUC-derived vector. In order to examine whether the low level expression was due to the large size of pRT45 plasmid, a smaller plasmid pRT41 (Figure 4) carrying the identical chimeric gene was electroporated into tobacco protoplasts and transient activity of the introduced gene was measured after 24 hours incubation (8). The results of this experiment are shown in Figure 5. It is clear that the small vector, pRT41, is a

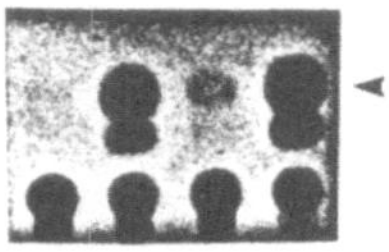

Figure 5. CAT activity in tobacco protoplasts after electroporation of the vectors pRT45 (#45); pRT50 (#50) and pRT41 (#41) and incubation of the protoplasts for 24 hr prior to assaying for CAT activity. The product of the assay, 3-acetyl chloramphenicol, is indicated by the arrow.

much more effective molecule for expressing the Inhibitor IIK-CAT gene via electroporation tobacco protoplasts. In this experiment, CAT expression, driven from the Inhibitor IIK control regions, was nearly equivalent to that of nos, which is considered a fairly strong constitutive plant promoter.

The strong expression of the Inhibitor IIK-CAT fusion gene having an Inhibitor IIK terminator, provided a convenient means of comparing this expression with a nos-CAT fusion gene and a CaMV-CAT fusion gene, both terminated with either the 1000 bp of the Inhibitor IIK 3' terminator or the 6b terminator. In Table 1 is shown the results of these experiments. The Table presents a comparison of the % conversion

Table 1. Effects of Inhibitor IIK 3' terminator and the 6b 3' terminator by the 5' region of Inhibitor IIK, nos and CaMV genes.

5' REGION		3" REGION	% CONVERSION
	cat		
INH II		INH II	11.1
INH II		6b Ti	< 1.0
nos		6b Ti	8.6
nos		INH II	22.7
CaMV		6b Ti	28.7
CaMV		INH II	68.0

of chloramphenicol to acetylated chloramphenicols by the CAT enzyme, whose level is directly propotional to the strength of the various control elements. The 3' of the Inhibitor IIK gene enhanced the expression of CAT gene driven from either the nos promoter or the CaMV 35S promoter by three fold over the levels expressed from chimeric genes, terminated with the 6b terminator of the Ti plasmid. These results are not only consistent with the earlier experiments, indicating that the 3' control region of the Inhibitor IIK gene contains DNA sequence which strongly influences the expression of CAT gene under the Inhibitor IIK 5' regions, but the Inhibitor IIK terminator can enhance activity of heterologous genes which are under the regulation of constitutive promoters.

The results in Table 1 indicate that strengths of constitutive promoters such as CaMV 35S and nos, in driving expression of the CAT gene, can be increased several fold by an element present in the Inhibitor IIK 3' region. Such an enhancement may have considerable significance in evaluating the capabilities for expression of plant genes regulated by constitutive promoters.

The Inhibitor IIK 3' control region is under intensive investigation to identify the specific sequences that are responsible for the regulation of the wound-inducibility of the Inhibitor IIK gene and for the enhancement of promoter strengths of constitutive promoters. The identification and characterization of these elements should serve as a basis to seek transacting factors that regulate wound-inducibility of proteinase inhibitor proteins in plants.

SUMMARY

Tobacco plants transformed with a chimeric gene containing the 1000 bp of the 5' and 1000 bp of the 3' flanking regions of proteinase Inhibitor IIK gene from potato, fused with the CAT gene, exhibit a systemically-mediated CAT expression when wounded. Although the wound-induced expression is relatively weak *in planta*, the fused gene is strongly expressed in excised leaf tissues or when introduced into tobacco protoplasts by electroporation. A gene fusion in which the 1000 bp Inhibitor IIK 3' terminator was replaced with a Ti-plasmid terminator sequence did not exhibit wound-inducible CAT gene expression *in planta*. This chimeric gene is also not expressed when electroporated into tobacco protoplasts. The 1000 bp 3' region from the Inhibitor IIK gene, was used to terminate transcription initiated from the constitutive nos and CaMV promoters fused with the CAT structural gene. When electroporated into tobacco protoplasts, CAT expression was increased three fold over its expression in constructs containing the constitutive terminator. The 3' region of the Inhibitor IIK gene therefore contains sequences that strongly influence the expression of the CAT gene under regulation of the 5' regions of both wound-inducible and constitutive genes.

ACKNOWLEDGEMENTS

This research was supported in part by NSF grant DCB-8702538 (CAR) and NSF grant DCB 86016018 (GA). The authors acknowledge Mr. Greg Wichelns for growing the plants and Mr. Mark Seymour and Mike Costa for excellent technical assistance.

REFERENCES

1. C.A. Ryan. Proteolytic enzymes and their inhibitors in plants, Ann. Rev. Plant Physiol. 24, 173 (1973).

2. M. Richardson. The Proteinase inhibitors of plants and microorganisms. Phytochemistry 16, 159.

3. C.A. Ryan. Proteinase inhibitors in plant leaves: A biochemical model for pest-induced natural plant protection, Trends in Biochemical Sci. 5, 148 (1978).

4. W. Brown and C.A. Ryan. Isolation and characterization of a wound-induced trypsin inhibitor from alfalfa leaves, Biochemistry 23, 3418 (1984).

5. M.E. Kraemer, M. Rangappa, W. Gade and Benepal, P.S. Induction of Trypsin inhibitors in soybean leaves by Mexican Bean Beetle (coleoptera: Corcinellidae) defoliation. J. Econ. Ent. 80, 237.

6. C.A. Ryan, P. Bishop, G. Pearce, A.G. Darvill, M. McNeil and P. Albersheim. A sycamore cell wall polysaccharide and a chemically related tomato leaf polysaccharide possess similar proteinase inhibitor-inducing activities, Plant Physiol. 68, 616 (1981).

7. Laskowski, M. Jr. and I. Kato. Protein inhibitors of proteinases. Ann. Rev. Biochem. 49, 593 (1980).

8. G. Gustafson, and C.A. Ryan. The specificity of protein turnover in tomato leaves: The accumulation of proteinase inhibitors, induced with the wound hormone, PIIF, J. Biol. Chem. 251, 7004 (1976).

9. J.C. Melville and C.A. Ryan. Chymotrypsin inhibitor I from potatoes: A multi-site inhibitor composed of subunits, Arch. Biochem. Biophys. 138, 700 (1970).

10. J. Bryant, T. Green, T. Gurusaddaiah and C.A. Ryan. Proteinase inhibitor II from potatoes: Isolation and characterization of the isoinhibitor subunits, Biochemistry 15, 3418 (1976).

11. C.A. Ryan, T. Kuo, G. Pearce and R. Kunkel. Variability in the concentration of three heat stable proteinase inhibitor proteins in potato tubers, Am. Potato J. 53, 443 (1976).

12. W.E. Brown, K. Takio, K. Titani and C.A. Ryan. Wound-induced trypsin inhibitor in alfalfa leaves: Identity as a member of the Bowman-Birk inhibitor family, Biochemistry 24, 2105 (1985).

13. J.S. Lee, W.E. Brown, G. Pearce, T.W. Dreher, J.S. Graham, K.G. Ahern, G.D. Pearson and C.A. Ryan. Molecular characterization and phylogenetic studies with a wound-inducible proteinase inhibitor gene in *Lycopersicum* species, Proc. Natl. Acad. Sci. (USA), 83, 7277 (1986).

14. T.E. Cleveland, R. Thornburg and C.A. Ryan. Molecular characterization of a wound-inducible proteinase inhibitor gene from potato and the processing of the mRNA and protein, Plant Molecular Biol., 8, 199 (1987).

15. R.W. Thornburg, G. An, T.E. Cleveland and C.A. Ryan. Characterization and expression of a wound-inducible proteinase inhibitor II gene from potato, Proc. Conf. Tayloring Plant Genes, T. Kosuge and G. Bruening, Eds., Proc. Natl. Acad. Sci. (USA), 84, 744 (1987).

16. G. An. Development of plant promoter expression vectors and their use for analysis of differential activity of nopaline synthase promoter in transformed tobacco cells, Plant Phys., 81, 86 (1986).

17. G. An, P.R. Ebert, B.-Y. Yi and C.-H. Choi. Both TATA box and upstream regions are required for the nopaline synthase promoter activity in transformed tobacco cells, Mol. Gen. Genet. 203, 245 (1986).

18. T. Murashige and F. Skoog. A revised medium for rapid growth and bioassays with tobacco tissue cultures, Phys. Plant. 15, 473 (1962).

HEAT STRESS: EXPRESSION AND STRUCTURE OF HEAT SHOCK PROTEIN GENES[1]

Joe L. Key[2], Ron T. Nagao[2],
Eva Czarnecka[3], and William B. Gurley[3]

[2]Botany Department
University of Georgia
Athens, Georgia 30602

[3]Department of Microbiology and Cell Science
University of Florida
Gainesville, Florida 32611

INTRODUCTION

There are several features of the HS response that make it an attractive system for study. The HS response is one of the most highly conserved genetic systems known in the biological world; it appears to be universal, having been observed in organisms ranging from eubacteria to archebacteria, from lower eukaryotes to plants and man. It occurs in almost every cell and tissue type and in cultured cells of animals and plants. The rapidity of the induction of HS gene expression, the magnitude of the response in terms of the levels of HS mRNAs and HS proteins that accumulate over a short period of time (Nagao et al., 1986), the ease of manipulation of the stimulus, and the great reproducibility of results make it an excellent model system for studies on the mechanisms operative in the regulated expression of genes. The HS response has great physiological importance. It appears to be critical in protecting cells from thermal damage by excessively high (otherwise lethal) temperatures and possibly even to other stresses. The system may have significant potential for genetic manipulations.

The HS response is characterized by a number of interesting regulatory phenomena. Transcription of HS genes is rapidly activated and the HS mRNAs accumulate to high levels (in soybean seedlings up to 20,000 or more copies per cell for some HS mRNAs). The initial transcriptional activation slows and cessation of HS gene transcription occurs after two to three hours of continuous HS. This shutdown of HS mRNA transcription under HS conditions has been referred to as self-regulation or autoregulation (DiDomenico et al., 1982; Key et al., 1983). There is a rapid slowing of normal mRNA translation (Ashburner and Bonner, 1979; Key et al., 1981); the normal mRNAs persist during HS from near normal levels to much reduced levels depending upon the specific mRNA (Mirault et al., 1978; Key et al., 1985a). These kinds of regulatory phenomena are much better defined in Drosophila (Lindquist, 1986), but these general features are observed in plants.

[1]This work was supported by U.S. Department of Energy Contract DE-FG09-86ER13602 to JLK and Lubrizol Genetics contracts to JLK and WBG.

There are more HS proteins from soybean and other plants investigated to date than have been observed in Drosophila and most other organisms. Different plant species accumulate from 12 to 27 or more 15 to 27 kD HS proteins (Mansfield, 1986) while Drosophila, for example, synthesizes only 4 prominent 22 to 27 kD low molecular weight HS proteins and these represent only a small fraction of hsp70 synthesis. The pattern and complexity of high molecular weight HS proteins (e.g., HS proteins of 70± kD and 84± kD) of plants and other organisms are much more similar and are much more conserved at the nucleotide and amino acid sequence level than are the low molecular weight HS proteins. Accordingly, we have concentrated much of our work on the relatively unique low molecular weight group of HS proteins of soybean.

The following discussion focuses on several aspects of the heat shock (HS) response in higher plants, in particular in the soybean (Glycine max) seedling: (1) thermotolerance; (2) aspects of expression of the HS genes relative to inductive temperature regimes; (3) analysis of the structure of some HS genes/proteins; and (4) experiments on transgenic expression of a soybean HS gene in sunflower tumors.

THERMOTOLERANCE

As noted above, thermotolerance likely represents the fundamental basis of the high evolutionary conservation of the HS response. Thermotolerance is used here to define the ability of an organism to withstand an otherwise lethal heat treatment if it has been pretreated with some appropriate nonlethal HS or some agent which induces a HS-like response (e.g., arsenite). Several kinds of HS treatments lead to the acquisition of thermotolerance in plants: (a) a HS at or near the optimum HS temperature (e.g., 40°C for soybean seedlings) for 1 to 2 hr prior to the shift to an otherwise lethal treatment (e.g., 2 hr at 45°C); (b) a HS at 40°C for 1 to 2 hr followed by several hours at 28°C to 30°C prior to the 45°C exposure; (c) a gradual increase in temperature (e.g., 3°C per hr) from the normal growth temperature up to as high as 47.5°C for soybean seedlings; (d) a 5 to 10 min treatment at 45°C followed by 2 hr or more at 30°C prior to an otherwise lethal treatment at 45°C for 2 hr; and (e) treatment with 50 to 100 μM arsenite for about 4 hr, which is itself a somewhat toxic or growth inhibitory treatment, prior to HS at 45°C for 2 hr. All of these treatments share in common the induction and accumulation to high levels of HS proteins. If, however, cycloheximide is included to inhibit protein synthesis during any of these inductive treatments which lead to the acquisition of thermotolerance, the subsequent 45°C, 2-hr HS treatment is lethal (C.-Y. Lin and J. L. Key, unpubl. data). In animal systems, the inclusion of amino acid analogues and their incorporation into protein during the tolerant HS prevent the acquisition of thermotolerance to lethal HS (Lindquist, 1986). These correlative data suggest that it is the accumulation, stability, and selective localization of normal HS proteins (i.e., no amino acid analogue substitutions) during permissive HS which allows organisms to survive otherwise lethal HS treatments (Key et al.,1985b; Lindquist, 1986), although other mechanisms may also contribute to the development of thermotolerance (see Key et al., 1987a).

Thermotolerance may be assessed by measuring growth and survival of seedlings, viability and/or growth of cells in culture, and at the cell structural/ultrastructural level. HS treatments which lead to the acquisition of thermotolerance cause the stabilization of the basic architectural features of cells, including membrane and organellar integrity. There are, however, some changes (based on electron microscopy analyses) in the ultrastructural features of the cell and its organelles in response to a permissive HS and to 45°C treatments that would otherwise be lethal. Membrane integrity and cell architecture are completely disrupted by a 45°C treatment without a pretreatment that permits the acquisition of thermotolerance in soybean seedlings (Mansfield et al., 1987; Lin and Key, unpubl. data). These structures survive a 45°C HS if a tolerant HS treatment precedes the otherwise lethal HS.

While some fundamental parameters of thermotolerance have been established, much remains to be learned about the development of thermotolerance and the role(s) of HS proteins in the process(es). There seems to be little question, however, that at least some HS proteins play critical roles in the maintenance of cell structure and the acquisition of thermotolerance during HS and in the ability of cells to resume normal functions following HS upon return to normal growth conditions. The stage of development and metabolic state of cells and tissues seem also to have modulating influences on the HS response and the acquisition of thermotolerance. These remain fruitful areas for experimentation and significant discovery along with investigations into the heat sensitivity of reproductive tissues (Altschuler and Mascarenhas, 1982; Cooper et al., 1984; Lindquist, 1986). There are many manifestations of thermotolerance and some of these will be noted below in terms of HS mRNA accumulation.

REGULATORY FEATURES OF HS mRNA PRODUCTION

Each plant species has a characteristic optimum or inductive breakpoint temperature at which HS mRNA and HS protein synthesis are maximal during a continuous HS. At that temperature (about 40°C for 2-day old soybean seedlings), HS mRNAs accumulate for 1 to 2 hr to a near steady state level; after about 2 hr the level of the mRNAs gradually declines (Nagao et al., 1986). These levels of HS mRNAs reflect the rapid and transient activation of HS gene transcription followed by a turn-off of transcription after 2 to 3 hr (Kimpel and Key, unpublished). The accumulated HS mRNAs gradually decay, but are still detectable after continuous 40°C HS for 12 hr. However, when seedlings are returned to 30°C, the HS mRNAs decay to essentially non-detectable levels within 3 to 4 hr (Nagao et al., 1986; Key et al., 1987b). These observations demonstrate that HS mRNAs are much more stable during continuous 40°C HS than at the normal growing temperature and that there is additional regulation of these genes during continuous HS beyond their activation during the shift to elevated temperatures. HS proteins continue to be synthesized for several hr after HS mRNA synthesis ceases. There is suggestive evidence that the level of the hsp70 protein of Drosophila (DiDomenico et al., 1982; Lindquist and DiDomenico, 1985) may relate to the autoregulation of HS gene transcription. The dnaK protein of E. coli, a HS protein that is considered the bacterial counterpart of hsp70 of Drosophila, yeast, etc., based on significant nucleotide/amino acid homologies, is a negative regulator of the HS response (Neidhardt et al., 1984).

The nature of this autoregulatory response is not understood in plants. Soybean seedlings respond to a sequential series of HS treatments (e.g., cycles of 2 to 3 hr at 40°C followed by 3 to 4 hr at 30°C) by induction of HS gene transcription during each HS cycle (Nagao et al., 1986). The level of HS mRNA which accumulates decreases somewhat during each successive cycle. This ability to undergo successive rounds of HS raises a question about the role of HS proteins per se in the autoregulation noted above. Most of the HS proteins of soybean are generally stable for many hours following either a single HS or successive rounds of HS. Thus, the mere presence of large amounts of HS proteins in the cell is not sufficient to prevent a subsequent typical HS response. These observations may be reconciled with the putative role of HS proteins (e.g., dnaK and hsp70 as noted above) in autoregulating HS gene transcription by assuming that (a) a specific regulatory HS protein(s) has a much shorter half-life than the majority of the HS proteins, (b) the loss of autoregulatory function could be the result of changes in state of one or more of the HS proteins during an intervening 30°C treatment, or (c) that HS proteins do not function directly in autoregulation. Given the recent genetic evidence from E. coli (Tilly et al., 1983) which shows that mutants in the dnaK locus over produce HS proteins indicating a loss of negative control, and from yeast where deletion of two hsp70 genes results in the constitutive expression of other HS proteins (Craig, 1985), it seems likely that hsp70-like proteins or some other functional counterpart in plants may play a role in autoregulation.

Another interesting feature associated with the regulation of the HS response relates to the inductive temperature which elicits a full HS response. As noted above, a 2 hr or continuous HS at 40°C elicits an optimal HS response. An exposure of seedlings to 45°C for 1 to 2 hr is lethal. On the other hand a 5 to 10 min 45°C treatment elicits a near normal HS response; in contrast a 5 to 10 min 40°C HS elicits only a minimal HS response (Nagao et al., 1986). The short-term 45°C HS raises several interesting questions. How do the cells/tissues perceive short-term 40°C and 45°C HS differently? How is it that HS mRNAs induced by the 5 to 10 min 45°C HS are stable and accumulate to high levels, and that the tissue completes a full cycle of HS protein production at 30°C while HS mRNAs induced during 40°C HS rapidly decay upon return to 30°C? The answers to these and other related questions are presently unknown.

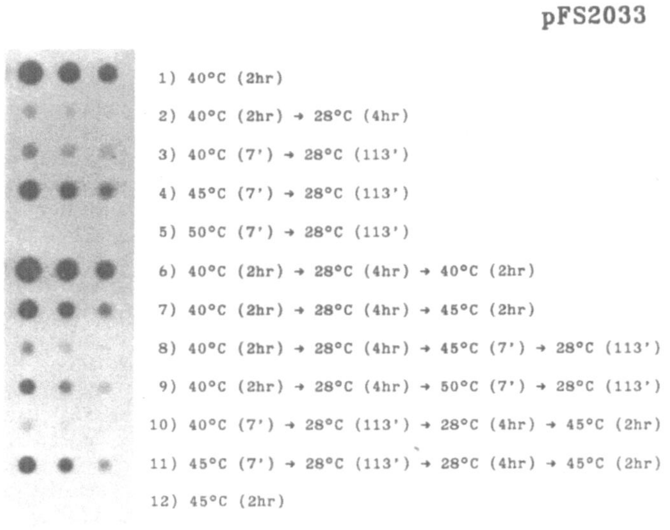

Figure 1. Dot blot hybridization analysis of HS mRNA levels following HS, recovery, and re-HS at various temperatures and lengths of time. Poly(A) RNA was isolated from soybean seedlings incubated under the indicated regimes, blotted and hybridized with ^{32}P-labeled HS cDNA clone pFS2033. Poly(A) RNA concentrations decreasing from left to right are 62.5, 31.25, 15.625 ng, respectively.

Another intriguing question is raised when a second HS treatment is administered following a full HS response (e.g., 2 to 3 hr at 40°C or 5 to 10 min at 45°C followed in each case by 3 to 4 hr at 30°C). The tissue is now thermotolerant, and a second treatment of 2 hr, 40°C HS or a 2 hr, 45°C HS induces a near normal HS response as noted above (Nagao et al., 1986). However, a 5 to 10 min, 45°C HS on the second cycle produces only a minimal HS response, similar to an initial 5 to 10 min, 40°C HS. Additionally, an initial 10 min, 50°C HS is very toxic and results in essentially no HS mRNA and HS protein accumulation over the next 2 hr; yet a second cycle, 10 min, 50°C HS produces a near normal HS response. In support of the general statements made above a summary set of data for pFS2033 mRNA accumulation during these various HS regimes is presented in Figure 1.

Clearly, the cells/tissues perceive 40°C, 45°C, and 50°C very differently depending upon (a) the length of HS at the various temperatures and (b) whether the seedlings have been exposed to an initial thermoprotective HS prior to the second HS. It appears that the thermoprotected state plays an important role in the response to varying temperatures during the second HS. Accordingly, the response to the initial HS is a corollary to that in that the treatment which produces a full HS response is also one which leads to a thermoprotected state while not being lethal. While we have no definitive answers to the intriguing regulatory questions raised by these observations, a framework has been established to permit the design of experiments which should provide answers to these questions and others on the mechanisms operative in the regulation of the HS response.

HEAT SHOCK GENE STRUCTURE

Significant progress in understanding the HS response has been made through the analyses and subsequent manipulation of the HS protein genes. The Drosophila genes have been the most extensively investigated, but many genes for HS proteins have been cloned and sequenced from several organisms (Nover et al., 1984; Craig, 1985; Lindquist, 1986; Nagao et al., 1986). One obvious parameter that determines gene expression is the presence of regulatory sequences in the HS genes. The identification of sequences necessary for induction of HS genes was initially accomplished by introduction of the genes into cells of other organisms, but investigations have been extended to homologous systems. A consensus Heat Shock Element (HSE; CTnGAAnnTTCnAG) is located about 20 bp upstream of the TATA-motif which is responsible for heat activation of the gene (Pelham, 1982). Basically the activation of HS genes is thought to be mediated by a transcription factor which after activation by HS binds tightly to the HSE regulatory sequence element in the promoter. While this mechanism of activation is consistent with heat induction, additional regulatory regions influence the level of transcription. For review of this literature see reviews noted above and those by Pelham (1985), Bienz (1985) and Voellmy (1985).

DNA sequence analyses of plant HS protein genes have identified sequences homologous to the Drosophila consensus HSE (Nagao et al., 1986). A comparison of potential proximal HSEs from plants with hsp70, hsp27 and the Drosophila consensus HSE are summarized in Table 1. Transgenic gene expression in heterologous systems has established that an 8 out of 10 match to the core inverted repeat of the HSE is sufficient for heat inducibility. The hsp27 gene is included to illustrate an exception where a 6 out 10 match can function when additional copies of the consensus are present. This is relevant to genes such as Gmhsp26-A where multiple HSEs can be identified but for which no match is greater than 6 out of 10 (Czarnecka et al., unpublished data). Most soybean genes sequenced to date have multiple HSEs with the proximal HSE consisting of an overlapping tandem element. The Arabidopsis HSEs are unusual in that they appear to consist of 13 bp rather than 14 bp dyad elements.

Comparison of the upstream 5' flanking region of four soybean HS protein genes in the 17 kD range shows relatively high nucleotide sequence homology (Nagao et al., 1985). A TATA-like motif, TTTAAATA, is present in each of these HS genes positioned 27 to 31 bp upstream of the transcription start site. Upstream, 31 and 41 bp from the 5' end of the TTTAAATA motif, are the 5' ends of two overlapping HSEs with additional HSEs located at various positions further upstream in each of these genes (Nagao et al., 1985).

Sequence comparisons of the 5' flanking regions of Gmhsp18.5-V and Gmhsp22-K which represent HS protein genes of differing molecular weight showed different TATA-like motifs between these two genes and the 17 kD group. Overlapping HSEs are present in both genes but the distance between the TATA-motif and the HSEs is greater for Gmhsp22-K than for the 17 to 18.5 kD group (Nagao et al., unpublished data). Additional HSEs are found

Table 1. Comparison of soybean and Drosophila sequences: Potential proximal heat shock promotor or regulatory elements (HSEs)

Sequence	Source	Reference
CTnGAAnnTTCnAG or CnnGAAnnTTCnnG	Drosophila consensus	Pelham, H. (1985)
(-61)CTcGAAtgTTCgcG(-48)	Drosophila hsp70	Pelham, H. (1982)
(-57)CTaGtgtgTgtgAG(-44)	Drosophila hsp27	Ingolia, T. D. and E. A. Craig (1981)
(-60)CTcGAAagTaCtAt(-47)	Soybean hs6871	Schoffl, F. et al. (1984)
(-62)CTgGAAcaTaCaAG(-49)	Soybean Gmhsp17.5-E	Czarnecka, E. et al. (1985)
CTaGAAacTTCtAG	Soybean Gmhsp17-BR	Nagao, R. T. et al., unpubl. data
(-62)CTgGAAcgTaCacG(-49)	Soybean Gmhsp17.5-M	Nagao, R. T. et al., (1985)
(-62)CTgGAAcaTaCaAG(-49)	Soybean Gmhsp17.6-L	Nagao, R. T. et al., (1985)
(-65)CTgGAAaacaCcAG(-52)	Soybean Gmhsp18.5-V	Nagao, R. T. et al., unpubl. data
(-87)aTgGAttcTTCaAG(-74)	Soybean Gmhsp22-K	Nagao, R. T. et al., unpubl. data
aTgGAccgTTtaAa	Soybean Gmhsp26-G	Nagao, R. T. et al., unpubl. data Czarnecka, E. et al., unpubl. data
(-63)CCcGAAtcTTCtgg(-50)	Maize HSP70	Rochester, D. E. et al., (1986)
(-61)aTgGAAacTctAG(-49)	Arabidopsis HSP70-1	Somerville, C. et al., unpubl. data
CTgGAAacTcaAG	Arabidopsis HSP70-2	Somerville, C. et al., unpubl. data

Proximal HSE represents the sequence having the greatest homology to the Drosophilia consensus positioned just upstream of the TATA box; most soybean genes sequenced to data have multiple HSEs positioned 5' to this proximal element and often overlapping HSEs.

upstream in each of these genes. Only one gene sequenced thus far, Gmhsp17-BR, does not have multiple HSEs within 900 bp of upstream flanking region; the HSE of this gene has perfect homology to the consensus, however (Nagao et al., unpublished data). Experiments to compare and evaluate the importance of the presence, location and sequence of upstream HSEs are in progress and discussed in the Transgenic Gene Expression section.

Heavy metals such as cadmium induce the synthesis of HS mRNAs and HS proteins in many organisms including soybean. These metals also induce the synthesis of metal-binding proteins, the metallothioniens, and specific regulatory elements known as metal responsive elements (MREs) (Karen et al., 1984) have been identified in the 5′ flanking regions of these genes. Homology searches have identified regions homologous to MREs in the 5′ flanking region of the soybean HS protein genes (Czarnecka et al., 1985; Nagao et al., 1985). Homology searches for other sequences which potentially could have regulatory significance have identified SV40 enhancer-like sequences, alternating purine-pyrimidine stretches having potential Z-DNA configuration, sequences of dyad symmetry, and hormone responsive elements (e.g. ecdysterone). The hormone responsive element is relevant because the expression of subsets of HS protein genes during development of Drosophila suggests that there would be additional regulatory elements such as hormone-receptor (e.g. ecdysterone) binding regions and corresponding specific factors or secondary promoter elements controlling these genes (Cohen and Meselson, 1985). Given the high conservation of the HS response, homology identification of these regulatory regions within the soybean genes is suggestive of function. As summarized below many of these regions are currently being tested for their functional significance in transformed tissues of sunflower.

HEAT SHOCK PROTEIN STRUCTURE

Characterization of HS cDNA and genomic clones has demonstrated that the soybean low molecular weight HS protein genes represent multigene families (Nagao et al., 1985; Nagao et al., 1986) with similarities to the low molecular weight group of HS proteins first characterized in Drosophila. Hybrid-selection translation and DNA sequence analyses indicate that the largest soybean family consists of thirteen proteins represented by cDNA clones pFS2109, pCE53, pFS2005 and pFS1968. The corresponding genomic clones which have been sequenced include Gmhsp17.5-E (Czarnecka et al., 1985), Gmhsp17.5-M, Gmhsp17.6-L (Nagao et al., 1985), HS6871 (Schoffl et al., 1984) and Gmhsp18.5-V and Gmhsp17-BR (Nagao et al., unpubl. data). Amino acid sequence comparisons of four soybean HS proteins of 17.3 to 17.6 kD revealed homologies of greater than 90% with approximately two-thirds of the nucleotide changes being silent substitutions (Nagao et al., 1985). Comparison of an 18.5 kD HS protein gene sequence with the four 17 kD sequences showed approximately 75% amino acid identity among the five sequences and 96% amino acid identity of Gmhsp18.5-V to at least one of the four 17 kD proteins. Of eleven amino acid changes unique to Gmhsp18.5-V, nine are located in one region close to the amino terminus of the protein. This is interestingly the same region that a single amino acid deletion is predicted in genes Gmhsp17.5-M and HS6871 in order to maintain amino acid alignment. This would suggest that this region is less significant functionally and therefore less conserved evolutionarily.

Based on hybrid-select translation and DNA sequence analysis data, additional families of small HS proteins include cDNA clones pCE75 (15 kD class), pFS2033, pEV1, pEV2 and pEV6 (21 to 24 kD proteins). Alignment of the deduced amino acid sequence of the 21 to 24 kD group with the 15 to 18 kD group shows lower overall homology, but distinct regions are conserved especially towards the carboxyl terminal regions of the proteins.

The 26 to 28 kD HS proteins represent additional multigene families of HS proteins. One gene represented by cDNA clone pEV3 encodes a 27 kD nuclear encoded HS protein which is transported into the chloroplast and processed to a 22 kD protein (Vierling et al., 1986). A class of general stress proteins is represented by cDNA clone pCE54. This cDNA hybrid-select translated five proteins of approximately 27 kD (Czarnecka et al., 1984). Sequence analysis of one genomic clone of the pCE54 group, Gmhsp26-A, gave a deduced amino acid sequence yielding a 26 kD polypeptide (Czarnecka et al., unpublished data).

Another means to evaluate common structural features of proteins is by hydropathy plots (Kyte and Doolittle, 1982). Comparison of the Gmhsp17.6-L as a representative of the soybean low molecular weight HS proteins with small HS proteins from other organisms shows a structural similarity of these proteins from such diverse organisms as Drosophila, Xenopus, and C. elegans (Nagao et al., 1985). Comparison of various soybean low molecular weight HS proteins ranging from 15 to 27 kD also shows similarity of structural features among these families of HS proteins.

A comparison of the base changes of several small HS protein genes with regard to composition and location suggests a duplication mechanism for the origin of these HS genes (Nagao et al., 1985). The very high homology among the coding regions of the low molecular weight HS genes of soybean and the conservation of hydropathic domains of HS proteins of differing molecular weight and species of origin suggests evolutionary conservation of function among these different proteins (Nagao et al., 1986).

ANALYSES OF THE PROMOTER STRUCTURE OF A HS GENE

Several approaches are being used to analyze the cis and trans regulatory elements involved in HS gene expression. Some results from these early experiments are summarized here.

a. Deletion analysis

The soybean Gmhsp17.5-E gene has been used in these experiments. This gene encodes a low molecular weight HS protein of 17.5 kDa (Czarnecka et al., 1985; Nagao et al., 1985) and is one of the most actively expressed HS genes in soybean (Schoffl and Key, 1982; Czarnecka et al., 1984). The 5' flanking region (Figure 2) contains three TATA-proximal HSEs that show 9, 6, and 7 out of 10 bp matches with the Drosophila consensus. Multiple HSEs (6 to 8 out of 10 bp matches) are located TATA-distal from position -358 to -900. The imperfect pentameric dyad (5'-AGAA_TTCT-3') is present in several Drosophila HS genes and is also found in Gmhsp17.5-E adjacent to a TATA-like sequence at position -221 (TATA/dyad). Other regions of interest include an 11 out of 14 bp match with the SV40 enhancer core sequence which overlaps the third TATA-proximal HSE at -87, two short stretches of potential Z-DNA upstream of the SV40 enhancer-like sequence, and homology to the steroid binding site in a Drosophila HS gene (Bienz, 1985).

A series of 5' deletions of Gmhsp17.5-E were analyzed in crown gall tumors of sunflower using a T-DNA vector that contained a single copy of the HS gene (Gurley et al., 1986). Results of this study showed that basal transcription was increased by removal of sequences far upstream, from position -3250 to -1175. Deletion of sequences -1175 to -95 removed several clusters of TATA-distal HSEs, the TATA/pentameric dyad and a portion of the third TATA-proximal HSE (-100 to -87). As a result, the amplitude of HS-inducible transcription was severely reduced to approximately 5% of the wild type level. The remaining activity, however, was still responsive to HS activation indicating that while the sequences involved in thermal induction reside downstream of position -95, sequences upstream determine the extent of transcription (see below).

These initial observations have been extended by analyzing more deletions and by incorporating a reference copy of Gmhsp17.5-E into the vector to serve as an internal standard for transcript quantification (Czarnecka, unpublished). By constructing a small duplication and substitution mutation in the reference gene leader sequence, RNA derived from the test gene could be distinguished from the longer reference gene transcripts by S1 nuclease hybrid protection mapping (Bruce and Gurley, 1987). The use of a T-DNA vector ensured a relatively low copy number per cell of the introduced gene, thereby eliminating the potential problems of template saturation sometimes encountered in transient expression assays. The variability in expression often seen between

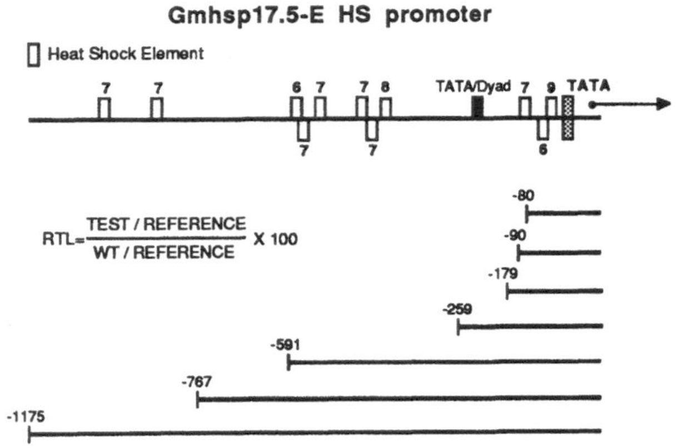

Figure 2. 5'-Deletion analysis of the soybean HS gene Gmhsp17.5-E in sunflower tumors. RTLs were determined by S1 nuclease mapping of poly(A) RNA using a homologous reference gene incorporated into the T-DNA-based vector as an internal standard (from Gurley et al., 1987).

individual tumors and regenerated plants was minimized by pooling 200 to 300 tumors for each mutant analyzed. Actual variability was ±10% of wild type activity between S1 nuclease mapping experiments using the same poly (A) RNA or between assays performed on RNA obtained from independently inoculated tumor batches (Gurley et al., 1987). The combination of an internal standard and the pooling of large numbers of tumors ensures reproducibility in results sufficient to resolve promoter fine structure.

Although HSEs seem to be the primary sequence involved in thermoinducibility of HS genes, the deletion results using the reference gene vector suggest that other sequences in the TATA-distal regions of the Gmhsp17.5-E promoter region may play a very important role in expression of this gene. A diagrammatic representation of the deletions analyzed using the reference gene vector is shown in Figure 2. The contribution of the TATA-distal HSEs to thermoinducibility is not clear from the results of these deletion analyses. For example, removal of sequences from -1175 to -767 containing two HSEs resulted in only a 10% loss in promoter activity, and deletion of a second group of six HSEs located between -591 and -259 had no apparent effect on the level of heat inducible transcription.

In contrast to the questionable role of the TATA-distal HSEs in thermoinduction, several other upstream regions of the Gmhsp17.5-E promoter are required for full transcriptional expression. Deletion of sequences from -767 to -591 resulted in a 14% drop (90% to 76%) in promoter activity. The only distinguishing feature of this region is its relatively high AT content. A second TATA-distal region that contributes to promoter activity is located within the area from -259 to -179. Removal of this region by sequential 5'-deletions reduced HS inducible transcription from 76% of wild type levels with the -259 deletion to 48% activity with the -179 deletion mutant. The only obvious homology with other sequences found in HS genes is the redundant TATA and associated pentameric dyad (5'-TATAAAGAA_TTCT-3'; TATA/dyad) located from -226 to -213. Although this sequence has not been shown to affect transcription in animal HS genes (Pelham, 1982), its conservation between Drosophila and soybean, together with these deletion results, suggest that it may contribute to Gmhsp17.5-E activity.

b. Factor binding

The formation of specific protein:DNA complexes has been analyzed by gel mobility retardation assays ("band shift") (Singh et al., 1986) and by DNAse I footprinting (Czarnecka et al., unpublished). The band shift assay works well in the testing of crude nuclear extracts and column fractions that may have endogenous DNAse. Using the band shift technique, bands of reduced mobility are seen representing the formation of specific protein complexes with the 5'-flanking DNA of Gmhsp17.5-E. The specificity of the interaction is seen by the competition of binding with unlabeled probe DNA and the lack of competition by nonspecific DNA (M13 RF DNA). DNAse I footprinting studies (Czarnecka et al., unpublished) with Gmhsp17.5-E indicate that the TATA-proximal domain of the promoter (-40 to -153) is bound in a DNA:protein complex in vitro by nuclear factors from soybean plumules. This protected region not only contains three HSEs, but other sequences as well which do not conform to known consensus motifs. The potential complexity of the TATA-proximal domain suggests that more than one type of transcription factor may interact in this area to modulate thermoinducible expression. A summary diagram of our preliminary DNAse I footprinting experiments is presented in Figure 3.

c. Summary of promoter structure

The overall view of the structure of the Gmhsp17.5-E promoter is based on in vivo expression of deletion mutants, preliminary DNAse I footprinting studies, and by analogy with the models developed for the role of HSEs and the HSTF (heat shock transcription factor) in Drosophila HS genes (Parker and

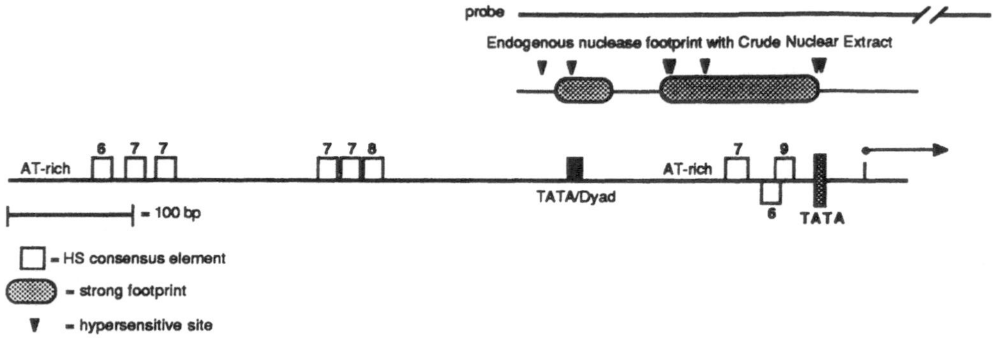

Figure 3. Summary of DNAse I footprint mapping of the Gmhsp17.5-E promoter. Nuclear extracts (S-100) were prepared from dark grown soybean plumules and incubated *in vitro* with 3′ end-labeled DNA restriction fragments from the 5′-flanking region of the gene and varying concentrations of DNAse I.

Topol, 1984; Wu, 1984a, 1984b, 1985; Topol et al., 1985; Shuey and Parker, 1986). The HS genes of soybean have regulatory sequences in the promoter that show a close homology to the HS consensus element (HSE) (5′-CTnGAAnnTTCnAG-3′) of *Drosophila*. As with animal HS genes, the dominant regulatory elements seem to be located in the TATA-proximal domain. The Gmhsp17.5-E promoter has three regions of homology to the HSE located between -49 and -99. Disruption of the distal HSE (7 out of 10 homology in wild type gene) by 5′-deletion, or substitution, significantly reduces thermoinducible activity suggesting that multiple HSEs within the TATA-proximal domain contribute strongly to induction (Gurley et al., 1986; Gurley et al., 1987). Removal of HSEs located far upstream had little or no effect on transcription. Additional sequence motifs such as the SV40 enhancer core (11 out of 14 bp match) and an inverted CCAAT box-like sequence are also found in the TATA-proximal domain but their role is still undetermined.

Nuclear factor preparations bind specifically to the AT-rich region from -153 to -119 suggesting that this area may be involved in the transcriptional regulation of the Gmhsp17.5-E promoter. A second AT-rich region located further upstream between deletions -767 and -591 also seems to be required for full activity of this promoter (Figure 2). The contribution of AT-rich sequences to promoter activity is consistent with the results of Baumann et al. (1987) who have identified a similar AT-rich region (14 consecutive adenosine residues) located from position -250 to -274 in the 5′-flanking region of hs6871. The possibility of multiple classes of *cis*-acting elements within the TATA-proximal domain is not unlikely since the analogous region of the human hsp70 promoter has been shown to contain binding sites for three types of *trans*-acting factors (Wu et al., 1987). In this regard, the unique aspect of the AT-rich areas of the plant HS genes is their apparent involvement as accessory elements contributing to maximum thermoinducibility. It is interesting to speculate that these types of regions may be responsible for the very high abundance of low molecular weight HS proteins found in plants.

Sequences immediately upstream of TATA seem to constitute a discrete functional domain. This TATA-proximal domain includes several subelements and extends from around -40 to approximately -160. The entire TATA-proximal element appears to bind proteins *in vitro* using nuclear extracts (S-100) from either control (28°C) or HS (40°C) tissues. The proximal portion of this region contains three HSEs with an inverted CCAAT-like motif between the second and third. From our DNAse I footprinting studies, an AT-rich region (-153 to -119) upstream of the TATA-proximal HSEs also seems to be a part of this domain. One interpretation of these results is that proteins bound to the AT-rich region may interact with HSTFs bound TATA-proximal to form a preinitiation complex with the TATA-binding factor.

The TATA-distal region seems to be comprised of at least three types of cis-acting elements: HSEs, an AT-rich region (-767 to -591), and the TATA/pentameric dyad at -231. Some of the TATA-distal HSEs and the TATA/dyad show some evidence of specific protein binding from the results of our preliminary DNAse I footprinting studies. Although the TATA-distal AT-rich region has not been examined by footprinting, the deletion analysis indicates that it also contributes to heat inducible transcription. The TATA/dyad seems to interact with a protein in vitro, and has been shown to contribute to promoter activity by these deletion studies. Due to the stepwise reduction in promoter activity obtained with sequential 5'-deletions, and the variety of consensus sequences and protein binding sites, the TATA-distal region is probably best viewed as a collection of separate promoter elements rather than as a single highly organized functional domain. Additional experiments using linker scan and site directed mutagenesis techniques should assist in further identifying and defining these promoter elements.

CONCLUSIONS

Plants respond to HS by producing a relatively complex group of HS proteins which appear to be functionally involved in providing thermotolerance. The expression of these genes is controlled at the transcriptional and translational levels. Plants can measure the severity and extent of heat stress to self-regulate or autoregulate HS mRNA and HS protein synthesis. The plant HS genes show homology to those of other systems. Based upon DNA sequence analyses and hybrid-select translation analyses, the soybean low molecular weight HS protein genes are complex multigene families. There is less homology among the low molecular weight HS proteins between plants and Drosophila. However, there is striking conservation of some structural features or domains among the corresponding low molecular weight HS proteins.

There appears to be very high conservation of at least one or more transcriptional regulatory regions of these genes (e. g. HSEs) among organisms. Transgenic gene expression of a series of 5' promoter deletions, band shift, and DNAse I footprinting experiments are being used to identify and evaluate regions of regulatory significance. The removal of a region containing a relatively high AT content or a redundant TATA with an associated pentameric dyad reduced heat inducible expression. These constitute TATA-distal regions in addition to HSEs which modulate the expression of HS genes. Whether these regions are unique to a few plant genes or conserved as accessory regulatory elements across kingdoms is not known. While substantial progress has been made toward the understanding of the HS response in plants, considerable research effort is still required to (a) define and understand the mechanism(s) (cis- and trans-acting factor interactions) operative in regulating the expression of HS genes in response to heat stress and other inducing stimuli, (b) identify the transducing mechanism(s) to perceive and measure the severity of stress, (c) define the mechanism(s) functioning in the self-regulating or autoregulation of the HS response, (d) identify the mechanism of preferential translation of HS mRNAs during heat stress, (e) identify the mechanism of preferential degradation of HS mRNAs during recovery at control temperatures, (f) define the function of HS proteins in the acquisition of thermotolerance, and (g) ascertain the function of constitutive and developmental expression of certain HS protein genes. This list of unanswered aspects of the HS response is by no means complete, but rather presented as perspective foci for future research in the important area of stress biology.

REFERENCES

Altschuler, M., and Mascarenhas, J. P., 1982, The synthesis of heat-shock and normal proteins at high temperatures in plants and their possible roles in survival under heat stress, in: "Heat Shock from Bacteria to Man," M. J. Schlesinger, M. Ashburner, and A. Tissieres, eds., Cold Spring Harbor Laboratory, Cold Spring Harbor, New York, pp. 321-327.

Ashburner, M., and Bonner, J. J., 1979, The induction of gene activity in Drosophila by heat shock, Cell, 17:241-254.

Baumann, G., Raschke, E., Bevan, M., and Schoffl, F., 1987, Functional analysis of sequences required for transcriptional activation of a soybean heat shock gene in transgenic tobacco plants, EMBO J., 6:1161-1166.

Bienz, M., 1985, Transient and developmental activation of heat-shock genes, Trends Biochem. Sci., 10:157-161.

Bruce, W. B., and Gurley, W. B., 1987, Functional domains of a T-DNA promoter active in crown gall tumors, Mol. Cell. Biol., 7:59-67.

Cohen, R. S. and Meselson, M., 1985, Separate regulatory elements for the heat-inducible and ovarian expression of the Drosophila hsp26 gene, Cell, 43:737-746.

Cooper, P., Ho, T.-H. D., and Hauptmann, R. M., 1984, Tissue specificity of the heat shock response in maize, Plant Physiol., 75:431-441.

Craig, E. A., 1985, The heat shock response, CRC Crit. Rev. Biochem., 18:239-280.

Czarnecka, E., Edelman, E., Schoffl, F., and Key, J. L., 1984, Comparative analysis of physical stress responses in soybean seedlings using cloned heat shock cDNAs, Plant Mol. Biol., 3:45-58.

Czarnecka, E., Gurley, W. B., Nagao, R. T., Mosquera, L., and Key, J. L., 1985, DNA sequence and transcript mapping of a soybean gene encoding a small heat shock protein, Proc. Natl. Acad. Sci. USA, 82:3726-3730.

DiDomenico, B. J., Bugaisky, G. E., and Lindquist, S., 1982, The heat shock response is self-regulated at both the transcriptional and posttranscriptional levels, Cell, 31:593-603.

Gurley, W. B., Czarnecka, E., Nagao, R. T., and Key, J. L., 1986, Upstream sequences required for efficient expression of a soybean heat shock gene, Mol. Cell. Biol., 6:559-565.

Gurley, W. B., Bruce, W. B., Czarnecka, E., Bandyopadhyay, R., Nagao, R. T., and Key J. L., 1987, Cis-regulatory elements in heat shock and T-DNA promoters, in: "Plant Gene Systems and Their Biology," L. McIntosh and J. L. Key, eds., UCLA Symposia on Molecular and Cellular Biology, New Series, Alan R. Liss, New York (in press).

Ingolia, T. D. and Craig, E. A., 1981, Primary sequence of the 5′ flanking region of the Drosophila heat shock genes in chromosome subdivision 67B, Nucleic Acids Res., 9:1627-1642.

Karin, M., Haslinger, A., Holtgreve, H., Richards, R. I., Krauter, P., Westphal, H. M., and Beats, M., 1984, Characterization of DNA sequences through which cadmium and glucocorticoid hormones induce human metallothionein-IIa gene, Nature, 308:513-519.

Key, J. L., Lin, C.-Y., and Chen, Y.-M., 1981, Heat shock proteins of soybean, Proc. Natl. Acad. Sci. USA, 78:3526-3530.

Key, J. L., Czarnecka, E., Lin, C.-Y., Kimpel, J., Mothershed, C. and Schoffl, F., 1983, A comparative analysis of the heat shock response in crop plants, in: "Current Topics in Plant Biochemistry and Physiology," D. D. Randall, D. G. Blevins, R. L. Larson, and B. J. Rapp, eds., University of Missouri, Columbia Press, pp.107-118.

Key, J. L., Kimpel, J. A., Lin, C.-Y., Nagao, R. T., Vierling, E., Czarnecka, E., Gurley, W. B., Roberts, J. K., Mansfield, M. A., and Edelman, L., 1985a, The heat shock response in soybean, in: "Cellular and Molecular Biology of Plant Stress," J. L. Key and T. Kosuge, eds., UCLA Symposia on Molecular and Cellular Biology, New Series, Alan R. Liss, New York, pp. 161-179.

Key, J. L., Kimpel, J., Vierling, E., Lin, C.-Y., Nagao, R. T., Czarnecka, E., and Schoffl, F., 1985b, Physiological and molecular analyses of the heat shock response in plants, in: "Changes in Eukaryotic Gene Expression in Response to Environmental Stress," B. G. Atkinson and D. B. Walden, eds., Academic Press, Inc., New York, pp. 327-348.

Key, J. L., Czarnecka, E., Gurley, W. B., and Nagao, R. T., 1987a, An analysis of physiological and molecular aspects of heat shock gene expression, in: "Tailoring Genes for Crop Improvement," G. Bruening, J. Harada, T. Kosuge, and A. Hollaender, eds., Plenum Press, New York, pp. 103-109.

Key, J. L., Kimpel, J., and Nagao, R. T., 1987b, Heat shock gene families of soybean and the regulation of their expression, in: "Plant Gene Systems and Their Biology," L. McIntosh and J. L. Key, eds., UCLA symposia on Molecular and Cellular Biology, New Series, Alan R. Liss, New York, (in press).

Kyte, J. and Doolittle, R. F., 1982, A simple method for displaying the hydropathic character of a protein, J. Mol. Biol., 157:105-132.

Lindquist, S. and DiDomenico, B., 1985, Coordinate and noncoordinate gene expression during heat shock: A model for regulation, in: "Changes in Eukaryotic Gene Expression in Response to Environmental Stress", B. G. Atkinson and D. B. Walden, eds., Academic Press, pp. 72-89.

Lindquist, S., 1986, The heat-shock response, Ann. Rev. Biochem., 55:1151-1191.

Mansfield, M. A., 1986, The heat shock response of soybean: Synthesis, accumulation, and distribution of the low molecular weight heat shock proteins, Ph.D. dissertation, University of Georgia, Athens, Georgia.

Mansfield, M. A., Lingle, W. L., and Key, J. L., 1987, An ultrastructural analysis of the effects of lethal heat shock on non-adapted and thermotolerant root cells of Glycine max, J. Ultrastruct. Res. (submitted for publication).

Mirault, M.-E., Goldschmidt-Clermont, Moran, M., Arrigo, A. P., and Tissieres, A., 1978, The effect of heat shock on gene expression in Drosophila melanogaster, Cold Spring Harbor Symp. Quant. Biol., 42:819-827.

Nagao, R. T., Czarnecka, E., Gurley, W. B., Schoffl, F., and Key, J. L., 1985, Genes for low-molecular-weight heat shock proteins of soybeans: Sequence analysis of a multigene family, Mol. Cell. Biol., 5:3417-3428.

Nagao, R. T., Kimpel, J. A., Vierling, E., and Key, J. L., 1986, The heat shock response: A comparative analysis, in: "Oxford Surveys of Plant Molecular and Cell Biology," B. J. Miflin, ed., Oxford University Press, Oxford, England, Vol. 3, pp. 384-438.

Neidhardt, F. C., Van Bogelen, R. A., and Vaughn, V., 1984, The genetics and regulation of heat-shock proteins, Ann. Rev. Genet., 18:295-329.

Nover, L., Hellmund, D., Neumann, D., Scharf, K.-D., and Serfling, E., 1984, The heat shock response of eukaryotic cells, Biol. Zbl., 103:357-435.

Parker, C. S. and Topol, J., 1984, A Drosophila RNA polymerase II transcription factor binds to the regulatory site of an hsp 70 gene, Cell, 37:273-283.

Pelham, H. R. B., 1982, A regulatory upstream promoter element in the Drosophila hsp70 heat-shock gene, Cell, 30:517-528.

Pelham, H. R. B., 1985, Activation of heat-shock genes in eukaryotes, Trends Gen. Sci., 1:31-35.

Rochester, D. E., Winter, J. A., Shah, D. M., 1986, The structure and expression of maize genes encoding the major heat shock protein, hsp70, EMBO J, 5:451-458.

Schoffl, F., and Key, J. L., 1982, An analysis of mRNAs for a group of heat shock proteins of soybean using cloned cDNAs, J. Mol. Appl. Genet., 1:301-314.

Schoffl, F., Raschke, E., and Nagao, R. T., 1984, The DNA sequence analysis of soybean heat shock genes and identification of possible regulatory promoter elements, EMBO J., 3:2491-2497.

Shuey, D. J. and Parker, C. S., 1986, Binding of Drosophila heat-shock gene transcription factor to the hsp70 promoter, J. Biol. Chem., 261:7934-7940.

Singh, H., Sen, R., Baltimore, D., Sharp, P. A., 1986, A nuclear factor that binds to a conserved sequence motif in transcriptional control elements of immunoglobulin genes, Nature, 319:154-158.

Tilly, K., McKittrick, N., Zylicz, M., and Georgopoulos, C., 1983, The dnaK protein modulates the heat-shock response of Escherichia coli, Cell, 34:641-646.

Topol, J., Ruden, D. M. and Parker, C. S., 1985, Sequences required for in vitro transcriptional activation of a Drosophila hsp70 gene, Cell, 42:527-537.

TISSUE SPECIFICITY AND DYNAMICS OF DISEASE RESISTANCE RESPONSES IN PLANTS

Klaus Hahlbrock, Claude Cretin, Beate Cuypers, Karl-Heinz Fritzemeier, Karl-Dietrich Hauffe, Willi Jahnen, Erich Kombrink, Frauke Rohwer, Dierk Scheel, Elmon Schmelzer, Martin Schröder, and Janet Taylor

Max-Planck-Institut für Züchtungsforschung, D-5000 Köln 30, FRG

INTRODUCTION

Plants showing biochemical responses when challenged with potential pathogens can be classified into two groups: host and nonhost plants. Host plants are operationally defined by the occurrence of both resistant and susceptible varieties depending on the genotype relation to a given race of the pathogen (pathotype). In the case of resistance, the two organisms undergo an incompatible interaction and the pathotype is avirulent; a susceptible plant undergoes a compatible interaction with a virulent pathotype. To emphasize the active participation of both organisms in the interplay of attempted invasion and defense, the terms *compatibility* and *imcompatibility* will be used mostly in the following sections. Plants not occurring in the form of a susceptible variety with respect to a given microorganism are called nonhost plants by that definition. Consequently, if the two organisms interact at all, this interaction is always incompatible.

Compatible plant-pathogen interactions are rare cases when compared with the large number of either incompatible interactions (most often involving nonhost plants) or indifferent combinations with no detectable interaction at all. However, the relatively few compatible interactions include those of greatest economical importance, *e. g.*, the devastating late-blight infections of potato that caused malnutrition, death and large emigration waves from Europe last century. Although the introduction of appropriate genes for resistance by conventional breeding of potato as well as many other crop plants greatly improved the situation for several decades, new problems have arisen in modern agriculture. For example, the large-scale propagation of genetically near-homogeneous cultivars has rendered most of our major crop plants highly vulnerable to the rapid spread of new, or previously less successful, virulent pathogen races, - with the increasingly adverse consequences of the wide application of pesticides.

One possible solution of this problem might be to drastically shorten the breeding procedure for new resistant cultivars and circumvent the time-consuming back-crossing in conventional breeding by transferring individual resistance genes by new gene transfer techniques to previously susceptible plants. However, far more desirable would be the complete change from host susceptibility to nonhost resistance, which so far has not been possible. To find out whether this goal can principally be reached, we are comparing the molecular mechanisms involved in nonhost- and host-plant interactions with potentially pathogenic fungi.

The various presently known biochemical reactions constituting the average defense response of plants to pathogens have been summarized elsewhere (Hahlbrock et al., 1986; Hahlbrock and Scheel, 1987). Using parsley (*Petroselinum crispum*) and potato (*Solanum tuberosum*) as especially suitable plants for studies of nonhost and host responses to infections with *Phytophthora megasperma* f. sp. *glycinea* (*Pmg*) and *P. infestans* (*Pi*), we have concentrated mainly on three defense reactions: The accumulation of phytoalexins, of wall-bound phenolics, and of lytic enzymes. Recent studies on the involvement of gene activation and on the structural organization of some of the responsible genes in parsley are outlined in the chapter by Scheel et al. Here, we report on the tissue-specific occurrence of these compounds as well as some of their biosynthetic enzymes; on the functional identification of several 'pathogenesis-related' (PR) proteins in potato; and on some dynamic aspects of the biochemical reactions investigated. The comparison of data from the two different plant systems intends to emphasize the variability of the response and possible deviations from the average combination of defense reactions.

THE INTERACTION OF PARSLEY WITH *P. MEGASPERMA*

Sites of Phytoalexin Synthesis and Accumulation

In parsley, one component of the nonhost resistance response to *Pmg* is the rapid accumulation of furanocoumarin phytoalexins at infection sites (Scheel et al., 1986). The same compounds are also present in uninfected tissue (Knogge et al., 1987), where they occur at high concentrations in oil ducts (Jahnen and Hahlbrock, unpublished results). S-Adenosyl-L-methionine:bergaptol O-methyltransferase (BMT), an enzyme specific for the furanocoumarin pathway (Hauffe et al., 1986) was localized by indirect immunoperoxidase staining both in oil-duct epithelial cells of uninfected parsley seedlings (Fig. 1) and at *Pmg* infection sites (Jahnen and Hahlbrock, unpublished results).

Similar mixtures of structurally related furanocoumarin derivatives accumulate at the two sites of BMT localization, indicating that essentially the same biosynthetic pathway is expressed in oil ducts and at infection sites. Thus, in this system phytoalexins are synthesized both transiently in response to infection and constitutively in other parts of the tissue.

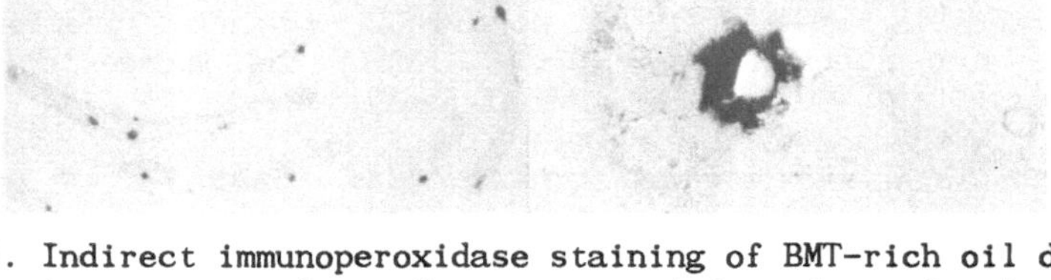

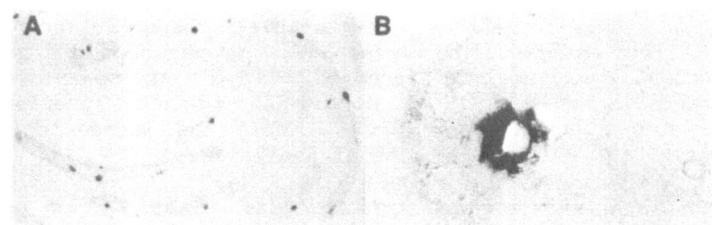

Fig. 1. Indirect immunoperoxidase staining of BMT-rich oil ducts in cross sections of a young, uninfected parsley leaf at low (A) and high (B) magnification.

Gene Activation

Probes to measure BMT transcription rates *in situ* are being generated (Hauffe, unpublished results) but have not been used for such studies. Indirect evidence from experiments with *Pmg* elicitor-treated cell suspension cultures of parsley (see chapter by Scheel et al.) suggests that stimulation of BMT and other enzyme activities is a result of gene activation. To test this in whole plants, preliminary studies on *in situ* mRNA hybridization were carried out using cross-sections of *Pmg*-infected leaves and a cDNA probe specific for PR protein 1 mRNA (Somssich et al., 1986), one of the various elicitor-induced mRNAs in cultured parsley cells. Figure 2 demonstrates that this mRNA accumulates rapidly at infection sites and has reached a high level 12 h post inoculation. Further work will concentrate on the relative timing of induction and the cellular localization of mRNAs synthesized *de novo* in the course of the pathogen defense response. These studies will include mRNAs encoding phytoalexin-biosynthetic enzymes.

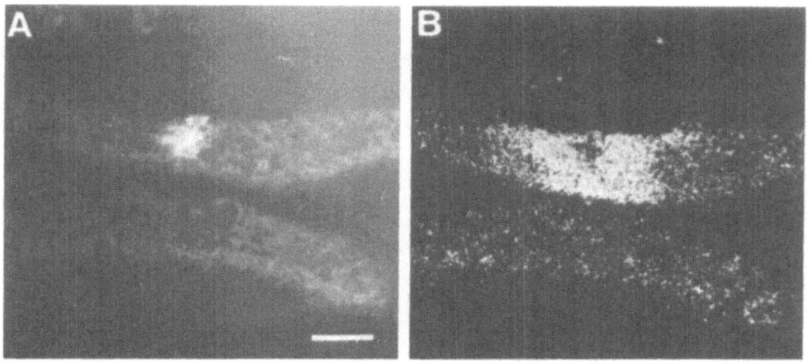

Fig. 2. *In situ* RNA hybridization in a *Pmg*-infected, formaldehyde/glutaraldehyde-fixed parsley leaf section 12 h post inoculation. A, UV fluorescence of infection site. B, Dark-field photomicrograph of the same section hybridized with "oligo"-^{3}H-labeled PR protein 1 cDNA (10^8 dpm/μg). Bar = 100 μm.

INTERACTIONS OF POTATO WITH *P. INFESTANS* AND *P. MEGASPERMA*

Limited Occurrence of Sesquiterpenoid Phytoalexins

In contrast to parsley, where the furanocoumarin phytoalexins were found both at *Pmg* infection sites and in oil ducts of axenically grown plants, the characteristic sesquiterpenoid phytoalexins of potato were detected only in infected tuber tissue. They were observed neither in uninfected tissue nor in infected leaves (Rohwer et al., 1987) and hence represent a striking example of organ-specific occurrence of a class of compounds previously thought to be invariably associated with incompatible interactions of plants with pathogens.

Functional Analysis of PR Proteins

In all cases so far investigated, PR proteins have been found to accumulate in incompatible, often also in compatible, plant-pathogen interactions (van Loon, 1985) and in elicitor-treated plant cell cultures (Somssich et al., 1986; Kombrink et al., 1986). So far, these proteins have been defined merely in operational terms (van Loon, 1985). Their

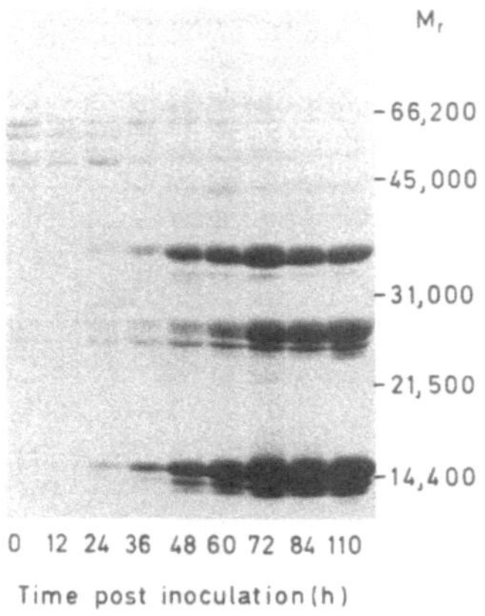

Fig. 3. Time course of changes in the patterns of proteins from the intercellular fluid of *Pi*-infected potato leaves. Separation was by polyacrylamide gel electrophoresis under denaturing conditions.

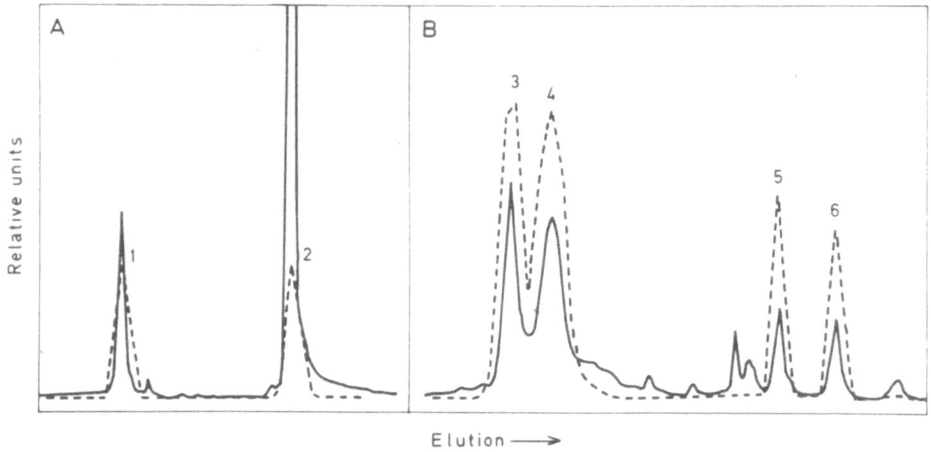

Fig. 4. Proteins (solid line) with chitinase activity (broken line) obtained by cation-exchange FPLC following partial purification on chitinase-affinity and sizing columns. A and B, resolution of first and second of the two peaks with chitinase activity eluting from the sizing column.

biochemical functions have not been reported. We have isolated several PR proteins from the intercellular fluid of *Pi*-infected (Fig. 3) or elicitor-treated potato leaves. They were purified to apparent homogeneity and identified as lytic enzymes. Altogether, six distinct proteins with chitinase activity (Fig. 4) and two with 1,3-β-glucanase activity were obtained in pure form. In both cases, the combined acitivities of the separated isoforms accounted for most of the respective enzyme acitivity extracted from the tissue. These results suggest that several major PR proteins in potato constitute the chitinase and 1,3-β-glucanase activities whose stimulation in infected tissue has been observed previously in this and many other plants (Hahlbrock and Scheel, 1987).

Immunohistochemical studies using antibodies to *Pmg* and *Pi* surface molecules (Cuypers, Jahnen and Hahlbrock, unpublished results) have enabled the localization of fungal hyphae in the invaded plant tissue (Fig. 5). The results of a detailed analysis of several *Pmg* and *Pi* infection sites in potato leaves are schematically outlined in Fig. 6. They are in agreement with independent results obtained using cDNA probes to determine the timing of changes in the levels of mRNAs encoding the two key enzymes of general phenylpropanoid metabolism, phenylalanine ammonia-lyase (PAL) and 4-coumarate:CoA ligase (4CL), in the course of *Pi* infections of potato leaves (Fig. 7). The changes in the amounts of these mRNAs were probably due to rapid, transient activation of the respective genes (Fritzemeier et al., 1987). These genes are of special interest for further studies of pathogen defense responses in plants. When measured, PAL and 4CL activities have always been found to increase rapidly in association with the hypersensitive cell death in incompatible interactions. Although biochemical details of the involvement of

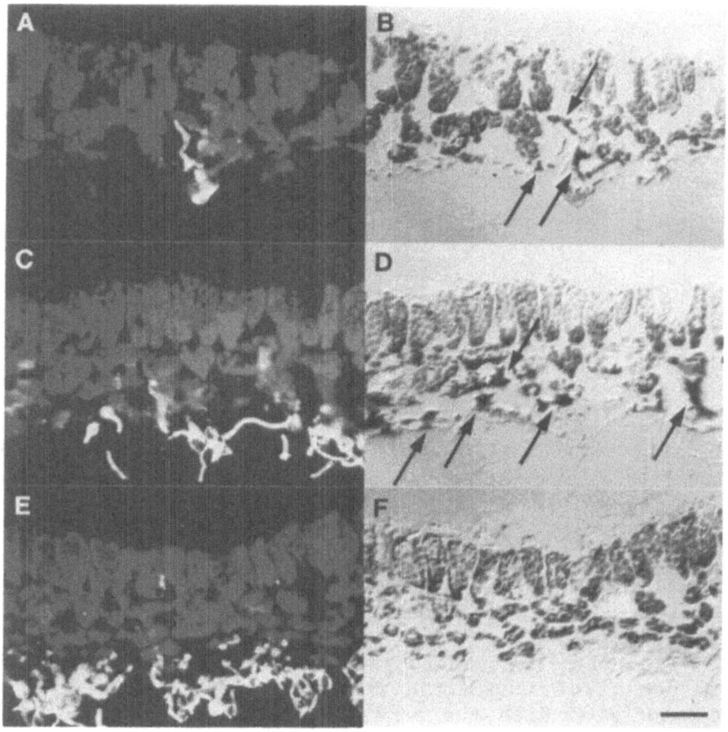

Fig. 5. Localization of fungal hyphae (A, C, E: light structures) and hypersensitive cell death (B, D, F: dark, shrunken cells) in cross sections of potato leaves infected with *Pmg* (upper row, 6 h post inoculation) or *Pi* (center row: incompatible interaction, lower row: compatible interaction; both 8 h post inoculation). A, C, E: yellow (light) fluorescence of antibody-decorated fungus and red (dark) fluorescence of chlorophyll in mesophyll cells under blue light. B, D, F: the same sections seen in visible light. Dark cells in B and D (arrows) are brown and dead. Bar: 50 µm.

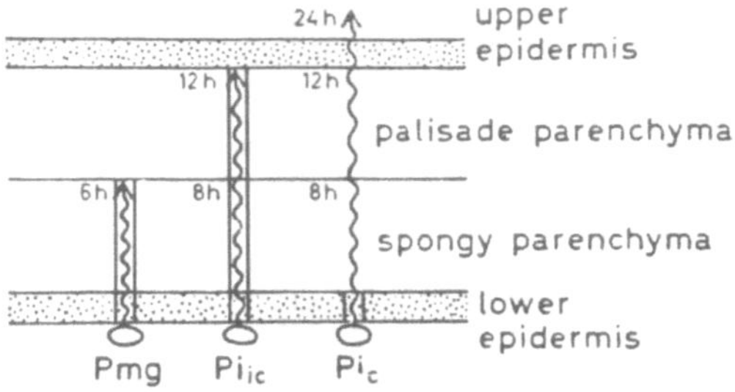

Fig. 6. Schematic representation of results from a series of immunocytological analyses analogous to those shown in Fig. 5. Arrows in boxes: fungal penetration is associated with hypersensitive cell death in the affected tissue. ic = incompatible, c = compatible.

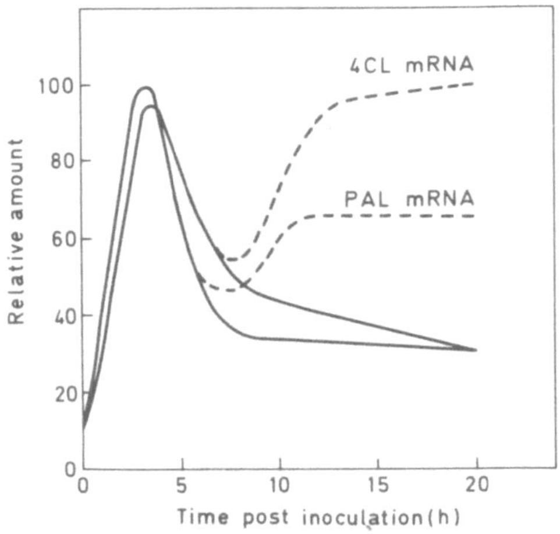

Fig. 7. Time courses of changes in hybridizable PAL and 4CL mRNA amounts in incompatible (solid lines) and compatible (lines ending as broken curves) interactions of potato with *Pi* (data taken from Fritzemeier et al., 1987)

phenylpropanoid metabolism in this process are still unknown, the rapid accumulation of substrates for the incorporation of phenolic compounds into the cell walls in necrotic areas is a likely explanation.

Taken together, the results shown in Fig. 5-7 might be interpreted as follows. The first, rapid and transient increase in PAL and 4CL mRNA levels is associated with the penetration of fungal hyphae through the inoculated epidermal cell layer. This causes hypersensitive cell death in the affected tissue, irrespective of whether the overall interaction is compatible or incompatible. Subsequent penetration of the mesophyll follows similar time courses in both types of host-pathogen interaction, but evokes rapid browning and hypersensitive cell death only in the incompatible interaction. In this case, the fungal hyphae appear to have largely detereorated by the time they either reach the opposite epidermis (host response to *Pi*) or have been confined to smaller areas in the mesophyll (nonhost response to *Pmg*). The upper epidermis is not penetrated in either case. In the compatible interaction, hyphal growth continues through this last cell layer, possibly again causing hypersensitive death of epidermal cells and thereby the second, large increase in the PAL and 4CL mRNA levels (broken lines in Fig. 7). Final steps in this process are sporulation of the fungus and browning and collapse of the whole leaf.

The so far hypothetical, causal relationship between phenylpropanoid metabolism and hypersensitive cell death is now being tested in more detail by *in situ* mRNA hybridization in individual cells at infection sites.

CONCLUSIONS

With regard to the occurrence of the most extensively studied pathogen defense compounds, the antibiotically active phytoalexins, our results indicate that plants can differ greatly from one another. We have presented two extreme cases. In the case of parsley, furanocoumarins meet the widely accepted definition (Paxton, 1982) of low-molecular-weight compounds with antibiotic activity accumulating upon infection, but also occur in uninfected plants elsewhere in the tissue. In the case of potato, the classical sesquiterpenoid phytoalexins, whose accumulation in *Pi*-infected tubers originally formed the basis for the phytoalexin theory of Müller and Börger (1940), were absent in other parts of the tissue as well as *Pi*-infected leaves even in strong incompatible interactions. These observations demonstrate the possible variability in the response of plants to potential pathogens.

The assignement of chitinase and 1,3-β-glucanase activities to PR proteins in potato, together with similar, unpublished results recently obtained with TMV-infected tobacco leaves (Legrand and Fritig, personal communication), suggests that these two lytic enzymes are important components of the pathogen defense response. It will be interesting to see whether all incompatible plant-pathogen interactions involve chitinase and 1,3-β-glucanase activity increases similar to those shown here. These two lytic activities are likely to play a major role in the degradation of fungal and bacterial cell walls, particularly because some, or perhaps all, plant chitinases have lysozyme activity (Boller et al., 1983).

In potato, as well as in parsley (Jahnen and Hahlbrock, unpublished results), much evidence has been accumulated suggesting that the activation of phenylpropanoid pathways is important or even essential for hypersensitive cell death in infected plant tissue to occur. Genes encoding PAL and 4CL are therefore prime candidates for studies of transcription activation mechanisms in relation to this key phenomenon in incompatible plant-pathogen interactions.

All data obtained so far with these two systems support the notion (Bailey and Deverall, 1983) that tissue specificity and the dynamics of the individual reactions are important aspects of disease resistance in plants. Work is in progress involving methods sensitive enough to study both aspects in infected tissue at the cellular level.

REFERENCES

Bailey, J. A., and Deverall, B. J., 1983, "The Dynamics of Host Defence", Academic Press, Sydney, New York, London.

Boller, T., Gehri, A., Mauch, F., and Vögeli, U., 1983, Chitinase in bean leaves: Induction by ethylene, purification, properties, and possible function, *Planta* 157:22-31.

Fritzemeier, K.-H., Cretin, C., Kombrink, E., Rohwer, F., Taylor, J., Scheel, D., and Hahlbrock, K., 1987, Transient induction of phenylalanine ammonia-lyase and 4-coumarate:CoA ligase mRNAs in potato leaves infected with virulent or avirulent races of *Phytophthora infestans*, Plant Physiol., in press.

Hahlbrock, K., Cuypers, B., Douglas, C., Fritzemeier, K.-H., Hoffmann, H., Rohwer, F., Scheel, D., and Schulz, W., 1986, Biochemical interactions of plants with potentially pathogenic fungi, *in*: "Recognition in Microbe-Plant Symbiotic and Pathogenic Interactions", B. Lugtenberg, ed., Springer-Verlag, Heidelberg.

Hahlbrock, K., and Scheel, D., 1987, Biochemical responses of plants to pathogens, *in*: "Innovative Approaches to Plant Disease Control", I. Chet, ed., Wiley & Sons, Inc.

Hauffe, K. D., Hahlbrock, K., and D. Scheel, 1986, Elicitor-stimulated furanocoumarin biosynthesis in cultured parsley cells: S-adenosyl-L-methionine:bergaptol and S-adenosyl-L-methionine:xanthotoxol O-methyltransferase, Z. *Naturforsch.* 41c:228-239.

Knogge, W., Kombrink, E., Schmelzer, E., and Hahlbrock, K., 1987, Phytoalexins and other putative defense-related materials are present in uninfected parsley plants, Planta, in press.

Kombrink, E., Bollmann, J., Hauffe, K. D., Knogge, W., Scheel, D., Schmelzer, E., Somssich, I., and Hahlbrock, K., 1986, Biochemical responses of non-host plant cells to fungi and fungal elicitor, *in*: "Biology and Molecular Biology of Plant-Pathogen Interactions", J. A. Bailey, ed., Plenum Publishing Corporation.

Müller, K. O., and Börger, H., 1940, Experimentelle Untersuchungen über die *Phytophthora*-Resistenz der Kartoffel. *Arb. Biol. Reichsanst. Land Forstwirtsch. Berlin-Dahlem* 23:189-231.

Paxton, J. E., 1982, Phytoalexins, *in*: "Active Defense Mechanisms in Plants", R. K. S. Wood, ed., Plenum Press, New York.

Rohwer, F., Fritzemeier, K. H., Scheel, D. and Hahlbrock, K., 1987, Biochemical reactions of different tissues and cultivars of potato (*Solanum tuberosum*) to zoospores or elicitors from *Phytophthora infestans*. Accumulation of sesquiterpenoid phytoalexins. *Planta* 170:556-561.

Scheel, D., Hauffe, K. D., Jahnen, W., and Hahlbrock, K., 1986, Stimulation of phytoalexin formation in fungus-infected plants and elicitor-treated cell cultures in parsley, *in*: "Recognition in Microbe-Plant Symbiotic and Pathogenic Interactions", B. Lugtenberg, ed., Springer Verlag, Heidelberg.

Somssich, I. E., Schmelzer, E., Bollmann, J., and Hahlbrock, K., 1986, Rapid activation by fungal elicitor of genes encoding 'pathogenesis-related' proteins in cultured parsley cells. *Proc. Natl. Acad. Sci. USA* 83:2427-2430.

Van Loon, L. C., 1985, Pathogenesis-related proteins. *Plant Mol. Biol.* 4:111-116.

THE RESPONSE TO ANAEROBIC STRESS: TRANSCRIPTIONAL REGULATION OF GENES FOR ANAEROBICALLY INDUCED PROTEINS

E.S. Dennis, J.C. Walker, D.J. Llewellyn, J.G. Ellis, K. Singh, J.G. Tokuhisa, D.R. Wolstenholme, and W.J. Peacock

CSIRO Division of Plant Industry, G.P.O. Box 1600, Canberra, A.C.T. 2601, Australia

Plants subjected to anaerobic stress, such as occurs during flooding, respond by switching from oxidative to fermentative carbohydrate metabolism. Under hypoxia, protein synthesis ceases and then there is synthesis of approximately 20 new polypeptides, some of which are known to be proteins catalysing the fermentative reactions. We have examined the DNA sequences responsible for the transcriptional regulation in Adh1 and aldolase. The critical segments of these two genes and the upstream regions of other anaerobic genes have an invariant core hexanucleotide, TGGTTT, in a longer consensus sequence $\substack{TC\\CG\\GT}TGGTTT\substack{TC\\GT\\CG}$. This suggests the anaerobic regulation of transcription of the different anaerobic polypeptide genes is mediated by binding of a common *trans*-acting factor.

Key words: Anaerobiosis, alcohol dehydrogenase, aldolase, sucrose synthase, anaerobic regulatory element, maize, electroporation.

Plants, like many other organisms, respond to conditions of low oxygen tension, such as occurs during flooding, by switching carbohydrate metabolism from an oxidative to a fermentative pathway (Davies, 1980). To enhance survival over long periods of low oxygen-stress, proteins synthesised during aerobic conditions cease being made and a specific set of polypeptides, the anaerobic polypeptides (ANPs) are synthesised (Sachs *et al*., 1980). When the ANPs are analysed by two dimensional polyacrylamide gel electrophoresis approximately 20 major components are seen; these components account for about 70% of the total protein synthesis occurring during hypoxia. This suggests that energy available for general cellular processes is limited and implies that the ANPs have a particular role in cell survival.

When intracellular conditions become anaerobic, lactate is used initially as the terminal electron acceptor. Lactate dehydrogenase reduces pyruvate to lactate. As lactate accumulates, cellular pH drops and there is a shift to the production of ethanol involving pyruvate decarboxylase and alcohol dehydrogenase (Davies, 1980; Roberts *et al*., 1984a,b). The ability of plants to utilise mainly ethanolic fermentation rather than be reliant on lactic fermentation is partly responsible for their ability to withstand hypoxic conditions for periods of up to several days. In vertebrates, prolonged periods of hypoxia lead to lactic acid accumulation, a substantial fall in cytoplasmic pH and eventual cell death. In maize root tips the cytoplasmic pH falls to a stable value, about 0.5 pH unit below the aerobic value (Roberts et al., 1984a,b), and the subsequent switch to ethanolic fermentation does not lead to any further acid accumulation. Mutants of maize (Schwartz, 1969) or barley (Haberd and Edwards, 1982) which are Adh1-null i.e., do not produce any active ADH1 enzyme, have a lower survival rate if flooded during germination than do Adh1-containing lines (Schwartz, 1969; Harberd

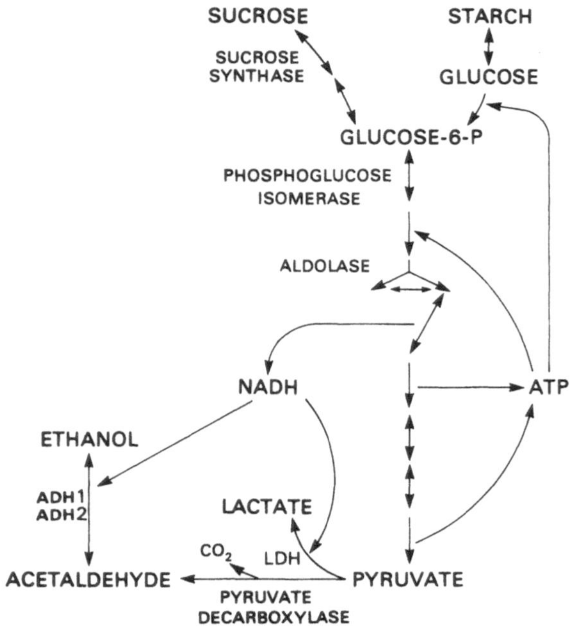

Fig. 1. Diagram of the pathways of anaerobic fermentation. The enzymes which have been identified as ANPs are shown. Lactate dehydrogenase and pyruvate decarboxylase have been shown to increase in amount under anaerobic conditions, but the corresponding genes have not been isolated.

and Edwards, 1982) and are unable to control the cytoplasmic pH drop (Roberts et al., 1984b). Adh2 is not required for viability of the flooded seedling except if the plant is Adh1-null (Dloughy, 1980).

In maize, the alcohol dehydrogenases, ADH1 and ADH2, have been studied extensively and together with aldolase (Kelley and Freeling, 1984b), phosphohexose isomerase (Kelley and Freeling, 1984a) and sucrose synthase (Springer et al., 1986) have been shown to be some of the polypeptides induced by hypoxia (Fig. 1). Also, in maize, pyruvate decarboxylase (Lazlo and St. Lawrence, 1983) increases in activity under anaerobic conditions. During severe hypoxia in barley, a 20-fold increase in lactate dehydrogenase activity has been demonstrated (Hoffman et al., 1986).

The maize genes corresponding to some of the ANPs, Adh1 and Adh2 (Dennis et al., 1984,1985), sucrose synthase (Werr et al., 1985), and aldolase (Dennis et al., 1987a)) have been cloned. The increase in synthesis of anaerobic polypeptides has been shown to be accompanied by an increased level of mRNA (Gerlach et al., 1982; Hake et al., 1985) and at least for Adh1 and Adh2 the increase results directly from an increased transcription rate (Rowland and Strommer 1986; Vayda and Freeling, 1986; Dennis et al., 1987b). The level of each specific mRNA increases 20- to 50-fold. There is also preferential translation of the ANP mRNAs under anaerobic conditions (Sachs et al., 1980).

These mechanisms of gene regulation, both transcriptional and translational, ensure a rapid and efficient switch in metabolism to cope with the altered environmental conditions. The presumption is that the increased synthesis of the ANPs is needed to increase the flux of carbohydrate through the glycolytic pathway to compensate for the lower efficiency of ATP production of the fermentative pathway compared to the oxidative pathway.

The specific synthesis of the ANPs should not be thought of as a response solely to external stress such as flooding. During normal development of the maize plant, some, and possibly all, of the ANPs are synthesised at various stages. In immature endosperm Adh1, Adh2 and sucrose synthase are expressed in the absence of external anaerobic conditions. This tissue specific expression could result from low oxygen tension within the endosperm tissue, the degree of hypoxia increasing as a function of distance from the outside of the developing kernel. While this resembles an anaerobic response in that these three anaerobic enzymes are synthesised, it is not known whether the other features usually associated with the response - cessation of aerobic protein synthesis and preferential translation of anaerobic mRNAs - are present in this developing endosperm. In the dry kernel, the aleurone and all the tissues of the embryo contain ADH1 activity, but the cell layers outside the aleurone (the pericarp) and the starchy endosperm layers beneath the aleurone layers lack ADH1 (Peacock et al., 1987). Whether the signals which bring about the developmentally regulated expression of these components of the anaerobic response are the same as those that respond to lower oxygen tension induced by the environment is not clear.

In one case of tissue specific expression, regulation of the ANPs must operate in a different way. In pollen there is expression of the Adh1 gene which appears not to be due to anaerobic induction since other ANPs such as Adh2 are not expressed (Freeling, 1976). In keeping with this suggestion is the behaviour of a mutant of Adh1 (Adh1-3F1124) isolated by Freeling and colleagues, which is the result of an insertion of the Robertson's mutator element in the promoter of the Adh1 gene. This mutant Adh1 is expressed at normal levels in pollen, but anaerobic expression in the roots and scutellum is only 0.5% of normal (reviewed in Freeling and Bennett, 1985). It is also clear that anaerobic conditions per se are not always sufficient to cause an anaerobic response since mature leaves do not show expression of the ANPs even after prolonged anaerobic induction (Okimoto et al., 1980).

We can envisage a model that involves intermediate steps leading from the external stimulus of low oxygen tension and results in expression of the ANP genes. Tissue specific expression may share some or all of these steps, but has the potential to bypass or block some steps for certain of the ANPs.

How is transcription of the anaerobic polypeptides increased by anaerobiosis?

We have studied transcriptional control of the *Adh1* gene under anaerobic conditions; *Adh1* is the best characterised of the ANPs. Using a transient expression system in which transcription from the endogenous *Adh1* promoter is regulated by oxygen tension (Howard *et al.*, 1987), we have introduced *in vitro* mutagenised promoter constructs into maize protoplasts and measured the activity of the promoter to determine the extent of anaerobic regulation (Howard *et al.*, 1987). The system involved the introduction of chimeric *Adh*CAT genes containing the *Adh1* promoter linked to the bacterial chloramphenicol acetyl transferase (*cat*) structural gene into maize suspension cell protoplasts by electroporation. These transfected cells were then incubated in various concentrations of oxygen and the promoter activity was assayed by measuring CAT activity 24 hours after introduction of the construct. The anaerobic regulation of the introduced *Adh*CAT gene paralleled that of the endogenous *Adh1* gene demonstrating that this system was suitable for dissecting the sequences responsible for anaerobically induced gene expression (Howard *et al.*, 1987).

The upstream boundary of the sequences necessary for anaerobic induction was defined by introducing constructs in which deletions from the 5' end of the promoter had been made. The sequences downstream of -140 were shown to be sufficient for anaerobic regulation of the gene (Walker *et al.*, 1987). Further deletions abolished expression in two steps and suggested the presence of two critical regions in the presumed anaerobic regulatory element (ARE). When the ARE sequences were joined to a truncated non-functional Cauliflower Mosaic Virus (CaMV) 35S gene promoter expression of the 35S gene was anaerobically inducible. The downstream boundary of sequences essential for expression was shown to be -81. Linker-scanner (clustered point) mutations created by the substitution of oligonucleotide linkers for sequences in the *Adh1* promoter supported the idea of two subregions within the ARE: loss of expression resulted when linkers were inserted into two different segments (Fig. 2) (Walker *et al.*, 1987). Full activity was found when linkers were used to substitute within the nucleotide sequence separating these two segments. The sequences of this anaerobic control region - the anaerobic regulatory element (ARE), and the effect of linker substitutions are shown in Fig. 2.

Another anaerobically induced gene of maize, aldolase, has also been subjected to functional analysis. Deletion analysis showed that the region upstream of -110 is sufficient for expression but removal of the sequences to -60 abolishes expression, (E.S. Dennis unpublished). Functional studies of the other anaerobically induced genes of maize have not been made but the promoter containing regions of *Adh2* (Dennis *et al.*, 1985) and sucrose synthase (Werr *et al.*, 1985) have been isolated and sequenced. If all anaerobically induced genes share a common regulatory mechanism it might be expected that the sequences shown to be functionally important for the *Adh1* gene would also be present in the upstream regions of the other anaerobically inducible genes. The *Adh2* upstream region has sequences homologous to the *Adh1* ARE sequences and these are located in a similar position relative to the start point of

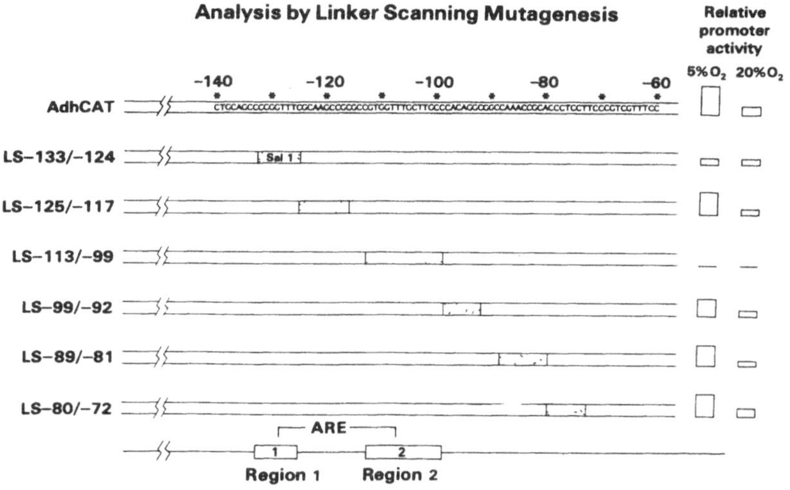

Fig. 2. The sequence of the ARE of maize with its two subregions. The effects of linker-scanner (clustered point) mutations in the ARE on expression of AdhCAT are shown on the right hand side of the figure.

transcription (Fig. 3) (Walker et al., 1987; Dennis et al., 1985). A sequence in the 5' region of the aldolase gene is homologous to a portion of the Adh1 ARE sequence, is located in a similar position, and is in the region shown to be functionally important for expression of the gene. Springer et al., (1986) found, in the sucrose synthase gene, a sequence in the first intron homologous to region II (Fig. 2) of the ARE of Adh1. We have detected a further region of homology located at -162 to -170, a position corresponding more closely to those in the Adh1, Adh2 and aldolase genes. When these homologies are examined in detail a consensus sequence emerges, $\begin{smallmatrix}TC\\CG\\GT\end{smallmatrix}$TGGTTT$\begin{smallmatrix}TC\\GT\\CG\end{smallmatrix}$, located in the region between -170 and -70 relative to the transcription start.

Adh1 and Adh2 mRNAs have higher steady state levels and different rates of accumulation relative to aldolase mRNA, under anaerobic conditions. This could relate to their different positions in the metabolic pathways and possibly to numbers of AREs. In the promoter region of the Adh1 gene there are two copies of the consensus anaerobic sequence whereas there is only one copy in the promoter of the aldolase gene. In sucrose synthase, another highly expressed ANP, there are at least two copies of the consensus sequence (Fig. 4); there are further homologous sequences at position -800 to -850.

In another environmentally induced response, that of heat shock, a single consensus element is sufficient to confer heat inducible transcription on exogenous genes in a high copy number system but most heat shock genes contain multiple elements. One of the Drosophila hsp70 genes has four consensus elements. Genes containing the two most proximal sites were transcribed more efficiently, in vitro, than those containing either one (reviewed in Linquist, 1986). It is possible that in the anaerobic response the number of consensus elements plays an analogous role in determining the level of transcriptional induction.

Organism	Gene	Position	Sequence
Maize	Adh1-1S (Dennis *et al.*, 1984)	-130 * -120 * -110 * -100 *	CTGCAGCCCCGGTTTCGCAAGCCGCGCCGTGGTTTGCTTGC
	Ferl & Nick (1987)#	-185 * -195 *	GGACGTGGTTTTCGCTCGT
	Adh2-N (Dennis *et al.*, 1985)	-140 * -130 *	CTCCCTGGTTTCTAACCGCG
	Sucrose synthase (Werr *et al.*, 1985)	+431 * +441 *	CGTGGTTTGCTTG
		-170 * -160 *	AATTCTGGTTTTGAGTAA
	Aldolase (Dennis *et al.*, 1987)	-70 * -60 *	TTTCGCTGGTTTCTTTCCCCTT
Pea	*Adh* (Llewellyn *et al.*, 1987)#	-100 * -110 *	CCGCTTTTGGTTTGTTT
Arabidopsis	*Adh* (Chang & Myerowitz, 1986)#	-150 * -160 *	GTATTTGGTTTTGCCTT

CONSENSUS SEQUENCE

TC TC
CGTGGTTTGT
GT CG

Fig. 3. Sequences conserved in anaerobically induced genes. The position of the bases in the gene sequence are given. Sequences which occur on the complementary strand are marked (#). The core hexanucleotide is underlined.

Is the anaerobic response conserved?

The ability to respond to anaerobic conditions by a switch in metabolism is widespread in plants and some, at least, of the enzymes which are induced are common to all plant species e.g. at least one anaerobically inducible Adh gene has been shown to be present in maize, barley, wheat, pea, tobacco and carrot. An interesting question is: are the signals responsible for anaerobic induction conserved? This can be examined by looking for homologous sequences in the promoters of anaerobically induced genes in a number of different species. Another, functional approach, is to introduce an anaerobically induced gene from one species into other species (for example maize Adh1 into tobacco) and determine if it is regulated anaerobically in the new host.

In order to look for homologous upstream sequences we cloned and sequenced an anaerobically induced Adh gene from pea (Pisum sativum) and analysed the promoter in a tobacco expression system using a CAT reporter gene (Llewellyn et al., 1987). Functional analysis indicated that the region between -120 and -51 is critical for anaerobic expression (Llewellyn, unpublished). We searched for sequence homologies to the ARE and in particular for homology to the 'core' anaerobic sequence which we suspect to be critical for anaerobic induction. No homologous sequence was found but the complement of the 'core' sequence, AAACCA, is located in the region shown to be functionally important (-120 to -51, see Fig. 4). This suggests that the anaerobic sequence is conserved and that it may work in either orientation, a property of many enhancer sequences.

Another anaerobically induced Adh gene which has been analysed is that of Arabidopsis thaliana (Chang and Myerowitz, 1986). In the upstream region of this gene (at -162 to -153) the sequence GCAAAACCAAA, containing the complement of the 'core', TGGTTT, occurs. This is another example of a potential ARE in a similar position in the promoter.

When the maize Adh1 promoter is introduced into tobacco in chimeric CAT constructs using the Agrobacterium tumefaciens stable transformation system, it is expressed at only a very low level under anaerobic conditions (Ellis et al., 1987a). In order to increase expression to a detectable level a 176 bp fragment (from -292 to -116) from the upstream region of the A. tumefaciens octopine synthase gene containing a transcriptional enhancer, (Ellis et al., 1987b) was substituted for the maize Adh1 sequences 5' of -140. This construct was found to be expressed at high levels in tobacco and was anaerobically inducible. When the region from -140 to -100 of the maize Adh1 gene, linked to the ocs enhancer, was substituted for a similar region of the nopaline synthase (nos) promoter coupled to a CAT gene, the nosCAT chimeric gene became anaerobically inducible (Ellis, unpublished). These results showed that the region of the maize Adh1 gene between -140 and -100 which contains the ARE is sufficient for anaerobic inducibility in tobacco. They also show that the maize ARE is recognised in tobacco. This argues that the factors responsible for transcription induction in response to anaerobic stimuli are sufficiently similar in maize and tobacco to function across species barriers.

How does the anaerobic response work?

The transcription of a number of genes has been shown to be regulated by the binding of protein factors to regulatory regions of the gene (e.g. Briggs et al., 1986). Different trans-acting factors bind to different promoter elements e.g. in the heat shock response there are proteins that bind to the heat shock consensus regulatory element as well as factors that bind to the TATA box. The regulation of the anaerobically inducible genes probably occurs by a similar mechanism; protein factors binding to the ARE and to the TATA box allow transcription to occur under anaerobic conditions.

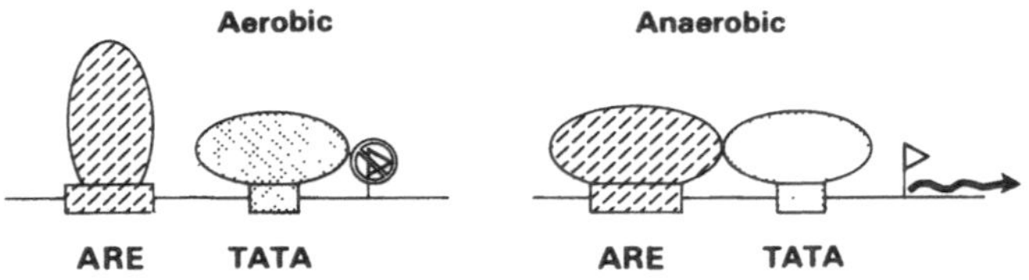

Fig. 4. Model of how the maize Adh1 gene is regulated by anaerobic conditions. ARE denotes the DNA sequence containing the ARE and TATA the equivalent TATA box sequence. The protein binding to the ARE is cross hatched, that to the TATA box, stippled. Transcription is indicated by an arrow.

Our analysis of the maize Adh1 promoter suggests a model. The deletion, linker-scanner and hybrid promoter experiments all indicate a positive control of gene expression (Walker et al., 1987). Disruption of the ARE sequence always leads to loss of activity, not to a high level of constitutive activity. Because of the results we obtained using a hybrid promoter with Adh1 and CaMV 35S elements, we conclude that protein factors are bound to the ARE under both aerobic and anaerobic conditions. The protein (or protein complex) may undergo a change, perhaps in conformation, in response to anaerobic conditions, thus allowing transcription to occur (Fig. 4) (Peacock et al., 1987). Experiments to determine whether factors that bind to the ARE are present in aerobic as well as anaerobic conditions are in progress. We are attempting to determine whether the ARE binding proteins are present in all tissues.

Another approach to studying the binding of trans-acting factors has been the analysis of the maize promoter in vivo, both for DNaseI hypersensitive sites and protection of the DNA against methylation by dimethyl sulphate. An anaerobically inducible DNaseI hypersensitive site from position -32 to -92 and a constitutively exposed region with several DNAaseI hypersensitive sites from position -150 to -400 have been reported (Ferl, 1985). Some of the sites are in regions which we have not shown to be functionally important. A number of regions have been found to be protected against dimethyl sulphate methylation (Ferl and Nick, 1987) suggesting the binding of protein factors. Ferl and Nick (1987) also suggested that two regulatory molecules bound to the maize Adh1 promoter only when the gene was induced by anaerobic conditions. These putative factors were associated with the regions -92 to -100 and -178 to -182. The sequence at -178 to -182 is included in a complement of the consensus anaerobic sequence, GAAAACCACG, extending from -187 to -176 (shown in Fig. 3 as -195 to -184 (Ferl and Nick, 1987)). Two other regulatory factors were deduced to be bound under both aerobic and anaerobic conditions and protected precisely the regions which we have identified as critical for anaerobic expression in our functional analysis of the promoter (positions -109 to -112 and -130 to -137). Ferl and Nick (1987) have argued that the factor bound at -130 to -137 alters its binding characteristics when the gene was induced by anaerobiosis.

Our data predict that since there are common upstream sequences in all the maize ANP genes there will be regulatory factors common to all these genes which bind to the conserved sequences. Two pieces of data suggest these factors are relatively abundant. Firstly, increasing concentrations of AdhCAT DNA (up to 2 mg/ml), in the electroporation experiments, did not saturate the anaerobic signal (Howard et al., 1987).

Secondly, addition of a 100-fold molar excess of a synthetic oligonucleotide encompassing the ARE did not lessen the anaerobic response of the AdhCAT chimeric gene when assayed transiently following electroporation (Howard, pers. comm.).

DISCUSSION

A common response to flooding, or other conditions that produce anaerobic stress, is widespread throughout the plant kingdom and our results suggest that the factors mediating the anaerobic response (the upstream regulatory sequences and sequences encoding the *trans*-acting regulatory factors) may also be conserved. We have not addressed the question of the mechanisms by which aerobic protein synthesis is inhibited and preferential translation of the aerobic mRNAs occurs. Nor have we attempted to determine the nature of the immediate effector of the transcriptional induction of the ANP genes. The anaerobic control of gene expression can be considered to be tripartite: 1) induction of ANP genes; 2) cessation of aerobic protein synthesis; 3) preferential translation of ANP mRNAs. It appears that response 1 can occur independently of responses 2 and 3, probably when conditions are not extreme, e.g. during normal development or under mildly hypoxic conditions.

We suggest that there is a consensus sequence $\begin{smallmatrix}TC\\CG\\GT\end{smallmatrix}TGGTTT\begin{smallmatrix}TC\\GT\\CG\end{smallmatrix}$, present in all anaerobically regulated genes; we have not yet tested whether this sequence contains sufficient information to cause a gene to be anaerobically induced. We are presently attempting to analyse the effect of changing each base in the ARE to identify the nucleotides critical to the anaerobic response.

Gel retardation has proved a convenient tool to assay for *trans*-acting regulatory factors. Such assays should confirm whether the same *trans*-acting factors bind to the other ANP genes, whether they are present in both aerobic and anaerobic tissue, and whether they are present in all tissues or only in tissue capable of anaerobic response.

Since plants do differ in their ability to withstand hypoxic conditions, e.g. peas are much less tolerant of hypoxia than maize (Roberts *et al.*, 1984), an understanding of the response to anaerobic stress and the way in which ANP genes are regulated may enable the strength of anaerobic response to be manipulated. It may thus prove possible to increase the flooding tolerance of specific plant species.

ACKNOWLEDGEMENTS

Research described in this paper was supported by Agrigenetics Research Corporation.

REFERENCES

Briggs, M.R., Kadonaga, J.T., Bell, S.P., and Tjian, R., 1986, Purification and biochemical characterization of the promoter-specific transcription factor, Spl. *Science*, 234:47.

Davies, D.D., 1980, Anaerobic metabolism and the production of organic acids, *in*: 'The Biochemistry of Plants,' D.D. Davies, ed., Vol.2. Academic Press, New York, pp 581-611.

Dennis, E.S., Gerlach, W.L., Pryor, A.J., Bennetzen, J.L., Inglis, A., Llewellyn, D., Sachs, M.M., Ferl, R.J., and Peacock, W.J., 1984, Molecular analysis of the alcohol dehydrogenase (*Adh1*) gene of maize. *Nucleic Acids Res.*, 12,9:3983.

Dennis, E.S., Gerlach, W.L., Walker, J.C., and Peacock, W.J., 1987, An anaerobically regulated aldolase gene of maize. EMBO J. (in press)

Dennis, E.S., Sachs, M.M., Gerlach, W.L., Beach, L., and Peacock, W.J., 1987, The Ds1 transposable element generates an additional intron in the Adh1 mutant allele Adh1-Fm335. Nucleic Acids Res., (in press).

Dennis, E.S., Sachs, M.M., Gerlach, W.L., Finnegan, E.J., and Peacock, W.J., 1985, Molecular analysis of the alcohol dehydrogenase 2 (Adh2) gene of maize. Nucleic Acids Res., 13,3: 727.

Dloughy, S.R., 1980, Ph.D. thesis, Indiana University.

Ellis, J.G., Llewellyn, D.J., Dennis, E.S., and Peacock, W.J., 1987a, Maize Adh1 promoter sequences control anaerobic regulation: addition of upstream promoter elements from constitutive genes is necessary for expression in tobacco. EMBO J., 6:11.

Ellis, J.G., Llewellyn, D.J., Walker, J.C., Dennis, E.S., and Peacock, W.J., 1987b, The OCS element : A 16 bp palindrome essential for activity of the octopine synthase enhancer. (submitted).

Ferl, R.J., 1985, Modulation of chromatin structure in the regulation of the maize Adh1 gene. Mol. Gen. Genet., 200:207.

Ferl, R.J., and Nick, H., 1987, In vivo detection of regulatory factor binding sites in the 5' flanking region of maize Adh1. J. Biol. Chem., (in press).

Freeling, M., 1976, Intragenic recombination in maize: pollen analysis methods and the effect of parental Adh1+ isoalleles. Genetics, 83:701.

Freeling, M., and Bennett, D.C., 1985, Maize Adh1. Ann Rev Genet. 19:297.

Gerlach, W.L., Pryor, A.J., Dennis, E.S., Ferl, R., Sachs, M.M., and Peacock, W.J., 1982, cDNA cloning and induction of the alcohol dehydrogenase gene from maize. Proc. Natl Acad. Sci., USA, 79:2981.

Hake, S., Kelley, P.M., Taylor, W.C., and Freeling, M., 1985, Co-ordinate induction of dehydrogenase 1, aldolase and other anaerobic RNA's in maize. J. Biol. Chem., 260:5050.

Harberd, N.P., and Edwards, K.J.R., 1982, The effect of a mutation causing alcohol dehydrogenase deficiency on flooding tolerance in barley. New Phytol., 90:631.

Howard, E.A., Walker, J.C., Dennis, E.S., and Peacock, W.J., 1987, Regulated expression of an alcohol dehydrogenase 1 chimeric gene introduced into maize protoplasts. Planta, 170:535.

Hoffman, N.E., and Hanson, A.D., 1986, Purification and properties of hypoxically induced lactate dehydrogenase from barley roots. Plant Physiol., 82:664.

Kelley, P.M., and Freeling, M., 1984a, Anaerobic expression of maize glucose phosphate isomerase 1. J. Biol. Chem., 259:673.

Kelley, P.M., and Freeling, M., 1984b, Anaerobic expression of maize fructose-1,6-diphosphate aldolase. J. Biol. Chem., 259:14180.

Lazlo, A., and St. Lawrence, P., 1983, Parallel induction and synthesis of PDC and ADH in anoxic maize roots. Molec. Gen. Genet., 192:110.

Lindquist, S., 1986, The heat shock response. Ann. Rev. Biochem., 55:1151.

Llewellyn, D.J., Finnegan, E.J., Ellis, J.G., Dennis, E.S., and Peacock, W.J., 1987, Structure and expression of an alcohol dehydrogenase 1 gene from Pisum sativum (cv. Greenfeast). J. Mol. Biol., 195:115.

Okimoto, R., Sachs, M., Porter, E., Freeling, M., 1980, Patterns of polypeptide synthesis in various maize organs under anaerobiosis. Planta, 150:89.

Peacock, W.M., Wolstenholme, D.R., Walker, J.C., Singh, K., Llewellyn, D., Ellis, J.G., and Dennis, E.S., 1987, Developmental and environmental regulation of the maize alcohol dehydrogenase 1 (Adh1) gene: promoter-enhancer interactions. UCLA, Alan R. Liss, New York. (in press).

Roberts, J.K.M., Callis, J., Jardetzky, O., Walbot, V., and Freeling, M., 1984a, Cytoplasmic acidosis as a determinant of flooding intolerance in plants. Proc. Natl Acad. Sci., USA, 81:6029.

Roberts, J.K.M., Callis, J., Wemmer, D., Walbot, V., and Jardetzky, O., 1984b, Mechanism of cytoplasmic pH regulation in hypoxic maize root tips and its role in survival under hypoxia. Proc. Natl Acad. Sci., USA, 81:3379.

Rowland, L.J., and Strommer, J.N., 1986, Anaerobic treatment of maize roots affects transcription of Adh1 and transcript stability. Mol. Cell Biol., 6:3368.

Sachs, M.M., Freeling, M., and Okimoto, R., 1980, The anaerobic proteins of maize. Cell, 20:761.

Schwartz, D., 1969, An example of gene fixation resulting from selective advantage in suboptimal conditions. Am. Nat., 103:479.

Springer, B., Werr, W., Starlinger, P., Clark Bennett, D., Zokolica, M., and Freeling, M., 1986, The Shrunken gene on chromosome 9 of Zea mays L is expressed in various plant tissues and encodes an anaerobic protein. Mol. Gen. Genet. 205:461.

Vayda, M., and Freeling, M., 1986, Insertion of the Mu1 transposable element into the first intron of maize Adh1 interferes with transcript elongation but does not disrupt chromatin structure. Plant Mol. Biol., 6:441.

Walker, J.C., Howard, E.A., Dennis, E.S., and Peacock, W.J., 1987, DNA sequences required for anaerobic expression of the maize alcohol dehydrogenase 1 gene. Proc. Natl Acad. Sci., (in press).

Werr, W., Frommer, W.B., and Starlinger, P., 1985, The structure of the sucrose synthase gene on chromosome 9 of Zea mays L. EMBO J., 4:1373.

MOLECULAR BIOLOGY AND MOLECULAR GENETICS OF PLANT BROMOVIRUSES

Paul Ahlquist

Institute for Molecular Virology
and Department of Plant Pathology
University of Wisconsin-Madison
Madison, WI 53706, U.S.A.

INTRODUCTION

The overwhelming majority of known plant viruses, including 21 of 26 formally recognized virus groups (Matthews, 1982), encapsidate their genomes in the form of messenger-sense single stranded RNA. These different groups of positive strand RNA viruses display an amazingly wide variety in genetic organization and virion design, having from one to four different viral RNAs encapsidated in from one to three physically separate particles of icosahedral, bacilliform, rigid rod or flexuous rod morphology. Despite this considerable outward diversity, nucleic acid sequencing has revealed in recent years that many of these viruses share extensive homology in one or more noncapsid proteins. Moreover, for a number of different virus genes, amino acid sequence conservation extends not only across plant virus groups but also between plant and animal viruses. As discussed below, the conserved proteins are implicated in central steps of virus replication, suggesting that many of the known plant viruses may have arisen by reassortment of an array of different coat, host- and vector-adaptation genes with a limited number of core replication genes.

One set of positive strand RNA viruses sharing homologous protein domains includes the plant comoviruses (discussed elsewhere in this volume), potyviruses, and animal picornaviruses such as poliovirus (Franssen et al., 1984; Allison et al., 1986; Domier et al., 1986). Some common features of these viruses include 5' genome-linked proteins on the viral RNA and extensive proteolytic processing of large primary translation products. Another important set of viruses sharing a distinguishable but extensive set of protein homologies includes, among others, the plant bromo-, cucumo-, almo-, tobamo-, and tobra-viruses, as well as the animal alphaviruses (Cornelissen and Bol, 1984; Haseloff et al., 1984; Rezaian et al., 1984; Ahlquist et al., 1985; Boccara et al., 1986). These viruses share three extensively conserved domains of amino acid sequence within virus-coded noncapsid proteins, as well as 5' caps on the viral RNA and the use of subgenomic mRNAs (see below) to express capsid proteins and, in some cases, other genes.

We have examined the molecular biology of brome mosaic virus (BMV) and some related viruses as useful examples of this second group of positive strand RNA viruses. The work described below includes studies of the organization of virus genes and their expression, the contribution of virus-encoded proteins and cis-acting RNA recognition signals to viral RNA replication, and the development of RNA virus-based expression vectors for foreign genes. Among other benefits, detailed understanding of the function of the conserved protein domains in virus replication should ultimately reveal useful targets for antiviral agents. Such agents, being directed against central conserved features of the virus replication machinery, might be effective against a broad spectrum of viruses within this large group of important plant and animal pathogens. In addition, just as DNA viruses have proved to be valuable sources of gene expression elements, it appears that the RNA-directed gene expression pathways which have been so evolutionarily successful in these widespread RNA viruses can also be useful in engineered genetic manipulation.

ORGANIZATION OF THE BMV GENOME

BMV is a mechanically transmissable virus which infects a number of cereal grains, typically causing mosaic symptoms and mild stunting (Lane, 1981). Several million virions accumulate in each infected cell (Loesch-Fries and Hall, 1980), leading to a yield of one to three mg of virus per g of infected tissue. The virions are nonenveloped 28 nm icosahedra whose outer capsid is composed of 180 copies of a single 20 kD coat protein (Lane, 1981). Density gradients fractionate the virions into three classes containing different species of single stranded RNA. Each heavy particle contains one copy of a 3.2 kilobase (kb) species denoted RNA1, each light particle contains one copy of a 2.9 kb RNA denoted RNA2, and each medium density particle contains one copy each of a 2.1 kb RNA (RNA3) and a 0.9 kb RNA (RNA4). Productive infection of whole plants requires co-inoculation with either all three particle types or the three largest RNA species, RNAs 1-3.

The BMV genome has been completely sequenced (Ahlquist et al., 1981b; 1984b), and a schematic view of some of its features is shown in Fig. 1. RNAs 1-3 encode the entire genome, since the sequence of "subgenomic" RNA4 is contained in RNA3. Each of the four RNAs bears a 5' m7GpppG... cap, but unlike most eukaryotic mRNAs does not bear a 3' poly(A) tract. Instead, each bears a highly conserved heteropolymeric sequence of around 200 bases (Ahlquist et al. 1981a; 1984b) possessing a variety of intriguing properties. This conserved 3' sequence folds into an extensively base-paired structure with at least two alternative conformational states, which are conserved in RNAs of other bromoviruses and also cucumoviruses (Ahlquist et al. 1981a; Rietveld et al., 1983). In vitro, a subset of this 3' terminal homology is sufficient to direct initiation of (-) strand RNA synthesis by a highly template selective polymerase extract from BMV-infected cells (Ahlquist et al., 1984a; Miller et al., 1986). In addition, this region interacts with several factors from uninfected cells, including tRNA nucleotidyl transferase, aminoacyl tRNA synthetase, and a translation elongation factor (Bujarski et al., 1986; Bastin and Hall, 1976). Although the 3' ends of a number of other plant viral RNAs are similarly recognized and aminoacylated by the normal cellular tRNA charging enzymes, the function(s) of this charging in the virus life cycle remain a subject of speculation (Haenni et al., 1982). Among other possibilities, alterations in the conformational state of the conserved 3' region, binding of viral or cellular proteins, and/or aminoacylation might regulate the participation of viral RNA in translation, replication and encapsidation, which all utilize BMV (+) strand RNA as a substrate (Ahlquist et al., 1984a,b).

EXPRESSION OF RNA VIRUS INFECTION FROM CLONED cDNA

Although it is possible to directly construct some recombinant RNA molecules in vitro from RNA precursors (Miele et al., 1983), the potential for laboratory manipulations at the RNA level is currently very limited. For this reason, the recent development of approaches to express complete, infectious RNA genomes from cloned viral cDNA has greatly enhanced prospects for further study of gene function and other properties of RNA viruses such as BMV. The resulting biologically active cDNA clones provide an artificial DNA stage in the virus life cycle, making the RNA genome accessible to recombinant DNA technology.

Complete cDNA clones of several RNA-based replicons, including bacteriophage Qß (Taniguchi et al., 1978), poliovirus and some related picornaviruses (Racaniello and Baltimore, 1981; Kandolf and Hofschneider, 1985), viroids (Cress et al., 1983), and some plant virus satellite RNAs (Gerlach et al., 1986), are directly infectious to normal hosts of these agents. In a number of other cases, complete viral cDNA clones have not displayed any direct biological activity, although in vitro transcripts derived from such clones are highly infectious. To date, such cases include BMV (Ahlquist et al., 1984c), two strains of TMV (Dawson et al., 1986; Meshi et al., 1986), some animal viruses (Mizutani et al., 1985; Levis et al., 1986), and a virulent satellite RNA of turnip crinkle virus (Simon and Howell, 1987).

Several factors influence the infectivity of transcripts from BMV cDNA. Independently isolated cDNA clones frequently contain sequence variations from the viral consensus, either representing variants in the virus population or reverse transcription errors (Fields and Winter, 1981; Domingo et al., 1978). Some BMV cDNA clones have been shown to contain lethal mutations which block the infectivity of their in vitro transcripts, and similar variation in biological activity of transcripts from separate clones was seen with TMV cDNA (Ahlquist et al., 1984c; Dawson et al., 1986). The possible presence of lethal mutations should thus be considered in attempts to express infection from any newly isolated viral cDNA clone.

Secondly, for BMV transcripts, 5' extensions containing nonviral bases generally reduce transcript infectivity. A single extra 5' G residue, e.g., reduces BMV transcript infectivity about threefold, while seven- or 16-base 5' extensions reduce infectivity 100-fold or more (Janda et al., 1987). A similar effect has been reported for transcripts of TMV cDNA (Dawson et al., 1986). To minimize or eliminate such 5' extensions, plasmid vectors have now been constructed with blunt end restriction/cloning sites at or within two bases of the transcription start of promoters for E. coli, T7 and SP6 RNA polymerases (Ahlquist and Janda, 1984; van der Werf et al., 1986; Bujarski and Kaesberg, 1987). A 5' cap structure greatly enhances the infectivity of in vitro transcripts from both BMV and TMV cDNA clones, possibly by stabilizing the transcript in vivo (Ahlquist et al. 1984c; Dawson et al., 1986; Meshi et al. 1986; Janda et al., 1987).

In contrast to the sensitivity of the 5' end, additional nonviral bases at the 3' ends of both BMV and TMV transcripts have less effect on infectivity (Ahlquist et al. 1984c; Meshi et al. 1986). With BMV cDNA clones, transcripts with 6-7 extra 3' bases are highly infectious, as well as shorter transcripts whose ends correspond to a natural intermediate in the replication process (French et al., 1986). Transcripts from BMV cDNA templates cleaved 2.7 kb downstream of the cDNA 3' end were not detectably infectious, however, and transcripts from uncleaved TMV cDNA plasmids were at least 20 times less infectious than transcripts with only three additional 3' bases (Ahlquist et al., 1984c; Meshi et al., 1986).

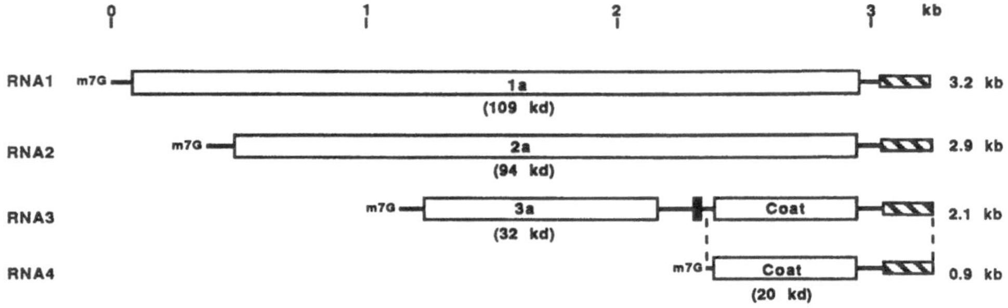

Fig. 1 - Schematic map of the four encapsidated BMV RNAs. The open boxes denote the open reading frames for the various BMV proteins discussed in the text. The crosshatched boxes represent the conserved 3' region of each RNA and the small filled box maps the length-heterogeneous oligo(A) sequence in RNA3 (Ahlquist et al. 1981b).

BMV RNA4 is an efficient mRNA for coat protein, while genomic RNAs 1-3 are mRNAs for noncapsid proteins 1a (109 kD), 2a (94 kD) and 3a (32 kD), respectively (Shih and Kaesberg, 1973). The precise functions of these noncapsid proteins are also under continuing investigation. As noted in the introduction, proteins 1a and 2a are extensively homologous to proteins encoded by a range of other plant and animal RNA viruses (Fig. 2), and evidence discussed below shows that these genes contribute required trans-acting functions to the viral RNA replication process. Comparisons of BMV with tobacco mosaic virus (TMV) (Haseloff et al., 1984) have led to the suggestion that the 3a protein may be functionally analogous to TMV p30, which has been implicated in facilitating cell to cell spread of TMV infection (Leonard and Zaitlin, 1982; Ohno et al., 1983).

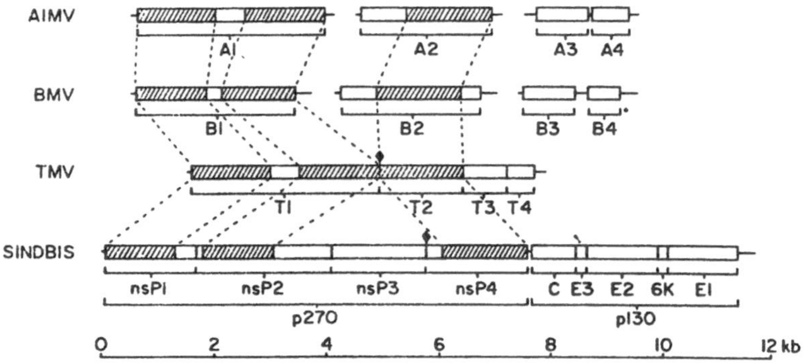

Fig. 2 - Location of coding regions for the homologous noncapsid protein domains encoded in the genomic RNAs of alfalfa mosaic virus (AlMV), brome mosaic virus (BMV), tobacco mosaic virus (TMV), and Sindbis virus (Ahlquist et al., 1985). Enlarged boxes correspond to expressed open reading frames, with the crosshatched regions and dotted lines indicating the three domains of conserved amino acid sequence homologous among all four viruses. Open reading frames B1-4 encode BMV proteins 1a, 2a, 3a, and coat, respectively. The filled diamonds denote the leaky termination codons read through at a significant frequency in TMV and Sindbis.

In most cases the 3' ends of viral cDNA transcripts have been defined by linearizing the template DNA by cleavage at a selected restriction site. An interesting alternative has recently been developed by Jozef Bujarski, who fused selected cDNA sequences from the satellite RNA of tobacco ringspot virus to the 3' end of a biologically active BMV cDNA clone (personal communication). As with the natural satellite RNA and other transcripts derived from its cDNA clones (Gerlach et al., 1986), the satellite RNA sequences added to the BMV transcript undergo self cleavage in the presence of magnesium ions. BMV transcripts synthesized from circular plasmids with this structure thus autolytically process to yield infectious transcripts with 15 extra 3' bases.

REPLICATION OF BMV RNA

Synthesis of BMV virion RNA in protoplasts continues in the presence of actinomycin D, showing that DNA templates are not required (Loesch-Fries and Hall, 1980). The independence of viral RNA synthesis from DNA templates is explained by the observations that infected cells contain both full-length RNA templates complementary to the virion RNAs and a virus-induced RNA-dependent RNA polymerase activity (Bastin and Kaesberg, 1976; Hardy et al., 1979; Loesch-Fries and Hall, 1980). The BMV nucleic acid replication cycle thus involves successive copying of complementary RNA strands. The physical state of (-) strand RNA templates in the cell is not known. While usually isolated as hybrids with the virion RNAs, this structure may form from largely single-stranded replicative intermediates during the isolation process (Bastin and Kaesberg, 1976; Richards et al., 1984).

The RNA-dependent RNA polymerase extractable from BMV-infected but not healthy cells can be used to synthesize (-) strand copies of BMV virion RNAs in vitro (Hardy et al., 1979; Bujarski et al., 1982; Miller and Hall, 1983). This polymerase extract is highly template-specific, copying RNAs from BMV and, to a lesser extent, the closely related cowpea chlorotic mottle virus (CCMV), but not other viral and cellular RNAs. Such selectivity is to be expected since successful initiation of infection requires the polymerase to efficiently recognize and copy a small number of incoming virion RNAs among the total cellular RNA population. While notable success has been achieved in copying BMV virion RNAs in vitro, the synthesis of (+) strand RNAs from (-) strand RNA templates has not yet been duplicated in vitro, and efforts in this direction continue.

In addition to (-) strands of the three genomic RNAs, (-) strands of subgenomic RNA4 can also be isolated from infected cells as part of a dsRNA4 complex (Bastin and Kaesberg, 1976). RNA4 is also a template for (-) strand RNA synthesis in vitro in the BMV RNA polymerase system (Hardy et al., 1979). Since present results (see below) suggest that these RNA4 (-) strands are not functional templates for (+) strand RNA4 synthesis, at least two hypotheses regarding them might be considered. First, production of RNA4 (-) strands may be an non-essential consequence of the presence in RNA4 of the same 3' terminal sequences which direct synthesis of (-) strands from genomic RNA3. Although possibly a nonproductive side reaction, the minor metabolic cost of producing a small population of RNA4 (-) strands might well be insufficient to select for more sophisticated regulation of viral (-) strand synthesis. Alternatively, at least some of the dsRNA4 "replicative form" isolated from BMV-infected cells may arise during extraction. I.e., full length RNA3 (-) strands engaged in (+) strand RNA4 synthesis might undergo product/template hybridization, followed by degradation of the remaining ssRNA portion of the RNA3 (-) strand (Bastin and Kaesberg, 1976).

Functional Analysis of Trans-Acting Factors in Viral RNA Replication

Inoculations of protoplasts with various combinations of fractionated BMV virion RNAs or infectious in vitro transcripts show that RNAs 1 and 2 are both necessary and together sufficient to induce viral RNA replication (Kiberstis et al., 1981; French et al., 1986). RNA3 is dispensable for replication of RNAs 1 and 2, and further work shows that expression of both RNA3 genes can be blocked without affecting synthesis of either RNA3 or subgenomic RNA4 (French and Ahlquist, 1987). Thus, all trans-acting factors contributed by the virus to RNA replication are encoded by RNAs 1 and 2. This suggests that the observed strong conservation of the 1a and 2a protein sequences (Fig. 2) arises from their required participation in the RNA replication process.

Despite their implied involvement in RNA replication, the precise enzymatic functions of the RNA 1- and 2-encoded 1a and 2a proteins have not been conclusively defined. Several lines of evidence suggest that one or both of these proteins may be multifunctional. In Sindbis virus, which shares common features of RNA replication with BMV and which encodes proteins homologous to BMV 1a and 2a, four different genetic complementation groups affecting viral RNA synthesis have been found (Strauss and Strauss, 1980). Evidence also exists for two separate complementation groups in the BMV 2a protein homologue of alfalfa mosaic virus (Roosien and van Vloten-Doting, 1982). Moreover, two distinct regions of amino acid sequence, separated by a region of variable length and amino acid sequence, are conserved among the BMV 1a protein homologues and may encode separate functions (Fig. 2). The Sindbis-encoded polypeptides homologous to these two BMV 1a protein domains are separated after translation by proteolytic cleavage (Fig. 2, segments nsP1 and nsP2).

Some clues about possible functions provided by the 1a-like and 2a-like proteins have come from comparisons with other proteins of known function. Kamer and Argos (1984) have pointed out sequence homologies between the 2a homologues and several known RNA-dependent polymerases, including the RNA polymerases of picornaviruses and bacteriophage Qß and retroviral reverse transcriptase. Gorbalenya et al. (1985) note that the C-terminal conserved region of protein 1a and its homologues contain a sequence motif common to a number of known ATP-utilizing proteins, which may be a nucleotide binding site. Such a nucleotide binding site might function in any of several aspects of viral RNA synthesis, including capping of the viral RNA. Genetic studies suggest additional RNA replication functions which may be virus-encoded. Among viruses encoding proteins homologous to BMV 1a and 2a, mutants have previously been isolated with RNA replication defects specifically in synthesis of genomic (-) strands or subgenomic RNA, suggesting that specific virus-encoded factors may exist to regulate synthesis of these classes of RNA independently from synthesis of virion (+) strand RNA (Keranen and Kaarianien, 1979; Sarachu et al., 1983).

Because of the central importance of RNA replication in the virus life cycle and the wide dispersal of homologues to the BMV RNA replication genes in a variety of important plant and animal viruses, we have recently begun a molecular genetic analysis of 1a and 2a gene functions using the biologically active BMV cDNA clones as the substrate for mutagenesis (P. Kroner, D. Richards, and P. Ahlquist, unpublished results). This cDNA-based approach bypasses or overcomes most of the difficulties which have previously hindered genetic analysis of RNA viruses. Thus, any of a wide variety of preselected mutant types can be reliably targeted to preselected regions or sites in the genome; multiple mutations can be avoided; muta-tions can be mapped unambiguously; and mutants can be isolated and tested without reference to phenotype, allowing ready identification of changes which are either phenotypically neutral or lethal. In preliminary results from this study, BMV mutants have been isolated with a broad range of phenotypic effects on RNA replication, and the prospects for future progress with this approach seem very promising.

Analysis of Cis-Acting Viral RNA Replication Signals

Successful replication of BMV RNAs depends not only on the formation of the necessary trans acting replication machinery by the 1a and 2a proteins, but also on the presence of suitable cis-acting signals which direct interaction of this machinery with its appropriate viral RNA substrates. To identify and characterize such cis-acting replication requirements, systematic deletions and sequence rearrangements were constructed in BMV RNA3 and their effects on RNA3 replication in protoplasts co-infected with wt BMV RNAs 1 and 2 was analyzed (French and Ahlquist, 1987). RNA3 was selected as the initial substrate for such analysis because, as noted above, it contributes no trans-acting information to the RNA replication process. Since BMV RNA replication in protoplasts is independent of virus packaging and intercell spread, any alterations influencing accumulation of RNA3 derivatives in such tests must reflect changes affecting cis-acting signals important in either synthesis or stability of progeny RNA.

One important principle established by this study is that considerable tolerance exists for the structure of RNAs which can be amplified in vivo by the BMV RNA replication machinery. A wide variety of essentially arbitrary RNA3 deletions, rearrangements, and substitutions are replicated effectively in BMV-infected cells (French et al., 1986; French and Ahlquist, 1987). Competence for viral replication thus does not seem to involve any global restrictions on RNA structure. This flexibility is in keeping with the apparent ability of members of the 1a and 2a gene family to reassort with new capsid and host adaptation genes in the course of virus evolution (Fig. 2).

Although many RNA3 sequence alterations are tolerated in replication, three specific regions of the RNA3 sequence were found to be required in cis for efficient viral amplification. As expected from in vivo results showing its role in (-) strand RNA initiation and tRNA-like interactions (see above), these requirements included the presence of a 3' terminal portion of the 3' noncoding region. However, the amount of 3' terminal sequence required in vivo, 163-200 bases, exceeds the 134 bases previously implicated in the above functions in in vitro tests. The additional sequence required in vivo may either provide a previously unrecognized function or contribute to known interactions in the more complex in vivo environment. 5'-terminal sequences were also found to be required for RNA3 replication (J. Bujarski and P. Ahlquist, unpublished results), as was expected since, by analogy to the (-) strand initiation at the 3' end of RNA3, these sequences might be involved in initiating (+) strand RNA3 synthesis.

In addition to these expected terminal sequence requirements, the deletion studies revealed that an approximately 150-base segment of the intercistronic region was also needed for efficient RNA3 amplification (French and Ahlquist, 1987). Interestingly, this region contains a conserved sequence element GGUUCAAYCCCU (Y=pyrimidine) also present in the intercistronic region of cucumber mosiac virus (CMV) RNA3 and the 5' untranslated regions of both BMV and CMV RNAs 1 and 2 (Rezaian et al., 1984). While this homology to 5' ends suggests that the required central region in BMV RNA3 may participate in initiation of (+) strand RNA synthesis, its actual role in replication remains undetermined. Nevertheless, the observed requirement for this region in replication, together with the studies on subgenomic promoter structure discussed below, largely explain the 250 base length of the intergenic region of BMV RNA3, which is unusually long for plant viruses.

EXPRESSION OF NATURAL AND FOREIGN GENES IN BMV

Production of Subgenomic Viral mRNAs

An important pathway for gene expression for all of the viruses in Fig. 2 is the production of one or more subgenomic RNAs, such as BMV RNA4, to serve as messengers for translation of capsid proteins and other internal genes. Production of such subgenomic mRNAs allows a virus to provide translational access to internal cistrons in the genomic RNA, by linking them to a newly created 5' end.

The subgenomic mRNAs of the viruses of Fig. 2 and their relatives are all 3' coterminal subsets of their parent genomic RNA. Early speculation included the possibility that these subgenomic RNAs were generated by site-specific cleavage of genomic RNA. However, as shown in vitro with the BMV RNA polymerase system, subgenomic BMV RNA4 is produced by partial transcription of (-) strand RNA3 template (Miller et al., 1985). This strategy not only allows multiple genes to be expressed from a given genomic RNA, but in principle also allows their expression to be differentially regulated in time and amount.

Because of the importance of subgenomic RNA synthesis in the replication of many viruses and its potential for study as a special case of RNA-directed RNA synthesis, we have examined the structure and function of the cis-acting sequences in BMV RNA3 which direct and regulate synthesis of subgenomic RNA4 in vivo. We find that a relatively small fragment of RNA3 sequence functions as a subgenomic promoter cassette, which will direct specific initiation in vivo of subgenomic RNA synthesis from any of a number of new insertion sites in RNA3 (R. French and P. Ahlquist, unpublished results). The contribution of various portions of sequence within the fully active promoter cassette has been tested by deletion analysis, and some operationally distinct promoter subdomains defined. Moreover, we find that multiple subgenomic promoters can be inserted and made to function in a single RNA3 derivative, allowing the engineered expression of many novel subgenomic mRNAs from a single genomic RNA (manuscript in preparation). This enhanced understanding and use of the subgenomic promoter will provide significant additional flexibility in vector applications as discussed below.

Expression of Foreign Genes in RNA Virus Vector

Although they depart from the DNA-based genetic pathways more generally familiar to molecular biologists, it would be simplistic to imagine that the RNA-directed pathways of replication and gene expression which have proven so successful in plant, animal, bacterial and fungal viruses do not offer new opportunities and tools which could enhance abilities for designed genetic manipulation. The greatest potential contribution of RNA-directed pathways would not be simply as substitutes for abilities already achievable by DNA-dependent means alone. Rather, RNA-based pathways should also be studied to provide new capabilities which can be used together with existing DNA-based approaches, expanding the total repertoire of techniques for gene expression and regulation. While a number of approaches can be envisioned, the future forms in which RNA-dependent gene expression may be applied cannot be completely predicted at this early stage. Neverthess, RNA viruses have already been used to direct engineered expression of foreign genes in eukaryotic cells, providing an early illustration of their potential.

Because, as discussed above, some of the basic cis-acting requirements for replication and gene expression in BMV RNAs are understood, it is possible to use BMV as an episomal gene expression vector in plant cells. Thus, for some purposes, BMV can be considered to function as an extraordinarily high copy number plasmid, which happens to exist in an RNA rather than DNA form. One of the foreign genes whose replication and expression in BMV-derived vectors has been studied is the popular chloramphenicol acetyl transferase or CAT gene (French et al., 1986; Ahlquist et al., 1987). In a variety of constructs, the CAT gene was inserted under control of the BMV subgenomic promoter, and the resulting hybrid RNA3 inoculated into barley cells along with wild type RNAs 1 and 2. The hybrid RNAs all replicate in vivo and produce the expected subgenomic RNAs, and those hybrids with the CAT gene in the same orientation as the natural viral genes express high levels of CAT gene activity in vivo. Not surprisingly in light of the high mRNA levels achieved, the CAT activity expressed compares very favourably to the highest levels of CAT expression reported from DNA transformation of plant cells. Moreover, the level of expression from BMV vectors can be readily controlled at the translational level by altering the 5' leader sequence of the subgenomic mRNA (French et al., 1986).

We have also inserted and expressed a number of additional foreign genes in plant cells via BMV-derived vectors (R. Sacher, A. Gatenby and P. Ahlquist, unpublished results). Experience with these genes shows that there is nothing remarkable about the CAT gene which facilitates its expression in the virus, and the CAT gene results should generally serve as a reasonable guide for expression of new genes. Moreover, in work analogous to that performed with BMV, the CAT gene has now been expressed in plant cells by derivatives of tobacco mosaic virus (Takamatsu et al., 1987) and in animal cells by derivatives of Sindbis virus (Schlesinger et al., 1987). Significantly, CAT expression is retained in repeated passages of CAT-expressing Sindbis virus derivatives, showing that the fidelity of RNA replication in such viruses is sufficient to maintain function even in unselected foreign genes over multiple rounds of virus replication (S. Schlesinger, personal communication).

CONCLUDING REMARKS

The recent development of several new approaches for studies of RNA viruses has led to rapid progress in investigations of the structure, function and expression of genes in this important and widespread class of viruses. Recognition of fundamental similarities in superficially dissimilar RNA viruses and important new observations from many laboratories have further accelerated the pace of this progress, creating exciting prospects for the field as a whole. Several RNA viruses have already been used as expression vectors in plant and animal cells, and further advanced applications of RNA viruses in engineered gene expression can be envisioned. Assays and experimental designs are being developed to extend molecular analysis to more broadly defined virus characteristics related directly to virus-host interactions, including virus specificity for particular hosts, virus pathogenicity, and host resistance.

ACKNOWLEDGMENTS

We are indebted to the experimental contributions and insight of many colleagues, including Roy French, Michael Janda, Richard Allison, Philip Kroner, Radiya Pacha, Robert Sacher, Pat Traynor and Douglas Richards. Research on these topics in the author's laboratory was supported by grants from Agrigenetics Research Associates, NIH (Grant GM35072) and NSF (Presidential Young Investigator Award DMB-8451884).

REFERENCES

Ahlquist, P., Bujarski, J., Kaesberg, P., and Hall, T., 1984a, Localization of the replicase recognition site within brome mosaic virus RNA by hybrid-arrested RNA synthesis, Plant Mol. Biol., 3:37.

Ahlquist, P., Dasgupta, R., and Kaesberg, P., 1981a, Near identity of 3' RNA secondary structure in bromoviruses and cucumber mosaic virus, Cell, 23:183.

Ahlquist, P., Dasgupta, R., and Kaesberg, P., 1984b, Nucleotide sequence of the brome mosaic virus genome and its implications for viral replication, J. Mol. Biol., 172:369.

Ahlquist, P., French, R., and Bujarski, J., 1987, Molecular studies of brome mosaic virus using infectious transcripts from cloned cDNA, Adv. Virus Res., 32:215.

Ahlquist, P., French, R., Janda, M., and Loesch-Fries, L. S., 1984c, Multicomponent RNA plant virus infection derived from cloned viral cDNA, Proc. Natl. Acad. Sci. USA, 81:7066.

Ahlquist, P., and Janda, M., 1984, cDNA cloning and in vitro transcription of the complete brome mosaic virus genome, Mol. Cell. Biol., 4:2876.

Ahlquist, P., Luckow, V., and Kaesberg, P., 1981b, Complete nucleotide sequence of brome mosaic virus RNA3, J. Mol. Biol., 153:23.

Ahlquist, P., Strauss, E., Rice, C., Strauss, J., Haseloff, J., and Zimmern, D., 1985, Sindbis virus proteins nsP1 and nsP2 contain homology to nonstructural proteins from several RNA plant viruses, J. Virol., 53:536.

Allison, R., Johnston, R., and Dougherty, W., 1986, The nucleotide sequence of the coding region of tobacco etch virus genomic RNA: evidence for the synthesis of a single polyprotein, Virology, 154:9.

Bastin, M., and Hall, T., 1976, Interaction of elongation factor 1 with aminoacylated brome mosaic virus and tRNAs, J. Virol., 20:117.

Bastin, M., and Kaesberg, P., 1976, A possible replicative form of brome mosaic virus RNA4, Virology, 72:536.

Boccara, M., Hamilton, W., and Baulcombe, D., 1986, The organization and interviral homologies of genes at the 3' end of tobacco rattle virus RNA1, EMBO J., 5:223.

Bujarski, J., Ahlquist, P., Hall, T., Dreher, T., and Kaesberg, P., 1986, Modulation of replication, aminoacylation and adenylation in vitro and infectivity in vivo of BMV RNAs containing deletions within the multifunctional 3' end, EMBO J., 5:1769.

Bujarski, J., Hardy, S., Miller, W. A., and Hall, T., 1982, Use of dodecyl-ß-D-maltoside in the purification and stabilization of RNA polymerase from brome mosaic virus-infected barley, Virology, 119:465.

Bujarski, J., and Kaesberg, P., 1987, DNA inserted two bases down from the initiation site of a SP6 polymerase transcription vector is transcribed efficiently in vitro, Nucl. Acids Res., 15:1337.

Cornelissen, B., and Bol, J., 1984, Homology between the proteins encoded by tobacco mosaic virus and two tricornaviruses, Plant Mol. Biol., 3:379.

Cress, D., Kiefer, M., and Owens, R., 1983, Construction of infectious potato spindle tuber viroid cDNA clones, Nucl. Acids Res., 11:6821.

Dawson, W., Beck, D., Knorr, D., and Grantham, G., 1986, cDNA cloning of the complete genome of tobacco mosaic virus and production of infectious transcripts, Proc. Natl. Acad. Sci. USA, 83:1832.

Domier, L., Franklin, K., Shahabuddin, M., Hellman, G., Overmeyer, J., Hiremath, S., Siaw, M., Lomonosoff, G., Shaw, J., and Rhoads, R., 1986, The nucleotide sequence of tobacco vein mottling virus RNA, Nucl. Acids Res., 14:5417.

Domingo, E., Sabo, D., Taniguchi, T., and Weissman, C., 1978, Nucleotide sequence heterogeneity of an RNA phage population, Cell, 13:735.

Fields, S., and Winter, G., 1981, Nucleotide sequence heterogeneity and sequence rearrangements in influenza virus cDNA, Gene, 15:207.

French, R., Janda, M., and Ahlquist, P., 1986, Bacterial gene inserted in an engineered RNA virus: efficient expression in monocotyledenous plant cells, Science, 231:1294.

French, R., and Ahlquist, P., 1987, Intercistronic as well as terminal sequences are required for efficient amplification of brome mosaic virus RNA3, J. Virol., 61:1457.

Franssen, H., Leunissen, J., Goldbach, R., Lomonosoff, G., and Zimmern, D., 1984, Homologous sequences in non-structural proteins from cowpea mosaic virus and picornaviruses, EMBO J., 3:855.

Gerlach, W., Buzayan, J., Schneider, I., and Bruening, G., 1986, Satellite tobacco ringspot virus RNA: biological activity of DNA clones and their in vitro transcripts, Virology, 151:172.

Gorbalenya, A. E., Blinov, V. M., and Kunin, E. V., 1985, Predictions of nucleotide-binding properties of virus-specific proteins from their primary structure, Molek. Genetika, 11:30.

Haenni, A., Joshi, S., and Chapeville, F., 1982, tRNA-like structures in the genomes of RNA viruses, Prog. in Nucl. Acid Res. and Mol. Biol., 27:85.

Hardy, S., German, R., Loesch-Fries, L. S., and Hall, T., 1979, Highly active template-specific RNA-dependent RNA polymerase from barley leaves infected with brome mosaic virus, Proc. Natl. Acad. Sci. USA, 76:4956.

Haseloff, J., Goelet, P., Zimmern, D., Ahlquist, P., Dasgupta, R., and Kaesberg, P., 1984, Striking similarities in amino acid sequence among nonstructural proteins encoded by RNA viruses that have dissimilar genomic organization, Proc. Natl. Acad. Sci. USA, 81:4358.

Janda, M., French, R., and Ahlquist, P., 1987, High efficiency T7 polymerase synthesis of infectious RNA from cloned brome mosaic virus cDNA and effects of 5' extensions on transcript infectivity, Virology, 158:259.

Kamer, G., and Argos, P., 1984, Primary structural comparison of RNA-dependent polymerases from plant, animal and bacterial viruses, Nucl. Acids Res., 12:7269.

Kandolf, R., and Hofschneider, P., 1985, Molecular cloning of the genome of a cardiotropic Coxsackie B3 virus: full-length reverse-transcribed recombinant cDNA generates infectious virus in mammalian cells, Proc. Natl. Acad. Sci. USA, 82, 4818.

Keranen, S., and Kaariainen, L., 1979, Functional defects of RNA-negative temperature sensitive mutants of Sindbis and Semliki Forest viruses, J. Virol., 32:19.

Kiberstis, P., Loesch-Fries, L.S., and Hall, T., 1981, Viral protein synthesis in barley protoplasts inoculated with native and fractionated brome mosaic virus RNA, Virology, 112:804.

Lane, L., 1981, Bromoviruses, in: "Handbook of Plant Virus Infections and Comparative Diagnosis," E. Kurstak, ed., Elsevier/North-Holland Biomedical Press, Amsterdam.

Leonard, D., and Zaitlin, M., 1982, Temperature sensitive strain of tobacco mosaic virus defective in cell-to-cell movement generates an altered viral-coded protein, Virology, 117:416.

Levis, R., Weiss, B., Tsiang, M., Huang, H., and Schlesinger, S., 1986, Deletion mapping of Sindbis virus DI RNAs derived from cDNAs defines the sequences essential for replication and packaging, Cell, 44:137.

Loesch-Fries, L., and Hall, T., 1980, Synthesis, accumulation, and encapsidation of individual brome mosaic virus RNA components in barley protoplasts, J. Gen. Virol., 47:323.

Matthews, R. E. F., 1982, Classification and nomenclature of viruses, Intervirology, 17:no.1-3.

Meshi, T., Ishikawa, M., Motoyoshi, F., Semba, K., and Okada, Y., 1986, In vitro transcription of infectious RNAs from full-length cDNAs of tobacco mosaic virus, Proc. Natl. Acad. Sci. USA, 83:5043.

Miele, E., Mills, D. and Kramer, F., 1983, Autocatalytic replication of a recombinant RNA, J. Mol. Biol., 171:281.

Miller, W. A., Bujarski, J., Dreher, T., and Hall, T., 1986, Minus-strand initiation by brome mosaic virus replicase within the 3' tRNA-like structure of native and modified RNA templates, J. Mol. Biol., 187:537.

Miller, W. A., Dreher, T., and Hall, T., 1985, Synthesis of brome mosaic virus subgenomic RNA in vitro by internal initiation on (-) sense genomic RNA, Nature, 313:68.

Miller, W. A., and Hall, T., 1983, Use of micrococcal nuclease in the purification of highly template dependent RNA-dependent RNA polymerase from brome mosaic virus-infected barley, Virology, 125:236.

Mizutani, S., and Collono, R., 1985, In vitro synthesis of an infectious RNA from cDNA clones of human rhinovirus type 14, J. Virol., 56:628.

Ohno, T., Takamatsu, N., Meshi, T., Okada, Y., Nishiguchi, M., and Kiho, Y., 1983, Single amino acid substitution in 30k protein of TMV defective in virus transport function, Virology, 131:255.

Racaniello, V., and Baltimore, D., 1981, Cloned poliovirus cDNA is infectious in mammalian cells, Science, 214, 916.

Rezaian, M., Williams, R., Gordon, K., Gould, A., and Symons, R., 1984, Nucleotide sequence of cucumber mosaic virus RNA2 reveals a translation product significantly homologous to corresponding proteins of other viruses, Eur. J. Biochem., 143:277.

Rietveld, K., Pleij, C., and Bosch, L., 1983, Three-dimensional models of the tRNA-like 3' termini of some plant viral RNAs, EMBO J., 2:1079.

Richards, O., Martin, S., Jense, H., and Ehrenfeld, E., 1984, Structure of poliovirus replicative intermediate RNA, J. Mol. Biol., 173:325.

Roosien, J., and van Vloten-Doting, L., 1982, Complementation and interference of ultraviolet-induced Mts mutants of alfalfa mosaic virus, J. Gen. Virol., 63:189.

Sarachu, A., Nassuth, A., Roosien, J., Van Vloten-Doting, L., and Bol, J., 1983, Replication of temperature sensitive mutants of alfalfa mosaic virus in protoplasts, Virology, 125:64.

Schlesinger, S., Levis, R., Weiss, B., Tsaing, M., and Huang, H., 1987, Replication and packaging sequences in defective interfering RNAs of Sindbis virus, in: "Positive Strand RNA Viruses," M. Brinton and R. Rueckert, eds., Alan R. Liss Co., New York.

Shih, D. S., and Kaesberg, P., 1973, Translation of brome mosaic viral RNA in a cell-free system derived from wheat embryo, Proc. Natl. Acad. Sci. USA, 70:1799.

Simon, A., and Howell, S., 1987, Synthesis in vitro of infectious RNA copies of the virulent satellite of turnip crinkle virus, Virology, 156:146.

Strauss, E., and Strauss, J., 1980, Mutants of alphaviruses: genetics and physiology, in: "The Togaviruses," R. Schlesinger, ed., Academic Press, Orlando.

Takamatsu, N., Ishikawa, M., Meshi, T., and Okada, Y., 1987, Expression of bacterial chloramphenicol acetyltransferase gene in tobacco plants mediated by TMV-RNA, EMBO J., 6:307.

Taniguchi, T., Palmieri, M., and Weissman, C., 1978, Qß DNA-containing hybrid plasmids giving rise to Qß phage formation in the bacterial host, Nature, 274:223.

van der Werf, S., Bradley, J., Wimmer, E., Studier, F. W., and Dunn, J., 1986, Synthesis of infectious poliovirus RNA by purified T7 RNA polymerase, Proc. Natl. Acad. Sci. USA, 83:2330.

THE USE OF FULL-LENGTH DNA COPIES IN THE STUDY OF THE EXPRESSION AND REPLICATION OF THE RNA GENOME OF COWPEA MOSAIC VIRUS

Ab van Kammen, Joan Wellink, Rob Goldbach, Rik Eggen and Pieter Vos
Department of Molecular Biology, Wageningen Agricultural University, De Dreijen 11, 6703 BC Wageningen. The Netherlands

INTRODUCTION

Cowpea mosaic virus (CPMV) is the type member of the comovirus group, which contains fourteen different plant viruses that have the same structural organisation of genomic RNAs and virus particles and use the same mechanism for expression and replication of the viral RNAs. CPMV is the best-studied virus of the group and in the following we shall mainly deal with CPMV. For recent reviews on comoviruses see ref. 1. and 2. CPMV has a genome consisting of two positive-strand RNA molecules separately encapsidated in icosahedral particles each with a diameter of 28 nm. The two nucleoprotein particles denoted B and M components have similar capsids composed of 60 copies of each of two different coat proteins but differ in nucleic acid content. B components contain a single RNA molecule (B-RNA) with a mol. weight of 2.04×10^6 and M components a RNA molecule (M-RNA) with a mol. weight of 1.22×10^6. Both B-RNA and M-RNA have a small protein VPg covalently linked to their 5' ends and a poly(A) tail at their 3'ends. Both RNAs are required for infectivity in plants but B-RNA can replicate independently in cowpea protoplasts. On the other hand, M-RNA, which codes for the two capsid proteins is completely dependent on B-RNA for its replication. This demonstrates that B-RNA must carry information for viral RNA replication.

Expression of CPMV RNAs

Both B- and M-RNA are translated *in vitro* and *in vivo* into large polyproteins, which are subsequently cleaved through a number of steps into functional polypeptides.
B-RNA, 5889 nucleotides long excluding the poly(A) tail is translated into a single 200K (= kilodalton) protein that is rapidly cleaved into 32K and 170K polypeptides. The 170K protein is further processed via two alternative routes (Fig.1): either the 170K is cleaved into 60K and 110K proteins or, by an other cleavage into 84K and 87K proteins. The 110K and 84K proteins can undergo additional cleavages to give the 87K and 60K proteins respectively together with a 24K protein. The 60K protein is the direct precursor of VPg. All five final cleavage products generated from the 200K polyprotein as well as the different intermediates, are present in CPMV infected protoplast, if in varying amounts. Free VPg has not been detected *in vivo* but it occurs either in the precursor form or linked to the 5' end of the terminal uridyl-residue of B-and M-RNA.

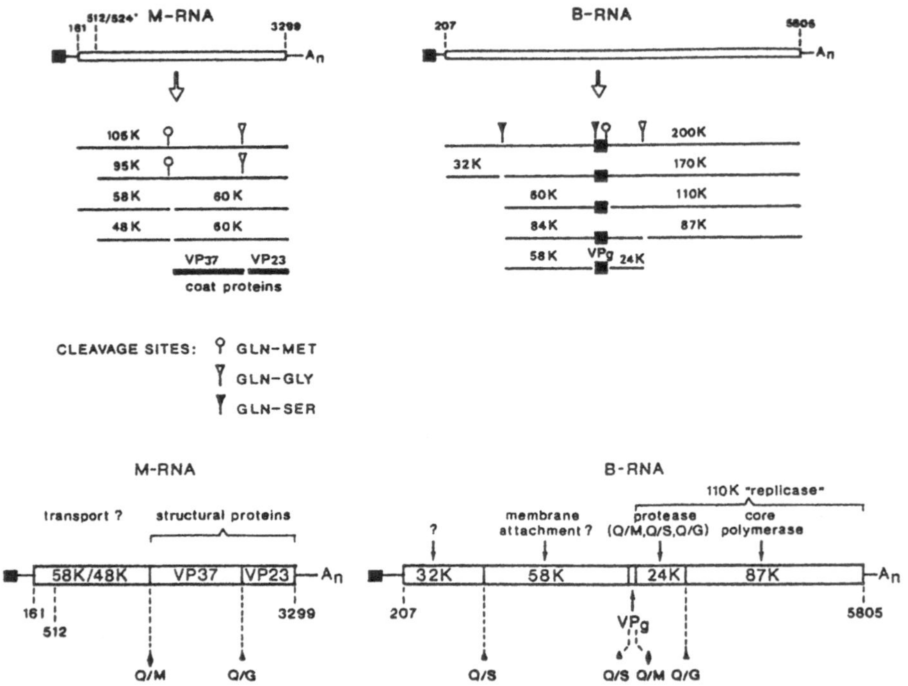

Fig.1.: Organization and expression of the genome of CPMV. the two genomic RNAs, M-RNA and B-RNA, contain each a single open reading frame represented by the open bars. The positions of the translational start and stop codons are indicated. B-RNA is translated into a 200K polyprotein and M-RNA into C-terminal overlapping 105K and 95K polyproteins, which are subsequently processed by proteolytic cleavages at the indicated sites. Translation of M-RNA in vitro starts at position 161, but more efficiently at the AUG codon at position 512. In vivo the 105K M-RNA encoded polyprotein has not been detected. In the genetic map in the lower part of the figure is the position of the coding regions of the different functional domains in the reading frames indicated drawn to scale. The same three types of cleavage sites used in the proteolytic processing are indicated with different abbreviations and marks in the upper and lower drawing.

The cleavage sites used in the proteolytic processing have been determined by determining partial amino-terminal sequences of the various B-RNA encoded proteins and locating the coding regions for these proteins on the B-RNA sequence (3). Three types of sites are used: a glutamine-serine pair (2x), a glutamine-methionine pair (1x) and a glutamine-glycine pair (1x). The order of the cleavage products in the 200K B-RNA encoded polyprotein is NH_2-32K-58K-4K(VPg)-24K-87K-COOH. (Fig.1.).
M-RNA, 3481 nucleotides long not including the poly(A) tail is translated into two C-terminal overlapping polypeptides of 105K and 95K which are cleaved into overlapping 58K and 48K proteins respectively and a 60K capsid protein, which is the precursor of the two capsid proteins VP37 and VP23. The proteolytic cleavage sites used in the processing are a glutamine-methionine pair and a glutamine-glycine pair. The order of the proteins in the M-RNA encoded polyprotein is NH_2-58K/48K-VP-37-VP23-COOH.
In the genetic map of CPMV shown in Fig.1. functions assigned to several of

the viral coded proteins are indicated. B-RNA encodes functions for viral RNA replication but, although B-RNA does not require M-RNA for replication, it cannot move to adjacent cells in plants in the absence of M-RNA. It has therefore been suggested that M-RNA encodes in addition to the two capsid proteins a protein to effect cell-to-cell transport. The 48K and/or 58K protein are possible candidates for such function. In vivo the capsid proteins are the only M-RNA encoded products easily detected but using specific antibodies it has recently been possible to detect in CPMV-infected protoplasts small amounts of the 60K capsid precursors and the 48K protein. The 58K protein has not been found (4). The 48K protein appears further to be excreted in the culture medium of CPMV-infected protoplasts. Therefore, if a specific protein is involved in cell-to-cell transport of CPMV, the 48K protein seems the most plausible candidate.
The detection in vivo of the 48K and 60K M-RNA encoded proteins is good evidence that M-RNA is translated in vivo in a 95K polyprotein but it is unclear if the expression of M-RNA into a 105K polypeptide as found upon in vitro translation, has a role in vivo. In this connection it is worth noting that M-RNA of several other comoviruses direct upon in vitro translation also the synthesis of two large polypeptides of approximately the same size as those produced by CPMV M-RNA. The occurrence of two AUG codons which give rise to translation into two large-sized proteins is apparently a common feature of comovirus M-RNAs which rather suggests that it may have biological significance.

Replication of CPMV-RNA

The replication of CPMV RNA takes place in the cytoplasm of infected cells associated with vesicular membrane structures the proliferation of which is induced upon virus infection. Accordingly, CPMV RNA replication complexes have been isolated and purified from membrane fractions of infected cells (see 2. and 1. for review). Highly purified replication complexes, still able to elongate in vitro nascent viral RNA chains, already initiated in vivo, to full-length viral RNA molecules, contain three major polypeptides: 110K, 68K and 57K. Using antisera against various viral proteins the 110K protein was proven to be the 110K polypeptide encoded by B-RNA (Fig.1.) while the 68K and 57K polypeptides are presumably host proteins which either have a function in the replication complex or are contaminating proteins. The 110K polypeptide (see Fig.1.) contains the sequence of both the 24K and B-RNA encoded proteins. The polymerase activity resides in the 87K polypeptide and the 24K polypeptide bears proteolytic activity (see next section). Since RNA replication appears to take place at membranes, other B-RNA encoded proteins may be important in attaching the replication to the plasma-membranes. The B-RNA encoded 58K protein and its direct 60K precursor which also contains the sequence of VPg are distinct membrane proteins and may perform this function. It then appears that four non-structural proteins with functions in viral RNA replication occur in the order 58K (membrane binding of replication complex) -VPg - 24K (protease) - 87K (RNA-dependent) on the 170K part of the 200K encoded polyprotein form which they originate by specific proteolytic cleavages.
A similar organization of corresponding functions in viral RNA replication is found in animal picornaviruses and plant viruses of the potyvirus and nepovirus group (5,6). All these viruses have positive-strand RNA genomes which are 3' poly-adenylated and carry a 5' terminal genome-linked protein VPg and are expressed through translation into large polyproteins from which native viral proteins arise by specific proteolytic cleavages. The analogy among these viruses furthermore includes significant amino acid sequence homology between the corresponding four proteins wich presumably have similar functions in viral RNA replication. The maintenance of the organization of functions and the conservation of their amino acid sequen-

ces strongly suggest that these plant and animal viruses use a similar mechanism for viral RNA replication jointedly carried out by the membrane protein, VPg, protease and replicase in a single complex.
An intriguing, unresolved question is the occurrence of VPg at the 5' end of the genomic RNAs. The release of VPg from the 60K precursor may be an important event in the initiation of viral RNA replication where it may have a role in priming RNA synthesis. A possible role of VPg in an early stage of CPMV-RNA replication is suggested by the finding of VPg at the 5' end of both negative and positive strands in the replicative form RNA isolated from virus-infected leaves (7).
The occurrence of VPg at the 5' end of each progeny viral RNA strand raises still further questions on the viral RNA replication mechanism. VPg is encoded by B-RNA that is translated into a 200K polyprotein which has to complete several successive processing steps to make VPg available and this then has to happen each time a B-RNA or a M-RNA molecule is produced. In this way protein processing appears once more to have a dominant role in the replication of CPMV-RNA as replication and processing seem to be closely connected.

For further elucidating the role of the virus-encoded proteins in the expression and to resolve the speculations about the mechanism of viral RNA replication and the possible regulating role of processing in this process, fullsize DNA copies of CPMV B- and M-RNA have been constructed and cloned. These clones can be used to produce site specific mutations the effects of which can subsequently be tested (8, 9, 10, 11). Some first results are reported in the following sections.

II PROTEOLYTIC PROCESSING OF THE CPMV POLYPROTEINS BY THE 24K PROTEIN

It has been thought for some years that CPMV B-RNA codes for two different proteases to achieve the cleavages at the three different types of sites employed during the processing of the B- and M-RNA encoded proteins (1). Inhibition studies using antiserum against the B-RNA encoded 32K polypeptide suggested that the 32K polypeptide represented a protease involved in cleaving the glutamine-methionine site releasing the 60K capsid precursor from the M-RNA encoded polyproteins (12). It appeared then also probable that this protein was involved in the glutamine-methionine cleavage in the B-RNA encoded polyprotein (Fig.1.). In the meantime it was proven that the 24K polypeptide encoded by B-RNA represents a viral protease capable of the primary cleavage at a glutamine-serine site in the B-RNA encoded polyprotein (8). This conclusion was based on experiments which used a full-length DNA copy synthesized from B component RNA and cloned in an appropriate vector downstream of a phage SP6 or T7 promoter. RNA molecules transcribed <u>in vitro</u> from this full-size DNA copy using SP6 or T7 RNA polymerase were efficiently translated in rabbit reticulocyte lysates into a 200K polypeptide similar to the polyprotein obtained by translation of viral B-RNA. This 200K polypeptide was rapidly cleaved in the lysate into 32K and 170K polypeptides exactly like the 200K polyprotein from viral B-RNA that is cleaved into these products at a glutamine-serine dipeptide sequence. <u>In vitro</u> transcription and translation of a DNA copy in which a 87 bp deletion in the coding sequence of the 24K polypeptide was introduced resulted in a primary translation product that was not processed which demonstrated the 24K polypeptide bears the enzyme activity involved in the primary cleavage of the 200K polyprotein at a glutamine-serine site. Since at that time the 32K protein released by the primairy cleavage from the B-RNA encoded 200K polyprotein was thought to be a viral protease involved in the cleavage of the M-RNA encoded polyproteins at a glutamine-methionine site (12) (see Fig.1.) it was of interest to examine the effect

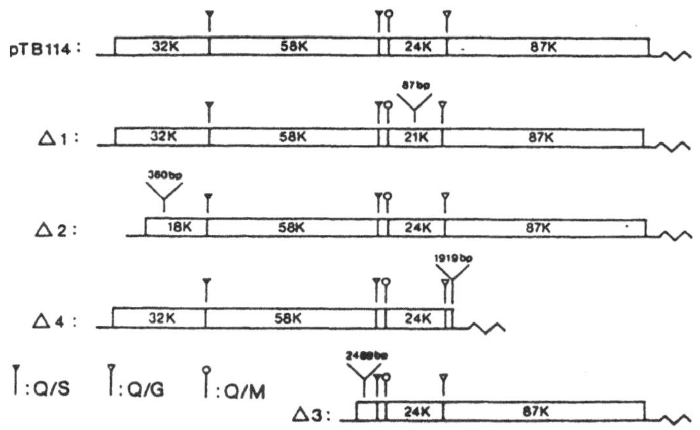

Fig.2.: Schematic representation of various cDNA clones of CPMV B-RNA. Clone pTB114 contains full-length cDNA of B-RNA whereas the clones pTB114 Δ1, -Δ2, -Δ3 and -Δ4 have deletions in the B cDNA at the indicated sites. The single open reading frames in the cDNAs are indicated by the open bars and the non-coding sequences as single lines. The different cDNAs were linked to a phage T7 promoter in a way that upon transcription after linearising of the vector DNA run-off transcripts are obtained wich, in comparison with viral RNAs, have two extra G residues at the 5' end and 4-5 non viral nucleotides at the 3' end of the poly(A) tails the cleavage sites are indicated as are the sizes of the final cleavage products.

of the deletion in the 24K protease i.e. blocking of the release of the 32K protein on the processing of the M-RNA encoded proteins. Surprisingly the small deletion in the coding region of the 24K polypeptide also prevented the processing of the M-RNA encoded polyproteins at a glutamine-methionine site. This prompted us to study in further detail the processing of the polyproteins produced by the CPMV-RNAs.

Proteolytic processing of B-RNA encoded polyprotein

The proteolytic processing of the B-polyprotein has further been elucidated by constructing a series of specific deletion mutants starting from a clone of full-length copy DNA of viral B RNA (10). RNA transcripts of the full-length cDNA in pTB114 (Fig.2.) was translated into a 200K polyprotein which was not only rapidly cleaved into 170K and 32K polypeptides but, upon longer incubation, for 20 hours, the 170K was further processed either by a glutamine-methionine cleavage into 110K and 60K polypeptides or by a glutamine-glycine cleavage into 87K and 84K polypeptides (cf Fig.1.). The 60K polypeptide was furthermore cleaved at a glutamine-serine site into 58K polypeptide and VPg.
The *in vitro* processing of the 200K polyprotein translated from B-RNA transcripts precisely imitated the generation by specific cleavages of B-encoded proteins *in vivo* as observed for example in cowpea protoplasts infected with CPMV B components. This can obviously make the study of the effect of various mutations on the processing *in vitro* of interest for understanding the processing *in vivo* . As already mentioned, a small deletion in the 24K polypeptide region blocked the primary cleavage of the 200K polyprotein at a glutamine-serine site. Even upon incubation for 20 hrs the 200K polyprotein remained stable and no specific cleavages were observed.

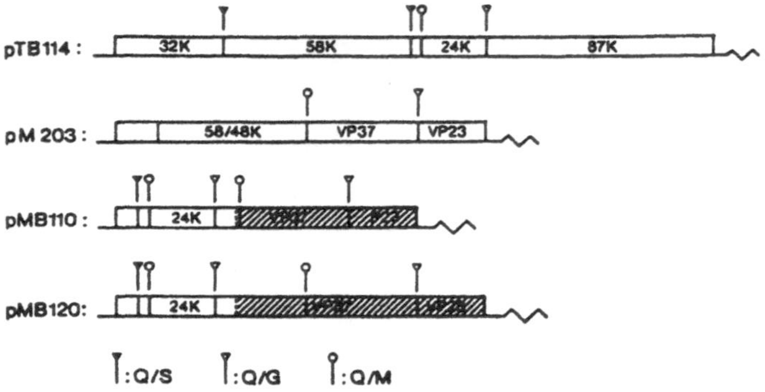

Fig.3.: Schematic representation of full-length cDNAs of B-RNA (pTB114) and M-RNA (pM203) and two hybrid cDNAs (pMB110 and pMB120) containing sequences derived from the cDNAs of M- and B-RNA. The single open reading frames are indicated by double-lined bars, the non-coding sequences as single lines. The shaded parts are derived from M cDNA and the open parts from B-cDNA. In vitro transcripts from the closed cDNA were obtained as described in the legend of Fig.2. The position of the proteolytic cleavage sites in the polyproteins are indicated.

Translation of RNA transcribed from a mutant cDNA clone with a 375 bp deletion in the coding sequence of the 32K polypeptide on the other hand did not change the processing of the translation product produced by this mutant RNA (Fig.2.) In that case the primary translation product of 190K was rapidly cleaved into a truncated 32K polypeptide of about 18K and the 170K polypeptide which was further processed precisely as the 170K polypeptide encoded by viral B-RNA or by B-RNA transcribed from the intact B DNA in pTB114. This processing proves that the 32K protein is not involved in any of the cleavages in the 170K polypeptide and that the 24K seems capable of cleaving all three types of cleavage sites in the 200K polyprotein without a second protease being necessary.

Proteolytic processing of M-RNA encoded polyproteins

For a closer examination of the proteolytic cleavages in the M-RNA encoded polyproteins hybrid DNA copies were constructed, which contained sequences of both M and B cDNA (10) (Fig.3.). These clones had in their open readingframe the coding sequence of the B-encoded 24K protease and the two M-RNA encoded capsid proteins including the glutamine-methionine and glutamine-glycine cleavage sites to release the coat proteins.
The polyprotein translated from the transcripts of these clones were rapidly processed into smaller proteins which were identified by immunoprecipitation using antisera against 24K polypeptide, VP37 and VP23. The results demonstrated that the 24K protein itself was released from the large primary translation product by cleaving at glutamine-methionine and glutamine-glycine cleavage sites bordering the 24K sequence in the hybrid polyprotein. In addition the 60K capsid precursor and the two capsid proteins VP37 and VP23 were found demonstrating that cleavage at the glutamine-methionine site and the glutamine-glycine site occurring in the M polyprotein part of the hybrid protein had also taken place. These cleavages did not happen with the polyprotein translated from a hybrid clone with the 89 bp deletion in the coding sequence of the 24K protein which proves that the cleavage in the hybrid proteins are indeed due to the protease activity of the 24K polypeptide. Thus it is clear that the B-encoded 24K protease is capable of cleaving both the glutamine-methionine and the methionine-glycine site in the M polyproteins.

Since in vivo the cleavage sites on the 105K and 95K M-RNA encoded polyproteins are not located on the same protein molecule as the 24K protease the proteolytic processing at the M polyprotein cleavage sites was also examined "in trans" by adding to polyproteins produced by viral M-RNA (see Fig.1.) unlabeled translation products from RNA transcripts of the full-size B cDNA and of the mutants made thereof. Addition of translation products from RNA transcribed from the intact fullsize B cDNA resulted in processing of the M-RNA encoded polyproteins into 58K/48K polypeptides and the 60K capsid precursor.

If now translation products obtained with mutant B cDNA lacking an intact 32K polypeptide but having an intact 24K protein were added to M-RNA encoded polyproteins no cleavage at the glutamine-methionine site occurred and no 60K capsid precursor was released, but cleavage of the glutamine-glycine site resulting in release of the small capsid protein VP23, was observed instead. Further experiments confirmed that only translation products from RNA transcripts of B cDNA clones with intact 32K and 24K polypeptides could achieve cleavage at the glutamine-methionine site in the M-RNA encoded polyproteins to release the 60K capsid precursor.

The results of all the processing experiments indicate that the B-RNA encoded 24K protease is responsible for all cleavages in the CPMV polyproteins. There is no specific proteolytic activity associated with the 32K polypeptide encoded by B-RNA as thought previously. For efficient cleavage of the glutamine-methionine site in the M-RNA encoded polyproteins the presence of the 32K polypeptide however is essential.

III INFECTIOUS RNA TRANSCRIPTS FROM FULL LENGTH cDNA CLONES OF CPMV RNAs

A set of full-length DNA copies of B-RNA and M-RNA were cloned downstream of a phage T7 promoter so that RNA transcripts from the inserted DNA had only two extra G residues at the 5' end as compared to viral RNA. The run-off transcripts from the linearised vector DNA further had a poly(A) tail of more than 60 nucleotides with four or five non-viral nucleotides at the 3' end of the poly(A) tail originating from the Cla I restriction site used in linearising the template DNA. The transcripts were tested for infectivity by inoculation of cowpea mesophyll protoplast. First different B-RNA transcripts were tested alone and after that, mixtures of B- and M-RNA transcripts.

Upon inoculation of cowpea protoplasts with 100 µg RNA transcribed from B cDNA it was demonstrated that various B-RNA encoded proteins were synthesized. Using antiserum raised against the 24K polypeptide (13) the 170K, 110K and 84K B-RNA specified polypeptides, all containing the 24K polypeptide domain could be detected by immunoprecipitation. The 24K protein itself was not found as it occurs only in very low amounts in virus-infected protoplasts (13). The synthesis of the B-RNA encoded proteins in the inoculated protoplasts indicated that the B-RNA transcripts are infectious since replication of RNA has to occur before viral protein synthesis becomes detectable in protoplasts. The amounts of B-RNA -specific proteins detected upon inoculation with transcripts from different B cDNA clones varied significantly.

Protoplasts inoculated with RNA transcripts from a B cDNA clone that produced a relatively large amount of B-RNA specific proteins were then analysed for the presence of viral RNA at different times after inoculation. Dot blot analysis of nucleic acids extracted from inoculated protoplasts showed that the amount of B-specific RNA decreased rapidly in the first 24 hours and then increased again indicating that after a lag period replication of transcripts became detectable.

Finally, using antiserum against the 24K polypeptide, it was possible to determine the number of protoplasts infected with B-RNA transcripts by fluorescent antibody staining. When $5x10^6$ protoplasts were inoculated with 100 µg of B-RNA transcripts about 1% of the protoplast had produced B-RNA

Table I. Infectivity of RNA transcribed from full-length cDNA clones of CPMV RNAs

RNA transcripts	μg RNA per 5x10^6 protopl.	% fluorescent cells with α-24K serum	% fluorescent cells with α-capsid serum
B114	100	0.8	0
B114 *	100	2.0	0
M203 *	100	0	0
B114 + M203 *	200	1.5	0.1
B114 Δ1 *	100	0	0
B114 Δ2 *	100	0	0
CPMV-RNA			
B RNA+M RNA	2	1.5	1.4
"	4	4.0	3.9
"	20	20	20
"	50	50	50

*) capped transcripts

In the first column are B114, B114 Δ1 and B114 Δ2 the same clones as pTB114, pTB114 Δ1 and pTB114 Δ2 depicted in Fig.2. M203 is the same as pM203 described in Fig.3.

specific proteins and were thus infected. (Table I)

Transcripts of cDNA clones of M-RNA were also tested for infectivity in cowpea protoplasts. As M-RNA is not replicated and expressed in the absence of B-RNA, infectious transcripts of B cDNA clones were used for co-infection to allow for replication of M cDNA transcripts. Since viral RNA from B-components is always contaminated with trace amounts of M-RNA irrespective of the isolation procedure and very low numbers of infected cells were to be expected, the use of infections transcripts of B cDNA clones was preferred as these transcripts are obviously free of any contaminating M-RNA.

In protoplasts inoculated with mixtures of infectious B cDNA transcripts and RNA transcribed from each of several different M cDNA clones no capsid protein synthesis could be detected by immunoprecipitation using antiserum against virus particles. But by fluorescent antibody staining it was demonstrated that 0,05% of the protoplasts showed the characteristic fluorescence of virus infected cells. This low number indicates that only approximately 5% of the protoplasts which were infected with B cDNA transcripts, were simultaneously succesfully infected with M cDNA transcripts (Table 1).

One reason for the low infectivity of the transcripts could be that the RNA is rapidly degraded upon inoculation of the protoplasts. The presence of a cap structure at the 5' end increases the stability of RNA (14) and therefore transcripts supplied with a cap were also tested for infectivity. With capped RNA the number of infected protoplasts was about two times larger than uncapped RNA, but still very low in comparison to viral RNA.

In table I are also recorded the results of infectivity tests with RNA transcribed from two mutant cDNAs of B-RNA. B114 Δ1 contained a deletion in the coding sequence of the 24K protease (8, 10) and B114 Δ2 a deletion in the coding sequence of the 32K protein with yet undefined function. The transcripts from both mutant cDNAs were not infectious. For mutant B114 Δ1 that is not surprising as this mutant has a defective 24K protein and the 24K protease is known to be essential for processing of the viral polyproteins. For the 32K protein there are however no indications of a specific function in the expression of B-RNA. The lack of infectivity of the RNA transcribed from the mutant B114 Δ2 cDNA might indicate that the 32K protein has a role in the replication of B-RNA.
The low infectivity of the RNA transcribed from the full-length cDNAs is a serious shortcoming of the present system. It seems plausible to blame the low infectivity of the RNA transcripts on the differences between the RNA transcripts and viral RNAs. As compared to viral B- and M-RNA the in vitro transcripts of the cDNAs of M- and B-RNA lack the 5' covalently linked VPg but have instead a 5' end extended with two guanylyl residues. Judged from the experiences with other viruses this can greatly affect the infectivity. The in vitro synthesized RNA transcripts of full-length cDNA clones of tobacco mosaic virus (TMV) (15, 16) and of brome mosaic virus BMV (17) displayed a relatively high infectivity if these transcripts had exactly the same 5' capped termini as the viral RNAs. The occurrence of a few extra nucleotides at the 5' ends decreased the infectivity of the transcripts dramatically (15, 17, 18). With these other viruses extra sequences at the 3' end of in vitro transcripts have no significant effect on the infectivity. Therefore we suppose that the 4 to 5 extra nucleotides at the 3' end of the poly(A) tail of the transcripts synthesized in vitro from the CPMV cDNAs are not the cause of the low infectivity.
Another reason for the low infectivity of the in vitro transcripts in the case of CPMV, might be the absence of VPg in the transcripts although it is not known whether the VPg at the 5' end has a specific function, it might protect the 5' ends against degradation by nucleases and contribute to the stability of the RNAs upon inoculation of protoplasts. The higher infectivity of the capped transcripts in comparison to uncapped transcripts (Table 1) could indicate that some kind of shielding of the 5' end is beneficial.
If the synthesis of infectious in vitro RNA transcripts for CPMV is an important step forward the system clearly needs improvement to increase the infectivity of the RNAs transcribed from the cDNAs before it will be optimum useful for further molecular genetics studies of CPMV multiplication.

Acknowledgements

The authors thank Hedy Adriaansz for typing the manuscript. This work was supported by the Netherlands Foundation for Chemical Research (SON with financial aid from the Netherlands Organization of Pure Research (ZWO).

REFERENCES

1. R. Goldbach, Comovirusses: Molecular biology and replication, in: Plant viruses: viruses with bipartite genomes and isometric particles, B.D. Harrison & A.F. Murant, eds. Plenum, New York, in press.
2. R. Eggen and A. van Kammen, RNA replication in comoviruses, in: RNA Genetics, Vol.I RNA replication, P. Ahlquist, J. Holland and E. Domingo, eds. CRC Press, Boca Raton, Florida, chapter 3, in press.
3. J. Wellink, G. Rezelman, R. Goldbach and K. Beyreuther, Determination of the proteolytic processing sites in the polyprotein encoded by the bottom-component RNA of cowpea mosaic virus, J. of Virol. 59:50 (1986).

4. J. Wellink, M. Jaegle, H. Prinz, A. van Kammen and R. Goldbach, Expression of the middle component RNA of cowpea mosaic virus in vivo, J. gen. Virol. 68 (1987), in press.
5. R.W. Goldbach, Molecular evolution of plant RNA viruses, Ann. Rev. Phytopathol. 24:289 (1986).
6. L.L. Domier, J.G. Shaw and R.E. Rhoads, Potyviral proteins share amino acid sequence homology with picorna-, como-, and caulimoviral proteins.
7. G. Lomonossoff, M. Shanks and D. Evans, The structure of cowpea mosaic virus replicative form RNA. Virology 144:351 (1985).
8. J. Verver, R. Goldbach, J.A. Garcia and P. Vos, In vitro expression of a full-length DNA copy of cowpea mosaic virus B-RNA: identification of the B-RNA encoded 24-kd protein as a viral protease, The EMBO Journal 6:549 (1987).
9. J.A. Garcia, L. Schrijvers, A. Tan, P. Vos, J. Wellink and R. Goldbach, Proteolytic activity of the cowpea mosaic virus encoded 24K protein synthetised in Escherichia coli, Virology 158, in press, (1987).
10. P. Vos, J. Verver, M. Jaegle, J. Wellink, A. van Kammen and R. Goldbach, Studies on the proteolytic processing of the cowpea mosaic virus polyproteins by in vitro expression of cDNA clones, submitted for publication, 1987.
11. P. Vos, M. Jaegle, J. Wellink, J. Verver, A. van Kammen and R. Goldbach, Infectious RNA transcripts derived from full-length DNA copies of the genomic RNAs of cowpea mosaic virus, submitted for publication, 1987.
12. H. Franssen, M. Moerman, G. Rezelman and R. Goldbach, Evidence that the 32.000-dalton protein encoded by bottom-component RNA of cowpea mosaic virus is a proteolytic processing enzyme. J. Virol. 50:183, 1984.
13. J. Wellink, M. Jaegle and R. Goldbach, Detection of a novel protein encoded by the bottom-component RNA of cowpea mosaic virus using antibodies raised against a synthetic peptide. J. Virol. 61:236 (1987).
14. M. Green, T. Maniatis and D. Melton, Human β-globin pre-mRNA synthesized in vitro is accurately spliced in Xenopus oocyte nuclei, Cell 32:681 (1983).
15. W.O. Dawson, D.L. Beck, D.A. Knorr and G.L. Grantham, cDNA cloning of the complete genome of tobacco mosaic virus and production of infectious transcripts. Proc. Natl. Acad. Sci. USA, 83:1832 (1986)
16. T. Meshi, M. Ishikawa, F. Motoyoshi, K. Sembla and Y. Okada, In vitro transcription of infectious RNAs from full-length cDNAs of tobacco mosaic virus. Proc. Natl. Acad. Sci. USA 83:5043 (1986).
17. P. Ahlquist, R. French, M. Janda and L.S. Loesch-Fries, Multi-component RNA plant virus infection derived from cloned viral cDNA. Proc. Natl. Acad. Sc. USA, 81:7066 (1984).
18. S. van der Werf, J. Bradley, E. Wimmer, F. W. Studier and J.J. Dunn, Synthesis of infectious poliovirus RNA by purified T7 RNA polymerase. Proc. Natl. Acad. Sci. USA 83:2330 (1986).

THE LIFE CYCLE OF CAULIFLOWER MOSAIC VIRUS

Pierre PFEIFFER [(1)], Karl GORDON [(2)], Johannes FÜTTERER [(2)] and Thomas HOHN [(2)]

(1) Institut de Biologie Moléculaire des Plantes,12 rue du Général Zimmer, F-67000 Strasbourg (France)

(2) Friedrich Miescher Institut, P.O. Box 2543, CH-4002 BASEL (Switzerland)

INTRODUCTION

Ever since the discovery that Cauliflower Mosaic Virus (CaMV) is one of the very few plant viruses with DNA instead of RNA as genetic material, the career of this unusual plant virus has been a series of ups and downs. At first, it was hailed in the late seventies as "the SV40 of plants" which should provide both virologists and the emerging plant molecular biologists with a vector for the introduction and expression of foreign genes in plants: although RNA viruses with subgenomic or satellite RNAs can also be used for the same purpose (see the paper on Brome Mosaic Virus by P. Ahlquist), CaMV had the obvious advantage of being a DNA virus and thus directly amenable to genetic engineering. Indeed, the identification of dispensable open reading frames permitted insertion of foreign DNA payloads which led, for instance, to the successful expression of a bacterial enzyme in plants (Brisson et al., 1984). In parallel, CaMV had again made the headlines in 1983 when several groups suggested independently that this plant DNA virus replicates by reverse transcription (Guilley *et al.*; Hull and Covey; Pfeiffer and Hohn). While being quite interesting in terms of molecular biology, this finding came as a frustration to those who wished to use CaMV as a gene vector: CaMV could be still considered as an episomal vector capable of replicating to high copy number, but now care had to be exercised not to incorporate DNA sequences which could interfere with the reverse transcription process, e.g. splicing sites which would lead to deletions.

Another intriguing question about CaMV concerns the expression of the genetic information borne by the DNA. Indeed, it was recognized very early that only two capped and polyadenylated CaMV-specific RNAs exist in detectable amounts in infected plants: one is a 19S monocistronic subgenomic messenger coding for a 62 kDa protein known to assemble into cytoplasmic inclusion bodies, and the other a 35S genomic RNA which contains all open reading frames (ORFs). However, the reverse transcription step uses this 35S RNA as a template: how could it then also be used for translation?

The purpose of this paper is to try to summarize our current knowledge of the biology of CaMV. Further details on specific points may also be found in several recent review articles (Dixon and Hohn, 1985; Maule, 1985, Bonneville *et al.*, 1987).

THE VIRUS AND ITS RELATION WITH THE HOST

CaMV virions appear under the electron microscope as isometric particles about 5O nm in diameter with no distinctive morphological features. The virions are highly resistant to dissociating agents, and accumulate in the cytoplasm of infected cells within amorphous inclusion bodies called viroplasms which are the hallmark of infection by CaMV and related viruses that constitute the group of the "Caulimoviruses".

The viroplasms are at least the sites of accumulation of the virions, if not the sites of synthesis and assembly of the viral components. Neither these viroplasms nor the virus particles are bounded by a membrane.

In the lab, CaMV can be transmitted by mechanical inoculation of either virus or viral DNA, but in nature, it is spread by vegetative propagation and by aphids. Aphid transmissibility requires a virus-encoded transmission factor associated with the viroplasms: naturally occuring non-transmissible strains of CaMV have a deficient gene II, and inactivation of gene II by mutagenesis causes the loss of aphid transmissibility (Givord *et al.*, 1984).

Al Ani *et al.* (1979) showed that the apparent multiplicity of the CaMV polypeptides arose from extensive proteolytic degradation and aggregation processes. They then postulated that the viral capsid probably comprised a single polypeptide species with a possible minor component. Furthermore, they demonstrated that the virion did not contain internal polypeptides forming a core with the nucleic acid, a view confirmed by neutron scattering studies (Chauvin *et al.*, 1979).

The genomic DNA also possessed an intriguing structure: it is found within the virion as a relaxed circle able to form variable numbers of intramolecular knots (Ménissier *et al.*, 1983), linearizes very easily at three specific points (rather than randomly), and generates upon electrophoresis under denaturing conditions three single-stranded molecules dubbed α, β, and γ (Volovitch *et al.* , 1978). This behaviour is due to the presence of three S_1 nuclease-sensitive single-stranded interruptions in the genomic DNA, one in the coding strand and two in the non-coding strand. However, sequencing revealed that these so-called "gaps" were regions of triple-stranded structure corresponding obviously to a limited strand displacement. Since CaMV DNA cloned in a plasmid is infectious, these gaps are not required for infectivity, but they reappear in the progeny virus and are thus somehow

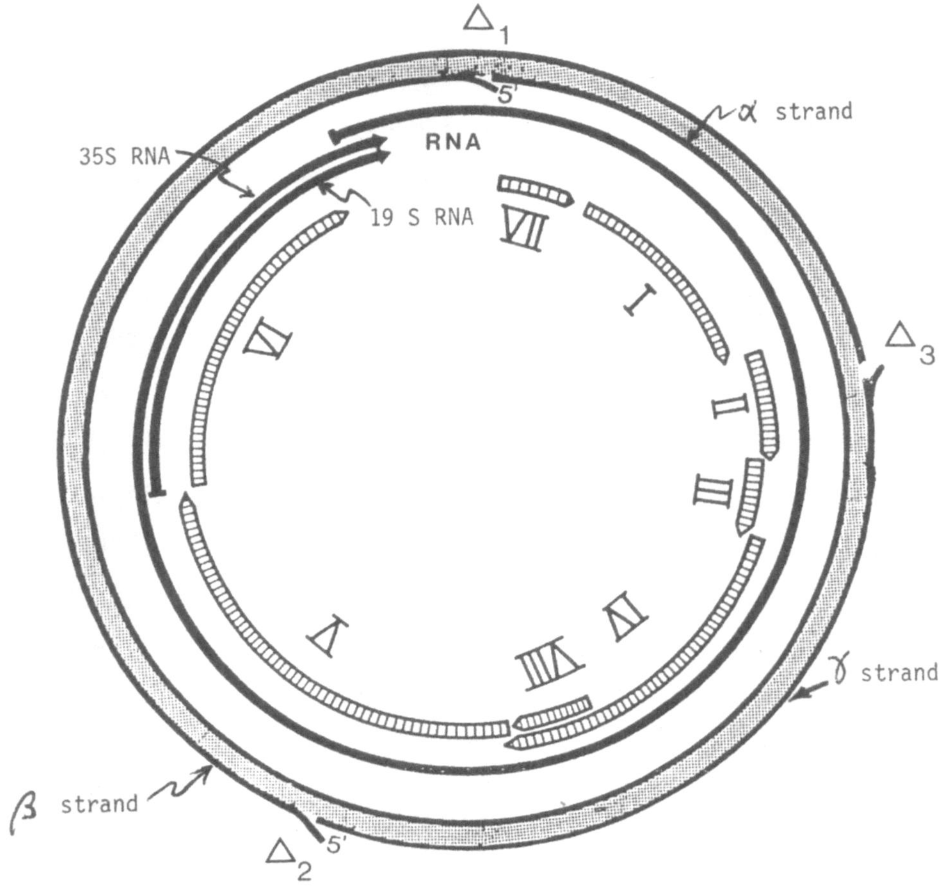

Figure 1 : Genetic organization of Cauliflower Mosaic Virus. The 8 kbp viral DNA features three single-stranded interruptions, one (Δ_1) in the α (or coding) strand, and two (Δ_2 and Δ_3) in the non-coding strand, defining the β and γ DNA species. The capped and polyadenylated 19S and 35S RNAs have different promoters, but share the same 3' termini. The DNA encodes 8 potential ORFs, of which two (ORFs II and VII) are dispensable. The functions of ORFS I and III and unknown, ORFIV codes for the capsid protein, ORFV for the viral reverse transcriptase, and ORF VI for the inclusion body protein.

connected with the replication of the virus. Finally, ribonucleotides are often attached 5' to the first deoxyribonucleotide, suggesting the use of stretches of RNA for the priming of DNA synthesis.

THE VIRAL GENOME AND ITS TRANSCRIPTION

A milestone in molecular studies on CaMV was the determination of the sequence of the viral DNA by Frank *et al.* in 1980. Figure 1 shows that the viral DNA contains six major ORFs and two minor ones that are tightly packed on a single coding strand, and separated by two intergenic regions, a smaller one between genes V and VI, and a larger one between genes VI and VII. The coding strand (or α strand) always features a single interruption (taken as the start of the sequence), whereas the complementary strand has two interruptions, thus generating the β and γ chains.

Transcription of the α strand is directed by the host RNA polymerase II and is asymmetric resulting in the production of the 19S RNA and of a long 35S RNA which covers the whole genome plus a terminal 180 nucleotide (NT) direct repeat. No discrete processing products of the 35S RNA have been detected, and the only other CaMV-specific RNA species detected reproducibly in infected plants is a 19S subgenomic mRNA covering ORFVI, the viroplasm protein gene.

The generation of the 35S RNA requires that the RNA polymerase involved crosses the Δ_1 interruption, and the unencapsidated form of covalently closed supercoiled CaMV DNA discovered by Ménissier *et al.* (1982) was a good candidate to serve as a template for this purpose. This was indeed confirmed by Olzewski *et al.* (1982) who showed that the supercoiled DNA associates with host histones to form a minichromosome transcribed by the host's RNA polymerase II in the nucleus; therefore, even though it ends up in cytoplasmic inclusion bodies, CaMV spends at least part of its life cycle in the nucleus. Accurate mapping of the two major transcripts enabled Guilley *et al.* (1982) to define their respective regulatory sequences: since then, the promoter regions of the 19S and especially of the 35S CaMV RNAs have been widely used as strong constitutive promoters in plant genetic engineering. An unexplained feature remains how the polyadenylation signal - common to the 19S and 35S transcripts - is only recognized by the RNA polymerase at the 3' end of the full length 35S RNA -which includes a terminal repeat- and not at the start of the transcript.

CAMV REPLICATION

The mechanism of CaMV replication has remained for many years more a matter of speculation than of experimental work. The lack of a proper experimental system permitting synchronous infection and labelling of replicating DNA led to tackle this problem via indirect approaches . The existence of supercoiled CaMV DNA as a minichromosome suggested that like in SV40, this structure could serve as a template for both transcription and replication. Pfeiffer and coworkers (1983, 1984) set out to isolate CaMV replicative complexes by hypotonic leaching of organelle suspensions that contained a mixture of nuclei and viroplasms. Such preparations did indeed support *in vitro* the synthesis of viral RNA and DNA, but the surprise came upon fractionation of these replicative complexes on sucrose gradients: CaMV-specific RNA synthesis comigrated with the minichromosomes as expected, but DNA synthesis was associated with much slower sedimenting structures near the top of the gradient rather than at a position where replicating minichromosomes were expected to migrate. Supercoiled DNA was found only in the minichromosome fractions, while free viral DNA and heterodisperse RNA were detected associated with the peak of DNA synthesis activity. Aphidicolin, a potent and specific

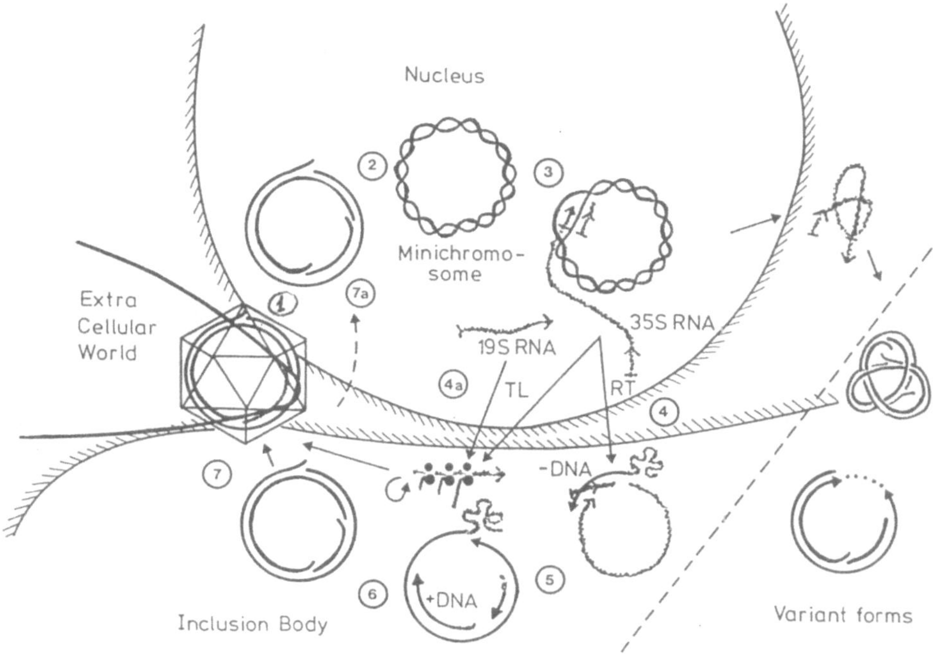

Figure 2: Current model of the life cycle of CaMV. Upon infection, the CaMV virions release their DNA (1) which gets repaired and sealed in the nucleus (2) where it associates with histones to generate a transcribing minichromosome (3). The 19S and 35S RNA are exported to the cytoplasm for translation (4a). Part of the 35S RNA goes to the viroplasms where it associates with a met tRNA molecule (4) to get reverse transcribed into DNA. After digestion of the RNA template and synthesis of the second strand of DNA (5), the viral DNA (6) is packaged into virions (7).

Two pathways may coexist: the 35S RNA may be encapsidated as early as step 4 and DNA synthesis occur in (pre)virions, or produce replication complexes that generate free viral DNA that is directed back into the nucleus for amplification (shunted pathway 7a).

inhibitor of eukaryotic DNA polymerase α, had no effect on CaMV DNA synthesis, while Actinomycin D seemed to inhibit it only partially. Furthermore, RNase treatment of the replicative complexes depleted incorporation by about half, and hybridization to single-stranded CaMV DNA probes showed that RNase affected only markedly the synthesis of the α strand. Meanwhile, workers at the John Innes Institute in Norwich had concentrated on the study of the various forms of unencapsidated DNA found in infected cells, and particularly of truncated DNA molecules: they were thus able to demonstrate the existence of CaMV DNA molecules containing a partial or complete α strand and a growing complementary strand, indicating that the α strand was synthesized first. Very often, the second DNA strand of incomplete duplexes spanned up to Δ_2 or Δ_3 , which confirmed the relevance of these structures to the replication cycle of CaMV. Another discrete single-stranded DNA molecule, dubbed α_1, was detected within virus particles by Guilley *et al.* (1983): the 5' end of this species maps close to the Δ_1 discontinuity and its 3' end is located in the region of the cap site of the 35S RNA. Other groups studied the structure of this DNA in detail and could show that it is in fact a composite molecule of about 600 NT DNA attached to a molecule of methionyl tRNA (Covey *et al.*, 1983).

This set of seemingly unrelated features led these three groups to postulate independently in 1983 that CaMV replicates by reverse transcription of an RNA intermediate, the 35S RNA. Figure 2 summarizes our current view of the replication cycle of CaMV.

The starting point of CaMV DNA replication is the pairing over 14 bases of the 3' end of the plant methionyl tRNA molecule with the 35S RNA immediately downstream from the Δ_1 discontinuity. A virus-specific reverse transcriptase activity (RTase) then commences DNA synthesis. One critical and rate-limiting step is the template switch required when the enzyme reaches the 5' end of the 35S RNA: this switch is rendered possible by the presence of the 180 NT direct repeat. If the template switch does not occur, a small piece of DNA results, functionally equivalent to the "minus strand strong stop DNA" found in the case of animal retroviruses. This small DNA has been detected within CaMV particles, suggesting that CaMV replicative complexes may be encapsidated at a very early stage of the replication cycle. Failure to perform the template switch properly can also generate an abortive "snapback", hairpin-shaped DNA species (Marco and Howell, 1984). Once the template switch is completed, reverse transcription of the 35S RNA resumes up to the tRNA primer which is displaced and now degraded. To allow second strand synthesis, the used RNA template has to be removed by an RNase H activity possibly borne by the reverse transcriptase itself. It is not known whether digestion is concomittant with the advance of the enzyme on the RNA template, or occurs at random, but in any case two purine-rich tracts of RNA located in the vicinity of the Δ_2 and Δ_3 are spared and used as primers for second strand synthesis. Here again, a critical step is the passage of the Δ_1 discontinuity by the growing plus strand which requires a second template switch, and indeed truncated DNA with the corresponding structure is found in infected cells.

ENZYMATICS OF THE CAMV REPLICATION CYCLE

The finding that CaMV DNA is converted into a covalently closed circular molecule to allow transcription of the 35S RNA points to early involvement of the host plant DNA repair enzymes. S. Howell (UCSD, La Jolla, personal communication) found that turnip leaf extracts indeed contain high levels of recombination/repair activities capable of sealing the interruptions present in viral DNA, recircularize cloned CaMV DNA excised from its plasmid vector, and of resolving infectious DNA from tandem dimers cloned into a plasmid. Unfortunately, these activities also recombine complementary mutants into wild type virus at a furious rate, thus tolling the bell for any hope to set up a CaMV-based helper virus system for the expression of foreign genes.

Another enzyme which plays an important - and in some respects unexpected - role in the life cycle of CaMV is the host RNA polymerase II whose duty is to supply the translation system with mRNA to produce sufficient amounts of the various viral gene products, but also to generate enough 35S RNA to keep the synthesis of viral DNA going: this is indeed a demanding process, since the generation of each copy of genomic DNA calls upon the consumption of a disposable 35S RNA molecule. At that stage, the viral reverse transcriptase comes into action: its duties are to position and anneal the tRNA primer to the 35S RNA, copy the latter into minus strand DNA, chew away the exhausted RNA template sparing only the two polypurine tracts earmarked to serve as primers for second strand synthesis, and to synthesise this second strand.

Evidently, this requires a multifunctional enzyme which has, in a first step, been indirectly characterized: the first hint that it may be virus-encoded rather than a host enzyme came from studies by Volovitch et al. (1983) who showed the expression in CaMV-infected turnips of a novel enzyme with a template preference typical for RTases, and pointed out the sequence homologies between the *pol* gene in animal retroviruses and the CaMV ORFV also noticed by Toh *et al.* (1983). Pfeiffer and coworkers (1983, 1984) studied CaMV replication complexes and could show by activity gel electrophoresis that they contain a DNA polymerase activity migrating with an Mr of about 79kDa, i.e. approximately the MW expected for the ORFV expression product. Later, it was confirmed that this is this enzyme is found only in CaMV-infected turnips and that it can use poly rC-oligo dG as a template (Hohn et al.,1985). However, the analysis by activity gel electrophoresis of RNase H activities present in uninfected and CaMV-infected turnips (using ^{32}P-labelled RNA hybridized to unlabelled DNA as a substrate) revealed no such activity comigrating with the RTase: instead, there was a massive stimulation of several RNase H activities in infected plants, a feature which can account for early failures to detect RTase activity in unfractionated extracts using RNA-DNA templates in which the RNA moiety gets rapidly digested (P.Pfeiffer, unpublished). Recently, Takatsuji *et al.* (1986) reported cloning of the CaMV RTase in yeast and also pointed out its apparent lack of RNase H activity: removal of the exhausted RNA template may therefore be a function of host enzyme(s).

Comparison of the arrangement of the coat protein and RTase genes in CaMV with the *gag* and *pol* genes in animal retroviruses suggests the possibility that in CaMV too the active RTase may be derived from a *gag-pol* fusion protein subsequently processed, a contention reinforced by the présence of sequence homologies in the vicinity of the N-terminus of ORFV with the protease region found in other retroviruses. Direct conclusive evidence is still lacking, but several groups are presently engaged into this line of research.

EXPRESSION OF THE CAMV GENES

The analogy between CaMV and other retroviruses and retroid elements allows the anticipation that it may use a similar strategy of gene expression. Indeed, the six or more ORFs present on the viral DNA are unlikely to be expressed from subgenomic messengers .Indeed, this would require the presence of promoter sequences upstream from each potential gene, and would generate a nested set of 3' coterminal mRNAs ending at the single polyadenylation site common to the 19S and 35S RNAs: neither of these have been detected. Alternatively, a series of monocistronic messengers could be derived from the 35S RNA by alternative splicing or separate polyadenylation sites, but none of these have been convincingly demonstrated to exist. Polyproteins subsquently processed into smaller species may exist in the case of ORFIV-ORFV, but are unlikely in the case of the upstream genes: ORFVII does not abut ORF I, and mutants with a truncated ORFII are viable, excluding an ORFI/II/III fusion protein.

Apparently, the synthesis of the various ORF products is not directed by monocistronic messengers, whether subgenomic or generated by splicing: this, together with the polar effect of mutations in CaMV DNA, led Sieg and Gronenborn (1982) to postulate that the CaMV 35S RNA be translated in a "relay-race" fashion, with the ribosomes continuing to scan the RNA after reaching the stop codon and reinitiating translation at the next ORF without disengaging from the 35S RNA. The ratio between translatable RNA and RNA to be reverse transcribed into genomic RNA could then be controlled by packaging (Fütterer and Hohn, 1987).

Testing this hypothesis *in vitro* with naturally occuring 35S RNA would have been a tricky business, because this would have entailed lengthy purification steps (e.g. hybrid selection procedures) with a high risk of fragmentation of the 8 kbp genome-length molecule - thus generating artifactual subgenomic RNAs. The development of RNA expression vectors was a quantum leap in this respect, making possible the *in vitro* synthesis of custom-taylored RNAs encoding each ORF either alone or with its neighbours downstream an adjustable leader sequence.

Using this approach, K.Gordon and coworkers (manuscript in preparation) were able first to express each CaMV gene taken individually both in the wheat germ and rabbit reticulocyte systems, and to study the sequential expression of the various ORFs. This approach proved very fruitful and enabled us to show that the long leader sequence of the 35S RNA reduces, but does not abolish, translational efficiency . Furthermore, a set of the various gene products could be generated as standards for a correlation with the *in vivo* situation. This revealed for instance surprising shifts in electrophoretic mobility of the ORFIV gene product which migrates at 80 kDa rather than at 57kDa as computed from its aminoacid sequence. In addition, the availability of radioactive translation products permitted an assessment of the virion-associated polypeptides by immune precipitation with antisera directed against dissociated virions. Attempts to detect an enzymatic activity associated with the ORFV translation product were unsuccessful, but showed clearly that the RTase activity detected in infected plants does not comigrate with the primary translation product of ORFV: the production of active RTase therefore requires post-translational modifications either of the ORFV translation product itself, or of an ORFIV/ORFV fusion protein akin to the *gag-pol* product of animal retroviruses. Finally, the amounts of translation product obtained are sufficient to generate antibodies to the various gene products, an approach already used successfully to detect *in vivo* non-structural proteins of RNA plant viruses (Berna et al., 1985).

A disappointment came however with the study of the relay-race model using dicistronic or polycistronic messengers: initially, our hope was that the head-to-tail arrangement of ORFs I through III would permit their sequential expression. However, the behaviour of the translation systems tested was just like with other polycistronic RNAs from RNA plant viruses (in which the various genes are separated by intergenic regions rather than abutting), i.e. only the 5' proximal gene was significantly expressed while translation of the downstream gene(s) was very limited, and most probably corresponded to internal initiation rather than proper reinitiation.

This *in vitro* approach has thus shown its limitations, and the group of T. Hohn has now shifted to the study of the relay-race model *in vivo* by quantitating the transient expression of various genes positioned like in CaMV downstream from various leader sequences, and by monitoring the *cis* and *trans* effects that the various CaMV gene products may have on CaMV gene expression.

REFERENCES

Al Ani, R., Pfeiffer, P., and Lebeurier, G., 1979, The structure of Cauliflower Mosaic Virus. II: Identity and location of the viral polypeptides. Virology, 93:188.

Berna, A., Godefroy-Colburn, T., and Stussi-Garaud, C., 1985, Preparation of an antiserum against an *in vitro* translation product of Alfalfa Mosaic Virus RNA 3, J. Gen. Virol. 66:1669.

Bonneville, J.M., Hohn, T., and Pfeiffer, P., 1987, Reverse transcription in the plant virus Cauliflower Mosaic Virus, in "RNA genetics", P. Ahlquist, ed., CRC Press, Boca Raton, Florida.

Brisson, N., Paszkowski, J., Penswick, J. R., Gronenborn, B., Potrykus, I., and Hohn, T., 1984, Expression of a bacterial gene in plants using a viral vector, Nature 310:511.

Chauvin, C., Jacrot, B., Lebeurier, G., and Hirth, L., 1979, The structure of Cauliflower Mosaic Virus. A neutron diffraction study. Virology, 96:640.

Covey, S. N., Turner, D., and Mulder, G., 1983, A small DNA molecule containing covalently-linked ribonucleotides originates from the large intergenic region of Cauliflower Mosaic Virus genome, Nucleic Acids Res. 11:251.

Dixon, L., and Hohn, T., 1985, Cloning and manipulating Cauliflower

Mosaic Virus, in "Recombinant DNA Research and Virus", Becker, ed., Martinus Nijhoff Publishing, Boston.

Frank., A., Guilley, H., Jonard, G., Richards, K., and Hirth, L., 1980, Nucleotide sequence of Cauliflower Mosaic Virus DNA, Cell, 29:285.

Fütterer, J., and Hohn, T., 1987, Involvement of nucleocapsids in reverse transcription: a general phenomenon?, TIBS,12:92.

Givord, L., Xiong, C., Giband, M., Koenig, I., Hohn, T., Lebeurier, G., and Hirth, L., A second Cauliflower Mosaic Virus gene product influences the structure of the viral inclusion body, EMBO J., 3:1423.

Guilley, H., Dudley, R. K., Jonard, G., Balazs, E., and Richards, K., 1982, Transcription of Cauliflower Mosaic Virus DNA: detection of promoter sequences and characterization of transcripts, Cell 30:763.

Guilley, H, Richards, K. E., and Jonard, G., 1983, Observations concerning the discontinuous DNAs of Cauliflower Mosaic Virus, EMBO J., 77:282.

Hull, R. , and Covey, S. N., 1983, Does Cauliflower Mosaic Virus replicate by reverse transcription? TIBS, 8:119.

Marco, Y., and Howell, S. N., 1984, Intracellular forms of viral DNA consistent with a model of reverse transcriptional replication of the Cauliflower Mosaic Virus, Nucleic Acids Res. , 12:1517.

Maule, A. J., 1985, Replication of Caulimoviruses in plants and in protoplasts, in "Molecular Plant Virology", Vol. II, J. W. Davies, ed., CRC press, Boca Raton, Florida.

Ménissier, J., Lebeurier, G., and Hirth, L., 1982, Free Cauliflower Mosaic Virus supercoiled DNA in infected plants, Virology, 117:322.

Olszewski,N., Hagen, G., and Guilfoyle, T., 1982, A transcriptionally active, covalently closed minichromosome of Cauliflower Mosaic Virus DNA isolated from infected turnip leaves, Cell, 29:395.

Pfeiffer, P., and Hohn, T., 1983, Involvement of reverse transcription in the replication of Cauliflower Mosaic Virus: a detailed model and test of some aspects, Cell, 33:781.

Pfeiffer, P., Laquel, P., and Hohn, T., 1984, Cauliflower Mosaic Virus replication complexes: characterization of the associated enzymes and of the polarity of the DNA synthesized in vitro, Plant Mol. Biol., 3:261.

Sieg, K., and Gronenborn, B., 1982, Evidence for polycistronic messenger RNA encoded by Cauliflower Mosaic Virus, Abstract NATO Advanced Studies Inst., Advanced Course 1982, 154.

Takatsuji, K., Hirochika, H., Fukushi, T., Ikeda, J. E., 1986, Expression of CaMV reverse transcriptase in yeast, Nature, 319:241.

Toh, H., Hayashida, H., and Miyata, T., 1983, Sequence homologies between retroviral reverse transcriptases and the putative polymerases of Hepatitis B virus and Cauliflower Mosaic Virus, Nature, 305:827.

Volovitch, M., Drugeon, G., and Yot, P., 1978, Studies on the single-stranded discontinuities in Cauliflower Mosaic Virus genome, Nucleic Acids Res., 5:2913.

Volovitch, M., Modjtahedi, N., Yot, P., and Brun, G., 1984, RNA-dependent DNA polymerase activity in Cauliflower Mosaic Virus infected plant leaves, EMBO J., 3:309.

AGROINFECTION OF *ZEA MAYS* WITH MAIZE STREAK VIRUS DNA

Barbara Hohn, Thomas Hohn, Margaret I.Boulton*, Jeffrey W. Davies* and Nigel Grimsley

Friedrich Miescher-Institut, PO Box 2543, CH-4002 Basel, Switzerland. * John Innes Institute, Colney Lane, Norwich NR4 7UH, U.K.

Agrobacterium tumefaciens is a gram-negative soil bacterium which can cause neoplastic proliferations on wounded plants. These galls also produce compounds called opines which are absent from normal plant cell but which can be used as carbon - and nitrogen sources by the bacteria. Virulent bacteria carry a plasmid, the Ti plasmid, which upon stimulation by the plant releases a portion of its DNA. This part, the T-DNA, is transferred in an unknown fashion to the plant cell, where eventually it is integrated in the plant genome. Requirements for T- DNA transfer are functional virulence regions on the Ti plasmid and the bacterial chromosome as well as short (25 bp long) imperfect direct repeat sequences which delineate the T-DNA. The Ti plasmid located virulence region is not active in vegetative cells but becomes activated upon receipt of (a) signal molecule(s) released from wounded plants (see reviews on *Agrobacterium*-mediated transformation by Gheysen et al., 1985 and Koukolíková-Nicola et al., 1987).

This unique plant transformation occurs with many dicotyledonous as well as some monocotyledonous plants (DeCleene, 1985; Hernalsteens et al, 1984; Hooykaas- Van Slogteren et al, 1984; Graves and Goldman, 1986; Graves and Goldman, 1987). However, the presence of T-DNA in the recipient plant has not been confirmed in most monocotyledonous genera, including the grasses. The possibility therefore remained that the *Agrobacterium* - plant interaction was not compatible with events involving early recognition, attachment, virulence gene activation, DNA transfer, DNA integration, DNA expression or plant response. A sensitive and efficient assay for detecting the expression of genes *known* to function in the host plant in question was required to test whether at least transfer and expression of *Agrobacterium* delivered DNA occurs. We report here on such a test, called agroinfection.

The term agroinfection, in its broadest sense, is used to describe the delivery of viral sequences into plants *via* *Agrobacterium* (Grimsley et al., 1986 a ; Grimsley and Bisaro, 1987). Here we will confine ourselves to the introduction of infectious viral DNA which requires a certain configuration of partial or complete viral genomes in the T-DNA of the agrobacterial donor. Inoculation of a plant susceptible to *Agrobacterium* and sensitive to the virus to be introduced leads to escape of infectious viral nucleic acid from the transforming T-DNA. Thus one or a few copies of successfully

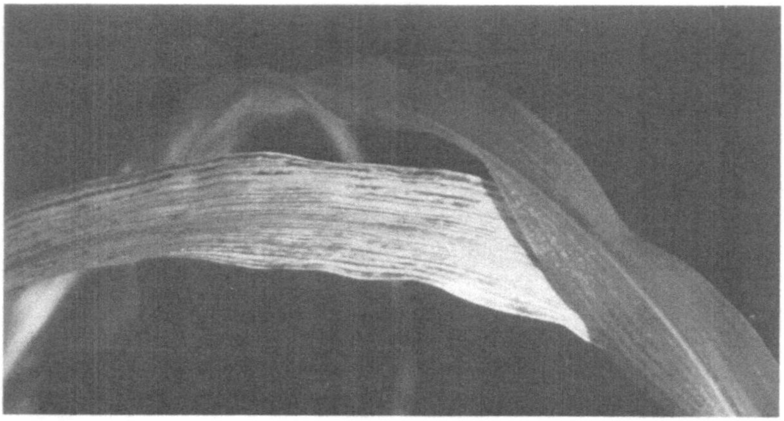

Figure 1. Symptoms of maize streak virus on Zea mays.

transforming DNA can be amplified by replication, spread systemically and cause development of symptoms on the whole organism. Agroinfection of this kind has been described and studied for cauliflower mosaic virus (CaMV, Grimsley et al., 1986 a, b; Hille et al., 1986), and potato spindle tuber viroid (PSTV, Gardner and Knauf, 1986; Gardner et al., 1986). In the latter case cDNA copies have been employed for agroinfection. In these studies important information on viral replication and the agrobacterial delivery process could be accumulated. In addition, using a dilution series, the sensitivity of the assay could be estimated: whereas 2µg CaMV DNA per inoculated turnip leaf was necessary to reproducibly yield symptoms 10^{-7}µg equivalents of CaMV were sufficient per inoculated plant if introduced by the agrobacterial route (Grimsley and Bisaro, 1987).

This assay then seemed the method of choice for reinvestigation of the question of the host range of Agrobacterium. The geminivirus maize streak virus (MSV) was chosen to test this approach, using maize as a host plant. Geminiviruses are single stranded DNA viruses with genomes consisting of one or two circles of 2.5 - 3.0 kb (see review by Davies et al, 1987). Members of the group here have been identified on the basis of their bipartite particle morphology. Distinctions can be made on the basis of their insect transmission vector (leafhopper or whitefly) and of their host range (mono- or dicotyledons). All viruses infecting monocotyledonous plants require leafhoppers as vectors and appear to have one DNA circle, whereas viruses infecting dicotyledonous plants use whiteflies as carriers and require two single stranded circular DNAs for infectivity and systemic spread. The only exception to this rule so far is beet curly top virus (BCTV) which is a monopartite virus transmitted by leafhoppers (Stanley et al, 1986). Cloned versions of several of the viruses infecting dicotyledonous plants have been shown to be infectious whereby coinoculation with both units of the bipartite viruses is required for symptom formation but the single unit of BCTV DNA is infectious by itself (reviewed in Davies et al, 1987).

Table 1.
Agroinfection of maize plants with different MSV constructions.

Ti plasmid	Binary plasmid	No.infected/no. inoculated plants
pTiC58	pEAP37	38/43
pTiC58	pBIN19-MSV	58/67
pTiC58	pEAP40	51/58
pGV3850::pEAP25		18/26
Naked DNA		
p3547		0/15

pEAP37 is described in Grimsley et al., 1987. pBIN19-MSV was constructed by inserting a MSV BamHI dimer into the Bam HI site located between the border sequences of pBIN19 (Bevan, 1984). p3547 is a MSV 1.6 mer (BglII - BamHI 0.6 + BamHI - BamHI monomer; see Fig.2) cloned into the BamHI site of pTZ19R (Mead et al., 1986). This plasmid was inserted, using its EcoRI site, into EcoRI cut pCIB200 (Rothstein et al, 1987), thus placing the MSV sequences into the T-region of pCIB200. pGV3850::pEAP25 was created by in vivo recombination of pEAP25, a pBR322 derivative carrying a MSV - BamHI dimer, with pGV3850, in Agrobacterium C58. Mobilizations into C58 were done by triparental mating (Rogers et al., 1986), and constructions in Agrobacterium were verified by Southern analysis. Inoculations of maize plants were done as described, by injecting agrobacterial suspension into the stems of 10-day old maize plants. DNA injections were done similarly, using 20µg DNA per plant. Symptom formation was scored three weeks after inoculation.

However, the cloned and sequenced isolates of MSV could not be demonstrated to be infectious. In fact, neither native viral DNA nor virus particles could be shown to infect maize plants. The only successful method for infection remained the natural one, using the insect vector Cicadulina mbila, a leafhopper resident in Africa and some Asian countries. The characterised clones of MSV therefore may not have represented the complete biologically active viral genome. Agroinfection of maize plants with dimer constructions therefore posed two questions: (1) Is maize a host for Agrobacterium? (2) Is the MSV isolate used for the experiment infectious?

The positive experimental result, as described below, gave both answers. Strains of Agrobacterium tumefaciens were constructed in which dimers of MSV DNA were inserted in the T-DNA of a binary vector. When such bacteria were inoculated on the stem and leaves of young maize plants, symptoms consisting of yellow-white streaks appeared (Fig.1; Grimsley et al, 1987). Control experiments involving binary vectors lacking the T-DNA border sequences or using an agrobacterial virulence mutant, both of which gave negative results, indicated that most probably the T-DNA transfer

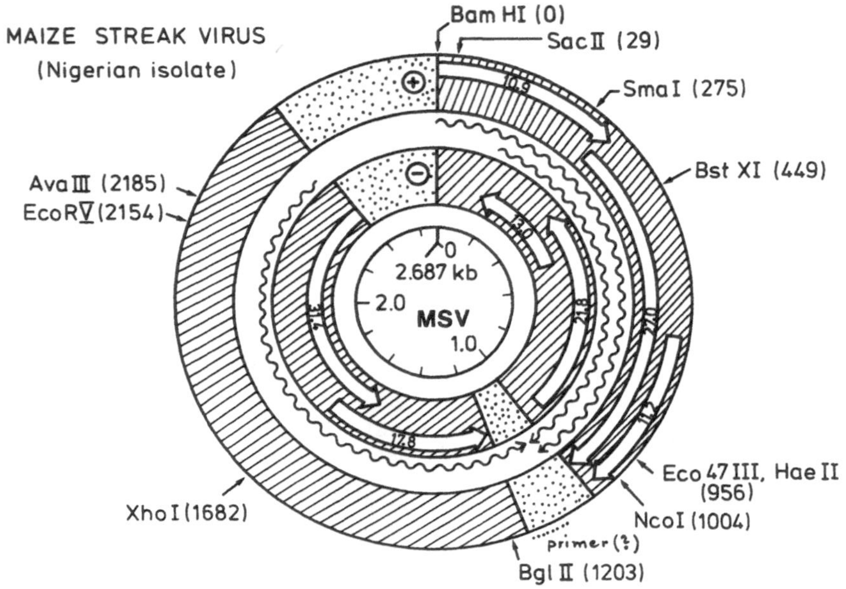

Figure 2. Map of MSV genome (Nigerian isolate), in its double stranded form. Compiled after Mullineaux et al. (1984), Donson et al. (1984) and Morris-Krisinich et al. (1985).

mechanism, as described for dicotyledonous plants, was operating in the maize system. This conclusion was further strengthened by the fact that naked DNA of MSV dimer used for the experiment was not infectious when inoculated in a similar fashion. Thus, lysis of the employed agrobacterial strains and infection of maize plants with the released DNA cannot have been the route of infection (Grimsley et al, 1987).

Also another binary vector, pBIN19 (Bevan, 1984), could be used to introduce MSV DNA into plants, via agroinfection (Table 1). In the constructions described so far complete tandem dimers of MSV, linearized in the BamHI site, (see Fig. 2) were used . In order to assess the requirements for duplications a partial dimer was tested: pEAP 40 contains, in a binary vector, one complete BamHI unit linked to the BglII to BamHI (clockwise) fragment. This clone was also agroinfectious. So, whatever the mechanism by which MSV DNA frees itself from the transforming T-DNA, two complete units of viral DNA are not required. Table 1 also shows that a concentrated solution of the 1.6mer DNA did not yield symptoms when applied to maize plants the way agrobacterial cells were applied, confirming our earlier observations. Thus, so far, only leafhoppers and Agrobacterium, the natural plant vectors, successfully introduce MSV particles and DNA, respectively, to plants.

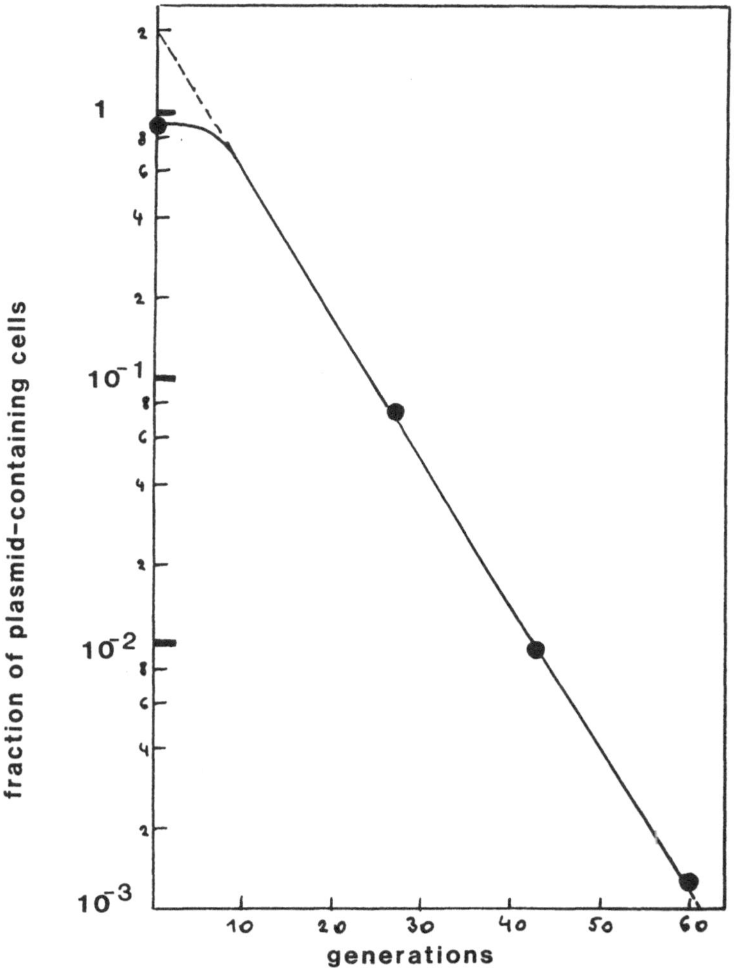

Figure 3. Maintenance of pEAP37 in C58, under nonselective conditions. Serial passages of C58(pTiC58,pEAP37) were conducted in L-broth containing 100µg/ml rifampicin, at 28°C. At intervals the culture was titred on plates containing rifampicin only or rifampicin and 25µg/ml kanamycin. The segregation kinetics indicates a plasmid copy number of about 2 per cell, under these conditions.

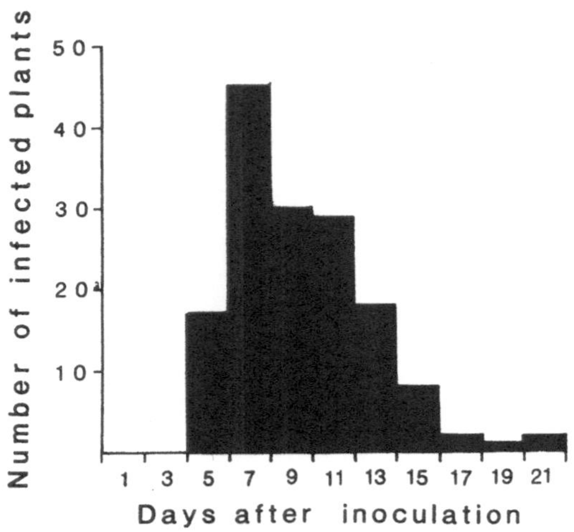

Figure 4. Kinetics of symptom appearance of MSV, following agroinfection. The number of new plants showing symptoms, scored at 2-day intervals, is shown.

Insertion of MSV tandem genomes into the T-DNA of pGV3850 yielded agroinfectious DNA too (Table 1). In this Ti plasmid the largest part of the original T-DNA genes has been replaced by the plasmid pBR322 (Zambryski et al, 1983). This experiment therefore demonstrates that transfer of intact natural T-DNA genes is not required to accompany transfer of MSV containing T- DNA. For biosafety reasons, however, for agroinfection we prefer the use of binary vectors which are unstable in the absence of the selecting antibiotic. Fig. 3 demonstrates this instability for pEAP37, a pRK derived plasmid, in C58, under defined conditions. An average rate of plasmid loss of 1.7% per generation can be calculated. Experiments have also been conducted in soil and plasmid instability was demonstrated, although quantitation is difficult. Another level of biological containment will be the employment of MSV coat protein mutants or packaging mutants in the constructions for agroinfection, since non-encapsidated DNA is not insect transmissible (Mullineaux et al., 1984). Whether such mutants will be able to systemically spread in the plant remains to be tested.

Symptoms usually appear between 4 and 15 days after inoculation with Agrobacterium. The distribution of the earliest time of symptom appearance, following inoculation by agrobacterial cells, is shown in Fig. 4. It is not known whether the spread in time of appearance of the earliest symptoms is due to (1) differences in the capacity of infection target cells to support release of viral DNA from the infecting T-DNA molecule, virus DNA replication and transport of viruses, or due to (2) differences in the amount of inoculum actually persisting at the point of inoculation or reaching the point most susceptible for DNA transfer, or due to (3)

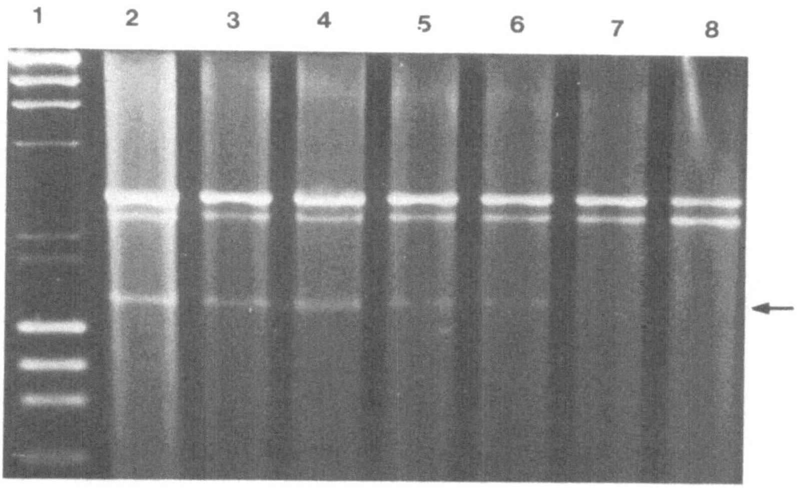

Figure 5. Intracellular MSV-DNA. After DNA extraction (Grimsley et al., 1987), total cellular DNA was loaded on a gel, either in undigested form (lanes 2, 3 and 4), or digested with 5, 10, 30 or 150 units of nuclease S1 (lanes 5, 6, 7, and 8, respectively). An arrow indicates the position of the nuclease-sensitive viral single-stranded DNA; the higher molecular weight nuclease-insensitive bands are double-stranded viral DNA. Molecular weight standards of HindIII cut lambda DNA + HaeIII cut ØX174 DNA are run in lane 1.

differences in the physiological state of the plant. A detailed analysis relating the point of virus injection by the leafhopper with place and date of symptom appearance has yet to be done.

Agroinfection of maize plants with MSV DNA results in a high copy number of MSV specific double-stranded DNA (Grimsley et al, 1987). The DNA isolation procedure used for this analysis enriched for double-stranded unencapsidated DNA. Later analyses revealed the presence of single-stranded DNA (Fig. 5). Single-stranded molecules, diagnosed by their sensitivity to the single-strand nuclease S1, are marked by an arrow. Since the life cycle of this virus requires a single stranded DNA for packaging, a search was undertaken for particles. Electron microscopic examination revealed the characteristic twinned particles (Fig. 6). Furthermore, leafhoppers allowed to feed on agroinfected plants were able to transmit the virus to healthy maize plants with an efficiency equal to that obtained when using insect-infected host plants (P. G. Markham, pers. comm.).

The experiments described here and published previously (Grimsley et al, 1987) demonstrate that the MSV clone employed contains all information necessary for replication, encapsidation and systemic spread. Moreover, the particles produced as a result are accepted by the normal insect vector

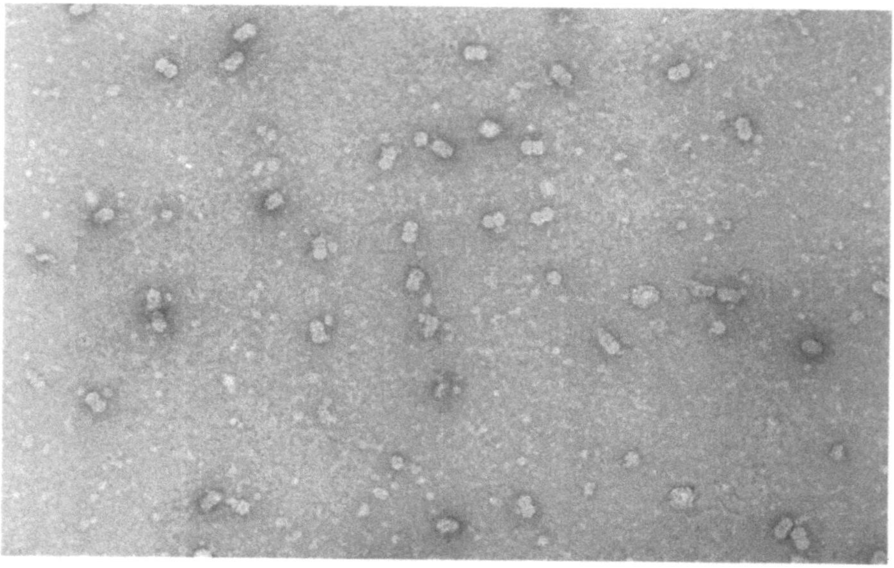

Figure 6. Electron micrograph of maize streak virus particles, extracted from an agroinfected maize plant.

and delivered to a new host where they elicit a new round of infection. It is not clear from the available data why viral DNA, single- or double-stranded, has not been infective when inoculated at the same position of maize plants as used for agroinfection. Is only the specialised feeding apparatus of a leafhopper or the sophisticated invasion mechanism of an Agrobacterium able to deliver the virus or DNA to plant cells competent for virus multiplication or transport? Whatever the reason, now that cloned MSV has been shown to be biologically active exploration and exploitation of the viral genome can be started. Mutation analysis will define functionally competent sequences. The possibility of transporting foreign genetic information into maize, using MSV, can be tested.

Agroinfection constitutes a sensitive assay for the interaction of Agrobacterium with plants. It is independent of the functional expression of the natural T-DNA genes. It should be kept in mind, however, that agroinfection is not an assay for integration of T-DNA into a plant genome. MSV-DNA carrying T-DNA molecules entering a plant cell nucleus have to undergo replicative and/or recombinational processes, the result of which are viable virus nucleic acids. Integration of the T-DNA molecules prior to this event may not be required. On the contrary, DNA molecules entering a plant cell might be more active in recombination, replication and transcription than integrated versions of the same sequences that could be repressed for the same activities by, for example, methylation or histone binding. However it is likely that the T-DNA enters the nucleus in agroinfection, since this is where the host enzymes for replication and recombination exist.

Now that maize has been established as a recipient for T-DNA transfer, more detailed questions can be asked: if T-DNA molecules are able to enter graminaceous plants like maize, can they also integrate into the genome? If they integrate, are the natural T-DNA genes properly expressed? If they are, why does the so well known phenotype of tumorous growth not result, as it does in dicotyledonous plants? It will be important to answer these questions in order to understand the possible generalities of the parasitic nature of *Agrobacterium* and in order to be able to use this knowledge for cereal manipulation.

ACKNOWLEDGEMENTS

We thank Cynthia Ramos (FMI), Marion Pinner (JII), and Kitty Plaskit (JII) for expert technical help, Michael Schläppi (FMI) for help in some constructions, and J. Schell for supply of biological material. Work at the JII was supported by the Agrigenetics Corporation.

REFERENCES

Bevan, M., Nucleic Acids Res.**12**, 8711-8721 (1984).

Davies, J. W., Townsend, R. & Stanley, J. in "Plant DNA Infectious Agents," pp. 31-52 eds. Hohn, T. & Schell, J. Springer, New York, in press (1987).

DeCleene, M. Phytopath. Z. **113**, 81-89 (1985).

Donson, J., Morris-Krisinich, B. A. M., Mullineaux, P. M., Boulton, M. I., Davies, J. W., EMBO J. **3**, 3069-3073 (1984).

Gardner, R. C., Chanoles, K. R., Owens, R. A., Plant Mol. Biol. **6**, 221-228 (1986).

Gardner, R. & Knauf, V., Science **231**, 725-727 (1986).

Gheysen, G., Dhaese, P., Van Montagu, M. & Schell, J. in "Genetic Flux in Plants" eds. Hohn, B. & Dennis, E. S., Springer, N.Y., 12-27 (1985).

Graves, A. C. F. & Goldman, S. L. Plant molec. Biol. **7**, 43-50 (1986).

Graves, A. C. F. & Goldman, S. L. J. Bact. **169**, 1745-1746 (1987).

Grimsley, N., Hohn, B., Hohn, T. & Walden, R. Proc. natn. Acad. Sci. U.S.A. **83**, 3282-3286 (1986a) .

Grimsley, N., Hohn, T. & Hohn, B. EMBO J. **5**, 641-646 (1986b).

Grimsley, N. & Bisaro, D. in "Plant DNA Infectious Agents," pp. 87-107, eds. Hohn, T. & Schell, J., Springer, New York, in press (1987).

Grimsley, N., Hohn, T. Davies, J. W. & Hohn, B. Nature **325**, 177-179 (1987).

Hernalsteens, J. P., Thia Toong, L., Schell, J. & Van Montagu, M. EMBO J. **3**, 3069-3042 (1984).

Hille, J., Dekker, M., Luttighuis, H. O., Van Kammen, A. & Zabel, P. Mol. Gen. Genet. **205**, 411-416 (1986).

Hooykaas-Van Slogteren, G. M. S., Hooykaas, P. J. J. & Schilperoot, R.A. Nature **311**, 763-764 (1984).

Koukolíková-Nicola, Z., Albright, L., and Hohn, B., in "Plant DNA Infectious Agents" pp. 109-148, eds. Hohn, T., and Schell, J., Springer, New York, in press (1987)

Mead, D. A., Szczesna-Skorupa, E. & Kemper, B. Protein Engineering 1, 67-74 (1986).

Morris-Krisinich, B. A. M., Mullineaux, P. M., Donson, J., Boulton, M. I., Markham, P. G., Short, M. N., Davies, J. W. Nucl. Acid Res. 3, 7237-7256 (1985).

Mullineaux, P. M., Donson, J., Morris-Krisinich, B. A. M., Boulton, M. I. & Davies, J. W. EMBO J. 3, 3063-3068 (1984).

Rogers, S. G., Horsch, R. B., and Fraley, R. T., Meth. Enzym., 118, 627-640 (1986).

Rothstein, S.J. Lahners, M., Lotstein, R.J., Carozzi, N. B., Jayne, S. M., and Rice, D., Gene, in the press (1987) .

Stanley, J., Markham, P. G., Callis, R. J. & Pinner, M. S. EMBO J. 5, 1761-1767 (1986).

Zambryski, P. J., Joos, H., Leemans, J., Van Montagu, M. & Schell, J. EMBO J. 2, 2143-2150 (1983).

STRUCTURAL PROPERTIES OF VIROID REPLICATIVE INTERMEDIATES AND OF SATELLITE RNAS OF CMV

Gerhard Steger

Institut für Physikalische Biologie
Universität Düsseldorf
Düsseldorf, F.R.G.

INTRODUCTION

Viroids form a unique class of plant pathogens which are distinguished from viruses by the absence of a protein coat and by their small size (for reviews see 1,2,3). They consist merely of a single-stranded, circular ribonucleic acid of a few hundred nucleotides. Viroids do not code for a translation product. Therefore, one has to assume that viroid replication depends completely on the enzyme systems of the host cell. DNA-dependent RNA polymerase II of plant origin is able to bind strongly to viroids (4) and to transcribe it into full length linear molecules of (-) polarity *in vitro* (5). The involvement of polymerase II *in vivo* was inferred from inhibition experiments in protoplasts with α-amanitin (6). A rolling-circle type of replication mechanism has been proposed (7,8) on the basis of the type of molecules found in infected plants: a (-) strand oligomer is transcribed from a circular viroid and, in a second step, a (+) strand oligomer is generated from the (-) strand oligomer. The oligomeric (+) strand intermediates require site-specific cleaving in order to yield monomeric units, which then ligate forming covalently closed circular molecules.

A very distinct feature of circular viroids is the well-known secondary structure (1,3): as a result of intramolecular base pairing, viroids form a unique rod-like secondary structure which is characterized by a series of double helical sections connected by small internal loops

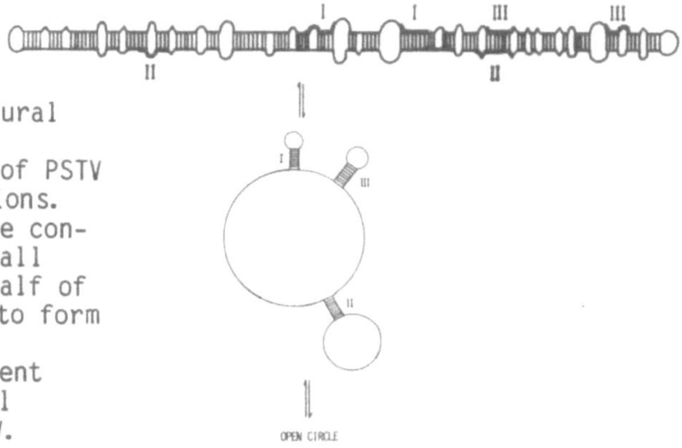

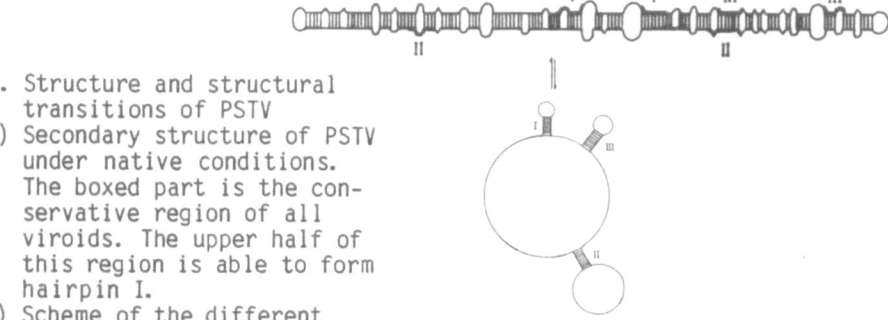

Fig. 1. Structure and structural transitions of PSTV
(a) Secondary structure of PSTV under native conditions. The boxed part is the conservative region of all viroids. The upper half of this region is able to form hairpin I.
(b) Scheme of the different stages during thermal denaturation of PSTV.

(Fig. 1a). Several experiments indicate that there is no additional tertiary structure. During denaturation, all native base pairs of viroids dissociate in one highly co-operative transition. In the same process, very stable hairpins not present in the native structure are newly formed. The mechanism is shown in Fig. 1b. The observed close similarity between different viroids is due mainly to a similar overall structure and to similar thermodynamic and functional domains, than to similarities in the primary sequence.

Since viroid DNA clones are widely available and since such DNA may be transcribed into RNA by means of the SP6 or T7 system, oligomeric viroid RNA of both polarities can be synthesized (9) in the amounts needed for physico-chemical experiments (10) and infectivity studies (11,12). From both a structural and functional model of viroid intermediates was derived (10). The most stable secondary structure of monomeric and oligomeric linear viroid molecules of both polarities is almost identical to that of circular mature viroids. However, it could clearly be shown that the intermediates may be trapped in a secondary structure which is bifurcated and deviates essentially from the one of mature viroids. This trapped structure contains three successive helices in the region of nucleotides 79-110 and is formed intermolecularly in the case of monomers (i.e. complex formation) or intramolecularly in the case of oligomers. This region, the central conserved region responsible for forming hairpin I, is made up of 28 base pairs with a G:C content of 71%. In Fig. 6 the tri-helical structure and the location of the corresponding sequence in the native viroid are depicted. In Fig. 5g is shown how the dimeric complexes are thought to be arranged. The dissociation of the three successive helices was observed by optical denaturation in the highest temperature transition (near 90°C in 1 M NaCl) of the (+) strand transcripts (cf. Fig. 2). The complex formation of monomers was verified by analytical ultracentrifugation. In the case of the (-) strands, two of the three helices are interrupted by A:C discontinuities resulting in a less stable complex. This difference between (+) and (-) strands may count for their different biological activities. The involvement of the region 79-110 in

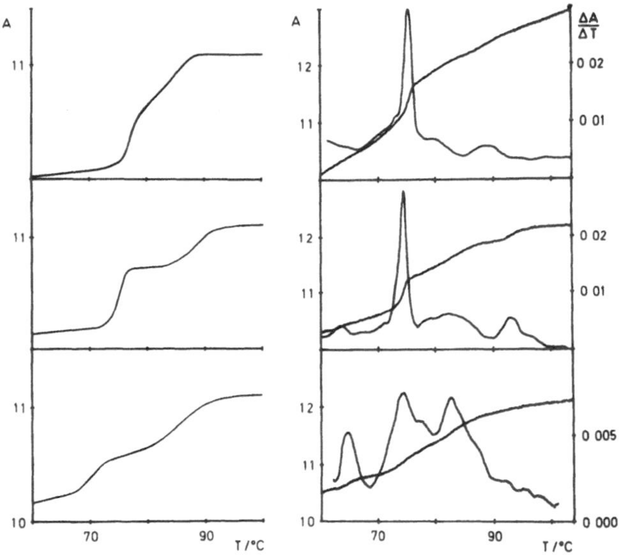

Fig. 2. Calculated (top) and experimental (bottom) optical melting curves of circular PSTV (left), monomeric SP6 PSTV (+) (middle), and (-) (right) (10). Experimental curves are extrapolated to 1 M NaCl. Special attention is called to the additional transition of the (+) intermediate at 93°C which is absent in the corresponding calculated curve. This transition is due to complex formation of the monomers.

processing of viroid intermediates was also concluded from biological studies with the shortest longer-than-unit molecules which are still infectious (11,12,13).

In contrast to the linear structures, which may assume structural alternatives and form bimolecular complexes, the circular viroids always have an unique structure and are not able to form intermolecular dimers. The change in free energy, when a viroid detaches from the oligomeric intermediate and circularises, is close to zero. Once the covalently closed circle of the unit-viroid has been formed, the completely extended structure is energetically more favorable than the branched structure. A backreaction, in which a viroid circle recombine with an oligomer, is impropable due to the very low free concentration of viroids and of intermediates. Hence the main "driving force" for the accumulation of circular viroids as end-product is the low rate of the backreaction. As may be seen from the model of the oligomeric transcripts in Fig. 6, an appropriate cut in the region forming the three successive helices releases a unit viroid molecule. Ligation of both ends of the unit viroid results in the circular mature viroid; ligation of the open ends left at the oligomer (due to the detachment of the monomer) leaves an oligomer containing one unit less than the original molecule. Obviously, the whole process may repeat producing further circular unit viroid molecules.

In this paper, experimental studies on synthetic oligomers of potato spindle tuber viroid (PSTV) by the method of temperature-gradient gel electrophoresis (TGGE), which was developed recently in this laboratory as a routine procedure (14), as well as theoretical treatment of their sequences are described. The results of these investigations and their implications for the model of viroid replicative intermediates will be described.

Cucumber mosaic virus (CMV) associated RNA 5 (CARNA 5) has been discovered as a fifth major RNA component among the RNAs encapsidated within CMV. CARNA 5 is a satellite, i.e. can replicate only in conjunction with the "helper" virus and does not share any significant sequence homology with the "helper" viral nucleic acids. CARNA 5 variants of which the sequences have been reported to date range in size from 334 to 368 nucleotides (15,16,17,18,19). The recent increasing interest in CMV satellites can be attributed in part to the fact that they have been shown capable of significantly modulating the disease symptoms of the helper viruses (20). At the present time the possibility to utilize CARNA 5 as a control agent of diseases caused by CMV in the field has only been exploited in China (21,22). In this biological control procedure seedlings are preventively inoculated with CMV and a symptom-attenuating CARNA 5. This establishes a steady-state low-level infection with virus, CARNA 5 and its double-stranded (ds) form, dsCARNA 5, spreading throughout the plant (23). It is important, however, to verify that during this process no CARNA 5 variants capable of causing lethal tomato necrosis are emerging (24). While during infection small quantities of viral dsRNA are produced, as expected, dsCARNA 5 accumulates in proportions far greater than needed for a replicative function, and sometimes exceeds the amounts of ssCARNA 5 within virions.

Due to stability, relative ease of detection and isolation, dsCARNA 5 seems to be suitable for diagnostic purposes. Its presence in relatively high copy-numbers in CMV infected tissue would increase diagnostic sensitivity. Furthermore, the physico-chemical properties of dsCARNA 5, which are known in some detail (25), show that their thermal stability (and thus their optical melting behavior) may be altered by changing a single or a few base pairs in their nucleotide sequence (25,26). This would mean that with appropriately simplified experimental technique it should be possible to differentiate dsCARNA 5 nucleotide sequence variants, including those derived from necrogenic or non-necrogenic CARNA 5 variants. In this report it will be shown that dsCARNA 5 isolated from CMV-infected plant tissues can be analyzed with respect to the presence of different nucleotide sequence variants by the method of TGGE.

MATERIALS AND METHODS

Transcripts of PSTV: Synthetic monomeric and oligomeric PSTV replication intermediates of both polarities were made by SP6 transcription of corresponding DNA clones as described earlier (9,10).

One to four DNA copies (connected 'head-to-tail') of PSTV (Diener strain, obtained from plasmid pST-B14, kindly provided from R. Owens) were inserted downstream of the Φ 10 promoter for the T7 RNA polymerase in a pBR322 derivative. *In vitro* transcription yielded the corresponding monomeric or oligomeric single stranded linear PSTV molecules of (+) and (-) polarity. Short terminal sequences were derived from the vector: at the 5' end 2 additional nucleotides (GG) and at the 3' end 7 additional nucleotides (GGGAAUU). In the case of the (+) transcripts, the sequence consists of nucleotides 282-359/1-281, in the case of the (-) species of nucleotides 281-1/359-282. The synthesized RNA was used directly or purified by a single HPLC run on Nucleogen-DEAE 4000-7 (Diagen, Düsseldorf, FRG) anion exchanger (manuscript in prep.). The purity of the transcripts was tested by gel electrophoresis and silver staining. The ethanol precipitates were washed twice with ethanol and dissolved in the appropriate buffers without dialysis.

dsCARNA 5: dsCARNA 5 was purified as described previously (27) from plant tissues infected with different isolates or strains of CMV. The dsCARNA 5 variants mentioned in this paper are: WT-dsCARNA 5 (Wisconsin tobacco isolate of CMV (28); necrogenic to tomato), D-dsCARNA 5 ((29); necrogenic to tomato), Yn-dsCARNA 5 ((30); necrogenic to tomato), 1-dsCARNA 5 ((31); non-necrogenic to tomato), Val 24-dsCARNA 5 (Valencia 24 isolate (24); non-necrogenic to tomato), S-dsCARNA 5 ((19); non-necrogenic to tomato), S52-dsCARNA 5 ((32,33); not necrogenic in pepper, squash, and tobacco; used as biological control agent in China (33)). After two passages in squash and about ten passages in tobacco, S52 induces necrosis in tomato. S52-dsCARNA 5 as used in this work was isolated from tobacco after ten passages.

Optical transition curves: The equilibrium thermal denaturation curves were measured by a two-wavelength spectrophotometer (SIGMA-ZWS 11, SIGMA Instrumente, Berlin, FRG) as described earlier (34). The samples of dsCARNA 5 were composed of about 0.25 A_{260} RNA/ml in 10 mM NaCl, 0.1 mM EDTA, 1 mM Na-cacodylate, and had a pH of 6.8; 50 µl were needed for each measurement. The experimental curves were extrapolated to the conditions of the calculation, i.e. 1 M ionic strength, with 14.3°C per factor of 10 in salt conditions. This value for the ionic strength dependence of T_m was determined experimentally for WT-dsCARNA 5.

Temperature-gradient gel electrophoresis: The technique of temperature-gradient gel electrophoresis is described in detail elsewhere (14). A horizontal slab gel (200 mm x 210 mm x 1 mm) on a film support (FMC Corp. Rockland, Maine, USA) is in thermal contact with a copper plate, but electrically insulated from the copper. Using two independent thermostats, the copper plate was cooled on one side and heated on the other. This permits establishment of a linear temperature gradient between 10°C and 80°C in the gel. In reference runs, linearity of the temperature gradient inside the gel was tested by thermoresistors; the deviation from linearity was smaller than 0.3°C everywhere in the gel. The device for the temperature-gradient was adapted to a LKB-Multiphor (LKB, Sweden). For electrophoresis perpendicular to the temperature gradient the sample was pipetted into a single slot (170 mm x 4 mm). Gels contained 5% acrylamide, 0.12% bisacrylamide, 0.08% TEMED, 8.9 mM TRIS, 8.9 mM boric acid, 0.25 mM EDTA, 8 M urea and 0.06% ammonium peroxodisulfate for starting the polymerization. The RNA samples were in 8.9 mM TRIS, 8.9 mM boric acid, 0.25 mM EDTA. For an electrophoretic analysis between 1 µg RNA (e.g. Fig. 4) if several heavy bands are present, and 200 ng (e.g. Fig. 3b) if only one component was contained, were used. The silver-staining of the gel originally described for proteins by Sammons et al. (35) was used in a modified form (36,37).

Calculation of thermal transition Curves: Calculation of the most stable secondary structures of single-stranded RNAs and of their possible alternative states (which were defined by fixing certain base pairs) was performed using a slightly modified Zuker-Nussinov-program (38) on a VAX 11/750 and a Siemens 7.580. The calculation of thermal denaturation curves, from a knowledge of the secondary structures existing at different temperatures, was performed as described earlier (38).

Denaturation curves of double-stranded RNAs were calculated according to the algorithm of Poland (39). Enthalpy and entropy parameters of helix growth were taken from Pörschke et al. (40,38); for interior loops containing n base pairs a loop weighting function $\delta(n) = \sigma \cdot n^{-1.75}$ was chosen with σ between 1E-3 and 1E-5 (41); For the product of the nucleation constant ß and the total single strand concentration c_0 a value between 1E-6 and 1E-8 was used. The calculation of one denaturation curve with 70 different temperature points needed less then 50 s on a Siemens 7.570. For calculation of relative mobility μ/μ_o of double-stranded and partially denatured molecules in the gel electrophoresis the semi-empirical equation of Lerman et al. (42,43) was used.

RESULTS and DISCUSSION

Analysis of dsCARNA 5 Variants

Experimentally determined optical melting curves: Optically determined thermal denaturation curves were performed for WT-, S52- and 1-dsCARNA 5. An example of such a melting curve is given in Fig. 3a. The curves of all dsCARNA 5s consist of at least three clearly separated transitions. The temperature difference between the transitions and the relative hypochromicities of the transitions are characteristic for each RNA species. Data derived from these curves are summarized in Table 1. The transition at highest temperature is irreversible because strand seperation is irreversible in the low ionic strength of these experiments.

Experimentally determined TGGE: All available dsCARNA 5 variants were analysed on temperature-gradient gels. Fig. 4 represents the gel electrophoretic analysis of WT-dsCARNA 5. At low and high temperatures the bands are moving much faster than at intermediate temperatures. Whereas the bands appear fairly homogeneous at low and high temperatures, they clearly split up in the region of the transition from the fast moving to the slow moving forms. This transition is called retardation transition. At least six different bands are separated. It is clearly demonstrated that WT-dsCARNA 5 contains at least six different RNA-species. Below the band of the dsRNAs a faster moving band is visible in all gels. This band is from an impurity in the gel matrix because it was also present if the sample slot was filled only with electrophoresis buffer without a nucleic acid sample.

The temperature-gradient gel analysis of 1-dsCARNA 5 (Fig. 2b) exhibits only one band although the gel had been overloaded on purpose. In clear contrast to WT-dsCARNA there is no indication for the presence of more than one molecular species. In order to compare the single component with the 6 components of WT-dsCARNA 5, a mixed sample of both was analysed: the heavy band of 1-dsCARNA does not coincide with any of the 6 of WT-dsCARNA 5. The temperatures of the retardation transition are summarised for all analysed variants in Table 2. From this table it is obvious that there is a correlation between necrogenicity and stability of the retardation transition of the variants: all non-necrogenic variants fall into the temperature range $T_m \leq 43.2°C$, the necrogenic variants in the temperature range $T_m \geq 43.8°C$.

Calculation of the helix-coil transition with the Poland algorithm: In order to obtain a quantitative thermodynamic treatment of the experimental data, the helix-coil transitions of certain dsCARNA 5s with known sequence were simulated. The first step in the theoretical treatment is

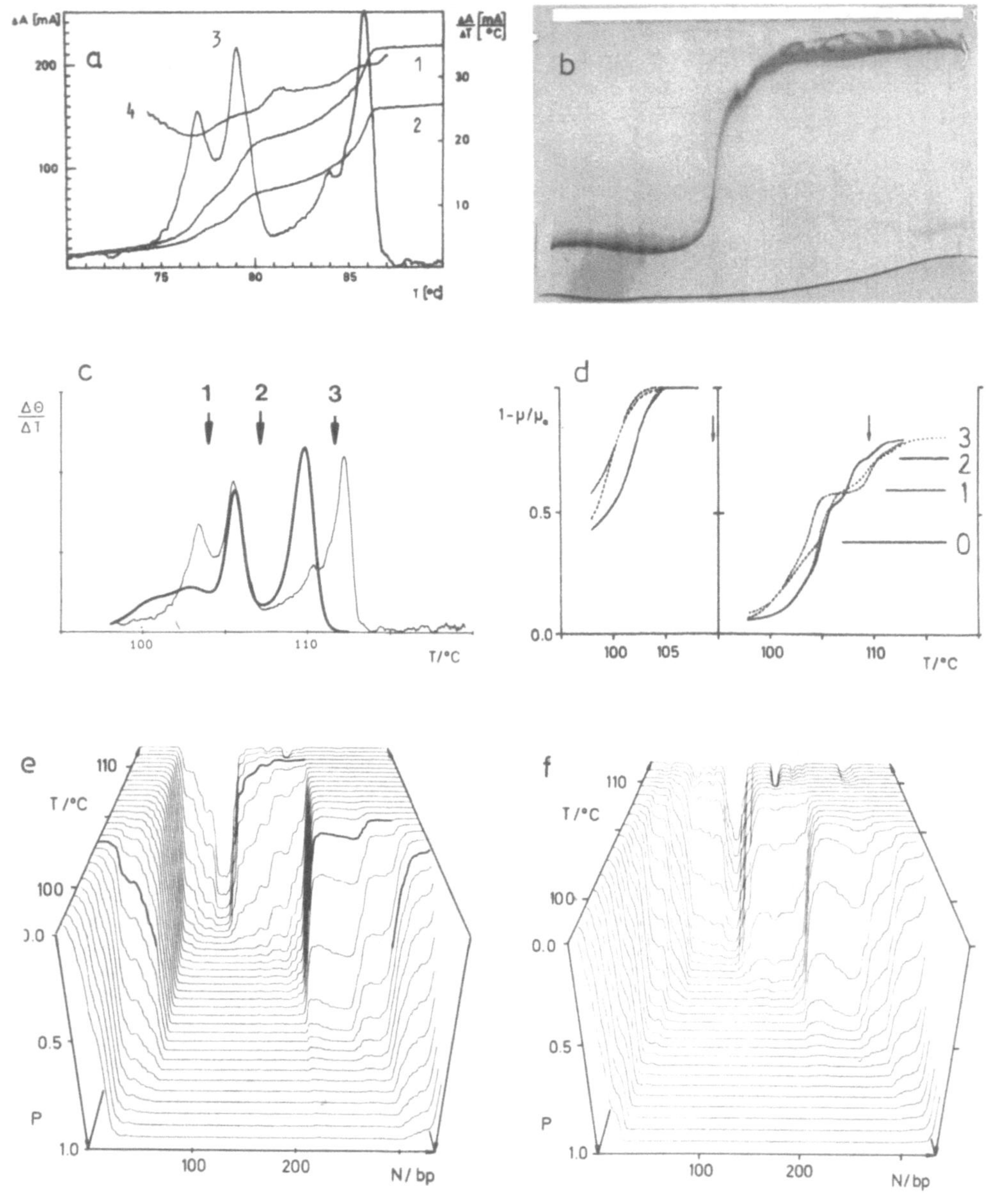

Fig. 3. (a) Experimental optical melting curve of 1-dsCARNA 5 in differentiated ($\Delta A_{260}/\Delta T$ (3)) and integrated (ΔA_{260} (1) and ΔA_{280} (2)) form. Scale for G:C-content (4) along ordinate is from 0 to 100%.
(b) TGGE pattern of 1-dsCARNA 5. Temperature-gradient is from 35° to 60°C.
(c) Superimposition of calculated and experimental optical denaturation curve of 1-dsCARNA 5 (extrapolated to 1 M NaCl with 14.7°C/log(c_{Na^+}). $\beta \cdot c_0$=1E-8; σ=1E-4.
(d) calculated mobility of dsCARNA 5 variants (—— D; ··· Yn; --- 1; -·- S). Left panel: L_r=20, right panel: L_r=200. Calculated temperatures of dissociation reaction are marked with arrows. Inserted levels 0 to 3 correspond to transitions marked in (c) and (e).
(e,f) Denaturation map of S-dsCARNA 5 (e) and D-dsCARNA 5 (f). Ordinate represents probability for a given base pair to be in the closed state (1 := closed base pair, 0 := open base pair). σ = 1E-4.

Table 1. Parameters of dsCARNA 5 variants determined by optical denaturation curves. All RNAs except S52-dsCARNA 5 were extensively dialysed against 10 mM NaCl, 0.1 mM EDTA, 1 mM Na-cacodylate, pH 6.8.
f_{GC}: determined from hypochromicities at 260 and 280 nm (10).
n_{Hy}: number of denaturing base pairs calculated from hypochromicity (10). Because of relative extinction changes, values are normalized to length of the nucleic acid, for WT- and S52-dsCARNA 5 a length of 335 was used.
$n(\Delta T_{1/2})$: number of denaturing base pairs calculated from $\Delta T_{1/2}$

RNA	transition	T_m/°C	$\Delta T_{1/2}$/°C	Hy/%	f_{GC}/%	$n_{\Delta T_{1/2}}$/Bp	n_{Hy}/Bp
1-dsCARNA 5	1	75.0	1.3	5	53	70	78
	2	77.0	1.1	8	55	80	128
	3	83.1	0.8	6	80	160	128
WT-dsCARNA 5	1	78.2	1.5	17	50	60	136
	2	82.2	1.6	12	59	55	105
	3	84.2	0.9	10	66	140	94
S52-dsCARNA 5	1	79.6	1.1	13.5	48	85	141
	2	83.5	1.8	9.0	56	50	102
	3	85.2	1.0	4.5	63	90	55
	4	86.1	1.0	3.0	66	130	40

Table 2. Midpoint temperatures of the retardation transitions of dsCARNA 5 variants determined by TGGE. *: major band

CARNA 5 variants	symptoms	T_m of retardation transition (°C)
WT	necrogenic	42.0, 42.4, 42.7, 43.7, 44.1*, 45.2
D		44.1*, 44.8
Yn		44.2
S52		43.9*
1	non-necrogenic	42.9
S		43.2
Val 24		40.5

the calculation of the state of each base pair of the double strand at a given temperature, i.e. the probability to be in the closed state. Such three-dimensional plots or denaturation maps are shown in Fig. 3e and f. The plots do not include the final dissociation step of the double strands. The plots of all examined satellite RNAs have the following steps in common: At low temperature the double strand opens up from its ends, mainly from the left end. With increasing temperatures three areas (marked with thick lines in Fig. 3f) of the double strand become successively destabilised. From the denaturation maps of the other variants (not shown) it is evident that all necrogenic variants of dsCARNA 5, examined so far, differ from the non-necrogenic mutants in the area with low stability at the left end of the molecules: This region ranges from nucleotides 1 to about 50 in the case of the non-necrogenic mutants whereas it ranges only to about nucleotide 20 in the case of the necrogenic ones.

Denaturation curves (calculated from the probability plots with inclusion of dissociation step) with several transitions were obtained for all variants of dsCARNA 5. The shape of the calculated curves corresponds qualitatetively to the measured curves (cf. Fig. 3a,c).

The experimental results of the TGGE were simulated by an exponential relation between electrophoretic mobility and degree of helix-coil transition (42,43). An example is given in Fig. 3d. The calculation does not take into account the final step of dissociation, and therefore a priori does not describe the increase in mobility at high temperatures. It follows from the exponential relationship that the base pairs which open up first contribute the most to the total decrease in mobility. In contrast, optical melting curves show comparable hypochromicities for all three transitions.

The agreement of experiments and calculations should be related to the following problems: a) the ionic strength dependence was only measured for WT-dsCARNA 5 and is probably not the same for the transitions of different dsCARNA 5s with different G:C contents; b) co-operativity of segmental transitions may change by shifting from low salt to high salt conditions; c) thermodynamic parameters and other constants are known with low accuracy. In the light of these uncertainties the agreement looks much better, particularly if one keeps in mind that on the whole temperature scale the very narrow transition range was hit within 1-2 Kelvin.

Assignment of the transitions: A comparison of the calculated gel electrophoretic curves (Fig. 3d) and the denaturation maps (Fig. 3e,f) at corresponding temperatures shows that solely the opening of base pairs at the left and at the right end of the satellite RNAs contributes to the retardation transition. In the optical melting curve (Fig. 3a,c) these base pairs denature in the low temperature shoulder (T ≤ 102°C). It follows that the retardation transition must correspond to the low-temperature shoulder of the first transition of the optical melting curve. However, if L_r is 200 (Fig. 3d), all transitions of the optical melting curve also contribute to the gel electrophoretic retardation steps. This effect is also visible in experimental TGGE if low voltage conditions are used (experiments not shown).

Effect of single base pair mutations: Between the non-necrogenic CARNA 5 variants on the one hand and the necrogenic ones on the other there are - among others - base exchanges at positions located in sequence domains which, according to our thermodynamic calculations, should be distinguishable in the simulated optical melting curves or in the temperature-gradient electrophoresis curves. For example, exchanges in the segment belonging to the highly co-operative transition 3 give rise to shifts in the calculated melting curves up to about 1°C if a G:C pair is changed to a A:U base pair or vice versa. Base pair 24 (numbering of 1-dsCARNA 5) is located in the small region which opens up first from the left end and

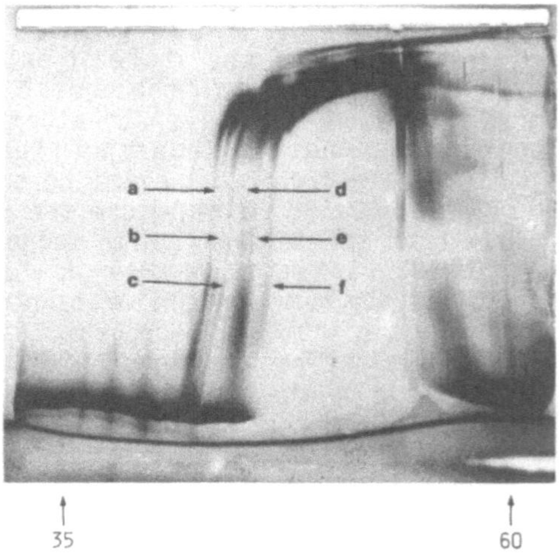

Fig. 4. TGGE pattern of WT-dsCARNA 5. At least six different RNA species, denoted a - f, are resolved in the retardation transition. The very broad bands in the region of low mobility are possibly an artifact from overloading. The arrows indicate the region of the linear temperature-gradient.

which in the optical melting curve gives rise to the broad shoulder at the low temperature side of the first transition. Its exchange leads to a deviation too small to be experimentally detectable in optical melting curves. This single base change gives rise, however, to a shift of about 0.8°C of the retardation-transition in a mobility curve. With gels the effect is larger than with optical melting curves because the mobility of the molecules is exponentially related to the opening of base pairs. Therefore opening up of small regions at the ends of the double strands leads to experimentally observable effects only with the temperature-gradient gels. Here, a shift of 0.8°C in a retardation-transition would be clearly detectable, particularly if another variant is used as an internal standard.

Sequence, gel electrophoretic mobility, and necrogenicity: Dissociation of the first 20 to 30 base pairs from the ends results in the largest retardation in gel electrophoresis. Consequently, minor sequence variations in these base pairs may result in clearly detectable shifts of the gel curves. The sequences of all non-necrogenic variants on the one hand and all necrogenic variants on the other are uniquely different in one position (24: U:A → C:G). This change leads to a less stable left end of the non-necrogenic dsCARNA 5 variants. In addition to this unique change a few other changes in the left as well as in the right ends of the molecule also shift the electrophoretic curves of the non-necrogenic variants in the same direction. It follows that exchanges in the ends of the molecule have the highest influence on the gel mobility. Therefore, the correlation between the stability (of the retardation transition) and necrogenicity of all CARNA 5 variants used in this work is based on special mutation of their sequences. We have to emphasize, however, that from such a correlation no mechanistic interpretation of necrogenicity can be derived. To do this one would have to assume that the double-stranded form of the satellite RNA is not an end product of replication but in itself in the cell might have a function in symptom modulation. In that case one could speculate that symptom modulation is connected with the ability of the double-stranded RNA to open up its ends. While the different "end stabilities" of necrogenic and non-necrogenic dsCARNA 5s could be coincidental, it seems more attractive to take them as an experimental hint for a possible mechanistic involvement of that region in symptom expression. One should note, however, that there are alternative models regarding more the single-stranded form of CARNA5 (reviewed in 44).

A more extensive treatment of the analysis of dsCARNA 5 variants is in preparation (45,46).

Analysis of viroid intermediates

Calculation of secondary structures: The structure of linear (+) monomers was calculated to be almost identical to that of native circular PSTV (10). Because the nucleotide sequence of PSTV is self-complementary except in the G:U base pairs, the structure of (-) monomers is very similar to that of (+) monomers. This type of unbranched secondary structure was also calculated to be the most stable structure for the oligomeric transcripts. There are at least two further secondary structures of oligomeric linear viroid molecules which are nearly as stable as the rod-like structure: i) a bifurcated structure containing all stable hairpins and ii) a structure with one bifurcation due to intramolecular base pairing of two hairpin I-regions. Schematic pictures of all three structures are given in Fig. 5g.

Temperature-Gradient gel electrophoresis: Mature circular PSTV as well as monomeric and oligomeric transcripts have been analysed by TGGE. The analysis of purified circular PSTV (Fig. 5a) shows a highly co-operative transition, the midpoint-temperature and the half width of which are very similar to those of the main peak of optical denaturation curves of PSTV. The transition is seen as a drastic reduction in mobility to less

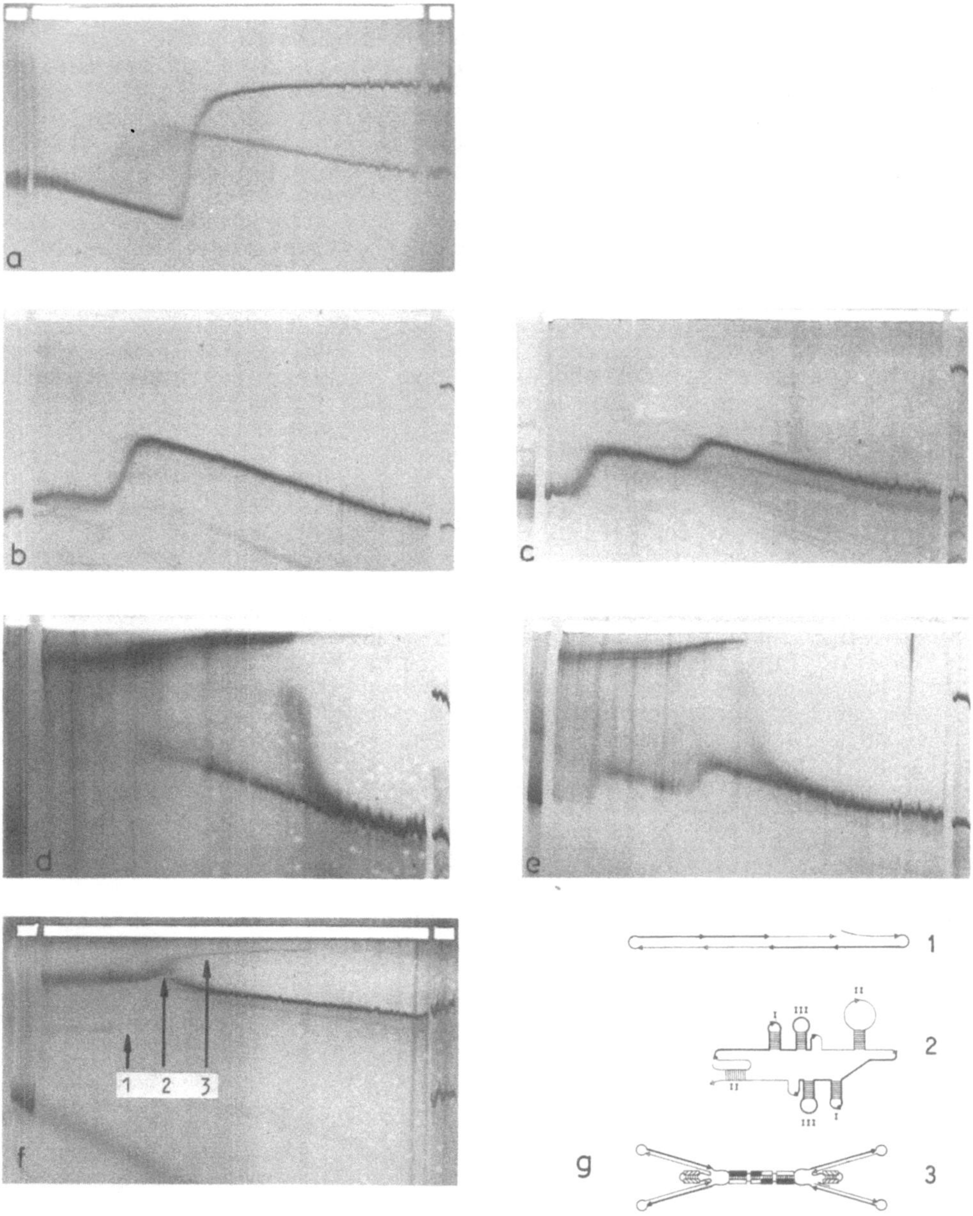

Fig. 5. Temperature-gradient gels of PSTV and PSTV-transcripts with a gradient from 25° to 70°C.
The reference slots contain always natural linear and circular PSTV.
(a) Circular PSTV from infected plants. A low amount of different natural linear molecules may also be seen.
(b,c) Monomeric T7 transcripts of (+) (b) and (-) (c) PSTV.
(d,e) Same samples as in (b,c) but denatured up to the temperature of the arrows in Fig. 2 and renatured prior to electrophoresis.
(f) TGGE of a transcription mix of dimeric (+) PSTV transcript.
(g) Structural models (compare with Fig. 6) of the RNAs of Fig. 5f. The numbers in the gel (Fig. 5f) refer to these structures. Roman numbers depict the stable hairpins of PSTV (cf. Fig. 1).

than a third. In addition to the band of circular viroid several transition curves of linear viroids, which are derived from the circular molecules by single nicks, are observed as faint curves. Their lower T_m values are also in accordance with optical data ((10), cf. Fig. 2), but with this technique only the superimposition of all transition curves could be measured. As only distinct curves appear, one has to assume that the nicks are not randomly distributed but restricted to very few sites.

The TGGE of the (+) strand transcript exhibits one co-operative transition (Fig. 5b), which corresponds to the main transition of the optical melting curve (Fig. 2). In the transition curve of the (-) strand two well-resolved transitions are seen (Fig. 5c), in accordance with the optical melting curve (Fig. 2).

If monomeric transcripts were denatured up to a temperature just below the last transition of the (+) strands (cf. arrows in Fig. 2), slowly renatured, and afterwards analysed with TGGE, the gels depicted in Fig. 5d and e were obtained. The transition curves, known from monomeric transcripts without heating and cooling (cf. Fig. 5b,c), are barely visible. The slow migration of the new band at low temperature indicates a conversion of the major part of all transcripts to bimolecular complexes. At very high temperature a discontinuous transition into the curves of the non-complexes monomers occurs. This transition is at higher temperatures in (+) strand (~ 65°C) than in (-) strand intermediates (~ 53°C). If the differing conditions are corrected for, the transition-temperature of the (+) strand complex agrees with the transition at highest temperature in the optical melting curve, yielding clear evidence that the interpretation of complex-dissociation was correct. The lower temperature of complex-dissociation of (-) strand transcripts explains why this transition could not be detected as a well resolved transition in optical melting.

The result of the following experiment is seen if Fig. 5f: an ongoing transcription of dimeric molecules at 37°C was stopped by adding EDTA, the mixture cooled to room temperature, transferred to the gel electrophoresis buffer by dilution and analysed without any additional treatment. From the gel it follows that the transcripts are present dominantly in a single structure right after their synthesis. The structure leads to slow migration at a temperature somewhat above that of the transition of the rod-like molecules. This is exactly the behaviour one does expect for the metastable structure as shown in Fig. 5g (middle part): i) this structure is more bulky than the rod-like structure, ii) after denaturing of less stable parts the structure is identical to that of the circular molecule

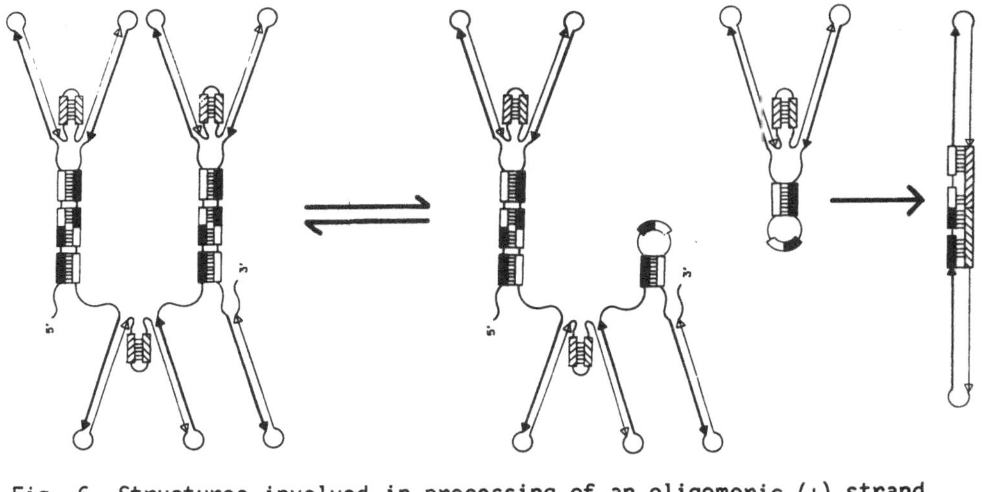

Fig. 6. Structures involved in processing of an oligomeric (+) strand intermediate to a circular viroid and an oligomeric strand containing one unit less than the original molecule.

after the main transition, i.e. it contains the three stable hairpins. The transition appears discontinuous, probably because the opening of hairpin II at the left side of the molecule is an irreversible process in the low ionic strength of the gel electrophoresis. One should note that to a minor extent also the rod-like structure and the structure with the three helices may be detected in the transcription mixture. One has to conclude, that the transient structure is able to switch over into the other structures under conditions which are close to physiological ones.

Conclusions: The analysis of synthetic transcripts of cloned PSTV cDNA with the method of TGGE was carried out under the assumption that these RNAs are models for the replicative intermediates of viroids. The results, which will be published in more detail elsewhere (47), confirm the earlier results (10) about the ability of viroid intermediates to adopt at least one special structure which may not be formed by mature viroids. These special structures may be the prerequisite for the recognition and processing of the intermediates to mature viroids by host enzymes. Because we know now the very distinct features of viroid intermediates *in vitro*, we would like to transfer the TGGE analysis to natural intermediates and their *in vivo* structures. Non-denaturing isolation of crude RNAs extracts from infected plants and TGGE analysis of these extracts with detection of the viroid species by hybridization (14) are the key experiments which are in progress.

SUMMARY

The examples given in this paper show that temperature-gradient gel electrophoresis may be applied to resolve quite different questions. In some cases, similar results have also been obtained by other methods. TGGE, however, bears the advantage that amounts of only 100 ng or less are needed because of the very sensitive staining techniques (silver or hybridization) and that the biopolymers have not to be purified for the study of their conformational transitions because electrophoresis is basically a separation technique.

ACKNOWLEDGEMENTS

I thank Prof. D. Riesner for support of the work and V. Rosenbaum, R. Hecker, Dr. Bewerunge, Dr. J. Kaper, and Prof. Tien Po for stimulating discussions. The work was supported by grants fom the Deutsche Forschungsgemeinschaft, the Minister für Wissenschaft und Forschung des Landes Nordrhein-Westfalen, and Fonds der Chemischen Industrie.

LITERATURE

1. "The Viroids", T.O. Diener, ed., Plenum Press, in press
2. H.L. Sänger *in*: "Encyclopedia of Plant Physiology", New Series, Vol. 14B, B. Parthier and D. Boulter, eds., Springer-Verlag, NY (1982).
3. D. Riesner and H.J. Gross, Viroids, *Ann. Rev. Biochem.* 54:531 (1985).
4. T.C. Goodman, L. Nagel, W. Rappold, G. Klotz, and D. Riesner, *Nucleic Acids Res.* 12:6231 (1984).
5. H.-R. Rackwitz, W. Rohde, and H.L. Sänger, *Nature* 291:297 (1981).
6. H.P. Mühlbach and H.L. Sänger, *Nature* 278:185 (1979).
7. A.D. Branch, H.D. Robertson, and E. Dickson, *Proc. Natl. Acad. Sci. USA* 78:6381 (1981).
8. R.A. Owens and T.O. Diener, *Proc. Natl. Acad. Sci. USA* 79:113 (1982).
9. M. Tabler and H.L. Sänger, *EMBO J.* 4:2191 (1985).
10. G. Steger, M. Tabler, W. Brüggemann, M. Colpan, G. Klotz, H.L. Sänger, and D. Riesner, *Nucleic Acids Res.* 14:9613 (1986).
11. M. Tabler and H.L. Sänger, *EMBO J.* 3:3055 (1984).
12. T. Meshi, M. Ishikawa, Y. Watanabe, J. Yamaya, Y. Okada, T. Sano, and E. Shikata, *Mol. Gen. Genet.* 200:199 (1985).

13. T.O. Diener, Proc. Natl. Acad. Sci. USA 83:58 (1986).
14. V. Rosenbaum and D. Riesner, Biophys. Chem. 26:235 (1987).
15. K.E. Richards, G. Jonard, M. Jacquemond, and H. Lot, Virology 89:395 (1978).
16. C.W. Collmer, M.E. Tousignant, and J.M. Kaper, Virology, 127:230 (1983).
17. K.H.J. Gordon and R.H. Symons, Nucleic Acids Res., 11:947 (1983).
18. S. Hidaka, K. Ishikawa, Y. Takanami, S. Kubo, and K. Miura, FEBS Lett., 174:38 (1984).
19. M.J. Avila-Rincon, C.W. Collmer, and J.M. Kaper, Virology, 152:446 (1986).
20. J.M. Kaper and M.E. Tousignant, Endeavour New Series, 8:194 (1984).
21. P. Tien and X.H. Chang, Seed Sci. Tech., 2:969 (1983).
22. P. Tien and X.H. Chang, Proc. 6th Int. Congr. Virol. Sendai, p. 379 (1984).
23. N. Habili and J.M. Kaper, Virology, 112:250 (1981).
24. I. Garcia-Luque, J.M. Kaper, J.R. Diaz-Ruiz, and M. Rubio-Huertos, J. Gen. Virol., 65:539 (1984).
25. G. Steger, Ph.D. thesis, T.H. Darmstadt, F.R.G. (1983).
26. J.R. Diaz-Ruiz and J.M. Kaper, Biochim. Biophys. Acta, 564:275 (1979).
27. J.E. Diaz-Ruiz and J.M. Kaper, Virology, 80:204 (1977).
28. J.M. Kaper and M.E. Tousignant, Virology, 85:323 (1978).
29. H. Lot, G. Marchoux, J. Marrou, J.M. Kaper, C.K. West, L. van Vloten-Doting, and R. Hull, J. Gen. Virol., 22:81 (1974).
30. J.M. Kaper, A.S. Duriat, and M.E. Tousignant, J. Gen. Virol., 67:2241 (1986).
31. J.M. Kaper, M.E. Tousignant, and S.M. Thompson, Virology, 114:526 (1981).
32. H. Lot and J.M. Kaper, Virology, 74:209 (1976).
33. P. Tien, X. Zhan, B. Qiu, B. Qin, and G. Qu, Annal. Appl. Biol., in press (1987).
34. H. Werntges, G. Steger, D. Riesner, and H.-J. Fritz, Nucleic Acids Res., 14:3773 (1986).
35. D.W. Sammons, L.D. Adams, and E.E. Nishizawa, Electrophoresis, 2:135 (1982).
36. J. Schumacher, N. Meyer, D. Riesner, and H.L. Weidemann, J. Phytopathol., 115:332 (1986).
37. E.A.C. Follett and U. Desselberger, J. Med. Virol., 11:39 (1983).
38. G. Steger, H. Hofmann, J. Förtsch, H.J. Gross, J.W. Randles, H.L. Sänger, and D. Riesner, J. Biomolec. Struct. Dyn., 2:543 (1984).
39. D. Poland, Biopolymers, 13:1859 (1974).
40. D. Pörschke, O.C. Uhlenbeck, and F.H. Martin, Biopolymers, 12:1313 (1973).
41. S. Ueno, H. Tachibana, Y. Husimi, and A. Wada, J. Biochem., 84:917 (1978).
42. S.G. Fischer and L.S. Lerman, Proc. Natl. Acad. Sci. USA, 80:1579 (1982).
43. L.S. Lerman, S.G. Fischer, I. Hurley, K. Silverstein, and N. Lumelsky, Ann. Rev. Biophys. Bioeng., 13:399 (1984).
44. J.M. Kaper and C.W. Collmer, in "RNA Genetics" Vol. II, E. Domingo, J. Holland, and P. Ahlquist, eds., CRC Press, in press (1987).
45. P. Tien, G. Steger, V. Rosenbaum, J.M. Kaper, and D. Riesner, Nucleic Acids Res. in press (1987).
46. G. Steger, P. Tien, J.M. Kaper, and D. Riesner, Nucleic Acids Res. in press (1987).
47. D. Riesner, R. Hecker, and G. Steger, J. Biomolec. Struct. Dyn. in press (1987).

MUTATIONAL ANALYSIS OF POTATO SPINDLE TUBER VIROID

Robert A. Owens, Rosemarie W. Hammond, and T. O. Diener

Microbiology and Plant Pathology Laboratory
Agricultural Research Service
U. S. Department of Agriculture
Beltsville, MD 20705

INTRODUCTION

Viroids are small (0.8-1.3 x 10^{-5} Mr), covalently circular RNAs that can be isolated from certain higher plant species afflicted with specific diseases. After introduction into healthy individuals of the same species, they replicate autonomously and cause the appearance of the characteristic disease syndrome. Although most viroids have been discovered because of their ability to cause readily recognizable disease symptoms in certain hosts, they may replicate in other species without causing obvious damage.

Since the discovery that viroids differ from conventional viruses in several fundamental respects (1), considerable effort has been devoted toward understanding the essential features of these unique pathogenic agents. Despite detailed knowledge (reviewed in 2) of their primary and secondary structures in vitro, relatively little is known about the relationship(s) between the structure of a viroid and its various biological functions. One approach to the identification of such relationships has involved nucleotide sequence comparisons of naturally occuring isolates of potato spindle tuber viroid (PSTV) and citrus exocortis viroid (CEV) which differ in the severity of the symptoms that they incite. Such comparisons (3,4) suggest that pathogenesis-modulating functions are localized within two regions of the native structure. When this approach was extended to a comparison of seven viroid species, Keese and Symons (5) were able to propose a model that correlates the functional domains of viroids with five structurally distinguishable regions--a conserved central core, flanking pathogenicity and variable domains, and two terminal loops (see Figure 1).

A second approach for dissecting structure-function relationships involves the introduction of site-directed mutations into infectious cDNA copies of viroids. Visvader et al. (6) have reported that a single nucleotide change in the upper central conserved region of CEV abolishes cDNA infectivity, and Owens et al. (7) reported similar results from chemical mutagenesis within the corresponding region of infectious PSTV cDNAs. These results are consistent with the suggestion (6,8,9) that this region contains a processing site for the longer-than-unit-length RNAs that are intermediates in viroid replication. BAL-31 deletion mutagenesis of an infectious dimer of hop stunt viroid (HSV) cDNA suggested that the corresponding region of HSV also contains a structure recognized by a putative endonuclease that produces unit-length linear molecules from (+) strand RNA multimers (10).

We have previously shown that a single C->T transition at the left border of the central conserved region abolishes PSTV cDNA infectivity (7), and various small insertions and deletions that alter the thermodynamic stability of HSV abolish the infectivity of both HSV cDNAs and their in vitro RNA transcripts (11). Because such changes must have significant effects upon essential nucleotide interactions within the viroid, we have begun to systematically study the effect of both single and multiple nucleotide changes upon the infectivity of PSTV cDNAs. In certain instances, we have introduced changes in regions of PSTV that exhibit natural sequence variation in hopes that these regions may tolerate minor alterations. The results presented below suggest that conservation of its native structure appears to be essential for viroid function. In at least one case, viroid viability can be restored by construction of a "pseudorevertant" in which the likely secondary structure has been restored.

SITE-SPECIFIC MUTAGENESIS OF PSTV

At present, site-specific mutagenesis of cloned viroid cDNAs involves two successive steps - introduction of the desired mutation into an infectious viroid cDNA and subsequent bioassay to detect phenotypic variation. Because naturally occuring variants are unlikely to contain alterations that severely affect essential functions, site-specific mutagenesis techniques complement comparative sequence analysis of naturally occurring variants.

Figure 1 shows the nucleotide sequence of PSTV (Intermediate strain); nucleotides altered by oligonucleotide-directed mutagenesis are denoted by asterisks. Results of infectivity trials with the altered PSTV cDNAs have been grouped according to their location within the molecule and are presented in Table 1.

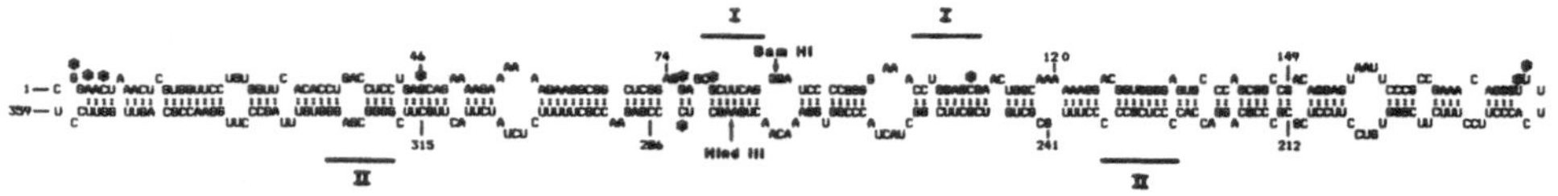

Figure 1. Schematic diagram of PSTV illustrating the nucleotide sequence and the highly base-paired, thermodynamically stable conformation of the intermediate strain. Asterisks denote the location of site-directed mutations within PSTV, BamHI and HindIII are restriction endonuclease recognition sites present in cloned PSTV cDNAs, and I and II indicate nucleotides comprising the stems of secondary hairpins formed during thermal denaturation of purified PSTV (2). The borders between the five proposed functional domains in PSTV (5) are indicated by arabic numerals.

Mutations Within the Terminal Loops

Mutations in the left and right terminal loops of PSTV were designed to introduce restriction endonuclease recognition sites into these domains. PSTV-P was constructed using an oligonucleotide containing mismatches at positions 2, 4, and 6, thereby introducing a PstI site. In the right terminal loop, changes at positions 177 and 178 created an EcoRI site. Both mutations abolished the infectivity of PSTV cDNA (Table 1), and extension of the rod-like structure of PSTV by the introduction of an 18-nucleotide insertion into the EcoRI site did not restore cDNA infectivity (data not shown).

Table 1. Infectivity of selected PSTV cDNAs containing site-specific mutations.

Plasmid	Modification: Substitution	Modification: Position	Infectivity
pST-B5	Wild type		5/5
PSTV-P	G->U	2	0/5
	A->C	4	
	C->G	6	
PSTV-R	U->A	177	0/5
	U->A	178	
PSTV-Ia	G->A	80	0/5
PSTV-Ib	C->U	109	0/5
PSTV-Ia+Ib	G->A	80	0/5
	C->U	109	
PSTV-H	C->U	284	0/5
PSTV-H'	G->A	76	0/5
PSTV-H+H'	C->U	284	2/5
	C->U	76	

Intact double-stranded plasmid DNA (1 ug/ul; 10 ul) was used as inoculum, and 10 plants were inoculated with each DNA to be tested. Plants were paired before hybridization analysis, and infectivity is expressed as the number of pairs containing at least one infected plant / total number of pairs inoculated. Modified from Hammond and Owens (12).

Mutations Within the Central Conserved Region

Quantitative thermodynamic and kinetic studies of viroid denaturation indicate that these molecules can assume alternative secondary structures that may control their various biological functions (2,5). During the process of denaturation, secondary hairpins involving distant complementary bases form - e.g., hairpins I and II (Figure 1 and Table 2). Although this reassociation occurs experimentally and can be visualized by electron microscopy, the biological significance of these structures is unknown.

Table 2. Mutations within the central conserved region of PSTV.

Native Structure

```
            *
 76        GC          GGA              G      A  U     *  111
      GAGC   UUCAG         UCC CCGGG          CC GGAGCGA
      ::::   :::::         ::: :::::          :: :::::::
      CUCG   AAGUC         AGG GGCCC          GG CUUCGCU
284                A   A    U      A      C               247
                    ACA                UCAU
```

Secondary Hairpin I

```
87        102
  G - C          o - o         o - o         o - o
  G - C          o - o         o - o         o - o
  A - U          o - o         o - o         o - o
  C - G          o - o         o - o         o - o
  U - G          o - o         o - o         o - o
  U - A          o - o         o - o         o - o
  C - G          o - o         o - o         o - o
 *G - C*        A     C        G   U         A - U
  C - G          o - o         o - o         o - o
79        110

PSTV-Int       PSTV-Ia       PSTV-Ib       PSTV-Ia+Ib
```

Infectivity

Viable	Nonviable	Nonviable	Nonviable

Positions of oligonucleotide-directed mutations PSTV-Ia and PSTV-Ib within the central conserved region and secondary hairpin I are indicated by asterisks. To emphasize the potential structural effects of the mutations, nucleotides which are identical in wild-type and mutant sequences are indicated by an "o".

Because the pair of inverted repeats within the upper central conserved region that form hairpin I may play a critical role in viroid replication (8), we decided to determine the effect of disrupting its base pairing by site-specific mutagenesis. As shown in Table 2, a G->A transition at residue 80 (PSTV-Ia) disrupts G:C base pairing in secondary hairpin I but not in the native structure, while the corresponding C->T transition at residue 109 in the right side of the hairpin converts a G:C base-pair to a G:U pairing in both the native structure and hairpin I. Both individual mutations abolished cDNA infectivity. The simultaneous presence of both mutations restores base-pairing within hairpin I but not the native structure; this pseudorevertant cDNA (PSTV-Ia+Ib) was not infectious.

A Pseudorevertant Within Premelting Loop 2 Restores PSTV cDNA Infectivity

We have previously reported (7) that a single C->T transition at position 284 abolishes PSTV cDNA infectivity, even though this change seems unlikely to destabilize the native structure of PSTV (Table 3). Because comparative sequence analysis (5) has shown that the uridine present at the corresponding positions in tomato apical stunt (TASV), CEV, and chrysanthemum stunt (CSV) viroids is base-paired with an adenosine residue in their native structures, we decided to investigate the effect of introducing a G->A transition at position 76 in PSTV. This transition alone (PSTV-H') also rendered PSTV cDNA noninfectious, but the simultaneous presence of both mutations (PSTV-H+H') replaces the G:C base pair with an A:U base pair and partially restored the impaired function(s) (Table 3). The extent of this functional restoration and the identity of the replicating progeny were studied further.

Table 3. Mutations within PSTV premelting loop 2.

```
----------------------------------------------------------------

     Native Structure                          Infectivity


   69      AG  GC        86
     CUCGG  GA  GCUUCAG
     :::::  ::  :::::::                          Viable
     GAGCC  CU  CGAAGUC
  289                    276

        PSTV-Int

           oo  oo
     ooooo  Go  ooooooo
     :::::   :  :::::::                          Nonviable
     ooooo  Uo  ooooooo


           PSTV-H

           ooA oo
     ooooo   o  ooooooo
     :::::   :  :::::::                          Nonviable
     ooooo   o  ooooooo
             C

            PSTV-H'

           oo  oo
     ooooo  Ao  ooooooo
     :::::  ::  :::::::                          Viable
     ooooo  Uo  ooooooo                       (attenuated)

           PSTV-H+H'
----------------------------------------------------------------
```

Positions of mutations just to the left of the central conserved region are shown. To emphasize the potential structural effects of the mutations, nucleotides which are identical in the wild-type and mutant sequences are indicated by an "o".

The specific infectivity of SP6 transcripts of PSTV-H+H' is very similar to that of PSTV-Int transcripts (data not shown), but appearance of progeny at the apex was delayed, and symptom severity was reduced. Six weeks after inoculation, the concentration of PSTV in the apex of tomatoes inoculated with SP6 RNA transcripts of PSTV-H+H' was approximately 4-fold lower than that in plants inoculated with corresponding wild-type transcripts; this difference was maintained when the plants were cut back and the axillary buds allowed to grow out. As expected, denaturing polyacrylamide gel electrophoresis of progeny RNA revealed two species with the same mobilities as the circular and linear forms of wild-type PSTV.

Having established that the infectious agent in plants inoculated with the double mutant was correctly processed viroid RNA, low molecular weight tomato RNA samples were analyzed by hybridization to three synthetic DNA probes. Hybridization with 5'-GGCTTCAGTTGTTTCCACC-3', a probe complementary to nucleotides 265-282 of both wild-type and putative mutant PSTV RNAs, was used to estimate the relative viroid concentration in each sample, and the mutagenesis primers PSTV-H and PSTV-H' were used to detect the C->U transition at position 284 and the G->A transition at position 76 which were expected to be present in the pseudorevertant PSTV. The results of such an analysis are presented elsewhere (12) and provide strong evidence that the replicating progeny in plants inoculated with the double mutant does indeed contain both mutations present in the cDNA inoculum. Sequence analysis of several cDNA clones produced from the PSTV-H+H' progeny RNA confirmed the presence of both mutations and the absence of additional mutations.

The results from oligonucleotide-directed mutagenesis of infectious PSTV cDNAs presented above allow several tentative conclusions about the relative importance of the native structure and secondary structural interactions for PSTV infectivity. As yet, we have been unable to confirm the participation of various conserved sequences (e.g., hairpin I) in viroid replication. Although the secondary mutation in PSTV-Ia+Ib restores base pairing within secondary hairpin I, it failed to restore cDNA infectivity. Furthermore, the dramatic and deleterious effect of comparatively minor sequence modifications in this and other viroid domains (i.e., PSTV-H and PSTV-Ib) indicate an important role for the native structure in viroid infectivity and strongly suggest the presence of previously unsuspected interactions within PSTV.

Comparison of the relative stabilities of the region beginning at position 68 and ending at position 86 in infectious (PSTV-Int and PSTV-H+H') and noninfectious PSTV molecules (PSTV-H and PSTV-H') illustrates the potential effect of perturbing the native structure. Even though the progeny from PSTV-H+H' would appear less stable than those from either the intermediate strain or the individual PSTV-H and PSTV-H' mutations (-19.3 kJ/mol vs. -20.2 or -19.7 kJ/mol, respectively), this cDNA was infectious. Characteristics of the viroid native structure such as base-stacking, number of helical turns, or the ability to participate in specific protein-nucleic acid interactions must also make important contributions to viroid infectivity.

All these factors, together with the observation by Schnolzer et al. (3) that many of the individual sequence differences among naturally occuring PSTV isolates would destabilize their rod-like secondary structures in the absence of compensating changes, suggest that more sophisticated approaches to mutagenesis of viroid cDNAs may be required. The value of one such approach, constructing pseudorevertant cDNAs containing compensating mutations, is illustrated by the restoration of infectivity observed in PSTV-H+H'. Saturation mutagenesis of infectious viroid cDNAs (see below) is a second possible approach.

EMERGING APPROACHES TO STRUCTURE/FUNCTION ANALYSIS

Conventional bioassay techniques for viroid cDNA infectivity are based upon the ability of viroid infections to rapidly become systemic and may be unable to differentiate between mutations that are actually lethal and those that severely inhibit replication or cell-to-cell spread. The use of oligonucleotide-directed mutagenesis techniques should lead to rapid refinements in our understanding of the molecular mechanisms involved in biological characteristics that can be studied using a systemic bioassay (e.g., replication and symptom expression), but the selective pressure in a systemic bioassay favors the propagation of wild-type viroid generated by sequence reversion at the RNA level rather than propagation of the presumably less fit mutant.

Alternative Assay Systems

Bioassay systems where the Ti plasmid of *Agrobacterium tumefaciens* is used to mediate viroid infection (13) can provide large amounts of tissue in which viroid infection need not be systemic to be detectable. We (Salazar et al., manuscript in preparation) have used this approach to investigate the factors which determine viroid host range. We found that although PSTV appears stable and able to undergo translocation, it is unable to undergo normal RNA-directed replication in nonhost species. Inoculation of tomatoes with *Agrobacterium* strains whose Ti plasmids contain PSTV cDNAs with lethal mutations may also be useful to characterize the nature of their defects (Hammond, unpublished data). Nevertheless, quantitative *in vitro* assay systems for specific viroid functions will be required for the detailed analysis of structure/function relationships by site-specific mutagenesis.

Hall and his collaborators (14) have used the ability of mutant brome mosaic virus (BMV) RNAs synthesized *in vitro* to act as substrates for both tyrosyl-tRNA synthetase and BMV RNA replicase in order to study RNA structure/function relationships in the 3'-terminal region. The ability of certain RNA transcripts of group I introns (see 15 for review) or plant satellite RNA sequences (16,17) to undergo spontaneous *in vitro* splicing has been used to characterize the effects of various site-directed mutations in these RNAs. In the case of the rRNA intron of *Tetrahymena thermophila*, an *in vivo* assay for RNA splicing that depends upon the ability of the intron to excise itself from B-galactosidase mRNA (18,19) allows the analysis of structure/function relationships in an RNA enzyme without the influence of preexisting models and the detection of long-range tertiary interactions between nucleotides that cannot be predicted by secondary structure calculations. As assay systems similar to those described above become available for mutant characterization and the questions posed become more detailed, rapid and efficient methods for generating and isolating large numbers of random, single-base substitutions will be required.

Saturation mutagenesis

Maniatis and coworkers have recently described a protocol for random chemical mutagenesis of single-stranded DNA that allows the isolation and characterization of large numbers of such mutations in the absence of phenotypic selection (20,21). Double-stranded DNA fragments containing base substitutions are physically separated from wild-type fragments by gel electrophoresis through a gradient of denaturants (formamide and urea). The protocol is applicable to viroid cDNAs (unpublished results), and mutations can be mapped either by direct sequence analysis or by pancreatic RNase cleavage of single-base mismatches present in RNA:DNA duplexes formed by annealing unlabeled single-stranded mutant DNAs to [^{32}P] labeled RNA probe complementary to the wild-type sequence. More than 50 percent of all possible single-base substitutions can be mapped by electrophoretic analysis of the resulting RNA cleavage products (21).

If the particular region of interest is sufficiently small and flanked by convenient restriction endonuclease recognition sites, a second approach to saturation mutagenesis is also possible. Both the position and number of nucleotide substitutions in a synthetic DNA can be controlled by adjusting the concentration of "dopant" nucleoside phosphoramidate added to one or more monomer reservoirs of the DNA synthesizer. Several groups (e.g., 22) have described cloning procedures for such mixed oligodeoxynucleotides whose efficiencies justify direct sequence analysis of the transformant clones.

Applications of Existing Mutant Viroid cDNAs

Apparent similarities between the mechanisms of viroid replication and intron excision from eukaryotic mRNAs have led Zimmern (23) to propose that viroids and RNA viruses may have originated from a novel class of eukaryotic RNAs whose primary function is the intercellular exchange of genetic information. Furthermore, he proposed that the origin, amplification, and extracellular exchange of these hypothetical regulatory RNAs depend upon RNA recombination, a phenomenon known to occur during the replication of poliovirus (24, 25) and brome mosaic virus (26). Kirkegaard and Baltimore (24) have presented genetic evidence which strongly supports a copy choice mechanism for RNA recombination in which the poliovirus RNA polymerase switches templates during negative strand synthesis. Keese and Symons (5) have proposed that viroid evolution has involved intermolecular recombination between distinct viroid species coinfecting a common host, an event that led to exchange of their terminal domains. We have previously speculated (see Figure 6 in reference 27) about how RNA recombination following Agrobacterium-mediated inoculation of tomato plants with a mixture of two mutant PSTV cDNAs might lead to the appearance of wild-type progeny. Figure 2 illustrates the essential features of several preliminary experiments that we have conducted to test our hypothesis.

As shown above (see Table 1), the two mutations chosen for use in this experiment, PSTV-P and PSTV-R, appear to completely abolish systemic viroid replication. The locations of these particular mutations within the terminal loops should also provide the greatest distance over which RNA recombination can occur. When tomato seedlings were inoculated with SP6 RNA transcripts of these two PSTV cDNAs, either singly or as equimolar mixtures, no evidence for systemic viroid replication could be obtained by nucleic acid spot hybridization analysis of the upper leaves. Even when the transcript mixtures were heated and annealed under conditions similar to those described by Steger et al. (28) to allow intramolecular formation of the palindromic structure possibly involved in the conversion of multimeric linear viroid RNAs to their mature circular forms, no progeny were produced.

The infectivity of SP6 transcripts prepared from both possible mixed tandem cDNA dimers containing these mutants was also tested. Our original motivation for testing these constructions was the belief that by fusing the transcripts of the two mutant cDNAs at the DNA level, we would leave only one intramolecular cleavage/ligation event involving nucleotides 79-110 to be accomplished in vivo. The result of such a cleavage/ligation event would be precisely the sort of circular RNA dimer that Keese and Symons (5) have postulated to be the template for RNA recombination. We hoped, therefore, to detect wild-type PSTV progeny in only a small percentage of the inoculated plants at best and, even then, only after comparatively long incubation periods. The results of our infectivity trials did not completely match these expectations.

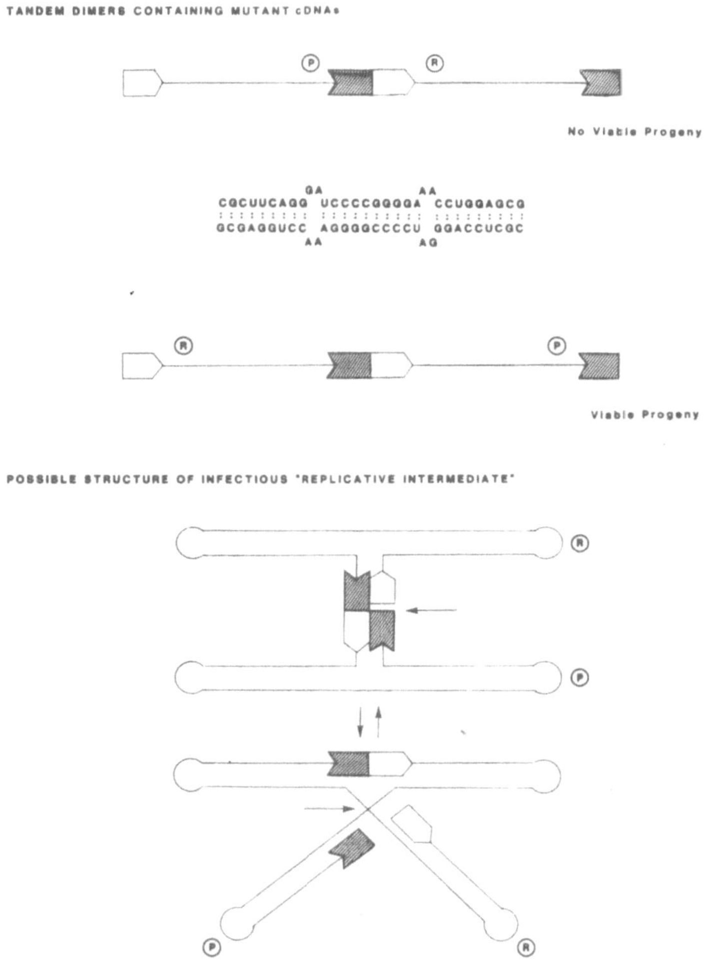

Figure 2. Infectivity trials with genetically marked PSTV RNA multimers. The relative positions of the PSTV-P and PSTV-R mutations within mixed tandem dimers of PSTV cDNA and their respective RNA transcripts are indicated by P and R , respectively. Only the construction in which the two mutations are separated by more than 359 nucleotides was infectious, and only two alternative conformations of the infectious RNA transcript are shown (lower). The open and shaded boxes indicate the positions of nucleotides comprising the putative palindromic processing site (shown between the two cDNA multimers), while possible cleavage/ligation sites are indicated by the arrows.

As expected, transcripts from the tandem cDNA dimer in which the two mutations were separated by only 180 nucleotides were noninfectious. To our initial surprise, however, the specific infectivity of transcripts from the cDNA dimer in which the mutations were separated by ~540 nucleotides appeared equal to that of transcripts from the wild-type cDNA dimer. When low molecular weight RNA isolated from infected plants was characterized by hybridization analysis with the PSTV-P and PSTV-R mutagenesis primers, the progeny appeared indistinguishable from the wild-type (PSTV-Int) strain. As indicated in the lower portion of Figure 2, efficient cleavage/ligation of multimeric PSTV (+)-strand RNAs can apparently occur at sites other than the palindromic processing site postulated by Diener (8). Additional experiments involving partial multimers of PSTV, tomato apical stunt (TASV), and tomato "planta macho" (TPMV) cDNAs are in progress to clarify this point.

CONCLUDING REMARKS

The results presented above are part of an extensive and ongoing analysis of the effects of mutation on viroid viability. We have detailed the effects of several single and multiple nucleotide changes on the infectivity of PSTV cDNAs or their respective RNA transcripts. Although conservation of the characteristic rod-like native structure appears to be essential for viroid function, we can only speculate about the molecular mechanisms mediating the effects we have observed. Minor sequence alterations could affect the specific transfer of viroids in and out of the nucleus, the site of viroid replication. Viroids spread throughout their hosts by translocation through the vascular system, and cell-to-cell spread may involve induction of a host component that aids in the spread of infection. Either of these processes could be disturbed by even minor changes in viroid sequence which alter the interaction between viroid-related RNAs and as-yet unidentified host components.

Although improved in vivo and in vitro assays systems will be required for complete characterization of the complex relationships between the structure of PSTV and its various biological functions, it is important to note that certain nonviable mutations could be rescued by introducing a second, compensating mutation elsewhere in the molecule. Site-directed mutagenesis of PSTV can also be used to study phenomena other than viroid replication and pathogenesis. In particular, we are trying to use genetically marked PSTV variants to assess the role of RNA recombination in viroid replication and evolution.

ACKNOWLEDGEMENTS

One of us (R.W.H.) has been supported by funds provided by U.S. Department of Agriculture Competitive Research Grants 81-CRCR-1-0719 and 85-CRCR-1-1738. We thank S.M. Thompson and M. Greyerbiehl for excellent technical assistance.

REFERENCES

1. T. O. Diener, Potato spindle tuber "virus". IV. A replicating, low molecular weight RNA, Virology 45:411 (1971).
2. D. Riesner and H. J. Gross, Viroids, Ann. Rev. Biochem. 54:531 (1985).
3. M. Schnolzer, B. Haas, K. Ramm, H. Hofmann, and H. L. Sanger, Correlation between structure and pathogenicity of potato spindle tuber viroid (PSTV), EMBO J. 4:2181 (1985).
4. J. E. Visvader and R. H. Symons, Eleven new sequence variants of citrus exocortis viroid and the correlation of sequence with pathogenicity, Nuc. Acids Res. 13:2907 (1985).

5. P. Keese and R. H. Symons, Domains in viroids: Evidence of intermolecular RNA rearrangements and their contribution to viroid evolution, Proc. Natl. Acad. Sci. USA. 82:4582 (1985).
6. J. E. Visvader, A. C. Forster, and R. H. Symons, Infectivity and in vitro mutagenesis of monomeric cDNA clones of citrus exocortis viroid indicates the site of processing of viroid precursors, Nuc. Acids Res. 13:5843 (1985).
7. R. A. Owens, R. W. Hammond, R. C. Gardner, M. C. Kiefer, S. M. Thompson, and D. E. Cress, Site-specific mutagenesis of potato spindle tuber viroid cDNA: Alterations within premelting region 2 that abolish infectivity, Plant Mol. Biol. 6:179 (1986).
8. T. O. Diener, Viroid processing: A model involving the central conserved region and hairpin I, Proc. Natl. Acad. Sci. USA. 83: 58 (1986).
9. J. Hashimoto and Y. Machida, The sequence in the potato spindle tuber viroid required for its cDNA to be infective: A putative processing site in virooid replication, J. Gen. Appl. Microbiol. 31:551 (1985).
10. T. Meshi, M. Ishikawa, Y. Watanabe, J. Yamaya, Y. Okada, T. Sano, and E. Shikata, The sequence necessary for the infectivity of hop stunt viroid cDNA clones, Mol. Gen Genet. 200:199 (1985).
11. M. Ishikawa, T. Meshi, Y. Okada, T. Sano, and E. Shikata, In vitro mutagenesis of infectious viroid cDNA clone, J. Biochem. 98:1615 (1985).
12. R. W. Hammond and R. A. Owens, Mutational analysis of potato spindle tuber viroid reveals complex relationships between structure and infectivity, Proc. Natl. Acad. Sci. USA. 84:(in press).
13. R. C. Gardner, K. Chonoles, and R. A. Owens, Potato spindle tuber viroid infections mediated by the Ti plasmid of Agrobacterium tumefaciens, Plant Mol. Biol. 6:221 (1986).
14. T. W. Dreher, J. J. Bujarski, and T. C. Hall, Mutant viral RNAs synthesized in vitro show altered aminoacylation and replicase activities, Nature 311:171 (1984).
15. T. R. Cech and B. L. Bass, Biological catalysis by RNA, Ann. Rev. Biochem. 5:599 (1986).
16. G. A. Prody, J. T. Bakos, J. M. Buzayan, I. R. Schneider, and G. Bruening, Autolytic processing of dimeric plant virus satellite RNA, Science 231:1577 (1986).
17. A. C. Forster and R. H. Symons, Self-cleavage of plus and minus RNAs of a virusoid and a structural model for the active sites, Cell 49: 211 (1987).
18. J. V. Price and T. R. Cech, Coupling of Tetrahymena ribosomal RNA splicing to B-galactosidase expression in Escherichia coli, Science 228:719 (1985).
19. R. B. Waring, J. A. Ray, S. W. Edwards, C. Scazzocchio, and R. W. Davies, The Tetrahymena rRNA intron self-splices in E. coli: In vivo evidence for the importance of key base-paired regions of RNA for RNA enzyme function, Cell 40:371 (1985).
20. R. M. Myers, L. S. Lerman, and T. Maniatis, A general method for saturation mutagenesis of cloned DNA fragments, Science 229: 242 (1985).
21. R. M. Myers, Z. Larin, and T. Maniatis, Detection of single base substitutions by ribonuclease cleavage at mismatches in RNA:DNA duplexes, Science 230:1242 (1985).
22. K. M. Derbyshire, J. J. Salvo, and N. D. F. Grindley, A simple and efficient procedure for saturation mutagenesis using mixed oligodeoxynucleotides, Gene 46:145 (1986).
23. D. Zimmern, Do viroids and viruses derive from a system that exhanges genetic information between eukaryotic cells, Trends Biochem. Sci. 7:205 (1982)

24. K. Kirkegaard and D. Baltimore, The mechanism of RNA recombination in poliovirus, Cell 47:433 (1986).
25. L. I. Romanova, V. M. Blinov, E. A. Tolskaya, E. G. Victorova, M. S. Kolesnikova, E. A. Guseva, and V. I. Agol, The primary structure of crossover regions of intertypic poliovirus recombinants: A model of recombination between RNA genomes, Virology 155:202 (1986).
26. J. J. Bujarski and P. Kaesberg, Genetic recombination between RNA components of a multipartite plant virus, Nature 321:528 (1986).
27. R. A. Owens and R. W. Hammond, Molecular biology of viroid-host interaction, in: "The Viroids," T. O. Diener, ed., Plenum, New York (1987).
28. G. Steger, M. Tabler, W. Bruggemann, M. Colpan, G. Klotz, H. L. Sanger, and D. Riesner, Structure of viroid replicative intermediates: Physico-chemical studies on SP6 transcripts of cloned oligomeric potato spindle tuber viroid, Nuc. Acids Res. 15:9613 (1986).

NON-ENZYMATIC CLEAVAGE AND LIGATION OF A PLANT SATELLITE RNA

George Bruening, Jamal Buzayan, Wayne Gerlach+) and
Arnold Hampel++)
Dept. Plant Pathology, Univ. Calif.-Davis, Ca 95616
+)Division of Plant Industry, Commonwealth Scientific and Industrial Research Organisation, Canberra, A.C.T. 2601, Australia, ++)Plant Molecular Biology Center, Northern Ill. Univ. DeKalb, Il. 60115

INTRODUCTION

Small satellite RNAs associated with certain plant viral infections require the virus for replication (1). They replicate extensively and become encapsidated in mixed infections in which both the satellite RNA and associated virus are present. The satellite RNA becomes encapsidated in virus like particles protected by the coat protein of the supporting virus. The satellite RNA may be 80% to 95% of the encapsidated RNA (1,2,3,4).

The 359 base long satellite RNA of tobacco ringspot virus (STobRV RNA) has been studied extensively in this laboratory and is the satellite about which the most information presently exists. STobRV RNA propagates in the presence of tobacco ringspot virus (TobRV). There is no extensive nucleotide sequence relationship between the satellite RNA and the two genomic RNAs of the supporting virus and as far as is known the satellite RNA has no translational product for transcription into DNA. Furthermore, STobRV acts as a parasite of the virus by ameliorating the symptoms that tobRV alone induces (1,5).

The fact that this and other small RNAs can replicate without the participation of DNA templates and without coding for a protein, argues for a number of specific functions for this RNA. This satellite RNA is known to have at least four functions: 1. recognition as template by the supporting virus and/or host/cell replication machinery. 2. encapsidation by the supporting virus coat protein. 3. attenuation of the viral infection, and 4. self processing of the satellite RNA itself (6).

The self-processing of the satellite itself would appear central to its biological role. The phenomenon of RNA itself mediating self-processing or autolytic cleavage and ligation is not a newly observed phenomenon. RNA autolytic cleavage was first observed in the site specific hydrolysis of tRNA in the presence of Mg^{++} and Zn^{++} in 1973 (7), Pb^{++} (8), or Eu^{++} (9). The mechanism of the RNA auto-cleavage in the presence of Pb^{++} was shown by direct crystallization observation to be carried out by the Pb^{++} ion held in a cavity between the T and D loops (10,11) which carried out the autolytic cleavage of phosphorous #18 in an acid/base type catalyzed reaction.

RNA autolytic reactions were later observed by Cech and colleagues (12) in the splicing reaction of a ribosomal RNA precursor (a gene of the group I precursor RNA class) in Tetrahymena. A 413 nucleotide intervening sequence is excised from the RNA using guanosine as a co-factor followed by ligation of the two flanking sequences (exons) and circularization of the intervening sequence. The RNA itself mediated the reaction. Other RNA autolytic reactions involved intron removal and exon splicing for other group I precursor RNAs (13,14,15); and group II precursor RNA intron removal and exon splicing (16,17,18). All of these self-processing RNA reactions involved transphosphorylation reactions through 5' phosphates and 3' or 2' OH groups. These reactions were all quasi-catalytic since the catalyst itself was consumes in the reaction. True RNA catalysis was described for the Ribonuclease P cleavage of transfer RNA by the RNA subunit M1 (19) and for the Tetrahymena intron catalytic activities (20).

NON-ENZYMATIC PROCESSING OF (+) STobRV RNA

Tissue infected with STobRV RNA and TobRV accumulates multimeric forms of STobRV RNA of both the (+) and (-) polarity (21,22) as well as the circular form of the (+) polarity which is the same form as found encapsidated in the viral coat protein (23). Virus particles also have in addition to the (+) STobRV RNA monomer, small amounts of multimeric (+) forms (21). No circular satellite RNA can be found associated with the virus like particles.

The dimeric form of STobRV RNA appears to be a simple direct repeat of the monomer with a normal 3'-5' nucleotide linkage between since reverse transcriptase reads through this region (5,24,25). The isolated dimer of (+)STobRV RNA was shown to autolytically cleave to form monomers (6). This reaction proceeded at room temperature at physiological ionic strengths with only a general polyvalent cation requirement. The monomer was reverse transcribed, cloned in a circular permuted form and sequenced (5,22). The RNA transcript of the clones sequence, has two junction sites (Figure 1).

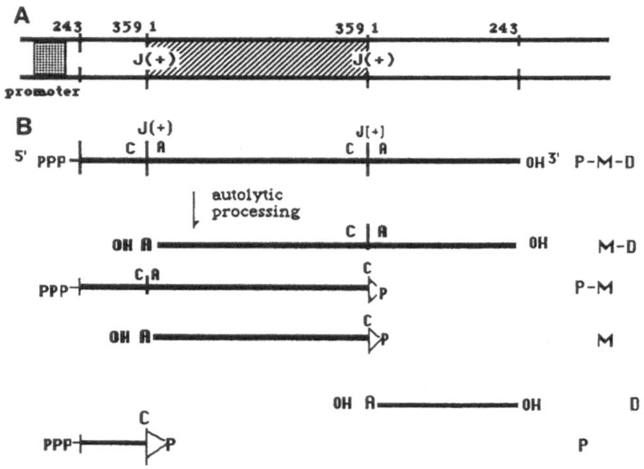

Figure 1. Transcription and self-processing of (+) STobRV RNA (5). (A) Plasmid pSP951 permuted dimeric clone of the cDNA of (+) STobRV RNA in plasmid pSP65 which contains the SP6 promoter. (B) After transcription by SP6 RNA polymerase the primary transcript P-M-D autolytically processes by cleaving at the CpA in the junction site J (+) to yield the various cleavage products shown. P is the fragment proximal to the promoter (5' fragment), M is linear monomer, and D is the fragment distal to the promoter (3' fragment).

Autolytic processing occurred at both of these junctions to generate biologically active monomeric STobRV RNA (5). Apparently, the RNA itself is the only molecule required for the autolytic cleavage at the junction site. The junction site is at a 359CpA1 bond. The autolytic products have a 3' terminus of a 2',3' cyclic phospho-cytidylate and a 5: OH terminus on the adenosine (22). Only 67 bases of the satellite sequence is sufficient to carry out this reaction (26, Buzayan et al. in preparation).

NON ENZYMIC CLEAVAGE AND LIGATION OF (-) STobRV RNA

When the circularly permuted dimeric STobRV cDNA sequence was inserted in the reverse orientation relative to promoter in the plasmid, the transcript was the complement to the (+) STobRV RNA sequence. This complement was designated (-) STobRV RNA. The primary transcript of (-) STobRV RNA (designated P-M-D) was shown to autolytically cleave to generate the M fragment (linear monomer), the P fragment (promoter proximal fragment) and the D fragment (promoter distal fragment) (5,27). The site of cleavage was at 49ApG48 with the numbering corresponding to that of the (+) sequence. Thus the site of cleavage was at a site corresponding to 48/49 bases away from the 359/1 cleavage site for the (+) strand. The termini of the autolytically cleaved RNA were a 2'3' cyclic phosphate on the 3' adenosine and a 5' OH on the 5' guanosine. The reaction occurred at neutral pH, room temperature, and with only a general polyvalent cation requirememt. No protein was required to carry out the reaction.

The autolytic reaction of (-) STobRV RNA was shown to be reversible (27). The newly formed P and D fragments ligated to forn P-D and the linear monomer M autoligated to form cM. The circularization of M is an efficient reaction with 40-50% of the linear form being converted to cM in one hour at 37deg in the presence of Mg^{++} and spermidine. Furthermore, this circularization could occur in the absence of polyvalent cation and in the presence of EDTA. The P-D ligation had a polyvalent cation requirement. Thus the circularization may be less dependent on the stabilization of structure than the ligation of two half molecules.

Presumably, these ligation reactions require an attack of the 5' OH upon a 2',3' cyclic phosphate to form the linear phosphodiester bond. The sequence of the newly ligated RNA in the junction region for both P-D and cM were identical to that of the uncleaved RNA showing ligation proceeds

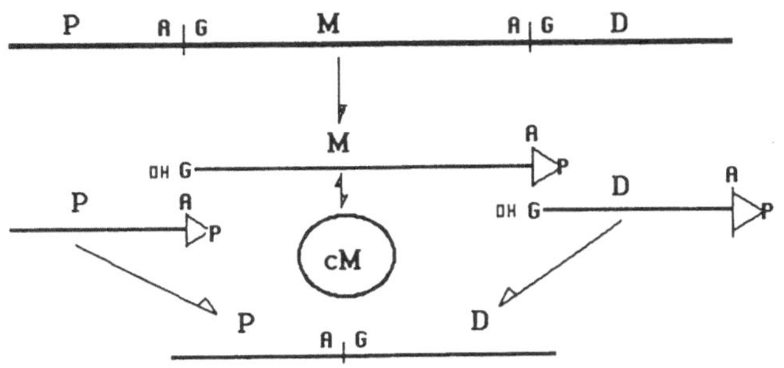

Figure 2. Autolytic cleavage and ligation of (-) STobRV RNA (5,27). Shown is a primary transcript (P-M-D) from a permuted dimeric clone of the cDNA of (-) STobRV RNA autolytically cleaving to P (promoter proximal), D (promoter distal), and M (linear monomer) fragments. The cleavage site is at 49ApG48. The cleavage reaction generates 2',3' cyclic phosphate and a 5' OH. The reaction is reversible, giving ligation products P-D and cM (circular monomer).

without addition or loss of bases. Furthermore, the sequencing reactions were done on DNA which was generated by reverse transcription of the newly ligated RNA - showing that reverse transcriptase could read through the junction region. This is evidence the junction ApG bond does not contain in addition a 2' phosphate (24) or branch as in mRNA splicing (25). The involvement of 5' phosphate was further eliminated by 5' phosphorylation of either D or M which prevented the ligation from occurring (27,28). Figure 2 summarizes these autolytic cleavage and ligation reactions.

THE NATURE OF THE NEWLY FORMED BOND

To analyze the exact nature of the newly formed ApG bond a 34mer oligodeoxynucleotide complementary to the junction region was hybridized to both cM and newly ligated P-D and the DNA/RNA hybrid digested with ribonuclease T1 (28). This yielded a junction fragment with the T1 protected ApG in a 39base long RNA. This RNA was sequenced by standard RNA sequencing methods and shown to be the junction containing RNA with no deletion or addition of bases.

The junction ApG was isolated by complete digestion of the junction containing 39mer with ribonuclease T1 which cleaves RNA at guanylate residues and pancreatic ribonuclease A which cleaves RNA at pyrimidine residues. This produces digestion products termining in 3' phosphate. These were then treated with phosphatase to remove the 3' phosphates and yield the junction ApG. This was chromatographed on thin layer chromatography, autoradiographed and shown to be bonafide 3'-5' ApG. No digestion products corresponding to 2'-5' ApG were seen. The isolated junction ApG was digestable with ribonuclease T2 and ribonuclease U2 as well, showing that it was 3'-5' ApG.

To show that this methodology indeed confirmed the existence of a normal 3'-5' ApG at the junction we did one more control experiment. Since the recovery of ApG from the junction was not quantitative the possibility existed that some junction ApG could have escaped detection if it were not 3'-5' ApG. It might have sufficiently distorted the DNA/RNA hybrid during protection to allow ribonuclease T1 digestion even under high salt conditions; whereupon we would not have isolated it from the junction 39mer RNA. Also the junction phosphodiester bond is known to be especially reactive so a 2' to 3' phosphate shift is conceivable. The confirmatory experiment was to digest the mixture of non-protected M and cM with ribonucleases T1 and A and subsequently with calf intestinal alkaline phosphatase. The digest should yield the dinucleoside phosphate ApG not only from the presumed junction in the ligates RNA but also from each of the remaining ApG sequences in the entire (-) satellite RNA. Thus the junction ApG should represent about 5% of the ApGs from cM, which is in turn about half of the cM and M mixture. Our detection limits in the experiment were within the required 2% and showed no 2'-5' ApG; confirming that the junction ApG is indeed normal 3'-5' ApG (28).

AN RNA SELF-CLEAVAGE MODEL

Recently, a model was proposed for the 2-dimensional structure of the processing site for (+) STobRV RNA, (+,-) forms of avocado sunblotch (ASBV) (29) and the (+,-) forms of the virusoid found associated with lucerne transient streak virus (vLTSV) (30). The latter authors named this the hammerhead structure and hypothesized the sequence of (+) STobRV RNA around the site of cleavage could also fit this model. All these RNAs are small (less than 400 nucleotides) and the cleavage products all terminate in 2',3' cyclic phosphates and 5' OH.

Figure 3 shows the smallest self-processing RNA to date (a 67mer) from (+) STobRV RNA drawn in the hammerhead configuration compared to the (+) model of ASBV as proposed by Forster and Symons (30). Shown are the 17 conserved bases and the similar location of the site of cleavage for the structures, CpA for (+) STobRV RNA and CpU for (+) ASBV. The sequence around the site of cleavage and ligation of (-) STobRV RNA cannot be put in the hammerhead configuration (30) which argues that the active site and RNA mediated cleavage mechanisms are likely different from that of (+-) ASBV, (+-) vLTSV, and (+) STobRV RNA.

(+) STobRV RNA

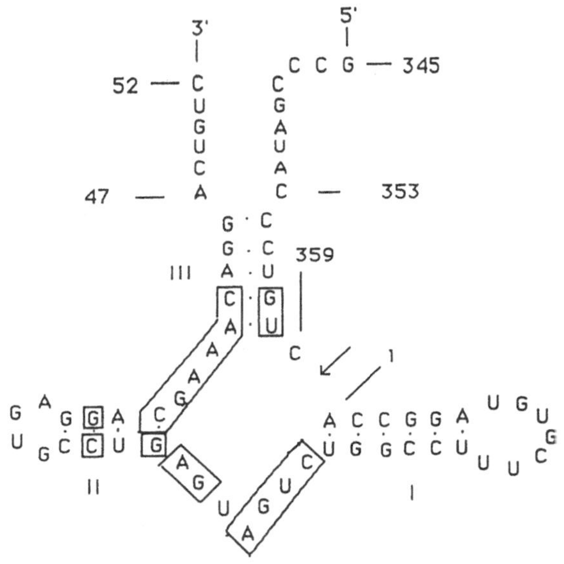

(+) ASBV

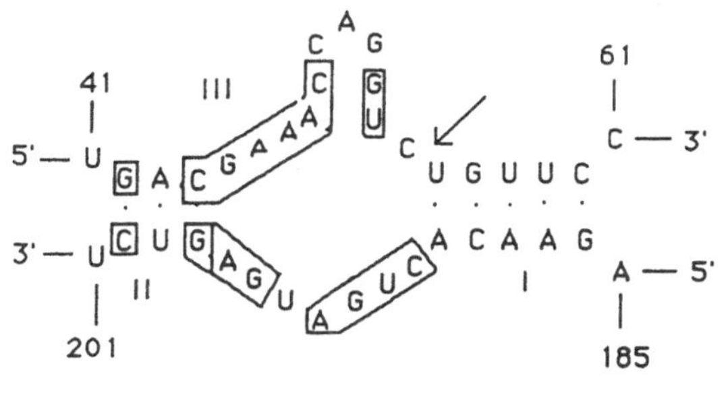

Figure 3. Hammerhead configuration of self-processing RNA. Shown is the self-cleaving 67mer of (+) STobRV RNA drawn in the hammerhead configuration of (+) ASBV as shown by Forster and Symons (30).

The (-) STobRV RNA ligation reaction is also likely different than the reaction of the (+) strand of the satellite RNA. The (+) STobRV RNA does show a small amount of reversibility to form the ligated bond (6) to give about 1% of the RNA in the ligated form. This is in contrast to the extensive reversibility of (-) STobRV RNA which can give as much as 40-50% of the RNA in the ligated form. The very different sequences and structures around the active site likely indicate the influence of nucleotide sequence on RNA mediated activity. For the (-) RNA strand, how a particular phosphodiester bond is activated for autolysis and how the 5' terminal hydroxyl of D and M can so readily attack the 2',3' cyclic phosphodiester group of P and M, are unknown. The particular RNA configuration must favor a trigonal bipyramid conformation about the reactive phosphorous atom (31).

COMPARISON WITH OTHER RNA MEDIATED LIGATION REACTIONS

Other RNA mediated ligation reactions involved a 5' phosphoryl transfer: self-splicing group I precursor RNA (20) and self-splicing group II precursor RNA (17,18). STobRV (-) RNA ligation involves a 2',3' phosphate reaction with a 5' OH. The cleavage reaction generates a 2',3' cyclic phosphate and a 5' OH. Thus the reactions of this molecule are a 3' phosphoryl transfer. This is similar to the simple acid/base catalysis of RNA which also generates 2',3' phosphate groups and a 5' OH, but it differs in three major respect - it is specific, reversible, and occurs at neutral pH.
These results show several characteristics of the non-enzymic cleavage and ligation reaction of STobRV (-) RNA. Neither the cleavage nor ligation reaction requires the full length monomeric 359 base RNA to be on both sides of the J (-) junction. We were able to achieve efficient ligation with a P RNA fragment of 199 bases and a D fragment of 130 bases (27,28). Furthermore, since the existence of 2',3' cyclic phosphate ends on RNA are not unusual whether generated from RNA hydrolysis or by a cyclase which converts 3' phosphates to 2',3' cyclic phosphates (32) it may not be surprising if this type of RNA ligation were to be found in other RNAs. This leads to the exciting possibility that such a mechanism might be used to generate chimeric RNAs which would give the cell a whole new mechanism for generating genetic diversity.

Enzyme catalyzed RNA ligations using a 2',3' cyclic phosphate group have been reported for wheat germ ligase (33). The other reactant has a 5' phosphoryl group, and the product has a junction with a 3',5' phosphodiester bond and a 2' phosphoryl group. This bond is very resistant to hydrolysis and is also found in the circular satellite RNAs of velvet tobacco mottle virus and Solanum nodiflorum mottle virus (24) as well as yeast (34) and plant (35) tRNA splicing reactions. Other enzymic RNA ligation reactions have as reactants or intermediates RNA molecules with 3' OH and 5' phosphoryl groups (bacteriophage T4 ligase) (36) and nuclear mRNA splicing (37). Another example of a ligation using a 2',3' phosphate ligating to a 5' OH is the enzyme catalyzed ligation of the two tRNA half molecules in Hela cells (38).

A MODEL FOR THE REPLICATION OF STobRV RNA

Our observations on the in vitro reactions of STobRV RNAs and the presence in infected tissues of multimeric and circular STobRV RNA suggest a replication model for STobRV RNA that has just six principal steps (39). These steps require ligation, rolling circle transcription and autolytic processing. The first three steps of the replication scheme as proposed for STobRV RNA are: 1. circularization of (+) RNA; 2. transcription of circular (+) RNA to generate multimeric (-) RNA; and 3. autolytic processing of multimeric (-) RNA. Step 4 is circularization of (-) RNA; 5. transcription of (-) RNA to generate multimeric (+) RNA and 6 is processing of (+) RNA to generate linear (+) monomer which is packaged in the viral like particles. The reactions are summarized in Figure 4. Four of these reactions proceed at least to a limited extent by RNA mediated reactions. These are reactions 1, 3, 4, and 6.

It can be hypothesized that the greater tendency of (-) RNA to form circles than (+) RNA may be part of a mechanism for controlling the relative extent of synthesis of the (+) and (-) forms. Only linear (+) RNA forms have been found in viral capsids (23). If the principal function of (-) RNA is to serve as template, the circular form should form easily. If most of the (+) form is to be encapsidated, then the (+) form should occur principally as the linear form.

The model shown in Figure 4 has four non-enzymatic steps based on in vitro reactions. The possibility exists of course that in vivo enzymatic reactions could be involved in these steps. As presented, however, the only enzymatic steps required are the two RNA polymerase reactions.

ACKNOWLEDGEMENTS

Research in the laboratory of George Bruening on STobRV RNA has been supported by USDA grant 84-CRCR-1-1449 and by the Agricultural Experiment Station of the University of California.

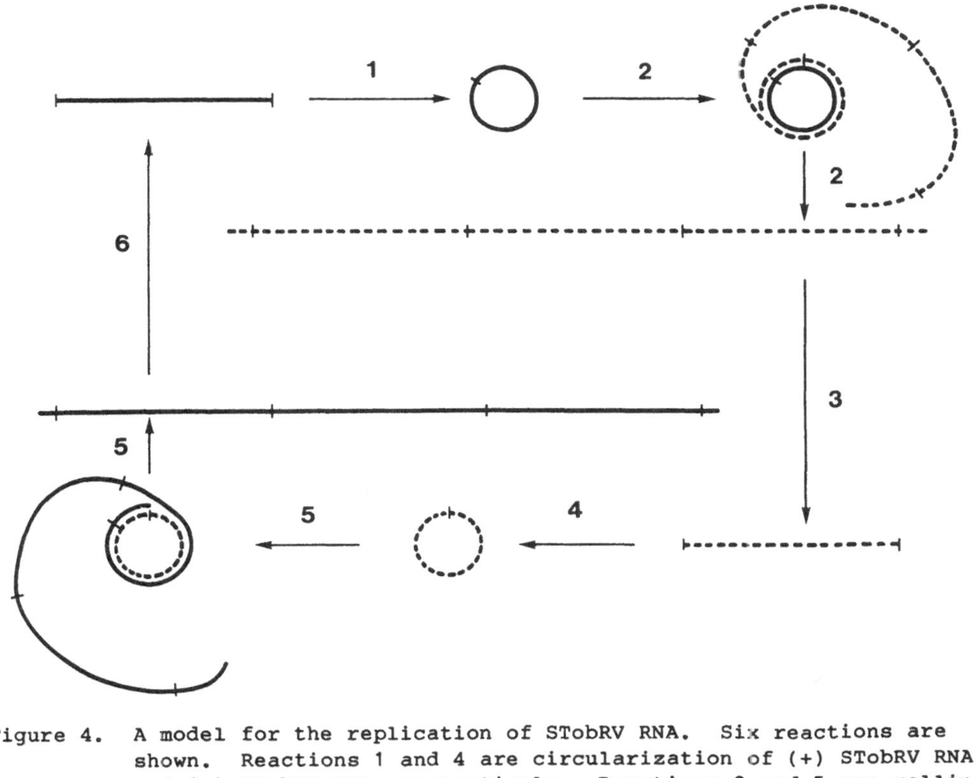

Figure 4. A model for the replication of STobRV RNA. Six reactions are shown. Reactions 1 and 4 are circularization of (+) STobRV RNA and (-) STobRV RNA, respectively. Reactions 2 and 5 are rolling circle transcription steps to generate (-) and (+) transcripts, respectively. The autolytic processing reactions 3 and 6 complete the cycle. RNA of (+) polarity is represented by the solid line and RNA of (-) polarity is represented by the dashed line.

REFERENCES

1. I. R. Schneider, Defective Plant Viruses. In Beltsville Symposia in Agricultural Research. I Virology in Agriculture, Romberger, J.A. ed., Allenheld, Osmun & Co. Montclair, New Jersey 201 (1977).
2. D. Davies and M. Clark, Ann. App. Biol. 103:439 (1983).
3. P. Piazzola and L. Rubino, Phytopath. Z. 111:199 (1984).
4. R. Francki, J. Randles, P. Chu, J. Rohozinski and T. Hatta, in Subviral Pathogens of Plants and Animals: Viroids and Prions, K. Maramorosch and J. McKelvey, eds., Academic Press, New York 265 (1985).
5. W. Gerlach, J. Buzayan, I. Schneider and B. Bruening, Virology 151:172 (1986).
6. G. Prody, J. Bakos, J. Buzayan, I. Schneider and G. Bruening, Science 231:1577 (1986).
7. W. Winterheimer and H. Zachau, Biochim. Biophys. Acta 299:82 (1973).
8. C. Werner, B. Krebs, G. Keith and G. Dirheimer, Biochim. Biophys. Acta 432:161 (1976).
9. B. Rordord and D. Kearns, Biopolymers 15:1491 (1976).
10. J. Ruben and M. Sundaralingam, J. Biomolecular Struc. and Dynamics 1:639 (1983).
11. R. Brown, B. Higerty, J. Dewan and A. Klug, Nature 303:543 (1983).
12. T. Cech, A. Zaug and P. Grabowski, Cell 27:487 (1981).
13. G. Gerriga and A. Lambowitz, Cell 39:631 (1984).
14. G. van der Horst and H. Tabak, Cell 40:759 (1985).
15. F. Chua, G. Maley, G. West, D. Belfort and F. Maley, Cell 45:157 (1986).
16. C. Schmelzer and R. Schweyen, Cell 46:557 (1986).
17. R. van der Veen, A. Arnberg, G. van der Horst, L. Bonen, H. Tabak and L. Grivell, Cell 44:225 (1986).
18. C. Peebles, P. Perlman, K. Mecklenburg, J. Petrillo, J. Tabor, K. Jarrell and H. Cheng, Cell 44:213 (1986).
19. C. Guerrier-Takada, K. Gardiner, T. Marsh, N. Pace and S. Altman, Cell 35:849 (1983).
20. T. Cech and B. Bass, Ann. Rev. Biochem. 55:599 (1986).
21. M. Kiefer, S. Daubert, I. Schneider and G. Bruening, Virology 121:262-273 (1982).
22. J. M. Buzayan, W.L. Gerlach, S. Bruening, P. Keese and A. Gould, Virology 151:186-199 (1986).
23. H. Linthorst and J. Kaper, Virology 137:206 (1984).
24. P. Kiberstis, J. Hazelhoff and D. Zimmern, EMBO J. 817 (1985).
25. B. Rushkin, A. Krainer, T. Maniatis and M. Green, Cell 38:317 (1984).
26. J. M. Buzayan, W. Gerlach and G. Bruening, Proc. Natl. Acad. Sci. USA 83:8859 (1986).
27. J. M. Buzayan, W. Gerlach and G. Bruening, Nature 323:349 (1986).
28. J. M. Buzayan, A. Hampel and G. Bruening, Nuc. Acid Res. 14:9729 (1986).
29. C. Hutchins, P. Rathjen, A. Forster and R. Symons, Nucl. Acids Res. 14:3627 (1986).
30. A. Forster and R. Symons, Cell 49:211 (1987).
31. F. Westheimer, Acc. Chem. Res. 1:70 (1968).
32. A. Reinberg, J. Arenas and J. Hurwitz, J. Biol. Chem. 260:6088 (1985).
33. M. Konarska, W. Filipowicz, H. Domdey and H. Gross, Nature 293:112 (1981).
34. C. Greer, C. Peebles, P. Gegenheimer, J. Abelson, Cell 32:537 (1983).
35. K. Tyc, Y. Kikuchi, M. Konarska, W. Filipowicz and H. Gross, EMBO J. 2:605 (1983).
36. P. Romaniuk and O. Uhlenbeck, Meth. Enzymol. 100:52 (1983).
37. M. Konarska, P. Grabowski, Padgett and P. Sharp, Nature 313:552 (1985).
38. W. Filipowicz and A. Shatkin, Cell 32:547 (1983(.
39. G. Bruening, J. Buzayan, A. Hampel and W. Gerlach, in RNA Genetics, Book I: RNA Replication, J. Holland, E. Domingo and P. Ahlquist, Eds. CRC Press, Inc. Boca Raton, Florida (1987).

EVOLUTION OF THE LEGHEMOGLOBINS

Kjeld A. Marcker and Niels N. Sandal

Department of Molecular Biology and Plant Physiology
University of Aarhus
DK-8000 Aarhus C, Denmark

Introduction

Leghemoglobins (Lbs) are monomeric hemoproteins synthesized in the root nodules which develop through the symbiotic association of Rhizobium species with leguminous plants. In the legumes so far investigated the Lbs are encoded in the plant genome as a small family of genes. Thus in soybean the Lb gene family consists of four functional genes (LbA, LbC1, LbC2 and LbC3), one pseudogene (PS1) and two or three truncated genes. The structure of all Lb genes contains three intervening sequences (IVS1, IVS2 and IVS3) which interrupt the coding sequences in identical positions. The coding sequences of all vertebrate globin genes are interrupted by two intervening sequences. When the amino sequences of globins and Lbs are aligned to maximize structural homology the splicing points of IVS1 and IVS3 in the Lbs coincide precisely with the two splicing points found in globins[1]. This finding supports the notion that all globin genes including the Lb genes are derived from a common ancestral gene.
The Lb genes are arranged in two independent clusters in the soybean genome. One cluster contains four genes in the order 5'Lba-LbC1-LbPS1-LbC3', while the other cluster contains two genes in the order 5'LbPS2-LbC2 3' where LbPS2 is a truncated gene[2].
During nodule development the Lb genes are sequentially activated in the opposite order to which they are arranged on the genome. In contrast to vertebrate globin genes the Lb genes are not regulated by a developmental switching mechanism, but all functional Lb genes remain active through most of the lifetime of the nodule.[3]
When a chimaeric soy bean Lb gene was introduced into the genomes of the legumes Lotus corniculatus and Trifolium repens, nodule specific expression of the chimaeric gene was found in root nodules formed on roots inoculated with the respective microsymbionts Rhizobium loti and Rhizobium leguminosarum[4]. Introduction of a complete Lb gene into Lotus corniculatus results in a nodule specific expression of the gene. The gene is expressed to a level which is comparable to its expression in soybean nodules[5]. These results are taken to imply that

the molecular mechanism responsible for Lb gene activation most likely is conserved through the various forms of legume-Rhizobium associations.

<u>Topology of evolutionary tree</u>

The amino acid sequences from several legume and Parasponia Lbs have been determined. In some cases the amino acid sequences were determined by amino acid sequencing[2-13], while

```
Parasp.      SSSEVNKVFTEEQEALVVKAWAVMKKNSAELGLQFFLKIFEIAPSAKNLFSYLKDSPVPLEQNP
soybean A       VAFTEKQDALVSSSFEAFKANIPQYSVVFYTSILEKAPAAKDLFSFL--ANGVDPTNP
soybean C1      GAFTEKQEALVSSSFEAFKANIPQYSVVFYNSILEKAPAAKDLFSFL--ANGVDPTNP
soybean C2      GAFTEKQEALVSSSFEAFKANIPQYSVVFYTSILEKAPAAKDLFSFL--SNGVDPSNP
soybean C3      GAFTDKQEALVSSSFEAFKTNIPQYSVVFYTSILEKAPVAKDLFSFL--ANGVDPTNP
soybean PS1     GAFTEKQEALVNSSFEAFKANLPHHSVVFFNSILEKAPAAKNMFSFL--GDAVDPKNP
kidney bean     GAFTEKQEALVNSSWEAFKGNIPQYSVVFYTSILEKAPAAKNLFSFL--ANGVDPTNP
pea             G-FTDKQEALVNSSSE-FKQNLPGYSVLFYTIILEKAPAAKGLFSFLKDTAGVEDS-P
broad bean      G-FTEKQEALVNSSSQLFKQNPSNYSVLFYTIILQKAPTAKAMFSFLKDSAGVVDS-P
Lotus corn.     G-FTAQQDALVGSSYEAFKQNLP
alfalfa         G-FTDKQEALVNSSWESFKQN-PGNSVLFYTIILEKAPAAKGMFSFLKDSAGVQDS-P
lupin 1         GVLTDVQVALVKSSFEEFNANIPKNTHRFFTLVLEIAPGAKDLFSFLKGSSEVPQNNP
lupin 2         GALTESQAALVKSSWEEFNANIPKHTHRFFILVLEIAPAAKDLFSFLKGTSEVPQNNP
                  .  . ...        .         .         .* ..  *. .         .

Parasp.         KLKPHATTVFVMTCESAVQLRKAGKVTVKESDLKRIGAIHFKTGVVNEHFEVTRFALL
soy A           KLTGHAEKLFALVRDSAGQLKASGTVVA-DAAL---GSVHAQKAVTDPEFVVVKEALL
soy C1          KLTGHAEKLFALVRDSAGQLKTNGTVVA-DAAL---VSIHAQKAVTDPQFVVVKEALL
soy C2          KLTGHAEKLFGLVRDSAGQLKANGTVVA-DAAL---GSIHAQKAITDPQFVVVKEALL
soy C3          KLTGHAEKLFGLVRDSAGQLKASGTVVI-DAAL---GSIHAQKAITDPQFVVVKEALL
soy PS1         KLAGHAEKLFGLVRDSAVQLQTKGLVVA-DATL---GPIHTQKGVTDLQFAVVKEALL
k. bean         KLTAHAESLFGLVRDSAAQLRANGAVVA-DAAL---GSIHSQKGVSNDQFLVVKEALL
pea             KLQAHAEQVFGLVRDSAAQLRTKGEVVLGNATL---GAIHVQKGVTNPHFVVVKEALL
b. bean         KLQAHAEKVFGMVRDSAVQLRATGEVVL-DGKD---GSIHIQKGVLDPHFVVVKEALL
alfalfa         KLQSHAEKVFGMVRDSAAQLRATGGVVLGDATL---GAIHIQKGVVDPHFAVVKEALL
lupin 1         DLQAHAGKVFKLTYEAAIQLEVNGAVAS-DATLKSLGSVHVSKGVVDAHFPVVKEAIL
lupin 2         ELQAHAGKVFKLVYEAAIQLEVTGVVVS-DATLKNLGSVHVSKGVADAHFPVVKEAIL
                .  *.   .      . ..  . .               *            * .   . .

Parasp.         ETIKEAVPEMWSPEMKNAWGVAYDQLVAAIKFEMKPSST
soy A           KTIKAAVGDKWSDELSRAWEVAYDELAAAIKKA
soy C1          KTIKEAVGGNWSDELSSAWEVAYDELAAAIKKA
soy C2          KTIKEAVGDKWSDELSSAWEVAYDELAAAIKKAF
soy C3          KTIKEAVGDKWSDELSSAWEVAYDELAAAIKKAF
soy PS1         KTIKEAVGDKWSEELSNAWEVAYDEIAAAIKKAMAIGSLV
k. bean         KTLKQAVGDKWTDQLSTALELAYDELAAAIKKAYA
pea             QTIKKASGNNWSEELNTAWEVAYDGLATAIKKAMKTA
b. bean         KTIKEASGDKWSEELSAAWEVAYDGLATAIKAA
alfalfa         KTIKEVSGDKWSEELNTAWEVAYDALATAIKKAMV
lupin 1         KTIKEVVGDKWSEELNTAWTIAYDELAIIIKKEMKDAA
lupin 2         KTIKEVVGAKWSEELNSAWTIAYDELAIVIKKEMDDAA
                    .      .     .   ...     ..
```

Fig.1. Alignment of Lb amino acid sequences. *: amino acid conserved in most of the globin sequences[21]. .: amino acid conserved in all Lbs. At a few positions in the pea and broad bean Lb sequences, ambiguities exist, but only one of the possibilities is shown here. References: Parasponia (Landsmann et al., 1986[13]), soybean A and soybean C1 (Hyldig-Nielsen et al.[14], 1982), soybean C2 and C3 (Wiborg et al., 1982[16]), soybean PS1 (Wiborg et al., 1983[17]), kidney bean (Lehtovaara and Ellfolk, 1975[7]), pea (Lehtovaara et al., 1980[12]), broad bean (Richardson et al., 1974[8]), Lotus corn. (Stougaard et al., 1987[5]), alfalfa (Kiss et al., 1987[20]), lupin 1 (Jegorov et al., 1976[9]), lupin 2 (Jegorov et al., 1978[10]).

in others the sequences were derived from cDNA or genomic clones[14-20]. We have used the available amino acid sequences to align the Lbs (fig.1). When positions with deleted amino acids are neglected in the rest of the aligned Lbs, the degree of homology between pairs of Lb sequences can be calculated. From such values the topology of an evolutionary tree was constructed using the UPGMA method. The lengths of the branches and the standard error were calculated as described by Nei et al., 1985[22]. The evolutionary tree thus obtained is shown in fig.2. Parasponia Lb is clearly separated from the different legume Lbs. Furthermore the lupin Lbs are very different from the rest of the legume Lbs. Kidney bean and soybean Lbs constitute one subgroup, while alfalfa, pea and broad bean constitutes another. It seems that Sesbania Lb is more similar to broad bean, pea and alfalfa Lbs than to soybean and kidney bean Lbs, but the difference is not statistically significant. An argument for the close relationship of Sesbania Lb to broad bean, pea and alfalfa Lbs is the common deletion of the second amino acid.
A simplified representation of the supposed taxonomical relationships of tribes belonging to Papilionideae is shown in fig. 3. (redrawn from Advances in Legume Systematics, 1981[23]). Inspection of the evolutionary trees presented in figs 2 and 3 reveals a close similarity. Thus in both cases lupin is distantly related to the other tribes from which Lb sequences are known. Soybean (Glycine max) and kidney bean (Phaseolus vulgaris) are closely related as they both belong to the Phaseoleae tribe. Similarly Alfalfa (Medicago sativa) belonging to Trifolieae, pea (Pisum sativum) and broad bean

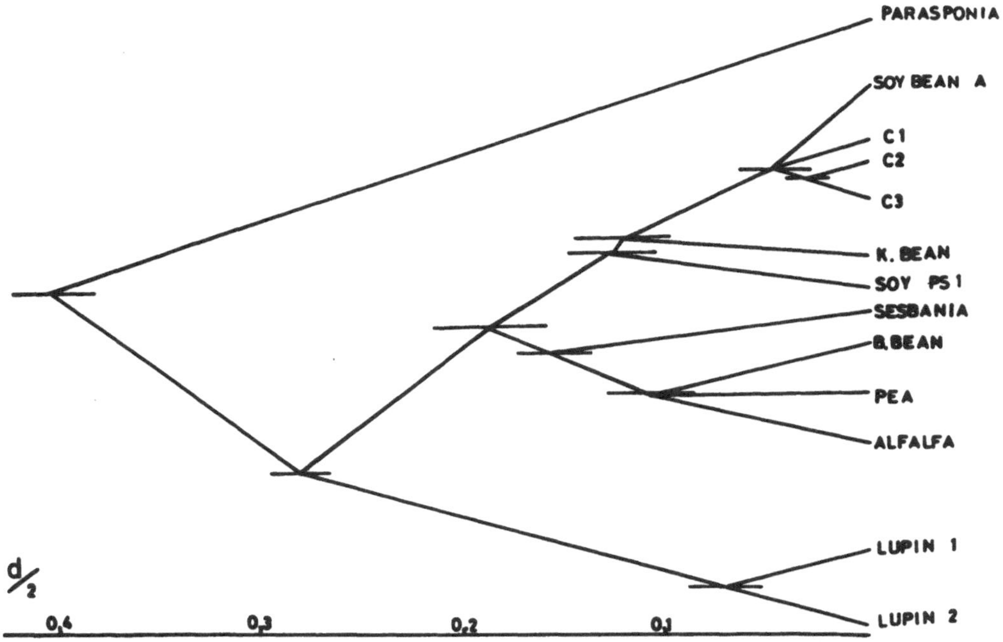

Fig.2. Evolutionary tree obtained from the alignment in fig. 1. The horizontal line at each branching point represents one standard error on each side. The Sesbania amino acid sequence was obtained from C. Appleby (pers. comm.).

(Vicia faba) belonging to Vicieae and Lotus corniculatus belonging to Loteae are related in both cases (trees). Only the N-terminal amino acid sequence of a Lotus Lb is known, but it has two features common to pea, broad bean and alfalfa: deletion of amino acid 2 and Q at position 20. The deletion of amino acid 48-49 in soy bean and kidney bean shows the close relationship of these species. The deletion of amino acid 92-95 seems to have occurred in a common ancestor to Phaseoleae, Vicieae, Trifolieae and Sesbanieae.

Evolution of the soybean Lb genes

The distances between soybean LbA, C1, C2 and C3 are about the same in fig. 2. The closest similarity is found between LbC2 and LbC3. The evolutionary history of the soybean Lb genes has been described by Jensen et al., 1983[24]. It is postulated that gene duplications resulted in a four gene cluster (A-C1-PS1-C3 ancestor). Soybean is tetraploid and it is therefore reasonable to assume that two four gene clusters arose as a result of a genome duplication in an ancestral soybean species. Subsequently a deletion event occurred in one of the clusters resulting in a two gene cluster. This event resulted in the formation of a hybrid LbC2 gene with the composition 5'LbC1-3'LbC3 with a division somewhere in the second intervening sequence. Furthermore the A like gene was truncated to give the LbPS2 gene. The pseudogene PS1 is included here because the distribution of mutations in the gene indicates that it was functional until recently (J. Hein, unpublished). It is not possible to determine if the gene duplication creating the A and PS1 ancestors happened before the ancestors of soybean and kidney bean separated.

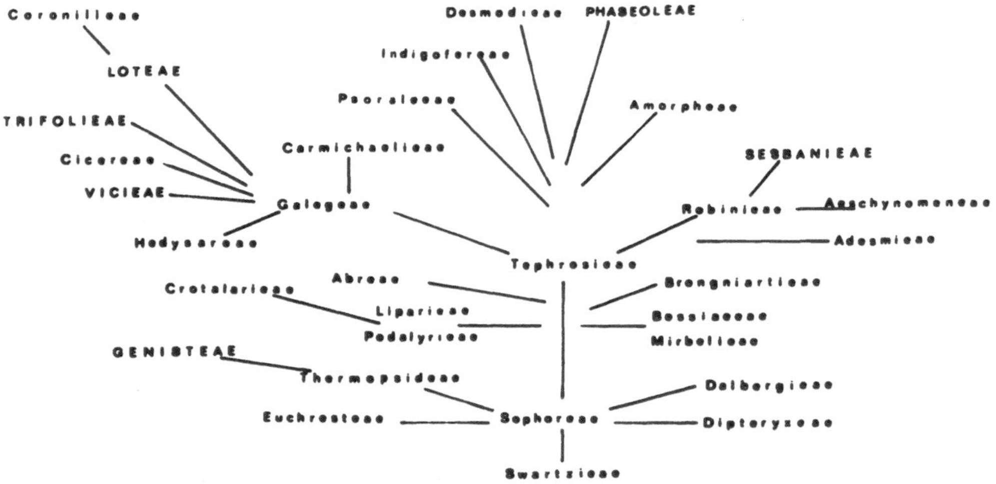

Fig. 3. Supposed taxonimical relationship between the tribes of Papilionideae. Redrawn from Advances in Legume Systematics, 1981[23]. The tribes from which Lb amino acid sequences are known are shown with block letters.

Time scale of Lb evolution

The lower axis in fig.2 corresponds to a time axis. Parasponia belongs to the Ulmaceae and Ulmaceae pollen are known from about 90 mill. years ago, which therefore is the minimum age of their divergence from Leguminoseae. As a consequence the minimum age of the branching point to lupin is about 63 mill. years. The branching point between Phaseoleae and the Trifolieae, Vicieae group would have a minimum age of about 43 mill. years. The evolutionary tree constructed here indicates a considerable age of the Leguminoseae, which is in agreement with the fact that the Papilionideae and Mimosoideae subfamilies were well developed at the Paleocene-Eocene boundary (55 mill. years ago), (Crepet and Taylor, 1985). The Papilionideae subfamily contains the legumes used in this investigation.

As previously mentioned a monophyletic origir of all plant and animal hemoglobins seems probable, because Lb genes are present in nonlegume plants. This raises the possibility that Lb genes or Lb like genes are present in all plants. Horizontal gene transfer is of course still a possibility, but if it occurred it must predate the diversification of angiosperm plants.

Acknowledgements

This investigation was supported by EEC contract BAP-0173-DK, the Danish State Biomolecular Engineering Programme and De Danske Sukkerfabrikker A/S.

References

1 Jensen, E.Ø., Paludan, K., Hyldig-Nielsen, J.J., Jørgensen, P. and Marcker, K.A. (1983), Nature 291, 677-679.

2 Bojsen, K., Abildsten, D., Jensen, E.Ø., Paludan, K. and Marcker, K.A. (1983), EMBO J. 2, 1165-1168.

3 Marcker, A., Lund, M., Jensen, E.Ø. and Marcker, K.A. (1984), EMBO J. 3, 1691-1695.

4 Stougaard, J., Marcker, K.A., Otten, L. and Schell, J. (1986), Nature 321, 667-674. 5 Ellfolk, N. and Sievers, G. (1971), Acta Chem. Scand. 25, 3532-3534.

5 Stougaard, J., Petersen, T.E. and Marcker, K.A, PNAS submitted.

6 Lehtovaara, P. and Ellfolk, N. (1975), Eur. J. Biochem. 54, 577-584.

7 Richardson, M., Dilworth, M.J. and Scarven, M.D. (1975), FEBS Lett. 51, 33-37.

8 Jegorov, C.A., Feigina, M.I., Kazakov, V.K., Shahparonov, M., Mitaleva, S.U. and Ovchinnikov, I.A. (1976), Bioorg. Khim. 2, 125-128.

9 Jegorov, C.A., Kazakov, V.K., Shahparonov, M., Feigina, M.I. and Kostetsky, P.V. (1978), Bioorg. Khim. 4, 476-480.

10 Sievers, G., Huhtula, M.L. and Ellfolk, N. (1978), Acta Chem. Scand. 32, 330-336.

11 Lehtovaara, P., Lappalainen, A. and Ellfolk, N. (1980), Biochim. Biophys. Acta 623, 98-106.

12 Kortt, A.A., Burns, J.E., Trinick, M.J. and Appleby, C.A. (1985), FEBS Lett. 180, 55-60.

14 Hyldig Nielsen, J.J., Jensen, E.Ø., Paludan, K., Wiborg, O., Garrett, R., Jørgensen, P. and Marcker, K.A. (1982) Nucl. Acids Res. 10, 689-701.
15 Wiborg, O., Hyldig-Nielsen, J.J., Jensen, E.Ø., Paludan, K. and Marcker, K.A. (1982), Nucl. Acids Res. 10, 3487-3494.
16 Brisson, N. and Verma, D.P.S (1982), Proc. Natl. Acad. Sci. USA 79, 4055-4059.
17 Wiborg, O., Hyldig-Nielsen, J.J., Jensen, E.Ø., Paludan, K. and Marcker, K.A. (1983), EMBO J. 2, 449-452.
18 Lee, J.S., Verma, D.P.S. (1984), EMBO J. 3, 2745-2752.
19 Landsmann, J., Dennis, E.S., Higgins, T.J.V., Appleby, C.A. Kortt, A.A. and Peacock, W.S. (1986), Nature 324, 166-169.
20 Kiss, G.B., Végh, Z. and Vincze, E. (1987), Nucl. Acids Res. 15, 3620.
21 Hunt, L.T., Hurst-Calderone, S. and Dayhoff, M.O. (1978), Atlas of Protein Sequence and Structure vol. 5, supp. 3, 229-234.
22 Nei M., Stephens, J.C. and Saitou, N. (1985), Mol. Biol. Evol. 2, 66-85.
23 Advances in Legume Systematics, Polhill, R.M. and Raven, P.H. Eds. (1981), Royal Botanic Gardens, Kew.
24 Jensen, E.Ø., Hein, J., Paludan, K. and Marcker, K.A. (1983), Goldberg, R. (ed.), Plant Molecular Biology, Alan R. Liss Inc. NY, 367-379.
25 Muller, J. (1981), Bot. Review 47, 1-144.
26 Crepet, W.,L. and Taylor, T.W. (1985), Science 228, 1087-1089.

TOWARDS NODULIN FUNCTION AND NODULIN GENE REGULATION

Jan-Peter Nap, Albert van Kammen, and Ton Bisseling

Department of Molecular Biology
Agricultural University
De Dreijen 11
6703 BC Wageningen
The Netherlands

INTRODUCTION

Since the introduction of the term 'nodulin' in 1979[1], the concept of nodulins[2] has been firmly established. The occurrence of nodule-specific plant gene products has been demonstrated in many *Rhizobium*-legume symbioses[3-6] and it has become clear that a differential expression of nodulin genes accompanies the development of a root nodule[6-11]. Most of the nodulin genes studied in any detail so far are expressed just prior to or concomitantly with the onset of nitrogen fixation, after the nodule structure has been formed. Consequently, their gene products are most likely involved in establishing an environment within the nodule that allows nitrogen fixation and ammonia assimilation. The leghemoglobin genes, at present the most extensively studied nodulin genes[12-14], can be considered the type member for this class of nodulin genes, a class denoted as late nodulin genes. Studies of these late nodulin genes add to our understanding of the functioning of a root nodule. Recently we have identified another class of nodulin genes, which we term early nodulin genes. These genes encode nodulins that are detectable during the formation of a nodule structure[15-19]. The analysis of this second class of nodulins is the first attempt to study genes and gene products most likely involved in the plant differentiation processes underlying the establishment of a nodule structure.

Detailed information on individual nodulins and nodulin genes is accumulating rapidly. Now, the elucidation of the functions of nodulins and the modes of regulation of nodulin genes is becoming a major issue in understanding the mechanisms of nodule differentiation and functioning. What are the regulatory elements that guide the spatial and temporal expression of the plant developmental program that results in the unique and highly specialized plant organ the root nodule is ? There is no doubt that *Rhizobium* is at least one of these 'regulatory elements'. This involvement of a bacterium in a plant differentiation process raises equally intriguing questions as to how a bacterial genome induces nodule formation and nodulin gene expression, and at the same time manages to escape the plant defense mechanism. The genetic origin, specificity, and succession of bacterial signals in the *Rhizobium*-legume interaction can be experimentally approached with the help of the identified nodulins.

1. *RHIZOBIUM* GENES INVOLVED IN THE INDUCTION OF NODULIN GENE EXPRESSION

Previously we have deduced that a 10 kilobases (kb) *nod* region from the *R. leguminosarum* sym plasmid in a cured *Rhizobium* strain is sufficient to elicit nodulin gene expression in pea[15]. The aim of the present study is to discriminate between the contribution of the *Rhizobium* chromosome on the one hand, and of the plasmid-borne *nod* region on the other, in the induction of nodulin gene expression. Therefore, the cloned *nod* region was introduced in an *Agrobacterium* chromosomal background. An attractive host plant for nodulation studies with such an engineered *Agrobacterium* strain is *Vicia sativa* subsp *nigra* (common vetch)[19], as root nodule formation is much more efficient on *Vicia* than on pea. However, nodulin gene expression had not been studied in *Vicia* nodules before. Thus, nodulin gene expression in wild-type *Vicia sativa* root nodules was studied to allow the subsequent analysis of nodulin gene expression in nodules formed on *Vicia sativa* by the *Agrobacterium* transconjugant harbouring the *Rhizobium nod* region.

Nodulin gene expression in wild-type nodules formed on *Vicia sativa* by *R. leguminosarum* PRE was examined by analyzing RNA on Northern blots using cDNA clones as probes, and by *in vitro* translation followed by two-dimensional (2D) gel electrophoresis[18]. The cDNA clones used in our studies were pENOD2[17] and pPsLb101[20]. Clone pENOD2 was isolated from a soybean nodule cDNA library and represents a soybean early nodulin gene[17]. This pENOD2 strongly cross-hybridizes to a *Vicia* nodulin mRNA, the gene of which is expressed well before the Lb genes[18]. This indicates that in addition to pea[15], also in *Vicia* nodule development an ENOD2-like early nodulin gene operates. In addition, the pea leghemoglobin cDNA clone pPsLb101 was used to follow *Vicia* Lb gene expression. The 2D pattern of *in vitro* translation products of *Vicia* root and nodule RNA revealed, after comparison, several nodulins[18]. Four of the major nodulins have an apparent molecular weight of 14,000. These are most probably the *Vicia* leghemoglobins. Two other major nodulins have molecular weights of 40,000 and 65,000 respectively. These nodulins are referred to as Nvs-40 and Nvs-65. Nvs-40 is an early nodulin; Nvs-65 and the Lbs belong to the class of late nodulins.

Through the use of the nodulins thus identified and classified as early (ENOD2, Nvs-40) and late (Nvs-65, Lb), we can now study nodulin gene expression in the *Vicia* nodules formed by the *Agrobacterium* transconjugant carrying the *Rhizobium nod* region. This study proposes to discover whether the *nod* region of the *Rhizobium* sym plasmid is exclusively involved in causing nodulin gene expression; consequently, the strategy of introducing the *nod* region in an *Agrobacterium* chromosomal background demands that the *Agrobacterium* chromosome itself does not harbour genes involved in the induction of nodulin gene expression. We examined the ability of the *Agrobacterium* genome to induce in tumors the expression of the genes identified as nodulin genes. No *Vicia* nodulins were detectable in tumors formed on the stem of *Vicia* plants after infection with *Agrobacterium*[18]. Thus, it is unlikely that the *Agrobacterium* genome itself codes for signals involved in the induction of nodulin gene expression. The *Agrobacterium* chromosome is, therefore, a suitable chromosomal background for the introduction of the *nod* region from the *Rhizobium* sym plasmid. Nodulin genes expressed in nodules formed by an *Agrobacterium* harbouring the *nod* region should be subject to signals derived from that *nod* region.

1.1 *Rhizobium nod* genes and early nodulin gene expression

Nodulin gene expression was studied in nodules formed by the Ti plasmid-cured *Agrobacterium* LBA4301[21] containing pMP104. Plasmid pMP104 contains a 12 kb *nod* region, carrying several *nod* genes from the *R. leguminosarum* sym plasmid pRL1JI, cloned in a low-copy number *Inc* P vector[22]. In the *Vicia* nodules formed by this *Agrobacterium* transconjugant the two early nodulin genes Nvs-40 and ENOD2 are expressed, but no late nodulin mRNAs are detectable (Nap, in preparation). This result establishes conclusively that the presence of only the *nod* region in an *Agrobacterium* background is sufficient to induce early nodulin gene expression. As the *nod* region is the only *Rhizobium* DNA required for the induction of early nodulin gene expression, this *nod* region itself must be involved in the induction of early nodulin gene expression. In fact, the number of genes involved may be reduced even further. Mutations in the *nodE, F, I* or *J* genes on the total sym plasmid cause delayed root nodule formation[23]. However, all nodulin genes are most probably expressed in these nodules, as they fix nitrogen. Consequently, it is not very likely that any of the *nodE, F, I* or *J* genes encode a signal essential for the induction of early nodulin gene expression. Thus, the other *nod* genes, *nodD, A, B,* and *C*, are probably involved in inducing early nodulin gene expression. Of these genes, *nod D* has been shown to induce the expression of other *nod* genes on the sym plasmid, in the presence of flavones or flavanones excreted by the root[24,25]. The *nod A, B,* and *C* genes are thus the most probable candidates for producing the signal responsible for the induction of early nodulin gene expression. It has been shown that these *nod* genes are absolutely essential for root hair curling[23], the infection process, and the induction of cortical cell division[26]. Because the early nodulin genes we have analyzed are first expressed when the nodule primordia have been formed[17], the induction of their expression is part of a developmental stage after the induction of cortical cell division. So if the *Rhizobium nodA, B,* and *C* genes are also involved in the induction of early nodulin gene expression, it appears that these genes play a role at many levels of the interaction between *Rhizobium* and its host plant (Fig. 1.).

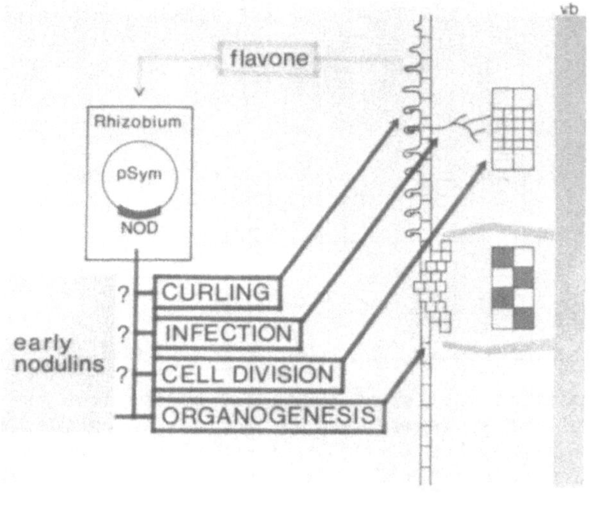

Figure 1. Schematic representation of the relationships between the *nod* genes (NOD) of *Rhizobium*, early nodulins, and steps in nodule development.
vb, vascular bundle; ?, relation unclear.

1.2 *Rhizobium nod* genes and leghemoglobin gene expression

The results described above also show that the 12 kb *nod* region on pMP104 in an *Agrobacterium* chromosomal background is not sufficient for the induction of the expression of late nodulin genes in *Vicia* nodules. Introduction of pMP104 in *R. leguminosarum* 248[27], cured of its sym plasmid (strain 248^c), and analysis of nodulin gene expression in the nodules formed by this 248^c(pMP104) showed that all nodulin genes are expressed in these nodules (Table 1.). The same is true for strain ANU845(pMP104), the sym plasmid-cured *R. trifolii* ANU845[28] harbouring pMP104. Thus, the 12 kb *nod* region in a *Rhizobium* chromosomal background is sufficient for induction of the expression of all nodulin genes examined. Because the engineered *Agrobacterium* is not able to induce the expression of late nodulin genes, whereas an engineered *Rhizobium* is, it seems that besides the *nod* region, chromosomal genes or non-sym plasmid genes are also required for the induction of late nodulin genes. However, an alternative explanation could be a different reaction of the plant defense mechanism towards the *Agrobacterium* transconjugant. It is conceivable that the *Agrobacterium* transconjugant is recognized by the host plant as an "*Agrobacterium*" at a certain stage of development. Then the apparent involvement of *Rhizobium* chromosomal genes might merely reflect the requirement of a distinct surface of the bacterium to prevent an interaction with the plant defense mechanism. The importance of the outer cell surface of *Rhizobium* is illustrated by the finding that mutations resulting in an altered outer surface lead to a disturbed infection process, in which the reaction of the plant resembles a hypersensitive response[29]. Because LBA4301(pMP104) manages to "fool" the plant defense mechanism for quite a while in terms of development, the differences between the outer surfaces of LBA4301(pMP104), 248^c(pMP104) and ANU845(pMP104) may be very minor.

Due to the limitations imposed by the use of engineered *Agrobacterium* strains, we cannot decide whether the *Agrobacterium* transconjugant has the genetic information for inducing late nodulin gene expression. If the defense mechanism counteracts nodulin gene expression before the induction of expression of these genes, it is possible that the *nod* region from the sym plasmid is involved in inducing late nodulin gene expression after all. As an experimental approach to this possibility, we compared nodulin gene expression in *Vicia* nodules formed by ANU845(pMP104) and ANU845(pRt032)(*nodE* K11::Tn5).

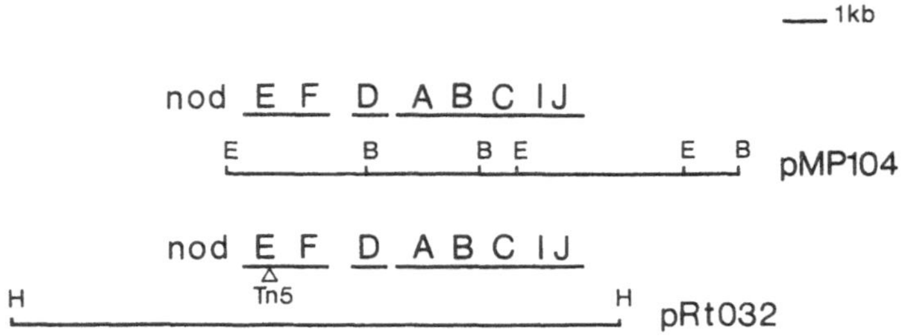

Figure 2. Simplified maps showing the *nod* region from the sym plasmid pRL1JI of *R. leguminosarum*, present in pMP104, and the *nod* region from the *R. trifolii sym* plasmid, present in pRt032, with the Tn5 in *nodE*. The maps are aligned to stress their similarities. Nomenclature of genes according to ref.22. H, *Hin*dIII; E, *Eco*R1, B, *Bam*H1.

Table 1. Pattern of early and late nodulin gene expression in nodules induced on *Vicia sativa* by the strains indicated.

bacterial strain	chromosomal background	EARLY		LATE	
		ENOD2	Nvs-40	Nvs-65	Lb
PRE	*R. leguminosarum*	+	+	+	+
248[c](pMP104)	*R. leguminosarum*	+	+	+	+
ANU843(*nodE* K11::Tn5)	*R. trifolii*	+	+	+	+
ANU845(pMP104)	*R. trifolii*	+	+	+	+
ANU845(pRt032)(*nodE* K11::Tn5)	*R. trifolii*	+	+	+	-
LBA4301(pMP104)	*A. tumefaciens*	+	+	-	-
hypothetical (see text)	?	+	-	-	-

Plasmid pRt032, which has a low copy number in *Rhizobium* (Drs. M.A. Djordjevic and R.W. Innes, personal communication), contains a 14 kb *nod* region from the sym plasmid of *R. trifolii* ANU843[29]. A Tn5 insertion in the *nodE* gene of pRt032 at position K11, yielding plasmid pRt032(*nodE* K11::Tn5), extends the host range of the recipient ANU845 to the pea/vetch cross-inoculation group[30]. In the *Vicia* nodules formed by ANU845(pMP104) all nodulin genes identified sofar are expressed (Table 1.) In nodules formed by ANU845(pRt032)(*nodE* K11::Tn5), the early nodulin genes ENOD2 and Nvs-40 are expressed, as well as the late nodulin gene Nvs-65. However, no Lb gene transcripts are detectable. Apparently there is a marked difference between the two nodule types with respect to Lb gene expression. The two bacterial strains used for inoculation differ only in the cloned *nod* region (Table 1.), implying that in some way the genetic information present on pMP104 is involved in Lb gene expression (see also section 2.1). The involvement of the *nod* genes in a stage of root nodule development associated with the expression of late nodulin genes is supported by the demonstration that the *nodA* and *nodC* genes remain expressed in *R. meliloti* bacteroids[31].

Since the genetic maps of pMP104 and pRt032(*nodE* K11::Tn5) are similar (Fig. 2.), the difference in the expression pattern of nodulin genes must be related to the presence of either plasmid. The major difference seems to be the presence of the Tn5 in *nodE*, indicating that *nodE* might be the cause for the observed difference. *R. trifolii* strain ANU843(*nodE* K11::Tn5)[30] however, in which the *nodE* K11 mutation is introduced in the complete sym plasmid, forms nodules on *Vicia* in which the Lb genes are expressed (Table 1.). No *nodE* (-like) gene is present on the *Rhizobium* chromosome (Dr. M.A. Djordjevic, personal communication), *nodE* cannot be required for the induction of expression of the Lb genes. Other distinguishing characteristics of pMP104 versus pRt032(*nodE* K11::Tn5) should therefore be responsible for the observed difference in Lb gene expression. Whether this reflects a (small) difference in copy number of the two plasmids, or a difference in the presence of essential genetic information -genes on pMP104 that are not present on pRt032(*nodE* K11::Tn5)- is not clear. The importance of the copy number of the *nod* genes has been demonstrated[32]. When a high-copy number plasmid with the *nodABC* genes, transcribed constitutively from a vector promotor, was introduced in a *R. leguminosarum* with a complete sym plasmid, nodulation ability was abolished completely. Slight differences in copy number between pMP104 and pRt032(*nodE* K11::Tn5) may therefore explain the observed differences in the nodulin gene expression pattern.

2. FUNCTIONS OF NODULINS

Turning towards the plant half of the *Rhizobium*-legume symbiosis, we are interested in the functions that nodulins may serve in the developmental program of nodule differentiation. A combined histological/molecular biological study of nodules disturbed in development may provide clues on the involvement of nodulin gene products in developmental processes. Alternatively, analysis of a nodulin cDNA clone at the nucleotide sequence level may reveal information about the functions of that nodulin, by allowing the protein structure of the nodulin to be deduced.

2.1 Nodulin gene expression and the developmental program of the nodule

The pattern of nodulin gene expression in nodules induced by the various strains listed in Table 1. show that we have obtained a collection of strains that induces a series of nodule types in which the induction of nodulin gene expression is more and more disturbed. This collection of strains may be useful in understanding the roles of different nodulin gene products in the developmental program of the nodule. When we combine our analyses of nodulin gene expression with histological studies on the nodule structures that are formed by the different strains, we may find correlations that reveal where in the developmental program of the nodule a particular nodulin has its function.

R. leguminosarum PRE induces on *Vicia* roots the formation of indeterminate[33] nodules, the development of which is similar to the development of pea root nodules (Van de Wiel, unpublished observations). The *Vicia* nodules have an apical meristem next to the early symbiotic zone. In this early symbiotic zone, the meristematic cells are differentiating into infected and uninfected cells, the bacteria are released from the infection threads and become surrounded by a peribacteroid membrane. In the distal part of the nodule the late symbiotic zone is located, in which fully differentiated infected and uninfected cells are present and bacteria have developed into bacteroids.

The strains 248^{c}(pMP104), ANU845(pMP104) and ANU843(*nodE* K11::Tn5) form nodules that have a similar organization as *Vicia* wild-type nodules. Also in all three nodule types, all nodulin genes identified are expressed. This shows that the whole developmental program of the nodule is maintained, although all *nif* and *fix* genes are absent in strains ANU845(pMP104) and 248^{c}(pMP104).

ANU845(pRt1032)(*nodE* K11::Tn5) forms nodules that deviate in the late symbiotic zone. This zone contains two to four layers of infected and uninfected cells, while in the distal part a large area of senescing tissue is present in which hardly any bacteria are left. In these nodules both early nodulin genes are expressed. Of the late nodulins. only the Nvs-65 nodulin gene is transcribed, while the Lb RNAs are not detectable. Apparently the class of late nodulins must be divided in two subclasses that are regulated in a different manner, as is found for the soybean late nodulins[10]. The absence of Lb transcripts in these nodules suggests that the Lb genes are first expressed when the nodule meristem cells are -according to cytological criteria- fully differentiated into infected cells. Immunocytological studies on pea wild-type nodules support this suggestion. In these nodules Lb is not detectable in the early symbiotic zone nor in the first two cell layers of the late symbiotic zone. (Van de Wiel, in preparation).

LBA4301(pMP104) forms nodules in which bacteria are released from the infection threads, and become surrounded by a peribacteroid membrane. However, upon release from the infection thread, despite the presence of a peribacteroid membrane, the bacteria are degraded and they never develop into bacteroid shaped structures. It appears that in the plant cytoplasm of these "infected" cells disintegration of cell components also takes place. This may indicate the activation of a plant defense response as soon as the bacteria are released from the infection threads. The uninfected cells look normal, as judged by electron microscopical observations. In these nodules the early nodulin genes ENOD2 and Nvs-40 are expressed, while none of the late nodulins are detectable. Since the differentiation into uninfected cells seems normal, the *Vicia* late nodulin genes Nvs-65 and Lb are apparently not expressed in uninfected cells. The absence of late nodulin Nvs-65 in the nodules formed by LBA4301(pMP104) suggests that release of the bacteria from the infection thread is not sufficient for the induction of expression of the Nvs-65 gene. Comparison between the nodules formed by ANU845(pRt1032)(*nodE* K11::Tn5) and LBA4301(pMP104) indicates that the Nvs-65 gene is probably first expressed in the youngest cells of the late symbiotic zone.

No bacterial strains are available that induce the formation of root nodules on *Vicia* roots without release of the bacteria from the infection thread. On pea roots the *Agrobacterium* transconjugant LBA2712, containing a complete *R. leguminosarum* sym plasmid instead of a Ti plasmid, forms nodules in which bacteria are not released from the infection threads[15]. Because the *Vicia* Nvs-40 can be immunoprecipitated with an antiserum raised against the pea early nodulin N-40'[16], and because the soybean ENOD2 cDNA clone cross-hybridizes with both a pea and a *Vicia* ENOD2-like early nodulin mRNA, we can study the expression of these two early nodulin genes in pea nodules as substitute for *Vicia*. The analyses showed that only the ENOD2 gene is expressed while the N-40' gene and all pea late nodulin genes are not transcribed. For the sake of illustration, a 'hypothetical strain' with this nodulin gene expression pattern on *Vicia* is added in Table 1. Assuming that the Nvs-40 and N-40' genes are regulated in the same way, the difference in Nvs-40/N-40' gene expression observed between LBA4301(pMP104) on *Vicia* and LBA2712 on pea suggests that the expression of the Nvs-40 gene may be related to the release of the bacteria from the infection threads and/or to the subsequent differentiation into infected and uninfected cells.

The pea nodules formed by the *Agrobacterium* transconjugant LBA2712 contain infection threads, so it cannot be determined whether the expression of the early nodulin ENOD2 is involved in the infection process or in root nodule organogenesis. To discriminate between these two possibilities we analyzed nodulin gene expression in nodule-like structures formed by *R. fredii* USDA257 on soybean roots. These structures lack intracellular bacteria as well as infection threads. Since ENOD2 is expressed in these nodule-like structures, we have concluded that ENOD2 is involved in nodule organogenesis and not in the infection process[17].

PHEKTP	PEYLP	PPHEKPP	PEYL
PPHEKPP	PEYQ	PPHEKPP	
HENPP	PEHQ	PPHEKPP	EHQ
PPHEKPP	PEYE	PPHEKPP	EYQ
PPHEKPP	PEYQ	PPHEKPP	PEYQ
PPHEKPP	PEHQ	PPHEKPP	EHQ
PPHEKPP	PEYQ	PPHEKPP	PEYQ
PPQEK		PPHEKPP	PEYQ
PPHEKPP	PEHQ	PPHEKPP	PVYP
PPYEKPP	PVYE	PPYEKPP	PVVYP
PPHEKPP	IYEP	PPLEKPP	VYNPPPYGRYPPSKKN

Figure 3. Amino acid sequence of the soybean early nodulin N-75 derived from the nucleotide sequence of the cDNA clone pENOD2. The sequence should be read left to right from top to bottom. The spacing is arranged to facilitate recognition of the repetitive structure of this protein.

2.2 The early nodulin ENOD2 is extremely (hydroxy)proline-rich

The soybean early nodulin cDNA clone pENOD2 hybrid-selects RNAs for two nodulins with an molecular weight of 75,000, the soybean N-75 nodulins[17]. DNA sequence analysis of pENOD2 revealed that proline accounts for 45% of the 240 amino acids deduced from the DNA sequence for the N-75 nodulin (Fig. 3). The amino acid sequence is organized in highly repetitive units, with a 17 times repeated heptapeptide sequence, consisting of Pro-Pro-His-Glu-Lys-Pro-Pro as part of a larger, but less conserved, repetitive sequence of 10 or 11 amino acids (Fig. 3). Although it remains to be established whether the prolines of the N-75 nodulins actually become hydroxylated and glycosylated *in vivo*, we assume this to be the case since four classes of hydroxyproline-rich glycoproteins occur in plants and no proline-rich proteins have been described. These four classes are the[34,35] :

(1) cell wall structural hydroxyproline-rich glycoproteins (HRGPs) or extensins,

(2) arabinogalactan proteins (AGPs),

(3) *Solanaceae* lectins,

(4) hydroxyproline-rich agglutinins.

These classes are characterized by their overall amino acid composition. In Table 2., the amino acid composition of a typical representative of each of these classes is given and compared to the (partial) amino acid composition derived for N-75. Also the amino acid composition, deduced from the DNA sequence, of two proline-rich proteins with unknown functions is added in this table. From Table 2., it can be concluded that the amino acid composition of none of the (hydroxy)proline-rich (glyco)proteins resembles the amino acid composition derived for N-75. Also the characteristic repetitive sequence of the extensins, Ser-Hyp-Hyp-Hyp-Hyp, is not found in N-75. The unique features of root nodule development seem therefore to be embodied by the occurrence of an unique class of (hydroxy)proline-rich proteins, a class that is characterized by its low serine, low alanine, and extremely high glutamic acid content. Both the high proline content and the repetitive nature of the protein suggests that the N-75 nodulins are structural proteins. This function derived from the structure of the nodulin is in perfect accordance with the role in nodule organogenesis deduced from analysis of nodules disturbed in development (see above).

Although the repetitive sequences of N-75 and extensin are different, the processes in which extensins are thought to play a role may also occur in root nodule development. Extensins are associated with the cell walls of most dicotelydenous plants and may be important in controlling growth and development[34,35]. They are assumed to be involved in host defense responses and have been shown to accumulate upon wounding[41] and pathogen attack[43]. We investigated extensin gene expression in soybean nodules and ENOD2 gene expression in wounded soybean roots. In preliminary studies using a sunflower extensin cDNA clone (the kind gift of Dr. Terry Thomas) as probe, we did not find a significant increase of extensin-related RNA in nodules compared to uninoculated roots. And, *vice versa*, we could not demonstrate ENOD2 gene expression in wounded soybean roots. These results tentatively indicate that different mechanisms may be involved in ENOD2 gene and stress-related extensin gene expression, suggesting that there may be no stress involved in root nodule formation; an illustration of the ingenuity of *Rhizobium* to bypass the plant defense mechanism[44].

Table 2. Comparison of the amino acid composition of typical representatives of hydroxyproline-rich glycoproteins, and of two proline-rich proteins, with the amino acid composition deduced for N-75. The amino acid composition, expressed as molar percentage, was obtained from either protein analysis or DNA sequencing. All amino acids not mentioned in this table comprise less than 2 molar percent each in all cases. Putative signal peptides are not included in the amino acid composition.

		amino acid													
protein	plant species	Pro + Hyp	Ser	Glu	Gln + Glu	Gln	His	Lys	Tyr	Ala	Val	Gly	Leu	Cys	Ref.
extensin	*Daucus*	45.7	14.2		0.4		11.6	6.7	11.0	0.4	5.9	0.4	0.4	0.0	36
extensin*	*Daucus*	42.2	10.9		2.6		9.5	11.7	12.0	1.4	4.0	0.0	0.0	0.0	37
AGP	*Phaseolus*	29.6	18.2		4.5		0.6	2.6	0.6	16.2	3.2	6.5	3.2	0.6	38
lectin	*Solanum*	21.9	12.6		6.9		0.0	3.7	3.3	4.1	0.4	12.2	1.2	10.6	39
agglutinin	*Solanum*	50.9	9.4		1.1		5.1	15.9	6.2	0.9	3.8	1.1	0.2	0.1	40
P33*	*Daucus*	30.8	2.8	4.7		1.4	12.8	9.5	4.3	4.2	10.0	1.4	1.4	0,0	41
"protein4"*	*Glycine*	40.0	0.0	3.9		0.0	0.0	20.0	16.0	0.0	17.4	0.0	0.0	0.0	42
N-75*	*Glycine*	44.5	0.4	16.6		5.4	9.5	9.5	7.5	0.0	2.1	0.4	1.2	0.4	17

*Amino acid sequence derived from the DNA sequence

3. PROMOTOR OF A PEA LEGHEMOGLOBIN GENE

In nodules formed by *R. trifolii* ANU843(*nodE* K11::Tn5) on *Vicia* roots, a host not belonging to the clover cross-inoculation group, all nodulin genes are expressed, including the Lb genes. Also some naturally occurring, broad host range rhizobia nodulate a variety of legumes as well as, beyond a taxonomical limit, the non-legume *Parasponia* [45]. In the nodules formed on all these different plants (leg)hemoglobin gene expression is induced. Since signals from the invading *Rhizobium* are involved in the induction of nodulin gene expression, similar mechanisms should regulate the expression of nodulin genes in different legumes. Convincing evidence for the apparent conservation of the regulation of Lb gene expression is the nodule specific expression of a chimeric soybean Lb gene in transgenic *Lotus corniculatus* plants[46]. These experiments have shown that a 2 kb region upstream of the start of transcription of the soybean Lbc gene, when present in a chimeric gene, is sufficient for a developmentally correct, nodule-specific expression of that chimeric gene in a different legume host.

To add to our understanding of the regulation of Lb gene expression and to obtain a strong nodule specific promotor that may be of use in future studies, we isolated a pea Lb gene. From a *Sau*3A partial pea genomic library constructed in EMBL3[47], one clone was isolated that hybridized with a pea Lb cDNA clone. The nucleotide sequence of the Lb encoding region indicates the presence of a complete Lb gene. Three intervening sequences interrupt the coding sequence at positions that correspond precisely to the positions of the introns found in all plant (leg)hemoglobin genes sequenced (from soybean[12,13], kidney bean[48], and the non-legume *Parasponia* [49]) so far.

As indicated above, the region of interest from the viewpoint of gene regulation is the 5'-flanking sequence. Analysis of the first 400 bp in front of the initiation codon shows that the presumptive promotor elements CCAAG and TATAAA, characteristic of eucaryotic genes in general, are found at the expected positions in this upstream region of the pea Lb gene. Because also the sequence of the coding regions of this Lb gene is identical to the sequence of a pea Lb-cDNA clone, the isolated gene is likely to be a functional Lb gene.

Extensive sequence homologies have been found between the promotor regions of the different soybean Lb genes[50], and between the promotors of the soybean Lb genes and a kidney bean Lb gene[48]. Also a *circa* 30-base-pair sequence surrounding the cap-site has been identified, that is conserved between soybean, kidney bean, and the animal globin genes[50] However, neither this "globin box" nor the other homologous sequences are found in the pea Lb gene promotor. The same applies to the *Parasponia* Lb gene promotor[49]. This indicates that the homologies found between soybean and kidney bean may reflect more the taxonomic relatedness of *Glycine* and *Phaseolus*, than reflect that this homology is correlated with or essential for Lb gene expression. *Glycine* and *Phaseolus* are both members of the tribe *Phaseoleae* (subfamily *Papilionoideae*) and form the same type of determinate nodules[33]. Although *Lotus* is rather distantly-related to soybean, it forms exactly the same type of desmodioid nodules, a type of nodules that is a subdivision of the determinate nodules thought to have a taxonomical and evolutionary relevance[33]. On the other hand, the regulation of Lb gene expression may be correlated with the observed homologies between soybean and kidney bean Lb, but then it seems probable that the regulatory sequences used for Lb gene expression are specific for the desmodioid type of nodule.

As stated in our introduction, the Lb genes can be considered the type member for the group of late nodulin genes, genes that are concomitantly induced around the onset of nitrogen fixation. It is likely, therefore, that these late nodulin genes share common regulatory elements. Comparison of the 5'-flanking regions of the soybean Lbc gene with the same region of the N-24 and N-23 genes has led to the identification of three consensus sequences that are putative tissue-specific regulatory sequences[51]. These sequences occur at distances up to 600 bp upstream from the cap site, so it is not yet possible to conclusively decide whether these consensus sequences are also present in the pea Lb promotor sequence. Recently a small soybean gene family of late nodulins comprising the genes for N-20, N-22, N-23, and N-44 has been sequenced and their 5'-flanking regions have been compared for a distance of approximately 200 nucleotides in front of the cap site[52]. The consensus sequences identified previously are reported to be lacking. Instead two sequences, of five and six nucleotides respectively, are identified that occur in front of the initiation codon in the 5'-flanking regions of all sequenced late nodulin genes, including all Lb genes. These consensus motifs are also found in the pea Lb promotor sequence, at presumably the appropriate positions. This may indicate that rather small regulatory sequences are involved in the regulation of the nodule-specific expression of the Lb genes.

CONCLUDING REMARKS

Development of a root nodule is an ongoing sequence of interactions involving two genomes. As a first approach to understand the process of nodule development, we attempted to dissect out the bacterial genes that underlie the control of these interactions. We have demonstrated that a fragment from the *R. leguminosarum* sym plasmid of only 12 kb, when present in an *Agrobacterium* chromosomal background, is sufficient for the induction of the expression of early nodulin genes. This 12 kb fragment, that carries the *nod* genes, is also in some way involved in inducing late nodulin gene expression, although in this case *Rhizobium* chromosomal or non-sym plasmid genes may contribute as well. In view of the available data on the *nod* genes present on the 12 kb fragment, the *nodABC* genes are the most likely candidates for providing the signal(s) that induce the expression of early nodulin genes. These *Rhizobium nod* genes are also absolutely required for root hair curling[23], the infection process, and cortical cell divisions[26]. Thus, an amazingly limited number of *Rhizobium* genes appear to elicit a whole range of different responses from the host plant. At present, little is known of the mode of action of the *nod* gene products. *Nod* gene-derived signals may change during the successive steps of nodule development as might be concluded from the different processing of the *nodC* gene product in bacteroids compared to bacteria[31]. Such a strategy would enlarge the genetic potentials of a limited fragment of DNA. Alternatively, the plant's responses may depend on the concentration of either the *nod* gene products themselves, or of factors that result from a cascade of reactions initiated by the *nod* gene products. In this case, analogies might be found with oligosaccharins, plant oligosaccharides that have been shown to influence a plant differentiation program in a profound manner[53].

Apart from the intimate relationship between the *Rhizobium nod* region and nodulin gene expression, we have also demonstrated that the developmental program of root nodule formation can be analyzed by manipulating the bacterial partner. With a set of bacterial strains that induce nodules more and more disturbed in development on the one hand, and the expression of nodulin genes as marker of development on the other, it proved possible to unravel the developmental program of the root nodule in its successive steps. In combination with structural analyses of the nodules formed by the various strains, these studies provided an entry to the elucidation of the functions nodulins may serve in the development of a root nodule. The relative ease of experimental approach by virtue of the advancement of bacterial genetics (compared to plant genetics), contributes greatly to the attractiveness of root nodule development as a system to study plant differentiation.

Although the most specific aspects of differentiation into a nodule will depend on the expression of nodule-specific genes, we have no evidence that the nodulins we have identified are actually essential in the developmental stage they correlate with. Plant transformation with antisense gene constructs[54] to "mutate" (*i.e.* reduce expression of) a specific nodulin gene *in vivo*[55] would seem a promising approach to ascertain the prerequisite of those nodulins to the successive stages of nodule development. We anticipate that the strong nodule-specific promotor of the Lb gene will be of invaluable use in these future studies.

ACKNOWLEDGEMENTS

We are greatly indebted to Herman Spaink and Michael Djordjevic for providing the constructs and strains used in this study. We gratefully acknowledge the support from the collegues of our research group and thank Jeremy Weinman for reading this manuscript. J.P.N. was financially supported by the Netherlands Organization for the Advancement of Pure Research.

REFERENCES

1. R.P. Legocki, and D.P.S. Verma, A nodule-specific plant protein (Nodulin-35) from soybean, *Science* 205:190 (1979).
2. A. van Kammen, Suggested nomenclature for plant genes involved in nodulation and symbiosis, *Plant Mol. Biol. Rep.* 2:43 (1984).
3. R.P. Legocki, and D.P.S. Verma, Identification of "nodule-specific" host proteins (nodulins) involved in the development of *Rhizobium*-legume symbiosis, *Cell* 20:153 (1980).

4. T. Bisseling, C. Been, J. Klugkist, A. van Kammen, and K. Nadler, Nodule-specific host proteins in effective and ineffective root nodules of *Pisum sativum*, *EMBO J.* 2: 961 (1983).
5. N. Lang-Unnasch, and F.M. Ausubel, Nodule-specific polypeptides from effective alfalfa root nodules and from ineffective nodules lacking nitrogenase, *Plant Physiol.* 77:833 (1985).
6. D.P.S. Verma, and N. Brisson, eds., "Molecular Genetics of Plant-Microbe Interactions", Nijhoff, Dordrecht (1987), various reports.
7. F. Füller, and D.P.S. Verma, Appearance and accumulation of nodulin mRNAs and their relationship to the effectiveness of root nodules, *Plant Mol. Biol.* 3:21 (1984).
8. F. Govers, T. Gloudemans, M. Moerman, A. van Kammen, and T. Bisseling, Expression of plant genes during the development of pea root nodules, *EMBO J.* 4:861 (1985).
9. C. Sengupta-Gopalan, J.W. Pitas, D.V. Thompson, and L.M. Hoffman, Expression of host genes during nodule development in soybean, *Mol. Gen. Genet.* 203:410 (1986).
10. T. Gloudemans, S.C. de Vries, H-J. Bussink, N.S.A. Malik, H.J. Franssen, J. Louwerse, and T. Bisseling, Nodulin gene expression during soybean (*Glycine max*) nodule development, *Plant Mol. Biol.* 8:395 (1987).
11. F. Govers, J.P. Nap, M. Moerman, H.J. Franssen, A. van Kammen, and T. Bisseling, cDNA Cloning and developmental expression of pea nodulin genes, *Plant Mol. Biol.* 8:425 (1987).
12. D. Sullivan, N. Brisson, B. Goodchild, D.P.S. Verma, and D.Y. Thomas, Molecular cloning and organization of two leghemoglobin genomic sequences of soybean, *Nature* 289:516 (1981).
13. J. Hyldig-Nielsen, E.Ø.Jensen, K. Paludan, O. Wiborg, R. Garret, P. Jørgensen, and K. Marcker, The primary structure of two leghemoglobin genes of soybean. *Nucl. Acids Res.* 10:689 (1982).
14. J.S. Lee, G.G. Brown, and D.P.S. Verma, Chromosomal arrangement of leghemoglobin genes in soybean. *Nucl. Acids Res.* 11:5541 (1983).
15. F. Govers, M. Moerman, J.A. Downie, P. Hooykaas, H.J. Franssen, J. Louwerse, A. van Kammen, and T. Bisseling, *Rhizobium nod* genes are involved in inducing an early nodulin gene, *Nature* 323:564 (1986).
16. J.P. Nap, M. Moerman, A. van Kammen, F. Govers, T. Gloudemans, H.J. Franssen, and T. Bisseling, Early nodulins in root nodule development, *in:* "Molecular Genetics of Plant-Microbe Interactions", D.P.S. Verma, and N. Brisson, eds., Nijhoff, Dordrecht (1987), p. 96.
17. H.J. Franssen, J.P. Nap, T. Gloudemans, W. Stiekema, H. van Dam, F. Govers, J. Louwerse, A. van Kammen, and T. Bisseling, Characterization of cDNA for nodulin-75 of soybean: a gene product involved in early stages of root nodule development, *Proc. Natl. Acad. Sci. USA* in press.
18. M. Moerman, J.P. Nap, F. Govers, R. Schilperoort, A. van Kammen, and T. Bisseling, *Rhizobium nod* genes are involved in the induction of two early nodulin genes in *Vicia sativa* root nodules. *Plant Mol. Biol.* in press.
19. A.A.N. van Brussel, T. Tak, A. Wetselaar, E. Pees, and C.A. Wijffelman, Small leguminosae as test plants for nodulation of *Rhizobium leguminosarum* and other rhizobia and agrobacteria harbouring a *leguminosarum* sym plasmid, *Plant Sci. Lett.* 27:317 (1982).
20. T.Bisseling, F. Govers, R. Wyndaele, J.P. Nap, J.W. Taanman, and A. van Kammen, Expression of nodulin genes during nodule development of effective and ineffective root nodules, *in* : "Advances in Nitrogen Fixation Research", C. Veeger, and W.E. Newton, eds., Nijhoff/Junk, The Hague, (1984) p. 579.
21. P.J.J. Hooykaas, F.M. Snijdewint, and R. Schilperoort, Identification of the sym plasmid of *Rhizobium leguminosarum* strain 1001 and its transfer to and expression in other rhizobia and *Agrobacterium tumefaciens*, *Plasmid* 8:73 (1982).

22. H.P. Spaink, R.J.H. Okker, C.A. Wijffelman, E. Pees, and B. Lugtenberg, Promotors and operon structure of the nodulation region of the *Rhizobium leguminosarum* symbiosis plasmid pRL1JI, *in* : "Recognition in Microbe-Plant Symbiotic and Pathogenic Interactions", B. Lugtenberg, ed., Springer-Verlag, Berlin (1986), p. 55.
23. J.A. Downie, C.D. Knight, A.W.B. Johnston, and L. Rossen, Identification of genes and gene products involved in the nodulation of peas by *Rhizobium leguminosarum*, *Mol. Gen. Genet.* 198:255 (1985).
24. J.L. Firmin, K.E. Wilson, L. Rossen, and A.W.B. Johnston, Flavonoid activation of nodulation genes in *Rhizobium* reversed by other compounds present in plants, *Nature* 324:90 (1986).
25. C. Wijffelman, B. Zaat, H. Spaink, I. Mulders, T. van Brussel, R. Okker, E. Pees, R. de Maagd, and B. Lugtenberg, Induction of *Rhizobium nod* genes by flavonoids : differential adaptation of promotor, *nodD* gene and inducers for various cross-inoculation groups, *in* : "Recognition in Microbe-Plant Symbiotic and Pathogenic Interactions", B. Lugtenberg, ed., Springer-Verlag, Berlin (1986), p. 123.
26. M.E. Dudley, T.W. Jacobs, and S.R. Long, Microscopic studies of alfalfa root cell divisions induced by *Rhizobium meliloti*, *Planta* in press.
27. D.P. Josey, J.L. Beynon, A.W.B. Johnston, and J. Beringer, Strain identification in *Rhizobium* using intrinsic antibiotic resistance, *J. Appl. Microbiol.* 46:343 (1979).
28. P.R. Schofield, R.W. Ridge, B.G. Rolfe, J. Shine, and J.M. Watson, Host-specific nodulation is encoded on a 14 kb DNA fragment in *Rhizobium trifolii*, *Plant Mol. Biol.* 3:3 (1984).
29. B.G. Rolfe, J.W. Redmond, M. Batley, H. Chen, S.F. Djordjevic, R.W. Ridge, B.J. Bassam, C.L. Sargent, F.B. Dazzo, and M.A. Djordjevic, Intercellular communication and recognition in the *Rhizobium*-legume symbiosis, *in* : "Recognition in Microbe-Plant Symbiotic and Pathogenic Interactions", B. Lugtenberg, ed., Springer-Verlag, Berlin (1986), p. 39.
30. M.A. Djordjevic, P.R. Schofield, and B.G. Rolfe, Tn5 mutagenesis of *Rhizobium trifolii* host-specific nodulation genes result in mutants with altered host-range ability, *Mol. Gen. Genet.* 200:463 (1985).
31. J. Schmidt, M. John, U. Wieneke, H-D. Krüssmann, and J. Schell, Expression of the nodulation gene *nodA* in *Rhizobium meliloti* and localization of the gene product in the cytosol, *Proc. Natl. Acad. Sci. USA* 83:9581 (1986).
32. C.D. Knight, L. Rossen, J.G. Robertson, B. Wells, and J.A. Downie, Nodulation inhibition by *Rhizobium leguminosarum* multicopy *nodABC* genes and analysis of early stages of plant infection, *J. Bact.* 166:552 (1986).
33. J.G. Bergersen, "Root Nodules of Legumes : Structure and Functions", John Wiley, Chichester, 1981.
34. M. McNeil, A.G. Darvill, S.C. Fry, and P. Albersheim, Structure and function of the primary cell walls of plants, *Ann. Rev. Biochem.* 53:625 (1984).
35. J.B. Cooper, J.A. Chen, G-J. van Holst, and J.E. Varner, Hydroxyproline-rich glycoproteins of plant cell walls, *Trends Biochem. Sci* 12:24 (1987).
36. G-J. van Holst, and J.E. Varner, Reinforced polyproline II conformation in a hydroxyproline-rich cell wall glycoprotein from carrot root, *Plant Physiol.* 74:247 (1984).
37. J. Chen, and J.E. Varner, An extracellular matrix protein in plants : characterization of a genomic clone for carrot extensin, *EMBO J.* 9:2145 (1985).
38. G-J. van Holst, F.M. Klis, P.J.M. de Wildt, C.A.M. Hazenberg, J. Buijs, and D. Stegwee, Arabinogalactan protein from a crude cell organelle fraction of *Phaseolus vulgaris L.*, *Plant Physiol.* 68:910 (1981).

39. A.K. Allen, N.N. Desai, A. Neuberger, and J.M. Creeth, Properties of potato lectin and the nature of its glycoprotein linkages, *Biochem. J.* 171:665 (1978).
40. J.E. Leach, M.A. Cantrell, and L. Sequeira, Hydroxyproline-rich bacterial agglutinin from potato; extraction, purification, and characterization, *Plant Physiol.* 70:1353 (1982).
41. J. Chen, and J.E. Varner, Isolation and characterization of cDNA clones for carrot extensin and a proline-rich 33-kDa protein, *Proc. Natl. Acad. Sci. USA* 82:4399 (1985).
42. J.C. Hong, R.T. Nagao, and J.L. Key, Sequence analysis of an auxin-responsive developmentally regulated soybean gene, *in* : "Abstracts First International Congress of Plant Molecular Biology", G.A. Galau, ed., Athens, Georgia, (1985), p. 91, abstract PO-1-117.
43. A.M. Showalter, J.N. Bell, C.L. Cramer, J.A. Bailey, J.E. Varner, and C.J. Lamb, Accumulation of hydroxyproline-rich glycoprotein mRNAs in response to fungal elicitor and infection, *Proc. Natl. Acad. Sci. USA* 82:6551 (1985).
44. M.A. Djordjevic, D.W. Gabriel, and B.G. Rolfe, *Rhizobium* : the refined parasite of legumes, *Ann. Rev. Phytopathol.* in press.
45. M.J. Trinick, and J. Galbraith, The *Rhizobium* requirements of the non-legume *Parasponia* in relation to the cross-inoculation concept of legumes, *New Phytol.* 86:17 (1980).
46. J.S. Jensen, K.A. Marcker, L. Otten, and J. Schell, Nodule-specific expression of a chimaeric soybean leghaemoglobin gene in transgenic *Lotus corniculatus*, *Nature* 321:669 (1986).
47. A.M. Frischauf, H. Lehrach, A. Poustka, and N. Murray, Lambda replacement vectors carrying polylinker sequences, *J. Mol. Biol.* 170:827 (1983).
48. J.S. Lee, and D.P.S. Verma, Structure and chromosomal arrangement of leghemoglobin genes in kidney bean suggest divergence in soybean leghemoglobin gene loci following tetraploidization. *EMBO J.* 3:2745 (1984).
49. J. Landsmann, E.S. Dennis, T.J.V. Higgins, C.A. Appleby, A.A. Kortt, and W.J. Peacock, Common evolutionary origin of legume and non-legume plant haemoglobins, *Nature* 324:166 (1986).
50. G.G. Brown, J.S. Lee, N. Brisson, and D.P.S. Verma, The evolution of a plant globin gene family, *J. Mol. Evol.* 21:19 (1984).
51. V.P. Mauro, T. Nguyen, P. Katinakis, and D.P.S. Verma, Primary structure of the soybean nodulin-23 gene and potential regulatory elements in the 5'-flanking region of nodulin and leghemoglobin genes, *Nucl. Acids Res.* 13:239 (1985).
52. N.N. Sandal, K. Bojsen, and K.A. Marcker, A small family of nodule specific genes from soybean, *Nucl. Acids Res.* 15:1507 (1987).
53. K. Tran Tranh Van, P. Toubart, A. Cousson, A.G. Darvill, D.J. Gollin, P. Chelf, and P. Albersheim, Manipulation of the morphogenetic pathways of tobacco explants by oligosaccharins, *Nature* 314:615 (1985).
54. J.R. Ecker, and R.W. Davies, Inhibition of gene expression in plant cells by expression of antisense RNA, *Proc. Natl. Acad. Sci. USA* 83:5372 (1986).
55. D.P.S. Verma, A.J. Delauney, and T. Nguyen, A strategy towards antisense regulation of plant gene expression, *in* : "Tailoring Genes for Crop Improvement, an agricultural perspective", G. Bruening, J. Harada, T. Kosuge, and A. Hollaender, eds., Plenum Press, New York (1987), p. 155.

ANALYSIS OF THE MECHANISM OF nod GENE REGULATION IN Rhizobium leguminosarum

G.-F. Hong*, J.L. Burn and A.W.B. Johnston

AFRC Institute of Plant Science Research
John Innes Institute
Colney Lane
Norwich NR4 7UH, U.K.

INTRODUCTION

Soil bacteria in the genera Rhizobium and Bradyrhizobium have the particular and important ability of recognizing and infecting specific legumes [different rhizobial strains infect different legume species] and inducing the formation of root nodules in which the bacteria fix N_2 and the resultant ammonia is made available to the host allowing many crop legumes to be grown without the need of nitrogenous fertilizer. Genes required for the early stages of infection and for the determination of the particular host-range specificity of a strain of Rhizobium are on a relatively small region of DNA, which, in the fast-growing Rhizobium species examined, is on a large, "symbiotic" [sym] plasmid [reviewed by Long,1984]. Although the fine-scale structure of these nodulation [nod] genes has been determined [see Rossen et al.,1987], the precise biochemical role of individual nod genes in the infection process remains to be established. There has been considerable progress in the understanding of the regulation of expression of these nod genes and this topic is the subject of this paper.

In R.leguminosarum biovar [bv.] viciae, which nodulates peas, vetches and lentils, eight contiguous genes, in three transcriptional units [nodFE, nodD and nodABCIJ] were identified [Rossen et al.,1985; Downie et al.,1985; Evans and Downie,1986; Shearman et al.,1986; unpublished observations]. Also, two other open reading frames, corresponding to genes termed nodL and M, located downstream of nodE were identified by sequence analysis [Downie et al.,1986].

By constructing fusions between different nod genes and lacZ of Escherichia coli in a wide host-range translational fusion vector and measuring β-galactosidase activities in strains of R.l. bv. viciae differing in their genetic background and grown in the presence or absence of extracts from the roots of peas, it was found that nodD is transcribed at relatively high levels whether root exudate was present or not and that transcription of nodD is auto-regulated; ie the higher the copy number of nodD, the lower the expression of the nodD-lacZ fusion [Rossen et al.,

*Present address: Institute of Biochemistry, Shanghai, People's Republic of China.

1985]. In contrast, in cells grown in normal growth media, nodFE and nodABCIJ were transcribed at low levels, if at all, but in strains exposed to pea root exudate, nodFE and nodABCIJ were actively transcribed, this activation requiring the presence of nodD. Thus nodD has at least two regulatory functions in this biovar [Rossen et al., 1985; Shearman et al., 1986]. Similar observations that nodD is transcribed constitutively but that transcription of other nod genes requires both nodD plus factor[s] from legume root exudate have been described in analogous studies on nod gene regulation in R.l. bv. trifolii and R. meliloti which respectively nodulate clover and alfalfa [Innes et al., 1985; Mulligan and Long, 1985]. In R. meliloti however, there was no evidence for autoregulation of nodD [Mulligan and Long, 1985], in contrast to the case in R.l. bv. viciae. This difference may be due to the fact that, unlike R.l. bv. viciae, strains of R. meliloti contain multiple copies of nodD on the sym plasmid [Gottfert et al., 1986]. Also, nodD mutations in R.l. bv. viciae completely abolish nodulation of peas [Downie et al., 1985], but mutations in the copy of the nodD gene of R. meliloti which lies upstream of nodA have only a slight effect on nodulation ability; again, this difference can be attributed to the duplication of nodD in the latter species.

Upstream of the transcriptional units that require nodD plus factors in legume root exudate for their expression, is a conserved sequence spanning approximately 35 bp's [the nod box], which is likely to be involved in nodD-mediated activation [Rossen et al., 1985; Rostas et al., 1986; Schofield and Watson, 1986; Shearman et al., 1986]. This sequence also exists upstream of other nod genes [nodG in R. meliloti and nodM in R.l. bv. viciae] and it seems likely that these two genes also require nodD plus root exudate for their transcription though this has not formally been demonstrated [Downie et al., 1986; Horvath et al., 1986; Schofield and Watson, 1986]. In initial reports, it was found that the nodD gene of one species of Rhizobium was funtionally equivalent to that of a strain with a different host-range; for example, Fisher et al., [1985] showed that the Nod⁻ defect of a nodD mutant strain of R.l. bv. viciae was corrected by the cloned nodD of R. meliloti. Further, extracts from the roots of R. meliloti can activate the nod genes of R.l. bv. viciae, a species which does not nodulate this host. However, it is now emerging that it will not be possible to generalize about whether nodD is a "common" nod gene or not; recently, Horvath et al., [1987] found that the ability of a wide-host range strain of Rhizobium to nodulate the tropical legume Macroptillium [siratro] was due to its possession of a version of nodD which responded to (so far unidentified) inducer molecule(s) in the roots of siratro but which is absent from root exudate of alfalfa. Transfer of this nodD gene to R. meliloti conferred on the recipient the ability to nodulate siratro; ie., in this case, the differences in host-range specificity could be attributed to the nature of the nodD genes in the two strains.

The inducer molecules in the root exudates of legumes which are responsible for nodD-dependent transcription of other nod genes have been identified as being particular flavone or flavanone molecules [Firmin et al., 1986; Peters et al., 1986; Redmond et al., 1986]. A representative of a different class of aromatic compound, the phytohormone trigonelline has also been found to activate the nodA gene of R. meliloti [Schmidt et al., 1986]. As well as exuding molecules that induce nod gene transcription, plants make compounds that antagonize this induction. Firmin et al., [1986] noted that in strains of R.l. bv. viciae exposed to certain flavonols, iso-flavonoids or aceto-phenone analogues which shared some structural features with the inducer flavone and flavanones, the ability of pea root exudate or defined flavone or flavanone inducers to activate transcription of the nodC-lacZ or nodF-lacZ fusions of this biovar was inhibited by as much as 90%. Thus there may be competition between the inducers and the "anti-inducers" which could occur at the level of uptake

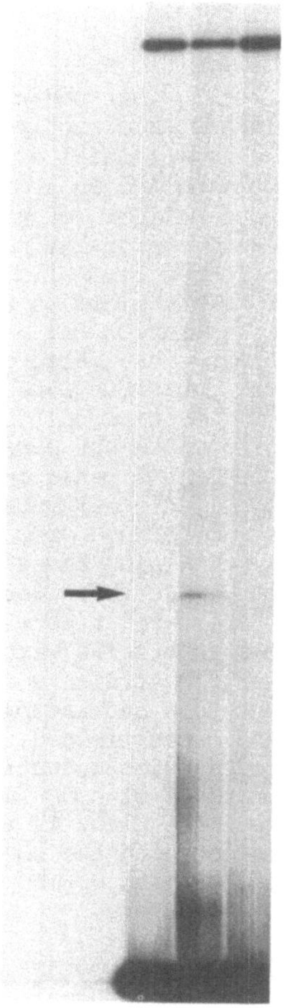

Fig. 1. Demonstration of complex formation between the *nodD-nodA* intergenic region and extracts from *Rhizobium* strains containing *nodD*. 2 µl *BamHI-* and *EcoRI*-cut DNA that spans the *nodD-nodA* intergenic region was end-labelled with ^{32}PdCTP and was preincubated for 15 mins at 20°C with 2 ul salmon sperm DNA (1 mg/ml^{-1}) plus cell extract obtained from the cured *R. leguminosarum* strain 8401 which lacks a *sym* plasmid, or from a derivative of 8401 with pIJ1518 which contains the cloned *nodD* gene of pRL1JI. Cell extracts were prepared from cells grown in 100 ml cultures. These were washed in 150 mM NaCl and then twice with buffer which contains 25 mM TES, pH 7.5, 5 mM EDTA, 5 mM mercaptoethanol, 10% glycerol, 150 mM NaCl). Cells were broken by sonication. Aliquots were loaded onto 8% non-denaturing polyacrylamide gel and following electrophoresis, the gel was examined by autoradiography. Note that the extra slowly migrating band, indicated by the arrow, was only seen when extract came from strain 8401pIJ1518 (central track). No extra band occurred when extract was from the cured strain 8401 (right-hand track) nor in the control in which the labelled DNA was not exposed to any extract (left track). In this Figure, cells had been grown in the absence of inducer; no further differences were seen when the cells were grown in the presence of the inducer naringenin (1 µM), the inducer also being present in the reaction mix and in the gel itself.

of these molecules or at the interaction of the flavonoids with the positively acting regulatory protein [presumed, but not yet proved to be, the product of nodD].

It has not been shown directly that the product of nodD [NodD] binds to the nod box nor that it interacts directly with the flavonoid inducer molecules. However, strong circumstantial evidence that both propositions are true comes from work involving the analysis of mutant forms of nodD of R.l. bv. viciae. In these studies [Burn et al., 1987], the cloned nodD of this biovar was mutagenized with hydroxylamine and the mutagenized plasmids were examined for any that differed from wild type nodD in their regulatory properties. Four Classes of mutant plasmids were isolated. Those of Class I were defective both in autoregulation and in induction of other nod genes in the presence of inducer flavonoids, those of Class II activated transcription of the other nod genes but were defective in autoregulation and those of Class III were the "reciprocal" of those of Class II; they retained the ability to auto-regulate but were unable to activate, to high level, transcription of the other nod genes in the presence of flavonoid inducers. Finally, two [probably clonal] mutant derivatives of nodD were identified which activated the other nod genes even in the absence of inducer flavonoids. Strains containing these Class IV mutants still responded to inducer molecules; when these were added, there was a "hyper-induction" of the other nod genes, their levels of expression being higher, by a factor of two, than those in strains with wild-type nodD grown in the presence of inducers. Further, in strains with Class IV mutant alleles of nodD, the "anti-inducers" genestein and kaempferol did not inhibit the transcription of the other nod genes; indeed, these molecules, which in strains with wild-type nodD antagonize induction, acted as inducers in strains containing Class IV mutants of nodD [Burn et al., 1987]. The sites of the mutations that conferred the Class IV mutant phenotype were located in the 3' end of nodD, consistent with the observation [Horvath et al., 1987] that the carboxy-terminal region of NodD may be responsible for the interaction with the flavonoid inducers.

Each of these four Classes of nodD mutant plasmids were mobilized into a Nod- strain of R.l. bv. viciae containing a nodD::Tn5 mutation in the sym plasmid pRL1JI. In all cases, the transconjugants formed nodules on peas; it was perhaps surprising, though, that the derivatives of the nodD::Tn5 mutant strain containing the Class I and the Class III plasmids could nodulate, since these mutants only conferred a slight induction of transcription of nodFE and nodABCIJ above background [3-fold, instead of the approximately 100-fold obtained in strains with the wild-type copy of nodD]. Further, derivatives containing either of the two Class IV mutant plasmids induced only about 20% the number of nodules compared to normal and those that were formed failed to fix nitrogen showing that continued, high level expression of nod genes during the infection process is detrimental to the development of normal nodules.

In this paper, we describe further experiments designed to test if the nodD gene product binds to the DNA that has been implicated in nodD-mediated nod gene expression. We also examined the effects of exposure of the strains to inducer molecules on any such binding and determined the influence of various types of nodD mutations on any DNA/protein complex formation.

RESULTS

In R.l. bv. viciae, as in other strains of Rhizobium, nodD is transcribed divergently from nodABCIJ and, in the intergenic space between nodA and nodD is a region which includes the nod box required for nodD-determined activation of transcription of nodABCIJ [Rossen et.al., 1985; Scott, 1986; Egelhoff et al., 1985; Shearman et al., 1986; Spaink et al., 1986; Rostas et al., 1986; Schofield & Watson, 1986]. This inter-genic region also contains the sequences which, in R.l. bv. viciae are involved in the repression of nodD by its own gene product [Rossen et al., 1985; Shearman et al., 1986]. One of the derivatives [termed IJ487] of bacteriophage mp8 which had been used for sequencing this region of the sym plasmid pRL1JI [Rossen et al., 1985; Shearman et al., 1986] contains this intergenic region. To see if there was any specific binding of proteins to this cloned DNA, the following was done.

The insert DNA was removed from the replicative form of mp8 clone IJ487 by digesting the double-stranded replicative form DNA with BamHI plus EcoRI [cleavage sites for these enzymes are in the linker in mp8, to either side of the insert, and do not cut in the insert] and the insert was end-labelled with ^{32}PdCTP by filling in the overhangs of the BamHI-cut terminii. The labelled fragment was then mixed with extracts obtained from sonicated Rhizobium cells [see legend to Fig. 1], these extracts being obtained from strains that differed with regard to the status of the nodD gene that they contained and which had been grown in the presence or absence of a nod gene inducer [naringenin, 1uM]. Aliquots of the reaction mixes were subjected to electrophoresis in non-denaturing, 8% polyacrylamide gels; if the fragment bound to any protein[s] in the extract, this would be detected [following autoradiography] by a retardation in the migration of the labelled fragment due to its complexing with polypeptide [Freid and Crothers, 1981; Garner and Revzin, 1981]. As shown in Fig.1, when the extract was obtained from strain 8401, which lacks any sym plasmid, or from strain 8401 containing the cloning vector pKT230, a single band was seen which had the same mobility as that of the control in which the labelled DNA had not been exposed to any cell-free extract. In contrast, with extracts obtained from strains 8401 pRL1JI [a native sym plasmid of R.l. bv. viciae] or 8401 pIJ1518 [which contains the nodD of pRL1JI cloned in pKT230], an additional, more slowly migrating band was observed [Fig.1]. This extra band had the same intensity and the same mobility whether the inducer flavanone naringenin was present or not [note that in the "plus inducer" treatments, the inducer was present in the growth media of the cells, in the reaction mix and was also incorporated into the gel]. It seemed likely that the slowly migrating band comprised a complex between the DNA that spanned the nodD-nodA intergenic region and protein[s] that are specifically present in strains containing a functional nodD gene. Evidence that this is the case came from the observation that when protease was added to the reaction mix, no new, slowly migrating band was obtained.

Extracts from representatives [two of each] of Rhizobium strains containing each of the four Classes of hydroxylamine-induced nodD mutants [see above] were also prepared and the labelled fragment was added to each of these extracts. Significantly, with the extracts obtained from the two Classes of nodD mutants which failed to autoregulate, the slower migrating band was either absent [Class II mutants] or was reduced in its intensity relative to that obtained with strains with wild-type nodD [Class I mutants]. With extracts from the Class III mutants [which autoregulate but do not activate], the mobility and intensity of the signal of the "complex" band was the same as was seen with extracts from strains with wild-type nodD. Interestingly, when extracts from the two "constitutive" Class IV mutants were used, a "complex" band was obtained with a similar intensity but with a significantly slower mobility than was found for the wild type control.

Upstream of nodF in pRL1JI is another copy of the nod box which is implicated in the activation of the nodFE genes by nodD in the presence of the inducer flavonoids [Shearman et al., 1986]. A fragment that spanned this region upstream of nodF was obtained from a second mp8 derivative [termed clone IJ1549] and was labelled [in a way similar to that described for the labelling of the clone IJ487 {see above}] and was exposed to extracts obtained from the Rhizobium strains described above. No new "complex" band was found whether the strains had been grown in the presence or absence of inducer or whether the strain contained a functional nodD or not.

DISCUSSION

In the presence of flavonoid inducer molecules, nodD is required for the activation of other nod genes and the transcripts that are so activated are preceded by [or include] a conserved region of approximately 35 bps, the nod box. Also, in R.l. bv. viciae, nodD is autoregulatory, repressing its own transcription. The simplest, but [as yet] unproven explanation for these observations is that in the absence of inducer molecules, the nodD gene product [NodD] binds to a region of the nod box and prevents [at least in R.l. bv. viciae] transcription of nodD. In the presence of inducer flavonoids, NodD might undergo an allosteric shift due to an interaction, either direct or indirect, with inducers which opens up the promoters of the genes whose transcription it directs. Even when inducers are present, nodD retains its autoregulatory properties suggesting that when its gene product is in an altered state due to its proposed interaction with inducers, it retains the ability to block transcription of itself. Although it has not formally been proved that NodD binds to inducers, strong circumstantial evidence for this has been provided by the finding that a single mutation in nodD [in the Class IV mutant strains] alters the effects both of inducers and anti-inducers on nodD-dependant nod gene regulation [Burn et al., 1987]. Such a general scheme for regulation in prokaryotes has been described for other regulons. For example, in E. coli, the regulatory gene araC is autoregulatory both in the presence and the absence of arabinose and, in cells exposed to the sugar, is responsible for the activation of other "structural" genes required for the catabolism of arabinose. The araC gene product binds to specific regions of DNA in the intergenic region between araC and araB [Lee et al., 1981] araC being located close to and transcribed divergently from the catabolic araBAD genes, an organization similar to that of nodD relative to that of nodABCIJ.

The results presented here show that the intergenic region between nodD and nodA specifically binds a protein [or proteins] if the cell-free extract to which this DNA is exposed is obtained from a Rhizobium strain that contains a functional nodD gene. In contrast, the region of DNA preceding nodF, which also contains a nod box, does not bind to protein under the conditions used in this study. Significantly, the DNA between nodD and nodA must also contain sequences that are involved in the repression of nodD transcription, whereas the DNA upstream of nodF does not [Shearman et al., 1986]. This indicates, therefore, that the DNA/protein complex that was observed here specifically reflects the binding of a protein to form a structure that is responsible for the repression of transcription of nodD. This conclusion is substantiated by the finding that extracts from the strains containing the Class I and Class II mutant forms of nodD, both of which failed to autoregulate, formed no or reduced amounts of the complex between protein and the DNA in the nodD-nodA intergenic region. In contrast, addition of extracts from strains with the Class III nodD mutants, which autoregulate but which fail to activate transcription of other nod genes, generated a DNA/protein complex that was indistinguishable from that found using extracts from wild-type strains.

The fact that the extracts from strains with the Class IV nodD mutants caused the fragment containing the nodD-nodA intergenic region to migrate more slowly than was the case with wild type extracts is significant since it provides strong [though not definitive] evidence that NodD binds directly to this piece of DNA. The reason for this additional retardation in migration is not known but some possible explanations are as follows. Perhaps additional molecules of the mutant NodD bind to the DNA or else the mutation causes the effective volume of NodD to increase and thus slow the migration of the complex. Thirdly, it may be relevant that the mutations in the two Class IV mutants of nodD lead to the generation of an asparagine residue in the mutants compared to the aspartate that is present in the corresponding position in the wild type [Burn et al., 1987]. This would cause the NodD to be more basic and this difference in charge might cause a decrease in the electrophoretic mobility under the conditions used here. However, unless a very large number of molecules of NodD were bound to this DNA, it is unlikely that this single charge difference would be sufficient to account for the extent of the change in the mobility of the complex which was observed.

In conclusion, the results presented here provide strong evidence that the product of the regulatory gene nodD does bind to the DNA between nodD and nodA and that this binding is responsible for the autoregulation of transcription of nodD. However, these results did not demonstrate that nodD-dependant activation of the other nod genes in the presence of flavonoid inducers involves binding of NodD to the nod box. However,this possibility cannot be ruled out since it is possible that such binding is less stable or requires particular conditions which we have been unable to mimic in the in vitro conditions employed here. Further studies are required to resolve this point, to establish formally if NodD does bind to the nodD-nodA intergenic region and to locate the contact points between the DNA and the regulatory protein.

ACKNOWLEDGMENTS

This work was funded by the Agricultural and Food Research Council via a grant-in-aid to the John Innes Institute. J.L.B. was supported by the Agricultural Genetics Company and G-F.H. received support from the Rockefeller Foundation.

REFERENCES

Burn, J.E., Rossen, L. and Johnston, A.W.B., 1987, Four classes of mutations in the nodD gene of Rhizobium leguminosarum which affect its ability to autoregulate and/or to activate other nod genes in the presence of flavonoid inducers, Genes and Development, in the press.

Debelle, F. and Sharma, S.B. 1986, Nucleotide sequence of Rhizobium meliloti RCR2011 genes involved in host-range specificity, Nucl. Acids Res., 14:7453, 1986.

Downie, J.A., Knight, C.D., Johnston, A.W.B., and Rossen, L., 1985 Identification of genes and gene products involved in the nodulation of peas by Rhizobium leguminosarum, Molec. Gen. Genet. 190:255.

Downie, J.A., Surin, B.P., Evans, I.J., Rossen, L., Firmin, J.L., Shearman C.A. and A.W.B. Johnston, 1986, Nodulation genes of Rhizobium leguminosarum, in "Proceedings Of The III International Symposium On Plant-Microbe Interactions" D.P.S. Verma, ed., Academic Press, N.Y.

Egelhoff, T.T., Fisher, R.F., Jacobs, T.W., and S.R. Long, 1985, Nucleotide

sequence of Rhizobium meliloti 1021 nodulation genes: nodD is read divergently from nodABC, DNA 4:241.
Evans, I.J. and Downie, J.A., 1986, The nodI gene product of Rhizobium leguminosarum is closely related to ATP-binding bacterial transport proteins; nucleotide sequence of the nodI and nodJ genes, Gene, 42:95.
Firmin, J.L., Wilson, K.E., Rossen, L. and Johnston, A.W.B., 1986, Flavonoid activation of nodulation genes in Rhizobium reversed by other compounds present in plants, Nature, 324:90.
Fisher, R.F., Tu, J.K. and Long S.R., 1985, Conserved nodulation genes in R. meliloti and R. trifolii, Appl. Env. Microbiol., 49:1432.
Fried, M.G. and Crothers, D.M., 1981, Equilibria and kinetics of lac repressor-operator interactions by polyacrylamide gel electrophoresis. Nucl. Acid Res., 5:3157.
Garner, M.M. and Revzin, A., 1981, A gel electrophoresis method for quantifying the binding of proteins to specific DNA regions. Applications to components of the E. coli lactose operon regulatory system. Nucl. Acids Res., 9:3047.
Gottfert, M., Horvath, B., Kondorosi, E., Putnocky, P., Rodriguez-Quinones, F. and Kondorosi, A., 1986, At least two nodD genes are necessary for efficient nodulation of alfalfa by Rhizobium meliloti, J. Mol. Biol., 191:411.
Horvath, B., Bachem, C.W.B., Schell, J. and Kondorosi, A., 1987, Host-specific regulation of nodulation genes in Rhizobium is regulated by a plant-signal, interacting with the nodD gene product, EMBO J,6:841.
Horvath, B., Kondorosi, E., John, M., Schmidt, J., Torok, I., Gyorgypal, Z., Barabas, I., Weineke, U., Schell, J., and Kondorosi, A., 1986 Organization, structure and symbiotic function of Rhizobium meliloti nodulation genes determining host specificity for alfalfa, Cell, 46:335.
Innes, R.W., Kuempel, P.L., Plazinski, J., Canter-Cramers, H., Rolfe, B.G. and Djordjevic, M.A., 1985, Plant factors induce expression of nodulation and host-range genes in Rhizobium trifolii, Mol. Gen. Genet.,207:426.
Lee, N.L., Gielow, W.O. and Wallace, R.G., 1981, Mechanism of araC auto-regulation and the domains of the two overlapping promoters pC and pBAD, in the L-arabinose regulatory region of Escherichia coli, Proc. Nat'l Acad. Sci. USA, 78:752.
Long, S.R., 1984, Genetics of Rhizobium nodulation, in "Plant Microbe Interactions, Vol I, Molecular and Genetic Perspectives", T. Kosuge and E.W. Nester, eds., MacMillan, N.Y., pp. 265-306.
Mulligan, J.T. and Long, S.R., 1985, Induction of Rhizobium meliloti nodC expression by plant exudate requires nodD, Proc. Nat'l. Acad. Sci. USA, 82:6609.
Peters, N.K., Frost, J.W. and Long, S.R., 1986, A plant flavone, luteolin, induces expression of Rhizobium meliloti nodulation genes, Science, 233:977.
Redmond, J.W., Batley, M., Djordjevic, M.A., Innes, R.W. Kuempel, P.L. and Rolfe, B.G., 1986, Flavones induce expression of nodulation genes in Rhizobium, Nature, 323:632.
Rossen, L., Davis, E.O., Burn, J.E., Latchford, J.W., Hong, G-F. and Johnston, A.W.B., 1987, Plant-induced Rhizobium genes involved in host specificity and early stages of nodulation, T.I.B.S., in the press.
Rossen, L., Shearman, C.A., Johnston, A.W.B. and Downie, J.A., 1985, The nodD gene of Rhizobium leguminosarum is autoregulatory and in the presence of plant exudate induces the nodABC genes, EMBO J., 4:3369.
Rostas, K., Kondorosi, E., Horvath, B., Simoncsits, A. and Kondorosi, A., 1986, Conservation of extended promoter regions of nodulation genes in Rhizobium, Proc. Nat'l. Acad. Sci. USA, 83:1757.
Schmidt, J., John, M., Weineke, U., Krussmann, H-D. and Schell, J., 1986, Expression of the nodulation gene nodA in Rhizobium meliloti and localization of its gene product in the cytosol, Proc. Nat'l. Acad.

PLANT-BACTERIAL SIGNALLING IN THE Rhizobium-LEGUME SYMBIOSIS

Frederick M. Ausubel, Mary A. Honma, Rebecca Dickstein, Wynne W. Szeto, B. Tracy Nixon and Clive W. Ronson

Department of Molecular Biology
Massachusetts General Hospital
Boston, MA 02114 USA

INTRODUCTION

Several projects in our laboratory concern the mechanism(s) of plant-bacterial communication in the Rhizobium meliloti - alfalfa symbiosis. In the symbiosis, nitrogen fixation occurs in root nodules -- complex, highly differentiated organs formed by the interaction of bacteria and plant. Within the nodule, individual plant cells become infected with bacteria that differentiate into specialized symbiotic forms called bacteroids. We have been investigating several stages of the symbiotic process during which signals are transmitted between the symbiotic partners.

THE R. Meliloti nodD PRODUCT PLAYS A HOST-SPECIFIC ROLE IN NODULATION

At the early stages of the symbiosis, the R. meliloti nodulation (nod) genes (Fig. 1) are involved in both receiving and sending signals to the plant host. R. meliloti nod genes can be divided into three broad groups: At least three of the nod genes, nodABC, appear to be functionally and structurally conserved in all Rhizobium and Bradyrhizobium species and are referred to as the "common" nod genes (Kondorosi and Kondorosi, 1986). In contrast, nodFEG and nodH (also called hsnFEG and hsnH for "host specific nodulation") are specifically required for nodulation of R. meliloti hosts (Debelle et al., 1986). Finally, the nodD product has been shown to be a positive transcriptional regulator which activates the remaining nod genes in the presence of specific flavone exudates from the host root (Mulligan and Long, 1985). In R. leguminosarum and R. trifolii, mutations in the nodD gene lead to a Nod⁻ phenotype (Djordjevic et al., 1985; Downie et al., 1985). However, R. meliloti strains containing mutations in the nodD gene still elicit nodules (Jacobs et al., 1985).

The apparent discrepancy between the phenotypes of R. meliloti nodD mutants and the phenotypes of R. trifolii and R. leguminosarum nodD mutants has been resolved by our finding that R. meliloti contains three copies of the nodD gene. When the R. meliloti nodD gene (adjacent to nodABC; see Fig. 1) was used as a hybridization probe against EcoRI digested R. meliloti DNA, two bands hybridized in addition to the fragment containing the nodABC genes. We refer to the two additional copies of nodD as nodD2 and nodD3 and to the nodD gene that is adjacent to nodABC as nodD1. NodD2 and nodD3 map to an 80 kilobase region of the R. meliloti symbiotic plasmid as shown in Fig. 1.

To determine the roles of the three R. meliloti nodD genes, we have cloned nodD2 and nodD3 and have constructed R. meliloti strains that contain mutations in all possible combinations of the three R. meliloti nodD genes. This was accomplished using a combination of homogenotization-mediated gene replacement technology (Ruvkun and Ausubel, 1981) and phage M12-mediated generalized transduction (Finan et al., 1984). We have assayed the phenotypes of the nodD mutants on two R. meliloti hosts, Medicago sativa (alfalfa) and Melilotus alba (sweet clover).

A triple R. meliloti nodD mutant (nodD1-nodD2-nodD3) failed to form nodules on either alfalfa or sweet clover, indicating that nodD is an essential nodulation gene for R. meliloti as well as for R. leguminosarum and R. trifolii. Single and double R. meliloti nodD mutants displayed varying phenotypes depending on the particular nodD genes mutated and the host plant. A nodD2 mutant had no discernible defect in nodulation on either alfalfa or sweet clover and nodD3 mutants exhibited a delayed nodulation phenotype of two to three days when inoculated onto either of these two hosts. In contrast, the nodulation phenotype of the nodD1 mutant depended on the host: on alfalfa nodules appeared five to six days after wild type nodules; on sweet clover nodules appeared two to three days after wild type nodules. Similarly, nodD1-nodD2 and nodD2-nodD3 double mutants formed nodules with the same delay as single nodD1 and nodD3 mutants. The nodD1-nodD3 double mutant also had a host specific phenotype: it formed only a few nodules on alfalfa after more than a one week delay and no nodules at all on sweet clover.

We have drawn the following conclusions from the nodulation tests with the nodD mutants: 1) All three copies of nodD are functional. 2) In the case of sweet clover nodulation, nodD1 and nodD3 have equivalent roles but nodD2 has no detectable role. 3) In the case of alfalfa nodulation, nodD1 plays a more important role than nodD3 which is more important than nodD2.

The three nodD genes in R. meliloti may have evolved to optimize the interaction between R. meliloti and its different host plants. One possibility is that each of the nodD products differs in its ability to respond to particular nod gene inducing factors exuded by different hosts. We are currently testing this latter possibility.

THE R. meliloti exo AND ndv GENE PRODUCTS ARE REQUIRED FOR NODULIN INDUCTION

When the R. meliloti nodABCD and nodFEGH genes are transferred on plasmid pMH36 (Fig. 1) to Agrobacterium tumefaciens, a plant pathogen related to R. meliloti, A. tumefaciens acquires the ability to elicit defective nodules on alfalfa roots. These pseudonodules have an overall morphology similiar to wild-type nodules but lack infection threads and intracellular bacteria (Hirsch et al., 1984; Truchet et al., 1985). These experiments indicated that bacterial invasion of the plant host is not required for nodule construction and that nod gene products are the signal to construct a nodule.

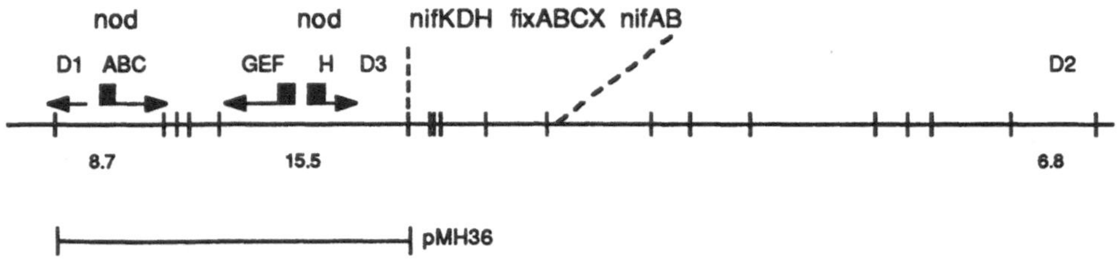

Figure 1. Genetic map of R. meliloti symbiotic region

Recently, we have investigated nodule-specific proteins (nodulins) that are expressed in A. tumefaciens-elicited pseudonodules by extracting RNA from pseudonodules and then subjecting the RNA to in vitro translation in a rabbit reticulocyte system. Two-dimensional electrophoretic analysis of the translation products showed that a very limited subset of nodulins are synthesized in pseudonodules. Interestingly, this same subgroup of nodulins was also synthesized in pseudonodules elicited by R. meliloti mutants that fail to synthesize a particular exopolysaccharide (exo mutants), or a β-1,2 glucan (ndv mutants). Like A. tumefaciens containing the R. meliloti nod genes, R. meliloti exo and ndv mutants elicit pseudonodules that lack infection threads and intracellular bacteria (Finan et al., 1985; Leigh et al., 1985; Dylan et al., 1986; Geremia et al., 1987). These results suggest that R. meliloti cell-surface polysaccharides act in the developmental pathway at the stage following the action of the nod gene products and that they may be important signals required for bacterial invasion of plant cells.

R. meliloti ntrA BUT NOT ntrC IS REQUIRED FOR nif ACTIVATION

A major unanswered question in the Rhizobium-legume symbiosis is the mechanism by which the R. meliloti nitrogen fixation (nif) genes are activated. In K. pneumoniae, a free-living nitrogen fixing species, nif genes are subject to two levels of positive regulation in response to limiting ammonia (see Gussin et al., 1986 for a review). The first level of regulation is nif-specific and is mediated by the product of the nifA gene. The nifA product is a transcriptional activator that is required for the expression of all nif operons, except its own. The second level of nif regulation is mediated by a centralized regulatory system (the ntr system) that controls the expression of a variety of nitrogen assimilatory genes in enteric bacteria. Under conditions of NH_4^+ starvation, the ntrC + ntrA products activate the nifLA operon and also activate other operons involved in nitrogen assimilation. Recent data have indicated that the ntrA protein is a sigma factor that directs core RNA polymerase to initiate transcription at ntrC activated promoters (Hirschmann et al., 1985; Hunt and Magasanik, 1985). Promoters activated by NtrC + NtrA do not contain canonical -35 and -10 sequences but instead have the consensus -26 CTGGYAYR-N_4-TTGCA -10 (Gussin et al., 1986).

In enteric bacteria, the ntr system responds to ammonia through the action of three additional genes, ntrB, glnD and glnB (see Kustu et al., 1986 for a review). Ninfa and Magasanik (1986) have shown recently that under conditions of limiting ammonia, NtrB is a protein kinase that phosphorylates NtrC to activate it. The functional state of NtrB, in turn, is determined by the products of glnB and glnD which respond to the intracellular 2-ketoglutarate to glutamine ratio (Bueno et al., 1985). Thus NtrC activity is modulated by a cascade system that is exquisitely sensitive to changes in ammonia availability.

To determine whether the same nif control circuitry operates in R. meliloti as in K. pneumoniae, we have identified and mutated the R. meliloti ntrA and ntrC genes. The R. meliloti ntrC gene was identified on the basis of interspecies homology with the E. coli ntrC gene (Szeto et al., 1987). As in enteric bacteria, the R. meliloti ntrC gene is located directly downstream of the ntrB gene. R. meliloti ntrC::Tn5 mutants failed to utilize nitrate as a sole nitrogen source. Unlike enteric ntrC mutants, however, R. meliloti ntrC mutants displayed no difficulty utilizing a variety of amino acids as nitrogen sources.

Cloned ntrA genes from K. pneumoniae and E. coli failed to cross-hybridize to R. meliloti DNA. Therefore, a genetic strategy was devised to identify the R. meliloti ntrA gene (Ronson et al., 1987a). This strategy was based on a recent study of the genetic regulation of C4-dicarboxylate transport in R. leguminosarum (Ronson and Astwood, 1985; C.W. Ronson, P.M. Astwood, B.T. Nixon and F.M. Ausubel, in preparation), where it was observed that dctA, a structural gene required for the transport of succinate, fumarate and malate, contained an NtrA-like promoter sequence. Two positive regulatory genes, dctB and dctD, required for activation of dctA transcription, were also identified. Interestingly, it was found that the C-terminus of dctB gene product (DctB) is homologous to the C-terminus of NtrB and the N-terminus of dctD product (DctD) is homologous to the N-terminus of NtrC. Furthermore, DctD also shares strong homology with NtrC and NifA in a central domain of these proteins that is postulated to interact with NtrA. These observations suggested that NtrA might be required for the expression of dctA; therefore, we formulated a selection scheme for mutations in a putative R. meliloti ntrA gene based on the supposition that NtrA is required for dctA expression.

To monitor dctA expression in R. meliloti, we utilized an in-frame dctA-lacZ fusion present on a broad host range plasmid that also contained the entire dct regulon from R. leguminosarum. R. meliloti containing the dctA-lacZ fusion plasmid formed white colonies on defined medium containing 5-bromo-4-chloro-3-indolyl β-D-galactopyranoside (X-gal) with glucose as sole carbon source, and blue colonies on defined medium containing X-gal with glucose plus succinate as carbon sources. Because of this latter result and because the dctA-lacZ fusion plasmid also contained the dctB and dctD genes, we reasoned that any R. meliloti mutant that contained the dctA-lacZ fusion plasmid and formed white colonies when plated on medium containing X-gal, glucose and succinate, should be mutated in the R. meliloti ntrA gene. We therefore conjugated the dctA-lacZ fusion plasmid into a culture of R. meliloti that had previously been mutagenized with transposon Tn5 and selected white colonies on defined medium containing X-gal, glucose, succinate, streptomycin and neomycin. Three white colonies were analyzed in detail and all three were subsequently shown to have Tn5 inserted in an ntrA-like gene as shown by DNA sequence analysis (Ronson et al., 1987a).

Consistent with the hypothesis that NtrA activates transcription in conjunction with either NtrC, DctD or NifA, R. meliloti ntrA mutants exhibited a pleiotropic phenotype: they failed to grow on nitrate as sole nitrogen source and succinate as sole carbon source and they elicited Fix$^-$ nodules (Ronson et al., 1987a). In contrast to the R. meliloti ntrA mutants, however, R. meliloti ntrC mutants elicited Fix+ nodules (Szeto et al., 1987). These results indicated that the R. meliloti nif genes are not activated via the centralized ammonia control system utilizing ntrC and ntrB and is consistent with the model that the R. meliloti nifA-like gene is activated by a centralized symbiotic regulatory system. Such a centralized symbiotic control sytem could be responsible for activating many symbiotic genes -- for example, genes whose products are involved in the early steps in nodulation or in nodule construction and maintenance. Thus the R. meliloti nifA regulatory gene would be only one of several different targets for a centralized regulatory system. We are currently attempting to identify the nature of the plant signal that we have hypothesized to be responsible for the activation of nifA.

Nif REGULATORY GENES ARE RELATED TO A VARIETY OF GENES THAT RESPOND TO ENVIRONMENTAL STIMULI

As indicated in the previous section, nucleotide sequence analysis carried out in our laboratory has revealed an unexpected degree of conservation among proteins involved in nif regulation and other bacterial proteins that are members of two-component regulatory systems (Nixon et al., 1986). One component of each system is thought to act as an environmental sensor (sensor component) that transmits a signal to the second component (regulatory component) which then effects the response, usually at the transcriptional level. The systems include those responding to osmolarity (envZ/ompR), nitrogen limitation (ntrB/ntrC), phosphate limitation (phoR/phoB) and toxic compounds (changes in membrane proteins controlled by cpxA/sfrA) in E. coli, genes controlling the virulence of

A. tumefaciens in response to plant exudate (*virA*/*virG*), and genes activating the expression of the C4-dicarboxylate transport structural gene in response to C4-dicarboxylates in *R. leguminosarum* (*dctB*/*dctD*) (see Fig. 2).

As illustrated in Fig. 2 for several such systems, about 200 amino acids of the C-terminal region is conserved among the proteins that act as sensors; about 120 amino acids of the N-terminal region is conserved among the proteins that act as regulators. The hydropathic profiles of members of the sensor class suggest that many contain an N-terminal periplasmic domain defined by two hydrophobic transmembrane regions (black boxes in Fig. 2) and a C-terminal cytoplasmic domain, a structure similar to that of the chemosensory receptor proteins of enteric bacteria (Krikos et al., 1983). NtrB clearly does not contain a transmembrane domain; as described above, it is also the only member of the sensor class that is thought to respond to an intracellular signal (Bueno et al., 1985).

The homology among the transcriptional activators shown in Fig. 2 is not limited to the N-terminal domain; the activators also share one of two C-terminal regions. Furthermore, the C-terminal region shared by OmpR, PhoB, SfrA and VirG also occurs at the N-terminus in ToxR, the activator of the cholera toxin operon (Miller at al., 1987) and the C-terminal region shared by NtrC and DctD also occurs in NifA (Buikema et al., 1985).

The domain conservation in the C-terminal portions of the regulator class of proteins most likely reflects common mechanisms of transcriptional activation. A helix-turn-helix motif, characteristic of prokaryotic DNA bining proteins, is found within the C-terminal 40 amino acids of NtrC, NifA and DctD and binding sites located 100-160 base pairs upstream of the transcriptional initiation site have been identified in NtrC- and NifA-activated promoters (Hirschman et al., 1985; Reitzer and Magasanik, 1986; Buck et al., 1986).

In addition to sharing a helix-turn-helix motif at their C-terminal ends and a common central domain, NtrC, NifA and DctD all require the alternate sigma factor NtrA (RpoN) as a co-activator. This suggests that

Environmental Stimulus	Sensor	Regulator	System Activated
Osmolarity Changes	envZ	ompR	ompC, ompF
Phosphate Limitation	phoR	phoB	phoA, phoE, etc.
Plant Exudate	virA	virG	virB, virC, etc.
Toxic Compounds, Dyes	cpxA	sfrA	traJ, etc.
Nitrogen Limitation	ntrB	ntrC	glnA, nifLA, etc.
4C-Dicarboxylic Acids	dctB	dctD	dctA
Oxygen/Ammonia ?	nifL?	Kp nifA	nifHDK etc.
Oxygen ?	?	Rm nifA	nif, fix

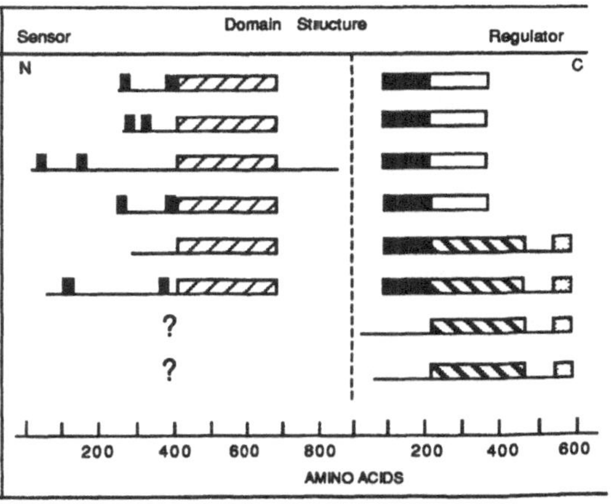

Figure 2. Conservation of two-component regulatory systems that respond to environmental stimuli. References: *E. coli envZ/ompR*, Hall and Silhavy, 1981; Comeau et al., 1985; Nara et al., 1986; *E. coli phoR/phoB*, Makino et al., 1986a; 1986b; *A. tumefaciens virA/virG*, Stachel and Zambryski, 1986; Winans et al., 1986; Melchers et al., 1986; LeRoux et al., 1987; *E. coli cpxA/sfrA*, Silverman, 1985; Albin et al., 1986; Drury et al., 1985; *K. pneumoniae* and *Bradyrhizibium* sp. (*Parasponia*) *ntrB/ntrC*, Buikema et al., 1985; Drummond et al., 1986; MacFarlane and Merrick, 1985; Nixon et al., 1986; *R. leguminosarum dctB/dctD*, Ronson et al., 1987a.

transcriptional activation by these proteins may be mediated by a protein-protein interaction of the conserved central domain with NtrA (Ronson et al., 1987b).

The lack of conservation between the N-terminal domains of the *K. pneumoniae* *ntrC* and *ntrA* products may account for their different sensitivities to environmental stimuli. Whereas *ntrB* product modulates the activity of *ntrC* product in response to nitrogen status, *K. pneumoniae* *nifL* product is hypothesized to inhibit *nifA* product activity in response to oxygen tension or low amounts of combined nitrogen (Buchanan-Wollaston and Cannon, 1984). It seems likely that in this species the N-terminal part of the *nifA* protein interacts with *nifL* protein and that the *nifL* protein will not contain the *ntrB*-set conserved sequence. Furthermore, although there is similarity between the N-terminal portions of *K. pneumoniae* and *R. meliloti* *nifA* gene products, it is possible that the low degree of conservation reflects evolution of the N-terminal region of the rhizobial *nifA* product to respond to some aspect of the symbiotic environment (Nixon et al., 1986).

Recently, we have proposed a model for the transduction of environmental signals by the two-component regulatory systems (Nixon et al., 1986; Ronson et al., 1987b). The N-terminal domain of the environmental sensor class perceives an environmental stimulus, and transmits a signal to its conserved cytoplasmic domain through an allosteric alteration much as envisaged for the chemosensory receptors. The activated C-terminal portion of the sensor protein then interacts with and modifies the conserved N-terminal portion of the regulator protein. This modified N-terminal domain is then able to effect the response through an interaction with its C-terminal domain, the interaction causing a switch in the conformation of the C-terminus between inactive and active forms or repressor and activator forms.

A clue to the mechanism by which the N-terminal domain of the regulator may be activated has been obtained recently with the demonstration by Ninfa and Magasanik described above that NtrC responds to NtrB by undergoing phosphorylation or dephosphorylation. This raises the possibility that all members of the sensor class modify their respective partners by the same mechanism.

The pattern of conserved domains found among the proteins that comprise two-component regulatory systems strongly suggests that these proteins have evolved by recruiting domains from various progenitor genes. Diverse bacterial species apparently employ common strategies for sensing specific changes in their environment and transducing that information to the transcriptional apparatus.

ACKNOWLEDGEMENTS

Unpublished work was supported by a grant from Hoechst A.G.

REFERENCES

Albin, R., Weber, R., and Silverman, P.M., 1986, The Cpx proteins of *Escherichia coli* K12, *J. Biol. Chem.* 261:4698-4705.

Buchanan-Wollaston, V. and Cannon, F., 1984, Regulation of *nif* transcription in *Klebsiella pneumoniae*, *in*: "Advances in Nitrogen Fixation Research," C. Veeger and W.E. Newton, eds., Nijhoff/Junk, The Hague, p. 732.

Buck, M., Miller, S., Drummond, M., and Dixon, R., 1986, Upstream activator sequences are present in the promoters of nitrogen fixation genes, *Nature* 320:374-378.

Bueno, R., Pahel, G., and Magasanik, B., 1985, Role of *glnB* and *glnD* gene products in regulation of the *glnALG* operon of *Escherichia coli*, *J. Bacteriol.* 164:816-822.

Buikema, W.J., Szeto, W.W., Lemley, P.V., Orme-Johnson, W.H., and Ausubel, F.M., 1985, Nitrogen fixation specific regulatory genes of Klebsiella pneumoniae and Rhizobium meliloti share homology with the general nitrogen regulatory gene ntrC of K. pneumoniae, Nucleic Acids Res. 13:4539-4555.
Comeau, D.E., Ikenaka, K., Tsung, K., and Inouye, M., 1985, Primary characterization of the protein products of the Escherichia coli ompB locus: structure and regulation of synthesis of the OmpR and EnvZ proteins, J. Bacteriol. 164:578-584.
Debelle, F., Rosenberg, C., Vasse, J., Maillet, F., Martinez, E., Denarie, J., and Truchet, G., 1986, Assignment of symbiotic developmental phenotypes to common and speciic nodulation (nod) genetic loci of Rhizobium meliloti, J. Bacteriol., 168:1075-1086.
Djordjevic, M.A., Schofield, P.R., and Rolfe, B.G., 1985, Tn5 mutagenesis of Rhizobium trifolii host-specific nodulation genes results in mutants with altered host-range ability, Molec. gen. Genet. 200:463-471.
Downie, J.A., Knight, C.D., Johnston, A.W.B., and Rossen, L., 1985, Identification of genes and gene products involved in the nodulation of peas by Rhizobium leguminosarum, Molec. gen. Genet. 198:255-262.
Drummond, M., Whitty, P., and Wootten, J., 1986, Sequence and domain relationships of ntrC and nifA from Klebsiella pneumoniae: homologies to other regulatory proteins, EMBO J. 5:441-447.
Drury, L.S. and Buxton, R.S., 1985, DNA sequence analysis of the dye gene of Escherichia coli reveals amino acid homology between the Dye and OmpR proteins, J. Biol. Chem. 260:4263-4242.
Dylan, T., Ielpi, L., Stanfield, S., Kashyap, L., Douglas, C., Yanofsky, M., Nester, E., Helinski, D.R., and Ditta, G., 1986, Rhizobium meliloti genes required for nodule development are related to chromosomal virulence genes in Agrobacterium tumefaciens, Proc. Natl. Acad. Sci. USA 83:4403-4407.
Finan, T.M., Hartwieg, E., Lemieux, K., Bergman, K., Walker, G.C., and Signer, E.R., 1984, General transduction in Rhizobium meliloti, J. Bacteriol. 159:120-124.
Finan, T.M., Hirsch, A.M., Leigh, J.A., Johansen, E., Kulday, G.A., Deegan S., Walker G.C., and Signer E.R., 1985, Symbiotic mutants of Rhizobium meliloti that uncouple plant from bacterial differentiation, Cell 40:869-877.
Geremia, R.A., Cavaignac, S., Zorreguieta, A., Toro, N., Olivares, J., and Ugalde, R.A., 1987, A Rhizobium meliloti mutant that forms ineffective pseudonodules in alfalfa produces exopolysaccharide but fails to form β-(1,2) glucan, J. Bacteriol. 169:880-884.
Gussin, G.N., Ronson, C.W., and Ausubel, F.M., 1986, Regulation of nitrogen fixation genes, Annu. Rev. Genet 20:567-591.
Hall, M.N. and Silhavy, T.J., 1981, Genetic analysis of the ompB locus in Escherichia coli K12, J. Mol. Biol. 151:1-15.
Hirsch, A.M., Wilson, K.J., Jones, J.D.G., Bang, M., Walker, V.V., and Ausubel, F.M., 1984, Rhizobium meliloti nodulation genes allow Agrobacterium tumefaciens and Escherichia coli to form pseudonodules on alfalfa. J. Bacteriol. 158:1133-1143.
Hirschmann, J., Wong, P.-K., Sei, K., Keener, J., and Kustu, S., 1985, Products of nitrogen regulatory genes ntrA and ntrC of enteric bacteria activate glnA transcription in vitro: evidence that the ntrA product is a sigma factor, Proc. Natl. Acad. Sci. USA 82:7525-7529.
Hunt, T.P., and Magasanik, B., 1985, Transcription of glnA by purified Escherichia coli components: core RNA polymerase and the products of glnF, glnG, and glnL, Proc. Natl. Acad. Sci. USA 82:8453-8457.
Jacobs, T.W., Egelhoff, T.T., and Long, S.R., 1985, Physical and genetic map of a Rhizobium meliloti nodulation region and nucleotide sequence of nodC, J. Bacteriol. 162:469-476.
Kondorosi, E. and Kondorosi, A., 1986, Nodule induction on plant roots by Rhizobium, TIBS 11:296-299.
Krikos, A., Mutoh, N., Boyd, A., and Simon, M.I., 1983, Sensory transducers of E. coli are composed of discrete structural and functional domains, Cell 33:615-622.

Leigh, J.A., Signer, E.R., and Walker, G.C., 1985, Exopolysaccharide-deficient mutants of Rhizobium meliloti that form ineffective nodules. Proc. Natl. Acad. Sci. USA 82:6231-6235.

LeRoux B., Yanofsky, M.F., Winans, S.C., Ward, J.E., Ziegler, S.F., and Nester, E.W., 1987, Characterization of the virA locus of Agrobacterium tumefaciens: a transcriptional regulator and host range determinant, EMBO J. 6:849-856.

MacFarlane, S.A. and Merrick, M., 1985, The nucleotide sequence of the nitrogen regulation gene ntrB and the glnA-ntrBC intergenic region of Klebsiella pneumoniae, Nucleic Acids Res. 13:7591-7606.

Makino , K., Shinagawa, H., Amemura, M., and Nakata, A., 1986a, Nucleotide sequence of the phoB gene, the positive regulatory gene for the phosphate regulon of Escherichia coli K-12, J. Mol. Biol. 190:37-44.

Makino, K., Shinagawa, H., Amemura, M., and Nakata, A., 1986, Nucleotide sequence of the phoR gene, a regulatory gene for the phosphate regulon of Escherichia coli, J. Mol. Biol. 192:549-556.

Melchers, L.S., Thompson, D.V., Idler, K.B., Shilperoort, R.A., and Hooykaas, P.J.J., 1986, Nucleotide sequence of the virulence gene virG of the Agrobacterium tumefaciens octopine Ti plasmid: significant homology between virG and the regulatory genes ompR, phoB and dye of E. coli, Nucleic Acids Res. 14:9933-9942.

Miller, V.L., Taylor, R.K., and Mekalanos, J.J., 1987, Cholera toxin transcriptional activator ToxR is a transmembrane DNA binding protein, Cell 48:271-279.

Mulligan, J.T. and Long, S.R., 1985, Induction of Rhizobium meliloti nodC expression by plant exudate requires nodD, Proc. Natl. Acad. Sci. USA 82:6609-6613.

Nara, F., Matsuyama, S., Mizuno, T., and Mizoshima, S., 1986, Molecular analysis of mutant ompR genes exhibiting different phenotypes as to osmoregulation of the ompF and ompC genes of Escherichia coli, Mol. gen. Genet. 202:194-199.

Ninfa, A.J. and Magasanik, B., 1986, Covalent modification of the glnG product NR_I, by the glnL product NR_{II}, regulates the transcription of the glnALG operon in Escherichia coli, Proc. Natl. Acad. Sci. USA 83:5909-5913.

Nixon, B.T., Ronson, C.W., and Ausubel, F.M., 1986, Two-component regulatory systems responsive to environmental stimuli share strongly conserved domains with the nitrogen assimilation regulatory genes ntrB and ntrC, Proc. Natl. Acad. Sci. USA 83:7850-7854.

Reitzer, L.J. and Magasanik, B., 1986, Transcription of glnA in E. coli is stimulated by activator bound to sites far from the promoter, Cell 45:785-792.

Ronson, C.W. and Astwood, P.M., 1985, Genes involved in the carbon metabolism of bacteroids, in: "Nitrogen Fixation Research Progress," P.J. Bottomley and W.E. Newton, eds., Martinus Nijhoff, Dordrecht, pp. 201-207.

Ronson, C.W., Nixon, B.T., Albright, L.M., and Ausubel, F.M., 1987a, Rhizobium meliloti ntrA (rpoN) gene is required for diverse metabolic functions, J. Bacteriol. 169:2424-2431.

Ronson, C.W., Nixon, B.T., and Ausubel, F.M., 1987b, Conserved domains in bacterial regulatory proteins that respond to environmental stimuli, Cell 49:579-581.

Ruvkun, G.B. and Ausubel, F.M., 1981, A general method for site directed mutagenes in prokaryotes: Construction of mutations in symbiotic nitrogen fixation genes of Rhizobium meliloti, Nature 289:85-88.

Silverman, P.M., 1985, Host cell plasmid interactions in the expression of DNA donor activity by F+ strains of Escherichia coli K12, Bioessays 2:254-259.

Stachel, S.E. and Zambryski, P.C., virA and virG control the plant-induced activation of the T-DNA transfer process of A. tumefaciens, Cell 46:325-333.

Szeto, W.W., Nixon, B.T., Ronson, C.W., and Ausubel, F.M., 1987, Identification and characterization of the Rhizobium meliloti ntrC gene: R. meliloti has separate regulatory pathways for activation of nitrogen fixation genes in free-living and symbiotic cells. J. Bacteriol. 169:1423-1432.

Truchet, G., Debelle, F., Vasse, J., Terzaghi, B., Garnerone, A.-M., Rosenberg, C., Batut, J., Maillet, F., and Denarie, J., 1985, Identification of a Rhizobium meliloti pSym2011 region controlling the host specificity of root hair curling and nodulation, J. Bacteriol. 164:1200-1210.

Winans, S.C., Ebert, P.R., Stachel, S.E., Gordon, M.P., and Nester, E.W., 1986, A gene essential for Agrobacterium virulence is homologous to a family of positive regulatory loci, Proc. Natl. Acad. Sci. USA 83: 8278-8282.

THE AGROBACTERIUM SYSTEM AND ITS APPLICATIONS

M. J. J. van Haaren, P. J. J. Hooykaas
and R. A. Schilperoort
Department of Plant Molecular Biology
Leiden University
The Netherlands

GENERAL INTRODUCTION

The family of the Rhizobiaceae embraces two important genera, viz Rhizobium and Agrobacterium, which have both been studied thoroughly because of their interactions with plants(Allen and Holding, 1974).
Rhizobium species are able to induce root nodules on leguminous plants; they can invade plant cells. This leads to a symbiosis between plant and bacterium, in which the bacterium provides fixed nitrogen to the plant and the plant provides carbon sources to the bacterium(Vincent, 1974). The Agrobacterium species do not invade plant cells, but are nevertheless able to induce tumor formation on the infected plant by transferring a piece of DNA, which is called the T-DNA, into cells of the infected plants. Tumor formation only occurs if fresh wounds are infected(Lipetz, 1965). In these wound sites the bacterium is able to penetrate the plant and to attach to plant cell walls, which is an essential step in the infection process(Lippincott and Lippincott, 1969; Schilperoort, 1969). The neoplastic plant disease Crown Gall(figure 1) is induced by Agrobacterium tumefaciens(Smith and Townsend, 1907), while Cane Gall is induced by A. rubi(Riker et al., 1946) and the Hairy Root disease by A. rhizogenes(Riker, 1930). The host range of these agrobacteria is different but can be very wide among dicotyledonous plants(De Cleene and De ley, 1976). Especially Crown Gall is a wide-spread disease that causes important losses in agriculture and arboriculture. For a long time it was thought that Agrobacterium could only transform dicotyledonous plants, but in 1984 Hooykaas-Van Slogteren et al. proved that also monocotyledonous plants are susceptible to Agrobacterium transformation(Hooykaas-Van Slogteren et al., 1984). Because the host cell response after infection of monocots by agrobacteria is much less pronounced -or absent entirely- than after infection of dicots, this phenomenon was not observed before. However, other characteristics than tumor formation proved that the plants do acquire T-DNA from the bacterium. Further results by Hernalsteens et al. (1984), Graves and Goldman (1986;1987) and Grimsley et al. (1987) confirmed and extended these initial findings. Evidence is now available for transfer of T-DNA to the monocot families Amaryllidaceae, Gramineae, Liliaceae and Iridaceae by Agrobacterium. These findings increased the importance of the research on agrobacteria and its developed vector systems, because of their potential application now also for monocotyledonous crops.

Fig. 1. A crown gall tumor induced by Agrobacterium tumefaciens on Kalanchoe daigremontiana.

Ti- AND Ri-PLASMIDS

Virulent agrobacteria contain a large plasmid which is responsible for its virulence(Zaenen et al., 1974). Strains cured of this plasmid have lost their oncogenicity. Reintroduction of the plasmid in a non-oncogenic strain leads to a restoration of its oncogenicity(Van Larebeke et al., 1974; Watson et al., 1975; Albinger and Beiderbeck, 1977). For A. tumefaciens and A. rubi this plasmid is called the tumor inducing(Ti)-plasmid and for A. rhizogenes the root inducing(Ri)-plasmid. A distinct part of these large(200-500kbp) plasmids, the T-region, is mobilized from the bacterium to the plant cell during tumor and hairy root induction(for a review see: Bevan and Chilton, 1982). This piece of DNA appears to be integrated at random into the genome of the transformed plant cells. Expression of the T-DNA leads to a transformation of the cells into tumor cells. Crown gall tumors and hairy roots are characterized by their ability to grow in the absence of phytohormones and by the presence of certain specific amino-acid derivatives which are called opines. The Ti- and Ri-plasmids are divided in groups which are classified by the kind of opines that are synthesized in the tumors or roots. The most detailed studies have been done on octopine and nopaline Ti-plasmids.

Comparison of different Ti-plasmids revealed four regions of extensive homology(Drummond and Chilton, 1978; Engler et al., 1981)(figure 2). By transposon insertion and deletion mutagenesis the properties determined by these homologous regions were elucidated. One of the homologous regions is the T-region, which is the part of the plasmid that is transferred to the plant cell. The origin of replication together with the incompatability functions of the plasmids form the second region that is conserved between the two plasmids(Koekman et al., 1982). The third conserved region encodes bacterial transfer functions of the plasmid and genes responsible for the exclusion of phage AP1(Hooykaas et al., 1977; Klapwijk et al., 1978). The

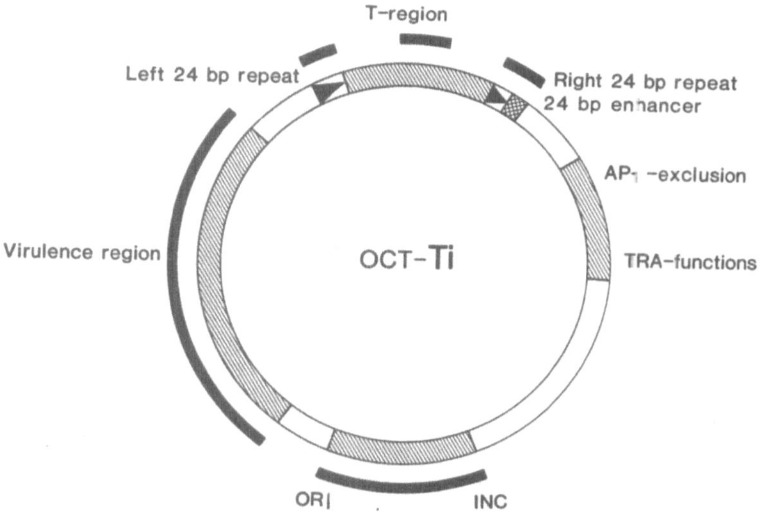

Fig. 2. A schematic map of an octopine type Ti-plasmid.
The dashed bars show the homologous regions with the nopaline Ti-plamsid and the black bars the less stringent homology with an agropine type Ri- plasmid. ORI = origin of replication; INC = incompability functions.

last and fourth region of homology(40kbp) codes for proteins involved in the T-DNA transfer process and was called the Virulence(Vir)-region(Ooms et al., 1980; Garfinkel and Nester, 1980). The homology between Ti- and Ri-plasmids is less extensive(Huffman et al., 1984; Jouanin, 1984; Lahners et al., 1984).

The non-homologous areas of different types of Ti- and Ri-plasmids determine functions such as the catabolism of specific opines or may consist of transposable elements that were picked up by these plasmids during evolution.

From extensive mutagenesis studies of the entire Ti-plasmid it was concluded that only two regions present on a Ti-plasmid are directly involved in tumorigenicity, the Vir-region and the T-region. however, besides genes present in the Vir- and T-region of the Ti-plasmid a number of chromosomal genes are involved in virulence. Some of these chromosomal virulence(chv)-genes are necessary for attachment of the bacterium to the plant cell wall, which is essential for transformation(Douglas et al., 1985; Matthysse, 1987).

THE VIRULENCE REGION

General Properties of the Virulence Region

To understand the process of plant transformation by agrobacteria much effort is spent on the elucidation of the functions that are encoded by genes directly involved in the transfer process. The virulence region present on the Ti-and Ri-plasmid consists of several complementation groups that are directly or indirectly involved in the T-DNA transfer process, but are

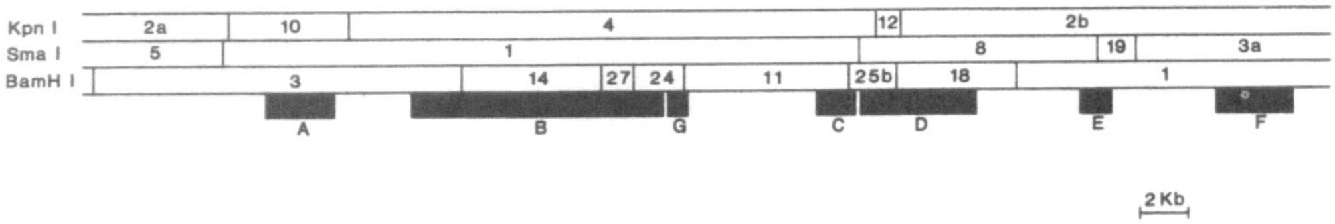

Fig. 3. A KpnI, SmaI and BamHI restriction map of the virulence region of the octopine Ti-plasmid pTi15955.
The black bars indicate the loci VirA to VirG.

themselves not transferred to the plant genome(Hille et al., 1982; Klee et al., 1982). A great extent of homology can be found among the Vir-regions of different Ti-plasmids but also to some extent between Ti- and Ri-plasmids, suggesting that they encode similar functions.

Six complementation groups are present on both the octopine and the nopaline Ti-plasmid. These are called VirA, VirB, VirC, VirD, VirE and VirG(figure 3). On the octopine Ti-plasmid a seventh locus was found and called VirF(Hooykaas et al., 1984), which is not present on the nopaline Ti-plasmid(Otten et al., 1985). Instead the nopaline Ti-plasmid contains a host range virulence gene called tzs, which is situated on the left hand site of virA and is involved in the production of trans-zeatin(Akiyoshi et al., 1985; Beaty et al., 1986; R. Shields, pers. commun.).

Mutations introduced in the octopine Ti Vir-region showed that four operons are absolutely essential for tumor induction(VirA, VirB, VirD and VirG) whereas the other three(VirC, VirE and VirF) are only necessary for tumor induction on certain plant species(Hooykaas et al., 1984; Yanofsky et al., 1985; Stachel and Nester, 1986). Mutations in the vir-genes can be complemented in trans by introducing the wild-type vir-operon in the mutant(Hille et al., 1982). Not only the vir-operon of the homologous Ti-plasmid can be used for complementation, but also that from a heterologous Ti- or Ri-plasmid(Hooykaas et al., 1984). In agreement with this observation it was found that transfer of the octopine Ti T-region to plant cells is

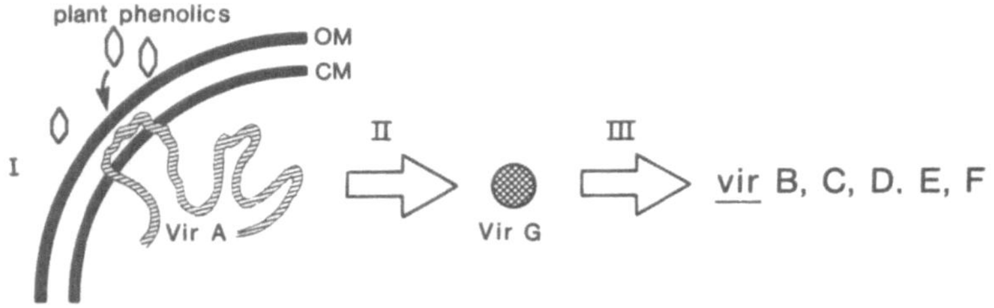

Fig. 4. Regulation of vir-genes via the VirA and VirG proteins.
OM = outer membrane; CM = cytoplasmic membrane.

mediated by the octopine Ti Vir-region, but also by that of the nopaline Ti-plasmid and the Ri-plasmid(Hoekema et al., 1984). These experiments confirmed that the functions encoded by the Vir-regions of different Ti- and Ri-plasmids are indeed similar. Large sections of the Vir-regions of both the nopaline and octopine Ti-plasmid have been sequenced by now. From the sequence information some insight was obtained into the possible functions of certain vir-genes.

The Plant Induced Vir-Expression Regulated by VirA and VirG

Promoter studies using the E.coli lacZ gene as a reporter gene revealed that some of the vir-promoters are plant inducible(Okker et al., 1984; Stachel et al., 1986a). It was found that virA is constitutively expressed and non-inducible, whereas virB, virC, virD and virE are expressed only upon activation by plant exudates and virG is both constitutively expressed and plant inducible(Stachel and Nester, 1986). The inducibility of virG was however questioned by Veluthambi et al. (1987). Whether virF contains a plant inducible promotor is not yet known. Sequence analysis of the promoter regions of the inducible genes virB, virC, virD and virE did not immediately show common recognition sites that might be involved in the regulation of vir-expression. Two boxes of hexamers(CGAGTA and GCAPuTT) present at different positions in the 5' upstream regions of the plant inducible vir-genes were mentioned as possible candidates for sequences involved in regulation(Das et al., 1986).

The activation of the vir-genes is a direct result of the recognition of specific metabolites in the wound sap of plants by Agrobacterium(Okker et al., 1984; Stachel et al., 1986a)(figure 4). The major inducing agents in the plant exudates of tobacco were purified and identified as acetosyringone(AS) and alpha-hydroxy-acetosyringone(OH-AS), but a wide range of other plant compounds were also found to be able to induce vir-gene expression(Stachel et al., 1985a). The identified plant signal molecules are also able to induce vir-gene expression in the bacterium when administrated in a pure form without any other plant material. It was reported however, that a cocktail of seven different inducing agents induced vir-expression to a higher level than each of these compounds individually(Bolton et al., 1986). It has to be noted however, that both AS and OH-AS were not present in this cocktail.

Mutations in the constitutively expressed virA or virG genes resulted in a loss of the stimulating effect of a plant exudate or AS on all vir-genes, indicating that the vir-genes are organized as a single regulon, the induction of which is controlled by virA and virG(Stachel and Zambryski, 1986).

The DNA sequence of both virA(Leroux et al., 1987; Melchers et al., 1987) and virG(Melchers et al., 1986; Winans et al., 1986a) were determined recently. Sequence comparison shows important homology between the VirA protein and the E.coli trans-membrane proteins encoded by the genes envZ, cpxA and phoR, while VirG was found to be closely related to the E.coli positive regulatory proteins OmpR, SfrA and PhoB. Nixon et al. (1986) proposed a model for such two factor regulatory systems, wherein one protein functions as a chemosensor, which becomes active after contact with a specific type of compound and then is able to transform the second protein into an active positive regulator of transcription. There are indications that (de)activation is achieved by the reversible (de)phosphorylation of the activator protein(Ninfa and Magasanik, 1986). Recent work with VirA specific antibodies showed that the 91,6Kda VirA protein is located in the cytoplasmic membrane of Agrobacterium(Leroux et al., 1987; Melchers et al., 1987). This is perfectly in line with the proposed function of VirA as a chemosensor for plant signal molecules such as acetosyringon(figure 4).

Role of VirB and VirD in T-region Transfer

The virD locus is an essential vir-locus, which was partly sequenced by Yanofsky et al. (1986). Recently the sequence of the complete virD locus -as defined by avirulent mutants- was determined(Thompson and Melchers, unpublished). It turned out that the virD locus (potentially) encodes four proteins of 16,2; 47,5; 21,4 and 72,4Kda. The first two proteins of the virD operon, called VirD1 and VirD2, are involved in the recognition of the T-region boundaries and the production of T-DNA intermediates. The functions of the other two essential VirD proteins have not yet been elucidated. From sequence analysis it can be concluded that the VirD1 protein has DNA binding properties(helix-turn-helix structure) and probably binds at the 24bp direct repeats of the T-region, although no experimental conformation of this hypothesis has been obtained till now. Proteins VirD1 and VirD2 together however are able to nick the 24bp T-region border repeats at a specific site and are the only Vir-proteins necessary for the formation of T-strands(Stachel et al., 1986b; Albright et al., 1987; Stachel et al., 1987; Wang et al., 1987). These VirD proteins are also essential for the induction of double stranded breaks in the 24bp repeats and for the formation of double stranded T-region circles(Alt Moerbe et al., 1986; Machida et al., 1986; Yanofsky et al., 1986; Veluthambi et al., 1987; Yamamoto et al., 1987). In a later section of this paper we will discuss these different observations in more detail.

The VirB locus is very large and encodes nine different proteins(Thompson and Melchers, unpublished). Some of the VirB proteins were found to be present in large quantities in the membranes of induced bacteria(Engstrom, cited in Stachel et al., 1987), which is in agreement with the hydrophilic character of many of the VirB proteins(Thompson and Melchers, unpublished). Therefore it might be that the VirB proteins are involved -as part of some transport apparatus- in the physical interaction between the bacteria and plant cells.

The Vir-loci VirC, VirE and VirF

VirC mutants display an attenuated or avirulent phenotype on certain host plants, but remain fully virulent on other host plants(Hooykaas et al., 1984; Yanofsky et al., 1985). The sequence of the VirC region of the octopine Ti-plasmid and part of the sequence of the nopaline Ti-plasmid was determined; the locus appeared to consist of two open reading frames encoding proteins of 25.7Kd and 22.7Kd(Close et al., 1987; Yanofsky and Nester, 1986). From experiments with small binary vectors, in which the vir-genes act in trans on the T-region(Horsch et al., 1986) as well as from experiments involving the isolation of T-region intermediates(Alt Moerbe et al., 1986), it was suggested that the VirC proteins might play a stimulating role in T-DNA intermediate formation, for instance by facilitating the interaction between the VirD proteins and the 24bp border repeats. Without the stimulating action of the VirC proteins it is possible that the process of T-DNA transfer becomes too inefficient to induce a tumorous response on certain host plants, that are less susceptable to the activity of the T-DNA onc-genes, which explains the host range effect of virC mutations.

The VirC and VirD operon are negatively regulated by the product of a chromosomal locus, called Ros(Close et al., 1985). Mutations in the ros locus resulted in elevated expression levels of both the virC and the virD operon. This negative control system operates independently of the positive regulation by the VirA and VirG proteins(Rogowsky et al., 1987). The function of this additional regulation system is not known.

Both virE and virF mutations affect the host range of the bacterium for tumor induction(Hooykaas et al., 1984; Hirooka and Kado, 1986). The virF locus is present in the octopine Ti-plasmid, the leucinopine Ti-plasmid and the Ri-plasmid, but not in the nopaline Ti-plasmid(Hooykaas et al., 1984; Otten et al., 1985). The virE locus was sequenced, which showed the presence of two -in the case of nopaline Ti three- open reading frames encoding proteins of 7.1 and 63.5Kd. The largest protein appeared to be very hydrophylic(Hirooka and Kado, 1986; Winans et al., 1986b). Mixed inoculation experiments revealed that both the virE and virF determined products are excreted by the bacterium(Otten et al., 1984; Otten et al., 1985). It could be that they play a role in the physiological conditioning of plant cells at the wound site of the plant or function in optimizing the transfer process from the bacterium to the plant cell. These optimal conditions might be more important during the interaction of the bacterium with certain plants than with others, explaining the fact that these vir-genes like the virC genes are only nessecary during the infection of certain host plants.

THE T-REGION

The T-region Genes

After the T-DNA is transferred and integrated at a proper location in the plant genome, its genes are expressed. The T-regions of the Ti- and Ri-plasmid contain several genes. Sequence analysis of the T-region of the octopine Ti-plasmid(Barker et al., 1983) and that of the agropine Ri-plasmid(Slightom et al., 1986) revealed that the genes present in the T-region contain promoter and terminator structures that have been reported to be nessecary for the expression of plant genes. More detailed genetic analysis of the promoter and terminator structures of certain T-region genes provided insight into the precise sequences that are nessecary for expression in plant cells(Shaw et al., 1984a; An et al., 1986; Bruce and Gurley, 1987; De Pater, 1987).

Three of the genes present in the T-region of the octopine and nopaline Ti-plasmids are directly involved in the biosynthesis of the plant hormones auxin(aux-genes) and cytokinin(cyt-gene) in the transformed plant cells(Akiyoshi et al., 1984; Barry et al., 1984; Thomashow et al., 1984; Buchmann et al., 1985; Thomashow et al., 1986). The acquired ability to produce these phyto-hormones continuously gives the tumor tissue the property to grow on hormone free media(Braun, 1958). Mutations in one or more of these onc-genes resulted in bacteria that were avirulent on certain plant species and which formed tumors of an altered morphology on other plant species(Ooms et al., 1981). Infection with mutants with a mutation in one or both of the genes involved in auxin production led to the formation of shoots at the infection site, while infection with mutants with a disruption in the cytokinin gene resulted in root formation(Garfinkel et al., 1981; Ooms et al., 1981). Mutations in the aux-locus as well as the cyt-locus resulted in strains that had lost their oncogenicity on several plant species(Leemans et al., 1982; Hille et al., 1983).

One or more of the T-DNA genes codes for enzymes involved in the synthesis of tumor specific compounds(opines). The opines are excreted by these plant cells via an enzyme encoded by the T-DNA(Messens et al., 1985a; Messens et al., 1985b). Opines can be used by *Agrobacterium* as a sole source of carbon and/or nitrogen(Bomhoff et al., 1976). The opine catabolic functions are encoded by the Ti- and Ri-plasmids.

Mutagenesis of the T-region revealed that disruption or deletion of T-region genes does not lead to a loss of the ability to transfer the T-DNA to plant cells(Leemans et al., 1982; Hille et al., 1983). Therefore, none of the T-region genes is essential for the T-region transfer process, which is in line with findings that the essential functions are encoded by the Vir-region and the bacterial chromosome.

The length of the T-region(8-20Kbp) seems not to be of influence on the DNA-transfer process. The T-regions that are present on the different types consensus : GGCAGGATATATxxxxxTGTAAxx

LB oct. TL-DNA: GGCAGGATATATTCAATTGTAAAC

RB oct. TL-DNA: GGCAGGATATATACCGTTGTAATT

LB oct. TR-DNA: GGCAGGATATATCGAGGTGTAAAA

RB oct. TR-DNA: GGCAGGATATATGCGGTTGTAATT

LB nop. T -DNA: GGCAGGATATATTGTGGTGTAAAC

RB nop. T -DNA: GACAGGATATATTGGCGGGTAAAC

LB Ri TL-DNA: GGCAGGATATATTGTGATGTAAAC

RB Ri TL-DNA: GACAGGATATATGTTCTTGTCATG

Fig. 5. The 24bp Border repeat sequences of the octopine Ti, TL- and TR-region, the nopaline Ti, T-region and the Ri, TL-region.

of Ti- and Ri-plasmids differ in length but are always bordered by 24bp imperfect direct repeats(Simpson et al., 1982; Yadav et al., 1982; Barker et al., 1983)(figure 5). From deletion mutagenesis it is known that these 24bp border repeats are important in the T-DNA transfer process(Ooms et al., 1982; Shaw et al., 1984b; Wang et al., 1984). Therefore the only two elements present on Ti- and Ri-plasmids nessecary to transfer DNA to the plant cell are the Vir-region and the border repeats of the T-region. After deletion of the complete sequence between the two border repeats surrounding the T-region, foreign DNA with a large size can be inserted between the borders, which subsequently can be transferred to the plant cell via Agrobacterium.

The Role of the Border Repeats in the Transfer Process

A lot of experiments have been done to get some more insight in the mechanism of T-DNA transfer. Initially these experiments focussed on the role of the 24bp border repeats in the transfer process. Deletion mutants of the border repeats showed a clear difference in function between the left and right border repeat. Deletion of the left border repeat did not lead to a diminished T-region transfer to the plant cell(Hille et al., 1983; Joos et al., 1983a), whereas deletion of the right border repeat almost completely abolished the transfer process(Ooms et al., 1982; Shaw et al., 1984b; Wang et al., 1984). It was, however, also observed that a T-region lacking a right border repeat can, although very inefficiently still be transferred to the plant cell, which could be monitored only on very sensitive plants(Koekman et al., 1979; Ooms et al., 1982; Hepburn and White, 1985). The result that a left border repeat is less important in the tranfer process than a right border repeat, might also be concluded from the fact that in transformed plant cells the integration site of the T-region in the plant genome is more accurate at the right junction than at the left junction(Simpson et al., 1982; Yadav et al., 1982; Zambryski et al., 1982; Holsters et al., 1983; Kwok et al., 1985; Slightom et al., 1985; Jorgenson et al., 1987).

The strong homology between right and left 24bp repeats at the sequence level made it unlikely that the minor differences in these border repeats were the cause of the observed functional difference(figure 5). In order to be able to examine the role of the border repeats independently of each other, experiments were performed in which border repeats were tested on small binary plant vectors, on which the Vir-functions act *in trans*. These experiments confirmed that right borders are more efficient in T-DNA transfer than left borders(Horsch and Klee, 1986; Jen and Chilton, 1986; Rubin, 1986).

However, when synthetic left and right border repeats were used in such studies it turned out that both types of border repeats were equally inefficient to transfer T-DNA to the plant cell(Peralta et al., 1986; Van Haaren et al., in prep.). This proved that the 24bp border repeats are sufficient for the transfer of the T-region to the plant cell, although the efficiency of transfer mediated by the synthetic border repeats was much lower than with the cloned larger border fragments. This directly indicated that a second sequence, next to the border repeats, is involved in the transfer process. The difference between the left and right border repeat, apart from their different position relative to the *onc*-genes, turned out to be due to a 24bp sequence that is present directly at the right hand side of the right border repeat, but not present near the left border repeat. This sequence enhances T-region transfer mediated by the right border repeat, but can not mediate T-DNA transfer by itself and therefore was called "enhancer"(Van Haaren et al., 1986) or "overdrive"(Peralta et al., 1987). Sequence comparison with other right border regions revealed that this 24bp enhancing sequence is not strongly conserved between the different Ti- and Ri- plasmids. However 75-100% homology was found with an 8bp core sequence 5'TGTTTGTT3' of the enhancer.

Experiments in which the right border repeat was replaced by a left border repeat, which was now present next to the enhancer, proved that left and right border repeats are functionally equivalent indeed(Van Haaren et al., 1987).

A general property of enhancers is that they can stimulate a process when present in different positions and even at a large distance from the site on the DNA where the process takes place. This is true for enhancers stimulating recombination and DNA-inversion (Plasterk et al., 1983; Plasterk and Van De Putte, 1985; Huber et al., 1985; Johnson and Simon, 1985), transcription (Goodbourn et al., 1985; Dyan and Tjian, 1985) and replication(Delucia et al., 1986). In order to test whether the T-region transfer enhancer shares these properties, we did experiments in which the position of the enhancer sequence relative to the border repeat was varied. These experiments demonstrated that even when the enhancer sequence was present on the other side of the border repeat, *i.e.* inside the T-region, it could stimulate the T-region transfer mediated by the border repeat(Van Haaren et al., 1987). The distance between the enhancer and the border repeat could also be enlarged considerably. Recent experiments showed that the enhancer sequence still retains its stimulating effect on T-DNA transfer if it is present at a distance of up to 6-7Kbp at the right hand site of the border repeat(Van Haaren et al., in prep.). From these experiments it can be concluded that the T-region transfer enhancer has the same characteristics as described for other enhancer sequences.

The orientation of the border repeat versus the T-region turned out to be important in the transfer process. When the right border repeat was inverted relative to the *onc*-genes, T-region transfer became almost undetectable(Wang et al., 1984; Caplan et al., 1985; Rubin, 1986; Peralta et al., 1986; Van Haaren et al., 1987). This indicated that the transfer of DNA to the plant cell is an orientation dependent process, starting at the right border repeat and ending at the left border repeat. This is in agreement with the fact that right borders are much more important than left borders for T-DNA transfer. However, with low efficiency also sequences on the other side of the border repeat are transferred to the plant cell. Also in this case the frequency is higher if an enhancer sequence is present in the vicinity of the border repeat(Van Haaren et al., 1987). Joos et al. (1983b) found that *onc*-genes that had been inserted in the Ti-plasmid somewhere outside the T-region were still transferred to the plant cell. This might have been due to transfer via the left border repeat or via the right border repeat in the reverse orientation, although the presence of pseudo borders elsewhere in the Ti-plasmid cannot be excluded. Pseudo borders having strong homology with the consensus border repeat are present in the T-region(Van Haaren et al., 1986).

Future experiments will be focussed on the proteins that act on the 24bp border repeats and the 24bp enhancer sequence. Mutagenesis studies will reveal the nucleotides in the border and enhancer sequences that are essential for protein binding on these elements and will give more insight in the molecular mechanism of the first steps in the T-region transfer process.

Mechanism of T-DNA Transfer

The detection of T-DNA intermediate structures in Agrobacterium might help us to get a better idea about the steps preceeding the actual transfer of the T-region to the plant cell. Recently several different approaches were used to isolate such intermediate structures from induced bacterial cultures.

The first succesfull report on the isolation of T-region intermediates was published by Koukolikova-Nicola et al. (1985). After induction by plant exudates, DNA was isolated from an Agrobacterium strain with the E.coli plasmid pBR322 between its T-region border repeats and used this DNA for transformation of E.coli with selection for the markers of pBR322. Transformants were obtained with a low frequency, which turned out to contain plasmids originating from the T-region. The plasmids had a single border repeat which proved to be the result of a fusion between the left and right border repeat. This could either be due to a recombination event between the border repeats, or of a cleavage-rejoining reaction(Koukolikova-Nicola et al., 1985). These results were confirmed and extended by Machida et al. (1986) and Alt-Moerbe et al. (1986), who used genetic techniques to select directly for T-region circles in Agrobacterium and E.coli(containing portions of the Vir-region), respectively. The results showed that T-region circle formation is dependent on the presence of two T-region border repeats in a direct orientation(Machida et al., 1986) and on the presence of a portion of the VirD operon(Alt-Moerbe et al., 1986; Yamamoto et al., 1987).

More recent experiments showed that the induction of agrobacteria by acetosyringon(AS) generates single stranded linear T-DNA molecules, called T-strands(Stachel et al., 1986b). The single stranded molecules could be visualized on Southern blots of gels loaded with undigested total DNA isolated from the induced bacterium. By using radioactive probes of both the bottom and upper strand of the T-region, it was shown that the T-strands correspond to the bottom strand of the T-region. Single stranded molecules corresponding to the upper strand were not be detected in these experiments. Double stranded linear or circular molecules with the size of the T-region could not be detected either. Only S1-nuclease treatment of the isolated DNA resulted in the appearence of bands corresponding to double stranded T-region molecules. This was interpreted to mean that nicks are present at the border repeats, which make the opposite strand sensitive to cleaveage by S1-nuclease(Stachel et al., 1986b). Nicks could indeed be detected in both the left and right 24bp border repeats of AS induced agrobacterial cultures(Yanofsky et al., 1986; Albright et al., 1987; Wang et al., 1987) and also in E.coli cells containing border constructs(Yanofsky et al., 1986). By using primer extension and RNA protection the major nick site was localized just after the third nucleotide of the bottom strand in both 24bp border repeats, although a number of minor nick sites were detected as well(Albright et al., 1987; Wang et al., 1987). The fact that T-DNA plant junctions have never been found to the right of the nick site at the right border repeat and never to the left of the left border repeat indicates that nicking indeed is a step in T-DNA processing. The production of bottom strand nicks is also in line with the production of single strand intermediates that correspond with the bottom strand of the T-region. Nicking of the border repeats seems to be independent of the presence of an enhancer sequence next to the border repeat, because nicking was found to take place on both left and right border repeats with the same efficiency(Wang et al., 1987). Besides bottom strand

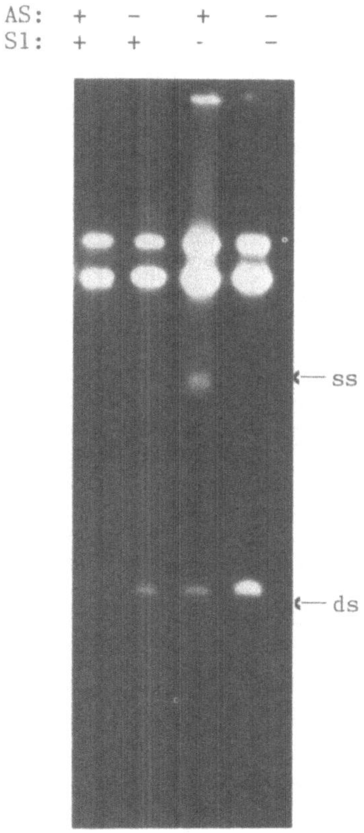

Fig. 6. Southern blot of a gel loaded with HindIII digested total DNAs from induced(AS+) and non-induced(AS-) Agrobacterium strain LBA5207. Half of the samples were digested with S1-nuclease before loading the gel. As a probe the HindIII fragment 18c was used which contains the left border of the octopine Ti-plasmid. Indicated with arrows are the T-strands(ss) and the fragments that are formed after a double stranded break in the left border repeat(ds).

nicks with low frequency also top strand nicks were detected, which suggests that with low frequency also double strand breaks might occur in the border repeats(Albright et al., 1987).

At the time of detection of single stranded border nicking, double stranded cleavages were reported to occur in the border repeats after induction of the bacterium(Veluthambi et al., 1987). Remarkably, double stranded border cleavage proved to occur with higher efficiency in left border repeats than in right border repeats. In these experiments detection of double strand cleavages in the DNA of induced bacteria was not dependent on treatment with S1-nuclease, whereas this was nessecary in the experiments of Stachel et al. (1986b) for the detection of double stranded T-region molecules.

The fact that both single stranded intermediates, as well as double stranded intermediates are present together in AS-induced bacteria was established in our recent experiments(figure 6). Total DNA was isolated from AS-induced bacteria and digested, before gel electrophoresis and Southern blotting. If the blot was hybridized with a left border specific probe, both the single stranded intermediates of the size of the T-region(sensitive for S1-nuclease) and the double stranded fragments that occur after cleavage on the left border repeat were detected.

Future experiments must reveal whether all of the events discussed above are indeed intermediate steps in T-DNA formation. Bottom nicking of right border repeats followed by 5'-3' synthesis -with the top strand as a template- towards the nicked left border could easily be envisaged to lead to the displacement of T-strands corresponding to the bottom strand of the T-region. Likewise double stranded breaks at both left and right border repeats would result directly in the release of double stranded linear T-region molecules. The double stranded T-region circles that were found with low frequency after strong selection most likely are due to a rare recombination reaction between the border repeats. Because of their rare occurence, it is doubtful whether these circular molecules are true T-DNA intermediates. They might represent products of an unusual side-reaction. An important question is whether the linear single stranded and double stranded molecules are both true T-DNA intermediates which form part of the same pathway or of different pathways leading to T-DNA transfer. If part of the same pathway the process requires the conversion of the single stranded molecule into a double stranded one. That this may occur to some extent is suggested by the fact that in experiments, in which border cleavage is visualized, the new bands representing the part inside the T-region, were more prominent than the "outside" bands(Stachel et al., 1987; Van Haaren et al., in prep.). Moreover, since synthesis of the top strand would start at the left border, the left part of the T-DNA would become double stranded first, which would lead to the detection of putative left border cleavages with higher frequency than right border cleavages.

Interestingly all discussed intermediate structures are formed after induction of only the first two genes of the virD operon(Alt-Moerbe et al., 1986; Machida et al., 1986; Yanofsky et al., 1986; Stachel et al., 1987; Veluthambi et al., 1987). However besides proteins nessecary for the nicking and cleavage of the border repeats, other functions like for instance a polymerase and helicase function are also likely to by involved in the initial steps of T-strand formation in the bacterium. These functions should be encoded by the bacterial chromosome, because the two VirD genes alone are not enough to fulfil all of these functions.

The role of the enhancer sequence in these processes is not yet very clear. It might stimulate either double stranded or single stranded intermediate formation by facilitating the recognition of the border repeat binding site by Vir-products, or by putting the border repeats in the right position with respect to each other, like was suggested for the Gin-mediated inversion process(Kanaar and Van De Putte, 1987). Experiments of Stachel et al. (1987) showed that both left and right border repeats can function as the initiation and termination sites for T-strand formation. However T-DNA structures with as a right terminus a left border repeat have never been found in plant cells. This might suggest that the enhancer present next to the right border repeat is involved in the production of double stranded T-molecules, after a T-strand has been formed.

PLANT VECTORS

The properties of the Agrobacterium make this bacterium a suitable vehicle for the transformation of plant cells. The development of plant vectors is of course important not only for basic research in plants, but also for applications in the field of plant genetic engineering.

Apart from Agrobacterium several other plant transforming agents -such as DNA and RNA viruses- have been studied as potential vectors. Examples are

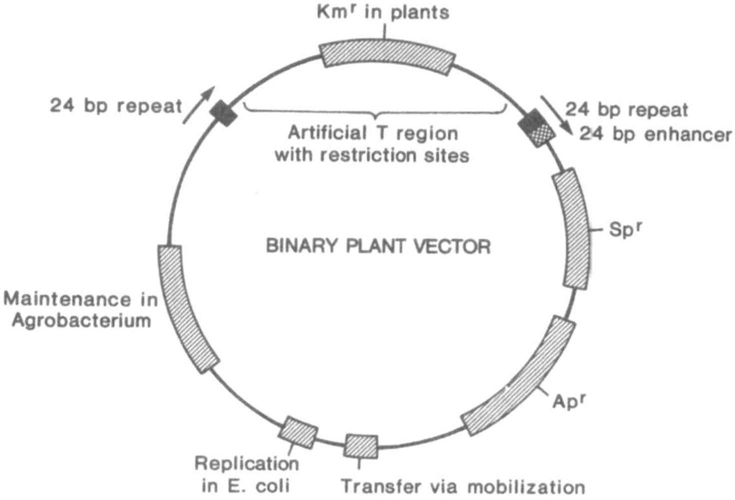

Fig. 7. A schematic map of a binary plant vector. Ap= Ampicillin resistance; Sp= spectinomycin resistance; Km= Kanamycin resistance.

cauliflower mosaic virus(CaMV), a circular dsDNA virus in which a non essential open reading frame can be replaced by foreign DNA(Paszkowski et al., 1986), and the bromo mosaic virus(BMV), a tripartite RNA virus in which the gene for the coat protein in RNA3 can be replaced by a foreign gene(French et al., 1986). Disadvantages of these virus vectors are that the host range is usually limited and that the amount of foreign DNA with which they can be loaded is rather small. In addition, the DNA introduced will not become integrated in the chromosome.

Advantages of the Agrobacterium system are that it can be manipulated easily, that the host range is wide -including both monocots and dicots- and that a large piece of DNA(up to at least 20Kbp) can be transferred as T-DNA to the plant cell, where it is stably integrated in the nuclear genome. Moreover, coinfection with two different Agrobacterium strains leads to cotransformation of the two T-DNA's with high frequency(Depicker et al., 1985; Petit et al., 1986). Furthermore plant cells which already received a "T-DNA" remain recipients of T-DNA in future transformation experiments, which makes it possible to further engineer already transformed cells.

Two different strategies are in use to apply Agrobacterium as a sophisticated DNA micro-injector for the transformation of plant cells. In both strategies the onc-genes must be deleted from the T-region first, since their presence would inhibit the regeneration of the transformed cells into normal plants.

A major problem hindering the use of the Ti-plasmid as a plant vector was its enormous size. This size made it impossible to manipulate the T-region directly. The strategy initially used to circumvent this problem was the use of homologous recombination to introduce foreign DNA between the border repeats of an intact Ti-plasmid(e.g. Zambryski et al., 1983; Fraley et al., 1985). An alternative strategy was devised later, which was based on the finding that a physical separation of the T-region from the rest of the Ti-plasmid does not interfere with virulence(De Framond et al., 1983; Hoekema et al., 1983). In this strategy which is called the binary vector system, the vir-genes act in trans to mediate T-region transfer from a vector present within the same bacterium. Preferably small wide host range plasmids are used as ("binary") vectors in the binary vector system, since these can easily be manipulated in E.coli before transfer to Agrobacterium. Vectors that can be used in the binary vector system have been described by several groups(Bevan, 1984; An et al., 1985; Hoekema et al., 1985; Klee et al., 1985; Van Der Elzen et al., 1985; An, 1986; Koncz and Schell, 1986; Simoens et al., 1986).

The ideal plant vector consists of a wide host range replicon, which enables it to be replicated in E.coli as well as in Agrobacterium, markers for selection in both bacterial hosts and the 24bp border repeats with the 24bp enhancer sequence. Moreover, between the border repeats several unique restriction sites should be present for the introduction of the genes that are to be introduced into the plant(figure 7). Many different chimaeric genes are already available that can serve as selection markers in plant cells, viz. for kanamycin(Bevan and Flavell, 1983), hygromycin(Van Der Elzen et al., 1985), methotrexate(Eichholtz et al., 1987) and bleomycin(Hille, et al., 1986). But also markers which can be used for quantitative assays are available for plants viz. based on cloramphenicol acetyl transferase(Herrera-Estrella et al., 1983), luciferase gene of the firefly(Ow et al., 1986;), luciferase gene of the bacterium Vibrio harveyi(Koncz et al., 1987) and E.coli beta-glucuronidase(Jefferson et al., 1986). The expression signals of T-region genes such as those of ocs and nos or those of CaMV are often used for the expression of these plant selectable markers(Sanders et al., 1987), but of course also tissue specific regulatory signals can be used. By introducing genes in the plant vector that areequiped with a transit peptide of a gene such as rbcS, it is possible to direct foreign proteins to the plant chloroplasts(Herrera-Estrella et al., 1984; Schreier et al., 1985; Van Den Broeck et al., 1985).

Since the T-DNA inserts rather randomly in the plant genome, this property is used to develop plant vectors with which promoters can be isolated from the plant genome. Such vectors consist of a selectable marker without plant promoter, so that the expression only occurs after insertion of the T-DNA behind a plant promoter. The vectors are equiped with a T-region with a bacterial selective marker and a cos-site, to facilitate cloning of the T-region from the plant genome(Andre et al., 1986;Teeri et al., 1986).

T-DNA Integration and Expression in the Plant Cell

The transfer of the T-region to the plant cell mediated by Agrobacterium is not always as accurate as was expected initially. Aberrant T-DNA structures can be found in some transformed plant tissues(Peerbolte et al., 1986; Van Lijsebettens et al., 1986; Spielman and Simpson, 1986). In some cases the transferred T-DNA turned out to be truncated or to have internal deletions. Tandem and inverted repetitions of T-DNA heve also been encountered(Kwok et al., 1985; Jorgensen et al., 1987). But even when the T-region is inserted properly into the plant genome, the expression of the inserted genes varies in different transformed lines. This is due to the position of the genes in the plant genome, where expression can be influenced by neighbouring plant expression signals or by modification of the DNA such as methylation which can inhibit expression of genes(Hepburn et al., 1983; Van Slogteren et al., 1984). Position effects on expression of introduced genes should be taken into account if Agrobacterium, or other tranformation systems, are used to insert foreign DNA in the plant genome. The transformed plant lines must therefore always be examined individually for their (artificial) T-DNA content and on the copy number and expression level of the inserted T-DNA.

Other Transformation Systems

Apart from tranformation of plant cells via Agrobacterium a number of alternative methods have been developed, which require the isolation of protoplasts. Such plant protoplasts can be transformed by microinjection of

DNA directly into the plant cell nucleus(Crossway et al., 1986), or by incubation with DNA in the presence of PEG/Ca(Krens et al., 1982;Lorz et al., 1985) or by a method called electroporation, in which DNA is taken up by the protoplasts after an electric pulse (Langridge et al., 1985; Fromm et al., 1985; Fromm et al., 1986; Riggs and Bates., 1986).

When DNA containing the T-region border repeats was introduced into plant protoplasts via one of these methods, integration of the DNA did not occur on the border repeats. This shows that the border repeats are not specifically recognized by the plant system, which is in agreement with its proposed role as a signal for transfer in the bacterium(Krens et al., 1985; Peerbolte et al., 1985).

CONCLUDING REMARKS

The Agrobacterium transfer system is being applied for stable transformation of plant cells with foreign DNA, and transformation by A.rhizogenes can more specifically be applied for the controlled induction of root formation. The transformation system can be used for a wide range of dicotyledonous plants and monocotyledonous plants. Whether there will be a limit to the host range of agrobacteria remains an open question, but more and more plants appear to be susceptible to transformation via Agrobacterium. The reason why some plants are not or only weakly susceptible to transformation by Agrobacterium might be the result of the absence of Vir-inducing compounds in the plant exudate(Usami et al., 1987). Addition of a Vir-inducer during infection can lead to optimized transformation, as was shown to be the case for Arabidopsis(Sheikholeslam et al., 1987). However for most plant species transformation itself is not a problem anymore, but rather the regeneration of the transformed plant cells into complete plants, especially when starting from protoplasts.

It should be noted here that the site of integration of the foreign DNA in the plant genome is important for the level of expression of the newly inserted genes. Therefore, when using a plant transformation system, one should always carefully check the level of expression of the introduced genes and then select the lines which give optimal expression.

To optimize the Agrobacterium system it is nessecary to obtain more information about the mechanism of T-region transfer. Some of the early steps of this process are currently under study, but much is still unknown. For instance the precise role of the enhancer sequence has not yet been elucidated. A better insight into the transfer mechanism might result in the production of more efficient plant vector systems.

Studies on the plant inducible promoters of the vir-genes in Agrobacterium are also of importance for application. For instance these promoters might be used to construct bacteria that produce certain products(like insecticides), only when they are in the rhizosphere or phyllosphere of the plant.

LITERATURE

Akiyoshi, D. E., Klee, H., Amasino, R. M., Nester, E. W., and Gordon, M. D., 1984, T-DNA of Agrobacterium tumefaciens encodes an enzyme of cytokinin biosynthesis, Proc. Natl. Acad. Sci. USA, 81:5994.

Akiyoshi, D. E., Regier, D. A., Jen, G. and Gordon, M. P., 1985, Cloning and nucleotide sequence of the tzs gene from Agrobacterium tumefaciens strain T37, Nucleic Acids Res., 13:2773.

Albinger, G. and Beiderbeck, R., 1977, Ubertragung der fahigkeit zur wurzelinduktion von Agrobacterium rhizogenes auf A.tumefaciens, Phytopathol. Z., 90:306.

Albright, L. M., Yanofsky, M. F., Leroux, B., Ma, D. and Nester, E. W., 1987, Processing of the T-DNA of Agrobacterium tumefaciens generates border

nicks and linear, single-stranded T-DNA, J. Bacteriol., 169:1046.
Allen, O. N. and Holding, A., 1974, Genus Agrobacterium, in: "Bergey's Manual of Determinative Bacteriology", R. E. Buchanan and N. E. Gibbons, eds., Williams and Wilkins, Baltimore.
Alt-Moerbe, J., Rak, B. and Schroder, J., 1986, A 3,6Kbp segment from the vir region of Ti-plasmids contains genes responsible for border sequence-directed production of T-region circles in E.coli, EMBO J., 5:1129.
An, G., Watson, B. D., Stachel, S., Gordon, M. P. and Nester, E. W., 1985, New cloning vehicles for transformation of higher plants, EMBO J., 4:277.
An, G., 1986, Development of plant promoter expression vectors and their use for analysis of differential activity of nopaline synthase promoter in transformed tobacco cells, Plant physiol., 81:86.
An, G., Ebert, P. R., Yi, B-Y. and Choi, C-H., 1986, Both TATA box and upstream regions are required for the nopaline synthase promotor activity in transformed tobacco cells, Mol. Gen. Genet., 203:245
Andre, D., Colau, D., Schell, J., Van Montagu, M. and Hernalsteens, J. P., 1986, Gene tagging in plants by a T-DNA insertion mutagen that generates APH(3') II plant gene fusions, Mol. Gen. Gent., 204:512.
Barker, R. F., Idler, K. B., Thompson, D. V. and Kemp, J. D., 1983, Nucleotide sequence of the T-DNA region from Agrobacterium tumefaciens octopine Ti plasmid pTi15955, Plant Mol. Biol., 2:335.
Barry, G. F., Rogers, S. G., Fraley, R. T. and Brand, L., 1984, Identification of a cloned cytokinin biosynthetic gene, Proc. Natl. Acad. Sci. USA, 81:4776.
Beaty, J. S., Powell, G. K., Lica, L., Regier, D. A., MacDonald, E. M. S., Hommes, N. G. and Morris, R. O., 1986, Tzs, a nopaline Ti-plasmid gene from Agrobacterium tumefaciens associated with trans-zeatin biosynthesis, Mol. Gen. Genet., 203:274.
Bevan, M. W. and Chilton, M. D., 1982, T-DNA of the Agrobacterium Ti and Ri plasmids, Ann. Rev. Genet. 16:357.
Bevan, M. W. and Flavell, R. B., 1983, A chimaeric antibiotic resistance gene as a selectable marker for plant cell transformation, Nature, 304:184.
Bevan, M., 1984, Binary Agrobacterium vectors for plant transformation, Nucleic Acids Res., 12:8711.
Bolton, G. W., Nester, E. W. and Gordon, M. P., 1986, Plant phenolic compounds induce expression of the Agrobacterium tumefaciens loci needed for virulence, Science, 232:983.
Bomhoff, G., Klapwijk, P. M., Kester, H. C. M., Schilperoort, R. A., Hernalsteens, J. P. and Schell, J., 1976, Octopine and nopaline synthesis and breakdown genetically controled by a plasmid of Agrobacterium tumefaciens, Mol. Gen. Genet., 145:177.
Braun, A. C., 1958, A physiological basis for autonomous growth of crown gall tumor cells, Proc. Natl. Acad. Sci. USA, 44:344.
Bruce, W. B. and Gurley, W. B., 1987, Functional domains of a T-DNA promoter active in crown gall tumors, Mol. Cell. Biol., 7, 59.
Buchmann, I., Marner, F-J., Schroder, G., Waffer-Schmidt, S. and Schroder, J., 1985, Tumour genes in plants: T-DNA encoded cytokinin biosynthesis, EMBO J., 4:853.
Caplan, A. B., Van Montagu, M. and Schell, J., 1985, Genetic analysis of integration mediated by single T-DNA borders, J. Bacteriol., 161:664.
Close, T. J., Tait, R. C. and Kado, C. I., 1985, Regulation of Ti plasmid virulence genes by a chromosomal locus of Agrobacterium tumefaciens, J. Bacteriol., 164:774.
Close, T. J., Tait, R. C., Rempel, H. C., Hirooka, T, Kim, L. and Kado, I., 1987, Molecular characterization of the virC genes of the Ti plasmid, J. Bacteriol., submitted.
Crossway, A., Oakes, J. V., Irvine, J. M., Ward, B., Knauf, V. C. and Shewmaker, C. K., 1986, Integration of foreign DNA following microinjection of tobacco mesophyll protoplasts, Mol. Gen. Genet., 202:179.

Das, A., Stachel, S., Ebert, P., Allenza, P., Montoya, A. and Nester, E., 1986, Promotors of Agrobacterium tumefaciens Ti-plasmid virulence genes, Nucleic Acids Res., 14:1355.
Deblaere, R., Bytebier, B., De Greve, H., Deboeck, F., Schell, J., Van Montagu, M. and Leemans, J., 1985, Efficient octopine Ti plasmid derived vectors for Agrobacterium mediated gene transfer to plants, Nucleic Acids Res., 13:4777.
De Cleene, M. and De ley, J., 1977, The host range of crown gall, Bot. Rev., 42:389.
De Framond, A. J., Barton, K. A. and Chilton, M. D., 1983, Mini-Ti: a new vector strategy for plant genetic engineering, Bio/Technology, 1:262.
Delucia, A. L., Deb, S., Partin, K and Tegtmeyer, P., 1986, Functional interactions of the simian virus 40 core origin of replication with flanking regulatory sequences, J. Virology, 57:138.
De Pater, B. S., 1987, "Plant expression signals of the Agrobacterium T-cyt gene", PhD. thesis, Leiden University.
Depicker, A., Herman, L., Jacobs, A., Schell, J. and Van Montagu, M., 1985, Frequencies of simultaneous transformation with different T-DNA's and their relevance to the Agrobacterium/plant cell interaction, Mol. Gen. Genet., 201:477.
Douglas, C. J., Staneloni, R. J., Rubin, R. A. and Nester, E. W., 1985, Indentification and genetic analysis of an Agrobacterium tumefaciens chromosomal virulence region, J. Bacteriol., 161:850.
Drummond, M. H. and Chilton, M-D., 1978, Tumor-inducing(Ti) plasmids of Agrobacterium share extensive regions of DNA homology, J. Bacteriol., 136:1178.
Dynan, W. S. and Tjian, R., 1985, Control of eukaryotic messenger RNA synthesis by sequence-specific DNA-binding proteins, Nature, 316:774.
Eichholz, D. A., Rogers, S. G., Horsch, R. B., Klee, H. J., Hayford, M., Hoffmann, N. L., Braford, S. B., Fink, C., Flick, J., O'Connell, K. M. and Fraley, R. T., 1987, Expression of mouse dihydrofolate reductase gene confers methotrexate resistance in transgenic petunia plants, Som. Cell. Mol. Genet., 13:67.
Engler, G., Depicker, A., Maenhaut, R., Villarroel, R., Van Montagu, M. and Schell, J., 1981, Physical mapping of DNA base sequence homologies between an octopine and a nopaline Ti-plasmid of Agrobacterium tumefaciens, J. Mol. Biol., 152:183.
Fraley, R. T., Rogers, S. G., Horsch, R. B., Eichholtz, D. A., Flick, J. S., Fink, J. S., Hoffmann, N. L. and Sanders, P. R., 1985, The SEV system: a new disarmed Ti-plasmid vector system for plant transformation, Bio/Technology, 3:629.
French, R., Janda, M. and Ahlquist, P., 1986, Bacterial gene inserted in an engineerd RNA virus: efficient expression in monocotyledonous plant cells, Science, 231:1294.
Fromm, M. E., Taylor, L. P. and Walbot, V., 1986, Stable transformation of maize after gene transfer by electroporation, Nature, 319:793.
Garfinkel, D. J. and Nester, E. W., 1980, Agrobacterium tumefaciens mutants affected in crown gall tumorigenesis and octopine catabolism, J. Bacteriol. 144:732.
Garfinkel, D. J., Simpson, R. B., Ream, L. W., White, F. F., Gordon, M. P. and Nester, E. W., 1981, Genetic analysis of crown gall: fine structure map of the T-DNA by site directed mutagenesis, Cell, 27:143.
Goodbourn, S., Burstein, H. and Maniatis, T., 1986, The human beta-inteferon gene enhancer in under negative control, Cell, 45:601.
Graves, A. C. F. and Goldman, S. L., 1986, The transformation of Zea mays seedlings with Agrobacterium tumefaciens, Plant Mol. Biol., 7:43.
Graves, A. C. F. and Goldman, S. L., 1987, Agrobacterium tumefaciens-mediated transformation of the monocot genus Gladiolus: detection of expression of T-DNA-encoded genes, J. Bacteriol., in press.
Grimsley, N., Hohn, T., Davies, J. W. and Hohn, B., 1987, Agrobacterium-mediated delivery of infectious maize streak virus into maize plants,

Nature, 325:177.
Hepburn, A. G., Clarke, L. E., Pearson, L. and White, J., 1983, The role of cytosine methylation in the control of nopaline synthase gene expression in a plant tumor, J. Mol. Appl. Genet., 2:315.
Hepburn, A. G. and White, J., 1985, The effect of right terminal repeat deletion on the oncogenicity of the T-region of pTiT37, Plant Mol. Biol., 5:3.
Hernalsteens, J. P., Thia-Toong, L., Schell, J. and Van Montagu, M., 1984, An Agrobacterium tumefaciens-transformed cell culture from the monocot Asparagus officinalis, EMBO J., 3:3039.
Herrera-Estrella, L., De Block, M., Messens, E., Hernalsteens, J. P., Van Montagu, M. and Schell, J., 1983, Chimeric genes as dominant selectable markers in plant cells, EMBO J., 2:987.
Herrera-Estrella, L., Van Den Broeck, G., Maenhaut, R., Van Montagu, M., Schell, J., Timko, M. and Cashmore, A., 1984, Light-inducible and chloroplast-associated expression of a chimaeric gene introduced into Nicotiana tabacum using a Ti plasmid vector, Nature, 310:115.
Hille, J., Klasen, I. and Schilperoort, R. A., 1982, Construction and application of R-prime plasmids, carrying different segments of an octopine Ti plasmid from Agrobacterium tumefaciens, for complementation of vir genes, Plasmid, 7:107.
Hille, J., Wullems, G. and Schilperoort, R., 1983, Non-oncogenic T-region mutants of Agrobacterium tumefaciens do transfer T-DNA into plant cells, Plant Mol. Biol., 2:155.
Hille, J., Verheggen, F., Roelvink, P., Franssen, H., Van Kammen, A. and Zabel, P., 1986, Bleomycin resistance: a new dominant selectable marker for plant transformation, Plant Mol. Biol., 7:171.
Hirooka, T. and Kado, C. I., 1986, Lacation of the right boundary of the virulence region on Agrobacterium tumefaciens plasmid pTiC58 and a host-specifying gene next to the boundary, J. Bacteriol., 168:237.
Hoekema, A., Hirsch, P. R., Hooykaas, P. J. J. and Schilperoort, R. A., 1983, A binary plant vector strategy based on separation of Vir- and T-region of the Agrobacterium tumefaciens Ti-plasmid, Nature, 303:179.
Hoekema, A., Hooykaas, P. J. J. and Schilperoort, R. A., 1984, Transfer of the octopine T-DNA segment to plant cells mediated by different types of Agrobacterium tumor- or root-inducing plasmids: generality of virulence systems, J. Bacteriol., 158:383.
Hoekema, A., Van Haaren, M. J. J., Fellinger, A. J., Hooykaas, P. J. J. and Schilperoort, R. A., 1985, Non-oncogenic plant vectors for use in the Agrobacterium binary system, Plant Mol. Biol., 5:85.
Holsters, M., Villarroel, R., Gielen, J., Seurinck, J., De Greve, H., Van Montagu, M. and Schell, J., 1983, An analysis of the boundaries of the octopine TL-DNA in tumors induced by Agrobacterium tumefaciens, Mol. Gen. Genet., 190:35.
Hooykaas, P. J. J., Klapwijk, P. M., Nuti, M. P., Schilperoort, R. A. and Rorsch, A., 1977, Transfer of the Agrobacterium tumefaciens Ti-plasmid to avirulent agrobacteria and to Rhizobium ex planta, J. Gen. Microbiol., 98:477.
Hooykaas, P. J. J., Hofker, M., Den Dulk-Ras, H. and Schilperoort, R.A., 1984, A comparison of virulence determinants in an octopine Ti-plasmid, a nopaline plasmid, and an Ri-plasmid by complementation analysis of Agrobacterium tumefaciens mutants, Plasmid, 11:195.
Hooykaas-Van Slogteren, G. M. S., Hooykaas, P. J. J. and Schilperoort, R. A., 1984, Expression of Ti-plasmid genes in monocotyledonous plants infected with Agrobacterium tumefaciens, Nature, 311:763.
Horsch, R. B. and Klee, H. J., 1986, Rapid assay of foreign gene expression in leaf discs transformed by Agrobacterium tumefaciens: role of T-DNA borders in the transfer process, Proc. Natl. Acad. Sci. USA, 83:4428.
Horsch, R .B., Klee, H. J., Stachel, S., Winans, S. C., Nester, E. W., Rogers, S. G. and Fraley, R. T., 1986, Analysis of Agrobacterium tumefaciens virulence mutants in leaf discs, Proc. Natl. Acad. Sci.

USA, 83:2571.
Huber, H. E., Iida, S., Arber, W. and Bickle, T. A., 1985, Site specific DNA inversion is enhanced by a DNA sequence element in cis, Proc. Natl. Acad. Sci. USA, 82:3776.
Huffman, G. A., White, F. F., Gordon, M .P. and Nester, E. W., 1984, Hairy root inducing plasmid: physical map and homology to tumor-inducing plasmids, J. Bacteriol., 157:269.
Jefferson, R. A., Bevan, M. and Kavanagh, T., 1986, The use of the Escherichia coli beta-glucuronidase gene as a gene fusion marker for studies of gene expression in higher plants, Biochem. Soc. Trans., 15:17.
Jen, G. C. and Chilton, M .D., 1986, Activity of T-DNA borders in plant cell transformation by mini-T plasmids, J. Bacteriol., 166:491.
Johnson, R. C. and Simon, M. I., 1985, Hin-mediated site-specific recombination requires two 26bp recombination sites and a 60bp recombinational enhancer, Cell, 41:781.
Joos, H., Inze, D., Caplan,A., Sormann, M., Van Montagu, M. and Schell, J., 1983a, Genetic analysis of T-DNA transcripts in nopaline crown galls, Cell, 32:1057.
Joos, H., Timmerman, B., Van Montagu, M. and Schell, J., 1983b, Genetic analysis of transfer and stabilization of Agrobacterium DNA in plant cells, EMBO J., 2:2151.
Jorgensen, R., Snyder, C. and Jones, J. D. G., 1987, T-DNA is organized predominantly in inverted repeat structures in plants transformed with Agrobacterium tumefaciens C58 derivatives, Mol. Gen. Genet., 207:471.
Jouanin, L., 1984, Restriction map of an agropine-type plasmid and its homology with Ti-plasmids, Plasmid, 12:91.
Kanaar, R. and Van De Putte, P., 1987, "Topological aspects of site specific DNA-inversion", Bioassays, in press.
Klapwijk, P. M., Scheulderman, T. and Schilperoort, R. A., 1978, Coordinated regulation of octopine degradetion and conjugative transfer of Ti-plasmids in Agrobacterium tumefaciens: evidence for a common regulatory gene and separate operons, J. Bacteriol., 136:775.
Klee, H. J., Gordon, M. P. and Nester, E. W., 1982, Complementation analysis of Agrobacterium tumefaciens Ti-plasmid mutations affecting oncogenicity, J. Bacteriol., 150:327.
Klee, H. J., Yanofsky, M. F. and Nester, E. W., 1985, Vectors for transformation of higher plants, Bio/Technology, 3:637.
Koekman, B. P., Ooms, G., Klapwijk, P. M. and Schilperoort, R. A., 1979, Genetic map of an octopine Ti-plasmid, Plasmid, 2:347.
Koekman, B. P., Hooykaas, P. J. J. and Schilperoort, R. A., 1982, A functional map of the replicator region of the octopine Ti-plasmid, Plasmid, 7:119.
Koncz, C. and Schell, J., 1986, The promotor of TL-DNA gene 5 controls the tissue-specific expression of chimaeric genes carried by a novel type of Agrobacterium binary vector, Mol. Gen. Genet., 204:383.
Koncz, C., Olsson, O., Langridge, W. H. R., Schell, J. and Szalay, A. A., 1987, Expression and assembly of functional bacterial luciferase in plants, Proc. Natl. Acad. Sci. USA, 84:131.
Koukolikova-Nicola, Z., Shillito, R. D., Hohn, B., Wang, K., Van Montagu, M. and Zambryski, P., 1985, Involvement of circular intermediates in the transfer of T-DNA from Agrobacterium tumefaciens to plant cells, Nature, 313:191.
Krens, F. A., Molendijk, L., Wullems, G. J. and Schilperoort, R. A., 1982, In vitro transformation of plant protoplasts with Ti-plasmid DNA, Nature, 296:72.
Krens, F. A., Mans, R. M. W., Hoge, J. H. C., Wullems, G. J. and Schilperoort, R. A., 1985, Structure and expression of DNA transferred to tobacco via transformation of protoplasts with Ti-plasmid DNA: co-transfer of T-DNA and non T-DNA sequences, Plant Mol. Biol., 5:223.
Kwok, W. W., Nester, E. W. and Gordon, M. P., 1985, Unusual plasmid DNA

organization in an octopine crown gall tumor, Nucleic Acids Res., 13:459.
Lahners, K., Byrne, M. C. and Chilton, M. D., 1984, T-DNA fragments of hairy root plasmid pRi8196 are distantly related to octopine and nopaline Ti-plasmid T-DNA, Plasmid, 11:130.
Langridge, W. H. R., Li, B. J. and Szalay, A. A., 1985, Electric field mediated stable transformation of carrot protoplasts with naked DNA, Plant Cell Reports, 4:355.
Leemans, J., Deblaere, R., Willmitzer, L., De Greve, H., Hernalsteens, J. P., Van Montagu, M. and Schell, J., 1982, Genetic identification of functions of TL-DNA transcripts in octopine crown galls, EMBO J., 1:147.
Leroux, B., Yanofsky, M .F., Winans, S. C., Ward, J. E., Ziegler, S. F. and Nester, E. W., 1987, Characterization of the virA locus of Agrobacterium tumefaciens: a transcriptional regulator and host range determinant, EMBO J., 6:849.
Lipetz, J., 1965, Crown gall tumorigenesis: effect of temperature on wound healing and conditioning, Science, 149:865.
Lippincott, B. B. and Lippincott, J. A., 1969, Bacterial attachment to a specific wound site as an essential stage in tumor initiation by Agrobacterium tumefaciens, J. Bacteriol., 97:620.
Lorz, H., Baker, B. and Schell, J., 1985, Gene transfer to cereal cells mediated by protoplast transformation, Mol. Gen. Genet., 199:178.
Machida, Y., Usami, S., Yamamoto, A., Niwa, Y. and Takebe, I., 1986, Plant-inducible recombination between the 25bp border sequences of T-DNA in Agrobacterium tumefaciens, Mol. Gen. Genet., 204:374.
Matthysse, A., 1987, Characterization of nonattaching mutants of Agrobacterium tumefaciens, J. Bacteriol. 169, 313.
Melchers, L. S., Thompson, D. V., Idler, K. D., Schilperoort, R. A. and Hooykaas, P. J. J., 1986, Nucleotide sequence of the virulence gene virG of the Agrobacterium tumefaciens octopine Ti plasmid: significant homology between virG and the regulatory genes ompR, phoB and dye of E.coli, Nucleic Acids Res., 14:9933.
Melchers, L. S., Thompson, D. V., Idler, K. D., Neuteboom, S. T. C., De Maagd, R. A., Schilperoort, R. A. and Hooykaas, P. J. J., 1987, Molecular characterization of the virulence gene virA of the Agrobacterium tumefaciens octopine Ti-plasmid, submitted.
Messens, E., Lenaerts, A., Hedges, R.W. and Van Montagu, M., 1985a, Agrocinopine A, a phosphorylated opine is secreted from crown gall cells, EMBO J., 4:571.
Messens, E., Lenaerts, A., Van Montagu, M. and Hedges, R.W., 1985b, Genetic basis for opine secretion from crown gall tumour cells, Mol. Gen. Genet., 199:344.
Ninfa, A. J. and Magasanik, B., 1986, Covalent modification of the glnG product, NRI, by the glnL product, NRII, regulates the transcription of the glnALG operon in Escherichia coli, Proc. Natl. Acad. Sci. USA, 83:5909.
Nixon, B. T., Ronson, C. W. and Ausubel, F. M., 1986, Two-component regulatory systems responsive to environmental stimuli share strongly conserved domains with the nitrogen assimilation regulatory genes ntrB and ntrC, Proc. Natl. Acad. Sci. USA, 83:7850.
Okker, R. J. H., Spaink, H., Hille, J., van Brussel, T. A. N., Lugtenberg, B. and Schilperoort, R. A., 1984, Plant-inducible virulence promotor of the Agrobacterium tumefaciens Ti-plasmid, Nature, 312:564.
Ooms, G., Klapwijk, P. M., Poulis, J. A. and Schilperoort, R. A., 1980, Characterization of Tn904 insertions in octopine Ti plasmid mutants of Agrobacterium tumefaciens, J. Bacteriol., 144:82.
Ooms, G., Hooykaas, P. J. J., Molenaar, G. and Schilperoort, R. A., 1981, Crown gall plant tumors of abnormal morphology, induced by Agrobacterium tumefaciens carrying mutated octopine Ti plasmids; analysis of T-DNA functions, Gene, 14:33.

Ooms, G., Hooykaas, P. J. J., Van Veen, R. J. M., Van Beelen, P., Regensburg-Tuink, A. J. G. and Schilperoort, R. A., 1982, Octopine Ti-plasmid deletion mutants of Agrobacterium tumefaciens with emphasis on the right side of the T-region, Plasmid, 7:15.
Otten, L., De Greve, H., Leemans, J., Hain, R., Hooykaas, P. and Schell, J., 1984, Restoration of virulence of vir region mutants of Agrobacterium tumefaciens strain B6S3 by coinfection with normal and mutant Agrobacterium strains, Mol. Gen. Genet., 195:159.
Otten, L., Piotrowiak, G., Hooykaas, P., Dubois, M., Szegedi, E. and Schell, J., 1985, Identification of an Agrobacterium tumefaciens pTiB6S3 vir region fragment that enhances the virulence of pTiC58, Mol. Gen. Genet., 199:189.
Ow, D. W., Wood, K. V., Deluca, M., De Wet, J. R., Helinski, D. R. and Howell, S. H., 1986, Transient and stable expression of the firefly luciferase gene in plant cells and transgenic plants, Science, 234:856.
Paszkowski, J., Pisan, B., Shillito, R. D., Hohn, T., Hohn, B. and Potrykus, I., 1986, Genetic transformation of Brassica campestris var. rapa protoplasts with an engineered cauliflower mosaic virus genome, Plant Mol. Biol., 6:303.
Peerbolte, R., Krens, F. A., Mans, R. M. W., Floor, M., Hoge, J. H. C., Wullems, G. J. and Schilperoort, R. A., 1985, Transformation of plant protoplasts with DNA: cotransformation of non-selected calf thymus carrier DNA and meiotic segregation of transforming DNA sequences, Plant Mol. Biol., 5:235.
Peerbolte, R., Leenhouts, K., Hooykaas-Van Slogteren, G. M. S., Hoge, J. H. C., Wullems, G. J. and Schilperoort, R. A., 1986, Clones from a shooty tobacco crown gall tumor I: deletions, rearrangements and amplifications resulting in irregular T-DNA structures and organizations, Plant Mol. Biol., 7:265.
Peralta, E. G., Hellmiss, R. and Ream, W., 1986, Overdrive, a T-DNA transmission enhancer on the A.tumefaciens tumour-inducing plasmid, EMBO J., 5:1137.
Petit, A., Berkaloff, A. and Tempe, J., 1986, Multiple transformation of plant cells by Agrobacterium may be responsible for the complex organization of T-DNA in crown gall and hairy root, Mol. Gen. Genet., 202:388.
Plasterk, R. H. A., Brinkman, A. and Van De Putte, P., 1983, DNA inversions in the chromosome of E.coli and in bacteriophage Mu: relationship to other site specific recombination systems, Proc. Natl. Acad. Sci. USA, 80:5355.
Plasterk, R. H. A. and Van De Putte, P., 1985, The invertable P-DNA segment in the chromosome of Escherichia coli, EMBO J., 4:237.
Riggs, S. D. and Bates, G. W., 1986, Stable transformation of tobacco by electroporation: evidence for plasmid concatenation, Proc. Natl. Acad. Sci. USA, 83:5602.
Riker, A. J., 1930, Studies on infectious hairy root of nursery apple trees, J. Agric. Res., 41:30.
Riker, A. J., Sproel, E. and Gutsche, A. E., 1946, Some comparisons of bacterial plant galls and of their causal agents, Bot. Rev., 12:57.
Rogowsky, P., Close, T. J. and Kado, C. I., 1987, Dual regulation of virulence genes of Agrobacterium tumefaciens plasmid pTiC58, in: "Moleculars genetics of plant-microbe interactions", D. P. S. Verma and N. Brisson, eds., M. Nijhoff, Dordrecht.
Rubin, R. A., 1986, Genetic studies on the role of octopine T-DNA border regions in crown gall tumor formation, Mol. Gen. Genet., 202:312.
Sanders, P. R., Winter, J. A., Barnason, A. R., Rogers, S. G. and Fraley, R. T., 1987, Comparison of cauliflower mosaic virus 35S and nopaline synthase promoters in transgenic plants, Nucleic Acids Res., 15:1543.
Schilperoort, R. A., 1969, "Investigations on plant tumors-crown gall. On the biochemistry of tumor induction by Agrobacterium tumefaciens" PhD. thesis, Leiden University.

Schreier, P. H., Seftor, E. A., Schell, J. and Bohnert, H. J., 1985, The use of nuclear-encoded sequences to direct the light-regulated synthesis and transport of a foreign protein into plant chloroplasts, EMBO J., 4:25.
Shaw, C. H., Watson, M. D., Carter, G. H. and Shaw, C. H., 1984a, The right hand copy of the nopaline Ti-plasmid 25bp repeat is required for tumour formation, Nucleic Acids Res., 12:6031.
Shaw, C. H., Carter, G. H., Watson, M. D. and Shaw, C. H., 1984b, A functional map of the nopaline synthase promotor, Nucleic Acids Res., 12:7831.
Sheikholeslam, S. N. and Weeks, D. P., 1987, Acetosyringon promotes high efficiency transformation of Arabidopsis thaliana explants by Agrobacterium, Plant Mol. Biol., 8:291.
Simoens, C., Alliotte, Th., Mendel, D., Muller, A., Schiemann, J., Van Lijsebettens, M., Schell, J., Van Montagu, M. and Inze, D., 1986, A binary vector for transferring genomic libraries to plants, Nucleic Acids Res., 14:8073.
Simpson, R. B., O'Hara, P. J., Kwok, W., Montoya, A. L., Lichtenstein, C., Gordon, M. P. and Nester, E. W., 1982, DNA from the A6S/2 crown gall tumor contains scrambled Ti-plasmid sequences near its junctions with plant DNA, Cell, 29:1005.
Slightom, J. L., Jouanin, L., Leach, F., Drong, R. F. and Tepfer, D., 1985, Isolation and identification of TL-DNA/plant junctions in Convolvulus arvensis transformed by Agrobacterium rhizogenes strain A4, EMBO J., 4:3069.
Slightom, J. L., Durand-Tardif, M., Jouanin, L. and Tepfer, D., 1986, Nucleotide sequence analysis of TL-DNA of Agrobacterium rhizogenes agropine type plasmid: identification of open reading frames, J. Biol. Chem., 261:108.
Smith, E. F. and Townsend, C. O., 1907, A plant tumor of bacterial origin, Science, 25:671.
Spielmann, A. and Simpson, R. B., 1986, T-DNA structure in transgenic tobacco plants with multiple independent integration sites, Mol. Gen. Genet., 205:34.
Stachel, S. E., Messens, E., Van Montagu, M. and Zambryski, P., 1985a, Identification of the signal molecules produced by wounded plant cells that activate T-DNA transfer in Agrobacterium tumefaciens, Nature, 318:624.
Stachel, S. E., An, G., Flores, C. and Nester, E. W., 1985b, A Tn3 LacZ transposon for the random generation of beta-galactosidase gene fusions: application to the analysis of gene expression in Agrobacterium, EMBO J., 4:891.
Stachel, S. E., Nester, E. W. and Zambryski, P. C., 1986a, A plant cell factor induces Agrobacterium tumefaciens vir gene expression, Proc. Natl. Acad. Sci. USA, 83:379.
Stachel, S. E., Timmerman, B. and Zambryski, P., 1986b, Generation of single-stranded T-DNA molecules during the initial stages of T-DNA transfer from Agrobacterium tumefaciens to plant cells, Nature, 322:706.
Stachel, S. E. and Nester, E. W., 1986, The genetic and transcriptional organization of the vir-region of the A6 Ti-plasmid of Agrobacterium tumefaciens, EMBO J., 5:1445.
Stachel, S. E. and Zambryski, P. C., 1986, VirA and VirG control the plant-induced activation of the T-DNA transfer process of A.tumefaciens, Cell, 46:325.
Stachel, S. E., Timmerman, B. and Zambryski, P., 1987, Activation of Agrobacterium tumefaciens vir gene expression generates multiple single-stranded T-strand molecules from the pTiA6 T-region: requirement for 5' virD gene products, EMBO J., 6:857.
Teeri, T. H., Herrera-Estrella, L., Depicker, A., Van Montagu, M. and Palva, E. T., 1986, Identification of plant promotors in situ by T-DNA mediated transcriptional fusions to the NPT-II gene, EMBO J., 5:1755.

Thomashow, L. S., Reeves, S. and Thomashow, M. F., 1984, Crown gall oncogenesis: evidence that a T-DNA gene from the Agrobacterium Ti plasmid pTiA6 encodes an enzyme that catalyzes synthesis of indoleacetic acid, Proc. Natl. Acad. Sci. USA, 81:5071.

Thomashow,M.F., Hugly, S., Buchholz, W. G. and Thomashow, L. S., 1986, Molecular basis for the auxin-independent phenotype of crown gall tumor tissues, Science, 231:616.

Usami, S., Morikawa, S., Takebe, I. and Machida, Y., 1987, Absence in monocotyledonous plants of the plant factors inducing T-DNA circularization and vir gene expression in Agrobacterium, Mol. Gen. Genet., submitted.

Van Den Broeck, G., Timko, M. P., Kausch, A. P., Cashmore, A. R., Van Montagu, M. and Herrera-Estrella, L., 1985, Targeting of a foreign protein to chloroplasts by fusion to the transit peptide from the small subunit of ribulose 1,5-biphosphate carboxylase, Nature, 313:358.

Van Der Elzen, P., Lee, K. Y., Townsend, J. and Bedbrook, J., 1985, Simple binary vectors for DNA transfer to plant cells, Plant Mol. Biol., 5, 149-154.

Van Haaren, M. J. J., Pronk, J. T., Schilperoort,R.A. and Hooykaas, P. J. J., 1986, T-region transfer from Agrobacterium tumefaciens to plant cells: functional charactarization of border repeats, in: "Recognition in microbe-plant symbiotic and pathogenic interactions", Lugtenburg, ed., NATO ISI series, Cell Biology, 4:203.

Van Haaren, M. J. J., Pronk, J. T., Schilperoort,R.A. and Hooykaas, P. J. J., 1987, Functional analysis of the Agrobacterium tumefaciens octopine Ti-plasmid left and right T-region border fragments, Plant Mol. Biol., 8:95.

Van Larebeke, N., Engler, G., Holsters, M., Van Den Elsacker, S., Zaenen, I., Schilperoort, R. A. and Schell, J., 1974, Large plasmid in Agrobacterium tumefaciens essential for crown gall inducing ability, Nature, 252:169.

Van Lijsebettens, M., Inze, D., Schell, J. and Van Montagu, M., 1986, Transformed cell clones as a tool to study T-DNA integration mediated by Agrobacterium tumefaciens, J. Mol. Biol., 188:129.

Van Slogteren, G. M. S., Hooykaas, P. J. J. and Schilperoort, R. A., 1984, Silent T-DNA genes in plant lines transformed by Agrobacterium tumefaciens are activated by grafting and by 5-azacytidine treatment, Plant Mol. Biol. 3:333.

Veluthambi, K., Jayaswal, R. K. and Gelvin, S. B., 1987, Virulence genes A, G, and D mediate the double-stranded border cleavage of T-DNA from the Agrobacterium Ti plasmid, Proc. Natl. Acad. Sci. USA, 84:1881.

Vincent, J. M., 1974, "The biology of nitrogen fixation", Ed. Quispel, A., North Holland publishing Co., Amsterdam.

Wang, K., Herrera-Estrella, L., Van Montagu, M. and Zambryski, P., 1984, Right 25bp terminal sequence of the nopaline T-DNA is essential for and determines direction of DNA transfer from Agrobacterium to the plant genome, Cell, 38:455.

Wang, K., Stachel, S. E., Timmerman, B., Van Montagu, M. and Zambryski, P. C., 1987, Site specific nick in the T-DNA border sequence as a result of Agrobacterium vir gene expression, Science, 235:587.

Watson, B., Currier, T. C., Gordon, M. P., Chilton, M .D. and Nester, E. W., 1975, Plasmid required for virulence of Agrobacterium tumefaciens, J. Bacteriol. 123:255.

Winans, S. C., Ebert, P. R., Stachel, S. E., Gordon, M. P. and Nester, E. W., 1986a, A gene essential for Agrobacterium virulence is homologous to a family of positive regulatory loci, Proc. Natl. Acad. Sci. USA, 83:8278.

Winans, S. C., Allenza, P., Stachel, S. E., McBride, K. E. and Nester, E. W., 1986b, Characterization of the virE operon of the Agrobacterium Ti plasmid pTiA6, Nucleic Acids Res., 15:825.

Yadav, N. S., Vanderleyden, J. Bennet, D. R., Barnes, W. M. and Chilton, M.

D., 1982, Short direct repeats flank the T-DNA on a nopaline Ti-plasmid, Proc. Natl. Acad. Sci. USA, 79:6322.
Yamamoto, A., Iwahashi, M., Yanofsky, M. F., Nester, E. W., Takebe, I. and Machida, Y., 1987, The promoter proximal region in the virD locus of Agrobacterium tumefaciens is necessary for the plant inducible circularization of T-DNA, Mol. Gen. Genet., 206:174.
Yanofsky, M., Lowe, B., Montoya, A., Rubin, R., Krul, W., Gordon, M. and Nester, E., 1985, Molecular and genetic analysis of factors controlling host range in Agrobacterium tumefaciens, Mol. Gen. Genet., 201:237.
Yanofsky, M. F. and Nester, E. W., 1986, Molecular characterization of a host range determining locus from Agrobacterium tumefaciens, J. Bacteriol., 168:244.
Yanofsky, M. F., Porter, S. G., Young, C., Albright, L. M., Gordon, M. P. and Nester, E. W., 1986, The virD operon of Agrobacterium tumefaciens encodes a site specific endonuclease, Cell, 47:471.
Zaenen, I., Van Larebeke, N., Teuchy, H., Van Montagu, M. and Schell, J., 1974, Supercoiled circular DNA in crown gall inducing Agrobacterium strains, J. Mol. Biol., 86:109.
Zambryski, P., Depicker, A., Kruger, K. and Goodman, H. M., 1982, Tumor induction Agrobacterium tumefaciens: analysis of the boundaries of T-DNA, J. Mol. Apll. Genet., 1:361.
Zambryski, P., Joos, H., Genetello, C., Leemans, J., Van Montagu, M. and Schell, J., 1983, Ti-plasmid vector for the introduction of DNA into plant cells without alteration of their normal regeneration capacity, EMBO J., 2:2143.

Ri T-DNA FROM *Agrobacterium rhizogenes*, A SEMIOCHEMICAL THAT ALTERS MORPHOLOGICAL PLASTICITY

D. Tepfer

The Rhizosphere Research Laboratory
I.N.R.A.
78000 Versailles, France

INTRODUCTION

In most instances plant development relies on strategies fundamentally different from those employed in animal development. Plants are morphologically plastic (see Trewavas and Jennings, 1986, for a general discussion). Plants retain embryonic tissues in their meristems, where cell division and enlargement continuously extend roots and shoots. In addition, new meristems form within existing organs and produce lateral ramifications. The plant's form is the product of differential growth and development that occur according its needs. The plant is sessile and its plasticity is thought to be an adaptation to the need to forage for water and minerals (on the part of the roots) and for air and light (on the part of the aerial parts) (Grime et al., 1986). Plasticity allows the plant to adapt to heterogeneity in its immediate environment and to respond to infection, wounding or excision of a stem or root, through the differentiation and growth of lateral or adventitious organs (a plastic response). The plant thus isolates the affected organ and attempts to establish an alternative. One of the consequences of morphological plasticity is the absence of a germ line in plants. The transition from a vegetative to a floral meristem may take place in one or often a number of meristems, each a different ramification of the main stem.

The molecular genetics of plant development are not understood. It would be useful and informative to describe the genetic basis for plasticity, in order to understand its novelty with respect to animal systems, to control plant morphology in agricultural applications and to stimulate the regeneration of shoots from cells in biotechnological uses of genetic transformation. Genes which control development are likely to be expressed at low levels, and would therefore be difficult to clone. We have taken an indirect approach to isolating and studying genes which alter development, relying on the fact that plants have co-evolved with microorganisms, such that plant growth and development are influenced not just by the physical environment, but also by these associated microorganisms.

THE EFFECTS OF Ri T-DNA FROM *Agrobacterium rhizogenes* ON PLANT DEVELOPMENT

Microorganisms influence plant development through the production of growth substances such as auxins and cytokinins (see review Morris, 1986). One of the best documented cases is Agrobacterium tumefaciens, which alters plant development by genetically transforming the plant cell (see reviews, Davey et al., 1986). In the case of A. tumefaciens, the transferred DNA (T-DNA) induces tumor formation and is referred to as Ti (tumor-inducing) T-DNA. A. rhizogenes uses a similar transformation mechanism (Hoekema et al., 1984; Slightom et al., 1985) to transmit a T-DNA that induces roots (Ri T-DNA). In both instances genetic transformation alters development. In the second case this alteration is responsible for the formation of an adventitious organ. The first question addressed here is, does Ri T-DNA therefore carry information which alters plasticity? Affirmative answers are given below, based on morphological observations: first, of the infected wound and, secondly, of the organs and plants derived from cells transformed in the wound. The second question is, If Ri T-DNA confers plasticity, what could be its role in rhizosphere ecology?

Inoculation of Agrobacterium rhizogenes on stems produces a variety of morphological reactions. Although root formation is the most common response, certain species react by forming tumors (White et al., 1982). Inoculation of leaves may produce shoots (Tepfer, 1984). Adventitious organ development is a manifestation of plasticity. Ri T-DNA is incorporated into the organs produced. Since both root and shoot differentiation can be induced by the same inoculation, a simple interpretation is that A. rhizogenes brings about a general increase in plasticity, or the propensity to differentiate. Analysis of the infected wound which produces transformed roots or shoots is complicated by the presence of the inciting bacteria. It is thus difficult to separate the effects of the bacterium and effects of the T-DNA inserted into the plant genome. A better answer to the question of changes in plasticity is obtained from decontaminated root cultures and plants (and their descendants), containing Ri T-DNA, but free of A. rhizogenes.

Fig. 1 gives the general scheme used to study the phenotype of organs and plants decontaminated of A. rhizogenes, but transformed by Ri T-DNA. Inoculation by A. rhizogenes results in the transfer of Ri T-DNA to plant cells exposed through wounding of a stem. The T-DNA inserts into the host genome and cell division occurs in the wound. Root meristems form from transformed cells. The roots that appear at the site of inoculation are excised, and used to establish a collection of genetically transformed, axenic root clones. Bacterial growth is suppressed by antibiotics and the growth of the roots is stimulated by the Ri T-DNA, such that the root tip outgrows the bacterial contamination. The organ clones so produced represent insertions of Ri T-DNA into the host genome in a variety of sites and configurations.

The phenotype of transformed roots is easily recognized and constitutes a marker for the presence of Ri T-DNA. These roots are characterized by rapid growth, frequently 40 (or more) times that of the controls (Jung and Tepfer, 1987), by altered geotropic behavior, and by a high degree of branching (Tepfer and Tempé, 1981; Tepfer, 1983a, 1984). Thus, apical dominance is reduced, more laterals form and the general propensity to differentiate root meristems is increased. In some clones of transformed fennel roots the tendency to regenerate stems is also increased (C. Attal and D. Tepfer, unpublished results). In other words morphological plasticity is augmented. If transformed tobacco roots are converted into callus by the application of auxins and cytokinins, the callus reforms roots after the growth substances are removed. Normal callus stops growing after removal of growth substances (Spano et al., 1981). Thus this increased plasticity is inherent to the transformed genome, not just to the transformed root.

Plants regenerate spontaneously from the roots of many species. Transformed roots produce plants that have the altered phenotype in their roots that is observed in transformed root cultures. Their aerial parts are also phenotypically altered (Tepfer, 1982, 1984) in ways that are indicative

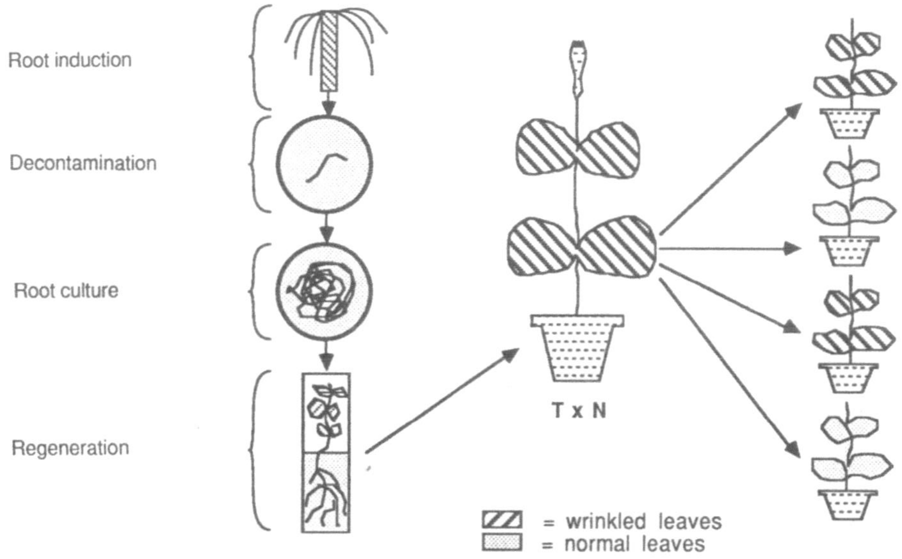

Figure 1: Flow diagram for producing transgenic plants having modified plasticity. Bacterial inoculation at day 0 produces roots after 2 weeks. Individual roots are decontaminated by culture on medium containing antibiotics. Root cultures serve to regenerate whole plants. The process typically takes two months, depending on species. The transformed phenotype segregates in the progeny after crossing a transformed plant (T) with the wild type (N).

of increased developmental plasticity. The transformed phenotype is best studied in the descendants of plants regenerated from transformed roots, thus avoiding chimeric variability and providing a large number of individuals (Fig. 1). The controls are root cultures established from excised roots and the plants derived from them. The following list summarizes the transformed phenotype, beginning with attributes that indicate increased plasticity.

THE TRANSFORMED PHENOTYPE

Evidence for increased morphological plasticity in whole plants:

1) Apical dominance is reduced in tobacco and carrot (stems and roots are highly branched) (Tepfer, 1984).

2) Adventitious roots form at high frequency on stems in tobacco (Ackermann, 1977; Tepfer, 1984).

3) Adventitious stems form on roots in tobacco (Tepfer, 1984).

4) Biennials (carrot and endive) become annuals: the propensity to differentiate flowers is increased (Tepfer, 1984; G. Touraud, pers. comm.).

Evidence for increased physiological plasticity:

1. Transformed roots of many species grow <u>in vitro</u>, while normal roots grow poorly, or not at all (Tepfer and Tempé, 1981, Tepfer, 1984; observations for about 50 species of dicots--published and unpublished results from various laboratories).

2. Transformed tobacco plants can grow in an environment where gas exchange is limited (Tepfer, 1984).

Other changes:

1) Leaves are wrinkled in nearly all of the 22 species so far examined, apparently due to differential growth of the veins and the leaf blade. Their overall shape changes (see for examples Tepfer, 1982, 1983, 1984; Ooms et al., 1985)

2) Flower morphology is altered in tobacco (Tepfer, 1984).

3) Internodal distance is shortened in tobacco (Tepfer, 1984; Durand-Tardif, 1985).

4) Sterility is sometimes observed in tobacco (Tepfer, 1984) and oilseed rape (Ooms et al., 1985; Guerche et al., 1987).

DISCUSSION

The first question was whether Ri T-DNA could be responsible for a general increase in morphological plasticity. The positive evidence presented above consists of the observation that, although the transformed phenotype differs somewhat from species to species, its general attribute is increased morphological (and physiological) plasticity. The source of the transformed phenotype is clearly the right hand portion of the Ri TL-DNA (Durand-Tardif et al., 1985; Vilaine and Casse-Delbart, 1987), comprising ORFs 10-18 (Slightom et al., 1986). Molecular hybridization (Huffman et al., 1984; Jouanin, 1984) and sequence comparison (Levesque et al., man. in prep.) have been used to search this region for sequence homology with genes shown to be responsible for the synthesis of plant growth substances, with negative results. Thus the mechanism behind the developmental effects of Ri T-DNA is completely unknown. One of the most surprising aspects of the transformed phenotype is that it is quite similar in widely different species (Tepfer, 1982, 1984). Ri T-DNA might interfere in a conserved mechanism, such that the same information has the same effect in different species. Alternatively, it might contain a collection of different kinds of information leading to the same phenotypic effect through different pathways in different species. Another possibility would be a compromise: a family of genes diverged from a common ancestor, affecting the same mechanism, each having diverged to be functional within a subset of the range of host species.

The possible evolutionary significance of Ri T-DNA has been discussed (Tepfer, 1982, 1983b; White et al., 1983). Parts of it are found in a number of species (White et al., 1983), possibly as a result of spontaneous transformation. Parasexual genetic exchange is not normally thought to be a factor in evolution. Nevertheless, Ri T-DNA could certainly provide a certain type of variability, and in species such as Convolvulus arvensis (morning glory) the transformed phenotype is not incompatible with an adaptive advantage. Transformed C. arvensis plants are generally more vigorous (unpublished observations). This species propagates by regenerating from roots and root fragments generated when the soil is disturbed. Increased plasticity may be a useful trait that can be acquired rapidly from a rhizosphere microorganism when the need arises.

The rhizosphere is rich in organisms living in intimate contact. The regulation of their growth is not understood, but surely depends upon semiochemicals, e.g. molecules that signal the presence of other organisms. For instance, small molecules produced in plant wounds apparently indicate to Agrobacterium and Rhizobium the proximity of a wounded host (Stachel et al., 1985; Peters et al., 1986; Ashby et al., 1987).

The Ri T-DNA can also be regarded as a semiochemical, a signal that is transmitted from the bacterium to the plant genome, where it becomes the starting point for a series of events. It instructs the plant to form organs that produce opines (Tepfer and Tempé, 1981) and have a more plastic pattern of growth. These events may be reply signals. The opines are believed to stimulate the growth of divers soil bacteria, such as the Agrobacteria carrying the T-DNA which initiated the roots (see review Tempé and Petit, 1982), as well as others (Rossignol and Dion, 1985; Bouzar and Moore, 1987). Secondly, the phenotype of the transformed roots renders them particularly sensitive to interactions with certain rhizosphere microorganisms. They have been used to culture obligate parasites, such as Polymyxa betae (the fungal vector for the rhizomania virus of sugar beet, see Tepfer and Yacoub, 1986). Their increased plasticity causes the formation of laterals at high frequency. Thus the bulk of the transformed root population is young, and the number of meristems is high: the preferred state for infection by fungal symbionts (Smith and Walker, 1981).

The T-DNA signal is amplified through meiosis. It is transmitted as a Mendelian dominant (Tepfer, 1982, 1984), and its level of phenotypic expression varies with the level of genotypic expression (Durand-Tardif et al., 1985) and with copy number. (In tobacco the level of phenotypic expression is reduced in homozygotes and increased in heterozygotes-- L. Heisler and D. Tepfer, unpublished observations.)

Against the background of general genetics, parasexual genetic exchange would seem aberrant. It becomes more acceptable, if not logical, in the context of the rhizosphere, where organisms co-evolve in close contact in a semi-liquid medium. If the general significance of Ri T-DNA is to increase plasticity, one particularly interesting consequence could be that it creates a germ line. Cells carrying and expressing Ri T-DNA would be more likely to form roots and shoots: roots more likely to regenerate and shoots more likely to flower. Genes incorporated into Ri T-DNA would be carried along and amplified through meiosis. One could thus expect that Ri T-DNA should carry "plast" genes insuring germ line formation in a variety of circumstances of insertion site, copy number and species, interspersed with other genes that do not directly influence plasticity, but benefit from a free ride through meiosis.

REFERENCES

Ackermann, C., 1977, Pflanzen aus Agrobacterium rhizogenes tumoren an Nicotiana tabacum, Plant. Sc. Lett., 8:23-30.

Ashby, A., Watson, M., and Shaw, C., 1987, A Ti-plasmid determined function is responsible for chemotaxis of Agrobacterium tumefaciens towards the plant wound product acetosyringone, FEMS Microbiol. Lett., 41:189-192.

Bouzar, H., and Moore, L., 1987, Isolation of different Agrobacterium biovars from a natural oak savanna and tallgrass prairie, Appl. Environ., Microbiol., 53:717-721.

Davey, M. Gartland, K., and Mulligan, B., 1986, Transformation of the genomic expression of plant cells, in: "Plasticity in Plants," A. Trewavas and D. Jennings, eds., The Company of Biologists Limited, Cambridge.

Durand-Tardif, M., Broglie, R., Slightom, J., and Tepfer, D., 1985, Structure and expression of Ri T-DNA from Agrobacterium rhizogenes in Nicotiana tabaccum. Organ and phenotypic specificity, J. Mol. Biol., 186:557-564.

Grime, J., Crick, J., and Rincon, J., 1986, The ecological significance of plasticity, in: "Plasticity in Plants," A. Trewavas and D. Jennings eds., The Company of Biologists Limited, Cambridge.

Guerche, P., Jouanin, L., Tepfer, D. and Pelletier, G., 1987, Genetic transformation of oilseed rape (Brassica napus) by the Ri T-DNA of Agrobacterium rhizogenes and analysis of inheritance of the transformed phenotype, Mol. Gen. Genet., 206:382-386.

Hoekema, A., Hooykass, P. and Schilperoort, R., 1984, Transfer of the octopine T-DNA segment to plant cells mediated by different types of Agrobacterium tumor or root-inducing plasmids: Generality of virulence systems, J. Bacteriol., 158:383-385.

Huffman,G., White,F., Gordon, M., and Nester, E., 1984, Hairy root inducing plasmid: Physical map and homology to tumor-inducing plasmids, J. Bacteriol., 157:269-276.

Jouanin, L., 1984, Restriction map of an agropine-type Ri plasmid and its homologies with Ti plasmids, Plasmid, 12:91-102.

Jung, G. and Tepfer, D., 1987, Use of genetic transformation by the Ri T-DNA of Agrobacterium rhizogenes to stimulate biomass and tropane alkaloid production in Atropa belladonna and Calystegia sepium roots grown in vitro, Plant Science, 50: (in press).

Morris, R., 1986, Genes specifying auxin and cytokinin biosynthesis in phytopathogens, Ann. Rev. Plant Physiol., 37:509-538.

Ooms, G., Bains, A., Burrell, M., Karp, A., Twell, D., and Wilcox, E., 1985, Genetic manipulation of cultivars of oilseed rape (Brassica napus) using Agrobacterium, Theor. Appl. Genet., 7:325-329.

Peters, N., Frost, J. and Long S., 1986, A plant flavone, luteolin, induces expression of Rhizobium meliloti nodulation genes, Science, 233:977-979.

Rossignol, G. and Dion P., 1985, Octopine, noapline and octopinic acid utilization in Pseudomonas, Can. J. Microbiol., 31:68-74.

Slightom, J., Durand-Tardif, M., Jouanin, L., and Tepfer, D., 1986, Nucleotide sequence analysis of TL-DNA of Agrobacterium rhizogenes agropinetype plasmid: identification of open-reading frames, J. Biol. Chem., 261:108-121.

Slightom, J., Jouanin, L., Leach, F., Drong, R. and Tepfer, D., 1985, Isolation and identification of TL-DNA/plant junctions in Convolvulus arvensis transformed by Agrobacterium rhizogenes strain A4, EMBO J., 4:3069-3077.

Smith, S., Walker, N., 1981, A quantitative study of mycorrhizal infection in Trifolium: separate determination of the rates of infection and mycellial growth, New Phytol., 89:225-240.

Spano, L., Wullems, G., Schilperoort, R. and Costantino, P., 1981, Hairy

root: in vitro growth properties of tissues induced by A. rhizogenes on tobacco, Plant. Sc. Lett., 23: 299-305.
Stachel, S., Messens, E., Van Montagu, M. and Zambryski, P., 1985, Identification of the signal molecules produced by wounded plant cells that activate T-DNA transfer in Agrobacterium tumefaciens, Nature, 318:624-629.
Tempé, J. and Petit, A., 1982, Opine utilization by Agrobacterium, in "The Molecular Biology of Plant Tumors," G.Kahl and J. Schell (eds.), Academic Press, New York.
Tepfer, D., 1982, La transformation génétique de plantes supérieures par Agrobacterium rhizogenes, in: 2è Colloque sur les Recherches Fruitières, I.N.R.A.-C.T.I.F.L., Bordeaux
Tepfer, D., 1983a, The potential uses of Agrobacterium rhizogenes in the genetic engineering of higher plants: nature got there first, in "Genetic Engineering in Eukaryotes," P. Lurquin and A. Kleinhofs, eds., Plenum, New York, London.
Tepfer, D., 1983b, The biology of genetic transformation of higher plants by Agrobacterium rhizogenes, in "Molecular Genetics of the Bacteria-Plant Interaction," A. Pühler, ed., Springer-Verlag, Berlin, Heildelberg.
Tepfer, D., 1984, Transformation of several species of higher plants by Agrobacterium rhizogenes: sexual transmission of the transformed genotype and phenotype, Cell, 37:959-967.
Tepfer, D. and Tempé, J., 1981, Production d'agropine par des racines formées sous l'action d'Agrobacterium rhizogenes, souche A4, C. R. Acad. Sc. Paris, 292:153-156.
Tepfer, D., and Yacoub, A. (1986), Modèle de rhizosphère utile notamment pour l'étude des parasites des plantes. Fr. pat. no. 2575759, Europ. pat. no. 190161.
Trewavas, A., and Jennings, D. 1986, Introduction, in: "Plasticity in Plants," D. Jennings and A. Trewavas (eds.), The Company of Biologists Limited, Cambridge.
Vilaine, F. and Casse-Delbart, F., 1987, Independent induction of transformed roots by the TL and TR regions of the Ri plasmid of agropine type Agrobacterium rhizogenes, Mol. Gen. Genet., 206:17-23.
White, F., Ghidossi, G., Gordon, M., and Nester, E., 1982, Tumor induction by Agrobacterium rhizogenes involves the transfer of plasmid DNA to the plant genome, Proc, Nat. Acad. Sci. USA, 79:3193-3197.
White, F., Garfinkel, D., Huffman, G., Gordon, M., and Nester, E., 1983, Sequences homologous to Agrobacterium rhizogenes T-DNA in the genomes of uninfected plants, Nature, 301:348-350.

THE ROLE OF VIRULENCE REGULATORY LOCI IN DETERMINING *AGROBACTERIUM* HOST RANGE

S.C. Winans, S. Jin, T. Komari, K.M. Johnson and E.W. Nester

University of Washington

Department of Microbiology, SC-42, Seattle, WA 98195, USA

Agrobacterium tumefaciens provides a useful model to study how pathogenic bacteria perceive a susceptible host and initiate responses appropriate to infection. This organism responds to cocultivation with plant suspension cells by inducing the transcription of 5 of its 6 *vir* transcriptional units encoded by the Ti plasmid. Induction of these genes is also observed by incubating bacteria in broth containing a class of substituted phenolic compounds such as acetosyringone (Stachel et al., 1985). The transcription of *virB*, *virC*, *virD*, *virE* and *virG* are all induced at least 100 fold by either treatment whereas the expression of *virA* is not affected. The *pinF* gene, which does not play an obligatory role in tumorigenesis, is also strongly inducible. These *vir* loci encode a total of 21 genes, most or all of which are thought to play a direct role in transferring a discrete segment of Ti plasmid DNA (the T-DNA) from the bacterium into the nucleus of an infected plant cell where it becomes covalently integrated into genomic DNA and directs the overproduction of several phytohormones, resulting in neoplastic growth. In this article, we will explore the regulatory network which controls *vir* gene expression. We will then examine two Ti plasmids (pTiAg162 and pTiBo542) which confer upon their host a host range different from that of the octopine plasmid (pTiA6) most widely studied in our lab. We will provide evidence that this altered host range conferred by these plasmids is due in part to distinctive properties of the plasmids' *vir* regulatory systems.

THE virA AND virG GENES ENCODE PROTEINS NECESSARY FOR TRANSCRIPTIONAL INDUCTION OF THE vir REGULON

Among the Ti plasmid-encoded *vir* genes are two genes, *virA* and *virG*, whose products are necessary and probably sufficient for *vir* gene induction by plant-derived phenolic compounds. Evidence that these genes encode positive regulatory loci is exemplified by data shown in Figure 1, which shows the induction of a *virD*::*lacZ* transcriptional fusion in wild type bacteria and in bacteria containing a mutation in *virA* or in *virG*. As can be seen, in wild type bacteria, the *virB* promoter is strongly induced by incubation with 100 μM acetosyringone, while in strains containing a mutation in *virA* or in *virG*, induction was not detected. Insertion mutations in other *vir* genes, or in the chromosomal *chvA* or *chvB* genes had no effect on induction (Stachel and Zambryski, 1986). Similar data has been obtained for the regulation of the *virB*, *virC*, and *virE* promoters

(Stachel and Zambryski, 1986, Winans et al., 1986, Winans, unpublished results). Full induction of virG itself also depends upon VirA and VirG (i.e., virG is autoregulatory). However, virG is partially inducible even in strains containing a mutation in either gene (Stachel and Zambryski, 1986). The kinetics of induction is consistently rather slow, compare to many other bacterial genes. Little or no induction is observed until about 4 hours after addition of the inducer, and intracellular levels of beta-galactosidase continue to increase for at least the next 24 hours. A chromosomal gene (ros) has also been identified which causes an elevated level of synthesis of the virC and virD promoters (Close et al., 1985; Close et al., 1987), but which does not affect the transcription of other promoters.

The nucleotide sequence of the virA and virG genes has been determined (Winans et al., 1986; Melchers et al., 1986; Leroux et al., 1987). virA is monocistronic and encodes a 92,000 dalton protein. The amino terminus of the predicted VirA protein contains a stretch of hydrophobic amino acids preceded by 4 basic residues. This sequence is similar to the so called "signal sequences" which have been found in a variety of secreted and transmembrane proteins. About 250 amino acids toward the carboxy terminus, VirA contains a second region of hydrophobic residues followed by 6 basic amino acid residues. Similar sequences have been found in several transmembrane proteins and have been denoted "stop transfer" sequences. If VirA does indeed have a functional signal sequence at the amino terminus and a functional stop transfer sequence 250 amino acids distal, then the amino-terminal 250 amino acids would be located in the periplasmic space and the remainder of the protein would lie in the cytoplasm. Consistent with this model, we have found by immunoblotting of fractionated cells with anti-VirA antisera that VirA is indeed located exclusively in the inner membrane (Leroux et al., 1987). The virG locus is also monocistronic, encoding a protein having a molecular weight of between 27,000 and 30,000 daltons, depending on which codon is utilized as the initiation codon. This protein is rather hydrophilic and does not contain a recognizable signal sequence, suggesting that it may be cytoplasmically localized.

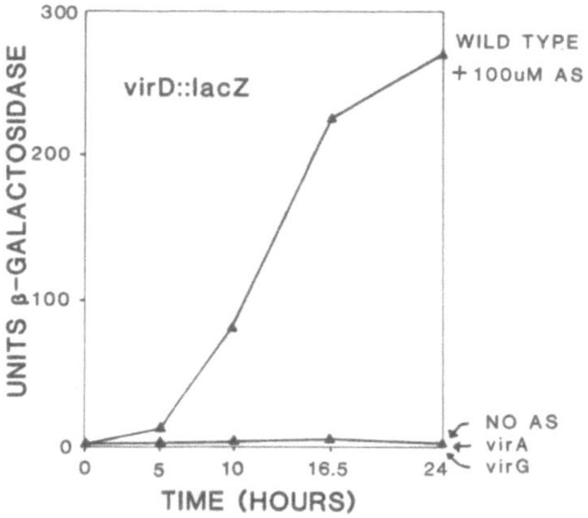

Figure 1. Transcriptional induction of a virD::lacZ transcriptional and translational fusion by incubation in MS medium (Murashige and Skoog, 1962) in the presence of 100 μM acetosyringone. All strains contain plasmid pSW162, which contains the same virD::lacZ fusion as pSM400 (Stachel and Nester, 1986) subcloned as a BglII fragment into the gap created by digestion of pVK102 with BglII. Strains also contain a wild type Ti plasmid, or Ti plasmids containing the virA226 or the virG19 mutation, as indicated.

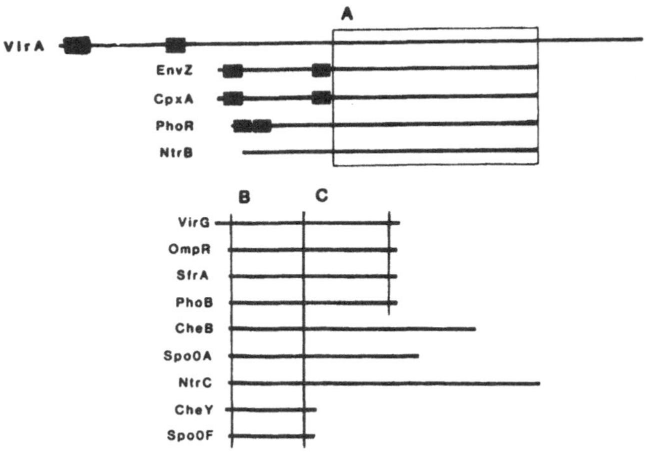

Figure 2. Possible domains of the VirA and VirG proteins and their homologues.

The PIR protein database of the National Biomedical Research Foundation was searched for proteins homologous to VirA and VirG (Leroux et al., 1987; Winans et al., 1986). These proteins were found to be strongly homologous to a family of two component regulatory systems (Leroux et al., 1987; Winans et al., 1986; Nixon et al., 1986). VirA is homologous to at least 4 known proteins, including the EnvZ and NtrB proteins, while VirG is homologous to at least 8 proteins, including the OmpR and NtrC proteins (these four homologoues have been found in E. coli and several other bacterial species). The EnvZ and OmpR proteins together regulate a number of genes whose transcription is influenced by extracellular osmotic pressure (Hall and Silhavy, 1981). They are also required for the regulation of genes which are not osmoregulated (Gibson et al., 1987). The NtrB and NtrC genes control a set of genes which are induced by nitrogen starvation including nifAL and glnA (Dixon, 1984).

Figure 2 shows a schematic representation of the regions of homology. As can be seen, the homology between VirA and its homologues lies over a discrete 250 amino acid region (domain A). In contrast, VirG is homologous to four proteins over its entire length and to an additional five proteins over its amino terminal half. The fact that these homologies tend to fall within rather discrete segments of each protein suggests that each segment may represent a functional domain having some degree of functional autonomy.

It seems reasonable that any property demonstrated of one member of this family may be conserved among some or all other members. Therefore, by drawing analogies between the VirA/VirG system and its homologues, we can gain some degree of understanding about how VirA and VirG may regulate the transcription of vir genes. We point out two recent studies which make strong predictions about the modes of action of VirA and VirG: (i) Ninfa and Magazanik (1987) have shown that purified NtrB protein is capable of converting NtrC between active and inactive states via phosphorylation and dephosphorylation. By analogy, we propose that VirA may convert VirG from an inactive state to an active state via phosphorylation. (ii) Norioka et al., 1987) have shown using the footprinting technique, that OmpR binds in vitro to a consensus sequence within the promoters of the genes it regulates. Since OmpR and VirG are strongly homologous, we believe that VirG may also physically bind to the promoters of other vir genes. Indeed, we do find a dodecanucleotide consensus sequence (TNCAATTGAAAPy) upstream of all vir promoters, as shown in Figure 3. Note that most vir promoters

virA	T	T	C	A	C	T	T	G	A	A	A	C	-68
virB	T	T	C	A	A	T	T	G	A	A	A	T	-57
virB	A	G	C	A	A	T	T	G	A	A	A	A	-38
virC	T	A	T	A	A	T	T	G	C	T	A	C	-57
virC	T	A	C	A	A	T	T	G	C	A	A	T	-41
virC	T	T	T	A	A	T	T	A	T	A	A	C	-37
virC	C	G	A	A	T	T	T	G	A	A	A	T	-26
virD	T	T	A	A	A	T	T	G	C	A	A	T	-76
virD	T	G	C	A	A	T	T	G	T	A	G	C	-70
virD	A	G	C	A	A	T	T	A	T	A	T	T	-61
virE	T	G	C	A	G	T	T	G	A	A	A	C	-55
virG	T	T	C	A	C	T	T	G	T	A	A	C	-52
virG	A	A	C	G	A	T	T	G	A	G	A	A	-40
virG	T	A	A	A	A	T	T	G	A	A	A	T	-19
G	0	5	0	1	1	1	0	13	0	1	1	0	
A	3	4	3	14	11	0	0	2	8	13	13	2	
T	11	6	2	0	1	14	15	0	4	1	1	6	
C	1	0	10	0	2	0	0	0	3	0	0	7	
CONSENSUS	T	N	C	A	A	T	T	G	A	A	A	Py	
ompF/C	N	C	N	T	T	T	T	G	A	A	A	C	
ctx				(T	T	T	T	G	A	T)$_8$			

Figure 3. A conserved dodecadeoxynucleotide found within all vir promoters. Numbers in the extreme right column indicate the position of the base in the second column with respect to the transcription initiation site (Das et al., 1986). Numbers in lines 15 to 18 indicate the number of sequences containing a particular base at that position. The last two lines show similar sequences in promoters regulated by proteins homologous to VirG.

contain multiple copies of this sequence and that these repeats are found at intervals of approximately 10 or 20 nucleotides. If VirG does bind these sequences, then perhaps several monomers may bind on the same face of the DNA helix in a cooperative manner. Figure 3 also shows that the sequences bound by two VirG homologues, OmpR and ToxR, show some degree of similarity with these putative VirG binding sites. We have presented elsewhere (Winans et al., 1986) a model that it is domain C of VirG (Figure 3) which physically associates with vir promoter DNA and with RNA polymerase to activate transcription, while domain B receives information from VirA and uses it to interconvert domain C between active and inactive states. Our model of the mechanism of VirA and VirG is shown in Figure 4.

THE LIMITED HOST RANGE OF THE GRAPE VINE ISOLATE A856 IS DUE IN PART TO ITS virA GENE

Most isolates of Agrobacterium are capable of inducing tumors on a wide variety of dicotyledonous and some monocotyledonous plants (DeCleene and DeLay, 1976; Graves and Goldman, 1987). Some Agrobacterium isolates, however, are tumorigenic on only one or a few plant species (Panagopoulos et al., 1978). The Ti plasmid is the primary determinant of host range, while chromosomal factors have been found not to contribute greatly to host specificity (Knauf et al., 1982; Loper and Kado, 1979; Thomashow et al., 1980). We have investigated the causes of the limited host range (LHR) of a strain of Agrobacterium which was originally isolated from grape vines. When the Ti plasmid of this strain (pTiAg162) is introduced into the C58 chromosomal background, the resulting strain (A856) had the same limited

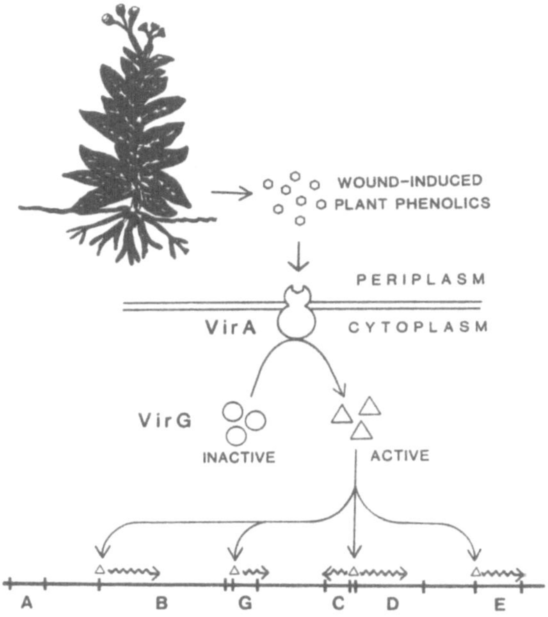

Figure 4. Model describing how VirA may sense the presence of plant-released phenolics and transform VirG, via phosphorylation, from an inactive form to a form capable of binding and activating vir promoters.

host range as the plasmid donor. The presence of three Ti plasmid encoded loci from the wide host range plasmid pTiA6 were required to widen the host range of A856: (i) the tmr gene of the T-DNA, (ii) the virC gene, and (iii) the virA gene (Buchholtz et al., 1984; Hoekema et al., 1984; Yanofsky, et al., 1985). The LHR plasmid (pTiAg162) was shown to carry a partially deleted and therefore defective tmr gene. Complementation experiments showed that pTiAg162 did contain a functional counterpart to the WHR virA (pTiA6) gene, even though no homology was observed between these loci in Southern hybridization analyses. Moreover, the LHR virA gene was only able to complement WHR virA mutants on plants normally infected by A856 (Yanofsky, et al., 1985). That is, the WHR strain containing a virA mutation, and complemented by the LHR virA gene, had acquired the more limited host range of the LHR strain (Yanofsky et al., 1985).

The fact that the host range of the limited host range strain can be extended by providing the WHR virA gene suggests that the LHR VirA protein may be ineffective during infections of plants not normally infected by this strain. Since VirA plays a critical role in the transcriptional induction of the vir regulon, we compared the ability of the wide and limited host range virA genes to induce the transcription of a virB::lacZ transcriptional fusion. Plasmid pSM243 contains a virB::lacZ fusion and also contains the WHR virA gene, while pSM243cd contains the identical virB::lacZ fusion, but does not contain virA (Stachel and Zambryski, 1986). These plasmids were introduced into the LHR strain and induction of virB was monitored during cocultivation with tobacco suspension cells. As shown in Figure 5, efficient induction occurred in the strain containing the WHR virA gene but was barely detectable in the strains containing the LHR virA gene. The same result was obtained when acetosyringone was used in place of suspension cells to induce virB (data not shown). None of these strains was induced during co cultivation with N. glauca or Vitus labruscana vc. Steuben suspension cells even though both strains infect these plants (data not shown).

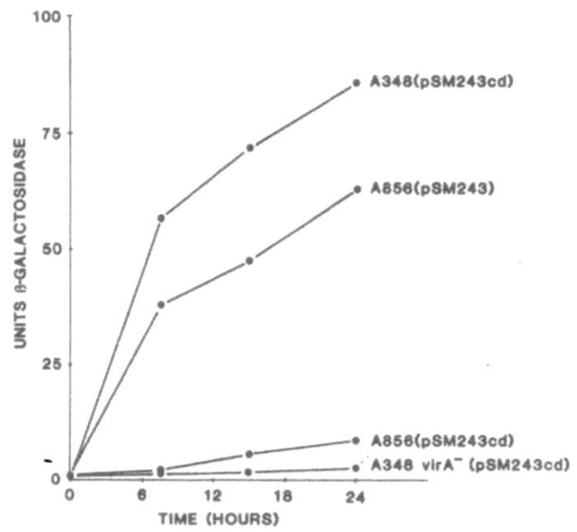

Figure 5. Induction of a virB::lacZ transcriptional fusion by strains containing WHR (A348) or LHR (A856) plasmids during cocultivation with NTIIB cells. Also included was a derivative of A348 bearing the non-inducible insertion mutation virA226::Tn3HoHo1. Plasmid pSM243cd contains a virB::lacZ fusion, while pSM243 contains the same fusion and also carries the WHR virA gene.

To determine the molecular basis for this host range phenotype, we determined the nucleotide sequences of the pTiAg162 virA locus. The LHR virA locus encodes a protein similar in size and overall features to the WHR virA gene. The predicted protein is 835 amino acids in length, and has a putative signal sequence and stop transfer sequence found in the WHR protein. The two proteins are recognizably homologous over their entire lengths and have identical residues at about 375 positions. Figure 6 illustrates the strength of homology of the two proteins as a function of amino acid position. It can be seen that the two proteins are most strongly conserved at their carboxy-terminal halves with greater divergence at the amino-terminal halves.

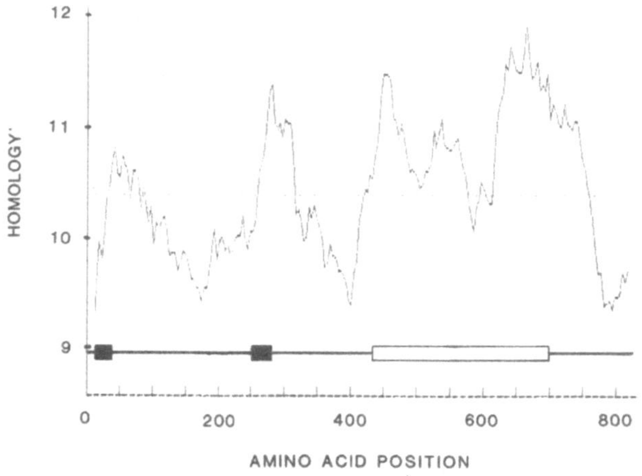

Figure 6. Strength of homology between WHR and LHR VirA proteins. The degree of homology of the two VirA proteins was plotted as a function of amino acid position. Above the x-axis is shown a schematic representation of VirA with solid boxes indicating hydrophobic regions and an open box indicating a region homologous to other bacterial proteins.

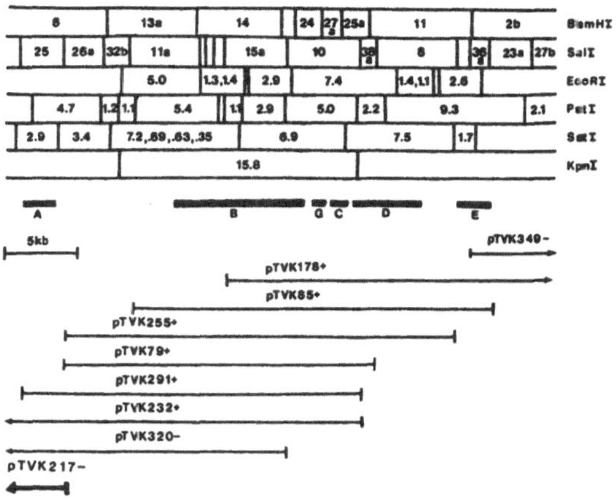

Figure 7. Clones of pTiBo542 which enhanced (+) or did not enhance (-) the virulence of A348. Regions of DNA shown below the physical and genetic map of the pTiBo542 *vir* region cloned into the cosmid vector pVK102.

A comparison between the *E. coli* aspartate and serine chemoreceptors has shown that these proteins show greater divergence at their presumed periplasmic regions than in the remainder of the proteins (Krikos et al., 1983). This is consistent with the proposal that the periplasmic region contains the binding site for the two chemoattractants. We find that the two VirA proteins also have diverged more strongly in their putative periplasmic domains than elsewhere (Figure 6). It is tempting to speculate therefore, that the LHR VirA protein may recognize grapevine-specific inducer molecules, and is unable to interact efficiently with inducers present in plants not normally infected by the limited host range strain. Experiments are necessary to determine whether plants normally infected by the LHR strain synthesize compounds which are recognized by the LHR VirA protein. We have previously shown that the *virA* locus is in part responsible for the narrow host range conferred by the plasmid pTiAg162 (Yanofsky et al., 1985). It is interesting to note that host range is reflected in one of the earliest steps of tumorigenesis, i.e., the recognition by *Agrobacterium* of a susceptible plant host.

THE SUPERVIRULENCE AND EXTENDED HOST RANGE OF STRAIN A281 IS DUE IN PART TO ITS *virG* GENE

A. tumefaciens strain A281 carrying the plasmid pTiBo542 induces large, early-appearing tumors and can promote tumors on at least one plant species (*Glycine max* [cv. Beeson]) not efficiently infected by A348 (Jin et al., 1986). Strains A281 and A348 contain the same chromosomal background and cryptic plasmids, and differ only in their Ti plasmids, indicating that this "supervirulent" phenotype and extended host range is due o plasmid encoded genes. We have determined which genes carried by pTiBo542 are necessary and sufficient to confer these properties (Jin et al., 1987), by introducing subcloned pTiBo542 DNA into strain A348 and screening for the supervirulent phenotype, that is, the ability to incite large, early appearing tumors, using both intact plants and tobacco leak discs (Horsch et al., 1986).

We created a clone bank of pTiBo542 DNA in the broad host range cosmid vector pVK102, introduced a total of 23 cosmids into A348, and screened for the ability of each cosmid to enhance the virulence of its host. Figure 7 shows that six cosmids were able to do so: pTVK79, pTVK85, pTVK178, pTVK232, pTVK255, and pTVK291. All of these plasmids share a 9 kb region containing the 3' half of virB, all of virG, all of virC, and the 5' third of virD, suggesting that the supervirulent phenotype may be encoded by one or more of these genes. Cosmids of pTiA6 DNA containing the same region slightly enhanced virulence, but far less than cosmids containing pTiBo542 DNA.

Two of the cosmids conferring supervirulence on A348 (pTVK79 and pTVK85) were mutagenized using Tn3HoHo1 (Stachel et al., 1985) and introduced into A348 to identify genes required for supervirulence. The 24 insertion mutations analyzed fell into 3 classes. The first class of mutations completely abolished the ability of the cosmid to confer supervirulence, while the second class caused a partial decrease in the ability to confer supervirulence, and the third class did not affect supervirulence. All mutations in the first class (represented by 7 insertions) were located in the virG gene. This was determined by restriction mapping and by finding that each insertion derivative had lost the ability to complement a pTiA6 virG mutant. All mutants in the second class (represented by 4 insertions) were found to have mutations in the 3' half of the virB operon. As expected, these mutants abolished the ability of the cosmid to complement pTiA6 virB mutations. The third class (represented by 13 mutants) included insertions in the 5' half of virB, in virC, and in virD. These results suggest that (i) the virG alone of pTiBo542 can confer partial supervirulent phenotype on A348, and (ii) the 3' half of the virB operon as well as virG are required to confer a full supervirulent phenotype.

All 24 insertions were introduced into the pTiBo542 plasmid by the "marker exchange" technique (Ruvkun and Ausubel, 1981) and assayed for virulence on tobacco leaf discs. 20 of these strains were completely avirulent, while 4 strains were retained a strongly attenuated ability to incite virulence. Two of these (M14 and M31) were located in virC, and showed a level of virulence similar to virC mutants of pTiA6. The other two insertions (M1 and M7) map at the 3' extreme of virB. Both of there strains could be complemented to full supervirulence by introducing pTVK85 derivatives containing virG::Tn3HoHo1 insertion mutations, confirming that they do not lie in the virG gene.

The pTiBo542 DNA required to confer supervirulence was further defined by subcloning fragments of the region containing the 3' end of virB and virG into broad host range cloning vectors and screening each transconjugant for supervirulence. The SalI fragment 10 was cloned into the SalI site of pVK102 in both possible orientations, to give pToK9 and pSG648. pToK9 conferred supervirulence on A348, while pSG648 gave an intermediate response. Since the only difference between these fragments is the orientation of the inserted fragment, we believe that a promoter in the vector is required for expression of a gene or genes in the fragment. The SalI site of pVK102 lies within the tet gene, and the SalI-10 fragment of pToK9 is oriented such that the virB and virG genes could be expressed from the tet promoter, while the insertion in pSG648 is inserted in the opposite orientation.

Since virG plays a central role in the supervirulence phenotype, and since virG controls the transcriptional induction of the vir regulon, we tested the notion that pTiBo542 might cause greater levels of induction of vir genes than pTiA6. We measured the efficiency of induction of a virB::lacZ fusion (pSM243cd) and a virE::lacZ fusion (pSM358cd) in A281 and in A348 after induction with acetosyringone. The B-galactosidase levels of both fusions were significantly higher in the A281 background than in the A348 background. Similar results were obtained for a virD::lacZ fusion (data not shown).

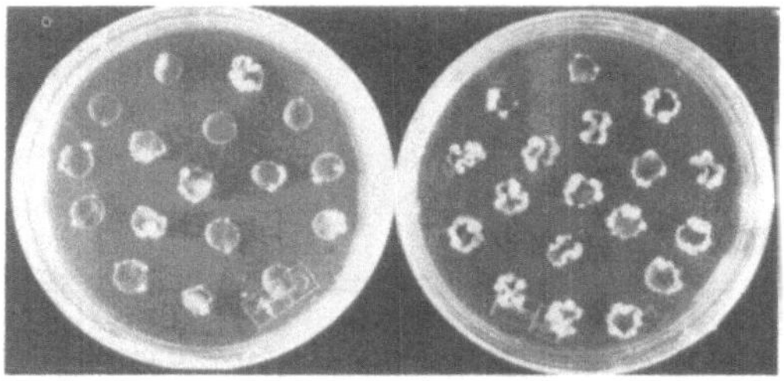

Figure 8. Comparison of tumor formation by A281 (right) and A281(pTVK291) (left) on N. *glauca* leaf disks. Leaf disks were cocultivated with *Agrobacterium* for 24 hours and pictures were taken 10 days after inoculation.

The level of basal and induced levels of transcription of the pTiBo542 *virG* gene were assayed directly using dot blot DNA. RNA samples from A348 and A281 after 16 hours of cocultivation with tobacco suspension cells were hybridized with ^{32}P-labelled pTiA6 *virG* DNA probe. Clearly, *virG*, like *virB*, *virD*, and *virE*, is expressed at much higher levels in A281 than in A348. Strikingly, the basal level of transcription of the pTiBo542 *virG* promoter was easily detectable in this assay, while that of the pTiA6 *virG* promoter was not detected. This result indicates that the uninduced levels of expression of the *virG* gene of pTiBo542 is significantly higher than that of the pTiA6 gene.

Since pToK9 was able to confer supervirulence upon A348 we tested its ability to increase the ability of this plasmid to enhance the already supervirulent phenotype of strain A281. This plasmid did not do so, since A281 and A281(pKo9) incited tumors with the same efficiency (data not shown). However, three different cosmids containing the entire pTiBo542 *virB* and *virG* operons were found to enhance the virulence of A281 (Figure 8). We have also constructed a derivative of pUCD2 containing a 15.8 kb *Kpn*I fragment of pTiBo542 DNA which includes all of *virB* and *virG*. This plasmid also enhanced the virulence of A281. Since the origin of this plasmid was derived from the W incompatibility group plasmid pSa, it is compatible with all *incP* and *incQ* plasmids most widely used as binary vectors to create transgenic plants. This plasmid could therefore be of value in increasing the efficiency of plant transformation.

REFERENCES

Buchholtz, W.G., and M.F. Thomashow. 1984. J. Bacteriol. 160:327-332.

Close, T.J., R.C. Tait, and C.I. Kado. 1985. J. Bacteriol. 164:774-781.

Close, T.J., P.M. Rogowsky, C.I. Kado, S.C. Winans, M.F. Yanofsky, E.W. Nester. 1987. J. Bacteriol. (in press).

Das, A., S. Stachel, P. Ebert, P. Allenza, A. Montoya, and E. Nester. 1986. Nucl. Acids Res. 14:1335-1364.

DeCleene, M., and J. De Lay. 1976. Bot. Rev. 42:389-466.

Dixon, R.A. 1984. J. Gen. Microbiol. 130:2745-2755.

Graves, A.C., and S.L. Goldman. 1987. J. Bacteriol. 169:1745-6

Hall, M.N. and T.J. Silhavy. 1981. Ann. Rev. Genet. 15:91-142.

Hoekema, A., B.S. DePater, A.J. Fellinger, P.J.J. Hooykaas, and R.A. Schilperoort. 1984. EMBO J. 3:3043-3047.

Hood, E.E., W.S. Chilton, M.-D. Chilton, and R.T. Fraley. 1986. J. Bacteriol. 168:1283-1290.

Horsch, R.B., H.J. Klee, S. Stachel, S.C. Winans, E.W. Nester, S.G. Rogers, and R.T. Fraley. 1986. Proc. Natl. Acad. Sci. USA 83:2571-2575.

Jin, S., T. Komari, M.P. Gordon, and E.W. Nester. Submitted.
Knauf, V.C., C.G. Panagopoulos, and E.W. Nester. 1982. Phytopathology 72:1545-1549.
Krikos, A., N. Mutoh, A. Boyd, and M.I. Simon. 1983. Cell 33:615-622.
Leroux, B., M.F. Yanofsky, S.C. Winans, J.E. Ward, S.F. Zeigler, and E.W. Nester. 1987. EMBO J. 6:849-856.
Lipman, D.J., and W.R. Pearson. 1985. Science 227:1434-1441.
Loper, J.E., and C.I. Kado. 1979. J. Bacteriol. 139:591-596.
Melchers, L.S., D.V. Thompson, K.B. Idler, R.A. Schilperoort, P.J. Hooykaas. 1986. Nucl. Acids Res.14:9933-42
Ninfa, A.J., and B. Magasanik. 1986. Proc. Natl. Acad. Sci. USA 83:5909-5913.
Nixon, T.B., C.W. Ronson, and F.M. Ausubel. 1986. Proc. Natl. Acad. Sci. USA 83:7850-7854.
Panagopoulos, C.G., P.G. Psallidas, and A.S. Alivizatos. 1978. Proc. 4th Int. Conf. Pl. Path. Bact. Angers pp. 221-228.
Ruvkun, G., and F. Ausubel. 1981. Nature(London) 289:85-88.
Stachel, S.E., and E.W. Nester. 1986. EMBO J. 5:1445-1454.
Stachel, S.E., E. Messens, M. Van Montagu, and P.C. Zambryski. 1985. Nature 318:624-629.
Stachel, S.E., and P.C. Zambryski. 1986. Cell 46:325-333.
Thomashow, M.F., C.G. Panagopoulos, M.P. Gordon, and E.W. Nester. 1980. Nature 283:794-796.
Winans, S.C., P.R. Ebert, S.E. Stachel, M.P. Gordon, and E.W. Nester. 1986. Proc. Natl. Acad. Sci. USA 83:8278-8282.
Yanofsky, M., B. Lowe, A. Montoya, R. Rubin, W. Krul, M. Gordon, and E.W. Nester. 1985. Mol. Gen. Genet. 201:237-246.
Yanofsky, M., S. Porter, C. Young, L. Albright, M.P. Gordon, and E.W. Nester. 1986. Cell 47:471-477.

MULTIPLE POSITIVE AND NEGATIVE SEQUENCE ELEMENTS MEDIATE THE LIGHT-INDUCED EXPRESSION OF A PLANT GENE

Robert Fluhr[1], Cris Kuhlemeier[2], Pamela J. Green[2]
Steve A. Kay[2], Gunter Strittmatter[2], Maria Cuozzo[2],
Ferenc Nagy[2], and Nam-Hai Chua[2]
[1]Department of Plant Genetics, Weizmann Institute of Science, Rehovot
[2] The Rockefeller University, Laboratory of Plant Molecular Biology, 1230 York Avenue, New York 10021-6399, U.S.A.

INTRODUCTION

Light so crucial to photosynthesis plays a decisive role in plant development. Plants have the necessary photoreceptor and signal transduction mechanisms to remain in cue with the quality and intensity of ambient radiation, and as expected, the activities of a large number of plant genes are modulated by light. We have studied some genes encoding the small subunit (rbcS) polypeptide of ribulose 1,5-bisphosphate carboxylase-oxygenase. Although its exact function is not clear the expression of rbcS genes is controlled by light as well as cell-type specific factors. This paper summarizes recent results from our laboratory on the selective expression of rbcS *in vivo* and in transgenic plants.

DIFFERENTIAL EXPRESSION IN A MULTIGENE FAMILY

The small subunit of ribulose 1,5-bisphosphate carboxylase (rbcS), is encoded by a small multigene family in pea (Coruzzi et al; 1984). The gene family is composed of at least five closely linked members (Polans et al., 1985). In all five pea rbcS genes, the non-coding region shows overall conservation but contains short regions of sequence divergence. We took advantage of this dishomology and developed gene specific tools using S1-nuclease derived transcript sizing (Fluhr et al., 1986a). Fig. 1 summarizes the differential expression of rbcS transcripts in pea leaves. While rbcS-3A and -3C together account for nearly 80% of rbcS transcripts in the mature leaf, rbcS -E9 and -3.6 are particularly low. The large differences in the steady state levels of the transcripts may indicate varied levels of transcriptional activity or be a result of specific mRNA stability. Two lines of evidence would suggest that the former is the case: 1. Transcriptional fusions involving highly active rbcS gene promoters (e.g. rbcS-3A) and nieve reporter genes are generally more active than similar constructs with less active rbcS genes (e.g. E9). 2. All rbcS transcripts decline in the dark at similar rates (Fluhr et al., 1986a), suggesting similar stability.

ORGAN-SPECIFIC DISTRIBUTION OF THE rbcS GENES

A quantitative comparison of a specific rbcS transcripts rbcS-3A in various pea organs is summarized in Fig. 2. rbcS transcripts such as

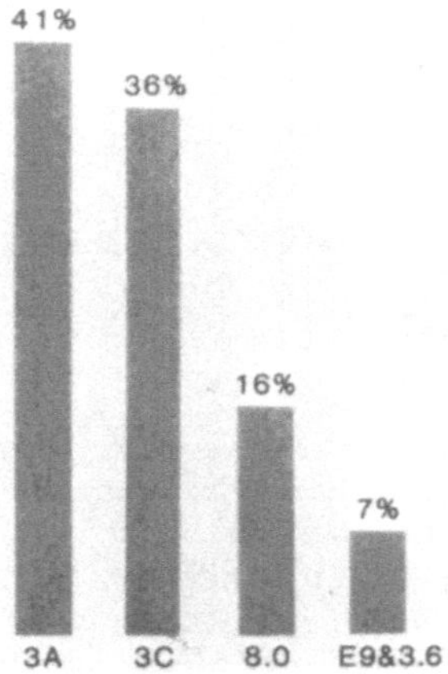

Fig. 1. Distribution of rbcS transcripts in pea leaves based on quantitative 3' S1 analysis and quantitative slot blot hybridization using gene specific oligonucleotides (Fluhr et al., 1986a).

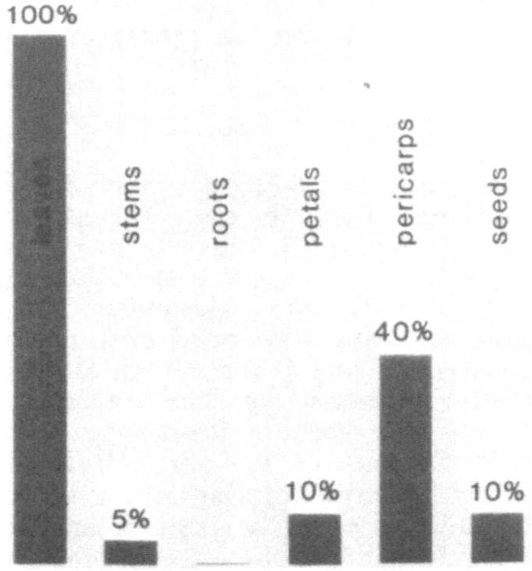

Fig. 2. Abundance of rbcS-3A transcripts in various pea organs relative to green leaves as determined by quantitative S1-nuclease analysis and quantative slot blot hybridization using gene specific oligonucleotides (Fluhr et al. 1986a).

rbcS-3A and also 3C are highest in chloroplast containing tissue, much lower in tissues containing plastids of alternate differentiation pathways (petals and seeds) and very low or not at all detectable in roots. In contrast the rbcS-E9 gene is virtually absent from any but chlorophyllus tissue. This is shown clearly in an experiment in which variable amounts of organ-specific RNA was analysed by the S1-nuclease method (Fig. 3). The relative ratio of transcripts of rbcS-E9,-3A and-3C are unchanged in leaves stems and pericarps. However in petals and seeds almost no rbcS-E9 transcript is detectable. Thus this particular gene is keyed to the developmental stage of the chloroplast. This experiment illustrates the sensitivity of the quantitative S1-nuclease approach. The varied nuances in tissue specific control of rbcS expression may be due to subtle differences in the arrangement of cis-acting regulatory elements.

LIGHT INDUCTION OF rbcS GENES

Genes for rbcS are among a group of 46 enzymes whose activity has been reported to increase upon illumination of etiolated plants. The photocontrol of rbcS mRNA abundance in the etiolated system has been shown to be phytochrome linked (Fluhr and Chua 1986b, Kaufman et al., 1984). Due to high levels of the rbcS gene product accumulating in green leaves much classical analysis of rbcS photocontrol was carried out on etiolated leaf material. In such experiments parallel photomorphogenic events can complicate interpretation. Analysis of rbcS-gene photocontrol in the mature leaf, carried out at the individual transcript level shows photoresponses to be more sophisticated than those in etiolated tissues. Flashes of red-light cannot induce rbcS transcripts in dark adapted mature leaves but effective conditions to restore rbcS transcript levels requires continuous illumination of light in the blue wavelength region.

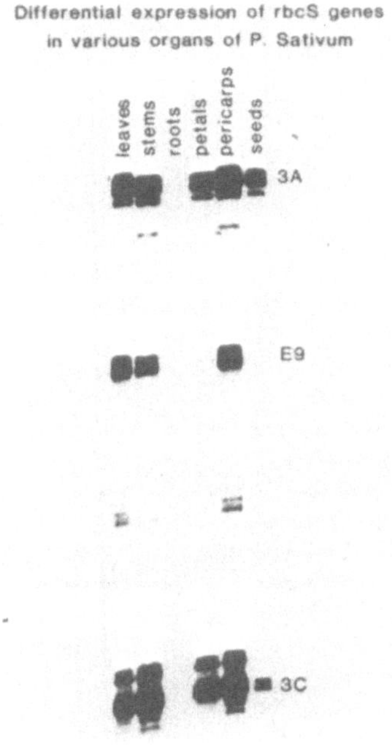

Fig. 3. Quantitative 3' S1-nuclease analysis of rbcS transcripts from various organs of pea plants. Varying amounts of total RNA were analyzed with an rbcS-3A probe so as to obtain a signal of equal strength with respect to the rbcS-3A transcript. The identification of each transcript type using this probe was ascertained by comparison of DNA sequence homology and with the use of transgenic petunia plants expressing each specific pea rbcS transcripts (Fluhr et al 1986a).

The enhancing effect of blue light can be reversed by far-red illumination. (Fluhr and Chua, 1986b). On the basis of these results we have proposed that photocontrol of rbcS gene expression is sensitive to the leafs developmental stage. Induction of rbcS mRNA in etiolated immature seedlings is mediated by phytochrome, in green mature tissue the induction is regulated by a blue sensitive photoreceptor working in concert with phytochrome. The resolution of the two photoreceptors for rbcS regulation raises the question whether signal transductions in these two systems involve the same cis-acting regulatory elements.

EXPRESSION OF rbcS IN TRANSGENIC PLANTS-LOCATION OF POSITIVE AND NEGATIVE CONTROL ELEMENTS

Transgenic tobacco and petunia plants carrying full length (>2kB) pea rbcS-3A, -3C and -E9 genes have been shown to faithfully recapitulate light inducibility and organ specifity of these transferred genes.(Nagy et al. 1985, Fluhr and Chua 1986b). The major drawback in this experimental system as a method of testing in vitro mutagenized genetic elements is clonal variance in the resultant transgenic plants. Well over fifty fold variation can be obtained in comparing the activity of any one construct, making it difficult to draw any but qualitative conclusions. A method which we developed to circumvent this problem includes co-transforming plants with a rbcS reference gene together with the gene under study. The gene products can then be differentiated by a specific quantitative S1 probe. We have used this system to delineate cis-regulatory DNA elements.

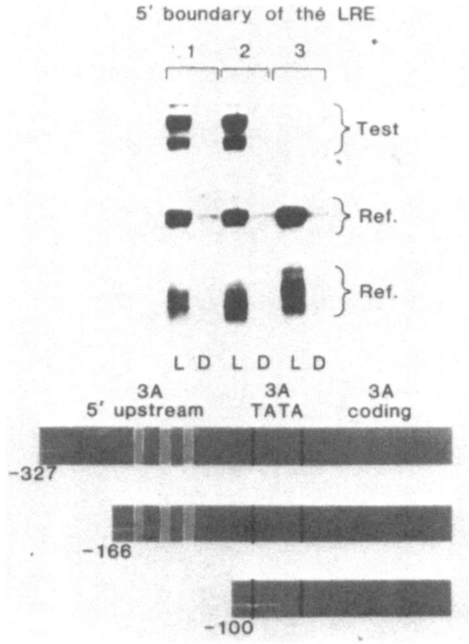

Fig. 4. Effect of 5' deletions on the transcript levels of the rbcS-3A gene. The deletions were formed by Bal 31 nuclease digestion. The test gene containing the deletion was inserted approximately 5 kilobase downstream from a rbcS-3A reference gene in the pMON 200 vector. S1 probe analysis produces a 160 nucleotide protected fragment for the reference gene and a 231 nucleotide fragment for the test gene transcripts.

When a series of BAL 31 deletions is performed on the rbcS-3A gene, full light-dark regulation is maintained to 166 base pairs upstream from the mRNA start site (Fig. 4). Further deletion to -149 severely lowers activity. However, even at -100 a very low amount of light regulated expression remains. From this we conclude that sequences downstream of -166 are sufficient for light regulated expression at wild type levels in transgenic tobacco and that the 5' end of a positive activating element is located between -166 and -149. (Kuhlemeier et al., 1987).

In order to identify more clearly, elements important for gene activation we constructed an enhancer trap vector. It consists of the 35S CaMV TATA box placed upstream from the bacterial chloramphenicol acetyl transferase gene which together have no intrinsic activity. Constructs containing DNA sequence of rbcS-3A from -160 to -50 activate this construct in light induced manner, however further 3' end deletion of -169 to -112 does not. This would place the 3' border of the active element between -50 and -112. (Kuhlemeier, unpublished).

Inspection of the upstream region of the activating elements (-169 to -50) in this regions reveals runs of sequence conserved in all the rbcS pea genes inspected. Prominent among them are the sequences GTGTGGTTAATATG from -151 to -138 (Box II) and ATCATTTTCAC from -125 to -114 (Box III) the former is homologus to the SV 40 core enhancer consensus and the latter to the β-interferon constitutive element. However, these elements by themselves or even as a unit from -169 to -112 are inactive. They cannot motivate transcription of a construct containing the 35S CaMV TATA box together with a CAT reporter gene. As these elements are unable to enhance transcription, we decided to combine them with a constitutive enhancer to determine whether it might modulate transcription in a negative way. To this end we made use of a chimeric gene containing the 35S CaMV upstream (-943 to -31) sequence juxtaposed to the otherwise inactive elements of rbcS-3A sequence. When Box II is present between the 35S enhancer and promoter a modest down-regulation of activity occurs in the dark. However three copies of the element lead to distinct light modulation. When a small segment of DNA (-136 -50) containing box III is used similar down regulation of the 35S enhancer is obtained. From many individual plants analyzed we have gained confidence that these two light-responsive elements (LREs) do not significantly influence mRNA levels in the light but act to decrease transcription in the dark, and hence act as negative regulatory units.

THE LIGHT RESPONSIVE ELEMENTS ARE PRESENT IN MULTIPLE COPIES

Deletion of the LRE elements from -162 to -125 (boxes II and III) in a construct which includes the upstream region of rbcS-3A from -410 does not effect expression of the rbcS-3A gene (Kuhlemeier et al, 1987). These results indicate that compensating enhancing and regulatory LRE functions exist in the more distal 5' end region. Indeed, inspection of this region reveals DNA sequences with homology to box II and box III. It would seem that in the mature green leaf under the conditions of light induction used in these experiments to LREs act in a redundant manner. It is conceivable that in other developmental stages or environmental conditions both sets of LREs are indispensible.

TRANS-ACTING FACTORS BIND TO LIGHT RESPONSIVE ELEMENTS

Gel retardation of DNA fragments by bound protein is one useful approach in the study of gene activation. Using pea nuclear extracts it was found that a DNA fragment containing sequences of box II and III

(-170 to -50) bind a nuclear protein species and are subsequently retarded in a gel assay. DNA fragments from a more distal portion (-330 to -171) of rbcS-3A compete for the same factor (P.J. Green, submitted). These results parallel the finding of redundant LREs established by in vivo analysis of mutant rbcS-3A promoter elements. In addition direct DNA foot printing experiment have also targeted the binding activity to the specific LREs defined by in-vivo experiments. The same binding activity can be detected in both light and dark grown plants. Thus, one can eliminate de novo synthesis of transactive factor as a mechanism in its function (P.J. Green submitted).

CONCLUSION

An upstream segment of pea rbcS-3A behaves like a transcriptional enhancer because it can motivate transcription of the heterologous 35S CaMV promoter. The transcription enhancer is made up of redundant components within which are elements that functionally behave as silencer sequences. Together this complex array of cis-acting DNA elements determine, in part, the differential response to light activation and organ specificity that each rbcS gene has been shown to possess.

The work summarized in this paper was supported by a grant from the Monsanto Company

REFERENCES

1. Coruzzi, G., Broglie, R., Edwards C, Chua N.-H), 1984 Tissue specific and light regulated expression of a pea nuclear gene encoding the small subunit of ribulose - 1,5-bisphosphate carboxylase. EMBO J 3: 1671-1679
2. Fluhr R, Kuhlemeier, C., Nagy, F, Chua N.H. (1986b) Organ-specific and light induced expression of plant genes. Science. 232: 1106-1112.
3. Fluhr R, Moses P, Morelli G, Coruzzi, G, Chua N.H., (1986a) Expresison dynamics of the pea rbcS multigene family and organ distribution of the transcripts. EMBO J. 5: 2063-2071.
4. Fluhr R, and Chua N.H., (1986b), Developmental regulation of two genes encoding ribulose-bisphosphate carboxylase small subunit in pea and transgenic petunia plants: Phytochrome response and blue-light induction. Proc. Natl. Acad. Sci. 83: 2358-2362, (1986b).
5. Kaufman L.S., Thompson W.F., Briggs W.R., (1984), Different red-light requirements for phytochrome-induced accumulation of Cab RNA and rbcS RNA. Science 226: 1447-1449.
6. Kuhlemeier C, Fluhr R, Green P.J., Chua N.H. (1987), Sequences in the pea rbcS-3A gene have homology to constitutive mammalian enhancers but function as negative regulatory elements, Genes and Development 1: 247-255.
7. Nagy F, Morelli G, Fraley R.T., Rogers S.G, Chua, N.H. (1985), Photo-regulated expression of a pea rbcS gene in leaves of transgenic plants. EMBO J 12: 3063-3068.
8. Polans, N.O., Weeden, N.F. and Thompson W.F., (1985), Inheritance organization and mapping of rbcS and Cab multigene families in pea. Proc. Natl. Acad. Sci (USA) . 82: 5083-5087.

STRUCTURE AND REGULATION OF THE MAIZE Suppressor-mutator TRANSPOSABLE ELEMENT

P.Masson, J.Banks, R.Surosky, J.Kingsbury and N.Fedoroff.

Carnegie Institution of Washington
Department of Embryology
115 W. University Parkway
Baltimore, MD 21210

ABSTRACT

The maize a gene encodes a product involved in the anthocyanin biosynthetic pathway, responsible for kernel pigmentation. The a-m2 allele, isolated by McClintock (1951), originated by insertion of an autonomous Suppressor-mutator (Spm or En) transposable element. Several new alleles (designated "states" by McClintock) were subsequently derived from the original one. The uniqueness of these a-m2 alleles resides in the fact that a gene expression is under the positive control of Spm. To gain insight into the structure and regulation of Spm, we have done genetic and molecular studies on certain a-m2 derivatives. We present evidence that both the termini and an internal GC-rich region of the element are involved in the determination of both Spm transposition frequency and Spm-dependent a gene expression. We also show that Spm is subject to both positive and negative control of expression. The Spm element can exist in both an unstably inactive state, as well as a stably inactive state, both of which are associated with methylation of element sequences.

INTRODUCTION

The Suppressor-mutator (Spm or En\I) family of maize transposable elements comprises autonomous elements, capable of promoting their own transposition, as well as transposition-defective (dSpm) elements. Autonomous En and Spm elements have been cloned from both the waxy (Pereira et al.,1986) and a loci (Masson et al.,1987); they are almost identical, 8.3-kb elements whose general structure is shown in Figure 1. dSpm elements are generally derived from autonomous ones by internal deletions (Schiefelbein et al., 1985; Schwarz-Sommer et al.,1987; Masson et al.,1987).

The interaction between an autonomous element and a defective one located elsewhere permitted McClintock to identify two major trans-active element functions: the mutator and the suppressor functions. The mutator is a transposition function defined by the Spm's ability to trans-activate transposition of a dSpm. The suppressor function is defined by the ability of an Spm element to suppress or inhibit the basal level of expression of a locus with a dSpm insertion.

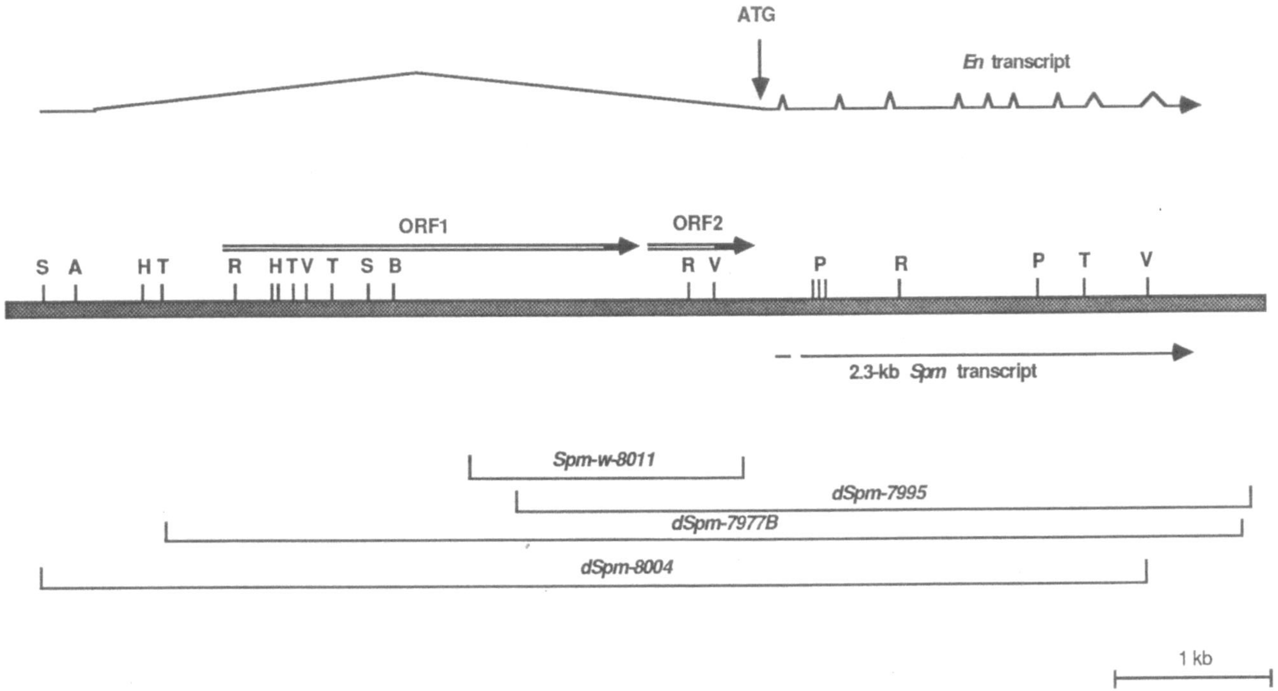

FIGURE 1. A diagrammatic representation of the Spm sequence.

The size, extent and orientations of ORF's 1 and 2 are represented by open arrows over the diagram. The approximate extent of the Spm's major 2.3-kb transcript is indicated below the diagram. Also represented by the lines below the diagram are the regions of the element deleted in each of the mutant elements analyzed. The extended arrow over the diagram represents the sequence of the mRNA of the closely related En element (Pereira et al., 1986), with horizontal portions of the arrow corresponding to exons. Several restriction sites within the element are indicated by the letters designating the following enzymes: A = AvaI; B = BamHI; H = HindIII; S = SalI; R = EcoRI; P = PvuII; T = PstI; and V = EcoRV.

There is a different type of interaction between the Spm element and the dSpm insertion in the a-m2 alleles (McClintock, 1951). In these alleles, the a gene (which encodes an enzyme involved in the anthocyanin biosynthetic pathway) is silent in the absence of an autonomous Spm element, but is expressed in its presence. Different derivatives of the original a-m2 insertion have been isolated and characterized genetically and at the molecular level (McClintock, 1962; Masson et al., 1987; Banks and Fedoroff, unpublished). We summarize the results of these studies and discuss their implications for understanding the regulation of Spm expression.

RESULTS AND DISCUSSION

Molecular Characterization of the Spm Insertion in the a-m2 Alleles

The a-m2 alleles analysed in this study contain an Spm insertion 0.1 kb upstream from the transcription start site at the a locus (Masson et al., 1987; Schwarz-Sommer et al., 1987). The fully functional autonomous Spm element present in the a-m2-7991A1 allele was sequenced and found to be almost identical to the En-1 element isolated from the wx locus (Pereira et al., 1986; Masson et al.,1987). The 8.3-kb Spm element contains 13-bp inverted terminal repeats, as well as subterminal regions with numerous short direct and inverted repeats extending for several hundred base pairs at each element end. (Schwarz-Sommer et al., 1984). The element contains two large open reading frames, and encodes a major 2.3-kb transcript. (Figure 1; Pereira et al.,1986; Masson et al., 1987)

The dSpm elements in several a-m2 derivatives have also been cloned and analysed (Masson et al.,1987). Almost all of them arose from the original Spm insertion by internal deletions (Figure 1). A careful analysis of the modified phenotype exhibited by each a-m2 derivative state has permitted us to identify sequences that determine, in cis and in trans, the freqency of Spm transposition and to develop hypotheses concerning Spm regulation.

Trans-determinants of Spm Transposition Frequency

The observation that an autonomous Spm element can promote or activate the transposition of a dSpm implies that Spm encodes a transposase. A comparison of the molecular structures of the Spm elements present in the different a-m2 alleles suggests that the transposase is encoded by the right half of Spm. The deletion of a 1.7-kb fragment in the middle of the element (a-m2-8011) decreases, but does not eliminate, its ability to transpose and to trans-activate the transposition of dSpm elements. A deletion partially overlapping the a-m2-8011 deletion on the left and extending to the right subterminal region completely abolishes the trans-active transposition function (dSpm-7995).

Cis-determinants of Spm Transposition Frequency

The dSpm elements of the different a-m2 alleles excise at different frequencies in the presence of a standard Spm element (Masson et al.,1987). The structural analysis of several of the dSpm elements has permitted the identification of internal cis-acting element sequences that determine the excision frequencies. Two different internal sequences have been identified: the subterminal repetitive domain and a sequence located at the extreme left end of the element, but distinct from the left subterminal domain.

The importance of a sequence at the right end of the element, probably the subterminal domain, has been deduced from the observation that both the dSpm-7995 and dSpm-7977B elements from which part of the right subterminal domain has been deleted (Masson et al.,1987), excise at a lower frequency than the Spm-w-8011 element in the presence of a standard Spm. The dSpm-8004 element has intact termini and subterminal regions, but excises at a very low frequency in the presence of an Spm element. A comparison between the structures of the dSpm-8004 element, and the dSpm-7977B and dSpm-I8 elements isolated from three different loci from which it can excise at a high frequency (Fedoroff et al., 1984; Schwarz-Sommer et al., 1984, 1985; Schiefelbein et al., 1985) indicates that an internal sequence extending from nucleotide 275 to nucleotide 860 at the element's left end is also an important cis-determinant of excision frequency. This conclusion is supported by the fact that the full-length defective dSpm-8167B element, which transposes at low frequency in the presence of Spm but has unaltered termini (Masson et al., 1987), is methylated at the SalI site located in this region, while the Spm-s-7991A1 element is not methylated (J.Banks and N.Fedoroff, unpublished).

Inactive and Cryptic Spm Elements

The existence of a positive control of Spm expression can be deduced from the interaction between active and inactive Spm elements. An inactive element has been isolated from the original a-m2-7991A1 allele (N.Fedoroff, unpublished data). That it is an inactive element and not a defective one is shown by its ability to undergo spontaneous reactivation in a fraction of the progeny. The inactive Spm element can be reactivated in trans by the introduction of an active Spm element (N.Fedoroff, unpublished data). Similarly, McClintock (1957-59,1971) reported that an inactive element can be transiently activated by an active one, suggesting that Spm encodes a positive regulatory gene product that can overcome the inactivating mechanism. Little is known about the molecular mechanism of element inactivation. However preliminary data indicate that methylation of the element may be correlated with inactivation (J.Banks and N.Fedoroff, unpublished data).

Cryptic copies of Spm also exist in the corn genome. Cryptic elements are in a stably inactive state. Spontaneous reactivation of a cryptic element occurs at a frequency of about 10-5, although there is evidence that it can be increased in several ways, including the introduction of an active Spm element (N.Fedoroff, 1986).We have identified an Spm element in the a-m2-8167B allele which is structurally indistinguishable from the Spm element of the a-m2-7991A1 allele, but exhibits the genetic behavior of a dSpm element. We have observed that this element is internally methylated (Masson et al., 1987; J. Banks and N.Fedoroff, unpublished data). One attractive hypothesis is that the Spm element of the a-m2-8167B allele is a cryptic element and not a defective one. We are currently testing this hypothesis by trying to reactivate the Spm element of the a-m2-8167B allele (J.Banks and N.Fedoroff, unpublished data).

Regulation of a Gene Expression in the a-m2 Alleles

The Spm insertion site is just upstream of the a gene's transcription start site in the a-m2 alleles. In those with a dSpm insertion, the gene is not expressed in the absence, but is expressed in the presence of an Spm element elsewhere in the genome. Hence a gene expression is under the positive control of the element and is mediated by an interaction between an element-encoded gene product and the dSpm element adjacent to the a gene.

As indicated in Figure 2, a gene expression is dependent on the Spm element in several a-m2 alleles with extensively deleted dSpm elements. The ability to mediate a gene expression therefore resides in element termini. The absence of deletions completely removing the right end of the element precludes assessment of its importance in mediating Spm-dependent a gene expression; however a gene expression is observed in a-m2 alleles in which the distance between the right element end and the gene varies from about 1 kb to more than 8 kb. Sequences at the left end of the element do appear to be more important. This conclusion follows from the observation that the a gene is expressed in the a-m2-8004 and a-m2-7977B alleles, but more weakly in the former than in the latter (Figure 2). This indicates that the sequence extending from nucleotide 275 to nucleotide 1075 of Spm influences the level of Spm-dependent a gene expression. About half of this sequence corresponds to the element's GC-rich first exon (Pereira et al.,1986; Masson et al.,1987). We also have preliminary evidence that reduced a gene expression results from methylation of this sequence. We have observed that the a gene in the a-m2-8167B allele is only weakly expressed in the presence of Spm. The inserted element is structurally indistinguishable from a standard Spm element, but is internally methylated.

The ability of the Spm element to trans-activate a gene expression parallels its ability to reactivate an inactive Spm element, suggesting that the a gene has come under the control of the Spm's regulatory mechanism. The element sequences implicated in regulating expression of the adjacent a gene are in the vicinity of the element's own transcription start site, identified at nucleotide 209 by Pereira et al. (1986). We have suggested that the ability of these element sequences to mediate a gene expression is consequent on their normal function in the activation of Spm expression in response to an element-encoded positive regulatory gene product (Masson et al.,1987).

CONCLUSIONS

The molecular and genetic analyses of several of the a-m2 alleles isolated by McClintock has established the molecular structure of an autonomous element and identified sequences that are important cis-determinants of excision frequency. Evidence has been presented that an Spm element can reactivate an inactive one, suggesting the existence of a positive control of Spm expression. The parallelism that exists between activation of a gene expression in the a-m2 alleles and reactivation of an inactive Spm element by an autonomous Spm element suggests that the a gene is under the control of Spm element's positive regulatory mechanism. We are now investigating the mechanisms of Spm regulation using a combination of molecular analysis of new a-m2 states derived by the classical genetic means and expression of different Spm constructs in a heterologous system (tobacco).

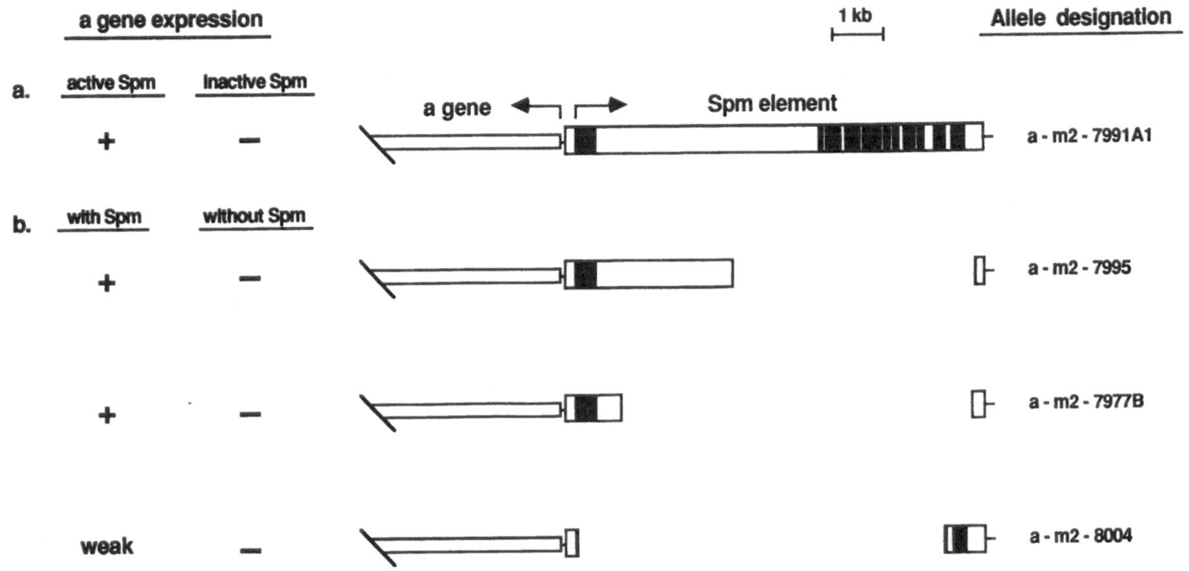

FIGURE 2. Structure of the _a_ locus in several _a-m2_ alleles.

The complete 8.3-kb _Spm_ element is inserted 0.1 kb upstream of the _a_ gene. As indicated by the arrows, the gene and the element are transcribed divergently. The _Spm_ element's intron/exon structure is assumed to be identical to that of the similar _En_ element (Pereira _et al._, 1986) and is represented by the unshaded (intron) and shaded blocks (exons) within the element). The structure of the element in several derivatives of the original _Spm_ insertion shown in part a is represented in part b of the Figure.

REFERENCES

Fedoroff N., Shure M., Kelly S., Johns M., Furtek D., Schiefelbein J. and Nelson O., 1984, Isolation of Spm controlling elements from maize, Cold Spring Harbor Symp. Quant. Biol., 49: 339-345.

Fedoroff N., 1986, Activation of Spm and modifier elements, Maize Gen. Coop. Newslet., 60: 18-20.

Masson P., Surosky R., Kingsbury J. and Fedoroff N., 1987, Genetic and molecular analysis of the Spm-dependent a-m2 alleles of the maize a locus, Genetics, in press.

McClintock B., 1951, Mutable loci in maize, Carnegie Inst. Wash. Yrbk., 50: 174-181.

McClintock B., 1957, Genetic and cytological studies of maize, Carnegie Inst. Wash.Yrbk.,56: 393-401.

McClintock B., 1958, The Suppressor mutator system of control of gene action in maize, Carnegie Inst. Wash. Yrbk.,57: 415-429.

McClintock B., 1959, Genetic and cytological studies of maize,Carnegie Inst. Wash. Yrbk., 58: 452-456.

McClintock B., 1962, Topographical relations between elements of control systems in maize, Carnegie Inst. Wash. Yrbk., 61:448-461.

McClintock B., 1975, The contribution of one component of a control system to versatility of gene expression, Carnegie Inst. Wash. Yrbk.,70: 5-17.

Pereira A., Cuypers H., Gierl A., Schwarz-Sommer Zs. and Saedler H., 1986, Molecular analysis of the En\Spm transposable element system of Zea mays, EMBO J., 5: 835-841.

Schiefelbein J., Raboy V., Fedoroff N. and Nelson O., 1985, Deletions within a defective Suppressor-mutator element in maize affect the frequency and developmental timing of its excision from the bronze locus, Proc. Natl. Acad. Sci. USA, 82: 4783-4787.

Schwarz-Sommer Zs., Gierl A., Klosgen R.B., Wienand U., Peterson P. and Saedler H., 1984, The Spm (En) transposable element controls the excision of a 2-kb DNA insert at the wx-m8 allele of Zea mays, EMBO J. 3: 1021-1028.

Schwarz-Sommer Zs., Gierl A., Berndtgen R. and Saedler H., 1985, Sequence comparison of "states" of al-ml suggests a model of Spm (En) action, EMBO J., 4: 2439-2443.

Schwarz-Sommer Zs., Shepherd N., Tacke E. Gierl A., Rohde W., Leclercq L., Mattes M., Berndtgen R., Peterson P. and Saedler H., 1987, Influence of transposable elements on the structure and function of the Al gene of Zea mays, EMBO J., 6: 287-294.

THE PHOTOSYNTHETIC REACTION CENTER FROM THE PURPLE BACTERIUM RHODOPSEUDOMONAS VIRIDIS AND ITS RELEVANCE TO PHOTOSYSTEM II

Hartmut Michel and Johann Deisenhofer

Max-Planck-Institut fuer Biochemie
D-8033 Martinsried, West Germany

INTRODUCTION

Photosynthesis is not confined to the world of plants and algae, but it is also wide-spread among prokaryotes. In the procaryotic, the bacterial, world oxygen evolution due to splitting of water ("water oxidation") is restricted to cyanobacteria and prochloron-type organisms, whereas the "purple" and "green photosynthetic bacteria" can oxidize reduced sulphur compounds or reduced organic materials in a so-called anoxygenic photosynthesis, or they derive energy from a cyclic electron-flow which is coupled to photophosphorylation.

How water is split in the oxygen evolving complex of photosystem II from chloroplasts and cyanobacteria belongs to the most important questions to be answered. It is clear that a direct photolysis of water does not take place. The primary charge separation occurs in the reaction center of photosystem II. This reaction center is a complex consisting of integral membrane proteins, several chlorophyll A molecules, two pheophytin A molecules, two or three plastoquinones and one non-heme-iron atom. Upon absorption of light an electron is transfered from the primary electron donor, which is a chlorophyll A molecule or a pair of chlorophyll A molecules, via a pheophytin to the plastoquinones. The photo-oxidized primary electron donor itself oxidizes a manganese compound in the oxygen evolving complex via Z. Z is presumably a plastoquinol. The manganese complex can accumulate four positive charges, which then extract four electrons from water leading to the release of protons and oxygen (for recent reviews see Dekker and van Gorkom, 1987; Brudvig, 1987). At the electron-accepting site two plastoquinones ("Q_A" and "Q_B") act in series. The electron is transfered first to the more firmly bound Q_A and then to Q_B. The Q_B plastoquinone receives two electrons successively, becomes protonated, and is then replaced by another oxidized plastoquinone. Q_B is the site of action of commercially used herbicides which presumably act by displacing the plastoquinone.

The location of the photosystem II reaction center is still a matter of debate. Nakatani *et al.* (1984) concluded from fluorescence measurements that a protein of apparent molecular weight 47000 (CP47) is the apoprotein of photosystem II reaction center. A different view emerged from work with the photosynthetic reaction centers from the purple bacteria. Williams *et al.* (1983) obtained the amino acid sequence of the L(ight) subunit of the reaction center from *Rhodopseudomonas* (*Rps.*) *sphaeroides* (now called *Rhodobacter* (*Rb.*) *sphaeroides*) and discovered sequence homologies with the D1 protein from spinach (Zurawski *et al.*, 1982). They suggested conservation of functionally analogous regions of the two proteins. D1 had been shown to become covalently labeled by the herbicide azido-atrazin in photoaffinity experiments (Pfister *et al.*, 1981). Later on, it became apparent that the M(edium) subunit of the bacterial reaction center possesses sequence homology with the L subunits and the D1 proteins (Youvan *et al.*, 1984, Williams *et al.*, 1984, Michel *et al.*, 1986). In parallel, the sequence of another photosystem II protein, D2, became known (Rasmussen *et al.*, 1984, Rochaix *et al.*, 1984, Alt *et al.*, 1984, Holschuh *et al.*, 1984) which also possesses sequence homologies to the D1 proteins and the L and M subunits. It was then evident that gene duplications gave rise to the L and M subunits as well as to the D1 and D2 proteins.

A substantial amount of structural information could be obtained with the reaction center from *Rps. viridis* which could be crystallized (Michel, 1982). The crystallographic analysis yielded a complete picture of the pigment arrangement and the structure of the protein subunits (Deisenhofer *et al.*, 1984, 1985). The most striking feature was the symmetric arrangement of the pigments and the L and M subunits. Both subunits are needed to establish the binding sites for the primary electron donor and the electron accepting quinone-iron complex. These characteristics and the sequence homologies led to the proposal that the D1 and D2 proteins form the core of photosystem II reaction center. D1 was proposed to correspond to the L subunit and D2 to the M subunit (Michel and Deisenhofer, 1986, Deisenhofer *et al.*, 1985). Independently, Hearst (1986) came to the same conclusion based on the particular good sequence homology between the D1 and D2 protein and the L and M subunits from *Rb. capsulatus*. This view received argumentative support (Trebst and Depka, 1985, Trebst, 1986). Strong experimental support was provided by the recent isolation of a complex of the D1 and D2 proteins together with cytochrome b559 (Nanba and Satoh, 1987, Satoh *et al.*, 1987). Here we discuss the structure of the photosynthetic reaction center from the purple bacterium *Rps. viridis* and describe the role of those amino acids which are conserved between the bacterial and the plant photosynthetic reaction center.

RESULTS AND DISCUSSION

The Structure of the Reaction Center Core from *Rps. viridis*

The core of the reaction center is made up by the L and M subunits. Both proteins possess a primarily helical secondary structure. Each subunit contains five long membrane-spanning helices and several short ones. The major helices of both subunits and the ring systems of the photosynthetic pigments are related by an internal twofold symmetry axis. The major helices and the pigments are shown in figure 1, the symmetry axis runs vertically in the centre of the figure. The structural differences between the L and M subunits are mainly at the amino-terminus on the

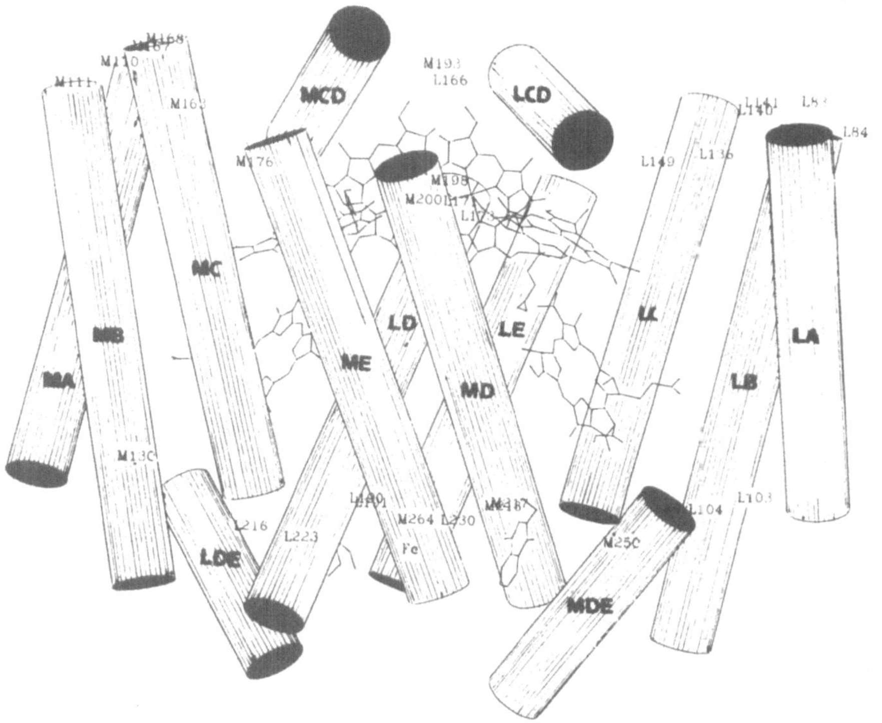

Fig. 1: Schematic drawing of the major helices and the pigments in the reaction center from *Rhodopseudomonas viridis*. The helices are represented as columns. The transmembrane helices of the L (M) subunit are labeled LA-LE (MA-ME), the major helices in the connections LCD (MCD) and LDE (MDE). The helix connections are omitted for clarity. The view is parallel to the plane of the membrane. The special pair bacteriochlorophylls are at the interface of the L and M subunits between the D and E helices, the accessory bacteriochlorophylls below helices LCD and MCD, and the bacteriopheohytins near the C helices. The binding site for Q_A (menaquinone in *Rps. viridis*) is between the MDE and MD helices, that for Q_B between the LDE and LD helices. The location of amino acids conserved between all L and D1, or M and D2 (see test) is shown.

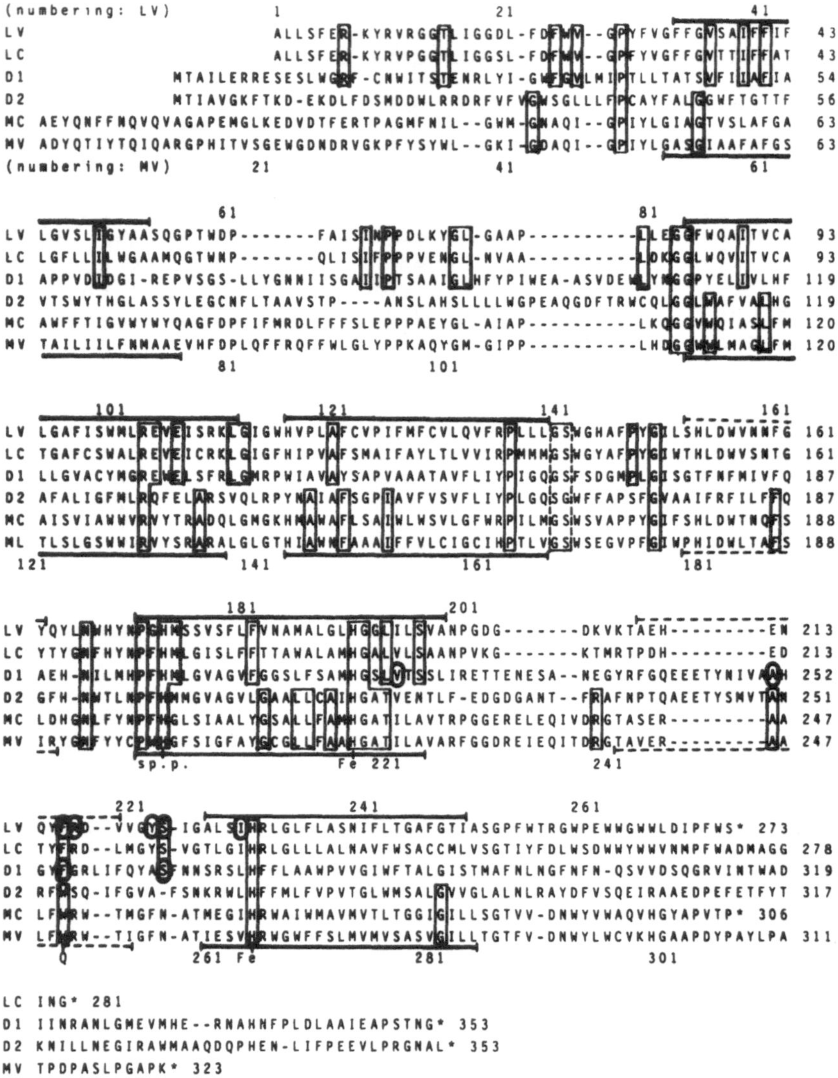

Fig. 2: The amino acid sequences of the L and M subunits from the purple bacteria *Rps. viridis* (LV, MV; Michel *et al.*, 1986a) and *Rb. capsulatus* (LC, MC; Youvan *et al.*, 1984) compared with those of the D1 and D2 proteins from spinach chloroplasts (Zurawski *et al.*, 1982, Alt *et al.*, 1984, Holschuh *et al.*, 1984). Amino acids common to all six subunits, or the L subunits and D1, or the M subunits and D2 are boxed. The position of the transmembrane helices in the *Rps. viridis* reaction center is indicated by bars above the sequences of the L subunits and below the sequences of the M subunits. The position of the short helices in the connections of transmembrane helices C and D , as well as D and E, is indicated by dashed lines. The histidine ligands of the special pair bacteriochlorophylls and of the non-heme-iron atom are marked by sp.p. or Fe. Circles show amino acids known to be mutated in herbicide resistant reaction centers from the purple bacteria or from photosystem II.

cytoplasmic side of the photosynthetic membrane (M has a longer amino-terminus), in the connection of the first and second transmembrane helices (M shows an in insertion of 7 amino acids), in the connection of the fourth and fifth transmembrane helices (M possesses an additional loop providing glu M232 as a ligand to the ferrous non-heme-iron atom) and at the carboxy-terminus, where the M subunit is considerably longer. At the interface of both subunits the primary electron donor, the "special pair", is found at the periplasmic side, whereas the ferrous non-heme-iron connects both subunits at the cytoplasmic side via histidine ligands of the D and E helices. The "accessory" bacteriochlorohylls, the bacteriopheophytins and the quinones form two structurally equivalent branches. One branch is more closely associated with the L subunit, the other with the M subunit. Only the branch associated with the L subunit is used for light-driven electron transfer across the photosynthetic membrane. The electron transfer from Q_A to Q_B occurs parallel to the membrane.

The Role of Conserved Amino Acids

Figure 2 shows the aligned amino acid sequences of the L and M subunits from *Rps. viridis*, *Rb. capsulatus* and the D1 and D2 proteins from spinach chloroplasts. The sequence homology seems not to be significant during the first quarter of the sequences. The D1 and D2 proteins are considerably longer, and even the existence in this sequence region of two transmembrane helices in D1 and D2 instead of only one in L and M is possible. However, this can be ruled out from the results of topological experiments using site-specific antibodies against peptides of the D1 protein (Sayre *et al.*, 1986). They are only compatible with a total of five transmembrane helices. The sequence homology starts to become significant with a gly-gly pair at position L83, L84 (M110, M111) which is at the start of the transmembrane helices B. Within the B helices arginine L103 (M130) is conserved. The function of its side chain seems to be to neutralize the partial negative charge due to the electric dipole moment of the short helix MDE (LDE), which intrudes partly into the membrane. The existence of such arginines in the B helices of D1 and D2 is a strong argument in favor of the occurence of such short helices in the connections of the D and E helices in photosystem II. The conserved pro L186 (M163) gives rise to a kink of 30-40° in the C helices. Glycine L149 (M176) is close to the start of helix LCD (MCD) and allows an abrupt turn of the peptide chain. Similarly, proline L171 (M198) at the start of transmembrane helix LD (MD) allows the necessary turn of the peptide chain to be made. Histidine L173 (M200) is the protein ligand to the magnesium atom of the special pair bacteriochlorophyll, and histidines L190 and M217 the ligand to the ferrous non-heme-iron atom. The other histidine ligands to the irons are L230 and M264 which are also found in D1 and D2 according to the alignment of figure 2.

In summary, primarily glycines and prolines of structural importance at or close to ends of helices are conserved, as well as the histidine ligands to the special pair bacteriochlorophylls and the non-heme-iron atom. Specifically conserved between L subunits and D1 proteins are phenylalanine L216 and serine L223 which form part of the binding site of Q_B (see figure 3) and the herbicide terbutryn (Michel *et al.*, 1986b). In D2 and the M subunits phenylalanine L216 is replaced by trp M250 which forms a major part of the Q_A binding site and may be involved in electron transfer from the bacteriopheophytin to Q_A. Another critical amino acid side chain seems to be that of glu L104. Apparently it undergoes a hydrogen bond with the bacteriopeophytin in L branch (Michel *et al.*, 1986b). It is not found in the M subunits and D2.

The role of all these conserved residues is in excellent agreement with the proposal that D1 and D2 constitute the core of photosystem II reaction center. However, there are also two interesting differences. From the sequence homologies there is no evidence for the presence of binding sites for the accessory chlorophylls in D1 and D2. This may mean that accessory chlorophylls in photosystem II do not exist.. It could also mean that they are bound to the D1 and D2 proteins in a different manner. The D1 and D2 proteins are also more homologous to one another than the L and M subunits. Especially their lengths in the connections of transmembrane helices D and E is nearly equal. An insertion in this region of seven amino acids as in M compared to L is missing in either subunit. In the bacteria this insertion gives rise to glu M232 as fifth protein ligand to the ferrous non-heme-iron atom. On the basis of the higher homology between D1 and D2 a deviation from the symmetric protein folding allowing either D1 or D2 to donate a fifth protein ligand to the ferrous non-heme-iron atom is unlikely. Having in mind the well-known effects of bicarbonate at the electron accepting site of photosystem II (for review see Vermaas and Govindjee, 1982)) we consider bicarbonate as a likely candidate for the fifth iron ligand.

The Q_B (and Herbicide) Binding Sites

The reaction centers from purple bacteria and photosystem II are both susceptible to herbicides of the *s*-triazin type like atrazin or terbutryn. the binding site of terbutryn and its mode of binding to the *Rps. viridis* reaction center has been established by X-ray crystallographic analysis (Michel *et al.*, 1986b). Apart from numerous van der Waals interactions terbutryn forms two hydrogen bonds with the protein matrix. One is found between a ring nitrogen as acceptor and the backbone N-H group of L224 as donor, the other between the serine side chain of L223 as acceptor and the aminoethyl side chain of terbutryn as donor.

Several mutations have been described leading to herbicide resistance. These mutations are indicated in the sequences of figure 2 and their spatial location is given in figure 3. In *Rps. viridis* three different mutants were obtained: (i) Phenylalanine L216 mutates to serine. This phenylalanine forms a major part of the terbutryn binding site. This mutation may change the overall polarity of the binding site, or the hydrogen bond capability of the side chain causes structural rearrangements. In the D1 protein of *Chlamydomonas* the corresponding, conserved phenylalanine

255 mutates to tyrosine (Erickson *et al.*, 1985) which is also likely to cause structural changes. (ii) Tyrosine L222 mutates to phenylalanine in terbutryn resistant *Rps. viridis* (Sinning and Michel, unpublished), and to glycine in *Rb, sphaeroides* (Paddock *et al.*, 1987). Since tyrosine L222 does not participate directly in terbutryn binding again a structural rearrangement is likely. (iii) Serine L223 mutates to alanine in *Rps. viridis* (Sinning and Michel, 1987) and to proline in *Rb. sphaeroides* (Paddock *et al.*, 1987). This mutation can be expected from the mode of terbutryn binding. The absence of one out of two hydrogen bonds between terbutryn and the protein has to decrease the binding by several orders of magnitude. In *Rps. viridis* this mutation is always accompanied by a second mutation: arginine L217 is replaced by a histidine. Since arginine L217 does not participate in terbutryn binding we consider this mutation as a secondary one somehow compensating a detrimentous effect of the first one. The corresponding serine 264 in D1 mutates to alanine (Erickson *et al.*, 1985) or glycine (Hirschberg *et al.*, 1984) in different species. These mutations are in line with the observed mode of terbutryn binding in *Rps. viridis*, and the formation of a hydrogen bond between serine 264 and the herbicide has actually been proposed by Hirschberg *et al.*.

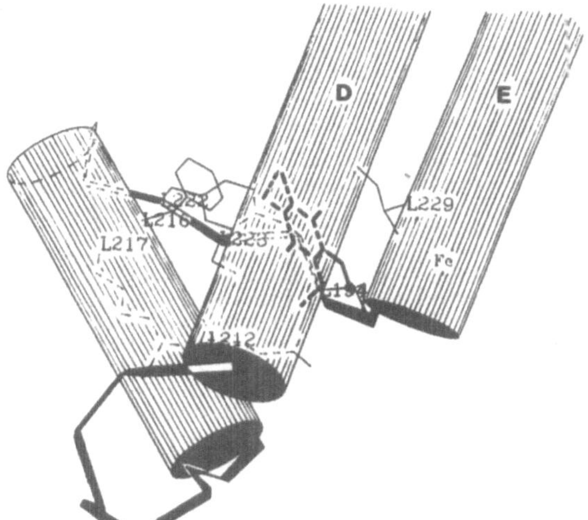

Fig. 3: The Q_A binding site in the photosynthetic reaction center from *Rps. viridis*. The location of amino acids which are mutated in herbicide resistant reaction centers from purple bacteria is shown, as well as the location of amino acids corresponding to mutated amino acids in photosystem II (see figure 2). Q_B (dashed) is seen behind the D helix of the L subunit.

Mutations so far not observed in *Rps. viridis* are the following: In *Rb. sphaeroides* ile L229 mutates to methionine (Paddock *et al.*, 1987). In these mutants the space available for terbutryn may be decreased due to the larger side chain.

Two other mutations have been described in D1 proteins: (i) Valine 219 mutates to ile (Erickson *et al.*, 1985). The mutation influences the binding of atrazine only slightly, but causes DCMU resistance. DCMU, however, has no effect on photosynthetic bacteria. Ile L194, corresponding to val 219 in the alignment of figure 2, is rather close to Q_B (see figure 3). (ii) Alanine 251 from *Chlamydomonas* mutates to valine (Johanningmeier *et al.*, 1987). The corresponding glu L212 (see figure 2) is part of the binding site of Q_B. Again it is hard to understand the detailed molecular mechanism of herbicide resistance due to this mutation. A considerable problem in this region is the low sequence homology between L and D1, especially the fact that the connection between the D and E helices is considerably longer in D1. Most likely these amino acids give rise to an additional loop of the peptide chain in the intrathylakoidal space of chloroplasts at the beginning of the small helix in the DE connection.

REFERENCES

Alt, J., Morris, J., Westhoff, P., and Herrmann, R.G., 1984, Nucleotide Sequence of the Clustered Genes for the 44kd Chlorophyll *a* Apoprotein and the "32kd"-like Protein of the Photosystem II Reaction Center in the Spinach Plastid Chromosome, Current Genetics, 8:597-606.

Brudvig, G.W., 1987, The Tetranuclear Manganese Complex of Photosystem II, J. Bioenergetics, 19:91-104.

Deisenhofer, J., Epp, O., Miki, K., Huber, R., and Michel, H., 1984, X-ray Structure Analysis of a Membrane Protein Complex: Electron Density Map at 3 Å Resolution and a Model of the Chromophores of the Photosynthetic Reaction Center from *Rhodopseudomonas viridis*, J. Mol. Biol., 180:385-398.

Deisenhofer, J., Epp, O., Miki, K., Huber, R., and Michel, H., 1985, Structure of the protein subunits in the photosynthetic reaction centre of *Rhodopseudomonas viridis* at 3Å resolution, Nature, 318:618-624.

Dekker, J.P., and Gorkum van, H.J., 1987, Electron Transfer in the Water-Oxidizing Complex of Photosystem II, J. Bioenergetics, 19:125-142.

Erickson, J.M., Rahire, M., and Rochaix, J.-D., 1985, Herbicide Resistance and Cross-Resistance: Changes at Three Distinct Sites in the Herbicide-Binding Protein, Science, 228:204-207.

Golden, S.S., Brusslan, J., and Haselkorn, R., 1986, Expression of a Family of *psbA* Genes Encoding a Photosystem II Polypeptide in the Cyanobacterium *Anacystis nidulans* R2, EMBO J., 5:2789-2798.

Hearst, J.E., 1986, Primary Structure and Function of the Reaction Center Polypeptides of *Rhodopseudomonas capsulata* - the Structural and Functional Analogies with the Photosystem II Polypeptides of Plants, in: "Photosynthesis III", L.A. Staehelin and C.J. Arntzen, eds., Springer-Verlag, Berlin.

Hirschberg, J., and McIntosh, L., 1983, Molecular Basis of Herbicide Resistance in *Amaranthus* hybridus, Science, 222:1346-1349.

Hirschberg, J., Bleecker, A., Kyle, D.J., McIntosh, L., and Arntzen, Ch.J., 1984, The Molecular Basis of Triazine-Herbicide Resistance in Higher-Plant Chloroplasts, Z. Naturforsch., 39c:412-420.

Holschuh, K., Bottomley, W., and Whitfield, P.R., 1984, Structure of the spinach chloroplast genes for the D2 and 44 kd reaction-centre proteins of photosystem II and for $tRNA^{Ser}$ (UGA), Nucleic Acids Res., 12:8819-8834

Johanningmeier, U., Bodner, U., and Wildner, G.F., 1987, A new Mutation in the Gene Coding for the Herbicide-Binding Protein in *Chlamydomonas*, FEBS Lett., 211:221-224.

Metz, J.G., Pakrasi, H.B., Seibert, M., and Arntzen, Ch.J., 1986, Evidence for a Dual Function of the Herbicide-Binding D1 Protein in Photosystem II, FEBS Lett., 205:269-274.

Michel, H., 1982, Three-dimensional Crystals of a Membrane Protein Complex: The Photosynthetic Reaction Centre from *Rhodopseudomonas viridis*, J. Mol. Biol., 158:567-572.

Michel, H., and Deisenhofer, J., 1986, X-Ray Diffraction Studies on a Crystalline Bacterial Photosynthetic Reaction Center: A Progress Report and Conclusions on the Structure of Photosystem II Reaction Centers, in: "Photosynthesis III", L.A. Staehelin and C.J. Arntzen, eds., Springer-Verlag, Berlin.

Michel, H., Weyer, K.A., Gruenberg, H., Dunger, I., Oesterhelt, D., and Lottspeich, F., 1986a, The 'light' and 'medium' subunits of the photosynthetic reaction centre from *Rhodopseudomonas viridis*: Isolation of the genes, nucleotide and amino acid sequence, EMBO J., 5:1149-1158.

Michel, H., Epp, O., and Deisenhofer, J., 1986b, Pigment-protein interactions in the photosynthetic reaction centre from *Rhodopseudomonas viridis*, EMBO J., 5:2445-2451.

Nakatani, H.Y., Ke, B., Dolan, E., and Arntzen, C.J., 1984, Identity of the Photosystem II Reaction Center Polypeptide, Biochim. Biophys. Acta, 765:347-352.

Nanba, O., and Satoh, K., 1987, Isolation of a photosystem II reaction center consisting of D-1 and D-2 polypeptides and cytochrome *b*-559, Proc. Natl. Acad. Sci. USA, 84:109-112.

Paddock, M.L., Williams, J.C., Rongey, S., Abresch, E.C., Feher, G., and Okamura, M.Y., 1987, Characterization of three herbicide resistant mutants of *Rhodopseudomonas sphaeroides* 2.4.1: Structure - Function Relationship, in: "Progress in Photosynthesis Research", J. Biggins, ed., Martinus Nijholt-Dordrecht-The Netherlands.

Pfister, K., Steinback, K.E., Gardner, G., and Arntzen, Ch.J., 1981, Photoaffinity labeling of an herbicide receptor protein in chloroplast membranes, Proc. Natl. Acad. Sci. USA, 78:981-985.

Rasmussen, O.F., Bookjans, G., Stummann, B.M., and Henningsen, K.W., 1984, Localization and Nucleotide Sequence of the Gene for the Membrane Polypeptide D2 from pea Chloroplasts DNA, Plant Mol. Biol., 3:191-199.

Rochaix, J.-D., Dron, M., Rahire, M., and Malnoe, P., 1984, Sequence Homology between the 32K dalton and the D2 Chloroplast Membrane Polypeptides of *Chlamydomonas reinhardii*, Plant Mol. Biol., 3:363-370.

Satoh, K., Fujii, Y., Aoshima, T., and Tado, T., 1987, Immunological identification of the polypeptide bands in the SDS-polyacrylamide gel electrophoresis of photosystem II preparations, FEBS Lett., 216:7-10.

Sayre, R.T., Andersson, B., and Bogorad, L., 1986, The Topology of a Membrane Protein: The Orientation of the 32 kd Qb-Binding Chloroplast Thylakoid Membrane Protein, Cell, 47:601-608.

Sinning, I., and Michel H., 1987, Herbicide Resistant *Rhodopseudomonas viridis* Mutants, Z. Naturforsch., submitted.

Trebst, A., and Depka, B., 1985, The Architecture of Photosystem II in Plant Photosynthesis Which Peptide Subunits Carry the Reaction Center of PS II?, in: "Antennas and Reaction Centers of Photosynthetic Bacteria", M.E. Michel-Beyerle, ed., Springer-Verlag, Berlin.

Trebst, A., 1986, The Topology of the Plastoquinone and the Herbicide Binding Peptides of Photosystem II in the Thylakoid Membrane, Z. Naturforsch., 41c:240-245.

Vermaas, W.F.J., and Govindjee, 1982, Development, Carbon Metabolism and Plant Poductivity, Photosynthesis, 2:541-558.

Williams, J.C., Steiner, L.A., Ogden, R.C., Simon, M.I., and Feher, G., 1983, Primary structure of the M subunit of the reaction center from *Rhodopseudomonas sphaeroides*, Proc. Natl. Sci. USA, 80:6505-6509.

Williams, J.C., Steiner, L.A., Feher, G., and Simon, M.I., 1984, Primary structure of the L subunit of the reaction center from *Rhodopseudomonas sphaeroides*, Proc. Natl. Sci. USA, 81:7303-7307.

Youvan, D.C., Bylina, E.J., Alberti, M., Begusch, H., and Hearst, J.E., 1984, Nucleotide and Deduced Polypeptide Sequences of the Photosynthetic Reaction-Center, B870 Antenna, and Flanking Polypeptides from R. capsulata, Cell, 37:949-957.

Zurawski, G., Bohnert, H., Whitfield, P.R., and Bottomley, W., 1982, Nucleotide Sequence of the Gene for the M_r 32,000 thylakoid Membrane Protein from *Spinacia oleracea* and *Nicotiana debneyi* Predicts a totally Conserved primary Translation Product of M_r 38,950, Proc. Natl. Sci. USA, 79:7699-7703.

MISSING LINKS FROM MOLECULAR BIOLOGY TO USEFUL PLANTS

Diter von Wettstein

Department of Physiology, Carlsberg Laboratory
Gamle Carlsberg Vej 10
DK-2500 Copenhagen Valby, Denmark

Useful plants are to be grown by the farmer, the gardener, the forester, the producer of vegetables and fruits. Human endeavour has created wheat more than 5000 years ago. Unfortunately we do not know how the three genomes of *Triticum monococcum*, *Triticum tauschii* and of a third species belonging to the genus *Aegilops* have come to be combined into bread wheat. Were they spotted and selected as spontaneous hybrids or was man involved in artificial crossings ? Hand-pollination of date palms was practiced by Assyrians 2000 B.C. and it is thus not excluded that some of our amphiploid crop plants are based on ancient artificial hybridizations. Selection for improved weight, disease resistance and climatic adaptation had and has an enormously successful impact in the breeding of our crop plants. As our knowledge in plant pathology, genetics and agricultural technology has improved, this has been successively exploited by plant breeders. Use of disease resistance genes, of spontaneous and induced mutants, chromosome translocations, cytoplasmic male sterility, hybrid vigour or new auto- and amphiploids can be traced in many of our highly bred crop plant varieties. A closer study reveals that these tools made available by genetics over the last 80 years only had an impact, if they were used in combination with the classical hybridization selection and evaluation procedures which require many years of field testing.

Triticale, a new amphihexaploid or amphioctoploid crop plant combining wheat genomes with the rye genome and providing grain of wheat quality on less demanding soil has existed for decades but selection and adaptation for making it a useful crop plant is still in progress. Excitement has been elicited, when hybrids between potato and tomato were produced by protoplast fusion[1,2], but it turned out that the two genomes have a hard time cooperating and that one of the original aims of moving chilling resistance genes via the hybrid from potato into the chilling sensitive domestic tomato[3] is not feasible without an enormous effort in experimental work and time.

Transfer of individual genes or groups of genes by genetic transformations, as examplified in many papers of this Advanced Study Institute is a more promising strategy for breeding useful plants, if we know of genes that encode proteins or enzymes supplying a valuable property for a crop plant. Two recent examples can serve as models for such an approach. (1) M. Vaeck and coworkers[4] have engineered insect resistant tobacco plants using a gene encoding a toxin found in Bacillus thuringensis and (2) De Block and coworkers[5] have engineered tobacco, potato and tomato plants resistant to a herbicide using a gene from Streptomyces hygroscopicus encoding a detoxifying enzyme.

(1) Bacillus thuringensis strain berliner contains a gene on a plasmid encoding a 130 kd protein which forms crystals in the spores of the bacterium. Proteases in the midgut of larvae of Pieris brassicae and Manduca sexta will cleave the protein into a 60 kd fragment which is toxic to the cells and kills the larvae. The gene was cloned and expressed in E. coli. By Bal 31 exonuclease deletion mapping the minimal protein fragment killing the larvae was determined to comprise 579 amino acids. The corresponding DNA fragment was inserted into a Ti-plasmid derived vector and the gene transferred into tobacco plants. The transformed plants expressed the gene as shown by immunochemistry of plant extracts and toxicity tests. Freshly hatched larvae of Manduca sexta were placed on leaves of 40 cm high control and transgenic tobacco plants. After 10-15 days the leaves of the control plants were completely eaten by the larvae. In contrast, the larvae on transgenic plants died within 4 days and leaf damage was restricted to a few small holes.

(2) The tripeptide bialaphos consisting of two alanine residues and an analogue of glutamate called phosphinothricin ($CH_3-POOH-CH_2-CH_2-CHNH_2-COO^-$) is produced by Streptomyces hygroscopicus. In bacteria and plants peptidases remove the alanine residues and release phosphinothricin which then blocks glutamine synthetase, the only enzyme which can detoxify ammonia released by nitrate reduction, amino acid catabolism and photorespiration. Phosphinotricin is thus a very potent herbicide. The Streptomyces species protects itself against phosphinothricin by an acetyltransferase enzyme (phosphinothricin acetyltransferase), an enzyme actually used by the streptomycete in the biosynthesis of bialaphos. The enzyme is encoded by the bar gene[6], which has been placed under the control of the 35S promoter of the cauliflower mosaic virus and transferred to plant cells using a Ti-plasmid derived vector[5]. Transgenic plants of tobacco, potato and tomato revealed resistance towards commercially used doses of phosphinotricin or bialaphos. The expressed enzyme was able to detoxify the herbicide and protect the plants from ammonia poisening.

To what extent these transgenic crop plants providing improved insect and herbicide resistance will be successful in commercial varieties is to be assessed.

Genetic transformation of wheat, barley, rice and maize cannot yet be employed in breeding of these commercially most important crop plants, but the achievement of transgenic plants in rye providing resistance to the antibiotic kanamycin[7] has brought this possibility within reach.

In the present paper I would like to discuss some goals in the improvement of the quality of barley in which molecular biology plays an important role in both defining these goals and reaching them by eliminating missing links.

THE CASE OF 1→3,1→4-β-GLUCANASE

Different forms of the enzyme (1→3,1→4)-β-D-glucan 4-glucanohydrolase are synthesized during germination respectively malting in the aleurone and the epithelial layer of the scutellum and released into the endosperm to degrade the 1→3,1→4-β-glucans of the endosperm cell walls[8,9,10]. Following this degradation amylolytic and proteolytic enzymes can reach and degrade the endosperm starch and protein storage components. Insufficient β-glucanase activity during malting gives rise to high levels of residual β-glucans in the wort[11], which can create problems during filtration of wort[12] and beer[13,14]. High β-glucan content may lead to precipitates in the finished beer[15]. Unmalted barley as adjunct in the brewhouse increases these problems[16]. The barley β-glucanase is almost completely inactivated during kilning of the malt and mashing. However, (1→3,1→4)-β-glucanases from microbial sources (Bacillus subtilis, Aspergillus niger or Trichoderma reesii) are less heat labile than barley β-glucanase. Such preparations may be added during mashing[17,18] or at a later stage during fermentation[13] to reduce the content of high molecular weight β-glucan.

With the isolation of a cDNA gene for barley β-glucanase and its expression in and secretion from yeast, a number of new approaches to alleviate these problems have become feasible. The complete amino acid sequence of 1→3,1→4-β-glucanase isoenzyme II comprising 306 amino acids was determined by sequencing nine tryptic peptides and by isolating from a cDNA library a clone encoding the 291 aminoterminal residues and nucleotide sequencing it[19]. The clone was selected by molecular hybridization with a mixed pentadecameric oligonucleotide probe synthesized with the genetic code dictionary from a piece of the amino acid sequence. With an additional oligonucleotide probe the cDNA clone encoding the aminoterminal part was isolated and the β-glucanase cDNA fragments spliced together[20]. The cDNA gene was supplied with the nucleotide sequence for the signal peptide of mouse α-amylase to permit transport of the barley β-glucanase through the secretory pathway of an eukaryotic cell. This construction was inserted in a yeast-E.coli shuttle plasmid behind the gene promoter of yeast alcohol dehydrogenase I. A laboratory yeast strain was transformed with the plasmid and the transformed strain secreted barley β-glucanase[20] in amounts which permitted during fermentation for one week at 10°C to degrade concentrations of β-glucans which are found troublesome in wort and beer filtration. Transformants of lager brewing strains also secreted the barley β-glucanase and effectively lowered β-glucan contents during fermentation. Presently experiments are directed towards the stable integration of the barley β-glucanase gene into the genome of brewing yeast strains and to test their effect on wort and beer.

While a lager yeast which synthesizes and secretes β-glucanase during fermentation and maturation of beer alleviates the problems of beer filtration and formation of precipitates in the finished beer, this will not help to solve problems of wort filtration. Here yeast strains which synthesize and secrete barley β-glucanase will provide a source of the natural enzyme which can be added during mashing. If the brewery has available a lager yeast strain secreting barley β-glucanase, it can produce the enzyme with its normal fermentation equipment, possibly even during the production of green beer. Such β-glucanase containing green beer might be added during mashing instead of microbial enzymes.

Determination of the primary structure of the β-glucanase and its comparison with a (1→3)-β-glucanase from tobacco indicates which regions of the enzyme contain the catalytic site for the hydrolysis of the glycosidic bonds[19]. This can be tested by making site-directed nucleotide changes in the cDNA gene cloned and expressing the protein with substituted amino acids in yeast. The secreted protein may be tested for enzyme activity and loss of activity upon specific amino acid changes will delineate the active sites. One can thereafter produce amino acid changes in other regions of the molecule and select for enzymes, which in analogy with the microbial enzymes are more heat stable; even construction of a barley β-glucanase which survives the kilning process may be feasible. Since genetic transformation of barley (see below) seems possible in the near future, the gene encoding a heat stable β-glucanase with an appropriate pH optimum might be incorporated into future malting barleys.

Other cloned genes encoding barley enzymes of importance for germination and malting are those encoding the two α-amylase isoenzymes[21,22,23,24] and one specifying β-amylase[25]. Of equal interest for plant breeders are the genes encoding chalcone synthase[26], an inhibitor of chymotrypsin[27], an α-amylase-protease inhibitor[28] and starch synthase[26]. Expression of these enzymes and proteins in yeast and their secretion permit to recognize the different properties of the isoenzymes and provide clues for breeding of improved enzyme characteristics.

THE PROSPECTS OF GENETIC TRANSFORMATION OF BARLEY

A. de la Peña, H. Lörz and J. Schell[7] have succeeded in transforming rye with a gene providing resistance to kanamycin. This antibiotic causes albinism, when grains of cereals including barley are germinated on media containing the antibiotic. The seedlings die after 10 days. A plasmid was used which carries the gene encoding aminoglycosidphosphotransferase II, an enzyme which inactivates kanamycin by its phosphorylating activity and results in green seedlings, if the enzyme is produced in the plant. The gene was coupled to the promoter of the gene for nopalinsyntase from Agrobacterium, a promoter strongly expressed in plant cells. The plasmid carried further genes to provide selection and replication in bacteria. An aqueous solution containing 100 μg/ml plasmid was injected in the culm above each node until all internodes are filled with the DNA solution (~ 300 μl). Injection has to take place about 14 days before meiotic metaphase in the pollen mother cells and ovules, so the DNA is incorporated in the nuclei of the germ cells. After pollination the kernels are ripened and sown on a medium containing kanamycin. Seven green plants were obtained among 3023 white seedlings. Extracts of leaf pieces from the green seedlings contained the aminoglycosidphosphotransferase and by molecular hybridization of isolated DNA from seedlings, the presence of the gene encoding the enzyme could be demonstrated.

There is reason to consider that this technique is applicable to barley and circumvents the problems which have been experienced in regenerating protoplasts of cereals into plants. The technique requires quite large amounts of plasmid and the number of grains per tiller is limited. There is, however, the possibility to amplify the number of germ cells, which can be tested for the presence of the selectable marker gene in this case kanamycin resistance. It is possible to regenerate embryos from pollen grains of barley[29,30,31,32,33,34]. A single anther may yield a 100 embryos which can be regenerated into seedlings. Since there are 3 anthers per flower and some 25 suitable flowers per spike, one may be able to screen more than a thousand embryos or regenerated seedlings for kanamycin resistance and thus transformants from a single spike. Once this system is established, one can investigate if co-transformation with non-selectable genes is feasible. In this respect the hor 2ca mutant line lacking the Hor-2 locus coding for B hordein polypeptides is a suitable host[35,36]. Transformation with the cloned Hor-2 genes[37,38] together with the canamycin resistance producing gene could be attempted and the expression of the Hor-2 gene analyzed in kanamycin resistant plants grown to maturity. The expression of the introduced Hor-2 gene would be monitored with fluorescing antibodies against B hordein in abraded grains with exposed endosperm[39].

Agrobacteria carrying a Ti-plasmid will transfer the T-DNA of this plasmid into cells of stems and leaves of maize plants when inoculated by injection. This was shown by N. Grimsley, T. Hohn, J.W. Davies and B. Hohn with a Ti-plasmid in which the T-DNA between the two 25 base pair borders was substituted for by tandemly repeated copies of the DNA of maize streak virus[40]. The virus multiplied and caused systemic infections even reaching the reproductive organs of the plant. It is therefore conceivable that genes inserted between the T-DNA borders in the Ti-plasmid can be used to transfect wounded embryos regenerated from barley pollen grains in another culture.

THE USE OF MONOCLONAL AND FLUORESCING ANTIBODIES IN BREEDING OF BARLEY

An efficient selection technique for identification of grains expressing newly introduced protein genes is required. Such a selection technique employing fluorescing antibodies has recently been developed. The increased lysine content of Hiproly (lys 1) is primarily due to an elevated production of lysine rich proteins such as chymotrypsin inhibitor CI-2 (M.W.9000, 10% lysine), protein Z and β-amylase[41,42,43]. U. Rasmussen has shown that monospecific antibodies directed against chymotrypsin inhibitor 2 can be conjugated with fluorescein isothiocyanate and reacted with the paraformaldehyde fixed surface of abraded grains in which the endosperm has been exposed[44]. The emitted fluorescence is proportional to the amount of chymotrypsin inhibitor present and Hiproly kernels with a high content can be picked out by ocular inspection from wild type kernels because of their higher fluorescence. The embedding of 100 seeds at a time in a block of thermoplastic clay[45,46] in such a manner that the embryo is protected allows one to identify a single kernel among a large population of kernels and to germinate it thereafter. High intensity light is projected through an appropriate interference filter onto the block of clay containing the embedded and abraded kernels to excite the fluorescence of the antibody with the appropriate wave length. The kernels are viewed through the interference filter appropriate for the emission maximum of the fluorescence employing a special video camera and a television screen.

This procedure is applicable to any protein for which antibodies are available[47]. Mutants, rare recombinants or transformants elevating the content of a desired protein or lowering the amount of an undesired protein can be screened out in such a way and it is estimated that two persons can screen 20,000 grains per day. Monoclonal antibodies have been produced against chymotrypsin inhibitor 2 and against individual families of hordein polypeptides and found to be useful in the same way as the rabbit elicited antibodies. Monoclonal antibodies have two advantages for the plant breeder: They can be produced in unlimited amounts once an appropriate hybridoma cell line has been isolated. They recognize single epitopes on a protein and can therefore be used to distinguish closely homologous polypeptides produced by individual members of multigene families. Screening with monoclonal antibodies in this non-destructive way facilitates breeding for increases or decreases of individual polypeptides. For the introduction of tailored Hor-2 genes the mutant Risø 56 (hor 2ca) with the deleted Hor-2 locus will be a suitable host and monoclonal antibodies recognizing the B hordein polypeptides can be used to detect transgenic grains expressing the introduced gene.

THE CASE OF PROANTHOCYANIDIN-FREE BARLEY

Proanthocyanidins of barley grains consist of dimers and trimers of (+)- catechin and (+) gallocatechin units forming the procyanidins and prodelphinidins. All flavanoids stay in the kernels during malting. During mashing these compounds dissolve in the wort and are retained during the following processes of primary and secondary fermentation of the wort. In the finished beer proanthocyanidins precipitate proteins and thereby give rise to undesirable non-biological haze. In order to avoid this haze formation several technologies are applied which aim at removing the proanthocyanidins (e.g. by filtering through polyvinylpolypyrrolidone containing sheets) or degrading the proteins with proteolytic enzymes or precipitating both compounds. While these proanthocyanidins are prominent constituents in grains of barley and millet, they are absent from kernels of maize or wheat, thus illustrating that they are not required for the survival of a crop plant. It was therefore desirable to explore if genetic blocking of the synthesis of proanthocyanidins would yield malting barley varieties which avoid the non-biological haze formation in beer. Proanthocyanidins are synthesized by the flavonoid pathway but by a different branch than the red anthocyanin pigments, the branch point being located at the 2,3-trans-3,4-cis-leucocyanidin intermediate[48,49]. If the pathway is blocked by a mutation prior to the branch point, plants result which lack the red anthocyanin pigmentation on the stem, auricles, lemma and awns as well as the colourless proanthocyanidins in the kernels. Accordingly anthocyanin-free mutants were tested for the presence or absence of proanthocyanidins in their kernels. Among the first 42 induced anthocyanin-less mutants tested one, ant 13-13, was found to be also blocked in the biosynthesis of proanthocyanidins. After propagation of the mutant pilot and full scale brewing tests were performed with malt prepared from this mutant[50]. Excellent haze stability was achieved in the beers. It was then possible with this and other mutants to test barley, malt and beer with an intrinsic absence of these polyphenols. Earlier hypotheses which regarded proanthocyanidins an important part of beer flavour and taste could be disproven[51,52].

In terms of agronomic performance, mutant ant 13-13 yielded about 25% less than its mother variety and was highly susceptible to infection by powdery mildew. Over the past years, an extensive program of mutation and recombination breeding has been carried out in order to obtain high yielding proanthocyanidin-free malting barley varieties adapted to various climates[51]. Up to now 569 proanthocyanidin-free mutants have been isolated after mutagenesis in 92 different spring and winter barley varieties comprising cultivars from Europe, North-America and Japan as well as from New Zealand and Australia. By diallelic crosses, 388 of the recessive mutants have been localized to 7 gene loci. The gene loci are designated ant 13, ant 17, ant 18, ant 19, ant 21, ant 22, ant 25. (Genes ant 1 to 12, ant 14 to 16, ant 23 and ant 24 are controlling the branch leading to anthocyanins).

The pathway is briefly as follows: The flavan skeleton of naringenin is synthesized from coumaroyl CoA and malonyl CoA. Naringenin is converted into eriodictyol, which is hydroxylated by a flavanone-3-hydroxylase to (+) dihydroquercetin. Dihydroquercetin reductase synthesizes (+)-2,3-trans-3,4-cis-leucocyanidin which is reduced to (+) catechin. The latter is used for the synthesis of oligomeric procyanidins in an as yet unknown enzymatic reaction. Three approaches have been used to locate the steps in the synthesis of proanthocyanidins which are blocked in the mutants[48,49,54]. 1) After establishing the time course for catechin and procyanidin synthesis in developing barley grains, ant mutants were analyzed for an accumulation of precursors. 2) Radioactive dihydroquercetin was prepared, fed to barley wild type and mutant testa-pericarp tissue at a time optimal for proanthocyanidin synthesis and the radioactive products analyzed. 3) Dihydroquercetin reductase and leucocyanidin reductase activities were measured in enzyme assays.

Mutants in the genes ant 17 and ant 22 accumulate the flavanone homoeriodictyol in the grains and contain the enzymes needed to convert (+)-dihydroquercetin into 3,4-cis-leucocyanidin, (+)-catechin and procyanidin B-3. Most likely these two genes control the step between eriodictyol and (+)-dihydroquercetin (flavanone-3-hydroxylation).

Mutants in the gene ant 18 accumulate dihydroquercetin and lack dihydroquercetin reductase activity. This gene is a candidate for the structural gene of this enzyme. A barley cDNA clone with partial homology to a cDNA clone of the a_1 gene in maize which encodes the maize dihydroquercetin reductase has been isolated[26]. It will thus be possible to analyze the molecular nature of the ant 18 mutants.

Mutants in the gene ant 19 (and ant 25) are proanthocyanidin-free but synthesize anthocyanin in the vegetative part of the plant. Mutant ant 19-

109 contains an active dihydroquercetin reductase. It is concluded that ant 19 controls the reduction of (+)-2,3-trans-3,4-cis-leucocyanidin to (+)-catechin, i.e. a step after the branch-point to anthocyanidin formation.

The modes of action of the two genes ant 13 and ant 21 remain unknown. Mutants in these genes do not accumulate flavonoid precursors of proanthocyanidins and they lack or have low flavanone-3-hydroxylase and dihydroquercetin reductase activities. Mutants in ant 13 are deficient in flavones and mutants in ant 21 contain an increased amount of caffeic acid[55]. These genes either control early steps in the biosynthesis of flavonoids or are regulatory genes for the pathway.

In general, the majority of the mutants show more or less severe defects in yield, fertility and other characteristics of agronomic significance. It is of great importance to establish the reason for this at the molecular level. The agronomic performance of some of the raw mutants and of various recombinant lines has been encouraging[51]. Yields of 7 to 8 t/ha achievable with modern varieties have also been obtained with proanthocyanidin-free lines. Resistance to diseases have been attained which are in line with those of modern varieties. The mutant line ant 17-148 isolated in the malting barley variety Triumph is on the Danish National list of approved varieties under the name Galant, but attempted marketing had to be discontinued because of insufficient modification during malting in commercial scale resulting in uneconomically low extract yields and high β-glucan contents of the wort[56]. New lines which give sufficient malt modification (e.g. Ca 737606; ant 13-13 x Dram) are in an advanced stage of field testing.

Brewing tests have shown that beer produced with proanthocyanidin-free barley has a better haze stability and shelf-life than that achievable with present technology[57]. This has two important consequences. If the brewer uses proanthocyanidin-free hops extract, thus avoiding the addition of proanthocyanidins from hops, he can blend proanthocyanidin-free malt with malt containing proanthocyanidins in quantities up to 50% and still achieve a shelf life of the beer corresponding to present day standards. If he wishes to use hop pellets or hop cones this can be done with varieties which have a low proanthocyanidin to α-acid ratio, preferably one below 0.4[51].

The malster and the brewer who have bought proanthocyanidin-free barley would like to sell off as feed the screenings comprising thin grains (below 2.5 mm width) or barley that happens not to meet his malting or brewing specifications. Chick performance trials have therefore been carried out on diets composed of proanthocyanidin-free barley. It was found that broiler chickens fed on proanthocyanidin-free barleys gained equivalent or better weight than on check varieties containing these polyphenols[51]. The feeding quality of barley is not influenced negatively by the absence of proanthocyanidins. Therefore this property can in the future be a standard characteristic of all barley varieties. The farmer, the trade and the malsters would like to be able to determine in a short time the purity of a lorry with proanthocyanidin-free grain or malt. The finding that the proanthocyanidins are located in the seed coat (testa) of the mature grain[58] has permitted the development of a procedure which can determine the frequency of proanthocyanidin containing grains in 15 minutes[59]. In this method grains are fixed in a block of thermoplastic clay, half the kernel sanded off, and the sanded grains stained with a 1% vanillin-6M-HCl solution. The testa of proanthocyanidin containing grains stains red and can be recognized by ocular inspection. Recently this procedure has been refined to allow the staining of testa tissue without damaging the embryo[46]. Non-destructive screenings for proanthocyanidin-free mutants in M_3 populations have so far yielded 50 mutants which form anthocyanins in the vegetative parts of the plant but lack the proanthocyanidins in the testa. It will be interesting to see, if these mutants blocking the last steps in proanthocyanidin synthesis will be equally pleiotropic and defective in their agronomic properties as the raw mutants with earlier blocks.

Characterization of the ant genes at the molecular level is feasible. If a transformation system in barley can be developed, the next missing link is a procedure to inactivate relevant genes without causing detrimental pleiotropic effects either by transposon insertion or by gene replacement through homologous recombination as is routine in yeast but not demonstrated in higher plants.

Financial support by the Biotechnology Action Programme of the Commission of the European Communities with contract No. BAP-0091-DK (B) is gratefully acknowledged.

REFERENCES

1. G. Melchers, M.D. Sacristan and A.A. Holder, Somatic hybrid plants of potato and tomato regenerated from fused protoplasts, Carlsberg Res. Commun. 43:203 (1978).
2. C. Poulsen, D. Porath, M.D. Sacristan and G. Melchers, Peptide mapping of the ribulose bisphosphate carboxylase small subunit from the somatic hybrid of tomato and potato, Carlsberg Res. Commun. 45:249 (1980).
3. R.M. Smillie, G. Melchers and D. von Wettstein, Chilling resistance of somatic hybrids of tomato and potato, Carlsberg Res. Commun. 44:127 (1979).
4. M. Vaeck, H. Höfte, A. Reynaerts, J. Leemans, M. van Montagu and M. Zabean, Engineering of insect resistant plants using a B.thuringiensis gene, in: Molecular Strategies of Crop Protection, Proc. A. du Pont-UCL Symposium, C.J. Arntzen and C.A. Ryan, eds., Steamboat Springs, Colorado (April 1986).
5. M. de Block, J. Botterman, M. Vandewiele, J. Dockx, C. Thoen, V. Gosselé, R. Movva, C. Thompson, M. van Montagu and J. Leemans, Engineering herbicide resistance in plants by expression of a detoxifying enzyme, EMBO J., in press (1987).
6. C.J. Thompson, N.R. Movva, R. Tizard, R. Crameri and J.E. Davies, Characterization of the herbicide resistance gene bar from Streptomyces hygroscopicus, EMBO J., in press (1987).
7. A. de la Peña, H. Lörz and J. Schell, Transgenic rye plants obtained by injecting DNA into young floral tillers, Nature 325:274 (1987).
8. J.R. Woodward and G.B. Fincher, Purification and chemical properties of two 1→3,1→4-β-glucan endohydrolases from germinating barley, Eur. J. Biochem. 121:663 (1982).
9. J. Mundy, A. Brandt and G.B. Fincher, Messenger RNAs from the scutellum and aleurone of germinating barley encode 1→3,1→4-β-D-glucanase, α-amylase and carboxypeptidase, Plant Physiol. 79:867 (1985).
10. I.M. Stuart, L. Loi and G.B. Fincher, Development of 1→3,1→4-β-D-glucan endohydrolase isoenzymes in isolated scutella and aleurone layers of barley (Hordeum vulgare), Plant Physiol. 80:310 (1986).
11. G.N. Bathgate and C.E. Dalgliesh, The diversity of barley and malt β-glucans, Proc. Am. Soc. Brew. Chem. 33:32 (1975).
12. R.W. Scott, The viscosity of worts in relation to their content of β-glucan, J. Inst. Brewing 78:179 (1972).
13. P.A. Leedham, D.J. Savage, D. Crabb and G.T. Morgan, Materials and methods of wort production that influence beer filtration, Proc. Eur. Brew. Conv. Congress, Nice:201 (1975)

14. L. Narziss, Betaglucan and betaglucanases, Proc. Eur. Brew. Conv. Barley and Malt Sympos., Helsinki:99 (1981).
15. S. Takayanagi, M. Amaha, K. Satake, Y. Kuroiwa, H. Igarashi and A. Murata, Studies on frozen beer precipitates. I. Formation and general characteristics, J. Inst. Brewing 75:284 (1969).
16. D.T. Bourne, M. Jones and J.S. Pierce, Beta-glucan and betaglucanases in malting and brewing, Mast. Brew. Assoc. Am. Tech. Quarterly 13:3 (1976).
17. C.W. Bamforth, Barley β-glucans, their role in malting and brewing, Brewers Digest 57:22 (1982).
18. B. Stentebjerg-Olesen, Application of beta-glucanases in the brewing industry, Proc. 16th Conv. Inst. Brew. (Austr. Section) Sydney:127 (1980).
19. G.B. Fincher, P.A. Lock, M.M. Morgan, K. Lingelbach, R.E.H. Wettenhall, J.F.B. Mercer, A. Brandt and K.K. Thomsen, Primary structure of the (1→3,1→4)-β-D-glucan 4-glucanohydrolase from barley aleurone, Proc. Natl. Acad. Sci. 83:2081 (1986).
20. E.A. Jackson, G.M. Ballance and K.K. Thomsen, Construction of a yeast vector directing the synthesis and release of barley (1→3,1→4)-β-glucanase, Carlsberg Res. Commun. 51:445 (1986).
21. S. Muthukrishman, G.R. Chandra and E.S. Maxwell, Hormonal control of α-amylase gene expression in barley. Studies using a cloned cDNA probe, J. Biol. Chem. 258:2370 (1983).
22. J.-K. Huang, M. Swegle, A.M. Dandekar and S. Muthukrishman, Expression and regulation of α-amylase gene family in barley aleurons. J. Mol. Genet. 2:579 (1984).
23. J.C. Rogers and C. Milliman, Coordinate increase in major transcripts from the high pI α-amylase multigene family in barley aleurone cells stimulated with gibberellic acid. J. Biol. Chem. 259:12234 (1984).
24. R.F. Whittier, D.A. Dean and J.C. Rogers, Nucleotide sequence analysis of alpha amylase and thiol protease genes that are hormonally regulated in barley aleurone cells, in press (1987).
25. M. Kreis, M. Williamson, B. Buxton, J. Pywell, J. Hejgaard and I. Svendsen, Primary structure and differential expression of β-amylase in normal and mutant barleys. in press (1987).
26. W. Rohde, E. Barzen, A. Marocco, Zs. Schwarz-Sommer, H. Saedler and F. Salamini, Isolation of genes that could serve as traps for transposable elements in Hordeum vulgare, in: Barley Genetics V. Proc. Fifth Intl. Barley Genet. Symp., Okayama, in press (1987).
27. M. Williamson, J. Forde, B. Buxton and M. Kreis, Nucleotide sequence of barley chymotrypsin inhibitor-2 (CI-2) and its expression in normal and high-lysine barley. Eur. J. Biochem. in press (1987).
28. J. Mundy and J.C. Rogers, Selective expression of a probable amylase/protease inhibitor in barley aleurone cells: Comparison to the barley amylase/subtilisin inhibitor, Planta 169:51 (1986).
29. B. Foroughi-Wehr and W. Friedt, Rapid production of recombinant barley yellow mosaic virus resistant Hordeum vulgare lines by anther culture, Theor. Appl. Genet. 67:377 (1984).
30. B. Huang, J.M. Dunwell, W. Powell, A.M. Hayter and W. Wood, The relative efficiency of microspore culture and chromosome elimination as methods of haploid production in Hordeum vulgare L., Z. Pflanzenzüchtg. 92:22 (1984).
31. R.L. Lyne, R.F. Benneth and C.P. Hunter, Embryoid and plant production from cultured barley anthers, in: Intern. Sympos. Genet. Manipulation in Crops, Beijing, China (1984).
32. C.P. Hunter, The effect of anther orientation on the production of microspore-derived embryoids and plants of Hordeum vulgare cv. Sabarlis, Plant Cell Rep. 4:267 (1985).
33. P.R.M. Shannon, A.E. Nickolson, J.M. Dunwell and D.R. Davies, Effect of anther orientation on microspore-callus production in barley (Hordeum vulgare L.), Plant Cell Tissue Organ Culture 4:271 (1985).

34. F.L. Olsen, Carlsberg Res. Commun. in preparation (1987).
35. E.H. Hopp, S.K. Rasmussen and A. Brandt, Organization and transcription of B1 hordein genes in high lysine mutants of barley. Carlsberg Res. Commun. 48:201 (1983).
36. M. Kreis, P.R. Shewry, B.G. Forde, S. Rahman and B.J. Miflin, Molecular analysis of a mutation confering the high-lysine phenotype on the grain of barley (Hordeum vulgare), Cell 34:161 (1983).
37. A. Brandt, A. Montembault, V. Cameron-Mills and S.K. Rasmussen, Primary structure of a B1 hordein gene from barley. Carlsberg Res. Commun. 50: 333 (1985).
38. B.G. Forde, A. Heyworth, J. Pywell and M. Kreis, Nucleotide sequence of a B1 hordein gene and the identification of possible upstream regulatory elements in endosperm storage protein genes from barley, wheat and maize. Nucl. Acids Res. 13:7327 (1985).
39. S.E. Ullrich, U. Rasmussen, G. Høyer-Hansen and A. Brandt, Monoclonal antibodies to hordein polypeptides. Carlsberg Res. Commun. 51:381 (1986).
40. N. Grimsley, Th. Hohn, J.W. Davies and B. Hohn, Agrobacterium-mediated delivery of infectious maize streak virus into maize plants. Nature 325:177 (1987).
41. J. Hejgaard and S. Boisen, High-lysine proteins in Hiproly barley breeding: Identification, nutritional significance and new screening methods. Hereditas 93:311 (1980).
42. I. Svendsen, B. Martin and I. Jonassen, Characteristics of Hiproly barley III. Amino acid sequences of two lysine-rich proteins. Carlsberg Res. Commun. 45:79 (1980).
43. J. Hejgaard, S.K. Rasmussen, A. Brandt and I. Svendsen, Sequence homology between barley endosperm protein Z and protease inhibitors of the α_1-antitrypsin family. FEBS Lett. 180:89 (1985).
44. U. Rasmussen, Immunological screening for specific protein content in barley seeds. Carlsberg Res. Commun. 50:83 (1985).
45. F. Heltved, S. Aastrup, O. Jensen, G. Gibbons and L. Munck, Preparation of seeds for mass screening. Carlsberg Res. Commun. 47:291 (1982).
46. H. Kristensen and S. Aastrup, A non-destructive screening method for proanthocyanidin-free barley mutants. Carlsberg Res. Commun. 51:509 (1986).
47. U. Rasmussen, The application of immunofluorescence for detecting specific proteins in barley seeds, in: Modern Methods of Plant Analysis. H.F. Linskens and J.F. Jackson, eds., Springer, Heidelberg, in press (1987).
48. K.N. Kristiansen, Biosynthesis of proanthocyanidins in barley: Genetic control of the conversion of dihydroquercetin to catechin and procyanidins, Carlsberg Res. Commun. 49:503 (1984).
49. K.N. Kristiansen, Conversion of (+)-dihydroquercetin to (+)-2,3-trans-3,4-cis-leucocyanidin and (+)-catechin with an enzyme extract from maturing grains of barley, Carlsberg Res. Commun. 51:51 (1986).
50. D. von Wettstein, B. Jende-Strid, B. Ahrenst-Larsen and J.A. Sørensen, Biochemical mutant in barley renders chemical stabilization of beer superfluous, Carlsberg Res. Commun. 42:341 (1977).
51. D. von Wettstein, R.A. Nilan, B. Ahrenst-Larsen, K. Erdal, J. Ingversen, B. Jende-Strid, K. Nyegaard Kristiansen, J. Larsen, H. Outtrup and S.E. Ullrich, Proanthocyanidin-free barley for brewing: Progress in breeding for high yield and research tool in polyphenol chemistry, MBAA Technical Quarterly 22:41 (1985).
52. K. Erdal, Proanthocyanidin-free barley - malting and brewing, J. Inst. Brew. 92:220 (1986).
53. B. Jende-Strid, Coordinator's Report: Anthocyanin genes, Barley Genetics Newsletter 14:76 (1985).
54. B. Jende-Strid and K. Nyegaard Kristiansen, Genetics of flavonoid biosynthesis in barley, in: Barley Genetics V. Proc. Fifth Intl. Barley Genet. Symp., Okayama, in press (1987).

55. B. Jende-Strid, Phenolic acids in grains of wild-type barley and proanthocyanidin-free mutants, Carlsberg Res. Commun. 50:1 (1985).
56. R. Schildbach, H. Pfenninger and F. Schur, Proanthocyanidinfreie Gerste Galant, Brauwelt 126:618 (1986).
57. B. Ahrenst-Larsen and K. Erdal, Anthocyanogen-free barley - A key to natural prevention of beer haze, Proc. Europ. Brewery Conv. Congr. Berlin (West):631 (1979).
58. S. Aastrup, H. Outtrup and K. Erdal, Location of the proanthocyanidins in the barley grain, Carlsberg Res. Commun. 49:105 (1984).
59. S. Aastrup, A test for presence or absence of proanthocyanidins in barley and malt, Carlsberg Res. Commun. 50:37 (1985).

IMPACT OF GENETICS AND GENETIC ENGINEERING ON IMPROVEMENT OF NITROGEN FIXATION IN FREE LIVING AND SYMBIOTIC BACTERIA

K.A. Ahmed

Genetics Dept., Fac. Agric.
Cairo University
Egypt

Applying conventional genetics methods (i.e. mutation and selection) using antibiotics, insectides, herbicides and phage resistance as well as physical and chemical mutagens on *Azotobacter* spp and *Rhizobium* spp resulted in improving nitrogen fixation ability of tested strains. Host specificity was also altered by intragenic transformation and intragenic conjugation methods.

Application of genetic engineering has resulted in obtaining *E.coli* and *B.subtilis* capable of nitrogen fixation as well as *Rhizobium* capable of free nitrogen fixation.

GENETICS AND MOLECULAR BIOLOGY OF DCMU AND ATRAZINE RESISTANCE IN CYANOBACTERIA, STRAINS SYNECHOCYSTIS 6714 AND 6803

G. Ajlani, I. Meyer and C. Astier

Laboratorie de Photosynthèse
ER 307, CNRS
91190 Gif sur Yvette, France

Mutations in the psbA gene encoding D_1, a photosystem II protein, have been correlated with DCMU and atrazine resistance in several organism which are able to perform oxygenic photosynthesis, including higher plants, green algae and cyanobacteria (1-4). Herbicide binding sites and the binding site of the secondary electron acceptor of the photosystem II, the quinone Q_B, might be located on this D_1 protein. Another photosystem II protein, the D_2 protein coded by the psbD gene exhibits homology with D_1 (5). In the green alga Chlamydomonas reinhardtii, the D_2 protein plays a role in the post transcriptional regulation of D_1 and in PSII complex stabilization (6). The sequence of this psbD gene is also highly conserved in the different organisms performing oxygenic photosynthesis. The exact functions of these two proteins are not yet completely known. More and more evidence suggests that they constitute the core of the PSII center. This hypothesis is also supported by the homology between D_1/D_2 and the subunits L/M which constitute the bacterial photoreaction center (7, 8). Photosystem II studies can be developed using mutants resistant to herbicides which are known to block PSII electron transfer.

From our collection of herbicide resistant mutants of Synechocystis, we have chosen two mutants having extreme phenotypes. The $DCMU^r$-II_B is highly resistant to DCMU (x 300) and still sensitive to atrazine. It was selected in two steps and has a good chance of being a double mutant. The Ar^r-V mutant is resistant to atrazine (x 67) but is little modified in its sensitivity to DCMU. In the two mutants, the acquisition of high level of resistance to one of the two herbicides did not induce any alteration in Q_B binding (9-11). We are cloning the different copies of the psbA and psbD genes from these two mutants and the wild type. The determination of their respective roles in the phenotype will be studied by transformation. We have constructed genomic libraries of these two mutants and of the wild type in the λEMBL4 vector. We have used the psbA gene cloned from Amaranthus and the psbD gene cloned from Chlamydomonas reinhardtii as probes to screen these 3 libraries. Several recombinant phages hybridizing with the probes were isolated and are being studied by transformation.

References

1. Erickson, J.M., Rahire, M., Bennoun, P., Delepelaire, P., Diner, B., and Rochaix, J.D., 1984, Herbicide resistance in Chlamydomonas reinhardtii results from a mutation in the chloroplast gene for the 32 kD protein of photosystem II. Proc. Natl. Acad. Sci. USA, 81:3617-3621.

2. Hirschberg, J., and McIntosh, L., 1983, Molecular basis of herbicide resistance in Amaranthus hybridus. Science, 122:1346-1349.

3. Golden, S.S., Brusslan, J.A., and Haselkorn, R., 1985, Herbicide resistance and the expression of psbA in Anacystis nidulans R2. V. International symposium on photosynthetic prokaryotes.

4. Erickson, J.M., Rahire, M., Rochaix, J.D., and Metz, L., 1985, Herbicide resistance and cross-resistance: changes at three distinct sites in the herbicide-binding protein. Science, 228:204-207.

5. Rochaix, J.D., Dron, M., Rahire, M., and Malnoe, P., 1984, Sequence homology between the 32 kD and the D_2 chloroplast membrane polypeptides in Chlamydomonas reinhardtii. Plant Mol. Biol., 3:363-370.

MOLECULAR ANALYSIS OF A RUBISCO ASSEMBLY MUTANT IN NICOTIANA TABACUM

Adi Avni, Ivan Rachailovich, Marvin Edelman,
Dvora Aviv, Esra Galun and Robert Fluhr

The Weizmann Institute of Science
Department of Plant Genetics
Rehovot, Israel

In photosynthetic eukaryotes, the enzyme ribulose-1,5-bisphosphate carboxylase/oxygenase (RUBISCO) is composed of eight small and eight large subunits. Newly-synthesized, chloroplast-coded large subunits can be found in oligomeric forms or in association with polypeptides of 60-61 kDa to form a 650-750 kDa pre-assembly complex. We have isolated a lethal plastid -coded mutant in Nicotiana tabacum which accumulates the 650-750 kDa complex but contains no RUBISCO holoenzyme or enzyme activity. The 650-750 kDa complex from both mutant and wild type sources disassociate in the presence of ATP. Processing of the nuclear-coded small subunit appears unchanged in the mutant however both large and small subunits are subsequently degraded. The large subunit gene (rbcL) of the mutant and wild type plants were cloned. The nucleotide sequence of the two genes revealed a single amino acid difference between them. We suggest that the mutant rbcL gene codes for a modified large-subunit protein with an altered binding site, which prevents the final assembly process with small-subunit.

THE FATE OF THE *AGROBACTERIUM TUMEFACIENS* T-DNA IN THE PLANT EARLY AFTER TRANSFER

Guus Bakkeren, Nigel Grimsley, Zdena Koukolikova-Nicola
and Barbara Hohn
Friedrich Miescher-Institut
P.O. Box 2543
CH-4002 Basel

Aim:

An assay-system has been set up to recover T-DNA border sequences *in planta* as to study their role in the T-DNA intermediate formation and transfer/integration process.

Approach:

The so-called 'Agroinfection'-system is used as a tool by means of which *Agrobacterium tumefaciens* can deliver very efficiently viral genome(s) that are cloned in between the 25 bp direct repeat border-sequences of the Ti-plasmid, into plant cells, whereupon infection occurs presuming the compatible virus/host combination is used.

In my system Cauliflower Mosaic Virus (CaMV) serves as a plant-replicon which upon transfer to *Brassica rapa* (Turnip) amplifies the hitch-hiking border-remnants of the transfer-event. We have sequenced these border-remnants from different, independently transferred virus and have already obtained nice data which show the power of this very clean system.

ONSET OF DESICCATION TOLERANCE DURING BARLEY EMBRYO DEVELOPMENT

D. Bartels, M. Singh, and F. Salamini

Max-Planck-Institut für Züchtungsforschung
Abteilung Pflanzenzüchtung und Ertragsphysiologie
D-5000 Köln 30, BRD

The response of excised barley embryos to rapid water stress is being investigated. Embryos from different developmental stages have been isolated, desiccated for 16 hrs to a water content of 2-3% and then put on germination medium (GM), on which excised, non-desiccated embryos of all developmental stages tested germinate precociously. At a precise developmental stage (16-18 days after pollination (DAP) under our conditions) the embryo acquires the ability to germinate after desiccation. The ability of the embryo to germinate precociously is distinguished from its ability to acquire desiccation tolerance. We are studying the metabolic changes which take place during the transition-phase from the desiccation intolerant embryo to the desiccation tolerant embryo. In order to study differentially expressed genes during embryogenesis, _in vivo_ labelled proteins were separated by 2-dimensional gel electrophoresis and compared with _in vitro_ synthesized proteins derived from mRNA of specific developmental stages.

Qualitative and quantitative changes between different developmental and physiological stages have been observed in the patterns of the synthesized proteins. Proteins expressed at specific developmental stages have been identified: e.g. at least five proteins (M_r between 30.000-40.000) are increasingly expressed in the protein pattern of embryos 17 DAP compared to embryos 12 DAP. Mainly quantitative differences can be seen between _in vitro_ synthesized proteins from fresh embryos (17 DAP) and dried embryos (17 DAP). ABA induced proteins can be identified among _in vitro_ translation products. A cDNA clone bank is being constructed in order to isolate cDNA clones specific for proteins appearing in the transition to desiccation tolerance. If embryos which would otherwise readily germinate but are not desiccation tolerant are cultured on GM medium containing ABA the embryos increase in size and also acquire the ability to germinate after subsequent desiccation.

SOLUBLE TRANSCRIPTION SYSTEMS FROM PLANTS

Alfred Batschauer and Eberhard Schäfer

Biologisches Institut der
Albert-Ludwigs-Universität Freiburg
West Germany

We are interested in the light regulation of different genes from barley and parsley. As we have shown (Batschauer and Apel 1984), the levels of mRNAs encoding the light-harvesting chlorophyll a/b proteins and the protochlorophyllide oxidoreductase are influenced by light via the phytochrome system.

Not only mRNA levels but also transcription rates could be modified by activating the phytochrome system in vivo (Mösinger et al. 1985). Besides phytochrome, other factors are involved in the induction of LHCP genes in barley depending on an intact plastid compartment (Batschauer et al. 1986).

In parsley cell cultures we have investigated the light regulation of genes involved in flavone and flavonol-glycoside biosynthesis. For example the chalcone synthase mRNA levels are influenced by three different photoreceptors (UV-B-, blue-light-receptors and phytochrome); while blue-light and red-light seem to modify the expression of this gene, UV-B-light is absolutely necessary for its induction (Bruns et al. 1986).

With the kind of analysis described above, even with measurements of transcription rates we get no information about the factors which are directly involved in the regulation of gene expression. We decided to isolate such factors to test them in vitro transcription systems. Different methods which were used for animal systems (for example Manley et al. 1980, Dignam et al. 1983) were tested and modified to obtain high incorporation of labeled UTP into RNA using soluble transcription systems prepared from barley and parsley cell cultures. Highest incorporation rates were obtained with a system according to Dignam et al. 1983 with some modifications using chromatin as the source of transcription factors. Tested with specific cloned genomic DNA, our results indicate that there is still high unspecific initiation beside specific one. We are planning further experiments to find conditions for transcription assays which enable us to gain specific initiation in a higher rate.

Literature

Batschauer, A., Apel, K., 1984, Eur. J. Biochem., 143:593-597.
Mösinger et al., 1985, Eur. J. Biochem., 147:137-142.
Batschauer, A. et al., 1986, Eur. J. Biochem., 154:625-634.
Bruns, B. et al., 1986, Planta, 169:393-398.
Manley, J.L. et al., 1980, Proc. Natl. Acad. Sci. USA, 77:3855-3859.
Dignam, J.D. et al., 1983, Nucl. Acids Res., 11:1475-1489.

TRANSFORMATION OF TOMATO AND TOBACCO WITH PLASTOCYANIN AND FERREDOXIN GENES

A.D. de Boer, R. Teertstra, F. Cremers, J. Hille,
P. Hooykaas, L. Smits, S. Smeekens and P. Weisbeek

Instituut voor Moleculaire Biologie en Medische
Biotechnologie, Rijksuniversiteit de Utrecht
3508 TB Utrecht, The Netherlands

Tomato and tobacco plants were transformed with Silene plastocyanin and ferredoxin genes or with fusions of both genes cloned in the vector Bin 19, by means of the binary cloning system developed for Agrobacterium tumefaciens. Regenerated plants that were kanamycine resistant, were tested for their ability to produce RNA that reacts with a Silene plastocyanin or ferredoxin probe under stringent conditions in RNA blotting experiments. Plants that reacted positively were analyzed in protein blots, for their ability to produce the Silene protein. Plants with the strongest protein production were also tested in DNA blotting experiments. The plastocyanin and ferredoxin genes were cloned behind the 35S promotor of the Cauliflower mosaic virus and therefore they are also expressed in non-photosynthetic tissue, as was confirmed with blotting experiments and electron microscopy. As concluded from the blotting experiments, the selected plants have the genes for Silene plastocyanin and ferredoxin integrated in their genome in an uninterrupted way and both genes are expressed in a normal way. The gene products are processed to the mature form properly. Presently, we are investigating whether the proteins are translocated to their proper location and we are examining the effects of interchanging the presequences of plastocyanin and ferredoxin on their transport. We also study transport in non-photosynthetic tissue, both with protein blotting and electron microscopical techniques. Initial experiments indicate import into plastids other than chloroplasts.

GENE REGULATION ASSOCIATED WITH THE SYNTHESIS OF CELL WALL COMPONENTS IN CULTURED CELLS OF PHASEOLUS VULGARIS

G. Paul Bolwell

Department of Biological Sciences
City of London Polytechnic
London, UK

Treatment of cell suspension cultures of *Phaseolus vulgaris* with plant growth regulators or with elicitor molecules from cell walls of the pathogen *Colletrotrichum lindemuthianum* brings about changes in cell wall components, the ultimate basis of most of which resides in differential gene expression. Thus, changes such as the cessation of pectin synthesis and increased hemicellulose deposition accompanied by the onset of lignification and characteristic of xylogenesis, can be brought about by application of plant growth regulators. In contrast, elicitor treatment results in the accumulation of wall-bound lignin-like phenolics and hydroxyproline-rich glycoproteins. The increases in extractable activities/levels of synthetic components, as part of a highly selective pattern of metabolic changes, have been followed to varying degrees of technical sophistication. Differential patterns of enzyme induction related to end-product accumulation are observed in response to the various treatments. Thus polysaccharide synthases and glycosyl transferases, enzymes of phenylpropanoid metabolism, hydroxyproline-rich glycoproteins and the enzymes of their post-translational modification have all been identified as targets for gene modulation. The regulation of phenylalanine ammonia-lyase at the transcriptional and post-translational level has been the most studied at the molecular level in this system.

POST-TRANSCRIPTIONAL REGULATION: CAULIFLOWER MOSAIC VIRUS GENE I EXPRESSION IS ENHANCED BY A VIRAL TRANS ACTING FACTOR

J.M. Bonneville, J.R. Penswick, H. Sanfacon,
B. Pisan and T. Hohn

Friedrich Miescher-Institut
CH-4002 Basel, Switzerland

We are studying the gene expression of cauliflower mosaic virus (CaMV) using gene fusions transfected into host plant protoplasts. The chloramphenicol acetyl transferase (CAT) reporter gene was fused to ORF I, the second reading frame on the genome-length viral RNA (35 S RNA). This dicistronic construct, when transfected alone, drives the synthesis of only barely detectable amounts of CAT activity. However, co-transfection of this plasmid with viral DNA results in a dramatic enhancement of the reporter gene expression. A control, indicating that this helping effect is a post-transcriptional one, is provided by monocistronic CAT fusions that use the same 35 S promoter: they are efficiently translated (and hence, transcribed) in the absence as well as in the presence of viral DNA. Use of mutants of both the dicistronic CAT plasmid and the helper viral DNA reveals that this transactivation phenomenon: i) is independent of the presence of the long viral untranslated leader on the chimeric mRNA, ii) is independent of the product of the first cistron, and iii) does not require a full-length, infectious viral helper DNA. Work is in progress to map the trans-acting factor gene(s), to define the sequences required in cis for transactivation and to elucidate the molecular level of this interaction, keeping in mind the relay-race translation model proposed for CaMV (see Dixon & Hohn, EMBO J. (1984) 3: 2731, and Sieg & Gronenborn as cited therein).

MITOCHONDRIAL DNA - ANALYSIS OF WHEAT ALLOPLASMIC LINE

Adina Breiman

Department of Botany
The George S. Wise Faculty of Life Sciences
Tel Aviv University, Tel Aviv 69978, Israel

Wheat alloplasmic lines, in which cytoplasms of diploid wild wheat relatives have been introduced by backcrosses into the nuclear background of Triticum aestivum, have been analysed for the organisation of the mtDNA. Mitochondrial DNA prepared from Aegilops squarrosa, Ae. variabilis, Ae. speltoides, T. timopheevi, and Ae. longissima alloplasmic lines, was compared to the mtDNA of the euplasmic lines, in order to detect polymorphism apparent in natural populations and to detect a nuclear effect on mtDNA organisation.

Whereas no major changes could be observed in the mtDNA or RNA transcripts of various genes in the alloplasmic fertile lines, the mtDNA of the alloplasmic male sterile T. timopheevi showed major variation in the organisation of the COX I ATPase 9' and ATPase 6 mt coding genes. The changes could also be detected at the RNA level by Northern analysis. Studies and further investigation of the T. timopheevi CMS alloplasmic line are being performed.

INSERTION OF SIGNALS FOR AUTOLYTIC CLEAVAGE IN VIRAL cDNAs PROVIDES NEARLY CORRECT 3' ENDS OF VIRAL RNA TRANSCRIPTS

Jozef J. Bujarski and Paul Kaesberg

Biophysics Laboratory and Department of Biochemistry
University of Wisconsin
Madison, WI 53706

Transcription of viral cDNAs integrated into nuclear genome of transgenic plants usually generates viral RNAs containing extraneous sequences at their termini. While the size of these sequences at the 5' end can be reduced relatively easily by reconstruction of the promoter junction it is difficult to avoid some spacer sequences and a polyA stretch at the 3' end. Existence of these extra sequences interferes with such processes as viral RNA replication or encapsidation.

We are using the autolytic cleavage activity of sequences found in satellite RNA of tobacco ringspot virus (STobRV RNA) in an attempt to overcome this problem (1). A DNA fragment, isolated from a STobRV cDNA clone (generously supplied by Dr. A. Hampel at Northern Illinois University) and containing all the sequences required for self-cleavage was inserted just downstream from the 3' end of an infectious cDNA clone of brome mosaic virus (BMV) RNA3 (2). Its self-cleavage activity was checked in vitro after transcription with SP6 RNA polymerase.

In cooperation with Drs. D. Merlo and S. Loesch-Fries at Agrigenetics, Madison, WI, we are generating transgenic tobacco plants which would transcribe BMV RNAs and self-regenerate their almost correct 3' ends. We expect that as in nonpolyadenylated mRNAs of histone genes, the removal of the polyA stretch (and most of the other 3' end heterologous sequences), will produce transcripts operational in the cytoplasm. We plan to use this system to analyze in vivo processes of recombination between BMV RNA components without replication and encapsidation and symptom generation without replication.

(1) Prody, G. et al., 1986, Science, 231:1577-1580.
(2) Ahlquist, P. and Janda, M., 1984, Molec. Cell. Biol., 4:2876-2882.

BACTERIAL EXOPOLYSACCHARIDES INVOLVED IN THE INTERACTIONS OF AGROBACTERIUM TUMEFACIENS AND RHIZOBIUM MELILOTI WITH PLANT HOSTS

Gerard A. Cangelosi, Lynn Hung, John A. Leigh,
David A. Ozga, V. Puvanesarajah* and Eugene W. Nester

Dept. of Microbiology, Univ. of Washington, Seattle, WA 98195; and *Dept. of Microbiology, Univ. of Tennessee, Knoxville, TN 37996, USA.

Agrobacterium tumefaciens and Rhizobium meliloti produce a high molecular weight, capsular exopolysaccharide (succinoglycan), as well as a smaller periplasmic β-1,2-glucan. Mutants of R. meliloti lacking succinoglycan (Exo$^-$), representing at least five different genetic complementation groups, do not form effective nodules on alfalfa roots. We have isolated analogous mutants of A. tumefaciens. Genetic complementation revealed that the Agrobacterium and Rhizobium Exo loci are functionally interchangable. With the exception of mutants belonging to the ExoC complementation group, all of the A. tumefaciens Exo mutants form normal tumors on four different plant hosts. The ExoC mutants are avirulent on all four plants and do not attach to plant cells in vitro. These mutants synthesize neither β-1,2-glucan nor succinoglycan, whereas the other Exo mutants lack only succinoglycan. A. tumefaciens strains with mutations in the chvB locus also do not synthesize β-1,2-glucan, do not attach to plant cells, and do not form tumors. The chvB and ExoC loci both map on the Agrobacterium chromosome, but are unlinked to each other. R. meliloti has a chromosomal locus (ndvB) which is homologous and functionally interchangable with chvB, and is required for effective nodulation. These data suggest that succinoglycan plays a role in nodulation by Rhizobium but not in crown gall tumor formation, whereas β-1,2-glucan is involved in both types of infections, perhaps at the level of attachment to plant cells.

CHARACTERISATION OF THREE STRAINS OF POTATO VIRUS X AND THEIR REPLICATION IN POTATO PROTOPLASTS CONTAINING GENES FOR COMPATIBILITY, INCOMPATIBILITY OR RESISTANCE

Robert H.A. Coutts[1] and Sally E. Adams[2]

[1]Dept. of Pure and Applied Biology, Imperial College, London, UK.
[2]Dept. of Biochemistry, University of Oxford.

Studies on aspects of resistance of potato protoplasts of differing genotype, and different reaction at the whole plant level with a range of PVX strains are now complete, as is the characterisation of a resistance-breaking and two other strains of the virus. As in other virus/host combinations while the whole plant reaction in the case of certain potato cultivars and PVX strains is an incompatable hypersensitive one, no such reaction was noted at the protoplast level. We have devised a new method of producing potato leaf material for the reproducible isolation and viral infection of protoplasts (Adams et al., 1985). Using this system we have investigated different potato cultivars in combination with the PVX strains which are incompatible, compatible or operationally field immune (Adams et al., 1986). The replication of the various PVX strains has been monitored by assaying infectious virus and viral RNA, and the production of capsid by ELISA and Western blotting using monoclonal antibodies raised against the different strains.

These studies show that the replication of the strains in compatible and incompatible combinations is similar and that large numbers of protoplasts (up to 70% of the cells) are infected which produce similar amounts of virus. However, in the case of protoplasts derived from plants which are operationally immune to common strains of PVX at the whole plant level, only a limited replication of the virus is tolerated. In contrast the resistance-breaking strain replicates in protoplasts of these plants to levels similar to those achieved in compatible or incompatible situations. The mode of operation of this limited resistance at the protoplast level is being investigated further with particular reference to the synthesis of viral sense and antisense RNA. The various strains used in the study have been characterised in detail with respect to their RNAs and capsids. Capsid proteins were characterised by V8 protease digestion, Western blotting and probing with monoclonal antibodies. Viral RNAs have been characterised at the gross levels of size and complexity and are currently being sequenced. The in vitro capabilities of the PVX RNAs have also been studied, and successful in vitro translation of viral RNA has been achieved with one strain of the virus using a novel extraction procedure of RNA from infected plants.

References: Adams, S.E., Jones, R.A.C., and Coutts, R.H.A., 1985, The Journal of General Virology, 66:1341-1346; Adams, S.E., Jones, R.A.C., and Coutts, R.H.A., 1986, The Journal of General Virology, 67:2341-2347.

ISOLATION OF L-PHOSPHINOTHRICIN RESISTANT PLANTS

Peter Eckes

Hoechst
Abteilung Pflanzenschutzforschung
D-6230 Frankfurt am Main 80

Glutamine synthetase (GS) plays a central role in nitrogen metabolism of higher plants. GS is responsible for the primary assimilation of ammonia produced by nitrate reduction and nitrogen fixation in roots as well as for the reassimilation of ammonia released by photorespiration in leaves. The enzyme is encoded by several organ-specifically expressed genes in plants (1,2).

Because of the central role of GS in the detoxification of ammonia, plants are very susceptible to the amino acid analog L-phosphinothricin (L-PPT), a potent inhibitor of GS. Recently an alfalfa cell line could be selected, which is resistant to L-PPT (3). This resistance was due to an amplification of the GS-gene and an overproduction of the corresponding GS-enzyme. But this cell line could by no means be regenerated to plants.

In the poster a way will be described to obtain L-PPT resistant plants by GS-overproduction. We fused an alfalfa GS-gene, which has previously been isolated (4), to different regulatory sequences, which can be recognized by the plant transcription apparatus. These constructions were integrated into the tobacco genome by *Agrobacterium tumefaciens* mediated gene transfer methods. GS expression in regenerated transgenic plants was monitored by RNA- and protein analyses. A significant increase in GS-specific RNA and protein could be observed. This GS-overproduction resulted in an increase in L-PPT resistance of the transgenic plants compared to wild type tobacco plants. Data upon DNA-, RNA- and protein analyses as well as some physiological parameters will be presented and discussed.

1) Cullimore, J.V. et al., 1984, *J. Mol. Appl. Gen.*, 2:589-599.
2) Tingey, S.V. et al., 1987, *EMBO J.*, 6:1-9.
3) Donn, G. et al., 1984, *J. Mol. Appl. Gen.*, 2:621-635.
4) Tischer, E. et al., 1986, *Mol. Gen. Genet.*, 203:221-229.

CHARACTERIZATION AND TRANSFER TO PLANT CELLS OF DIFFERENT GENES OF A. RHIZOGENES 1855 TL-DNA

M. Cardarelli[1], D. Mariotti[1], M. Pomponi[2], L. Spanò[2], M. Trovato[2], M.L. Mauro[2], P. Filetici[1], G. Vitali[2], I. Capone[2], and P. Costantino[2]

[1]Centro Acidi Nucleici, CNR, Roma, Italy
[2]Dip. Genetica e Biologia Molecolare, Università di Roma "La Sapienza", Roma, Italy

Agropine-type Ri plasmids of Agrobacterium rhizogenes contain two distinct T-regions, denominated TL and TR. Genes encoding synthesis of agropine and of the plant hormone auxin have been localized on the TR-DNA. Very little is known, on the contrary, on the nature of the genes residing on the TL-DNA. These genes confer to otherwise unresponsive plant cell the capability to differentiate roots upon stimulation by auxin (in press); moreover they confer characteristic morphological traits to plants regenerated from hairy roots. In an effort to identify TL genes relevant for hairy root induction and plant morphology we have cloned several segments of the TL-region of the agropine-type Ri plasmid pRi1855 in the binary vector system Bin19. These constructions were utilized for infections on carrot discs and tobacco leaf disc regeneration of transformed plants.

Plants transformed with a TL-DNA segment encompassing ORFs 10, 11 and 12 and even just ORF 11 (rolB) show the characteristic hairy root phenotype (wrinkled leaves, excessive root system); hairy roots can be induced on tobacco stems by ORF 11 and, more abundantly, by ORFs 10, 11 and 12. On the contrary, rooting on carrot discs requires a more complex complement of TL-genes, comprising ORF 10 to 14.

Morphological and biochemical analysis of regenerants containing different portions of the TL-DNA of pRi1855 as well as expression in vitro and in yeast of different ORFs is currently underway in our laboratory.

TOLERANCE OF CYANOBACTERIA TO THE HERBICIDE SULFOMETURON METHYL: A NATURAL OCCURRENCE AND ISOLATION OF SM RESISTANT MUTANTS

Devorah Friedberg

Institute of Life Sciences, Division of Microbial and Molecular Ecology, The Hebrew University of Jerusalem
91904 Jerusalem, Israel

The herbicide sulfometuron methyl (SM) inhibits the growth of the cyanobacterium *Anacystis nidulans* R2K, while *Aphanocapsa* 6714 is tolerant. Two types of SM resistant mutants were isolated from *Anacystis*. The activity of acetolactate synthase (ALS), a key enzyme in the biosynthesis of branched chain amino acids was found to be apparently 1000 fold more resistant to SM in cell free extracts of the mutants compared to that of the wild type. Unlike *Anacystis* the ALS activity of the cyanobacterium *Aphanocapsa* 6714 was found to tolerate high concentrations of the herbicide. SM growth retardation of *Anacystis* wild type was alleviated by the addition of both valine and leucine. The specific activity, of mutant SM2/32 ALS was markedly lower than that of other strains. This mutant growth was extremely susceptible to valine inhibition, though the sensitivity of its ALS activity to valine was similar to that of the other strains. Leucine and isoleucine alleviated the valine induced growth inhibition of this strain as well as that of R2K. The data suggests that SM inhibits cyanobacterial growth via inhibition of ALS and that cyanobacteria should be useful for the isolation of altered ALS gene which confer resistance to herbicides.

PROTEIN BINDING DOMAINS IN THE 5'-UPSTREAM REGION OF RIBULOSE BISPHOSPHATE CARBOXYLASE SMALL SUBUNIT GENES

Thomas F. Gallagher

Department of Botany
University College Dublin
Belfield, Dublin 4

The binding of proteins, isolated from nuclei, to the 5'-upstream region of rubisco SSU genes has been studied by gelshift and DNA footprinting techniques. The relevance of these results to the study of the sequences required for light regulated and tissue specific expression of SSU genes will be discussed.

References

Ellis, R.J., Gallagher, T.F., Jenkins, G.I., and Lennox, C.R., 1984, Photoregulation of the biosynthesis of ribulose bisphosphate carboxylase. J. Embryo. Exptl. Morph., 83:163-174.

Gallagher, T.F., Jenkins, G.I., and Ellis, R.J., 1985, Rapid modulation of transcription of nuclear genes encoding chloroplast proteins by light. FEBS Letters, 186:241-245.

FUNCTIONAL MAPPING OF PEA LEGUMIN UPSTREAM REGULATORY ELEMENTS USING Ti-PLASMID VECTORS

C.S. Garrett and C.H. Shaw

Department of Botany
University of Durham, Science laboratories
South Road, Durham DH1 3LE, England

The expression of legumin genes from pea using Agrobacterium Ti-plasmid vectors is currently being investigated.

A number of plasmids containing a maximum of 1.2 kb of 5' upstream sequences from a legumin gene of Pisum sativum L. ligated to the coding region of the nopaline synthase (nos) gene have been constructed. The production of nopaline in various plant tissues is being used as a quick initial assay of gene expression from the legumin promoter. The use of smaller promoter fragments and the insertion of spacer DNA within the promoter region have been employed in an effort to localize the regions of 5' flanking sequence which may play a role in tissue specific expression.

These constructs have been mobilised to Agrobacterium tumefaciens using pRN3 and homologous recombination with pDUB1006, an oncogenic Ti-plasmid, which contains the nos coding region and the 25 base pair repeat from the right border. These vectors have been screened for nopaline production in callus tissue obtained from Kalanchoe. Tumours derived from strains carrying the smallest promoter fragment (which includes the TATA box and upstream to -95) do not show nopaline production but a small proportion of tumours with the full length promoter did show low levels of nopaline production. This could indicate that the legumin promoter is not completely inactive in undifferentiated tissue from Kalanchoe giving nopaline levels on the limits of detection. Tumour tissue from Nicotiana tabacum and Pisum sativum L. will also be tested.

In parallel these chimaeric legumin-nos genes have been inserted into Bin 19 to give a series of disarmed vectors which have allowed the regeneration of whole plants from infected Nicotiana tabacum leaf discs. No nopaline has been detected in leaf tissue from regenerated plants transformed with a chimaeric legumin-nos gene containing 0.8 kb of upstream sequences.

Additional promoter deletions spanning from 200-600 base pairs upstream of the legumin transcription start site are currently being prepared. DNA from transformed plant tissue will be analysed for unrearranged insertion of the chimaeric leg-nos genes. Northern blots will be carried out to assay for nopaline synthase mRNA production from a variety of plant tissues as they become available.

LIGHT CONTROL AND DEVELOPMENTAL REGULATION OF EXPRESSION OF NUCLEAR GENES IN YOUNG BARLEY SEEDLINGS

Rosa Barkardottir, Birgit F. Jensen, Jette D. Kreiberg, Lene H. Madsen, Peter S. Nielsen and Kirsten Gausing
Department of Molecular Biology and Plant Physiology
University of Aarhus, C.F. Møllers Allé 130
DK-8000 Århus C

Taking advantage of the natural developmental gradient in graminaceous leaves, the levels of several messengers encoded by nuclear genes have been measured as a function of cell age. RNA from sections (0.5 and 1.0 cm) of the first leaf from seven day old barley seedlings grown under carefully controlled diurnal conditions was isolated and analysed by Northern hybridizations using different barley cDNA clones as probes. Semi-quantitative estimates of relative messenger levels were obtained by scanning of autoradiograms of the Northern hybridizations. The influence of light on messenger levels was studied in seven day old green and etiolated leaves, divided into a basal section (one third of the leaves, cell age up to 36 hrs) and a top section (most cells between 1.5 and 5 day old). In these experiments, RNA from coleoptiles and roots were included to evaluate the tissue specificity of gene expression. Results for three main groups of genes will be reported:

Genes positively regulated by light and with leaf specific expression. This group includes the two multigene families encoding the small subunit of RuBPCase and chlorophyll a/b binding protein and the possibly single gene encoding plastocyanin.

Genes negatively regulated by light and with leaf specific expression. This group includes the multigene families encoding leaf specific thionin and, presumably, protochlorophyllide oxidoreductase.

Genes relatively unaffected by light and expressed both in leaves and roots. Two quite different types of genes belong to this group. In one gene family, most genes (distinguished by their different size messenger) give rise to nearly identical levels of messenger in all tissue. The other type shows a very interesting pattern of expression: the gene(s) is predominantly expressed in leaf and root meristem (root tips) and the messenger disappears very rapidly in the slightly older leaf sections just above the meristematic zone.

THE MOBILISATION OF THE T-DNA FROM AGROBACTERIUM TO THE PLANT CELLS INVOLVES A SINGLE STRANDED DNA-BINDING PROTEIN

Christine Gietl

Friedrich Miescher-Institut, Postfach 2543, CH-4002 Basel
Permanent address: Technical University of Munich
Institute of Botany, Arcisstr. 21, D-8030 Munich 2

During infection of a plant, Agrobacterium tumefaciens transfers a portion of the Ti-plasmid DNA (the T-DNA) to plant cells. Genes of the T-region are responsible for tumor induction, but not for gene transfer itself; transacting functions which mediate the process of the T-region transfer, are specified in the bacterium by its chromosomal virulence loci (chvA and chvB) and the Ti-plasmid virulence loci (vir A, B, G, C, D, E). Whereas chv expression is constitutive and nonregulated, transcription of the vir loci is induced by plant synthesized phenolic compounds such as acetosyringone. As one of the early steps the T-DNA is cut out by a site specific endonuclease coded by the vir D-region. Both transport of the T-DNA in a single stranded or in a double stranded form is discussed.

It was therefore interesting to find out, if acetosyringone treated Agrobacteria contain single stranded and/or double stranded DNA binding proteins.

Extracts of induced and uninduced Agrobacteria strains were tested with synthetic, radioactively labeled oligonucleotide probes for the presence of DNA-binding proteins using polyacrylamide gel retardation assays. One inducible and several constitutively synthesized, sequence unspecific, single stranded DNA-binding proteins determined by the Ti-plasmid, the vir-region and the Agrobacterium chromosome are identified. The inducible ssDNA-binding protein is absent in mutants of the vir E locus, as well as in mutants of the A and G loci, in which induction of transcription of the entire vir-region is abolished. A minimum ssDNA probe length is required for complex formation: 36 nucleotides in the case of the induced protein, 30 nucleotides in the case of the constitutively synthesized proteins. Isolation of complexes by sucrose gradient centrifugation was achieved.

The function of the inducible, sequence unspecific, ssDNA-binding protein most likely determined by the vir E-region during bacteria/plant conjugation is discussed.

FUNCTIONAL ORGANIZATION OF THE POLYPEPTIDE CHAINS OF THE NICOTINIC ACETYLCHOLINE RECEPTOR

J. Giraudat, M. Dennis, T. Heidmann, J.P. Changeux (Inst. Pasteur, Paris), F. Kotzyba-Hibert, M. Goeldner, C. Hirth Univ. Pasteur, Strasbourg), P.Y. Haumont, F. Lederer (Hop. Necker, Paris), and Y. Chang (Ciba-Geigy, Basel)

The nicotinic acetylcholine receptor from Torpedo electric organ is a heterologous pentamer of four different transmembrane polypeptides in the ratio $\alpha_2\beta\gamma\delta$. This complex contains all the functional elements of the receptor, including the ion channel, and possesses several classes of ligand binding sites: two sites for nicotinic agonists (and competitive antagonists) carried, at least in part, by the two α-chains, plus a unique high affinity site for noncompetitive blockers of the permeability response.

The cDNAs coding for the constitutive four chains of the receptor have been cloned and sequenced in various laboratories. The four chains display striking homologies in amino acid sequence and similar hydrophobicity profiles suggesting an organization of the receptor molecule into a transmembrane bundle of five subunits symmetrically organized around a central ionic channel. On the basis of these sequence data, models of transmembrane topology common to all four subunits have been proposed. Each chain would contain: a large amino terminal hydrophilic domain located in the extracellular space, a compact core of three uncharged transmembrane α-helices (MI-MIII), a cytoplasmic hydrophilic domain, and a fourth uncharged transmembrane helix (MIV) close to the C-terminus. Some authors propose the presence of one or two additional transmembrane segments.

In order to test several features of these various structural models, we decided to covalently label specific sites on the receptor with radioactive irreversible ligands and then to identify the modified amino acid(s) using protein chemical techniques.

The acetylcholine binding site was labeled with the irreversible antagonist DDF (p(N,N-dimethylamino) benzene diazonium fluoroborate). The nicotinic receptor was labeled by (^{3}H)DDF. The α-chain was purified by polyacrylamide gel electrophoresis, cleaved with cyanogen bromide and the generated peptides were analysed by HPLC. Three distinct radioactive peaks were observed, which were all decreased when the agonist carbamylcholine was present during irradiation. Sequence analysis demonstrated that the predominant (^{3}H)DDF labeled species corresponds to the fragment α 179-207, and permitted the identification of several modified amino acids clustered around the residues Cys 192 and Cys 193. In the native receptor this region of the α-chain is thus involved in the formation of the acetylcholine binding site and must therefore be located in the extracellular domain of the receptor molecule. The existence of other fragments specifically labeled by (^{3}H)DDF suggests that distant regions of the primary structure of the α-chain might participate in the formation of the acetylcholine binding site.

The unique high affinity site for noncompetitive blockers was probed using chlorpromazine. Chlorpromazine is a well-characterized noncompetitive blocker (NCB) which covalently labels all four receptor chains when irradiated (254 nm) under conditions where it occupies the high affinity NCB site. The nicotinic receptor was labeled by (^{3}H)chlorpromazine at equilibrium in the presence of agonist and the various subunits were subsequently purified by polyacrylamide gel electrophoresis. We have studied so far the δ and β subunits. The purified chain was digested by trypsin and/or cyanogen bromide, and the generated fragments analysed by HPLC. Sequence analysis of the radioactive species enabled us to identify residues Ser 262 of the δ-chain and Ser 254, Leu 257 of the β-chain as sites of (^{3}H)chlorpromazine incorporation. Sequence analysis of peptides purified in parallel from a control batch demonstrated that the labeling of these particular residues was inhibited by phencyclidine, a highly specific ligand for the high affinity NCB site.

The identified labeled residues are located in highly homologous regions of the δ and β chains. These results provide experimental evidence that, at least for the two chains analysed so far, homologous domains of the receptor subunits are involved in the formation of the unique high affinity NCB site. Several lines of evidence support the view that the high affinity NCB site is located within the ionic channel. Our data suggest that, at least in the subunits characterized so far, the uncharged segment MII (to which belong the residues labeled by chlorpromazine) is part of the transmembrane portion of the ionic channel.

MOLECULAR GENETICS OF PHOTOSYNTHESIS IN CHLAMYDOMONAS

M. Goldschmidt-Clermont, Y. Choquet, M. Kuchka, J. Girard-Bascou*, P. Bennoun*, V. Kück+ and J.-D. Rochaix
Departments of Plant Biology and Molecular Biology, Univ. of Geneva; *Institut de Biologie Physico-Chimique, Paris; +Lehrstuhl für Allgemeine Botanik, Ruhr-Universität Bochum

We are investigating the molecular mechanisms of the assembly of the photosynthetic apparatus in the green alga *Chlamydomonas reinhardtii*, a process that involves the genetic elements of the chloroplasts as well as those of the nucleus. Photosynthetic mutants in both genomes can be obtained and used to analyze the function, the coordinate assembly and the stability of these large multimolecular complexes.

Rubisco (ribulose bisphosphate carboxylase/oxygenase) is a simple example of this cooperation, being composed of two types of subunits. The large subunit is encoded in the chloroplast, while the small subunits are encoded in the nucleus as precursors which are imported, processed and assembled in the chloroplast. There are only two small subunit genes in *C. reinhardtii* which produce two distinct mRNAs that accumulate to different levels depending on the growth conditions (illumination and carbon source). In chloroplast mutants that lack the large subunit, small subunit genes are still expressed and the precursors are synthesized and processed, but are then rapidly degraded: protein stability or degradation thus appear to play a role in the coordinate accumulation of the holoenzyme (Spreitzer *et al*., 1985).

In a similar approach of a multi-component membrane complex, we are studying the assembly of photosystem II. Chloroplast and nuclear mutations are being analyzed for their effects on the synthesis and stability of the PSII core and of the oxygen evolving complex.

One of the two chloroplast genes for the apoproteins of CPI (psa A1) consists of three exons that are dispersed at widely separate locations in the *C. reinhardtii* chloroplast DNA. We have obtained chloroplast and nuclear mutants that are deficient in the maturation of the mRNA. New RNAs of abnormal size accumulate in these mutant strains and their structure suggests that they may be intermediates in the trans-splicing of the precursor molecules.

Reference

Spreitzer, R., Goldschmidt-Clermont, M., Rahire, M., and Rochaix, J.-D., 1985, *Proc. Natl. Acad. Sci. USA*, 82:5460-5464.

GEMINIVIRUS DNA IS PRESENT IN PLASTIDS

B.R. Gröning, A. Abouzid, and H. Jeske

Institut für Allgemeine Botanik
Abt. Genetik, Universität Hamburg
D-2000 Hamburg 52

The characteristic leaf mosaic of infected Abutilon sellovianum is caused by abutilon mosaic virus (AbMV), a geminivirus with single-stranded (ss) DNA. Although pathological symptoms can clearly be observed in the plastids the virus particles have never been detected in these organelles. Unbroken plastids were isolated from AbMV infected and uninfected plants in order to determine the localization of the viral genome.

As judged by the binding of the fluorochrome DAPI plastids of infected plants showed a larger DNA content than plastids of uninfected plants. Furthermore their thylakoid system was degenerated and substituted by amorphous electron-dense material. Southern blot hybridization experiments indicated that the viral ssDNA of AbMV was present in plastids of infected plants. Control plants gave no hybridization signals. DNase treatment of isolated plastids excluded an unspecific adsorption of viral DNA to the plastids. This is the first evidence for the presence of viral DNA in plastids.

HIGH LEVEL EXPRESSION OF MAIZE 15 KD ZEIN POLYPEPTIDE IN TRANSGENIC TOBACCO SEEDS

Leslie M. Hoffman, Roger Bookland, Kay Rashka and Debra D. Donaldson

Agrigenetics Advanced Science Company
5649 East Buckeye Road
Madison, WI 53716, U.S.A.

The maize 15 kD zein structural gene was placed under the transcriptional regulation of French bean β-phaseolin gene flanking regions. The entire RNA coding region of the zein gene was sandwiched between 850 bp of the 5' flank and 1100 bp of the 3' flank of the phaseolin gene. *Agrobacterium tumefaciens*-mediated transformation was used to insert the chimeric phaseolin-zein gene carried on a micro Ti-plasmid into the genome of tobacco leaf cells. Regenerated transgenic plants are capable of synthesizing zein in a tissue specific manner during the latter half of seed development. Zein accumulates to high levels in seeds, reaching 1.6% of the total seed protein in several plants. Among 19 transformed plants, there is an 80 fold variation in the level of zein polypeptide detected with rabbit polyclonal anti-15 kD zein sera. No zein protein is found above background ELISA levels in leaves, stems, roots, sepals, or petioles of plants expressing zein in their seeds. The maize storage protein is also detected in roots, hypocotyls, and cotyledons of germinating transgenic tobacco seeds. In developing seeds, zein is correctly processed by the removal of a 20 amino acid signal peptide. Mung bean nuclease mapping shows that transcription of the chimeric gene is initiated in phaseolin-derived sequences, and transcription is most likely terminated within the phaseolin gene 3' flanking region. Polyadenylation signals derived from both the zein and phaseolin genes are used interchangeably in the processing of zein RNA in transgenic plant seeds.

MOLECULAR BASIS FOR THE OVERPRODUCTION OF 5-ENOLPYRUVYLSHIKIMATE 3-PHOSPHATE SYNTHASE IN A GLYPHOSATE-TOLERANT CELL SUSPENSION CULTURE OF *CORYDALIS SEMPERVIRENS*

H. Holländer-Czytko, D. Johänning, and N. Amrhein

Lehrstuhl für Pflanzenphysiologie
Ruhr-Universität Bochum
D-4630 Bochum

Cultured *Corydalis sempervirens* cells have been adapted to increasing concentrations of glyphosate in the medium. In cells grown routinely in the presence of 5 mM glyphosate, the tolerance was correlated with a 30 to 40 fold overproduction of 5-enolpyruvylshikimate 3-phosphate (EPSP) synthase, the target enzyme of the herbicide (1,2).

Translatable mRNA-activity for EPSP-synthase in glyphosate-adapted, as compared to non-adapted, cells was, however, increased only 10 fold. The protein was synthesized as a higher molecular weight precursor of M_r 53,900, while the mature enzyme had a M_r 45,500 (3). This observation is in agreement with the known localization of the enzyme in plastids (4).

In Northern blots, a single transcript of 1.8 kb was detected by an oligonucleotide probe deduced from the N-terminal amino acid sequence of the protein. Levels of the transcript for EPSP-synthase in the cells corresponded to the translatable mRNA activities.

Probing of EcoRI-, BamHI-, or HindIII-digested total DNA with the labelled oligonucleotide revealed in each case only a single band identical for adapted and non-adapted cells.

Southern and dot blot analyses showed no significant differences in the relative proportion of EPSP-synthase specific sequences in DNA from either adapted or non-adapted cells. Thus, amplification of the EPSP-synthase gene, which was recently demonstrated for a glyphosate-tolerant tissue culture of *Petunia hybrida* (5), is not the basis for glyphosate tolerance in our culture of *Corydalis sempervirens*. Evidence for other mechanism(s) will be discussed.

References

1. Amrhein, N., Johänning, D., Schab, J., and Schulz, A., 1983, *FEBS Lett.*, 157:191.
2. Smart, C.C., Johänning, D., Müller, G., and Amrhein, N., 1985, *J. Biol. Chem.*, 260:30, 16338.
3. Holländer-Czytko, H., and Amrhein, N., 1987, *Plant Physiol.*, in press.
4. Smart, C.C. and Amrhein, N., 1987, *Planta*, 170:1.
5. Shah, D.M., Horsch, R.B., Klee, H.J., Kishore, G.M., Winter, J.A., Tumer, N.E., Hironaka, C.M., Sanders, P.R., Gasser, C.S., Aykent, S., Siegel, N.R., Rogers, S.G., and Fraley, R.T., 1986, *Science*, 233:478.

INSERTION ELEMENTS WHICH CAN GENERATE TARGET DUPLICATIONS OF VARIABLE LENGTH

Shigeru Iida

Biozentrum der Universität Basel, CH-4056 Basel
Present address: Institute for Plant Sciences,
ETH Zürich, CH-8092 Zürich

One of the characteristics of transposable elements is the generation of a target duplication upon integration. This had first been observed with the prokaryotic element IS1. IS1 residing in the *Escherichia coli* chromosome is known to integrate into AT-rich regions and to generate 9-bp target duplications. We have shown that IS1 can intrinsically generate not only 9-bp but also 8-bp or 10-bp target duplications. IS186, another IS element on the *E. coli* chromosome, transposes preferentially into GC-rich sequences and generates duplications varying in length between 8- and 12-bp, even at the same target site. Comparisons with other IS elements known to generate duplications of variable length and implications to transposition mechanisms are presented.

References

Iida, S., Hiestand-Nauer, R., and Arber, W., 1985, *Proc. Natl. Acad. Sci. USA*, 82:839-843.

Iida, S. and Hiestand-Nauer, R., 1986, *Gene*, 45:233-235.

Sengstag, C., Iida, S., Hiestand-Nauer, R., and Arber, W., 1986, *Gene*, 49:153-156.

TRANSACTING REGULATORY FACTORS FOR THE SOYBEAN LEGHEMOGLOBIN GENES

Erik Ø. Jensen[1], Frans de Bruijn[2] and Kjeld A. Marcker[1]

[1]Department of Molecular Biology and Plant Physiology, University of Aarhus, DK-8000 Århus C.
[2]Max Planck Institut für Züchtungsforschung, D-5000 Köln 30

Recently we have shown that 1 kb of the soybean leghemoglobin C_3 (lbc_3) gene promotor contributes to its nodule specific expression (Jens Stougaard Jensen et al., 1986, Nature, 321:669-674). The organ specificity of the cis-acting lbc_3 regulatory sequences is thought to be due to their interaction with nodule specific trans-acting factors. We have developed a novel method to identify two nodule specific factors, which bind to the lbc_3 promotor. The binding sites are located on a segment spanning from -245 bp to -150 bp upstream of the start of transcription. No other binding sites have been detected downstream of these two binding sites. The regulatory factors seem to be highly expressed in nodules and may comprise about 0.05% of crude nuclear extract, whereas no binding can be detected using nuclear extracts from leaves and roots. A synthetic oligonucleotide containing 19 bp homologous to the proximal binding region does also bind the specific factor. This region is also highly conserved in all four soybean leghemoglobin genes, whereas the distal binding site seems to be specific for the lbc_3 gene. Work is in progress to purify the regulatory factors and to determine their functions in control of transcription.

LOCALIZATION OF THE REGION OF Ri TL-DNA RESPONSIBLE FOR THE A. RHIZOGENES TRANSFORMED PHENOTYPE

L. Jouanin, F. Vilaine, J. Tourneur, V. Pautot, C. Tourneur, J.F. Muller, M. Caboche and F. Casse-Delbart

Laboratoire de Biologie Cellulaire, INRA
Route de Saint Cyr
F-78000 Versailles

Agrobacterium rhizogenes provokes hairy roots on sensitive plants. Plants regenerated from these roots display a particular phenotype, such as wrinkled leaves, reduced internodal distance, root plagiotropism (1). Two non-contiguous regions (TL and TR) of the Ri plasmid of strain A4 can be integrated into the plant genome (2). The TL-DNA (approximatively 20 kb) is responsible for the hairy root transformed phenotype (2,3). A 4.3 kb Eco RI fragment of the TL region was introduced into the genome of tobacco using two strategies: a direct transfer method and a *Agrobacterium*-mediated transformation.

In the liposome-mediated procedure (4), the plasmid containing the Eco RI fragment was co-transferred with a plasmid conferring kanamycin resistance. Among the selected kanamycin resistant plants, one displayed a phenotype similar to that of pRiA4 transformed plants. Unlike leaves of a normal tobacco plant, leaf fragments of this plant are able to form hairy roots *in vitro*. The expression of sequences from the plasmids and their transmission to progeny were studied.

The second procedure is based on a binary vector system using a micro-Ri plasmid derived from *A. rhizogenes* plasmids (5). It contains a kanamycin resistance gene and the Eco RI-15 fragment. This plasmid is able to induce proliferation of transformed roots on tobacco leaf disks. The phenotype of the plants regenerated from these roots is similar to the phenotype of the plant obtained with the first technique. The three ORFs (6) contained in this Eco RI fragment were subcloned in the micro-Ri plasmid in order to determine which one is responsible for the root formation.

These experiments allowed us to localize the part of the TL-DNA responsible for most of the characteristics of the hairy root phenotype, in particular the ability of the transformed cells to differentiate into roots and to regenerate plants with a phenotype related to the pRi-transformed phenotype.

(1) Tepfer, D., 1984, *Cell*: 959-967.
(2) Jouanin, L., Guerche, P., Pamboukdjan, N., Tourneur, C., Casse-Delbart, F., Tourneur, J., 1987, *Mol. Gen. Genet.*, 206, in press.
(3) Vilaine, F., Casse-Delbart, F., 1987, *Mol. Gen. Genet.*, 206:17-23.
(4) Deshayes, A., Herrera-Estrella, L., Caboche, M., 1985, *EMBO J.*, 4: 2731-2737.
(5) Vilaine, F. and Casse-Delbart, F., submitted.
(6) Slightom, J., Jouanin, L., Leach, F., Drong, R., Tepfer, D., 1986, *J. Biol. Chem.*, 261:108-121.

DIURNAL AND CIRCADIAN CONTROL IN THE EXPRESSION OF LIGHT- AND HEAT-INDUCIBLE CHLOROPLAST PROTEINS IN PEA (PISUM SATIVUM)

Klaus Kloppstech, Bernhard Grimm, Peter Ottersbach
and Beate Otto
Institut für Botanik, Universität Hannover
Herrenhäuser Strasse 2
D-3000 Hannover 21

When pea plants grown under a normal light-dark regime were analyzed for their mRNA composition during a 24 h day it was found that the levels for three light inducible proteins vary considerably. The maxima for the mRNAs of the early light-inducible protein (ELIP), the light-harvesting chlorophyll a/b protein (LHCP) and the small subunit of ribulose-1,5-bisphosphate carboxylase (ssRuBPCase) are reached in the early morning whereas the minima for these proteins occur during the dark period. The amplitude of the fluctuation is between 5- and 10-fold for particular proteins.

It has been shown before that active phytochrome is involved in the expression of light-inducible genes in etiolated plants. The fact that we observe an increase in the mRNA levels already at the end of the night, indicates that this phytochrome action should not be a direct one. We propose an indirect coupling, e.g. via a second messenger.

A similar change in the mRNA abundancy has been found for the heat-shock proteins. In this case the ability of the plants to transcribe heat-shock genes varies throughout the day. The maxima for the inducibility by a 2 hour heat-shock are found in the second half of the night and the morning. The minima are reached during the afternoon. The degree of fluctuation is between 3- and 5-fold. There is no difference in the expression for mRNAs of cytosolic and nuclear coded plastid heat-shock proteins.

A CHIMERIC TRANSCRIPT CONTAINING A 16S rRNA AND A POTENTIAL mRNA IN CHLOROPLASTS OF EUGLENA GRACILIS

Barbara Koller*, Etienne Roux, Paul-Etienne Montandon, and Erhard Stutz
*European Molecular Biology Laboratory, Heidelberg, FRG
Laboratoire de Biochimie, Universite de Neuchatel, Neuchatel, Switzerland

The ribosomal RNA operons (rrn operons) in the chloroplast genome of Euglena gracilis are arranged in tandemly repeated units. The number of these units can fluctuate between one and five in different strains of the alga (1,2). Genes for supplementary 16S and 5S rRNAs (s16S and s5S rrn genes), and in some strains also a partial 23S rRNA sequence, were found upstream of the complete rrn operons (3,4,6). In the Z strain an open reading frame (ORF 406) and an inverted repeat (IR1) of about 100 bp were localized between the s16S rrn gene and the first complete rrn operon (5,6).

It was not known, whether the supplementary rrn genes are transcribed, and whether the ORF 406 codes for a mRNA. To answer this question, cloned fragments of chloroplast DNA, containing this region of the genome, were hybridized with homologous RNA. The DNA-RNA hybrids were analysed by electron microscopy, by northern hybridization and by nuclease S1 protection mapping.

An unusual transcript, consisting at its 5'-end of the s16S rRNA and at its 3'-end of a potential mRNA, coding for the ORF 406, was found. The transcript has a size of about 3.6 kb and starts near, at at, the 5'-end of the s16S rrn gene. It includes the entire ORF 406. Termination of transcription occurs at multiple sites downstream of the ORF 406 within a region of about 400 bp. Since the RNA contains both sequences of the inverted repeat IR1, it forms a stable stem loop structure. The abundance of the transcript is comparable to the abundance of chloroplast mRNAs of thylakoid membrane proteins. It is not known, whether this chimeric transcript can function as ribosomal RNA, or as mRNA, or both.

References

1) Wurtz, E.A. and Buetow, W., 1981, Current Genetics, 3:181-187.
2) Koller, B. and Delius, H., 1982, Mol. gen. Genet., 188:305-308.
3) Jenni, B. and Stutz, E., 1979, FEBS Letters, 102:95-99.
4) Koller, B., Delius, H., and Helling, R.B., 1984, Plant Molec. Biol., 3:127-136.
5) Roux, E., Graf, L., and Stutz, E., 1983, Nucleic Acid Res., 11: 1957-1968.
6) Roux, E. and Stutz, E., 1985, Current Genetics, 9:221-227.

ISOLATION OF PUTATIVE PETUNIA HYBRIDA CHLOROPLAST AND MITOCHONDRIAL REPLICATION ORIGINS AND ANALYSIS OF THE INITIATION OF DNA SYNTHESIS

Jan M. de Haas, Frank Kors, Ad J. Kool and
H. John J. Nijkamp

Department of Genetics, Free University
De Boelelaan 1087
1081 HV Amsterdam

In addition to the nucleus there are two types of plant cell organelles which contain their own DNA, transcription-, translation- and replication machinery, the mitochondria and chloroplasts. The genetic organization of the DNA and the regulation of expression are intensively studied in order to get insight into the biogenesis and the regulation of this process, in which both the nuclear and organelle genes are involved. Until now only little attention has been paid to the process of organelle DNA replication and the regulation of this replication. Therefore, as a first step to elucidate the process of organelle DNA replication we have isolated potential chloroplast (cp) and mitochondrial (mt) replication origins. Because replication origins contain AT-rich stretches of nucleotides we isolated ARS sequences (which originate from AT-rich regions of the DNA) of the Petunia hybrida chloroplast (cp) and mitochondrial (mt) DNA. Subcloning and sequence analysis of the functional ARS and surrounding regions followed by searching for replication origin characteristics in the sequence is the first step to identify potential replication origins.

Three cpDNA-ARS-sequences (ARS A, B and C, (1)) have been isolated and sequenced, whereas four mtDNA-ARS-sequences have been isolated (ARS I, II, III and IV) of which ARS I has been sequenced. All sequenced cp and mt ARS's show stretches with a high AT-content, numerous direct and inverted repeats and the presence of at least one yeast ARS consensus (5'A/TTTTATPuTTTA/T), essential for ARS activity in yeast. Part of the ARS B region (which carries in yeast no functional ARS activity) shows almost all known origin of replication characteristics and extensive sequence homology with the possible replication origin of the Euglena gracilis cpDNA (2). In order to test this putative replication origin for functioning we have developed an in vitro DNA synthesizing lysate system, isolated from purified chloroplasts. This specific part of the ARS B region, inserted in a vector, is labeled more intensively in an in vitro DNA synthesizing system than the vector or an insert of a control cpDNA insert. This suggests that in this fragment, DNA synthesis is initiated. The observed DNA synthesis proceeds towards the ribosomal RNA genes (located in the inverted repeat), but up till now bidirectional DNA synthesis can not be excluded. Experiments are in progress to answer this possibility. Our chloroplast work demonstrates that the strategy described above is probably a good one for isolating potential replication origins. The functioning in vivo will be confirmed both in Chlamydomonas reinhardii CW15 (as a possible model system for replication) and in the P. hybrida chloroplasts by means of transformation with the cell organelle vectors (3).

The four mtDNA ARS fragments range in size from 6.5 to 0.6 kbp. They have been subcloned and the smallest fragment has been sequenced. The sequence of ARS I (0.6 kbp) shows some origin of replication characteristics such as a potential stem-and-loop structure, two GC-rich stretches and a potential gyrase binding site. The sequence of all mt ARS's will be compared (in analogy with the chloroplast work) with all characterized replication origins and mt "plasmid-like" DNA molecules which replicate autonomously in mitochondria.

(1) Overbeeke, N., Haring, M.A., Nijkamp, H.J.J., and Kool, A.J., 1984, Plant Mol. Biol., 3:235.

(2) de Haas, J.M., Boot, K.J.M., Haring, M.A., Kool, A.J., and Nijkamp, H.J.J., 1986, Mol. Gen. Genet., 202:48.

(3) van Grinsven, M.Q.J.M., Haring, M.A., de Haas, J.M., Nijkamp, H.J.J., and Kool, A.J., 1985, in: Genetic manipulation in plant breeding, Eucarpia. W. Horn, C.J. Jensenen, W. Odenbach and O. Schieder, eds., Walter de Gruyter Publ., Berlin, New York, pp. 859.

EASY DETECTION OF SUB-PICOGRAM QUANTITIES OF DNA BY IMPROVEMENT IN THE USE OF NON-RADIOACTIVE DNA PROBES. APPLICATION TO THE STUDY OF PLANT GENOME

Philippe Lebacq, Martine Duchenne, Driss Squalli and Martine Joannes
Laboratoire de Biologie Moléculaire Végétale
Université Paris-Sud
F-91405 Orsay

The use of non-radioactive systems to detect target DNA or RNA displays many advantages:

- Possible use in under-equipped areas.
- No need for radioactive confinement regulation and authorization.
- No exposure to health hazards which can exist with the use of radioactivity or the non-isotopic system using carcinogenic products as acetoxy-acetyl-amino-fluorene for instance.
- Sensitivity at least equal to that achievable with radiolabelled probes.
- Possible long storage of the probe (one year or more).

The system we use has been developed by Orgenics Ltd. and improved by us in France and in Israel. The tagging is accomplished by inserting an antigenic sulfone group into cytosine moities of the denatured DNA by simple chemical modification (1). Visualization of the tagged DNA is carried out by inserting an enzymatic sandwich immune reaction. A specific monoclonal antibody binds to the sulfone group of the tagged DNA. Then an enzyme anti-immune-immunoglobulin-antibody conjugate binds to the monoclonal antibody. The enzyme converts a soluble chromogenic substrate into an insoluble dye which precipitates at the exact location of the immune reaction. So sulfonation is a rapid procedure which, in contrast with nick-translation protocol, does not require enzymatic cocktails and special temperature conditions.

Improvement in the use of non-radioactive sulfonated probes has been carried out by our group and now allows:

- Use of the sulfonated procedure in routine and without special care in delicate target DNA detection such as spacer rDNA sequences in a wheat nuclear DNA smear.
- The detection of sub-picogram quantities of target DNA. For instance a single copy gene such as the ras gene is routinely detected by using our blocking solution with heparin (2). Minicircle DNAs, present in very small amounts in sugar-beet mitochondrial DNA have been easily visualized on nylon membranes by our technique.
- Possible use of nylon membranes (which improve DNA retention and have an important mechanical resistance) with very low colour background if any.

(1) Sverdlov, E.D., Monastyrskaya, G.S., Guskova, L.I., Sheichenko, V.J., and Budowski, E.J., 1974, Biochim. Biophys. Acta, 340:153-165.
(2) Lebacq, P., Pouletty, P., and Joannes, M., 1986, article submitted.

THE TRANSCRIPTIONALLY ACTIVE CHROMOSOME (TAC) OF THE SPINACH CHLOROPLAST: COMPARISON OF ITS PROPERTIES AFTER ISOLATION AT HIGH OR LOW IONIC STRENGTH

Michel Lebrun*, Jean-Francois Briat and Jean-Pierre Laulhere

Laboratoire de Biologie Moléculaire Végétale, Université de Grenoble, F-38402 Saint Martin d'Hères
*Present address: Laboratoire de Biologie Moléculaire et Cellulaire Végétale, Rhone-Poulenc Agrochimie, F-69263 Lyon

We describe the isolation of the TAC from spinach chloroplasts at high ionic strength, with minimum mechanical shearing. The integrity of the TAC, its influence on the transcriptional activity and the transcription steps performed by this high salt prepared TAC (1,2) were investigated, and compared to its low salt isolated counterpart.

The TAC isolated at high ionic strength is heterogenous in size. The transcription activity is found in three different regions of the gradient which could correspond to simple or multiple DNA copies of replicating structures of the chloroplast DNA.

As shown by CsCl density gradients, the TAC isolated at high salt concentration and by glycerol gradient centrifugation is more homogenous in density than the TAC isolated by gel filtration at low ionic strength. The proportion of DNA in the TAC isolated at low ionic strength is 65 to 79% while it is 91% in the TAC prepared at high salt. The protein moiety of both TAC preparations have been analysed for polypeptide composition. Many proteins, among which the basic histone-like proteins, cannot be detected in the high salt TAC.

The *in vitro* transcripts of both preparations are hybridized to Southern blots of chloroplast DNA. The two preparations show about the same transcriptional activity, in particular, transcripts hybridizing to the fragments which contain the rDNA, are as abundant in both preparations. The specificity of the transcription is not altered by the high salt extraction. The high salt isolated TAC transcribes the structural part of the ribosomal RNA genes, but does not transcribe the 5' region of the rDNA operon at a detectable level. Only the elongating RNA chains are labelled *in vitro*. Mechanical alteration was imposed on the TAC preparation by passing it through a syringe needle. There is no significant change in the transcriptional pattern after this mechanical shearing of the TAC but a general decrease in the transcriptional activity.

To determine if RNA affects the transcriptional activity, we isolated the TAC at high salt after treatment with RNAse A for various times. A 16 fold increase of the transcriptional activity was found after a 5 min RNAse pretreatment of the chloroplast lysates.

Contrary to the *Euglena* chloroplast TAC (3), the spinach high salt TAC is able to transcribe non-ribosomal genes.

References

(1) Briat, J.F., Laulhère, J.P., and Mache, R., 1979, Eur. J. Biochem., 98:285-292.
(2) Blanc, M., Briat, J.F., and Laulhère, J.P., 1981, Biochim. Biophys. Acta, 655:374-382.
(3) Rushlow, K.E., Orozco, E.M., Lipper, C., and Hallick, R.B., 1980, J. Biol. Chem., 255:3786-3792.

GENE EXPRESSION IN NICOTIANA TABACUM IN RESPONSE TO COMPATIBLE AND INCOMPATIBLE RACES OF PSEUDOMONAS SOLANACEARUM

F. Ragueh, N. Lescure, D. Roby and Y. Marco

Laboratoire de Biologie Moléculaire des Relations
Plantes-microorganismes, Chemin de Borde Rouge,
Auzeville, F-31326 Castanet-Tolosan

The molecular mechanisms involved in the plant response to compatible and incompatible pathogens are poorly understood. However, several host genes have been already implicated in the reaction of plants infected by a variety of pathogens. The changes in the gene expression of *Nicotiana tabacum* in response to challenge by *Pseudomonas solanacearum* (P.S.) is being studied in our laboratory. This bacterium is the causative agent of bacterial wilt of several economically important plants. Based on their behaviour on differential hosts, several pathovars have been defined: whereas the K60 strain (compatible) induces the development of the disease on tobacco plants, the GMI1000 strain (incompatible) elicits an hypersensitive response (H.R.) within 24 hours after infiltration in leaves. A deletion mutant of the GMI1000 strain, GMI1178 has lost its ability to induce the H.R. and is used as a control in our experiments. Total and poly A^+ RNA were purified from tobacco leaves infiltrated with K60, GMI1000, GMI1178 and H_2O at selected time points over 18 hours. *In vitro* translation of polyA^+ RNA was performed using rabbit reticulocyte lysate in the presence of S^{35} methionine and the products analyzed by both 1D and 2D dimensional electrophoresis (PAGE). 2D gel analysis demonstrates that up to 13 hours after infiltration, the expression of many plant genes is altered in a similar way in leaves infiltrated with the K60 and GMI1178 strains. Both protein patterns differ markedly from the control (leaves infiltrated with water and frozen immediately) and show some differences with the protein pattern from leaves infiltrated with water and incubated for the same time. Most of these changes (increase and/or appearance and decrease and/or disappearance of new mRNA activities) observed in leaves infiltrated with the K60 and GMI1178 strains are not specific of the plant response to *Pseudomonas solanacearum* since very similar protein patterns have been obtained in leaves infiltrated with saprophytic bacteria (*E. coli*). However, 6 hours after infiltration, the expression of plant genes in leaves undergoing an H.R. differs markedly from the ones observed with leaves infiltrated with the K60 and GMI1178 strains. After 13 hours, the changes are often more pronounced. These results suggest that the hypersensitive response involves a cascade of very early and specific molecular events. Moreover, up to 13 hours, the gene expression in leaves infiltrated with the compatible strain of P.S. is altered although the observed changes are probably not pathogen specific.

POTATO SPINDLE TUBER VIROID REPLICATION IN CELL SUSPENSION CULTURES AND PROTOPLASTS OF POTATO

H.-P. Mühlbach, R. Kern and R. Luckinger

Max-Planck-Institut für Biochemie
Abteilung Viroidforschung
Am Klopferspitz 18, D-8033 Martinsried

Viroids are small single-stranded circular RNA molecules which cause disease in higher plants. The potato spindle tuber viroid (PSTV) and related viroids are known to replicate in the nucleus of infected cells via concatameric RNA intermediates. To study the organisation of PSTV-specific RNA complexes occurring during replication, RNA was extracted from PSTV-infected fermenter grown potato cells and separated by column chromatography. Subsequent biochemical analysis revealed the existence of different classes of PSTV RNA: (i) free monomeric PSTV(+)RNA, which represents the viroid proper; (ii) free multimeric PSTV(+)RNA forms; (iii) a variety of partly double-stranded PSTV RNA complexes of different size. The majority of these complexes consists of circular monomeric PSTV(+)RNA and a heterogeneous population of multimeric PSTV(-)RNAs of various length. All these complexes are supposed to represent different intermediates of PSTV replication, which are formed during the subsequent steps of RNA-RNA transcription. Of particular interest are the multimeric PSTV(+)RNA forms, which appear in the fraction of free, non-complexed RNA. They consist of single-strand circular and linear PSTV dimers, which most probably represent precursors of the finally accumulating circular PSTV monomer.

The temporal order of synthesis of PSTV(+) and (-)RNA was studied in protoplasts, which were isolated from healthy potato cells and inoculated in vitro via liposomes. Using dot blot and Northern blot hybridisation, the synthesis of PSTV(-)RNA was first detectable 24 hours after inoculation. It was followed by a drastic increase of PSTV(+)RNA synthesis. During prolonged culture of the protoplasts the PSTV(+)RNA synthesis continued, while PSTV(-)RNA synthesis slowed down. This finding clearly substantiates the presumed role of multimeric PSTV(-)RNA forms as the transient intermediates of PSTV replication.

STRUCTURE AND EXPRESSION OF GENES CODING FOR THE SWEET POTATO TUBEROUS ROOT STORAGE PROTEIN SPORAMIN

Kenzo Nakamura

Lab. of Biochemistry, Fac. of Agriculture
Nagoya University, Chikusa-ku
Nagoya 464, Japan

About 80% of the total soluble protein in mature tuberous roots of the sweet potato is accounted for by sporamin, a group of related proteins with mol. wt. of 20 kD, deposited in vacuole. Very little, if any, sporamin is present in other tissues of the plant. Genes encoding sporamin exist as a multigene family in the sweet potato genome. The family consists of sporamin A and sporamin B subfamilies based on their sequence homology with intra-subfamily homologies (>94%) being much higher than inter-subfamily homologies (82-84%), and each subfamily consists of more than 6 members. Comparison of the nucleotide sequence of 5 full-length sporamin cDNAs suggest unique features in the evolution of sporamin multigene family. The nucleotide sequence of one of the sporamin A genomic clones indicates that this gene does not contain intron and a perfect consensus TATA box sequence is present 23 bp 5' to the 5'-terminus of the full-length sporamin A cDNA sequence. In order to define the sequence elements important for the regulation of gene expression during tuberous root differentiation, we are currently analyzing the structure of sporamin B genomic clone.

Sporamin is synthesized by membrane-bound polysomes as larger precursors carrying N-terminal extra sequences of 35 to 37 amino acids. Co-translational processing in vitro by dog pancreas membrane or in vivo in E. coli cells removes only the N-terminal signal peptide of 19 to 21 amino acids. The following 16 amino acid segment enriched with charged amino acids is probably removed by post-translational cleavage during or after its transport to vacuole. Several lines of evidence suggest that at least part of sporamin precursors expressed in yeast cells are transported into vacuole. We have joined a full-length sporamin cDNA downstream of the CaMV 35S promoter, and this chimeric gene was introduced into tobacco by Agrobacterium/Ti plasmid vector system. The transformed tobacco callus cells accumulated sporamin with the mature size. Intracellular localization of sporamin in transformed tobacco cells is now under investigation.

THE MAIZE TRANSPOSABLE ELEMENT Ds AS A SITE FOR CHROMOSOME BREAKAGE

Birgit Nelsen, R. Garber*, E. Tillmann, and Hans-Peter Döring

Institut für Genetik, Universität zu Köln
Weyertal 121, D-5000 Köln 41
*Present address: Cornell University, Dept. of Plant Physiology, Ithaca, NY, USA

One of the maize Ds (Dissociation) transposable elements isolated by Barbara McClintock exhibits an unusual behaviour compared to most other elements. It is able to promote chromosome breakage at specific sites and to create chromosomal rearrangements upon transposition.

This Ds element and some derivatives associated with several unstable alleles of the shrunken (sh) locus, have peculiar structures. At the alleles displaying chromosome breakage, two, or one and a half 2 kbp unit Ds elements are present. One element apparently was transposed into the approximate center of an identical copy of itself.

Structural and genetic analysis of these elements in comparison to other Ds and Ac elements that do not induce chromosome breakage led us to the following conclusions: The presence of one and a half or two Ds elements is required for chromosome breakage. In addition, the Ds elements must be oppositely orientated with respect to each other. A single Ds element is unable to induce chromosome breakage.

EXPRESSION OF A PROTEIN ENCODED BY AN OPEN READING FRAME OF CHLOROPLAST tRNA LYS GENE FROM MUSTARD

H. Neuhaus, J. Hughes and G. Link

AG Pflanzliche Zellphysiologie
Ruhr-Universität Bochum, Universitätsstr. 150
D-4630 Bochum

The chloroplast gene for tRNA lys is located 213 bp upstream of the psbA gene (1), i.e. the gene for the major light inducible plastid transcript. Sequence analysis has shown a 2.6 kb intron in the anticodon loop which contains a long open reading frame of 524 amino acids (2). The intron derived polypeptide appears structurally related to mitochondrial maturases, proteins known to be involved in RNA splicing.

Transcript levels of these two adjacent chloroplast genes during early seedling development until 84 h after sowing are under differential temporal and light control (3). Our data are consistent with a hypothetical mechanism which implies a role of the trnK gene product(s) in the photocontrol of plastid gene expression.

Our initial approach to provide evidence for this hypothesis is to investigate the putative intron encoded polypeptide. We inserted three different fragments comprising the entire ORF sequence in phase into the C-terminal BamHI site of the lac Z gene. The hybrid gene was expressed in *E. coli* under control of the lac promoter and the purified hybrid protein was used to raise antibodies against it. These antibodies were tested against chloroplast proteins and *in vitro* translation products of the trnK gene transcripts.

References

1. Link, G. and Langridge, U., 1984, *Nucleic Acids Res.*, 12:945-957.
2. Neuhaus, H. and Link, G., 1987, *Curr. Genet.*, 11:251-257.
3. Hughes, J., Neuhaus, H., and Link, G., 1987, *Plant Physiol.*, submitted.

THE ROLE OF LIGHT AND NITRATE ON THE INDUCTION OF NITRATE REDUCTASE IN CORN LEAVES

A. Oaks, H. Deising and L.A. Cass

McMaster University
Hamilton
Ontario, L8S 4K1

It has long been known that both light and NO_3^- are required for the appearance of nitrate reductase activity in leaves (Beevers et al., Plant Physiology (1965) 40:691-698). Recently we have re-examined this question using monospecific antibodies prepared against a nitrate reductase purified from corn leaves according to the method of Nakagawa et al. (Plant Physiol. (1984) 75:285). We have grown corn seedlings hydroponically, on washed sand or on Kimpack in the light in the presence or absence of NO_3^- or in the dark in the presence or absence of NO_3^-. SDS-PAG electrophoresis followed by western blotting showed that crude extracts prepared from light grown plants had a polypeptide of about 116 kD (the size of the subunit for nitrate reductase) whether nitrate was present or not. Crude extracts prepared from dark grown plants did not have this polypeptide, although smaller polypeptides which reacted with the NR-IgG were found at the gel front. These results suggest that the mRNA coding for nitrate reductase is present in the light grown plants and possibly in dark grown plants as well. In the dark the NR subunits appear to be unstable. In the light NO_3^- appears to be responsible for the activation of a nitrate reductase precursor protein. We are currently examining the effect of NH_4Cl additions on the appearance of the NR protein. (Research supported from NSERC, Canada).

GENE TARGETING IN PLANTS

J. Paszkowski*, M. Baur, A. Bogucki* and I. Potrykus*
Friedrich Miescher-Institut, CH-4002 Basel
*Present address:
Swiss Federal Institute of Technology Zurich
Institute of Plant Sciences, ETH-Zentrum, CH-8092 Zurich

Co-transformation of different DNA molecules into protoplasts during direct DNA mediated transformation was used to construct several lines of tobacco plants containing genome integrated copies of a mutated selectable marker gene. These mutated copies served as the targets for site directed integration of complementing DNA molecules which upon homologous recombination should be able to restore function of a marker gene.

The results of experiments leading to site specific homologous recombination and frequencies of gene targeting in plants will be discussed.

DISSECTION OF UPSTREAM SEQUENCES INVOLVED IN REGULATED EXPRESSION OF THE *NICOTIANA PLUMBAGINIFOLIA* *rbc*S-8B GENE IN HOMOLOGOUS NUCLEAR BACKGROUND

Carsten Poulsen and Nam-Hai Chua

Laboratory of Plant Molecular Biology
The Rockefeller University, 1230 York Avenue
New York, N.Y. 10021-6399, U.S.A.

We have previously investigated the expression of the pea *rbc*S-3A gene in a heterologous nuclear background (transgenic tobacco) and have identified several positive and negative elements involved in light response. To see if similar elements also mediate the regulated expression of *rbc*S genes in a homologous system, we have analyzed the expression of the *N. plumbaginifolia* *rbc*S-8B promoter in the same nuclear background. An *rbc*S promoter with 1 kbp of 5' upstream sequence (-1038 to +32) was chosen as the wild type from which a series of 3' and 5' deletion mutants with breakpoints 50 to 100 bp apart were constructed. We have also assembled chimeric promoters comprising various 5' upstream fragments joined to the CaMV 35S promoter fragments previously used in this laboratory (-105/+9, -46/+9). The mutant promoters were fused to the coding sequence of the bacterial chloramphenicol acetyltransferase (CAT) which is used as a reporter gene. Analysis of hundreds of transgenic *N. plumbaginifolia* plants by CAT assays and by S1 nuclease protection assays, reveal the following points: (1) Light-regulated and organ-specific expression requires only 312 bp of 5' upstream sequence. Deletion breakpoints at -257, -192 and -94 yield 50%, 10% and 0% CAT mRNA levels, respectively. The fraction of plants which show properly regulated expression decreases with decreasing length of the 5' flanking sequence. (2) 3' deletion to -10, i.e. removal of the cognate *rbc*S-8B mRNA start site, does not affect expression or regulation. New mRNA start site can be demonstrated. (3) Additional 3' deletion of the TATA-box (to -41) results in a 10-fold decrease in activity, although fully regulated expression is maintained. (4) Fusion of other 3' deleted fragments (-52 and -92) to the CaMV 35S promoter fragments (-46/+9 and -105/+9) results in full restoration of promoter capability. (5) Removal of 0.7 kbp of upstream sequence from the chimeric promoters has no deleterious effect on the expression levels. However, transgenic plants containing these constructs show some level of expression in roots and in the dark.

When inserted in a dual promoter system, where another reporter gene, *nos*, was located divergently upstream of the *rbc*S-8B CAT test gene, the following observations were made by 5' *nos* mRNA S1 nuclease assays: (6) The *rbc*S-8B promoter fragment contains a light responsive enhancer(s) which can confer light regulation on the *nos* promoter. The enhancer function is maximal with the -312 mutant, i.e. when in a proximal position. *Nos* activity with the -192 mutant is about 5-fold lower. (7) When *rbc*S-8B sequences downstream of -92 are removed (see points (2), (3) and (4) above), full light-dark enhancer response is reacquired.

We conclude that the major *rbcS*-8B regulated enhancer is *not* upstream of -312 and *not* downstream of -192. Regulated expression of CAT and *nos* mRNAs can be conferred by sequence elements between -312 and -102. This is in accordance with results obtained with the pea *rbcS*-3A gene, and is substantiated by the finding of homologous sequence elements probably serving for binding of similar DNA binding factors.

SYNTHESIS AND PROCESSING OF RAPE-SEED STORAGE

Lars Rask, Mats Ericson, Joakim Rödin, Marit Lenman, Mats Ellerström, Hans-Olof Gustafsson, Eva Muren and Lars-Göran Josefsson
Department of Cell Research, Uppsala Biomedical Center
University of Agricultural Sciences, Uppsala, Sweden

In rape-seed, two types of storage proteins are synthesized, napin (formerly denoted 1.7S) and cruciferin (formerly called 12S). Napin which is encoded by a small gene family consisting of about ten members displays only limited heterogeneity. It is composed of two disulfide-linked peptide chains of 86 and 29 amino acid residues, respectively, processed from a common precursor. During the processing two highly negatively charged peptide structures are removed.

Cruciferin, which has a molecular weight of approximately 300,000, probably consists of twelve peptide chains processed pairwise from three different precursors. Two pairs of these chains apparently are similar whereas the other pair only displays 65% amino acid identity to one of the other pairs. The topography of each chain within the heterohexamer is not known.

Napin and cruciferin are synthesized on rough endoplasmic reticulum during days 21 to 45 post-anthesis. They accumulate rapidly in the same protein bodies of the embryo cells. The synthesis of both proteins is promoted by abscisic acid.

A DEFECTIVE TRANSPOSITION EVENT IN *ANTIRRHINUM MAJUS* - IMPLICATIONS FOR THE NORMAL TRANSPOSITION PROCESS

Tim Robbins, Rosemary Carpenter and Enrico Coen

Department of Genetics
John Innes Institute
Colney Lane, Norwich NR4 7UH

An insertion of the transposable element Tam3 in the promoter of the *pallida* gene of *Antirrhinum majus* blocks the synthesis of anthocyanin in the flowers. A derived allele with an altered phenotype has exchanged a sequence that was originally 5' to the element, with a sequence 3 recombination map units away from *pallida*, probably by inversion. A short sequence is duplicated at either end of the rearrangement. A model put forward to explain these events proposes that during the normal process of transposition, the excision and integration sites are brought together in a complex. The consequences of the rearrangement for *pallida* gene expression are discussed.

INDUCTION OF CRASSULACEAN ACID METABOLISM IN MESEMBRYANTHEMUM: MASS INCREASE AND DE-NOVO SYNTHESIS OF PEP CARBOXYLASE

J.M. Schmitt, R. Höfner, C.B. Michalowski, S.W. Olson, L. Vazquez-Moreno, J.A. Ostrem and H.J. Bohnert
Botanisches Institut der Universität Würzburg, Mittlerer Dallenbergweg, D-8700 Würzburg and Department of Biochemistry University of Arizona, Tucson, AZ 85721, USA

Exposure of Mesembryanthemum crystallinum to high salinity in the soil causes a shift in the mode of carbon assimilation from that typical of C_3 plants to that typical of Crassulacean acid metabolism (CAM) plants, which are able to exhibit substantial net carbon gain at night. Intact plants were induced to exhibit CAM by irrigation with nutrient solution containing 500 mM NaCl. During the induction period, the extractable activity of phosphoenol-pyruvate carboxylase (PEPcase) increased approximately 40-fold. This increase was linearly correlated with a mass increase of PEPcase protein as measured by single radial immunodiffusion. De-novo synthesis of PEPcase protein was shown by immunoprecipitation of the newly synthesized, radioactively labelled protein in leaf-discs from salt-treated plants. Non-treated plants were characterized by a low level of the enzyme and low rates of PEPcase synthesis. Synthesis of this enzyme in leaf-discs was correlated with the concentration of NaCl during induction of CAM.

Clones for the enzymes PEPcase, pyruvate, Pi dikinase (PPDK) and NADP malic enzyme have been isolated from a cDNA library (lambda gt11) by immunological screening. The increase in enzymatic activity and protein is preceded by an accumulation of mRNA for PEPcase and PPDK while mRNAs for e.g. rbcS and cab genes decrease.

GENETIC MANIPULATION OF RICE (O. SATIVA) BY USING PROTOPLAST CULTURE

K. Shimamoto

Plantech Research Institute
1000 Kamoshida-cho, Midori-ku
Yokohama, Japan

1. Plant regeneration from protoplasts

Novel nurse culture methods were developed for high frequency plant regeneration from rice protoplasts. With these methods, protoplasts isolated from five cultivars of rice formed colonies with a frequency of 3-9% and plants were regenerated from 20-50% of protoplast-derived calli. The field test of 131 protoplast-derived plants showed that 80-90% of those plants were normal in 10 agronomic characters examined and this offers a rationale for using protoplasts for genetic manipulation of rice.

2. Somatic hybridization

To incorporate economically important traits of other Gramineae species into rice, somatic hybridization of rice and several Gramineae species was performed. The fusion method employs electrofusion and a selection scheme using metabolic inhibitor was developed. To date, somatic hybrid plants, 1) rice + barnyard grass (wild millet), 2) rice + wild *Oryza* species, have been obtained. Analyses of isozymes and chromosomes confirmed the hybrid nature of these hybrids.

3. Gene transfer by electroporation

A number of parameters influencing transformation frequency have been examined to obtain stable transformants by electroporation procedure and NPT positive clones were obtained with kanamycin constructs having CaMV promoter. Transformed calli were transferred onto regeneration medium to obtain transgenic rice plants. Further optimization of electroporation procedure and introduction of plant genes are now in progress.

GENE EXPRESSION IN RESISTANCE OF FRENCH BEAN TO *PSEUDOMONAS SYRINGAE* PV. *PHASEOLICOLA*

Alan Slusarenko, Christopher Knight and Sharon Alldrick

Department of Plant Biology & Genetics
The University
Hull, HU6 7RX, U.K.

The plant pathogenic bacterium *Pseudomonas syringae* pv. *phaseolicola* causes halo blight of French bean (*Phaseolus vulgaris*). Currently, three races of the pathogen may be identified using different bean cultivars. The French bean cv. Red Mexican is resistant to race 1 of the pathogen but susceptible to races 2 and 3 while the cv. Tendergreen is resistant to race 3 but susceptible to races 1 and 2. The resistant trait in both cultivars is associated with a hypersensitive reaction (HR) and is inherited as though it were conditioned by a single dominant gene.

The products of such resistance genes may function as part of a recognition system which, when triggered by the pathogen, sets into motion a series of responses in the host which lead, ultimately to resistance. These host responses may be mediated at the level of transcription, translation, post-translational activation of holoenzymes or inactivation of enzymes.

The hypersensitive response is typified by a rapid, localized necrosis of host cells at the site of infection, beyond which the pathogen does not spread. Inhibitors of translation in eukaryotes have been used to show that a period of host protein synthesis is a pre-requisite for tissues to be able to respond hypersensitively to pathogens. However, very little is known of what happens in the plant cell between the induction of the HR and the appearance of cellular and tissue necroses. The work with protein synthesis inhibitors suggests that a controlled sequence in the disorganization of the cellular machinery, i.e. 'apoptosis', might be involved in the HR rather than a simple destruction of host cells by the pathogen. The responses of the surrounding healthy tissue to hypersensitive necrosis are also uncertain except that it is thought to be the site of much of the phytoalexin synthesis. Accumulation of antibacterial isoflavonoid phytoalexins in hypersensitively responding Red Mexican leaves has been correlated with cessation of bacterial multiplication *in vivo*. The mRNA activities for many of the enzymes on the biosynthetic pathway of isoflavonoid phytoalexins have been shown to increase in infected hypocotyls and elicitor-treated cell suspension cultures of bean.

Total cellular and polysomal RNAs were isolated from leaves of the French bean cultivar Red Mexican which had been inoculated with virulent or avirulent races of *Pseudomonas syringae* pv. *phaseolicola*. The mRNAs were translated *in vitro* and the polypeptide fingerprints obtained from the various treatments indicated specific changes in host mRNA activities in leaf tissue expressing a hypersensitive reaction. The mRNAs coded for several high M_r polypeptides, and coordinated decreases and transient increases in activity were observed, defining a cascade of changes in gene expression in tissues undergoing the HR. Some changes were apparent as early as 2 h after inoculation. Messenger RNA for phenylalanine ammonia lyase (PAL), an early enzyme involved in phytoalexin biosynthesis in bean, was estimated by probing Northern blots with a cDNA probe for PAL (a kind gift from C.J. Lamb). PAL mRNA began to increase between 6 to 9 hours and peaked around 12 hours after inoculation of leaves with avirulent bacteria; this was some hours after the earliest changes in mRNA activity which we correlated with the expression of resistance. We have constructed cDNA libraries from mRNA from inoculated tissues and we are attempting to identify the early-resistance specific clones by differential screening.

DIFFERENTIAL GENE EXPRESSION DURING TUBER DEVELOPMENT AND CLONING OF CIS-ACTING REGULATORY SEQUENCES INVOLVED IN TUBER-SPECIFIC GENE EXPRESSION OF POTATO

Willem J. Stiekema, Freek Heidekamp, Wim. G. Dirkse and Jeanine Louwerse
Research Institute Ital
P.O. Box 48
6700 AA Wageningen, The Netherlands

The molecular mechanisms involved in the regulation of organ-specific gene expression in potato are still obscure. We have focussed our attention upon those genes specifically expressed in the tuber.

A number of cDNA recombinants were isolated containing copies of tuber mRNAs. Among these full-length patatin cDNAs were identified by hybrid-select translation and DNA sequence analysis. Three additional tuber mRNAs were cloned encoding polypeptides of respectively 11.5, 16 and 21.5 kD.

The expression of these tuber mRNAs in other potato tissues and during tuber development has been studied by Northern blot analysis. All four mRNAs were predominantly present in tubers (100-1000 x more abundant in tuber compared to stolons, stems, leaves or roots). During tuber development three of the four, including the patatin mRNA could be detected 5 days after tuber initiation while the fourth was already present 2 days after tuber initiation. These results will be discussed.

To study the cis-acting regulatory elements involved in the tuber-specific expression of genes, a DNA library was screened for the presence of patatin gene sequences. Three genomic clones were isolated and regions containing putative cis-acting controlling elements were characterized by DNA sequence analysis. These sequences will be fused to reporter genes and transferred to potato, to establish the regulatory character of these elements in transgenic potato plants.

IDENTIFICATION AND ANALYSIS OF EXPRESSION OF CHLOROPLAST GENES INVOLVED IN TRANSCRIPTION, DNA REPLICATION AND REPAIR IN *CHLAMYDOMONAS REINHARDTII*

Stefan J. Surzycki, Steven Fong, Tsai-Hsia Hong and Timothy Opperman
Department of Biology
Indiana University
Bloomington, IN 47405

Both the nuclear and chloroplast genomes of plants encode proteins required for photosynthesis and chloroplast DNA expression and replication. Most of the genes encoding chloroplast proteins studied to date have been those involved in photosynthesis and translation. Relatively little is known about plant genes encoding polypeptides involved in chloroplast DNA transcription, replication and repair.

To gain more information in this area, we have been identifying and characterizing nuclear and chloroplast sequences in *Chlamydomonas reinhardtii* that bear homology to *Escherichia coli* genes encoding polypeptides of the transcription, DNA replication and DNA repair apparatus. Using heterologous hybridization analysis, we have found chloroplast sequences in *C. reinhardtii* that are homologous to *E. coli* genes involved in transcription: RNA polymerase subunits alpha (*rpo*A), beta (*rpo*B), beta' (*rpo*C), and transcription termination factor *rho*. We have also found chloroplast sequences that are homologous to *E. coli* genes involved in DNA replication: beta subunit of DNA polymerase III (*dna*N), single-stranded DNA binding protein (*ssb*), and chromosomal replication origin binding protein (*dna*A).

Additionally, we have found chloroplast sequences homologous to bacterial genes encoding proteins of the SOS response: regulatory proteins *rec*A and *lex*A; proteins of the DNA error-less repair system *uvr*A, *uvr*B, *uvr*D; and a protein involved in UV induced mutagenesis (error-prone repair), *umu*DC.

Data will be presented on the location of all of these homologous sequences on the *C. reinhardtii* chloroplast genome. We will also present data on the sequence, intron structure and expression of the chloroplast *rpo*C gene. The results of experiments examining the cellular response to DNA damage produced by ultraviolet light (UV) and the UV-mimetic agent 4-Nitroquinoline-N-oxide (NQO) in wild type and UV sensitive (UVS) strains of *Chlamydomonas* will be presented. Data indicate that both error-prone and error-less DNA repair mechanisms operate in the nuclear and plastid compartments in *Chlamydomonas*. An analysis of UV-inducible transcripts and proteins will be presented.

ANALYSIS OF VIROID FUNCTIONS BY SITE-SPECIFIC MUTAGENESIS

Martin Tabler, Mina Tsagris, Isolde Günther and Heinz L. Sänger
Max-Planck-Institut für Biochemie
Abteilung Viroidforschung
D-8033 Planegg-Martinsried

The viroids are autonomously replicating pathogens of higher plants. They consist of a single-stranded covalently closed RNA with a chain length between 240-380 nucleotides and lack mRNA capacity. It has been found that cloned viroid cDNA and also in vitro synthesized RNA transcripts thereof are able to induce viroid infections when mechanically inoculated onto the appropriate host plants. This provides the experimental basis for analysing the functional domains of the viroid genome by introducing mutations into the viroid cDNA and studying their effects on infectivity and on the in vitro processing of the RNA transcripts of the corresponding mutants.

We manipulated the potato spindle tuber viroid (PSTV) in the so-called "central conserved region" which is part of all viroids except the avocado sunblotch viroid and supposed to be involved in processing. Secondly we mutated the "virulence-modulating region" (VM-region) which influences the severity of symptom expression. The SP6 RNA transcripts of more than a dozen mutants were assayed for the in vitro processing in a nuclear extract prepared from potato suspension cells and for infectivity. Those mutations which affected the correct processing were found to be of lower specific infectivity or non-infectious at all. With respect to infectivity we were able to classify three categories of mutations (i) lethal mutations, which result in a complete loss of infectivity although some of these mutated PSTV RNAs are still correctly processed to the circular monomer in vitro; (ii) unstable mutations, where the mutated RNA causes infections, but the corresponding viroid RNA progeny reverts to the wild type viroid; (iii) viable mutations, where the mutated PSTV RNA is infectious and accumulating as viroid RNA progeny, thus leading to a new PSTV isolate. Viable mutations could thus far be only obtained when the mutation was introduced into the VM-region. This finding corroborates previous observations that sequence variations between different naturally occuring PSTV isolates are mainly restricted to this domain. The thermodynamic instability of this region could be correlated with the severity of disease symptoms induced by the corresponding isolates. In line with these empiric observations our viable PSTV mutant VM2 induced significantly reduced disease symptoms as compared with the "wild type" isolate KF440-2 into which the mutation had been introduced. Thus mutations into the VM region should allow to develop a "disarmed" PSTV mutant which might be used to "cross-protect" host plants against infections with virulent viroids.

STRUCTURE, BIOSYNTHESIS AND TURNOVER OF THE CELL WALL OF THE UNICELLULAR GREEN ALGA CHLAMYDOMONAS REINHARDII

Jürgen Voigt and Hans-Peter Vogeler

Institut für Allgemeine Botanik
Universität Hamburg
Ohnhorststr. 18, D-2000 Hamburg 52

The cell wall of the unicellular green alga Chlamydomonas reinhardii consists of several layers of hydroxyproline-containing glycoproteins. The insoluble cell wall component was found to be located within the outer wall layers. The innermost wall layer consists of soluble glycoproteins, which can be extracted from intact cells by treatment with aqueous LiCl. Pulse-labelling and pulse-chase experiments performed during the cell enlargement period revealed that the LiCl-soluble wall glycoproteins are precursors of the insoluble cell wall component (1). The turnover products of the insoluble wall layer were found to be accumulated in the culture medium (2).

Several putative cell wall glycoproteins were purified from the LiCl-extracts of intact cells (3) and used to raise antibodies. All the obtained antibodies showed cross-reactivity with the different antigens. Two of these antibody preparations also recognized chemically deglycosylated cell wall components.

The polypeptide subunits of the insoluble wall layer were found to be solubilized completely during chemical deglycosylation with HF/pyridine. Several polypeptides of different molecular weights were released from the insoluble wall component by this procedure, most of which were recognized by the antibodies raised against soluble wall glycoproteins.

When in-vitro translation products of poly(A)-containing RNA from C. reinhardii were immunologically analysed using these antibodies, some polypeptides were specifically precipitated, whose molecular weights differed considerably (31 kD - 150 kD). These findings indicate that the cell wall of C. reinhardii contains several different, but immunologically related polypeptides.

The amino acid compositions of the different cell wall polypeptides were found to be very similar, but differed with respect to the proportion of hydroxyproline.

The antibodies raised against soluble cell wall glycoproteins from C. reinhardii cross-reacted with soluble cell wall glycoproteins of some other green alga and also with three glycoproteins which were extracted from isolated cell walls of tobacco pollen tubes.

References:

(1) Voigt, J., 1986, Z. Naturforsch., 41c:885-896.
(2) Voigt, J., 1985, Biochem. J., 226:259-268.
(3) Voigt, J., 1985, Planta, 164:379-389.

DNA THAT MAPS OUTSIDE OF THE BORDERS OF THE T DNA OF THE Ti PLASMID IS NOT TRANSFERRED TO THE PLANT CELL DURING TUMOR FORMATION BY *AGROBACTERIUM TUMEFACIENS*

R. Walden

Leicester Biocentre, Leicester University
University Road, Leicester LE1 7RH, U.K.

Agrobacteria containing dimers of the cauliflower mosaic virus (CaMV) genome cloned into the T DNA of the Ti plasmid can be used to initiate CaMV infection following the inoculation of turnip. This process has been termed "agroinfection" and provides a sensitive method by which the transfer of DNA from *Agrobacterium* to the plant cell can be investigated. I have used agroinfection to ask the question: "Does *Agrobacterium* transfer all of its DNA, or a specific portion of it, to the plant cell during tumor formation ?". In order to do this dimers of the CaMV genome were cloned into different regions of the Ti plasmid, the chromosome of *Agrobacterium* and a broad host range plasmid. The dimeric DNA is stable when introduced into the Ti plasmid and the chromosomal DNA of *Agrobacterium* but is unstable when cloned in an broad host range plasmid and introduced into *Agrobacteria* by conjugation. *Agrobacteria* containing the dimeric CaMV genome cloned into both different regions of the Ti plasmid and the chromosomal DNA have been used to inoculate turnip plants and demonstrate that during tumor formation only DNA between the T-DNA border sequences is transferred from the *Agrobacterium* to the plant cell. A sequence located within the kanamycin resistance gene of Tn903 has been shown to be able to act as a pseudo-right border and direct transfer of DNA from the Ti plasmid to the plant genome.

FIELD BEAN STORAGE PROTEIN GENES: STRUCTURE AND EVOLUTION OF THE LEGUMIN B GENE SUBFAMILY AND EXPRESSION OF LEGUMIN GENE LeB4 OR DERIVED CHIMAERIC CONSTRUCTS IN TRANSGENIC TOBACCO AND XENOPUS OOCYTES

U. Wobus, H. Bäumlein, D. Helbing, D. Hellmund, U. Heim, R. Manteuffel, A. Müller and J. Schiemann
Zentralinstitut für Genetik und Kulturpflanzenforschung
Akademie der Wissenschaften der DDR
DDR-4325 Gatersleben

Field bean (Vicia faba var. minor) seed globulins contain 70% legumin, a polymorphic protein coded for by two main subfamilies of genes. We have determined nucleotide sequences of five legumin genes belonging to the B subfamily. Four of them differ from gene LeB4 (proven to be expressed) by deletions, insertions and/or point mutations and are classified as pseudogenes. The unequal distribution of mutations together with high homologies reaching far into the 3'-flanking regions suggest frequent recombination events during evolution.

We have transferred the legumin gene LeB4 (coding plus 2.7 kb 5'- and 0.4 kb 3'-flanking sequences) via a Ti-derived vector into tobacco and found the encoded polypeptide in the transgenic seeds in amounts up to 0.6% of the salt-extractable proteins. There is an only about 6-fold variation in the amount of Vicia polypeptide accumulated in the seeds of 10 different transgenic plants. This level is further reduced by calculating the amount of LeB4 protein per integrated copy of the LeB4 gene which varies from one to at least five. The low plant-to-plant variation found will aid in the analysis of cis-acting elements like the "legumin box" defined earlier (Bäumlein et al., 1986, Nucl. Acids Res. 14, 2707).

In a first step towards this analysis we fused the chloramphenicol acetyl transferase (CAT) gene to LeB4 5'-flanking regions of different length and transferred the constructs into tobacco or in Xenopus oocytes where they result in different levels of CAT activity.

PHYTOCHROME REGULATION OF mRNA LEVELS OF RIBULOSE-1,5-BISPHOSPHATE CARBOXYLASE AND LIGHT-HARVESTING CHLOROPHYLL a/b PROTEIN IN ETIOLATED RYE SEEDLINGS (Secale cereale)

Dietrich Ernst[1], Friedhelm Pfeiffer[1], Petra Eilfeld[2], Peter Eilfeld[2], Wolfhart Rüdiger[2] and Dieter Oesterhelt[1]

[1]Max-Planck-Institut für Biochemie
8033 Martinsried, FRG

[2]Botanisches Institut der Universität
Menzingerstr. 67
8000 München 19, FRG

INTRODUCTION

Phytochrome is a protein with a linear tetrapyrol as chromophore and acts as one of the main photoreceptors in plants controlling photomorphogenesis. Over the past years it became obvious, that via phytochrome red light controls the stationary level of a number of mRNA species in plants (Tobin and Silverthorne, 1985 and articles cited herein). Further studies have shown, that the transcription of genes coding for the small subunit (SSU) and the large subunit (LSU) of ribulose-1,5-bisphosphate carboxylase (Rubisco), for the light-harvesting chlorophyll a/b protein (LHCP) and for the protochlorophyllide oxidoreductase are good probes for the analysis of the phytochrome-mediated response of gene expression to red light (Gallagher and Ellis, 1982; Silverthorne and Tobin, 1984; Berry-Lowe and Meagher, 1985; Mösinger et al., 1985). However, even in closely related species the details of phytochrome controlled gene expression might be different (Tobin and Silverthorne, 1985). Therefore our studies reported here are exclusively concerned with rye.

Light or phytochrome induced gene regulation is complex and the molecular mechanisms involved in the signal transduction chain are far from being understood. The first report on a simplifying approach to this problem was given by Ernst and Oesterhelt (1984), who demonstrated that the addition of purified rye P_{fr} to nuclei isolated from dark grown rye seedlings increased the total UMP incorporation into RNA, suggesting a direct interaction of phytochrome with nuclei without the necessity of cytoplasmic factors, participating in the signal chain. For the demonstration of gene specific effects, first the regulation of the two genes for SSU and LHCP were studied in vivo and then the appropriate conditions applied to an in vitro system. In the present review we will describe our results on phytochrome mediated gene regulation in vivo and in vitro.

Isolation of RNA, hybridization, preparation of cDNA clones and labelled riboprobes, as well as nuclei isolation and in vitro transcription experiments have been described in detail (Ernst et al., 1987b). Irradiation of dark grown rye seedlings with white light resulted in a strong increase of the stationary message levels for both subunits of Rubisco, although in the dark a relatively high level was present already (Fig.1). The increase is mediated by phytochrome, as a pulse of red light resulted in a transient increase for both mRNAs lasting the following 24 h dark period (Fig.1,2). The simultaneous increase indicates a coordinated synthesis for both mRNA species, although the kinetics of induction are different. This may reflect a different regulation mechanism in the nucleus and the plastid (Sasaki et al., 1986; Ernst et al., 1987b). A smaller increase in both mRNA species is also seen after a pulse of far-red light and due to very low amounts of P_{fr}, established by such a far-red pulse (Schäfer et al., 1975; Colbert et al., 1983). The effect of red light can be reduced by a subsequent far-red light treatment to the level of the control, demonstrating the classical phytochrome response (Fig.1,2). Because in the dark both mRNAs are already present we conclude, that phytochrome is not essential for de novo induction, but rather for influencing the level of the two mRNAs.

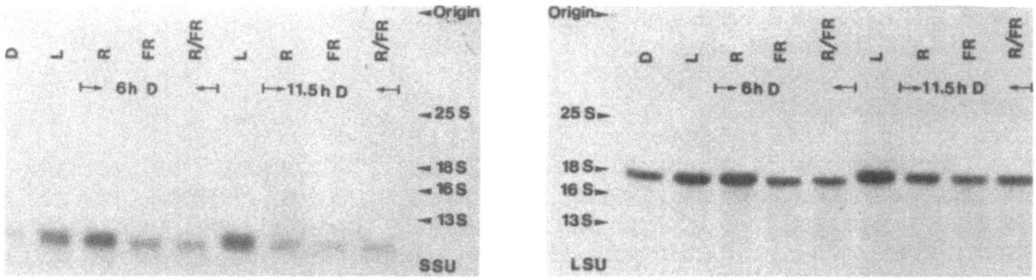

Figure 1. Northern blotting of total RNA (10 µg) of different light treated seedlings. Hybridization was carried out with labelled SSU 'anti-sense' RNA and LSU 'anti-sense' RNA. Seedlings were grown for 5 d in the dark, exposed to different light treatment and RNA was isolated. D=dark; R=5 min red, FR=5 min far-red, R/FR=1 min red/4 min far-red and then darkness for the time indicated; L=white light.

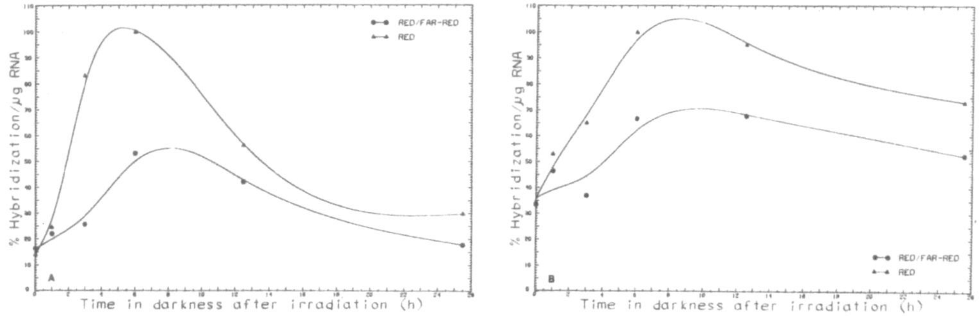

Figure 2. Time course of change in SSU (A) and LSU (B) of Rubisco mRNA levels in response to different light treatment.

In order to discriminate between transcriptional and post-transcriptional control of SSU mRNA, we carried out in vitro transcription studies. Nuclei were isolated from seedlings which were (i) grown for 5 d in the dark, (ii) grown for 4 d in the dark, then for 1 d in white light and (iii) grown for 5 d in the dark, irradiated with a pulse of red light and kept in the dark afterwards, up to 24 h. After incubation of the nuclei for 30 min at 25° C in the presence of ^{32}P-UTP, RNA was isolated and hybridized to gene specific probes (Fig.3). From corresponding slot blots an enhanced transcription rate is evident for white light-grown seedlings (24 h light). In seedlings pulsed with red light, a time course of the transcriptional rate was measured and as a result a transient increase of SSU mRNA transcription observed (Fig.3). Quantitation revealed a 2.5 fold increase in the transcription rate, whereas the mRNA abundance (level) was increased by a factor of 7-10 (Fig.4). If mRNA levels are considered as a stationary state, composed

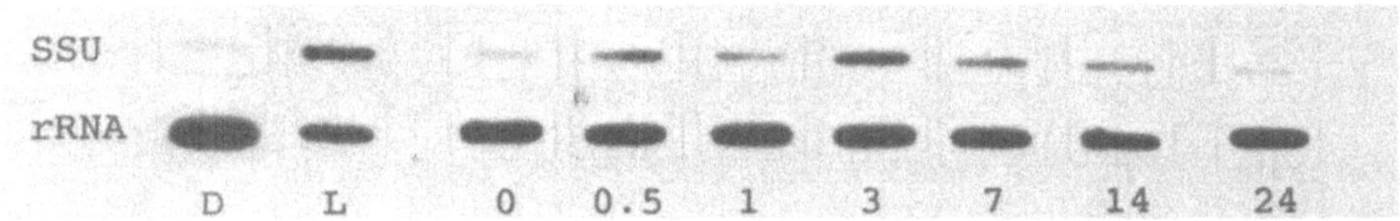

Figure 3. Autoradiogram of slot blots of plasmid DNA, containing specific sequences for SSU and rRNA, probed with in vitro labelled RNA transcripts in nuclei, isolated from light-treated rye seedlings. Seedlings were grown for 5 d in the dark (D), for 4 d in the dark and 1 d in white light (L), or for 5 d in the dark, illuminated with red light for 5 min and then returned to darkness for the time indicated.

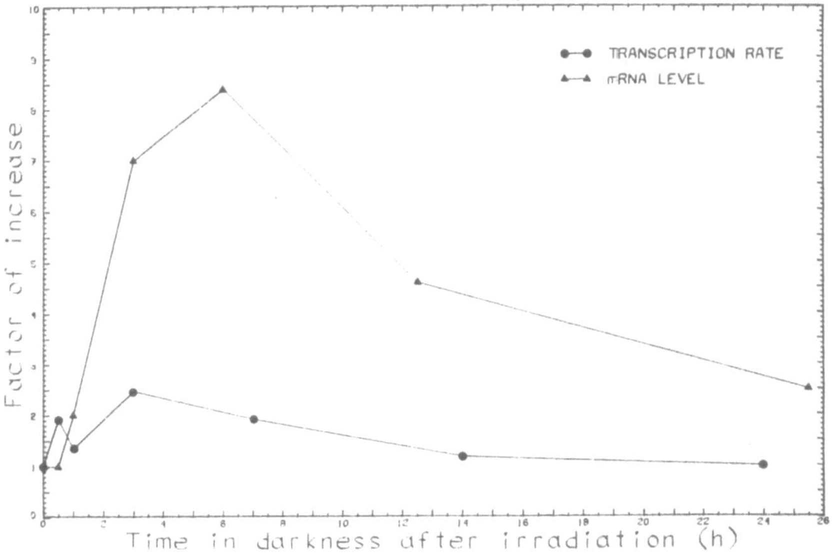

Figure 4. Comparison of change in the transcription rate of SSU genes and of SSU mRNA levels in response to red light treatment. Slot blots of autoradiograms were scanned and the dark value set as 1.

of mRNA synthesis and degradation (modification would still allow hybridization), any increase in level must reflect the changing ratio of synthesis and degradation:

$$\text{mRNA level} = \frac{\text{rate of synthesis}}{\text{rate of degradation}}$$

The results in Fig.4 clearly demonstrate that not only the rate of synthesis for the SSU message is increased, but the rate of its degradation decreased at the same time. The result can be quantitate by the following equation:

$$\frac{dm}{dt} = {}^{o}k_{s} - {}^{1}k_{d}m$$

where ${}^{o}k_{s}$ is the zero-order rate constant of synthesis, ${}^{1}k_{d}$ the first order rate constant of degradation and m the amount of mRNA at time t (Schimke, 1975; Chappell and Hahlbrock, 1984; Quail et al., 1986). The areas at various times t under the two curves for level and rate should be related by a constant factor, if transcription alone is involved in gene regulation. For the SSU mRNA this factor is 0.45 at 1 h, 4.5 at 6 h and 7 at 24 h. This indicates post-transcriptional regulatory phenomena involved in SSU gene expression.

IN VIVO GENE REGULATION OF LIGHT-HARVESTING CHLOROPHYLL a/b PROTEIN

Besides Rubisco, LHCP is one of the best studied plant proteins, including the regulation of its gene (Apel et al., 1983; Tobin et al., 1984; Bennett et al., 1984; Tobin and Silverthorne 1985; Mösinger et al., 1985). From a cDNA bank prepared from rye poly A mRNA, clones were screened with a heterologous probe from spinach (R. Herrmann, München). First labelled LHCP-specific riboprobes were prepared and then hybridization carried out as described for Rubisco genes (Ernst et al., 1987b). One positive clone (pFPA 641) was obtained and partially sequenced, revealling 60% sequence identy with the known LHCP-sequence from pea. The complete insert of the LHCP-clone pFPA 641 was inserted in pGEM 2 and clones pFPB 302 and pFPB 303 prepared, which contained the insert in the respective orientations. The preparation of labelled riboprobe RNA and all other hybridizations were as described by Ernst et al. (1987b).

In etiolated rye seedlings the LHCP mRNA is present in very low concentrations, compared with the SSU mRNA. After a red light pulse the mRNA level increased by a factor of 14 within additional 6 h darkness, indicating the involvement of phytochrome in the regulation (Fig.5). After longer times in darkness, the value declined to that of the dark control. This behaviour is similar to that found for SSU mRNA. However, the factor of increase is greater and the decrease is more complete after 1 d.

In contrast to the low but detectable levels of LHCP mRNA, transcriptional studies with isolated nuclei from dark grown seedlings showed no detectable transcription for the LHCP mRNA, when compared with the SSU mRNA under identical conditions. Only prolonged exposure of the slot blot preparations allowed the determination of a transcriptional rate. It is about 1/30 in the dark than that for the SSU mRNA. A red light pulse resulted in an increased transcriptional rate of the LHCP mRNA by a factor of 18 as maximum at 3 h, followed by a decline in rate almost to the dark control within 1 d (Fig.6). Treatment of the data in Fig.6 as described above for the SSU mRNA reveals less contribution of post-transcriptional regulatory elements in LHCP, as already evident from the closer match of enhancements in rate and level. Table 1 summarizes the data about mRNA regulation for SSU and LHCP in etiolated rye seedlings and includs rRNA synthesis.

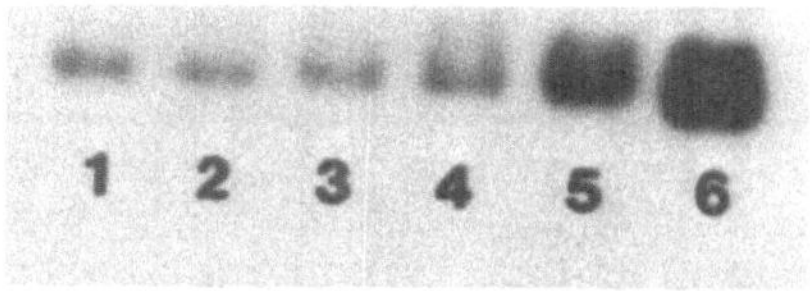

Figure 5. Northern blotting of total RNA (5 µg) of rye seedlings. Hybridization was carried out with labelled LHCP 'anti-sense' RNA. Seedlings were grown in the dark for 5 d, irradiated with red light for 5 min and returned to darkness for different times. 1=dark control; 2=15 min; 3=30 min; 4=60 min; 5=3 h; 6=6 h.

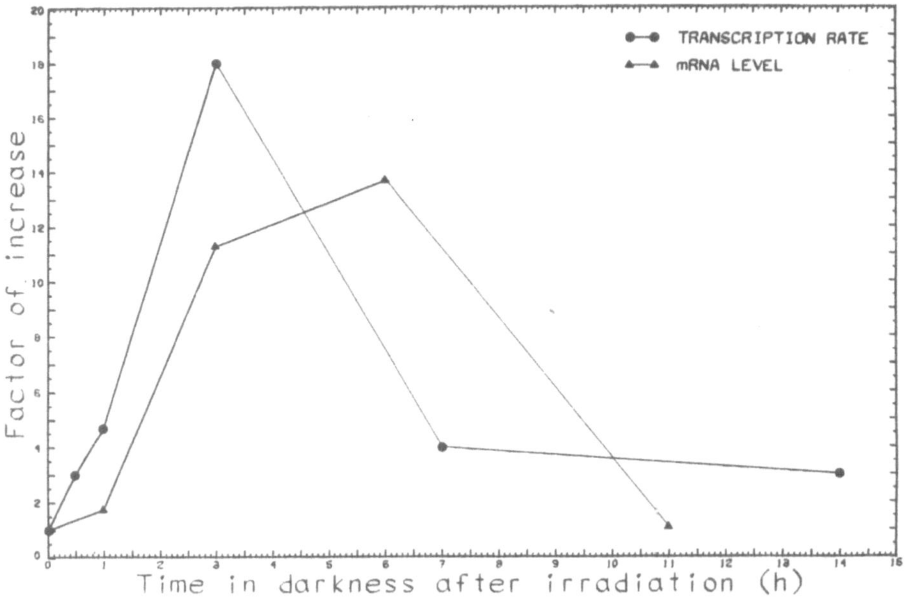

Figure 6. Comparison of change in the transcription rate of LHCP genes and LHCP mRNA levels in response to red light treatment. Slot blots of autoradiograms were scanned and the dark value set as 1.

IN VITRO GENE REGULATION IN ISOLATED NUCLEI

Addition of FPLC-purified rye P_{fr} (Ernst et al., 1987a) to nuclei, isolated from dark grown rye seedlings, increased the total incorporation of ^{32}P-UMP into RNA by an average of 65 % in 4 independent experiments. Addition of P_r increased also the transcription rate, resulting in about 3/4 of the P_{fr} induced enhancement. This confirmed results obtained with smaller amounts of nuclei (Ernst and Oesterhelt,

Table 1. In vivo gene regulation in etiolated rye seedlings.

RNA	R level	R rate	L level	L rate
SSU	+	+	+	+
LSU	+	n.d.	+	n.d.
LHCP	+	+	+	+
rRNA	n.d.	o	n.d.	-

Seedlings were grown for 5 d in the dark and then illuminated. R=5 min red, 6 h dark; L=24 h white light; +=increase; -=decrease; o=no change; n.d.=not determined

1984). In order to exclude any effect of weak light, a X-ray film was exposed in the growth chamber of the seedlings during the growth phase of 5 d and the fact, that no darkening of the developed film was observed, was taken as an indication that total darkness was guaranteed. In contrast to our results Schäfer et al. (1986), using a heterologous system with oat nuclei and rye P_{fr}, could only detect an increase of overall transcription, if nuclei from red light pre-irradiated seedlings were used. Wether these discrepances are due to the different systems used, is not known. Nevertheless, also nuclei from pre-irradiated rye seedlings show higher rates of ^{32}P-UMP incorporation than nuclei isolated from the dark grown seedlings. Therefore the experiments shown in Fig.7 were carried out with such nuclei. Addition of oat P_{fr}, purified according to Eilfeld et al. (1986), to rye nuclei showed also an increase in the overall transcription rate. However, there was no significant difference if instead of P_{fr}, oat P_r was added. Furthermore, addition of modified oat phytochrome (obtained by adding 1-10 equivalents fluorescein mercuric acetate) also increased the incorporation of ^{32}P-UMP into RNA (Fig.7). Identical results were obtained, even if the modified phytochrome showed no longer photoreversibility. The alteration of the protein induced by fluorescein mercuric acetate, leading to reduced or lost photoreversibility is irreversible, even if the modifying reagent is removed by dithioerythritol under incubation conditions (Eilfeld et al., 1987). This behaviour rises the question, whether the application of phytochrome mimics the in vivo situation, although other proteins showed no significant increase in the overall transcription (Fig.7). Additionally there was no photoreversibility of the in vitro system, as irradiation with far-red light, after addition of rye P_{fr} revealed no decrease in the P_{fr} stimulated overall transcription.

In further experiments we tried to demonstrate an influence of exogenous phytochrome on specific gene transcription and used the measurement of SSU and LHCP mRNA, after application of rye P_{fr} to isolated nuclei, as an assay. Two independent experiments with nuclei isolated from dark grown and red light pre-irradiated seedlings showed no increase in SSU and LHCP mRNA transcription (Fig.8). This result together with the non photoreversibility of the effect of P_{fr} on the incorporation of ^{32}P-UMP into RNA make its unlikely, that the phenomena observed reflect the in vivo action of P_{fr}. It should, however, be mentioned that a recent report by Mösinger et al. (1987) claimed, that the addition of exogenous oat P_{fr} to nuclei isolated from red light pre-irradiated barley seedlings increased the LHCP mRNA transcription and reduced the protochlorophyllide oxidoreductase mRNA transcription, relative to the overall transcription. Wether these discrepances are due to different nuclei preparations or to different phytochrome preparations is not clear. Nevertheless all these reports stress the implication of additional factor(s), in addition to phytochrome, in the first steps of phytochrome mediated gene regulation.

As plastidic factor(s) may be involved in nuclear gene regulation (Mayfield and Taylor, 1984; Oelmüller and Mohr, 1986; Batschauer et al., 1986) we analyzed the influence of purified chloroplasts and P_{fr} on the transcription rate of SSU and LHCP mRNA. Chloroplasts were isolated

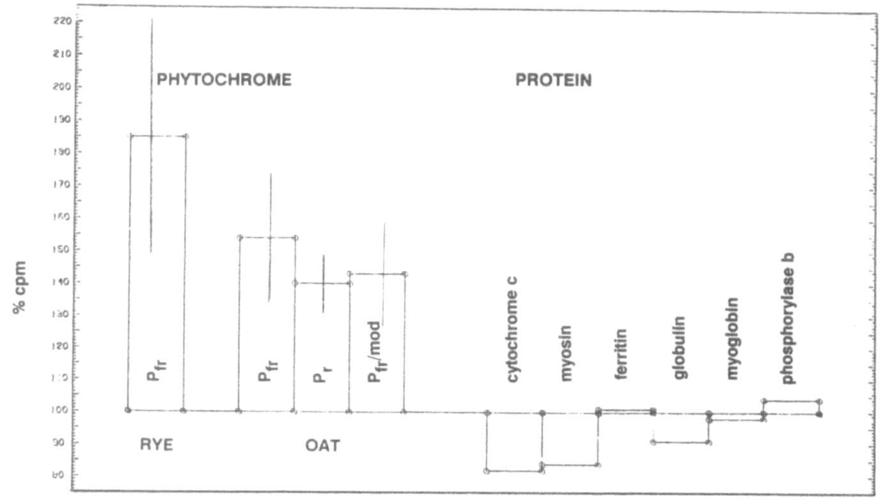

Figure 7. Overall transcription in isolated rye nuclei, in the presence of phytochrome or other proteins. Values are expressed as percentage of control. 10^6 nuclei/300 µl were incubated in the presence of 12 µg protein. Proteins were added to the nuclei 10 min before start of transcription and kept on ice. For phytochrome data points are the means of at least 5 experiments +/- SD.

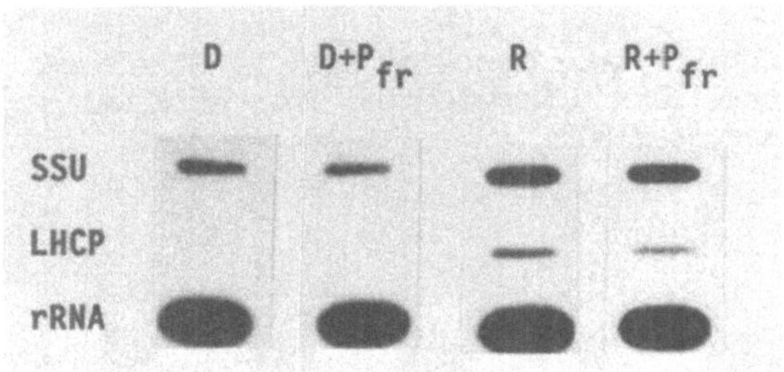

Figure 8. Autoradiogram of slot blots of in vitro transcription. Seedlings were grown in the dark for 5 d (D), or in the dark for 5 d, illuminated with red light and brought back to the dark for 6 h (R). Nuclei were isolated and incubated ($7x10^6$/300 µl) in the absence or presence of purified P_{fr} (20 µg). P_{fr} was added to the nuclei 10 min before start of transcription and kept on ice. RNA was isolated and probed with plasmids, containing specific sequences for SSU, LHCP and rRNA.

Table 2. Effect of exogenous factors on transcription rate in isolated rye nuclei.

Factor	Increase of SSU/LHCP mRNAs over control
Dark + P_{fr}	-
Dark + crude extract$_R$	-
Dark + crude extract$_L$	-
Dark + chloroplasts	-
Dark + chloroplasts + P_{fr}	-
Red	+
Red + P_{fr}	+ (≙Red)

Nuclei were isolated from dark grown or from red light pre-irradiated seedlings. R=5 min red; 5 h dark; L=22 h white light. ($7x10^6$ nuclei/300 µl; 20 µg P_{fr}; about 40-80 chloroplasts / nucleus; crude extract: seedlings were homogenized with transcription buffer, containing 2 mM PMFS, centrifuged and the supernatant was used)

according to Herrmann (1982) and purified, using a Percoll gradient (Berkowitz and Gibbs, 1985). Intactness of chloroplasts was analyzed, using ferricyanide as a Hill oxidant (Leegood and Walker, 1983). Incubation of chloroplasts with nuclei in the in vitro transcription system, revealed a strong hybridization signal for LSU mRNA (data not shown), indicating active chloroplast transcription. However, no significant increase was found for SSU and LHCP mRNA, both in the absence and presence of P_{fr} (Table 2). In addition, crude extract from red light pre-irradiated and from white light grown seedlings showed no increase in transcription of these two mRNA species (Table 2). In conclusion, the signal chain: light - phytochrome - gene regulation remains a puzzle and needs further analysis.

ACKNOWLEDGEMENTS

The excellent technical assistance of Mrs. K. Schefbeck, R. Vojacek and C. Weyrauch is gratefully acknowledged. This work was supported by the Deutsche Forschungsgemeinschaft.

REFERENCES

Apel, K., Gollmer, I., and Batschauer, A., 1983, The light-dependent control of chloroplast development in barley (Hordeum vulgare L), J. Cell. Biochem., 23:181.

Batschauer, A., Mösinger, E., Kreuz, K., Dörr, I., and Apel, K., 1986, The implication of a plastid-derived factor in the transcriptional control of nuclear genes encoding the light-harvesting chlorophyll a/b protein, Eur. J. Biochem, 154:625.

Bennett, J., Jenkins, G. I., and Hartley, M. R., 1984, Differential regulation of the accumulation of the light-harvesting chlorophyll a/b complex and ribulose bisphosphate carboxylase/oxygenase in greening pea leaves, J. Cell. Biochem., 25:1.

Berkowitz, G. A., and Gibbs, M., 1985, Chloroplasts as a whole, in: "Cell Components, Modern Methods of Plant Analysis New Series Vol. 1," H. F. Linskens and J. F. Jackson, eds., Springer, Berlin Heidelberg NewYork Tokyo.

Berry-Lowe, S. L., and Meagher, R. B., 1985, Transcriptional regulation of a gene encoding the small subunit of ribulose-1,5-bisphosphate carboxylase in soybean tissue is linked to the phytochrome response, Mol. Cell. Biol., 5:1910.

Chappell, J., and Hahlbrock, K., 1984, Transcription of plant defence genes in response to UV light or fungal elicitor, Nature, 311:76.

Colbert, J. T., Hershey, H. P., Quail, P. H., 1983, Autoregulatory control of translatable phytochrome mRNA levels, Proc. Natl. Acad. Sci. USA, 80:2248.

Eilfeld, P., Eilfeld, P., and Rüdiger, W., 1986, On the primary photoprocess of 124-kdalton phytochrome, Photochem. Photobiol., 44:761.

Eilfeld, P., Widerer, G., Malinowski, H., Eilfeld, P., and Rüdiger, W., 1987, Topography of the phytochrome molecule as determined from chemical modification of SH-groups, Biosciences, submitted.

Ernst, D., and Oesterhelt, D., 1984, Purified phytochrome influences in vitro transcription in rye nuclei, EMBO J., 3:3075.

Ernst, D., Vojacek, R., and Oesterhelt, D., 1987a, Purification of phytochrome from rye by fast protein liquid chromatography, Photochem. Photobiol., 45:859.

Ernst, D., Pfeiffer, F., Schefbeck, K., Weyrauch, C., and Oesterhelt, D., 1987b, Phytochrome regulation of mRNA levels of ribulose-1,5-bisphosphate carboxylase in etiolated rye seedlings (Secale cereale), Plant. Mol. Biol., submitted.

Gallagher, T. F., and Ellis, R. J., 1982, Light-stimulated transcription of genes for two chloroplast polypeptides in isolated pea leaf nuclei, EMBO J., 1:1493.

Herrmann, R. G., 1982, The preparation of circular DNA from plastids, in: "Methods in Chloroplast Molecular Biology," M. Edelmann, R. B. Hallick and N.-H. Chua, eds., Elsevier Biomedical Press, Amsterdam NewYork Oxford.

Leegood, R. C., and Walker, D. A., 1983, Chloroplasts, in: "Membranes and Organelles from Plant Cells," J. L. Hall and A. L. Moore, eds., Academic Press, London NewYork Paris SanDiego SanFrancisco SaoPaulo Sydney Tokyo Toronto.

Mayfield, S. P., and Taylor, W. C., 1984, Carotenoid-deficient maize seedlings fail to accumulate light-harvesting chlorophyll a/b binding protein (LHCP) mRNA, Eur. J. Biochem., 144:79.

Mösinger, E., Batschauer, A., Schäfer, E., and Apel, K., 1985, Phytochrome control of in vitro transcription of specific genes in isolated nuclei from barley (Hordeum vulgare), Eur. J. Biochem., 147:137.

Mösinger, E., Batschauer, A., Vierstra, R., Apel, K., and Schäfer, E., 1987, Comparison of the effects of exogenous native phytochrome and in-vivo irradiation on in-vitro transcription in isolated nuclei from barley (Hordeum vulgare), Planta, 170:505.

Oelmüller, R., Dietrich, G., Link, G., and Mohr, H., 1986, Regulatory factors involved in gene expression (subunits of ribulose-1,5-bisphosphate carboxylase) in mustard (Sinapis alba L.) cotyledons, Planta, 169:260.

Quail, P. H., Colbert, J. T., Peters, N. K., Christensen, A. H., Sharrock, R. A., and Lissemore, J. L., 1986, Phytochrome and the regulation of the expression of its genes, Phil. Trans. R. Soc. Lond. B., 314:469.

Sasaki, Y., Nakamura, Y., and Matsuno, R., 1986, Phytochrome-mediated accumulation of chloroplast DNA in pea leaves, FEBS Lett., 196:171.

Schäfer, E., Lassig, T. U., Schopfer, P., 1975, Photocontrol of phytochrome destruction in grass seedlings. The influence of wavelenght and irradiance, Photochem. Photobiol., 22:193.

Schäfer, E., Apel, K., Batschauer, A., and Mösinger, E., 1986, Phytochrome The molecular biology of action, in: "Photomorphogenesis in Plants", R. E. Kendrick and G. H. M. Kronenberg, eds., Martinus Nijhoff Publishers, Dordrecht Boston Lancaster.

Schimke, R. T., 1975, Methods for analysis of enzyme synthesis and degradation in animal tissues, in: "Methods in Enzymology Vol. 40", B. W. O'Malley and J. G. Hardman, eds., Academic Press, NewYork SanFrancisco London.

Silverthorne, J., and Tobin, E. M., 1984, Demonstration of transcriptional regulation of specific genes by phytochrome action, Proc. Natl. Acad. Sci. USA, 81:1112.

Tobin, E. M., and Silverthorne, J., 1985, Light regulation of gene expression in higher plants, Ann. Rev. Plant. Physiol., 36:569.

Tobin, E. M., Wimpee, C. F., Silverthorne, J., Stiekama, W. J., Neumann, G. A., and Thornber, J. P., 1984, Phytochrome regulation of the expression of two nuclear-coded chloroplast proteins, in: "Biosynthesis of the Photosynthetic Apparatus Molecular Biology, Development and Regulation," J. P. Thornber, L. A. Staehelin and R. B. Hallick, eds., Alan R. Liss Inc., New York.

AUTHOR INDEX

SUBJECT INDEX